BIOLOGY

FIFTH EDITION

Biology

HELENA CURTIS

N. SUE BARNES

WORTH PUBLISHERS, INC.

Male wood duck (Aix sponsa) *in breeding plumage, drying its wings, photographed on the St. Croix River, Minnesota. Animals, such as the wood duck, in which the males and females differ in appearance are said to be sexually dimorphic ("having two forms"). In such species, the males are characteristically polygamous. (© Scott Nielsen)*

BIOLOGY, FIFTH EDITION

COPYRIGHT © 1968, 1975, 1979, 1983, 1989 BY WORTH PUBLISHERS, INC.

LIBRARY OF CONGRESS CATALOG CARD NUMBER: 88-51041

ISBN: 0-87901-394-X

PRINTED IN THE UNITED STATES OF AMERICA

PRINTING: 3 4 5 6 7   YEAR: 0 1 2 3 4

EDITOR: SALLY ANDERSON

PRODUCTION: SARAH SEGAL

ART DIRECTOR: GEORGE TOULOUMES

LAYOUT DESIGN: PATRICIA LAWSON, DAVID LOPEZ

DESIGN: MALCOLM GREAR DESIGNERS

ILLUSTRATOR: SHIRLEY BATY

PICTURE EDITORS: DAVID HINCHMAN, ANNE FELDMAN, ELAINE BERNSTEIN

TYPOGRAPHY: NEW ENGLAND TYPOGRAPHIC SERVICE, INC.

COLOR SEPARATION: CREATIVE GRAPHIC SERVICES CORPORATION

PRINTING AND BINDING: VON HOFFMANN PRESS, INC.

COVER PHOTOGRAPH: WOOD DUCK (© ROD WILLIAMS/BRUCE COLEMAN LTD.)

FRONTISPIECE: TROPICAL FOREST, COSTA RICA (© GARY BRAASCH)

ILLUSTRATION ACKNOWLEDGMENTS BEGIN ON PAGE A–1, CONSTITUTING

AN EXTENSION OF THE COPYRIGHT PAGE.

WORTH PUBLISHERS, INC.

33 IRVING PLACE

NEW YORK, NEW YORK 10003

This book is dedicated to those whose creative and painstaking studies have contributed to our understanding of biology.

Contents in Brief

Giraffe with oxpeckers

Contents

Ladybug on flower of shepherd's needle

SECTION 5 Biology of Plants

CHAPTER 29

The Flowering Plants: An Introduction

CHAPTER 30

The Plant Body and Its Development

CHAPTER 31

Transport Processes in Plants

CHAPTER 32

Plant Responses and the Regulation of Growth

Preface

In the twenty years since the first edition of *Biology* appeared, the science of biology has been characterized by ever accelerating change, including not only a flood of new information but also new ideas and unifying concepts. Some areas of biology have undergone metamorphoses before our very eyes, while others have attained a new maturity. This has presented us—now at our word processors—and you—in the classroom and laboratory—with new challenges and opportunities as we work together to provide students with a solid foundation in the principles of biology while simultaneously sharing with them the excitement of the contemporary science.

With this, the Fifth Edition, *Biology* becomes both one of the oldest and one of the newest introductory biology textbooks. One of our principal goals in preparing this edition has been to maintain a balance between the old and the new. This has required not only a willingness to discard material, but also considerable care that we not eliminate or slight material that, although not new, is essential if students are to be adequately prepared to understand current and future developments in biology. Simultaneously, we have, of course, wanted to be as up to the minute as possible, without becoming merely trendy. At a time when important discoveries are published almost continuously, there is a temptation to become so engrossed in the new that we lose sight of the fact that the majority of today's students are, like their predecessors, coming to the formal study of contemporary biology for the first time. The clear explication of the basic principles of biology, with pertinent and readily understood examples, has become increasingly important with each passing year. Thus, specific topics for detailed treatment have been selected on the basis of their centrality to modern biology, their utility in illuminating basic principles, their importance as part of the requisite store of knowledge of an educated adult as we approach a new century, and their inherent interest and appeal to students. Throughout, we have tried to provide the underlying framework and arouse student curiosity so that a foundation is laid for those areas—very diverse—in which you may wish to give more extensive coverage in the classroom or laboratory than is possible in any introductory textbook, regardless of its length.

The central, essential foundation of biology is, of course, evolution, the major organizing theme of this text as of all modern biology texts. As in previous editions, the stage is set in the Introduction, which focuses on the development of the Darwinian theory. New to the Introduction is a section previewing the other major unifying principles of modern biology that are also recurring themes throughout the text; also new is a brief overview of the diversity of life. Both are designed to provide students with a broad framework before they begin their study of the details on which modern biology is built. In this edition, we have also strengthened the introductory discussion of the nature of science, and, throughout the text, we have included more information about how biologists know what they know and how scientists in general go about their business.

After the Introduction, this edition, like previous editions, follows the levels-of-organization approach. Part 1 deals with life at the subcellular and cellular levels, Part 2 with organisms, and Part 3 with populations, ending with a survey of the distribution of life on earth. Each part is divided into two or three sections. A significant amount of restructuring has occurred in the sequence of chapters within certain sections and within the chapters themselves.

One of the most striking aspects of the enormous burst of new discoveries in molecular and cell biology since the Fourth Edition is the power of these discoveries to explain processes that previously could only be described—and in the most general terms. The immune response, olfaction and color vision, events at the synapse, the summing of information by individual neurons, and differentiation and morphogenesis in animal development are just a few of the many phenomena whose secrets are being revealed by studies at the molecular and cellular level. These revelations depend, in large part, on what is now a flood tide of reports identifying specific membrane proteins, their amino acid sequences, their three-dimensional structures, and, in many cases, the nucleotide sequences of the genes coding for the proteins and the location of these genes within the genome. Because these discoveries, although fascinating in and of themselves, are of such value in explaining organismal phenomena, we have generally chosen to defer their discussion to later sections of the text, where their significance will be most readily grasped by students. Molecular and cell biology have, in many ways, come of age, and it seems to us that the essential task in these early sections of the book has become the clear communication of the underlying principles on which so much is now being built—rather than a catalog of the latest new discoveries, which will soon be superseded by even more exciting ones.

The extraordinary pace of discovery in genetics, principally as a result of recombinant DNA technology, requires, with each new edition, a major rethinking of Section 3. Responses to our surveys indicate that, however tempting it might be to begin the section with molecular genetics and to reduce the coverage of classical genetics, doing so could make this most exciting area of modern biology less accessible to students. Thus, as in previous editions, we begin our consideration of genetics with Mendel and take an essentially historical approach to the development of the powerful science we know today. Within that overall framework, however, there has been the addition of a significant amount of new material, coupled with a number of internal reorganizations that we believe provide greater clarity and a smoother conceptual development in our coverage of molecular genetics.

Part 2, Biology of Organisms, has also undergone many changes, particularly in the early chapters of Section 4 and in Sections 5 and 6. In the previous edition, Section 4, The Diversity of Life, was significantly expanded. The enthusiasm with which the revised section was received—plus our own continuing awe at the incredible variety of living organisms—led to our decision to retain the expanded section intact. We have, however, made major revisions in the first chapter of the section, dealing with the classification of organisms, and minor revisions throughout the section.

The organization of Section 5, the Biology of Plants, has long been problematic. It has been difficult to find a sequence that would flow logically, coordinate well with laboratory programs, and—most important—captivate students with the beauty and biological accomplishments of plants without overwhelming them with the vocabulary necessary for an accurate description of the living plant. In this edition we have chosen to begin the section with the familiar—the flower—and with the dynamic process of plant reproduction, a sequence that flows directly from the discussion of plant evolution and diversity in Section 4. In the following chapter, the anatomy of the plant body is considered in conjunction with another dynamic process, the development of the embryo into the mature sporophyte. The two chapters on plant hormones and plant responses in the Fourth Edition have now been merged into one integrated chapter; new understandings of the physiological processes of plants have made such a separation increasingly artificial.

In Section 6, the Biology of Animals, we have retained the overall organizational scheme and problem-solving approach of the Fourth Edition, while significantly revising many chapters. As noted previously, animal physiology is one of the principal areas in which enormous and rapid progress is being made as a result of new discoveries at the molecular and cellular level, and we have tried to capture and share with students as much of the current excitement as possible. Although we have continued to use the human animal—inherently fascinating to most students—as our representative organism in these chapters, we have strengthened the comparative thread and made explicit much comparative material that was previously implicit.

Part 3, the Biology of Populations, covers what G. E. Hutchinson aptly described as "the ecological theater and the evolutionary play." Modern evolutionary theory and ecology are so intertwined that any separation of the two is arbitrary. We believe, however, that the student's understanding of modern ecology is deepened and enriched if it is preceded by a knowledge of the mechanisms of evolution.

In Section 7, Evolution, the five chapters of the previous edition have been reworked into four. As in the Fourth Edition, the section begins with a chapter that reviews the key points of Darwin's theory, examines the types of evidence that support evolution, and considers the changes that have occurred in evolutionary theory since Darwin's original formulation. This is followed by extensively revised chapters on the genetic basis of evolution, natural selection, and the origin of species. Then follow two chapters, also heavily revised, on the evolution of the hominids and on animal behavior and its evolution. Many of you have told us that you prefer to cover human evolution while the discussion of evolutionary mechanisms is still fresh in students' minds, and we have accordingly shifted that chapter to this section from the end of the book. Behavior is a topic for which little, if any, time is available in many courses, but it holds great interest for students, professors, and these authors alike. We have tried to provide students with a solid introduction to the contemporary study of behavior and then to focus on topics that we think are most likely to be of immediate interest and appeal to them.

Section 8, Ecology, has also been extensively revised, as we attempt to track the continual shifts, rethinkings, and controversies that characterize this most vibrant science. As in the Fourth Edition, the section moves from population dynamics, through the interactions of populations in communities and ecosystems, to the overall organization and distribution of life on earth. The text ends with a consideration of the tropical forests—the most complex and most seriously threatened of all ecological systems.

Each section ends with suggestions for further reading. Scientifically speaking, the selections are arbitrary. They were chosen not as documentation for statements in the book or as fuller presentations of difficult subjects, but rather because of their accessibility to students. Our hope is that at least some students will continue reading on their own, preferably reports not yet published about discoveries just now being dreamed of.

A number of new supplements accompany this edition of *Biology.* Of particular interest is *More Biology in the Laboratory,* by Doris R. Helms of Clemson University, an expanded version of *Biology in the Laboratory,* which accompanies the Fourth Edition of *Invitation to Biology.* A detailed Preparator's Guide accompanies the lab manual. Other supplements include *BioBytes,* a series of computer simulations by Robert Kosinski of Clemson University, a Study Guide and a Test

Bank by David J. Fox of the University of Tennessee, a new computerized test-generation system, a new and greatly expanded Instructor's Resource Manual by Debora Mann of Clemson University, and an extensive set of acetate transparencies, most of them in color.

As with previous editions, we have been deeply dependent on the advice of consultants and reviewers. In addition to her work on the new laboratory manual, Dori Helms played a major role in the revisions of the genetics section, the plant section, and the development chapter in animal physiology. She has generously shared with us her extensive knowledge, her wealth of experience in the classroom and laboratory, and her enthusiasm—all of which have been marvelous resources that we have greatly appreciated.

We are also deeply indebted to Rita Calvo of Cornell University, who reviewed a series of revisions of the genetics section; to Jacques Chiller of the Lilly Research Laboratories, who has been an invaluable source on contemporary immunology; to Mark W. Dubin of the University of Colorado, who guided us through our revision of the integration and control chapters of animal physiology; and to Manuel C. Molles, Jr., of the University of New Mexico, and Andrew Blaustein of Oregon State University, both of whom made major contributions to our revision of the evolution and ecology sections.

In addition, we have been greatly assisted by advice and counsel from the following reviewers:

BRUCE ALBERTS, University of California Medical School, San Francisco
WILLIAM E. BARSTOW, University of Georgia
CHARLES J. BIGGERS, Memphis State University
WILLIAM L. BISCHOFF, University of Toledo
ROBERT BLYSTONE, Trinity University
LEON BROWDER, University of Calgary
RALPH BUCHSBAUM, Pacific Grove, California
JAMES J. CHAMPOUX, University of Washington
JAMES COLLINS, Arizona State University
JOHN O. CORLISS, University of Maryland
MICHAEL CRAWLEY, Imperial College at Silwood Park, Ascot, England
CHARLES CURRY, University of Calgary
FRED DELCOMYN, University of Illinois, Urbana-Champaign
RUTH DOELL, San Francisco State University
RICHARD DUHRKOPF, Baylor University
DAVID DUVALL, University of Wyoming
JUDI ELLZEY, University of Texas, El Paso
ROBERT C. EVANS, Rutgers University, Camden
RAY F. EVERT, University of Wisconsin
KATHLEEN FISHER, University of California, Davis
ROBERT P. GEORGE, University of Wyoming
URSULA GOODENOUGH, Washington University, St. Louis
PATRICIA GOWATY, Clemson University
LINDA HANSFORD, Baltimore, Maryland
JEAN B. HARRISON, University of California, Los Angeles
STEVEN HEIDEMANN, Michigan State University
MERRILL HILLE, University of Washington
GERALD KARP, San Francisco, California
JOHN KIRSCH, University of Wisconsin
ROBERT M. KITCHIN, University of Wyoming
KAREL LIEM, Harvard University
JANE LUBCHENCO, Oregon State University
R. WILLIAM MARKS, Villanova University

LARRY R. McEDWARD, University of Washington
SUE ANN MILLER, Hamilton College
RANDY MOORE, Wright State University
BETTE NICOTRI, University of Washington
JAMES PLATT, University of Denver
FRANK E. PRICE, Hamilton College
EDWARD RUPPERT, Clemson University
TOM K. SCOTT, University of North Carolina, Chapel Hill
LARRY SELLERS, Louisiana Tech University
DAVID G. SHAPPIRIO, University of Michigan
JOHN SMARRELLI, Loyola University, Chicago
GILBERT D. STARKS, Central Michigan University
IAN TATTERSALL, American Museum of Natural History
ROBERT VAN BUSKIRK, State University of New York, Binghamton
ERIC WEINBERG, University of Pennsylvania
JOHN WEST, University of California, Berkeley
ARTHUR WINFREE, University of Arizona

As always, the preparation of a new edition is a staggering and complex task, and its successful completion has depended on the efforts of many highly talented individuals. In particular, we wish to thank Shirley Baty, who, in addition to preparing many new illustrations for this edition, has reworked virtually all of the Fourth Edition art as we converted the book to full color throughout; David Hinchman, Anne Feldman, and Elaine Bernstein, who have located an enormous number of marvelous new photographs and micrographs; John Timpane, who prepared the comprehensive index; George Touloumes and the members of his staff who are responsible for the design and layout of each page of the book; Sarah Segal, who has managed the production process and somehow kept us all on course; and Sally Anderson, our extraordinary editor, and her capable assistant, Lindsey Bowman. Sally's editorial expertise, her thorough knowledge of biology in general and of this text in particular, and her long experience in working with us both have played an incalculable role in the successful completion of this revision. And, a special thank you to Bob Worth, whose vision and constant support have made it all possible.

Finally, we want to thank all of the professors and students who have written to us, some with criticisms, some with suggestions, some with questions, and some simply because they enjoyed the book. These letters serve to remind us of how privileged we are to be writing for young people. We continue to appreciate their curiosity, their energy, their imaginativeness, and their dislike of the pompous and pedantic. We hope we serve them well.

New York   Helena Curtis

December, 1988   N. Sue Barnes

An added note: As you may have noticed, with this edition of *Biology,* N. Sue Barnes is listed as coauthor. This is a recognition long overdue. Sue has been a member of the team for eleven years now. Over this period of time, she has assumed increasing responsibility for the revisions of both *Biology* and *Invitation to Biology* (on which she has been listed as coauthor for the last two editions). The Fifth Edition of *Biology* would have been impossible without her. In addition, I wish to express my personal gratitude for her integrity, patience, fortitude, and good spirits—and for the fact that she always comes through.

                                                                  H.C.

BIOLOGY

Introduction

In 1831, the young Charles Darwin set sail from England on what was to prove the most consequential voyage in the history of biology. Not yet 23, Darwin had already abandoned a proposed career in medicine—he described himself as fleeing a surgical theater in which an operation was being performed on an unanesthetized child—and was a reluctant candidate for the clergy, a profession deemed suitable for the younger son of an English gentleman. An indifferent student, Darwin was an ardent hunter and horseman, a collector of beetles, mollusks, and shells, and an amateur botanist and geologist. When the captain of the surveying ship H.M.S. *Beagle*, himself only a little older than Darwin, offered passage for a young gentleman who would volunteer to go without pay, Darwin eagerly seized this opportunity to pursue his interest in natural history. The voyage, which lasted five years, shaped the course of Darwin's future work. He returned to an inherited fortune, an estate in the English countryside, and a lifetime of independent work and study that radically changed our view of life and of our place in the living world.

THE ROAD TO EVOLUTIONARY THEORY

That Darwin was the founder of the modern theory of evolution is well known. Although he was not the first to propose that organisms **evolve**—or change—through time, he was the first to amass a large body of supporting evidence and the first to propose a valid mechanism by which evolution might occur. In order to understand the meaning and significance of Darwin's theory, it is useful to look at the intellectual climate in which it was formulated.

Aristotle (384–322 B.C.), the first great biologist, believed that all living things could be arranged in a hierarchy. This hierarchy became known as the *Scala Naturae*, or ladder of nature, in which the simplest creatures had a humble position on the bottommost rung, man occupied the top rung, and all other organisms had their proper places in between. Until the late nineteenth century, many biologists believed in such a natural hierarchy. But whereas to Aristotle living organisms had always existed, the later biologists (at least those of the Occidental world) believed, in harmony with the teachings of the Old Testament, that all living things were the products of a divine creation. They believed, moreover, that most were created for the service or pleasure of mankind.

That each type of living thing came into existence in its present form—specially and specifically created—was a compelling idea. How else could one explain the astonishing extent to which every living thing was adapted to its environment and to its role in nature? It was not only the authority of the Church but also, so it seemed, the evidence before one's own eyes that gave such strength to the concept of special creation.

I–1 *When Charles Darwin visited the Galapagos archipelago, he found that each major island had its own variety of tortoise, so distinct from the others that it was easily recognized by local sailors and fishermen. This was one of the clues that led him to the formulation of the theory of evolution.*

The Galapagos consists of 13 volcanic islands that pushed up from the sea more than a million years ago. The major vegetation is thornbush and cactus, and the original black basaltic lava is often visible, as it is beneath the lumbering feet of this tortoise on Hood Island—"what we might imagine the cultivated parts of the Infernal regions to be," young Darwin wrote in his diary.

Among those who believed in divine creation was Carolus Linnaeus (1707–1778), the great Swedish naturalist who devised our present system of nomenclature for species, or kinds, of organisms. In 1753, Linnaeus published *Species Plantarum*, which described, in two encyclopedic volumes, every species of plant known at the time. Even as Linnaeus was at work on this massive project, explorers were returning to Europe from Africa and the New World with previously undescribed plants and animals and even, apparently, new kinds of human beings. Linnaeus revised edition after edition to accommodate these findings, but he did not change his opinion that all species now in existence were created by the sixth day of God's labor and have remained fixed ever since. During Linnaeus's time, however, it became clear that the pattern of creation was far more complex than had been originally envisioned.

Evolution before Darwin

The idea that organisms might evolve through time, with one type of organism giving rise to another type of organism, is an ancient one, predating Aristotle. A school of Greek philosophy, founded by Anaximander (611–547 B.C.) and culminating in the writings of the Roman Lucretius (99–55 B.C.), developed not only an atomic theory but also an evolutionary theory, both of which are strikingly similar to modern conceptions. The work of this school, however, was largely unknown in Europe at the time that the science of biology, as we know it today, began to take form.

In the eighteenth century, the French scientist Georges-Louis Leclerc de Buffon (1707–1788) was among the first to propose that species might undergo changes in the course of time. He suggested that, in addition to the numerous creatures that were produced by divine creation at the beginning of the world, "there are lesser families conceived by Nature and produced by Time." Buffon believed that these changes took place by a process of degeneration. In fact, as he summed it up, ". . . improvement and degeneration are the same thing, for both imply an alteration of the original constitution." Buffon's hypothesis, although vague as to the way in which changes might occur, did attempt to explain the bewildering variety of creatures in the modern world.

Another doubter of fixed and unchanging species was Erasmus Darwin (1731–1802), Charles Darwin's grandfather. Erasmus Darwin was a physician, a gentleman naturalist, and a prolific writer, often in verse, on both botany and zoology. He suggested, largely in asides and footnotes, that species have historical connections with one another, that animals may change in response to their environment, and that their offspring may inherit these changes. He maintained, for instance, that a polar bear is an "ordinary" bear that, by living in the Arctic, became modified and passed the modifications along to its cubs. These ideas were never clearly formulated but are interesting because of their possible effects on Charles Darwin, although the latter, born after his grandfather died, did not profess to hold his grandfather's views in high esteem.

The Age of the Earth

It was geologists, more than biologists, who paved the way for modern evolutionary theory. One of the most influential of these was James Hutton (1726–1797). Hutton proposed that the earth had been molded not by sudden, violent events but by slow and gradual processes—wind, weather, and the flow of water—the same processes that can be seen at work in the world today. This theory of Hutton's, which was known as uniformitarianism, was important for three reasons. First, it implied that the earth has a long history, which was a new idea to eighteenth-century Europeans. Christian theologians, by counting the successive generations since Adam (as recorded in the Bible), had calculated the maximum

I–2 *Charles Darwin in 1840, four years after he returned from his five-year voyage on H.M.S. Beagle. In his later book,* The Voyage of the *Beagle, Darwin made the following comments about his selection for the voyage: "Afterwards, on becoming very intimate with Fitz Roy [the captain of the Beagle], I heard that I had run a very narrow risk of being rejected on account of the shape of my nose! He . . . was convinced that he could judge of a man's character by the outline of his features; and he doubted whether anyone with my nose could possess sufficient energy and determination for the voyage. But I think he was afterwards well satisfied that my nose had spoken falsely."*

I-3 *While the* Beagle *sailed up the west coast of South America, Darwin explored the Andes on foot and horseback. He saw geological strata such as these, discovered fossil sea shells at about 3,700 meters (12,000 feet), and was witness to the upheaval of the earth produced by a major earthquake that occurred while he was there. In 1846, he published a book on his geological observations in South America. Strata are now seen as pages in evolutionary history.*

I-4 *Particular strata, even though widely separated geographically, have characteristic assemblages of fossils. These fossil trilobites from the Devonian period (360 to 408 million years ago) were found in strata in (***a***) Ohio, (***b***) Oklahoma, and (***c***) upstate New York.*

age of the earth at about 6,000 years. As far as we know, no one since the followers of Anaximander (whose school maintained that the earth was infinitely old) had thought in terms of a longer period. Yet 6,000 years is far too short for major evolutionary changes to take place, by any theory. Second, the theory of uniformitarianism stated that change is itself the *normal* course of events, as opposed to a static system interrupted by an occasional unusual event, such as an earthquake. Third, although this was never explicit, uniformitarianism suggested that there might be alternatives to the literal interpretation of the Bible.

The Fossil Record

During the latter part of the eighteenth century, there was a revival of interest in fossils, which are the preserved remains of organisms long since deceased. In previous centuries, fossils had been collected as curiosities, but they had generally been regarded either as accidents of nature—stones that somehow looked like shells—or as evidence of great catastrophes, such as the Flood described in the Old Testament. The English surveyor William Smith (1769–1839) was among the first to study the distribution of fossils scientifically. Whenever his work took him down into a mine or along canals or cross-country, he carefully noted the order of the different layers of rock, known as geological strata, and collected the fossils from each layer. He eventually established that each stratum, no matter where he came across it in England, contained characteristic kinds of fossils and that these fossils were actually the best way to identify a particular stratum in a number of different geographic locations. (The use of fossils to identify strata is still widely practiced, for instance, by geologists looking for oil.) Smith did not interpret his findings, but the implication that the present surface of the earth had been formed layer by layer over the course of time was an unavoidable one.

Like Hutton's world, the world seen and described by William Smith was clearly a very ancient one. A revolution in geology was beginning; earth science was becoming a study of time and change rather than a mere cataloging of types of rocks. As a consequence, the history of the earth became inseparable from the history of living organisms, as revealed in the fossil record.

Catastrophism

Although the way was being prepared by the revolution in geology, the time was not yet ripe for a parallel revolution in biology. The dominating force in European science in the early nineteenth century was Georges Cuvier (1769–1832). Cuvier was the founder of vertebrate paleontology, the scientific study of the fossil record of vertebrates (animals with backbones). An expert in anatomy and zoology, he applied his knowledge of the way in which animals are constructed to the study of fossil animals, and he was able to make brilliant deductions about the

(a)

(b)

(c)

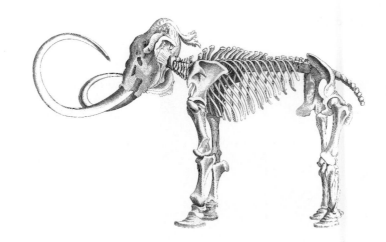

I–5 *A drawing by Georges Cuvier of a mastodon. Although Cuvier was one of the world's experts in reconstructing extinct animals from their fossil remains, he was a powerful opponent of evolutionary theories.*

I–6 *According to Lamarck's hypothesis —now known to be in error—as giraffes stretched to reach the high branches, their necks lengthened, and this acquired characteristic was transmitted to their offspring.*

form of an entire animal from a few fragments of bone. Today we think of paleontology and evolution as so closely connected that it is surprising to learn that Cuvier was a staunch and powerful opponent of evolutionary theories. He recognized the fact that many species that had once existed no longer did. (In fact, according to modern estimates, considerably less than one percent of all species that have ever lived are represented on the earth today.) Cuvier explained the extinction of species by postulating a series of catastrophes. After each catastrophe, the most recent of which was the Flood, new species filled the vacancies.

Cuvier hedged somewhat on the source of the new animals and plants that appeared after the extinction of older forms; he was inclined to believe they moved in from parts unknown. Another major opponent of evolution, Louis Agassiz (1807–1873), America's leading nineteenth-century biologist, was more straightforward. According to Agassiz, the fossil record revealed 50 to 80 total extinctions of life, followed by an equal number of new, separate creations.

The Concepts of Lamarck

The first modern scientist to work out a systematic concept of evolution was Jean Baptiste Lamarck (1744–1829). "This justly celebrated naturalist," as Darwin himself referred to him, boldly proposed in 1801 that all species, including *Homo sapiens,* are descended from other species. Lamarck, unlike most of the other zoologists of his time, was particularly interested in one-celled organisms and invertebrates (animals without backbones). Undoubtedly it was his long study of these forms of life that led him to think of living things in terms of constantly increasing complexity, each species derived from an earlier, less complex one.

Like Cuvier and others, Lamarck noted that older rocks generally contained fossils of simpler forms of life. Unlike Cuvier, however, Lamarck interpreted this as meaning that the more complex forms had arisen from the simpler forms by a kind of progression. According to his hypothesis, this progression, or evolution, to use the modern term, is dependent on two main forces. The first is the inheritance of acquired characteristics. Organs in animals become stronger or weaker, more or less important, through use or disuse, and these changes, according to Lamarck's proposal, are transmitted from the parents to the progeny. His most famous example was the evolution of the giraffe. According to Lamarck, the modern giraffe evolved from ancestors that stretched their necks to reach leaves on high branches. These ancestors transmitted the longer necks—acquired by stretching—to their offspring, which stretched their necks even longer, and so on.

The second, equally important force in Lamarck's concept of evolution was a universal creative principle, an unconscious striving upward on the *Scala Naturae* that moved every living creature toward greater complexity. Every amoeba was on its way to man. Some might get waylaid—the orangutan, for instance, had been diverted from its course by being caught in an unfavorable environment—but the will was always present. Life in its simplest forms was constantly emerging by spontaneous generation to fill the void left at the bottom of the ladder. In Lamarck's formulation, Aristotle's ladder of nature had been transformed into a steadily ascending escalator powered by a universal will.

Lamarck's contemporaries did not object to his ideas about the inheritance of acquired characteristics, which we, with our present knowledge of genetics, know to be false. Nor did they criticize his belief in a metaphysical force, which was actually a common element in many of the concepts of the time. But these vague, untestable postulates provided a very shaky foundation for the radical proposal that more complex forms evolved from simpler forms. Moreover, Lamarck personally was no match for the brilliant and witty Cuvier, who relentlessly attacked his ideas. As a result, Lamarck's career was ruined, and both scientists and the public became even less prepared to accept any evolutionary doctrine.

DEVELOPMENT OF DARWIN'S THEORY

The Earth Has a History

The person who most influenced Darwin, it is generally agreed, was Charles Lyell (1797–1875), a geologist who was Darwin's senior by 12 years. One of the books that Darwin took with him on his voyage was the first volume of Lyell's newly published *Principles of Geology,* and the second volume was sent to him while he was on the *Beagle.* On the basis of his own observations and those of his predecessors, Lyell opposed the theory of catastrophes. Instead, he produced new evidence in support of Hutton's earlier theory of uniformitarianism. According to Lyell, the slow, steady, and cumulative effect of natural forces had produced continuous change in the course of the earth's history. Since this process is demonstrably slow, its results being barely visible in a single lifetime, it must have been going on for a very long time. What Darwin's theory needed was time, and it was time that Lyell gave him. In the words of Ernst Mayr of Harvard University, the discovery that the earth was ancient "was the snowball that started the whole avalanche."

The Voyage of the *Beagle*

This, then, was the intellectual climate in which Charles Darwin set sail from England. As the *Beagle* moved down the Atlantic coast of South America, through the Strait of Magellan, and up the Pacific coast, Darwin traveled the interior. He explored the rich fossil beds of South America (with the theories of Lyell fresh in his mind) and collected specimens of the many new kinds of plant and animal life

I–7 (a) *A reproduction of the* Beagle, *sailing off the coast of South America.* (b) *Cutaway view of the ship. Only 28 meters in length, this "good little vessel" set sail on its five-year voyage with 74 people aboard. Darwin shared the poop cabin with a midshipman and 22 chronometers belonging to Captain Fitz Roy, who had a passion for exactness. Darwin's sleeping space was so confined that he had to remove a drawer from a locker to make room for his feet.*

(a)

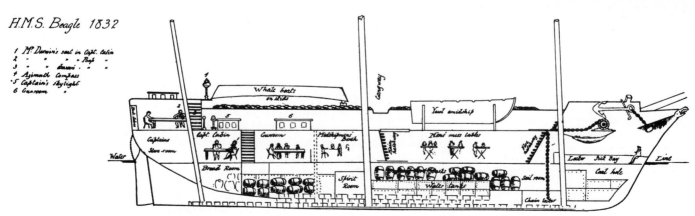

(b)

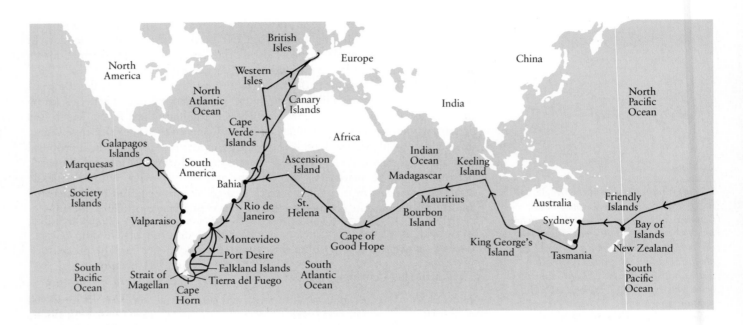

I–8 *The* Beagle's *voyage. The ship left England in December of 1831 and arrived at Bahia, Brazil, in late February of 1832. About 3¹/₂ years were spent along the coast of South America, surveying and making inland explorations. The stop at* the Galapagos Islands was for slightly more than a month, and, during that brief time, Darwin made the wealth of observations that were to change the course of the science of biology. The remainder of the voyage, across the Pacific to New Zealand and Australia, across the Indian Ocean to the Cape of Good Hope, back to Bahia once more, and at last home to England, occupied another year.

I–9 *A distinguishing feature of the Galapagos tortoise is the shape of its carapace, or shell, which varies according to its island of origin. The tortoises found on the islands with comparatively lush vegetation are characterized by a domed shell, shown here, which affords protection of the tortoise's soft parts as it makes its way through the thick undergrowth. The high arch at the front of the saddleback shell (see Figure I–1) enables the tortoise to reach upward in search of food; such shells are typical of tortoises living on arid islands where food may be scarce.*

he encountered. He was impressed most strongly during his long, slow trip down one coast and up the other by the constantly changing varieties of organisms he saw. The birds and other animals on the west coast, for example, were very different from those on the east coast, and even as he moved slowly up the western coast, one species would give way to another.

Most interesting to Darwin were the animals and plants that inhabited a small, barren group of islands, the Galapagos, which lie some 950 kilometers off the coast of Ecuador. The Galapagos were named after the islands' most striking inhabitants, the tortoises (*galápagos* in Spanish), some of which weigh 100 kilograms or more. Each island has its own type of tortoise; sailors who took these tortoises on board and kept them as convenient sources of fresh meat on their sea voyages could readily tell which island any particular tortoise had come from. Then there was a group of finchlike birds, 13 species in all, that differed from one another in the sizes and shapes of their bodies and beaks, and particularly in the type of food they ate. In fact, although clearly finches, they had many characteristics seen only in completely different types of birds on the mainland. One finch, for example, feeds by routing insects out of the bark of trees. It is not fully equipped for this, however, lacking the long tongue with which the true woodpecker flicks out insects from under the bark. Instead, the woodpecker finch uses a small stick or cactus spine to pry the insects loose.

From his knowledge of geology, Darwin knew that these islands, clearly of volcanic origin, were much younger than the mainland. Yet the plants and animals of the islands were different from those of the mainland, and in fact the inhabitants of different islands in the archipelago differed from one another. Were the living things on each island the product of a separate special creation? "One might really fancy," Darwin mused at a later date, "that from an original paucity of birds in this archipelago one species had been taken and modified for different ends." This problem continued, in his own word, to "haunt" him.

(a)

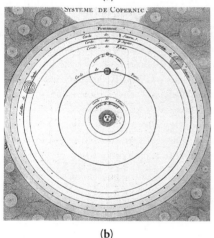

(b)

I–10 (a) *A view of the universe first proposed by the early Greeks and accepted throughout the Middle Ages. In this colored woodcut from Martin Luther's Bible, dated 1534, earth is in the center of the universe, surrounded by a layer of air containing clouds, stars, planets, the sun, and the moon. Beyond this is an outer layer of fire.* (b) *The solar system, as proposed by Nicholas Copernicus. In 1543, Copernicus set forth in* De Revolutionibus *the new concept that the sun, not the earth, is the center of the solar system. His theory was supported by the German astronomer Johannes Kepler (1571–1630), who discovered the laws of planetary motion, and by the Italian Galileo Galilei (1564–1642). The latter spent the last 10 years of his life confined to his home for heresy because of his advocacy of Copernican beliefs.*

The Darwinian Theory

Darwin was an assiduous and voracious reader. Not long after his return, he came across a short but much talked about sociological treatise by the Reverend Thomas Malthus that had first appeared in 1798. In this essay, Malthus warned, as economists have warned ever since, that the human population was increasing so rapidly that it would soon be impossible to feed all the earth's inhabitants. Darwin saw that Malthus's conclusion—that food supply and other factors hold populations in check—is true for all species, not just the human one. For example, Darwin calculated that a single breeding pair of elephants, which are among the slowest reproducers of all animals, would, if all their progeny lived and reproduced the normal number of offspring over a normal life span, produce a standing population of 19 million elephants in 750 years, yet the average number of elephants generally remains the same over the years. So, although a single breeding pair could have, in theory, produced 19 million descendants, it did, in fact, produce an average of only two. But why these particular two? The process by which the two survivors are "chosen" Darwin called **natural selection.**

Natural selection, according to Darwin, was a process analogous to the type of selection exercised by breeders of cattle, horses, or dogs. In artificial selection, we humans choose individual specimens of plants or animals for breeding on the basis of characteristics that seem to us desirable. In natural selection, the environment takes the place of human choice. As individuals with certain hereditary characteristics survive and reproduce and individuals with other hereditary characteristics are eliminated, the population will slowly change. If some horses were swifter than others, for example, these individuals would be more likely to escape predators and survive, and their progeny, in turn, might be swifter, and so on.

According to Darwin, inherited variations among individuals, which occur in every natural population, are a matter of chance. They are not produced by the environment, by a "creative force," or by the unconscious striving of the organism. In themselves, they have no goal or direction, but they often have positive or negative adaptive values; that is, they may be more or less useful to an organism as measured by its survival and reproduction. It is the operation of natural selection —the interaction of individual organisms with their environment—over a series of generations that gives direction to evolution. A variation that gives an organism even a slight advantage makes that organism more likely to leave surviving offspring. Thus, to return to Lamarck's giraffe, an animal with a slightly longer neck may have an advantage in feeding and thus be likely to leave more offspring than one with a shorter neck. If the longer neck is an inherited characteristic, some of these offspring will also have long necks, and if the long-necked animals in this generation have an advantage, the next generation will include more long-necked individuals. Finally, the population of short-necked giraffes will have become a population of longer-necked ones (although there will still be variations in neck length).

As you can see, the essential difference between Darwin's formulation and that of any of his predecessors is the central role he gave to variation. Others had thought of variations as mere disturbances in the overall design, whereas Darwin saw that variations among individuals are the real fabric of the evolutionary process. Species arise, he proposed, when differences among individuals within a group are gradually converted into differences between groups as the groups become separated in space and time.

The Origin of Species, which Darwin pondered for more than 20 years after his return to England, is, in his own words, "one long argument." Fact after fact, observation after observation, culled from the most remote Pacific island to a neighbor's pasture, is recorded, analyzed, and commented upon. Every objection is weighed, anticipated, and countered. *The Origin of Species* was published on November 24, 1859, and the Western world has not been the same since.

Darwin's Long Delay

Darwin returned to England with the *Beagle* in 1836. Two years later, he read the essay by Malthus, and in 1842 he wrote a preliminary sketch of his theory, which he revised in 1844. On completing the revision, he wrote a formal letter to his wife requesting her, in the event of his death, to publish the manuscript (which was some 230 pages long). Then, with the manuscript and letter in safekeeping, he turned to other work, including a four-volume treatise on barnacles. For more than 20 years following his return from the Galapagos, Darwin mentioned his ideas on evolution only in his private notebooks and in letters to his scientific colleagues.

In 1856, urged on by his friends Charles Lyell and botanist Joseph Hooker, Darwin set slowly to work preparing a manuscript for publication. In 1858, some 10 chapters later, Darwin received a letter from the Malay archipelago from another English naturalist, Alfred Russel Wallace, who had corresponded with Darwin on several previous occasions. Wallace presented a theory of evolution that exactly paralleled Darwin's own. Like Darwin, Wallace had traveled extensively and also had read Malthus's essay. Wallace, tossing in bed one night with a fever, had a sudden flash of insight. "Then I saw at once," Wallace recollected, "that the ever-present variability of all living things would furnish the material from which,

by the mere weeding out of those less adapted to the actual conditions, the fittest alone would continue the race." Within two days, Wallace's 20-page manuscript was completed and in the mail.

When Darwin received Wallace's letter, he turned to his friends for advice, and Lyell and Hooker, taking matters into their own hands, presented the theory of Darwin and Wallace at a scientific meeting just one month later. (Darwin described Wallace as "noble and generous," as indeed he was.) Lyell and Hooker read four papers from Darwin's notes of 1844, excerpts from two letters written by Darwin, and Wallace's manuscript. Their presentation received little attention, but for Darwin the floodgates were opened. He finished his long treatise in little more than a year, and the book was finally published. The first printing was a mere 1,250 copies, but they were sold out the same day.

Why Darwin's long delay? His own writings, voluminous though they are, shed little light on this question. But perhaps his background does. He came from a conventionally devout family, and he himself had been a divinity student. Perhaps most important, his wife, to whom he was deeply devoted, was extremely religious. It is difficult to avoid the speculation that Darwin, like so many others, found the implications of his theory difficult to confront.

Alfred Russel Wallace (1823–1913). As a young man, Wallace explored the Malay archipelago for eight years, covering about 22,500 kilometers (14,000 miles) by foot and native canoe. During his stay there, he collected 125,000 specimens of plants and animals, many of them previously unknown. His book about his Malay travels bears this inscription: "To Charles Darwin, Author of 'The Origin of Species,' I dedicate this book, not only as a token of personal esteem and friendship but also to express my deep admiration for his genius and his works."

(a)

(b)

I–11 *According to biochemical tests, made possible by new genetic engineering techniques, there is a close evolutionary link between (a) the woolly mammoth, a creature that roamed North America, Asia, and Europe thousands of years ago, and (b) the modern elephant. Several years ago, a baby woolly mammoth that died about 40,000 years ago was found frozen in Siberia. Its tissues were so perfectly preserved that the exact structure of certain key molecules could be determined and compared with the structure of the same molecules in living elephants.*

Acceptance of Darwin's argument revolutionized the science of biology. "The theory of evolution," in the words of Ernst Mayr, "is quite rightly called the greatest unifying theory in biology." As we shall see throughout this text, it is the thread that links together all the diverse phenomena of the living world. It also deeply influenced our way of thinking about ourselves. With the possible exception of the new astronomy of Copernicus and Galileo in the sixteenth and seventeenth centuries, no revolution in scientific thought has had as much effect on human culture as this one. One reason is, of course, that evolution is in contradiction to the literal interpretation of the Bible. Another difficulty is that it seems to diminish human significance. The new astronomy had made it clear that the earth is not the center of the universe or even of our own solar system. Then the new biology asked us to accept the proposition that, as far as science can show, we are not fundamentally different from other organisms in either our origins or our place in the natural world.

Challenges to Evolutionary Theory

Today, with almost no exceptions, modern biologists are convinced by a vast body of accumulated evidence that the earth has a long history and that all living organisms, including ourselves, arose in the course of that history from earlier, more primitive forms. This accumulated evidence consists of an interlocking fabric of thousands upon thousands of pieces of data concerning past and present organisms, including not only anatomical structure but also physiological and biochemical processes, patterns of embryonic development, patterns of behavior, and most recently, the sequences of genetic information encoded in the DNA molecules of the chromosomes.

Yet, as everyone who reads a newspaper or watches television knows, evolutionary theory remains a matter of lively public controversy. Moreover, the advocates of special creation—who maintain that each species was created separately—seek to lend strength to their arguments from the fact that scientists ask many questions about evolution. They point out that even among scientists, evolution is "only a theory," and that even leading scientists do not agree on this "theory." Much of the confusion surrounding this controversy stems from the very definition of the word "theory," and from a misunderstanding of the nature and limitations of the scientific process, topics that we shall consider later in this Introduction. Among biologists, there is almost unanimous agreement that evolution has occurred in the past and continues to occur today. As we shall see in Section 7, however, the details and relative importance of the different processes involved in evolutionary change are currently the subject of intense research and discussion among biologists.

UNIFYING PRINCIPLES OF MODERN BIOLOGY

The foundations of modern biology include not only evolution but also three other principles that are so well established that biologists seldom discuss them among themselves. You could read widely in the current biological literature without seeing any of them mentioned, yet it is impossible to understand either the ideas or the data of contemporary biology without being aware of them. These principles, like evolution, will be discussed in greater detail in the course of this text and will recur as major themes, but you should have them in mind from the outset.

All Organisms Are Made Up of Cells

One of the fundamental principles of biology is that all living organisms are composed of one or more similar units, known as **cells.** This concept is of tremendous and central importance to biology because it emphasizes the basic sameness of all living systems. It therefore brings an underlying unity to widely varied studies involving many different kinds of organisms.

The word "cell" was first used in a biological sense some 300 years ago. In the seventeenth century, the English scientist Robert Hooke, using a microscope of his own construction, noticed that cork and other plant tissues are made up of small cavities separated by walls. He called these cavities "cells," meaning "little rooms." However, "cell" did not take on its present meaning—the basic unit of living matter—for more than 150 years.

In 1838, Matthias Schleiden, a German botanist, came to the conclusion that all plant tissues consist of organized masses of cells. In the following year, zoologist Theodor Schwann extended Schleiden's observations to animal tissues and proposed a cellular basis for all life. In 1858, the idea that all living organisms are composed of one or more cells took on an even broader significance when the great pathologist Rudolf Virchow generalized that cells can arise only from preexisting cells: "Where a cell exists, there must have been a preexisting cell, just as the animal arises only from an animal and the plant only from a plant. . . . Throughout the whole series of living forms, whether entire animal or plant organisms or their component parts, there rules an eternal law of continuous development."

From the perspective provided by Darwin's theory of evolution, published in the following year, Virchow's concept takes on an even larger significance. There is an unbroken continuity between modern cells—and the organisms they compose—and the primitive cells that first appeared on earth more than 3 billion years ago.

All Organisms Obey the Laws of Physics and Chemistry

Until fairly recently, many prominent biologists believed that living systems are qualitatively different from nonliving ones, containing within them a "vital spirit" that enables them to perform activities that cannot be carried on outside the living organism. This concept is known as vitalism and its proponents as vitalists.

In the seventeenth century, the vitalists were opposed by a group known as the mechanists. The French philosopher René Descartes (1596–1650) was a leading proponent of this point of view. The mechanists set about showing that the body worked essentially like a machine; the arms and legs move like levers, the heart like a pump, the lungs like a bellows, and the stomach like a mortar and pestle. Although these simple mechanical models provided much insight into the functioning of the animal body, by the nineteenth century the debate about the distinctiveness of living systems had moved beyond them. The argument became centered on whether or not the chemistry of living organisms was governed by the

(a)

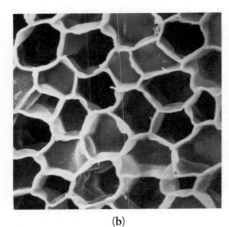

(b)

I–12 (a) *Robert Hooke's drawings of two slices of a piece of cork, reproduced from his* Micrographica, *published in 1665, and* (b) *a scanning electron micrograph of a slice of cork. Hooke was the first to use the word "cells" to describe the tiny compartments that together make up an organism. The cells in these pieces of cork have died—all that remain are the outer walls. As we shall see in subsequent chapters, the living cell is filled with a variety of substances, organized into distinct structures and carrying out a multitude of essential processes.*

I-13 *Wöhler was a student in this laboratory at Giessen, in what is now West Germany. It was one of the first where practical work in chemistry could be done.*

same principles as the chemistry performed in the laboratory. The vitalists claimed that the chemical operations performed by living tissues could not be carried out experimentally in the laboratory, categorizing reactions as either "chemical" or "vital." Their new opponents, known as reductionists (since they believed that the complex operations of living systems could be reduced to simpler and more readily understandable ones), achieved a partial victory when the German chemist Friedrich Wöhler (1800–1882) converted an "inorganic" substance (ammonium cyanate) into a familiar organic substance (urea). On the other hand, the claims of the vitalists were supported by the fact that, as chemical knowledge improved, many new compounds were found in living tissues that were never seen in the nonliving, or inorganic, world.

In the late 1800s, the leading vitalist was Louis Pasteur, who claimed that the changes that took place when fruit juice was transformed to wine were "vital" and could be carried out only by living cells—the cells of yeast. In spite of many advances in chemistry, this phase of the controversy lasted until almost the turn of the century. However, in 1898, the German chemists Eduard and Hans Büchner showed that a substance extracted from yeast cells could produce fermentation outside the living cell. (This substance was given the name enzyme, from *zyme*, the Greek word meaning "yeast" or "ferment.") A "vital" reaction was demonstrated to be a chemical one, and the subject was eventually laid to rest. Today it is generally accepted that living systems "obey the rules" of chemistry and physics, and modern biologists no longer believe in a "vital principle."

The realization that living systems obey the laws of physics and chemistry opened a new era in the history of biology. Increasingly organisms were studied in terms of their chemical composition and the chemical reactions that take place within their bodies. These studies, which continue today at an extraordinary pace, have yielded a vast amount of information and provide an essential foundation for contemporary biology. Perhaps their greatest test came about 40 years ago. One of the most striking characteristics of living things is their capacity to reproduce, to generate faithful copies of themselves. By about 1950, this capacity had been shown to reside in a single type of chemical molecule, deoxyribonucleic acid (DNA). The race to discover the structure of this molecule began, and the question in everyone's mind was whether or not the structure of this one "simple" molecule could possibly explain the mysteries of heredity. As it turned out, and as we shall discuss in Section 3, it could.

I–14 *The acquisition of energy resources to power life processes is of fundamental importance for all living organisms. These cheetahs, photographed in Kenya, have caught and killed a springbok, on which they will feed until sated. The remainder of the carcass—and the stored chemical energy it contains—will then be abandoned by the cheetahs, which rarely return to a previous day's kill. Cheetahs typically hunt in small groups, stealthily stalking their prey and then rushing at the intended victim at high speed. Over short distances, the cheetah is the fastest land animal in the world. It can attain speeds of 110 kilometers per hour but can maintain such speeds for little more than 30 seconds (covering a distance of about half a kilometer).*

All Organisms Require Energy

Among the laws of physics that are pertinent to biology are the laws of thermodynamics. They state simply that (1) energy can be changed from one form to another but it cannot be created or destroyed, that is, the total energy of the universe remains constant; and (2) all natural events proceed in such a way that concentrations of energy tend to dissipate or become random. A heated object, which is one example of concentrated energy, loses its heat to its surroundings.

A living system, which is a concentration of energy of another kind, can maintain itself in the face of this tendency only by a constant intake of energy. Living organisms are experts at energy conversion. The energy they take in— whether in the form of sunlight or chemical energy stored in food—is transformed and used by each individual cell to do the work of the cell. This work includes powering not only the numerous processes that constitute the activities of the organism but also the synthesis of an enormous diversity of molecules and cellular structures. In the course of the cell's work, the energy may be further transformed to energy of motion, to heat energy, or even back to light energy again. It is ultimately dissipated, and the organism must take in more energy.

This flow of energy is the essence of life. A cell can be best understood as a complex of systems for transforming energy. At the other end of the biological scale, the structure of the biosphere—that is, the entire living world—is determined by the energy exchanges occurring among the groups of organisms within it. Similarly, evolution may be viewed as a competition among organisms for the most efficient use of energy resources.

THE FORMS OF LIFE

One of the principal consequences of the evolutionary competition is an incredible diversity in the living world. It is estimated that we share this planet with more than 5 million different species of organisms. These different organisms exhibit great variety in the organization of their bodies, in their patterns of reproduction, growth, and development, and in their behavior.

I–15 (a) *The smallest and most numerous living organisms known to biology are the prokaryotes, which include the bacteria and closely related forms. Despite their small size, bacteria, like all living things, are highly organized and contain a variety of structures.* Neisseria gonorrhoeae (*dark spheres*), *the causative agent of gonorrhea, is shown here being ingested by white blood cells.*

(b) *Although significantly more complex than prokaryotes, protists are generally quite small. This drop of water, from a freshwater lake, contains at least four different kinds of protists. The bright yellow-green organism is an alga, which manufactures its own food by photosynthesis. The larger, blue organisms are protozoans; they feed on bacteria and algae.*

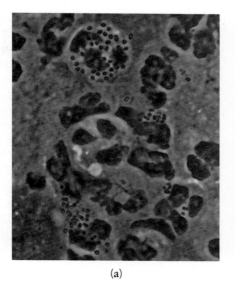

(a)

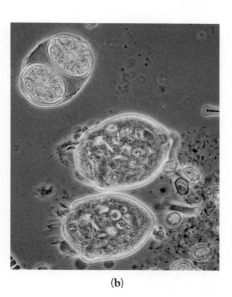

(b)

I–16 *This mushroom, growing on a forest floor, is the reproductive structure of a fungus. The bulk of the body of the fungus consists of underground filaments through which nutrients are absorbed.*

Despite the seemingly overwhelming diversity of living organisms, it is possible to group them in ways that reveal not only patterns of similarity and difference but also historical relationships among different groups. In Section 4, we shall consider these patterns and relationships in some detail. Before reaching Section 4, however, we shall encounter a marvelous variety of organisms. Thus you will find it helpful to be aware of the five major categories, or kingdoms, into which we group organisms in this text.

The first of these, which includes the earliest life forms known to have appeared on this planet, is the kingdom Monera. It comprises the smallest and simplest of all organisms, the bacteria and their relatives (Figure I–15a). Each individual organism consists of a single structural unit—a cell. This group of one-celled organisms makes up the prokaryotes. The term "prokaryote" means "before the nucleus" and refers to the internal organization of the cells, which have neither a clearly defined nucleus nor the other structures that can be found in all other kinds of cells. The first prokaryotes made their appearance at least 3.5 billion years ago, when the earth was very different from the green planet we know today, and prokaryotes were the earth's sole inhabitants for more than 2 billion years.

The second major category of life is known as the kingdom Protista. Protists are also mostly one-celled, but the cells are structurally very different from the prokaryotes. They are known as eukaryotes, meaning "truly nucleated." The cells of all the organisms in the other three kingdoms are also eukaryotes. Many biologists believe that the transition from the prokaryotic to the eukaryotic cell was the largest and most significant event in the history of life, second only in biological importance to the first appearance of living systems. The protists are an extremely varied collection of organisms (Figure I–15b), and this kingdom includes the most structurally complex and versatile of all cells. Examples of protists are amoebas, paramecia, and the many forms of algae.

The third kingdom is made up of the fungi, including such organisms as molds, yeasts, and mushrooms (Figure I–16). Their mode of existence is very different from that of all other living things. Fungi digest complex macromolecules, which they may find in soil, water, cotton, leather, or even on the surface of the human skin, into smaller molecules. They then absorb these smaller molecules into their bodies, typically composed of masses of fine filaments, the surfaces of which are in direct contact with the nutrient source.

I-17 *Among the largest and oldest of all living things are the giant sequoias of North America. Members of the plant kingdom, sequoias typically have trunks more than 10 meters in circumference and are sometimes more than 75 meters tall. These particular trees were already growing a thousand years ago, long before Europeans settled in North America.*

I-18 *Among the animals, the insects are noted for both their numbers and their variety. A major reason for their success, biologically speaking, is the diversity of their life styles. This is a cetonid beetle, a resident of the tropical forests of western Africa. Note its spiny legs, which are raised in defense. Adults of this species are thought to feed on the pollen and nectar of flowers. The eggs are laid in rotting wood, and the larvae (the immature insects) develop there.*

The other two kingdoms are, of course, the plants and the animals, Plantae and Animalia. Plants are most concisely defined as many-celled organisms that collect energy from light—sunshine. They then transform this energy into the complex molecules of which their bodies are composed (Figure I–17). These molecules, which include sugars, proteins, and oils, are the energy sources for animal life.

The fifth kingdom, the animals, includes those life forms that are many-celled and that depend on other forms—mostly plants or other animals—for their sustenance (Figure I–18). From our anthropocentric point of view, animal usually means mammal, but actually most animals are invertebrates. More than one and a half million different kinds of animals have been recorded, of which 95 percent are invertebrates and more than a million of these are insects.

In little more than a century, our knowledge of the diversity of organisms, past and present, of the processes occurring within their bodies, and of their interrelationships with one another has rapidly outstripped that gained in all the previous centuries of human inquiry. This explosion of knowledge, which continues at an ever-accelerating pace, is the direct consequence of that particular form of inquiry that we call science (from the Latin *scientia*, "having knowledge").

THE NATURE OF SCIENCE

Science, whether biological or other, is a way of seeking principles of order. Art is another way, as are religion and philosophy. Science differs from these others in that it limits its search to the natural world, the physical universe. Also, and perhaps even more significant, it differs from them in the central value it gives to observation (particularly that structured kind of observation called experimentation). Scientists begin their search by accumulating data—evidence—and trying to fit these data into systems of order, conceptual schemes that organize the data in some meaningful way. Accumulating and ordering data are not two steps; they go on simultaneously. Or to put it another way, the accumulation of data is undertaken by scientists as a way of answering a question, or of supporting or rejecting an idea. The data may be generated by systematic observation, including deliberate, planned experiments (of which we shall see many examples in the chapters that follow); they may also be gleaned retrospectively from reevaluation of one's earlier systematic observation or from reevaluation of verifiable information recorded by others. (Darwin, like many biologists before and since, made copious use of all of these methods.)

The great discoveries in science are not merely the addition of new data but the perception of new relationships among the available data—in other words, the development of new ideas. The ideas of science are categorized in ascending order of validity as hypotheses, theories, and principles or laws. Lower on the scale than the hypothesis is the hunch, or educated guess, which is how most hypotheses begin. A hunch becomes a hypothesis—and therefore an idea that can be investigated scientifically—when and only when it is stated in such a way that it is potentially testable, even if the test cannot be done immediately. The testing of a hypothesis can often be done quite promptly, but sometimes it is, of necessity, long delayed. For example, some current hypotheses about the interactions that determine the structure of tropical forests cannot be tested until many more data have been gathered by biologists working in the forests. Similarly, a number of hypotheses about the organization of the cell could not be tested until after the development of the electron microscope.

A hypothesis can sometimes be tested directly. For example, one can determine whether caterpillars are repelled by a particular substance, isolated from plant leaves, that is hypothesized to protect the leaves from the caterpillars' predation.

I-19 *In a swamp in Peru, two biologists, Terry Erwin and Linda Sims of the Smithsonian Institution, are collecting data on the population structure of the tropical forest. Erwin is shooting a line high into the treetops as the first step in insect collection. Many of the specimens they have found are wholly new to science.*

This type of test frequently involves a controlled experiment, in which two groups of organisms (or cells) are exposed to conditions that are identical in all respects except the one being tested. Often, however, the most important tests of a hypothesis are indirect. A basic assumption of science is that the universe is consistent and that its component parts interact with and affect one another in understandable and predictable ways. An essential part of the testing of a hypothesis is the logical deduction of what other things should follow if the hypothesis is correct. The deduction constitutes a prediction of what should be observed if the hypothesis is correct and indicates the type of data that need to be gathered in order to test it.

Although a key predictive test may demonstrate that a hypothesis is false or may indicate that it must be modified, such a test can never definitely prove, once and for all, that a hypothesis is true—simply because we can never be certain that we have examined all of the relevant evidence. However, repeated successful tests of a hypothesis—either directly or in terms of the consequences that would follow if the hypothesis were correct—provide strong evidence in favor of the hypothesis.

When a scientist has collected sufficient data to support a particular hypothesis, he or she then reports the results to other scientists; such a report usually takes place at a scientific meeting (such as the meeting at which Darwin's and Wallace's papers were read), or in a scientific publication, such as a journal or a book (for example, *The Origin of Species*). If the data are sufficiently interesting or the hypothesis important, the observations or experiments will be repeated in an attempt to confirm, deny, or extend them. Hence, scientists always report the methods that they used in gathering and analyzing their data as well as their conclusions.

When a hypothesis of broad, fundamental importance has survived a number of independent tests, involving a diversity of data, it is generally referred to as a theory. Thus, a theory in science has a somewhat different meaning from the word "theory" in common usage, in which "just a theory" carries with it the implication of a flight of fancy, a hunch, or an abstract notion, rather than a carefully formulated, well-tested proposition. A theory that has withstood repeated testing over a period of time becomes elevated to the status of a law or principle, although not always identified as such. The "theory" of evolution, which has been tested and retested, directly and indirectly, for the past 130 years, is an example. As far as scientists are concerned, it is a basic principle of biology, just as is the cell "theory." However, our knowledge of many of the details of cellular structure and function and of the details of the evolutionary process is in the stage of theory, or even hypothesis.

Because the subject matter of biology is enormously diverse, biologists utilize a wide variety of approaches in their studies. Careful and systematic observation, of the sort practiced by the nineteenth-century naturalists, Darwin among them, remains a cornerstone. It is now supplemented by an enormous array of technological innovations that began with the microscope. The experimental procedures of chemistry are essential for studying the physiological processes occurring within organisms and their constituent cells. The study of populations of organisms and their interactions depends on the same kind of statistical mathematics employed by economists and is enhanced by computers that can analyze large quantities of data very rapidly. Charting the past course of evolution depends not only on the work of paleontologists in the field and in the laboratory but also on the intellectual tools of the historian and the homicide detective. As we shall see in the course of this text, there is no single "scientific method" in biology; there are instead a multiplicity of methods, with the particular method used in any given instance being determined by the question to be answered.

Some Comments on Science and Scientists

Science and Theory

The power of theories is that they combine many generalizations and other theories into networks of interlocking ideas that point to the future.

J. Bronowski, *The Common Sense of Science,* Random House, Inc., New York, 1959.

On principle, it is quite wrong to try founding a theory on observable magnitudes alone. It is the theory which decides what we can observe.

A. Einstein, from J. Bernstein, "The Secrets of the Old Ones, II," *The New Yorker,* March 17, 1973.

The Scientific Method

Indeed, scientists are in the position of a primitive tribe which has undertaken to duplicate the Empire State Building, room for room, without ever seeing the original building or even a photograph. Their own working plans, of necessity, are only a crude approximation of the real thing, conceived on the basis of miscellaneous reports volunteered by interested travelers and often in apparent conflict on points of detail. In order to start the building at all, some information must be ignored as erroneous or impossible, and the first constructions are little more than large grass shacks. Increasing sophistication, combined with methodical accumulation of data, make it necessary to tear down the earlier replicas (each time after violent arguments), replacing them successively with more up-to-date versions. We may easily doubt that the version current after only 300 years of effort is a very adequate restoration of the Empire State Building; yet, in the absence of clear knowledge to the contrary, the tribe must regard it as such (and ignore odd travelers' tales that cannot be made to fit).

E. J. DuPraw, *Cell and Molecular Biology,* Academic Press, Inc., New York, 1968.

Scientists at Work

Scientists are like pickpockets. God has all the secrets in his pockets, and we try to pick them. You make an assumption in science—and it *is* an assumption—that there are fundamental laws you can find out. You have an idea you think can be proved and you try to prove it. Depending on how it goes, you make a step forward or you make a fool of yourself. Nature doesn't care whether you're right or wrong. Nature is the way it is, and you had better be smart enough to get a little glimpse.

Abraham Pais, Rockefeller University

Scientists at work have the look of creatures following genetic instructions; they seem to be under the influence of a deeply placed human instinct. They are, despite their efforts at dignity, rather like young animals engaged in savage play. When they are near to an answer their hair stands on end, they sweat, they are awash in their own adrenalin. To grab the answer, and grab it first, is for them a more powerful drive than feeding or breeding or protecting themselves against the elements.

It sometimes looks like a solitary activity, but it is as much the opposite of solitary as human behavior can be. There is nothing so social, so communal, so interdependent. An active field of science is like an immense intellectual anthill; the individual almost vanishes into the mass of minds tumbling over each other, carrying information from place to place, passing it around at the speed of light.

There are special kinds of information that seem to be chemotactic. As soon as a trace is released, receptors at the back of the neck are caused to tremble, there is a massive convergence of motile minds flying upwind on a gradient of surprise, crowding around the source. It is an infiltration of intellects, an inflammation.

There is nothing to touch the spectacle. In the midst of what seems a collective derangement of minds in total disorder, with bits of information being scattered about, torn to shreds, disintegrated, reconstituted, engulfed, in a kind of activity that seems as random and agitated as that of bees in a disturbed part of the hive, there suddenly emerges, with the purity of a slow phrase of music, a single new piece of truth about nature. . . .

There is something like aggression in the activity, but it differs from other forms of aggressive behavior in having no sort of destruction as the objective. While it is going on, it looks and feels like aggression: get at it, uncover it, bring it out, grab it, halloo! It is like a primitive running hunt, but there is nothing at the end of it to be injured. More probably, the end is a sigh. But then, if the air is right and the science is going well, the sigh is immediately interrupted, there is a yawping new question, and the wild, tumbling activity begins once more, out of control all over again.

Lewis Thomas, *The Lives of a Cell: Notes of a Biology Watcher,* Viking Press, Inc., New York, 1974.

Science and Human Values

The raw materials of science are our observations of the phenomena of the natural universe. Science—unlike art, religion, or philosophy—is limited to what is observable and measurable and, in this sense, is rightly categorized as materialistic. Hunches are abandoned, hypotheses superseded, theories revised—and occasionally, shattered—but the observations endure, and, moreover, they are used over and over again, sometimes in wholly new ways. It is for this reason that scientists stress and seek objectivity. In the arts, by contrast, the emphasis is on subjectivity —experience as filtered through the individual consciousness.

Because of this emphasis on objectivity, value judgments cannot be made in science in the way that such judgments are made in philosophy, religion, and the arts, and indeed in our daily lives. Whether something is good or beautiful or right in a moral sense, for example, cannot be determined by scientific methods. Such judgments, even though they may be supported by a broad consensus, are not subject to scientific testing.

(a)

(b)

I-20 *In science, great stress is placed on objectivity. As a consequence, scientists, speaking as scientists, refrain from making value judgments. Thus, for example, they would not identify any organism as the ugliest or the most beautiful. As a private individual, however, a scientist may have his or her own opinions about (a) wart hogs and (b) butterflies and flowers.*

At one time, sciences, like the arts, were pursued for their own sake, for pleasure and excitement and satisfaction of the insatiable curiosity with which we are both cursed and blessed. In the twentieth century, however, the sciences have spawned a host of giant technological achievements—the hydrogen bomb, polio vaccine, pesticides, indestructible plastics, nuclear energy plants, perhaps even ways to manipulate our genetic heritage—but have not given us any clues about how to use them wisely. Moreover, science, as a result of these very achievements, appears enormously powerful. It is thus little wonder that there are many people who are angry at science, as one would be angry at an omnipotent authority who apparently has the power to grant one's wishes but who refuses to do so.

The reason that science cannot and does not solve the problems we want it to is inherent in its nature. Most of the problems we now confront can be solved only by value judgments. For example, science gave us nuclear weapons and can give us predictions as to the extent of the biological damage that their use might cause. Yet it cannot help us, as citizens, in weighing the risk of damage from a nuclear exchange against the desire for a strong national defense.

Science has produced the knowledge that makes possible not only the construction of kidney dialysis machines but also the surgical replacement of a diseased kidney by a healthy one. There are, however, many more patients who need kidney transplant operations than there are healthy kidneys available for transplanting. Scientific methods cannot help us decide who should be on the waiting list for a kidney transplant and who should remain dependent on repeated dialysis treatments. Similarly, scientists can predict the possible extent of damage to the plants and animals of a particular area from the use of pesticides; they can also predict the reduction in food crops or the increase in malaria that would occur were pesticides prohibited. But, scientists, in their capacity as scientists, cannot make the choice as to whether we should or should not use pesticides.

It is one of the ironies of this so-called "age of science and materialism" that probably never before have ordinary individual men and women, including scientists, been confronted with so many moral and ethical dilemmas. In this text, we shall discuss some of the dilemmas that have grown out of the achievements of modern science and technology. Our greater concern, however, is to provide you with the biological knowledge necessary to understand the relevant data as you make your own value judgments regarding the problems that confront us now and that will do so in the future.

Science as Process

You are fortunate to be studying biology now, in what many consider to be its "golden age." New ideas and unexpected discoveries have opened up exciting frontiers in many different areas—cell biology, genetics, immunology, neurobiology, evolution, ecology, to mention just a few. Because there is so much to tell, most biology texts, and this one is no exception, tend to stress what is known at the present time, rather than what is not known or how we came to know what we do. This tendency, although understandable, distorts the nature of biology and, indeed, of science in general.

A modern science is not a static accumulation of facts organized in a particular way but a somewhat amorphous body of knowledge that constantly grows, developing new bulges and unpredictable appendages. It may also suddenly change its entire shape—as biology did in the nineteenth century with the realization of the quantity and diversity of evidence supporting evolution. Science is not information contained within textbooks or libraries or information-retrieval centers, but rather it is a dynamic process taking place in the minds of living scientists. In our enthusiasm for telling you what biologists have learned thus far about living organisms—their history, their properties, and their activities—do not let us convince you that all is known. Many questions are still unanswered. More important, many good questions have not yet been asked. Perhaps you will be the one to ask them.

You may have been persuaded to study biology because of current environmental problems or because of a desire to know more about the mechanisms of your own body or an interest in genetic engineering or a career in medicine—in short, because it is "relevant." The study of biology is, indeed, pertinent to many aspects of our day-to-day existence, but do not make this your main focus as you embark on the study of biology. Above all other considerations, study biology because it is "irrelevant"—that is, study it for its own sake, because, like art and music and literature, it is an adventure for the mind and nourishment for the spirit.

QUESTIONS

1. What is the essential difference between Darwin's theory of evolution and that of Lamarck?

2. The chief predator of an English species of snail is the song thrush. Snails that inhabit woodland floors have dark shells, whereas those that live on grass have yellow shells, which are less clearly visible against the lighter background. Explain, in terms of Darwinian principles.

3. The phrase "chance and necessity" has been used to describe the Darwinian theory of evolution. Relate this to the fact that snails living on grass do not have green shells, but there are, for example, green frogs and green insects.

4. The largest terrestrial organisms known are the giant sequoias (Figure I–17) of North America, which dwarf even an elephant. The largest animals, members of the whale family, are all aquatic. Can you explain why this should be the case?

5. When scientists report new findings, they are expected to reveal their methods and raw data as well as their conclusions. Why is such reporting considered essential?

SUGGESTIONS FOR FURTHER READING

BATES, MARSTON, and PHILIP S. HUMPHREY (eds.): *The Darwin Reader*, Charles Scribner's Sons, New York, 1956.*

A collection of Darwin's writings, including The Autobiography, *and excerpts from* The Voyage of the Beagle, The Origin of Species, The Descent of Man, *and* The Expression of the Emotions. *Darwin was a fine writer, and you can discover here the wide range of his interests and concerns at different periods of his life.*

BRONOWSKI, J.: *The Ascent of Man*, Little, Brown and Company, Boston, 1973.*

An informal and illuminating history of the sciences, originally prepared as a television series. The emphasis is on science's relation to human culture. Well designed and illustrated.

DARWIN, CHARLES: *The Origin of Species by Means of Natural Selection, or The Preservation of Favored Races in the Struggle for Life*, W. W. Norton & Company, New York, 1975.*

Darwin's "long argument." Every student of biology should, at the very least, browse through this book to catch its special flavor and to begin to understand its extraordinary force.

DARWIN, CHARLES: *The Voyage of the Beagle*, Doubleday & Company, Inc., Garden City, N.Y., 1962.*

Darwin's own chronicle of the expedition on which he made the discoveries and observations that eventually led him to his theory of evolution. The sensitive, eager, young Darwin that emerges from these pages is very unlike the solemn image many of us have formed of him from his later portraits.

LEWIN, ROGER: *Thread of Life: The Smithsonian Looks at Evolution*, W. W. Norton & Company, New York, 1982.

An absorbing account of the historical development of evolutionary thought, with current applications from biochemistry, paleontology, and geology. Well written and beautifully illustrated with a rich assortment of color photographs.

MAYR, ERNST: *The Growth of Biological Thought: Diversity, Evolution, and Inheritance*, Harvard University Press, Cambridge, Mass., 1982.*

This is the first of two volumes on the history of biology and its major ideas, written by one of the leading figures in the study of evolution. The introductory chapters provide an outstanding analysis of the philosophy and methodology of the biological sciences. This book, like Darwin's masterpiece, should at least be sampled by every serious student of biology.

MOOREHEAD, ALAN: *Darwin and the Beagle*, Harper & Row, Publishers, Inc., New York, 1969.*

A delightful narrative of Darwin's journey, beautifully illustrated with contemporary or near-contemporary drawings, paintings, and lithographs.

---

* Available in paperback.

P A R T **1**

Biology of Cells

S E C T I O N **1**

The Unity of Life

This explosion in the sky, a supernova in astronomers' terms, was caused by the death of a star. About 170,000 years ago, the star, then 10 million years old, exhausted its fuel. During its life span, thermonuclear reactions, such as those now taking place in our own sun, had turned hydrogen to helium, and helium to carbon and oxygen, which themselves fused to even heavier elements. Once twenty times the size of the sun, the star cooled and, under the force of gravity, exploded inward. In this way, in the death of stars, all of the atoms of which our planet and its inhabitants are formed had their beginnings. This supernova, the first to be recorded in 383 years, was initially seen on February 24, 1987, by astronomers at a remote observatory in Chile.

Atoms and Molecules

Our universe began, according to current theory, with an explosion that filled all space, with every particle of matter hurled away from every other particle. The temperature at the time of the explosion—some 10 to 20 billion years ago—was about 100,000,000,000 degrees Celsius (10^{11} °C). At this temperature, not even atoms could hold together; all matter was in the form of subatomic, elementary particles. Moving at enormous velocities, even these particles had fleeting lives. Colliding with great force, they annihilated one another, creating new particles and releasing more energy.

As the universe cooled, two types of stable particles, previously present only in relatively small amounts, began to assemble. (By this time, several hundred thousand years after the "big bang" is believed to have taken place, the temperature had dropped to a mere 2500°C, about the temperature of a white-hot wire in an incandescent light bulb.) These particles—protons and neutrons—are very heavy as subatomic particles go. Held together by forces that are still incompletely understood, they formed the central cores, or nuclei, of atoms. These nuclei, with their positively charged protons, attracted small, light, negatively charged particles—electrons—which moved rapidly around them. Thus, atoms came into being.

It is from these atoms—blown apart, formed, and re-formed over the course of several billion years—that all the stars and planets of our universe are formed, including our particular star and planet. And it is from the atoms present on this planet that living systems assembled themselves and evolved. Each atom in our own bodies had its origin in that enormous explosion 10 to 20 billion years ago. You and I are flesh and blood, but we are also stardust.

This text begins where life began, with the atom. At first, the universe aside, it might appear that lifeless atoms have little to do with biology. Bear with us, however. A closer look reveals that the activities we associate with being alive depend on combinations and exchanges between atoms, and the force that binds the electron to the atomic nucleus stores the energy that powers living systems.

ATOMS

All matter, including the most complex living organisms, is made up of combinations of **elements**. Elements are, by definition, substances that cannot be broken down into other substances by ordinary chemical means. The smallest particle of an element is an **atom**. There are 92 naturally occurring elements, each differing from the others in the structure of its atoms (Table 1–1).

The atoms of each different element have a characteristic number of positively charged particles, called protons, in their nuclei. For example, an atom of hydrogen, the lightest of the elements, has 1 proton in its nucleus; an atom of carbon has 6 protons in its nucleus. The number of protons in the nucleus of a

1–1 *"From so simple a beginning,"* Darwin wrote in Origin of Species, *"endless forms most beautiful and most wonderful have been, and are being evolved."* Among them is this oceanic worm, about one-tenth of a centimeter in length. Its tentacles, which are green, can be withdrawn into its body. The yellow mass is its digestive gland, and the adjacent coiling orange-yellow tube is its gut. Like the water that surrounds it, this tiny worm, formally known as* Poeobius meseres, *is made up of atoms created in the death of stars.*

TABLE 1-1   **Atomic Structure of Some Familiar Elements**

ELEMENT	SYMBOL	NUCLEUS NUMBER OF PROTONS	NUMBER OF NEUTRONS*	NUMBER OF ELECTRONS
Hydrogen	H	1	0	1
Helium	He	2	2	2
Carbon	C	6	6	6
Nitrogen	N	7	7	7
Oxygen	O	8	8	8
Sodium	Na	11	12	11
Phosphorus	P	15	16	15
Sulfur	S	16	16	16
Chlorine	Cl	17	18	17
Potassium	K	19	20	19
Calcium	Ca	20	20	20

* In most common isotope.

particular atom is called its **atomic number.** The atomic number of hydrogen is therefore 1, and the atomic number of carbon is 6.

Outside the nucleus of an atom are negatively charged particles, the electrons, which are attracted by the positive charge of the protons. The number of electrons in an atom equals the number of protons in its nucleus. The electrons determine the chemical properties of atoms, and chemical reactions involve changes in the numbers and energy of these electrons.

Atoms also contain neutrons, which are uncharged particles of about the same weight as protons. These, too, are found in the nucleus of the atom, where they seem to have a stabilizing effect. The **atomic weight** of an element is essentially equal to the number of protons plus neutrons in the nuclei of its atoms. The atomic weight of carbon is 12, whereas that of hydrogen, which contains no neutrons, is 1. Electrons are so light by comparison to protons and neutrons that their weight is usually disregarded. When you weigh yourself, only about 30 grams—approximately 1 ounce—of your total weight is made up of electrons.

Isotopes

All atoms of a particular element have the same number of protons in their nuclei. Sometimes, however, different atoms of the same element contain different numbers of neutrons. These atoms, which therefore differ from one another in their atomic weights but not in their atomic numbers, are known as **isotopes** of the element. For example, three different isotopes of hydrogen exist (Table 1–2). The common form of hydrogen, with its one proton, has an atomic weight of 1 and is

TABLE 1-2   **Isotopes of Hydrogen**

ISOTOPE NAME	SYMBOL	ATOMIC NUMBER	ATOMIC WEIGHT	NUMBER OF PROTONS	NUMBER OF NEUTRONS	NUMBER OF ELECTRONS
Hydrogen	^1H	1	1	1	0	1
Deuterium	^2H	1	2	1	1	1
Tritium	^3H	1	3	1	2	1

1–2 *The dating of fossils by determining the relative proportions of different isotopes in the nearby or surrounding volcanic rocks is an important tool in tracing the course of evolutionary history. About 50 million years ago, this perch (Mioplosus), unable to swallow or dislodge the herring (Knightia) it had voraciously attacked, suffocated and sank to the bottom of the lake in which it lived. Lake sediments built up year after year and eventually dried, preserving the bones in limestone. These fossils were found in the Green River Formation in southwestern Wyoming in an area that is now a dry, rocky basin but for millions of years was covered by a lake surrounded by lush vegetation.*

symbolized as ^{1}H, or simply H. A second isotope of hydrogen, known as deuterium, contains one proton and one neutron and so has an atomic weight of 2; this isotope is symbolized as ^{2}H. Tritium, ^{3}H, a third, extremely rare isotope, has one proton and two neutrons and so has an atomic weight of 3. The chemical behavior of the two heavier isotopes is essentially the same as that of ordinary hydrogen—all three isotopes have only one electron each, and it is the electrons that determine chemical properties.

Most elements have several isotopic forms. The differences in weight, although very small, are sufficiently great that they can be detected with modern laboratory apparatus. Moreover, many, but not all, of the less common isotopes are radioactive. This means that the nucleus of the atom is unstable and emits energy as it changes to a more stable form. The energy given off by the nucleus of a radioactive isotope may be in the form of rapidly moving subatomic particles, electromagnetic radiation, or both. It can be detected with a Geiger counter or on photographic film.

Isotopes have a number of important uses in biological research and in medicine. They can be used, for example, to determine the age of fossils and of the rocks in which fossils are found (Figure 1–2). Each type of radioactive isotope emits energy and changes into another kind of isotope at a characteristic and fixed rate. As a result, the relative proportions of different isotopes in a rock sample give a good indication of how long ago that rock was formed.

Another use of radioactive isotopes is as "tracers." Since isotopes of the same element all have the same chemical properties, a radioactive isotope will behave in an organism just as its more common nonradioactive isotope does. As a result, biologists have been able to use isotopes of a number of elements—especially hydrogen, carbon, nitrogen, oxygen, and phosphorus—to trace the course of many essential processes in living organisms.

Isotopes play a role in the treatment of many forms of cancer, and they also have a number of diagnostic uses in medicine. For example, an isotope of the element thallium, which is unreactive in the human body, can be used to identify blocked blood vessels in persons with symptoms of heart disease. The isotope is first injected into the bloodstream. Then, while the patient exercises on a treadmill, the movement of the radioactive isotope is detected by a Geiger counter that is connected to a computer. The result is a "picture" of the distribution of the isotope in the heart muscle. If a blood vessel is blocked by fatty deposits, the isotope cannot penetrate the region of heart muscle supplied by that blood vessel. This procedure, which has no known side effects, provides an extremely reliable indication of the presence or absence of a common type of heart disease (see page 71).

The Signs of Life

What do we mean when we speak of "the evolution of life," or "life on other planets," or "when life began"? Actually, there is no simple definition. Life does not exist in the abstract; there is no "life," only living things. Moreover, there is no single, simple way to draw a sharp line between the living and the nonliving. There are, however, certain properties that, taken together, distinguish animate (that is, living) objects from inanimate ones.

(a) Living things are highly organized, as in this cross section of a stem of a sycamore sapling. This stem reflects the complicated organization of many different kinds of atoms into molecules and of molecules into complex structures. Such complexity of form, which is never found in inanimate objects of natural origin, makes possible the specialization of different parts of a living organism for different functions.

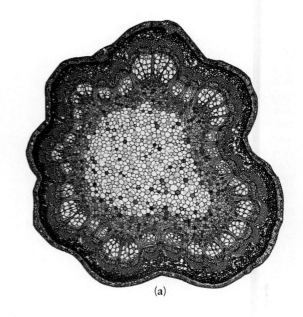

(a)

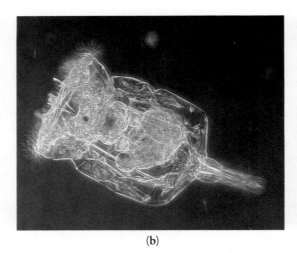

(b)

(c)

(b) Living things are homeostatic, which means simply "staying the same." Although they constantly exchange materials with the outside world, they maintain a relatively stable internal environment quite unlike that of their surroundings. Even this tiny, apparently fragile animal, a rotifer, has an internal chemical composition that differs from its changing environment.

(c) Living things reproduce themselves. They make more of themselves, generation after generation, with astonishing fidelity (and yet, as we shall see, with just

enough variation to provide the raw material for evolution). Flowers, the familiar symbols of spring and romance, are the reproductive structures of the largest and most diverse group of plants.

(d) Living organisms grow and develop. Growth and development are the processes by which, for example, a single living cell, the fertilized egg, becomes a tree, or an elephant, or as shown here, a newborn zebra.

(e) Living things take energy from the environment and change it from one form to another. The pro-

(d)

(e)

(f)

(g)

cesses of energy conversion are highly specialized and remarkably efficient. This American bald eagle has converted chemical energy stored in its body to energy of motion used in catching a salmon. After the eagle has eaten and digested the salmon, the chemical energy stored in the body of the salmon will be available for the eagle's use.

(f) Living organisms respond to stimuli. When this fish jostled the tentacles hanging from the rim of the jellyfish's bell-shaped body, the response—injection of a paralyzing substance into the fish's body—was immediate. As the fish lies helpless, attached to sticky strands from the tentacles, the fluttering mouth of the jellyfish responds to its presence.

(g) Living things are adapted. Moles, for instance, live underground in tunnels shoveled out by their large forepaws. Their eyes are small and almost sightless. Their noses, with which they sense the worms and other small invertebrates that make up their diet, are fleshy and enlarged.

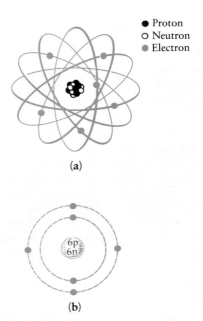

- ● Proton
- ○ Neutron
- ● Electron

(a)

(b)

1–3 *Two models of the carbon atom:* **(a)** *a planetary model and* **(b)** *a Bohr model.*

1–4 **(a)** *The energy used to push a boulder to the top of a hill (less the heat energy produced by friction between the boulder and hill) becomes potential energy, stored in the boulder as it rests at the top of the hill. This potential energy is converted to kinetic energy (energy of motion) as the boulder rolls downhill.*

(b) *When an atom, such as the hydrogen atom diagrammed here, receives an input of energy, an electron may be boosted to a higher energy level. The electron thus gains potential energy, which is released when the electron returns to its previous energy level.*

Models of Atomic Structure

The concept of the atom as the indivisible unit of the elements is almost 200 years old; however, our ideas about its structure have undergone many changes and may well undergo further changes in the future. These ideas, or hypotheses, are usually presented in the form of models, as are many other scientific hypotheses.

The earliest model of the atom, emphasizing its indivisibility, portrayed the atom as a sphere like a billiard ball. When it was realized that electrons could be removed from the atom, the billiard-ball model gave way to the plum-pudding model, in which the atom was represented as a solid, positively charged mass with negatively charged particles, the electrons, embedded in it. Subsequently, however, physicists found that an atom is, in fact, mostly empty space. The distance from electron to nucleus, experiments indicated, is about 1,000 times greater than the diameter of the nucleus; the electrons are so exceedingly small that the space is almost entirely empty. Thus the more familiar planetary model of the atom came into being, in which the electrons were depicted as moving in orbits around the nucleus (Figure 1–3a). Later, it was replaced by the Bohr model, named after physicist Niels Bohr. The Bohr model (Figure 1–3b) emphasized the fact that different electrons of an atom have different amounts of energy and are at different distances from the nucleus. As we shall see, neither the planetary model nor the Bohr model gives an accurate "picture" of an atom, and they have been superseded by another model, the orbital model (see Figure 1–6). However, the Bohr model can help us to understand certain properties of atoms that are of great importance in the chemistry of living systems.

ELECTRONS AND ENERGY

The distance of an electron from the nucleus is determined by the amount of **potential energy** (often called "energy of position") the electron possesses. The greater the amount of energy possessed by the electron, the farther it will be from the nucleus. Thus, an electron with a relatively small amount of energy is found close to the nucleus and is said to be at a low **energy level;** an electron with more energy is farther from the nucleus, at a higher energy level.

An analogy may be useful. A boulder resting on flat ground neither gains nor loses potential energy. If, however, you change its position by pushing it up a hill, you increase its potential energy. As long as it sits on the peak of the hill, the rock once more neither gains nor loses potential energy. If it rolls down the hill, however, potential energy is converted to energy of motion and released (Figure 1–4a). Similarly, water that has been pumped up to a water tank for storage has potential energy that will be released when the water runs back down.

The electron is like the boulder, or the water, in that an input of energy can move it to a higher energy level—farther away from the nucleus. As long as it

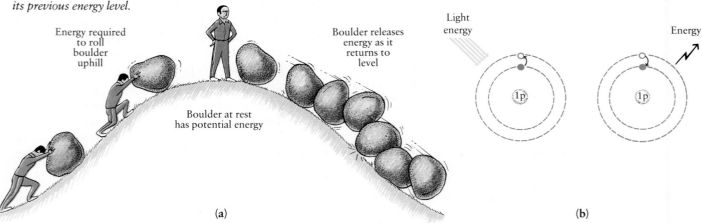

Energy required to roll boulder uphill

Boulder at rest has potential energy

Boulder releases energy as it returns to level

Light energy

Energy

(a)

(b)

1-5 *The leaves of these corn plants contain chlorophyll, which gives them their green color. When a packet of light energy—a photon—strikes a molecule of chlorophyll, electrons in the molecule are raised to higher energy levels. As each electron returns to its previous energy level, a portion of the energy released is captured in the bonds of carbon-containing molecules.*

1-6 *The most accurate representation of our knowledge of atomic structure is provided by orbital models.* (**a**) *The two electrons at the first energy level of an atom occupy a single spherical orbital. The nucleus is at the intersection of the axes.*

(**b**) *At the second energy level, there are four orbitals, each containing two electrons. One of these orbitals is spherical and the other three are dumbbell-shaped. The axes of the dumbbell-shaped orbitals are perpendicular to one another. The orbitals are shown individually in this diagram. In reality, the spherical orbital of the second energy level surrounds the orbital of the first energy level, and portions of the dumbbell-shaped orbitals pass through the two spherical orbitals. The orbitals influence one another and determine the overall shape of the atom.*

remains at the higher energy level, it possesses the added energy. And, just as the rock is likely to roll, and the water to run, downhill, the electron also tends to go to its lowest possible energy level.

It takes energy to move a negatively charged electron farther away from a positively charged nucleus, just as it takes energy to push a rock up a hill. However, unlike the rock on the hill, the electron cannot be pushed partway up. With an input of energy, an electron can move from a lower energy level to any one of several higher energy levels, but it cannot move to an energy state somewhere in between. For an electron to move from one energy level to a higher one, it must absorb a discrete amount of energy, equal to the difference between the two particular energy levels. When the electron returns to its original energy level, that same amount of energy is released (Figure 1–4b). The discrete amount of energy involved in the transition between two energy levels is known as a quantum. Thus the study of electron movements is known as quantum mechanics, and the term "quantum jump," which has invaded our everyday discourse, refers to an abrupt, discontinuous movement from one level to another.

In the green cells of plants and algae, the radiant energy of sunlight raises electrons to a higher energy level. In a series of reactions, which will be described in Chapter 10, these electrons are passed "downhill" from one energy level to another until they return to their original energy level. During these transitions, the radiant energy of sunlight is transformed into the chemical energy on which life on earth depends.

The Arrangement of Electrons

At a given energy level, an electron moves around the nucleus at almost the speed of light. The electron is so small and moves so rapidly that it is theoretically impossible to determine, at any given moment, both its precise location and the exact amount of energy it possesses. As a result of this difficulty, the current model of atomic structure describes the pattern of the electron's motion rather than its position. The volume of space in which the electron will be found 90 percent of the time is defined as its **orbital.**

In any atom, the electrons at the lowest energy level—the first energy level—occupy a single spherical orbital, which can contain a maximum of two electrons (Figure 1–6a). Thus, for instance, hydrogen's single electron moves about the nucleus—90 percent of the time—within this single spherical orbital. Similarly, the two electrons of helium (atomic number 2) move within the single spherical orbital of the first energy level.

Atoms of higher atomic number than helium have more than two electrons. Since the first energy level is filled by two electrons, the additional electrons must occupy higher energy levels, farther from the nucleus. At the second energy level, there are four orbitals, each of which can hold a maximum of two electrons (Figure 1–6b). Thus, the second energy level can contain a total of eight electrons—and so can the third energy level of elements through atomic number 20 (calcium).

First energy level:

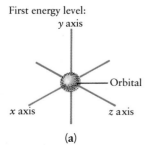

(a)

Second energy level:

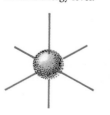  

(b)

The way an atom reacts chemically is determined by the number and arrangement of its electrons. An atom is most stable when all of its electrons are at their lowest possible energy levels. Therefore, the electrons of an atom fill the energy levels in order—the first is filled before the second, the second before the third, and so on. Moreover, an atom in which the outermost energy level is completely filled with electrons is more stable than one in which the outer energy level is only partially filled. For example, helium (atomic number 2) has two electrons at the first energy level, which means that its outer energy level (in this case, also its lowest energy level) is completely filled. Helium is therefore extremely stable and tends to be unreactive. Similarly, neon (atomic number 10) has two electrons at the first energy level and eight at the second energy level; both energy levels are completely filled, and neon is unreactive. Helium, neon, and argon (atomic number 18) are called the "noble" gases because of their disdain for reacting with other elements.

In the atoms of most elements, however, the outer energy level is only partially filled (Table 1–3). These atoms tend to interact with other atoms in such a way that after the reaction both atoms have completely filled outer energy levels. Some atoms lose electrons; others gain electrons; and, in many of the most important chemical reactions that occur in living systems, atoms share their electrons with each other.

TABLE 1-3   Electron Arrangements in Some Familiar Elements

ELEMENT	ATOMIC NUMBER	NUMBER OF ELECTRONS IN EACH ENERGY LEVEL*			
		FIRST	SECOND	THIRD	FOURTH
Hydrogen (H)	1	1	—	—	—
Helium (He)	2	2	—	—	—
Carbon (C)	6	2	4	—	—
Nitrogen (N)	7	2	5	—	—
Oxygen (O)	8	2	6	—	—
Neon (Ne)	10	2	8	—	—
Sodium (Na)	11	2	8	1	—
Phosphorus (P)	15	2	8	5	—
Sulfur (S)	16	2	8	6	—
Chlorine (Cl)	17	2	8	7	—
Argon (Ar)	18	2	8	8	—
Potassium (K)	19	2	8	8	1
Calcium (Ca)	20	2	8	8	2

* The first energy level can hold a maximum of 2 electrons; the second energy level can hold a maximum of 8 electrons, as can the third energy level of the elements through atomic number 20 (calcium). In elements of higher atomic number, the third energy level has additional, inner orbitals that can hold a maximum of 10 more electrons.

BONDS AND MOLECULES

When atoms interact with one another, resulting in filled outer energy levels, new, larger particles are formed. These particles, consisting of two or more atoms, are known as **molecules,** and the forces that hold them together are known as **bonds.** There are two principal types of bonds: ionic and covalent.

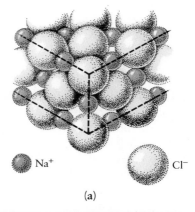

Na⁺   Cl⁻

(a)

(b)

1–7 **(a)** *Oppositely charged ions, such as the sodium and chloride ions depicted here as spheres, attract one another. Table salt is crystalline NaCl, a latticework of alternating Na⁺ and Cl⁻ ions held together by their opposite charges. Such bonds between oppositely charged ions are known as ionic bonds.*

(b) The regularity of the latticework is reflected in the structure of salt crystals, magnified here about 14 times.

1–8 *Ostriches, racing across the Etosha Pan in Namibia. Although these birds, the largest in the world, cannot fly, they are extremely graceful runners. Speeds of 40 to 60 kilometers per hour are not unusual. The movements of ostriches—like those of all complex animals—are the result of muscle contractions triggered by nerve impulses. Sodium, potassium, and calcium ions are involved in producing and propagating nerve impulses, and calcium ions are required for the contraction of muscle fibers.*

Ionic Bonds

For many atoms, the simplest way to attain a completely filled outer energy level is either to gain or to lose one or two electrons. For example, chlorine (atomic number 17) needs one electron to complete its outer energy level (see Table 1–3). By contrast, sodium (atomic number 11) has a single electron in its outer energy level. This electron is strongly attracted by the chlorine atom and jumps from the sodium to the chlorine. As a result of this transfer, both atoms have outer energy levels that are completely filled, and all the electrons are at the lowest possible energy levels. In the process, however, the original atoms have become electrically charged. Such charged atoms are known as **ions.** The chlorine atom, having accepted an electron from sodium, now has one more electron than proton and is a negatively charged chloride ion: Cl⁻. Conversely, the sodium ion has one less electron than proton and is positively charged: Na⁺.

Because of their charges, positive and negative ions attract one another. Thus the sodium ion (Na⁺) with its single positive charge is attracted to the chloride ion (Cl⁻) with its single negative charge. The resulting substance, sodium chloride (NaCl), is ordinary table salt (Figure 1–7). Similarly, when a calcium atom (atomic number 20) loses two electrons, the resulting calcium ion (Ca²⁺) can attract and hold two Cl⁻ ions. Calcium chloride is identified in chemical shorthand as $CaCl_2$, with the subscript 2 indicating that two chloride ions are present for each ion of calcium.

Bonds that involve the mutual attraction of ions of opposite charge are known as **ionic bonds.** Such bonds can be quite strong, but, as we shall see in the next chapter, many ionic substances break apart easily in water, producing free ions. Small ions such as Na⁺ and Cl⁻ make up less than 1 percent of the weight of most living matter, but they play crucial roles. Potassium ion (K⁺) is the principal positively charged ion in most organisms, and many essential biological processes occur only in its presence. Calcium ion (Ca²⁺), K⁺, and Na⁺ are all involved in the production and propagation of the nerve impulse. In addition, Ca²⁺ is required for the contraction of muscles and for the maintenance of a normal heartbeat. Magnesium ion (Mg²⁺) forms a part of the chlorophyll molecule, the molecule in green plants and algae that traps radiant energy from the sun.

Hydrogen molecule (H₂)

1-9 *In a molecule of hydrogen, each atom shares its single electron with the other atom. As a result, both atoms effectively have a filled first energy level, containing two electrons—a highly stable arrangement. This type of bond, in which electrons are shared, is known as a covalent bond.*

Covalent Bonds

Another way for an atom to complete its outer energy level is by sharing electrons with another atom. Bonds formed by shared pairs of electrons are known as **covalent bonds.** In a covalent bond, the shared pair of electrons forms a new orbital (called a molecular orbital) that envelops the nuclei of both atoms (Figure 1–9). In such a bond, each electron spends part of its time around one nucleus and part of its time around the other. Thus the sharing of electrons both completes the outer energy level and neutralizes the nuclear charge.

Atoms that need to gain electrons to achieve a filled, and therefore stable, outer energy level have a strong tendency to form covalent bonds. Thus, for example, a hydrogen atom forms a single covalent bond with another hydrogen atom. It can also form a covalent bond with any other atom that needs to gain an electron to complete its outer energy level.

Of extraordinary importance in living systems is the capacity of carbon atoms to form covalent bonds. A carbon atom has four electrons in its outer energy level (see Table 1–3). It can share each of those electrons with another atom, forming covalent bonds with as many as four other atoms (Figure 1–10). The covalent bonds formed by a carbon atom may be with different atoms (most frequently hydrogen, oxygen, and nitrogen) or with other carbon atoms. As we shall see in Chapter 3, this tendency of carbon atoms to form covalent bonds with other carbon atoms gives rise to the large molecules that form the structures of living organisms and that participate in essential life processes.

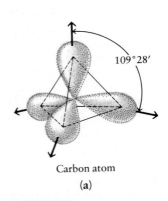

Carbon atom

(a)

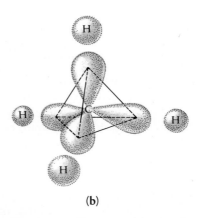

(b)

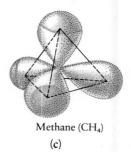

Methane (CH₄)

(c)

1-10 **(a)** *When a carbon atom forms covalent bonds with four other atoms, the electrons in its outer energy level form new orbitals. These new orbitals, which are all the same shape, are oriented toward the*

four corners of a tetrahedron. Thus the four orbitals are separated as far as possible. **(b)** *When a carbon atom reacts with four hydrogen atoms, each of the electrons in its outer energy level forms a covalent*

bond with the single electron of one hydrogen atom, producing a methane molecule **(c)**. *Each pair of electrons moves in a new, molecular orbital. The molecule has the shape of a tetrahedron.*

Polar Covalent Bonds

The atomic nuclei of different elements have different degrees of attraction for electrons. Factors that determine the strength with which a nucleus attracts its outer electrons include the number of protons it contains, the closeness of the outer electrons to the nucleus, and the number of other, "shielding" electrons between the nucleus and the outer electrons. In covalent bonds formed between atoms of different elements, the electrons are not shared equally between the atoms involved—instead, the shared electrons tend to spend more time around the nucleus with the greater attraction. The atom around which the electrons

Slightly
positive
charge

Slightly
negative
charge

Hydrogen chloride (HCl)

1–11 In a polar molecule, such as hydrogen chloride (HCl), the shared electrons tend to spend more time around one atom, in this case, the chlorine atom, than around the other atom. As a result, the atom that attracts the electrons more strongly (chlorine) has a slightly negative charge, and the atom that attracts the electrons less strongly (hydrogen) has a slightly positive charge.

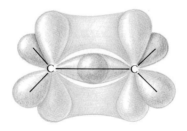

1–12 An orbital model of a carbon-carbon double bond. One pair of electrons occupies the inner orbital between the two carbon atoms. The other pair of electrons occupies the outer orbital, which has two phases, one above the plane and one below. This creates a rigid bond, about which the atoms cannot rotate. Each of the two smaller orbitals (shown in gray) extending from each carbon atom contains one electron and can form a covalent bond with another atom. Thus two covalent bonds can be formed at each end of this structure.

spend more time has a slightly negative charge; the other atom has a slightly positive charge, since the electrons spend less time around it and thus its nuclear charge is not entirely neutralized (Figure 1–11).

Covalent bonds in which electrons are shared unequally are known as **polar covalent bonds,** and the molecules containing these bonds are said to be **polar molecules.** Such molecules often contain oxygen atoms, to which electrons are strongly attracted. The polar properties of many oxygen-containing molecules have very important consequences for living things. For example, many of the special properties of water (H_2O), upon which life depends, derive largely from its polar nature, as we shall see in the next chapter.

Ionic, polar covalent, and covalent bonds actually may be considered different versions of the same type of bond. The differences depend on the differing attractions of the combining atoms for electrons. In a wholly nonpolar covalent bond, the electrons are shared equally; such bonds can exist only between identical atoms, H_2, Cl_2, O_2, and N_2, for example. In polar covalent bonds, electrons are shared unequally, and in ionic bonds, there is an electrostatic attraction between the negatively and positively charged ions as a result of their having previously gained or lost electrons.

Double and Triple Bonds

There are various ways in which atoms can participate in covalent bonds and fill their outer energy levels. Oxygen, for example, has six electrons in its outer energy level. Four of these electrons are grouped into two pairs and are generally unavailable for covalent bonding; the other two electrons are unpaired, and each can be shared with another atom in a covalent bond. In the water molecule (H_2O), one of these electrons participates in a covalent bond with one hydrogen atom, and the other in a covalent bond with a different hydrogen atom. Two single bonds are formed, and all three atoms have filled outer energy levels.

The bonding situation is different in another familiar substance, carbon dioxide (CO_2). In this molecule, the two available electrons of each oxygen atom participate with two electrons of a *single* carbon atom in the formation of *two* covalent bonds. Each oxygen atom is joined to the central carbon atom by two pairs of electrons (four electrons). Such bonds are called **double bonds,** and they are symbolized in a structural formula by two lines connecting the atomic symbols: O═C═O. Carbon atoms can form double or even triple bonds (in which three pairs of electrons are shared) with each other as well as with other atoms, and so the variety of kinds of molecules that carbon can form is very large.

Electrons shared in double and triple bonds form orbitals that differ in shape from the orbitals filled by single electron pairs. For instance, when four single bonds satisfy the electron requirements of carbon, they will be directed toward the four corners of a tetrahedron that has the carbon atom at its center, as we saw in Figure 1–10. When two bonds are replaced by a double bond, the remaining single bonds form the arms of a Y with the double bond as its leg (Figure 1–12). When two double bonds are made by a single carbon atom, as in carbon dioxide, the three bonded atoms lie in a straight line.

The symmetry of the carbon dioxide molecule has an important consequence. The double covalent bonds in carbon dioxide, like all covalent bonds between nonidentical atoms, are polar. However, because the molecule is perfectly symmetrical, the electrons are pulled in opposite directions by the two oxygen atoms, cancelling out the unequal distribution of charge. As a result, the carbon dioxide molecule is nonpolar. Similarly, the symmetry of the methane molecule (see Figure 1–10) produces a nonpolar molecule, even though the individual bonds between the atoms are polar.

Single bonds are flexible, leaving atoms free to rotate in relation to one another. Double and triple bonds hold the atoms relatively rigid in relation to one another. The presence of double bonds in a molecule can make a significant difference in its properties. For example, both fats and oils are composed of carbon and hydrogen atoms covalently bonded together, but in fats the bonds are all single and in oils some of the bonds are double. The rigidity caused by these double bonds prevents the oil molecules from packing together, and, as a consequence, oils are liquids at room temperature. In fats, by contrast, the molecules can bend and twist, fitting closely together in a solid structure at room temperature.

CHEMICAL REACTIONS

Chemical reactions—exchanges of electrons among atoms—can be compactly described by chemical equations. For example, the equation for the formation of sodium chloride is

$$Na^+ + Cl^- \longrightarrow NaCl$$

The arrow in the equation designates "forms" or "yields" and shows the direction of chemical change. Like algebraic equations, chemical equations "balance"; that is, there are the same number of atoms in the products of the reaction as in the original reactants. To take a slightly more complex example, hydrogen gas can combine with oxygen gas to produce water. Hydrogen gas is H_2, and oxygen gas is O_2. However, we know that each molecule of water contains two atoms of hydrogen and one of oxygen, and therefore the proportions must be two to one:

$$2H_2 + O_2 \longrightarrow 2H_2O$$

Two molecules of H_2 plus one molecule of O_2 yield two molecules of water. The equation for a chemical reaction thus tells the kinds of atoms that are present, their proportions, and the direction of the reaction.

A substance consisting of molecules that contain atoms of two or more different elements, held together in a definite and constant proportion by chemical bonds, is known as a chemical **compound.** Examples of chemical compounds include water (H_2O), sodium chloride (NaCl), carbon dioxide (CO_2), methane (CH_4), and glucose ($C_6H_{12}O_6$).

Types of Reactions

The multitude of chemical reactions that occur in both the animate and the inanimate worlds can be classified into a few general types. One type of reaction is a simple combination, represented by the expression:

$$A + B \longrightarrow AB$$

Examples of this type of reaction are the combination of sodium ions and chloride ions to form sodium chloride and the combination of hydrogen gas with oxygen gas to produce water.

A reaction may also take the form of a dissociation:

$$AB \longrightarrow A + B$$

For example, the earlier equation showing the formation of water can be reversed:

$$2H_2O \longrightarrow 2H_2 + O_2$$

This means that water molecules yield hydrogen and oxygen gases.

A reaction may also involve an exchange, taking the form:

$$AB + CD \longrightarrow AD + CB$$

An example of such an exchange occurs when the chemical compounds sodium hydroxide (NaOH) and hydrochloric acid (HCl) react, producing table salt and water:

$$NaOH + HCl \longrightarrow NaCl + H_2O$$

As we pursue our study of living organisms, we shall encounter numerous examples of these three general types of chemical reactions.

THE BIOLOGICALLY IMPORTANT ELEMENTS

Of the 92 naturally occurring elements, only six make up some 99 percent of all living tissue (Table 1–4). These six elements are carbon, hydrogen, nitrogen, oxygen, phosphorus, and sulfur, conveniently remembered as CHNOPS. These are not the most abundant of the elements of the earth's surface. Why, as life assembled and evolved from stardust, were these of such importance? One clue is that the atoms of all of these elements need to gain electrons to complete their outer energy levels (see Table 1–3). Thus they generally form covalent bonds. Because these atoms are small, the shared electrons in the bonds are held closely to the nuclei, producing very stable molecules. Moreover, with the exception of hydrogen, atoms of these elements can all form bonds with two or more atoms, making possible the formation of the large and complex molecules essential for the structures and functions of living systems.

TABLE 1-4 **Atomic Composition of Three Representative Organisms**

ELEMENT	HUMAN	ALFALFA	BACTERIUM
Carbon	19.37%	11.34%	12.14%
Hydrogen	9.31	8.72	9.94
Nitrogen	5.14	0.83	3.04
Oxygen	62.81	77.90	73.68
Phosphorus	0.63	0.71	0.60
Sulfur	0.64	0.10	0.32
CHNOPS total:	97.90%	99.60%	99.72%

LEVELS OF BIOLOGICAL ORGANIZATION

One of the fundamental principles of biology is that living things obey the laws of physics and chemistry. Organisms are made up of the same chemical components —atoms and molecules—as nonliving things. This does not mean, however, that organisms are "nothing but" the atoms and molecules of which they are composed. As we have seen, there are recognizable differences between living and nonliving systems. To understand the basis of these differences, let us consider the most thoroughly studied of all living things, the bacterium *Escherichia coli.**

---

* Biologists use a binomial ("two-name") system for designating organisms. Every different kind of organism has a unique two-part name. The first part of the name refers to the genus (plural, genera) to which the organism belongs; the second part, in combination with the first part, refers to the particular species, a subdivision of the genus category. In this name, for example, *Escherichia* denotes the genus, while *coli* designates a particular kind, or species, of *Escherichia,* distinguished from all others by certain characteristics. By convention, with the second mention of a scientific binomial, it is permissible to abbreviate the first (genus) name. This is fortunate, particularly when dealing with such names as *Escherichia.*

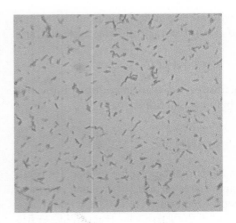

1–13 *Cells of* Escherichia coli, *as photographed through a light microscope. They have been stained with a dye that adheres to their surface, making them easier to see. Although these cells, magnified 450 times, are minute, their structure is quite complex and they exhibit all of the properties that characterize a living system.*

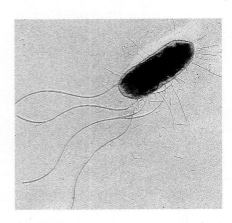

1–14 *An* E. coli *cell, magnified 11,280 times by the electron microscope. The short, rigid structures extending from the surface are used by the cell to attach to a food source or, when exchanging genetic information, to another* E. coli *cell. The longer, more flexible structures, which rotate at about 40 revolutions per second, propel the cell through the surrounding medium. When they rotate in a counterclockwise direction, the cell moves forward; when they rotate in a clockwise direction, it tumbles through the water and then, with the return of counterclockwise rotation, moves forward again.*

The atoms constituting this bacterium (see Table 1–4) are combined with each other in very specific ways. Much of the hydrogen and oxygen is present in the form of water, which accounts for most of the weight of *E. coli*. In addition to water, each bacterium contains about 5,000 different kinds of macromolecules (very large molecules). Some of these play structural roles, others regulate cell function, and nearly 1,000 of them are involved in decoding the genetic information. Some of the macromolecules interact with water to form a delicate, pliant film that encloses all the other atoms and molecules of which *E. coli* is composed. So enclosed, they constitute, remarkably, a cell, a living entity.

An *E. coli* cell is very small, appearing no bigger than a hyphen even when magnified by the most powerful light microscope, but it possesses some astonishing capacities. Like other living things, it can transform energy, taking molecules from the medium and using them to fuel its processes of growth and reproduction. It can exchange genetic information with other *E. coli* cells. It can move, propelling itself by the rotation of thin, flexible fibers attached to a structure that resembles—but considerably predates—an automobile transmission. The direction of motion is not random; *E. coli*, small as it is, has a number of different sensing devices, enabling it to detect and move toward nutrients and away from noxious substances.

E. coli is one of the most common of microscopic organisms. Its preferred residence is the human intestinal tract, where it lives in close association with the cells that form the lining of that tract. These human cells resemble *E. coli* in many important ways: they contain about the same proportions of the same six kinds of atoms, and, as in *E. coli*, these atoms are organized into macromolecules. However, the human cells are also very different from *E. coli*. For one thing, they are much larger; for another, they are much more complex. Most important, they are not independent entities, like the *E. coli* cells; each is part of a larger organism. Individual cells are specialized for particular functions that are subservient to the function of the organism as a whole. Each intestinal lining cell lives for only a few days; the organism, with any luck, will live for many decades.

E. coli, the cells of its human host, and other microorganisms living in the intestinal tract all interact with one another. Usually this takes place so uneventfully that we are unaware of the interactions, but occasionally we are reminded of the delicate balance. For example, many of us have had the experience of taking an antibiotic to cure one type of infection and ending up with another infection—usually caused by a type of yeast cell. What has happened is that the antibiotic has killed not only the bacteria causing our initial infection but also *E. coli* and the other normal inhabitants of our intestinal tract. Yeast cells are not susceptible to the antibiotic, and so they take over the territory, in much the same way that certain species of plants will quickly take over any patch of countryside from which the usual vegetation is removed.

E. coli and the cells with which it interacts illustrate what are known as levels of organization. The first level of organization with which biologists are usually concerned is subatomic—the particles that form atoms. The organization of these particles into atoms represents a second level, and the organization of atoms into molecules represents a third level. Although each level consists of components from the preceding level, the new organization of the components at a given level results in the emergence of new properties that are quite different from those of the preceding level. For example, hydrogen and oxygen are gases at ordinary temperatures, yet water—composed of hydrogen and oxygen—is a liquid with properties very different from those of either gas.

At a fourth level of organization, the most remarkable property of all emerges —life, in the form of a cell. Other properties emerge when individual, specialized cells are organized at a still higher level, in a multicellular organism. Organized in

1–15 *A gallery of cells.* (**a**) *Amoeba proteus, a single-celled organism named for Proteus, a Greek god capable of changing his shape at will. Extensions of the cell, known as pseudopodia, enable amoebas to move and to capture prey.*

(**b**) *This simple organism, called Pandorina, is made up of 32 cells, most of which are visible here, held together by a jellylike substance, whose outlines you can also see. Each of these cells can survive independently of the others. To reproduce, each cell divides, producing a new cell inside, and then the parent colony breaks apart.*

(**c**) *The embryo of a sea urchin at the two-celled stage. Within each cell is a nucleus that carries all the genetic information needed for every cell in the mature sea urchin.*

(**d**) *These cells are from the cerebral cortex of a human brain—the most highly organized structure on earth. The actions of the cells of the cerebral cortex and the interconnections among them are responsible for consciousness, intelligence, dreams, and memory.*

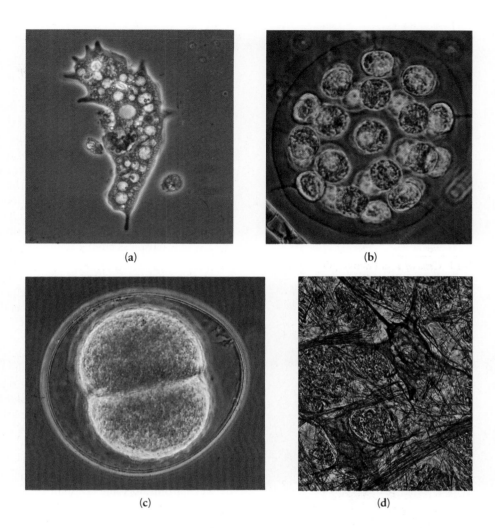

(a)

(b)

(c)

(d)

one way, the cells form a liver; in another way, the intestinal tract; in yet another, the human brain, which represents an extraordinary level of organizational complexity. Yet it, in turn, is only part of a larger entity whose characteristics are different from those of the brain, although they depend on the characteristics of the brain. Nor is the individual organism the ultimate level of biological order. Living organisms interact with each other, and finally, groups of living organisms are themselves part of an even vaster system of organization. This ultimate level of organization, the **biosphere,** involves not only the great diversity of plants and animals and microorganisms and their interactions with each other but also the physical characteristics of the environment and the planet earth itself.

The organization of this text parallels the levels of biological organization. In Part 1, we are beginning with atoms and molecules, and we will move on to examine the structure and activities of the living cell. In Part 2, our focus will be on individual organisms as we first consider their diversity and then, in more detail, the essential characteristics of plants and animals. Our view will expand further in Part 3, as we look at the interactions of organisms with each other. Over long periods of time these interactions give rise to evolutionary change; on a shorter time scale, they determine the organization of the communities of living organisms we find around us.

SUMMARY

Matter is composed of atoms, the smallest units of chemical elements. Atoms are made up of smaller particles. The nucleus of an atom contains positively charged protons and (except for hydrogen, ^1H) neutrons, which have no charge. The atomic number of an atom is equal to the number of protons in its nucleus. The atomic weight of an atom is the sum of the number of protons and neutrons in its nucleus. The chemical properties of an atom are determined by its electrons—small, negatively charged particles found outside the nucleus. The number of electrons in an atom equals the number of protons and thus the atomic number.

The nuclei of different isotopes of the same element contain the same number of protons but different numbers of neutrons. Thus the isotopes of an element have the same atomic number but different atomic weights.

The electrons of an atom have differing amounts of energy. Electrons closer to the nucleus have less energy than those farther from the nucleus and thus are at a lower energy level. An electron tends to occupy the lowest available energy level, but with an input of energy, it can be boosted to a higher energy level. When the electron returns to a lower energy level, energy is released.

The chemical behavior of an atom is determined by the number and arrangement of its electrons. An atom is most stable when all of its electrons are at their lowest possible energy levels and those energy levels are completely filled with electrons. The first energy level can hold two electrons, the second energy level can hold eight electrons, and so can the third energy level of the small atoms of greatest interest in biology. Chemical reactions between atoms result from the tendency of atoms to reach the most stable electron arrangement possible.

Particles consisting of two or more atoms are known as molecules, which are held together by chemical bonds. Two common types of bonds are ionic and covalent. Ionic bonds are formed by the mutual attraction of particles of opposite electric charge; such particles, formed when an electron jumps from one atom to another, are known as ions. In covalent bonds, pairs of electrons are shared between atoms; in some covalent bonds, known as polar covalent bonds, pairs of electrons are shared unequally, giving the molecule regions of positive and negative charge. Covalent bonds in which two atoms share two pairs of electrons (four electrons) are known as double bonds, and those in which they share three pairs of electrons (six electrons) are known as triple bonds.

Chemical reactions—exchanges of electrons among atoms—can be represented by chemical equations. Three general types of chemical reactions are (1) the combination of two or more substances to form a different substance, (2) the dissociation of a substance into two or more substances, and (3) the exchange of atoms among two or more substances. Substances that consist of the atoms of two or more different elements, in definite and constant proportions, are known as chemical compounds.

Living things are made up of the same chemical and physical components as nonliving things, and they obey the same chemical and physical laws. Six elements (CHNOPS) make up 99 percent of all living matter. The atoms of all of these elements are small and form tight, stable covalent bonds. With the exception of hydrogen, they can all form covalent bonds with two or more atoms, giving rise to the complex molecules that characterize living systems.

The properties of a complex molecule depend upon the organization of the atoms within the molecule. Similarly, the properties of a living cell depend upon the organization of molecules within the cell, and the properties of a multicellular organism depend upon the organization of the cells within its body. The ultimate level of biological organization, the biosphere, results from the interactions of the plants, animals, and microorganisms of the earth with each other and with physical factors in the environment.

QUESTIONS

1. Describe the three types of particles of which atoms are composed. What is the atomic number of an atom? The atomic weight?

2. For each of the following isotopes, determine the number of protons and neutrons in the nucleus: (a) ^{11}C, ^{12}C, ^{14}C; (b) ^{31}P, ^{32}P, ^{33}P; (c) ^{32}S, ^{35}S, ^{38}S.

3. Consider the isotopes of phosphorus listed in Question 2. Would you expect all three of these isotopes to exhibit the same chemical properties in a living organism? Why or why not?

4. Although no model of the atom gives us an exact "picture," different models can help us to understand important characteristics of atoms. What characteristic of the atom was stressed by the planetary model? What important characteristic of electrons is emphasized by the Bohr model? What additional information about electrons is provided by the orbital model?

5. The street lights in many cities contain bulbs filled with sodium vapor. When electrical energy is passed through the bulb, a brilliant yellow light is given off. What is happening to the sodium atoms to cause this?

6. What is the difference between an energy level and an orbital? How many electrons can the first energy level of an atom hold? The second energy level? The third energy level?

7. Determine the number of protons, the number of neutrons, the number of energy levels, and the number of electrons in the outermost energy level in each of the following atoms: oxygen, nitrogen, carbon, sulfur, phosphorus, chlorine, potassium, and calcium.

8. How many electrons does each of the atoms in Question 7 need to share, gain, or lose to acquire a completed outer energy level?

9. Magnesium has an atomic number of 12. How many electrons are in its first energy level? Its second energy level? Its third energy level? How would you expect magnesium and chlorine to interact? Write the formula for magnesium chloride.

10. Explain the differences between ionic, covalent, and polar covalent bonds. What tendency of atoms causes them to interact with each other, forming bonds?

11. Molecules that contain polar covalent bonds typically have regions of positive and negative charge and thus are polar. However, some molecules containing polar covalent bonds are nonpolar. Explain how this is possible.

12. Knowing that chemical reactions have to be balanced, fill the appropriate numbers into the underlined spaces (*hint*: from 1 to 6 in all cases):

(a) _____H_2CO_3 $\longrightarrow$ _____H_2O + _____CO_2
   Carbonic
   acid

(b) _____H_2 + _____N_2 $\longrightarrow$ _____NH_3
                            Ammonia

(c) _____$NaOH$ + _____H_2CO_3 $\longrightarrow$ _____Na_2CO_3 + _____H_2O
   Sodium   Sodium
   hydroxide   carbonate

(d) _____CH_3OH + _____O_2 $\longrightarrow$ _____CO_2 + _____H_2O
   Methyl
   alcohol

(e) _____O_2 + _____$C_6H_{12}O_6$ $\longrightarrow$ _____H_2O + _____CO_2
                Glucose

13. What six elements make up the bulk of living tissue? What characteristics do the atoms of these six elements share?

Water

In this chapter and the next, we are going to examine the molecules of which living things are composed. By far the most abundant of these molecules is water, which makes up 50 to 95 percent of the weight of any functioning living system.

Life on this planet began in water, and today, wherever liquid water is found, life is also present. There are one-celled organisms that eke out their entire existence in no more water than can cling to a grain of sand. Some kinds of algae are found only on the melting undersurfaces of polar ice floes. Certain bacteria can tolerate the near-boiling water of hot springs. In the desert, plants race through an entire life cycle—seed to flower to seed—following a single rainfall. In the tropical rain forest, the water cupped in the leaves of a plant forms a microcosm in which a myriad of small organisms are born, spawn, and die.

Water is the most common liquid on earth. Three-fourths of the surface of the earth is covered by water. In fact, if the earth's land surface were absolutely smooth, all of it would be 2.5 kilometers* under water. But do not mistake "common" for "ordinary"; water is not in the least an ordinary liquid. Compared with other liquids it is, in fact, quite extraordinary. If it were not, it is unlikely that life on earth could ever have evolved.

---

* A metric table with English equivalents is provided in Appendix A.

2–1 *The first living systems came into being, according to present hypotheses, in the warm primitive seas, and for many organisms, ourselves included, each new individual begins life bathed and cradled in water. These are salamander larvae.*

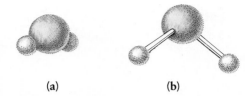

2-2 *The structure of the water molecule (H₂O) can be depicted in several different ways.* (**a**) *In the space-filling model, the oxygen atom is represented by the red sphere and the hydrogen atoms by the blue spheres. Because of its simplicity, this model is often used as a convenient symbol of the water molecule.*

(**b**) *The ball-and-stick model emphasizes that the atoms are joined by covalent bonds; it also gives some indication of the geometry of the molecule. A more accurate description of the molecule's shape is provided by the orbital model in Figure 2–3a.*

THE STRUCTURE OF WATER

In order to understand why water is so extraordinary and how, as a consequence, it can play its unique and crucial role in relation to living systems, we must look again at its molecular structure. Each water molecule is made up of two atoms of hydrogen and one atom of oxygen (Figure 2–2). Each of the hydrogen atoms is linked to the oxygen atom by a covalent bond; that is, the single electron of each hydrogen atom is shared with the oxygen atom, which also contributes an electron to each bond.

The water molecule as a whole is neutral in charge, having an equal number of electrons and protons. However, the molecule is polar (page 33). Because of the very strong attraction of the oxygen nucleus for electrons, the shared electrons of the covalent bonds spend more time around the oxygen nucleus than they do around the hydrogen nuclei. As a consequence, the region near each hydrogen nucleus is a weakly positive zone. Moreover, the oxygen atom has four additional electrons in its outer energy level. These electrons are paired in two orbitals that are not involved in covalent bonding to hydrogen. Each of these orbitals is a weakly negative zone. Thus, the water molecule, in terms of its polarity, is four-cornered, with two positively charged "corners" and two negatively charged ones (Figure 2–3a).

When one of these charged regions comes close to an oppositely charged region of another water molecule, the force of attraction forms a bond between them, which is known as a **hydrogen bond.** Hydrogen bonds are found not only in water but also in many large molecules, where they help maintain structural stability. They are, however, very specific. A hydrogen bond can form only between a hydrogen atom that is covalently bonded to an atom that has a strong attraction for electrons and a strongly electron-attracting atom in another molecule. In the molecules found in living systems, hydrogen bonding typically occurs between a hydrogen atom covalently bonded to either oxygen or nitrogen and the oxygen or nitrogen atom of another molecule. In water, a hydrogen bond forms between a negative "corner" of one water molecule and a positive "corner" of another. Every water molecule can establish hydrogen bonds with four other water molecules (Figure 2–3b).

Any single hydrogen bond is significantly weaker than either a covalent or an ionic bond. Moreover, it has an exceedingly short lifetime; on an average, each hydrogen bond in liquid water lasts approximately 1/100,000,000,000th of a second. But, as one is broken, another is made. All together, the hydrogen bonds have considerable strength, causing the water molecules to cling together as a liquid under ordinary conditions of temperature and pressure.

Now let us look at some of the consequences of these attractions among water molecules, especially as they affect living systems.

2-3 *The polarity of the water molecule and its consequences.* (**a**) *As shown in this model, four orbitals branch off from the oxygen nucleus of a water molecule. Two of the orbitals are formed by the shared electrons bonding the hydrogen atoms to the oxygen atom. They have a slightly positive charge. The other two orbitals have a slightly negative charge.*

(**b**) *As a result of these positive and negative zones, each water molecule can form hydrogen bonds (dashed lines) with four other water molecules. Under ordinary conditions of pressure and temperature, the hydrogen bonds are continually breaking and re-forming in a shifting pattern. Thus water is a liquid.*

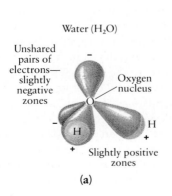

Water (H₂O)

Unshared pairs of electrons— slightly negative zones

Oxygen nucleus

Slightly positive zones

(a)

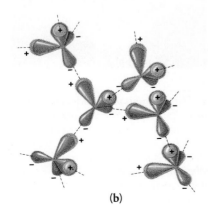

(b)

2–4 *This remarkable photo shows a kingfisher, just as it breaks the surface of the water in a dive for food. Note the many droplets surrounding the diving bird and the continuous sheet formed by the surface of the water. These are both results of the surface tension of water.*

CONSEQUENCES OF THE HYDROGEN BOND

Surface Tension

Look at water dripping from a faucet. Each drop clings to the rim and dangles for a moment by a thread of water; then just as the tug of gravity breaks it loose, its outer surface is drawn taut, to form a sphere as the drop falls free. Gently place a needle or a razor blade flat on the surface of the water in a glass. Although the metal is denser than water, it floats. Look at a pond in spring or summer; you will see water striders and other insects walking on its surface almost as if it were solid. These phenomena are all the result of **surface tension.** Surface tension is a result of the cohesion, or clinging together, of the water molecules. (**Cohesion** is, by definition, the holding together of molecules of the same substance. **Adhesion** is the holding together of molecules of different substances.)

The only liquid with a surface tension greater than that of water is mercury. Atoms of mercury are so greatly attracted to one another that they tend not to adhere to anything else. Water, however, because of its negative and positive charges, adheres strongly to any other charged molecules and to charged surfaces. The "wetting" capacity of water—that is, its ability to coat a surface—results from its polar structure, as does its cohesiveness.

Capillary Action and Imbibition

If you hold two dry glass slides together and dip one corner in water, the combination of cohesion and adhesion will cause water to spread upward between the two slides. This is **capillary action.** Capillary action similarly causes water to rise in very fine glass tubes, to creep up a piece of blotting paper, or to move slowly through the minute spaces between soil particles and so become available to the roots of plants.

Imbibition ("drinking up") is the capillary movement of water molecules into substances such as wood or gelatin, which swell as a result. The pressures developed by imbibition can be astonishingly great. It is said that stone for the ancient Egyptian pyramids was quarried by driving wooden pegs into holes drilled in the rock face and then soaking the pegs with water. The swelling of the wood created a force great enough to break the stone slab free. Seeds imbibe water as they begin to germinate, swelling and bursting their seed coats (Figure 2–5).

2–5 *The germination of seeds begins with changes in the seed coat that permit a massive uptake of water. The embryo and surrounding structures then swell, bursting the seed coat. In this acorn, photographed on the forest floor, the embryonic root has emerged through the tough outer layers of the fruit.*

TABLE 2-1 **Comparative Specific Heats (The quantity of heat, in calories, required to raise the temperature of 1 gram through 1°C)**

SUBSTANCE	SPECIFIC HEAT
Liquid ammonia	1.23
Water	1.00
Ethyl alcohol (ethanol)	0.60
Sugar (sucrose)	0.30
Chloroform	0.24
Salt (NaCl)	0.21
Glass	0.20
Iron	0.10
Lead	0.03

TABLE 2-2 **Comparative Heats of Vaporization (The quantity of heat, in calories, required to convert 1 gram of liquid to 1 gram of gas)**

LIQUID	HEAT REQUIRED
Water (at 0°C)	596
Water (at 100°C)	540
Hydrofluoric acid	360
Ammonia	295
Ethyl alcohol (ethanol)	236.5
Nitric acid	115
Carbon dioxide	72.2
Chlorine	67.4
Ether	9.4

Resistance to Temperature Change

If you go swimming in the ocean or a lake on one of the first hot days of summer, you will quickly be aware of a striking difference between the air temperature and the water temperature. This difference occurs because a greater input of energy is required to raise the temperature of water than to raise the temperature of air. The amount of heat a given amount of a substance requires for a given increase in temperature is its **specific heat** (also called heat capacity). One calorie* is defined as the amount of heat that will raise the temperature of 1 gram (1 milliliter or 1 cubic centimeter) of water 1°C. The specific heat of water is about twice the specific heat of oil or alcohol; that is, approximately 0.5 calorie will raise the temperature of 1 gram of oil or alcohol 1°C. It is four times the specific heat of air or aluminum and 10 times that of iron. Only liquid ammonia has a higher specific heat (Table 2–1).

Heat is a form of energy—the **kinetic energy,** or energy of motion, of molecules. Molecules are always moving; they vibrate, rotate, and shift position in relation to other molecules. Heat, which is measured in calories, reflects the *total* kinetic energy in a collection of molecules; it includes both the magnitude of the molecular movements and the mass and number of moving molecules present. By contrast, temperature, which is measured in degrees, reflects the *average* kinetic energy of the molecules. Thus, heat and temperature are not identical. For example, a lake may have a lower temperature than does a bird flying over it, but the lake contains more heat because it has many more molecules in motion.

The high specific heat of water is a consequence of hydrogen bonding. The hydrogen bonds in water tend to restrict the movement of the molecules. In order for the kinetic energy of water molecules to increase sufficiently for the temperature to rise 1°C, it is necessary first to rupture a number of the hydrogen bonds holding the molecules together. When you heat a pot of water, much of the heat energy added to the water is used in breaking the hydrogen bonds between the water molecules. Only a relatively small amount of heat energy is therefore available to increase molecular movement.

What does the high specific heat of water mean in biological terms? It means that for a given rate of heat input, the temperature of water will rise more slowly than the temperature of almost any other material. Conversely, the temperature will drop more slowly as heat is removed. Because so much heat input or heat loss is required to raise or lower the temperature of water, organisms that live in the oceans or large bodies of fresh water live in an environment where the temperature is relatively constant. Also, the high water content of terrestrial plants and animals helps them to maintain a relatively constant internal temperature. This constancy of temperature is critical because biologically important chemical reactions take place only within a narrow temperature range.

Vaporization

Vaporization—or evaporation, as it is more commonly called—is the change from a liquid to a gas. Water has a high **heat of vaporization**. At water's boiling point (100°C at a pressure of 1 atmosphere), it takes 540 calories to change 1 gram of liquid water into vapor, almost 60 times as much as for ether and twice as much as for ammonia (Table 2–2).

Hydrogen bonding is also responsible for water's high heat of vaporization. Vaporization comes about because some of the most rapidly moving molecules of a liquid break loose from the surface and enter the air. The hotter the liquid, the more rapid the movement of its molecules and, hence, the more rapid the rate of

---

* Nutritional calories are actually kilocalories (kcal); 1 kilocalorie equals 1,000 calories.

"Ammonia! Ammonia!"

[Drawing by R. Grossman, © 1962
The New Yorker Magazine, Inc.]

2–6 *Ammonia is very similar to water in its molecular structure, and biologists have speculated about whether it might substitute for water in life processes. The ammonia molecule (NH_3) is made up of hydrogen atoms covalently bonded to nitrogen, which, like the oxygen in the water molecule, retains a slight negative charge. However, because there are three hydrogens to one nitrogen, the charge difference between the positive and negative zones in the ammonia molecule is not as great as in the water molecule, and the hydrogen bonds formed by ammonia are slightly weaker than those formed by water. Moreover, the 3:1 ratio of hydrogen to nitrogen makes it difficult for ammonia molecules to form an interlocking network. As a consequence, ammonia does not have the cohesive power of water and evaporates much more quickly. Perhaps this is why no form of life based on ammonia has been found, although NH_3 may have been very common in the primitive atmosphere.*

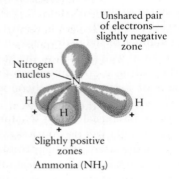

Unshared pair
of electrons—
slightly negative
zone

Nitrogen
nucleus

–

N

H
+

H

H
+

+

Slightly positive
zones

Ammonia (NH_3)

evaporation. But, whatever the temperature, so long as a liquid is exposed to air that is less than 100 percent saturated with the vapor of that liquid, evaporation will take place, right down to the last drop.

In order for a water molecule to break loose from its fellow molecules—that is, to vaporize—the hydrogen bonds have to be broken. This requires heat energy. As a consequence, when water evaporates, as from the surface of your skin or a leaf, the escaping molecules carry a great deal of heat away with them. Thus evaporation has a cooling effect. Evaporation from the surface of a land-dwelling plant or animal is one of the principal ways in which these organisms "unload" excess heat and so stabilize their temperatures.

Freezing

Water exhibits another peculiarity when it undergoes the transition from a liquid to a solid (ice). In most liquids, the **density**—that is, the weight of the material in a given volume—increases as the temperature drops. This greater density occurs because the individual molecules are moving more slowly and so the spaces between them decrease. The density of water also increases as the temperature drops, until it nears 4°C. Then the water molecules come so close together and are moving so slowly that *every one* of them can form hydrogen bonds simultaneously with four other molecules—something they could not do at higher temperatures. However, the geometry of the water molecule is such that, as the temperature drops below 4°C, the molecules must move slightly apart from each other to maintain the maximum number of hydrogen bonds in a stable structure. At 0°C, the freezing point of water, this creates an open latticework (Figure 2–7) that is the most stable structure for an ice crystal. Thus water as a solid takes up more volume than water as a liquid. Ice is less dense than liquid water and therefore floats in it.

This increase in volume has occasional disastrous effects on water pipes but, on the whole, turns out to be enormously beneficial for life forms. If water continued to contract as it froze, ice would be heavier than liquid water. As a result, lakes and ponds and other bodies of water would freeze from the bottom up. Once ice began to accumulate on the bottom, it would tend not to melt, season after season. Spring and summer might stop the freezing process, but laboratory experiments have shown that if ice is held to the bottom of even a relatively shallow tank, water can be boiled on the top without melting the ice. Thus if water did not expand when it froze, it would continue to freeze from the bottom up, year after year, and never melt again. Eventually, the body of water would freeze solid and any life in it would be destroyed. By contrast, the layer of floating ice that actually forms tends to protect the organisms in the water. The ice layer effectively insulates the liquid water beneath it, keeping its temperature at or above the freezing point of water.

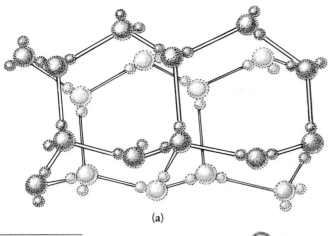

(a)

(b)

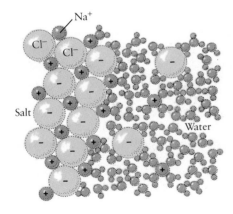

○ Oxygen

○ Hydrogen

2–7 (a) *In the crystalline structure of ice, each water molecule is hydrogen-bonded to four other water molecules in a three-dimensional open latticework. The bond angles in some of the water molecules are distorted as they link up in a hexagonal arrangement. This arrangement, shown here in a small section of the latticework, is repeated throughout the crystal and is responsible for the beautiful patterns seen in snowflakes and frost. The water molecules are actually farther apart in ice than they are in liquid water.*

(b) *When water freezes in the cracks and crevices of rock, the force created by its expansion splits the rock. Over long periods of time, this process breaks up masses of rock and contributes to the formation of soil.*

2–8 *Because of the polarity of water molecules, water can serve as a solvent for ionic substances and polar molecules. This diagram shows sodium chloride (NaCl) dissolving in water as the water molecules cluster around the individual sodium and chloride ions, separating them from one another. Notice the difference between the way the water molecules are arranged around the sodium ions and the way they are arranged around the chloride ions.*

The melting point of water is 0°C, the same temperature as the freezing point. To make the transition from a solid to a liquid, water requires 79.7 calories per gram, a quantity known as the **heat of fusion.** As ice melts, it draws this much heat from its surroundings, thereby cooling the surroundings. The heat energy absorbed by the ice breaks the hydrogen bonds of the latticework. Conversely, as water freezes, it releases the same amount of heat into its surroundings. In this way, ice and snow also serve as temperature stabilizers, particularly during the transition periods of fall and spring. Moderation of sudden changes in temperature gives organisms time to make seasonal adjustments essential to survival.

The presence of dissolved substances in water lowers the temperature at which water freezes, which is why salt is thrown on icy sidewalks and used in ice-cream freezers. The "hardening" process in several species of winter-hardy plants, by which they prepare themselves for cold weather, includes the breakdown of starch (which is insoluble in the fluids of the plant cell) into simple sugars (which are soluble). Freshwater fish, whose body fluids are salty compared to the pond or lake in which they live, do not freeze when the temperature of water is at or near 0°C. However, logically speaking, saltwater fish, whose body fluids are less salty than the ocean water surrounding them, should freeze at the below-zero temperatures of Arctic water. They do not, however, and animal physiologists investigating this phenomenon have discovered that at least one species, the ghost fish, produces a complex protein named, appropriately, antifreeze protein. This protein, which is secreted into the bloodstream, appears to interfere with the formation of the crystalline structure of ice. Recently, studies of several species of terrestrial frogs that spend the winter hibernating beneath leaf litter have revealed that their body fluids contain a high concentration of glycerol, one of the ingredients sometimes used in automobile antifreeze.

WATER AS A SOLVENT

Many substances within living systems are found in aqueous solution. (A **solution** is a uniform mixture of the molecules of two or more substances. The substance present in the greatest amount—usually a liquid—is called the **solvent,** and the substances present in lesser amounts are called **solutes.**) The polarity of water molecules is responsible for water's capacity as a solvent. The polar water molecules tend to separate ionic substances, such as sodium chloride (NaCl), into their constituent ions. As shown in Figure 2–8, the water molecules cluster around and segregate the charged ions.

The Seasonal Cycle of a Lake

As we have seen, water increases in density as its temperature drops, until it reaches 4°C, the temperature of maximum density. Water either colder or warmer than 4°C is less dense and floats above water at 4°C. As a result, the water of temperate-zone lakes is stratified in the summer and winter but undergoes considerable mixing in the fall and spring. The stratifications of summer and winter enable lake-dwelling organisms to avoid life-threatening temperature extremes, while the mixing that occurs in fall and spring provides nutrients and oxygen to organisms at all levels of the lake.

In the summer, the top layer of water, called the epilimnion, is heated by the sun and the surrounding air, becoming warmer than the lower layers. Since it becomes less dense as it becomes warmer, this water remains at the surface. Only the water in the epilimnion circulates. In the middle layer, there is an abrupt drop in temperature, known as the thermocline. Since the water in this layer is progressively more dense, it does not mix with the lighter water above. The water of the middle layer effectively cuts off the circulation

of oxygen from the surface into the third layer, the hypolimnion. As the organisms of the hypolimnion gradually use up the available oxygen, the summer stagnation results (a).

In the fall, the temperature of the epilimnion drops until it is the same as that of the hypolimnion. The warmer water in the middle layer then rises to the surface, producing the fall overturn (b). Aided by the fall winds, water begins to circulate throughout the lake (c); oxygen is returned to the depths, and nutrients released by the activities of bottom-dwelling bacteria are carried to the upper layers of the lake.

As winter deepens, the surface water cools below 4°C, becoming lighter as it expands. This water remains on the surface and, in many areas, freezes. The result is winter stratification (d).

In the spring, as the ice melts and the water on the surface warms to 4°C, it sinks to the bottom, producing the spring overturn (e). Another thorough mixing of the water in the lake follows, comparable to that shown in (c).

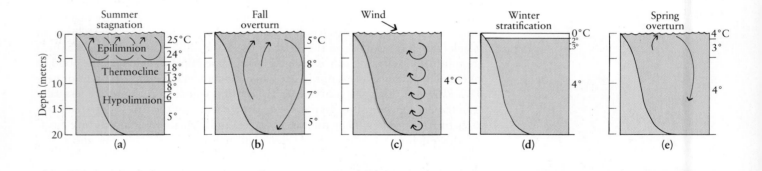

Many of the covalently bonded molecules important in living systems—such as sugars—have regions of partial positive and negative charge. (Such polar regions arise, as you might expect, in the neighborhood of covalently bonded atoms whose nuclei exert differing degrees of attraction for electrons.) These molecules therefore attract water molecules and also dissolve in water. Polar molecules that readily dissolve in water are often called **hydrophilic** ("water-loving"). Such molecules slip into aqueous solution easily because their partially charged regions attract water molecules as much as or more than they attract each other. The polar water molecules thus compete with the attraction between the solute molecules themselves.

Molecules, such as fats, that lack polar regions tend to be very insoluble in water. The hydrogen bonding between the water molecules acts as a force to exclude the nonpolar molecules. As a result of this exclusion, nonpolar molecules tend to cluster together in water, just as droplets of fats tend to coalesce, for example, on the surface of chicken soup. Such molecules are said to be **hydrophobic** ("water-fearing"), and the clusterings are known as hydrophobic interactions.

We will encounter these properties of hydrophilic and hydrophobic molecules again in later chapters. These weak forces—hydrogen bonds and hydrophobic forces—play crucial roles in shaping the architecture of large, biologically important molecules and, as a consequence, in determining their properties.

IONIZATION OF WATER: ACIDS AND BASES

In liquid water, there is a slight tendency for a hydrogen atom to jump from the oxygen atom to which it is covalently bonded to the oxygen atom to which it is hydrogen-bonded (Figure 2–9). In this reaction, two ions are produced: the hydronium ion (H_3O^+) and the hydroxide ion (OH^-). In any given volume of pure water, a small but constant number of water molecules will be ionized in this way. The number is constant because the tendency of water to ionize is offset by the tendency of the ions to reunite; thus even as some molecules are ionizing, an equal number of others are forming, a state known as dynamic **equilibrium.**

2–9 *When water ionizes, a hydrogen nucleus (that is, a proton) shifts from the oxygen atom to which it is covalently bonded to the oxygen atom to which it is hydrogen-bonded. The resulting ions are the negatively charged hydroxide ion and the positively charged hydronium ion. In this diagram, the large spheres represent oxygen and the small spheres represent hydrogen.*

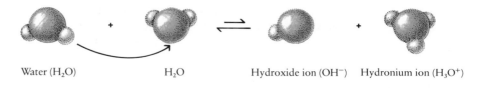

Water (H_2O) H_2O Hydroxide ion (OH^-)   Hydronium ion (H_3O^+)

Although the positively charged ion formed when water ionizes is the hydronium ion (H_3O^+), rather than the hydrogen ion (H^+), by convention the ionization of water is expressed by the equation:

$$HOH \rightleftharpoons H^+ + OH^-$$

The arrows indicate that the reaction goes in both directions. The fact that the arrow pointing toward HOH is longer indicates that, at equilibrium, most of the H_2O is not ionized. As a consequence, in any sample of pure water, only a small fraction exists in ionized form.

In pure water, the number of H^+ ions exactly equals the number of OH^- ions. This is necessarily the case since neither ion can be formed without the other when only H_2O molecules are present. However, when an ionic substance or a substance with polar molecules is dissolved in water, it may change the relative numbers of H^+ and OH^- ions. For example, when hydrogen chloride (HCl) dissolves in water, it is almost completely ionized into H^+ and Cl^- ions; as a result, an HCl solution (hydrochloric acid) contains more H^+ ions than OH^- ions. Conversely, when sodium hydroxide (NaOH) dissolves in water, it forms Na^+ and OH^- ions; thus, in a solution of sodium hydroxide in water, there are more OH^- ions than H^+ ions.

A solution acquires the properties we recognize as acidic when the number of H^+ ions exceeds the number of OH^- ions; conversely, a solution is basic (alkaline) when the number of OH^- ions exceeds the number of H^+ ions. Thus, an **acid** is a substance that causes an increase in the relative number of H^+ ions in a solution, and a **base** is a substance that causes an increase in the relative number of OH^- ions.

Strong and Weak Acids and Bases

Strong acids and bases are substances, like HCl and NaOH, that ionize almost completely in water, resulting in relatively large increases in the concentrations of H^+ and OH^- ions, respectively. Weak acids and bases, by contrast, are those that ionize only slightly, resulting in relatively small increases in the concentration of H^+ or OH^- ions.

Because of the strong tendency of H^+ and OH^- ions to combine and the weak tendency of water to ionize, the concentration of OH^- ions will always decrease as the concentration of H^+ ions increases (as, for example, when HCl is added to water), and vice versa. If HCl is added to a solution containing NaOH, the following reaction will take place:

$$H^+ + Cl^- + Na^+ + OH^- \longrightarrow H_2O + Na^+ + Cl^-$$

In other words, if an acid and a base of comparable strength are added in equivalent amounts, the solution will not have an excess of either H^+ or OH^- ions.

Many of the acids important in living systems owe their acidic properties to a group of atoms called the carboxyl group, which includes one carbon atom, two oxygen atoms, and a hydrogen atom (symbolized as —COOH). When a substance containing a carboxyl group is dissolved in water, some of the —COOH groups dissociate to yield hydrogen ions:

$$—COOH \rightleftharpoons —COO^- + H^+$$

Thus compounds containing carboxyl groups are hydrogen-ion donors, or acids. They are weak acids, however, because, as indicated by the arrows, the —COOH ionizes only slightly.

Among the most important bases in living systems are compounds that contain the amino group (—NH_2). This group has a weak tendency to accept hydrogen ions, thereby forming —NH_3^+:

$$—NH_2 + H^+ \rightleftharpoons —NH_3^+$$

As hydrogen ions are removed from solution by the amino group, the relative concentration of H^+ ions decreases and the relative concentration of OH^- ions increases. Groups, such as —NH_2, that are weak hydrogen-ion acceptors are thus weak bases.

The pH Scale

Chemists express degrees of acidity by means of the **pH scale.** The symbol "pH" is derived from the French *pouvoir hydrogène* ("hydrogen power"). It stands for the negative logarithm of the concentration of hydrogen ions in moles per liter. Although this sounds complicated, in practice it is relatively simple. As you may recall from your mathematics courses, the logarithm is the exponential power to which a specified number (commonly 10) must be raised to equal a given number. For example, the logarithm of 100 is 2, since 100 equals 10^2 (that is, 10×10). The logarithm of 1/100 is -2, since 1/100 equals 10^{-2} (that is, $1/10 \times 1/10$). The numbers whose logarithms are of interest to us are the concentrations of hydrogen ions in solutions, expressed in moles per liter.

A **mole** is the amount of an element equivalent to its atomic weight expressed in grams, or the amount of a substance equivalent to its molecular weight expressed in grams. (The **molecular weight** of a substance is the sum of the atomic weights of the atoms constituting the molecule.) Thus, a mole of atomic hydrogen (atomic weight 1) is 1 gram of hydrogen atoms; a mole of atomic oxygen (atomic weight 16) is 16 grams of oxygen atoms; and a mole of water (molecular weight 18) is 18 grams of water molecules. The most interesting thing about the mole is that a mole—of any substance—contains the same number of particles as any other mole. This number, known as Avogadro's number, is 6.02×10^{23}. Thus, a mole of water molecules (18 grams) contains exactly the same number of molecules as a mole of hydrogen chloride molecules (36.5 grams). Use of the mole in specifying quantities of substances involved in chemical reactions makes it possible for us to consider comparable numbers of reacting particles.

The ionization that occurs in a liter of pure water results in the formation, at equilibrium, of 1/10,000,000 mole of hydrogen ions (and, as we noted earlier, of exactly the same quantity of hydroxide ions). In decimal form, this concentration of hydrogen ions is written as 0.0000001 mole per liter. This same concentration of hydrogen ions can be written even more conveniently in exponential form as 10^{-7} mole per liter. The logarithm is the exponent, -7, and the negative logarithm is 7; in terms of the pH scale, it is referred to simply as pH 7 (see Table 2–3). At pH 7, the concentrations of free H^+ and OH^- are exactly the same, as they are in pure water. This is a neutral state. Any pH below 7 is acidic, and any pH above 7 is basic. The lower the pH number, the higher the concentration of hydrogen ions. Thus pH 2 means 10^{-2} mole of hydrogen ions per liter of water, or 1/100 mole per liter (0.01 mole per liter)—which is, of course, a much larger figure than 1/10,000,000 (0.0000001). Since the pH scale is logarithmic, a difference of one pH unit represents a tenfold difference in the concentration of hydrogen ions; for example, a solution at pH 3 has 1,000 times as many H^+ ions as a solution at pH 6.

We can now define "acid" and "base" more fully:

1. An acid is a substance that causes an increase in the number of H^+ ions and a decrease in the number of OH^- ions in a solution. Most acids are hydrogen-ion donors, but some acids function by removing OH^- ions from the solution. A solution with a pH below 7 (with more than 10^{-7} mole of H^+ ions per liter) is acidic.

2. A base is a substance that causes a decrease in the number of H^+ ions and an increase in the number of OH^- ions in a solution. Some bases, such as NaOH, donate OH^- ions to the solution; others, such as the $-NH_2$ group, are hydrogen-ion acceptors, removing H^+ ions from the solution. A solution with a pH above 7 (with less than 10^{-7} mole of H^+ ions per liter) is basic.

TABLE 2–3 The pH Scale

		CONCENTRATION OF H⁺ IONS (MOLES PER LITER)	pH	CONCENTRATION OF OH⁻ IONS (MOLES PER LITER)
Increasing H⁺ / Decreasing OH⁻	Acidic	$1.0 = 10^{0}$	0	10^{-14}
		$0.1 = 10^{-1}$	1	10^{-13}
		$0.01 = 10^{-2}$	2	10^{-12}
		$0.001 = 10^{-3}$	3	10^{-11}
		$0.0001 = 10^{-4}$	4	10^{-10}
		$0.00001 = 10^{-5}$	5	10^{-9}
		$0.000001 = 10^{-6}$	6	10^{-8}
	Neutral	$0.0000001 = 10^{-7}$	7	$10^{-7} = 0.0000001$
Decreasing H⁺ / Increasing OH⁻	Basic	10^{-8}	8	$10^{-6} = 0.000001$
		10^{-9}	9	$10^{-5} = 0.00001$
		10^{-10}	10	$10^{-4} = 0.0001$
		10^{-11}	11	$10^{-3} = 0.001$
		10^{-12}	12	$10^{-2} = 0.01$
		10^{-13}	13	$10^{-1} = 0.1$
		10^{-14}	14	$10^{0} = 1.0$

Acid Rain

The average pH of normal rainfall is about 5.6 (mildly acidic), a result of the combination of carbon dioxide with water vapor to produce carbonic acid. In the 1920s, however, the pH of rain and snow in Scandinavia began to drop, and by the 1950s, similar phenomena were observed elsewhere in Europe and in the northeastern United States. As more data were collected, it was found that, in certain geographic areas, the average annual pH of precipitation was between 4.0 and 4.5. Occasional storms would release rain with a pH as low as 2.1, which is extremely acidic.

The low pH was traced primarily to two acids found in the rainfall: sulfuric (H_2SO_4) and nitric (HNO_3), both of which ionize almost completely in aqueous solution, releasing hydrogen ions. These acids are formed when the gaseous oxides of sulfur and nitrogen react with water vapor and other gases in the air. Sulfur and nitrogen oxides are released into the atmosphere by some natural processes (for example, volcanic eruptions), but far greater quantities are released as a result of human activities. Sulfur oxides are produced by the combustion of high-sulfur coal and oil and by the smelting of sulfur-containing ores. Nitrogen oxides are by-products of gasoline combustion in automobile engines and of some generating processes for electricity.

That sulfur oxides could be damaging to vegetation was evident as early as the turn of the century, when a large copper smelter was opened in a mountainous area in Tennessee. Within a few years, all vegetation had been killed in the formerly luxuriant forest surrounding the smelter. The solution devised for this problem—still used today—was to build very tall smokestacks, so the wind would carry the pollutants away from the immediate area. It was assumed that they would be so widely dispersed that they would be rendered harmless. By the 1960s, accumulating evidence indicated that sulfur oxides released from tall smokestacks are transported hundreds or thousands of miles by the prevailing winds (generally west to east in the Northern Hemisphere) and are then returned to the earth in rain and snow. Nitrogen oxides released from automobiles are also carried off by the wind. What was once a local problem has become an international problem, in which the pollutants respect no boundaries.

The biological consequences of acid rain depend in part on the characteristics of the soil and underlying rock on which it falls. In areas where the principal rock is limestone (calcium carbonate), the buffering action of the H_2CO_3–HCO_3^- system (see page 52) can generally prevent acidification of the soil, lakes, and streams. In other areas, where the soil and bodies of water do not contain such a natural buffer, the pH drops gradually but steadily as a result of repeated additions of acid rain. The drop in pH is often quite sudden and extreme when the spring melt brings an infusion of acid accumulated in the winter snows. Although the low pH resulting from the spring melt is usually temporary, it can be particularly devastating to salamanders and frogs, many of which lay their eggs in small ponds and puddles formed by the melt water.

Lakes in mountainous regions are especially vulnerable to acid rain. A 1977 Cornell University study of the lakes at high elevations (above 600 meters) in the western Adirondack Mountains of New York found that 51 percent had a pH below 5.0, of which 90 percent were devoid of fish life. By contrast, a similar study performed between 1929 and 1937 found that only 4 percent of the lakes were acidic and without fish. More recent studies indicate increasingly acidic conditions in lakes at lower levels in the Adirondacks (which were previously unaffected) and in many lakes of the Cascade Mountains in the Pacific Northwest. The effects of low pH on fish include the depletion of calcium in their bodies, leading to weakened and deformed bones; the failure of many eggs to hatch and deformed fish from those that do hatch; and the clogging of the gills by aluminum, which is released from the soil by acid.

The effects of acid rain on plants depend on both the species and the soil conditions. Among the observed effects are reduced germination of seeds, a decrease in the number of seedlings that mature, reduced growth, and lowered resistance to disease. If the soil is not adequately buffered, essential nutrients are leached from it and are thus unavailable to the plants. Recently, it has become apparent that the forests of the eastern United States, from Maine through Georgia, are in serious decline. Detailed studies reveal a dramatic slowing of growth during the past 20 years, and in some locations at high altitude,

(a) *This map, based on estimates made by the Environmental Protection Agency, shows the sensitivity of different areas of the continental United States to acid rain. It takes into account such factors as major sources of sulfur and nitrogen oxides, weather patterns, altitude, and soil characteristics.*

(b) *In the mountains of New England, red spruce trees are dying at a high rate. Only their skeletons remain, standing silent sentinel over the forest.*

(c) *In a forest in Vermont, a student sets up an apparatus to collect rainwater that will be tested for acidity.*

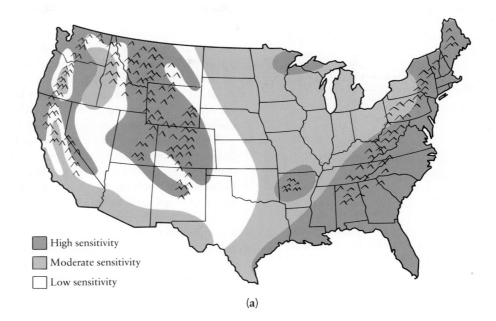

High sensitivity

Moderate sensitivity

Low sensitivity

(a)

trees are dying in large numbers and reproduction of some plants has come to a halt. At the present time, it is not known whether the decline is a result of acid rain, other pollutants, disease, subtle climate changes, or, most likely, some combination of these factors.

The accumulating evidence indicates that acid rain is one of the most serious worldwide pollution problems confronting us today. The potential consequences of its effects on biological systems are immense: lowered crop yields, decreased timber production, the need for greater amounts of increasingly expensive fertilizer to compensate for nutrient leaching, the loss of important freshwater fishing areas, and, possibly, of the eastern forests as well. The monetary and social costs of allowing the conditions that create acid rain to continue (or even to increase) are potentially very great, as are the costs of available processes to remove the sulfur and nitrogen oxides at the source, before they enter the air.

Scientists from many fields are presently engaged in research to gain a greater understanding of the causes and effects of acid rain and the likely consequences of proposed solutions. Although scientists can provide information on which decisions can be based, the choices that lie ahead are essentially social and economic, to be made through political processes.

(b)

(c)

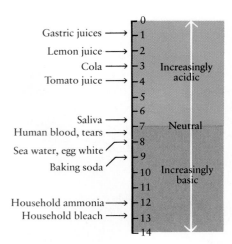

Gastric juices → 1
Lemon juice → 2
Cola → 3
Tomato juice → 4
5
6
Saliva → 7
Human blood, tears →
Sea water, egg white → 8
Baking soda → 9
10
11
Household ammonia → 12
Household bleach → 13
14

0
Increasingly acidic
Neutral
Increasingly basic

2-10 *pH values of various common solutions. A difference of one pH unit reflects a tenfold difference in the H+ ion concentration. Cola, for instance, is 10 times as acidic as tomato juice, and gastric juices are about 100 times more acidic than cola drinks.*

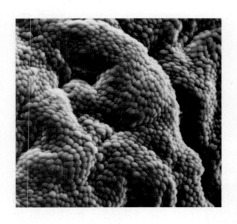

2-11 *Surface of the lining of the stomach, as shown in a scanning electron micrograph (magnified approximately 185 times). The numerous indentations are the openings into gastric pits, in which acid-secreting cells are located. Mucus, also secreted by cells of the stomach lining, coats the surface of the stomach and protects it from the acid.*

Buffers

Solutions more acidic than pH 1 or more basic than pH 14 are possible, but these are not included in the scale because they are almost never encountered in biological systems (Figure 2–10). In fact, almost all the chemistry of living things takes place at pH's between 6 and 8. Notable exceptions are the chemical processes in the stomach of humans and other animals, which take place at a pH of about 2 (Figure 2–11). Human blood, for instance, maintains an almost constant pH of 7.4 despite the fact that it is the vehicle for a large number and variety of nutrients and other chemicals being delivered to the cells, as well as for the removal of wastes, many of which are acids and bases.

The maintenance of a constant pH—an example of homeostasis (see page 26)—is important because the pH greatly influences the rate of chemical reactions. Organisms resist strong, sudden changes in the pH of blood and other fluids by means of **buffers,** which are combinations of H^+–donor and H^+–acceptor forms of weak acids or bases.

Buffers help maintain constant pH by their tendency to combine with H^+ ions and thus remove them from solution as the H^+ ion concentration begins to rise and to release them as it falls. The capacity of a buffer system to resist changes in pH is greatest when the concentrations of its H^+–donor and H^+–acceptor forms are equal. As the concentration of one form increases and that of the other decreases, the buffer becomes less effective. A variety of buffers function in living systems, each most effective at the particular pH at which its H^+–donor and H^+–acceptor concentrations are equal.

The major buffer system in the human bloodstream is the acid-base pair H_2CO_3–HCO_3^-. The weak acid H_2CO_3 (carbonic acid) dissociates into H^+ and bicarbonate ions as shown in the equation below.

$$H_2CO_3 \rightleftharpoons H^+ + HCO_3^-$$
$$\text{H}^+\text{ donor} \qquad \text{H}^+\text{ acceptor}$$

The H_2CO_3–HCO_3^- buffer system resists the changes in pH that might result from the addition of small amounts of acid or base by "soaking up" the acid or base. For example, if a small amount of H^+ is added to the system, it combines with the H^+ acceptor HCO_3^- to form H_2CO_3. This reaction removes the added H^+ and maintains the pH near its original value. If a small amount of OH^- is added, it combines with the H^+ to form H_2O; more H_2CO_3 tends to ionize to replace the H^+ as it is used.

Control of the pH of the blood is rendered even "tighter" by the fact that the H_2CO_3 is in equilibrium with dissolved carbon dioxide (CO_2) in the blood:

$$H_2O + CO_2 \rightleftharpoons H_2CO_3$$

As the arrows indicate, the two reactions are in equilibrium, and the equilibrium favors the formation of CO_2; in fact, the ratio is about 100 to 1 in favor of CO_2 formation.

Dissolved CO_2 in the blood is, in turn, in equilibrium with the CO_2 in the lungs. By changing your rate of breathing, you can change the HCO_3^- concentration in the blood and thus adjust the pH of your internal fluids.

Obviously, if the blood should be flooded with a very large excess of acid or base, the buffer would fail, but normally it is able to adjust continuously and very rapidly to the constant small additions of acid or base that normally occur in body fluids.

2-12 *The water cycle.*

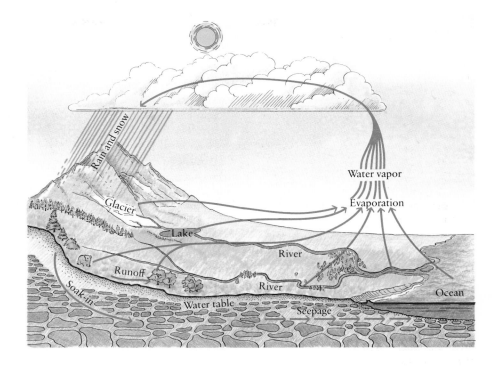

THE WATER CYCLE

Most of the water on earth—almost 98 percent—is in liquid form, in the oceans, lakes, and streams. Of the remaining 2 percent, some is frozen in polar ice and glaciers, some is in the soil, some is in the atmosphere in the form of vapor, and some is in the bodies of living organisms.

Water is made available to land organisms by processes powered by the sun. Solar energy evaporates water from the oceans, leaving the salt behind. Water is also evaporated, but in much smaller amounts, from moist soil surfaces, from the leaves of plants, and from the bodies of other organisms. These molecules—now water vapor—are carried up into the atmosphere by air currents. Eventually they fall to the earth's surface again as snow or rain. Most of the water falls on the oceans, since these cover most of the earth's surface. The water that falls on land is pulled back to the oceans by the force of gravity. Some of it, reaching low ground, forms ponds or lakes and streams or rivers, which pour water back into the oceans.

Some of the water that falls on the land percolates down through the soil until it reaches a zone of saturation. In the zone of saturation, all pores and cracks in the rock are filled with water (groundwater). The upper surface of the zone of saturation is known as the water table. Below the zone of saturation is solid rock, through which the water cannot penetrate. The deep groundwater, moving extremely slowly, eventually also reaches the ocean, thereby completing the water cycle.

As we have seen in this chapter, water, essential for life, is a most extraordinary substance. The earth's supply of water is the permanent possession of our planet, held to its surface by the force of gravity. Through the movements of the water cycle, it is perpetually available to living organisms.

SUMMARY

Water, the most common liquid on the earth's surface and the major component, by weight, of all living things, has a number of remarkable properties. These properties are a consequence of its molecular structure and are responsible for water's "fitness" for its roles in living systems.

Water is made up of two hydrogen atoms and one oxygen atom held together by covalent bonds. The water molecule is polar, with two weakly negative zones and two weakly positive zones. As a consequence, weak bonds form between water molecules. Such bonds, which link a somewhat positively charged hydrogen atom that is part of one molecule to a somewhat negatively charged oxygen atom that is part of another molecule, are known as hydrogen bonds. Each water molecule can form hydrogen bonds with four other water molecules. Although individual bonds are weak and constantly shifting, the total strength of the bonds holding the molecules together is very great.

Because of the hydrogen bonds holding the water molecules together (cohesion), water has a high surface tension and a high specific heat (the amount of heat that a given amount of the substance requires for a given increase in temperature). It also has a high heat of vaporization (the heat required to change a liquid to a gas) and a high heat of fusion (the heat required to change a solid to a liquid). Just before water freezes, it expands; thus ice has a lower density and a larger volume than liquid water. As a result, ice floats in water.

The polarity of the water molecule is responsible for water's adhesion to other polar substances and hence its tendency for capillary movement. Similarly, water's polarity makes it a good solvent for ions and polar molecules. Molecules that dissolve readily in water are known as hydrophilic. Water molecules, as a consequence of their polarity, actively exclude nonpolar molecules from solution. Molecules that are excluded from aqueous solution are known as hydrophobic.

Water has a slight tendency to ionize, that is, to separate into H^+ ions (actually H_3O^+, hydronium ions) and OH^- ions. In pure water, the number of H^+ ions and OH^- ions is equal at 10^{-7} mole per liter. A solution that contains more H^+ ions than OH^- ions is acidic; one that contains more OH^- ions than H^+ ions is basic. The pH scale reflects the proportion of H^+ ions to OH^- ions. An acidic solution has a pH lower than 7.0; a basic solution has a pH higher than 7.0. Almost all of the chemical reactions of living systems take place within a narrow range of pH around neutrality. Organisms maintain this narrow pH range by means of buffers, which are combinations of H^+–donor and H^+–acceptor forms of weak acids or bases.

Through the water cycle, the water above, on, and below the earth's surface is recirculated. As a result, it is continuously available to living organisms.

QUESTIONS

1. (a) Sketch the water molecule and label the zones of positive and negative charge. (b) What are the major consequences of the polarity of the water molecule? (c) How are these effects important to living systems?

2. The trick with the razor blade (see page 42) works better if the blade is a little greasy. Why?

3. Surfaces such as glass or raincoat cloth can be made "nonwettable" by application of silicone oils or other substances that cause water to bead up instead of spread flat. What do you suppose is happening, in molecular terms, when a surface becomes nonwettable?

4. Can you explain maple sugar production in terms of its value to the sugar maple tree?

5. Generally, coastal areas have more moderate temperatures (not as cold in winter, nor as hot in summer) than inland areas at the same latitude. What reasonable explanation can you give for this phenomenon?

6. What is vaporization? Describe the changes that take place in water as it vaporizes. What is heat of vaporization? Why does water have an unusually high heat of vaporization?

7. As we have seen, the digestive processes in the human stomach take place at a pH of about 2. When the food being digested reaches the small intestine, sodium bicarbonate ($NaHCO_3$) is released from the pancreas into the small intestine. What effect would you expect this to have on the pH of the partially digested food mass?

Organic Molecules

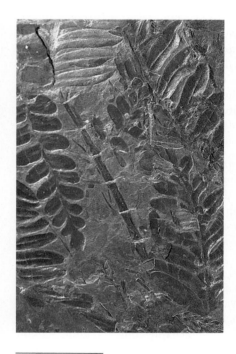

3–1 *In the process of photosynthesis, carbon from carbon dioxide in the atmosphere is incorporated into organic molecules by plants. These molecules provide the energy that powers living systems and are also used to build the larger structural molecules of which living organisms are composed. Some 300 million years ago, conditions on the earth were such that the carbon-containing dead bodies of vast numbers of organisms did not decay but were instead converted into coal and petroleum. Coal deposits are rich with the fossilized remains of plants that lived at the time, such as the leaves of the fern* Alethopteris *and the branch of the giant horsetail* Calamites *shown here.*

In this chapter, we present some of the types of **organic molecules**—molecules containing carbon—that are found in living things. As you will see, the molecular drama is a grand spectacular with, literally, a cast of thousands; a single bacterial cell contains some 5,000 different kinds of molecules, and an animal or plant cell has about twice that many. These thousands of molecules, however, are composed of relatively few elements (CHNOPS). Similarly, relatively few kinds of molecules play the major roles in living systems. As we noted previously, water makes up from 50 to 95 percent of a living system, and small ions such as K^+, Na^+, and Ca^{2+} account for no more than 1 percent. Almost all the rest, chemically speaking, is composed of organic molecules.

Four different kinds of organic molecules are found in large quantities in organisms. These four are **carbohydrates** (composed of sugars), **lipids** (nonpolar molecules, many of which contain fatty acids), **proteins** (composed of amino acids), and **nucleotides** (complex molecules that play key roles in energy exchanges and that can also combine to form very large molecules known as nucleic acids). All of these molecules—carbohydrates, lipids, proteins, and nucleotides—contain carbon, hydrogen, and oxygen. In addition, proteins contain nitrogen and sulfur, and nucleotides, as well as some lipids, contain nitrogen and phosphorus.

It has been said that it is necessary only to be able to recognize about 30 molecules for a working knowledge of the biochemistry of cells. Two of these are the sugars glucose and ribose; another is a fatty acid; 20 are the biologically important amino acids; and five are nitrogenous bases, nitrogen-containing molecules that are key constituents of nucleotides. If you bear with us, you will find that you readily learn to recognize the players and their roles and to distinguish the stars from the members of the chorus. Consider this, if you will, an introduction to the principal characters; the plot begins to unfold in Chapter 4.

THE CENTRAL ROLE OF CARBON

The Carbon Backbone

As you will recall from Chapter 1, a carbon atom has six protons and six electrons, two electrons in its first energy level and four in its second energy level. Thus carbon can form four covalent bonds with as many as four different atoms. Methane (CH_4), which is natural gas, is an example (Figure 1–10, page 32). Even more important, in terms of carbon's biological role, carbon atoms can form bonds with each other. Ethane, for example, contains two carbons; propane, three; butane, four; and so on, forming long chains (Figure 3–2). In general, an organic molecule derives its overall shape from the arrangement of the carbon atoms that form the backbone, or skeleton, of the molecule. The shape of the molecule, in turn, determines many of its properties and its function within living systems.

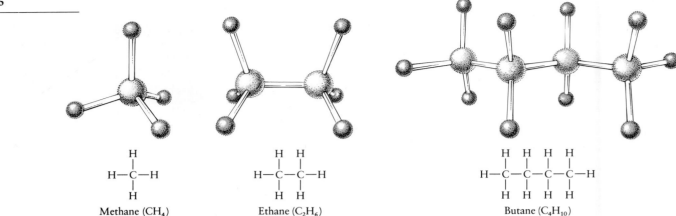

Methane (CH₄) Ethane (C₂H₆) Butane (C₄H₁₀)

3–2 *Ball-and-stick models and structural formulas of methane, ethane, and butane. In the models, the gray spheres represent carbon atoms and the smaller blue spheres represent hydrogen atoms. The sticks in the models—and the lines in the structural formulas—represent covalent bonds, each of which consists of a pair of shared electrons. Note that every carbon atom forms four covalent bonds.*

In the molecules shown in Figure 3–2, every carbon bond that is not occupied by another carbon atom is taken up by a hydrogen atom. Such compounds, consisting of only carbon and hydrogen, are known as **hydrocarbons**. Structurally, they are the simplest kind of organic compounds. Although most hydrocarbons are derived from the remains of organisms that died millions of years ago, they are relatively unimportant in living organisms. They are, however, of great economic importance; the liquid fuels upon which we depend—gasoline, diesel fuel, and heating oil—are all composed of hydrocarbons.

Functional Groups

The specific chemical properties of an organic molecule derive principally from groups of atoms known as **functional groups**. These groups are attached to the carbon skeleton, replacing one or more of the hydrogens that would be present in a hydrocarbon. An —OH (hydroxyl) group is an example of a functional group.* When one hydrogen and one oxygen are bonded covalently, one outer electron of the oxygen is left over, unpaired and unshared; it can be shared with a similarly available outer electron of a carbon atom, thereby forming a covalent bond with the carbon. A compound with a hydroxyl group in place of one or more of the hydrogens in a hydrocarbon is known as an alcohol. Thus methane (CH_4), with the replacement of one hydrogen atom by a hydroxyl group, becomes methanol, or wood alcohol (CH_3OH), a pleasant-smelling, poisonous compound noted for its ability to cause blindness and death. Ethane similarly becomes ethanol, or grain alcohol (C_2H_5OH), which is present in all alcoholic beverages. Glycerol, $C_3H_5(OH)_3$, contains, as its formula indicates, three carbon atoms, five hydrogen atoms, and three hydroxyl groups.

Table 3–1 illustrates the functional groups that will be of greatest interest to us in our exploration of living systems. A knowledge of functional groups makes it easy to recognize particular molecules and to predict their properties. For example, the carboxyl group (—COOH), mentioned in the previous chapter, is a functional group that gives a molecule the properties of an acid. Alcohols, with their polar hydroxyl groups, tend to be soluble in water, for instance, whereas hydrocarbons, such as butane, with only nonpolar functional groups (such as methyl groups), are highly insoluble in water. Aldehyde groups are often associated with pungent odors and tastes. Smaller molecules with aldehyde groups, such as formaldehyde, have unpleasant odors, whereas larger ones, such as the chemicals that give vanilla, apples, cherries, and almonds their distinctive flavors, tend to be pleasing to the human sensory apparatus.

As you can see, most of the functional groups in Table 3–1 are polar and so have regions of positive and negative charge in aqueous solution. Thus they confer water-solubility and local electric charge to the molecules that contain them.

---

* —OH, the functional group, is called hydroxyl; OH⁻, the ion, is called hydroxide.

TABLE 3-1 **Some Biologically Important Functional Groups**

GROUP	NAME	BIOLOGICAL SIGNIFICANCE
—OH	Hydroxyl	Polar, thus water-soluble; forms hydrogen bonds
—C(=O)OH	Carboxyl	Weak acid (hydrogen donor); when it loses a hydrogen ion, it becomes negatively charged: —C(=O)O$^-$ + H$^+$
—N(H)(H)	Amino	Weak base (hydrogen acceptor); when it accepts a hydrogen ion, it becomes positively charged: —N$^+$(H)(H)—H
H—C=O	Aldehyde	Polar, thus water-soluble; characterizes some sugars
C=O	Ketone (or carbonyl)	Polar, thus water-soluble; characterizes other sugars
H—C(H)(H)—H	Methyl	Hydrophobic (insoluble in water)
(O=)P(OH)—OH	Phosphate	Acid (hydrogen donor); in solution, usually negatively charged: (O=)P(O$^-$)—O$^-$ + 2H$^+$

Some of the polar functional groups tend to become fully ionized, depending on the pH of the solution. Many functional groups participate directly in the chemical reactions of greatest interest in biological systems.

The Energy Factor

Covalent bonds—the bonds commonly found in organic molecules—are strong, stable bonds consisting of electrons moving in orbitals about two or more atomic nuclei. These bonds have different characteristic strengths, depending on the configurations of the orbitals. You will recall from the last chapter that molecules are always in motion—vibrating, rotating, and shifting position in relation to other molecules. The atoms within molecules are also in motion—vibrating and, often, rotating about the axes of their bonds. If this motion becomes great enough (that is, if the atoms possess enough kinetic energy), the bond will "break" and the atoms will become separated from each other. Bond strengths are conventionally expressed in terms of the energy, in kilocalories per mole, that must be supplied to break the bond under standard conditions of temperature and pressure (Figure 3–3).

Kilocalories per mole

93.4	N	0.102 nm	H
98.8	C	0.109	H
171	C	0.123 (double)	O
147	C	0.127 (double)	N
147	C	0.133 (double)	C
84	C	0.143	O
69.7	C	0.148	N
83.1	C	0.154	C

3–3 *A chemical bond is a force holding atoms together. The strength of the bond is measured in terms of the energy required to break it. The figures at the left indicate the number of kilocalories that will break the bonds between the pairs of atoms shown. The lines connecting the atoms represent the bonds; the figures above the lines represent the characteristic center-to-center distances between the atoms, expressed in nanometers (1 nanometer, abbreviated nm, equals 10^{-9} meter). Double lines indicate double bonds, which, as you can see, hold the atoms closer together and are stronger.*

Why Not Silicon?

Silicon (atomic number 14) is more abundant than carbon (atomic number 6). As you can tell from its atomic number, silicon also requires four electrons to complete its outer energy level. Why then is it found so rarely in living systems? Because silicon atoms are larger than carbon atoms, the distance between two silicon atoms is much greater than the distance between two carbon atoms. As a result, the bonds between the more tightly held carbon atoms are almost twice as strong as those between silicon atoms. Thus carbon can form long stable chains and silicon cannot.

Carbon's capacity to form double bonds is also crucial to its central role in biology. As we saw in Chapter 1, a carbon atom can combine with two oxygen atoms by means of two double bonds; the carbon dioxide molecule, all its electron requirements satisfied, floats in air as a gas, free and independent. It also dissolves readily in water and so is available to living systems. In silicon dioxide, by contrast, a silicon atom forms single bonds to two oxygen atoms, leaving two unpaired electrons on the silicon atom and one on each oxygen. As a consequence, the silicon atom needs to gain two electrons to fill its outer energy level, and each oxygen atom needs to gain one electron. Thus the unpaired electrons are readily

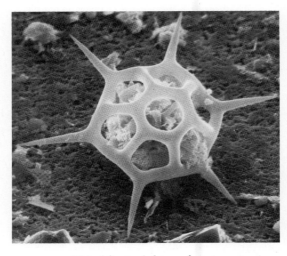

This delicate skeleton of a microorganism, from Narragansett Bay, Rhode Island, is composed of silicon dioxide. The material within the skeleton is organic debris.

shared with the unpaired electrons on neighboring SiO_2 molecules, forming, eventually, grains of sand, rocks, or with biological intervention, the shells of microscopic marine organisms.

When a covalent bond breaks, atoms (or, in some cases, groups of atoms) are released, and each atom usually takes its own electrons with it. This results in atoms whose outer energy levels are only partially filled with electrons. For example, when the atoms of a methane molecule are vibrating and rotating so rapidly that the four carbon-hydrogen bonds break, one carbon atom and four hydrogen atoms are produced—and each of these atoms needs to gain electrons to complete its outer energy level. Thus, the atoms tend to form new covalent bonds quite rapidly, restoring the stable condition of filled outer energy levels. Whether the new bonds that form are identical to those that were broken or are different depends on a number of factors—the temperature, the pressure, and, most important, what other atoms are available in the immediate vicinity.

Chemical reactions in which new combinations are formed always involve a change in electron configurations and therefore in bond strengths. Depending on the relative strengths of the bonds broken and the bonds formed in the course of a chemical reaction, energy will either be released from the system or will be taken up by it from the surroundings. Consider, for example, the burning of methane, represented by the following equation:

$$CH_4 + 2O_2 \longrightarrow CO_2 + 2H_2O$$

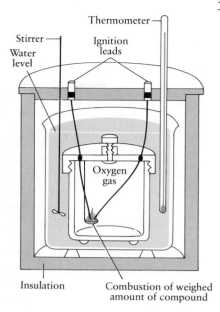

3–4 *A calorimeter is used to measure the amount of energy stored in an organic compound. A known quantity of the compound is ignited electrically. As it burns, the rise in the temperature of the surrounding water is measured. Using the specific heat of water and the known weight of water in the calorimeter, one can then calculate the number of calories released by the burning of the sample.*

This reaction, which can be set in motion by a spark, is often the cause of explosions in coal mines; when it occurs, it releases energy in the form of heat. The amount of energy released can be measured quite precisely, as shown in Figure 3–4. It turns out to be 213 kilocalories per mole of methane. This can be expressed by a simple equation:

$$\Delta H^\circ = -213 \text{ kcal/mole}$$

The Greek letter delta (Δ) stands for change, H for heat, and the superscript $^\circ$ indicates that the reaction occurs under certain standard conditions of temperature and pressure. The minus sign indicates that energy has been released.

Similarly, changes in energy occur in the chemical reactions that take place in organisms. However, as we shall see in Section 2, living systems have evolved strategies for minimizing not only the energy required to set a reaction in motion but also the proportion of energy released as heat. These strategies involve, among other factors, specialized protein molecules known as **enzymes,** which are essential participants in the chemical reactions of living systems. (The word "strategy" in its ordinary meaning is a deliberate plan to achieve a specified goal. Biologists use it to mean a group of related traits, evolved by organisms under the influence of natural selection, that solves particular problems encountered by living systems.)

CARBOHYDRATES: SUGARS AND POLYMERS OF SUGARS

Carbohydrates are the primary energy-storage molecules in most living things. In addition, they form a variety of structural components of living cells; the walls of young plant cells, for example, are about 40 percent cellulose, which is the most common organic compound in the biosphere.

Carbohydrates are formed from small molecules known as **sugars.** There are three principal kinds of carbohydrates, classified according to the number of sugar molecules they contain. **Monosaccharides** ("single sugars"), such as ribose, glucose, and fructose, contain only one sugar molecule. **Disaccharides** consist of two sugar molecules linked covalently. Familiar examples are sucrose (table sugar), maltose (malt sugar), and lactose (milk sugar). **Polysaccharides,** such as cellulose and starch, contain many sugar molecules linked together. Large molecules, such as polysaccharides, that are made up of similar or identical subunits are known as **polymers** ("many parts"), and the subunits are called **monomers** ("single parts").

Monosaccharides: Ready Energy for Living Systems

Monosaccharides are organic compounds composed of carbon, hydrogen, and oxygen. They can be described by the formula $(CH_2O)_n$, where n may be as small as 3, as in $C_3H_6O_3$, or as large as 8, as in $C_8H_{16}O_8$ (Figure 3–5). These proportions gave rise to the term carbohydrate ("hydrate of carbon") for sugars and the larger molecules formed from sugar subunits.

As you can see by studying Figure 3–5, monosaccharides are characterized by hydroxyl groups and an aldehyde or ketone group. These functional groups make sugars highly soluble in aqueous solution and, in molecules containing more than five carbon atoms, lead to an internal reaction that dramatically changes the shape of the molecule. When these monosaccharides are in solution, the aldehyde or ketone group has a tendency to react with one of the hydroxyl groups, producing a ring structure. In glucose, for example, the aldehyde group on the first carbon atom reacts with the hydroxyl group on the fifth carbon atom, producing a six-membered ring structure, as shown in Figure 3–6. When the ring forms, it may close in one of two ways, with the hydroxyl group now on the first carbon positioned either above or below the plane of the ring. The form in which the hydroxyl group is below the plane is known as alpha-glucose, and the form in which it is above the plane is known as beta-glucose. As we shall see, this small difference between the alpha and beta forms of glucose can lead to very significant differences in the properties of larger molecules formed by living systems from glucose.

3–5 *Two different ways of classifying monosaccharides: according to the number of carbon atoms and according to the functional groups, indicated here in color. Glyceraldehyde, ribose, and glucose contain, in addition to hydroxyl groups, an aldehyde group, indicated in green; they are called aldose sugars (aldoses). Dihydroxyacetone, ribulose, and fructose each contain a ketone group, indicated in brown, and are called ketose sugars (ketoses).*

Number of carbon atoms

	Trioses (3 carbons)	Pentoses (5 carbons)	Hexoses (6 carbons)
Aldoses:	Glyceraldehyde ($C_3H_6O_3$)	Ribose ($C_5H_{10}O_5$)	Glucose ($C_6H_{12}O_6$)
Ketoses:	Dihydroxyacetone ($C_3H_6O_3$)	Ribulose ($C_5H_{10}O_5$)	Fructose ($C_6H_{12}O_6$)

^{6}CH$_2$OH

^{5}C—O H

H | H | ^{1}C

^{4}C OH H

OH ^{3}C ^{2}C OH

H OH

alpha-Glucose

H
|
^{1}C=O

H—^{2}C—OH

HO—^{3}C—H

H—^{4}C—OH

H—^{5}C—OH

^{6}CH$_2$OH

Glucose, straight-chain form

^{6}CH$_2$OH

^{5}C—O OH

H | H | ^{1}C

^{4}C OH H

OH ^{3}C ^{2}C H

H OH

beta-Glucose

3–6 *In aqueous solution, the six-carbon sugar glucose exists in two different ring structures, alpha and beta, that are in equilibrium with each other. The molecules pass through the straight-chain form to get from one structure to the other. The sole difference in the two ring structures is the position of the hydroxyl group attached to carbon atom 1; in the alpha form it is below the plane of the ring, and in the beta form it is above the plane.*

Like hydrocarbons, monosaccharides can be burned, or oxidized, to yield carbon dioxide and water:

$$(CH_2O)_n + nO_2 \longrightarrow (CO_2)_n + (H_2O)_n$$

This reaction, like the burning of methane, releases energy, and the amount of energy released as heat can be calculated by burning sugar molecules in a calorimeter. The same amount of energy is released—although not nearly so wastefully—when the equivalent amount of carbohydrate is oxidized in a living cell as when it is burned in a calorimeter.

This statement comparing the oxidation of food molecules with that of fuel molecules is not a metaphor; it is a fact. For example, the energy cost of transporting a kilogram of body weight the distance of a kilometer is 0.95 kilocalorie for a pigeon, 0.73 kcal for a person, and 0.83 kcal for a Cadillac.

A principal energy source for humans and other vertebrates is the monosaccharide glucose. It is in this form that sugar is generally transported in the animal body. A patient receiving an intravenous feeding in a hospital is getting glucose dissolved in a salt solution that approximates the ionic composition of body fluids. The dissolved glucose is carried through the bloodstream to the cells of the body where the energy-releasing reactions are carried out. As measured in a calorimeter, the oxidation of a mole of glucose releases 673 kilocalories:

$$C_6H_{12}O_6 + 6O_2 \longrightarrow 6CO_2 + 6H_2O$$
$$\Delta H° = -673 \text{ kcal}$$

Disaccharides: Transport Forms

Although glucose is the common transport sugar for vertebrates, sugars are often transported in other organisms as disaccharides. Sucrose, commonly called cane sugar, is the form in which sugar is transported in plants from the photosynthetic cells (mostly in the leaves), where it is produced, to other parts of the plant body. Sucrose is composed of the monosaccharides glucose and fructose. Sugar is transported through the blood of many insects in the form of another disaccharide, trehalose, which consists of two glucose units linked together. Another common disaccharide is lactose, a sugar that occurs only in milk. Lactose is made up of glucose combined with another monosaccharide, galactose.

3–7 *A male calliope hummingbird drinking sugary syrup—nectar—from the flower of an Oregon grape. Many animals have sensitive detection mechanisms for sugar and apparently find its taste pleasurable. In the course of consuming sugary plant products, animals obtain not only a rich energy supply but also other essential nutrients, such as plant proteins, lipids, vitamins, and minerals.*

Representations of Molecules

As we saw in Chapters 1 and 2, chemists have developed various models to represent the structures of atoms and molecules. Each of these models is a way of organizing a particular set of scientific data and of focusing attention on particular characteristics of atoms and molecules.

Because the properties of a molecule depend on its three-dimensional characteristics, physical models are often the most useful. For example, ball-and-stick models of the kind shown in Figure 3–2 emphasize the geometry of a molecule and, in particular, the bonds between atoms. But these models fail to suggest the overall shape of the molecule created by the movement of electrons within their orbitals.

A closer approximation of molecular shape is provided by space-filling models. Each atom is represented by the edge of its outermost orbitals. Space-filling models are misleading, however, in that molecules do not fill space in the same way that we think of a table or a rock as filling space. The atoms that make up molecules consist mostly of empty space. If the perimeter of the outer orbitals of the electrons in an oxygen atom were the size of the perimeter of the Astrodome in Houston, the nucleus would be a ping-pong ball in the center of the stadium. What "fills" the space in molecules are regions of charge, associated with the movements of the electrons around the nuclei. One molecule "sees" another molecule in terms of these regions of charge. As a consequence, for instance, a protein that transports glucose molecules into the living cell will not transport fructose molecules because of the differences in the shape of the regions of charge. All the intricate biochemistry that goes on in the cell is based on this ability of molecules to "recognize" one another.

Ball-and-stick and space-filling models are often used in the laboratory, but they are less useful on paper because it is necessary to see them from all angles to see all of the atoms and their bonds. The most accurate two-dimensional representations of molecular structure are orbital models, such as those shown in Figure 2–3 (page 41). For molecules containing more than a few atoms, however, orbital models become extremely complicated. Thus, when representing complex molecules, such as those found in living systems, chemists usually use molecular formulas or structural formulas. A molecular formula indicates the number of atoms of each kind within the molecule, while a structural formula shows how the atoms are bonded to one another.

Space-filling models of the sugars glucose and fructose. The gray spheres, almost completely hidden at the center of each molecule, represent the carbon atoms. The red spheres at the surface of each molecule represent oxygen atoms, while the blue spheres represent hydrogen atoms.

Glucose   Fructose

Glucose, for example, has 6 carbon atoms, 12 hydrogen atoms, and 6 oxygen atoms. Its molecular formula is $C_6H_{12}O_6$. However, fructose also contains 6 carbons, 12 hydrogens, and 6 oxygens and has a similar structure—a chain of carbon atoms to which hydrogen and oxygen atoms are attached. The differences between glucose and fructose are determined by which carbon atoms the other atoms are attached to. The molecules can therefore be distinguished by their structural formulas:

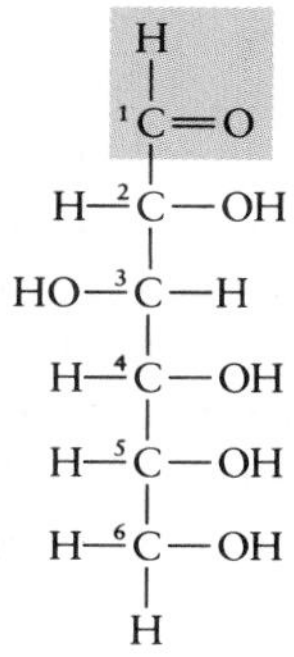

Glucose
(chain)

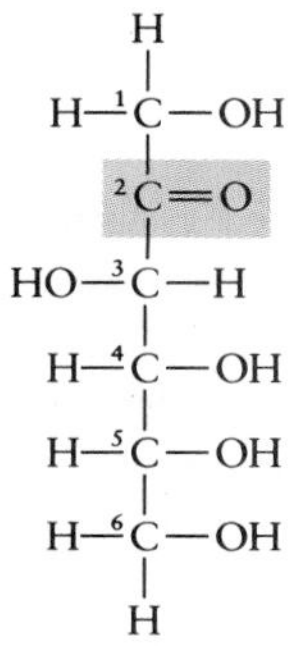

Fructose
(chain)

Or, in the ring forms:

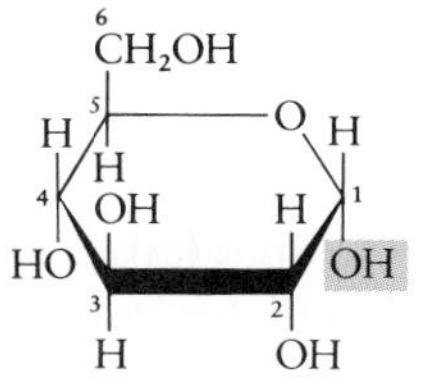

alpha-Glucose

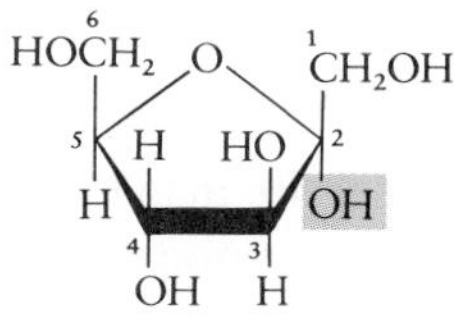

alpha-Fructose

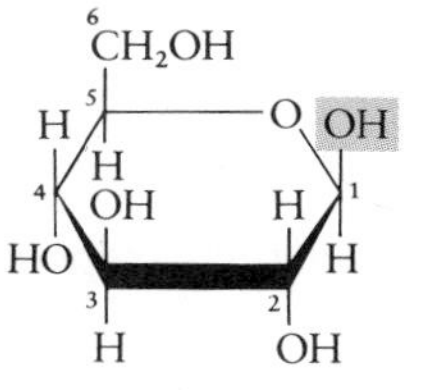

beta-Glucose

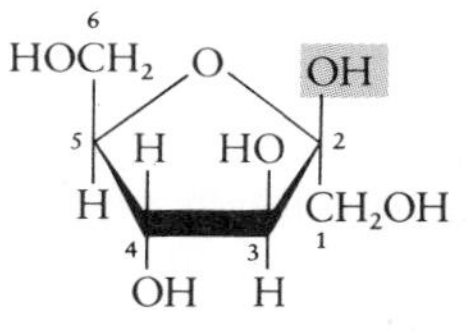

beta-Fructose

The lower edges of the rings are made thicker to hint at a three-dimensional structure. By convention, the carbon atoms at the intersections of the links in an organic ring structure are "understood" and not labeled. Although it is not necessary to number the carbon atoms, doing so often makes it easier to interpret the structural formula.

Notice that the reaction leading to the formation of the ring structures of fructose involves the ketone group, which is on carbon 2, and the hydroxyl group on carbon 5. The result is a five-membered ring, with an overall shape quite unlike that of the six-membered ring formed by glucose. The hydroxyl group whose position above or below the plane of the ring determines whether the molecule is alpha-fructose or beta-fructose is on carbon 2.

In sugars that form ring structures in solution, the positions of the hydroxyl groups not involved in ring formation are easily determined: —OH groups that appear on the left side of the structural formula for the straight-chain form go above the plane of the ring, and —OH groups that appear on the right side of that structural formula go below the plane of the ring.

Although structural formulas do not give us exact information about the shapes of the regions of charge that are so critical in biological reactions, they do give us more information than is apparent at first glance. You will find them a convenient tool as we examine the molecules involved in the structures and processes of living systems.

alpha-Glucose   alpha-Glucose   alpha-Glucose   beta-Fructose

Maltose   Sucrose
(a)   (b)

3–8 *The condensation reactions producing two common disaccharides.* (**a**) *Maltose is a disaccharide made up of two alpha-glucose units, joined in what is known as a 1→4 linkage (the bonding between the two rings involves the 1-carbon of one glucose subunit and the 4-carbon of the other).* (**b**) *Sucrose is a disaccharide formed from an alpha-glucose unit and a beta-fructose unit, joined in a 1→2 linkage (the bonding between the two rings involves the 1-carbon of glucose and the 2-carbon of fructose). In order to represent this bond on paper, we must rotate the structural formula for beta-fructose 180° (right to left), which has the disconcerting effect of turning everything upside down to our eyes. In the three-dimensional world in which the molecules actually exist, however, formation of this 1→2 linkage creates no problems.*

As you can see, the condensation reactions producing these disaccharides involve the removal of a molecule of water. Splitting them back into their constituent monosaccharides requires the addition of a water molecule (hydrolysis).

In the synthesis of a disaccharide molecule from two monosaccharide molecules, a molecule of water is removed in the process of forming the new bond between the two monosaccharides (Figure 3–8). This type of chemical reaction, which occurs in the synthesis of most organic polymers from their subunits, is known as **condensation.** Thus, only the free monomers of carbohydrates actually have a CH_2O ratio because of the removal of two atoms of hydrogen and one of oxygen every time such a bond is formed.

When a disaccharide is split into its monosaccharide units, which happens when it is used as an energy source, the molecule of water is added again. This splitting is known as **hydrolysis,** from *hydro,* meaning "water," and *lysis,* meaning "breaking apart." Hydrolysis is an energy-releasing reaction. The hydrolysis of sucrose, for example, releases 5.5 kilocalories per mole. Conversely, the formation of sucrose from glucose and fructose requires an energy input of 5.5 kilocalories per mole of sucrose.

Storage Polysaccharides

Polysaccharides are made up of monosaccharides linked together in long chains. Some of them are storage forms of sugar. **Starch,** for instance, is the principal food storage form in most plants. A potato, for example, contains starch produced from the sugar formed in the green leaves of the plant; the sugar is transported underground and accumulated there in a form suitable for winter storage, after which it will provide for new growth in the spring. Starch occurs in two forms, amylose and amylopectin. Both consist of glucose units linked together (Figure 3–9).

Glycogen is the principal storage form for sugar in higher animals. Glycogen has a structure much like that of amylopectin except that it is more highly branched, with branches occurring every eight to ten glucose units. In vertebrates, glycogen is stored principally in the liver and in muscle tissue. When there is an excess of glucose in the bloodstream, the liver forms glycogen. When the concentration of glucose in the blood drops, the hormone glucagon, produced by the pancreas, is released into the bloodstream; glucagon stimulates the liver to hydrolyze glycogen to glucose, which then enters the bloodstream.

Formation of polysaccharides from monosaccharides requires energy. However, when the cell needs energy, these polysaccharides can be hydrolyzed, releasing monosaccharides that can, in turn, be oxidized to provide energy for cellular work.

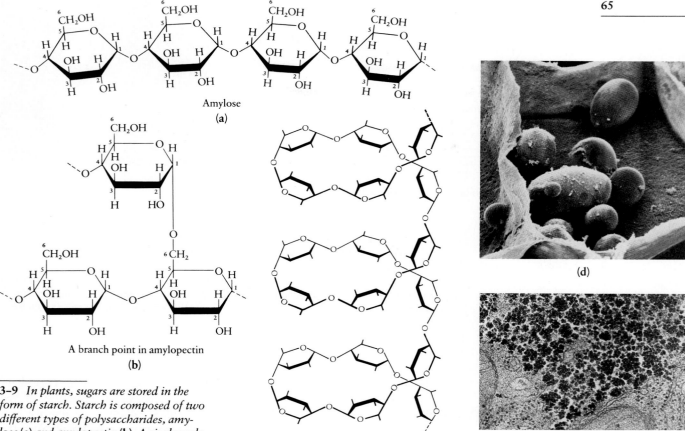

Amylose
(a)

A branch point in amylopectin
(b)

(c)

(d)

(e)

3-9 *In plants, sugars are stored in the form of starch. Starch is composed of two different types of polysaccharides, amylose (a) and amylopectin (b). A single molecule of amylose may contain 1,000 or more alpha-glucose units with carbon 1 of one glucose ring linked to carbon 4 of the next in a long, unbranched chain, which coils to form a helix (c). A molecule of amylopectin may contain from 1,000 to 6,000 alpha-glucose units; short chains containing about 24 to 36 alpha-glucose units periodically branch off from the main chain.*

(d) Starch molecules, perhaps because of their helical nature, tend to cluster into granules. In this scanning electron micrograph of a single storage cell of a potato, the spherical and egg-shaped objects are starch granules. They are magnified about 1,000 times.

(e) Glycogen, which is the common storage form for sugar in vertebrates, resembles amylopectin in its general structure except that each branch contains only 16 to 24 alpha-glucose units. The dark granules in this liver cell, magnified about 55,000 times, are glycogen. When glucose is needed, it is provided by the hydrolysis of glycogen.

Structural Polysaccharides

A major function of molecules in living systems is to form the structural components of cells and tissues. The principal structural molecule in plants is **cellulose.** In fact, half of all the organic carbon in the biosphere is contained in cellulose. Wood is about 50 percent cellulose, and cotton is nearly pure cellulose.

Cellulose molecules form the fibrous part of the plant cell wall. The cellulose fibers, embedded in a matrix of other kinds of polysaccharides, form an external envelope around the plant cell. When the cell is young, this envelope is flexible and stretches as the cell grows, but it becomes thicker and more rigid as the cell matures. In some plant tissues, such as the tissues that form wood and bark, the cells eventually die, leaving only their tough outer walls.

Cellulose is a polymer composed of monomers of glucose, just as starch and glycogen are. Starch and glycogen can be readily utilized as fuels by almost all kinds of living systems, but only a few microorganisms—certain bacteria, protozoa, and fungi—can hydrolyze cellulose. Cows and other ruminants, termites, and cockroaches can use cellulose for energy only because of microorganisms that inhabit their digestive tracts.

To understand the differences between structural polysaccharides, such as cellulose, and energy-storage polysaccharides, such as starch or glycogen, we must look again at the glucose molecule. You will remember that the molecule is basically a chain of six carbon atoms and that when it is in solution, as it is in the cell, it assumes a ring form. The ring may close in either of two ways (see Figure 3–6). One ring form is known as alpha, and the other as beta. The alpha and beta forms are in equilibrium, with a certain number of molecules changing from one form to the other all the time, going through the open-chain structure to reach the other form. Starch and glycogen are both made up entirely of alpha units.

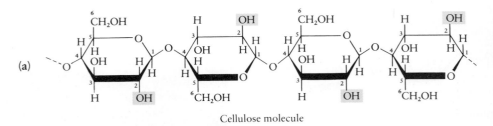

Cellulose molecule

(c)

Model of cross-linked cellulose molecules

3–10 **(a)** *Cellulose consists of beta-glucose monomers, joined in 1 ⟶ 4 linkages. (Note that the structural formulas for alternating beta-glucose units have been rotated 180° to show the bonding.) In cellulose, the —OH groups (indicated in color), which project from both sides of the chain, form hydrogen bonds with neighboring —OH groups, resulting in* the formation of bundles of cross-linked parallel chains **(b)**. *By contrast, in the starch molecule (Figure 3–9), most of the —OH groups capable of forming hydrogen bonds face toward the exterior of the helix, making it more readily soluble in the surrounding water.*

(c) *The wall of a young plant cell is about 40 percent cellulose. Each of the* microfibrils you can see here (magnified about 30,000 times) is a bundle of hundreds of cellulose strands, and each strand is a chain of beta-glucose monomers **(a)**. *The microfibrils, as strong as an equivalent amount of steel, are embedded in other polysaccharides, one of which is pectin.*

Cellulose, however, consists entirely of beta units (Figure 3–10). This slight difference has a profound effect on the three-dimensional structure of the molecules, which align in parallel, forming crystalline cellulose microfibrils. As a result, cellulose is impervious to the enzymes that so successfully break down the storage polysaccharides.

Chitin, which is a major component of the exoskeletons of arthropods, such as insects and crustaceans, and also of the cell walls of many fungi, is a tough, resistant, modified polysaccharide (Figure 3–11). At least 900,000 different species of organisms can synthesize chitin, and it has been estimated that the individuals belonging to a single species of crab produce several million tons of chitin a year.

3–11 **(a)** *Chitin is a polymer consisting of repeated modified monosaccharides. As you can see, the monomer is a six-carbon sugar, like glucose, in which a nitrogen-containing group has replaced the —OH group on carbon 2.* **(b)** *A cicada molting. The relatively hard outer coverings, or* exoskeletons, of insects contain chitin. Because exoskeletons do not grow as the insect grows, they must be molted periodically. The discarded exoskeleton is at the top, above the insect, which is drying out and waiting for its new exoskeleton to harden.

Chitin

(b)

LIPIDS

Lipids are a general group of organic substances that are insoluble in polar solvents, such as water, but that dissolve readily in nonpolar organic solvents, such as chloroform, ether, and benzene. Typically, lipids serve as energy-storage molecules—usually in the form of fats or oils—and for structural purposes, as in the case of phospholipids, glycolipids, and waxes. Some lipids, however, play major roles as chemical "messengers," both within and between cells.

Fats and Oils: Energy in Storage

Unlike many plants, such as the potato, animals have only a limited capacity to store carbohydrates. In vertebrates, sugars in excess of what can be stored as glycogen are converted into fats. Some plants also store food energy as oils, especially in seeds and fruits. Fats and oils contain a higher proportion of energy-rich carbon-hydrogen bonds (see Figure 3–3) than carbohydrates do and, as a consequence, contain more chemical energy. On the average, fats yield about 9.3 kilocalories per gram* as compared to 3.79 kcal per gram of carbohydrate, or 3.12 kcal per gram of protein. Also, because fats are nonpolar, they do not attract water molecules and hence are not "weighted down" by them, as glycogen is. Taking into account the water factor, fats store six times as much energy, gram for gram, as glycogen, which is undoubtedly why in the course of evolution they came to play a major role in energy storage.

An example of the value of this concentrated energy storage is provided by hummingbirds. A male ruby-throated hummingbird has a fat-free weight of 2.5 grams (about 1/10 ounce). It migrates every fall from Florida to Yucatan, some 2,000 kilometers. Before doing so, it accumulates 2.0 grams of body fat, an amount almost equal to its original weight. However, if it were to carry the same energy reserves in the form of glycogen, it would have to carry 5 grams, twice its own fat-free weight.

A fat molecule consists of three molecules of fatty acid joined to one glycerol molecule. Glycerol, as we noted previously, is a three-carbon alcohol that contains three hydroxyl groups. A fatty acid consists of a long hydrocarbon chain that terminates in a carboxyl group (—COOH); the nonpolar chain is hydrophobic, whereas the carboxyl group gives one portion of the molecule the properties of an acid. As with the disaccharides and polysaccharides, each bond between glycerol and a fatty acid is formed by the removal of a molecule of water (condensation), as shown in Figure 3–12. Fat molecules, which are also known as triglycerides, are said to be neutral because they contain no polar groups. As you would expect, they are extremely hydrophobic.

Fatty acids, which are seldom found in cells in a free state (that is, not as part of another molecule), consist of chains containing an even number of carbon atoms, typically between 14 and 22. About 70 different fatty acids are known. They differ in their chain lengths, in whether the chain contains any double bonds (as in oleic acid) or not (as in stearic acid), and in the position in the chain of any double bonds (see Figure 3–12). A fatty acid, such as stearic acid, in which there are no double bonds is said to be **saturated** because the bonding possibilities are complete for all the carbon atoms of the chain (that is, each carbon atom has formed bonds to four other atoms). A fatty acid, such as oleic acid, that contains carbon atoms joined by double bonds is said to be **unsaturated** because those carbon atoms have the potential to form additional bonds with other atoms.

Unsaturated fats, which tend to be oily liquids, are more common in plants than in animals; examples are olive oil, peanut oil, and corn oil. Animal fats, such as butter and lard, contain saturated fatty acids and usually have higher melting temperatures.

* 1,000 grams = 1 kilogram = 2.2 pounds, so oxidation of a pound of fat would yield about 4,200 kilocalories, more than the 24-hour requirement for a moderately active adult.

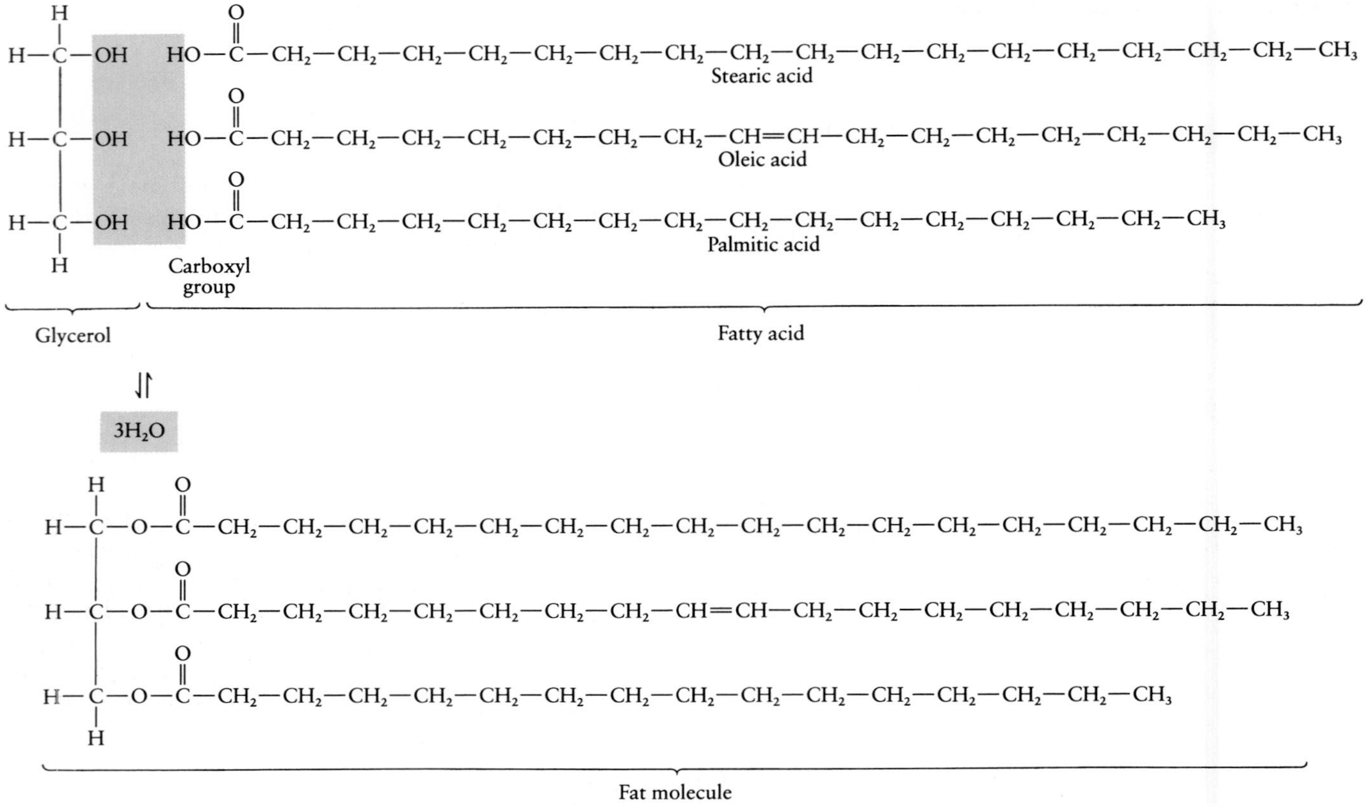

3–12 *A fat molecule consists of three fatty acids joined to a glycerol molecule (hence the term "triglyceride"). The long hydrocarbon chains of which the fatty acids are composed terminate in carboxyl (—COOH) groups, which become covalently bonded to the glycerol molecule. Each bond is formed when a molecule of water (color) is removed (condensation). The physical properties of a fat—such as its melting point—are determined by the lengths of its fatty acid chains and by whether the chains are saturated or unsaturated. Three different fatty acids are shown here. Stearic acid and palmitic acid are saturated, and oleic acid is unsaturated, as you can see by the double bond in its structure.*

Sugars, Fats, and Calories

As we noted earlier, when carbohydrates are taken into the body in excess of the body's energy requirements, they are stored temporarily as glycogen or, more permanently, as fats. Conversely, when the energy requirements of the body are not met by its immediate intake of food, glycogen and, subsequently, fat are broken down to fill these requirements. Whether or not the body uses up its own storage molecules has nothing to do with the molecular form in which the energy comes into the body. It is simply a matter of whether these molecules, as they are broken down, release sufficient numbers of calories.

Insulators and Cushions

In general, fat stored in fat cells can be mobilized for energy when caloric intake is less than caloric expenditures. Some types of fat, however, seem to be protected from such mobilization. Large masses of fatty tissue, for example, surround mammalian kidneys and serve to protect these precious organs from physical shock. For reasons that are not understood, these fat deposits remain intact even at times of starvation. Another mammalian characteristic is a layer of fat under the skin, which serves as thermal insulation. This layer is particularly well developed in seagoing mammals.

Among humans, females characteristically have a thicker layer of subdermal ("under-the-skin") fat than males. This capacity to store fat, although not much admired in our present culture, was undoubtedly very valuable 10,000 or more years ago. At that time, as far as we know, there was no other reserve food supply, and this extra fat not only nourished the woman but, more important, the unborn child and the nursing infant, whose ability to fast without damage is much less than that of the adult. Thus many of us are strenuously dieting off what millennia of evolution have given us the capacity to accumulate.

3–13 *This harp seal, resting on an ice floe in the Gulf of St. Lawrence, is well insulated by a thick layer of fat under the skin, which serves the same function that a wet suit serves for a diver.*

Phospholipids and Glycolipids

Lipids, especially phospholipids and glycolipids, also play extremely important structural roles. Like fats, both phospholipids and glycolipids are composed of fatty acid chains attached to a glycerol backbone. In the **phospholipids,** however, the third carbon of the glycerol molecule is occupied not by a fatty acid but by a phosphate group (Figure 3–14) to which another polar group is usually attached. Phosphate groups are negatively charged. As a result, the phosphate end of the molecule is hydrophilic, whereas the fatty acid portions are hydrophobic. The consequences are shown in Figure 3–15. As we shall see in Chapter 5, this arrangement of phospholipid molecules, with their hydrophilic heads exposed and their hydrophobic tails clustered together, forms the structural basis of cellular membranes.

In the **glycolipids** ("sugar lipids"), the third carbon of the glycerol molecule is occupied not by a phosphate group but by a short carbohydrate chain. Depending on the particular glycolipid, this chain may contain anywhere from 1 to 15 monosaccharide monomers. Like the phosphate head of a phospholipid, the carbohydrate head of a glycolipid is hydrophilic, and the fatty acid tails are, of course, hydrophobic. In aqueous solution, glycolipids behave in the same fashion as phospholipids, and they are also important components of cellular membranes.

3–14 *A phospholipid molecule consists of two fatty acids linked to a glycerol molecule, as in a fat, and a phosphate group (indicated by color) linked to the glycerol's third carbon. It also usually contains an additional chemical group, indicated by the letter R. The fatty acid "tails" are nonpolar and therefore insoluble in water (hydrophobic); the polar "head" containing the phosphate and R groups is soluble (hydrophilic).*

Polar head | Nonpolar tails

$$R-O-P-O-{}^3CH_2$$

$$H-{}^2C-O-C-CH_2CH_2CH_2CH_2CH_2CH_2CH_2CH=CHCH_2CH_2CH_2CH_2CH_2CH_2CH_2CH_3$$

$$H-{}^1C-O-C-CH_2CH_2CH_2CH_2CH_2CH_2CH_2CH_2CH_2CH_2CH_2CH_2CH_2CH_2CH_2CH_3$$

Glycerol

(a)

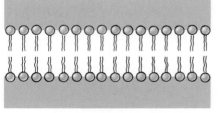

(b)

3–15 (a) *Because phospholipids have water-soluble heads and water-insoluble tails (Figure 3–14), they tend to form a thin film on a water surface with their tails extending above the water.* (b) *Surrounded by water, they spontaneously arrange themselves in two layers with their hydrophilic (water-loving) heads extending outward and their hydrophobic (water-fearing) tails inward. This arrangement forms the structural basis of cellular membranes.*

3–16 *A scanning electron micrograph of waxy deposits on the upper surface of a eucalyptus leaf. The deposits are magnified 10,800 times. All groups of land plants synthesize waxes, which protect exposed plant surfaces from water loss.*

Waxes

Waxes are also a form of structural lipid. They form protective coatings on skin, fur, feathers, on the leaves and fruits of land plants (Figure 3–16), and on the exoskeletons of many insects.

Cholesterol and Other Steroids

Cholesterol belongs to an important group of compounds known as the **steroids** (Figure 3–17). Although steroids do not resemble the other lipids structurally, they are grouped with them because they are insoluble in water. All the steroids have four linked carbon rings, like cholesterol, and several of them, like cholesterol, have a tail. In addition, many of them have the —OH functional group, which makes them alcohols.

Cholesterol is found in cell membranes (with the exception of bacterial cells); about 25 percent (by dry weight) of the cell membrane of a red blood cell is cholesterol. It is also a major component of the myelin sheath, the lipid membrane that wraps around fast-conducting nerve fibers, speeding the nerve impulse. Cholesterol is synthesized in the liver from saturated fatty acids and is also obtained in the diet, principally in meat, cheese, and egg yolks. High concentrations of cholesterol in the blood are associated with atherosclerosis, in which cholesterol is found in fatty deposits on the interior lining of diseased blood vessels (see essay).

Sex hormones and the hormones of the adrenal cortex (the outer portion of the adrenal glands, which lie atop the kidneys) are also steroids. These hormones are formed from cholesterol in the ovaries, testes, adrenal cortex, and other glands that produce them. Prostaglandins are a group of lipids with hormonelike actions, which are derived from fatty acids. Both the steroid hormones and the prostaglandins will be discussed more fully in Section 6.

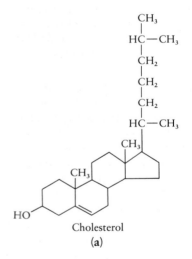

3–17 *Two examples of steroids. (a) The cholesterol molecule consists of four carbon rings and a hydrocarbon chain. (b) Testosterone, a male sex hormone synthesized from cholesterol by cells in the testes, also has the characteristic four-ring structure but lacks the hydrocarbon tail.*

PROTEINS

Proteins are among the most abundant organic molecules; in most living systems they make up 50 percent or more of the dry weight. Only plants, with their high cellulose content, are less than half protein. There are many different protein molecules: enzymes; hormones; storage proteins, such as those in the eggs of birds and reptiles and in seeds; transport proteins, such as hemoglobin; contractile proteins of the sort found in muscle; immunoglobulins (antibodies); membrane proteins; and many different types of structural proteins (Table 3–2). In their functions, their diversity is overwhelming. In their structure, however, they all follow the same simple blueprint: they are all polymers of amino acids, arranged in a linear sequence.

Regulation of Blood Cholesterol

Although cholesterol plays essential roles in the animal body, it is also a principal villain in heart disease. Deposits containing cholesterol can narrow the arteries carrying blood to the heart muscle, and people with unusually large amounts of cholesterol in their blood have a high risk of heart attacks. How does the body regulate cholesterol levels? What goes wrong to cause elevated levels? How does cholesterol cause heart attacks? Given the fact that heart disease is the major cause of death in this country, these questions are not only of biological interest but are also of importance to just about every one of us.

The key organ in cholesterol regulation is the liver, which not only synthesizes needed cholesterol from saturated fatty acids but also degrades excess cholesterol circulating in the blood—as a result, for example, of a diet rich in milk, cheese, and egg yolks. Cholesterol is transported to and from body cells, including those of the liver, by way of the bloodstream. Like other lipids, however, it is insoluble in water, and thus in plasma, the fluid portion of the blood. It is carried in particles consisting of a cholesterol interior and a lipid "wrapper" that has water-soluble proteins embedded in its outer surface. These large complexes exist in two principal forms: low-density lipoproteins (LDLs) and high-density lipoproteins (HDLs). LDLs function as the delivery trucks of the system, carrying dietary cholesterol and newly synthesized cholesterol to various destinations in the body, including both the liver and the hormone-synthesizing organs. HDLs, however, function more like garbage trucks, carrying excess cholesterol on a one-way trip to the liver for degradation and excretion.

Normally the system is in balance, and the liver synthesizes or degrades cholesterol depending on the body's current needs and the amount of circulating cholesterol. It can, however, be thrown out of balance by a number of factors. If, for example, the dietary intake of cholesterol is high, the liver becomes swamped and cannot degrade all of the excess. If the dietary intake of saturated fats is high, even in the absence of a high intake of cholesterol itself, the liver increases its synthesis of cholesterol. Current evidence indicates that the liver monitors the level of cholesterol in the blood through its uptake of LDLs, for which cell surfaces have specialized receptors. If these receptors are absent or damaged, liver cells continue

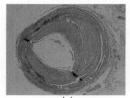 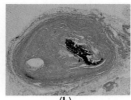

(a)   (b)

In one type of heart disease, atherosclerosis, cholesterol and other fatty substances accumulate in the walls of the coronary arteries, which supply the heart muscle. This accumulation triggers abnormal growth and the production of fibrous tissues by the cells of the walls. (a) A cross section of a coronary artery in which moderate atherosclerosis has developed. Fatty deposits have formed, and the space left for blood flow is significantly decreased. (b) A coronary artery in which the deposits have become so great that only a very narrow channel remains open. Such a narrow channel can be completely blocked by a blood clot. The result is a heart attack and the death of the heart muscle supplied by the artery.

to synthesize and export cholesterol in the form of LDLs, even when blood cholesterol levels are high.

When the quantities of circulating LDLs are greater than can be taken up by the liver and hormone-synthesizing organs, they are taken up by the cells lining the arteries supplying the heart. This can ultimately lead to total blockage of an artery and thus to a heart attack.

Heart disease often runs in families, suggesting that hereditary factors are involved in some cases. In one type of hereditary heart disease, the cells of the body have no LDL receptors. Individuals with this disease have six to eight times the normal amount of cholesterol in their blood, usually have their first heart attack in childhood, and die of heart disease in their early twenties. Other families seem to be protected against heart disease, apparently because the individuals' bodies synthesize large quantities of HDLs, ensuring that all excess cholesterol makes a speedy one-way trip to the liver. For most of us, however, the degree of risk depends on our behavior: whether or not we exercise regularly, which seems to increase HDL levels and thus protect against cholesterol buildup; whether or not we smoke cigarettes, which seems to decrease HDL levels; and the quantities of cholesterol and saturated fats that we consume.

Protein molecules are large, often containing several hundred amino acids. Thus the number of different amino acid sequences, and therefore the possible variety of protein molecules, is enormous—about as enormous as the number of different sentences that can be written with our own 26-letter alphabet. Organisms, however, have only a very small fraction of the proteins that are theoretically possible. The single-celled bacterium *Escherichia coli,* for example, contains 600 to 800 different kinds of proteins at any one time, and the cell of a plant or animal has several times that number. In a complex organism, there are at least several thousand different proteins, each with a special function and each, by its unique chemical nature, specifically fitted for that function.

TABLE 3–2   **Biological Functions of Proteins**

TYPES OF PROTEINS*	EXAMPLES
Structural proteins	Collagen, silk, virus coats, microtubules
Regulatory proteins	Insulin, ACTH, growth hormones
Contractile proteins	Actin, myosin
Transport proteins	Hemoglobin, myoglobin
Storage proteins	Egg white, seed protein
Protective proteins in vertebrate blood	Antibodies, complement
Membrane proteins	Receptors; membrane-transport proteins; antigens
Toxins	Botulism toxin, diphtheria toxin
Enzymes	Sucrase, pepsin

* Many of the proteins listed here will be discussed in other sections of the book, particularly in Section 6.

Amino Acids: The Building Blocks of Proteins

Every amino acid has the same fundamental structure, which consists of a central carbon atom bonded to an amino group ($-NH_2$), to a carboxyl group ($-COOH$), and to a hydrogen atom (Figure 3–18a). In every amino acid there is also another atom or group of atoms (designated as $-R$) bonded to the central carbon. As we saw in Chapter 2 (page 48), the amino group is a weak base and the carboxyl group is a weak acid. Depending on the pH of the surrounding solution, a free amino acid may be uncharged, negatively charged (if the $-COOH$ group is ionized to $-COO^-$ and H^+), or positively charged (if the $-NH_2$ group has acquired a hydrogen ion, becoming $-NH_3^+$).

A large variety of different amino acids is theoretically possible, but only 20 different kinds are used to build proteins (Figure 3–18b). And it is always the same 20, whether in a bacterial cell, a plant cell, or a cell in your own body. The only differences in these 20 amino acids lie in their side ($-R$) groups. In eight of the molecules, the side group consists of short chains or rings of carbon and hydrogen atoms; as you would expect, such groups are nonpolar and thus hydrophobic. The side groups in seven of the amino acids have polar regions; in acidic or basic solution, these regions can become charged. The remaining five amino acids have side groups that are either weak acids or weak bases; depending on the particular side group and the pH of the solution, they may be negatively or positively charged.

In yet another example of a condensation reaction, the amino "head" of one amino acid can be linked to the carboxyl "tail" of another by the removal of a

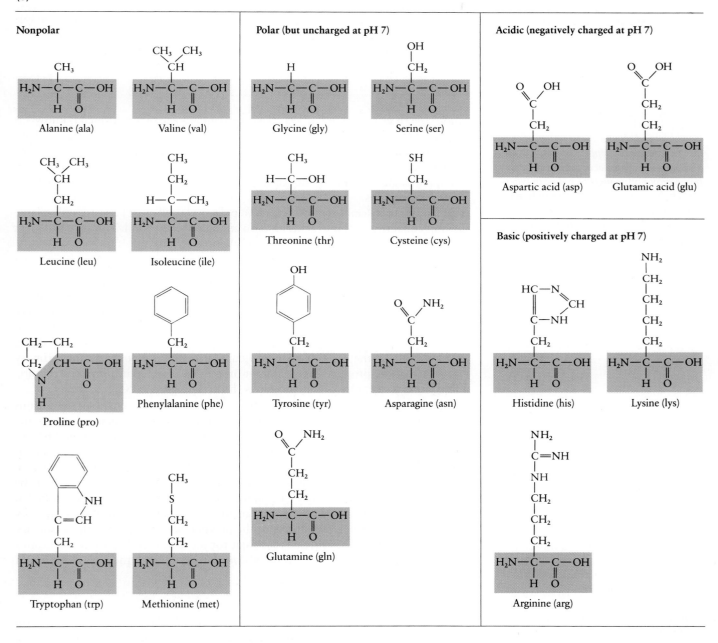

(a)

(b)

3–18 **(a)** *Every amino acid contains an amino group (—NH₂) and a carboxyl group (—COOH) bonded to a central carbon atom. A hydrogen atom and a side group are also bonded to the same carbon atom. This basic structure is the same in all amino acids. The "R" stands for the* side group, which is different in each kind of amino acid. **(b)** *The 20 different kinds of amino acids used in making proteins. As you can see, the essential structure is the same in all 20 molecules, but the side groups differ. These groups may be nonpolar (with no difference in charge* between one zone and another), polar but with the charges balancing one another out so that the side group as a whole is uncharged, negatively charged, or positively charged. The nonpolar side groups are not soluble in water, whereas the polar and charged side groups are water-soluble.

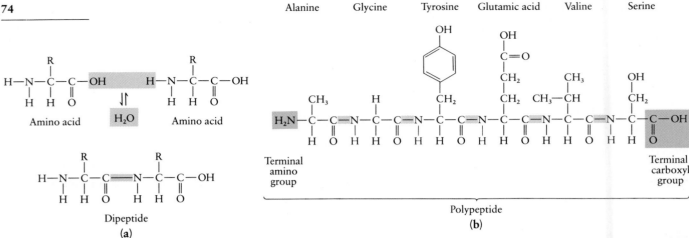

3–19 (a) *A peptide bond is a covalent bond formed by condensation.* (b) *Polypeptides are polymers of amino acids linked together by peptide bonds, with the amino group of one acid joining the carboxyl group of its neighbor. The polypeptide chain shown here contains only six amino acids, but some chains may contain as many as 1,000 linked amino acid monomers.*

molecule of water (Figure 3–19a). The covalent linkage that is formed is known as a **peptide bond,** and the molecule that is formed by the linking of many amino acids is called a **polypeptide** (Figure 3–19b). The sequence of amino acids in the polypeptide chain determines the biological character of the protein molecule; even one small variation in the sequence may alter or destroy the way in which the protein functions.

In order to assemble amino acids into proteins, a cell must have not only a large enough quantity of amino acids but also enough of every kind. This fact is of great importance in human nutrition (see essay).

The Levels of Protein Organization

In a living system, a protein is assembled in a long polypeptide chain, one amino acid at a time. In this process (to be described in some detail in Chapter 15), the amino group of one amino acid is linked to the carbonyl* of another, like a line of boxcars. The linear sequence of amino acids, which is dictated by the hereditary information in the cell for that particular protein, is known as the **primary structure** of the protein. Each different protein has a different primary structure. The primary structure of one protein is shown in Figure 3–20.

As the chain is assembled, interactions begin to take place among the various amino acids along the chain. Linus Pauling and coworker Robert Corey discovered that hydrogen bonds could form between the slightly positive amino hydrogen of one amino acid and the slightly negative carbonyl oxygen of another amino acid. They elucidated two structures that could result from these hydrogen bonds. One of these they called the alpha helix, because it was the first to be discovered, and the second, the beta pleated sheet. These structures are shown in Figure 3–21. Biochemists refer to the regular, repeated configurations caused by hydrogen bonding between atoms of the polypeptide backbone as the **secondary structure** of a protein. Proteins that exist for most of their length in a helical or pleated-sheet form are known as fibrous proteins, and they play important structural roles in organisms.

---

* When a peptide bond forms, the OH of the carboxyl group and an H of the amino group split out to form a water molecule. All that remains of the carboxyl group is the $\rangle$C$=$O group, which, in this context, we refer to as "carbonyl" (see Table 3–1).

3–20 *Primary structure of a relatively small protein, human adrenocorticotropic hormone (ACTH). This was one of the first proteins for which the primary structure was determined. As you can see, it consists of a single polypeptide chain containing 39 amino acids. This hormone, secreted by the pituitary gland, stimulates production of cortisol and related steroid hormones by the adrenal cortex.*

ser—tyr—ser—met—glu—his—phe—arg—trp—gly—lys—pro—val—gly—lys—lys—arg—arg—pro—val—lys—val—tyr—pro—asp—ala—gly—glu—

asp

phe—glu—leu—pro—phe—ala—glu—ala—ser—gln

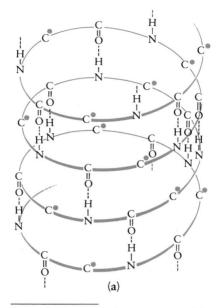

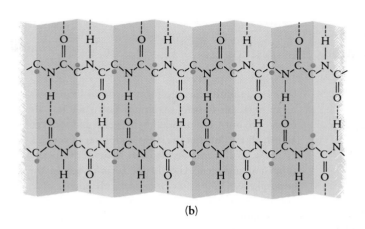

(b)

75

3–21 *Protein secondary structures.* (a) *The alpha helix. The helix is held in shape by hydrogen bonds, indicated by the dashed lines. The hydrogen bonds form between the oxygen atom of the carbonyl group in one amino acid and the hydrogen atom of the amino group in another amino acid that occurs four amino acids farther along the chain. The R groups, not shown in this diagram, are attached to the carbons indicated by the red dots. The R groups extend out from the helix.* (b) *The beta pleated sheet. The pleats are formed by hydrogen bonding between atoms of the backbone of the polypeptide; the R groups, which are attached to the carbons indicated by the red dots, extend above and below the folds of the pleat.*

Other forces, which involve the nature of the R groups in the individual amino acids, are also at work on the polypeptide chain, and these counteract the formation of the hydrogen bonds just described. For instance, an R group such as that of isoleucine is so bulky that it interrupts the turn of the helix, making the hydrogen bonding impossible. When the —SH portion of the R group of a cysteine encounters the same portion of another cysteine, two hydrogen atoms may split out, resulting in the formation of a covalent bond between the sulfur atoms of the two amino acids. This bond, known as a disulfide bridge, locks the molecule in that position. R groups with unlike charges are attracted to each other, and those with like charges are mutually repelled. As the molecule is twisted and turned in solution, the hydrophobic R groups tend to cluster together in the interior of the molecule and the hydrophilic R groups tend to extend outward into the aqueous solution. Hydrogen bonds form, linking together segments of the amino acid backbone. The intricate three-dimensional structure that results from these interactions among R groups is called the **tertiary structure** of a protein. Figure 3–22a shows the various types of bonds that are involved in forming the tertiary structure.

● = Amino acid
N = Nonpolar (hydrophobic) interactions
S — S = Disulfide bridges
– – – = Hydrogen bonds
P⁺, P⁻ = Polar (hydrophilic) groups

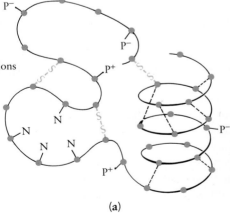

(a)

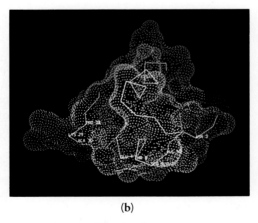

(b)

3–22 (a) *Types of bonds that stabilize the tertiary structure of a protein molecule. These same types of bonds also stabilize the structure of protein molecules that consist of more than one polypeptide chain.* (b) *A computer-generated model of the insulin molecule, which consists of*

two short polypeptide chains, folded together in an intricate three-dimensional structure. The two lines represent the backbones of the chains, and the dots represent the atoms on the surface of the molecule that are accessible to the surrounding solvent. In the insulin mole-

cule, as in all molecules, atoms are constantly vibrating and rotating. The atoms shown in red and orange are most likely to undergo slight shifts of position in the insulin crystal, whereas those shown in green and blue are least likely to shift position.

Amino Acids and Nitrogen

Like fats, amino acids are formed within living cells using sugars as starting materials. But while fats are made up only of carbon, hydrogen, and oxygen atoms, all available in the sugar and water of the cell, amino acids also contain nitrogen. Most of the earth's supply of nitrogen exists in the form of gas in the atmosphere. Only a few organisms, all microscopic, are able to incorporate nitrogen from the air into compounds—ammonia, nitrites, and nitrates—that can be used by living systems. Hence, the proportion of the earth's nitrogen supply available to the living world is very small.

Plants incorporate the nitrogen in ammonia, nitrites, and nitrates into carbon-hydrogen compounds to form amino acids. Animals are able to synthesize some of their amino acids, using ammonia as a nitrogen source. The amino acids they cannot synthesize, the so-called essential amino acids, must be obtained either directly or indirectly from plants. For adult human beings, the essential amino acids are lysine, tryptophan, threonine, methionine, phenylalanine, leucine, valine, and isoleucine.

People who eat meat usually get enough protein and the correct balance of amino acids. People who are vegetarians, whether for philosophical, esthetic, or economic reasons, have to be careful that they get enough protein and, in particular, the essential amino acids.

Until recently, agricultural scientists concerned with the world's hungry people concentrated on developing plants with a high caloric yield. Increasing recognition of the role of plants as a major source of amino acids for human populations has led to emphasis on the development of high-protein strains of food plants and of plants with essential amino acids, such as "high-lysine" corn.

Another approach to the right balance of amino acids is to combine certain foods. Beans, for instance, are likely to be deficient in tryptophan and in the sulfur-containing amino acids, but they are a good-to-excellent source of isoleucine and lysine. Rice is deficient in isoleucine and lysine but provides an adequate amount of the other essential amino acids. Thus rice and beans in combination make just about as perfect a protein menu as eggs or steak, as some nonscientists seem to have known for quite a long time.

In many proteins, the tertiary structure produces an intricately folded, globular shape for the molecule as a whole; these proteins are called globular proteins. Enzymes—proteins that regulate chemical reactions in living systems—are globular proteins, as are the membrane receptors for an enormous variety of molecules. Antibodies, important components of the immune system, are also globular proteins. As we shall see in subsequent chapters, the three-dimensional structures of all of these molecules are of critical importance in determining their biological functions.

Many proteins are composed of more than one polypeptide chain. These chains may be held to each other by hydrogen bonds, disulfide bridges, hydrophobic forces, attractions between positive and negative charges, or, most often, by a combination of these types of interactions. Such proteins are often called multimeric; a protein containing two polypeptide chains is termed a dimer, one containing three chains is a trimer, and one containing four chains is a tetramer. The hormone insulin, for example, is a dimer; it is composed of two polypeptide chains (Figure 3–22b). This level of organization of proteins, which involves interaction of two or more polypeptides, is called a **quaternary structure.**

The secondary, tertiary, and quaternary structures of a protein all depend on the primary structure—the sequence of amino acids—and on the local chemical environment.

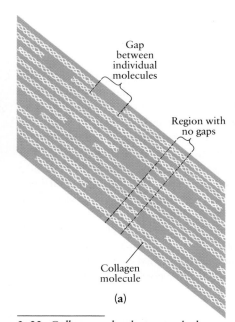

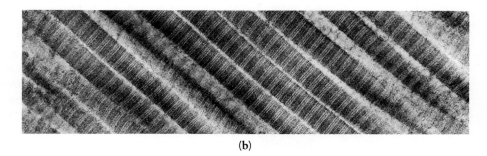

(b)

3–23 *Collagen molecules are packed together in fibrils that are a major constituent of skin, tendon, ligament, cartilage, and bone. Within an individual fibril (a), the collagen molecules are arranged in a staggered pattern, with gaps between individual molecules. This arrangement strengthens the fibrils, making them resistant to shearing forces. In electron micrographs of collagen fibrils, such as (b), a striated pattern is observed because the stain used in preparing the specimen is concentrated in the gaps between the molecules, causing the regions with no gaps to appear as lighter bands. These fibrils are magnified 23,500 times.*

Structural Uses of Proteins

Fibrous Proteins

In general, fibrous proteins have a regular, repeated sequence of amino acids and therefore a regular, repetitious structure. An example is collagen, which makes up about one-third of all the protein in vertebrates. The basic collagen molecule is composed of three very long polypeptides—about 1,000 amino acids per chain. These three polypeptides, which are made up of repeating groups of amino acids, are held together by hydrogen bonds linking amino acids of different chains into a tight coil. The molecules can coil so tightly because every third amino acid is glycine, the smallest of the amino acids. The collagen molecules are packed together to form fibrils (Figure 3–23), which are, in turn, associated into larger fibers.

Collagen is actually a family of proteins. Different types of collagen molecules contain polypeptides with slightly different amino acid sequences. The larger structures formed from the different types of molecules perform a variety of functions in the body. Consider a cow: Tendons, which link muscle to bone, are made up of collagen fibers in parallel bundles; thus arranged, they are very strong but do not stretch. The cow's hide, by contrast, is made up of collagen fibrils arranged in an interlacing network laid down in sheets. Even its corneas—the transparent coverings of the eyeballs—are composed of collagen. When collagen is boiled in water, the polymers are dispersed into shorter chains, which we know as gelatin.

Other fibrous proteins include keratin (Figure 3–24), silk, and elastin, present in the elastic tissue of ligaments.

3–24 *The fibrous protein keratin is found in all vertebrates. It is the chief component of scales, wool, nails, and feathers. (a) The horn of a rhinoceros consists of tightly packed strands of keratin. Solid rhino horn is used for dagger handles, and powdered horn as an aphrodisiac. A single horn can net a poacher far more than the average annual wage in many parts of Africa. (b) A feather, such as this spectacularly colored peacock feather, is made up of a shaft to which thousands of barbs—each with many tiny barbules—are attached.*

(a)

(b)

α tubulin

β tubulin

Soluble tubulin dimer

Microtubule

(a)

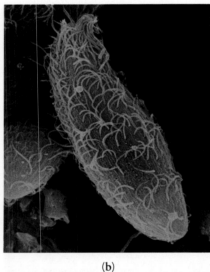

(b)

3–25 (a) *Microtubules are hollow tubes, so small that they cannot be visualized by a light microscope. They are composed of subunits, each of which is a globular protein. The subunits are of two types, alpha tubulin and beta tubulin, which first come together to form a soluble dimer. The dimers then self-assemble into insoluble hollow tubules.* (b) *Among their many functions, microtubules make up the internal structure of cilia, the small, hairlike appendages found on the surface of many eukaryotic cells, such as the protist Dileptus. The organism is magnified 1,000 times.*

Globular Proteins

Some structural proteins are globular. For example, microtubules, which function in a variety of ways inside the cell, are made up of globular proteins. These proteins associate to form long, hollow tubes—so long that their entire length can seldom be traced in a single microscopic section. Microtubules play a critical role in cell division, as we shall see in Chapter 7. They also participate in the internal skeleton that stiffens parts of the cell body and also seem to function as a kind of scaffolding for cellular construction projects. For example, the formation of a new cell wall in a plant can be predicted by the appearance at the site of large numbers of microtubules; when a plant cell wall is forming or growing and cellulose fibrils are being laid down outside the cell membrane, microtubules can be detected inside the cell, aligned in the same direction as the fibrils outside.

Chemical analysis shows that each microtubule consists of a very large number of subunits. There are two types of subunits, each of which is a globular protein formed from one polypeptide chain. Because of their complementary configurations, the two subunits fit together, forming approximately dumbbell-shaped dimers. The dimers assemble themselves into tubules (Figure 3–25), adding on length as required. When their job is over, they separate. The way in which the cell controls the assembly and disassembly of microtubules is the subject of a great deal of current research.

Hemoglobin: An Example of Specificity

Fibrous proteins, like polysaccharides, are usually molecules with a relatively small variety of monomers in a repetitive sequence. Many globular proteins, by contrast, have extremely complex, irregular amino acid sequences, as complex and irregular as the sequence of letters in a sentence on this page. Just as these sentences make sense (if they do) because the letters are the right ones and in the right order, the proteins make sense, biologically speaking, because their amino acids are the right ones in the right order.

Hemoglobin, for example, is a protein that is manufactured and carried in the red blood cells. Its molecules have the special property of being able to combine loosely with oxygen, collecting it in the lungs and releasing it in the tissues. The hemoglobin molecule has a quaternary structure that consists of four polypeptide chains, each of which is combined with an iron-containing group known as **heme**. In heme, an iron atom is held by nitrogen atoms that are part of a larger structure known as a porphyrin ring (Figure 3–26). Hemoglobin has two identical alpha chains and two identical beta chains, each with a unique primary structure containing about 150 amino acids, for a total of about 600 amino acids in all (Figure 3–27).

Sickle cell anemia is a disease in which the hemoglobin molecules are defective. When oxygen is removed from them, these molecules change shape and combine with one another to form stiffened rodlike structures. Red blood cells containing large proportions of such hemoglobin molecules become stiff and deformed, taking on the characteristic sickle shape (Figure 3–28). The deformed cells may clog the smallest blood vessels (capillaries). This causes blood clots and deprives vital organs of their full supply of blood, resulting in pain, intermittent illness, and, in many cases, a shortened life span.

Analysis of the hemoglobin molecules reveals that the only difference between normal and sickle cell hemoglobin is that in a precise location in each beta chain, one glutamic acid is replaced by one valine. In the quaternary structure of the molecule, this particular location is on the outer surface, and valine, unlike glutamic acid, contains a nonpolar R group. The result is a hydrophobic, "sticky" region that can interact with hydrophobic regions on neighboring hemoglobin molecules, producing the observed clumping. When one considers that this

3–26 *The heme group of hemoglobin. It contains an iron atom (Fe) held in a porphyrin ring. The porphyrin ring consists of four nitrogen-containing rings, which are numbered in the diagram. Each heme group is attached to a long polypeptide chain that wraps around it. The oxygen molecule is held flat against the heme.*

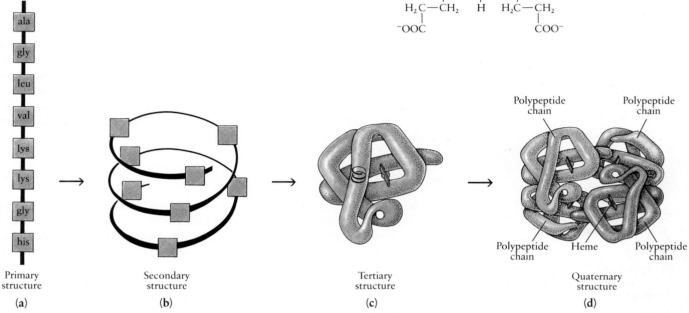

Primary structure	Secondary structure	Tertiary structure	Quaternary structure
(a)	**(b)**	**(c)**	**(d)**

3–27 *Levels of organization in the hemoglobin molecule. (a) The sequence of the amino acids in each chain is its primary structure. (b) The helical form assumed by any part of the chain as a consequence of hydrogen bonding between nearby* $\diagup$C$=$O *and* —NH *groups is its secondary structure. (c) The folding of the chains in three-dimensional shapes is the tertiary structure, and (d) the combination of the four chains into a single functional molecule is the quaternary structure. The outside of the molecule and the hole through the middle are lined by charged amino acids, and the uncharged amino acids are packed inside. Each of the four chains surrounds a heme group (red), which can hold a single oxygen molecule. A hemoglobin molecule is therefore capable of transporting four oxygen molecules.*

3–28 *Scanning electron micrographs of (a) human red blood cells containing normal hemoglobin, and (b) a red blood cell containing the abnormal hemoglobin associated with sickle cell anemia. When the oxygen concentration in the blood is low, the abnormal hemoglobin molecules stick together, distorting the shape of the cells. As a result, the cells cannot pass readily through the capillaries. These cells are magnified about 7,000 times.*

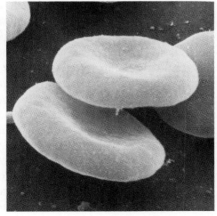

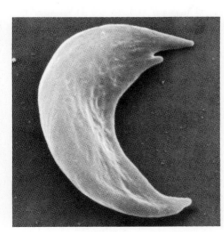

(a)   **(b)**

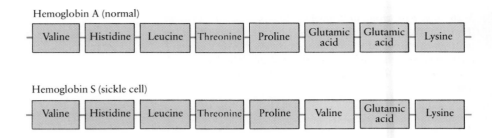

3-29 *An example of the remarkable precision of the "language" of proteins. Portions of the beta chains of the hemoglobin A (normal) molecule and the hemoglobin S (sickle cell) molecule are shown. The entire structural difference between the normal molecule and the sickle cell molecule (literally, a life-and-death difference) consists of one change in the sequence of each beta chain: one glutamic acid is replaced by one valine.*

difference of two amino acids in a total of almost 600 can make such a profound difference in the properties of the molecule as a whole—indeed, can be the difference between life and death—one begins to get an idea of the precision and the importance of the arrangement of amino acids in a particular sequence in a protein. In living systems, which must perform many different activities simultaneously, the specificity of function that results from the structural precision of different protein molecules is of crucial importance.

NUCLEOTIDES

The information dictating the structures of the enormous variety of protein molecules found in living organisms is encoded in and translated by molecules known as **nucleic acids.** Just as proteins consist of long chains of amino acids, nucleic acids consist of long chains of nucleotides. A nucleotide, however, is a more complex molecule than an amino acid. As shown in Figure 3–30, it consists of three subunits: a phosphate group, a five-carbon sugar, and a **nitrogenous base**—a molecule that has the properties of a base and contains nitrogen.

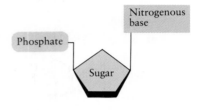

3-30 *A nucleotide is made up of three different subunits: a phosphate group, a five-carbon sugar, and a nitrogenous base. As we shall see in Chapter 14, nucleotides can be linked together in long chains by condensation reactions involving the hydroxyl groups of the phosphate and sugar subunits.*

The sugar subunit of a nucleotide may be either ribose or deoxyribose, which contains one less oxygen atom than ribose (Figure 3–31). Ribose is the sugar subunit in the nucleotides that form **ribonucleic acid (RNA),** and deoxyribose is the subunit in the nucleotides that form **deoxyribonucleic acid (DNA).** Five different nitrogenous bases are found in the nucleotides that are the building blocks of nucleic acids. Two of these, adenine and guanine, have a two-ring structure and are known as **purines** (Figure 3–32a). The other three, cytosine, thymine, and uracil, have a single-ring structure and are known as **pyrimidines** (Figure 3–32b). Adenine, guanine, and cytosine are found in both DNA and RNA, while thymine is found only in DNA and uracil only in RNA. As we shall see in Chapter 8, adenine and ribose sugar are also found in the nucleotides that are essential participants in the chemical reactions occurring within living systems.

Although their chemical components are very similar, DNA and RNA play very different biological roles. DNA is the primary constituent of the chromosomes of the cell and is the carrier of the genetic message. The function of RNA is to transcribe the genetic message present in DNA and translate it into proteins. The discovery of the structure and function of these molecules is undoubtedly the greatest triumph thus far of the molecular approach to the study of biology. In Section 3, we shall trace the events leading to the key discoveries and shall consider in some detail the marvelous processes—the details of which are still being worked out—by which these molecules perform their functions. First, however, we must turn our attention to the living cell—its origins, its structure, and the activities by which it maintains itself as an entity distinct from the nonliving world surrounding it.

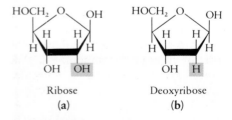

Ribose
(a)

Deoxyribose
(b)

3-31 *The sugar subunit of a nucleotide may be either (a) ribose or (b) deoxyribose. As you can see, the structural difference between the two sugars is slight. In ribose, carbon 2 bears a hydrogen atom above the plane of the ring and a hydroxyl group below the plane; in deoxyribose, the hydroxyl group on carbon 2 is replaced by a hydrogen atom.*

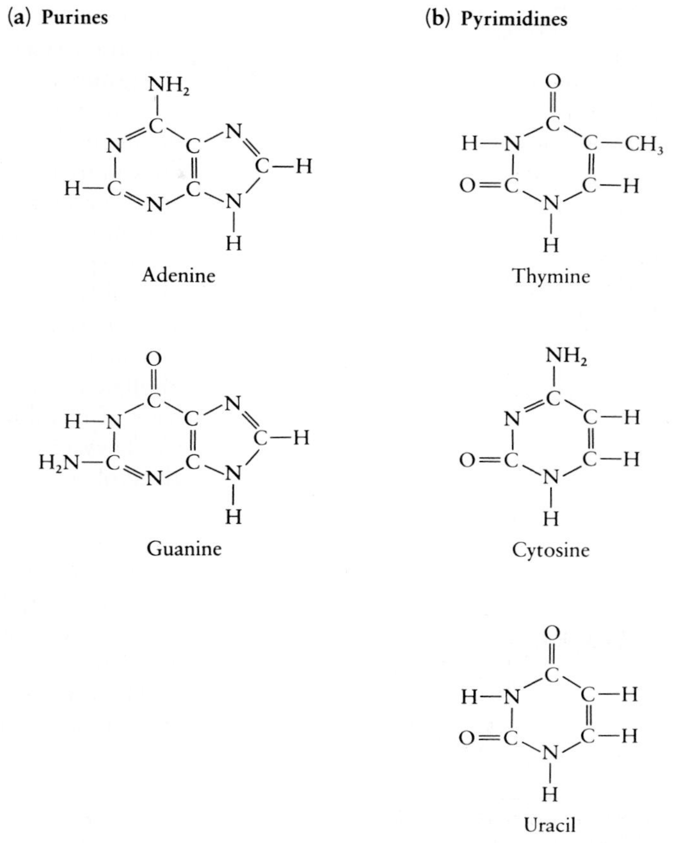

3–32 *The five nitrogenous bases of the nucleotides that form nucleic acids. (a) Adenine and guanine, the purines, occur in both DNA and RNA, as does cytosine, one of the pyrimidines (b). Thymine, also a pyrimidine, is found in DNA but not in RNA, and uracil, a third pyrimidine, is found in RNA but not in DNA. As we shall see in Section 3, the sequence of these simple molecules within long chains of nucleic acids is responsible for the transmission and translation of hereditary information, generation after generation.*

(a) Purines

Adenine

Guanine

(b) Pyrimidines

Thymine

Cytosine

Uracil

SUMMARY

The chemistry of living organisms is, in essence, the chemistry of organic compounds—that is, compounds containing carbon. Carbon is uniquely suited to this central role by the fact that it is the lightest atom capable of forming multiple covalent bonds. Because of this capacity, carbon can combine with carbon and other atoms to form a great variety of strong and stable chain and ring compounds. Organic molecules derive their three-dimensional shapes primarily from their carbon skeletons. Many of their specific properties, however, are dependent on functional groups. A general characteristic of all organic compounds is that they release energy when oxidized. Among the major types of organic molecules important in living systems are carbohydrates, lipids, proteins, and nucleotides.

Carbohydrates serve as a primary source of chemical energy for living systems. The simplest carbohydrates are the monosaccharides ("single sugars"), such as glucose and fructose. Monosaccharides can be combined to form disaccharides ("two sugars"), such as sucrose, and polysaccharides (chains of many monosaccharides). The polysaccharides starch and glycogen are storage forms for sugar, whereas cellulose, another polysaccharide, is an important structural material in plants. Disaccharides and polysaccharides are formed by condensation reactions in which monosaccharide units are covalently bonded with the removal of a molecule of water. They can be broken apart again by hydrolysis, with the addition of a water molecule.

Lipids are hydrophobic organic molecules that, like carbohydrates, play important roles in energy storage and as structural components. Compounds in this group include fats and oils, phospholipids, glycolipids, waxes, and cholesterol and other steroids. Fats are the chief energy-storing lipids. A fat molecule consists of one molecule of glycerol bonded to three fatty acids. Fats are designated as unsaturated or saturated depending on whether or not their fatty acids contain any double bonds. Unsaturated fats, which tend to be oily liquids, are more commonly found in plants.

Phospholipids are major structural components of cellular membranes. Phospholipids consist of one unit of glycerol, two fatty acids (instead of the three fatty acids present in fats), and a phosphate group to which another polar group may be attached. Because of their hydrophilic "heads" and hydrophobic "tails," phospholipids spontaneously orient in water to form films and clusters that are the basis of membrane structure. Glycolipids, which consist of one unit of glycerol, two fatty acids, and a short carbohydrate chain attached to the third carbon of the glycerol, are also important components of cellular membranes.

Proteins are very large molecules composed of long chains of amino acids; these are known as polypeptide chains. The 20 different amino acids used in making proteins vary according to the properties of their side (R) groups. From these relatively few amino acids, an extremely large variety of different kinds of protein molecules can be synthesized, each of which has a highly specific function in living systems.

The sequence of amino acids is known as the primary structure of the protein. Depending on the amino acid sequence, the molecule may take on any of a variety of forms. Hydrogen bonds between $>$C$=$O and $>$NH groups tend to fold the chain into a repeating secondary structure, such as the alpha helix or the beta pleated sheet. Interactions between the R groups of the amino acids may result in further folding into a tertiary structure, which is often an intricate, globular form. Two or more polypeptides may interact to form a quaternary structure.

In fibrous proteins, the long molecules interact with other similar, or identical, long polypeptide chains to form cables or sheets. Collagen and keratin are fibrous proteins that play a variety of structural roles. Globular proteins may also serve structural purposes. Microtubules, which are important cell components, are composed of repeating units of globular proteins assembled helically into a hollow tubule. Other globular proteins have regulatory, transport, and protective functions.

Because of the variety of amino acids, proteins can have a high degree of specificity. An example is hemoglobin, the oxygen-carrying molecule of the blood, which is composed of four (two pairs of) polypeptide chains, each attached to an iron-containing (heme) group. Substitution of one amino acid for another in one of the pairs of chains alters the surface of the molecule, producing a serious and sometimes fatal disease known as sickle cell anemia.

Nucleotides are complex molecules consisting of a phosphate group, a five-carbon sugar, and a nitrogenous base. They are the building blocks of the nucleic acids deoxyribonucleic acid (DNA) and ribonucleic acid (RNA), which transmit and translate the genetic information. Nucleotides also play key roles in the energy exchanges accompanying chemical reactions within living systems.

QUESTIONS

1. Distinguish among the following: hydrocarbon/carbohydrate; glucose/fructose/sucrose; monomer/polymer; glycogen/starch/cellulose; saturated/unsaturated; phospholipid/glycolipid; polysaccharide/polypeptide; peptide bond/disulfide bridge/hydrophobic interaction; primary structure/secondary structure/tertiary structure/quaternary structure; heme/hemoglobin; nitrogenous base/nucleotide/nucleic acid.

2. Identify the functional groups in the compounds below. Which of these is hydrophilic? Hydrophobic?

(a) CH_3COOH

Major component
of vinegar

(b) HCOOH

Active ingredient
in an ant's sting

(c) CH_2-CH_2
      $\;\;|\qquad\;\;|$
      OH   OH

Automobile
antifreeze

(d)
$$\begin{array}{c} H \\ | \\ H-C=O \end{array}$$

Preservative used
for biological
specimens

(e)
$$\begin{array}{c} CH_3-C-CH_3 \\ \| \\ O \end{array}$$

Nail-polish
remover

(f) NH₂

Used in manufacture
of commercial dyes

3. Draw a structural formula for (a) a monosaccharide; (b) a fatty acid; (c) an amino acid.

4. Butyric acid, $CH_3CH_2CH_2COOH$, gives rancid butter its odor and flavor. Draw its structural formula.

5. Many of the synthetic reactions in living systems take place by condensation. What is a condensation reaction? What types of molecules undergo condensation reactions to form disaccharides and polysaccharides? To form fats? To form proteins?

6. Disaccharides and polysaccharides, as well as lipids and proteins, can be broken down by hydrolysis. What is hydrolysis? What two types of products are released when a polysaccharide such as starch is hydrolyzed? How are these products important for the living cell?

7. What do we mean when we say that some polysaccharides are "energy-storage" molecules and that others are "structural" molecules? Give an example of each. In what sense should any polysaccharide be regarded as an "energy-storage" molecule?

8. Plants usually store energy reserves as polysaccharides, whereas, in most animals, lipids are the principal form of energy storage. Why is it advantageous for animals to have their energy reserves stored as lipids rather than as polysaccharides? (Think about the differences in "life-style" between plants and animals.) What kinds of storage materials would you expect to find in seeds?

9. Sketch the arrangement of phospholipids when they are surrounded by water.

10. In pioneer days, soap was made by boiling animal fat with lye (potassium hydroxide). The bonds linking the fatty acids to the glycerol molecule were hydrolyzed, and the potassium hydroxide reacted with the fatty acid to produce soap. A typical soap available today is sodium stearate. In water, it ionizes to produce sodium ions (Na^+) and stearate ions:

$$CH_3(CH_2)_{16}C{<}{\begin{array}{c}O\\O^-\end{array}}$$

Explain how soap functions to trap and remove particles of dirt and grease.

11. Silk is a protein in which polypeptide chains are arranged in a beta pleated sheet. In these chains, the peptide sequence glycine-serine-glycine-alanine-glycine-alanine occurs repeatedly. (a) Draw the structural formula for this hexapeptide, and show the peptide bonds in color. (b) Explain how a peptide bond is formed.

C H A P T E R **4**

Cells: An Introduction

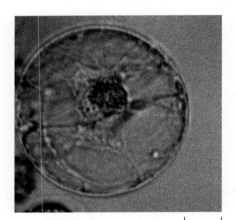

⊢——⊣ 500 μm

4–1 *A single living cell from the leaf of a poplar tree. Made up mostly (more than 95 percent) of only four kinds of atoms, it is nevertheless capable of carrying out a wide variety of intricate chemical reactions. This cell's capacity to exhibit the properties, such as energy conversion, homeostasis, reproduction, and growth, that are characteristic of living organisms depends on the complex organization of its constituent parts.*

The short, straight line at the bottom of this micrograph and those that follow provide a reference for size; a micrometer, abbreviated μm, is 1/1,000,000 meter. The same system is used to indicate distances on a road map.

In the last three chapters, we have progressed from subatomic particles through atoms and molecules to complex macromolecules, such as proteins and nucleic acids. At each level of organization, new properties appear. For instance, water, as we have seen, is not the sum of the properties of elemental hydrogen and oxygen; it is something more and also something different. In proteins, amino acids become organized into polypeptides, and polypeptide chains are arranged in a new level of organization, the tertiary or quaternary structure of the complete protein molecule. Only at this level of organization do the complex properties of the protein emerge, and only then can the molecule assume its function.

The characteristics of living systems, like those of atoms or of molecules, do not emerge gradually as the degree of organization increases. They appear quite suddenly and specifically, in the form of the living cell—something that is more than and different from its constituent atoms and molecules. No one knows exactly when or how this new level of organization—the living cell—first came into being. However, increasing knowledge of the history of our planet and the results of numerous laboratory experiments provide evidence for the hypothesis that living cells spontaneously self-assembled from molecules present in the primitive seas.

THE FORMATION OF THE EARTH

About 5 billion years ago, cosmologists calculate, the star that is our sun came into being. According to current theory, it formed, like other stars, from an accumulation of particles of dust and hydrogen and helium gases whirling in space among the older stars.

The immense cloud that was to become the sun condensed gradually as the hydrogen and helium atoms were pulled toward one another by the force of gravity, falling into the center of the cloud and gathering speed as they fell. As the cluster grew denser, the atoms moved more rapidly. More atoms collided with each other, and the gas in the cloud became hotter and hotter. As the temperature rose, the collisions became increasingly violent until the hydrogen atoms collided with such force that their nuclei fused, forming additional helium atoms and releasing nuclear energy. This thermonuclear reaction is still going on at the heart of the sun and is the source of the energy radiated from its glowing surface.

The planets, according to current theory, formed from the remaining gas and dust moving around the newly formed star. At first, particles would have collected at random, but as each mass grew larger, other particles began to be attracted by the gravity of the largest masses. The whirling dust and forming spheres continued to revolve around the sun until finally each planet had swept its own path clean, picking up loose matter like a giant snowball. The orbit nearest the sun was swept

4–2 *The tremendous amounts of energy released by the thermonuclear reactions at the heart of the sun give rise to an envelope of extremely hot gases surrounding its surface. The layer of gases may extend as far as 64,000 kilometers from the surface of the sun—a distance about five times greater than the diameter of the earth.*

by Mercury, the next by Venus, the third by earth, the fourth by Mars, and so on out to Neptune and Pluto, the most distant of the planets. The planets, including earth, are calculated to have come into being about 4.6 billion years ago.

During the time earth and the other planets were being formed, the release of energy from radioactive materials kept their interiors very hot. When earth was still so hot that it was mostly liquid, the heavier materials collected in a dense core whose diameter is about half that of the planet. As soon as the supply of stellar dust, stones, and larger rocks was exhausted, the planet ceased to grow. As earth's surface cooled, an outer crust, a skin as thin by comparison as the skin of an apple, was formed. The oldest known rocks in this layer have been dated by isotopic methods as about 4.1 billion years old.

Only 50 kilometers below its surface, the earth is still hot—a small fraction of it is even still molten. We see evidence of this in the occasional volcanic eruption that forces lava (molten rock) through weak points in the earth's skin, or in the geyser, which spews up boiling water that has trickled down to the earth's interior.

The biosphere is the part of the planet within which life exists. It forms a thin film on the outermost layer, extending only about 8 or 10 kilometers up into the atmosphere and about as far down into the depths of the sea.

THE BEGINNING OF LIFE

Until very recently, the earliest fossil organisms known were a mere 600 million years old, and for a long time after the publication of *The Origin of Species,* biologists regarded the earliest events in the history of life as chapters that would probably remain forever closed to scientific investigation.

Two developments, however, have greatly improved our long-distance vision. The first was the formulation of a testable hypothesis about the events preceding life's origins. This hypothesis generated questions for which answers could be sought experimentally. The results of the initial experimental tests led to the formulation of further hypotheses and to additional experiments, a process that continues today as scientists in many laboratories explore the question of life's origins. The second development was the discovery of fossilized cells more than 3 billion years old.

The Question of Spontaneous Generation

Most of the early biologists, from the time of Aristotle, believed that simple living things, such as worms, beetles, frogs, and salamanders, could originate spontaneously in dust or mud, that rodents formed from moist grain, and that plant lice condensed from a dewdrop. In the seventeenth century, Francesco Redi performed a famous experiment in which he put out decaying meat in a group of wide-mouthed jars—some with lids, some covered by a fine veil, and some open—and demonstrated that maggots arose only where flies were able to lay their eggs.

By the nineteenth century, no scientist continued to believe that complex organisms arose spontaneously. The advent of microscopy, however, led a vigorous renewal of belief in the spontaneous generation of very simple organisms. It was necessary only to put decomposing substances in a warm place for a short time and tiny "live beasts" appeared under the lens, before one's very eyes. By 1860, the controversy had become so spirited that the Paris Academy of Sciences offered a prize for experiments that would throw new light on the question. The prize was claimed in 1864 by Louis Pasteur, who devised experiments to show that microorganisms appeared only as contaminants from the air and not "spontaneously" as his opponents claimed. In his experiments he used swan-necked flasks, which permitted the entrance of oxygen, thought to be necessary for life, but which, in their long, curving necks, trapped bacteria, fungal spores, and other microbial life and thereby protected the contents of the flasks from contamination. He showed that if the liquid in the flask was boiled (which killed microorganisms already present) and the neck of the flask was allowed to remain intact, no microorganisms would appear. Only if the curved neck of the flask was broken off, permitting contaminants to enter the flask, did microorganisms appear. (Some of his original flasks, still sterile, remain on display at the Pasteur Institute in Paris.)

"Life is a germ, and a germ is Life," Pasteur proclaimed at a brilliant "scientific evening" at the Sorbonne before the social elite of Paris. "Never will the doctrine of spontaneous generation recover from the mortal blow of this simple experiment."

In retrospect, Pasteur's well-planned experiments were so decisive because the broad question of whether or not spontaneous generation had *ever* occurred was reduced to the simpler question of whether or not it occurred under the specific conditions claimed for it. Pasteur's experiments answered *only* the latter question, but the results were so dramatic that for many years very few scientists were able to entertain the possibility that under quite different conditions, when the earth was very young, some form of "spontaneous generation" might indeed have taken place. The question of the origin of the first living systems remained unasked until well into the twentieth century.

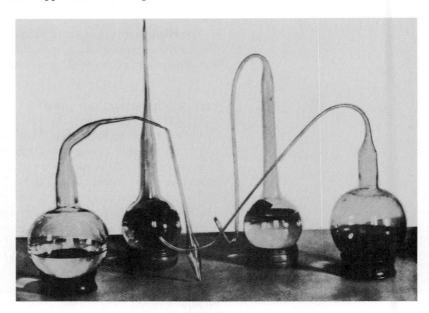

Pasteur's swan-necked flasks, which he used to counter the argument that spontaneous generation failed to occur in sealed vessels because air was excluded. These flasks permitted the entrance of oxygen, thought to be essential for life, but their long, curving necks trapped spores of microorganisms and thereby protected the liquids in the flasks from contamination.

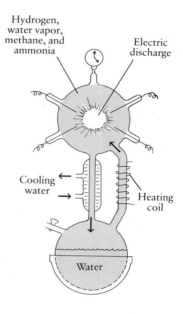

4–3 *Miller's experiment. Conditions believed to have existed on the primitive earth were simulated in the apparatus diagrammed here. Methane (CH$_4$) and ammonia (NH$_3$) were continuously circulated between a lower "ocean," which was heated, and an upper "atmosphere," through which an electric discharge was transmitted. At the end of 24 hours, about half of the carbon originally present in the methane gas was converted to amino acids and other organic molecules. This was the first test of Oparin's hypothesis.*

The testable hypothesis was offered by the Russian biochemist A. I. Oparin. According to Oparin, the appearance of life was preceded by a long period of what is sometimes called chemical evolution. The identity of the substances, particularly gases, present in the primitive atmosphere and in the seas during this period is a matter of controversy. There is general agreement, however, on two critical issues: (1) Little or no free oxygen was present, and (2) the four elements—hydrogen, oxygen, carbon, and nitrogen—that make up more than 95 percent of living tissues were available in some form in the atmosphere and waters of the primitive earth.

In addition to these raw materials, energy abounded on the young earth. There was heat energy, both boiling (moist) heat and baking (dry) heat. Water vapor spewed out of the primitive seas, cooled in the upper atmosphere, collected into clouds, fell back on the crust of the earth, and steamed up again. Violent rainstorms were accompanied by lightning, which provided electrical energy. The sun bombarded the earth's surface with high-energy particles and ultraviolet light, another form of energy. Radioactive elements within the earth released their energy into the atmosphere. Oparin hypothesized that under such conditions organic molecules were formed from the atmospheric gases and collected in a thin soup in the earth's seas and lakes. Because there was no free oxygen present to react with and degrade these organic molecules to simple substances such as carbon dioxide (as would happen today), they tended to persist. Some of these molecules might have become locally more concentrated by the drying up of a lake or by the adhesion of the molecules to a solid surface.

Oparin published his hypothesis in 1922, but at that time biochemists were so convinced by Pasteur's demonstration disproving spontaneous generation (see essay) that the scientific community ignored his ideas. In the 1950s, the first test of Oparin's hypothesis was performed by Stanley Miller, then a graduate student at the University of Chicago (Figure 4–3). Experiments of this sort, now repeated many times, have shown that almost any source of energy—lightning, ultraviolet radiation, or hot volcanic ash—would have converted molecules believed to have been present on the earth's surface into a variety of complex organic compounds. With various modifications in the experimental conditions and in the mixture of gases placed in the reaction vessel, almost all of the common amino acids have been produced, as well as the nucleotides that are the essential components of DNA and RNA.

These experiments have not proved that such organic compounds were formed spontaneously on the primitive earth, only that they could have formed. The accumulated evidence is nevertheless very great, and most biochemists now

4–4 *Volcanic eruptions, such as the one shown here off the coast of Iceland, occurred frequently on the restless surface of the young earth. The intense fields of electrical, thermal, and shock-wave energy generated by such eruptions of steam and lava could have been a major factor in the formation of organic molecules. This eruption, which took place in 1963, resulted in the birth of the island of Surtsey.*

believe that, given the conditions existing on the young earth, chemical reactions producing amino acids, nucleotides, and other organic molecules were inevitable.

As the concentrations of such molecules increased, bringing them into closer proximity to each other, they would have been subject to the same chemical forces that act on organic molecules today. As we saw in the last chapter, small organic molecules react with each other, typically in condensation reactions, to form larger molecules; moreover, such forces as hydrogen bonds and hydrophobic interactions cause these molecules to assemble themselves into more complex aggregates. In modern chemical systems—either in the laboratory or in the living organism—the more stable molecules and aggregates tend to survive and the least stable are transitory. Similarly, the compounds and aggregates that had the greatest chemical stability under the prevailing conditions on the primitive earth would have tended to survive. Hence a form of natural selection played a role in chemical evolution as well as in the biological evolution that was to follow.

The First Cells

From a biochemical perspective, three characteristics distinguish the living cell from other chemical systems: (1) the capacity to replicate itself, generation after generation; (2) the presence of enzymes, the complex proteins that are essential for the chemical reactions on which life depends; and (3) a membrane that separates the cell from the surrounding environment and enables it to maintain a distinct chemical identity. Which of these characteristics appeared first—and made possible the development of the others—remains an open question. However, as we shall see in Chapter 18, recently discovered functions of RNA suggest that the starting point may well have been the self-assembly of RNA molecules from nucleotides produced by chemical evolution.

In other studies, simulating conditions during the earth's first billion years, Sidney W. Fox and his coworkers at the University of Miami have produced membrane-bound protein structures that can carry out a few chemical reactions analogous to those of living cells. These structures are produced through a series of chemical reactions, beginning with dry mixtures of amino acids. When the mixtures are heated at moderate temperatures, polymers (known as thermal proteinoids) are formed, each of which may contain as many as 200 amino acid monomers. When these polymers are placed in an aqueous salt solution and maintained under suitable conditions, they spontaneously form proteinoid microspheres (Figure 4–5). The microspheres grow slowly by the addition of

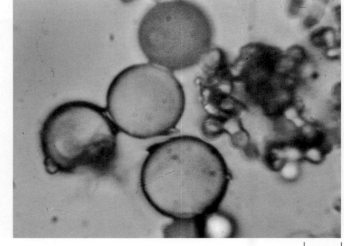

4–5 *Proteinoid microspheres, which form spontaneously when thermal proteinoids are maintained in aqueous solution under suitable conditions. As you can see, the microspheres are separated from the surrounding solution by a membrane that appears to be two-layered.*

1μm

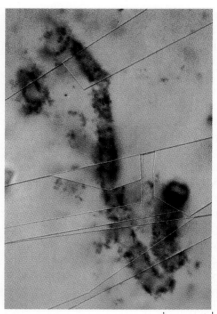

├── 50 µm ──┤

4-6 *This microfossil of a filament of bacteria-like cells was found in Western Australia in a deposit of a flintlike rock known as black chert. Dated at 3.5 billion years of age, it is one of the oldest fossils known.*

proteinoid material from the solution and eventually bud off smaller microspheres. These microspheres are not living cells. Their formation, however, suggests the kinds of processes that could have given rise to self-sustaining protein entities, separated from their environment and capable of carrying out the chemical reactions necessary to maintain their physical and chemical integrity.

It is not known when the first living cells appeared on earth, but we can establish some sort of time scale. The earliest fossils found so far (Figure 4-6), which resemble present-day bacteria, have been dated at 3.4 and 3.5 billion years—about 1.1 billion years after the formation of the earth itself. Although the fossils are so small that their structure can be made visible only by electron microscopy, they are sufficiently complex that it is clear that some little aggregation of chemicals had moved through the twilight zone separating the living from the nonliving millions of years before.

Why on Earth?

On the basis of astronomical studies and the explorations carried out by unmanned space vehicles, it appears that earth alone among the planets of our solar system supports life. The conditions on earth are ideal for living systems based on carbon-containing molecules. A major factor is that earth is neither too close to nor too distant from the sun. The chemical reactions on which life—at least as we know it—depends require liquid water, and they virtually cease at very low temperatures. At high temperatures, the complex chemical compounds essential for life are too unstable to survive.

Earth's size and mass are also important factors. Planets much smaller than earth do not have enough gravitational pull to hold a protective atmosphere, and any planet much larger than earth is likely to have so dense an atmosphere that light from the sun cannot reach its surface. The earth's atmosphere blocks out many of the most energetic radiations from the sun, which are capable of breaking the covalent bonds between carbon atoms. It does, however, permit the passage of visible light, which made possible one of the most significant steps in the evolution of complex living systems.

HETEROTROPHS AND AUTOTROPHS

The energy that produced the first organic molecules came from a variety of sources on the primitive earth and in its atmosphere—heat, ultraviolet radiations, and electrical disturbances. When the first primitive cells or cell-like structures evolved, they required a continuing supply of energy to maintain themselves, to grow, and to reproduce. The manner in which these cells obtained energy is currently the subject of lively discussion.

Modern organisms—and the cells of which they are composed—can meet their energy needs in one of two ways. **Heterotrophs** are organisms that are dependent upon outside sources of organic molecules for both their energy and their small building-block molecules. (*Hetero* comes from the Greek word meaning "other," and *troph* comes from *trophos*, "one that feeds.") All animals and fungi, as well as many single-celled organisms, are heterotrophs. **Autotrophs,** by contrast, are "self-feeders." They do not require organic molecules from outside sources for energy or to use as small building-block molecules; they are, instead, able to synthesize their own energy-rich organic molecules from simple inorganic substances. Most autotrophs, including plants and several different types of single-celled organisms, are **photosynthetic,** meaning that the energy source for their synthetic reactions is the sun. Certain groups of bacteria, however, are **chemosynthetic;** these organisms capture the energy released by specific inorganic reactions to power their life processes, including the synthesis of needed organic molecules.

Both heterotrophs and autotrophs seem to be represented among the earliest microfossils. It has long been postulated that the first living cell was an extreme heterotroph. As the primitive heterotrophs increased in number, according to this hypothesis, they began to use up the complex molecules on which their existence depended and which had taken millions of years to accumulate. As the supply of these molecules decreased, competition began. Under the pressure of this competition, cells that could make efficient use of the limited energy sources now available were more likely to survive and reproduce than cells that could not. In the course of time, other cells evolved that were able to synthesize organic molecules out of simple inorganic materials.

Recent discoveries, however, have raised the possibility that the first cells may have been either chemosynthetic or photosynthetic autotrophs rather than heterotrophs. First, several different groups of chemosynthetic bacteria have been found that would have been well-suited to the conditions prevailing on the young earth (Figure 4–7). Some of these bacteria are the inhabitants of swamps, while others have been found in deep ocean trenches in areas where gases escape from fissures in the earth's crust. There is evidence (to be discussed in Chapter 21) that these bacteria are the surviving representatives of very ancient groups of unicellular organisms. Second, organic molecules that are, in plants, the chemical precursors of chlorophyll have been produced in experiments analogous to that performed by Miller. When these molecules are mixed with simple organic molecules in an oxygen-free environment and illuminated, primitive photosynthetic reactions occur. These reactions resemble the reactions that occur in some types of photosynthetic bacteria.

Although biologists are presently unable to resolve the question of whether the earliest cells were heterotrophs or autotrophs, it is certain that without the evolution of autotrophs, life on earth would soon have come to an end. In the more than 3.5 billion years since life first appeared on earth, the most successful autotrophs (that is, those that have left the most offspring and diverged into the greatest variety of forms) have been those that evolved a system for making direct use of the sun's energy in the process of photosynthesis. With the advent of photosynthesis, the flow of energy in the biosphere came to assume its dominant modern form: radiant energy from the sun channeled through photosynthetic autotrophs to all other forms of life.

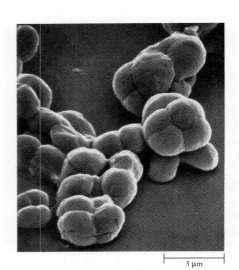

5 μm

4–7 *Methanogens, such as the cells shown here, are chemosynthetic bacteria that produce methane from carbon dioxide and hydrogen gas. They can live only in the absence of oxygen—a condition prevailing on the young earth but occurring today only in isolated environments, such as the muck and mud at the bottom of swamps.*

PROKARYOTES AND EUKARYOTES

The cell theory, as we noted in the Introduction, is one of the foundations of modern biology. This theory states simply that (1) all living organisms are composed of one or more cells; (2) the chemical reactions of a living organism, including its energy-releasing processes and its biosynthetic reactions, take place within cells; (3) cells arise from other cells; and (4) cells contain the hereditary information of the organisms of which they are a part, and this information is passed from parent cell to daughter cell. All available evidence indicates that there is an unbroken continuity between modern cells—and the organisms they compose—and the first primitive cells that appeared on earth.

All cells share two essential features. One is an outer membrane, the **cell membrane** (also known as the **plasma membrane**), that separates the cell from its external environment. The other is the genetic material—the hereditary information—that directs a cell's activities and enables it to reproduce, passing on its characteristics to its offspring.

The organization of the genetic material is one of the characteristics that distinguish two fundamentally distinct kinds of cells, **prokaryotes** and **eukaryotes.**

In prokaryotic cells, the genetic material is in the form of a large, circular molecule of DNA, with which a variety of proteins are loosely associated. This molecule is known as the **chromosome.** In eukaryotic cells, by contrast, the DNA is linear, forming a number of distinct chromosomes; moreover, it is tightly bound to special proteins known as **histones,** which are an integral part of the chromosome structure. Within the eukaryotic cell, the chromosomes are surrounded by a double membrane, the **nuclear envelope,** that separates them from the other cell contents in a distinct **nucleus** (hence the name, *eu,* meaning "true," and *karyon,* meaning "nucleus" or "kernel"). In prokaryotes ("before a nucleus"), the chromosome is not contained within a membrane-bound nucleus, although it is localized in a distinct region known as the **nucleoid.**

The remaining components of a cell (that is, everything within the cell membrane except the nucleus or nucleoid and its contents) constitute the **cytoplasm.** The cytoplasm contains a large variety of molecules as well as formed bodies called **organelles.** These specialized structures carry out particular functions within the cell. Both prokaryotes and eukaryotes contain very small organelles called **ribosomes,** on which protein molecules are assembled. In addition, eukaryotes contain a variety of more complex organelles, which are often enclosed within membranes.

The cell membrane of prokaryotes is surrounded by an outer **cell wall** that is manufactured by the cell itself. Some eukaryotic cells, including plant cells and fungi, have cell walls, although their structure is different from that of prokaryotic cell walls. Other eukaryotic cells, including those of our own bodies and of other animals, do not have cell walls. Another feature distinguishing eukaryotes and prokaryotes is size: eukaryotic cells are usually larger than prokaryotic cells.

Modern prokaryotes include the bacteria (Figure 4–8) and the cyanobacteria (Figure 4–9, on the next page), a group of photosynthetic prokaryotes that were formerly known as the blue-green algae.* According to the fossil record, the

4–8 *Cells of* Escherichia coli, *the heterotrophic prokaryote that is the most thoroughly studied of all living organisms. The genetic material (DNA) is in the lighter-appearing area in the center of each cell; this region, which is not enclosed by a membrane, is known as the nucleoid. The small, dense bodies in the cytoplasm are ribosomes. The two cells in the center have just finished dividing and have not yet separated completely.*

---

* The Latin term *alga,* plural *algae,* means "seaweed." Algae is a general term aplied to eukaryotic single-celled photosynthetic organisms and to many simple multicellular forms. Until recently, it was also applied to some prokaryotes.

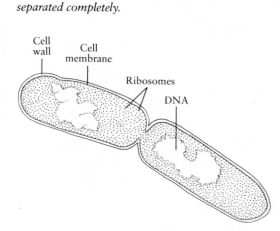

Cell wall   Cell membrane   Ribosomes   DNA

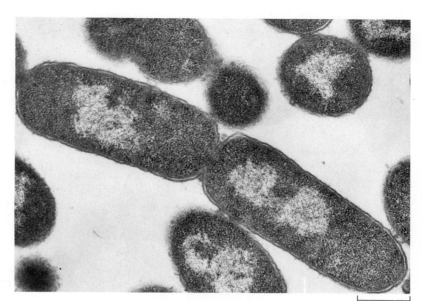

0.5 μm

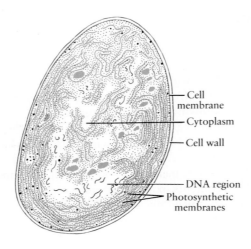
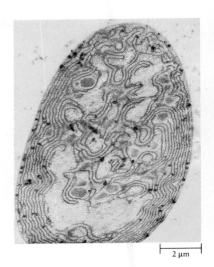

Cell
membrane

Cytoplasm

Cell wall

DNA region

Photosynthetic
membranes

2 µm

4-9 *Electron micrograph and diagram of a photosynthetic prokaryotic cell, the cyanobacterium* Anabaena azollae. *In addition to the genetic material, this cell contains a series of membranes in which chlorophyll and other photosynthetic pigments are embedded.* Anabaena *synthesizes its own energy-rich organic compounds in chemical reactions powered by the radiant energy of the sun.*

earliest living organisms were comparatively simple cells, resembling present-day prokaryotes. Prokaryotes were the only forms of life on this planet for almost 2 billion years until eukaryotes evolved. (The evolutionary relationship between prokaryotes and eukaryotes—quite an interesting subject—will be explored in Chapter 22.)

Figure 4–10 gives an example of a single-celled photosynthetic eukaryote, the alga *Chlamydomonas*. It is a common inhabitant of freshwater ponds and aquariums. These organisms are small and bright green (because of their chlorophyll), and they move very quickly with a characteristic darting motion. Being photosynthetic, they are usually found near the water's surface.

The Origins of Multicellularity

The first multicellular organisms, as far as can be told by the fossil record, made their appearance a mere 750 million years ago (Figure 4–11). The major groups of multicellular organisms—such as the fungi, the plants, and the animals—are thought to have evolved from different types of single-celled eukaryotes.

4-10 *Electron micrograph of* Chlamydomonas, *a photosynthetic eukaryotic cell, which contains a membrane-bound ("true") nucleus and numerous organelles. The most prominent organelle is the single, irregularly shaped chloroplast that fills most of the cell. It is surrounded by a double membrane and is the site of photosynthesis. Other membrane-bound organelles, the mitochondria, provide energy for cellular functions, including the flicking movements of the two flagella (one of which is visible in the micrograph). These movements propel the cell through the water. The organism's food reserves are in the form of starch granules. The cytoplasm is surrounded by a cell membrane, outside of which is a cell wall composed of polysaccharides.*

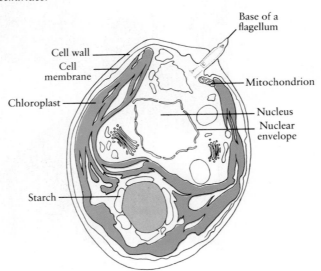

Base of a
flagellum

Cell wall

Cell
membrane

Chloroplast

Mitochondrion

Nucleus

Nuclear
envelope

Starch

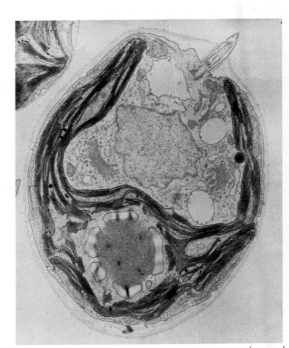

1 µm

4–11 *The clockface of biological time. Life first appears relatively early in the earth's history, before 6:00 A.M. on a 24-hour time scale. The first multicellular organisms do not appear until the early evening of that 24-hour day, and* Homo, *the genus to which humans belong, is a late arrival—at about 30 seconds to midnight.*

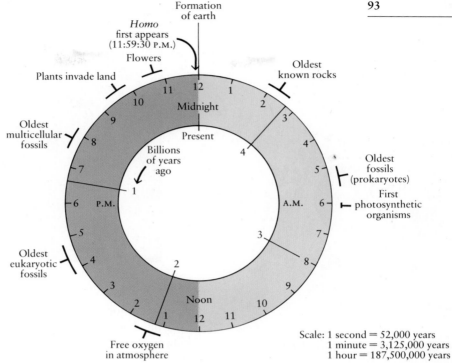

Formation of earth

Homo first appears (11:59:30 P.M.)

Flowers

Plants invade land

Oldest known rocks

Oldest multicellular fossils

Midnight

Present

Billions of years ago

Oldest fossils (prokaryotes)

First photosynthetic organisms

P.M.

A.M.

Oldest eukaryotic fossils

Noon

Free oxygen in atmosphere

Scale: 1 second = 52,000 years
1 minute = 3,125,000 years
1 hour = 187,500,000 years

4–12 *Electron micrograph of cells from the leaf of a corn plant. The nucleus can be seen on the right side of the central cell. The granular material within the nucleus is chromatin; it contains DNA associated with histone proteins. Note the many mitochondria and chloroplasts, all enclosed by membranes. The vacuole and cell wall are characteristic of plant cells but are generally not found in animal cells. As you can see, this cell closely resembles* Chlamydomonas, *shown in Figure 4–10.*

The cells of modern multicellular organisms closely resemble those of single-celled eukaryotes. They are bound by a cell membrane identical in appearance to the cell membrane of a single-celled eukaryote. Their organelles are constructed according to the same design. The cells of multicellular organisms differ from single-celled eukaryotes in that each type of cell is specialized to carry out a relatively limited function in the life of the organism. However, each remains a remarkably self-sustaining unit.

Notice how similar a cell from the leaf of a corn plant (Figure 4–12) is to *Chlamydomonas*. This plant cell is also photosynthetic, supplying its own energy needs from sunlight. However, unlike the alga, it is part of a multicellular organism and depends on other cells for water, minerals, protection from desiccation (drying out), and other necessities.

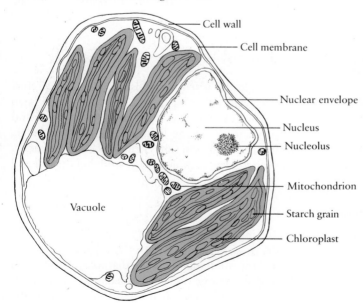

Cell wall

Cell membrane

Nuclear envelope

Nucleus

Nucleolus

Mitochondrion

Starch grain

Chloroplast

Vacuole

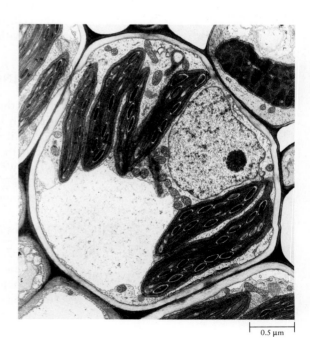

0.5 µm

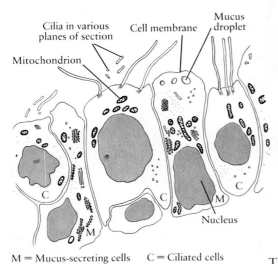

Cilia in various planes of section

Mitochondrion

Cell membrane

Mucus droplet

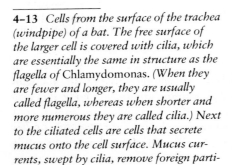

C

M

C

M

C

Nucleus

M = Mucus-secreting cells   C = Ciliated cells

4-13 *Cells from the surface of the trachea (windpipe) of a bat. The free surface of the larger cell is covered with cilia, which are essentially the same in structure as the flagella of* Chlamydomonas. *(When they are fewer and longer, they are usually called flagella, whereas when shorter and more numerous they are called cilia.) Next to the ciliated cells are cells that secrete mucus onto the cell surface. Mucus currents, swept by cilia, remove foreign particles from the surface of the trachea. Note the mitochondria located near the base of the cilia.*

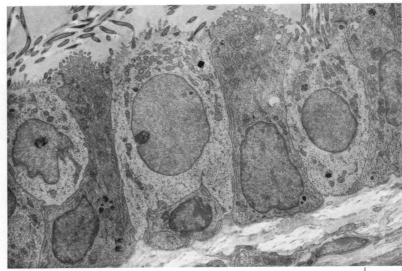

2.5 μm

The human body, made up of trillions of individual cells, is composed of at least 200 different types of cells, each specialized for its particular function but all working as a cooperative whole. Figure 4–13 shows cells from an animal trachea (windpipe). These cells are epithelial cells, the sort of cells that line all the internal and external body surfaces of animals. The epithelial cells of the trachea are part of an elaborate organ system involved in delivering oxygen to other cells in the body.

As we discussed in the Introduction, prokaryotes, protists, fungi, plants, and animals constitute the five major categories, or kingdoms, into which organisms are classified in this text. The prokaryotes (kingdom Monera) are essentially unicellular, although in some types the cells form clusters, threads, or chains; this kingdom includes chemosynthetic, photosynthetic, and heterotrophic forms. Protists are a diverse group of eukaryotic one-celled organisms and some simple multicellular ones; the protists include both heterotrophs and photosynthetic autotrophs. The fungi, plants, and animals are all multicellular eukaryotic organisms. All animals and fungi are heterotrophs, whereas all plants, with a few curious exceptions (such as Indian pipe and dodder, which are parasites) are photosynthetic autotrophs. Within the multicellular plant body, however, some of the cells are photosynthetic, such as the cells of a leaf, and some are heterotrophic, such as the cells of a root. The photosynthetic cells supply the heterotrophic cells of the plant with sucrose.

VIEWING THE CELLULAR WORLD

In the three centuries since Robert Hooke first observed the structure of cork through his simple microscope (see page 10), a wealth of knowledge has been accumulated both about the structure of cells and their component parts and about the dynamic processes that characterize the living cell. This knowledge, which we shall begin to examine in the next chapter, has generally come in bursts, following the development of new and better techniques for studying the cell and its contents.

Types of Microscopes

Unaided, the human eye has a resolving power of about 1/10 millimeter, or 100 micrometers (Table 4–1). Resolving power is a measure of the capacity to distinguish objects from one another; it is the minimum distance that must be between two objects for them to be perceived as separate objects. For example, if you look at two lines that are less than 100 micrometers apart, you will see a single,

TABLE 4-1 **Measurements Used in Microscopy**

1 centimeter (cm) = 1/100 meter = 0.4 inch*
1 millimeter (mm) = 1/1,000 meter = 1/10 cm
1 micrometer (μm)† = 1/1,000,000 meter = 1/10,000 cm
1 nanometer (nm) = 1/1,000,000,000 meter = 1/10,000,000 cm
1 angstrom (Å)‡ = 1/10,000,000,000 meter = 1/100,000,000 cm
or
1 meter = 10^2 cm = 10^3 mm = 10^6 μm = 10^9 nm = 10^{10} Å

* A metric-to-English conversion table is found in Appendix A.
† Micrometers were formerly known as microns (μ), and nanometers as millimicrons (mμ).
‡ The angstrom is not an accepted measurement in the International System of Units; in the past, however, it was widely used in microscopy, and you will occasionally encounter it in your reading.

somewhat thickened line. Similarly, two dots less than 100 micrometers apart look like a single blurry dot. Conversely, if you look at two lines (or two dots) that are 120 micrometers apart, you can easily distinguish them from each other.

Most eukaryotic cells are between 10 and 30 micrometers in diameter—some 3 to 10 times below the resolving power of the human eye—and prokaryotic cells are smaller still. In order to distinguish individual cells, to say nothing of examining the structures of which they are composed, we must use instruments that provide greater resolution. Most of our current knowledge of cell structure has been gained with the assistance of three different types of instruments: the light microscope, the transmission electron microscope, and the scanning electron microscope (Figure 4–14).

The best light microscopes have a resolving power of 0.2 micrometer, or 200 nanometers, and so improve on the naked eye about 500 times. It is theoretically

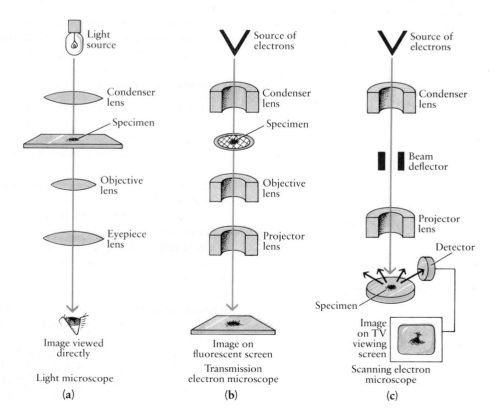

4–14 *A comparison of (a) the light microscope, (b) the transmission electron microscope, and (c) the scanning electron microscope. The light microscope is shown upside-down, to emphasize its similarities with the electron microscopes. The focusing lenses in the light microscope are glass or quartz; those in the electron microscopes are magnetic coils. In both the light microscope and the transmission electron microscope, the illuminating beam passes through the specimen; in the scanning electron microscope, it is deflected from the surface.*

impossible to build a light microscope that will do better than this. The limiting factor is the wavelength of light, which ranges from about 0.4 micrometer for violet light to about 0.7 micrometer for red light. With the light microscope, we can distinguish the larger structures within eukaryotic cells and can also distinguish individual prokaryotic cells. We cannot, however, visualize the internal structure of prokaryotic cells or distinguish between the finer structures of eukaryotic cells.

Notice that resolving power and magnification are two different things. If you take a picture through the best light microscope of two lines that are less than 0.2 micrometer, or 200 nanometers, apart, you can enlarge that photograph indefinitely, but the two lines will continue to blur together. By using more powerful lenses, you can increase magnification, but this will not improve resolution.

With the transmission electron microscope, resolving power has been increased about 1,000 times over that provided by the light microscope. This is achieved by using "illumination" of a much shorter wavelength, consisting of electron beams instead of light rays. Areas in the specimen that permit the transmission of more electrons—"electron-transparent" regions—show up bright, and areas that scatter electrons away from the image—"electron-opaque" regions—are dark. Transmission electron microscopy at present affords a resolving power of about 0.2 nanometer, roughly 500,000 times greater than that of the human eye. This is about twice the diameter of a hydrogen atom.

Although the resolving power of the scanning electron microscope is only about 10 nanometers, this instrument has become a valuable tool for biologists. In scanning electron microscopy, the electrons whose imprints are recorded come from the surface of the specimen, rather than from a section through it. The electron beam is focused into a fine probe, which is rapidly passed back and forth over the specimen; complete scanning from top to bottom usually takes a few seconds. Variations in the surface of the specimen affect the pattern in which the electrons are scattered from it; holes and fissures appear dark, and knobs and ridges are light. The scattered electrons are amplified and transmitted to a television monitor, producing a visual image of the specimen. Scanning electron microscopy provides vivid three-dimensional representations of cells and cellular structures that compensate, in part, for its limited resolution.

Preparation of Specimens

In both the light microscope and the transmission electron microscope, the formation of an image with perceptible contrast requires that different parts of the cell differ in their transparency to the beam of illumination—either light rays or electrons. Parts of the specimen that readily permit the passage of light or of electrons appear bright, whereas parts that block the passage of the illuminating beam appear dark. In the scanning electron microscope, areas that appear light are those that deflect electrons back into the image; parts that appear dark are those that are not well illuminated by the electron beam or that deflect electrons away from the detector.

Living cells and their component parts are, however, almost completely transparent to light. By weight, cells are about 70 percent water, through which light passes easily. Moreover, water and the much larger molecules that form cellular structures are composed of small atoms of low atomic weight (CHNOPS). These atoms are relatively transparent to electrons, which are strongly deflected only by atoms of high atomic weight, such as those of heavy metals. To create sufficient contrast for the light microscope, cells must be treated with dyes or other substances that differentially adhere to or react with specific subcellular components, producing regions of differing opacity. For the electron microscope, the specimens are similarly treated with compounds of heavy metals.

4–15 *Rabbit sperm cells, as seen in* (**a**) *a light micrograph,* (**b**) *a transmission electron micrograph, and* (**c**) *a scanning electron micrograph. Note the dramatic increase in the resolution of ultrastructural detail in the electron micrographs.*

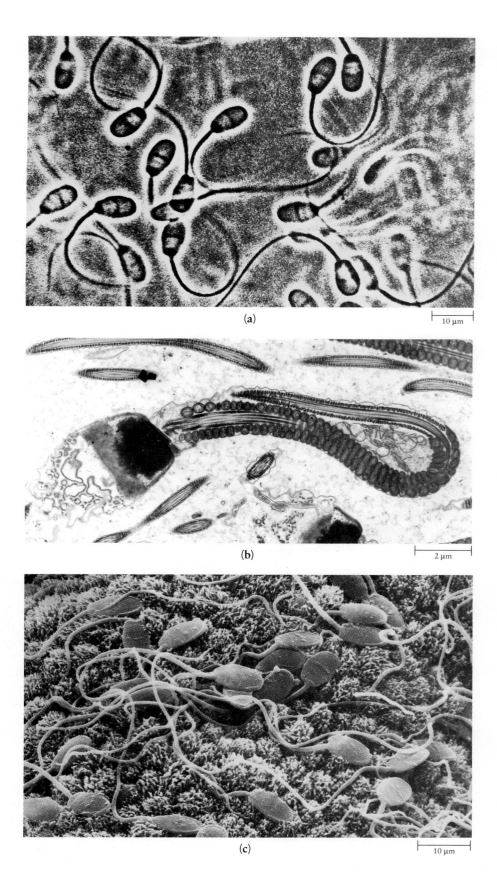

(a)

10 μm

(b)

2 μm

(c)

10 μm

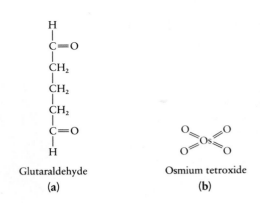

Glutaraldehyde
(a)

Osmium tetroxide
(b)

4–16 (a) *Glutaraldehyde and* (b) *osmium tetroxide, two compounds frequently used as fixatives in preparing specimens for microscopic examination. Note that glutaraldehyde has two aldehyde functional groups; each of these groups can react with a different protein molecule, linking the molecules together. The oxygen atoms of osmium tetroxide react with lipids, leading to a similar cross-linking.*

After a specimen has been treated with a staining substance, all of the stain that has not adhered to specific structures must be washed away. Cells, however, are quite fragile and any kind of rough treatment disrupts their structure. To solve this problem, biological specimens are generally "fixed" before staining. This procedure involves treatment with compounds that bind cell structures in place, usually through the formation of additional covalent bonds between molecules. Aldehydes, for instance, react with the amino groups of protein molecules, linking adjacent protein molecules together in a fairly rigid structure. Osmium tetroxide, a compound frequently used in preparing specimens for electron microscopy, interacts with lipids, binding the molecules together. Fixation has the added advantage of making cells more permeable to staining substances and to the solutions used to wash away excess stain.

Fixation and staining procedures are usually carried out with groups of cells as in, for example, a piece of liver tissue. Such specimens are not transparent to an illuminating beam—they are simply too thick to allow the passage of light rays or electrons. Before examination under the light microscope or the transmission electron microscope, they must be sliced into sections so thin that the unstained regions are transparent. After fixation and staining, the specimens are usually embedded in wax or a plastic resin to provide enough firmness to allow a clean slicing.

With the scanning electron microscope, as we noted earlier, electrons do not pass through the sample but are instead deflected from its surface. Usually, the surface of the specimen is first coated with metal, a process known as shadowing (Figure 4–17). Often, the organic material of the original specimen is removed by chemical treatment, leaving only a metallic replica of the surface, which is reinforced with a carbon film. Depending on its thickness, the replica can be examined under either the transmission electron microscope or the scanning electron microscope.

In addition to these rather drastic treatments, specimens for the electron microscope must also be dehydrated, either by chemical methods or by freeze-dry methods similar to those used in preparing freeze-dried coffee. This step is necessitated by the properties of the electrons forming the illuminating beam. If they pass through a chamber containing gaseous molecules, the electrons are deflected by the molecules and cannot be focused into a beam. Thus all air must be evacuated from the inner, working chamber of an electron microscope, creating a vacuum. If the sample were not first dehydrated, water molecules would evaporate from the sample into the chamber, destroying the vacuum and the focused electron beam.

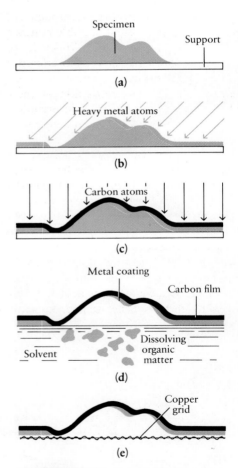

4–17 *"Shadowing" and the preparation of a replica.* (a) *The specimen is placed on a support and* (b) *is "shadowed" by heavy metal atoms evaporated from a heated filament at the side. Because the atoms are deposited from an angle, the metal coating is thicker on raised areas of the specimen.* (c) *A uniform film of carbon atoms is deposited from above, reinforcing and strengthening the replica,* (d) *which is then floated to the surface of a solvent that dissolves away the organic material. The completed replica* (e) *is washed and picked up on a copper grid.*

If the replica is thin enough, it can be examined under the transmission electron microscope as well as under the scanning electron microscope.

The procedures required to prepare most specimens for either the light microscope or the electron microscope usually result in the death of the constituent cells. Moreover, they raise serious questions about whether the structures we see in micrographs are "real" or are distortions introduced by the preparation process. One way of reducing the likelihood that microscopic observations are in error is to prepare similar samples using different techniques. If a feature appears repeatedly in different types of preparation, the probability is great that it exists within the living cell. Another approach has been the development of a variety of new techniques for viewing living cells.

Observation of Living Cells

When light waves are emitted from a coherent source (such as a laser), the waves are in phase, that is, the peaks and troughs of the waves match (Figure 4–18a). This has the effect of reinforcing the waves and creating a greater amplitude, which we perceive as increased brightness. When light waves are out of phase, however, they interfere with one another, reducing amplitude and brightness (Figure 4–18b).

As light waves pass from one material through another, they are bent, or diffracted, and their paths are slightly changed. This diffraction also alters the phase relationships of the light waves, resulting in varying amounts of interference. The amount of interference produced when light passes through the different structures of a cell is not great and provides little detectable contrast when the cell is viewed through an ordinary light microscope—which is, of course, why cells must be stained. In phase-contrast and differential-interference microscopes, however, specially designed optical systems enhance the small amounts of interference, providing greater contrast. The resolution of these microscopes is limited, as in ordinary light microscopes, but they do provide a different perspective on the living cell, revealing features difficult to detect with other systems.

Another technique frequently used with living cells is dark-field microscopy. The illuminating beam strikes the specimen from the side and the lens system detects light reflected from the sample, which appears as a bright object against a dark background. Features of the cell that are invisible in other micrographs often come into sharp relief in dark-field micrographs.

At the present time, work is progressing rapidly on other microscopic techniques. For example, video cameras are being coupled with light microscopes, producing a display on a screen and a videotape record. By adjusting the controls, as you do on a television set, the background "noise" can be reduced, contrast improved, and particular features enhanced. Video techniques, as applied to the study of the living cell, are in their infancy, but they are generating great excitement as they reveal previously unseen processes within the cell.

4–18 *The brightness of light depends on the amplitude of the light waves. (a) When two light waves are in phase, they reinforce one another, resulting in greater amplitude and brightness. (b) When the waves are completely out of phase, as shown here, they cancel each other out and no light is perceived.*

Waves that have passed through different structures within the unstained cell are partially out of phase. The resulting interference produces slight differences in contrast that can be enhanced by special optical systems, such as those of phase-contrast and differential-interference microscopes.

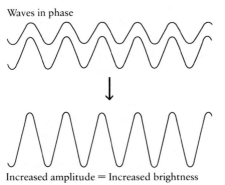

Waves in phase

Increased amplitude = Increased brightness

(a)

Waves out of phase

Decreased amplitude = Darkness

(b)

4–19 *Four views of a living fibroblast, a type of connective tissue cell that can be propagated in the laboratory in a culture dish. This cell has been photographed through (a) a conventional light microscope, (b) a phase-contrast microscope, (c) a differential-interference microscope, and (d) a dark-field microscope.*

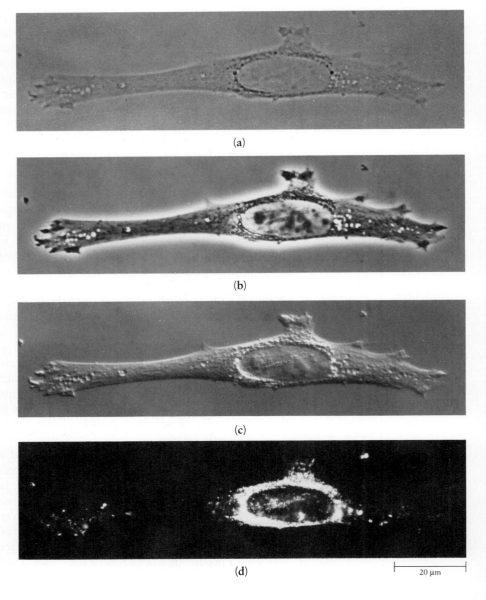

(a)

(b)

(c)

(d)

20 μm

SUMMARY

The properties associated with living systems emerge at the cellular level of organization. One of the fundamental principles of biology is the cell theory, which states that (1) all living organisms consist of one or more cells; (2) the chemical reactions of a living organism, including energy-releasing processes and biosynthetic reactions, take place within cells; (3) cells arise from other cells; and (4) cells contain the hereditary information of the organisms of which they are a part, and this information is passed from parent cell to daughter cell.

The age of the earth is estimated at 4.6 billion years. Microfossils of bacteria-like cells have been discovered that are 3.5 billion years old. The complexity of these cells suggests that the first primitive cells arose very early in the earth's existence—sometime during the first billion years.

The primitive atmosphere held the raw materials of living matter—hydrogen, oxygen, carbon, and nitrogen—combined in water vapor and other simple gases. The energy required to break apart the molecules of these gases and re-form them into more complex molecules was present in heat, lightning, radioactive elements, and high-energy radiation from the sun. Laboratory experiments have shown that under these conditions, the types of organic molecules characteristic of living systems can be formed. Other experiments have suggested the kinds of processes by which aggregations of organic molecules could have formed cell-like structures, separated from their environment by a membrane and capable of maintaining their structural and chemical integrity.

The earliest cells may have been heterotrophs (organisms that depend on outside sources for their energy-rich organic molecules) or autotrophs (organisms that can make their own organic molecules from inorganic substances). The first autotrophs may have been chemosynthetic (using the energy released by specific inorganic reactions to synthesize their own organic molecules) or photosynthetic (using the sun's energy to power their synthetic reactions). With the advent of photosynthesis, the flow of energy through the biosphere assumed its dominant modern form—radiant energy from the sun is captured by photosynthetic autotrophs and channeled through them to heterotrophic organisms. Modern heterotrophs include fungi and animals, as well as many types of single-celled organisms; modern autotrophs include other types of single-celled organisms and, most important, the green plants.

There are two fundamentally distinct types of cells—prokaryotes, which include only the bacteria and cyanobacteria, and eukaryotes, which include the protists, fungi, plants, and animals. Prokaryotic cells lack membrane-bound nuclei and most of the organelles found in eukaryotic cells. Prokaryotes were the only form of life on earth for almost 2 billion years, and then, about 1.5 billion years ago, eukaryotic cells evolved. Multicellular organisms, which are composed of eukaryotic cells specialized to perform particular functions, evolved comparatively recently—only about 750 million years ago.

Because of the small size of cells and the limited resolving power of the human eye, microscopes are required to visualize cells and subcellular structures. The three principal types are the light microscope, the transmission electron microscope, and the scanning electron microscope. Specimens to be studied using a conventional light microscope or a transmission electron microscope must be fixed, stained, dehydrated (for the electron microscope), embedded, and sliced into thin sections. Surface replicas are generally prepared for study with the scanning electron microscope. Concern about the distortions that may be introduced by these preparation procedures has led to the development of other microscopic techniques. The special optical systems of phase-contrast, differential-interference, and dark-field microscopes make it possible to study living cells. An important new development is the use of video cameras with microscopes.

QUESTIONS

1. Distinguish among the following: heterotroph/autotroph; chemosynthetic autotroph/photosynthetic autotroph; prokaryote/eukaryote; light microscope/transmission electron microscope/scanning electron microscope.

2. Why would energy sources have been necessary for the synthesis of simple organic molecules on the primitive earth?

3. Although there is some uncertainty as to the exact mixture of gases that constituted the early atmosphere, there is general agreement that free oxygen was not present. What properties of oxygen would have made chemical evolution unlikely in an atmosphere containing O_2?

4. A key event in the origin of life was the formation of a membrane that separated the contents of primitive cells from their surroundings. Why was this so critical?

5. Some scientists think that other planets in our galaxy may well contain some form of life. If you were seeking such a planet, what characteristics would you look for?

6. Return to Chapter 2 and add approximate scale markers to Figures 2–1, 2–5, and 2–11.

7. What are the advantages and disadvantages of studying cells with the transmission electron microscope and the scanning electron microscope? With special optical microscopes such as phase-contrast, differential-interference, and dark-field?

CHAPTER 5

How Cells Are Organized

There are many, many different kinds of cells. In a drop of pond water, you are likely to find a variety of protists, and in even a small pond, there are probably several hundred different kinds of protists, plus a variety of prokaryotes. Our own tissues and organs are constructed of at least 200 different and distinct types of somatic ("body") cells. Plants are composed of cells that appear quite different from those of our bodies, and insects have many cells of kinds not found either in plants or in vertebrates. Thus, the first remarkable fact about cells is their diversity.

The second, even more remarkable, fact is their similarity. Every cell is a self-contained and at least partially self-sufficient unit, surrounded by a membrane that controls the passage of materials into and out of the cell. This makes it possible for the cell to differ biochemically and structurally from its surroundings. All cells also have an information and control center in which the genetic material is localized. As we noted in the last chapter, this region in prokaryotic cells is known as the nucleoid; in eukaryotic cells, it is the nucleus. Many eukaryotic cells also have a variety of internal structures, the organelles, which are similar or identical from one cell to another through a wide range of cell types. And, all cells are composed of the same remarkably few kinds of atoms and molecules.

CELL SIZE AND SHAPE

Most of the cells that make up a plant or animal body are between 10 and 30 micrometers in diameter. A principal restriction on cell size is that imposed by the relationship between volume and surface area. As Figure 5–1 shows, as volume decreases, the ratio of surface area to volume increases rapidly. Materials—such as oxygen, carbon dioxide, ions, food molecules, and waste products—entering and leaving the cell must move through its membrane-bound surface. These substances are the raw materials and products of the cell's **metabolism,** which is the total of all of the chemical activities in which it is engaged. The more active the cell's metabolism, the more rapidly materials must be exchanged with the environment if the cell is to continue to function. In smaller cells, the ratio of surface area to volume is higher than in larger cells, and thus proportionately greater quantities of materials can move into, out of, and through smaller cells in a given period of time. A larger cell, by contrast, requires the exchange of greater quantities of materials in order to meet the needs of the larger volume of living matter—and yet, the larger the cell, the smaller the ratio of surface area to volume.

A second limitation on cell size appears to involve the capacity of the nucleus, the cell's control center, to provide enough copies of the information needed to regulate the processes occurring in a large, metabolically active cell. The exceptions seem to "prove" the rule. In certain large, complex one-celled protists—the ciliates, of which *Paramecium* is an example—each cell has two or more nuclei, the additional ones apparently copies of the original. Other organisms, such as the

5-1 *The single 4-centimeter cube, the eight 2-centimeter cubes, and the sixty-four 1-centimeter cubes all have the same total volume. As the volume is divided up into smaller units, however, the total amount of surface area increases, as does the ratio of surface area to volume. For example, the sixty-four 1-centimeter cubes have four times the total surface area of the single 4-centimeter cube, and the ratio of surface area to volume in each 1-centimeter cube is four times that in the 4-centimeter cube. Similarly, smaller cells have a higher ratio of surface area to volume than larger cells. This means not only more membrane surface through which materials can move into or out of the cell but also less living matter to be serviced and shorter distances through which materials must move within the cell.*

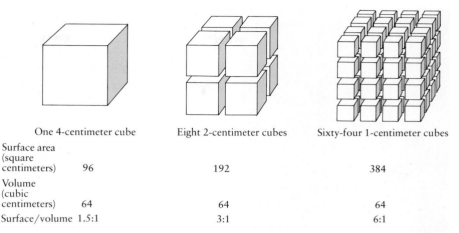

	One 4-centimeter cube	Eight 2-centimeter cubes	Sixty-four 1-centimeter cubes
Surface area (square centimeters)	96	192	384
Volume (cubic centimeters)	64	64	64
Surface/volume	1.5:1	3:1	6:1

slime molds, are made up, in effect, of one giant cell but have thousands of nuclei. Such organisms are also, frequently, very thin and spread out, thereby avoiding the problem of the ratio of surface area to volume.

It is not surprising that the most metabolically active cells are usually small. The relationship between cell size and metabolic activity is nicely illustrated by egg cells. Many egg cells are very large. A frog's egg, for instance, is 1,500 micrometers in diameter. Some egg cells are several centimeters across—for example, the cell, or yolk, of a chicken's egg. Most of this mass consists of stored nutrients for the developing embryo. When the egg cell is fertilized and begins to be active metabolically, many nuclear divisions occur and the cell divides many times before there is any actual increase in volume or mass. Thus the total mass is cut into cellular units small enough for efficient transfer and control processes.

Like drops of water and soap bubbles, cells have a tendency to be spherical. As we have seen, however, cells often have other shapes. This is because of cell walls, found in plants, fungi, and many one-celled organisms; because of attachments to and pressure from other neighboring cells or surfaces (as in intestinal epithelial cells); or because of arrays of microtubules and other structural filaments within the cell (Figure 5–2).

5-2 *(a) Numerous fine strands of cytoplasm (axopods) extend radially from the body of the protist* Actinosphaerium. *Each axopod is bounded by an extension of the cell membrane and contains many microtubules, arranged longitudinally, which stiffen and extend the axopods. (b) An axopod in cross section. The microtubules are arranged in two interlocking spirals that form a twelvefold pattern.*

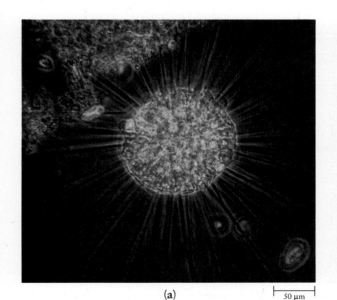

(a)

50 μm

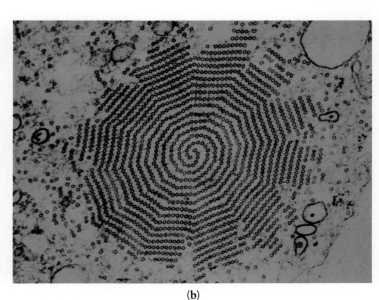

(b)

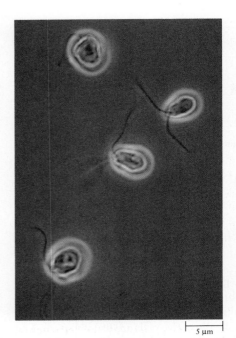

⊢————⊣
5 µm

5-3 *Cells of the green alga* Chlamydomonas. *Note the pair of flagella with
which each cell propels itself through the
water.*

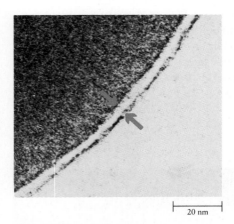

⊢————⊣
20 nm

5-4 *Electron micrograph showing a cross
section of the cell membrane of a human
red blood cell. The cell membrane is indicated by the arrows. The "molecular
sandwich" structure of the membrane is
believed to consist of two electron-opaque
(dark) layers of phospholipid molecules
arranged with their hydrophobic tails
pointing inward, forming the electron-
transparent (light) inner "filling," with
globular proteins embedded throughout.
The darker material at the left of the
micrograph is hemoglobin, which fills the
red blood cell.*

SUBCELLULAR ORGANIZATION

Antony van Leeuwenhoek discovered the protists some 300 years ago. "This was
for me," he wrote, "among all the marvels that I have discovered in nature, the
most marvelous of all." As Leeuwenhoek and his successors observed the thousands of living creatures that they found "all alive in a drop of water," they were
able to see, but only barely, structures within them. They interpreted these
structures as miniature hearts and stomachs and lungs, in other words, tiny
organs, or organelles.

Modern microscopic techniques have confirmed that eukaryotic cells do
indeed contain a multitude of structures. They are not, of course, organs such as
those found in multicellular organisms, but they are in some ways comparable:
they are specialized in form and function to carry out particular activities required
in the cellular economy. Just as the organs of multicellular animals work together
in organ systems, the organelles of cells engage in a number of cooperative and
interdependent functions.

Every cell must carry out essentially the same processes—acquire and assimilate food, eliminate wastes, synthesize new cellular materials, and in many cases,
be able to move and to reproduce. Just as the various organs of your body have a
structure that suits them for the specific functions they carry out (the kidney for
elimination of wastes from the blood, the intestine for food absorption, and so
on), so all cells have an internal architecture that includes organelles suited to the
functions they perform. It is important to realize that a cell is not a random
assortment of parts but a dynamic, integrated entity.

Also, remember that although we can look at only one structure or process at a
time, most activities of a cell go on simultaneously and influence one another.
Chlamydomonas, for instance, is swimming, photosynthesizing, absorbing
nutrients from the water, building its cell wall, making proteins, converting sugars
to starch (or vice versa), and oxidizing food molecules for energy, all at the same
time. It is also likely to be orienting itself in the sunlight, it is probably preparing to
divide, it is possibly "looking" for a mate, and it is undoubtedly carrying out at
least a dozen or more other important activities, many of which may be still
unknown.

CELL BOUNDARIES

The Cell Membrane

All cells, as we stated previously, are basically very similar. They all have DNA as
the genetic material, they perform the same types of chemical reactions, and they
are all surrounded by an external cell membrane that conforms to the same
general design in both prokaryotic and eukaryotic cells. The living matter
bounded by the membrane consists, in eukaryotes, of the nucleus and the
cytoplasm, which contains the organelles.

A cell can exist as a distinct entity because of the cell membrane, which
regulates the passage of materials into and out of the cell. The cell membrane (also
called, as we noted previously, the plasma membrane) is only about 7 to 9
nanometers thick and cannot be resolved by the light microscope. With the
electron microscope, it can be visualized as a continuous, thin double line (Figure
5–4).

The eukaryotic cell membrane is formed from a phospholipid bilayer, that is, a
double layer of phospholipid molecules arranged with their hydrophobic fatty
acid tails pointing inward (Figure 5–5). Cholesterol molecules are embedded in the
hydrophobic interior of the bilayer, in which numerous protein molecules are also

suspended. These proteins, known as integral membrane proteins, generally span the bilayer and protrude on either side; the portions embedded in the bilayer have hydrophobic surfaces, whereas the surfaces of those portions that extend beyond the bilayer are hydrophilic.

The two surfaces of the cell membrane differ considerably in chemical composition. The two layers of the bilayer generally have different concentrations of specific types of lipid molecules. In many types of cells, the outer layer is particularly rich in glycolipid molecules. The carbohydrate chains of these molecules are, like the phosphate heads of the phospholipid molecules, exposed on the surface of the membrane; the hydrophobic fatty acid tails are within the membrane. The protein composition of the two layers also differs. The integral membrane proteins have a definite orientation within the bilayer, and the portions extending on either side are completely different in both amino acid composition and tertiary structure. On the cytoplasmic side of the membrane, additional protein molecules, known as peripheral membrane proteins, are bound to some of the integral proteins protruding from the bilayer. On the outside of the membrane, short carbohydrate chains are covalently linked to the protruding proteins. These chains, along with the carbohydrate chains of the glycolipids, form a carbohydrate coat on the outer surface of the membranes of many types of cells. The carbohydrates are thought to play a role in the adhesion of cells to one another and in the "recognition" of molecules that interact with the cell (such as hormones, antibodies, and viruses).

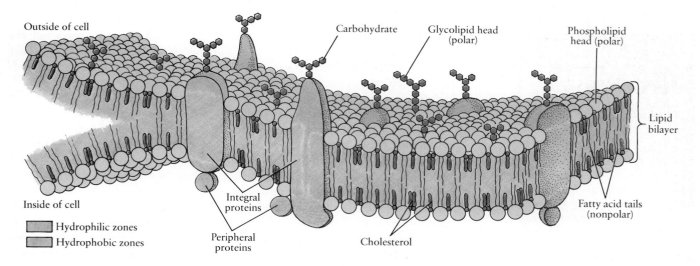

5-5 Model of a cell membrane, as determined from electron micrographs and biochemical data. The basic membrane structure is formed from a network of phospholipid molecules, in which cholesterol molecules and large protein molecules are embedded. The phospholipid molecules are arranged in a bilayer with their hydrophobic tails pointing inward and their hydrophilic phosphate heads pointing outward. Cholesterol molecules nestle among the hydrophobic tails.

The proteins embedded in the bilayer are known as integral membrane proteins. On the cytoplasmic face of the membrane, peripheral membrane proteins are bound to some of the integral proteins. The portion of a protein molecule's surface that is within the lipid bilayer is hydrophobic; the portion of the surface that is exposed outside the bilayer is hydrophilic. It is believed that pores with hydrophilic surfaces pass through some of the protein molecules.

Interspersed among the phospholipid molecules of the outer layer of the bilayer are glycolipid molecules. Their carbohydrate chains and the carbohydrate chains attached to the proteins protruding on the outside of the membrane are thought to be involved in the adhesion of cells to each other and with the "recognition" of molecules at the membrane surface.

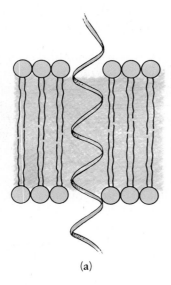

(a)

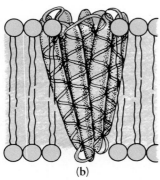

(b)

5-6 *Two principal configurations that have been determined for membrane proteins are (a) an alpha helix, and (b) a globular tertiary structure, formed by repeating segments of alpha helix that zig-zag through the membrane. The helical segments are linked by irregular, hydrophilic segments of the polypeptide chain that extend on either side of the membrane.*

Although many of the integral proteins appear to be anchored in place, either by peripheral proteins or by cytoplasmic protein filaments that are concentrated near the membrane, the structure of the bilayer is generally quite fluid. The lipid molecules and at least some of the protein molecules can move laterally within it, forming different patterns that vary from time to time and place to place. Consequently, this widely accepted model of membrane structure is known as the **fluid-mosaic model.**

Recent studies have revealed the detailed structure of a number of membrane proteins. Two basic configurations have been identified among the integral proteins studied thus far (Figure 5–6). One is a relatively simple rodlike structure, consisting of an alpha helix embedded in the hydrophobic interior of the membrane, with less regular, hydrophilic portions extending on either side of the membrane. These hydrophilic portions are often extensively folded into an intricate tertiary structure. The other configuration is found in large, globular molecules with complex tertiary or quaternary structures that result from repeated "passes" through the membrane. The portions of these proteins embedded in the hydrophobic interior of the bilayer consist of segments of tightly coiled alpha helix. In globular proteins formed from a single polypeptide chain, these regular helical segments alternate with segments of the polypeptide chain that have an irregular structure. The irregular segments, which are hydrophilic, are exposed on either side of the membrane, while the helical segments zig-zag back and forth through the membrane. Although the embedded surfaces in contact with the lipid bilayer are always hydrophobic, the interior portions of some of the globular proteins are apparently hydrophilic, creating "pores" through which certain polar substances can cross the membrane.

The cell membrane of bacterial cells is much the same in basic composition as the cell membrane of eukaryotic cells, except that, with a few exceptions, bacterial cell membranes do not contain cholesterol. In eukaryotes, all the membranes of a cell, including those surrounding the various organelles, also have the same general structure. There are, however, differences in the types of lipids and, particularly, in the number and types of proteins and carbohydrates, which vary from membrane to membrane and also from place to place on the same membrane. These differences give the membranes of different types of cells and of the different organelles unique properties that can be correlated with differences in function. Most membranes are about 40 percent lipid and 60 percent protein, though there is considerable variation. The proteins, which are extremely diverse structurally, perform a variety of essential functions. Some are enzymes, regulating particular chemical reactions; others are receptors, involved in the recognition and binding of signalling molecules, such as hormones; and still others are transport proteins, playing critical roles in the movement of substances across the membrane. As we shall see repeatedly in the course of this text, discoveries concerning the structure and function of specific membrane proteins are shedding new light on a diversity of processes, ranging from the navigation of bacterial cells to photosynthesis to the transmission of the nerve impulse.

The Cell Wall

A principal distinction between plant and animal cells is that plant cells are surrounded by a cell wall. The wall is outside the membrane and is constructed by the cell. As a plant cell divides, a thin layer of gluey material forms between the two new cells; this becomes the **middle lamella** (Figure 5–7a). Composed of pectins (the compounds that make jellies gel) and other polysaccharides, it holds adjacent cells together. Next, on either side of the middle lamella, each plant cell constructs its primary cell wall. The primary wall contains cellulose molecules bundled together in microfibrils that are laid down in a matrix of gluey polymers.

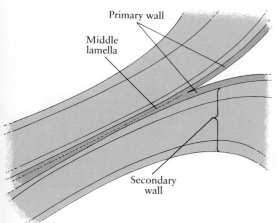

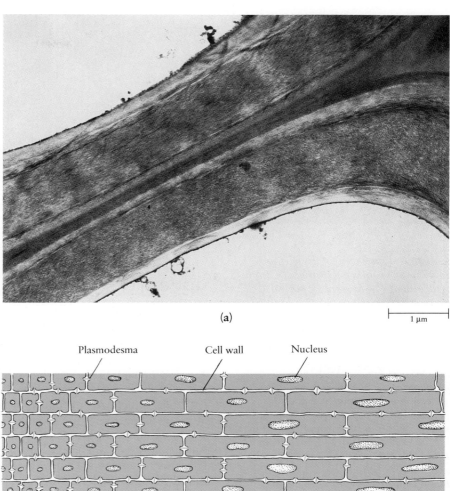

(a)

1 μm

5–7 (a) *Electron micrograph of two adja-cent cell walls of tracheids, cells through which water is conducted in plants. You can see the middle lamella, the primary walls, and the layered secondary walls, deposited inside the primary wall. The cells, which are from the wood of a ground hemlock, have died. The electron-trans-parent areas in the top left and lower right portions of the micrograph represent empty space, once filled by the living mat-ter of the cells.*

(b) *Growth of plant cells is limited by the rate at which the cell walls expand. The walls control both the rate of growth and its direction; they do not expand in all directions but elongate in a single dimension. Cells at the left are newly formed; cells farther to the right are older and have started to elongate. Plasmodes-mata (singular, plasmodesma) are chan-nels connecting adjacent cells.*

(b)

As you can see in Figure 3–10c on page 66, successive layers of cellulose microfibrils are oriented at right angles to one another in the completed cell wall. (Those of you familiar with building materials will note that the cell wall thus combines the structural features of both fiberglass and plywood.)

In plants, growth takes place largely by cell elongation. Studies have shown that the cell adds new materials to its walls throughout this elongation process. The cell, however, does not simply expand in all directions; its final shape is deter-mined by the structure of its cell wall (Figure 5–7b).

As the cell matures, a secondary wall may be constructed. This wall is not capable of expansion, as is the primary wall. It often contains other molecules, such as lignin, that have stiffening properties. In such cells, the living material of the cell often dies, leaving only the outer wall, a monument to the cell's architec-tural abilities (see Figure I–12, page 10).

Cellulose-containing cell walls are also found in many algae. Fungi and prokar-yotes also have cell walls, but they usually do not contain cellulose. The cell walls of fungi are composed principally of chitin (Figure 3–11a, page 66). Prokaryotic cell walls contain polysaccharides and complex polymers known as peptidogly-cans, which are formed from amino acids and sugars. We shall examine the structure of prokaryotic cell walls in Chapter 21.

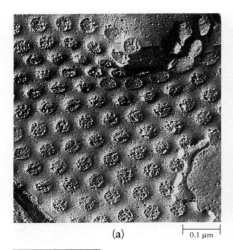

(a)   0.1 μm

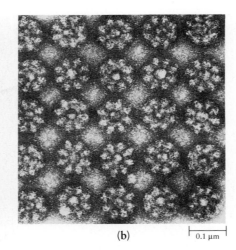

(b)   0.1 μm

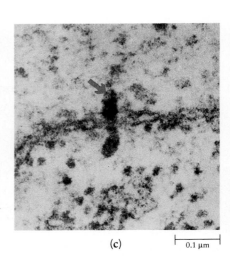

(c)   0.1 μm

5–8 (a) *A surface view of the nuclear envelope of a guinea pig sperm cell. Clearly visible on this surface are the nuclear pores. Biochemical studies and electron micrographs of sections through the plane of the envelope have revealed that the structure of each nuclear pore consists of eight protein-containing granules* (b). *The opening of the pore is a very narrow channel in the center of each octagonal array.*

(c) A granule of protein and RNA, indicated by the arrow, moves from the nucleus (top of the micrograph), through a nuclear pore, and into the cytoplasm (bottom). This cell is from the salivary gland of the midge Chironomus, *a delicate insect that resembles a mosquito but does not feed on humans or other mammals.*

THE NUCLEUS

In eukaryotic cells, the nucleus is a large, often spherical body, usually the most prominent structure within the cell. It is surrounded by the nuclear envelope, which is made up of two concentric membranes, each of which is a lipid bilayer. These two membranes are separated by a gap of about 20 to 40 nanometers. At frequent intervals, however, they are fused together, creating small **nuclear pores** through which materials pass between the nucleus and cytoplasm (Figure 5–8). The pores, which are surrounded by large protein-containing granules arranged in an octagonal pattern, form a narrow channel through the fused lipid bilayers.

The chromosomes are found within the nucleus. When the cell is not dividing, the chromosomes are visible only as a tangle of fine threads, called **chromatin.** The most conspicuous body within the nucleus is the **nucleolus.** There are typically two nucleoli per nucleus, although often only one is visible in a micrograph. As we shall see in Chapter 18, the nucleolus is the site at which ribosomal subunits are constructed. Viewed with the electron microscope, the nucleolus appears to be a collection of fine granules and tiny fibers (Figure 5–9). These are thought to be parts of ribosomal subunits and threads of chromatin.

5–9 *Electron micrograph of the nucleus of the alga* Chlamydomonas. *The dark body in the center of the nucleus is the nucleolus. The major components of ribosomes are produced in the nucleolus; you can see partially formed ribosomes around its periphery. Notice also the nuclear envelope with its nuclear pores, two of which are indicated by arrows. Above the nucleus is a portion of a chloroplast containing starch grains. A Golgi complex is to the upper left of the nucleus, and mitochondria are visible below.*

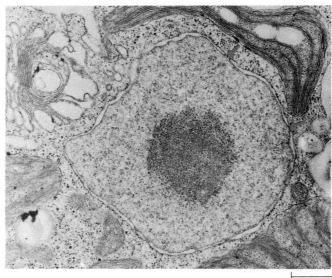

0.5 μm

(a)

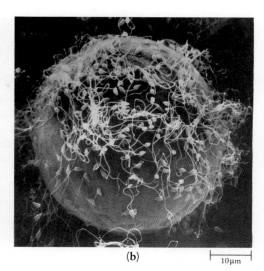

(b) |—| 10μm

5–10 (**a**) *Sea urchins, readily identified by their spiny body surfaces, are common inhabitants of rocky seashores.* (**b**) *This egg from a sea urchin is surrounded by sperm cells. Despite the great differences in size of egg and sperm, both contribute equally to the hereditary characteristics of the individual. Since the nucleus is approximately the same size in both cells, the early microscopists postulated that this part of the cell must be the carrier of the hereditary information. Sea urchin eggs and sperm cells have been used in many studies because sea urchins are relatively easy to obtain and fertilization, which is external, can be easily observed in the laboratory.*

5–11 *Drawings made by Walther Flemming in 1882 of chromosomes in dividing cells of salamander larvae. Flemming's observations were dependent upon the development of new cytological staining techniques.*

The Functions of the Nucleus

Our current understanding of the role of the nucleus in the life of the cell began with some early microscopic observations. One of the most important of these observations was made more than a hundred years ago by a German embryologist, Oscar Hertwig, who was observing the eggs and sperm of sea urchins (Figure 5–10). Sea urchins produce eggs and sperm in great numbers. The eggs are relatively large and so are easy to observe. They are fertilized in the open water, rather than internally, as is the case with land-dwelling vertebrates such as ourselves. Watching the eggs being fertilized under his microscope, Hertwig observed that only a single sperm cell was required. Further, when the sperm cell penetrated the egg, its nucleus was released and fused with the nucleus of the egg. This observation, confirmed by other scientists and in other kinds of organisms, was important in establishing the fact that the nucleus is the carrier of the hereditary information: the only link between father and offspring is the nucleus of the sperm.

Another clue to the importance of the nucleus came about as the result of the observations of Walther Flemming, also about 100 years ago. Flemming observed the "dance of the chromosomes" that takes place when eukaryotic cells divide (a process to be described in Chapter 7), and he painstakingly pieced together the sequence of events. (The fact that Hertwig and Flemming made their observations at about the same time was no coincidence; enormous improvements had just occurred in the light microscope and in techniques of microscopy.)

Since Flemming's time, a number of experiments have explored the role of the nucleus of the cell. In one simple experiment, the nucleus was removed from an amoeba by microsurgery. The amoeba stopped dividing and, in a few days, it died. If, however, a nucleus from another amoeba was implanted within 24 hours after the original one was removed, the cell survived and divided normally.

In the early 1930s, Joachim Hämmerling studied the comparative roles of the nucleus and the cytoplasm by taking advantage of some unusual properties of the marine alga *Acetabularia*. The body of *Acetabularia* consists of a single huge cell 2 to 5 centimeters in height. Individuals have a cap, a stalk, and a "foot," all of which are differentiated portions of the single cell. If the cap is removed, the cell will rapidly regenerate a new one. Different species of *Acetabularia* have different kinds of caps. *Acetabularia mediterranea*, for example, has a compact umbrella-shaped cap, and *Acetabularia crenulata* has a cap of petal-like structures.

Hämmerling took the "foot," which contains the nucleus, from a cell of *A. crenulata* and grafted it onto a cell of *A. mediterranea*, from which he had first

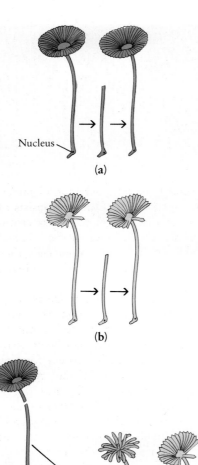

5–12 (a) *One species of* Acetabularia *has an umbrella-shaped cap, and* (**b**) *another has a ragged, petal-like cap. If the cap is removed, a new cap forms, similar in appearance to the amputated one. However, if the "foot" (containing the nucleus) is removed at the same time as the cap and a new nucleus from the other species is transplanted, the cap* (**c**) *that forms will have a structure with characteristics of both species. If this cap is removed, the next cap* (**d**) *that grows will be characteristic of the cell that donated the nucleus, not of the cell that donated the cytoplasm.*

removed the "foot" and the cap. The cap that then formed had a shape intermediate between those of the two species. When this cap was removed, the next cap that formed was completely characteristic of *A. crenulata* (Figure 5–12).

Hämmerling interpreted these results as meaning that certain cap-determining substances are produced under the direction of the nucleus. These substances accumulate in the cytoplasm, which is why the first cap that formed after nuclear transplantation was of an intermediate type. By the time the second cap formed, however, the cap-determining substances present in the cytoplasm before the transplant had been exhausted, and the form of the cap was completely under the control of the new nucleus.

We can see from these experiments that the nucleus performs two crucial functions for the cell. First, it carries the hereditary information that determines whether a particular cell will develop into (or be a part of) a *Paramecium*, an oak, or a human—and not just any *Paramecium*, oak, or human, but one that resembles the parent or parents of that particular, unique organism. Each time a cell divides, this information is passed on to the two new cells. Second, as Hämmerling's work indicated, the nucleus exerts a continuing influence over the ongoing activities of the cell, ensuring that the complex molecules that the cell requires are synthesized in the number and of the kind needed. The way in which the nucleus performs these functions will be described in Section 3.

THE CYTOPLASM

Not long ago, the cell was visualized as a bag of fluid containing enzymes and other dissolved molecules along with the nucleus, a few mitochondria, and occasional other organelles that could be seen by special microscopic techniques. With the development of electron microscopy, however, an increasing number of structures have been identified within the cytoplasm, which is now known to be highly organized and crowded with organelles. On page 112, the interior of a typical animal cell is shown in Figure 5–15; Figure 5–16 shows a corresponding view of a typical plant cell.

The Cytoskeleton

As we noted in the previous chapter, extremely thin sections are required for study with the transmission electron microscope. With the recent development, however, of the high-voltage electron microscope, which produces a beam of electrons with greater penetration, it has become possible to use much thicker specimens—in some cases, whole cells. The resulting visualization of the interior of the cell in three dimensions has revealed previously unsuspected interconnections among filamentous protein structures within the cytoplasm. These structures form an internal **cytoskeleton** that maintains the shape of the cell, enables it to move, anchors its organelles, and directs its traffic (Figure 5–13). Three different types of filaments have been identified as major participants in the cytoskeleton: **microtubules, actin filaments** (formerly known as microfilaments), and **intermediate filaments.**

Microtubules, as we saw on page 78, are long, hollow tubes, assembled from dimers of the globular proteins alpha and beta tubulin. They are about 22 nanometers in diameter, but their length varies. In many cells, the microtubules extend outward from an "organizing center" near the nucleus, ending near the cell surface (Figure 5–14a). As noted previously, microtubules play an important role in cell division and seem to provide a temporary scaffolding for the construction of other cellular structures. As we shall see later in this chapter, they are also key components of cilia and flagella, permanent structures used for locomotion by many types of cells.

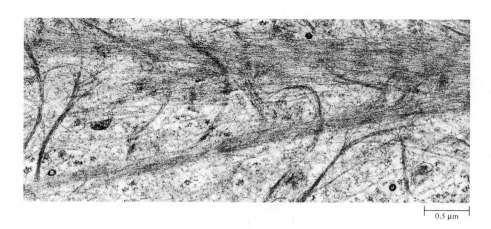

0.5 µm

5–13 *The three principal elements of the cytoskeleton are visible in this electron micrograph of an epithelial cell from a rat kangaroo. The thick bundles of relatively straight fibers running horizontally across the micrograph are actin filaments. Microtubules are the somewhat thicker individual fibers that resemble railroad tracks as seen in aerial photographs taken at high altitude. The bundles of fibers that curve vertically through the micrograph consist of intermediate filaments.*

Actin filaments are fine protein threads, averaging 6 nanometers in diameter, formed from molecules of the globular protein actin. Each filament consists of many globular actin molecules, linked together in a helical chain. Like microtubules, actin filaments can be readily assembled and disassembled by the cell, and they also play important roles in cell division and cell motility. In some cells, they are concentrated in bundles, known as stress fibers, near the cell membrane (Figure 5–14b).

Intermediate filaments, as their name implies, are intermediate in size between microtubules and actin filaments, with a diameter of 7 to 11 nanometers. Unlike microtubules and actin filaments, which consist of globular protein subunits, intermediate filaments are composed of fibrous proteins and cannot be as easily disassembled by the cell once they are formed. The specific protein forming intermediate filaments varies according to the cell type; in different types of epithelial cells, for example, these filaments are composed of different types of keratin. Each of the protein molecules making up an intermediate filament has a rodlike portion of constant length, with terminal regions that vary in length and amino acid composition. In many cells, the intermediate filaments radiate out from the nuclear envelope and are closely associated with the microtubules (Figure 5–14c); in epithelial cells, they are also anchored at specific points in the cell membrane. The function of intermediate filaments in the life of the cell is still poorly understood, but they are found in the greatest density in cells subject to mechanical stress.

5–14 *The distribution of elements of the cytoskeleton in whole cells is dramatically revealed by immunofluorescence microscopy. The cell is treated with specially prepared fluorescent antibodies to the protein of interest. The antibodies attach to the protein, and the pattern of their fluorescence indicates the location of the protein. These micrographs of epithelial cells from the rat kangaroo reveal (a) microtubules radiating from the cell center, (b) actin filaments, bundled together in stress fibers, and (c) intermediate filaments, extending throughout the cytoplasm.*

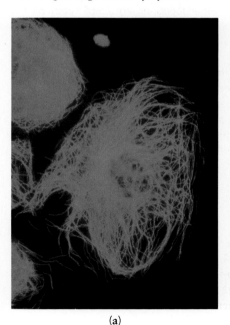

(a)

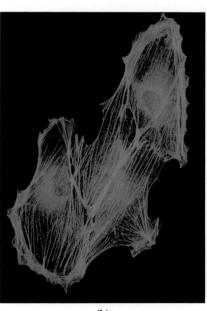

(b)

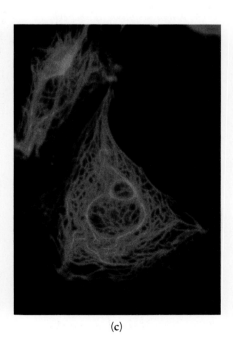

(c)

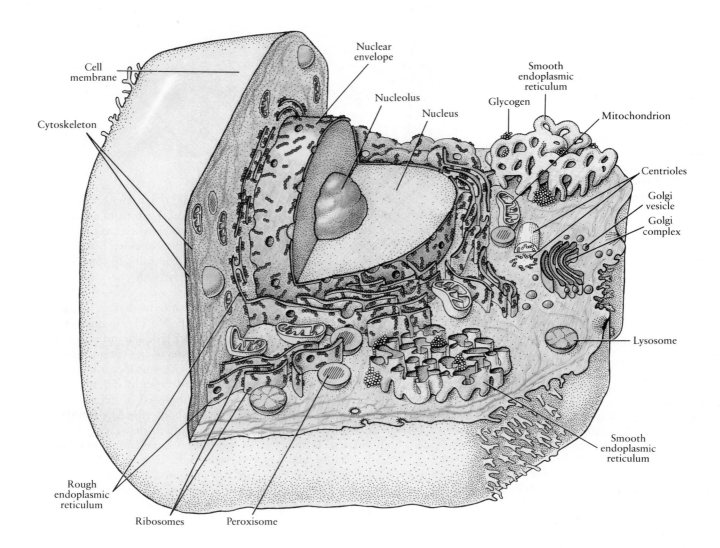

5–15 *A representative animal cell, as interpreted from electron micrographs. Like all cells, this one is bounded by a cell membrane (the plasma membrane), which acts as a selectively permeable barrier to the surrounding environment. All materials that enter or leave the cell, including food, wastes, and chemical messages, must pass through this barrier.*

Within the membrane is found the cytoplasm, which contains the enzymes and other solutes of the cell. The cytoplasm is traversed and subdivided by an elaborate system of membranes, the endoplasmic reticulum, a portion of which is shown here. In some areas, the endoplasmic reticulum is covered with ribosomes, the special structures on which amino

acids are assembled into proteins. Ribosomes are also found elsewhere in the cytoplasm.

Golgi complexes are packaging centers for molecules synthesized within the cell. Lysosomes and peroxisomes are vesicles in which a number of different types of molecules are broken down to simpler constituents that can either be used by the cell or, in the case of waste products, safely removed from it. The mitochondria are the sites of the chemical reactions that provide energy for cellular activities.

The largest body in the cell is the nucleus. It is surrounded by a double membrane, the nuclear envelope, the outer membrane of which is continuous with

the endoplasmic reticulum. Within the nuclear envelope is a nucleolus, the site where the ribosomal subunits are formed, and the chromatin, which is the material of the chromosomes in an extended form.

An elaborate, highly structured network of protein filaments, the cytoskeleton, pervades the cytoplasm. Among its components are microtubules, which have a rodlike appearance, and intermediate filaments, threadlike structures concentrated near the cell membrane. Other elements of the cytoskeleton are too fine to be visible at this magnification. The filaments of the cytoskeleton maintain the cell's shape, anchor its organelles, and direct the intracellular molecular traffic.

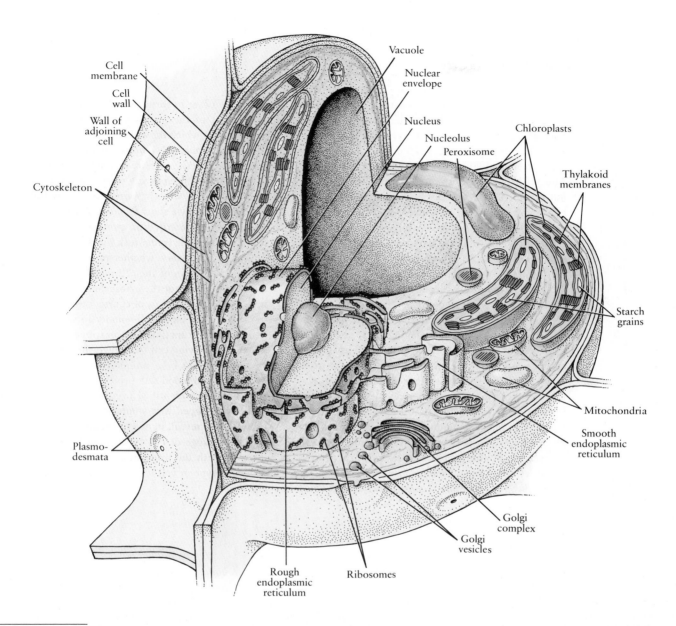

Cell
membrane

Cell
wall

Wall of
adjoining
cell

Cytoskeleton

Plasmo-
desmata

Vacuole

Nuclear
envelope

Nucleus

Nucleolus

Peroxisome

Chloroplasts

Thylakoid
membranes

Starch
grains

Mitochondria

Smooth
endoplasmic
reticulum

Golgi
complex

Golgi
vesicles

Rough
endoplasmic
reticulum

Ribosomes

5–16 *A relatively young plant cell, as interpreted from electron micrographs. Like the animal cell, it is bounded by a cell membrane. Surrounding the cell membrane is a cellulose-containing cell wall. Plasmodesmata, which are channels through the cell walls, provide a cytoplasmic connection between adjacent cells.*

The most prominent structure in many plant cells is a large vacuole, filled with a solution of salts and other substances. In mature plant cells, the vacuole often occupies the bulk of the cell, and the other cel-

lular contents are squeezed into a narrow region next to the cell membrane. As we shall see in the next chapter, the vacuole plays a key role in keeping the cell wall stiff and the plant body crisp.

Chloroplasts, the large organelles in which photosynthesis takes place, are generally concentrated near the surface of the cell. Molecules of chlorophyll and the other substances involved in the capture of light energy from the sun are located in the thylakoid membranes within the chloroplasts.

Like the animal cell, the living plant cell contains a prominent nucleus, extensive endoplasmic reticulum, and many ribosomes and mitochondria. Especially numerous in the growing plant cell are Golgi complexes, which play an important role in the assembly of materials for the expanding cell wall. The orientation of cellulose microfibrils as they are added to the cell wall is determined by the orientation of microtubules in portions of the cytoskeleton close to the cell membrane.

Spectrin and the Red Blood Cell

As the human red blood cell matures, it synthesizes large quantities of hemoglobin and then extrudes its nucleus, organelles, and other cytoplasmic constituents. The resulting mature red blood cell is essentially a bag of hemoglobin, enclosed by the cell membrane. And yet it has a definite shape, resembling a doughnut in which the center has been compressed but not removed (see Figure 3–28a, page 79). Moreover, as it is carried along in the bloodstream, the cell can twist, turn, bend, and fold as it makes its way through the smallest of the blood vessels, the capillaries. Many of the capillaries are so narrow that a single red blood cell can barely squeeze through to deliver its precious load of oxygen to the body tissues.

Until recently, the question of how the red blood cell maintains its shape, springing back unchanged from its contortions in the capillaries, remained a mystery. Studies of the cell membrane and the cytoskeleton have now revealed that the secret lies in several different kinds of protein molecules that together form a supporting meshwork just inside the cell membrane. The major component of this meshwork is a protein known as spectrin, which constitutes about 30 percent of all the protein associated with the red blood cell membrane. Each spectrin molecule consists of two long polypeptide chains, loosely intertwined with each other to form a dimer. Another important component of the meshwork is actin, the globular protein from which the actin filaments of the cytoskeleton are formed. As shown in the diagram, each spectrin dimer is anchored at one end to the cell membrane by a peripheral protein known as ankyrin. The other end of each spectrin dimer is linked to another dimer by a short filament of actin monomers and an actin-spectrin link protein. The result is a secure, yet flexible framework for the red blood cell.

It is hard to know whether to consider spectrin a peripheral protein of the cell membrane or an additional constituent of the cytoskeleton. This difficulty in definition—which is, of course, of no consequence to the living red blood cell—underlines once more the structural and functional integration of all the components of a living cell. Although the spectrin meshwork appears to be a unique feature of the highly specialized red blood cell, proteins related to spectrin have now been identified in a number of other cell types. It is hoped that studies of these spectrin-like proteins will reveal, in more general terms that apply to a wide range of cell types, how the cytoskeleton interacts with the proteins of the cell membrane to give each type of cell its particular overall structure and shape.

The supporting framework of the mature red blood cell. Each spectrin dimer is attached at one end to a peripheral membrane protein known as ankyrin; another spectrin dimer is attached to a second, nearby molecule of ankyrin. The ankyrin molecules, in turn, are linked to integral membrane proteins. At the opposite end, each spectrin dimer is attached to another spectrin dimer by way of a short filament of globular actin monomers and still another protein, the actin-spectrin link protein. The resulting meshwork, securely anchored to the cell membrane, is flexible yet strong, enabling the red blood cell to move efficiently through the capillaries.

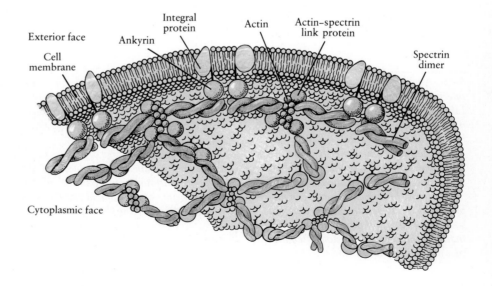

A dense network of wisplike fibers interconnects all of the other structures within the cytoplasm. These fibers, which are formed from accessory proteins of the cytoskeleton, link the cytoskeletal filaments together in specific ways. Although the resulting network gives the cell a highly ordered three-dimensional structure, it is neither rigid nor permanent. The cytoskeleton is a dynamic framework, changing and shifting according to the activities of the cell.

Vacuoles and Vesicles

In addition to organelles and the cytoskeleton, the cytoplasm of many cells, especially plant cells, contains **vacuoles.** A vacuole is a space in the cytoplasm filled with water and solutes; it is surrounded by a single membrane, known in plant cells as the **tonoplast.** Immature plant cells characteristically have many vacuoles, but as a plant cell matures, the numerous smaller vacuoles coalesce into one large, central, fluid-filled vacuole that then becomes a major supporting element of the cell (Figure 5–17). This vacuole also increases the size of the cell, including the amount of surface exposed to the environment, with a minimal investment in structural materials by the cell.

Vesicles, which are found in all metabolically active eukaryotic cells, have the same general structure as vacuoles. They are distinguished by size, function, and composition. Vesicles are usually less than 100 nanometers in diameter, whereas vacuoles are larger. One of the principal functions of vesicles is transport; as we shall see, vesicles participate in the transport of materials both within the cell and into and out of the cell.

Ribosomes

Ribosomes are the most numerous of the cell's many organelles. A rapidly growing *E. coli* cell has approximately 15,000 ribosomes, and a eukaryotic cell may have many times that number. Ribosomes, which are not enclosed by a membrane, are of similar construction in both prokaryotic and eukaryotic cells; the ribosomes of eukaryotic cells, however, are somewhat larger than those of prokaryotes. As we noted in Chapter 3, proteins are chains of amino acids, assembled in a specific sequence. The ribosomes are the sites at which this assembly takes place, a process that will be explored in some detail in Chapter 15. The more protein a cell is making, the more ribosomes it has.

The way in which the ribosomes are distributed in the eukaryotic cell is related to the way the newly synthesized proteins are utilized. Some proteins, such as collagen, digestive enzymes, hormones, or mucus, are released outside the cell, sometimes carrying out their functions at a great distance, on a cellular scale, from their source. Other proteins are essential components of cellular membranes. And, still others—hemoglobin and some enzymes, for example—are used within the cytoplasm. In cells, such as immature red blood cells, that are making cytoplasmic proteins for their own use, the ribosomes are distributed throughout the cytoplasm. In cells that are making new membrane material or proteins that are to be exported from the cell, large numbers of ribosomes are found attached to a complex system of internal membranes, the endoplasmic reticulum.

Endoplasmic Reticulum

The **endoplasmic reticulum** is a network of interconnecting flattened sacs, tubes, and channels found in eukaryotic cells. The amount of endoplasmic reticulum in a cell is not fixed but increases or decreases depending on the cell's activity.

There are two general categories of endoplasmic reticulum, rough (with ribosomes attached) and smooth (without ribosomes), which are, however, continuous with each other. Rough endoplasmic reticulum is present in all eukaryotic cells and predominates in cells making large amounts of proteins for export. It is

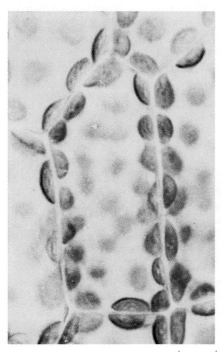

├── 10 μm ──┤

5–17 In this cell from the photosynthetic structure of a moss, the vacuole has expanded until it almost entirely fills the cell. The small amount of living cytoplasm, containing the chloroplasts, has been forced to the edges of the cell, up against the cell membrane.

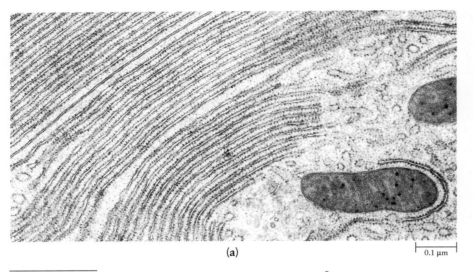

(a)

`|—— 0.1 μm ——|`

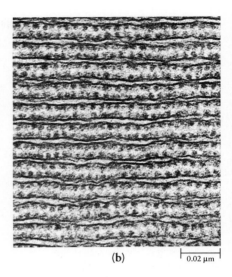

(b)

`|— 0.02 μm —|`

5–18 *(a) Rough endoplasmic reticulum, which fills most of this micrograph, is a system of membranes that separates the cell into channels and compartments and provides surfaces on which chemical activities take place. The dense objects on the membrane surfaces are ribosomes. This cell is from a pancreas, an organ extremely active in the synthesis of digestive enzymes, which are "exported" to the upper intestine, where most digestion takes place. In the lower right-hand corner of the micrograph are one mitochondrion and, above it, a portion of another. (b) Rough endoplasmic reticulum at higher magnification, showing the individual ribosomes. The compartments formed by the membranes of the endoplasmic reticulum are filled with newly synthesized proteins. (c) An interpretation of the rough endoplasmic reticulum based on electron micrographs.*

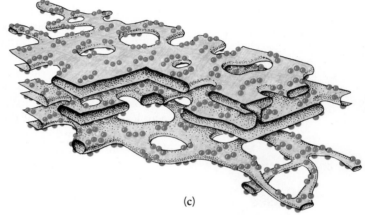

(c)

continuous with the outer membrane of the nuclear envelope, which also has ribosomes attached (see Figures 5–15 and 5–16). Rough endoplasmic reticulum often includes large, flattened sacs called cisternae. If cells engaged in protein synthesis are permitted to take up radioactive amino acids, the radioactive labels are first detected in the cytoplasm, then at the membrane of the rough endoplasmic reticulum, and then, slightly later, within its cisternae.

The synthesis of a protein destined for export from the cell, for incorporation into a specific organelle or its membrane, or for incorporation into the cell membrane itself begins in the cytoplasm with the synthesis of a "leader" of hydrophobic amino acids. This portion of the molecule, known as the **signal sequence,** is thought to direct the newly forming protein and the ribosomes that are participating in its synthesis to a specific region of the rough endoplasmic reticulum. The ribosomes attach to the endoplasmic reticulum, and the hydrophobic amino acids of the signal sequence assist in the transport of the protein through the lipid bilayer to the interior cavity, or **lumen,** of the endoplasmic reticulum. As synthesis of the protein proceeds, the growing polypeptide chain continues to move into the lumen. The newly synthesized protein molecule then moves from the rough endoplasmic reticulum through a special transitional endoplasmic reticulum, in which it is packaged in a transport vesicle destined for the Golgi complex. In the course of this progression from the endoplasmic reticulum to the Golgi complex, and subsequently to its ultimate destination in the cell, the protein molecule undergoes further processing. This processing includes cleavage of the signal sequence and, often, the addition of carbohydrate groups to the protein.

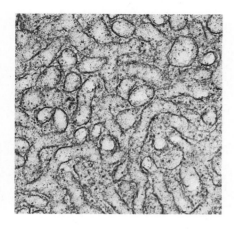

5-19 *Smooth endoplasmic reticulum from the testicle of an opossum. These membranes participate in the synthesis of the steroid hormone testosterone.*

Only cells specialized for the synthesis or metabolism of lipids—such as the gland cells that make steroid hormones—have large amounts of smooth endoplasmic reticulum. Smooth endoplasmic reticulum is also found in liver cells, where it appears to be involved in various detoxification processes (one of the many functions of the liver). For instance, in experimental animals fed large amounts of phenobarbital, the amount of smooth endoplasmic reticulum in the liver cells increases severalfold. A specialized transitional endoplasmic reticulum also seems to be active in the liver's breakdown of glycogen to glucose. As more of its functions are discovered, it seems likely that smooth endoplasmic reticulum actually represents a number of quite different specializations of endoplasmic reticulum, resembling one another only in their lack of ribosomes.

Golgi Complexes

Each **Golgi complex** consists of flattened, membrane-bound sacs stacked loosely on one another and surrounded by tubules and vesicles (Figure 5–20). The function of the Golgi complex is to accept vesicles from the endoplasmic reticulum, to modify the membranes and contents of the vesicles, and to incorporate the finished products in transport vesicles that deliver them to other parts of the cell and, especially, to the cell surface. Thus Golgi complexes serve as packaging and distribution centers. They are found in almost all eukaryotic cells. Animal cells usually contain 10 to 20 Golgi complexes, and plant cells may have several hundred.

One of the most critical products processed, packaged, and distributed by the Golgi complexes is new material for the membranes of the cell and its organelles. Membrane lipids and proteins, synthesized in the endoplasmic reticulum, are delivered to the Golgi complex in vesicles that fuse with it. Within the cisternae of the Golgi complex, the final assembly of carbohydrates with proteins (forming glycoproteins) and with lipids (forming glycolipids) occurs; as we noted previously, these carbohydrate combinations found on the surface of cell membranes are thought to play key roles in membrane function. Current evidence indicates that different stages in this chemical processing occur in different cisternae of the Golgi complex, and that materials are transported from one cisterna to the next via vesicles. After the chemical processing is completed, the new membrane material is packaged in vesicles that are targeted to the correct location, whether it

5-20 *Graphic interpretation and electron micrograph of a Golgi complex. A Golgi complex consists of four or more membrane-bound cisternae, arranged in a loose stack. Materials are packaged in membrane-enclosed vesicles at the Golgi complexes and are distributed within the cell or shipped to the cell surface. Note the vesicles pinching off from the edges of the flattened cisternae.*

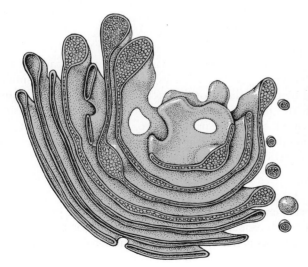

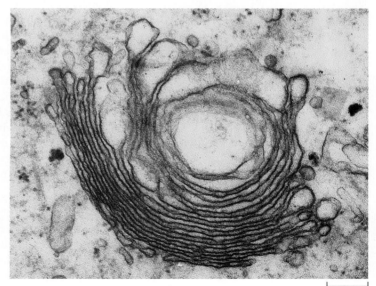

0.25 μm

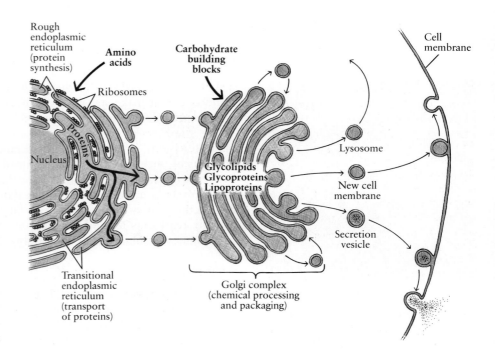

5-21 *Diagram illustrating the interaction of ribosomes, the endoplasmic reticulum, and the Golgi complex and its vesicles. These organelles work together in the synthesis, chemical processing, packaging, and distribution of macromolecules and new membrane material. As proteins are synthesized on the ribosomes, they are fed into the rough endoplasmic reticulum. They then move through a specialized transitional region of endoplasmic reticulum and are released in vesicles that fuse with the sacs of the Golgi complex. These vesicles incorporate in their membranes lipids newly synthesized in the endoplasmic reticulum. In the Golgi complex, carbohydrates are added to some of the proteins and lipids, producing glycoproteins and glycolipids; these macromolecules are common components of membranes. In some types of cells, lipids are added to other proteins in the Golgi complex, producing lipoproteins. Molecules destined for export from the cell also undergo chemical processing in the Golgi complex. Vesicles containing the finished molecules and macromolecules are released from the Golgi complex and move to other locations within the cell or to its exterior surface.*

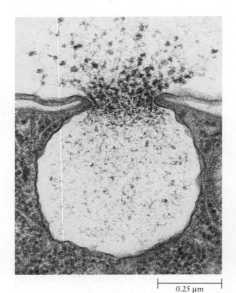

0.25 μm

5-22 *A secretion vesicle, formed by a Golgi complex of the protist* Tetrahymena furgasoni, *discharges mucus at the cell surface. Notice how the membrane enclosing the vesicle has fused with the cell membrane.*

be the cell membrane or the membrane of a particular organelle. In plant cells, Golgi complexes also bring together some of the components of the cell walls and export them to the cell surface where they are assembled.

In addition to their function in the assembly of cellular membranes, Golgi complexes have a similar function in the processing and packaging of materials that are released outside the cell. Figure 5–21 summarizes the way in which the ribosomes, the endoplasmic reticulum, and the Golgi complex and its vesicles interact to produce new material for the cell membrane and macromolecules for export.

Lysosomes

One type of relatively large vesicle commonly formed from the Golgi complex is the **lysosome.** Lysosomes are essentially membranous bags that enclose hydrolytic enzymes, thereby separating them from the rest of the cell; these enzymes are involved in breaking down proteins, polysaccharides, and lipids. If the lysosomes break open, the cell itself will be destroyed, since the enzymes they carry are capable of hydrolyzing all the major types of macromolecules found in a living cell. The tenderness and inflammation associated with rheumatoid arthritis and gout appear to be related to the escape of hydrolytic enzymes from lysosomes.

An example of the function of lysosomes is seen among white blood cells, which engulf bacteria in the human body. As the bacteria are taken up by the cell, they are wrapped in a membrane-enclosed sac, a vacuole. (This process is known as phagocytosis; we shall discuss it further in the next chapter.) When this occurs, the lysosomes within the cell fuse with the vacuoles containing the bacteria and release their hydrolytic enzymes. The bacteria are quickly digested. In similar fashion, the lysosomes of heterotrophic protists, such as *Paramecium, Didinium,* and amoebas, fuse with the phagocytic vacuoles containing food organisms. Hydrolytic enzymes released by the lysosomes into the vacuoles digest the contents. Why the enzymes do not destroy the membranes of the lysosomes that carry them is a pertinent question yet to be answered.

Peroxisomes

Another type of relatively large vesicle containing lytic enzymes is the **peroxisome.** Peroxisomes are vesicles in which purines (one of the two major categories of nitrogenous bases) and several other types of compounds are broken down by the cell. In plants, peroxisomes are also the site of a series of reactions that occur in

5-23 *Lysosomes and peroxisomes are vesicles within which different types of molecules are broken down.* (a) *In this portion of a cell from the adrenal gland, the dark oval bodies are lysosomes. They may contain 40 or more different hydrolytic enzymes.* (b) *A peroxisome. The crystalline material in the center is a peroxide-producing enzyme involved in the breakdown of purines. Another enzyme destroys the peroxide, preventing its escape into the cytoplasm.*

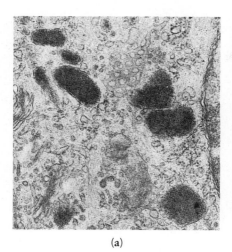

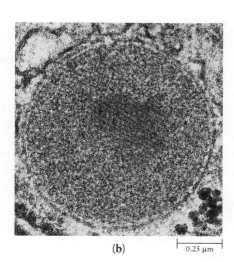

(a)

(b)

0.25 μm

sunlight when the cell contains relatively high concentrations of oxygen. These reactions and the breakdown of purines both produce hydrogen peroxide (H_2O_2), a compound that is extremely toxic to living cells. The peroxisomes, however, contain another enzyme that immediately breaks hydrogen peroxide into water and oxygen, preventing any damage to the cell.

Mitochondria

Mitochondria (singular, mitochondrion) are among the largest organelles in a cell. In the mitochondria, energy-yielding organic molecules are broken down and their energy is repackaged into smaller units, convenient for most cellular processes. The higher the energy requirements of a particular eukaryotic cell, the more mitochondria it is likely to have. A liver cell, for example, which has modest energy requirements, has about 2,500, making up about 25 percent of the volume of the cell, whereas a heart muscle cell has several times as many very large mitochondria. Mitochondria are often found clustered in areas in the cell where energy requirements are high.

Mitochondria vary in shape from almost spherical, to potato-shaped, to greatly elongated cylinders. As shown in Figure 5–24, they are always surrounded by two membranes, the inner one of which folds inward; these folds, known as **cristae,** are working surfaces for mitochondrial reactions. The more active a mitochondrion, the more cristae it is likely to have. In Chapter 9, we shall examine in more detail the structure of mitochondria and the processes that occur within these important organelles.

5-24 *A mitochondrion is surrounded by two membranes. The inner membrane folds inward to make a series of shelves, or cristae. The membrane forming these shelves plays a crucial role in the energy-releasing chemical reactions that occur in the mitochondria.*

Outer membrane

Inner membrane

Cristae

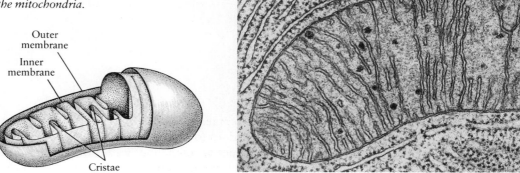

0.25 μm

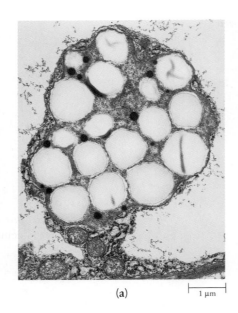

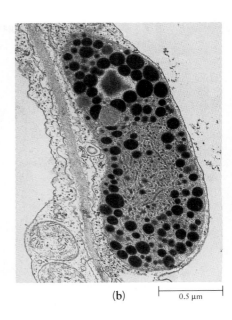

5-25 (a) *A leucoplast from the embryo sac of a soybean. The embryo sac is the structure in the flower in which the egg is fertilized and the embryonic plant begins its development. The large round, clear bodies are starch granules, and the smaller, dark bodies are lipid droplets.*

(b) *Chromoplast from a forsythia petal. The large, dark granules contain the orange and yellow pigments characteristic of certain flowers and fall leaves. To the left of the chromoplast are the faintly visible cell wall and two mitochondria. To the left and right of the micrograph are portions of two vacuoles.*

(a)　　　　　1 μm

(b)　　　　　0.5 μm

Plastids

Plastids are membrane-bound organelles found only in the cells of plants and algae. They are surrounded by two membranes, like mitochondria, and have an internal membrane system that may be folded intricately. Mature plastids are of three types: leucoplasts, chromoplasts, and chloroplasts.

Leucoplasts (*leuco* means "white") store starch or, sometimes, proteins or oils. Leucoplasts are likely to be numerous in storage organs such as roots, as in a turnip, or tubers, as in a potato.

Chromoplasts (*chromo* means "color") contain pigments and are associated with the bright orange and yellow colors of fruits, flowers, fall leaves, and carrots.

Chloroplasts (*chloro* means "green") are the chlorophyll-containing plastids in which photosynthesis takes place. Like other plastids, they are surrounded by two membranes; the third, internal membrane of chloroplasts forms an elaborate series of internal compartments and work surfaces. The sequence of events that lead to the formation of a mature chloroplast are shown in Figure 5-26. In Chapter 10, we shall consider the structure of the chloroplast and its functional significance in more detail.

5-26 *Chloroplast development.* (a) *The immature plastid contains small crystalline structures (top).* (b) *In the presence of light, these structures begin to break up into elongated vesicles.* (c) *The vesicles flatten into membranes that are grouped together in stacks. These internal membranes are known as thylakoids, and the stacks they form as grana.* (d) *A mature chloroplast. Like the mitochondrion, it is surrounded by two membranes; in addition, it contains an elaborate internal membrane system in which the light-capturing reactions of photosynthesis take place.*

Chromoplasts and leucoplasts develop from plastids similar to (a).

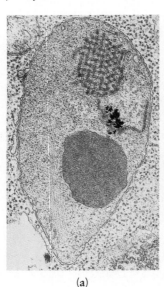

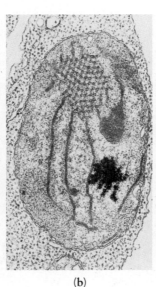

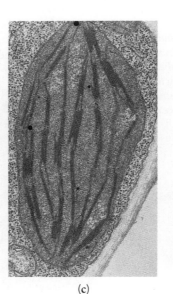

(a)　　　　　　(b)　　　　　　(c)　　　　　　(d)　　　0.5 μm

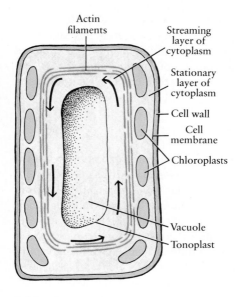

Actin filaments
Streaming layer of cytoplasm
Stationary layer of cytoplasm
Cell wall
Cell membrane
Chloroplasts
Vacuole
Tonoplast

5-27 *The role of actin in cytoplasmic streaming. In the cell of a green alga, the portion of the cytoplasm containing the chloroplasts is located against the cell membrane. Laid down across the chloroplasts are bundles of actin filaments that form distinct "tracks"; these are thought to direct the regular streaming movement that occurs in the portion of the cytoplasm located between the actin filaments and the vacuole. The outer portion of the cytoplasm, by contrast, is stationary.*

HOW CELLS MOVE

All cells exhibit some form of movement. Even plant cells, encased in a rigid cell wall, exhibit active cytoplasmic streaming (movement of cytoplasm within the cell) as well as chromosomal movements and changes in shape during cell division. As we saw in Figure 4–13 (page 94), cilia beat along the surface of the tracheal cells of animals. Embryonic cells migrate in the course of animal development. Differentiating and regenerating nerve cells send out axons, which are long, slender extensions that may be a meter or more in length. Amoebas pursue and engulf their prey, and little *Chlamydomonas* cells dart toward a light source. Two different molecular mechanisms of cellular movement have been identified: (1) assemblies of protein filaments, in which actin filaments (page 111) are present in large quantity, and (2) assemblies of microtubules in cilia and flagella.

Actin and Associated Proteins

As you will recall, the actin filaments of the cytoskeleton consist of helical chains composed of subunits of the globular protein actin. Actin filaments are present in a great variety of cells, including plant cells. They participate not only in the maintenance of cytoplasmic organization but also in cell motility and the internal movement of cellular contents. Interacting with the actin filaments in producing cellular motion are bundles of another protein, known as myosin. Additional proteins, thought to play regulatory roles, are associated with the actin and myosin molecules.

Actin, myosin, and their associated proteins are involved in a variety of different cellular processes. For example, they are found in the large, multinucleate organisms known as slime molds, which move like giant amoebas and exhibit vigorous cytoplasmic streaming. Actin has also been found in a true amoeba, *Amoeba proteus*. In such cells, which move by gradual changes of shape, the actin filaments are found concentrated in bundles or a meshwork near the moving edge. These filaments have also been shown to act as a sort of "purse string" in animal cells during cell division, pinching off the cytoplasm to separate the two daughter cells. In algal cells, actin filaments occur in bundles wherever cytoplasmic streaming is taking place (Figure 5–27). The way in which actin and its associated proteins bring about amoeboid movement and cytoplasmic streaming is currently under active investigation.

5-28 *A macrophage, a type of phagocytic white blood cell, on the move across the surface of a culture dish. Its motility is made possible by the interaction of actin filaments with another protein known as myosin. As we shall see in Chapter 39, macrophages perform a number of vital functions in the body's response to invading microorganisms. They are also one of the principal cell types attacked by the AIDS virus.*

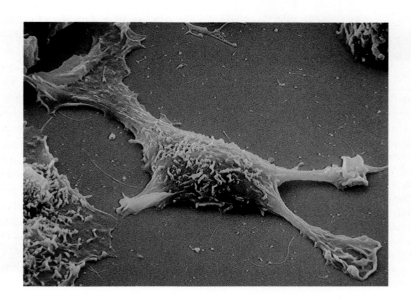

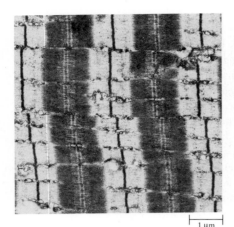

5–29 *Contractile assemblies of protein filaments, principally actin and myosin, in a vertebrate skeletal muscle. Each unit is known as a sarcomere, approximately 18 of which are visible in this electron micrograph. When the muscle contracts, the distance from left to right in this micrograph shortens.*

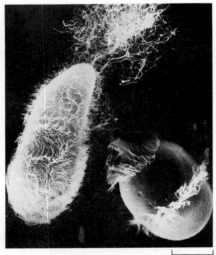

5–30 *Two ciliates, one-celled protists distinguished by their many cilia. On the left,* Paramecium; *on the right,* Didinium. Didinium *is stalking* Paramecium. *Paramecium, in defense, has discharged a barrage of barbs (visible as a cloud at the top of the micrograph).* Didinium *is about to eject a bundle of slender, poisonous strands (not visible), which will paralyze* Paramecium *in a matter of seconds. In* Paramecium, *the cilia are distributed fairly evenly over the cell surface. In* Didinium, *they form two wreaths that circle the organism's barrel-shaped body.*

Actin and myosin are also the principal components of the elaborate contractile assemblies (Figure 5–29) found in the muscle cells of vertebrates and many other animals. This specialized organization of actin and myosin (to be described in detail in Chapter 42) makes possible the rapid, coordinated movements that give animals, ranging from insects to fishes to birds to race horses to ourselves, their great mobility.

Cilia and Flagella

Cilia and **flagella** are long, thin (0.2 micrometer) structures extending from the surface of many types of eukaryotic cells. They are essentially the same except for length (the names were given before their basic similarity was realized). When they are shorter and occur in larger numbers, they are more likely to be called cilia; when they are longer and fewer, they are usually called flagella. (Prokaryotic cells also have flagella, but they are so different in construction from those of eukaryotes that it would be useful if they had a different name. We shall examine their structure and mechanism of movement in Chapter 21.)

In one-celled protists and some small animals (such as a few types of flatworms), cilia and flagella are associated with movement of the organism. For example, one species of *Paramecium* has approximately 17,000 cilia, each about 10 micrometers long, which propel it through the water by beating in a coordinated fashion. Other protists, such as the members of the genus *Chlamydomonas*, have only two whiplike flagella, which protrude from the anterior end of the organism and move it through the water (see Figure 5–3, page 104). The motile power of the human sperm cell comes from its single powerful flagellum, or "tail."

Many of the cells that line the surfaces within our bodies are also ciliated. These cilia do not move the cells but, rather, serve to sweep substances across the cell surface. For example, cilia on the surface of cells of the respiratory tract beat upward, propelling a current of mucus that sweeps bits of soot, dust, pollen, tobacco tar—whatever foreign substances we have inhaled either accidentally or on purpose—to our throats, where they can be removed by swallowing. Human egg cells are propelled down the oviducts by the beating of cilia that line the inner surfaces of these tubes. Cilia and flagella are found extensively throughout the living world, on the cells of invertebrates, vertebrates, the sex cells of ferns and other plants, as well as on protists. Only a few large groups of eukaryotic organisms, such as red algae, fungi, flowering plants, and roundworms (nematodes), have no cilia or flagella on any cells.

Almost all eukaryotic cilia and flagella, whether on a *Paramecium* or a sperm cell, have the same internal structure. Nine pairs of fused microtubules form a ring that surrounds two additional, solitary microtubules in the center (Figure 5–31). Microtubules, you will recall, are composed of identical globular protein units assembled in a hollow helix. The movement of cilia and flagella comes from within the structures themselves; if cilia are removed from cells and placed in a medium containing energy-yielding chemicals, they beat or swim through the medium. The movement, according to the generally accepted hypothesis, is caused by each outer pair of microtubules moving tractor-fashion with respect to its nearest neighbor. The two "arms" that you can see on one of each pair of outer tubules (Figure 5–31a) have been shown to be enzymes involved in energy-releasing chemical reactions. Other proteins are involved in the formation of spokes connecting the nine pairs of outer microtubules to the central pair, and still other proteins form more widely spaced links, rather like the hoops of a barrel, connecting the nine outer pairs to each other. The spokes are thought to play a role in coordinating the tractorlike movements of the microtubules, whereas the links limit the amount of sliding possible and thus convert it into a bending motion.

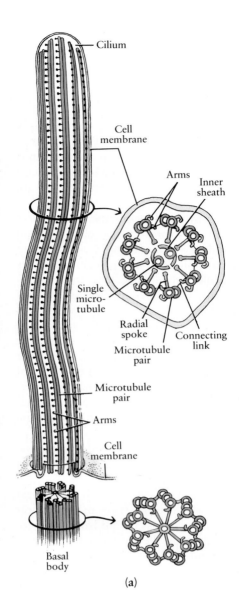

Cilium

Cell
membrane

Arms

Inner
sheath

Single
micro-
tubule

Radial
spoke

Connecting
link

Microtubule
pair

Microtubule
pair

Arms

Cell
membrane

Basal
body

(a)

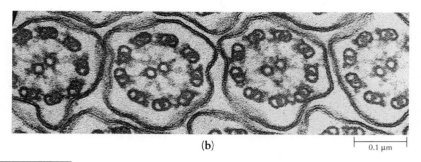

(b)

0.1 μm

5–31 (a) *Diagram of a cilium with its underlying basal body. Virtually all eukaryotic cilia and flagella, whether they are found on protists or on the surfaces of cells within our own bodies, have this same internal structure, which consists of an outer ring of nine pairs of microtubules surrounding two additional microtubules in the center. The "arms," the radial spokes, and the connecting links are formed from different types of protein. The basal bodies from which cilia and flagella arise have nine outer triplets, with* no microtubules in the center. The "hub" of the wheel in the basal body is not a microtubule, although it has about the same diameter.

(b) *Cross section of cilia from a gill cell of a mussel. The beating of cilia on the gills of mussels, clams, oysters, and other two-shelled mollusks sweeps water through the sievelike gills. Small organisms and particles of food are trapped in mucus on the gill surface and are then swept toward the mouth by the cilia.*

Basal Bodies and Centrioles

Underlying each cilium is a structure known as a **basal body,** which has the same diameter as a cilium, about 0.2 micrometer. It consists of microtubules arranged in nine triplets (rather than pairs) around the periphery. Unlike the cilium, it has no microtubules in the center, and none of the microtubules in the basal body have arms. Cilia and flagella arise from basal bodies. For instance, as a sperm cell takes form, a basal body moves near the cell membrane, and the sperm's flagellum arises from it through the assembly of microtubules.

Many types of eukaryotic cells contain **centrioles.** Centrioles, which typically occur in pairs, are small cylinders, about 0.2 micrometer in diameter, containing nine microtubule triplets (Figure 5–32). Their structure is identical to that of basal bodies; however, their distribution in the cell is different. Until recently, it appeared that their function was also different, which is why they are called by different names even though electron microscopy has revealed their identical structures. Centrioles usually lie in pairs with their long axes at right angles to one another in the region of the cytoplasm, near the nuclear envelope, from which the microtubules of the cytoskeleton radiate. They are found only in those groups of organisms that also have cilia or flagella (and, therefore, basal bodies). There is evidence that centrioles play a role in organizing a structure known as the spindle, which appears at the time of cell division and is involved in chromosome movements. The spindle, it has been shown, also contains numerous microtubules; thus centrioles and basal bodies both appear to be organizers of microtubules. However, as we shall see in Chapter 7, cells that have no centrioles—such as the cells of the flowering plants—are also able to organize microtubules into a spindle.

The discovery of the complex internal structure of cilia and flagella, basal bodies and centrioles, repeated over and over again throughout the living world, was one of the spectacular revelations of electron microscopy. For biologists, it is another glimpse down the long corridor of evolution, providing overwhelming evidence, once again, of the basic unity of earth's living things.

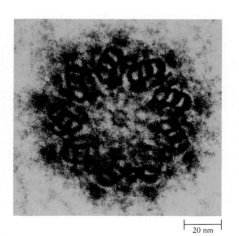

20 nm

5–32 *Cross section of a centriole from a cell of the fruit fly* Drosophila. *Centrioles are structurally identical to basal bodies.*

SUMMARY

Cells are the basic units of biological structure and function. The size of cells is limited by proportions of surface to volume; the greater a cell's surface area in proportion to its volume, the greater the quantity of materials that can move into and out of the cell in a given period of time. Cell size is also limited by the capacity of the nucleus to regulate cellular activities. Cells that are active metabolically are likely to be small.

Cells are separated from their environment by a cell membrane that restricts the passage of materials into and out of the cell and so protects the cell's structural and functional integrity. According to the fluid-mosaic model, cell membranes are formed from phospholipid bilayers in which cholesterol and protein molecules are embedded. The embedded protein molecules, which typically span the membrane, are known as integral membrane proteins. Different integral proteins perform different functions; some are enzymes, others are receptors, and still others are transport proteins. The two faces of the membrane differ in chemical composition. The cytoplasmic face is characterized by peripheral membrane proteins, attached to the integral proteins embedded in the bilayer. The exterior face of the membrane is characterized by short carbohydrate chains. Some of these chains are the hydrophilic heads of glycolipid molecules that are interspersed among the phospholipid molecules of the outer layer of the bilayer; other carbohydrate chains are covalently linked to the protruding portions of integral membrane proteins.

The cells of plants, most algae, fungi, and prokaryotes are further separated from the environment by a cell wall constructed by the cell itself. Cellulose is an important constituent of the cell walls of plants and many algae.

The nucleus of eukaryotic cells is separated from the cytoplasm by the nuclear envelope, which consists of two concentric lipid bilayers. Pores in the nuclear envelope provide channels through which molecules pass to and from the cytoplasm. The nucleus contains the hereditary material, the chromosomes, which, when the cell is not dividing, exist in an extended form called chromatin. The nucleolus, visible within the nucleus, is involved in the formation of ribosomes. Interacting with the cytoplasm, the nucleus helps to regulate the cell's ongoing activities.

The cytoplasm of the cell is a concentrated aqueous solution containing enzymes, many other dissolved molecules and ions, and also, in the case of eukaryotic cells, a variety of membrane-bound organelles with specialized functions in the life of the cell. The eukaryotic cytoplasm has a supporting cytoskeleton that includes three principal types of structures: microtubules, actin filaments, and intermediate filaments. The cytoskeleton maintains the shape of the cell, enables it to move, anchors its organelles, and directs its traffic. Vacuoles, which are bounded by a single membrane, are also present in the cytoplasm of many cells, particularly plant cells. The vacuoles of plant cells are storage reservoirs and play a role in cell support. Vesicles, present in all metabolically active eukaryotic cells, are usually smaller than vacuoles. They perform a variety of functions, of which transport is one of the most important.

Eukaryotic cells contain many organelles, most of which are not found in prokaryotic cells (see Table 5–1). The most numerous organelles (in both prokaryotes and eukaryotes) are the ribosomes, the sites of assembly of proteins. The cytoplasm of eukaryotic cells is subdivided by a network of membranes known as the endoplasmic reticulum, which serves as a work surface for many of the cell's biochemical activities. In eukaryotic cells, many ribosomes are bound to the surface of the endoplasmic reticulum, producing the rough endoplasmic reticulum. Rough endoplasmic reticulum is especially abundant in cells producing proteins for export. Smooth endoplasmic reticulum, which lacks ribosomes, is

TABLE 5-1 A Comparison of Cell Characteristics

KINGDOM	MONERA	PROTISTA	FUNGI	PLANTAE	ANIMALIA
Cell type	Prokaryotic	Eukaryotic	Eukaryotic	Eukaryotic	Eukaryotic
Cell membrane	Present	Present	Present	Present	Present
Cell wall	Noncellulose (polysaccharide and peptidoglycan)	Present in some forms, various types	Chitin and other noncellulose polysaccharides	Cellulose and other polysaccharides	Absent
Nuclear envelope	Absent	Present	Present	Present	Present
Chromosomes	Single, continuous DNA molecule	Multiple, consisting of DNA and histone proteins	Multiple, consisting of DNA and histone proteins	Multiple, consisting of DNA and histone proteins	Multiple, consisting of DNA and histone proteins
Ribosomes	Present (smaller)	Present	Present	Present	Present
Endoplasmic reticulum	Absent	Present	Present	Present	Present
Mitochondria	Absent	Usually present	Present	Present	Present
Plastids	Absent	Present in some forms	Absent	Present	Absent
Golgi complexes	Absent	Present	Present	Present	Present
Lysosomes	Absent	Often present	Often present	Similar structures (lysosomal compartments present)	Often present
Peroxisomes	Absent	Often present	Present in some forms	Often present	Often present
Vacuoles	Absent	Present	Present	Usually large single vacuole in mature cell	Small or absent
9 + 2 cilia or flagella	Absent	Often present	Absent	Absent (in flowering plants)	Often present
Centrioles	Absent	Often present	Absent	Absent (in flowering plants)	Present

abundant in cells specialized for lipid synthesis or metabolism. Golgi complexes, also composed of membranes, are processing and packaging centers for materials being moved through and out of the cell. Lysosomes, which contain hydrolytic enzymes, are involved in intracellular digestive activities in some cells. The enzymes for cellular reactions that produce hydrogen peroxide as a by-product are sequestered in the peroxisomes, along with an enzyme that breaks down the toxic peroxide into water and oxygen.

Mitochondria are membrane-bound organelles in which energy-yielding organic molecules are broken down and the released energy repackaged into smaller units. Plastids are membrane-bound organelles found only in photosynthetic organisms. Leucoplasts are storage compartments, chromoplasts contain pigments, and chloroplasts are the sites of photosynthesis in plants and algae.

Assemblies of protein filaments, principally actin and myosin, are associated with internal cellular movement, whereas cilia and flagella are associated with the external movement of cells or the movement of materials along cell surfaces. These whiplike appendages are found on the surface (yet within the cell membrane) of many types of eukaryotic cells. They have a highly characteristic 9 + 2 structure, with nine pairs of microtubules forming a ring surrounding two central microtubules. One of each pair of outer microtubules contains enzymes involved in the chemical reactions that release energy for ciliary motion.

Cilia and flagella arise from basal bodies, which are cylindrical structures containing nine microtubule triplets with no inner pair. Centrioles have the same internal structure as basal bodies and are found in those groups of organisms that also have cilia or flagella. They typically occur in pairs, lying near the nuclear envelope, and may play a role in the formation of the spindle during cell division.

QUESTIONS

1. Distinguish between the following: cell membrane/cell wall; nucleus/nucleolus; rough endoplasmic reticulum/smooth endoplasmic reticulum; lysosomes/peroxisomes; chloroplasts/mitochondria; cilia/flagella; basal body/centriole.

2. Describe the structure of the cell membrane. How do the two faces of the membrane differ? What is the functional significance of these differences?

3. (a) Sketch an animal cell. Include the principal organelles and label them. (b) Prepare a similar, labeled sketch of a plant cell. (c) What are the major differences between the animal cell and the plant cell?

4. Why is the secondary wall of a plant cell *inside* the primary cell wall? Where is the cell membrane in relation to the two cell walls?

5. What are the functions of the cytoskeleton? Describe the similarities and differences between microtubules, actin filaments, and intermediate filaments.

6. Explain the functions of each of the following structures: ribosomes, endoplasmic reticulum, vesicles, and Golgi complexes. How do they interact in the synthesis and delivery of new membrane material and in the export of proteins from the cell?

7. Use a ruler and the scale marker at the bottom of each micrograph on pages 104 and 123 to determine: (a) the thickness (roughly) of a cell membrane, (b) the diameter of a cilium, and (c) the diameter of a microtubule within a cilium. (This is how the sizes of cellular components are determined by microscopists.) Would a cilium be resolvable in a light microscope (that is, is its diameter more than 0.2 μm)?

8. (a) Sketch a cross section of a cilium. (b) Sketch a cross section of the basal body of a cilium. (c) What are the differences between the two structures?

9. On the basis of what you know of the functions of each of the structures in Table 5–1, what components would you expect to find most prominently in each of the following cell types: muscle cells, sperm cells, green leaf cells, red blood cells, white blood cells?

10. Two brothers were under medical treatment for infertility. Microscopic examination of their semen showed that the sperm were immotile and that the little "arms" were missing from the microtubular arrays. The brothers also had chronic bronchitis and other respiratory difficulties. Can you explain why?

C H A P T E R **6**

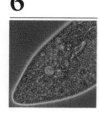

How Things Get into and out of Cells

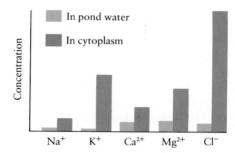

6–1 *The relative concentrations of different ions in pond water and in the cytoplasm of the green alga* Nitella. *Differences such as these indicate that cells regulate their exchanges of materials with the surrounding environment.*

One of the criteria by which we identify living systems is that living matter, although surrounded on all sides by nonliving matter with which it constantly exchanges materials, is different from that nonliving matter in the kinds and amounts of chemical substances it contains (Figure 6–1). Without this difference, of course, living systems would be unable to maintain the organization and structure on which their existence depends.

In all living systems, ranging from prokaryotes to the most complex multicellular eukaryotes, regulation of exchanges of substances between the living system and the nonliving world occurs at the level of the individual cell and is accomplished by the cell membrane. In multicellular organisms, the cell membrane has the additional task of regulating exchanges of substances among the various specialized cells that constitute the organism. Control of these exchanges is essential to protect each cell's integrity, to maintain those very narrow conditions of pH and ionic concentrations at which its metabolic activities can take place, and to coordinate the activities of the different cells. In addition to the cell membrane, which controls the passage of materials between the cell and its environment, internal membranes, such as those surrounding mitochondria, chloroplasts, and the nucleus, control the passage of materials among intracellular compartments (Figure 6–2). This makes it possible for the cell to maintain the specialized chemical environments necessary for the processes occurring in the different organelles.

Maintenance of the internal environment of the cell and its constituent parts requires that cell membranes perform a complex double function: they must keep

6–2 *Electron micrograph of a portion of a cell from the pancreas. The cell has a large, central nucleus with scattered chromatin, many mitochondria, large quantities of rough endoplasmic reticulum, and many small vesicles. Not only is the cell itself surrounded by a membrane and the nucleus by a double membrane system (the nuclear envelope), but its organelles are also surrounded by membranes. The membranes of the endoplasmic reticulum further divide the cell into membrane-bound compartments. Collectively, all of these different membranes regulate the movements of substances into and out of cells and restrict their passage from one part of the cell to another.*

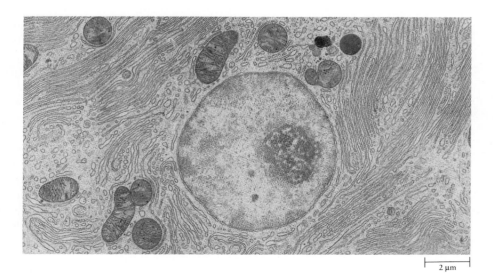

127

certain substances out while letting others in, and, conversely, they must keep certain substances in while letting others out. The capacity of a membrane to accomplish this function depends not only on the physical and chemical properties that result from its lipid and protein structure, but also on the physical and chemical properties of the substances—ions, molecules, and aggregations of molecules—that interact with the membrane. Of the many kinds of molecules surrounding and contained within the cell, by far the most common is water. Further, the many other molecules and ions important in the life of the cell are carried in an aqueous solution. Therefore, let us begin our consideration of transport across cell membranes by looking again at water, focusing our attention this time on how it moves.

THE MOVEMENT OF WATER AND SOLUTES

In both the animate and inanimate worlds, water molecules move from one place to another because of differences in potential energy, usually referred to as the **water potential.** Water moves from a region where water potential is greater to a region where water potential is lower, regardless of the reason for the water potential. A simple example is water running downhill in response to gravity. Water at the top of a hill has more potential energy (that is, a greater water potential) than water at the bottom of a hill. As the water runs downhill, its potential energy is converted to kinetic energy; this, in turn, can be converted to mechanical energy doing useful work if, for example, a water mill is placed in the path of the moving water.

Pressure is another source of water potential. If we fill a rubber bulb with water and then squeeze, water will squirt out of the nozzle. Like water at the top of a hill, this water has been given a high water potential and will move to a lower one. Can we make the water that is running downhill run uphill by means of pressure? Obviously we can. But only so long as the water potential produced by the pressure exceeds the water potential produced by gravity.

In solutions, water potential is affected by the concentration of dissolved particles (solutes). As the concentration of solute particles (that is, the number of solute particles per unit volume of solution) increases, the concentration of water molecules (that is, the number of water molecules per unit volume of solution) must necessarily decrease, and vice versa. In the absence of other factors (such as pressure), the water potential of a solution is directly related to the concentration of water molecules—the higher the concentration of water molecules, the greater the water potential. Conversely, the higher the concentration of solute particles, the lower the water potential. Water molecules move from regions of higher water potential to regions of lower water potential, a fact of great importance for living systems.

The concept of water potential is a useful one because it enables us to predict the way that water will move under various combinations of circumstances. Measurements of water potential are usually made in terms of the pressure required to stop the movement of water—that is, the hydrostatic (water-stopping) pressure—under the particular circumstances. The unit usually used to measure this pressure is the atmosphere. One atmosphere is the average pressure of the air at sea level, about 1 kg/cm² (or 15 lb/in²).

Two mechanisms are involved in the movement of water and solutes: bulk flow and diffusion. In living systems, bulk flow moves water and solutes from one part of a multicellular organism to another part, whereas diffusion moves molecules and ions into, out of, and through cells. A particular instance of diffusion—that of water across a membrane that separates solutions of different concentration—is known as osmosis.

6–3 *Water at the top of a falls, like a boulder on a hilltop, has potential energy. The movement of water molecules as a group, as from the top of the falls to the bottom, is referred to as bulk flow.*

Bulk Flow

Bulk flow is the overall movement of a fluid. The molecules move all together and in the same direction. For example, water runs downhill by bulk flow in response to the differences in water potential at the top and the bottom of a hill. Blood moves through your body by bulk flow as a result of the water potential (blood pressure) created by the pumping of your heart. Sap—an aqueous solution of sucrose and other solutes—moves by bulk flow from the leaves of a plant to other parts of the plant body.

Diffusion

Diffusion is a familiar phenomenon. If you sprinkle a few drops of perfume in one corner of a room, the scent will eventually permeate the entire room even if the air is still. If you put a few drops of dye in one end of a glass tank full of water, the dye molecules will slowly become evenly distributed throughout the tank. The process may take a day or more, depending on the size of the tank, the temperature, and the relative size of the dye molecules.

Why do the dye molecules move apart? If you could observe the individual dye molecules in the tank (Figure 6–4), you would see that each one of them moves individually and at random. Looking at any single molecule—at either its rate of motion or its direction of motion—gives you no clue at all about where the molecule is located with respect to the others. So how do the molecules get from one end of the tank to the other? Imagine a thin section through the tank, running from top to bottom. Dye molecules will move in and out of this section, some moving in one direction, some moving in the other. But you will see more dye molecules moving from the side of greater dye concentration. Why? Simply because there are more dye molecules at that end of the tank. If there are more dye molecules on the left, more dye molecules, moving at random, will move to the right, even though there is an equal probability that any one molecule of dye will move from right to left. Consequently, the *net* movement of dye molecules will be from left to right. Similarly, if you could see the movement of the individual water molecules in the tank, you would see that their *net* movement is from right to left.

Substances that are moving from a region of higher concentration of their own molecules to a region of lower concentration are said to be moving *down a gradient*. (A substance moving in the opposite direction, toward a higher concentration of its own molecules, moves *against a gradient*, which is analogous to being pushed uphill.) Diffusion occurs only down a gradient. The steeper the downhill gradient—that is, the larger the difference in concentration—the more rapid the diffusion. In our imaginary tank, there are two gradients; the dye molecules are moving down one of them, and the water molecules are moving down the other in the opposite direction. In each case, the molecules are moving from a region of higher potential energy to a region of lower potential energy.

What happens when all the molecules are distributed evenly throughout the tank? The even distribution does not affect the behavior of the molecules as individuals; they still move at random. And, since the movements are random, just as many molecules go to the left as to the right. But because there are now as many molecules of dye and as many molecules of water on one side of the tank as on the other, there is no *net* movement of either. There is, however, just as much random motion as before, provided the temperature has not changed. When the molecules have reached a state of equal distribution, that is, when there are no more gradients, they are said to be in dynamic equilibrium.

The essential characteristics of diffusion are (1) that each molecule or ion moves independently of the others and (2) that these movements are random. The net result of diffusion is that the diffusing substances become evenly distributed.

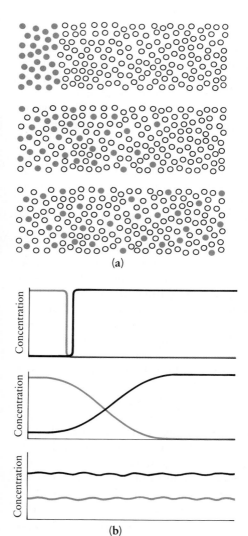

6–4 (a) *Diagram of the diffusion process. Diffusion is the result of the random movement of individual molecules (or ions), which produces a net movement from a more concentrated to a less concentrated region. Notice that as one type of molecule (indicated by color) diffuses to the right, the other diffuses in the opposite direction. The result will be an even distribution of both types of molecules. Can you see why the net movement of molecules will slow down as equilibrium is reached?* (b) *Graphs showing the concentration gradients of dye and water.*

Cells and Diffusion

Water, oxygen, carbon dioxide, and a few other simple molecules diffuse freely across cell membranes. Carbon dioxide and oxygen, which are both nonpolar, are soluble in lipids and move easily through the lipid bilayer of the membrane. Despite their polarity, water molecules also move through the membrane without hindrance, apparently through hydrophilic apertures. These may be either permanent pores created by the tertiary structure of some of the integral membrane proteins or momentary openings resulting from movements of the lipid molecules. Other polar molecules, provided they are small enough, also diffuse through these openings. The permeability of the membrane to these solutes varies inversely with the size of the molecules, indicating that the apertures are small and that the membrane acts like a sieve in this respect.

Diffusion is also a principal way in which substances move within cells. One of the major factors limiting cell size is this dependence upon diffusion, which is essentially a slow process, except over very short distances. As you can see by studying Figure 6–4, the process becomes increasingly slower and less efficient as the distance "covered" by the diffusing molecules increases. The rapid spread of a substance through a large volume, such as perfume through the air of a room, is due not primarily to diffusion but rather to the circulation of air currents. Similarly, in many cells, the transport of materials is speeded by active streaming of the cytoplasm, a process in which actin filaments of the cytoskeleton play a key role (see Figure 5–27, page 121).

Efficient diffusion requires not only a relatively short distance but also a steep concentration gradient. Cells maintain such gradients by their metabolic activities, thereby hastening diffusion. For example, carbon dioxide is constantly produced as the cell oxidizes fuel molecules for energy. As a result, there is a higher concentration of carbon dioxide inside the cell than out. Thus a gradient is maintained between the inside of the cell and the outside, and carbon dioxide diffuses out of the cell down this gradient. Conversely, oxygen is used up by the cell in the course of its activities, so oxygen present in air or water or blood tends to move into the cells by diffusion, again down a gradient. Similarly, within a cell, molecules or ions are often produced at one place and used at another. Thus a concentration gradient is established between the two regions, and the substance diffuses down the gradient from the site of production to the site of use.

Countercurrent Exchange

Although diffusion is efficient only over short distances, it nevertheless plays a key role in the transport of substances into and out of multicellular organisms, as well as between different compartments within the organism. For example, oxygen enters the bloodstream of an animal by diffusing through cells that are in contact with the environment. Unless the animal is quite small or leads a sedentary existence, a high rate of diffusion is required to supply it with adequate oxygen. Organisms have developed a number of anatomical arrangements that maintain steep concentration gradients and thus maximize diffusion rates. One of the most common types of arrangement is found in fish gills. Gills are divided into filaments, which are further divided into a number of flat, densely packed lamellae—platelike structures containing many blood vessels (Figure 6–5). Water, containing dissolved oxygen, flows over the lamellae, and oxygen moves by diffusion into the blood vessels just below the thin surfaces. In the lamellae, the direction of blood flow is opposite to the direction of water flow. This countercurrent arrangement maintains a constant concentration gradient between the bloodstream and the water, and oxygen can diffuse in through the entire surface of the lamellae. As we shall see in Section 6, this principle of countercurrent exchange is used in several different anatomical systems.

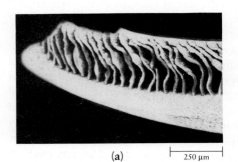

(a)   250 μm

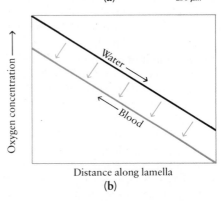

Distance along lamella
(b)

6–5 (a) Micrograph of a gill filament from a trout, showing the densely packed lamellae, which contain the blood vessels. Water, carrying dissolved oxygen, flows between the lamellae in one direction; blood flows through them in the opposite direction. Thus the blood carrying the most oxygen (that is, the blood leaving the lamellae) meets the water carrying the most oxygen, and the blood carrying the least oxygen (the blood entering the lamellae) meets the water carrying the least oxygen. (b) In this way, a constant concentration gradient is maintained along the lamella, and the transfer of oxygen to blood by diffusion (blue arrows) takes place all across its surface.

Sensory Responses in Bacteria: A Model Experiment

Concentration gradients are important not only in the diffusion of substances into, out of, and through cells but also, in the case of many single-celled organisms, in the movement of the cell itself through the surrounding medium. Bacterial cells, as we noted on page 36, are able to swim toward a food source or away from a noxious chemical. They accomplish this by moving along a concentration gradient, from a lower concentration of a particular type of molecule to a higher one, or vice versa. Such directed movements in bacteria are extremely sensitive and highly specific: the bacteria can sense only certain molecules and can sense them at very low concentrations. The sensory abilities are due to receptor sites in the cell membrane that detect the molecules in question.

If you watch flagellated bacteria swimming freely, you will see two types of movement. When the flagella are rotating, they drive the cell through the water in much the same way a propeller drives a boat. When the flagella stop, the cell tumbles wildly for perhaps a tenth of a second. Then the propellerlike motion begins again, and the cell moves off in a new direction. When the concentration of chemicals in the water is uniform, the cell tumbles often, changing direction every time. By contrast, when the cell is moving along a gradient, there are fewer tumbles, so the cell continues longer in the same direction.

How do bacterial cells "decide" to move in a particular direction? How do they know there *is* a concentration gradient? For many years, the most widely held hypothesis was that a bacterial cell could detect the difference in concentration between its front end and its rear end. However, when Daniel E. Koshland, Jr., of the University of California, calculated the concentrations of molecules to which a cell could respond, he began to question this concept. A bacterial cell is so small that, in a gradient steep enough to produce a strong response, the difference in concentration between the front end of the cell and the rear end would be only on the order of one molecule in 10,000. Also, no gradient would be exactly uniform. In short, the analytical task confronting the cell on its journey would seem virtually impossible.

Koshland then formulated an alternative hypothesis and, more important, figured out a way to test between the two. Koshland's hypothesis was that the bacterial cells were making a comparison not in space —between their front end and their rear end—but in time—from one microsecond to the next as they moved along the gradient. In order to choose between the alternatives, he formulated an ingeniously simple experiment. Using a strain of the common bacterium *Salmonella,* Koshland set up an apparatus with which he could transfer cells almost instantaneously from one liquid medium to another and compare their motility. First, he put *Salmonella* in a medium that contained no chemical attractants. The cells exhibited their normal tumble-and-run pattern of behavior. He transferred them to a new medium, also containing no chemical attractants. They did not change their pattern of movement. This part of the experiment— known as a control—showed that moving the cells, by itself, did not affect their motility.

Next, in the crucial part of the experiment, he placed the bacteria in a medium containing a uniform concentration of the amino acid serine, an attractant for *Salmonella.* The cells behaved just as they had when no attractant was present. Then he transferred the cells to a medium with a slightly higher concentration of serine. There was an immediate change: for a few seconds, the cells ran more than they tumbled. Then he transferred them to a medium with a lower concentration of serine; for a few seconds, they tumbled more and ran less. In other words, although the bacteria were actually moving from one uniform concentration to another, they behaved as if they were moving up or down a gradient. Koshland had tricked the bacteria into revealing their secret and so was able to choose between the alternative hypotheses. The bacteria were analyzing differences in time, not space.

This experiment is a minilesson in how scientists go about their business. They formulate a testable hypothesis and then they challenge it. The test of the hypothesis can take the form of a clever, well-designed experiment, as in this example, of accumulated observations, or of the analysis of reports made by other observers. However, two components are always necessary: the testable hypothesis and the data with which to test it.

Of course, many questions remain. Exactly how do the receptor sites on the cell membrane recognize particular substances? How does the cell "remember" the concentration from one moment to the next? How does the sensory response (the detection of the chemical) trigger the motor response (the movement of the flagella)? Here again is a characteristic of the scientific process: the answer to one question nearly always raises still more questions.

Osmosis: A Special Case of Diffusion

A membrane that permits the passage of some substances, while blocking the passage of others, is said to be **selectively permeable.** The movement of water molecules through such a membrane is a special case of diffusion, known as **osmosis.** Osmosis results in a net transfer of water from a solution that has higher water potential to a solution that has lower water potential. In the absence of other factors that influence water potential (such as pressure), the movement of water in osmosis will be from a region of lower solute concentration (and therefore of higher water concentration) into a region of higher solute concentration (lower water concentration). The presence of solute decreases the water potential and so creates a gradient of water potential along which water diffuses.

The diffusion of water is not affected by *what* is dissolved in the water, only by *how much* is dissolved—that is, the concentration of particles of solute (molecules or ions) in the water. The word **isotonic** was coined to describe two or more solutions that have equal numbers of dissolved particles per unit volume and therefore the same water potential. There is no net movement of water across a membrane separating two solutions that are isotonic to one another, unless, of course, pressure is exerted on one side. In comparing solutions of different concentration, the solution that has less solute (and therefore a higher water potential) is known as **hypotonic,** and the one that has more solute (a lower water potential) is known as **hypertonic.** (Note that *iso* means "the same"; *hyper* means "more"—in this case, more particles of solute; and *hypo* means "less"—in this case, fewer particles of solute.) In osmosis, water molecules diffuse from a hypotonic solution (or from pure water), through a selectively permeable membrane, into a hypertonic solution (Table 6–1).

TABLE 6–1   **The Direction of Water Movement in Osmosis**

WATER MOVES ACROSS A SELECTIVELY PERMEABLE MEMBRANE FROM	TO
Region of higher water potential	Region of lower water potential
Higher water concentration	Lower water concentration
Lower solute concentration	Higher solute concentration
Hypotonic solution (less solute)	Hypertonic solution (more solute)
Region of lower osmotic potential	Region of higher osmotic potential

Osmosis and Living Organisms

The osmotic movement of water across the selectively permeable cell membrane causes some crucial problems for living systems. These problems vary according to whether the cell or organism is hypotonic, isotonic, or hypertonic in relation to its environment. One-celled organisms that live in the seas, for example, are usually isotonic with the salty medium they inhabit, which is one way of solving the problem. The cells of most marine invertebrates are also isotonic with sea water. Similarly, the cells of vertebrate animals are isotonic with the blood and lymph that constitute the watery medium in which they live.

Many types of cells, however, live in a hypotonic environment. In all single-celled organisms that live in fresh water, such as *Paramecium,* the interior of the cell is hypertonic to the surrounding water; consequently, water tends to move into the cell by osmosis. If too much water were to move into the cell, it could dilute the cell contents to the point of interfering with function and could even eventually rupture the cell membrane. This is prevented by a specialized organelle

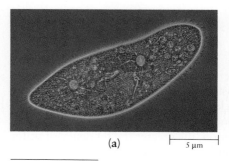

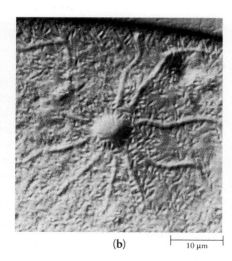

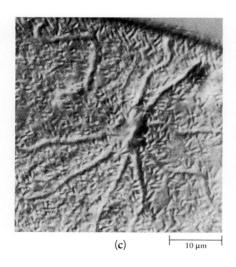

(a)   5 µm

(b)   10 µm

(c)   10 µm

6–6 *A* Paramecium *is hypertonic in relation to its environment, and hence water tends to move into the cell by osmosis. Excess water is expelled through its contractile vacuoles. (a) Phase-contrast micrograph of a living* Paramecium *showing the position of its two rosette-like contractile vacuoles. As revealed by the scanning electron microscope, (b) collecting tubules converge toward the vacuole, filling it. (c) Then it contracts, emptying outside the cell membrane by way of a small central pore. Actin filaments (page 121) are involved in the contraction of the vacuole.*

known as a contractile vacuole, which collects water from various parts of the cell and pumps it out with rhythmic contractions (Figure 6–6). As you might expect, this bulk transport process requires energy.

Osmotic Potential

The water potential on the two sides of a selectively permeable membrane will become equal if enough water moves from the hypotonic solution into the hypertonic solution to equalize the water concentrations (and therefore the solute concentrations)—that is, to make the solutions isotonic. If, however, physical barriers prevent the expansion of the hypertonic solution as water moves into it by osmosis, there will be increasing resistance as water molecules continue to move across the membrane. This resistance creates a buildup of pressure that gradually increases the water potential of the hypertonic solution, decreasing the gradient of water potential between the two solutions. As the pressure increases, the *net* flow of water molecules will slow and then cease as the gradient of water potential disappears. (Individual water molecules, of course, continue to move back and forth across the membrane, but these movements are in equilibrium and there is no net movement of water.) The pressure that is required to stop the osmotic movement of water into a solution is called the osmotic pressure. It is a measure of the **osmotic potential** of the solution—that is, of the tendency of water to move across a membrane into the solution.

The measurement of the osmotic potential of a solution is illustrated in Figure 6–7. The beaker contains distilled water, and within the tube is the solution. Across the mouth of the tube is a selectively permeable membrane; such a membrane is freely permeable to water but not to the particles (ions or molecules) in the solution. Water moving from the beaker through the membrane into the solution causes the solution to rise in the tube. As it does so, the pressure created by the force of gravity acting on the column of solution gradually increases, thus increasing the water potential of the solution. The solution rises in the tube until

6–7 *Osmosis and the measurement of osmotic potential. (a) The tube contains a solution and the beaker contains distilled water. (b) The selectively permeable membrane permits the passage of water but not of solute. The diffusion of water into the solution causes the solution to rise in the tube until the tendency of water to move into a region of lower water concentration is counterbalanced by the pressure resulting from the force of gravity acting on the column of solution. This hydrostatic (water-stopping) pressure is proportional to the height, h, and density of the column of solution. (c) The pressure that must be applied to the piston to force the column of solution back to the level of the water in the beaker provides a quantitative measure of the osmotic potential of the solution—that is, of the tendency of water to diffuse across a membrane into the solution.*

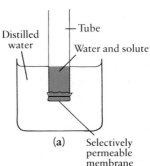

Distilled water

Tube

Water and solute

(a)   Selectively permeable membrane

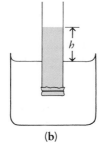

h

(b)

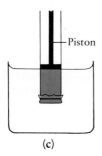

Piston

(c)

equilibrium is reached—that is, until the water potential on both sides of the membrane is equal. The amount of pressure that must then be applied to the piston to force the solution in the tube back to the level of the water in the beaker provides a quantitative measure of the osmotic potential.

The lower the water potential of a solution, the greater the tendency of water molecules to move into it by osmosis and, therefore, the greater its osmotic potential. Since solutes decrease the water potential of a solution, a higher solute concentration means a greater osmotic potential.

As we shall see in Chapter 37, the osmotic potential of solutions within the vertebrate kidney plays an important part in determining the composition of the excreted urine.

Turgor

Plant cells are usually hypertonic to their surrounding environment, and so water tends to diffuse into them. This movement of water into the cell creates pressure within the cell against the cell wall. The pressure causes the cell wall to expand and the cell to enlarge. The elongation that occurs as a plant cell matures (see Figure 5–7b, page 107) is a direct result of the osmotic movement of water into the cell.

As the plant cell matures, the cell wall stops growing. Moreover, mature plant cells typically have large central vacuoles that contain solutions of salts and other materials. (In citrus fruits, for example, they contain the acids that give the fruits their characteristic sour taste.) Because of these concentrated solutions, plant cells have a high osmotic potential—that is, water has a strong tendency to move into the cells. In the mature cell, however, the cell wall does not expand further. Its resistance to expansion results in an inward directed pressure, analogous to the pressure exerted by the depressed piston in Figure 6–7. This pressure, known as the wall pressure, prevents the net movement of additional water into the cell. Consequently, equilibrium of water concentration is not reached and water continues to "try" to move into the cell, maintaining a constant pressure on the cell wall from the inside (Figure 6–8). This internal pressure on the cell wall is known as **turgor,** and it keeps the cell walls stiff and the plant body crisp. When turgor is reduced, as a consequence of water loss, the plant wilts.

CARRIER-ASSISTED TRANSPORT

As we noted in Chapter 2, water and other polar or charged (hydrophilic) molecules exclude lipids and other hydrophobic molecules. Conversely, hydrophobic molecules exclude hydrophilic ones. This behavior of molecules, determined by the presence or absence of polar or charged regions, is of fundamental importance in the capacity of cellular membranes to regulate the passage of materials into and out of cells and organelles. As we have seen, cell membranes are formed from a lipid bilayer, the interior of which is filled by the hydrophobic tails of the lipid molecules. This interior lipid sea is a formidable barrier to ions and most hydrophilic molecules, but it does allow easy passage of hydrophobic molecules, such as steroid hormones. (It was, in fact, the observation that hydrophobic molecules diffuse readily across cell membranes that provided the first evidence of the lipid nature of the membrane.)

Most organic molecules of biological importance, however, have polar functional groups and are therefore hydrophilic; unlike carbon dioxide, oxygen, and water, they cannot move freely through the lipid barrier by simple diffusion. Similarly, the ions that are of crucial importance in the life of the cell cannot diffuse through the membrane. Although individual ions, such as sodium (Na^+) and chloride (Cl^-) are quite small, in aqueous solution they are surrounded by water molecules (see Figure 2–8, page 45); both the size of the resulting aggrega-

Movement of water
Solutes

Cell wall   Cytoplasm   Vacuole

(a)

(b)

(c)

6–8 (a) *A turgid plant cell. The central vacuole is hypertonic in relation to the fluid surrounding it and so gains water. The expansion of the cell is held in check by the cell wall.* (b) *A plant cell begins to wilt if it is placed in an isotonic solution, so that water pressure no longer builds up within the vacuole.* (c) *A plant cell in a hypertonic solution loses water to the surrounding fluid and so collapses, with its membrane pulling away from the cell wall. Such a cell is said to be plasmolyzed.*

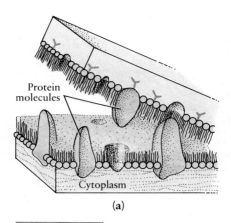

Protein molecules

Cytoplasm

(a)

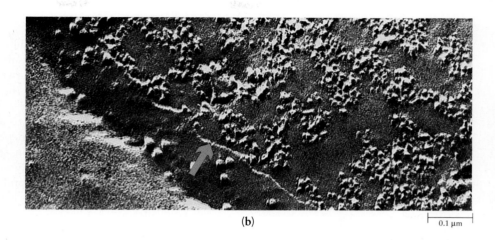

(b)

0.1 μm

6–9 (a) *The interior surface of cellular membranes can be revealed when specimens are prepared for the electron microscope by the freeze-fracture technique. In this procedure, the cell is frozen and then fractured with a sharp blow. The fracture line generally runs between the two lipid layers of a membrane, exposing the embedded proteins. A replica of the exposed surface is prepared for examination, using the procedures described on page 98.*

(b) The interior surface of the membrane of a red blood cell, prepared by the freeze-fracture technique. The arrow indicates the edge of the fracture line. The numerous particulate structures visible in the micrograph are integral membrane proteins, many of which are thought to be transport proteins.

tions and their charges prevent ions from slipping through the apertures that allow the passage of water molecules (see page 130). Transport of these aggregations and of all but the very smallest hydrophilic molecules depends upon integral membrane proteins that act as carriers, ferrying molecules back and forth.

The transport proteins of cell and organelle membranes are highly selective; a particular protein may accept one molecule while it excludes a nearly identical one. It is the configuration of the protein molecule—that is, its tertiary or, in some cases, quaternary structure—that determines what molecules it can transport. Although the protein typically undergoes temporary changes in configuration in the course of the transport process, it is not permanently altered. As we shall see in Chapter 8, enzymes are also highly selective in the molecules with which they interact and are not permanently altered by those interactions. Enzymes are often given names that end in "-ase," and to emphasize the similarities, transport proteins are sometimes referred to as permeases. Unlike enzymes, however, transport proteins do not necessarily produce chemical change in the molecules with which they interact.

Some transport proteins can move substances across the membrane only if there is a favorable concentration gradient; such carrier-assisted transport is known as **facilitated diffusion**. Other proteins can move molecules against a concentration gradient, a process known as **active transport**. Facilitated diffusion, like the simple diffusion discussed earlier, is a passive process, requiring no energy outlay by the cell; active transport, by contrast, requires the expenditure of cellular energy (Figure 6–10).

6–10 *Modes of transport through the cell membrane. In simple diffusion and facilitated diffusion, molecules or ions move down a concentration gradient. The potential energy of the concentration gradient drives these processes, which are, from the standpoint of the cell, passive. In active transport, by contrast, molecules or ions are moved against a concentration gradient. Energy, released by cellular reactions, is required to power active transport. Both facilitated diffusion and active transport require the presence of integral membrane proteins, specific for the substance being transported.*

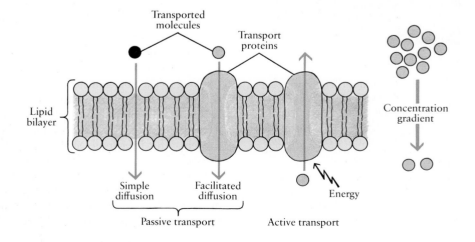

Transported molecules

Transport proteins

Lipid bilayer

Simple diffusion

Facilitated diffusion

Energy

Concentration gradient

Passive transport

Active transport

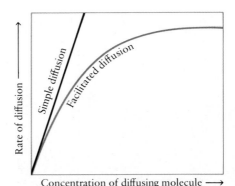

6–11 *A comparison of the rates of simple diffusion and facilitated diffusion across the cell membrane. In simple diffusion, the rate increases steadily as the concentration of diffusing molecules (or ions) increases. In facilitated diffusion, by contrast, the rate increases only as long as there are additional transport protein molecules, specific for the diffusing substance, available. When all of the protein molecules are in use, the rate levels off and does not increase further.*

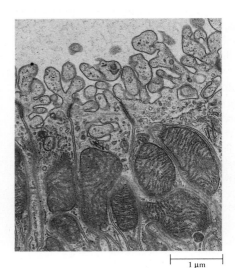

6–12 *Mitochondria clustered near the surface of kidney cells of a bat. These cells are concerned with pumping out sodium ions against a concentration gradient. The mitochondria provide the energy for this active-transport process.*

Facilitated Diffusion

Both facilitated diffusion and simple diffusion are driven by the potential energy of a concentration gradient. Molecules move down the gradient from a region of higher concentration to a region of lower concentration. Ions and hydrophilic molecules, however, can move across the lipid barrier of a cell membrane, from the region of higher concentration to the region of lower concentration, only if a specific transport protein is available to allow them passage. The rate at which they can diffuse across the membrane depends not only upon the steepness of the concentration gradient but also upon the number of their specific transport protein molecules present in the membrane.

Glucose, for example, is a hydrophilic molecule that enters most cells by facilitated diffusion. Because glucose is rapidly broken down when it enters a cell, a steep concentration gradient is maintained between the inside and outside. However, when very large numbers of glucose molecules are present in the surrounding medium, the rate of entry does not increase beyond a certain point; it reaches a peak and then remains steady at that level (Figure 6–11). This limitation on the rate of entry is a result of the limited number of specific glucose-transporting protein molecules in the cell membrane.

Active Transport

Depending on the direction of the concentration gradient, a molecule may be transported across the cell membrane by either facilitated diffusion or active transport. In active transport, molecules or ions are moved against a concentration gradient, an energy-requiring process analogous to pushing a boulder up a hill. For example, glucose is transported into liver cells, where it is stored as glycogen, even though the concentration of glucose is higher inside the liver cells than in the bloodstream. This active-transport process presumably involves different membrane proteins than those used in facilitated diffusion.

The Sodium-Potassium Pump

One of the most important and best-understood active-transport systems is the sodium-potassium pump. Most cells maintain a differential concentration gradient of sodium ions (Na^+) and potassium ions (K^+) across the cell membrane: Na^+ is maintained at a lower concentration inside the cell, and K^+ is kept at a higher concentration. This concentration gradient is exploited by nerve cells to propagate electrical impulses, as you will learn in Chapter 41. The sodium-potassium pump requires energy made available by a molecule known as ATP (adenosine triphosphate), which is the form in which most of a cell's ready energy currency is carried. A measure of the importance of the sodium-potassium pump to the organism is that more than a third of the ATP used by a resting animal is consumed by this one ion-pumping mechanism.

The pumping of Na^+ and K^+ ions is accomplished by a transport protein thought to exist in two alternative configurations. One configuration has a cavity opening to the inside of the cell, into which Na^+ ions can fit; the other has a cavity opening to the outside, into which K^+ ions fit. As shown in Figure 6–13, Na^+ within the cell binds to the transport protein. At the same time, an energy-releasing reaction involving ATP results in the attachment of a phosphate group to the protein. This triggers its shift to the alternative configuration and the release of the Na^+ to the outside of the membrane. The transport protein is now ready to pick up K^+, which results in the release of the phosphate group from the protein, thus causing it to return to the first configuration and to release the K^+ to the inside of the cell. As you can see, this process will generate a gradient of Na^+ and K^+ ions across the membrane.

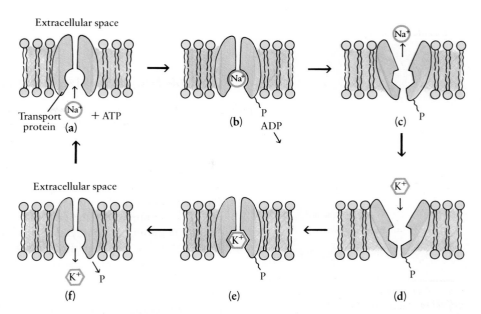

6–13 *A model of the sodium-potassium pump.* (a) *An Na$^+$ ion in the cytoplasm fits precisely into the transport protein.* (b) *A chemical reaction involving ATP then attaches a phosphate group (P) to the protein, releasing ADP (adenosine diphosphate). This process results in* (c) *a change of shape that causes the Na$^+$ to be released outside the cell.* (d) *A K$^+$ ion in the extracellular space is bound to the transport protein* (e), *which in this form provides a better fit for K$^+$ than for Na$^+$.* (f) *The phosphate group is then released from the protein, inducing conversion back to the other shape, and the K$^+$ ion is released into the cytoplasm. The protein is now ready once more to transport Na$^+$ out of the cell.*

For clarity, only single ions are shown in this diagram. Quantitative studies have shown, however, that each complete pumping sequence transports three Na$^+$ ions out of the cell and two K$^+$ ions into the cell.

6–14 *Three types of transport molecules. In the simplest, known as a uniport, one particular solute is moved directly across the membrane in one direction. In the type of cotransport system known as a symport, two different solutes are moved across the membrane, simultaneously and in the same direction. Often, a concentration gradient involving one of the transported solutes powers the transport of the other solute; for example, a concentration gradient of Na$^+$ ions frequently powers cotransport of glucose molecules. In another type of cotransport system, known as an antiport, two different solutes are moved across the membrane, either simultaneously or sequentially, in opposite directions. The Na$^+$–K$^+$ pump is an example of a cotransport system involving an antiport.*

Types of Transport Molecules

Many ingenious models have been proposed to show how transport proteins, such as the sodium-potassium pump, might accept and eject their passengers. One of the first suggested that the protein rotated, like a revolving door. More recent evidence indicates that although membrane proteins can move laterally, they are not free to flip-flop through the membrane as that model required. A current model hypothesizes that transport proteins have hydrophilic cores, which the transported molecules are squeezed through, propelled by changes in the configuration of the protein. These changes in configuration may be triggered directly, by the binding of the molecule to be transported, or indirectly, by the interaction of a receptor on the protein surface with some other molecule or ion that is not actually transported through the membrane. The cell membrane of most nerve cells, for example, contains a complex protein molecule that is a receptor for a molecule known as acetylcholine. When acetylcholine binds to this receptor, a channel is opened in another protein molecule, closely associated with the acetylcholine receptor; this channel allows sodium ions (Na$^+$) to flow into the cell, down the concentration gradient created and maintained by the action of the sodium-potassium pump.

Current evidence indicates that there are at least three general types of transport proteins (Figure 6–14). The simplest transfers one particular kind of molecule or ion directly across the membrane. More complex proteins function as cotransport systems, in which the transport of a particular molecule or ion depends upon the simultaneous or sequential transport of a different molecule or ion. In some cotransport systems, both solutes are transported in the same direction; in others, exemplified by the sodium-potassium pump, the two different solutes are transported in opposite directions.

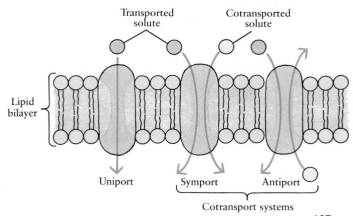

137

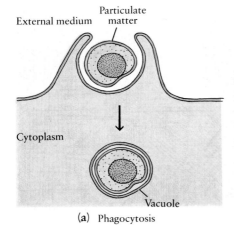

External medium — Particulate matter

Cytoplasm

Vacuole

(a) Phagocytosis

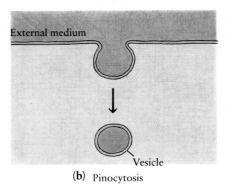

External medium

Vesicle

(b) Pinocytosis

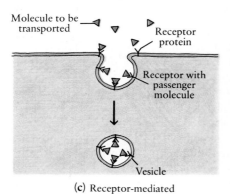

Molecule to be transported

Receptor protein

Receptor with passenger molecule

Vesicle

(c) Receptor-mediated endocytosis

6–15 *Three types of endocytosis.* (a) *In phagocytosis, contact between the cell membrane and particulate matter causes the cell membrane to extend around the particle, engulfing it in a vacuole. If the particle is a food item, lysosomes fuse with the vacuole, spilling their digestive enzymes into it.* (b) *In pinocytosis, the cell membrane pouches inward, forming a vesicle around liquid from the external medium that is to be taken into the cell.* (c) *In receptor-mediated endocytosis, the substances to be transported into the cell must first bind to specific receptor molecules. The receptors are either localized in indented areas of the cell membrane, known as pits, or migrate to such areas after binding the molecules to be transported. When filled with receptors carrying their particular substance, the pit buds off as a vesicle.*

VESICLE-MEDIATED TRANSPORT

Although crossing the cell membrane, with or without the assistance of transport proteins, is one of the principal ways substances get into and out of the cell, it is not the only way. Another type of transport process involves vesicles or vacuoles that are formed from or that fuse with the cell membrane. For example, many substances are exported from cells in vesicles formed by the Golgi complexes. As we saw in Figure 5–21 (page 118), vesicles move from the Golgi complexes to the surface of the cell. When a vesicle reaches the cell surface, its membrane fuses with the membrane of the cell, thus expelling its contents to the outside. This process is known as **exocytosis.** Transport by means of vesicles or vacuoles can also work in the opposite direction. In **endocytosis,** material to be taken into the cell induces the membrane to bulge inward, producing a vesicle enclosing the substance. This vesicle is released into the cytoplasm. Three different forms of endocytosis are known: **phagocytosis** ("cell-eating"), **pinocytosis** ("cell-drinking"), and **receptor-mediated endocytosis** (Figure 6–15).

When the substance to be taken into the cell in endocytosis is a solid, such as a bacterial cell, the process is usually called phagocytosis. Many heterotrophic protists, such as amoebas, feed in this way; similarly, macrophages (Figure 5–28, page 121) and other types of white blood cells in our own bloodstreams engulf bacteria and other invaders in phagocytic vacuoles. Often lysosomes fuse with these vacuoles, emptying their enzymes into them and so digesting or destroying their contents.

The taking in of liquids, as distinct from particulate matter, is given the special name of pinocytosis, although it is the same in principle as phagocytosis. Pinocytosis occurs not only in single-celled organisms but also in multicellular animals. One type of cell in which it has been frequently observed is the human egg cell. As the egg cell matures in the ovary of the female, it is surrounded by "nurse cells." These cells apparently transmit dissolved nutrients to the egg cell, which takes them in by pinocytosis.

In receptor-mediated endocytosis, currently the subject of a great deal of research, particular membrane proteins serve as receptors for specific molecules that are to be transported into the cell. Cholesterol, for example, is carried into animal cells by receptor-mediated endocytosis. As we noted earlier (page 71), cholesterol circulates in the bloodstream in the form of LDL particles, which interact with specific receptors on the cell surface. Binding of LDL particles to the receptor molecules triggers the formation of a vesicle that transports the cholesterol molecules into the cell. The protein forming the LDL receptor is a complex molecule that includes three different functional regions: a large segment that projects outside the cell, to which the LDL particles bind; a single, short segment that crosses the membrane; and a "tail" segment of about 50 amino acids that projects into the cytoplasm of the cell.

Membrane receptors for some substances, such as the hormone insulin, are apparently free to move laterally in the membrane and, when unoccupied, are scattered at random locations on its surface. As the molecules to be transported into the cell bind to the receptors, the receptors move close together. A vesicle forms, and the hormone-laden receptors are carried into the cell. Receptors for other substances, such as LDL particles, appear to be localized in groups in specific areas of the cell membrane even before binding of the substance to be transported. The segment of the LDL receptor that projects into the cytoplasm is thought to play a key role in this clustering of the receptors.

In the areas where specific receptors are localized—or to which they migrate, as in the case of insulin receptors—the inner, or cytoplasmic, face of the cell membrane is characterized by a peripheral membrane protein known as clathrin. These areas, which are slightly indented, are known as coated pits. The vesicles that form from them, containing receptor molecules and their passengers, thus

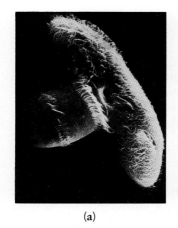
(a)

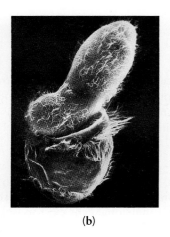
(b)

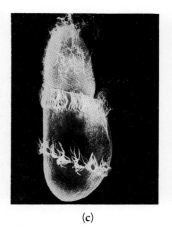
(c)

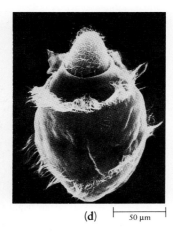
(d)

|—| 50 μm

6–16 *Phagocytosis of* Paramecium *by* Didinium. *(See Figure 5–30, on page 122, for the preamble to their encounter.)* (**a**) *Ingestion of the* Paramecium *has begun. The concave area just above the oral rim of the* Didinium *is the oral groove of the* Paramecium. Paramecium, *a heterotroph, feeds largely on bacteria.* (**b**) *Because the* Paramecium *is larger than the* Didinium, *folding helps.* (**c**) *The* Paramecium *is half*

"swallowed"; the part that is within the Didinium *is surrounded by a membrane composed of the cell membrane of the* Didinium. *The process of compression has begun, as you can see in the tip of the* Paramecium *protruding from the oral rim of the* Didinium. *This compression is largely a matter of squeezing out water.* (**d**) *Once the* Paramecium *is completely*

inside, the cell membrane of the Didinium *will fuse over it, forming a food vacuole. The* Paramecium, *however, must provide the means for its own demolition: the* Didinium *apparently lacks a crucial digestive enzyme that the* Paramecium *supplies. A* Didinium *can eat a dozen* Paramecium, *each larger than itself, in a single day.*

acquire an external, cagelike coating of clathrin. The formation of such a coated vesicle is illustrated in Figure 6–17.

As you can see by studying Figures 6–15 and 6–17, the surface of the membrane facing the interior of a vesicle or vacuole is equivalent to the surface facing the exterior of the cell; similarly, the surface of the vesicle or vacuole membrane facing the cytoplasm is equivalent to the cytoplasmic surface of the cell membrane. As we noted in the last chapter, new material needed for expansion of the cell membrane is transported, ready-made, from the Golgi complexes to the

6–17 *The formation of a coated vesicle in the developing egg cell of a hen. The vesicles are much larger than those of smaller cells and are thus easier to visualize.* (**a**) *A coated pit in the cell membrane is covered on the cytoplasmic face with a latticework of clathrin molecules. The large particles clustered in the shallow pit on the external face are lipoprotein molecules, gathered from the surrounding medium and bound to specific membrane receptors that are associated with the underlying clathrin layer.* (**b**) *The pit deepens, and then* (**c**) *the cell membrane closes around the pit to form the vesicle.* (**d**) *The completed vesicle with its outer coating of clathrin buds off and moves into the cell. The lipoproteins carried by this coated vesicle will be incorporated into the egg yolk.*

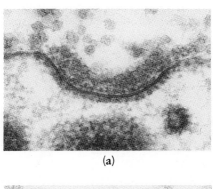
(a)

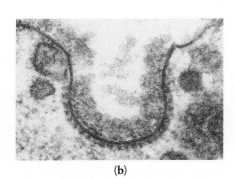
(b)

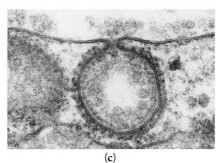
(c)

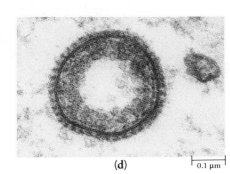
(d)

|—| 0.1 μm

139

membrane by a process similar to exocytosis. Current evidence indicates that the portions of the cell membrane used in forming endocytic vesicles or vacuoles are also returned to the membrane in exocytosis, thus recycling the membrane lipids and proteins, including the specific receptor molecules.

CELL-CELL JUNCTIONS

Thus far in our consideration of the transport of substances into and out of cells, we have assumed that individual cells exist in isolation, surrounded by a watery environment. In multicellular organisms, however, this is generally not the case. Cells are organized into **tissues,** groups of specialized cells with common functions. For example, in animals, the four principal types of tissues are muscle, nerve, connective, and epithelial (covering). Tissues are further organized in concert to form **organs,** such as the heart, brain, or kidney, each of which, like a subcellular organelle, has a design that suits it for a specific function.

As you might imagine, in multicellular organisms it is essential that individual cells communicate with one another so that they can collaborate to create a harmonious tissue or organ. Nerve impulses are transmitted from neuron to neuron, or from neuron to muscle or gland. Cells in the body of a plant or animal release hormones that travel over a distance and affect other cells in the same organism. In the course of development, embryonic cells influence the differentiation of neighboring cells into tissues and organs. These communications are all accomplished by means of chemical signals—that is, by substances that are transported out of one cell and travel to another cell. When they reach the cell membrane of the target cell, they may actually be transported into the cell, by any one of the processes we have considered, or they may bind to specific membrane receptors at the surface of the target cell, thereby triggering chemical reactions within the cell.

Often, however, the cells within a tissue or organ are tightly packed, allowing for direct and intimate contacts of various types between the cells. Among plant cells, which are separated from one another by cell walls, channels called **plasmodesmata** traverse the walls, directly connecting the cytoplasm of adjacent cells (Figure 6–18). Plasmodesmata, which are from 30 to 60 nanometers in diameter, appear to be lined by the cell membrane; in addition, they generally contain tubular extensions of the endoplasmic reticulum known as desmotubules.

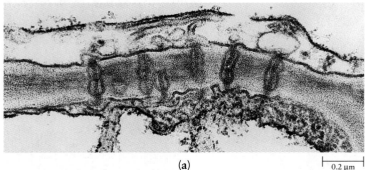

(a) |—————| 0.2 μm

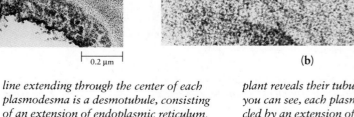

(b) |———| 50 nm

6–18 **(a)** *Plasmodesmata connecting two leaf cells from a corn plant. The wide gray area running horizontally in this electron micrograph is the cell wall, which is traversed by the plasmodesmata. The dark line extending through the center of each plasmodesma is a desmotubule, consisting of an extension of endoplasmic reticulum.* **(b)** *A cross section through the plasmodesmata connecting root cells in a lettuce plant reveals their tubular structure. As you can see, each plasmodesma is encircled by an extension of the cell membrane and contains a desmotubule in its center.*

Communication in the Cellular Slime Mold

A cellular communication system of particular interest to biologists, because of the comparative ease with which it can be studied, is seen among a group of organisms known as the cellular slime molds. The slime mold *Dictyostelium discoideum* is an example. At one stage in its life cycle, it exists as a swarm of small individual amoebas (a), which divide and grow and feed, amoeba-fashion, until their food supply (mostly bacteria) gives out. At this point, the cells alter both their shape and behavior: they become sausage-shaped (b) and begin to migrate toward the center of the group. (The direction in which the stream is moving is indicated by the arrow.) Eventually, they pile up in a heap; the heap gradually takes on the form of a multicellular mass somewhat resembling a garden slug (c), slowly migrating and depositing a thick slime sheath that collapses behind it. The sluglike mass soon stops its migration, gathers itself into a mound, and sends up a long stalk at the tip of which a fruiting

body forms (d, e, f). The fruiting body matures (g) and eventually bursts open, releasing a new swarm of tiny amoebas, and the cycle begins again.

The chemical that spreads from cell to cell to initiate this remarkable sequence of events was first called acrasin, after Acrasia, the cruel witch in Spenser's *Faerie Queene* who attracted men and turned them into beasts. Acrasin was later identified as the chemical compound cyclic AMP (adenosine monophosphate). In recent years, it has become clear that many of the communications among cells in the human body also involve cyclic AMP. As we shall see in Chapter 40, the interaction of a number of vertebrate hormones (principally hormones that are proteins or amino acid derivatives) with their specific receptors in the cell membrane triggers a sequence of events within the cell in which cyclic AMP plays a central role.

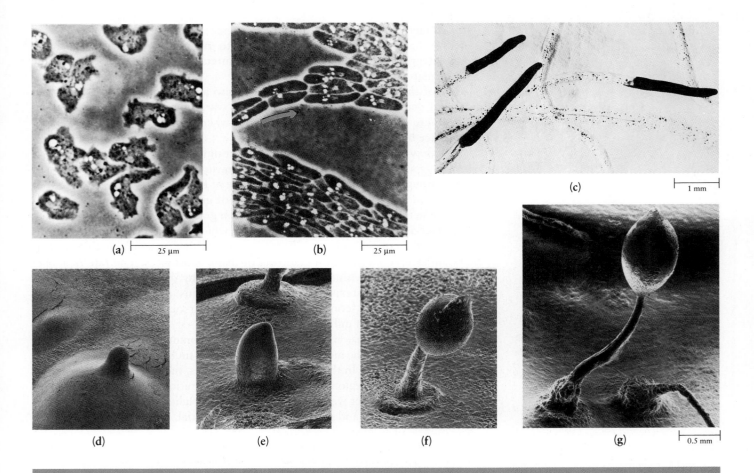

(a) ├─── 25 μm ───┤   (b) ├─── 25 μm ───┤   (c) ├─── 1 mm ───┤

(d)   (e)   (f)   (g) ├─── 0.5 mm ───┤

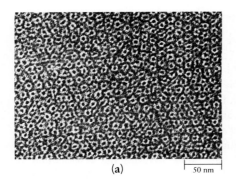

(a)   50 nm

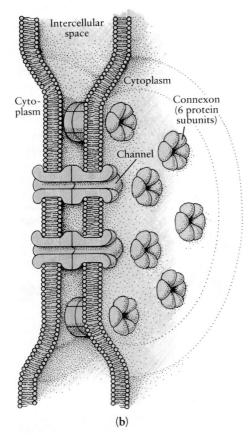

Intercellular space

Cytoplasm

Cyto-plasm

Connexon (6 protein subunits)

Channel

(b)

6–19 (a) *A portion of a gap junction between two liver cells, as seen in cross section. In this electron micrograph, the narrow channels through which small molecules can pass between the cells are filled with an electron-opaque staining substance and appear as black dots. (b) A model of a gap junction. Embedded in the cell membranes are structures that have been appropriately named "connexons." Each connexon consists of six identical membrane protein subunits, arranged in a hexagonal pattern with a space through the center. Connexons in the adjacent cell membranes abut each other in perfect alignment, providing a channel connecting the cytoplasm of the two cells.*

In animal tissues, structures known as **gap junctions** permit the passage of materials between cells. These junctions appear as fixed clusters of very small channels (about 2 nanometers in diameter) surrounded by an ordered array of proteins (Figure 6–19). Experiments with radioactively labeled molecules have shown that small messenger molecules pass through these channels. Gap junctions also serve to transmit electrical signals in the form of ions. For example, the contractions of muscle cells in the heart are synchronized by the flow of sodium ions (Na^+) through gap junctions.

The transport of materials into and out of cells through the channels of plasmodesmata or gap junctions, through integral membrane proteins, and by means of endocytosis and exocytosis appear superficially to be three quite different processes. They are fundamentally similar, however, in that they all depend on the precise, three-dimensional structure of a great variety of specific protein molecules. These protein molecules not only form channels through which transport can occur but also endow the cell membrane with the capacity to "recognize" particular molecules. This capacity is the result of billions of years of an evolutionary process that began, as far as we are able to discern, with the formation of a fragile film around a few organic molecules. This film separated the molecules from their external environment and permitted them to maintain the particular kind of organization that we recognize as life. It is one of the many critical capacities transmitted from parent to offspring each time a cell divides, a process we shall examine in the next chapter.

SUMMARY

The cell membrane regulates the passage of materials into and out of the cell, a function that makes it possible for the cell to maintain its structural and functional integrity. This regulation depends on interactions between the membrane and the materials that pass through it.

One of the principal substances passing into and out of cells is water. Water potential determines the direction in which water moves; that is, water moves from where the water potential is higher to where it is lower. Water movement takes place by bulk flow and diffusion.

Bulk flow is the overall movement of water molecules and dissolved solutes as a group, as when water flows in response to gravity or pressure. The circulation of blood through the human body is an example of bulk flow.

Diffusion involves the random movement of individual molecules or ions and results in net movement down a concentration gradient. It is most efficient when the surface area is large in relation to volume, when the distance involved is short, and when the concentration gradient is steep. By their metabolic activities, cells maintain steep concentration gradients of many substances. The rate of movement of substances within cells is also increased by cytoplasmic streaming. In several anatomical systems of multicellular organisms, countercurrent exchange maintains a constant concentration gradient over a large surface area, thus maximizing the rate of diffusion.

Osmosis is the diffusion of water through a membrane that permits the passage of water but inhibits the movement of most solutes; such a membrane is said to be selectively permeable. In the absence of other forces, the net movement of water in osmosis is from a region of lower solute concentration (a hypotonic medium), and therefore of higher water potential, to one of higher solute concentration (a hypertonic medium), and so of lower water potential. Turgor in plant cells is a consequence of osmosis.

Molecules cross the cell membrane by simple diffusion or are transported by carrier proteins embedded in the membrane. If carrier-assisted transport is driven by the concentration gradient, the process is known as facilitated diffusion. If the transport requires the expenditure of energy by the cell, it is known as active transport. Active transport can move substances against their concentration gradients. One of the most important active-transport systems is the sodium-potassium pump, which maintains sodium ions at a relatively low concentration and potassium ions at a relatively high concentration in the cytoplasm.

Controlled movement into and out of a cell may also occur by endocytosis or exocytosis, in which the substances are transported in vacuoles or vesicles composed of portions of the cell membrane. Three forms of endocytosis are phagocytosis, in which solid particles are taken into the cell; pinocytosis, in which liquids are taken in; and receptor-mediated endocytosis, in which molecules or ions to be transported into the cell are bound to specific receptors in the cell membrane.

In multicellular organisms, communication among cells is essential for coordination of the different activities of the cells in the various tissues and organs. Much of this communication is accomplished by chemical agents that either pass through the cell membrane or interact with receptors in its surface. Communication may also occur directly, through the channels of plasmodesmata (in plant tissues) or gap junctions (in animal tissues).

QUESTIONS

1. Distinguish among the following: bulk flow/diffusion/osmosis; water potential/hydrostatic pressure/osmotic potential; hypotonic/hypertonic/isotonic; endocytosis/exocytosis; phagocytosis/pinocytosis/receptor-mediated endocytosis; coated pit/coated vesicle; plasmodesmata/gap junctions.

2. What is a concentration gradient? How does a concentration gradient affect diffusion? How does a concentration gradient affect osmosis?

3. When diffusion of dye molecules in a tank of water is complete, random movement of molecules continues (as long as the temperature remains the same). However, net movement stops. How do you reconcile these two facts?

4. Why is diffusion more rapid in gases than in liquids? Why is it more rapid at higher temperatures than at lower temperatures?

5. Three funnels have been placed in a beaker containing a solution (see the figure below). What is the concentration of the solution? Explain your answer.

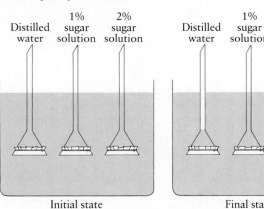

Initial state

Final state

| Distilled water | 1% sugar solution | 2% sugar solution |

6. How does countercurrent exchange affect diffusion?

7. Imagine a pouch with a selectively permeable membrane containing a saltwater solution. It is immersed in a dish of fresh water. Which way will the water move? If you add salt to the water in the dish, how will this affect water movement? What living systems exist under analogous conditions? How do you think they maintain water balance?

8. When you forget to water your house plants, they wilt and the leaves (and sometimes the stems) become very limp. What has happened to the plants to cause this change in appearance and texture? Within a few hours after you remember to water your plants, they resume their normal, healthy appearance. What has occurred within the plants to cause this restoration? Sometimes, if you wait too long to water your plants, they never revive. What do you suppose has happened?

9. What limits the passage of water and other polar molecules and ions through the cell membrane? How do such molecules get into and out of the cell? Describe four possible routes.

10. In what three ways does active transport differ from simple diffusion? How does it differ from facilitated diffusion?

11. Justify the conclusion that differences in ion concentration between cells and their surroundings (see Figure 6–1) indicate that cells regulate the passage of materials across membranes.

12. In Figure 6–16d, the *Paramecium* is sinking below the oral rim of *Didinium*. What will happen next? Complete the scenario, giving as many details as possible. (You might want to end your account with the fact that *Didinium* divides once for every two *Paramecium* consumed.)

How Cells Divide

Cell division is the process by which cellular material is divided between two new daughter cells. In one-celled organisms, it increases the number of individuals in the population. In many-celled plants and animals, it is the means by which the organism grows, starting from one single cell, and also by which injured or worn-out tissues are replaced and repaired. An individual cell grows by assimilating materials from its environment and synthesizing these materials into new structural and functional molecules. When a cell reaches a certain critical size and metabolic state, it divides. The two daughter cells, each of which has received about half of the mass of the parent cell, then begin growing again. A bacterial cell may divide every six minutes. In a one-celled eukaryote such as *Paramecium*, cell division may occur every few hours.

The new cells produced are structurally and functionally similar both to the parent cell and to one another. They are similar, in part, because each new cell usually receives about half of the parent cell's cytoplasm and organelles. More important, in terms of structure and function, each new cell inherits an exact replica of the hereditary information of the parent cell.

7-1 *Single-celled eukaryotic organisms typically reproduce by simple cell division. In this scanning electron micrograph of the ciliated protist* Opisthonecta, *the separation of the two daughter cells is almost complete. Each cell has received not only an exact replica of the parent cell's hereditary information but also approximately half of its organelles and cytoplasm.*

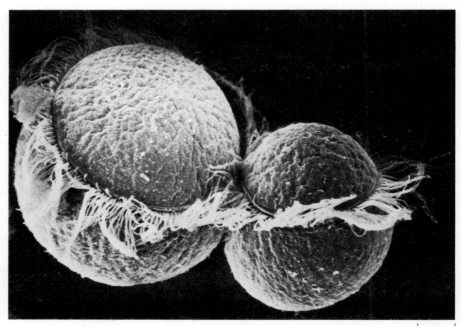

10 μm

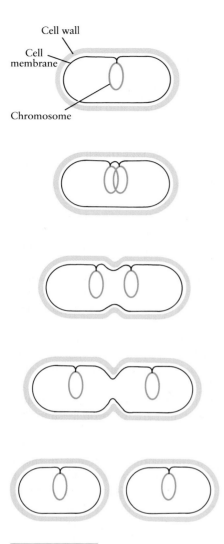

Cell wall

Cell membrane

Chromosome

7–2 *Schematic diagram of cell division in a bacterium. Attachment of the chromosome to an inward fold of the cell membrane ensures that one chromosome replicate is distributed to each daughter cell as the cell membrane elongates.*

CELL DIVISION IN PROKARYOTES

The distribution of exact replicas of the hereditary information is comparatively simple in prokaryotic cells. In such cells, most of the hereditary material is in the form of a single, long, circular molecule of DNA, with which a variety of proteins are associated. This molecule, the cell's chromosome, is replicated before cell division. According to present evidence, each of the two daughter chromosomes is attached to a different spot on the interior of the cell membrane. As the membrane elongates, the chromosomes move apart (Figure 7–2). When the cell has approximately doubled in size and the chromosomes are separated, the cell membrane pinches inward, and a new cell wall forms that separates the two new cells and their chromosome replicas.

The prokaryotic chromosome has been the subject of an enormous amount of research. In Section 3, we shall consider this marvelous structure, its replication and its functions, in more detail.

CELL DIVISION IN EUKARYOTES

In eukaryotic cells, the problem of exactly dividing the genetic material is much more complex. A typical eukaryotic cell contains about a thousand times more DNA than a prokaryotic cell, and this DNA is linear, forming a number of distinct chromosomes. For instance, human somatic ("body") cells have 46 chromosomes, each different from the others; when these cells divide, each daughter cell has to receive one copy, and only one copy, of each of the 46 chromosomes. Moreover, as we have seen, eukaryotic cells contain a variety of organelles, and these must also be apportioned between the daughter cells.

The solutions to these problems are, as you will see, ingenious and elaborate. In a series of steps called, collectively, **mitosis,** a complete set of chromosomes is allocated to each of two daughter nuclei. Mitosis is usually followed by **cytokinesis,** a process that divides the cell into two new cells, each of which contains not only a nucleus with a full chromosome complement but also approximately half of the cytoplasm and organelles of the parent cell.

Although mitosis and cytokinesis are the culminating events of cell division in eukaryotes, they represent only two stages of a larger process.

7–3 *A human cell divides. The long, dark bodies are chromosomes, the carriers of the hereditary information. The chromosomes have replicated and moved apart. Each set of chromosomes is an exact copy of the other. Thus the two new cells, the daughter cells, will contain the same hereditary material.*

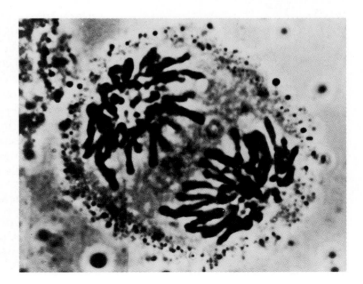

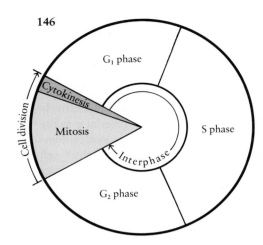

7–4 *The cell cycle. Cell division, which consists of mitosis (the division of the nucleus) and cytokinesis (the division of the cytoplasm), takes place after the completion of the three preparatory phases that constitute interphase. During the S (synthesis) phase, the chromosomal material is replicated. Separating cell division and the S phase are two G (gap) phases. The first of these (G₁) is a period of general growth and replication of cytoplasmic organelles. During the second (G₂), structures directly associated with mitosis and cytokinesis begin to assemble. After the G₂ phase comes mitosis, which is usually followed immediately by cytokinesis. In cells of different species or of different tissues within the same organism, the different phases occupy different proportions of the total cycle.*

0.2 μm

7–5 *A centriole pair from a dividing cell of the fruit fly* Drosophila, *as seen in longitudinal section. The many thin fibers also visible in this micrograph are microtubules.*

Centrioles are formed either from preexisting centrioles, with the newly formed centriole appearing at right angles to the previously existing one, or from basal bodies. The structure of a centriole, as seen in cross section, is identical to that of a basal body (see Figure 5–32, page 123).

THE CELL CYCLE

Dividing cells pass through a regular sequence of cell growth and division, known as the **cell cycle** (Figure 7–4). The cycle consists of five major phases: G₁, S, G₂, mitosis, and cytokinesis. Completion of the cycle requires varying periods of time from a few hours to several days, depending on both the type of cell and external factors, such as temperature or available nutrients.

Before a cell can begin mitosis and actually divide, it must replicate its DNA, synthesize more of the histones and other proteins associated with the DNA in the chromosomes, produce a supply of organelles adequate for two daughter cells, and assemble the structures needed to carry out mitosis and cytokinesis. These preparatory processes occur during the G₁, S, and G₂ phases of the cell cycle, which are known collectively as **interphase.**

The key process of DNA replication occurs during the S (synthesis) phase of the cell cycle, a time in which many of the histones and other DNA-associated proteins are also synthesized. G (gap) phases precede and follow the S phase; during the G phases, no DNA synthesis can be detected in the nucleus of the cell.

The G₁ phase, which follows cytokinesis and precedes the S phase, is a period of intensive biochemical activity. The cell increases in size, and its enzymes, ribosomes, mitochondria, and other cytoplasmic molecules and structures also increase in number. Some of the cellular structures can be synthesized entirely *de novo* ("from scratch") by the cell; these include microtubules, actin filaments, and ribosomes, all of which are composed, at least in part, of protein subunits. Membranous structures, such as the Golgi complexes, lysosomes, vacuoles, and vesicles, are all apparently derived from the endoplasmic reticulum, which is renewed and enlarged by the synthesis of lipid and protein molecules. In those cells that contain centrioles (that is, virtually all eukaryotic cells except those of fungi, flowering plants, and nematodes), the two centrioles begin to separate from each other and to replicate. Each member of the original centriole pair gives rise to a smaller daughter centriole by a copying process that is not yet understood. Mitochondria and chloroplasts, which are produced only from previously existing mitochondria and chloroplasts or plastids (page 120), also replicate. Each of these organelles has its own chromosome, which is organized much like the single chromosome of the bacterial cell. (These are two of the reasons that many biologists hypothesize that mitochondria and chloroplasts originated as separate organisms and then took up a new way of life inside early eukaryotic cells, more than a billion years ago. This question will be discussed further in Chapter 22.)

During the G₂ phase, which follows the S phase and precedes mitosis, the final preparations for cell division occur. The newly replicated chromosomes, which are dispersed in the nucleus in the form of threadlike strands of chromatin, slowly begin to coil and condense into a compact form; this condensation appears to be necessary for the complex movements and separation of the chromosomes that will occur in mitosis. Replication of the centriole pair is completed, with the two mature centriole pairs lying just outside the nuclear envelope, somewhat separated from each other (Figure 7–5). Also during this period, the cell begins to assemble the special structures required for the allocation of a complete set of chromosomes to each daughter cell during mitosis and for the separation of the two daughter cells during cytokinesis.

Regulation of the Cell Cycle

Dividing cells of different species show characteristic variations in the pattern of the cell cycle. In the common bean, for example, the complete cycle requires about 19 hours, of which 7 hours are taken up by the S phase; G₁ and G₂ are of equal length (about 5 hours each), and mitosis lasts 2 hours. By contrast, in mouse fibroblast cells, the cell cycle is approximately 22 hours, of which mitosis is less than 1 hour, S is almost 10 hours, G₁ is 9 hours, and G₂ is a little more than 2 hours.

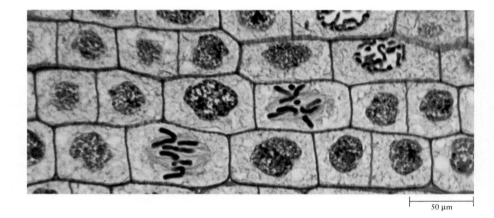

7–6 *Cross section of the tip of an onion root. The cells in this region undergo repeated divisions, providing new cells for the growth of the root. Different cells in this tissue section are in different phases of the cell cycle. The cells that are significantly larger than the others and that contain dark, rodlike structures are in mitosis.*

50 µm

7–7 *An experimental demonstration of density-dependent inhibition, which is also known as contact inhibition. (a) When isolated cells are grown in a nutrient medium, they divide until they form a continuous layer, one cell thick, across the surface of the culture dish. (b) If several rows of cells are removed, for example, by scraping them off, the adjacent cells ruffle their borders and flatten out (c). These cells then begin dividing, stopping once more when the dish is completely covered by a single layer of cells (d).*

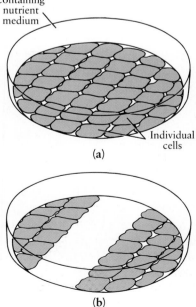

Culture dish containing nutrient medium

Individual cells

(a)

Some cell types pass through successive cell cycles throughout the life of the organism. This group includes the one-celled organisms and certain cells in growth centers of both plants and animals (Figure 7–6). An example is the cells in the human bone marrow that give rise to red blood cells. The average red blood cell lives only about 120 days, and there are about 25 trillion (2.5×10^{13}) of them in an adult. To maintain this number, about 2.5 million new red blood cells must be produced by cell division each second. At the other extreme, some highly specialized cells, such as nerve cells, lose their capacity to replicate once they are mature. A third group of cells retains the capacity to divide but does so only under special circumstances. Cells in the human liver, for example, do not ordinarily divide, but if a portion of the liver is removed surgically, the remaining cells (even if as few as a third of the total remain) continue to replicate themselves until the liver reaches its former size. Then they stop. All told, about 2 trillion (2×10^{12}) cell divisions occur in an adult human every 24 hours, or about 25 million per second.

In a multicellular organism, it is of critical importance that cells of the various different types divide at a sufficient rate to produce as many cells as are needed for growth and replacement—and only that many. If any particular cell type divides more rapidly than is necessary, the normal organization and functions of the organism may be disrupted as specialized tissues are invaded and overwhelmed by the rapidly dividing cells. Such is the course of events in cancer. In both multicellular and unicellular organisms, it is also important that cells divide only when they have reached a size large enough to ensure that the resulting daughter cells will contain all of the metabolic machinery needed for survival.

A number of environmental factors, including the depletion of nutrients and changes in temperature or pH, can cause cells to stop growing and dividing. In multicellular organisms, contact with adjacent cells can have the same effect. If normal cells from vertebrates (including humans) are isolated from one another and grown in a nutrient medium on a smooth glass surface, they move, amoeba-like, ruffling their cell borders until they encounter another cell, at which time they stop. More significantly, they undergo repeated cell cycles until enough cells have been produced that each cell is touching another cell; then cell division ceases (Figure 7–7). This phenomenon, known as density-dependent inhibition, does not occur in cancer cells; they pile on top of one another, moving, multiplying, crowding each other until all of the nutrients are used up.

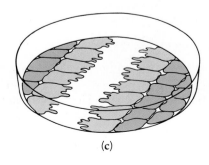

(b)   **(c)**

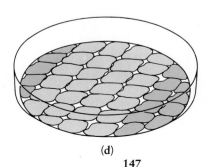

(d)

When normal cells stop growing, as a result of depletion of nutrients, density-dependent inhibition, or other factors, they stop at a point late in the G_1 phase. This point is known as the R ("restriction") point of the cell cycle. Once a cell passes the R point, it is committed to follow through the remaining parts of the cycle and then to divide. The G_1 phase is rapidly completed, synthesis of DNA and histone proteins in the S phase begins, and the cell moves steadily through the remaining phases of the cycle. The nature of the control or controls that act at the R point is currently the subject of intense research, not only because of its biological interest but also because of its potential importance in the control of cancer. One hypothesis suggests that passing the R point of the cycle requires a specific concentration of a particular protein synthesized in small quantities during the G_1 phase; only when this protein reaches the necessary concentration, thereby signaling that the cell has attained a suitable size and metabolic state, do the subsequent events of the cycle occur. Other hypotheses suggest that regulation involves a variety of stimulatory and inhibitory growth factors, some of which may be synthesized by the cell itself, and some of which may be synthesized and released into the surrounding medium by neighboring cells. A number of factors that influence the growth and division of cells in tissue culture have been discovered in the last few years; work is now under way characterizing their structure, the structure of the membrane receptors to which they bind, and the events triggered within the cell in response to that binding.

MITOSIS

The function of mitosis is to maneuver the replicated chromosomes so that each new cell gets a full complement—one of each. The capacity of the cell to accomplish this distribution depends on the condensed state of the chromosomes during mitosis and on an assembly of microtubules known as the **spindle.** Let us examine these structures before considering the "dance of the chromosomes."

The Condensed Chromosomes

As we noted previously, the threadlike chromosomes slowly begin to condense after their synthesis in the S phase of the cell cycle. By the beginning of mitosis, they are sufficiently condensed to become visible under the light microscope. Each chromosome can be seen to consist of two replicas, called **chromatids** (Figure 7–8a), joined together by a constricted area common to both chromatids.

7–8 (a) *A fully condensed chromosome. The chromosomal material was replicated during the S phase of the cell cycle, and each chromosome now consists of two identical parts called chromatids. The centromere, the constricted area at the center, is the site of attachment of the two chromatids. The kinetochores are protein-containing structures, one on each chromatid, associated with the centromere. Attached to the kinetochores are microtubules that form part of the spindle.*

(b) *In this electron micrograph of a portion of a dividing green alga, spindle fibers can be seen extending from the kinetochores. The dark material is the centromere region of the chromosome, most of which is out of the plane of the thin section prepared for the micrograph. The kinetochores are the two disk-shaped areas at either side of the chromosome.*

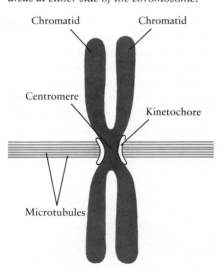

A replicated, condensed chromosome

(a)

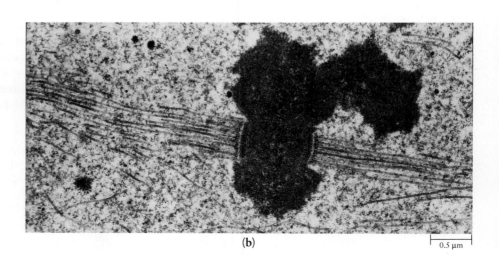

(b)   0.5 μm

(a)

⊢————⊣
10 µm

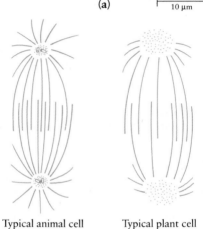

Typical animal cell
(b)

Typical plant cell
(c)

7–9 (a) *This micrograph of a dividing cell of the lung epithelium of an Oregon newt, an amphibian, illustrates the spindle's three-dimensional quality. The red fibers are the spindle microtubules. The large blue bodies near the equator of the spindle are the chromosomes.*

The basic framework of the spindle in (b) *an animal cell and* (c) *a plant cell. In the animal cell, a centriole pair is present at each pole; the polar fibers, which form the bulk of the spindle, are sharply focused on the centrioles, and additional fibers radiate outward from the centrioles, forming the aster. In plant cells, by contrast, centrioles are absent, the spindle is less sharply focused at the poles, and no aster is formed. Not shown in these diagrams are the replicated chromosomes and the spindle fibers attached to their kinetochores.*

This region of attachment is known as the **centromere.** Within the constricted region are disk-shaped protein-containing structures, the **kinetochores,** to which microtubules of the spindle are attached (Figure 7–8b).

The Spindle

When completely formed, the spindle (Figure 7–9) is a three-dimensional football-shaped structure, consisting of at least two groups of microtubules: (1) polar fibers, which reach from each pole of the spindle (analogous to the ends of the football) to a central region midway between the poles, and (2) kinetochore fibers, which are attached to the kinetochores of the replicated chromosomes. As we shall see, these two groups of spindle fibers are responsible for separating the sister chromatids during mitosis.

In those cells that contain centrioles, each pole of the spindle is marked by a pair of newly replicated centrioles. Such cells also contain a third group of shorter spindle fibers, extending outward from the centrioles. These additional fibers are known collectively as the **aster.** It has been hypothesized that the fibers of the aster may brace the poles of the spindle against the cell membrane during the movements of mitosis; in cells that lack centrioles and asters, the rigid cell wall may perform a similar function.

Most of the tubulin dimers from which the microtubules of the spindle are formed (see Figure 3–25, page 78) are apparently borrowed from the cytoskeleton. Immunofluorescence micrographs have shown that the network of cytoskeletal microtubules that radiate outward from the center of a nondividing cell (see Figure 5–14a, page 111) is disassembled at the beginning of mitosis. As a consequence, dividing cells take on a characteristic rounded appearance. Following cell division, the spindle is disassembled, the cytoskeletal network of microtubules is reassembled, and the cell assumes its nondividing shape.

Centrioles and the Microtubule Organizing Center

As we noted in Chapter 5, basal bodies and centrioles are the same structure used, perhaps, for different purposes. Basal bodies organize the microtubules of flagella and cilia, and centrioles have long been thought to play a role in organizing the microtubules of the spindle fibers—spinning them out somewhat as a spider spins out silk. A striking example of the interchangeability of basal bodies and centrioles is provided by the alga *Chlamydomonas.* At the beginning of mitosis, its two flagella are reabsorbed by the cell, and the basal bodies move near the nucleus—to the same location occupied in other cells by the centrioles. During mitosis, they behave exactly like centrioles, appearing to organize the spindle. When mitosis is complete, they migrate to the ends of the daughter cells, giving rise to new flagella. Despite this evidence for the role of the centrioles in spindle formation, cells that do not have centrioles or basal bodies also form spindles with microtubules. In some animal cells, moreover, it is possible to remove the centrioles from the cells—yet spindle formation proceeds normally.

The explanation for these seemingly contradictory observations may lie in a densely staining region seen around the centrioles in many electron micrographs. Such a densely staining region is also present in cells without centrioles and is the area from which both the spindle fibers and the microtubules of the cytoskeleton originate. The material in this region, rather than the centrioles themselves, is now thought to be the microtubule organizing center. It has also been suggested that the spindles, instead of forming from the centrioles, separate them, pushing them apart and so ensuring that each daughter cell receives an adequate supply of basal bodies from which to construct flagella or cilia.

The Phases of Mitosis

The process of mitosis is conventionally divided into four phases: prophase, metaphase, anaphase, and telophase. Of these, prophase is usually by far the longest. If a mitotic division takes 10 minutes (which is about the minimum time required), during about six of these minutes the cell will be in prophase. The schematic drawings that follow show mitosis as it takes place in an animal cell. (Similar schematic drawings of mitosis in a plant cell are shown in Figure 7–16 on page 154.)

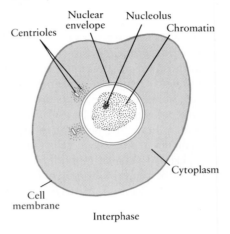
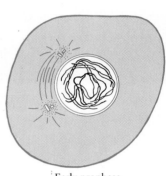

Centrioles Nuclear envelope Nucleolus Chromatin

Cytoplasm

Cell membrane

Interphase   Early prophase

During the interphase portions of the cell cycle, little can be seen in the nucleus. By early **prophase,** however, the chromatin has condensed sufficiently that the individual chromosomes become visible under the light microscope. Each chromosome consists of two duplicate chromatids pressed closely together longitudinally and connected at the centromere. In the cells of most organisms (fungi, flowering plants, and nematodes are the principal exceptions), two centriole pairs can be seen at one side of the nucleus, outside the nuclear envelope. As we noted previously, replication of the original centriole pair began during the G_1 phase of the cell cycle and was completed late in the G_2 phase. The cell becomes more spheroid and the cytoplasm more viscous at this stage, as the microtubules of the cytoskeleton are disassembled in preparation for the formation of the spindle.

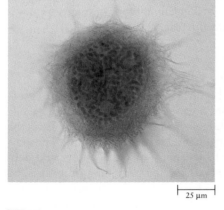

├─── 25 μm ───┤

7–10 *Early prophase in a cell from a seed of the African globe lily,* Haemanthus katherine. *The microtubules of the cytoskeleton are still in a meshwork surrounding the nucleus and have not yet become reorganized into a spindle. At this stage, the chromosomes are condensing. Their threadlike appearance when they first become visible under the microscope is the source of the name mitosis.* Mitos *is the Greek word for "thread."*

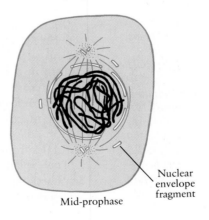

Nuclear envelope fragment

Mid-prophase

During prophase, the centriole pairs move apart. Between the centriole pairs, forming as they separate (or, more likely, separating them as they form), are the microtubules that become the polar fibers of the spindle. In those cells that have centrioles, the microtubules that form the aster radiate outward from the centrioles. By this time the nucleoli usually have disappeared from view. As the chromosomes continue to condense, the nuclear envelope breaks down, dispersing into membranous fragments similar to fragments of endoplasmic reticulum.

By the end of prophase, the chromosomes are fully condensed and are no longer separated from the cytoplasm. The centriole pairs have reached the poles of the cell, and the members of each pair are of equal size. The polar fibers of the spindle are fully formed, and the kinetochore fibers, attached to the kinetochores of the chromosomes, have also formed.

Early metaphase

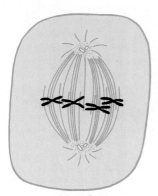
Metaphase

During early **metaphase,** the chromatid pairs move back and forth within the spindle, apparently maneuvered by the spindle fibers, as if they were being tugged first toward one pole and then the other. Finally the chromatid pairs become arranged precisely at the midplane (equator) of the cell. This marks the end of metaphase.

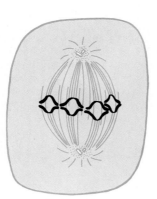
Early anaphase

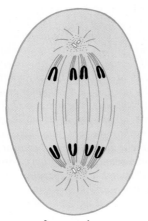
Late anaphase

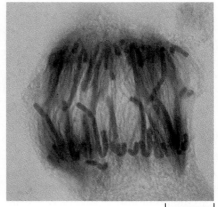

20 μm

7–11 *Middle anaphase in a cell from the seed of an African globe lily. The chromosomes have moved halfway toward the poles.*

At the beginning of **anaphase,** the most rapid stage of mitosis, the centromeres separate simultaneously in all the chromatid pairs. The chromatids of each pair then move apart, each chromatid becoming a separate chromosome, each apparently drawn toward the opposite pole by the kinetochore fibers. The centromeres move first, while the arms of the chromosomes seem to drag behind. In most cells, the spindle as a whole also elongates, with the poles appearing to be pushed farther apart. As anaphase continues, the two identical sets of newly separated chromosomes move rapidly toward the opposite poles of the spindle.

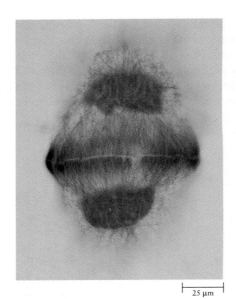

25 µm

7-12 *Middle telophase in a cell from a seed of an African globe lily. Note the unstained region at the equator of the cell. It is here that the cell plate will form during cytokinesis.*

7-13 *Mitosis in embryonic cells of a whitefish. (a) Prophase. The chromosomes have become visible, the nuclear envelope is breaking down, and the spindle apparatus is forming. Note the prominent asters. (b) Metaphase. The chromatid pairs are lined up at the equator of the cell. (c) Anaphase. The two sets of chromosomes are moving apart. (d) Telophase. The chromosomes are completely separated, the spindle apparatus is disappearing, and a new cell membrane is forming that will complete the separation of the two daughter cells.*

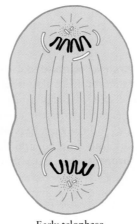

Early telophase

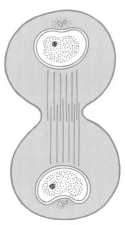

Late telophase

By the beginning of **telophase,** the chromosomes have reached the opposite poles and the spindle begins to disperse into tubulin dimers. During late telophase, nuclear envelopes re-form around the two sets of chromosomes, which once more become diffuse. In each nucleus, the nucleoli reappear. Often, a new centriole begins to form adjacent to each of the previous ones. As we saw earlier, replication of the centrioles continues during the subsequent cell cycle, so that each daughter cell has two centriole pairs by prophase of the next mitotic division.

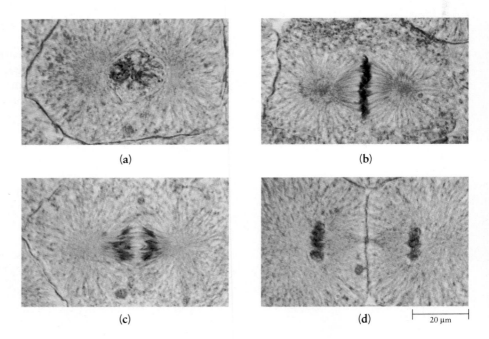

(a)

(b)

(c)

(d)

20 µm

The Mechanism of Chromosome Movement

Although there is little doubt that the movement of the chromosomes toward the poles and the separation of the poles from each other is the result of interactions among the kinetochore fibers and the polar fibers of the spindle, the exact mechanism or mechanisms remain unknown. One possibility is suggested by the fact that the kinetochore fibers lengthen during prophase and then shorten during anaphase—without getting either thinner or thicker. This indicates that the fibers probably do not contract but that material is added to or removed from them during different phases of mitosis, leading to the pushing apart or pulling together of the structures to which the fibers are attached. Another mechanism is suggested by the fact that the microtubules of the spindle fibers have little protein

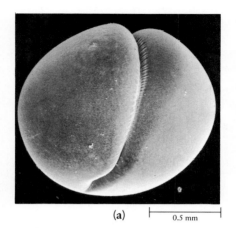

(a) 0.5 mm

(b) 50 μm

7-14 *Cytokinesis in an animal cell, a frog egg.* (a) *The egg is dividing in two;* (b) *close-up showing the constriction furrows.*

"arms," analogous to those seen in the microtubules of cilia and flagella (see Figure 5–31a, page 123). These "arms," like those of cilia and flagella, contain an enzyme involved in energy-releasing chemical reactions. It is thus hypothesized that the spindle microtubules may also "walk" along each other, tractor-fashion, powered by the energy released in the enzymatic reactions. Although microtubules constitute the bulk of the spindle fibers, there are other proteins associated with the spindle; it remains possible that these proteins, about which very little is presently known, also play key roles in the movements of mitosis.

CYTOKINESIS

Cytokinesis, the division of the cytoplasm, usually but not always accompanies mitosis, the division of the nucleus. The visible process of cytokinesis generally begins during telophase of mitosis, and it usually divides the cell into two nearly equal parts. Although the spindle does not seem to be directly involved in the division of the cytoplasm, the cleavage always occurs at the midline of the spindle, in the region where the polar fibers overlap. There is evidence that these spindle microtubules play a role in positioning the other structures that are responsible for the actual division of the cytoplasm. If, for example, the spindle is pushed to the side of the cell shortly after its formation is complete—so that its equator extends only halfway across the cell—only the cytoplasm of that half of the cell is ultimately divided in cytokinesis. The result is a two-lobed cell, with a daughter nucleus in each lobe.

Cytokinesis differs significantly in plant and animal cells. In animal cells, during early telophase, the cell membrane begins to constrict along the circumference of the cell in the plane of the equator of the spindle. At first, a furrow appears on the surface, and this gradually deepens into a groove (Figure 7–14). Eventually, the connection between the daughter cells dwindles to a slender thread, which soon parts. Actin filaments (page 110), seen in large numbers near the furrows, are thought to play a role in the constriction. They are believed to act as a sort of "purse string," gathering in the membrane of the parent cell at its midline, thus pinching apart the two daughter cells.

In plant cells, the cytoplasm is divided at the midline by a series of polysaccharide-containing vesicles produced from the Golgi complexes (Figure 7–15). The

7-15 *In plants, the separation of the two cells is effected by the formation of a structure known as the cell plate. Vesicles appear across the equatorial plane of the cell and gradually fuse, forming a flat, membrane-bound space, the cell plate, which extends outward until it reaches the wall of the dividing cell. The large, dark forms on either side of the micrograph are the chromosomes.*

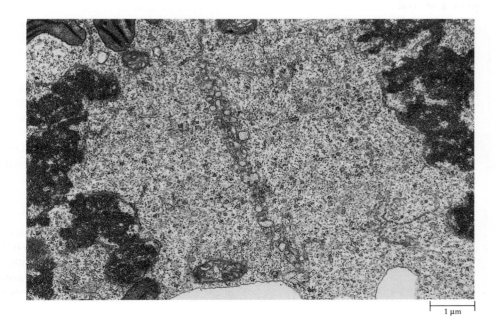

1 μm

7–16 *Mitosis in a plant cell with four chromosomes. Note that a spindle forms, although no centrioles are present and no aster is visible.*

The plane of cell division is established late in the G₂ phase of the cell cycle when the microtubules of the cytoskeleton are reorganized into a circular structure, known as the preprophase band, just inside the cell wall. Although this band disappears early in prophase, it deter-mines the future location of the equator and the cell plate. The microtubules of the band are later reassembled into the spindle in a clear zone that develops around the nucleus in the course of prophase.

In cytokinesis, which begins during telophase, the cell plate gradually extends outward until it meets the exact region of the cell wall occupied earlier by the pre-prophase band. The vesicles that give rise to the cell plate are apparently guided into position by the spindle fibers remaining between the daughter nuclei.

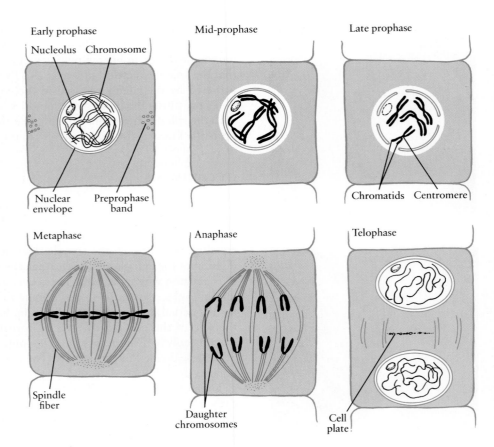

Early prophase
Nucleolus   Chromosome
Nuclear envelope   Preprophase band

Mid-prophase

Late prophase
Chromatids   Centromere

Metaphase
Spindle fiber

Anaphase
Daughter chromosomes

Telophase
Cell plate

vesicles eventually fuse to form a flat, membrane-bound space, the **cell plate.** As more vesicles fuse, the edges of the growing plate fuse with the membrane of the cell. In this way, a layer of polysaccharides is established between the two daughter cells, completing their separation. This layer becomes impregnated with pectins, ultimately forming the middle lamella (see page 107). Each new cell then constructs its own cell wall, laying down cellulose and other polysaccharides against the outer surface of its cell membrane.

When cell division is complete, two daughter cells are produced, smaller than the parent cell but otherwise indistinguishable from it and from each other.

Although we have focused in this chapter on the processes directly involved in cell division, it is important to realize that the other activities characteristic of the living cell are also occurring throughout the cell cycle. The cell is synthesizing the macromolecules necessary to maintain its structure, it is degrading other mole-cules, it is regulating both the internal movement of substances and the movement of substances between the cell interior and the exterior environment, and it is responding to a variety of stimuli. All of these activities—as well as cell division itself—require a steady expenditure of energy by the cell. In the next section, we shall see how cells provide themselves with this essential energy.

SUMMARY

Cell division in prokaryotes is a relatively simple process, in which two daughter chromosomes are attached to different spots on the interior of the cell membrane. As the membrane elongates, the chromosomes are separated. The cell membrane then pinches inward, and a new cell wall forms, completing the division of the daughter cells.

Cell division is a more complex process in eukaryotes, which contain a vast amount of genetic material, organized into a number of different chromosomes. Dividing cells pass through a regular sequence of cell growth and cell division known as the cell cycle. The cycle consists of a G_1 phase, during which the cytoplasmic molecules and structures increase; an S phase, during which the chromosomes are replicated; a G_2 phase, during which condensation of the chromosomes and assembly of the special structures required for mitosis and cytokinesis begin; mitosis, during which the replicated chromosomes are apportioned between two daughter nuclei; and cytokinesis, during which the cytoplasm is divided, separating the parent cell into two daughter cells. The first three phases of the cell cycle are known, collectively, as interphase. Regulation of the cell cycle occurs late in the G_1 phase and may involve a number of interacting factors.

When the cell is in the interphase portions of the cycle, the chromosomes are visible only as thin strands of threadlike material (chromatin) within the nucleus. As mitosis begins, the condensing chromosomes, previously replicated during the S phase, become visible under the light microscope. At these early stages of mitosis, the chromosomes consist of pairs of identical replicas, called chromatids, held together at the centromere. Simultaneously, the spindle is forming. In animal cells, it forms between the centrioles as they separate. In both animal and plant cells, the framework of the spindle is formed by fibers that extend from the poles to the equator of the cell; other fibers are attached to the chromatids at their kinetochores, protein-containing structures associated with the centromeres. Prophase ends with the breakdown of the nuclear envelope and disappearance of the nucleoli. During metaphase, the chromatid pairs, maneuvered by the spindle fibers, move toward the center of the cell. At the end of metaphase, they are arranged on the equatorial plane. During anaphase, the sister chromatids separate, and each chromatid, now an independent chromosome, moves to an opposite pole. During telophase, a nuclear envelope forms around each group of chromosomes. The spindle begins to break down, the chromosomes uncoil and once more become extended and diffuse, and the nucleoli reappear.

Cytokinesis in animal cells results from constrictions in the cell membrane between the two nuclei. In plant cells, the cytoplasm is divided by the coalescing of vesicles to form the cell plate, within which the cell wall is subsequently laid down. In both cases, the result is the production of two new, separate cells. As a result of mitosis, each has received an exact copy of the enormous skein of genetic material of the parent cell, and, as a result of cytokinesis, approximately half of the cytoplasm and organelles.

QUESTIONS

1. Distinguish among the following terms: cell cycle/cell division; mitosis/cytokinesis; chromatid/chromosome; centriole/ centromere/kinetochore.

2. Describe the activities occurring during each phase of the cell cycle and the role of each phase in the overall process of cell division.

3. What is a chromosome? How is it related to chromatin?

4. Why do we often refer to chromatids as sister chromatids? When are sister chromatids formed? How? When do they first become visible under the microscope?

5. Describe the structure of the spindle in a typical animal cell. What are thought to be the functions of each of the different groups of spindle fibers?

6. In what ways does cell division in plant cells differ from that in animal cells?

7. What is the function of cell division in the life of an organism? Suppose you, as an organism, were made up of a large single cell rather than trillions of small ones. How would you differ from your present self?

SUGGESTIONS FOR FURTHER READING

Books

ALBERTS, BRUCE, DENNIS BRAY, JULIAN LEWIS, MARTIN RAFF, KEITH ROBERTS, and JAMES D. WATSON: *Molecular Biology of the Cell*, 2d ed., Garland Publishing, Inc., New York, 1989.

> *Progressing from the molecules of which cells are composed, through an examination of cellular structure and function, to the interactions of cells within tissues, this outstanding text describes not only our most current knowledge and how it was attained but also the many areas still to be explored. It is clearly written and filled with wonderful micrographs and explanatory diagrams. Highly recommended.*

DARNELL, JAMES, HARVEY F. LODISH, and DAVID BALTIMORE: *Molecular Cell Biology*, W. H. Freeman and Company, New York, 1986.

> *A comprehensive treatment of modern cell biology, richly illustrated with diagrams and micrographs. This text places particular emphasis on molecular genetics, membrane structure and function, cytoplasmic organelles, and the cytoskeleton, as well as on the techniques used in the contemporary study of cell biology.*

DE DUVE, CHRISTIAN: *A Guided Tour of the Living Cell*, Scientific American Library, W. H. Freeman and Company, New York, 1984.*

> *In this beautifully illustrated two-volume set, de Duve, one of the pioneers of modern cell biology, takes the reader—imagined to be a "cytonaut," a bacterium-sized tourist—on a journey through the eukaryotic cell. The first portion of the journey explores the cellular membranes; the second, the cytoplasm and its organelles; and the third, the nucleus. At the conclusion of the journey, de Duve considers such key questions of modern biology as the origin of life and the mechanisms of evolution.*

LEDBETTER, M. C., and KEITH R. PORTER: *Introduction to the Fine Structures of Plant Cells*, Springer-Verlag, New York, 1970.

> *An excellent atlas of electron micrographs of plant cells, with detailed explanations.*

LEHNINGER, ALBERT L.: *Principles of Biochemistry*, Worth Publishers, Inc., New York, 1982.

> *This introductory text is outstanding both for its clarity and for its consistent focus on the living cell.*

OPARIN, A. I.: *The Origin of Life*, Dover Publications, Inc., New York, 1938.*

> *Oparin, a Russian biochemist, was the first to argue that life arose spontaneously in the oceans of the primitive earth. Although his concepts have been somewhat modified in detail, they form the basis for the present scientific theories on the origin of living things.*

PORTER, KEITH R., and MARY A. BONNEVILLE: *An Introduction to the Fine Structures of Cells and Tissues*, 4th ed., Lea & Febiger, Philadelphia, 1973.

> *An atlas of electron micrographs of animal cells; detailed commentaries accompany each. These are magnificent micrographs, and the commentaries describe not only what the pictures show but also the experimental foundations of our knowledge of cell ultrastructures.*

PRESCOTT, DAVID M.: *Cells: Principles of Molecular Structure and Function*, Jones and Bartlett Publishers, Boston, 1988.

> *An up-to-date, yet concise, textbook of cell biology, written for a first course at the undergraduate level. It is a wonderful introduction for any reader wishing to gain an overview of the exciting developments in contemporary cell biology.*

SCIENTIFIC AMERICAN: *The Molecules of Life*, W. H. Freeman and Company, New York, 1986.*

> *This reprint of the October 1985 issue of* Scientific American *includes 11 articles on molecules that play key roles in the living cell. The articles on proteins, the molecules of the cell membrane, and the molecules of the cytoskeleton are of particular interest at this point in your study of biology. You will find the other articles in this collection useful at later points in the course.*

SCIENTIFIC AMERICAN: *Molecules to Living Cells*, W. H. Freeman and Company, New York, 1980.*

> *A collection of outstanding articles from* Scientific American. *Chapters 1, 2, 4, 10, and 11 cover the origin of life, evolution of the earliest cells, the cell cycle, and cell membranes and their assembly. Highly recommended.*

SILK, JOSEPH: *The Big Bang: The Creation and Evolution of the Universe*, 2d ed., W. H. Freeman and Company, New York, 1988.*

> *A discussion of modern evidence concerning the formation of the solar system and the planet earth. An excellent, well-written introduction to cosmology.*

STRYER, LUBERT: *Biochemistry*, 3d ed., W. H. Freeman and Company, New York, 1988.

> *An introductory text, with many examples of medical applications of biochemistry. Handsomely illustrated.*

THOMAS, LEWIS: *The Lives of a Cell: Notes of a Biology Watcher*, Viking Press, Inc., New York, 1974.*

THOMAS, LEWIS: *The Medusa and the Snail: More Notes of a Biology Watcher*, Viking Press, Inc., New York, 1979.*

> *Thomas, a physician and medical researcher, reveals the extent to which science can tune our intellectual antennae, broaden our perceptions, and extend our appreciation of ourselves and of the world around us. Anyone who wants to refute the contention that science destroys human values need look no further than these short, sensitive essays.*

WEINBERG, STEVEN: *The Discovery of Subatomic Particles*, Scientific American Library, W. H. Freeman and Company, New York, 1984.

> *In this handsome book, an introduction to the structure of the atom is combined with a lively history of twentieth-century physics. This revolution in physics profoundly influenced modern biology.*

* Available in paperback.

WEINBERG, STEVEN: *The First Three Minutes: A Modern View of the Origin of the Universe*, Basic Books, Inc., New York, 1977.*

A wonderful story, written for the intelligent nonscientist (characterized by the author as a smart old attorney who expects to hear some convincing arguments before he makes up his mind).

Articles

ALBERSHEIM, PETER: "The Walls of Growing Plant Cells," *Scientific American*, April 1975, pages 81–95.

ALLEN, ROBERT DAY: "The Microtubule as an Intracellular Engine," *Scientific American*, February 1987, pages 42–49.

BONNER, JOHN TYLER: "Chemical Signals of Social Amoebae," *Scientific American*, April 1983, pages 114–120.

BRETSCHER, MARK S.: "Endocytosis: Relation to Capping and Cell Locomotion," *Science*, vol. 224, pages 681–686, 1984.

BRETSCHER, MARK S.: "How Animal Cells Move," *Scientific American*, December 1987, pages 72–90.

BROWN, MICHAEL S., and JOSEPH L. GOLDSTEIN: "A Receptor-Mediated Pathway for Cholesterol Homeostasis," *Science*, vol. 232, pages 34–47, 1986.

DAUTRY-VARSAT, ALICE, and HARVEY F. LODISH: "How Receptors Bring Proteins and Particles into Cells," *Scientific American*, May 1984, pages 52–58.

DE DUVE, CHRISTIAN: "Microbodies in the Living Cell," *Scientific American*, May 1983, pages 74–84.

DUSTIN, PIERRE: "Microtubules," *Scientific American*, August 1980, pages 67–76.

ELGSAETER, ARNLJOT, BJORN T. STOKKE, ARNE MIKKELSEN, and DANIEL BRANTON: "The Molecular Basis of Erythrocyte Shape," *Science*, vol. 234, pages 1217–1223, 1986.

FAUL, HENRY: "A History of Geologic Time," *American Scientist*, vol. 66, pages 159–165, 1978.

GRIFFITHS, GARETH, and KAI SIMONS: "The *trans* Golgi Network: Sorting at the Exit Site of the Golgi Complex," *Science*, vol. 234, pages 438–443, 1986.

KASTING, JAMES F., OWEN B. TOON, and JAMES B. POLLACK: "How Climate Evolved on the Terrestrial Planets," *Scientific American*, February 1988, pages 90–97.

KELLY, REGIS B.: "Pathways of Protein Secretion in Eukaryotes," *Science*, vol. 230, pages 25–32, 1985.

KOSHLAND, DOUGLAS E., T. J. MITCHISON, and MARC W. KIRSCHNER: "Polewards Chromosome Movement Driven by Microtubule Depolymerization *in vitro*," *Nature*, vol. 331, pages 499–504, 1988.

LAZARIDES, ELIAS, and JEAN PAUL REVEL: "The Molecular Basis of Cell Movement," *Scientific American*, May 1979, pages 100–113.

MARX, JEAN L.: "A Potpourri of Membrane Receptors," *Science*, vol. 230, pages 649–651, 1985.

MERTZ, WALTER: "The Essential Trace Elements," *Science*, vol. 213, pages 1332–1338, 1981.

MILLER, JULIE ANN: "Cell Communication Equipment: Do-It-Yourself Kit," *Science News*, April 14, 1984, pages 236–237.

MOHNEN, VOLKER A.: "The Challenge of Acid Rain," *Scientific American*, August 1988, pages 30–38.

PETERSON, IVARS: "A Biological Antifreeze," *Science News*, November 22, 1986, pages 330–332.

RACKER, EFRAIM: "Structure, Function, and Assembly of Membrane Proteins," *Science*, vol. 235, pages 959–961, 1987.

ROTHMAN, JAMES E.: "The Compartmental Organization of the Golgi Apparatus," *Scientific American*, September 1985, pages 74–89.

SATIR, BIRGIT: "The Final Steps in Secretion," *Scientific American*, October 1975, pages 28–37.

SATIR, PETER: "How Cilia Move," *Scientific American*, October 1974, pages 44–52.

SCHINDLER, D. W.: "Effects of Acid Rain on Freshwater Ecosystems," *Science*, vol. 239, pages 149–157, 1988.

SHARON, NATHAN: "Carbohydrates," *Scientific American*, November 1980, pages 90–116.

SLOBODA, ROGER D.: "The Role of Microtubules in Cell Structure and Cell Division," *American Scientist*, vol. 68, pages 290–298, 1980.

SPORN, MICHAEL B., and ANITA B. ROBERTS: "Peptide Growth Factors Are Multifunctional," *Nature*, vol. 332, pages 217–219, 1988.

STAEHELIN, L. ANDREW, and BARBARA E. HULL: "Junctions between Living Cells," *Scientific American*, May 1978, pages 141–152.

STILLINGER, FRANK H.: "Water Revisited," *Science*, vol. 209, pages 451–457, 1980.

UNWIN, NIGEL, and RICHARD HENDERSON: "The Structure of Proteins in Biological Membranes," *Scientific American*, February 1984, pages 78–94.

WEISSKOPF, VICTOR F.: "The Origin of the Universe," *American Scientist*, vol. 71, pages 473–480, 1983.

WICKNER, WILLIAM T., and HARVEY F. LODISH: "Multiple Mechanisms of Protein Insertion Into and Across Membranes," *Science*, vol. 230, pages 400–407, 1985.

---

* Available in paperback.

SECTION **2**

Energetics

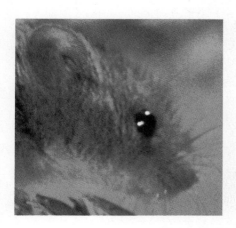

The energy of the summer sun is stored in these wheat plants, ready for harvest. Of the radiant energy of the sun striking the wheat field, less than 10 percent is converted into stored chemical energy.

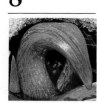

The Flow of Energy

Life here on earth depends on the flow of energy from the thermonuclear reactions taking place at the heart of the sun. The amount of energy delivered to the earth by the sun is about 13×10^{23} (the number 13 followed by 23 zeros) calories per year. It is a difficult quantity to imagine. For example, the amount of solar energy striking the earth every day is about 1.5 billion times greater than the amount of electricity generated in the United States each year.

About one-third of this solar energy is reflected back into space as light. Much of the remaining two-thirds is absorbed by the earth and converted to heat. Some of this absorbed heat energy serves to evaporate the waters of the oceans, producing the clouds that, in turn, produce rain and snow. Solar energy, in combination with other factors, is also responsible for the movements of air and of water that help set patterns of climate over the surface of the earth.

A small fraction—less than 1 percent—of the solar energy reaching the earth becomes, through a series of operations performed by the cells of plants and other photosynthetic organisms, the energy that drives all the processes of life. Living systems change energy from one form to another, transforming the radiant energy from the sun into the chemical and mechanical energy used by everything that is alive.

This flow of energy is the essence of life. As we noted in the Introduction, evolution may be viewed as a competition among organisms for the most efficient use of energy resources. A cell can be best understood as a complex of systems for transforming energy. At the other end of the biological scale, the structure of an ecosystem or of the biosphere itself is determined by the energy exchanges occurring among the groups of organisms within it.

In this chapter, we shall look first at the general principles governing all energy transformations and then at the characteristic ways in which cells regulate the energy transformations that take place within living systems. In the chapters that follow, the principal and complementary processes of energy flow through the biosphere will be examined—glycolysis and respiration in Chapter 9 and photosynthesis in Chapter 10.

8–1 *Harvest mice, feeding on wheat kernels. Of the chemical energy stored in the wheat kernels, less than 10 percent will be converted to chemical energy stored in the body tissues of the mice. Life runs downhill and is sustained only by a constant flow of radiant energy from the sun.*

8–2 *Another downhill step. Less than 10 percent of the chemical energy stored in the body tissues of the mouse will be converted to chemical energy stored in the body tissues of its predator, the red rat snake.*

THE LAWS OF THERMODYNAMICS

Energy is such a common term today that it is surprising to learn that the word was coined less than 200 years ago, at the time of the development of the steam engine. It was only then that scientists and engineers began to understand that heat, motion, light, electricity, and the forces holding the atoms together in molecules are all different forms of the same capacity to cause change, or, as it is often expressed, to do work. This new understanding led to the study of **thermodynamics**—the science of energy transformations—and to the formulation of its laws.

The First Law

The **first law of thermodynamics** states, quite simply: *Energy can be changed from one form to another, but it cannot be created or destroyed.* The total energy of any system plus its surroundings thus remains constant, despite any changes in form.

Electricity is a form of energy, as is light. Electrical energy can be changed to light energy (for example, by letting an electric current flow through the tungsten wire in a light bulb). Conversely, light energy can be changed to electrical energy, a transformation that is the essential first step of photosynthesis, as we shall see in Chapter 10.

Energy can be stored in various forms and then changed into other forms. In automobile engines, for example, the energy stored in the chemical bonds of gasoline is converted to heat (kinetic energy), which is then partially converted to mechanical movements of the engine parts. Some of the energy is converted back to heat by the friction of the moving engine parts, and some of it leaves the engine in the exhaust products. Similarly, when organisms oxidize carbohydrates, they convert the energy stored in chemical bonds to other forms. On a summer evening, for example, a firefly converts chemical energy to mechanical energy, to heat, to flashes of light, and to electrical impulses that travel along the nerves of its body. Birds and mammals convert chemical energy into the heat necessary to maintain their body temperature, as well as into mechanical energy, electrical energy, and other forms of chemical energy. According to the first law of thermodynamics, in these energy conversions, and in all others, energy is neither created nor destroyed.

In all energy conversions, however, some useful energy is converted to heat and dissipates. In an automobile engine, for example, the heat produced by friction and lost in the exhaust, unlike the heat confined in the engine itself, cannot produce work—that is, it cannot drive the pistons and turn the gears—because it is dissipated into the surroundings. But it is nevertheless part of the total equation. In a gasoline engine, about 75 percent of the energy originally present in the fuel is transferred to the surroundings in the form of heat—that is, it is converted to increased motion of atoms and molecules in the air. Similarly, the heat produced by the metabolic processes of animals is dissipated into the surrounding air or water.

It was in the course of studies of engine efficiency that the notion of potential energy was first developed. A barrel of gasoline or a ton of coal could be assigned a certain amount of potential energy, measured in terms of the amount of heat it would liberate when burned. The efficiency of the conversion of the potential energy to "useful" energy depended on the design of the system.

Although these concepts were formulated in terms of engines running on heat energy, they apply to other systems as well. As we saw in Chapter 1, a boulder pushed to the top of a hill gains energy—potential energy. Given a little push, it rolls down the hill again, converting that potential energy to motion and the heat produced by friction. Water, as we saw in Chapter 6, may also possess potential energy. As it moves by bulk flow from the top of a waterfall or over a dam, it can turn waterwheels that turn gears and, for example, grind corn. Thus the potential

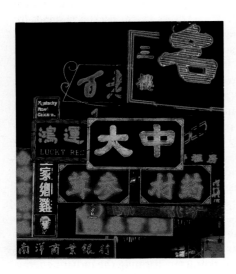

8–3 *Electrical energy can be converted to light energy, as in these Hong Kong signs, for example. The energy emitted as an electron falls from one energy level to another is a discrete amount, characteristic for each atom. When electricity is passed through a tube of gas, electrons in the atoms of the gas are boosted to higher energy levels. As they fall back, light energy is emitted, producing, for example, the red glow characteristic of neon and the yellow glow characteristic of sodium vapor.*

$E = mc^2$

Protons and neutrons, as we noted in Chapter 1, are arbitrarily assigned an atomic weight of 1. One would therefore expect that an element with, for example, twice as many protons and neutrons as another element would weigh twice as much. This assumption is true—almost. If the weights of nuclei are measured with great accuracy, as they can be by instruments developed by modern physics, small nuclei always have proportionately slightly greater weights than larger nuclei. For example, the most common isotope of carbon, as you know, has a combined total of 12 protons and neutrons, and carbon, by convention, is assigned an atomic weight of 12. The hydrogen atom, however, has an atomic weight not exactly of 1, as would be expected, but of 1.008. Helium has two protons and two neutrons. It does not have a weight of exactly 4, however, or of 4.032 (four times the weight of hydrogen); it weighs, in relation to carbon, 4.0026. Similarly, oxygen, with a combined total of 16 protons and neutrons, has an atomic weight—in relation to that of carbon—of 15.995. In short, when protons and neutrons are assembled into an atomic nucleus, there are slight changes in weight, which reflect changes in mass.

One of the oldest and most fundamental concepts of chemistry is the law of conservation of mass—that mass is never created or destroyed. We assume the law of conservation of mass every time we write a chemical equation. Yet under conditions of extremely high temperature, atomic nuclei fuse to make new elements and there is a measurable decrease in mass. What happens to the mass "lost" in the course of this fusion? This is the question answered by Einstein's fateful equation $E = mc^2$. E stands for energy, m stands for mass, and c is a constant equal to the speed of light. Einstein's equation means simply that under certain extreme and unusual conditions, mass is turned into energy.

The sun consists largely of hydrogen nuclei. At the extremely high temperatures at the core of the sun, hydrogen nuclei strike each other with enough velocity to fuse. In a series of steps, four hydrogen nuclei fuse to form one helium nucleus. In the course of these steps, energy is released, enough to keep the fusion reaction going and to emit tremendous amounts of radiant energy into space. Life on this planet depends on energy emitted by the sun in the course of this fusion reaction. This same reaction provides, of course—as Einstein foresaw—the energy of the hydrogen bomb.

Albert Einstein in 1905, the year he published his paper on the theory of relativity. He was 26 years old and working at the Swiss Patent Office in Bern as a technical expert third class.

energy of water, in this system, is converted to the mechanical energy of the wheels and gears and to heat, produced by the movement of the water itself and also by the turning wheels and gears. Molecules also contain potential energy, stored in the chemical bonds between their constituent atoms. When these bonds are broken in chemical reactions, the energy they contain can be used to form other chemical bonds or can be released as heat.

The first law of thermodynamics states that in energy exchanges and conversions, wherever they take place and whatever they involve, the total energy of the system and its surroundings after the conversion is equal to the total energy before the conversion. In the case of chemical reactions, this means that the energy of the products of the reaction plus the energy released in the reaction is equal to the initial energy of the reactants.

The Second Law

The energy that is dissipated as heat as the result of an energy conversion has not been destroyed—it is still present in the random motion of atoms and molecules —but it has been "lost" for all practical purposes. It is no longer available to do useful work. This brings us to the **second law of thermodynamics,** which is the more interesting one, biologically speaking. It predicts the direction of all events involving energy exchanges; thus it has been called "time's arrow." The second law states that *in all energy exchanges and conversions, if no energy leaves or enters the system under study, the potential energy of the final state will always be less than the potential energy of the initial state.* The second law is entirely in keeping with everyday experience. A boulder will roll downhill but never uphill. Heat will flow from a hot object to a cold one and never the other way. A ball that is dropped will bounce—but not back to the height from which it was dropped.

A process in which the potential energy of the final state is less than that of the initial state is one that releases energy (otherwise it would be in violation of the first law). An energy-releasing process is called an **exergonic** ("energy-out") reaction. As the second law predicts, only exergonic reactions can take place spontaneously—that is, without an input of energy from outside the system. (Spontaneously, though the word has an explosive sound to it, says nothing about the rate of the reaction, just whether or not it can take place at all.) By contrast, a process in which the potential energy of the final state is greater than that of the initial state is one that requires energy. Such energy-requiring processes are known as **endergonic** ("energy-in") reactions, and in order for them to proceed, an input of energy is required that is greater than the difference in energy between the products and the reactants.

One important factor in determining whether or not a reaction is exergonic is already familiar to us: ΔH, the change in heat content of a system. As we noted in Chapter 3, the energy change that takes place when glucose, for instance, is oxidized can be measured in a calorimeter and expressed in terms of ΔH. The oxidation of a mole of glucose yields 673 kilocalories. Or,

$$C_6H_{12}O_6 + 6O_2 \longrightarrow 6CO_2 + 6H_2O$$
$$\Delta H = -673 \text{ kcal/mole}$$

Generally speaking, an exergonic chemical reaction is also an exothermic reaction —that is, it gives off heat and thus has a negative ΔH. However, there are exceptions. One of the most dramatic is found with a substance known as dinitrogen pentoxide, which decomposes spontaneously and with explosive force to nitrogen dioxide and oxygen, and in so doing, absorbs heat:

$$2N_2O_5 \longrightarrow 4NO_2 + O_2$$
$$\Delta H = +26.18 \text{ kcal/mole}$$

In short, another factor besides the gain or loss of heat affects the change in potential energy and thus the direction of the process. This factor is given the formal name of **entropy,** and it is a measurement of the disorder or randomness of a system.

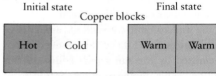

Initial state — Final state
Copper blocks

Heat flows from warm body to cool body

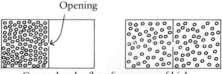

Opening

Gas molecules flow from zone of high pressure to zone of low pressure

Order becomes disorder

8–4 *Some illustrations of the second law of thermodynamics. In each case, a concentration of energy—in the hot copper block, in the gas molecules under pressure, and in the neatly organized books—is dissipated. In nature, processes tend toward randomness, or disorder. Only an input of energy can reverse this tendency and reconstruct the initial state from the final state. Ultimately, however, disorder will prevail, since the total amount of energy in the universe is finite.*

Before we examine more closely why the decomposition of dinitrogen pentoxide is exergonic despite its positive ΔH, let us return to the more familiar example of water. The change from ice to liquid water and the change from liquid water to water vapor are both endothermic processes—a considerable amount of heat is removed from the surrounding air as they take place. Yet, under the appropriate conditions, they proceed spontaneously. The key factor in all three of these examples is the increase in entropy. In the case of the dinitrogen pentoxide, a solid is being changed into two gases, and two molecules are being converted into five. In the case of ice and liquid water, a solid is being turned into a liquid, and some of the bonds that hold the water molecules together in a crystal (ice) are being broken. As the liquid water turns to vapor, the rest of the hydrogen bonds are ruptured as the individual water molecules dance off, one by one. In every case, the disorder of the system has increased.

The notion that there is more disorder associated with more numerous and smaller objects than with fewer, large ones is in keeping with our everyday experience. If I have 20 papers on my desk, the possibilities for disorder are greater than if I have 2 or even 10. If I cut each of the 20 in half, the entropy of the system—the capacity for randomness—increases. Also, the relationship between entropy and energy is a commonplace idea. If you were to find your room tidied up and your books in alphabetical order on the shelf, you would recognize that someone had been at work—that energy had been expended. For me to organize the papers on my desk similarly requires that I expend energy. Furthermore, it would be feasible to measure the energy expenditure in calories.

Now let us return to the question of the energy changes that determine the course of chemical reactions. Both the change in the heat content of the system (ΔH) and the change in entropy (which is symbolized as ΔS) contribute to the overall change in energy. This total change—the one that takes into account both heat and entropy—is called the **free energy change** and is symbolized as ΔG, after the American physicist Josiah Willard Gibbs (1839–1903), who was one of the first to put all of these ideas together.

The relationship between ΔG, ΔH, and entropy is given in the following equation:

$$\Delta G = \Delta H - T\Delta S$$

It states that the free energy change is equal to the change in heat (a negative value in exothermic reactions, remember) minus the change in entropy multiplied by the absolute temperature, T. In exergonic reactions, ΔH may be zero or may even be positive, but ΔG is always negative. As you can see in the equation, $T\Delta S$ is preceded by a minus sign. The greater the increase in entropy, the more negative ΔG will be; that is, the more exergonic the reaction will be.

With ΔG in mind, let us examine once more the combustion of glucose. The ΔH of that reaction is −673 kcal/mole. The free energy change, ΔG, is −686 kcal/mole. The increase in entropy has contributed 13 kcal/mole to the free energy change of the process. Thus both the change in heat and the change in entropy contribute to the lower energy state of the products of the reaction.

ΔG can also enable one to predict processes that occur when ΔH is zero or even positive, as with dinitrogen pentoxide. For instance, it confirms our earlier observations that heat will flow spontaneously from a hot object to a cold one, that dye molecules will diffuse spontaneously in a beaker of water, or that my desk will revert to disorder. In each of these processes, the final state has more entropy—and therefore less potential energy—than the initial state.

Previously we stated the second law in terms of the energy change between the initial and final states of a process. The second law can also be stated in another, simpler way: *All natural processes tend to proceed in such a direction that the disorder or randomness of the universe increases.*

(a)

(b)

(c)

(d)

8-5 *Although energy is dissipated in every conversion from one form to another, considerable work can be accomplished in the process. For example, the transformation of stored chemical energy to mechanical energy can be used by an organism to move to a more satisfactory position.* (a) *A queen conch, a mollusk, resting on the surface of the sand.* (b, c) *Vigorous muscular movements of the conch's spade foot create a depression of just the right size to reorient the animal in relation to its world* (d).

Living Systems and the Second Law

The universe, according to the present model, is a closed system—that is, neither matter nor energy either enters or leaves the system. The matter and energy present in the universe at the time of the primordial explosion (see page 23) are all the matter and energy it will ever have. Moreover, after each and every energy exchange and transformation, the universe as a whole has less potential energy and more entropy than it did before. In this view, of course, the universe is running down. The stars will flicker out, one by one; life—any form of life on any planet—will come to an end. Finally, even the motion of individual molecules will cease. However, even the most pessimistic among us do not believe this will occur for another 20 billion years or so.

In the meantime, life can exist *because* the universe is running down. Although the universe as a whole is a closed system, the earth is not. As we noted at the beginning of this chapter, it is receiving an energy input of 13×10^{23} calories per year from the sun. Photosynthetic organisms are specialists at capturing the light energy released by the sun as it slowly burns itself out. They use this energy to organize small, simple molecules (water and carbon dioxide) into larger, more complex molecules (sugars). In the process, the captured light energy is stored in the chemical bonds of sugars and other molecules. Living cells—including photosynthetic cells—can convert this stored energy into motion, electricity, light, and, by shifting the energy from one type of chemical bond to another, more convenient forms of chemical energy. At each transformation, energy is lost to the surroundings as heat. But before the energy captured from the sun is completely dissipated, organisms use it to create and maintain the complex organization of structures and activities that we know as life.

OXIDATION-REDUCTION

You will recall from Chapter 1 that electrons possess differing amounts of potential energy depending on their distance from the atomic nucleus and the attraction of the nucleus for electrons. An input of energy will boost an electron to a higher energy level, but without added energy an electron will remain at the lowest energy level available to it.

Chemical reactions are essentially energy transformations in which energy stored in chemical bonds is transferred to other, newly formed chemical bonds. In such transfers, electrons shift from one energy level to another. In many reactions, electrons pass from one atom or molecule to another. These reactions, which are of great importance in living systems, are known as oxidation-reduction (or redox) reactions. The *loss* of an electron is known as **oxidation,** and the atom or molecule that loses the electron is said to be oxidized. The reason electron loss is called oxidation is that oxygen, which attracts electrons very strongly, is most often the electron acceptor.

Reduction is, conversely, the *gain* of an electron. Oxidation and reduction always take place simultaneously because an electron that is lost by the oxidized atom is accepted by another atom, which is reduced in the process.

Redox reactions may involve only a solitary electron, as when sodium loses an electron and becomes oxidized to Na^+, and chlorine gains an electron and is reduced to Cl^-. Often, however, the electron travels with a proton, that is, as a hydrogen atom. In such cases, oxidation involves the removal of hydrogen atoms, and reduction the gain of hydrogen atoms. For example, when glucose is oxidized, hydrogen atoms are lost by the glucose molecule and gained by oxygen:

$$C_6H_{12}O_6 + 6O_2 \longrightarrow 6CO_2 + 6H_2O + \text{Energy}$$

The electrons are moving to a lower energy level, and energy is released.

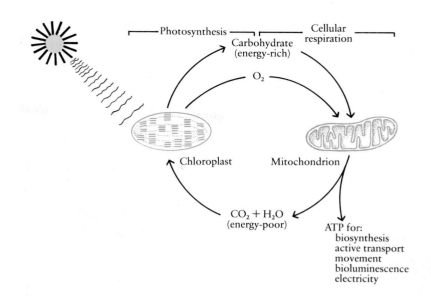

8–6 *The flow of biological energy. Chloroplasts, present in all photosynthetic eukaryotic cells, capture the radiant energy of sunlight and use it to convert water and carbon dioxide into carbohydrates, such as glucose, starch, and other foodstuff molecules. Oxygen is released as a product of the photosynthetic reactions.*

Mitochondria, present in all eukaryotic cells, carry out the final steps in the breakdown of these carbohydrates and capture their stored energy in ATP molecules. This process, cellular respiration, consumes oxygen and produces carbon dioxide and water, completing the cycling of the molecules.

With each transformation, some energy is dissipated to the environment in the form of heat. Thus the flow of biological energy is one-way and can continue only so long as there is an input of energy from the sun.

Conversely, in the process of photosynthesis, hydrogen atoms are transferred from water to carbon dioxide, thereby reducing the carbon dioxide to form glucose:

$$6CO_2 + 6H_2O + Energy \longrightarrow C_6H_{12}O_6 + 6O_2$$

In this case, the electrons are moving to a higher energy level, and an energy input is required to make the reaction occur.

In living systems, the energy-capturing reactions (photosynthesis) and energy-releasing reactions (glycolysis and respiration) are oxidation-reduction reactions. As we have seen, the complete oxidation of a mole of glucose releases 686 kilocalories of free energy (conversely, the reduction of carbon dioxide to form a mole of glucose stores 686 kilocalories of free energy in the chemical bonds of glucose). If this energy were to be released all at once, most of it would be dissipated as heat. Not only would it be of no use to the cell, but the resulting high temperature would be lethal. However, mechanisms have evolved in living systems that regulate these chemical reactions—and a multitude of others—in such a way that energy is stored in particular chemical bonds from which it can be released in small amounts as the cell needs it. These mechanisms generally involve sequences of reactions, some of which are oxidation-reduction reactions. Although each reaction in the sequence represents only a small change in the free energy, the overall free energy change for the sequence can be considerable.

METABOLISM

In any living system, energy exchanges occur through thousands of different chemical reactions, many of them taking place simultaneously. As we noted in Chapter 5, the sum of all these reactions is referred to as metabolism (from the Greek *metabole*, meaning "change"). If we were merely to list the individual chemical reactions, it would be difficult indeed to understand the flow of energy through a cell. Fortunately, there are some guiding principles that lead one through the maze of cell metabolism. First, virtually all the chemical reactions that take place in a cell involve enzymes—large protein molecules that play very specific roles. Second, biochemists are able to group these reactions in an ordered series of steps, commonly called a pathway; a pathway may have a dozen or more sequential reactions or steps. Each pathway serves a function in the overall life of the cell or organism. Furthermore, certain pathways have many steps in common —for instance, those that are concerned with the synthesis of the different amino acids or the various nitrogenous bases. Some pathways converge; for example, the pathway by which fats are broken down to yield energy leads to the pathway by which glucose is broken down to yield energy.

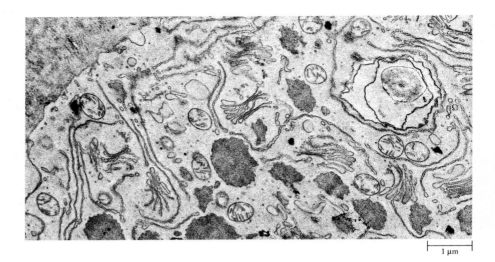

1 µm

8-7 *A chemical factory. Part of a cell from the root of a wheat plant. Cells such as this contain thousands of different organic molecules. These molecules form the many intricate structures of which the cell is composed and carry out the multitude of chemical reactions necessary for the life of the cell, its maintenance and growth, and its interactions with the cells around it. Many of the chemical activities of the cell are compartmentalized within the organelles, vesicles, and vacuoles, on the surfaces of the membranes, and in different regions of the cytoplasmic solution.*

Many types of living systems have pathways unique to them. Plant cells, for example, expend much of their energy building their cell walls, an activity not engaged in by animal cells. Red blood cells specialize in the synthesis of hemoglobin molecules, not made anywhere else in the animal body. It is not surprising that the distinctive differences in function among cells and organisms are correlated not only with their forms but also with their biochemistry. What *is* surprising, however, is that much of the metabolism of even the most diverse of organisms is exceedingly similar; the differences in many of the metabolic pathways of humans, oak trees, mushrooms, and jellyfish are very slight. Some pathways—for example, those of glycolysis and respiration—are virtually universal, found in almost all living systems.

The magnitude of the chemical work carried out by a cell—and its consequent energy expenditure—can be understood if one recognizes that, for the most part, the thousands of different molecules, large and small, found within a cell are synthesized there. The total of chemical reactions involved in synthesis is called **anabolism.** Cells also are constantly involved in the breakdown of larger molecules; these activities are known, collectively, as **catabolism.** Catabolism serves two purposes: (1) it releases the energy for anabolism and other work of the cell, and (2) it provides raw materials for anabolic processes.

Not only do living systems carry out this multitude of chemical activities, but they do so under what might seem, at first glance, extraordinarily difficult conditions. Most of their chemical reactions are carried out within individual living cells, and in cells, not two or three but thousands of different kinds of molecules are present. As we saw in Chapter 5, however, the cytoplasm of the living cell is highly structured, with the organelles, endoplasmic reticulum, vesicles, vacuoles, and the cytoskeleton itself effectively compartmentalizing the cell into different "work areas" (Figure 8-7). This segregates different reaction pathways from one another, enabling different chemical reactions to take place without mutual interference.

However, for any particular molecules to react with one another, it is not enough that they be in the same general region of the cell; they must be in extremely close proximity and, moreover, must collide with sufficient force to overcome the mutual repulsion of their electron clouds. The force required varies with the nature of the molecules; the more stable their initial state, the more forceful the collision must be. The force with which molecules collide depends on their kinetic energy, and the average kinetic energy of the molecules in a cell is quite moderate, as reflected in the moderate temperatures of living systems. In a group of molecules, it is likely that some proportion is moving with sufficient energy to cause a reaction to occur, but often this proportion is so small that the reaction, for all practical purposes, does not take place. How, then, can the complex chemical work of a cell be accomplished? The question can be answered in a single word: enzymes.

8-8 *In order to react, molecules must possess enough energy—the energy of activation—to collide with sufficient force to overcome their mutual repulsion and to weaken existing chemical bonds. An uncatalyzed reaction requires more activation energy than a catalyzed one, such as an enzymatic reaction. The lower activation energy in the presence of the catalyst is often within the range of energy possessed by the molecules, and so the reaction can occur at a rapid rate with little or no added energy. Note, however, that the overall energy change (ΔG) from the initial state to the final state is the same with and without the catalyst.*

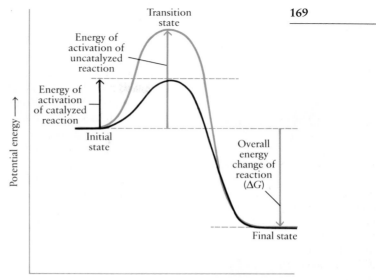

ENZYMES

To proceed at a reasonable rate, most chemical reactions require an initial input of energy to get started. This is true even for exergonic reactions such as the oxidation of glucose or the burning of natural gas (methane). The added energy increases the kinetic energy of the molecules, enabling a greater number of them to collide with sufficient force not only to overcome their mutual repulsion but also to break existing chemical bonds within the molecules. The energy that must be possessed by the molecules in order to react is known as the **energy of activation.** Sometimes, as in the case of natural gas, a spark is all that is needed to supply enough energy. Once the reaction begins, it liberates energy that is transferred to the other methane molecules until all are moving rapidly enough to react almost simultaneously with explosive force.

In the laboratory, the energy of activation is usually supplied as heat. But, in a cell, many different reactions are going on at the same time, and heat would affect all of these reactions indiscriminately. Moreover, heat would break hydrogen bonds and would have other generally destructive effects on the cell. Cells get around this problem by the use of enzymes, globular proteins that are specialized to serve as catalysts. A **catalyst** is a substance that lowers the activation energy required for a reaction by forming a temporary association with the molecules that are reacting (Figure 8-8). This temporary association brings the reacting molecules close to one another and may also weaken the existing chemical bonds, making it easier for new ones to form. The lower activation energy in the presence of the catalyst is within the range of energy possessed by a greater proportion of the reacting molecules; as a result, the reaction goes more rapidly than it would in the absence of a catalyst. The catalyst itself is not permanently altered in the process, and so it can be used over and over again.

Because of enzymes, cells are able to carry out chemical reactions at great speed and at comparatively low temperatures. For instance, the combination of carbon dioxide with water,

$$CO_2 + H_2O \rightleftharpoons \underset{\substack{\text{Carbonic} \\ \text{acid}}}{H_2CO_3}$$

can take place spontaneously, as it does in the oceans. In the human body, however, this reaction is catalyzed by an enzyme, carbonic anhydrase. This is one of the fastest enzymes known, with each enzyme molecule catalyzing the production of 6×10^5 (600,000) molecules of carbonic acid per second. The catalyzed reaction is 10^7 times faster than the uncatalyzed one. In animals, this reaction is

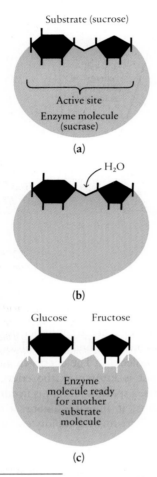

Substrate (sucrose)

Active site
Enzyme molecule
(sucrase)

(a)

H_2O

(b)

Glucose   Fructose

Enzyme
molecule ready
for another
substrate
molecule

(c)

8-9 A model of enzyme action. (a)
Sucrose, a disaccharide, (b) is hydrolyzed
to yield (c) a molecule of glucose and a
molecule of fructose. The enzyme involved
in this reaction, sucrase, is specific for this
process; as you can see, the active site of
the enzyme fits the opposing surface of
the sucrose molecule. The fit is so exact
that a molecule composed, for example, of
two subunits of glucose would not be
affected by this enzyme.

essential in the transfer of carbon dioxide from the cells, where it is produced, to the bloodstream, which transports it to the lungs. As carbonic anhydrase illustrates, enzymes are typically effective in very small amounts.

Almost 2,000 different enzymes are now known, each of them capable of catalyzing a specific chemical reaction. However, different types of cells are able to manufacture different types of enzymes—no cell contains all the known enzymes. The particular enzymes that a cell can manufacture are a major factor in determining the biological activities and functions of that cell. A cell can carry out a given chemical reaction at a reasonable rate only if it has a specific enzyme that can catalyze that reaction. The molecule (or molecules) on which an enzyme acts is known as its **substrate**. For example, in the reaction diagrammed in Figure 8–9, sucrose is the substrate and sucrase is the enzyme. (Note that enzyme names often end in "-ase.")

Enzyme Structure and Function

Enzymes are large to very large,* complex globular proteins consisting of one or more polypeptide chains. They are folded so as to form a groove or pocket on the surface into which the reacting molecule or molecules—the substrate—fit and where the reactions take place. This portion of the enzyme is known as the **active site.** Only a few amino acids of the enzyme are involved in any particular active site. Some of these may be adjacent to one another in the primary structure, but often the amino acids of the active site are brought close to one another by the intricate folding of the amino acid chain that produces the tertiary structure. In an enzyme with a quaternary structure, the amino acids of the active site may even be on different polypeptide chains (Figure 8–10).

The active site not only has a three-dimensional shape complementary to that of the substrate, but it also has a complementary array of charged or uncharged, hydrophilic or hydrophobic, areas on the binding surface. If a particular portion of the substrate has a negative charge, any corresponding feature on the active site is likely to have a positive charge, and so on. Thus the active site not only recognizes and confines the substrate molecule but also orients it in a particular direction.

The Induced-Fit Hypothesis

When the existence of active sites was first postulated by Emil Fischer in 1894, he compared the relationship between the active site and the substrate to that between a lock and a key. Within the last few years, however, studies of enzyme structure have suggested that the active site is considerably more flexible than a keyhole. The binding between enzyme and substrate appears to alter the conformation of the enzyme, thus inducing a close fit between the active site and the substrate (Figure 8–11). It is believed that this induced fit may put some strain on the reacting molecules and so further facilitate the reaction.

Cofactors in Enzyme Action

The catalytic activity of some enzymes appears to depend only upon the physical and chemical interactions between the amino acids of the active site and the substrate. Many enzymes, however, require additional nonprotein low-molecular-weight substances in order to function. Such nonproteins that are essential for enzyme function are known as **cofactors.**

---

* Different enzymes vary in their molecular weights from about 12,000 to more than 1 million. Amino acids have an average molecular weight of about 120, and this figure is used to estimate the number of amino acids in a polypeptide of known molecular weight.

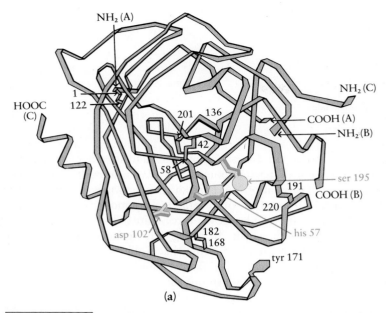

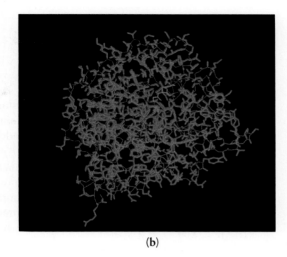

8–10 (a) A model of the digestive enzyme chymotrypsin. This enzyme is composed of three polypeptide chains. The amino (NH₂) and carboxyl (COOH) ends of each are labeled. The numbers represent the positions of particular amino acids in the chains. Five disulfide bridges connect amino acids 1 and 122, 42 and 58, 136 and 201, 168 and 182, and 191 and 220. The three-dimensional shape of the molecule is a result of a combination of disul-

fide bridges and of interactions among the chains and between the chains and the surrounding water molecules. These interactions are based on the positive or negative charges or the polarity of the various amino acids. As a result of the bending and twisting of the polypeptide chains, particular amino acids come together in a highly specific configuration to form the active site of the enzyme. Three amino acids known to be part of the active site

are shown in red. (b) A computer-generated model of the digestive enzyme carboxypeptidase. The backbone of the polypeptide chain is shown in red, and the side (R) groups of the amino acids are in blue. The yellow-green structure represents the substrate, a dipeptide, nestled in the active site. The small white star burst just below the substrate is the zinc ion (Zn²⁺), an essential cofactor for this enzyme.

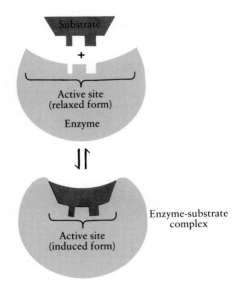

8–11 The induced-fit hypothesis. The active site is believed to be flexible and to adjust its conformation to that of the substrate molecule. This induces a close fit between the active site and the substrate and may also put some strain on the substrate molecule.

Ions as Cofactors

Certain ions are cofactors for particular enzymes. For example, the magnesium ion (Mg^{2+}) is required in all enzymatic reactions involving the transfer of a phosphate group from one molecule to another. As you will recall (see Table 3–1, page 57), the phosphate group is usually negatively charged in solution; its two negative charges are attracted by the two positive charges of the magnesium ion, holding it in position at the active site of the enzyme. K^+, Ca^{2+}, and other ions play similar roles in other reactions. In some cases, bonds between ions and the R groups of particular amino acids help to maintain certain folds in the tertiary structure or to hold a quaternary structure together.

Coenzymes and Vitamins

Nonprotein organic molecules may also function as cofactors in enzyme-catalyzed reactions. Such molecules are called **coenzymes;** they are bound, either temporarily or permanently, to the enzyme, usually fairly close to the active site. Some coenzymes function as electron-acceptors in oxidation-reduction reactions, receiving electrons—often a pair of electrons accompanied by a hydrogen ion (that is, a proton)—and then passing them on to another molecule. There are several different kinds of electron-accepting coenzymes in any given cell, each capable of holding electrons at a slightly different energy level.

One of the most frequently encountered coenzymes is known as nicotinamide adenine dinucleotide (NAD). A nucleotide is, as we saw in Chapter 3 (page 80), a complex molecule composed of three subunits: a phosphate group, a five-carbon sugar, and a nitrogenous base. A dinucleotide, such as NAD, is a molecule that consists of two nucleotides. In NAD, the two nucleotides forming the molecule

8-12 *The nitrogenous bases* **(a)** *nicotinamide and* **(b)** *adenine. Each of the nitrogen atoms in these molecules has an unshared pair of electrons that exerts a weak attraction for hydrogen ions (H⁺). Thus nicotinamide and adenine are bases, combining with hydrogen ions and thereby increasing the relative number of hydroxide ions in a solution (see page 48).*

Ribose

8-13 *The five-carbon sugar ribose. Ribose is an essential component of many biologically important molecules. These include not only many coenzymes but also ATP and related compounds and, as we shall see in Chapter 15, several types of RNA molecules involved in protein synthesis.*

contain two different nitrogenous bases, nicotinamide and adenine (Figure 8–12). Adenine is a molecule we shall be mentioning frequently in subsequent chapters; in addition to its role as a subunit of NAD and other coenzymes, it is part of the ATP molecule (see page 180) and is one of the principal components of the nucleic acids DNA and RNA. (The use of adenine for three quite different purposes is an example of the economy with which the living cell operates.) Nicotinamide occurs as a component of the vitamin niacin. Niacin, like other vitamins, is a compound, required in small quantities, that we and other animals cannot synthesize ourselves and so must obtain in our diets (see Table 34–3). Thus we must eat foods containing niacin (which includes both nicotinamide and nicotinic acid but should not be confused with nicotine, found in tobacco). When nicotinamide is present, our cells can use it to make NAD. A number of other coenzymes are also vitamins or contain vitamins as subunits of the molecule.

The other components of NAD are two molecules of the five-carbon sugar ribose (Figure 8–13) and two phosphate groups. In the NAD molecule, the ribose subunits are linked together by the phosphate groups. One of the ribose units is linked, in turn, to the nicotinamide subunit, and the other to the adenine subunit (Figure 8–14).

The nicotinamide ring is the business end of NAD, the part that accepts—and subsequently releases—the electrons. In its oxidized, electron-accepting state, the molecule has a positive charge and is written as NAD⁺. When it accepts two electrons and one proton, it is reduced to NADH. Like other coenzymes, this

NAD⁺ $H^+ + 2e^-$ ⇌   NADH

8-14 *Nicotinamide adenine dinucleotide (NAD) in its oxidized form, NAD⁺, and its reduced form, NADH. Notice how the bonding within the nicotinamide ring shifts as the molecule changes from the oxidized to the reduced form, and vice versa.*

As indicated above the arrows, reduction of NAD⁺ to NADH requires two electrons and one hydrogen ion (H⁺). The two elec-

trons, however, generally travel as components of two hydrogen atoms; thus, there is one hydrogen ion "left over" when NAD⁺ is reduced. As we shall see in the next chapter, H⁺ ions released into the surrounding solution when NAD⁺ is reduced play a critical role in powering vital cellular processes.

Auxotrophs

Sometimes, as a result of a genetic mutation, an organism is unable to synthesize a particular enzyme in an active form. When this occurs, the reactions of the pathway in which that enzyme participates cannot proceed to completion. The end product, which may be of critical importance to the organism, is not formed, and there may be, moreover, an accumulation of the substrate of the defective or missing enzyme. As we shall see in Chapter 19, such accumulations can lead to serious human disease or even death.

Although mutations resulting in defective or missing enzymes are no less serious for microorganisms, such as bacteria and fungi, they do provide biologists with a valuable tool for elucidating enzymatic pathways. Cells with a defect in a biosynthetic pathway

are known as auxotrophs. They grow normally only if they are supplied with either the end product of the entire pathway or the product of the specific reaction normally catalyzed by the defective or missing enzyme. Auxotrophs can be produced by irradiating cells with x-rays; they are identified and isolated by their inability to grow on chemical media that will support the growth of normal cells. Studies of the exact chemical requirements of different auxotrophs have revealed the details of many biosynthetic pathways, particularly those involved in the synthesis of amino acids. As we shall see in Chapter 15, auxotrophs have also made a major contribution to our understanding of the mechanism by which hereditary information is converted into the structures and processes of the living cell.

One group of auxotrophs of the red bread mold Neurospora crassa *have defective enzymes (indicated in red) at different points in the biosynthesis of the amino acid arginine (arg). These mutants, each deficient in one enzyme, retain the activity of the other enzymes of the arginine pathway. Four of the enzymes (E_1 through E_4) and four of the intermediate products (A through D) are shown in the diagram. The substance accumulated (indicated in gray) by each auxotroph and its requirements for growth reveal the sequence of the enzymes and the intermediates in the reaction pathway.*

$$A \xrightarrow{E_1} B \xrightarrow{E_2} C \xrightarrow{E_3} D \xrightarrow{E_4} arg \qquad \text{Normal cell}$$

$$A \xrightarrow{E_1} B \xrightarrow{E_2} C \xrightarrow{E_3} D - \boxed{E_4} \; arg \qquad \text{Auxotroph I: accumulates D and requires arg for growth}$$

$$A \xrightarrow{E_1} B \xrightarrow{E_2} C - \boxed{E_3} \; D \xrightarrow{E_4} arg \qquad \text{Auxotroph II: accumulates C and requires either D or arg for growth}$$

$$A \xrightarrow{E_1} B - \boxed{E_2} \; C \xrightarrow{E_3} D \xrightarrow{E_4} arg \qquad \text{Auxotroph III: accumulates B and requires C, D, or arg for growth}$$

molecule is recycled—that is, NAD^+ is regenerated when NADH passes its two electrons and one proton on to another electron acceptor. Thus, although this coenzyme is involved in many cellular reactions, the actual number of $NAD^+/NADH$ molecules required is relatively small.

Enzymatic Pathways

Enzymes typically work in series—the pathways we referred to earlier.

$$\text{Product 1} \xrightarrow{\text{Enzyme 1}} \text{Product 2} \xrightarrow{\text{Enzyme 2}} \text{Product 3} \xrightarrow{\text{Enzyme 3}}$$
$$\text{Product 4} \xrightarrow{\text{Enzyme 4}} \text{Product 5} \xrightarrow{\text{Enzyme 5}} \text{End product}$$

Cells derive several advantages from this sort of arrangement. First, the groups of enzymes making up a common pathway can be segregated within the cell. Some are found in solution, as in the lysosomes, whereas others are embedded in the membranes of particular organelles. The enzymes located in membranes appear to be lined up in sequence, so the product of one reaction moves directly to the adjacent enzyme for the next reaction of the series. A second advantage is that there is little accumulation of intermediate products, since each product tends to

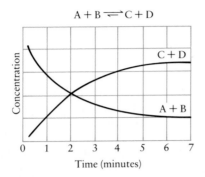

$$A + B \rightleftharpoons C + D$$

8–15 *The changes in concentration of products and reactants in a reversible reaction. At first, only molecules of A and B are present. The reaction begins when A and B start to yield products C and D. At the end of two minutes, the concentrations of A + B and C + D are equal. As the reaction proceeds, the concentration of C + D will continue to increase to the point of chemical equilibrium (at about the sixth minute) and will thereafter remain greater than the concentration of A + B. This is the proportion at which the rates of forward and reverse reactions are the same.*

be used up in the next reaction along the pathway. A third and most important advantage can be understood by considering the nature of chemical equilibrium.

As we have noted previously (page 47), chemical reactions can go in either direction. When net change ceases, the reaction is said to be at equilibrium. In the reaction

$$A + B \rightleftharpoons C + D$$

the point of equilibrium is reached when as many molecules of C and D are being converted to molecules of A and B as molecules of A and B are being converted to molecules of C and D.

The concentration of reactants does *not* have to equal the concentration of products in order for equilibrium to be established; only the *rates* of the forward and reverse reactions must be the same. Consider the reaction shown above. The different lengths of the arrows indicate that at equilibrium there is more C + D present than A + B. If only A and B molecules are present initially, the reaction occurs at first to the right, with A and B molecules converting into C and D molecules. Figure 8–15 shows the relative changes in concentration as the reaction continues. As C and D accumulate, the rate of the reverse reaction increases, and at the same time, the rate of the forward reaction decreases because of the decreasing concentrations of A and B. At about minute 6, the rates of the forward and reverse reactions equalize and no further changes in concentration take place. The proportions of A + B and C + D will remain the same. There will always be more C + D molecules in the system than A + B molecules, but the reaction will not go to completion—that is, not all of the A + B molecules will be converted to C + D molecules.

The relative proportions of A + B and C + D at equilibrium are determined by the free energy change (ΔG) of the reaction. Only if there were no net change in the free energy (that is, $\Delta G = 0$) would the concentrations of A + B and C + D be equal at equilibrium. The fact that the concentration of C + D molecules is greater than that of A + B molecules at equilibrium tells us that the potential energy of C + D is less than the potential energy of A + B. The reaction proceeding from A + B to C + D is exergonic (that is, it has a negative ΔG); conversely, the reaction from C + D to A + B is endergonic (positive ΔG). In any reversible reaction, the point of equilibrium will lie in the direction for which ΔG is negative. The larger the negative value of ΔG, the more strongly the reaction will be pulled in that direction.

This has important consequences in the sequential reactions that occur within living cells. If A + B and C + D molecules are in a closed system, equilibrium will be attained and there will be no subsequent change in their concentrations. If, however, the system is open and C + D molecules are continually removed from it, equilibrium will never be attained, and the conversion of A + B molecules into C + D molecules will continue. The sequential reactions of enzymatic pathways have the effect of removing the product of each reaction from the system, so that equilibrium is not attained. If, for example, product 2 in the enzymatic pathway shown on page 173 is used up (by being converted into product 3) almost as rapidly as it is formed, the reaction product 1 $\rightarrow$ product 2 can never reach equilibrium. If the eventual end product is also used up, the whole series of reactions will move toward completion. Moreover, if any of the reactions along the pathway are highly exergonic (large negative ΔG), they will rapidly use up the products of the preceding reactions, pulling those reactions forward; similarly, the accumulation of the products from the exergonic reactions will push the subsequent reactions forward by increasing the concentrations of the reactants. The linking of reactions in enzymatic pathways, with exergonic reactions moving the entire series forward, is a key factor in the remarkable efficiency with which living organisms carry out their chemical activities.

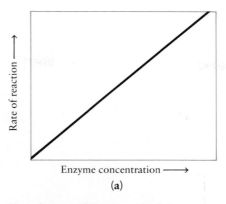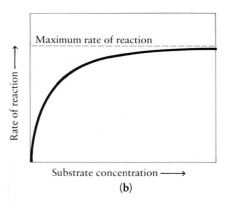

(a) (b)

8–16 *The effect of enzyme and substrate concentrations on the rate of a typical enzymatic reaction for which all necessary cofactors are available in ample supply.* **(a)** *In the presence of excess substrate, the reaction rate increases in direct proportion to the concentration of enzyme. Regulation of enzyme concentrations is a principal means by which cells regulate the rate of chemical reactions.* **(b)** *If the enzyme concentration remains constant as the substrate concentration increases, the rate of reaction increases until it approaches a maximum. Further increases in the substrate concentration have no additional effect on the reaction rate. The maximum rate is attained when all of the enzyme molecules are occupied by substrate molecules.*

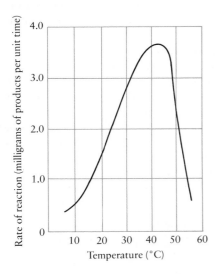

8–17 *The effect of temperature on the rate of an enzyme-controlled reaction. The concentrations of enzyme and substrate molecules were kept constant. As you can see, the rate of the reaction, as in most chemical reactions, approximately doubles for every 10°C rise in temperature. In the enzymatic reactions of humans (body temperature 37°C) and other mammals, the maximum rate of reaction is attained at about 40°C. Above this temperature, the rate decreases, and at about 60°C the reaction stops altogether, presumably because the enzyme is denatured. Although the shape of the curve is similar for all enzymatic reactions, the temperature range through which an enzyme is active varies with the type of organism and the particular enzyme.*

Regulation of Enzyme Activity

Another remarkable feature of the metabolic activity of cells is the extent to which each cell regulates the synthesis of the products necessary to its well-being in the amounts and at the rates required, while avoiding overproduction, which would waste both energy and raw materials. This regulation depends, in turn, on the regulation of enzyme activity.

The concentrations of enzyme and substrate molecules, as well as the availability of cofactors, are principal factors in limiting enzyme action (Figure 8–16). Most enzymes probably work at a rate well below their maximum because of these limitations. Moreover, many enzymes are broken down rapidly, characteristically by other enzymes that hydrolyze peptide bonds. A highly efficient means of regulation of these quickly degraded enzymes is for the cell to produce them only when they are needed. The way in which bacterial cells turn on and off their enzyme production at the source is described in some detail in Chapter 16.

Some enzymes are produced only in an inactive form and, just when they are needed, are activated, usually by another enzyme. Chymotrypsin, a digestive enzyme, is controlled in this way. It is synthesized by cells in the pancreas in the form of chymotrypsinogen, which consists of a single, very long polypeptide chain and is inactive. When this molecule is released into the small intestine, where it performs its digestive work, the enzyme trypsin snips out dipeptides at two points in the chain. The resulting three segments constitute the active chymotrypsin molecule (Figure 8–10a, page 171). In this way, chymotrypsin molecules (and other digestive enzymes) are prevented from digesting the proteins in the cells in which they are synthesized.

Living systems also have several other ways of turning enzyme activity on and off and of regulating its level.

Effects of Temperature and pH

As we noted earlier, an increase in temperature increases the rate of uncatalyzed chemical reactions. This temperature effect also holds true for enzyme-catalyzed reactions—but only up to a point. As you can see in Figure 8–17, the rate of most enzymatic reactions approximately doubles for each 10°C rise in temperature and then drops off very quickly above 40°C. The increase in reaction rate occurs because, at higher temperatures, more of the substrate molecules possess sufficient energy to react; the decrease in the reaction rate occurs as the movement and vibration of the enzyme molecule itself increases, disrupting the hydrogen bonds and other relatively fragile forces that maintain its tertiary structure. A molecule that has lost its characteristic three-dimensional structure in this way is said to be **denatured.** Partially denatured enzymes (in which the structure is only slightly distorted) regain their activity on being cooled, indicating that their polypeptide chains have regained their necessary shape. If the denaturation is sufficiently severe, however, it is irreversible, leaving the polypeptide chains permanently tangled and inactivated.

8–18 *The body of a man with a noose around his neck was found in a Danish peat bog in 1950. He died some 2,000 years ago. The remarkable preservation of the body is the result of the extremely acidic pH of the peat bog, which almost completely inhibited the enzymatic activities of the microorganisms that customarily decompose organic molecules.*

The pH of the surrounding solution also affects enzyme activity. The conformation of an enzyme depends, among other factors, on attractions and repulsions between negatively charged (acidic) and positively charged (basic) amino acids. As the pH changes, these charges change, and so the shape of the enzyme changes until it is so drastically altered that it is no longer functional. More important, probably, the charges of the active site and the substrate are changed so that the binding capacity is affected. The optimum pH of one enzyme is not the same as that of another. The digestive enzyme pepsin, for example, works at the very low (highly acidic) pH of the stomach (page 52), in an environment where most other proteins would be permanently denatured. Some enzymes are usually found at a pH that is not their optimum, suggesting that this discrepancy may not be an evolutionary oversight but a way of damping enzyme activity.

Allosteric Interactions

An ingenious mechanism by which an enzyme may be temporarily activated or inactivated is known as **allosteric interaction.** Allosteric interactions occur among enzymes that have at least two binding sites, one the active site and another, into which a second molecule, known as an allosteric effector, fits. The binding of the effector changes the shape of the enzyme molecule and either activates or inactivates it (Figure 8–19).

8–19 *An allosteric ("other shape") effector can bind to an enzyme and, by altering the bonds determining its tertiary structure, change the conformation of the active site. As a consequence, the enzyme may be altered so that it cannot interact with its substrate, as shown here, or it may be activated.*

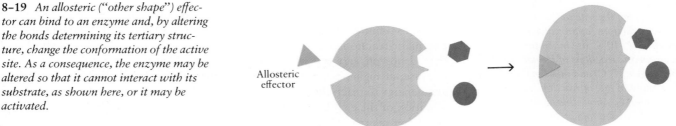

Allosteric effector

Allosteric interactions are frequently involved in **feedback inhibition,** which is a common means of biological control. A familiar nonbiological example of feedback inhibition is a thermostat that turns off the furnace when the room temperature reaches a desired level. In feedback inhibition of enzymatic reactions,

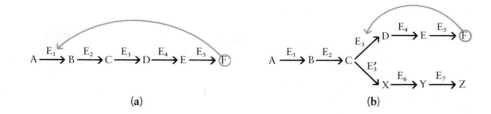

(a)   **(b)**

8-20 *Feedback inhibition.* (a) *In this series of reactions, each step (the black arrows) is catalyzed by a specific enzyme. Enzyme E_1, which converts A to B, is allosteric, and the allosteric inhibitor is the product F. Thus, enzyme E_1 will be more active when amounts of F are low.* (b) *A branched metabolic pathway, A being converted to B, B being converted to C, and C being converted to both D and X. In this case, the enzyme at the branch, E_3, is allosteric. As the quantity of the allosteric inhibitor F increases, the upper branch of the pathway will become less active, and the reaction series will be primarily shunted to the production of Z.*

one of the products, often the last in the series, acts as an allosteric effector, inhibiting the function of one of the enzymes, often the first in the series (Figure 8–20a). Or, in a reaction that may take one of two directions, the effector may act to shunt the reactions along another pathway (Figure 8–20b).

Competitive Inhibition

Some compounds inhibit enzyme activity by temporarily occupying the active site of the enzyme; regulation in this way is known as competitive inhibition, because the regulatory compound and the substrate compete with each other for binding to the active site. Competitive inhibition is completely reversible; the result of the competition at any particular time depends on how many of each kind of molecule are present. For example, in the reaction series

$$A \xrightarrow{E_1} B \xrightarrow{E_2} C \xrightarrow{E_3} D \xrightarrow{E_4} E \xrightarrow{E_5} F$$

the final product F might be rather similar in structure to product D. F could occupy the active site of enzyme E_4, preventing D, the normal substrate, from binding to the enzyme. As F was used up by the cell, the active site of enzyme E_4 would once more become available to D.

Competitive inhibition is the mechanism of action of some drugs used to treat bacterial infections in animals. For instance, bacteria make the vitamin folic acid, which animal cells do not make (animals obtain folic acid from their food). One of the compounds in the metabolic pathway leading to folic acid is *para*-aminobenzoic acid (PABA). As you can see in Figure 8–21, the drug sulfanilamide has a structure very similar to that of PABA. The two structures are so similar, in fact, that the enzyme involved in converting PABA to folic acid combines with the drug rather than with the PABA. Without folic acid, the bacterial cell dies, leaving the animal cell, which lacks this enzyme, unharmed.

para-Aminobenzoic acid
(PABA)

Sulfanilamide

8-21 para-*Aminobenzoic acid (PABA) is one of the compounds in the metabolic pathway to folic acid in bacterial cells. Sulfanilamide, a drug, has a similar structure. It can combine with the enzyme that converts PABA to folic acid, thereby blocking the synthesis of folic acid, without which the bacterial cell cannot live.*

Noncompetitive Inhibition

In noncompetitive inhibition, the inhibitory chemical, which need not resemble the substrate, binds with the enzyme at a site on the molecule other than the active site. Lead, for instance, forms covalent bonds with sulfhydryl (SH) groups. Many enzymes contain cysteine, which has a sulfhydryl group. The binding of lead to such enzymes disrupts their tertiary structure and deactivates them, producing the symptoms associated with lead poisoning. Like competitive inhibition, noncompetitive inhibition is often reversible, but such reversal is not accomplished by increased concentrations of substrate. In the case of lead, for example, the inhibition can be reversed by treatment with other sulfhydryl-containing compounds that bind the lead atoms more tightly than cysteine does.

Irreversible Inhibition

Some substances inhibit enzymes irreversibly, either binding permanently with key functional groups of the active site or so thoroughly denaturing the protein that

8–22 *The bacterial cell wall has polysaccharide backbones formed by alternating molecules of two sugars, NAG and NAM. The backbones (indicated by the thick black lines) are cross-linked by short peptide chains. One of the amino acids in these chains, lysine (shown in red), forms three peptide bonds—one with the amino acid above it, one with the amino acid below it, and another with the adjacent amino acid. These bonds are formed in an unusual way—by the enzymatic transfer of a peptide bond from one molecule to another. Penicillin blocks this transfer. Its structure mimics that of a dipeptide, and it binds to the active site of the enzyme, irreversibly inhibiting it. Without cross-links, the cell wall cannot hold the cell together. Thus penicillin acts specifically on growing bacterial cell walls.*

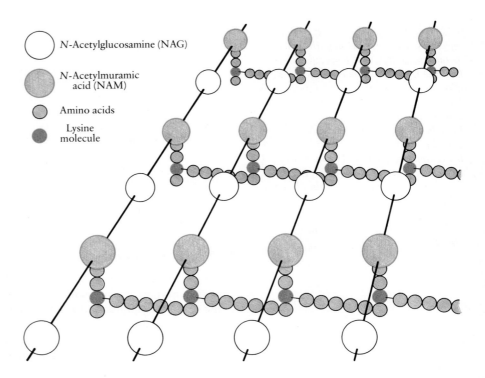

○ N-Acetylglucosamine (NAG)

● N-Acetylmuramic acid (NAM)

· Amino acids

· Lysine molecule

its tertiary structure cannot be restored. The nerve gases, used widely during World War I and now banned, are among the most potent poisons known. They irreversibly inhibit enzymes involved in transmission of the nerve impulse, resulting in paralysis and death. Many useful drugs, including the antibiotic penicillin (Figure 8–22), are also irreversible inhibitors of enzyme activity.

Membrane Transport Proteins and Receptors Revisited

In Chapter 6, we noted that the integral membrane proteins that transport molecules and ions across the lipid bilayer are sometimes called permeases because of their similarities to enzymes. We are now in a position to see just how similar they are. Both enzymes and membrane transport proteins are large, globular protein molecules with complex tertiary or quaternary structures. In both types of molecules, the tertiary or quaternary structure produces a precisely configured region on the surface—an active site or a binding site—into which another specific molecule or ion fits. In enzymes, association of a substrate molecule with the active site leads to a chemical change in the substrate; in transport proteins, the association of a molecule or ion with the binding site leads to the movement of that molecule or ion across the membrane. The changes in protein conformation that result from the binding play an important role in either catalyzing the chemical reaction or propelling the molecule or ion through the protein to the other side of the membrane.

These obvious similarities are, however, just the beginning. The rate at which a specific membrane protein transports its particular substance across the lipid bilayer—like the rate at which an enzyme catalyzes its particular reaction—is affected by the number of molecules of that specific protein in the membrane, the concentration of the molecules or ions to be transported, the temperature, and the pH. Moreover, the transport functions of integral membrane proteins are similarly regulated by allosteric interactions, reversible inhibitions (both competitive and noncompetitive), and irreversible inhibitions. An example of allosteric

Some Like It Cold

Enzyme action is exquisitely dependent on the tertiary and quaternary structure of the globular protein, particularly as it affects the active site. As a consequence, many enzymes are thermolabile—that is, they do not function at higher temperatures, even within the normal physiological range. One such enzyme is responsible for color in Siamese cats. It functions adequately in the cooler, peripheral areas of the body, such as the ears, nose, paws, and tip of the tail, but becomes inactive in the warmer areas of the body. For similar reasons, Himalayan rabbits are all black when raised at temperatures of about 5°C; white with black ears, forepaws, noses, and tails when raised at normal room temperatures; and all white when raised at temperatures above 35°C.

Thermolability has its advantages. In the northern seal, the newborns are white, as a result of their developing at a warm (internal) temperature. The newborns cannot swim and so are restricted to the ice floes, where their white coats provide them with color camouflage. By the time they are able to swim, their coats have turned brown and so blend with the dark Arctic waters. The thermolability of this same enzyme (tyrosinase) provides similar camouflage for the Arctic fox. During the summer it develops a white coat, which provides color protection during the winter months. During the winter, its dark coat develops, which is revealed when the white coat is shed in the spring. All of which illustrates that, as we shall see often, evolution is opportunistic.

(a)

(b)

(a) *The characteristic color pattern of the Siamese cat is a result of the thermolability of an enzyme affecting coat color. This enzyme, involved in controlling the synthesis of a dark pigment, is active only in the cooler, more peripheral areas of the* body. (b) *Similarly, the enzyme controlling dark pigmentation in northern seals is active only at low temperatures. The newborns, which, like other mammals, develop within the warm bodies of their mothers, are white. New fur, growing in* after the seals are exposed to the colder external environment, is brown. Although the white fur protects the newborn seals against most predators, it makes them highly desirable to human hunters, now their principal predators.

regulation of a transport protein is provided by the acetylcholine receptor. As we noted previously (page 137), when acetylcholine binds to this particular protein in the cell membrane of a nerve cell, a channel is opened in a closely associated protein through which sodium ions (Na^+) move down a concentration gradient into the cell.

Some of the protein receptors located on the surface of cell membranes are not involved in transport at all. They are instead the allosteric binding sites of enzyme molecules that are integral membrane proteins. In at least some cases, the allosteric effector binds to its receptor—located in the portion of the protein exposed on the external face of the membrane—and triggers a change in the

8-23 *Adenosine triphosphate (ATP) is the cell's chief energy currency. The bonds between the three phosphate groups in the molecule are important in ATP function. In order to represent more accurately the position of the adenine subunit in the ATP molecule, we have rotated it 180° (right to left) from the orientation depicted in Figures 8-12 and 8-14.*

Adenine

Phosphates

Ribose

Adenosine triphosphate (ATP)

conformation of the active site of the enzyme—located on the cytoplasmic face of the membrane. The result is a chemical reaction within the cell. In other cases, such as the cell membrane receptor for the hormone adrenaline, the receptor and the enzyme in which activity is triggered are apparently two distinct proteins. In Chapter 40, we shall examine the sequence of enzymatic reactions that result from the binding of adrenaline to its receptor.

Transport processes across cell membranes are, like chemical reactions, governed by the laws of thermodynamics. Substances can move spontaneously only from a state of higher potential energy to a state of lower potential energy. Such spontaneous movement in diffusion—either simple or facilitated—is driven by differences in potential energy that result from differences in concentration. Often, however, cells must transport substances against a concentration gradient—that is, from a state of lower potential energy to a state of higher potential energy. For such active transport processes, as for many chemical reactions, an input of energy is required.

THE CELL'S ENERGY CURRENCY: ATP

All of the biosynthetic activities of the cell, many of its transport processes, and a variety of other activities require energy. A large proportion of this energy is supplied by a single substance, **adenosine triphosphate,** or ATP. Glucose and other carbohydrates are storage forms of energy and also forms in which energy is transferred from cell to cell and organism to organism. In a sense, they are like money in the bank. ATP, however, is like the change in your pocket—it is the cell's immediately spendable energy currency.

At first glance, ATP, as shown in Figure 8-23, appears to be a complex, unfamiliar molecule. If you look at it more closely, however, you will find that you recognize all of its component parts. It is made up of the nitrogenous base adenine, the five-carbon sugar ribose, and three phosphate groups. These three phosphate groups with strong negative charges are covalently bonded to one another; this is an important feature in ATP function. Three linked phosphate groups are also characteristic of other molecules that play a role similar to that of ATP in certain cellular reactions; for example, guanosine triphosphate (GTP), another energy-carrying molecule, differs from ATP only in the substitution of the nitrogenous base guanine for adenine.

To understand the role of ATP and related triphosphate compounds, we must return briefly to the concept of the chemical bond. Because a chemical bond is a stable configuration of electrons, reacting molecules must possess a certain amount of energy in order to collide with sufficient force to overcome their mutual repulsion and to weaken existing chemical bonds, allowing the formation of new bonds. This energy is the energy of activation (Figure 8–8). Because of enzymes, which reduce the required energy of activation to a level already possessed by a significant proportion of the reacting molecules, the reactions essential to life are able to proceed at an adequate rate. As we have seen, however, the *direction* in which a reaction proceeds is determined by the free energy change, ΔG. Only if the reaction is exergonic (negative ΔG), will it proceed to any significant extent. Yet many cellular reactions, including synthetic reactions such as the formation of a disaccharide from two monosaccharide molecules, are endergonic (positive ΔG). In such a reaction, the electrons forming the chemical bonds of the product are at a higher energy level than the electrons in the bonds of the starting materials—that is, the potential energy of the product is greater than the potential energy of the reactants, an apparent violation of the second law of thermodynamics. Cells circumvent this difficulty by **coupled reactions** in which endergonic reactions (or transport processes, such as the active transport of a substance against a concentration gradient) are linked to exergonic reactions that provide a surplus of energy, making the entire process exergonic and thus able to proceed spontaneously. The molecule that most frequently supplies energy in such coupled reactions is ATP.

The internal structure of the ATP molecule makes it unusually suited to this role in living systems. In the laboratory, energy is released from the ATP molecule when the third phosphate is removed by hydrolysis, leaving ADP (adenosine diphosphate) and a phosphate:

$$ATP + H_2O \longrightarrow ADP + Phosphate$$

In the course of this reaction, about 7 kilocalories of energy are released per mole of ATP. Removal of the second phosphate produces AMP (adenosine monophosphate) and releases an equivalent amount of energy.

$$ADP + H_2O \longrightarrow AMP + Phosphate$$

The covalent bonds linking these two phosphates to the rest of the molecule are symbolized by a squiggle, $\sim$, and were, for many years, called "high-energy" bonds—an incorrect and confusing term. These bonds are not strong bonds, like the covalent bonds between carbon and hydrogen, which have a bond energy of 98.8 kilocalories per mole. They are instead bonds that are easily broken, releasing an amount of energy—about 7 kilocalories per mole—adequate to drive many of the essential endergonic reactions of the cell. Moreover, the energy released does not arise entirely from the movement of the bonding electrons to lower energy levels. It is also a result of a rearrangement of the electrons in other orbitals of the ATP or ADP molecules. The phosphate groups each carry negative charges and so tend to repel each other. When a phosphate group is removed, the molecule undergoes a change in electron configuration that results in a structure with less energy.

ATP in Action

In living cells, ATP is sometimes directly hydrolyzed to ADP plus phosphate, releasing energy for a variety of activities. ATP hydrolysis provides, for example, a means for producing heat, as in those animals, such as birds and mammals, that generally maintain a high and constant body temperature. Enzymes catalyzing the hydrolysis of ATP are known as ATPases; a variety of different ATPases have been identified. The protein "arms" in cilia and flagella (page 123) and on microtubules of the mitotic spindle (page 153), for example, are ATPase molecules, catalyzing the energy release that causes the microtubules to move past one another. Many of the proteins that move molecules and ions through cell membranes against a concentration gradient are not only transport proteins but also ATPases, releasing energy to power the transport process.

(a)

(b)

8–24 *Living organisms use the energy stored in the phosphate bonds of ATP for a variety of purposes. (a) In the skunk cabbage, a common plant in bogs and marshy areas of the northeastern United States, ATP is hydrolyzed to produce heat. The plant produces enough heat to melt surrounding snow or ice, while maintaining a nearly constant internal temperature of about 22°C (72°F). (b) Certain organisms, such as these bioluminescent mushrooms, transform ATP energy into light energy, thereby glowing in the dark.*

Usually, however, the terminal phosphate group of ATP is not simply removed but is transferred to another molecule. This addition of a phosphate group is known as **phosphorylation;** enzymes that catalyze such transfers are known as kinases. Phosphorylation reactions transfer some of the energy of the phosphate group in the ATP molecule to the phosphorylated compound, which, thus energized, participates in a subsequent reaction.

For example, in the reaction

$$W + X \longrightarrow Y + Z$$

if the potential energy of W plus the potential energy of X were less than that of Y plus Z, the reaction would not take place to any significant extent. Chemists could drive the reaction forward by supplying outside energy, probably in the form of heat. The cell might handle it in a two-step process. First:

$$W + ATP \longrightarrow W\text{-}P + ADP$$

The potential energy of the products is less than that of the reactants, so this reaction will take place. However, much of the energy made available when the phosphate group was removed from ATP is conserved in the new compound W–phosphate, or W–P. The next step in the process becomes

$$W\text{-}P + X \longrightarrow Y + Z + P$$

With the release of the phosphate from W, this second reaction also becomes one in which the potential energy of the products is less than the potential energy of the reactants and which, therefore, can take place.

Take, for instance, the formation of sucrose in sugarcane.

$$Glucose + Fructose \longrightarrow Sucrose + H_2O$$

In this reaction, the potential energy of the products is 5.5 kilocalories per mole greater than the potential energy of the reactants (that is, $\Delta G = +5.5$ kcal/mole). However, the sugarcane plant carries out this synthesis through a series of reactions coupled to the breakdown of ATP and the accompanying phosphorylation of the glucose and fructose molecules. The overall reaction is

$$Glucose + Fructose + 2ATP \longrightarrow Sucrose + 2ADP + 2P$$

Since the potential energy of 2 ADPs is about 14 kilocalories per mole less than the potential energy of 2 ATPs, the overall difference in products and reactants becomes 8.5 kilocalories per mole (that is, $\Delta G = -8.5$ kcal/mole). The coupling of reactions permits sugarcane to form sucrose.

Where does the ATP come from? As we shall see in the next chapter, energy released in the cell's catabolic reactions, such as the breakdown of glucose, is used to "recharge" the ADP molecule to ATP. Thus the ATP/ADP system serves as a universal energy-exchange system, shuttling between energy-releasing reactions and energy-requiring ones.

One marvels at the process of evolution as exemplified by the intricate flower of an orchid, the shell of a chambered nautilus, or the opposable thumb and forefinger of the human hand. Remember that ATP, NAD, and indeed the place of each amino acid in the polypeptide chain of an enzyme are also the products of evolution and also, you must admit, quite marvelous. Perhaps even beautiful.

SUMMARY

Living systems convert energy from one form to another as they carry out essential functions of maintenance, growth, and reproduction. In these energy conversions, as in all others, some useful energy is lost to the surroundings at each step.

The laws of thermodynamics govern transformations of energy. The first law states that energy can be converted from one form to another but cannot be created or destroyed. The potential energy of the initial state (or reactants) is equal to the potential energy of the final state (or products) plus the energy released in the process or reaction. The second law of thermodynamics states that in the course of energy conversions, the potential energy of the final state will always be less than the potential energy of the initial state. The difference in potential energy between the initial and final states is known as the free energy change and is symbolized as ΔG. Exergonic (energy-releasing) reactions have a negative ΔG, and endergonic (energy-requiring) reactions have a positive ΔG. Factors that determine ΔG include ΔH, the change in heat content, ΔS, the change in entropy, which, multiplied by the absolute temperature (T), is a measure of randomness or disorder:

$$\Delta G = \Delta H - T\Delta S$$

Another way of stating the second law of thermodynamics is that all natural processes tend to proceed in such a direction that the entropy of the universe increases. To maintain the organization on which life depends, living systems must have a constant supply of energy to overcome the tendency toward increasing disorder. The sun is the original source of this energy.

The energy transformations in living cells involve the movement of electrons from one energy level to another and, often, from one atom or molecule to another. Reactions in which electrons move from one atom to another are known as oxidation-reduction reactions. An atom or molecule that loses electrons is oxidized; one that gains electrons is reduced.

Metabolism is the total of all the chemical reactions that take place in cells. Reactions resulting in the breakdown or degradation of molecules are known, collectively, as catabolism; biosynthetic reactions are known, collectively, as anabolism. Metabolic reactions take place in series, called pathways, each of which serves a particular function in a cell. Each step in the pathway is controlled by a specific enzyme. The stepwise reactions of enzymatic pathways enable cells to carry out their chemical activities with remarkable efficiency in terms of both energy and materials.

Enzymes serve as catalysts, lowering the energy of activation and thus enormously increasing the rate at which reactions take place. They are large globular protein molecules folded in such a way that particular groups of amino acids form an active site. The reacting molecules, known as the substrate, fit precisely into this active site. Although the conformation of an enzyme may change temporarily in the course of a reaction, it is not permanently altered. Many enzymes require cofactors, which may be simple ions, such as Mg^{2+} or Ca^{2+}, or nonprotein organic molecules known as coenzymes. Many coenzymes, such as NAD, function as electron carriers, with different coenzymes holding electrons at slightly different energy levels. Many vitamins are parts of coenzymes.

Enzyme-catalyzed reactions are under tight cellular control. Principal factors in the rate of enzymatic reactions are the concentrations of enzyme and substrate and the availability of required cofactors. Many enzymes are synthesized by the cell or activated only when they are needed. The rate of enzymatic reactions is also affected by temperature and pH, which affect the attractions among the amino acids of the protein molecule and also between the active site and substrate.

A precise means of enzyme control is allosteric interaction. Allosteric interaction occurs when a molecule other than the substrate combines with an enzyme at a site other than the active site and in so doing alters the shape of the active site to render it either functional or nonfunctional. Feedback inhibition occurs when the product of an enzymatic reaction at the end or at a branch of a particular pathway acts as an allosteric effector, temporarily inhibiting the activity of an enzyme earlier in the pathway and thus temporarily stopping the series of chemical reactions.

Enzymes may also be regulated by competitive inhibition, in which another molecule, similar to the normal substrate, competes for the active site. Competitive inhibition can be reversed by increased concentrations of the substrate. Noncompetitive inhibitors bind elsewhere on the molecule, altering the tertiary structure so that the enzyme cannot function. Noncompetitive inhibition is usually reversible, but not by the substrate. Irreversible inhibitors bind permanently to the active site or irreparably disrupt the tertiary structure.

The transport proteins of cell membranes resemble enzymes in their complex protein structures, their specificity, and in the variety of ways in which their activity is regulated. The protein receptors found on the surface of the cell membrane are often allosteric binding sites, regulating the conformation—and thus the activity—of either transport proteins or membrane-bound enzymes.

ATP participates as an energy carrier in most series of reactions that take place in living systems. The ATP molecule consists of the nitrogenous base adenine, the five-carbon sugar ribose, and three phosphate groups. The three phosphate groups are linked by two covalent bonds that are easily broken, each yielding about 7 kilocalories of energy per mole. Cells are able to carry out endergonic reactions and processes (such as biosynthetic reactions, active transport, or the movement of microtubules) by coupling them with exergonic reactions that provide a surplus of energy. Such coupled reactions usually involve ATP or related triphosphate compounds.

QUESTIONS

1. Distinguish among the following terms: the first law of thermodynamics/the second law of thermodynamics; $\Delta H/\Delta S/\Delta G$; exergonic/endergonic; oxidation/reduction; metabolism/catabolism/anabolism; active site/substrate; competitive inhibition/noncompetitive inhibition/irreversible inhibition; ATP/ADP/AMP.

2. At present, at least four types of energy conversions are going on in your body. Name them.

3. All natural processes proceed with an increase in entropy. How then do you explain the freezing of water?

4. The laws of thermodynamics apply only to closed systems, that is, to systems into which no energy is entering. Is an aquarium ordinarily a closed system? Could you convert it to one? A space-ship may or may not be a closed system, depending on certain features of its design. What would these features be? Is the earth a closed system?

5. Explain why it is that living systems, despite appearances, are not in violation of the second law of thermodynamics.

6. What is there about the orderliness of a living organism that most significantly distinguishes it from the orderliness of a machine, such as a computer or the telephone system?

7. What is the basis for the specificity of enzyme action? What is the advantage to the cell of such specificity? What might be its disadvantages to the cell?

8. Turn back to Figure 3–18, in which all the amino acids are shown, and try to make some educated guesses about which amino acids can substitute for one another in the structure of an enzyme, and which substitutions would have drastic effects.

9. When a plant does not have an adequate supply of an essential mineral, such as magnesium, it is likely to become sickly and may die. When an animal is deprived of a particular vitamin in its diet, it too is likely to become ill and may die. What is a reasonable explanation of such phenomena?

10. Most organisms cannot live at high temperatures. Explain at least one way in which high temperatures are harmful to organisms. Some bacteria and algae, however, live in hot springs, at temperatures far above those that can be tolerated by most organisms. How might such bacteria and algae differ from most other organisms?

11. In enzyme regulation by allosteric interaction, the inhibitor often works on the first enzyme of the series. In regulation by competitive inhibition, it often works on the last. Why this difference?

12. In a series of experiments with an enzyme that catalyzes a reaction involving substrate A, it was found that a particular substance X inhibited the enzyme. When the concentration of A was high and the concentration of X was low, the reaction proceeded rapidly; as the concentration of X was increased and that of A was decreased, the reaction slowed down; when the concentration of X was high and that of A was low, the reaction stopped. If the concentration of A was again increased, the reaction resumed. How can you explain these results?

13. When a sulfa drug, such as sulfanilamide, is prescribed for a bacterial infection, it is very important to remember to take the drug at the prescribed times and in the prescribed quantity. Why is this essential? Suppose you were instructed to take two tablets every three hours, and instead you took only one tablet every five hours. What do you think would happen?

14. Some human societies use the barter system for exchange of goods and services. However, all complex societies have some form of monetary exchange. What are the advantages of a monetary exchange? Relate your answer to the ADP/ATP system.

15. Why, in the accompanying photograph, are there more plants than zebras and more zebras than lions? (Explain in terms of thermodynamics.)

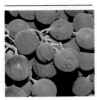

How Cells Make ATP: Glycolysis and Respiration

ATP is the principal energy carrier in living systems. It participates in a great variety of cellular events, from chemical biosyntheses, to the flick of a cilium, the twitch of a muscle, or the active transport of a molecule across a cell membrane. It is involved in the propagation of an electric impulse along a nerve or, in some remarkable organisms, the electrocution of prey (Figure 9–1). In the following pages, we shall show in some detail how a cell breaks down carbohydrates and captures and stores a portion of their potential energy in the terminal phosphate bonds of ATP. The oxidation of glucose (or other carbohydrates) is complicated in detail—so go slowly—but simple in its overall design.

AN OVERVIEW OF GLUCOSE OXIDATION

Oxidation, as you will recall (page 166), is the loss of an electron. Reduction is the gain of an electron. Since, in spontaneous oxidation-reduction reactions, electrons go from higher to lower energy levels, a molecule usually releases energy as it is oxidized. In the oxidation of glucose, carbon-carbon bonds, carbon-hydrogen bonds, and oxygen-oxygen bonds are exchanged for carbon-oxygen and hydrogen-oxygen bonds, as oxygen atoms attract and hoard electrons. The summary equation for this process is

$$\text{Glucose} + \text{Oxygen} \longrightarrow \text{Carbon dioxide} + \text{Water} + \text{Energy}$$

Or,

$$C_6H_{12}O_6 + 6O_2 \longrightarrow 6CO_2 + 6H_2O$$

$$\Delta G = -686 \text{ kcal/mole}$$

9–1 (a) *The electric ray converts the chemical energy of ATP—provided by the oxidation of glucose and other organic compounds—to electrical energy, stunning and immobilizing its prey with electric discharges.* (b) *The prey shown here, a small reef fish, is then moved to the mouth by the pectoral fins and* (c) *is swallowed. The reef fish will be converted to chemical energy, which will, in turn, be converted to mechanical and electrical energy for the capture of additional prey. Only a fraction of the potential energy is transferred at each passage.*

Some other types of electric fish produce discharges of lower voltage that are used in establishing territories, in locating objects (including both potential prey and potential predators), and, perhaps, in communicating with members of the same species.

(a)

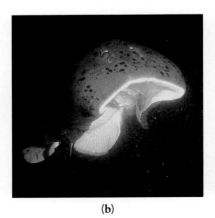

(b)

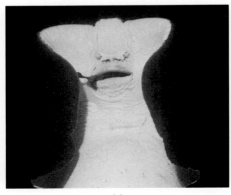

(c)

Living systems are experts at energy conversions. They are organized to trap this free energy so that it will not be dissipated randomly but can be used to do the work of the cell. About 40 percent of the free energy released by the oxidation of glucose is conserved in the conversion of ADP to ATP. As you will recall, about 75 percent of the energy in gasoline is "lost" as heat in an automobile engine, and only 25 percent is converted to useful forms of energy. The living cell is significantly more efficient.

In living systems, the oxidation of glucose takes place in two major stages. The first is known as **glycolysis.** The second is **respiration,** which, in turn, consists of two stages: the **Krebs cycle** and terminal **electron transport.** Glycolysis occurs in the cytoplasm of the cell, and the two stages of respiration take place within the mitochondrion.

In glycolysis, the six-carbon glucose molecule is split into two molecules of a three-carbon compound, pyruvic acid (Figure 9–2). Four hydrogen atoms (that is, four electrons and four protons) are removed from the glucose molecule in this process. The electrons and two of the protons are accepted by NAD^+ molecules (page 172), while the two other protons remain in solution as hydrogen ions (H^+). The free energy change, ΔG, of this stage is -143 kcal/mole; this represents a relatively small proportion of the potential energy stored in the glucose molecule.

9–2 *In glycolysis, the six-carbon glucose molecule is split into two molecules of a three-carbon compound known as pyruvic acid.*

Glucose ⟶ 2 Pyruvic acid

In respiration, the remaining hydrogen atoms are removed from the pyruvic acid molecules, and the carbon atoms are oxidized to carbon dioxide. The hydrogen atoms, in the form of electrons and protons, are initially accepted by NAD^+ and a related electron acceptor. Ultimately, all of the electrons and protons removed from the carbon atoms of the original glucose molecule are transferred to oxygen, forming water. The free energy change, ΔG, of respiration is -543 kcal/mole, a comparatively large energy yield.

In the course of glycolysis and respiration, about 38 molecules of ATP are regenerated from ADP in the breakdown of each molecule of glucose. As we shall see, the exact number of ATP molecules produced depends on at least two variables.

GLYCOLYSIS

Glycolysis—the lysis (splitting) of glucose—exemplifies the way the biochemical processes of a living cell proceed in small sequential steps. It takes place in a series of nine reactions, each catalyzed by a specific enzyme.

As we examine the details of glycolysis, notice how the carbon skeleton is dismembered and its atoms rearranged step by step. Note especially the formation of ATP from ADP and of NADH and H⁺ from NAD⁺. ATP and NADH represent the cell's net energy harvest from this reaction pathway.

Step 1. The first steps in glycolysis require an input of energy, which is supplied by coupling these steps to the ATP/ADP system. The terminal phosphate group is transferred from an ATP molecule to the carbon in the sixth position of the glucose molecule, to make glucose 6-phosphate. (The symbol Ⓟ represents a phosphate group.) The reaction of ATP with glucose to yield glucose 6-phosphate and ADP is an exergonic reaction. Some of the free energy is conserved in the chemical bond linking the phosphate to the glucose molecule, which then becomes energized. This reaction is catalyzed by a specific enzyme (hexokinase), and each of the reactions that follows is also catalyzed by a specific enzyme.

Step 2. The molecule is reorganized, again with the help of a particular enzyme. The six-sided ring characteristic of glucose becomes the five-sided fructose ring. (As you know, glucose and fructose both have the same number of atoms— $C_6H_{12}O_6$—and differ only in the arrangement of these atoms.) This reaction can proceed approximately equally well in either direction; it is pushed forward by the accumulation of glucose 6-phosphate and the removal of fructose 6-phosphate as the latter enters Step 3.

Step 3. In this step, which is similar to Step 1, fructose 6-phosphate gains a second phosphate by the investment of another ATP. The added phosphate is bonded to the first carbon, producing fructose 1,6-diphosphate, that is, fructose with phosphates in the 1 and 6 positions. Note that in the course of the reactions thus far, two molecules of ATP have been converted to ADP and no energy has been recovered.

The enzyme catalyzing this step, phosphofructokinase, is an allosteric enzyme, and ATP is an allosteric effector inhibiting its activity. The allosteric interaction between them is the chief regulatory mechanism of glycolysis. If the ATP concentration in the cell is high—that is, if ATP is present in quantities more than adequate to meet the various needs of the cell—ATP inhibits the activity of phosphofructokinase. Glycolysis, and thus ATP production, cease, and glucose is conserved. As the cell uses up its supply of ATP and the concentration drops, the enzyme is released from inhibition, and the breakdown of glucose resumes. This is one of the major control points of ATP production.

Step 4. The six-carbon sugar molecule is split into two three-carbon molecules, dihydroxyacetone phosphate and glyceraldehyde phosphate. The two molecules

9–3 *The steps of glycolysis.*

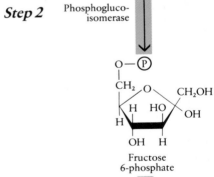

Glucose

Step 1 Hexokinase

Glucose 6-phosphate

Step 2 Phosphogluco-isomerase

Fructose 6-phosphate

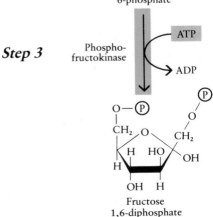

Step 3 Phospho-fructokinase

Fructose 1,6-diphosphate

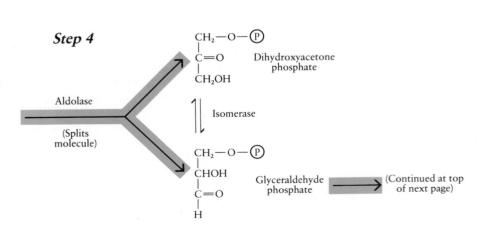

Step 4 Aldolase (Splits molecule)

Dihydroxyacetone phosphate

Isomerase

Glyceraldehyde phosphate

(Continued at top of next page)

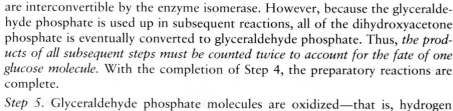

are interconvertible by the enzyme isomerase. However, because the glyceraldehyde phosphate is used up in subsequent reactions, all of the dihydroxyacetone phosphate is eventually converted to glyceraldehyde phosphate. Thus, *the products of all subsequent steps must be counted twice to account for the fate of one glucose molecule.* With the completion of Step 4, the preparatory reactions are complete.

Step 5. Glyceraldehyde phosphate molecules are oxidized—that is, hydrogen atoms with their electrons are removed—and NAD^+ is reduced to NADH and H^+ (a total of two molecules of NADH and two H^+ ions per molecule of glucose). This is the first reaction from which the cell harvests energy. Some of the energy from this oxidation reaction is also conserved in the attachment of a phosphate group to what is now the 1 position of the glyceraldehyde phosphate molecule. (The designation P_i represents inorganic phosphate available as a phosphate ion in solution in the cytoplasm.) The properties of this bond are similar to those of the phosphate bonds of ATP, as indicated by the squiggle.

Step 6. This phosphate is released from the diphosphoglycerate molecule and used to recharge a molecule of ADP (a total of two molecules of ATP per molecule of glucose). This is a highly exergonic reaction (ΔG is negative and large), and so it pulls all the preceding reactions forward.

Step 7. The remaining phosphate group is enzymatically transferred from the 3 position to the 2 position.

Step 8. In this step, a molecule of water is removed from the three-carbon compound. This internal rearrangement of the molecule concentrates energy in the vicinity of the phosphate group.

Step 9. The phosphate is transferred to a molecule of ADP, forming another molecule of ATP (again, a total of two molecules of ATP per molecule of glucose). This is also a highly exergonic reaction and thus pulls forward the preceding two reactions (Steps 7 and 8).

Summary of Glycolysis

The complete sequence begins with one molecule of glucose. Energy is invested at Steps 1 and 3 by the transfer of a phosphate group from an ATP molecule—one at each step—to the sugar molecule. The six-carbon molecule splits at Step 4, and from this point onward, the sequence yields energy. At Step 5, a molecule of NAD^+ is reduced to NADH and H^+, storing some of the energy from the oxidation of glyceraldehyde phosphate. At Steps 6 and 9, molecules of ADP take energy from the system, becoming phosphorylated to ATP.

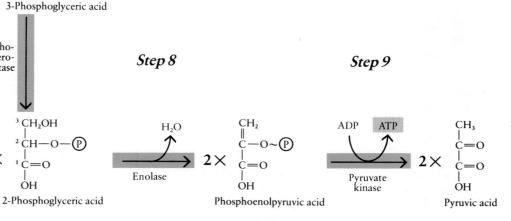

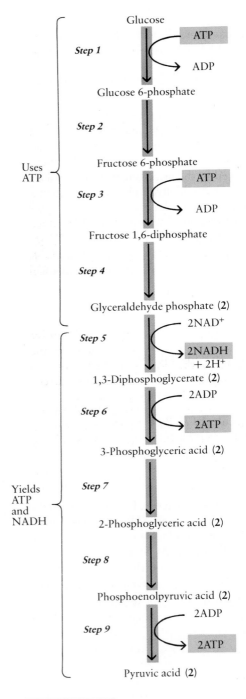

Uses ATP

Step 1

Glucose

ATP

ADP

Glucose 6-phosphate

Step 2

Fructose 6-phosphate

Step 3

ATP

ADP

Fructose 1,6-diphosphate

Step 4

Glyceraldehyde phosphate (2)

2NAD⁺

Yields ATP and NADH

Step 5

2NADH + 2H⁺

1,3-Diphosphoglycerate (2)

Step 6

2ADP

2ATP

3-Phosphoglyceric acid (2)

Step 7

2-Phosphoglyceric acid (2)

Step 8

Phosphoenolpyruvic acid (2)

Step 9

2ADP

2ATP

Pyruvic acid (2)

9–4 *Summary of the two stages of glycolysis. The first stage utilizes 2ATP; the second stage yields 4ATP and 2NADH. Compounds other than glucose, such as the hexose galactose and a number of pentoses, as well as glycogen and starch, can undergo glycolysis once they have been converted to glucose 6-phosphate.*

To sum up: The energy from the phosphate bonds of two ATP molecules is needed to initiate the glycolytic sequence. Subsequently, two NADH molecules are produced from two NAD⁺ and four ATP molecules from four ADP:

$$Glucose + 2ATP + 4ADP + 2P_i + 2NAD^+ \longrightarrow$$
$$2 \text{ Pyruvic acid} + 2ADP + 4ATP + 2NADH + 2H^+ + 2H_2O$$

Thus one glucose molecule has been converted to two molecules of pyruvic acid. The net harvest—the energy recovered—is two molecules of ATP and two molecules of NADH per molecule of glucose. The two molecules of pyruvic acid still contain a large amount of the potential energy that was stored in the original glucose molecule. This series of reactions is carried out by virtually all living cells—from prokaryotes to the eukaryotic cells of our own bodies.

ANAEROBIC PATHWAYS

Pyruvic acid can follow one of several pathways. One pathway is aerobic (with oxygen), and the others are anaerobic (without oxygen). We shall briefly discuss two of the most interesting anaerobic pathways and then follow the aerobic one, which is the principal pathway of energy metabolism for most cells in the presence of oxygen.

In the absence of oxygen, pyruvic acid can be converted to ethanol (ethyl alcohol) or to one of several different organic acids, of which lactic acid is the most common. The reaction product depends on the type of cell. For example, yeast cells, present as a "bloom" on the skin of grapes, can grow either with or without oxygen. When the sugar-filled juices of grapes and other fruits are extracted and stored under anaerobic conditions, the yeast cells turn the fruit juice to wine by converting glucose into ethanol (Figure 9–5). When the sugar is exhausted, the yeast cells cease to function; at this point, the alcohol concentration is between 12 and 17 percent, depending on the variety of the grapes and the season at which they were harvested.

The formation of alcohol from sugar is called **fermentation.** Because of the economic importance of the wine industry, fermentation was the first enzymatic process to be intensively studied. In fact, before their effects were known to be so diverse, enzymes were commonly referred to as "ferments."

Lactic acid is formed from pyruvic acid by a variety of microorganisms and also by some animal cells when O_2 is scarce or absent (Figure 9–6). It is produced, for example, in vertebrate muscle cells during strenuous exercise, as by an athlete during a sprint. We breathe hard when we run fast, thereby increasing the supply of oxygen, but even this increase may not be enough to meet the immediate needs of the muscle cells. These cells, however, can continue to work by accumulating what is known as an oxygen debt. Glycolysis continues, using glucose released from glycogen stored in the muscle, but the resulting pyruvic acid does not enter the aerobic pathway of respiration. Instead, it is converted to lactic acid, which, as it accumulates, lowers the pH of muscle and reduces the capacity of the muscle fibers to contract, producing the sensations of muscle fatigue. The lactic acid diffuses into the blood and is carried to the liver. Later, when oxygen is more abundant (as a result of the deep breathing that follows strenuous exercise) and ATP demand is reduced, the lactic acid is resynthesized to pyruvic acid and back again to glucose or glycogen.

Why is pyruvic acid converted to lactic acid, only to be converted back again? The function of the initial conversion is simple: it uses NADH and regenerates the NAD⁺ without which glycolysis cannot go forward (see Step 5, page 189). Even though the overall process seems to be wasteful in terms of energy consumption, the regeneration of NAD⁺ may be all-important in the economy of the organism, spelling the difference between life and death when an animal "out of breath" needs one last burst of ATP to escape from a predator or catch a prey.

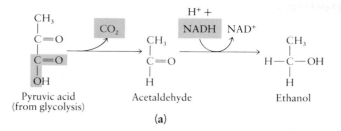

(a)

9–5 (a) *The steps by which pyruvic acid, formed by glycolysis, is converted anaerobically to ethanol (ethyl alcohol). In the first step, carbon dioxide is released. In the second, NADH is oxidized, and acetaldehyde is reduced. Most of the potential energy of the glucose remains in the alcohol, which is the end product of the sequence. However, by regenerating NAD^+, these steps allow glycolysis to continue, with its small but sometimes vitally necessary yield of ATP.*

(b) Yeast cells on the skins of these grapes give them their dust-like "bloom." When the grapes are crushed, the yeast cells mix with the juice. Storing the mixture under anaerobic conditions causes the yeast to break down the glucose in the grape juice to alcohol.

(b)

9–6 *The enzymatic reaction that produces lactic acid from pyruvic acid anaerobically in muscle cells. In the course of this reaction, NADH is oxidized and pyruvic acid is reduced. The NAD^+ molecules produced in this reaction and the one shown in Figure 9–5 are recycled in the glycolytic sequence. Without this recycling, glycolysis cannot proceed. Lactic acid accumulation results in muscle soreness and fatigue.*

An oxygen debt is usually accumulated during a short burst of intensive exercise—for example, a rapid sprint. During sustained moderate exercise, the intake of oxygen by the lungs and the circulation of the blood supplying oxygen to the muscle tissues often catch up with the oxygen consumption in the muscle, thus producing the "second wind" phenomenon well known to runners.

The fact that glycolysis does not require oxygen suggests that the glycolytic sequence evolved early, before free oxygen was present in the atmosphere. Presumably, primitive one-celled organisms used glycolysis (or something very much like it) to extract energy from the organic compounds they absorbed from their watery surroundings. Although anaerobic glycolysis generates only two molecules of ATP for each glucose molecule processed (a small fraction—about 5 percent—of the ATP that can be generated through aerobic processes), it was and is adequate for the needs of many organisms.

RESPIRATION

In the presence of oxygen, the next stage in the breakdown of glucose involves the stepwise oxidation of pyruvic acid to carbon dioxide and water—the process known as respiration. Respiration has two meanings in biology. One is the breathing in of oxygen and breathing out of carbon dioxide; this is also the ordinary, nontechnical meaning of the word. The second meaning of respiration is the oxidation of food molecules by cells. This process, sometimes qualified as cellular respiration, is what we are concerned with here.

As we noted previously, cellular respiration takes place in two stages: the Krebs cycle and terminal electron transport. In eukaryotic cells, these reactions take place within the mitochondria. Mitochondria, as we saw in Chapter 5, are surrounded by two membranes. The outer one is smooth, and the inner one folds inward. The folds are called cristae. Within the inner compartment of the mitochondrion, surrounding the cristae, is a dense solution known as the **matrix.** The matrix contains enzymes, coenzymes, water, phosphates, and other molecules involved in respiration. The outer membrane is permeable to most small molecules, but the inner one permits the passage of only certain molecules, such as pyruvic acid and ATP, and restrains the passage of others. As we shall see, this selective permeability of the inner membrane is critical to the ability of the mitochondria to harness the power of respiration to the production of ATP.

Dissecting the Cell

In every living cell, many hundreds of chemical reactions proceed simultaneously. Molecules are continuously synthesized via certain enzymatic pathways, and molecules are broken down by other enzymatic pathways. Many of these reactions are mutually incompatible, as can be demonstrated by destroying the structure of cells and mixing their enzymes in a test tube. Chemical chaos results, and the enzymes are soon inactivated. In the cell, however, anabolic and catabolic pathways operate in harmony because biochemical reactions are spatially localized and compartmentalized within specific subcellular organelles. A living cell is the most intensely concentrated set of chemical reactions known. A cell carries out many more chemical reactions than any apparatus devised by chemical engineers, and all within the space of a few cubic micrometers. The cell's unique chemical versatility results from the compartmentalization of biochemical pathways within organelles.

In order to study the specific functions of any organelle type, the organelle must be dissected free from all other cell structures and collected in large quantities. Mitochondria, lysosomes, and other organelles are, of course, far too small for hand dissection, but cell biologists can prepare pure samples of any organelle type by the technique of preparative centrifugation.

Small particles, ranging in size from cells to macromolecules, can be separated by centrifugation if the particle types differ in size and density. Particles suspended in fluid and then subjected to strong gravitational force will move through the fluid at varying rates, the largest, densest particles settling most rapidly. Forces up to 400,000 times the force of gravity (400,000 g) can be generated in a test tube of suspended particles by rotating the tube at very high speeds in an ultracentrifuge. Thus, subcellular structures, such as mitochondria, nuclei, and intracellular membranes, can be separated into purified fractions by spinning fragmented cells at appropriate centrifugal forces.

For instance, in order to determine which enzymatic pathways are present in mitochondria, a tissue, such as rat liver, is minced into small pieces and homogenized—that is, the cells are gently broken up by grinding the tissue, for example, in a glass tube fitted with a Teflon pestle. The tube contains a sucrose solution that is isotonic with intracellular fluid. The resulting suspension of cell organelles is then placed in an unbreakable test tube and spun in the centrifuge at low speed (700 g) for 10 minutes, so as to drive the bulkiest structures, the nuclei, to the bottom of the tube. All the lighter organelles remain suspended in the fluid, which is called the supernatant. The supernatant is transferred to another centrifuge tube and spun at a higher speed (10,000 g for 20

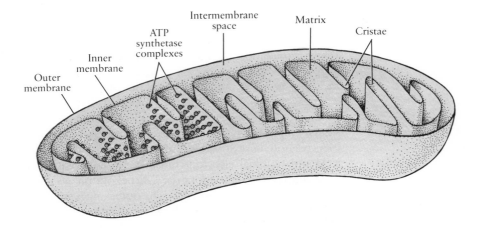

9-7 *A mitochondrion is surrounded by two membranes. The inner membrane folds inward to make a series of shelves, or cristae. Many of the enzymes and electron carriers involved in cellular respira-* *tion are built into these internal membranes. Among the enzymes are the ATP synthetase complexes, which, as we shall see, play a critical role in the formation of ATP in the final stage of cellular* *respiration. The matrix is a dense solution containing enzymes involved in earlier stages of cellular respiration, plus coenzymes, phosphates, and other solutes.*

minutes), which sediments particles such as mitochondria and lysosomes into a pellet at the bottom of the tube. The supernatant, containing ribosomes and various membranes, is discarded, and the pellet is retained.

At this point, the mitochondria have been partially purified by means of differential centrifugation. This type of centrifugation separates particles of quite different size and density by "spinning down" the large particles to form a pellet. To separate particles of rather similar size, such as mitochondria and lysosomes, the more subtle technique of zonal centrifugation is used. To return to the experiment we have outlined, the pellet containing the mitochondria and lysosomes is resuspended and gently layered atop a sucrose density gradient in a centrifuge tube. A sucrose density gradient is prepared by layering sucrose solutions of differing densities one above the other so that the densest solution is at the bottom of the tube and the least dense solution is at the top of the tube. When organelles are centrifuged in a density gradient (120,000 g for 8 hours), each organelle type moves through the gradient at a different rate, depending on the organelle's density. Following centrifugation, mitochondria will occupy one zone in the gradient, lysosomes another, and other organelles will be found in other zones. By puncturing the bottom of the tube and removing the contents drop by drop, we

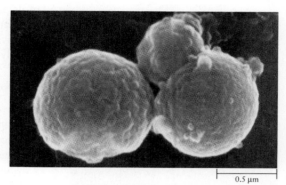

Scanning electron micrograph of isolated, intact liver mitochondria.

can collect a pure sample of mitochondria. The purity can be verified by the electron microscope. If the procedure has been performed correctly, the isolated mitochondria will emerge with membranes intact and all enzymatic pathways functioning. It is then possible to test for the activities of specific enzymes, thus determining which biochemical functions are compartmentalized within mitochondria. Similarly, other cell constituents can be isolated and their biochemical activities determined.

Some of the enzymes of the Krebs cycle are in solution in the matrix. Other Krebs cycle enzymes and the enzymes and other components of the electron transport chains are built into the membrane of the cristae. These inner membranes of the mitochondria are about 80 percent protein and 20 percent lipid. In the mitochondria, pyruvic acid from glycolysis is oxidized to carbon dioxide and water, completing the breakdown of the glucose molecule. Ninety-five percent of the ATP generated by heterotrophic cells is produced in the mitochondria.

A Preliminary Step: The Oxidation of Pyruvic Acid

Pyruvic acid passes from the cytoplasm, where it is produced by glycolysis, and crosses the outer and inner membranes of the mitochondria. Before entering the Krebs cycle, the three-carbon pyruvic acid molecule is oxidized (Figure 9–8). The carbon and oxygen atoms of the carboxyl group are removed in the form of carbon dioxide, and a two-carbon acetyl group (CH_3CO) remains. In the course of this exergonic reaction, the hydrogen of the carboxyl group reduces a molecule of NAD^+ to NADH. The original glucose molecule has now been oxidized to two

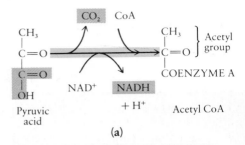

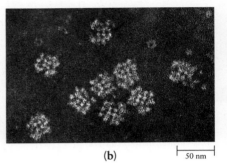

(a)

(b) $\overline{\quad}$ 50 nm

9-8 (a) *The three-carbon pyruvic acid molecule is oxidized to the two-carbon acetyl group, which is combined with coenzyme A to form acetyl CoA. The oxidation of the pyruvic acid molecule is coupled to the reduction of NAD^+. Acetyl CoA enters the Krebs cycle.* (b) *Electron micrograph showing the enzymes involved in the oxidation of pyruvic acid to acetyl CoA. Each of the complexes visible here represents multiple copies of the three different enzymes required for this reaction sequence.*

CO_2 molecules and two acetyl groups, and, in addition, four NADH molecules have been formed (two in glycolysis and two in the oxidation of pyruvic acid).

Each acetyl group is momentarily accepted by a compound known as coenzyme A. Like the coenzymes we have examined previously, coenzyme A is a large molecule, a portion of which is a nucleotide and a portion of which is a vitamin (pantothenic acid, one of the B complex vitamins). The combination of the acetyl group and CoA is abbreviated acetyl CoA. Its formation is the link between glycolysis and the Krebs cycle.

The Krebs Cycle

Upon entering the Krebs cycle (Figure 9–9), the two-carbon acetyl group is combined with a four-carbon compound (oxaloacetic acid) to produce a six-carbon compound (citric acid). In the course of the cycle, two of the six carbons are oxidized to CO_2, and oxaloacetic acid is regenerated—making this series literally a cycle. Each turn around the cycle uses up one acetyl group and regenerates a molecule of oxaloacetic acid, which is then ready to begin the sequence again.

In the course of these steps, some of the energy released by the oxidation of the carbon-hydrogen and carbon-carbon bonds is used to convert ADP to ATP (one molecule per cycle), and some is used to produce NADH and H^+ from NAD^+

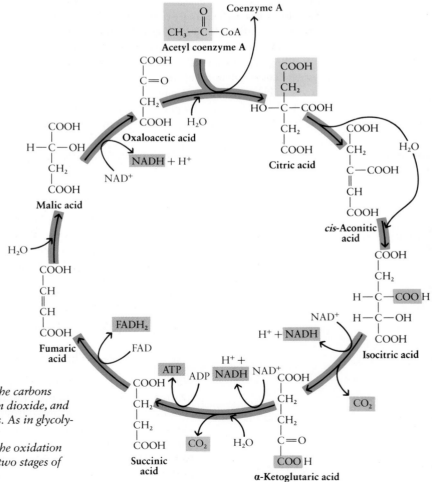

9-9 *The Krebs cycle. In the course of the cycle, the carbons donated by the acetyl group are oxidized to carbon dioxide, and the hydrogen atoms are passed to electron carriers. As in glycolysis, a specific enzyme is involved at each step.*

Coenzyme A shuttles back and forth between the oxidation of pyruvic acid and the Krebs cycle, linking these two stages of respiration.

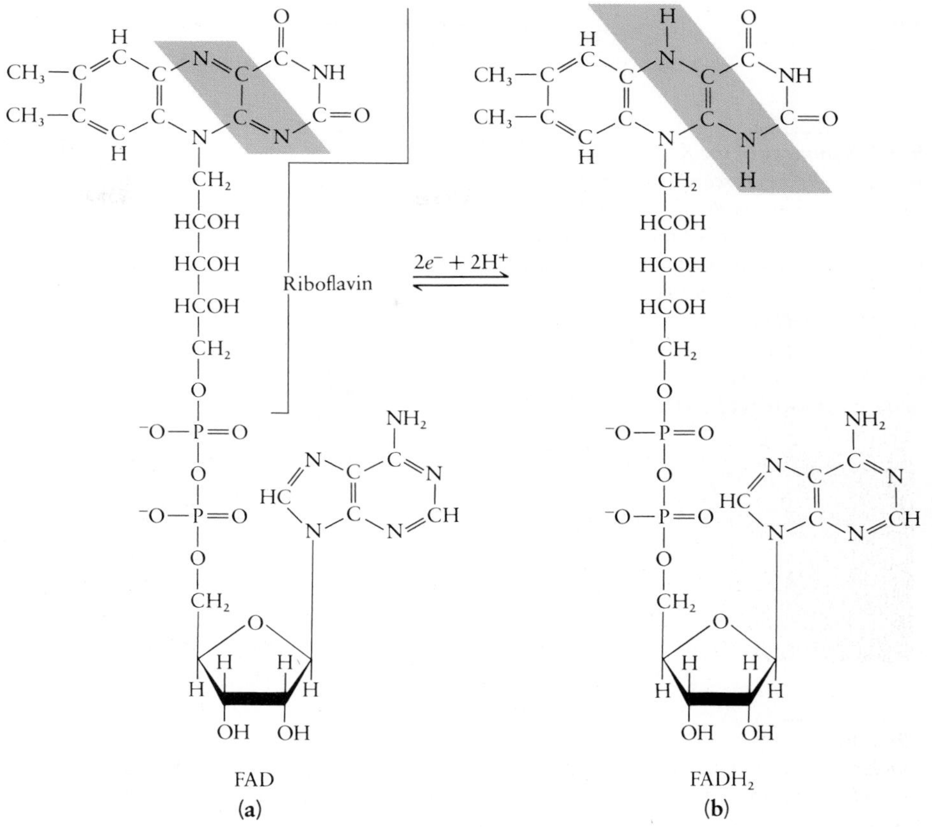

9–10 *Flavin adenine dinucleotide, an electron acceptor, in (a) its oxidized form (FAD) and (b) its reduced form (FADH₂). Riboflavin is a vitamin made by all plants and many microorganisms. It is also known as vitamin B₂. It is a pigment; in its oxidized form it is a bright yellow.*

A related electron acceptor, flavin mononucleotide (FMN), consists of riboflavin and the first phosphate group shown here. It accepts electrons from NADH in the electron transport chain.

In the living cell, both flavin adenine dinucleotide and flavin mononucleotide are bound to specific proteins, forming macromolecules known as flavoproteins.

(three molecules per cycle). In addition, some energy is used to reduce a second electron carrier, flavin adenine dinucleotide, abbreviated FAD (Figure 9–10). One molecule of FADH₂ is formed from FAD per turn of the cycle. No O_2 is required for the Krebs cycle; the electrons and protons removed in the oxidation of carbon are accepted by NAD^+ and FAD.

Oxaloacetic acid + Acetyl CoA + ADP + P_i + 3NAD$^+$ + FAD $\longrightarrow$
Oxaloacetic acid + 2CO$_2$ + CoA + ATP + 3NADH + FADH$_2$ +
$$3H^+ + H_2O$$

Note that the oxaloacetic acid molecule with which the cycle ends is not the same molecule with which the cycle began. If one begins with a glucose molecule in which the carbon atoms are radioactive, radioactive carbon atoms will appear among the four carbons of the oxaloacetic acid.

Electron Transport

The carbon atoms of the glucose molecule are now completely oxidized. Some of the potential energy of the glucose molecule has been used to produce ATP from ADP. Most of the energy, however, remains in electrons removed from the C—C and C—H bonds and passed to the electron carriers NAD^+ and FAD. These electrons are still at a high energy level.

In the final stage of respiration, these high-energy-level electrons are passed step-by-step to the low energy level of oxygen. The energy they yield in the course of this passage is ultimately used to regenerate ATP from ADP. This step-by-step passage is made possible by a series of electron carriers, each of which holds the electrons at a slightly lower level.

These carriers make up what is known as an **electron transport chain.** At the top of the energy hill the electrons are held by NADH and FADH₂. Most of the potential energy of the glucose molecule now resides in these electron acceptors.

195

9–11 *Summary of the Krebs cycle. One molecule of ATP, three molecules of NADH, and one molecule of FADH$_2$ represent the energy yield of the cycle. Two turns of the cycle are required to complete the oxidation of one molecule of glucose. Thus the total energy yield of the Krebs cycle for one glucose molecule is two molecules of ATP, six molecules of NADH, and two molecules of FADH$_2$.*

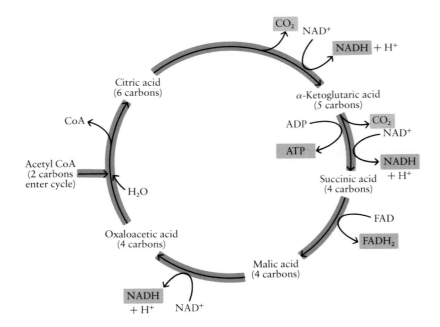

9–12 *Cytochromes are molecules that participate in electron transfer in the mitochondria. A cytochrome molecule consists of a heme group held in an intricate protein structure. (a) The heme group of cytochrome c. The iron (Fe) atom, enclosed in a porphyrin ring (a nitrogen-containing ring), combines with and then releases electrons. (b) The overall structure of the cytochrome c molecule, showing the position of the heme group (color) within the globular protein.*

The Krebs cycle yielded two molecules of FADH$_2$ and six molecules of NADH for each molecule of glucose. The oxidation of pyruvic acid to acetyl CoA yielded two molecules of NADH. Also, you will recall, two molecules of NADH were produced in glycolysis. In the presence of oxygen, the electrons held by these two NADH molecules are also transported into the mitochondrion where they are fed into the electron transport chain. In the process, NAD$^+$ is regenerated in the cytoplasm, allowing glycolysis to continue.

Among the principal components of the electron transport chain are molecules known as cytochromes. These molecules consist of a protein and a heme group, analogous to that of hemoglobin, in which an atom of iron is enclosed in a porphyrin ring (Figure 9–12). Although similar, the protein structures of the individual cytochromes differ enough to enable them to hold electrons at different energy levels. The iron atom of each cytochrome alternately accepts and releases an electron, passing it along to the next cytochrome at a slightly lower energy level until the electrons, their energy spent, are accepted by oxygen (Figure

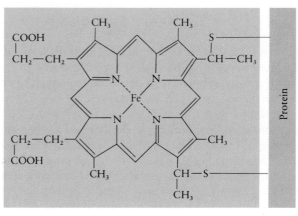

Cytochrome *c*

(a)

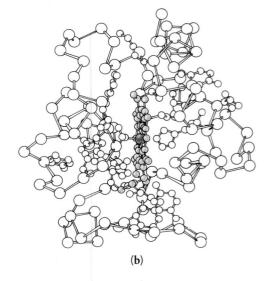

(b)

9-13 *The principal electron-carrier molecules of the electron transport chain. At least nine other carrier molecules function as intermediates between the carriers shown here.*

Flavin mononucleotide (FMN) and coenzyme Q (CoQ) transfer electrons and protons. The cytochromes transfer only electrons. The electrons carried by NADH enter the chain when they are transferred to FMN; those carried by FADH$_2$ enter the chain farther down the line at CoQ. The electrons are ultimately accepted by oxygen, which combines with protons (hydrogen ions) in the solution to form water.

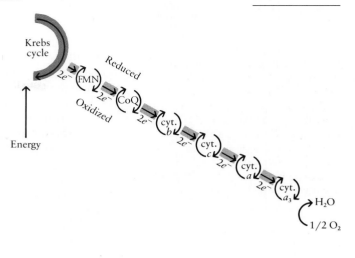

9–13). The energy released in this downhill passage of electrons is harnessed, as we shall see, to form ATP molecules from ADP. Such ATP formation is known as **oxidative phosphorylation.** At the end of the chain, the electrons are accepted by oxygen, which then combines with protons (H$^+$ ions) from the solution to produce water.

Quantitative measurements show that for every two electrons that pass from NADH to oxygen, three molecules of ATP are formed from ADP and phosphate. For every pair of electrons that passes from FADH$_2$, which holds them at a slightly lower energy level than NADH, two molecules of ATP are formed. In oxidative phosphorylation, the electron transfer potential of NADH and FADH$_2$ is converted to the phosphate transfer potential of ATP.

The Mechanism of Oxidative Phosphorylation: Chemiosmotic Coupling

For many years, the mechanism of oxidative phosphorylation—that is, the way in which ATP is formed from ADP and phosphate as electrons pass down the electron transport chain—was a puzzle. A major breakthrough occurred when a British biochemist, Peter Mitchell, proposed that the process is powered by a gradient of protons (H$^+$ ions) established across the inner mitochondrial membrane. Continuing studies have revealed many details of this mechanism, known as **chemiosmotic coupling,** although much remains to be learned.

The term "chemiosmotic" reflects the fact that the production of ATP in oxidative phosphorylation includes both chemical processes and transport processes across a selectively permeable membrane. Two distinct events take place in chemiosmotic coupling: (1) a proton gradient is established across the inner membrane of the mitochondrion, and (2) potential energy stored in the gradient is released and captured in the formation of ATP from ADP and phosphate.

The proton gradient is established as electrons move down the electron transport chain. At three transition points in this chain, significant drops occur in the amount of potential energy held by the electrons. As a consequence, a relatively large amount of free energy is released at each of these steps—as the electrons move from FMN to coenzyme Q, as they move from cytochrome *b* to cytochrome *c*, and as they move from cytochrome *a* to cytochrome *a$_3$*. This energy powers the pumping of protons from the mitochondrial matrix through the inner membrane to the intermembrane space—that is, the space between the inner and outer membranes of the mitochondrion. From the intermembrane space, some of the protons pass through the outer membrane into the cytoplasm (the outer membrane, you will recall, is freely permeable to most ions and small molecules).

Exactly how the pumping of protons is accomplished is a matter of current investigation. One hypothesis proposes that the electron carriers in the chain are

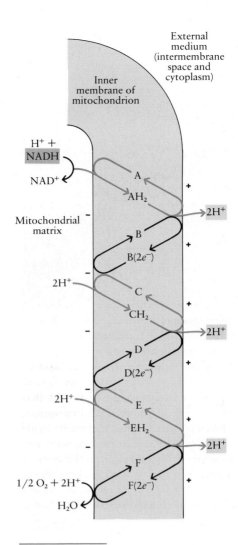

H⁺ + NADH → NAD⁺ (A / AH₂) — labels in diagram:

External medium (intermembrane space and cytoplasm)

Inner membrane of mitochondrion

H⁺ + NADH

NAD⁺

Mitochondrial matrix

A
AH₂
2H⁺
B
B(2e⁻)
2H⁺
C
CH₂
2H⁺
D
D(2e⁻)
2H⁺
E
EH₂
2H⁺
F
F(2e⁻)
1/2 O₂ + 2H⁺
H₂O

9–14 *According to one hypothesis, the chemiosmotic proton gradient is established as the electrons contributed by NADH follow a looping pattern across the inner mitochondrial membrane during their transit down the electron transport chain. With each passage of two electrons from the matrix side of the membrane to its outer side, two protons are transported by carrier molecules (loops indicated in blue) and then discharged into the external medium. After discharge of the protons, the two electrons are transported back to the matrix side of the membrane by other carrier molecules (black loops). This hypothesis assumes that the carrier molecules (here designated as A through F) are fixed in the membrane in the necessary positions.*

positioned so that the electrons travel a zigzag course from the inner to the outer surface of the inner membrane, traversing it three times in the course of their passage from NADH to oxygen. According to this hypothesis, each time two electrons travel from the inside of the membrane to the outside, they pick up two protons from the matrix and release them to the outside (Figure 9–14). Another hypothesis proposes that the energy released by the electrons triggers conformational changes in specific membrane transport proteins that enable them to carry protons against the concentration gradient from the matrix into the intermembrane space. The precise number of protons transported as each electron pair moves down the chain is uncertain; it is thought to be at least six, but it may be more.

Like the boulder at the top of the hill, the water at the top of the falls, or the chemical energy in a stick of dynamite, the difference in the concentration of protons between the matrix and the outside represents potential energy. This potential energy results not only from the difference in pH (more H⁺ ions outside than inside) but also from the difference in electric charge (Figure 9–15). Because the inner membrane is impermeable to virtually all charged particles, other positive ions cannot move into the matrix to neutralize the negative charge created when the protons are pumped out. The movement of negative ions out of the matrix, which would also neutralize the charge difference, is similarly blocked. The result is potential energy, available to power any process that provides a channel allowing the protons to flow down the electrochemical gradient back into the matrix.

Such a channel is provided by a large enzyme complex known as **ATP synthetase.** This complex consists of two major portions, or factors, known as F_O and F_1 (Figure 9–16a). F_O is embedded in the inner mitochondrial membrane, traversing the membrane from outside to inside. It is thought to have an inner channel through which protons can pass. F_1 is a large globular structure, consisting of nine polypeptide subunits, that is attached to F_O on the matrix side of the membrane.

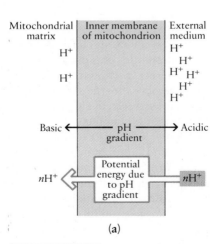

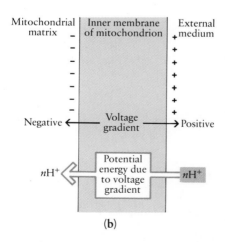

9–15 *The potential energy stored by the pumping of protons from the mitochondrial matrix has two components: (**a**) a chemical gradient resulting from the difference in pH between the matrix and the external medium, and (**b**) a voltage gradient, resulting from the difference in electric charge on the inside and outside of the membrane. When a channel is provided that allows protons to reenter the matrix, they descend both gradients simultaneously. The contribution of the voltage gradient to the total potential energy is, however, greater than that of the chemical gradient.*

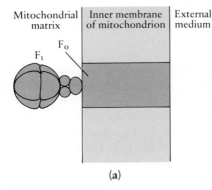

Mitochondrial matrix	Inner membrane of mitochondrion	External medium

F_1 F_O

(a)

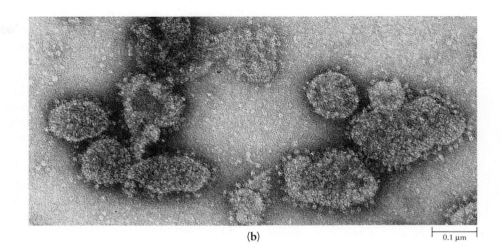

(b)

0.1 μm

9–16 **(a)** *Diagram of the ATP synthetase complex. The F_O portion is contained within the inner membrane of the mitochondrion, and the F_1 portion, which consists of nine subunits, extends into the mitochondrial matrix.*

(b) *The knobs protruding from the membrane of these vesicles are the F_1 portions of ATP synthetase complexes. The F_O portions to which they are attached are embedded in the membrane and are not visible in this electron micrograph. These vesicles were prepared by disrupting the inner mitochondrial membrane with ultrasonic waves. When the membrane is disrupted in this fashion, the membrane fragments immediately reseal, forming closed vesicles. These vesicles are, however, inside-out; the outer surface here is the surface that faces the matrix in the intact mitochondrion. Such inside-out vesicles are an important tool in the continuing study of oxidative phosphorylation.*

In electron micrographs of the inner mitochondrial membrane, the F_1 units appear as protruding knobs (Figure 9–16b). With proper chemical treatment, the F_1 unit can be removed from the mitochondrial membrane and subjected to detailed study. It has been shown to have binding sites for ATP and ADP, and, in solution, it catalyzes the hydrolysis of ATP to ADP, thus functioning as an ATPase. Its usual function when attached to the F_O unit in the intact mitochondrion is, however, the reverse. As protons flow down the electrochemical gradient from the outside into the matrix, passing through the F_O unit and then the F_1 unit, the free energy released powers the synthesis of ATP from ADP and phosphate. It is not certain whether the phosphorylation of one molecule of ADP to ATP requires the passage of two, three, or four protons through the ATP synthetase complex. Figure 9–17 summarizes the chemiosmotic coupling of oxidative phosphorylation.

9–17 *According to the chemiosmotic theory, protons are pumped out of the mitochondrial matrix as electrons are passed down the electron transport chain, which forms a part of the inner mitochondrial membrane. The movement of the protons down the electrochemical gradient as they pass through the ATP synthetase complex provides the energy by which ATP is regenerated from ADP and phosphate. The exact number of protons pumped out of the matrix as each electron pair moves down the chain is still to be determined, as is the number that must flow through ATP synthetase for each molecule of ATP formed.*

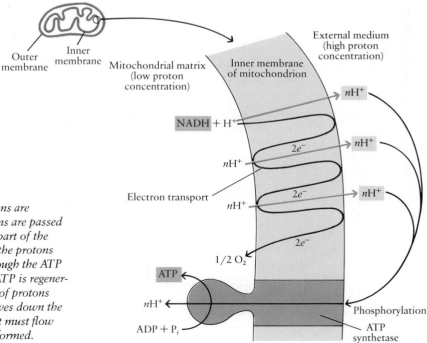

Chemiosmotic power also has other uses in living systems. For example, it provides the power that drives the rotation of bacterial flagella, which we shall discuss in Chapter 21. In photosynthetic cells, as we shall see in the next chapter, it is involved in the formation of ATP using energy supplied to electrons by the sun. And, it can be used to power other transport processes. In the mitochondrion, the energy stored in the proton gradient is used not only to drive the synthesis of ATP but also to carry other substances through the inner membrane via cotransport systems (page 137). Both phosphate and pyruvic acid are carried into the mitochondrion by integral membrane proteins that simultaneously transport protons down the gradient.

You will recall that we mentioned earlier that "about" 38 molecules of ATP are formed for each molecule of glucose oxidized to carbon dioxide and water. One of the reasons for this vagueness is that the exact amount of ATP formed depends on how the cell apportions the energy made available by the proton gradient. When more of this energy is used in other transport processes, less of it is available for ATP synthesis. The needs of the cell vary according to the circumstances, and so does the amount of ATP synthesized.

Control of Oxidative Phosphorylation

Electrons continue to flow along the electron transport chain, providing energy to create and maintain the proton gradient, only if ADP is available to be converted to ATP. Thus oxidative phosphorylation is regulated by supply and demand. When the energy requirements of the cell decrease, fewer molecules of ATP are used, fewer molecules of ADP become available, and electron flow is decreased.

OVERALL ENERGY HARVEST

We are now in a position to see how—and how much of—the potential energy originally present in the glucose molecule has been recovered in the form of ATP. Bear in mind that because the proton gradient in the mitochondrion can be used for purposes other than ATP synthesis, the figures we give represent the maximum energy harvest.

9–18 *Summary of glycolysis and respiration. Glucose is first broken down to pyruvic acid, with a yield of two ATP molecules and the reduction (dashed arrows) of two NAD⁺ molecules to NADH. Pyruvic acid is oxidized to acetyl CoA, and one molecule of NAD⁺ is reduced. (Note that this and subsequent reactions occur twice for each glucose molecule; this electron passage is indicated by solid arrows.) In the Krebs cycle, the acetyl group is oxidized and the electron acceptors NAD⁺ and FAD are reduced. NADH and FADH₂ then transfer their electrons to the series of cytochromes and other electron carriers that make up the electron transport chain. As the electrons are passed "downhill," relatively large amounts of free energy are released during the passage from FMN to CoQ, from cytochrome b to cytochrome c, and from cytochrome a to cytochrome a₃. These bursts of free energy transport protons through the inner mitochondrial membrane, establishing the proton gradient that powers the synthesis of ATP from ADP.*

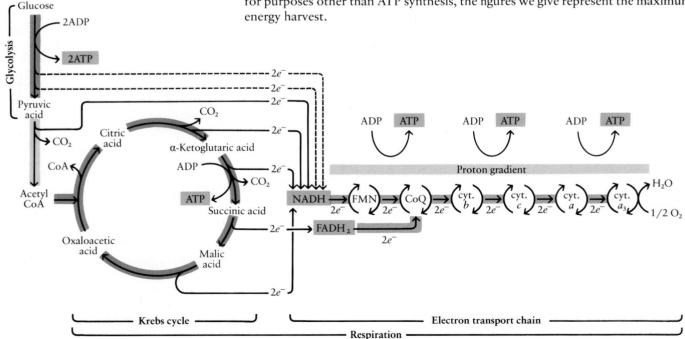

Glycolysis, in the presence of oxygen, yields two molecules of ATP directly and two molecules of NADH. These NADH molecules, however, cannot cross the inner membrane of the mitochondrion, and the electrons they carry must be "shuttled across" the membrane. In most cells, the energy cost of this process is quite low, and each NADH formed in glycolysis ultimately results in the synthesis of three molecules of ATP. In these cells, the total gain from glycolysis is 8 ATP. In other cells, including those of brain, skeletal muscle, and insect flight muscle, the energy cost of the shuttle is higher; the electrons are at a lower energy level when they reach the electron transport chain, and they enter the chain at coenzyme Q, rather than at FMN. Thus, like the electrons carried by $FADH_2$ from the Krebs cycle, they yield only two molecules of ATP per electron pair. In such cells, the total gain from glycolysis is only 6 ATP. (This was the second factor in our earlier hedging about the number of ATPs formed.)

The conversion of pyruvic acid to acetyl CoA, which occurs inside the mitochondrion, yields two molecules of NADH for each molecule of glucose and so produces six molecules of ATP.

The Krebs cycle, which also occurs inside the mitochondrion, yields two molecules of ATP, six of NADH, and two of $FADH_2$, or a total of 24 ATP, for each molecule of glucose.

As a balance sheet (Table 9–1) shows, the complete yield from a single molecule of glucose is a maximum of 38 molecules of ATP. Note that all but 2 of the 38 molecules of ATP have come from reactions taking place in the mitochondrion, and all but 4 result from the passage down the electron transport chain of electrons carried by NADH or $FADH_2$.

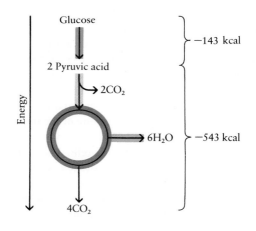

9-19 *Energy changes in the oxidation of glucose. The complete respiratory sequence (glucose + $6O_2 \longrightarrow 6CO_2$ + $6H_2O$) proceeds with an energy drop of 686 kcal/mole. Of this, almost 40 percent (266 kilocalories) is conserved in 38 ATP molecules. In anaerobic glycolysis (glucose $\longrightarrow$ lactic acid), by contrast, only 2 ATP molecules are produced, representing only about 2 percent of the available energy of glucose.*

TABLE 9–1 **Summary of Maximum Energy Yield from the Oxidation of One Molecule of Glucose**

In the Cytoplasm					
Glycolysis:		2 ATP	$\longrightarrow$		2 ATP
In the Mitochondria					
From glycolysis:	2 NADH	$\longrightarrow$ 6 ATP		$\longrightarrow$	6 ATP*
From respiration:					
Pyruvic acid $\longrightarrow$ Acetyl CoA:	1 NADH	$\longrightarrow$ 3 ATP	($\times$ 2)	$\longrightarrow$	6 ATP
Krebs cycle:		1 ATP			
	3 NADH	$\longrightarrow$ 9 ATP	($\times$ 2)	$\longrightarrow$	24 ATP
	1 $FADH_2$	$\longrightarrow$ 2 ATP			
Total:					**38 ATP**

* In some cells, the energy cost of transporting the electrons from the NADH molecules formed in glycolysis across the inner mitochondrial membrane lowers the net yield from these 2 NADH to 4 ATP; thus the total maximum yield in these cells is 36 ATP.

The free energy change (ΔG) that occurs during glycolysis and respiration is −686 kilocalories per mole. About 266 kilocalories per mole (7 kilocalories per mole of ATP $\times$ 38 moles of ATP) have been captured in the phosphate bonds of the ATP molecules, an efficiency of almost 40 percent.

The ATP molecules, once formed, are exported across the membrane of the mitochondrion by a shuttle system that simultaneously brings in one molecule of ADP for each ATP exported.

Ethanol, NADH, and the Liver

The human body can dispose fairly readily of most toxic products of its own manufacture, such as carbon dioxide and nitrogenous wastes. In contrast, most ingested toxic substances, such as ethanol (beverage alcohol), must first be broken down by the liver, which possesses special enzymes not present in other tissues.

It has been known for many years that heavy drinkers are at great risk for severe, and often fatal, liver disease. Studies conducted by Charles S. Lieber and his colleagues at the Bronx Veterans Administration Hospital and the Mount Sinai School of Medicine in New York City, have demonstrated that the origin of the problem lies in the simple chemical steps involved in the breakdown of ethanol. Enzymes in the liver first oxidize ethanol (CH_3CH_2OH) to acetaldehyde (CH_3CHO), removing two hydrogen atoms and reducing a molecule of NAD^+; this is the reverse of the second reaction shown in Figure 9–5a on page 191. The acetaldehyde is then oxidized to acetic acid, which is, in turn, oxidized to carbon dioxide and water and eliminated from the body.

Although the intoxicating effects of alcohol are due mostly to the acetaldehyde, which stimulates the release of adrenalinelike agents, the chief culprits in the development of liver disease are the hydrogen atoms (electrons and protons) removed from ethanol. These "extra" hydrogens—carried by NADH—follow two principal pathways within the cell. Most are fed directly into the electron transport chain, producing water and ATP. Because of the high levels of NADH present in the cell from the oxidation of ethanol, the production of NADH by glycolysis and the Krebs cycle is reduced. As a result, sugars, amino acids, and fatty acids are not broken down but are instead converted to fats. The fats accumulate in the liver. The mitochondria also swell, presumably as a result of the distortion of their normal function—the electron transport chain is doing very heavy duty, while the Krebs cycle is effectively shut down.

Other hydrogen atoms are used in the synthesis of fatty acids from the carbohydrate skeletons that are not being processed in glycolysis and the Krebs cycle. More fats accumulate. It does not take long. In human volunteers fed a good high-protein, low-fat diet, six drinks (about 10 ounces) a day of 86 proof alcohol produced an eightfold increase in fat deposits in the liver in only 18 days. Fortunately, these early effects are completely reversible.

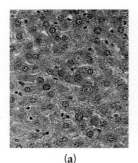 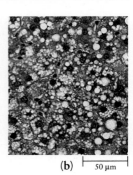

(a)   (b) ⊢ 50 μm

(a) *Normal liver tissue from a rat fed a balanced liquid diet for 24 days.* (b) *In this liver tissue from another rat fed a liquid diet in which ethanol provided 36 percent of the total calories, many globular fat droplets have accumulated. This rat was also maintained on its special diet for 24 days.*

The liver cells work hard to get rid of the excess fats, which are not soluble in blood plasma. Before being released into the bloodstream, the fats are coated with a thin layer of protein in a process carried out on the membranes of the endoplasmic reticulum. The liver cells of heavy drinkers show enormous proliferation of the endoplasmic reticulum.

After a few years—depending on how much alcohol is consumed—liver cells, engorged with fat, begin to die, triggering the inflammatory process known as alcoholic hepatitis. Liver function becomes impaired. Cirrhosis is the next step; it is the formation of scar tissue, which interferes with the function of the individual cells and also with the supply of blood to the liver. This leads to the death of more cells. The liver can no longer carry out its normal activities—such as breaking down nitrogenous wastes—which is why cirrhosis is a cause of death. In fact, cirrhosis of the liver is the ninth leading cause of death in the United States.

Not so long ago, it was commonly believed that a good diet was all that was required to protect even a heavy drinker from the deleterious effects of alcohol. In fact, if one were just to add a few vitamins to the alcohol itself, some sophisticates maintained, most of the long-term physical damage of alcohol would disappear. This new evidence refutes these comforting notions, and it comes at a time when alcohol is enjoying a resurgence of popularity among persons of high school and college age. (In populations of postgraduate age, as in other human societies the world over, it never lost its status as the drug of choice for abuse.)

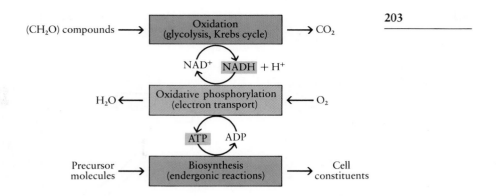

9–20 *The strategy of energy metabolism. Organisms extract energy from compounds by oxidizing them to carbon dioxide and water. NAD⁺ is the major oxidizing agent. The NADH gives up its high-energy electrons and associated protons to electron acceptors in the electron transport chain. The electrons are passed down the chain, ultimately to oxygen. The energy released in this process is used to phosphorylate ADP, converting it to ATP, which is used to drive the endergonic reactions of the organism. Note that both NADH and ATP are used in a cyclic fashion. Although they are critically important compounds, they are present in very small amounts, accomplishing their work as a result of constant and rapid turnover. The need for a constant supply of these molecules explains why a deprivation of oxygen brings about death in a very few minutes.*

OTHER CATABOLIC PATHWAYS

Most organisms do not feed directly on glucose. How do they extract energy from, for example, fats or proteins? The answer lies in the fact that the Krebs cycle is a Grand Central Station for energy metabolism. Other foodstuffs are broken down and converted into molecules that can feed into this central pathway.

Polysaccharides, such as starch, are broken down to their constituent monosaccharides and phosphorylated to glucose 6-phosphate; in this form, they enter the glycolytic pathway. Fats are first split into their glycerol and fatty acid components. The fatty acids are then chopped up into two-carbon fragments and slipped into the Krebs cycle as acetyl CoA. Proteins are broken down into their constituent amino acids. The amino acids are deaminated (the amino groups removed), and the residual carbon skeleton is either converted to an acetyl group or to one of the larger carbon compounds of the glycolytic pathway or the Krebs cycle so that it can be processed at this stage of the central pathway. The amino groups, if not reutilized, are eventually excreted as urea or other nitrogen-containing wastes. These various degradative pathways are, collectively, catabolism.

BIOSYNTHESIS

The pathways of glucose breakdown, central to catabolism, are also central to the biosynthetic, or anabolic, processes of life. These processes are the pathways of synthesis of the various molecules and macromolecules that make up an organism.

Since many of these substances, such as proteins and lipids, can be broken down and fed into the central pathway, you might guess that the reverse process can occur—namely, that the various intermediates of glycolysis and the Krebs cycle can serve as precursors for biosynthesis. This is in fact the case. However, the biosynthetic pathways, while similar to the catabolic ones, are distinctive. Different enzymes control the steps, and various critical steps of anabolism differ from those of the catabolic processes. These general pathways, which are followed in the cells of virtually all living organisms, are outlined in Figure 9–21.

In order for the reactions of the catabolic and anabolic pathways to occur, there must be a steady supply of organic molecules that can be broken down to yield energy and building block molecules. Without a supply of such molecules, the metabolic pathways cease to function and the life of the organism ends. Heterotrophic cells (including the heterotrophic cells of plants, such as the cells of the roots) are dependent on external sources—specifically, autotrophic cells—for the organic molecules that are essential to life. Autotrophic cells, however, are able to synthesize monosaccharides from simple inorganic molecules and an external energy source. These monosaccharides are then used not only to supply energy but also as building blocks for the variety of organic molecules synthesized in the anabolic pathways. By far the most important autotrophic cells are the photosynthetic cells of algae and plants. In the next chapter, we shall examine how these cells capture the energy of sunlight and use it to synthesize the monosaccharide molecules on which life on this planet depends.

9–21 *Major pathways of catabolism and anabolism in the cell.*

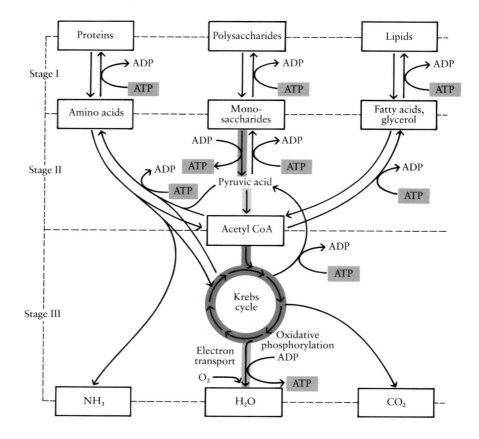

SUMMARY

The oxidation of glucose is a chief source of energy in most cells. As the glucose is broken down in a series of small enzymatic steps, a significant proportion of the potential energy of the molecule is repackaged in the phosphate bonds of ATP molecules.

The first phase in the breakdown of glucose is glycolysis, in which the six-carbon glucose molecule is split into two three-carbon molecules of pyruvic acid. A net yield of two molecules of ATP (from ADP) and two of NADH (from NAD^+) results from the process. Glycolysis takes place in the cytoplasm of the cell.

The second phase in the breakdown of glucose and other fuel molecules is respiration. It requires oxygen and, in eukaryotic cells, takes place in the mitochondria. It occurs in two stages: the Krebs cycle and terminal electron transport. (In the absence of oxygen, the pyruvic acid produced by glycolysis is converted to either ethanol or lactic acid by the process of fermentation. NAD^+ is regenerated, allowing glycolysis to continue, producing a small but vital supply of ATP for the organism.)

In the course of respiration, the three-carbon pyruvic acid molecules from glycolysis are broken down to two-carbon acetyl groups, which then enter the Krebs cycle. In a series of reactions in the Krebs cycle, the two-carbon acetyl group is oxidized completely to carbon dioxide. In the course of the oxidation of each acetyl group, four electron acceptors (three NAD^+ and one FAD) are reduced, and another molecule of ATP is formed.

The final stage of respiration is terminal electron transport, which involves a chain of electron carriers and enzymes embedded in the inner membrane of the mitochondrion. Along this series of electron carriers, the high-energy electrons carried by NADH from glycolysis and by NADH and $FADH_2$ from the Krebs cycle pass downhill to oxygen. At three points in their passage down the complete electron transport chain, large quantities of free energy are released that power the pumping of protons (H^+ ions) out of the mitochondrial matrix. This creates an electrochemical gradient of potential energy across the inner membrane of the mitochondrion. When protons pass through the ATP synthetase complex as they flow down the electrochemical gradient back into the matrix, the energy released is used to form ATP molecules from ADP and phosphate. This mechanism, by which oxidative phosphorylation is accomplished, is known as chemiosmotic coupling.

In the course of the breakdown of the glucose molecule, a maximum of 38 molecules of ATP can be formed. The exact number of ATP molecules formed depends on how much of the energy of the proton gradient is used to power other mitochondrial transport processes and on the shuttle mechanism by which the electrons from the NADH molecules formed in glycolysis are brought into the mitochondrion. Generally, almost 40 percent of the free energy released in glucose oxidation is retained in the form of newly synthesized ATP molecules.

Other food molecules, including fats, polysaccharides, and proteins, are utilized by being degraded to compounds that can enter these central pathways at various steps. The biosynthesis of these substances also originates with precursor compounds derived from intermediates in the respiratory sequence and is driven by the energy derived from those processes.

QUESTIONS

1. Distinguish among the following: oxidation of glucose/glycolysis/respiration/fermentation; aerobic pathways/anaerobic pathways; FAD/FADH$_2$; Krebs cycle/electron transport.

2. Describe the process of fermentation. What conditions are essential if it is to occur? With some strains of yeast, fermentation stops before the sugar is exhausted, usually at an alcohol concentration in excess of 12 percent. What is a plausible explanation?

3. If aerobic (oxygen-utilizing) organisms are so much more efficient than anaerobes in converting energy, why are there any anaerobes left on this planet? Why didn't they all become extinct long ago?

4. Sketch the structure of a mitochondrion. Describe where the various stages in the breakdown of glucose take place in relation to mitochondrial structure. What molecules and ions cross the mitochondrial membranes during these processes?

5. Each FADH$_2$ molecule produced in the Krebs cycle results in the formation of only two molecules of ATP when its electrons pass down the electron transport chain. It was long thought that because these electrons enter the chain at coenzyme Q rather than at FMN they "missed" one of the sites of phosphorylation. What is a more accurate explanation?

6. Cyanide can combine with—and deactivate—cytochrome a and cytochrome a_3. In our bodies, however, cyanide tends to react first with hemoglobin and to make it impossible for oxygen to bind to the hemoglobin. Either way, cyanide poisoning has the same effect: it inhibits the synthesis of ATP. Explain how this is so.

7. When the F$_1$ portion of the ATP synthetase complex is removed from the mitochondrial membrane and studied in solution, it functions as an ATPase. Why does it not function as an ATP synthetase?

8. Certain chemicals function as "uncoupling" agents when they are added to respiring mitochondria. The passage of electrons down the chain to oxygen continues, but no ATP is formed. One of these agents, the antibiotic valinomycin, is known to transport K^+ ions through the inner membrane into the matrix. Another, 2,4-dinitrophenol, transports H^+ ions through the membrane. How do these substances prevent the formation of ATP? Which would you expect to have the most profound effect on ATP formation? Why?

9. In the cells of a specialized tissue known as brown fat, the inner membrane of the mitochondrion is permeable to H^+ ions. These cells contain large stores of fat molecules, which are gradually broken down and the resulting acetyl groups are fed into the Krebs cycle. The electrons captured by NADH and FADH$_2$ are, in turn, fed into the electron transport chain and ultimately accepted by oxygen. No ATP is synthesized, however. Why not? Brown fat tissue is found in some hibernating animals and in mammalian infants that are born hairless, including human infants. What do you suppose the function of brown fat tissue is?

10. (a) As we have seen, a cell can obtain a maximum of 38 molecules of ATP from each molecule of glucose that is completely oxidized. Account for the production of each molecule of ATP. (b) In the course of glycolysis, the Krebs cycle, and electron transport, 40 molecules of ATP are actually formed. Why is the net yield for the cell only 38 molecules? (c) What other factors can reduce the yield of ATP?

11. Describe how the processes of the cell are adapted to the efficient use of a variety of foodstuffs, and to the efficient production of the variety of materials that the cell needs to manufacture for its own use.

12. In terms of the cell's economy, what do anabolic processes provide for the cell? What do catabolic processes provide? How are they dependent on each other?

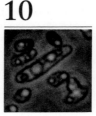

Photosynthesis, Light, and Life

The first photosynthetic organism probably appeared 3 to 3.5 billion years ago. Before the evolution of photosynthesis, the physical characteristics of earth and its atmosphere were the most powerful forces in shaping the course of natural selection. With the evolution of photosynthesis, however, organisms began to change the face of our planet and, as a consequence, to exert strong influences on each other. Organisms have continued to change the environment, at an ever-increasing rate, up to the present day.

As we saw earlier, the atmosphere in which the first cells evolved lacked free oxygen. These earliest organisms were, of course, adapted to living in an environment without free oxygen; in fact, oxygen, with its powerful electron-attracting capacities, would have been poisonous to them (as it is to many modern anaerobes). Their energy came from anaerobic processes, most likely glycolysis and fermentation, which would have resulted in a gradual accumulation of carbon dioxide in the atmosphere. The organic molecules they used as fuel may have been formed by nonbiological processes (page 87), or they may have been produced by chemosynthetic autotrophs (page 89) or by primitive photosynthetic cells that, like some modern photosynthetic bacteria, did not release oxygen to the environment.

Then, it is hypothesized, there slowly evolved photosynthetic organisms that used carbon dioxide as their carbon source and released oxygen, as do most modern photosynthetic forms. As these photosynthetic organisms multiplied, they provided a new supply of organic molecules, and free oxygen began to accumulate. In response to these changing conditions, cell species arose for which oxygen was not a poison but rather a requirement for existence. (It has been proposed that the original function of the electron transport chain found in the cell membrane of aerobic bacteria—thought to be the forerunner of the mitochondrial electron transport chain—was to protect the cell from oxygen.)

As we saw in the last chapter, oxygen-utilizing organisms have an advantage over those that do not use oxygen. A higher yield of energy can be extracted per molecule from the aerobic breakdown of carbon-containing compounds than from anaerobic processes, in which fuel molecules are not completely oxidized. Energy released in cells by reactions using oxygen made possible the development of increasingly active, increasingly complex organisms. Without oxygen, the complex forms of life that now exist on earth could not have evolved.

Life on earth continues to be dependent on photosynthesis both for its oxygen and for its carbon-containing fuel molecules. Photosynthetic organisms capture light energy and use it to form carbohydrates and free oxygen from carbon dioxide and water, in a complex series of reactions. The overall equation for photosynthesis can be summarized as:

$$CO_2 + H_2O + \text{Light energy} \longrightarrow \underset{\text{Carbohydrate}}{(CH_2O)} + O_2$$

To understand how organisms are able to capture light energy and convert it into stored chemical energy, we must first look at the characteristics of light itself.

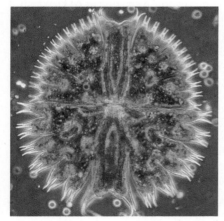

| 20 μm |

10–1 *A freshwater green alga of the genus* Micrasterias. *In this alga and other unicellular algae, each cell is an individual, self-sufficient photosynthetic organism. The green color is due to chlorophyll, in which the radiant energy of sunlight is converted to chemical energy.*

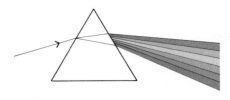

10–2 *White light is actually a mixture of different colors, ranging from violet at one end of the spectrum to red at the other. It is separated into its component colors when it passes through a prism—"the celebrated phaenomena of colors," as Newton referred to it.*

THE NATURE OF LIGHT

Over 300 years ago, the English physicist Sir Isaac Newton (1642–1727) separated visible light into a spectrum of colors by passing it through a prism (Figure 10–2). Then by passing the light through a second prism, he recombined the colors, producing white light once again. By this experiment, Newton showed that white light is actually made up of a number of different colors, ranging from violet at one end of the spectrum to red at the other. Their separation is possible because light of different colors is bent at different angles in passing through the prism. Newton believed that light was a stream of particles (or, as he termed them, "corpuscles"), in part, because of its tendency to travel in a straight line.

In the nineteenth century, through the genius of James Clerk Maxwell (1831–1879), it became known that what we experience as light is in truth a very small part of a vast continuous spectrum of radiation, the electromagnetic spectrum (Figure 10–3). As Maxwell showed, all the radiations in this spectrum act as if they travel in waves. The wavelengths—that is, the distances from one wave peak to the next—range from those of gamma rays, which are measured in nanometers (1 nanometer = 10^{-9} meter), to those of low-frequency radio waves, which are measured in kilometers (1 kilometer = 10^3 meters). Within the spectrum of visible light, red light has the longest wavelength, violet the shortest. Another feature that these radiations have in common is that, in a vacuum, they all travel at the same speed—300,000 kilometers per second.

By 1900, it had become clear, however, that the wave model of light was not adequate. The key observation, a very simple one, was made in 1888: when a zinc plate is exposed to ultraviolet light, it acquires a positive charge. The metal, it was soon deduced, becomes positively charged because the light energy dislodges electrons, forcing them out of the metal atoms. Subsequently, it was discovered that this photoelectric effect, as it is called, can be produced in all metals. Every metal has a critical wavelength for the effect; the light (visible or invisible) must be of that wavelength or a shorter wavelength for the effect to occur.

With some metals, such as sodium, potassium, and selenium, the critical wavelength is within the spectrum of visible light, and as a consequence, visible light striking the metal can set up a moving stream of electrons (such a stream is an electric current). Burglar alarms, exposure meters, television cameras, and the electric eyes that open doors for you at supermarkets or airline terminals all operate on this principle of turning light energy into electrical energy.

10–3 *Visible light is only a small portion of the vast electromagnetic spectrum. For the human eye, the visible spectrum ranges from violet light, which is made up of comparatively short rays, to red light, the longest visible rays.*

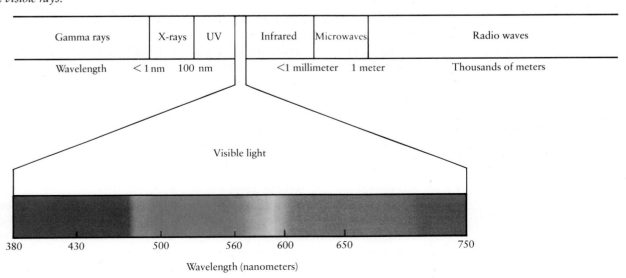

No Vegetable Grows in Vain

Until about 350 years ago, observers of the biological world, noting that the life processes of animals were dependent on the food they ate, thought that plants derived their food from the soil in a similar way. This concept was widely accepted until the Belgian physician Jan Baptista van Helmont (1577–1644) offered the first experimental evidence to the contrary.

Van Helmont grew a small willow tree in an earthenware pot for five years, adding only water to the pot. At the end of five years, the willow had increased in weight by 74 kilograms, while the earth in the pot had decreased in weight by only 57 grams. On the basis of these results, van Helmont concluded that all the substance of the plant was produced from the water and none from the soil. (This experiment is of general interest to those concerned with tracing the history of science, because it is one of the first carefully designed biological experiments ever reported. Van Helmont's conclusions, however, were too broad.)

The next stage in our understanding of plant nutrition resulted from studies of combustion, a topic that intrigued not only the medieval alchemists but also their successors who laid the foundations of modern chemistry. One of the fascinating problems about combustion was that it in some way "injured" air. For example, if a candle was burned in a closed container, it would soon go out; if a mouse was then put in the container, it would die.

One of those concerned with the changes produced in air by burning was Joseph Priestley (1733–1804), an English clergyman and chemist. On August 17, 1771, Priestley "put a sprig of mint into air in which a wax candle had burned out and found that, on the 27th of the same month, another candle could be burned in this same air." Priestley believed, as he reported, that he had "accidentally hit upon a method of restoring air that has been injured by the burning of candles." The "restorative which nature employs for this purpose," he stated, is "vegetation." Priestley extended his observations and soon showed that air "restored" by vegetation was not "at all inconvenient to a mouse." These experiments offered the first logical explanation of how the air remained "pure" and able to support life despite the burning of countless fires and the breathing of many animals. When Priestley was presented with a medal for his discovery, the citation read in part: "For these discoveries we are assured that no vegetable grows in vain ... but cleanses and purifies our atmosphere."

Priestley's reports that plants purify the air were of great interest to his fellow chemists, but they soon attracted criticism because the experiments could not be confirmed. In fact, when Priestley tried to do the experiments again himself, he did not get the same results. (We think now that he may have moved his equipment to a dark corner of his laboratory.) It was a Dutch physician, Jan Ingenhousz (1730–1799), who

The wave model of light would lead you to predict that the brighter the light—that is, the stronger, or more intense, the beam—the greater the force with which the electrons would be dislodged. But as we have already seen, whether or not light can eject the electrons of a particular metal depends not on the brightness of the light but on its wavelength. A very weak beam of the critical wavelength or a shorter wavelength is effective, while a stronger beam of a longer wavelength is not. Furthermore, as was shown in 1902, increasing the brightness of the light increases the number of electrons dislodged but not the velocity at which they are ejected from the metal. To increase the velocity, one must use a shorter wavelength of light. Nor is it necessary for energy to be accumulated in the metal. With even a dim beam of a critical wavelength, electrons may be emitted the instant the light hits the metal.

To explain such phenomena, the particle model of light was resurrected by Albert Einstein in 1905. According to this model, light is composed of particles of energy called **photons**. The energy of a photon is not the same for all kinds of light but is, in fact, inversely proportional to the wavelength—the longer the wave-

was finally able to confirm Priestley's work with an important addition. He found that the purification takes place only in sunlight. Plants at night or in the shade, he reported, "contaminate the air which surrounds them, throwing out an air hurtful to animals." He also observed that only the green parts of plants restore the air and, on the basis of control experiments, that "the sun by itself has no power to mend air without the concurrence of plants."

While Ingenhousz was performing his experiments on plants, Antoine Lavoisier (1743–1794) was carrying out the experiments that put chemistry on a modern basis. Among Lavoisier's many discoveries, those that had the most impact on studies of plant processes concerned the exchanges of gases that take place when animals breathe. Working with the mathematician P. S. Laplace (1749–1827), Lavoisier confined a guinea pig for about 10 hours in a jar containing oxygen and measured the carbon dioxide produced. He also measured the amount of oxygen used by a man active and at rest. In these experiments, he was able to show that the combustion of carbon compounds with oxygen is the true source of animal heat and that oxygen consumption increases during physical work. "Respiration is merely a slow combustion of carbon and hydrogen, which is similar in every respect to that which occurs in a lighted lamp or candle, and, from this point of view, animals that

breathe are really combustible bodies which burn and are consumed."

The work of Ingenhousz spanned the prematurely terminated career of Lavoisier, who was guillotined on May 8, 1794, during the French Revolution. (The judge presiding over the case is reported to have said, "The Republic has no need of savants.") Quick to adopt Lavoisier's ideas about gases, Ingenhousz hypothesized that the plant was not just exchanging "good air" for "bad" and so making the world habitable for animal life. In the sunshine, he suggested, a plant absorbs the carbon from carbon dioxide, "throwing out at that time the oxygen alone, and keeping the carbon to itself as nourishment."

Nicholas Theodore de Saussure (1767–1845) later showed that equal volumes of CO_2 and O_2 are exchanged during photosynthesis and that the plant does indeed retain the carbon. He also showed that more weight was gained by the plant during photosynthesis than could be accounted for by the carbon taken in as carbon dioxide. In other words, the carbon in the dry matter of plants comes from carbon dioxide, but equally important, the rest of the dry matter, with the exception of minerals from the soil, comes from water. Thus, all the components—carbon dioxide, water, and light—were identified, and it became possible to write the overall photosynthetic equation, as shown on page 206.

length, the lower the energy. Photons of violet light, for example, have almost twice the energy of photons of red light, the longest visible wavelength.

The wave model of light permits physicists to describe certain aspects of its behavior mathematically, and the photon model permits another set of mathematical calculations and predictions. These two models are no longer regarded as opposed to one another; rather, they are complementary, in the sense that both—or a totally new model—are required for a complete description of the phenomenon we know as light.

The Fitness of Light

Light, as Maxwell showed, is only a tiny band in a continuous spectrum. From the physicist's point of view, the difference between radiations we can see and radiations we cannot see—so dramatic to the human eye—is only a few nanometers of wavelength, or, expressed differently, a small amount of energy. Why does this particular group of radiations, rather than some other, make the leaves

(a)

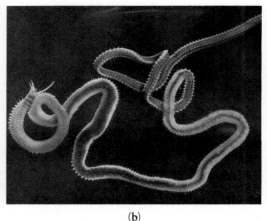

(b)

10–4 (a) *Gathering palolo worms at dawn in Western Samoa. Once a year, following the first full moon after the autumnal equinox, Pacific palolo worms leave their undersea burrows and swarm. The rear portion of each worm, filled with eggs or sperm, breaks off and, attracted to the* moonlight, *rises to the water's surface. There, in the early morning hours, before daylight has stimulated the release of eggs and sperm, people gather the wriggling worms for a communal feast. The sea is said to resemble vermicelli soup. Because* it occurs so accurately, the day of the "big rising" marked the beginning of the Samoan New Year in premissionary days. (b) A palolo worm. Its narrower posterior region is filled with sperm and will break off when the worm swarms.

grow and the flowers burst forth, cause the mating of fireflies and the spawning of palolo worms (Figure 10–4), and, when reflecting off the surface of the moon, excite the imagination of poets and lovers? Why is it that this tiny portion of the electromagnetic spectrum is responsible for vision, for the rhythmic day-night regulation of many biological activities, for the bending of plants toward the light, and also for photosynthesis, on which life depends? Is it an amazing coincidence that all these biological activities are dependent on these same wavelengths?

George Wald of Harvard, an expert on the subject of light and life, says no. He thinks that if life exists elsewhere in the universe, it is probably dependent on the same fragment of the vast spectrum. Wald bases this conjecture on two points. First, living things, as we have seen, are composed of large, complicated molecules held in special configurations and relationships to one another by hydrogen bonds and other weak bonds. Radiation of even slightly higher energies than the energy of violet light breaks these bonds and so disrupts the structure and function of the molecules. Radiations with wavelengths less than 200 nanometers—that is, with still higher energies—drive electrons out of atoms. On the other hand, light of wavelengths longer than those of the visible band—that is, with less energy than red light—is absorbed by water, which makes up the great bulk of all living things on earth. When this light is absorbed by molecules, its lower energy causes them to increase their motion (increasing heat) but does not trigger changes in their electron configurations. Only those radiations within the range of visible light have the property of exciting molecules—that is, of moving electrons into higher energy levels—and so of producing chemical and, ultimately, biological changes.

The second reason that the visible band of the electromagnetic spectrum has been "chosen" by living things is that it, above all, is what is available. Most of the radiation reaching the surface of the earth from the sun is within this range.

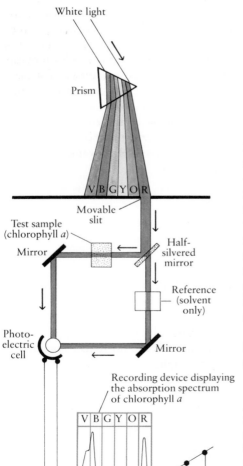

White light

Prism

V B G Y O R

Movable slit

Test sample (chlorophyll *a*)

Mirror

Half-silvered mirror

Reference (solvent only)

Photo-electric cell

Mirror

Recording device displaying the absorption spectrum of chlorophyll *a*

V B G Y O R

10–5 *The absorption spectrum of a pigment is measured with a spectrophotometer. This device directs a beam of light of each wavelength at the substance to be analyzed and records what percentage of light of each wavelength is absorbed by the pigment sample as compared to a ref-* erence sample. Because the mirror is lightly (half) silvered, half of the light is reflected and half is transmitted. The photoelectric cell is connected to an electronic device that automatically records the percentage absorption at each wavelength.

Higher-energy wavelengths are screened out by the oxygen and ozone high in the modern atmosphere. Much infrared radiation is screened out by water vapor and carbon dioxide before it reaches the earth's surface.

This is an example of what has been termed "the fitness of the environment"; the suitability of the environment for life and that of life for the physical world are exquisitely interrelated. If they were not, life could not exist.

CHLOROPHYLL AND OTHER PIGMENTS

In order for light energy to be used by living systems, it must first be absorbed. A pigment is any substance that absorbs light. Some pigments absorb all wavelengths of light and so appear black. Some absorb only certain wavelengths, transmitting or reflecting the wavelengths they do not absorb. Chlorophyll, the pigment that makes leaves green, absorbs light in the violet and blue wavelengths and also in the red; because it reflects green light, it appears green. Different pigments absorb light energy at different wavelengths. The absorption pattern of a pigment is known as the **absorption spectrum** of that substance (Figure 10–5).

Different groups of plants and algae use various pigments in photosynthesis. There are several different kinds of chlorophyll that vary slightly in their molecular structure (Figure 10–6). In plants, chlorophyll *a* is the pigment directly involved in the transformation of light energy to chemical energy. Most photosynthetic cells also contain a second type of chlorophyll—in plants, it is chlorophyll *b*—and a representative of another group of pigments called the carotenoids. One of the carotenoids found in plants is beta-carotene. The carotenoids are red, orange, or yellow pigments. In the green leaf, their color is masked by the chlorophylls, which are more abundant. In some tissues, however, such as those of a ripe tomato, the carotenoid colors predominate, as they do also when leaf cells stop synthesizing chlorophyll in the fall.

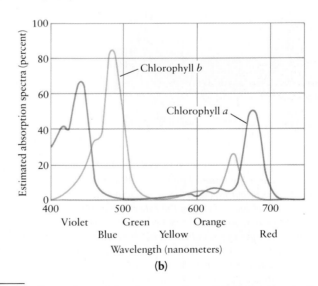

(b)

10–6 **(a)** *Chlorophyll* a *is a large molecule with a central atom of magnesium held in a porphyrin ring. Attached to the ring is a long, hydrophobic carbon-hydrogen chain that may help to anchor the molecule in the internal membranes of the chloroplast. Chlorophyll* b *differs from chlorophyll* a *in having an aldehyde (CHO) group in place of the CH₃ group indicated in green. Alternating single and double bonds, such as those in the chlorophylls, are common in pigments.* **(b)** *The estimated absorption spectra of chlorophyll* a *and chlorophyll* b *within the chloroplast. (Prepared by Govindjee.)*

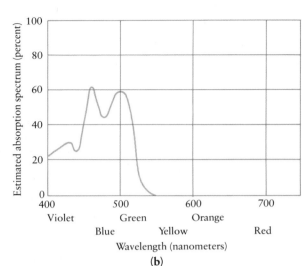

beta-Carotene

Vitamin A

Retinal

(a)

10–7 **(a)** *Related carotenoids. Cleavage of the beta-carotene molecule at the point indicated by the arrow yields two molecules of vitamin A. Oxidation of vitamin A yields retinal, the pigment involved in vision. Absorption of light energy changes the electron configuration of retinal and triggers a nerve impulse in the retina. All of this explains why you were told to eat your carrots.* **(b)** *The estimated absorption spectrum of carotenoids in the chloroplast. (Prepared by Govindjee.)*

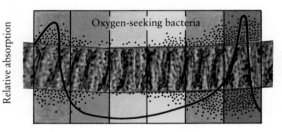

10–8 *Results of an experiment performed in 1882 by T. W. Englemann revealing the action spectrum of photosynthesis in a filamentous alga. Like more recent investigators, Englemann used the rate of oxygen production to measure the rate of photosynthesis. Unlike his successors, however, he lacked sensitive devices for detecting oxygen. As his oxygen indicator, he chose motile bacteria that are attracted by oxygen. In place of the mirror and diaphragm usually used to illuminate objects under view in his microscope, he substituted a "microspectral apparatus," which, as its name implies, produced a tiny spectrum of colors that it projected upon the slide under the microscope. Then he arranged a filament of algal cells parallel to the spread of the spectrum. The oxygen-seeking bacteria congregated mostly in the areas where the violet and red wavelengths fell upon the algal filament. As you can see, the action spectrum for photosynthesis Englemann revealed in this elegant experiment paralleled the absorption spectrum of chlorophyll. He therefore concluded that photosynthesis depends on the light absorbed by chlorophyll.*

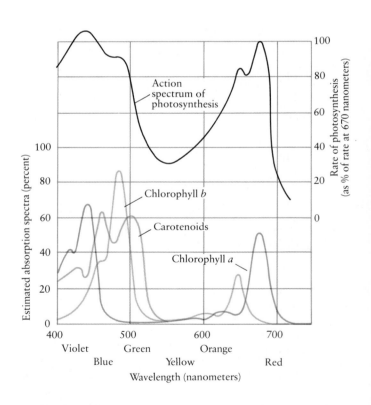

10–9 *The upper curve shows the action spectrum for photosynthesis and the lower curves, absorption spectra for chlorophyll a, chlorophyll b, and carotenoids in the chloroplast. Note that the action spectrum of photosynthesis indicates that chlorophyll a, chlorophyll b, and carotenoids all absorb light used in photosynthesis. (Prepared by Govindjee.)*

The other chlorophylls and the carotenoids are able to absorb light at wavelengths different from those absorbed by chlorophyll *a*. They apparently can pass the energy on to chlorophyll *a*, thus extending the range of light available for photosynthesis (Figure 10–7).

An **action spectrum** defines the relative effectiveness (per number of incident photons) of different wavelengths of light for light-requiring processes, such as photosynthesis, flowering, phototropism (the bending of a plant toward light), and vision. Similarity between the absorption spectrum of a pigment and the action spectrum of a process is considered evidence that that particular pigment is responsible for that particular process (Figures 10–8 and 10–9).

When pigments absorb light, electrons within the pigment molecules are boosted to a higher energy level. Three possible consequences are: (1) the energy may be dissipated as heat; (2) it may be re-emitted immediately as light energy of a longer wavelength, a phenomenon known as fluorescence; (3) the energy may trigger a chemical reaction, as happens in photosynthesis. Whether or not a chemical reaction occurs depends not only on the structure of the particular pigment but also on its relationship with neighboring molecules. For example, if chlorophyll molecules are isolated in a test tube and light is permitted to strike them, they fluoresce. In other words, the molecules absorb light energy, and the electrons are momentarily raised to a higher energy level and then fall back again to a lower one. As they fall to a lower energy level, they release much of this energy as light. None of the light absorbed by isolated chlorophyll molecules is converted to any form of energy useful to living systems. Chlorophyll can convert light energy to chemical energy only when it is associated with certain proteins and embedded in a specialized membrane.

PHOTOSYNTHETIC MEMBRANES: THE THYLAKOID

The structural unit of photosynthesis is the **thylakoid**, which usually takes the form of a flattened sac, or vesicle. In the photosynthetic prokaryotes, thylakoids may form a part of the cell membrane, or they may occur singly in the cytoplasm, or, as in the cyanobacteria, they may be part of an elaborate internal membrane structure (see Figure 4–9, page 92). In eukaryotes, the thylakoids form a part of the internal membrane structure of specialized organelles, the chloroplasts (Figure 10–10). The alga *Chlamydomonas*, for instance, has a single very large chloroplast; the cell of a leaf characteristically has 40 to 50 chloroplasts, and there are often 500,000 chloroplasts per square millimeter of leaf surface.

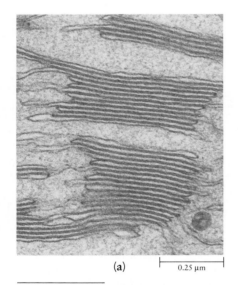

(a)   0.25 μm

10–10 *The unit of photosynthesis is the thylakoid, a flattened sac, whose membranes contain chlorophyll and other pigments. In plants and algae, thylakoids are part of an elaborate membrane system enclosed in a special organelle, the chloroplast. (a) Stacks of thylakoids (grana) from a plant cell. The inner compartments of the thylakoids are interconnected, forming the thylakoid space, which contains a solution whose composition differs from that of the stroma and the cytoplasm. (b) A chloroplast, showing the elaborate system of internal membranes comprising interconnected stacks of thylakoids.*

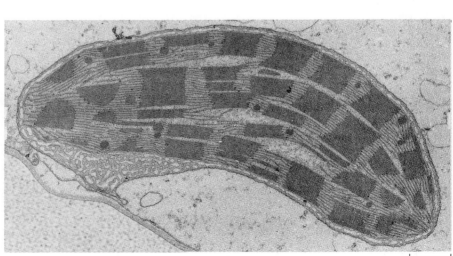

(b)   1 μm

The Structure of the Chloroplast

Chloroplasts, like mitochondria, are surrounded by two membranes that are separated by an intermembrane space. The inner membrane, unlike that of the mitochondrion, is smooth. The thylakoids, in the interior of the chloroplast, constitute a third membrane system. Surrounding the thylakoids, and filling the interior of the chloroplast, is a dense solution, the **stroma,** which (like the matrix of the mitochondrion) is different in composition from the cytoplasm. The thylakoids enclose an additional compartment, known as the thylakoid space, which contains a solution of still different composition. Thus, whereas the mitochondrion has two membrane systems (the outer and the inner) and two compartments (the intermembrane space and the matrix), the chloroplast has three membrane systems (outer, inner, and thylakoid) and three compartments (intermembrane space, stroma, and thylakoid space).

With the light microscope under high power, it is possible to see little spots of green within the chloroplasts of leaves. The early microscopists called these green specks **grana** ("grains"), and this term is still in use. Under the electron microscope, it can be seen that the grana are stacks of thylakoids. Some of the thylakoid membranes have extensions that interconnect the grana through the stroma that separates them.

All the thylakoids in a chloroplast are oriented parallel to each other. Thus, by swinging toward the light, the chloroplast can simultaneously aim all of its millions of pigment molecules for optimum reception, as if they were miniature electromagnetic antennae (which, of course, they are).

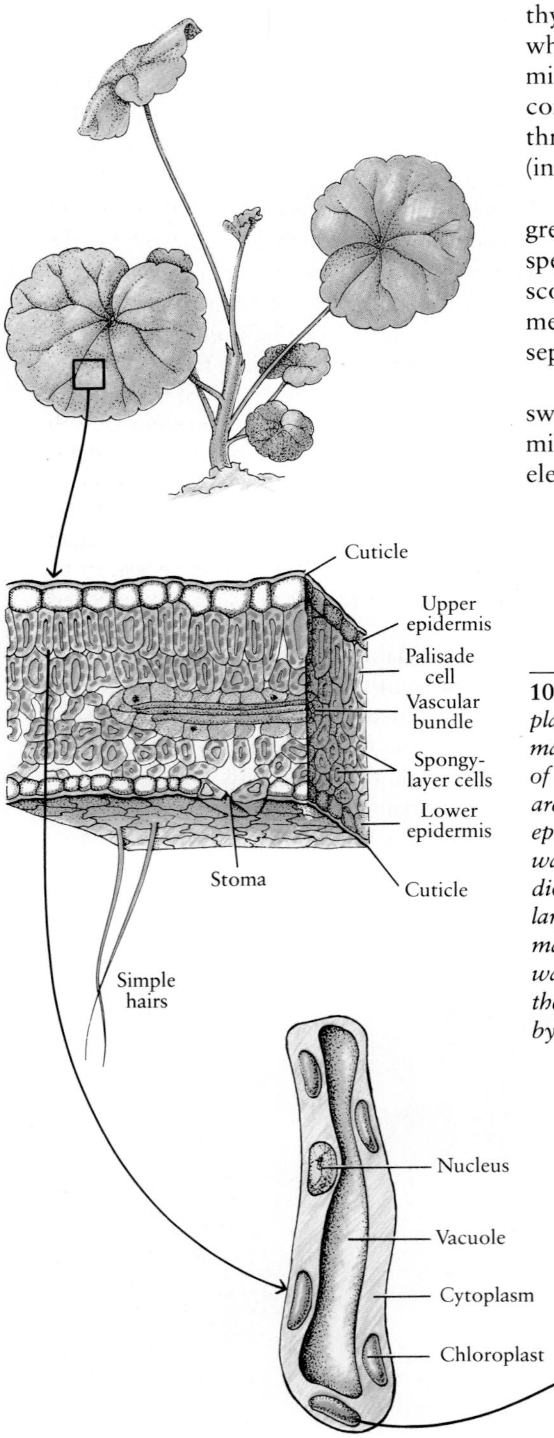

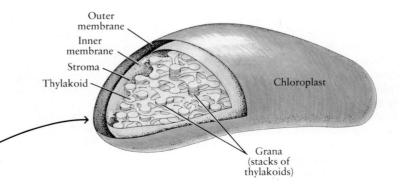

10–11 *Journey into a chloroplast. The plant shown is a geranium, which you may recognize by the characteristic shape of its leaves. The inner tissues of the leaf are completely enclosed by transparent epidermal cells that are coated with a waxy layer, the cuticle. Oxygen, carbon dioxide, and other gases enter the leaf largely through special openings, the stomata (singular, stoma). These gases and water vapor fill the spaces between cells in the spongy layer, leaving and entering cells by diffusion. Water, taken up by the roots,* enters the leaf by way of the vascular bundle, and sugars, the products of photosynthesis, leave the leaf by this route, traveling to nonphotosynthetic parts of the plant. Much of the photosynthesis takes place in the palisade cells, elongated cells directly beneath the upper epidermis. They have a large central vacuole and numerous chloroplasts that move within the cell, orienting themselves with respect to the light. Light is captured in the membranes of the disk-shaped thylakoids within the chloroplast.*

Cuticle

Upper epidermis

Palisade cell

Vascular bundle

Spongy-layer cells

Lower epidermis

Stoma

Cuticle

Simple hairs

Nucleus

Vacuole

Cytoplasm

Chloroplast

Outer membrane

Inner membrane

Stroma

Thylakoid

Chloroplast

Grana (stacks of thylakoids)

THE STAGES OF PHOTOSYNTHESIS

As we noted earlier (page 208), it was demonstrated about 200 years ago that light is required for the process we know as photosynthesis. It is now known that photosynthesis actually takes place in two stages, only one of which requires light. Evidence for this two-stage mechanism was first presented in 1905 by the English plant physiologist F. F. Blackman, as the result of experiments in which he measured the rate of photosynthesis under varying conditions.

Blackman first plotted the rate of photosynthesis at various light intensities. In dim to moderate light, increasing the light intensity increased the rate of photosynthesis, but at higher intensities, a further increase in light intensity had no effect. He then studied the combined effects of light and temperature on photosynthesis. In dim light, an increase in temperature had no effect. However, Blackman found that if he increased the light and also increased the temperature, the rate of photosynthesis was greatly accelerated (Figure 10–12). As the temperature increased above 30°C, the rate of photosynthesis slowed and finally the process ceased.

On the basis of these experiments, Blackman concluded that more than one set of reactions was involved in photosynthesis. First, there was a group of light-dependent reactions that were temperature-independent. The rate of these reactions could be accelerated in the dim-to-moderate light range by increasing the amount of light, but it was not accelerated by increases in temperature. Second, there was a group of reactions that were dependent not on light but rather on temperature. Both sets of reactions seemed to be required for the process of photosynthesis. Increasing the rate of only one set of reactions increased the rate of the entire process only to the point at which the second set of reactions began to hold back the first (that is, it became rate-limiting). Then it was necessary to increase the rate of the second set of reactions in order for the first to proceed unimpeded.

In Blackman's experiments, the temperature-dependent reactions increased in rate as the temperature was increased, but only up to about 30°C, after which the rate began to decrease. From this evidence it was concluded that these reactions were controlled by enzymes, since this is the way enzymes are expected to respond to temperature (see Figure 8–17, page 175). This conclusion has since proved to be correct.

10–12 (a) *An increase in light intensity beyond about 1,200 candelas does not produce a corresponding increase in the rate of photosynthesis. A curve such as the one shown here indicates that some other factor—known as a rate-limiting factor—is involved in the process under study. Under field conditions, CO_2 concentration is commonly the rate-limiting factor.* (b) *At a low intensity of light, an increase in temperature does not increase the rate of photosynthesis. At a high intensity, however, an increase in temperature has a very marked effect. From these data, Blackman concluded that photosynthesis includes both light-dependent and light-independent reactions.*

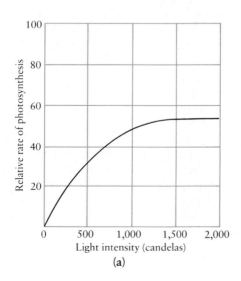

(a)

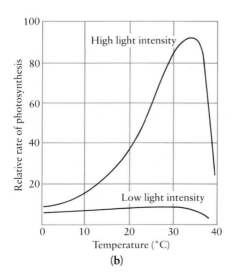

(b)

10–13 *Although NAD⁺ (Figure 8–14) and NADP⁺ resemble one another very closely, their biological roles are distinctly different. NADH generally transfers its electrons to other electron carriers, which continue to pass them on down to successively lower energy levels in discrete steps. In the course of this electron transfer, ATP molecules are formed. NADPH, by contrast, provides energy directly to biosynthetic processes of the cell that require large energy inputs.*

Photosynthesis was thus shown to have both a light-dependent stage, the so-called "light" reactions, and an enzymatic, light-independent stage, the "dark" reactions. The terms "light" and "dark" reactions have created much confusion, for although the "dark" reactions do not require light as such—only the chemical products of the "light" reactions—they can occur in either light or darkness. Moreover, recent work has shown that the enzyme controlling one of the key "dark" reactions is indirectly stimulated by light. As a result, these terms are now falling into disfavor. They are being replaced by terms that more accurately describe the processes occurring during each stage of photosynthesis.

In the first stage of photosynthesis—the energy-capturing reactions—light strikes chlorophyll *a* molecules that are packed in a special way in the thylakoid membranes. Electrons from the chlorophyll *a* molecules are boosted to higher energy levels, and, in a series of reactions, their added energy is used to form ATP from ADP and to reduce an electron-carrier molecule known as NADP⁺ (Figure 10–13). NADP⁺ closely resembles NAD⁺, and it too is reduced by the addition of two electrons and a proton, forming NADPH. Water molecules are also broken apart in this stage of photosynthesis, supplying electrons that replace those boosted from the chlorophyll *a* molecules.

In the second stage of photosynthesis, the ATP and NADPH formed in the first stage are used to reduce the carbon in carbon dioxide to a simple sugar. Thus the chemical energy temporarily stored in ATP and NADPH molecules is transferred to molecules suitable for transport and storage in the algal cell or plant body. At the same time, a carbon skeleton is formed from which other organic molecules can be built. This incorporation of CO_2 into organic compounds is known as the **fixation of carbon**. The steps by which it is accomplished, called the carbon-fixing reactions, occur in the stroma of the chloroplast.

THE ENERGY-CAPTURING REACTIONS

The Photosystems

In the thylakoids, chlorophyll and other molecules are, according to the present model, packed into units called photosystems. Each unit contains from 250 to 400 molecules of pigment, which serve as light-trapping antennae. Once a photon of light energy is absorbed by one of the antenna pigments, it is bounced around (like a hot potato) among the other pigment molecules of the photosystem until it reaches a special form of chlorophyll *a*, which is the reaction center. When this particular chlorophyll molecule absorbs the energy, an electron is boosted to a higher energy level from which it is transferred to another molecule, a primary electron acceptor. The chlorophyll molecule is thus oxidized (minus an electron) and positively charged.

Present evidence indicates that there are two different photosystems. In Photosystem I, the reactive chlorophyll *a* molecule is known as P_{700} (P is for pigment) because one of the peaks of its absorption spectrum is at 700 nanometers, a slightly longer wavelength than the usual chlorophyll *a* peak. When P_{700} is oxidized, it bleaches, which is how it was detected. No one has managed to isolate pure P_{700}. Recent evidence indicates that P_{700} is not a different kind of chlorophyll but rather a dimer of two chlorophyll *a* molecules; its unusual properties result from its association with special proteins in the thylakoid membrane and its position in relation to other molecules. Photosystem II also contains a reactive chlorophyll *a* molecule, which passes its electron on to a different primary electron acceptor. The reactive chlorophyll *a* molecule of Photosystem II is P_{680}.

Van Niel's Hypothesis

For more than 100 years after the completion of the work of Ingenhousz (page 208), it was generally assumed that in the equation

$$CO_2 + H_2O + Light \longrightarrow (CH_2O) + O_2$$

the carbohydrate (CH_2O) resulted from the combination of carbon atoms with water molecules and that the oxygen was released from the carbon dioxide molecule. This entirely reasonable hypothesis was widely accepted. But, as it turned out, it was wrong.

The investigator who upset this long-held assumption was the late C. B. van Niel of Stanford University. Van Niel, then a graduate student, was investigating photosynthesis in different types of photosynthetic bacteria. In their photosynthetic reactions, bacteria reduce carbon to carbohydrates, but they do not release oxygen. Among the types of bacteria van Niel was studying were the purple sulfur bacteria, which require hydrogen sulfide for photosynthesis. In the course of photosynthesis, globules of sulfur (S) are excreted or accumulated inside the bacterial cells. In these bacteria, van Niel found that this reaction takes place during photosynthesis:

$$CO_2 + 2H_2S \xrightarrow{\text{Light}} (CH_2O) + H_2O + 2S$$

This finding was simple enough and did not attract much attention until van Niel made a bold extrapolation. He proposed that the generalized equation for photosynthesis is

$$CO_2 + 2H_2A \xrightarrow{\text{Light}} (CH_2O) + H_2O + 2A$$

In this equation, H_2A stands for some oxidizable substance such as hydrogen sulfide, free hydrogen, or any one of several other compounds used by photosynthetic bacteria—or water. In the cyanobacteria, the algae, and the green plants, H_2A is water. In short, van Niel proposed that it was the water that was the source of oxygen in photosynthesis, *not* the carbon dioxide.

This brilliant speculation, first proposed in the early 1930s, was not proved until many years later. Eventually, investigators, using a heavy isotope of oxygen (^{18}O), traced the oxygen from water to oxygen gas:

$$CO_2 + 2H_2^{18}O \xrightarrow{\text{Light}} (CH_2O) + H_2O + {}^{18}O_2$$

This confirmed van Niel's hypothesis. The overall concept of photosynthesis has remained unchanged from the time of van Niel's proposal. However, many of its details have subsequently been worked out, and more are still under active investigation.

Purple sulfur bacteria. In these cells, hydrogen sulfide plays the same role as water does in the photosynthetic process of plants. The hydrogen sulfide (H_2S) is split, and the sulfur accumulates as globules, visible within the cells.

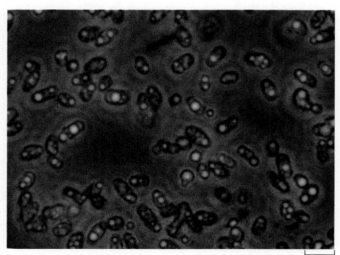

5 μm

10–14 *Light energy trapped in the reactive chlorophyll* a *molecule of Photosystem II boosts electrons to a higher energy level. These electrons are replaced by electrons pulled away from water molecules, releasing protons (H⁺ ions) and oxygen gas. The electrons are passed from the primary electron acceptor along an electron transport chain to a lower energy level, the reaction center of Photosystem I. As they pass along this electron transport chain, some of their energy is packaged in the form of ATP. Light energy absorbed by Photosystem I boosts electrons to another primary electron acceptor. From this acceptor, they are passed via other electron carriers to NADP⁺ to form NADPH. The electrons removed from Photosystem I are replaced by those from Photosystem II.*

ATP and NADPH represent the net gain from the energy-capturing reactions. To generate one molecule of NADPH, two electrons must be boosted from Photosystem II and two from Photosystem I. Two molecules of water are split into protons and oxygen gas, making available the two replacement electrons needed by Photosystem II. One molecule of water is regenerated in the formation of ATP.

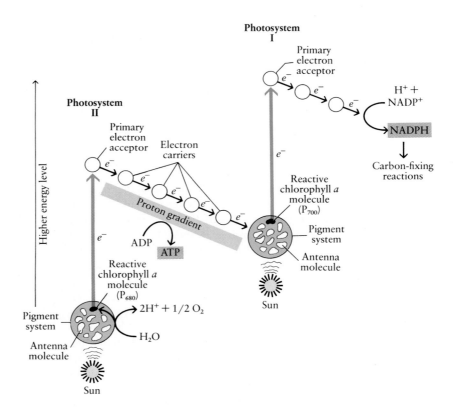

The Light-Trapping Reactions

The two photosystems probably evolved separately, with Photosystem I coming first. As we shall see, Photosystem I can operate independently. In general, however, the two systems work together simultaneously and continuously, as shown in Figure 10–14. According to the current model, light energy enters Photosystem II, where it is trapped by the reactive chlorophyll *a* molecule P_{680}. An electron from the P_{680} molecule is boosted to a higher energy level from which it is transferred to a primary electron-acceptor molecule. The electron then passes downhill along an electron transport chain to Photosystem I. As electrons pass along this transport chain, a proton gradient is established across the thylakoid membrane; the potential energy of this electrochemical gradient is used to form ATP from ADP in a chemiosmotic process similar to that in the mitochondrion. This process is known as **photophosphorylation.**

Three other events are taking place simultaneously:

1. The P_{680} chlorophyll molecule, having lost its electron, is avidly seeking a replacement. It finds it in the water molecule, which, while bound to a manganese-containing protein, is stripped of an electron and then broken into protons and oxygen gas.

2. Additional light energy is trapped in the reactive chlorophyll molecule (P_{700}) of Photosystem I. The molecule is oxidized, and an electron is boosted to a primary electron acceptor from which it goes downhill to NADP⁺.

3. The electron removed from the P_{700} molecule of Photosystem I is replaced by the electron that moved downhill from the primary electron acceptor of Photosystem II.

Thus in the light there is a continuous flow of electrons from water to Photosystem II to Photosystem I to NADP⁺. In the words of the late Nobel laureate Albert Szent-Györgi: "What drives life is . . . a little electric current, kept up by the sunshine."

The energy harvest from these steps is represented by an ATP molecule (whose formation releases a water molecule) and NADPH, which then become the chief sources of energy for the reduction of carbon dioxide. To generate one molecule of NADPH, four photons must be absorbed, two by Photosystem II and two by Photosystem I.

Cyclic Electron Flow

As we mentioned previously, there is also evidence that Photosystem I can work independently. When this occurs, no NADPH is formed. In this process, called **cyclic electron flow,** electrons are boosted from P_{700} to the primary electron acceptor of Photosystem I. They do not, however, pass down the series of electron carriers leading to $NADP^+$. Instead, they are shunted to the electron transport chain that connects Photosystems I and II and pass downhill through that chain back into the reactive P_{700} molecule (Figure 10–15). ATP is produced in the course of this passage. In the absence of $NADP^+$ (as, for example, when all of the available $NADP^+$ has been reduced to NADPH but has not yet been reoxidized in the carbon-fixing reactions), or when the cell needs additional supplies of ATP but not of NADPH, photosynthetic eukaryotic cells are able to synthesize ATP using the energy of sunlight to power cyclic electron flow. However, no oxygen is released and no carbon dioxide is reduced.

It is believed that the most primitive photosynthetic mechanisms worked by cyclic electron flow. This is also apparently the way in which some photosynthetic bacteria carry out photosynthesis. What is an alternative shunt pathway in eukaryotes is, in these bacteria, the principal pathway of electron flow in photosynthesis.

Photosynthetic Phosphorylation

The photophosphorylation of ADP to ATP as electrons pass down the electron transport chain from Photosystem II to Photosystem I (or, as they pass down a portion of that chain during cyclic electron flow) is a chemiosmotic process, similar in many ways to oxidative phosphorylation in the mitochondria. In both the mitochondria and the chloroplasts, the electron transport chains contain cytochromes, and the electron carriers and enzymes of these chains are embedded

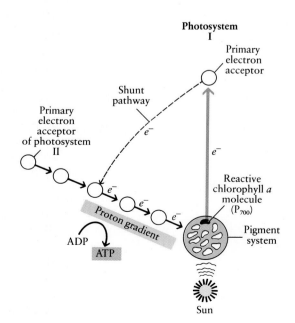

10–15 *When cyclic electron flow occurs in eukaryotic photosynthetic cells, Photosystem II is bypassed. Only Photosystem I and a portion of the electron transport chain between the two photosystems are utilized. ATP is produced from ADP, but oxygen is not released and $NADP^+$ is not reduced. In some photosynthetic bacteria, which have only Photosystem I, cyclic electron flow is the principal photosynthetic mechanism.*

10–16 *The chemiosmotic mechanism of photophosphorylation. In this process, electrons from chlorophyll a—boosted to a high energy level by sunlight—flow down an electron transport chain in the thylakoid membrane. The energy they release as they move to a lower energy level is used to pump protons from the stroma into the thylakoid space, creating an electrochemical gradient of potential energy. As the protons flow down the gradient from the thylakoid space back into the stroma, ADP is phosphorylated to ATP by ATP synthetase. The chemical structures of the electron carriers and enzymes of the thylakoid membrane (including ATP synthetase) are only slightly different from those of the mitochondrial membrane.*

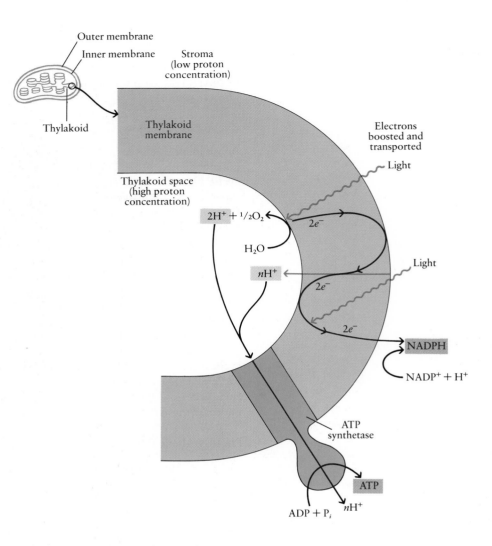

in membranes (the inner membrane of the mitochondrion and the thylakoid membrane of the chloroplast) that are impermeable to protons (H^+ ions). In both organelles, an electrochemical gradient of potential energy is established as protons are pumped through the membrane using energy released as electrons pass down the chain. Moreover, ADP is phosphorylated to ATP as protons flow down the potential energy gradient through ATP synthetase complexes consisting of two factors.

The topography of this chemiosmotic process is, however, slightly different in chloroplasts. In mitochondria, as we saw on page 197, protons are pumped out of the matrix into the external medium; they flow down the gradient from the exterior back into the matrix. Similarly, in chloroplasts the protons are pumped out of the stroma (analogous to the mitochondrial matrix); they are not, however, pumped out of the organelle itself but rather *into* the thylakoid space (Figure 10–16). The potential energy gradient is between this internal, third compartment and the stroma. When protons flow down the gradient, they move from the thylakoid space back into the stroma, where the ATP is synthesized.

Many of the details of photophosphorylation, like those of oxidative phosphorylation, remain to be worked out. For example, it is clear that protons are released into the thylakoid space when water is split into protons, free oxygen, and electrons at the reaction center of Photosystem II. Additional protons are pumped into the thylakoid space as electrons flow down the transport chain, but

Photosynthesis without Chlorophyll

Halobacteria are rod-shaped cells, quite similar in appearance to *Escherichia coli*. They grow best in very salty water, about seven times as salty as sea water. If the salt concentration is reduced to only about three times that of sea water, the cell wall falls apart, and as the concentration is reduced still further, the cell membrane begins to break up. Walther Stoeckenius, then at Rockefeller University, separated the membrane fragments by centrifugation. One of the fractions, it turned out, was purple, and this, though the investigators did not know it at the time, was the first clue to a major energy source of the salt-loving bacteria.

Halobacteria are aerobes, and when suitable substrates and adequate oxygen are available, they oxidize organic molecules, producing ATP by oxidative phosphorylation. Oxygen, however, is often unavailable in the salty waters in which the halobacteria live. The secret of their success, it has now been shown, lies in the purple patches of the cell membrane, which provide an alternative, photosynthetic mechanism for producing ATP. The photosynthetic pigment of the halobacteria is not a form of chlorophyll, as in all other photosynthetic organisms, but retinal (Figure 10–7, page 212), which is also the visual pigment of the vertebrate eye. The membrane of the halobacteria contains molecules of retinal plus protein—the complex is called bacteriorhodopsin. When bacteriorhodopsin is excited by light and then returns to its original energy level, the energy released pumps protons across the membrane, out of the cell. This pumping establishes a proton gradient that drives the phosphorylation of ADP to ATP, thus providing additional support that the chemiosmotic mechanism is indeed a universal one for the regeneration of ATP.

0.25 μm

A freeze-fracture preparation of the membrane of a halobacterium. The fine-grained regions with a hexagonal pattern are patches of purple membrane.

Darwin himself confessed a certain uneasiness when called upon to explain how an organ as complex as the eye might have arisen by the slow, cumulative steps of evolution. Was retinal "invented" twice? Or do these purple fragments of membrane hold clues both to the mechanisms of human vision and also to its origins?

the exact number pumped and the mechanism are not yet certain. A similar uncertainty concerns the number of protons that must flow through the ATP synthetase complex to synthesize one molecule of ATP. The best estimate at the present time is three protons for each ATP synthesized.

The proton gradient that drives photophosphorylation is established by the release of energy that originally entered the system from sunlight. Phosphorylation by isolated chloroplasts can also be driven by a proton gradient established artificially across the thylakoid membrane. In such experimental systems, phosphorylation proceeds in the dark, providing impressive evidence that the key factor in photophosphorylation, as in oxidative phosphorylation, is the establishment of a proton gradient.

10–17 *Scanning electron micrograph of open stomata on the lower surface of a cottonwood leaf. The carbon dioxide used in photosynthesis reaches the photosynthetic cells through these openings.*

20 μm

THE CARBON-FIXING REACTIONS

The reactions that we have just described are the energy-capturing reactions of photosynthesis. In the course of these reactions, as we saw, light energy is converted to electrical energy—the flow of electrons—and the electrical energy is converted to chemical energy stored in the bonds of NADPH and ATP. In the second stage of photosynthesis, this energy is used to reduce carbon. Carbon is available to photosynthetic cells in the form of carbon dioxide. Algae, such as the cell shown in Figure 10–1, obtain dissolved carbon dioxide directly from the surrounding water. In plants, carbon dioxide reaches the photosynthetic cells through specialized openings in leaves and green stems, called **stomata** (Figure 10–17).

The Calvin Cycle: The Three-Carbon Pathway

The reduction of carbon takes place in the stroma in a cycle named after its discoverer, Melvin Calvin. The Calvin cycle is analogous to the Krebs cycle (page 194) in that, in each turn of the cycle, the starting compound is regenerated. The starting (and ending) compound is a five-carbon sugar with two phosphates attached, ribulose bisphosphate (RuBP).

The cycle begins when carbon dioxide is bound to RuBP, which then splits to form two molecules of phosphoglycerate, or PGA (Figure 10–18). (Each PGA molecule contains three carbon atoms, hence the name, the three-carbon pathway.) The enzyme catalyzing this crucial reaction, RuBP carboxylase, is very abundant in chloroplasts, making up more than 15 percent of the total chloroplast protein. RuBP carboxylase, said to be the most abundant protein in the world, is located on the surface of the thylakoid membranes.

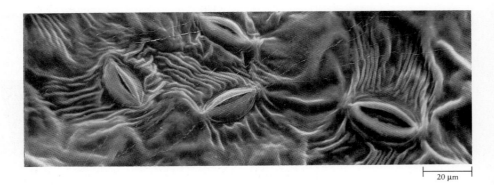

10–18 *Calvin and his collaborators briefly exposed photosynthesizing algae to radioactive carbon dioxide ($^{14}CO_2$). They found that the radioactive carbon is first bound to ribulose bisphosphate (RuBP), which then immediately splits to form two molecules of phosphoglycerate (PGA). The radioactive carbon atom, indicated in color, appears in one of the two molecules of PGA. This is the first step of the Calvin cycle.*

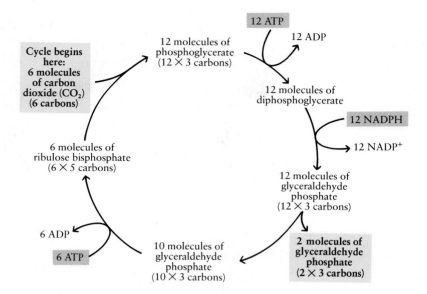

10–19 *Summary of the Calvin cycle. At each full "turn" of the cycle, one molecule of carbon dioxide enters the cycle. Six turns are summarized here—the number required to make two molecules of glyceraldehyde phosphate, the equivalent of one molecule of a six-carbon sugar. Six molecules of ribulose bisphosphate (RuBP), a* *five-carbon compound, are combined with six molecules of carbon dioxide, yielding twelve molecules of phosphoglycerate, a three-carbon compound. These are reduced to twelve molecules of glyceraldehyde phosphate. Ten of these three-carbon molecules are combined and rearranged to form six five-carbon molecules of RuBP.* *The two "extra" molecules of glyceraldehyde phosphate represent the net gain from the Calvin cycle. The energy that drives the Calvin cycle is in the form of ATP and NADPH, produced by the energy-capturing reactions in the first stage of photosynthesis.*

The complete cycle is diagrammed in Figure 10–19. As in the Krebs cycle, each step is catalyzed by a specific enzyme. At each full turn of the cycle, a molecule of carbon dioxide enters the cycle, is reduced, and a molecule of RuBP is regenerated. Three turns of the cycle introduce three molecules of carbon dioxide, the equivalent of one three-carbon sugar, and produce one molecule of glyceraldehyde phosphate, which is the immediate product of the Calvin cycle. This same three-carbon sugar-phosphate molecule is formed when the fructose diphosphate molecule is split at the fourth step in glycolysis (page 188).

Six revolutions of the cycle, with the introduction of six molecules of carbon dioxide, are necessary to produce the equivalent of a six-carbon sugar, such as glucose. These six revolutions of the cycle produce two molecules of glyceraldehyde phosphate, which can subsequently react to produce one molecule of a six-carbon sugar. The overall equation for the series of reactions required for the synthesis of glucose is

$$6RuBP + 6CO_2 + 18ATP + 12NADPH + 12H^+ + 12H_2O \longrightarrow$$
$$6RuBP + Glucose + 18P_i + 18ADP + 12NADP^+$$

The Four-Carbon Pathway

In most plants, the first step in the fixation of carbon is the binding of carbon dioxide to RuBP and its entrance into the Calvin cycle. Some plants, however, first bind carbon dioxide to a compound known as phosphoenolpyruvate (PEP) to form the four-carbon compound oxaloacetic acid. (Oxaloacetic acid, you may recall, is also an intermediate in the Krebs cycle.) The carbon dioxide incorporated into oxaloacetic acid is ultimately transferred to RuBP and enters the Calvin cycle,

10–20 *Carbon dioxide fixation by the* C₄ *pathway. Carbon dioxide is bound to phosphoenolpyruvate (PEP) by the enzyme PEP carboxylase. The resulting oxaloacetic acid is converted either to malic acid or aspartic acid. These steps later will be reversed, releasing carbon dioxide for use in the Calvin cycle.*

but not until it has passed through a series of reactions that transport it more deeply into the leaf. Plants that utilize this pathway, also known as the Hatch-Slack pathway, are commonly called C₄, or four-carbon, plants, as distinct from the C₃ plants in which carbon is bound first to RuBP to form the three-carbon compound phosphoglycerate (PGA).

In C₄ plants, the binding of carbon dioxide to PEP is catalyzed by the enzyme PEP carboxylase (Figure 10–20). The resulting oxaloacetic acid is then reduced to malic acid or converted (with the addition of an amino group) to aspartic acid. These steps take place in mesophyll cells, whose chloroplasts are characterized by an extensive network of thylakoids, organized into well-developed grana. The next step is a surprise: the malic acid (or aspartic acid, depending on the species) is transported to bundle-sheath cells. The chloroplasts of these cells, which form tight sheaths around the vascular bundles of the leaf, have poorly developed grana and often contain large grains of starch (Figure 10–21). In the bundle-sheath cells, the malic (or aspartic) acid is decarboxylated to yield CO₂ and pyruvic acid. The CO₂ then enters the Calvin cycle. This process, summarized in Figure 10–22, physically separates the capture of CO₂ by the plant from the reactions of the Calvin cycle.

One might well ask why C₄ plants should have evolved such an energetically expensive and seemingly clumsy method of providing carbon dioxide to the Calvin cycle. To answer this question, we must consider both the function of the leaf as a whole and the properties of PEP carboxylase and RuBP carboxylase, the enzymes that catalyze the first step of the carbon-fixing reactions in C₄ and C₃ plants, respectively.

10–21 *In* C₄ *plants, the chloroplasts of mesophyll cells differ from those of bundle-sheath cells. As shown in this micrograph of portions of chloroplasts in two adjacent cells of a corn leaf, the mesophyll chloroplast (top) contains well-developed grana, while the grana of the bundle-sheath chloroplast are poorly developed. Notice the plasmodesmata that connect the two cells, providing channels through which substances can flow from one cell to the other.*

0.5 µm

10–22 *A pathway for carbon fixation in C₄ plants. CO₂ is first fixed in mesophyll cells as oxaloacetic acid. It is then transported to bundle-sheath cells, where the carbon dioxide is released. The CO₂ thus formed enters the Calvin cycle. Pyruvic acid returns to the mesophyll cell, where it is phosphorylated to PEP.*

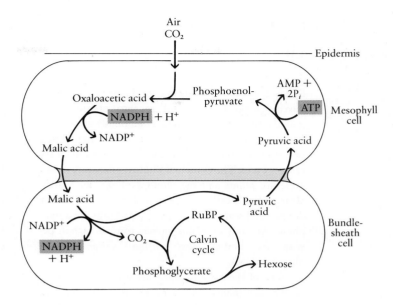

Carbon dioxide is not continuously available to the photosynthesizing cells. It enters the leaf by way of the stomata, specialized pores that open and close depending on, among other factors, water stress. Moreover, when plants are growing in close proximity to one another, the air surrounding the leaves may be quite still, with little gas exchange between the immediate environment and the atmosphere as a whole. Under such conditions, the concentration of carbon dioxide in the air closest to the leaves may be rapidly reduced to low levels by the photosynthetic activity of the plants. PEP carboxylase, the enzyme that catalyzes the formation of oxaloacetic acid in C₄ plants, has a higher affinity for carbon dioxide than does RuBP carboxylase. Even at low concentrations of carbon dioxide, the enzyme works rapidly to bind it to PEP. Compared with RuBP carboxylase, PEP carboxylase fixes carbon dioxide faster, at lower levels, keeping the CO₂ concentration lower within the cells near the surface of the leaf. This maximizes the gradient of carbon dioxide between these cells and the outside air. Thus, when the stomata are open, carbon dioxide readily diffuses down the concentration gradient into the leaf. If the stomata must be closed much of the time—as they must be to conserve water in a hot, dry climate—the plant with C₄ metabolism will take up more carbon dioxide with each gasp (so to speak) than the plant that has only C₃ metabolism. Hence, the C₄ plant is at a distinct advantage in drought-ridden areas.

In the presence of ample carbon dioxide, RuBP carboxylase fixes carbon dioxide efficiently, feeding it into the Calvin cycle. However, when the carbon dioxide concentration in the leaf is low in relation to the oxygen concentration, this same enzyme catalyzes a reaction of RuBP with oxygen, rather than with carbon dioxide. This reaction leads to the formation of glycolic acid, the substrate for a process known as photorespiration. Photorespiration, which occurs in the peroxisomes (page 118) of photosynthetic cells, is the oxidation of carbohydrates in the presence of light and oxygen. Unlike mitochondrial respiration, however, it yields neither ATP nor NADH. Under normal atmospheric conditions, as much as 50 percent of the carbon fixed in photosynthesis by a C₃ plant may be reoxidized to CO₂ during photorespiration. Thus photorespiration greatly reduces the photosynthetic efficiency of C₃ plants.

The Carbon Cycle

By photosynthesis, living systems incorporate carbon dioxide from the atmosphere into organic compounds. In respiration, these compounds are broken down again into carbon dioxide and water. These processes, viewed on a worldwide scale, result in the carbon cycle. The principal photosynthesizers in this cycle are plants and the phytoplankton, the marine algae. They synthesize carbohydrates from carbon dioxide and water and release oxygen into the atmosphere. About 100 billion metric tons of carbon per year are bound into carbon compounds by photosynthesis.

Some of the carbohydrates are used by the photosynthesizers themselves. Plants release carbon dioxide from their roots and leaves, and marine algae release it into the water where it maintains an equilibrium with the carbon dioxide of the air. Some 500 billion metric tons of carbon are "stored" as dissolved carbon dioxide in the seas, and some 700 billion metric tons in the atmosphere. Some of the carbohydrates are used by animals that feed on the living plants, on algae, and on one another, releasing carbon dioxide. An enormous amount of carbon is contained in the dead bodies of plants and other organisms plus discarded leaves and shells, feces, and other waste materials that settle into the soil or sink to the ocean floors where they are consumed by small invertebrates, bacteria, and fungi. Carbon dioxide is also released by these processes into the reservoir of the air and oceans.

Another, even larger store of carbon lies below the surface of the earth in the form of coal and oil, deposited there some 300 million years ago.

The natural processes of photosynthesis and respiration generally balance one another out. Over the long span of geologic time, the carbon dioxide concentration of the atmosphere has varied, but for the last 10,000 years it has remained relatively constant. By volume, it is a very small proportion of the atmosphere, only about 0.03 percent. It is important, however, because carbon dioxide, unlike most other components of the atmosphere, absorbs heat from the sun's rays. Since 1850, carbon dioxide concentrations in the atmosphere have been increasing, owing in large part to our use of fossil fuels, to our plowing of the soil, and to our destruction of forest land, particularly in the tropics. A recent study by the Environmental Protection Agency predicts that this increase in the carbon dioxide "blanket" will significantly increase the average temperatures here on earth, beginning within the next 20 years. Average increases of about 2°C by the year 2040 and 5°C by the year 2100 are expected. The consequences of these temperature increases cannot be known with certainty. In some parts of the world, there may be lengthened growing seasons, increased precipitation, and, in conjunction with the increased levels of carbon dioxide available to plants, greater agricultural productivity. In other parts of the world, however, it is thought that

High CO_2 and low O_2 concentrations limit photorespiration. Consequently, C_4 plants have another distinct advantage over C_3 plants. First, the RuBP carboxylase is sequestered in the bundle-sheath cells, in the interior of the leaf, where it is to some extent protected from atmospheric oxygen. Second, because the CO_2 fixed by the C_4 pathway is essentially "pumped" from the mesophyll cells into the bundle-sheath cells, it is delivered to RuBP carboxylase in a concentrated form. The ratio of CO_2 to O_2 is high enough that the enzyme catalyzes the reaction involving CO_2 and not the reaction leading to photorespiration. Any carbon dioxide that is released by photorespiration is immediately recaptured by PEP carboxylase in the mesophyll cells and fed back to RuBP and into the Calvin cycle. As a result of all of these factors, the net rates of photosynthesis in C_4 grasses, such as corn, sugarcane, and sorghum, can be two to three times the rates in C_3 grasses, such as wheat, oats, and rice.

C_4 plants evolved primarily in the tropics and are especially well adapted to high

precipitation will be significantly reduced, lowering crop yields and, in already arid areas, accelerating the spread of the great deserts of the world. Most biologists are greatly concerned that although we do not know the consequences—either for ourselves or for other organisms—of what we are doing, we keep right on doing it.

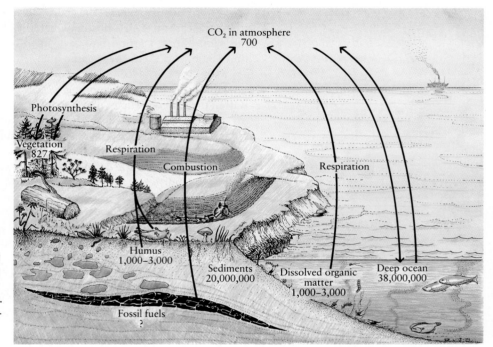

The carbon cycle. The arrows indicate the movement of carbon atoms. The numbers are all estimates of the amount of carbon stored, expressed in billions of metric tons. The amount of carbon released by respiration and combustion has begun to exceed the amount fixed by photosynthesis.

light intensities, high temperatures, and dryness. The optimal temperature range for C_4 photosynthesis is much higher than that for C_3 photosynthesis, and C_4 plants flourish even at temperatures that would eventually be lethal to many C_3 species. Because of their more efficient use of carbon dioxide, C_4 plants can attain the same photosynthetic rate as C_3 plants but with smaller stomatal openings and, hence, with considerably less water loss.

Perhaps the most familiar example of the competitive capacity of C_4 plants is seen in lawns in the summertime. In most parts of the United States, lawns consist mainly of C_3 grasses such as Kentucky bluegrass. As the summer days become hotter and drier, these dark green, fine-leaved grasses are often overwhelmed by rapidly growing crabgrass, which disfigures the lawn as its yellowish-green, broader-leaved plants slowly take over. Crabgrass, you will not be surprised to hear, is a C_4 plant.

The list of plants known to utilize the four-carbon pathway has grown to over 100 genera, at least a dozen of which include both C_3 and C_4 species. This pathway has undoubtedly arisen independently many times in the course of evolution and is another example of the exquisite adaptation of living systems to their environment.

THE PRODUCTS OF PHOTOSYNTHESIS

Glyceraldehyde phosphate, the three-carbon sugar produced by the Calvin cycle, may seem an insignificant reward, both for all the enzymatic activity on the part of the cell and for our own intellectual stress. However, this molecule and those derived from it provide (1) the energy source for virtually all living systems, and (2) the basic carbon skeleton from which the great diversity of organic molecules can be synthesized. Carbon has been fixed—that is, it has been brought from the inorganic world into the organic one.

Molecules of glyceraldehyde phosphate may flow into a variety of different metabolic pathways, depending on the activities and requirements of the cell. Often they are built up to glucose or fructose, following a sequence that is in many of its steps the reverse of the glycolysis sequence described in the previous chapter. (At some steps, the reactions are simply reversed and the enzymes are the same. Other steps—the highly exergonic ones of the downhill sequence—are bypassed.) Plant cells use these six-carbon sugars to make starch and cellulose for their own purposes and sucrose for export to other parts of the plant body. Animal cells store them as glycogen. All cells use sugars, including glyceraldehyde phosphate and glucose, as the starting point for the manufacture of other carbohydrates, fats and other lipids, and, with the addition of nitrogen, amino acids and nitrogenous bases. Finally, as we saw in the preceding chapter, oxidation of the carbon fixed in photosynthesis is the source of ATP energy for heterotrophic organisms and for the heterotrophic cells of plants.

SUMMARY

In photosynthesis, light energy is converted to chemical energy, and carbon is fixed into organic compounds. The generalized equation for this reaction is

$$CO_2 + 2H_2A + \text{Light energy} \longrightarrow (CH_2O) + H_2O + 2A$$

in which H_2A stands for water or some other substance from which electrons can be removed.

Light energy is captured by the living world by means of pigments. The pigments involved in photosynthesis in eukaryotes include the chlorophylls and the carotenoids. Light absorbed by the pigments boosts their electrons to higher energy levels. Because of the way the pigments are packed into membranes, they are able to transfer this energy to reactive molecules, probably chlorophyll *a* packed in a particular way.

Photosynthesis takes place within organelles known as chloroplasts, which are surrounded by two membranes. Contained within the membranes of the chloroplast are a solution of organic compounds and ions known as the stroma and a complex internal membrane system consisting of fused membranes that form sacs called thylakoids. The pigments and other molecules responsible for capturing light are located in and on the thylakoid membranes.

Photosynthesis takes place in two stages, as summarized in Table 10–1. In the currently accepted model of the energy-capturing reactions, light energy strikes antennae pigments of Photosystem II, which contains several hundred molecules of chlorophyll *a* and chlorophyll *b*. Electrons are boosted uphill from the reactive chlorophyll *a* molecule P_{680} to a primary electron acceptor. As the electrons are

TABLE 10–1 Summary of the Stages of Photosynthesis

	CONDITIONS	WHERE	WHAT APPEARS TO HAPPEN	RESULTS
Energy-capturing reactions	Light	Thylakoids	Light striking Photosystem II boosts electrons uphill. (These electrons are replaced by electrons from water molecules, which release O.) Electrons then pass downhill along an electron transport chain to Photosystem I; in the process, ATP is formed by a chemiosmotic process. Light hits Photosystem I, boosts electrons uphill to an electron acceptor from which they are passed to $NADP^+$ to form NADPH. Electrons are replaced in Photosystem I by electrons from Photosystem II.	Energy of light is converted to chemical energy stored in bonds of ATP and NADPH.
Carbon-fixing reactions	Do not require light, although some enzymes stimulated by light	Stroma	Calvin cycle. NADPH and ATP formed in energy-capturing reactions are used to reduce carbon dioxide. The cycle yields glyceraldehyde phosphate from which glucose or other organic compounds can be formed.	Chemical energy of ATP and NADPH is used to incorporate carbon into organic molecules.

removed, they are replaced by electrons from water molecules, with the simultaneous production of free O_2 and protons (H^+ ions). The electrons then pass downhill to Photosystem I along an electron transport chain; this passage generates a proton gradient that drives the synthesis of ATP from ADP (photophosphorylation). Light energy absorbed in antennae pigments of Photosystem I and passed to chlorophyll P_{700} results in the boosting of electrons to another primary electron acceptor. The electrons removed from P_{700} are replaced by the electrons from Photosystem II. The electrons are ultimately accepted by the electron carrier $NADP^+$. The energy yield from this sequence of reactions is contained in the molecules of NADPH and in the ATP formed by photophosphorylation.

Photophosphorylation also occurs as a result of cyclic electron flow, a process that bypasses Photosystem II. In cyclic electron flow, the electrons boosted from P_{700} in Photosystem I do not pass to $NADP^+$ but are instead shunted to the electron transport chain that links Photosystem II to Photosystem I. As they flow down this chain, back into P_{700}, ADP is phosphorylated to ATP.

Like oxidative phosphorylation in the mitochondria, photophosphorylation in the chloroplasts is a chemiosmotic process. As electrons flow down the electron transport chain from Photosystem II to Photosystem I, protons are pumped from the stroma into the thylakoid space, creating an electrochemical gradient of potential energy. As protons flow down this gradient from the thylakoid space back into the stroma, passing through ATP synthetase complexes, ATP is formed.

In the carbon-fixing reactions, which take place in the stroma, NADPH and ATP produced in the energy-capturing reactions are used to reduce carbon dioxide to organic carbon. This is accomplished by means of the Calvin cycle. In the Calvin cycle, a molecule of carbon dioxide is combined with the starting material, a five-carbon sugar called ribulose bisphosphate. At each turn of the cycle, one carbon atom enters the cycle. Three turns of the cycle produce a three-carbon molecule, glyceraldehyde phosphate. Two molecules of glyceraldehyde phosphate (six turns of the cycle) can combine to form a glucose molecule. At each turn of the cycle, RuBP is regenerated. The glyceraldehyde phosphate can also be used as a starting material for other organic compounds needed by the cell.

In C_4 plants, carbon dioxide is initially accepted by a compound known as PEP (phosphoenolpyruvate) to yield the four-carbon oxaloacetic acid. Following a series of reactions that transport it to cells more deeply within the leaf, the carbon dioxide is released, binding with RuBP and entering the Calvin cycle. Although C_4 plants use more energy to fix carbon, their net photosynthetic efficiency can be higher than that of C_3 plants. The greater affinity of PEP carboxylase for carbon dioxide enables C_4 plants to capture ample CO_2 with minimal water loss. Also, photorespiration, a process in which fixed carbon is reoxidized to carbon dioxide and lost to the plant, is limited in C_4 plants. Under conditions of intense sunlight, high temperatures, or drought, C_4 plants are more efficient than C_3 plants.

QUESTIONS

1. Distinguish between the following: absorption spectrum/action spectrum; grana/thylakoid; stroma/thylakoid space; NAD^+/$NADP^+$; energy-capturing reactions/carbon-fixing reactions; Photosystem I/Photosystem II; C_3 photosynthesis/C_4 photosynthesis.

2. Why is it plausible to argue, as the Nobel laureate George Wald does, that wherever in the universe we find living organisms, we will find them (or at least some of them) to be colored?

3. Predict what colors of light might be most effective at stimulating plant growth. (Such is the principle of the special light bulbs used for plants.)

4. Sketch a chloroplast and label all membranes and compartments. In what ways does the structure of a chloroplast resemble that of a mitochondrion? In what ways is it different?

5. Describe in general terms the events of photosynthesis. Compare your description with Table 10–1.

6. In what ways are photophosphorylation and oxidative phosphorylation alike? In what ways are they different?

7. The experiment shown in Figure 4–3 on page 87 was attempted in Stockton, California, in 1973, but instead of making amino acids, the electric sparks caused the apparatus to explode.

(No one was hurt, fortunately.) What is present in today's atmosphere that was not present in the primitive atmosphere and that would account for the explosion?

8. Given the scarcity of high-salt environments and the difficulties of surviving in them, explain in evolutionary terms why the halobacteria (page 221) are found there.

9. Return to Figure 10–21 and describe the biochemical events taking place in each of the chloroplasts.

10. Consider the anatomy of the leaf of a C_4 plant. In which type of cell—mesophyll or bundle-sheath—would you expect the oxygen-releasing reaction of photosynthesis to occur? Why? Does this localization of oxygen release increase or decrease photorespiration?

11. Over 100 genera of plants have acquired C_4 photosynthesis in the course of evolution. How is this adaptation advantageous to these plants? Many plants, however, have not evolved C_4 photosynthesis. Why is it advantageous to such plants *not* to have C_4 photosynthesis?

12. Trace a carbon atom through a series of biological events, such as those illustrated on pages 158, 160, and 161.

SUGGESTIONS FOR FURTHER READING

Books

ALBERTS, BRUCE, DENNIS BRAY, JULIAN LEWIS, MARTIN RAFF, KEITH ROBERTS, and JAMES D. WATSON: *Molecular Biology of the Cell*, 2d ed., Garland Publishing, Inc., New York, 1989.

This outstanding cell biology text includes a clear, up-to-date discussion of our current understanding of the processes that occur in mitochondria and chloroplasts. The authors' discussion of the experimental procedures used to study these processes is especially helpful.

CONANT, JAMES BRYANT (ed.): *Harvard Case Histories in Experimental Science*, vol. 2, Harvard University Press, Cambridge, Mass., 1964.

Case #5, Plants and the Atmosphere, *edited by Leonard K. Nash, describes the early work on photosynthesis, often presented in the words of the investigators themselves. The narrative illuminates the historical context in which the discoveries were made.*

DARNELL, JAMES, HARVEY F. LODISH, and DAVID BALTIMORE: *Molecular Cell Biology*, W. H. Freeman and Company, New York, 1986.

A comprehensive treatment of modern cell biology, richly illustrated with diagrams and micrographs. Chapter 20 of this outstanding text is devoted to a thorough explication of our current understanding of the processes that occur in the mitochondria and chloroplasts.

LEHNINGER, ALBERT L.: *Principles of Biochemistry*, Worth Publishers, Inc., New York, 1982.

This introductory text is outstanding both for its clarity and for its consistent focus on the living cell. Lehninger was one of the foremost experts on cellular energetics, and this text is enriched by his vast experience and thorough understanding of the processes by which cells provide themselves with energy.

NEWSHOLME, ERIC, and TONY LEECH: *The Runner: Energy and Endurance*, Fitness Books, Roosevelt, N.J., 1984.*

> *A lively introduction to the energetics of the working human body. Equally useful as a primer for runners who want to know more about their own physiology and as an introduction to the biochemistry of carbohydrate and fat metabolism.*

PRESCOTT, DAVID M.: *Cells: Principles of Molecular Structure and Function*, Jones and Bartlett Publishers, Boston, 1988.

> *An up-to-date, yet concise, textbook of cell biology, written for a first course at the undergraduate level. Chapter 4 is devoted to energy flow and metabolism.*

RABINOWITCH, EUGENE, and GOVINDJEE: *Photosynthesis*, John Wiley & Sons, Inc., New York, 1969.*

> *Although now out of date in many respects, this book remains a lucid introduction, suitable for undergraduate students, to the processes of photosynthesis and to related physical and chemical concepts, such as entropy and free energy.*

SCIENTIFIC AMERICAN: *Molecules to Living Cells*, W. H. Freeman and Company, New York, 1980.*

> *A collection of articles from* Scientific American. *Chapters 5 and 6 are concerned with the structure and function of enzymes, Chapter 12 is an excellent presentation of the chemiosmotic synthesis of ATP in both mitochondria and chloroplasts, and Chapter 13 examines the photosynthetic membrane.*

STRYER, LUBERT: *Biochemistry*, 3d ed., W. H. Freeman and Company, New York, 1988.

> *A good introduction, handsomely illustrated, to cellular energetics.*

Articles

AHERN, TIM J., and ALEXANDER M. KLIBANOV: "The Mechanism of Irreversible Enzyme Inactivation at 100°C," *Science*, vol. 228, pages 1280–1284, 1985.

BARBER, JIM: "Signals from the Reaction Centre," *Nature*, vol. 332, pages 111–112, 1988.

BJORKMAN, O., and J. BERRY: "High-Efficiency Photosynthesis," *Scientific American*, October 1973, pages 80–93.

CLOUD, PRESTON: "The Biosphere," *Scientific American*, September 1983, pages 176–189.

DETWILER, R. P., and C. A. S. HALL: "Tropical Forests and the Global Carbon Cycle," *Science*, vol. 239, pages 42–47, 1988.

DICKERSON, RICHARD E.: "Cytochrome *c* and the Evolution of Energy Metabolism," *Scientific American*, March 1980, pages 136–153.

DICKINSON, ROBERT E., and RALPH J. CICERONE: "Future Global Warming from Atmospheric Trace Gases," *Nature*, vol. 319, pages 109–115, 1986.

KARPLUS, MARTIN, and J. ANDREW MCCAMMON: "The Dynamics of Proteins," *Scientific American*, April 1986, pages 42–51.

KERR, RICHARD A.: "Is the Greenhouse Here?" *Science*, vol. 239, pages 559–561, 1988.

KLIBANOV, ALEXANDER M.: "Immobilized Enzymes and Cells as Practical Catalysts," *Science*, vol. 219, pages 722–727, 1983.

KNOWLES, JEREMY R.: "Tinkering with Enzymes: What Are We Learning?" *Science*, vol. 236, pages 1252–1258, 1987.

KOLATA, GINA: "How Do Proteins Find Mitochondria?" *Science*, vol. 228, pages 1517–1518, 1985.

LANE, M. DANIEL, PETER L. PEDERSEN, and ALBERT S. MILDVAN: "The Mitochondrion Updated," *Science*, vol. 234, pages 526–527, 1986.

LEWIN, ROGER: "A Downward Slope to Greater Diversity," *Science*, vol. 217, pages 1239–1240, 1982.

NASSAU, KURT: "The Causes of Color," *Scientific American*, October 1980, pages 124–154.

NEURATH, HANS: "Evolution of Proteolytic Enzymes," *Science*, vol. 224, pages 350–357, 1984.

REVELLE, ROGER: "Carbon Dioxide and World Climate," *Scientific American*, August 1982, pages 35–43.

SCHOPF, J. WILLIAM: "The Evolution of the Earliest Cells," *Scientific American*, September 1978, pages 110–138.

SRIVASTAVA, D. K., and SIDNEY A. BERNHARD: "Metabolite Transfer via Enzyme-Enzyme Complexes," *Science*, vol. 234, pages 1081–1086, 1986.

STOECKENIUS, WALTHER: "The Purple Membrane of Salt-loving Bacteria," *Scientific American*, June 1976, pages 38–46.

WAGGONER, PAUL E.: "Agriculture and Carbon Dioxide," *American Scientist*, vol. 72, pages 179–184, 1984.

WEBB, A. DINSMOOR: "The Science of Making Wine," *American Scientist*, vol. 72, pages 360–367, 1984.

WESTHEIMER, F. H.: "Why Nature Chose Phosphates," *Science*, vol. 235, pages 1173–1178, 1987.

WOODWELL, G. M., et al.: "Global Deforestation: Contribution to Atmospheric Carbon Dioxide," *Science*, vol. 222, pages 1081–1086, 1983.

YOUVAN, DOUGLAS C., AND BARRY L. MARRS: "Molecular Mechanisms of Photosynthesis," *Scientific American*, June 1987, pages 42–48.

---

* Available in paperback.

SECTION 3

Genetics

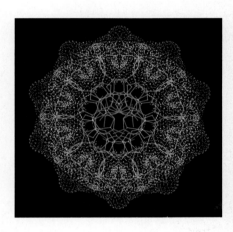

"*Of everything that creepeth on the earth, there went in two and two . . .*" *Knowledge that like begets like is at least as old as recorded history. The way in which this continuity of life is preserved from generation to generation, and the way that male and female each contribute to the characteristics of their offspring were fundamental questions of biological science. These ancient questions have now been answered in a few extraordinary decades of modern genetic research.*

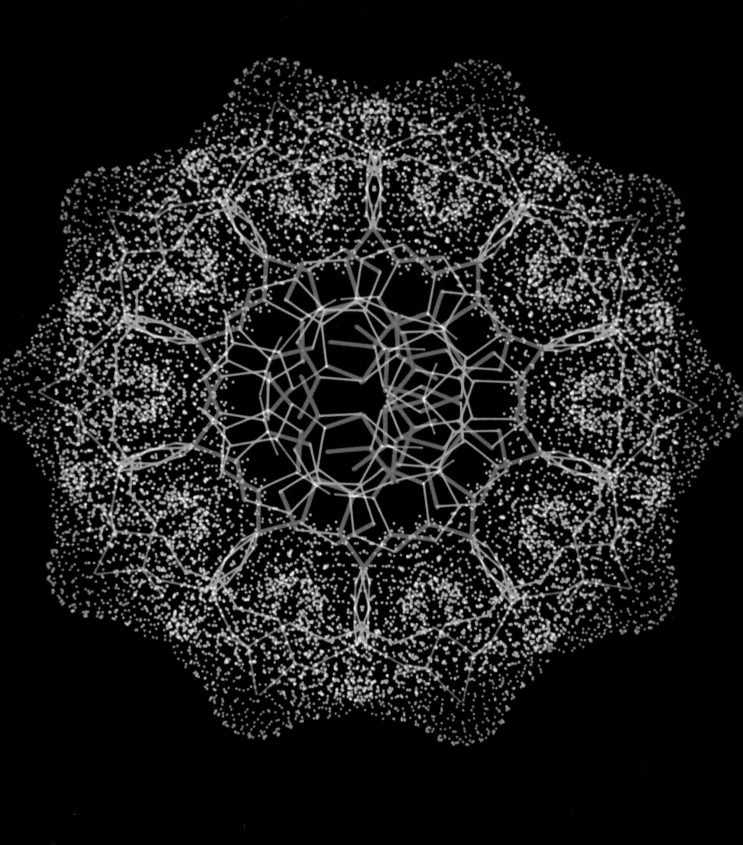

C H A P T E R **11**

From an Abbey Garden: The Beginning of Genetics

11–1 *The answers to the age-old questions of the nature of heredity were found in the structure of a remarkable molecule, deoxyribonucleic acid, or DNA. This molecular "rose window" is a computer-generated image of one complete turn of the DNA helix, with red representing oxygen, green for carbon, yellow for phosphorus, and blue for nitrogen. In this image, the helix has been collapsed into a single plane.*

Among all the symbols in biology, perhaps the most widely used and most ancient are the hand mirror and comb of Venus (♀) and the shield and spear of Mars (♂), the scientific shorthand for female and male. Ideas about the biological roles of male and female are even older than these familiar symbols. Very early, it must have been noticed that male and female were both necessary to produce children and that both transmitted characteristics—hair color, for example, a large nose, or a small chin—to their children. And throughout history, biological inheritance has been an important factor in human social organization, often determining the distribution of wealth, power, land, and royal privileges.

Sometimes a family characteristic is so distinctive that it can be traced through many generations. A famous example of such a characteristic is the Hapsburg lip (Figure 11–2), which has appeared in Hapsburg after Hapsburg, over and over again since at least the thirteenth century. Examples such as this have made it easy to accept the importance of inheritance, but it is only comparatively recently that we have begun to understand something about how this process works. In fact, heredity—the transmission of characteristics from parent to offspring—was not really studied as a science until the second half of the nineteenth century. Yet the questions addressed in this study are among the most fundamental in biology, since self-replication is the essence of the hereditary process and one of the principal properties of living systems.

(a)

(b)

(c)

11–2 *The protruding lip of the Hapsburgs is a famous example of an inherited characteristic. These portraits of members* *of the Hapsburg family encompass a period of about 200 years:* **(a)** *Ferdinand I (1503–1564), Holy Roman Emperor,* **(b)** *Rudolph II (1552–1612), Holy Roman Emperor, and* **(c)** *Charles II (1661–1700), King of Spain.*

235

11–3 *Only fairly recently has it been realized that living things come only from other living things of the same species. This picture from an old Turkish history of India shows a wakwak tree, which bears human fruit. According to the account, the tree is to be found on an island in the South Pacific.*

EARLY IDEAS ABOUT HEREDITY

Far back in human history, people learned to improve domestic animals and crops by selective breeding of individuals with desirable characteristics. The ancient Egyptians and Babylonians, for example, knew how to produce fruits by artificial fertilization, crossing male flowers borne on one date palm tree with female flowers from another tree. The nature of the difference between the male and female flowers was understood by the Greek philosopher and naturalist Theophrastus (371–287 B.C.). "The males should be brought to the females," he wrote, "for the male makes them ripen and persist." In the days of Homer, breeding a male donkey to a mare was known to produce a mule, although little explanation could be given for the manner in which the beast came by its unusual appearance.

Many legends were based on bizarre possibilities of matings between individuals of different species. The wife of Minos, according to Greek mythology, mated with a bull and produced the Minotaur. Folk heroes of Russia and Scandinavia were traditionally the sons of women who had been captured by bears, from which these men derived their great strength and so enriched the national stock. The camel and the leopard also mated from time to time, according to the early naturalists, who were otherwise unable—and it is hard to blame them—to explain an animal as improbable as the giraffe (the common giraffe still bears the scientific name of *Giraffa camelopardalis*). Thus folklore reflected early and imperfect glimpses into the nature of hereditary relationships.

The first scientist known to have pondered the mechanism of heredity was Hippocrates (460?–377? B.C.). He hypothesized that specific particles, or "seeds," are produced by all parts of the body and are transmitted to offspring at the time of conception, causing certain parts of the offspring to resemble those parts of the parents. A century later, Aristotle rejected the ideas of Hippocrates. Children often seem to inherit characteristics of their grandparents or even their great-grandparents rather than their parents, observed Aristotle. How could these distant relatives have contributed the "seeds" of flesh and blood that were transmitted from parent to offspring? To resolve the conflict, Aristotle postulated that the male semen was made up of imperfectly blended ingredients, some of which were inherited from past generations. At fertilization, he proposed, the male semen mixed with the "female semen," the menstrual fluid, giving form and power *(dynamis)* to the amorphous substance. From this material, flesh and blood formed as the offspring developed.

For two thousand years no one had a better idea. Indeed, there were not many new ideas at all. Seventeenth-century medical texts continued to show various stages in the coagulation of the embryo from the mixture of maternal and paternal semens. In fact, many scientists as well as laymen did not believe that such mixtures were even always necessary; they held that life, at least the "simpler" forms of life, could arise by spontaneous generation. Worms, flies, and various crawling things, it was commonly believed, took shape from putrid substances, ooze, or mud, and a lady's hair dropped in a rain barrel could turn into a snake. Jan Baptista van Helmont, a seventeenth-century physician known for his experiments on the growth of plants (see page 208), actually published his personal recipe for the production of mice: One need only place a dirty shirt in a pot containing a few grains of wheat, and in 21 days mice would appear. He had performed the experiment himself, he said. The mice would be adults, both male and female, he added, and would be able to produce more mice by mating. As we saw in Chapter 4 (page 86), spontaneous generation did not lose its grip on the imagination until Pasteur's decisive disproof in 1864.

11-4 *What the animalculists, or spermists, of the seventeenth and eighteenth centuries believed they saw when they looked through a microscope at sperm cells. This is a homunculus ("little man"), a future human being in miniature, in a sperm cell.*

THE FIRST OBSERVATIONS

In 1677, the Dutch lens maker Anton van Leeuwenhoek discovered living sperm—"animalcules," as he called them—in the seminal fluid of various animals, including man. Enthusiastic followers peered through Leeuwenhoek's "magic looking glass" (his homemade microscope) and believed they saw within each human sperm a tiny creature—the homunculus, or "little man" (Figure 11–4). This little creature was thought to be the future human being in miniature. Once implanted in the womb of the female, the future human being would be nurtured there, but the mother's only contribution would be to serve as an incubator for the growing fetus. Any resemblance a child might have to its mother, these theorists held, was because of the "prenatal influences" of the womb.

During the very same decade (the 1670s), another Dutchman, Régnier de Graaf, described for the first time the ovarian follicle, the structure in which the human egg cell (the ovum) forms. Although the actual human egg was not seen for another 150 years, the existence of a human egg was rapidly accepted. In fact, de Graaf attracted a school of followers, the ovists, who were as convinced of their opinions as the animalculists, or spermists, were of theirs and who soon contended openly with them. It was the female egg, the ovists said, that contained the future human being in miniature; the animalcules in the male seminal fluid merely stimulated the egg to grow. Ovists and spermists alike carried the argument one logical step further. Each homunculus was thought to have within it another perfectly formed but smaller being, and in that was still another one, and so on—children, grandchildren, and great-grandchildren, all stored away for future use. Some ovists even went so far as to say that Eve had contained within her body all the unborn generations yet to come, each egg fitting closely inside another like a child's hollow blocks. Each female generation since Eve had contained one fewer than the previous generation, they explained, and after 200 million generations, all the eggs would be spent and human life would come to an end.

BLENDING INHERITANCE

By the middle of the nineteenth century, the concepts of the ovists and spermists began to yield to new data. The facts that challenged these earlier hypotheses came not so much from scientific experiments as from practical attempts by master gardeners to produce new ornamental plants. Artificial crossings of such plants showed that, in general, regardless of which plant supplied the pollen—which contains the sperm cells—and which plant contributed the egg cells, both contributed to the characteristics of the new variety. But this conclusion raised even more puzzling questions: What exactly did each parent plant contribute? How did all the hundreds of characteristics of each plant get combined and packed into a single seed?

The most widely held hypothesis of the nineteenth century was that of blending inheritance. According to this concept, when the sperm and egg cells, or **gametes** (from the Greek word *gamos,* meaning "marriage"), combine, there is a mixing of hereditary material that results in a blend, analogous to a blend of two different-colored inks. On the basis of such a hypothesis, one would predict that the offspring of a black animal and a white animal would be gray, and their offspring would also be gray because the black and white hereditary material, once blended, could never be separated again.

You can see why this concept was unsatisfactory. It ignored the phenomenon of characteristics skipping a generation, or even several generations, and then reappearing. To Charles Darwin and other proponents of the theory of evolution, it presented particular difficulties. As we noted in the Introduction (page 7), evolution, according to Darwin, takes place as natural selection acts on existing hereditary variations—that is, variations that can be inherited. If the hypothesis of blending inheritance were valid, the hereditary variations would disappear, like a single drop of ink in the many-colored mixture. Sexual reproduction would eventually result in complete uniformity, natural selection would have no raw material on which to act, and evolution would not occur.

THE CONTRIBUTIONS OF MENDEL

At about the same time that Darwin was writing *The Origin of Species,* an Austrian monk, Gregor Mendel, was beginning a series of experiments that would lead to a new understanding of the mechanism of inheritance. Mendel, who was born into a peasant family in 1822, entered a monastery in Brünn (now Brno, Czechoslovakia), where he was able to receive an education. He attended the University of Vienna for two years, pursuing studies in both mathematics and science. He failed his tests for the teaching certificate he was seeking and so retired to the monastery, of which he eventually became abbot. Mendel's work, carried on in a quiet monastery garden and ignored until after his death, marks the beginning of modern genetics.

Mendel's great contribution was to demonstrate that inherited characteristics are carried by discrete units that are parceled out separately (reassorted) in each generation. These discrete units, which Mendel called *Elemente,* eventually came to be known as **genes.**

Mendel's Experimental Method

For his experiments in heredity, Mendel chose the common garden pea. It was a good choice. The plants were commercially available, easy to cultivate, and grew rapidly. Different varieties had clearly different characteristics that "bred true," appearing unchanged from one crop to the next. For instance, a variety with tall plants always produced tall offspring, and one with yellow seeds always produced yellow seeds, generation after generation. Moreover, the reproductive structures of the pea flower are entirely enclosed by petals, even when they are mature (Figure 11–5). Consequently, the flower normally self-pollinates; that is, sperm cells from the flower's own pollen fertilize its egg cells. Although the plants could be crossbred experimentally, accidental crossbreeding could not occur to confuse the experimental results. As Mendel said in his original paper, "The value and utility of any experiment are determined by the fitness of the material to the purpose for which it is used."

Mendel's choice of the pea plant for his experiments was not original. However, he was successful in formulating the fundamental principles of heredity—where others had failed—because of his approach to the problem. First, he tested a very specific hypothesis in a series of logical experiments. He planned his experiments carefully and imaginatively, choosing for study only clear-cut, measurable hereditary differences. Second, he studied the offspring of not only the first generation but also the second and subsequent generations. Third, and most important, he counted the offspring and then analyzed the results mathematically. Even though his mathematics was simple, the idea that a biological problem could be studied quantitatively was startlingly new. Finally, he organized his data in such a way that his results could be evaluated simply and objectively. The experiments themselves were described so clearly that they could be repeated and checked by other scientists, as eventually they were.

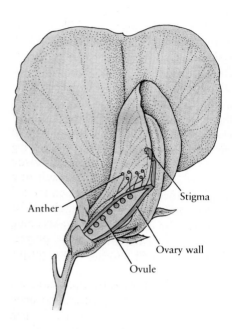

11–5 *In a flower, pollen develops in the anthers and the egg cells in the ovules. Pollination occurs when pollen grains, trapped on the stigma, germinate and grow down to the ovules, where they release sperm. Egg and sperm nuclei unite, and the fertilized eggs develop within the ovules, which are attached to the ovary wall. In the garden pea (Pisum sativum), the ovules with their enclosed embryos form the peas (the seeds), while the ovary wall becomes the pod.*

Pollination in most species of flowering plants involves the pollen from one plant (often carried by an insect) being caught on the stigma of another plant. This is called cross-pollination.

In the pea flower, however, the stigma and anthers are completely enclosed by petals, and the flower, unlike most, does not open until after fertilization has taken place. Thus the plant normally self-pollinates. In his crossbreeding experiments, Mendel pried open the bud before the pollen matured and removed the anthers with tweezers, preventing self-pollination. Then he artificially pollinated the flower by dusting the stigma with pollen collected from another plant.

Labels on figure: Anther, Stigma, Ovary wall, Ovule

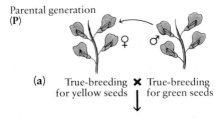

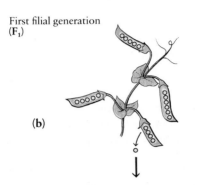

(a) True-breeding ✗ True-breeding
for yellow seeds for green seeds

First filial generation
(F₁)

(b)

Second filial generation
(F₂)

(c)

11–6 *An outline of one of Mendel's experiments.* **(a)** *A true-breeding yellow-seeded pea plant was crossed with a true-breeding green-seeded plant by removing pollen from the anthers of flowers on one plant and transferring it to the stigmas of flowers on the other plant.* **(b)** *Pea pods containing only yellow seeds developed from the fertilized flowers. These peas (seeds) were planted, and the resulting plants were allowed to self-pollinate.* **(c)** *Pea pods developing from the self-pollinated flowers contained both yellow and green peas in an approximate ratio of 3:1, that is, about 3/4 were yellow and 1/4 were green.*

The Principle of Segregation

Mendel began with 32 different types of pea plants, which he studied for several years before he began his quantitative experiments. As he said later in his report on this work, he did not want to experiment with traits in which the difference could be "of a 'more or less' nature, which is often difficult to define." As a result of his preliminary observations, Mendel selected for detailed study seven traits that appeared in two conspicuously different forms in different varieties of plants. One variety of plant, for example, always produced yellow peas (seeds), while another always produced green ones. In one variety, the seeds, when dried, had a wrinkled appearance; in another variety they were smooth. The complete list of traits is given in Table 11–1.

Mendel performed experimental crosses, removing the pollen-containing anthers from flowers and dusting their stigmas with pollen from a flower of another variety. He found that in every case in the first generation (now known in biological shorthand as the F_1, for "first filial generation"), all of the offspring showed only one of the two alternative characteristics; the other characteristic disappeared completely. For example, all of the plants produced as a result of a cross between true-breeding yellow-seeded plants and true-breeding green-seeded plants were as yellow-seeded as the yellow-seeded parent. Similarly, all of the flowers produced by plants resulting from a cross between a true-breeding purple-flowered plant and a true-breeding white-flowered plant were purple. Characteristics that appeared in the F_1 generation, such as yellow seeds and purple flowers, Mendel called **dominant**.

The interesting question was: What had happened to the alternative characteristic—the greenness of the seed or the whiteness of the flower—that had been passed on so faithfully for generations by the parent stock? Mendel let the pea plant itself carry out the next stage of the experiment by permitting the F_1 plants to self-pollinate (Figure 11–6). The characteristics that had disappeared in the first generation reappeared in the second, or F_2, generation. In Table 11–1 are the results of Mendel's actual counts. These characteristics, which were present in the parent generation and reappeared in the F_2 generation, must also have been present somehow in the F_1 generation, although they were not apparent there. Mendel called these characteristics **recessive**.

Looking at the results in Table 11–1, you will notice, as Mendel did, that the dominant and recessive characteristics appear in the second, or F_2, generation in ratios of about 3:1. How do the recessives disappear so completely and then reappear again, and always in such constant proportions? It was in answering this question that Mendel made his greatest contribution. He saw that the appearance and disappearance of alternative characteristics, as well as their constant proportions in the F_2 generation, could be explained if hereditary characteristics are

---

TABLE 11–1 **Results of Mendel's Experiments with Pea Plants**

	ORIGINAL CROSSES		SECOND FILIAL GENERATION (F_2)			
TRAIT	DOMINANT	× RECESSIVE	DOMINANT	RECESSIVE	TOTAL	RATIO
Seed form	Round	× Wrinkled	5,474	1,850	7,324	2.96:1
Seed color	Yellow	× Green	6,022	2,001	8,023	3.01:1
Flower position	Axial	× Terminal	651	207	858	3.14:1
Flower color	Purple	× White	705	224	929	3.15:1
Pod form	Inflated	× Constricted	882	299	1,181	2.95:1
Pod color	Green	× Yellow	428	152	580	2.82:1
Stem length	Tall	× Dwarf	787	277	1,064	2.84:1

11–7 *A pea plant homozygous for purple flowers is represented as WW in genetic shorthand. The allele for purple flowers is designated W because of a convention by which geneticists use the first letter of the less common form (white) of the gene. The capital indicates the dominant allele, the lowercase the recessive. A WW plant can produce only gametes with a purple-flower (W) allele. The female symbol ♀ indicates that this flower contributed the egg cells, or female gametes.*

♀   WW

A white-flowered pea plant (ww) can produce only gametes with a white-flower (w) allele. The male symbol ♂ indicates that this flower contributed the sperm cells, or male gametes.

♂   ww

When a w sperm cell fertilizes a W egg cell, the result is a Ww, which, since the W allele is dominant, will produce purple flowers. However, this Ww plant can produce gametes with either a W or a w allele.

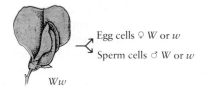

Egg cells ♀ W or *w*
Sperm cells ♂ W or *w*

Ww

And so, if the plant self-pollinates, four possible combinations can occur:

♀ W × ♂ W ⟶ *purple flowers*
♀ W × ♂ w ⟶ *purple flowers*
♀ w × ♂ W ⟶ *purple flowers*
♀ w × ♂ w ⟶ *white flowers*

determined by discrete (separable) factors. These factors, Mendel realized, must have occurred in the F_1 plants in pairs, with one member of each pair inherited from the maternal parent, the other from the paternal parent. The paired factors separated again when the mature F_1 plants produced sex cells, resulting in two kinds of gametes, with one member of the pair in each.

The hypothesis that every individual carries pairs of factors for each trait and that the members of the pair segregate (separate) during the formation of gametes has come to be known as Mendel's first law, or the **principle of segregation.**

Consequences of Segregation

We now recognize that any given gene, for instance, the gene for seed color, can exist in different forms. These different forms of a gene are known as **alleles.** For example, yellow seededness and green seededness are determined by different alleles—that is, different forms—of the gene for seed color. The alleles are represented in biological shorthand by letters; the allele for yellow seededness is represented by *Y*, and the allele for green seededness by *y*.

How a given trait is expressed in an organism is determined by the particular combination of two alleles for that trait carried by the organism. If the two alleles are the same (for example, *YY* or *yy*), then the organism is said to be **homozygous** for that particular trait. If the two alleles are different from one another (for example, *Yy*), then the organism is **heterozygous** for the trait.

When gametes are formed, alleles are passed on to them, but each gamete contains only one allele for any given gene. When two gametes combine in the fertilized egg, the alleles occur in matched pairs again. If the two alleles in a given pair are the same (a homozygous state), the characteristic they determine will be expressed. If they are different (a heterozygous state), one may be dominant over the other; a dominant allele is one that produces its particular characteristic in the heterozygous as well as in the homozygous state. The outward appearance and other observable characteristics of an organism constitute its **phenotype,** a term that is also used to describe a single characteristic, such as yellow or green seeds. Even though a recessive allele may not be expressed in the phenotype, each allele of a matched pair still exists independently and as a discrete unit in the genetic makeup, or **genotype,** of the organism. The two alleles of a pair will separate from each other when gametes are again formed. Only if two recessive alleles come together in the fertilized egg—one from the female gamete and one from the male gamete—will the phenotype then show the recessive characteristic.

When pea plants homozygous for purple flowers are crossed with pea plants having white flowers, only pea plants with purple flowers are produced. Each plant in this F_1 generation, however, carries both an allele for purple and an allele for white. Figure 11–7 shows what happens in the F_2 generation if the F_1 generation self-pollinates. One of the simplest ways to predict the types of offspring that might be produced from such a cross is to diagram it using a Punnett square, as shown in Figure 11–8. Notice that the result would be the same if an F_1 individual were cross-fertilized with another F_1 individual, which is how these experiments are performed with animals and with plants that are not self-pollinating.

In order to test the hypothesis that alleles occur in pairs and that the two alleles of a pair segregate during gamete formation, it is necessary to perform an additional experiment: cross purple-flowering F_1 plants (the result of a cross between purple- and white-flowering plants) with white-flowering plants. To the casual observer, it would appear as if this were simply a repeat of Mendel's first experiment, crossing plants having purple flowers with plants having white flowers. But if Mendel's hypothesis is correct, the results will be different from those of his first experiment. Can you predict the results of such a cross? Stop a moment and think about it.

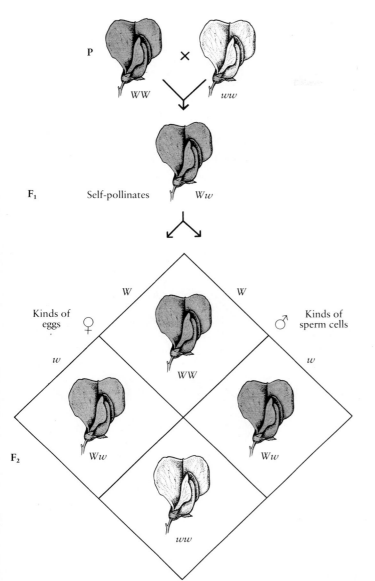

P

WW × ww

F₁   Self-pollinates   Ww

Kinds of eggs ♀   W   W   ♂   Kinds of sperm cells

w   w

F₂

WW   Ww

Ww   ww

P

Ww × ww

♀   W   w   ♂

w   w

F₁   Ww

ww   Ww

ww

11–8 *Mendel's hypothesis, as exemplified in the F₁ and F₂ generations following a cross between a parent (P) pea plant with two dominant alleles for purple flowers (WW) and one with two recessive alleles for white flowers (ww). The phenotype of the offspring in the F₁ generation is purple, but note that the genotype is Ww. The F₁ heterozygote produces four kinds of gametes, ♀ W, ♂ W, ♀ w, ♂ w, in equal proportions. When this plant self-pollinates, the W and w sperm cells and eggs combine randomly to form, on the average, 1/4 WW, 2/4 (or 1/2) Ww (purple), and 1/4 ww offspring. It is this underlying 1:2:1 genotypic ratio that accounts for the phenotypic ratio of 3 dominants (purple) to 1 recessive (white). The distribution of characteristics in the F₂ is shown by a Punnett square, named after the English geneticist who first used this sort of checkerboard diagram for the analysis of genetically determined traits.*

11–9 *A testcross. In order for a pea flower to be white, the plant must be homozygous for the recessive allele (ww). But a purple pea flower can be produced by a plant with either a Ww or a WW genotype. How could you tell such plants apart? Geneticists solve this problem by breeding such plants with homozygous recessives. This sort of experiment is known as a testcross. As shown here, a phenotypic ratio in the F₁ generation of one purple to one white indicates that the purple-flowering parent used in the testcross must have been heterozygous.*

The easiest way to analyze the possible result of this cross is again to use a Punnett square, as in Figure 11–9. This type of experiment, which reveals the genotype of the parent with the dominant phenotype, is known as a **testcross**. A testcross is an experimental cross between an individual having the dominant phenotype (genotype unknown) for a given trait with another individual that is known to be homozygous for the recessive allele. Whether or not two different phenotypes are produced indicates whether the individual with the dominant phenotype is heterozygous or homozygous for the trait being studied. The testcross shown in Figure 11–9 reveals that the genotype of the plant being tested is *Ww* rather than *WW*. What would have been the results of the testcross if the plant tested had been homozygous for the purple-flower allele?

The Principle of Independent Assortment

In a second series of experiments, Mendel studied crosses between pea plants that differed in two characteristics; for example, one parent plant produced peas that were round and yellow, and the other had peas that were wrinkled and green. The

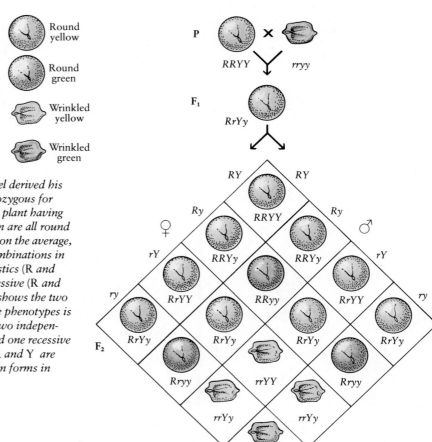

11-10 *One of the experiments from which Mendel derived his principle of independent assortment. A plant homozygous for round (RR) and yellow (YY) peas is crossed with a plant having wrinkled (rr) and green (yy) peas. The F₁ generation are all round and yellow, but notice how the characteristics will, on the average, appear in the F₂ generation. Of the 16 possible combinations in the offspring, 9 show the two dominant characteristics (R and Y), 3 show one combination of dominant and recessive (R and y), 3 show the other combination (r and Y), and 1 shows the two recessives (r and y). This 9:3:3:1 distribution of the phenotypes is always the expected result from a cross involving two independently assorting genes, each with one dominant and one recessive allele in each of the parents. (Note that the letters R and Y are used because round and yellow are the less common forms in nature.)*

round and yellow characteristics, you will recall (see Table 11–1), are dominant, and the wrinkled and green are recessive. As you would expect, all the seeds produced by a cross between the true-breeding parental types were round and yellow. When these *F₁* seeds were planted and the resulting flowers allowed to self-pollinate, 556 seeds were produced. Of these, 315 showed the two dominant characteristics, round and yellow, but only 32 combined the recessive characteristics, green and wrinkled. All the rest of the seeds were unlike either parent; 101 were wrinkled and yellow, and 108 were round and green. Totally new combinations of characteristics had appeared.

This experiment did not contradict Mendel's previous results. If the two traits, seed color and seed shape, are considered independently, round and wrinkled still appeared in a 3:1 ratio (423 round to 133 wrinkled), and so did yellow and green (416 yellow to 140 green). But the seed shape and the seed color characteristics, which had originally been combined in a certain way in one plant (round only with yellow and wrinkled only with green), behaved as if color factors and shape factors were entirely independent of one another (yellow could now be found with wrinkled and green with round). From this, Mendel formulated his second law, the **principle of independent assortment.** This principle states that when gametes are formed, the alleles of a gene for one trait segregate independently of the alleles of a gene for another trait.

Figure 11–10 diagrams Mendel's interpretation of these results. It shows why, in a cross involving two independently assorting genes, with each gene having one dominant and one recessive allele, the phenotypes in the offspring will, on the

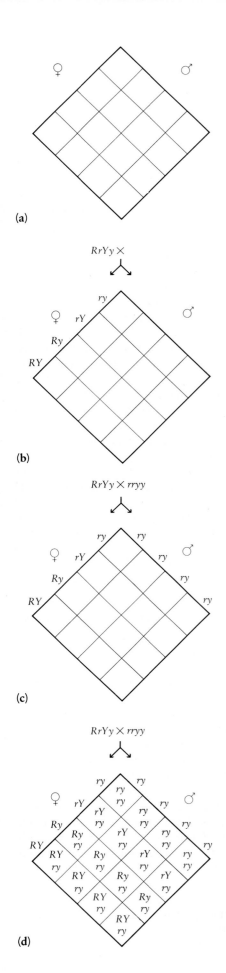

(a)

$RrYy \times$

(b)

$RrYy \times rryy$

(c)

$RrYy \times rryy$

(d)

11-11 *A testcross involving two gene pairs.*

average, be in the ratio of 9:3:3:1. Nine out of the total 16, or 9/16, represents the proportion of F_2 offspring that will show the two dominant characteristics, 1/16 the proportion that will show the two recessive characteristics, and 3/16 and 3/16 the proportions that will show the two alternative combinations of dominants and recessives. In terms of probability (see essay), a seed has a 3/4 chance of being yellow and a 3/4 chance of being round and thus a 9/16 chance (3/4 × 3/4) of being both yellow and round. It has only a 1/4 chance of being wrinkled, however, so its chance of being yellow and wrinkled is 3/4 × 1/4, or 3/16.

The 9:3:3:1 ratio holds true when one of the original parents is homozygous for both recessive characteristics and the other homozygous for both dominant ones, as in the experiment just described *(RRYY × rryy)*, as well as when each original parent is homozygous for one recessive and also homozygous for one dominant characteristic *(rrYY × RRyy)*. The F_1 progeny from either of these crosses will always be heterozygous for both traits *(RrYy)*; crossing these heterozygotes produces the F_2 generation with the expected 9:3:3:1 ratio of phenotypes.

A Testcross

Can you predict the outcome of a testcross between an individual homozygous for the recessive alleles for each of two traits and an individual who is heterozygous for both traits? Such a cross is similar to the one analyzed in Figure 11–9 but involves alleles of two independently assorting genes instead of one. For simplicity, let us again study the distribution of the alleles for seed shape, round versus wrinkled *(R versus r)*, and the alleles for seed color, yellow versus green *(Y versus y)*, in a cross between a heterozygote and a homozygous recessive. Draw a Punnett square with 16 squares (Figure 11–11a). Put the female symbol on one side and the male on the other. Assume that the heterozygote contributes the female gametes. With a genotype of *RrYy*, the heterozygote can produce four kinds of gametes: *RY, Ry, rY,* and *ry*. At the head of each column at the left, where the female symbol is, put one of these possible combinations (Figure 11–11b). Notice that in this step we are assuming, as Mendel did, that each of the possible kinds of gametes is produced in equal numbers.

The homozygous recessive can produce only one type of gamete in terms of the traits being studied: *ry*. Put *ry* at the head of each column on the right (Figure 11–11c). The advantage of using a Punnett square is that it makes it impossible to overlook any combination of gametes.

Now, starting with the column on the far left, begin to fill in the squares. You are less likely to make a mistake if you fill in all the female gametes first (or all male gametes), a column at a time, than if you try to work with both male and female gametes at the same time.

Next, fill in the symbols for the alleles carried in the male gametes. Your square will then look like the one in Figure 11–11d.

Now count the phenotypes that this square predicts. Every capital *R* indicates a round seed (since *R* is dominant); there are eight capital *R*'s and, hence, eight round seeds. Conversely, you can count eight wrinkled *(rr)* seeds. Similarly, there are eight yellow *(Y)* seeds and eight green *(yy)* seeds.

Notice that each of the possible combinations of characteristics, round green, round yellow, wrinkled green, and wrinkled yellow, appears in equal proportions. This ratio of 1:1:1:1 is the typical result of a testcross involving an individual that is heterozygous for both traits.

Mendel and the Laws of Probability

In applying mathematics to the study of heredity, Mendel was asserting that the laws of probability apply to biology as they do to the physical sciences. Toss a coin. The probability that it will turn up heads is fifty-fifty, that is, one chance in two, or 1/2. The probability (or chance) that it will turn up tails is also fifty-fifty, or 1/2. The chance that it will turn up one or the other is certain, or one chance in one. Now toss two coins. The chance that one will turn up heads is again 1/2. The chance that the second will turn up heads is also 1/2. The chance that both will turn up heads is $1/2 \times 1/2$, or 1/4. The probability of two independent events occurring simultaneously is simply the probability of one occurring alone multiplied by the probability of the other occurring alone. This is known as the product rule of probability. The probability that both coins will turn up tails is similarly $1/2 \times 1/2$. The probability of the first coin turning up tails and the second coin turning up heads is $1/2 \times 1/2$, and the probability of the second turning up tails and the first heads is also $1/2 \times 1/2$.

We can diagram this in a Punnett square (see figure), which indicates that the combination in each square has an equal chance of occurring. Similarly, in Mendel's experiment diagrammed in Figure 11–8, the probability that a gamete produced by an F_1 plant of Ww genotype will carry the W allele is 1/2, and the probability that it will carry the w allele is 1/2. The probability, therefore, of any specific combination of the two alleles in the offspring—that is, WW, Ww, wW, or ww—is $1/2 \times 1/2$, or 1/4. It was undoubtedly the observation that one-fourth of the offspring in the F_2 generation showed the recessive phenotype that indicated to Mendel that he was dealing with a simple case of the laws of probability.

Returning to our coin toss, if there were three coins involved, the probability of any given combination would be simply the product of all three individual possibilities: $1/2 \times 1/2 \times 1/2$, or 1/8. Similarly, with four coins, the probability of any specific combination is $1/2 \times 1/2 \times 1/2 \times 1/2$, or 1/16. The Punnett square on page 245 expresses the probability of each of any one of four possible phenotype combinations.

When there is more than one possible arrangement of the events producing the specified outcome, however, the individual probabilities are added. For instance, what is the probability of throwing a head and a tail, in either order? There are two ways you could do this: by throwing a head first and then a tail (HT), or a tail first and then a head (TH). The probability of throwing a tail and a head or a head and a tail is the sum of their individual probabilities: $(1/2 \times 1/2) + (1/2 \times 1/2) = 1/4 + 1/4 = 1/2$. This is

11–12 *A flower of one species of evening primrose.*

MUTATIONS

In 1902, a Dutch botanist, Hugo de Vries, reported results of his studies on Mendelian inheritance in the evening primrose. Heredity in the primrose, he found, was generally orderly and predictable, as in the garden pea. Occasionally, however, a characteristic appeared that was not present in either parent or indeed anywhere in the lineage of that particular plant. De Vries hypothesized that such characteristics came about as the result of abrupt changes in genes and that the characteristic produced by a changed gene was then passed along like any other hereditary characteristic. De Vries called these abrupt hereditary changes **mutations,** and organisms exhibiting such changes came to be known as **mutants.** Different alleles of a gene, de Vries proposed, arose as a result of mutations. For example, the allele for wrinkled peas is thought to have arisen as a mutation of the gene for round peas.

As it turned out, only about 2 of some 2,000 changes in the evening primrose observed by de Vries were actually mutations. The vast majority were due to new

known as the sum rule of probability. In the cross diagrammed in Figure 11–8, a heterozygote is produced by either *Ww* or *wW*. The probability of a heterozygote in the F_2 generation is the sum of the probability of each of the two possible combinations: $1/4 + 1/4 = 1/2$.

The sum rule of probability, like the product rule, applies in more complex cases as well. For example, if you were asked the probability of throwing two heads and a tail, the answer would be 3/8. Three combinations are possible: HHT, HTH, and THH. For each of these combinations, the probability is $1/2 \times 1/2 \times 1/2 = 1/8$, that is, the product of three independent throws. Thus, the probability of throwing two heads and a tail is $1/8 + 1/8 + 1/8 = 3/8$.

Notice that in planning his experiments, Mendel made several assumptions: (1) of the male gametes produced, one-half contain one paternal allele and one-half contain the other paternal allele for each gene; (2) of the female gametes produced, one-half contain one maternal allele and one-half contain the other maternal allele for each gene; (3) the male and female gametes combine at random. Thus, the laws of probability could be employed—an elegant marriage of biology and mathematics.

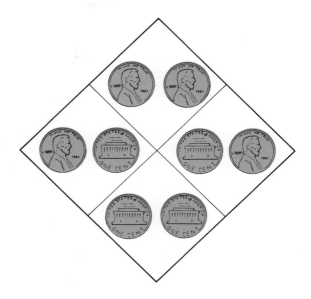

If you toss two coins 4 times, it is unlikely that you will get the precise results diagrammed above. However, if you toss two coins 100 times, you will come close to the proportions predicted in the Punnett square, and if you toss two coins 1,000 times, you will be very close indeed. As Mendel knew, the ratio of dominants to recessives in the F_2 generation might well not have been so clearly visible if he had been dealing with a small sample. The larger the sample, however, the more closely it will conform to results predicted by the laws of probability.

combinations of alleles, about which we shall have more to say in subsequent chapters, rather than to actual changes in any particular gene. However, de Vries's concept of mutation as the source of genetic variation proved of great importance, even though most of his examples were not valid.

Mutations and Evolutionary Theory

An important gap in Darwin's theory of evolution, first published in 1859, was the lack of explanation as to how variations can persist in populations. Mendel's work filled this gap. Segregation of alleles explained how variation is maintained from generation to generation. Independent assortment explained how individuals could have characteristics in combinations not present in either parent and so perhaps be better adapted, in evolutionary terms, than either parent. However, Mendelian principles presented new problems to the early evolutionists. If all hereditary variations were to be explained by the reshuffling process proposed by Mendel, there would be little or no opportunity for the kind of change in organisms envisioned by Darwin. As a result of mutations, however, there is a wide range of variability in natural populations. In a complex or shifting environment, a particular variation may give an individual or its offspring a slight edge. Although mutations seldom, if ever, determine the direction of evolutionary change, they are now recognized as the ultimate—and continual—source of the hereditary variations that make evolution possible.

THE INFLUENCE OF MENDEL

Mendel's experiments were first reported in 1865 before a small group of people at a meeting of the Brünn Natural History Society. None of them, apparently, understood what Mendel was talking about. But his paper was published the following year in the *Proceedings* of the Society, a journal that was circulated to libraries all over Europe. In spite of this, his work was ignored for 35 years, during most of which he devoted himself to the duties of an abbot, and he received no scientific recognition until after his death. He was, to use DuPraw's phrase (page 16), an odd traveler whose tale could not be made to fit.

(a)   **(b)**

11–13 **(a)** *Gregor Mendel, holding a fuchsia, is third from the right in this photograph of members of the Augustinian monastery in Brünn in 1862. In his experiments carried out in the monastery garden, Mendel showed that hereditary determinants are carried as separate units* *from generation to generation. His discoveries explained how inherited variations can persist for generation after generation.*

Although Mendel published only one other scientific paper during his lifetime, he continued his breeding experiments with a variety of plants until his election *as abbot of the monastery in 1871. Unfortunately, almost all of Mendel's papers relating to his scientific work were destroyed shortly before or after his death in 1884.* **(b)** *A facsimile of a page from one of Mendel's remaining handwritten manuscripts.*

It was not until 1900 that biologists were finally prepared to accept Mendel's findings. Within a single year, his paper was independently rediscovered by three scientists, one of whom was de Vries, each working in a different European country. Each of these scientists had done similar experiments and was searching the scientific literature for confirmation of his results. And each found, in Mendel's brilliant analysis, that much of his own work had been anticipated.

During the 35 years that Mendel's work remained in obscurity, great improvements were made in microscopy and, as a consequence, in the study of the structure of the cell (cytology). It was during this period that chromosomes were discovered and their movements during mitosis, which we described in Chapter 7, were first observed and recorded. Also discovered during this period was the process by which the gametes—the bearers of the hereditary information from one generation to the next—are formed; we shall examine this process in the next chapter.

SUMMARY

We have begun our consideration of genetics with the earliest ideas about biological inheritance, tracing the gradual development of these ideas into a science. The first question with which this new science was concerned was the mechanism of inheritance. How are hereditary characteristics passed from generation to generation?

By the middle of the nineteenth century, it was recognized that ova and sperm are specialized cells and that the ovum and sperm both contribute to the hereditary characteristics of the new individual. But how are these special cells, called gametes, able to pass on the many hundreds of characteristics involved in inheritance? Blending inheritance, which held that the characteristics of the parents blended in the offspring, like a mixture of two fluids, was one hypothesis. This explanation, however, did not allow for the inheritance of variations, which clearly occurred.

The revolution in genetics came when the blending concept was replaced by a unit concept. According to Mendel's principle of segregation, hereditary characteristics are determined by discrete factors (now called genes) that occur in pairs—one of each pair inherited from each parent. The members of the pair may be the same, in which case the individual is homozygous for the trait determined by the gene, or they may be different, in which case the individual is heterozygous for that trait. Different forms of the same gene are known as alleles.

The genetic makeup of an organism is known as its genotype. Its outward, observable characteristics are known as its phenotype. An allele that is expressed in the phenotype of a heterozygous individual to the exclusion of the other allele is a dominant allele; one whose effects are concealed in the phenotype is a recessive allele. In crosses involving two individuals heterozygous for the same gene, the ratio of dominant to recessive in the phenotypes of the offspring is 3:1.

A testcross, in which an individual with a dominant phenotypic characteristic but an unknown genotype is crossed with an individual homozygous for the recessive allele, reveals the genotype of the individual. If, in a testcross involving one gene, the two possible phenotypes appear in the offspring, the tested individual is heterozygous; if only the dominant phenotype appears in the offspring, the individual is homozygous for the dominant allele.

Mendel's other great principle, the principle of independent assortment, applies to the behavior of two or more different genes. This principle states that the alleles of one gene segregate independently of the alleles of another gene. When organisms heterozygous for each of two independently assorting genes are crossed, the expected phenotypic ratio in the offspring is 9:3:3:1.

Mutations are abrupt changes in the genotype—the ultimate source of the genetic variations studied by Mendel. Different mutations of a single gene increase the diversity of alleles of that gene in a population. As a consequence, mutation provides the variability among organisms that is the raw material for evolution.

QUESTIONS

1. Distinguish between the following terms: gene/allele; dominant/recessive; homozygous/heterozygous; genotype/phenotype; the F_1/the F_2; mutation/mutant.

2. In the experiments summarized in Table 11–1, which of the alternative characteristics appeared in the F_1 generation?

3. Why is a homozygous recessive always used in a testcross?

4. (a) What is the genotype of a pea plant that breeds true for tall? (Use the symbol T for tall, t for dwarf.) What possible gametes can be produced by such a plant? (b) What is the genotype of a pea plant that breeds true for dwarf? What possible gametes can be produced by such a plant? (c) What will be the genotype of the F_1 generation produced by a cross between a true-breeding tall pea plant and a true-breeding dwarf pea plant? (d) What will be the phenotype of this F_1 generation? (e) What will be the probable distribution of characteristics in the F_2 generation? Illustrate with a Punnett square.

5. The ability to taste a bitter chemical, phenylthiocarbamide (PTC), is due to a dominant allele. In terms of tasting ability, what are the possible phenotypes of a man both of whose parents are tasters? What are his possible genotypes?

6. If the man in Question 5 marries a woman who is a nontaster, what proportion of their children could be tasters? Suppose one of the children is a nontaster. What would you know about the father's genotype? Explain your results by drawing Punnett squares.

7. A taster and a nontaster have four children, all of whom can taste PTC. What is the probable genotype of the parent who is a taster? Is there another possibility?

8. What is the probability of drawing two aces out of a deck of 52 playing cards, one the ace of hearts and the other the ace of spades?

9. You have just flipped a coin five times and it has turned up heads every time. What is the chance that the next time you flip it, it will turn up tails?

10. (a) Suppose you would like to have a family consisting of two girls and a boy. What are your chances, assuming you have no children now? (b) If you already have one boy, what are your chances of completing your family as planned? (c) If you have two girls, what is the probability that the next child will be a boy?

11. PKU, phenylketonuria, is a disease caused by the presence of two recessive alleles of a particular gene. Individuals who are homozygous for the dominant allele of that gene or who are heterozygous show no signs of the disease. If two healthy parents have a child with PKU, what are their genotypes with respect to PKU? What are their chances of having another child with the same disease?

12. A pea plant that breeds true for round, green seeds *(RRyy)* is crossed with a plant that breeds true for wrinkled, yellow seeds *(rrYY)*. Each parent is homozygous for one dominant characteristic and for one recessive characteristic. (a) What is the genotype of the F_1 generation? (b) What is the phenotype? (c) The F_1 seeds are planted and their flowers are allowed to self-pollinate. Draw a Punnett square to determine the ratios of the phenotypes in the F_2 generation. How do the results compare with those of the experiment shown in Figure 11–10?

CHAPTER 12

Meiosis and Sexual Reproduction

Mendelian genetics is concerned with the way in which hereditary traits are passed from parents to offspring. Most eukaryotic organisms—including, for example, pea plants, sea urchins, and human beings—reproduce sexually. Sexual reproduction generally requires two parents, and it always involves two events: **fertilization** and **meiosis.** Fertilization is the means by which the different genetic contributions of the two parents are brought together to form the new genetic identity of the offspring. Meiosis is a special kind of nuclear division that is believed to have evolved from mitosis and uses much of the same cellular machinery. However, as you will see, meiosis differs from mitosis in some important respects.

HAPLOID AND DIPLOID

12–1 (a) *Wake-robin* (Trillium erectum) *flowers in the early spring.* (b) *Meiosis leading to the formation of pollen grains occurs in the anthers. The clearly visible chromosomes shown here are almost completely separated. Each diploid nucleus has divided to form four haploid sets of chromosomes.*

To understand meiosis, we must look again at the chromosomes. Every organism has a chromosome number characteristic of its particular species. A mosquito has 6 chromosomes per somatic (body) cell; a cabbage, 18; corn, 20; a sunflower, 34; a cat, 38; a human being, 46; a potato, 46; a plum, 48; a dog, 78; and a goldfish, 94. However, in these organisms and most other familiar plants and animals, the sex cells—or gametes—have exactly half the number of chromosomes that is characteristic of the somatic cells of the organism. The number of chromosomes in the gametes is referred to as the **haploid** ("single set") number, and the number in the somatic cells as the **diploid** ("double set") number. Cells that have more than two sets of chromosomes are known as **polyploid** ("many sets").

(a)

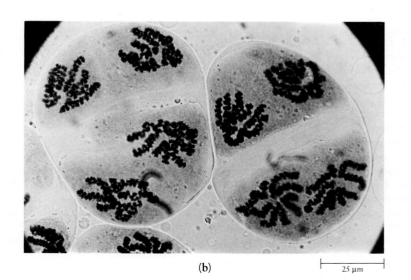

(b)

$\vdash\!\!-\!\!\dashv$ 25 μm

249

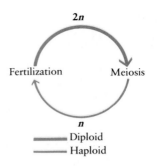

12-2 *Sexual reproduction is characterized by two events: the coming together of the gametes (fertilization) and meiosis. Following meiosis, there is a single set of chromosomes, that is, the haploid number (n). Following fertilization, there is a double set of chromosomes, that is, the diploid number (2n).*

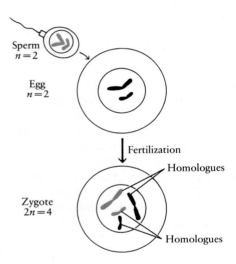

12-3 *During gamete formation, individual homologues (members of a homologous pair of chromosomes) are parceled out by meiosis so that a haploid (n) gamete, which is produced from a diploid (2n) cell, carries only one member of each homologous pair. At fertilization, the chromosomes in the sperm and egg nuclei come together in the zygote, producing, once again, pairs of homologous chromosomes. Each pair consists of one homologue from the father (paternal chromosome) and one from the mother (maternal chromosome). Here, and in subsequent diagrams, red and black are used to indicate the paternal and maternal chromosomes of a homologous pair.*

For brevity, the haploid number is designated *n* and the diploid number as *2n*. In humans, for example, $n = 23$ and therefore $2n = 46$. When a sperm fertilizes an egg, the two haploid nuclei fuse, $n + n = 2n$, and the diploid number is restored (Figure 12–2). A diploid cell produced by the fusion of two gametes is known as a **zygote.**

In every diploid cell, each chromosome has a partner. These pairs of chromosomes are known as homologous pairs, or **homologues.** The two resemble each other in size and shape and also, as we shall see, in the kinds of hereditary information (genes) each contains. One homologue comes from the gamete of one parent, and its partner is from the gamete of the other parent. After fertilization both homologues are present in the zygote (Figure 12–3).

In the special kind of nuclear division called meiosis, the diploid set of chromosomes, which contains the two homologues of each pair, is reduced to a haploid set, which contains only one homologue of each pair. Meiosis thus counterbalances the effects of fertilization. Cytologists predicted the existence of this so-called "reduction division" before it was actually observed; as they realized, without such a division, fertilization would double the chromosome number with each succeeding generation. In addition to maintaining a constant number of chromosomes from generation to generation, meiosis is, as we shall see shortly, a source of new combinations within the chromosomes themselves.

MEIOSIS AND THE LIFE CYCLE

Meiosis occurs at different times during the life cycle of different organisms (Figure 12–4). In many protists and fungi, such as the alga *Chlamydomonas* and the mold *Neurospora*, it occurs immediately after fusion of the mating cells (Figure 12–5). The cells are ordinarily haploid, and meiosis after fertilization restores the haploid number.

In plants, such as ferns, a haploid phase typically alternates with a diploid phase (Figure 12–6). The common and conspicuous form of a fern is the sporophyte, the diploid organism. By meiosis, fern sporophytes produce spores, usually on the undersides of their fronds (leaves). These spores have the haploid number of chromosomes. They germinate to form much smaller plants (gametophytes), typically only a few cell layers thick. In these plants, all the cells are haploid. The

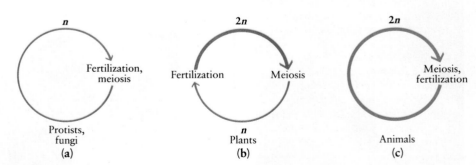

12-4 *Fertilization and meiosis occur at different points in the life cycle of different organisms.* (a) *In many—but not all—protists and fungi, meiosis occurs immediately after fertilization. Most of the life cycle is spent in the haploid state (signified by the thin line).* (b) *In plants, fertilization and meiosis are separated in time. The life cycle of the organism consists of both a diploid phase and a haploid phase.* (c) *In animals, completion of meiosis is immediately followed by fertilization. As a consequence, during most of the life cycle the organism is diploid (signified by the thick line).*

12–5 *The life cycle of* Chlamydomonas *is of the type shown in Figure 12–4a. The organism is haploid for most of its life cycle (thin arrows). Fertilization, the fusion of cells of different mating strains (indicated here by + and −), temporarily produces the diploid zygote (thick arrows). The zygote produces a thick coat that allows it to remain dormant during harsh conditions. Following dormancy, the diploid zygote divides meiotically, forming four new haploid cells. Each haploid cell can reproduce asexually (by mitosis and cytokinesis) to form either more haploid cells or, in periods of stress, haploid cells of a particular mating type. These cells can fuse with cells of the opposite mating type, and another sexual cycle is underway.*

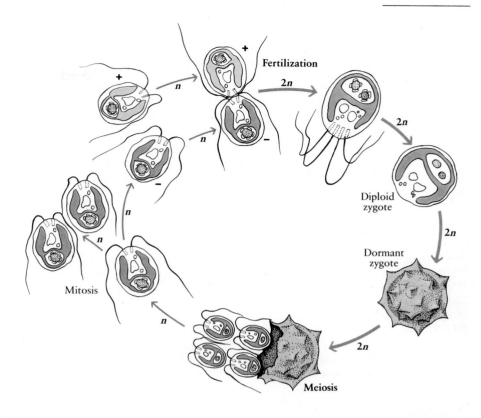

12–6 *The life cycle of a fern is of the type shown in Figure 12–4b. Following meiosis, spores, which are haploid, are produced in the sporangia and then are shed (extreme right). The spores develop into haploid gametophytes. In many species, the gametophytes are only one layer of cells thick and are somewhat heart-shaped, as shown here (bottom). From the lower surface of the gametophyte, filaments, the rhizoids, extend downward into the soil.*

On the lower surface of the gametophyte are borne the flask-shaped archegonia, which enclose the egg cells, and the antheridia, which enclose the sperm. When the sperm are mature and there is an adequate supply of water, the antheridia burst, and the sperm cells, which have numerous flagella, swim to the archegonia and fertilize the eggs. From the zygote, the diploid (2n) sporophyte develops, growing out of the archegonium within the gametophyte. After the young sporophyte becomes rooted in the soil, the gametophyte disintegrates. The sporophyte matures, develops sporangia, in which meiosis occurs, and the cycle begins again.

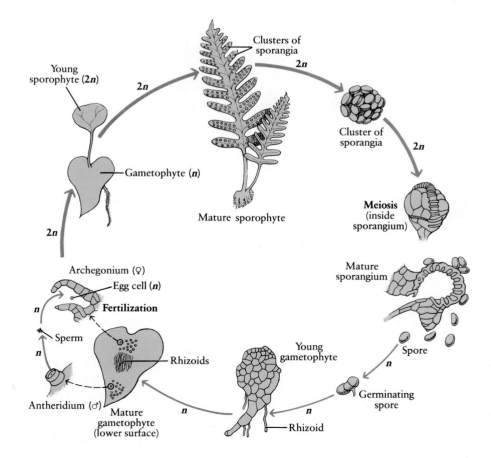

small, haploid gametophytes produce gametes by mitosis. The gametes fuse and then develop into a new, diploid sporophyte. This process, in which a haploid phase is followed by a diploid phase and again by a haploid phase, is known as **alternation of generations.** As we shall see, alternation of generations occurs in all sexually reproducing plants, although not always in the same form.

Human beings have the typical animal life cycle, in which the diploid individual produces haploid gametes by meiosis immediately preceding fertilization. Fusion of male and female gametes at fertilization restores the diploid chromosome number, and virtually all of the life cycle is spent in the diploid state (Figure 12–7).

12–7 *The life cycle of* Homo sapiens. *Gametes—egg cells and sperm cells—are produced by meiosis. At fertilization, the haploid gametes fuse, restoring the diploid number in the fertilized egg. The zygote develops into a mature man or woman, who again produces haploid gametes. As is the case with most other animals, the cells are diploid during almost the entire life cycle, the only exception being the gametes. This is the type of life cycle diagrammed in Figure 12–4c.*

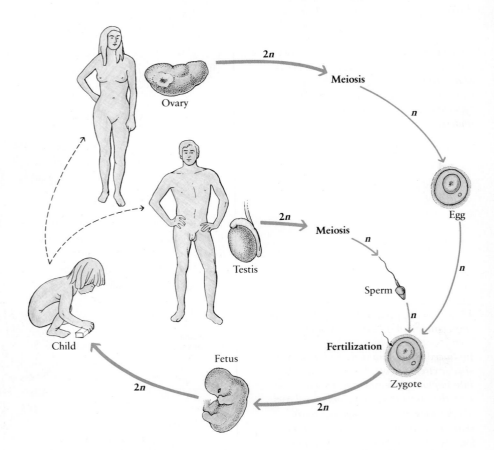

Note that although meiosis in animals produces gametes, meiosis in plants produces **spores.** A spore is a haploid reproductive cell that, unlike a gamete, can develop into a haploid organism without first fusing with another cell. With the formation of either gametes and spores, however, meiosis has the same result: at some point during the life cycle of a sexually reproducing organism, it reduces the diploid chromosome set to the haploid chromosome set.

MEIOSIS VS. MITOSIS

As we saw in Chapter 7, the events that occur in mitosis result in the formation of two daughter nuclei, each of which receives an exact copy of the parent cell's chromosomes. The events that take place during meiosis resemble those of mitosis, but there are some important differences:

1. Mitosis can occur in either haploid or diploid cells, whereas meiosis occurs only in cells with the diploid (or polyploid) number of chromosomes.

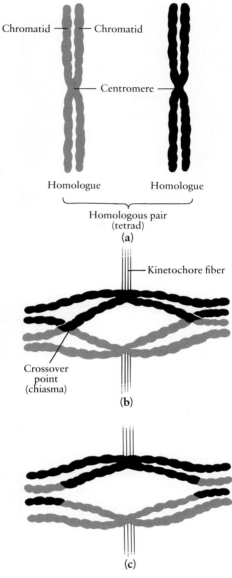

Chromatid — — Chromatid

Centromere

Homologue   Homologue

Homologous pair
(tetrad)
(a)

Kinetochore fiber

Crossover
point
(chiasma)
(b)

(c)

12–8 (a) *During prophase I of meiosis, chromosomes become arranged in homologous pairs. Each homologous pair consists of four chromatids and is therefore also known as a tetrad (from the Greek tetra meaning "four"). (b) Crossing over. During meiosis I, homologues, paired together as tetrads, connect at crossover points, where exchanges of segments of the chromosomes—crossing over—take place. Homologues remain associated at the crossover points, or chiasmata, until the end of prophase I. (c) Then, as the chromosomes separate slightly, the chiasmata appear to slip off. As you can see, crossing over results in a recombination of the genetic material of the two homologues.*

2. During meiosis, each diploid nucleus divides twice, producing a total of four nuclei. The chromosomes, however, replicate only once—prior to the first nuclear division.

3. Thus, each of the four nuclei produced contains half the number of chromosomes present in the original nucleus.

4. The haploid nuclei produced by meiosis contain new combinations of chromosomes. That is, the homologous chromosomes, originally derived from the organism's parents, are assorted randomly among the four new haploid nuclei. (For example, whether a given chromosome in a particular gamete produced by your body is the homologue derived originally from your mother or from your father is purely a matter of chance.)

Before looking at meiosis, you might wish to review the events of mitosis (pages 150–153).

THE PHASES OF MEIOSIS

Meiosis consists of two successive nuclear divisions, conventionally designated meiosis I and meiosis II. In meiosis I, homologous chromosomes pair and then separate from one another; in meiosis II, the chromatids of each homologue separate. In the following discussion we shall describe meiosis in a plant cell in which the diploid number is 6 *(n = 3)*. Three of the six chromosomes were originally derived from one parent and three from the other parent. For each chromosome from one parent there is a homologous chromosome, or homologue, from the other parent.

During interphase preceding meiosis, the chromosomes are replicated, so that by the beginning of meiosis each chromosome consists of two identical sister chromatids held together at the centromere region (Figure 12–8a). The first of the two nuclear divisions in meiosis then proceeds through the stages of prophase, metaphase, anaphase, and telophase (all of these are given the designation I to indicate that they are substages of meiosis I).

At the beginning of meiosis, **prophase I,** the chromatin condenses and the chromosomes come into view. By this time, an event has occurred for which the mechanism is completely unknown—the homologous chromosomes have come together in pairs. Once contact is made at any point between the two homologues, pairing extends, zipperlike, along the length of the chromatids in a process called **synapsis.** Since each chromosome consists of two identical chromatids, the pairing of the homologous chromosomes actually involves four chromatids; this complex of paired homologous chromosomes is known as a **tetrad.**

At this point, a crucial process occurs that can alter the genetic makeup of the chromosomes. This process, known as **crossing over,** involves the exchange of segments of one chromosome with corresponding segments from its homologous chromosome (Figure 12–8b). At the sites of crossing over, portions of the chromatids of one homologue are broken and exchanged with the corresponding portions of one or the other of the chromatids of the second homologue. The breaks are resealed, and the result is that the sister chromatids of a single homologue no longer contain identical genetic material (Figure 12–8c). The maternal homologue now contains portions of the paternal homologue, and vice versa. Thus crossing over is an important mechanism for recombining the genetic material from the two parents.

As prophase progresses, the homologues begin to pull away from each other—except in the areas of crossing over. Here, at the crossover points, or **chiasmata** (singular, chiasma), the homologues remain in close association until the end of prophase. Then the chiasmata seem to slip off the ends of the chromosomes. (If

12–9 (a) *Early prophase I in the forma-*
tion of a sperm cell in a grasshopper. The
homologous chromosomes are now
paired; the individual chromatids are not
visible, however, so each chromosome
appears as a single structure, and the
tetrads appear double-stranded (rather
than four-stranded). The dark area in the
upper right is a sex chromosome, which is
very prominent in the grasshopper. (b)
Late prophase I. All four chromatids can
be seen in some of the tetrads, and chias-
mata are evident.

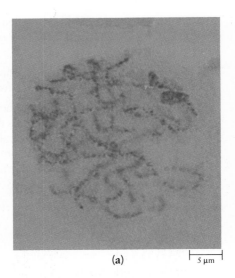
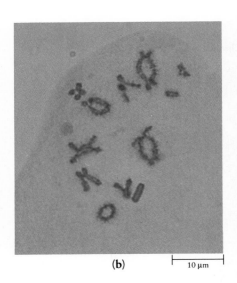

(a)   5 μm

(b)   10 μm

you cross your arms in front of you and then slide them apart, the crossed area will eventually reach your hands and then slide off your fingertips, much like the chiasmata.)

Although the homologous chromosomes have moved slightly apart by the end of prophase I, the homologues are still paired. Spindle microtubules can be seen radiating out from the two poles of the cell. The nucleoli and nuclear envelope disappear toward the end of this stage.

If you remember one important point—that the homologous chromosomes are arranged in pairs at this stage of meiosis—you will be able to remember all of the subsequent events with little difficulty.

In **metaphase I,** the homologous pairs (as you will recall, three pairs in this example) line up along the equatorial plane of the cell. (By contrast, in metaphase of mitosis, replicated chromosomes line up in single file with no sign of pairing of homologues.) The centromere region of each homologue has doubled by the end of metaphase, and spindle fibers have become associated with the kinetochores (see page 148). In an animal cell, centrioles and asters are also present.

During **anaphase I,** the homologues, each consisting of two sister chromatids, separate, as if pulled apart by the spindle fibers attached to the kinetochores. However, the two sister chromatids of each homologue do not separate as they did in mitosis.

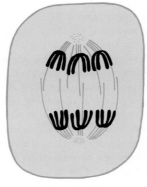

Late prophase I   Metaphase I   Anaphase I

By the end of the first meiotic division, **telophase I,** the homologues have moved to the poles. Each chromosome group now contains only half the number of chromosomes as the original nucleus.* Moreover, these chromosomes may be different from any of those present in the original cell because of exchanges that took place during crossing over. Depending on the species, new nuclear envelopes may or may not form, and cytokinesis may or may not take place. In some animal cells, but not all, the centrioles also divide at this stage.

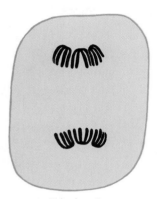

Telophase I

Meiosis, however, does not end here. Although two haploid nuclei have been formed, each nucleus contains double the haploid amount of hereditary material. Why? Because each chromosome consists of two chromatids.

Meiosis II resembles mitosis except that it is not preceded by replication of the chromosomal material. A short interphase may occur, during which the chromosomes partially unfold, but meiosis in many species proceeds from telophase directly to prophase II.

At the beginning of the second meiotic division, the chromosomes, if dispersed, condense fully again. Remember, in this example there are three chromosomes in each nucleus (the haploid number), and each is still in the form of two chromatids held together at the centromere region.

During **prophase II,** the nuclear envelopes, if present, disintegrate, and new spindle fibers begin to appear.

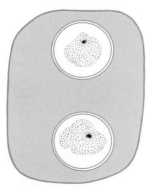

Interphase II

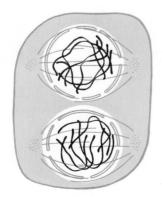

Prophase II

---

* In counting, it is often difficult to know whether to count a chromosome that has replicated but has not divided as 1 or 2 chromosomes. It is customary to count such a chromosome as 1. The trick is to count centromeres. If a chromatid has its own centromere, not associated with the centromere of its sister chromatid, then it can be called a chromosome.

During **metaphase II,** the three chromatid pairs in each nucleus line up on the equatorial plane. Spindle fibers are once again associated with the kinetochores, and other spindle fibers extend from the poles.

At **anaphase II,** as in anaphase of mitosis, the sister chromatids separate from one another. Each chromatid, which can now be called a chromosome, moves toward one of the poles.

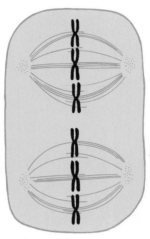

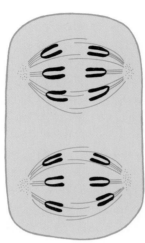

Metaphase II   Anaphase II

During **telophase II,** the spindle microtubules disappear and a nuclear envelope forms around each set of chromosomes. There are now four nuclei in all, each containing the haploid number of chromosomes. Cytoplasmic division (cytokinesis) proceeds as it does following mitosis. Cell walls form, dividing the cytoplasm, and these haploid plant cells begin to differentiate into spores.

Thus, beginning with one cell containing six chromosomes (three homologous pairs), we end with four cells, each with three chromosomes (no homologous pairs). The chromosome number has been reduced from the diploid to the haploid number.

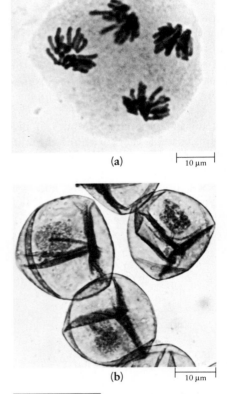

(a) | 10 μm

(b) | 10 μm

12–10 (a) *Anaphase II in the royal fern* Osmunda regalis. *The chromatids have separated, and the daughter chromosomes are moving to the opposite poles of the spindles.* (b) *The end of spore formation in* Osmunda regalis. *Each of these haploid cells can germinate to produce a gametophyte, the haploid phase in the life cycle.*

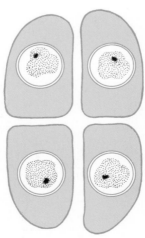

Telophase II   Four haploid cells

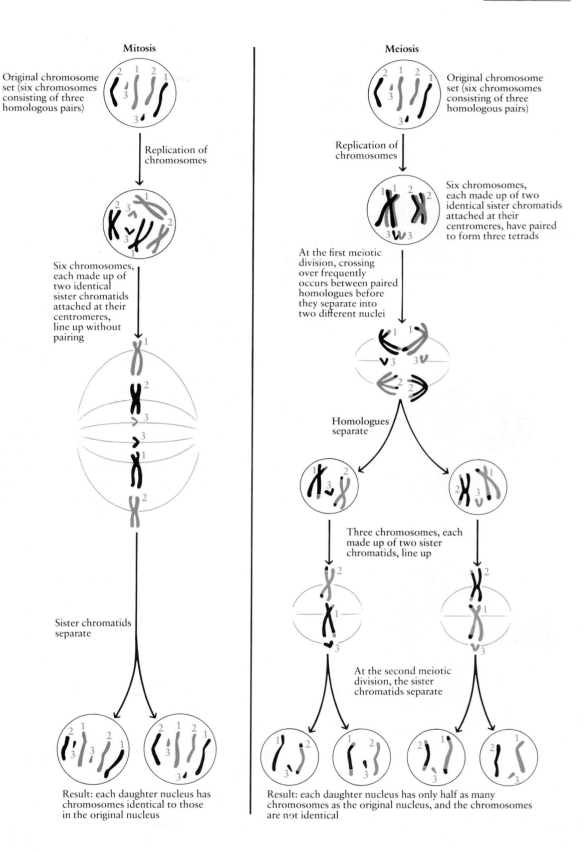

Mitosis

Original chromosome set (six chromosomes consisting of three homologous pairs)

Replication of chromosomes

Six chromosomes, each made up of two identical sister chromatids attached at their centromeres, line up without pairing

Sister chromatids separate

Result: each daughter nucleus has chromosomes identical to those in the original nucleus

Meiosis

Original chromosome set (six chromosomes consisting of three homologous pairs)

Replication of chromosomes

Six chromosomes, each made up of two identical sister chromatids attached at their centromeres, have paired to form three tetrads

At the first meiotic division, crossing over frequently occurs between paired homologues before they separate into two different nuclei

Homologues separate

Three chromosomes, each made up of two sister chromatids, line up

At the second meiotic division, the sister chromatids separate

Result: each daughter nucleus has only half as many chromosomes as the original nucleus, and the chromosomes are not identical

12–11 *A comparison of mitosis and meiosis. In these examples, each diploid cell has six chromosomes (2n = 6).*

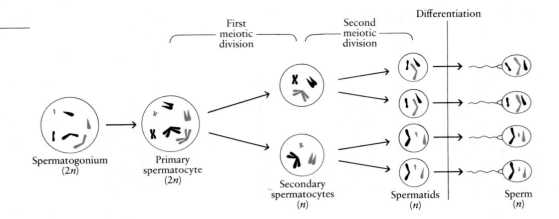

12–12 *The series of changes resulting in the formation of sperm cells begins with the growth of spermatogonia into primary spermatocytes. For simplicity, only six (n = 3) chromosomes are shown. At the first meiotic division, each primary spermatocyte divides into two haploid secondary spermatocytes. The second* *meiotic division results in the formation of four haploid spermatids, which differentiate into functional sperm. This process occurs continuously; the normal ejaculate of an adult human male contains between 300 and 400 million sperm cells.*

Meiosis in the Human Species

In all vertebrates, including humans, meiosis takes place in the reproductive organs, the testes of the male and the ovaries of the female. In the male, a cell known as a primary spermatocyte undergoes the two divisions of meiosis to produce four haploid spermatids, each of which then differentiates into a sperm cell (Figure 12–12).

In females, the meiotic divisions also produce haploid nuclei but the cytoplasm is apportioned unequally during cytokinesis in both meiosis I and II. One egg cell (the ovum) is produced, along with two or three polar bodies (Figure 12–13). The polar bodies contain the other post-meiotic nuclei and usually disintegrate. As a result of this unequal division, the ovum is well supplied with cytoplasmic materials, such as ribosomes, mitochondria, enzymes, and stored nutrients, important for the development of the embryo.

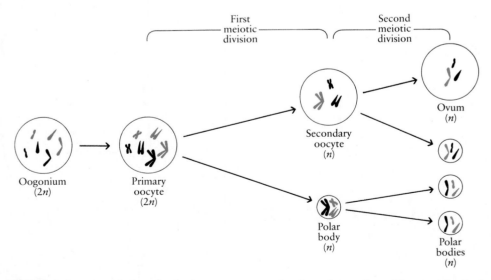

12–13 *Formation of the ovum begins with the growth of an oogonium into a primary oocyte. In the first meiotic division, this cell divides into a secondary oocyte and a polar body. The first meiotic division begins in the human female during the third month of fetal development and ends at ovulation, which may* *take place 50 years later. The second meiotic division, which produces the egg cell and a second polar body, does not take place until after the fertilizing sperm cell has penetrated the secondary oocyte. The first polar body may also divide.*

The Consequences of Sexual Reproduction

Many organisms can reproduce both asexually (by mitosis and cytokinesis) and sexually. Most unicellular eukaryotes have a life cycle similar to that of *Chlamydomonas* (Figure 12–5), in which either sexual or asexual reproduction may take place, often depending on environmental circumstances. Some unicellular eukaryotes, such as amoebas, reproduce only asexually. Many plants can also reproduce asexually; many grasses, for instance, spread by means of horizontal stems (rhizomes), growing either just above or just below the surface of the soil. In animals, asexual reproduction can take place by budding, as in *Hydra,* or by the breaking off of a fragment of the parent animal, as occurs in sponges, sea anemones, and certain types of worms. Because of the careful copying process of mitosis, asexually produced individuals are genetically identical to their parents.

By contrast, the potential for genetic variability in sexually produced individuals is enormous. The diagram below shows the possible distributions of chromosomes at meiosis in organisms with relatively few chromosomes. The red chromosomes were originally of paternal origin, and the black chromosomes of maternal origin. In the course of meiosis, these chromosomes are distributed among the haploid cells. As you can see, chromosomes of maternal or paternal origin do not stay together but are assorted independently. (At metaphase I of meiosis, the orientation of the homologous pairs is random, with no "rule" as to how many paternal or maternal homologues should be on either side of the equator.)

The number of possible chromosome combinations in the gametes is 2^n; 2 is the number of homologues in a pair, and n equals the haploid chromosome number. For example, (a) if the original number of chromosomes is 4 $(n = 2)$, the number of possible combinations of chromosomes is 2^2, or 4. (b) If the original number is 6 $(n = 3)$, the number of possible combinations is 2^3, or 8. (c) If there are 8 chromosomes $(n = 4)$, 16 different combinations (2^4) are possible.

A human male with his 46 chromosomes is capable of producing 2^{23} kinds of sperm cells—8,388,608 different combinations of chromosomes, about equal in number to the population of New York City. Similarly, a human female is capable of producing 2^{23} kinds of egg cells—8,388,608 different combinations of chromosomes. And this does not take into account the additional variations that may be introduced by crossing over.

Sexual reproduction, however, is expensive and inefficient, since a sexual population wastes half of its reproductive capacity on producing males and can therefore reproduce only half as fast as an asexual population. This raises one of the most intriguing questions of modern evolutionary biology: Given that sexual reproduction is expensive and inefficient, why is it so widely practiced throughout the multicellular world? The answer to this question is not known, but, as we shall see in Chapter 47, several different possibilities are the subject of current discussion and debate.

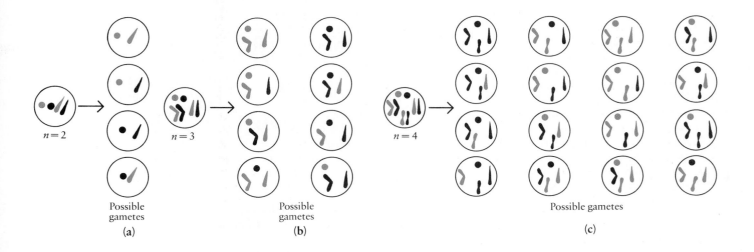

Possible gametes (a)

Possible gametes (b)

Possible gametes (c)

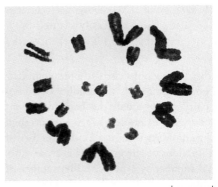

12–14 *Chromosomes from a diploid cell of a grasshopper, during metaphase of a mitotic division. Each chromosome consists of two closely aligned chromatids. Note that even though these chromosomes are not paired, it is possible to pick out some of the homologues. The observation that chromosomes come in homologous pairs was one of Sutton's clues to the meaning of meiosis.*

12–15 *The chromosome distributions in Mendel's cross of round yellow and wrinkled green peas, according to Sutton's hypothesis. Although the pea has 14 chromosomes (n = 7), only 4 are shown here, the two carrying the alleles for round or wrinkled and the two carrying the alleles for yellow or green. (This selection of specific chromosomes is analogous to Mendel's selection of specific traits to study.) As you can see, one parent is homozygous for the recessives, one for the dominants. Therefore, the only gametes they can produce are RY and ry. (Remember, R now stands not just for the allele but also for the chromosome carrying the allele, as do the other letters.) The F_1 generation, therefore, must be Rr and Yy. When a cell of this generation undergoes meiosis, R is separated from r and Y from y when the respective homologues separate at anaphase I. These alleles assort independently. Four different types of haploid egg nuclei are possible, as the diagram reminds us, and also four different types of sperm nuclei. These can combine in 4 × 4, or 16 different ways, as illustrated in the Punnett square.*

CYTOLOGY AND GENETICS MEET: SUTTON'S HYPOTHESIS

In 1902, shortly after the rediscovery of Mendel's work, Walter S. Sutton, a graduate student at Columbia University, was studying the formation of sperm cells in male grasshoppers. Observing the process of meiosis, Sutton noticed that the chromosomes were paired at the beginning of the first meiotic division. He also noticed that the two chromosomes of any one pair had physical resemblances to one another. In diploid cells, he noted, chromosomes apparently come in pairs. The pairing was obvious only at meiosis, although the discerning eye might also find the matching, but unpaired, homologues during metaphase of mitosis (Figure 12–14).

Sutton was struck by the parallels between what he was seeing and the first principle of Mendel—the principle of segregation. Suddenly the facts fell into place. Suppose chromosomes carried genes, the *Elemente* described by Mendel. This idea may not seem very startling to you now, but remember that at the turn of

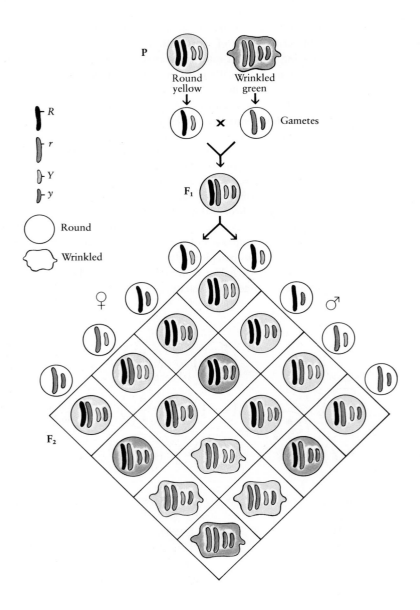

the century the gene was just an abstract idea or mathematical unit to the geneticist, and to the cytologist the chromosome was just an easily stained body with unknown function. Suppose, Sutton reasoned, alleles occurred on homologous chromosomes. Then the alleles would always remain independent and so would be separated at meiosis I as homologous chromosomes separated. New combinations of alleles would be formed as gametes fused at fertilization. Mendel's principle of the segregation of alleles could thus be explained by the segregation of the homologous chromosomes at meiosis.

What about Mendel's second principle in relation to the movement of the chromosomes at meiosis? This principle, as you will recall, states that the alleles of different genes assort independently. How one pair of alleles is segregated has no effect on the segregation of another pair of alleles, provided—and this is an important point—the two different pairs of alleles are on different pairs of chromosomes. The fact that Mendel chose traits for which the genes were found on different pairs of homologous chromosomes was essential to the success of his work (Figure 12–15).

As occurs often in the history of science, two other biologists recognized, at about the same time, the correlation between the behavior of Mendel's *Elemente* and the observed movement of the chromosomes. Young Sutton's paper appeared first, however, and his presentation was by far the most convincing. Nonetheless, much more evidence was required before, more than a decade later, most biologists were ready to concede that the little "colored bodies" performing their stereotyped, repetitive dance within the cell's nucleus actually held the secrets of the most ancient mysteries of heredity.

SUMMARY

Sexual reproduction involves a special kind of nuclear division called meiosis. Meiosis is the process by which the chromosomes are reassorted and cells are produced that have the haploid chromosome number *(n)*. The other principal component of sexual reproduction is fertilization, the coming together of haploid cells to form the zygote; fertilization restores the diploid number *(2n)*. There are characteristic differences among major groups of organisms as to when in the life cycle these events take place.

At the start of meiosis, the chromosomes arrange themselves in pairs. The members of the pairs are known as homologues. One homologue of each pair is of maternal origin, and one is of paternal origin. Each homologue consists of two identical sister chromatids, held together at the centromere. Early in meiosis, crossing over occurs between homologues, resulting in exchanges of chromosomal material.

In the first stage of meiosis, meiosis I, the homologues are separated. Two nuclei are produced, each with a haploid number of chromosomes, which, in turn, consist of two chromatids each. The nuclei enter interphase, but the chromosomal material is not replicated. In the second stage of meiosis, meiosis II, the sister chromatids of each chromosome separate as in mitosis. When the two nuclei divide, four haploid cells result.

Each of the haploid cells produced by meiosis contains a unique assortment of chromosomes due to crossing over and random assortment of homologues. Thus meiosis is a source of variation in the offspring.

Sutton was among the first to notice the analogy between the behavior of the chromosomes at meiosis and the segregation and assortment of the factors described by Mendel. On the basis of this observation, Sutton proposed that genes are carried on chromosomes.

QUESTIONS

1. Distinguish among the following: haploid/diploid/polyploid; sporophyte/gametophyte; gamete/zygote; meiosis I/meiosis II; homologues/tetrad; crossing over/chiasmata.

2. Dogs have a diploid chromosome number of 78. How many chromosomes would you expect to find in a gamete? In a liver cell? Plums have a haploid chromosome number of 24. How many chromosomes would you expect to find in a stem cell? In the nucleus of a pollen grain?

3. Draw a diagram of a cell with six chromosomes $(n = 3)$ at meiotic prophase I. Label each pair of chromosomes differently (for example, label one pair A^1 and A^2, and another B^1 and B^2, etc.).

4. Diagram the eight possible gametes resulting from meiosis in a plant cell with six chromosomes $(n = 3)$. Label each chromosome differently, as in Question 3. Assume that crossing over does not occur.

5. Identify the stages of meiosis in *Lilium* shown in the micrographs below. What stage of meiosis is visible in Figure 12–1?

6. (a) Compare metaphase of mitosis and metaphase II of meiosis. (b) Compare anaphase of mitosis with anaphase I and anaphase II of meiosis. In your answers, consider both the positions and composition of the chromosomes, as well as the consequences.

7. Compare and contrast the overall processes and the genetic consequences of meiosis and mitosis.

8. In our bodies and in those of most other animals, both mitosis and meiosis occur. What are the end products of these two processes? Where in our bodies do these processes occur?

9. Is sexual reproduction—that is, fertilization and meiosis—possible with only one parent? Explain.

10. Mendel did not know of the existence of chromosomes. Had he known, what change might he have made in his second principle?

11. You do not look exactly like your mother or your father. Why is this so? Explain how you might have inherited some of your maternal grandfather's characteristics. (Start with a gamete produced by your grandfather, and end with one of your somatic cells.)

(a)   (b)

(c)

(d)

Genes and Gene Interactions

When Mendel's work was rediscovered in Europe in 1900 by Hugo de Vries and others, it attracted wide attention throughout the world and stimulated many studies by investigators seeking to confirm and extend Mendel's observations. Prominent among these were the English scientists Reginald Punnett, the geneticist immortalized in the Punnett square, and William Bateson, a zoologist.

In 1909, Thomas Hunt Morgan, a biologist from the United States who had visited de Vries's laboratory in Holland and been impressed by his work, abandoned his previous work in embryology and began a study of genetics. At Columbia University, he founded what was to be the most important laboratory in the field for several decades. The wealth of data that emerged from these studies was so impressive that this period in genetics research, which lasted until World War II, has been characterized as the "golden age" of genetics (though some would argue that the golden age is now).

By a remarkable combination of insight and good fortune, Morgan selected the fruit fly *Drosophila melanogaster* as his experimental organism. Biologists have often used for their experiments "insignificant" plants and animals—such as Mendel's pea plants, for instance, or Hertwig's sea urchins (page 109). Underlying this approach is the assumption that basic biological principles are universal, applying equally to all living things. As it turned out, the little fruit fly proved to be a "fit material" for a wide variety of genetic investigations. In the decades that followed, *Drosophila* was to become famous as the biologist's principal tool in studying animal genetics.

Drosophila means "lover of dew," although actually this useful animal is not attracted by dew but feeds on the fermenting yeast that it finds in rotting fruit. The fruit fly was an excellent choice for genetic studies since it is easy to breed and maintain. Only 3 millimeters long, these tiny flies can produce a new generation every two weeks. Each female lays hundreds of eggs during her adult life, and very large numbers of flies can be kept in a half-pint bottle, as they were in Morgan's laboratory, familiarly known as the "Fly Room."

13–1 *Thomas Hunt Morgan at work in Columbia University's "Fly Room" in 1917. The camera-shy Morgan was photographed surreptitiously by a colleague who concealed a camera under a pile of milk bottles on his own desk. The half-pint milk bottles were used to house experimental* Drosophila.

THE REALITY OF THE GENE

Perhaps the most important of the principles established by Morgan and his colleagues was that Mendel's factors—genes—are located on chromosomes. Early in the twentieth century, at the beginning of the "golden age," this idea, so commonplace to us now, evoked raging controversy. At this stage of genetics research, the gene still had no physical reality. It was a pure abstraction. The work of Sutton and other cytologists was known, but it seemed irrelevant to studies of inheritance. As late as 1916, Bateson wrote, "The supposition that particles of chromatin, indistinguishable from each other and indeed almost homogeneous under any known test, can by their material nature confer all the properties of life, surpasses the range of even the most convinced materialism."

Sex Determination

One line of cytological observations, however, was providing further evidence linking "particles of chromatin" with heredity. As early as the 1890s, cytologists noticed that male and female organisms often show chromosomal differences, and they began to speculate that these differences are related to sex determination.

As Sutton had observed, the chromosomes of a diploid organism come in pairs. In all of the pairs except one, the chromosomes in both males and females appear to be the same; these are called **autosomes.** The structure of one pair, however, may differ between males and females. The chromosomes of this pair are known as the **sex chromosomes.** In many species, the two sex chromosomes are identical in the female but are dissimilar in the male, with one male sex chromosome the same as the female sex chromosomes and the other usually smaller. The sex chromosome that is the same in the cells of both males and females is called the *X* chromosome, and the unlike chromosome characteristic of the cells of males is called the *Y* chromosome. Thus we can characterize the two sexes as *XX* (female) and *XY* (male). In some insects, such as the grasshopper, which Sutton studied, there is no *Y* chromosome. In such cases, the females are characterized as *XX* and the males as *XO* (the *O* does not represent a chromosome but indicates the absence of one). In species in which the male is *XY* or *XO*, the male is said to be **heterogametic,** since he can produce two types of gametes, and the female is said to be **homogametic.**

Not all organisms, however, have heterogametic males and homogametic females. In birds, moths, and butterflies (and in occasional species in other groups), the sex chromosomes are reversed; the male has the two *X* chromosomes, and the female only one. The *Y* chromosome may or may not be present. In these organisms, it is the female that is heterogametic.

Human beings have 22 pairs of autosomes, which are structurally the same in both sexes. Females have a twenty-third matching pair, the sex chromosomes, *XX*. Human males, as their twenty-third pair, have one *X* and one *Y* (see Figure 19–2, page 383). During meiosis, as each diploid spermatocyte undergoes meiotic division into four haploid sperm cells, two of the sperm cells receive *X* chromosomes and two receive *Y* chromosomes. The ovum always contains an *X* chromosome, since a human female does not normally possess the *Y* in any of her cells. Thus the zygote will be *XX* or *XY*, depending on whether an *X*-bearing sperm or a *Y*-bearing sperm fertilizes the egg (Figure 13–2). It is in this way that the sperm cell contributed by the male determines the sex of the offspring, and it is the process of meiosis that governs the almost equal production of male and female offspring.

The correlation of the appearance of chromosomes with a particular trait—sex—gave strength to the gene-chromosome hypothesis. As we shall see, however, a large body of evidence, from a variety of different observations and experiments, was required before biologists became convinced that genes are on the chromosomes.

Sex Linkage

One of the advantages of *Drosophila melanogaster* for genetic studies is that it has only four pairs of chromosomes. Three of these pairs are autosomes, and the fourth is a pair of sex chromosomes, an *XX* pair in the female and an *XY* pair in the male (Figure 13–3). This feature of the fruit fly turned out to be particularly useful, although Morgan could not have foreseen that when he first selected *Drosophila* as his experimental organism.

When he began his investigations in 1909, Morgan intended to use *Drosophila* for breeding experiments similar to those Mendel had carried out with the pea

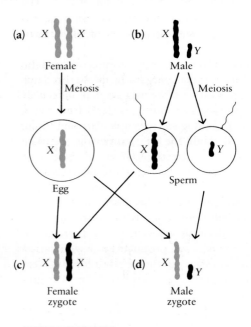

13–2 *Sex determination in an organism (such as humans) in which the male is heterogametic.* **(a)** *At meiosis, every egg cell receives an* X *chromosome from the mother.* **(b)** *A sperm cell may receive either an* X *chromosome or a* Y *chromosome.* **(c)** *If a sperm cell carrying an* X *chromosome fertilizes the egg, the offspring will be female* (XX). **(d)** *If a sperm cell carrying a* Y *chromosome fertilizes the egg, the offspring will be male* (XY).

13–3 *The fruit fly* (Drosophila melanogaster) *and its chromosomes. Fruit flies have only four pairs of chromosomes (2n = 8), a fact that simplified Morgan's experiments. Six of the chromosomes (three pairs) are autosomes (including the two dot-like chromosomes in the center), and two are sex chromosomes.*

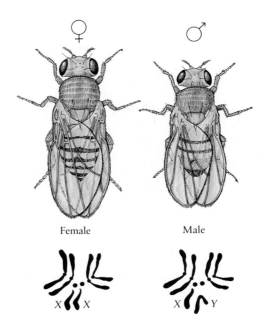

Female Male

plant. He was looking, as Mendel had, for patterns of inheritance. Such experiments involved examining under a magnifying lens hundreds—eventually tens of thousands—of individual fruit flies.

Initially, the investigators in Morgan's laboratory were looking for genetic differences among individual flies that they could then study in the breeding experiments. A year after Morgan established his colony, such a difference appeared. One of the prominent and readily visible characteristics of fruit flies is their brilliant red eyes. One day, a white-eyed fly, a mutant, was observed in the colony (Figure 13–4). This fly, a male, was mated with a red-eyed female, and all of the F_1 offspring had red eyes.

Morgan then crossbred the F_1 offspring, just as Mendel had done in his pea experiments. However, instead of the expected 3:1 ratio of dominant to recessive phenotypes (that is, of red-eyed to white-eyed individuals), the ratio was closer to 4:1, and, moreover, all of the white-eyed individuals were males:

Red-eyed females	2,459
White-eyed females	0
Red-eyed males	1,011
White-eyed males	782

Why were there no white-eyed females? To explore the situation further, Morgan crossed the original white-eyed male with one of the F_1 females. The following results were obtained from this testcross:

Red-eyed females	129
White-eyed females	88
Red-eyed males	132
White-eyed males	86

In other words, females can be white-eyed. The characteristic behaves pretty much like a typical recessive (the expected ratio is 1:1:1:1). So why were there no white-eyed females in the F_2 generation? Stop here a moment and see if you can answer this question. Morgan was able to do so.

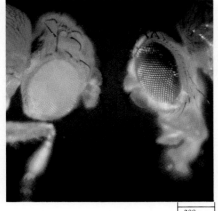

200 µm

13–4 *A mutant white-eyed fruit fly (left) and a normal, or wild-type, red-eyed fruit fly (right). While looking for genetic differences among* Drosophila, *Morgan discovered a single white-eyed fruit fly in his population of thousands. This chance occurrence launched an avalanche of studies and established a new concept of the gene.*

P $X^{w+}X^{w+}$ × X^wY
Red-eyed   White-eyed
female   male

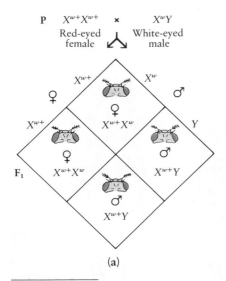

X^{w+} X^w
♀   ♂

X^{w+} Y

$X^{w+}X^w$ $X^{w+}X^w$ $X^{w+}Y$
♀   ♀   ♂

F₁

$X^{w+}Y$
♂

(a)

$X^{w+}X^w$ × $X^{w+}Y$
Red-eyed   Red-eyed
female from F₁   male from F₁

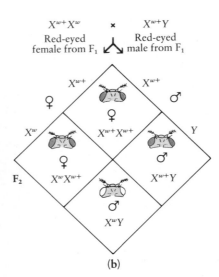

X^{w+} X^{w+}
♀   ♂

X^w Y

X^wX^{w+} $X^{w+}X^{w+}$ $X^{w+}Y$
♀   ♀   ♂

F₂

X^wY
♂

(b)

$X^{w+}X^w$ × X^wY
Red-eyed   White-eyed
female from F₁   male

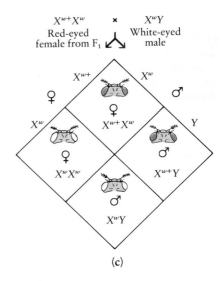

X^{w+} X^w
♀   ♂

X^w Y

X^wX^w $X^{w+}X^w$ $X^{w+}Y$
♀   ♀   ♂

X^wY
♂

(c)

13–5 *Punnett square diagrams of the experiments Morgan performed after discovery of the white-eyed male Drosophila. The least common characteristic, white eyes, is represented by w; w⁺ symbolizes the wild-type allele for red eyes. Alleles located on the sex chromosomes are commonly designated by superscripts to X and Y. (a) Morgan first mated a true-*

breeding red-eyed female to the white-eyed male. All of the offspring had red eyes. (b) Next, he mated an F₁ red-eyed female to an F₁ red-eyed male. Although both red-eyed and white-eyed males were produced in the F₂ generation, all F₂ females had red eyes, suggesting a relationship between the inheritance of eye color and the behavior of the sex chromosomes.

(c) A testcross between a red-eyed F₁ female and the original white-eyed male produced both red-eyed and white-eyed flies of both sexes. This led to the conclusion that the gene for eye color must be carried on the X chromosome. The allele for red eyes (w⁺) is dominant, and the allele for white eyes (w) is recessive.

On the basis of these experiments, outlined in Figure 13–5, Morgan and his coworkers formulated the following hypothesis: The gene for eye color is carried only on the X chromosome. (In fact, as it was later shown, the Y chromosome of *Drosophila* carries very little genetic information.) The allele for white eyes must indeed be recessive, since all of the F₁ flies had red eyes. Thus a heterozygous female would have red eyes—which is why there were no white-eyed females in the F₂ generation. However, a male that received an X chromosome carrying the allele for white eyes would always be white-eyed since no other allele would be present.

Further experimental crosses, such as the one shown in Figure 13–6, confirmed Morgan's hypothesis. They also revealed that white-eyed fruit flies are more likely to die before they reach adulthood than are red-eyed fruit flies, which explains their lower-than-expected number in the F₂ generation and the testcross.

These experiments introduced the concept of **sex-linked traits,** which are, as we shall see in Chapter 19, important in the genetics of human beings as well as of fruit flies. With regard to the alleles for sex-linked traits, members of the heterogametic sex—most often, the male—are neither homozygous nor heterozygous; instead, they are said to be **hemizygous.** Moreover, with sex-linked traits, the terms dominant and recessive apply, strictly speaking, only in the case of the homogametic sex—most often, the female. In the heterogametic male, any allele carried on the X chromosome will be expressed in the phenotype. As a consequence, sex-linked "recessives" appear much more frequently in males.

The results of the breeding experiments with white-eyed and red-eyed fruit flies convinced Morgan, and many other geneticists as well, that Sutton's hypothesis was correct: Genes *are* on chromosomes. Conclusive demonstration of the physical location of the gene, however, depended on subsequent experiments, to which we shall return later in this chapter.

Tortoiseshell Cats, Barr Bodies, and the Lyon Hypothesis

A dark spot of chromatin—called a Barr body—can be seen at the outer edge of the nucleus of female mammalian somatic cells in interphase. This dark spot is an inactivated X chromosome. According to the Lyon hypothesis—named after the British geneticist Mary Lyon, who proposed it—early in the development of the embryo of the female mammal, one or the other X chromosome is inactivated in each somatic cell already formed. This inactivation occurs randomly, with the result that the embryo becomes a mosaic of cells, some with one X chromosome inactivated, and others with the other X chromosome inactivated. Thus, all the somatic cells of a female mammal are not identical but are one of two types, depending on which of the X chromosomes is active and which is inactive. Once an X chromosome is inactivated, all the daughter cells of that cell will have the same X chromosome inactivated. In the germ cells from which the egg cells will ultimately be produced by meiosis, one X chromosome appears to be inactivated early in development, but it is reactivated prior to meiosis.

In human females heterozygous for certain sex-linked traits, it has been found that the recessive is expressed to varying degrees, with some populations of cells expressing the recessive phenotype and other populations not expressing it. Whether a sex-linked characteristic is expressed depends upon which X chromosome (and, consequently, which allele for a particular trait) was inactivated in the embryonic cell from which the population of cells descended. A striking example is provided by color blindness, a human sex-linked characteristic that we will consider in more detail in Chapter 19. Women who are heterozygous for color blindness are sometimes color blind in one eye but not the other.

In cats, the alleles for black or yellow coat color are carried on the X chromosome. Male cats, having only a single X chromosome with one or the other of these alleles, are either black or yellow. Tortoiseshell cats have coats with patches of both black and yellow. As you would expect, they are almost always female—neatly fitting the predictions of the Lyon hypothesis.

13–6 *Offspring of a cross between a white-eyed female fruit fly and a red-eyed male fruit fly, illustrating what happens when a recessive allele is carried on an X chromosome. The F₁ females, with one X chromosome from the mother and one from the father, are all heterozygous (X^wX^{w+}) and so will be red-eyed. But the F₁ males, with their single X chromosome received from the mother carrying the recessive (w) allele, will all be white-eyed because the Y chromosome carries no gene for eye color. Thus the recessive allele on the X chromosome inherited from the mother will be expressed in the male offspring.*

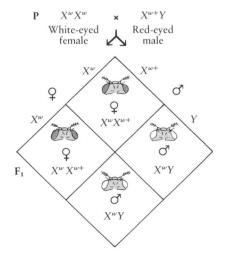

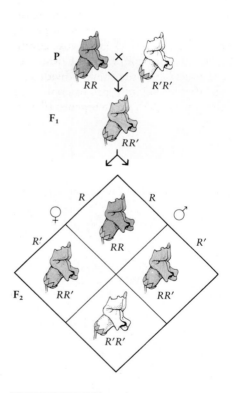

13–7 *Many human characteristics are governed by simple Mendelian inheritance of dominant alleles. Among these are dimples and cleft chin, as illustrated by the actor Kirk Douglas and his sons.*

13–8 *A cross between a red* (RR) *snapdragon and a white* (R′R′) *snapdragon. This looks very much like the cross between a purple- and a white-flowering pea plant shown in Figure 11–8, but there is a significant difference because in this case neither allele is dominant. The flower of the heterozygote is a blend of the two colors.*

BROADENING THE CONCEPT OF THE GENE

At the same time that the early work on *Drosophila* was proceeding in Morgan's laboratory, experiments and observations in other laboratories were also confirming and extending Mendel's principles. Many additional dominant and recessive alleles were demonstrated in a variety of other organisms, including humans (Figure 13–7). Alleles, even sex-linked ones, segregate and assort independently according to the principles of Mendel. However, as the golden age of genetics progressed, new studies, while confirming Mendel's work in principle, showed that the patterns of inheritance are not always as simple and direct as Mendel's reported results had indicated. (This fact is not surprising when you recall that Mendel had carefully selected certain traits for study, using only those that showed clear-cut differences.) The phenotypic effects of a particular gene, it was found, are influenced not only by the alleles of that gene present in the organism, but also by other genes and by the environment. And many, indeed most traits are influenced by more than one gene, just as most genes can influence more than a single trait. Some examples follow.

Allele Interactions

Incomplete Dominance and Codominance

Dominant and recessive characteristics are not always as clear-cut as in the seven traits studied by Mendel in the pea plant. Some characteristics appear to blend. For instance, as Bateson and Punnett showed in 1906, a cross between a homozygous red-flowering snapdragon *(RR)* and a homozygous white-flowering snapdragon *(R′R′)* produces heterozygotes that are pink, a phenotype intermediate between those of the homozygotes (Figure 13–8). This phenomenon is known as **incomplete dominance.** As we shall see in Chapter 15, it is a result of the combined effects of gene products. When the heterozygous pink snapdragons are allowed to self-pollinate, red and white characteristics sort themselves out once again, showing that the alleles themselves, as Mendel had asserted, remain discrete and unaltered.

In other cases, alleles may act in a **codominant** manner, with heterozygotes expressing not an intermediate phenotype but rather both homozygous phenotypes simultaneously. A familiar example is found in the human blood type AB, in which the distinctive characteristics of red blood cells of both type A and type B are expressed in the phenotype.

Multiple Alleles

Although any individual diploid organism can have only two alleles of any given gene, it is possible that more than two forms of a gene—**multiple alleles**—may be present in a population of organisms. Multiple alleles result from different mutations of a single gene.

The members of the "set" of alleles may have different dominance relationships with one another. For instance, coat color in rabbits is determined by a series of four alleles: C (wild type, or agouti), c^{ch} (light gray, or chinchilla), c^h (albino with black extremities, or Himalayan), and c (albino). Different combinations of any two of these four possible alleles produce different coat colors (Figure 13–9).

In humans, the three alleles—A, B, and O—that determine the principal blood groups (to be discussed in more detail in Chapter 39) are probably the best known example of multiple alleles.

Gene Interactions

In addition to the interactions that occur between alleles of the same gene, interactions also occur among the alleles of different genes. Indeed, most of the

(a)

(b)

(c)

(d)

13-9 Coat color in rabbits is determined principally by a single gene, of which four different alleles are known. Any individual rabbit, of course, carries only two alleles for this gene in its body cells. Different combinations of alleles produce (a) the wild-type, or agouti, rabbit (CC, Cc^{ch}, Cc^h, or Cc), (b) the chinchilla rabbit (c^{ch}c^{ch}, c^{ch}c^h, or c^{ch}c), (c) the Himalayan rabbit (c^{h}c^h or c^{h}c), and (d) the albino rabbit (cc).

characteristics (both structural and chemical) that constitute the phenotype of an organism are the result of the interaction of many distinct genes.

Appearance of Novel Phenotypes

Sometimes, when one trait is affected by two (or more) different genes, a completely novel phenotype may appear. For instance, as demonstrated by Bateson and Punnett, comb shape in chickens is determined by two different genes, rose and pea, each with two alleles *(R, r* and *P, p)*. *RR* or *Rr* results in rose comb, whereas *rr* produces single comb. *PP* or *Pp* produces pea comb, and *pp* produces single comb. However, when *R* and *P* occur together in the same individual, a novel phenotype, walnut comb, results. Thus, four different types of combs are possible depending on the interaction of the alleles of these two genes (Figure 13–10).

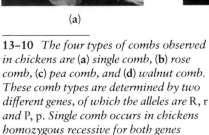

(a)

(b)

(c)

(d)

13-10 The four types of combs observed in chickens are (a) single comb, (b) rose comb, (c) pea comb, and (d) walnut comb. These comb types are determined by two different genes, of which the alleles are R, r and P, p. Single comb occurs in chickens homozygous recessive for both genes (rrpp). Rose comb results from the presence of at least one dominant R allele coupled with two recessive p alleles (Rrpp or RRpp). Pea comb, by contrast, is produced by at least one dominant P allele coupled with two recessive r alleles (rrPp or rrPP). When at least one dominant allele of each gene is present in the same individual, a novel phenotype, walnut, is produced. Genotypically, chickens with walnut comb may be RRPP, RRPp, RrPP, or RrPp.

Epistasis

In other cases, a novel phenotype does not appear when genes interact. Instead, different genes interact so that one gene interferes with or modifies the effect of the other. This type of interaction is called **epistasis** ("standing upon"). If gene *A* masks the effects of gene *B*, then *A* is said to be epistatic to *B*.

A classic example of epistatic gene interaction was reported by Bateson. Scientists in his laboratory crossed two pure-breeding white-flowered varieties of sweet pea *(Lathyrus odoratus)* and found that the progeny all had purple petals. When these F_1 plants were allowed to self-pollinate, of 651 plants that flowered in the F_2 generation, 382 had purple petals and 269 were white. At first, these figures may seem meaningless, but if you examine them closely, you will see that they fit a 9:7 ratio. In a cross demonstrating independent assortment of two nonallelic genes in a ratio of 9:3:3:1, 9/16, you will remember, is the proportion of offspring that show the effects of the two dominant alleles. So we can conclude that only a plant that has received at least one dominant allele from each gene (that is, allele *P* and allele *C*) can make the purple pigment (Figure 13–11).

In this case, either gene in the recessive homozygous condition is epistatic to, or hides, the effects of the other gene. When gene *C* is homozygous recessive *(cc)*, flowers will be white even if a dominant *P* is present (as in the phenotypes of *ccPp* and *ccPP*). Similarly, when gene *P* is homozygous recessive *(pp)*, flowers are also white (as in the phenotypes of *Ccpp* and *CCpp*).

13–11 *Epistasis in sweet peas. When two different varieties of white-flowered sweet pea plants are crossed, all of the* F_1 *plants have purple flowers. In the* F_2 *generation, the ratio of purple- to white-flowered plants is 9:7. Purple color is due to the presence of both dominant alleles* P *and* C*; the homozygous recessive of either gene masks, or is epistatic to, the effects of the other gene.*

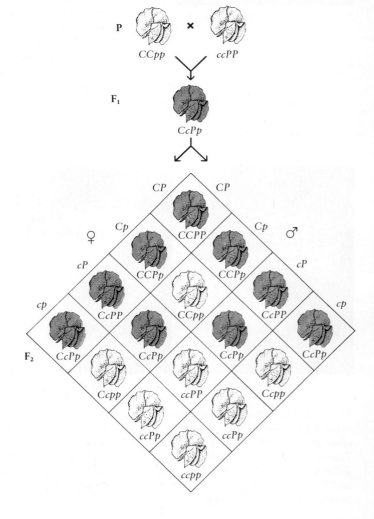

The phenomenon of epistasis indicates once again that what we see at the phenotypic level is actually the product of a very complicated series of events that influence one another during the development and lifetime of the organism. No gene works alone. The expression of all genes is in some way influenced by many other genes.

Genes and the Environment

The expression of a gene is always the result of its interaction with the environment. To take a common example, a seedling may have the genetic capacity to be green, to flower, and to fruit, but it will never turn green if it is kept in the dark, and it may not flower and fruit unless certain precise environmental requirements are met.

The water buttercup, *Ranunculus peltatus*, is a more striking example. It grows with half the plant body submerged in water. Although the leaves are genetically identical, the broad, floating leaves differ markedly in both form and physiology from the finely divided leaves that develop under water (Figure 13–12).

Temperature often affects gene expression. Primrose plants that are red-flowered at room temperature are white-flowered when raised at temperatures above 30°C (86°F). Similarly, as we noted in Chapter 8, Himalayan rabbits are white at high temperatures and black at low temperatures. In addition, Siamese cats raised at room temperature are black in their cooler peripheral areas, such as ears, nose, and tail tip.

These are extreme examples of a universal verity: The phenotype of any organism is the result of interaction between genes and environment.

Expressivity and Penetrance

When the expression of a gene is altered by environmental factors or other genes, two outcomes are possible. First, the degree to which a particular genotype is expressed in the phenotype of an individual may vary. This variable **expressivity** is seen for polydactyly, the presence of extra fingers and toes, which is caused by a dominant allele. Often, there is great variability in expressivity among members of a family, with the result that some individuals have extra digits on both hands and feet, while others may have only a portion of an extra toe on one foot.

Second, the proportion of individuals that show the phenotype ascribed to a particular genotype may be less than expected; the genotype shows incomplete **penetrance.** For example, individuals known to carry the allele for polydactyly may have absolutely normal hands and feet.

Examples of variable expressivity and incomplete penetrance abound among human genetic characteristics, often making it difficult to analyze the patterns of inheritance for certain genetic diseases or abnormalities.

Polygenic Inheritance

Some traits, such as size or height, shape, weight, color, metabolic rate, and behavior, are not the result of interactions between one, two, or even several genes; instead, they are the cumulative result of the combined effects of many genes. This phenomenon is known as **polygenic inheritance.**

A trait affected by a number of genes, or polygenes, does not show a clear difference between groups of individuals—such as the differences tabulated by Mendel. Instead, it shows a gradation of small differences, which is known as **continuous variation.** If you make a chart of differences among individuals for any trait affected by a number of genes, you get a curve such as that shown in Figure 13–13.

13–12 *The water buttercup. Leaves growing above water are broad, flat, and lobed. The genetically identical under-water leaves are thin and finely divided.*

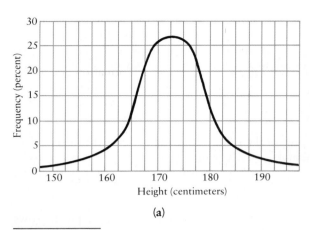

(a)

1	0	0	1	5	7	7	22	25	26	27	17	11	17	4	4	1
4:10	4:11	5.0	5:1	5:2	5:3	5:4	5:5	5:6	5:7	5:8	5:9	5:10	5:11	6:0	6:1	6:2

(b)

13–13 (a) *Height distribution of males in the United States. Height is an example of polygenic inheritance; that is, it is affected by a number of genes. Such genetic traits are characterized by small gradations of difference. A graph of the distribution of such traits takes the form of a bell-shaped curve, as shown, with the mean, or average, usually falling in the center of the curve. The larger the number of genes involved, the smoother the curve.* (b) *A company of student recruits at the Connecticut Agricultural College about 80 years ago. The number of men in each group and their height (in feet and inches) are shown below the photograph.*

Males in the United States are taller, on the average, than they were fifty years ago, due to better nutrition and other environmental factors. However, the shape of the curve is the same; in other words, the great majority fall within the middle range, and the extremes in height are represented by only a few individuals. Some of these height variations are produced by environmental factors, such as diet, but even if all the men in a population were maintained from birth on the same type of diet, there would still be a continuous variation in height in the population. This is due to genetically determined differences in hormone production, bone formation, and numerous other factors.

Table 13–1 illustrates a simple example of polygenic inheritance, color in wheat kernels, which is controlled by two genes, the four alleles of which exhibit cumulative quantitative effects. Human skin color is believed to be under a similar kind of genetic control (although involving more than two genes), as are many other traits. In fact, most normal human traits are believed to be polygenic.

TABLE 13–1 The Genetic Control of Color in Wheat Kernels (Polygenic inheritance*)

Parents:	$R_1R_1R_2R_2$ $\times$ $r_1r_1r_2r_2$		
	(Dark red) (White)		

F_1:	$R_1r_1R_2r_2$ (Medium red)		

F_2:	**Genotype**		**Phenotype**	
1	$R_1R_1R_2R_2$		Dark red	
2 } 4	$R_1R_1R_2r_2$		Medium-dark red	
2	$R_1r_1R_2R_2$		Medium-dark red	
4	$R_1r_1R_2r_2$		Medium red	15 red
1 } 6	$R_1R_1r_2r_2$		Medium red	to
1	$r_1r_1R_2R_2$		Medium red	1 white
2 } 4	$R_1r_1r_2r_2$		Light red	
2	$r_1r_1R_2r_2$		Light red	
1	$r_1r_1r_2r_2$		White	

* Two genes are involved, each with two alleles: R_1 and r_1 for gene 1, and R_2 and r_2 for gene 2.

Pleiotropy

It is also possible for a single gene to affect more than one characteristic, a phenomenon known as **pleiotropy.** The frizzle trait in fowl (Figure 13–14) is an example of pleiotropy. In these animals, a change in feather formation due to the abnormal action of a single gene leads to drastic changes in many other aspects of their physiology. Similarly, the gene for white coat color in cats has a pleiotropic effect on the eye and ear. Cats that are all white and have blue eyes are often deaf. Some white cats have one blue eye and one yellow-orange eye. These too are deaf, but only on the side where the blue eye is present.

In rats, a single mutation affecting a gene that produces a protein involved in the formation of cartilage causes a whole complex of congenital deformities. These include thickened ribs, a narrowing of the tracheal passage (through which air moves to and from the lungs), blocked nostrils, a blunt snout, a loss of elasticity in the lungs, thickening of the heart muscle, and, needless to say, a greatly increased mortality. Since cartilage is one of the most common structural substances of the body, the widespread consequences of such a mutation are not difficult to understand. In fact, it is very likely that Mendel's allele for wrinkled peas, for example, also affected other structural characteristics of the pea plant. Similarly, the higher mortality rate of white-eyed fruit flies was a result of pleiotropic effects.

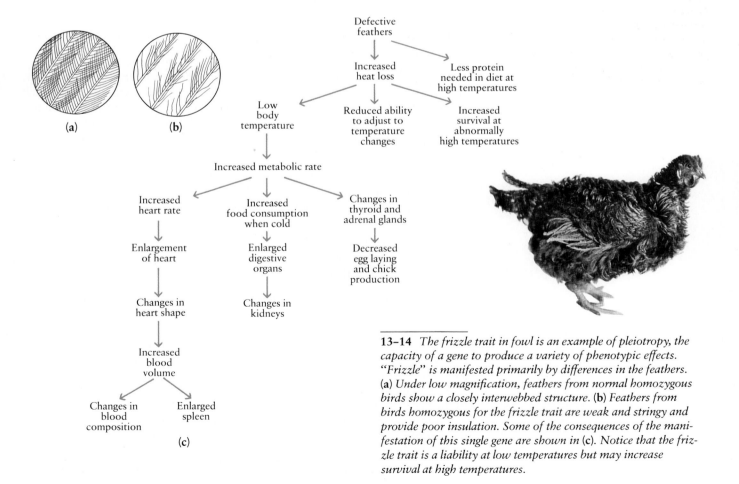

13–14 *The frizzle trait in fowl is an example of pleiotropy, the capacity of a gene to produce a variety of phenotypic effects. "Frizzle" is manifested primarily by differences in the feathers.* (a) *Under low magnification, feathers from normal homozygous birds show a closely interwebbed structure.* (b) *Feathers from birds homozygous for the frizzle trait are weak and stringy and provide poor insulation. Some of the consequences of the manifestation of this single gene are shown in* (c). *Notice that the frizzle trait is a liability at low temperatures but may increase survival at high temperatures.*

GENES AND CHROMOSOMES

Linkage

Mendel showed that certain pairs of alleles, such as those for round and wrinkled peas, assort independently of other pairs, such as those for yellow and green peas. However, as we noted previously, the alleles of two different genes will always assort independently if the genes are on different pairs of homologous chromosomes. If the alleles of the two genes are on the same pair of homologous chromosomes, then segregation of the alleles of one gene will not be independent of the segregation of the alleles of the other gene. In other words, if the alleles of two different genes are on the same chromosome, they should both be transmitted to the same gamete at meiosis. Genes that tend to stay together because they are on the same pair of homologous chromosomes are said to be **linked,** or in the same **linkage group.**

In 1927, an important research tool became available when H. J. Muller, one of Morgan's collaborators, found that exposure to x-rays greatly increases the rate at which mutations occur in *Drosophila*. Other forms of radiation, such as ultraviolet light, and certain chemicals were also shown to act as **mutagens,** or agents that produce mutations. As increasing numbers of mutants were found in Columbia University's *Drosophila* collection, as a result of the application of Muller's discovery, the mutations began to fall into four linkage groups, in accord with the four pairs of chromosomes visible in the cells. Indeed, in all organisms that have been studied in sufficient genetic detail, the number of linkage groups and the number of pairs of chromosomes have been the same, providing further support for Sutton's hypothesis that genes are on the chromosomes.

Recombination

Large-scale studies of linkage groups soon revealed some unexpected difficulties. For instance, most fruit flies have light tan bodies and long wings, both of which are dominant characteristics. When individuals homozygous for these characteristics were bred with mutant fruit flies having black bodies and short wings (both recessive characteristics), all the F_1 offspring had light tan bodies and long wings, as would be expected. Then the F_1 generation was inbred. Two outcomes seemed possible:

1. The genes for body color and wing length would be assorted independently, giving rise to Mendel's 9:3:3:1 ratio in the phenotypes and indicating that the genes for these two traits were on different chromosomes.
2. The genes for the two traits would be linked. In this case, 75 percent of the flies would be tan with long wings and 25 percent, homozygous for both recessives, would be black with short wings.

In the case of these particular traits, the results closely resembled the second possibility, but they did not conform exactly. In a few of the offspring, the genes for these traits seemed to assort independently; that is, some few flies appeared that were tan with short wings, and some that were black with long wings. How could this be? Somehow genes that were presumed to be on the same chromosome had become separated.

To find out what was happening, Morgan tried a testcross, breeding a member of the F_1 generation with a homozygous recessive. If black and tan, long and short, assorted independently—that is, if they were on different chromosomes—25 percent of the offspring of this cross should be black with long wings, 25 percent tan with long wings, 25 percent black with short wings, and 25 percent tan with

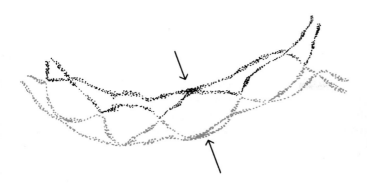

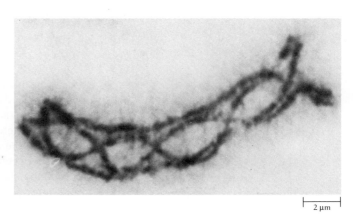

├─┤ 2 μm

13–15 *Homologous chromosomes of a grasshopper, as seen in prophase I. All four chromatids are visible. Crossing over—that is, the exchange of genetic material—has probably occurred at* the points at which these chromatids intersect, the chiasmata. The arrows indicate the position of the centromeres.

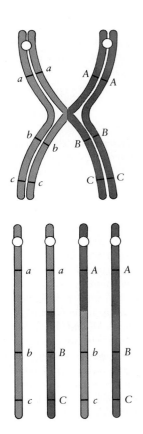

13–16 *Crossing over takes place when breaks occur in the chromatids of homologous chromosomes during prophase I, when the chromosomes are paired. The broken end of each chromatid joins with the chromatid of a homologous chromosome. In this way, alleles are exchanged between chromosomes. The white circles symbolize centromeres.*

short wings. On the other hand, if the genes for color and wing size were on the same chromosome and so moved together, half of the testcross offspring should be tan with long wings and half should be black with short wings. But actually, as it turned out, over and over, in counts of hundreds of fruit flies resulting from such crosses, 41.5 percent were tan with long wings, 41.5 percent were black with short wings, 8.5 percent were tan with short wings, and another 8.5 percent were black with long wings.

Morgan was convinced by this time that genes are located on chromosomes. It now seemed clear that the genes for the two traits, body color and wing length, were located on a single pair of homologous chromosomes, since the characteristics did not show up in the 1:1:1:1 ratio of independently assorted alleles. The only way in which the observed figures could be explained, Morgan reasoned, was if alleles could sometimes be exchanged between homologous chromosomes, that is, be recombined.

As we noted in Chapter 12, it now has been established that exchange of portions of homologous chromosomes—crossing over—takes place at the beginning of meiosis (Figure 13–15). If crossing over takes place between the positions at which two different genes are located on the same pair of homologues, then the alleles of the two genes can become separated as chromatids of the two homologues break and rejoin with each other (Figure 13–16).

Mapping the Chromosome

With the discovery of crossing over, accumulating evidence clearly supported not only the premise that genes are carried on chromosomes, but also that they must be positioned at particular spots, or **loci** (singular, locus), on the chromosomes. It followed that the alleles of any given gene must occupy corresponding loci on homologous chromosomes. Otherwise, the exchange of sections of chromosomes would result in genetic chaos rather than in an exact exchange of alleles.

As other traits were studied, it became clear that the percentage of recombinations between any two genes, such as those for body color and wing length, was different from the percentage of recombinations between two other genes, such as those for body color and leg length. In addition, as Morgan's experiments had shown, these percentages were fixed and predictable. It occurred to A. H. Sturtevant, who was an undergraduate working in Morgan's laboratory at this time, that the percentage of recombinations probably had something to do with the physical distances between the gene loci, or, in other words, with their spacing along the chromosome. This concept opened the way to the "mapping" of chromosomes.

Sturtevant postulated (1) that genes are arranged in a linear order on chromosomes, like beads on a string; (2) that genes that are close together will be separated by crossing over less frequently than genes that are farther apart; and (3) that it should therefore be possible, by determining the frequencies of recombinations, to plot the sequence of the genes along the chromosome and the relative distances between them. In Figure 13–16, for example, you can see that in a crossover, the chance that a strand would break and rejoin with its homologous strand somewhere between *B* and *C* should be less likely than if this were to happen somewhere between *A* and *B*, simply because the distance between *B* and *C* is less and there is less room (fewer chances) for crossovers to occur. Similarly, the chance of a crossover between *A* and *B* is less than the chance of a crossover between *A* and *C*.

In 1913, Sturtevant began constructing chromosome maps using data from crossover studies in fruit flies. As a standard unit of measure, he arbitrarily defined one map unit as equal to the distance that would give (on the average) one recombinant organism per 100 fertilized eggs (1 percent recombination). Genes with 10 percent recombination would be 10 map units apart; those with 8 percent recombination would be 8 map units apart (Figure 13–17). The fewer map units between genes, the less likely they were to be separated, while genes more than 50 map units apart on the same chromosome assorted independently.

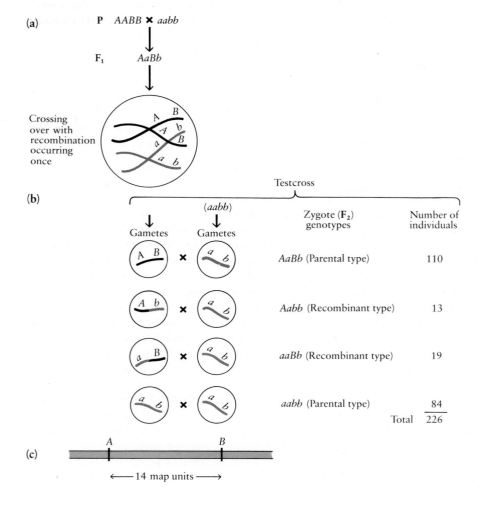

13–17 *Determining the map distance between two genes on the same chromosome.* (a) *When an individual homozygous dominant* (AABB) *for two genes located on the same pair of homologous chromosomes is crossed with one that is homozygous recessive* (aabb), *the F₁ offspring will all be heterozygous for both genes* (AaBb). *If crossing over occurs during meiosis in the heterozygote, alleles on the chromatids of the two homologues can be exchanged, and four different types of gametes can result from the recombination: parental-type gametes AB and ab, along with recombinant-type gametes Ab and aB.* (b) *The heterozygote is then mated to a homozygous recessive individual (a testcross).* (c) *The number of recombinant offspring divided by the total number of offspring yields a recombination percentage* (32/226 = 0.14) *that is defined as the map distance between the genes. Genes A and B are 14 map units apart.*

13–18 *Mapping a chromosome. Alleles A and B recombine with a and b in 4 percent of the offspring, and alleles A and C with a and c in 9 percent of the offspring. Use as the map unit the distance that will give (on the average) one recombinant per 100 fertilized eggs. (a) Start with the highest recombination percentage and establish the relative positions of A and C on the chromosome. (b) A and B, as you know from the data, are 4 units apart. Thus B could theoretically be either to the left of A or to the right. (c) However, if it were to the left, B and C would be 13 units apart, a distance that does not conform to the data. B must therefore be between A and C.*

Cross	Offspring		
$AB \times ab$	$AB + ab$	(Parental)	96%
	$Ab + aB$	(Recombinant)	4%
$AC \times ac$	$AC + ac$	(Parental)	91%
	$Ac + aC$	(Recombinant)	9%
$BC \times bc$	$BC + bc$	(Parental)	95%
	$Bc + bC$	(Recombinant)	5%

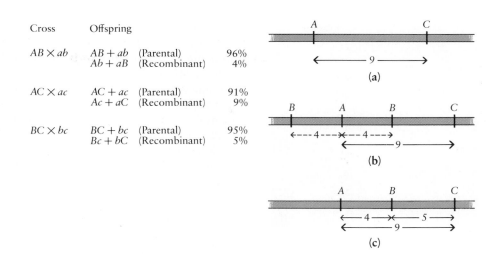

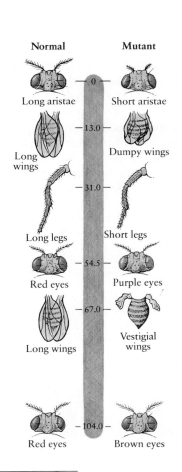

13–19 *A portion of one chromosome map of* Drosophila melanogaster, *showing relative positions of some of the genes on chromosome 2, as calculated by the frequency of recombinations. As you can see, more than one gene may affect a single trait, such as eye color.*

By correlating recombination frequencies (that is, crossover percentages) with the relative distance between genes, Sturtevant and other geneticists located a variety of genes on the chromosome maps of *Drosophila* (Figure 13–19). These map distances, however, do not necessarily accurately reflect physical distances because breaking and rejoining of chromosome segments are more likely to occur at some sites on the chromosomes than at others.

These important studies confirmed that not only are genes located on chromosomes, as Sutton had hypothesized, but they also have fixed positions in a linear sequence.

ABNORMALITIES IN CHROMOSOME STRUCTURE

Recombination does not affect the order of genes on a chromosome. However, it is possible, under certain circumstances, for chromosomes to break apart and rejoin in a different orientation on the same chromosome or even join with another chromosome. The outcome for both possibilities is a change in the sequence of genes on the affected chromosomes.

Cytological evidence for such changes in chromosome structure, giving rise to changes in gene order and altered hereditary patterns, came from study of the giant chromosomes discovered in 1933 in the salivary glands of *Drosophila* larvae. In *Drosophila*, as in many other insects, certain cells do not divide during the larval stages of the insect. In such cells, however, the chromosomes continue to replicate, over and over again, but since daughter chromosomes do not separate from one another after replication, they simply become larger and larger until they are composed of thousands of copies. Also, salivary gland homologous chromosomes pair tightly along their entire lengths, adding to their size; for example, a salivary gland cell of *Drosophila*, in which $2n = 8$, appears to have only four giant chromosomes. As you can see in Figure 13–20, when stained, these giant chromosomes are characterized by very distinctive dark and light bands.

These banding patterns became another useful tool for geneticists, enabling them to detect structural changes in the chromosomes themselves. In the giant chromosomes, geneticists can actually locate the positions where changes in chromosome structure have occurred by observing changes in the banding pattern. Sometimes a whole segment is lost; usually the effect of such a loss—known as a **deletion**—is lethal. In some cases, the "lost" segment becomes incorporated into its homologue, in which the segment then appears twice; this phenomenon is known as **duplication**. Sometimes a portion is transferred from one chromosome to another, nonhomologous chromosome, a process known as **translocation**. In

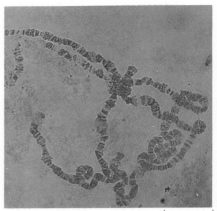

13–20 *Chromosomes from the salivary gland of a* Drosophila *larva. These chromosomes are 100 times larger than the chromosomes in ordinary body cells, and their details are therefore much easier to see. (This micrograph was taken with a light microscope, not an electron microscope.) Because of the distinctive banding patterns, it is possible in some cases to associate genes with specific regions in particular chromosomes. Each visible band represents dozens of genes.*

other cases, a double break in a chromosome occurs and a segment is turned 180° and then reincorporated in the chromosome; this phenomenon is known as **inversion.** The existence of such chromosomal aberrations had been hypothesized by Morgan's group on the basis of mapping studies. The studies of the giant chromosomes of *Drosophila* confirmed the existence of these chromosomal aberrations (Figure 13–21), and it became possible to assign genes to physical locations on the *Drosophila* chromosome.

Perhaps more important, almost a quarter of a century after the *Drosophila* work had begun, it was no longer just supposition that little, seemingly homogeneous, particles of chromatin could serve as the repositories of the mysteries of heredity.

SUMMARY

The rediscovery of Mendel's work in 1900 was the catalyst for many new discoveries in genetics, leading to the identification of chromosomes as the carriers of heredity and to the modification and extension of some of Mendel's conclusions.

Strong support for the hypothesis that genes are on the chromosomes came from studies by Morgan and his group on the fruit fly *Drosophila melanogaster*. Because it is easy to breed and maintain, *Drosophila* has been used in a wide variety of genetic studies. It has four pairs of chromosomes; three pairs (the autosomes) are structurally the same in both sexes, but the fourth pair, the sex chromosomes, is different. In fruit flies, as in many other species (including humans), the two sex chromosomes are *XX* in females and *XY* in males.

At the time of meiosis, the sex chromosomes, like the autosomes, segregate. Each egg cell receives an *X* chromosome, but half the sperm cells receive an *X* chromosome and half receive a *Y* chromosome. Thus, in fruit flies, humans, and many (but not all) other organisms, it is the paternal gamete that determines the sex of the offspring.

In the early 1900s, breeding experiments with *Drosophila* showed that certain traits are sex-linked, that is, their genes are carried on the sex chromosomes.

13–21 *A chromosomal inversion occurs when a segment of a chromosome breaks off, is turned 180°, and rejoins by the "wrong" ends. (a) An inversion results in a reversal of the sequence of genes, as shown in the chromosome on the right. (b) When one member of a homologous pair contains a large inverted segment, that chromosome must loop inside the other for close pairing to occur. (c) In giant chromosomes, such loops are greatly magnified, making it possible to readily identify regions with inverted segments.*

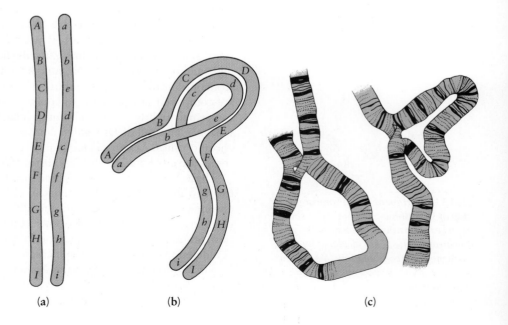

(a)   (b)   (c)

Because the *X* chromosome carries genes that are not present on the *Y* chromosome, a single recessive allele in the male, if carried on the *X* chromosome, will result in a recessive phenotype since no other allele is present. By contrast, a female heterozygous for a sex-linked trait will show the dominant characteristic.

Many traits are inherited according to the patterns revealed by Mendel. However, in others—perhaps the majority—the patterns are more complex. These complexities are caused by interactions among alleles, interactions among genes, and interactions with the environment.

Although many alleles interact in a dominant-recessive manner, some show varying degrees of incomplete dominance and codominance. In a population of organisms, a single gene may occur as multiple alleles, as the result of a series of different mutations of that gene. However, only two alleles can be present in any particular diploid individual.

Different genes can also interact with one another. Novel phenotypes may result from these interactions, or genes may affect one another in an epistatic manner, such that one hides the effect of the other. As a result, the expected phenotypic ratio is altered. Moreover, variable expressivity or reduced penetrance often results from the effects of other genetic influences, environmental conditions, or both.

The phenotypic expression of many traits is influenced by a number of genes. This phenomenon is known as polygenic inheritance. Such traits typically show continuous variation, as represented by a bell-shaped curve. Conversely, a single gene can affect two or more superficially unrelated traits; this property of a gene is known as pleiotropy.

Some genes assort independently in breeding experiments, and others tend to remain together. Genes that do not assort independently (because they are on the same chromosome) are said to be linked; a linkage group consists of a pair of homologous chromosomes.

Alleles are sometimes exchanged between homologous chromosomes as a result of crossing over in meiosis. Such recombinations can take place because: (1) the genes are arranged in a fixed linear array along the length of the chromosomes, and (2) the alleles of a given gene are at the corresponding sites (loci) on homologous chromosomes. Chromosome maps, showing the relative positions of gene loci along the chromosomes, have been developed from recombination data provided by breeding experiments.

Genetic studies have shown that chromosome breaks other than those resulting in crossovers may sometimes occur. A portion of a chromosome may be lost, or deleted, it may be duplicated, it may be translocated to a nonhomologous chromosome, or it may be inverted. Studies of the giant chromosomes of *Drosophila* larvae provided visual confirmation of these changes, as well as the final, conclusive evidence that the chromosomes are the carriers of the genetic information.

QUESTIONS

1. Distinguish among the following: sex chromosome/autosome; heterogametic/homogametic; incomplete dominance/codominance/epistasis; polygenic inheritance/pleiotropy; variable expressivity/incomplete penetrance; inversion/deletion/duplication/translocation.

2. Draw a diagram similar to Figure 13–2 indicating sex determination in a robin. Note that in birds the sex chromosomes are labeled *Z* and *W*; the genotype for the homogametic sex is *ZZ* and for the heterogametic sex is *ZW*.

3. With regard to sex-linked traits, the heterogametic sex, usually the male, is said to be hemizygous. What does this term mean? What two types of individuals can be used as testcross organisms for sex-linked traits?

4. The genes for coat color in cats are carried on the X chromosome. Black *(b)* is the recessive and yellow *(B)* is the dominant. What coat colors would you expect in the offspring of a cross between a black female and a yellow male? What coat colors would you expect in the sons of a tortoiseshell female?

5. Judging from the colors of the tortoiseshell cat (page 267), at what stage in development does the X chromosome become inactivated?

6. The so-called "blue" (really gray) Andalusian variety of chicken is produced by a cross between the black and white varieties. Only a single pair of alleles is involved. What color chickens (and in what proportions) would you expect if you crossed two blues? If you crossed a blue and a black? Explain.

7. In snapdragons, the allele that produces tall stems is completely dominant to the allele for dwarf stems, while the allele that produces red flowers is only partially dominant to that for white flowers. Describe the phenotype (height and flower color) of the F_1 plants resulting from a cross between a homozygous tall, red-flowered plant and a homozygous dwarf, white-flowered plant. If one of these F_1 plants self-pollinates, what will be the appearance and proportions of phenotypes in the resulting F_2 generation? Which two of these phenotypes will breed true?

8. In chickens, as we have seen, two pairs of alleles determine comb shape. *RR* or *Rr* results in rose comb, whereas *rr* produces single comb. *PP* or *Pp* produces pea comb, and *pp* produces single comb. When *R* and *P* occur together, they produce a new type of comb: walnut. What would be the genotype of the F_1 generation resulting from *RRpp* × *rrPP?* The phenotype? If F_1 hybrids were crossbred, what would be the probable distribution of genotypes? Of phenotypes? (Illustrate this cross with a Punnett square.)

9. In Duroc-Jersey pigs, coat color is determined by two genes, *R* and *S*. The homozygous recessive condition, *rrss*, produces a white coat. The presence of at least one copy each of *R* and *S* produces red. The presence of one or the other allele (either *R* or *S)* produces a new phenotype, sandy. Give the phenotypes of the following genotypes:

RRSS	*rrss*
RrSs	*rrSs*
RRSs	*rrSS*
RrSS	*RRss*

10. Mating a red Duroc-Jersey boar to sow A (white) gave pigs in the ratio of 1 red: 2 sandy: 1 white. Mating this same boar to sow B (sandy) gave 3 red: 4 sandy: 1 white. When this boar was mated to sow C (sandy), the litter had equal numbers of red and sandy piglets. Using the information presented in Question 9, give the possible genotypes of the boar and the three sows.

11. You and a geneticist are looking at a mahogany-colored Ayrshire cow with a newly born red calf. You wonder if it is male or female, and the geneticist says it is obvious from the color which sex the calf is. She explains that in Ayrshires the genotype *AA* is mahogany and *aa* is red, but the genotype *Aa* is mahogany in males and red in females. What is she trying to tell you—that is, what sex is the calf? What are the possible phenotypes of the calf's father?

12. In one strain of mice, skin color is determined by five different pairs of alleles. The colors range from almost white to dark brown. Would it be possible for some pairs of mice to produce offspring darker or lighter than either parent? Explain.

13. The size of an egg laid by one variety of hens is determined by three pairs of alleles; hens with the genotype *AABBCC* lay eggs weighing 90 grams, and hens with the genotype *aabbcc* lay eggs weighing 30 grams. Each of the alleles *A*, *B*, or *C* adds 10 grams to the weight of the egg. When a hen from the 90-gram strain is mated with a rooster from the 30-gram strain, the hens of the F_1 generation lay eggs weighing 60 grams. If a hen and rooster from this F_1 generation are mated, what will be the weight of the eggs laid by hens of the F_2?

14. Height and weight in animals follow a distribution similar to that shown in Figure 13–13. By inbreeding large animals, breeders are usually able to produce some increase in size among their stock. But after a few generations, increase in size characteristically stops. Why?

15. In Jimson weed, the allele that produces violet petals is dominant over that for white petals, and the allele that produces prickly capsules is dominant over that for smooth capsules. A plant with white petals and prickly capsules was crossed with one that had violet petals and smooth capsules. The F_1 generation was composed of 47 plants with white petals and prickly capsules, 45 plants with white petals and smooth capsules, 50 plants with violet petals and prickly capsules, and 46 plants with violet petals and smooth capsules. What were the genotypes of the parents?

16. A diploid organism has 42 chromosomes per cell. How many linkage groups does it have?

17. Segregation of alleles can occur at either of two stages of meiosis. Name the two stages and explain what happens in each of them.

18. Does crossing over necessarily result in a recombination of alleles? Explain.

19. In a series of breeding experiments, a linkage group composed of genes *A, B, C, D,* and *E* was found to show approximately the recombination frequencies in the chart below. Using Sturtevant's standard unit of measure, "map" the chromosome.

		Gene				
		A	*B*	*C*	*D*	*E*
	A	—	8	12	4	1
	B	8	—	4	12	9
Gene	*C*	12	4	—	16	13
	D	4	12	16	—	3
	E	1	9	13	3	—

Recombinations per 100 fertilized eggs

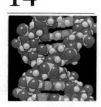

The Chemical Basis of Heredity: The Double Helix

By the early 1940s, the existence of genes and the fact that they were in chromosomes were no longer in doubt. But what were the genes? What did they really do? A turning point in genetics came when scientists focused on the question of how it was possible for these little lumps of matter—the chromosomes—to be the bearers of what they had come to realize must be an enormous amount of extremely complex information.

THE CHEMISTRY OF HEREDITY

The chromosomes, like all the other parts of a living cell, are composed of atoms arranged into molecules. As we noted previously, some scientists, a number of them eminent in the field of genetics, thought it would be impossible to understand the complexities of heredity in terms of the structure of "lifeless" chemicals. Others thought that if the chemical structure of the chromosomes was understood, we could then come to understand how chromosomes function as the bearers of the genetic information. This thinking marked the beginning of the vast range of investigations that we know as "molecular genetics."

The Language of Life

Early chemical analyses of the hereditary material revealed that the eukaryotic chromosome consists of both deoxyribonucleic acid (DNA) and protein, in about equal amounts. Thus, both were candidates for the role of the genetic material. Proteins seemed the more likely choice because of their greater chemical complexity. (As you will recall from Chapter 3, proteins are polymers of amino acids, of which there are 20 different types found in living cells; DNA, by contrast, is a polymer formed from only four different types of nucleotides.) Speculative thinkers in the field of biology were quick to point out that the amino acids, the number of which is so provocatively close to the number of letters in our own alphabet, could be arranged in a variety of different ways. The amino acids were seen as making up a sort of language—"the language of life"—that spelled out the directions for all the many activities of the cell. Many prominent investigators, particularly those who had been studying proteins, believed that the genes themselves were proteins. They thought that the chromosomes contained master models of all the proteins that would be required by the cell and that enzymes and other proteins active in cellular life were copied from these master models. This was a logical hypothesis, but, as it turned out, it was wrong.

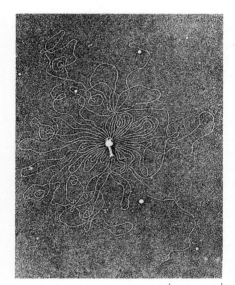

├─ 0.5 µm ─┤

14–1 *Molecular genetics is concerned with the chemical basis of heredity. The genetic information has been shown to be contained in a large, complex molecule known as deoxyribonucleic acid (DNA). This electron micrograph shows a bacteriophage, a virus that attacks bacterial cells. It is surrounded by its genetic material, a long, continuous molecule of DNA. (The molecule shown here, however, has been broken apart at one point—note the free ends at the top and bottom.) In the center of the micrograph is the outer coat of the virus, made of protein, from which the DNA has been released.*

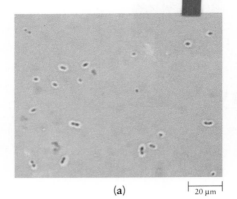

(a)   ⊢__20 μm__⊣

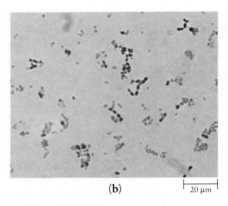

(b)   ⊢__20 μm__⊣

14–2 (**a**) *Encapsulated and* (**b**) *nonencapsulated forms of pneumococci. The capsule is made up of polysaccharides deposited outside the cell wall. The encapsulated form, which is resistant to phagocytosis by white blood cells, produces pneumonia; the mutant, nonencapsulated form is harmless.*

THE DNA TRAIL

Sugar-Coated Microbes and the Transforming Factor

To trace the beginning of the other hypothesis—the one that ultimately proved correct—it is necessary to go back to 1928 and pick up an important thread in modern biological history. In that year, an experiment was performed that seemed at the time to have little relevance to the field of genetics. Frederick Griffith, a public health bacteriologist in England, was studying the possibility of developing vaccines against *Streptococcus pneumoniae,* a type of bacterium that causes one form of pneumonia. In those days, before the development of antibiotics, bacterial pneumonia was a serious disease, the grim "captain of the men of death."

As Griffith knew, these bacteria, commonly called pneumococci, come in either virulent (disease-causing) forms with polysaccharide capsules or nonvirulent (harmless) forms without capsules (Figure 14–2). The production of the capsule and its composition are both genetically determined—that is, they are inherited properties of the bacteria. (It is now known that the nonencapsulated pneumococcus is a mutant form; in Griffith's time, however, the term mutant was not applied to bacteria.) Griffith was interested in finding out whether injections of heat-killed virulent pneumococci, which do not cause disease, could be used to immunize against pneumonia. In the course of various experiments, he performed one that gave him very puzzling results. He injected mice simultaneously with heat-killed virulent bacteria and with living nonvirulent bacteria, each of which was harmless—but all the mice died. When Griffith performed autopsies on them, he found their bodies filled with living encapsulated (and therefore virulent) bacteria (Figure 14–3). Had the dead virulent cells come back to life, or had something been passed from them to the living, nonvirulent cells that, in turn, endowed the living cells with the capacity to make capsules and therefore to be virulent?

Within the next few years, it was shown that the same phenomenon could be reproduced in the test tube and these questions could be answered. It was found that extracts from the killed encapsulated bacteria, when added to the living harmless bacteria, could convert them to the virulent type with the capacity to make capsules. Furthermore, once converted, they could transmit this characteristic to their progeny. This phenomenon was known as **transformation,** and the "something" in the extract that caused the conversion was called the **transforming factor.**

In 1943, after almost a decade of patient chemical isolation and analysis, O. T. Avery and his coworkers at Rockefeller University demonstrated that the transforming factor was DNA. Subsequent experiments showed that a variety of genetic factors could be passed from bacterial cells of one strain to bacterial cells of another, similar strain by means of isolated DNA.

The Nature of DNA

DNA had first been isolated by a German physician named Friedrich Miescher in 1869—in the same remarkable decade in which Darwin published *The Origin of Species* and Mendel presented his results to the Brünn Natural History Society. The substance Miescher isolated was white, sugary, slightly acidic, and contained phosphorus. Since he found it only in the nuclei of cells, he called it "nuclein." This name was later amended to nucleic acid, and then, still later, to deoxyribonucleic acid, to distinguish it from a similar chemical also found in cells, ribonucleic acid (RNA).

Almost 50 years later, in 1914, another German, Robert Feulgen, discovered that DNA had an unusually strong attraction for a red dye called fuchsin.

14–3 *Discovery of the transforming factor, a substance that can transmit genetic characteristics from one cell to another, resulted from studies of pneumococci, which are pneumonia-causing bacteria. One strain of these bacteria has polysaccharide capsules (protective outer layers); another does not. The capacity to make capsules and cause disease is an inherited characteristic, passed from one bacterial generation to another as the cells divide. (a) Injection into mice of encapsulated pneumococci killed the mice. (b) The nonencapsulated strain produced no infection. (c) If the encapsulated strain was heat-killed before injection, it too produced no infection. (d) If, however, heat-killed encapsulated bacteria were mixed with live nonencapsulated bacteria and the mixture was injected into mice, the mice died. (e) Blood samples from the dead mice revealed live encapsulated pneumococci. Something had been transferred from the dead bacteria to the live ones that endowed them with the capacity to make polysaccharide capsules and cause pneumonia. This "something" was later isolated and found to be DNA.*

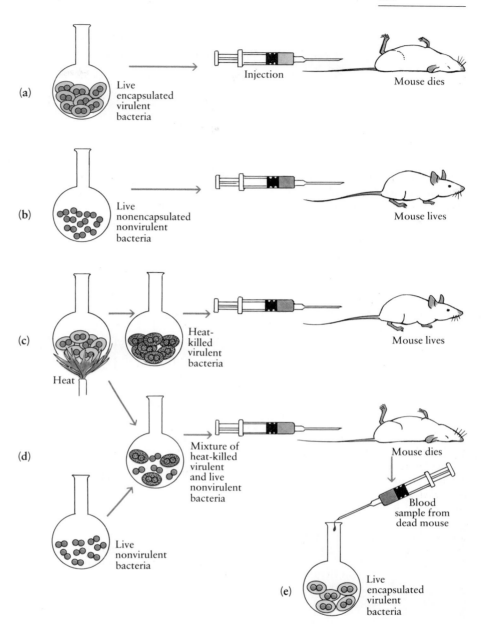

However, he considered this finding so unimportant that he did not trouble to report it for a decade. Feulgen staining, as it was called when it finally made its way into use, revealed that DNA was present in all cells and was characteristically located in the chromosomes.

For the next few decades, however, there was no particular interest in DNA since no role had been postulated for it in cellular metabolism. During the 1920s, most of the work on its chemistry was carried out in a single laboratory by the eminent biochemist P. A. Levene. He showed that DNA could be broken down into a five-carbon sugar, a phosphate group, and four nitrogenous bases—adenine and guanine (the purines) and thymine and cytosine (the pyrimidines). From the proportions of these components he made two deductions, one correct and one incorrect:

1. Each nitrogenous base is attached to a molecule of sugar, which is, in turn, attached to a phosphate group to form a single molecule, a nucleotide (Figure 14–4). This deduction was correct.

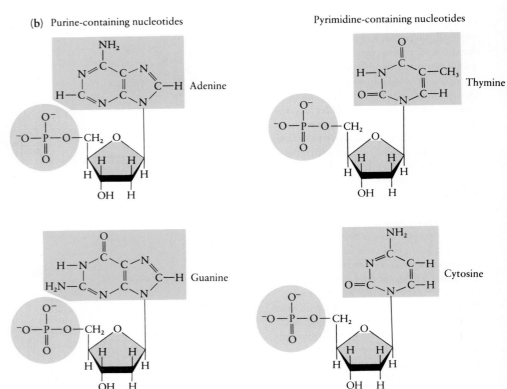

(a)

(b) Purine-containing nucleotides   Pyrimidine-containing nucleotides

Adenine

Thymine

Guanine

Cytosine

14–4 (a) *A nucleotide is made up of three different components: a nitrogenous base, a five-carbon sugar, and a phosphate group.* (b) *The four types of nucleotides found in DNA. Each nucleotide consists of one of four possible nitrogenous bases, a deoxyribose sugar, and a phosphate group.*

14–5 *Max Delbrück and Salvador Luria at the Cold Spring Harbor Laboratory of Quantitative Biology in 1953. They shared the Nobel Prize with A. D. Hershey in 1969 for "their discoveries concerning the replication mechanism and the genetic structure of viruses."*

2. Since, in all the samples he measured, the proportions of the nitrogenous bases were approximately equal, Levene concluded that all four nitrogenous bases must be present in nucleic acid in equal quantity. Furthermore, he hypothesized that these molecules must be grouped in clusters of four—a tetranucleotide, he called it—that repeated over and over again along the length of the molecule. Although this deduction was incorrect, it dominated scientific thinking about the nature of DNA for more than a decade.

Because Levene's "tetranucleotide theory" was given great weight by his renown as a biochemist, biologists were generally slow to recognize the importance of Avery's demonstration that the transforming factor in bacteria is DNA. This was partly because bacteria, which are, of course, prokaryotes, were considered "lower" and "different" and partly because the DNA molecule—made up of only four components—seemed too simple for the enormously complex task of carrying the hereditary information. Avery, like Mendel before him, was a traveler bearing an odd tale that did not fit.

The Bacteriophage Experiments

In 1940, Max Delbrück and Salvador Luria, both of whom had left Europe in the mass intellectual exodus of the 1930s, initiated a series of studies with another "fit material," destined to become as important to genetic research as the garden pea and the fruit fly. The fit material was a group of viruses that attack bacterial cells and are therefore known as **bacteriophages** ("bacteria eaters"), or phage for short. Every known type of bacterial cell is preyed upon by its own type of bacterial virus, and many bacteria are host to many different kinds of viruses. Delbrück, Luria, and the group that joined them in these studies agreed to concentrate on a series of seven related viruses that attack *Escherichia coli*, the familiar bacterium

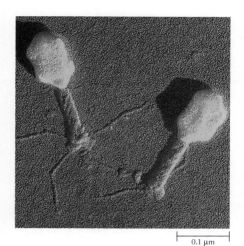

⊢———⊣ 0.1 μm

14-6 *T4 bacteriophages, as revealed by a transmission electron micrograph of a replica prepared by "shadowing" (page 98). Notice their highly distinctive "tadpole" shape. Each bacteriophage consists of a head, which appears hexagonal in electron micrographs, and a complex tail assembly. These bacteriophages attach to* E. coli *cells by means of the thin fibers extending from the tail assembly.*

that inhabits the healthy human intestine. These viruses were numbered T1 through T7, with the T standing simply for "type." As it turned out, most of the early work was done on T2 and T4, which became known as the T-even bacteriophages.

These viruses were inexpensive to work with, easy to maintain in the laboratory, and demanded little space or equipment. Furthermore, they were phenomenal at reproducing themselves. Twenty-five minutes from the time a single virus infected a bacterial cell, that cell would burst open, releasing a hundred or more new viruses, all exact copies of the original virus. Another advantage (which was not discovered until after the research was begun) was that this group of bacteriophages has a highly distinctive shape (Figure 14-6), and so can be readily identified with the electron microscope.

According to electron-microscope studies of infected *E. coli* cells (broken open at regular intervals after infection), the bacteriophages do not multiply like bacteria. Except for a few fragments, they disappear the moment after infection, and for the first 10 to 11 minutes of the infection cycle, not a single virus can be seen within the bacterial cell. Then, depending on when the cell is opened during the course of the infection, increasing numbers of completed bacteriophages can be seen and, mixed with them, odds and ends that appear to be bits of incomplete bacteriophages.

Chemical analysis of the bacteriophages revealed that they consist quite simply of DNA and of protein, the two leading contenders in the 1940s for the role of the genetic material. The chemical simplicity of the bacteriophage offered geneticists a remarkable opportunity. The viral genes—the hereditary material that directed the synthesis of new viruses within the bacterial cell—had to be carried either on the protein or on the DNA. If it could be determined which of the two it was, then the chemical identity of the gene would be known.

In 1952, a set of simple but ingenious experiments were carried out by Alfred D. Hershey and his colleague, Martha Chase.* They prepared two separate samples of viruses, one in which the DNA was labeled with a radioactive isotope of phosphorus, ^{32}P, and the other in which the protein was labeled with a radioactive isotope of sulfur, ^{35}S. Each type of virus was produced by growing the *E. coli* host on a medium that contained the appropriate radioactive isotope. After a cycle of multiplication, the newly formed viruses all contained some of the radioactive isotope in place of the common nonradioactive isotope. If you recall the chemical structure of nucleic acids and proteins, you will note that DNA contains phosphorus but no sulfur, while the amino acid components of proteins contain no phosphorus, although two amino acids (methionine and cysteine) contain sulfur. Thus ^{32}P and ^{35}S can serve as specific radioactive labels that distinguish DNA from protein.

One culture of bacteria was infected with ^{32}P-labeled phage and another with ^{35}S-labeled phage (Figure 14-7). After the infection cycle had begun, the cells were agitated in a blender and then spun down in a centrifuge to separate them from any viral material remaining outside the cells. The two samples—one containing extracellular material and the other intracellular material—were then tested for radioactivity. Hershey and Chase found that the ^{35}S had remained outside the bacterial cells with the empty viral coats and the ^{32}P had entered the cells, infected them, and caused the production of new virus progeny. It was therefore concluded that the genetic material of the virus is DNA rather than protein.

---

* We are including the names of the scientists involved in these experiments, not only to give credit where it is due, but also because the names have become synonymous with the work. What we are describing now are the Hershey-Chase experiments.

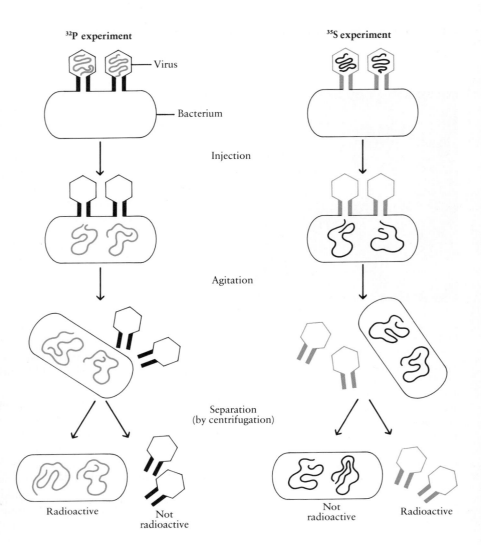

14-7 *A summary of the Hershey-Chase experiments demonstrating that DNA is the hereditary material of a virus. Radioactively labeled molecules are shown in color.*

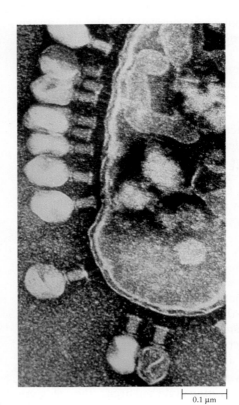

14-8 *Electron micrograph of T4 bacteriophages attacking a cell of* E. coli. *The viruses attach to the bacterial cell by their tail fibers; viral DNA, contained within the head of the virus, is injected through the tail and into the cell. As you can see, the heads of some of the viruses are empty, indicating that the injection process has already occurred. A complete cycle of virus infection takes only about 25 minutes. At the end of that period about a hundred new virus particles are released from the cell.*

Electron micrographs have now confirmed that the T4 bacteriophage attaches to the bacterial cell wall by its tail fibers. They also indicate that the phage injects its DNA into the cell, leaving the empty protein coat on the outside (Figure 14–8). In short, the protein is just a container for the bacteriophage DNA. It is the DNA of the bacteriophage that enters the cell and carries the complete hereditary message of the virus particle, directing the formation of new viral DNA and new viral protein.

Further Evidence for DNA

The role of DNA in transformation and in viral replication formed very convincing evidence that DNA is the genetic material. Two other lines of experimental work also helped to lend weight to the argument. First, Alfred Mirsky, in a long series of careful studies conducted at Rockefeller University, showed that, in general, the somatic cells of any given species contain equal amounts of DNA and the gametes contain just half as much DNA as is found in the somatic cells. This is consistent with the observed result of meiosis, in which the diploid chromosome number is reduced to the haploid number.

Chargaff's Results

A second important series of contributions was made by Erwin Chargaff of Columbia University's College of Physicians and Surgeons. Chargaff analyzed the purine and pyrimidine content of the DNA of many different kinds of living things and found that, in contradiction to Levene's conclusions (page 283), the nitrogenous bases do *not* always occur in equal proportions. The proportions of the four nitrogenous bases are the same in all cells of all individuals of a given species, but they vary from one species to another. Therefore variations in base composition could very well provide a "language" in which the instructions controlling cell growth could be written. Some of Chargaff's results are reproduced in Table 14–1. Can you, by examining these figures, notice anything interesting about the proportions of purines and pyrimidines?

TABLE 14–1   **Percentage Composition of DNA in Several Species**

SOURCE	PURINES		PYRIMIDINES	
	ADENINE	GUANINE	CYTOSINE	THYMINE
Human being	30.4%	19.6%	19.9%	30.1%
Ox	29.0	21.2	21.2	28.7
Salmon sperm	29.7	20.8	20.4	29.1
Wheat germ	28.1	21.8	22.7	27.4
E. coli	24.7	26.0	25.7	23.6
Sea urchin	32.8	17.7	17.3	32.1

The Hypothesis Is Confirmed

Taken together, all of the studies we have traced thus far provided convincing evidence that DNA is the genetic material. Nonetheless, a critical question remained unanswered: *How* is the genetic information contained in the DNA? The answer to this question was to be found in the structure of the DNA molecule itself.

To fulfill its biological role, the genetic material has to meet at least four requirements:

1. It must carry the genetic information from parent cell to daughter cell and from generation to generation. Further, it must carry a great deal of information. Consider how many instructions must be contained in the set of genes that directs, for example, the development of an elephant, or a tree, or even a *Paramecium*.

2. It must contain information for producing a copy of itself, for it is copied with every cell division and with great precision.

3. It must be chemically stable; otherwise, it would not carry identical information from generation to generation and offspring would not resemble their parents.

4. On the other hand, it must be capable of mutation. When a gene changes, that is, when a "mistake" is made, the "mistake" must be copied as faithfully as was the original. This is a most important property, for without the capacity to replicate "errors," there would be no genetic variation. As a consequence, there would also be no evolution by natural selection.

It was when the DNA molecule was found to have the size, the configuration, and the complexity necessary to meet these requirements that DNA became universally accepted as the genetic material. The scientists primarily responsible for working out the structure of DNA were James Watson and Francis Crick, and their feat is one of the milestones in the history of science.

THE WATSON-CRICK MODEL

In the early 1950s, a young American scientist, James Watson, went to Cambridge, England, on a research fellowship to study problems of molecular structure. There, at the Cavendish Laboratory, he met physicist Francis Crick. Both were interested in DNA, and they soon began to work together to solve the problem of its molecular structure. They did not do experiments in the usual sense but rather undertook to examine all the data about DNA and to unify them into a meaningful whole.

The Known Data

By the time Watson and Crick began their studies, quite a lot of information on the subject had already accumulated:

1. The DNA molecule was known to be very large, and also very long and thin, and to be composed of nucleotides containing the nitrogenous bases adenine, guanine, thymine, and cytosine.

2. According to Levene's interpretation of his data, these nucleotides were assembled in repeating units of four.

3. Linus Pauling, in 1950, had shown that a protein's component chains of amino acids are often arranged in the shape of a helix and are held in that form by hydrogen bonds between successive turns of the helix (see page 75). Pauling had suggested that the structure of DNA might be similar.

4. X-ray diffraction photographs of DNA (Figure 14–9) from the laboratories of Maurice Wilkins and Rosalind Franklin at King's College, London, showed patterns that almost certainly reflected the turns of a giant helix.

5. Also crucial were the data of Chargaff indicating, as you perhaps noticed in Table 14–1, that (within experimental error) the amount of adenine is the same as the amount of thymine, and the amount of guanine is the same as the amount of cytosine: $A = T$ and $G = C$.

Building the Model

From these data, some of them contradictory, Watson and Crick attempted to construct a model of DNA that would fit the known facts and explain the biological role of DNA. In order to carry the vast amount of genetic information, the molecules should be heterogeneous and varied. Also, there must be some way for them to replicate readily and with great precision so that faithful copies could be passed from cell to cell and from parent to offspring, generation after generation.

On the other hand, Watson and Crick could not be sure that the chemical structure of DNA would actually reveal its biological function. After all, this idea had never really been tested rigorously. "In pessimistic moods," Watson has recalled, "we often worried that the correct structure might be dull—that is, that it would suggest absolutely nothing."

It turned out, in fact, to be unbelievably "interesting." By piecing together the various data, they were able to deduce that DNA is an exceedingly long, entwined double helix.

If you were to take a ladder and twist it into a helix, keeping the rungs perpendicular, you would have a crude model of the DNA molecule (Figure 14–10). The two rails, or sides, of the ladder are made up of alternating sugar and phosphate molecules. The perpendicular rungs of the ladder are formed by the nitrogenous bases—adenine (A), thymine (T), guanine (G), and cytosine (C). Two bases form each rung, with each base covalently bonded to a sugar-phosphate

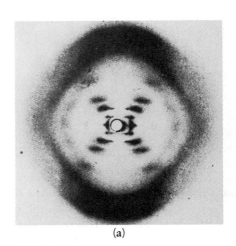

(a)

(b)

14–9 (a) *The critical x-ray diffraction photograph of DNA, taken by Rosalind Franklin. The reflections crossing in the middle indicate that the molecule is a helix. The heavy dark regions at the top and bottom are due to the closely stacked bases perpendicular to the axis of the helix.* (b) *Rosalind Franklin, photographed while vacationing in France in 1950 or 1951. She died of cancer in 1958, at the age of 37.*

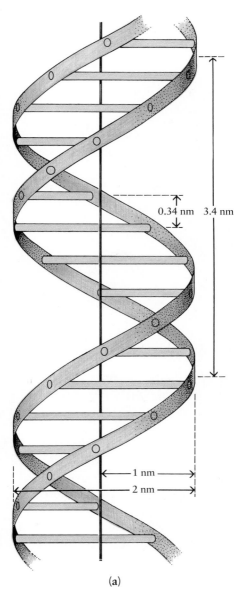

(a)

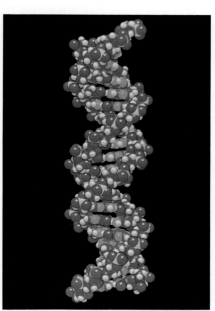

(b)

14–10 (a) *The double-stranded helical structure of DNA, as first presented in 1953 by Watson and Crick. The framework of the helix is composed of the sugar-phosphate units of the nucleotides. The rungs are formed by the four nitrogenous bases adenine and guanine (the purines) and thymine and cytosine (the pyrimidines). Each rung consists of two bases. Knowledge of the distances between the atoms was crucial in establishing the structure of the DNA molecule. The dis-tances were determined from x-ray diffraction photographs of DNA taken by Rosalind Franklin and Maurice Wilkins.* **(b)** *A computer-generated space-filling model of DNA. The paired bases form the rungs connecting the sugar-phosphate side rails. In this model, the atoms are color-coded as follows: white = hydrogen (H), red = oxygen (O), yellow = phosphorus (P), dark blue = carbon (C), and turquoise = nitrogen (N).*

unit. The paired bases meet across the helix and are joined together by hydrogen bonds, the relatively weak bonds that Pauling had demonstrated in his studies of protein structure.

The distance between the two sides, or railings, according to x-ray measurements, is 2 nanometers. Two purines in combination would take up more than 2 nanometers, and two pyrimidines would not reach all the way across. But if a purine paired in each case with a pyrimidine, there would be a perfect fit and the molecule would be the same width along its entire length. The paired bases—the "rungs" of the ladder—would therefore always be purine-pyrimidine combinations.

As Watson and Crick analyzed the data, they assembled actual tin-and-wire models of the molecules (see essay, page 291), testing where each piece would fit into the three-dimensional puzzle. As they worked with the models, they realized that the nucleotides along any one strand of the double helix could be assembled in any order: for example, TTCAGTACATTGCCA, and so on (Figure 14–11a). Since a DNA molecule may be thousands of nucleotides long, there is a possibility for great variety in the sequence of bases, and variety is one of the primary requirements for the genetic material. Note also that the strand has direction: each phosphate group is attached to one sugar at the 5′ position (the fifth carbon in the sugar ring) and to the other sugar at the 3′ position (the third carbon in the sugar ring). Thus, the strand has a 5′ end and a 3′ end.

The most exciting discovery came, however, when Watson and Crick set out to construct the matching strand. They encountered another interesting and important restriction. Not only could purines not pair with purines and pyrimidines not pair with pyrimidines, but because of the structures of the bases, adenine could pair only with thymine, forming two hydrogen bonds, (A=T), and guanine only with cytosine, forming three hydrogen bonds (G≡C). The paired bases were **complementary.** Look at Table 14–1 again and see how well these chemical requirements explain Chargaff's data.

The double-stranded structure of a DNA molecule is shown in Figure 14–11b. As you can see, the two strands run in opposite directions; that is, the direction from the 5′ end to the 3′ end of each strand is opposite. The strands are said to be **antiparallel.** Although the nucleotides along one chain of the double helix can occur in any order, their sequence then determines the order of nucleotides in the other chain. This is necessarily the case, since the bases are complementary (G with C and A with T). Thus, for example, the complementary strand of (5′)-TTCAGTACATTGCCA-(3′) must have the nucleotide sequence (3′)-AAGTCATGTAACGGT-(5′).

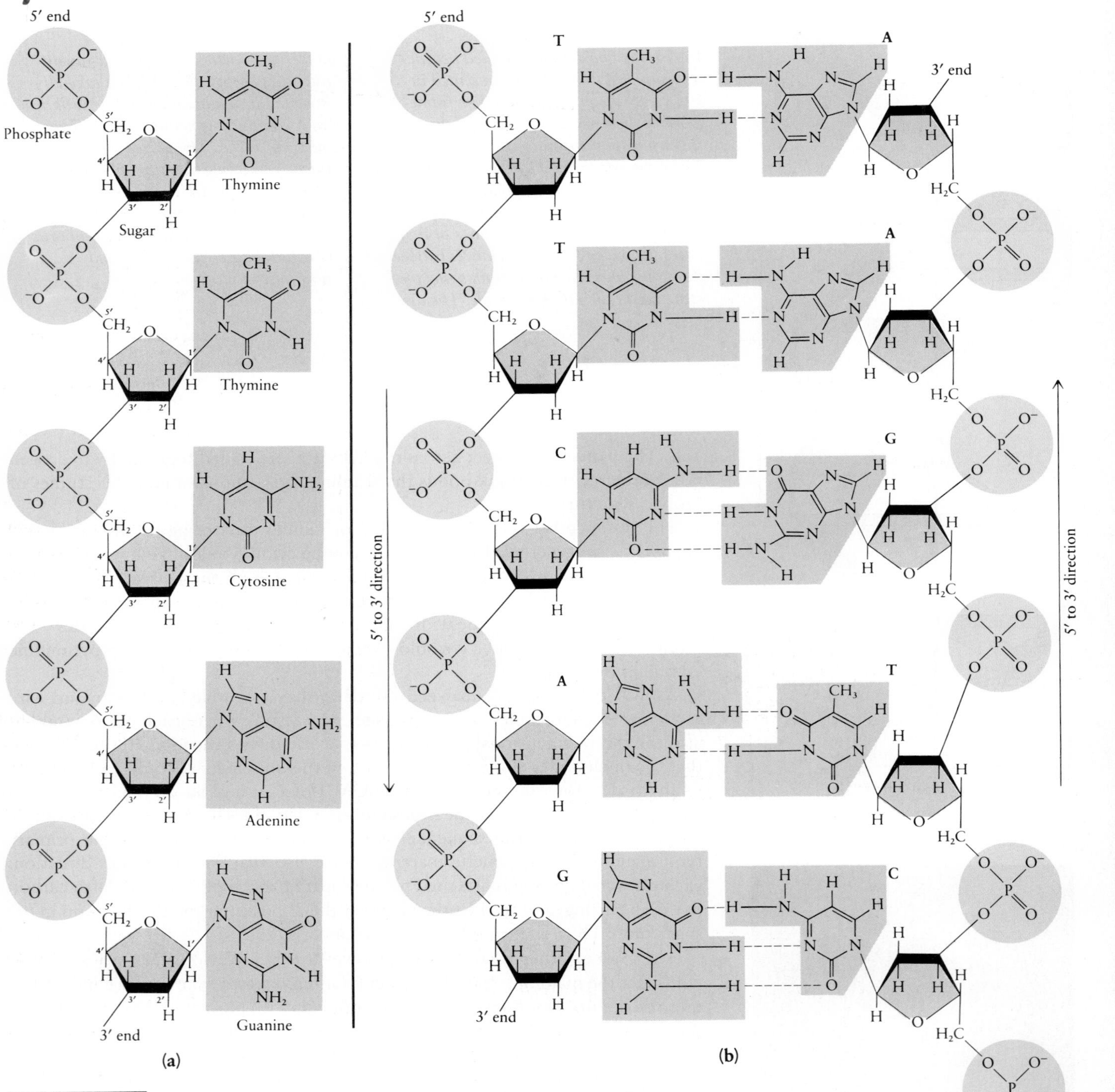

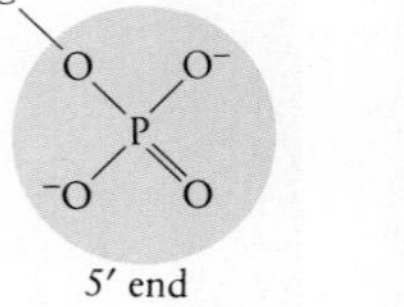

(a)

(b)

14–11 **(a)** *The structure of a portion of one strand of a DNA molecule. Each nucleotide consists of a deoxyribose sugar, a phosphate group, and a purine or pyrimidine base. Note the repetitive sugar-phosphate-sugar-phosphate sequence that forms the backbone of the molecule. Each phosphate group is attached to the 5' carbon of one sugar subunit and to the 3' carbon of the sugar subunit in the adjacent nucleotide. The DNA strand thus has a 5' end and a 3' end, determined by these 5'*

and 3' carbons. The sequence of bases varies from one DNA molecule to another. Here, the order of the nucleotides, which is customarily written in the 5' to 3' direction, is TTCAG. **(b)** *The double-stranded structure of a portion of a DNA molecule. The strands are held together by hydrogen bonds (represented here by dashes) between the bases. Notice that adenine and thymine can form two hydrogen bonds, whereas guanine and cytosine can form three. Because of these bonding require-*

ments, adenine can pair only with thymine and guanine only with cytosine. Thus the order of bases along one strand determines the order of bases along the other strand. The strands are antiparallel; that is, the direction from the 5' to the 3' end of one strand is opposite to that of the other.

Who Might Have Discovered It?

Then there is the question, what would have happened if Watson and I had not put forward the DNA structure? This is "iffy" history which I am told is not in good repute with historians, though if a historian cannot give plausible answers to such questions I do not see what historical analysis is about. If Watson had been killed by a tennis ball I am reasonably sure I would not have solved the structure alone, but who would? Olby has recently addressed himself to this question. Watson and I always thought that Linus Pauling would be bound to have another shot at the structure once he had seen the King's College x-ray data, but he has recently stated that even though he immediately liked our structure it took him a little time to decide finally that his own was wrong. Without our model he might never have done so. Rosalind Franklin was only two steps away from the solution.

Watson (left) and Crick in 1953 with one of their models of DNA. "DNA, you know, is Midas' gold," said Maurice Wilkins, with whom they shared the Nobel Prize. "Everyone who touches it goes mad."

She needed to realise that the two chains must run in opposite directions and that the bases, in their correct tautomeric forms, were paired together. She was, however, on the point of leaving King's College and DNA, to work instead on TMV [tobacco mosaic virus] with Bernal. Maurice Wilkins had announced to us, just before he knew of our structure, that he was going to work full time on the problem. Our persistent propaganda for model building had also had its effect (we had previously lent them our jigs to build models but they had not used them) and he proposed to give it a try. I doubt myself whether the discovery of the structure could have been delayed for more than two or three years.

There is a more general argument, however, recently proposed by Gunther Stent and supported by such a sophisticated thinker as Medawar. This is that if Watson and I had not discovered the structure, instead of being revealed with a flourish it would have trickled out and that its impact would have been far less. For this sort of reason Stent had argued that a scientific discovery is more akin to a work of art than is generally admitted. Style, he argues, is as important as content.

I am not completely convinced by this argument, at least in this case. Rather than believe that Watson and Crick made the DNA structure, I would rather stress that the structure made Watson and Crick. After all, I was almost totally unknown at the time and Watson was regarded, in most circles, as too bright to be really sound. But what I think is overlooked in such arguments is the intrinsic beauty of the DNA double helix. It is the molecule which has style, quite as much as scientists. The genetic code was not revealed all in one go but it did not lack for impact once it had been pieced together. I doubt if it made all that difference that it was Columbus who discovered America. What mattered much more was that people and money were available to exploit the discovery when it was made. It is this aspect of the history of the DNA structure which I think demands attention, rather than the personal elements in the act of discovery, however interesting they may be as an object lesson (good or bad) to other workers.

FRANCIS CRICK: "The Double Helix: A Personal View," *Nature*, vol. 248, pages 766–769, 1974.

DNA REPLICATION

An essential property of the genetic material is the ability to provide for exact copies of itself. Does the Watson-Crick model satisfy this requirement? In their published account, Watson and Crick wrote, "It has not escaped our notice that the specific pairing we have postulated immediately suggests a possible copying mechanism for the genetic material." Implicit in the double and complementary structure of the DNA helix is a method by which it can reproduce itself. At the time of chromosome replication, the molecule "unzips" down the middle, the paired bases separating at the hydrogen bonds. As the two strands separate, they act as **templates,** or guides; each directs the synthesis of a new complementary strand along its length, using the raw materials in the cell (Figure 14–12). This mechanism of DNA replication, hypothesized by Watson and Crick on the basis of their model of DNA structure, is called **semiconservative replication,** since half the molecule is conserved. Each old strand forms a template for the production of a new one. If a T is present on the old strand, only an A can fit into place in the new strand; a G will pair only with a C, and so on. In this way, each strand forms a copy of the original partner strand, and two exact replicas of the molecule are produced. The age-old question of how hereditary information is duplicated and passed on, generation after generation, had apparently been answered.

A Confirmation of Semiconservative Replication

The Watson-Crick model of DNA replication, however, was not the only possible mechanism. Matthew Meselson and Franklin W. Stahl, working at the California Institute of Technology, devised an elegant experiment to choose among three possible models (Figure 14–13).

In designing their experiment, they took advantage of the availability of a heavy isotope of nitrogen (^{15}N) and an extremely sensitive method of separating macro-molecules on the basis of density. The method, which had been devised by Meselson while he was a graduate student, involves placing a solution of cesium chloride (CsCl) in a tube and spinning it in an ultracentrifuge. The small, dense CsCl molecules form a continuous density gradient, less concentrated at the top of the tube and more concentrated at the bottom. When DNA molecules are centrifuged in this solution, they will form a band at the point in the gradient at which the DNA and the CsCl solution have equal densities (Figure 14–14a). CsCl was selected for this procedure because the range of densities in the gradient it forms includes that of DNA.

Meselson and Stahl grew *E. coli* for several generations in a medium in which the sole nitrogen source contained ^{15}N, the heavy isotope of nitrogen. At the end of this period, the DNA of the bacterial cells contained a large proportion of heavy nitrogen. Although the density of this DNA was only about 1 percent greater than that of normal DNA, it formed a separate and distinct band in the cesium chloride gradient (Figure 14–14b and c).

They then placed a sample of cells containing heavy nitrogen in a medium containing ^{14}N; the cells were left in this medium only long enough for the DNA to replicate once (as determined by a doubling of the number of cells). A sample of DNA from these cells was spun in the ultracentrifuge (Figure 14–14d). Then, a second sample of cells was grown in the ^{14}N medium for two generations. Their DNA was also ultracentrifuged (Figure 14–14e).

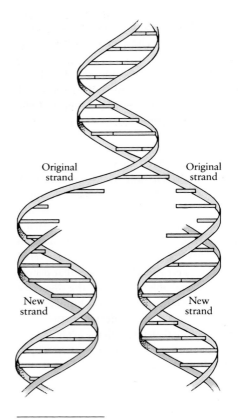

Original strand   Original strand

New strand   New strand

14–12 *Replication of the DNA molecule, as predicted by the Watson-Crick model. The strands separate down the middle as the paired bases separate at the hydrogen bonds. Each of the original strands then serves as a template along which a new,* complementary strand forms from nucleotides available in the cell. Subsequent research has resulted in modification of some of the details of this process, as we shall see shortly, but the underlying principle is unchanged.

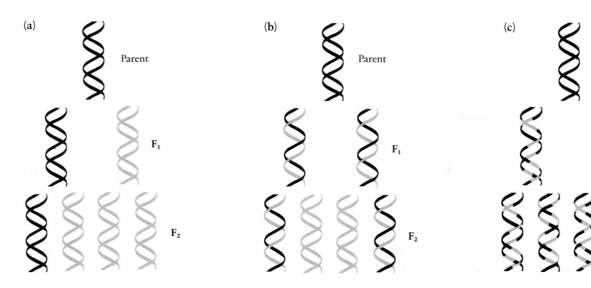

(a)   (b)   (c)

Parent   Parent   Parent

F_1 F_1 F_1

F_2 F_2 F_2

14–13 *Three possible mechanisms of DNA replication. In this diagram the original strands are shown in black and the newly replicated strands are shown in color. (a) Conservative replication. Each of the two strands of parent DNA is replicated, without strand separation. In the first generation, one daughter is all old DNA and one daughter is all new. The second generation contains one helix com-*

posed of two old strands and three made up entirely of new strands. (b) Semiconservative replication. The two parental strands separate, and each forms a template for a new strand. In the first generation, each daughter is half old and half new. The second generation comprises two hybrid DNAs (half old, half new) and two DNAs made up entirely of new strands.

(c) Dispersive replication. During replication, parent chains break at intervals, and replicated segments are combined into strands with segments from parent chains. All daughter helixes are part old, part new. The Meselson-Stahl experiment (Figure 14–14) was undertaken to determine which of these three possibilities was correct. Watson and Crick had predicted (b).

Each sample of DNA contained more light DNA, as would be expected, because newly formed DNA had to incorporate the available ^{14}N. Moreover—and this was of crucial importance—the density of the first generation DNA was exactly halfway between that of heavy parent DNA and that of ordinary light DNA, as it should be if each molecule contained one old (heavy) strand and one new (light) strand. The second generation contained one-half half-heavy DNA and one-half light DNA, which again, exactly and ingeniously, confirmed the Watson-Crick hypothesis of semiconservative replication (Figure 14–13b).

14–14 *The Meselson-Stahl experiment. (a) Normal light DNA forms a precisely located band when ultracentrifuged in a density gradient of cesium chloride. (b) E. coli cells cultured in a medium containing heavy nitrogen (^{15}N) accumulate a heavy DNA, which forms a separate and distinct band. (c) When a mixture of heavy and light DNA is centrifuged in a cesium chloride density gradient, the two types of DNA separate into two distinct bands. (d) When cells grown in a medium containing heavy nitrogen (^{15}N) are permitted to multiply for one generation in a medium containing ordinary light nitrogen (^{14}N), their DNA forms a band in the cesium chloride density gradient that is located midway between the bands of the heavy and light DNAs. (e) When cells containing heavy DNA are grown for two generations in the ^{14}N medium, their DNA forms two bands in the density gradient—one band of light DNA and one band of half-heavy DNA. The column on the right shows the investigators' interpretations. As you can see, this experiment confirmed the Watson-Crick hypothesis of semiconservative replication.*

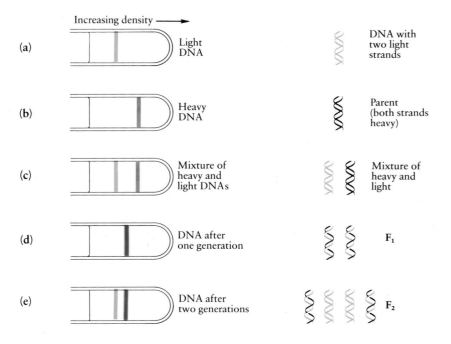

Increasing density ⟶

(a)   Light DNA   DNA with two light strands

(b)   Heavy DNA   Parent (both strands heavy)

(c)   Mixture of heavy and light DNAs   Mixture of heavy and light

(d)   DNA after one generation F_1

(e)   DNA after two generations F_2

293

The Mechanics of DNA Replication

Replication of DNA is a process that occurs only once in each cell generation, during the S phase of the cell cycle (page 146); it is the essential event in the replication of the chromosomes. In most eukaryotic cells, DNA replication leads ultimately to mitosis, but in the primary spermatocytes and oocytes it leads instead to meiosis. It is a remarkably rapid process; for example, in humans and other mammals, the rate of synthesis is about 50 nucleotides per second. In prokaryotes, it is even faster—about 500 nucleotides per second.

The principle of semiconservative replication, in which each strand of the DNA double helix serves as a template for the formation of a new strand, is relatively simple and easy to understand. However, the actual process by which the cell accomplishes replication is considerably more complex. Like other biochemical reactions of the cell, DNA replication requires a number of different enzymes, each catalyzing a particular step of the process. The identification of the principal enzymes, their precise functions, and the sequence of events in replication has required a number of years and the efforts of many scientists working in different laboratories. Although our understanding is still incomplete, the general outlines of the process are now clear.

Initiation of DNA replication always begins at a specific nucleotide sequence known as the **origin of replication.** It requires special initiator proteins and, in addition, enzymes known as **helicases.** These enzymes break the hydrogen bonds linking the complementary bases at the replication origin, opening up the helix so replication can occur. However, as the strands of the helix separate, the adjacent portions of the double helix are in danger of becoming more and more tightly coiled, or supercoiled. Other enzymes, the **topoisomerases,** break and reconnect one or both strands of the helix, allowing swiveling to occur and thus relieving strain on adjacent portions of the molecule.

Once the two strands of the DNA double helix are separated, additional proteins, known as single-strand binding proteins, attach to the individual strands, holding them apart and preventing kinking. This makes possible the next stage, the actual synthesis of the new strands, catalyzed by a group of enzymes known as **DNA polymerases.**

14–15 *An overview of DNA replication. The two strands of the DNA molecule (a) separate at the origin of replication as a result of the action of initiator proteins and enzymes known as helicases. (b, c) The two replication forks move away from the origin of replication in opposite directions, forming a replication bubble that expands bidirectionally. (d) When synthesis of the new DNA strands is complete, the two double-stranded chains separate into two new double helixes. The DNA has been replicated semiconservatively; that is, each new double helix consists of one old strand and one new strand.*

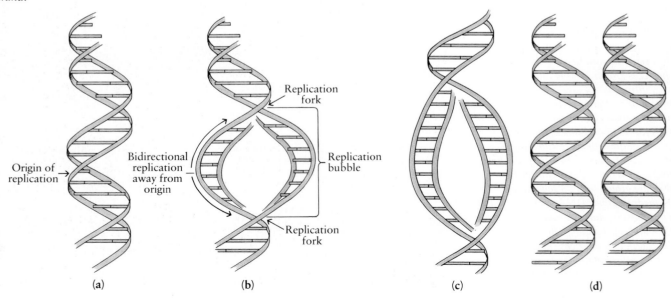

(a)   (b)   (c)   (d)

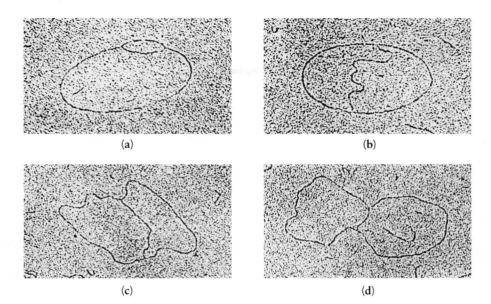

(a)　　　　　　　　　(b)

(c)　　　　　　　　　(d)

14–16 *Replication of the circular* E. coli *chromosome, as revealed by the electron microscope. The process begins in a specific sequence of nucleotides, the origin of replication, and proceeds in opposite directions from that point.* **(a)** *In the early stages of bidirectional replication, the enlarging replication bubble resembles an eye.* **(b)** *Subsequently, the replicating chromosome takes on the shape of the Greek letter theta.* **(c)** *Bidirectional replication continues until the entire chromosome has been replicated* **(d)**. *The two newly formed chromosomes then separate from each other (see Figure 7–2, page 145).*

If replicating DNA is viewed under the electron microscope, the region of synthesis appears as an "eye," or replication bubble. At either end of the bubble, where the old strands are being separated by helicase and the complementary strands are being synthesized, the molecule appears to form a Y-shaped structure. This is known as a **replication fork**. Replication is bidirectional, with two replication forks moving in opposite directions away from the origin (Figure 14–15).

In prokaryotes, there is a single replication origin, located within a specific nucleotide sequence about 300 base pairs long. As the replicated strands begin to separate, a structure forms that resembles the Greek letter theta (θ), eventually giving rise to two circular DNAs (Figure 14–16). In eukaryotes, by contrast, there are many replication origins; replication proceeds along the linear chromosomes as each bubble expands bidirectionally until it meets an adjacent bubble (Figure 14–17). The entire length of the chromosome is replicated as these bubbles merge.

14–17 *Replication of eukaryotic chromosomes is initiated at multiple origins. The individual replication bubbles spread until ultimately they meet and join. In this electron micrograph of an embryonic cell of* Drosophila melanogaster, *the centers of the replication bubbles are indicated by the arrows.*

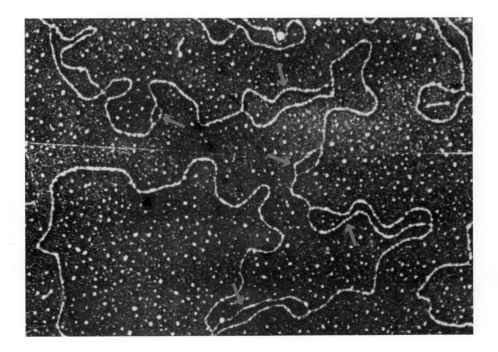

RNA Primers and the Direction of Synthesis

In order for synthesis of a new, complementary strand of DNA to occur, it is not enough that the old strand—serving as the template for the new strand—be present. There must also be the beginning of the new strand. This beginning is provided by a **primer,** formed of nucleotides of ribonucleic acid (RNA). As you will recall from Chapter 3 (page 80), RNA is a closely related nucleic acid in which the five-carbon sugar is ribose; in addition, it contains the nitrogenous base uracil instead of thymine. RNA nucleotides can form hydrogen bonds with the nucleotides of a DNA strand, following a similar principle of complementarity. Guanine pairs with cytosine, the adenine of RNA pairs with the thymine of DNA, and the uracil of RNA pairs with the adenine of DNA.

In eukaryotes, RNA primers are typically about 10 nucleotides long. Their synthesis on the exposed single strand of DNA is catalyzed by an enzyme known as **RNA primase.** With the RNA primers in place, DNA polymerases begin to synthesize new complementary DNA strands along the template strands, adding nucleotides one by one to the growing strands.

The first DNA polymerases discovered were found to synthesize new DNA strands only in the 5′ to 3′ direction; that is, incoming nucleotides were added only at the 3′ end of the chain. This posed a major problem. Because of the antiparallel structure of the DNA double helix, replication of the two new DNA strands on the two arms of the Y-shaped replication fork seemed to require not only synthesis in the 5′ to 3′ direction but also in the 3′ to 5′ direction. For a number of years, researchers sought unsuccessfully to identify another DNA polymerase that would function in the 3′ to 5′ direction. The cell's solution to this problem was eventually revealed by the Japanese scientist Reiji Okazaki, who found that, although the 5′ to 3′ strand is synthesized continuously as a single unit, the 3′ to 5′ strand is synthesized discontinuously, as a series of fragments, each synthesized in the 5′ to 3′ direction (Figure 14–18). The strand that is synthesized continuously is known as the **leading strand,** and the strand synthesized as a series of fragments is known as the **lagging strand.** The fragments that

14–18 *A close-up of the leading and lagging strands near the replication fork. The addition of nucleotides to both of these strands is catalyzed by the enzyme DNA polymerase, which works only in the 5′ to 3′ direction. In order for DNA polymerase to begin adding nucleotides, an RNA primer must be present, hydrogen bonded to the template strand; synthesis of the RNA primer is catalyzed by RNA primase. (The RNA primer for the leading strand is located at the origin of replication, not visible in this diagram.) The leading strand is synthesized continuously in the 5′ to 3′ direction. The lagging strand, by contrast, is synthesized discontinuously, in the form of Okazaki fragments. These fragments are synthesized in the 5′ to 3′ direction, which is, however, opposite to the overall direction of replication of this strand. When an Okazaki fragment has become long enough to encounter the RNA primer ahead of it, other enzymes replace the RNA nucleotides of the primer with DNA nucleotides. DNA ligase then connects the fragment with the adjacent newly synthesized fragment of the strand.*

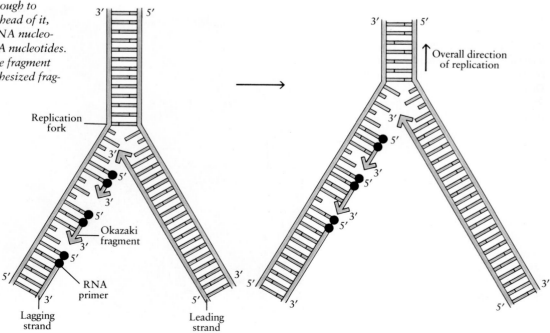

form the lagging strand, the **Okazaki fragments,** are typically 1,000 to 2,000 nucleotides long in prokaryotes and 100 to 200 nucleotides long in eukaryotes.

The role of the RNA primers, on both the leading and lagging strands, is to provide correctly paired strands of nucleotides with exposed 3′ OH groups to which the DNA polymerase can begin attaching DNA nucleotides in sequence. DNA polymerase, beginning at an RNA primer, moves along a DNA strand, adding nucleotides to the 3′ end of the growing strand, until it encounters the 5′ end of another strand of RNA primer. On the lagging strand, in which synthesis moves away from the replication fork, this primer represents the beginning of another Okazaki fragment. On the leading strand, in which synthesis moves in the same direction as the replication fork, this primer represents the beginning of an Okazaki fragment of the next replication bubble. When DNA polymerase makes contact with the 5′ end of an RNA primer, other enzymes are activated; these enzymes remove the RNA nucleotides and replace them with DNA nucleotides. Another enzyme, **DNA ligase,** then connects the newly synthesized DNA segment to the growing DNA strand by catalyzing the condensation reaction that bonds adjacent phosphate and sugar groups.

This complex process, which occurs prior to every cell division, is summarized in Figure 14–19.

Proofreading

One of the essential features of DNA replication is that DNA polymerase can add nucleotides to the 3′ end of a strand only if the nucleotides previously added to that strand are correctly paired with their complementary nucleotides on the template strand. As we have seen, the function of the RNA primer is to provide a correctly paired sequence of nucleotides at which DNA synthesis can begin. In the

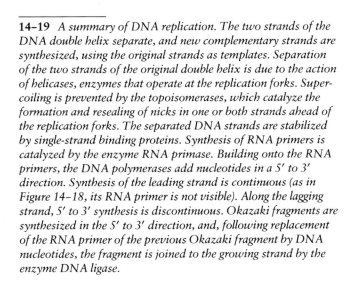

14–19 *A summary of DNA replication. The two strands of the DNA double helix separate, and new complementary strands are synthesized, using the original strands as templates. Separation of the two strands of the original double helix is due to the action of helicases, enzymes that operate at the replication forks. Super-coiling is prevented by the topoisomerases, which catalyze the formation and resealing of nicks in one or both strands ahead of the replication forks. The separated DNA strands are stabilized by single-strand binding proteins. Synthesis of RNA primers is catalyzed by the enzyme RNA primase. Building onto the RNA primers, the DNA polymerases add nucleotides in a 5′ to 3′ direction. Synthesis of the leading strand is continuous (as in Figure 14–18, its RNA primer is not visible). Along the lagging strand, 5′ to 3′ synthesis is discontinuous. Okazaki fragments are synthesized in the 5′ to 3′ direction, and, following replacement of the RNA primer of the previous Okazaki fragment by DNA nucleotides, the fragment is joined to the growing strand by the enzyme DNA ligase.*

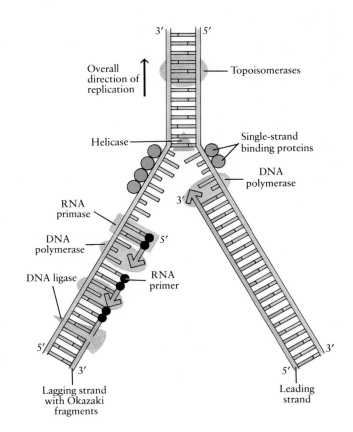

course of synthesis, however, errors are sometimes made, and the wrong nucleotide is added to the newly forming strand—that is, the nucleotide added to the strand is not complementary to the nucleotide on the template strand. When this occurs, DNA polymerase backtracks, removing nucleotides until it encounters a correctly paired nucleotide. At that point, the enzyme stops its reverse movement and resumes moving in the 5′ to 3′ direction, adding nucleotides to the growing strand as it goes. This ability to remove incorrectly paired nucleotides provides an important proofreading that ensures the accuracy of DNA replication.

In addition to the proofreading that occurs during DNA replication, other enzymes constantly monitor the DNA double helixes of the cell. Whenever an incorrectly paired nucleotide is encountered, DNA repair enzymes move in to snip it out and replace it with the correct nucleotide. This process is essential in maintaining the integrity of the genetic machinery.

The Energetics of DNA Replication

The nucleotides required for DNA synthesis are assembled by the same biosynthetic pathways as the nucleotides required for other functions of the cell. They are assembled not in the form of the monophosphates shown in Figure 14–4 but rather as triphosphates; that is, adenine is provided as deoxyadenosine triphosphate (dATP), guanine as the analogous deoxyguanosine triphosphate (dGTP), and so forth. The $P \sim P$ groups of the triphosphates provide the energy to power the reactions catalyzed by DNA polymerase. As each nucleotide is attached to the growing DNA strand, the two "extra" phosphates are removed and released. Almost immediately, another enzyme breaks the bond between the two phosphates, releasing them as inorganic phosphates.

Why does the cell do it this way? The release of the energy-yielding $P \sim P$ group and the subsequent breaking of the bond between the two phosphates seem like a waste of carefully stored chemical energy. Is the cell really so profligate?

Measurement of the energy changes involved reveals an interesting point. The reaction in which the activated nucleotide is attached to the growing DNA strand is only slightly exergonic. Therefore, it could go in either direction. Under certain equilibrium conditions, the DNA strand could come apart about as fast as it was synthesized, with the enzyme working both ways. However, with removal of the $P \sim P$ fragment and its immediate degradation, the reaction becomes highly exergonic. Thus, the reverse reaction—which would necessitate reforging the $P \sim P$ group—becomes highly endergonic and so, for all practical purposes, does not occur. This is yet another example of the way in which the living cell exercises tight control over its biochemical activities.

DNA AS A CARRIER OF INFORMATION

You will recall that a necessary property of the genetic material is the capacity to carry information. The Watson-Crick model showed that the DNA molecule is able to do this. The information is carried in the sequence of the bases, and *any* sequence of bases is possible. Since the number of paired bases ranges from about 5,000 for the simplest known virus up to an estimated 5 billion in the 46 human chromosomes, the number of possible variations is astronomical. The DNA from a single human cell—which if extended in a single thread would be almost 2 meters long—can contain information equivalent to some 600,000 printed pages of 500 words each, or a library of about a thousand books. Obviously, the DNA structure can well account for the endless diversity among living things.

SUMMARY

Classical genetics had been concerned with the mechanics of inheritance—how the units of heredity are passed from one generation to the next and how changes in the hereditary material are expressed in individual organisms. In the 1930s, new questions arose and geneticists began to explore the nature of the gene—its structure, composition, and properties.

During the 1940s, many investigators believed that genes were proteins, but others were convinced that the hereditary material was deoxyribonucleic acid (DNA). Important, although not widely accepted, evidence for the genetic role of DNA was presented by Avery in his experiments to identify the transforming factor of pneumococci. Confirmation of Avery's hypothesis came from studies with bacteriophages (bacterial viruses) showing that DNA and not protein is the genetic material of the virus.

Further support for the genetic role of DNA came from two more sets of data: (1) Almost all somatic cells of any given species contain equal amounts of DNA, and (2) the proportions of nitrogenous bases are the same in the DNA of all cells of a given species, but they vary in different species.

In 1953, Watson and Crick proposed a structure for DNA. The DNA molecule, according to their model, is a double-stranded helix, shaped like a twisted ladder. The two sides of the ladder are composed of repeating subunits consisting of a phosphate group and the five-carbon sugar deoxyribose. The "rungs" are made up of paired nitrogenous bases, one purine base pairing with one pyrimidine base. There are four bases in DNA—adenine and guanine, which are purines, and thymine and cytosine, which are pyrimidines. Adenine (A) can pair only with thymine (T), and guanine (G) only with cytosine (C). The four bases are the four "letters" used to spell out the genetic message. The paired bases are joined by hydrogen bonds.

When the DNA molecule replicates, the two strands come apart, separating at the hydrogen bonds. Each strand acts as a template for the formation of a new complementary strand from nucleotides available in the cell. The semiconservative (one strand conserved) nature of this process was confirmed by studies using heavy isotopes.

DNA replication begins at a particular nucleotide sequence on the chromosome, the origin of replication. It proceeds bidirectionally, by way of two replication forks that move in opposite directions. Helicases unwind the double helix at each replication fork, and single-strand binding proteins stabilize the separated strands. Topoisomerases relieve supercoiling of the helix ahead of the replication forks by nicking, which allows the chains to swivel freely, followed by resealing. A strand of RNA primer, correctly base-paired to the template strand, is required before replication can begin; attachment of the nucleotides that form the primer is catalyzed by the enzyme RNA primase. Addition of DNA nucleotides to the strand is catalyzed by DNA polymerases. These enzymes synthesize new strands in the 5′ to 3′ direction only, adding nucleotides, one by one, to the 3′ end of the growing strand. Replication of the leading strand is continuous, but replication of the lagging strand is discontinuous. On the lagging strand, segments known as Okazaki fragments are synthesized in the 5′ to 3′ direction. DNA ligase catalyzes the condensation reaction that links adjacent Okazaki fragments together. In the course of DNA synthesis, DNA polymerase proofreads, backtracking when necessary to remove nucleotides that are not correctly paired with the template strand.

The nucleotides incorporated into the growing DNA strands are supplied in the form of triphosphates. The energy required to power replication is supplied by removal of the two "extra" phosphates and degradation of the $P \sim P$ bond.

With the elucidation of the structure of the DNA double helix by Watson and Crick, the role of DNA as the carrier and transmitter of the genetic information was universally accepted. With the discovery of the complex—and extremely precise—mechanism by which the living cell replicates its DNA, the question of how the hereditary information is faithfully transmitted from parent cell to daughter cell, generation after generation, was answered.

QUESTIONS

1. Distinguish among the following: purine/pyrimidine; origin of replication/replication bubble/replication fork; helicases/topoisomerases; RNA primase/DNA polymerases; leading strand/lagging strand; RNA primers/Okazaki fragments/DNA ligase.

2. What are the steps by which Griffith demonstrated the existence of the transforming factor? Can you think of any implications of Griffith's discovery for modern medicine?

3. What characteristics of bacteriophages make them a useful experimental tool?

4. One of the chief arguments for the erroneous hypothesis that proteins constitute the genetic material was that proteins are heterogeneous. Explain why the genetic material must have this property. What feature of the Watson-Crick model of DNA structure is important in this respect?

5. When the structure of DNA was being worked out, it became apparent that one purine base must be paired with a pyrimidine base, and that the other purine base must be paired with the other pyrimidine base. The evidence for this requirement came from two types of data. What were the data, and how did they indicate this structural requirement?

6. Further consideration of the structures of the four nitrogenous bases indicated that adenine could pair only with thymine, and cytosine only with guanine. What feature of the structures of the bases imposed this requirement on the structure of the DNA molecule?

7. Suppose you are talking to someone who has never heard of DNA. How would you support an argument that DNA is the genetic material? List at least five of the strong points in such an argument.

8. Suppose the Meselson-Stahl experiment were extended to the third generation. What would be the proportion of light to half-heavy DNA?

9. Eukaryotic cells are grown for a number of generations in a medium containing thymine labeled with ^{3}H. They are then removed from the radioactive medium, placed in an ordinary medium, and allowed to divide. Studies of the distribution of the radioactive isotope are made after each generation to determine the presence or absence of radioactive material in the chromatids. Before the cells are placed in the nonradioactive medium, all the chromatids contain ^{3}H. After one generation in the nonradioactive medium, all of the chromatids still contain radioactive ^{3}H. Assuming each chromatid contains a single DNA molecule, explain the results. Is this consistent with the Watson-Crick hypothesis? What would be the distribution of the ^{3}H after two generations in the nonradioactive medium? Why?

10. When 5-bromodeoxyuridine (BrdU) is added to the medium in which cells are cultured, it is used in place of thymidine (thymine plus deoxyribose). All DNA synthesized after the cells have been placed in the medium will contain BrdU. After the cells have divided once in the BrdU-containing medium, the chromatids all look alike, as you can see in micrograph (a) below. Both sister chromatids of each chromosome contain parent strands of original DNA combined with new strands of BrdU-substituted DNA. The original DNA stains darkly, so both sides are dark. Micrograph (b) shows the appearance of the chromosomes after the second cell division in the BrdU-containing medium. As you will notice, one sister chromatid is dark, and the other is light. Does this confirm or refute the results of the Meselson-Stahl experiment? Explain your answer.

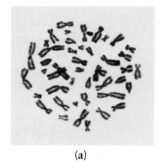   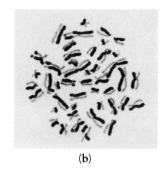

(a)   (b)

11. In the DNA replication bubble diagrammed below, label all 5′ and 3′ ends on each nucleotide strand and identify the leading and lagging strands. Describe what will happen next.

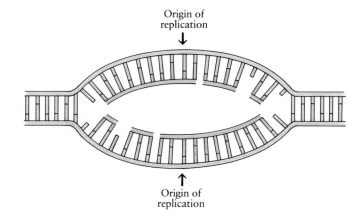

Origin of
replication
↓

↑
Origin of
replication

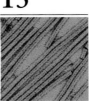

The Genetic Code and Its Translation

15–1 Neurospora, *the red bread mold, growing on a corn tortilla. Colonies of unidentified black and gray molds are also present.* Neurospora *is another "fit material" that has played a key role in the history of genetics. Studies of* Neurospora *mutants by George Beadle and Edward Tatum in the 1940s provided the first definitive demonstration that genes contain the information specifying particular protein molecules.*

The Watson-Crick model, which both established the structure of DNA and indicated the mechanism for its replication, led to the virtually universal acceptance of DNA as the repository of the hereditary information. But how does a DNA molecule—"a lump of inert matter"—embody the instructions that specify a bacterium or a fruit fly or a student of biology? In the same year (1953) that their model was first published, Watson and Crick speculated—quite correctly as it turned out—that "it . . . seems likely that the precise sequence of the bases is a code which carries the genetical information." And so began an avalanche of studies to decipher the code and, even more interesting, to discover how the instructions it contains are decoded, transmitted, and carried out. These studies had their beginnings in a concept first proposed early in this century, shortly after Mendel's work was rediscovered.

GENES AND PROTEINS

Inborn Errors of Metabolism

In 1908, an English physician, Sir Archibald Garrod, presented a series of lectures in which he set forth a new concept of human diseases, which he called "inborn errors of metabolism." Garrod postulated that certain diseases that are caused by the body's inability to perform particular chemical processes are hereditary in nature.

One such disease was described in 1649:

The patient was a boy who passed black urine and who, at the age of fourteen years, was submitted to a drastic course of treatment which had for its aim the subduing of the fiery heat of his viscera, which was supposed to bring about the condition in question by charring and blackening his bile. Among the measures prescribed were bleedings, purgation, baths, a cold and watery diet, and drugs galore. None of these had any obvious effect, and eventually the patient, who tired of the futile and superfluous therapy, resolved to let things take their natural course. None of the predicted evils ensued, he married, begat a large family, and lived a long and healthy life, always passing urine black as ink.

Sir Archibald hypothesized that this condition, alkaptonuria, is the result of an enzyme deficiency and is hereditary in nature. Implicit in his hypothesis was the idea that genes act by influencing the production of enzymes.

One Gene–One Enzyme

By the 1940s, biologists had come to realize that all of the biochemical activities of the living cell, including the multitude of synthetic reactions that produce all of its constituent molecules—carbohydrates, lipids, and proteins—depend upon different specific enzymes. Even the synthesis of enzymes depends on enzymes.

301

Further, it was becoming clear that the specificity of the different enzymes is a result of their primary structure, the linear sequence of amino acids in the molecule.

Meanwhile, George Beadle, a geneticist, was working with the eye-color mutants of *Drosophila* that had been discovered in Morgan's laboratory (page 265). As a result of his studies, Beadle formulated the hypothesis that each of the various eye colors observed in the mutants is the result of a change in a single enzyme in a biosynthetic pathway. To test this notion—that genes control enzymes—on a broader scale, he teamed up in 1941 with Edward L. Tatum, a biochemist. Instead of picking a genetic characteristic and working out its chemistry, they decided to begin with step-by-step chemical reactions—controlled by enzymes—to see if mutations affected these reactions.

The organism they chose for their studies was the red bread mold *Neurospora crassa* (Figure 15–1), which has since become almost as famous a research tool in genetics as the fruit fly. This fungus has several obvious advantages for genetic research:

1. Its life cycle is brief.

2. It can be grown in vast quantities in the laboratory.

3. Throughout most of its life cycle, it is haploid. As a consequence, when a mutation occurs, its effects are detectable immediately. The lack of a homologous chromosome rules out the masking of any mutation by a dominant allele.

4. Meiosis takes place in saclike reproductive structures known as asci (singular, ascus), in which the products of meiosis are not only packaged but also are replicated by mitosis and lined up neatly for inspection (Figure 15–2).

5. Many chromosome mapping studies had already been done with *Neurospora*, which facilitated further genetic analysis.

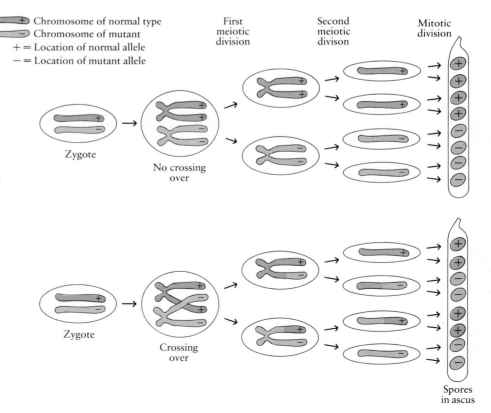

15–2 *Meiosis in* Neurospora crassa, *showing only one of its seven chromosomes. The zygotes shown here were produced by crossing a mutant strain with a normal strain (lacking that particular mutation). As a result of meiosis and the single mitotic division that follows it, eight spores are produced, lined up in a single, narrow spore case, the ascus. Four spores will be normal, and four will be mutants. The order in which the spores are lined up in the ascus depends on whether crossing over has or has not occurred.*

Chromosome of normal type
Chromosome of mutant
+ = Location of normal allele
− = Location of mutant allele

First meiotic division

Second meiotic divison

Mitotic division

Zygote

No crossing over

Zygote

Crossing over

Spores in ascus

15–3 *How Beadle and Tatum tested the mutants of* Neurospora. *In these experiments, they were able to show that a change in a single gene results in a change in a single enzyme. (**a**) Asci are removed from the fruiting bodies (ascocarps) of* Neurospora, *and the spores are dissected out. (**b**) Each spore is transferred to an enriched medium, containing all that* Neurospora *normally needs for growth plus supplementary amino acids. (**c**) A fragment of the mold is tested for growth on the minimal medium. If no growth is observed on the minimal medium, it may mean that a mutation has occurred that renders this mutant incapable of making a particular amino acid. (**d**) Subcultures of molds that grow on the enriched medium but not on the minimal medium are tested for their ability to grow on minimal media supplemented with only one of the amino acids. In the example shown here, a mold that has lost its capacity to synthesize the amino acid proline is unable to survive in a medium that lacks that amino acid. Further tests are then made to discover which enzymatic step in the synthesis of proline has been impaired.*

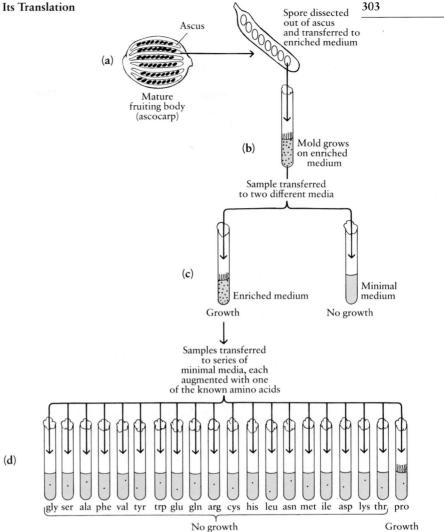

Another important feature of *Neurospora,* from the point of view of the investigators, was the fact that it can be grown on a very simple medium—a minimal medium—containing any one of several sugars as a carbon and energy source, one vitamin (biotin), and a few minerals. This undemanding fungus is able to make for itself all the amino acids, other vitamins, polysaccharides, and other substances essential for its growth and reproduction.

The synthesis of an amino acid or a vitamin requires a series of chemical reactions, each of which is catalyzed by a particular enzyme. If, as a result of a mutation, *Neurospora* were to lose any one of the enzymes involved, for example, in making the amino acid arginine, it could no longer grow on the minimal medium. The mutant could, however, grow on a medium supplemented with arginine.

Beadle and Tatum x-rayed *Neurospora* spores to increase the mutation rate, allowed them to germinate and grow on a medium enriched with all of the amino acids, and then crossed these irradiated strains with normal strains. The spores produced as a result of these crosses were then used in the experimental tests, ensuring that any changes observed were indeed genetic ones. On the basis of their experiments, summarized in Figure 15–3, Beadle and Tatum were able to demonstrate that different mutations resulted in loss of the ability to synthesize different amino acids. Moreover, as a result of the genetic analyses they performed, they showed that, in a number of cases, several different mutations could result in loss of the ability to synthesize the same amino acid. This indicated that the different mutations were affecting enzymes catalyzing different steps of the reaction pathway for the synthesis of that amino acid. For example, as we saw on page 173, three different mutations, affecting three different enzymes, can render *Neurospora* unable to synthesize arginine.

On the basis of their studies, Beadle and Tatum set forth the then-daring (and later, Nobel Prize–winning) proposal that a single gene specifies a single enzyme; in other words, one gene–one enzyme.

This formulation turned out to be an oversimplification, however, for although enzymes are indeed proteins, not all proteins are enzymes. Some proteins, for instance, are hormones, like insulin, and others are structural proteins, like collagen. These proteins, too, are specified by genes. This led to an expansion of the original concept, but did not modify it in principle. "One gene–one enzyme," as it was first formulated, was simply amended to "one gene–one protein." Subsequently, with the realization that many proteins consist of more than one polypeptide chain, it was modified once more, to the less memorable but more precise "one gene–one polypeptide chain." (As we shall see, this, too, has now been further amended.)

The Structure of Hemoglobin

What causes the change or loss of function in an enzyme or protein specified by a gene that has undergone a mutation? Linus Pauling was one of the first to see some of the implications of the work of Beadle and Tatum with regard to this question. Perhaps, Pauling reasoned, human diseases involving hemoglobin, such as sickle cell anemia, could be traced to a variation from the normal protein structure of the hemoglobin molecule. To test this hypothesis—which, at this stage, was pure speculation—he took samples of hemoglobin from people with sickle cell anemia (a homozygous recessive condition), from others heterozygous for the allele, and from still others homozygous for the normal allele. To try to detect differences in these proteins, Pauling used a technique known as electrophoresis, in which organic molecules dissolved in a solution are separated in a weak electric field.

As we noted in Chapter 3, individual amino acids may have a positive charge, a negative charge, or no electric charge at all. Therefore, a substitution of one amino acid for another may change the total charge of the protein molecule, and the normal and the variant protein molecules will, as a result, move at different rates in an electric field.

Figure 15–4 shows the results of Pauling's experiment. A person who has sickle cell anemia makes a different sort of hemoglobin than a person who does not have the disease. A person who is heterozygous (carrying one copy of the allele for sickling and one copy of the allele for normal hemoglobin) makes both kinds of hemoglobin molecules; however, enough normal molecules are produced to prevent anemia. (Notice that the terms "dominant" and "recessive" are beginning to have a less definite meaning.)

A few years later, Vernon Ingram, of Cambridge University in England, was able to show that the actual difference between the normal and the sickle cell hemoglobin molecules is a change in a single amino acid out of 300, as we noted previously (page 80).

The Virus Coat

Additional evidence that DNA specifies the structure of proteins came from the studies of bacteriophages described in Chapter 14. You will recall that the introduction of viral DNA into a bacterial cell results in the production not only of more viral DNA, but also of the proteins of the virus coat. Clearly, the viral DNA carries the information for the synthesis of the coat proteins.

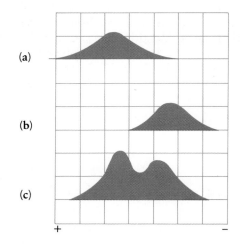

(a)

(b)

(c)

+   −

15–4 *Results from electrophoresis of* (a) *normal hemoglobin,* (b) *the hemoglobin of a person with sickle cell anemia, and* (c) *the hemoglobin of a person who is heterozygous for the sickle cell allele. The height of the curves indicates the amount of hemoglobin at each point. Because of slight differences in electric charge, normal and sickle cell hemoglobins move at different rates in an electric field. The normal hemoglobin is more negatively charged; hence it has moved closer to the positive pole than has the sickle cell hemoglobin. The hemoglobin from the heterozygote separates into the two different positions, indicating that it consists of both normal and sickle cell hemoglobins.*

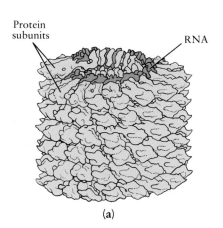

Deoxyribose   Ribose

(a)

Thymine   Uracil

(b)

15–5 *Chemically, RNA is very similar to DNA, but there are two differences in its nucleotides. (a) One difference is in the sugar component; instead of deoxyribose, RNA contains ribose, in which a hydroxyl group replaces a hydrogen on the 2' carbon. (b) The other difference is that instead of thymine, RNA contains the closely related pyrimidine uracil (U). Uracil, like thymine, pairs only with adenine.*

A third, and very important, difference between the two nucleic acids is that most RNA is single-stranded and does not form a regular helical structure.

FROM DNA TO PROTEIN: THE ROLE OF RNA

As a result of all of these studies, there was general agreement that the DNA molecule is a code that contains instructions for biological structure and function. Moreover, these instructions are carried out by proteins, which also contain a highly specific biological "language." As we saw in Chapter 3, the linear sequence of amino acids in a polypeptide chain determines the three-dimensional structure of the completed protein molecule, and it is the three-dimensional structure that determines function. The question thus became one of translation. How did the order of bases in DNA specify the sequence of amino acids in a protein molecule?

The search for the answer to this question led to ribonucleic acid (RNA), the close chemical relative of DNA that we have encountered previously. (The chemical differences between the two molecules are reviewed in Figure 15–5.) There were several clues that RNA might play a role in the translation of genetic information from DNA into a sequence of amino acids. First, there was some circumstantial evidence provided by eukaryotic cells. Unlike DNA, which is found primarily in the nucleus, RNA is found mostly in the cytoplasm, and it is there that most protein synthesis takes place. Embryologists noted that the cells of developing embryos of many different kinds contain high levels of RNA. Second, both prokaryotic and eukaryotic cells making large amounts of protein have numerous ribosomes. A rapidly growing *Escherichia coli* cell, for instance, contains about 15,000 ribosomes, constituting about one-half of the total mass of the cell. And ribosomes are two-thirds RNA and one-third protein.

Additional evidence came from experiments with viruses. When a bacterial cell is infected by a DNA-containing bacteriophage, RNA is synthesized from the viral DNA before viral protein synthesis begins. Also, some viruses contain no DNA—only RNA and protein; tobacco mosaic virus (TMV) is an example (Figure 15–6). When a tobacco leaf is infected with RNA purified from TMV, new viruses are produced, protein coats and all. In other words, RNA as well as DNA seemed to contain information about proteins.

Protein
subunits

RNA

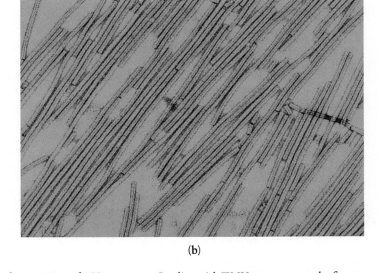

(a)   **(b)**

15–6 (a) *Diagram of a very small portion of tobacco mosaic virus (TMV). This virus has a central core not of DNA but rather of RNA (ribonucleic acid). Its outer coat is composed of 2,150 identical pro-* *tein molecules, each consisting of 158 amino acids. If the RNA is separated from its protein coat and rubbed into scratches on a tobacco leaf, new TMV particles are formed, complete with new protein coats.* *Studies with TMV were among the first to suggest that RNA also could direct the assembly of proteins. (b) When viewed under the electron microscope, TMV particles appear as rod-shaped structures.*

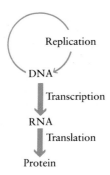

15-7 *The central dogma of molecular genetics: information flows from DNA to RNA to protein. Replication of the DNA occurs only once in each cell cycle, during the S phase prior to mitosis or meiosis. Transcription and translation, however, occur repeatedly throughout the interphase portions of the cell cycle.*

The Central Dogma

Based on early evidence, Crick proclaimed what he called the "central dogma," illustrated in Figure 15–7. DNA specifies RNA, which, in turn, specifies proteins. Note that the arrows go in only one direction. The genotype (DNA) determines the phenotype by dictating the composition of proteins. However, proteins do not alter the genotype—that is, proteins do not send instructions back to the DNA.

Crick called his proposal "dogma" because, at the time, there was little supporting evidence for it. A diversity of experiments have since, however, showed it to be true, almost without exception.*

The direction of information flow from DNA to RNA to proteins provided an important confirmation of Darwin's theory of evolution, in which natural selection acts on inherited variations—and a refutation of Lamarck's contention (page 4) that acquired characteristics could be inherited.

RNA as Messenger

As it turned out, not one but three kinds of RNA play roles as intermediaries in the steps that lead from DNA to protein. At this point in our story, we shall describe just one of them: **messenger RNA** (mRNA).

Messenger RNA molecules are copies (transcripts) of DNA sequences. Unlike DNA molecules, however, RNA molecules are usually single-stranded. Each new mRNA molecule is copied, or transcribed, from one of the two strands of DNA (the template strand) by the same base-pairing principle that governs DNA replication (Figure 15–8). Like a strand of DNA, each RNA molecule has a 5' end and a 3' end. As in the synthesis of DNA, the ribonucleotides, which are present in

---

* As we shall see in Chapter 17, the principal exception to the central dogma is a process known as reverse transcription, in which the information encoded by some RNA viruses is transcribed into DNA.

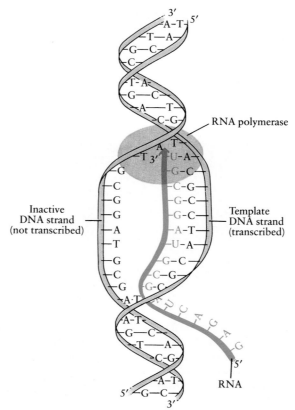

15-8 *A schematic representation of RNA transcription. At the point of attachment of the enzyme RNA polymerase, the DNA opens up, and, as the RNA polymerase moves along the DNA molecule, the two strands of the molecule separate. Nucleotide building blocks are assembled into RNA in a 5' to 3' direction as the enzyme reads the template DNA strand in a 3' to 5' direction. Note that the RNA strand is complementary—not identical—to the template strand from which it is transcribed; its sequence is, however, identical to that of the inactive (untranscribed) DNA strand, except for the replacement of thymine (T) by uracil (U).*

the cell as triphosphates, are added, one at a time, to the 3′ end of the growing RNA chain. The process, known as **transcription,** is catalyzed by the enzyme RNA polymerase. This enzyme operates in the same manner as DNA polymerase, moving in a 3′ to 5′ direction along the template DNA strand, synthesizing a new complementary strand of nucleotides—in this case, ribonucleotides—in a 5′ to 3′ direction. Thus, the mRNA strand is antiparallel to the DNA template strand from which it is transcribed.

Messenger RNA is the working copy of the genetic information. Incorporating the instructions encoded in DNA, mRNA dictates the sequence of amino acids in proteins.

THE GENETIC CODE

The identification of mRNA as the working copy of the genetic instructions still left the big question unresolved. Scientists from many disciplines were intrigued by the puzzle of the genetic code. One of these was George Gamow, the astronomer who was the father of the "big bang" theory of cosmology (page 23).

Proteins contain 20 different amino acids (Figure 3–18, page 73), but DNA and RNA each contain only four different nucleotides. As Gamow pointed out, if a single nucleotide "coded" for one amino acid, only four amino acids could be specified by the four bases. If two nucleotides specified one amino acid, there could be a maximum number, using all possible arrangements, of 4 × 4, or 16—still not quite enough to code for all 20 amino acids. Therefore, following the code analogy, at least three nucleotides in sequence must specify each amino acid. This would provide 4 × 4 × 4, or 64, possible combinations, or **codons**—clearly, more than enough.

The three-nucleotide, or triplet code, was widely adopted as a working hypothesis. Its existence, however, was not actually demonstrated until the code was finally broken, a decade after Watson and Crick first presented their DNA structure. The scientists who performed the initial, crucial experiments toward breaking the code were Marshall Nirenberg and his colleague Heinrich Matthaei, both of the National Institutes of Health.

Breaking the Code

Messenger RNA, then newly discovered, gave Nirenberg the tool he needed. He broke apart *E. coli* cells, extracted their contents, and added to them radioactively labeled amino acids and crude samples of RNA from a variety of cell sources. All of the RNA samples stimulated protein synthesis; the amounts of protein produced were small but measurable. In other words, the material extracted from the *E. coli* cells would start producing protein molecules even when the RNA "orders" it received were from a "complete stranger." Even the RNA from tobacco mosaic virus, which naturally multiplies only in cells of the leaves of tobacco plants, could be read as an mRNA by the machinery of the bacterial cell.

Nirenberg and Matthaei then tried an artificial RNA. Perhaps if the cell-free extracts could read a foreign message and translate it into protein, they could read a totally synthetic message, one dictated by the scientists themselves. Severo Ochoa of New York University had developed an enzymatic process for linking ribonucleotides into a long strand of RNA. With this process, carried out in a test tube, he had produced an RNA molecule that contained only one nitrogenous base, uracil, repeated over and over again. It was called "poly-U."

Nirenberg and Matthaei prepared 20 different test tubes, each of which contained cell-free extracts of *E. coli* that included ribosomes, ATP, the necessary enzymes, and all of the amino acids. In each test tube, one of the amino acids, and only one, carried a radioactive label. Synthetic poly-U was added to each test tube.

The Elusive Messenger

The cytoplasm of cells that are synthesizing proteins is full of RNA. This observation was the major clue that RNA plays a role in directing the assembly of proteins. Although the existence of RNA molecules that carry the genetic information from DNA to protein was hypothesized, confirmation of the hypothesis required the detection and isolation of the messenger molecules. The problem was complicated not by a shortage of RNA but by the fact that most of any cell's RNA is bound into ribosomes, and, as we shall see shortly, ribosomal RNA is not heterogeneous. It was therefore an unlikely candidate for a carrier of

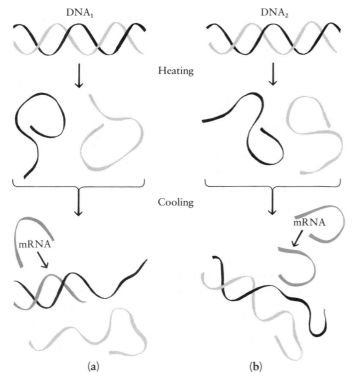

(a)   (b)

RNA-DNA hybrids can be used to show the complementarity in nucleotide sequence between an RNA molecule and the DNA molecule from which it was transcribed. (a) The formation of a hybrid molecule between a radioactive RNA molecule (color) and its template DNA strand (black). (b) If the same RNA molecule is mixed with unrelated DNA, no hybrid molecules containing radioactivity are formed.

genetic information. This paradox puzzled molecular biologists for almost a decade.

Escherichia coli and its phages once again provided the tools of discovery. As you know, the genetic material of the bacteriophages is DNA, and their coats are made of protein. When these viruses infect a bacterial cell, new coat proteins are synthesized. If the messenger hypothesis were true, new RNA should be formed between the time of cell infection and the making of new virus particles. To test the hypothesis, *E. coli* cells were infected with phage and then exposed briefly to uracil labeled with radioactive carbon. Short-lived RNA molecules with radioactive labels were detected in the cells; they were associated with ribosomes but they were not a part of the ribosomes. Were they the long-sought messengers?

To answer this second question it was necessary to show that the radioactive RNA was complementary to the phage DNA. The method used was simple but ingenious. If dissolved DNA molecules are heated gently, the hydrogen bonds break apart and the two strands of the double helix separate. Subsequently, when the solution is slowly cooled, complementary strands pair up again and the hydrogen bonds reform. If the newly formed, radioactive RNA were complementary to the phage DNA, the investigators reasoned, it too should form a duplex with that DNA. As a control, the radioactive RNA molecules were first mixed in a beaker with a solution of *E. coli* DNA and heated; when this mixture was cooled, no duplexes containing radioactivity could be detected. No hybrid RNA-DNA molecules had formed. Then the radioactive RNA molecules were mixed with a solution of phage DNA; when this mixture was heated and then cooled, the results were clear-cut. Duplexes containing radioactivity had formed, indicating that the RNA had bound to its complementary strand of viral DNA. The messenger had been disclosed.

Hybridization of DNA with DNA and of RNA with DNA has since become an enormously powerful tool in both molecular genetics and evolutionary taxonomy. It is used in a great variety of studies, ranging from the detection of specific genes responsible for particular human diseases (page 397) to the resolution of evolutionary enigmas (page 422).

In 19 of the test tubes, no radioactive polypeptides were produced, but in the twentieth tube, to which radioactive phenylalanine had been added, the investigators were able to detect newly formed, radioactive polypeptide chains. When the polypeptides were analyzed, they were found to consist only of phenylalanines, one after another. Nirenberg and Matthaei had dictated the message "uracil . . . uracil . . . uracil . . . uracil . . . uracil . . . uracil . . . ," and a clear answer had come back, "phenylalanine . . . phenylalanine. . . ." The experiment not only defined the first code word (UUU = phe) but also made available a method for defining the others.

As a result of further experiments (see essay on page 310), the mRNA codons for all of the amino acids were worked out (Figure 15–9). Of the 64 possible triplet combinations, 61 specify particular amino acids and three are termination codons. With 61 combinations coding for 20 amino acids, you can see that there must be more than one codon for many of the amino acids; thus the genetic code is said to be **degenerate**.*

15–9 *The genetic code, consisting of 64 triplet combinations (codons) and their corresponding amino acids. The codons shown here are the ones that would appear in the mRNA molecule. Of the 64 codons, 61 specify particular amino acids. The other three codons are stop signals, which cause the chain to terminate.*

Since 61 triplets code for 20 amino acids, there are "synonyms," as many as six different codons for leucine, for example. Most of the synonyms, as you can see, differ only in the third nucleotide. Each codon, however, specifies only one amino acid.

Second letter

First letter (5′ end)	U	C	A	G	Third letter (3′ end)
U	UUU UUC } phe UUA UUG } leu	UCU UCC UCA UCG } ser	UAU UAC } tyr UAA stop UAG stop	UGU UGC } cys UGA stop UGG trp	U C A G
C	CUU CUC CUA CUG } leu	CCU CCC CCA CCG } pro	CAU CAC } his CAA CAG } gln	CGU CGC CGA CGG } arg	U C A G
A	AUU AUC AUA } ile AUG met	ACU ACC ACA ACG } thr	AAU AAC } asn AAA AAG } lys	AGU AGC } ser AGA AGG } arg	U C A G
G	GUU GUC GUA GUG } val	GCU GCC GCA GCG } ala	GAU GAC } asp GAA GAG } glu	GGU GGC GGA GGG } gly	U C A G

PROTEIN SYNTHESIS

With a knowledge of the genetic code, we can turn our attention to the question of how the information encoded in the DNA and transcribed into mRNA is subsequently translated into a specific sequence of amino acids in a polypeptide chain. The answer to this question is now understood in great detail. The basic principles of protein synthesis are the same in both prokaryotic and eukaryotic cells, but there are some differences in detail, which will be described in subsequent chapters. Here we shall focus on the process as it takes place in prokaryotes, particularly *E. coli*.

Instructions for protein synthesis are encoded in sequences of nucleotides in the DNA molecule. Semiconservative replication of DNA transmits these instructions from parent cell to daughter cell and from generation to generation. Thus, each new cell and each new organism inherits the necessary information for synthesizing the specific proteins that determine its particular structure and functions.

* Degeneracy, in this context, does not imply a moral judgment. It is a term used by physicists to describe multiple states that amount to the same thing. The word persists in biology as a testimony to the roles of physicists Delbrück, Wilkins, Crick, Gamow, and others in the research that ultimately led to the breaking of the genetic code.

AGA-GAG-AGA

Among the many experiments that contributed to breaking the genetic code were those of H. G. Khorana at the University of Wisconsin. Khorana synthesized an artificial messenger in which two nucleotides were repeated over and over again in a known sequence: AGAGAGAGAG, UCUCUCUCUC, ACACACACAC, and UGUGUGUGUG. Each of these RNA chains, when used as a messenger in the cell-free system, produced polypeptide chains of alternating amino acids. Poly-AG produced arginine and glutamic acid over and over again; poly-UC, serine and leucine; poly-AC, threonine and histidine; and poly-UG, cysteine and valine. This is, of course, what you would expect from a triplet code. A poly-AG message would be read AGA. . .GAG. . .AGA. . . , for instance.

Khorana also synthesized artificial messengers in which three nucleotides were repeated over and over again. These messengers could produce three different polypeptides, each consisting of only one amino acid, repeated over and over. Which polypeptide was produced depended on where the reading process began.

These studies provided the first clear demonstration (1) that mRNA is read sequentially (that is, one codon after another); (2) that how it is read depends on the reading frame—that is, the nucleotide at which translation starts; and (3) that the codon consists of an uneven number of nucleotides, lending support to the triplet hypothesis.

mRNA base sequence	Read as	Amino acid sequence obtained
(AG)$_n$	···AGA GAG AGA GAG···	···arg-glu-arg-glu···
(AGC)$_n$	··· AGC AGC AGC ···	···ser-ser-ser···
	··· GCA GCA GCA ···	···ala-ala-ala···
	··· CAG CAG CAG ···	···gln-gln-gln···

In Khorana's experiments, an artificial mRNA in which two nucleotides alternated over and over produced a polypeptide chain of alternating amino acids. An artificial mRNA with three different nucleotides produced three different polypeptides, each consisting of only one type of amino acid.

As we have seen, these instructions are transcribed—that is, copied—into an mRNA molecule following the same base-pairing rules that govern DNA replication; the only difference is that in mRNA uracil substitutes for thymine. Specific nucleotide sequences of the DNA, called **promoters,** are the start signals for RNA synthesis, and others, called **terminators,** are the stop signals for RNA synthesis. The RNA is transcribed in only one direction—5′ to 3′—and along only one strand of the DNA duplex. The mRNA molecules are long—500 to 10,000 nucleotides—and single-stranded. These molecules, as we have noted previously, are the working copies used in protein synthesis to determine the amino acid sequences.

The synthesis of proteins requires, in addition to mRNA molecules, two other types of RNA: **ribosomal RNA** (rRNA) and **transfer RNA** (tRNA). These molecules differ both structurally and functionally from mRNA. In most cells, ribosomal RNA is by far the most abundant type, a fact that hampered the search for mRNA, which typically has only a very transitory existence in an *E. coli* cell. Ribosomes consist of two subunits (Figure 15–10) and are, by weight, about two-thirds RNA and one-third protein. In the ribosomes of *E. coli*, the smaller (30S) subunit has one type of rRNA, 1,542 nucleotides in length (commonly referred to as 16S rRNA), and a single molecule each of 21 different proteins. The larger (50S) subunit has two types of rRNA, one consisting of 120 nucleotides (5S rRNA) and the other of 2,904 nucleotides (23S rRNA), and 34 different proteins.

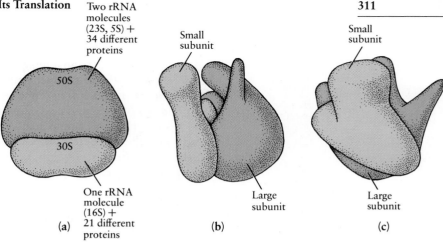

15–10 *Ribosomes, in both prokaryotes and eukaryotes, consist of two subunits, one large and one small. Each subunit is composed of specific rRNA and protein molecules. Studies of rates of sedimentation in the ultracentrifuge have revealed that the size and density of both the subunits and the whole ribosome differ in prokaryotes and eukaryotes. (a) The subunits of the E. coli ribosome have sedimentation values of 50S and 30S; these values, which are a function of both molecular weight and the shape of the molecule, are not additive. The prokaryotic 30S and 50S ribosomal subunits combine to form a ribosome that has a sedimentation value of only 70S.*

(b, c) Two views of the three-dimensional structure of the E. coli ribosome, as revealed by electron micrographs.

Two rRNA molecules (23S, 5S) + 34 different proteins

50S

30S

One rRNA molecule (16S) + 21 different proteins

(a)

Small subunit

Large subunit

(b)

Small subunit

Large subunit

(c)

The smaller subunit has a binding site for the mRNA molecule; in *E. coli* and other prokaryotes, the leading (5′) end of the mRNA molecule attaches to this binding site even as the rest of the molecule is still being transcribed. The larger subunit has two binding sites for the third type of RNA, transfer RNA.

Transfer RNA molecules are, in effect, the dictionary by which the language of nucleic acids is translated into the language of proteins. These molecules are comparatively small, ranging from 75 to 85 nucleotides in length. There are more than 20 different kinds in every cell, at least one for each of the kinds of amino acids found in proteins. Each tRNA molecule has two important attachment sites. One such site, known as the **anticodon,** binds to the codon on the mRNA molecule. The other, on the 3′ end of the tRNA molecule, attaches to a particular amino acid. Thus, tRNA molecules provide the crucial link between nucleic acids and proteins, the two languages of the living cell.

All tRNA molecules have approximately the same cloverleaf shape shown in Figure 15–11. The 3′ end of the molecule—the one that attaches to the amino acid—always terminates in a (5′)-CCA-(3′) sequence. The sequence of the other nucleotides, however, varies according to the particular type of tRNA. Attachment of tRNA molecules to their amino acids is brought about by a group of enzymes known as aminoacyl-tRNA synthetases. There are at least 20 different aminoacyl-tRNA synthetases, one or more for each amino acid. Each of these enzymes has a binding site for a particular amino acid and for its matching tRNA molecule.

15–11 (a) *The structure of a tRNA molecule. Such molecules consist of about 80 nucleotides linked together in a single chain. The chain always terminates in a (5′)-CCA-(3′) sequence. An amino acid can link to its specific tRNA at this end. Some nucleotides are the same in all tRNAs; these are shown in gray. The other nucleotides vary according to the particular tRNA. The symbols D, γ, ψ, and T represent unusual modified nucleotides characteristic of tRNA molecules.*

Some of the nucleotides are hydrogen-bonded to one another, as indicated by the dashed lines. In some regions, the unpaired nucleotides form loops. The loop on the right in this diagram, known as the TψC loop, is thought to play a role in binding the tRNA molecule to the surface of the ribosome. Three of the unpaired nucleotides in the loop at the bottom of the diagram (indicated in color) form the anticodon. They serve to "plug in" the tRNA molecule to an mRNA codon.

(b) The molecule folds over on itself, producing this three-dimensional structure. This is a photograph of a model based on x-ray analysis.

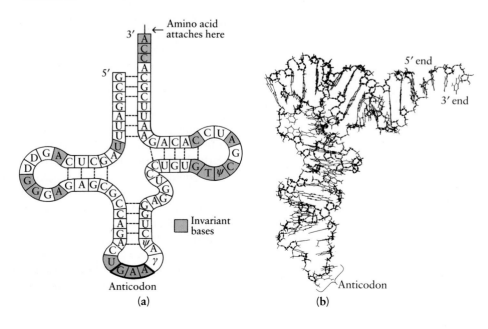

Amino acid attaches here

Invariant bases

Anticodon

(a)

5′ end

3′ end

Anticodon

(b)

The enzymatic reaction linking an amino acid to its tRNA molecule takes place in two steps. In the first, which supplies the energy required for the reaction, an ATP molecule is cleaved, two phosphates are released, and a complex is formed that consists of an amino acid, a molecule of AMP (adenosine monophosphate), and the enzyme. This amino acid–AMP–enzyme complex remains intact until it encounters the appropriate tRNA molecule (out of the 20 or more different types in the cell). Work is presently underway to identify the portion of the tRNA molecule that is recognized by the enzyme and to determine the means by which recognition occurs.

The second step of the reaction takes place when the amino acid–AMP–enzyme complex meets its tRNA molecule. The AMP molecule is released from the enzyme, a bond is formed between the amino acid and the 3′ end of the tRNA molecule, and then the amino acid–tRNA complex is also released. Subsequently, when the tRNA molecule is hydrogen-bonded to the mRNA molecule, anticodon to codon, it thus brings the specified amino acid into place. Only then is the bond broken between the tRNA and the amino acid, as a new bond is formed—a peptide bond that links the newly arrived amino acid to the growing polypeptide chain. Simultaneously, the tRNA molecule is released, once again free to pick up another molecule of its amino acid and to repeat the cycle.

Translation

The synthesis of proteins is known as **translation,** since it is the transfer of information from one language (nucleotides) to another (amino acids). It takes place in three stages: initiation, elongation, and termination (Figure 15–12).

The first stage, **initiation,** begins when the smaller ribosomal subunit attaches to a strand of mRNA near its 5′ end, exposing its first, or initiator, codon. (As we noted earlier, in *E. coli* the 3′ end of the mRNA is still bound to the DNA helix, with transcription continuing even as translation begins at the 5′ end.) Next, the first tRNA comes into place to pair with the initiator codon of mRNA. This initiator codon, which is usually (5′)-AUG-(3′), pairs in an antiparallel fashion with the tRNA anticodon (3′)-UAC-(5′). The incoming initiator tRNA, which binds to the AUG codon, carries a modified form of the amino acid methionine, *N*-formylmethionine, or fMet, as its amino acid (Figure 15–13). This fMet will be the first amino acid in the newly synthesized polypeptide chain, but it may later be removed. The combination of the small ribosomal subunit, mRNA, and the initiator tRNA is known as the **initiation complex.** The larger ribosomal subunit then attaches to the smaller subunit, and the initiator tRNA becomes locked into the P (peptide) site of the larger subunit—one of two sites for binding tRNA molecules. The energy for this step is provided by the hydrolysis of guanosine triphosphate (GTP).

At the beginning of the **elongation** stage, the second codon of the mRNA is positioned opposite the A (aminoacyl) site of the large subunit. A tRNA with an anticodon complementary to the second mRNA codon plugs into the mRNA molecule and, with its amino acid, occupies the A site of the ribosome. When both the P and A sites are occupied, an enzyme, peptidyl transferase, which is part of the larger subunit of the ribosome, forges a peptide bond between the two amino acids, attaching the first amino acid (fMet) to the second. The first tRNA is released. The ribosome moves one codon down the mRNA chain; consequently, the second tRNA, to which is now attached fMet and the second amino acid, is transferred from the A to the P position. A third tRNA–amino acid moves into the A position opposite the third codon on the mRNA, and the step is repeated. The P position accepts the tRNA bearing the growing polypeptide chain; the A position accepts the tRNA bearing the new amino acid that will be added to the chain. As the ribosome moves along the mRNA chain, the initiator portion of the

15–12 *Three stages in protein synthesis.* (a) *Initiation. The smaller ribosomal subunit attaches to the 5′ end of the mRNA molecule. The first tRNA molecule, bearing the modified amino acid fMet, plugs into the AUG initiator codon on the mRNA molecule. The larger ribosomal subunit locks into place, with the tRNA occupying the P (peptide) site. The A (aminoacyl) site is vacant. The initiation complex is now complete.*

(b) *Elongation. A second tRNA with its attached amino acid moves into the A site, and its anticodon plugs into the mRNA. A peptide bond is formed between the two amino acids brought together at the ribosome. At the same time, the bond between the first amino acid and its tRNA is broken. The ribosome moves along the mRNA chain in a 5′ to 3′ direction, and the second tRNA, with the dipeptide attached, is moved to the P site from the A site as the first tRNA is released from the ribosome. A third tRNA moves into the A site, and another peptide bond is formed. The growing peptide chain is always attached to the tRNA that is moving from the A site to the P site, and the incoming tRNA bearing the next amino acid always occupies the A site. This step is repeated over and over until the polypeptide is complete.*

(c) *Termination. When the ribosome reaches a termination codon (in this example, UGA), the polypeptide is cleaved from the last tRNA and the tRNA is released from the P site. The A site is occupied by a release factor that triggers the dissociation of the two subunits of the ribosome.*

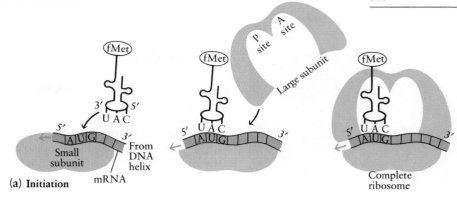

(a) Initiation

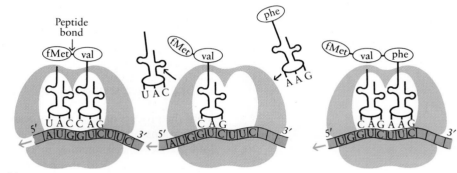

(b) Elongation

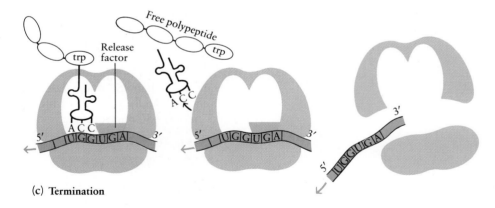

(c) Termination

15–13 *The structures of methionine and N-formylmethionine (fMet), the modified amino acid carried by the initiator tRNA in prokaryotes. Note that the attachment of the formyl group (shown in color) prevents a peptide bond from forming on the "wrong" (amino) end of the amino acid and ensures that its peptide bond with the second amino acid in the sequence will occur at the carboxyl end. The mRNA codon for both methionine and fMet is (5′)-AUG-(3′). It is not known how AUG can code for fMet in the initiator position and methionine elsewhere in the polypeptide chain. Two different tRNAs are involved, but the anticodons are the same.*

Methionine
(met)

N-Formylmethionine
(fMet)

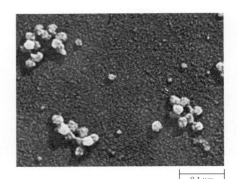

|__ 0.1 µm

15–14 *Clusters of ribosomes reading the same mRNA strand. Such groups are called polyribosomes, or polysomes.*

mRNA molecule is freed, and another ribosome can form an initiation complex with it. A group of ribosomes reading the same mRNA molecule is known as a **polysome** (Figure 15–14).

Toward the end of the coding sequence of the mRNA molecule is a codon that serves as a **termination** signal. Three termination codons are known (UAG, UAA, and UGA), and often more than one is present. No tRNAs exist with anticodons that "match" these codons, and so no tRNAs will enter the A site in response to them. When a termination codon is reached, translation stops, the polypeptide chain is freed, and the two ribosomal subunits separate. It is estimated that *E. coli* can synthesize as many as 3,000 proteins, each different and each assembled in this same way.

The elucidation of the details of this precise and elegant process of translation was an awe-inspiring achievement. Even more awe-inspiring is the knowledge that at this very moment a similar process is taking place in virtually every cell of our own bodies.

REDEFINING MUTATIONS

With the process of protein synthesis in mind, let us consider some of the broader implications of the genetic code and its translation. For example, take another look at sickle cell anemia, in the light of Figure 15–9. Normal hemoglobin contains glutamic acid at a particular position; sickle cell hemoglobin contains valine in the same position. The difference between the mRNA codons for glutamic acid and valine is a single nucleotide. In mRNA, GAA or GAG specifies glutamic acid (glu), and GUU, GUC, GUA, or GUG specifies valine (val). So the difference between the two is the replacement of one adenine by one uracil. In the template strand of DNA from which the mRNA was transcribed, the difference is the replacement of one thymine by one adenine in a sequence of nucleotides that, since it dictates a polypeptide that contains more than 150 amino acids, must contain more than 450 nucleotides. In other words, the tremendous functional difference—literally a matter of life and death—can be traced to a single "misprint" in over 450 nucleotides.

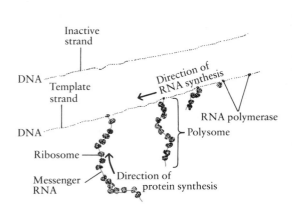

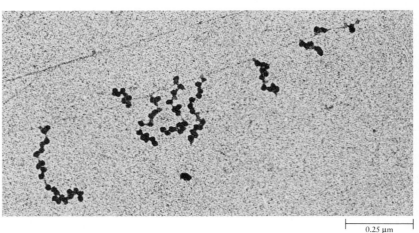

|__ 0.25 µm

15–15 *A bacterial gene in action. In the micrograph, you can see several different mRNA strands (shown in color in the diagram) being transcribed simultaneously from the same DNA template. The longest one, at the left, was the first one synthe-* *sized. As each mRNA strand peels off the DNA molecule, ribosomes attach to the mRNA, translating its encoded information into protein. You can also see molecules of RNA polymerase, the enzyme that catalyzes the transcription of RNA* *from DNA. The RNA polymerase molecule at the far right is approximately at the point where transcription begins. The protein molecules are not visible in this micrograph.*

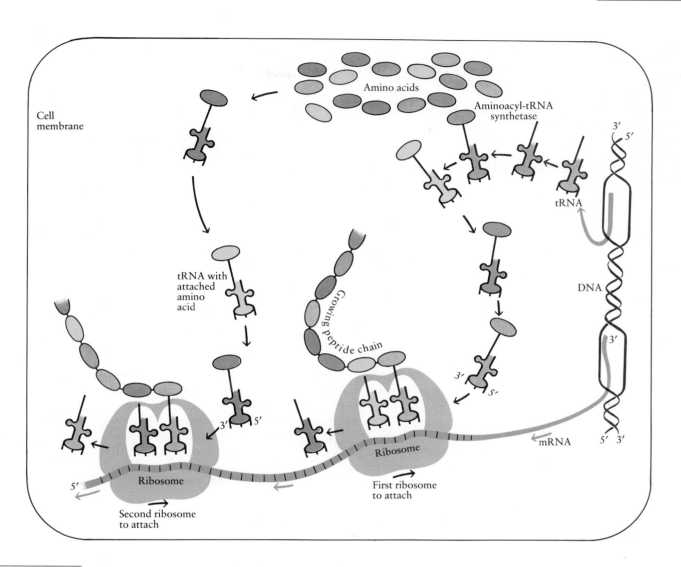

15–16 *A summary of protein synthesis in a bacterial cell. At least 32 different kinds of tRNA molecules are transcribed from the DNA of the bacterial cell. These molecules are so structured that each can be attached at one end (by an aminoacyl-tRNA synthetase) to a specific amino acid. Each contains an anticodon that is complementary to an mRNA codon for that particular amino acid.*

The process begins when an mRNA strand is transcribed from a DNA template. At the point of attachment to the ribosome, the matching tRNA molecule, with its amino acid, plugs in momentarily to the codon in the mRNA. As the ribosome moves along the mRNA strand, a tRNA linked to its particular amino acid fits into place and the first tRNA molecule

is released, leaving behind its amino acid, now enzymatically linked to the second amino acid by a peptide bond. As the process continues, the amino acids are brought into line one by one, following the exact order originally dictated by the DNA from which the mRNA was transcribed.

De Vries, almost 90 years ago, defined mutation in terms of characteristics appearing in the phenotype. In the light of current knowledge, the definition is somewhat different: A mutation is a change in the sequence or number of nucleotides in the nucleic acid of a cell. Mutations that occur in gametes or the cells that give rise to gametes are transmitted to future generations. Mutations that occur in somatic cells are transmitted to the daughter cells produced by mitosis and cytokinesis.

15–17 *Deletion or addition of nucleotides within a gene leads to changes in the protein produced. The original DNA molecule, the mRNA transcribed from it, and the resulting polypeptide are shown in (a).*

In (b) we see the effect of the deletion of a nucleotide pair (T-A), as indicated by the arrow. The reading frame for the gene is altered, and a different sequence of amino acids occurs in the polypeptide.

A similar change results from the addition of a nucleotide pair (brown), as shown in (c).

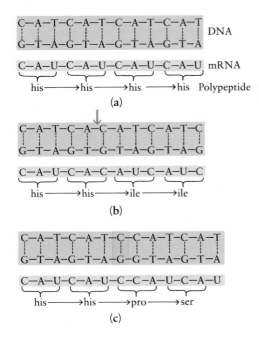

Most mutations involve only a single nucleotide substitution and are called **point mutations.** As in the case of sickle cell anemia, such a substitution can lead to changes in the protein produced by a gene. Other changes in the amino acid sequence of a protein can result from deletion or addition of nucleotides within a gene (Figure 15–17). When this occurs, the reading frame of the gene may shift—that is, the way in which the nucleotides are grouped into triplets changes resulting in the production of an entirely new protein. These **frame-shifts,** as they are known, almost invariably lead to "bad" proteins.

UNIVERSALITY OF THE GENETIC CODE

In the decades since the genetic code was broken, the DNA and proteins of many more organisms have been examined. The evidence is now overwhelming: for virtually all organisms, from *E. coli* to *Homo sapiens,* the genetic code is universal.* It evolved early, remained constant, and determines the underlying unity of all living things.

SUMMARY

Genetic information is coded in the sequence of nucleotides in molecules of DNA, and these, in turn, determine the sequence of amino acids in molecules of protein. That genes exert their effects by influencing the production of specific protein molecules was demonstrated by the experiments of Beadle and Tatum with *Neurospora* mutants. As a result of their experiments, they formulated the principle of "one gene–one enzyme," subsequently amended to "one gene–one polypeptide." Studies of normal and sickle cell hemoglobin molecules demonstrated that a change in a single amino acid of a polypeptide chain can cause a dramatic change in the function of the resulting protein.

---

* The only apparent exceptions to this statement, as we shall see in Chapter 18, involve mitochondria. These organelles contain their own DNA, transcribe their own mRNA, rRNA, and tRNA molecules, and carry out some protein synthesis. In several instances, the mitochondrial code differs from that carried in the chromosomes of both prokaryotes and eukaryotes.

The way in which DNA is translated into protein has been worked out in considerable detail. The information is transcribed from one strand of the DNA (the template strand) into a long, single strand of RNA (ribonucleic acid). This type of RNA molecule is known as messenger RNA, or mRNA. An enzyme, RNA polymerase, catalyzes the transcription process. The mRNA is synthesized in the 5′ to 3′ direction, following the principles of base pairing first suggested by Watson and Crick. It is therefore complementary to the template DNA strand. Each group of three nucleotides in the mRNA molecule is the codon for a particular amino acid.

The genetic code has been deciphered, that is, it is now known which amino acid is called for by a given mRNA codon. Of the 64 possible triplet combinations of the four-letter nucleotide code, 61 combinations have been identified with one of the 20 amino acids that make up protein molecules. The other three triplets serve as "stop signals," terminating protein synthesis. The code is universal.

Protein synthesis—translation—takes place at the ribosomes. A ribosome is formed from two subunits, one large and one small, each consisting of characteristic ribosomal RNAs (rRNAs) complexed with specific proteins. Also required for protein synthesis are another group of RNA molecules, known as transfer RNA (tRNA), each of which is folded into a cloverleaf configuration. These small molecules can carry an amino acid on one end and have a triplet of bases, the anticodon, on a central loop at the opposite end of the molecule. The tRNA molecule is the adapter that pairs the correct amino acid with each mRNA codon during protein synthesis. There is at least one kind of tRNA molecule for each kind of amino acid found in cells. Enzymes known as aminoacyl-tRNA synthetases catalyze the binding of each amino acid to its specific tRNA molecule.

In *E. coli* and other prokaryotes, even as the 3′ end of an mRNA strand is being transcribed, ribosomes are attaching near its 5′ end. At the point where the strand of mRNA is in contact with a ribosome, tRNAs are bound temporarily to the mRNA strand. This bonding takes place by complementary base pairing between the mRNA codon and the tRNA anticodon. Each tRNA molecule carries the specific amino acid called for by the mRNA codon to which the tRNA attaches. Thus, following the sequence originally dictated by the DNA, the amino acid units are brought into line one by one and, as peptide bonds form between them, are linked into a polypeptide chain.

Mutations are now defined as changes in the sequence or number of nucleotides in the nucleic acid of a cell or organism. Point mutations may take the form of substitutions of one nucleotide for another, or deletions or additions of nucleotides.

QUESTIONS

1. Distinguish among the following: mRNA/tRNA/rRNA; code/codon/anticodon; transcription/translation; P site/A site; initiation/elongation/termination.

2. A person heterozygous for the allele for sickle cell anemia makes the variant hemoglobin molecules but does not suffer from anemia. In what respect is this situation different from the definitions of "dominant" and "recessive" given in Chapter 11?

3. Define the "central dogma" and describe its importance in evolutionary theory.

4. Most of the bacterial DNA codes for mRNA, and most of the RNA produced by the cell is mRNA. Yet analysis of the RNA content of a cell reveals that, typically, rRNA is about 80 percent of the cellular RNA and tRNA makes up most of the rest. Only about 2 percent is normally mRNA. How do you explain these findings? What do you think the functional explanation might be?

5. Even before the genetic code was deciphered, it was believed to be degenerate. Explain why.

6. Transcription and translation in prokaryotes are "linked" processes but, as we shall see in Chapter 18, this is not the case in eukaryotes. On the basis of your knowledge of cell structure, propose a likely explanation.

7. Given the details of protein synthesis, what further amendment would you make to the principle "one gene–one polypeptide"?

8. In a hypothetical segment of one strand of a DNA molecule, the sequence of bases is (3′)-AAGTTTGGTTACTTG-(5′). What would be the sequence of bases in an mRNA strand transcribed from this DNA segment? What would be the sequence of amino acids coded by the mRNA? Does it matter at what point on the template strand the transcription from DNA to mRNA begins? Explain your answer.

9. Suppose you have the peptide arg-lys-pro-met, and you know that the tRNA molecules used in its synthesis had the following anticodons:

(3′)-GGU-(5′)
(3′)-GCU-(5′)
(3′)-UUU-(5′)
(3′)-UAC-(5′)

Determine the DNA nucleotide sequence for the template strand of the gene that codes for this peptide.

10. Deletion or addition of nucleotides within a gene leads to changes in the protein produced. The original DNA molecule, the mRNA transcribed from it and the resulting polypeptide are:

```
T–C–G–T–C–G–T–C–G–T–C–G
|  |  |  |  |  |  |  |  |  |  |  |   DNA
A–G–C–A–G–C–A–G–C–A–G–C

U–C–G–U–C–G–U–C–G–U–C–G   mRNA
 ser ——→ ser ——→ ser ——→ ser   Polypeptide
```

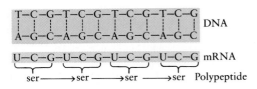

Deleting the second T-A pair yields the following DNA molecule:

```
T–C–G–C–G–T–C–G–T–C–G–T
|  |  |  |  |  |  |  |  |  |  |  |
A–G–C–G–C–A–G–C–A–G–C–A
```

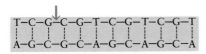

How is the resulting amino acid sequence altered?
How does the addition of a C-G pair to the original molecule,

```
T–C–G–T–C–G–C–T–C–G–T–C
|  |  |  |  |  |  |  |  |  |  |  |
A–G–C–A–G–C–G–A–G–C–A–G
```

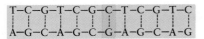

affect the amino acid sequence?

The Molecular Genetics of Prokaryotes and Viruses

As we have seen in the two previous chapters, many of the important early advances in molecular genetics resulted from experiments with prokaryotes and viruses. Investigations with pneumococci, *Escherichia coli*, bacteriophages, and TMV contributed to the identification of DNA as the genetic material, to the breaking of the genetic code, and to the elucidation of both the principles and the details of transcription and translation. This work, however, is not just of historical interest. Studies with these and other viruses and bacteria have laid the foundation for much of the current work in genetics.

Biologists are now able to manipulate genes in ways never before imagined—modifying and recombining portions of DNA molecules from different sources and inserting these altered molecules into other cells where they are expressed. This new technology, known as **recombinant DNA,** has generated an avalanche of studies and information, some of which will be discussed in the following chapters. In order to understand this rapidly advancing frontier of genetics, however, we must look more closely both at the ways in which bacteria and viruses modify, recombine, and exchange genetic material entirely on their own, without any human intervention whatsoever, and at the ways in which they regulate the expression of their genes.

THE *E. COLI* CHROMOSOME

The chromosome of *E. coli* is a single, continuous (circular) thread of double-stranded DNA, approximately 1 millimeter long when fully extended but only 2 nanometers in diameter (Figure 16–1). It contains some 4.7 million base pairs. The

16–1 *An* E. coli *cell that has been lysed (broken apart), releasing its circular chromosome. Even when freed from the cell, the chromosome appears as a highly folded structure, composed of many loops. In the intact cell, the chromosome, which is some 500 times longer than the cell itself, is tightly packed into the nucleoid region.*

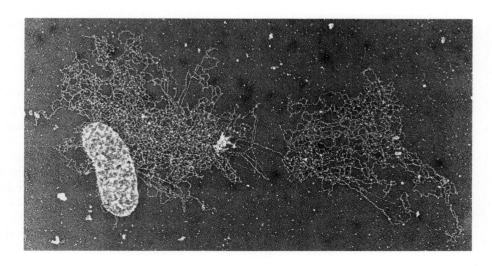

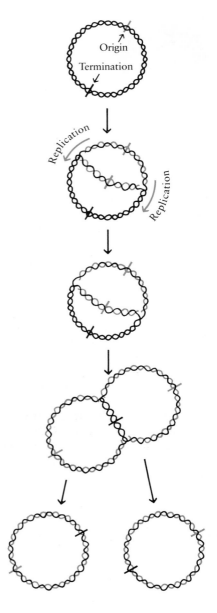

16–2 *A schematic representation of bidirectional (theta) replication of the DNA of the prokaryotic chromosome. The points of origin and termination of replication are indicated in red and black, respectively, and the newly synthesized DNA strand is shown in red. The replication forks are moving away from the origin in opposite directions.*

bacterial cell itself is less than 2 micrometers long, about 1/500th the extended length of its chromosome. Within the cell, the chromosome is compacted into an irregularly shaped body known as the nucleoid (see Figure 4–8, page 91).

As we saw in Chapter 14, bidirectional replication of the chromosome in *E. coli* and other bacteria begins at one specific nucleotide sequence known as the origin of replication. As the two replication forks move away from the origin in opposite directions, DNA polymerase adds nucleotides, one by one, to the 3′ ends of both the leading strands and the Okazaki fragments of the lagging strands. When the circular bacterial chromosome is replicating, it forms a structure resembling the Greek letter θ (theta); hence, its replication is known as **theta replication** (Figure 16–2).

TRANSCRIPTION AND ITS REGULATION

Transcription in *E. coli* and other prokaryotes takes place, as we saw in the last chapter, by the synthesis of a molecule of mRNA along a template strand of DNA. The process begins when the enzyme RNA polymerase attaches to the DNA at a specific site known as the promoter. The RNA polymerase molecule binds tightly to the promoter and causes the DNA double helix to open, initiating transcription. The growing RNA strand remains hydrogen-bonded to the DNA template briefly—only about 10 or 12 ribonucleotides are bonded to the DNA at any one time—and then it peels off in a single strand.

A segment of DNA that codes for a polypeptide (a protein) is known as a **structural gene**. Often structural genes coding for polypeptides with related functions occur together in sequence on the bacterial chromosome. Such functional groups might include, for instance, two polypeptide chains that together constitute a particular enzyme or three enzymes that work in a single enzymatic pathway. Groups of genes coding for such molecules are typically transcribed into a single mRNA strand. Thus, a group of polypeptides that are needed by the cell at the same time can be synthesized simultaneously, a simple and efficient inventory-control system.

The newly synthesized mRNA molecule (Figure 16–3) has a short "leader" sequence at its 5′ end, part of which may assist in binding mRNA to the ribosome. The coding region of the molecule is a linear sequence of nucleotides that precisely dictates the linear sequence of amino acids in particular polypeptide chains. There may be several stop and start codons within the mRNA molecule, marking the end of one structural gene and the beginning of the next. An additional nucleotide sequence at the 3′ end is known as a "trailer." As we have seen previously, ribosomes attach to the mRNA molecule even before transcription is complete, initiating the sequence of events shown in Figure 15–12 (page 313).

The Need for Regulation

In the course of their long evolutionary history, *E. coli* and other prokaryotes have evolved ways to maximize their utilization of nutrients for cellular growth. If a bacterial cell could be said to have a purpose or function, it would be to grow and multiply as rapidly as possible. And, bacteria are excellent at achieving this; a culture of *E. coli* cells, for example, can double in number every 20 minutes.

One reason for *E. coli*'s effectiveness in using nutrients is its versatility; it can make at least 1,700 enzymes and other proteins, enabling it to utilize a wide range of potential nutrients. A second reason is that the cell is highly efficient in its synthetic activities. It does not make all of its possible proteins all of the time, but only when they are needed and only in the amounts needed. For example, cells of *E. coli* supplied with the disaccharide lactose as a carbon and energy source

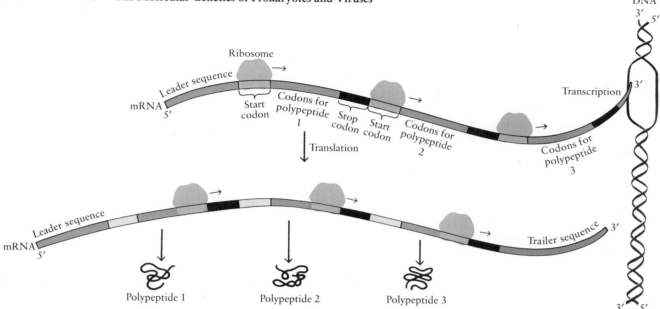

16–3 *In prokaryotes, transcription often results in an mRNA molecule that contains coding sequences for several different polypeptide chains, with the sequences separated by stop and start codons. In this diagram, the stop and start codons are adjacent, but sometimes they are separated by as many as 100 to 200 nucleotides. The 5' end of the mRNA molecule has a short leader sequence, and the 3' end has a trailer sequence; neither of these sequences codes for protein. Translation generally begins at the leading end of the mRNA while the rest of the molecule is still being transcribed.*

require the enzyme beta-galactosidase to split the disaccharide (Figure 16–4). Cells growing on lactose have approximately 3,000 molecules of beta-galactosidase per cell. In the absence of lactose, there is an average of one molecule of the enzyme per cell. In short, the presence of lactose induces the production of the enzyme molecules needed to break it down (Figure 16–5a). Such enzymes are said to be **inducible**.

Conversely, the presence of a particular nutrient may inhibit the transcription of a group of structural genes. *E. coli*, like other bacteria, can synthesize each of its amino acids from ammonia and a carbon source. The structural genes for the enzymes needed for the biosynthesis of the amino acid tryptophan, for instance,

16–4 *Lactose (milk sugar) is an important energy source for* E. coli. *The splitting of lactose into galactose and glucose requires the enzyme beta-galactosidase. Normal* E. coli *cells synthesize beta-galactosidase only when lactose is present in the medium in which they are growing.*

16–5 *Inducible and repressible enzymes.* (a) *The rate of synthesis of beta-galactosidase, an inducible enzyme produced by E. coli, increases dramatically when lactose is added to the surrounding growth medium. As long as lactose is abundant in the medium, enzyme production continues at its maximum rate. However, when lactose is removed from the medium, the rate of synthesis of beta-galactosidase immediately plummets.* (b) *In the absence of an essential substance, such as the amino acid tryptophan, the enzymes required for its production are synthesized at a maximum rate. If, however, tryptophan is added to the medium, synthesis of these enzymes is rapidly repressed.*

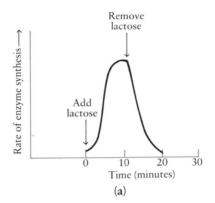

(a)

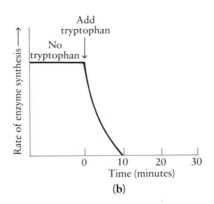

(b)

are grouped together and are transcribed into a single mRNA molecule. This mRNA is produced continuously in growing cells—unless tryptophan is present. In the presence of tryptophan, production of the enzymes ceases (Figure 16–5b). Such enzymes, the synthesis of which is reduced by the presence of the products of the reactions they catalyze, are said to be **repressible**.

Mutants of *E. coli* sometimes occur that are unable to regulate enzyme production. These cells produce beta-galactosidase even in the absence of lactose, for example, or the enzymes that synthesize tryptophan even when tryptophan is present. These and similar mutants are generally at a disadvantage because they are squandering their energies and resources. Normal *E. coli* cells rapidly outmultiply them.

Although regulation of protein synthesis could theoretically take place at many points in the biosynthetic process, in prokaryotes it occurs mostly at the level of transcription. Regulation involves interactions between the chemical environment of the cell and special regulatory proteins, coded by regulatory genes. These proteins can work either as negative controls, repressing mRNA transcription, or as positive controls, enhancing transcription. The fact that mRNA is translated into protein so immediately (before transcription is even completed) and broken down so rapidly further increases the efficiency of this strategy of regulation.

The Operon

It was the detection of the mutants described above that led to our current understanding of the regulation of transcription in prokaryotes. This understand-

16–6 *François Jacob, André Lwoff, and Jacques Monod in Paris in 1966, shortly after they shared the Nobel Prize for their discoveries concerning the genetics of prokaryotes. Jacob and Monod were honored for their work on the operon model of genetic regulation, and Lwoff for his work on a phenomenon known as lysogeny, which we shall consider later in this chapter. Most of the research of these three scientists was carried out at the Pasteur Institute, beginning at the time of World War II. During that war Jacob served in the French army, while Lwoff and, most notably, Monod were active in the French Resistance.*

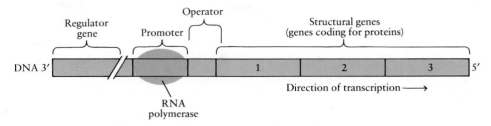

16–7 *A schematic representation of an operon. An operon consists of a promoter, an operator, and structural genes (that is, genes that code for proteins, often enzymes that work sequentially in a particular reaction pathway). The promoter, which precedes the operator, is the binding site for RNA polymerase. The operator is the site at which a repressor protein can bind; it may overlap the promoter, the first structural gene (as shown here), or both. Another gene involved in operon function is the regulator, which codes for the repressor. Although the regulator may be adjacent to the operon, in most cases it is located elsewhere on the bacterial chromosome.*

ing rests upon a model, known as the **operon** model, proposed some years ago by the French scientists François Jacob and Jacques Monod, who shared the Nobel Prize in 1965 with their colleague André Lwoff. According to the model formulated by Jacob and Monod, groups of genes coding for proteins with related functions are arranged in units known as operons. An operon (Figure 16–7) comprises the promoter, the structural genes, and another DNA sequence known as the **operator.** The operator is a sequence of nucleotides located between the promoter and the structural gene or genes; the operator may overlap the promoter, the adjacent structural gene, or both.

Transcription of the structural genes often depends on the activity of still another gene, the **regulator,** which may be located anywhere on the bacterial chromosome. This gene codes for a protein called the **repressor,** which binds to the operator. When a repressor is bound to the operator, it obstructs the promoter. As a consequence, RNA polymerase either cannot bind to the DNA molecule or, if bound, cannot begin its movement along the molecule. The result in either case is the same: no mRNA transcription occurs. However, when the repressor is removed, transcription may begin. Evidence for the existence of the regulator gene was derived from studies of *E. coli* cells that could not stop making beta-galactosidase. In these cells, a mutation in the regulator gene for the lactose *(lac)* operon provided the essential clue that such a gene existed in normal cells.

The capacity of the repressor to bind to the operator and thus to block protein synthesis depends, in turn, on another molecule that functions as an effector. Depending on the operon, an effector can either activate or inactivate the repressor for that particular operon (Figure 16–8). For example, when lactose is present in the growth medium, the first step in its metabolism produces a closely related sugar, allolactose, that binds to and inactivates the repressor, removing it

16–8 *In an operon system, the synthesis of proteins is regulated by interactions involving either a repressor and an inducer or a repressor and a corepressor. (a) In inducible systems, such as the* lac *operon, the repressor molecule is active until it combines with the inducer (in this case, allolactose). (b) In repressible systems, such as the* trp *operon, the repressor is not active until it combines with the corepressor.*

(a) Active repressor + Inducer (allolactose) = Inactive repressor-inducer complex

(b) Inactive repressor + Corepressor (tryptophan) = Active repressor

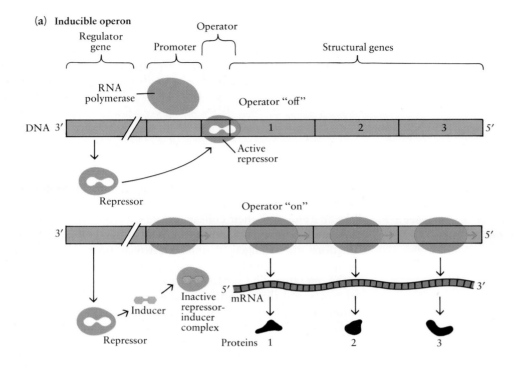

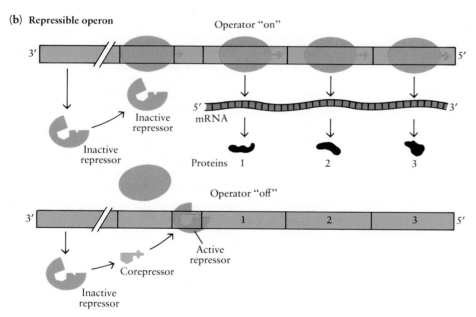

16–9 *Inducible and repressible operons are both turned off by repressor proteins that are coded by regulator genes. The repressor binds to DNA at the operator and so prevents RNA polymerase from initiating transcription. (a) In inducible operons, the inducer counteracts the effect of the repressor by binding with it and maintaining it in an inactive form. Thus, when the inducer is present,* *the repressor can no longer attach to the operator, permitting transcription and translation to proceed.*

(b) In repressible operons, the repressor can bind to the operator only when it is combined with a corepressor. Thus transcription and translation proceed until a corepressor is produced.

from the operator of the *lac* operon. As a consequence, RNA polymerase can begin its movement along the DNA molecule, transcribing the structural genes of the operon into mRNA (Figure 16–9a). In the case of the tryptophan *(trp)* operon, the presence of the amino acid activates the repressor, which then binds to the operator and blocks the synthesis of the unneeded enzymes (Figure 16–9b). Both allolactose and tryptophan—as well as the molecules that interact with the repressors of other operons—are allosteric effectors (page 176), exerting their effects by causing a change in the configuration of the repressor molecule.

Some 75 different operons have now been identified in *E. coli*, comprising 260 structural genes. Some are, like the *lac* operon, inducible, while others are, like the *trp* operon, repressible. Note, however, that both inducible and repressible systems are examples of negative control, since both involve repressors that turn off transcription.

The CAP–Cyclic AMP System

Catabolite activator protein, or CAP, is a regulatory protein that exerts positive control on the operon. Like the operon itself, the CAP system was initially investigated in relation to lactose metabolism and is now known to be of much wider significance. CAP combines with a molecule known as cyclic AMP (cAMP), and this combination binds to the promoter region of the operon. Only then, when the CAP-cAMP complex is bound to the promoter, does maximum transcription take place (Figure 16–10). As we have seen, the operon is under the negative control of the repressor: no transcription takes place unless the repressor is removed. It is also under the positive control of the CAP-cAMP complex, which enhances transcription when it is bound to the operon.

16–10 *Negative and positive regulation of the* lac *operon.* **(a)** *In the* lac *operon (and other operons regulated by the CAP-cAMP system), the promoter includes two distinct regions: a binding site for the CAP-cAMP complex and an entry site for RNA polymerase molecules. In order for RNA polymerase to bind efficiently to the promoter, the CAP-cAMP complex must be in place on its binding site.* **(b)** *In the absence of the inducer (allolactose), the repressor binds to the operator, which, in the* lac *operon, overlaps the first structural gene. Although RNA polymerase can bind to the promoter, it cannot move past the repressor to begin transcription.* **(c)** *In the presence of the inducer, the repressor is inactivated and can no longer attach to the operator. If, under these circumstances, the CAP-cAMP complex is in place at its binding site, RNA polymerase molecules immediately begin transcription of mRNA molecules that direct the synthesis of three proteins: the enzyme beta-galactosidase, a transport protein that brings lactose from the external medium into the cell, and the enzyme transacetylase, which transfers an acetyl group from acetyl CoA (page 194) to galactose.*

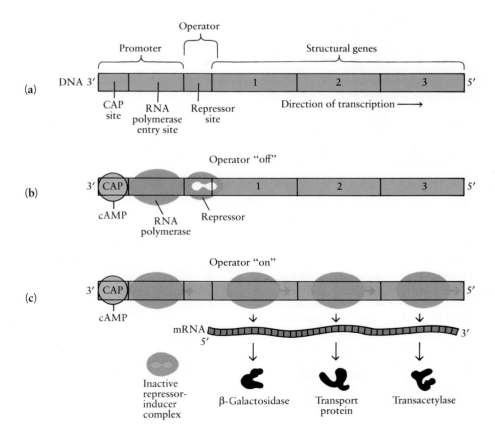

Discovery of this control system came about because of the observation that *E. coli* will not use lactose as an energy source if glucose is present. In other words, in the presence of glucose, the *lac* operon remains repressed, even though lactose is present in the cell. The intermediary in this regulatory process is cAMP. When the supply of glucose in the cell decreases, the level of cAMP increases, more CAP-cAMP complexes form and become available to bind to the *lac* operon, more of the proteins coded by the *lac* operon are produced, and more lactose is broken down. The process by which a decrease in the concentration of glucose leads to an increase in the concentration of cAMP remains a mystery.

These mechanisms are, in themselves, further examples of the precision with which the living cell regulates its biochemical activities. Their manipulation is, as we shall see in the next chapter, an essential component in the scientific trick of inducing bacterial cells to synthesize mammalian proteins of medical importance, such as human insulin.

PLASMIDS AND CONJUGATION

Although the bacterial chromosome contains all of the genes necessary for growth and reproduction of the cell, virtually all types of bacteria have been found to carry additional DNA molecules known as **plasmids.** Plasmids, which are much smaller than the bacterial chromosome, may carry as few as two genes or as many as thirty. Certain plasmids can move into and out of the bacterial chromosome; a plasmid that is incorporated into the chromosome is known as an **episome.**

Like the bacterial chromosome, plasmids are circular and self-replicating (Figure 16–11). Some plasmids replicate in synchrony with the chromosome, and each daughter cell has only one copy of the plasmid. Other plasmids replicate asynchronously, with the result that the cell may contain multiple copies. In the case of some small plasmids, as many as 50 copies have been detected in a single cell. Alternatively, if the plasmid replicates less frequently than the chromosome, some daughter cells may not receive any copies of the plasmid. The DNA of an episome, as you would expect, replicates when the chromosome itself replicates.

About a dozen different kinds of plasmids have been described in *E. coli* alone. Two of the most important types are sex factor, or F, plasmids, and drug resistance, or R, plasmids.

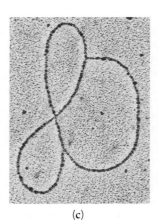
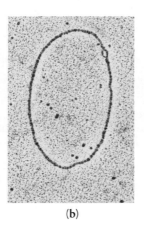
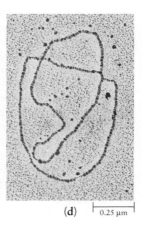

(a)   0.5 μm

16–11 (a) *Plasmids from* Neisseria gonorrhoeae, *the bacterium that causes gonorrhea. The two pairs of connected plasmids (indicated by the arrows) are probably just completing replication.* (b) *A plasmid from* E. coli *replicating. Its replication, like that of the bacterial chromosome, is bidirectional. At almost 2 o'clock in the plasmid, you can see the replication "eye," where the two DNA double helixes (each consisting, as you will remember, of one old strand and one new strand) are beginning to separate from each other.* (c) *The replication process is more than half-completed, and* (d) *the two plasmids are almost at the point of separation.*

(b)   **(c)**   **(d)**   0.25 μm

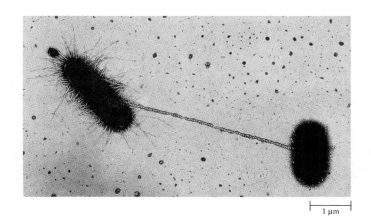

16–12 *Electron micrograph of conjugating* E. coli *cells. The elongated F⁺ (male) cell at the top of the micrograph is connected to the more rotund F⁻ (female) cell by a long pilus. Genes on the F plasmid are responsible for the production of these specialized pili, which are necessary for conjugation. Numerous shorter pili are visible on the F⁺ cell.*

1 μm

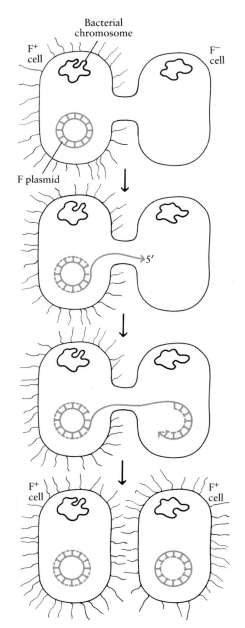

Bacterial chromosome

F⁺ cell

F⁻ cell

F plasmid

5′

F⁺ cell

F⁺ cell

The F Plasmid

The first plasmid to be recognized as such was the F (for fertility) factor of *E. coli*. This F factor, or F plasmid as it is sometimes called, contains some 25 genes, many of which control the production of F pili. F pili are long, rod-shaped protein structures that extend from the surface of cells containing the F plasmid, which are known as male (donor), or F⁺, cells. Cells that lack the F plasmid are known as female (recipient), or F⁻, cells. The F⁺ cells can attach themselves to F⁻ cells by the pili (Figure 16–12) and transfer the F plasmid to them through cytoplasmic bridges. Transfer of the F factor gives the recipient cells the capacity to produce F pili and to transfer the F plasmid (that is, the recipient cells become F⁺ cells). In a mixed bacterial culture, all F⁻ cells quickly become F⁺ cells. In bacteria, maleness is thus a highly contagious condition, a phenomenon fortunately without parallel in humans. Transfer of DNA from one cell to another by cell-to-cell contact is known as **conjugation.**

The transfer of an F plasmid is diagrammed in Figure 16–13. Note that it involves a mode of replication, known as **rolling-circle replication,** that differs significantly from the bidirectional (theta) replication of the bacterial chromosome (see Figure 16–2).

The F factor, like many other plasmids, can become integrated into the bacterial chromosome. A bacterial cell that contains the F factor as part of its chromosome—that is, as an episome—is known as an Hfr (high frequency of recombination) cell. An Hfr cell has an astonishing property: when it attaches to an F⁻ cell, the replicating bacterial chromosome itself (or a portion thereof) can be transferred from the Hfr cell to the F⁻ cell. In other words, the genes in the bacterial chromosome can be passed from one cell to another, resulting in a new gene combination in the recipient cell. Conjugation is thus, in effect, a form of sexual recombination.

In the Hfr cell at conjugation, a break occurs in the F-factor sequence in the chromosome, and rolling-circle replication begins (Figure 16–14). Leading by its 5′ end, a single strand of DNA passes from the Hfr cell to the F⁻ cell. Recombination may then occur between the chromosome of the recipient cell and portions

16–13 *Transfer of an F plasmid from an F⁺ cell to an F⁻ cell via rolling-circle replication. In these diagrams, the plasmid is shown greatly enlarged; in reality, it is much smaller than the bacterial chromosome and contains far fewer nucleotide pairs.*

A single strand of DNA moves into the recipient cell, where its complementary strand is synthesized (dashed lines). As the DNA strand is transferred, the donor

strand "rolls" counterclockwise, exposing the unpaired nucleotides. These serve as a template for the synthesis of a complementary DNA strand (dashed lines). As a result, the plasmid in the donor cell continues to be a circle of double-stranded DNA. The transferred plasmid converts the recipient cell to an F⁺ cell. Transfer of DNA by cell-to-cell contact is known as conjugation.

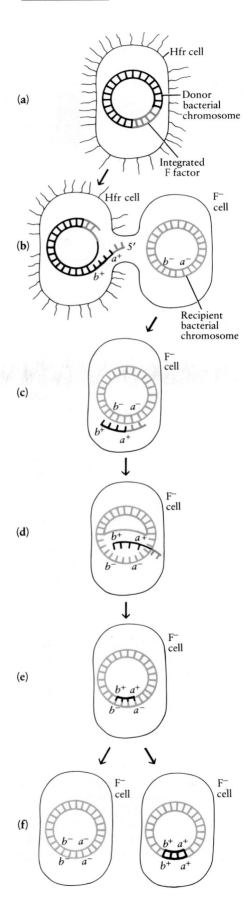

16-14 *Transfer of a portion of the bacterial chromosome during conjugation.* (a) *An F⁺ cell is converted into an Hfr cell when an F plasmid becomes inserted into its chromosome.* (b) *A break occurs in the F-factor sequence in the chromosome, and rolling-circle replication begins. Leading by its 5′ end, a single strand of the DNA, containing a portion of the F-factor sequence followed by genes a⁺ and b⁺, moves into the recipient F⁻ cell. In this example, only a portion of the chromosome is transferred before the cells separate from each other.* (c) *The single strand of transferred DNA is complementary to the part of the recipient chromosome that carries the same genes. The "match" is not exact, however, since genes a⁻ and b⁻ are alternate forms of genes a⁺ and b⁺. They differ in nucleotide sequence as a result of mutations that have caused, in this example, a⁻ and b⁻ to be nonfunctional—that is, they do not result in the synthesis of products a and b.* (d) *Recombination occurs between the donor DNA and the recipient chromosome. The displaced recipient DNA strand and the non-complementary portion of the donor strand (that is, the portion containing part of the F-factor sequence) are degraded by enzymes. The structure shown here exists only briefly.* (e) *The chromosome containing the integrated recipient DNA immediately replicates. This allows each of the "almost-matched" strands to act as a template for the formation of a new complementary strand in which all bases can be paired exactly.* (f) *The daughter cell containing genes a⁻ and b⁻ will not be able to synthesize products a and b, but the daughter cell that contains transferred genes a⁺ and b⁺ will be able to synthesize these products. This offers a means by which conjugation can be demonstrated. Note that the donor cell remains Hfr, and the recipient cell is still F⁻, as are its daughter cells.*

of the chromosome of the donor cell, with new material replacing old in the recipient cell's chromosome. Such recombinations can be detected by working with suitable auxotrophic strains of F⁻ bacteria—that is, mutant strains unable to synthesize particular molecules, which must be supplied in the growth medium for the bacteria to survive. Gaining the ability to synthesize those particular molecules demonstrates that the deficient F⁻ cells have received the necessary genes from the normal donor cells.

Chromosome Mapping

Studies of conjugation revealed that the genes of a bacterium are arranged in a regular linear order around the circular chromosome. The essential laboratory tools in these conjugation studies were a kitchen blender and a timer. Stop here a moment and see if you can figure out the role of these two comparatively humble instruments in mapping the bacterial chromosome. Here are two clues: (1) movement of a strand of DNA from the donor cell into the recipient cell during conjugation proceeds at a constant rate, and (2) transmission of a copy of the entire *E. coli* chromosome requires about an hour and a half at 37°C.

During the approximately 90 minutes required for the transfer of a complete chromosome copy from an Hfr cell to an F⁻ recipient cell, the newly synthesized strand of DNA is making its way into the recipient cell. By breaking the physical contact between the conjugating cells at various times in the process (which can be accomplished by whirling them at high speed in a blender) and then analyzing which genes have been transferred to the F⁻ recipient, it is possible to construct a map of the chromosome. The mapping process confirms that the genes are carried in a linear array on the chromosome: *A* is followed by *B*, *B* by *C*, and so on, down to *XYZ*.

As different Hfr strains were studied, it was found that insertion sites for the F plasmid differed from strain to strain and that the chromosome could be transferred in either direction, depending on the orientation of the inserted F plasmid. It was these studies that also first gave a clue that the *E. coli* chromosome is circular; that is, Y followed X, and Z followed Y, but next would come A and then B. The circularity of the chromosome—the fact that it has no ends—has now been confirmed by electron microscopy.

R Plasmids

In 1959, a group of Japanese scientists discovered that resistance to certain antibiotics and other antibacterial drugs can be readily transferred from one bacterial cell to another. Under experimental conditions, 100 percent of a population of drug-sensitive cells can become resistant within an hour after being mixed with suitable drug-resistant bacteria. It was subsequently found that the genes conveying drug resistance are often carried on plasmids, which have come to be known as R plasmids.

Resistance genes can also be transferred from one R plasmid to another. A single plasmid may collect as many as 10 resistance genes, making the cell it inhabits (and any cell to which it is transferred) resistant to as many as 10 different antibiotics (Figure 16–15). Resistance genes can also be transferred from plasmids to the bacterial chromosome, to viruses, and, most disturbing of all, to bacteria of other species. Thus, the usually innocuous *E. coli* can pick up R plasmids by conjugation and transfer them to *Shigella*, a bacterium capable of causing a sometimes fatal form of dysentery. Infectious drug resistance has now been found among an increasing number of types of pathogens, including those responsible for typhoid, gastroenteritis, plague, undulant fever, meningitis, and gonorrhea.

Typically, only a few copies of these large plasmids exist in a single cell. They are passed from mother to daughter cells at cell division, are transferred by conjugation, or, in another example of bacterial transformation (page 282), they may be simply passed from cell to cell through the cell membranes.

Drug resistance in bacterial cells is often the result of the synthesis of enzymes that break down the drug or that set up a new enzymatic pathway, circumventing the effects of the drug. Thus, resistance may depend on the synthesis of specific enzymes in high concentrations. The fact that resistance genes are on plasmids may allow for many copies of those genes to be produced very rapidly within a single cell.

16–15 *This plasmid, known as R6, carries genes that confer resistance to six different drugs, including the antibiotics tetracycline, neomycin, and streptomycin.*

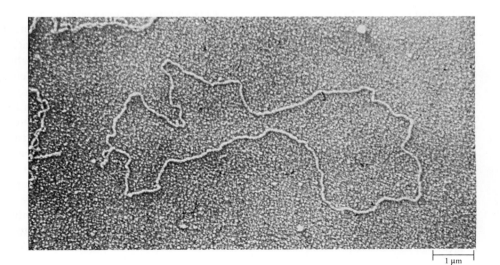

1 μm

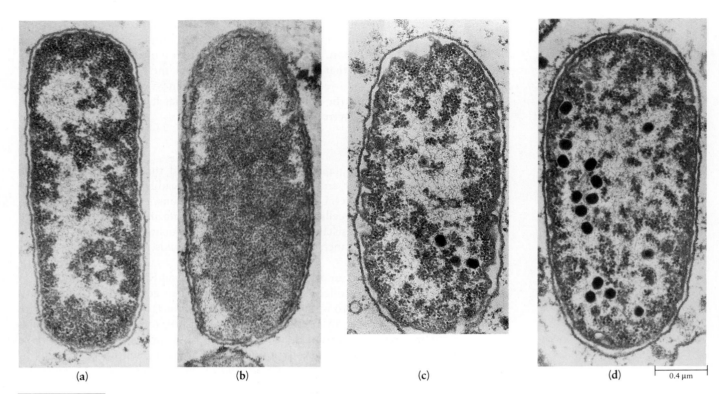

0.4 μm

(a)   (b)   (c)   (d)

16–16 *Stages in the replication of T4
bacteriophage in cells of* E. coli. *(a) As the
viral infection begins, the DNA of the
bacterial cell is visible in the electron-
transparent nucleoid region. (b) After 5
minutes, the bacterial DNA has changed
in appearance and has moved toward the
cell membrane. (c) After 15 minutes, the
bacterial DNA has disappeared, replaced
by vacuoles containing threads of replicating phage DNA; transcription from phage
DNA is occurring simultaneously. You
can also see proteins assembling themselves into the distinctive outer coat of the
bacteriophage. (d) After 30 minutes, many
phages, both complete and incomplete, are
present.*

VIRUSES

Viruses consist essentially of a molecule of nucleic acid enclosed in a protein coat,
or **capsid.** They contain no cytoplasm or ribosomes or other cellular machinery.
However, as we saw in Chapter 14, they can move from cell to cell and, within a
host cell, utilize its enzyme systems and organelles to replicate their nucleic acid
and synthesize new coat proteins. The coat may consist of one protein molecule
repeated over and over, as in tobacco mosaic virus (page 305), or a number of
different kinds of proteins, as in the T-even bacteriophages with their complex tail
assemblies (page 285). The composition of the protein coat determines the
attachment of the virus to the membrane of the host cell and the subsequent entry
of the viral nucleic acid into the cell.

Once within the host cell, the viral nucleic acid directs the production of new
viruses (Figure 16–16). This is accomplished using the raw materials of the
cell—such as nucleotides, amino acids, and the cell's ATP and other energy
sources—and also the cell's metabolic machinery. Thus viruses are obligate
parasites; they cannot multiply outside the host cell.

The nucleic acid of a virus—the viral chromosome—may be either DNA or
RNA, single-stranded or double-stranded, circular or linear. Viral chromosomes
vary greatly in size from some 5,400 nucleotides for a small, single-stranded DNA
bacteriophage, φX174, to 180,000 for the T-even bacteriophages. The viral
chromosome always codes for the coat protein or proteins, and also for one or
more enzymes involved in replication of the viral chromosome. These enzymes
ensure the rapid replication of viral nucleic acid in preference to the nucleic acid
of the host cell. The viral chromosome also codes for an enzyme or enzymes that,
once the new virus particles are assembled, enable them to lyse the host cell and
escape. The infection cycle is complete when the viral nucleic acid molecules are
packaged into the newly synthesized protein coats and the virus particles break
out of the host cell.

16–17 *When certain types of viruses—known as temperate bacteriophages—infect bacterial cells, one of two events may occur. (**a**) The viral DNA may enter the cell and set up an infection, or lytic, cycle, or (**b**) the viral DNA may become part of the bacterial chromosome, replicating with it and being passed on to daughter cells. Bacteria harboring such viruses are known as lysogenic, because from time to time, such a virus, called a prophage, becomes activated and sets up a new lytic cycle.*

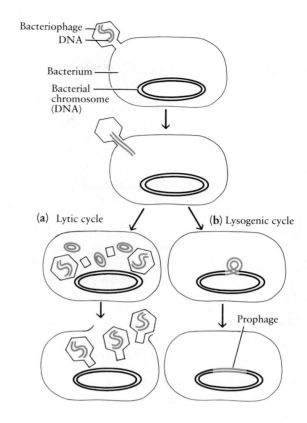

(**a**) Lytic cycle   (**b**) Lysogenic cycle

Prophage

Viruses as Vectors

The genetic makeup of the DNA of bacterial cells can be altered, as we have seen, by the introduction of DNA from other bacterial cells. Both transformation (page 282) and conjugation can result in recombination between donor and host DNA. Similarly, the transfer of plasmids and their insertion into the chromosome of a bacterial cell can change the DNA composition of the recipient cell. Viruses can also play a role as vectors (carriers) that move pieces of DNA from one bacterium to another.

Temperance and Lysogeny

Early in the study of bacteriophages, it was noted that a virus infection could suddenly erupt in a colony of apparently uninfected bacterial cells. Such cells were termed **lysogenic,** because of their capacity to generate a cycle of cell lysis that spread through neighboring cells. The cause of lysogeny, it was discovered, was the capacity of certain viruses to set up a long-term relationship with their host cell, remaining latent for many cellular generations before initiating an infection cycle. Such viruses became known as **temperate bacteriophages.** The DNA of temperate phages, like that of the F plasmid, may become integrated at specific sites in the host chromosome, replicating along with the chromosome. Such integrated bacteriophages are known as **prophages.** Prophages break loose from the host chromosome spontaneously about once in every 10,000 cell divisions, triggering a lytic cycle (Figure 16–17). In the laboratory, lysis may be triggered by ultraviolet light (see essay), x-rays, or other agents that damage nucleic acids.

Temperate phages resemble plasmids in that (1) they are autonomously replicating molecules of DNA, and (2) they may become integrated into the bacterial cell chromosome. They differ from plasmids both in their capacity to manufacture a protein coat and thus to exist (though not to replicate) outside the cell and in their capacity to lyse the host cell.

"Sir, I Am Entirely Lysed"

André Lwoff, who appears in the photograph on page 322, wrote the following account of his discovery of a technique for inducing lysogenic bacteria to undergo lysis:

Our aim was to persuade the totality of the bacterial population to produce bacteriophage. All our attempts—a large number of attempts it was—were without result. . . . Yet I had decided that extrinsic factors must induce the formation of bacteriophage. Moreover, the hypothesis had been published already [1949], and when one publishes an hypothesis, one is sentenced to hard labor. . . .

Our experiments consisted in inoculating exponentially growing bacteria into a given medium and following bacterial growth by measuring optical density [that is, the turbidity of the culture, which provided an indirect measure of the number of intact bacterial cells in the culture]. Samples were taken every fifteen minutes, and the technicians reported the results. They (the technicians, that is) were so involved that they had identified themselves with the bacteria, or with the growth curves, and they used to say, for example: "I am exponential," or "I am slightly flattened. . . ."

So negative experiments piled up, until after months and months of despair, it was decided to irradiate the bacteria with ultraviolet light. This was not rational at all, for ultraviolet radiations kill bacteria and bacteriophages, and on a strictly logical basis the idea still looks illogical in retrospect. Anyhow, a suspension of lysogenic bacilli was put under the UV lamp for a few seconds.

The Service de Physiologie Microbienne is located in an attic, just under the roof of the Pasteur Institute, with no proper insulation. The thermometer sometimes rises in a manner that leaves no conclusion other than that the temperature is high. It was a very hot summer day and the thermometer was unusually high. After irradiation, I collapsed in an armchair, in sweat, despair, and hope. Fifteen minutes later, Evelyne Ritz, my technician, entered the room and said: "Sir, I am growing normally." After another quarter of an hour, she came again and reported simply that she was normal. After fifteen more minutes, she was still growing. It was very hot and more desperate than ever. Now sixty minutes had elapsed since irradiation; Evelyne entered the room again and said very quietly, in her soft voice: "Sir, I am entirely lysed." So she was: the bacteria had disappeared! As far as I can remember, this was the greatest thrill—molecular thrill—of my scientific career.

From André Lwoff, "The Prophage and I," in *Phage and the Origins of Molecular Biology,* a collection of essays compiled in 1966 by the Cold Spring Harbor Laboratory and dedicated to Max Delbrück on his sixtieth birthday.

Transduction

The process known as **transduction** is the transfer of cellular DNA from one host cell to another by means of viruses. In the course of a lytic cycle, as we have noted, viruses exploit the resources of the host cell. During the lytic cycle of many viruses, the host DNA becomes fragmented; when these viruses leave the cell, some of them may contain DNA fragments from the host chromosome. Since the amount of DNA that can be packaged within the protein coat is limited, such viruses lack some or all of their own necessary genetic information. Although they may be able to infect a new host cell, they are not be able to complete a lytic cycle. However, the genes they carry from their previous host may become incorporated into the chromosome of the new host. Depending on the genes that are transferred, these recombinations may be detected. This process is called general transduction (Figure 16–18a) because virtually any gene can be transferred by this mechanism.

16–18 *The two types of viral transduction.* **(a)** *General transduction occurs when a nontemperate bacteriophage infects a bacterial cell. The viral DNA enters the bacterial cell and undergoes a lytic cycle. In the course of this cycle, the DNA of the host cell is broken apart, and some of the fragments are accidentally incorporated into newly formed virus particles. When released, a virus particle containing bacterial DNA may infect another bacterial cell. Although such a virus is defective and unable to set up a lytic cycle, the bacterial DNA it has introduced may recombine with the DNA of the new host cell.*

(b) *Specialized, or restricted, transduction occurs when a temperate bacteriophage infects a bacterial cell and enters a lysogenic cycle. The viral DNA is incorporated into the host chromosome, where it may remain as a prophage for many generations. When the prophage leaves the bacterial chromosome, it often takes with it a piece of the bacterial DNA. In this case, only DNA adjacent to the insertion site of the prophage is picked up with the viral DNA. The linked bacterial and viral DNA is replicated and incorporated into new virus particles that are released from the cell when it is lysed. These particles infect other bacterial cells, and the genes of the first host cell may recombine with those of the new host cell. Viral DNA may also become integrated into the DNA of the new host cell.*

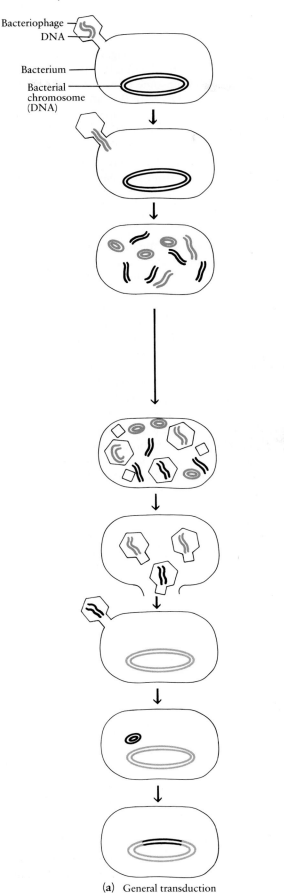

Bacteriophage
DNA

Bacterium

Bacterial
chromosome
(DNA)

(a) General transduction

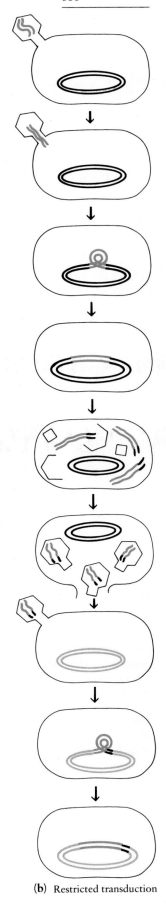

(b) Restricted transduction

When prophages break loose from a host chromosome to initiate a lytic cycle, they may, similarly, take a fragment of the host chromosome with them. The chromosome of each newly formed phage then consists of both host DNA and viral DNA. In this situation, the host DNA is not picked up at random, as is the case with nontemperate phages, but is, quite specifically, restricted to the portions of the host chromosome adjoining the insertion site of the prophage. Hence, this process is known as specialized, or restricted, transduction (Figure 16–18b).

Transduction resembles conjugation in that it involves the transfer of bacterial genes from one bacterial cell to another. It differs from conjugation in that in transduction the genes are carried by viruses.

Introducing Lambda

Lambda, the best studied of the temperate bacteriophages, has several interesting and instructive features. When the double-stranded DNA of the virus is packaged in its protein coat, it is linear—that is, it has two free ends. Once the viral chromosome is released into the host cell, however, it forms a circle. This closing of the circle takes place because a single strand of 12 nucleotides protrudes on the 5′ end of each strand of the DNA molecule (Figure 16–19). These strands are exactly complementary to one another and are said to be "sticky." Hydrogen bonds form between the complementary base pairs, joining the ends of the molecule together. Such a cohesive nucleotide sequence, which has since been found in other DNAs, is known in shorthand as a COS region.

When the DNA of lambda's circular chromosome is replicated, it is rolled out in numerous copies, joined in a single long molecule. A special enzyme then cleaves the molecule over and over at the COS sites and packages the individual lambda chromosomes into the waiting protein coats.

Lambda's integration into the E. coli chromosome takes place because the E. coli chromosome contains a short nucleotide sequence identical to a sequence on the lambda chromosome. Attachment of the bacteriophage chromosome to the

16–19 (a) A portion of the lambda chromosome, showing the nucleotide sequences of the single-stranded "sticky" ends. (b) In the host cell, the DNA molecule forms a circle when these single strands come together and their complementary bases are paired. The enzyme DNA ligase catalyzes the condensation reaction that links the phosphate subunit at each 5′ end to the deoxyribose subunit at each 3′ end, closing the "breaks."

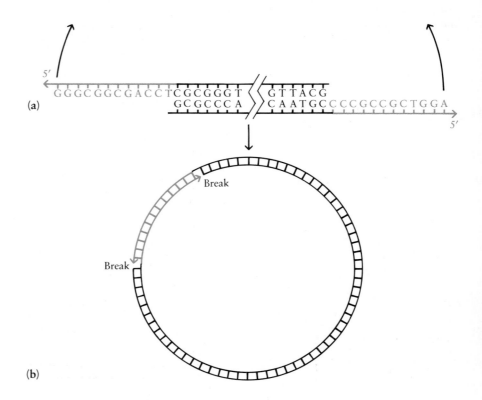

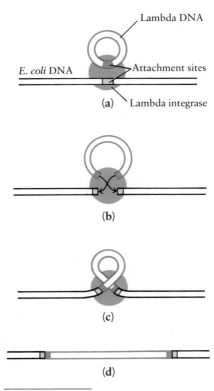

16-20 *Integration of lambda DNA into the chromosome of* E. coli. **(a)** *Lambda integrase recognizes the attachment sites on the DNA molecules of lambda and* E. coli *and brings them into close proximity.* **(b)** *This enzyme catalyzes the breakage of both circular DNA helixes and* **(c)** *their joining together, resulting in the incorporation of the lambda prophage into the* E. coli *chromosome* **(d)**.

bacterial chromosome is brought about by a special enzyme that recognizes both sequences, brings the two circular DNA helixes together, and initiates the cutting and sealing reactions (Figure 16-20). This enzyme, lambda integrase, is coded by the lambda chromosome. When lambda leaves the bacterial chromosome to begin a lytic cycle, another enzyme frees it from the bacterial chromosome by making a staggered cut that leaves "sticky" ends protruding. These ends rapidly rejoin one another, forming a circular chromosome once more.

As we shall see in the next chapter, specific recognition sites, staggered cuts, and "sticky" ends, the everyday equipment of lambda, were to prove equally essential to molecular geneticists.

TRANSPOSONS

More recently, another type of movable genetic element has been found; it is known as a **transposon.** Like episomes and prophages, transposons are segments of DNA that are integrated into the chromosomal DNA. However, they differ from episomes and prophages in that they contain a gene that codes an enzyme, transposase, that catalyzes their insertion into a new site. Also, at each end they have a repeated nucleotide sequence. This sequence may consist of direct repeats —such as ATTCAG and ATTCAG—or of inverted repeats—such as ATTCAG and GACTTA. The repeated sequences are typically 20 to 40 nucleotides in length. At the time of insertion, the target site on the host chromosome—the site at which the transposon becomes inserted—is duplicated. The target sequence, which is 5 to 10 base pairs in length, then flanks the transposon (Figure 16-21). In some cases, the transposons do not actually move—that is, they do not disappear from their initial site when they appear at a new location. Instead the original, parental transposon gives rise to a new copy that becomes inserted elsewhere.

Two kinds of transposons are known, simple and complex. Simple transposons, also called **insertion sequences,** are only about 600 to 1,500 base pairs in length and do not carry any genes beyond those essential for the process of transposition. At least six different simple transposons have been found in E. coli. They are detectable because they cause mutations. If one of these transposons becomes inserted into a gene, it inactivates it. Simple transposons also contain promoter sequences, which may lead to the inappropriate initiation of transcription of previously inactive genes of the host chromosome. Simple transposons appear to have no function but to duplicate themselves; thus they are among that group of molecules known as "selfish DNA."

16-21 *Insertion of a transposon into recipient DNA.* **(a)** *The nucleotide sequence at which insertion occurs is known as the target site.* **(b)** *Staggered cuts are made in the target site, and* **(c)** *the transposon is attached to the protruding ends of the cuts.* **(d)** *As the gaps are filled by complementary nucleotides, identical repeats are formed on the two sides of the inserted transposon. These are often used as "landmarks" in the identification of DNA sequences that have been transposed.*

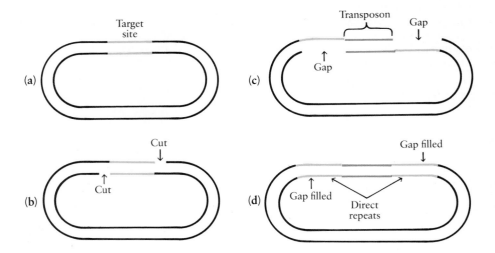

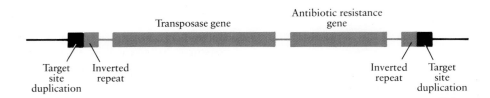

16-22 *The structure of a complex transposon. The transposable element shown here consists of the gene that codes for transposase, a gene that codes for an enzyme that confers resistance to an antibiotic, and, at each end, inverted repeats. The duplicated target site sequence is part of the host chromosome.*

Complex transposons are much larger and carry genes that code for additional proteins (Figure 16–22). As is the case with simple transposons, complex transposons may cause mutations, but they are also detectable because of their gene products. Genes that are part of a complex transposon can move from place to place on a chromosome or from chromosome to chromosome, and are therefore known as "jumping genes." Drug-resistance genes are often part of transposons, and so can be transferred readily from plasmid to plasmid and from plasmid to bacterial chromosome to plasmid again. Complex transposons are often found to have simple transposons flanking them—one at each end—which suggests that complex transposons might have come about by two simple transposons jumping at the same time, taking with them everything in between.

RECOMBINATION STRATEGIES

We have now described four different ways that new, information-carrying DNA can be introduced into a bacterial cell: transformation (page 282), which is the uptake of fragments of DNA; conjugation, which is the direct transfer of DNA from one cell to another; viral infection, with the injection of viral nucleic acid; and transduction, the transfer by viruses of nonviral genetic material from one cell to another.

We have also, as you may have noticed, described two different ways in which genetic recombination can take place. One involves exchange between homologous segments of DNA. When two such segments of double-stranded DNA are aligned with one another, exchanges occur between the molecules in such a way that genes may be transferred from one molecule to the other. This phenomenon occurs in eukaryotic cells during meiosis as a result of crossing over; it also takes place during conjugation, transformation, and transduction in bacterial cells.

16-23 *Transposons can carry genes conferring antibiotic resistance from R plasmid to R plasmid and also into and out of the bacterial chromosome. This micrograph shows three transposons that carry a gene conferring resistance to the antibiotic ampicillin; they are from a gonorrhea-causing bacterium.*

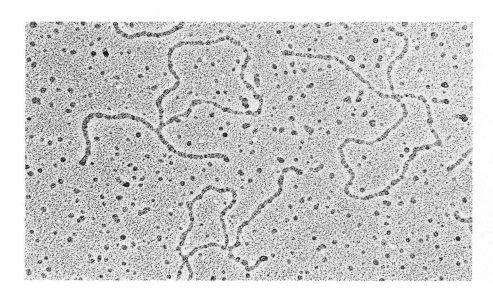

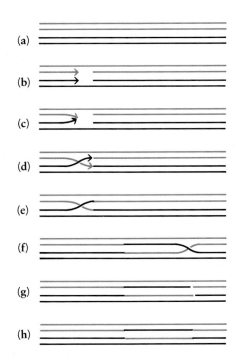

16–24 *The "single-strand switch" model of genetic recombination between two homologous strands of DNA. (a) The homologous parental DNAs are indicated in black and color. (b) One strand of each DNA molecule is broken, (c, d) switched over to the other molecule, and (e) joined to the opposite switched strand. (f) The exchange of strands between DNAs proceeds along the chromosome, and (g) at a specific point the switched strands are again broken and (h) resealed, completing the exchange and recombination of the genes.*

Several models of how homologous recombination takes place have been proposed; one of them, the "single-strand switch," is shown in Figure 16–24.

A second kind of recombination involves the insertion of movable (and removable) genetic elements. These elements can enter or leave the DNA of a chromosome without the occurrence of homologous recombination. The F plasmid, lambda phage, and transposons are all examples of this phenomenon.

Genetic recombination clearly was not invented by late twentieth-century molecular biologists; it has undoubtedly been taking place for billions of years, long before *Homo sapiens* was even a twinkle in the eye of evolution.

SUMMARY

The essential genetic information of prokaryotes, of which *E. coli* is the best-studied example, is coded in a circular double-stranded molecule of DNA. It is tightly packed in the bacterial cell in the nucleoid region. Replication begins at a particular site on the chromosome and proceeds bidirectionally (theta replication).

A principal means of genetic regulation in bacteria is the operon system. An operon is a linear sequence of genes coding for a group of functionally related proteins plus the promoter and operator. The structural genes of the operon are transcribed as a single mRNA molecule. Transcription from the operon is controlled by the promoter and operator sequences, which are adjacent to the structural genes and bind specific proteins. The promoter contains the binding site for RNA polymerase and may contain a binding site for the CAP-cyclic AMP complex. The operator is the binding site for a repressor, a protein coded by another gene, the regulator, which may be located some distance away on the bacterial chromosome. The operator overlaps the promoter, the first structural gene, or both; when the repressor is attached to the DNA molecule at the operator, RNA polymerase cannot initiate transcription of mRNA. When the repressor is not present, RNA polymerase can attach to the DNA and begin its movement along the chromosome, permitting transcription and protein synthesis to take place.

The *lac* operon is an example of an inducible operon. It is turned from "off" to "on" when an inducer binds to and inactivates the repressor. Other operons, such as the *trp* operon, are repressible. These are turned from "on" to "off" by the action of a corepressor that binds to an inactive repressor. This activates the repressor and it binds to the operator. Both induction and repression are forms of negative regulation.

Positive regulation of some operons is provided by the binding of the CAP-cAMP complex. For example, when glucose is present in the cell, cyclic AMP levels are low, and the CAP-cAMP complex does not form. As glucose is depleted, cAMP levels rise, and CAP-cAMP complexes form and then bind to the promoter. With lactose present (and the repressor thus inactivated) and the CAP-cAMP complex in place, RNA polymerase also binds to the promoter, and transcription from the operon takes place.

In addition to the genes carried on the bacterial chromosome, the bacterial cell contains other genes carried in plasmids, which are much smaller, also circular, double-stranded DNA molecules. Most plasmids can be transferred from cell to cell. Such transfer of DNA by cell-to-cell contact is known as conjugation. Some plasmids can become reversibly integrated into the bacterial chromosome, in which case they are known as episomes. Plasmids often carry genes for drug resistance; as many as 10 such genes have been located on a single plasmid.

The F (fertility) factor of *E. coli* is a plasmid present in donor F$^+$ (male) cells that can be transferred to recipient F$^-$ (female) cells; such cells then become F$^+$ and can transfer the F factor. When the F factor becomes integrated into the chromosome of an *E. coli* cell (making it an Hfr cell), part or (rarely) all of the chromosome can be transferred to another *E. coli* cell. At the time of transfer, the chromosome replicates by the rolling-circle mechanism, and a single-stranded copy of the DNA enters the recipient cell linearly so that the bacterial genes enter the cell one after another in a fixed sequence. The complementary strand is then synthesized from nucleotides available in the recipient cell. Because the rate at which the bacterial genes enter the recipient cell is constant at a given temperature, separation of conjugating cells at regular intervals provides a means for mapping the bacterial chromosome.

Viruses consist of either DNA or RNA wrapped in a protein coat. Within a host cell, the viral nucleic acid can make use of the cell's metabolic resources to synthesize more viral nucleic acid molecules and more viral proteins. Packaged in their protein coats, the virus particles can then break out of the cell to start a new infection cycle.

The DNA of some viruses, known as temperate viruses, can become integrated into the host chromosome where, like an episome, it replicates along with the chromosome. When integrated into a host chromosome, the DNA of a bacterial virus is known as a prophage. From time to time, prophages break loose from the chromosome and set up a new infection (lytic) cycle.

Viruses can serve as vectors, transporting genes from cell to cell, a process known as transduction. General transduction occurs when host DNA, fragmented in the course of the viral infection, is incorporated into new virus particles that carry these fragments to a new host cell. Specialized transduction occurs when a prophage, on breaking away from the host chromosome, carries with it—as part of the viral chromosome—host genes, which are then transported to a new host cell.

Lambda is a temperate bacteriophage of *E. coli*. The lambda chromosome is linear when it is in the viral protein coat, but when it is released into the cytoplasm of *E. coli*, it forms a circle. The closing occurs because of the existence of "sticky" ends—single-stranded complementary DNA sequences on either end of the molecule. Lambda becomes integrated into the bacterial chromosome at a specific attachment site bearing a nucleotide sequence identical to a sequence in lambda itself.

Transposons are movable genetic elements that differ from plasmids and viruses in several respects: (1) they carry a gene for the enzyme transposase, which catalyzes their integration into the host chromosome; (2) a repeated sequence, either direct or inverted, is present at each end of the transposon; (3) the target sequence on the host chromosome is duplicated when the transposon is inserted, with the result that a transposon is flanked on each end by the target sequence. Transposons may cause mutations by interfering with the normal expression of host-cell genes. Simple transposons contain only genes involved in their transposition; complex transposons carry additional structural genes.

Genetic recombinations can take place either because of exchanges between homologous sequences of DNA—substitution of one sequence for another—or by insertion of additional, new DNA into a recipient chromosome. Exchanges of host-cell genes in transformation, conjugation, and transduction all take place by the former mechanism (substitution of one sequence of DNA for another similar sequence). The second type of genetic recombination (the insertion of additional, new DNA) is characteristic of transposons, of the viral DNA of prophages, and of plasmids, such as the F factor, that can be added to or subtracted from the bacterial chromosome.

QUESTIONS

1. Distinguish among the following: inducible enzyme/repressible enzyme; operon/promoter/operator; inducer/repressor/corepressor; transformation/conjugation/transduction; plasmid/virus/transposon; F plasmid/R plasmid; F^+ cell/F^- cell/Hfr cell; theta replication/rolling-circle replication; prophage/lysogenic bacterium; general transduction/specialized transduction.

2. Compounds can be used by cells in two different ways. One type is broken down (usually as an energy source). Another is used as a building block for a larger molecule. Which type of compound would you expect to function as a corepressor? As an inducer? Do the examples given in the text conform to your expectations?

3. A culture of bacterial cells is grown in a medium in which both glucose and lactose are present in fixed amounts as the sole carbon sources. Describe the series of events that take place in the operon as the sugars are metabolized.

4. Gene expression theoretically can be regulated at the level of transcription, translation, or activation of the protein. In the latter case, the polypeptide produced by protein synthesis is in an inactive form. It undergoes some structural modification (catalyzed by enzymes) before it can perform its function in the cell. What would be the advantages of each type of regulation, in terms of the cell? Under what circumstances might one type be more useful than another? Which is the more economical?

5. How did the mapping experiments with *E. coli* demonstrate that the bacterial chromosome is both linear and circular?

6. You are trying to map the DNA of a strange new bacterium. It is a circular duplex similar to that of *E. coli*. It contains an F plasmid and can transfer DNA during conjugation. After allowing different strains of the bacterium to conjugate, two by two, for different lengths of time, the process is interrupted. In each experiment, you test whether a certain gene (for example, met^-) that was previously nonfunctional in the recipient strain is now functional in that strain—that is, you want to determine if transfer of the gene from the donor chromosome, followed by recombination with the recipient chromosome, has occurred. To do this, you plate the recipient bacteria onto a minimal medium and see if they grow. Only cells containing genetic recombinants will grow. From the results given below, determine the order of genes on the chromosome.

		GROWTH ON MINIMAL MEDIUM			
DONOR	RECIPIENT	5 MIN	10 MIN	15 MIN	20 MIN
phe^+met^- ×	phe^-met^+	no	no	yes	yes
met^+leu^- ×	met^-leu^+	yes	yes	yes	yes
leu^+phe^- ×	leu^-phe^+	no	yes	yes	yes
leu^-phe^+ ×	leu^+phe^-	no	no	yes	yes
met^-pro^+ ×	met^+pro^-	no	no	no	yes

7. As bacterial conjugation indicates, it is possible to separate the production of new genetic combinations from reproduction. Why do you think these two processes are combined in eukaryotic cells?

8. Describe the two possible outcomes of infection of a bacterial cell by a temperate virus.

Recombinant DNA: The Tools of the Trade

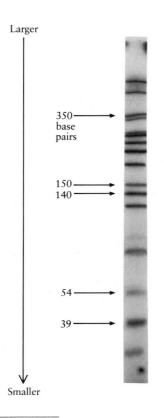

Larger

350 base pairs

150
140

54

39

Smaller

17–1 *Electrophoresis, the technique used by Pauling to separate normal and sickle cell hemoglobins (page 304) can also be used to separate fragments of DNA. In electrophoresis, the electrical field can separate molecules not only on the basis of their charge (the feature utilized by Pauling) but also on the basis of their size. Smaller molecules move faster than larger ones. As shown in this photograph of an actual electrophoretic gel, DNA fragments containing different numbers of base pairs can be cleanly separated from one another. The separated, purified fragments can subsequently be washed out of the gel unharmed. This separation procedure is important in many aspects of recombinant DNA work.*

The revelation of the many ways in which cells process, add, delete, and transfer genetic information opened the way for molecular biologists to carry out their own genetic manipulations. This field of human technological activity is often referred to simply as recombinant DNA—a term that overlooks the fact that the recombining of DNA was, by all evidence, going on long before the first amoeba appeared, much less a curious primate.

Recombinant DNA technology has made possible further investigations of the structure and function of genes, especially the otherwise inaccessible genes of eukaryotes. The results of some of these studies will be described in the next chapter. It has also suddenly and dramatically opened the way to new understandings of human genetics (the subject of the final chapter of this section), leading to the accurate diagnosis of many human genetic diseases, and, very likely, someday, the successful treatment of such diseases.

In this chapter we shall describe some of the basic techniques of recombinant DNA—the tools of the trade. The most important ones are (1) methods for obtaining specific, uniform DNA sequences—that is, segments of DNA molecules of a size suitable for analysis and manipulation; (2) DNA cloning, which makes it possible to produce such segments of DNA in large quantities; (3) nucleic acid hybridization, a method for identifying specific segments of DNA and RNA and for estimating similarities between nucleic acids from different sources; and (4) DNA sequencing, the determination of the exact order of nucleotides in a DNA segment—in effect, allowing a direct "reading" of the encoded genetic information.

ISOLATION OF SPECIFIC DNA SEGMENTS

When researchers first confronted the size and complexity of the DNA of even the simplest virus, the possibility of ever deciphering the encoded genetic information seemed virtually hopeless. Chemical analysis depends upon being able to obtain uniform samples of a manageable size and, for a while, that seemed impossible. However, as has occurred repeatedly in recombinant DNA research, the tools were provided by the organisms themselves. In this case, the tools were **restriction enzymes,** which are synthesized by certain bacterial cells, and another enzyme, known as **reverse transcriptase,** which is encoded by the nucleic acid of certain RNA viruses.

Restriction Enzymes: gDNA

Restriction enzymes were discovered in the early 1970s in a number of bacterial species, including *Escherichia coli.* The function of these enzymes is to cleave foreign DNA—that is, DNA from other bacterial strains or from viruses. For example, if DNA from *E. coli* strain B is introduced into cells of *E. coli* strain C, it is broken into fragments by restriction enzymes of the strain C cells.

The essential feature of restriction enzymes, which has made them indispensable for recombinant DNA technology, is that they cleave the DNA not at random but only at very specific nucleotide sequences, some four to eight base pairs in length. These sequences are referred to as **recognition sequences,** since they are "recognized" by specific restriction enzymes. The bacteria—for example, *E. coli* cells of strain C—protect their own DNA from their restriction enzymes by adding a methyl group ($-CH_3$) to one or more of the nucleotides in the recognition sequences of their DNA. This methylation, which occurs during DNA replication, is accomplished by special enzymes that are commonly found with the restriction enzymes. For example, one restriction enzyme of *E. coli,* called *Eco*RI, cleaves DNA only at the sequence GAATTC. Cells that produce *Eco*RI also produce a specific methylating enzyme that adds a methyl group to one of the adenines in the GAATTC sequence, thus protecting their own DNA from recognition and cleavage.

A second important feature of restriction enzymes, as shown in Figure 17–2, is that not all of them make straight cuts through both strands of the DNA molecule; some, including *Eco*RI, cut through the strands a few nucleotides apart, leaving "sticky" ends, as in lambda (page 334). As we have seen, such "sticky" ends can rejoin when hydrogen bonds form spontaneously between complementary bases and the enzyme DNA ligase forges a sugar-phosphate bond linking the ends of each strand together. Also—and this is most important—these "sticky" ends can join with any other segment of DNA that has been cleaved by the same restriction enzyme and, as a result, has complementary "sticky" ends. This discovery marked the beginning of recombinant DNA technology.

More than 200 different restriction enzymes have now been isolated, making it possible to cleave a DNA molecule at any one of more than 90 recognition sequences, thus producing multiple uniform fragments of DNA molecules. The DNA fragments produced by restriction enzymes acting on the DNA of a cell or an organism are known as **genomic DNA,** or gDNA. Because of the number of different restriction enzymes and recognition sequences, it is now possible to splice together segments of gDNA from a limitless variety of sources.

It is interesting to note that certain viruses have enzymes analogous to the restriction enzymes and methylating enzymes of bacteria. For instance, the DNA of the T-even bacteriophages codes for enzymes that break the DNA of *E. coli* cells into fragments; the fragments are then recycled by the bacteriophages to make new bacteriophage DNA. These viral enzymes, like restriction enzymes, cleave DNA only at specific nucleotide sequences. The bacteriophages protect their own DNA from cleavage by methylation of the cytosines that occur within the sequences recognized by the enzymes.

17–2 *The DNA nucleotide sequences recognized by three widely used restriction enzymes:* (a) Hpa*I,* (b) Eco*RI, and* (c) Hind*III. Recognition sequences are frequently, as for these enzymes, six base pairs long, and, when read in the 5′ to 3′ direction, the two strands of the sequence are identical. EcoRI and HindIII cleave the DNA so that "sticky" ends result. Restriction enzymes are generally obtained from bacteria:* Hpa*I is from* Hemophilus parainfluenzae, *EcoRI is from* E. coli, *and* Hind*III is from* Hemophilus influenzae.

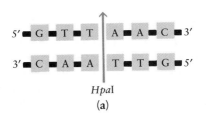

(a)

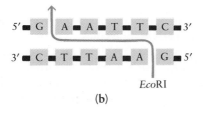

(b)

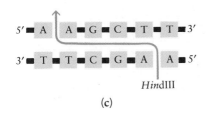

(c)

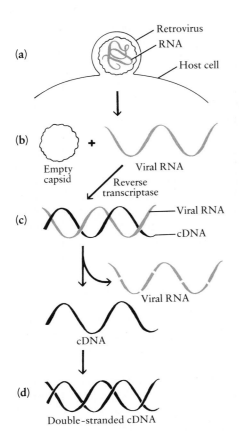

(a) Retrovirus
RNA
Host cell

(b) Empty capsid + Viral RNA
Reverse transcriptase

(c) Viral RNA
cDNA
Viral RNA

cDNA

(d) Double-stranded cDNA

17–3 *Infection of an animal cell by a retrovirus.* **(a)** *The capsid of a retrovirus is typically surrounded by an outer lipoprotein envelope formed from elements of the cell membrane of its previous host. This envelope can fuse with the cell membrane of a new host, allowing the virus to enter the cell.* **(b)** *Once the retrovirus has gained entry to the cell, the viral RNA is released from the capsid and* **(c)** *transcribed into a single strand of complementary DNA (cDNA).* **(d)** *Synthesis of the matching DNA strand follows immediately, producing a double-stranded molecule of cDNA. These reactions, as well as the degradation of the original viral RNA molecule, are all catalyzed by reverse transcriptase.*

As we shall see in the next chapter, the double-stranded cDNA can become integrated into the host-cell chromosome. Subsequently, new viral RNA molecules are transcribed from the cDNA, as are mRNA molecules that direct the synthesis of viral proteins.

Reverse Transcriptase: cDNA

A second method of obtaining specific DNA segments for cloning and manipulation became available with the unexpected discovery of a type of animal virus known as a **retrovirus.** Eukaryotic cells, including the cells of plants and animals, are host to both DNA and RNA viruses. In the case of DNA viruses, the viral DNA is both replicated, forming more viral DNA, and transcribed into messenger RNA, directing the synthesis of viral proteins. With most RNA viruses, the RNA is similarly replicated, forming new viral RNA, and also serves as messenger RNA. However, some RNA viruses have a different method of replication: in these viruses, the RNA serves first as a template for synthesizing DNA, using a viral enzyme, reverse transcriptase (Figure 17–3). Both new viral RNA and mRNA for the synthesis of viral proteins are subsequently transcribed from this DNA.

The discovery of retroviruses, as you might expect, has led to a revision in the "central dogma" of molecular genetics (Figure 17–4). These viruses are also of great interest to medical scientists; a number of retroviruses have been shown to cause cancer in animals, and the virus responsible for the devastating disease AIDS (to be discussed in Chapter 39) is a retrovirus. In addition, retroviral reverse transcriptase has proved to be a valuable tool in recombinant DNA studies.

DNA molecules synthesized by reverse transcriptase from an RNA template are known as **complementary DNA,** or cDNA. cDNA molecules can be spliced into other DNA molecules by means of artificial "sticky" ends. These are constructed by the addition of a single strand of a single nucleotide—TTTTTT, for instance—to the end of the cDNA sequence. The cDNA molecule can then combine with any other DNA molecule to which a complementary single strand has been added—AAAAAA, for instance.

17–4 *The revision of the "central dogma" of molecular genetics required by the discovery of retroviruses. Note that although information can flow from RNA to DNA, as well as from DNA to RNA, the essential feature of the "central dogma" remains intact: information does not flow from the phenotype (protein) to the genotype (nucleic acid).*

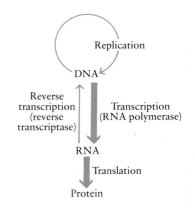

Replication
DNA
Reverse transcription (reverse transcriptase)
Transcription (RNA polymerase)
RNA
Translation
Protein

An advantage of using reverse transcriptase to produce DNA segments is that the cDNA molecules represent genes, rather than the fragments produced by restriction enzymes. A disadvantage, however, is that it requires the isolation of a specific mRNA molecule from which the cDNA can be transcribed. This is usually feasible only in cells producing large quantities of a particular protein—such as immature red blood cells making hemoglobin or lymphocytes (a type of white blood cell) producing antibodies—a serious limitation on the cDNA molecules that can be prepared.

17–5 *Preparation of a synthetic oligonucleotide.* (a) *Organic molecules that function as blocking groups are added chemically to either the 5′ or 3′ ends of mononucleotides. Some blocking groups can be removed by treatment with an acid, and others by treatment with a base. Typically, one type of blocking group is used on the 3′ end and another type on the 5′ end.* (b) *The blocking groups prevent a condensation reaction at the "wrong" ends of the nucleotides, but allow the desired reaction to occur. In this example, treatment with acid then removes the blocking group from the 5′ end of the newly formed dinucleotide,* (c) *allowing it to undergo a condensation reaction with another nucleotide that is blocked at the 5′ end but free to react at the 3′ end. The result is a trinucleotide* (d). *These steps are repeated over and over until the desired oligonucleotide is synthesized.*

Synthetic Oligonucleotides

Scientists have also developed methods for synthesizing short sequences of DNA and RNA in the laboratory. These molecules, called **synthetic oligonucleotides** (from *oligo,* meaning "few"), are a third source of uniform DNA or RNA segments.

As shown in Figure 17–5, the preparation of synthetic oligonucleotides depends on the capacity to selectively block either the 3′ end or the 5′ end of nucleotides. This ensures that the condensation reactions forming the sugar-phosphate bonds occur in the proper sequence to link the nucleotides in the desired order. When this procedure was first developed, synthesis of a segment containing from 12 to 20 nucleotides required several days. It has now become possible to attach one end of the chain to a resin bead in a column through which nucleotides and the appropriate chemical reagents can be passed sequentially, leading to automation of the procedure. As a result, longer oligonucleotides can be synthesized, in a much shorter period of time.

CLONES AND VECTORS

The next requirement for the detailed study of DNA was a methodology for obtaining gDNA, cDNA, or synthetic oligonucleotide molecules in large quantities. The machinery for duplicating DNA was already present in *E. coli* and other bacterial cells. What were needed were vectors that could carry the DNA molecules of interest into these cells and initiate replication. Again, prokaryotes and viruses provided the solution.

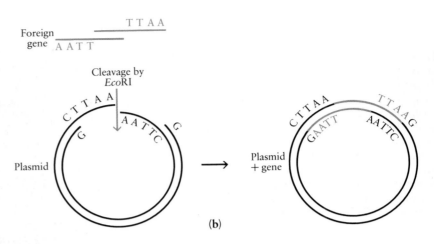

(b)

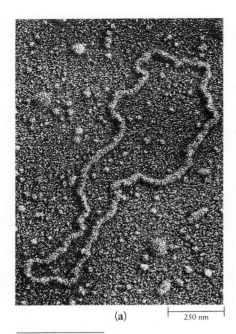

(a)

|_____|
                    250 nm

17-6 (a) *The plasmid pSC101. (b) The*
restriction enzyme EcoRI cleaves this
plasmid at the sequence GAATTC, leaving
"sticky" ends exposed. These ends, con-
sisting of TTAA and AATT sequences, can
join with any other segment of DNA that
has been cleaved by the same enzyme.
Thus it is possible to splice a foreign gene
into the plasmid. (In this diagram, the
length of the GAATTC sequences is exag-
gerated and the lengths of the other por-
tions of both the foreign gene and the
plasmid are compressed.)

When such plasmids, incorporating a
foreign gene, are released into a medium
in which bacteria are growing, they are
taken up by some of the bacterial cells. As
these cells multiply, the plasmids also rep-
licate; the result is an increasing number
of cells, all making copies of the same
plasmid. The plasmids can then be sepa-
rated from the other contents of the cells
and treated with EcoRI to release multiple
copies of the cloned gene.

Plasmids as Vectors

Not long after the discovery of the restriction enzyme *Eco*RI, Stanley Cohen and
his coworkers from the Stanford University Medical School isolated a small
plasmid of *E. coli,* designated pSC101 (note the initials of its discoverer), that
makes the bacteria resistant to the antibiotic tetracycline (Figure 17–6). pSC101
has only one GAATTC sequence in its entire molecule, and, as a consequence, it is
cleaved at only one site by *Eco*RI. The insertion of a small segment of foreign
DNA into the plasmid does not affect either the uptake of the plasmid by *E. coli,*
its capacity to make the recipient cells tetracycline-resistant, or its ability to
replicate. Typically, the plasmid replicates several times in its host cell, producing
about 10 new plasmids per cell. Because the pSC101 plasmids containing the
foreign DNA segment are larger than those that do not contain it, after replication
they can be readily isolated and collected. Treatment of the isolated plasmids by
*Eco*RI releases the foreign DNA, which can then be separated from the pSC101
DNA by electrophoresis (see Figure 17–1).

The discovery that pSC101 could be used as a vector for foreign DNA opened
the way to the production of uniform, identical segments of DNA in large enough
amounts to be analyzed by biochemical means. Such multiple copies are known as
clones, a term also applied to genetically identical bacteria and other organisms
produced asexually from a single parent cell or organism.

Lambda and Cosmids

Plasmids are useful as vectors because they multiply rapidly and are easily taken up
by bacteria through the cell membrane. Their chief shortcoming is that the life of
a plasmid is very competitive, and those that multiply the most rapidly have an
evolutionary advantage. And, as you might expect, the larger the plasmid, the
more time it requires in order to replicate. Thus, although short sequences of
foreign DNA are tolerated, longer segments tend to be eliminated over the
generations. Plasmids are reliable vectors only for segments of up to about 4,000
base pairs in length.

Specially modified strains of bacteriophage lambda can be used to clone larger
DNAs, segments of up to 20,000 base pairs in length. Such preparations are made
by removing the central section of the phage DNA, using strains of lambda that
have appropriately located recognition sites for restriction enzymes. This central
section contains genes that are involved solely with the integration of lambda into
the bacterial host chromosome (see Figure 16–20, page 335); they are not required
for infection of the cell or multiplication within it. This large section of the
lambda genome is replaced with foreign DNA. If the foreign DNA is about the
same length as the deleted DNA—that is, some 20,000 base pairs—it can be
introduced into the *E. coli* cell and will multiply as the virus undergoes its usual
cycles of lytic infection.

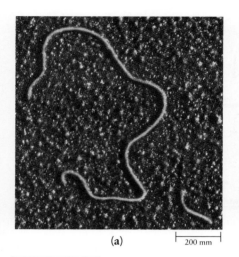

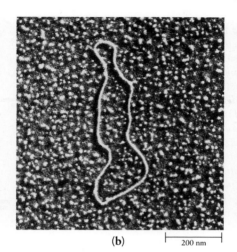

(a)   200 mm

(b)   200 nm

(c)   200 mm

17-7 *Insertion of a foreign gene into a bacterial cell.* (a) *A plasmid has been cut open with a restriction enzyme, leaving two "sticky" ends. A small segment of foreign DNA (lower right) also has "sticky" ends that can join with the ends of the plasmid by base pairing.* (b) *With the aid of DNA ligases, the foreign DNA has been spliced into the plasmid.* (c) *The plasmid, now containing the foreign DNA, is about to enter a bacterial cell. (Huntington Potter and David Dressler,* LIFE *Magazine, 1980, Time, Inc.)*

In order to clone still larger DNA segments, biologists have taken further advantage of the unusual and marvelous features of lambda, in this case, its cohesive ends. As we noted in the last chapter, the COS regions of lambda are the recognition sites for the enzyme that cuts the newly synthesized DNA into individual viral chromosomes and packs them into their protein coats. As it turns out, all that is required for the packaging to take place are two COS regions 35,000 to 40,000 base pairs apart. Using restriction enzymes, biologists are now able to construct DNA segments flanked by appropriately placed COS regions; these segments are then conveniently packaged into lambda protein coats by the viral enzyme. The protein coat gains them entry into the bacterial cell; once inside, the introduced DNA segments, like the normal lambda chromosome, assume a circular form (see Figure 16–19, page 334) and begin to multiply like plasmids. These vectors are appropriately known as **cosmids.**

NUCLEIC ACID HYBRIDIZATION

One of the earliest—and still one of the most useful—methods for studying DNA and RNA molecules is nucleic acid hybridization. This technique takes advantage of the base-pairing properties of nucleic acids. If DNA is heated, the hydrogen bonds holding the two strands together are broken and the strands separate; the three-dimensional structure of the molecule is lost, and the molecule is said to be denatured. When the solution is cooled, the hydrogen bonds re-form, reconstituting the double helix.

When denatured DNAs from different sources are mixed together, they undergo random collisions. If two strands with nearly complementary sequences (the matching need not be exact) find each other, they will form a hybrid double helix. The extent to which the segments from two samples reassociate and the speed with which they do so provide an estimate of the similarity between their nucleotide sequences. As we shall see in Chapter 20, this method is being used with increasing success to determine the evolutionary relationships among organisms.

When denatured DNA is mixed with single-stranded RNA, DNA-RNA hybrids can be formed. As we saw in Chapter 15 (page 308), this technique made possible the first identification and isolation of mRNA. Now, mRNA molecules are routinely used to identify and isolate corresponding DNA segments, and vice versa.

17–8 *The use of a radioactive probe to locate a DNA segment of interest.* (**a**) *Multiple copies of the DNA of a vector, such as a modified lambda chromosome, are cut with a suitable restriction enzyme, such as EcoRI. Foreign DNA containing the segment of interest is cut with the same restriction enzyme.* (**b**) *The restriction fragments of the two DNAs are mixed under conditions that allow the foreign DNA to become incorporated into the lambda chromosomes.* (**c**) *The resulting DNA molecules are enclosed in protein capsids, in a reaction catalyzed by an enzyme obtained from lambda. This produces phages capable of infecting bacterial cells.* (**d**) *Colonies, each consisting of a few identical bacterial cells, are established in small wells on a culture plate and infected with the phages. On the average, only one phage is added to each colony. After the cells have multiplied a number of times, which allows the vectors to multiply also, a replica of the bacterial colonies is obtained by blotting them with a specially prepared filter.* (**e**) *Chemical treatments release, denature, and attach the DNA to the filter. The filter is then incubated with a single-stranded radioactive probe, complementary to the DNA segment of interest.* (**f**) *After incubation, all single-stranded nucleic acid is washed from the filter, and the colony replica containing radioactive hybrid molecules is identified. The corresponding original colony is then treated to release the vectors, which can be purified and cloned in other bacteria to produce multiple copies of the DNA of interest.*

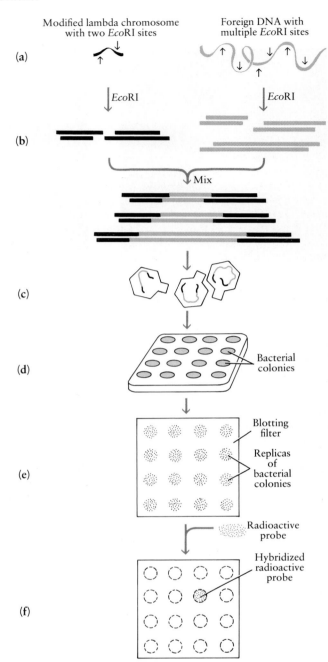

Radioactive Probes

Before a particular DNA or mRNA segment of interest can be cleaved, cloned, sequenced, or otherwise manipulated, it must first be located and isolated. One of the most important tools in accomplishing such location and isolation is an application of nucleic acid hybridization, using radioactive probes. These probes are short segments of single-stranded DNA or RNA labeled with a radioactive isotope. They may be mRNA or synthetic oligonucleotides or cloned fragments of gDNA or cDNA. Such probes may be used in a variety of ways.

For example, suppose that you are interested in studying a particular gene. One procedure that you could follow to locate and isolate DNA containing the gene is illustrated in Figure 17–8. First, a restriction enzyme is used to fragment DNA molecules known to contain the gene, and the fragments are incorporated into

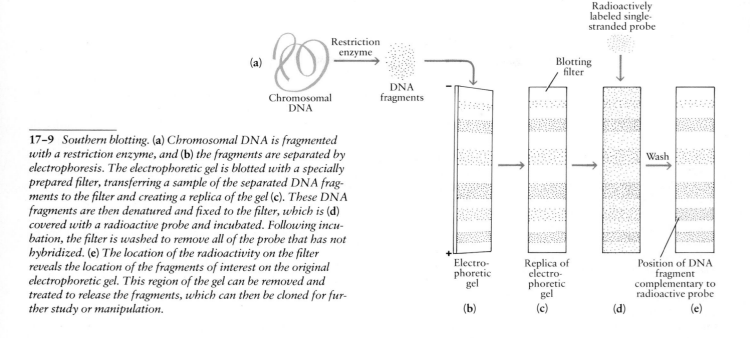

17-9 *Southern blotting.* (a) *Chromosomal DNA is fragmented with a restriction enzyme, and* (b) *the fragments are separated by electrophoresis. The electrophoretic gel is blotted with a specially prepared filter, transferring a sample of the separated DNA fragments to the filter and creating a replica of the gel* (c). *These DNA fragments are then denatured and fixed to the filter, which is* (d) *covered with a radioactive probe and incubated. Following incubation, the filter is washed to remove all of the probe that has not hybridized.* (e) *The location of the radioactivity on the filter reveals the location of the fragments of interest on the original electrophoretic gel. This region of the gel can be removed and treated to release the fragments, which can then be cloned for further study or manipulation.*

appropriate vectors. Some of the vectors carry fragments containing the gene of interest (or, more likely, portions of that gene), while others carry other fragments, which are, for your present purposes, not of particular interest. The vectors are then introduced into a series of identical bacterial colonies, each containing a very small number of cells. Approximately one vector is added to each colony, with the result that—if the vector is taken up by a cell—as each colony multiplies, an increasing number of copies of that one vector will be present. Using a special type of blotting filter, a replica of the colonies is made by transferring a portion of each colony to the filter, which can be treated in such a way that the DNA is released, denatured, and chemically attached to the filter. Next, the replica is exposed to a radioactive probe, previously prepared from, for example, a segment of corresponding mRNA. The probe is allowed to hybridize with any complementary DNA it encounters, and then all of the single-stranded nucleic acid is washed out of the filter; any double-stranded molecules that have formed remain in place on the filter. The colony replicas that are radioactive after this treatment identify the original colonies containing the vectors with the DNA segment you wish to study. Once the colony or colonies are identified, the vectors bearing the desired fragment can be collected, introduced into other bacterial cells, and cloned to produce large quantities of that DNA segment.

An analogous technique, illustrated in Figure 17–9, is known as Southern blotting (named after molecular biologist E. M. Southern, who developed it). DNA is first digested with one or more restriction enzymes. The resulting restriction fragments are separated by size on an electrophoretic gel and then blotted onto a filter. The filter thus contains a replica of the fragments on the gel, just as the filter in the previously described method contained a replica showing the positions of the bacterial colonies. The DNA fragments on the filter are then denatured, attached, and exposed to a radioactive probe, which, following the treatment outlined above, reveals the location of fragments containing a sequence complementary to the probe. Once the positions of the fragments of interest are identified, those fragments can be purified from the electrophoretic gel and used for subsequent study or manipulation. Southern blotting provides a very sensitive method for detecting specific nucleotide sequences of interest.

DNA SEQUENCING

The development of techniques for cleaving DNA molecules into smaller pieces and cloning them into multiple copies now makes it possible, in principle, to determine the nucleotide sequence of any isolated DNA molecule. One of the most important features of restriction enzymes is that different enzymes cleave DNA molecules at different sites (Figure 17–10). Cleavage of a DNA molecule with one restriction enzyme produces one particular set of short DNA fragments; cleavage of an identical DNA molecule with a different restriction enzyme produces a different set of short DNA fragments. The fragments of each set can be separated from each other by electrophoresis on the basis of their lengths and then cloned into multiple copies for use in sequencing studies.

One of the earliest methods for determining the nucleotide sequence of short DNA segments—and also one of the easiest to understand—was worked out by Allan Maxam and Walter Gilbert. It uses radioactive labeling techniques and electrophoresis. Multiple copies of single-stranded fragments are labeled on one end by removing a phosphate and substituting a radioactive phosphate in its place. Then, the mixture of identically labeled DNA molecules is separated into four portions. Each portion is treated chemically in such a way that one base—C, for example—and no other, is damaged and removed from the molecule, breaking the strand at that site. Crucial to the method is the fact that the chemical treatment is regulated so that not all the C's are damaged and those that are damaged are hit at random. Take, for example, a hypothetical 10-nucleotide sequence in which the base at the 5′ end bears a radioactive label, as indicated by the color:

$$5'3'$$
$$\boxed{G}-A-T-C-A-G-C-T-A-G$$

Random excision of the C's produces the following mixture:

$$\boxed{G}-A-T-C-A-G$$
$$T-A-G$$
$$\boxed{G}-A-T$$
$$A-G-C-T-A-G$$
$$A-G$$

Of these fragments, two are radioactively labeled at the 5′ end.

17–10 (a) *Simian virus 40, or SV40, has an icosahedral (20-sided) protein coat enclosing its chromosome, a circular molecule of double-stranded DNA. SV40, originally isolated from monkeys, is of particular interest because it can produce cancers in baby hamsters and other laboratory animals.* (b) *Because the SV40 chromosome has only one GAATTC sequence, EcoRI cleaves the SV40 DNA at only one point, which becomes the reference point for other analyses and manipulations.* (c) *The restriction enzyme HindIII cleaves the SV40 DNA at six points, yielding six fragments.* (d) *Electrophoresis separates the fragments according to size. Lane 1 shows uncut DNA, and lane 2 shows SV40 DNA cleaved by HindIII.*

The use of different restriction enzymes to cleave DNA molecules, such as the SV40 chromosome, into different sets of specific fragments is an essential component in determining the complete nucleotide sequence of the molecules.

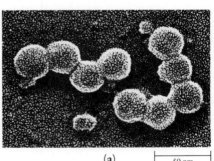

(a)   50 nm

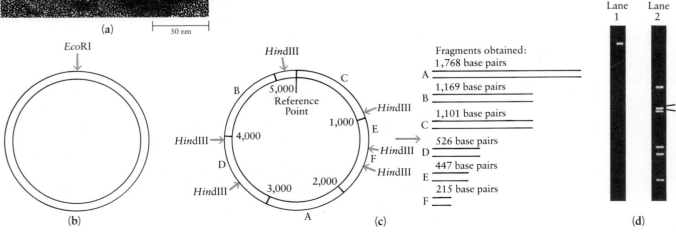

(b)   (c)   (d)

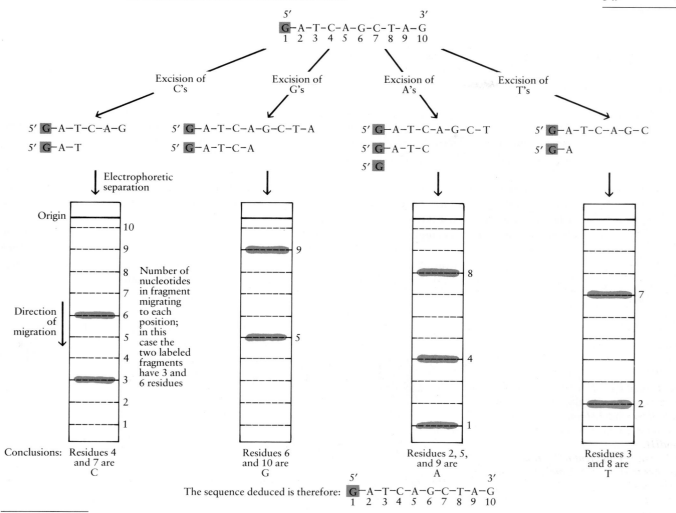

17–11 *The sequencing of a segment of a DNA molecule by the method of Maxam and Gilbert. The single-stranded segment (present in multiple copies) is radioactively labeled at the 5′ end. The solution containing the labeled DNA is divided into four portions, each of which is subjected to a different chemical treatment to break the molecule at only one of the four bases. The resulting fragments are then separated by electrophoresis on lanes that are calibrated to indicate the positions occupied by fragments of different lengths. Thus the location of the radioactivity reveals the number of nucleotides contained in the labeled fragments. By combining the information gained from each procedure, the sequence of the complete segment can be read directly off the lanes.*

The mixture of nucleotide fragments is then placed on a gel and subjected to electrophoresis, which separates them on the basis of their lengths. The shorter fragments move farther through the gel than do the longer fragments. The positions to which the two radioactive fragments move, which is revealed by their radioactive label, can be used to determine the number of nucleotides the labeled fragments contain. This same procedure, repeated for all four bases, results in a sequence ladder from which the order of nucleotides can be read off directly (Figure 17–11).

Because the sets of fragments produced by different restriction enzymes overlap, the information obtained from sequencing the fragments can be pieced together like a puzzle to reveal the entire sequence of the DNA molecule or isolated gene.

Another, widely used method for determining the nucleotide sequences of short DNA segments, using enzymatic rather than chemical procedures, was devised by Frederick Sanger. Using this method, Sanger provided the first complete sequence of a genome, that of the bacteriophage φX174 (see essay on the next page). Gilbert and Sanger were honored with the Nobel Prize in 1980 for their accomplishments in nucleic acid sequencing. It was Sanger's second; his first, received 22 years earlier, was for the first sequencing of a protein, insulin.

Today, the sequencing of genes from sources that range from the smallest viruses to human cells is going forward in hundreds of laboratories around the world.

Bacteriophage φX174 Breaks the Rules

The first complete nucleotide sequence to be worked out was that for the DNA of a small bacteriophage known as φX174. The single-stranded DNA that forms the chromosome of this virus was known to code for nine proteins, and the number of amino acids in each of these proteins was also known. This knowledge raised a serious and interesting question, even before the analysis was complete. The DNA (which contains 5,375 nucleotides) was insufficient to code for the nine proteins by the triplet code hypothesis. It simply was not long enough. Were the concepts established by a quarter-century of intensive effort to be overthrown by a submicroscopic particle?

When the nucleotide sequence became known, its secret was revealed. The investigators had originally assumed that each gene was physically separate along the DNA molecule. However, it turns out that there are pairs of overlapping genes. In other words, different genes in φX174 are coded by the same regions of DNA but using different reading frames. More recently, several additional examples of overlapping genes have been found in other viruses.

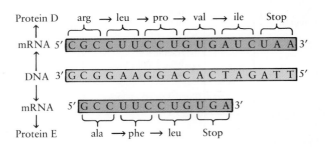

A segment of the DNA of φX174, showing a portion of the genes for proteins D and E. Both are coded by the same stretch of DNA; their reading frames, however, are different. The gene for protein E is completely contained within the gene for D, an entirely unrelated protein.

BIOTECHNOLOGY

Early in the course of recombinant DNA research, biologists realized that if the segments of DNA coding for certain proteins (particularly those of medical or agricultural importance) could be transferred into bacteria and expressed, the bacteria could function as "factories," providing a virtually limitless source of the proteins.

The first synthesis of a mammalian protein in a bacterial cell was reported by Keiichi Itakura and his associates at the City of Hope Medical Center. They selected the gene for the hormone somatostatin because it was a small protein (only 14 amino acids) and it could be detected in very small amounts.

The amino acid sequence of somatostatin is alanine-glycine-cysteine-lysine-asparagine-phenylalanine-phenylalanine-tryptophan-lysine-threonine-phenylalanine-threonine-serine-cysteine. Knowing this sequence, the investigators determined what one of the possible sequences of nucleotides in the DNA could be (you can do this, too) and then prepared a synthetic gene, including an initiation codon, bonding together the nucleotides one by one (see page 343).

Then, using a restriction enzyme, the synthetic gene was spliced into plasmids carrying genes for drug resistance, and the plasmids were supplied to *E. coli* cells. Some cells took up the plasmids (as evidenced by the fact that they were now drug-resistant), but there was no evidence of somatostatin synthesis. Itakura's

17–12 *The somatostatin project. A gene for somatostatin, synthesized artificially, was fused to the beta-galactosidase gene in a bacterial plasmid. Introduced into* E. coli, *this plasmid directs the synthesis of a hybrid protein that begins as beta-galactosidase but ends as somatostatin. Cyanogen bromide cleaves the protein at methionine, thus releasing the hormone intact (together with many fragments of beta-galactosidase, from which it can be separated).*

Reading in a clockwise direction, the nucleotide sequence shown in this diagram is the 5' to 3' sequence of the inactive strand of the DNA double helix. It is complementary to the template strand from which the mRNA is transcribed, and, with the exception of the substitution of thymine for uracil, it is identical to the 5' to 3' sequence of the transcribed mRNA molecule.

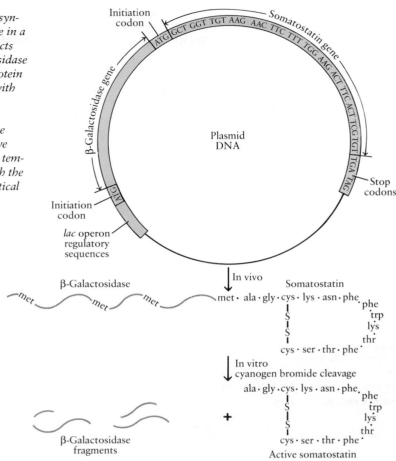

17–13 *Crystals of human insulin, known as "Humulin," produced by bacteria that have been modified by genetic engineering.*

group needed some way to turn the gene on. They inserted the regulatory sequences for the *lac* operon into the plasmid upstream from the somatostatin gene. When they turned the operon on, they were able to detect somatostatin, but only in very small amounts. Cells degrade foreign proteins, and the hormone was being destroyed almost as fast as it was being produced. Finally, to protect the newly synthesized somatostatin from the bacterial enzymes, the scientists spliced the somatostatin gene onto the beta-galactosidase gene, the first structural gene of the *lac* operon. At the point of the splice, for reasons that we shall see shortly, they retained the initiation codon of the synthetic gene; as you will recall, this is also the codon for methionine.

These plasmids were reintroduced into host cells, and, at last, clones of bacteria were obtained that were able to synthesize the hybrid beta-galactosidase-somatostatin protein. The protein was then isolated and treated with the chemical cyanogen bromide, which cleaved it exactly at the methionine insert, releasing the somatostatin (Figure 17–12). The somatostatin, tested in laboratory animals, was found to have the biological activity of the natural hormone. In other words, it worked!

More recently, genes for other medically useful proteins have been successfully introduced into bacterial cells and have functioned in protein synthesis. One example is the gene for human insulin (Figure 17–13). Another is the gene for somatotropin, or growth hormone, which is used to treat some forms of dwarfism in children. Somatotropin had previously been available only in very small amounts, being extracted from human pituitary glands. Several cases of contamination of these pituitary extracts by viruses causing human neurological diseases made the bacterial synthesis of this hormone of crucial medical importance.

On the economic front, the bacterial synthesis of other proteins is of increasing importance. For example, the enzyme rennin, extracted from calves' stomachs and used by the dairy industry for cheesemaking, has been produced by recombinant DNA technology. More recently, scientists at Cornell University have succeeded in inducing bacteria to synthesize the enzyme cellulase, which is produced in nature by certain fungi. This enzyme converts cellulose, which is indigestible by most organisms, into glucose, a food molecule of major importance.

Vaccines against viral disease are another important product of the new biotechnology. All viruses, as you know, consist of nucleic acid wrapped in a protein coat. It is the exterior proteins of the capsid that determine whether or not the virus can attach itself to and enter a target cell. In the animal bloodstream, these proteins of the virus, recognized by cells of the immune system as foreign, evoke the formation of antibodies, molecules that play a crucial role in future immunity against the virus. Most vaccines are made using killed or altered forms of the virus particles. Vaccines produced from synthetic protein coats alone are safer, since without the viral nucleic acid, no contamination of the vaccine by infectious particles can occur.

GENE TRANSFER: THE CASE OF THE GLOWING TOBACCO PLANT

Agrobacterium tumefaciens is a common soil bacterium that infects plants, producing a lump, or tumor, of tissue known as a crown gall (Figure 17–14). Even if the bacteria are destroyed with antibiotics, the gall, once started, continues to grow. Investigations have revealed that the cause of the gall is not *Agrobacterium tumefaciens* itself but rather a large (200,000 base pairs) plasmid of the bacterium. A portion of this plasmid, which is known as Ti (for tumor-inducing), becomes integrated into the DNA of the host plant cell. Crown gall disease is the only naturally occurring genetic recombination yet recorded between prokaryotic and eukaryotic cells.

To try to understand how Ti exerts its effects, recombinant DNA technology has been used to examine the genes of the Ti plasmid. Three of its genes, it has been found, direct the synthesis of plant hormones that act directly on the gall cells to promote their growth. One or more additional genes subvert the cell's machinery to produce unusual amino acids, called opines, which can be used by the gall cells but not by normal cells. Moreover, opines act as molecular aphrodisiacs, increasing bacterial conjugation and thus promoting the spread of the Ti plasmid to uninfected bacterial cells. In effect, Ti takes over and directs the activities of both of its hosts—the bacterial cells and the plant cells—to promote its own multiplication.

The Ti plasmid has attracted considerable attention not only because of its remarkable powers but also because of its potential as a vector to ferry useful genes into crop plants. Among the genes that are candidates for such transfer are those conferring resistance to major plant diseases and those required for C_4 photosynthesis (page 223) and for nitrogen fixation (page 664).

In a notable recent experiment, investigators from the University of California at San Diego isolated the gene for the enzyme luciferase from fireflies. The substrate of luciferase is a protein called luciferin; in the presence of oxygen, luciferin plus luciferase plus ATP produces bioluminescence, as seen in the flash of the firefly. The luciferase gene was cloned in *E. coli* and then spliced into the chromosome of a plant virus, which provided a regulatory sequence for the gene.

17–14 *Crown galls growing on a tobacco stem.*

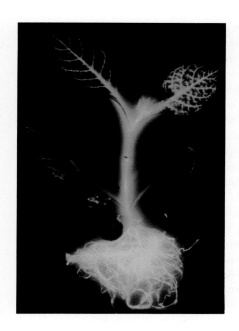

17-15 *Genes for the production of the enzyme luciferase are inserted into cells isolated from the normal tobacco plant* (Nicotiana tabacum), *using* Agrobacterium *as a vector. After the undifferentiated callus cells develop into a mature plant, the cells that have incorporated the luciferase gene into their DNA are luminescent in the presence of luciferin, ATP, and oxygen.*

The altered viral chromosome was then inserted into Ti plasmids, the plasmids were transferred to the bacteria, and the bacteria were incubated with tobacco leaf cells. The cells formed a mass of tissue, known as a callus, from which, in a suitable growth medium, new plants were produced. The new plants were watered with a solution containing luciferin. You can see the results in Figure 17–15: the plants shone!

One of the technical problems encountered in attempts at gene transfer is knowing whether a particular gene has actually been introduced into a new host cell and, if transferred, whether it is directing the synthesis of protein. Luciferase clearly provides an extraordinary signal that gene transfer has taken place, and experiments are now underway to combine it with other genes that are candidates for transfer. In the meantime, these experiments are a glowing example of the ingenuity of both molecular geneticists and the Ti plasmid.

SUMMARY

Recombinant DNA technology includes four basic techniques: (1) methods for obtaining specific, uniform DNA segments of a size suitable for analysis and manipulation; (2) DNA cloning, which makes it possible to produce identical DNA segments in large quantities; (3) nucleic acid hybridization, a method for identifying specific segments of DNA and RNA and for estimating similarities between nucleic acids from different sources; and (4) DNA sequencing, the determination of the exact order of nucleotides in a DNA segment. Another important tool used in recombinant DNA work is electrophoresis, which provides a means of separating DNA segments from one another on the basis of their size.

Specific, uniform DNA segments may be produced by cleaving DNA molecules with restriction enzymes, by transcribing mRNA into DNA with the enzyme reverse transcriptase, or through the laboratory synthesis of oligonucleotides. Restriction enzymes are found in nature in bacterial cells. They cleave DNA molecules at specific recognition sequences that are typically four to eight nucleotides in length. Their function in the bacterial cells in which they occur is the degradation of foreign DNA molecules. The DNA of the bacterial cell is protected from its own restriction enzymes by the methylation of nucleotides at the recognition sequences. Some restriction enzymes produce straight cuts through the DNA molecule. Others cut unevenly, leaving "sticky" ends that can then join by complementary base pairing with other fragments produced by the same enzyme. This makes it possible to combine DNA segments from different sources. The DNA segments produced by cleavage with restriction enzymes are known as genomic DNA, or gDNA.

Reverse transcriptase is an enzyme produced by certain RNA viruses, known as retroviruses. When these viruses infect a host cell, reverse transcriptase catalyzes the synthesis of DNA from the viral RNA template; viral mRNA, coding for viral proteins, is transcribed from this DNA, as is the viral RNA to be packaged into new virus particles. In the laboratory, reverse transcriptase can be used to synthesize DNA from an RNA template, such as an mRNA molecule. DNA segments produced in this manner are known as complementary DNA, or cDNA.

Clones, in molecular genetics, are multiple copies of the same DNA sequence. The sequences to be cloned are introduced into bacterial cells by means of vectors. Plasmids and bacteriophages, particularly bacteriophage lambda, are used as vectors; cosmids are synthetic vectors that combine the cohesive ends (COS regions) of lambda with the DNA segment to be cloned. Once in the bacterial cell, the vector and the foreign DNA it carries are replicated, and the multiple copies can be harvested from the cells.

Nucleic acid hybridization techniques depend upon the capacity of a single strand of RNA or DNA (which can be released from the double helix by heating) to combine, or hybridize, with another strand with a complementary nucleotide sequence. The greater the similarity between the nucleotide sequences of the two strands, the more rapid and more complete the hybridization. This technique makes possible a range of procedures, including estimating the evolutionary affinities of different organisms, determining relationships between DNAs and transcribed RNAs, and the use of radioactive probes to identify specific nucleotide sequences of interest. Radioactive probes can be used, for example, to identify bacterial colonies in which a particular DNA sequence, introduced by a vector, is being cloned, or to identify segments of interest on an electrophoretic gel.

DNA sequencing is the determination, nucleotide by nucleotide, of the sequence of bases in a molecule of DNA. Two principal techniques of sequencing are in current use, one involving enzymatic methods and the other chemical methods. Sequencing depends on the availability of multiple copies of uniform DNA segments, cloned from the DNA fragments produced by restriction enzymes. By combining sequencing information for sets of short segments produced by different restriction enzymes, molecular biologists can determine the complete sequence of a long DNA segment (such as an entire gene).

Techniques have now been developed for incorporating specific genes into suitable vectors, introducing them into bacterial cells, and inducing the bacterial cells to synthesize the proteins coded by the genes. Human insulin and human growth hormone are two medically important proteins now produced in this way.

The Ti plasmid of *Agrobacterium tumefaciens*, the cause of crown gall disease in plants, is being used as a vector for introducing genes into plant cells. Incorporation of the gene for the enzyme luciferase into the Ti plasmid makes it possible to determine visually if the genes carried by the plasmid have been successfully transferred to the plant cells and are being expressed.

QUESTIONS

1. Distinguish between the following: gDNA/cDNA; restriction enzyme/methylating enzyme; retrovirus/reverse transcriptase.

2. Identify the four techniques that form the basis of recombinant DNA technology and give a brief description of each.

3. What is electrophoresis? Why is it of such great value in recombinant DNA studies?

4. How is the DNA of a bacterial cell protected from the action of its own restriction enzymes? Why does this protection not extend to foreign DNA introduced into the cell?

5. Describe the role of "sticky" ends in recombinant DNA technology. What enzyme is required to complete the recombination?

6. Compare and contrast the features of plasmids, bacteriophage lambda, and cosmids as vectors. What are the advantages of the Ti plasmid as a vector for introducing genes into plant cells?

7. Suppose you treated a DNA molecule with a particular restriction enzyme and obtained five fragments, which you separated and cloned into multiple copies. Using the multiple copies, you then sequenced the five fragments. What would you do next to establish the order of the five fragments in the original molecule?

8. Suppose you wish to locate on the chromosome the gene coding for a small protein molecule. You know the amino acid sequence of the protein, and you have the technical skill to synthesize an artificial mRNA molecule with any nucleotide sequence you choose. How would you go about locating the gene? Suppose you wish to separate the gene from the rest of the chromosome. How would you proceed?

9. Why, in the somatostatin project, did the investigators link the synthetic somatostatin gene to regulatory elements of the *lac* operon? Why did they also include the first structural gene of the *lac* operon (the gene coding for beta-galactosidase) in the plasmid that was introduced into the *E. coli* cells?

10. *E. coli* cells used in recombinant DNA studies are "disabled"; that is, they lack the capacity to synthesize key components of their cell walls, for example, or to make a nitrogenous base, such as thymine. Consequently, they are able to survive only in an enriched laboratory medium. Why is such a precaution taken by molecular biologists?

C H A P T E R 18

The Molecular Genetics of Eukaryotes

The discovery, early in the history of molecular genetics, that the genetic code is apparently universal—the same in *Escherichia coli, Homo sapiens,* and all other organisms—is awesome evidence that all living things are descended from a common ancestor. It originally tempted molecular biologists to think that the eukaryotic chromosome would turn out to be simply a large-scale version of the *E. coli* chromosome. This has not proved to be the case. As studies of the molecular genetics of eukaryotes have progressed, increasingly aided by the tools of recombinant DNA technology, the gulf between eukaryotes and prokaryotes has widened rather than narrowed. It is now clear that, at the molecular level, there are many important differences in the genetics of eukaryotes and prokaryotes—some expected and some very surprising. As we shall see, these differences include (1) a far greater quantity of DNA in the eukaryotic cell; (2) a great deal of repetition in this DNA, with much of it lacking any apparent function; (3) a close association of the DNA with proteins that play a major role in chromosome structure; and (4) considerably more complexity in the organization of the protein-coding sequences of the DNA and the regulation of their expression.

THE EUKARYOTIC CHROMOSOME

DNA is an "exquisitely thin filament," in the words of E. J. DuPraw, who calculated that a length sufficient to reach from the earth to the sun would weigh

18–1 *In eukaryotic cells, the DNA of the genome is always found in association with protein. Chromosomes are made up of a protein-DNA combination called chromatin. At mitosis and meiosis, the chromatin condenses and becomes visible under the microscope. Each chromatid, according to present evidence, contains a single molecule of double-stranded DNA and is, by weight, about 60 percent protein. Shown here is human chromosome 12 at metaphase of mitosis.*

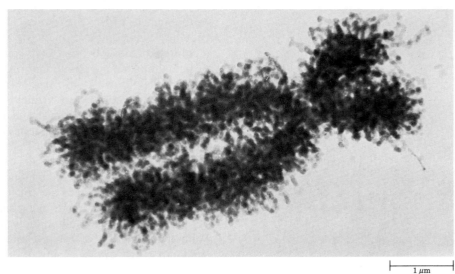

1 μm

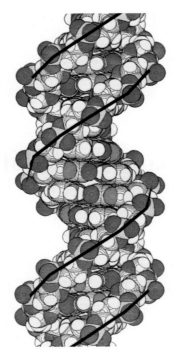 

B-DNA   Z-DNA

18–2 *Computer-generated diagrams of B-DNA and Z-DNA. B-DNA is the form described by Watson and Crick. It consists of two strands of nucleotides twisted around each other, with the backbones forming a smooth, right-handed helix. In Z-DNA, the backbones of the nucleotide strands form a zig-zag, left-handed helix. The two forms are reversible, suggesting that the changes in structure might reflect some regulatory role.*

only half a gram. The DNA of each eukaryotic chromosome is believed to be in the form of a single molecule. In a human chromosome, each of these molecules is believed to be from 3 to 4 centimeters long. Each diploid cell, with its 46 chromosomes, thus contains about 2 meters of DNA, and the entire human body contains some 25 billion kilometers of DNA double helix.

Double-stranded DNA is always a helix, and the helix is usually the tightly coiled right-handed helix, known as the B form, first described by Watson and Crick. However, x-ray diffraction studies have shown that DNA can assume other helical conformations: A-DNA, which is also a right-handed helix, less tightly coiled than the B form; and Z-DNA, which is a left-handed helix (Figure 18–2). The left-handed coiling results in zigs and zags of the sugar-phosphate backbone, hence the name Z-DNA. It is hypothesized that changes in the configuration of the DNA molecule may affect the binding of proteins to the molecule, and that this, in turn, affects gene expression.

Structure of the Chromosome

In the nucleus of the eukaryotic cell, the DNA is always found combined with proteins. This combination, as we noted in Chapter 5, is known as chromatin ("colored threads") because of its staining properties. Chromatin is more than half protein, and the most abundant proteins, by weight, belong to a class of small polypeptides known as histones. Histones are positively charged (basic) and so are attracted to—and, in turn, attract—the negatively charged (acidic) DNA. They are always present in chromatin and are synthesized in large amounts during the S phase of the cell cycle. The histones are primarily responsible for the folding and packaging of DNA. In a human cell, for instance, the approximately 2 meters of DNA are packed into 46 cylinders that, when condensed at metaphase, have a combined length of only 200 nanometers.

There are five distinct types of histones, known as H1, H2A, H2B, H3, and H4. They are present in enormous quantities—about 30 million molecules of H1 per cell, and about 60 million molecules of each of the other four types per cell. With the exception of H1, the amino acid sequences of the histones are very similar in widely diverse groups of organisms. The H3 molecule of the garden pea, for instance, differs from the H3 molecule of the cow by only four amino acids out of a total of 135.

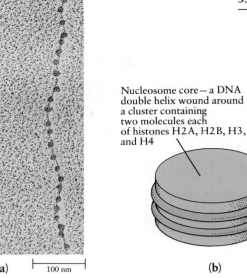

18–3 (a) *Chromatin that has been decondensed to reveal the beadlike nucleosomes. The distance between nucleosomes is about 10 to 11 nanometers, and the diameter of each bead is about 7 nanometers. The core of each nucleosome is composed of about 140 base pairs of DNA and an assembly of eight histone molecules. The strand of linker DNA between the nucleosome cores contains another 30 to 60 base pairs.*

(b) The structure of a nucleosome. The negatively charged DNA coils twice around the protein core, composed of eight positively charged histone molecules. An H1 histone molecule (also positively charged) binds to the outer surface of the nucleosome.

Nucleosome core — a DNA double helix wound around a cluster containing two molecules each of histones H2A, H2B, H3, and H4

H1 histone

Linker DNA

(a) 100 nm

(b)

The fundamental packing unit of chromatin is the **nucleosome** (Figure 18–3), which is composed of a core of two molecules each of histones H2A, H2B, H3, and H4—eight molecules in all—around which the DNA filament is wrapped twice, like thread around a spool. Each nucleosome contains, in addition to the eight histone molecules, about 140 pairs of nucleotides, and the strand of DNA between the nucleosomes contains another 30 to 60 nucleotide pairs. The fifth type of histone, H1, lies on this strand, outside the nucleosome core. When a fragment of DNA is tied up in a nucleosome, it is about one-sixth the length it would be if fully extended.

In electron micrographs, such as Figure 18–3a, the nucleosomes and the in-between, linking strands of DNA resemble beads on a string. In fact, electron micrographs provided the first clue to this remarkable structure, which has since been analyzed in detail by biochemical techniques.

Chromatin isolated at the next level of condensation is shown in Figure 18–4a. The details of this structure (which is 30 nanometers in diameter and is thus known as the 30-nanometer fiber) are not known. Two models that have been proposed are shown in Figure 18–4b and c.

18–4 (a) *Electron micrograph of a chromatid strand that is more tightly condensed than the beads-on-a-string form shown in Figure 18–3a. Because of its diameter, this form of chromatin is known as a 30-nanometer fiber. (b, c) Two proposed models for the way in which the beads-on-a-string form is packed into the 30-nanometer fiber.*

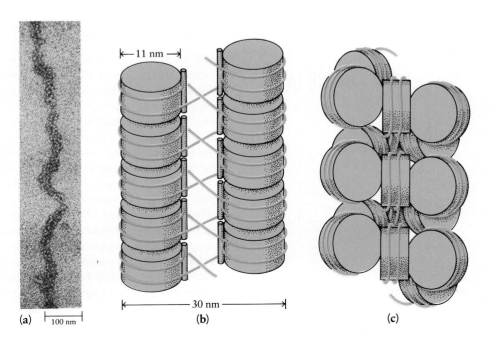

(a) 100 nm

11 nm

30 nm

(b)

(c)

18–5 *The looped domains of a chromosome can be revealed by chemical treatment that removes most of the histone proteins. In this electron micrograph of a single chromatid of an insect chromosome at mitotic anaphase, the DNA can be seen looping out from a nonhistone protein scaffolding. This scaffolding is essentially the same size and shape as the original anaphase chromosome.*

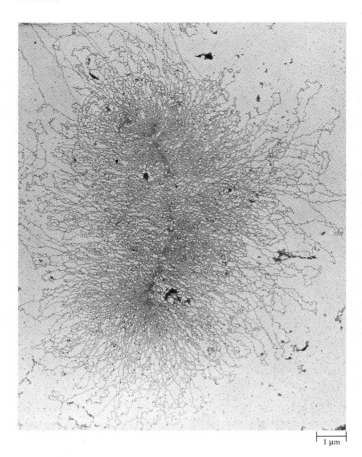

1 μm

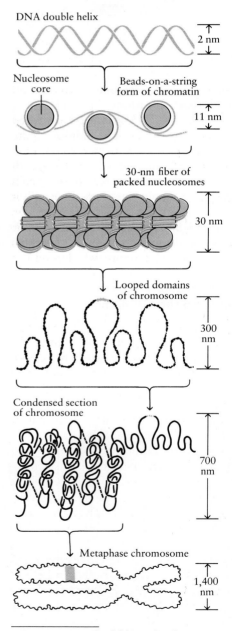

DNA double helix

2 nm

Nucleosome core

Beads-on-a-string form of chromatin

11 nm

30-nm fiber of packed nucleosomes

30 nm

Looped domains of chromosome

300 nm

Condensed section of chromosome

700 nm

Metaphase chromosome

1,400 nm

18–6 *Stages in the folding of a chromosome, according to various models.*

Further condensation occurs as the 30-nanometer fiber forms a series of loops, known as **looped domains** (Figure 18–5). Each looped domain coils until, ultimately, clusters of neighboring looped domains condense into the compact chromosomes that become visible during mitosis and meiosis (Figure 18–6).

Other proteins associated with the chromosome are the enzymes concerned with DNA and RNA synthesis, along with regulatory proteins, plus a large number and variety of unidentified molecules. Unlike the histones, these proteins vary from one cell type to another, both in their abundance and in their identity.

Replication of the Chromosome

As we saw in Chapter 14, replication of the DNA of eukaryotes is the same, in principle, as replication of the DNA of prokaryotes. Nucleotides in the energy-yielding form of triphosphates are assembled along a template DNA strand in the semiconservative manner first proposed by Watson and Crick and subsequently confirmed by Meselson and Stahl (page 293). As in prokaryotes, DNA polymerases operate only in the 5′ to 3′ direction; the 3′ to 5′ strand is synthesized as a series of Okazaki fragments, which are then joined together by the enzyme DNA ligase to form the complementary strand (see Figure 14–19, page 297).

In the comparatively small, circular prokaryotic chromosome, replication begins at a single replication origin and proceeds bidirectionally along two replication forks (page 295). In eukaryotic chromosomes, there are many replication origins, and bidirectional synthesis takes place until the replication forks merge (page 295). Replication is much slower in eukaryotes than in prokaryotes; in human cells, for example, the rate is about 50 base pairs per second per replication fork. As it is synthesized, eukaryotic DNA becomes complexed with histones and other proteins.

REGULATION OF GENE EXPRESSION IN EUKARYOTES

As we saw in Chapter 16, regulation of gene expression in prokaryotes largely involves the fine tuning of the metabolic machinery of the cell in response to changes in available nutrients in the environment. In eukaryotes, especially multicellular eukaryotes, the problems of regulation are very different. A multicellular organism usually starts life as a fertilized egg, the zygote. The zygote divides repeatedly by mitosis and cytokinesis, producing many cells. At some stage these cells begin to differentiate, becoming, for example, muscle cells, nerve cells, blood cells, intestinal cells, and so forth. Each cell type, as it differentiates, begins to produce characteristically different proteins that distinguish it from other types of cells. This is nicely illustrated by mammalian red blood cells. In the early stages of fetal life, developing red blood cells synthesize one type of fetal hemoglobin; red blood cells produced at later stages contain a second type of fetal hemoglobin; then, sometime after the birth of the organism, the developing red blood cells begin to produce the alpha and beta chains characteristic of adult hemoglobin. Thus the genes are expressed in a carefully controlled sequence, one after the other. The DNA segments that code for these hemoglobin molecules are expressed only in developing red blood cells.

There is evidence, however, that all of the genetic information originally present in the zygote is also present in every diploid cell of the organism (Figure 18–7). In other words, the DNA segments that code for hemoglobin (both the fetal types and the adult type) are present in skin cells and heart cells and liver cells and nerve cells and, indeed, in every one of the nearly 200 different types of cells in the body. Similarly, the DNA sequence that codes for the hormone insulin is present not only in the specialized cells of the pancreas that manufacture insulin but also in all the other cells. Since each type of cell produces only its characteristic proteins—and not the proteins characteristic of other cell types—it becomes apparent that differentiation of the cells of a multicellular organism depends on the inactivation of certain groups of genes and the activation of others.

Condensation of the Chromosome and Gene Expression

Many lines of evidence indicate that the degree of condensation of the DNA of the chromosome, as shown by chromatin staining, plays a major role in the regulation of gene expression in eukaryotic cells. Staining reveals two types of chromatin: **euchromatin,** the more open chromatin, which stains weakly, and **heterochromatin,** the more condensed chromatin, which stains strongly (Figure 18–8). During interphase, heterochromatin remains condensed, but euchromatin becomes dispersed. Transcription of DNA to RNA takes place only during interphase, when the euchromatin is dispersed.

Some regions of heterochromatin are constant from cell to cell and are never expressed. An example is the highly condensed chromatin located in the centromere region of the chromosome. This region, which does not code for protein, is believed to play a structural role in the movement of the chromosomes during mitosis and meiosis. Similarly, little or no transcription takes place from Barr bodies (page 267), which are X chromosomes that are tightly condensed and irreversibly inactivated.

Other regions of condensed chromatin, by contrast, vary from one type of cell to another within the same organism, reflecting, it is believed, the biosynthesis of different proteins by different types of cells. Also, as a cell differentiates during embryonic development, the proportion of heterochromatin to euchromatin increases as the cell becomes more specialized.

Additional evidence linking the degree of chromosome condensation to gene expression comes from studies of the giant chromosomes of insects (see page 277).

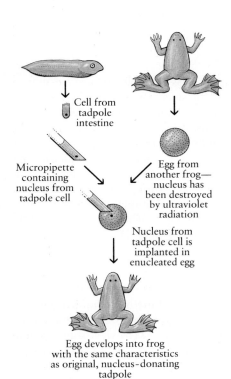

18–7 *A number of experiments have shown that early development does not result in permanent inactivation of genes or the loss of functional DNA. In experiments by J. B. Gurdon, illustrated here, nuclei were removed from intestinal cells of a tadpole and implanted into egg cells in which the nucleus had been destroyed. In some cases, the egg developed normally, indicating that the tadpole intestinal cell nucleus contained all the information required for all the cells of the organism. In other experiments, F. C. Steward demonstrated that under certain conditions, a single differentiated cell of a carrot can be persuaded to reconstitute an entire carrot plant.*

Figure labels:
Cell from tadpole intestine

Micropipette containing nucleus from tadpole cell

Egg from another frog— nucleus has been destroyed by ultraviolet radiation

Nucleus from tadpole cell is implanted in enucleated egg

Egg develops into frog with the same characteristics as original, nucleus-donating tadpole

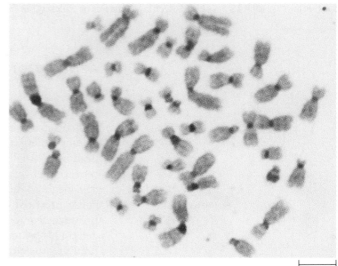

5 µm

18-8 *Human chromosomes at metaphase, stained to distinguish the more tightly condensed heterochromatin from the less tightly condensed euchromatin. Notice the deeply stained heterochromatin in the region of the centromere.*

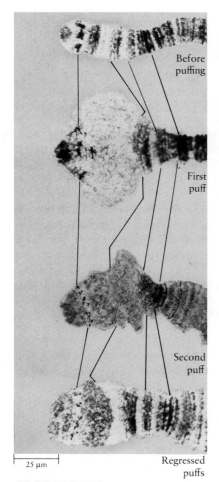

Before puffing

First puff

Second puff

25 µm

Regressed puffs

18-9 *Observations of chromosome puffs support the concept that the DNA is somehow unwound to make it available for RNA transcription. These puffs were observed in chromosomes of the Brazilian gnat, which, like the fruit fly, has giant chromosomes in some of its cells. Puffs occur normally but can also be induced experimentally. These puffs occurred in response to a hormone that causes molting. As the micrographs indicate, the puffs occurred sequentially along one chromosome.*

At various stages of larval growth in insects, it is possible to observe diffuse thickenings, or "puffs," in various regions of these chromosomes (Figure 18–9). The puffs are open loops of DNA, and studies with radioactive isotopes indicate that these loops are sites of rapid RNA synthesis. When ecdysone, a hormone that produces molting in insects, is injected, the puffs occur in a definite sequence that can be related to the developmental stage of the animal. For example, in one species of *Drosophila*, ecdysone initiates three new puffs and causes increases in 18 other puffs within 20 minutes after it is injected; during this same period, 12 other puffs decrease in size. After 4 to 6 hours, five additional puffs can be seen. The looping out of the DNA occurs before RNA synthesis is initiated. The mechanism of this spinning out of the chromosomal DNA is still poorly understood.

Methylation and Gene Expression

Once the DNA helix is formed, specific enzymes add methyl groups to nucleotides of cytosine. Methylation in eukaryotes is hypothesized to inhibit gene expression and so to provide a form of gene regulation. Methylcytosine is found almost exclusively at the complementary sequences (5′)-C-G-(3′) and (3′)-G-C-(5′). In birds and mammals, 50 to 70 percent of cytosines in such sequences are methylated. Z-DNA (page 356) contains a large proportion of methylated cytosines, and Z-DNA is believed to be inactive. On the other hand, some eukaryotes —insects, for example—have no methylated bases in their DNA and yet they regulate the expression of their genes with no apparent loss of efficiency.

Regulation by Specific Binding Proteins

In eukaryotes, as in prokaryotes, transcription is also regulated by proteins that bind to specific sites on the DNA molecule. To date, only a few of these proteins and their binding sites have been identified. However, it is increasingly clear that this level of transcriptional control is far more complex in eukaryotes, particularly multicellular eukaryotes, than in prokaryotes. A gene in a multicellular organism appears to respond to the sum of many different regulatory proteins, some tending to turn the gene on and others to turn it off. The sites at which these regulatory proteins bind may be hundreds or even thousands of base pairs away from the promoter sequence at which RNA polymerase binds and transcription begins. This, as you might expect, adds to the difficulty of identifying the regulatory molecules and also of understanding exactly how they exert their effects. Recent research suggests that changes in the activities of some of these regulators are linked with the development of cancers, a matter we shall explore later in this chapter.

The DNA of the Energy Organelles

Two important generalizations have emerged from our study of genetics in the preceding chapters. One is that the genome of the offspring of sexually reproducing organisms is a new combination of maternal and paternal genes. The second, emphasized in the last three chapters, is that the genetic code is universal. As it turns out, neither of these generalizations is quite true.

The exceptions are found in the eukaryotic cell's energy organelles, the mitochondria and the chloroplasts. These organelles contain their own DNA, which, like that of prokaryotes, is not associated with histones. This DNA is replicated within the organelle, and new mitochondria and chloroplasts are formed by simple division, much as in *E. coli*. Because they are self-replicating and because their DNA resembles that of prokaryotes, mitochondria and chloroplasts are hypothesized to have evolved from prokaryotes that became parasites in primitive eukaryotic cells early in evolutionary history, a topic that we shall explore further in Chapter 22.

The DNA of mitochondria and chloroplasts is transcribed and translated, although most of the proteins of these organelles are coded by the nuclear DNA and are synthesized in and imported from the cytoplasm. The relatively small DNA molecule (16,569 base pairs) of human mitochondria has now been sequenced, making possible a variety of studies. Comparison of this DNA sequence with the RNAs and the few proteins made by the mitochondria has revealed that the genetic code is not, strictly speaking, universal. For instance, in human mitochondria, UGA codes for tryptophan and not termination, and AUA codes for methionine and not isoleucine (see page 309). Also, mitochondria have many fewer tRNAs than either *E. coli* or eukaryotic cells, not enough to translate all of the possible codons by conventional base pairing. The origin of these discrepancies remains unknown and puzzling, for despite the differences in the mitochondrial code, the code employed in the translation of the information encoded in the DNA of the nucleus is the same in all eukaryotic organisms in which it has been checked and is the same as the code employed in prokaryotic cells.

In about two-thirds of all plant species, the male gamete contributes neither chloroplasts nor mitochondria to the zygote. Such non-Mendelian inheritance was first noted some 80 years ago in plants in which deficient chloroplasts produced mottled leaves —but only if the defect was present in the plant in which the egg cell developed. Similarly, in humans and other animal species, the sperm contributes almost no cytoplasm to the fertilized egg and, consequently, all of the mitochondria are maternal in origin. As we shall see in Chapter 50, the maternal transmission of the mitochondrion—coupled with the capacity to rapidly sequence the mitochondrial DNA of different individuals—is shedding new light on the evolution of *Homo sapiens*.

THE EUKARYOTIC GENOME

Examination of the DNA of eukaryotic cells revealed four major surprises. First, with some few exceptions, the amount of DNA per cell is the same for every diploid cell of any given species (which is not surprising), but the variations among different species are enormous. *Drosophila* has about 1.4×10^8 base pairs per haploid genome, only about 70 times more than *E. coli*. Humans (with approximately 3.5×10^9 base pairs) have 25 times as much as *Drosophila*, somewhat more than a mouse, but about the same amount as a toad (3.32×10^9 base pairs). The largest amount of DNA found so far has been located in a salamander with 8×10^{10} base pairs per haploid genome.

Second, in every eukaryotic cell, there is what appears to be a great excess of DNA, or at least of DNA whose functions are unknown. It is estimated that in eukaryotic cells less than 10 percent of all the DNA codes for proteins; in humans, it may even be as little as 1 percent. By contrast, prokaryotes, as we have seen, use their DNA very thriftily; viruses even more so. Except for regulatory or signal sequences, virtually all of their DNA is expressed.

Third, almost half of the DNA of the eukaryotic cell consists of nucleotide sequences that are repeated hundreds, even millions, of times. This was a particularly startling discovery. In *E. coli,* long the model for molecular geneticists, each chromosomal DNA molecule typically contains only one copy of any given gene. (The principal exceptions are the genes coding for the ribosomal RNAs.) Moreover, according to Mendelian genetics, a gene should be present only twice per diploid eukaryotic cell, not in a multitude of copies.

Introns

The fourth surprise—and perhaps the most unexpected of all—was that the protein-coding sequences of eukaryotic genes are usually not continuous but are instead interrupted by noncoding sequences. These noncoding interruptions within the gene are known as intervening sequences, or **introns,** and the coding sequences, the sequences that are expressed, are called **exons.**

Introns were discovered in the course of hybridization experiments, when investigators found that there was not a perfect match between eukaryotic messenger RNA molecules and the genes from which they were transcribed. The nucleotide sequences of the genes were much longer than their complementary mRNA molecules that were found in the cytoplasm. Subsequently, introns and exons were actually visualized in electron micrographs (Figure 18–10).

18–10 *This electron micrograph reveals the results of an experiment in which a single strand of DNA containing the gene coding for ovalbumin was hybridized with the messenger RNA for ovalbumin. The complementary sequences of the DNA and mRNA are held together by hydrogen bonds; there are eight such sequences, the exons labeled L and 1 through 7 in the accompanying diagram. Some segments of the DNA do not have corresponding mRNA segments and so loop out from the hybrid; these are the seven introns, labeled A through G. Only the exons are translated into protein.*

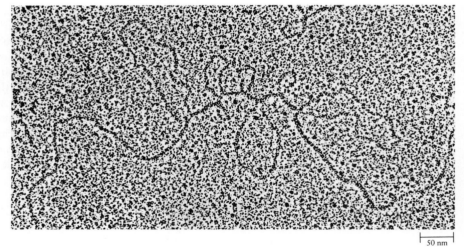

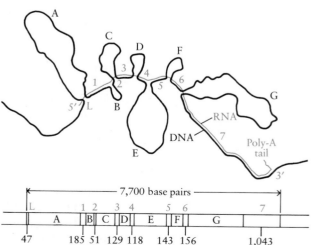

It is now known that most, but not all, structural genes of multicellular eukaryotes contain introns. The introns are transcribed onto RNA molecules and excised before translation. The number of introns per gene varies widely. For example, the gene for ovalbumin, a protein found in large quantities in vertebrate egg cells, has seven introns. By contrast, the mammalian gene for beta globin, one of the polypeptides of the hemoglobin molecule, has only two introns, one large and one small. In chickens, the gene for collagen, a very common protein (Figure 3–23, page 77), has 50 introns. Introns have also been found in genes coding for transfer RNAs and ribosomal RNAs, and even in some viruses.

In general, the more complex an organism and the more recently it has evolved, the larger and more abundant are its introns. It is not known which came first—continuous genes lacking introns or interrupted genes containing introns. It has been suggested that perhaps the latter came first, but that in bacteria and other unicellular organisms that are highly selected for rapid growth, any unneeded DNA has been eliminated in the course of their evolution.

The Function of Introns

Are introns accidents, or do they have a function? One suggestion for their continued existence in multicellular eukaryotes is that they promote recombination; crossing over during meiosis is more likely in genes containing introns than in genes lacking introns, just because of the distances involved. There are also indications that, in some cases, different exons code for different structural and functional segments, or domains, of the finished protein (Figure 18–11). For example, the central exon of the beta-globin gene codes for the domain of the polypeptide that holds the heme group, and the other two exons code for domains of the molecule that fold around this central portion (see Figure 3–27, page 79). It is hypothesized that new combinations of such domains, brought about by the reshuffling of exons, might foster the rapid evolution of new proteins.

18–11 *An attractive hypothesis: Exons code for discrete functional regions in the protein for which the entire gene codes.*

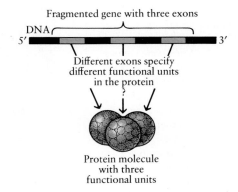

Fragmented gene with three exons

DNA

5′   3′

Different exons specify
different functional units
in the protein
?

Protein molecule
with three
functional units

Classes of DNA: Repeats and Nonrepeats

The repetitious nature of much of the DNA in the eukaryotic cell was first revealed in hybridization studies carried out before the discovery of restriction enzymes. In these studies, DNA was broken by chemical means into fragments about 1,000 base pairs in length, denatured, and allowed to reassociate. When *E. coli* DNA is treated in this way, hybridization takes place at a uniform rate; every strand present has an equal chance of finding a partner strand because each is present in equal numbers. However, if eukaryotic DNA is treated in this way, up to 30 percent of the DNA, depending on the species, reassociates very rapidly, indicating that multiple copies of the same sequence are present in each genome. This class of DNA has come to be called simple-sequence DNA. Another fraction—known as intermediate-repeat DNA—reassociates more slowly. A third fraction forms hybrids at a still slower rate, indicating that only one or a few copies of each sequence are present. This latter fraction contains most of the protein-coding genes.

Simple-Sequence DNA

Simple-sequence DNA proved easy to analyze because it consists of short sequences, as its name implies, arranged in tandem, head to tail. These sequences are typically 5 to 10 base pairs in length, although a few are as long as 200 to 300 base pairs.

Simple-sequence DNA is present in enormous quantities. Half of the DNA in one species of crab, for instance, is made up of ATATATAT and so on; *Drosophila virilis* has ACAAACT repeated 12 million times. About 10 percent of the DNA of the mouse, and about 20 to 30 percent of human DNA, is made up of short, highly repetitive sequences. It is interesting to note that repetitive sequences, especially those in which purines (A or G) and pyrimidines (T or C) alternate, are especially likely to form Z-DNA.

Simple-sequence DNA is thought by many investigators to be vital to chromosome structure. Long blocks of short repetitive sequences have been found around the centromere (Figure 18–12) and, indeed, may *be* the centromere. More recently, the tips of all human chromosomes have been found to consist of some 1,500 to 6,000 nucleotides in which a simple sequence—TTAGGG—is repeated over and over. This same simple, repeated sequence has also been found at the chromosome tips in a wide range of other mammals, birds, reptiles, and even protists. The "caps" formed by this repeated sequence are hypothesized to play a role in chromosome integrity and stability.

18–12 *Salamander chromosomes in which the concentration of simple-sequence DNA in the centromere regions was revealed by the use of a radioactive probe. The probe—RNA labeled with tritium—was prepared in a test tube, using as the template simple-sequence DNA that had been separated from other chromosomal DNA in a density gradient. When the salamander chromosomes were exposed to the radioactively labeled RNA, the regions of the chromosomes complementary to the RNA hybridized with it. After the RNA that had not hybridized was washed from the preparation, a photographic emulsion was placed over it. The dark spots indicate the location of the radioactivity. These chromosomes are in meiotic metaphase I.*

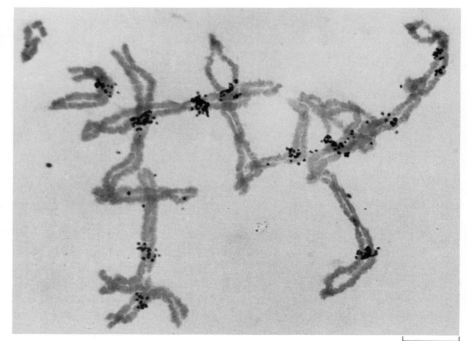

10 μm

Intermediate-Repeat DNA

In the hybridization experiments described earlier, intermediate-repeat sequences reassociated more slowly than simple sequences and more rapidly than single-copy DNA. About 20 to 40 percent of the DNA of multicellular organisms consists of this class. Intermediate-repeat DNA differs in several features from simple-sequence DNA. First, the sequences are longer, generally about 150 to 300 nucleotides. Second, they are similar but not identical to one another (hence they are sometimes referred to as "families"). Third, with the exception of the rRNA and histone genes, they are scattered throughout the genome. Fourth, some of the intermediate-repeat sequences, though only a small proportion, have known functions.

18–13 *A single gene codes for three types of rRNA molecules (18S, 5.8S, and 28S) found in the eukaryotic ribosome. This gene occurs in multiple copies, repeated in tandem (head to tail). (a) Electron micrograph of five copies of this gene separated by nontranscribed spacer sequences. (b) An enlargement of one of the genes and its transcribed RNA precursor molecules (the fine fibrils perpendicular to the DNA molecule). Enzymatic cleavage of the RNA precursor molecules and removal of the transcribed spacer sequences yields individual 18S, 5.8S, and 28S rRNA molecules. (c) A map of the three-rRNA gene. The portions of the DNA coding for the three rRNA molecules are shown in color.*

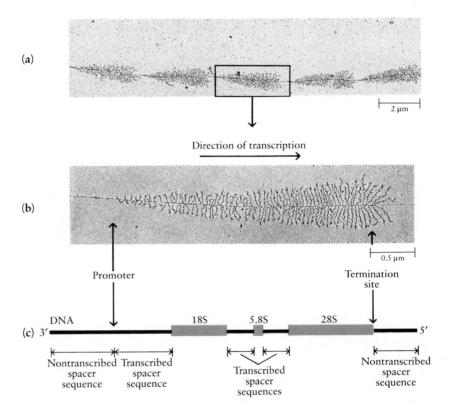

Among the most thoroughly studied intermediate-repeat sequences are the genes coding for histones and for the ribosomal RNAs (Figure 18–13). The histone genes are present in multiple copies (from 50 to 500) in the cells of all multicellular eukaryotes. Cells of multicellular eukaryotes, which may contain some 10 million ribosomes per cell, also have from 50 to 5,000 copies of the rRNA genes. The rRNA genes occur in tandem, head to tail; the chromosome regions in which they are located form the structure we recognize as the nucleolus (see essay).

Most intermediate-repeat sequences are more mysterious in nature. One of the most common families of intermediate repeats, for example, is the *Alu* sequence. Although all members of this family are not identical, they typically contain a recognition sequence for a restriction enzyme known as *Alu*I. About 5 to 10 percent of the entire human genome is made up of *Alu* sequences, often located within introns.

The Nucleolus

As you know, the most prominent feature in the nucleus of a cell in interphase is the nucleolus. During mitosis and meiosis, however, the nucleolus disappears, only to reappear following telophase. Because of its prominence in the cell and its puzzling behavior, the nucleolus has long been an object of scrutiny by cytologists. (A review of observations on the nucleolus published in 1898 contained some 700 references.) Although some details are still missing, its structure and function are now known in broad outline.

Structurally, the nucleolus is not actually a distinct entity, but rather a cluster of loops of chromatin, often from different chromosomes. For example, 10 of the 46 human chromosomes contribute chromatin loops to the nucleolus. The loops that form the basic structure of the nucleolus are the DNA segments that contain copies of the gene coding for three of the four types of rRNA molecules found in eukaryotic ribosomes. During the condensation of the chromosomes at the beginning of meiosis or mitosis, these loops are reeled back into their respective chromosomes, and the nucleolus disappears.

Functionally, the nucleolus is a ribosome factory in which rRNA molecules are transcribed from the chromatin loops and ribosomal subunits are assembled. Ribosomal proteins, themselves synthesized on ribosomes in the cytoplasm of the cell, are transported into the eukaryotic nucleus and assembled into subunits. The nearly completed ribosomal subunits, containing both rRNAs and protein, are then shipped back into the cytoplasm where, after a few finishing touches, they begin to perform their essential functions in the assembly of amino acids into proteins.

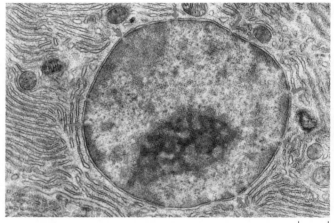

Electron micrograph of the nucleus of a type of pancreatic cell that produces and exports many of the enzymes used in digestion. The dark body in the center is the nucleolus, where the RNAs of the ribosomes are synthesized and the ribosomal subunits are assembled. You can see partially formed ribosomal subunits around its periphery. Notice also the nuclear envelope, with its many nuclear pores (arrows). Surrounding the nucleus are membranes of the endoplasmic reticulum, as well as a few mitochondria.

1 µm

Single-Copy DNA

The rest of the genome (anywhere from 50 to 70 percent, depending on the species) is made up of sequences that are not repeated or are only repeated a few times. With the exception of the histone genes, all of the known protein-coding genes belong to this fraction of the DNA. However, only a small proportion of the single-copy DNA—perhaps as little as 1 percent of the total—appears to be translated into protein. Transcription units, which consist of exons plus introns, are separated by great distances of nontranscribed spacer DNA. Moreover, introns are often longer than exons; for example, in vertebrates, introns may make up more than 80 percent of the DNA within transcription units.

"Protein-coding genes," in the words of molecular biologist James Darnell, "seem to be islands floating in a sea of meaningless DNA."

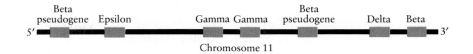

Chromosome 11

18–14 *One important group of globin genes is the beta globin family, which is located in a long nucleotide sequence in human chromosome 11. The epsilon (ε) gene is expressed early in embryonic life, followed by the two gamma (γ) genes (which code for the gamma chains that combine with alpha chains to produce fetal hemoglobin). The delta (δ) gene, found only in primates, and the beta (β) gene are expressed in adult life. This gene family also includes two pseudogenes—DNA sequences similar to the structural genes, but which are not expressed.*

Gene Families

As we have seen, some genes occur in multiple identical copies. These include the genes coding for the ribosomal RNAs and also the genes coding for histones, the only multiple repeats that are known to code for proteins. Both of these families of genes code for molecules needed in large quantities.

Other protein-coding genes are found in gene families made up of similar but not identical genes. The best studied of these is the globin family. Adult hemoglobin, as we saw in Chapter 3, is a complex of four polypeptide chains—two alpha (α) globin chains and two beta (β) globin chains—each carrying an identical heme group. In humans, the beta branch of the globin gene family is clustered together on one chromosome (chromosome 11). The cluster contains five different protein-coding genes, spaced out along the chromosome (Figure 18–14). They differ slightly in their nucleotide sequences, and all consist of three exons and two introns, one large and one small, in the same position in each gene.

As we noted previously, these genes are expressed one after the other in the course of embryonic development, producing polypeptide molecules that differ very slightly. Combined with alpha chains, they form hemoglobins with higher affinities for oxygen than the alpha-beta hemoglobin of the mother; hence the developing fetus can compete successfully with the mother for oxygen, wresting O₂ molecules from the heme in her bloodstream.

Figure 18–15 summarizes the evolutionary steps that are believed to have led to the present-day human globin gene family. The ancestral molecule is believed to have resembled myoglobin, which is a relatively small protein (153 amino acids) found in muscle cells; myoglobin consists of a single polypeptide chain and holds a single oxygen-binding heme group. It is believed that the ancestral gene—coding for the ancestral protein—was accidentally duplicated several times in the course of evolutionary history and that these duplicates were preserved by unequal cross-over events. Once the genes were duplicated, mutations led to their divergence, eventually giving rise to the present family of genes. A corresponding divergence of function occurred in the proteins for which the genes coded, leading to modern myoglobin and to the forerunners of the alpha and beta chains of hemoglobin.

18–15 *Proposed evolutionary relationships among some of the globin genes. The solid dots represent duplications of genes. The divergence of the genes following duplication is the result of mutations.*

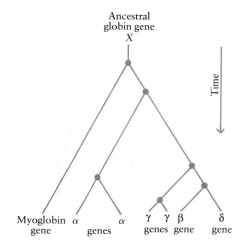

In addition to the protein-coding genes shown in Figure 18–14, there are two other nucleotide sequences that closely resemble members of the beta group but are not expressed. These sequences, known as beta pseudogenes, are believed to be copies that have been disabled by their accumulated mutations. It remains to be seen—if there are human descendants and molecular biologists in future millennia—whether these will be discarded in the course of subsequent evolution or if, as the result of further mutations, they will eventually become active family members.

Several other gene families are known that are also made up of closely related genes coding for proteins with slightly different properties. The actin family, for example, codes for the various forms of the contractile protein actin (page 121), a component of the cytoskeleton (page 110) and also one of the principal components of animal muscle fibers. Slightly different forms of actin are present in the mammalian fetus and in the adult, as is the case with beta globin; also, slightly different forms of actin are present in different types of muscle, for example, skeletal muscle and cardiac muscle.

TRANSCRIPTION AND PROCESSING OF mRNA IN EUKARYOTES

Transcription in eukaryotes is the same, in principle, as in prokaryotes. It begins with the attachment of a special enzyme, an RNA polymerase, to a particular nucleotide sequence, the promoter, on one strand of the DNA double helix. This strand then functions as a template for the assembly of ribonucleotides, as shown in Figure 15–8 (page 306). The transcribed RNA molecules (rRNAs, tRNAs, and mRNA) then play their various roles in the translation of the encoded genetic information into protein.

Despite this basic similarity, there are some significant differences between transcription in prokaryotes and eukaryotes. One difference is that eukaryotic genes are not grouped in operons in which two or more structural genes are transcribed onto a single RNA molecule, as they often are in prokaryotes. In eukaryotes, each structural gene is transcribed separately, and its transcription is under separate controls.

There are also differences in the enzymes involved in transcription. Most notably, in prokaryotes, a single RNA polymerase catalyzes the biosynthesis of the three types of RNA—messenger, transfer, and ribosomal. In eukaryotes, there are three different RNA polymerases: one transcribes the genes that will be translated into proteins, a second transcribes the genes for the large ribosomal RNAs, and the third transcribes a variety of small RNAs, including the tRNAs and the small RNAs of the ribosome.

mRNA Modification and Editing

In prokaryotes, as you will recall, ribosomes attach to an mRNA molecule and begin its translation into protein even before transcription is completed. In eukaryotes, however, transcription and translation are separated in both time and space. After transcription is completed in the nucleus, the mRNA transcripts are extensively modified before they are transported to the cytoplasm—the site of translation.

Even before transcription is completed, while the newly forming RNA strand is only about 20 base pairs long, a "cap" of an unusual nucleotide, 7-methylguanine, is added to the 5′ end of the messenger. This cap, it is now known, is necessary for the binding of the mRNA to the ribosome. After transcription has been com-

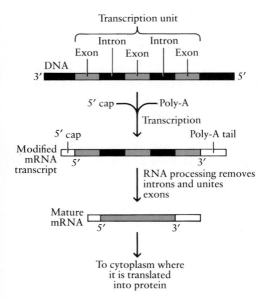

18–16 *A summary of the stages in the processing of mRNA transcribed from the structural genes of eukaryotes. The genetic information encoded in the DNA is transcribed into an RNA copy. This copy is then edited, with the addition of the 5′ cap and a poly-A tail, the excision of the introns, and the splicing together of the exons. The mature mRNA then goes to the cytoplasm, where it is translated into protein.*

pleted and the molecule released from the DNA template, special enzymes add a string of adenine nucleotides to the 3′ end of the molecule. This added segment, known as the poly-A tail, may contain as many as 200 nucleotides. Although its function in the cell has not been determined, its utility to molecular biologists is clear: by synthesizing a poly-T strand and anchoring it to a suitable support, they can catch mRNA molecules by their poly-A tails.

Before the modified mRNA molecules leave the nucleus, the introns are excised, and the exons spliced together to form a single, continuous molecule (Figure 18–16). The exact way that excision and splicing take place is not known, but it must be very exact, since the slightest error would cause a frame-shift in the transcribed message (see page 316).

Several instances have now been found in which identical mRNA transcripts are processed in more than one way. Such alternative splicing can result in the formation of more than one functional polypeptide from RNA molecules that were originally identical (Figure 18–17). In such cases, an intron may become an exon, or vice versa. Thus, as you can see, the more that is learned about eukaryotic DNA and its expression, the more difficult it becomes to define "gene" or "intron" or "exon."

The mRNAs that are transported to the cytoplasm are associated with proteins in ribonucleoprotein particles (mRNPs). The associated proteins may aid in transporting the mRNA molecules through pores in the nuclear envelope and may also help to bind the mRNAs to ribosomes.

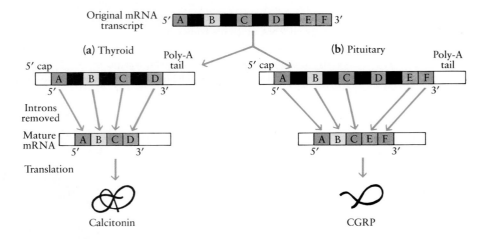

18–17 *Alternative splicing of identical mRNA transcripts results in the synthesis of different polypeptides from the information encoded by a single gene. For example, when the mRNA transcript shown here is processed in (**a**) the thyroid gland, the transcript is cut and the poly-A tail is added to the 3′ end of exon D. The introns are then removed, and the mature mRNA molecule is translated into the*

*peptide hormone calcitonin. (**b**) In the pituitary gland, however, the poly-A tail is added to the 3′ end of exon F. Five introns are removed from this transcript, including segment D, which in the thyroid was retained as an exon. The mature mRNA, composed of five exons, is translated into a different hormone, known as calcitonin-gene-related protein (CGRP).*

RNA and the Origin of Life

The discovery of the role of DNA in heredity launched a debate among biologists as to whether life had its molecular beginnings in protein, the original candidate, or in DNA. DNA was a likely choice because it is the repository of the genetic information and provides the template for its own precise replication; proteins have neither of these qualifications. Proponents of proteins, however, have noted that virtually all of the chemical reactions of the cell depend on the catalytic activities of proteins. Even the so-called self-replicating properties of DNA are actually protein-dependent.

An important clue toward the resolution of this "chicken-and-egg" dilemma has come from an unexpected source. T. C. Cech and his coworkers at the University of Colorado were studying the excision of introns and the splicing together of exons. This process must be carried out with exquisite precision; a mistake of one nucleotide could render the entire molecule nonfunctional. By a happy coincidence, the biological system the investigators were using was the unicellular protist *Tetrahymena*. In order to isolate the catalysts required for the reaction, Cech and his coworkers set up two cell-free systems. One contained not only an RNA molecule from which an intron was to be excised but also proteins that were potential catalysts; the other system, the control, was protein-free. The intron was neatly excised in the first system,

as expected, but, to everyone's surprise, the excision and splicing process also took place in the control. It was subsequently shown that the intron itself—a 400-nucleotide sequence of RNA—has an enzyme-like catalytic activity that carries out the excision and splicing. This sequence folds up to form a complex surface that functions like an enzyme. Although RNA catalysts are not common, other examples have now been found both in other types of reactions and in exon-splicing in other types of cells.

The discovery that RNA can act as a catalyst makes it easier to imagine how life had its beginnings. According to Bruce M. Alberts, "One suspects that a crucial early event was the evolution of an RNA molecule that could catalyze its own replication." These molecules then diversified into a collection of catalysts that could, for example, assemble ribonucleotides in RNA synthesis or accumulate lipid-like molecules to form the first primitive cell membranes. Gradually, other RNAs evolved and assembled the first proteins, which, because they were better catalysts, gradually took over the enzymatic functions. In the third step, DNA appeared on the scene, and its more stable double-stranded structure became the ultimate repository of the genetic information. Thus, researchers speculate, the catalytic intron can be regarded as a living fossil, a provocative clue to the events of almost 4 billion years ago.

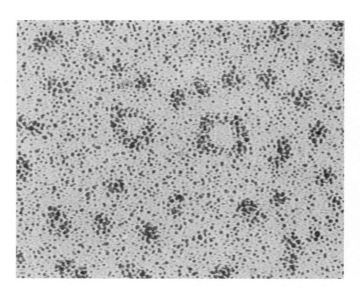

The small circles visible in this electron micrograph are introns removed from a transcribed RNA molecule of the protist Tetrahymena. *These introns have the capacity to catalyze their own excision and to splice together the exons.*

GENES ON THE MOVE

As we saw in Chapter 16, studies of the prokaryotic chromosome revealed, unexpectedly, the existence of a number of genetic elements—transforming factors, plasmids, bacteriophages, and transposons—that move into and out of the bacterial genome, affecting both structure and function. Analysis of the eukaryotic chromosome has revealed that it too is subject to rearrangements, deletions, and additions. We shall discuss a few examples, notably antibody-coding genes, viruses, and transposons.

Antibody-Coding Genes

Antibodies are complex globular proteins produced in large quantities by specialized white blood cells (lymphocytes) in response to the presence of foreign molecules. A substance that evokes the production of antibodies is known as an **antigen;** virtually all foreign proteins and most foreign polysaccharides can act as antigens. An antibody recognizes and combines with its particular antigen in much the same way, and as specifically, as an enzyme combines with its substrate. Antibodies immobilize or destroy foreign proteins, virus particles, bacterial cells, and other invaders. The problem with antibodies, from the geneticist's perspective, is that a single organism—a mouse, for instance—is capable of making at least 10 million different kinds of antibodies, and there are not enough protein-coding genes in the entire mammalian genome to account for this many different proteins.

Analyses of the amino acid sequences of antibody molecules have shown that each is made up of two heavy (long) polypeptide chains and two light (short) chains (Figure 18–18). Each type of chain has a constant region that is characteristic of the species of organism and the type of antibody, and each has variable regions. The variable regions are responsible for the highly specific reaction between antigen and antibody. These variable regions consist of only about 100 amino acids.

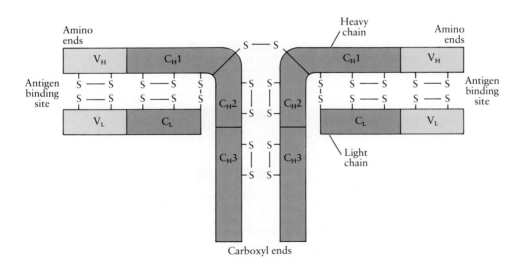

18–18 *Structure of an antibody molecule. It has two light (L) and two heavy (H) polypeptide chains, each of which has a variable (V) region (yellow) and a constant* (C) *region (brown). The polypeptide chains are connected to each other by disulfide bridges. The variable regions of the light and heavy chains form a complex three-dimensional structure that is the portion of the antibody molecule that recognizes and binds to foreign antigens.*

More than 20 years ago it was hypothesized that the constant and variable regions of antibody molecules might be encoded by separate genes. Variability could thus be generated by combining a single constant sequence with different variable sequences. With the discovery of restriction enzymes and the refinement of hybridization techniques, it became possible to test this hypothesis. Studies conducted by Susumu Tonegawa, now of the Massachusetts Institute of Technology, demonstrated conclusively that separate DNA segments code for the variable portions of the antibody molecules. Detailed analysis has shown that the DNA for the variable regions of the heavy chain consists of at least 400 different variable (V) sequences, about 12 diversity (D) sequences, and four joining (J) sequences, which can be assembled in many millions of different ways.

Moreover, by comparing the nucleotide sequences of mature mouse lymphocytes with those of embryonic mouse cells, Tonegawa was able to show that segments that code for the variable regions of an antibody molecule are actually moved into a new place on the chromosome during the differentiation of a lymphocyte (Figure 18–19). This phenomenon—the rearrangement of gene fragments in somatic cells to produce functional genes—is known to occur only in the immune system. Tonegawa's work, for which he received the 1987 Nobel Prize, laid the foundation for a virtual avalanche of new discoveries concerning the cells of the immune system, about which we shall have a great deal more to say in Chapter 39.

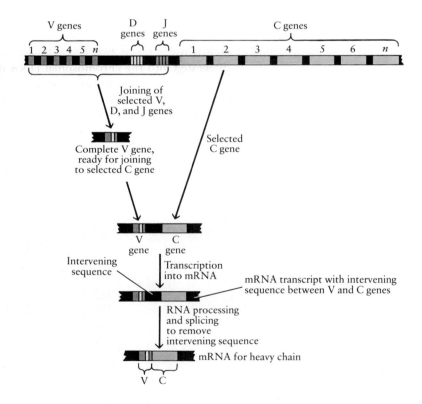

18–19 *A schematic representation of the assembly of the genes coding for an antibody heavy chain. Selected variable (V), diversity (D), and joining (J) genes are transposed from different regions of the chromosome to form a complete variable gene that then joins to one of the constant (C) genes. The intervening sequence between the variable and constant sequences is removed from the RNA transcript to yield the finished mRNA molecule.*

18-20 *Susumu Tonegawa, the first Japanese scientist to receive the Nobel Prize, was inundated with congratulatory telephone calls at his Massachusetts home when the award was announced in October of 1987. For his nine-month-old son, Hidde, however, the focus of attention was the photographer and his equipment.*

Viruses

The viruses of eukaryotes, like those of prokaryotes, consist essentially of nucleic acid enclosed in a protein capsid. Like the bacteriophages described in Chapter 16, eukaryotic viruses may be either DNA or RNA. Moreover, like lambda and other temperate bacteriophages, certain eukaryotic viruses can also become integrated into the chromosomal DNA of the host cell. When integrated, these viruses are known as **proviruses**. They are, in effect, mobile genetic elements. In eukaryotes, such viruses are of two general types: DNA viruses (analogous to the temperate bacteriophages) and RNA retroviruses.

A number of DNA viruses are known that can, depending on the type of cell they infect, either initiate an infection cycle, damaging the cell, or insert themselves into the chromosomal DNA of the host cell. An example is simian virus 40, or SV40 (Figure 17–10, page 348). SV40 is a virus of monkeys that was first discovered in cells, growing in tissue culture, that were being used for the development of polio vaccines. SV40 was subsequently found to cause cancers in newborn hamsters, though not in the monkeys that are its normal hosts. The cancers are caused by specific growth-promoting proteins produced in the cells by the viral genes. In short, SV40 can introduce new, functional genes into the DNA of the host cell, as can a number of other DNA viruses.

The second group of eukaryotic viruses that can become integrated into host cell chromosomes are the RNA retroviruses. The integration of an RNA virus into a DNA chromosome poses some special problems that are, as we saw in the previous chapter, solved by the enzyme reverse transcriptase. Molecules of reverse transcriptase are carried within the capsid of an RNA retrovirus, along with its RNA. Once within the host cell, the viral RNA is copied by reverse transcriptase to produce, after a complex series of events, a double-stranded DNA molecule. In the course of these events, reverse transcriptase also directs the duplication of sequences at the ends of the virus, producing repeated sequences known as long terminal repeats (LTRs). LTRs are distinctive features of retroviruses.

Once integrated into a host chromosome, the DNA derived from the viral RNA utilizes the RNA polymerases and other resources of the host cell to produce new viral RNA and protein molecules, which are packaged into new virus particles. Depending on the site of insertion into the host chromosome, DNA derived from a retrovirus may cause mutations by interfering with the expression of host cell genes, either inhibiting them or releasing them from repression. Characteristically, however, most retroviral insertions do not damage or destroy their host cells but become permanent additions to the host cell genome. If germ cells (the cells destined to become eggs and sperm) are infected with such a retrovirus, its genetic information will be transmitted to the next generation.

In mice, it is estimated that from 0.5 to 1.0 percent of the total DNA is of retroviral origin. Moreover, retroviruses are so efficient at promoting their own transcription that as much as 10 percent of the total mRNA in a cell may be of retroviral origin.

Eukaryotic Transposons

Transposons in prokaryotes, as you will recall, are genetic elements—nucleotide sequences—that can move, either directly or in the form of replicas, from one place to another in the bacterial genome. Analogous transposable elements have also been identified in eukaryotic cells. In fact, they were first reported in plants some 40 years ago (see essay). More recently, transposons have been identified in yeast and in *Drosophila,* and there is evidence that many of the repeated DNA sequences in these and other organisms originated as transposons.

Eukaryotic transposons resemble their bacterial counterparts in structure (Figure 18–21), and, like bacterial transposons, they can cause mutations when they become inserted into structural genes or promoter regions. They differ from bacterial transposons in one significant and interesting feature: in eukaryotes, many transposons are first copied into RNA—and then back into DNA—before their insertion into a new location in the chromosomal DNA. This discovery was surprising, for it was previously thought that reverse transcription was unique to retroviruses. Moreover, DNA sequencing has uncovered pseudogenes (nonfunctional genes) that lack introns and, in some cases, even have poly-A tails—clear evidence that they were originally copied from messenger RNA rather than DNA. If this were a mystery story, this would be the devastating clue incriminating the perpetrator—the enzyme reverse transcriptase.

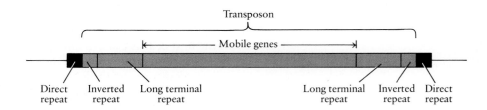

18–21 *Elements typically found in eukaryotic transposons. The central region of the transposon is a DNA sequence, often several thousand base pairs in length, that contains the gene coding for transposase (the enzyme that inserts the transposon into the host chromosome by cutting and splicing the host DNA). This segment may also carry other protein-coding genes. The insertion sequence, which also contains a promoter sequence, consists of a* long terminal repeat, usually several hundred base pairs in length and the same at both ends of the transposon. The inverted repeats are also identical at each end, except that they are mirror images; if one, for example, is AATG, the other will be GTAA. Inverted repeats are the recognition sequences for transposases. The direct repeats, outside the transposon itself, are copies of a short segment (5 to 10 base pairs) of host DNA.

"It Was Fun . . ."

Forty years ago, Barbara McClintock, working in the Cold Spring Harbor Laboratories on Long Island, New York, was studying the genetics of corn *(Zea mays)*. Working with corn kernels (each of which is an embryonic corn plant), she was performing genetic analyses of color differences and other variations, similar to those carried out in Morgan's laboratory with *Drosophila*. In the course of these studies, she encountered instances of unexplained sudden gene inactivation. On the basis of mapping and cytological studies, she was able to deduce that these changes in gene function were not due to mutations but rather came about as a consequence of the movement of genetic elements—"controlling elements," she called them—from place to place on the chromosome. These elements, she reported, actually "jumped" from one site to another on a chromosome and even from one chromosome to another.

Her findings, first published in 1951, were largely ignored, another odd traveler's tale that did not fit into the scheme of things as then understood. "Fiercely independent, beholden to no one," in the words of James Watson, she stubbornly pursued her research, sometimes working without pay. "It was fun," she is reported to have said, "I could hardly wait to get up in the morning."

In the last decade, with the discovery of a host of movable genetic elements, McClintock has received a

Barbara McClintock, photographed in 1983 with one of her colleagues, Stephen Dellaporta, at the Cold Spring Harbor Laboratories.

barrage of accolades. In 1981, at the age of 79, she was given eight separate awards, including a lifetime grant, and was hailed as a scientific prophet. In 1983, she became a Nobel laureate. Her life hasn't changed much, however; she is still at work in her laboratories in Cold Spring Harbor.

Although nucleotide sequences that resemble defective genes for reverse transcriptase have been identified in the genome of some eukaryotes, there is, thus far, no evidence of reverse transcriptase activity in cells uninfected by retroviruses. It is, however, widely accepted that the action of reverse transcriptase in the course of retroviral infections has played an important role in the evolution of the eukaryotic genome and, in particular, in the evolution of transposons.

GENES, VIRUSES, AND CANCER

Cancer is a disease in which cells escape the factors, still largely unknown, that regulate normal cell growth. As a consequence, the cells multiply out of control, crowding out, invading, and destroying other tissues. Cancer is often considered a group of diseases rather than a single disease because, with few exceptions, any one of the 200 or more cell types in the human body can become malignant. The behavior of the cells and the prognosis of the illness depend on the type of cells that have become malignant.

Three lines of evidence have long linked the development of cancer with changes in the genetic material. First, once a cell has become cancerous, all of its daughter cells are cancerous; in other words, cancer is an inherited property of cells. Second, gross chromosomal abnormalities, such as deletions and translocations, are often visible in cancer cells. Third, most carcinogens—agents known to cause cancer, such as x-rays, ultraviolet radiation, tobacco smoke, and a variety of chemicals—are also mutagens.

As long ago as 1911, a cancer-causing virus, the Rous sarcoma virus, was isolated from chicken tumors. (In those days, a virus was defined simply as "a cell-free extract that produces a disease when it is injected into a suitable host.") Even though other cancer-causing viruses, particularly viruses affecting laboratory mice, were gradually discovered, a viral theory of cancer was slow to emerge. For one thing, viruses could not be shown to be important as causes of human cancer. (Even today, after years of searching, only a few rare human cancers have been linked to viruses.) For another, the "viral theory" of cancer seemed to be at odds with the "mutation theory." In addition, the fact that most of the known cancer-causing viruses, including the Rous sarcoma virus, are RNA viruses rather than DNA viruses also seemed to make these two hypotheses incompatible.

Eventually, however, evidence emerged that viruses, like mutagens, can bring about changes in the cell's genetic makeup and that, furthermore, all known cancer-causing viruses are viruses that introduce information into host cell chromosomes. These include both DNA viruses, like SV40, and RNA retroviruses. The discovery of the role of reverse transcriptase forged the crucial link between retroviruses and the chromosomes of eukaryotic cells. The Rous sarcoma virus has been shown to be a retrovirus, and DNA segments produced by reverse transcriptase from its RNA have been located in host cell chromosomes.

Recombinant DNA techniques have enabled molecular biologists to study some of the changes in eukaryotic chromosomes that lead to cancer. These studies are usually carried out in cells growing in tissue culture. When such cells are exposed to a cancer-causing agent, such as a virus, they may undergo characteristic changes in their growth patterns and in their shape. Such cells are said to be **transformed** (Figure 18–22). Transformed cells can produce cancers when they are transplanted into laboratory animals. (Note that transformation has two meanings in biology: one is the introduction of new characteristics into a cell by means of DNA from another cell, as in the pneumococcus experiments of some 60 years ago; the other is the induction of cancer.)

Studies of transformed cells have uncovered a group of genes known as **oncogenes** (from the Greek word *onkos*, meaning "tumor"). Oncogenes closely resemble normal genes of the eukaryotic cells in which they are found. According to the oncogene hypothesis, cancer is caused when something goes wrong in the expression of these normal cellular genes, as a result of mutations in the genes themselves, changes in gene regulation, or both. Thus, viruses can cause cancer in three different ways. First, simply by their presence in the chromosome, viruses may disrupt the function of normal genes. Second, viruses may encode proteins needed for viral replication that also affect the regulation of cellular genes. Third, and most interesting of all, viruses may serve as vectors of oncogenes. In fact, oncogenes were first discovered when genetic analyses of cancer-causing retroviruses revealed the presence of genes that were not required by the viruses for their own multiplication. It was subsequently found that the nucleotide sequences of these genes not only closely resemble those of normal genes of the host cell but also cause malignant transformation of the cells.

With these discoveries, the "viral theory" and the "mutation theory" of cancer are no longer regarded as incompatible but rather as mutually supportive. About

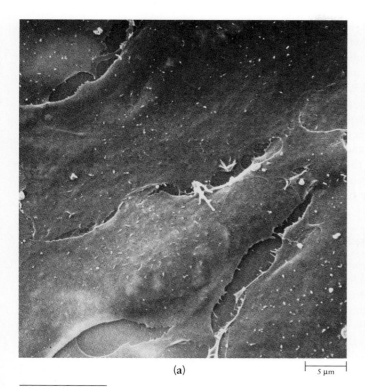

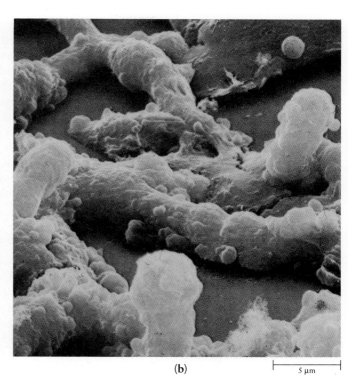

(a)   5 μm   (b)   5 μm

18–22 *Scanning electron micrographs of* (a) *normal cells growing in tissue culture and* (b) *the same type of cells after transformation with a cancer-causing virus. Observe that the cancer cells not only show striking surface changes but also have piled up on top of one another. The normal cells are inhibited by cell-to-cell contact and stop multiplying (see page 147), whereas the cancer cells do not.*

50 oncogenes have been discovered so far. Their gene products that have been identified all seem to be regulatory proteins of some sort, involved with the control of either cell growth or cell division. Thus, this work not only is bringing us closer to the control of one of our oldest and ugliest enemies but also is yielding new information on the fundamental question of the regulation of cell growth.

TRANSFERS OF GENES BETWEEN EUKARYOTIC CELLS

The most extravagant hope with regard to the applications of recombinant DNA technology is that at some future time it may be possible to correct genetic defects by substituting "good" genes for "bad" ones. This is an enormously complex undertaking. It requires, first, the preparation of a gene that will be taken up by a eukaryotic cell, become incorporated into a chromosome, and be expressed there—but this is only the beginning. The new gene must be established in a large number of cells of the appropriate type (blood cells should not produce somatostatin, for example) and be subject to the complicated—and as yet largely unknown—regulatory controls of the normal gene.

To Cells in Test Tubes

The first stage of the project has proved easier than expected; foreign genes will undergo recombination in eukaryotic cells growing in test tubes. In the first such experiment, SV40 virus was used as a vector to insert a rabbit gene for the beta-globin polypeptide into monkey cells. The recipient cells produced rabbit beta globin. The advantage of using a virus as a vector is that not only can viruses gain access to target cells but also they characteristically possess strong promoters, with the result that the gene is efficiently expressed.

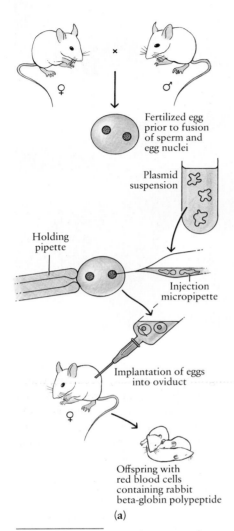

Subsequently, it was found that exposing cells in tissue culture to purified DNA precipitated with calcium ion (Ca^{2+}) stimulates the uptake of DNA; apparently some of the cells—about one in a million—actually phagocytize the precipitated granules and incorporate the DNA. To identify those cells that have taken up the foreign DNA, a marker is needed. To solve this problem, a line of mouse cells has been developed that lacks the enzyme thymidine kinase (TK). These TK^- cells are exposed to DNA molecules containing both the gene for thymidine kinase and other genes. Cells that have taken up the DNA molecules can be identified by their ability to grow in a medium in which the TK^- cells cannot.

Studies such as these demonstrate that eukaryotic cells have, in principle, the same capacities for incorporating foreign DNA as do prokaryotic cells. Moreover, they make possible further analysis of the regulation of expression of eukaryotic genes.

To Fertilized Mouse Eggs

Foreign genes have also been introduced into fertilized eggs and expressed in the organisms that developed from the eggs. Jon W. Gordon and Frank H. Ruddle of Yale University were the first to successfully insert a DNA sequence into the fertilized eggs of mice. When the eggs were injected with a rabbit beta-globin gene, the mice derived from the eggs were found to contain the rabbit beta globin in their red blood cells (Figure 18–23). The fact that the gene was expressed only in the red blood cells—and not in other tissues of the mouse—indicated that it had been incorporated in the "right" place and so had come under cellular control mechanisms. The gene was also passed, in a Mendelian distribution, to subsequent generations.

Using the same technique, Ralph Brinster of the University of Pennsylvania and Richard Palmiter of the University of Washington combined the human gene for the growth hormone somatotropin with the regulatory portion of a mouse gene and injected it into fertilized mouse eggs. The resultant "transgenic" mice—mice receiving the transferred human gene—grew to twice the normal size (Figure 18–24) indicating that the human gene was incorporated into the mouse genome and was producing growth hormone. Because the DNA was injected into the egg cells, it was found in all cells, including the germ cells, and thus could be passed on to the next generation.

This type of procedure has now been carried out with a number of cloned genes. From 10 to 30 percent of the eggs survive the manipulation and, of these, the foreign gene functions in up to 40 percent.

18–23 (a) *The procedure by which Gordon and Ruddle inserted the gene for rabbit beta globin into mice. The gene was spliced into plasmids, which were then injected into fertilized eggs before the egg and sperm nuclei fused. After injection, the fertilized eggs were implanted in a female mouse, who gave birth. The rabbit beta-globin polypeptide was present in the red blood cells of the offspring, and hybridization techniques revealed that the gene had been incorporated into their DNA.*

(b) Injection of the plasmid suspension into a fertilized egg. The diameter of the micropipette tip is only about 0.5 micrometer, but Gordon says that the injection "is equivalent to your being speared by a telephone pole." Nevertheless, a significant number of the eggs survive.

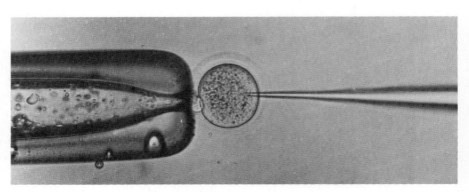

(b)

18-24 *The two female mice shown here are litter mates, approximately 24 weeks old. The fertilized egg from which the mouse on the left developed was injected with a gene consisting of the promoter and regulator sequences of a mouse gene combined with the structural gene for human growth hormone. Following integration of the new gene into the genome of the female mouse, it is passed on to her offspring. On the average, mice that express the new gene grow two to three times as fast as mice lacking the gene and, as adults, they are twice the normal size. (Those unable to see the utility of a giant mouse may be interested to learn that a similar procedure has now been successfully performed in fish.)*

18-25 *The red-eyed fruit fly at the right is the offspring of the brown-eyed fly at the left.* Drosophila *transposons bearing a gene for red eyes were injected into the brown-eyed fly when it was an early embryo. Transposons with the gene for red eyes were incorporated into chromosomes of the cells that ultimately formed its gametes. The gene for red eyes was therefore passed on to its offspring.*

To *Drosophila* Embryos

More recently, in a most elegant experiment, Allan C. Spradling and Gerald M. Rubin at the Carnegie Institution of Washington used naturally occurring transposons of *Drosophila* to ferry genes into embryonic fruit flies. They first demonstrated that if the transposons were injected into early embryos, they could become incorporated into the cells destined to become gametes. They then spliced a gene for red eyes into the transposons, cloned the transposons bearing the inserted gene, and injected them into early embryos of mutant brown-eyed flies. About 8 percent of the injected embryos developed into fertile adults. Although these adults had brown eyes, 39 percent of them produced offspring with normal red eyes (Figure 18–25). Subsequent generations of offspring from these red-eyed flies also had red eyes, indicating that the gene had been incorporated into the *Drosophila* chromosomes in a stable fashion. Early indications are that transposons may prove to be the vectors of choice.

The hope is, of course, that it may someday be possible to correct human genetic defects. The immediate goal is the development of more reliable vectors to carry genes into somatic cells and, most important, increased knowledge of the regulatory factors that control their functions. Scientists in this field are cautious —but optimistic.

SUMMARY

The eukaryotic chromosome differs in many ways from the chromosome of prokaryotes. Its DNA may take the form of B-DNA (the right-handed helix described by Watson and Crick), A-DNA (a less tightly coiled right-handed helix), or Z-DNA (a left-handed helix). Eukaryotic DNA is always associated with proteins, which constitute more than half the weight of the chromosome. Most of these proteins are histones, which are relatively small, positively charged molecules. The DNA molecule wraps around cores made of eight molecules of histone to form nucleosomes, which are the basic packaging units of eukaryotic DNA.

Biologists are beginning to understand some aspects of the regulation of gene expression in eukaryotes. During embryonic development, different groups of genes are activated or inactivated in different types of cells. According to several lines of evidence, gene expression is correlated with the degree of condensation of the chromosome. Condensed chromatin may take the form of either euchromatin, which is loosely packed, or heterochromatin, which is tightly packed. Another factor thought to be involved in gene regulation is the methylation of cytosine nucleotides, which takes place after replication. A variety of specific regulatory proteins, still very poorly understood, are also thought to play key roles in the regulation of gene expression.

Eukaryotes have far more DNA than prokaryotes. The amount of DNA is constant in the cells of any given species but is not correlated with the size, complexity, or position on the evolutionary scale of the organism. Eukaryotic cells seem to have a great excess of DNA, much of which appears to be "meaningless."

In complex multicellular eukaryotes, the coding sequence of most structural genes is not continuous but contains introns, which are also known as intervening sequences. Although introns are transcribed into RNA in the nucleus, they are not present in the mRNA in the cytoplasm and thus are not translated into protein. The segments that are present in the cytoplasmic mRNA and are translated into protein are known as exons.

Hybridization and sequencing studies have revealed three classes of eukaryotic DNA. Multiple repeats of short nucleotide sequences, characteristically arranged in tandem, are known as simple-sequence DNA. Simple-sequence DNA is associated with the tightly coiled heterochromatin in the region of the centromere. Longer repeats, usually dispersed throughout the chromosomes, are known as intermediate-repeat DNA. Intermediate-repeat DNA includes multiple copies of the genes coding for the rRNAs, tRNAs, and histones. The third class, single-copy DNA, makes up about 70 percent of the chromosomal DNA in humans. Current data indicate that as little as 1 percent of the human genome may be translated into protein.

Some structural genes, such as those coding for the polypeptide chains of hemoglobin molecules, form gene families. The individual genes of the family differ slightly in their nucleotide sequences; as a consequence, the proteins for which they code differ slightly in structure and biological properties. Some members of gene families are not expressed, presumably because of deleterious mutations; these DNA sequences are known as pseudogenes. Gene families are believed to have their origins in gene duplications that occurred as a result of recombination "errors," followed by different mutations in different copies of the gene.

Transcription in eukaryotes differs from that in prokaryotes in a number of respects. Several different RNA polymerases are involved, as well as a multiplicity of regulatory proteins. Also, in eukaryotes, structural genes are not grouped in operons as they often are in prokaryotes; the transcription of each gene is regulated separately, and each gene produces an RNA transcript containing the encoded information for a single product. RNA transcripts are processed in the nucleus to produce the mature messenger RNA molecules that move through the nuclear pores into the cytoplasm. This processing includes addition of a methyl-guanine cap to the 5′ end of the molecule, addition of a poly-A tail to the 3′ end, and removal of the introns. Alternative splicing of identical RNA transcripts in different types of cells can produce different mRNA molecules and different polypeptides.

As is the case with prokaryotes, the eukaryotic genome contains a surprising array of movable genetic elements. Functional antibody genes are formed during the differentiation of lymphocytes by the rearrangement of gene sequences coding for different parts of the antibody molecule. Viruses, including both DNA viruses and RNA retroviruses, can become integrated into the eukaryotic chromosome; when integrated into the eukaryotic chromosome, they are known as proviruses. The incorporation of genetic information carried by RNA retroviruses depends on the transcription of the RNA into DNA, which is catalyzed by reverse transcriptase. Eukaryotic transposons resemble those of prokaryotes in that, by becoming inserted in the genome, they activate or inactivate genes, either by disrupting the coding sequences or by interfering with regulation. As in prokaryotes, the transposon may not actually move but instead may generate a copy that becomes integrated elsewhere in the genome. Many eukaryotic transposons differ from those of prokaryotes most notably in that the transposon is first transcribed to RNA and then back to DNA before insertion elsewhere in the chromosome. Current evidence suggests that the evolutionary origin of eukaryotic transposons is to be found in retroviruses.

Cancer, according to present evidence, is caused by alterations in the function of some normal cellular genes; such genes are known as oncogenes. These alterations in gene function may be caused by mutations, by changes in the regulation of gene expression, or both. Viruses can cause cancer by inserting an oncogene into a chromosome or by disrupting gene regulation. Elucidation of this role of viruses has served to unify conflicting hypotheses concerning the causes of cancer.

Limited success has been achieved in the transfer of genes to eukaryotic cells growing in test tubes, to fertilized eggs of the mouse, and to *Drosophila* embryos. In each of these cases, foreign genes have been incorporated and expressed in the new host. Such studies are leading to increased understanding of the regulatory factors governing the expression of eukaryotic genes.

QUESTIONS

1. Distinguish among the following: B-DNA/Z-DNA; nucleosome/nucleolus; euchromatin/heterochromatin; intron/exon; gene/pseudogene/oncogene.

2. In what ways are the chromosomes of prokaryotes and eukaryotes similar? In what ways are they different?

3. Describe the three classes of eukaryotic DNA. What are some of the functions that have been identified for each class?

4. What might be the advantage to an organism of having multiple copies of the genes for the rRNAs, tRNAs, and histones?

5. Although a human cell contains considerably fewer than 1 million structural genes, a human individual is, according to immunologists, capable of making at least 100 million different kinds of antibodies. How is this possible?

6. What is the function of the intron in the assembly of an mRNA molecule coding for a polypeptide chain of an antibody molecule?

7. Most bacteriophages lack introns, but introns have been found in the DNA viruses of eukaryotes. The RNA viruses of eukaryotes, however, lack introns. What do these findings suggest about the origin of viruses?

8. When DNA is heated, the hydrogen bonds between paired bases break and the complementary strands separate, denaturing the molecule. As shown in the electron micrograph below, when a DNA molecule containing the five histone genes was heated slightly, only the spacer sequences were denatured, revealing the location of the histone-coding regions. The genes coding for histones are rich in guanine and cytosine, while the spacers between them are rich in adenine and thymine. How does this explain the appearance of the DNA molecule? (*Hint:* Think about the chemistry of the DNA molecule and, if necessary, review Figure 14–11b on page 290.)

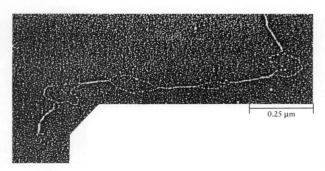

0.25 μm

C H A P T E R 19

Human Genetics: Past, Present, and Future

The principles of genetics are, of course, the same for humans as they are for members of any other diploid, eukaryotic species. In practice, however, there are some important differences. With the exception of those of us who belong to royal families (Figure 19–1), most people do not have information about their forebears that extends back over more than three generations. By contrast, any individual fruit fly in Morgan's laboratory had a pedigree that went back many generations. Breeding experiments, so readily performed with pea plants, are not possible with humans and, even if they were, the small numbers of offspring and the long generation time would make such investigations impractical. Consequently, in the past most of our knowledge about human genetics has come from the observation of abnormalities with a hereditary pattern—some trivial curiosities, such as the man with the urine black as ink (page 301), and some life-threatening, such as sickle cell anemia. Thus, until very recently, the flow of information has been from medical science to basic genetics. Now, however, the advances in molecular genetics made possible by recombinant DNA technology have reached the point of practical application in human genetics. They have revolutionized the understanding of many genetic defects while simultaneously providing new means for their diagnosis and new hopes for cure and prevention.

19–1 *Queen Victoria (seated center) and some of her immediate family. Seventeen of the people in this photograph, which was taken in 1894, are her direct descendants. These include Princess Irene of Prussia, standing to the right of Victoria and wearing a feather boa, and, to the left of Victoria, Alexandra (also wearing a boa), the future Tsarina of Russia. Nicholas II, to become the last Tsar of Russia, is standing beside Alexandra. Both Irene and Alexandra were carriers of a sex-linked recessive allele for hemophilia, a blood-clotting disorder.*

19-2 *The normal diploid chromosome number of a human being is 46, 22 pairs of autosomes and two sex chromosomes. The autosomes are grouped by size (A, B, C, etc.), and then the probable homologues are paired. A normal woman has two X chromosomes and a normal man, shown here, an X and a Y.*

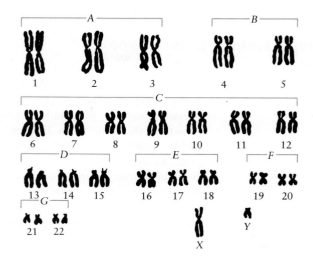

THE HUMAN KARYOTYPE

The number of chromosomes in the human species is 46, 44 autosomes and two sex chromosomes. Although cytologists began counting human chromosomes in the 1890s, it took a long time for them to arrive at the correct figure. The nucleus is small and the number of chromosomes large (as compared, for instance, to a fruit fly's four). Counts were made with tissues taken from corpses, often executed criminals; after death, the chromosomes tend to clump together, resulting in falsely low counts. In the 1920s, a cytologist was able to obtain fresh tissue, from the the testes of three patients at a state mental institution who were castrated for "excessive self-abuse." Working with this preparation, he reported finding a diploid number of 48. So seemingly authoritative was his statement and so technically difficult the problem that other cytologists also reported 48—for more than 30 years. It was not until techniques became available for growing cells in tissue culture and spreading them out for observation (see essay) that the correct number was revealed; this was in 1956, three years after Watson and Crick published their structure of DNA.

A graphic (or photographic) representation of the chromosomes present in the nucleus of a single somatic cell of a particular organism is known as a **karyotype.** From a karyotype, as shown in Figure 19–2, we can determine the number, size, and shape of the chromosomes and identify the homologous pairs. As you can see, however, some of the smaller chromosomes are quite similar in appearance. In these cases, staining to reveal banding patterns, as shown in Figure 19–3, makes it possible to distinguish similarly sized chromosomes and to identify homologues.

19-3 *A standard map of the banding patterns of chromosomes 8 through 11 in the human karyotype as determined both at the metaphase stage (black bands) and at the early prophase stage of mitosis (colored bands). The early prophase chromosomes are much longer and thinner than the metaphase chromosomes, and many more bands can therefore be detected. All of the bands shown here are those that stain with a specific reagent. Note how these chromosomes, which are similar in size and shape, can readily be distinguished by their banding patterns.*

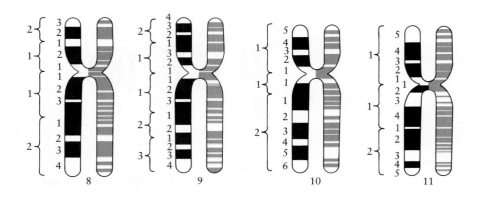

Preparation of a Karyotype

Chromosome typing for the identification of gross chromosomal abnormalities is being carried out at an increasing number of genetic counseling centers throughout the United States. The result of the procedure is a graphic display of the chromosome complement, known as a karyotype. The chromosomes shown in a karyotype are mitotic metaphase chromosomes, each consisting of two sister chromatids held together at their centromeres. To prepare a karyotype, cells in the process of dividing are interrupted at metaphase by the addition of colchicine, a drug that prevents the subsequent steps of mitosis from taking place by interfering with the spindle microtubules. After treating and staining, the chromosomes are photographed, enlarged, cut out, and arranged according to size. Chromosomes of the same size are paired according to centromere position, which results in different "arm" lengths. From the karyotype, certain abnormalities, such as an extra chromosome or piece of a chromosome, can be detected.

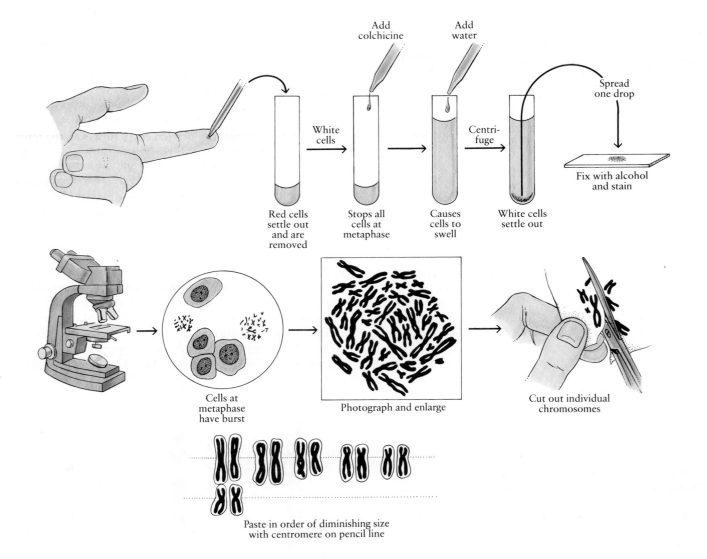

Add colchicine

Add water

White cells

Red cells settle out and are removed

Stops all cells at metaphase

Causes cells to swell

Centri-fuge

White cells settle out

Spread one drop

Fix with alcohol and stain

Cells at metaphase have burst

Photograph and enlarge

Cut out individual chromosomes

Paste in order of diminishing size with centromere on pencil line

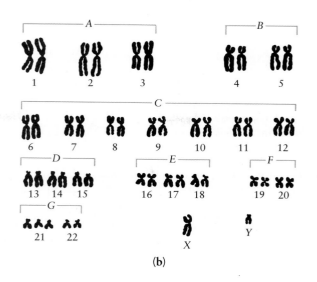

(b)

19–4 (a) *Although children with Down's syndrome share certain physical characteristics, there is a wide range of mental capacity among these individuals.* **(b)** *The karyotype of a male with Down's syndrome caused by nondisjunction. Note that there are three chromosomes 21.*

Chromosome Abnormalities

Certain genetic diseases are caused by abnormalities in the number or structure of chromosomes so severe that they can be detected in the karyotype. For example, from time to time, usually because of "mistakes" during meiosis or mitosis, homologous chromosomes or their chromatids may not separate. This phenomenon is known as **nondisjunction**. In meiosis, the results of nondisjunction are gametes with one or more chromosomes too many and other gametes with one or more chromosomes too few. A gamete with too few chromosomes (unless the missing chromosome is a sex chromosome) cannot produce a viable embryo. Sometimes, although rarely, a cell with too many chromosomes can produce a viable embryo; the result is an individual with one or more extra chromosomes in every cell of his or her body. In the vast majority of such cases, however, the fetus is spontaneously aborted early in pregnancy, an event that occurs in 15 to 20 percent of recognized pregnancies.

Individuals with additional autosomal chromosomes always have widespread abnormalities; with the exception of those with Down's syndrome, those that are not stillborn typically survive only a few months. Among the few who survive, most are mentally retarded and those who survive to maturity are usually sterile. They frequently have abnormalities of the heart and other organs as well.

Other abnormalities that may be visible in the karyotype are deletions and translocations. A deletion is simply the loss of a portion of a chromosome. A translocation (page 277) occurs when a deleted portion of one chromosome is transferred to and becomes part of another, nonhomologous chromosome.

Down's Syndrome

One of the most familiar conditions resulting from an abnormality in an autosomal chromosome is Down's syndrome, named after the physician who first described it. Because it usually involves more than one defect, it is referred to as a syndrome, a group of disorders that occur together. Down's syndrome includes, in most cases, a short, stocky body type with a thick neck; mental retardation, ranging from mild to severe in different individuals; a large tongue, resulting in speech defects; an increased susceptibility to infections; and, often, abnormalities of the heart and other organs. Individuals with Down's syndrome who survive into their thirties or forties also have a high probability of developing a form of senility similar to Alzheimer's disease (to be discussed in Chapter 43).

Down's syndrome arises when an individual has three, rather than two, copies of chromosome 21. In about 95 percent of the cases, the cause of the genetic abnormality is nondisjunction during formation of a parental gamete, resulting in 47 chromosomes, with an extra copy of chromosome 21 in the cells of the affected individual (Figure 19–4).

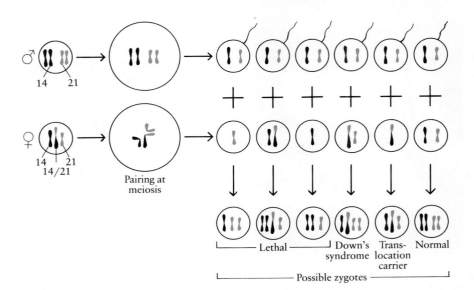

19–5 *Transmission of translocation Down's syndrome. The father, top row, has normal pairs of chromosomes 21 and 14, and each of his sperm cells will contain a normal 21 and a normal 14. The mother (shown here as the translocation carrier) has one normal 14, one normal 21, and a translocation 14/21. She herself appears normal, but her chromosomes cannot pair normally at meiosis. There are six possibilities for the offspring of these parents; the infant will (1) die before birth (three of the six possibilities), (2) have Down's syndrome, (3) be a translocation carrier like the mother, or (4) be normal. Tests for the chromosomal abnormality can be made in prospective parents and in the fetus before birth.*

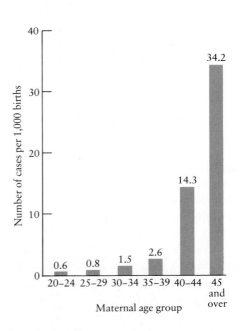

19–6 *The frequencies of births of infants with Down's syndrome in relation to the ages of the mothers. The number of cases shown for each age group represents the occurrence of Down's syndrome in every 1,000 live births by mothers in that group. As you can see, the risk of having a child with Down's syndrome increases rapidly after the mother's age exceeds 40. An increased risk is also thought to occur after the father's age exceeds 55.*

Down's syndrome may also result from a translocation in the chromosomes of one of the parents. The person with Down's syndrome caused by translocation usually has a third chromosome 21 (or, at least, most of it) attached to a larger chromosome, most often chromosome 14. Such an individual, although he or she has only 46 chromosomes, has the functional equivalent of a third chromosome 21.

When cases of Down's syndrome due to translocation are studied, it is usually found that one parent, although phenotypically normal, has only 45 separate chromosomes. One chromosome is usually composed of most of chromosomes 14 and 21 joined together. The possible genetic makeups of the offspring of this parent are diagrammed in Figure 19–5. Three out of the six possible combinations are lethal. One of the remaining three will produce Down's syndrome, one will be normal, and one will be an asymptomatic carrier of the 14/21 translocation. Thus, parents who have a child with Down's syndrome are advised to have their karyotypes prepared. If either parent has the translocation, they are warned that they are at a high risk of having another child with Down's syndrome and that half of their normal children will be carriers of the translocation.

It has been known for many years that Down's syndrome and a number of other defects involving nondisjunction are more likely to occur among infants born to older women (Figure 19–6); the reasons for this are not known. Recent studies, however, have also indicated that in about 25 percent of the cases of Down's syndrome due to nondisjunction, the extra chromosome comes from the father rather than the mother.

Abnormalities in the Sex Chromosomes

Nondisjunction may also produce individuals with unusual numbers of sex chromosomes. An *XY* combination in the twenty-third pair, as you know, produces maleness, but so do *XXY, XXXY,* and even *XXXXY.* These latter males, however, are usually sexually underdeveloped and sterile. *XXX* combinations sometimes produce normal females, but many of the *XXX* women and almost all *XO* women (women with only one X chromosome) are sterile.

Chromosome Deletions

Small chromosome deletions can also result in congenital defects or other illnesses. For example, a small deletion on the short arm of chromosome 11 is

19–7 (a) *A chromosomal abnormality associated with cancer. The chromosomes shown here have been stained to reveal banding patterns. The chromosome on the left is normal. The one on the right has a deletion, shown by the smaller size of the bracket. Such deletions have been found in children with Wilms' tumor. (b) The left eye of a 15-year-old boy who has this chromosomal deletion and who developed Wilms' tumor in infancy. Note the absence of an iris. An older half-brother and a maternal aunt also had aniridia and developed Wilms' tumor at an early age. Another brother and the boy's mother are phenotypically normal. Analysis of the mother's chromosomes revealed that although she carries the deletion in chromosome 11, the missing segment is present in her cells in chromosome 2. Almost all other chromosomal abnormalities associated with cancer have occurred only in somatic cells and are not inherited.*

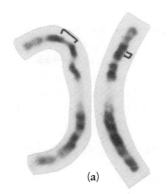

(a)

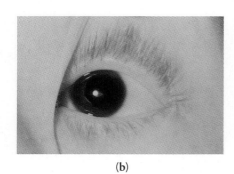

(b)

associated with Wilms' tumor, a cancer of the kidney found in infants and young children. Associated with this same deletion is a condition known as aniridia, the congenital absence of the iris of the eye (Figure 19–7). Not everyone with the deletion develops Wilms' tumor, but children with both aniridia and the chromosome deletion are at a high risk. By testing children with aniridia for the chromosome deletion, medical researchers can detect and treat Wilms' tumor at an early stage before it has caused any symptoms.

Prenatal Detection

A procedure known as **amniocentesis** makes possible the prenatal detection of Down's syndrome and a number of other genetic conditions in the fetus. A thin needle is inserted through the mother's abdominal wall and through the membranes that enclose the fetus, and a sample of the amniotic fluid surrounding the fetus is withdrawn (Figure 19–8). While the procedure is being performed, a sonogram of the fetus is displayed on a monitor, enabling the physician to identify its precise location. Although amniocentesis must be done with great care, it is simple, quick, and usually harmless. The amniotic fluid contains living cells sloughed off by the fetus. These cells, grown in tissue culture, can provide mitotic cells from which a karyotype can be made.

19–8 (a) *Amniocentesis. The position of the fetus is first determined by ultrasound. Then a needle is inserted into the amniotic cavity, and fluid containing fetal cells is withdrawn into a syringe. The cells are grown in tissue culture and then are analyzed for chromosomal abnormalities and other genetic defects. The procedure is usually not performed until the sixteenth week of pregnancy, to ensure both that there are enough fetal cells in the amniotic cavity to make detection possible and that there is sufficient amniotic fluid that removal of the small amount necessary for the test will not endanger the fetus.*

(b) A sonogram of the uterus of a pregnant woman carrying a four-month-old fetus. The solid black regions enclosed by the muscular uterine walls are the fluid surrounding the fetus. The fetus is lying on its back, with its head at the left and its right arm in the foreground. Toward the right, you can see portions of the umbilical cord extending from the abdomen of the fetus up to the thickened tissues of the placenta.

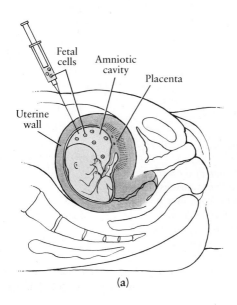

Fetal cells
Amniotic cavity
Placenta
Uterine wall

(a)

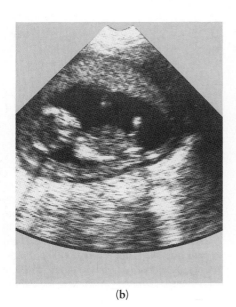

(b)

More recently, a technique has been developed for collecting cells from the chorion, one of the fetal membranes. The advantage of this method is that the test can be performed as early as the eighth week of pregnancy.

In most cases, at this stage of our knowledge, treatment of conditions detected by prenatal testing is not possible. Parents are faced with the difficult decision of whether or not to abort the affected fetus.

PKU, SICKLE CELL ANEMIA, AND OTHER RECESSIVES

Many inherited disorders are, like the white flowers in Mendel's pea plants, the result of the coming together of two recessive alleles. Individuals heterozygous for the gene are usually symptom-free; in the heterozygote enough of the particular protein is produced from the normal allele to make up for the recessive allele, which codes for either a defective, poorly functioning polypeptide or for none at all. Thus, the clinical symptoms show up only in the recessive homozygote.

Phenylketonuria

One of the best-studied examples of a genetic disorder inherited as a Mendelian recessive is phenylketonuria, or PKU, as it is commonly known. Individuals with PKU lack the enzyme that normally converts the amino acid phenylalanine to tyrosine (Figure 19–9). When this enzyme is missing or deficient, phenylalanine and its abnormal breakdown products accumulate in the bloodstream and urine. These breakdown products are harmful to the cells of the developing nervous system and can result in profound mental retardation.

19–9 *Some steps in the pathway for the breakdown of the amino acids phenylalanine and tyrosine. If the enzyme that catalyzes Step 4, the conversion of homogentisate into 4-maleylacetoacetate, is missing, alkaptonuria results (page 301). If the enzyme that catalyzes Step 1, the conversion of phenylalanine to tyrosine, is defective, the result is an accumulation of phenylalanine and the disease known as phenylketonuria (PKU). If one of the enzymes that converts tyrosine to melanin is defective, albinism results.*

PKU is caused by a recessive allele in the homozygous state. About 1 in every 15,000 infants born in the United States is homozygous for this allele. Such infants usually appear healthy and normal at birth, but after the first few months the symptoms of the disease set in, and without treatment severe mental retardation usually results. Many never learn to walk or talk and are subject to periodic convulsions and seizures. Most afflicted individuals must be hospitalized for their entire lives, which, in untreated persons, is seldom more than 30 years.

It is not yet known how the high levels of phenylalanine and derivative compounds bring about the neurological symptoms. However, the knowledge we do have is enough to effectively treat infants with PKU and prevent the symptoms from appearing. Most states now require routine tests of all newborn babies in order to detect PKU homozygotes. Those identified at birth are put on a special diet containing low amounts of phenylalanine—enough to supply dietary needs but not enough to permit toxic accumulations. As a result, they are able to develop normally.

Albinism

Albinism—the lack of pigmentation in skin, hair, and eyes—is due to an inability to make the brown pigment melanin. Melanin is produced in pigment cells via an enzymatic pathway from the amino acid tyrosine; as shown in Figure 19–9, it is a product of the same pathway associated with the breakdown of phenylalanine. Most albinos lack one of the enzymes necessary to produce melanin. Other albinos have the enzyme, which is, however, unable to enter the pigment cells; as a consequence, the tyrosine within these cells is not acted upon by the enzyme and melanin is not produced. Both forms of albinism are inherited as autosomal recessives.

Tay-Sachs Disease

Tay-Sachs disease is an autosomal recessive condition resulting in degeneration of the nervous system. As with PKU, Tay-Sachs homozygotes appear normal at birth and through the early months. However, by about eight months, symptoms of severe listlessness become evident. Blindness usually occurs within the first year. Afflicted children rarely survive past their fifth year. The biochemical basis of this disease is now, at least in part, understood. Homozygous individuals lack an enzyme, N-acetyl-hexosaminidase, which breaks down a lipid known as GM_2 ganglioside. The enzyme is normally found in the lysosomes of brain cells and plays a crucial role in keeping GM_2 ganglioside from accumulating. In the child lacking the enzyme, the lysosomes of the brain cells fill with this lipid and swell, and the cells die (Figure 19–10). There is no therapy yet available for Tay-Sachs disease.

While Tay-Sachs disease is a rare disorder in the general population (1 in 300,000 births), until recently it has had a much higher incidence (1 in 3,600 births) among Jews of Eastern and Central European extraction (Ashkenazic Jews), who make up more than 90 percent of the American Jewish population. It is estimated that among this population approximately 1 in 28 individuals is a heterozygous carrier of the Tay-Sachs allele. The development of a blood test measuring levels of the enzyme that is deficient in Tay-Sachs disease has made it possible for prospective parents to determine if they are carriers (heterozygotes have half the normal levels of the enzyme). Since its development, this test has been so extensively utilized by prospective parents in the American Jewish population that the incidence of Tay-Sachs births has now dropped dramatically; in fact, in the United States today, the majority of the Tay-Sachs babies born are to non-Jewish parents.

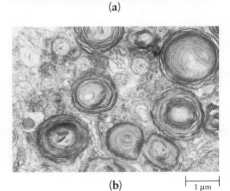

(a)

(b) $\vdash$ 1 μm $\dashv$

19–10 (a) *In this 17-month-old child with Tay-Sachs disease, deterioration of the brain, already begun, progressed rapidly. The child died before his sixth birthday.* (b) *Tay-Sachs disease is caused by the absence of an enzyme involved with lipid metabolism. Without the enzyme, harmful lipid deposits accumulate in the lysosomes of brain cells, as shown in this micrograph.*

Sickle Cell Anemia

The allele that, in the homozygous recessive condition, is responsible for sickle cell anemia apparently originated in Africa. For reasons that we shall discuss in Chapter 47, it has been maintained by natural selection at a very high frequency in the populations of certain regions of Africa. As a consequence, sickle cell anemia occurs at a high frequency among blacks. In the United States about 9 percent of blacks are heterozygous for the sickle cell allele, and about 0.2 percent are homozygous for the allele and therefore have the symptoms of sickle cell anemia.

Sickle cell anemia, you will recall, is due to a single amino acid substitution in the beta chains of the hemoglobin molecule (page 80). When the oxygen concentration is low, sickle cell hemoglobin becomes insoluble and forms bundles of stiff fibers. These fibers distort the shape of the red blood cells, making them more fragile; premature degradation of the red blood cells causes the anemia. Also, the loss of flexibility of the red blood cells (which are normally very flexible) makes it difficult for them to make their way through small blood vessels. Blocking of the blood vessels in the joints and in vital organs by these abnormal red cells is both painful and life-threatening.

Individuals heterozygous for the sickle cell allele are generally symptomless. Their hemoglobin, however, contains both normal and sickle cell beta chains. As with PKU, the "good" allele makes enough normal hemoglobin that the effects of the "bad" allele are not discernible. However, if blood samples are treated in ways that remove oxygen from all the hemoglobin molecules, some of the blood cells of a heterozygote will sickle (Figure 19–11). Thus it is possible to detect heterozygotes quite easily. If two heterozygotes have children, there is 1 chance in 4 that they will have a child with sickle cell anemia, and a fifty-fifty chance that they will have a child who is, like themselves, a heterozygote and so a carrier of the allele.

19–11 *Scanning electron micrograph of deoxygenated blood from an individual heterozygous for the sickle cell allele. As you can see, some—but not all—of the red blood cells have sickled.*

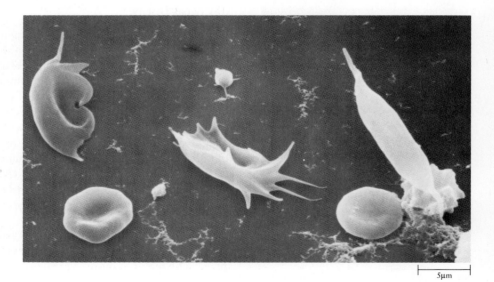

Discovering the cause of a disease, conventional wisdom dictates, should lead us directly to a cure. The cause of sickle cell anemia is known down to the last nucleotide, but no effective treatments are yet available.

Sickle cell hemoglobin is not the only known genetic alteration of the hemoglobin molecule. More than 100 hereditary variants have now been found, and about 20 of these cause disease.

19-12 *In the past, dwarfs were frequently attendants at the royal courts of Europe. This 1656 painting by Velasquez,* Las Meninas, *shows the Infanta attended by her maids of honor (*las meninas*), including, at the right, the dwarfs Maribarbola and Pertusato. Also in this most famous of Velasquez's works can be seen the artist himself at the left and the images of Philip IV and his queen caught in the mirror above the Infanta's head. Notice how the viewer is drawn into the scene; it is as if you had just opened the door and everyone looked up.*

DWARFS AND OTHER DOMINANTS

In terms of the numbers of individuals affected, serious medical problems caused by autosomal dominants are rare, simply because severely afflicted individuals are typically not able to reproduce. One of the more common disorders caused by a dominant allele is achondroplastic dwarfism (Figure 19–12). Although achondroplastic dwarfs are less likely to have children than are other individuals, there is apparently a high rate of mutation in the gene involved, causing the condition to reappear.

Perhaps the most familiar autosomal dominant is Huntington's disease. Huntington's is the disease that afflicted the folk singer and songwriter Woodie Guthrie. It is progressive, involving the destruction of brain cells; death usually occurs 10 to 20 years after the onset of symptoms. Huntington's disease is caused by a single dominant allele. As you can readily calculate, any child who has a parent with Huntington's disease has a 50 percent chance of inheriting the disease. Victims of Huntington's, however, usually have no symptoms of brain cell damage until they are past 30 years of age, and, until recently, there was no clinical way to tell who among those at risk would develop the disease and who would not. By the age at which symptoms first appear, individuals with the disease often have already had children who might then, as with the Guthrie family, spend years waiting to see whether they, too, would develop the disease. We shall have more to say about this later in the chapter.

SEX-LINKED TRAITS

Color Blindness

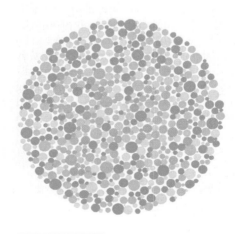

19-13 *A simple diagram used to detect red-green color blindness. Individuals with normal color vision can easily read the two-digit number embedded in the pattern of the dots. Individuals with red-green color blindness cannot distinguish the number.*

Can you distinguish the number in Figure 19–13? Eight percent of human males and 0.04 percent of human females cannot do so. The difference in these percentages is due to the fact that, in humans as in fruit flies (page 266), the Y chromosome carries much less genetic information than the X chromosome. Among the genes carried in humans on the X chromosome and not on the Y are genes affecting color discrimination.

The ability to perceive color depends on three genes, coding for three different visual pigments, each responsive to light in a different region of the spectrum of visible light. One of these pigments is responsive to red wavelengths, another to green, and the third to blue. The gene coding for the pigment responsive to blue light is on an autosome, but the genes coding for the pigments responsive to red and green light are both on the X chromosome. In males, if the gene for green is defective, green cannot be distinguished from red, and, conversely, a defect in the gene for red results in red appearing as green. (The former is about three times more common than the latter.) In heterozygous females, the defective alleles are recessive to the normal alleles on the other X chromosome, and so vision is usually normal. However, as we noted in Chapter 13 (page 267), there are occasional cases in which a heterozygous female is color-blind in one eye but has normal color vision in the other eye. Complete red-green color blindness in females occurs only in those rare instances in which both X chromosomes carry the same defective allele.

If a woman carrying a defective allele on one X chromosome transmits that X chromosome to a daughter, the daughter will also have normal color vision if she receives an X chromosome with the normal allele from her father (that is, if he is not color-blind). If, however, the X chromosome with the defective allele is transmitted from mother to son, he will be color-blind since, lacking a second X chromosome, he has only the defective allele (Figure 19–14). As we noted in Chapter 13, traits such as eye color in *Drosophila* or color discrimination in humans, which are controlled by genes on the X chromosome, are said to be sex-linked.

Hemophilia

Another classic example of a sex-linked characteristic is the hemophilia that has afflicted some royal families of Europe since the nineteenth century. Hemophilia is a group of diseases in which the blood does not clot normally. Clotting occurs through a complex series of reactions in which each reaction depends on the presence of certain protein factors in the blood plasma. Failure to produce one essential plasma protein, known as Factor VIII, results in the most common form of hemophilia, hemophilia A, which is associated with a recessive allele of a gene carried on the X chromosome. In this type of hemophilia, even minor injuries carry the risk of the patient's bleeding to death. Persons with hemophilia A can be treated with Factor VIII extracted from normal human blood, but the cost is very high—an estimated $6,000 to $26,000 per year—and carries the risk of transmission of infectious diseases, including AIDS. Work is currently underway, using recombinant DNA techniques, to develop a genetically engineered Factor VIII that can be synthesized in bacteria, eliminating the risk of contamination with infectious agents.

19–14 *The pedigree of a family in which the mother has inherited one normal and one defective allele for red-green color discrimination. The normal allele is dominant, and she has normal color vision. However, half of her eggs (on the average) will carry the defective allele and half will carry the normal allele—and it is a matter of chance which kind is fertilized. Since her husband's Y chromosome, the one that determines a son rather than a daughter, carries no gene for color discrimination, the single allele the wife contributes (even though it is a recessive allele) will determine whether or not a son is color-blind. Therefore, half of her sons (on the average) will be color-blind. Assuming that her children marry individuals with X chromosomes with the normal alleles, the expected distribution of the trait among her grandchildren will be as shown in the F_2 generation. Note that all the daughters of a color-blind man will be carriers of the defective allele, and all his sons will be normal with respect to color discrimination, unless their mother is a carrier or is herself color-blind.*

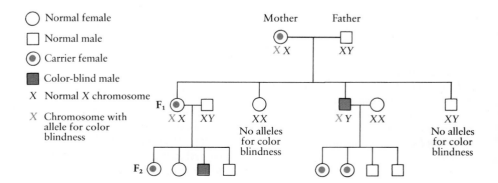

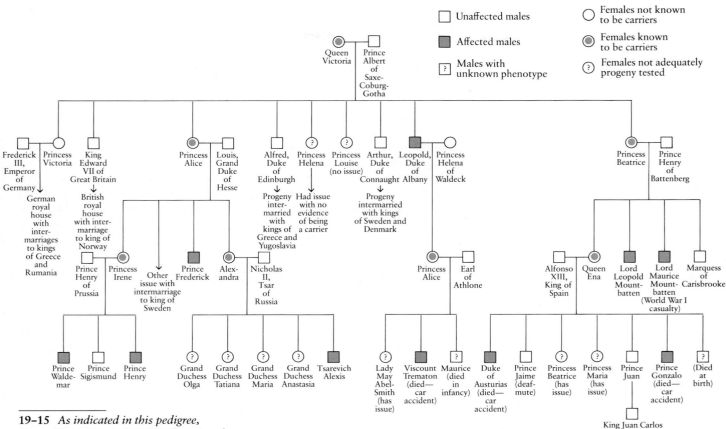

19–15 *As indicated in this pedigree, Queen Victoria was the original carrier of the allele for hemophilia that has afflicted male members of the royal families of Europe since the nineteenth century. The British royal family escaped the disease because King Edward VII, and consequently all his descendants, did not inherit the defective allele.*

Queen Victoria was probably the original carrier in her family (Figure 19–15). Prince Albert, Victoria's husband, could not have been the source because male-to-male inheritance of the disease, as with all X-linked characteristics, is impossible. Because none of Victoria's relatives other than her descendants were afflicted, we conclude that the mutation occurred on an X chromosome in one of her parents or in the cell line from which her own eggs were formed. One of her sons, Leopold, Duke of Albany, died of hemophilia at the age of 31. At least two of Victoria's daughters were carriers, since a number of *their* descendants were hemophiliacs. And so, through various intermarriages, the disease spread from throne to throne across Europe. Tsarevitch Alexis, the only son of Nicholas II and Alexandra, the last Tsar and Tsarina of Russia, inherited the allele for hemophilia from his mother. His parents' great concern for his health, which apparently distracted them from affairs of state, contributed to the turbulent events surrounding the Russian revolution.

Muscular Dystrophy

Muscular dystrophy is the name given to a group of diseases characterized by muscle wasting. The most common and severe type, Duchenne muscular dystrophy, affects cardiac muscle as well as skeletal muscle and is accompanied by mental retardation in about 30 percent of the cases. It is X-linked, occurring almost exclusively in males, with an incidence of about 1 in 3,500 newborn boys. The first symptoms usually develop in affected boys between the ages of 2 and 6, and most die by their early twenties.

In December of 1987, Louis Kunkel of the Harvard University Medical School reported that he had isolated the protein that is defective in muscular dystrophy patients. Called dystrophin, it accounts for only 0.002 percent of the protein in the muscles of normal individuals. It was totally absent in the two patients with Duchenne muscular dystrophy tested so far.

Although the role of this protein in normal muscle is not yet clear, some of the events that accompany its absence are. One of the most critical is a hardening of the muscles, a condition known as fibrosis. As a result of fibrosis, blood supply to the muscle cells is restricted, and they die. This poorly understood phenomenon is thought to cause the weakness and eventual death of patients with Duchenne muscular dystrophy.

Researchers are now in the process of cloning and sequencing the gene coding for dystrophin. This undertaking is made more difficult by the fact that it has proved to be the largest human gene ever known, with about 2 to 3 million base pairs, including some 60 exons and huge introns. Duchenne muscular dystrophy appears to be associated with deletions in this gene.

DIAGNOSIS OF GENETIC DISEASES: RFLPs

Sickle Cell Anemia

One of the first rewards of recombinant DNA technology, in terms of human genetics, has been the ability to diagnose many of the hereditary disorders we have just discussed. Sickle cell anemia, the first to be so diagnosed, is an instructive example. The DNA coding for beta globin, as we have seen, is one of the most extensively studied of human genes and was the first human disease-related gene to be cloned. In order to develop a diagnostic test for sickle cell anemia, radioactive copies of portions of the beta-globin nucleotide sequence were prepared as a probe. DNA from persons with normal hemoglobin and DNA from persons with sickle cell anemia were cleaved with the restriction enzyme *Hpa*I. The fragments produced were then exposed to the radioactive beta-globin probe. In persons with normal hemoglobin, it was found, the probe consistently hybridized with a fragment that was either 7,000 or 7,600 nucleotides long. By contrast, in 87 percent of persons with sickle cell anemia, the probe hybridized with a much longer fragment—some 13,000 nucleotides in length (Figure 19–16). The same result is seen with cells obtained by amniocentesis, thus providing the first prenatal screening test for one of the most common serious genetic disorders.

These markers are known as RFLPs, pronounced "rif-lips" and translated as "restriction-fragment-length polymorphisms." This means simply that inherited

19–16 *A test, using RFLPs, to detect the presence of the sickle cell allele. Treating human DNA with the restriction enzyme* Hpa*I produces three possible restriction fragments containing the gene for the beta chain of hemoglobin. In persons with the normal allele (gray) for beta globin, the fragments are either 7,000 or 7,600 nucleotides long. Among blacks in the United States with the sickle cell allele (color), the fragments are 13,000 nucleotides in length. A recognition sequence for the restriction enzyme* Hpa*I—present in the DNA of persons with the normal beta-globin allele—is missing in individuals carrying the sickle cell allele. The linkage between the sickle cell allele and the loss of the* Hpa*I site holds true only for populations in or originally from West Africa. Among blacks in or from East Africa, the sickle cell allele is associated with the 7,600-nucleotide fragment.*

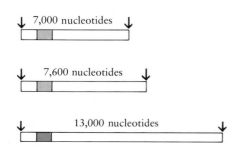

variations—mutations—in the nucleotide sequences of different individuals lead to differences in the lengths of the fragments produced by restriction enzymes. This occurs because the recognition sequence for the restriction enzyme has been eliminated or otherwise altered by the mutation. The principle involved here is gene linkage, the same principle that made possible the genetic mapping studies described in Chapter 13. Those studies in *Drosophila* confirmed that two genes that are close together on the same chromosome tend to stay together. The updated version is that two nucleotide sequences close together on the same DNA molecule tend to stay together, just as purple eyes and vestigial wings were inherited together in the fruit fly. In the *Drosophila* days, a marker gene was one that produced a detectable phenotypic change, such as a different wing shape or eye color. In the days of recombinant DNA technology, a marker may be "anonymous"—detectable only in the pattern produced by a restriction enzyme and a molecular probe. In the case of sickle cell RFLPs, the allele for sickle cell beta globin has become linked in some populations with another nucleotide sequence that alters the site for *Hpa*I recognition. The closer the allele and its marker are, the more accurate the diagnosis.

In the first family studied with this technique, both parents were known carriers of the sickle cell allele and had previously had a baby with sickle cell anemia. The woman was pregnant again, and the parents did not want to have another child with this extremely painful disease. DNA tests from each of the parents yielded both the short and the long fragments, as would be expected. Tests from their child with sickle cell anemia yielded only the long fragments. Tests of the fetal cells yielded both short and long fragments; the unborn child would be a carrier like the parents, but would not have the disease.

Huntington's Disease

James F. Gusella of the Massachusetts General Hospital, with a large team of coworkers, set out to find a diagnostic test for Huntington's disease based on the RFLP technique. Huntington's presented a new problem because neither the defective gene nor its normal allele had been identified—nor have they yet. What was available was a large library of cloned restriction fragments from human DNA. Gusella and coworkers assigned themselves the monumental task of working their way through this vast collection to find a Huntington's disease marker. They had astonishing good fortune. They found their marker within the first dozen restriction fragments tested, in a polymorphic restriction fragment produced by cleaving human DNA with the restriction enzyme *Hin*dIII. Four different patterns of fragments had been identified, known simply as A, B, C, and D, which were present in varying proportions in the normal population. Working with an American family in which some members had Huntington's disease, scientists screened the DNA from affected and normal individuals and found that, in all cases, the presence of Huntington's was associated with pattern A. This was a good start, but pattern A is the most common pattern, occurring in some 60 percent of the population, and the family was not a large one. More data were needed.

Gusella and his coworkers next turned their attention to Lake Maracaibo, Venezuela, where there is a family of more than 3,000 individuals, all apparently descended from one German sailor with Huntington's (Figure 19–17). For three years, the team in Venezuela, led by Nancy Wexler, interviewed family members to obtain the information needed to construct a pedigree, performed neurological examinations of family members, and collected skin and blood samples from 570 people. As the samples were collected, they were flown to Gusella's laboratory where a marker could be identified by correlation with the family pedigree and the results of the neurological examinations.

(a)

(b)

19-17 *The search for the Huntington's disease marker involved meticulous study of an isolated, inbred Venezuelan family. Pedigree data and neurological examinations were correlated with genetic studies of DNA. (a) Team leader Nancy Wexler (right) and other researchers recording pedigree information supplied by individuals of the afflicted family. (b) Two family members: a man in his forties with Huntington's disease and his daughter, who has a 50 percent chance of developing it.*

Witness for the Prosecution

An individual's DNA is as distinctive as a fingerprint and, in certain types of violent crime, more likely to be obtainable. The method for "DNA fingerprinting," which was devised by Alec Jeffreys of the University of Leicester in England, is basically simple. As we noted in Chapter 18, the eukaryotic genome contains many regions of simple-sequence DNA, identical short nucleotide sequences lined up in tandem and repeated thousands of times. Jeffreys found that the number of repeated units in such regions differs distinctively from individual to individual. (The only exceptions, as with fingerprints, are in the case of identical twins.) These regions can be excised from the total DNA by the use of appropriate restriction enzymes, placed on an electrophoretic gel, separated by length, denatured, and identified by a radioactive probe. When the process is completed, the end result, visible on x-ray film, looks like the bar code on a supermarket package.

Such a DNA bar code helped to convict Randall Jones, now on Death Row in Florida. Jones's car got stuck in the mud. In search of a tow, he found a young couple asleep in a pickup truck parked by a fishing ramp. He shot each of them in the head with a high-powered rifle, dragged their bodies into the woods, used the truck to pull out his car, and then went back and raped the woman. In such cases, standard blood or semen analysis can identify a suspect with a certainty of about 90 to 95 percent, leaving some room for argument. However, Jones's DNA pattern, which

matched the sperm found in the victim's body, could occur in only one person out of 9.34 billion—about double the present population of the world. (Note that this test would also completely exonerate a defendant who was innocent.)

Often only very small samples of biological evidence are found at a crime scene. A gene amplification method known as PCR (polymerase chain reaction) has been developed that can take a minute fragment of DNA and, in a few hours, synthesize millions of copies. This method involves adding a short primer at each end of a selected DNA sequence, separating the two strands of the double helix by heating, and exposing them to a bacterial DNA polymerase that recognizes the primers. This enzyme is from a bacterium that thrives at high temperatures in hot springs and is not denatured by the temperatures used to denature the DNA. In the polymerase chain reaction, both strands of the DNA are copied simultaneously. If this is repeated for 20 cycles, the amount of DNA present is increased about one million times.

Gene amplification has made it possible to obtain DNA fingerprints from trace amounts of blood and semen and even from the root of a single hair. It has been used to speed up prenatal diagnosis of genetic disease and to detect latent virus infections. It also made possible the analysis of mitochondrial DNA from a woolly mammoth that died some 40,000 years ago (see Figure I–11, page 9).

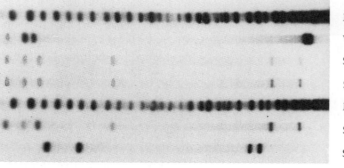

An x-ray film comparing the patterns produced by electrophoresis of restriction fragments of simple-sequence DNA from a rape victim, from the semen of the rapist, and from two suspects. The evidence revealed here by the use of a single radioactive probe strongly suggests that suspect 1 was, in fact, the rapist. In practice, three or four different probes are used, to provide conclusive identification.

Markers

Victim

Semen sample

Semen sample

Markers

Suspect 1

Suspect 2

This group, it was found, differed from the American family: in this family, the presence of Huntington's disease correlated with pattern C. Pattern C is much less common than pattern A, being present in only about 20 percent of the general population. Thus, the scientists were able to predict, in some cases, which individuals in the Venezuelan family would develop Huntington's later in life. For example, if a young person's mother did not have Huntington's and the father did, and the RFLP analysis of the mother did not reveal pattern C and the father's did, then the presence or absence of pattern C in the DNA of the person under study would predict with great certainty whether or not that person would be afflicted.

Actually, as it turns out, many members of the Venezuelan family do not want to know their future in terms of the disease. Similarly, Arlo Guthrie and his sisters have elected not to be tested. In the case of Huntington's, the most significant development may not be the possibility of earlier detection but rather the identification of the gene responsible. Work is now underway to locate the gene, with the hope of ultimately isolating and characterizing both the abnormal allele that causes Huntington's disease and its normal allele. It is hoped that this work will not only reveal the underlying abnormality that causes the brain cells to deteriorate some three to five decades after birth, but will also illuminate fundamental questions of the normal development and aging of the nervous system.

RFLPs have now been found for other hereditary diseases, including cystic fibrosis, the most common genetic disease in Caucasians.

DIAGNOSIS OF GENETIC DISEASES: RADIOACTIVE PROBES

As we noted earlier, the most common form of hemophilia is caused by defects in a plasma protein known as Factor VIII. It is possible to diagnose hemophilia in the fetus by testing for Factor VIII, but only after 20 weeks of gestation. A better test was sought, one that could be performed much earlier. The location of the gene coding for the factor was completely unknown, but, in this case, the protein had been isolated. From the known amino acid sequence of the protein, a short (36-nucleotide) fragment of RNA was synthesized and used as a radioactive probe to screen restriction fragments of the human genome. Starting with the few fragments to which the probe hybridized, it was possible to reconstruct the entire gene. It turned out to consist of 186,000 base pairs—26 exons separated by 25 introns. Once the gene was identified, it was found that mutations in the gene could be detected by RFLPs. Thus, it is now possible to identify those women who are carriers of the disease and also to detect hemophilia in fetuses at a very early stage of development.

Another example of the use of radioactive probes is a new and extremely precise test for sickle cell anemia. This test utilizes short synthetic radioactive probes that detect the single base-pair difference between the alleles for normal beta globin and sickle cell beta globin.

Until recently the only prenatal diagnosis for Duchenne muscular dystrophy was a test to determine the sex of the fetus. Parents who had a child with muscular dystrophy could elect to abort a male fetus. Now this form of muscular dystrophy is detected by a somewhat different use of a radioactive probe. A DNA probe has been synthesized that hybridizes with a portion of the nucleotide sequence that codes for dystrophin—the portion that is missing in children with the disease. Thus, the failure of the probe to find a complementary sequence among restriction fragments is diagnostic of the disease. Prospective parents, knowing whether their male fetus will be normal or will develop Duchenne muscular dystrophy, can therefore make an informed decision.

Some Ethical Dilemmas

Although the rapid advances in molecular genetics are contributing greatly to our understanding of human genetic diseases, the capacity to identify individuals either at risk of developing certain diseases or of transmitting them to their children is creating perplexing ethical dilemmas both for individuals and for our society as a whole. The four questions below cover situations that have either arisen already or are likely to arise by the year 2000. They were posed by Dr. Eric Lander, a human geneticist at the Whitehead Institute for Biomedical Research and Harvard University.

1. While the constitutional right to abortion is absolute, the choice is never easy. Imagine that you learned very early in a pregnancy that the child would certainly:

   (a) die within nine months from spinal muscular atrophy, a fatal genetic disorder;

   (b) suffer throughout life from cystic fibrosis, a painful chronic disease, and die at about age 20;

   (c) suffer from Huntington's disease at age 40 and die at about 50;

   (d) suffer from Alzheimer's disease at about 60;

   (e) be congenitally deaf;

   (f) be a dwarf, but otherwise healthy;

   (g) be predisposed to severe manic depression, which could be partially controlled by medication.

Would you choose to abort? (Assume that you are young enough that you may reasonably expect to have more children if you wish.) Regardless of your own choice, would you consider it *unethical* for another couple to do so? What principles underlie your choices?

2. Suppose you could learn with certainty whether you would suffer:

   (a) from Huntington's disease at about 40;

   (b) from Alzheimer's disease at about 60.

Would you want to know?

3. Should insurance companies have the right to:

   (a) charge higher premiums to individuals with higher risk of inherited disease?

   (b) know the results of testing for genetic predispositions?

   (c) refuse to cover a child whom prenatal tests show will suffer a severe genetic disease?

4. Suppose that 10 percent of the work force is particularly prone to cancer induced by an industrial chemical. Should an employer:

   (a) have the obligation to make the work place safe for this minority?

   (b) have the right to require pre-employment genetic screening?

   (c) have the right to refuse employment to such workers?

   (d) have the right to require such workers to pay for supplementary insurance?

Would your answer differ if the minority involved were 1 percent of the work force or 40 percent of the work force?

Although correction of the genetic defects is not possible at this time and treatment by gene replacement seems far distant, the development of radioactive probes for the detection of abnormal alleles has revolutionized medical genetic diagnosis. The use of radioactive probes is also revolutionizing forensic medicine, making it possible to conclusively identify the perpetrators of certain types of violent crime, while exonerating innocent suspects (see essay, page 396).

THE "BOOK OF MAN"

Plans are now underway for the largest and most extravagant enterprise ever undertaken in medical or biological research: the mapping and sequencing of the entire human genome. It has been compared in magnitude to the Manhattan Project, which brought forth the atomic bomb, and the space program, which culminated with a human footprint on the moon. Although initially a subject of some controversy and dissent, which was also true of the projects to which it is compared, it now has gained a momentum of its own and is clearly going forward.

There is action on two fronts. One multi-institutional enterprise is focused on mapping the human chromosomes. As in *Drosophila*, this involves finding marker genes—genes with different and detectable alleles—and documenting their rate of recombination in breeding populations. Ray White and his colleagues at the Howard Hughes Medical Center at the University of Utah, for example, have been tracing such markers through three generations of 60 Mormon families. (Mormons were chosen not only because they are in Utah, but also because their families are large and cohesive and, moreover, their genealogical records are meticulously maintained.) His group has mapped almost 500 genes. Another group, led by Helen Donis-Keller of Collaborative Research, Inc., a biotechnology company, has located several hundred markers in some 20 additional families. Her group has been conducting an intensive search for the gene responsible for cystic fibrosis and has identified some 60 markers on chromosome 7 alone, the chromosome to which this gene has been traced.

The second, and more controversial, effort involves the sequencing of the entire genome. When it was first proposed years ago, many molecular biologists were against it, mainly because of the great expense that would be involved—several billion dollars, it was estimated at the time. Much of the genome is "senseless," they pointed out, most funds available for research would have to be channeled into this one project, it would take decades to complete, and, perhaps most important of all, the creative talent of almost an entire generation of young scientists in the field would be wasted on this routine enterprise. However, the prevailing winds of opinion have now clearly shifted. The crucial change came about as a result of the development of methods for automatic sequencing of nucleotides, which is greatly decreasing the estimates of time and especially money required for the project. Also, as it has come closer to realization, there is mounting excitement about what reading this "book of man," as it has been called, might reveal. What is the mysterious role of the "senseless" DNA? How are genes regulated? (The answer to this question may hold the key to understanding and treating cancers and also to gene replacement therapy.) What further evolutionary relationships will be revealed among the human genes themselves? And among humans and other species? At this writing, the way in which this enterprise will be organized has not yet been decided, but a completion date has been set: the year 2000. That is less than a hundred years after the rediscovery of Mendel's work, which marks the beginning of genetics as a science, and less than 50 years after the announcement by Watson and Crick that opened up the field of molecular biology. And you will see it happen.

SUMMARY

The normal diploid number of human chromosomes is 46: 44 autosomes and two sex chromosomes, XX in females and XY in males. A karyotype is a graphic representation of a set of chromosomes. In preparing a karyotype, metaphase chromosomes are paired in homologues and the homologues are grouped by size. Nonhomologous chromosomes of similar size and shape can be distinguished by staining techniques that reveal chromosome banding.

Visible chromosomal abnormalities include extra chromosomes (usually a result of nondisjunction—the failure of two homologues to separate at the time of meiosis), translocations, and deletions. Down's syndrome is among the disorders associated with an extra chromosome; it may be caused either by nondisjunction or, less commonly, by translocation. Extra sex chromosomes can also result from nondisjunction and are usually associated with sterility. Aniridia and Wilms' tumor are defects associated with deletion of one small region of chromosome 11. A number of genetic defects can now be detected in the fetus by the use of amniocentesis, the collection of fetal cells from the amniotic fluid.

Many genetic diseases are the result of deficiencies or defects in enzymes or other critical proteins. These are caused, in turn, by mutations in the genes coding for the proteins. When the alleles resulting from the mutations are recessive, the diseases are apparent only in the homozygote. Such diseases include phenylketonuria, Tay-Sachs disease, sickle cell anemia, and a number of other disorders associated with variations in the hemoglobin molecules. Genetic conditions caused by autosomal dominants include a form of dwarfism and Huntington's disease.

Genetic defects expressed much more frequently in males than in females are caused by mutant alleles on the X chromosome. The mutant alleles responsible for sex-linked characteristics are usually recessive to the normal alleles and are thus not expressed in heterozygous females, who can, however, transmit them to their children. Human sex-linked characteristics include color blindness, hemophilia, and Duchenne muscular dystrophy.

Recombinant DNA technology is providing new means for early diagnosis of hereditary diseases. Among the most important tools for such diagnosis are RFLPs (restriction-fragment-length polymorphisms) and radioactive probes. RFLPs are the result of natural variations—mutations—that eliminate or alter the recognition sequence for a restriction enzyme. When such a mutation is associated with an allele causing a genetic disease, it can provide a diagnostic marker for that allele. Radioactive probes, which bind to either the normal or the mutant allele, can be used for detection and diagnosis when the nucleotide sequence of the allele is known or can be deduced.

A massive effort is now beginning to map and sequence the entire human genome. This project, which is expected to be completed by the year 2000, may answer many puzzling questions. If past experience is any guide, it is also likely to raise many new questions.

QUESTIONS

1. Nondisjunction can occur at the first meiotic division or the second. How do the effects differ? Include diagrams with your answers.

2. Describe the two types of chromosomal abnormalities that can cause Down's syndrome. With which type is it possible to identify unequivocally prospective parents who are at a higher-than-average risk of having a child with Down's syndrome? How?

3. If two healthy parents have a child with sickle cell anemia, what are their genotypes with respect to this allele? Having had one such child, what are their chances of having another child with the same disease?

4. What proportion of the children of the parents in Question 3 will be carriers of the sickle cell allele (that is, heterozygous)? What proportion of their children will not carry the allele? (Draw a Punnett square to diagram this problem.)

5. The first child of a normally pigmented man and woman is an albino. They plan to have three more children and want to know the probability that all will be normal. Calculate the probability that three normal children will be born.

6. In 1952 it was reported that two albinos, who had met at a school for the partially sighted, had married and had three children, all of whom had normal pigmentation. Assuming the children are not illegitimate, how do you explain that the children are normal?

7. In humans, either of two recessive alleles (a or b), when homozygous, can cause congenital deafness. Hence, two people who are congenitally deaf could marry and have children, all of whom are normal. What would be the genotypes of the parents in such a situation?

8. A woman homozygous for the dominant alleles A and B, necessary for normal hearing, marries a man who is congenitally deaf. What are the possible genotypes of the man? What is the probability that this couple will have a deaf child?

9. A man with a particular hereditary disease marries a normal woman. They have six children, three girls and three boys. The girls all have the father's disease but the boys do not. What type of inheritance pattern is suggested? (Although this situation was not discussed in the text, you should be able to deduce the answer.)

10. Why is male-to-male inheritance of color blindness impossible? Under what conditions would color blindness be found in a woman? If she married a man who was not color-blind, what proportion of her sons would be color-blind? Of her daughters?

11. What is the probability that a woman with normal color vision whose father was color-blind but whose husband has normal color vision will have a color-blind son? A color-blind daughter?

12. How do we know that Prince Albert was not the source of the hemophilia in Queen Victoria's descendants?

13. A woman whose maternal grandfather was hemophiliac has parents who are clinically normal. She too seems normal, as does her husband. What are the chances that her first son will be normal? (*Hint:* Determine the genotype of the woman's mother and then the possible genotypes of the woman herself.)

SUGGESTIONS FOR FURTHER READING

Books

ALBERTS, BRUCE, DENNIS BRAY, JULIAN LEWIS, MARTIN RAFF, KEITH ROBERTS, and JAMES D. WATSON: *Molecular Biology of the Cell,* 2d ed., Garland Publishing, Inc., New York, 1988.

Chapters 5, 9, and 10 of this outstanding cell biology text provide an excellent discussion of contemporary molecular genetics.

BODMER, WALTER F., and L. L. CAVALLI-SFORZA: *Genetics, Evolution, and Man,* W. H. Freeman and Company, San Francisco, 1976.

Although now somewhat dated, this excellent undergraduate text, with an emphasis on human genetics, remains a valuable resource.

DARNELL, JAMES, HARVEY LODISH, and DAVID BALTIMORE: *Molecular Cell Biology,* Scientific American Books, New York, 1986.

A thorough review of gene structure and function. Chapter 7 discusses the techniques used in molecular biology.

GOODENOUGH, URSULA: *Genetics,* 3d ed., Saunders College/Holt, Rinehart and Winston, New York, 1984.

An up-to-date, general introductory text with emphasis on molecular genetics. Outstanding for its clarity of explanation.

HAWKINS, JOHN D.: *Gene Structure and Expression,* Cambridge University Press, New York, 1985.

A short synopsis on DNA and the organization of prokaryotic and eukaryotic genes.

JACOB, FRANÇOIS: *The Logic of Life: A History of Heredity,* Pantheon Books, New York, 1973.

Jacob's principal theme concerns the changes in the way people have looked at the nature of living beings. These changes, which are part of our total intellectual history, determine both the pace and direction of scientific investigation. The opening chapters are particularly brilliant.

JUDSON, HORACE F.: *The Eighth Day of Creation: Makers of the Revolution in Biology,* Simon and Schuster, New York, 1979.*

A comprehensive study of the human and scientific aspects of molecular biology from the 1930s through the mid-1970s. As events unfold, the story is told from each participant's point of view. We are treated to an enlightening and personal glimpse of the development of scientific thought.

LEHNINGER, ALBERT L.: *Principles of Biochemistry,* Worth Publishers, Inc., New York, 1982.

This introductory text is outstanding both for its clarity and for its consistent focus on the living cell. There are numerous medical and practical applications throughout.

LEWIN, BENJAMIN: *Genes III,* John Wiley & Sons, New York, 1987.

Revised frequently, Lewin's classic text provides an excellent update on the structure and function of genes. Chapters 29 and 30 present a concise discussion of movable genetic elements in both prokaryotes and eukaryotes.

NOSSAL, G. J. V.: *Reshaping Life,* Cambridge University Press, New York, 1985.

A short, excellent, easily read introduction to recombinant DNA techniques and the biotechnology industry.

OLBY, ROBERT: *The Path to the Double Helix,* University of Washington Press, Seattle, 1975.

An account, written by a professional historian of science, of twentieth-century genetics. Olby is interested not only in the scientific concepts and experiments but also in the various personalities involved and their effects on one another and on the course of scientific discovery.

OLIVER, STEPHEN G., and JOHN M. WARD: *A Dictionary of Genetic Engineering,* Cambridge University Press, Cambridge, 1985.

Short descriptions of the techniques and vocabulary used in modern molecular genetics are a help to any reader. This book is good to have as a reference when reading original articles in this field.

PETERS, JAMES A. (ed.): *Classic Papers in Genetics,* Prentice-Hall, Inc., Englewood Cliffs, N.J., 1959.*

Includes papers by most of the scientists responsible for the important developments in genetics: Mendel, Sutton, Morgan, Beadle and Tatum, Watson and Crick, and so on. You should find this book very interesting; the authors are surprisingly readable, and the papers give a feeling of immediacy that no modern account can achieve.

* Available in paperback.

RUSSELL, PETER J.: *Genetics,* Little, Brown, and Company, Boston, 1986.

> *This clearly written text covers the principles of genetics, using a problem-solving approach.*

SCIENTIFIC AMERICAN: *Molecules to Living Cells,* W. H. Freeman and Company, San Francisco, 1980.*

> *More than half of the articles in this collection from* Scientific American *are accounts of major breakthroughs in our understanding of the nucleic acids and their functions. Written for the general reader by the scientists who made the key discoveries, they are highly recommended.*

SCIENTIFIC AMERICAN: *Recombinant DNA,* W. H. Freeman and Company, San Francisco, 1978.*

> *The articles in this collection from* Scientific American *include discussions of the critical experiments that paved the way for modern molecular genetics.*

STRICKBERGER, MONROE W.: *Genetics,* 3d ed., The Macmillan Company, New York, 1986.

> *A cohesive account of the science, this book provides a broad coverage of classical genetics.*

STRYER, LUBERT: *Biochemistry,* 3d ed., W. H. Freeman and Company, New York, 1988.

> *An introductory text, with many examples of medical applications of biochemistry. Handsomely illustrated.*

WATSON, JAMES D.: *The Double Helix,* Atheneum Publishers, New York, 1968.*

> *"Making out" in molecular biology. A brash and lively book about how to become a Nobel laureate.*

WATSON, JAMES D., NANCY H. HOPKINS, JEFFREY W. ROBERTS, JOAN A. STEITZ, and ALAN M. WEINER: *Molecular Biology of the Gene,* 4th ed., vols. I and II, The Benjamin/Cummings Publishing Company, Menlo Park, Calif., 1987.

> Molecular Biology of the Gene *is a classic; now in its Fourth Edition, it continues to serve as a celebration of the extraordinary achievements of biology during the second half of the twentieth century. Well written and richly illustrated, the book covers an incredible array of topics, ranging from molecular biology of prokaryotes to that of eukaryotes and multicellular organisms.*

WATSON, JAMES D., JOHN TOOZE, and DAVID T. KURTZ: *Recombinant DNA: A Short Course,* W. H. Freeman and Company, New York, 1983.*

> *A short course in genetics, organized around the central theme of recombinant DNA. Clearly written and handsomely illustrated, it is accessible to anyone who enjoys reading* Scientific American.

Articles

ALBERTS, BRUCE M.: "The Function of the Hereditary Materials: Biological Catalyses Reflect the Cell's Evolutionary History," *American Zoologist,* vol. 26, pages 781–798, 1986.

ANDERSON, W. FRENCH: "Prospects for Human Gene Therapy," *Science,* vol. 226, pages 401–409, 1984.

ANDERSON, W. R., and E. G. DIACHMAKOS: "Genetic Engineering in Mammalian Cells," *Scientific American,* July 1981, pages 106–121.

ANGIER, NATALIE: "A Stupid Cell with all the Answers," *Discover,* November 1986, pages 71–83.

BISHOP, J. MICHAEL: "Oncogenes," *Scientific American,* March 1982, pages 80–92.

BROWN, DONALD D.: "Gene Expression in Eukaryotes," *Science,* vol. 211, pages 667–674, 1981.

CAMPBELL, A. M.: "How Viruses Insert Their DNA into the DNA of the Host Cell," *Scientific American,* December 1976, pages 102–113.

CECH, T. R.: "The Generality of Self-Splicing RNA: Relationship to Nuclear nRNA Splicing," *Cell,* January 1986, pages 207–210.

CHAMBON, PIERRE: "Split Genes," *Scientific American,* May 1981, pages 60–71.

CHILTON, MARY-DELL: "A Vector for Introducing New Genes into Plants," *Scientific American,* June 1983, pages 51–59.

COHEN, S. N., and J. A. SHAPIRO: "Transposable Genetic Elements," *Scientific American,* February 1980, pages 40–49.

DARNELL, JAMES E., JR.: "The Processing of RNA," *Scientific American,* October 1983, pages 90–100.

DARNELL, JAMES E., JR.: "RNA," *Scientific American,* October 1985, pages 68–87.

DAVIS, BERNARD D.: "Frontiers of the Biological Sciences," *Science,* vol. 209, pages 78–79, 1980.

DELISI, CHARLES: "The Human Genome Project," *American Scientist,* vol. 76, pages 488–493, 1988.

DE ROBERTIS, E. M., and J. B. GURDON: "Gene Transplantation and the Analysis of Development," *Scientific American,* December 1979, pages 74–82.

DICKERSON, RICHARD E.: "The DNA Helix and How It is Read," *Scientific American,* December 1983, pages 94–111.

EIGEN, MANFRED, WILLIAM GARDINER, PETER SCHUSTER and RUTHILD WINKLER-OSWATITSCH: "The Origin of Genetic Information," *Scientific American,* April 1981, pages 88–118.

FEDOROFF, NINA V.: "Transposable Genetic Elements in Maize," *Scientific American,* June 1984, pages 85–98.

FELSENFELD, GARY: "DNA," *Scientific American,* October 1985, pages 58–67.

FIDDES, J. C.: "The Nucleotide Sequence of a Viral DNA," *Scientific American,* December 1977, pages 54–67.

FISHER, ARTHUR: "Sinistral DNA," *Mosaic,* September 1983, pages 1–7.

GALLO, ROBERT C.: "The First Human Retrovirus," *Scientific American,* December 1986, pages 88–98.

GILBERT, WALTER: "DNA Sequence and Gene Structure," *Science*, vol. 214, pages 1305–1312, 1981.

GILBERT, WALTER, and LYDIA VILLA-KAMAROFF: "Useful Proteins from Recombinant Bacteria," *Scientific American*, April 1980, pages 74–94.

GORDON, JON W., and FRANK H. RUDDLE: "Integration and Stable Germ Line Transmission of Genes Injected into Mouse Pronuclei," *Science*, vol. 214, pages 1344–1345, 1981.

GOULD, STEPHEN JAY: "Linnaean Limits," *Natural History*, August 1986, pages 16–23.

HALL, B. D.: "Mitochondria Spring Surprises," *Nature*, vol. 282, pages 129–130, 1979.

HOLZMAN, DAVID: "Ribosomal RNA: Where the Action Is," *Mosaic*, vol. 16, pages 32–39, 1985.

HOLZMAN, DAVID: "RNA: Messenger, Self-splicer, Catalyst . . .," *Mosaic*, vol. 15, pages 16–21, 1984.

KOLATA, GINA: "Two Disease-Causing Genes Found," *Science*, vol. 224, pages 669–670, 1986.

KORNBERG, ROGER D., and AARON KLUG: "The Nucleosome," *Scientific American*, February 1981, pages 52–64.

LAKE, JAMES A.: "The Ribosome," *Scientific American*, August 1981, pages 84–97.

LAWN, RICHARD M., and GORDON A. VEHAR: "The Molecular Genetics of Hemophilia," *Scientific American*, March 1986, pages 48–54.

LEWIN, ROGER: "Biggest Challenge Since the Double Helix," *Science*, vol. 212, pages 28–32, 1981.

LEWIN, ROGER: "Do Jumping Genes Make Evolutionary Leaps?" *Science*, vol. 213, pages 634–636, 1981.

LEWIN, ROGER: "On the Origin of Introns," *Science*, vol. 217, pages 921–922, 1982.

MARX, JEAN L.: "A Movable Feast in the Eukaryotic Genome," *Science*, vol. 211, pages 153–155, 1981.

MARX, JEAN L.: "Restriction Enzymes: Prenatal Diagnosis of Genetic Disease," *Science*, vol. 202, pages 1068–1069, 1978.

MILLER, JULIE ANN: "Building a Better Mouse," *BioScience*, February 1987, pages 103–106.

MILLER, JULIE ANN: "Switch-on Genes in Development," *Science News*, November 1985, page 205.

MOSES, PHYLLIS B., and NAM-HAI CHUA: "Light Switches for Plant Genes," *Scientific American*, April 1988, pages 88–93.

MURRAY, ANDREW W., and JACK W. SZOSTAK: "Artificial Chromosomes," *Scientific American*, November 1987, pages 62–68.

OLDROYD, DAVID: "Gregor Mendel: Founding-father of Modern Genetics," *Endeavour*, vol. 8, pages 29–31, 1984.

PATTERSON, DAVID: "The Causes of Down's Syndrome," *Scientific American*, April 1987, pages 52–60.

PINES, MAYA: "In the Shadow of Huntington's," *Science 84*, May 1984, pages 30–39.

PTASHNE, MARK, ALEXANDER D. JOHNSON, and CARL O. PALO: "A Genetic Switch in a Bacterial Virus," *Scientific American*, November 1982, pages 128–140.

RADMAN, MIROSLAV, and ROBERT WAGNER: "The High Fidelity of DNA Duplication," *Scientific American*, August 1988, pages 40–46.

RENSBERGER, BOYCE: "Cancer, the New Synthesis," *Science 84*, September 1984, pages 28–40.

RICH, A., and S. H. KIM: "The Three-Dimensional Structure of Transfer RNA," *Scientific American*, January 1978, pages 52–62.

SCHMID, CARL W., and WARREN R. JELINEK: "The *Alu* Family of Dispersed Repetition Sequences," *Science*, vol. 216, pages 1065–1070, 1982.

SMITH, MICHAEL: "The First Complete Nucleotide Sequencing of an Organism's DNA," *American Scientist*, vol. 67, pages 47–67, 1979.

STAHL, FRANKLIN W.: "Genetic Recombination," *Scientific American*, February 1987, pages 90–101.

STEITZ, JOAN A.: "Snurps," *Scientific American*, June 1988, pages 56–63.

TANGLEY, LAURA: "Gearing Up for Gene Therapy," *BioScience*, vol. 35, pages 8–10, 1985.

VARMUS, HAROLD: "Reverse Transcription," *Scientific American*, September 1987, pages 56–64.

WALTERS, LEROY: "The Ethics of Human Gene Therapy," *Nature*, vol. 320, pages 225–227, 1986.

WEINBERG, ROBERT A.: "Finding the Anti-Oncogene," *Scientific American*, September 1988, pages 44–51.

WEINBERG, ROBERT A.: "A Molecular Basis of Cancer," *Scientific American*, November 1983, pages 126–142.

WHITE, RAY, and JEAN-MARC LALOUEL: "Chromosome Mapping with DNA Markers," *Scientific American*, February 1988, pages 40–48.

* Available in paperback.

PART **2**

Biology of Organisms

SECTION **4**

The Diversity of Life

Suspended in the canopy of a tropical rain forest, a biologist encounters a vast variety of new species of plants and animals. According to a recent estimate, there may be as many as 30 million species of insects alone in the tropical rain forest, the great majority of them still unknown and unclassified. At present, the tropical rain forests are being destroyed so rapidly that, in 25 years, most of the inhabitants of this rich environment will have been lost forever.

CHAPTER 20

The Classification of Organisms

Part 1 of this book dealt largely with cellular and subcellular aspects of living things. In Part 2, we shall be concerned with a higher level of organization, the whole organism. The chief focus will be on the plants, particularly the flowering plants, and on the more complex animals, particularly the vertebrates. These are the subjects of Sections 5 and 6, respectively. By way of introduction to these organisms, in the next few chapters we shall examine the tremendous diversity of life on our planet. Our discussion begins, in Chapter 21, with the simplest of single-celled organisms, the bacteria (which, as we have seen, are astonishingly gifted and complex), and ends, in Chapter 28, with the most familiar multicellular organisms, the mammals. At the same time, we shall trace the major developments in the course of evolution that have given rise to this diversity.

THE NEED FOR CLASSIFICATION

Most people have a limited awareness of the natural world and are concerned chiefly with the organisms that influence their own lives. For example, gauchos, the cowboys of Argentina, who are famous for their horsemanship, have some 200 names for different colors of horses but generally divide plants into four groups: *pasto*, or fodder; *paja*, bedding; *cardo*, wood; and *yuyos*, everything else.

Most of us are like the gauchos. Once beyond the range of common plants and animals, and perhaps a few uncommon ones that are of special interest to us, we usually run out of names and categories. Biologists, however, face the task of systematically identifying, studying, and exchanging information about the vast diversity of organisms—more than 5 million different species—with which we relative newcomers share this planet. In order to do this, they must have a system for naming all these organisms and for grouping them together in orderly and logical ways. The problems of developing such a system are immensely complicated and begin with the basic unit of biological classification, the species.

WHAT IS A SPECIES?

Species in Latin simply means "kind," and so species, in the simplest sense, are different kinds of organisms. A more rigorous definition of species was set forth in 1940 by Ernst Mayr of Harvard University, who said that species are "groups of actually or potentially interbreeding natural populations which are reproductively isolated from other such groups." The phrase "actually or potentially" allows for the fact that although members of the human population of Greenland are not likely to interbreed with those of Patagonia, they are still members of the human species; similarly, transporting a group of insects to some remote island does not automatically make them members of another species. The words "groups" and "populations" are important in this definition also. The possibility that single

20-1 *An inhabitant of the canopy. This masked puddle frog, like many other animals that dwell in the tropical treetops, seldom descends to the forest floor. Active at night, puddle frogs sleep during the day, preferably wedged into a tight place, such as an unfurling leaf, where they are well camouflaged. The great diversity of life on earth is a result of a long evolutionary history in which different kinds—species —of organisms have become adapted to different environments and different ways of life.*

individuals of different species may have occasional offspring—such as by the crossing of lions and tigers in a zoo—is unimportant in terms of the group. Mayr's definition conforms to common sense: if members of one species freely exchanged genes with members of another species, they could no longer retain those unique characteristics that identify them as different kinds of organisms.

This definition works well for animal species and is generally accepted by zoologists. Many plants, however, can reproduce asexually and also can form fertile hybrids with other species. Bacteria, with their variety of forms of genetic exchange, do not fit this definition neatly, nor do the many unicellular eukaryotes that reproduce by cell division, forming clones of identical cells. Thus, although botanists and microbiologists use the term "species," they are more likely to consider it a category of convenience, existing rather in the human mind than in the natural world.

For most practical purposes, a species is a category into which is placed an individual organism that conforms to certain fairly rigid criteria concerning its structure and other characteristics. From an evolutionary perspective, however, a species is a group or population of organisms, reproductively united but very probably changing as it moves through space and time. Splinter groups, reproductively isolated from the population as a whole, can undergo sufficient change that they become new species. This process is known as **speciation.** Occurring repeatedly in the course of more than 3.5 billion years, it has given rise to the diversity of organisms that have lived in the past and that live today.

The Naming of Species

A group of closely related species, presumably derived by speciation from a common ancestor, constitute a **genus** (plural, genera). According to the binomial system of nomenclature devised by the Swedish naturalist Linnaeus in the eighteenth century and still in use today, the scientific name of an organism consists of two parts—the name of the genus plus a specific epithet (an adjective or modifier). The genus name is always written first, as in *Drosophila melanogaster,* and it may be used alone when one is referring to members of the entire group of species making up that genus, such as *Drosophila* or *Paramecium.*

20–2 *These plants are all members of the genus* Viola: (**a**) *the common blue violet,* Viola papilionacea, (**b**) *the wild pansy,* Viola tricolor, *and* (**c**) *the long-spurred violet,* Viola rostrata. *Although there is an overall similarity among all three species, there are clear-cut differences in leaf shape, flower color and size, and other characteristics.*

(a)

(b)

(c)

A specific epithet is meaningless when written alone, however, because many different species in different genera may have the same specific epithet. For example, *Drosophila melanogaster* is the fruit fly that has played such an important role in genetics; *Thamnophis melanogaster,* however, is a semiaquatic garter snake. Thus, by itself, the specific epithet *melanogaster* ("black stomach") would not identify either organism. For this reason, the specific epithet is always preceded by the genus name, or, in a context where no ambiguity is possible, the genus name may be abbreviated to its initial letter. Thus *Drosophila melanogaster* may be designated *D. melanogaster.*

Whoever describes a genus or species first has the privilege of naming it. It may not be named after oneself, but often it is named after a friend or colleague. *Escherichia,* for example, is named after Theodor Escherich, a German physician (*coli* simply means intestinal); and *Rhea darwinii,* an ostrichlike bird found in Patagonia, is named after Charles Darwin.

Names may be descriptive. The first discovered early fossil of a horse—which does not superficially resemble a modern horse at all—was named *Hyracotherium,* the "hyrax-like beast." (A hyrax looks like a big guinea pig and is now thought to be a distant relative of the elephant.) When O. C. Marsh of Yale University began, in the 1870s, to study fossil horses, he recognized that the little dog-sized *Hyracotherium* was an early equine and gave it the charming name Eohippus ("dawn horse"). However, *Hyracotherium* remains its official designation because it was published first.

Some names are heartfelt. Thus, members of various mosquito genera have been given the specific epithets *punctor, tormentor, vexans, horrida, perfidiosus, abominator,* and *excrucians.* Others are frivolous. An English entomologist coined a whole series of generic names based on the pseudo-Greek ending *chisme,* pronounced "kiss me." Thus, there are squash bugs, stink bugs, and seed bugs known variously as *Polychisme, Peggichisme, Dolichisme,* and the promiscuous *Ochisme.* A new species of wasp in the genus *Lalapa* was recently given the specific epithet *lusa;* there is a (presumably treacherous) beetle named *Ytu brutus,* and a horsefly called *Tabanus balazaphyre.* So, although species names may appear formidable and be unpronounceable, they are not necessarily as pompous (or even as informative) as they may seem.

These binomials are a necessary tool for clear and unambiguous communication among biologists. Many species lack common names, and, even when common names do exist for a kind of organism, more than one name may be given to the same species, such as groundhog and woodchuck, gnu and wildebeest, pill bug and sow bug and wood louse. Or names may vary from place to place. As shown in Figure 20–3, a robin in North America is distinctly different from the

20–3 *Though they are both called robins,* (a) *the North American robin,* Turdus migratorius *("thrush, of the migratory habit"), is a distinctly different bird from* (b) *the English robin,* Erithacus rubecula.

(a)

(b)

20–4 *The eighteenth-century Swedish professor, physician, and naturalist Carolus Linnaeus (1707–1778), who developed the binomial system for naming species of organisms and established the major categories that are used in the hierarchical system of biological classification. (Linnaeus was born Carl von Linné but latinized his name in the scholarly fashion of the time.) When he was 25, Linnaeus spent five months exploring Lapland for the Swedish Academy of Sciences; he is shown here wearing his Lapland collector's outfit, which you might want to compare with those of modern-day collectors, as shown in the photograph on page 15.*

English bird of the same name. A yam in the southern United States is a totally different vegetable from a yam several hundred kilometers away in the West Indies. When different languages are involved, the problems of communication would be virtually insurmountable without a system of nomenclature universally recognized and agreed upon by biologists.

HIERARCHICAL CLASSIFICATION

One of the fundamental goals of observers of the natural world, from well before the time of Aristotle, has been to perceive order in the diversity of life. One way of achieving order is through **taxonomy,** which is the classification of organisms (or, for that matter, of any other item, such as books in the library or groceries on the shelf).

The taxonomy of organisms is a hierarchical system—that is, it consists of groups within groups, with each group ranked at a particular level. In such a system, a particular group is called a **taxon** (plural, taxa), and the level at which it is ranked is called a **category.** For example, in political geography, nation, state or province, and city are categories, whereas Canada, Ontario, and Toronto are taxa within those categories. Similarly, genus and species are categories, and *Homo* and *Homo sapiens* are taxa.

In the time of Linnaeus, three categories were in common use: the species, the genus, and a category of much higher level, the **kingdom.** Naturalists recognized three kingdoms: plant, animal, and mineral. Kingdom is still the highest category used in biological classification. Between the level of genus and the level of kingdom, however, Linnaeus and subsequent taxonomists have added a number of other categories. Thus, genera are grouped into **families,** families into **orders,** orders into **classes,** and classes into **phyla** or **divisions.** (The categories of division

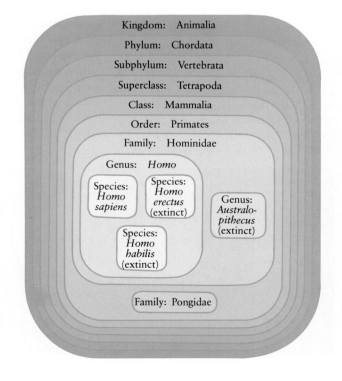

20–5 *The hierarchical nature of biological classifications, consisting of groups within groups, can be represented visually in what is known as a Venn diagram. This Venn diagram shows the classification of the genus* Homo.

and phylum are equivalent. The term "division" is generally used in the classification of prokaryotes, algae, fungi, and plants, whereas "phylum" is used in the classification of protozoa and animals.) These categories may be further subdivided or aggregated into a number of less frequently employed categories such as subphylum or superfamily. By convention, generic and specific names are written in italics, while the names of families, orders, classes, and other taxa whose categories rank above the genus level are not, although they are capitalized.

This system of classification makes it possible to generalize. For example, Table 20–1 shows the classification of two different organisms. Notice how the classification of an animal as a mammal, or a plant as Anthophyta, provides an index to a vast amount of information. Notice also that, in progressing downward from kingdom to species, there is an increase in detail, proceeding from the general to the particular. In short, hierarchical classification is a highly useful means of storing and retrieving information.

TABLE 20–1 Biological Classifications

Red Maple (*Acer rubrum*)

CATEGORY	TAXON	CHARACTERISTICS
Kingdom	Plantae	Multicellular organisms primarily adapted for life on land; usually have rigid cell walls and chlorophylls *a* and *b* contained in chloroplasts
Division	Anthophyta	Vascular plants (plants with conducting tissues) with seeds and flowers; ovules enclosed in ovary; seeds enclosed in fruit; the flowering plants
Class	Dicotyledones	Embryo with two seed leaves (cotyledons)
Order	Sapindales	Soapberry order; usually woody plants
Family	Aceraceae	Maple family; characterized by watery, sugary sap; opposite leaves; winged fruit; chiefly trees of temperate regions
Genus	*Acer*	Maples and box elder
Species	*Acer rubrum*	Red maple

Human Being (*Homo sapiens*)

CATEGORY	TAXON	CHARACTERISTICS
Kingdom	Animalia	Multicellular organisms requiring complex organic substances for food; food usually ingested
Phylum	Chordata	Animals with notochord, dorsal hollow nerve cord, gill pouches in pharynx at some stage of life cycle
Subphylum	Vertebrata	Spinal cord enclosed in a vertebral column, body basically segmented, skull enclosing brain
Superclass	Tetrapoda	Land vertebrates, four-limbed
Class	Mammalia	Young nourished by milk glands, skin with hair, body cavity divided by a muscular diaphragm, red blood cells without nuclei, three ear bones (ossicles), high body temperature
Order	Primates	Tree dwellers or their descendants, usually with fingers and flat nails, sense of smell reduced
Family	Hominidae	Flat face; eyes forward; color vision; upright, bipedal locomotion
Genus	*Homo*	Large brain, speech, long childhood
Species	*Homo sapiens*	Prominent chin, high forehead, sparse body hair

The fundamental category in the hierarchical classification of organisms is the species, which, despite the difficulties in its definition, may be considered a biological reality. The other categories, however, exist only in the human mind. To take a familiar group as an example, some taxonomists, "lumpers," would combine all the cats except one into the single genus *Felis*, excluding only the cheetah *(Acinonyx)*, because of its nonretractable claws. Others, "splitters," would reserve the designation *Felis* for the smaller cats, such as the cougar, the ocelot, and the domestic cat, and divide the others into the larger cats *(Panthera)*, including the lion, tiger, and leopard, and the bobtail cats *(Lynx)*. More extreme splitters favor separate genera for the clouded leopard *(Neofelis)* and the snow leopard *(Uncia)*. No one disagrees about the characteristics of the animals themselves but only about the weighing of similarities and differences. One taxonomist's family can easily be another's order.

EVOLUTIONARY SYSTEMATICS

For Linnaeus and his immediate successors, the goal of taxonomy was the revelation of the grand, unchanging design of creation. After 1859, however, differences and similarities among organisms came to be seen as products of their evolutionary history, or **phylogeny.** Thus genera came to be regarded as, ideally, groups of recently diverged (and therefore closely related) species, families as less recently divergent genera, and so on. Biologists now wanted their taxonomies to be not only convenient and useful but also an accurate reflection of the historical relationships among organisms. Such taxonomies are, in effect, hypotheses about evolutionary history. Like other hypotheses, they can be tested (through detailed study of the fossil record and of the structure and other characteristics of living organisms) and revised as necessary. This study of the historical relationships among organisms is known as evolutionary **systematics.**

The Monophyletic Ideal

In a classification scheme that accurately reflects evolutionary history, every taxon is, ideally, **monophyletic.** This means that the members of a taxon, at whatever

(a)

20–6 *A sampling of members of the class Mammalia, order Carnivora, family Felidae. According to some taxonomic schemes, all of these felines would be considered members of the genus* Felis. *Alternatively, the genus* Felis *is reserved for small cats, such as* (a) *the domestic cat,* (b) *the mountain lion, or cougar, and* (c) *the ocelot. The larger cats, including* (d) *the leopard, are classified as* Panthera, *and the bobtail cats, such as* (e) *the lynx, as* Lynx.

(b)

(c)

categorical level, should all be descendants of the nearest common ancestral species. In other words, taxa should be real historical units. Thus, a genus should consist of species descended from the most recent common ancestor—and only of species descended from that ancestor. Similarly, a family should consist of genera descended from a more distant common ancestor—and only of genera descended from that ancestor.

Although this ideal sounds relatively straightforward, it is often difficult to attain. In many cases, biologists do not know enough about the evolutionary history of the organisms to establish taxa that are, with a reasonable degree of certainty, monophyletic. In other cases, the two different functions of biological classification come into conflict, and convenience and utility may be deemed more important than an accurate reflection of phylogeny. Thus, as we shall see, some generally accepted taxa contain organisms descended from more than one ancestral line. Such taxa are said to be **polyphyletic.**

Homology and Phylogeny

The grouping of organisms into taxa from the categorical levels of genus through phylum or division is based on similarities in structure and other phenotypic characteristics. From Aristotle on, however, biologists have recognized that superficial similarities are not useful criteria for taxonomic decisions. To take a simple example, birds and insects should not be grouped together simply because both have wings. A wingless insect (such as an ant) is still an insect, and a flightless bird (such as the kiwi) is still a bird on the basis of overall structure. Linnaeus classified whales with mammals and not with fish, despite their external similarities.

A key question in evolutionary systematics is the origin of a similarity or difference. Does a similarity reflect inheritance from a common ancestor, or does it reflect adaptation to similar environments by organisms that do not share a common ancestor? A related question arises with differences between organisms: Does a difference reflect separate phylogenetic histories, or does it reflect instead the adaptations of closely related organisms to very different environments? A classic example is the vertebrate forelimb. The wing of a bird, the flipper of a whale, the foreleg of a horse, and the human arm have quite different functions

(d)

(e)

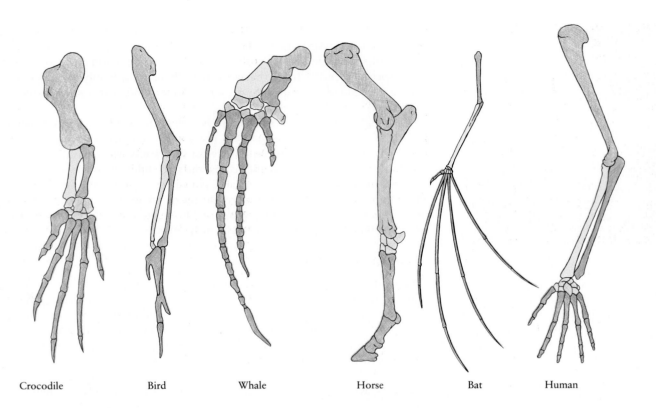

Crocodile   Bird   Whale   Horse   Bat   Human

20-7 *The bones in these forelimbs are color-coded to indicate fundamental similarities of structure and organization. The crocodile is placed first because it is the closest to the ancestral type—the form from which all the others arose. (Note also the similarity between the forelimb of the crocodile and that of the human.) Structures that have a common origin but not necessarily a common function are known as homologues. Analogous structures, by contrast, are superficially similar but have an entirely different evolutionary background—for example, the spine of a cactus (a modified leaf) and the thorn of a rose (a modified branch).*

and appearances. Detailed study of the underlying bones reveals, however, the same basic structure (Figure 20–7). Such structures, which have a common origin but not necessarily a common function, are said to be **homologous.** These are the features upon which evolutionary classification systems are ideally constructed.

By contrast, other structures, which may have a similar function and superficial appearance, have an entirely different evolutionary background. Such structures are said to be **analogous.** Thus the wings of a bird and the wings of an insect are analogous, not homologous.

Decisions as to homology and analogy are seldom so simple. In general, the features most likely to be homologous—and thus useful in determining phylogenetic relationships—are those that are complex and detailed, consisting of a number of separate parts. This is true whether the similar feature is anatomical, as in the bones of the vertebrate forelimb, or is a biochemical pathway or a behavioral pattern. The more separate parts involved in a feature shared by several species, the less likely it is that the feature evolved independently in each.

TAXONOMIC METHODS

The traditional means of determining the classification of a newly discovered organism requires several different steps, involving different kinds of appraisals. First, the organism is tentatively assigned to a particular taxon on the basis of its overall outward similarities to other members of that taxon. Then these similarities are tested for homologies. Fossils are taken into account when possible. For instance, hares and rabbits (collectively known as lagomorphs) were long believed to be rodents, but the earliest fossil remains of lagomorphs and rodents show that the two groups had quite different origins. Conversely, bears, once considered a very distinct group of carnivores, are now known, on the basis of paleontology, to have diverged relatively recently from dogs. Various stages in the life cycle and the patterns of embryonic development are also compared; as you will see in the chapters that follow, some major decisions concerning the phylogenetic relationships of animal groups are based on similarities in early development.

Taxonomies constructed by the traditional methods reflect the consideration and weighing of a large number of factors. Some of these factors provide evidence of the genealogy, or branching patterns, that have characterized the evolutionary history of the organisms, while others reflect the degree to which the organisms have diverged since they began to travel separate evolutionary paths. Thus, traditional taxonomies contain information about both the sequence in which branchings occurred and the extent of the subsequent biological changes. Such taxonomies can be summarized in the form of phylogenetic trees, as shown in Figure 20–9.

Given the large number of factors that are taken into account in constructing an evolutionary taxonomy and the fact that different biologists have different views of the importance of various factors, it is not surprising that radically different classifications have sometimes been proposed for the same organisms.

Alternative Methodologies

Two alternative methodologies—numerical phenetics and cladistics—have been proposed as replacements for the traditional methods of evolutionary systematics. Both pheneticists and cladists have sought to develop a truly objective taxonomic method that would eliminate the subjectivity that seems unavoidable with the traditional methods. Both groups have also noted that it is not really possible for a

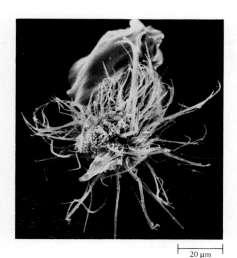

20–8 *Occasionally, newly discovered organisms cannot be classified in existing taxa and require the creation of new taxa. This tiny animal,* Nanaloricus mysticus, *dredged from the ocean bottom off the coast of France in 1982, is a case in point. With a plate-covered body, numerous spines projecting from its head, and a retractable tube for a mouth, it is unlike any previously known animal. It is the first member of a newly established phylum,* Loricifera *("girdle-wearer"). Closely related species, also assigned to the new phylum, have now been found in the Coral Sea, in Greenland, and off the coast of Florida.*

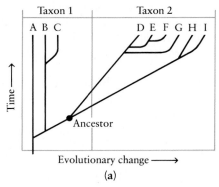

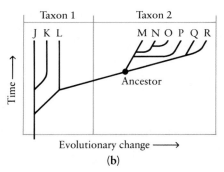

20–9 *The evolutionary history of a group of related organisms can be represented by a phylogenetic tree. The vertical locations of the branching points indicate when particular taxa diverged from one another; the horizontal distances indicate how much the taxa have diverged, taking into account a number of different characteristics.*

The two diagrams shown here represent the evolutionary histories of two different groups of taxa, labeled A through I in (a) *and J through R in* (b). *Each group has been classified using traditional methods. In* (a), *the ancestor of taxa D through I is included in taxon 1 because of its close resemblance to taxa B and C.*

In (b), *the ancestor of taxa M through R is placed in taxon 2, because of its close resemblance to taxon M. In each case, taxa 1 and 2 would themselves be members of a taxon at a higher categorical level, which would probably include other taxa as well.*

single classification scheme to indicate both overall similarity (the concern of the earliest taxonomists) and genealogy (the additional concern of taxonomists since the time of Darwin). They point out that some long-separated lineages have evolved in parallel and so continue to resemble one another more closely than organisms that have diverged rapidly from a recent common ancestor. Not only are the traditional methods suspect, but the goals are unattainable, according to this joint analysis. The remedies proposed by the two groups are, however, exactly opposite.

Numerical Phenetics

Numerical phenetics is based only on a species' observable characteristics. To begin, the characteristics of the species being studied are divided into unit characters, that is, characters of two or more states that cannot logically be subdivided further. These unit characters are assigned numbers and coded as plus or minus or 0 (data not available). As many different characters as possible—at least 100—are taken into consideration. The data are then processed by computer, which scores the taxa according to the number of unit characters they share.

Each character is given equal weight by this system with no subjective evaluation or prior knowledge taken into consideration. For instance, the possession of five fingers in a phenetic analysis would signify that lizards are more similar to humans than to snakes. The difference between homology and analogy is disregarded. Characters known to be subject to environmental pressures—such as the shape of a leaf—are weighted equally with more constant characters—for example, the structure of a flower. Pheneticists maintain that such problems are resolved if enough characters are taken into consideration. Thus, for instance, despite the fact that one has five fingers and one does not, the relationship between lizard and snake would emerge when the other characters are taken into account.

Cladistics

In contrast to numerical phenetics, which bases classification exclusively on degree of overall similarity, cladistics ignores overall similarity and is based exclusively on phylogeny. Cladists maintain that the branching of one lineage from another in the course of evolution is the one event that can be determined objectively. Such points are marked by the appearance of evolutionary novelties—that is, characteristics that were not present in the ancestral, or **primitive,** condition (see essay).

The goal of the cladists is the construction of **holophyletic** taxa. Whereas a monophyletic taxon includes only organisms descended from a common ancestor (but not necessarily all such organisms nor the ancestor itself), a holophyletic taxon must include all of the descendants from the common ancestor, plus the ancestor. None of the taxa in Figure 20–9, determined by traditional methods, would satisfy these requirements.

The substitution of cladistics for traditional methods of classification can produce revolutionary changes. Figure 20–10 shows a traditional phylogenetic tree of the principal land-dwelling vertebrates. Four of the modern groups (B through E), which include organisms such as crocodiles, turtles, snakes, and lizards, are placed in class Reptilia according to conventional schemes. Two of the groups, the birds and the mammals, are placed in separate classes because of their obvious biological differences. According to the cladist scheme, however, taxa must conform strictly to branching patterns, forming "nested sets" within the hierarchy of the more inclusive taxa. Thus, crocodiles (B) end up with the birds (A), rather than with the turtles, snakes, and lizards.

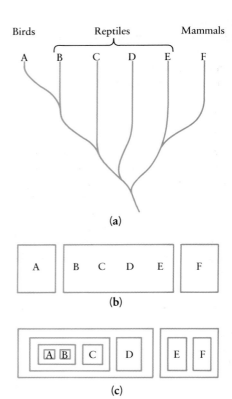

20–10 (a) *The phylogeny of the living representatives of the dominant terrestrial vertebrates. Branches that have become extinct, such as the dinosaurs, are not shown.* (b) *According to traditional classification, taxa B through E are grouped together (class Reptilia). Taxa A and F, the birds and the mammals, are placed in class Aves and class Mammalia because of their extensive biological differences from the other groups.* (c) *According to cladistic methodology, classification must be based solely on genealogy (branching patterns), with the more recent branches assigned lower categorical ranks in the hierarchical system. The biological changes that have occurred since the groups branched from one another are not taken into account.*

How to Construct a Cladogram

A cladogram is a hypothesis of branching sequences. It looks, at first glance, like a phylogenetic tree, but it is not one. It contains no ancestors, only branching points as determined by the appearance of evolutionary novelties. For example, consider the following organisms: lizard, mouse, trout, cow, shark. They have some shared characteristics—a nerve cord on the dorsal (top) side of the body, a chambered heart, four appendages, and jaws—that set them off, as a group, from other groups, such as clams or grasshoppers, for instance. Four of them have bony skeletons and one, the shark, does not. Three of them are characterized by an amniote egg (that is, an egg that contains its own water supply). Two of them have mammary glands. Thus, in terms of evolutionary novelties, the branching sequence shown in (a) is constructed.

The hypothesis can now be tested. For instance, cow and mouse not only both have mammary glands, but both also have hair, another evolutionary novelty. If the lizard and mouse had hair, but the cow lacked hair, the hypothesis would have to be reexamined. Or, if the mouse and trout had three ear bones, but the lizard did not, the cladogram would have to be reconstructed. Thus, every cladogram fulfills the scientific ideal of presenting a testable hypothesis.

From the cladogram, a hierarchical classification can be constructed:

Group: shark, trout, lizard, mouse, cow

Subgroup 1a: shark
Subgroup 1b: trout, lizard, mouse, cow

Subgroup 2a: trout
Subgroup 2b: lizard, mouse, cow

Subgroup 3a: lizard
Subgroup 3b: mouse, cow

Subgroup 4a: mouse
Subgroup 4b: cow

Cladists view such a hierarchical listing as representing "nested sets," that is, groups within groups, as shown in (b). Examination of this figure reveals that, following the logical steps outlined, we have been led to put sharks in one high-ranking taxon and trout in another, along with mammals. This sort of result has led traditionalists to note that cladists cannot tell a fish from a cow. In this we see the essence of the disagreement.

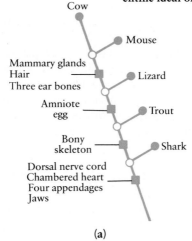

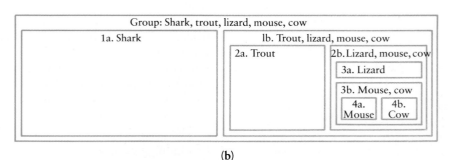

(a) A cladogram for five types of vertebrates. Each branch point is defined by one or more characteristics interpreted as evolutionary novelties. (b) From the cladogram, the cladistic hierarchy, represented as "nested sets," can be constructed.

Traditional systematists point out that such a classification ignores an important aspect of the evolutionary history of these organisms: the striking biological changes that occurred as the early birds diversified into a new life zone—the air—and as the early mammals diversified into a terrestrial environment made available by the demise of the dinosaurs. Given the obvious differences between birds and crocodiles and the similarity, for instance, of crocodiles and lizards (neither of which have changed significantly from the ancestral condition), a system that groups birds with crocodiles, instead of crocodiles with lizards, makes no biological sense. Cladists, however, are concerned with the lack of consistency of the traditional system; class Reptilia, for the cladist, simply should not exist. It is analogous to a political taxon made of the United States minus all states admitted since 1912. Such a taxon, which consists of a holophyletic group from which strikingly divergent members, and perhaps the common ancestor as well, have been removed, is said to be **paraphyletic.** Even in political taxonomy, however, paraphyletic taxa—such as the contiguous 48 states—make sense and are useful.

It has become apparent in recent years that no one approach to biological classification is adequate for all purposes. In their search for consistency and objectivity, both pheneticists and cladists must exclude a portion of the available knowledge about organisms. Moreover, neither phenetics nor cladistics completely escapes subjectivity; decisions about the characters to be tabulated or the identification of evolutionary novelties must still be made by fallible human beings on the basis of incomplete knowledge. These two methodologies, however, have provided important new perspectives in taxonomy and triggered many new questions about organisms and their evolutionary relationships. The answers to some of these questions are beginning to emerge as taxonomists apply the tools of molecular biology to their studies.

MOLECULAR TAXONOMY

Taxonomy by any methodology has been based largely on anatomy, and this will probably continue to be true in the future since a large body of data on comparative anatomy has been accumulated. As we have seen, however, it is often difficult to determine the appropriate weights to assign to different kinds of similarities. Moreover, such differences are of little help in the study of structurally dissimilar organisms—fish and fungi, for instance. New biochemical techniques are, however, becoming increasingly important in evolutionary systematics. They offer two distinct advantages: the results are objectively quantifiable, and very diverse organisms can be compared. Biochemical studies can reveal, for example, similarities and differences in enzymes, reaction pathways, hormones, and important structural molecules. With the development of techniques for sequencing the amino acids in proteins and the nucleotides in DNA and RNA molecules, it has become possible to compare organisms at the most basic level of all—the gene.

Amino Acid Sequences

One of the first proteins to be analyzed in taxonomic studies was cytochrome *c*, one of the carriers of the electron transport chain (page 196). Cytochrome *c* molecules from a great variety of organisms were sequenced, making it possible to determine the number of amino acids by which the molecules of the various organisms differ. Presumably, the greater the number of amino acid differences between any two organisms, the more distant their evolutionary relationship; conversely, the smaller the number of differences, the closer their relationship. Figure 20–11 illustrates a phylogeny based on cytochrome *c* data. The results conform well, but not perfectly, with phylogenies constructed by more traditional methods.

As data on protein variations accumulated, it became clear that although protein structure is a useful parameter of evolutionary relationships, there are difficulties in interpreting the results. Some biologists maintain that differences in

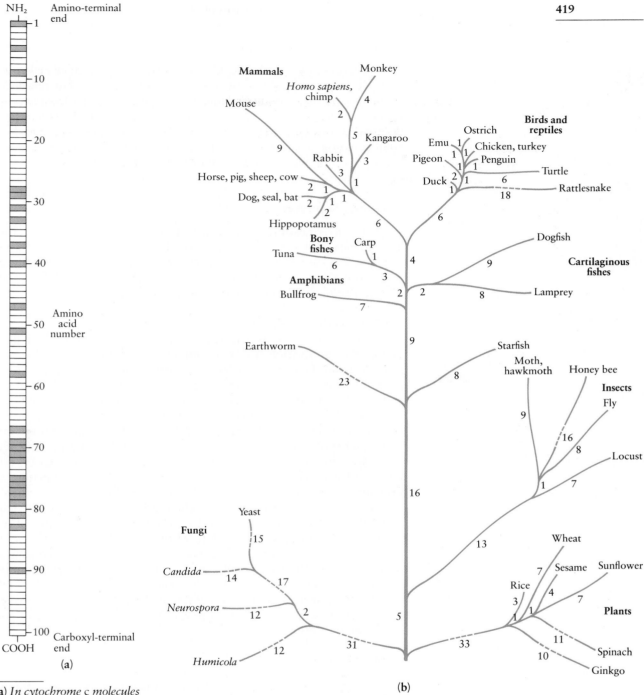

(a)

(b)

20–11 (a) *In cytochrome c molecules from the more than 60 species that have been studied, 27 of the amino acids are identical (color). (b) The main branches of a phylogenetic tree based on comparisons of the amino acid sequences of cytochrome c molecules. The numbers indicate the number of amino acids by which each cytochrome c differs from the cytochrome c at the nearest branch point. Dashes indicate that a line has been shortened and is therefore not to scale. Although based on comparisons of a single type of protein molecule, this tree is in fairly good agreement with phylogenies constructed by more conventional means.*

protein structure represent functional differences among the molecules, just as differences in the structure of the beaks of birds represent adaptations to different food sources. Other biologists believe that amino acid changes occur at random —as the result of random mutations—and that they do not represent the result of a selection process but merely mark off the passage of time, like grains of sand trickling through an hour glass or the decay of radioisotopes. From this viewpoint, the amino acid differences in the homologous proteins of different groups of organisms do not represent functional differences; instead, they are **molecular clocks** and can be used to determine the time at which various groups diverged.

In support of the "random-tick" hypothesis, biological clock makers point out that two frog species that diverged millions of years ago, but remained similar enough in appearance to be included in a single genus, differ from one another in amino acid substitutions as much as a bat does from a whale. *Homo sapiens* and the chimpanzee, on the other hand, two species that differ anatomically and in a number of other characteristics but that diverged recently, according to paleontological evidence, have identical amino acid sequences in cytochrome *c* and some other proteins as well.

Nucleotide Sequences

Now that it is possible to sequence nucleic acids (see page 348), the use of homologous proteins to estimate evolutionary relationships has been largely abandoned. One reason is that nucleic acid sequencing is technically far easier than protein sequencing, dealing, as it does, with only four different nucleotides as compared to 20 amino acids. Also, it is more sensitive, since changes in nucleotides may not be reflected in changes in amino acids because of the many synonyms in the genetic code.

As the sequences of nucleic acids from a variety of species have been determined, the information has been entered into computer data banks, making possible detailed comparisons. Such comparisons have demonstrated the value of nucleic acid sequences in taxonomic studies. For example, analyses of the rRNA and tRNA molecules of prokaryotes have made it possible, for the first time, to begin determining the evolutionary relationships among these organisms, which are extremely difficult to distinguish on the basis of structural features alone.

A number of studies have shown, however, that there are a variety of different molecular clocks, ticking at different rates. As you might expect, nucleotide changes that result in amino acid substitutions affecting the function of critical proteins are weeded out by natural selection; thus, clocks based on such proteins are a bit slow. Conversely, nucleotide changes in segments of DNA that are never translated—such as introns (page 361)—appear to be relatively free of functional constraints, leading to a faster ticking of the clock. In addition, for reasons that are only poorly understood, different portions of the DNA of an organism appear to be subject to different rates of mutation. Theoretically, these difficulties would be eliminated (or at least minimized) if many different homologous DNA segments were sequenced. Limitations of time and resources, however, make such an approach impractical and have led to the search for simpler ways of comparing large portions of the genome.

DNA-DNA Hybridization

One of the earliest techniques used in the study of large nucleic acid molecules was hybridization. As we saw earlier (page 345), when a solution of DNA is heated, the hydrogen bonds break and the double-stranded molecules separate into single strands. Upon cooling, the single strands reassociate with each other—or with homologous strands of DNA or RNA from another source. This property of nucleic acid molecules has been invaluable in locating particular genes of interest for further study. More recently, Charles G. Sibley and Jon E. Ahlquist of Yale University have devised an elegant taxonomic adaptation of the technique.

In their studies, Sibley and Ahlquist break organismal DNA into fragments of approximately 500 nucleotides each and remove the repeated DNA segments (page 363) that characterize the eukaryotic genome. In groups of two, solutions of the resulting single-copy DNA are mixed, heated, and cooled, allowing hybridization of homologous sequences to occur (Figure 20–12). The DNA from one source is radioactively labeled, and the other is unlabeled; the amount of unlabeled DNA is approximately 1,000 times the amount of labeled DNA. Because of

20–12 *In DNA-DNA hybridization experiments, radioactively labeled DNA* (**a**) *is mixed with unlabeled DNA* (**b**) *from the same species or a different species. The amount of unlabeled DNA is approximately 1,000 times the amount of labeled DNA. The solution is boiled to dissociate the double-stranded molecules and then incubated to allow the single strands to reassociate. Most of the unlabeled strands reassociate with other unlabeled strands* (**c**), *and some strands fail to reassociate at all* (**d**). *However, because of the vast excess of unlabeled DNA, about 99 percent of the radioactive strands reassociate with unlabeled strands, forming hybrid molecules* (**e**). *The more similar the nucleotide sequences of the strands forming a hybrid molecule, the more tightly they are held together—and the higher the temperature required to dissociate them again. As you can see, one of the hybrid molecules is a perfect match, while the other two have differing degrees of mismatch.*

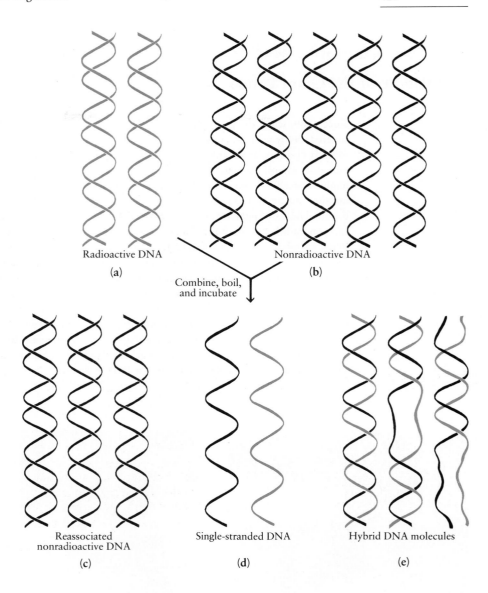

Radioactive DNA
(a)

Nonradioactive DNA
(b)

Combine, boil,
and incubate

Reassociated
nonradioactive DNA
(c)

Single-stranded DNA
(d)

Hybrid DNA molecules
(e)

this vast excess of unlabeled DNA, most of the reassociated double-stranded molecules that contain radioactivity are hybrids. As the solution is then gradually reheated, the molecules that dissociate into single strands are removed and tested for radioactivity. The temperature at which 50 percent of the hybrid molecules dissociate is a measure of how tightly the strands are bonded, which is, in turn, a measure of the similarity of the DNA sequences. The higher the temperature, the more similar the DNAs.

The 50-percent-dissociation temperature is determined individually for the DNA of each species to be studied (using radioactive and nonradioactive samples from that species). This provides a base line for each species, against which the 50-percent-dissociation temperatures of the hybrids its DNA forms with the DNA of other species can be compared. A 1 °C lowering from the single-species 50-percent-dissociation temperature corresponds to a 1 percent difference in the nucleotide sequences of the two species being compared; this, in turn, corresponds to an evolutionary separation of approximately 4.5 million years. By correlating hybridization results with known dates of specific evolutionary and geologic events, as determined from the fossil record, it is possible to establish, in real time, the branching points in evolutionary lineages.

The Riddle of the Giant Panda

Since 1869, when the giant panda of China was discovered, its true identity has been a riddle. It was initially classified as a member of the bear family, but biologists almost immediately began to wonder if it might instead be most closely related to another unusual mammal indigenous to China, the lesser panda. The lesser panda was clearly a member of the raccoon family, even though there were no other living members of that family in the Old World—unless, perhaps, the giant panda was also a raccoon, albeit a very odd one. The two pandas share many unusual anatomical and behavioral characteristics, and although the giant panda does resemble the bears in certain features, it differs significantly in others. Over the years, biologists debated the question, opting in roughly equal numbers for "bear" or "raccoon." The answer seemed to be at hand in 1964 with the publication of a detailed anatomical study of the giant panda. This study demonstrated, to the satisfaction of most biologists, that the giant panda is a bear and that the features in which it resembles the lesser panda are adaptations to feeding on its exclusive energy source, bamboo.

This conclusion has now been confirmed by the application of four different techniques: DNA-DNA hybridization, comparison of the charge and size of homologous protein molecules (as determined by electrophoretic studies), comparison of the antibody-binding properties of homologous proteins, and detailed study of the banding patterns of the chromosomes. Each procedure provided the same answer: the giant panda is a bear, as children the world over have known all along.

(a)

Although both native to China and sharing many unusual anatomical and behavioral adaptations, the lesser panda (a) and the giant panda (b) are not each other's closest relatives. The results of DNA-DNA

(b)

hybridization studies (c), as well as studies using other techniques, have demonstrated that the lesser panda is most closely related to the raccoons, whereas the giant panda is most closely related to the bears. The

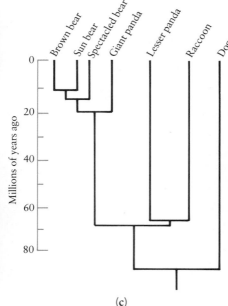

(c)

hybridization results also confirm the fossil evidence indicating that the lineage that gave rise to the bears is a branch of the lineage that gave rise to the dogs (see page 417).

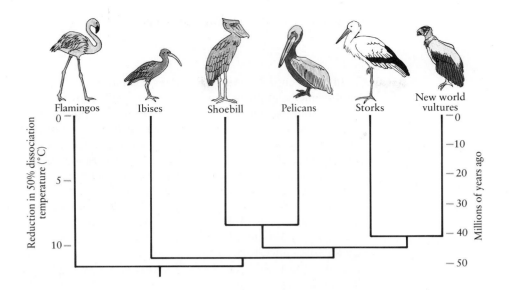

20-13 *The family tree of six closely related groups of living birds, as determined by DNA-DNA hybridization studies. The DNA used in these studies was extracted from red blood cells, which, in birds, contain nuclei. This particular set of experiments not only answered a long-standing question about the proper classification of the flamingos but also provided surprising new information about the history of the New World vultures, such as the California condor. On the basis of structural and behavioral similarities, these birds have traditionally been classified with the Old World vultures; it now appears that those similarities are the result of adaptation to similar life styles. As shown here, the New World vultures are most closely related to the storks. The Old World vultures, by contrast, belong to a completely different evolutionary lineage and are more closely related to the hawks and the eagles.*

Using this method, Sibley and Ahlquist have compared—two by two—more than 25,000 DNA samples from about 1,600 species of birds. In the process, they have solved a number of puzzles in bird taxonomy. For example, flamingos have long been classified with the storks by some authorities, with the geese and ducks by others; hybridization studies have now demonstrated unequivocally that flamingos are more closely related to the storks (Figure 20–13). To cite another example, the starlings were long classified with the crows. However, data from DNA-DNA hybridization studies have revealed that these two lineages separated some 60 million years ago and that starlings are much more closely related to mockingbirds, catbirds, and thrushes, from which they diverged a mere 25 million years ago.

Recent evidence indicates that the DNA of some widely separated groups (such as, for example, insects and mammals) evolves at quite different rates. Similarly, mitochondrial DNA appears to evolve much more rapidly than nuclear DNA. Thus even the molecular clock of DNA-DNA hybridization, which is based on the entire single-copy genome, is not the answer to all taxonomic problems. For groups of organisms that are not too distantly related, however, it is the molecular technique of choice. In conjunction with other methods of analysis, it provides taxonomists with the opportunity not only to answer long-standing questions about the relationships of organisms (see essay) but also to ask many new questions. The answers to these will undoubtedly upset some traditional classifications and generate additional questions, providing new insights into the biology of both familiar and unfamiliar organisms. Laboratory taxonomists are engaged in an adventure equal in excitement—if not in physical danger—to that of the tropical biologists who are almost daily discovering organisms previously unknown to science.

A QUESTION OF KINGDOMS

One problem of biological classification currently unresolved—and probably unresolvable, even with the battery of new methods available—is the placement of taxa in the category of kingdom. In Linnaeus's time, as we mentioned earlier, three kingdoms were recognized—animals, plants, and minerals—and until very recently it was common to classify every living thing as either an animal or a plant. Kingdom Animalia included those organisms that moved and ate things, and whose bodies grew to a certain size and then stopped growing. Kingdom Plantae comprised all living things that did not move or eat and that grew indefinitely. Thus the fungi, algae, and bacteria were grouped with the plants, and the protozoa—the one-celled organisms that ate and moved—were classified with the animals.

In the twentieth century, new data began to emerge. This was partly a result of improvements in the light microscope and, subsequently, the development of the electron microscope, and partly because of the application of biochemical techniques to studies of differences and similarities among organisms. As a result, the number of groups recognized as constituting different kingdoms has increased. The new techniques revealed, for example, the fundamental differences between prokaryotic and eukaryotic cells—differences sufficiently great to warrant placing the prokaryotes in a separate kingdom, Monera.

Other studies have provided new information about the evolutionary history of the major types of organisms. As we shall see in Chapter 22, there is strong evidence that different lineages of eukaryotes arose independently from different prokaryotic ancestors. Moreover, different lineages of unicellular eukaryotes appear to have given rise to the multicellular plants, fungi, and animals, and, at least in the case of photosynthetic organisms, multicellularity has arisen several times. This history makes it impossible, on the basis of current knowledge, to establish monophyletic kingdoms and still have the kingdoms reflect similarities and differences among major groups of living organisms.

Most contemporary proposals concerning kingdoms are based not on evolutionary history, but rather on the cellular organization and the mode of nutrition of the organisms. The proposal we shall follow recommends five kingdoms: Monera, Protista, Fungi, Plantae, and Animalia. In Appendix C, the major groups of organisms are classified in these five kingdoms. The members of kingdom Monera, the prokaryotes, are identified on the basis of their unique cellular organization and biochemistry. Members of the kingdom Protista are eukaryotes, both autotrophs and heterotrophs, and most are unicellular. A few groups of relatively simple multicellular organisms are also included in this kingdom, because they are more closely related to unicellular forms than to superficially similar fungi, plants, or animals. All the other multicellular eukaryotes are divided into three kingdoms, based primarily on their mode of nutrition: fungi absorb organic molecules from the surrounding medium, plants manufacture them by photosynthesis, and animals ingest them in the form of other organisms. These three groups of organisms have distinct ecological roles—plants are generally producers, animals are consumers, and fungi are decomposers. Table 20–2 summarizes some of the essential similarities and differences among these five kingdoms of organisms. We shall discuss each group in turn in the chapters that follow.

No system of kingdoms is completely satisfactory. For instance, as we shall see in Chapter 24, there is a clear evolutionary sequence leading from certain one-celled photosynthetic eukaryotes to the flowering plants. So if we group all of the one-celled eukaryotes together in kingdom Protista, as we do in this text, we break up what would appear to be a direct evolutionary line to the plants. However, at the one-celled level of organization, the capacity for photosynthesis is not always a useful criterion for determining the degree of relationship among organisms. Two species of single-celled, motile organisms may be virtually identical in most respects, except that one has chloroplasts and the other does not. In some cases, the one that has chloroplasts can lose them from time to time and still continue to survive and reproduce indefinitely. To separate these two closely related forms at the level of kingdom seriously misrepresents biological reality.

The number of kingdoms into which organisms are classified and the decisions made in assigning problematic taxa to particular kingdoms do not, fortunately, affect the genus and species designations of those organisms nor most of the other hierarchical categories in which they are now classified. Most important, the unique characteristics and evolutionary history of each group of organisms are totally unaffected by any difficulties we may have in ordering them to suit our purposes.

20–14 *Representatives of the five kingdoms. (a) Monera. Cells of the bacterium* Pseudomonas aeruginosa, *which thrives in standing water, such as that in wooden hot tubs. This bacterium produces a characteristic rash on the skin of its human cohabitants. It is also the cause of "swimmer's ear." (b) Protista.* Vorticella, *like most protists, is a single cell. It attaches itself to a substrate by a long stalk. A contractile fiber, visible in this micrograph, runs through the stalk. (c) Fungi. Inky cap mushroom. Fungi are characterized by a multicellular underground network and also spores and sporangia, which, in mushrooms, are borne in these familiar structures. (d) Plantae. California poppies and lupines. Flowers, attractive to pollinators, are among the principal reasons for the evolutionary success of the plants of division Anthophyta. (e) Animalia. A luna moth. A nervous system with a variety of sense organs is a chief characteristic of the animal kingdom.*

TABLE 20-2 **Characteristics of the Five Kingdoms**

	MONERA	PROTISTA	FUNGI	PLANTAE	ANIMALIA
Cell type	Prokaryotic	Eukaryotic	Eukaryotic	Eukaryotic	Eukaryotic
Nuclear envelope	Absent	Present	Present	Present	Present
Mitochondria	Absent	Present	Present	Present	Present
Chloroplasts	Absent (photosynthetic membranes in some types)	Present (some forms)	Absent	Present	Absent
Cell wall	Noncellulose (polysaccharide and peptidoglycan)	Present in some forms, various types	Chitin and other noncellulose polysaccharides	Cellulose and other polysaccharides	Absent
Means of genetic recombination	Conjugation, transduction, transformation, or none	Fertilization (syngamy) and meiosis, conjugation, or none	Fertilization and meiosis, dikaryosis (page 481), or none	Fertilization and meiosis	Fertilization and meiosis
Mode of nutrition	Autotrophic (chemosynthetic or photosynthetic) or heterotrophic	Photosynthetic or heterotrophic, or combination of these	Heterotrophic, by absorption	Photosynthetic	Heterotrophic, by ingestion
Motility	Bacterial flagella, gliding, or nonmotile	9 + 2 cilia and flagella, amoeboid, contractile fibrils	Nonmotile	9 + 2 cilia and flagella in gametes of some forms, none in most forms	9 + 2 cilia and flagella, contractile fibrils
Multi-cellularity	Absent	Absent in most forms	Present	Present	Present
Nervous system	Absent	Primitive mechanisms for conducting stimuli in some forms	Absent	Absent	Present, often complex

SUMMARY

More than 5 million different kinds of organisms are known to inhabit the earth, and many times that number await discovery. Scientists have sought order in this vast diversity of living things by classifying them, that is, by grouping them in meaningful ways.

The basic unit of classification is the species, most easily defined as an interbreeding group of organisms that does not breed with other groups of organisms. Species are named by a binomial system that includes the name of the genus (written first) and the specific epithet, a modifier that identifies the particular species within the genus.

Taxonomy, the science of classifying groups (taxa) of organisms in formal groups, is hierarchical. Genera are groups of similar species, presumably species that have recently diverged. Genera are grouped into families, families into orders, orders into classes, classes into phyla or divisions, and divisions or phyla into kingdoms. The grouping of divisions or phyla into kingdoms is based on cellular organization and mode of nutrition. The classification system followed in this text consists of five kingdoms: Monera, Protista, Fungi, Plantae, and Animalia.

In classifying organisms in the categories of genus through phylum or division, evolutionary systematists seek to group the organisms in ways that reflect their phylogeny (evolutionary history). In a phylogenetic system, every taxon should be, ideally, monophyletic; that is, every taxon should consist only of organisms descended from a common ancestor. A major principle of such classification is that the similarities taken into account should be homologous—that is, the result of common ancestry rather than of adaptation to similar environments (analogous). Among the types of data used by evolutionary systematists are structural and biochemical characteristics, fossil evidence, stages in the life cycle, and patterns of embryonic development.

Two alternative methodologies used in classification are numerical phenetics and cladistics. Numerical phenetics relies solely on the scoring of equally weighted, objectively observable similarities and differences among groups of organisms, without regard to homology and analogy. Cladistics, by contrast, is based entirely on branching sequences (genealogy), determined by evolutionary novelties, and ignores overall similarities. The goal of cladistics is the creation of holophyletic taxa, in which the common ancestor and all of its descendants are included.

New techniques in molecular taxonomy are providing objective, numerical comparisons of organisms at the most basic level of all, the gene. A large body of evidence indicates that protein and nucleic acid molecules are molecular clocks, with changes in composition reflecting the time that has elapsed since different groups of organisms diverged from one another. Amino acid sequencing, nucleotide sequencing, and DNA-DNA hybridization are all making valuable contributions to more accurate classification schemes and, most important, to our understanding of organisms and their evolutionary history.

QUESTIONS

1. Distinguish among the following: taxonomy/evolutionary systematics; category/taxon; genus/species; division/phylum; homology/analogy; numerical phenetics/cladistics; monophyletic/holophyletic/paraphyletic/polyphyletic.

2. Identify which of the following are categories and which are taxa: undergraduates; the faculty of the University of Tennessee; the Washington Redskins; major league baseball teams; the U.S. Marine Corps; Mozart's symphonies.

3. The use of the terms "division" and "phylum" for the same category in biological classification is an accident of history, resulting from the human tendency to place different subjects of study in separate compartments. What compartmentalization produced these two terms?

4. It is generally thought that nucleic acid molecules are more reliable molecular clocks than protein molecules. Give at least three reasons why this should be the case.

5. Based on your knowledge of DNA structure, the genetic code, and protein structure, what sorts of random mutations would you expect to persist in a lineage of organisms, generation after generation, unaffected by natural selection? What sorts of mutations would you expect to be harmful to the organisms and thus suppressed by natural selection?

6. What are the major identifying characteristics of each of the five kingdoms?

C H A P T E R **21**

The Prokaryotes and the Viruses

In the next three chapters, we are going to describe three diverse kingdoms of organisms: Monera (the prokaryotes), Protista (unicellular and some simple multicellular eukaryotes), and Fungi. Many of the organisms in these three kingdoms are major agents of disease not only in *Homo sapiens,* but also in the plants and animals on which we depend. Thus, although they are dissimilar in many important ways, and some multicellular protists and fungi are quite large, these organisms have conventionally been studied together in the discipline known as microbiology. Because of their disease-causing properties, viruses, which are, by most definitions, not living organisms at all and thus are not classified in any of the five kingdoms, have also traditionally been studied as a part of microbiology. We shall therefore examine their diversity in the latter part of this chapter.

The prokaryotes are, in evolutionary terms, the oldest group of organisms on earth. And, despite their relative simplicity, contemporary prokaryotes are the most abundant organisms in the world. Although there are sometimes difficulties in defining prokaryotic species unambiguously, about 2,700 distinct species are currently recognized. Prokaryotes are the smallest cellular organisms; a single gram (about 1/28 of an ounce) of fertile soil can contain as many as 2.5 billion individuals.

The success of the prokaryotes, biologically speaking, is undoubtedly due to their great metabolic diversity and their rapid rate of cell division. Growing under

21–1 *Thermophilic ("heat-loving") bacteria thrive at 92°C, a temperature close to the boiling point of water. Shown here is a boiling hot spring, Grand Prismatic Spring, in Yellowstone National Park in Wyoming. Carotenoid pigments of the thickly growing thermophilic bacteria and cyanobacteria color the run-off channels a brownish orange.*

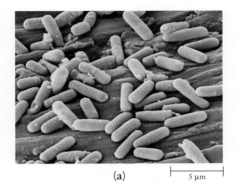

(a) |⎯⎯ 5 µm ⎯⎯|

(b) |⎯ 1 µm ⎯|

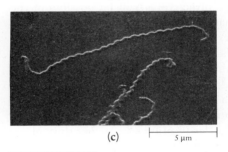

(c) |⎯⎯ 5 µm ⎯⎯|

21–2 *The cells of many familiar genera of bacteria have one of three readily distinguished shapes:* (a) *straight rods,* (b) *spheres, or* (c) *long, spiral rods. The rod-shaped bacteria (bacilli) include those microorganisms that cause lockjaw (Clostridium tetani), diphtheria (Corynebacterium diphtheriae), and tuberculosis (Mycobacterium tuberculosis), as well as the familiar E. coli. Among the spherical cocci, which may form pairs, clusters, or chains, are* Streptococcus pneumoniae, *a cause of bacterial pneumonia;* Streptococcus lactis, *which is used in the commercial production of cheese; and* Nitrosococcus, *soil bacteria that oxidize ammonia to nitrates. The helically coiled spirilla are less common. Cell shape is a relatively constant feature in many species of bacteria.*

optimum conditions, a population of *Escherichia coli,* probably the best-known prokaryote, can double in size every 20 minutes. Prokaryotes can survive in many environments that support no other form of life. They have been found in the icy wastes of Antarctica, the dark depths of the ocean, and even in the near-boiling waters of natural hot springs. Some prokaryotes are among the very few modern organisms that can survive without free oxygen, obtaining their energy by anaerobic processes (see page 190). Oxygen is lethal to some types (obligate anaerobes), whereas others can exist with or without oxygen (facultative anaerobes).

When conditions are unfavorable, some types of prokaryotes can form thick-walled spores. These spores are inactive, resistant forms that enable the cells to survive for long periods of time without water or nutrients or in conditions of extreme heat or cold. They may stay dormant for years, and some remain viable even when boiled in water for several hours.

From an ecological point of view, prokaryotes are most important as decomposers, breaking down organic material to forms in which it can be used by plants. They also play a major role in the process known as nitrogen fixation, by which nitrogen gas (N_2) is reduced to ammonia (NH_3) or ammonium ion (NH_4^+). Although nitrogen is abundant in the atmosphere, eukaryotes are not able to use atmospheric nitrogen, and so the crucial first step in the incorporation of nitrogen into organic compounds depends largely on certain species of prokaryotes; some of these species are free-living, while others are found only in close association with plants. Some prokaryotes are photosynthetic, and a few species are both photosynthetic and nitrogen-fixing.

THE CLASSIFICATION OF PROKARYOTES

Until quite recently, the possibility of classifying the prokaryotes on an evolutionary basis seemed remote. Most of the characteristics used to determine phylogenetic relationships among eukaryotes—for example, intricate anatomical structures composed of interconnected parts and complex patterns of reproduction, development, and growth—simply do not exist in prokaryotes. Biologists studying the prokaryotes were forced to rely upon differences in the size and shape of individual cells, the general appearance of the colonies formed by some prokaryotes, type of movement, mode of nutrition, the presence or absence of spores, the presence or absence of a cell wall, and the characteristics of the cell wall, as revealed by its ability to retain certain chemical dyes. Many of these characteristics do not reflect phylogenetic relationships. The prokaryotes have been in existence—and evolving in response to diverse selection pressures—for a very long time. Certain features, such as cell shape and colony form, have probably evolved over and over again. In contrast, other features, such as a cell wall, the capacity for photosynthesis, or the ability to form spores, have been lost independently in a number of lineages. As a result of these difficulties, the classification schemes for the prokaryotes that have long been in use are not hierarchical; the only category above the level of genus is division, with the number of divisions varying in different schemes.

In recent years, studies of cell ultrastructure and biochemistry (particularly the details of metabolic pathways) have enabled biologists to begin unraveling the evolutionary relationships of the prokaryotes. Crucial breakthroughs have come with the development of the molecular techniques described in the previous chapter: amino acid sequencing, nucleotide sequencing, and DNA-DNA hybridization. In general, as with eukaryotes, the smaller the number of differences in the nucleotide sequences of two organisms, the more recent their common ancestor. An additional technique, determination of the percentage of guanine and cytosine (G + C) and of adenine and thymine (A + T) in the DNA has also proved valuable in identifying the degree of relationship among prokaryotic species.

These techniques are rapidly leading to a revolution in our understanding of prokaryotic phylogeny. One of the most striking discoveries thus far has been that a few genera of prokaryotes, not previously classified together, differ very dramatically from all other organisms. These genera include the methanogens (bacteria that synthesize methane from carbon dioxide and hydrogen gas), the salt-loving halobacteria (see page 221), and the thermoacidophiles (bacteria that thrive in very acidic environments at high temperatures). They differ from all other prokaryotes in several significant ways. First, their cell walls have a unique composition. Second, they do not use the Calvin cycle for carbon reduction. Third, they require several unique coenzymes that serve the function of NAD$^+$ and FAD as electron acceptors. Fourth, and perhaps most important, the nucleotide sequences of their transfer RNAs and ribosomal RNAs are markedly different from those in all other organisms. A particularly interesting feature of the methanogens is that they would have been excellently suited to the conditions prevailing on earth during the earliest stages of biological evolution. It appears that the methanogens, extreme halophiles, and thermoacidophiles may be the few surviving representatives of a lineage that diverged very early from the common ancestor that gave rise to the other prokaryotes. Recently it has been proposed that they be placed in a new kingdom, Archaebacteria; all of the other prokaryotes, collectively known as the Eubacteria, would remain in kingdom Monera. (A corollary of this proposal recommends the establishment of two superkingdoms: Prokaryota, to include kingdoms Archaebacteria and Monera, and Eukaryota, to include kingdoms Protista, Fungi, Plantae, and Animalia.)

At the present time, molecular taxonomic studies of the prokaryotes are proceeding rapidly in a number of laboratories, and the probable evolutionary relationships of the different groups are being worked out. The outlines of the emerging prokaryotic phylogeny are shown in Figure 21–3. It is anticipated that a detailed, evolutionary classification of the prokaryotes will be possible in the relatively near future, but it is unlikely that the older classification schemes will become obsolete. Although those schemes do not necessarily reflect phylogeny, they do group prokaryotes on the basis of similarities that are of immense practical importance in medical and veterinary diagnosis and treatment, agriculture, and industrial microbiology. Two different schemes, each with different uses, each illuminating different aspects of the incredible diversity of prokaryotes, are likely to exist side by side. For our purposes, however, the best introduction to this diversity is not through the details of a classification scheme but through an examination of the characteristics found among the prokaryotes.

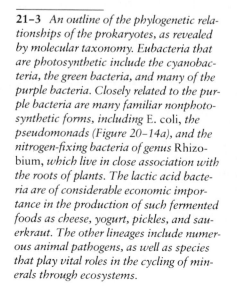

21–3 *An outline of the phylogenetic relationships of the prokaryotes, as revealed by molecular taxonomy. Eubacteria that are photosynthetic include the cyanobacteria, the green bacteria, and many of the purple bacteria. Closely related to the purple bacteria are many familiar nonphotosynthetic forms, including E. coli,* the *pseudomonads (Figure 20–14a), and the nitrogen-fixing bacteria of genus* Rhizobium, *which live in close association with the roots of plants. The lactic acid bacteria are of considerable economic importance in the production of such fermented foods as cheese, yogurt, pickles, and sauerkraut. The other lineages include numerous animal pathogens, as well as species that play vital roles in the cycling of minerals through ecosystems.*

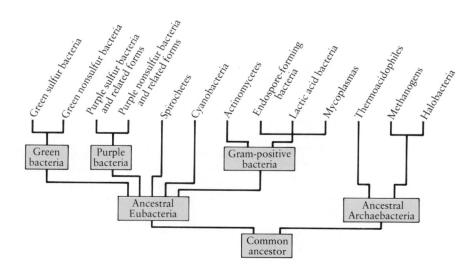

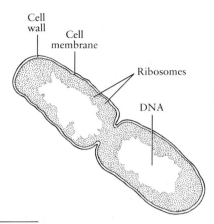

21–4 *Cells of* Escherichia coli, *a modern prokaryote that is a common inhabitant of the human digestive tract. The DNA is localized in the nucleoid—the less dense (lighter-appearing) region in the center of each cell. The small, dense bodies in the cytoplasm are ribosomes. The two cells in the center have just finished dividing and have not yet separated completely.*

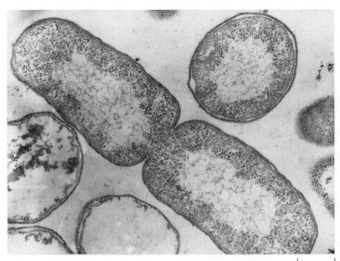

0.25 µm

THE PROKARYOTIC CELL

The essential features of a prokaryotic cell were outlined in Chapter 4 and are reviewed in Figure 21–4. The cell shown here is the familiar *Escherichia coli,* the most thoroughly studied of the prokaryotes.

The most prominent feature within the prokaryotic cell is the nucleoid, the region in which the chromosome is localized. All prokaryotic chromosomes analyzed so far have proved to consist of one single, continuous ("circular") molecule of DNA associated with a small amount of RNA and non-histone proteins. A prokaryotic cell may also contain one or more plasmids (page 326). As we saw in Section 3, studies of the prokaryotic chromosome have contributed greatly to our understanding of genetic mechanisms.

The cytoplasm of most prokaryotes is relatively unstructured, although it often has a fine granular appearance due to its many ribosomes. These are somewhat smaller than eukaryotic ribosomes but have the same general shape. Generally the cytoplasm is not divided or compartmentalized by membranes and does not contain any membrane-bound organelles. The principal exception occurs in the cyanobacteria, which contain an extensive membrane system bearing chlorophyll and other photosynthetic pigments.

The Cell Membrane

The membrane enclosing the cytoplasm of a prokaryotic cell is formed from a lipid bilayer; it is similar in chemical composition to that of a eukaryotic cell (see page 104). However, except in the mycoplasmas (the smallest free-living cells known), the membranes of prokaryotes lack cholesterol or other steroids. In the aerobic prokaryotes, the cell membrane incorporates the electron transport chain found in the mitochondrial membrane of eukaryotic cells. In the photosynthetic green and purple bacteria (but not in the cyanobacteria), the sites of photosynthesis are found in the cell membrane. In the photosynthetic purple bacteria and in aerobes with large energy requirements, the membrane is often extensively convoluted, with folds extending into the interior. These, of course, greatly increase the working surface of the membrane. Also, as we noted on page 145, the membrane appears to contain specific attachment sites for the DNA molecules; these sites are believed to play a role in ensuring the separation of the replicated chromosomes at cell division.

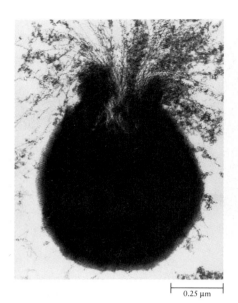

0.25 μm

21–5 *This unusual micrograph, taken by Victor Lorian at the Bronx-Lebanon Hospital Center in New York, shows a bacterial cell exploding. This bacterium is a member of the species* Staphylococcus aureus, *the cause of many human infections. Because water tends to move into the cell by osmosis, the contents of bacterial cells are under pressure. When this cell was treated with an antibiotic that damaged the cell wall, it exploded.*

The Cell Wall

Almost all prokaryotes are surrounded by a cell wall, which gives the different types their characteristic shapes. Many prokaryotes have rigid walls, some have flexible walls, and only the mycoplasmas have no cell walls at all. Because most bacterial cells are hypertonic in relation to their environment, they would burst without their walls. (The mycoplasmas live as intracellular parasites in an isotonic environment.)

Chemical Structure of the Cell Wall

The cell walls of prokaryotes are complex and contain many kinds of molecules not present in eukaryotes. Except for the Archaebacteria (the methanogens and their close relatives), the walls of prokaryotes contain complex polymers known as peptidoglycans, which are primarily responsible for the mechanical strength of the wall.

Prokaryotic cell walls occur in two different configurations, which are readily distinguished by their capacity to combine firmly with such dyes as gentian violet. Those that combine with the dyes are known as gram-positive, whereas those that do not combine with them are known as gram-negative, after Hans Christian Gram, the Danish microbiologist who discovered the distinction. In gram-positive cells (Figure 21–6a), the wall consists of a homogeneous layer of peptidoglycans and polysaccharides that ranges from 10 to 80 nanometers in thickness. In gram-negative cells (Figure 21–6b), by contrast, the wall consists of two layers: an inner peptidoglycan layer, only 2 to 3 nanometers thick, and an outer layer of lipoproteins and lipopolysaccharides. These molecules are arranged in the form of a bilayer, about 7 to 8 nanometers in thickness, similar in structure to the cell membrane. Thus, the boundary of a gram-negative cell is actually a sandwich within a sandwich: an inner cell membrane, a thin peptidoglycan layer, and an outer membrane.

Gram staining is widely used as a basis for classifying bacteria, since it reflects a fundamental difference in the architecture of the cell wall. As shown in Figure 21–3, all gram-positive bacteria are thought to constitute a distinct phylogenetic lineage. The architecture of the cell wall, in turn, affects various other characteristics of the bacteria, such as their patterns of susceptibility to antibiotics (see Figure 8–22, page 178). Gram-positive bacteria, such as *Staphylococcus,* are much more susceptible to some types of antibiotics than are gram-negative bacteria, such as *E. coli.* Their walls are also more readily digested by lysozyme, an enzyme found in the nasal secretions, saliva, and other body fluids of many animals.

21–6 *Electron micrographs of sections through the cell walls of (**a**) Bacillus polymyxa, an endospore-forming gram-positive bacterium, and (**b**) E. coli, a gram-negative bacterium. The wall of a gram-positive bacterium consists of a homogeneous layer of peptidoglycans and polysaccharides, seen here as the lower dark band. The upper dark band represents a layer of surface proteins. In a gram-negative bacterium, a layer of peptidoglycan (visible here as a light band) is sandwiched between the cell membrane and an outer membrane, similar in composition to the cell membrane.*

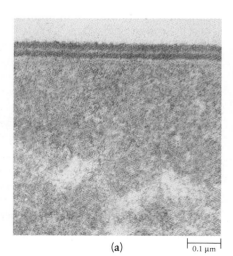

(a)   0.1 μm

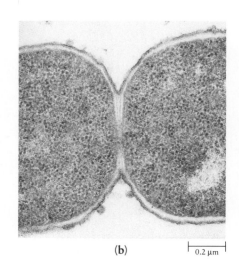

(b)   0.2 μm

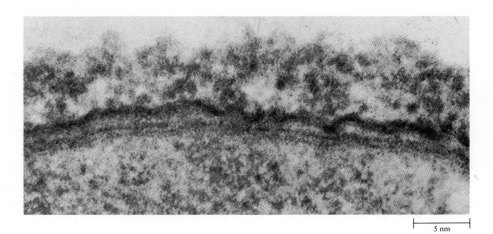

21-7 *Most forms of meningitis, an infection of the membranes covering the brain and spinal cord, are caused by the gram-negative bacterium* Neisseria meningitidis. *In this electron micrograph of* N. meningitidis, *you can distinguish the cell membrane (visible here as a bilayer), the peptidoglycan layer, the outer membrane, and a fuzzy-appearing polysaccharide capsule. Different strains of* N. meningitidis *are characterized by polysaccharide capsules of different composition. Scientists at the Walter Reed Army Institute of Research and Rockefeller University have used purified capsule polysaccharides to develop effective vaccines against two of the three major strains of this bacterium.*

5 nm

In some bacteria, a gluey polysaccharide capsule, which is secreted by the bacterium, is present outside the cell wall (Figure 21–7). The function of the capsule is not entirely clear, but its presence is associated with pathogenic activity in certain organisms. For example, as shown in Figure 14–2, the encapsulated form of *Streptococcus pneumoniae* is virulent, whereas the nonencapsulated form is generally nonvirulent. It appears that the capsule may interfere with phagocytosis by host white blood cells.

Flagella and Pili

Some types of bacteria have long, slender extensions, known as flagella and pili. Each bacterial flagellum is made up of monomers of a small globular protein, flagellin, assembled into chains that are wound in a triple helix (three chains) with a hollow central core. Bacterial flagella grow from the tip; the individual flagellin molecules pass down a channel in the center of the helix and are assembled onto the ends of the chains. The flagella of different species differ slightly in diameter (12 to 18 nanometers), probably due to slight differences in the composition of their flagellin. In some species, the flagella are distributed over the entire surface of the cell (Figure 21–8); in others, they occur in tufts at one end, or pole, of the cell.

21-8 *A bacterial cell* (Proteus mirabilis) *with numerous flagella—176, to be exact.*

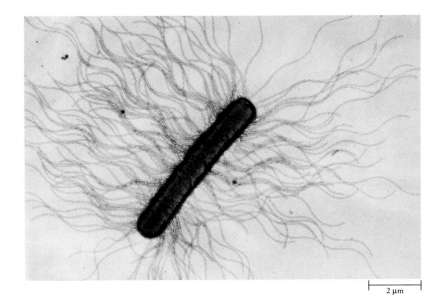

2 µm

21–9 (a) *A flagellum from a gram-negative bacterium, showing the basal end.* (b) *Diagram of a flagellum from* E. coli. *The basal body, which serves to anchor the flagellum, consists of two pairs of rings surrounding a rod. The M ring is integrated in the cell membrane, the S ring in the peptidoglycan layer, and the P and L rings in the outer membrane. The filament is made up of several protein chains that form a helix with a hollow core.*

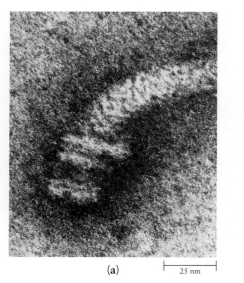

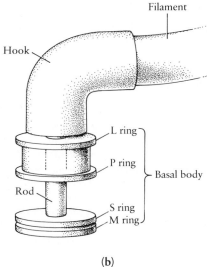

(a)   25 nm   (b)

The bacterial flagellum is not enclosed within the cell membrane, as eukaryotic flagella are, but protrudes from the cell as a naked, corkscrew-shaped protein filament. It is anchored into the cell membrane and wall by a complicated assembly (Figure 21–9). The filament (the helix of flagellin monomers with its hollow core) terminates in a hook made up of a different protein. The hook is inserted into a basal body that consists of a rod and, in gram-negative bacteria, two pairs of rings, one pair of which is embedded in the inner membrane and one pair in the outer membrane. In gram-positive bacteria, only the inner pair of rings is present, embedded in the cell membrane.

Bacterial flagella beat with a rotary movement, but they are so fine and the beat so fast that the motion of the flagellum itself cannot be seen. (The flagella of *Spirillum serpens,* for instance, have been clocked at 2,400 rpm.) Ingenious methods have been devised by which the cells can be tethered in place by the flagella. As a result, the cells rotate instead of the flagella, and the movement can be observed. The basal end of the flagellum, with its complex structure, apparently acts as a motor, running on chemiosmotic power.

Flagellated bacteria typically move rapidly in gently curved lines, or "runs," each lasting approximately a second. In species in which the flagella originate at sites all over the cell surface, a series of runs is interrupted frequently by periods of tumbling, lasting about a tenth of a second. After each tumble, the bacterium starts off in a new direction on its next run. The frequency of tumbling controls the length of the runs and hence the overall direction of movement. Whether such a cell moves in a run or tumbles is determined by the direction in which its flagella rotate. When the flagella rotate in a counterclockwise direction, they work together, producing sustained swimming in one direction. But when the flagella reverse direction and rotate in a clockwise manner, they work independently, and the cell tumbles. In species with tufts of polar flagella, clockwise rotation does not produce tumbling but rather a 180° reversal in the direction of the cell's movement. As we saw in Chapter 6 (see page 131), concentration gradients of attractive and noxious chemicals in the surrounding medium are the critical stimuli affecting the frequency with which the flagella reverse their direction of rotation. Receptors in the cell membrane measure changes in the concentration of specific molecules from one moment to the next; in some way, as yet unknown, this information is translated into counterclockwise or clockwise rotation of the flagella.

Navigation by the Poles

For heterotrophic prokaryotes, as for heterotrophic eukaryotes, adequate supplies of digestible organic molecules are essential. As we have seen, flagellated prokaryotes actively swim toward food sources, typically guided by concentration gradients of attractive chemicals in the surrounding medium. In 1975, however, the navigation of a bacterial species from a Massachusetts swamp was shown to be guided not by concentration gradients but rather by the magnetic field of the earth. All of the Massachusetts bacteria observed were found to swim toward magnetic north.

Detailed studies revealed that each bacterium accumulates about 20 opaque, roughly cubic granules of magnetite (Fe_3O_4), the iron ore from which magnets are made. The granules are arranged in a linear fashion, forming a single magnet with a north-seeking pole and a south-seeking pole. The interaction of the earth's magnetic field and the bacterium's internal magnet orients it toward the north; the rotation of its flagella, which are in a cluster at the end of the bacterium opposite the north-seeking pole of its magnet, then propel it in that direction. Because the earth's magnetic field has a vertical component as well as a horizontal component, this interaction also orients the bacterium downward—toward sediments rich in decaying matter.

After the discovery of these bacteria, scientists predicted that if similar magnet-containing bacteria existed in the Southern Hemisphere, their polarity would be reversed. In 1980, such bacteria were found in New Zealand and Tasmania, and, as predicted, their internal magnets are oriented so that the south-seeking poles are opposite the end at which the flagella are attached. Similar bacteria have also been isolated from sediments near the earth's magnetic equator. These bacteria are equally divided between north-seeking and south-seeking forms. There is virtually no vertical component to the earth's magnetism at the equator, so neither form has an advantage over the other. Their internal magnets, however, do prevent them from swimming upward, away from rich sources of food.

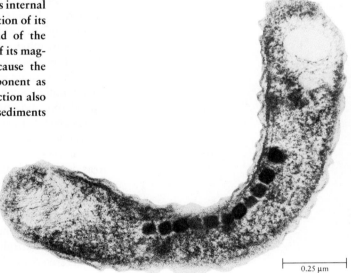

0.25 µm

A chain of tiny magnets is clearly visible in this swamp-dwelling bacterium. The capacity to manufacture a magnetic chain is inherited. When such a cell divides, the magnetic particles are divided equally between the two daughter cells. Each daughter cell then manufactures additional particles. Although not visible in this micrograph, a membrane surrounds the magnetic chain and holds the particles in place.

Pili (singular, pilus) are assembled from protein monomers in much the same way that the filaments of flagella are. (You will not be surprised to learn that the protein is called pilin.) They are rigid, cylindrical rods that extend out from the cell, sometimes to a considerable distance. Pili range from 4 to 35 nanometers in diameter, and they are often present in large numbers (hundreds on a single cell). They serve to attach bacteria to a food source, to the surface of a liquid (where oxygen is present), or, in the case of conjugating bacteria, to one another (see Figure 16–9, page 324).

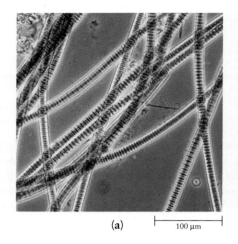

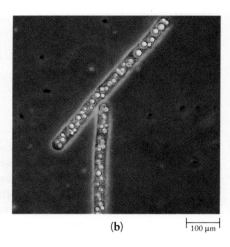

(a)   ⊢————⊣ 100 μm   (b)   ⊢—⊣ 100 μm

21–10 (a) Oscillatoria, *a filamentous cyanobacterium. Although many cyano-bacteria are nonmotile, the filamentous forms typically glide on a slime secreted by the cells. As you may have guessed from the green color, all cyanobacteria are photosynthetic.*

(b) Beggiatoa, *a genus of gliding bacteria, consists of relatively large cells that form filaments and move in a fashion similar to that of many cyanobacteria. The members of this genus are all chemosynthetic autotrophs. The conspicuous granules in the cells are sulfur, produced by the oxidation of hydrogen sulfide—a process utilized by the bacteria for the production of energy. Other types of gliding bacteria are heterotrophic.*

DIVERSITY OF FORM

The oldest method of identifying microorganisms is by their physical appearance. The shape of individual prokaryotes is a result, as we noted previously, of their cell wall. Bacteria exhibit considerable diversity of form, but many of the most familiar species fall into one of three form-groups (see Figure 21–2). Straight, rod-shaped forms like *E. coli* are known as **bacilli;** spherical ones are called **cocci;** and long, spiral rods are called **spirilla.** A fourth form-group, the **vibrios,** consists of short, curved rods that are thought to be incomplete spirals.

Different types of bacteria have characteristic patterns of growth, producing filaments, clusters, or colonies that also have a distinctive shape. For example, cocci may stick together in pairs after division (diplococci), they may occur in clusters (staphylococci), or they may form chains (streptococci). One bacterium that causes pneumonia is a streptococcus, while the staphylococci are responsible for many serious infections characterized by boils or abscesses. The rod-shaped bacilli usually separate after cell division. When they do remain together, they spread out end to end in filaments, since they always divide in the same plane (transversely). In some genera, these filaments are funguslike in appearance and the combining form *myco-* (from the Greek word for "fungus") is part of the generic name. *Mycobacterium tuberculosis,* for example, the cause of tuberculosis, is a bacillus that forms a filamentous, funguslike growth.

Many of the cyanobacteria (Figure 21–10a) as well as members of a nonphotosynthetic group, the gliding bacteria (Figure 21–10b), also form filamentous structures consisting of numerous individual cells. These cells typically secrete a slime or mucus that attaches to a solid surface and provides a pathway along which they can glide from one place to another. Other types of bacteria are characterized by distinctive sheaths or coatings outside their cell walls (Figure 21–11).

Spirochetes are among the easiest microorganisms to identify (Figure 21–12). They are very long (5 to 500 micrometers) and slender (about 0.5 micrometer in diameter) and have an unusual structure, known as an axial filament, made up of

21–11 Sphaerotilus natans, *a filamentous bacterium enclosed in a sheath. The sheath is encrusted with particles of iron oxide, formed by the bacterium from soluble iron compounds. Filaments of Sphaerotilus are responsible for the brownish scum often seen on the surface of polluted streams.*

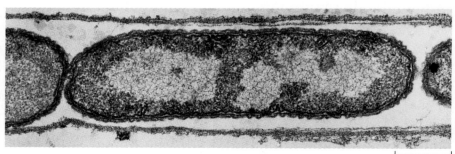

⊢——⊣ 0.5 μm

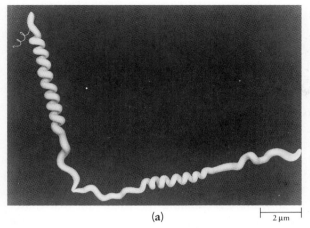

(a)

2 μm

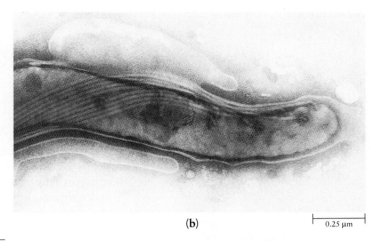

(b)

0.25 μm

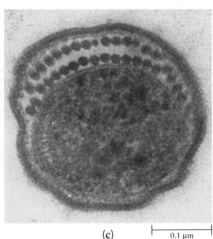

(c)

0.1 μm

21–12 **(a)** *The spirochetes range in size up to 500 micrometers long, which is an enormous size for prokaryotes.* Treponema pallidum, *two of which are shown here, is the causative agent of syphilis.* **(b)** *The end of a spirochetal cell that has* been stained to show the insertion points of two fibrils of the axial filament. **(c)** *Cross section of a spirochete showing the fibrils of the axial filament between the cell membrane and the cell wall.*

two sets of fibrils attached at each end of the cell. The fibrils are identical to flagella in structure and so are recognized as modified flagella, or endoflagella. They are wrapped around the cell between the cell membrane and the delicate wall, with the fibrils of each set overlapping at the middle of the cell. Rotation of the axial filament is thought to produce the corkscrew movement characteristic of the spirochetes. All members of this group are thought to belong to the same phylogenetic lineage.

Rickettsiae, which are the smallest cells known, constitute still another group of prokaryotes distinguishable by their form (Figure 21–13).

REPRODUCTION AND RESTING FORMS

Most prokaryotes reproduce by simple cell division (page 145), also called **binary fission.** In some forms, reproduction is by budding or by the fragmentation of filaments of cells. As they multiply, these prokaryotes, barring mutations, produce clones of genetically identical cells. Mutations do occur, however; it has been estimated that in a culture of *E. coli* that has divided 30 times, about 1.5 percent of the cells are mutants. Mutations, combined with the rapid generation time of prokaryotes, are responsible for their extraordinary adaptability. Further adaptability is provided by the genetic recombinations that take place as a result of conjugation, transformation, transduction, and exchanges of plasmids. It is not known how common such genetic recombinations are in nature or whether they occur in all types of prokaryotes.

Many prokaryotes have the capacity to form spores, which are dormant, resting cells. This process has been studied most extensively in the bacilli. It characteristically occurs when a population of cells, growing very rapidly, begins to use up its food supply. Each cell, at the beginning of sporulation (spore formation), contains two duplicate chromosomes. A cell membrane grows around one of the chromosomes, separating it from the rest of the cell, which then engulfs the newly formed cell. Thus the spore-to-be is now surrounded by two membranes, its own and that of the larger cell (these events are shown in Figure 21–14). A spore coat, consisting of two layers, then forms around this smaller cell. The inner layer (the cortex) contains a peptidoglycan that is completely different from that present in the bacterial cell wall; the outer layer consists of proteins that are composed largely of hydrophobic amino acids. The mature spore is released from

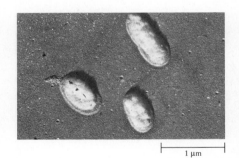

1 μm

21–13 *The tiny cells of* Rickettsia prowazekii, *shown here, are the cause of typhus. They are transmitted by human body lice, and under crowded conditions, large numbers of people can be infected in a very short time. More human lives have been taken by rickettsial diseases than by any other infection except malaria. At the siege of Granada in 1489, 17,000 Spanish soldiers were killed by typhus, but only 3,000 in combat. In the Thirty Years War, the Napoleonic campaigns, and the Serbian Campaign during World War I, typhus was also the decisive factor.*

437

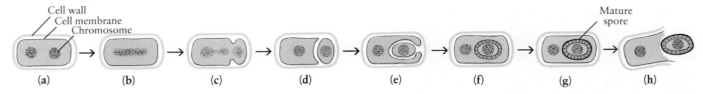

Cell wall
Cell membrane
Chromosome

Mature spore

(a)   (b)   (c)   (d)   (e)   (f)   (g)   (h)

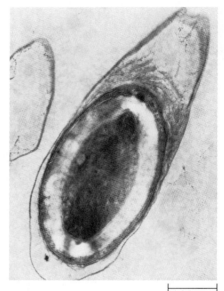

0.5 μm

21–14 *Endospore formation in* Bacillus cereus. (**a, b**) *The two chromosomes within the vegetative cell have condensed into a rod-shaped form.* (**c**) *The transverse wall begins to form,* (**d**) *cutting off the spore material from the vegetative cell.* (**e, f**) *The vegetative cell grows around the spore, and the spore coat is formed.* (**g, h**) *The spore matures and is released from the cell. The micrograph shows an endospore of* Clostridium.

the cell in this protected state, and it remains dormant until appropriate events trigger its germination. Germination takes place rapidly, with the uptake of water, the dissolution of the spore coat, and the formation of a new cell wall.

Spore formation greatly increases the capacity of prokaryotic cells to survive. For example, the spores of *Clostridium botulinum,* the bacterium that causes botulism, are not destroyed by boiling for several hours. Genetic studies of the spore-making *Bacillus subtilis* indicate that some 50 genes, clustered in about five segments of the chromosome, are involved in spore formation.

The myxobacteria, a type of gliding bacteria, form fruiting bodies, which are brightly colored collections of spores and slime large enough to be seen by the unaided eye (Figure 21–15). Although the cells of the myxobacteria retain their independent function, their association represents a rudimentary form of multicellularity, described by one whimsical biologist as "multicellularity by committee."

PROKARYOTIC NUTRITION

Heterotrophs

Although the members of kingdom Monera exhibit tremendous metabolic diversity, most prokaryotes are heterotrophs. Of these, the vast majority are **saprobes** (from the Greek *sapros,* "rotten" or "putrid"), feeding on dead organic matter.

0.25mm

21–15 *A fruiting body of* Chondromyces crocatus, *a myxobacterium. Each fruiting body, which may contain as many as 1 million cells, consists of a central stalk that branches to form clusters of oval structures that each contain a large number of single-cell spores.*

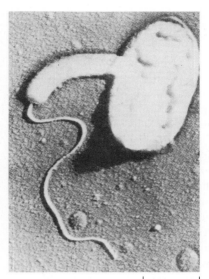

—————

21–16 *Pathogenic bacteria are themselves vulnerable to other pathogenic bacteria. This electron micrograph captured a cell of* Bdellovibrio bacteriovirus *(left) attacking a cell of* Erwinia amylovora *(upper right), the bacterium that causes fire blight of pears and apples.* Bdellovibrio bacteriovirus, *which is abundant in soil and sewage, moves by means of a single posterior flagellum. Like other members of the genus* Bdellovibrio *(from* bdello, *the Greek word for "leech"), it makes a hole in the host cell wall and multiplies between the wall and the cell membrane, digesting the host cell as it multiplies.*

Bacteria and other microorganisms are responsible for the decay and recycling of organic material in the soil. Typically, different groups of bacteria play different, specific roles—such as the digestion of cellulose, starches, or other polysaccharides, or the hydrolysis of specific peptide bonds, or the breakdown of amino acids. Because of their high degree of nutritional specialization, bacteria are able to live in large numbers in the same small area with little competition and, indeed, with mutual assistance, as the activities of one group make food molecules available to another group. These combined activities release the nutrients and make them available to plants and, through plants, to animals. Thus, the bacteria are an essential part of ecological systems.

Some heterotrophic bacteria live in close association with other organisms. Some of these are parasites that break down organic material in the bodies of living organisms (Figure 21–16). The disease-causing (pathogenic) bacteria belong to this group, as do a number of nonpathogenic forms. Some of these bacteria have little effect on their hosts, and some are actually beneficial. Cows and other ruminants can utilize cellulose only because their stomachs contain bacteria and protists that have cellulose-digesting enzymes. Our own intestines contain a number of types of generally harmless bacteria (including *E. coli*). Some supply vitamin K, which is necessary for blood clotting. Others prevent us from developing serious infections. When the normal bacterial inhabitants of the human intestinal tract are destroyed—as can happen, for example, following prolonged antibiotic therapy—our tissues are much more vulnerable to disease-causing microorganisms.

Chemosynthetic Autotrophs

Chemosynthetic autotrophs obtain their energy from the oxidation of inorganic compounds. Only prokaryotes are able to use inorganic compounds as an energy source.

An unusual group of chemosynthetic prokaryotes are the methanogens (page 90). Although their existence was long known, they were not studied extensively until quite recently because they are obligate anaerobes—that is, they are poisoned by oxygen—and hence are difficult to isolate and impossible to grow under ordinary laboratory conditions. The methanogens are the final participants in decomposition processes involving organic matter in anaerobic environments, including marshes, lake sediments, and the digestive tracts of animals. They convert CO_2 and H_2 formed by the fermentation processes of other anaerobes to methane (CH_4).

Certain chemosynthetic bacteria are essential components of the nitrogen cycle, the process by which nitrogen compounds are cycled and recycled through ecosystems. One group oxidizes ammonia or ammonium (derived from the breakdown of organic materials, the activities of nitrogen-fixing prokaryotes, or, to a minor extent, from lightning or volcanic activities). The products of this reaction are nitrite (NO_2^-) and energy. Another group oxidizes nitrites, producing nitrate (NO_3^-) and energy. Nitrate is the form in which nitrogen moves from the soil into the roots of plants.

Sulfur is also required by plants for amino acid synthesis. Like nitrogen, it is converted to the form in which it is taken up by plant roots by the activities of chemosynthetic bacteria that oxidize elemental sulfur to sulfate:

$$2S + 2H_2O + 3O_2 \longrightarrow 2H_2SO_4$$

Other sulfur bacteria, such as *Thiothrix* and *Beggiatoa,* obtain energy by oxidizing hydrogen sulfide.

These important biogeochemical cycles and the roles of microorganisms in them will be discussed further in Chapter 54.

(a)

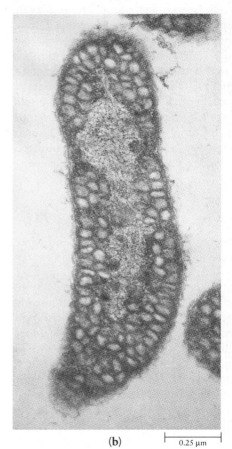

(b)

0.25 μm

21–17 *Purple bacteria. (a) Mats of the purple sulfur bacterium* Thiopedia roseopersicinia *growing in a spring in Wisconsin. The bacteria grow below the surface of the water. Gas vesicles within the cells cause them to rise to the surface when disturbed. (b) A purple nonsulfur bacterium,* Rhodospirillum rubrum. *The structures that resemble vacuoles are intrusions of the cell membrane, containing the photosynthetic pigments. This cell, which has a very high content of bacteriochlorophyll, is from a culture that was grown in dim light; in cells grown in bright light, the membrane intrusions are less extensive.*

Photosynthetic Autotrophs

Green and Purple Bacteria

Among the eubacteria are five photosynthetic types, which are thought to represent three distinct phylogenetic lineages: the cyanobacteria, the green bacteria, and the purple bacteria (see Figure 21–3). The green bacteria and the purple bacteria include both sulfur and nonsulfur forms. A number of nonphotosynthetic eubacteria, including the familiar *Escherichia*, are believed to belong to the same lineage as the purple bacteria.

The distinctive colors of the green and purple bacteria result, of course, from the pigments they contain. The reaction centers of the photosynthetic green bacteria contain bacteriochlorophyll *a*, which differs only slightly from the chlorophyll *a* of eukaryotes (page 211). The antennae pigments are bacteriochlorophylls *c, d,* or *e*, which differ more significantly. In the two groups of photosynthetic purple bacteria, the reaction centers contain bacteriochlorophylls *a* or *b*, and the same molecules also function as antennae pigments. Bacteriochlorophyll *b*, which differs chemically in several details from chlorophyll *a*, is a pale blue-gray. The colors of the purple bacteria are due to the presence of several different yellow and red carotenoids, which function as additional accessory pigments. In the purple nonsulfur bacteria, the colors may actually range from purple to red or brown.

In the photosynthetic sulfur bacteria, as we noted in Chapter 10, the sulfur compounds are the electron donors, playing the same role in bacterial photosynthesis that water does in photosynthesis in eukaryotes.

$$CO_2 + 2H_2S \xrightarrow{\text{Light}} (CH_2O) + H_2O + 2S$$

In the photosynthetic nonsulfur bacteria, other compounds, including alcohols, fatty acids, and a variety of other organic substances, serve as electron donors for the photosynthetic reactions.

Photosynthesis by green and purple bacteria is carried out anaerobically, and it never results in the production of molecular oxygen. However, in all species except the extreme halophiles (which, as we have seen, are members of a completely different phylogenetic lineage), carbon is fixed by means of the Calvin cycle.

Cyanobacteria

Unlike other photosynthetic prokaryotes, but like all photosynthetic eukaryotes, the cyanobacteria contain chlorophyll *a* and lyse water during photosynthesis, producing molecular oxygen. For this reason, they were classified with the eukaryotic algae until biochemical studies and electron microscopy revealed their prokaryotic nature. Although the name has no taxonomic status, they are still often referred to as the "blue-green algae." Cyanobacteria thrive in freshwater environments, and they are a principal component of the bluish-green scum seen on many ponds in late summer.

Cyanobacteria have several kinds of accessory pigments, including xanthophyll, which is a yellow carotenoid, and several other carotenoids. The cells also contain one or two pigments known as phycobilins: phycocyanin, a blue pigment, which is always present, and phycoerythrin, a red one, which is often present. Chlorophyll and the accessory pigments are not enclosed in chloroplasts, as they are in eukaryotic cells, but are part of a membrane system distributed in the peripheral portion of the cell (Figure 21–18).

Cells of the cyanobacteria, like those of many other prokaryotes, have an outer polysaccharide sheath, or coating. In some species, particularly those that spread up onto the land, this outer sheath is deeply pigmented; in most species, however,

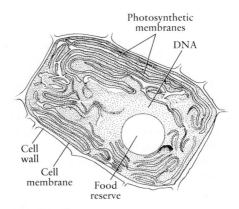

Photosynthetic
membranes

DNA

Cell
wall

Cell
membrane

Food
reserve

21–18 *Electron micrograph of the cyanobacterium* Anabaena cylindrica. *Cyanobacteria have no chloroplasts and no
membrane-bound nucleus, such as are
found in eukaryotic photosynthetic cells.
Photosynthesis takes place in chlorophyll-
containing membranes within the cell, and
the chromosome is a single molecule of
DNA. The three-dimensional quality of
this electron micrograph is due to freeze-
fracturing (page 135).*

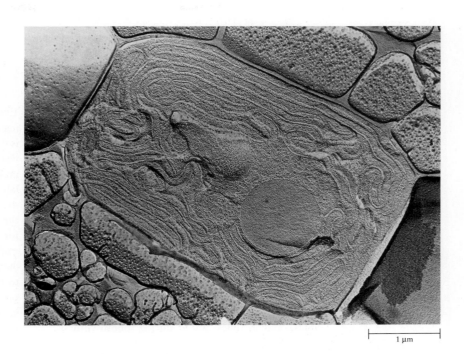

1 μm

it is not pigmented, and the colors of the cells result from the carotenoids and
phycobilins that they contain. Different species of cyanobacteria are golden
yellow, brown, red, emerald green, blue, violet, or blue-black. The Red Sea was so
named because of the dense concentrations, or "blooms," of red-pigmented
cyanobacteria that float on its surface.

Some species of cyanobacteria are capable of nitrogen fixation. The photosynthetic, nitrogen-fixing cyanobacteria are among the most self-sufficient of organisms. They have the simplest nutritional requirements of any living thing, needing
only nitrogen and carbon dioxide, which are always present in the atmosphere, a
few minerals, and water. On a worldwide scale, the ecological importance of the
nitrogen-fixing cyanobacteria appears to be less than that of the nitrogen-fixing
bacteria, at least for agriculture. However, in Southeast Asia, such cyanobacteria
make the productivity of rice paddies about 10 times greater than that of other
types of farmland.

21–19 *Women planting rice in a paddy in Perak, Malaysia. In
Southeast Asia, rice can often be grown on the same land continuously without the addition of fertilizers because of nitrogen-
fixing cyanobacteria. The cyanobacteria are found not in the
water but rather in the tissues of a small water fern* (Azolla) *that
grows in the rice paddies.*

Two Unusual Photosynthetic Prokaryotes

Virtually all of the contemporary photosynthetic eubacteria can be readily identified as members of one of three lineages: cyanobacteria, green bacteria, or purple bacteria, all of which are gram-negative. Moreover, many nonphotosynthetic gram-negative bacteria, including *E. coli,* are thought to be descended from purple bacteria that lost the capacity for photosynthesis. The recent serendipitous discoveries of two photosynthetic prokaryotes with unique combinations of characteristics suggest the existence of other lineages.

The first of these unusual prokaryotes, discovered in 1975, exhibits some important similarities to the cyanobacteria. Its principal photosynthetic pigment is chlorophyll *a,* it contains the same types of carotenoids, and it lyses water during photosynthesis, releasing molecular oxygen. However, unlike the cyanobacteria, it contains no phycobilins. The principal accessory pigment is chlorophyll *b,* found elsewhere only in the chloroplasts of green algae and plants. As in the cyanobacteria, the photosynthetic membranes are not invaginations of the cell membrane but are instead separate structures. They consist, however, of two membranous sacs pressed together—a structure intermediate between the individual membranes of the cyanobacteria (Figure 4–9, page 92) and the stacks of thylakoids in the chloroplasts of the green algae and the plants (Figure 10–10, page 213). This constellation of characteristics has led to the name *Prochloron.* As we shall see in the next chapter, there is reason to believe that *Prochloron* is a living representative of an ancient lineage of prokaryotes that gave rise to the chloroplasts of the green algae and, ultimately, of the plants.

The second unusual prokaryote was discovered in 1981. An undergraduate at the University of Indiana made a mistake in a laboratory exercise designed to isolate photosynthetic bacteria from samples of pond water and soil, and none of the desired organisms multiplied in the experimental flasks. However, as the laboratory instructors were discarding the flasks at the end of the exercise, they noticed a green film on the bottom of one of the "botched" flasks. Happily for biological knowledge (and, undoubtedly for the student as well), they saved the contents and set about determining the source of the green film. It turned out to be a photosynthetic bacterium quite unlike any previously known. This organism, which has been named *Heliobacterium chlorum,* is a photosynthetic, nitrogen-fixing bacterium that is an obligate anaerobe. It contains a distinct type of bacteriochlorophyll that, unlike all other types of bacteriochlorophyll and chlorophyll, contains no oxygen atoms. This suggests that *Heliobacterium chlorum* is a representative of a lineage that evolved very early, well before the presence of even small amounts of free oxygen in the earth's atmosphere. Although *Heliobacterium chlorum* is gram-negative, analyses of its rRNA molecules indicate that it is most closely related to the group of gram-positive bacteria that includes *Clostridium.* This raises the possibility that the contemporary gram-positive bacteria (all of which are nonphotosynthetic) are the descendants of photosynthetic ancestors. And, where had the student obtained the soil sample that contained *Heliobacterium chlorum?* Right in front of the biology building.

Because of their nutritional independence, the cyanobacteria are able to colonize bare areas of rock and soil. A dramatic example of such colonization was seen on the island of Krakatoa in Indonesia, which was denuded of all visible plant life by a cataclysmic volcanic explosion in 1883. Filamentous cyanobacteria were the first living things to appear on the pumice and volcanic ash; within a few years they had formed a dark-green gelatinous growth. The layer of cyanobacteria eventually became thick enough to provide a substrate for the growth of plants. It is very probable that the cyanobacteria were similarly the first colonizers of land in the course of biological evolution.

A single cell of the genus Prochloron. *Analyses of* Prochloron *DNA and rRNA molecules indicate that its closest prokaryotic relatives are the cyanobacteria. The natural habitat of* Prochloron *cells is the body surface of sea squirts, soft-bodied animals that live along seashores and that belong to the lineage from which vertebrate animals arose.*

1 μm

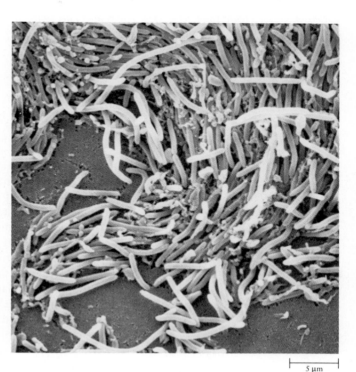

Scanning electron micrograph of Heliobacterium chlorum *cells. These rod-shaped cells form motile masses similar to those of the gliding bacteria. The cell membrane of* Heliobacterium *is not amplified and intruded into the cytoplasm, which suggests a photosynthetic process more primitive than that of the green and purple bacteria.*

5 μm

VIRUSES: DETACHED BITS OF GENETIC INFORMATION

Viruses do not fit easily into any of the kingdoms of living organisms. Because of their small size and infectious capacities, however, they have usually been studied with the prokaryotes. As we saw in Section 3, viruses consist of a nucleic acid core—either DNA or RNA—surrounded by a protein coat, or capsid. They reproduce only within living cells, commandeering the enzymes and other metabolic machinery of their hosts. Without this machinery, they are as inert as any other macromolecule, lifeless by most criteria. As our knowledge of viruses has

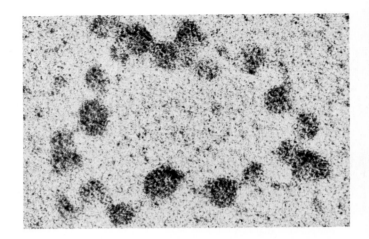

21-20 *One of the more striking evidences of the cellular origin of viruses is provided by the chromosome of simian virus 40 (SV40), shown here. As in the eukaryotic chromosome, its double-stranded DNA is complexed with histone proteins to form nucleosomes (see Figure 18-4, page 357). The histones are those of the host cell; SV40 not only uses host-cell machinery for its own synthetic activities but also directly incorporates products of the host cell's synthetic activities.*

grown, and particularly as the nucleotide sequences of increasing numbers of viruses have been determined—and shown to be very similar to sequences in the chromosomes of cellular organisms—biologists have come to regard viruses as cellular fragments that have set up a partially independent existence (Figure 21-20). Once this partial independence is gained, however, viruses appear to pursue their own evolutionary course, distinct, in many important aspects, from that of the cells from which they were originally derived.

In size, viruses range from about 17 nanometers (a hemoglobin molecule is 6.4 nanometers in diameter) to about 300 nanometers, larger than small bacteria. The larger ones are at the limits of resolution of the light microscope. Viruses can be characterized and classified on the basis of their usual host cells, their nucleic acid content (DNA or RNA, single-stranded or double-stranded), and their specific shapes, which are determined by the structure of the protein capsid (Figure 21-21). The proteins of the capsid may take the form of a helix, as in tobacco mosaic virus (see Figure 15-6, page 305) and influenza virus, or the form of triangular plates arranged in a polyhedron, as in adenovirus and the T-even bacteriophages. The capsid may be surrounded by additional layers or have other complex protein structures attached to it.

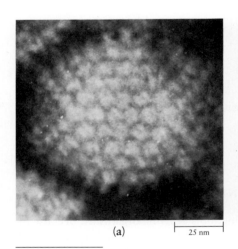

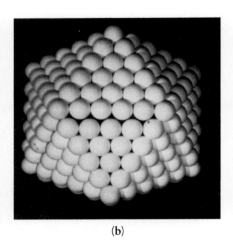

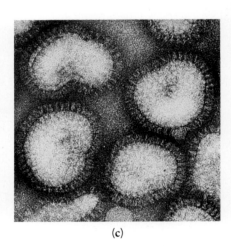

| (a) | 25 nm | (b) | (c) |

21-21 *Representative viral structures.* (a) *Adenovirus, one of the many viruses that cause colds in humans. This virus is an icosahedron. Each of its 20 sides is an equilateral triangle composed of identical protein subunits. Many viruses, and also Buckminster Fuller's geodesic domes, are constructed on this principle. There are 252 subunits in all. Within the icosahedron is a core of dou-* ble-stranded DNA. (b) *A model of the adenovirus, made up of 252 tennis balls.*

(c, d) *Electron micrograph and diagram of influenza virus. The virus is composed of a core of RNA surrounded by a helical protein capsid and a lipoprotein membrane through which protrude stubby protein spikes. The influenza virus, for reasons*

The proteins of the capsid determine the specificity of a virus; a cell can be infected by a virus only if viral protein can fit into one of the specific receptor sites in the cell membrane of that type of cell. Thus bacteriophages attack bacterial cells; tobacco mosaic virus infects leaf cells of the tobacco plant; adenoviruses and rhinoviruses, causes of the common cold, invade cells in the mucous membranes of the respiratory tract; and polio virus infects cells of the upper respiratory tract, the intestinal lining, and sometimes the nervous system. Apparently all types of cells—both prokaryotic and eukaryotic—are susceptible to infection by specific viruses capable of interacting with their membrane receptors.

In some virus infections, the protein coat is left outside the cell while the nucleic acid enters (see Figure 14–7, page 286); in others, the intact virus enters the cell, but once inside, the protein is destroyed by enzymes, freeing the viral nucleic acid. In the DNA viruses, the DNA of the virus replicates and is also transcribed into messenger RNA. The mRNA codes for viral enzymes, viral coat protein, and probably, at least in some cases, repressors and other regulatory chemicals. For its synthetic activities, the virus uses the equipment of the host cell, including ribosomes, transfer RNA molecules, amino acids, and nucleotides. Many viruses use host enzymes as well as those coded by their own nucleic acids, and some break up host DNA and recycle the nucleotides as viral DNA. In most RNA viruses, the viral RNA both replicates and serves directly as messenger RNA. In other RNA viruses, however, the viral RNA is transcribed into DNA, from which mRNA is later transcribed. This phenomenon of **reverse transcription** is characteristic of the cancer-causing viruses and the virus responsible for AIDS (acquired immune deficiency syndrome, to be discussed further in Chapter 39).

Virus particles are assembled within the host cell. In viruses with helical capsids, the protein subunits of the capsid come together around the newly synthesized nucleic acid. In other types of viruses, the capsid is formed separately, and then the nucleic acid is inserted into it, apparently drawn in by a protein functioning as an internal spool. Some viruses, such as influenza virus, code for additional proteins that are inserted into the membrane of the host cell; the newly formed viruses bud off from portions of the host cell membrane containing the viral proteins, and, in so doing, become wrapped in fragments of it (Figure 21–22).

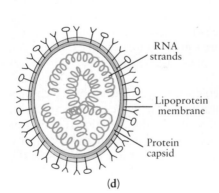

RNA strands

Lipoprotein membrane

Protein capsid

(d)

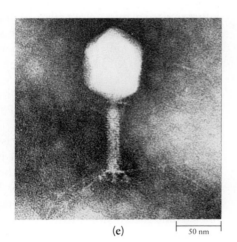

50 nm

(e)

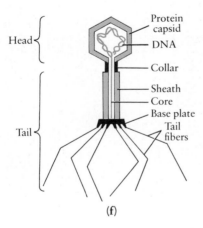

Head

Protein capsid

DNA

Collar

Sheath

Core

Base plate

Tail fibers

Tail

(f)

that are not understood, mutates frequently. The changes in its nucleic acid alter the proteins of the outer envelope and, hence, previously formed antibodies no longer "recognize" it. New strains of influenza viruses are likely to arise more rapidly than new vaccines can be produced to combat them.

(e, f) Electron micrograph and diagram of a T-even bacterio-

phage, showing its many different structural components. The DNA of the virus codes for all of the necessary proteins. The capsid head, the major structures of the tail, and the tail fibers are assembled separately. After the DNA has been inserted into the capsid head, the pre-formed tail assembly is attached to it. Addition of the tail fibers completes the viral particle.

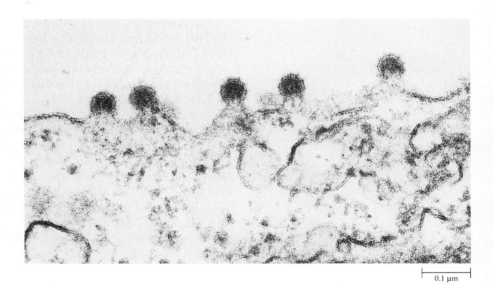

0.1 μm

21–22 *In this electron micrograph, particles of Semliki Forest virus are budding from the cell membrane of an animal cell, acquiring a lipoprotein envelope in the process. Semliki Forest virus is named for the rain forest in Uganda where it was first isolated from the tissues of mosquitoes. It infects a variety of animals and is related to the virus that causes yellow fever.*

When assembly of the viral particles is completed, they are released from the host cell, often lysing its membrane in the process. Each new viral particle is capable of setting up a new infection cycle in an uninfected cell.

Viroids and Prions: The Ultimate in Simplicity

For many years, the agents causing certain serious diseases of plants and animals remained mysterious. These diseases exhibit the characteristics of viral infection and yet no viruses could be identified. Recent studies have now demonstrated that even simpler infectious agents are at work: **viroids,** which are naked RNA molecules found in plants, and **prions,** which are small proteinaceous particles found in animals. Both viroids and prions replicate in susceptible cells.

The first viroid to be characterized, known as PSTV, is the causative agent of potato spindle-tuber disease; in this disease, the plant produces elongated, gnarled potatoes (tubers) that sometimes have deep crevices in the surface. PSTV can also infect tomato plants—close relatives of potato plants—producing stunted growth and twisted leaves. PSTV is a single-stranded RNA molecule containing 359 nucleotides; the molecule can be either linear or a closed circle. The linear form folds back on itself in a hairpin-shaped structure, whereas the circular form is flattened. In each case, complementary base pairs are joined by hydrogen bonds, resulting in a double-stranded RNA structure similar to that of DNA. Under the electron microscope, both forms of PSTV appear as rods about 50 nanometers long (Figure 21–23).

The way in which viroids replicate is an enigma, as is the means by which they cause disease. They are found almost exclusively in the nuclei of infected cells, and experiments have shown that the RNA of viroids does not function as messenger RNA. Unlike the DNA or RNA of viruses, it is not translated into enzymes that participate in its own replication. The location of viroids in the nucleus and their inability to act as messenger RNAs has led to the hypothesis that they cause their symptoms by interfering with gene regulation in the infected host cells. This hypothesis is supported by the fact that certain plant proteins, found in healthy cells, are present in significantly larger quantities in infected cells. New evidence suggests that the interference may be at the stage of mRNA editing, particularly the removal of introns and the splicing together of exons (see page 369). All viroids studied thus far share important similarities with a group of introns that includes "self-splicing" members—that is, introns that can remove

21–23 *The short, rodlike structures indicated by arrows in this electron micrograph are potato spindle-tuber viroids. The much larger, looping structure is a portion of the DNA molecule from a bacteriophage, included to show the tremendous difference in size between a viroid and a virus. Viroids, unlike viruses, have no protein coat. Other viroids have now been identified that cause diseases in citrus trees, cucumbers, and chrysanthemums.*

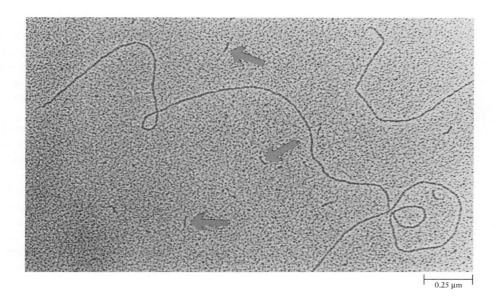

0.25 μm

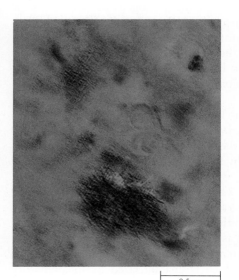

0.5 μm

21–24 *A collection of prions in a section of the brain of a hamster infected with scrapie. This section is from a region of the brain near the hippocampus, a structure associated with memory (see Chapter 43). Prions are known to be the infectious agent of Creutzfeldt-Jakob disease, a degenerative neurological disease of humans, but whether they play a role in Alzheimer's disease remains an open question.*

themselves from an RNA molecule, splicing together the exons, without the assistance of any enzymes. Viroids have certain nucleotide sequences that are identical to sequences found in these introns; among the identical segments are nucleotide sequences that play a major role in determining the configuration of the excised intron. Are viroids actually "escaped" introns that have established a partially independent existence, analogous to that of viruses? The present evidence suggests that this may be the case; it also suggests that excised introns may have previously unsuspected functions in the cell.

The discovery of viroids in plant cells led to the suspicion that they were also the cause of scrapie, an infectious neurological disease in sheep that is similar in its symptoms to several devastating human diseases, including Alzheimer's disease. The infectious agent of scrapie, however, has now been identified as a prion—a protein particle not associated with any detectable nucleic acid (Figure 21–24). Amino acid sequencing of the prion has made it possible to develop DNA probes to search for its gene. Although no gene with a corresponding sequence has been identified in any purified preparation of the infectious agent itself, such a gene has been located in the chromosomes of both infected and healthy animals. Moreover, mRNA transcribed from the gene has been identified in the cells of both types of animals. The normal function of this gene and its product, the effect of prions on either the gene or its product, and the way in which the prions are replicated remain totally mysterious—but perhaps not for long, given the intense interest in the human diseases similar to scrapie.

MICROORGANISMS AND HUMAN ECOLOGY

Symbiosis

Symbiosis ("living together") is a close and long-term association between organisms of different species. Although there is some disagreement as to precisely what constitutes a symbiotic relationship, and the details of the relationship between two closely associated species are often difficult to determine, symbiotic relationships are generally considered to be of three kinds. If the relationship is beneficial to both species, it is called **mutualism.** If one species benefits from the association while the other is neither harmed nor benefited, it is called **commensalism.** If one species benefits and the other is harmed, the relationship is known as **parasitism.**

An enormous variety of microorganisms live symbiotically with human beings, but the lines of demarcation between the different categories of symbiosis are not clear-cut. A bacterium such as *E. coli,* which lives in the lower intestine, may be commensal, depending on its host for food and shelter and neither helping nor harming. If it produces a needed vitamin or digestive enzyme, the relationship is mutualistic. If it gets into the bloodstream and causes septicemia (blood poisoning), it is a parasite and a pathogen.

Evolutionary success is measured in terms of surviving progeny. A symbiotic microorganism that destroys its host before the reproduction and dispersal of its progeny to new hosts is less likely to be successful by the evolutionary criterion than one that enjoys a long and comfortable relationship with its protector. Thus disease is generally the result of a sudden change in the microorganism, in the host, or in their relationship. For instance, many people harbor small numbers of *Mycobacterium tuberculosis* without any symptoms of disease; however, factors such as malnutrition, fatigue, or other diseases may weaken host defenses so that the signs of tuberculosis appear. Similarly, the herpes simplex viruses that cause fever blisters or genital lesions may remain latent for months or years at a time, with an outbreak occurring only in response to some change in the condition of the host.

How Microbes Cause Disease

The pathogenic effects of microbes are produced in a variety of ways. Viruses, as we have seen, enter particular types of cells and often destroy them. Bacteria may produce cell destruction also. Frequently, however, the effects we recognize as disease are caused not by the direct action of the pathogens but by toxins, or poisons, produced by them. For instance, diphtheria is caused by a bacillus, *Corynebacterium diphtheriae.* The organisms are inhaled and establish infection in the upper respiratory tract, where they produce a powerful toxin that is transported through the bloodstream to body cells. This toxin, which is made only when the bacterium is harboring a particular prophage (page 331), inhibits protein synthesis.

Some diseases are the result of the body's reaction to the pathogen. In pneumonia caused by *Streptococcus pneumoniae,* the infection causes a tremendous outpouring of fluid and cells into the air sacs of the lungs, thus interfering with breathing. The symptoms caused by fungus infections of the skin similarly result from inflammatory responses.

A single disease agent can cause a variety of diseases. Skin infections of *Streptococcus pyogenes* cause the disease known as impetigo. Throat infections by the same bacterium are the familiar strep throat. Throat infections with strains of the bacterium that produce toxins (again as a result of a bacteriophage) are known as scarlet fever. Among persons with untreated strep throat or scarlet fever, about 0.5 percent develop rheumatic fever, which is characterized by inflammatory changes in the joints, heart, and other tissues, apparently as a result of reactions involving the body's own immune system. Conversely, many agents may cause the same symptoms; the "common cold" can result from infection with any one of a large number of viruses.

Prevention and Control of Infectious Disease

Although microorganisms were first seen and depicted with remarkable accuracy by Antony van Leeuwenhoek in the late seventeenth century, they were not generally recognized as a cause of disease until about 100 years ago. This recognition opened the way to control measures, among the most important of which was the introduction of sterile procedures in hospitals. Even more important than improvement in medical practices was the institution of public health

21–25 *Until the 1950s, poliomyelitis was one of the most feared of all childhood diseases. It was brought under control by immunization, initially with a killed-virus vaccine (Salk vaccine) that was subsequently replaced by a live-virus vaccine (Sabin vaccine), the one now in wide general use. The possible aftereffects of polio are depicted in this Egyptian hieroglyph, dated at approximately 1400 B.C.*

measures—for example, the eradication of fleas, lice, mosquitoes, and other agents that carry disease; disposal of sewage and other wastes; protection of public water supplies; pasteurization of milk; and quarantine. Not long ago, for instance, infant mortality during the first years of life was often as high as 50 percent in some localities owing to infant diarrhea caused by contaminated milk and water; in some regions in the developing countries, it still remains quite high.

Many infectious diseases, both bacterial and viral, can be prevented by immunization (to be discussed in Chapter 39). Bacteria, in particular, are also susceptible to antimicrobial drugs, such as sulfa and penicillin, for which the battlefields of World War II were the proving grounds. Penicillin, which is synthesized by the fungus *Penicillium,* was the first known antibiotic—by definition, a chemical that is produced by a living organism and is capable of inhibiting the growth of microorganisms. Many antibiotics are produced by bacteria, especially the actinomycetes; some are formed by fungi. Many, including penicillin, can now be synthesized in the laboratory. Antibiotics and other chemotherapeutic agents are effective because they interfere with some essential process of the pathogen without affecting the cells of the host.

Antimicrobial drugs made possible not only the treatment of battlefield wounds and of common infectious diseases but also the widespread and often life-saving use of major surgery as a treatment for cancer and other diseases. The tremendous decrease in deaths from infectious disease over the last decades is a chief cause of the present population explosion. Ironically, with the advent of strains of bacteria resistant to these drugs (page 329) and endemic in hospitals, the latter are once more becoming reservoirs of serious bacterial diseases.

With a few exceptions, viruses have been impervious to attack by chemotherapeutic agents; drugs that successfully disrupt viral replication generally have devastating effects on cellular processes. However, recent advances in determining the three-dimensional structure of macromolecules are providing the first

(a)

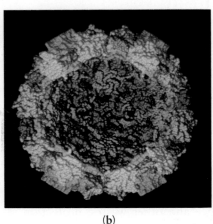

(b)

21–26 *Computer-generated images of the protein capsid of type I polio virus, as seen (a) from the outer surface and (b) in a cut-away view. The capsid is formed from four protein subunits, each of which is repeated 60 times. Three of the subunits are visible on the exterior surface, and one of them occurs only on the inner surface of the hollow capsid. The structure of the capsid of a human rhinovirus was also determined recently, and it is likely that the details of other capsid structures will follow in the near future.*

realistic hopes for the development of antiviral agents. Membrane receptors and viral capsids (Figure 21–26) are beginning to yield the structural secrets that explain their interactions, raising the possibility of devising molecules that will block either the relevant receptors or the portions of the capsid that fit into them.

New approaches to the control of both bacterial and viral diseases are also likely to emerge from the wealth of information being generated by recombinant DNA studies. Almost as a by-product, these studies are shedding new light on disease processes, for example, by identifying bacterial and viral genes that affect virulence. If previous patterns of biological history hold, it is reasonable to expect that as molecular biologists learn more about the organisms themselves—and about the genes dictating their properties—medical scientists will apply that knowledge in practical ways.

SUMMARY

The prokaryotes are both the most ancient group of organisms and the most abundant. The application of molecular techniques is enabling biologists to begin determining the phylogenetic relationships in this diverse kingdom. Two distinct lineages of prokaryotes, the Archaebacteria and the Eubacteria, have been identified. Among the major lineages of Eubacteria are green bacteria, purple bacteria and related forms, spirochetes, cyanobacteria, and gram-positive bacteria.

The prokaryotes are the only organisms with cells in which the DNA is not associated with histones and in which there is no membrane-bound nucleus or membrane-bound organelles. Their cell membrane is formed from a lipid bilayer that does not contain cholesterol and that often incorporates enzyme systems such as—in aerobic forms—the electron transport system. The varied forms of prokaryotes are imparted by their cell walls, which may be rigid or flexible. In gram-positive bacteria, the cell wall consists of a relatively thick layer of polysaccharides and complex polymers known as peptidoglycans; in gram-negative bacteria, a thin layer of peptidoglycans is enclosed by an outer membrane, similar in composition to the cell membrane. A polysaccharide capsule, surrounding the cell wall, is also present in many strains of bacteria.

Many species have long, slender extensions, flagella and pili, extending from the surface of the cell. Flagella are locomotor structures, whereas pili serve to attach the cells to surfaces or to each other. Bacterial flagella, which have a complex structure quite different from that of eukaryotic flagella, propel the cell through the medium by a rotary movement. In spirochetes, the flagella are modified to form an axial filament between the cell wall and the cell membrane; rotation of this filament imparts a corkscrew motion to the cells. Nonflagellated forms are nonmotile or move by gliding on the surface of a secreted mucus or slime.

Reproduction of prokaryotes is usually by binary fission, but some forms reproduce by budding or fragmentation. Genetic variability is introduced by mutation and by genetic exchanges and recombinations. Many types of prokaryotes form tough, resistant spores.

Prokaryotes comprise both heterotrophs and autotrophs; the autotrophs include both chemosynthetic and photosynthetic forms. Chemosynthetic autotrophs are found only among prokaryotes. The photosynthetic green and purple bacteria use a variety of substances, including sulfur compounds, as electron donors and do not produce oxygen gas. The cyanobacteria have a type of photosynthesis essentially like that in plants. Photosynthetic prokaryotes lack chloroplasts, and their photosynthetic pigments are embedded in the cell membrane or in internal membranes.

Viruses consist of nucleic acid (DNA or RNA) enclosed in a protein capsid.

They are now thought to be cellular fragments that have set up a partially independent existence. Viral reproduction can occur only within a host cell; the nucleic acid of the virus replicates and directs the formation of new protein capsids, utilizing the host cell's enzymes and other metabolic equipment. Infectious agents even simpler than viruses have recently been isolated and identified: viroids (small molecules of RNA with which no protein is associated) and prions (proteinaceous particles with which no nucleic acid is associated). Viroids are the causative agent of certain plant diseases, and prions transmit scrapie, a disease of sheep. The mechanisms by which viroids and prions exert their pathogenic effects remain unknown.

Microorganisms and human beings live in a multitude of symbiotic relationships, which may take the form of (1) commensalism, in which one partner benefits and the other is neither harmed nor benefited, (2) mutualism, in which both benefit, and (3) parasitism, in which one partner benefits and the other is harmed. A symbiotic relationship may change from one type to another as a result of changes in the host or, sometimes, in the microorganism. Pathogenic microorganisms may cause disease by direct tissue destruction, by producing toxins, and by provoking host defenses.

QUESTIONS

1. Distinguish among the following: Archaebacteria/Eubacteria; gram-positive/gram-negative; eukaryotic flagella/bacterial flagella/pili; axial filament/fibril; pathogen/toxin; virus/viroid/prion.

2. Label the drawings below.

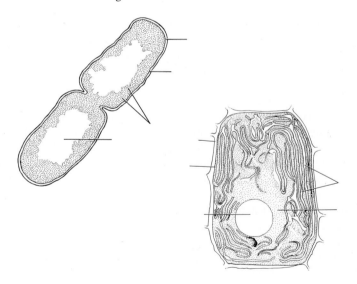

3. Identify the types of bacteria shown in the drawings below.

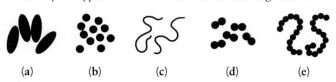

(a)   (b)   (c)   (d)   (e)

4. Many microorganisms produce antibiotics. What do you think their function might be for the organisms that produce them?

5. Five types of Eubacteria are photosynthetic: the green sulfur bacteria, the green nonsulfur bacteria, the purple sulfur bacteria, the purple nonsulfur bacteria, and the cyanobacteria. Describe the similarities and differences among these groups of organisms and in their photosynthetic processes.

6. In addition to the photosynthetic Eubacteria, a few genera of salt-loving prokaryotes are photosynthetic. How do these cells differ from the other photosynthetic prokaryotes? (In thinking about this question, you may wish to review the essay on page 221 as well as material in this chapter.)

7. Before the structure of any virus was known, Crick and Watson predicted that the protein coats of viruses would prove to be made up of large numbers of identical subunits. Can you explain the basis of their prediction?

8. Some biologists consider viruses to be living organisms. By what criteria might viruses be considered alive?

9. Which of the three types of symbiotic association would, in your opinion, apply to the following relationships: dog and flea; human being and dog; maggot and wound; *E. coli* and human being; diphtherial microorganism and human being?

10. Prokaryotes are considered more primitive than eukaryotes. Does this mean they are all identical to forms of life that existed before eukaryotes arose? Explain your answer.

C H A P T E R **22**

The Protists

100 μm

22–1 Stentor, *a ciliated protist.*
*Extended, it looks like a trumpet (this
genus is named after "bronze-voiced"
Stentor of the* Iliad, *"who could cry out in
as great a voice as 50 other men"). Stentor
is crowned with a wreath of membranelles. These beat in rhythm, creating a
powerful vortex that draws edible particles up to and into the funnel-like oral
groove leading to the buccal cavity
("mouth"). Elastic protein threads, myonemes ("little muscles"), run the length of
the cell. When contracted,* Stentor *is an
almost perfect sphere.*

The protists are all eukaryotic. The majority of protist species are unicellular, and
those that are not have relatively simple multicellular or multinucleate structures.
Apart from these two characteristics, they are an extremely diverse group of
organisms. The kingdom Protista includes heterotrophs, among them a number of
parasitic forms; photosynthetic autotrophs; and, as you will see, a few versatile
organisms that are both heterotrophic and photosynthetic.

Protists are far from simple. Indeed, among one group, the ciliates, are found
probably the most complex of all cells, with an astonishing variety of highly
specialized structures (Figure 22–1). That single cells should be so complicated is
actually not surprising. Among the protists, each cell is a self-sufficient organism,
as capable of meeting all of the requirements of living as any multicellular plant or
animal.

THE EVOLUTION OF THE PROTISTS

The microfossil record indicates that the first eukaryotes evolved at least 1.5
billion years ago. Eukaryotes, as we saw in Section 1, are distinguished from
prokaryotes by their larger size, the separation of nucleus from cytoplasm by a
nuclear envelope, the association of the DNA with histone proteins and its
organization into a number of distinct chromosomes, and complex organelles,
among which are chloroplasts and mitochondria.

The step from the prokaryotes to the first eukaryotes (the protists) was one of
the major evolutionary transitions, second in importance only to the origin of life
itself. The question of how it came about is a matter of current and lively
discussion. One interesting hypothesis, gaining increasing acceptance, is that
larger, more complex cells evolved when certain prokaryotes took up residence
inside other cells.

As we noted previously, oxygen began slowly to accumulate in the atmosphere
about 2.5 billion years ago, as a result of the photosynthetic activity of the
cyanobacteria. Those prokaryotes that were able to use oxygen in ATP production gained a strong advantage, and so such forms began to prosper and increase.
Some of these cells evolved into modern forms of aerobic bacteria. Others,
according to this hypothesis, became symbionts within larger cells and evolved
into mitochondria.

Several lines of evidence support the idea that mitochondria are descended
from specialized bacteria. Mitochondria contain their own DNA, and this DNA is
present in a single, continuous ("circular") molecule, like the DNA of bacteria.
Many of the enzymes contained in the cell membranes of bacteria are found in
mitochondrial membranes. Mitochondria contain ribosomes that resemble those
of bacteria both in their small size and in details of their chemical composition,
including some of the nucleotide sequences in the constituent rRNA molecules.

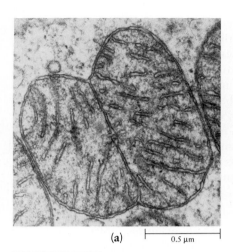

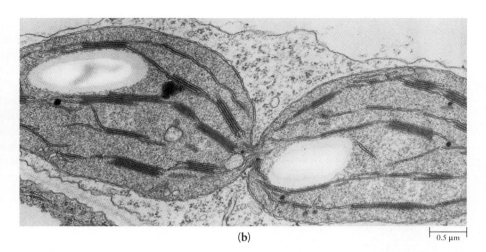

(a) 0.5 μm

(b) 0.5 μm

22–2 (a) Mitochondria and (b) chloroplasts may have originated as symbiotic prokaryotes. This hypothesis is supported by the facts that both contain their own DNA and both, as shown here, replicate by binary fission.

Further, mitochondria appear to be produced only by other mitochondria, which divide within their host cell (Figure 22–2). However, to make the situation more complex, there is not nearly enough DNA in mitochondria to code for all the mitochondrial proteins. Host-cell DNA is also required for the synthesis of mitochondrial enzymes and structural proteins.

It is thought that mitochondria are most likely derived from a lineage of purple nonsulfur bacteria in which the capacity for photosynthesis had been lost. Little is known, however, about the original cells in which these bacteria first set up housekeeping—or, indeed, if they actually existed. Contemporary prokaryotes provide few clues about the origins of two key features of all eukaryotic cells—a nuclear envelope and chromosomes containing both DNA and histone proteins. On the basis of nucleotide sequences in transfer and ribosomal RNAs, it has been suggested that very early in the history of life a common ancestral lineage gave rise not only to the two lineages (Archaebacteria and Eubacteria) shown in Figure 21–3 (page 430) but also to a third lineage. Members of this third lineage, which have been dubbed the "urkaryotes," are hypothesized to have been the host cells in the evolution of the eukaryotes. If such cells existed, they probably had no means of using oxygen for cellular respiration. Thus they were dependent entirely on glycolysis and fermentation, anaerobic processes that are, as we have seen (page 190), relatively inefficient. Cells with oxygen-utilizing respiratory assistants would have been more efficient than those lacking them and so would have reproduced at a faster rate.

In an analogous fashion, photosynthetic prokaryotes ingested by larger, nonphotosynthetic cells are believed to be the forerunners of chloroplasts. Such symbioses are thought to have occurred independently in a number of lineages, giving rise to the various groups of modern photosynthetic eukaryotes.

The endosymbiotic hypothesis accounts for the presence in eukaryotic cells of complex organelles not found in the far simpler prokaryotes. It gains support from the fact that many modern organisms contain intracellular symbiotic bacteria, cyanobacteria, or photosynthetic protists, indicating that such associations are not difficult to establish and maintain. Different species of cyanobacteria, for example, are found living within other cyanobacteria—and within protists, fungi, plants, and animals. These symbiotic cyanobacteria typically lack cell walls and are, functionally, chloroplasts. Moreover, they divide at the same time as the host cell, by a process similar to chloroplast division. A number of species of photosynthetic protists also form symbiotic associations with invertebrate animals (Figure 22–3). In such symbioses, whether ancient or modern, the smaller cells gain nutrients and protection, and the larger cells are given a new energy source.

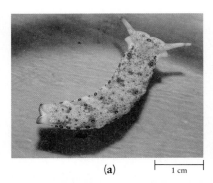

(a)

1 cm

(b)

22–3 Plakobranchus, *a marine mollusk. The tissues of this animal contain chloroplasts, which it obtains by eating certain green algae.*

(a) Ordinarily, structures called parapodia are folded over the animal's back, hiding the chloroplast-containing tissues. (b) When the parapodia are spread apart, however, the deep green tissues become visible. The chloroplasts carry on photosynthesis so efficiently that within a 24-hour cycle of light and darkness some individuals produce more oxygen than they consume.

The greater complexity of the eukaryotic cell brought with it a number of advantages that ultimately made possible the evolution of multicellular organisms. The eukaryotic cell is capable of carrying vastly more genetic information than the prokaryotic cell—enough, for example, to specify an oak tree or a human being. Because of the compartmentalization of functions by membranes, eukaryotic cells are more efficient metabolically and can be larger. Although a larger cell requires more energy, it can obtain it more easily, either because of an increased surface area for photosynthesis or absorption of food molecules or because of a greater ability to catch and subdue prey organisms. Moreover, larger cells are generally better able to make necessary adjustments to potentially life-threatening environmental changes, for example, in temperature or available water.

Classification of Protists

Most of the modern protists probably bear little or no resemblance to the first eukaryotes. As their complexity reminds us, they are not beginnings. They, like the modern prokaryotes, are the end products of millions of years of evolution, and in that time, they have undergone tremendous diversification. Those that have not changed much—amoebas might be an example—have survived over the millennia because they are exquisitely adapted to the relatively unchanging environment in which they live. Biologists generally agree (1) that the modern protists represent a number of quite different evolutionary lineages, and (2) that all other eukaryotic organisms—fungi, plants, and animals—originated from primitive protists (Figure 22–4).

The classification of the protists, like that of the prokaryotes, is currently undergoing a major upheaval, as the electron microscope and modern biochemical and molecular techniques make available a vast amount of new information about these organisms. Such questions as the number of divisions or phyla constituting the kingdom, the relationships among the various divisions and phyla, the proper placement of particular species, and the criteria to be used in making these decisions are all topics of discussion and, in some cases, of heated

22–4 *A simplified scheme of the possible evolutionary relationships among the major groups of organisms. According to current hypotheses, the fungi and different groups of modern protists evolved separately from single-celled eukaryotes. The origins of the plants have been traced to one group of photosynthetic protists, the Chlorophyta, or green algae. The animals are presumed to have arisen from a single-celled eukaryotic heterotroph, although direct evidence is lacking.*

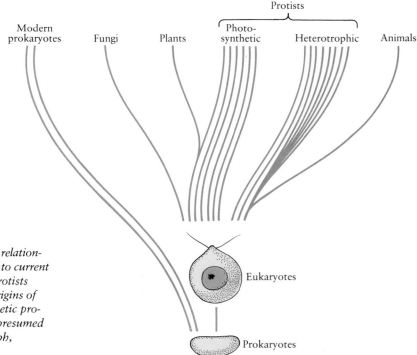

TABLE 22–1 **The Kingdom Protista**

Photosynthetic Autotrophs

Division Euglenophyta	*Euglena* and related algae, all unicellular, mostly freshwater
Division Chrysophyta	Diatoms and related algae, all unicellular, marine and freshwater
Division Dinoflagellata	Dinoflagellates, all unicellular, mostly marine
Division Chlorophyta	Green algae, including unicellular, colonial, and multicellular forms; freshwater and marine
Division Phaeophyta	Brown algae, including the kelps; all multicellular; almost all marine
Division Rhodophyta	Red algae, all multicellular, mostly marine

Heterotrophs

Division Myxomycota	Plasmodial slime molds
Division Acrasiomycota	Cellular slime molds
Division Chytridiomycota	Water molds (chytrids)
Division Oomycota	Water molds (oomycetes)
Phylum Mastigophora	Flagellates, parasitic or free-living
Phylum Sarcodina	Amoeba-like organisms, free-living or parasitic
Phylum Ciliophora	Ciliates, free-living or parasitic
Phylum Opalinida	Opalinids, intestinal parasites of lower vertebrates
Phylum Sporozoa	Sporozoans, parasitic

controversy. At present, no comprehensive classification of kingdom Protista exists that is widely accepted by biologists concerned with these organisms. For our purposes, we shall consider 15 principal divisions and phyla, which are grouped informally in Table 22–1 on the basis of their mode of nutrition.

In the past, it was common to regard the photosynthetic protists (the algae) as "lower plants" and the unicellular heterotrophs (the protozoa) as "lower animals." Similarly, the slime molds and water molds (unusual heterotrophic organisms that resemble the fungi in some features but differ from them in others) were considered "lower fungi." However, it is now increasingly apparent that, with one notable exception, the contemporary protists represent lineages that are not closely related to the lineages that gave rise to the members of the three multicellular kingdoms. (The notable exception is to be found among the green algae of division Chlorophyta, from which the plants are thought to have arisen.) Moreover, as we shall see, certain divisions of algae appear to be more closely related to certain phyla of protozoa than to other divisions of algae, and vice versa. Thus "algae" and "protozoa" have been abandoned as formal terms in modern classification; they remain, however, convenient informal terms and will be used as such in this text.

PHOTOSYNTHETIC AUTOTROPHS

By the time of the earliest known eukaryotes—about 1.5 billion years ago—a number of distinct lines of simple photosynthetic eukaryotes had already evolved. The approximately 30,000 described species now in existence can be grouped into six divisions. Three of these divisions (Euglenophyta, Chrysophyta, and Dinoflagellata) consist almost entirely of unicellular organisms. The other three divisions (Chlorophyta, Rhodophyta, and Phaeophyta) include groups that are multicellular.

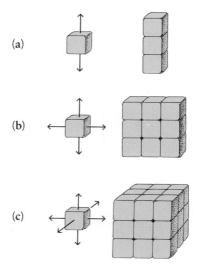

22–5 *Different planes of cell division can produce different patterns of growth, all represented among the algae. (**a**) Cell divisions in one plane produce a filament. (**b**) Divisions in two planes produce a one-celled layer. (**c**) Divisions in three planes produce a three-dimensional solid body. The arrows indicate the direction of growth.*

22–6 *Rocks along the coast of Cornwall, in Britain, showing the zonation of algal species from lower levels to higher levels. Zonation is based on a variety of factors, including not only light requirements and resistance to desiccation but also predation and competition.*

Nearly all members of these divisions are photosynthetic. In general, they have a relatively simple structure, which may be a single cell, a filament of cells, a plate of cells, or a solid body that may begin to approach the complexity of a plant body (Figure 22–5).

The unicellular algae are usually found floating near the surface of the oceans and inland waters, where light is abundant. Each cell is a totally self-sufficient individual, dependent only on light from the sun and carbon dioxide and minerals from the water that surrounds it. These cells may live in colonies, and sometimes the cells of a single colony may be attached to one another as the gliding bacteria, for instance, are held together by sticky secretions; occasionally there is even a division of labor among the cells. Together with small invertebrates and immature forms of larger animals they form the **plankton** (from *planktos,* the Greek word for "wanderer"). The photosynthetic members of the plankton community—sometimes called the **phytoplankton**—carry out most of the photosynthesis that takes place in the oceans. They are the ultimate energy source for most of the oceans' other inhabitants and, as a by-product of their photosynthetic activities, contribute enormous amounts of oxygen to the atmosphere.

Most of the multicellular algae are adapted to living in shallow waters and along the shores. Here the waters are usually rich in nutrients, washed down from the land or swept up in currents from the deeper waters. Living conditions are otherwise difficult, however. Along a rocky shore, for instance, multicellular algae—the seaweeds—are subject to great fluctuations of humidity, temperature, salinity, and light; to the pounding of the surf; and to the abrasive action of sand particles churned up by the waves. Life close to the shore is also crowded and competitive. Under these pressures, different groups of algae have become specialists at exploiting particular areas along tidal shores. The result is the zonation of life forms visible in many coastal areas (Figure 22–6).

Characteristics of the Photosynthetic Protists

The six divisions of algae vary widely in their biochemical characteristics, especially pigmentation, nature of stored food reserves, and cell-wall components, and in the number and position of their flagella, when present (Table 22–2). Typically, the cell walls of algae have a cellulose matrix, often with massive amounts of other polysaccharides that give certain algae a mucilaginous consistency.

The names of some divisions are derived from the colors of the predominant accessory pigments, which mask the bright green of the chlorophylls. A wide variety of carotenoids is found in the chloroplasts of algae. The xanthophylls are yellowish-brown carotenoids; the xanthophyll fucoxanthin gives the brown algae their characteristic color and name. It is also found in the golden-brown algae and diatoms (division Chrysophyta). The red algae (Rhodophyta) owe their colors to several kinds of phycobilins, accessory pigments that are also characteristic of the cyanobacteria. In the green algae the color of the chlorophylls is usually not masked by accessory pigments.

The wide variety of pigments found in the chloroplasts of the various divisions of algae suggests that (1) different types of oxygen-producing prokaryotes were in existence prior to the development of eukaryotic cells, and (2) the various divisions of algae may have evolved as a result of the establishment of symbiotic relationships with these different photosynthetic prokaryotes, which then evolved into modern chloroplasts. The chloroplasts of the red algae (Rhodophyta) are clearly derived from primitive cyanobacteria. Similarly, the chloroplasts of the Chlorophyta and Euglenophyta are thought to be derived from a prokaryotic lineage represented by the newly discovered *Prochloron* (page 442). It has also been suggested that the chloroplasts of the Chrysophyta, Dinoflagellata, and Phaeophyta may be derived from a lineage represented by *Heliobacterium chlorum* (page 442), although many authorities regard this as highly unlikely.

TABLE 22-2 Comparative Summary of Characteristics of the Photosynthetic Protists

DIVISION	NUMBER OF SPECIES	PHOTOSYNTHETIC PIGMENTS	FOOD RESERVE	FLAGELLA	CELL WALL COMPONENT	REMARKS
Euglenophyta (euglenoids)	1,000	Chlorophylls *a* and *b*, carotenoids	Paramylon and fats	1, 2, or 3 per cell, apical (at end of cell)	No cell wall; have a proteinaceous pellicle	Mostly freshwater; sexual reproduction unknown
Chrysophyta (diatoms, golden-brown algae, and yellow-green algae)	13,000	Chlorophylls *a* and *c*, carotenoids, including fucoxanthin	Laminarin and oils	None, 1, or 2, apical, equal or unequal	Cellulose, frequently impregnated with silica; cell wall lacking in some	Marine and freshwater
Dinoflagellata (dinoflagellates)	2,000	Chlorophylls *a* and *c*, carotenoids	Starch and oils	None or 2, lateral	Cellulose	Mostly marine; sexual reproduction rare
Chlorophyta (green algae)	9,000	Chlorophylls *a* and *b*, carotenoids	Starch	Usually 2 per cell, identical	Polysaccharides, cellulose in some	Freshwater and marine
Phaeophyta (brown algae)	1,500	Chlorophylls *a* and *c*, carotenoids, including fucoxanthin	Laminarin and oils	2, lateral, in reproductive cells only	Cellulose and algin	Almost all marine, flourish in cold ocean waters
Rhodophyta (red algae)	4,000	Chlorophyll *a*, carotenoids, phycobilins	Floridean starch	None	Cellulose, pectin compounds, impregnated with calcium carbonate in some	Mostly marine; complex sexual cycle; many species tropical

A diversity of storage products is found among the algae, with most having distinctive carbohydrate food reserves and many containing lipids as well. The green algae and dinoflagellates store their carbohydrates as starch, as do the plants. In the brown algae, golden-brown algae, and diatoms, another glucose polymer, laminarin (Figure 22–7), takes the place of starch. The carbohydrate reserves of the red algae are biochemically similar to starch.

When algal cells divide, the cell membrane generally pinches inward from the margin of the cell (furrowing), just as in heterotrophic protists, fungi, and animals (see page 153). Cell plates, like those of plants (also on page 153), are formed during cell division only in one brown alga and a few genera of filamentous green algae. Unlike plant cells, most algal cells—except those of the red algae—have centrioles.

22-7 *Laminarin, the principal storage product in the protists of divisions Chrysophyta and Phaeophyta. Like starch, it is made up of glucose residues, but unlike starch, there are only about 15 to 30 glucose units per molecule, and their linkage is different (see page 65).*

Laminarin

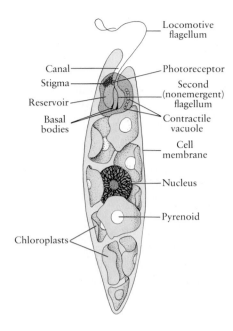

Locomotive
flagellum

Canal
Stigma
Reservoir
Basal
bodies

Photoreceptor
Second
(nonemergent)
flagellum
Contractile
vacuole
Cell
membrane

Nucleus

Pyrenoid

Chloroplasts

22–8 *Euglena is one of the most versatile of all one-celled organisms. Containing numerous chloroplasts, it is photosynthetic, but it can also absorb organic nutrients from the surrounding medium and can live without light. The pyrenoid is believed to be a food storage body. Euglena is pulled through the water by the whiplike motion of the single locomotive flagellum.*

Division Euglenophyta: Euglenoids

Euglenophyta are a small group of unicellular organisms (about 1,000 species), most of which are found in fresh water. They are named for the genus *Euglena*, the most common member of the group (Figure 22–8). About a third of the approximately 40 genera of euglenoids contain chloroplasts. These chloroplasts, which contain chlorophylls *a* and *b*, are very similar to those of the green algae but are enclosed by a triple—rather than a double—membrane. This suggests that euglenoids originally acquired chloroplasts by ingesting green algal cells; over long periods of time, most elements of the symbiotic green algae were lost, ultimately leaving only the chloroplast and the cell membrane. An alternative hypothesis is that euglenoid chloroplasts were derived independently from symbiotic photosynthetic prokaryotes resembling *Prochloron*. The nonphotosynthetic euglenoids are frequently able to absorb dissolved organic substances, and certain forms can ingest living prey, including other euglenoids. Euglenoids store their food as paramylon, an unusual polysaccharide that is found almost exclusively in this group.

Euglenoid cells are complex. *Euglena* is characteristically an elongated cell with a nucleus, two flagella (one of which is short and inactive), and numerous small emerald-green chloroplasts that give the cell its bright color. The flagella are attached at the base of a flask-shaped opening, the reservoir, at the anterior end of the cell. Emptying into the reservoir is the contractile vacuole, which collects excess water from all parts of the cell. Contractile vacuoles, which are found in flagellated cells of other algal groups as well, are also common features among the ciliates and other protozoa (see page 133). Euglenoid cells lack a cell wall but have a flexible series of protein strips, which make up the pellicle, inside the cell membrane (Figure 22–9). Unlike the stiff wall of other algal cells, the flexible pellicle permits *Euglena* to change its shape, providing an alternative form of locomotion—wriggling—for mud-dwelling forms.

If one leaves a culture of *Euglena* near a sunny window, a clearly visible green cloud will form in the water, and this will follow the light as it moves. If the light is too bright, however, the cells will swim away from it. *Euglena* is probably able to orient with respect to the light because of a pair of special structures: the stigma, or eyespot, which is a patch of pigment on the reservoir, and a photoreceptor on the locomotive flagellum. When the cell is in certain orientations with respect to light, the photoreceptor is shaded by the stigma. The amount of light striking the

22–9 (a) *A living cell of* Euglena, *and* (b) *a cell broken open, showing the flexible protein strips that make up the pellicle. The large organelle at the upper right is a chloroplast. The spongy network below the chloroplast is endoplasmic reticulum. At the bottom left is a broken mitochondrion, identifiable by its characteristic cristae.*

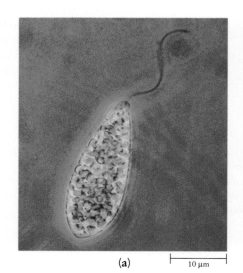

(a)   10 μm

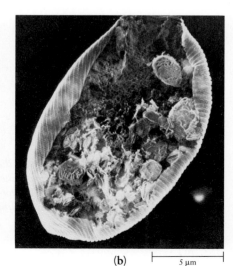

(b)   5 μm

photoreceptor influences the action of the flagellum and therefore directs the movement of the cell.

Euglenoids reproduce asexually, dividing longitudinally to form two new cells that are mirror images of one another. If some strains of *Euglena* are kept at an appropriate temperature and in a rich medium, the cells may divide faster than the chloroplasts, producing nonphotosynthetic cells, which can survive indefinitely in a suitable medium containing a carbon source. It is tempting to speculate that some modern heterotrophs, including the nonphotosynthetic euglenoids, arose from autotrophs in an analogous manner.

Division Chrysophyta: Diatoms and Golden-Brown Algae

This division includes two major classes, the diatoms (almost 10,000 species) and the golden-brown algae (about 3,000 species), and one minor class, the yellow-green algae (about 600 species). The chrysophytes have several identifying characteristics: (1) their photosynthetic pigments are chlorophyll *a* and chlorophyll *c*, which closely resembles the chlorophyll *b* of green algae and plants; (2) all species except the yellow-green algae contain a yellow-brown carotenoid, fucoxanthin, which functions as an accessory pigment and gives them their characteristic color; (3) their cell walls, which contain cellulose, are often impregnated with silicon compounds and thus are very rigid; (4) they characteristically store food in the form of oils rather than starch. Because of these oil reserves, fresh water containing large numbers of chrysophytes may have an unpleasant oily taste, as may fish caught in such waters. It has, in fact, been suggested that chrysophytes might be a potential source of oil for fuel.

Diatoms are a major component of the plankton and are an important source of food for small marine animals. Recently, the golden-brown algae have been found to be of extraordinary importance in the nanoplankton—components of the plankton so small that they pass through the fine mesh of an ordinary plankton net. It is now thought that the golden-brown algae may be the major food-producing organisms of the oceans.

Diatoms are enclosed in a fine double shell, the two halves of which fit together, one on top of the other, like a carved pillbox (Figure 22–10). Electron microscopy has shown that the fine tracings in diatom shells actually represent minute, intricately shaped pores or passageways connecting the interior of the cell to the exterior environment. The piled-up silica shells of diatoms, which have

22–10 *Diatoms (division Chrysophyta).* (a) *Side view showing the characteristic intricately marked shell.* (b) *Pillbox type seen from above and from the side. Notice that one cell is dividing. Each new cell will get half of the "pillbox" and then construct the other half. When the parent cell divides, one half is always smaller than the other. When this smaller cell divides, one of its daughter cells is, in turn, still smaller. When the individual cells are reduced in size to about 30 percent of the maximum diameter characteristic of the species, a cycle of sexual reproduction may be triggered. The offspring, which are full size, then undergo a new series of cell divisions.*

(a)   50 μm

(b)   100 μm

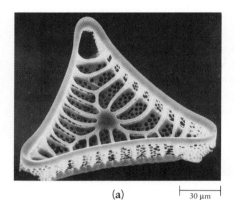

(a)   ├─────┤ 30 μm

(b)   ├──┤ 1 μm

22–11 *Scanning electron micrographs of*
(a) the silicon-containing shell of a diatom
and (b) the shell of a golden-brown alga,
which is made up of scales containing
calcium carbonate. Each diatom species
has its characteristic pattern of perfora-
tions in the walls. The delicate markings
of these shells, by which the species are
identified, were traditionally used by
microscopists to test the resolving power
of their lenses.

22–12 *Dinoflagellates.* (a) Ceratium
tripos, *an armored dinoflagellate.* (b) Noc-
tiluca scintillans, *a bioluminescent marine*
dinoflagellate. Yellow-brown diatoms that
have been ingested can be seen inside the
cell.

collected over millions of years, form the fine, crumbly substance known as "diatomaceous earth," used as an abrasive in silver polish and toothpaste and for filtering and insulating materials.

Some other members of this division lack cell walls and are amoeboid. Except for the presence of chloroplasts, the amoeboid chrysophytes are indistinguishable from some species of phylum Sarcodina (page 470), and the two groups may be closely related.

Reproduction in the chrysophytes usually takes place by cell division, an asexual process, but diatoms sometimes reproduce sexually. Gametes are produced by meiosis and then undergo fusion, or **syngamy.** (Syngamy in plants and animals is usually referred to as fertilization.) The resulting zygote expands to the full size characteristic of the species and undergoes cell division; the daughter cells then produce new silica shells. Diatoms are diploid, except for the gametes.

Division Dinoflagellata: "Spinning" Flagellates

The Dinoflagellata, of which about 2,000 different species are known, are also largely composed of single-celled algae, most of them marine forms. Like the chrysophytes, the dinoflagellates are important components of the phytoplankton. Other members of the division are heterotrophs clearly related to the photosynthetic forms.

Many of the dinoflagellates are bizarre in appearance, with a stiff cellulose wall (theca) that often looks like a strange helmet or an ancient coat of armor. There are also "naked" forms, including some parasitic species. Dinoflagellates usually have two flagella that beat within grooves; one encircling the body like a belt and a second lying perpendicular to the first. The beating of the flagella in their respective grooves causes the cells to spin like tops as they move through the water. Many dinoflagellates are bioluminescent.

Dinoflagellates are often red in color, and the infamous red tides, in which thousands of fish die, are caused by great blooms of red dinoflagellates. The poison in these red tides has been traced to several species of dinoflagellates, including the armored *Gessnerium catenellum*. It is such an extraordinarily powerful nerve toxin that 1 gram of it would be enough to kill 5 million mice in 15 minutes. Blooms of *Gessnerium catenellum* appear regularly on the Pacific Coast and the Gulf of Mexico and have been reported off the New England coast. Mussels can also ingest the algae responsible for the red tides, concentrating the poison; these mollusks then become dangerous for consumption by vertebrates, including humans.

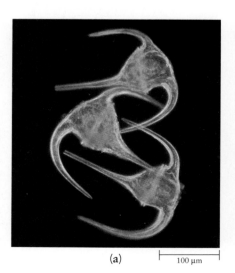

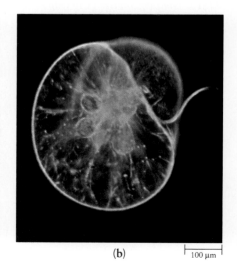

(a)   ├─────┤ 100 μm   (b)   ├─────┤ 100 μm

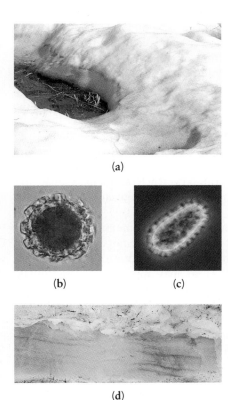

(a)

(b)   (c)

(d)

22–13 *Snow algae.* (a) *In many parts of the world, the presence of large numbers of snow algae produces "red snow" in high-altitude snow banks during the summer. This photograph was taken near Cedar Breaks National Monument in Utah.* (b) *Dormant zygote of the snow alga* Chlamydomonas nivalis. *The red color results from pigments that serve to protect the chlorophyll within the zygote following fertilization (see Figure 12–5, page 251).* (c) *Resting zygote of the alga* Chloromonas brevaspina, *found in green snow.* (d) *Green snow occurs just below the surface, usually near tree canopies. It is widespread, occurring as far south as Arizona and as far north as Alaska.*

Division Chlorophyta: Green Algae

The chlorophytes are the most diverse of all the algae, comprising about 9,000 species. They are thought to represent four distinct evolutionary lineages, one with only a few living species and three with the vast majority of contemporary species. The three principal classes of Chlorophyta, corresponding to these three lineages, are Chlorophyceae, Charophyceae, and Ulvophyceae. Although most green algae are aquatic, others occur in a wide variety of habitats, including the melting surface of snow (Figure 22–13), as green patches on tree trunks, and as symbionts in a variety of organisms. Of the aquatic species, most of the Ulvophyceae are marine, but the great majority of the Chlorophyceae and Charophyceae are found in fresh water. Many green algae are unicellular and microscopic in size. For example, *Chlamydomonas* (Figure 5–3, page 104), a member of class Chlorophyceae, is visible only as a fleck of green under the light microscope. Some of the Ulvophyceae are large; *Codium magnum* in the Gulf of Mexico, for example, sometimes attains a breadth of 25 centimeters and a length of more than 8 meters.

The plants are believed to have originated from the same lineage of green algae as the contemporary Charophyceae. Both the plants and the green algae—but no other group of algae—contain chlorophylls *a* and *b* and beta-carotene as their photosynthetic pigments and store their foods as starch. In addition, most of the Charophyceae also have cellulose-containing cell walls, and cell division in a few members of this class is characterized by the formation of a cell plate.

The Increase of Complexity

The green algae are of particular interest to students of evolution not only because of their relationship to the plants but also because of the wide range of complexity they exhibit. Whereas the three divisions previously discussed contain only unicellular forms and the members of the two divisions that follow are almost entirely multicellular, division Chlorophyta contains both unicellular and multicellular organisms, as well as a number of intermediate forms.

One form intermediate between the unicellular and the multicellular involves the association of individual cells in colonies. Colonies differ from true multicellular organisms in that, in colonies, the individual cells preserve a high degree of independent function. The cells are often connected by cytoplasmic strands, which integrate the colony sufficiently that it may be regarded as a single organism. The classic example of increasing complexity among colonial organisms is called the volvocine line, after one of its more spectacular members, *Volvox*. It is based on the cohesion of *Chlamydomonas*-like cells in motile colonies. The simplest member of the volvocine line is *Gonium* (Figure 22–14a). The *Gonium* colony consists of 4 to 32 separate cells (depending on the species) arranged in a shield-shaped disk. The flagella of each cell beat separately, pulling the entire colony forward. Each cell in *Gonium* divides to produce an entire new colony.

A closely related colonial organism is *Pandorina* (Figure 22–14b), which consists of 16 or 32 cells in a tightly packed ovoid or ellipsoid shape. The colony is polar; that is, one end is different from the other: the eyespots are larger in the cells at one pole of the colony. Each cell has two flagella, and because all the flagella point outward, *Pandorina* rolls through the water like a ball. When the cells attain their maximum size, the colony sinks to the bottom, and each of the cells divides to form a daughter colony. The parent colony then breaks open like Pandora's box (which suggested its name), releasing new daughter colonies.

Volvox (Figure 22–14c), the culmination of the volvocine line, is a hollow sphere that is made up, according to species, of a single layer of 500 to 60,000 tiny biflagellate cells. The flagella of each cell beat in such a way as to turn the entire colony around its axis, spinning it majestically through the water. Like *Pandorina*, *Volvox* is polar; it orients its anterior end toward the light but moves away from

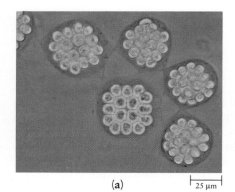

(a)   `25 µm`

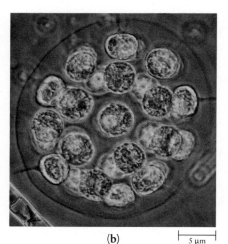

(b)   `5 µm`

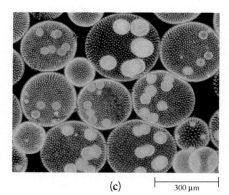

(c)   `300 µm`

22–14 *The volvocine line.* **(a)** *Gonium. Each colony is composed of a shield-shaped mass of* Chlamydomonas-*like cells, held together in a gelatinous matrix.* **(b)** *Pandorina,* in which the individual *cells form an egg-shaped colony.* **(c)** *Vol-vox, the largest and most complex member of this evolutionary line. Each colony is made up of hundreds or thousands (depending on the species) of individual bright green biflagellate cells, attached to one another by fine threads of cytoplasm. Each colony forms a hollow sphere that spins through the water as a result of coordinated beating of the flagella. Daughter colonies form inside the mother colony. Many such mother-daughter combinations are visible in this micrograph.*

light that is too strong. Most of the cells are vegetative and do not take part in reproduction. Only some of the cells of the lower hemisphere can form daughter colonies; in other words, there is specialization of function. These cells, which appear identical to others in the young colony, become larger, greener, and structurally distinct as the colony matures. Then they enlarge and divide, forming new spheres of hundreds or thousands of cells. Daughter colonies remain inside the mother colony until the mother colony eventually breaks apart. In some species, the reproductive cells produce gametes and reproduction is sexual.

In these four extant genera of class Chlorophyceae, *Chlamydomonas, Gonium, Pandorina,* and *Volvox,* there is a steady progression in size and complexity and also a trend toward specialization of function. Nevertheless, this particular line represents an evolutionary "dead end," in that it has not given rise to a more complex group of organisms.

Another type of intermediate form, characteristic of some members of class Ulvophyceae, results from repeated nuclear divisions that are not accompanied by a corresponding division of the cytoplasm and the formation of cell walls. An entire organism, such as *Valonia* (Figure 22–15a), may appear to be unicellular, but, in fact, it consists of many nuclei within a common cytoplasm. This type of organization is neither truly unicellular nor multicellular but is known as **coenocytic** (pronounced "see-no-sit-ik"; from the Greek *koinos,* "shared in common"). Some coenocytic green algae, such as *Cladophora,* are filamentous, while others, such as *Valonia* and *Codium magnum,* form more massive structures. A number of common seaweeds, especially tropical ones, are coenocytic.

True multicellularity is exemplified by a number of green algae in classes Charophyceae and Ulvophyceae, including *Spirogyra* and *Ulva* (Figure 22–15b and c). In these organisms, nuclear division is followed by cytokinesis and the formation of cell walls, but the daughter cells do not separate from one another. As we saw in Figure 22–5, such cell division can lead to the formation of filaments, sheets, or a three-dimensional body.

Life Cycles in the Green Algae

A variety of different life cycles are observed among the green algae. In Figure 12–5 (page 251), we examined the life cycle of the unicellular green alga *Chlamydomonas.* The haploid cells of this alga usually reproduce asexually, with each cell undergoing two successive mitotic divisions, resulting in four new haploid cells. However, when essential nutrients, particularly nitrates, are in short supply, *Chlamydomonas* reproduces sexually. Haploid cells of different mating types function as gametes—they fuse, forming a diploid zygote around which a protective coat forms. This cell, known as a **zygospore,*** remains dormant until conditions are once more favorable for growth. It then undergoes meiosis, producing four haploid cells. The correlation between nutrient deficiency and sexual reproduction in *Chlamydomonas* and in many other organisms suggests that the rate of genetic recombination is subject to environmental influences.

---

* Note that the word "spore" has more than one meaning in biology. Among the prokaryotes, spores are produced by a transformation of the original cell contents. Among the protists and fungi, spores may contain haploid nuclei produced by meiosis or diploid nuclei produced by syngamy ("zygospores"); such diploid nuclei typically undergo meiosis, producing haploid nuclei, before they are released from the protective coat. Among the plants, spores are always haploid cells produced by meiosis. All spores, however, have some essential features in common: they are tiny, light, and able to withstand desiccation and extremes of temperature, and, under favorable circumstances, they are able to germinate without combining with another cell. Thus, in four of the five kingdoms, spores serve as resting and dispersal forms, providing the organism with a means of waiting out hostile environmental conditions and of colonizing new territories.

(a)　　　　　 10 mm

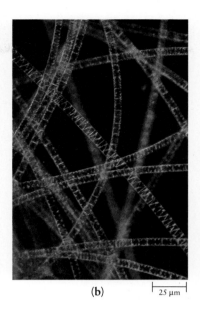

(b)　　　 25 μm

(c)

22–15 *Division Chlorophyta includes a variety of multinucleate and multicellular organisms in addition to colonial forms. (a) Valonia, a coenocytic green alga. About the size of a hen's egg, Valonia contains many nuclei but has no partitions separat-* ing them. *It is common in tropical waters.* **(b)** Spirogyra *is a freshwater alga in which the cells all elongate and then are divided by transverse cross walls so that they are strung together in long, fine filaments. The chloroplasts form spirals that look like* strips of green tape within each cell. **(c)** Ulva, *or sea lettuce, is a marine alga in which the cells divide both longitudinally and laterally, with a single division in the third plane. This produces a broad thallus (vegetative body) two cells thick.*

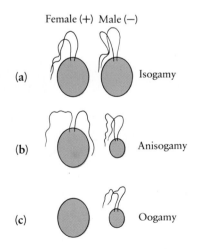

Female (+)　Male (−)

(a)　　　　　　Isogamy

(b)　　　　　　Anisogamy

(c)　　　　　　Oogamy

22–16 *Evolution of gametes.* (a) *Isogamy. The gametes are similar in size and shape.* (b) *Anisogamy. One gamete, conventionally termed male, is smaller than the other.* (c) *Oogamy. One gamete, usually the larger, is not motile. This is where sexual differentiation, with its mixed blessings and many complications, all began.*

In most species of *Chlamydomonas,* the cells of the two different mating types, conventionally designated as plus (+) and minus (−), are identical in size and structure; this condition is known as **isogamy** (Figure 22–16). In other species of *Chlamydomonas,* one gamete is larger than the other, but both are motile, a condition known as **anisogamy.** In still other species, one gamete, usually the larger, is not motile, a condition known as **oogamy.** The larger, nonmotile gametes are specialized for storing nutrients for the zygote, whereas the smaller gametes are specialized for seeking out and finding the first type of gamete. An individual that produces gametes that are nonmotile and (usually, though not always) larger is known as female. The entire range of differences between gametes that occurs among the algae (and among other types of organisms, as well) is exhibited in the various species of the single genus *Chlamydomonas.*

A more complex life cycle, characterized by alternation of generations, is found in some multicellular green algae, as well as in all plants. As we saw in Chapter 12, in this type of life cycle, a diploid, spore-producing generation alternates with a haploid, gamete-producing generation. The gamete-producing form is known as the **gametophyte;** the individual cells of which it is composed are haploid *(n)* and so are the gametes it produces. The gametes fuse to form a diploid *(2n)* zygote. The zygote develops into the spore-producing form, the **sporophyte,** in which all the cells are diploid. Spores, which are produced in specialized structures of the sporophyte by meiosis, are always haploid. Spores differ from gametes in that when a spore germinates it forms a new organism, while a gamete must first unite with another gamete before further development occurs. The gametophyte and the sporophyte are always genetically different, since one is composed of haploid cells and the other of diploid cells. Although all gametes from a given gametophyte are genetically identical (since they are produced by mitosis from haploid cells), genetic recombination occurs both in the fusion of

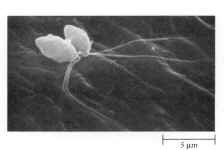

5 μm

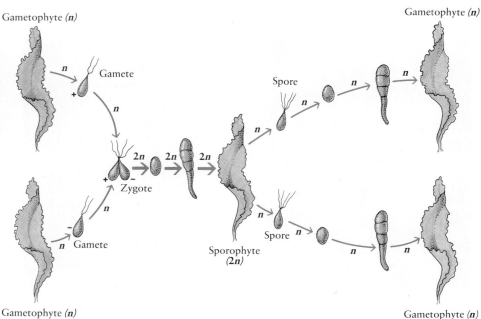

Gametophyte (*n*)

Gamete

n

n

n

n

Gamete

n

Gametophyte (*n*)

2*n*

+

−

Zygote

2*n*

2*n*

2*n*

n

Sporophyte
(2*n*)

n

Spore

n

Spore

n

n

Gametophyte (*n*)

n

n

n

n

Gametophyte (*n*)

Gametophyte (*n*)

22–17 *In the sea lettuce,* Ulva, *we can see the reproductive pattern known as alternation of generations, in which one generation produces spores, the other gametes. The haploid (n) gametophyte produces haploid isogametes, and the gametes fuse to form a diploid (2n) zygote. A sporophyte, a multicellular body in which all the cells are diploid, develops from the zygote. The sporophyte produces haploid spores by meiosis. The haploid spores develop into haploid gametophytes, and the cycle begins again. The micrograph shows isogametes of* Ulva *before cytoplasmic fusion.*

gametes of different types to form the zygote and in the meiotic divisions that produce spores. Moreover, the spores, with their new genetic combinations, may be dispersed quite widely from the parent sporophyte, enabling the organism to spread into new environments.

In some green algae, such as the sea lettuce, *Ulva* (Figure 22–17), the two generations look alike and are said to be **isomorphic** (from the Greek word *morphe,* meaning "shape" or "form"). In other species, the sporophyte and the gametophyte do not resemble one another, and the generations are said to be **heteromorphic**. In fact, in some cases, the two generations of the same alga were once considered to represent two entirely different genera until the organism was studied in the laboratory (Figure 22–18).

22–18 *In some algae the alternating generations are so different that they once were believed to be different genera. At one time, the gametophyte of the* Valonia-*like alga shown here was called* Halicystis *and the sporophyte* Derbesia. *As you can see, this alga exhibits anisogamy.*

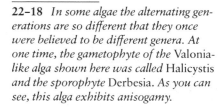

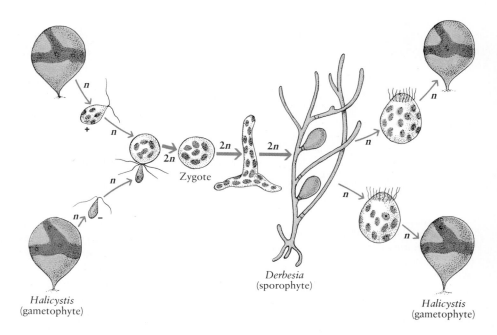

n

+

n

2*n*

Zygote

2*n*

2*n*

n

n

n

−

Halicystis
(gametophyte)

Derbesia
(sporophyte)

n

Halicystis
(gametophyte)

n

Halicystis
(gametophyte)

Division Phaeophyta: Brown Algae

The 1,500 species of brown algae are the principal seaweeds of temperate and polar regions. An almost exclusively marine group, they dominate the rocky shores throughout the cooler regions of the world, and some, like the kelps, often form extensive beds offshore. The brown algae contain chlorophylls *a* and *c* and fucoxanthin, as do the Chrysophyta. They store their food as an unusual polysaccharide (laminarin), or sometimes as oil, but never as starch, as do green algae and plants. Their cell walls contain cellulose.

The brown algae are often very large, and many have a variety of specialized tissues. Some of the giant kelps are nearly 60 meters long; many of these are annuals, reaching their full size in a single season. The body of a protist (or a plant) that is multicellular, but relatively unspecialized, is known as a **thallus**. In kelps, the thallus is well differentiated into **holdfast** ("root"), **stipe** ("stalk"), and **blade** ("leaf"). The words are put into quotation marks because, although the corresponding parts of the alga superficially resemble the structures of plants, they are not really comparable in their internal organization. However, most kelps have strands of elongated conducting cells in the center of the stipe that are similar to the cells that conduct sugars in the vascular plants. Carbohydrates produced in the blades, which are exposed to sunlight, are thus transported to the stipe and holdfast, which may be far below the surface of the water.

There are no modern unicellular forms in this group, except for the gametes. The life cycles of most brown algae involve alternation of generations. In some species the two generations are isomorphic, whereas in others they are heteromorphic. In the larger kelps, the gametophyte is much smaller than the conspicuous, familiar sporophyte. In the rockweed *Fucus* and its relatives, the life cycle is superficially similar to that in higher animals (see Figure 12–4c, page 250). Meiosis occurs in specialized cells of the diploid organism, producing haploid cells that immediately undergo mitosis, giving rise to gametes.

Division Rhodophyta: Red Algae

Most of the seaweeds of the world are red algae, of which there are some 4,000 species. They are most commonly found in warm marine waters; fewer than 2 percent of the species are freshwater forms. Red algae usually grow attached to rocks or other algae. Their red color indicates that they absorb blue light, which is the color with the greatest penetration in water. As a result, red algae can grow at greater depths than other algae; some have been found attached 175 meters below the ocean surface in the clear water of the tropics. Although some grow to lengths of several meters, red algae never attain the size of the largest of the brown algae.

The red algae contain chlorophyll *a* and carotenoids, and also certain phycobilins, which give them their distinctive colors. The cell walls of most forms include an inner layer of cellulose and an outer layer of mucilaginous carbohydrates, from which agar, used in culturing bacteria in the laboratory, is derived. In addition, some red algae have the capacity to deposit calcium carbonate in their cell walls. Such algae are called coralline algae, and they play an important role in building coral reefs.

The basic life cycle of the red algae involves alternation of generations. In most species, the gametophyte and sporophyte are isomorphic, but an increasing number of heteromorphic life cycles are being discovered. The gametes are unusual in that neither type is motile; the male gamete is carried to the fixed female reproductive cell by the movement of the water. The red algae are one of the few groups of organisms that have no flagellated cells; they also lack centrioles.

(a)

(b)

22–19 (a) *A brown alga,* Laminaria, *showing holdfasts, stipes, and portions of the blades.* (b) *Unlike the brown algae, most red algae are made up of filaments. The branched filaments of this red alga are hooked, enabling it to cling to other seaweeds.*

MULTINUCLEATE AND MULTICELLULAR HETEROTROPHS

The Slime Molds

The slime molds are a group of curious organisms, usually classified with the protists because of their similarity to the amoebas. Two main groups are known, the plasmodial slime molds (division Myxomycota), with about 550 species, and the cellular slime molds (division Acrasiomycota), with 65 species.

Most slime molds live in cool, shady, moist places in the woods—on decaying logs, dead leaves, or other damp organic matter. One of the common plasmodial species *(Physarum cinereum),* however, is sometimes found creeping across city lawns. Plasmodial slime molds come in a variety of colors and can be spectacularly beautiful. The function of the pigments is not known with certainty, but they are probably photoreceptors because only pigmented species require light for spore production.

During their nonreproductive stages, the plasmodial slime molds are thin, streaming masses of protoplasm, which creep along in amoeboid fashion. As one of these plasmodia travels, it engulfs bacteria, yeast, fungal spores, and small particles of decayed plant and animal matter, which it digests. It may grow to weigh as much as 50 grams or more, and, since slime molds are spread thinly, this mass can cover an area more than a meter in diameter. The plasmodium is coenocytic; as it grows, the nuclei divide repeatedly and, in the early stages, synchronously.

Plasmodial growth continues as long as an adequate food supply and moisture are available. When either of these is in short supply, the plasmodium separates into many mounds of protoplasm, each of which develops into a mature **sporangium** (a structure in which spores develop) borne at the tip of a stalk. Meiosis takes place and cell walls form around the individual haploid nuclei, producing the spores.

The spores germinate under favorable conditions, and each spore, depending on the species, produces one to four haploid, flagellated cells. Some of these cells fuse to form a zygote, from which a new plasmodium develops.

22-20 (a) *The plasmodium—a streaming mass of protoplasm—of a slime mold. Such a plasmodium, with its multiple nuclei, can pass through a piece of silk or filter paper and come out the other side apparently unchanged.* (b) *Sporangia of a plasmodial slime mold on a rotting log.*

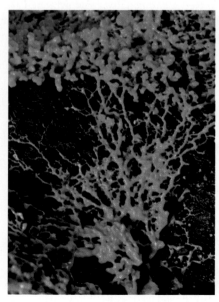

(a)

(b)

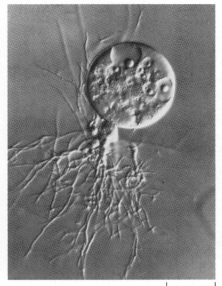

25 μm

22-21 Chytridium confervae, *a common chytrid, as photographed through a differential-interference microscope. Note the slender, nutrient-absorbing rhizoids extending downward from the growing sporangium. The sporangium of this coenocytic organism contains numerous nuclei, vacuoles, and organelles, such as mitochondria. The rhizoids contain vacuoles and organelles but no nuclei.*

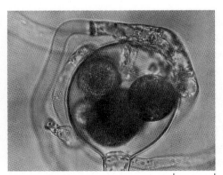

25 μm

22-22 *Mating in the oomycete* Achyla ambisexualis. *The large spherical structure is the female gametangium. The dark bodies within it contain eggs. Encircling the female gametangium is the male gametangium. Fertilization tubes extending from the male into the female gametangium are barely visible. Sperm nuclei pass through these tubes to the egg nuclei. Development of the male gametangia and their attraction to the female are controlled by the production of a steroid hormone remarkably similar in structure to the human sex hormones.*

The life cycle of an acrasiomycete, or cellular slime mold, is shown on page 141. These slime molds also begin as amoeba-like organisms but differ from the plasmodial slime molds in that the amoebas, on swarming together, do not lose their cell membranes but retain their identity as individual cells.

The Water Molds

The water molds are coenocytic organisms, many of which resemble fungi in their basic structure. Until quite recently, they were usually classified with the fungi, but their biochemical characteristics and the presence of flagellated reproductive cells (which do not occur in any of the fungi) indicate that they represent distinct lineages. The two principal divisions are Chytridiomycota, with about 900 species, and Oomycota, with about 800 species.

Division Chytridiomycota

Most chytrids consist of a small vegetative body (a thallus), differentiated into a coenocytic sporangium and rootlike anchoring structures, called **rhizoids** (Figure 22–21). The rhizoids, which contain no nuclei, absorb dissolved nutrients from the substrate. When the chytrid is mature, the sporangium cleaves into flagellated spores, each of which contains a single nucleus. Following free-swimming and encysted stages, each spore can germinate to form a new thallus.

Other chytrids are much more complex in their structure and reproduction. Sexually reproducing forms have flagellated gametes, which may be similar or different in size and motility. In the genus *Allomyces*, the larger, less active gametes produce a hormone, appropriately named sirenin, that attracts the smaller, more active gametes. Like some algae and all plants, *Allomyces* and one other closely related genus have a life cycle characterized by alternation of generations.

Chytrids are found in both fresh and salt water and in moist soil. They live as parasites on algae, plants, and fungi, or as saprobes, feeding on dead algae, pollen grains, and other plant debris.

Division Oomycota

Structurally, most oomycetes resemble fungi, consisting of coenocytic filaments known as **hyphae**. As in the fungi, portions of the hyphae form specialized structures, the **gametangia**, in which gametes are produced. However, the cell walls of oomycetes, like those of many algae but unlike those of fungi, contain cellulose.

The oomycetes derive their name from *oion*, the Greek word for "egg." All oomycetes display oogamy—the differentiation of gametes into forms that are clearly male (sperm) and female (egg). However, the gametes of oomycetes, unlike those of chytrids, have no flagella and are nonmotile. In sexual reproduction, the sperm nucleus and the egg, each of which is borne in its own type of gametangium, fuse to produce a zygote (Figure 22–22). Oomycetes can also reproduce by forming asexual spores, each of which bears two flagella. Many oomycetes are aquatic (and thus they also are known as water molds), but even the terrestrial forms can produce flagellated spores that require free water.

Most oomycetes are saprobes, living on dead organic matter. Some forms are parasitic and pathogenic, however, and as mycologist C. J. Alexopoulos has said, "At least two of them have had a hand—or should we say a hypha!—in shaping the economic history of an important portion of mankind." The first of these is *Phytophthora infestans* (*phytophthora* literally means "plant destroyer"), the cause of the "late blight" of potatoes, which produced the great potato famines in Ireland (Figure 22–23). The second economically important member of this group

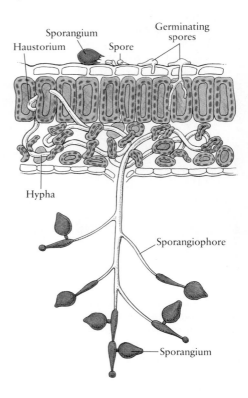

22–23 Phytophthora infestans, *cause of potato blight. Infection begins when an airborne sporangium alights on a leaf, releasing spores that move about in the film of water on the leaf's surface. These spores germinate, producing haustoria (specialized hyphae) that penetrate the epidermis and attack the photosynthetic cells in the interior of the leaf. Eventually aerial hyphae—sporangiophores—bear sporangia from which a new generation of asexual spores is released.*

is *Plasmopara viticola*, the cause of downy mildew of grapes. This mildew threatened the entire French wine industry during the latter part of the nineteenth century.

UNICELLULAR HETEROTROPHS

In addition to the multinucleate and multicellular heterotrophs we have just considered, kingdom Protista includes a vast number of unicellular heterotrophs (Table 22–3). These organisms, known informally as the protozoa, are classified in five principal groups. Three phyla, which contain both free-living and parasitic members, are distinguished on the basis of mode of locomotion: (1) by flagellar movement (phylum Mastigophora, the mastigophorans), (2) by pseudopodia (phylum Sarcodina, the sarcodines), and (3) by ciliary movements (phylum Ciliophora, the ciliates). Two phyla are entirely parasitic: (1) Opalinida, the opalinids, which have ciliary movement, and (2) Sporozoa, the sporozoans, in which the motility of the cells is much reduced.

Protozoans usually reproduce asexually, by binary fission. Many also have sexual cycles, involving meiosis and the fusion of gametes, which results in a diploid (2n) zygote. The zygote often takes the form of a thick-walled, resistant zygospore, especially during periods of drought or cold. Some protozoans, notably the ciliates, undergo conjugation, in which nuclei are exchanged between cells.

Phylum Mastigophora

The Mastigophora are regarded as the most primitive of the heterotrophic protists. Some authorities believe they are derived from photosynthetic flagellated cells, such as *Euglena*, that lost their chloroplasts; others think they may be the descendants of protists that never acquired photosynthetic symbionts. Almost all of the smaller cells have one or two flagella, and larger cells frequently have many. These flagella, like those of other protists, have the characteristic 9 + 2 structure (page 122). The mastigophorans multiply asexually by binary fission (mitosis and cytokinesis), and, in some forms, sexually, by syngamy. They generally have no

TABLE 22-3 **Comparative Summary of Characteristics of the Heterotrophic Protists**

DIVISION OR PHYLUM	NUMBER OF SPECIES	LOCOMOTOR STRUCTURES	MODE OF REPRODUCTION	REMARKS
Myxomycota (plasmodial slime molds)	550	Pseudopodia; flagella on reproductive cells	Asexual (spores) and sexual (fusion of germinated spores)	Coenocytic, except in reproductive stages
Acrasiomycota (cellular slime molds)	65	Pseudopodia	Asexual (spores); also sexual in some species	Unicellular, with multicellular aggregates forming prior to reproduction
Chytridiomycota (chytrids)	900	Flagella on spores and gametes	Asexual (spores); also sexual in some forms	Coenocytic; mostly aquatic; cell walls contain chitin
Oomycota (oomycetes)	800	Flagella on spores	Asexual (spores) and sexual; all exhibit oogamy	Coenocytic; many aquatic; cell walls contain cellulose; cause blights and mildews of plants
Mastigophora (flagellates)	1,500	Flagella; some also form pseudopodia	Asexual (binary fission); sexual (meiosis and syngamy) in some species	Unicellular; mostly parasitic; some free-living
Sarcodina (amoebas and related forms)	11,500	Pseudopodia; a few develop flagella at some stages of life cycle	Asexual (binary fission) or sexual	Unicellular; mostly free-living; many have outer shells, or tests; about 33,000 fossil species known
Ciliophora (ciliates)	8,000	Cilia	Asexual; genetic exchanges through conjugation	Unicellular; mostly free-living; have micronuclei and macronuclei
Opalinida (opalinids)	400	Cilia or flagella; flagellated gametes	Asexual or sexual	Unicellular; all intestinal parasites of lower vertebrates
Sporozoa (sporozoans)	5,000	None	Complex life cycle, involving both asexual and sexual reproduction	Unicellular; all parasitic; cause devastating diseases, including malaria, in humans and other animals

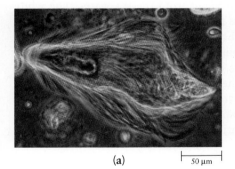

(a)   50 μm

outer wall, and some are able to form pseudopodia, which are, as you will recall, temporary extensions of the cell body that are used in locomotion and in engulfing food particles.

Most mastigophorans are parasitic, but some are free-living. The former include *Trypanosoma gambiense* and *Trypanosoma rhodesiense*, flagellates that cause African sleeping sickness, and members of the genus *Trichonympha*, complex and beautiful flagellates that live as symbionts in the digestive tracts of termites, where they digest the wood ingested by the termite. A number of other mastigophorans cause debilitating diseases in humans and their poultry and livestock; "hiker's diarrhea" is caused by members of the genus *Giardia*, which is also endemic in some day-care centers.

22–24 (a) Trichonympha. *This flagellate, which breaks down cellulose, is responsible for the well-known proficiency of its termite hosts to digest wood.* (b) *Three cells of* Giardia lamblia, *one of the many protists that cause diarrhea in humans. G. lamblia cells are characterized by four pairs of flagella and an adhesive disk with which they adhere to the lining of the small intestine. Using his primitive microscope, van Leeuwenhoek first identified* Giardia *about 300 years ago and hypothesized its pathogenic role; confirmation of his hypothesis has come only in recent years, with the development of methods to study this protist in the laboratory and in other animals. Encysted* Giardia *are transmitted from host to host in the feces.*

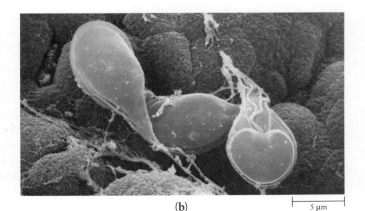

(b)   5 μm

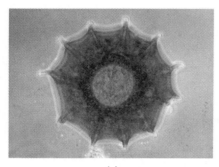

(a)

(b)   200 μm

22–25 *Outer shells, or tests, are characteristic of certain groups of sarcodines. (a)
The brilliantly colored test of* Arcella den-
tata *consists of a proteinaceous material
secreted by the organism. (b) The shell of a
foraminiferan. In addition to as many as
7,000 living species of foraminiferans,
there are about 30,000 extinct species,
known only from their fossilized shells.*

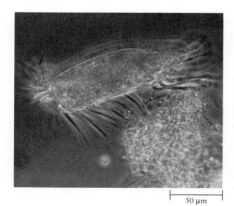

50 μm

22–26 *The cilia of some protists are
clumped to form cirri. In* Euplotes patella,
*shown here, the cirri move individually,
propelling the cell in a jerky motion.*

Phylum Sarcodina

The sarcodines include the amoebas and related forms. They have no coat or wall outside their cell membrane and generally move and feed by the formation of pseudopodia (see Figure 1–15a, page 37). They take their name from the word "sarcode," coined in the early nineteenth century to describe the "simple, glutinous, and homogeneous jelly" of which, at one time, simple cells were thought to be composed. Despite their uncomplicated appearance, they are complex cells and, as we shall see (page 474), are even capable of complex behavior patterns—for example, when sensing and pursuing prey organisms.

Sarcodines are thought to have originated from mastigophorans; some sarcodines may develop flagella during particular stages of their life cycle or under particular environmental conditions. They are found in both fresh and salt water. Some are parasites, such as those that cause amoebic dysentery in humans.

Reproduction may be asexual or sexual. Asexual reproduction takes place by cell division accompanied by mitosis in which the nuclear envelope usually does not break down. In sexual reproduction, the cells, which are diploid, undergo meiosis, forming gametes, which then fuse to form zygotes.

Many of the sarcodines have outer shells, or tests, sometimes brightly colored. Some, like *Arcella*, secrete a proteinaceous material that hardens on exposure. Others, like *Difflugia*, exude a sticky organic substance in which they deposit silicon-containing particles. These particles, which were previously ingested by the organism, are divided between the two daughter cells at cell division; each daughter cell then arranges them into an almost exact replica of the parental shell. Such shells, which are produced primarily by amoebas that live in sand or soil, on the mosses of bogs and forest floors, or in crevices in tree trunks, are thought to have evolved as a protection against abrasion or dehydration of the organism within. Other sarcodines, the heliozoans ("sun animals"), resemble pincushions, with fine pseudopodia stiffened with microtubules (page 78) radiating from their bodies; they are found in both salt and fresh water. *Actinosphaerium*, the protist shown in Figure 5–2 (page 103), may reach as much as a millimeter in diameter and may sometimes be seen as a tiny white speck floating on the surface of a pond.

Another group of sarcodines, the Foraminifera (numerically the most common of all unicellular heterotrophs), have snail-like shells and live in the sea. Their shells are made of calcium carbonate, extracted from the sea water. The white cliffs of Dover and similar chalky deposits throughout the world are the result of the long accumulation of these shells. The shells of the Foraminifera have been accumulating on the ocean bottom for millions of years, and in many areas, as a result of geologic changes, thick deposits of their skeletons (the "foraminiferan ooze") can be found on the surface of the land or under later rock formations. Since the skeletons have evolved over this long period of time, it is possible to date a particular stratum by the type of Foraminifera that it contains, a fact that has proved of immense practical value in locating oil-bearing strata.

Phylum Ciliophora

The ciliophores, or ciliates, are the most highly specialized and complicated of the protozoans. About 8,000 species of ciliates are known, both freshwater and saltwater forms; most are free-living. They are believed to have been derived from primitive mastigophorans, which, traveling in the opposite evolutionary direction from the sarcodines, developed elaborate ciliary systems (see Figure 5–30, page 122). In some species, the cilia adhere to each other in rows, forming brushlike structures called membranelles (see Figure 22–1) or clumps of cilia called cirri, which can be used for walking or jumping, as well as in feeding. Cilia, membranelles, and cirri move in a coordinated fashion, although the way in which they are coordinated is not entirely understood. Some ciliates also have myonemes, contractile threads. All have a complex "skin," the cortex, which is bound on the outer side by the cell membrane. In a few groups, the cortex contains small barbs known as trichocysts, which are discharged when the cell is stimulated in certain ways.

The ciliates have another unusual feature: they have two kinds of nuclei, micronuclei and macronuclei. One or more of each kind is present in all cells. They also have a complex process for exchange of genetic information, in which cells conjugate and the micronuclei undergo meiosis. Cells then exchange haploid micronuclei that fuse, so that each cell has a new diploid micronucleus, which then divides. The old macronucleus dissolves, and a new macronucleus develops from one of the daughter micronuclei. The macronucleus in certain ciliates contains 50 to 100 times as much DNA as the micronucleus and so is believed to represent multiple copies of it. This view is supported by the fact that, in many ciliates, a cell can survive indefinitely without a micronucleus if even a small portion of a macronucleus is left, although the cell cannot conjugate. However, it cannot live without a macronucleus, even if it has a micronucleus. The macronucleus does not divide mitotically but is apportioned approximately equally between dividing cells as they constrict and separate.

The complexity of the conjugation process among ciliates reminds us again of the expenditure of energy and other resources involved in effecting exchanges of genetic information among individuals of the same species, from bacteria to *Homo sapiens*. The high biological cost of these activities is an indication of the great survival value, in evolutionary terms, of the genetic variability produced by such exchanges.

22–27 *Drawing of* Paramecium, *a ciliate. The body of this protist is completely covered by 9 + 2 cilia, although only a relative few are shown here. Like other ciliates,* Paramecium *feeds largely on bacteria, smaller microorganisms, and other particulate matter. The beating of specialized cilia drives particles into the buccal cavity, where they are formed into food vacuoles that enter the cytoplasm. The food is digested in the vacuoles, and the undigested matter, still in vacuoles, is emptied out through the anal pore. The contractile vacuoles serve to eliminate excess water from the cell.*

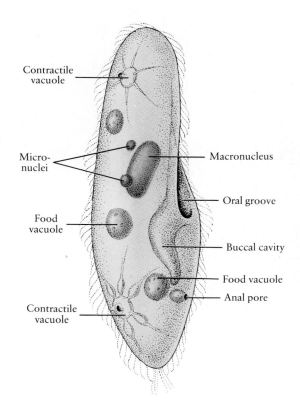

Contractile vacuole

Micro-nuclei

Macronucleus

Oral groove

Food vacuole

Buccal cavity

Food vacuole

Anal pore

Contractile vacuole

The Evolution of Mitosis

Among modern prokaryotes and protists, it is possible to observe varying patterns in the division of the genetic material. These patterns are believed to reflect evolutionary history. Among prokaryotes, you will recall, the replicated chromosomes (mostly DNA) attach to the cell membrane and so are separated as the newly formed cells divide. Among plant and animal cells, the nuclear envelope breaks down at mitosis, and the spindle fibers (composed of microtubules) are apparently involved in the mechanism of separation. Detailed analysis of some of the protists has revealed a number of intermediate stages between these two. In dinoflagellates, for instance, the chromosomes—which have no histones attached to the DNA and are always condensed—are anchored permanently to the nuclear envelope. The envelope remains intact during mitosis, and the chromosomes, as in bacteria, are separated as the enveloping membrane elongates. At the time of cell division, cytoplasmic channels containing bundles of microtubules form between the two sets of attached chromosomes; these microtubules are all oriented in the same direction and are thought to regulate the separation process.

The macronucleus of ciliated protozoa appears to be apportioned in just this same way, but in the micronucleus of ciliates, both kinetochore and polar spindle fibers are seen, as they are in slime molds. The spindle fibers appear to push the poles apart as the fibers grow, elongating the nucleus in preparation for its division. The nuclear envelope, however, does not break down.

As the history is reconstructed, in the earliest eukaryotes the nuclear envelope remained intact, and the chromosomes separated as it elongated. Next, microtubules, which first played an extranuclear role, entered the nuclear envelope, and, in an increasingly more organized way, aided in its elongation. Finally, the nuclear envelope, no longer useful in mitosis, began to fragment at prophase, and spindle fibers took over as a more efficient means for separating pairs of chromosomes in a diploid organism.

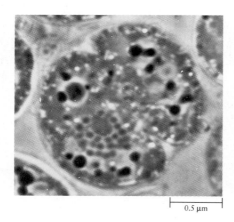

0.5 μm

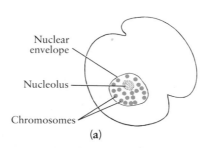

Nuclear envelope

Nucleolus

Chromosomes

(a)

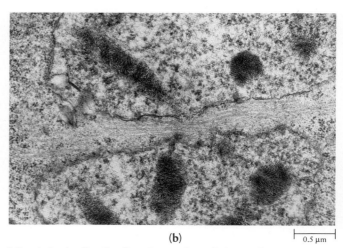

(b)   0.5 μm

Mitosis in the dinoflagellate Cryptothecodinium cohnii. (a) *Photomicrograph and diagram of interphase chromosomes.* (b) *Microtubules in the cytoplasmic channel separating the two nuclei. The dark areas are chromosomes.*

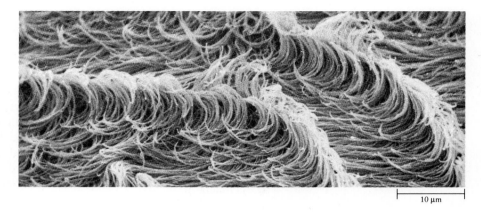

22-28 *The surface of* Opalina, *one of the parasitic opalinids, as seen through the scanning electron microscope. Its cilia beat continuously, about 40 to 60 times per second, in a coordinated beat that produces synchronous waves of motion.*

10 μm

Phylum Opalinida

The opalinids have been found most often in the digestive tracts of frogs and toads, and, occasionally, in fishes and reptiles. They are structurally simple, with a uniform covering of cilia or flagella (Figure 22–28). At least two nuclei are present; in some species, the number of nuclei increases as the cell gets larger. The nuclei are not, however, differentiated into macronuclei and micronuclei. The opalinids may be distant relatives of the ciliates, or they may represent a distinct evolutionary lineage. Like certain groups of mastigophorans, opalinids produce gametes that fuse to form a zygote; some authorities consider this an indication of a close relationship with those particular mastigophoran groups.

Phylum Sporozoa

All sporozoans are parasites. They are characterized by the lack of cilia or flagella and by complex life cycles. The best-known sporozoans are members of the genus *Plasmodium,* which cause malaria in many species of birds and mammals. A typical *Plasmodium* life cycle is shown in Figure 22–29. Malaria has generally been

22-29 *Life cycle of* Plasmodium vivax, *one of the sporozoans that cause malaria in humans. The cycle begins (a) when a female* Anopheles *mosquito "bites" a person with malaria, and, along with the blood, sucks up gametes (b) of the sporozoan. In the mosquito's digestive tract, the gametes unite (c) and form a zygote (d). From the zygotes, multinucleate structures called oocysts develop (e), which, within a few days, divide into thousands of very small, spindle-shaped cells, sporozoites (f), which then migrate to the mosquito's salivary glands. When the mosquito "bites" another victim (g), she infects the person with the sporozoites. These first enter liver cells (h), where they undergo multiple division (i). The products of these divisions (merozoites) enter the red blood cells (j), where again they divide repeatedly (k). They break out of the blood cells (l) at regular intervals of about 48 hours, producing the recurring episodes of fever characteristic of the disease. After a period of asexual reproduction, some of these merozoites become gametes, and, if they are ingested by a mosquito at this stage, the cycle begins anew.*

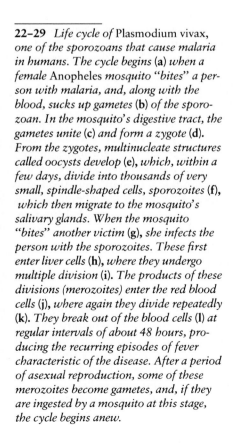
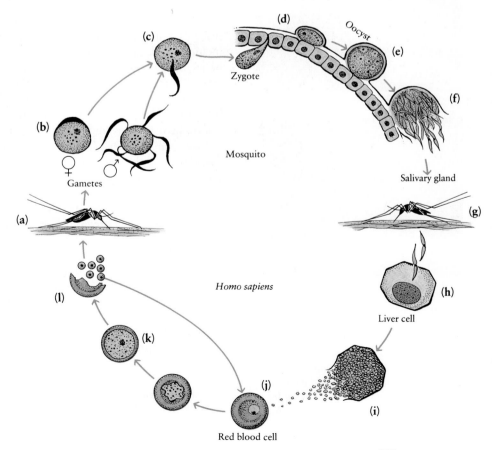

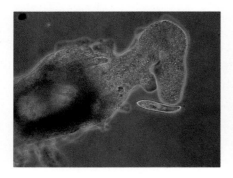

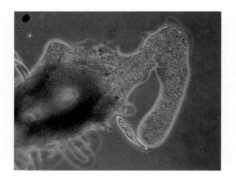

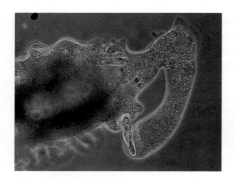

22–30 *The giant amoeba* (Chaos chaos) *capturing its prey. The initial stimulus produced by the prey—a* Paramecium— *induces pseudopodia from the amoeba and causes the amoeba to move toward the* Paramecium. *A large pseudopodium surrounds the* Paramecium, *drawing it toward the amoeba and engulfing it. A vacuole forms around the prey.*

controlled by drugs that act on the parasite at various stages of its life cycle and by insecticides that kill the mosquitoes that transmit it. Recently, however, both the parasite and the mosquitoes have begun to develop resistance to the chemicals used to attack them, and malaria remains a major cause of human death and disability. It is estimated that some 200 to 400 million people, principally in tropical regions of the world, are infected with the *Plasmodium* parasite. In Africa, about 10 percent of its victims die as a direct consequence of the infection and the remaining 90 percent experience repeated episodes of severe illness; mortality rates in young children often approach 50 percent.

With the advent of techniques for culturing *Plasmodium* in the laboratory, it has become possible to begin detailed studies of this parasite. Attention has been focused on the sporozoites, the merozoites, and the gametes—stages in which the life cycle is potentially most vulnerable to interruption—and particularly on the surface proteins of these cells. With recombinant DNA techniques, genes coding for the surface proteins are being identified, as are the regulatory mechanisms governing their expression. The hope is to use recombinant DNA techniques to prepare synthetic versions of key proteins that will function as effective vaccines, stimulating the immune system to attack the parasite. The development of such vaccines, however, is an extraordinarily complex task because, in the course of its evolution, the parasite has developed ways of altering these proteins, thereby evading attacks by the immune system. Despite the inherent difficulties and the relatively small amount of money devoted to research on malaria, considerable progress has been made in the last few years and there is hope that an effective vaccine—probably incorporating a variety of surface proteins—will eventually be available.

PATTERNS OF BEHAVIOR IN PROTISTS

As we have seen in previous chapters, even at the prokaryotic level of organization, organisms are capable of behavioral responses to environmental stimuli, including light, chemicals, and magnetic fields. Slightly more complex patterns of behavior are seen among the protists, at the lowest level of eukaryotic organization. Photosynthetic protists, such as *Euglena,* are especially sensitive to light intensity, moving into areas with optimal levels of light but out of areas where the light is too bright. Nonphotosynthetic protists, such as amoebas, may also exhibit **phototaxis,** or movement in response to light. If a bright pinpoint of light is focused on the advancing pseudopodium of an amoeba, the pseudopodium withdraws. If the entire body of the amoeba is exposed to bright light, the cell contracts suddenly and will even extrude any half-digested food. However, if it cannot escape from the light, the amoeba, after a few moments' hesitation, will resume its normal activities. This type of response, by which the stimulus comes to be ignored and a previous behavior pattern restored, is known as **habituation.** It is an important component of the behavior of multicellular eukaryotes, especially animals. (If our own cells were not capable of habituation, we would be continuously responding to and distracted by, for example, the touch of our own clothing or background noises.)

Amoebas are also **chemotactic,** that is, responsive to chemical stimuli. When something edible, an algal cell or a fellow protozoan, is in the vicinity of an amoeba, the amoeba can sense it at some distance, at least the length of its own body away. It then sends out a pseudopodium that is shaped quite specifically for its intended victim. A fine pincerlike projection will be formed for a small, quiet morsel; a much stronger and more massive pseudopodium will reach for a large ciliate or vigorously moving object (Figure 22–30). If the intended victim moves away, the amoeba will remain in pursuit so long as it is close enough to the prey to receive stimuli from it.

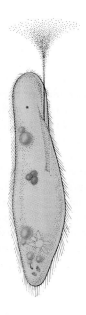

22–31 Paramecium *samples a drop of India ink. Currents created by the strongly beating cilia draw particles of the ink toward the protist's buccal cavity.*

Avoidance in *Paramecium*

Other patterns of protist behavior have been studied extensively in ciliates. *Paramecium,* for example, shows behavior that appears more complex than that of amoebas but is just as simple in principle. These protists are responsive to a variety of stimuli, including very subtle changes in temperature and chemistry that we can detect only by finely calibrated instruments. But though the detection methods are highly sensitive, the responses are fixed (stereotyped).

As a *Paramecium* swims, the strongly beating cilia around the oral groove create currents that constantly bring a sample of water to the buccal cavity, which seems to be the testing area. In this way, the *Paramecium* continuously explores the environment that lies ahead, shifting and sampling (Figure 22–31).

In general, *Paramecium* responds to changes in its environment by avoidance reactions. If it receives a negative stimulus, it turns away. If it turns, it always turns in the same direction, to its left, which is the aboral side—the side away from the oral groove. This occurs because of the relaxation of the beat of the body cilia. It does not matter which side the stimulus comes from. If the microscopist takes a blunt needle and jabs a *Paramecium* on the aboral side, it will still turn toward that side; and once it has turned, it will continue in this new direction indefinitely.

As shown in Figure 22–32, if the negative stimulus is powerful, such as a poisonous chemical, the *Paramecium* will stop short, reverse its ciliary beat and back up, turn toward the aboral side about 30°, and then start forward again, testing. If necessary, it will repeat this performance. Under a strong negative stimulus, it will turn a full 360° and will continue to turn until an avenue of escape is found. It can find its way around solid objects in the same way.

The end result of this, of course, is about the same as if the organisms were "attracted" to a favorable situation. For example, *Paramecium* is extremely temperature-sensitive. If you place the cells in a culture dish that can be made warmer at one end than another, you will find, by trying different combinations of temperatures, that they will tend to congregate in water that is about 27°C (80°F). If you place them on a slide with water that is below 27°C and then warm one little portion of it gently (you can do this by placing a hot needle on the cover slip), the cells will all gather in this warm spot. They reach it by chance as they move about through the water. But once they get into it, they literally cannot get out again, since as soon as they reach the edge of the warm water, they receive a sample of cooler water, which will turn them right back again. Similarly, on a slide that contains water that is too hot, they will trap themselves in an area that has been cooled slightly, even by only a few degrees.

22–32 *Avoidance behavior in* Paramecium. *The colored area at the top of the figure represents a drop of a toxic substance, and the arrows indicate the direction of movement. The* Paramecium (**a**) *moves up to the substance, (**b**) tests it, (**c**) backs up, (**d**) turns 30°, and (**e**) moves forward in the new direction. Many protists are capable of this sort of simple behavior.*

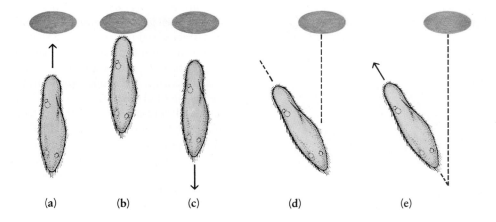

(a) (b) (c) (d) (e)

They have also been shown to congregate in the same way in a spot that is slightly acidic, by avoiding the neutral or alkaline areas. Bacteria, which are a primary food of *Paramecium,* create a slightly acidic environment by their metabolism, and indeed, *Paramecium* itself creates a slightly acidic environment by giving off carbon dioxide; so their avoidance of neutral or alkaline areas enhances the tendency of these protists to group together and also to gather where there is food. This tendency is strengthened further by the fact that they are more likely to cling to some other object—such as the leaf or stalk of a plant, a pile of debris, or even a shred of filter paper—if they are in a slightly acidic medium.

As described here, the behavior of *Paramecium,* helpless to escape from a drop of warm water or forced to cling by the presence of carbon dioxide, seems very unlifelike. Yet, as one watches the cell negotiating the changes in its environment, its actions seem purposeful. Both of these impressions are true. The individual *Paramecium* has no choice as to its behavior, but the behavior of these protists, in general, is a result of millions of years of choice, the choice by natural selection.

SUMMARY

The kingdom Protista comprises an enormous variety of eukaryotic organisms, mostly unicellular with some relatively simple multicellular forms. A major factor in the evolution of the eukaryotes may have been the establishment of symbiotic relationships with prokaryotic cells that, internalized, eventually became specialized as mitochondria and chloroplasts. It is thought that the protists represent a number of quite distinct phylogenetic lineages and, moreover, that all other eukaryotic organisms are derived from primitive protists. We have considered 15 divisions and phyla of protists, informally grouped into photosynthetic autotrophs (algae), multinucleate and multicellular heterotrophs (slime molds and water molds), and unicellular heterotrophs (protozoa).

The photosynthetic protists are classified into six divisions on the basis of their photosynthetic pigments, reserve food supply, and cell wall composition. Euglenophyta represent a small group of unicellular algae, mostly found in fresh water. They contain chlorophylls *a* and *b* and store carbohydrates in an unusual starch-like substance, paramylon. The cells lack a wall but have a flexible series of protein strips, which make up the pellicle, inside the cell membrane. The cells are highly differentiated, containing chloroplasts, a contractile vacuole, an eyespot, and flagella. No sexual cycle is known. This division also contains nonphotosynthetic forms.

The Chrysophyta—diatoms and golden-brown algae—are important components of freshwater and marine phytoplankton. They are unicellular. Diatoms are characterized by fine, double silicon-containing shells. They usually reproduce asexually, but syngamy also occurs.

The Dinoflagellata are unicellular biflagellates, many of which are marine. They are characterized by two flagella that beat in different planes, causing the organism to spin; dinoflagellates often have stiff, bizarrely shaped cellulose walls.

The Chlorophyta, or green algae, are thought to represent four distinct evolutionary lineages. The plants appear to have originated from the same lineage as the green algae of class Charophyceae. All green algae and plants have chlorophylls *a* and *b* and beta-carotene as their photosynthetic pigments and store their food reserves in the form of starch. Members of class Charophyceae, like plants, have cellulose-containing cell walls, and a few also form a cell plate during cell division.

Among the green algae are a variety of forms, representing different degrees of complexity: unicellular organisms, colonies, coenocytic (multinucleate) forms, and true multicellular organisms. The reproductive cycles of green algae are often quite complex. In species with sexual cycles, the gametes of different mating types may be similar in size and structure (isogamy), different in size but both motile (anisogamy), or different in size, with one, usually the larger, not motile (oogamy). Some multicellular green algae have a life cycle known as alternation of generations, in which a haploid phase alternates with a diploid phase. The haploid *(n)* generation, known as the gametophyte, produces haploid gametes. The gametes fuse to form the zygote, which develops into a diploid *(2n)* sporophyte. The sporophyte produces spores by meiotic division. A spore is a single cell that, unlike a gamete, can develop into an adult organism without combining with another cell. In organisms with alternation of generations, the spore, which is haploid, germinates to produce the haploid gametophyte.

The Phaeophyta (brown algae) and Rhodophyta (red algae) are the principal seaweeds. The brown algae, which include the kelps, are found more commonly in cooler water; the red algae, in the tropics. In many brown algae the thallus is differentiated into holdfast, stipe, and blade, analogous to the root, stem, and leaf of plants. Most kelps have tissues specialized for the conduction of sugar from the blades to nonphotosynthetic parts of the thallus.

The slime molds are heterotrophic, amoeba-like organisms that reproduce by the formation of spores. There are two principal divisions: Myxomycota (plasmodial slime molds), which are coenocytic during the nonreproductive stages, and Acrasiomycota (cellular slime molds), in which the aggregating amoeboid cells retain their individual identity. The water molds (divisions Chytridiomycota and Oomycota) are coenocytic heterotrophs that superficially resemble fungi. They reproduce both asexually and sexually. In chytrids, both the spores and the gametes are flagellated, whereas in oomycetes, only the spores are flagellated; all oomycetes exhibit oogamy.

Unicellular heterotrophic protists (the protozoa) are thought to have evolved from nonphotosynthetic, flagellated ancestors. Among the protozoa are some of the largest known cells and also the most complex. Three phyla—the Mastigophora (flagellates), the Sarcodina (amoebas and their relatives), and the Ciliophora (ciliates)—contain both free-living and parasitic species, and their members may be identified on the basis of their locomotor structures. Two phyla—Opalinida and Sporozoa—contain only parasitic forms. The opalinids are ciliated, but the sporozoans have no locomotor organelles. In the course of their complex life cycles, however, sporozoans are transported quite efficiently from host to host.

Protists exhibit a variety of simple behavioral responses that prefigure the complex behaviors characteristic of multicellular eukaryotes. Among the protist responses are phototaxis (in both photosynthetic and nonphotosynthetic organisms), chemotaxis, avoidance, and habituation.

QUESTIONS

1. Describe the similarities and differences among the Euglenophyta, the Chrysophyta, and the Dinoflagellata.

2. To which other divisions or phyla of protists may the Euglenophyta, the Chrysophyta, and the Dinoflagellata be related? Describe the evidence for the relationships you cite.

3. In some classification schemes, the Chlorophyta, the Phaeophyta, and the Rhodophyta are placed in the plant kingdom. What similarities between the plants and these three divisions of algae might justify such a placement? What differences between the plants and these divisions justify their placement in kingdom Protista? Which of these similarities and differences are most likely to be homologous and which analogous?

4. Describe three different pathways to multicellularity, as exemplified by organisms discussed in this chapter.

5. In the previous chapter, the fruiting bodies of the myxobacteria were described as "multicellularity by committee." To which protists could such a description also apply? Explain your answer.

6. Distinguish among the following: colonial organism/multicellular organism; syngamy/fertilization; isogamy/anisogamy/oogamy; isomorphic/heteromorphic; sporophyte/gametophyte; spore/gamete/zygospore.

7. Depending on the stage in the life cycle of an organism at which meiosis occurs, it may result in one of three types of haploid cells. All three types are found among the various protists.

Identify the three types of haploid meiotic products, and for each give an example of a protist in which it is formed.

8. Explain alternation of generations, using as your example the sea lettuce, *Ulva.*

9. Name the five phyla of unicellular heterotrophic protists, and give the distinguishing characteristics of each.

10. Label the drawing below.

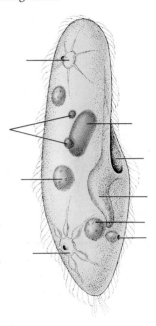

11. Consider the life cycle of *Plasmodium*. At what stages in the cycle do its numbers increase? Why might a parasite that requires several hosts find it advantageous to evolve a life cycle in which its numbers increase at several stages? Why might it be advantageous to a parasite to have a second host, such as the mosquito?

12. Early work on malaria vaccines has suggested that the *Plasmodium* life cycle may be most vulnerable to disruption at the stage of gamete fusion, which occurs in the *Anopheles* mosquito. A synthetic vaccine, based on surface proteins of the gametes, triggers the production of circulating antibodies in the blood of experimental animals with malaria. When a mosquito feeds on these animals, it ingests both *Plasmodium* gametes and the antibodies. In the mosquito, the antibodies cause the male gametes to clump together, preventing fusion of gametes to form zygotes. However, medical scientists involved in malaria research feel that a vaccine that had *only* this effect would be unethical, and they are also seeking vaccine components that will disrupt the life cycle at the sporozoite and merozoite stages as well. Why?

13. Given that light is essential for photosynthesis, it is easy to understand the adaptive significance of behavioral responses to light by photosynthetic protists. The significance of such responses for nonphotosynthetic organisms is not so obvious. What might be the advantages to an amoeba of a capacity to respond to light?

14. Arrange in order of evolutionary development: shell or skeleton, phagocytosis, active transport, multicellularity, symbiosis. (Like the rest of us, you will only be guessing, but be prepared to defend your guess.)

The Fungi

As we saw in the last chapter, the eukaryotic cell, by virtue of its size and complexity, has a number of properties that made possible the great diversification of the protists in both structure and mode of life. These properties include the capacity to carry a great deal of genetic information and to transmit it reliably from generation to generation; the compartmentalization and specialization of different parts of the cell for different functions, leading to greater efficiency; the ability to acquire more food; and greater adaptability to life-threatening environmental change.

There are, however, limits to the size that a single cell can attain and still function efficiently. One critical factor is the surface-to-volume ratio (page 102); the larger the cell, the greater the quantity of materials that must be moved in and out, but this movement can occur only through the cell surface. Another critical factor is the capacity of the nucleus to regulate a large amount of cytoplasm and the diverse functions of a complex cell. One solution to these problems is multicellularity: the repetition of individual units—cells—each with an efficient surface-to-volume ratio and each with its own nucleus. Another solution is the flattening or spreading out of the cells and the presence of multiple nuclei in a common cytoplasm, exemplified most strikingly in the fungi (Figure 23–1). Unlike true multicellularity, this solution does not open the way for great diversification of form, but it is nonetheless an evolutionary success. Fungi, like bacteria, are literally everywhere; thus far about 100,000 distinct species have been identified, and it is estimated that another 200,000 species await discovery.

23–1 *A fungus growing on a fallen tree trunk. The bulk of its body consists of masses of thin filaments through which nutrients are absorbed. In many fungi, the cells are incompletely separated by perforated cell walls; both cytoplasm and nuclei flow through the filaments.*

Chitin

$$HNCOCH_3 \quad CH_2OH \quad HNCOCH_3$$

(a)

Cellulose

(b)

23–2 *The cell walls of fungi contain the polysaccharide chitin (**a**) rather than cellulose (**b**), the polysaccharide found in the cell walls of plants. Chitin resembles cellulose in that it is tough, inflexible, and insoluble in water. As you can see, the two molecules are also structurally very similar; in chitin, the hydroxyl (OH) group on the carbon atom in position 2 of each glucose subunit is replaced by a nitrogen-containing group.*

23–3 *Gill fungi on the trunk of a dead tree in southern Ontario. The only portions of these fungi visible are the spore-producing structures, composed of tightly packed hyphae; the bulk of the mycelium is below the surface of the dead trunk. This fungus,* Flammulina velutipe, *is commonly known as velvet foot or as winter mushroom.*

CHARACTERISTICS OF THE FUNGI

The fungi are so unlike any other organisms that, although they were long classified with the plants, biologists now assign them to a separate kingdom. Although some fungi, including the yeasts, are unicellular, most species are composed of masses of coenocytic or multicellular filaments. A fungal filament is called a hypha, and all the hyphae of a single organism are collectively called a **mycelium.** The walls of the hyphae are composed primarily of chitin (Figure 23–2), a polysaccharide that is never found in plants. (It is, however, the principal component of the exoskeleton—the hard outer covering—of insects and other arthropods.) The visible structures of most fungi represent only a small portion of the organism; these structures, such as mushrooms, are tightly packed hyphae, specialized for the production of spores.

A mycelium normally arises by the germination and outgrowth of a single spore, with growth taking place only at the tips of hyphae. All fungi are nonmotile throughout their life cycle, although spores may be carried great distances by the wind. Growth of the mycelium substitutes for motility, bringing the organism into contact with new food sources and different mating strains. This growth can be quite rapid—some fungi can grow a mass of new hyphae in 24 hours that, if placed end-to-end, would total more than a kilometer in length.

All fungi are heterotrophs, either saprobes or parasites. Because of their filamentous form, each fungal cell is no more than a few micrometers from the soil, water, or other substance in which the fungus lives, and is separated from it by only a thin cell wall. Because their cell walls are rigid, fungi are unable to engulf small microorganisms or other particles. They obtain food by absorbing dissolved inorganic or organic materials. Typically a fungus will secrete digestive enzymes onto a food source and then absorb the smaller molecules that are released. The mycelium may appear as a mass on the surface of the food source or may be hidden beneath the surface. Parasitic fungi often have specialized hyphae, called **haustoria** (singular, haustorium), that absorb nourishment directly from the cells of the host organism. (As we shall see on page 691, some parasitic plants have analogous structures, also known as haustoria.)

The fungi, together with the bacteria, are the principal decomposers of organic matter. It is estimated that the top 20 centimeters of fertile soil contain, on the average, nearly 5 metric tons of fungi and bacteria per hectare (2.47 acres). As we shall see in Section 8, the activities of these organisms are as vital to the continued function of the earth's ecosystems as are those of the food producers. Some fungi are, from the human standpoint, destructive, attacking our foodstuffs, our domestic plants and animals, our shelter, our clothing, and even our persons. Others, however, are essential for the production of bread, cheese, and wine. Moreover, fungi are the source of a number of antibiotics and other life-saving medications.

10 μm

23-4 *A sporangium, an asexual reproductive structure of a fungus, from the black bread mold* Rhizopus. *The contents of the sporangium are segregated from the rest of the mycelium by a cell membrane and a cell wall.*

Reproduction in the Fungi

Most fungi reproduce both asexually and sexually. Asexual reproduction takes place either by the fragmentation of the hyphae (with each fragment becoming a new individual) or by the production of spores. In some of the fungi, spores are produced in sporangia, which are borne on specialized hyphae called **sporangiophores.** Fungal spores are often, but not necessarily, resting forms, surrounded by a tough, resistant wall. Like the spores of other organisms, these spores are able to survive during periods of drought or extreme temperatures. The airborne spores of some fungi are very small and therefore can remain suspended in the air for long periods and be widely dispersed. Often the sporangia are raised above the mycelium by the sporangiophores; thus the spores are easily caught up and transported by air currents. The bright colors and powdery textures associated with many types of molds are the colors and textures of the spores and sporangia.

Sexual reproduction in many fungi involves the specialization of portions of the hyphae to form gametangia (page 467). The contents of a gametangium, like those of a sporangium, are separated from the hypha from which it is formed by a cell membrane and a complete cell wall, known as a **septum.** Sexual reproduction can occur in a variety of ways: (1) by fusion of gametes that have been released from the gametangia, (2) by fusion of gametangia, or (3) by fusion of unspecialized hyphae.

Sometimes the fusion of fungal hyphae is not followed immediately by the fusion of nuclei. Thus strains of fungi may exist with two or more genetically distinct kinds of nuclei operating simultaneously. When such a combination contains two nuclei of complementary mating types, it is known as a **dikaryon.** Dikaryons are found uniquely among the fungi.

CLASSIFICATION OF THE FUNGI

The members of the kingdom Fungi are generally classified into three principal divisions: Zygomycota, the zygomycetes; Ascomycota, the ascomycetes; and Basidiomycota, the basidiomycetes. The criteria used in distinguishing these three divisions involve both features of basic structure and patterns of reproduction, particularly sexual reproduction (Table 23–1). An additional taxon, Deuteromycota, or the Fungi Imperfecti, includes fungi in which sexual reproduction is unknown, either because it has been lost in the course of evolution or because it has not been observed. Also included in this taxon for convenience are certain

---

TABLE 23–1 **The Kingdom Fungi**

DIVISION	NUMBER OF SPECIES	EXAMPLES	DISTINCTIVE CHARACTERISTICS	DISEASES	ECONOMIC USES
Zygomycota	About 600	Black bread mold	Formation of zygospores (tough, resistant spores resulting from a fusion of gametangia)	Few	None
Ascomycota	30,000	*Neurospora,* yeasts, morels, truffles	Formation of fine asexual spores (conidia); sexual spores in asci; hyphae divided by perforated septa; dikaryons	Powdery mildews of fruits, chestnut blight, Dutch elm disease, ergot	Food (morels, truffles); wine-, beer-, and bread-making (yeasts)
Basidiomycota	25,000	Toadstools, mushrooms, rusts, smuts	Sexual spores in basidia; hyphae divided by perforated septa; dikaryons	Rusts, smuts	Food (mushrooms)
Deuteromycota (Fungi Imperfecti)	25,000	*Penicillium*	Fungi with no known sexual cycles	Ringworm, thrush	Cheeses, antibiotics

23–5 *The perforated cross wall of an ascomycete showing a nucleus squeezing through the perforation. The fungus is* Neurospora crassa, *the red bread mold.*

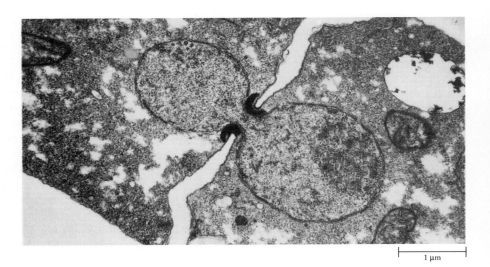

1 μm

other closely related fungi whose sexual stages are known. Taxonomists refer to groups like this as "waste baskets" because species are included here only because they do not fit in other taxa. Because of similarities between the patterns of asexual reproduction in many of the Fungi Imperfecti and the ascomycetes, this taxon is sometimes considered a class of division Ascomycota; most authorities, however, rank it at the level of division.

The ancestors of the fungi were probably unicellular eukaryotic organisms that apparently have no living counterparts. These organisms are thought to have given rise to three distinct lineages, one leading to the modern chytrids (page 467), a second leading to the oomycetes (page 467), and a third leading to the zygomycetes. All three of these groups are characterized by a coenocytic organization, and with the exception of the reproductive structures, there is no compartmentalization of the hyphae. However, the chytrids and the oomycetes differ so significantly from the fungi in a number of basic features that most authorities believe that they should be classified in kingdom Protista, a recommendation that we have followed in this text.

In both the ascomycetes and the basidiomycetes, the hyphae are septate—divided by transverse cell walls—but the walls are perforated, and the cytoplasm and even the nuclei (Figure 23–5) are able to flow through the septa. It is generally thought that the ascomycetes and the basidiomycetes are derived from a common ancestor, and that this ancestor and the zygomycetes evolved from an earlier common ancestor.

DIVISION ZYGOMYCOTA

The zygomycetes are terrestrial fungi, most of which are saprobes, living in the soil and feeding on dead plant or animal matter. Some are parasites of plants, insects, or small soil animals. Their sexual reproduction is characterized by the formation of zygospores, which, as we saw in the last chapter, are thick-walled, resistant spores that develop from a zygote.

One of the most common members of this division is *Rhizopus stolonifer,* the black bread mold. Infection begins when a spore germinates on the surface of bread, fruit, or some other organic matter and forms hyphae. Some of the hyphae extend rhizoids, which anchor the fungus to the substrate, secrete digestive enzymes, and absorb dissolved organic materials. Other specialized hyphae, the sporangiophores, push up into the air, and sporangia form at their tips. As the sporangia mature, they become black, giving the mold its characteristic color.

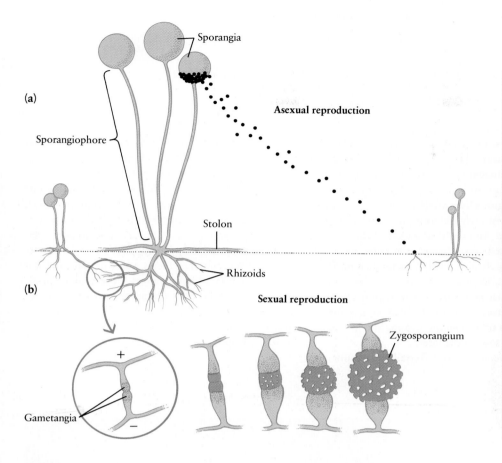

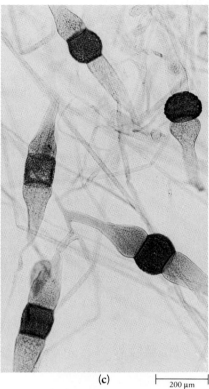

(c)   200 μm

23–6 *Asexual and sexual reproduction in a black bread mold. The mycelium consists of branched hyphae, including rhizoids, which anchor the organism and absorb nutrients; stolons, which run above the surface of the bread; and sporangiophores, which elevate the sporangia. (a) At maturity, the fragile wall of the sporangium disintegrates, releasing the asexual* spores, *which are carried away by air currents. Under suitable conditions of warmth and moisture, the spores germinate, giving rise to new masses of hyphae.* **(b)** *Sexual reproduction occurs when two hyphae from different mating strains come together, forming gametangia. After fusion, the gametangia develop into a thick-walled, resistant structure, the zygo-* sporangium, *which contains a number of zygotes. After a period of dormancy, the zygotes undergo meiosis, and the zygosporangium germinates, producing a new sporangium from which haploid spores are released.*

(c) Zygosporangia of the black bread mold Rhizopus stolonifer.

They eventually break open, releasing numerous airborne spores, each of which can germinate to produce a new mycelium.

Sexual reproduction in *Rhizopus* occurs when the specialized hyphae of two different mating strains meet and fuse, attracted toward one another by hormones that diffuse in the form of gases. The two strains are designated + and −, since there are no structural differences between them on which to base male and female designations. Septa, or cross walls, form behind the tips of the touching hyphae; the two tip cells thus formed are gametangia, one containing numerous + nuclei, the other containing numerous − nuclei. The two gametangia fuse, and then many pairs of + and − nuclei fuse, producing diploid nuclei; any unfused haploid nuclei degenerate. The resulting multinucleate cell, containing a number of zygotes, forms a hard, warty wall and becomes a dormant zygosporangium. During this dormant stage, the zygosporangium can survive periods of extreme heat or cold or desiccation. At the end of dormancy, just prior to germination, the diploid nuclei undergo meiosis. Half of the haploid nuclei resulting from meiosis are + nuclei, and half are − nuclei. Following meiosis, the zygosporangium germinates, producing a sporangium from which both + and − airborne spores are released.

Ready, Aim, Fire!

Over the millennia, the fungi have evolved a variety of methods that ensure wide dispersal of their spores. One of the most ingenious is found in *Pilobolus*, a zygomycete that grows on dung. The sporangiophores of this fungus, which attain a height of 5 to 10 millimeters, are positively phototropic—that is, they grow toward light. An expanded region of the sporangiophore located just below the sporangium (known appropriately as the subsporangial swelling) functions as a lens, focusing the rays of the sun on a photoreceptive area at its base. The region of the sporangiophore *away* from the focused light grows more rapidly than other regions, causing the sporangiophore to curve toward the light.

A vacuole in the subsporangial swelling contains a high concentration of solutes, with the result that water moves in by osmosis. Eventually the pressure becomes so great that the swelling splits, firing the intact sporangium in the direction of the light. The initial velocity approaches 50 kilometers per hour, and the sporangia often travel farther than 2 meters —an enormous distance, when you consider that mature sporangia are only 80 micrometers in diameter. Each sporangium adheres where it lands, and if it should land on a blade of grass—a reasonable possibility, since it was fired toward the light—it may be eaten by a grazing animal. It then passes through the digestive tract of the animal unharmed and is deposited in the dung, where the cycle begins anew.

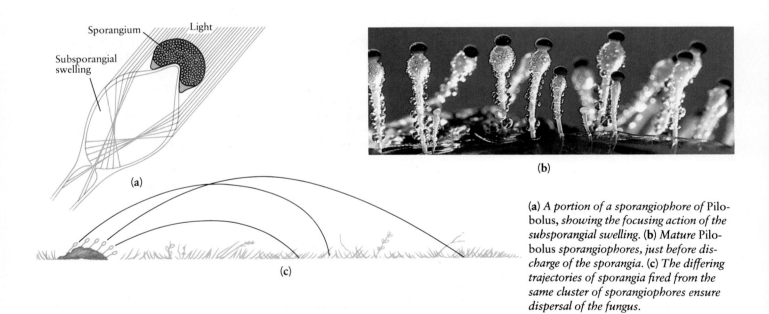

(a) *A portion of a sporangiophore of* Pilobolus, *showing the focusing action of the subsporangial swelling.* (b) *Mature* Pilobolus *sporangiophores, just before discharge of the sporangia.* (c) *The differing trajectories of sporangia fired from the same cluster of sporangiophores ensure dispersal of the fungus.*

DIVISION ASCOMYCOTA

The ascomycetes are the largest division of fungi, with some 30,000 species, plus an additional 25,000 species that are found only in lichens (page 488). Among the ascomycetes are the yeasts and powdery mildews, many of the common black and blue-green molds, and the morels and truffles prized by gourmets. Members of this group of fungi are the cause of many plant diseases, such as chestnut blight and Dutch elm disease, and are the source of many of the antibiotics. The red bread mold *Neurospora,* which played a major role in the history of modern genetics, is an ascomycete.

(a)   **(b)**

23–7 *Two ascomycetes.* (**a**) *A common morel,* Morchella esculenta. *These (and the truffles) are among the most prized of the edible fungi. The structure recognized as the morel is the ascocarp, in which asci and ascospores are produced.* (**b**) *Scarlet cup,* Sarcoscypha coccinea, *a harbinger of spring in hardwood forests throughout the United States. It is usually found arising from a fallen branch.*

In ascomycetes, as we noted earlier, the hyphae are divided by cross walls, or septa. Each compartment generally contains a separate nucleus, but the septa have pores in them through which the cytoplasm and the nuclei can move (see Figure 23–5). The life cycle of an ascomycete (Figure 23–8) typically includes both asexual and sexual reproduction. Asexual spores are commonly formed either singly or in chains at the tip of a specialized hypha. They are characteristically very fine and so are often called **conidia,** from the Greek word for "dust."

Sexual reproduction in the ascomycetes always involves the formation of an **ascus** ("little sac"), a structure that characterizes this division (Figure 23–9a).

23–8 *Life cycle of an ascomycete. An ascospore (upper left) germinates to produce a haploid monokaryotic mycelium, which reproduces through the formation of asexual spores (conidia). When monokaryotic mycelia of different mating strains form gametangia, the stage is set for sexual reproduction. A bridge forms between female (color) and male (black) gametangia, allowing the haploid male nuclei to enter the female gametangium. The hyphae that proliferate from this gametangium are dikaryotic—that is, each cell contains a pair of haploid nuclei, one of each parental type. These dikaryotic hyphae, with interspersed monokaryotic hyphae, give rise to the ascocarp (bottom). In the ascocarp, the dikaryotic hyphae grow and differentiate to form the asci, within which the haploid nuclei fuse. The resulting diploid nucleus undergoes meiosis, producing four new haploid nuclei. These nuclei then divide mitotically, and the mature ascus thus contains eight haploid ascospores. With the release and germination of the ascospores, the cycle begins once more.*

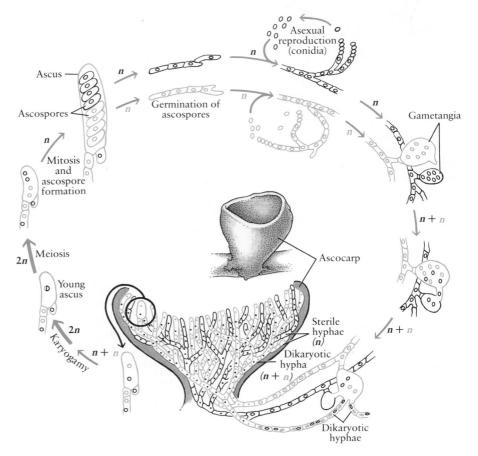

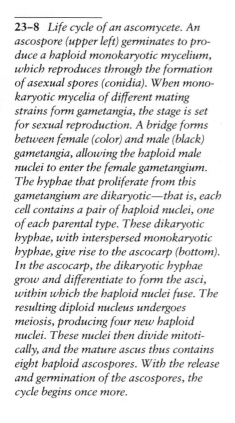

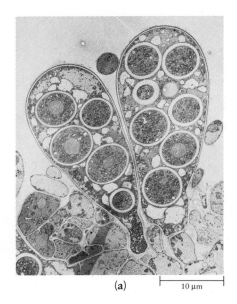

(a)   10 μm

(b)

23–9 (a) *Electron micrograph of two asci in which ascospores are maturing. The closed ascus within which the sexually produced spores develop is the "trademark" of the ascomycete.* (b) *Asci of the red bread mold* Neurospora. *Each ascus contains eight haploid ascospores, lined up in the order in which they were produced by meiosis and the subsequent mitotic division.*

Depending on the species, ascus formation is preceded by the fusion of gametes, gametangia, or unspecialized hyphae of different mating strains. The nuclei form pairs—dikaryons—that divide synchronously as the hypha grows. Eventually some of the nuclei fuse; this is the only truly diploid stage in the life cycle. The diploid nuclei immediately undergo meiosis, producing four haploid nuclei, which then usually divide mitotically, producing eight haploid nuclei (Figure 23–9b). Each of these nuclei becomes surrounded by a tough wall; a mature ascus contains eight of these spores (ascospores). In most ascomycetes, the asci are formed in complex structures called ascocarps. At maturity the asci become turgid and finally burst, releasing their ascospores explosively into the air.

Single-celled ascomycetes are known as yeasts. Many yeasts are adapted to environments with high sugar content, such as the nectar of flowers or the surface of fruits and, as we noted in Chapter 9, they are responsible for the fermentation of fruit juice to wine. Yeasts are characteristically small, oval cells that reproduce asexually by budding. Sexual reproduction in yeasts occurs when two cells (or two ascospores) unite and form a zygote. The zygote may produce diploid buds or may undergo meiosis to produce four haploid nuclei. There may be a subsequent mitotic division. Within the zygote wall, which is now an ascus, walls are laid down around the haploid nuclei, forming ascospores. The ascospores are liberated when the ascus wall breaks down.

Many ascomycetes are plant parasites. Ergot, for instance, one of the most famous fungus-produced diseases, is caused by *Claviceps purpurea*, a parasite of rye. Although ergot seldom causes serious damage to a crop of rye, it is dangerous because a small amount mixed with rye grains is enough to cause severe illness among domestic animals that eat the grain or among people who eat bread made with the flour. Ergotism is often accompanied by gangrene, nervous spasms, psychotic delusions, and convulsions. It occurred frequently during the Middle Ages, when it was known as St. Anthony's fire. In one epidemic in the year 994, more than 40,000 people died. Ergot, which causes muscles to contract and blood vessels to constrict, has various medical uses. It is also the initial source for the psychedelic drug lysergic acid diethylamide (LSD).

DIVISION BASIDIOMYCOTA

The most familiar basidiomycetes are mushrooms. The mushroom, or basidiocarp, which is the spore-producing body, is composed of masses of tightly packed hyphae. The mycelium from which the mushrooms are produced forms a diffuse mat, which may grow as large as 35 meters in diameter. Mushrooms usually form at the outer edges, where the mycelium grows most actively, since this is the area in which there is the most nutritive material. As a consequence, the mushrooms appear in rings, and as the mycelium grows, the rings become larger and larger in diameter. Such circles of mushrooms, which might appear in a meadow overnight, were known in European folk legends as "fairy rings." They can make such a rapid appearance because most of the new protoplasm is produced underground, in the mycelium. The protoplasm then streams into the new hyphae of the fruiting body as it forms above ground.

The basidiomycetes, like the ascomycetes, have hyphae subdivided by perforated septa. Sexual reproduction is initiated by the fusion of haploid hyphae to form a dikaryotic mycelium (Figure 23–10). The dikaryotic mycelium may persist for years, forming an elaborate structure. Eventually, some of the nuclei fuse to form diploid nuclei that immediately undergo meiosis. Fusion and meiosis always take place in a specialized hypha called a **basidium** (from the Greek word for "club"). The spores (basidiospores) are formed externally on the basidium. Many of the larger basidiomycetes seem to have lost the capacity to produce asexual spores.

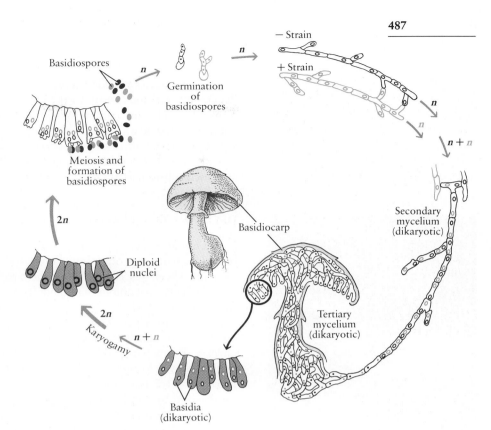

23–10 *Life cycle of a basidiomycete. Basidiospores (upper left) germinate to produce primary monokaryotic mycelia. Secondary dikaryotic mycelia are formed by the fusion of hyphae from different mating types. The secondary mycelia grow and differentiate to form the reproductive structures (basidia). In mushrooms, the basidia form within the gills. After the basidium enlarges, the two nuclei, one from each mating strain, fuse. Meiosis follows almost immediately, resulting in the formation of four nuclei, from each of which a basidiospore develops. After the basidiospores are released, the basidiocarp disintegrates.*

23–11 *Three basidiomycetes. (a) Corn smut, a common fungal disease of corn. The black, dusty-looking masses are spores. (b) A shelf fungus, which grows on decaying wood. (c) Amanita bisporigera. Members of the genus* Amanita *include the most beautiful and also the most poisonous of the mushrooms. The "skirt" near the top of the stalk is one of the identifying characteristics of this genus. One mushroom has been picked to show the gills, on which the sexual spores are formed. The toxin in* Amanita bisporigera *consists of two linked cyclopeptides, each containing eight amino acids. This protein binds to an RNA polymerase in liver cells and causes acute liver damage, resulting in death.*

The best-known mushrooms belong to the group known as the gill fungi. The spores of these fungi are found in the furrows, or "gills," under the cap. If you cut off the cap of a mature mushroom and place it on a piece of white paper, it will release fine spores that trace out a negative copy of the gill structure. The spores, which come in a wide range of colors, are a useful means of identifying various mushrooms. Varieties of the gill fungus *Agaricus campestris,* the common field mushroom, are among the most familiar of the mushrooms commercially cultivated in North America. Most of the known poisonous mushrooms are also gill fungi. Mushrooms of the genus *Amanita* are the most highly poisonous of all mushrooms; even one bite of the white *Amanita bisporigera,* a "destroying angel," can be fatal. Some species of toxic mushrooms, such as *Psilocybe mexicana* (the source of psilocybin), are eaten for their hallucinogenic effects.

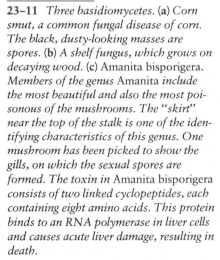

(a)

(b)

(c)

Other types of basidiomycetes include puffballs (a few of which are a meter in diameter), earthstars, stinkhorns, jelly fungi, and the parasitic rusts and smuts, some of which cause severe losses among cereal crops. Another basidiomycete, white rot fungus, a voracious destroyer of wood, is currently being investigated for its potential in destroying toxic wastes. This fungus contains a complex enzyme that is able to degrade not only the tough polymers found in wood but also such substances as DDT, dioxin, and a wide range of organic pollutants.

DIVISION DEUTEROMYCOTA

As we noted earlier, the deuteromycetes, or Fungi Imperfecti, are generally fungi in which sexual reproduction is unknown. About 25,000 species of deuteromycetes have been described, among them parasites that cause diseases of plants and animals. The most common human diseases caused by this group are infections of the skin and mucous membranes, such as ringworm (including "athlete's foot") and thrush (to which infants are particularly susceptible). A few deuteromycetes are of economic importance due to the part they play in the production of certain cheeses (Roquefort and Camembert, for example) and of antibiotics, including penicillin. Cyclosporin, a compound that suppresses the immune reactions involved in the rejection of organ transplants (to be discussed in Chapter 39), is synthesized by a soil-dwelling deuteromycete. Consisting of 13 amino acids, one of which appears to be synthesized only by this fungus, cyclosporin has made possible a dramatic increase in successful heart transplants.

SYMBIOTIC RELATIONSHIPS OF FUNGI

Although most fungi are saprobes, living on dead organic matter, a large number of fungi are parasitic on plants and animals, causing a variety of diseases. Fungi are also involved in other types of symbioses. Two of these—lichens and mycorrhizae—have been and are of extraordinary importance in enabling photosynthetic organisms to become established in previously barren terrestrial environments.

The Lichens

A lichen is a combination of a specific fungus and a green alga or a cyanobacterium. The product of such a combination is very different from either the photosynthetic organism or the fungus growing alone, as are the physiological conditions under which the lichen can survive. The lichens are widespread in nature. They occur from arid desert regions to the Arctic and grow on bare soil, tree trunks, sun-baked rocks, fence posts, and windswept alpine peaks all over the world; they have even been found growing in air spaces within Antarctic rocks, a few millimeters below the frigid rock surface. Lichens are often the first colonists of bare rocky areas; their activities begin the process of soil formation, gradually creating an environment in which mosses, ferns, and other plants can gain a foothold.

Lichens do not need an organic food source, as do their component fungi, and unlike many free-living algae and cyanobacteria, they can remain alive even when desiccated. They require only light, air, and a few minerals. They apparently absorb some minerals from their substrate (this is suggested by the fact that particular species are characteristically found on specific kinds of rocks or soil or tree trunks), but minerals reach the lichen primarily through the air and in rainfall. Because lichens rapidly absorb substances from rainwater, they are particularly susceptible to airborne toxic compounds. Thus, the presence or absence of lichens is a sensitive index of air pollution.

├─── 50 μm ───┤

23–12 Penicillium, *an "imperfect fungus," showing conidiophores ("conidia-bearers"), which have formed at the tips of hyphae. Conidia are dust-fine asexual spores, characteristic of the ascomycetes.*

Predaceous Fungi

Among the most highly specialized of the fungi are the predaceous fungi that have developed a number of mechanisms for capturing small animals that they use as food. Some secrete a sticky substance on the surface of their hyphae in which passing protists, rotifers, small insects, or other animals become glued. More than 50 species of Fungi Imperfecti capture small roundworms (nematodes) that abound in the soil. In the presence of a population of roundworms (or even of water in which the worms have been growing), the hyphae of the fungi produce loops that swell rapidly, closing the opening like a noose when a nematode rubs against its inner surface. The stimulation of the cell walls is thought to increase the amount of osmotically active material in the cells, causing water to enter and expand them rapidly.

(a) The predaceous imperfect fungus *Arthrobotrys dactyloides* has trapped a nematode. The traps consist of rings, each comprising three cells, which swell rapidly to about three times their original size and garrote the nematode. Once the worm has been trapped, fungal hyphae grow into its body and digest it. When triggered, the ring cells can expand completely in less than a tenth of a second. This species was appropriately called the "nefarious noose fungus" by the late W. H. Weston of Harvard University, who made vast contributions to our present knowledge of the fungi. (b) Another nematode-trapping fungus, *Dactylella drechsleri*. This species traps the worms with small adhesive knobs and was dubbed the "lethal lollipop fungus" by Weston.

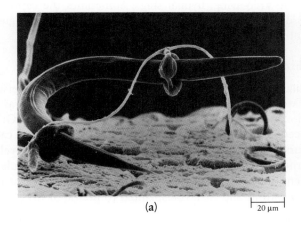

(a)　　　　　　　　20 μm

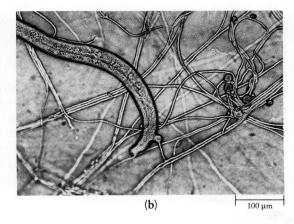

(b)　　　　　　　　100 μm

Many of the algae and cyanobacteria found in lichens are also commonly found as free-living species. Lichen fungi are thought to have free-living hyphal stages, but these fungi can generally be detected and identified only after they have encountered a suitable photosynthetic organism and formed a lichen. For these reasons, lichens are generally named and classified according to the species of the fungal component. There are about 25,000 species of lichens, in almost all of which the fungus is an ascomycete. Photosynthetic organisms from some 26 different genera are found in symbiotic association with these fungi. The most frequent are the green algae *Trebouxia* and *Trentepohlia* and the cyanobacterium *Nostoc*; members of one of these three genera are found in about 90 percent of all lichens.

(a)   (b)   (c)

23–13 (a) *Crustose ("encrusting") lichen growing on bare rock in central California.* (b) *Foliose ("leafy") lichen growing on a tree in northern Ontario.* (c) *British soldier lichen* (Cladonia cristatella), *a fruticose ("shrubby") lichen. Each soldier (so called because of the scarlet color) is 1 to 2 centimeters tall.*

Lichens reproduce most commonly by the breaking off of fragments containing both fungal hyphae and photosynthetic cells. New individuals can also be formed by the capture of an appropriate alga or cyanobacterium by a lichen fungus in its free-living, hyphal stage. Sometimes the captured photosynthetic cells are destroyed by the fungus, in which case the fungus also dies. If the photosynthetic cells survive, a lichen is produced.

Mycorrhizae

Mycorrhizae ("fungus-roots") are symbiotic associations between fungi and the roots of vascular plants. The importance of mycorrhizae was first recognized in connection with efforts to grow orchids in greenhouses. Orchids have microscopic seeds that germinate to form a tiny pad of tissue called a protocorm. Cultivators of orchids found that the plants seldom developed beyond the protocorm stage unless they were infected with a particular kind of fungus. Subsequently it was found that if seedlings of many forest trees are grown in nutrient solutions and then transplanted to prairie or other grassland soils, they

23–14 *Scanning electron micrographs showing the establishment of the British soldier lichen* (Cladonia cristatella) *in sterile laboratory culture.* (a) *An early interaction between fungal and algal components of the lichen.* (b) *Penetration of an algal cell by a fungal haustorium (arrow).*

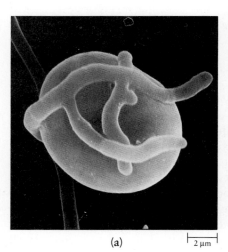

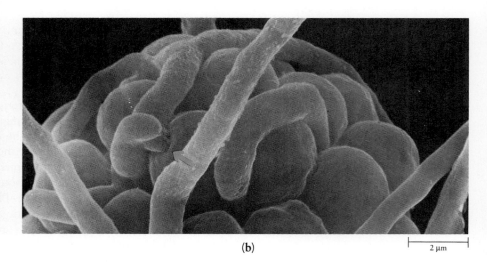

(a)   ⊢ 2 μm ⊣   (b)   ⊢ 2 μm ⊣

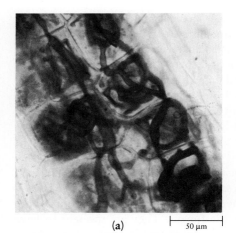

(a)   50 μm

(b)

23–15 (a) *Surface view of an endomy-corrhizal rootlet of fescue, a perennial grass, showing coiled hyphae in some of the cells. Mycorrhizal associations are especially important for grasses growing on nutrient-poor soils or at high eleva-tions.* (b) *Ectomycorrhizal rootlets from a western hemlock. Hormones secreted by the fungus cause the root to branch in a special pattern. This growth pattern and the swollen hyphal sheath impart a char-acteristic appearance to the ectomycorrhi-zae. The narrow strands extending from the mycorrhizae are bundles of hyphae that function as extensions of the root system.*

fail to grow. Eventually they may die from malnutrition, even if analysis shows that there are abundant nutrients in the soil. If a small amount (0.1 percent by volume) of forest soil containing fungi is added to the soil around the roots of the seedlings, however, they will grow promptly and normally. Mycorrhizae are now thought to occur in more than 90 percent of all families of plants.

In some mycorrhizal associations, known as endomycorrhizae, the fungal hyphae penetrate the cells of the root, forming coils, swellings, or branches (Figure 23–15a). The hyphae also extend out into the surrounding soil. Endomycorrhizae occur in about 80 percent of all vascular plants, and the fungal component is usually a zygomycete. In other associations, known as ectomycorrhizae, the hyphae form a sheath around the root but do not actually penetrate its cells (Figure 23–15b). Ectomycorrhizae are characteristic of certain groups of trees and shrubs, including pines, beeches, and willows; the fungus is usually a basidiomy-cete, but some associations involve ascomycetes, including truffles.

The exact relationship between roots and fungi is not known. Apparently the roots secrete sugars, amino acids, and possibly some other organic substances that are used by the fungi. Although the evidence is just now accumulating, it appears that the fungi convert minerals in the soil and decaying material into an available form and transport them into the root. It has been shown experimentally that mycorrhizae transfer phosphorus from the soil into roots, and there is evidence that water uptake is facilitated by the fungi. One of the most intriguing recent observations is that, under certain circumstances, mycorrhizae appear to function as a bridge through which phosphorus, carbohydrates, and probably other sub-stances pass from one host plant to another.

A study of the fossils of early vascular plants has revealed that mycorrhizae occurred as frequently in them as they do in modern vascular plants. This has led to the interesting suggestion that the evolution of mycorrhizal associations may have been the critical step allowing plants to make the transition to the bare and relatively sterile soils of the then-unoccupied land.

SUMMARY

Fungi are typically composed of masses of filaments called hyphae; the principal component of the hyphal walls is the polysaccharide chitin. Fungi are hetero-trophs, deriving their nutrition by absorption of organic compounds digested extracellularly by secreted enzymes.

Fungi form both asexual and sexual spores, although not all fungi form both kinds. The asexual spores may be formed in sporangia. The sexual cycle is initiated by the fusion of hyphae of different mating strains. In some fungi, the nuclei in the fused hyphae combine immediately, and a zygote is formed. In others, the two genetically distinct nuclei usually remain separate, forming pairs—known as dikaryons—that divide synchronously, sometimes over prolonged periods. Once the nuclei fuse, meiosis always follows immediately.

The kingdom Fungi includes three principal divisions—Zygomycota, Ascomy-cota, and Basidiomycota—as well as another major taxon, Deuteromycota, or Fungi Imperfecti, which is usually ranked at the level of division. The zygomy-cetes, which are thought to share a common ancestor with the lineage leading to the ascomycetes and basidiomycetes, are characterized by thick-walled zygo-spores. The hyphae of both ascomycetes and basidiomycetes are compartmenta-lized by perforated septa. In the ascomycetes, the sexual spores develop within an ascus (sac), and in the basidiomycetes, the sexual spores develop on a basidium (club). Most of the Deuteromycota have no known sexual cycle.

Fungi have an important ecological role as decomposers of organic material. They are also parasitic on many types of organisms, particularly plants, in which they often cause serious disease. Fungi participate in two additional types of symbioses that are of ecological significance—lichens and mycorrhizae.

Lichens are combinations of fungi and green algae or cyanobacteria that are structurally and physiologically different from either organism as it exists separately. They are able to survive under adverse environmental conditions where neither partner could exist independently. The lichen represents a symbiotic relationship in which a fungus encloses photosynthetic cells and is dependent upon them for nourishment.

Mycorrhizae ("fungus-roots") are associations between soil-dwelling fungi and plant roots. There are two principal types: endomycorrhizae and ectomycorrhizae. Mycorrhizal associations facilitate the uptake of minerals by the roots of the plant and provide organic molecules for the fungus. They are thought to have played a key role in enabling plants to make the transition to land.

QUESTIONS

1. Distinguish between the following: hypha/mycelium; chitin/cellulose; sporangia/gametangia; conidia/ascospores; ascus/basidium; monokaryotic/dikaryotic; endomycorrhizae/ectomycorrhizae.

2. In a multicellular organism, all of the individual cells must be supplied with food and water. What evolutionary solution to this problem is exemplified by the fungi? How have the fungi, which have no motile cells at any stage of the life cycle, solved the problem of obtaining new food supplies when they have exhausted a particular source?

3. Give the distinguishing characteristics of the Zygomycota, Ascomycota, and Basidiomycota.

4. Some authorities believe that the Chytridiomycota and Oomycota should be placed in kingdom Fungi, while others believe they are better placed in kingdom Protista. What characteristics do the members of these two divisions share with the fungi? In what ways do the chytrids and oomycetes differ from the fungi?

5. As you can see from the last two chapters, our method of classification into kingdoms does not produce entirely satisfactory results. What are some of the difficulties unresolved (or even created) by this system? Can you suggest alternative solutions?

6. Most of the fungi that have no sexual reproduction cannot currently be classified with their sexually-reproducing relatives. Is this likely to change in the future? Why or why not?

7. Coenocytic organisms, such as the zygomycetes, show little differentiation. When differentiation does occur, as in gametangium formation, it is preceded by construction of a septum. In your opinion, why?

8. Most fungi can reproduce both asexually and sexually. What are the advantages and disadvantages of each type of reproduction?

9. What type of symbiotic relationship would you say exists between the fungus and the alga or cyanobacterium of a lichen? Between the fungus and the plant roots of a mycorrhizal association?

10. How are fungi thought to have aided photosynthetic organisms in the transition to land?

The Plants

24-1 *Although plants are primarily adapted for life on land, some, such as the water lily,* Nymphaea odorata, *have returned to an aquatic existence. Like whales and dolphins,* Nymphaea *retains the traces of its ancestors' terrestrial sojourn. These include a water-resistant cuticle, stomata (openings through which gas exchange occurs), and a highly developed internal transport system.*

Plants are, quite simply, multicellular photosynthetic organisms primarily adapted for terrestrial life. Their characteristics are best understood in terms of the transition from water to land, an event that occurred some 500 million years ago. The land offered a wealth of advantages to photosynthetic organisms. On land, light is abundant from daybreak to dusk and is not blocked by turbulent water. Carbon dioxide, needed for photosynthesis, is plentiful in the atmosphere and circulates more freely in air than in water. And, most important, the land was then unoccupied by competing forms of life.

Terrestrial life, however, presented photosynthetic organisms with a new and major difficulty, that of obtaining and retaining adequate amounts of water. The solutions to this problem that gradually evolved in the plants depended on multicellularity, which made possible specialization on a broad scale. As we saw in Chapter 18, the selective activation or inactivation of particular genes during development results in cells or groups of cells specialized for particular functions. Specialization brings with it the potential for a vast increase in the size of the organism, which, in turn, creates additional problems to be solved: the organism must synthesize increased amounts of food to supply its numerous cells with energy; larger quantities of materials must be moved into and out of the organism and transported to the individual cells; a larger body requires more physical support; the activities of all of the cells must be integrated for the smooth functioning of the entire organism; and a greater length of time is required for the full development of the organism, often requiring protection and nourishment of the immature stages. The story of plant evolution represents a series of natural experiments at solving these problems; the experiments have been enormously successful, producing a diversity of multicellular photosynthetic organisms that, collectively, have been able to occupy most of the earth's land surface for at least 300 million years.

THE ANCESTRAL ALGA

As we noted in Chapter 22, all plants appear to have arisen from the green algae (division Chlorophyta). Like the plants, the green algae contain chlorophylls *a* and *b* and beta-carotene as their photosynthetic pigments, and they accumulate their food reserves in the form of starch. In plants and green algae—but in no other organisms—the starch is stored in plastids (page 120), rather than in the cytoplasm. Beyond these similarities, however, the green algae exhibit a great diversity of characteristics, some of which are shared with the plants and some of which are not. The constellation of characteristics that could have given rise to the plants is found among contemporary green algae only in some members of class Charophyceae, most strikingly in the genus *Coleochaete* (Figure 24–2).

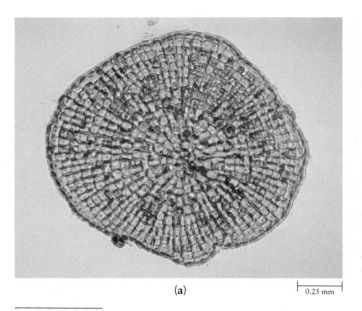

(a)

|⎯⎯⎯| 0.25 mm

(b)

24–2 (a) Coleochaete, *a green alga of class Charophyceae that grows on the surface of submerged freshwater plants in shallow water. This multicellular alga takes the form of a disk, generally one cell thick.* (b) *Fossils of* Parka decipiens, *an extinct green alga, dated at about 380 million years ago.* Parka decipiens *closely resembles* Coleochaete *in its shape, tissue structure, and chemistry. It has many features suggesting that it may be closely related to the alga that gave rise to the plants.*

Although *Coleochaete* itself does not seem to have been the alga from which the plants evolved, it is thought to be closely related to it. Several sets of data lead to this conclusion. The cells of *Coleochaete*, like those of plants, have cellulose in their walls and contain peroxisomes (page 119), in which the key enzyme involved in photorespiration (page 225) is synthesized. Additional evidence is found in its pattern of cytokinesis. In almost all other organisms, including most green algae, division of the cytoplasm takes place by constriction and pinching off of the cell membrane. In plants and in *Coleochaete*, the cytoplasm is divided by the formation of a cell plate at the equator of the spindle. Similarities are also found in the pattern of microtubules underlying the flagella in *Coleochaete* and in those plant cells that have flagella.

One important characteristic shared by all plants, but absent in *Coleochaete*, is a well-defined alternation of generations. As we saw in Chapter 22, this type of life cycle is found not only in many multicellular green algae but also in the red algae and the brown algae, which biochemical evidence indicates originated from totally different symbiotic events than the green algae. This suggests that alternation of generations has arisen independently on several occasions, and *Coleochaete* provides a significant clue as to how it may have occurred. *Coleochaete*, like *Chlamydomonas* and a number of other green algae, is haploid for most of its life cycle. It is oogamous, producing clearly differentiated egg and sperm that fuse to form a zygote that subsequently undergoes meiosis, producing haploid cells from which new individuals develop. In *Coleochaete*, however, fusion of the gametes occurs not in the open waters but rather on the surface of the parent organism, and neighboring cells grow around the zygote, enclosing and protecting it (Figure 24–3). Prior to meiosis, additional rounds of DNA replication occur in the zygote,

24–3 *The large, dark cells in this micrograph of* Coleochaete *are the diploid zygotes. As you can see, they are protected by a layer of the smaller, haploid cells of the parent organism. The hair cells that extend outward from the disk are ensheathed at its base and are the source of the organism's name;* Coleochaete *means "sheathed hair." The hairs are thought to discourage aquatic animals from feeding on the alga.*

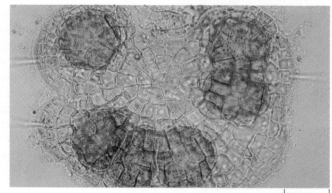

|⎯⎯⎯| 100 μm

24–4 *The aboveground surfaces of plants are characteristically covered with a waxy cuticle that retards water loss. Epidermal pores permit gas exchange. The pore shown in this scanning electron micrograph is from* Marchantia, *a liverwort. Unlike the stomata of vascular plants (Figure 10–17, page 222), it does not open and close in response to environmental cues.*

with the result that the number of haploid cells ultimately released is not a mere 4, as in *Chlamydomonas,* but anywhere from 8 to 32. Only slight modifications of this life cycle—division of the diploid cells of the zygote by mitosis and cytokinesis *prior to* meiosis, followed by meiotic division of some, but not all, of the diploid cells—would be required to produce alternation of generations.

THE TRANSITION TO LAND

By the time the immediate ancestor of the plants moved from shallow waters onto the land, it had apparently evolved a well-defined alternation of heteromorphic generations. After the transition to land, new adaptations began to evolve in the life cycle and in other features as well. These adaptations were critical to the ultimate success of the plants on land and must have occurred early in their evolutionary history—most modern plants, even though very diverse, share them. One such characteristic, clearly associated with the transition to land, is the protective cuticle that covers the aboveground surfaces of plants and retards the loss of water from the plant body. The cuticle is formed of a waxy substance called cutin, secreted by the epidermal cells. Associated with the cuticle—in fact, made necessary by it—are pores through which the gas exchanges necessary for photosynthesis can take place (Figure 24–4).

Another adaptation was the development of multicellular reproductive organs (gametangia and sporangia) that were surrounded by a protective layer of sterile (nonreproductive) cells. The gametangia are called **archegonia** if they give rise to egg cells and **antheridia** if they give rise to sperm (Figure 24–5). A correlated adaptation was the retention of the fertilized egg (the zygote) within the female gametangium (the archegonium) and its development there into an embryo. Thus, during its critical early stages of development, the embryo, or young sporophyte, is protected by the tissues of the female gametophyte.

24–5 *Multicellular gametangia of the liverwort* Marchantia, *a member of division Bryophyta. (a) Female gametangia, or archegonia, in various stages of development. The archegonia are flask-shaped, with a single egg developing in the base of the flask. (b) Antheridium developing in the male gametophyte. The spermatogenous tissue will give rise to sperm cells, which, when mature, will swim to the egg through the neck canal of the archegonium.*

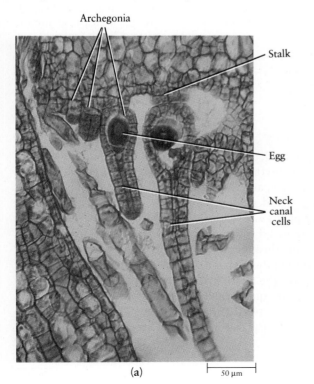

Archegonia

Stalk

Egg

Neck canal cells

(a)

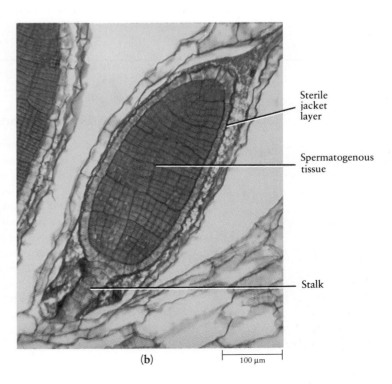

Sterile jacket layer

Spermatogenous tissue

Stalk

(b)

TABLE 24-1 **Major Physical and Biological Events in Geologic Time**

MILLIONS OF YEARS AGO	PERIOD	EPOCH	LIFE FORMS	CLIMATES AND MAJOR PHYSICAL EVENTS
Cenozoic Era				
0–2	Quaternary	Recent Pleistocene	Planetary spread of *Homo sapiens;* extinction of many large mammals and birds. Deserts on large scale.	Fluctuating cold to mild. Numerous glacial advances and retreats; uplift of Sierra Nevada.
2–5	Tertiary	Pliocene	Large carnivores. First known appearance of hominids (humanlike primates).	Cooler. Continued uplift and mountain building, with widespread glaciation in Northern Hemisphere. Uplift of Panama joins North and South America.
5–25		Miocene	Whales, apes, grazing mammals. Spread of grasslands as forests contract.	Moderate. Extensive glaciation begins again in Southern Hemisphere. Moderate uplift of Rockies.
25–38		Oligocene	Large, browsing mammals; monkey-like primates appear. Origin of many modern families of flowering plants.	Rise of Alps and Himalayas. Lands generally low. Volcanoes in Rockies. South America separates from Antarctica.
38–55		Eocene	Primitive horses, tiny camels, modern and giant types of birds. Formation of grasslands.	Mild to very tropical. Many lakes in western North America. Australia separates from Antarctica; India collides with Asia.
55–65		Paleocene	First known primitive primates and carnivores.	Mild to cool. Wide, shallow continental seas largely disappear.
Mesozoic Era				
65–144	Cretaceous		Extinction of dinosaurs at end of period. Marsupials, insectivores, and flowering plants become abundant.	Tropical to subtropical. Elevation of Rockies at end of period. Africa and South America separate.
144–213	Jurassic		Dinosaurs' zenith. Flying reptiles, small mammals. Birds appear. Gymnosperms, especially cycads, and ferns.	Mild. Continents low, with large areas covered by seas. Mountains rise from Alaska to Mexico.
213–248	Triassic		First dinosaurs. Primitive mammals appear. Forests of gymnosperms and ferns.	Continents mountainous and joined in one mass. Large areas arid. Eruptions in eastern North America. Appalachians uplifted and broken into basins.

Subsequent Diversification

Not long after the transition to land, the plants diverged into at least two separate lineages. One gave rise to the bryophytes, a group that includes the modern liverworts, hornworts, and mosses, and the other to the vascular plants, a group that includes all of the larger land plants. A principal difference between the bryophytes and the vascular plants is that the sporophytes of the latter, as their name implies, have a well-developed vascular system that transports water, minerals, sugars, and other nutrients throughout the plant body. The bryophytes first appear in the fossil record during the Devonian period, about 370 million years ago (see Table 24–1, which gives the major events in the evolution of plants and animals). These ancient fossils are quite similar to bryophytes living today. The oldest fossils of vascular plants are from the early part of the Silurian period, about 430 million years ago. The vascular plants, as we shall see, subsequently underwent a great diversification.

TABLE 24-1 **Major Physical and Biological Events in Geologic Time** *(Continued)*

MILLIONS OF YEARS AGO	PERIOD	EPOCH	LIFE FORMS	CLIMATES AND MAJOR PHYSICAL EVENTS
Paleozoic Era				
248–286	Permian		Reptiles diversify. Origin of conifers, cycads, and ginkgos; possible origin of flowering plants; earlier forest types wane.	Extensive glaciation in Southern Hemisphere. Seas drain from land; worldwide aridity. Appalachians formed by end of Paleozoic.
286–360	Carboniferous Pennsylvanian Mississippian		Age of amphibians. First reptiles. Variety of insects. Sharks abundant. Great swamps, forests of ferns, gymnosperms, and horsetails.	Warm; conditions like those in temperate or subtropical zones—little seasonal variation, water plentiful. Lands low, covered by shallow seas or great coal swamps. Mountain building in eastern U.S., Texas, Colorado.
360–408	Devonian		Age of fishes. Amphibians appear. Mollusks abundant. Lunged fishes. Extinction of primitive vascular plants. Origin of modern groups of vascular plants.	Europe mountainous with arid basins. Mountains and volcanoes in eastern U.S. and Canada. Rest of North America low and flat. Sea covers most of land.
408–438	Silurian		Rise of fishes and reef-building corals. Shell-forming sea animals abundant. Invasion of land by arthropods. Earliest vascular plants. Modern groups of algae and fungi.	Mild. Continents generally flat. Again flooded. Mountain building in Europe.
438–505	Ordovician		First primitive fishes. Invertebrates dominant. First fungi. Possible invasion of land by plants.	Mild. Shallow seas, continents low; sea covers U.S. Limestone deposits.
505–590	Cambrian		Shelled marine invertebrates. Explosive diversification of eukaryotic organisms.	Mild. Extensive seas, spilling over continents.
Precambrian Era				
590–4,500			Origin of life. Prokaryotes. Eukaryotic cells and multicellularity by close of era. Earliest known fossils, including soft-bodied marine invertebrates.	Dry and cold to warm and moist. Planet cools. Formation of earth's crust. Extensive mountain building. Shallow seas. Accumulation of free oxygen.

CLASSIFICATION OF THE PLANTS

According to the classification scheme we follow, the modern plants are placed in 10 separate divisions (Table 24–2). The liverworts, hornworts, and mosses of division Bryophyta are quite different from one another, and there is some question as to whether they represent three distinct lineages from the ancestral plant or subsequent branchings of one lineage from that ancestor. By contrast, each of the nine divisions of vascular plants is, by all available evidence, monophyletic—that is, all of its members are descended from a common ancestor.

The vascular plants are frequently grouped, for convenience, in ways that may or may not reflect evolutionary relationships. For instance, these plants, as a group, are often referred to as tracheophytes. They can be grouped into those without seeds (divisions Psilophyta, Lycophyta, Sphenophyta, and Pterophyta) and those with seeds. The seed plants also form two informal groups, the gymnosperms and the angiosperms. The gymnosperms are those with "naked," unprotected seeds (divisions Coniferophyta, Cycadophyta, Ginkgophyta, and Gnetophyta); the angiosperms, with enclosed, protected seeds, are, formally speaking, the Anthophyta, the flowering plants.

TABLE 24-2   **A Classification of Living Plants**

TAXON	COMMON NAME	NUMBER OF SPECIES
Division Bryophyta	Bryophytes	16,000
Class Hepaticae	Liverworts	6,000
Class Anthocerotae	Hornworts	100
Class Musci	Mosses	9,500
Division Psilophyta	Whisk ferns	Several
Division Lycophyta	Club mosses	1,000
Division Sphenophyta	Horsetails	15
Division Pterophyta	Ferns	12,000
Division Coniferophyta	Conifers	550
Division Cycadophyta	Cycads	100
Division Ginkgophyta	Ginkgos	1
Division Gnetophyta	Gnetophytes	70
Division Anthophyta	Flowering plants (angiosperms)	235,000
Class Monocotyledones	Monocots	65,000
Class Dicotyledones	Dicots	170,000

DIVISION BRYOPHYTA: LIVERWORTS, HORNWORTS, AND MOSSES

Lacking water-gathering roots and the kind of specialized tissues that transport water up the body of a vascular plant, the bryophytes must absorb moisture through aboveground structures. As a consequence, they grow most successfully in moist, shady places and in bogs. Some of them, like sphagnum (peat moss), are able to absorb and hold large amounts of water; in effect, they maintain a watery existence even on land. The majority of the bryophytes are tropical, but some species occur in temperate regions, and a few even reach the Arctic and Antarctic.

Most bryophytes are comparatively simple in their structure and are relatively small, usually less than 20 centimeters in length. A single moss plant may sprawl over a considerable area, but most liverworts are so small that they are noticeable only to a keen observer. In the damp environments frequented by the bryophytes, individual cells can absorb water and nutrients directly from the air or by diffusion from nearby cells. Like the lichens, they are sensitive indicators of air pollution.

Although bryophytes do not have true roots, they are generally attached to the substrate by means of rhizoids, which are elongate single cells or filaments of cells. Many bryophytes also have small leaflike structures in which photosynthesis takes place. These structures lack the specialized tissues of the "true" leaves of the vascular plants and are only one or a few cell layers thick. For these reasons, the leaflike structures of the bryophytes and the leaves of the vascular plants are believed to have evolved separately. As in other plants, the body of a bryophyte is specialized for support and food storage.

(a)

(b)

(c)

24–6 *Representative bryophytes, the only plants in which the gametophyte, which is haploid (n), is the dominant, nutritionally independent generation. (a) A young gametophyte of the liverwort* Marchantia, *growing on a rock. (b)* Anthoceros, *a hornwort. The "horns" are the diploid (2n) sporophytes, which are attached to a dish-shaped gametophyte. (c) A haircap moss with spore capsules. The lower green structures are the gametophytes; the non-photosynthetic stalks and capsules are the sporophytes. Depending on the species, the sporophytes of mosses growing in temperate regions take 6 to 18 months to reach maturity.*

Bryophyte Reproduction

Bryophytes, like all plants, have a life cycle with alternation of generations. In contrast to the vascular plants, however, the bryophytes are characterized by a haploid gametophyte that is usually larger than the diploid sporophyte.

The life cycle of a moss begins when a haploid spore germinates to form a network of horizontal filaments known as protonemata (singular, protonema) (Figure 24–7). Individual gametophytes grow up like branches from this network. The multicellular antheridia and archegonia are borne on the gametophyte. When sufficient moisture is present, the sperm, which are biflagellate, are released from the antheridium and swim to the archegonium, to which they are attracted chemically. Without free water in which the sperm can swim, the life cycle cannot be completed.

24–7 *Protonema of a moss, with a budlike structure from which the gametophyte will develop. Protonemata are characteristic of mosses and liverworts. They often resemble filamentous green algae. For the subsequent stages of the life cycle, see Figure 24–8.*

24–8 *The moss life cycle begins with the release of spores from a capsule, which opens when a small lid bursts (upper left). The spore germinates to form a branched, filamentous protonema, from which a leafy gametophyte develops. Sperm cells, which are expelled from the mature antheridium, are attracted into the archegonium, where one fuses with the egg cell to produce the zygote. The zygote divides mitotically to form the sporophyte and, at the same time, the base of the archegonium divides to form the protective calyptra. The mature sporophyte consists of a capsule, which may be raised on a stalk— also part of the sporophyte—and a foot. Meiosis occurs within the capsule, resulting in the formation of haploid spores.*

In this moss, the gametophytes bear both antheridia and archegonia. In other species, a single gametophyte may bear either antheridia or archegonia, but not both.

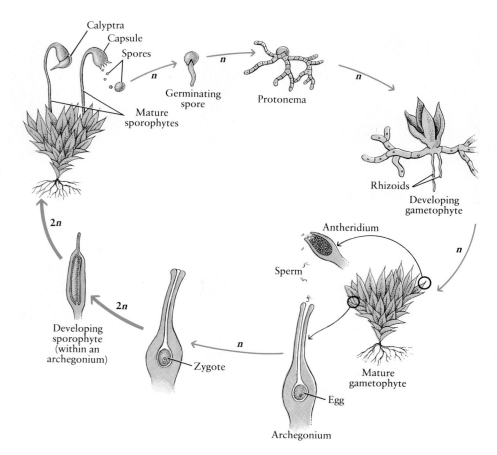

24–9 (a) *The bowl-shaped gemma cups visible on this gametophyte of the liverwort* Marchantia *contain minute bodies, the gemmae, which are splashed out by the rain and grow in the vicinity of the parent plant. Asexual reproduction by fragmentation or by gemmae is common among the liverworts.*

In Marchantia, *the elevated antheridia and archegonia are formed on different plants, the male (b) and the female (c) gametophytes. The zygote, formed in the archegonium, develops into the sporophyte, which remains attached to the female gametophyte. Gemma cups develop on both types of gametophyte.*

Fusion of sperm and egg takes place within the archegonium. Inside the archegonium, the zygote develops into a sporophyte, which remains attached to the gametophyte and is nutritionally dependent upon it. Typically the sporophyte consists of a foot, a stalk, and a single, large sporangium (or capsule), from which the spores are discharged.

Asexual reproduction, often by fragmentation, is also common. Many mosses and liverworts also produce minute bodies, known as gemmae, that can give rise to new plants (Figure 24–9).

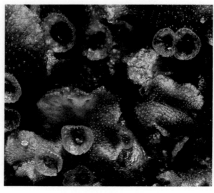

(a)

(b)

(c)

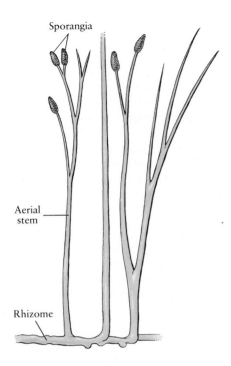

24–10 Rhynia major, *one of the earliest known vascular plants. It lacked leaves and roots. Its aerial stems, which were photosynthetic, were attached to an underground stem, or rhizome. The aerial stems were covered with a cuticle and contained stomata. The dark structures at the tips of the stems are sporangia, which apparently released their spores by splitting lengthwise.*

THE VASCULAR PLANTS: AN INTRODUCTION

Evolutionary Developments in the Vascular Plants

Rhynia major, now extinct, is an example of the earliest known vascular plants, dating back some 400 million years. As you can see in Figure 24–10, it is much more primitive in appearance than a bryophyte. However, it differs from the bryophytes in one important respect: within its stem is a central cylinder of vascular tissue, specialized for conducting water and dissolved substances up the plant body and products of photosynthesis down.

Beginning with a very simple vascular plant, such as *Rhynia,* it is possible to trace some major evolutionary trends, as well as several key innovations. One early innovation was the root, a structure specialized for anchorage and the absorption of water. Another was the leaf, a structure specialized for photosynthesis. Two distinct types of leaves evolved, the microphyll and the megaphyll. The microphyll contains only a single strand of vascular tissue, whereas the megaphyll typically contains a complex system of veins. These two types of leaves seem to have originated in different ways (Figure 24–11). Of the modern plants, only the Psilophyta, the Lycophyta, and the Sphenophyta have microphylls.

One of the most striking trends in plant evolution has been the development of increasingly efficient conducting systems between the two portions of the plant body. The conducting system in modern vascular plants consists of two distinct tissues: the **xylem,** which transports water and ions from the roots to the leaves, and the **phloem** (pronounced "flow-em"), which carries dissolved sucrose and other products of photosynthesis from the leaves to the nonphotosynthetic cells of the plant. The conducting elements of the xylem are **tracheids** and **vessel members,** and the conducting elements of the phloem are **sieve cells** or **sieve-tube members.** In stems, the longitudinal strands of xylem and phloem are side by side, either in vascular bundles or arranged in two concentric layers (cylinders), in which one tissue (typically the phloem) occurs outside the other. With the development of roots, leaves, and efficient conducting systems, the plants effectively solved the most basic problems confronting multicellular photosynthetic organisms on land—acquiring adequate supplies of water and food and delivering them to all of the cells making up the organism.

Another pronounced trend is the reduction in size of the gametophyte. In all vascular plants, the gametophyte is smaller than the sporophyte. However, in the more primitive vascular plants, the gametophyte is separate and nutritionally independent of the sporophyte. In the more recently evolved groups—the gymnosperms and the angiosperms—the gametophyte has been reduced to microscopic size and dependent status.

Also related to the reproductive cycle is a trend toward heterospory. The earliest vascular plants produced only one kind of spore (homospory) in one kind of sporangium. Upon germination, such spores typically produce gametophytes on which both antheridia and archegonia form. In plants that are heterosporous, two different kinds of gametophytes develop: one bearing archegonia and the other, antheridia.

24–11 *According to one widely accepted theory,* (a) *microphylls evolved as lateral outgrowths of the stem. Representatives of the three divisions of modern plants with microphylls are shown in Figure 24–12.*

(b) *Megaphylls evolved by fusion of branch systems and thus have a complex vascular network. The great majority of vascular plants have megaphylls.*

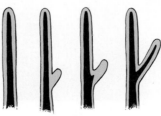

Evolution of microphylls
(a)

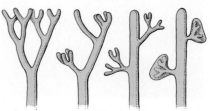

Evolution of megaphylls
(b)

24–12 *Representatives of three divisions of seedless vascular plants.* (a) *The whisk fern,* Psilotum, *one of the two living genera of division Psilophyta. The bulbous structures are the sporangia, which occur in fused groups of three.* Psilotum *is unique among living vascular plants in that it lacks roots and leaves. If you look closely, however, you can see small, scalelike outgrowths below the sporangia.*

(b) *The club mosses of the genus* Lycopodium *are the most familiar members of division Lycophyta. In this genus, the sporangia are borne on specialized leaves, sporophylls, which are aggregated into a cone at the apex (top) of the branches, as shown in this running ground pine,* Lycopodium complanatum. *The airborne, waxy spores give rise to small, independent subterranean gametophytes. The sperm, which are biflagellate, swim to the archegonium, where the young sporophyte, or embryo, develops.*

(c) *The horsetails, division Sphenophyta, of which there is only one living genus (*Equisetum*), are easily recognized by their jointed, finely ribbed stems, which contain silica. At each node, there is a circle of small, scalelike leaves. Spore-bearing structures are clustered into a cone at the apex of the stem. The gametophytes are independent, and the sperm are coiled, with numerous flagella.*

As the gametophytes became reduced, archegonia and antheridia decreased in size until, in the angiosperms, they disappeared altogether, made obsolete by the seed—perhaps the innovation most important in the enormous success of the vascular plants on land. The seed is a complex structure in which the young sporophyte, or embryo, is contained within a protective outer covering, the seed coat. The seed coat, which is derived from tissues of the parent sporophyte, protects the embryo while it remains dormant, sometimes for years, until conditions are favorable for its germination. The earliest known seeds were fossilized in late Devonian deposits some 360 million years ago.

THE SEEDLESS VASCULAR PLANTS

Four divisions of seedless vascular plants have living representatives: Psilophyta (whisk ferns), Lycophyta (club mosses), Sphenophyta (horsetails), and Pterophyta (ferns), the largest group.

Division Pterophyta: The Ferns

Ferns are vascular plants that can usually be distinguished from most other plants by their large feathery leaves, which, in most species, unroll from base to tip as they develop. According to the fossil record, ferns first appeared almost 400 million years ago, and they are still relatively abundant. Most of the 12,000 living species are found in the tropics, but many also occur in temperate and even arid regions. Because they have flagellated sperm and need free water for fertilization, those species growing in arid regions exploit the seasonal occurrence of water for sexual reproduction.

The stems of ferns are usually not as complex as those of gymnosperms and angiosperms, and they are often reduced to a creeping underground stem (rhizome). Although ferns do not exhibit secondary growth—the type of growth that results in increase of girth and formation of bark and woody tissue—some grow

(a)

(b)

(c)

24-13 *Ferns.* **(a)** *The immature sporophyte of many common ferns develops as a "fiddle head," which uncoils and spreads as it elongates.* **(b)** *A cinnamon fern. The sporangia are borne on separate stalks, visible in the center of the photograph.*

(a)

(b)

24-14 *Fern spores develop on the sporophyte in sporangia, which are usually found in clusters (sori) on the underside of a leaf (sporophyll), as shown here. Fern spores give rise to tiny gametophytes that, although photosynthetic, are barely visible to the naked eye.*

very tall. For instance, *Cyathea australis,* a tree fern found on Norfolk Island in the South Pacific, sometimes reaches 28 meters in height.

The leaves (fronds) of ferns are often finely divided into leaflets (pinnae). With a high surface-to-volume ratio, these widely spread, divided leaves are very efficient light collectors, well adapted to growing on the forest floor in diffuse light. The sporangia commonly are on the undersurface of the leaves or, sometimes, on specialized leaves. Sporangium-bearing leaves are called **sporophylls.** (As we shall see, the reproductive structures of flowers are also sporophylls.) The sporophylls may resemble the other green leaves of the plant or may be nonphotosynthetic stalks (modified leaves). The sporangia of ferns commonly occur in small clusters known as sori (singular, sorus).

In the ferns, as in all the vascular plants, the dominant generation is the sporophyte. The gametophyte of the homosporous ferns begins development as a small algalike filament of cells, each filled with chloroplasts, and then develops into a flat structure, often only one layer of cells in thickness. Although this gametophyte is small, it is nutritionally independent, as is the sporophyte. All but a few genera of ferns are homosporous, and a single gametophyte usually produces both antheridia and archegonia. The sperm are coiled and multiflagellate. The life cycle of a fern is shown in Figure 12–6 on page 251.

THE SEED PLANTS

The humid Carboniferous period, which ended some 286 million years ago, was the age when most of the earth's coal deposits were formed from lush vegetation that sank so swiftly into the warm, marshy soil that there was no chance for much of it to decompose. Seed plants were in existence by the close of this period; according to the fossil record, some of the fernlike plants and even some of the club mosses had seedlike structures.

In the Permian period (248 to 286 million years ago), there were worldwide changes of climate, with the advent of widespread glaciers and drought. Terrestrial plants and animals were under strong selection pressures for specialized structures that would enable them to survive during periods when no water was available.

Coal Age Plants

About 100 billion metric tons of carbon dioxide are fixed in photosynthesis every year. Almost the same amount is released into the atmosphere by living organisms in the course of respiration—less only 1 part in 10,000. This very slight imbalance is caused by the burying of organisms in sediment or mud under conditions in which oxygen is excluded and decay is only partial. This accumulation of partially decayed material is known as peat. The peat may eventually become covered with sedimentary rock and so be placed under pressure. Depending on time, temperature, and other factors, peat may become compressed into coal, petroleum, or natural gas—the so-called fossil fuels.

During certain periods in the earth's history, the rate of fossil fuel formation was greater than at other times. One such period, aptly called the Carboniferous, extended from 360 to 286 million years ago. What are now the temperate regions of Europe and North America were then tropical to subtropical, supporting year-round growth. The lands were low, covered by shallow seas or swamps. The dominant plants of the Carboniferous period were lycopod trees and giant horsetails; these disappeared during the Permian period, a time of worldwide drought and extensive glaciation. Other plants with surviving descendants include several families of ferns and one group of gymnosperms, the conifers. The flowering plants had not yet put in an appearance.

A reconstruction of a Carboniferous swamp forest, dominated by the lycopod tree Lepidodendron. *The young, unbranched trees that resemble bottle brushes matured into the many-branched trees that formed the canopy of the forest. Giant horsetails of the genus* Calamites *are at the left and in the center foreground. Seed plants also thrived in this environment; the plant at the far right with forked branches is the seed fern* Medullosa. *As you can see, these plants had shallow root systems, as do modern swamp plants, and so were easily toppled by the winds.*

(a)

(b)

(c)

(d)

24–15 *Representative gymnosperms.* (a) *Male and female plants of* Zamia pumila, *the only cycad native to the United States. It is common in the sandy woods of Florida. The stems are mostly or entirely underground and, along with the roots, were used by the Seminole Indians as food. The two large gray cones in the foreground are female cones; the smaller brown cones are male cones.*

(b) *Leaves and fleshy seeds of* Ginkgo biloba, *the only surviving species of the Ginkgophyta, a lineage that extends back to the late Paleozoic era. Ginkgo is especially resistant to air pollution and is commonly cultivated in urban parks and along city streets. The fleshy coating of the seeds has a putrid smell, similar to that of rancid butter. However, the inner "kernel" of the seed, which has a fishy taste, is a much-prized delicacy in the Orient.*

(c) *A branch of a conifer, the Ponderosa pine (also called western yellow pine), bearing a female cone. When the cone was mature, it opened and released its winged seeds, two of which have become caught between the scales. Like other pines, the Ponderosa has flexible, needlelike leaves, which are held together in a bundle (fascicle). Ponderosa pine, one of the principal forest trees of the Rockies from Canada to Mexico, is a staple of the Northwest lumber industry.*

(d) *A large seed-producing plant of* Welwitschia mirabilis, *a gnetophyte, growing in the Namib Desert of southern Africa.* Welwitschia *produces only two adult leaves, which continue to grow for the life of the plant. As growth continues, the leaves break off at the tips and split lengthwise; hence, older plants appear to have numerous leaves.*

Those organisms in which water-conserving, protective structures had previously evolved were at a great advantage. For example, the amphibians gave way to the reptiles with their scaly skins and shelled eggs, which may have been better suited to the harsh climate. Similarly, by the close of this period, which ended the Paleozoic era, plants with seeds gained a major evolutionary advantage and came to be the dominant plants of the land.

The two major modern groups of plants are both seed plants: the **gymnosperms,** which have naked seeds, and the **angiosperms** (from the Greek word *angio,* meaning "vessel"—literally, a seed borne in a vessel), which have protected seeds.

Gymnosperms

It was during the Permian period that the gymnosperms diversified. Four groups of gymnosperms have living representatives—three small divisions (Cycadophyta, Ginkgophyta, and Gnetophyta) and one large and familiar division (Coniferophyta). The conifers ("cone-bearers") include the pines, firs, spruces, hemlocks, cypresses, junipers, and the giant coastal redwoods of California and Oregon.

505

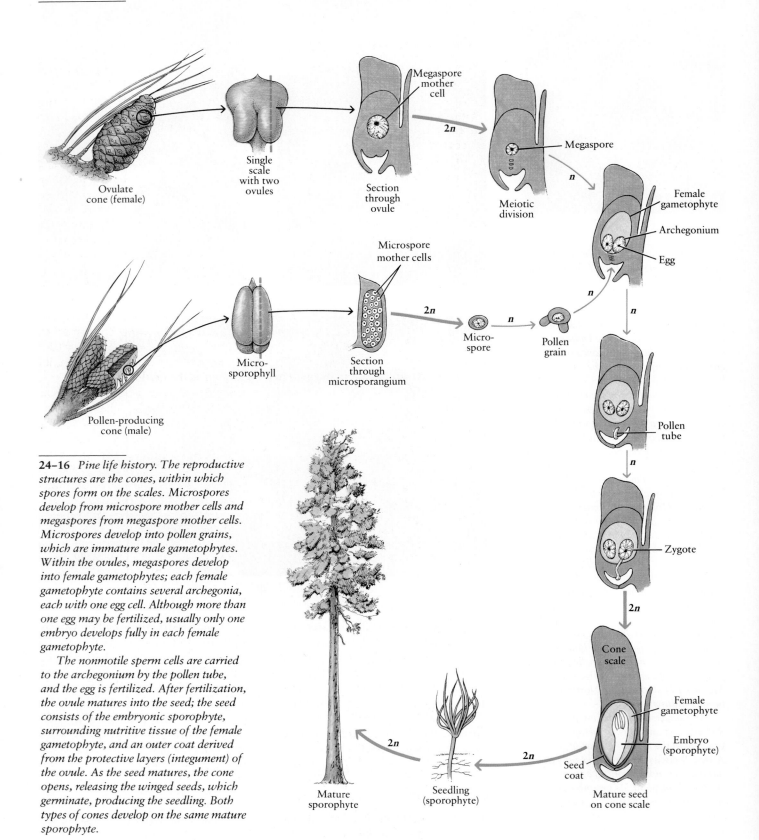

24–16 *Pine life history. The reproductive
structures are the cones, within which
spores form on the scales. Microspores
develop from microspore mother cells and
megaspores from megaspore mother cells.
Microspores develop into pollen grains,
which are immature male gametophytes.
Within the ovules, megaspores develop
into female gametophytes; each female
gametophyte contains several archegonia,
each with one egg cell. Although more than
one egg may be fertilized, usually only one
embryo develops fully in each female
gametophyte.*

*The nonmotile sperm cells are carried
to the archegonium by the pollen tube,
and the egg is fertilized. After fertilization,
the ovule matures into the seed; the seed
consists of the embryonic sporophyte,
surrounding nutritive tissue of the female
gametophyte, and an outer coat derived
from the protective layers (integument) of
the ovule. As the seed matures, the cone
opens, releasing the winged seeds, which
germinate, producing the seedling. Both
types of cones develop on the same mature
sporophyte.*

Labels in figure:

Ovulate cone (female)

Single scale with two ovules

Section through ovule

Megaspore mother cell

2n

Meiotic division

Megaspore

n

Female gametophyte

Archegonium

Egg

Microspore mother cells

Micro-sporophyll

Section through microsporangium

2n

Micro-spore

n

Pollen grain

n

n

Pollen-producing cone (male)

Pollen tube

n

Zygote

2n

Cone scale

Female gametophyte

Embryo (sporophyte)

Seed coat

Mature seed on cone scale

2n

Seedling (sporophyte)

2n

Mature sporophyte

Formation of the Seed

The seed is a protective structure in which the embryonic plant can be dispersed and lie dormant until conditions become favorable for its survival. Thus, in its functions, it parallels the spores of bacteria or the resistant zygotes of the freshwater algae. In structure it is far more elaborate, however. A seed includes the embryo (the young, dormant sporophyte), a store of nutritive tissue, and an outer protective coat.

To understand the structure of the seed, it is necessary to return for a moment to the alternation of generations found in all plants and in some of the green algae (see pages 463–464). In the green algae, the two generations, sporophyte and gametophyte, are independent and are generally about the same size. In the seedless vascular plants, including the ferns, the gametophytes, although still independent, are smaller than the sporophytes. In the seed plants, the gametophytic generation is reduced still further and is totally dependent on the sporophyte.

All gymnosperms are heterosporous, producing two different types of spores in two different types of sporangia. Spores that give rise to male gametophytes are known as **microspores,** and they are formed in structures known as microsporangia. Spores from which female gametophytes develop are **megaspores,** and they are formed in megasporangia. A megasporangium contains a single megaspore mother cell, which gives rise by meiosis to a megaspore, and it is surrounded by one or two layers of tissue, the integument. The entire structure—the megasporangium, its protective integument, and its contents—is known as the **ovule.**

With this information in mind, let us look at a specific example, the formation of a pine seed (Figure 24–16). A pine tree—the mature sporophyte—has two types of cones, which produce the two types of spores. The small, male cones superficially resemble those of the club mosses, but the large, female cones have ovule-bearing scales that are much thicker and tougher than the sporophylls of the male cones.

In the male cones, specialized microspore mother cells inside the microsporangia undergo meiosis to produce haploid microspores. Each microspore differentiates into a microscopic, windborne pollen grain, an immature male gametophyte. The wind is an unreliable messenger, disseminating the pollen grains at random, and wind-pollinated plants characteristically produce pollen in great quantities (Figure 24–17).

24–17 (a) *Male cones of Scotch pine* (Pinus sylvestris) *shedding pollen. The pollen grains are immature male gametophytes, which complete their maturation when they reach the ovules, embedded in the female cones. There they produce pollen tubes that carry the nonmotile sperm cells to the egg cells.*

(b) *Female cone of Scotch pine. The female gametophytes develop in ovules on the base of a scale of the cone, and the eggs are fertilized there. Each scale contains two ovules. When the seeds are mature, they drop from the cone.*

(a)

(b)

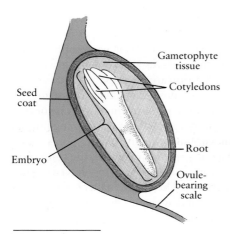

Seed coat

Gametophyte tissue

Cotyledons

Embryo

Root

Ovule-bearing scale

24–18 *Pine seed. The outer layers (integument) of the ovule have hardened into a seed coat, enclosing the female gametophyte and the embryo, which now consists of an embryonic root and a number of embryonic leaves, the cotyledons.*

When the seed germinates, the root will emerge from the seed coat and penetrate the soil. When the root absorbs water, the tightly packed cotyledons will elongate and swell with the moisture, rising above ground on the lengthening stem and forcing off the seed coat. During this period, the cotyledons absorb nutrients that are stored in the gametophyte tissue and are essential for the growth of the embryo into a seedling.

Within the ovules of the female cones, the megaspores are formed by meiosis. Of the four cells produced within the ovule by each meiotic sequence, three disintegrate and the remaining one—the megaspore—develops into a tiny female gametophyte. This haploid gametophyte grows within the ovule and develops two or more archegonia, each of which contains a single egg cell. The development from the megaspore into the gametophyte with its egg cells may take many months—slightly more than a year, for example, in some common pines.

As the ovule ripens, it secretes a sticky liquid. When the female cone becomes dusted with pollen, some of the pollen sifts down between the cone scales and comes into contact with the liquid. Pollen grains, caught in the sticky liquid, are drawn to the ovule as the liquid dries. Here, some three months later, the pollen grain develops into a mature male gametophyte. This gametophyte produces two nonmotile cells, the male gametes, or sperm. These are carried toward the egg within the pollen tube, which is produced by the male gametophyte and slowly grows through the tissues of the ovule—a process that takes almost a year.

Because the drought-resistant pollen is blown to the female cones by the wind, and the sperm are carried to the egg by the pollen tube, the pines and other conifers are not dependent on free water for fertilization. Thus they are able to reproduce sexually when (and where) ferns and bryophytes cannot.

Following fertilization, the zygote begins to divide and forms the embryo, or young sporophyte. As the ovule matures, its integument hardens into a seed coat, enclosing both the embryo and the female gametophyte (the latter provides food for the embryo when the seed germinates). After the cone matures, it opens, releasing its seeds; this typically occurs in the fall of the second year after the initial appearance of the cones. In most conifers the seeds are winged and are dispersed by the wind.

Figure 24–18 shows a longitudinal section of a pine seed, which has been described as "three generations under one roof." The seed coat and the wing on which the seed is carried arise from the hardened integument of the ovule, derived from the mother sporophyte. The seed coat surrounds the tissue of the female gametophyte; swollen and packed with stored food reserves, the gametophyte tissue grows and displaces the tissue of the original megasporangium. Innermost in the seed is the embryo with its several **cotyledons,** or seed leaves, which will appear as the first leaves of the shoot of the new sporophyte when the seed germinates. The lower part of the embryo will develop into the first root.

Under favorable conditions, the seed—the mature ovule and its contents—will germinate and give rise to a seedling. If conditions remain favorable, the seedling will ultimately become a mature pine tree, producing spores that give rise to gametophytes that, in turn, produce gametes.

The Conifer Leaf

Another feature commonly associated with conifers, although not with all gymnosperms, is a needlelike leaf. Figure 24–19 shows a cross section of a pine leaf, which may be 10 or more centimeters long but only 1 to 3 millimeters in diameter. In the center are the veins, which carry water in one set of conducting cells (the tracheids) and sugars in another (the sieve cells). Outside the region containing the veins are the cells in which photosynthesis takes place. The ducts on the flat sides of the needles carry resin, a substance that is released when the plant is wounded and apparently serves to close the break. The outer layer of cells, the epidermis, is very hard but contains stomata through which gas exchange takes place.

The needlelike leaf, despite its slender form, is a megaphyll (page 501). It is well adapted to long periods of low humidity, as in regions with seasonal rainfall or long, cold winters, and to moisture-losing, sandy soils. One or more of those are characteristic of many regions in which modern conifers are abundant.

24–19 *Cross section of a pine leaf stained to show the tissues and cell types. The hard outer covering and compact shape protect the leaf from water loss, an essential factor in the survival of these trees in areas in which there is little rainfall, or in which the water is locked in the ground as ice during many months of the year.*

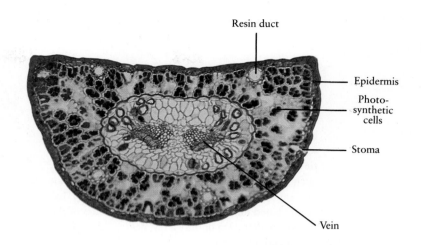

Resin duct

Epidermis

Photo-synthetic cells

Stoma

Vein

Angiosperms: The Flowering Plants

It is believed that the angiosperms—plants with enclosed, protected seeds—evolved from a now-extinct group of gymnosperms. They appear in the fossil record in abundance during the Cretaceous period, about 120 million years ago, as the dinosaurs were declining. Of the numerous angiosperm genera that appeared at that time, many seem to have been very similar to our modern genera.

Daniel Axelrod, a paleobotanist at the University of California, believes that angiosperms probably arose long before the Cretaceous, during the Permian period. They are likely to have originated on the less fertile hills and uplands of tropical areas, the richer lowlands being crowded with club mosses, ferns, and gymnosperms. Once established, they spread into the lowlands, where they became the dominant plant forms and were deposited as fossils. During this mid-Cretaceous period, the climate of the earth was warmer and more uniform than it is at present, and by the end of the Cretaceous period, much of the land was covered with a rich forest of angiosperms, reaching almost as far north as the Arctic Circle.

Angiosperms, like gymnosperms, have megaphylls, stomata, and a cuticle impervious to water. The modern forms, however, have a more highly evolved vascular system than is found in the gymnosperms (Figure 24–20). They also have two new, interrelated structures that distinguish them from all other plants: the flower and the fruit. Both are devices by which animals are induced, rewarded, tricked, and even seduced into carrying out the plants' reproductive strategies.

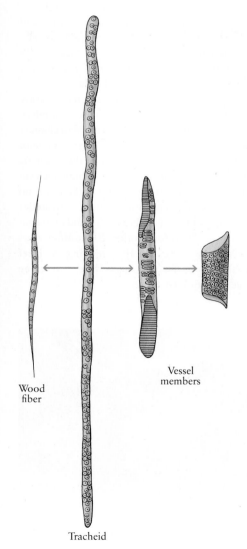

Wood fiber

Vessel members

Tracheid

24–20 *Evolutionary relationships among some cells of the xylem of vascular plants. Tracheids, which are the only water-conducting cells of the conifers, are believed to resemble more primitive cells. Tracheids are elongated cells with thin areas (pits) in their lateral walls through which water moves from one tracheid to another up the trunk from the roots. Tracheids also provide mechanical support. Wood fibers,* *specialized for support, and vessel members, specialized for conducting water, are presumed to have evolved from the primitive water-conducting and supporting tracheids of early vascular plants. In the most highly evolved vessels, the end walls of the individual cells (vessel members) disintegrate during development, and the members are stacked on top of one another, leaving a continuous tube.*

The Ice Ages

During most of our planet's history, its climate appears to have been warmer than it is at the present time. However, these long periods of milder temperatures have been interrupted periodically by Ice Ages, so called because they are characterized by glaciations, or persistent accumulations of ice and snow. Such glaciations occur whenever the summers are not hot enough and long enough to melt ice that has accumulated during the winter. In many parts of the world, an alteration of only a few degrees in temperature is enough to begin or end a glaciation.

An early Ice Age appears to have occurred at the beginning of the Paleozoic era, some 590 million years ago. Another, marked by extensive glaciations in the Southern Hemisphere, closed the Paleozoic, some 248 million years ago. The conifers evolved during this period and possibly the angiosperms, as older forest types disappeared. A more recent, less severe cold, dry period occurred at the end of the Mesozoic, about 65 million years ago, and may be associated with the events causing the extinction of the dinosaurs (to be discussed in Chapter 49).

The most recent Ice Age began during the Pleistocene epoch, about 1.5 million years ago. The Pleistocene has been marked by four extensive glaciations that have covered large areas of North America, Great Britain, and northern Europe. Between the glaciations there have been intervals, called interglacials, during which the climate has become warmer. In each of these four Pleistocene glaciations, sheets of ice, thicker than 3 kilometers in some regions, spread out locally from the poles, scraped their way over much of the continents—reaching as far south as southern Illinois in North America and covering Scandinavia, most of Great Britain, northern Germany, and north-

ern Russia—and then receded again. We are living at the end of the fourth glaciation, which completed its retreat only some 8,000 years ago.

The fossil record shows that during these periods of violent climatic change the populations of these regions were under extraordinary evolutionary pressures. Plant and animal populations moved, changed, or became extinct. In the interglacial periods, during which the average temperatures were at times warmer than those of today, the tropical forests and their inhabitants spread up through today's temperate zones. During the periods of glaciations, only animals of the northern tundra could survive in these same locations. Rhinoceroses, great herds of horses, large bears, and lions roamed Europe in the interglacial periods. In North America, as the fossil record shows, there were camels and horses, saber-toothed cats, and great ground sloths, one species as large as an elephant. In the colder periods, reindeer ranged as far south as southern France, while during the warmer periods, the hippopotamus reached England.

The reason for these large changes in temperature is one of the most controversial issues in modern science. They have been variously ascribed to changes in the earth's orbit, variations in the earth's angle of inclination toward the sun, migration of the magnetic poles, fluctuations in the solar energy, higher elevations of the continental masses, continental drift, and combinations of these and other causes. Also under debate is the question of whether this period of glaciations is over—perhaps for another 200 million years —or whether we are merely enjoying a brief interglacial before the ice sheets begin to creep toward the equator once more.

Glaciers are accumulations of snow and ice that flow across a land surface as a result of their own weight. This Alaskan glacier is flowing into the sea. As the glacier moves forward, its edges melt; the rocks carried within it have become concentrated in the dark band at the right.

About 235,000 different species of angiosperms are known. They dominate the tropical and temperate regions of the world, occupying well over 90 percent of the earth's vegetative surface. The angiosperms include not only the plants with conspicuous flowers but also the great hardwood trees, all the fruits, vegetables, nuts, and herbs, and grains and grasses that are the staples of the human diet and the basis of agricultural economy all over the world. These tremendously diverse plants are classified in two large groups: class Monocotyledones (the monocots), with about 65,000 species, and class Dicotyledones (the dicots), with about 170,000 species. Among the monocots are such familiar plants as the grasses, lilies, irises, orchids, cattails, and palms. The dicots include many of the herbs, almost all the shrubs and trees (other than conifers), plus many other plants. The major differences between these two classes are summarized in Table 24–3 and will be discussed further in Section 5.

TABLE 24–3 **Principal Differences between Monocots and Dicots**

CHARACTERISTIC	MONOCOTS	DICOTS
Flower parts	Usually in threes	Usually in fours or fives
Pollen grains	Have one furrow or pore	Have three furrows or pores
Cotyledons ("seed leaves")	One	Two
Leaf venation	Major veins usually parallel	Major veins usually netlike
Vascular bundles in young stem	Scattered	In a ring
Secondary (woody) growth	Absent	Usually present

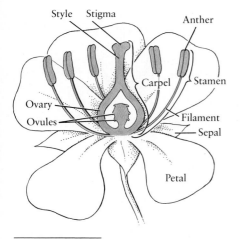

24–21 *Structure of a flower. The reproductive structures are shown in color. Some flowers have only the male structures, some only the female; such flowers are said to be imperfect. A flower that possesses both stamens and carpel, such as this flower, is known as a perfect flower.*

The petals and sepals are, like the stamens and carpel, modified leaves.

The Flower

Figure 24–21 is a diagram of a simple flower. The central structure is the **carpel,** the female reproductive structure, which is thought to be a modified sporophyll (sporangium-bearing leaf). (A single carpel or a group of fused carpels is also known as a pistil, because of its resemblance to an apothecary's pestle.) The swollen base of the carpel is the **ovary,** within which is the ovule, or ovules, in which the female gametophyte develops from a megaspore. The tip of the carpel is specialized as the **stigma,** a sticky surface to which pollen grains adhere. The stigma and ovary are connected by a slender column of tissue, the **style.**

The pollen grains (immature male gametophytes) develop in the **stamen,** which, like the carpel, is a sporophyll. It consists of the **anther,** which contains microsporangia in which the pollen grains develop, and a supporting **filament.**

As shown in Figure 24–22, pollen grains produced in the anthers are carried to the stigma of (usually) another flower, where they germinate, developing pollen tubes that grow through the style toward the ovule. The nonmotile sperm cells are carried by the pollen tubes to the female gametophyte, which typically consists of only seven cells. The extraordinary events of angiosperm fertilization, which give rise not only to an embryonic sporophyte but also to a special nutritive tissue, will be discussed in detail in Chapter 29 (page 615).

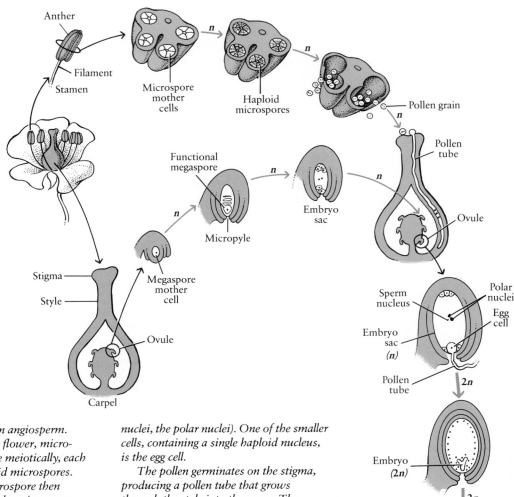

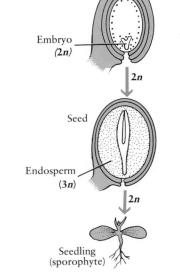

24–22 *Life history of an angiosperm. Within the anther of the flower, microspore mother cells divide meiotically, each giving rise to four haploid microspores. The nucleus in each microspore then divides mitotically, and the microspore develops into a two-celled pollen grain, which is an immature male gametophyte. One of the cells subsequently divides again, usually after germination, resulting in three haploid cells per pollen grain: two sperm cells and the tube (pollen tube) cell.*

Within the ovule, a megaspore mother cell divides meiotically to produce four haploid megaspores. Three of the megaspores disintegrate; the fourth divides mitotically, developing into an embryo sac—the female gametophyte—consisting of seven cells with a total of eight haploid nuclei (the large central cell contains two nuclei, the polar nuclei). One of the smaller cells, containing a single haploid nucleus, is the egg cell.

The pollen germinates on the stigma, producing a pollen tube that grows through the style into the ovary. The growing pollen tube enters the ovule through a small opening known as the micropyle. The two sperm cells pass through the tube into the embryo sac; one sperm nucleus fertilizes the egg cell, the other merges with the polar nuclei, forming a triploid (3n) cell that develops into a nutritive tissue, the endosperm. The embryo undergoes its first stages of development while still within the ovary of the flower, and the ovary itself matures to become a fruit. The seed, released from the mother sporophyte in a dormant form, eventually germinates, forming a seedling.

Evolution of the Flower

Plants are, generally speaking, immobile. Thus, uniting the sperm of one individual with the egg of another individual of the same species presents a problem, to which the flower—that apt symbol of spring and romance—is a solution. The early gymnosperms from which the angiosperms evolved were probably wind-pollinated, as are modern gymnosperms. And, as in the modern gymnosperms, the ovule probably exuded droplets of sticky sap in which pollen grains were caught and drawn to the female gametophyte. Insects, principally beetles and flies, that fed on plants must have come across the protein-rich pollen grains and the sticky, sugary droplets. As they began to depend on these new-found food supplies, they inadvertently carried pollen from plant to plant.

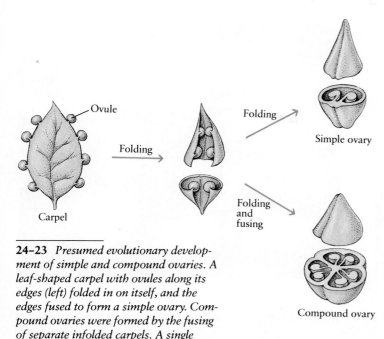

24–23 *Presumed evolutionary development of simple and compound ovaries. A leaf-shaped carpel with ovules along its edges (left) folded in on itself, and the edges fused to form a simple ovary. Compound ovaries were formed by the fusing of separate infolded carpels. A single flower may contain one or more carpels, which may be separate, as in blackberry flowers, or fused, as in the flowers of tomato plants and apple trees.*

24–24 *Flower of a round-leaved hepatica* (Hepatica americana). *Notice the open, bowl shape and the numerous and separate floral parts. The insect is a pollen-eating beetle.*

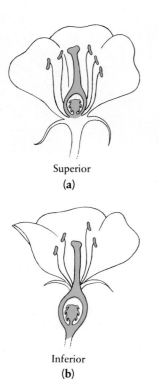

24–25 (a) *Superior and* (b) *inferior ovaries. In flowers with superior ovaries, the floral parts are attached below the ovaries. In flowers with inferior ovaries, the floral parts are attached above, and the ovaries are protected.*

Insect pollination must have been more efficient than wind pollination for some plant species because, clearly, selection began to favor those plants that had insect pollinators. The more attractive the plants were to the insects, the more frequently they would be visited and the more seeds they would produce. Any chance variations that made the visits more frequent or that made pollination more efficient offered immediate advantages; more seeds would be produced, and thus more offspring would be likely to survive. Nectaries (nectar-secreting structures) evolved, which lured the pollinators. Plants developed white or brightly colored flowers that called attention to the nectar and other food supplies. The carpel, originally a leaf-shaped structure, became folded on itself, enclosing and protecting the ovule from hungry pollinators (Figure 24–23). By the beginning of the Cenozoic era, some 65 million years ago, the first bees, wasps, butterflies, and moths had appeared. These are insects for which flowers are often the only source of nutrition for the adult forms. From this time on, flowers and certain insect groups have had a profound influence on one another's history, each shaping the other as they evolved together.

The primitive flower is believed to have resembled that of the modern hepatica, shown in Figure 24–24, which has numerous floral parts, each separate from the other. By comparing this type of flower with some of the more specialized ones shown on page 616, it is possible to see four main trends in flower evolution:

1. Reduction in number of floral parts. Most specialized flowers have few stamens and few carpels.

2. Fusion of floral parts. Carpels and petals, in particular, have become fused, sometimes elaborately so.

3. Elevation of free floral parts above the ovary. In the primitive flower, the floral parts arise at the base of the ovary (Figure 24–25a). Such ovaries are said to be superior. In the more advanced flowers, the free portions of the floral parts are above the ovary (Figure 24–25b); such ovaries are said to be inferior. This is an important adaptation by which the ovules are believed to be protected from foraging insects.

4. Changes in symmetry. The radial symmetry of the primitive flower has given way, in more advanced forms, to bilaterally symmetrical forms.

513

(a) (b)

(c) (d) (e)

24–26 *Pollinators.* **(a)** *A honey bee foraging in a flower of Gaillardia pulchella. Seen frequently along roadsides in the American Southwest, this flower is commonly known as firewheel because of its resemblance to a Fourth of July pinwheel. It is also known as Indian blanket.*

(b) *A longhorn beetle pollinating a lily. The pollen-covered head of the beetle is brushing against a stigma of the flower. Notice how precisely the pollen grains were deposited on the beetle's head by the flowers it visited previously.*

(c) *A gossamer-winged butterfly sipping nectar in a daisy. Notice the long, sucking tongue of the butterfly. The head of a daisy and other similar flowers consists of numerous separate florets.*

(d) *Female rufous hummingbird probing for nectar in a columbine. Her head is collecting pollen, which she will carry to another flower. Flowers pollinated by birds are scentless, bright red or orange, and have copious nectar that makes the visit worthwhile.*

(e) *By thrusting its face into the center of an organ-pipe cactus flower, a Leptonycteris bat is able to lap up nectar with its long, bristly tongue. Pollen grains clinging to its face and neck are transferred to the next flower visited by the bat. Bat-pollinated flowers have dingy colors and a musty scent (similar to that produced by bats to attract one another), and they open at night. Other mammalian pollinators include a few rodents, marsupials, and primates.*

A flower that attracts only a few kinds of animal visitors and attracts them regularly has an advantage over flowers visited by more promiscuous pollinators: its pollen is less likely to be wasted on a plant of another species. In turn, it is an advantage for an animal to have a "private" food supply that is relatively inaccessible to competing species. Many of the distinctive features of modern flowers are special adaptations that encourage regular visits (constancy) by particular pollinators. The varied shapes, colors, and odors allow sensory recognition by pollinators. The diverse, sometimes bizarre, structures such as deep nectaries and complex landing platforms that are found, for example, in orchids, snapdragons, and irises, represent ways of excluding indiscriminate pollinators and of increasing the precision of pollen deposits on the animal messengers.

Evolution of the Fruit

A fruit is the mature, ripened ovary of an angiosperm and contains the seeds. A great variety of fruits, adapted for many different dispersal mechanisms, have evolved in the course of angiosperm history (Figure 24–27). In most cases, the chief requirement is that the seed be transported some distance from the parent plant, where it is more likely to find open ground and sunlight. Many fruits, like flowers, evolved as a payment to an animal visitor for transportation services. A familiar example is provided by the edible fleshy fruits that become sweet and brightly colored as they ripen, attracting the attention of birds and mammals, including ourselves. The seeds within the fruits pass through the digestive tract hours later and are often deposited some distance away. In some species the seed coat's exposure to digestive juices is a prerequisite for germination. The seeds themselves may be bitter or toxic, as in apples, discouraging animals from grinding up and digesting them.

Biochemical Evolution

Another factor in the dominance of the flowering plants has been the evolution of characteristic bad-tasting or toxic compounds. The mustard family (Brassicaceae), for example, is characterized by the pungent taste and odor associated with cabbage, horseradish, and mustard. Milkweeds (family Asclepiadaceae) contain cardiac glycosides, substances that act as heart poisons in vertebrates, potential predators of this group of plants. Bitter-tasting quinine is derived from tropical trees and shrubs of the genus *Cinchona*. Nicotine and caffeine are plant products. Mescaline comes from the peyote cactus; tetrahydrocannabinol from *Cannabis sativa;* opium from a poppy; cocaine from the coca leaf. All of these substances were, at one time, thought of as by-products of plant metabolism and termed "secondary plant substances"; they are now recognized as products of angiosperm evolution that provide powerful defenses against animal predators.

Angiosperms represent the most successful of all plants in terms of numbers of individuals, numbers of species, and their effects on the existence of other organisms. In Section 5, we shall consider the structure and physiology of this dominant group of plants, paying particular attention to the processes of reproduction, growth, and development, the transport of materials by the vascular system, and the integration of the activities of the diverse types of cells forming the plant body.

(a)

(b)

24–27 *Angiosperms are characterized by fruits. As we shall see in Chapter 29, a fruit is a mature ovary, enclosing the seed or seeds. It often also includes accessory parts of the flower. The fruit aids in dispersal of the seed. Some fruits are borne on the wind, some are carried from one place to another by animals, some float on water, and some are even forcibly ejected by the parent plant.*

(a) In milkweed, the fruit bursts open when it is ripe, releasing seeds with tufts of silky hair that aid in their dispersal.

(b) The tough seeds in these blackberries will pass unharmed through the digestive tract of the harvest mouse. If they are deposited in a suitable environment, they will germinate, giving rise to new blackberry plants.

THE ROLE OF PLANTS

The only forms of life on land that do not depend on plants for their existence are a few kinds of autotrophic prokaryotes and protists. For all other terrestrial organisms, the chloroplast of the plant cell is the "needle's eye" through which the sun's energy is channeled into the biosphere. Even those animals that eat only other animals—the carnivores—could not exist if their prey, or their prey's prey, had not been nourished by plants.

24-28 *Poison ivy* (Toxicodendron radicans) *produces an organic alcohol that causes an irritating rash on the skin of many people. The ability to produce this compound is thought to have evolved under the selection pressure exerted by herbivorous animals. Fortunately, the plant is easily identified by its characteristic compound leaves with their three leaflets. Leaves growing in the sun are generally shiny, while those growing in the shade are dull.*

Moreover, plants are the channels by which many of the simple inorganic substances vital to life enter the biosphere. Carbon in the form of carbon dioxide is taken from the atmosphere and incorporated into organic compounds during photosynthesis. Elements such as nitrogen and sulfur are taken from the soil in the form of simple inorganic compounds and incorporated into proteins, vitamins, and other essential organic compounds within green plant cells. Animals cannot make these organic compounds from inorganic materials and so are entirely dependent on plants for these compounds as well as for their energy supply.

It was only after the plants had successfully invaded the land that members of a number of different animal groups could, in turn, carry out their own invasions of this vast new environment. As we shall see in subsequent chapters of this section, both invertebrates and vertebrates underwent great diversification as they took advantage of the variety of habitats and food supplies made available by the plants.

SUMMARY

Plants are multicellular photosynthetic organisms adapted for life on land. Among their adaptations are a waxy cuticle, pores through which gases are exchanged, protective layers of cells surrounding the reproductive cells, and retention of the young sporophyte within the female gametophyte during its embryonic development.

The ancestor of the plants is believed to have been a multicellular green alga of class Charophyceae, similar to the modern genus *Coleochaete*. It was oogamous (that is, its gametes were differentiated into egg and sperm), and it evolved a life cycle characterized by the alternation of heteromorphic generations, two features present in all plants. From this common ancestor, two principal lineages diverged: the bryophytes and the vascular plants.

The mosses and other members of division Bryophyta are relatively small plants usually found in moist locations. Most lack specialized vascular tissues and all lack true leaves, although the plant body is differentiated into photosynthetic, food-storing, and anchoring tissues. In sexual reproduction, the female gamete (egg) develops in the archegonium; the male gametes (sperm) develop in the antheridium. The sperm, which are flagellated, swim to the archegonium and fertilize the egg cell. The resulting zygote develops into an embryo within the archegonium. In the bryophytes—and in no other members of the plant kingdom—the gametophyte *(n)* is dominant and the sporophyte *(2n)* is smaller, attached, and often nutritionally dependent.

Although the bryophytes seem to have changed little in the course of their history, the vascular plants have undergone a great diversification. Major developments in their evolution include better conducting systems, a progressive reduction in the size of the gametophyte, and the "invention" of the seed. The nine divisions of vascular plants can be informally grouped into the seedless vascular plants (divisions Psilophyta, Lycophyta, Sphenophyta, and Pterophyta) and the seed plants. The seed plants can be grouped into the gymnosperms, or naked-seed plants (divisions Coniferophyta, Cycadophyta, Ginkgophyta, and Gnetophyta), and the angiosperms, or flowering plants (division Anthophyta).

Among living seedless vascular plants, the ferns (division Pterophyta) are the most numerous. They are characterized by large, often finely dissected leaves called fronds. These leaves are, like the leaves of the seed plants, megaphylls. The sporophyte is the dominant generation, but in most ferns the gametophytes are independent. The sperm are flagellated, and free water is needed for fertilization. Sporangia are typically formed on the underside of specialized leaves (sporophylls) of the sporophyte.

Gymnosperms and angiosperms are seed plants. The most numerous modern gymnosperms are the conifers ("cone-bearers"). On the scales of the smaller, male cones, microspores differentiate into male gametophytes, which are released in the form of windblown pollen. In the ovules that form on the scales of the larger, female cones, the female gametophytes develop from megaspores. Within the female gametophyte, archegonia form. The male gametophyte germinates and produces a pollen tube through which the nonmotile sperm enter the archegonium and fertilize the egg cell. The seed, or mature ovule, consists of the seed coat, the young embryo, and the female gametophyte, which serves as nutritive tissue. The seed, which is shed from the female cone, can remain dormant for long periods of time and hence is adapted to withstanding cold and drought.

The angiosperms, of which there are about 235,000 species, are characterized by the flower and the fruit. Flowers attract pollinators, and fruits enhance the dispersal of seeds. The reproductive structures of the flower are the stamens, composed of filament and anther, and the carpels, composed of ovary, style, and stigma. The stamens and the carpels are highly specialized sporophylls. The ovule or ovules are enclosed in the ovary, which is the base of the carpel or group of fused carpels. Pollen grains, the immature male gametophytes, are formed in the anthers and germinate on the sticky surface of the stigma, the tip of the carpel.

The various shapes and colors of flowers evolved under selection pressures for more efficient pollinating mechanisms. Major trends in flower evolution include reduction and fusion of floral parts, a change in the position of the ovary relative to the other flower parts to a more protected (inferior) position, and a shift from radial to bilateral symmetry.

In addition to the flower and the fruit, a third factor in the success of the angiosperms has been the evolution of bad-tasting or toxic chemicals that discourage the predations of foraging animals. Angiosperms are the dominant plants of the modern landscape, providing a diversity of habitats and foods for terrestrial animals.

QUESTIONS

1. Distinguish among the following: bryophytes/vascular plants; moss/club moss; spore/sporangium/sporophyte/sporophyll; homosporous/heterosporous; gametophyte/antheridium/archegonium; xylem/phloem; microphyll/megaphyll; gymnosperm/angiosperm; microspore/megaspore; carpel/stamen.

2. Consider the green alga *Coleochaete*. What advantages does the alga gain from the production of "extra" haploid cells, made possible by additional replication of the DNA of the zygote prior to meiosis? What might be the advantages—to *Coleochaete*—of retaining the zygote on the body of the parent alga?

3. Bryophytes, among the plants, and amphibia, among the animals, often live in habitats intermediate between fresh water and dry land, rather than between salt water and land. Propose a physiological argument (referring back to Chapter 6) to explain why invasions of the land were more likely by organisms previously adapted for life in fresh water.

4. Where did the liverworts get their name? (The answer is not in the text, but a little research should tell you.) What does this name tell you about the human perspective of the rest of the biosphere?

5. Vascular plants are characterized by the presence of lignin, which stiffens and supports the plant body as it grows upward toward the light. What is the primary source of support for the photosynthetic portions of the multicellular seaweeds?

6. Tall trees today are not appreciably taller than plants of the Devonian forests. What factors select for tallness in trees? What factors, by contrast, select against tallness? What kind of evolutionary innovation might be required to alter the optimal balance among these various factors?

7. In some areas on earth, large gymnosperms are either dominant or manage to coexist with angiosperms. List some advantages for a big plant's being a gymnosperm.

8. Describe the sporophylls of a club moss, a horsetail, a fern, a pine, and an angiosperm.

9. Sketch a pine seed and label it. Indicate the origin of each of its components, and whether the component is haploid or diploid. How many generations are represented in the tissues of the seed?

10. Sketch a flower and label the structures. What is the function of each floral part?

11. A major trend in the evolution of angiosperms is a reduction in the number of floral parts. Why should reduction, rather than greater elaboration, be so common?

12. How have plants solved the problem of obtaining adequate food and water? Of supplying food and water to the individual cells? Of protecting and nourishing the immature stages?

The Animal Kingdom I: Introducing the Invertebrates

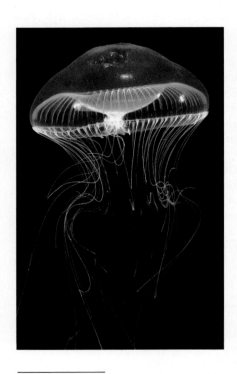

25-1 *Animals are characterized by their motility, usually a result of the contraction of assemblies of protein fibers within specialized (muscle) cells. Aequorea victoria, a jellyfish, is shown here swimming actively, its bell contracted by muscles around its margin.*

Animals are many-celled heterotrophs, and their principal mode of nutrition is ingestion. They depend directly or indirectly for their nourishment on photosynthetic autotrophs—algae or plants. Typically they digest their food in an internal cavity and store food reserves as glycogen or fat. Their cells, unlike those of most other eukaryotes, do not have walls. Generally, animals move by means of contractile cells (muscle cells) containing actin, myosin, and associated proteins (see page 122). The most complex animals—the octopuses, the insects, and the vertebrates—have many kinds of specialized tissues, including elaborate sensory and neuromotor mechanisms not found in any other kingdom.

For most of us, animal means mammal, and mammals are, in fact, the chief focus of attention in Section 6. However, the mammals, or even the vertebrates as a whole, represent only a small fraction of the animal kingdom. More than 1.5 million different species of animals have been described, of which more than 95 percent are invertebrates—that is, animals without backbones. Indeed, the enormous variety displayed by the invertebrates is partly why they are so endlessly fascinating to study. There are also two practical reasons for the study of invertebrates. First, by virtue of their variety and their sheer numbers, they are of great ecological importance. Second, the invertebrates demonstrate a wide spectrum of ingenious solutions to the biological problems confronting multicellular heterotrophic organisms. In this way, they illuminate the fundamental nature of these problems and so lead us to a deeper understanding of the physiology of the vertebrate animals, including ourselves.

THE DIVERSITY OF ANIMALS

The unifying characteristic among all the animals is their mode of nutrition. Unlike plants, which are passive recipients of energy from the sun, animals must seek out food sources or, alternatively, devise strategies for ensuring that the food comes to them. Thus motility—of the entire organism, or its parts, or both—is a requirement for animal survival. Motility in general and the hunting of prey in particular require efficient systems of integration and control of the multitude of cells that make up the organism. In short, muscle and nerve, distinguishing features of the animal kingdom, have their origin in heterotrophy.

Paradoxically, these unifying characteristics of the animal kingdom are also the key to its diversity. Given the properties of living tissue and the characteristics of our planet—particularly the force of gravity and the physical properties of water and air—there are only a few basic ways that locomotion, the capture of food, self-defense, and coordination can be accomplished. In the course of evolution, however, as animals have adapted to new or changing environments, those few basic ways have been "reinvented," refined, and elaborated upon, resulting in the great diversity of structural and functional detail that we see today.

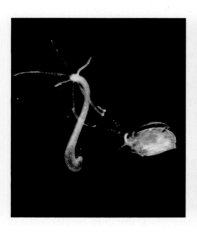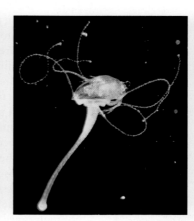

25–2 *Some animals actively seek their food, whereas others arrange to have it delivered (see Figure 25–3). Here a Hydra encounters a small crustacean (Daphnia), grasps it with its tentacles, engulfs it, and digests it. These activities are coordinated by the animal's simple but effective nervous system.*

25–3 *The adult barnacle is nonmotile, depending on its six pairs of bristly appendages to deliver its food supply. Louis Agassiz, America's leading nineteenth century naturalist, characterized a barnacle as "nothing more than a little shrimp-like animal, standing on its head in a limestone house and kicking food into its mouth."*

A similar diversification has occurred in reproductive patterns, providing mechanisms not only for genetic recombination but also for dispersal into and exploitation of habitats with abundant food supplies. Although sexual reproduction is the usual pattern in animals, many different types of animals are also capable of rapid asexual reproduction when conditions are suitable. Most animals are diploid, with gametes the only haploid stage in the life cycle (see Figure 12–4c, page 250). In the simplest and most primitive form of sexual reproduction, sperm and egg cells produced by different individuals are shed into the water, where they unite. Although external fertilization is retained in many animals that are otherwise highly evolved (for example, most frogs and toads), a variety of different methods of internal fertilization have also evolved. Additional versatility, particularly in feeding and dispersal, is provided by the immature forms, or **larvae,** that are characteristic of the life cycles of many different types of animals. As adults, most animals are fixed in size and shape (in contrast to plants, in which growth often continues for the lifetime of the organism); as we shall see, however, larvae often differ dramatically in size and shape from their adult counterparts.

In Chapter 21, we commented upon the tremendous diversity of the prokaryotes, as exemplified by the wide range of environments they inhabit and the many ways in which they satisfy their energy requirements. Among the more than 1.5 million species of animals that have been described, we see a similar pattern of adaptation to many different ways of life. Thus, for instance, on and around a single coral head, only a meter or two in diameter, one finds a dazzling array of different forms—reef fish, crustaceans, sponges, jellyfish, starfish, sea urchins, anemones, and the coral animals themselves. Similarly, to take a terrestrial example, a single spadeful of soil turns up earthworms, pillbugs, spiders, nematodes, and various other tiny animals. The branch of a single tree may harbor a dozen different kinds of insects, both adults and larvae, differing from one another in their nutritional specializations.

THE ORIGIN AND CLASSIFICATION OF ANIMALS

Animals, like plants, presumably had their origins among the protists. In the case of animals, however, we have fewer clues about which protists most closely resemble the ancestral ones, and multicellularity may well have arisen more than once. By the Cambrian period (590 to 505 million years ago), two major—and distinct—diversifications of animal life had already occurred, only one of which was to survive. The earliest evidences of animal life are provided by the fossilized

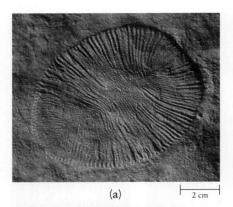

(a)   ⊢—⊣ 2 cm

(b)   ⊢—⊣ 250 mm

25–4 *Among the unusual organisms of the Ediacaran deposits are* **(a)** *Dickinsonia, a segmented animal that lacked a clearly defined head and appendages, and* **(b)** *Spriggina, a segmented animal with a prominent "head shield." Although these organisms superficially resemble some modern invertebrates, they appear to have lacked mouths, and their flattened bodies contained no tubular structures.*

animals found in greatest abundance in the Ediacaran Hills of Australia and a rock formation known as the Burgess Shale in the Canadian Rockies. The Ediacaran animals (Figure 25–4), which arose some 670 million years ago and flourished for about 100 million years, were quite unlike later animals and apparently were not their ancestors. As we shall see, the fundamental structure of virtually all living animals is a tube (or, in some cases, a cavity) within a tube, modified and elaborated in a variety of ways that maintain a high ratio of surface to volume. Ediacaran animals, by contrast, lacked tubular internal structures and were flat, leaflike, or, occasionally, quilted like an air mattress. Although this evolutionary solution to the problem of maintaining an adequate surface-to-volume ratio was ultimately unsuccessful, it produced a diversity of exquisitely beautiful organisms.

The Burgess Shale, which is dated at about 530 million years ago, includes members of at least 10 extinct phyla as well as fossil representatives of virtually all contemporary phyla. Some of these animals (Figure 25–5) are very similar indeed to animals living today. Although the fossils of the Burgess Shale and other Cambrian formations around the world attest to the ancient origins of the modern phyla of animals, they shed little light on the chronological order of their appearance. Most of the evidence for charts such as Figure 25–6 comes from studies of living animals.

Modern animals are classified in about 30 phyla, each of which is thought to be monophyletic. A number of criteria are used in classifying an animal and in attempting to elucidate the evolutionary relationships among the different phyla. Among the factors considered are the number of tissue layers into which the cells are organized, the basic plan of the body and the arrangement of its parts, the presence or absence of body cavities and the manner in which they form, and the pattern of development from fertilized egg to adult animal. Table 25–1 provides a short outline of animal classification and indicates the key features that are used in grouping the various phyla. In our examination of the diversity of animals, we shall devote most of our attention to the phyla indicated in boldface type, which exemplify the most significant developments and adaptations in animal evolution.

TABLE 25–1   **An Outline of Animal Classification***

I. Subkingdom Parazoa: phylum **Porifera**

II. Subkingdom Mesozoa: phylum Mesozoa

III. Subkingdom Eumetazoa

    A. Radially symmetrical animals: phyla **Cnidaria** and Ctenophora

    B. Bilaterally symmetrical animals

        1. Acoelomates (animals that lack a body cavity): phyla **Platyhelminthes,** Gnathostomulida, and **Rhynchocoela**

        2. Pseudocoelomates (animals with the type of body cavity known as a pseudocoelom): phyla **Nematoda,** Nematomorpha, Acanthocephala, Kinorhyncha, Gastrotricha, Loricifera, Rotifera, and Entoprocta

        3. Coelomates (animals with the type of body cavity known as a coelom)

            a. Protostomes (animals in which the mouth appears at or near the first opening that forms in the developing embryo): phyla **Mollusca, Annelida,** Sipuncula, Echiura, Priapulida, Pogonophora, Pentastomida, Tardigrada, Onychophora, and **Arthropoda**

            b. Lophophorates (protostomes but with some deuterostome characteristics): phyla Brachiopoda, Phoronida, and Bryozoa

            c. Deuterostomes (animals in which the anus appears at or near the first opening that forms in the developing embryo): phyla **Echinodermata,** Chaetognatha, Hemichordata, and **Chordata**

* A summary of the distinctive characteristics of each phylum can be found in Appendix C.

25–5 *Some animals of the Burgess Shale.*
(a) Amiskwia, *a gelatinous worm with prominent fins, and* (b) Dinomischus, *a stalked animal, represent phyla known only from fossil deposits of the Cambrian period. Members of living phyla include* (c) Hyolithes, *a mollusk distinguished by a cone-shaped shell with a protective cap;* (d) Aysheaia, *an onychophoran that bears a striking resemblance to the modern* Peripatus *(see Figure 26–22, page 561); and* (e) Pikaia, *an early chordate, readily identified by the stiffening rod, or notochord, that runs the length of its body and by the segmental pattern of its muscles (see Figure 28–10, page 591, for its modern counterpart,* Branchiostoma). *This specimen is 4 centimeters long.*

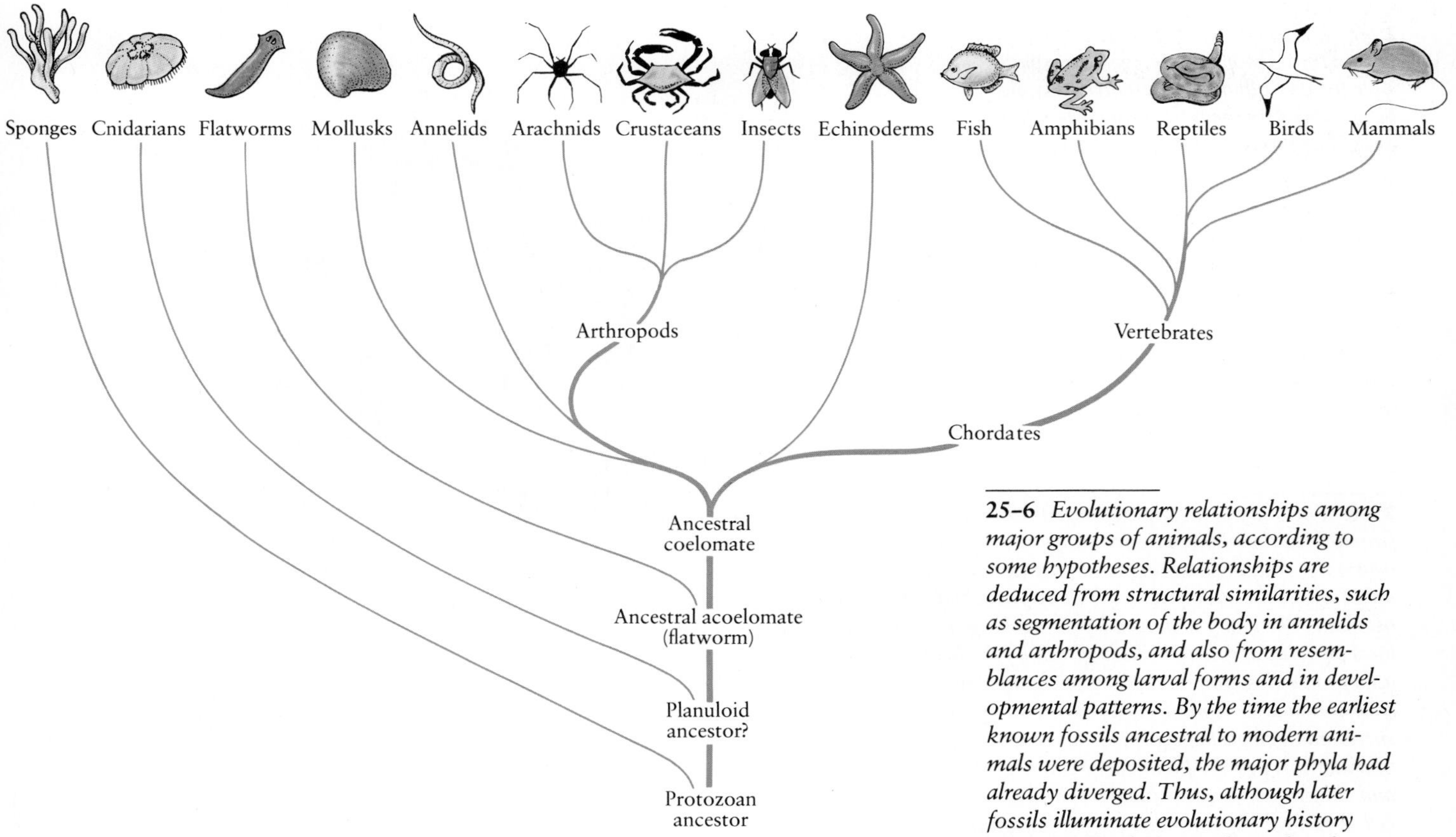

(a)

(b)

(c)

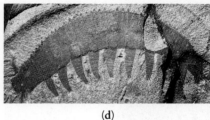

(d)

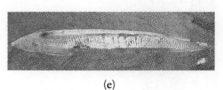

(e)

Sponges Cnidarians Flatworms Mollusks Annelids Arachnids Crustaceans Insects Echinoderms Fish Amphibians Reptiles Birds Mammals

Arthropods

Vertebrates

Chordates

Ancestral coelomate

Ancestral acoelomate (flatworm)

Planuloid ancestor?

Protozoan ancestor

25–6 *Evolutionary relationships among major groups of animals, according to some hypotheses. Relationships are deduced from structural similarities, such as segmentation of the body in annelids and arthropods, and also from resemblances among larval forms and in developmental patterns. By the time the earliest known fossils ancestral to modern animals were deposited, the major phyla had already diverged. Thus, although later fossils illuminate evolutionary history within phyla (for example, within the arthropods or the chordates), we lack clear documentary evidence of the relationships among the different phyla.*

In the remainder of this chapter, we shall look at the so-called lower inverte-brates, characterized by relatively simple body plans. In the three subsequent chapters, we shall consider the more complex animals, all of which possess the type of body cavity known as a coelom.

PHYLUM PORIFERA: SPONGES

Sponges may have had a different origin from other members of the animal kingdom and seem to have traveled a solitary evolutionary route. For this reason, they are often placed in a subkingdom of their own, the Parazoa ("beside the animals"). In fact, until the nineteenth century, the sponges were classified as plant-animals ("zoophytes"), since during their adult life they are all **sessile** (attached to a substrate). Sponges are common on ocean floors throughout much of the world. Most live along the coasts in shallow water, but some, such as the fragile glass sponges, are found at great depths, where currents are relatively slow. A few types are found in fresh water.

A sponge is essentially a water-filtering system, made up of one or more chambers through which water is driven by the action of numerous flagellated cells. Sponges are made up of a relatively few cell types, the most characteristic of which are the **choanocytes,** or collar cells, the flagellated cells that line the interior cavity of the sponge (Figure 25–8). The protozoan choanoflagellates have similar cells, and it is possible that sponges arose from colonial forms of such organisms.

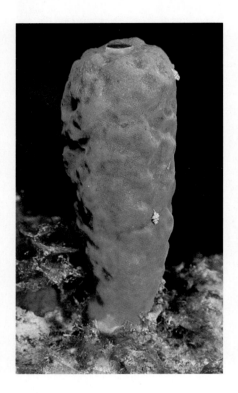

25–7 *A purple tube sponge. Pigments within the cells of this sponge are responsible for its brilliant color, which may also be enhanced by the refraction, or bending, of light as it passes through the water.*

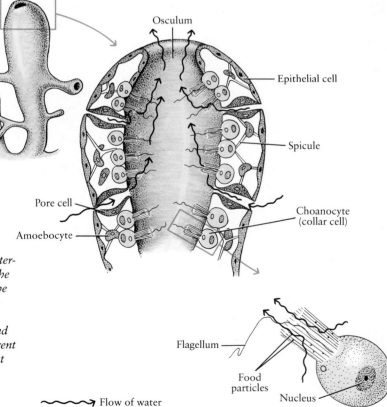

Flow of water

25–8 *The body of a simple sponge is dotted with tiny pores, from which the phylum derives its name (Porifera, or "pore-bearers"). Water containing food particles is drawn into the internal cavity of the sponge through these pores and is forced out the osculum. The water is moved by the sucking effect of flow of the local currents across the osculum and by the beating of the flagella protruding from the collars of choanocytes. The collar of each choanocyte is made up of about 20 retractile filaments and surrounds a single flagellum, the lashing of which directs a current of water through the filaments. Minute particles are filtered out and cling to one or more filaments and are then drawn into the cell. A sponge 10 centimeters high filters more than 20 liters of water a day.*

25–9 *A section of the skeleton of* Euplectella aspergillum, *a species of glass sponge picturesquely known as Venus's flower basket. These fragile sponges, with their delicate silica-containing skeletons, are usually found at great depths. According to the fossil record, glass sponges were present in the Ordovician period, which began 505 million years ago.*

A sponge represents a level of organization somewhere between a colony of cells and a true multicellular organism. The cells are not organized into tissues or organs, yet there is a form of recognition among the cells that holds them together and organizes them. If the body of a living sponge is squeezed through a fine sieve or cheesecloth, it is separated into individual cells and small clumps of cells. Within an hour, the isolated sponge cells begin to reaggregate, and as these aggregations get larger, canals, flagellated chambers, and other features of the body organization of the sponge begin to appear. This phenomenon has been used as a model for the analysis of cell adhesion, recognition, and differentiation, all of which are basic features of embryonic development in higher organisms.

The outer surface of a sponge is covered with epithelial cells, some of which contract in response to touch or to irritating chemicals, and in so doing, close up pores and channels. Each cell acts as an individual, however; there is little coordination among them. Between the epithelial cells and the choanocytes is a middle, jellylike layer, and in this layer are amoebocytes, amoeba-like cells that carry out various functions. Amoebocytes play several roles in reproduction, secrete skeletal materials, and, most important, carry food particles from the choanocytes to the epithelial and other nonfeeding cells. Because all the digestive processes of sponges are carried out within single cells, even a giant sponge—and some stand taller than a man—can consume nothing larger than microscopic particles.

The sponge shown in Figure 25–8 is a small and simple one. In larger sponges, the body plan, although essentially the same, looks far more complex. These sponges, which need correspondingly more food, have highly folded body walls that greatly increase the filtering and feeding surfaces. We have already encountered this evolutionary stratagem for increasing biological work surfaces at the cellular level—as in the inner membrane of the mitochondrion—and we shall be encountering it repeatedly in different animal structures.

The approximately 5,000 species of sponges are grouped into four classes, according to their skeletal structure, which serves for protection, stiffening, and support. In class Calcarea, the skeleton consists of individual spicules of calcium carbonate. Members of class Hexactinellida, the glass sponges, have spicules of silica fused in a continuous and often very beautiful latticework (Figure 25–9). The largest class, Demospongiae, has unfused silica spicules, or a tough, keratin-like protein called spongin, or a combination of the two. The cleaned and dried proteinaceous skeletons of this group are the "natural" sponges available commercially. Members of the fourth and smallest class, Sclerospongiae, have skeletons that contain all three kinds of material—calcium carbonate, silica, and spongin.

Reproduction in Sponges

The reproduction of sponges exhibits many of the features characteristic of sessile or slow-moving animals. Asexual reproduction is quite common, either by fragments that break off from the parent animal, or by gemmules, aggregations of amoebocytes within a hard, protective outer layer. Production of such resistant forms occurs most frequently in freshwater organisms. In the ocean, conditions are relatively unchanging, but the freshwater environment is much more variable. Invertebrates that live in freshwater are more likely to have protected embryonic forms than even closely related marine species.

Sexual reproduction in sponges is highly specialized. As we noted earlier, the simplest and most primitive form of fertilization is external, with sperm and egg cells shed into the water. In most sponges, however, fertilization is internal.

25-10 *A breadcrumb sponge* (Halichondria panacea) *showing some of the many openings (oscula) through which water leaves the animal. This sponge grows as a flat, encrusting mass on the undersurfaces of rocks where it is protected from direct sunlight at low tide. The oscula usually protrude from the rest of the animal. This arrangement allows the natural water flow in the habitat to draw water through the sponge.*

Gametes appear to arise from enlarged amoebocytes, but there are reports that choanocytes can also form gametes. The sperm cells, which resemble those of other animals, are carried by the water currents out of the osculum of one sponge and into the interior cavity of another sponge. There they are captured by choanocytes and transferred to amoebocytes, which then transfer them to ripe eggs—a method of fertilization unique to the sponges. Most sponges even provide a certain amount of maternal care, retaining the young during the early stages of development. The embryonic sponge develops into a flagellated, free-swimming larva that, after a short life in the plankton, locates an appropriate site, settles, and develops into an adult sponge.

Most kinds of sponges are **hermaphrodites** (from Hermes and Aphrodite); that is, the same individual has both male and female reproductive structures and produces both sperm and egg cells. This is a great advantage for animals with little or no motility. In animals with separate sexes, a given individual can mate only with members of the opposite sex, but for a hermaphrodite any partner—or the gametes of any partner—will suffice.

PHYLUM MESOZOA: MESOZOANS

Like the sponges, the mesozoans are so different from all other animals that they are placed in a subkingdom of their own. About 50 species of mesozoans are known; they are extremely simple wormlike animals that live as parasites inside a variety of marine invertebrates. The 20 to 30 cells that make up the body of a mesozoan are organized in two layers—a mass of reproductive cells surrounded by a single layer of ciliated cells (Figure 25–11). There are no organs or body cavities of any kind. As their name ("middle animals") suggests, the mesozoans may be primitive forms, transitional between protists and multicellular animals. Some authorities believe, however, that they are flatworms (phylum Platyhelminthes) that have become simplified as an adaptation to a parasitic way of life.

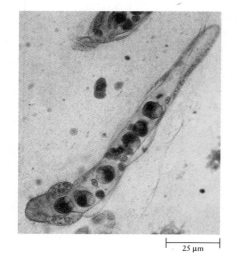

$25\ \mu m$

25-11 Dicyema typoides *is a mesozoan found in the kidneys of octopuses living in the Gulf of Mexico. Mesozoans typically reproduce asexually during one phase of the life cycle and sexually during another phase. This individual is a sexually reproducing adult, forming both sperm and egg cells.*

RADIALLY SYMMETRICAL ANIMALS

Two phyla, Cnidaria (jellyfishes, sea anemones, and corals) and Ctenophora (comb jellies and sea walnuts), consist of gelatinous animals in which the adult form is generally radially symmetrical. In radial symmetry, the body parts are arranged around a central axis, like spokes around the hub of a wheel (Figure 25–12).

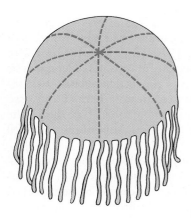

25–12 *In organisms with radial symmetry, any plane through the animal that passes through the central axis divides the body into halves that are mirror images of one another.*

The animals in these two phyla are also characterized by a **gastrovascular cavity,** which has only one opening. Within this cavity, enzymes are released that break down food, partially digesting it extracellularly, as our own food is digested within the stomach and intestinal tract. The food particles are then taken up by the cells lining the cavity; they complete the digestive process and pass the products on to the other cells of the animal. Water circulating through the gastrovascular cavity supplies dissolved oxygen to the lining cells and carries carbon dioxide, other waste products, and the inedible remains of food particles out through the single opening.

Phylum Cnidaria

The cnidarians are a large and often strikingly beautiful group of aquatic organisms. The cells of cnidarians, unlike those of sponges and mesozoans, are organized into distinct tissues, and their activities are coordinated by a nervous system.

As you can see in Figure 25–13a, the basic body plan is simple: the animal is essentially a hollow container, which may be either vase-shaped, the **polyp,** or bowl-shaped, the **medusa.** The polyp is usually sessile; the medusa, motile. Both consist of two layers of tissue: **epidermis** and **gastrodermis** (from the Greek *epi,* "on" or "over," and *gaster,* "stomach," plus *derma,* "skin"). Between the two layers is a gelatinous filling, the **mesoglea** ("middle jelly"), which is made of a collagenlike material. In the polyp form, the mesoglea is sometimes very thin, but in the medusa, it often accounts for the major portion of the body substance. The two tissues of cnidarians are derived from two embryonic tissues, the **ectoderm** and the **endoderm** (from the Greek *ektos,* "outer," and *endon,* "inner"). Because they have two embryonic tissue layers, cnidarians are said to be diploblastic.

25–13 (a) *Among cnidarians, there are two basic body forms: the vase-shaped polyp (left) and the bowl-shaped medusa (right). The gastrovascular cavity has a single opening. The cnidarian body has two tissue layers, epidermis and gastrodermis, with gelatinous mesoglea between them.*

(b) *Cnidocytes, specialized cells located in the tentacles and body wall, are a distinguishing feature of cnidarians. The interior of the cnidocyte is filled by a nematocyst, which consists of a capsule containing a coiled tube, as shown on the left. A trigger on the cnidocyte, responding to chemical or mechanical stimuli, causes the tube to shoot out, as shown on the right. The capsule is forced open and the tube turns inside out, exploding to the outside. The cnidocyte cannot be "reloaded"; it is absorbed and a new cell grows to take its place.*

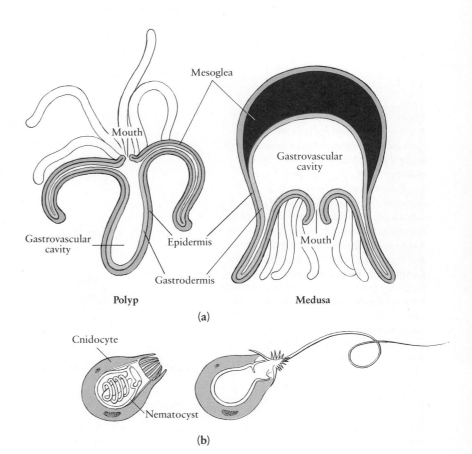

Mesoglea

Mouth

Gastrovascular cavity

Gastrovascular cavity

Epidermis

Mouth

Gastrodermis

Polyp

Medusa

(a)

Cnidocyte

Nematocyst

(b)

A distinctive feature of these animals, the **cnidocyte** (Figure 25–13b), gives the phylum its name. Cnidarians are carnivores. They capture their prey by means of tentacles that form a circle around the mouth. These tentacles are armed with cnidocytes, special cells that contain **nematocysts** (thread capsules). Nematocysts are discharged in response to a chemical stimulus or touch. The tubular nematocyst threads, which are often poisonous and may be sticky or barbed, can lasso prey, harpoon it, or paralyze it—or some useful combination of all three. The toxin apparently produces paralysis by attacking the lipoproteins of the nerve cell membranes of the prey.

Nematocysts occur only in this phylum, with some interesting exceptions. Certain other invertebrates, including nudibranchs (a kind of mollusk) and flatworms, can eat cnidarians without triggering the nematocysts. The nematocysts are then moved to the surface of the predator and can be fired in their new host's defense.

The cnidarian life cycle is characterized by an immature larval form, known as the **planula,** which is a small, free-swimming ciliated organism. Following this larval stage, some cnidarians go through both a polyp and a medusa stage in their life cycles (Figure 25–14). In such species, polyps reproduce asexually and medusas sexually. This sort of life cycle, in which the sexually reproductive form is distinctly different from the asexual form, superficially resembles alternation of generations in plants. There is, however, no alternation between haploid and diploid forms as there is in plants; the only haploid forms are the gametes. The cnidarian life cycle allows for rapid asexual reproduction (by the polyp), dispersal and genetic recombination (by the medusa), and habitat selection (by the planula larva).

The approximately 9,000 species of cnidarians are grouped in three major classes: Hydrozoa, in which the polyp is usually the dominant form; Scyphozoa, predominantly medusoid, exemplified by the large jellyfishes; and Anthozoa, which includes the sea anemones and the reef-building corals, and has only the polyp.

25–14 *The life cycle of the cnidarian Aurelia. Sperm and egg cells are released from adult medusas into the surrounding water. Fertilization takes place, and the resulting zygote develops first into a hollow sphere of cells, the blastula. It then elongates and becomes a ciliated larva called a planula. After dispersal, the planula settles to the bottom, attaches by one end to some object, and develops a mouth and tentacles at the other end, thus transforming into the polyp stage. The body of the polyp grows and, as it grows, begins to form medusas, stacked upside down like saucers. In a phase of rapid asexual reproduction, these bud off, one by one, and grow into full-sized jellyfish.*

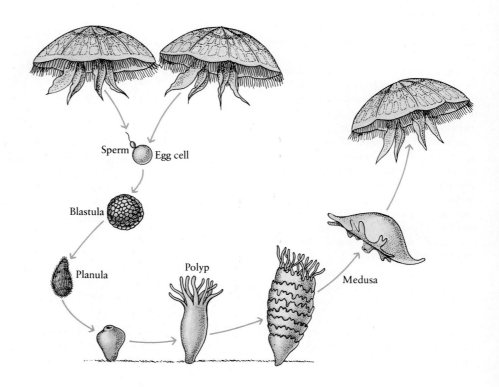

Sperm   Egg cell

Blastula

Planula

Polyp

Medusa

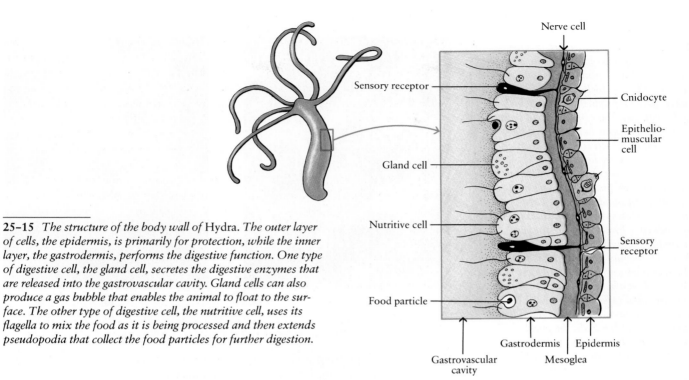

25–15 *The structure of the body wall of* Hydra. *The outer layer of cells, the epidermis, is primarily for protection, while the inner layer, the gastrodermis, performs the digestive function. One type of digestive cell, the gland cell, secretes the digestive enzymes that are released into the gastrovascular cavity. Gland cells can also produce a gas bubble that enables the animal to float to the surface. The other type of digestive cell, the nutritive cell, uses its flagella to mix the food as it is being processed and then extends pseudopodia that collect the food particles for further digestion.*

25–16 Hydra *has a nervous system that integrates the body into a functional whole, making possible a range of fairly complex activities. For instance,* Hydra *may float, glide on its base, or, as shown here, it may travel by a somersaulting motion.*

Class Hydrozoa

Among the most thoroughly studied of the cnidarians are species of *Hydra* and related genera, which are small, common freshwater forms, convenient to keep in the laboratory. Figure 25–15 shows a small section of the body wall of *Hydra.* The epidermis is composed largely of epitheliomuscular cells, which perform a covering, protective function and also serve as muscle cells. Each cell has contractile fibers, myonemes, at its base and so can contract individually, like the contractile epithelial cells of the sponge. The gastrodermis is mostly made up of cells concerned with digestion; these cells also contain contractile fibers. In *Hydra,* as in other polyps, the contractile fibrils of the epidermis run lengthwise in the animal and the fibrils of the gastrodermis run circularly, so the body wall can stretch or bulge, depending on which group contracts.

In addition to cnidocytes and epitheliomuscular cells, which are independent effectors—cells that both receive and respond to stimuli—*Hydra* contains two other types of nerve cells: sensory receptor cells and cells connected into a network, the nerve net. Sensory receptor cells are more sensitive than other epithelial cells to chemical and mechanical stimuli, and when stimulated they transmit their impulses to an adjacent cell or cells, which then respond. Note that this system is one step more complicated than the individual epitheliomuscular cell or cnidocyte, which acts as both receptor and effector. The nerve net, a loose connection of nerve cells lying at the base of the epidermal layer, is the simplest example of a nervous system that links an entire organism into a functional whole. It coordinates the muscular contractions of *Hydra,* making possible a wide variety of activities (Figures 25–2 and 25–16). However, there is no center of operations for the nervous system. This type of conducting system occurs in *Hydra* and certain other cnidarians.

Hydra, a relatively simple hydrozoan that lives as a solitary polyp and has no medusa stage, is thought to be descended from more complex ancestors. Most hydrozoans are colonial marine animals and have both hydroid (polyp) and medusoid forms at different times in their life cycles. Members of the genus *Obelia,* for example, spend most of their lives as colonial polyps (Figure 25–17a).

25–17 *Colonial hydrozoans. (a) Obelia is made up of two types of polyp, a Hydra-like form—shown here with tentacles extended—which is the feeding polyp, and a reproductive form, which lacks tentacles. You can see two of these reproductive forms in the axils of the branches. At the left is a newly released free-swimming medusa.*

(b) Cnidarians of the order Siphonophora are large, floating colonies made up of both polyps and medusas. The polyps are feeding forms and the medusas are reproductive forms; in some species, nonreproductive medusas are swimming bells. The colony produces a gas-filled float, or pontoon. In the Portuguese man-of-war, shown here with a captured fish, the large float serves also as a sail. The blue strands are composed of reproductive and feeding individuals; the purple strands, which may grow as long as 15 meters, are made up of stinging, food-gathering polyps armed with nematocysts. A large colony can inflict serious injury on a human being, leading, in some cases, to death by drowning or as a consequence of a severe allergic reaction.

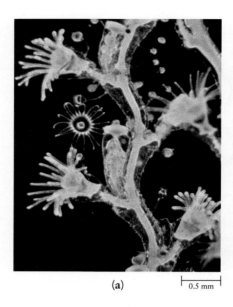

(a)   ⊢ 0.5 mm ⊣   (b)

The colony arises from a single polyp, which multiplies by budding. The new polyps do not separate but remain interconnected so that their gastrovascular cavities form a continuous channel, through which food particles and oxygen-bearing water are circulated. Within the colony are two types of polyps: feeding polyps with tentacles and cnidocytes, and reproductive polyps from which tiny medusas bud off. These medusas produce eggs or sperm that are released into the water and fuse to form zygotes. Many colonial hydrozoans, with their division of labor between feeding and reproductive forms, are very like a single organism (Figure 25–17b).

Class Scyphozoa

A second major class of cnidarians is Scyphozoa, or "cup animals," in which the medusa form is dominant. Although the fairly simple polyp form reproduces itself asexually, in addition to budding off medusas, elaborate colonies are not formed. Scyphozoan medusas, known more commonly as jellyfishes, range in size from less than 2 centimeters up to 4 meters in diameter; the largest jellyfishes trail tentacles as long as 10 meters. In the adult animal, the mesoglea is so firm that a large, freshly beached jellyfish can easily support the weight of a man. The mesoglea of some jellyfishes contains wandering, amoeba-like cells, which serve to transport food from the nutritive cells of the gastrodermis.

In scyphozoans, the epitheliomuscular cells, which are more specialized than those of *Hydra,* underlie the epidermis, contracting rhythmically to propel the medusa through the water (see Figure 25–1). The contractions are coordinated by concentrations of nerve cells in the margin of the bell. These nerve cells connect with fibers innervating (providing the nerve supply for) not only the musculature but also the tentacles and the sense organs.

The bell margin is liberally supplied with sensory receptor cells sensitive to mechanical and chemical stimuli. In addition, jellyfishes have two types of multicellular sense organs: statocysts and light-sensitive ocelli. **Statocysts** are specialized receptor organs that provide information by which an animal can orient itself with respect to gravity (Figure 25–19a). The statocyst, which seems to have been one of the first types of sensory organ to have evolved, has persisted apparently unchanged to the present day. This simple structure, providing essential information for motile organisms, is found in many animal phyla and is

25–18 *The moon jellyfish (Aurelia aurita) has four frilly mouth lobes that gather minute organisms into the gastrovascular cavity. It has short tentacles and usually grows to only about 30 centimeters in diameter.*

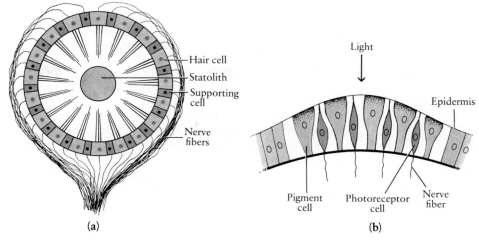

25-19 (a) *The statocyst is a specialized receptor organ that orients the jellyfish with respect to gravity. When the bell tilts, gravity pulls the statolith, a grain of hardened calcium salts, down against the hair cells. This stimulates the nerve fibers and signals the animal to right itself.* (b) *Ocellus (eyespot), the simplest type of photoreceptor organ. Ocelli of this sort are common among the cnidarians.*

thought to have arisen independently on a number of occasions. (As we shall see in Chapter 32, similar structures may also play a role in orienting the growth of plant roots with respect to gravity.) **Ocelli,** which may have evolved even earlier, are groups of pigment cells and photoreceptor cells (Figure 25–19b). They are typically located at the bases of the tentacles.

Class Anthozoa

Anthozoans ("flower animals")—the sea anemones and corals—are cnidarians that, like *Hydra,* have no medusa stage. These polyps reproduce both asexually, by budding, division, or fragmentation, and sexually, by the production of gametes. The zygote develops into a planula larva that may travel some distance from the parents before attaching itself to a suitable substrate and developing into an adult animal.

The gastrovascular cavity of anthozoans, unlike that of hydrozoan polyps, is divided by vertical partitions. In most corals, which are colonies of anthozoans, the epidermal cells secrete protective outer walls, usually of calcium carbonate (limestone), into which each delicate polyp can retreat. The limestone-forming polyps are the most ecologically important of the cnidarians. A coral reef is composed primarily of the accumulated limestone skeletons of the corals, covered by a thin crust occupied by the living colonial animals. A reef is both the structural and nutritional basis of the complex coral reef community (see essay).

25-20 *Ghost anemones* (Diadumene leucolena), *photographed in the waters off the coast of Rhode Island. The flowerlike appearance is deceptive; sea anemones are carnivorous animals belonging to a class of cnidarians that, like* Hydra, *have dropped the medusa stage. In common with other cnidarians, their tentacles are equipped with stinging nematocysts. The tentacles move food into the gastrovascular cavity, which is divided longitudinally by partitions.*

The Coral Reef

The coral reef is the most diverse of all marine communities. The reef structure itself is formed by colonial anthozoans. Each polyp in the colony secretes its own calcium-containing skeleton, which then becomes part of the reef. The photosynthetic activity of the reef is carried out almost entirely by symbiotic dinoflagellates and green algae living within the tissues of the corals; in fact, as much as half of the living substance of a coral reef may consist of green algae. Carbon, oxygen, and dissolved minerals flow over the reef as a result of the movement of waves and the ocean currents. The reef furnishes both food and shelter for other sea animals, including numerous species of reef fishes and a tremendous variety of invertebrates, such as sponges, sea urchins, marine worms, and crustaceans. The coral polyps and algae that form the reef can grow only in warm, well-lighted surface water, where the temperature seldom falls below 21°C.

The largest coral-created land masses in the world are the Marshall Islands in the Pacific and the 2,000-kilometer-long Great Barrier Reef, off the northeast shores of Australia. Other reefs are found throughout tropical waters and as far north as Bermuda, which is warmed by the Gulf Stream.

(a)

(b)

(a) *Living gorgonian coral, photographed in the Caribbean. Notice the individual polyps, extended from their limestone skeletons.*

(b) *A portion of a coral reef in the New Hebrides, islands in the Pacific, east of Australia. Staghorn coral is in the foreground.*

25–21 Mnemiopsis leidyi, *a ctenophore that is common along the Atlantic and Gulf coasts of the United States. In adults of* Mnemiopsis *and related genera, the two long tentacles characteristic of other ctenophores are greatly reduced in size. Muscular mouth lobes and short tentacles around the mouth work together in the capture and movement of prey organisms into the gastrovascular cavity.*

Phylum Ctenophora

About 90 species, commonly known as comb jellies and sea walnuts, make up phylum Ctenophora ("comb-bearers"). These small animals are readily identified by comblike plates of fused cilia that are arranged in eight longitudinal bands on the body surface (Figure 25–21). The coordinated beating of the cilia is their primary means of locomotion, although a few species also use muscular movements of the body wall for more rapid swimming. Ctenophores are abundant in the plankton of the open ocean; comb jellies and sea walnuts are bioluminescent, giving off bright flashes of light that are particularly visible at night.

Ctenophores, like cnidarians, are characterized by a gastrovascular cavity and two tissue layers, epidermis and gastrodermis, with a gelatinous mesoglea between. Amoeboid cells and muscle fibers are scattered through the mesoglea, which makes up most of the body substance. In many species, two long tentacles, containing specialized cells that secrete a sticky substance, are used to capture food.

Reproduction is sexual, and all individuals are hermaphrodites. The fertilized eggs develop into free-swimming larvae that gradually change into the adult form.

BILATERALLY SYMMETRICAL ANIMALS: AN INTRODUCTION

All of the remaining animals, including ourselves, are bilaterally symmetrical. The apparent exceptions, such as the echinoderms, which are radially symmetrical as adults, pass through a bilateral stage during their development. In bilaterally symmetrical animals, the body is organized along a longitudinal axis, with the right half an approximate mirror image of the left half (Figure 25–22). Bilateral symmetry makes possible more efficient locomotion than does radial symmetry, which is typically found in animals that move slowly or are sedentary.

A bilaterally symmetrical animal also has a top and a bottom or, in more precise terms, a **dorsal** and a **ventral** surface. These terms are applicable even when the organism is turned upside down, or, as with humans, stands upright, in which case dorsal means back and ventral means front. Most bilateral organisms also have distinct "head" and "tail" ends—**anterior** and **posterior.** Having one end that goes first is characteristic of actively moving animals. In such animals, many of the sensory cells are collected into the anterior end, enabling the animal to assess an area before entering it. With the aggregation of sensory cells, there came a concomitant gathering of nerve cells that is the forerunner of the brain. Structures useful in capturing and consuming prey are also generally located in the anterior region of the animal, whereas digestive, excretory, and reproductive structures

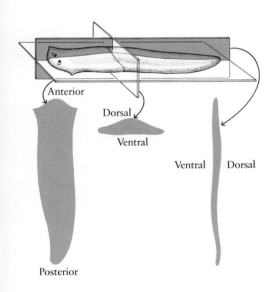

25–22 *In a bilaterally symmetrical organism, such as the planarian shown here, the right and left halves of the body are mirror images of one another. The upper and lower (or back and front) surfaces are known as dorsal and ventral. With some exceptions, the end that goes first is termed anterior and the one that brings up the rear, posterior.*

tend to be located toward the posterior. The concentration of sensory and nerve cells and of structures associated with feeding at the anterior end of an animal is known as **cephalization.**

The bilateral animals are all triploblastic; that is, they have three embryonic tissue layers. The third layer is the **mesoderm,** a layer of cells located between the ectoderm and the endoderm. These three layers can be detected very early in the development of bilateral animals and give rise to the various specialized tissues of the adult animal. Thus they are known as the "germ layers." Generally speaking, covering and lining tissues, as well as nerve tissues, are derived from ectoderm; digestive structures from endoderm; and muscles and most other parts of the body from mesoderm. This general pattern makes functional sense. When the sensory and nerve cells evolved, they did so as specializations of the outer layer, the ectoderm, where sensation was most important. Similarly, digestive structures evolved in the inner layer, the endoderm, which surrounds the food-containing cavity. As animals became more complex and additional structures evolved for locomotion, internal transport, excretion, and reproduction, they did so from the "new," middle layer, the mesoderm.

The triploblastic animals can be grouped in three categories, according to the presence or absence of a body cavity—a **coelom** (pronounced "see-loam")—in addition to the digestive cavity. In the simplest arrangement (Figure 25–23b), tissues derived from the three germ layers are packed together and there is no body cavity other than the digestive cavity. Animals of this type are known as **acoelomates.** A more complex arrangement is found in the **pseudocoelomates.** These animals have an additional cavity that develops between the endoderm and the mesoderm (Figure 25–23c); this cavity is known as a **pseudocoelom** because of its location and because it lacks the epithelial lining characteristic of a coelom. The **coelomates** (mollusks, annelids, and all of the more complex animals) have a true coelom, which is a fluid-filled cavity that develops *within* the mesoderm (Figure 25–23d). Within the coelom, the digestive tract ("gut") and other internal organs—lined with epithelium—are suspended by double layers of mesoderm known as mesenteries. The advantages of these increasingly complex arrangements will become apparent as we examine the animals themselves.

25–23 *Basic arrangements of animal tissue layers, as shown in cross section.* (a) *A body that consists of only two tissue layers is characteristic of cnidarians and ctenophores.*

(b) *Acoelomates, such as flatworms and ribbon worms, have three-layered bodies, with the layers closely packed on one another.*

(c) *Pseudocoelomates, such as nematodes, have three-layered bodies in which a pseudocoelom develops between the endoderm and mesoderm.*

(d) *Coelomates—mollusks, annelids, and most other animals, including vertebrates—have bodies that are three-layered with a cavity, the coelom, that develops within the middle layer (mesoderm). The mesodermal mesenteries suspend the gut within the body wall.*

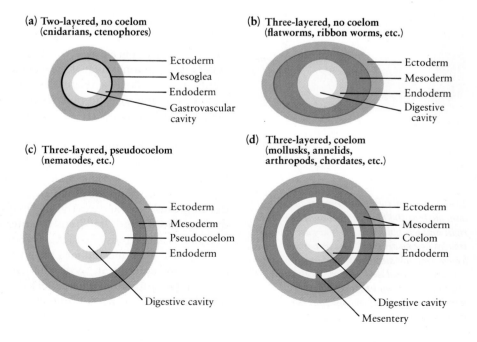

(a) **Two-layered, no coelom** (cnidarians, ctenophores)
— Ectoderm
— Mesoglea
— Endoderm
— Gastrovascular cavity

(b) **Three-layered, no coelom** (flatworms, ribbon worms, etc.)
— Ectoderm
— Mesoderm
— Endoderm
— Digestive cavity

(c) **Three-layered, pseudocoelom** (nematodes, etc.)
— Ectoderm
— Mesoderm
— Pseudocoelom
— Endoderm
— Digestive cavity

(d) **Three-layered, coelom** (mollusks, annelids, arthropods, chordates, etc.)
— Ectoderm
— Mesoderm
— Coelom
— Endoderm
— Digestive cavity
— Mesentery

PHYLUM PLATYHELMINTHES: FLATWORMS

The flatworms are the simplest animals with bilateral symmetry and three distinct germ layers. Moreover, not only are their tissues specialized for various functions, but also two or more types of tissue cells may combine to form organs. Thus, while sponges are made up of aggregations of cells and cnidarians are largely limited to the tissue level of organization, flatworms can be said to exemplify the organ level of complexity.

Like the cnidarians, but unlike most other bilaterally symmetrical animals, the flatworms have a digestive cavity with only one opening. Since the animal cannot feed, digest, and eliminate undigested residues simultaneously, food cannot be processed continuously.

Flatworms are acoelomates with solid bodies, and they have no circulatory system for the transport of oxygen and food molecules. Thus all cells must be within diffusion distance of sources of oxygen and of food. Flatworms have solved this problem in two ways. First, the body is flattened, which keeps the cells close to the external oxygen supply, and second, the digestive cavity is branched, carrying food particles to all regions of the body.

The flatworms are believed by some zoologists to have evolved from the cnidarians (or, perhaps, vice versa), not by way of either adult form, however, but from a ciliated planula larva that became sexually mature without going through polyp or medusa stages. Others are persuaded that the flatworms had independent origins among the ciliates. Still another group maintains that they are degenerate annelids. The new molecular techniques for determining phylogenetic relationships (page 418) may ultimately resolve this question.

About 13,000 species of flatworms have been described, and they are placed in three classes. Class Turbellaria contains mostly free-living forms, whereas the members of classes Trematoda (flukes) and Cestoda (tapeworms) are parasitic.

Class Turbellaria

The free-living flatworms form a large and varied group, and we shall single out just one type for examination, the freshwater planarian (Figure 25–24). The ectoderm of a planarian gives rise to cuboidal epithelial cells, many of which are ciliated, particularly those on the ventral surface. Ventral epithelial cells secrete mucus, which provides traction for the planarian as it moves by means of its cilia along its own slime trail. Planarians are among the largest animals that can use cilia for locomotion. The cilia in larger species are usually employed for moving water or other substances along the surface of the animal, as in the human respiratory tract, rather than for propelling the animal.

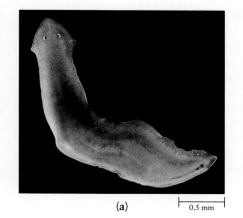

(a) |———| 0.5 mm

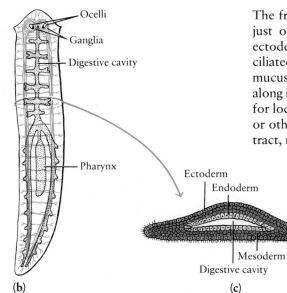

(b)

Ocelli
Ganglia
Digestive cavity
Pharynx

Ectoderm
Endoderm
Mesoderm
Digestive cavity

(c)

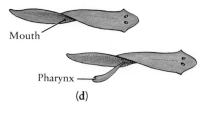

Mouth
Pharynx

(d)

25–24 (a) *Example of a flatworm: a freshwater planarian.* (b) *The nervous system is indicated in color. Note that some of the fibers are aggregated into two cords, one on each side of the body. Clusters of nerve cells in the head, known as* ganglia, *are the forerunner of a brain. The light-sensitive ocelli enable a planarian to see about as well as you can with your eyes closed.* (c) *Planarians, like other flatworms, but unlike cnidarians, have three* layers of body tissues, derived from the three germ layers. The only body cavity is the digestive cavity; thus flatworms are acoelomates. (d) *A carnivore, the planarian feeds by means of its extensible pharynx.*

25–25 *A free-living marine flatworm,* Pseudoceros zebra, *on the surface of a red sponge. The many species belonging to phylum Platyhelminthes, though widely varied in color and shape, are nearly all small and flat-bodied. This flatworm, a member of the order Polycladida, has a highly branched digestive cavity, extending into all parts of its body.*

Like other turbellarians, the planarian is carnivorous. It eats either dead meat or other slow-moving animals, including smaller planarians. It feeds by means of a muscular organ, the pharynx, which is free at one end. The free end can be stretched out through the mouth opening. Muscular contractions in the tubular pharynx cause strong sucking movements, which tear the meat into microscopic bits and draw them into the digestive cavity, where they are phagocytized by the cells of the lining. Because this digestive cavity has three main branches, planarians are placed in the order Tricladida.

Unlike the sponges or cnidarians, most flatworms have an excretory system, which is especially well developed in freshwater species. In the planarian, the system is a network of fine tubules that runs the length of the animal's body (Figure 25–26). Side branches of the tubules contain flame cells, each of which has a hollow center in which a tuft of cilia beats, moving water along the tubules to the exit pores between the epithelial cells. The flame-cell system appears to function largely to regulate water balance; most of the metabolic waste products probably leave the flattened body by diffusion.

25–26 *Flatworms have a tubular excretory system. The system usually consists of two or more branching tubules running the length of the body. In the planarian and its relatives, the tubules open to the body surface through a number of tiny pores. At the ends of the side branches are small bulblike structures known as flame cells. Within each of these cells, a tuft of cilia in constant motion resembles the flickering of a flame. Water and some waste materials from the tissue fluids are moved by the cilia through tubules to the excretory pores, where the collected liquid leaves the body.*

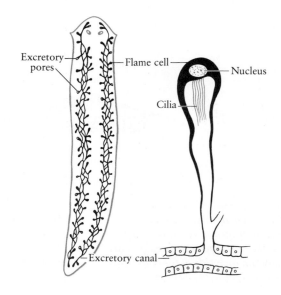

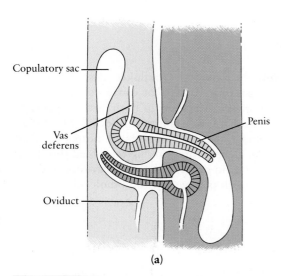

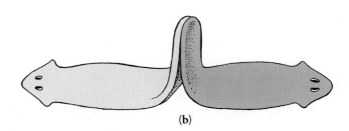

(a)   **(b)**

25-27 (a) *Planarians have both male and female reproductive structures, and mating involves a mutual exchange of sperm. The erect penis of each partner is inserted into the copulatory sac of the other. The* *vas deferens is a duct leading from the testes, where sperm are produced by meiosis, to the penis.* (b) *Two planarians mating.* *Planarians can also reproduce asexually, by fragmentation or by fission. Fragments of the tail, for example, may regenerate new heads.*

In most respects, the free-living flatworms are relatively simple, and the rest of the bilaterally symmetrical animals may have evolved from something quite like them. Their reproductive systems, however, are complex, and fertilization is internal. Planarians, for example, are hermaphrodites, like most other flatworms, and when they mate, each partner deposits sperm in the copulatory sac of the other (Figure 25–27). These sperm then travel along special tubes, the **oviducts,** to fertilize the eggs as they become ripe.

The Planarian Nervous System

The evolution of bilateral symmetry brought with it marked changes in the organization of the nervous system as well as of other systems. Even among primitive flatworms, the neurons (nerve cells) are not dispersed in a loose network, as in *Hydra,* but are instead condensed into longitudinal cords. In the planarians, this condensation is carried further, and there are two main conducting channels, one on each side of the flat, ribbonlike body. These channels carry impulses to and from the aggregation of nerve cells in the anterior end of the body (see Figure 25–24b). Such aggregations of nerve cell bodies are known as **ganglia** (singular, ganglion).

The ocelli of the planarian are usually inverted pigment cups (Figure 25–28). They have no lenses, and they cannot form an image. However, they can distinguish light from dark and can identify the direction from which the light is coming. Planarians are photonegative; if you shine a light from the side on a dish of planarians, they will move steadily away from the source of light.

Among the epithelial cells are receptor cells sensitive to certain chemicals and to touch. The head region in particular is rich in chemoreceptors. If you place a small piece of fresh liver in the culture water so that its juices diffuse through the medium, the planarians will raise their heads off the bottom and, if they have not eaten recently, will lope directly and rapidly (on a planarian scale) toward the meat, to which they then attach themselves to feed. The animal locates the food source by repeatedly turning toward the side on which it receives the stimulus more strongly until the stimulus is equal on both sides of its head. This can be demonstrated experimentally; if the chemoreceptor cells are removed from one side of the head, the animal will turn constantly toward the intact side.

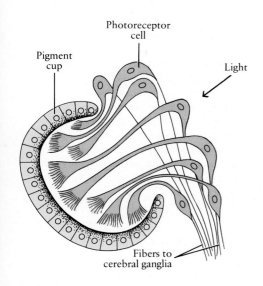

25-28 *Ocellus of the planarian. The light stimulus is received by the ends of the photoreceptor cells adjacent to the pigment cup. Thus the light travels first through the fibers carrying the signals to the cerebral ganglia. In this inside-out organization, the planarian ocellus resembles the vertebrate eye (see Figure 42–12).*

Classes Trematoda and Cestoda

Phylum Platyhelminthes includes also the trematodes, or flukes (class Trematoda), and the tapeworms (class Cestoda), parasitic forms that can cause serious and sometimes fatal diseases among vertebrates. Members of both of these parasitic classes have a tough outer layer of cells that is resistant to the body fluids of their hosts, particularly digestive fluids; most also have suckers or hooks on their anterior ends by which they fasten to their victims. Trematodes feed through a mouth, but the tapeworms—which have no mouths, digestive cavities, or digestive enzymes—merely hang on and absorb predigested food molecules through their skin (Figure 25–29).

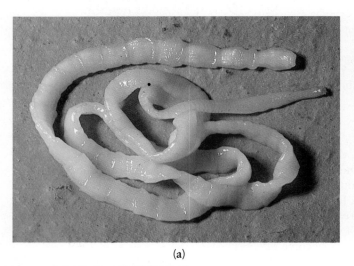

(a)

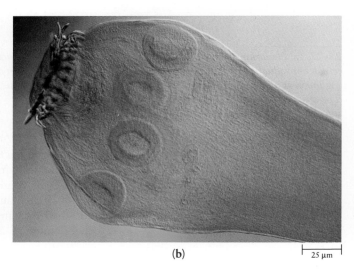

(b)

25 µm

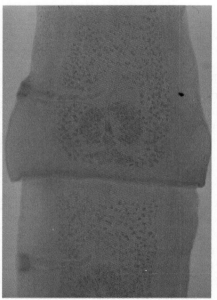

(c)   50 µm

25–29 (a) Dipylidium caninum, *a common tapeworm of dogs and cats. This specimen was removed from a cat. Tapeworms are intestinal parasites that lack any digestive system of their own. They cling by their heads, which, as shown in* (b) Taenia pisiformis, *another common tapeworm of dogs, are equipped with hooks and suckers. They absorb food molecules—digested by their hosts—through their body walls.* (c) *Their bodies, posterior to the heads, are divided into segments known as proglottids. In these* mature proglottids from Taenia pisiformis, *the dark areas are the reproductive structures; the opening is the genital pore. Each proglottid is a sexually complete hermaphroditic unit in which sperm and eggs are produced. Proglottids break off when the eggs are mature, and new ones are formed. Typically, the eggs develop into larvae in members of one species that eat food contaminated by the eggs, and the larvae develop into adult worms in members of another species that eat the flesh of the first species.*

Tapeworms are found in the intestines of many vertebrates, including humans, and may grow as long as 5 or 6 meters. They cause illness not only by encroaching on the food supply but also by producing wastes and by obstructing the intestinal tract. The most common human tapeworm, the beef tapeworm, infects people who eat the undercooked flesh of cattle that have grazed on land contaminated by human feces containing tapeworm segments.

All parasites, including parasitic flatworms, are believed to have originated as free-living forms and to have lost certain tissues and organs (such as the digestive cavity) as a secondary effect of their parasitic existence, while developing adaptations of advantage to the parasitic way of life. Such adaptations often include a complex life cycle involving two or more hosts (see essay).

The Politics of Schistosomiasis

Many of the parasitic platyhelminths have a complex life cycle involving two different hosts. In the case of three species of the trematode genus *Schistosoma,* the hosts are freshwater snails and humans. The life cycle begins when small, swimming larvae are released from a snail; a single infected snail can shed some 100,000 larvae in its six-month lifetime. The larvae, if successful, attach to human skin and penetrate it. They feed, mature, and mate in the bloodstream. The female lays her eggs, which have sharp spines on the surface, in the capillaries of the bladder wall or the intestine (depending on the species). The symptoms of schistosomiasis (also sometimes called bilharzia, or snail fever) are caused by the eggs; they lodge in the liver and spleen, blocking blood vessels, and their sharp spines tear the surrounding tissues, causing hemorrhages.

Eggs leave the human host by passing through the blood vessel wall into the bladder or intestinal tract, where they are flushed out with the urine or feces. If the eggs are deposited in fresh water, they hatch immediately into ciliated larvae that seek out the particular species of snail in which they multiply asexually.

Schistosomiasis now affects some 200 to 300 million people in 71 countries in Asia, Africa, and South America. Ironically, western-style progress has contributed to its spread. Snails thrive in the still waters of irrigation canals and artificial lakes. The disease spread rapidly through Upper Egypt following the construction of the Aswan High Dam and the digging of permanent irrigation canals. Before the creation of the huge man-made Lake Volta in Ghana, the schoolchildren of the villages along the riverbank had an incidence of schistosomiasis of 1 percent. Now the prevalence in some lakeside villages is 100 percent.

Medical progress has not followed technology into the tropics. When the colonial period ended, the major research programs into tropical diseases ended too. Western medical research is geared overwhelmingly to the conquest of diseases affecting mainly rich, urban, older men and women. For instance, more than $1.4 billion was spent on cancer research by the U.S. government alone in 1987, as compared to $8 million worldwide on schistosomiasis and $24 million on malaria. Cancer affects some 10 million people, schistosomiasis 200 million, malaria 200 to 400 million. The challenge of schistosomiasis and the other tropical parasitic diseases lies not so much in the disease process itself as in providing incentives (and therefore funding) to cure—or, better yet, prevent—the diseases of the politically powerless rural poor.

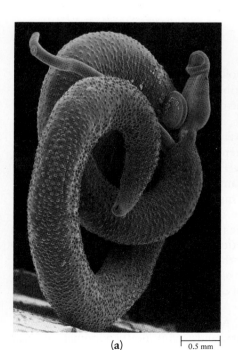

(a)
⊢——⊣ 0.5 mm

(b)

(a) *A scanning electron micrograph of male and female worms of* Schistosoma mansoni, *one of the three species that cause schistosomiasis in humans. The slender female is enveloped by the much larger body of the male. The two worms remain intertwined except when the female moves away to lay her eggs—about 300 per day. Adult worms live from 3 to 30 years, during which egg production continues unabated.* (b) *In the snail-infested waters near Aswan, Egypt, bathing children are rapidly infected by* Schistosoma *larvae. During the years that pass before the symptoms of disease become apparent, large quantities of eggs are shed by the children, continually infecting new generations of snails.*

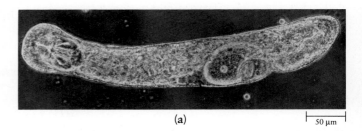

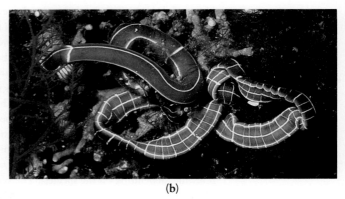

25–30 (a) *A living gnathostomulid,* Problognathia minima, *as seen through the phase-contrast microscope. This species lives in the intertidal sand flats along the coast of Bermuda. It moves by slowly gliding between the sand grains, moving its head rhythmically from side to side.*

(b) *Two ribbon worms of the species* Tubulanus superbus, *photographed in Scotland. Ribbon worms range in length from less than 2 centimeters to 30 meters and are virtually all the colors of the spectrum. All members of the phylum, however, have thin bodies (seldom more than 0.5 centimeter thick), a mouth-to-anus digestive tract, a circulatory system, and a long, muscular tube than can be thrust out to grasp prey.*

OTHER ACOELOMATES

Phylum Gnathostomulida

This small phylum contains about 80 species of tiny marine worms. Their distinguishing characteristic is a unique pair of hard jaws, from which the phylum derives its name (from the Greek *gnathos,* "jaw," plus *stoma,* "mouth"). Gnathostomulids are abundant along coastal shorelines, where they live in the spaces between particles of sand and silt, using their hard jaws to scrape bacteria and fungi from the particles.

These tiny worms have no coelom or pseudocoelom, and the digestive cavity has only one opening. Some zoologists believe that they are closely related to the free-living flatworms, whereas others think they are degenerate forms, derived from pseudocoelomates.

Phylum Rhynchocoela

Phylum Rhynchocoela ("snout" plus "hollow") consists of about 650 species of acoelomate worms, commonly called ribbon worms or nemertines. They are characterized by a long, retractile, slime-covered hollow tube (proboscis). The proboscis, sometimes armed with a barb, seizes prey and draws it to the mouth, where it is engulfed. Some inject a paralyzing poison into their prey.

The ribbon worms are of special interest to biologists attempting to reconstruct the evolution of the invertebrates. They appear to be closely related to the flatworms, but they exhibit two significant new features. Ribbon worms have a one-way digestive tract beginning with a mouth and ending with an anus. This is a far more efficient arrangement than the digestive cavity of the cnidarians and flatworms, with its single opening. In the two-opening tract, food moves assembly-line fashion, always in the same direction, with the consequent possibilities (1) that eating can be continuous and (2) that various segments of the tract can become specialized for different stages of digestion. These worms also have a circulatory system, typically consisting of one dorsal and two lateral blood vessels that carry the colorless blood.

Reproduction in the ribbon worms, however, is simpler than in most flatworms. Ordinarily the sexes are separate, and fertilization is external. Asexual reproduction by fragmentation of the body and regeneration of whole worms from the parts is also fairly common.

PSEUDOCOELOMATES

One major phylum (Nematoda) and seven minor phyla are characterized by a pseudocoelom, a body cavity that develops between the endoderm and the mesoderm (see Figure 25–23c). The pseudocoelom, which is essentially a sealed, fluid-filled tube, increases the effectiveness of the animal's muscular contractions. In addition to working against the water or a substrate, muscles must also have something to work against in an animal's body—otherwise the body just bends in the direction of contraction and a floppy, uncoordinated motion results. Because it resists bending, the pseudocoelom functions as a **hydrostatic skeleton** within the animal's body, causing the body to return to its original shape after the muscles have contracted. It makes possible a significant advance over the simple, rather flaccid movements of the acoelomate worms and most cnidarians.

All of the pseudocoelomates have a one-way digestive tract, but they lack a circulatory system. However, the movement of fluids within the pseudocoelom, enhanced by muscular contractions of the body wall, compensates for this.

Phylum Nematoda

About 12,000 species of nematodes (roundworms) have been described and named, but some authorities think that there may be as many as 400,000 to 500,000. Most are free-living, microscopic forms. It has been estimated that a spadeful of good garden soil usually contains about a million nematodes. Some are parasites; most species of plants and animals are parasitized by at least one species of nematode.

Nematodes (Figure 25–31) are cylindrical, unsegmented worms and are covered by a thick, continuous cuticle, which is molted periodically as they grow. An interesting, and unique, feature of nematode construction is the absence of circular muscles. The contraction of the longitudinal muscles acting against both the tough, elastic cuticle and the internal hydrostatic skeleton gives the worm its characteristic whipping motion in water. The mouth of a nematode has a muscular pharynx and often is equipped with piercing stylets. Reproduction is sexual, and the sexes are usually separate.

Humans are hosts to about 50 species of parasitic nematodes, which are a major cause of death and disability throughout the world. In North America, the most common parasitic nematodes are pinworm *(Enterobius),* whipworm *(Trichuris),* hookworm *(Ancylostoma),* intestinal roundworm *(Ascaris),* and *Trichinella.* The latter causes trichinosis, which is contracted by eating uncooked or undercooked pork, a single gram of which may contain 3,000 cysts (resting forms)

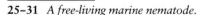

25–31 *A free-living marine nematode.*

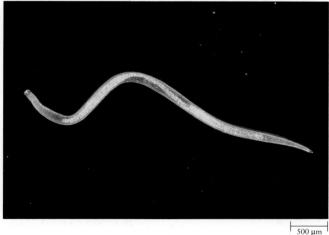

500 μm

of *Trichinella*. Ingestion of only a few hundred of these cysts can be fatal. On a worldwide basis, it is estimated that 650 million people are infected by *Ascaris*, 450 million by *Ancylostoma*, 250 million by *Filaria*, the cause of filariasis, and 50 million by *Onchocerca*, the cause of "river blindness" in fertile regions of West Africa and in the coffee-growing highlands of Mexico and Guatemala. Research is revealing that many of these worms have unique enzymes and metabolic pathways directly related to such features of the parasitic life style as migration within the body of the host, attachment to host tissues, and the massive production of eggs.

Other Pseudocoelomate Phyla

The members of seven other phyla, quite diverse, have body plans based upon the pseudocoelom. Phylum Nematomorpha consists of about 230 species of horse-hair worms that resemble nematodes. The adults, which do not feed, are free-living and reproduce in water (Figure 25–32a); the juveniles are parasites of arthropods. The 500 species of spiny-headed worms, phylum Acanthocephala, are parasitic throughout their life cycle. The larval forms develop in the tissues of arthropods, and the adults live and reproduce in the intestines of vertebrates. The adult worms, which superficially resemble tapeworms, lack a digestive tract and have a proboscis bearing hooks that attach to the intestinal wall of the host (Figure 25–32b and c).

The members of four pseudocoelomate phyla are often found, along with the acoelomate gnathostomulids, in the spaces between sand and silt particles along shorelines. These include the three species of the new phylum Loricifera (see Figure 20–8, page 415), discovered clinging tightly to sand grains in ocean waters at a depth of 25 to 30 meters. The larvae of these animals are free-swimming, but the adults are sedentary. By contrast, the adults of the other three phyla (Figure 25–33) are motile. Phylum Kinorhyncha contains about 100 species of tiny, burrowing marine worms that feed on diatoms in muddy ocean shores. Short-bodied, these worms are covered with spines and have a spiny, retractile proboscis. The sexes are separate. Most of the approximately 400 species in phylum Gastrotricha, however, are hermaphrodites. They live in sandy shores of both fresh and salt water, where they feed on protists and dead organic matter. Many of the 1,500 to 2,000 species of phylum Rotifera are found in the same environment in fresh water, as well as around the vegetation in ponds and along lake shores. Other members of the phylum are marine, and some are terrestrial. Rotifers are sometimes called "wheel animalcules" because the beating of a crown of cilia around the mouth causes them to spin through the water like tiny wheels.

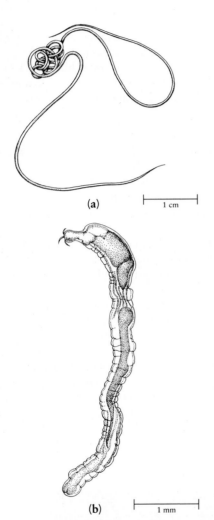

(a)

|———— 1 cm ————|

(b)

|—— 1 mm ——|

25–32 *Two phyla of pseudocoelomates consist of worms that are parasitic during at least one stage of the life cycle.* (a) *A horse-hair worm, phylum Nematomorpha. The specimen from which this drawing was prepared was found in a garden hose on Long Island, New York. It was probably carried into the hose by a cricket or a grasshopper that it parasitized during its juvenile stage. Because of the intricate knots into which they tie themselves, nematomorphs are also known as gordian worms, after Gordius, King of Phrygia, who tied the gordian knot severed by Alexander the Great.*

(b) *An adult male of phylum Acanthocephala. Known commonly as spiny-headed worms, acanthocephalans are characterized by a hook-bearing proboscis* (c).

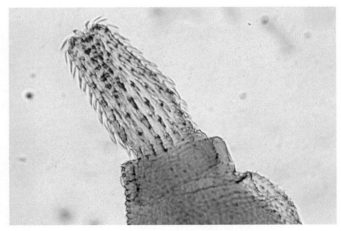

(c)

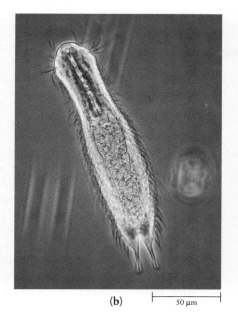

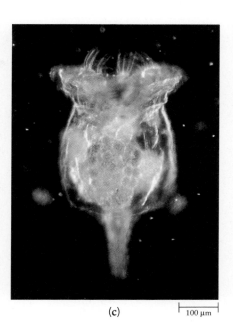

(a)   100 µm

(b)   50 µm

(c)   100 µm

25-33 *Among the residents of the sand and silt of shorelines are members of three pseudocoelomate phyla—Kinorhyncha, Gastrotricha, and Rotifera. (a) Centroderes spinosus, a kinorhynch. Unable to swim, a kinorhynch burrows by forcing fluid into its head; when the head is anchored in the mud by its spines, the animal can pull the rest of its body forward. (b) Chaetonotus, a common gastrotrich. Gastrotrichs can both swim and crawl, clinging to surfaces by means of adhesive tubes that project from the sides of their bodies. (c) A rotifer of the genus Brachionus, commonly found in freshwater plankton. The cilia that propel the rotifer through the water are visible at the anterior end of the organism.*

Rotifers have a muscular pharynx with hard jaws; they feed on protists, bits of vegetation, and other animals even smaller than themselves. The sexes are separate in rotifers, but in some groups the females produce eggs that can develop without fertilization, a phenomenon known as **parthenogenesis.**

Phylum Entoprocta contains about 75 species of stalked, usually sessile animals that superficially resemble both hydrozoans and a group of coelomate animals known as bryozoans, or ectoprocts (see page 562). Their internal structure, however, identifies them as pseudocoelomates. They have a U-shaped digestive tract, and both the mouth and anus are located within a circle of tentacles (Figure 25–34). (In the bryozoans, by contrast, the anus is located outside a circle of tentacles, and the tentacles are structurally different from those of the entoprocts.) Reproductive patterns vary among the entoprocts. Some species have separate sexes, some are simultaneous hermaphrodites, and in others a single reproductive organ produces sperm at one stage in the life cycle and eggs at a later stage. As we shall see in the next chapter, this phenomenon, known as sequential hermaphroditism, is not limited to the entoprocts.

25-34 *A marine entoproct,* Barentsia benederi. *Most entoprocts are colonial, but members of a few species live as solitary individuals.*

SUMMARY

Animals are multicellular heterotrophs that depend directly or indirectly on plants or algae as their source of food energy. Almost all digest their food in an internal cavity. Most are motile. Reproduction is usually sexual. More than 95 percent of the animal species are invertebrates, animals without backbones.

Modern animals are classified into about 30 phyla. Criteria used in classification include the number of embryonic tissue layers, the basic body plan and the arrangement of body parts, the presence or absence of body cavities, and the pattern of development from fertilized egg to adult.

Both the sponges (phylum Porifera) and the extremely simple, parasitic mesozoans (phylum Mesozoa) are so different from all other animals that they are placed in separate subkingdoms. Sponges are composed of a number of different cell types, including choanocytes, which are the feeding cells; epithelial cells, some of which are contractile; and amoebocytes, which perform a variety of functions in reproduction, the secretion of skeletal materials, and the transport of food particles within the animal. In the sponge, there is little coordination among the various cells. Sponges reproduce both asexually and sexually, with a highly specialized form of internal fertilization. Most sponges, like many other sessile or slow-moving animals, are hermaphrodites.

The animals in phylum Cnidaria and phylum Ctenophora share three major characteristics: (1) radial symmetry; (2) a gastrovascular cavity, in which food is partially digested extracellularly and oxygen-bearing water is circulated; and (3) a two-layered body plan, in which the two tissue layers, the epidermis and the gastrodermis, are separated by a noncellular jellylike substance, the mesoglea. The jellyfishes, sea anemones, and corals of phylum Cnidaria are distinguished from all other animals by their special stinging cells, the cnidocytes; the sea walnuts and comb jellies of phylum Ctenophora are identified by comblike plates of fused cilia, arranged in eight rows on the body surface. Adult cnidarians may take the form of either polyps or medusas; in many species, the life cycle includes an asexually reproducing polyp, a sexually reproducing medusa, and a ciliated planula larva. Cnidarians have a simple nervous system that coordinates the movements of the animal, a variety of sensory cells, and two types of specialized sensory organs, statocysts and ocelli.

The animals in all of the remaining phyla have primary bilateral symmetry, with distinct "head" and "tail" ends and a concomitant clustering of nerve cells in the anterior region (cephalization). They also have three distinct embryonic tissue layers—ectoderm, mesoderm, and endoderm. The simplest of the bilateral animals are the acoelomates (phyla Platyhelminthes, Gnathostomulida, and Rhynchocoela), which have no body cavity other than the digestive cavity.

The flatworms, phylum Platyhelminthes, are characterized by a flattened body, a branched digestive system with only one opening, and an excretory system, involving flame cells, that serves largely to maintain water balance. Ocelli are present, as are sensory cells responsive to touch and to various chemicals; the nerve cells are organized into longitudinal cords. Flatworms may be free-living (class Turbellaria) or parasitic (classes Trematoda and Cestoda).

Other acoelomates include the tiny marine worms of phylum Gnathostomulida and the ribbon worms of phylum Rhynchocoela. The ribbon worms, which are characterized by a retractile, prey-seizing proboscis, are the most primitive animals with a one-way (mouth-to-anus) digestive tract and a circulatory system.

Eight phyla of animals have body plans based on the pseudocoelom, a fluid-filled cavity that develops between the endoderm and the mesoderm. The pseudocoelom functions as a firm hydrostatic skeleton, enabling these animals to move more efficiently than the cnidarians or the acoelomate worms. The pseudocoelomates all have a one-way digestive tract but lack a circulatory system.

The roundworms (phylum Nematoda) are the largest and ecologically most important group of pseudocoelomates. Most are free-living, but many are parasitic, causing a variety of serious diseases in plants and animals, including humans. Unlike other types of worms, roundworms have only longitudinal muscles and move in a characteristic whipping manner.

Other pseudocoelomates include the horsehair worms (phylum Nematomorpha) and the spiny-headed worms (phylum Acanthocephala), which are both parasitic; the tiny marine and freshwater animals of phyla Loricifera, Kinorhyncha, Gastrotricha, and Rotifera, most of which are bottom-dwellers; and the Entoprocta, which superficially resemble the hydrozoans of phylum Cnidaria. Among the various pseudocoelomates, the entire range of sexual reproductive patterns in animals is seen: separate sexes, simultaneous hermaphrodites, sequential hermaphrodites, external fertilization, internal fertilization, and parthenogenesis.

QUESTIONS

1. Distinguish among the following: cnidocyte/nematocyst; polyp/medusa; diploblastic/triploblastic; endoderm/mesoderm/ectoderm; radial symmetry/bilateral symmetry; acoelomate/pseudocoelomate/coelomate.

2. In which phylum or phyla do you find each of the following and what is its function: choanocyte; gastrovascular cavity; cnidocyte; statocyst; planula larva; flame cell?

3. Describe the similarities and differences among the three classes of phylum Cnidaria.

4. How are the ocelli of jellyfish and planarians similar? How are they different?

5. What advantages does radial symmetry provide to sessile organisms? What advantages does bilateral symmetry provide to motile organisms?

6. A pseudocoelom (or, as we shall see in the next chapter, a coelom) provides hydrostatic support for an animal. Why is a gastrovascular cavity less likely to fill this same function?

7. How do the parasitic flatworms of phylum Platyhelminthes differ from the free-living flatworms? What general features are characteristic of adaptation to a parasitic way of life?

8. Virtually all of the animals in phyla Gnathostomulida, Loricifera, Kinorhyncha, and Gastrotricha and many animals in phylum Rotifera (as well as some animals we have yet to meet) live in the spaces between sand and silt grains along shorelines. What specializations enable such a large number of different animals to occupy the same microenvironment?

9. On the basis of your knowledge of sexual reproduction, what do you think might be the advantages to an organism of functioning as a male in the earlier stages of its life and then as a female in the later stages, or vice versa?

10. What might be the advantages of parthenogenetic reproduction?

11. What might be the advantages and disadvantages of internal fertilization as compared to external fertilization?

The Animal Kingdom II: The Protostome Coelomates

The coelom, a fluid-filled cavity that develops within the mesoderm, characterizes all the remaining phyla in the animal kingdom. Although it may seem less dramatic than other evolutionary innovations, the coelom is extremely important. Within this cavity, organ systems—suspended by the mesenteries—can bend, twist, and fold back on themselves, increasing their functional surface areas and filling, emptying, and sliding past one another, surrounded by lubricating coelomic fluid. Consider the human lung, constantly expanding and contracting in the chest cavity, or the 6 or 7 meters of coiled human intestine; neither of these could have evolved before the coelom. The coelom, like the pseudocoelom, also constitutes a hydrostatic skeleton, stiffening the body in somewhat the same way water pressure stiffens and distends a fire hose.

The various phyla of coelomate animals fall into two broad groups that roughly correspond to major branches of the phylogenetic tree. These groups are based on characteristic features of embryonic development. When a fertilized egg—the zygote—begins to divide, the early cell divisions usually follow one of two patterns. In mollusks, annelids, arthropods, and a number of other coelomate phyla (as well as in acoelomates and pseudocoelomates), the early cleavages are spiral, occurring in a plane oblique to the long axis of the egg. In echinoderms, chordates, and a few other coelomate phyla, the cleavage pattern is radial, parallel to and at right angles to the axis of the egg (Figure 26–1). With both types of cleavage, the embryo gradually develops to a stage known as the blastula, which is a hollow ball of cells. Next, an opening, the blastopore, appears. Among the animals with spiral cleavage, the mouth (stoma) develops at or near the blastopore, and this group is called the **protostomes**—"first the mouth." In the animals with radial cleavage, the anus forms at or near the blastopore and the mouth forms secondarily elsewhere; thus these animals are known as **deuterostomes**—"second the mouth." These differences are so fundamental that they are believed to have originated very early in animal evolution, before the branchings that gave rise to the modern coelomate phyla.

Another characteristic difference between the two groups of animals is the manner in which the coelom forms (Figure 26–2). In the protostomes, the coelom usually forms by a splitting of the mesoderm, and the process is said to be

26–1 *Egg cleavages, showing spiral and radial patterns at the third cleavage. Spiral cleavage is characteristic of protostomes (acoelomates, pseudocoelomates, mollusks, annelids, and arthropods); radial cleavage is characteristic of deuterostomes (echinoderms and chordates).*

First cleavage

Second cleavage

Third cleavage—spiral

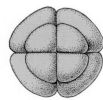

Third cleavage—radial

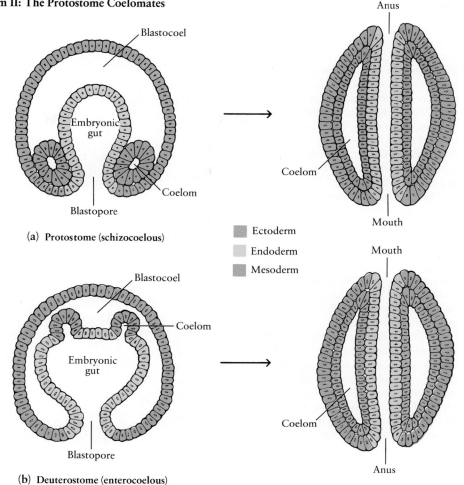

26–2 *Key features of embryonic develop-ment in coelomate animals.* **(a)** *In proto-stomes, the mesoderm takes shape between the endoderm and the ectoderm, in the region around the blastopore. The coelom arises from a splitting of the solid meso-derm. As the cells continue to multiply, the blastocoel—the original embryonic cavity of the blastula—is obliterated. The blasto-pore of protostomes becomes the mouth.*

   (b) *In deuterostomes, the mesoderm originates from outpocketings of the embryonic gut. These outpocketings create cavities within the mesoderm that become the coelom. The blastopore becomes the anus, and the mouth develops elsewhere.*

(a) Protostome (schizocoelous)

(b) Deuterostome (enterocoelous)

schizocoelous (from the Greek *schizo,* "split"). In the deuterostomes, however, the coelom is usually formed by an outpouching of the cavity of the embryonic gut, a process said to be **enterocoelous** (from the Greek *enteron,* "gut"). Entero-coelous formation of the coelom is thought to have arisen from the schizocoelous process several times in the course of animal evolution.

   In the next two chapters, we shall focus on the protostome coelomates—the mollusks, annelids, and a variety of smaller groups in this chapter, and the arthropods in the next. In the final chapter of this section, our attention will be devoted to the deuterostomes.

PHYLUM MOLLUSCA: MOLLUSKS

The mollusks constitute one of the largest phyla of animals, both in numbers of living species (at least 47,000, and perhaps many more) and in numbers of individuals. Their name is derived from the Latin word *mollus,* meaning "soft," and refers to their soft bodies, which are generally protected by a hard, calcium-containing shell. In some forms, however, the shell has been lost in the course of evolution, as in slugs and octopuses, or greatly reduced in size and internalized, as in squids.

   Mollusks exhibit a tremendous diversity of form and behavior. The three major classes range from largely sedentary or sessile filter-feeding animals, such as clams and oysters (class Bivalvia), through aquatic and terrestrial snails and slugs (class Gastropoda), to the predatory cuttlefish, squids, and octopuses (class Cephalopoda), which are not only the most active of the mollusks but also the most intelligent of all invertebrates.

26–3 *Mollusks are characterized by soft bodies composed of a head-foot, a visceral mass, and a mantle, which can secrete a shell. They exchange gases with the surrounding water through gills, except for the land snails, in which the mantle cavity has been modified for air breathing. A hypothetical primitive mollusk is shown in (a).*

The three major modern classes are the bivalves, the gastropods, and the cephalopods. (b) The bivalves, such as the clam shown here, are generally sedentary and feed by filtering water currents, created by beating cilia, through large gills. (c) In the gastropods, exemplified by the snail, the visceral mass has become coiled and rotated through 180° so that mouth, anus, and gills all face forward and the head can be withdrawn into the mantle cavity. (d) In the cephalopods, such as the squid, the head is modified into a circle of arms, and part of the head-foot forms a tubelike siphon through which water can be forcibly expelled, providing for locomotion by jet propulsion. The arrows indicate the direction of water movement.

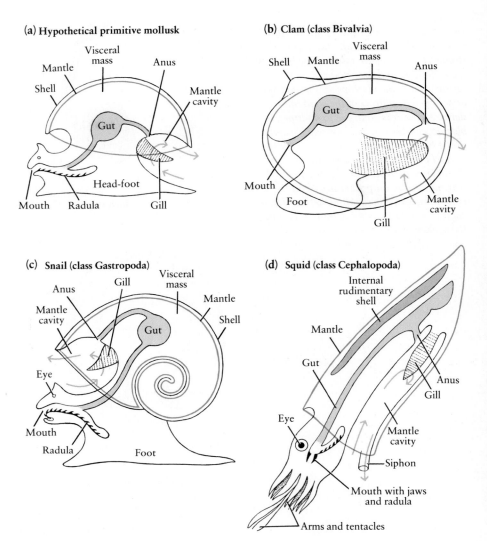

(a) Hypothetical primitive mollusk

(b) Clam (class Bivalvia)

(c) Snail (class Gastropoda)

(d) Squid (class Cephalopoda)

Characteristics of the Mollusks

Structurally, mollusks are quite distinct from all other animals (Figure 26–3). As you can see, the hypothetical primitive mollusk displayed a clear bilateral symmetry. Among modern mollusks, only the polyplacophorans (chitons) and monoplacophorans, both relatively small classes, bear any obvious resemblance to the primitive model. Modern mollusks, however, all have the same fundamental body plan. There are three distinct body zones: a **head-foot,** which contains both the sensory and motor organs; a **visceral mass,** which contains the well-developed organs of digestion, excretion, and reproduction; and a **mantle,** a specialized tissue formed from folds of the dorsal body wall, that hangs over and enfolds the visceral mass and that secretes the shell. The **mantle cavity,** a space between the mantle and the visceral mass, houses the gills; the digestive, excretory, and reproductive systems discharge into it. Water sweeps into the mantle cavity (usually propelled by cilia on the gills), passing across the surface of the gills and aerating them. Water leaving the mantle cavity carries excreta and, in season, gametes, both of which are discharged downstream from the gills.

A characteristic organ of the mollusk, found only in this phylum, and in all classes except the bivalves, is the **radula** (Figure 26–4), a movable, tooth-bearing strap of chitinous material suggestive of a tongue. The radular apparatus, which can be projected out of the mouth and then drawn back in with a licking

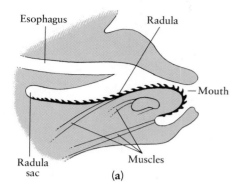

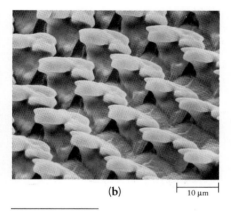

(b) |— 10 μm —|

26–4 (a) *A vertical section through the head of a land snail to show the radula. Teeth on the radula rasp and tear food materials and then convey them to the esophagus, a narrow tube leading to the stomach.* (b) *Scanning electron micrograph of radular teeth from a minute African land snail that feeds on bits of dead leaves. The large teeth (upper left) cut and tear the leaf; the smaller teeth (lower right) pull the pieces into the snail's mouth.*

movement, serves both to scrape off algae and other food materials and also to convey them backward to the digestive tract. In some species, it is also used in combat.

Supply Systems

In the protists and in the smaller and simpler animals, oxygen and food molecules are supplied to cells—and waste products removed from them—largely by diffusion, aided by the movement of external fluids. Internal transport may also be assisted, as we have seen, by wandering, amoeboid cells, as in the cnidarians, or by the movement of fluids in a body cavity, as in the pseudocoelomates. For larger, thicker animals, a more effective method of providing each cell with a direct and rapid line of supply is a circulatory system that propels extracellular fluid—blood—around the body in a systematic fashion.

The molluscan circulatory system consists of a muscular pumping organ, the heart, and vessels that carry the blood to and from the heart. The heart usually has three chambers: two of them (atria) receive blood from the gills, and the third (the ventricle) pumps it to the other body tissues. Except for the cephalopods, mollusks have what is known as an open circulation; that is, the blood does not circulate entirely within vessels—as it does in the annelids, for example—but is collected from the gills, pumped through the heart, and released directly into spaces in the tissues from which it returns to the gills and then to the heart. Such a blood-filled space is known as a **hemocoel** ("blood cavity"). In the mollusks, the hemocoel has largely replaced the coelom, which is reduced to a small area around the heart and to the cavities of the organs of reproduction and excretion. Cephalopods, whose vigorous activities require that the cells be supplied with large quantities of oxygen and food molecules, have a closed circulatory system of continuous vessels and accessory hearts that propel blood into the gills.

Oxygen enters the body of a mollusk through the moist surface of the mantle and the gills. A **gill** is an external structure with an increased amount of surface area, through which gases can diffuse, and a rich supply of blood for the transport of the gases to and from the rest of the body. Oxygen travels inward by diffusion, down the concentration gradient. The gradient exists because the surface film, exposed to the dissolved oxygen in the passing water, contains more oxygen than does the blood within the gills, which was depleted of oxygen as it passed through the body tissues. Carbon dioxide, produced by cellular respiration, moves out to the surface film and then into the surrounding water by the same mechanism. In fact, all gas exchange in animals, whether water-dwelling or land-dwelling, takes place by diffusion across moist membranes.

The digestive tract of all mollusks is extensively ciliated and has many different working areas. Food is taken up by cells lining the digestive glands arising from the stomach and the anterior intestine, and then is passed into the blood. Undigested materials are compressed into mucus-coated fecal pellets, which are discharged through the anus into the mantle cavity and are carried away from the animal in the water currents. This packaging of digestive wastes in solid form prevents fouling of the water passing over the gills.

Nitrogenous wastes produced by the metabolic activities of the cells are removed by one or two tubular structures known as **nephridia.** One opening of each nephridium is in the coelom surrounding the heart, and the other opening discharges into the mantle cavity. Coelomic fluid is forced, under pressure, into the opening near the heart; as the fluid passes through the tubule, water, sugar, salts, and other needed materials are returned to the tissues through its walls, while waste products are secreted into it for excretion. Thus the excretory system is concerned not only with the problem of water balance, as are the contractile vacuoles of *Paramecium* and the flame cells of planarians, but also with the regulation of the chemical composition of body fluids.

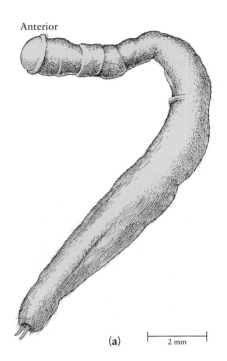

Anterior

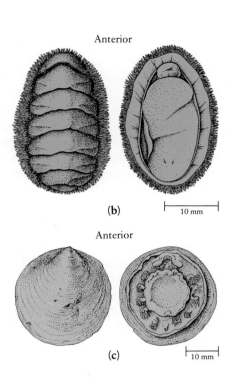

Anterior

(b) | 10 mm |

Anterior

(c) | 10 mm |

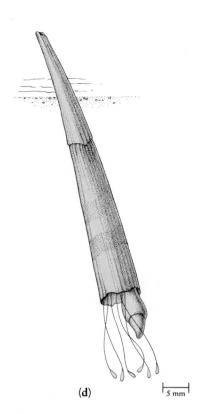

(a) | 2 mm |

(d) | 5 mm |

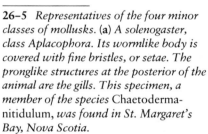

26–5 *Representatives of the four minor classes of mollusks.* (a) *A solenogaster, class Aplacophora. Its wormlike body is covered with fine bristles, or setae. The pronglike structures at the posterior of the animal are the gills. This specimen, a member of the species* Chaetoderma- nitidulum, *was found in St. Margaret's Bay, Nova Scotia.*

(b) *Dorsal (left) and ventral (right) views of a chiton, class Polyplacophora. The dorsal shell of eight plates is fringed by a girdle of hard spicules. As seen from the ventral surface, the bulk of the chiton's body is its foot, which has been pulled aside at the left to reveal the gills within the mantle cavity. At the anterior of the animal is the mouth, through which it extends the radula when grazing.*

(c) *Dorsal and ventral views of* Neopi- lina galatheae, *the first known living rep- resentative of class Monoplacophora. The large central structure of the ventral sur- face is the foot, which is surrounded by five pairs of gills.*

(d) *A tusk shell, class Scaphopoda. This mollusk burrows in sand or mud with the large foot that extends from the wide end of the shell. The specialized tentacles bring food particles, such as diatoms, to the mouth hidden within the shell.*

Minor Classes of Mollusks

Four of the seven classes of mollusks with living representatives are fairly small. About 250 species of wormlike marine animals, known as solenogasters (Figure 26–5a), constitute the class Aplacophora ("no plates"). Although they have no shell and the foot is greatly reduced, they are unequivocally identified as mollusks by the presence of a radula.

The approximately 600 species of chitons (Figure 26–5b) are placed in class Polyplacophora. As we noted previously, they bear some resemblance to the hypothetical primitive mollusk (Figure 26–3a). Common as grazers on surf-swept ocean shores, chitons have a somewhat flattened body, covered by a dorsal shell formed from a series of eight plates. On either side of the body, a series of gills are suspended between the mantle and the foot and are continuously aerated by water flowing through the mantle cavity from the anterior to the posterior of the animal.

Class Monoplacophora contains just eight living species. The first species, *Neopilina galatheae* (Figure 26–5c), was discovered in the 1950s in a deep ocean trench off the coast of Costa Rica. The only previously known representatives of the class were fossils from the Cambrian period, which ended 505 million years ago. *Neopilina,* which is little more than 2.5 centimeters long, also has an organization suggestive of the hypothetical primitive mollusk. It has a large, single dorsal shell, but is unusual in having five pairs of gills, six pairs of nephridia, and eight pairs of retractor muscles, with which it can pull the shell down securely over its soft body.

About 350 species of tusk or tooth shells (Figure 26–5d) make up class Scaphopoda. These familiar residents of the seashore have long, tubular shells, open at both ends. They live a sedentary life, with the wide end of the shell, containing the head (which bears structures used in feeding) and the foot, buried in the sand or mud. Water, carrying dissolved gases, enters and leaves the mantle cavity through the exposed, narrow end of the shell.

Class Bivalvia

The approximately 7,500 living species of bivalves include such common animals as clams, oysters, scallops, and mussels. They derive their name from the two parts, or valves, into which the shell is divided. The left and right valves are connected dorsally by a hinge with a flexible ligament. One or two large **adductor muscles**, familiar as the delectable portion of the scallop, are used to close the shell tightly in times of danger. These muscles are so strong in scallops that the animals can move swiftly through the water by clapping their shells together, thereby eluding predators such as starfish.

In this class of mollusks, the body has become flattened between the valves, and a distinct "head" has generally disappeared (see Figure 26–3b). The bivalves are sometimes called Pelecypoda—"hatchet foot"—because the muscular foot is often highly developed in this group. A clam, using its "hatchet foot," can dig itself into sand or mud with remarkable speed. However, many bivalves are sessile, and some of them secrete strands of protein by which they anchor themselves to rocks. The adhesive polymer synthesized by mussels, which consists of a repeated sequence of 10 amino acids, is as powerful as epoxy—and, as you might expect, it is resistant to damage by salt water. Work is currently underway, using recombinant DNA techniques, to produce sufficient quantities of this adhesive for use in dentistry and medicine, for example, in reattaching broken teeth and bones.

Abundant in both salt and fresh water, most bivalves are filter-feeding herbivores; they live largely on microscopic algae. Their gills, which are large and elaborate, collect food particles. Water is circulated through the sievelike gills by the beating of gill cilia. Small organisms and particles of food are trapped in mucus on the gill surface and swept toward the mouth by the cilia; the gills also sort particles by size, rejecting sand and other larger particles.

Throughout the molluscan phylum, there is a wide range of development of the nervous system. The bivalves have three pairs of ganglia of approximately equal size—cerebral, visceral, and pedal (supplying the foot)—and two long pairs of nerve cords interconnecting them. They have statocysts, usually located near the pedal ganglia, and sensory cells for discrimination of touch, chemical changes, and light. The scallop has quite complex eyes; a single individual may have a hundred or more eyes located among the tentacles on the fringe of the mantle (Figure 26–6). The lens of this eye cannot focus images, however, so it does not appear to serve for more than the detection of light and dark, including passing shadows cast by other moving organisms.

26–6 A bivalve, the calico scallop (Aequipecten gibbos). Its bright blue eyes are visible among its tentacles. This scallop lives in the waters off the coast of Florida.

In the earliest mollusks, the sexes were separate and fertilization was external. This primitive condition is retained in most bivalves, but internal fertilization has evolved in a number of different bivalve lineages. Specialized "brood pouches," in which the young are protected during their early development, are found in hermaphroditic species as well as in the females of some species with separate sexes.

Class Gastropoda

The gastropods, which include the snails, whelks, periwinkles, abalones, and slugs, are the largest group of mollusks (at least 37,500 living species). They have either a single shell or, as a secondary evolutionary development, no shell. Gastropods are common in both salt and fresh water and on land. The group is unusual in that some members are able to digest cellulose and other structural carbohydrates. In addition to herbivores, class Gastropoda includes omnivores, a wide variety of specialized carnivores, scavengers, and even some parasites.

Gastropods have lost the bilateral symmetry characteristic of other mollusks and have become asymmetrical through a curious anatomical rearrangement called **torsion.** Torsion, which is a separate phenomenon from the coiling of the shell, is a twisting through 180° of the rest of the body relative to the head-foot. It occurs at a specific stage of larval development and, in many cases, requires only a few minutes. Torsion is initiated by the contraction of a large muscle that runs from the right side of the shell to the left side of the head-foot; it is generally completed by asymmetrical growth in which one side of the larval gastropod's body grows more rapidly than the other side. As a result of torsion, the shell, mantle cavity, and visceral mass are moved so that parts that once were located at the rear of the body now lie over the head (see Figure 26–3c). Other events in gastropod development usually result in a spiral coiling of the visceral mass and the shell. In response to the displacement and consequent crowding of the internal organs, the gill and nephridium of the right side have been completely lost in many species. In cases where the shell has been lost in the course of subsequent evolution and snails have evolved into slugs, the mantle cavity is generally moved back toward its original position by a process of detorsion—but the missing gill and other organs have not been regained.

Land-dwelling snails do not have gills, but the area in their mantle cavities once occupied by gills is rich in blood vessels, and the snail's blood is oxygenated there.

26–7 *Land-dwelling gastropods.* (a) *A terrestrial snail. The shell, which is secreted by the mantle and grows as the soft body grows, covers and protects the visceral mass. The head contains sensory organs, including two eyes at the tips of the longer tentacles. Note the thick trail of slime secreted by the snail as it glides along.* (b) *A banana slug,* Ariolimax columbianus, *photographed at Point Reyes, California.*

(a)

(b)

Thus the mantle cavity has become, in effect, a lung. Moreover, as with all lungs, the opening is reduced to retard evaporation. Some snails that were probably once land dwellers have returned to the water, but they have not regained gills. Instead, they bob up to the surface at intervals to entrap a fresh bubble of air in their mantle cavities.

Gastropods, which lead a more mobile, active existence than bivalves, have a ganglionated nervous system with as many as six pairs of ganglia connected by nerve cords (Figure 26–8). There is a concentration of nerve cells at the anterior end of the animal, where the tentacles, which have chemoreceptors and touch receptors, are located. In some of the animals, the eyes are quite highly developed in structure; they appear, however, to function largely in the detection of changes in light intensity, like the eyes of the scallop.

In some gastropods, the primitive form of reproduction—separate sexes with external fertilization—is retained. In most gastropods, however, fertilization is internal, and hermaphroditism has evolved repeatedly in this group. The simultaneous hermaphroditism found in many snails and slugs probably resulted from the difficulties of such slow-moving animals in finding mates. In some species, the animals are sequential hermaphrodites: they are males when they are younger, then females when they are older and larger.

26–8 *The ganglionated nervous system of a gastropod. The cerebral ganglia supply the tentacles and eyes; the pleural ganglia, the mantle; the pedal ganglia, the foot muscles; and the subesophageal, supraesophageal, and visceral ganglia supply the visceral mass. Counterclockwise torsion of the visceral mass during development results in the figure-eight pattern seen here, in which one major nerve passes over and another under the esophagus, a tubular structure leading from the mouth to the stomach.*

The operculum, a hardened plate attached to the foot, functions as a trapdoor when the snail withdraws into its shell.

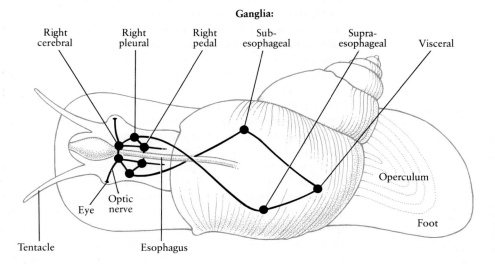

Ganglia:

Right cerebral   Right pleural   Right pedal   Sub-esophageal   Supra-esophageal   Visceral

Operculum

Foot

Eye   Optic nerve

Tentacle   Esophagus

Class Cephalopoda

The cephalopods (the "head-foots") are, in many respects, the most evolutionarily advanced animals to be found among the invertebrates. The 600 living species in this strictly marine class rival the vertebrates in complexity and, in some cases, in intelligence. Active predators, they compete quite successfully with fish.

Although obviously mollusks, the cephalopods have become greatly modified (see Figure 26–3d). The large head has conspicuous eyes and a central mouth surrounded by arms, some 90 in the chambered *Nautilus,* 10 in the squid, and 8 in the octopus. The octopus body seldom reaches more than 30 centimeters in diameter (except on the late late show), but giant squids sometimes attain sea-monster proportions. One caught in the Atlantic in 1861 was about 6 meters long, not counting the tentacles.

Nautilus, as the only modern shelled cephalopod, offers an indication of some of the steps by which this class disposed of the shell entirely. The animal occupies only the outermost portion of its elaborate and beautiful shell, the rest of which serves as a flotation chamber. In the squid and its relative, the cuttlefish, the shell has become an internal stiffening support, and in the octopus, it is lacking entirely.

Behavior in the Octopus

Behaviorally, the octopus is the best studied of the cephalopods. The octopus is a sea dweller that creeps about actively on its arms or swims rapidly through the water by strong rhythmic muscular contractions that expel water from the mantle cavity. The octopus, like other cephalopods, is carnivorous, living on smaller sea animals, usually crabs. It bites the crab, or other prey, with its parrotlike beak, injecting a toxin from its salivary glands. It then bundles up the paralyzed animal in its web and carries it home to eat. When it is not actively in pursuit of food, the octopus, lacking any protective shell, lives in small caves behind rocks or in reefs or wreckage.

Curious and able to use its arms with great dexterity, the octopus makes an extremely apt experimental subject. In a seawater tank, it will gather together bricks, shells, and any movable debris into a crude sort of house, and there it sits and watches, often bobbing its head up and down. Although the eye is remarkably similar to our own and equally acute, apparently the octopus does not have stereoscopic vision. Head bobbing seems to be the way it estimates distance, fixing on an object from two points, just as surveyors triangulate a distant landmark.

When confronted by an object larger than itself, an octopus literally pales. It flattens and turns almost white except for a dark area around the eyes and a dark trim along the edges of its web. This makes the octopus seem larger than it is and probably serves to deter would-be predators. By contrast, the sight of a crab can so excite an octopus that its arms weave about and its skin color darkens, breaking out in patches of bright blue, pink, or purple, depending on the species. These color changes are brought about by the contraction of muscles that draw out small sacs of pigment, the chromatophores, to form flat plates of color. When the muscles relax, the chromatophores return to their original size and the patches of color disappear.

By using a system of rewards (crabs, for example) and punishments (mild electric shock), the investigator can readily teach the octopus to seize certain objects and not to seize others. Such experiments approximate what an octopus must learn in nature—that some objects are edible, some are not, and some may bite or sting. A small crab, for example, is a meal, but if it is carrying a sea anemone on its shell, it had better be left alone. Because of the comparative ease with which such tests can be performed, owing to the octopus's natural curiosity and appetite, the animal has been the subject of a great many experiments on vision, touch, and learning.

The manipulatory powers of the octopus are great, and the arms are very sensitive to texture and are rich in chemoreceptors. However, the animal seems to have difficulty processing and coordinating sensory data. Martin Wells, of Churchill College, Cambridge, who has carried out many studies on octopus behavior, points out that the problems of handling data from eight extensible arms, each of which has several hundred suckers and can move separately and in any direction, are probably insurmountable. As we shall see in the next chapter, the severe limitations of movement imposed by an articulated skeleton prove, in contrast, to be a great advantage for many types of specialized activities.

A simplified view of the brain and nervous system of the octopus. The large lobes behind the eyes are the optic lobes, concerned with collecting and analyzing visual data. The octopus often finds its prey by reaching its arms into a crevice into which it cannot see. The suckers on the arms of an octopus are softer and more flexible than our fingertips and so can make fine distinctions among textures. They also contain chemoreceptors that can detect sugars, salts, and other chemicals in dilutions well below the range of discrimination of the human tongue.

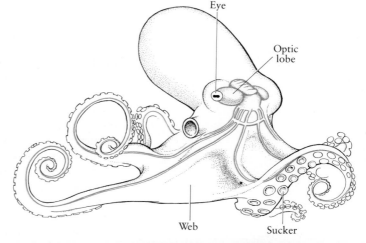

Eye

Optic lobe

Web

Sucker

Freedom from the external shell has given the mantle more flexibility. The most obvious effect of this is the jet propulsion by which cephalopods dart through the water. Usually, water taken into the mantle cavity bathes the gills and is then expelled slowly through a tube-shaped structure, the siphon; but when the cephalopod is hunting or being hunted, it can contract the mantle cavity forcibly and suddenly, thereby squirting out a sudden jet of water. Contraction of the mantle-cavity muscles usually shoots the animal backward, head last, but the squid and the octopus can turn the siphon in almost any direction they choose. Cephalopods also have sacs from which they can release a dark fluid that forms a cloud, concealing their retreat and confusing their enemies. These colored fluids were at one time a chief source of commercial inks. *Sepia* is the name of the genus of cuttlefish from which a brown ink used to be obtained.

26–9 *A cephalopod. Jet-propelled, this multicolored cuttlefish is moving to the right. You can see its siphon extending from the mantle, below its eye. On either side, an undulating fin runs the length of the broad, flattened body.*

26–10 Eledone cirrhosa, *the lesser octopus, from the British seacoast. Note its siphon, its well-developed eyes, and the suckers on the undersurface of its arms. The suckers, which contain a variety of sensory receptors, are also used to immobilize prey organisms, such as crabs, so the octopus can bite them and inject its paralyzing toxin.*

The cephalopods have well-developed brains, composed of many groups of ganglia, in keeping with their highly developed sensory systems and their lively, predatory behavior. These large brains are covered with cartilaginous cases. The rapid responses of the cephalopods are made possible by a bundle of giant nerve fibers that control the muscles of the mantle. Many of the studies on conduction of nerve impulses are made with the giant axons of squids, which are large enough to permit the insertion of electrodes.

In cephalopods, the sexes are always separate, and fertilization is internal. Courtship and mating behavior are complex, and males often fight for access to females. Fertilization occurs when the male uses one of his arms to transfer packets of sperm, called spermatophores, from his mantle cavity to the mantle cavity of the female. The female produces an intricate, gelatinous egg mass in which the developing embryos are protected until they hatch as miniature adults. In *Octopus* and some other cephalopod genera, the mother guards the egg masses, cleaning and aerating them.

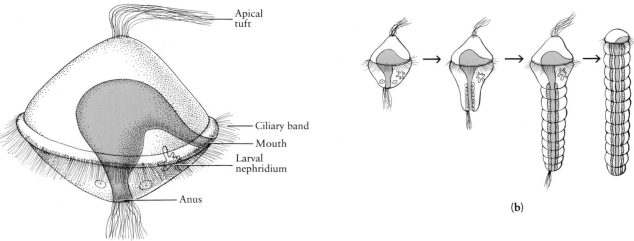

Apical
tuft

Ciliary band

Mouth

Larval
nephridium

Anus

(a)

(b)

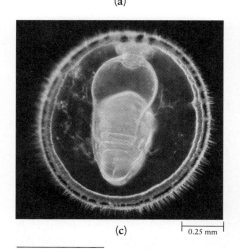

(c)   0.25 mm

26–11 **(a)** *Diagram of a trochophore larva. Although their adult forms are very different, certain annelids and mollusks have larvae of this type.* **(b)** *Development of the annelid trochophore into a segmented worm. The process begins with the elongation of the lower part of the trochophore. The elongated region then becomes constricted into segments, which soon develop bristles. The apical tuft disappears, and the upper part of the trochophore becomes the head. The worm's growth will continue throughout its lifetime by the addition of new segments just in front of the rear segment.*

(c) The trochophore larva of a marine annelid; this larva will develop into a polychaete worm.

Evolutionary Affinities of the Mollusks

Although the mollusks and the annelids (segmented worms) are, as we are about to see, quite different in their basic body plans, they have an important similarity that suggests an evolutionary link. This is the trochophore larva (Figure 26–11). Most marine mollusks (except the cephalopods) and marine annelids pass through this very distinct larval form during their development. With the exception of *Neopilina* and *Nautilus,* however, adult mollusks do not exhibit signs of the segmentation so characteristic of the annelids, and no traces of segmentation are seen in the larval development of any known mollusk. Most, but not all, authorities think that the segmental patterns seen in the gills, nephridia, and some other structures of *Neopilina, Nautilus,* and some fossil bivalves were late evolutionary developments, unrelated to the development of segmentation in the annelids. If so, the lineages giving rise to the mollusks and to the annelids diverged from an unsegmented coelomate ancestor that was characterized by the trochophore larva.

PHYLUM ANNELIDA: SEGMENTED WORMS

This phylum includes almost 9,000 species of marine, freshwater, and terrestrial worms, including the familiar earthworms. The term annelid is from the Latin for "ringed" and refers to the most distinctive feature of this group: the division of the body into segments, or **metameres.** The metameres are visible as rings on the outside and are separated by partitions (septa) on the inside. This segmented pattern is found in a modified form in arthropods, too, such as millipedes, crustaceans, and insects, which are thought to have evolved from the same ancestors that gave rise to modern annelids.

The annelids have a segmented coelom, a tubular gut, and a closed circulatory system that transports oxygen (diffused through the skin or through fleshy extensions of the skin) and food molecules (from the gut) to all parts of the body. The excretory system consists of paired nephridia, which typically occur in each segment of the body except the head. Annelids have a centralized nervous system and a number of special sensory cells, including touch cells, taste receptors, light-sensitive cells, and cells concerned with the detection of moisture. Some also have well-developed eyes and sensory antennae.

The three classes of annelids are Oligochaeta (terrestrial worms, with some freshwater and marine relatives), Polychaeta (mostly marine worms), and Hirudinea (leeches). It is generally agreed that the hirudineans are derived from oligochaetes, but the question of whether the polychaetes or the oligochaetes came first remains unresolved. However, we shall begin with class Oligochaeta, since its most familiar members—the earthworms—are such clear exemplars of annelid structure.

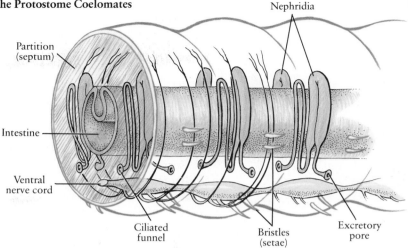

26–12 *Three segments of the earthworm, an annelid. On each segment are four pairs of bristles, which are extended and retracted by special muscles. These are used by the worm to anchor one part of its body while it moves another part forward. Two excretory tubes, or nephridia, are in each segment (except the first three and the last). Each nephridium really occupies two segments, since it opens externally by a pore in one segment and internally by a ciliated funnel in the segment immediately in front of it. The intestine, nephridia, and other internal organs are suspended in the large fluid-filled coelom, which also serves as a hydrostatic skeleton.*

Class Oligochaeta: The Earthworms

Figure 26–12 shows a portion of the body of an earthworm, a representative oligochaete. Note how the body is compartmentalized into regular segments. Most of these segments, particularly the central and posterior ones, are identical, each exactly like the one before and the one after. Each identical segment contains two nephridia, three pairs of nerves branching off from the central nerve cord running along the ventral surface, a portion of the digestive tract, and a left and right coelomic cavity; each segment also bears four pairs of bristles, or **setae.** The chief exceptions to this rule of segmented structure are found in the most forward segments. In these, specialized areas of the nervous, digestive, circulatory, and reproductive systems are found.

The tubelike body is wrapped in two sets of segmental muscles, one set running longitudinally and the other encircling the segments. When the earthworm moves, it anchors some of its segments by its setae, and the circular muscles of the segments anterior to the anchored segments contract, thus extending the body forward. Then its forward setae take hold, and the longitudinal muscles contract while the posterior anchor is released, drawing the posterior segments forward. The ultimate in a hydrostatic skeleton is provided by the coelom of the earthworm. Not only is the coelom partitioned by the septa between segments, but it is also divided into left and right compartments within each segment. This arrangement, in combination with the two sets of segmental muscles, allows exquisite control over movements of small parts of the body.

Digestion in Earthworms

The digestive tract of the earthworm (Figure 26–13) is a long, straight tube. The mouth leads into a strong, muscular **pharynx,** which acts like a suction pump, helping the mouth to draw in decaying leaves and other organic matter, as well as dirt, from which organic materials are extracted. The earthworm makes burrows in the earth by passing such material through its digestive tract and depositing it outside in the form of castings, a ceaseless activity that serves to break up, enrich, and aerate the soil. Darwin, in his study of earthworms (*The Formation of Vegetable Mould through the Action of Worms),* estimated 15 tons as the weight of castings thrown up annually on an acre of land—perhaps 20 ounces per worm per year, he calculated.

The narrow section of digestive tract posterior to the pharynx, the **esophagus,** leads to the **crop,** where food is stored. In the **gizzard,** which has thick, muscular walls lined with protective cuticle, the food is ground up with the help of the ever-present soil particles. The rest of the digestive tract is made up of a long intestine, which has a large fold along its upper surface that increases its surface area. The intestinal epithelium consists of enzyme-secreting cells and ciliated absorptive cells.

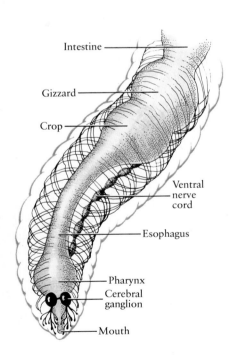

26–13 *The digestive tract of an earthworm. The mouth leads into a muscular pharynx, which sucks in decaying vegetation and other materials. These are stored in the crop and ground up in the gizzard with the help of soil particles. The rest of the tract is a long intestine in which food is digested and absorbed.*

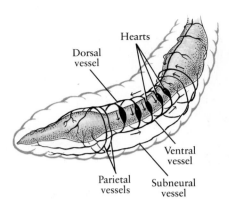

Hearts

Dorsal vessel

Ventral vessel

Parietal vessels

Subneural vessel

26–14 *The circulatory system of the earthworm is made up of longitudinal vessels running the entire length of the animal, one dorsal and several ventral. Smaller vessels (the parietal vessels) in each segment collect the blood from the tissues and from the subneural vessel and feed it into the muscular dorsal vessel, through which it is pumped forward. In the anterior segments are five pairs of hearts—muscular pumping areas in the blood vessels—whose irregular contractions force the blood downward to the ventral vessel, from which it returns to the posterior segments. The arrows indicate the direction of blood flow.*

Circulation in Earthworms

The circulatory system of the earthworm (Figure 26–14) is composed of longitudinal vessels running the entire length of the worm, one dorsal and several ventral. The largest ventral vessel underlies the intestinal tract, supplying blood to it, to the smaller ventral vessels that surround the nerve cord, and, by means of many small branches, to all the tissues of the body. Numerous small capillaries in each segment carry blood from the tissues to the dorsal vessel. Also in each segment are larger parietal ("along the wall") vessels transporting blood from the subneural vessel to the dorsal vessel, which also collects nutrients from the intestinal tract. The muscular dorsal vessel propels forward the fluids collected in this way from all over the animal's body.

Connecting the dorsal and ventral vessels, and so completing the circuit, are five pairs of hearts, muscular pumping areas in the blood vessels. Their irregular contractions force the blood down to the ventral vessels and also forward to the vessels that supply the more anterior segments. Both the hearts and the dorsal vessel have valves that prevent backflow. Note that annelids have a closed circulatory system in which the blood flows entirely through vessels. Evolution of a closed circulatory system made possible a degree of control, not previously feasible, over the composition of the circulating body fluid.

Respiration in Earthworms

Earthworms, unlike mollusks, have no special respiratory organs; respiration takes place by simple diffusion through the body surface. The gases of the atmosphere dissolve in the liquid film on the earthworm's body, which is kept moist by secreted mucus and by coelomic fluid released through dorsal pores. Oxygen diffuses inward to the network of capillaries just underlying the body surface and is consumed by body cells as the blood circulates. Carbon dioxide, which is picked up by the blood, diffuses out into the air through the body surface.

Excretion in Earthworms

In annelids, as in mollusks, the nephridia remove nitrogenous wastes from the coelomic fluid and regulate its chemical composition. The excretory system of the earthworm consists of one pair of nephridia for each segment (see Figure 26–12). Each nephridium consists of a long, convoluted tubule that begins with a ciliated funnel opening into the coelomic cavity of the anteriorly adjacent segment. Coelomic fluid is carried into the funnel by the beating of the cilia and is excreted through an outer pore. During its passage through the nephridium, needed water, other molecules, and ions are reabsorbed and waste products are secreted into the fluid.

The Nervous System of Earthworms

Earthworms have a variety of sensory cells. Touch cells, or **mechanoreceptors,** contain tactile hairs, which, when stimulated, trigger a nerve impulse. Patches of these hair cells are found on each segment of the earthworm. The hairs probably also respond to vibrations in the ground, to which the earthworm is very sensitive. The earthworm does not have ocelli or eyes—as one might expect, since it rarely emerges from underground—but it does have light-sensitive cells. Such cells are more abundant in its anterior and posterior segments, the parts of its body most likely to be outside of the burrow. These cells are not responsive to light in the red portion of the spectrum, a fact exploited by anglers who search for worms in the dark using red-lensed flashlights.

Among the earthworm's most sensitive cells are those that detect moisture. The cells are located on its first few segments. If an earthworm emerging from its burrow encounters a dry spot, it swings from side to side until it finds dampness; failing that, it retreats. However, when the anterior segments are anesthetized, the earthworm will crawl over dry ground. The animal also appears to have taste cells. In the laboratory, worms can be shown to select, for example, celery in preference to cabbage leaves and cabbage leaves in preference to carrots.

Each segment of the worm is supplied by nerves that receive impulses from sensory cells and by nerves that cause muscles to contract. The cell bodies for these nerves are grouped together in clusters (ganglia). The movements of each segment are directed by a pair of ganglia and are triggered by movement in the adjacent anterior segment; thus a headless earthworm can move in a coordinated manner. However, an earthworm without its cerebral ganglia moves ceaselessly; in other words, the cerebral ganglia inhibit and modulate activity.

There are also, as in planarians, conducting channels made up of nerve fibers bound together in bundles, like cables, which run lengthwise through the body. These nerve fibers are gathered together in a fused, double nerve cord that runs along the ventral surface of the body and then divides to encircle the pharynx at the anterior end of the animal (see Figure 26–13). The nerve cords contain fast-conducting fibers that make it possible for the earthworm to contract its entire body very quickly, withdrawing into its burrow when disturbed.

Reproduction in Earthworms

Earthworms are hermaphrodites, and some can reproduce parthenogenetically. In most species, however, two earthworms, held together by mucous secretions from the clitellum (a special collection of glandular cells), exchange sperm and separate (Figure 26–15). Two or three days later, the clitellum forms a second mucous sheath surrounded by an outer, tougher protective layer of chitin. This sheath is pushed forward along the animal by muscular movements of its body. As it passes over the female gonopores, it picks up a collection of mature eggs, and then, continuing forward, it picks up the sperm deposited in the sperm receptacles, or spermathecas. Once the mucous band is slipped over the head of the worm, its sides pinch together, enclosing the now fertilized eggs in a small capsule from which the infant worms hatch.

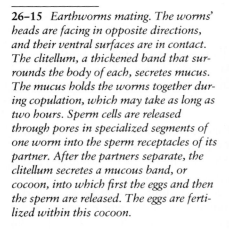

26–15 *Earthworms mating. The worms' heads are facing in opposite directions, and their ventral surfaces are in contact. The clitellum, a thickened band that surrounds the body of each, secretes mucus. The mucus holds the worms together during copulation, which may take as long as two hours. Sperm cells are released through pores in specialized segments of one worm into the sperm receptacles of its partner. After the partners separate, the clitellum secretes a mucous band, or cocoon, into which first the eggs and then the sperm are released. The eggs are fertilized within this cocoon.*

26–16 *A deep-sea polychaete worm. Unlike the more familiar earthworm, this annelid has a well-differentiated head with sensory appendages and lateral parapodia ("side feet") with many setae.*

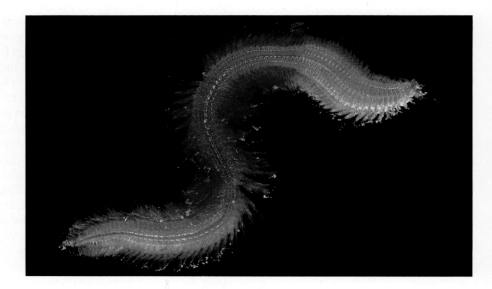

Class Polychaeta

The polychaetes, which are almost all marine, differ from the earthworms and other oligochaetes in a number of ways (Figure 26–16). The most striking difference is that they typically have a variety of appendages, including tentacles, antennae, and specialized mouthparts. Each segment contains two fleshy extensions, **parapodia,** which function in locomotion and also, because they contain many blood vessels, are important in gas exchange. In polychaetes, as in other segmented animals, there is a tendency for the division of labor between segments to lead to **tagmosis**—the formation of groups of segments into body regions with functional differences. This tendency is more pronounced in the polychaetes than in the oligochaetes and often results in distinct head, trunk, and tail regions.

Polychaetes have diverse life styles. Some are motile predators, using their strong jaws to feed on small animals. Others are more sedentary, feeding on materials suspended in the water or deposited in bottom sediments. Many polychaetes live in elaborately fashioned tubes constructed in the mud or sand of the ocean bottom. Usually the sexes are separate, fertilization is external, and there is a free-swimming trochophore larva.

Class Hirudinea

Hirudineans are the leeches, which have flattened, often tapered, bodies with a sucker at each end (Figure 26–17). In most species, the setae have been lost, and the animals either creep along with a loping movement or swim with undulating motions of the body. Like the earthworm and other oligochaetes, leeches are hermaphrodites.

Bloodsucking leeches attach themselves to their hosts by their posterior sucker, and then, using their anterior sucker, either slit the host's skin with their sharp jaws or digest an opening through the skin by means of enzymes. Finally, they secrete chemicals into the host's blood that prevent the formation of clots. One of these substances, appropriately called hirudin, is the most powerful natural anticoagulant known. The gene coding for hirudin, which consists of 65 amino acids, has recently been cloned, and it is expected that synthetic hirudin will soon be available for treating heart attack victims and patients with severe atherosclerosis. Another substance in leech saliva has been demonstrated to inhibit the spread of malignant cells from lung cancers. In keeping with their rather unpleasant image, however, leeches secrete no anesthetic in their saliva.

26–17 *Leeches are primarily bloodsuckers with digestive tracts specially adapted for storage of blood. They range in size from 1 to 30 centimeters and are found mostly in inland waters or in damp places on land, where they parasitize fish, turtles, and other vertebrates. A few species are predatory, feeding on worms and insects that they swallow whole.*

Shown here is a medicinal leech, genus Hirudo, of the type commonly used for bloodletting. Bloodletting, one of the most common forms of medical treatment as late as the 1800s, was used for patients suffering from whooping cough, gout, drunkenness, rheumatism, sore throat, and asthma, among other ailments. Patients were commonly "bled to faintness," losing as much as 1.5 liters of blood.

MINOR PROTOSTOME PHYLA

In addition to the three major phyla of protostome coelomates (Mollusca, Annelida, and Arthropoda), there are seven minor phyla with living representatives. Four of these phyla contain bottom-dwelling marine worms that show varying degrees of similarity to the annelids.

The peanut worms, phylum Sipuncula, range from 1 to more than 60 centimeters in length and are characterized by a long proboscis that can be retracted into the stout, bulblike body (Figure 26–18a). Although the approximately 300 species of this phylum have neither segmentation nor setae, their trochophore larvae are very similar to those of the polychaete annelids.

Phylum Echiura contains about 100 species that are sometimes called spoon worms. Most have a long proboscis that, unlike the proboscis of the peanut worms, cannot be retracted into the stout body; it can contract, however, forming a structure that resembles the bowl of a spoon (Figure 26–18b). As in the peanut worms, the sexes are separate, fertilization is usually external, and there is a trochophore larva. The adult worms have setae but show no traces of segmentation.

Fifteen species of burrowing worms, ranging from 0.5 millimeter to 20 centimeters in length, make up phylum Priapulida. They are characterized by a retractile proboscis bearing spines that are used to capture soft-bodied prey (Figure 26–18c). Priapulids are not segmented internally, but the external surface of the body may be marked by many superficial rings; setae are found only on the males of one genus. These worms most closely resemble the pseudocoelomate kinorhynchs (see page 541) but may have a true coelom. Very little is known about either their embryonic development or their phylogenetic relationships.

(a)

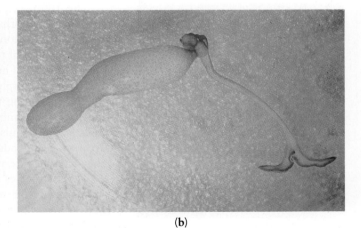

(b)

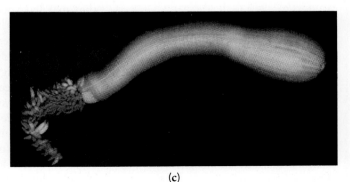

(c)

26–18 *Representatives of the smaller phyla of protostome worms.* (**a**) Dendrostomum pyroides, *a sipunculan. The retractile proboscis represents about one-third of the worm's total length. Food particles, trapped by the mucus-covered tentacles, are moved to the mouth by cilia. When disturbed, these worms contract into the shape of a peanut.* (**b**) *An echiuran,* Bonellia viridis, *from the bottom sediments off the Mediterranean coast of France. The algal cells on which this species feeds are caught in the cilia of the long, forked proboscis and are carried to the mouth.* (**c**) Priapulus caudatus, *the first priapulid described, with its anterior proboscis everted. The bushy tail at the posterior of the animal consists of tubular structures that are thought to have a respiratory function. In some species the tail is muscular and bears hooks; it presumably anchors the worm when it is burrowing.*

26-19 *The most stunning members of phylum Pogonophora are the giant tube worms that live near fissures in the earth's crust, deep in the Pacific Ocean. These worms, first discovered in 1977, sometimes reach 1.5 meters in length. Their bodies harbor large quantities of symbiotic chemosynthetic bacteria that provide them with organic molecules.*

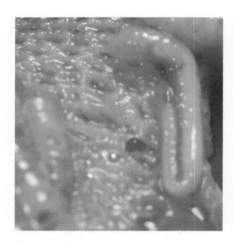

26-20 Porocephalus crotali, *a pentasto-mid. Adults of this genus live as parasites in the lungs of snakes, attaining lengths of 10 to 13 centimeters. Their life cycle includes two larval stages, each parasitic in a different host, which may be a snake or a mammal. Although not visible in this photograph, the mouth of a pentastomid is surrounded by four projections. At one time, these projections were erroneously thought to be additional mouths; hence the name "five mouths."*

The 100 species of beard worms, phylum Pogonophora, have a segmented posterior end with setae but have no mouth or digestive tract. These very slender worms live in long tubes buried in deep-sea sediments (Figure 26–19). The anterior region of the body bears a crown of tentacles that either provide surfaces for the uptake of nutrients or, in some species, contain symbiotic chemosynthetic bacteria that supply the worms with nutrients.

The remaining three phyla—Pentastomida, Tardigrada, and Onychophora—have varying combinations of annelid and arthropod characteristics. These animals are distinguished from all others by unjointed appendages bearing claws; arthropod appendages, by contrast, are jointed. Like the arthropods, however, they have an external cuticle that is molted periodically.

The 70 species of wormlike pentastomids (Figure 26–20) are all parasites of vertebrate respiratory systems. Pentastomids are so highly specialized for their parasitic existence—lacking, for example, circulatory, respiratory, and excretory organs—that it has been difficult to determine the primitive form from which they may have evolved. Although their larvae resemble young tardigrades, recent studies indicate that they are most closely related to the arthropods. Pentastomids have separate sexes and internal fertilization and, like many other parasites, produce tremendous numbers of eggs.

Phylum Tardigrada contains about 350 species of tiny segmented animals often called "water bears" (Figure 26–21). Common in fresh water and in the film of moisture on mosses, they lumber along on their four pairs of stubby legs. Their protective cuticle is thin, but they are able to survive drying out; when conditions are unfavorable, they enter a state of suspended animation and remain dormant—in some cases, for years—until moisture is once more available. The sexes are separate, but in some species males are unknown and the females produce eggs that develop parthenogenetically.

26-21 *A "water bear," phylum Tardigrada. These remarkable animals are able to survive extremes of temperature as well as extreme desiccation. They are found not only in temperate climates, moving in their slow, deliberate way over the surfaces of mosses and lichens, but also in the Arctic, in the tropics, and even in hot springs. Their mouths are equipped with piercing stylets, through which they suck the juices of plant cells or of such small animals as rotifers and nematodes.*

50 μm

The caterpillarlike animals of phylum Onychophora have a particularly striking combination of annelid and arthropod characteristics. The annelid characteristics shared by the 70 species in this phylum include relatively soft bodies, segmentally arranged nephridia, muscular body walls, and ciliated reproductive tracts. On the other hand, their jaws (derived from appendages), protective cuticle, relatively large brains, and circulatory and respiratory systems resemble those of arthropods. Their antennae and eyes are similar to those of both the polychaete annelids and the arthropods. Although it is reasonable to consider onychophorans primitive animals, suggestive of a stage in the early evolution of the arthropods, their reproduction is very advanced. Although some onychophorans lay eggs, most give birth to live young (Figure 26–22). Moreover, in some species the embryos develop within a uterus and are nourished through a placenta-like structure, analogous to that of the mammals. Modern onychophorans are all terrestrial, living inconspicuously in moist habitats, mainly in the Southern Hemisphere. Fossil evidence, however, indicates that the earliest species were marine (see Figure 25–5d, page 521). As onychophorans made the transition to land, they, like the plants (and, as we shall soon see, the arthropods and the vertebrates), successfully solved the problem of protecting and nourishing the developing young.

26-22 *The birth of an onychophoran. With a strong contraction of its body, this young* Peripatus *has just freed itself from its mother's genital orifice. Although a newborn* Peripatus *is able to walk immediately on its short, unjointed legs and to eject the "glue" with which these carnivorous animals ensnare prey, it usually remains with its mother for a few days.*

26–23 *The extensive lophophore of a phoronid worm,* Phoronopsis harmeri. *The lophophore is in the shape of a horseshoe, with each of the ends forming a spiral coil. Cilia on the tentacles of the lophophore direct water currents through the groove between the two coils. Food particles are trapped in mucus on the tentacles and carried by the cilia to the mouth, which is at the base of the groove. Although phoronids are not attached to the chitinous or leathery tubes in which they live, they never leave them. When disturbed, the animal can completely withdraw into the safety of the tube.*

26–24 *The brachiopod* Terebratulina septen, *with its shell closed and open. When the shell is open, the only structure visible is the large, spiral lophophore, which constitutes about two-thirds of the animal's body. Water is circulated through the shell by the moving tentacles of the lophophore. Its cilia sweep small organisms in the water toward the mouth, which is under the lophophore, toward the left. The brachiopod is anchored to the substrate by a stalk, or pedicel.*

THE LOPHOPHORATES

The animals in three additional phyla, known collectively as the lophophorates, are, strictly speaking, protostomes (that is, the mouth develops at or near the blastopore). Nevertheless, they possess other characteristics that lead many authorities to regard them as primitive deuterostomes. Lophophorates do not display the segmentation characteristic of annelids and arthropods, but the coelom of the adult forms of some species and of the embryos of most species is partitioned into several major compartments. A similar sort of partitioning is characteristic of many deuterostomes. Other deuterostome characteristics shared by the lophophorates include radial cleavage in the early stages of embryonic development and, in some species, enterocoelous formation of the coelom (see page 545).

The lophophorates, which are all aquatic, include the lamp shells (phylum Brachiopoda), the phoronid worms (phylum Phoronida), and the bryozoans, or "moss animals" (phylum Bryozoa, or Ectoprocta). Although the three phyla have diverged considerably, all of these animals have a characteristic food-gathering organ known as the **lophophore** (Figure 26–23). Located at the anterior end of the animal, the lophophore consists of a crown of hollow, ciliated tentacles used to capture small organisms and bits of organic debris suspended in the surrounding water. The cavity within the tentacles is an extension of the coelom, and gases are exchanged between the coelomic fluid and the external environment through the thin walls of the tentacles.

The most familiar of the lophophorates are the lamp shells, phylum Brachiopoda. Until well into the nineteenth century, these animals, which resemble a Greek or Roman oil-burning lamp (Figure 26–24), were classified with the bivalve mollusks. The lophophore, however, clearly distinguishes the brachiopods from the bivalves. Moreover, the two valves of the brachiopod shell are dorsal and ventral, rather than left and right as in the bivalves. There are about 250 species of brachiopods living today, but the fossil record reveals an additional 30,000 extinct species.

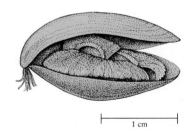

1 cm

Simpler in structure are the 18 species of phylum Phoronida (see Figure 26–23). These worms, which range from 4 to 25 centimeters in length, live in tubes in or on soft ocean bottoms in shallow waters. The lophophore projects above the top of the tube but can be withdrawn into the tube when the animal is disturbed. Although phoronids are often found clustered together, each animal is independent of the others. Most species are hermaphroditic, and one species is known to reproduce asexually.

In contrast to the phoronids, the tiny bryozoans, or ectoprocts, are colonial animals, often with a considerable division of labor among the members of the colony. Their colonies can be found on virtually any type of firm surface in salt water and, less frequently, in fresh water. Bryozoans secrete a hard, protective covering around themselves, from which only the individual lophophores extend (Figure 26–25). Their name derives from the superficial resemblance of their colonies to patches of moss. As we noted earlier (page 541), the pseudocoelomate

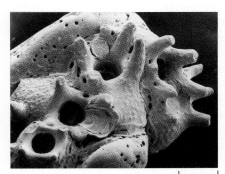

26–25 *Colonies of the bryozoan* Hippodiplosia insculpta, *photographed in Monterey Bay, California. Note the extended lophophores with which the colony members catch passing algal cells. Because of the protective covering that bryozoans secrete around themselves, their colonies are also known as "sea mat."*

200 μm

26–26 *The skeletal remains of a microscopic colony of the bryozoan* Trematooecia psammophila *on the edge of a sand grain. The stalklike extensions are spines. When the bryozoan was alive, the lophophores extended through the large openings visible in this scanning electron micrograph as dark holes. The thumbprint-like depression adjacent to the second hole is a developing brood chamber. Within each brood chamber an embryo develops into a motile larva.*

Entoprocta have a similar appearance, and before the differences in internal structure were fully understood, the entoprocts were also included in phylum Bryozoa. About 4,000 living species of bryozoans are known, and most are hermaphrodites; the freshwater forms, however, reproduce asexually as well as sexually.

The bryozoan life cycle includes a motile larva that, in many species, is such a poor swimmer that biologists have long wondered how new colonies can become established at relatively great distances from the parent colony. Recent studies have demonstrated that, in at least some species, the secret is a series of short-lived sexually reproducing generations. The larvae attach themselves to individual sand grains, rapidly mature, and reproduce again, without investing either the time or the resources to form an elaborate colony—for which there is, in any case, insufficient space (Figure 26–26). By jumping from sand grain to sand grain, one generation at a time, descendants of the original colony ultimately reach a distant site suitable for the establishment of a new colony. Among the many mechanisms of genetic recombination and dispersal in the animal kingdom, this may be one of the most ingenious.

SUMMARY

One of the most significant innovations in the course of animal evolution was the coelom, a fluid-filled cavity that develops within the mesoderm. The coelom not only functions as a hydrostatic skeleton but also provides space within which the internal organs can be suspended by the mesenteries.

The coelomate animals are divided into two broad groups on the basis of their embryonic development. In the protostomes, the cleavage of the fertilized egg is usually spiral, the mouth develops at or near the blastopore, and the coelom is formed by a splitting of the solid mesoderm. In the deuterostomes, the cleavage pattern is radial, the anus develops at or near the blastopore, and the coelom is formed by outpocketings of the primitive gut.

The soft-bodied animals of phylum Mollusca are classified into four relatively small classes and three major classes: Bivalvia (oysters and clams), Gastropoda (snails), and Cephalopoda (octopuses and their relatives). The basic body plan of all mollusks is the same—a head-foot, a visceral mass, and a shell-secreting mantle—but it has been modified in the course of adaptation to different ways of life. Shells are often present but may be reduced or absent. Mollusks are also characterized by a toothed tongue (the radula), an open circulatory system, and a greatly reduced coelom. In most mollusks, respiration is carried out by means of gills, thin-walled structures that are an extension of the epidermis and are located in the mantle cavity. Excretion is accomplished by special organs, the nephridia, tubular structures that collect fluids from the coelom and exchange salts and other substances with body tissues as the fluid passes along the tubules for excretion. Nervous systems and behavior vary among the species, reaching a zenith of complexity in the brainy octopus.

On the basis of similarities in their trochophore larvae, the mollusks and the segmented worms of phylum Annelida are thought to have diverged from a common coelomate ancestor. Annelids are characterized by bodies that are conspicuously segmented, both internally and externally; well-developed coeloms; tubular digestive tracts; paired nephridia in each segment; and closed circulatory systems, often with contractile vessels. The phylum includes terrestrial worms, such as the earthworms (class Oligochaeta), marine worms (class Polychaeta), and leeches (class Hirudinea). Unlike the oligochaetes, the polychaetes have a variety of appendages and exhibit tagmosis—the formation of groups of segments into body regions with functional differences. Annelids have relatively complex nervous systems, consisting of a ventral nerve cord and pairs of ganglia (clusters of nerve cells), one pair to each segment, which receive impulses from a variety of sensory receptors and trigger motor activities in each segment.

Of the seven minor phyla of protostome coelomates, four consist of bottom-dwelling marine worms that show varying similarities to the annelids. These groups include the peanut worms (phylum Sipuncula), the spoon worms (phylum Echiura), the priapulids (phylum Priapulida), and the beard worms (phylum Pogonophora). Three other phyla—Pentastomida, Tardigrada, and Onychophora—all characterized by unjointed, clawed appendages and an external cuticle, have combinations of annelid and arthropod features. The onychophorans are of particular interest, for they suggest a stage in the evolution of arthropods from a segmented, coelomate ancestor common to both the annelids and the arthropods.

Lophophorates, although protostomes, resemble deuterostomes in the radial cleavage of the fertilized egg and, in some species, the enterocoelous formation of the coelom. The lamp shells (phylum Brachiopoda), the marine worms of phylum Phoronida, and the colonial Bryozoa are all lophophorates. They are characterized by the lophophore, a crown of hollow tentacles bearing cilia that sweep food particles into the mouth.

QUESTIONS

1. Distinguish between the following terms: protostome/deuterostome; schizocoelous/enterocoelous; coelom/hemocoel; planula/trochophore.

2. What modifications of the basic molluscan body plan have occurred in bivalves, gastropods, and cephalopods?

3. Many mollusks have lost or may be in the evolutionary process of losing their shells. What are the advantages to an organism of having a shell? Of losing one?

4. Smallness may also be an advantage to an organism. What are some of the advantages of smallness?

5. Why does any heart truly worthy of the name have at least two chambers?

6. Label the drawing at the right.

7. Nematodes have only longitudinal muscles in their body walls but have very high internal fluid pressures; earthworms, with both longitudinal and circular muscles, have low fluid pressures. Can you suggest a mechanism by which internal fluid at high pressure can circumvent the need for certain muscles?

8. Describe the structure and function of a lophophore. How does it differ from the tentacles of a cnidarian, such as *Hydra?*

9. A larval or medusoid stage occurs in the life histories of many marine invertebrates. Yet in group after group of freshwater and terrestrial invertebrates, each presumably separately evolved from marine ancestors, the free-living larva or medusa has been lost. What might be the selective factor underlying the loss?

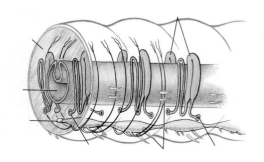

The Animal Kingdom III: The Arthropods

Phylum Arthropoda, the "joint-footed" animals, is by far the largest of the animal phyla. More than 1 million species of insects and other arthropods have been classified to date, and estimates of the total number are as high as 50 million. As to the number of individuals, it has been calculated that of insects alone, as many as 10^{18}—a billion billion—are alive at any one time. Arthropods are abundant in virtually all habitats. It has been estimated that over every square kilometer in the temperate zone there are, at certain seasons, some 20 million individual arthropods, layered in the atmosphere like plankton.

CHARACTERISTICS OF THE ARTHROPODS

Arthropods are characterized by their jointed appendages. In the more highly evolved members of the phylum, the number of appendages is reduced, but they are more specialized and efficient. Particularly among the insects and crustaceans, these specialized appendages include not only walking legs but also a kit of marvelously adapted tools—jaws, gills, tongs, egg depositors, sucking tubes, claws, antennae, paddles, and pincers.

All arthropods are segmented, a characteristic that strongly suggests a common ancestry with the annelids (Figure 27–2). In the course of arthropod evolution, however, the body has become shorter, and it has fewer segments, which have become fixed in number and more specialized. In many arthropods, tagmosis

27–1 *A harlequin beetle in flight. Arthropods are characterized by a variety of specialized appendages. Notice the three pairs of jointed legs, the jointed antennae, the intricately colored wings, and the segmented body.*

27–2 *Arthropods, like the annelids, are segmented. In the more primitive arthropods, such as this millipede, the segmented pattern remains clearly visible in the adult animal. However, unlike annelids, adult arthropods have rigid, jointed exoskeletons and appendages. In millipedes, which are members of class Diplopoda, most of the body segments are fused into "diplosegments," each of which has two pairs of legs.*

(page 558) has been carried much farther than in the polychaete annelids, with the fusion of segments to form distinct body regions—a head, a thorax (sometimes fused with the head to form a cephalothorax), and an abdomen. But the basic segmented pattern is often still clearly evident in the immature stages (witness the caterpillar) and can be discerned in the adult by examination of the appendages, the musculature, and the nervous system.

At some point well after the lineage leading to the arthropods diverged from that leading to the annelids, further major branchings occurred (see Figure 25–6, page 521). These branchings gave rise to three principal types of arthropods: chelicerates, aquatic mandibulates, and terrestrial mandibulates. The conspicuous differences in the appendages of these three types can be clearly distinguished by even an inexperienced eye. In both the aquatic mandibulates (class Crustacea) and the terrestrial mandibulates (class Insecta and four smaller classes), the most anterior appendages are one or two pairs of **antennae,** and the next are **mandibles** (jaws). Differences in the development and structure of the mandibles suggest that they evolved independently in the two groups and do not reflect a common mandibulate ancestor. The chelicerates, which include class Merostomata (horseshoe crabs), class Pycnogonida (sea spiders), and class Arachnida (spiders, mites, scorpions, and their relatives), have no antennae and no mandibles. Their first pair of appendages consist of **chelicerae** (singular, chelicera), which take the form of pincers or fangs. Chelicerates may also have **book lungs** or **book gills;** these respiratory structures, which are not present in mandibulates, derive their name from their resemblance to the leaves of a partially opened book.

Despite the huge number of arthropods and the richness of their diversity, there are a number of features shared by all members of this phylum.

The Exoskeleton

The most striking characteristic of all arthropods is their articulated (jointed) exoskeleton. This exoskeleton, or cuticle, is secreted by the underlying epidermis and is attached to it; it is made up of an outer, often waxy layer, composed of lipoprotein, a hardened middle layer, and an inner flexible one, both composed principally of chitin and proteins. The exoskeleton not only covers the surface of the animal but also extends inward at both ends of the digestive tract; in insects, it lines the tracheae (breathing tubes) as well. The cuticle serves as protection against predators, and it is often waterproof, keeping exterior water out and interior water in. It is used for food grinders in the foregut, for wings, and for tactile hairs. Cuticle even forms the lens of the arthropod eye.

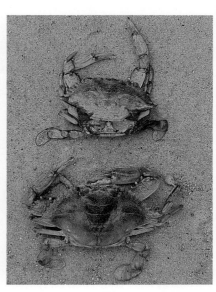

(a)

(b)

27–3 (a) *A many-jointed exoskeleton, as in this South American katydid, is characteristic of the arthropods. The slits in each foreleg are the insect's ears.*

(b) *Molting. An exoskeleton, once formed, does not grow and so must be periodically discarded. After shedding the old cuticle, an arthropod expands rapidly by taking in air or water, stretching the new exoskeleton before it hardens. The old cuticle is at the top, and below is the newly emerged, already expanded "soft-shelled" blue crab. Its new exoskeleton has begun to harden.*

The exoskeleton may form a veritable coat of armor, as it does in beetles and in some of the crustaceans (in which it is often infiltrated with calcium salts), but at the joints it is flexible and thin, permitting free movement. Muscles are attached to the various portions of the exoskeleton, just as they are attached to the various bones of the endoskeleton in vertebrates. When the muscles contract, the exoskeleton moves at its joints; such movements can be exquisitely precise because the force is brought to bear on very small areas, for example, on the different sections of an insect's leg.

The exoskeleton has certain disadvantages. It does not grow (as the bony vertebrate endoskeleton does), and so it must be discarded and re-formed many times as the animal grows and develops. At molting time, the animal secretes an enzyme that dissolves the inner layer of the exoskeleton, and a new skeleton, not yet hardened, is formed beneath the old one. Molting is dangerous; the newly molted animal is particularly vulnerable to predators and, in the case of terrestrial forms, subject to water loss. Many arthropods go into hiding until their new cuticle has hardened. Molting is also costly in terms of metabolic expenditures, although a number of insects and some freshwater crustaceans limit their losses by thriftily reabsorbing the dissolved inner layer and eating the outer layers of the old exoskeleton.

The fact that the exoskeleton can be waterproofed made possible the evolution of terrestrial forms among the arthropods, a process that was well underway by the late Ordovician. Fossil traces of animal burrows in relatively dry soil, thought to have been made by arthropods similar to millipedes, were recently discovered in deposits near Potters Mills, Pennsylvania; dated at 440 to 445 million years ago, these fossils are the earliest evidence of animals adapted to a semiarid or arid environment. Of all the invertebrate phyla, only Arthropoda contains a multitude of species well adapted to withstand the drying action of the air.

Internal Features

Arthropods, like annelids, have a tubular mouth-to-anus gut. They also have a coelom, but the arthropod coelom—like that of the mollusks—is markedly reduced, consisting only of the cavities of the reproductive and excretory organs.

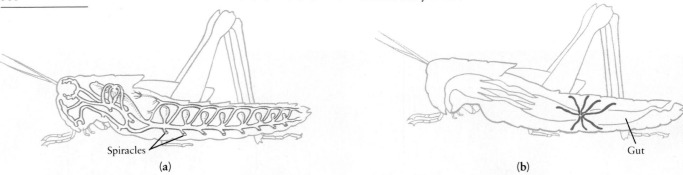

Spiracles

(a)

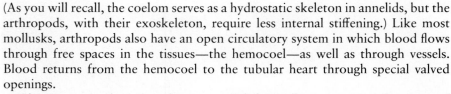

Gut

(b)

27–4 **(a)** *Respiration by means of a system of internal tubes (called tracheae) is found almost exclusively among terrestrial arthropods. Tracheae are usually branched, like those shown here, and open to the outside by spiracles that may be closed to conserve water. The tubes are lined with spiral rings and cuticle, which keeps them open. Because it delivers oxygen directly to the cells, a tracheal system is one of the most efficient respiratory systems in the animal kingdom. As the size of the animal increases, however, its efficiency decreases.*

(b) *Malpighian tubules represent another exclusively arthropod characteristic, although, like tracheae, they are not found in all classes. These tubules collect water and nitrogenous wastes from the hemocoel and empty them into the gut. The wastes (in the form of uric acid or guanine) are excreted with the feces.*

(As you will recall, the coelom serves as a hydrostatic skeleton in annelids, but the arthropods, with their exoskeleton, require less internal stiffening.) Like most mollusks, arthropods also have an open circulatory system in which blood flows through free spaces in the tissues—the hemocoel—as well as through vessels. Blood returns from the hemocoel to the tubular heart through special valved openings.

The insects and some other terrestrial forms have an unusual means of respiration, consisting of a system of cuticle-lined air ducts, known as **tracheae,** that pipe air directly into the various parts of the body (Figure 27–4a). Gases must pass through the tracheae largely by diffusion, thus placing a limit on the size of insects. Some terrestrial arthropods, such as spiders, have book lungs, either in addition to or instead of tracheae. Excretion in terrestrial forms is by means of **Malpighian tubules** (Figure 27–4b), which are attached to and empty into the midgut or hindgut. These tubules absorb wastes from the body cavities. Respiration by tracheae, book lungs, or book gills and excretion by Malpighian tubules are found almost exclusively in the arthropods (although not all arthropods have these features).

The Arthropod Nervous System

The bulk of the arthropod nervous system consists of a double chain of segmental ganglia running along the ventral surface (Figure 27–5). The double chains part to encircle the esophagus, ending in three fused pairs of dorsal ganglia. These fused dorsal ganglia, which perform a number of specialized functions, constitute a brain. However, many arthropod activities are controlled at the segmental level, as in annelids. For example, even after the brain has been removed, members of a number of species can move, eat, and carry on other functions normally. In fact, in the arthropods generally, the brain appears to act not so much as a stimulator of the action of the animal but as an inhibitor, as in the earthworms. The grasshopper, for example, can walk, jump, or fly with its brain removed; indeed, the brainless grasshopper responds to the slightest stimulus by jumping or flying.

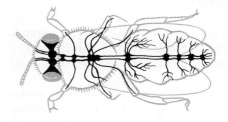

27–5 *The arthropod nervous system, as exemplified in the bee. The brain consists of three fused pairs of dorsal ganglia at the anterior end of a double chain of ganglia. The ganglia are interconnected by two bundles of nerve fibers running along the ventral surface. Because of the arthropod's segmental type of nervous system, many functions are controlled at a local level, and a number of species can carry on some of their normal activities after the brain has been removed.*

SUBDIVISIONS OF THE PHYLUM

The Chelicerates

As we noted previously, chelicerates have neither antennae nor mandibles. The first pair of appendages, the chelicerae, which bear pincers or are sharp and fanglike, are used for biting prey. The second pair are the **pedipalps,** which may bear pincers, be modified as walking legs, or serve as sensory organs. Posterior to the pedipalps are a series of jointed walking legs. The body segments have become fused into two tagmata—an anterior cephalothorax and a posterior abdomen, which is unsegmented in most chelicerates but conspicuously segmented in scorpions. Except for some mites, all chelicerates are carnivorous. The sexes are almost always separate.

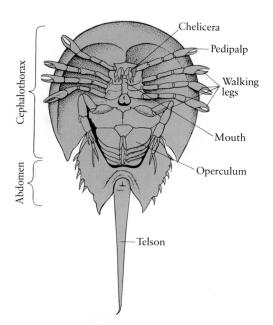

27–6 *Because the body of the horseshoe crab is largely covered with a heavy shield, or carapace, its segmented body plan is only evident from the underside. The operculum is a flat, movable plate that covers and protects the book gills. Horseshoe crabs are often called living fossils because remarkably similar organisms date back to the Silurian period (408 to 438 million years ago). There are now only four living species.*

Two classes of chelicerates are relatively small. Class Merostomata consists of only four species of horseshoe crabs (Figure 27–6). In viewing a horseshoe crab from below, one can see the chelicerae, pedipalps, and four pairs of walking legs. Posterior to the walking legs are a series of flaplike book gills, derived from modified limbs. *Limulus,* a bottom dweller that feeds upon such small animals as annelids and clams, is common in the shallow waters along the East Coast of the United States.

The approximately 500 species of sea spiders (class Pycnogonida) have slender bodies and four (or, rarely, five) pairs of legs, which are often very long (Figure 27–7). They feed on soft-bodied invertebrates—particularly cnidarians—by sucking juices from their bodies. Although sea spiders are common in coastal waters, most species are quite small and inconspicuous.

All of the remaining chelicerates belong to class Arachnida, with about 57,000 species.

27–7 *A male sea spider,* Nymphon gracile, *carrying two large masses of developing eggs. Among pycnogonids, the eggs are always carried by the males on a subsidiary pair of specialized legs. Sea spiders have neither respiratory nor excretory systems; gases and waste products are thought to diffuse through the large surface area provided by the long, thin legs.*

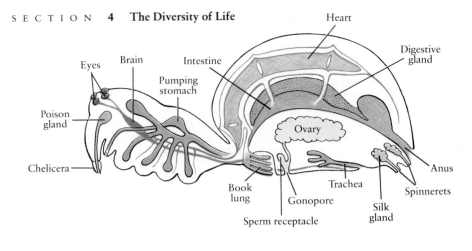

27–8 *A spider, an arachnid. Ducts from the poison glands open at or near the tips of the chelicerae. The flow of poison is voluntarily controlled by the spider. Only a few spiders are dangerous to human beings; perhaps the most dangerous are members of the species shown here, the black widow.*

(a)

27–9 *Some arachnids.* (a) *A giant desert hairy scorpion,* Hadrurus hirsutus, *photographed in Arizona. The southwestern United States and Mexico are also home to the highly toxic genus* Centruroides. *Scorpions sting to immobilize their prey and in self-defense, releasing a venom that attacks the nervous system of the victim.* (b) *An adult red velvet mite of the genus* Trombidium. (c) *A female hunting spider, photographed in a rain forest in Costa Rica as she guarded her cocoon of fertilized eggs. Note how the segments of the body have become fused into two regions —cephalothorax and abdomen.* (d) *An orb-weaving spider* (Neoscona oaxacensis) *has caught a grasshopper in its web. The prey, immobilized by venom, is being wrapped in silk. If you look closely, you can see the silk being spun out from the spider's abdomen. Webs, which may be as much as a meter in diameter, are reconstructed daily.* (e) *A daddy longlegs. These familiar arachnids are also known as harvestmen; most species mature in late summer, and they are seen in greatest numbers at harvest time.*

Class Arachnida

The arachnids, which include spiders, ticks, mites, scorpions, and daddy longlegs, are almost all terrestrial, except for a few species that have returned to the water. Arachnids have four pairs of walking legs, and their chelicerae and pedipalps are often highly specialized. In spiders, ducts from a pair of poison glands lead through the chelicerae, which are sharp and pointed and are used for biting and paralyzing prey (Figure 27–8). Scorpions use their pedipalps to handle and to tear food. Male spiders also use the pedipalps to transfer semen to the female.

Spiders, like most arachnids, live on a completely liquid diet. All are predatory. The prey is bitten and often paralyzed by the chelicerae, and then enzymes from the midgut are poured out over the torn tissues to produce a partially digested broth. The liquefied tissues of the prey are pumped into the stomach and then to the intestine, where digestion is completed and the juices absorbed. Arachnids respire by means of tracheae or book lungs, or both. Book lungs are a series of leaflike plates within a chitin-lined chamber into which air is drawn and expelled by muscular action.

On the posterior portion of the spider's abdominal surface is a cluster of spinnerets, modified appendages from which a fluid protein exudes that polymerizes into silk as it is exposed to air. Silk is used not only for the variety of

(b)

(c)

(d)

(e)

prey-snaring webs made by the different species but for a number of other purposes as well—such as a drop line, on which the spider can make a defensive dive, draglines for marking a course, gossamer threads for ballooning, hinges for trap doors, an egg case, lining for a burrow, the shroud of a victim, or a wrapping for an edible offering presented to the female of certain species by the courting male. Most spiders can spin several kinds and thicknesses of silk. Webs are species-specific, and web-building is a genetically programmed behavior.

The Aquatic Mandibulates: Class Crustacea

The 25,000 species of crustaceans include crabs, crayfish, lobsters, shrimp, prawns, barnacles, *Daphnia* (water fleas), and a number of smaller forms. Some crustaceans, such as the familiar pillbugs, or sowbugs, are adapted to life in moist land environments. Crustaceans differ from the terrestrial mandibulates, such as insects, in that they have legs or leglike appendages on the abdomen as well as the thorax and have two pairs of antennae as compared to the insects' one pair.

Among the crustaceans, the sexes are usually separate, but there are exceptions. Barnacles, which are sessile as adults, are simultaneous hermaphrodites. Some species of shrimp are sequential hermaphrodites; when the animal is small, it is male, but when it reaches a size that is effective for carrying eggs, it becomes female. Most marine crustaceans have larval stages that swim about before developing into the adult animals.

The Lobster

Figure 27–10 shows the structure of a lobster, a representative crustacean. A crayfish, which is a freshwater form, differs anatomically from the lobster only in minor respects.

27–10 *A representative crustacean, the American lobster,* Homarus americanus. *The lobster has 19 pairs of appendages, including antennae, mouthparts, and legs specialized for feeding, walking, and swimming.*

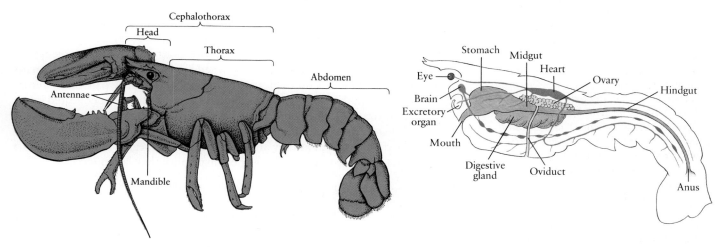

The lobster has 19 segments, the first 13 of which are united on the dorsal side in a combined head and thorax, or cephalothorax. A heavy shield, or carapace, arises from the head and covers the thorax. Carapaces are common among crustaceans. The abdomen consists of six distinct segments. (To lobster eaters, the abdomen is the "tail," but to biologists, the designation "tail" is usually reserved for areas of the body posterior to the anus.)

The various appendages have special functions. The antennae, of which there are two pairs, are sensory. The mandibles, or jaws, are used for crushing food; like all arthropod jaws, they move laterally, opening and closing from side to side like a pair of ice tongs rather than up and down like the jaws of vertebrates. Crustacean mandibles are thought to have arisen from a pair of limbs, the bases of which took

(a)

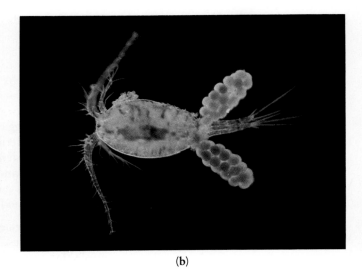

(b)

(c)

27-11 *Some crustaceans.* (a) *A cleaner shrimp of the Indian and Pacific Oceans, photographed near the Philippine Islands.* (b) *A female copepod, an inhabitant of the plankton. The two large structures at the posterior are egg sacs in which zygotes are developing; they will be released as free-swimming larvae. Copepods are the most numerous animals not only in the plankton but also in the world; the individuals of a single genus (Calanus) are thought to outnumber all other animals put together.* (c) *Pillbugs (also known as sowbugs or wood lice) are terrestrial crustaceans found in damp places.*

Other crustaceans include the barnacles (Figure 25–3, page 519) and the crabs (Figure 27–3b, page 567). Crabs resemble lobsters, but the relative proportions of the body have changed; the cephalothorax is broader and larger than the abdomen, which is tucked under the rest of the body.

on a chewing function. The thorax of the lobster bears five pairs of walking legs, the first pair of which are modified as claws. The claws are unequal in size in the full-grown lobster; the larger claw is used for defense and for crushing food, and the smaller claw, which has the sharper teeth, seizes and tears the prey. The next two pairs of walking legs have small pincers, which can also seize prey. The last pair of walking legs also serve to clean the appendages of the abdomen. These flattened appendages, or swimmerets, are used, like flippers, for swimming. The claws and abdomen are filled almost entirely with the large striated muscles many of us enjoy eating. These muscles are extremely powerful. A lobster can snap its abdomen ventrally with enough force to shoot backward through the water, and a large lobster can shatter the shell of a clam or oyster with its crushing claw. Any one of its appendages can be regenerated over a series of molts if lost, and a lobster will often drop off (autotomize) a claw or leg that is held by a predator, in order to escape. Severely damaged legs are also autotomized, reducing blood loss.

The anterior and posterior regions of the digestive tract, the foregut and hindgut, are lined with cuticle. As a consequence, most of the food is absorbed through the midgut and through the cells of the large digestive gland, an appendage of the midgut. Respiration is accomplished by the flow of water over 20 pairs of feathery gills attached on or near the bases of the legs. A pair of excretory organs is located in the head; wastes extracted from the blood are collected into a bladder and excreted from a pore at the base of each of the second antennae.

When lobsters mate, the sperm are deposited in or near the female gonopore; the eggs are then fertilized as they are laid. The fertilized eggs cling by means of a sticky secretion to the swimmerets of the female until they hatch.

Terrestrial Crustaceans

Unlike other arthropod groups, almost all crustaceans are aquatic; some crabs, however, are amphibious or terrestrial. Amphibious crabs continue to respire with gills, carrying water in their thoracic cavities with which to keep the gills wet. The true land crabs have lost some of the gill structures but have an area of highly vascularized epithelial tissue through which gases are exchanged. The land snails, you will recall, solved the respiratory problems involved in the transition from water to land in an analogous way.

The Terrestrial Mandibulates: Myriapods

Terrestrial mandibulates are identified by their single pair of antennae and by their mandibles, which, as we noted earlier, differ from those of the crustaceans. They respire through tracheae, and excretion is by means of Malpighian tubules. In addition to the insects, there are four smaller classes of relatively unspecialized terrestrial mandibulates. There has been little tagmosis in these arthropods, and

27–12 *A centipede of the genus* Scolopendra, *photographed in southern Africa. In centipedes, the appendages of the first segment are modified as poison claws. Prey is killed with the claws and then chewed with the mandibles.*

their bodies consist of a head region followed by an elongated trunk with many distinct segments, all more or less alike. With a few exceptions, all segments have paired appendages, and the members of these four classes are known collectively as myriapods ("many-footed").

The most familiar myriapods are the centipedes (class Chilopoda) and the millipedes (class Diplopoda). The approximately 3,000 species of centipedes prefer damp places—under logs or rocks or in basements. They are all carnivorous, feeding on cockroaches and other insects, as well as on soft-bodied annelids. A centipede (Figure 27–12) has one pair of appendages on each body segment, unlike a millipede (see Figure 27–2, page 566), which appears to have two pairs of appendages per segment. Each body ring of the millipede, however, actually represents two segments fused into a double unit. About 7,500 species of millipedes have been described, and they too prefer damp environments. Unlike centipedes, they are herbivorous, usually feeding on bits of decaying vegetation.

Less familiar are the approximately 300 species of class Pauropoda and the 130 species of class Symphyla. Unlike other arthropods, they are soft-bodied, living in moist soil, leaf litter, and other decaying matter. Although both pauropods and symphylans are abundant, they are quite small and, with the exception of one species of symphylan that is a common pest in greenhouses, they usually go about their business unnoticed by human eyes.

The Terrestrial Mandibulates: Class Insecta

The insects constitute the largest class, by far, of the arthropods. In fact, more than 70 percent of all animal species on earth are insects.

Insects are the only invertebrates capable of flight. When they began to fly—more than 240 million years ago—they were able to move into and exploit a life zone almost totally unoccupied by any other form of animal life. In terms of both numbers of species and numbers of individuals, they are the dominant terrestrial organisms of this planet.

There are about 30 orders of insects, of which the four largest are Diptera, Lepidoptera, Hymenoptera, and Coleoptera. The Diptera ("two-winged") include the familiar flies, gnats, and mosquitoes. The Lepidoptera ("scale-winged") are the moths and the butterflies. Hymenoptera ("membrane-winged") include ants, wasps, and bees, many species of which live in complex societies (to be described in Chapter 51). The Coleoptera ("shield-winged") are the beetles, most of which have a pair of hard protective forewings, which pivot forward out of the way during flight (see Figure 27–1, page 565), and a pair of membranous hindwings used for flying. Of the approximately 1 million classified species of insects, at least 300,000 are beetles.

(a)

(b)

(c)

27–13 *The late J. B. S. Haldane, who was noted for his crusty disposition, as well as for his scientific achievements, was once asked what his study of biology had revealed to him about the mind of God. "Madame," he replied, "only that He had an inordinate fondness for beetles."*

(a) Everybody's favorite beetle, a ladybird ("ladybug"), feeding on aphids. This is the seven-spot ladybird, Coccinella septempunctata. *(b) A stag beetle, so called because of the antlerlike mandibles characteristic of the pugnacious males. This species,* Durcus curvidens, *is common in the rain forests of Southeast Asia. (c) A scarab beetle,* Stephanorhina guttata, *photographed in the Congo.*

Insect Characteristics

Figure 27–14 shows a grasshopper. Here you can see many of the characteristic features of insects: three body regions—the head, the thorax, and the abdomen; three pairs of legs; one pair of antennae; and a set of complex mouthparts. In the less specialized insects, such as the grasshopper, the mouthparts are used for handling and masticating food (Figure 27–15), but in the more specialized groups, the mouthparts are often modified into sucking, piercing, slicing, or sponging organs. Some are exquisitely adapted to draw in nectar from the deep, tubular nectaries of specialized flowers.

Most adult insects have two pairs of wings made up of light, strong sheets of chitin; the veins in the wings are chitinous tubules that serve primarily for strengthening. In some primitive orders, wings never evolved, and in other orders, such as fleas and lice, wings were lost secondarily, returning the insect to the condition of its wingless ancestors. Other species may have short, nonfunctional wings in one or both sexes.

Digestive, Excretory, and Respiratory Systems

The foregut and hindgut of the insect digestive tract are lined with cuticle. Salivary gland fluids are carried with food into the crop, where digestion begins. The stomach, or midgut, which lies mainly in the abdomen, is the chief organ of absorption. Insects have digestive enzymes as specialized as their mouthparts; the structure of the enzymes depends on whether the insect dines on blood, seeds, other insects, eggs, flour, cereal, glue, wood, paper, or your woolen clothes.

27–14 *In the grasshopper, an insect, the head consists of six fused segments that have appendages specialized for tasting and biting. Each of the three segments of the thorax carries a pair of legs (three pairs in all), and two of them carry wings (in the grasshopper, the forewings are hardened as protective covers). The spiracles in the abdomen open into a network of chitin-lined tubules through which air circulates to various tissues of the body. This sort of tubular breathing system is found in the terrestrial mandibulates and some arachnids. Excretion takes place by Malpighian tubules that empty into the hindgut.*

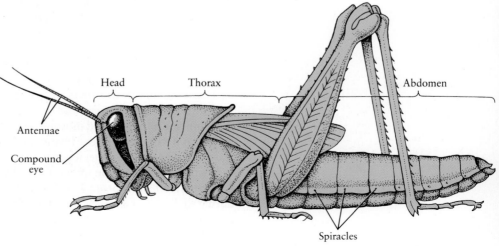

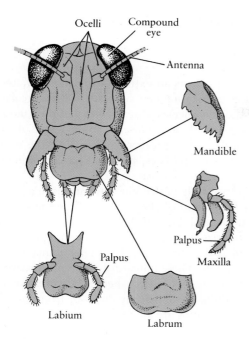

27–15 *Mouthparts of a grasshopper. The mandibles are crushing jaws. The labium and the labrum are the lower and upper lip. The maxillae move food into the mouth, and the palpi assist in tasting.*

Excretion is carried out through Malpighian tubules. In the grasshopper and many other insects, the nitrogenous wastes are eliminated in the form of nearly dry crystals of uric acid, an adaptation that promotes water conservation.

The respiratory system consists of a network of cuticle-lined tubules through which air circulates to the various tissues of the body, supplying each cell directly. Muscular movements of the animal's body improve the internal circulation of air. The amount of incoming air and also the degree of water loss is regulated by the opening and closing of the spiracles.

Insect Life Histories

The life histories of insects are fundamentally different from those of marine invertebrates, particularly forms that are sessile or slow-moving as adults. Although many marine invertebrate larvae feed, their most important biological function seems to be the invasion and selection of new habitats at some distance from the parental habitat. Such larvae may migrate great distances, resulting in the dispersal of the species over a wide area. The immature forms of most insects, by contrast, have limited mobility and usually pass through the various stages of their development very close to the location where the adult female originally laid the eggs—sometimes in the exact spot. Insect young are voracious feeders, acquiring the resources needed for their own growth and development and, in some species, storing up reserves for an adult stage in which they will not feed. The capacity to

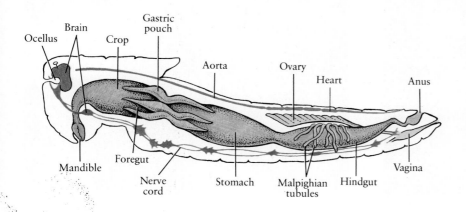

27–16 *Some immature insect forms.*
(a) Scarab beetle (white grub) larva in soil.
(b) The larva (caterpillar) of the cecropia
moth, Hyalophora cecropia. *(c) Mosquito*
larvae and a pupa (on right). Mosquito
larvae are aquatic, hanging onto the
undersurface of the water with respiratory
tubes. (d) Tent caterpillars on their protec-
tive web.

(a)

(b)

fly has, of course, given most adult insects an extraordinary mobility, and they have little difficulty finding mates or locating new habitats in which to lay their eggs—thereby ensuring dispersal of the species. One corollary of their mobility is that the sexes are separate in all insects. In many insects, some degree of parental care has evolved, with one or both parents protecting or feeding the young, or both. Among the termites, ants, wasps, and bees, such care has led ultimately to complex societies with a substantial division of labor.

Young growing insects change not only in size but often in form, a phenomenon known as **metamorphosis.** The extent of change varies. In some species, the young, although sexually immature, look like small adults; they grow larger by a series of molts until they reach full size. In others, like the grasshopper, the newly hatched young are wingless but may gain wingpads in later immature stages; otherwise they are similar to the adult. These immature, nonreproductive forms are known as **nymphs.** In almost 90 percent of insect species, however, a complete metamorphosis occurs, and the adults are drastically different from their immature forms. These immature feeding forms are all correctly referred to as larvae, although they are also commonly known as caterpillars, grubs, or maggots, depending on the species. Following the larval period, the insect undergoing complete metamorphosis enters a pupal stage, in which extensive remodeling of the organism occurs. The adult (sexually mature) insect emerges from the pupa. Both eggs and pupae (which are nonfeeding) can endure lengthy cold or dry seasons, a critical adaptation for terrestrial organisms.

An insect that undergoes complete metamorphosis exists in four different forms in the course of its life history. The first form is the egg and the embryo. The second form is the larva, the animal that hatches from the egg; larvae eat and grow. In many larvae, such as those of flies, growth takes place not by an increase in the number of cells, as in most animals, but by an increase in the size of the cells, in somewhat the same way that growth takes place in certain plant tissues. During the course of its growth, the larva molts a characteristic number of times—twice in the fruit fly, for example. The stages between molts are known as **instars.** Then, when the larva is full-grown, it molts to form the pupa. In many, but not all species, the pupa is enclosed within a cocoon or some other protective covering. During the outwardly lifeless pupal stage, many of the larval cells break down, and entirely new groups of cells, set aside in the embryo, begin to proliferate, using the degenerating larval tissue as a culture medium. These groups of cells develop into the complicated structures of the adult.

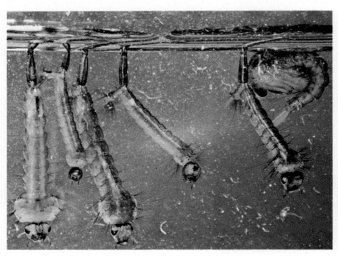

(c)

(d)

27-17 *Life history of* Heliconius ismenius, *a butterfly of Central America. The life history begins* (a) *with the mating of male and female butterflies. Fertilization is internal, and the fertilized eggs* (b) *undergo the early stages of their development within a rubbery egg shell. The newly hatched caterpillar* (c), *the larval form, eats, grows, and continues its development;* (d) *and* (e) *show the second and third instars—that is, the caterpillar after its first and second molts.* (f) *When ready to pupate, the caterpillar attaches itself upside down to a leaf with a patch of silk. Inside the pupa* (g), *metamorphosis into the butterfly occurs through a recycling of the tissues of the caterpillar. The emergent adult butterfly* (h) *is the sexually reproductive form, ready to start the cycle once more.*

(a)

(b)

(c)

(d)

(e)

(f)

(g)

(h)

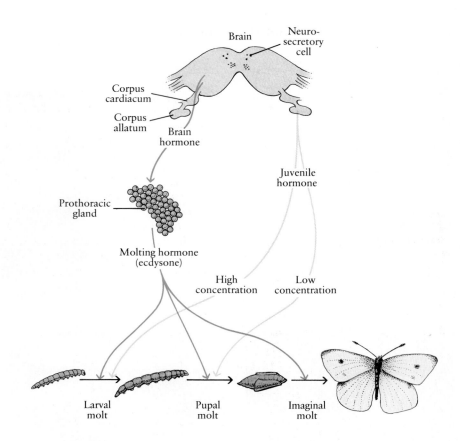

27–18 *Molting is under hormonal control. In insects, a hormone (brain hormone) produced by neurosecretory cells in the brain and released from the corpora cardiaca stimulates the prothoracic gland, which, in turn, produces molting hormone (ecdysone). Although all molts require ecdysone, whether or not metamorphosis occurs depends on a third hormone, juvenile hormone, produced by the corpora allata. Continued presence of juvenile hormone at high concentrations ensures that larval molts occur during the first portion of the life history. In later larval life, production of juvenile hormone declines, permitting adult structures to develop.*

Aristotle and his contemporaries thought of the adult insect as the imago, *the perfect form or ideal image that the immature form was "seeking to express." Although this concept has long since been abandoned by biologists, the molt giving rise to the adult is still known as the imaginal molt.*

Molting and metamorphosis are under the control of **hormones,** which are organic molecules secreted by one tissue of an organism that regulate the functions of another tissue or organ of the same organism. Like most hormone-controlled processes, molting and metamorphosis are the end result of an interplay of several substances: brain hormone, molting hormone (ecdysone), and juvenile hormone (Figure 27–18). At intervals during larval growth, brain hormone, produced by neurosecretory cells in the brain, is released into the blood. It stimulates the release, in turn, of molting hormone from a gland in the thorax. The molting hormone stimulates not only molting but also the formation of a pupa and the development of adult structures. The latter are held in check, however, by a third hormone, the juvenile hormone. Only when the production of juvenile hormone declines, in later larval life, can metamorphosis to the adult form take place.

REASONS FOR ARTHROPOD SUCCESS

Among all the invertebrates, why are the arthropods, in general, and the insects, in particular, so spectacularly successful? One important reason is undoubtedly the nature of the exoskeleton, which waterproofs, provides protection, and makes possible the evolution of the many finely articulated appendages characteristic of this phylum.

Another reason, which applies especially to insects, is their small size and the high specificity of diet and other requirements of each species. As a consequence, many different species can live in a single small environment—in a few cubic centimeters of soil, on a small plant, or on or within the egg or body of a single animal—without competing with one another. The varied and highly specialized mouthparts are a reflection of this specificity of diet.

The diversity of insects and the specificity of their requirements may be, in part, an evolutionary response to the great diversity of microenvironments provided by the vascular plants. Not only are there 235,000 species of angiosperms alone, but also the structure of each individual plant is so complex that it provides a variety of resources that can be exploited by insects with different requirements and adaptations. This is a much richer environment than open waters, the seashore, or the soil. And, virtually every cubic centimeter of it is available to insects because of the extraordinary mobility that results from their small size and their capacity for flight. The resources provided by vascular plants were, in effect, an evolutionary laboratory in which great numbers of variations could be tested, leading to the diversity of adaptations we see today. This process, of course, continues.

Another factor in the success of insects is the complete metamorphosis that occurs in the vast majority of insect species. In these insects, the adaptations for feeding and growth that are found in the larvae are separated from the adaptations for dispersal and reproduction found in the adults. The adaptations for each function can be more finely honed because compromises between conflicting needs are minimal. Another consequence of complete metamorphosis is that the dietary and other requirements of the larvae and the adults of the same species are so different that they do not compete with each other. In effect, they occupy different environments.

A final reason for success is undoubtedly the arthropod nervous system, with its fine control over the various appendages and the many highly sensitive sensory organs found in great diversity throughout the phylum. Sensory perception among arthropods, especially insects, is so important that we shall devote the remainder of this chapter to that topic and to several examples of arthropod behavior.

Arthropod Senses and Behavior

Vision: The Compound Eye

The most conspicuous sensory organ of the arthropods is the compound eye, which is an evolutionary development characteristic of this one phylum. The basic structural unit of this eye is the **ommatidium** (Figure 27–20), repeated over and over. A dragonfly, for example, has some 30,000 ommatidia. Each ommatidium is covered by a cornea, usually with a round or hexagonal surface; these are visible under low-power magnification as individual facets of the eye. Underlying the cornea is a group of eight retinular cells surrounded by pigment cells. The light-sensitive portion of the ommatidium is the **rhabdom,** which is the central core of the ommatidium. Nerve fibers carry the stimulus from each ommatidium to the brain. The pigment cells prevent light from traveling from one ommatidium to another. An ommatidium is much larger than a vertebrate photoreceptor, and so there are far fewer in an equivalent space. Hence, the image has less resolution, like a newspaper picture under high magnification.

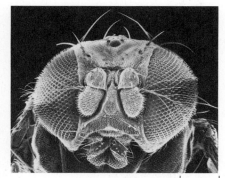

100 μm

27–19 *The head of* Drosophila, *as shown by the scanning electron microscope. Note the large compound eyes on either side of the head. Although insect eyes cannot change focus, they can define objects only a millimeter from the lens, a useful adaptation for an insect.*

27–20 *Structure of the compound eye. The eye is composed of a large number of structural and functional units called ommatidia. Each ommatidium has its own cornea, which forms one of the facets of the compound eye, and its own light-focusing lens. The light-sensitive part is the rhabdom, which is surrounded by the retinular cells, which transmit the stimulus. The ommatidium is surrounded by pigment cells that prevent light from traveling from one ommatidium to another.*

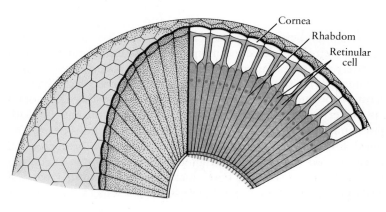

Cornea
Rhabdom
Retinular cell

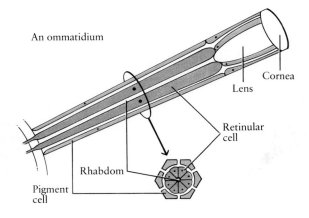

An ommatidium

Cornea
Lens
Retinular cell
Rhabdom
Pigment cell

Firefly Light: A Warning, an Advertisement, a Snare

Fireflies are beetles of the family Lampyridae. As is characteristic of most other insects, they undergo a complete metamorphosis. The eggs are laid in moist soil, hatching into larvae in about three weeks. They spend up to two years as larvae; the larvae are carnivorous, feeding on annelids, insects, and mollusks, which they subdue with poison from their mandibles.

Firefly larvae are luminescent, producing a glow that waxes and wanes over a period of seconds. (Even the embryos glow a little.) The survival value of larval luminescence is not clear. The best hypothesis is that the larvae taste bad and that the glow is a warning signal to predators. Compounds related to the skin poisons of toads have been found in some species. Experiments reported by Albert Carlson of the State University of New York at Stony Brook have shown that mice that bite into larvae reject them violently, flinging the carcass away and scrubbing their mouths with their forefeet. It has also been shown that mice can be readily trained to avoid glowing larvae.

After molting through a number of instars, the larvae pupate over a period of two to three weeks. During this time the adult luminescent organ—a lantern—develops on the ventral surface of the terminal abdominal segments. This luminescent organ is under neural control and is used to solve the all-important (for the firefly) problem of finding a mate and reproducing. The adult lives only one to four weeks, so time is of the essence. Each species of firefly has its own signal pattern composed of light flashes lasting only a fraction of a second. Males emit their signals in flight; the females are stationary, near or on the ground. Females flash in response to the signals from males of their own species, aiming their lantern toward the flashing male. When the male sees the

(a) *Two fireflies of the species* Photuris hebes *mating as they hang beneath a goldenrod leaf.* (b) *A female of the species* Photuris versicolor *devours a male of the species* Photinus tanytoxus.

answering flash, he alights and walks toward the female, still flashing. In some species, this courtship flashing is different from the previous flashing pattern of seeking. Copulation lasts minutes to hours, after which the female lays her eggs. The male immediately resumes his signaling and searching activities, looking for another mate. On any given summer evening, there are many more males than females seeking mates—perhaps a ratio of 50 to 1—and competition among males is very intense.

Unlike the larvae, most adult flies are not carnivorous, and some species do not feed at all. Females of the genus *Photuris,* however, are exceptions. They are predaceous carnivores, and their prey is the male firefly of other species. *Photuris versicolor* is one of the most extensively studied species. Within three days after mating, the female undergoes a behavioral

Although the compound eye is deficient in acuity, offering less detail than the vertebrate eye, it is better for detecting motion because each ommatidium is stimulated separately and so has a separate visual field. Also, each ommatidium responds to stimuli more rapidly than does a vertebrate photoreceptor. Ability to detect motion can be measured accurately in the laboratory by testing a phenomenon known as flicker fusion. In this test, a light is flicked on and off with increasing rapidity until the observer sees the flicker as a continuous beam. The beam is perceived as continuous because stimulation of any photoreceptor cell persists for a brief period even after the stimulus disappears. So, in effect, the flicker-fusion test is a measurement of how quickly the cell recovers from one stimulus and becomes sensitive to another. It is possible to test flicker-fusion rates

(a)　　　　　　　　　　　　　　　　　　　　　(b)

change. Instead of responding only to the flashing pattern of a male of her own species, she becomes an active hunter, flying to an area where the prey species is active. When a male firefly flashes nearby, she responds—not with her own signal, but rather with a signal characteristic of the female of the male's own species. A *Photuris versicolor* female can respond appropriately to the male signals of at least five different firefly species. The male's dilemma is acute. If he hesitates, he loses his chance to mate, given the 50 to 1 competition. If he rushes in, he risks being devoured by the aggressive female.

Males of the species *Photinus macdermotti* have responded to this dilemma by finding a means of stalling for time while holding off the competition. These tiny males are preyed upon by the females of at least three different *Photuris* species, so their incen-

tives are particularly strong. Their weapons are also false flashes. As they approach a female, they emit flashes that mimic the female predator of the other species, thereby deterring the rival males of their own species. Also, by injecting their flashes into the flash patterns of rival males, they can disrupt their signals and get the females to respond to them rather than to their rivals.

James Lloyd, of the University of Florida, who has been a leader in these studies for almost 20 years, reminds us that on the same summer evenings that the fireflies are engaged in these rituals of courtship, mating, deceit, treachery, and death, thousands of other species of insects are also occupied in similar activities. Are fireflies more complex than other insects, he asks, or are comparable but unilluminated dramas taking place throughout the insect world?

in animals by training experiments in which the animal learns to associate a flickering light with a reward (usually food) and a steady beam with no reward, or vice versa. Such tests have demonstrated that the compound eye greatly exceeds the camera eye of vertebrates in this respect. A bee would see in clear outline a moving figure that we would see as blurred, and if the bee went to the movies, the film seen by us as a continuous picture would jerk along from frame to frame for the bee. The ability to perceive motion is extremely important for an insect since it must be able to make out objects when it is flying at high speed (which, as far as the visual apparatus is concerned, presents the same problems as following a moving object).

27–21 *The hairs on the legs of this red and green tiger beetle* (Cicindela scutellaris) *are touch receptors, or sensilla. At the base of the hairs are sensory cells. When a hair is touched or bent, nerve impulses are initiated.*

In addition to, or instead of, compound eyes, many of the arthropods possess simple eyes, or ocelli, which seem generally to serve only for light detection. Most insects have two or three ocelli, and spiders, which do not have compound eyes, may have as many as eight ocelli, depending on the species.

Touch Receptors

The body surfaces of terrestrial arthropods are often covered with sensory-receptor units known as **sensilla** (singular, sensillum), or "little sense organs." Most of the sensilla take the form of fine spines, or setae, composed of hollow shafts of cuticle. At the bases of these shafts are sensory cells. In their simplest form, the sensilla are touch receptors. In these, when the spine is touched or bent, the sensory cell responds and initiates nerve impulses. Such receptors are found, in particular, on the antennae and the legs. In addition to being stimulated by direct contact, they can also be stimulated by vibrations and air currents. A spider monitors what is going on in its web by sensing vibrations transmitted through the threads when the web is touched. Soldier termites of certain species strike the ground or the walls of their nest with their heads when threatened or disturbed; the vibrations they produce warn their colony mates. A fly perceives the air currents from the movement of a hand or fly swatter and so escapes; a fly in a glass jar is much less likely to be disturbed by such movements.

Proprioceptors

Proprioceptors are sensory receptors that provide information about the position of various parts of the body and the stresses and strains on them. A type common in the arthropods is the campaniform sensillum (Figure 27–22a). Campaniform sensilla are located in thin, stretchable areas of the cuticle. When the cells are twisted or stretched, a nerve fiber signals the central nervous system.

Touch receptors can also serve as proprioceptors. The praying mantis, for example, is capable of making a lightning-swift strike at a moving object. When it sights a potential victim, the insect moves its entire head to bring it into binocular range, since the eyes themselves do not move (Figure 27–22b). Movement of the head results in the stimulation of proprioceptive hairs on the head and thorax of the insect. On the basis of the impulses received from these hairs, the position of the prey and the movement of its own legs are automatically coordinated by the brain and nervous system of the mantis. If these hairs are removed, the mantis can strike a moving object only if the object is directly in front of it.

27–22 (a) *Campaniform sensillum, a type of proprioceptor common in arthropods.*

(b) *Since its eyes do not move, the praying mantis must move its entire head to bring its victim into binocular range. The movement of its head sends impulses through proprioceptive hairs on its head and thorax to its legs. The movement of its legs is thus automatically coordinated with the position of its prey, giving the mantis the ability to strike at a victim swiftly and at the proper range.*

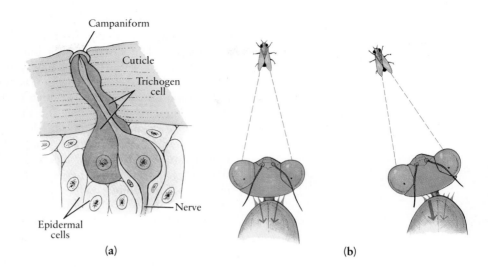

(a)

(b)

Communication by Sound

Arthropods, particularly insects, have developed complex forms of communication. A number of species, such as the locusts, grasshoppers, and crickets, call to one another with sounds made by rubbing their legs or wings together or against their bodies (Figure 27–23). Five distinct types of calls are known: (1) calling by males and (2) calling by females, both of which are long-range sounds; (3) courtship sounds by males and (4) aggressive sounds by males, both of which are short-range; and (5) alarm sounds, which may be given by either males or females. Recognition of and response to the sound may be based on its pattern, rhythm, or frequency (pitch). Some, but not all, insects are unable to distinguish frequencies (the differences between high and low notes) and so are essentially "tone deaf." The effectiveness of calling songs is often increased by group singing, such as the famous chorus of male seventeen-year cicadas, which can attract females from distances far greater than an individual "voice" would reach. Insects produce songs and respond to appropriate songs without ever having heard a song before.

27–23 *The katydid produces its characteristic loud, shrill sounds by rubbing the scraper at the base of its right wing against the file at the base of its left wing.*

Sound waves are set in motion by the vibration of some object, for example, a beating wing, a flexible area of an arthropod's skeleton, a human larynx, or a violin string. The vibrations not only cause the molecules of the surrounding medium (air or water) to travel away from the object in small bursts but also produce a series of changes in air (or water) pressure. The diverse sound receptors of arthropods may be sensitive either to the impact of the moving molecules or to the pressure changes, enabling the animal to detect the calls produced by other members of its species or the sounds of approaching predators.

The simplest of the arthropod sound receptors are sensilla with tactile hairs that vibrate when they are struck by moving molecules in the air; the vibrations of these hairs, in turn, trigger nerve impulses in the animal. The antennae of the male mosquito, for example, contain thousands of such hairs, which are sensitive to the sounds produced by the vibrating wings of the female mosquito in flight. The response of the male to these sounds serves to bring the sexes together. When the male mosquito first emerges from its pupal shell, it is sexually immature and also deaf, with its antennal hairs lying flat along the shafts. When the male matures sexually, some 24 hours later, the hairs almost simultaneously become erect and free to vibrate when a female approaches.

Other insects have developed special structures that respond to the pressure changes in sound waves; these structures, which may be located on the legs (see Figure 27–3a), the thorax, or the abdomen, are known as tympanic organs. In these organs, a fine membrane, the tympanum (or eardrum), is stretched across one or more closed air sacs (Figure 27–24). The tympanic membrane vibrates in response to the pressure differences in sound waves of certain frequencies, and this vibration is transmitted to underlying receptor cells. In many insects, the tympanic air sacs abut portions of the tracheal system and, in some species, are adjacent to the air sacs of the tympanic organ on the opposite side of the body. As a result, the pressure changes of the sound waves are received by the tympanic membrane not only on its outer surface but also on its inner surface, after transmission from the outside through the tracheae or the opposite tympanic organ. The differences in the pressures on the two sides of the membrane enable the insect to detect the direction from which the sound is coming. Many moths, in particular, have a remarkable ability to identify precisely the direction of approach of predatory bats, which locate their prey by emitting sounds that reflect off the prey, back to the bat.

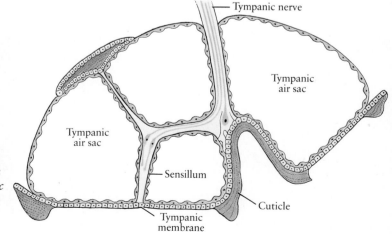

27–24 *The most elaborate of the insect sound receptors is the tympanic organ. The tympanic air sacs are covered by a membranous drum, and the sensory cells are so arranged in the organ that they are stimulated by movements of the drum or the air-sac walls. Tympanic organs respond to pressure changes in the air or in water. This is a cross section of the tympanic organ of a noctuid moth.*

27-25 *The two large appendages extending from either side of the head of this cricket are palpi, which contain special chemoreceptors (taste organs).*

27-26 *The lavishly plumed antennae of the male cecropia moth are receptors for the alluring pheromones released by the females.*

Communication by Pheromones

The use of chemicals for communication is common among organisms, and the substances employed range from the sex attractants of the little algal cell *Chlamydomonas* to Chanel No. 5. Many insects communicate by chemicals known as **pheromones.** These chemical messengers, usually produced in special glands, are discharged into the environment, where they act on other members of the same species.

Among the best studied of the pheromones are the mating substances of moths. Female gypsy moths, by the emission of minute amounts of a pheromone commonly known as disparlure, can attract male moths that are several kilometers downwind. One female produces about one-millionth (10^{-6}) of a gram of disparlure, enough to attract more than a billion males if it were distributed with maximum efficiency. The male moth characteristically flies upwind, and the pheromone, of course, disperses downwind. When a male moth detects the pheromone of a female of the species, he will fly toward the source. Since the male can detect as little as a few hundred molecules per milliliter of the attractant, disparlure is still potent even when it has become widely diffused. If the male loses the scent, he flies about at random until he either picks it up again or abandons the search. It is not until he is quite close to the female that he can fly "up the gradient" and use the intensity of the odor as a locating device. Figure 27–26 shows the antennae of a male cecropia moth by which the pheromone emitted by the female is detected.

Programmed Behavior

Complex patterns of unlearned, genetically transmitted behavior—such as web-building in spiders—are another arthropod characteristic. This relatively rigid programming of behavior may be a necessary correlate of the shortness of the life span of the smaller arthropods. It provides an interesting contrast to the more flexible behavior patterns of the mammals, with their comparatively long life spans, long periods of learning, and more complex brains.

SUMMARY

Arthropoda is the largest animal phylum in both number of species and of individuals. Arthropods are segmented animals with jointed chitinous exoskeletons and a variety of highly specialized appendages and sensory organs. In most groups, the segments are combined, forming a head, a thorax (sometimes fused with the head as a cephalothorax), and an abdomen. The arthropods are also characterized by an open circulatory system and a nervous system consisting of a series of ganglia, a pair per segment, interconnected by a double ventral nerve cord. Tracheae (cuticle-lined breathing tubes), book gills, book lungs, and Malpighian tubules (excretory ducts leading into the hindgut) are found almost exclusively among arthropods.

There are three major groups of arthropods: the chelicerates, characterized by chelicerae (fangs or pincers) and pedipalps; the aquatic mandibulates, with two pairs of antennae and a pair of mandibles (jaws); and the terrestrial mandibulates, with one pair of antennae and a pair of mandibles that differ from those of aquatic mandibulates. The chelicerates include the horseshoe crabs (class Merostomata), the sea spiders (class Pycnogonida), and the spiders, scorpions, mites, and ticks (class Arachnida). The aquatic mandibulates (some of which actually live in moist environments on land) all belong to class Crustacea and include such familiar animals as lobsters, crabs, shrimp, and barnacles. Terrestrial mandibulates include four relatively small classes (Chilopoda, Diplopoda, Pauropoda, and Symphyla) and the largest class in the animal kingdom, Insecta, with about 1 million species. The insects are the only invertebrates capable of flight.

In the life histories of most insects, dispersal and habitat selection are carried out by the highly mobile adult forms, whereas the immature forms have limited mobility and feed voraciously in a fairly restricted area. In the course of their development, most insects pass through a complete metamorphosis that is controlled by the interaction of at least three hormones. The stages in the life history are egg, larva, pupa, and adult. In a minority of species, including the grasshopper, the hatchling looks much like a miniature adult; in these, the immature form is known as a nymph.

Among the factors contributing to the extraordinary success of the arthropods are their exoskeleton, their generally small size, and their great specialization in both diet and habitat. Additional factors in the success of the insects are the capacity for flight and complete metamorphosis, which allows for greater refinement of adaptations for feeding and for reproduction and dispersal, as well as reducing competition between adults and immature forms.

The finely tuned arthropod nervous system, with its diverse sensory organs, has also been important in arthropod success. Among the most important sensory receptors are the compound eye, touch receptors, proprioceptors, and tympanic organs. Arthropods communicate with members of the same species by sound and also by pheromones—chemicals released by one individual that affect the behavior or physiology of another. Arthropod behavior is notable not only for its complexity and diversity but also for the extent to which it is programmed in the nervous system, that is, not learned.

QUESTIONS

1. Distinguish among the following terms: chelicerae/mandibles; nephridia/Malpighian tubules; nymph/larva/pupa; hormone/pheromone; brain hormone/molting hormone/juvenile hormone; ommatidium/sensillum.

2. Describe the following arthropod structures and explain their functions: book gills, tracheae, spiracles, spinnerets, compound eye, tympanic organ.

3. How would you distinguish an insect from an arachnid? From a crustacean?

4. Note that both the mollusks and the arthropods have a greatly reduced coelom. What two features—one obvious, the other not—are also shared by most members of these two groups? What is the correlation between these structural similarities?

5. Describe gas exchange in a clam, a terrestrial snail, an earthworm, a lobster, an arachnid, and an insect. How does gas exchange in these animals differ from that in a cnidarian? How is it similar?

6. Compare the nervous systems of a clam, an octopus, an annelid, and an arthropod. How do they differ from those of a *Hydra* and a planarian?

7. The arthropods, which include some of the most active animals, have open circulatory systems, often considered inefficient. Annelids, which may share a common ancestor with the arthropods, have closed circulatory systems. Presumably, half of the annelid system must have been lost. How could the acquisition of a relatively rigid exoskeleton make superfluous the vessels returning blood to the heart?

The Animal Kingdom IV: The Deuterostomes

On the basis of characteristic features of embryonic development, the coelomate animals fall into two broad groups. In the protostomes—the subject of the two preceding chapters—the cleavage pattern in the early cell divisions of the embryo is spiral, the mouth develops at or near the blastopore, and the coelom results from a splitting of the mesoderm (schizocoelous formation). By contrast, in the deuterostomes, the early cleavage pattern is radial, the anus develops at or near the blastopore, the mouth forms secondarily elsewhere, and the coelom is formed by outpocketings of the embryonic gut (enterocoelous formation).

These features of embryonic development are shared by four strikingly different phyla of animals: Echinodermata (starfish, sea urchins, and related forms), Chaetognatha (arrow worms), Hemichordata (acorn worms and their relatives), and Chordata. The largest subphylum of Chordata includes all the vertebrate animals—fishes, amphibians, reptiles, birds, and mammals, among them *Homo sapiens*.

PHYLUM ECHINODERMATA: THE "SPINY-SKINNED" ANIMALS

The echinoderms include the sea lilies and feather stars (class Crinoidea), starfish and brittle stars (class Stelleroidea), sea urchins and sand dollars (class Echinoidea), and sea cucumbers (class Holothuroidea). The majority of the echinoderm species are known only through fossils, but the 6,000 living species are abundant throughout the oceans of the world. Particularly in deep waters, echinoderms often make up the bulk of living tissue.

Most adult echinoderms are radially symmetrical, like most cnidarians, but the symmetry is imperfect with some traces of bilaterality. The larvae, however, are bilaterally symmetrical. During their development, different parts of the larval body grow at different rates, preparing the way for the change in body symmetry. At the time of metamorphosis, the larvae, which have drifted in the ocean currents for weeks, feeding all the while, temporarily attach to a solid surface; they then rapidly transform into adults with a fivefold radial symmetry. The echinoderms are believed to have evolved from an ancestral, bilateral, motile form that settled down to a sessile life and then became radially symmetrical. The feather stars (Figure 28–1) represent this second hypothetical stage. In the third evolutionary stage, some of the animals, as represented by the starfish and sea urchins, became motile again. Following this line of reasoning, one might expect an eventual return to bilateral symmetry in this group, and, in fact, this is seen to some extent in the soft, elongated bodies of sea cucumbers.

The characteristic features of this phylum are clearly visible in the most familiar of the echinoderms, the starfish.

28–1 *According to available evidence, echinoderms evolved from bilaterally symmetrical, motile animals into radially symmetrical, sessile forms, such as feather stars and sea lilies. This is the feather star* Nemaster rubiginosa, *photographed off South Caicos Island in the British West Indies.*

28–2 *Echinoderms are characterized by a calcium-containing skeleton of rods, plates, or spicules embedded just below the skin, a water vascular system of canals, and tube feet. The five-part body plan, clearly visible in the candy-cane sea star shown here, is characteristic of most living echinoderms.*

28–3 *The water vascular system of the starfish supports its locomotion. Five radial canals, one for each arm, connect the ring canal with many pairs of tube feet, which are hollow, thin-walled cylinders ending in suckers. At the other end of each tube foot is a rounded muscular sac, the ampulla. When the ampulla contracts, the water in it, prevented by a valve from flowing back into the radial canal, is forced under pressure into the tube foot. This stiffens the tube, making it rigid enough to walk on, and extends the foot until it attaches to the substrate by its sucker. The muscles of the foot then contract, forcing the water back into the sac and creating the suction that holds the foot to the surface.*

Class Stelleroidea: Starfish and Brittle Stars

The body of a starfish consists of a central disk from which radiate a number of arms. Most starfish have five arms, which was the ancestral number, but some have more. Like all echinoderms, a starfish has an interior skeleton that typically bears projecting spines, the characteristic from which the phylum derives its name. The skeleton is made up of tiny, separate calcium-containing plates held together by the skin tissues and by muscles.

The central disk of a starfish contains a mouth on the lower surface, above which is the stomach. A starfish has no head or brain, but rings of nerves around the mouth provide coordination of the arms, any one of which may lead the animal in its sluggish, creeping movements along the sea bottom. Each arm contains a pair of digestive glands and also a nerve cord, with an eyespot at the end. These eyespots are the only sensory organs, strictly speaking, of the starfish, but the epidermis contains thousands of neurosensory cells (as many as 70,000 per square millimeter) concerned with touch, photoreception, and chemoreception. The sexes are separate in most starfish, and each arm also has its own pair of sperm-producing or egg-producing organs, which open directly to the exterior through small pores. Although reproduction is usually sexual, with external fertilization, some starfish multiply asexually, regenerating whole animals after division of the central disk. A few species can reproduce asexually by regeneration of shed arms.

The coelom forms a complicated system of cavities and tubes and helps provide for circulation. Respiration is accomplished by many small, fingerlike projections, the skin gills, which are protected by spines. Waste removal is carried out by amoeboid cells that circulate in the coelomic fluid, picking up the wastes and then escaping through the thin walls of the skin gills, where they are ejected.

The **water vascular system** (Figure 28–3) is a unique feature of this phylum. This system, which is a modified coelomic cavity, creates hydrostatic support for the **tube feet,** unusual locomotor structures found only in the echinoderms. Each arm of a starfish contains two or more rows of fluid-filled tube feet, which are interconnected by radial canals and a central ring canal. At one end of each tube foot is a sucker, with which the foot attaches to the substrate, and at the other end, a rounded muscular sac, the ampulla. When the ampulla contracts, water is forced under pressure through a valve into the soft, hollow tube, extending the foot and making it rigid enough to walk on.

When the tube feet are planted on a hard surface, such as a rock or a clam shell, contractions of the muscles at the base of each tube foot collectively exert enough force to pull the starfish forward or to pull open a bivalve mollusk, a feat that will be appreciated by anyone who has ever tried to open an oyster or a clam. When attacking bivalves, which are its staple diet, the starfish everts its stomach through its mouth opening and then squeezes the stomach tissue through the opening that it has made between the bivalve shells. The stomach tissues can insinuate themselves through a slit as narrow as 0.1 millimeter to digest the soft tissue of the prey.

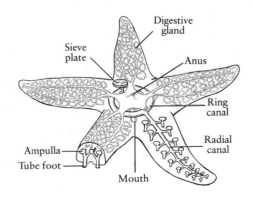

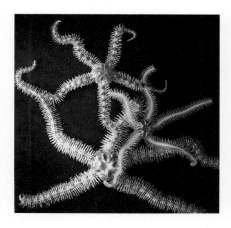

28-4 *Brittle stars,* Ophiothrix fragilis. *The echinoderm class Stelleroidea includes both the brittle stars (subclass Ophiuroidea) and the starfish (subclass Asteroidea).*

(a)

(b)

28-5 *Among the most familiar echinoderms are sea urchins and sand dollars.* **(a)** *The edible sea urchin,* Echinus esculentus, *resting on the surface of a brown alga of the genus* Laminaria. *Sea urchins are com-* mon inhabitants of tidal pools. **(b)** *A sand dollar,* Echinarachnius parma, *photographed on a beach on the Isles of Shoals, Maine. Sand dollars are often seen on sandy shores.*

Brittle stars (Figure 28–4), which are also known as serpent stars, look like starfish with particularly skinny arms. The arms bend from side to side in a snakelike motion, enabling the animal to crawl about on the ocean bottom. Their tube feet, which lack suckers, are used in gathering and handling food, rather than for locomotion. As you might suspect from their name, the brittle stars have arms that break easily. These echinoderms, like many crustaceans (see page 572), can autotomize body parts seized by a predator, thereby making good their escape.

Other Echinoderms

The members of the other classes of echinoderms show a number of variations on the body plan of the starfish. The sea lilies and feather stars (Figure 28–1), which are thought to be the most ancient of the echinoderms, spend most of their lives either attached or clinging to the substrate. As we have seen with so many sessile animals, the food-gathering structures (in this case, the arms, tube feet, and mouth) are directed upward, enabling the animal to gather food items from the surrounding water.

In the sea urchins and sand dollars (Figure 28–5) and in the tiny members of a newly discovered group of echinoderms (Figure 28–6), arms are not present. The five-part body plan of sea urchins and sand dollars is clearly visible in five paired rows of tube feet that extend through the strong skeleton. As in the starfish, the tube feet are used primarily for locomotion. These animals are formidably armed

28-6 **(a)** *Dorsal and* **(b)** *ventral views of* Xyloplax turnerae, *the "sea-daisy" from the Caribbean. This is the second described species of the Concentricycloidea, the first new class of living echinoderms to be established by taxonomists since 1821. The first "sea-daisy,"* Xyloplax medusiformis, *was found in New Zealand waters, living in crevices in waterlogged, rotting wood on the ocean bottom. Calcium-containing skeletal plates, a water vascular system, and the tube feet on the ventral surface identify these organisms as echinoderms. Adults of the Caribbean form have a sac-like stomach, but the New Zealand form lacks a stomach, prompting the conclusion that it obtains nutrients from the bacteria that are in large quantities in its preferred habitat. The Caribbean species appears to lay fertilized eggs and probably has a larval stage in its life history. In the New Zealand species, the offspring develop within the body of the parent and resemble miniature adults at the time of their release.*

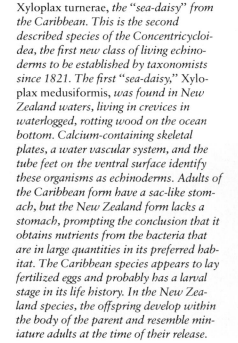

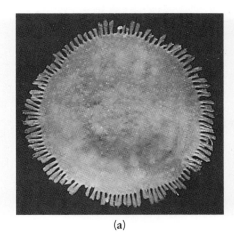

(a)

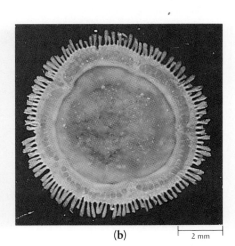

(b)

2 mm

589

28-7 *A sea cucumber,* Parastichopus californicus, *photographed at Point Loma, California. Depending on the species, sea cucumbers range in length from about 5 millimeters to more than 1 meter. Some tropical species attain a length of 3 meters.*

with movable spines and, in some species, with poison glands hidden among the spines. Typically they are also equipped with complex and powerful jaws that are used to graze on algae and other organic materials adhering to the substrate.

Sea cucumbers (Figure 28–7), with their relatively soft bodies and partial return to bilateral symmetry, bear little superficial resemblance to the other echinoderms, but they too have five rows of tube feet on the body surface. The calcium-containing skeleton is greatly reduced, usually consisting of ossicles in the skin. The body wall is often tough and leathery. Modified tube feet around the mouth look like tentacles and are used in gathering food. Some sea cucumbers feed on plankton in the surrounding water, while others feed primarily on organic matter in the bottom deposits. Some simply ingest the bottom sediment, extracting whatever nutrients it may contain as it passes through the digestive tract.

28-8 **(a)** *The predatory arrow worms are an important component of marine plankton, feeding on copepods and, occasionally, small fish. They usually feed near the surface at night, descending to deeper waters during the day.*

(b) *An arrow worm is marvelously equipped for its predaceous activities. The head bears two large eyes on the dorsal surface and numerous sharp spines, used to spear prey. Within the chamber leading to the mouth are strong teeth. This particular arrow worm, however, was captured by a medusa of the cnidarian* Obelia *(page 528). When photographed, it was dead, and most of its body was in the gastrovascular cavity of its captor.*

PHYLUM CHAETOGNATHA: ARROW WORMS

Although there are only about 60 species in phylum Chaetognatha ("bristle jaws"), these arrow-shaped animals are among the most abundant predators in marine plankton. They range from 1 to 10 centimeters in length and, with a flick of the tail, can shoot forward to capture other small planktonic animals (Figure 28–8). Arrow worms are unsegmented, but their bodies have three distinct regions—head, trunk, and tail. Their embryonic development identifies them as deuterostomes, but they do not seem to be closely related to any other deuterostome phylum. All arrow worms are hermaphroditic, with self-fertilization occurring in some species. The newly hatched young resemble miniature adults and, without passing through a dramatically different larval stage, soon begin a life of active predation.

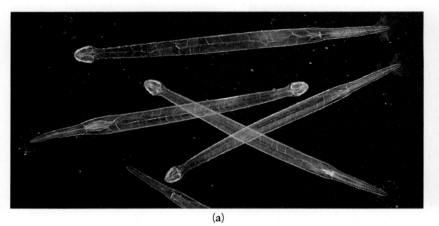

(a)

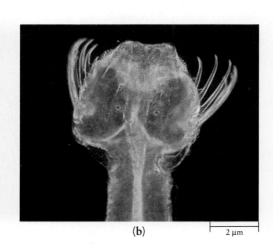

(b)

2 µm

28-9 *An acorn worm,* Glossobalanus sarniensis, *burrowing in shell gravel. Its body consists of three regions—a proboscis, a short collar, and a long trunk, only a portion of which is visible here. Notice the gill slits in the anterior portion of the trunk. Such slits are one of the identifying characteristics of the chordates.*

28-10 (a) Branchiostoma, *a lancelet, exemplifies the four distinctive chordate characteristics: (1) a notochord, the dorsal rod that extends the length of the body; (2) a dorsal, tubular nerve cord; (3) pharyngeal gill slits; and (4) a tail. Branchiostoma retains these four characteristics throughout its life; many chordates, however, have a notochord, pharyngeal gill slits or pouches, and a tail only during their immature stages.*

(b) Lancelets, with their anterior ends protruding from the substrate. Much of the body tissue consists of segmental blocks of muscle. The square, yellowish structures along the side of the body are the reproductive organs. The sexes are separate in lancelets, and sperm or egg cells are released into the atrium. The gametes pass out through the atriopore, with fertilization occurring externally.

PHYLUM HEMICHORDATA: ACORN WORMS

Most of the 80 species in phylum Hemichordata are acorn worms (Figure 28–9). As you can see, the body is divided into three regions—a proboscis, with which the animal burrows in ocean sediments in shallow waters, a short collar, and a long trunk. A few species of sessile hemichordates, known as pterobranchs, look more like bryozoans than acorn worms, but they have the same three body regions and share other hemichordate characteristics.

The hemichordates are of particular interest to evolutionary biologists because they have features characteristic of both the echinoderms and the chordates. Some of the hemichordates have ciliated larvae that are almost identical to the larvae of starfish. Moreover, the coelomic cavities of the tentacles of the pterobranchs provide a hydrostatic support similar to the water vascular system of echinoderms, although there are no tube feet. The hemichordate nervous system is, in some respects, similar to that of the echinoderms, but it includes both ventral and dorsal nerve cords that are joined by a ring at the posterior limit of the collar. Elsewhere in the animal kingdom, a principal nerve cord on the dorsal side of the body is found only among the chordates. The dorsal nerve cord of chordates is hollow, unlike the ventral nerve cords in other animals, which are solid; in some hemichordates, the anterior portion of the dorsal cord is also hollow. The strongest evidence of a close relationship between the hemichordates and the chordates is provided by the pharynx, a structure in the anterior portion of the trunk, which is perforated by holes known as pharyngeal gill slits. As we are about to see, such a pharynx is one of the identifying characteristics of the chordates.

PHYLUM CHORDATA: THE CEPHALOCHORDATES AND UROCHORDATES

The phylum Chordata includes some 43,000 species, grouped in three subphyla: the Cephalochordata, or lancelets; the Urochordata, or tunicates, of which the most familiar are the sea squirts; and the Vertebrata, or vertebrates. Thus the term "invertebrate" refers to all animals except the members of one subphylum of the Chordata.

Subphylum Cephalochordata includes only 28 species. The best-known cephalochordate is *Branchiostoma* (Figure 28–10), a small, blade-shaped, semitransparent animal found in shallow marine waters all over the warmer parts of the world. Although it can swim very efficiently, it spends most of its time buried in the sandy bottom, with only its mouth protruding above the surface. This animal exemplifies all four of the salient features of the chordates. The first is the **notochord,** a rod that extends the length of the body and serves as a firm but flexible axis. The notochord is a structural support. Because of it, *Branchiostoma* can swim with strong undulatory motions that move it through the water with a speed unattainable by the flatworms or aquatic annelids.

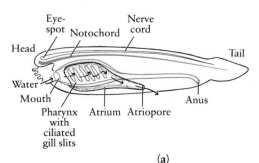

(a)

(b)

28-11 *Two stages in the life of a tunicate:* (**a**) *the larva, and* (**b**) *the adult form. In the larva, the tunic covers the mouth and atriopore, preventing the flow of water through the pharynx and its ciliated gill slits. Thus, even in species in which the larval pharynx is well developed, as shown here, the larva is unable to feed. After a brief free-swimming existence, it settles to the bottom and attaches at the anterior end. Metamorphosis then begins. The larval tail, with the notochord and dorsal nerve cord, disappears, and the animal's entire body is turned 180°. The mouth is carried backward to open at the end opposite that of attachment, and all the other internal organs are also rotated back. It has been hypothesized that the ancestral vertebrates arose from tunicate larvae that became sexually mature—and thus capable of reproducing—without undergoing metamorphosis.*

(**c**) *Living adult tunicates, or sea squirts, photographed in the U.S. Virgin Islands.*

The second chordate characteristic is the **dorsal, hollow nerve cord,** a tube that runs beneath the dorsal surface of the animal, above the notochord. (The principal nerve cords in other phyla, as we have noted, are solid and are almost always near the ventral surface.)

The third characteristic is a **pharynx with gill slits.** The pharyngeal apparatus becomes highly developed in fishes, in which it serves a respiratory function, and traces of the gill pouches remain even in the human embryo. In *Branchiostoma,* the pharynx serves primarily for collecting food. The cilia on the sides of the gill slits pull in a steady current of water, which passes through the slits into a chamber known as the atrium and then exits through the atriopore. Food particles are collected in the sievelike pharynx, mixed with mucus, and channeled along ciliated grooves to the intestine.

The fourth characteristic is a **tail,** posterior to the anus, consisting of blocks of muscle around an axial skeleton. Most of the body tissue of *Branchiostoma* is made up of blocks of muscles, the myotomes.

Although *Branchiostoma* usefully exemplifies the chordates, many biologists believe it is more likely to be a degenerate form of fish rather than a truly primitive member of the phylum. Other possible candidates for the ancestral form are found among the tunicates of the subphylum Urochordata. Although adult tunicates do not have all of the typical chordate features, the larvae, which resemble *Branchiostoma,* are clearly chordates possessing the four identifying characteristics (Figure 28–11). About 1,300 urochordate species are known, and they are found in the plankton and on ocean bottoms throughout the world in both shallow and deep waters. They are commonly known as tunicates because the body is covered by a firm, protective tunic that contains cellulose.

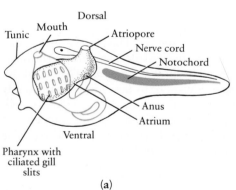

(a)

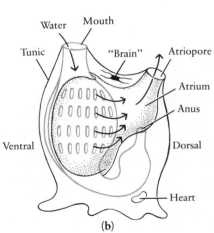

(b)

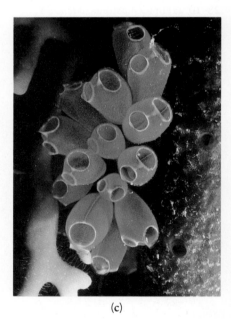

(c)

PHYLUM CHORDATA: THE VERTEBRATES

The vertebrates constitute the largest (about 41,700 species) and most familiar subphylum of the chordates. Vertebrates typically have a vertebral column, or backbone, as their structural axis. This is a flexible, usually bony support that develops around the notochord, supplanting it entirely in most species. Dorsal projections of the vertebrae encircle the nerve cord along the length of the spine.

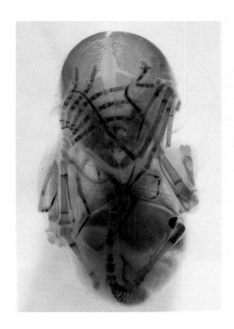

28–12 *The fetus of a long-legged bat. The bones have been stained red and the cartilage blue, so that you can see the extent to which the skeleton is still cartilaginous. Notice the legs, for example. Only the red areas are bone; these will gradually grow and replace the cartilage as the animal matures.*

The brain is similarly enclosed and protected by a cranium, usually consisting of bony skull plates. Between the vertebrae are cartilaginous disks, which give the vertebral column its flexibility. Associated with the vertebrae are segmental muscles by which sections of the vertebral column can be moved separately. This segmental pattern persists in the embryonic forms of higher vertebrates but is largely lost in the course of development.

One of the great advantages of a bony endoskeleton is that it is composed of living tissue that can grow with the animal. In the developing vertebrate embryo, the skeleton is largely cartilaginous; in most vertebrates, bone gradually replaces cartilage in the course of maturation (Figure 28–12). The growing portions of the bones characteristically remain cartilaginous until the animal reaches its full size.

There are seven living classes of vertebrates: the fishes (comprising three classes), the amphibians, the reptiles, the birds, and the mammals. Their evolution is clearly documented in the fossil record.

Classes Agnatha, Chondrichthyes, and Osteichthyes: Fishes

The first fishes were jawless and had a strong notochord running the length of their bodies. Today these jawless fishes (class Agnatha), once a large and diverse group, are represented only by the hagfish and the lampreys. They have a notochord throughout their lives, like *Branchiostoma*. Although their ancestors had bony skeletons, modern agnaths have a cartilaginous skeleton. Lacking true bones, they are very flexible; a hagfish can actually tie itself in a knot. Many cyclostomes ("round mouths"), as the agnaths are also called, are highly predatory, attaching to other fish by their suckerlike mouths (Figure 28–13), and rasping through the skin into the viscera of their hosts. The juvenile lamprey, which resembles *Branchiostoma*, however, feeds by sucking up mud containing microorganisms and organic debris—as, most probably, did the primitive Agnatha.

The sharks (including the dogfish) and skates, the Chondrichthyes, the second major class of fishes, also have a cartilaginous skeleton. Their ancestors, like those of the agnaths, were bony animals. Their skin is covered with small, pointed teeth (denticles), which resemble vertebrate teeth structurally and give the skin the texture and abrasive quality of coarse sandpaper.

28–13 *A sea lamprey (Petromyzon marinus), which normally attaches to other fish and feeds on their blood, has attached itself to a rock. The majority of lampreys are detritus feeders, eating decayed organic material. Respiration occurs through the prominent gill pouches, seen here as seven indentations.*

28–14 *Rays, like skates and sharks, are cartilaginous fish that have existed in their present form for about 350 million years. Their flattened body is an adaptation to bottom living. Shown here is a spotted eagle ray, which feeds on clams and oysters that it locates by rooting in the sand with its snout.*

28–15 *The evolution of the jaw.* (**a**) *In the earliest filter-feeding vertebrates, water currents were created by the beating of cilia around the mouth. As the water moved through the pharyngeal apparatus, food items were filtered out.* (**b**) *In some groups, the cilia were lost, and modified filters began to function in both feeding and respiration; they were not only pumps, moving food-bearing water through strainers, but also gills, across the surface of which gases were exchanged. At the same time, scales migrated to the region of the mouth. Subsequently,* (**c**) *the bones of the first gill arch evolved into the upper and lower jaws, and* (**d**) *the second gill arch moved forward, bracing the jaws at the back of the skull. The scales around the mouth evolved into teeth, and the muscles attached to the jaws became highly developed.*

The result was fish, such as (**e**) *the deep-sea viperfish (Chauliodus), able to feed on a wide variety of large prey organisms.*

(e)

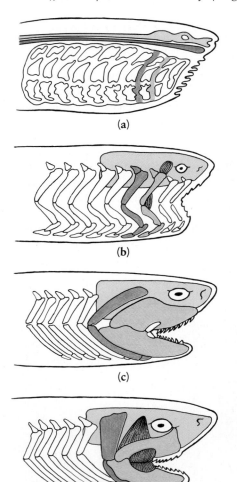

(a)

(b)

(c)

(d)

The third major class of fishes includes those with bony skeletons, the Osteichthyes. This group includes the trout, bass, salmon, perch, and many others—almost all of the familiar freshwater and saltwater fishes.

One of the major events in the evolution of fishes—and of the vertebrate groups descended from them—was the transformation of the anterior gill arches of filter-feeding fish into jaws (Figure 28–15). The development of powerful jaws, often armed with formidable teeth, greatly increased the range of other organisms on which the fish could feed. With more efficient feeding on larger and more concentrated sources of energy came the possibility of significant increases in size.

According to present evidence, fishes evolved in fresh water. The chondrichthyans moved to the sea early in their evolution, while the bony fishes went through most of their evolution in fresh water and spread to the seas at a much later period. Some still make this difficult physiological transition in each lifetime. Salmon, for example, hatch in fresh water, spend most of their lives in salt water, and return to fresh water to spawn. At breeding time, eels travel from the fresh waters of Europe and North America to the Sargasso Sea (an area of the south Atlantic), from which distant point the young begin the long, difficult journey, often lasting many years, back to the rivers and lakes.

The Transition to Land

Another characteristic of the bony fishes is that the early forms seem to have had lungs or lunglike structures, as well as gills. These lungs, however, were not efficient enough to serve as more than accessory structures to the gills. They were a special adaptation to fresh water, which, unlike ocean water, may become stagnant (depleted of oxygen) because of decay of organic matter or of algal bloom. Lunged fishes apparently evolved independently several times, and they were the most common fishes in the later Devonian period, a time of recurring drought. In most of them, the lung evolved into an air bladder, or swim bladder; many modern osteichthyans have gas-filled swim bladders that serve as flotation chambers or organs of sound production. A fish raises or lowers itself in the water by adding gases to or removing them from the air bladder via the bloodstream. Still other primitive fishes evolved into the modern lungfish (Figure 28–17). These fish can live in water that does not have sufficient oxygen to support other fish life. Lungfish surface and gulp air into their lungs in much the same way that certain aquatic but air-breathing snails bob to the surface to fill their mantle cavities.

28-16 *The heavily armored placoderms were the ancestors of two major classes of present-day fishes—the Chondrichthyes (cartilaginous fishes) and the Osteichthyes (bony fishes).*

28-17 *A modern lungfish,* Protopterus annectens. *When the dry seasons come, members of this African genus wriggle downward into the mud, which eventually hardens around them. Mucus glands under the skin secrete a watertight film around the body, preventing evaporation. Only the mouth is left exposed. During this period, the fish takes a breath only about once every two hours.*

In yet others, skeletal supports evolved that served to prop up the thorax of the fish. These fish could gulp air even when their bodies were not supported by water. It is thought that these osteichthyans could waddle, dragging their bellies on the ground, along the muddy bottom of a drying stream bed to seek deeper water or perhaps even make their way from one water source to another one nearby. Thus the transition to land may have begun as an attempt to remain in the water.

Class Amphibia

Amphibians descended from air-breathing lunged fish. Modern amphibians include frogs and toads (which typically lack tails as adults) and salamanders (which have tails throughout their lives). They can readily be distinguished from the reptiles by their thin, usually scaleless skins, which serve as respiratory organs. Adult frogs also have lungs, into which they gulp air, but some salamanders respire entirely through their skins and the mucous membranes of their throats. Because water evaporates rapidly through their skins, amphibians can die of desiccation in a dry environment. Those found in deserts spend the drier times of the day far below the surface of the sand.

Most frogs in cold climates have two life stages, one in water and the other on land (hence their name, from *amphi* and *bios*, meaning "both lives"). The eggs are laid in water and are fertilized externally. They hatch into gilled larvae (tadpoles). The tadpoles later develop into adults that lose their gills and develop lungs. The adults may live out of the water, at least in the summer. However, there are many variations on this theme. Some of the American salamanders fertilize their eggs internally; the males deposit sperm packets, either in water or on moist land, and these packets are picked up by the females. Many modern amphibians skip the free-living larval stage. The eggs, which may be laid on land, in a hollow log or cupped leaf, or may even be carried by the parent, hatch into miniature versions of the adult. Some salamanders, such as the mud puppy and the axolotl, never complete their metamorphosis, remaining essentially aquatic larvalike forms (see Figure 35–4, page 735), even as sexually mature adults. In some species, these larvalike forms can be induced to metamorphose into "adult" forms by administration of hormones, indicating that the genetic capacity for this later developmental stage has not been lost.

28-18 *Many toads are clearly fishlike in their larval (tadpole) stages. As adults, they require water to reproduce, and their moist skins are an important accessory respiratory organ. Toads, like all adult amphibians, are carnivores. They catch insects with a flick of their long tongues, which are attached at the front of their mouths and which have a sticky, flypaperlike surface. This common toad has just captured a beetle larva (a grub).*

595

28–19 *Amniote egg. The membranes, which are produced as outgrowths from the embryo as it develops, surround and protect the embryo and the yolk (its food supply). The egg shell and egg membrane, which are waterproof but permeable to gases, are added as the early embryo passes down the maternal reproductive tract.*

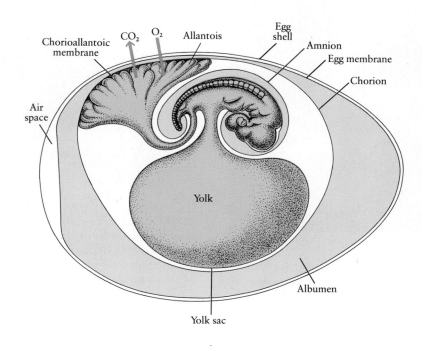

28–20 (a) *A painted turtle,* Chrysemys picth. *The dorsal carapace of turtles and tortoises is partly fused to the vertebral column and the ribs, with a mosaic of horny plates on the surface. Unlike other reptiles, most tortoises do not molt but add new epidermal scales on the undersurface. This results in the addition of a growth ring each year.*

(b) *Alligators and crocodiles, the largest modern reptiles, lay their eggs on land, and their skins are reinforced with horny epidermal scales. Crocodiles, which are essentially tropical animals, have more slender snouts than alligators. Alligators have jaws that are broader and more rounded anteriorly; they are also reported to be less aggressive. The animal shown here is the American alligator,* Alligator mississipiensis.

Class Reptilia

As you will recall, the vascular plants were freed from the water by the evolution of the seed. Analogously, the vertebrates became truly terrestrial with the evolution in the reptiles of the **amniote egg** (Figure 28–19), an egg that retains its own water supply and so can survive on land. The reptilian egg, which is much like the familiar hen's egg in basic design, contains a large yolk, the primary food supply for the developing embryo, and abundant albumen (egg white), which supplies additional nutrients and water. A membrane, the amnion, surrounds the developing embryo with a liquid-filled space that substitutes for the ancestral pond. A gill-like stage is passed in a shelled egg or in the maternal oviduct or uterus. In mammals also, although their eggs typically develop internally, the embryos are enclosed in water within the amnion and pass through a stage with gill pouches before birth.

Reptiles are characteristically four-legged, although the legs are absent in snakes and some lizards. They have a dry skin, usually covered by protective scales, that makes possible their terrestrial existence. Modern reptiles, of which there are about 6,000 species, include lizards, snakes, turtles, and crocodiles.

(a)

(b)

(a)

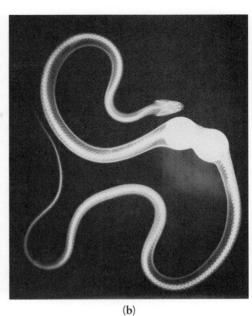

(b)

28–21 (a) *Eastern hognose snakes,* Heterodon playrhinos, *hatching from their leathery egg shells. Even as they hatch, these young snakes are sunning themselves, a behavior characteristic of ectotherms—animals that take in heat from the environment. Eastern hognose snakes, like all other snakes, are carnivorous, devouring their prey whole. Such behavior is made possible by the structure of the jaws; the upper jaw is loosely articulated with the skull, and the lower jaw consists of two bones joined only by muscle and skin, which can be easily stretched. Swallowing prey whole can, however, lead to untoward consequences, as befell one rattlesnake. As shown in its x-ray (b), taken at the hospital of the University of Florida, it swallowed two golf balls, which it apparently misidentified as eggs. Following surgery, the snake was returned to the wild, well away from the nearest golf course.*

Evolution of the Reptiles

By late in the Carboniferous period (see Table 24–1, page 496), the first reptiles had begun to evolve from closely similar amphibian ancestors. During the succeeding Permian period, there was an explosive increase in the number of reptilian species. (During this same period, conifers began to replace the ferns and other "amphibious" plants, suggesting that a drier climate may have been a primary selective force in both of these events.) Throughout the Permian and much of the Triassic, the dominant land vertebrates were mammal-like reptiles, an abundant and diverse group, members of which were later to give rise to the mammals. Around the beginning of the Triassic period, several of the more specialized reptile groups arose, including turtles, lizards, and the thecodonts, ancestors of the Archosauria, or ruling reptiles, perhaps the most spectacular of all the land's inhabitants so far.

The origin and rise of the Archosauria were associated with improvements in reptilian locomotion. Vertebrates first came to the land with all four limbs splayed far out to the side; turtles have retained this sprawling gait. Among the early Archosauria, there was a progressive tendency toward bipedalism, with a concomitant freeing of the front legs for other purposes—flight, for example. There were three major groups of archosaurs: the pterosaurs, or flying reptiles; the crocodilians, which, reverting to (or perhaps retaining) four-legged posture, became our modern crocodiles and alligators; and the dinosaurs, a varied and splendid group of reptiles.

Dinosaurs were long assumed to have been **ectothermic**—that is, to have maintained their body temperatures within broad limits by taking in heat from the environment or by giving it off to the environment. Some biologists contend, however, that at least some groups of dinosaurs were **endothermic**—that is, their body temperatures were maintained by heat generated internally, as are the body temperatures of birds and mammals. Only a few modern reptiles, such as leatherback turtles, show any degree of endothermy.

Fossil remains of the largest dinosaur known thus far, dubbed *Seismosaurus* ("earthshaker"), were discovered near Albuquerque, New Mexico, in 1985. It is estimated that this animal was 30 to 40 meters in length and weighed between 70 and 95 metric tons, far larger than any land animal that has succeeded it.

Throughout the long Mesozoic era, the dinosaurs dominated the life of the land, rulers of the earth for 150 million years. Then, about 65 million years ago, they vanished, leaving only a single line of descendants, the birds. The cause of their extinction has been the subject of speculation since the first dinosaur fossils were discovered in the nineteenth century, but it is only recently that new data have emerged, making possible the formulation of testable hypotheses. In Chapter 49, we shall consider these data and current hypotheses about the phenomenon of extinction and its role in the process of evolution. As we shall see, extinction is a common fate; it is now estimated that at least 99.9 percent of all the species that have ever lived have become extinct.

Class Aves: Birds

Birds are essentially reptiles specialized for flight (Figure 28–22). Their bodies contain air sacs, and their bones are hollow. The frigate bird, a large seagoing bird with a wingspread of more than 2 meters, has a skeleton that weighs only 110 grams (about 4 ounces). The most massive bone in the bird skeleton is the keel, or breastbone, to which are attached the huge muscles that operate the wings. Flying birds have jettisoned all extra weight; the female's reproductive system has been trimmed down to a single ovary, and even this becomes large enough to be functional only in the mating season.

Birds have feathers, which is their outstanding, unique physical characteristic. They are endothermic, generating heat by internal metabolic processes and maintaining a high and constant body temperature. In modern birds, feathers make flight possible and also serve as insulation. (Only animals that are endothermic require insulation; insulation would be a disadvantage for animals that warm their bodies by exposure to the environment.) Birds also have scales on their legs, a reminder of their reptilian ancestry. Many birds hatch at a very immature stage, and virtually all birds require a long period of parental care.

Evolution of Flight

How did flight evolve? Biologists agree that evolution occurs by a series of changes, each of which, to be conserved by natural selection, must be of survival value. Being able to fly not very well is a dubious advantage. The most widely accepted hypothesis for the origin of flight is that the ancestors of the birds were tree-dwelling reptiles and that flight evolved from gliding, as a way to extend or brake jumps from branch to branch.

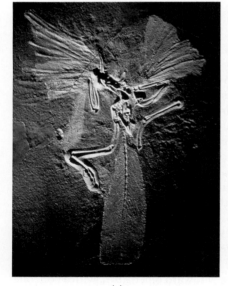

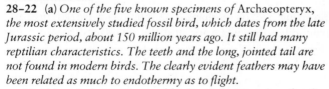

(a)

28–22 (a) *One of the five known specimens of* Archaeopteryx, *the most extensively studied fossil bird, which dates from the late Jurassic period, about 150 million years ago. It still had many reptilian characteristics. The teeth and the long, jointed tail are not found in modern birds. The clearly evident feathers may have been related as much to endothermy as to flight.*

(b) *Although clear evidence of feathers is lacking, these fossil bones, discovered in 1984 in western Texas, are thought to be those of a bird that lived 75 million years before* Archaeopteryx. *Tentatively named* Protoavis, *it has hollow bones, a well-developed "wishbone," and a breastbone with a keel, all characteristic of modern birds. Its skull is also similar to that of a bird, but its hind legs, bony tail, and pelvis resemble those of ground-dwelling dinosaurs.*

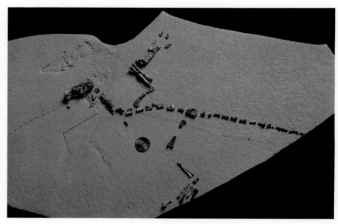

(b)

28–23 *A red-shouldered hawk capturing a mouse.*

However, John Ostrom of Yale University, having studied the anatomy of the known specimens of *Archaeopteryx* and many related forms as well, supports a second hypothesis: that the ancestors of birds were ground-dwelling reptiles. *Archaeopteryx,* according to the fossil evidence, is a close relative of small, bipedal, carnivorous dinosaurs known as theropods, one of the groups of dinosaurs now thought to have been endothermic. The only major distinctions between the fossil theropods and *Archaeopteryx* are that *Archaeopteryx* has feathers and fused collar bones (a "wishbone"), like modern birds. But why should feathers evolve in a ground-dwelling animal? According to Ostrom, feathers were originally an adaptation not for flight but for insulation.

Evidence that the original function of feathers was not flight is provided by anatomical studies showing that the wing feathers apparently were not attached to the bones of the "hand" in *Archaeopteryx,* as they are in modern birds, but were merely embedded in the skin. Also the breastbone and its keel are lacking, indicating that the wing muscles were not well developed. *Archaeopteryx* was clearly not a good flyer, if indeed it could fly at all. In Ostrom's reconstruction, the early stages of flight began with a feathered, endothermic, carnivorous dinosaur running after its prey, flapping its long feathered arms, and leaping. The fact that the feathered forelimbs of *Archaeopteryx* end in claws, as did the elongated arms of the theropods, lends support to this image. Natural selection may have favored the evolution of the long "wing" feathers because they increased the speed of the running predator or because they served as cagelike traps—natural nets—for capturing prey; some modern predatory birds use their wings in this way (Figure 28–23). Thus flight is seen, in this hypothesis, as the culmination of a long, successful predatory leap.

Ostrom's reconstruction, first set forth in 1974, focused new attention on the evolution of flight, which remains the subject of active inquiry. In addition to continuing study of the available fossils, including the newly discovered *Protoavis,* computer models are now being used to analyze the aerodynamic properties of wings of different sizes, shapes, and patterns of movement. Although the evolutionary question may not be soon resolved, this ferment should provide new understanding of the marvelous feats of aerial skill exhibited by contemporary birds.

Whether flight evolved from the "trees down" or from the "ground up," it made available to the birds a new and vast life zone, filled with a previously inaccessible food supply—airborne insects. An enormous diversification followed, producing not only the 9,000 species of living birds but also some 14,000 species that have become extinct.

Class Mammalia

Mammals, like the birds, descended from the reptiles. Although their ancestors predated the dinosaurs, their great diversification did not occur until after the demise of the dinosaurs. When the dinosaurs' long reign ended, previously occupied terrestrial habitats became available to the mammals. Characteristics distinguishing these animals from other vertebrates are that mammals (1) have hair, (2) provide milk for their young from specialized glands (mammary glands), and (3) like birds, but unlike other vertebrates, maintain a high body temperature by generating heat metabolically.

Nearly all mammalian species bear live young, as do some fish and reptiles, which retain the eggs in their bodies until they hatch. However, some very primitive mammals, the **monotremes,** such as the duckbilled platypus, lay eggs with shells but nurse their young after hatching. By contrast, the **marsupials,** which include the opossums and the kangaroos, bear live young. They differ from the largest group of mammals, however, in that the infants are born at a tiny and extremely immature stage and are often kept in a special protective pouch in which they suckle and continue their development (Figure 28–24). Most of the familiar mammals are **placentals,** so called because they utilize their efficient nutritive connection, the placenta, between the uterus and the embryo for a relatively long period of time. As a result, the young develop to a much more advanced stage before birth. Thus the young are afforded protection during their most vulnerable period.

The earliest placentals were small, secretive, and probably nocturnal, thus avoiding the carnivorous dinosaurs that were active during the daylight hours. They undoubtedly lived mostly on insects, grubs, worms, and eggs. Shrews, which are believed to closely resemble these primitive mammals, have retained their elusive habits.

28–24 *Marsupial infants are born at an immature stage and continue their development attached to a nipple in a special protective pouch of the mother.* (a) *This tiny kangaroo accidentally became dislodged from its mother's pouch. As you can see, it is still attached to the nipple. After the picture was taken, the baby was restored to the pouch, with no apparent ill effects from its premature introduction to the outside world.* (b) *Opossum infants spend about two weeks in the womb and about three months in their mother's pouch. A newborn opossum is much smaller than a honey bee.*

(a)

(b)

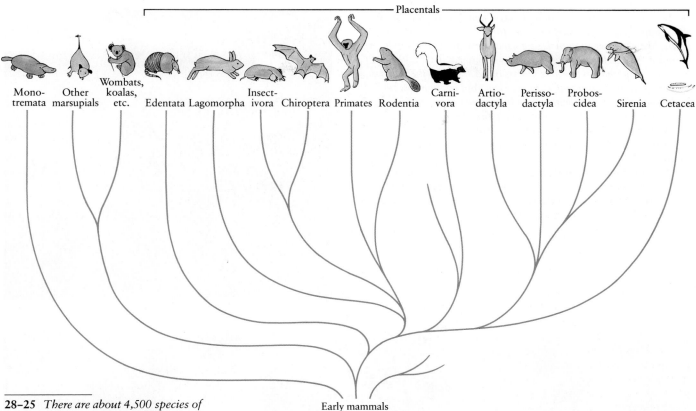

Placentals

Mono-tremata | Other marsupials | Wombats, koalas, etc. | Edentata | Lagomorpha | Insect-ivora | Chiroptera | Primates | Rodentia | Carni-vora | Artio-dactyla | Perisso-dactyla | Probos-cidea | Sirenia | Cetacea

Early mammals

28–25 *There are about 4,500 species of mammals, divided into three subclasses: the monotremes, the marsupials, and the placentals. Twelve of the sixteen orders of placentals are shown here.*

Mammals have fewer, but larger, skull bones than the fishes and reptiles, an example of the fact that "simpler" and "more primitive" may have quite opposite meanings. In the mammals, as in some other vertebrates, a bony platform or partition has developed that separates nasal and food passages far back in the throat, preventing food from entering the lungs. The lower mammalian jaw, unlike that of reptiles, consists of a single bone. Moreover, mammals, unlike snakes or lizards, cannot move the upper jaw in relation to the brain case. As a result, mammals lack the ability of snakes to swallow food items larger than themselves (see Figure 28–21b); mammals must either feed on organisms smaller than themselves or tear the food into pieces small enough to be swallowed.

The major evolutionary lines of mammals are summarized in Figure 28–25. The primates, the order to which we belong, are placental mammals that retain all four kinds of teeth (canines, incisors, premolars, and molars) and have opposable first digits (thumbs and usually big toes), two mammary glands in the chest (rather than in the abdomen), frontally directed eyes, and a relatively large brain with a convoluted cerebral cortex. We are distinguished from the other primates by our upright posture, long legs and short arms, high forehead and small jaw, and sparse body hair.

In many respects, humans are among the least specialized of the mammals. Unlike the carnivores, which are meat eaters, and the several orders of herbivores, we are omnivores, eating a wide variety of fruits, vegetables, and other animals. Our hands closely resemble those of a primitive reptile (see Figure 20–7, page 414), in contrast to the highly specialized forelimbs developed by, for example, whales, bats, and horses. We cannot see as well as an owl monkey. Our sense of smell is much less keen than a dog's, and our sense of taste far less sensitive than that of an enormous variety of other animals. Many animals can run faster, swim more powerfully, and climb trees with more agility (though few can do all three).

28–26 *An assortment of mammals. (a) A female three-toed sloth, photographed with her baby in a rain forest in Panama. Young sloths, like all young mammals, are dependent on mother's milk for nourishment. Unlike most members of order Edentata, the three-toed sloth has teeth—five molars on each side of the upper jaw and four on each side of the lower jaw. (b) Lagomorphs, such as the snowshoe hare shown here in its summer coat, have, in addition to molars and premolars, two pairs of upper incisors. By contrast, rodents, such the beaver (c), have only one pair of upper incisors. In both lagomorphs and rodents, the teeth grow continuously.*

(d) Carnivores, such as these two young male lions chasing a herd of zebras and springbok, are adapted to hunt and kill for food. The zebras are perissodactyls (odd-toed ungulates), and the springbok are artiodactyls (even-toed ungulates).

(a)

(b)

(c)

(d)

(e) *Hippopotamuses,* which are also artio-dactyls, *graze by night and spend most of the daylight hours resting in the water.*

(e)

(f) *Elephants, the* Proboscidea, *are the largest land mammals living today; some reach a weight of 7.5 metric tons.*

(f)

(g) *Among the mammals that have returned to the water are the whales and dolphins of order* Cetacea. *This is the common dolphin,* Delphinus delphis.

(g)

Humans have, however, one area of extreme specialization: the brain. Because of our brain, we are unique among all the other animals in our capacity to reason, to speak, to plan, to learn, and so, to some extent, to control our own future and that of the other organisms with which we share this planet.

SUMMARY

The four deuterostome phyla are Echinodermata, Chaetognatha, Hemichordata, and Chordata. Echinoderms include the sea lilies and feather stars, starfish and brittle stars, sea urchins and sand dollars, and sea cucumbers. Although echinoderms have bilaterally symmetrical larvae, the adult forms of most species are radially symmetrical, with a five-part body plan. Echinoderms have an internal, calcium-containing skeleton that typically bears spines. Their most unusual characteristic is the water vascular system, a modified coelomic cavity that provides hydrostatic support for the tube feet and creates suction for the clinging and pulling activities associated with locomotion and feeding.

The arrow worms, phylum Chaetognatha, have three distinct body regions and are among the most active predators in the marine plankton. Hemichordates, of which the majority are acorn worms, have a mixture of echinoderm and chordate characteristics. Their larvae, the coelomic cavities in some members of the phylum, and portions of the nervous system resemble those of echinoderms. In the hemichordates, however, the nerve cord is dorsal (rather than ventral), a feature found elsewhere only among the chordates. Hemichordates also have a pharynx with gill slits, another of the identifying features of the chordates.

The phylum Chordata comprises three subphyla: the Cephalochordata (lancelets), the Urochordata (tunicates), and the Vertebrata. The primary characteristics of the chordates are the notochord, a flexible longitudinal rod running just ventral to the nerve cord and serving as the structural axis of the body (present only in embryonic life in most vertebrates); the nerve cord, which is a hollow tube located dorsally; a pharynx with gill slits; and a tail. *Branchiostoma* best illustrates the basic chordate body plan. It is hypothesized that the tunicate larva, which also shows these characteristics, resembles the primitive chordate from which the vertebrates evolved.

The vertebrates, the largest subphylum of the chordates, are characterized by a vertebral column, a flexible and usually bony support that develops around and supplants the notochord and encloses the nerve cord, and a cranium enclosing the brain. The vertebrates include the fishes (three living classes), the amphibians, the reptiles, the birds, and the mammals.

Some primitive bony fishes, forerunners of the amphibians, were aided in the transition to land by the development of lungs or lunglike structures and by strong skeletal supports that could prop up the body of the fish out of water. Most modern amphibians remain incompletely adapted to life on land and must spend part of the life cycle in water. Vertebrates became truly terrestrial with the development, in the reptiles, of the amniote egg. The diversification of the reptiles that followed their conquest of the land ultimately gave rise not only to a great variety of reptiles (most of which have been extinct for about 65 million years) but also to their descendants, the birds and the mammals.

QUESTIONS

1. Distinguish among the following: exoskeleton/endoskeleton; tube foot/ampulla; ectotherm/endotherm; monotreme/marsupial/placental.

2. Describe the water vascular system of a starfish. What are its similarities to and differences from the hydrostatic skeleton of an earthworm?

3. Describe the identifying characteristics of the phylum Chordata. What is the functional significance of each?

4. How do the three living classes of fishes differ from one another?

5. Consider the graceful swimming of a squid, a lancelet, or a fish. What is the role of the semirigid beam running the length of the animal, whether the "pen" of a squid, the notochord of a *Branchiostoma*, or the vertebral column of a fish?

6. What anatomical features make possible the flight of birds?

7. Among fish and reptiles, some species are oviparous (that is, they lay eggs from which the young hatch) and some are viviparous (giving birth to live young). Name some of the advantages of each alternative. Why are all birds oviparous?

SUGGESTIONS FOR FURTHER READING

Classification

Books

AYALA, F. J., and J. A. KIGER, JR.: *Modern Genetics,* 2d ed., The Benjamin/Cummings Publishing Company, Menlo Park, Calif., 1984.

> *Contains a good chapter on phylogeny and the comparison of proteins and nucleotides.*

BARNES, R. S. K.: *A Synoptic Classification of Living Organisms,* Sinauer Associates, Inc., Sunderland, Mass., 1984.*

> *A concise presentation of the diversity of life on earth, organized according to a five-kingdom system. A comparison of the taxonomic decisions made in this book, which incorporates the judgment of many British systematists, with those made by Margulis and Schwartz (see below) reveals many of the current disagreements about the classification of particular groups of organisms.*

DOBZHANSKY, THEODOSIUS, et al.: *Evolution,* W. H. Freeman and Company, San Francisco, 1977.

> *Although now becoming somewhat dated, this text remains a good general introduction to evolution. It contains valuable chapters on taxonomy and phylogeny.*

KESSEL, R. G., and C. Y. SHIH: *Scanning Electron Microscopy in Biology: A Students' Atlas on Biological Organization,* Springer-Verlag, New York, 1976.

> *An atlas filled with marvelous scanning electron micrographs of specialized structures of prokaryotes, protists, fungi, plants, and animals, as well as of whole organisms.*

MARGULIS, LYNN, and KARLENE V. SCHWARTZ: *Five Kingdoms: An Illustrated Guide to the Phyla of Life on Earth,* 2d ed., W. H. Freeman and Company, New York, 1987.

> *A concise presentation of the diversity of life on earth, organized according to a five-kingdom system. Although many biologists disagree with some of the taxonomic decisions made by the authors, this book contains a wealth of fascinating information about organisms and their life styles, coupled with outstanding micrographs and diagrams.*

MAYR, ERNST: *The Growth of Biological Thought: Diversity, Evolution, and Inheritance,* Harvard University Press, Cambridge, Mass., 1982.*

> *This is the first of two projected volumes on the history of biology and its major ideas, written by one of the leading figures in the study of evolution. Following an outstanding introductory analysis of the philosophy and methodology of the biological sciences, the first major section of the book provides a thorough treatment of the history and current status of taxonomy.*

PERRY, DONALD: *Life Above the Jungle Floor,* Simon and Schuster, New York, 1986.*

> *Many, and perhaps most, of the earth's undiscovered organisms reside high in the trees of tropical forests. In a lively book, Perry, who devised the climbing and suspension apparatus that is making possible the exploration of this life zone, describes his adventures in the treetops. For a photograph of the author at work, see page 404.*

Articles

BRITTEN, R. J.: "Rates of DNA Sequence Evolution Differ Between Taxonomic Groups," *Science,* vol. 231, pages 1393–1398, 1986.

DIAMOND, JARED: "How Many Unknown Species Are Yet to Be Discovered?" *Nature,* vol. 315, pages 538–539, 1985.

LEWIN, DONALD A.: "The Nature of Plant Species," *Science,* vol. 204, pages 381–384, 1979.

LEWIN, ROGER: "Molecules vs. Morphology: Of Mice and Men," *Science,* vol. 229, pages 743–745, 1985.

LOWENSTEIN, J. M.: "Molecular Approaches to the Identification of Species," *American Scientist,* vol. 73, pages 541–547, 1985.

MAY, ROBERT M.: "How Many Species Are There?" *Nature,* vol. 324, pages 514–515, 1986.

MAYR, ERNST: "Biological Classification: Toward a Synthesis of Opposing Methodologies," *Science,* vol. 214, pages 510–516, 1981.

MAYR, ERNST: "Uncertainty in Science: Is the Giant Panda a Bear or a Raccoon?" *Nature,* vol. 323, pages 769–771, 1986.

O'BRIEN, STEPHEN J.: "The Ancestry of the Giant Panda," *Scientific American,* November 1987, pages 102–107.

SIBLEY, CHARLES G., and JON E. AHLQUIST: "Reconstructing Bird Phylogeny by Comparing DNA's," *Scientific American,* February 1986, pages 82–92.

WILSON, EDWARD O.: "Time to Revive Systematics," *Science,* vol. 230, page 1227, 1985.

* Available in paperback.

Prokaryotes and Viruses

Books

AUSTRIAN, ROBERT: *Life with the Pneumococcus: Notes from the Bedside, Laboratory, and Library,* University of Pennsylvania Press, Philadelphia, 1985.

> *The history of the efforts to understand, control, and prevent pneumococcal pneumonia is lucidly recounted in this collection of lectures, reviews, and research reports. Although the unexpected results of Griffith's experiments (see page 282) and the introduction of antibiotics diverted attention from the quest for an effective vaccine against pneumococcal pneumonia, work has gone forward, with Austrian one of the major participants. This book provides many insights into the challenges and rewards of a career devoted to one specific biological problem.*

BURNET, MacFARLANE, and DAVID O. WHITE: *Natural History of Infectious Disease,* 4th ed., Cambridge University Press, New York, 1972.*

> *A general introduction to the ecology of infectious diseases and their influence on human activities.*

STANIER, R. Y., J. L. INGRAHAM, M. L. WHEELIS, and P. R. PAINTER: *The Microbial World,* 5th ed., Prentice-Hall, Inc., Englewood Cliffs, N.J., 1986.

> *An introduction to the biology of microorganisms, with special emphasis on the properties of bacteria. It is widely considered one of the most authoritative accounts.*

ZINSSER, HANS: *Rats, Lice, and History,* The Atlantic Monthly Press/Little, Brown and Company, Boston, 1935.*

> *A classic popular account of the influence of infectious diseases and their animal vectors on the course of human history, reissued in 1984.*

Articles

BLAKEMORE, R. P., and R. B. FRANKEL: "Magnetic Navigation in Bacteria," *Scientific American,* December 1981, pages 58–65.

BROCK, T. D.: "Life at High Temperatures," *Science,* vol. 230, pages 132–138, 1985.

BUTLER, P. J. G., and A. KLUG: "The Assembly of a Virus," *Scientific American,* November 1978, pages 62–69.

COSTERTON, J. W., G. G. GEESEY, and K. J. CHENG: "How Bacteria Stick," *Scientific American,* January 1978, pages 86–95.

DIENER, T. O.: "The Viroid—A Subviral Pathogen," *American Scientist,* vol. 71, pages 481–489, 1983.

DWORKIN, M., and D. KAISER: "Cell Interactions in Myxobacterial Growth and Development," *Science,* vol. 230, pages 18–24, 1985.

FOX, G. E., et al.: "The Phylogeny of Prokaryotes," *Science,* vol. 209, pages 457–463, 1980.

GALLO, ROBERT C.: "The First Human Retrovirus," *Scientific American,* December 1986, pages 88–98.

GALLO, ROBERT C.: "The AIDS Virus," *Scientific American,* January 1987, pages 46–56.

HIRSCH, MARTIN S., and JOAN C. KAPLAN: "Antiviral Therapy," *Scientific American,* April 1987, pages 76–85.

HOGLE, J. M., M. CHOW, and D. J. FILMAN: "The Structure of Poliovirus," *Scientific American,* March 1987, pages 42–49.

LERNER, R. A.: "Synthetic Vaccines," *Scientific American,* February 1983, pages 66–74.

PENNY, DAVID: "What Was the First Living Cell?" *Nature,* vol. 331, pages 111–112, 1988.

PRUSINER, S. B.: "Prions," *Scientific American,* October 1984, pages 50–59.

SHAPIRO, JAMES A.: "Bacteria as Multicellular Organisms," *Scientific American,* June 1988, pages 82–89.

SIMONS, K., H. GAROFF, and A. HELENIUS: "How an Animal Virus Gets into and out of Its Host Cell," *Scientific American,* February 1982, pages 58–66.

WOESE, CARL R.: "Archaebacteria," *Scientific American,* June 1981, pages 98–122.

Protists

Books

BOLD, HAROLD C., and MICHAEL J. WYNNE: *Introduction to the Algae: Structure and Reproduction,* 2d ed., Prentice-Hall, Inc., Englewood Cliffs, N.J., 1985.

> *A detailed reference work on the algae that contains a wealth of information on all groups; taxonomically oriented.*

BONNER, JOHN T.: *The Cellular Slime Molds,* 2d ed., Princeton University Press, Princeton, N.J., 1968.

> *A record of experimental work with a small but fascinating group of organisms.*

CURTIS, HELENA: *The Marvelous Animals,* Natural History Press, Garden City, N.Y., 1968.

> *An informal introduction to one-celled eukaryotes.*

JURAND, A., and G. C. SELMAN: *The Anatomy of Paramecium aurelia,* The Macmillan Company, New York, 1964.

> *An exploration, mainly by electron microscopy, of the astonishing complexity of a single-celled organism.*

LEE, J. J., S. H. HUTNER, and E. C. BOVEE (eds.): *An Illustrated Guide to the Protozoa,* Society of Protozoologists, Lawrence, Kansas, 1985.

> *The most recent comprehensive reference on the protozoan protists. This well-illustrated guide to an enormous diversity of fascinating organisms also includes a glossary of terms used by students of the protists.*

MARGULIS, LYNN: *Symbiosis in Cell Evolution: Life and Its Environment on the Early Earth,* W. H. Freeman and Company, New York, 1981.*

> *A fascinating discourse proposing the origin of eukaryotic cells by serial symbiotic events.*

PICKETT-HEAPS, JEREMY D.: *Green Algae: Structure, Reproduction and Evolution in Selected Genera,* Sinauer Associates, Inc., Sunderland, Mass., 1975.

---

* Available in paperback.

A beautifully illustrated book providing much insight into the variety of form and function in the cells of the green algae.

ROUND, F. E.: *The Ecology of Algae*, Cambridge University Press, New York, 1981.

A comprehensive account of the ecology of both freshwater and marine algae.

Articles

CORLISS, JOHN O.: "The Kingdom Protista and Its 45 Phyla," *BioSystems*, vol. 17, pages 87–126, 1984.

COX, F. E. G.: "Malaria Vaccines: The Shape of Things to Come," *Nature*, vol. 333, page 702, 1988.

DONELSON, J. E., and M. J. TURNER: "How the Trypanosome Changes Its Coat," *Scientific American*, February 1985, pages 44–51.

FRIEDMAN, M. J., and W. TRAGER: "The Biochemistry of Resistance to Malaria," *Scientific American*, March 1981, pages 154–164.

GODSON, G. N.: "Molecular Approaches to Malaria Vaccines," *Scientific American*, May 1985, pages 52–59.

SCHOPF, J. W.: "The Evolution of the Earliest Cells," *Scientific American*, September 1978, pages 110–138.

SCHWARTZ, R. M., and M. O. DAYHOFF: "Origins of Prokaryotes, Eukaryotes, Mitochondria, and Chloroplasts," *Science*, vol. 199, pages 395–403, 1978.

VIDAL, G.: "The Oldest Eukaryotic Cells," *Scientific American*, February 1984, pages 48–57.

YATES, G. T.: "How Microorganisms Move through Water," *American Scientist*, vol. 74, pages 358–365, 1986.

Fungi

Books

AHMADJIAN, V., and S. PARACER: *Symbiosis: An Introduction to Biological Associations*, University Press of New England, Hanover, N.H., 1986.

In this small but fascinating book, Ahmadjian describes the nature of the relationship between the fungal and algal components of a lichen.

ALEXOPOULOS, C. J., and C. W. MIMS: *Introductory Mycology*, 3d ed., John Wiley & Sons, Inc., New York, 1979.

A thorough introduction to the fungi, the slime molds, and the water molds.

LARGE, E. C.: *The Advance of the Fungi*, Dover Publications, Inc., New York, 1962.*

A fascinating popular account of the closely interwoven histories of fungi and humans, first published in 1940.

SMITH, A. H.: *The Mushroom Hunter's Field Guide*, The University of Michigan Press, Ann Arbor, Mich., 1980.

A clear, concise, well-illustrated guide to edible mushrooms, enlivened with good advice and pertinent anecdotes.

Articles

AHMADJIAN, VERNON: "The Nature of Lichens," *Natural History*, March 1982, pages 30–37.

FRIEDMANN, E. I.: "Endolithic Microorganisms in the Antarctic Cold Desert," *Science*, vol. 215, pages 1045–1053, 1982.

LITTEN, W.: "The Most Poisonous Mushrooms," *Scientific American*, March 1975, pages 90–101.

MATOSSIAN, MARY K.: "Ergot and the Salem Witchcraft Affair," *American Scientist*, vol. 70, pages 355–357, 1982.

RUEHLE, J. L., and D. H. MARX: "Fiber, Food, Fuel, and Fungal Symbionts," *Science*, vol. 206, pages 419–422, 1979.

STROBEL, G. A., and G. N. LANIER: "Dutch Elm Disease," *Scientific American*, August 1981, pages 56–66.

Plants

Books

BARTH, FRIEDRICH G.: *Insects and Flowers: The Biology of a Partnership*, Princeton University Press, Princeton, N.J., 1985.

A well-written and beautifully illustrated introduction to the interrelationships of flowers and insects. Incorporating many recent discoveries, the text considers not only the diverse structures of flowers but also the sensory, navigational, and communication abilities of pollinating insects.

CONARD, HENRY S., and PAUL L. REDFEARN, JR.: *How to Know the Mosses and Liverworts*, 2d ed., William C. Brown Publishing Co., Dubuque, Iowa, 1979.*

A good beginner's guide for identifying many of the more common bryophytes. It includes a profusely illustrated key and an excellent glossary.

MILNE, DAVID, et al. (eds.): *The Evolution of Complex and Higher Organisms*, NASA Special Publication 478, U.S. Government Printing Office, 1985.

A concise summary of the history of multicellular life on earth, including an up-to-date review of the history of fossil plants.

RAVEN, PETER H., RAY F. EVERT, and SUSAN E. EICHHORN: *Biology of Plants*, 4th ed., Worth Publishers, Inc., New York, 1986.

This general botany text contains an excellent presentation of the evolution of plants and related organisms, as well as a wealth of information on prokaryotes, protists, and fungi.

Articles

CREPET, W. L.: "Ancient Flowers for the Faithful," *Natural History*, April 1984, pages 38–45.

DILCHER, D., and P. R. CRANE: "In Pursuit of the First Flower," *Natural History*, March 1984, pages 56–61.

GENSEL, PATRICIA G., and HENRY N. ANDREWS: "The Evolution of Early Land Plants," *American Scientist*, vol. 75, pages 478–489, 1987.

GRAHAM, LINDA E.: "The Origin of the Life Cycle of Land Plants," *American Scientist*, vol. 73, pages 178–186, 1985.

---

* Available in paperback.

MULCAHY, DAVID L.: "Rise of the Angiosperms," *Natural History,* September 1981, pages 30–35.

NIKLAS, KARL J.: "Aerodynamics of Wind Pollination," *Scientific American,* July 1987, pages 90–95.

NIKLAS, KARL J.: "Computer-simulated Plant Evolution," *Scientific American,* March 1986, pages 78–86.

NIKLAS, KARL J.: "Wind Pollination—A Study in Controlled Chaos," *American Scientist,* vol. 73, pages 462–470, 1985.

NORSTOG, KNUT: "Cycads and the Origin of Insect Pollination," *American Scientist,* vol. 75, pages 270–279, 1987.

ROSENTHAL, GERALD A.: "The Chemical Defenses of Higher Plants," *Scientific American,* January 1986, pages 94–99.

VALENTINE, JAMES W.: "The Evolution of Multicellular Plants and Animals," *Scientific American,* September 1979, pages 140–158.

WIENS, D.: "Secrets of a Cryptic Flower," *Natural History,* May 1985, pages 70–77.

Animals

Books

BARNES, ROBERT D.: *Invertebrate Zoology,* 5th ed., Saunders College/Holt, Rinehart and Winston, Philadelphia, 1987.

One of the best general introductions to protozoans and invertebrates.

BUCHSBAUM, RALPH, MILDRED BUCHSBAUM, JOHN PEARSE, and VICKI PEARSE: *Animals without Backbones,* 3d ed., University of Chicago Press, Chicago, 1987.*

A delightful introduction to the invertebrates, for the general reader, with a multitude of photographs.

DESMOND, ADRIAN J.: *The Hot-Blooded Dinosaurs: A Revolution in Paleontology,* The Dial Press, Inc., New York, 1976.*

Desmond, a historian of science, describes the development of evolutionary theories, particularly as they were influenced by the discovery of dinosaur fossils, and presents a lively review of the evidence that is leading an increasing number of paleontologists to the conclusion that dinosaurs were endothermic. (If you are not interested in the history of paleontology, you may want to begin in the middle.) The text is well written and the illustrations are wonderful.

EVANS, HOWARD E.: *Life on a Little-Known Planet,* University of Chicago Press, Chicago, 1984.*

Professor Evans is the author of many popular articles and books on insects. This book, originally published in 1966, profits from his wide knowledge, clarity, and humor.

---

* Available in paperback.

FEDUCCIA, ALAN: *The Age of Birds,* Harvard University Press, Cambridge, Mass., 1980.*

A history of the birds, tracing their evolution from reptilian ancestors through the diversifications that gave rise to the major groups of modern birds. Well-illustrated with many photographs and drawings.

FENTON, M. BROCK: *Just Bats,* University of Toronto Press, Toronto, 1983.*

A short, well-illustrated account of the fascinating lives of the bats, the only mammals capable of flight.

HANSON, EARL D.: *The Origin and Early Evolution of Animals,* Wesleyan University Press, Middletown, Conn., 1977.

Recommended to the serious student who is interested in how the experts make decisions concerning evolutionary relationships.

KLOTS, ALEXANDER B., and ELSIE B. KLOTS: *Living Insects of the World,* Doubleday & Company, Inc., Garden City, N.Y., 1975.

A spectacular gallery of insect photos. The text is informal but informative, written by experts for the general public.

MCMAHON, T. A., and J. T. BONNER: *On Size and Life, Scientific American* Library, W. H. Freeman and Company, New York, 1985.*

Written by an engineer and a biologist, this beautifully illustrated book explores the relationship of the size of an organism to its shape, its speed, its physiological functions, its evolution, and its ecology.

PEARSE, VICKI, JOHN PEARSE, MILDRED BUCHSBAUM, and RALPH BUCHSBAUM: *Living Invertebrates,* Blackwell Scientific Publications, Palo Alto, Calif., 1987.

An introduction to the invertebrates, for the more advanced student, by the authors of Animals without Backbones *(see above). Although this text is more detailed and contains a great deal more information than* Animals without Backbones, *it retains the same delightful writing style. A wonderful book for both serious study and browsing.*

ROMER, ALFRED: *The Procession of Life,* World Publishing Co., Cleveland, 1968.*

A history of evolution, written by an expert, but as readable as a novel.

WELLS, M. J.: *Brain and Behavior in Cephalopods,* Stanford University Press, Stanford, Calif., 1962.

Experimental analyses of behavior in the octopus and squid.

WHITTINGTON, H. B.: *The Burgess Shale,* Yale University Press, New Haven, Conn., 1985.

The Burgess Shale, a rock formation in the Canadian Rockies, contains one of the world's largest fossil assemblages of soft-bodied invertebrates from the Cambrian period. This beautifully illustrated book summarizes two decades of study of these fossilized animals, many of which were very different from the animals that succeeded them.

WILFORD, JOHN NOBLE: *The Riddle of the Dinosaur*, Alfred A. Knopf, Inc., New York, 1985.*

A well-written history of dinosaur discovery and research, with balanced discussions of the many controversies that surround these fascinating animals.

Articles

AUSTAD, STEVEN N.: "The Adaptable Opossum," *Scientific American*, February 1988, pages 98–104.

BUFFETAUT, ERIC: "The Evolution of the Crocodilians," *Scientific American*, October 1979, pages 130–144.

CAMERON, J. N.: "Molting in the Blue Crab," *Scientific American*, May 1985, pages 102–109.

CRACRAFT, JOEL: "Early Evolution of Birds," *Nature*, vol. 331, pages 389–390, 1988.

DOMNING, D. P.: "Sea Cow Family Reunion," *Natural History*, April 1987, pages 64–71.

FIELD, KATHARINE G., et al.: "Molecular Phylogeny of the Animal Kingdom," *Science*, vol. 239, pages 748–753, 1988.

GHISELIN, MICHAEL T.: "A Movable Feaster," *Natural History*, September 1985, pages 54–61.

GOSLINE, J. M., and M. E. DeMONT: "Jet-propelled Swimming in Squids," *Scientific American*, January 1985, pages 96–103.

GRIFFITHS, MERVYN: "The Platypus," *Scientific American*, May 1988, pages 84–91.

HADLEY, N. F.: "The Arthropod Cuticle," *Scientific American*, July 1986, pages 104–112.

HORN, MICHAEL H., and ROBIN N. GIBSON: "Intertidal Fishes," *Scientific American*, January 1988, pages 64–70.

HORNER, J. R.: "The Nesting Behavior of Dinosaurs," *Scientific American*, April 1984, pages 130–137.

HORRIDGE, G. A.: "The Compound Eye of Insects," *Scientific American*, July 1977, pages 108–122.

KIRSCH, JOHN A. W.: "The Six-Percent Solution: Second Thoughts on the Adaptedness of the Marsupialia," *American Scientist*, vol. 65, pages 276–288, 1977.

KOEHL, M. A. R.: "The Interaction of Moving Water and Sessile Organisms," *Scientific American*, December 1982, pages 124–134.

LaBARBERA, MICHAEL, and STEVEN VOGEL: "The Design of Fluid Transport Systems in Organisms," *American Scientist*, vol. 70, pages 54–60, 1982.

LENHOFF, HOWARD M., and SYLVIA G. LENHOFF: "Trembley's Polyps," *Scientific American*, April 1988, pages 108–113.

LENT, CHARLES M., and MICHAEL H. DICKINSON: "The Neurobiology of Feeding in Leeches," *Scientific American*, June 1988, pages 98–103.

LEWIN, ROGER: "On the Origin of Insect Wings," *Science*, vol. 230, pages 428–429, 1985.

LLOYD, JAMES E.: "Mimicry in the Sexual Signals of Fireflies," *Scientific American*, July 1981, pages 138–145.

MACURDA, D. B., JR., and D. L. MEYER: "Sea Lilies and Feather Stars," *American Scientist*, vol. 71, pages 354–365, 1983.

McMENAMIN, MARK A. S.: "The Emergence of Animals," *Scientific American*, April 1987, pages 94–102.

MICHELSEN, AXEL: "Insect Ears as Mechanical Systems," *American Scientist*, vol. 67, pages 696–706, 1979.

MORRIS, SIMON CONWAY: "The Search for the Precambrian-Cambrian Boundary," *American Scientist*, vol. 75, pages 156–167, 1987.

OSTROM, JOHN H.: "Bird Flight: How Did It Begin?" *American Scientist*, vol. 67, pages 46–56, 1979.

PANCHEN, ALEC L.: "In Search of the Earliest Tetrapods," *Nature*, vol. 333, page 704, 1988.

RICHARDSON, J. R.: "Brachiopods," *Scientific American*, September 1986, pages 100–106.

SEBENS, K. P.: "The Anemone Below," *Natural History*, November 1986, pages 48–53.

WEBB, P. W.: "Form and Function in Fish Swimming," *Scientific American*, July 1984, pages 72–82.

---

* Available in paperback.

SECTION 5

Biology of Plants

Branch of an apple tree, showing one of the major characteristics of the angiosperms, the most successful group of plants. Flowers lure the insects and other animals that carry pollen from plant to plant.

C H A P T E R   **29**

The Flowering Plants: An Introduction

For most of earth's history, the land was bare. A billion years ago, algae clung to the shores at low tide and may even have begun to cover a few moist surfaces farther inland. But, had anyone been there to observe it, the earth's surface would generally have appeared almost as barren and forbidding as the bleak Martian landscape. According to the fossil record, plants began to invade the land a mere half billion years ago. Not until then did the earth's surface truly come to life. As a film of green spread from the edges of the waters, other forms of life—heterotrophs—were able to follow. The shapes of these new forms and the ways in which they lived were determined by the plant life that preceded them. Plants supplied not only their food—their chemical energy—but also their nesting, hiding, stalking, and breeding places.

And so it is today. In all terrestrial communities except those created by human activities, the character of the plants still determines the character of the animals and other forms of life that inhabit a particular area. Even we members of the human species, who have seemingly freed ourselves from the life of the land and even, on occasion, from the surface of the earth, are still dependent on the photosynthetic events that take place in the green leaves of plants.

In Chapter 24, we traced the evolution of the plants; if you have not already read that chapter, we recommend that you do so now. In this section, we shall focus on the group of plants that evolved most recently, the angiosperms (division Anthophyta). The angiosperms are, by far, the most abundant plants on earth today, with about 235,000 living species. As we noted in Chapter 24, the group is divided into two large classes, Monocotyledones (the monocots), with about 65,000 species, and Dicotyledones (the dicots), with about 170,000 species. (See Table 24–3 on page 511 for the principal differences between these two classes.)

Angiosperms are characterized by specialized reproductive structures—flowers—in which sexual reproduction occurs, in which the seeds are formed, and from which the fruits develop. For these plants, a new cycle of life begins when a pollen grain—often brushed from the body of a foraging insect—comes into contact with the stigma of a flower of the same species. As we saw in Chapter 24, flowers are exquisitely adapted to achieve this act of pollination. Unlike the reproductive organs of animals, which are permanent structures that develop in the embryo, flowers are transitory. After fertilization, some parts of the flower become the fruit, protecting and enclosing the seed or seeds; other parts die and are discarded.

29–1 *The second major characteristic of the angiosperms is the fruit. Seeds are dispersed from the fruit, sometimes with the assistance of other organisms, such as this obliging porcupine.*

SEXUAL REPRODUCTION: THE FLOWER

Most flowers consist of four sets of floral parts—sepals, petals, stamens, and carpels (Figure 29–2). Each floral part is thought to be, evolutionarily speaking, a modified leaf (see page 511). The floral parts may be arranged spirally on a more or

613

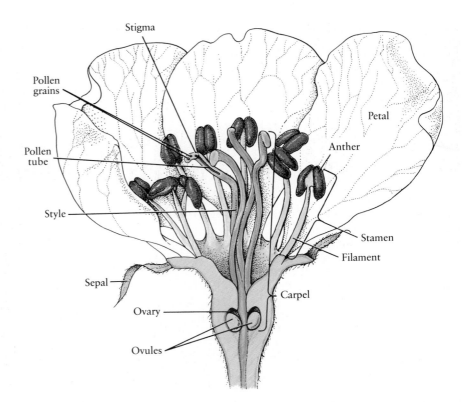

29–2 *The flower of a pear tree. This flower is a complete flower, which means that it contains all four floral parts—sepals, petals, stamens, and carpels. It is also a perfect flower, which means that it contains both male structures (stamens) and female structures (carpels). Each stamen consists of a pollen-bearing anther and its filament. Each carpel consists of a stigma, a style, and an ovary. In the pear flower, each set of floral parts is located at one level, in a whorl.*

Pollen grains have been deposited on the sticky surface of a stigma of this flower, and the pollen tube of one is growing down the style to an ovule.

Stigma

Pollen grains

Pollen tube

Style

Sepal

Ovary

Ovules

Petal

Anther

Stamen

Filament

Carpel

less elongated stalk, or similar parts—such as the petals—may be located at one level in a whorl. In monocots, each of the floral parts is usually present in multiples of three; in dicots, by contrast, the floral parts typically occur in multiples of four or five.

The outermost parts of the flower are the **sepals,** which are usually green and leaflike. The sepals, collectively known as the **calyx,** enclose and protect the developing flower bud. Next are the **petals,** which together are called the **corolla.** Petals may also be leaf-shaped, but they are often brightly colored. They advertise the presence of the flower among the green leaves, attracting insects or other animals that visit flowers for their nectar or for other edible substances. As these animals forage for food, they are likely to carry pollen from flower to flower (see Figure 24–26, page 514).

Within the corolla are the stamens. Each stamen consists of a single elongated stalk, called the filament, and at the end of the filament, the anther. The pollen grains, formed within the anther, are the immature male gametophytes. (For a review of alternation of generations in angiosperms, see Figure 24–22 on page 512). When ripe, the pollen grains are released, often in large numbers, through slits or pores in the anther.

The centermost parts of the flower are the carpels, which contain the female gametophytes. A single flower may have one carpel or several carpels, which may be separate or fused together. Typically a single carpel or fused carpels consist of a stigma, which is a sticky surface to which pollen grains adhere; a stalk, the style, through which the pollen tube grows; and a swollen base, the ovary. Within the ovary are one or more ovules, each of which encloses a female gametophyte, or **embryo sac,** containing a single egg cell. After the egg is fertilized, the ovule develops into a seed and the ovary into a fruit.

A flower that contains both stamens and carpels is known as a perfect flower. In some species, however, the flowers are imperfect—that is, they are either male (staminate) or female (carpellate). Male and female flowers may be present on the same plant, as in corn, squash, oaks, and birches; such plants are said to be monoecious ("one house"). Species in which the male and female flowers are on separate plants, such as the tree of heaven *(Ailanthus),* American mistletoe, and holly, are known as dioecious ("two houses"). As gardeners know, in order for a female holly plant to produce berries, a male holly—which never produces berries—must be planted nearby.

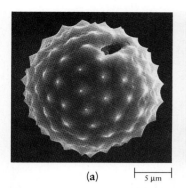
(a) |—— 5 μm ——|

(b) |—15 μm—|

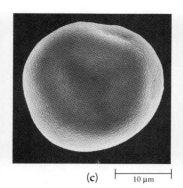
(c) |—— 10 μm ——|

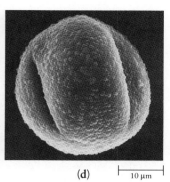
(d) |— 10 μm —|

29–3 *Pollen grains. The walls of the pollen grain protect the male gametophyte on its journey from the anther to a stigma. These outer surfaces, which are remarkably tough and resistant, are often elaborately sculptured. As you can see, the pollen grains of different species are distinctly different: (a) common ragweed (such spiny pollen grains trigger the misery of hay fever in susceptible individuals); (b) Geranium, a dicot; (c) timothy grass, a monocot (smooth pollen grains are found in most wind-pollinated plants); and (d) oak, a dicot.*

Upon germination, the pollen tube emerges through a pore or slit in the outer coat, clearly visible here in the ragweed and oak pollen grains. The pollen grains of monocots typically have one pore or slit, whereas those of dicots have three.

The Pollen Grain

By the time the pollen grain is released from the anther, it usually consists of three haploid cells—two sperm cells contained within a larger cell, known as a tube cell. The tube cell, in turn, is enclosed by the thick outer wall of the pollen grain. The pollen grain contains its own nutrients and has so tough an outer coating that intact grains thousands of years old have been found in peat bogs.

As you will recall, in many multicellular algae and in the bryophytes and seedless vascular plants, there is a distinct alternation of generations in which the sporophyte produces spores that produce gametophytes that produce gametes, with the gametophyte and sporophyte having separate existences. In the course of plant evolution, the gametophyte has been steadily reduced in size. In the angiosperms, all that remains of the male gametophyte is the tough, tiny pollen grain and the pollen tube that grows from it. The sperm cells are the gametes.

Fertilization

Once on the stigma, the pollen grain germinates, and, under the influence of the tube nucleus, a pollen tube grows through the style into an ovule (Figure 29–4).

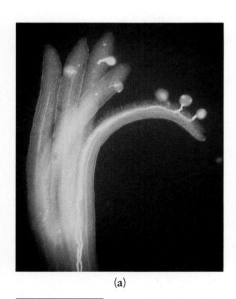
(a)

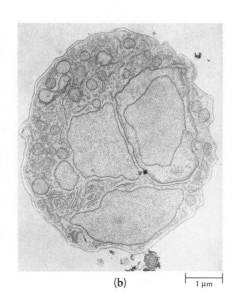
(b) |— 1 μm —|

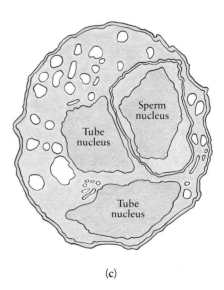
(c)

29–4 *(a) Pollen tube growth in* Geranium maculatum, *a species in which the stigma is divided into distinct lobes. Germination of pollen grains is thought to depend on species-specific recognition, perhaps involving the interaction of chemical substances on the sticky surface of the stigma with the pollen grain. It has also been sug-*

gested that sugary substances on the stigma are used by the pollen grain to provide energy for the rapid growth of the pollen tube. In Geranium, *the pollen tube from a grain 0.1 millimeter in diameter can grow about 1 centimeter in length within 20 minutes.*

(b, c) Electron micrograph and diagram

of a cross section of a pollen tube. Two lobes of the single tube nucleus are visible. This nucleus directs the formation of the pollen tube and eventually disintegrates. Numerous mitochondria are visible in the tube cell, as are several plastids. At the right is a sperm cell, which is actually a cell within a cell.

615

(a)

(b)

(c)

(d)

(e)

(f)

(g)

29–5 *Flowers—forms and variations.*
(a) *The sweet bay magnolia,* Magnolia virginiana. *The carpels form a cone-shaped receptacle from which the curved styles emerge. The cream-colored stamens, some of which have dropped off, surround the carpels. Magnolias are very primitive flowers with numerous separate floral parts arranged in a spiral pattern.* (b) *The lady's slipper orchid,* Paphiopedilum invincible. *Orchids are highly specialized flowers, with the style and filaments fused into a single structure that bears both a stigma and, a short distance away, an anther. The lip of the flower is a modified petal that can serve as a landing platform for insects. Orchidaceae, with at least 17,000 species, is the second largest family of flowering plants.* (c) *The sunflower,* Helianthus annuus *("annual flower of the sun"). In the sunflower and other composites, numerous individual flowers make up the head, which acts as a large, single flower in attracting insects. The central portion of the head consists of separate florets, each comprising a pair of fused carpels forming a single ovary and fused anthers enclosed in a small corolla of fused petals. This central portion is surrounded by ray flowers (with yellow petals), which are often sterile. Composites (family Asteraceae), with some 22,000 species, are the largest family of flowering plants.* (d) *In Hibiscus, a column of stamens is fused around the style.* (e) *Stamens and stigma of a purple crocus.* (f) *Corn* (Zea mays), *a monoecious species. Separate male and female flowers are borne on the same plant. The tassels, at the top of the stem, are the male (pollen-producing) flowers. Each thread of "silk," seen emerging from the ear of corn, is the combined stigma and style of a female flower.* (g) *Water hemlock, a member of the carrot family, provides an example of inflorescences, or flower clusters.*

This may be a long distance; in corn, for instance, the pollen tube may grow to a length of 40 centimeters. The number of grains reaching the stigma is often greater than the number of ovules available for fertilization, creating intense competition among pollen tubes, the race going to the swift.

Each ovule contains a female gametophyte, which has also become reduced in size in the course of evolution. In many species, the female gametophyte consists of seven cells, with a total of eight haploid nuclei (Figure 29–6). One of the seven cells is the egg, containing a single haploid nucleus. On either side of the egg is a small cell known as a synergid. At the opposite end of the gametophyte are three small cells, the antipodal cells, whose function, if any, is unknown. The large central cell contains two haploid nuclei, called the polar nuclei because they move to the center from each end, or pole, of the gametophyte.

The two synergids, one on each side of the egg cell, attract the growing pollen tube. When the pollen tube reaches one of the synergids, it fuses with it, releasing the tube nucleus and the two sperm nuclei into the synergid. From the synergid, one sperm nucleus enters the egg cell and unites with the egg nucleus. The fertilized egg, or zygote, develops into the embryo—the young diploid sporophyte. The second sperm nucleus is released into the large central cell, where it unites with the two polar nuclei in a process of triple fusion. From the resulting $3n$ (triploid) cell, a specialized tissue called the **endosperm** develops. The endosperm surrounds and nourishes the developing embryo. These extraordinary phenomena of fertilization and triple fusion—together called "double fertilization"—take place, in all the natural world, only in the flowering plants.

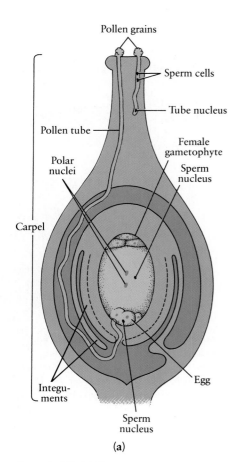

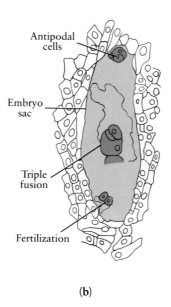

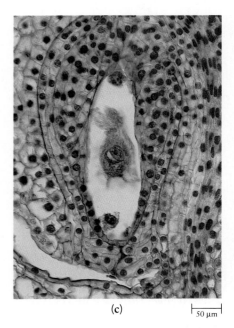

29–6 (a) *Fertilization in angiosperms. The pollen tube of the male gametophyte, or pollen grain, grows through the style and enters an ovule, which contains a seven-celled female gametophyte (the embryo sac). One of the sperm nuclei*
unites with the egg, forming the zygote. The other sperm nucleus fuses with the two polar nuclei that are present in a single large cell (which in the drawing fills most of the ovule). This triple fusion produces a triploid (3n) cell, from which the endo-
sperm will develop. The carpel shown here contains a single ovule.

(b, c) Diagram and micrograph of the embryo sac of a lily (Lilium) shortly after "double fertilization" has occurred.

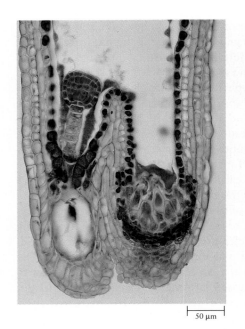

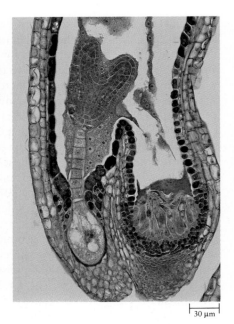

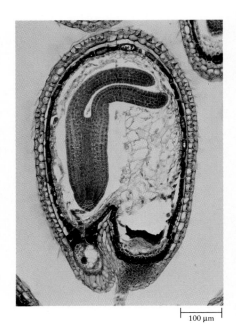

50 µm

30 µm

100 µm

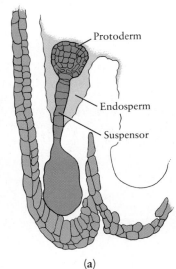

Protoderm

Endosperm

Suspensor

(a)

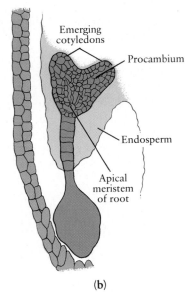

Emerging
cotyledons

Procambium

Endosperm

Apical
meristem
of root

(b)

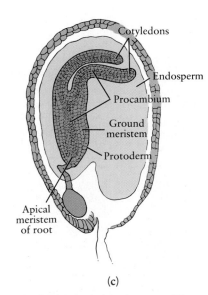

Cotyledons

Endosperm

Procambium

Ground
meristem

Protoderm

Apical
meristem
of root

(c)

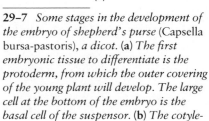

29–7 *Some stages in the development of
the embryo of shepherd's purse* (Capsella
bursa-pastoris), *a dicot.* (a) *The first
embryonic tissue to differentiate is the
protoderm, from which the outer covering
of the young plant will develop. The large
cell at the bottom of the embryo is the
basal cell of the suspensor.* (b) *The cotyle-*

*dons ("seed leaves") are beginning to
emerge. The procambium, a second
embryonic tissue, will later give rise to the
vascular tissues of the plant.* (c) *The cotyledons have developed further. Additional
differentiation has produced a third
embryonic tissue, the ground meristem,
from which the bulk of the tissue of the*

*young leaves, stems, and roots will be
derived. The three embryonic tissues,
known as the primary meristems, are continuous between the cotyledons and the
axis of the embryo.* (d) *The mature embryo
within its protective seed coat. The apical
meristems of the root and the shoot are
clearly differentiated.*

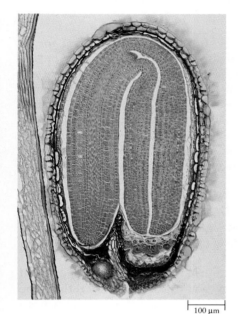

100 μm

THE EMBRYO

Following "double fertilization," the $3n$ cell divides mitotically to produce endosperm. The zygote also divides mitotically to form the embryo. As the embryo grows, its cells begin to **differentiate**—that is, they become different from one another. In its earliest stages, the embryo consists of a globular mass of cells on a stalk known as the suspensor. The cells of the suspensor, also formed from divisions of the fertilized egg, are actively involved in the delivery of nutrients to the embryo.

As development of the embryo proceeds, changes in its internal structure result in the formation of three distinct embryonic tissues. At the same time, or slightly later, emergence of the one cotyledon ("seed leaf") in monocots, or the two in dicots, occurs. Gradually the embryo takes on its characteristic form, a process known as **morphogenesis.** The developmental stages of a dicot embryo are shown in Figure 29–7.

In the early stages of embryonic growth, cell division takes place throughout the body of the young plant. As the embryo grows older, however, the addition of new cells becomes gradually restricted to certain parts of the plant body: the **apical meristems** (from the Greek *merizein,* "to divide"), located near the tips of the root and the shoot. During the rest of the life of the plant, primary growth—which chiefly involves the elongation of the plant body—originates in the apical meristems of roots and shoots.

THE SEED AND THE FRUIT

The seed consists of the embryo, which develops from the fertilized egg; the stored food, which consists of or derives from the endosperm; and the seed coat, which develops from the outermost layer or layers (integuments) of the ovule (Figure 29–8). At the same time, the fruit develops from the wall of the ovary (the base of the carpel). As the ovary ripens into fruit and the seeds form, the petals, stamens, and other parts of the flower may fall away.

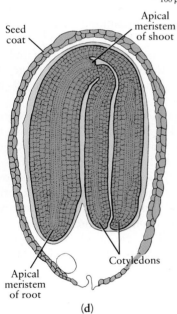

Seed coat

Apical meristem of shoot

Cotyledons

Apical meristem of root

(d)

29–8 *Seeds.* (a) *In dicots such as the common bean, the endosperm is digested as the embryo grows, and the food reserve is transferred to the fleshy cotyledons.* (b) *In corn and other monocots, the single cotyledon, known as the scutellum in corn and other grains, absorbs food reserves from the endosperm. The coleoptile is a sheath that encloses the apical meristem of the shoot; it is the first structure to appear above ground after the seed germinates.*

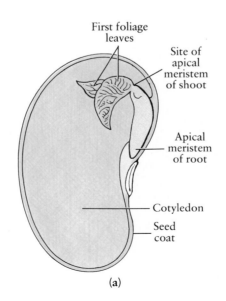

First foliage leaves

Site of apical meristem of shoot

Apical meristem of root

Cotyledon

Seed coat

(a)

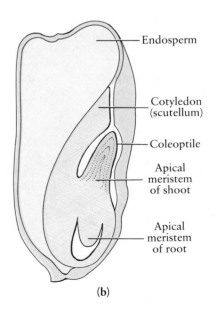

Endosperm

Cotyledon (scutellum)

Coleoptile

Apical meristem of shoot

Apical meristem of root

(b)

Peach

Pineapple

Raspberry

29–9 *A simple fruit, an aggregate fruit, and, on the right, a multiple fruit.*

29–10 *Development and structure of the pear, a pome. (a) Flower of the pear. The ovary is the basal portion of the carpel. (b) Older flower, after petals have fallen. (c) Longitudinal section and (d) cross section of the mature fruit. The core of the pear is the ripened ovary wall. The fleshy, edible part develops from the floral tube.*

Types of Fruits

As you know from your own observations, fruits take many different forms. These forms are adaptations for a variety of dispersal mechanisms, as we noted in Chapter 24. Fruits are generally classified as simple, aggregate, or multiple (Figure 29–9), depending on the arrangement of the carpels in the parent flower. Simple fruits develop from one carpel or the fused carpels of a single flower, whereas aggregate fruits, such as magnolia, raspberry, and strawberry, develop from several separate carpels of a single flower. Multiple fruits consist of the carpels of more than one flower. A pineapple, for example, is a multiple fruit formed from an inflorescence, or flower cluster. The ovaries of the individual flowers fuse as they mature.

Simple fruits are by far the most diverse. When ripe, they may be soft and fleshy or dry. The three principal types of fleshy fruit are the berry, the drupe, and the pome. In berries, such as tomatoes, dates, and grapes, there are one to several carpels, each of which may have one or many ovules. The inner layer of the fruit wall is usually fleshy.

In drupes, there are also one to several carpels, but usually only a single seed develops. The inner wall of the fruit is stony and usually adheres tightly to the seed. Some familiar drupes are peaches, cherries, olives, and plums. The peach is a typical drupe; the skin, the succulent, edible portion of the fruit, and the stone are three distinct layers of the wall of the mature ovary. The almond-shaped structure within the stone is the seed.

Pomes are highly specialized fleshy fruits, characteristic of the subfamily of roses that produces rose hips. A pome is derived from an inferior ovary (page 513) in which the fleshy portion comes largely from the floral tube (Figure 29–10). Apples and pears are pomes.

Dry fruits are classified into two groups, dehiscent and indehiscent. Mature dehiscent fruits break open while still attached to the parent plant, releasing the seeds; the seeds of indehiscent fruits, by contrast, are still within the fruit when it is shed from the parent plant (Figure 29–11). Among the most familiar dehiscent fruits are those of legumes (such as the pea family), in which the ovary splits down two sides; the pod is the mature ovary wall and the peas are the seeds. In other dehiscent fruits, the fruit is derived from a single carpel in which the ovary wall splits down one side; the fruits of columbines and milkweeds are examples.

Indehiscent fruits occur in many plant families. The most common is the achene, a small, single-seeded fruit; some achenes, such as those produced by the

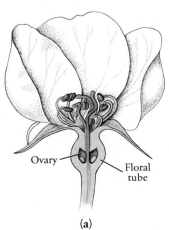

Ovary

Floral tube

(a)

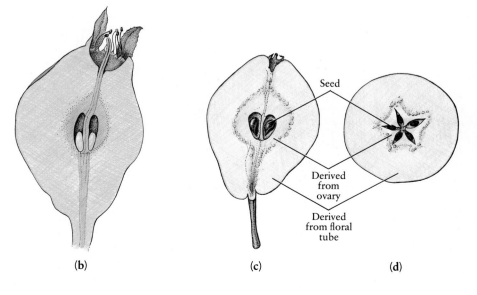

Seed

Derived from ovary

Derived from floral tube

(b)   **(c)**   **(d)**

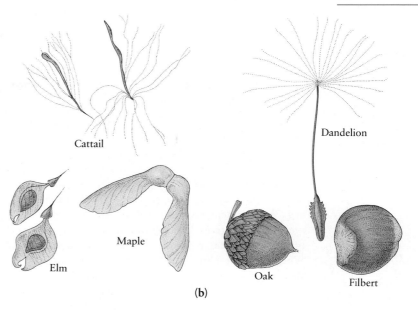

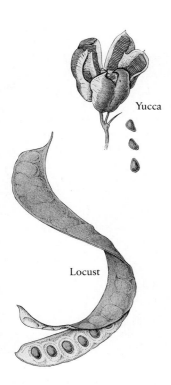

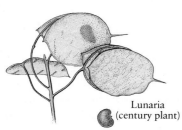

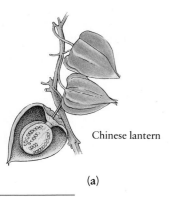

(a)

29–11 *Examples of (a) dehiscent and (b) indehiscent fruits. The distance that the seeds of dehiscent fruits are dispersed from the parent plant often depends on the force with which the fruit splits open. Many indehiscent fruits are carried a great distance from the parent plant, either by the wind or by hungry animals, before the seeds are released.*

elm and the ash, are winged. The most familiar kind of indehiscent fruit is the nut, which resembles the achene but has a stony coat and is derived from a compound ovary (an ovary formed from fused carpels). Examples of nuts are acorns and hazelnuts. Note that the word "nut" is used very indiscriminately in common speech: peanuts are the seeds of legumes; pine nuts are conifer seeds; almonds and coconuts are drupes.

ADAPTATIONS TO SEASONAL CHANGE

As we noted in Chapter 24, the angiosperms are thought to have evolved in upland regions of the tropics during a relatively mild period in the earth's history. As the climate grew colder, some angiosperms became extinct, while others were able to survive only near the equator. Some, presumably because of adaptations to drought (perhaps reflecting their highland origins), were able to survive in the cold, when water is locked up in ice. Chief among such adaptations is the capacity to remain dormant.

Dormancy and the Life Cycle

Depending on their characteristic patterns of active growth, dormancy, and death, modern plants are classified as annuals, biennials, and perennials. Among **annual** plants, the entire life cycle from seed to vegetative plant to flower to seed again takes place within a single growing season. Annuals include many familiar weeds, wild flowers, garden flowers, and vegetables. All vegetative organs (roots, stems, and leaves) die, and only the seeds, which are characteristically highly resistant to cold, desiccation, and other environmental hazards, bridge the gap between one generation and the next. Annual plants are usually soft-stemmed (nonwoody), or **herbaceous.** (An herb, botanically speaking, is a nonwoody seed plant, a definition that differs from the more common, culinary designation.)

In **biennial** plants, the period from seed germination to seed formation spans two growing seasons. The first season of growth often results in a short stem, a rosette of leaves near the soil surface, and a root. The root is often modified for food storage; sugar beets and carrots are examples of such storage roots. During the second growing season, the stored food reserves are mobilized for flowering, fruiting, and seed formation, after which the plant dies. Species that are characteristically biennials may, in some locations, complete their cycle in a single season or may require three or more years under adverse conditions.

The Staff of Life

Grains are the small, one-seeded fruits of grasses. Because they are relatively dry, they can be stored for long periods of time. The collecting and storing of grains from wild grasses is believed to have been an important impetus to the agricultural revolution of some 11,000 years ago (see Chapter 54). Today we are heavily dependent on cultivated wheat, rice, corn, rye, and other grains. In many countries, they constitute the principal component of the human diet. Wheat is about 9 to 14 percent protein. Its protein value is diminished, however, by its deficiency in certain essential amino acids, notably lysine (see page 73).

The grain of wheat, sometimes known as the kernel, is made up of the embryo, the endosperm, and, fused together, the mature ovary wall and the remains of the seed coat. More than 80 percent of the bulk of the wheat kernel and 70 to 75 percent of its protein are in the endosperm. White flour is made from the endosperm. Wheat germ, which is the embryo, forms about 3 percent of the kernel. It is usually removed as wheat is processed because it contains oil, which makes the flour more likely to spoil. Bran is the mature ovary wall, the remains of the seed coat, and the aleurone layer (the outer part of the endosperm); it constitutes about 14 percent of the kernel and is comprised mostly of cellulose. The bran is also removed when wheat is milled to make white flour. Bran somewhat decreases the caloric value of the wheat kernel, because we are unable to digest cellulose; bran therefore tends to speed the passage of food

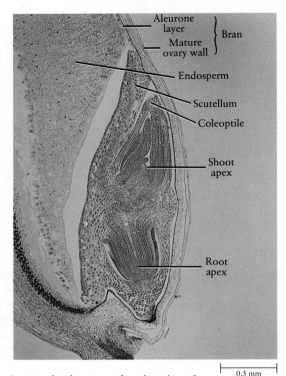

Longitudinal section of a wheat kernel.

through our intestinal tracts, resulting in decreased absorption. Until fairly recently, wheat germ and bran, which contain most of the vitamins, were sometimes used for human consumption but more often were fed to livestock. With new evidence of the importance of adequate amounts of fiber in the human diet, many more cereals and breads are made from whole grains in which the bran is retained.

Perennials are plants in which the vegetative structures persist year after year. The herbaceous perennials remain dormant as modified underground structures during unfavorable seasons, while the woody perennials, which include vines, shrubs, and trees, survive above ground. Woody perennials flower only when they become adult plants; a horse chestnut, for instance, may not flower until it is 25 years old.

The great advantage to being woody and perennial is that height can be added each growing season. Thus leaves can overtop and shade those of neighboring plants, flowers can be more conspicuously displayed, and seeds can be dispersed over a wider area. Woody perennials in favorable climates, such as the tropical rain forest, may live year after year with little change during the annual cycle, just as did the ancestral angiosperms. Perennials that live in areas where part of the year is unfavorable to growth have a variety of adaptations. Some, such as cacti and most gymnosperms, undergo little apparent change, although their rates of metabolism,

(a)

(b)

(c)

29–12 *Annual, biennial, and perennial plants.* (**a**) *An annual,* Linanthus dianthiflorus, *photographed in Baja California in the Sonoran desert. Many desert plants are annuals, racing through the entire life cycle from seed to flower to seed during a brief period following the seasonal rains.*

(**b**) *A biennial, the black-eyed Susan,* Rudbeckia hirta, *photographed on a July day in a Michigan field.*

(**c**) *An ancient perennial, the dawn redwood,* Metasequoia. *Twenty-five million years ago, the temperate zone of the Northern Hemisphere was covered with great forests of* Metasequoia. *Long thought to be extinct, living trees were discovered in western China in 1944. This tree in Philadelphia grew from seeds planted in 1947. Like the ginkgo, the bald cypress, and the larch, but unlike other gymnosperms, the dawn redwood is deciduous, shedding all its leaves in the fall.*

and therefore of growth, change with the seasons. Among dicots, many of the common vines, shrubs, and trees drop their leaves annually; such plants are said to be **deciduous.** They are typically found in regions where there is a marked seasonal variation in available water. The broad leaf of a deciduous dicot presents a much more efficient light-collecting surface than the needlelike leaf of a conifer (see page 508). The annual replacement of leaves is expensive, however, in terms of energy and other resources and can only take place in areas where the soil is fertile enough to supply the necessary nutrients and where the growing season is long enough to show an energy profit.

Seed Dormancy

The seeds of many wild plants require a period of dormancy before they will germinate. For example, the seeds of almost all plants growing in areas with marked seasonal temperature variations require a period of cold prior to germination. This physiological requirement ensures that the seed will "wait" at least until the next favorable growth period. The seeds of a few species can remain dormant and yet viable—with the embryo in a state of suspended animation—for hundreds of years. The record for dormancy, as far as is known, has been set by some seeds of East Indian lotus from a deposit at Pulantien, Manchuria. As determined by radioisotope dating, their age at the time of germination, in 1982, was 466 years.

The seed coat often plays a major role in maintaining dormancy. In some species, it acts primarily as a mechanical barrier, preventing the entry of water and gases, without which growth is not possible. In these plants, growth is initiated when the seed coat is worn away—for example, by being abraded by sand or soil, burned away in a forest fire, or partially digested as it passes through the digestive tract of a bird or other animal. In other species, dormancy is maintained chiefly by chemical inhibitors in the seed coat. These inhibitors undergo chemical changes in response to various environmental factors, such as light or prolonged cold or a sudden rise in temperature, which neutralize their effects, or they may be eroded or washed away by rainfall. Eventually, dormancy is broken and the stage is set for germination.

The dormancy requirement of seeds apparently evolved only recently, geologically speaking, among groups of plants subjected to the environmental stress of increasing winter cold characteristic of the most recent Ice Age. By this time— only 1.5 to 2 million years ago—the angiosperms were already a highly diversified group, and different populations responded to these pressures in different ways.

623

This explains why even closely related types of plants may have different mechanisms for maintaining and breaking dormancy.

A dormant seed is usually very dry—only about 5 to 20 percent of its total weight is water. Germination cannot begin until the seed has imbibed the water required for the metabolic activities of the growing embryo. Once germination is underway, the seed coat ruptures and the young sporophyte emerges. As we shall see in the next chapter, a series of growth processes immediately begin that give rise to the plant body and continue throughout its life.

SUMMARY

Flowers are the structures of sexual reproduction in the angiosperms. The pollen grains, which contain the male gametophytes, are produced by the anthers. Anther plus filament is known as the stamen. The carpel typically consists of the stigma (an area on which the pollen grains germinate), a style, and, at its base, the ovary. The ovary contains one or more ovules, and within each ovule is a female gametophyte that contains an egg cell.

A new cycle of life begins when a pollen grain germinates on the stigma of a flower of the same species, sending a pollen tube through the style and into an ovule. One of the sperm nuclei of the pollen grain fertilizes the egg cell in the female gametophyte; the other sperm nucleus unites with the two polar nuclei of the female gametophyte to form a triploid ($3n$) cell. Division of this triploid cell produces a special nutritive tissue, the endosperm. The phenomena of fertilization and triple fusion, called "double fertilization," occur only in the flowering plants. Following "double fertilization," the ovule develops into a seed and the ovary into a fruit.

The angiosperm seed, or mature ovule, consists of the embryo, the seed coat, and stored food. The petals, stamens, and other floral parts of the parent plant may fall away as the ovary ripens into fruit and the seeds form.

Dormancy—of the seed, of vegetative parts of the plant body, or of both— enables angiosperms to bridge periods of drought or cold unsuitable for plant growth. Angiosperms are classified as annuals, biennials, or perennials, depending on whether the plant body dies at the end of one growing season (annual) or after two seasons (biennial) or whether vegetative portions of the plant body persist from year to year (perennial). In regions where there is marked variation in the availability of water, perennials are frequently deciduous, dropping their leaves at the end of each growing season.

QUESTIONS

1. Distinguish among the following: ovary/ovule/egg cell/seed; male gametophyte/female gametophyte/embryo sac; pollination/fertilization; protoderm/procambium/ground meristem; annual/biennial/perennial.

2. In many plants, pollen production in the anthers occurs prior to or after the full development of the carpel of the same flower. What are the consequences of this shift in time frames?

3. J. B. S. Haldane, a mathematician who made major contributions to biology and whose pithy comment about beetles we quoted earlier (see page 574), once remarked: "A higher plant is at the mercy of its pollen grain." Explain.

4. Which of the following is a berry: blackberry, strawberry, mulberry, grape, pumpkin?

5. Seeds of the jack pine (a gymnosperm) are released from the female cones only after exposure to intense heat, as in a forest fire. Following release, the tough seed coats rupture, and the seeds germinate. What might be the advantage for the jack pine of this delay in seed release and germination?

6. Sketch a dicot embryo at the time of seed release. Identify each part in terms of the future development of the plant body.

7. What are the major adaptations that enable plants to survive the periodic drought (winter) of temperate regions?

The Plant Body and Its Development

As we discussed in Chapter 24, plants are multicellular photosynthetic organisms primarily adapted for life on land. The ancestor of the plants is thought to have been a multicellular green alga similar to *Coleochaete* (page 494). As in modern plants, its principal photosynthetic pigments were chlorophylls *a* and *b* and beta-carotene, all of which were contained in chloroplasts. Each of its cells had a membrane-bound nucleus and mitochondria and other organelles, as well as an external cell wall containing cellulose. Its energy source was sunlight, and it obtained oxygen, carbon dioxide, and minerals from the water in which it lived.

The photosynthetic cells of plants have the same few and relatively simple requirements: light, water, oxygen, carbon dioxide, and certain minerals. From these simple materials they, like their ancestors, make the sugars, fatty acids, amino acids, nucleotides, and other organic substances on which all plant and animal life depends. In algae and in small, simple plants (such as bryophytes) living in a moist environment, each of the required materials is immediately available to every cell. Most plants, however, live in a very different environment, and their photosynthetic cells require a complex life-support system in order to function. The plant body is, in effect, that life-support system.

Figure 30–1 diagrams the external structure of an economically important angiosperm, the potato plant. This plant, like other vascular plants, is characterized by a root system that anchors the plant in the ground and collects water and minerals from the soil; a stem or trunk that raises the photosynthetic parts of the plant toward the sun; and structures highly specialized for light capture and photosynthesis, the leaves. Its cells, unlike those of a multicellular alga, are not autonomous and can survive only through cooperation and a division of labor. This division of labor is made possible by processes of differentiation that begin during embryonic development and continue throughout the life of the plant.

THE CELLS AND TISSUES OF THE PLANT BODY

As we saw in the last chapter, the cells of the angiosperm embryo differentiate early in its development into three distinct tissues: the protoderm, the procambium, and the ground meristem (see Figure 29–7, page 618). These embryonic tissues, known as the primary meristems, give rise to three **tissue systems** that are continuous throughout the plant body. The protoderm, the first tissue to differentiate, is the origin of the **dermal tissue** system, which provides an outer protective covering for the entire plant body. The procambium, the next tissue to differentiate, gives rise to the **vascular tissue** system, composed of xylem and phloem. As we saw in Chapter 24, xylem transports water and dissolved minerals, while phloem transports dissolved sugars and other organic compounds from the photosynthetic (autotrophic) cells of leaves and green stems to the nonphoto-

30–1 *The body plan of a flowering plant, the potato* (Solanum tuberosum). *The aboveground shoot system consists of the stem, the leaves, whose primary function is photosynthesis, and the flowers, the reproductive structures. After fertilization, the flower petals fall away and the ovaries mature to form the fruits, as we saw in the last chapter. Leaves, which may be simple or, as in the potato, compound, appear at regions on the stem known as nodes. The portions of the stem between successive nodes are called internodes.*

In many plants, the only belowground structures are the roots, which supply water and minerals to the stem, leaves, flowers, and fruits. In the potato, however, the most conspicuous underground structures are the tubers, which are enlarged stems adapted for food storage.

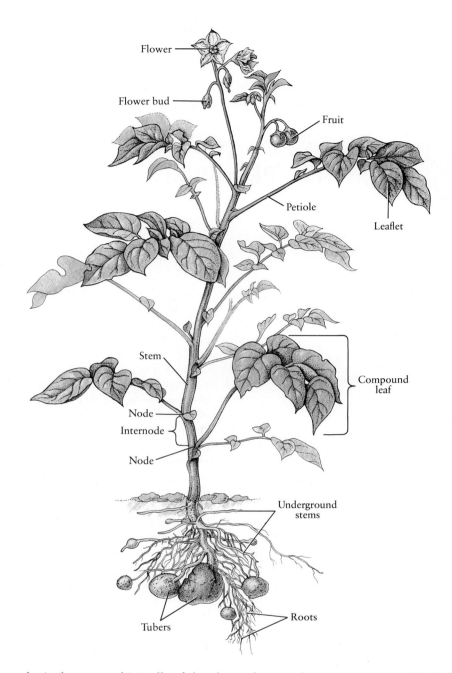

synthetic (heterotrophic) cells of the plant. The vascular tissues are embedded in the **ground tissue** system, which is derived from the embryonic ground meristem. As we shall see in the course of this chapter, the principal differences in the structure of leaves, stems, and roots lie in the relative distribution of the vascular and ground tissue systems.

The most frequently encountered cells in the plant body are of a type known as **parenchyma**. These cells, which occur in all three tissue systems and predominate in the ground tissues, are typically many-sided, with thin, flexible walls. In addition to a nucleus, mitochondria, and other organelles, parenchyma cells generally contain plastids, which, depending on the location of the cell, may be chloroplasts, leucoplasts, or chromoplasts (see page 120). The plant cell shown in Figure 5–16 (page 113) is a photosynthetic parenchyma cell, containing chloroplasts. In addition to photosynthesis, parenchyma cells perform a variety of essential functions in the plant, including respiration and storage of food and water. Each of the tissue systems also contains additional cell types, specialized for the particular functions of the tissue.

Photosynthesis is, of course, the most fundamental activity of a plant, on which all else depends. We shall therefore begin our examination of the plant body with the leaves, the primary photosynthetic organs, which are essentially the same in a seedling as in a mature plant. Following our consideration of leaves, we shall shift our attention back to the germinating seed, and examine the structure and development of roots and stems, the other two organs of the plant body, without which the leaves could not survive.

LEAVES

Leaf Structure

The structure of a leaf (Figure 30–2) is a compromise between three conflicting evolutionary pressures: to expose a maximum photosynthetic surface to sunlight, to conserve water, and, at the same time, to provide for the exchange of gases necessary for photosynthesis.

The photosynthetic cells of leaves are parenchyma cells of two types: **palisade parenchyma,** which are densely packed, columnar cells located just below the upper surface of the leaf, and **spongy parenchyma,** which are irregularly shaped cells in the interior of the leaf, often with large spaces between them. These spaces are filled with gases, including water vapor, oxygen, and carbon dioxide. Most of the photosynthesis occurs in the palisade cells, which are specialized for intercepting light.

Palisade and spongy parenchyma make up the ground tissue of the leaf, known as the **mesophyll,** or "middle leaf." The mesophyll is enclosed in an almost airtight wrapping of epidermal cells, which secrete a waxy substance called **cutin.** The cutin forms a coating, the **cuticle,** over the outer surface of the epidermis. The epidermal cells and the cuticle are transparent, permitting light to penetrate to the photosynthetic cells.

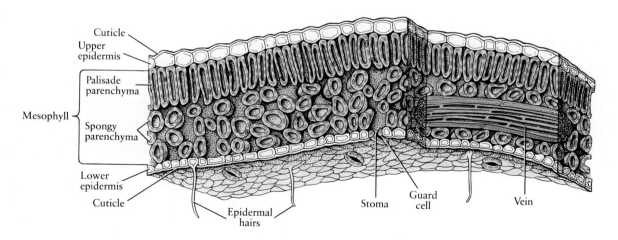

30–2 *The structure of a leaf. Photosynthesis takes place in the palisade cells and, to a lesser extent, in the spongy parenchyma. The chloroplasts are indicated in bright green. Note that the cytoplasm, which contains the chloroplasts, is concentrated near the cell surface, and the* centers of the cells are filled with large vacuoles. The chloroplasts move within the cytoplasm, orienting themselves to the sun. The veins carry water and solutes to and from these mesophyll cells. The interior of the leaf is enclosed by epidermal cells covered with a waxy layer, the cuti- cle. Openings in the epidermis, the stomata, permit the exchange of gases. The guard cells surrounding the stomata also have chloroplasts, which may (or may not) play a role in opening and closing the stomata (see pages 653–655).

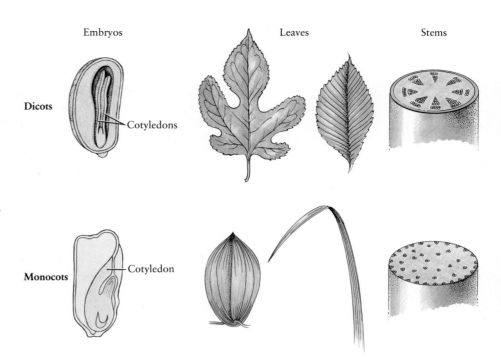

Embryos   Leaves   Stems

Dicots — Cotyledons

Monocots — Cotyledon

30-3 *As we have noted previously, the two classes of angiosperms are the dicots and the monocots. The names refer to the fact that the dicot embryo has two cotyledons ("seed leaves"), and the monocot embryo has one. Other characteristic differences are visible in the plant body. The principal veins of dicot leaves are usually netted; those of monocot leaves are usually parallel. In dicot stems, bundles of vascular tissue are arranged around a central core of ground tissue; in monocot stems, the vascular bundles are scattered throughout the ground tissue. As we saw in the last chapter, there are also characteristic differences in the number of floral parts and in the structure of the pollen grains.*

Substances move into and out of leaves through two quite different structures: vascular bundles and stomata. Water and dissolved minerals are transported into leaves—and the products of photosynthesis are transported out—by way of the vascular bundles. The vascular bundles, which are known in leaves as **veins,** pass through the petioles (leaf stalks) and are continuous with the vascular tissues of the stem and root. Veins form distinctive patterns in leaf blades, which are conspicuously different in monocots and dicots. In monocots, the major veins are usually all parallel; in dicots, the venation (vein pattern) is netted (Figure 30–3).

Gases—oxygen and carbon dioxide—move into and out of leaves by diffusion through stomata (plural of *stoma,* the Greek word for "mouth"). A stoma consists of a small opening, or pore; it is surrounded by two specialized cells in the leaf epidermis, called guard cells, that open and close the pore (Figure 30–4). The exchange of oxygen and carbon dioxide is, as we saw in Chapter 10, necessary for photosynthesis. However, as these gases are exchanged between the atmosphere and the leaf interior, water also escapes from the leaf. About 90 percent of the water loss from the plant body is through the stomata; the remaining 10 percent is through the epidermal cells.

30-4 *An open stoma. The stomata lead into air spaces within the leaf that surround the thin-walled spongy parenchyma cells. The air in these spaces, which make up 15 to 40 percent of the total volume of the leaf, is saturated with water vapor that has evaporated from the photosynthetic cells.*

10 μm

Stomata are typically most abundant on the undersurface of leaves. They may be very numerous. For example, on the lower surface of tobacco leaves there are about 19,000 stomata per square centimeter; on the upper surface there are about 5,000 stomata per square centimeter.

Leaf Adaptations and Modifications

Leaves come in a variety of shapes and sizes, ranging from broad fronds to tiny scales. Some of these differences can be correlated with the environments in which the plants live. Large leaves with broad surfaces are often found in plants that grow under the canopy in a tropical rain forest, where water is plentiful but competition for light is intense. Leaves of such plants sometimes have "drip tips," which facilitate the runoff of rainwater. Leaves with small surfaces are usually associated with dry climates. In conifers, for example, the photosynthetic surfaces are greatly reduced and there is an extra-thick layer of epidermis and cuticle (see page 508). In such plants, photosynthesis is reduced but so is water loss. The trees are thus able to survive long periods of drought, including the drought of winter, when water is frozen in snow or ice and is unavailable to plants. Similarly, angiosperms in dry habitats often have small, leathery leaves. This reduction in leaf surface reaches an extreme in those desert cacti in which the leaves are modified as spines—hard, dry, nonphotosynthetic structures. (The terms "spine" and "thorn" are often used interchangeably; however, thorns are technically modified branches.) In these plants, photosynthesis takes place in the fleshy stems, which are also water-storage organs.

In many plants, leaves are succulent; that is, they are adapted for water storage. Among the most interesting examples of this adaptation are the "window" plants native to the deserts of South Africa. Their leaves grow almost entirely underground, with only the transparent "window" tip of the leaf protruding above the soil surface. The transparent water-storage tissue of the leaf provides a conduit for light to reach the underground photosynthetic cells.

30–5 *Cross section of the leaf of an oleander, an angiosperm adapted to a dry climate. At the top, a very thick cuticle covers the multiple epidermis, so called because it consists of four layers of cells. The stoma is contained within a stomatal crypt, which is lined with epidermal hairs. The veins are seen in cross section, which is why they look different from the vein in Figure 30–2.*

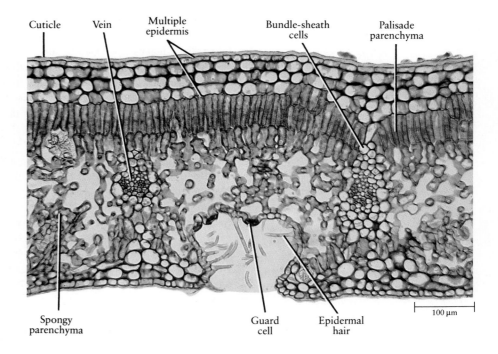

Cuticle Vein Multiple epidermis Bundle-sheath cells Palisade parenchyma

Spongy parenchyma Guard cell Epidermal hair

100 µm

(a)

(b)

(c)

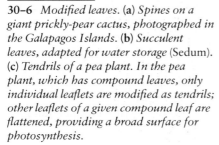

30–6 *Modified leaves.* **(a)** *Spines on a giant prickly-pear cactus, photographed in the Galapagos Islands.* **(b)** *Succulent leaves, adapted for water storage (Sedum).* **(c)** *Tendrils of a pea plant. In the pea plant, which has compound leaves, only individual leaflets are modified as tendrils; other leaflets of a given compound leaf are flattened, providing a broad surface for photosynthesis.*

Leaves may also be specialized for other functions, such as food storage or support. For example, a bulb, such as the onion, is a large bud consisting of a short stem with many leaves modified for storing food. The "head" of a cabbage also consists of a compressed stem bearing numerous thick, overlapping leaves. In some plants, the petioles become thick and fleshy: celery and rhubarb are two familiar examples. The tendrils of some climbing plants—the garden pea, for instance—are modified leaves or leaflets.

CHARACTERISTICS OF PLANT GROWTH

As we discussed in the last chapter, the embryos of many angiosperms pass through a dormant stage prior to germination of the seed. With germination, growth resumes, the seed coat ruptures, and the young sporophyte emerges. The first foliage leaves open to the sun and begin photosynthesizing, while internally the growth processes that give rise to the plant body continue.

The primary growth of the plant involves differentiation of the three tissue systems, elongation of roots and stems, and the formation of lateral roots and of branches. As we noted earlier, after development of the embryo is complete, subsequent primary growth originates in the apical meristems of the root and shoot.

The existence of such meristematic areas, which add to the body of the plant throughout its life, is one of the principal differences between plants and animals. Higher animals stop growing when they reach maturity, although the cells of certain "turnover" tissues, such as the skin or the lining of the intestine, continue to divide. Plants, however, continue to grow during their entire life span. Growth in plants is the counterpart, to some extent, of motility in animals. Plants "move" by extending their roots and shoots, both of which involve changes in size and form. As a result of these changes, a plant modifies its relationship with the environment, for example, by curving toward the light and extending its roots toward water. The sequence of growth stages in plants thus corresponds to a whole series of motor acts in animals, especially those associated with obtaining food and water. In fact, growth in plants serves many of the functions that we group under the term "behavior" in animals.

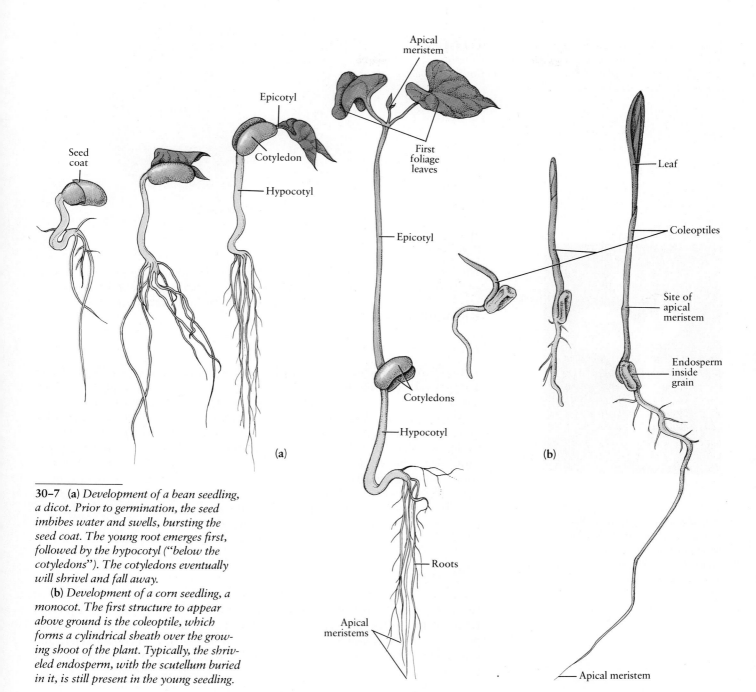

30-7 (a) *Development of a bean seedling, a dicot. Prior to germination, the seed imbibes water and swells, bursting the seed coat. The young root emerges first, followed by the hypocotyl ("below the cotyledons"). The cotyledons eventually will shrivel and fall away.*

(b) *Development of a corn seedling, a monocot. The first structure to appear above ground is the coleoptile, which forms a cylindrical sheath over the growing shoot of the plant. Typically, the shriveled endosperm, with the scutellum buried in it, is still present in the young seedling.*

ROOTS

Roots are specialized structures that anchor the plant and take up water and essential minerals. The embryonic root is the first structure to break through the seed coat, and, in an older plant, the root system may make up more than half of the plant body. The lateral spread of tree roots is usually greater than the spread of the crown of the tree. In a study made on a four-month-old rye plant, the total surface area of the root system was calculated to be 639 square meters, 130 times the surface area of the leaves and stem. Root growth is affected by soil conditions and availability of water. The deepest known roots were those of a mesquite (a desert shrub) found growing in a new open-pit mine near Tucson, Arizona, in 1960; they had penetrated to a depth of 53.3 meters.

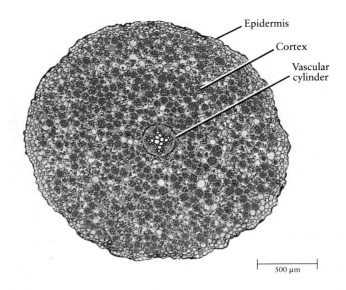

Epidermis

Cortex

Vascular
cylinder

500 μm

30-8 *Root of a buttercup, a dicot, in cross section. The plastids in the parenchyma cells of the cortex contain starch grains, stained purple in this preparation. The vascular cylinder is shown in more detail in Figure 30-11a.*

30-9 *A radish seedling. Note the discarded seed coat, the cotyledons, and the primary root with its numerous root hairs. Most of the uptake of water and minerals occurs through the root hairs, which form just behind the growing tip of the root.*

Root Structure

The internal structure of the angiosperm root is comparatively simple. In dicots and most monocots, the three tissue systems (dermal, ground, and vascular) are arranged in three concentric layers: the **epidermis,** the **cortex,** and the **vascular cylinder** (Figure 30-8).

The Epidermis

The epidermis, which covers the entire surface of the young root, absorbs water and minerals from the soil and protects the internal tissues. The cuticle is either absent or very thin compared with that found on the surface of a leaf.

The epidermal cells of the root are characterized by fine, tubular outgrowths, known as **root hairs** (Figure 30-9). Root hairs are slender extensions of the epidermal cells; in fact, the nucleus of the epidermal cell is often found within the root hair. In the rye plant previously mentioned, the roots were estimated to have some 14 billion root hairs. Placed end to end, they would have extended more than 10,000 kilometers. Most of the water and minerals that enter the root are absorbed by these delicate outgrowths of the epidermis. In the mature plants of many species, however, mycorrhizal associations (page 490) substitute for root hairs.

The Cortex

As you can see in Figure 30-8, the cortex occupies by far the greatest volume of the young root. The cells of the cortex are parenchyma cells, as in the ground tissue of the leaf. However, root parenchyma usually lacks functional chloroplasts; instead, the plastids are specialized for food storage and contain starch and other organic substances. (In some species, the roots are highly specialized for this function; beets and carrots are examples of roots with an abundance of storage parenchyma.) There are many air spaces in the cortex, and oxygen from the soil enters these spaces through the epidermal cells and is used by the cortical cells in respiration.

Unlike the rest of the cortex, the cells of the innermost layer, the **endodermis,** are compact and have no spaces between them. Each endodermal cell is encircled by a continuous band of wax, known as the **Casparian strip** (Figure 30-10). The Casparian strip, which is located within the cell wall and adheres tightly to the cell membrane, is not permeable to water. Therefore, water and dissolved substances, which can move freely around the other cortical cells and through their cell walls, must pass through the cell membranes of endodermal cells. As you will recall (page 130), water, oxygen, and carbon dioxide pass easily through cell membranes, but many ions and other substances do not. Thus, the membranes of the endodermal cells regulate the passage of such substances into the vascular tissues of the root, thereby determining what is transported to the rest of the plant body.

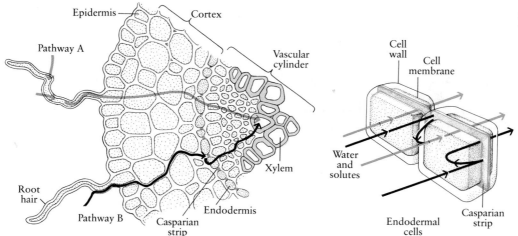

30-10 *Diagrammatic cross section of a root, showing the two pathways of uptake of water and dissolved substances. Most of the solutes and some of the water entering the root follow pathway A, indicated in color; the solutes move by active transport and diffusion and the water by osmosis through the cell membranes and plasmodesmata (page 140) of a series of living cells. Most of the water and some of the solutes entering the root follow pathway B, indicated in black, flowing through the cell walls and along their surfaces. Notice, however, the location of the Caspar-ian strip and how it blocks pathway B all around the vascular cylinder of the root. In order to pass the Casparian strip, both water and solutes must be transported through the cell membranes of the endodermal cells, as in pathway A. After the water and solutes have crossed the endodermis, most of the solutes continue along pathway A to the conducting cells of the xylem, and most of the water returns to pathway B for the remaining distance to the conducting cells.*

The Vascular Cylinder

The vascular cylinder of the root consists of the vascular tissues (xylem and phloem) surrounded by one or more layers of cells, the **pericycle,** from which branch roots arise. In most species, the vascular tissues of the root are grouped in a solid cylinder, similar to that shown in Figure 30–8. Figure 30–11a shows the details of the vascular cylinder of a buttercup root (a dicot). In some monocots, however, the vascular tissues form a cylinder around a **pith,** a central core of ground tissue (Figure 30–11b).

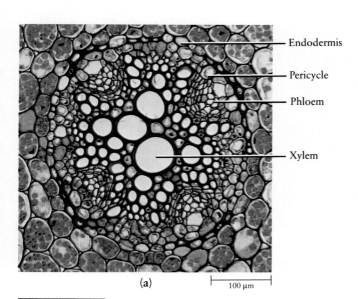

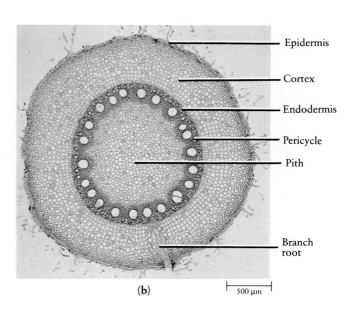

30-11 (a) *Details of the vascular cylinder of the buttercup root (a dicot) shown in Figure 30–8. The endodermis, which is outside the pericycle, is considered part of the cortex. The endodermis contains the Casparian strips.*

(b) *Cross section of the root of a corn plant (a monocot), showing the vascular cylinder enclosing the pith. Part of a branch root, which emerges from the pericycle, can be seen in the lower portion of the micrograph.*

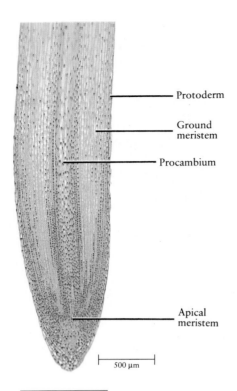

Protoderm

Ground
meristem

Procambium

Apical
meristem

500 μm

30-12 *Longitudinal section of the root tip of an onion, a monocot. The primary meristematic tissues—protoderm, ground meristem, and procambium—can be distinguished close to the apical meristem.*

Primary Growth of the Root

The first part of the embryo to break through the seed coat, in nearly all seed plants, is the embryonic root, or **radicle.** At its tip is the root cap, which protects the apical meristem as the root is pushed through the soil. Cells of the root cap wear away and are constantly replaced by new cells from the meristem.

Certain cells in the meristem retain the capacity to produce new cells and thus perpetuate the meristem. All the other cells in the root—which are the progeny of these relatively few meristematic cells—eventually differentiate, some becoming cells of the root cap and others forming the tissue systems of the root. The maximum rate of cell division occurs at a point well above the tip of the meristem. Then, just above the point where cell division is greatly reduced, the cells gradually elongate, growing to 10 or more times their original length, often within the span of a few hours. This elongation is the principal cause of primary growth in roots, although, of course, growth is ultimately dependent on the production of new cells that become part of the zone of elongation.

As the cells elongate, they begin to differentiate. The first cells to differentiate in roots are the conducting cells of the phloem, followed by the conducting cells of the xylem. In the region of the root where the xylem first forms, the endodermis also differentiates. To the inside of the endodermis, the pericycle forms. At approximately the same level in the elongating root, the epidermal cells differentiate and begin to extend root hairs into crevices between the soil grains.

This sequence of growth, as shown in Figure 30–14, occurs in the first root of a seedling and is repeated over and over again in all the growing root tips of a plant—even those of a tree 50 meters tall.

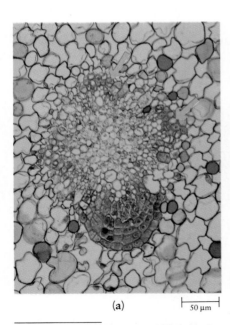

(a)   50 μm

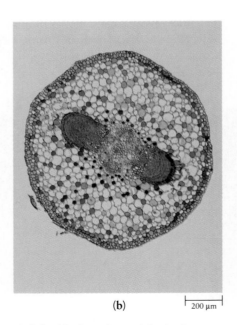

(b)   200 μm

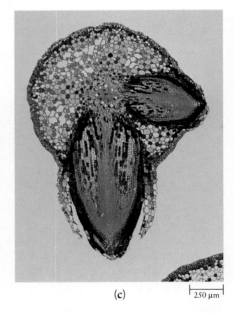

(c)   250 μm

30-13 *Sections from three different willow (Salix) roots, showing stages in the development of branch roots. In the upper portion of (a), the origin of two branch roots in the pericycle is marked by arrows; in the lower portion of the micrograph, a*

slightly older branch root is beginning to grow through the cortex. This is a very young root, and the vascular cylinder is not yet fully developed. In (b), two branch roots have traveled almost halfway across the cortex. In (c), one branch root has bro-

ken through the epidermis to the outside, and another is about to do so. As they grow, branch roots destroy the tissues in their path, partly by crushing and partly by digestion with enzymes.

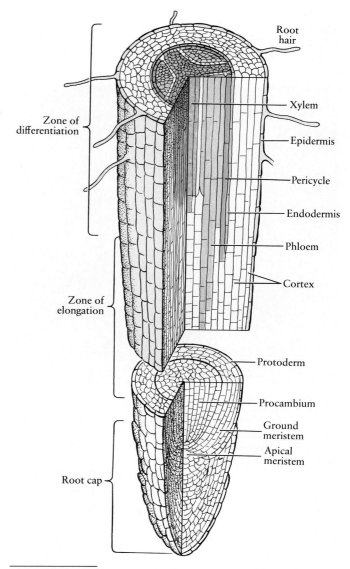

30–14 *The growth regions of a dicot root. New cells are produced by the division of cells within the apical meristem. The cells above the meristem undergo a characteristic series of changes as the distance increases between them and the root tip. First, there is a maximum rate of cell division, followed by cell elongation with little further division. As the cells elongate, they differentiate into the three primary meristems that give rise to the three tissue systems of the root. The protoderm becomes the epidermis, the ground meristem becomes the cortex, and the procambium becomes the primary xylem and primary phloem. Some of the cells produced by the apical meristem differentiate to form the protective root cap.*

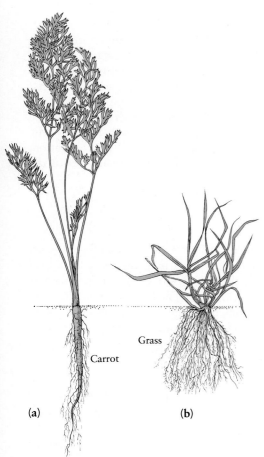

30–15 *Types of root systems. (a) Taproot system of a carrot (a dicot). (b) Fibrous root system of rye grass (a monocot).*

Patterns of Root Growth

In many dicots, the primary root develops into a large **taproot,** which, in turn, gives rise to lateral, or branch, roots (Figure 30–15a). In monocots, the primary root is usually short-lived and the final root system develops from the base of the stem; roots of this kind are called **adventitious roots.** ("Adventitious" describes any structure growing from other than its "usual" place.) These adventitious roots and their branches form a fibrous root system (Figure 30–15b).

635

(a)

(b)

30–16 (a) *Prop roots of corn. These are adventitious roots, arising from the stem.* (b) *Air roots (pneumatophores) of a white mangrove. The root tips grow up out of the mud in which these trees grow and take up oxygen needed by the roots for respiration.*

Aerial roots are adventitious roots produced from aboveground structures. Some aerial roots, such as those of English ivy, cling to vertical surfaces and thus provide support for the climbing stem. In other plants, such as corn, aerial roots serve as prop roots (Figure 30–16a). Trees that grow in swamps, such as the red mangrove and the bald cypress, often have prop roots.

In swampy areas the soil is usually low in oxygen. Some trees that grow in these habitats develop roots that grow out of the water and serve not only to anchor the plant but also to supply the root cells with the oxygen needed for respiration. White mangroves, for example, have roots whose tips grow upward out of the mud and serve this aerating function (Figure 30–16b).

STEMS

Stems display leaves to the light and are the pathway through which substances are transported between the roots and the leaves. They may also be adapted for storing food or water. As we saw in Figure 30–1, white potatoes are enlarged, underground stem tips, known as tubers, that are filled with starch. In some species of plants that grow in arid environments, water is stored in large parenchyma cells in the stems; the stem of a cactus, for example, may be 98 percent water by weight.

Stem Structure

The outer surface (dermal tissue) of young green stems, like that of leaves and roots, is made of epidermal cells. Like leaves, green stems are covered with a waxy cuticle, contain stomata, and are photosynthetic.

The bulk of the tissue in a young stem is ground tissue. As in leaves and roots, it is composed mostly of parenchyma cells. The turgor (page 134) of these cells provides the chief support for young green stems. The ground tissue of stems also may contain specialized supporting tissues known as **collenchyma** and **sclerenchyma**. Unlike the thin-walled parenchyma cells, collenchyma cells (Figure 30–

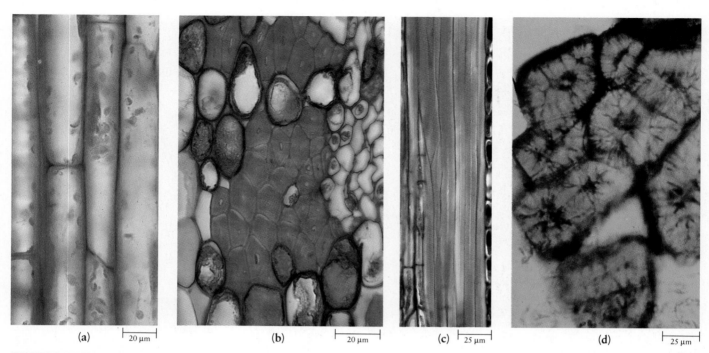

(a) ⊢ 20 μm ⊣   (b) ⊢ 20 μm ⊣   (c) ⊢ 25 μm ⊣   (d) ⊢ 25 μm ⊣

30–17 *Some types of cells found in the ground tissue of stems. (a) Collenchyma cells, viewed in longitudinal section. Their irregularly thickened cellulose walls are rich in pectin and contain much water. The walls are plastic, and so the cell can continue to grow. (b) Cross section and (c)*

longitudinal section of phloem fibers from the stem of a linden tree (Tilia americana). *Only a portion of the length of the fibers can be seen in* (c). *These sclerenchyma cells have thickened, often lignified cell walls that give them strength and rigidity. Many fibers, but not all, are dead at*

maturity. (d) Sclereids, another type of sclerenchyma cell, have very thick lignified walls. Sclereids are often found in seeds and fruits, as well as in stems. These sclereids, called stone cells, are from a pear; they give the fruit its characteristic gritty texture.

17a) have primary walls (see page 107) that are thickened at the corners or in some other uneven fashion. Their name derives from the Greek word *colla,* meaning "glue," which refers to their characteristic thick, glistening walls. Collenchyma cells are often located just inside the epidermis, forming either a continuous cylinder or distinct vertical strips of tissue. They provide support for the growing regions of young stems and branches.

Sclerenchyma cells are of two types: fibers and sclereids. Fibers, which are extremely elongated, somewhat elastic cells (Figure 30–17b and c), typically occur in strands or bundles arranged in a definite pattern characteristic of the particular species. They are often associated with the vascular tissues. Plant fibers such as flax, hemp, jute, sisal, and raffia have long been used in human artifacts, including baskets, rope, and cloth. Sclereids (Figure 30–17d), which are variable in form, are also common in stems. Layers of sclereids are found in seeds, nuts, and fruit stones as well, where they form the hard outer coverings.

The name sclerenchyma is derived from the Greek *skleros,* meaning "hard." This property of sclerenchyma cells is a result of the impregnation of their cell walls with lignin, a complex macromolecule that toughens and hardens cellulose. Sclerenchyma cells differ from collenchyma in three other respects: (1) they have secondary walls (page 107) in addition to primary walls; (2) the cells are often dead at maturity, with only their cell walls remaining; and (3) they usually occur in regions of the plant body that have completed primary growth.

30–18 *In angiosperms, the conducting elements of the phloem are sieve tubes, made up of individual cells, the sieve-tube members. These cells, which lack nuclei at maturity, are usually found in close association with companion cells, which do have nuclei. Sieve-tube members are joined to other sieve-tube members at their ends by sieve plates. (a) Longitudinal view of a sieve tube in the stem of the squash (Cucurbita maxima). P-protein lines the inner surface of the cell walls of the sieve-tube members. (b) Face view of a sieve plate between two mature sieve-tube members. In this electron micrograph, the pores of the sieve plate are open. In most cut sections of phloem tissue, however, the pores are plugged either with callose, a polysaccharide deposited by the sieve-tube members in response to injury, or with P-protein. Both callose and P-protein prevent leakage from the sieve tube. Callose is also deposited as part of the normal aging process. (c) Longitudinal section of phloem, showing mature and immature sieve-tube members. The arrows point to P-protein bodies in immature cells.*

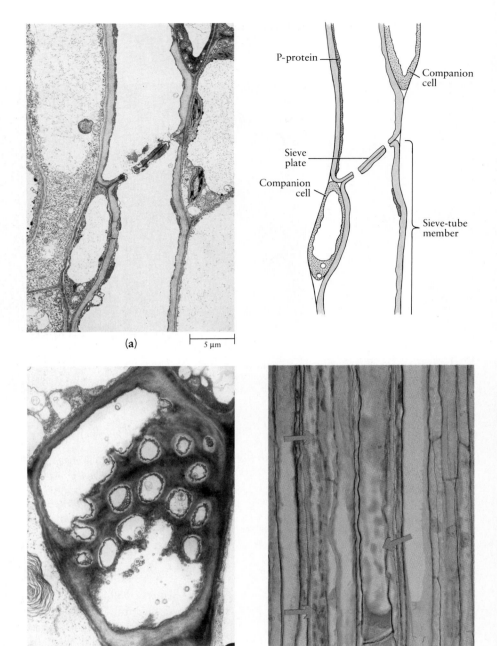

(a) 5 µm

(b) 2 µm

(c) 50 µm

Vascular Tissues

The vascular tissues, phloem and xylem, consist of specialized conducting cells, supporting fibers, and parenchyma cells, which store food and water. The conducting cells of the phloem, as we noted previously, transport the products of photosynthesis, chiefly in the form of sucrose, from the leaves to the nonphotosynthetic cells of the plant. In gymnosperms, these conducting cells are sieve cells; in angiosperms, they are sieve-tube members (Figure 30–18a). A **sieve tube** is a vertical column of sieve-tube members joined by their end walls. These end walls, called **sieve plates,** have pores leading from one sieve-tube member to the next (Figure 30–18b).

The sieve-tube members, which are alive at maturity, are filled largely with a watery fluid called sieve-tube sap. In dicots and some monocots, they also contain a proteinaceous substance called slime, or P-protein (the "P" stands for phloem). The function of P-protein is unknown, although some botanists believe that, along with the polysaccharide callose, it seals sieve-plate pores in response to injury. In mature sieve-tube members, P-protein lies along the inner surface of the cell wall and is continuous from one cell to the next through the sieve plates.

As a sieve-tube member matures, its nucleus and many of its organelles disintegrate. Sieve-tube members, however, are characteristically associated with specialized parenchyma cells called **companion cells,** which contain all of the components commonly found in living plant cells, including a nucleus. Companion cells may be responsible for the secretion of substances into and out of the sieve-tube members and are thought also to provide nuclear functions for the sieve tubes and fulfill their energy requirements. A sieve-tube member and its companion cell arise from the same mother cell.

Specialized cells in the xylem conduct water and minerals from the roots to other parts of the plant body. It is customary to think of xylem as transporting water up and phloem as transporting sugars down, but if you think of the various shapes of plants you can see that water must also often be transported laterally, as along a tendril, or even down, as to the branches of a weeping willow. Conversely, sugars must often go upward, as into a flower or fruit.

In angiosperms, the conducting cells of the xylem are tracheids and vessel members. Both of these cell types have thick secondary walls impregnated with lignin, and both are dead at functional maturity. Tracheids are long, thin cells that overlap one another on their tapered ends (Figure 30–19a). These overlapping surfaces contain thin areas, known as pits, where no secondary wall has been deposited. Water passes from one tracheid to the next through the pits. Vessel members, which are much shorter and wider, also differ from tracheids in that their end walls contain perforations or are entirely absent (Figure 30–19b and c). Thus, the vessel members form a continuous **vessel,** which is a more effective conduit than a series of tracheids. Seedless vascular plants and most gymnosperms have only tracheids; most angiosperms have both tracheids and vessel members.

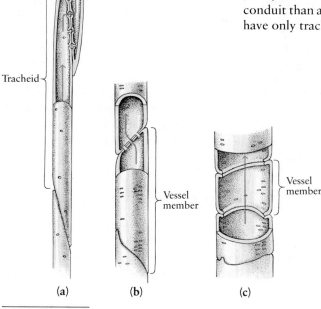

Tracheid

Vessel member

Vessel member

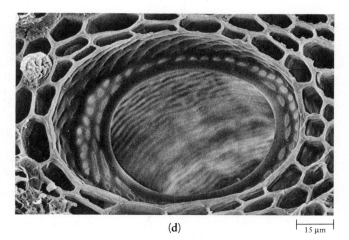

(a) (b) (c) (d) ⊢ 15 μm ⊣

30–19 *Tracheids and vessel members are the conducting cells of the xylem in angiosperms.* (a) *Tracheids are a more primitive and less efficient type of conducting cell. Water moving from one tracheid to another passes through pits. Pits are not perforations but simply areas in which there is no secondary cell wall. Water moving from one tracheid to another passes through two primary cell walls and the middle lamella (see page 106).*

Vessel members differ from tracheids in that the primary walls and middle lamellae of vessel members are perforated at the ends where they are joined with other vessel members. (b) *There may be numerous perforations in adjoining walls of vessel members, or* (c) *the adjoining walls may dissolve completely as the cells mature, forming a single opening. Vessel members are also characteristically shorter and wider than tracheids, and their*

adjoining walls are less oblique. Vessel members are connected with other vessel members and also with other cells by pits in the side walls.

(d) *A view into a vessel in the xylem of a prop root in a corn plant.*

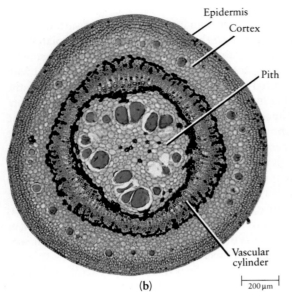

Epidermis
Cortex
Pith
Vascular cylinder
(b)
200 μm

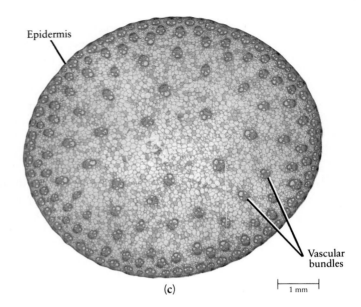

Epidermis
Vascular bundles
(c)
1 mm

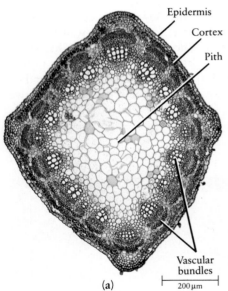

Epidermis
Cortex
Pith
Vascular bundles
(a)
200 μm

30–20 *Cross sections of two dicot stems and one monocot stem. (a) In alfalfa, a dicot, the vascular cylinder is made up of separate vascular bundles. (b) In this young stem of the linden, also a dicot, the vascular tissue forms a continuous cylinder. This stem contains mucilage ducts, which stain red. (c) In corn, a monocot, numerous vascular bundles are scattered throughout the ground tissue.*

Stem Patterns

In green stems, the xylem and the phloem are usually arranged in longitudinal parallel strands, the vascular bundles, which are embedded in the ground tissue. In young dicot stems, the vascular bundles form a ring, the vascular cylinder, around a central pith (Figure 30–20a and b). The cylinder of ground tissue outside the vascular bundles is the cortex. Within each bundle, the xylem is characteristically on the inside, adjacent to the pith, and the phloem is on the outside, adjacent to the cortex. In monocots, the vascular bundles are usually scattered throughout the ground tissue (Figure 30–20c).

The vascular tissues of the stem are continuous with those of the root, and yet, as we have seen, their arrangement in the stem is quite different from that in the root. The change from the patterns observed in the root to those found in the stem is a gradual one. The region of the plant axis between root and stem in which the change occurs is known as the transition region.

PRIMARY GROWTH OF THE SHOOT SYSTEM

The shoot system includes the stem and all of the structures that develop from it—typically, all of the aboveground parts of the plant. The pattern of growth of the developing shoot tip is similar to the pattern we saw earlier in the root: first, cell division takes place; next, cell elongation; and finally, differentiation. However, because of the regular occurrence of nodes and their appendages—leaves and buds—the growth zones are not as distinct in the shoot as they are in the root. Also, no covering analogous to the root cap is produced over the shoot tip.

As in the root, the outermost layer of cells develops into the epidermis. In the shoot, these cells are covered by a relatively conspicuous cuticle. Underlying cells differentiate to form the ground tissues and the primary vascular tissues—the primary xylem and the primary phloem. The pattern of development is more complicated, however, than in the root tip since the apical meristem of the shoot is the source of tissues that give rise to new leaves, branches, and flowers.

Figure 30–21 shows the shoot tip of the familiar house plant *Coleus*. In the center is the apical meristem, which is very small, with the beginnings—primordia

—of two leaves. Flanking the apical meristem are the two previously formed leaves, which are completing their growth and development. The leaves originate by cell division in localized areas along the side of the apical meristem. As growth progresses, the vascular tissue of the stem differentiates upward into the leaf primordia, becoming part of the general vascular system that connects the plant from root to leaf tip. Leaves are formed in an orderly sequence at the shoot tip. In some species, they arise simultaneously in pairs opposite one another, as in *Coleus*. In other species, they form spirally or in circles (whorls) at the nodes. As the internodes elongate, the young leaves become separated and thus spaced out along the stem of the plant.

As the shoot tip elongates, small masses of meristematic tissue are left just above the points at which the leaves are attached to the stem (the leaf axils). These new meristematic regions, the axillary buds, remain dormant until after growth of the adjacent leaf and internode is complete, or, in perennials, until the next growing season. In many species, including *Coleus*, development of the axillary buds is suppressed by the influence of the terminal bud. (This phenomenon, known as apical dominance, will be discussed further on page 675.) In some species, specific buds are destined to become lateral branches or specialized shoots, such as tubers or flowers. In other species, the fate of the buds is determined by environmental conditions, particularly day length, as we shall see in Chapter 32. When a flower forms—from either an axillary bud or a terminal bud—the meristem gives rise to the floral parts and ceases to exist as a source of new cells.

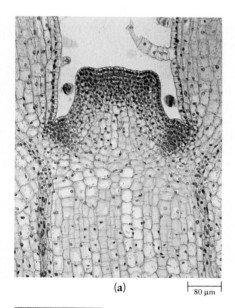

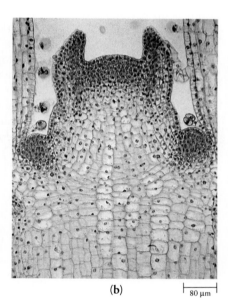

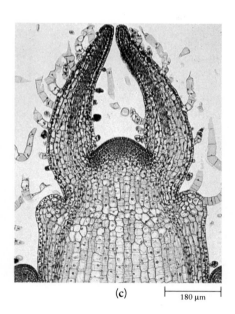

(a)   80 μm   (b)   80 μm   (c)   180 μm

30–21 *Stages in leaf development at the shoot apex of* Coleus, *as seen in longitudinal section.* Coleus *leaves develop in pairs, opposite one another, with each successive pair at right angles to the preceding pair. In these micrographs, the nuclei are stained purple; thus, the meristematic regions, densely packed with small, rapidly dividing cells, appear pur-* ple. **(a)** *Two small bulges, or leaf buttresses, appear on opposite sides of the stump-shaped apical meristem. In addition, buds are developing in the axils of the two previously formed leaves.* **(b)** *Two erect, peg-like leaf primordia have developed from the leaf buttresses. Strands of vascular tissue are extending upward into the leaf primordia.* **(c)** *As the leaf primordia elon-* gate, *the vascular tissues continue their upward differentiation. The epidermal hairs, visible on the outer surface of the developing leaves, originate from certain cells of the protoderm very early, long before the protoderm matures to become the epidermis.*

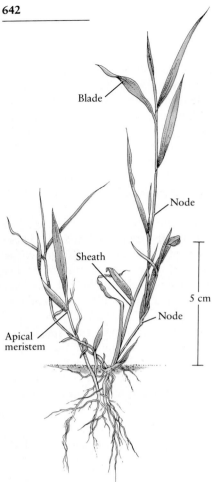

Blade

Node

Sheath

5 cm

Node

Apical
meristem

30–22 *The structure of a grass plant. The capacity of a grass leaf to grow from the sheath at its base is a useful adaptation to herbivore grazing. It has also resulted, over the eons, in the piteous spectacle of thousands of hominids vibrating behind their lawnmowers from spring to fall.*

Modifications in the Pattern of Shoot Growth

In most plants, primary growth of the shoot proceeds from its tip, the apical meristem, and from the axillary buds. Grass plants, however, also have meristems at the bases of the leaves. The grass leaf consists of a blade (the broad part) and a sheath that fits around the plant stem (Figure 30–22). When the blade is cut off, cells in the sheath, in response to some unknown signal, are activated, producing more blade.

The shoots of some climbing plants coil themselves around the structures on which they are growing, a process that we shall examine in Chapter 32. Others produce modified branches in the form of tendrils. (As we saw previously, tendrils may also be modified leaves or leaflets.) The tendrils of grape, English ivy, and Virginia creeper are all modified branches.

Vegetative Reproduction

Among the specialized shoots that originate from the axillary buds of many species are runners and rhizomes. Runners, such as those found in most varieties of strawberry, are long, slender stems that grow along the surface of the soil. Rhizomes are also horizontal stems, growing either along or below the surface of the soil. Both runners and rhizomes develop adventitious roots and are the source of new plants, genetically identical to the parent plant. Strawberries (Figure 30–23) are a familiar example of plants that propagate vegetatively by runners, as are spider plants, commonly grown as hanging plants. Plants that reproduce by rhizomes include potatoes, many flowering garden perennials, such as irises and lilies-of-the-valley, the sod-forming grasses of lawns and pastures, and bamboos and reed grasses (Figure 30–24). In grasses, for example, rhizomes form buds that produce upright stems bearing leaves and flowers.

The asexual production of genetically identical clones from runners and rhizomes is a very efficient way for a plant to spread quickly and invade new territory. The young plants that develop in this way have a continuous source of nourishment from the parent plant and a lower mortality than seedlings. The production of clones can also provide a mechanism for reproducing in environments that are only periodically hospitable to the germination of seeds. Creosote bushes, common in the North American deserts, produce clones that form a circle or an ellipse around a bare area where the parent plant first took root (Figure

30–23 *Strawberry plants reproduce asexually by means of runners, thin horizontal stems that grow along the surface of the soil. Roots and leaves develop at every second node along the modified stems. These plants also form flowers and reproduce sexually.*

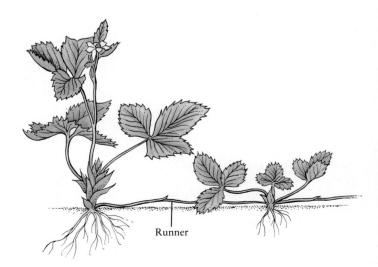

Runner

30-24 *Common reed grass* (Phragmites communis) *propagates asexually by means of rhizomes. Each stand of reed grass is, in effect, a single individual. The tassels at the tops of the stalks are flowers; the seeds produced through sexual reproduction give rise to new stands of reed grass, genetically different from the parent plants.*

30–25). At ages approaching 12,000 years, some creosote clones are the oldest living plants known; these extraordinary plants are thought to have started from seeds that germinated near the end of the last glacial expansion.

The capacity of plants to reproduce vegetatively has been exploited in the development of cultivated varieties of plants for food or ornamental use. Because such plants are genetically identical to the parent stock, vegetative reproduction is a way of preserving uniformity. Many plants are reproduced by stem cuttings, which involves placing young stems in soil or some other planting medium and protecting them from drying out until adventitious roots appear. Rooting can often be facilitated by treating the cuttings with hormones (see page 674). Many stem cuttings will also form roots when placed in water. Another artificial form of plant propagation is grafting, in which a stem cutting is inserted in a slit in the stem or trunk of a rooted woody plant. The wound is sealed with tape, and if the stem cutting and the rooted plant have not been damaged too much by the surgery, the grafted shoot will "take" and begin growing. Most fruit trees and roses are propagated in this way.

Many economically important plants are sterile and can only be propagated vegetatively; these include pineapples, bananas, seedless grapes, navel oranges, sugar cane, and numerous ornamental plants.

30-25 *Creosote bushes of the King Clone in the Mojave desert in California. The bushes form an elliptical ring that is about 23 meters by 8 meters. Standing just within the ring is Andrew Sanders of the University of California, Riverside.*

When a creosote seed germinates, it gives rise to a small stem, known as the crown. As the crown grows thicker, it produces many new branches at its base. Over the first 40 to 90 years of its life, the crown splits into lobes, each of which forms a new crown and produces more branches. Eventually the original crown

dies, leaving a central area of bare sand surrounded by the genetically identical bushes of the clone. Although a single crown has a life expectancy of a mere 80 to 100 years, the clone derived from the original seedling may live far longer. The parent crown of the King Clone has been located and dated using radioactive isotope methods. It is roughly 11,700 years old, indicating that the King Clone is, by far, the oldest living plant known. Recently, the Nature Conservancy purchased 17 acres of land surrounding the King Clone, ensuring its protection.

SECONDARY GROWTH

As you know from your own observations, many plants not only grow taller with age (or longer, as they expand into new territory by vegetative reproduction) but also become thicker. The process by which woody dicots increase the thickness of their trunks, stems, branches, and roots is known as secondary growth (Figure 30–26). The so-called "secondary tissues" are not derived from the apical meristems; instead, they are produced by **lateral meristems** known as the vascular cambium and the cork cambium.

The **vascular cambium** is a thin, cylindrical sheath of tissue between the xylem and the phloem. In stems, its cells are derived from the embryonic procambium and from parenchyma cells in the pith rays; in roots, the cells are derived from the procambium and the pericycle. In plants with secondary growth, the cambial cells divide continually during the growing season, adding secondary xylem toward the inside of the cambium and secondary phloem toward the outside. Some meristematic cells remain as a cylinder of undifferentiated cambium, in which cell division will resume at the start of the next growing season.

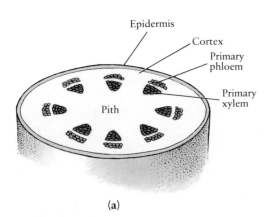

(a)

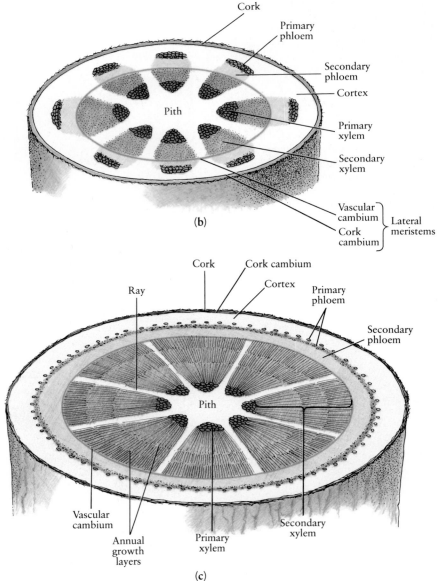

(b)

(c)

30–26 (a) *Stem of a dicot before the onset of secondary growth.*

(b) *Beginnings of secondary growth. Secondary xylem and secondary phloem are produced by the vascular cambium, a meristematic tissue formed late in primary growth. As the trunk increases in diameter, the epidermis is stretched and torn. This is accompanied by the formation of the cork cambium, from which cork is formed, replacing the epidermis.*

(c) *Cross section of a three-year-old stem, showing annual growth rings. Rays are rows of living cells that transport nutrients and water laterally (across the trunk). On the perimeter of the outermost growth layer of secondary xylem is the vascular cambium, encircled by a band of secondary phloem. The primary phloem and the cortex will eventually be sloughed off. In an older stem, the thin cylinder of active secondary phloem is immediately adjacent to the vascular cambium. The tissues outside the vascular cambium, including the phloem, constitute the bark.*

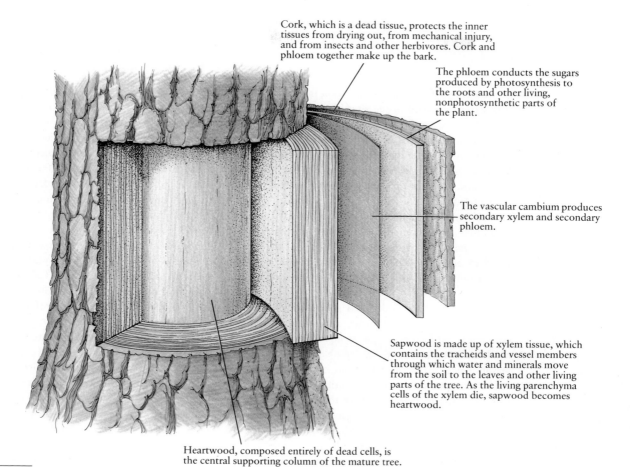

Cork, which is a dead tissue, protects the inner tissues from drying out, from mechanical injury, and from insects and other herbivores. Cork and phloem together make up the bark.

The phloem conducts the sugars produced by photosynthesis to the roots and other living, nonphotosynthetic parts of the plant.

The vascular cambium produces secondary xylem and secondary phloem.

Sapwood is made up of xylem tissue, which contains the tracheids and vessel members through which water and minerals move from the soil to the leaves and other living parts of the tree. As the living parenchyma cells of the xylem die, sapwood becomes heartwood.

Heartwood, composed entirely of dead cells, is the central supporting column of the mature tree.

30–27 *A tree trunk showing the relationships of the successive concentric layers. The heartwood is composed entirely of dead cells. The cortex and epidermis, which are outside the phloem in a green stem, are sloughed off during secondary growth.*

As secondary growth increases the girth of stems and roots, the epidermis becomes stretched and torn. In conjunction with this tearing process, a new type of cambium, the **cork cambium,** forms from the cortex. The cork cambium produces cork (phellem), which replaces the epidermis as the protective covering of woody stems and roots. Unlike the epidermis, cork is a dead tissue at maturity. The cork cambium often forms anew each year, moving further inward until, finally, there is no cortex left. The tissues outside the vascular cambium—a thin layer of phloem, the cork cambium, and the cork—constitute the bark.

As the plant grows older, the parenchyma cells of the xylem in the center of the stem and root die, and their neighboring vessels become clogged and cease to function. This nonconducting xylem, called heartwood, forms the center of the trunk and major roots of a tree, providing the support and anchorage required as the height of the tree continues to increase through primary growth. The living parenchyma cells and open vessels just inside the vascular cambium constitute the sapwood, through which water and minerals flow from the root tips to the leaves.

Season after season the new xylem forms visible growth layers, or rings. Each growing season leaves its trace, so that the age of a tree can be estimated by counting the number of growth rings in a section near its base. Since the growth rate of a tree depends on climatic conditions, the width of the annual growth layers can be used to estimate fluctuations in temperature and rainfall that occurred many years ago (see essay).

The Record in the Rings

As we have seen, each growing season in the life of a tree leaves its record in the ring of secondary xylem formed in the trunk. The growth rings are visible because of differences in the density of wood produced early in the growing season and that produced late in the season. The early wood has large cells with thin walls, whereas the late wood has smaller cells with proportionately thicker walls. Within a given growth layer, the change from early wood to late wood may be quite gradual, but a distinct change is visible where the small, thick-walled cells of the late wood of one growing season abut the larger, thin-walled cells of the early wood of the next growing season.

The width of the individual growth layers may vary greatly from year to year, depending on such environmental factors as light, temperature, rainfall, available soil water, and the length of the growing season. Under favorable conditions, the growth rings are wide; under unfavorable conditions, they are narrow. In semiarid regions, where there is very little rain, trees are sensitive rain gauges. An excellent example is the bristlecone pine *(Pinus longaeva)* of the western Great Basin. Each growth ring is different, and a study of the rings tells a story that dates back thousands of years.

The oldest known living specimen of bristlecone pine is 4,900 years old. Dendrochronologists—scientists who conduct historical research using the growth rings of trees—have been able to match samples of wood from living and dead trees. In this way, they have built a library containing a continuous series of rings dating back more than 8,200 years. This record of average ring width has provided a valuable guide to past conditions of precipitation and temperature. The widths of the growth rings of bristlecone pines at higher elevations (the upper tree line) have been found to be closely related to temperature changes. The rings of these trees have revealed that the summers in the White Mountains of California were relatively warm from 3500 B.C. to 1300 B.C., and the tree line was about 150 meters higher than it is at present. Summers were cool from 1300 B.C. to 200 B.C.

Information provided by the growth rings of both gymnosperms and angiosperms is being used not only to reconstruct past climatic conditions but also to help predict future conditions. With more accurate knowledge of past climates than is provided by human records—even those of the past few centuries—it is possible to determine cyclic patterns of temperature change and of rainfall and drought. Such information is of considerable importance, for example, in planning the wise management and allocation of our finite resources of fresh water.

(a)

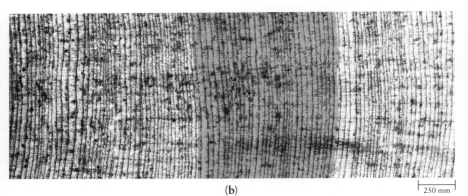

(b)

250 mm

(a) *A bristlecone pine. These pines, which grow near the timberline, are the oldest living trees (but not, as we saw in Figure 30–25, the oldest living plants).* (b) *A portion of a cross section of wood from a bristlecone pine, showing the variation in width of the annual rings. This section begins approximately 6,260 years ago; the band of rings in color represents the years from 4240 B.C. to 4210 B.C.*

By the continuous formation of secondary xylem and (to a lesser extent) phloem, woody dicots increase their diameter as primary growth increases their height. Moreover, newly formed tracheids, vessel members, and sieve-tube members provide fresh conduits—undamaged by the activities of the numerous parasites and herbivores that attack plants—for the transport of water and nutrients from one part of the plant body to another. In the next chapter, we shall examine the mechanisms by which materials move through these cells.

SUMMARY

Plants are multicellular photosynthetic organisms adapted to life on land. The plant body has specialized photosynthetic areas (leaves), conducting and supporting structures (stems), and organs that anchor the plant in the soil and absorb water and minerals from it (roots).

When a seed germinates, growth of the root and shoot proceeds from the apical meristems of the embryo. Certain cells within the meristem retain the capacity to divide. Others elongate and then differentiate, forming, according to their position, the various specialized cells of the plant (see Table 30–1). These

TABLE 30–1 Summary of Main Cell Types in Angiosperms

CELL TYPE	ORIGIN	LOCATION	CHARACTERISTICS	FUNCTION
Meristematic (in apical meristem)	Embryonic cells	Apices of shoots and roots	Many-sided, small, thin-walled cells; vacuoles usually small	Origin of primary meristematic tissues and of root cap cells
Meristematic (in vascular cambium)	Procambium and parenchyma of pith rays in stems; procambium and pericycle in roots	Lateral, between secondary phloem and xylem tissues	Elongate, often spindle-shaped	Produces secondary xylem and phloem
Epidermal	Protoderm	Surface of entire primary plant body	Flattened, variable in shape, overlaid by cuticle; some specialized as guard cells	Protective covering; prevents desiccation yet allows gas exchange
Parenchyma	Protoderm, ground meristem, procambium, vascular cambium, cork cambium, wound tissues	Everywhere, usually dominant in pith, cortex, mesophyll	Many-sided, usually thin-walled; abundant air spaces between cells	Photosynthesis, respiration, storage, wound healing, among others
Collenchyma	Ground meristem of leaf and stem	Peripheral in cortex of stem and in leaves	Elongate, with irregularly thickened primary walls	Support for young stems and leaves
Sclereid	Protoderm, ground meristem, procambium, vascular cambium, and cork cambium	In pith and cortex of stems; in leaves and flesh of fruits; seed coats	Irregular; massive secondary wall; alive or dead at maturity	Produces hard texture, mechanical support
Fiber	Procambium or vascular cambium; ground meristem	Primary and secondary xylem and phloem; cortex	Very long, narrow cell, with secondary cell wall; usually dead at maturity	Support
Tracheid	Procambium or vascular cambium	Primary or secondary xylem	Elongate, with pits in walls; dead at maturity	Conduction of water and solutes
Vessel member	Procambium or vascular cambium	Interconnected series (= vessels) in primary or secondary xylem	Elongate, with pits in walls and end walls perforated; dead at maturity	Conduction of water and solutes
Sieve-tube member	Procambium or vascular cambium	Primary or secondary phloem, usually with companion cells; form interconnected series (= sieve tubes)	Elongate, with specialized sieve plates; nucleus lacking at maturity	Conduction of organic solutes
Cork (phellem)	Cork cambium	Surface of stems and roots with secondary growth	Flattened cells, compactly arranged; dead at maturity, the cells often air-filled	Restricts gas exchange and water loss

cells, in varying combinations, form the three tissue systems—dermal, ground, and vascular—that are continuous throughout the plant body. Parenchyma cells, which are thin-walled and many-sided, are the most common cells in plants.

The ground tissue of the leaf is composed primarily of photosynthetic parenchyma cells. Palisade cells, in which most of the photosynthesis takes place, are elongated cells with large central vacuoles. Spongy parenchyma cells, also photosynthetic, are surrounded by large air spaces. The palisade and spongy parenchyma cells make up the mesophyll, or "middle leaf."

The upper and lower surfaces of the leaf consist of one or more layers of transparent epidermal cells covered with a waxy layer, the cuticle. Specialized pores, the stomata, open and close, regulating the exchange of gases and the release of water vapor. Veins, the vascular bundles of the leaf, conduct water and minerals to the mesophyll cells of the leaf (through the xylem) and transport sugars away from them (through the phloem). The vascular tissues of leaves are continuous with those of the stem and roots.

The embryonic root is the first structure to break out of the germinating seed. Cells produced by its apical meristem form a root cap, which protects the root tip as it is pushed through the soil. Young roots have an outer layer of epidermis and, at most, a very thin cuticle. Extensions of the epidermal cells form root hairs, which greatly increase the absorptive surface of the root. Inside the epidermis is the ground tissue of the root, the cortex, composed mostly of parenchyma cells, often modified for storage. The innermost layer of the cortex is the endodermis, a single layer of specialized cells whose walls contain a waterproof zone, the Casparian strip. Just inside the endodermis is another layer of cells, the pericycle, from which branch roots arise. Within the pericycle are the xylem and phloem.

Green stems, like leaves, have an outer layer of epidermal cells covered with a cuticle. The bulk of the young stem is ground tissue, which may be divided into an outer cylinder (the cortex) and an inner core (the pith). The ground tissue is composed largely of parenchyma cells but also may contain collenchyma cells and sclerenchyma cells (fibers and sclereids).

The vascular tissues consist of phloem and xylem. In angiosperms, the conducting cells of the phloem are the sieve-tube members, living cells with perforated end walls that form continuous sieve tubes. Associated closely with each sieve-tube member is a companion cell. The conducting tissue of the xylem is made up of a series of tracheids or vessel members. Tracheids and vessel members characteristically have thick secondary walls and are dead at functional maturity. Phloem and xylem also contain parenchyma cells and fibers.

The height (or length) of the aboveground parts of a plant is increased through primary growth of the shoot system. Leaf primordia arise from the apical meristem of the shoot. As the nodes are separated by elongation of the internodes, small masses of meristem (buds) form in the axils of the leaves. These axillary buds may remain dormant or may give rise to branches or specialized shoots. Among the specialized shoots that may form are tubers, runners, and rhizomes, through which many species reproduce asexually, producing populations of genetically identical individuals.

Secondary growth is the process by which woody plants increase their girth. Such growth arises primarily from the vascular cambium, a sheath of meristematic tissue completely surrounding the xylem and completely surrounded by phloem. The cambial cells divide during the growing season, adding new xylem cells (secondary xylem) on their inner surfaces and new phloem cells (secondary phloem) on their outer surfaces. As the trunk increases in girth, the epidermis is eventually ruptured and replaced by cork.

QUESTIONS

1. Distinguish among the following: dermal tissue/ground tissue/vascular tissue; epidermis/cuticle; parenchyma/collenchyma/sclerenchyma; palisade parenchyma/spongy parenchyma; sieve-tube member/tracheid/vessel member; apical meristem/lateral meristem; primary growth/secondary growth.

2. Describe the distinguishing characteristics of monocots and dicots.

3. Sketch the structure of a leaf, labeling the principal cells and tissues. Describe the function of each of the labeled parts.

4. Where might you expect to find a leaf with stomata only on the bottom? Only on the top?

5. Reexamine the oleander leaf in Figure 30–5, relating its various structural features to the ecology of the oleander.

6. Some leaves are specialized for functions other than photosynthesis. List three other functions for which leaves may be specialized, and give an example of each specialization.

7. We have seen that the tissue of the root cortex contains many air spaces. Also, we noted that some plants growing in swampy areas have aerial roots. You may have discovered that overwatering can kill house plants. What do these data indicate about roots?

8. Why are there no root hairs or lateral roots at the extreme tip of a root?

9. Sketch cross sections of (a) a root, (b) a dicot stem, and (c) a monocot stem. Label the principal cells and tissues in each sketch, and describe the function of each of the labeled parts.

10. What two cell layers are present in roots but absent in stems? What functions do the cells of these layers perform in the roots, and why are they necessary for the survival of the plant?

11. It has often been said that the principal differences among root, stem, and leaf are quantitative in nature rather than qualitative. Explain.

12. Suppose you carve your initials 1 meter above the ground on the trunk of a mature tree that is growing vertically at an average of 15 centimeters per year. How high will your initials be at the end of 2 years? At the end of 20 years?

13. As a result of secondary growth, most of our common trees increase in girth as they increase in height. What limitations would a lack of secondary growth impose on the form and mechanical support of a tree?

14. Sketch a tree trunk with secondary growth. Compare your sketch with Figure 30–26c.

15. Porcupines often eat all of the bark off young trees, at the height they can reach. When they do this, the tree dies. Why?

16. Consider a mature tree. Which parts of the tree are composed of living cells? Why might it be advantageous for living cells to make up such a small fraction of the mass of the tree? Similarly, why might it be advantageous for a deciduous tree to invest as little material as possible in making leaves?

C H A P T E R   **31**

Transport Processes in Plants

As we noted in Chapter 24, the plants that have come to dominate the modern landscape are those that have evolved the best solutions (so far) to two critical problems: conveying the sperm to the egg in the absence of free water, and transporting needed materials to the various cells and tissues of the plant body. In this chapter, we shall discuss first the movement of water and minerals in the xylem; second, the transport of dissolved sugars and other organic substances in the phloem; and third, a related subject, factors affecting the availability of nutrients to the plant. You may find it helpful to review the properties of water (especially pages 42 to 44) and the ways in which it moves (pages 128 to 134) before proceeding. The transport processes of plants depend on the extraordinary properties of this very common liquid.

THE MOVEMENT OF WATER AND MINERALS

Transpiration

A plant needs far more water than an animal of comparable weight. In an animal, most of its water remains in its body and recirculates; by contrast, more than 90 percent of the water entering a plant's roots is released into the air as water vapor. This loss of water vapor from the plant body is known as **transpiration.** It is a necessary consequence of the opening of the stomata that must occur for a plant to obtain the carbon dioxide required for photosynthesis (Figure 31–1), but the price is high. For every gram of carbon dioxide incorporated into organic matter, a plant utilizing the C_3 pathway of photosynthesis loses 400 to 500 grams of water, and a plant utilizing the C_4 pathway (pages 223 to 228) loses 250 to 300 grams of water.

Transpiration creates an enormous "thirst" in a plant. Thus, for example, a single corn plant requires 160 to 200 liters of water to grow from seed to harvest, and 1 hectare (2.47 acres) of corn requires almost 5 million liters of water a season. British ecologist John L. Harper describes the terrestrial plant as a "wick connecting the water reservoir of the soil with the atmosphere."

The Uptake of Water

As we saw in the last chapter, water enters most plants almost entirely through the roots, principally through the root hairs. During periods of rapid transpiration, water may be removed from around the roots so quickly that the soil water in the vicinity of the roots becomes depleted. Water then moves slowly by diffusion and capillary action through the soil toward the depleted region near the roots. Roots

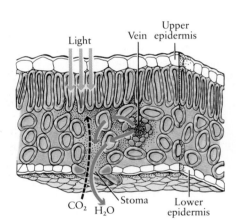

31–1 *As carbon dioxide, essential for photosynthesis, enters the leaf through the stomata, water vapor is lost. Although this water loss poses serious problems for plants, it provides both the motive force for the uptake of water by the roots and a mechanism for cooling the leaf. The temperature of a leaf may be as much as 10 to 15°C lower than that of the surrounding air, simply because evaporating water carries heat away (see page 44).*

31–2 *Guttation droplets on the edge of wild strawberry leaves. Guttation, the loss of liquid water, is a result of root pressure. The water escapes through specialized pores located near the ends of the principal veins of the leaves. Guttation, which is restricted to relatively small plants, usually occurs at night when the air is moist and transpiration is greatly reduced.*

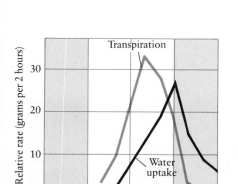

31–3 *Measurements in ash trees show that a rise in water uptake follows a rise in transpiration. These data suggest that the loss of water generates forces for its uptake.*

also obtain additional water by growing beyond the depleted region; the main roots of corn plants, for example, grow an average of 52 to 63 millimeters a day.

Root cells, like other living parts of the plant, contain a higher concentration of solutes (both organic and inorganic) than does soil water. As a consequence, water from the soil enters the roots by osmosis. The resulting pressure, known as root pressure, is sufficient to move water a short distance up the stem. Guttation, the loss of liquid water through the leaves (Figure 31–2), is a visible consequence of root pressure. But how can water reach 20 meters high to the top of an oak tree, travel three stories up the stem of a vine, or move 125 meters up in a tall redwood?

One important clue is the observation that during times when the most rapid transpiration is taking place—which is when the flow of water up the stem must be the greatest—xylem pressures are negative (less than atmospheric pressure). This negative pressure can be demonstrated. If you peel a piece of bark from a transpiring tree and make a cut in the xylem, no liquid runs out. In fact, if you place a drop of water on the cut, the drop will be drawn in.

What is the pulling force? It is not simple suction, as the negative pressure might indicate. Suction simply removes air from a system so that the water (or other liquid) is pushed up by atmospheric pressure. But atmospheric pressure is only enough to raise water (against no resistance) approximately 10 meters at sea level, and many trees are much taller than 10 meters.

The Cohesion-Tension Theory

According to the now generally accepted theory, the explanation for the movement of water is to be found not only in the properties of plants but also in the properties of water, to which plants have become exquisitely adapted. As we pointed out in Chapter 2, in every water molecule, two hydrogen atoms are covalently bonded to a single oxygen atom. Each hydrogen atom is also held to the oxygen atom of a neighboring water molecule by a hydrogen bond. The cohesion resulting from this secondary attraction is so great that the tensile strength in a thin column of water can be as much as 140 kilograms per square centimeter (2,000 pounds per square inch). This means that a negative pressure of more than 140 kilograms per square centimeter is required to pull the column of water apart.

In a leaf, water evaporates, molecule by molecule, from the walls of the parenchyma cells into the air spaces of the leaf. As the water potential of a leaf cell decreases, water from the vessels and tracheids moves, molecule by molecule, into the cell. But each water molecule in a tracheid or vessel is linked to other water molecules in the tracheid or vessel. They, in turn, are linked to others, forming a long, narrow, continuous stream of water reaching down all the way to a root hair and, in fact, into the surrounding soil solution. As a molecule of water moves through the stem and into the leaf, it tugs the next molecule along behind it. In plants with large vessels, this process moves water at a rate of 30 to 40 meters per hour; in plants with smaller vessels, the rate is about 5 to 10 meters per hour.

Because the diameter of the vessels is relatively small and because water molecules adhere to the cell walls of the vessel members (even as they are cohering to one another), gas bubbles, which could rupture the water column, do not usually form. The pulling action, molecule by molecule, causes the negative pressure observed in the xylem. The technical term for a negative pressure is tension, and this theory of water movement is known as the **cohesion-tension theory.**

The principle of cohesion-tension is illustrated in Figure 31–4. As indicated in the diagram, the power for this process comes not from the plant, which plays only a passive role in transpiration, but from the energy of the sun.

31–4 (a) *A simple model that illustrates the cohesion-tension theory. A piece of porous clay pipe, closed at both ends, is filled with water and attached to the end of a long, narrow glass tube also filled with water. The water-filled tube is placed with its lower end below the surface of a volume of mercury contained in a beaker. As water molecules evaporate from the pores in the pipe, they are replaced by water "pulled up" through the narrow glass tube in a continuous column. As the water evaporates, the mercury rises in the tube to replace it. (b) Transpiration from plant leaves results in sufficient water loss to create a similar negative pressure.*

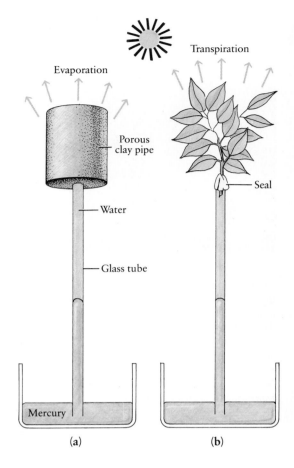

(a)   (b)

Factors Influencing Transpiration

Transpiration is, as we have seen, costly, especially when water supply is limited. Several factors affect the rate of water loss. One is temperature; the rate of evaporation doubles for every 10°C increase in temperature. Humidity is also important. Water is lost much more slowly into air already laden with water vapor. Air currents also affect transpiration. Wind blows away the water vapor from leaf surfaces, which makes the concentration gradient of water steeper and so hastens the evaporation of water molecules from the leaf. Leaves of plants that grow in exposed, windy areas are often hairy; these hairs are believed to protect the leaf surface from the wind and so retard transpiration (Figure 31–5). By far the most important factor affecting transpiration, however, is the regulatory effect exercised by the opening and closing of the stomata.

31–5 *Leaf hairs create a layer of still air over the epidermis, restricting gas exchange and water loss from the stomata (visible in the lower portion of the micrograph).*

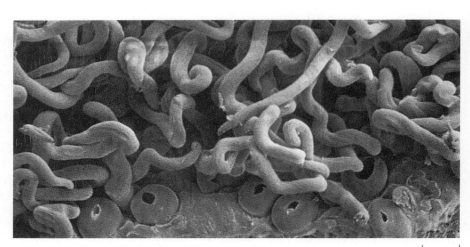

25 µm

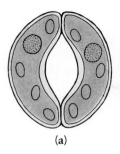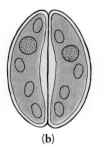

(a)　　　　(b)　　　　(c)　　　　(d)

31–6 *Mechanism of stomatal movement. A stoma is bordered by two guard cells that (**a**) open the stoma when they are turgid and (**b**) close it when they lose turgor. In many species, the inner walls of the guard cells are slightly thickened. For many years it was thought that as turgor increased, the thinner parts of the cell wall were stretched more than the thicker parts, causing the cells to bow out and the stoma to open. It is now known that the key to stomatal opening lies instead in* *cellulose microfibrils that are arranged in hoops around the circumference of the guard cells (**c**). The arrangement is similar to that of the belts in radial tires. When water enters the guard cells, the microfibrils prevent radial expansion, and the only direction in which the cells can expand is lengthwise. Because the two cells are attached to each other at the ends, this lengthwise expansion forces them to bow out and the stoma to open (**d**).*

The Mechanism of Stomatal Movements

It has long been known that the osmotic movement of water is involved in the opening and closing of the stomata. As shown in Figure 31–6, each stoma has two surrounding guard cells. Stomatal movements are caused by changes in the turgor of these cells. When the guard cells are turgid, they bow out, opening the stoma; when they lose water, they relax, and the stoma closes.

Turgor, as we saw on page 134, is maintained or lost due to the osmotic movement of water into or out of cells. As you will recall, water moves across the cell membrane from a solution of low solute concentration into a solution of high solute concentration. The active accumulation of solutes in the guard cells causes water to move into them by osmosis; conversely, a decrease in the solute concentration of guard cells results in the osmotic movement of water out of the cells.

Techniques that make it possible to measure ionic concentrations within individual guard cells have revealed that the critical solute affecting the osmotic movement of water into and out of these cells is the potassium ion (K^+). With an increase in the K^+ concentration, the stomata open, and with a decrease, they close. Current evidence indicates that potassium ions are actively transported between the guard cells and the reservoir provided by the surrounding epidermal cells (Figure 31–7). Water follows by osmosis, causing the observed changes in the turgor and shape of the guard cells.

In many, but not all, species, chloride ions, Cl^-, accompany the K^+ ions across the membrane, thus maintaining electrical neutrality. In other, perhaps all, species, H^+ ions are transported in the opposite direction, producing a decreased hydrogen ion concentration within the guard cells of open stomata.

The active transport of K^+ ions between the guard cells and the surrounding epidermal cells is, of course, an energy-requiring process. The energy source is not yet known, but, on the basis of present evidence, one of two possibilities seems likely. As you will recall, guard cells contain chloroplasts, and it may be that K^+ transport is powered by ATP produced by photophosphorylation reactions (page 218) occurring in these chloroplasts. Another possibility is that the transport of H^+ ions, in a direction opposite to that of the K^+ ions, establishes an electrochemical gradient down which the K^+ ions move. If this should be the case, the transport of K^+ ions into and out of the guard cells—and thus, the opening and closing of the stomata—would be yet another example of a vital process powered by a chemiosmotic mechanism (page 197).

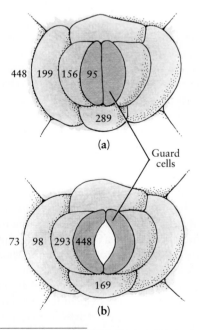

448 | 199 | 156 | 95

289

(a)

Guard cells

73 | 98 | 293 | 448

169

(b)

31–7 *Quantitative changes in potassium ion (K^+) concentrations in the guard cells and neighboring epidermal cells of (**a**) a closed stoma and (**b**) an open stoma of the dayflower (*Commelina communis*) leaf. The measurements were made with K^+-sensitive microelectrodes that were inserted into the individual cells.*

31–8 *Corn plants, photographed in a Wisconsin field in June of 1988, during a severe drought that affected many of the prime agricultural areas of North America. The leaves of corn plants have numerous stomata on both the upper and lower surfaces, with slightly more on the lower surface. The cuticle on the lower surface is significantly heavier than on the upper surface. When corn plants are under severe water stress, and closure of the stomata is inadequate to protect the plants against further water loss, the leaves roll up, as shown here. Leaf rolling reduces transpiration, protecting virtually all of the stomata on the upper surface, as well as many on the lower surface. All that remains exposed to the environment is a portion of the lower surface, with its heavy cuticle.*

Factors Influencing Stomatal Movements

A number of environmental factors affect—either directly or indirectly—the movement of potassium ions into and out of the guard cells and thus influence the opening and closing of the stomata. The principal one is the availability of water. When the water available to a leaf drops below a certain critical point (which varies from species to species), the stomata close, thereby limiting evaporation of the remaining water. This is not a direct effect; stomatal closure occurs before the leaf loses turgor and wilts.

A plant's capacity to anticipate water stress depends on the action of the hormone abscisic acid, about which we shall have more to say in the next chapter. Several lines of evidence link abscisic acid with stomatal movements. First, during the initial stage of water stress, levels of the hormone increase markedly in many species of plants. Second, application of abscisic acid to a leaf causes stomatal closure within a few minutes. Third, a strain of tomato plant, known as the "wilty mutant," produces very little abscisic acid and fails to close its stomata as the water supply decreases; application of abscisic acid to a "wilty mutant" tomato plant enables it to regulate its stomata normally. Abscisic acid acts by binding to specific receptors in the guard cell membrane; the receptor-hormone complex then triggers a change in the cell membrane, making it more permeable to potassium ions.

Stomatal movement also occurs quite independently of the loss or gain of water by the plant. Among the other factors that affect stomatal movement are the carbon dioxide concentration, temperature, and light. In most species, an increase in carbon dioxide concentration in the intercellular spaces of a leaf causes the stomata to close. The magnitude of this response varies greatly from species to species and with the degree of water stress that a particular plant has undergone. In corn, the stomata may respond to changes in the internal carbon dioxide concentration within seconds. The site for sensing the level of carbon dioxide is inside the guard cells.

Within normal ranges, temperature changes have little effect on the stomata, but temperatures higher than 35°C (95°F) can cause stomatal closure in some species. This temperature-induced closure can be prevented, however, by maintaining the plant in air that contains no carbon dioxide. This suggests that temperature changes affect stomatal movements primarily by altering the carbon dioxide concentration within the leaf. An increase in temperature results in an increase in respiration and thus in the concentration of intercellular carbon dioxide. Many plants in hot climates close their stomata regularly at midday, apparently because of both increased water stress and the effect of temperature on the concentration of carbon dioxide in the leaves.

In most species, the stomata close regularly in the evening, when photosynthesis is no longer possible, and open again in the morning. These movements in response to light occur even when there are no changes in the water available to the plant. Opening in the light and closing in the dark may be a result of changes in the carbon dioxide concentration—photosynthesis uses carbon dioxide and thus reduces the internal concentration, whereas respiration causes an increase in the carbon dioxide concentration. Light may, however, have another, more direct effect. Blue light has long been known to stimulate stomatal opening independently of the carbon dioxide concentration. When isolated guard cells from onion are illuminated with blue light in the presence of potassium ions, they swell. The blue-absorbing pigment (which is thought to be located in either the cell membrane or the vacuole membrane) promotes the uptake of K$^+$ ions by the guard cells. Recent experiments have demonstrated that this uptake of K$^+$ ions is a secondary effect, and that the primary effect of the interaction of blue light with the pigment is to stimulate the pumping of H$^+$ ions *out of* the guard cells. Although these experiments provide support for the hypothesis that an electro-

31–9 *Among the most familiar CAM plants are the stonecrops of genus* Sedum. *These stonecrops, flourishing on otherwise barren cliffs high above the surf, were photographed in Redwood National Park, California.*

chemical gradient of H^+ ions powers K^+ transport, they do not rule out the possible involvement of ATP. The blue-absorbing pigment, which would be yellow in color, is thought to be a flavin; as you may recall (page 195), flavin-containing nucleotides are among the electron carriers involved in cellular respiration in the mitochondria.

Crassulacean Acid Metabolism

Although the stomata of most plants are open during the day and closed at night, the stomata of some species close in the daytime and open at night. Not only is the temperature lower at night but also the humidity is usually higher; both factors reduce the rate of transpiration. The species that open their stomata only at night include a variety of plants adapted to hot, dry climates, for example, cacti, pineapples, and members of the stonecrop family (Crassulaceae). These plants take in carbon dioxide at night, converting it to malic and isocitric acids. During the day, when the stomata are closed, the carbon dioxide is released from these organic acids and used immediately in photosynthesis. This process is known as Crassulacean acid metabolism, and plants that employ it are known as CAM plants. It is analogous to the C_4 pathway in photosynthesis described on page 224, although it apparently evolved independently. CAM plants achieve a considerable savings in water; in contrast to C_3 and even C_4 plants (see page 650), they lose only 50 to 100 grams of water for each gram of carbon dioxide incorporated into organic matter.

The Uptake of Minerals

In addition to water and the energy-rich sugars and other compounds produced by photosynthesis, plant cells require a number of different chemical elements, which are found in the earth in the form of minerals. A **mineral** is a naturally occurring inorganic substance, usually solid, with a definite chemical composition. Some minerals, such as diamond, sulfur, and copper, consist of single elements. Others, such as quartz (SiO_2) and calcite ($CaCO_3$), are compounds. Of the 92 different elements found in rocks and soil, oxygen and silicon are by far the most abundant, followed by aluminum and iron.

Mineral ions are taken up in water solution by the roots and travel through the xylem in the transpiration stream. As you saw in Figure 30–10 (page 633), the cells of the endodermis play a major role in determining which substances enter the xylem. The ionic composition of plant cells is far different from the ionic composition of the medium in which the plant grows. For example, in one study, cells of pea roots were found to have a concentration of potassium ions (K^+) 75 times greater than that of the nutrient solution. Similarly, in another study, the vacuoles of rutabaga cells were shown to contain 10,000 times more K^+ than the external solution.

Thus it seems evident that mineral ions are brought into plant cells by active transport. Support for this hypothesis comes from observations indicating that the uptake of ions is an energy-requiring process. For instance, if roots are deprived of oxygen or poisoned so that respiration is curtailed, ion uptake is drastically decreased. Also, if a plant is deprived of light, it stops absorbing ions and eventually releases them back into the soil.

One way that roots move ions such as K^+ against a concentration gradient is by the action of transport proteins in the cell membrane (see page 136). Such proteins combine with the ions and then, utilizing the energy of ATP, carry them to the other side of the membrane and release them. Because the membrane is only slowly permeable to ions such as K^+, the continued action of the "pump proteins" results in differences in the concentration of the transported ions on one side of the membrane as compared to the other.

(a)

(b)

31–10 Some plants have special mineral requirements. (a) Plants of the mustard family use sulfur in the synthesis of the mustard oils that give the plants their characteristic sharp taste. These mustard plants were growing along the side of a highway near Lima, Ohio. (b) Horsetails incorporate silicon into their cell walls, making them indigestible to most herbivores but useful, at least in colonial America, for scouring pots and pans. These bushy vegetative shoots, as well as several stalk-like fertile shoots, of the horsetail Equisetum telmateia *were photographed in California.*

As a result of these differences in ion concentrations, there is often a difference not only of osmotic potential but also of electric potential between the inside and the outside of the membrane. As we have seen, such differences may play a role in the transport of K^+ ions into and out of the guard cells. They may also have other functions. For example, in the nerve cells of animals, differences in electric potential on either side of the membrane provide the basis for the nerve impulse. Stimulation of the nerve cell membrane results in an abrupt change in its permeability to ions, allowing ions to move rapidly down the gradient of potential energy. This rapid movement of ions across the membrane constitutes the nerve impulse. Similar events are apparently involved in certain plant responses, as we shall see in the next chapter.

Mineral Requirements of Plants

Plants require minerals for many different functions. One of the most critical, of course, is regulation of water balance. Because osmotic potential is determined by the number of solute particles in a solution, rather than by their specific chemical identity, several kinds of ions may serve interchangeably in this role. Thus, the requirement is described as nonspecific. On the other hand, a specific element may be an essential component of a critical biological molecule, which will not function properly in its absence. In such cases, the requirement is highly specific. For example, magnesium is an essential component of the chlorophyll molecule (see Figure 10–6, page 211). Some minerals—phosphorus and calcium, for example—are required constituents of cell membranes, and some control membrane permeability. Others are indispensable components of a variety of enzyme systems that catalyze chemical reactions in the cell. Still others provide a proper ionic environment in which such reactions can occur. Because mineral elements are involved in many fundamental processes, the effects of mineral deficiencies are typically very wide-ranging, affecting a number of structures and functions in the plant body.

Table 31–1 lists the mineral elements required by plants, the form in which they are usually absorbed, and some of the uses plants make of them. Depending on the concentration of a mineral element typically found in plants, it is classified as either a macronutrient or a micronutrient. A macronutrient may constitute as much as 0.5 to 3 or 4 percent of the dry weight of a plant; by contrast, only a few parts per million of a micronutrient may be present. For many elements, the amount present is determined by burning the plant completely—which permits the carbon, hydrogen, oxygen, nitrogen, and sulfur to escape as gases—and then analyzing the remaining ash. Proportions of each element vary in different species and in the same species grown under different conditions. Also, the ash often contains elements, such as silicon, which are present in the soil and are taken up by plants but which are not generally thought to be required for growth.

Mineral requirements can also be identified by studying the capacity of plants to grow in distilled water to which small amounts of various minerals are added. This sounds easier than it is. Sometimes it has been found that a substance—molybdenum, for example—is needed in such small amounts that it is almost impossible to set up experimental conditions that exclude it. Thus, it is difficult to prove that its absence is lethal.

You might anticipate that organisms make use of what is most readily available; indeed, they seem to have done this when life originated from elements in the gases of the primitive atmosphere. But Table 31–1 reveals some findings that you might not expect. Sodium, for instance, which is one of the most abundant of the elements, is apparently not required at all by most plants. The fact that plants evolved with no functions requiring sodium is even more striking when you consider that sodium is vital to animals. (Sodium is the principal osmoregulator in

TABLE 31-1 A Summary of Mineral Elements Required by Plants

ELEMENT	PRINCIPAL FORM IN WHICH ABSORBED	APPROXIMATE CONCENTRATION IN HEALTHY WHOLE PLANTS (AS % OF DRY WEIGHT)	SOME FUNCTIONS
Macronutrients			
Nitrogen	NO_3^- (or NH_4^+)	1–4%	Component of amino acids, proteins, nucleotides, nucleic acids, chlorophyll, and coenzymes
Potassium	K^+	0.5–6%	Involved in osmosis and ionic balance and in opening and closing of stomata; activator of many enzymes
Calcium	Ca^{2+}	0.2–3.5%	Component of cell walls; enzyme cofactor; involved in cell membrane permeability and transport of ions and hormones
Phosphorus	$H_2PO_4^-$ or HPO_4^{2-}	0.1–0.8%	Component of energy-carrying phosphate compounds (ATP and ADP), phospholipids, nucleic acids, and several essential coenzymes
Magnesium	Mg^{2+}	0.1–0.8%	Part of the chlorophyll molecule; activator of many enzymes
Sulfur	SO_4^{2-}	0.05–1%	Component of some amino acids, proteins, and coenzyme A
Micronutrients			
Iron	Fe^{2+}, Fe^{3+}	25–300 parts per million (ppm)*	Required for chloroplast development; component of cytochromes and nitrogenase
Chlorine	Cl^-	100–10,000 ppm	Involved in osmosis and ionic balance; essential in photosynthesis in the reactions in which oxygen is produced
Copper	Cu^{2+}	4–30 ppm	Activator or component of some enzymes
Manganese	Mn^{2+}	15–800 ppm	Activator of some enzymes; required for integrity of chloroplast membrane and for oxygen release in photosynthesis
Zinc	Zn^{2+}	15–100 ppm	Activator or component of many enzymes
Boron	BO^{3-} (borate) or $B_4O_7^{2-}$ (tetraborate)	5–75 ppm	Influences Ca^{2+} utilization, nucleic acid synthesis, and membrane integrity
Molybdenum	MoO_4^{2-}	0.1–5.0 ppm	Required for nitrogen metabolism
Elements Essential to Some Plants or Organisms			
Cobalt	Co^{2+}	Trace	Required by nitrogen-fixing microorganisms
Sodium	Na^+	Trace	Involved in osmosis and ionic balance; for many plants, probably not essential; required by some desert and salt-marsh species and may be required by all plants that utilize C_4 pathway of photosynthesis

* Parts per million (ppm) equal units of an element by weight per million units of oven-dried plant material; 1% equals 10,000 ppm.

animals and, as we shall see in Chapter 41, is necessary for transmission of the nerve impulse.) In the seas, where both plant and animal life seem to have originated, sodium is the most abundant mineral element and is far more available than potassium, which it closely resembles in its essential properties. Potassium, however, is the principal osmoregulator in plants. Similarly, silicon and aluminum are usually present in large amounts in soils, but only a few plants (specifically, grasses) require silicon and apparently none requires aluminum. Conversely, most plants need molybdenum, which is relatively rare.

THE MOVEMENT OF SUGARS: TRANSLOCATION

In addition to water and minerals, the cells of a plant also need energy. As we have seen, the photosynthetic cells of a plant, which are typically most abundant in the leaves, not only capture the energy of sunlight for their own use but also provide the organic molecules that are the energy source for all of the other cells of the plant. The process by which the products of photosynthesis are transported to other tissues is known as **translocation.**

Halophytes: A Future Resource?

Unlike most animals, most plants do not require sodium and, moreover, cannot survive in brackish waters or saline soils. In such environments, the solution surrounding the roots often has a higher solute concentration than the cells of the plant, causing water to move out of the roots by osmosis. Even if the plant is able to absorb water, it faces additional problems from the high level of sodium ions. If the plant takes up water and excludes sodium ions, the solution surrounding the roots becomes even saltier, increasing the likelihood of water loss through the roots. The salt may even become so concentrated that it forms a crust around the roots, effectively blocking the supply of water to the roots. Another problem is that sodium ions may enter the plant in preference to potassium ions, depriving the plant of an essential nutrient as well as inhibiting some enzyme systems.

Some plants, however—known as halophytes—can grow in saline environments such as deserts, salt marshes, and coastal areas. All of these plants have evolved mechanisms for dealing with high sodium concentrations, and for some of them sodium appears to be a required nutrient. The adaptations of halo-phytes vary. In many halophytes, the sodium-potassium pump (page 136) seems to play a major role in maintaining a low sodium concentration within the cells while simultaneously ensuring that a sufficient supply of potassium ions enters the plant. In some species, the pump operates primarily in the root cells, pumping sodium back to the environment and potassium into the root. The presence of calcium ions (Ca^{2+}) in the soil solution is thought to be essential for the effective functioning of this mechanism.

Other halophytes take in sodium through the roots but then either secrete it or isolate it from the living cytoplasm of the plant body. In *Salicornia* (pickleweed), a sodium-potassium pump (or a variant of it) operates in the membranes of the vacuoles of leaf cells. Sodium ions enter the cells but are immediately pumped into vacuoles and isolated from the cytoplasm. In such plants, the solute concentration of the vacuoles is higher than that of the environment, establishing the necessary osmotic potential for the movement of water into the roots. In other genera, the salt is pumped into the intercellular spaces of the leaves and then secreted from the plant. In *Distichlis*

(a) Atriplex *(saltbush) is one of several halophytes being evaluated as potential crop plants.*

(a)

palmeri (Palmer's grass), the salt exudes through specialized cells (not the stomata) onto the surface of the leaf. In *Atriplex* (saltbush), it is concentrated by special salt glands and pumped into bladders. The bladders expand as the salt accumulates, finally bursting. Rain or the passing tide washes the salt away.

Halophytes are of current interest not only because of the light they may shed on the osmoregulatory mechanisms of plants, but also because of their potential as crop plants. In a world with an ever-increasing need for food, vast areas are unsuitable for agricultural purposes because of the salinity of the soil. For example, there are over 30,000 kilometers of desert coastline and about 400 billion hectares of desert with potential water supplies that are, however, too salty for crop plants. Moreover, each year about 200,000 hectares of irrigated crop land become so salty that further agriculture is impossible. When arid land is heavily irrigated, as in large areas of the western United States, salts from the irrigation waters accumulate in the soil. This occurs because in both evaporation from the soil and transpiration from plants, essentially pure water is given off, leaving all solutes behind. Over many years, the salt concentration of the soil increases, eventually reaching levels that cannot be tolerated by most plants. It has been suggested that the ancient civilizations of the Near East ultimately fell because their heavily irrigated land became so salty that food could no longer be grown on it.

One way of extending the life of irrigated crop lands and of bringing presently barren areas into agricultural use would be to breed salt-tolerance into conventional crop plants. Thus far, however, such efforts have met with little success. Scientists at the University of Arizona's Environmental Research Laboratory are taking what appears to be a more promising approach. They have gathered halophytes from all over the world and are engaged in an extensive research program to determine optimal growing conditions, potential yields, and the nutritional value and palatability of the seeds and vegetative parts of the various species. Their results suggest that a number of halophyte species have great potential for use in livestock feed and, quite possibly, for human consumption as well.

(b) *The surface of a leaf of* Atriplex. *Salt is pumped from the leaf tissues through narrow stalk cells into the large, expandable bladder cells.*

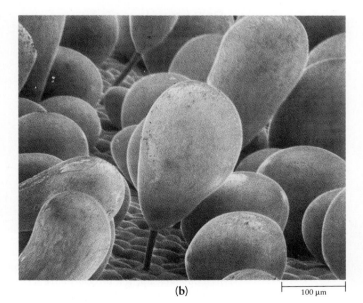

(b)

100 μm

Evidence for the Phloem

The fact that water is transported in the tracheids and vessels of the xylem was recognized by botanists 300 years ago. The role of the phloem in the movement of sugar was not generally agreed upon, however, until well into the twentieth century. Early evidence suggesting that the phloem is involved in sugar transport came from observations of trees that had a complete ring of bark removed from them. As we saw in the last chapter, bark contains the phloem but not the xylem. When a photosynthesizing tree is "girdled" in this manner, the tissue above the ring becomes swollen, indicating that fluid moving downward in the phloem from the photosynthesizing leaves has accumulated there. Convincing evidence for the role of the phloem was obtained when radioactive tracers became available. If plants carry out photosynthesis in air containing carbon dioxide made with carbon 14 ($^{14}CO_2$), the sugars produced will carry a radioactive tag. Studies of the passage of radioactivity through the plant have shown conclusively that sugars are transported in the sieve tubes (see essay).

Aphids, which are very small sap-sucking insects, have provided valuable information on the movement of substances through the sieve tubes (Figure 31–11). Data gathered with the assistance of aphids indicate that sieve-tube sap contains (by weight) 10 to 25 percent solutes, more than 90 percent of which are sugars, mostly sucrose. Low concentrations of amino acids and other nitrogen-containing substances are also present. Tracer stains indicate that the rate of movement of the solutes along the sieve tube is remarkably fast: in one set of experiments, for example, it was estimated that the sap was moving at a rate of about 100 centimeters per hour, far faster than could be accounted for by diffusion alone. At this rate, each individual sieve-tube member was emptying and completely refilling every two seconds.

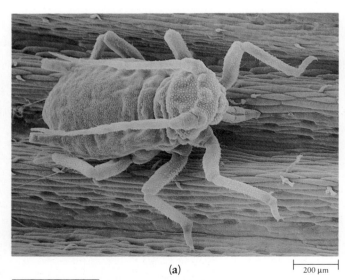

(a)   200 μm

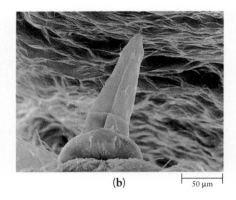

(b)   50 μm

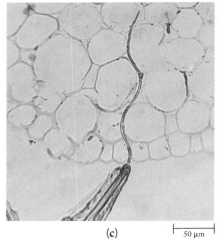

(c)   50 μm

31–11 *Assistance by aphids.* (a) *Aphids are small insects that feed on plant sap; you probably have seen them on rose bushes.* (b) *The aphid drives its sharp mouthparts, or stylets, like a hypodermic needle through the epidermis.* (c) *As this micrograph reveals, the stylets traverse the cortical or mesophyll cells and then tap the contents of a single sieve-tube member. If the aphid is anesthetized, it is possible to sever its body from the stylets, leaving the latter undisturbed in the cell. The sieve-tube sap often continues to exude through the stylets for many hours, and pure samples can be collected for analysis without damaging the sieve tube or interfering with its function.*

Radioactive Isotopes in Plant Research

Radioactive isotopes can be used in a number of ways to study the synthesis, transport, and use of materials within plants. Initially, the radioactive isotope must be incorporated into the plant. Radioactive carbon, for example, will be taken up by a plant if its leaves are exposed to carbon dioxide containing carbon 14. Or, radioactive phosphorus will be taken up if the roots are exposed to a solution containing phosphorus-32 ions.

The length of time the plant is exposed to the radioactive material is determined by the information the investigators hope to obtain. For example, in studies to determine the time required for carbon dioxide to be incorporated into the various products of photosynthesis, a sequence of exposure times would be used. In studies focusing on the location of a particular product formed from the radioactive substance, the length of exposure would depend on the time required for the chemical reactions under study.

After exposure to the radioactive substance, the plant is quick-frozen and freeze-dried. In whole-plant autoradiography, the plant is flattened and then pressed against a sheet of x-ray film. Radiation given off by the isotope exposes the film adjacent to the portions of the plant in which it is located. By comparing the flattened plant with the developed film, investigators can determine the location of the radioactive substance within the plant.

In tissue autoradiography (histoautoradiography), the freeze-dried plant tissues are embedded in paraffin, resin, or a similar material. Next, they are sliced into very thin sections, which are mounted on microscope slides. The tissue sections are then placed in contact with a photographic emulsion or film. As in whole-plant autoradiography, the radiation from the isotope exposes the film in contact with the portions of the tissue section containing radioactive material. After an appropriate interval of time, the film is developed. Comparison, under the microscope, of the developed film and the underlying tissue section reveals the exact location of the radioactive substance in the plant tissues.

In the study illustrated here, three leaflets of a bean plant were enclosed in a flask and were exposed to $^{14}CO_2$ and light for 35 minutes (a). During that time, the radioactive carbon dioxide was incorporated into sugars, which were then being transported to other parts of the plant. A cross section (b) and a longitudinal section (c) from the stem were placed in contact with autoradiographic film for 32 days. When the film was developed and compared with the underlying tissue sections, it was apparent that the radioactivity (visible as dark specks on the film) was confined almost entirely to the sieve tubes.

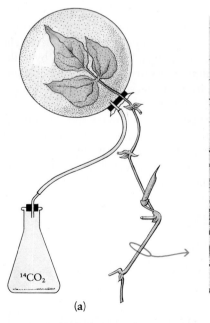

$^{14}CO_2$

(a)

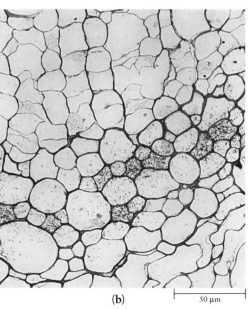

50 μm

(b)

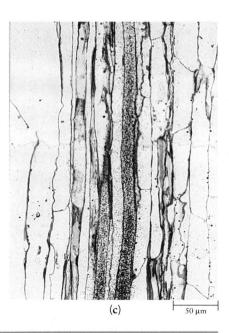

50 μm

(c)

The Pressure-Flow Hypothesis

The movement of sugars and other organic solutes in translocation follows what is known as a **source-to-sink pattern.** The principal sources of the solutes are the photosynthesizing leaves, but storage tissues may also serve as important sources. All plant parts unable to meet their own nutritional needs may act as sinks, that is, as importers of organic solutes. Thus, storage tissues act as sinks when they are importing solutes and as sources when they are exporting solutes.

The most widely accepted explanation for the source-to-sink movement in translocation is the **pressure-flow hypothesis.** According to this hypothesis, the solutes move in solutions that, in turn, move because of differences in water potential caused by concentration gradients of sugar. The principle underlying this hypothesis can be illustrated by a simple physical model consisting of porous bulbs permeable only to water and connected by glass tubes (Figure 31–12). The first bulb contains a solution in which there is a dissolved material, such as sucrose, and the second, to make the example as simple as possible, contains only water. When these interconnected bulbs are placed in distilled water, water will enter the first by osmosis. The entry of water will increase the hydrostatic pressure within this bulb and cause the water and the solutes in it to move along the tube to the second bulb, where the pressure again builds up. If the second bulb is connected with a third bulb containing water or a sucrose concentration lower than that now in the second bulb, the solution will flow from the second to the third by the same process, and so on indefinitely down the sucrose gradient. This hypothesis is supported by the fact that distinct gradients in the concentration of sucrose have been demonstrated along the phloem tissues. Moreover, it can account for the known rates of movement in the phloem.

As shown in Figure 31–13, sugars from the photosynthetic cells of the leaf are moved into sieve tubes against a concentration gradient. In sugar beets, it has been shown that sucrose molecules are moved from the mesophyll cells of the leaf to the phloem of the veins where they are "loaded" into the sieve tubes by active transport. This loading process appears to involve a cotransport of sucrose molecules and hydrogen ions by way of a specific transport protein on the sieve-tube membrane. The incoming sugar decreases the water potential in the sieve tube and causes water to move into the sieve tube from the xylem by osmosis. At a sink—for example, a storage root—sugar molecules leave the sieve tube. Water molecules then follow the sugar molecules out, again by osmosis. Thus the water flows in at one end of the sieve tube and out at the other. Between these two points, the water and its solutes, including sugar, move passively by bulk flow. The speed of transport depends on the differences in concentration between source and sink. At the sink, sugars may be either utilized or stored, but most of the water returns to the xylem and is recirculated in the transpiration stream.

Companion cells, because of their dense appearance and many mitochondria, are believed to be very active metabolically, and it is hypothesized that one of their functions is to meet the energy requirements of the sieve-tube members with which they are associated. Note, however, that the actual flow of the sugar solution is a passive process, requiring no metabolic activity on the part of the phloem cells.

FACTORS INFLUENCING PLANT NUTRITION

As we saw earlier, the mineral elements needed by plants are taken up by the roots in solution and are transported through the plant body in the transpiration stream. Although the availability of minerals depends principally on the nature of the surrounding soil, the activities of symbiotic fungi and bacteria also play a crucial role.

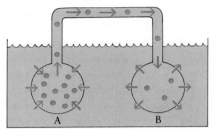

- ●   Sugar molecule
- →   Movement of water
- →   Movement of sugar solution

31–12 Model of the pressure-flow hypothesis. Bulbs A and B, which are interconnected and are permeable to water, are placed in a bath of distilled water. Bulb A contains a higher concentration of sucrose than bulb B. Water enters bulb A from the medium, increasing the hydrostatic pressure and pushing the solution to bulb B. If B were connected to a third bulb, C, with a still lower concentration of sucrose, as sieve-tube members are connected in a series, hydrostatic pressure building up in B would push the solution on to C, and so on.

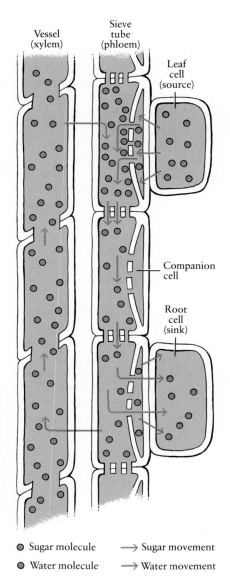

Sugar molecule   → Sugar movement

Water molecule   → Water movement

31–13 *The pressure-flow mechanism as it is thought to occur in the plant body. Sugar molecules enter a companion cell at the source by active transport and then move into the sieve tube through the many cytoplasmic connections in the common cell wall of the sieve-tube member and its companion cell. As a consequence of the increased concentration of sugar, the water potential is decreased, and water enters the sieve tube by osmosis. Sugar molecules leave the sieve tube at the sink, and the sugar concentration in the sieve tube falls; as a result, water moves out of the sieve tube by osmosis. Because of the active secretion of sugar molecules into the sieve tube at the source and their removal from the sieve tube at the sink, a flow of sugar solution takes place along the tube between source and sink.*

Soil Composition

Soil, the uppermost layer of the earth's crust, is composed of rock fragments associated with organic material, both living and in various stages of decomposition. It typically has three layers: the A horizon, the B horizon, and the C horizon. The A horizon, or topsoil, is the zone of maximum accumulation of organic matter (humus). The B horizon, or subsoil, consists of inorganic particles in combination with mineral nutrients that have leached (washed) down from the A horizon. The C horizon is made up of loose rock that extends down to the bedrock beneath it. As shown in Figure 31–14, the depth and composition of these three layers—and consequently, the fertility of the soil—vary considerably in different environments.

The mineral content of soil depends in part on the parent rock from which the soil was formed. These differences in mineral content can be extremely localized, with sharp lines of demarcation. Geologists sometimes use types of vegetation or changes in color or growth patterns of plants as indicators of mineral deposits.

In most soils, however, the mineral content is more dependent on biological factors. In an undisturbed environment, most of the mineral nutrients stay within the system—the soil itself and the plants, microorganisms, and small soil animals that it supports and contains. If, however, the vegetation is repeatedly removed, as when crops are harvested, grasslands are overgrazed, or the top, humus-rich layer of the A horizon is eroded, the soil rapidly becomes depleted of nutrients. It can then be used for agricultural purposes only if it is heavily fertilized.

Another factor influencing the mineral content of soil is the size of the soil particles. The smaller fragments of rock are classified as sand, silt, or clay (Table

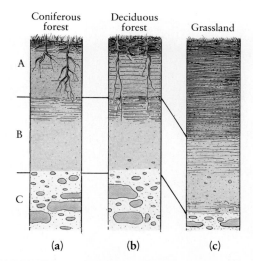

(a)   (b)   (c)

31–14 *Diagrams of soil layers of three major soil types. (a) The litter of the northern coniferous forest is acidic and slow to decay, and the soil has little accumulation of humus, is very acidic, and is leached of minerals. (b) In the cool, temperate deciduous forest, decay is somewhat more rapid, leaching less extensive, and the soil more fertile. Such soils have been widely used for agriculture, but they need to be prepared by adding lime (to reduce acidity) and fertilizer. (c) In the grasslands, almost all of the plant material above the ground dies each year, as do many of the roots, and thus large amounts of organic matter are constantly returned to the soil. In addition, the finely divided roots penetrate the soil extensively. The result is highly fertile soil, often black in color, with a topsoil sometimes more than a meter in depth.*

TABLE 31–2	Soil Classification
	DIAMETER OF PARTICLES (MICROMETERS)
Coarse sand	200–2,000 (0.2–2 millimeters)
Sand	20–200
Silt	2–20
Clay	Less than 2

31–2). Water and minerals drain rapidly through soil composed of large particles (sandy soil). Soil composed of small particles (clay) holds the water against gravity. Moreover, the small clay particles are negatively charged and therefore bind positively charged ions, such as calcium (Ca^{2+}), potassium (K^+), and magnesium (Mg^{2+}). However, a pure clay soil is poorly suited for plant growth because it is usually too tightly packed to let in enough oxygen for the respiration of plant roots, soil animals, and most soil microorganisms. Clay soils that contain enough particles of sand and silt to keep the soil from packing are known as loams, and these are generally the best soils for plant growth.

The pH of the soil also affects its capacity to retain minerals. In acidic soil, hydrogen ions replace other positively charged ions clinging to clay particles, and these nutrient ions leach out of the soil. Soil pH also affects the solubility of certain nutrient elements. Calcium, for example, is more soluble (and therefore more available to plant roots) as pH increases, whereas iron becomes less available to plants as pH increases. Crops such as alfalfa, sweet clover, and other legumes have high calcium requirements and therefore grow best in alkaline soil. Rhododendrons and azaleas, on the other hand, have high requirements for iron, which is abundant only when the soil is acidic.

Soils and plant life interact. Plants secrete hydrogen ions, which help to degrade rock surfaces and release positively charged ions from those surfaces. As they decay, plant parts constantly add to the humus, thereby changing not only the content of the soil but also its texture and its capacity to hold minerals and water. In turn, the plants are dependent upon the mineral content of the soil and its holding capacity. As these improve, plants increase in both number and size and also often change in kind, thereby producing further changes in the soil. Thus, under natural conditions, the soil—and the availability of nutrients and water to plant roots—are constantly changing.

The Role of Symbioses

Mycorrhizae

Two types of symbiotic relationships play important roles in plant nutrition. One of these is mycorrhizae ("fungus-roots"), the associations between fungi and roots that we described in Chapter 23. In these associations, the fungi extract nutrients from the soil and make them available to plants, thus enabling plants to prosper on nutrient-poor soils. Recent studies indicate that the fungi may also screen out chemicals, making it possible for plants to live in soils that would otherwise be toxic.

Rhizobia and Nitrogen Fixation

All but one of the elements listed in Table 31–1 are derived principally from the weathering of rocks. The exception is nitrogen, the nutrient for which plants have the greatest requirement. Although nitrogen constitutes about 78 percent of the air, most plants cannot use gaseous nitrogen, in which each molecule consists of two atoms of nitrogen held together by a triple covalent bond (page 33), an exceptionally strong bond. Plants are dependent upon nitrogen-containing ions— ammonium (NH_4^+) and nitrate (NO_3^-)—from the soil.

In plant cells, nitrate ions are reduced to ammonium ions, and the ammonium ions are then combined with carbon-containing compounds to form amino acids, nucleotides, chlorophyll, and other nitrogen-containing compounds. These nitrogen-containing compounds are then returned to the soil with the death of the plants (or of the animals that have eaten the plants) and are reprocessed by soil organisms, taken up by plant roots in the form of nitrate dissolved in soil water,

31–15 *Indian pipe* (Monotropa uniflora). *It was once believed that this nonphotosynthetic angiosperm, with its strange white waxy flowers, lived on decaying matter in the soil. It is now known, however, that it is dependent on mycorrhizal fungi that transfer nutrients from other plants to the parasitic Indian pipe.*

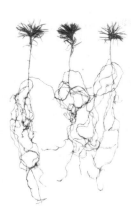

31–16 *Effects of mycorrhizae on tree nutrition. Nine-month-old seedlings of white pine were grown for two months in a sterile nutrient solution and then transplanted to prairie soil. The seedlings on the left were transplanted directly. The seedlings on the right were placed in forest soil containing fungi for two weeks before being transplanted to the prairie.*

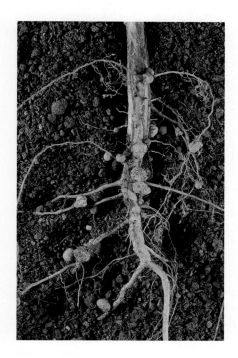

31–17 *Nitrogen-fixing nodules on the roots of a soybean plant, a legume. These nodules are the result of a symbiotic relationship between a soil bacterium (Rhizobium) and root cells.*

and reconverted to organic compounds. In the course of this cycle (to be discussed in more detail in Chapter 54), a certain amount of nitrogen is always "lost," in the sense that it becomes unavailable to plants.

The main source of nitrogen loss is the removal of plants from the soil. Soils under cultivation often show a steady decline of nitrogen content. Nitrogen may also be lost when topsoil is carried off by soil erosion or when ground cover is destroyed by fire. Nitrogen-containing ions are also leached away by water percolating down through the soil. In addition, numerous types of bacteria are present in the soil that, when oxygen is not available, break down nitrates, releasing nitrogen into the air and using the oxygen for cellular respiration.

If the nitrogen lost from the soil were not steadily replaced, virtually all life on this planet would finally flicker out. The "lost" nitrogen is returned to the soil by **nitrogen fixation,** the process by which atmospheric nitrogen is incorporated into organic nitrogen-containing compounds. On a worldwide basis, most nitrogen fixation is carried out by a few kinds of prokaryotes, including both free-living and symbiotic forms of cyanobacteria and heterotrophic bacteria. Of the various classes of nitrogen-fixing organisms, the symbiotic bacteria are by far the most important in terms of total amounts of nitrogen fixed. The most common of the nitrogen-fixing symbiotic bacteria is *Rhizobium,* which invades the roots of leguminous plants, such as clover, peas, beans, vetches, and alfalfa (Figure 31–17).

The beneficial effects to the soil of growing leguminous plants are so obvious that they have been recognized for hundreds of years. Where leguminous plants are grown, some of the "extra" nitrogen is usually released into the soil; it then becomes available to other plants. In modern agriculture, it is common practice to rotate a nonleguminous crop, such as corn, with a leguminous one, such as alfalfa. The leguminous plants are then either harvested, leaving behind the nitrogen-rich roots, or, better still, plowed back into the field. A crop of alfalfa that is plowed back into the soil may add as much as 350 kilograms of nitrogen to the soil per hectare, frequently enough to grow a crop of a nonleguminous plant without any additional fertilization.

Carnivorous Plants

More than 350 species of plants are meat-eaters; their diets include insects, other invertebrates, and even some vertebrates, such as small birds and frogs. Unlike carnivorous animals, they do not utilize their prey for energy but rather as a source of mineral elements, particularly nitrogen, phosphorus, and calcium. As Darwin noted in his book on this subject, published in 1875, these plants are usually found in swamps, bogs, and peat marshes where acids leach the soil of nutrients.

(a) The sundew is a tiny plant, often only 2 to 5 centimeters across, with club-shaped tentacles on the upper surface of its leaves. These tentacles secrete a clear, sticky liquid that attracts insects. When an insect is caught by one tentacle, the other tentacles bend toward it until the insect drowns, its air passages filled with a mucilage-like fluid. The tentacles also secrete digestive enzymes.

(b) Pitcher plants attract insects into their flower-like tubular leaves by means of nectar. Following the

(a)   (b)

The Symbiotic Relationship

The symbiosis between a species of *Rhizobium* and a legume is quite specific; for example, the species of bacteria that invade and induce nodule formation in clover roots will not induce nodules on the roots of soybeans. The recognition process on which this specificity depends is thought to involve the interaction of plant proteins known as lectins on the root surface with polysaccharides of the bacterial cell wall. Rhizobia enter the root-hair tips of legumes while the plants are still seedlings (Figure 31–18a). The bacteria usually induce these epidermal cells to produce internal cellulose tubes, or infection threads, through which the bacteria move to the cortical cells of the root (Figure 31–18b). Within the cortex, the infection threads branch, becoming populated with multiplying bacteria. Soon there is a proliferation of the surrounding cortical cells, presumably the result of

nectar over the rim, the insect finds itself on a carpet of fine, transparent hairs. When the nectar trail ends, and the insect turns to exit, it encounters the points of these fine hairs, pointing downward and blocking its way. If it moves inward, it encounters a slick waxed surface from which it slides or drops into a foul-smelling broth of rainwater, digestive enzymes, and bacteria at the pitcher's base.

(c, d) The Venus flytrap has leaves that close around any insect moving on the surface of the leaf.

The closing of the leaf is triggered by touching one or two of the three trigger hairs in the middle of each leaf lobe. Venus flytraps are found in nature only on the coastal plain of North and South Carolina, usually at the edges of wet depressions and pools. It has long been believed that the Venus flytrap lures its victims by exuding nectar, but recent studies indicate that insects visit the leaves by chance. Despite its name, a Venus flytrap's usual diet in the wild consists of crawling invertebrates, such as ants.

(c)

(d)

hormones released during the growth of the rhizobia. The infection threads typically pass close to the nuclei of the cortical cells and seem to cause them to degenerate. Near the nuclei, vesicles form in the threads and then break open, releasing the bacteria into the cytoplasm of the host cells (Figure 31–18c).

Soon after their release, the bacteria begin to grow, increasing in size some tenfold and synthesizing an enzyme complex called **nitrogenase.** Nitrogenase, which consists of two different polypeptides, one containing iron and the other molybdenum, catalyzes the following reaction:

$$\underset{\text{Nitrogen}}{N_2} + \underset{\substack{\text{Hydrogen} \\ \text{ions}}}{6H^+} + \underset{\text{Electrons}}{6e^-} \xrightarrow{\text{Nitrogenase}} \underset{\text{Ammonia}}{2NH_3}$$

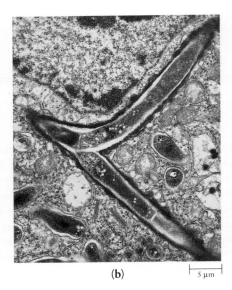

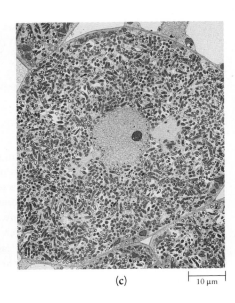

(a) 5 μm   (b) 5 μm   (c) 10 μm

31–18 (a) *Scanning electron micrograph of an emerging root hair of a soybean seedling, with several rhizobia (arrows) attached.* (b) *Two branches of an infection thread in a soybean nodule cell. As you can see, the thread is passing near the cell nucleus, at the top of the micrograph.* (c) *Cross section of a nodule, showing a cell containing numerous bacteria. The plant supplies the bacteria with an energy source; the bacteria supply the plant with fixed nitrogen.*

The energy required to break the strong triple bond of the nitrogen molecule is supplied by ATP; some 15 to 20 molecules of ATP are hydrolyzed to ADP for each molecule of gaseous nitrogen incorporated into ammonia. The ammonia produced by this reaction is combined with carbon compounds synthesized by the photosynthetic cells of the plant, producing amino acids.

The legumes are by far the largest group of plants that enter into nitrogen-fixing partnerships with symbiotic bacteria. There are, however, numerous nitrogen-fixing symbioses that involve plants other than legumes. Sweet fern (an angiosperm), for example, forms nodules that are induced by and contain nitrogen-fixing actinomycetes (moldlike bacteria), rather than *Rhizobium*. Like rhizobia, the actinomycetes enter the host plant by way of a root-hair infection. The resulting symbiosis allows the plants to carry out a pioneering role in the revegetation of barren land. Sweet fern, for instance, is commonly planted along highways in Massachusetts for this purpose. There is also evidence that the roots of some grasses, such as sorghum, are involved in less intimate symbiotic associations with nitrogen-fixing bacteria. The bacteria involved, the most prevalent of which is *Azospirillum,* usually reside in the transition zone between the root and the soil. Some of the nitrogen fixed by the bacteria is absorbed by the roots, and carbohydrates released from the roots provide energy for the bacteria.

31–19 *Sweet fern,* Comptonia peregrina, *growing on West Rattlesnake Mountain in New Hampshire. Sweet fern, which obtains its needed nitrogen from symbiotic nitrogen-fixing actinomycetes, is able to colonize nutrient-poor soils on which most other plants cannot grow.*

Recombinant DNA and Nitrogen Fixation

Genetic mapping of a free-living nitrogen-fixing bacterium has shown that the 17 genes known to be involved in nitrogen fixation are clustered on one portion of the chromosome. In experiments at the University of Sussex, biologists have succeeded in transferring this gene cluster to a plasmid and then introducing the plasmid into *Escherichia coli* cells. The *E. coli* cells were then able to synthesize nitrogenase and to fix nitrogen.

In other experiments at Cornell University, the gene cluster for nitrogenase has been transferred to yeast cells (eukaryotes) and has remained intact during repeated cell divisions. These experiments raise the hope, of course, that nitrogen-fixing genes can be transferred to the cells of plants such as corn, which could then, in effect, make their own nitrogen-containing compounds. Thus far, however, the yeast cells have not demonstrated the capacity to fix nitrogen, perhaps because of differences in start/stop signals or in the cytoplasmic makeup of prokaryotic and eukaryotic cells. An alternative way of conferring nitrogen-fixing capability on plants would be to transfer to nonlegumes the genes involved in the leguminous plant's contribution to the symbiotic association. This might be accomplished either by plasmids or by somatic cell hybridization. However, because of the number of genes involved and the complexity of the factors controlling nitrogen fixation, many botanists believe that such an achievement is, at best, many years away.

About 250 million metric tons of nitrogen are added to the soil each year, of which some 200 million tons are biological in origin. The other 50 million tons are largely in the form of chemical fertilizers, produced by commercial nitrogen-fixation processes in which the required energy is supplied by fossil fuels. More than one-third of the total amount of energy needed to produce a crop of corn is used in the manufacture, transportation, and application of chemical fertilizers. As the cost of the fossil fuels required for all of these processes fluctuates, so does the cost of producing the crop. From this perspective, the possible use of recombinant DNA techniques to increase biological nitrogen fixation takes on great practical significance. For, just as all animals, including ourselves, are ultimately dependent on photosynthesis for carbon-containing molecules, they— and we—depend on nitrogen fixation for the nitrogen-containing molecules without which living cells cannot function.

SUMMARY

Transpiration is the loss of water vapor by plants. As a consequence of transpiration, plants require large amounts of water. Water enters the plant from the soil through the roots and travels through the plant body in the conducting cells of the xylem (vessel members and tracheids). According to the cohesion-tension theory, water moves through the tracheids and vessels under negative pressure (tension). Because the molecules of water cling together (cohesion), a continuous column of water molecules is pulled from the soil solution and into the root, molecule by molecule, by the evaporation of water above.

Diffusion of gases, including water vapor, into and out of the leaf is regulated by the stomata. The stomata are opened and closed by the guard cells due to changes in turgor. Turgor is increased or decreased by the osmotic movement of water, which follows the movement of potassium ions into or out of the guard cells. The active transport of potassium ions is regulated by a variety of factors, including water stress, abscisic acid, carbon dioxide concentration, temperature, and light.

Minerals, which are naturally occurring inorganic substances, are brought into the plant from the soil and are carried with the transpiration stream in the xylem. They fulfill a variety of functions in plants, some of which are nonspecific, such as the effects on osmotic potential. Other functions are specific, such as the presence of magnesium in the chlorophyll molecule. A number of minerals are essential components of enzyme systems.

The movement of organic compounds from the photosynthetic parts of the plant is known as translocation. It takes place in the phloem and follows a source-to-sink pattern. According to the pressure-flow hypothesis, sugars are loaded into the sieve tubes in the leaf by active transport and are removed from them in other parts of the plant body where they are needed for growth and energy. Water moves into and out of the sieve tubes by osmosis, following the sugar molecules. These processes create a difference in water potential along the sieve tube, which causes water and the sugars dissolved in it to move by bulk flow along the sieve tube.

Characteristics of the soil affect the availability of minerals to plants. These characteristics include the rock from which the soil was formed, the size of the soil particles, the amount of humus present, and the soil pH.

Two types of symbiotic relationships are important in plant nutrition: mycorrhizae (discussed in Chapter 23) and relationships involving nitrogen-fixing bacteria. The associations between nitrogen-fixing bacteria, such as rhizobia, and the roots of certain plants, particularly legumes, result in the incorporation of gaseous nitrogen from the atmosphere into organic nitrogen-containing compounds.

QUESTIONS

1. Distinguish among the following: transpiration/translocation; cohesion-tension theory/pressure-flow hypothesis; source/sink; A horizon/B horizon/C horizon; rhizobia/mycorrhizae.

2. What properties of water discussed in Section 1 are important to the movement of water and solutes through plants?

3. Transpiration has often been described as a "necessary evil" to the plant. Why is it necessary? How is it evil?

4. Gardeners advise removing many leaves of a plant after transplanting. How does this help the plant to survive?

5. Consider a tree transpiring most rapidly at midday and an investigator with a sensitive instrument for measuring changes in the diameter of the trunk. If water is pulled up from the top (cohesion-tension theory), what changes in diameter should be observed from night to day?

6. In Figure 31–4, why doesn't air enter the tube through the top of the enclosed porous pipe or the leaf, even though water vapor can easily escape? How can the porous pipe (and, by analogy, the leaf) be permeable to air or water but effectively impermeable at an air-water interface?

7. Identify the cells and tissues through which a molecule of water travels from the time it enters the root until it is used in photosynthesis.

8. When K^+ ions move out of the guard cells, they move into adjacent epidermal cells. How would the influx of K^+ ions affect the epidermal cells? What effect would this have on the stomata?

9. Using the techniques described in this chapter for the analysis of sieve-tube sap, how would you measure the rate of movement? (*Hint:* Aphids are available.)

10. A particular sugar molecule was produced by photosynthesis in a leaf of a perennial plant late one August. Throughout the following winter, it was stored in a root of the plant. The next spring, it was oxidized in the process of respiration, providing energy for the growth of a new shoot tip. Identify the cells and tissues through which this sugar molecule traveled between its synthesis and its ultimate use.

11. As you will recall from Chapter 10, the immediate product of the Calvin cycle is glyceraldehyde phosphate, a three-carbon sugar (page 222). What chemical reactions did the sugar molecule in Question 10 probably undergo between its synthesis and its use?

12. (a) Experts in flower arranging advise recutting the stems of flowers while holding them under water. Explain why. (b) Some florists advise adding ordinary table sugar to the water in which cut flowers are placed. When this is done, certain types of flowers will remain fresh for several weeks. What is the explanation?

13. As we have seen, nitrogen fixation is an energy-requiring process. What is the source of the energy used by *Rhizobium?* If new symbiotic associations are developed in which the nitrogen-fixing bacteria produce greater yields of nitrogen-containing compounds, what price is the host plant likely to pay? Why?

Plant Responses and the Regulation of Growth

32-1 *This young bean seedling has just broken through the soil. Its growth into a mature plant will depend on light, water, temperature, and minerals from the soil and also on the interaction of many internal factors, among which are plant hormones.*

The capacity to respond to stimuli—both internal and external—is one of the essential properties of all living organisms (page 27). However, perhaps because of an inevitable bias based on our own experience, we tend to think of the perception of and active response to stimuli as phenomena limited to motile organisms, which rapidly and visibly alter their relationship with their environment when stimulated in various ways. Although plants cannot literally move from one place to another in response to a stimulus, they can significantly alter their relationship with the environment through their patterns of growth. As we noted in Chapter 30, growth in plants serves many of the same functions as motility in animals and is, to some extent, a counterpart of behavior. Many plant responses are slower than those of animals, but they are no less effective in solving the problems with which the environment confronts the organism.

We saw earlier that as a plant grows it does far more than simply increase its mass and volume. In a process that begins in the embryo and continues throughout the lifetime of a plant, cells produced in the meristematic tissues differentiate and elongate to form the specialized tissues and organs of the plant body. The rate at which this process occurs—in the plant as a whole and in different parts of the plant body—is affected by such environmental factors as temperature, water, sunlight, gravity, and physical contact with other objects, including other organisms. Growth rates, in turn, affect both the size of the whole plant and the relative sizes of its parts, thus determining its shape, or form. Moreover, many of a plant's activities—especially its cycles of active growth, reproduction, and dormancy—are finely tuned to the pattern of the changing seasons. Not only are plants able to sense and react to a range of factors in the immediate environment, but also, and perhaps more important, they are able to anticipate environmental changes and prepare for them.

Many of the details of how plants sense their environment and then respond with altered patterns of growth and development are, despite many years of research, poorly understood. Application of the techniques of molecular biology to plant physiology is beginning to yield new insights into many of the mechanisms underlying specific plant responses, but a complete and coherent picture of the functioning plant remains elusive. It is clear, however, that regulation of the physiological activities of a plant depends on the interplay of a number of internal and external factors. Chief among the internal factors are the plant hormones.

PHOTOTROPISM AND THE DISCOVERY OF PLANT HORMONES

A **tropism** is a growth response that involves the curvature of a plant part toward or away from an external stimulus that determines the direction of movement. If the plant part curves toward the stimulus, the tropism is said to be positive; if it curves away, the tropism is negative. One of the most obvious and useful

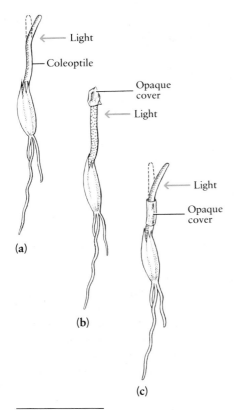

32-2 *The Darwins' experiment.* (a) *Light striking a growing coleoptile (such as the tip of this oat seedling) causes it to curve toward the light.* (b) *Placing an opaque cover over the tip of the seedling inhibits this curvature, but* (c) *an opaque collar placed below the tip does not. These experiments indicate that something produced in the tip of the seedling and transmitted down the stem causes the curvature.*

responses of plants is their positive **phototropism,** that is, their curvature toward light. It was with the study of phototropism that our knowledge of plant hormones and their effects on growth began, and it remains an appropriate starting point in the consideration of plant responses.

Charles Darwin and his son Francis performed some of the first experiments on phototropism. Working with grass seedlings, they noted that the curvature occurs below the tip, in a lower part of the coleoptile (the hollow sheath that surrounds the shoot tip in the embryos and seedlings of grasses). Then they showed that if they covered the tip of the coleoptile with a cylinder of metal foil or a hollow tube of glass blackened with India ink and exposed the plant to a light coming from the side, the characteristic curvature of the seedling did not occur. If, however, light was permitted to penetrate the cylinder, curving occurred normally. Curving also occurred normally when the lightproof cylinder was placed below the tip (Figure 32–2). "We must therefore conclude," they wrote, "that when seedlings are freely exposed to a lateral light some influence is transmitted from the upper to the lower part, causing the material to bend."*

In 1926, the Dutch plant physiologist Frits W. Went succeeded in separating this "influence" from the plants that produced it. Went cut off the coleoptile tips from a number of oat seedlings. He placed the tips on a slice of agar (a gelatinlike substance), with their cut surfaces touching the agar, and left them there for about an hour. He then cut the agar into small blocks and placed a block off-center on each stump of the decapitated plants, which were kept in the dark during the entire experiment. Within one hour, he observed a distinct curvature *away* from the side on which the agar block was placed (Figure 32–3).

Agar blocks that had not been previously in contact with a coleoptile tip produced either *no* curvature or only a slight curvature *toward* the side on which the block had been placed. Agar blocks that had been exposed to a section of coleoptile lower on the shoot produced no physiological effect.

Went interpreted these experiments as showing that the coleoptile tip exerted its effects by means of a chemical stimulus rather than a physical stimulus, such as

---

* Writing in 1881, the Darwins used the term "bend" to describe the phototropic response of a plant. Contemporary botanists, however, make a distinction between "bending," which does not involve growth, and "curvature," which does. As we shall see, the phototropic response is caused by differential cell elongation, and thus is, strictly speaking, a "curvature."

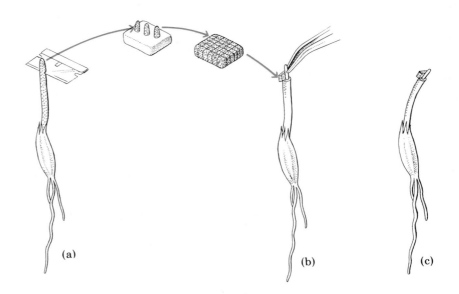

32-3 *Went's experiment.* (a) *He cut the coleoptile tips from oat seedlings and placed them on a slice of agar. After about an hour, he cut the agar in small blocks and* (b) *placed each block, off-center, on a decapitated seedling (the leaf is pulled up).* (c) *The seedlings curved away from the side on which the block was placed.*

an electrical impulse. This chemical stimulus came to be known as **auxin,** a term coined by Went from the Greek word *auxein,* "to increase." Auxin was one of the first plant hormones to be discovered.

The phototropism observed by the Darwins, it is now known, results from the fact that, under the influence of light, auxin migrates from the light side to the dark side of the shoot tip. The cells on the dark side, which contain more auxin, elongate more rapidly than those on the light side, causing the plant to curve toward the light—a response with high survival value for young plants. As with the influence of light on stomatal opening, only blue light—that is, light less than 500 nanometers in wavelength—is effective. In the phototropic response, as in the stomatal response, the blue-absorbing pigment is thought to be a flavin (see page 655); when stimulated, it apparently alters the permeability of the cell membrane, facilitating the movement of auxin to the shaded side of the stem.

HORMONES AND THE REGULATION OF PLANT GROWTH

A hormone, by definition, is a chemical substance that is produced in one tissue and transported to another, where it exerts one or more highly specific effects. Hormones integrate the growth, development, and metabolic activities of the various tissues of the plant. Typically they are active in very small quantities. In the shoot of a pineapple plant, for example, only 6 micrograms of auxin are found per kilogram of plant material. One enterprising plant physiologist calculated that the weight of the hormone in relation to that of the shoot is comparable to the weight of a needle in 20 metric tons of hay.

The term "hormone" comes from the Greek word *hormaein,* meaning "to excite." It is now clear, however, that many hormones also have inhibitory effects. So, rather than thinking of hormones as stimulators, it is perhaps more useful to consider them as chemical regulators. But this term also needs qualification. As we shall see, the response to a particular regulatory "message" depends not only on its content (that is, its chemical structure) but also on the identity of its recipient (that is, the specific tissue) and when and how it is received. Moreover, the response to any particular hormone is influenced by a variety of other factors in the internal environment of the plant, chief among which are likely to be other hormones.

Five principal types of plant hormones are known: auxins, cytokinins, ethylene, abscisic acid, and gibberellins. Recent evidence suggests that short carbohydrate chains, known as oligosaccharins, also function as plant hormones. Although we shall, for clarity, discuss these groups of hormones one at a time, keep in mind that we are always, in fact, dealing with interactions among them.

Auxins

The substance isolated by Went and given the name auxin is indoleacetic acid, or IAA. Several different substances with activity similar to that of IAA have now been isolated from plant tissues, and others have been synthesized in the laboratory. These substances are known collectively as auxins (Figure 32–4).

32–4 *Natural and synthetic auxins. IAA (indoleacetic acid), isolated from plant tissues, is the most common natural auxin. Naphthalenacetic acid, a synthetic auxin, is commonly used to induce the formation of adventitious roots in cuttings and to reduce fruit drop in orchards. 2,4-D, also a synthetic auxin, is used as an herbicide.*

IAA

Naphthalenacetic acid

2,4-D

32-5 *The stalk of the African violet leaf on the left was placed in a solution containing the synthetic auxin naphthalenacetic acid for 10 days before the picture was taken. The stalk of the leaf on the right was placed in pure water. Note the growth of adventitious roots on the stalk of the hormone-treated leaf.*

IAA is synthesized in the plant by enzymatic conversion of the amino acid tryptophan (see Figure 3–18, page 73). It is produced principally by the apical meristems of shoots and is transported to other parts of the plant, moving in one direction only, from shoot to root. (If a cut portion of a stem is turned upside down, IAA moves from bottom to top.) IAA causes cells in the growing region of the shoot to elongate; if the shoot apex is removed, growth stops. If the hormone is applied to the cut surface, growth resumes. If an auxin is applied to the intact plant, no growth effect on the stem occurs until relatively high concentrations are reached; then growth is inhibited. In short, it appears as if the apex produces the optimal concentration of auxin to which the shoot can respond positively.

Roots, which also receive their auxin from the shoot apex, are more sensitive to the hormone. Auxin in very small amounts is required for root growth; however, even a slight increase in auxin concentration inhibits root growth. As we shall see later in this chapter, such increases in auxin concentration along the lower side of horizontally oriented roots are thought to be involved in the downward growth of roots in response to gravity.

Synthetic auxins, unlike IAA, are not readily broken down by natural plant enzymes or by the enzymes of soil bacteria. Their long-lasting effects make them better suited for commercial purposes than IAA. The rooting preparations commonly used by gardeners to stimulate the growth of branch roots and adventitious roots in stem cuttings contain a synthetic auxin. Another synthetic auxin, 2,4-D, is used as an herbicide. For reasons not known, 2,4-D and related compounds are toxic to dicots at concentrations not harmful to monocots. Because of this selective effect and their low cost, synthetic auxins are commonly used on lawns to control broad-leaved weeds.

Mechanism of Action of Auxin

The stimulatory effects of auxin on the growth of shoots and roots are the result of a relatively rapid elongation of cells. Under the influence of auxin, the plasticity of the cell wall increases and the cell expands in response to the turgor exerted by the movement of water into the cell vacuole (see page 134). Current evidence indicates that IAA exerts its influence by activating a proton pump in the plant cell membrane that transports hydrogen ions (H^+) from the cell into the cellulose-containing cell wall. As a result, the cell wall becomes acidified and this, in turn, activates a pH-dependent enzyme in the cell wall that breaks the cross-links between the cellulose molecules. This allows the molecules to slide past one another as turgor acts against the cell wall. Subsequently, the cross-links re-form, and the wall becomes rigid once more. This response, known as "acid growth," is rapid, often beginning within 3 to 5 minutes, reaching a maximum in 30 minutes, and ending after 1 to 3 hours. As we shall see later in this chapter, a similar acid growth response is involved in the closing of Venus flytrap leaves.

The induction of acid growth, however, is only part of the auxin story. Auxin also stimulates long-term growth, triggering the expression of at least 10 specific genes. Long-term growth induced by auxin involves increased transcription of messenger RNA and ribosomal RNA, the assembly of many new ribosomes, and the biosynthesis of proteins different from those previously present in the cell. Although these proteins have not been identified with certainty, it is reasonable to believe, as a working hypothesis, that some of them are enzymes involved in the synthesis of the new cell-wall materials necessary for continued growth.

An increasing body of evidence indicates that calcium ion (Ca^{2+}) is intimately involved in the effects of auxin on growth. Inhibitors of auxin transport abolish not only auxin effects but also Ca^{2+} transport. Moreover, plants deficient in Ca^{2+} are often unresponsive to auxin. It has been hypothesized that auxin induces an increase in the amount of Ca^{2+} in treated cells, which, in turn, activates a regulatory substance known as calmodulin. Calmodulin, which has been found in a wide variety of eukaryotic organisms, affects membrane permeability, the transport of H^+ ions, and the activity of a number of enzymes.

Apical Dominance and Other Auxin Effects

In most dicot species, the growth of axillary buds is inhibited by auxin that moves down the stem from the shoot apex. This phenomenon is known as **apical dominance.** If you cut off the growing tip (apical meristem) of the stem of the house plant *Coleus,* for example, the axillary buds begin to grow vigorously, producing a plant with a bushier, more compact body and with more flowers. These derepressed buds can be repressed again by applying auxin to the cut surface of the shoot tip. Similarly, if you treat the "eyes" (actually axillary buds) of a potato with auxin, they will be inhibited from sprouting and the potato can therefore be stored longer.

Auxin was long thought to be directly responsible for apical dominance. However, recent experiments suggest that its effects are indirect. It now appears that auxin stimulates the production of ethylene in cells surrounding the axillary buds, and that this hormone, rather than auxin, inhibits bud growth. As we shall see shortly, cytokinins are also involved in apical dominance.

In woody plants, auxin also plays a role in the seasonal initiation of activity in the vascular cambium (page 644). When the meristematic region of the shoot begins to grow in the spring, auxin moving down from the shoot tip stimulates cambial cells to divide, forming secondary phloem and secondary xylem. Here, as elsewhere, its effects are modulated by other growth-regulating substances in the plant body.

Although auxin is produced principally in the apical meristems of shoots, it is also produced by young leaves, flowers, developing embryos, and fruits. Auxin promotes the maturation of the ovary wall and the development of fleshy fruits (Figure 32–6). Auxin treatment of the female floral parts of some species is used to produce fruits without fertilization, such as seedless tomatoes, cucumbers, and eggplants.

(a)

(b) (c)

32–6 *Auxin, apparently produced by developing seeds, promotes the growth of fruit. (a) Normal strawberry, (b) strawberry from which all seeds have been removed, and (c) strawberry in which three horizontal rows of seeds were left. If a paste containing auxin is applied to (b), the strawberry grows normally.*

Cytokinins

The isolation of auxin spurred the search for other growth-promoting factors in plants, particularly for a hormone that would stimulate cell division. Such a factor was first detected in coconut "milk," which is a liquid endosperm. It or related compounds have now been found in all plants examined, especially in actively dividing tissues, such as meristems, germinating seeds, fruits, and roots. They are called **cytokinins** (from "cytokinesis"). The most active naturally occurring cytokinin is zeatin, which was originally isolated from corn *(Zea mays)*; the synthetic cytokinin most commonly used in research is known as kinetin. Cytokinins resemble the purine adenine (Figure 32–7).

Studies of the way in which cytokinins promote cell division have shown that the hormone is required for some process that takes place after DNA replication is complete but before mitosis begins. Cytokinins increase the rate of protein synthesis, and it is thought that some of the resulting proteins may be necessary for cell division. The effect of cytokinins is not on the transcription of messenger RNA but rather on its translation into protein. The relationship, if any, between the mechanism of action of the cytokinins and their resemblance to adenine is not known.

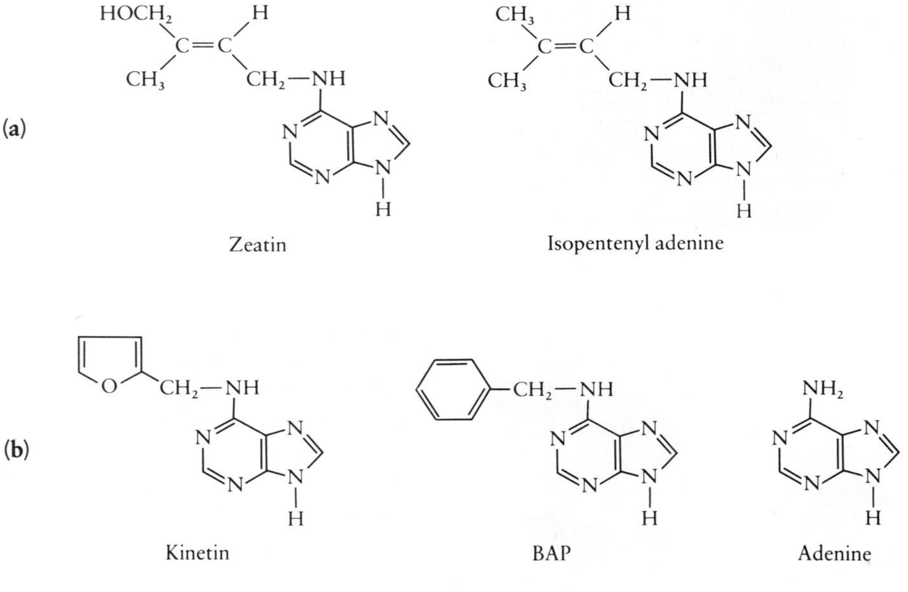

32–7 *Cytokinins.* **(a)** *Zeatin and isopentenyl adenine have been isolated from plant material.* **(b)** *Kinetin and 6-benzylamino purine (BAP) are commonly used synthetic cytokinins. Note the resemblances between the purine adenine and a portion of the molecule of each of these cytokinins. The significance of the resemblance between adenine and the hormones of this group is not known. It may merely be another example of biological economy: the use of a single major biosynthetic pathway to produce a number of functionally different products.*

(a)

Zeatin   Isopentenyl adenine

(b)

Kinetin   BAP   Adenine

Responses to Cytokinin and Auxin Combinations

Studies of responses to combinations of auxin and cytokinins are helping physiologists glimpse how plant hormones work together to produce the total growth pattern of the plant. Apparently, an undifferentiated plant cell—such as a meristematic cell—has two courses open to it: either it can enlarge, divide, enlarge, and divide again, or it can elongate without cell division. The cell that divides repeatedly remains essentially undifferentiated, whereas the elongating cell tends to differentiate and become specialized. In studies of tobacco stem cells, the addition of IAA to the culture medium on which the cells were growing produced rapid cell expansion, so that giant cells were formed. Adding kinetin alone had little or no effect. However, adding IAA plus kinetin resulted in rapid cell division, so that large numbers of relatively small cells were formed (Figure 32–8).

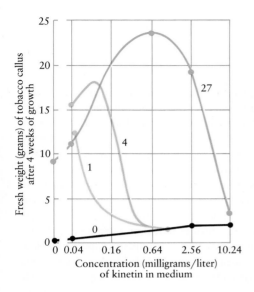

32–8 *The response of tobacco cells in tissue culture to combinations of auxin (IAA) and cytokinin (kinetin). The concentrations of IAA (in milligrams per liter) are indicated on the curves. Kinetin alone has little effect on the growth of undifferentiated tobacco tissue (a callus). IAA alone causes the culture to grow to a weight of about 10 grams, regardless of the concentration used. When both hormones are present, growth is greatly increased. Notice, however, that when optimum concentrations are exceeded, the growth rate declines.*

By slightly altering the relative concentrations of IAA and kinetin, investigators have been able to affect the development of undifferentiated cells growing in tissue culture. At roughly equal concentrations of the two hormones, the cells remain undifferentiated, forming a mass of tissue known as a **callus**. When a higher concentration of IAA is present, undifferentiated tissue gives rise to organized roots. With a higher concentration of kinetin, buds appear (Figure 32–9). Further, careful balancing of the two hormones can produce both roots and buds and, thus, an incipient plant.

However, lest you think this is simple, we shall describe another tissue culture study, in which tuber tissue of the Jerusalem artichoke was used. In this study, it was shown that a third factor, the calcium ion (Ca^{2+}), can modify the action of the auxin-cytokinin combination. Auxin plus low concentrations of kinetin favored cell enlargement, but as Ca^{2+} was added to the culture, there was a steady shift in the growth pattern from cell enlargement to cell division. High concentrations of Ca^{2+} apparently prevent the cell wall from relaxing and expanding, and at such concentrations the cell switches course and divides. Thus, not only do hormones modify the effects of hormones, but these combined effects may, in turn, be modified by nonhormonal factors, such as calcium ion and, undoubtedly, many others.

Other Cytokinin Effects

Cytokinins have also been shown to reverse the inhibitory effect of auxin in apical dominance. Local application of kinetin to repressed axillary buds releases them from inhibition. In intact plants, cytokinins are synthesized in the roots and travel upward, primarily through the xylem, reaching the lower buds first and in highest concentration. As the dominating apex of the plant grows and moves away from these lower buds, the influence of the cytokinins overcomes that of auxin and the buds begin to grow.

Another, apparently unrelated function of cytokinins is preventing the senescence (aging) of leaves. In many plants, the lower leaves turn yellow and drop off as the upper, new leaves develop; if kinetin is applied to the lower leaves, they remain green. Similarly, excised leaves remain green when maintained in a nutrient solution containing kinetin.

It is hypothesized that senescence in leaves, and probably in other plant parts as well, results from the progressive "turning off" of segments of DNA, with a consequent loss of messenger RNA production and protein synthesis. In experiments with excised leaves that contain radioactive amino acids and have been "spotted" with kinetin, the amino acids migrate to the kinetin-treated areas.

32–9 *Two buds forming on undifferentiated tissue (a callus) from a geranium following treatment with both an auxin and a cytokinin. Callus from some types of plants will continue to grow as undifferentiated tissue, or roots, or buds, depending on the relative proportions of auxin and cytokinins.*

Plants in Test Tubes

As long ago as the 1930s, scientists developed techniques for growing plant cells in test tubes. Tiny fragments of meristem are implanted under sterile conditions in a medium containing minerals and various combinations of organic compounds. Under these conditions, the meristematic cells proliferate to form clumps of undifferentiated cells. Subsequently, it was discovered that, by adjusting the hormone balance in the medium, it was possible to make these cells differentiate and grow into mature plants. Using these techniques, hundreds or even thousands of subcultures of meristematic tissue can be produced in a relatively short time and in a small space. Then, by altering the hormone balance, each of these can be turned into a small but perfect plant. This technique was initially employed in the culture of orchids and other hard-to-grow plants and is now being extended to many other plants.

Within the last 20 years, it has become possible, using similar techniques, to grow isolated protoplasts —plant cells without cell walls—in the laboratory. These protoplasts can be grown as isolated cells, like cultures of bacteria. Alternatively, if they are grown in a suitable medium, they will regenerate cell walls, multiply, and differentiate into whole plants. The young plants produced in this way are not only genetically uniform but also free of infectious disease.

Cultured cells are being used in a variety of ways. For instance, they are used in rapid screening tests to determine resistance to infectious diseases or to detect nutritional requirements. In this way, a scientist not only can work much more rapidly but can also do assays with millions of cells growing in a very small space, as compared to the far fewer number of plants that can be grown in a field or a greenhouse. Moreover, protoplasts from two different species of plant can be fused to create a hybrid. Protoplast fusion holds particular promise for combining the desirable characteristics of species that are sexually incompatible and therefore cannot be crossbred by conventional techniques.

The first plant created by a fusion of protoplasts was a hybrid of tobacco that had already been produced by conventional crossbreeding. (Producing a plant whose characteristics were known was a necessary preliminary test for the method.) Subsequently, protoplast fusion was used to create hybrids of species within the genera that include petunia *(Petunia),* carrot *(Daucus),* and potato *(Solanum).* More recently, hybrids involving species belonging to different genera have been obtained by protoplast fusion. Of particular interest is the fusion of the potato *(Solanum tuberosum)* and the tomato *(Lycopersicon esculentum),* both of which are members of the nightshade family (Solanaceae). Tomato plants are resistant to the oomycete that causes potato blight (see page 468), a disease that remains a serious threat to potato crops. It is hoped that, through protoplast fusion, it will be possible to develop potato plants that incorporate the tomato plants' resistance genes. Similar work, funded by the Campbell Soup Company, is under way with protoplast fusion involving tomato plants and strains of tobacco that are resistant to diseases to which tomato plants are particularly vulnerable.

The regeneration of whole plants from cultured cells, protoplast fusion procedures, and, most recently, the application of recombinant DNA techniques to plants (see page 352) together raise hopes for the production of entirely new "superplants." Such plants might be not only resistant to diseases that destroy a large proportion of the world's food supply each year, but also, for instance, capable of simultaneously carrying out C_4 photosynthesis and nitrogen fixation.

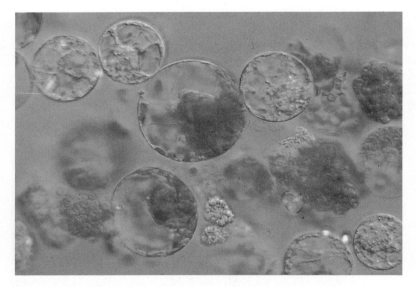

Protoplasts of wild tobacco (green) and commercial tobacco (clear). As you can see, some of the protoplasts have fused, giving rise to hybrid cells. With suitable treatment, the hybrid cells will regenerate cell walls, multiply, and differentiate into whole plants.

These experiments have led to the proposal that cytokinins prevent the DNA from being turned off and so promote continued enzyme synthesis and the production of such compounds as chlorophyll.

Ethylene

Ethylene is an unusual hormone in that it is a gas, a simple hydrocarbon, $H_2C = CH_2$. Its effects have been known for a long time. In the early 1900s, many fruit growers made a practice of improving the color and flavor of citrus fruits by "curing" them in a room with a kerosene stove. (Long before this, the Chinese used to ripen fruits in rooms where incense was being burned.) It was long believed that it was the heat that ripened the fruits. Ambitious fruit growers, who went to the expense of installing more modern heating equipment, found to their sorrow that this was not the case. As experiments later showed, the incomplete combustion products of kerosene were actually responsible for ripening the fruits. The most active gas was identified as ethylene. As little as 1 part per million of ethylene in the air will speed the ripening process. Subsequently, it was found that ethylene is produced by plants, fungi, and bacteria, as well as by kerosene stoves. The ethylene-synthesizing system is apparently located on the cell membrane, from which the hormone is released. Ethylene can be detected just before and also during fruit ripening and is responsible for a number of changes in color, texture, and chemical composition that take place as fruits mature. It also is involved in the senescence of floral parts that follows fertilization and precedes fruit development.

Auxin at certain concentrations causes a burst of ethylene production in some plants, and it is now believed that some of the effects on fruits and flowers generally attributed to auxin are related to the release of ethylene. As we noted earlier, ethylene is thought to be an effector of apical dominance. Auxin induces ethylene production in or near the axillary buds, whereas the cytokinins may inhibit it.

Ethylene and Leaf Abscission

Plants drop their leaves at regular intervals, either as a result of normal aging of the leaf or, in the case of deciduous shrubs and trees, in response to environmental cues. This process of **abscission** ("cutting off") is preceded by changes in the **abscission zone,** at the base of the petiole. In woody dicots, the abscission zone consists of two cell layers: (1) a structurally weak layer in which the actual abscission occurs, and (2) a protective layer that forms a leaf scar on the stem (Figure 32–10).

Once leaf senescence begins, ethylene produced in the abscission layer is the principal regulator of leaf drop. It acts by promoting the synthesis and release of cellulase, an enzyme that breaks down plant cell walls (and which is also involved in the ripening of fruits). Auxin inhibits abscission if applied to the leaf before senescence begins; however, once the abscission layer is formed, auxin promotes abscission by stimulating ethylene production.

Abscission layer   Protective layer

Petiole   Stem   (a)   1 mm

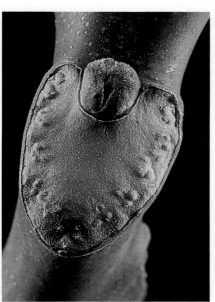

(b)

32–10 (a) *Abscission zone in a maple leaf, as seen in a longitudinal section. The abscission layer, which forms across the base of the petiole, consists of structurally weak cells. Under the influence of ethylene, enzymes are produced that cause the walls of these cells to dissolve. (b) After the leaf drops off, the protective layer forms a covering, the leaf scar, on the stem. This leaf scar is on a branch of* Ailanthus, *the tree of heaven.*

32–11 *Abscisic acid, an inhibitor that blocks the action of the growth-promoting hormones. In some species, it produces dormancy in buds and leaves. As we saw in the last chapter, it also plays a major role in stomatal closing.*

Abscisic acid

Abscisic Acid

Soon after the discovery of the growth-promoting hormones, plant physiologists began to speculate that growth-inhibiting hormones would be found, since it is clearly advantageous to the plant not to grow at certain times and in certain seasons. Not long afterwards, an inhibitory hormone, which was called dormin, was isolated from dormant buds. Subsequently, the same hormone was discovered in leaves, where it was thought to promote abscission. Thus it was called **abscisic acid,** or ABA (Figure 32–11). The latter name proved dominant, which is somewhat unfortunate since it is now known that, in most plants at least, ABA has little to do with abscission.

Abscisic acid may, however, induce dormancy. For example, application of ABA to vegetative buds changes them to winter buds, converting the outermost leaf primordia to bud scales (Figure 32–12). ABA is also present in the seeds of many species, where it is a major factor in maintaining seed dormancy. Moreover, as we saw in the last chapter, ABA brings about the closing of stomata under conditions of impending water shortage. Thus, ABA has come to be known as the stress hormone in recognition of its role as a protector of the plant against unfavorable environmental conditions.

32–12 *Winter buds of an American basswood. These buds, or young shoots, form in one growing season and elongate in the next. The buds contain small amounts of water and high concentrations of proteins and lipids, which prevent the formation of ice crystals that could kill the cells. Bud scales, which are small, tough modified leaves, protect the delicate tissues of the bud from mechanical injury and desiccation. As the first shoots emerge in the spring, the bud scales are forced off. Note the leaf scar at the base of each bud.*

Gibberellins

The **gibberellins** were discovered slightly earlier than auxin by a Japanese scientist who was studying a disease of rice plants called "foolish seedling disease." The diseased plants grew very rapidly but were spindly and tended to fall over under the weight of the developing grains. The cause of the symptoms, it was found, was a chemical produced by a fungus, *Gibberella fujikuroi*, which infected the seedlings. The substance, which was named gibberellin, and many closely related substances were subsequently isolated not only from the fungus but also from bacteria and many species of plants. More than 70 different gibberellins are now known (Figure 32–13); of these, gibberellic acid (GA_3) has been studied most thoroughly. Gibberellin and auxin (to a lesser extent) control elongation in mature trees and shrubs. The mechanism by which gibberellin affects elongation remains unknown; however, it does not involve the transport of H^+ ions or acid growth and thus would appear to be different from that of auxin.

Gibberellic acid (GA_3) GA_7 GA_4

32–13 *Three of the more than 70 gibberellins that have been isolated from natural sources. Gibberellic acid (GA_3) is the most abundant in fungi and the most biologically active in many tests. The minor structural differences that distinguish the other two gibberellins are indicated by arrows.*

In many species of plants, gibberellins characteristically produce hyperelongation of the stem, such as that seen in the "foolish seedlings." Particularly striking effects are seen in some plants that are genetic dwarfs. In some of these, application of gibberellin makes them indistinguishable from normal plants, suggesting that these dwarfs lack a gene coding for an enzyme needed for gibberellin production.

Gibberellins can also produce bolting, a phenomenon observed in many plants, especially biennials, that first grow as rosettes. Just before flowering, the flower stem elongates rapidly ("bolts"). Bolting and flowering, which normally occur only after an environmental cue (such as a cold season), occur in some plants following treatment with gibberellin (Figure 32–14). Abscisic acid inhibits the effects of gibberellin in promoting bolting; this inhibition is, in turn, reversed by cytokinins.

Gibberellins can also induce cellular differentiation. In woody plants, gibberellins stimulate the vascular cambium to produce secondary phloem. Both phloem and xylem develop in the presence of gibberellins and auxins. In intact plants, interactions between the two types of hormones are thought to determine the relative rates of production of secondary phloem and secondary xylem.

Gibberellins and Seed Germination

The highest concentrations of gibberellins have been found in immature seeds, although they are present in varying amounts in all parts of plants.

In grains (the one-seeded fruits of grasses), there is a specialized layer of cells, the aleurone layer, just inside the seed coat (see page 622). These cells are rich in protein. During the early stages of germination, following water uptake, the embryo produces gibberellin, which diffuses to the aleurone layer. In response to gibberellin, the aleurone cells produce and release enzymes that hydrolyze the starch, lipids, and proteins in the endosperm, converting them to sugars, fatty acids, and amino acids that the embryo and, later, the emerging seedling can use (Figure 32–15). In this way, the embryo itself calls forth the substances needed for its metabolism and growth at just the times they are required.

32–14 *Bolting and flowering in cabbage plants following gibberellin treatment. The plants on the left were not treated.*

32–15 *The action of gibberellin in a barley seed. The embryo releases gibberellic acid (GA), which diffuses to the aleurone layer. The gibberellin induces the aleurone cells to synthesize enzymes that digest the food reserves of the endosperm into smaller molecules. These molecules are absorbed by the scutellum and transported to the growing regions of the embryo.*

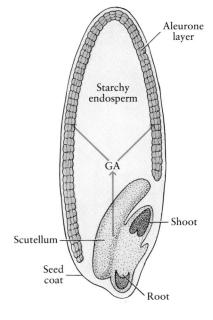

32–16 *An experiment investigating the action of gibberellin in barley seeds. Forty-eight hours before the picture was taken, each of these seeds was cut in half and the embryo removed. The seed at the bottom was treated with plain water, the seed in the center was treated with a solution of 1 part per billion of gibberellin, and the seed at the top was treated with 100 parts per billion of gibberellin. As you can see, digestion of the starchy storage tissue has begun to take place in the seeds treated with gibberellin.*

The production of the hydrolytic enzymes is the result of the synthesis of specific new messenger RNA molecules from the DNA template. For example, transcription of the gene coding for alpha-amylase, which digests starch in the endosperm, increases in response to gibberellin. Thus it is hypothesized that gibberellins, in this instance at least, are acting as derepressors, turning on certain genes.

Oligosaccharins

Like the gibberellins, the most recently discovered plant regulatory substances came to the attention of botanists in the course of studies of diseased plants. Plants infected by bacteria, fungi, and some protists defend themselves by producing antibiotics at the site of infection. These antibiotics are known as phytoalexins (from the Greek *alexein,* "to ward off"), having been originally discovered in plants infected by oomycetes of the genus *Phytophthora,* the cause of potato blight (see page 468).

Plant physiologists studying this defensive response hypothesized that some signal released by the pathogen was triggering the expression of the plant genes coding for the enzymes involved in antibiotic synthesis. Studies to test this hypothesis and to identify the signal were begun in the early 1970s by Peter Albersheim and his colleagues at the University of Colorado. They began with a fungus that attacks soybeans and found that the signal recognized by the plant was an oligosaccharide—a short chain of sugar molecules—released from the fungal cell wall. Related studies of antibiotic synthesis in response to bacterial infections also indicated that oligosaccharides were the signaling molecules, but, surprisingly, these oligosaccharides were released not from the bacterial cell wall but from the plant cell wall. Later investigations of plant responses to other pathogens and to wounding revealed a number of different, specific oligosaccharide molecules released from plant cell walls in response to different stimuli. In all cases, these molecules, dubbed **oligosaccharins,** were released from the noncellulose matrix of the primary cell wall and stimulated the transcription of messenger RNA.

These discoveries raised the possibility that oligosaccharins might have a more general role as plant regulatory substances. Could it be that the other plant hormones, such as auxin and gibberellin, work indirectly, by activating enzymes that release specific oligosaccharins from the cell wall, and that these molecules are the real regulators of the physiological processes of the plant? Studies published in 1985 by Kiem Tran Thanh Van and her colleagues in France indicate that oligosaccharins can indeed alter the growth and development of plants in cultures in which hormone levels and pH are carefully controlled (Figure 32–17). Her studies have generated considerable excitement, and the pace of research on oligosaccharins is quickening. There is hope that this new class of plant substances will provide the key to understanding what, as you have seen, is a bewildering diversity of responses to the five classic categories of plant hormones.

GRAVITROPISM

As we noted when we began our consideration of plant hormones, phototropism —curvature toward the light—has high survival value for a young plant. Another response with high survival value is the capacity of a young plant to respond to gravity, righting itself so that the shoot grows up and the root grows down. This response is known as **gravitropism;** like phototropism, gravitropism is thought to involve auxin.

When young monocot shoots are oriented horizontally, auxin migrates to the lower side of the coleoptile and calcium ion (Ca^{2+}) accumulates along the upper

32–17 *The effects of oligosaccharins on plant development have been studied using thin cell layers of stem from tobacco plants (a). The thin cell layers, which include the epidermis, the layer of cells immediately below the epidermis, and one to three layers of parenchyma cells, are maintained in a nutrient medium containing glucose and salts. The ratio of auxin to cytokinin, the pH, and the particular oligosaccharins added to the culture medium determine how the thin cell layer develops. Under the influence of different oligosaccharins, it may form (b) an undifferentiated callus, (c) roots, (d) vegetative shoots and leaves, or (e) floral shoots.*

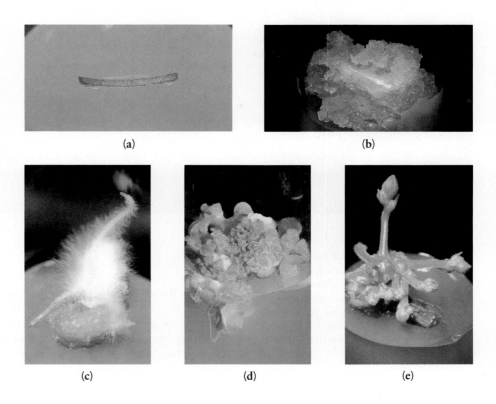

(a)

(b)

(c)

(d)

(e)

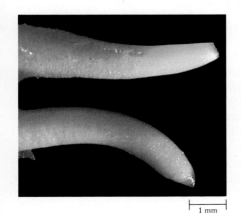

1 mm

32–18 *A demonstration of the importance of the root cap in gravitropism. The root cap of the corn root at the top of the micrograph has been removed; the root cap of the root at the bottom is intact. When the two roots were oriented horizontally, only the root with an intact cap responded with downward curvature. Application of a small amount of auxin to one side of a root from which the cap has been removed will cause it to curve toward the side on which the auxin is applied.*

side. The increased concentration of Ca^{2+} inhibits growth along the upper side, while auxin stimulates elongation of cells on the lower side. The result is an upward curvature of the shoot. When the shoot becomes vertical, the differences in auxin and Ca^{2+} concentrations disappear, and growth continues in an upright direction. Whether a similar mechanism is involved in the gravitropism of dicot shoots remains unknown. Thus far, experiments designed to determine concentrations of auxin and Ca^{2+} in different parts of horizontally oriented dicot shoots have yielded conflicting results.

As you will recall, roots are exquisitely sensitive to even slight increases in auxin concentration. Their downward growth is thought to result from the inhibitory effects of increased auxin concentrations along the lower side. As cells along the upper side of the root elongate more rapidly than those along the lower side, the root curves downward. The transport of auxin to the lower side of roots is dependent upon Ca^{2+}, but nothing is known of the mechanism by which it occurs.

Another major question concerns the way in which the influence of gravity is detected. How does a seedling "know" it is on its side? Hormones and ions are soluble, and so gravity itself should have no effect on their distribution. The answer to this question has long been thought to lie in specialized cells of the shoot and the root cap (Figure 32–18). The inner, or core, cells of the root cap seem to be analogous to the statocysts found in many animals (see page 528). Like statocysts, these cells contain statoliths—particles that move in response to gravity. In jellyfish, the statoliths are grains of hardened calcium salts; in the core cells of the root cap, they are starch-containing plastids, known as amyloplasts. When a root is growing vertically, the amyloplasts collect near the lower walls of the core cells (Figure 32–19a). If the root is placed in a horizontal position, however, the amyloplasts slide downward and come to rest near what were previously vertically oriented walls (Figure 32–19b). Within minutes, the root begins to curve down, gradually returning the amyloplasts to their original

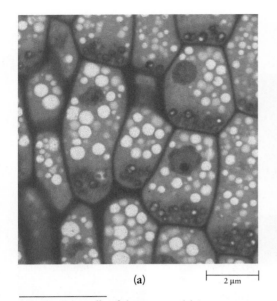 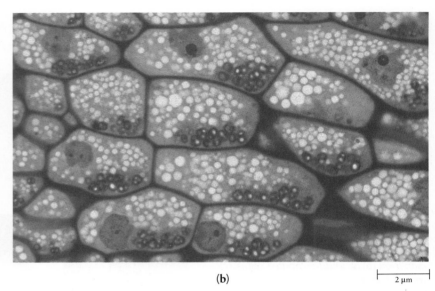

(a)   2 μm   (b)   2 μm

32–19 *Core cells of the root cap* **(a)** *in their normal vertical orientation and* **(b)** *after the root has been turned horizontally. The amyloplasts are the dark globular bodies containing white starch grains. Note how their position within the cells changes when the orientation of the root is changed.*

position. Roots from which the amyloplasts have been removed are unable to respond to gravity, suggesting that their movement is indeed critical in the detection of gravity. Recently, however, a mutant plant has been discovered in which the roots lack amyloplasts and yet exhibit nearly normal responses to gravity. Thus, the mechanism of gravity detection, like so many other mechanisms in plants, remains in question.

How the detection of gravity—by whatever means—is translated into chemical gradients of auxin and calcium ion is similarly unknown. Many of the investigators of root gravitropism think that both the detection of gravity and the translation of that information into chemical gradients probably involve alterations in both electrical properties and membrane permeability.

PHOTOPERIODISM

In many regions of the biosphere, the most important environmental changes affecting plants (and, indeed, land organisms in general) are those that result from the changing seasons. Plants are able to accommodate to these changes because of their capacity to anticipate the yearly calendar of events: the first frost, the spring rains, long dry spells, long growing periods, and even the time that nearby plants of the same species will flower. For many plants, all of these determinations are made in the same way: by measuring the relative periods of light and darkness. This phenomenon is known as **photoperiodism.**

Photoperiodism and Flowering

The effects of photoperiodism on flowering are particularly striking (Figure 32–20). Plants are of three general types: day-neutral, short-day, and long-day. Day-neutral plants flower without regard to day length. Short-day plants flower in early spring or fall; they must have a light period *shorter* than a critical length—for instance, the cocklebur flowers when exposed to 16 hours or *less* of light. Other short-day plants are poinsettias, strawberries, primroses, ragweed, and some chrysanthemums. Long-day plants, which flower chiefly in the summer, will flower only if the light periods are *longer* than a critical length. Spinach, potatoes, clover, henbane, and lettuce are examples of long-day plants. With both short-day and long-day plants, the photoperiodic initiation of flowering can take place only if the plant has passed from its juvenile stage into a phase designated as "ripeness to flower." In woody perennials, it may take decades for a plant to reach this stage.

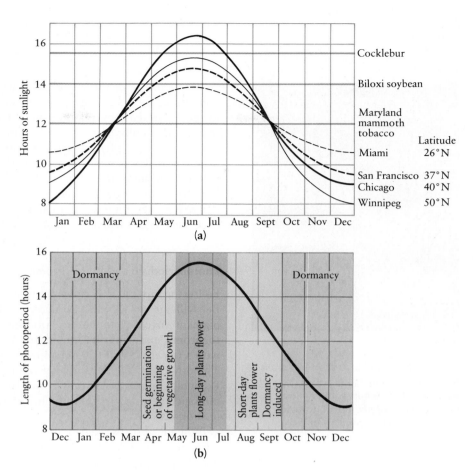

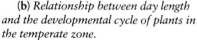

32–20 (a) *The relative length of day and night determines when many plants flower. The four curves depict the annual change in day length in four North American cities at four different latitudes. The green lines indicate the effective photoperiod of three different short-day plants. The cocklebur, for instance, requires 16 hours or less of light. In Chicago, it can flower as soon as it matures, but in Winnipeg the buds do not appear until early in August, so late that frost usually kills the plants before the seeds are mature.*

(b) *Relationship between day length and the developmental cycle of plants in the temperate zone.*

The discovery of photoperiodism explained some puzzling data about the distribution of common plants. Why, for example, is there no ragweed in northern Maine? The answer, investigators found, is that ragweed starts producing flowers when the day length is 14.5 hours. The long summer days do not shorten to 14.5 hours in northern Maine until August, and then there is not enough time for ragweed seed to mature before the frost. For somewhat similar reasons, spinach cannot produce seeds in the tropics. Spinach needs 14 hours of light a day for a period of at least two weeks in order to flower, and days are never this long in the tropics.

Note that ragweed and spinach will both bloom if exposed to 14 hours of daylight, yet one is designated as short-day and one as long-day. The important factor is not the absolute length of the photoperiod but rather whether it is longer or shorter than a particular critical interval for that variety. Detection of the interval can be very precise; in some varieties 5 or 10 minutes' difference in exposure can determine whether or not a plant will flower.

The first field studies on photoperiodism in plants were performed more than 50 years ago by scientists at the U.S. Department of Agriculture. These scientists have since become known as the Beltsville group, for the small town in Maryland where they carried out their studies.

Measuring the Dark

Subsequently, other investigators, Karl C. Hamner and James Bonner, began a laboratory study of photoperiodism. They used the cocklebur as the experimental organism. As we mentioned previously, the cocklebur is a short-day plant, requiring 16 hours or less of light per 24-hour cycle to flower. It is particularly useful for experimental purposes because a single exposure under laboratory conditions to a short-day cycle will induce flowering two weeks later, even if the plant is immediately returned to long-day conditions.

32-21 *The cocklebur, a short-day plant that can stand a lot of abuse (see essay on page 688), has been important in experimental studies of photoperiodism. Each "burr" is an inflorescence with two flowers.*

In the course of these studies, in which they tested a variety of experimental conditions, the investigators made a crucial and totally unexpected discovery. If the period of darkness is interrupted by as little as a 1-minute exposure to the light of a 25-watt bulb, flowering does not occur. Interruption of the light period by darkness has no effect whatsoever on flowering. Subsequent experiments with other short-day plants showed that they, too, required periods not of uninterrupted light but of uninterrupted darkness (Figure 32–22).

What about long-day plants? They also measure darkness. A long-day plant that will flower if it is kept in a laboratory in which there is light for 16 hours and dark for 8 hours will also flower on 8 hours of light and 16 hours of dark if the dark is interrupted by even a brief exposure to light.

Photoperiodism and Phytochrome

Following up on the clues from the Hamner and Bonner experiments, the Beltsville group was able to detect and eventually to isolate the light-detecting pigment involved in photoperiodism. This pigment, which they called **phytochrome,** exists in two different forms. One form, known as P_r, absorbs red light with a wavelength of 660 nanometers; phytochrome is synthesized in the P_r form. The other form, P_{fr}, absorbs far-red light with a wavelength of 730 nanometers. P_{fr}, which is the biologically active form of the pigment (that is, it can trigger a biological response), promotes flowering in long-day plants and inhibits flowering in short-day plants.

When P_r absorbs red light, it is converted to P_{fr} (Figure 32–23). This conversion takes place in sunlight or in incandescent light; in both of these types of light, red wavelengths predominate over far-red. The persistence of the effect of even a very brief red-light exposure is due to the fact that P_{fr}, the active form of phytochrome, is relatively stable in the dark. When P_{fr} absorbs far-red light, it is converted back to P_r. In the dark, P_{fr} is converted slowly to P_r, is degraded and replaced by newly synthesized P_r, or both.

32-22 *Experiments on photoperiodism showed that plants measure the length of darkness rather than that of light. Short-day plants flower only when the dark period exceeds some critical value. Thus, the cocklebur, for instance, will flower on 8 hours of light and 16 hours of darkness. If the 16-hour period of darkness is interrupted even very briefly, as shown on the right, the plant will not flower.*

The long-day plant, on the other hand, which will not flower on 16 hours of darkness, will flower if the dark period is interrupted. Long-day plants flower only when the dark period is less *than some critical value.*

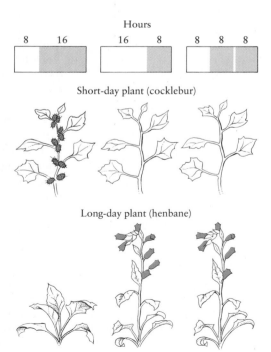

Hours

Short-day plant (cocklebur)

Long-day plant (henbane)

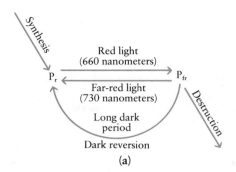

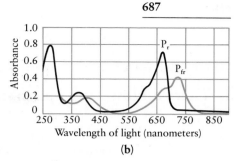

(a)

(b)

32–23 (a) *Phytochrome is synthesized (from amino acids) in the P_r form. P_r changes to P_{fr} when exposed to red light, which is present in sunlight. P_{fr} is the active form that induces the biological response. P_{fr} reverts to P_r when exposed to far-red light. In darkness, P_{fr} slowly reverts to P_r or is degraded. (b) Absorption spectra of the two forms of phytochrome. The similarities between the action spectra of the biological responses and the absorption spectra of the pigment provided important evidence that phytochrome is the pigment responsible for the responses. The reversible changes in absorbance provided the necessary clues both for detecting the pigment in plant extracts by spectrophotometry (page 210) and for isolating it.*

Other Phytochrome Responses

The $P_r \rightleftharpoons P_{fr}$ conversion, it is now known, acts as an off-on switch for a number of plant responses to light. Many kinds of small seeds, such as lettuce, germinate only when they are in loose soil, near the surface. This ensures that the seedling will reach the light before it runs out of stored food. Red light, a sign that sunlight is present, stimulates seed germination by converting phytochrome to the active form (P_{fr}). As with flower induction, exposure of the seeds to far-red light prevents the effect. The exposures can be alternated repeatedly and the seeds respond only to the final stimulus of the series. The inhibition of seeds by far-red light is interpreted as an adaptation that prevents them from germinating in the soil under a leaf canopy. As light filters through green leaves, the red is absorbed, leaving far-red predominant. Light-requiring plants have a better chance of survival if the seeds remain dormant, for years if necessary, until the required light is available.

Phytochrome is also involved in the early development of seedlings. When a seedling develops in the dark, as it normally does underground, the stem elongates rapidly, pushing the shoot up through the soil layers. A seedling grown in the dark will be elongated and spindly with small leaves (Figure 32–24). It will also be almost colorless, because the chloroplasts do not synthesize chlorophyll until they are exposed to light. Such a seedling is said to be etiolated. When the seedling tip reaches the light, normal growth begins. Phytochrome is involved in the switching from etiolated to normal growth. If a dark-grown bean seedling is exposed to only 1 minute of red (660 nanometers) light, it will respond with normal growth. If, however, the exposure to red light is followed by a 1-minute exposure to far-red light, thus negating the original exposure, etiolated growth continues.

The way in which phytochrome acts is not known. One recent suggestion is that it alters the permeability of the cell membrane, permitting particular substances to enter the cell, or, perhaps, inhibiting their entry, and that these substances, which probably include hormones, regulate the cell's activities. An increase in the concentrations of all the growth-promoting hormones can be detected almost immediately after phytochrome activation.

32–24 *Dark-grown seedlings, such as the bean plants on the left, are thin and pale with longer internodes and smaller leaves than the normal seedlings on the right. This group of characteristics, known as etiolation, has survival value because it increases the seedling's chances of reaching light before its stored energy supplies are used up.*

Is There a Flowering Hormone?

One of the most persistent—and still unanswered—questions in plant physiology concerns the existence of a flowering hormone. Since the 1930s, numerous experiments have suggested that such a hormone is formed in the leaves in response to an appropriate cycle of light and dark and then travels to the apical meristem of the plant and initiates flowering.

For example, (a) when certain plants, such as the cocklebur shown in the illustration below, are exposed to an appropriate light cycle, those with leaves flower and those without leaves do not. When even one-eighth of a leaf remains on a plant, flowering occurs, and the illumination of a single leaf—not necessarily the whole plant—suffices. These experiments indicate that chemical signals originating in the leaves cause the plant to flower. (b) This conclusion is supported by experiments on branched plants.

Exposure of one branch to the light induces flowering on the other branch as well, even when only a portion of a leaf is present on the lighted branch. (c) When two plants are grafted together, exposure of one of the plants to the light cycle induces flowering in both the lighted plant and the grafted one.

Despite the evidence of these and other experiments, no specific flowering hormone has ever been isolated and identified. However, both auxin and gibberellin have been shown to induce flowering in some plants under some circumstances. These results suggest that flowering does not involve a single, unique hormone but rather a combination of hormones, probably acting in conjunction with other, as yet unknown, chemical factors. Recent tissue culture studies indicate that among these factors may be oligosaccharins (see Figure 32–17).

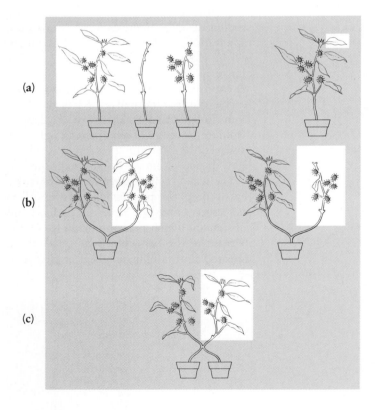

(a)

(b)

(c)

(a)

(b)

32–25 *Leaves of the oxalis plant, day (a) and night (b). Darwin believed that this folding of the leaves conserved heat energy. A more recent but also unproved hypothesis is that the "sleep movements" prevent the leaves from absorbing moonlight on bright nights, thus protecting photoperiodic phenomena.*

CIRCADIAN RHYTHMS

How can a spinach plant distinguish a 14-hour day from a 13.5-hour day? Seeking an answer to this question leads us to another group of readily observable phenomena. Some species of plants, for instance, have flowers that open in the morning and close at dusk. Others spread their leaves in the sunlight and fold them toward the stem at night (Figure 32–25). As long ago as 1729, the French scientist Jean-Jacques de Mairan noticed that these diurnal (daily) movements continue even when the plants are kept in constant dim light. More recent studies have shown that less evident activities, such as photosynthesis, auxin production, and the rate of cell division, also have daily rhythms. The rhythms continue even when all environmental conditions are kept constant. These regular day-night cycles are called **circadian rhythms,** from the Latin words *circa,* meaning "about," and *dies,* "day." They have been found throughout the four kingdoms of eukaryotic organisms.

Biological Clocks

Are these rhythms internal—that is, caused by factors within the organism itself—or is the organism keeping itself in tune with some external factor? For a number of years, biologists debated whether there might not be some environmental force, such as cosmic rays, the magnetic field of the earth, or the earth's rotation, that was setting the rhythms. Attempts to identify such an external factor have led to numerous expeditions under an extraordinary variety of conditions. Organisms have been taken down into salt mines, shipped to the South Pole, flown halfway around the world in airplanes, and, more recently, orbited in satellites—all without any detectable effect on their circadian rhythms.

Virtually all biologists now agree that circadian rhythms are endogenous—that is, they originate within the organism itself. The mechanism by which they are controlled is known as a **biological clock.** Strong evidence in support of an internal biological clock is that circadian rhythms are not exact. Different species and different individuals of the same species often have slightly different, but consistent, rhythms, often as much as an hour or two longer or shorter than 24 hours.

Biological clocks play a role in many aspects of plant and animal physiology, synchronizing internal and external events. For example, some flowers secrete nectar or perfume at certain specific times of the day or night. As a result, pollinators—which have their own biological clocks—have become programmed to visit these flowers at these times, thereby ensuring maximum rewards for both

32–26 *"Sleep movement" rhythms in the bean plant. Many legumes, such as the bean, orient their leaves perpendicular to the rays of the sun during the day and fold them up at night. These movements can easily be transcribed on a rotating chart by a delicately balanced pen-and-lever system attached to the leaf by a fine thread (a). The rhythm will persist for several days in continuous dim illumination. A representative recording is seen in (b).*

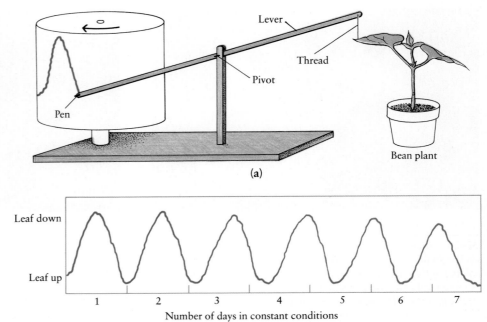

(a)

(b)

the pollinators and the flowers. However, for most organisms, the "use" of the clock for such purposes is thought to be a secondary development. The primary function of the clock appears to be in enabling organisms to recognize the changing seasons of the year by "comparing" external rhythms of the environment, such as changes in day length, to their own internal rhythms.

Resetting the Clock

Although circadian rhythms originate within the organisms themselves, they can be modified by external conditions—which is, of course, important in keeping organisms in tune with their environment. For instance, a plant whose natural daily rhythm shows a peak every 26 hours when grown under continuous dim light will reset its rhythm, as if the hands of its clock had been turned, when exposed to an unusual interval of light and darkness. For example, exposed to 14-hour intervals of light alternating with 10-hour intervals of darkness, the plant resets its clock repeatedly, eventually becoming synchronized with that cycle. It can also adjust to 11 hours of light and 11 hours of darkness (or 22 hours). Such adjustment to an externally imposed rhythm is known as **entrainment.** If the new rhythm is too far removed from the original one, however, the organism will "escape" the entrained rhythm and revert to its natural one. A plant that has been kept on an artificial or forced rhythm, even for a long period of time, immediately reverts to its normal internal rhythm when returned to continuous dim light.

The Nature of the Clockwork

The chemical nature of the biological clock—or indeed whether there is just one kind of clock or many—is still not known. When the on-off switch of phytochrome was first discovered, it was hypothesized that the reversion of P_{fr} to P_r might be the time-measuring system of the plant, functioning by the same principle as an hourglass. Experiments have disproved this hypothesis, however; for instance, in many plants, P_{fr} disappears within three or four hours of darkness.

32–27 *Tendrils of a wild grape plant. Twining is caused by varying growth rates on different sides of the tendril.*

32–28 (a) *Dodder* (Cuscuta), *a parasitic plant that wraps itself around and around its host. The vegetative parts of the plant are usually bright orange or yellow; the flowers are tiny, waxy, and absolutely colorless. One of the plants frequently parasitized by dodder is the halophyte Salicornia (see page 658), visible here as green stalks.* (b) *A salt marsh near Mill Valley, California, in which the host Salicornia plants are almost totally obscured by a sea of parasitic dodder.*

However, in some plants, the circadian rhythm of sleep movements can be rephased by conversion of P_r to P_{fr}. The best evidence to date indicates that the timing mechanism involves rhythmic changes in the cell membrane, either in the protein components, in the phospholipids, or both. Thus, phytochrome conversion might reset the clock by producing changes in membrane structure or permeability.

TOUCH RESPONSES

Twining and Coiling

Many plants respond to touch. One of the most common examples is seen in tendrils—the modified stems or leaves with which many plants support themselves and climb (Figure 32–27). Tendrils often move in a spiral as they grow; this process, called circumnutation, increases the tendrils' chances of finding a support. When the apex of a tendril touches any object, it responds to the touch by forming a tight coil. Cells touching the support shrink slightly, and those on the outer side elongate. In the garden pea, stroking the tendril with a glass rod for 2 minutes can induce a coiling response that lasts more than 48 hours. Coiling requires light, apparently for ATP production, since ATP will substitute for light in excised tendrils. Specialized epidermal cells of the tendrils are the sensors for touch, but the mechanism by which these cells induce coiling is unknown. Current evidence indicates that auxin and ethylene are probably involved; these hormones cause excised tendrils to coil even in the absence of touch.

Studies by M. J. Jaffe of Wake Forest University in North Carolina have shown that tendrils can store the "memory" of tactile stimulation. For example, if pea tendrils are kept in the dark for three days and then stroked, they will not coil, perhaps because of the requirement for ATP. If, however, they are illuminated within two hours of the stroking, they will show the coiling response.

A complex repertoire of coiling behavior is seen in the parasitic plant called dodder *(Cuscuta)*. A mature dodder plant looks like a tangle of cooked spaghetti (Figure 32–28). The plants, which have no leaves and no chlorophyll, range in color from almost white to yellow or orange and have a tiny waxy flower. When a dodder seed germinates, the seedling anchors itself with a small temporary root and begins to circumnutate. It can sense a host plant up to 8 centimeters away, presumably by chemical stimulus, and, in response, moves toward it. When it touches its new host (or any other object), it coils itself rapidly around it, pulling

(a)   (b)

up its roots behind it. It then develops haustoria (see page 480) that it sinks into the vascular tissues of the host plant, tapping its food and water supply. Once its feeding network is established, the tip begins to circumnutate again. Thus alternately winding and looping, a single dodder plant can crochet itself into a mat that covers as much as a kilometer.

Rapid Movements in the Sensitive Plant

A more rapid response to touch occurs in the sensitive plant, *Mimosa pudica*. A few seconds after a leaf is touched, the petiole droops and the leaflets fold (Figure 32–29). This response is a result of a sudden change in turgor in specialized "motor cells" of the jointlike thickenings called pulvini (singular, pulvinus) at the base of the leaflets and leaves. These same cells are involved with sleep movements, which also occur in *Mimosa*. Depending on the variety and the stimulus, a single leaf or all the leaves on the plant may be affected. A series of reactions appears to be involved in spreading the stimulus. The sensory stimulus is apparently rapidly translated into an electrical signal, similar to a nerve impulse in animals, that passes along the petiole and can be detected by microelectrodes placed in the plant. The electrical signal, in turn, triggers a chemical signal that makes the cell membrane of the motor cells more permeable to potassium (K^+) and chloride (Cl^-) ions. Movement of these ions out of the motor cells causes water to leave the cells by osmosis. As a result of this water loss, the motor cells collapse, thereby causing movement of the leaf or leaflet. Tannins are also secreted by the motor cells into extracellular spaces and may protect the plant against predation.

Rapid Movements in Carnivorous Plants

Rapid responses to touch also occur in the capture of prey by the carnivorous Venus flytrap and the sundew (see essay, pages 666 to 667). The leaves of the Venus flytrap have two hinged lobes, each of which is equipped with three sensitive hairs. When an insect alights on a leaf, it brushes against the hairs, setting off an electrical impulse that triggers the closing of the leaf. The toothed edges mesh like a bear trap, snapping shut in less than half a second. Once the insect is trapped, the leaf halves gradually squeeze closed, and the captive animal is pressed against digestive glands located on the inner surface of the trap. The movement of the insect stimulates the trap to close more tightly and glands on the upper leaf surface to secrete digestive enzymes.

32–29 *Sensitive plant* (Mimosa pudica). (**a**) *Normal position of leaves and leaflets.* (**b**) *Response to touch. It has been hypothesized that these reactions may prevent wilting (when they occur in response to strong winds), startle insects, or dismay larger herbivores. Collapse of the leaflets is caused by rapid turgor changes in cells of the petioles. These changes are accompanied by the release from the cells of substances known as tannins. Tannins have an astringent taste and are repellent to herbivores. In the evolution of the touch response in* Mimosa, *the release of tannins may have been more important than the rapid movements.*

(a)

(b)

It was long thought that the response of the Venus flytrap, like that of the sensitive plant, involved turgor changes. Recent research, however, has revealed that the response is instead another example of acid growth (page 675). The electrical impulse set off when an insect brushes against the trigger hairs activates an enzyme that, using ATP for energy, pumps H^+ ions into the walls of epidermal cells along the outer surface of the trap's hinge; almost 30 percent of the ATP in these cells is consumed in the 1 to 3 seconds required to close the leaf completely. The rapid, irreversible cell expansion that results from acidification of the cell walls closes the trap. During this process, the cells on the inner surface of the hinge continue to grow at their usual, slow rate; some 10 hours later their growth catches up with the earlier growth of the cells on the outer surface, and the trap opens. As a result of both growth processes, the leaf is slightly larger than it was before it closed.

Studies by investigators at Washington University in St. Louis have shown that an electrical impulse is also involved in the trapping of insects by the sundew. As you saw on page 666, the club-shaped leaves of the plant are covered with tiny tentacles. A sticky droplet surrounding the tip of each tentacle attracts insects. When an insect is caught on the tip of a tentacle, the surrounding tentacles bend in, carrying the prey to the center of the leaf, where it is digested. Microelectrodes placed in the tentacles (Figure 32–30) have revealed that the touch of an insect's feet triggers an electrical impulse that moves down the tentacle.

The electrical impulses that have been observed in plants are the same, in principle, as the nerve impulses of animals (see page 847). Many botanists expect that electrical signals will be found to coordinate a variety of activities involving cells in different parts of the plant body.

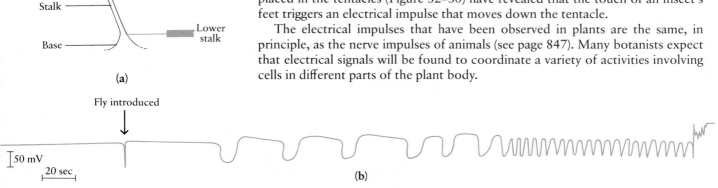

32–30 (a) *Diagram of a sundew tentacle with electrodes in place.* (b) *A record of electrical impulses produced by positioning a fruit fly so that its feet stroked the head of a tentacle. The recording ended when the contact was broken as the tentacle bent.*

Generalized Effects of Touch on Plant Growth

In addition to the specialized responses of some plants, touch and other mechanical stimuli may also have widespread effects on patterns of growth. Although botanists have long known that plants grown in a greenhouse tend to be taller and more spindly than plants of the same species grown outside, it was not until the 1970s that systematic studies revealed that regular rubbing or bending of stems inhibits their elongation and results in shorter, stockier plants. Plants in a natural environment are, of course, subjected to similar stimuli, in the form of wind, raindrops, and the movements of passing animals. This response also involves both electrical signals and a change in cell membrane permeability. The change in membrane permeability is believed to affect the relative proportions of hormones available to the plant cells.

CHEMICAL COMMUNICATION AMONG PLANTS

As we noted in Chapter 24 (page 515), many angiosperms produce toxic or bad-tasting compounds that function as powerful defenses against herbivorous animals. In some species, the production of such compounds is initiated or increased in response to damage inflicted on the plant by herbivores, for example, chewing insects. The resulting higher concentrations of these substances deter further predation and thereby protect the plant from severe damage. Studies by scientists at the University of Washington and at Dartmouth College have shown that the plants do more than protect themselves: they apparently also "warn" neighboring plants of the same species to mobilize their defenses prior to attack.

The effect was first observed in Sitka willows; three days after high levels of defensive chemicals were detected in the leaves of trees being directly attacked by tent caterpillars, high levels of the same chemicals were found in the leaves of trees that had not yet been touched by the caterpillars, some as far as 60 meters from the damaged trees. Presumably, the damaged leaves release an airborne substance that, upon reaching the leaves of another plant, triggers their synthesis of defensive chemicals. Similar effects have been observed in carefully controlled laboratory studies in which mechanical damage was inflicted on the leaves of poplar and maple seedlings. Studies currently in progress are designed to isolate and identify the airborne substance or substances involved in this communication.

As we saw in Chapter 27, chemical communication among individuals of the same species is a familiar phenomenon in animals, particularly insects. This recent research is the first hint that there may be many more interactions among individual plants than are apparent to even the well-trained eye.

SUMMARY

Plants respond to stimuli in both their internal and external environments. Such responses enable the plant to develop normally and to remain in touch with changing external conditions.

Hormones are important factors in plant responses. A hormone is a chemical that is produced in particular tissues of an organism and carried to other tissues of the organism, where it exerts one or more specific influences. Characteristically, it is active in extremely small amounts. Table 32–1 summarizes the five major groups of hormones that have been isolated from plants: auxins, cytokinins, ethylene, abscisic acid, and gibberellins. Other growth-regulating substances, such as the recently discovered oligosaccharins, may also be present.

Auxin is produced principally in rapidly dividing tissues, such as apical meristems. It causes lengthening of the shoot, chiefly by promoting cell elongation. In conjunction with cytokinin and ethylene, it is involved in apical dominance, in which the growth of axillary buds is inhibited, thus restricting growth principally to the apex of the plant. At low concentrations, auxin promotes the initiation of

TABLE 32–1   **Plant Hormones and Their Effects**

HORMONE	PHYSIOLOGICAL EFFECTS AND ROLES
Auxin	Stimulates cell elongation; involved in phototropism, gravitropism, apical dominance, and vascular differentiation; inhibits abscission prior to formation of abscission layer; stimulates ethylene synthesis; stimulates fruit development; induces adventitious roots on cuttings
Cytokinin	Stimulates cell division, reverses apical dominance, involved in shoot growth and fruit development, delays leaf senescence
Ethylene	Stimulates fruit ripening, leaf and flower senescence, and abscission; may be effector of apical dominance
Abscisic acid	Stimulates stomatal closure; may be necessary for abscission and dormancy in certain species
Gibberellin	Stimulates shoot elongation, stimulates bolting and flowering in biennials, regulates production of hydrolytic enzymes in grains

branch roots and adventitious roots; at higher concentrations, it inhibits the growth of the main root system. In developing fruits, auxin produced by seeds stimulates growth of the ovary wall; diminished production of auxin is correlated with abscission of fruits and leaves. The capacity of auxin to produce such varied effects appears to result from different responses of the various target tissues and the presence of other factors, including other hormones.

The cytokinins promote cell division. It is possible, by altering the concentrations of auxin and cytokinins, to alter patterns of growth in undifferentiated plant tissue in tissue culture.

Ethylene is a gas produced by fruits during the ripening process, which it promotes. It plays a major role in leaf abscission and is thought to be an effector of apical dominance. Abscisic acid, a growth-inhibiting hormone, affects stomatal closure and may be involved in the induction of dormancy in vegetative buds and the maintenance of dormancy in seeds.

Gibberellins stimulate shoot elongation, induce bolting and flowering in many plants, and are also involved in embryo and seedling growth. In grasses, they stimulate the production of hydrolyzing enzymes that act on the stored starch, lipids, and proteins of the endosperm, converting them to sugars, fatty acids, and amino acids, which nourish the seedling.

Oligosaccharins, short carbohydrate chains released from plant cell walls in response to various stimuli, may also be important growth regulators. Recent studies have shown that specific oligosaccharins can determine the course of development in tissue culture.

Plants respond to a number of environmental stimuli. Two responses with high survival value for young plants are phototropism, the curvature of a plant toward the light, and gravitropism, the capacity of the shoot to grow up and the root to grow down. Phototropism is mediated by auxin and gravitropism by auxin and calcium ion (Ca^{2+}).

Photoperiodism is the response of organisms to changing periods of light and darkness in the 24-hour day. Such a response controls the onset of flowering in many plants. Some plants, known as long-day plants, will flower only when the periods of light exceed a critical length. Other plants, short-day plants, flower only when the periods of light are less than some critical period. Day-neutral plants flower regardless of photoperiod. Experiments have shown that the dark period rather than the light period is the critical factor.

Phytochrome, a pigment commonly present in small amounts in the tissues of plants, is the receptor molecule for detecting transitions between light and darkness. The pigment exists in two forms, P_r and P_{fr}. P_r absorbs red light, whereas P_{fr} absorbs far-red light. P_{fr} is the biologically active form of the pigment; among its many known effects, it promotes flowering in long-day plants, inhibits flowering in short-day plants, promotes germination in lettuce seeds, and promotes normal growth in seedlings. Its mechanism of action appears to involve changes in permeability of the cell membrane.

Circadian rhythms are regular cycles of growth and activity occurring approximately on a 24-hour basis. Many of these rhythms are independent of the organism's environment and are controlled by some endogenous regulator—a biological clock. The principal function of the biological clock is apparently to provide the timing mechanism necessary for photoperiodic phenomena. Its chemical nature is not known.

Some species of plants exhibit specific, rapid movements in response to touch. Examples include the winding of tendrils, the collapse of the leaves of the sensitive plant *(Mimosa),* the triggering of the carnivorous Venus flytrap, and the bending of the tentacles of the sundew. In addition, all vascular plants appear to respond to touch and other mechanical stimuli with altered growth patterns, resulting in shorter, stockier plants. Plant responses to touch involve various combinations of electrical impulses, changes in turgor, and chemical changes that result in differential rates of growth.

Some species of plants respond to mechanical damage, such as may be caused by herbivorous animals, by synthesizing chemicals that deter further predation. They apparently also release airborne substances that "communicate" with other individuals of the same species, triggering in them the synthesis of defensive chemicals before any actual damage occurs.

QUESTIONS

1. Describe Went's experiments and the conclusions that can be drawn from them.

2. Describe the principal roles played by each of the following hormones: auxin, cytokinin, ethylene, abscisic acid, and gibberellin.

3. Describe the phenomenon of apical dominance and the role of each of the three hormones involved.

4. The distinctions we make among the plant hormones refer to general classes of their effects rather than to the precise chemical names of the substances actually produced by particular plants. Why might it be relatively easy to test chemicals for hormonal activity but be particularly difficult to determine just which chemical a plant actually produces and uses?

5. One bad apple can spoil the whole barrel. Explain.

6. Label the drawing at the right. In terms of the labeled structures, describe the action of gibberellin in initiating growth of the embryo.

7. Distinguish between the following: phototropism/photoperiodism; circadian rhythm/biological clock.

8. Why is a biological clock necessary for photoperiodism?

9. Plants must, in some manner, synchronize their activities with the seasons. Of the cues that they might use, day (or night) length seems to have been selected. What might be the advantage of using photoperiod rather than, say, temperature as a season detector?

10. Photoperiodic systems are not nearly as sensitive to low levels of illumination as are many visual systems. Why might extreme sensitivity be a disadvantage in a photoperiodic system? What might be the effect of the widespread use of bright lights for street illumination?

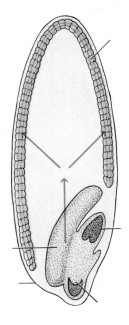

11. Suppose you were given a chrysanthemum plant, in bloom, one autumn, and you decided to keep it indoors as a house plant. What precautions would you need to take the following autumn to ensure that it would bloom again?

12. Carefully examine the graph in Figure 32–26. What is happening to the rhythm of the plant's "sleep movements" toward the end of the week? Speculate as to why this might be happening. It has also been observed that as the plant is maintained over time under the experimental conditions, it becomes rather sickly. What factors do you suppose might be involved in this change?

SUGGESTIONS FOR FURTHER READING

Books

BRADY, JOHN: *Biological Clocks, Studies in Biology, No. 104,* University Park Press, Baltimore, Maryland, 1979.*

An interesting and well-written introduction to the subject of biological clocks and their experimental study.

CUTLER, DAVID F.: *Applied Plant Anatomy,* Longman, Inc., New York, 1978.*

An interestingly written textbook on the fundamentals of plant anatomy showing some of the ways in which a knowledge of plant anatomy can be applied to solve many important everyday problems.

* Available in paperback.

GALSTON, ARTHUR W., PETER J. DAVIES, and RUTH L. SATTER: *The Life of the Green Plant*, 3d ed., Prentice-Hall, Inc., Englewood Cliffs, N.J., 1980.*

A comprehensive and up-to-date description of the functioning of the green plant, especially strong in plant physiology. Written for students without an advanced background in biology or chemistry.

GREULACH, VICTOR: *Plant Structure and Function*, 2d ed., The Macmillan Company, New York, 1983.

A short, lucid introduction to botany, with emphasis on physiology.

HEYWOOD, VERNON H. (ED.): *Flowering Plants of the World*, Prentice-Hall, Inc., Englewood Cliffs, N.J., 1985.

The best available guide for students to the families of flowering plants.

KENDRICH, R. E., and B. FRANKLAND: *Phytochrome and Plant Growth, Studies in Biology*, No. 68, 2d ed., Edward Arnold Publishers, London, 1983.*

A concise, well-illustrated introduction to the molecular properties and physiology of phytochromes.

LEDBETTER, M. C., and KEITH PORTER: *Introduction to the Fine Structure of Plant Cells*, Springer-Verlag, New York, 1970.

An excellent atlas of electron micrographs of plant cells, with a detailed explanation of each.

RAVEN, PETER H., RAY F. EVERT, and SUSAN E. EICHHORN: *Biology of Plants*, 4th ed., Worth Publishers, Inc., New York, 1986.

An up-to-date and handsomely illustrated general botany text, especially strong in evolution and ecology.

SALISBURY, FRANK B., and CLEON W. ROSS: *Plant Physiology*, 3d ed., Wadsworth Publishing Co., Inc., Belmont, Calif., 1985.

A good, modern plant physiology text for more advanced students.

SCHNELL, DONALD E.: *Carnivorous Plants of the United States and Canada*, John F. Blair, Winston-Salem, N.C., 1976.

Descriptions of 45 species of carnivorous plants with many color photographs. This book is not only interesting reading but also useful as a field guide, and it contains a chapter on growing techniques.

TROUGHTON, JOHN H., and F. B. SAMPSON: *Plants: A Scanning Electron Microscope Survey*, John Wiley & Sons, Inc., New York, 1973.*

A scanning electron microscope study of some anatomical features of plants and the relationship of these features to physiological processes.

ZIMMERMANN, MARTIN H.: *Xylem Structure and the Ascent of Sap*, Springer-Verlag, New York, 1983.

A delightfully written "idea" book on xylem structure and function by one who contributed a great deal to our understanding of functional xylem anatomy and sap movement.

ZIMMERMANN, MARTIN H., and CLAUDE L. BROWN: *Trees: Structure and Function*, Springer-Verlag, New York, 1975.

An up-to-date discussion of how trees work, with emphasis on structure as it relates to function.

Articles

ALBERSHEIM, PETER, and ALAN G. DARVILL: "Oligosaccharins," *Scientific American*, September 1985, pages 58–64.

BRILL, WINSTON J.: "Nitrogen Fixation: Basic to Applied," *American Scientist*, vol. 67, pages 458–466, 1979.

BURNHAM, CHARLES R.: "The Restoration of the American Chestnut," *American Scientist*, vol. 76, pages 478–487, 1988.

CHILTON, MARY-DELL: "A Vector for Introducing New Genes into Plants," *Scientific American*, June 1983, pages 50–59.

EVANS, MICHAEL L., RANDY MOORE, and KARL-HEINZ HASENSTEIN: "How Roots Respond to Gravity," *Scientific American*, December 1986, pages 112–119.

EVERT, RAY F.: "Sieve Tube Structure in Relation to Function," *BioScience*, vol. 32, pages 789–795, 1982.

FOLKERTS, GEORGE W.: "The Gulf Coast Pitcher Plant Bogs," *American Scientist*, vol. 70, pages 260–267, 1982.

GOODMAN, R. M., H. HAUPTLI, A. CROSSWAY, and V. C. KNAUF: "Gene Transfer in Crop Improvement," *Science*, vol. 236, pages 48–54, 1987.

HITCH, CHARLES J.: "Dendrochronology and Serendipity," *American Scientist*, vol. 70, pages 300–305, 1982.

MARX, JEAN L.: "How Rhizobia and Legumes Get It Together," *Science*, vol. 230, pages 157–158, 1985.

MULCAHY, DAVID L., and GABRIELLA B. MULCAHY: "The Effects of Pollen Competition," *American Scientist*, vol. 75, pages 44–50, 1987.

OLSEN, RALPH A., RALPH B. CLARK, and JESSE H. BENNETT: "The Enhancement of Soil Fertility by Plant Roots," *American Scientist*, vol. 69, pages 378–384, 1981.

PETTITT, JOHN, SOPHIE DUCKER, and BRUCE KNOX: "Submarine Pollination," *Scientific American*, March 1981, pages 134–143.

RICK, CHARLES M.: "The Tomato," *Scientific American*, August 1978, pages 76–87.

RUEHLE, JOHN L., and DONALD H. MARX: "Fiber, Food, Fuel, and Fungal Symbionts," *Science*, vol. 206, pages 419–422, 1979.

SHIGO, ALEX L.: "Compartmentalization of Decay in Trees," *Scientific American*, April 1985, pages 96–103.

TOMLINSON, P. B.: "Tree Architecture," *American Scientist*, vol. 71, pages 141–149, 1983.

TORREY, JOHN G.: "The Development of Plant Biotechnology," *American Scientist*, vol. 73, pages 354–363, 1985.

WALKER, DAN B.: "Plants in the Hostile Atmosphere," *Natural History*, June 1978, pages 74–81.

* Available in paperback.

SECTION 6

Biology of Animals

Blood carries the oxygen required for the energy-yielding reactions of the body—the reactions needed for maintenance, growth, reproduction, and the winning of gold medals. During exertion, heart output may increase tenfold, pumping up to 40 liters—about 10 gallons—of blood from the heart's left ventricle every minute, thus meeting the increased demand for oxygen.

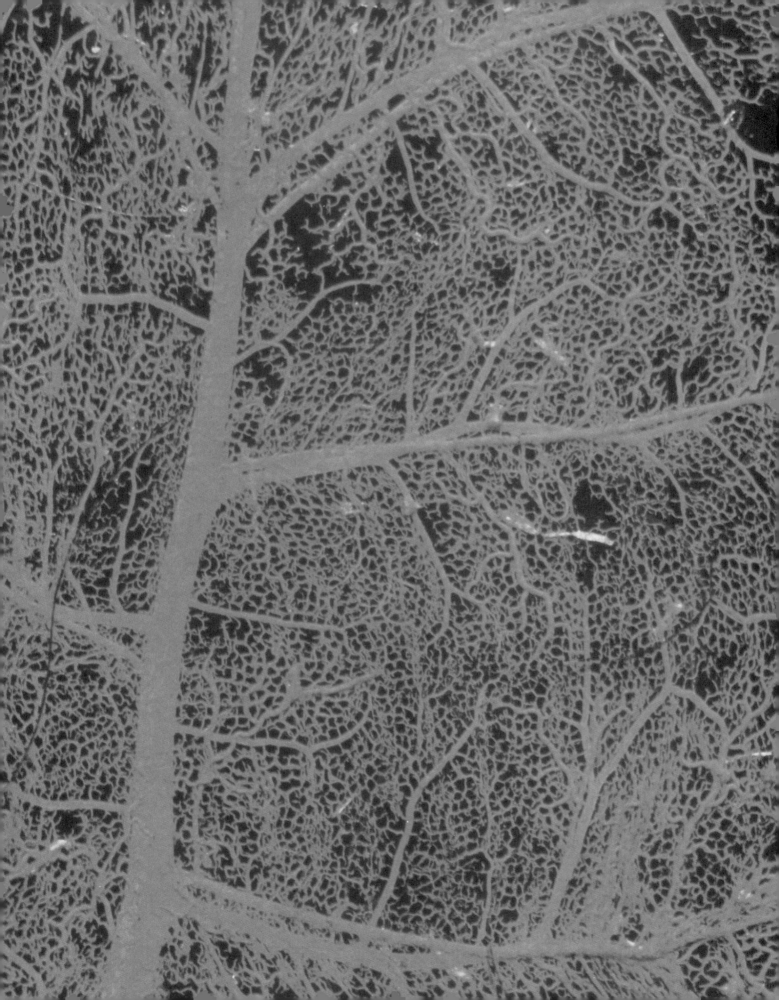

The Vertebrate Animal: An Introduction

33–1 *An 80,000-kilometer network of capillaries delivers the blood to the muscles and other tissues of the body. Oxygen and nutrients move by diffusion across the thin, permeable capillary walls to enter the body cells, no one of which is more than 0.01 centimeter from one of these minute blood vessels. Depleted of oxygen, the blood returns to the heart again, propelled by muscle contraction.*

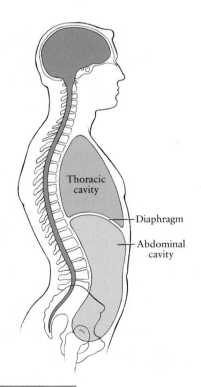

33–2 *Humans, like other vertebrates, are characterized by a dorsal central nervous system (spinal cord and brain) enclosed in vertebrae and the skull. As in other mammals, a muscular diaphragm divides the coelom into the thoracic cavity and the abdominal cavity.*

The animal kingdom, as we saw in Chapters 25 through 28, is made up of a vast array of organisms, ranging from microscopic aquatic forms to extremely complicated and highly organized multicellular creatures that are, by far, the most complex of all living systems. In this section of the book, we shall consider in more detail the principles of vertebrate anatomy and physiology, using *Homo sapiens* as our representative organism. Because we are part of a continuum of nature, it is logical to study the human animal (now one of the best understood) in order to gain a deeper understanding of animal life in general. This one species, however, will not command our full attention, for it is equally logical to study other animals in order to understand the human animal. Thus we shall rely on numerous examples from other organisms to gain insight into important physiological principles and adaptations.

Even those of us who have most marveled at the exquisite architecture of an orchid flower, or who are as content as Leeuwenhoek to watch the intricate and varied movements of a *Paramecium* and its neighbors, approach the subject of the biology of our own systems with a quickened interest. Indeed for some, student and scientist alike, the principal and perhaps the only reason for studying "lower forms" is the extent to which such studies bear directly on human welfare. But as *Escherichia coli*, T2 and T4 bacteriophages, and *Drosophila* remind us, there is no way to study only the human species, any more than it is possible to study only a fruit fly.

CHARACTERISTICS OF *HOMO SAPIENS*

The human being is a vertebrate and as such has a bony, articulated (jointed) endoskeleton that supports the body and grows as it grows. The dorsal nerve cord (the spinal cord) is surrounded by bony segments, the vertebrae, and the brain is enclosed in a protective casing, the skull.

As in other vertebrates, and most invertebrates as well, the human body contains a coelom—a cavity that forms within the mesoderm of the developing embryo (see page 532). In humans and other mammals, the coelom is divided into compartments, of which the two largest are the thoracic cavity and the abdominal cavity (Figure 33–2). These are separated by a dome-shaped muscle, the diaphragm. The thoracic cavity contains the heart, lungs, and esophagus (the upper portion of the digestive tract). The abdominal cavity contains a large number of organs, including the stomach, intestines, and liver.

Human beings are, of course, mammals. One of the most important characteristics of mammals is that they are warm-blooded. More precisely, they are endotherms (page 597), generating heat internally to maintain a high and relatively constant body temperature. As a consequence, mammals (and birds, which are also endotherms) are able to achieve and sustain levels of physical activity and mental alertness generally far greater than those of ectothermic animals, which can regulate their body temperature only within broad limits by absorbing heat from or releasing it to the environment. A concomitant of endothermy is a high metabolic rate, which requires relatively large and constant supplies of food (fuel) molecules and oxygen.

Mammals have other important characteristics. They have hair or fur rather than scales or feathers. In keeping with their high levels of activity and mental alertness, they have highly developed systems for receiving, processing, and reacting to information from the environment. All mammals (except the monotremes) give birth to live young, as distinct from laying eggs, which all birds and most fish, amphibians, and reptiles do. Mammals nurse their offspring, a process that involves a relatively long period of parental care and makes possible a long learning period. Contrast this, for example, with most insects and nearly all species of fish, amphibians, and reptiles, in which the young are independent from the moment they hatch from the egg. There is a tendency among the large mammals, in particular, toward fewer offspring per litter and prolongation of parental care. Humans, for instance, rarely have more than two surviving young per birth, only two mammary glands with which to nurse them, and an extraordinarily long period of infancy and childhood, with dependency on parents often lasting well past physical maturity. Finally, for better or worse, *Homo sapiens* is by far the most intelligent of all mammals.

CELLS AND TISSUES

The body of a vertebrate, like that of all complex multicellular organisms, is made up of a variety of different, specialized cells. Although these cells greatly resemble one-celled organisms in their requirements, they differ from one-celled organisms in that they develop and function as part of an organized whole. Cells are organized into tissues, which are groups of cells carrying out a unified function. Different kinds of tissues, united structurally and coordinated in their activities, form organs, such as the stomach.

Experts can distinguish about 200 different cell types in the human body, which are customarily classified into only four tissue types: (1) epithelial, (2) connective, (3) muscle, and (4) nerve.

Epithelial Tissues

Epithelial tissues consist of continuous sheets of cells that provide a protective covering over the whole body and contain sensory nerve endings. They also provide a protective wrapping for individual internal organs and form the interior lining membranes of organs, cavities, and passageways. As a moment's reflection will reveal, everything that goes into and out of the body and its various organs must pass through epithelium, which thus plays an important regulatory role in the movement of molecules and ions.

Epithelial tissues are classified according to the shape of the individual cells as squamous, cuboidal, or columnar (Figure 33–3). They may consist of only a single layer of cells (simple epithelium), as found in the inner lining of the circulatory system, or several layers (stratified epithelium), as found in the outer layer (epidermis) of the skin (Figure 33–4). One surface of the epithelial sheet is always attached to an underlying layer, called the basement membrane. This layer is composed of glycoproteins in which a fibrous protein derived from collagen is embedded; both the glycoproteins and the fibrous protein are produced by the epithelial cells themselves.

The epithelium of the body cavities and passageways frequently contains modified epithelial cells that secrete mucus, which lubricates the surfaces (see

Squamous

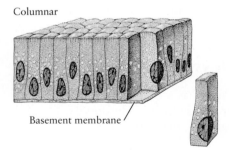

Cuboidal

Columnar

Basement membrane

33–3 *The three types of epithelial cells that cover the inner and outer surfaces of the body. Squamous cells, which usually perform a protective function, make up the outer layers of the skin and the lining of the mouth and other mucous membranes. There are usually several layers of these flat cells piled on top of one another. Cuboidal and columnar cells, which line many internal passageways, are often involved in active transport and other energy-requiring processes. They also perform much of the chemical work of the body.*

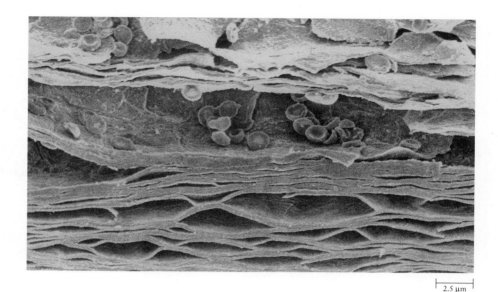

2.5 μm

33-4 *The outer surface, or epidermis, of the skin is composed of stratified squamous epithelium. In the lower portion of this scanning electron micrograph, the layers of epithelium are clearly visible. On the skin surface, some of the older cells are being sloughed off. This is a continuous process in epithelial tissue that is subjected to wear and tear. Replacement cells are produced by cell division in the underlying layers of the epidermis. A number of red blood cells can be seen just below the top layer of epithelial cells.*

Figure 4–13, page 94). Other epithelial cells, specialized for the synthesis and secretion of specific substances for export, are often clustered together to form **glands.** Among the substances produced by glands are perspiration, saliva, milk, hormones, and digestive enzymes. Glandular epithelium is composed of cuboidal or columnar cells.

Desmosomes and Tight Junctions

In epithelial tissues that function as coverings and linings, the physical integrity of the tissue as a whole—no tears, no leaks—is often of paramount importance. Two general types of cell-cell junctions play an essential role in maintaining this integrity: **desmosomes** and **tight junctions** (Figure 33–5). Desmosomes have often been compared to spot welds between cells. They consist of plaques of dense fibrous material between cells, with clusters of filaments from the cytoplasm of the neighboring cells looping in and out of them. Desmosomes attach cells to one another and give tissues mechanical strength. They are found in especially large numbers in tissues subjected to mechanical stress, such as the skin.

Tight junctions, which appear to involve the fusion of adjacent cell membranes, form a continuous seal around each cell in a layer of tissue, preventing leakage between cells. For example, intestinal epithelial cells are surrounded by tight junctions that keep the intestinal contents from seeping between the cells. Similar arrangements are found in the epithelial cells that form the linings of other internal organs, such as the bladder and the kidney. Tight junctions occur only in vertebrates, but slightly different structures, known as septate junctions, perform the same function in invertebrates.

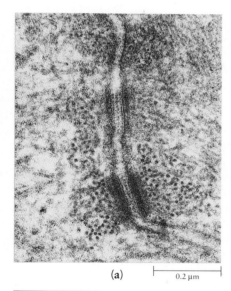

(a) |— 0.2 μm —|

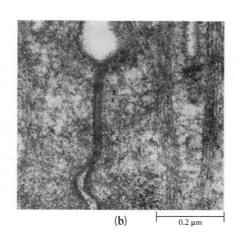

(b) |— 0.2 μm —|

33-5 *Junctions between epithelial cells maintain the structural integrity of the tissue.* (a) *Desmosomes cement adjacent cells together. Here three desmosomes connect the cell membranes of two adjacent epithelial cells of the large intestine of a rat. The upper structure is a belt desmosome, and the lower two structures are spot desmosomes. Filaments cut in cross section as the sample was prepared for the electron microscope appear as black dots. As you can see, the filaments within belt desmosomes have a smaller diameter than those within spot desmosomes; the belt desmosome filaments are thought to contain actin (page 121).* (b) *A tight junction (arrow) seals cells together and prevents materials from leaking between them. These cells are from the epithelium lining the small intestine of a rat.*

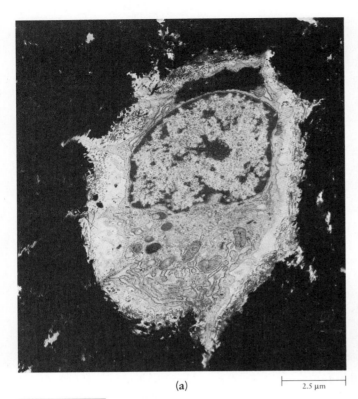

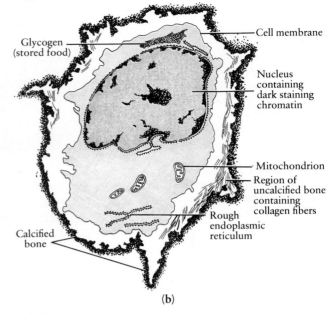

Glycogen
(stored food)

Cell membrane

Nucleus
containing
dark staining
chromatin

Mitochondrion

Region of
uncalcified bone
containing
collagen fibers

Rough
endoplasmic
reticulum

Calcified
bone

(b)

(a)

2.5 μm

(c)

33–6 *(a) Electron micrograph and *(b)* diagram of a bone cell (osteocyte). Young bone cells (osteoblasts) produce the extracellular matrix, an organic material consisting of collagen fibers and ground substance. This matrix gradually hardens as calcium compounds in the ground substance crystallize. Cytoplasmic processes extend from the cells through channels in the hardened bone, as shown in *(c).*

Connective Tissues

Connective tissues bind together, support, and protect the other three kinds of tissues. Unlike epithelial tissue cells, the cells of connective tissues are widely separated from one another by large amounts of extracellular material, the matrix, which anchors and supports the tissue. The matrix, synthesized by the cells of the tissue, consists of a ground substance, which is more or less fluid and amorphous (formless), and, in many connective tissues, fibers. There are several different types of fibers, which vary according to the particular tissue: (1) connecting and supporting fibers, such as collagen fibrils (Figure 3–23, page 77), which are a major component of skin, tendons, ligaments, cartilage, and bone; (2) elastic fibers, which are found, for example, in the walls of large blood vessels; and (3) reticular fibers, which form networks inside solid organs, such as the liver. While epithelial tissues are classified according to cell shape and arrangement, connective tissues are grouped by the characteristics of their extracellular matrix.

The principal connective tissues, by volume, in the human body are bone, blood, and lymph. In blood and lymph, the extracellular matrix is a watery fluid, called plasma, that contains numerous ions and molecules. A variety of specialized cells (to be discussed in Chapters 36 and 39) circulate through the body in this fluid matrix. The extracellular matrix of bone, by contrast, is impregnated with hard crystals of calcium compounds. Like other connective tissues, however, bone is living matter, consisting of cells, fibers, and ground substance (Figure 33–6). Bone tissue, despite its strength, is amazingly light; the human skeleton (Figure 33–7) makes up only about 18 percent of our weight.

(a)

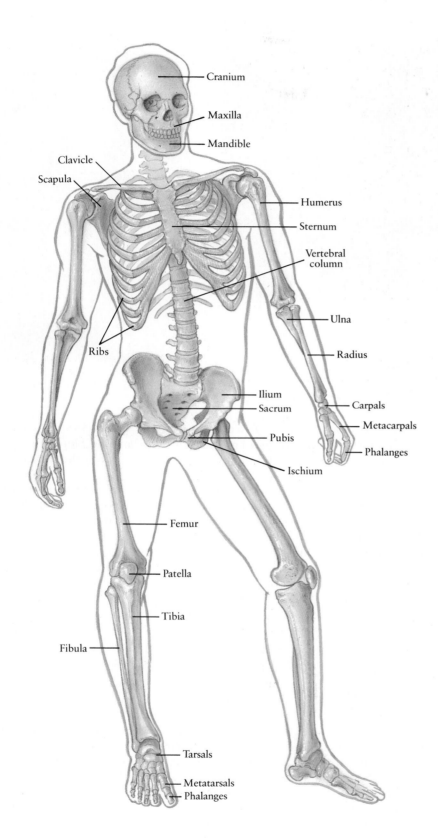

Cranium

Maxilla

Mandible

Clavicle

Scapula

Humerus

Sternum

Vertebral
column

Ulna

Radius

Ribs

Ilium

Sacrum

Carpals

Pubis

Metacarpals

Ischium

Phalanges

Femur

Patella

Tibia

Fibula

Tarsals

Metatarsals

Phalanges

(b)

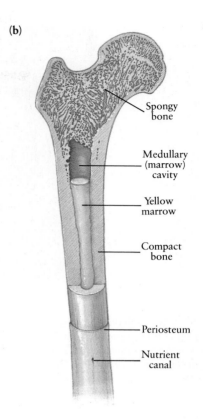

Spongy
bone

Medullary
(marrow)
cavity

Yellow
marrow

Compact
bone

Periosteum

Nutrient
canal

33–7 (a) *The skeleton of a human adult contains 206 bones. Twenty-nine are in the skull, including 14 face bones and six small bones (ossicles) of the ears. There are 27 bones in each hand and 26 in each foot. Bones are living organs, made up not only of connective tissue but of the other tissue types as well, including nerve, muscle, and the epithelial lining of the blood vessels that traverse the bones.*

(b) *The ends of long bones, such as this femur, consist of spongy bone, in which there are large spaces, surrounded by compact bone. The shaft, which is hollow, is composed of compact bone. A central cavity containing bone marrow extends through the center of the shaft. The marrow of long bones is yellow because of stored fat. The periosteum is a fibrous sheath, which contains the blood vessels that supply oxygen and nutrients to the bone tissues. Blood vessels emerge from bone through openings known as nutrient canals.*

(a)

2 µm

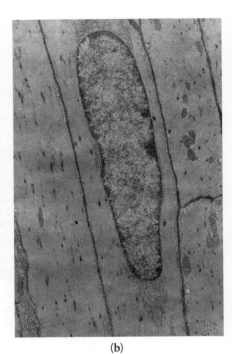

(b)

33–8 *Muscle cells are characterized by very thin fibrils of contractile proteins that run lengthwise through the cells and make up the bulk of the cytoplasm. These fibrils are arranged in a regular pattern in striated muscle, but they are irregular in smooth muscle and form no apparent pattern.*

(a) Electron micrograph of skeletal muscle, showing the striated pattern. A muscle fiber is a single large cell, containing many nuclei. The individual fibrils within the cell are separated from each other by a specialized endoplasmic reticulum, visible here as relatively wide, approximately horizontal, transparent bands.

(b) Electron micrograph of smooth muscle from the spermatic duct of a mouse. Smooth muscle is made up of long, spindle-shaped cells; portions of several cells are visible in this micrograph. Unlike the cells of skeletal muscle, but like most cardiac muscle cells, each smooth muscle cell contains only a single nucleus. One cell's nucleus, distinguishable by its granular texture, can be seen in the middle of the micrograph.

Muscle Tissue

Muscle cells are specialized for contraction. Every function of muscle—from running, jumping, smiling, and breathing to propelling the blood through the body and ejecting the fetus from the uterus—is carried out by the contraction of muscle cells in concert.

There are two general types of muscle tissue, as shown in Figure 33–8: striated muscle, which has a striped appearance under the microscope, and smooth muscle (no stripes). Smooth muscle surrounds the walls of internal organs, such as the digestive organs, uterus, bladder, and blood vessels. The muscles that move the skeleton are striated muscle and are sometimes called voluntary muscles since we can move them at will. A special type of striated muscle, cardiac muscle, makes up the wall of the heart. Smooth muscle and cardiac muscle are not, except in rare individuals, under direct voluntary control and are thus categorized as involuntary.

Contraction of muscle cells depends on the interaction of two proteins: actin and myosin (page 121). In skeletal and cardiac muscle, these proteins are arranged in regular, repeating assemblies, resulting in the characteristic striations.

Striated Muscle

Some 40 percent of a man's body, by weight, consists of the striated muscle that moves the skeleton; women characteristically have less, around 20 percent. A skeletal muscle is typically attached to two or more bones, either directly or, more often, indirectly, by means of the tough strands of connective tissue known as tendons. Some tendons, such as those that connect the finger bones with their muscles in the forearm, are very long. When the muscle contracts, the bones move around a joint, which is held together by ligaments and generally contains a lubricating fluid. Most of the skeletal muscles of the body work in antagonistic groups, one flexing, or bending, the joint, and the other extending, or straightening, it (Figure 33–9). Also, two antagonistic groups may contract together to stabilize a joint. Such muscle action makes it possible for us (and others) to stand upright.

The Injury-Prone Knee

A joint is a structure, usually movable, at which two bones are opposed. The knee is the hinge joint between two strong, inflexible bones, the femur and the tibia. It is covered by the "kneecap," or patella (removed in this drawing, but visible in Figure 33–7a). The patella acts as a stop so that the hinge can open to 180° but no farther.

As in most joints, the bones of the knee are bound together by ligaments. The collateral ligaments bind the joint externally; they are slack when the knee is flexed and taut when the knee is extended. The cruciate ligaments cross within the joint, stabilizing it. The menisci are C-shaped wedges of cartilage, resting on the upper surface of the tibia and cushioning the joint.

Knees are particularly susceptible to injury, as Billie Jean King, Joe Namath, Wilt Chamberlain, and others will sadly attest. The reason for the knee's vulnerability becomes clear when you contemplate the analogous problem of fastening together two match sticks end-to-end with several rubber bands, to produce a hinge that is both strong and flexible and can not only swing back and forth but also, to some extent, twist, bend, and rotate. Common knee injuries are torn ligaments, especially the collateral ligaments (as a result of the knee's being hit from the side), and

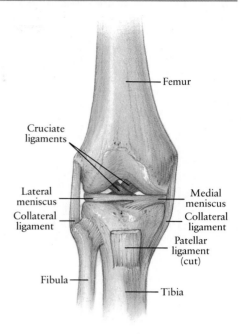

The right knee, viewed from the front, with the patella removed.

crushed menisci. When the shock-absorbing wedges of menisci are lost, bone chips begin to accumulate in the joint, making it less flexible and much more painful.

33–9 *Muscles attached to bone move the vertebrate endoskeleton. They often work in antagonistic pairs, with one relaxing as the other contracts. Muscles cannot lengthen spontaneously; they lengthen only when the joint moves in the opposite direction, due to contraction of the antagonistic muscles. For example, when you move your hand toward your shoulder, as shown here, the biceps contracts and the triceps relaxes. When you move your hand down again, the triceps contracts, while the biceps relaxes. The muscles that move the skeleton, such as those diagrammed here, are known as skeletal muscles. They are striated, as shown in Figure 33–8a.*

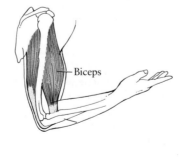
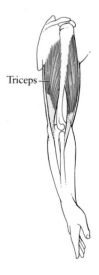

A skeletal muscle, such as the biceps, consists of bundles of muscle fibers—often hundreds of thousands of fibers—held together by connective tissue. Each fiber is a single cell, cylindrical or spindle-shaped (tapering at both ends), with many nuclei, formed by the fusion of a large number of small, mononucleate cells during embryonic development. These fibers are often very large cells—50 to 100 micrometers in diameter and many centimeters long. Their internal structure and the mechanism by which they contract will be described in Chapter 42.

Cardiac muscle resembles skeletal muscle in its assemblies of actin and myosin and, thus, in its striated appearance. However, its cells, which are shorter than those of skeletal muscle, are usually mononucleate and have branched, rather than tapered, ends. As we shall see in Chapter 36, certain cardiac muscle cells have the capacity to contract spontaneously some 70 times per minute, thus initiating the heartbeat hour after hour, day after day, throughout an entire lifetime.

Smooth Muscle

Although smooth muscle cells also contain actin and myosin, the molecules are not arranged in regular assemblies and do not form a striated pattern. Smooth muscle cells are spindle-shaped and mononucleate. Functionally, smooth muscle contracts much less rapidly than skeletal striated muscle, and its contractions are more prolonged. As we noted previously, it is generally not under voluntary control.

In most hollow organs, such as the intestines, smooth muscle fibers are organized into sheets arranged in two layers—an outer, longitudinal layer and an inner, circular layer—which can contract alternately. As you will recall, such a juxtaposition of circular and longitudinal muscles makes possible the locomotion of the earthworm (page 555). In the internal organs of a vertebrate, the changes in diameter or shape that result from alternating contraction of the two sets of muscles are utilized to propel fluids or semisolid materials (such as a food mass) through the organ. In blood vessels, bundles of smooth muscle fibers encircle the walls, constricting the vessels as the fibers contract. Depending on the location of the vessel and the strength of the contraction, these muscles either provide added propulsion to the moving blood or regulate its pressure and rate of flow.

Nerve Tissue

The fourth major tissue type is nerve tissue. The essential functional units of nerve tissue are the **neurons,** which transmit nerve impulses. Nerve tissue also contains cells of another type, known as glial cells, that physically support and insulate the neurons. These cells are called neuroglia when they are in the central nervous system (brain and spinal cord) and Schwann cells when they are outside the central nervous system. Glial cells are believed to supply the neurons with nutrients and other molecules and may play an important part in maintaining the ionic composition of nerve tissue, which is, as we shall see in Chapter 41, central to its function.

As shown in Figure 33–10, neurons come in a diversity of sizes and shapes. Typically, a neuron consists of a **cell body,** which contains the nucleus and much of the metabolic machinery of the cell; **dendrites,** usually numerous, short, threadlike cytoplasmic extensions—processes—that, with the cell body, receive stimuli from other cells; and an **axon,** a long process that is capable of rapidly conducting an electrochemical signal—the nerve impulse—over great distances. Axons are also known as nerve fibers.

Neurons are specialized to receive signals—from the external environment, the internal environment, or other neurons—to integrate the signals received, and to transmit the integrated information to other neurons, muscles, or glands. Func-

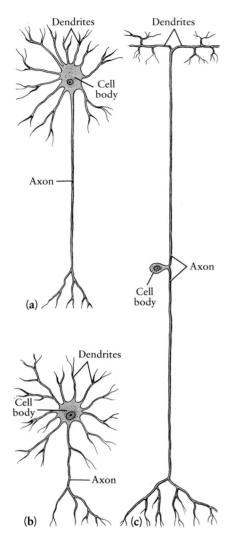

33–10 *Three of the many different forms characteristic of vertebrate neurons.* **(a)** *Motor neurons and relay neurons are characterized by a cell body with numerous dendrites and a long axon that travels without interruption to its terminal, where it branches.* **(b)** *Interneurons, which are found within localized regions of the central nervous system, typically have a complex system of dendrites and, as shown here, a short axon with branches —or no axon at all.* **(c)** *In sensory neurons, which transmit impulses from sensory receptors at the ends of the dendrite branches, the cell body is often off to one side of the long axon. All of these neurons form connections, known as synapses, with other neurons.*

33–11 *A representative reflex arc, showing the functions of the major types of neurons. Sensory receptor cells stimulate a sensory neuron, which relays a signal to an interneuron within a localized region of the central nervous system (often the spinal cord). From the interneuron, the signal is transmitted to a motor neuron, which, in turn, stimulates an effector, shown here as a muscle cell. Not shown in this diagram are relay neurons, which simultaneously transmit to other regions of the central nervous system (typically in the brain) the information received from the sensory neuron. These basic components of the reflex arc are found in all vertebrates, from the simplest to the most complex. Reflex arcs play an essential role in the regulation of many internal processes, as well as making possible almost instantaneous responses to numerous environmental stimuli.*

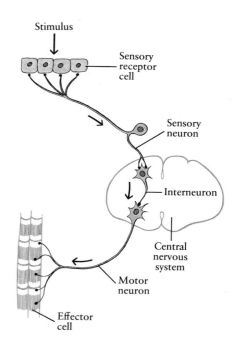

tionally, there are four classes of neurons: **sensory neurons,** which receive sensory information and relay it to the central nervous system; **interneurons,** which transmit signals within localized regions of the central nervous system; **relay neurons,** which relay signals between different regions of the central nervous system; and **motor neurons,** which transmit signals from the central nervous system to effectors, such as muscles or glands. These four types of neurons are linked in a variety of circuits, ranging from simple reflex arcs (Figure 33–11) to the extremely complex interconnections that characterize the human brain.

Neurons may reach astonishing lengths. For example, the axon of a single motor neuron may extend from the spinal cord down the whole length of the leg to the toe. Or the axon of a sensory neuron, the cell body of which is located just outside the spinal cord, may extend from the toe to the cell body and then up the entire length of the spinal cord to the lower part of the brain, where it terminates. In a giraffe, such a cell might be close to 5 meters long, and in an adult human, close to 2 meters.

Nerves are bundles of many axons from many neurons—usually hundreds and sometimes thousands. Each axon is capable of transmitting a separate message, like the wires in a telephone cable.

LEVELS OF ORGANIZATION

In animals, as in all other organisms, the basic structural unit is the living cell. As we have just seen, the vertebrate body comprises a variety of cells, organized into four types of tissues—groups of cells with a similar function. At the next level of organization, different kinds of tissues, united structurally and coordinated in their activities, form organs. The stomach, for example, is an organ made up of layers of glandular epithelium (in the stomach lining), connective tissue, nerves, and smooth muscle. The structure of the largest organ of the human body, the skin, is shown in Figure 33–12 on the next page.

33–12 *A section of human skin, the body's largest organ, showing the epidermis and the dermis, which together make up the skin, and the underlying subcutaneous fat.*

The epidermis is made up mostly of epithelial cells and consists of two layers: an inner layer of living cells, and an outer layer of dead cells filled with keratin (page 77). The epidermis is a turnover system: cells produced in the inner layer migrate toward the surface and die. At the base of the epidermis are pigment cells that produce the granules responsible for skin color.

The dermis, consisting mostly of connective tissue, contains sensory nerve endings, small blood vessels known as capillaries, erector muscles, which raise the hair when contracted, and sweat and sebaceous glands, which are composed of modified epithelial cells. The latter produce a fatty substance that lubricates the skin surface. Fatty tissue, which makes up the insulating layer below the dermis, is also a form of connective tissue, as is the blood found in the capillaries.

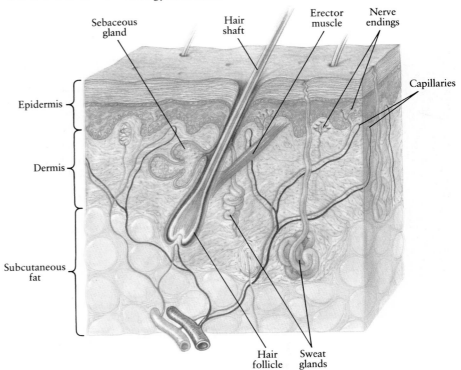

Organs that work together in an integrated fashion to perform a particular function make up **organ systems,** the next level of organization. The digestive system, for example, is composed of the stomach and a number of other organs, each of which carries out specific activities that contribute toward the overall process. The organ systems together constitute the complete animal, a living organism that interacts with its external environment, including not only the physical environment but also other individuals, both of its own species and of other species. As we examine the structure and physiology of the vertebrate animal in the subsequent chapters of this section, we shall do so principally in terms of organ systems and the functions they perform.

FUNCTIONS OF THE ORGANISM

In Chapter 30, we examined the structure of the plant body in terms of the challenges imposed by the transition from life in the water to life on the land. Similarly, the structures of the animal body "make sense" only when seen as adaptations that solve particular problems presented by the relationship between the organism and its environment.

Before we further employ this metaphor of problem and solution, however, we should stop for a moment to see what we really mean by biological problem solving. An organism confronts its "problems" with a set of genetic instructions. If they work, the organism lives and passes on these instructions to its offspring. If its instructions are better than those carried by its neighbor, its offspring will probably be more numerous. It is in this way that "problems" get "solved."

Energy and Metabolism

A major problem for any living system—indeed, *the* major problem—is that posed by the second law of thermodynamics: to maintain the high level of organization characteristic of such systems in the face of the universal trend toward disorder

33–13 *Members of the shrew family, such as this pygmy shrew, have a regular three-hour cycle of activity and rest. Throughout the day and night, winter and summer, they alternate three hours of rest with three hours of feeding on insects and other small invertebrates. Pygmy shrews are common in Europe and central Asia. Their bodies are about 5 centimeters long, and their tails are almost 4 centimeters long.*

(page 164). To do this, organisms need sources of energy and raw materials to maintain and operate their energy-extracting machinery. Animals, as we saw in Chapter 25, are multicellular heterotrophs that ingest their food, from which energy is ultimately released by the oxygen-requiring reactions of cellular respiration. In large animals, a critical problem is transforming the ingested food into molecules that can be used by individual cells and then providing those molecules, along with oxygen, to each of the multitude of cells that make up the organism's body. The complexities of the vertebrate digestive, respiratory, and circulatory systems, to be discussed in Chapters 34 through 36, represent particular evolutionary solutions to this problem.

Homeostasis

An equally critical but less obvious problem facing all animals is the maintenance of a relatively constant internal environment. This problem is also imposed by the laws of physics and chemistry. The chemical reactions of a living cell and the synthesis and maintenance of its constituent structures require a tightly controlled chemical environment, a fairly narrow range of temperature, and protection from foreign invaders, such as bacteria and viruses, that may feed on the substance of the cell, poison its enzymes with toxins, or subvert its genetic machinery for their own replication.

As we saw in Chapter 1, **homeostasis,** the maintenance of a relatively constant internal environment, is one of the identifying characteristics of living systems. Such constancy is, in many ways, more difficult for one-celled organisms and small multicellular organisms than it is for large animals, such as *Homo sapiens.* Small organisms are extremely vulnerable to changes in the temperature or chemical composition of the medium in which they live, and it is likely that the strongest evolutionary pressures toward larger size and multicellularity were related to homeostasis. Larger organisms, with their lower surface-to-volume ratios (page 103) and, often, protective outer coverings, have a greater resistance to external change and foreign invaders. The world in which your individual cells function and flourish is decidedly different from the world around you—but not so different from the warm primordial soup in which we all began.

Maintaining a relatively constant environment throughout the body of a large animal is a complex process. It involves not only continuous monitoring and regulation of many different factors but also ready defenses against an enormous diversity of microorganisms. As we shall see, virtually all of the organ systems participate in homeostasis. Particularly important roles are played by the urinary system (Chapter 37), the immune system (Chapter 39), and the two organ systems responsible for integration and control.

Integration and Control

The third problem faced by animals has two aspects. First, homeostasis requires coordinating the activities of the numerous cells that constitute the organism, so that tissues and organs respond to overall physiological needs, which change with fluctuations in the environment. Second, animals are typically quite active, moving about as they seek mates and search for and capture food, while simultaneously trying to avoid being captured by other animals. A life of active movement requires receiving and processing information from the external environment, followed by coordinated and appropriate contractions of the skeletal muscles.

There are two major control systems in animals, the endocrine system (the hormone-secreting glands and their products) and the nervous system. Speaking very generally, the endocrine system is responsible for changes that take place over

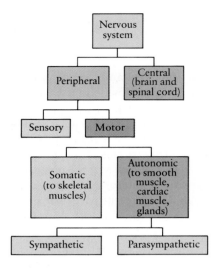

33–14 *The nervous system of vertebrates, including* Homo sapiens, *consists of a central nervous system (the brain and spinal cord) and a peripheral nervous system, a vast network of nerves connecting the central nervous system with all other parts of the body. Sensory neurons carry information to and motor neurons carry information from the central nervous system. The motor neurons are organized into the somatic and autonomic systems, and the autonomic system contains two divisions, the sympathetic and the parasympathetic.*

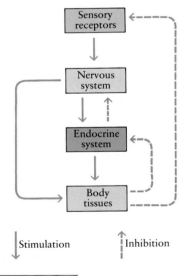

Stimulation   Inhibition

33–15 *A schematic representation of the feedback control pathways that regulate virtually all physiological processes. As we shall see in subsequent chapters, the details—the type of sensory stimuli perceived, the regions of the nervous system activated, the hormones released, and the target tissues of nervous and endocrine control—vary from process to process, but the underlying principle is always the same.*

a relatively long time period—minutes to months—whereas the nervous system is involved with more rapid responses—milliseconds to minutes. However, the more we learn about these systems, the more they are seen to be closely interrelated. For instance, the production of sex and other hormones was once believed to be under the control of the "master" pituitary gland; it is now known that the pituitary, though perhaps a master in some aspects of the endocrine world, is actually an executive secretary of the **hypothalamus,** a major brain center. And within little more than the last decade, it has been discovered that the embryonic development of certain major brain centers is profoundly influenced by hormones regulated by the pituitary.

Even anatomically, the endocrine and nervous systems are not distinct. One of the body's most important glands, the adrenal medulla, a major source of adrenaline (also known as epinephrine), is not, strictly speaking, a gland. That is, it is not modified epithelial tissue but is rather a large ganglion—a collection of nerve cell bodies—whose nerve endings secrete the hormone.

Within the nervous system, there are a number of structural and functional subdivisions (Figure 33–14). One subdivision, the **somatic** system, innervates skeletal muscle. Another, the **autonomic** ("involuntary") system, innervates smooth muscle, cardiac muscle, and glands. The autonomic system is further subdivided into the **sympathetic** and the **parasympathetic** divisions, which interact with one another in a finely tuned system of checks and balances. The sympathetic division is most active in times of stress or danger. Among the major effectors of the sympathetic division is the adrenal medulla, and the overall effects of general sympathetic stimulation are those we associate with a "rush of adrenaline." The parasympathetic division plays its major role in supporting everyday activities such as digestion and excretion. The functions of the endocrine and nervous systems will be described in greater detail in Chapters 40 through 43, but we introduce them now because every organ system to be discussed is under the influence of both the endocrine system and the sympathetic and parasympathetic divisions of the autonomic nervous system.

Feedback Control

The body's integration and control systems characteristically act through feedback loops, both negative and positive. The simplest example from everyday life of a negative feedback system is the thermostat that regulates your furnace. When the temperature in your house drops below the preset thermostat level, the thermostat turns the furnace on. When the temperature rises above the preset level, the thermostat turns the heat off. In a living organism, systems are seldom completely on or off, and homeostatic control is much more finely modulated. The principle, however, is the same: a deviation from a "preset" condition stimulates a response that reduces the deviation.

Feedback loops typically involve both the nervous and endocrine systems (Figure 33–15). One important homeostatic function of the body, for example, is keeping the blood volume constant. This is accomplished, in large part, by regulation of the rate at which water is removed from the bloodstream by the kidneys. A hormone known as antidiuretic hormone (ADH), which is produced by the hypothalamus and released from the pituitary gland, acts on the tubules of the kidney to decrease water excretion. The production of ADH is controlled by sensory receptors in the circulatory system, particularly in the heart, that measure blood pressure—an indirect measure of blood volume. When blood pressure goes up, firing of these receptors inhibits the release of ADH, and more water is excreted, reducing blood volume. As blood pressure goes down, the stimulus from the receptors decreases, ADH production increases, water is retained by the kidney, and blood pressure and blood volume increase.

Some feedback systems involve additional relay loops. The thyroid gland, for instance, produces the thyroid hormone, thyroxine, which, among other effects, increases the rate of cellular metabolism. The amount of thyroxine produced depends on the production of another hormone, thyroid-stimulating hormone (TSH) from the pituitary gland. A major factor regulating the production of TSH is the concentration of thyroid hormone in the circulating blood. Thus, although an additional step is involved, the principle is the same: the concentration of the hormone itself or the response to the hormone by a target tissue inhibits the synthesis or secretion of the hormone in question.

Many functions are controlled by a number of separate feedback control pathways. An example is temperature regulation, which we shall consider in Chapter 38. Core body temperature is controlled by a thermostat located in the hypothalamus. This thermostat collects information from a number of sensory receptors, integrates it, compares the results with its thermostat setting, and marshals a complex of appropriate responses.

Continuity of Life

The fourth challenge an organism faces—which may or may not be a problem—is to multiply. The biological imperative to reproduce, following the dictates of the genes, is enormous. As we shall see when we consider evolution and ecology later in this text, animals devote much of their energy and resources to meeting this challenge. Reproduction, as we saw in Section 4, may be carried out in a variety of ways, but in mammals it is always sexual and always involves formation of gametes, their union to form a zygote, or fertilized egg, and the development of the zygote into an adult. The last two chapters of this section will describe these processes, with particular emphasis on the formation of human gametes, their fusion in the fertilized egg, and the almost miraculous, though oft-repeated, phenomena by which this single cell develops into a human being.

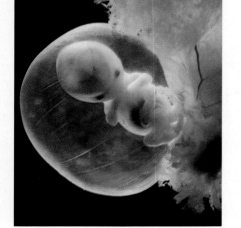

33–16 *Genus,* Homo; *species,* sapiens. *Embryo at 7 weeks.*

SUMMARY

Vertebrates, including *Homo sapiens,* are characterized by a bony, jointed endoskeleton that includes a skull and a vertebral column enclosing the central nervous system (the brain and spinal cord). The human body, like that of other mammals, contains a coelom that is divided by a muscle, the diaphragm, into two major compartments, the abdominal cavity and the thoracic cavity.

The cells of the vertebrate body are organized into tissues, groups of cells that carry out a unified function. Various types of tissues are grouped in different ways to form organs, and organs are grouped to form organ systems. The four principal tissue types of which the vertebrate body is made are epithelial tissue, connective tissue, muscle, and nerve. Epithelium serves as a covering or lining for the body and its cavities. Glands are composed of specialized epithelial cells; their secretions include mucus, perspiration, milk, saliva, hormones, and digestive enzymes. Connective tissues are characterized by their capacity to secrete substances, which often include collagen and other fibers, that make up the extracellular matrix. They serve to support, strengthen, and protect the other tissues of the body. Muscle cells are specialized for contraction, which is accomplished by assemblies of two proteins, actin and myosin. In striated muscle, which includes skeletal muscle and cardiac muscle, these assemblies form a striped pattern, visible under the microscope. In smooth muscle, no such pattern is apparent. Nerve cells, or neurons, are specialized for the reception, processing, and transmission of information. Neurons typically consist of a cell body, dendrites, and an axon. Signals, in the form of electrochemical impulses, can be conducted rapidly over great distances by the axon. Neurons are surrounded and supported by glial cells.

Four basic functions are essential to the on-going life of a multicellular animal. Food must be obtained and processed to yield molecules that can be used by individual cells, and these molecules, as well as oxygen, must be delivered to the cells. The internal environment must be closely regulated, a process known as homeostasis. The contractions of its skeletal muscles in locomotion and the physiological activities of its many tissues and organs must be coordinated in response to changes in both the internal and external environments. Integration and control involve both the endocrine and nervous systems and characteristically function through feedback loops. Finally, as dictated by its genes, the organism reproduces.

QUESTIONS

1. Distinguish among the following terms: animal/vertebrate/ mammal; coelom/diaphragm; endoskeleton/exoskeleton; muscle cell/muscle fiber; striated muscle/smooth muscle; skeletal muscle/cardiac muscle; nerve cell/neuron/glial cell; neuroglia/ Schwann cell; dendrite/axon; sensory neuron/interneuron/relay neuron/motor neuron; nerve fiber/nerve; somatic/autonomic; sympathetic/parasympathetic.

2. What is the functional significance of each of the four tissue types? Give an example of each type.

3. What are the three types of epithelial tissue? What is the basis for classification of epithelial cells?

4. In addition to protection, what is the other major function of epithelial tissue?

5. What is the major structural difference between connective tissue and epithelial tissue? What functions of connective tissue account for this structural difference?

6. What is meant by antagonistic paired muscles?

7. Give some examples of homeostasis. Why is it so important?

8. What are the major survival problems of an organism? Compare the solutions to these problems achieved by plants with those achieved by vertebrate animals.

9. Label the drawing below. Explain the process that is taking place, as signified by the four arrows.

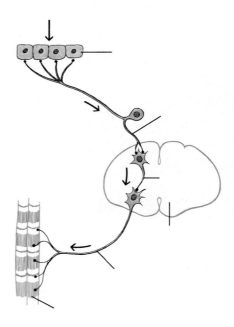

Energy and Metabolism I: Digestion

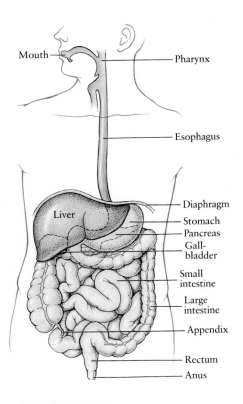

Mouth

Pharynx

Esophagus

Diaphragm

Liver

Stomach

Pancreas

Gall-
bladder

Small
intestine

Large
intestine

Appendix

Rectum

Anus

34–1 *The human digestive tract. Food passes from the mouth through the pharynx and esophagus to the stomach and small intestine, where most digestion takes place. Undigested materials pass through the large intestine, are stored briefly in the rectum, and are eliminated through the anus. Accessory organs of the digestive system are the salivary glands (shown in Figure 34–6), pancreas, liver, and gallbladder.*

Digestion is the breakdown of ingested food materials into molecules that can be delivered to and utilized by the individual cells of the animal body. These molecules serve a variety of functions. For example, they may be energy sources; they may provide essential chemical elements, such as calcium, nitrogen, or iron; or they may be molecules—such as certain amino acids, fatty acids, and vitamins —that cells need but cannot synthesize for themselves.

In Section 4, we traced the evolution of digestive systems from the feeding cells of sponges (page 522), through the gastrovascular cavity of jellyfish (page 525) and the digestive cavity of flatworms (page 533), to the major evolutionary break-through, exemplified by ribbon worms, the invention of the one-way digestive tract (page 538). As animals became larger and more complex, the tract became more convoluted, providing increased working surfaces, and different portions became specialized for different stages of the digestive process.

DIGESTIVE TRACT IN VERTEBRATES

In vertebrates, the digestive tract—known rather less elegantly but just as accurately as the gut—consists of a long convoluted tube, extending from mouth to anus (Figure 34–1). The digestive system also includes the salivary glands, the pancreas, the liver, and the gallbladder, accessory organs that provide the enzymes and other substances essential for digestion.

The inner surface of the gut is continuous with the outer surface of the body, and so, technically, the cavity of the gut is outside the body. Because the gut contents are thus sequestered, they can be subjected to the action of enzymes and bacteria and to conditions of pH that, although optimal for the breakdown of food, would rapidly destroy the living cells and tissues of the body proper. Nutrient molecules actually enter the body only when they pass through the epithelial lining of the digestive tract. Thus, the process of digestion involves two components: the breakdown of food molecules and their absorption into the body.

The digestive tract begins with the oral cavity and includes the mouth, pharynx, esophagus, stomach, small intestine, large intestine, and anus. Each of these areas is specialized for a particular phase in the overall process of digestion, but the fundamental structure of each is similar. As shown in Figure 34–2, the tube, from beginning to end, has four layers: (1) the innermost layer, the **mucosa,** which is made up of epithelial tissue, an underlying basement membrane, and connective tissue, with a thin outer coating of smooth muscle in some places; (2) the **submucosa,** which is made up of connective tissue and contains nerve fibers and blood and lymph vessels; (3) the **muscularis externa,** muscle tissue; and (4) the **serosa,** an outer coating of connective tissue.

34-2 *The layers of the digestive tract include (1) the mucosa, (2) the submucosa, which contains nerves and blood and lymph vessels, (3) the muscularis externa, and (4) the serosa (also known as the visceral peritoneum), which covers the outer surfaces of the organs in the abdominal cavity. A mesentery is a fold of the peritoneum that connects the digestive tract with the posterior abdominal wall. Glands outside the digestive tract, principally the pancreas and liver, discharge digestive enzymes and bile into the tract through various ducts.*

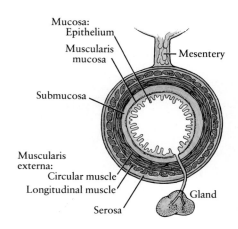

The epithelium of the digestive tract contains many mucus-secreting goblet cells and, in some parts of the gut, glands that secrete digestive enzymes. Along most of the digestive tract, the muscularis is made up of two layers of smooth muscle, an inner layer, in which the orientation is circular, and an outer layer, in which the cells are longitudinally arranged. Coordinated contractions of these muscles produce ringlike constrictions that mix the food, as well as the wavelike motions, known as **peristalsis,** that move food along the digestive tract (Figure 34–3). At several points the circular layer of muscle thickens into heavy bands, called **sphincters.** These sphincters, by relaxing or contracting, act as valves to control passage of food from one area of the digestive tract to another.

The Oral Cavity: Initial Processing

As we noted in Chapter 25, the great diversification of kingdom Animalia was determined, in large part, by the differing adaptations for obtaining food. In vertebrates, the seizing and ingesting of food is typically accomplished by the mouth, the structure in which the mechanical breakdown of food also begins.

Many vertebrates, including most mammals, have teeth, which are complex structures adapted for the tearing and grinding of food. (Modern birds, which are toothless, have gizzards containing particles of sand and gravel that serve the same tearing and grinding function.) The crown of a tooth—the visible part—is covered with enamel, the hardest substance in the body (mostly calcium phosphate); the root, within the gum, is covered with cement, a substance similar to bone. The bulk of the tooth is composed of dentine, another bonelike material, which forms slowly during the life of the tooth. The pulp cavity within the tooth contains the cells that produce dentine, as well as the nerve endings and blood vessels.

Figure 34–4 illustrates the pattern of dentition in five different mammalian orders. As you can see, the dentition of humans is relatively unspecialized. Children have 20 teeth, which are gradually lost and replaced by a second set of 32 teeth as the jaw grows larger. Of the 16 adult teeth in each jaw, four are incisors, flat chisel-like structures specialized for cutting; two are canines, used by carnivores for stabbing and tearing; four are premolars ("in front of the molars"), each of which has two cusps, or protuberances, and so are also called bicuspids; and six are molars, each of which has four or five cusps. The premolars and molars are used for grinding. Molars do not replace temporary teeth but are added as the jaw grows.

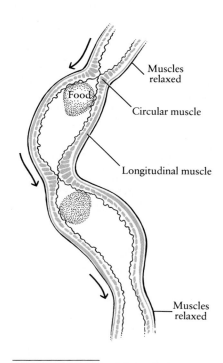

34-3 *Food is moved through the vertebrate digestive tract by several different patterns of contraction of the smooth muscle in its walls. One of the most important patterns, known as peristalsis, consists of progressive waves of contraction of the circular muscles.*

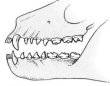

Generalized mammal

Rodent
(Rodentia)

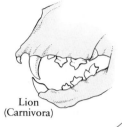

Lion
(Carnivora)

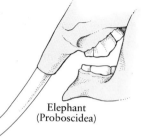

Elephant
(Proboscidea)

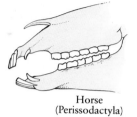

Horse
(Perissodactyla)

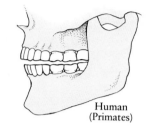

Human
(Primates)

34–4 *The basic pattern of dentition in mammals includes, in each quadrant, two incisors, one canine ("eye tooth"), two to four premolars, and three molars. The various orders of mammals have modifications of this pattern that reflect their dietary habits. Rodents (whose name comes from the Latin* rodere, *"to gnaw") are characterized by their sharp, chisel-shaped incisors that grow throughout their*

lives. Most rodents have no canines. Large predatory carnivores, such as the lion, have canines adapted for stabbing and slicing, and large molars, which can easily crush the bones of large herbivores. In elephants, tusks are modified incisors; they are used for attack and defense and for rooting food from the ground or breaking branches. (The tusks of most other mammals, such as walruses, are modified

canines.) The modern horse is a grazing animal; its incisors clip off grasses, which are then ground by the large, flat molars. Humans, with a relatively unspecialized diet, have a correspondingly unspecialized dentition. Similar, though less spectacular, differences in anatomy and physiology of other components of the digestive system also reflect adaptation to different diets.

The tongue, also a vertebrate development, serves largely to move and manipulate food in mammals. However, some vertebrates, such as the jawless hagfish and lampreys (page 593), have tongues equipped with horny "teeth." The sticky tongues of frogs and toads, which are attached at the front rather than the back of the mouth, flip out to catch insects (see Figure 28–18, page 595). Mammalian tongues carry taste buds (Figure 34–5). In humans, the tongue has acquired a secondary function of formulating sounds for communication.

As the food is being chewed, it is moistened by saliva, a watery secretion produced by three pairs of large salivary glands (Figure 34–6) plus numerous minute glands, the buccal glands, which are located in the mucosal lining of the mouth. The saliva, which contains mucus, lubricates the food so that it can be swallowed easily. Saliva is slightly alkaline, owing to the presence of sodium

34–5 (a) *The outer pore of a human taste bud, as revealed by a scanning electron micrograph of the surface of the tongue. The sensory receptor cells of the taste bud are just visible within the pore. Other taste buds are found on the roof of the mouth and the pharynx.* (b) *Longitudinal section of a taste bud. As we shall see in Chapter 42, different receptor cells respond to the molecules responsible for different tastes. These cells perform a critical function in enabling an animal to distinguish what is good to eat from what is not.*

(a)   5 µm

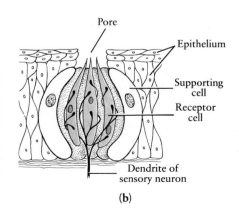

(b)

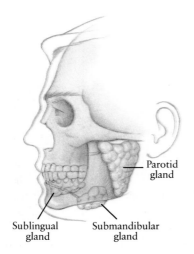

Parotid
gland

Sublingual
gland

Submandibular
gland

34–6 *The bulk of the saliva is produced by three pairs of salivary glands. Additional amounts are supplied by minute glands, the buccal glands, in the mucous membrane lining the mouth. The parotid glands are the sites of infection by mumps virus.*

bicarbonate. In humans and other mammals that chew their food, saliva also contains a digestive enzyme, salivary amylase, that begins the breakdown of starches. Like all digestive enzymes, amylase works by hydrolysis; that is, the breaking of each bond involves addition of a molecule of water (see page 64). Carnivores, such as dogs, which characteristically tear and gulp their food, have no digestive enzymes in their saliva.

The secretion of saliva is controlled by the autonomic nervous system. It can be initiated by the presence of food in the mouth, which triggers reflexes originating in taste buds and in the walls of the mouth, and also by the mere smell or anticipation of food. (Think hard, for a moment, about eating a lemon.) Fear inhibits salivation; at times of great danger or stress, the mouth may become so dry that speech is difficult. Conversely, the presence of a noxious substance in the mouth or stomach stimulates the copious production of a watery saliva as a protective reaction. On the average, we produce 1 to 1.5 liters of saliva every 24 hours.

The Pharynx and Esophagus: Swallowing

From the mouth, food is propelled backward toward the esophagus, a muscular tube about 25 centimeters long in adult humans. Swallowing is the passing of food to the esophagus and through the esophagus on to the stomach (Figure 34–7). It begins as a voluntary action but, once under way in humans, it continues involuntarily. In humans, the upper part of the esophagus is striated muscle, but the lower part is smooth muscle. In dogs, cats, and other animals that gulp their food, the entire length of the esophagus is striated muscle. Both liquids and solids are propelled along the esophagus by peristalsis. This process is so efficient that we can swallow water while standing on our heads.

34–7 *Separation of the digestive and respiratory systems in mammals makes it possible to keep food out of the lungs. (a) Face and neck, showing parts of the respiratory and digestive systems. The pharynx, the common passageway of the two systems, is at the back of the mouth and connects with the trachea (windpipe) and the esophagus. (b) Swallowing. As the food mass descends, the epiglottis tips downward, blocking the entrance to the trachea. The food mass then passes into the esophagus.*

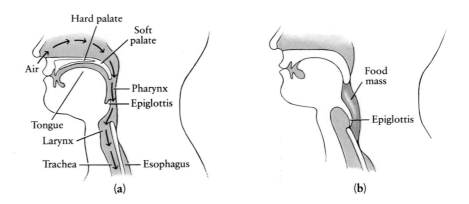

The esophagus passes through the diaphragm, which separates the thoracic and abdominal cavities, and opens into the stomach, which, with the remaining digestive organs, lies in the abdomen. The abdominal cavity is completely lined by the **peritoneum,** a thin layer of connective tissue covered by moist epithelium. (The portion of the peritoneum that covers the outer surface of the digestive tract is, as we noted earlier, known as the serosa.) The stomach, intestines, and other organs in the abdominal cavity are suspended by folds of peritoneum known as **mesenteries,** as shown in Figure 34–2. Mesenteries consist of double layers of tissue, with blood vessels, lymph vessels, and nerves lying between the two layers.

The Heimlich Maneuver

Accidental choking on food claims the lives of almost 3,000 persons a year in this country alone, more than accidents involving firearms or airplanes. It occurs when a food mass enters the trachea rather than the esophagus (Figure 34–7). If the food becomes lodged, the victim cannot speak or breathe and, if the airway is blocked for four or five minutes, the victim will die. (The fact that the victim cannot speak helps onlookers to distinguish such blockage from a heart attack. Although the symptoms are similar, heart attack sufferers can talk.)

The food can nearly always be dislodged by the Heimlich maneuver, a procedure so simple that it has been carried out successfully in at least two instances by eight-year-olds. There are three steps: (1) Stand behind the victim and wrap your arms around his or her waist. (2) Make a fist with one hand, grasp it with your other hand, and then place the fist against the victim's abdomen, slightly above the navel and below the rib cage. (3) Press your fist into the victim's abdomen with a quick upward thrust. The sudden elevation of the diaphragm compresses the lungs and forces air up the trachea, pushing the food out. Repeat several times if necessary.

If the victim is sitting, the procedure can be carried out in the same way. If the victim is lying on his back, face him, and kneel astride the hips. Put the heel of one hand on the abdomen above the navel and below the rib cage, put the other hand on top of the first hand, and press with a quick upward thrust. You can even do it on yourself: place your hands at your waist, make a fist, and press quickly upward.

The victim should be seen by a physician immediately after the emergency treatment because it is possible to break a rib or cause other internal injuries, especially if the movements are performed incorrectly. But, considering the alternative, it is well worth the risk.

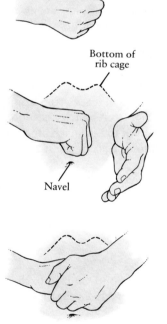

Bottom of rib cage

Navel

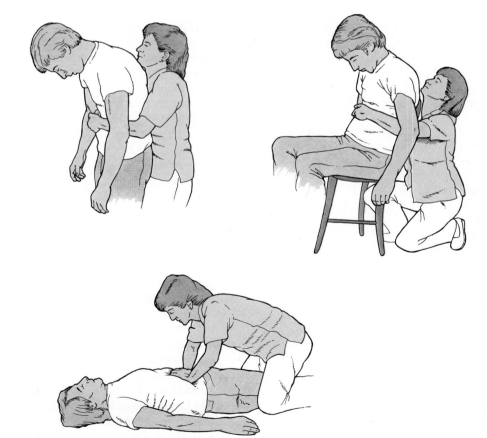

The maneuver can be carried out with the victim in a standing, sitting, or lying position. If the person is larger than you are, it is preferable to have him sitting or lying down. Note how the fist is made (you might want to practice this) and the position of the hands. These are both very important.

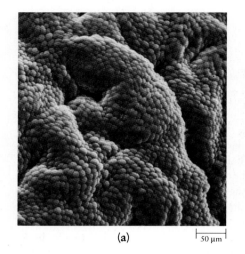

(a) 50 μm

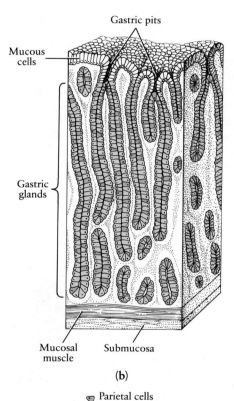

Gastric pits

Mucous cells

Gastric glands

Mucosal muscle   Submucosa

(b)

Parietal cells
(secrete HCl)

Chief cells
(secrete pepsinogen)

34–8 (a) *Surface of the stomach, as shown in a scanning electron micrograph. The numerous grooves and indentations are gastric pits.* (b) *A cross section of stomach mucosa. The parietal cells secrete hydrochloric acid, and the chief cells produce pepsinogen. Mucus, secreted by other cells of the epithelium, coats the surface of the stomach and lines the gastric pits, protecting the stomach surface from digestion.*

The Stomach: Storage and Liquefaction

The stomach is a collapsible, muscular bag, which, unless it is fully distended, lies in folds. Stomachs vary widely in capacity. A hyena's stomach can hold an amount equivalent to one-third of the animal's own body weight, for instance, whereas a mammal that eats small, regularly available, and highly nutritious food items—such as seeds or insects—needs only a small stomach. The human stomach, distended, holds 2 to 4 liters of food.

The mucosal layer of the stomach is very thick and contains numerous gastric pits (Figure 34–8). Mucus-secreting epithelial cells cover the surface of the stomach and line the gastric pits. Opening into the lower portions of the gastric pits are the gastric glands, whose walls contain the parietal cells, which produce hydrochloric acid (HCl), and the chief cells, which produce pepsinogen, the precursor of the digestive enzyme pepsin. These secretions, with the water in which they are dissolved, constitute gastric juice.

As a consequence of the HCl secretion, the pH of gastric juice is normally between 1.5 and 2.5, far more acidic than any other body fluid. The burning sensation you feel if you vomit is caused by the acidity of gastric juice acting on unprotected membranes. Normally, the mucus in the stomach forms a barrier between the epithelium and the gastric juices and so prevents the stomach from digesting itself. The HCl, which kills most bacteria and other living cells in the ingested food, loosens the tough, fibrous components of plant and animal tissues and erodes the cementing substances between cells. HCl also initiates the conversion of pepsinogen to its active form, pepsin, by splitting off a small portion of the molecule. Once pepsin is formed, it acts on other molecules of pepsinogen to form more pepsin. Pepsin, which breaks proteins down into peptides, is active only at the low pH of the normal stomach.

The stomach is influenced by both the nervous and endocrine systems. Anticipation of food and the presence of food in the mouth stimulate churning movements of the stomach and the production of gastric juice. Fear and anger decrease the stomach's motility. When protein-containing food reaches the stomach, its presence causes the release of a hormone, gastrin, from gastric cells into the bloodstream. This hormone acts on the epithelial cells of the stomach mucosa to increase their secretion of gastric juice and on the muscle cells of the stomach wall to increase their contractions.

In the stomach, food is converted into a semiliquid mass, which is gradually moved by peristalsis through the pyloric sphincter, separating the stomach and small intestine. The stomach is usually empty four hours after ingestion of a meal.

The Small Intestine: Digestion and Absorption

In the small intestine, the breakdown of food begun in the mouth and stomach is completed. The resulting nutrient molecules are then absorbed from the digestive tract into the circulatory system of the body, through which they are delivered to the individual cells.

Anatomically, the small intestine is characterized by circular folds in the submucosa; numerous microscopic fingerlike projections, **villi,** on the mucosa (Figure 34–9); and tiny cytoplasmic projections, **microvilli,** on the surface of the individual epithelial cells (Figure 34–10). All of these structural features increase the surface area of the small intestine. If the small intestine of an adult human were to be fully extended, it would be some 6 meters long; the total surface area of the human small intestine is approximately 300 square meters, about the size of a doubles tennis court. The **duodenum,** the upper 25 centimeters of the small intestine, is the most active in the digestive process; the rest is principally concerned with absorption of nutrients.

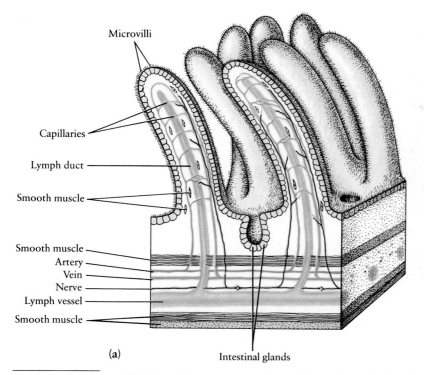

Microvilli

Capillaries

Lymph duct

Smooth muscle

Smooth muscle
Artery
Vein
Nerve
Lymph vessel
Smooth muscle

(a)

Intestinal glands

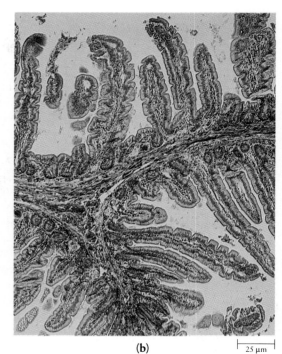

(b)

25 μm

34–9 **(a)** *Diagram and* **(b)** *photomicrograph of villi of the small intestine, as seen in longitudinal section. Nutrient molecules are absorbed through the walls of the villi and, with the exception of fat molecules, enter the bloodstream by means of the capillaries. Fats—hydrolyzed to fatty* acids and glycerol, resynthesized into new fats, and packaged into particles known as chylomicrons—are taken up by the lymphatic system. The villi can move independently of one another; their motion increases after a meal.

In the micrograph, the columnar epi- thelial cells that are responsible for the absorption of nutrient molecules are stained pink, and their nuclei are stained dark red. Mucus-secreting goblet cells are stained blue, as is the layer of smooth muscle at the base of the villi.

34–10 **(a)** *A columnar epithelial cell from the lining of the intestine. Notice, in particular, the tight junctions and desmosomes that bind these cells together in a continuous sheet.* **(b)** *Microvilli on the surface of two adjacent intestinal epithelial cells. These cellular extensions greatly increase the absorptive surface of the intestine. They also contain digestive enzymes, which are part of the cell membrane, and proteins involved in the transport of nutrient molecules across the cell membrane. Notice the tight junction at the surface where the two cells meet and, slightly lower, a desmosome. Although not visible in this micrograph, intestinal epithelial cells contain numerous mitochondria. Their abundance indicates that the cells have high energy requirements, probably due, in large part, to active transport processes.*

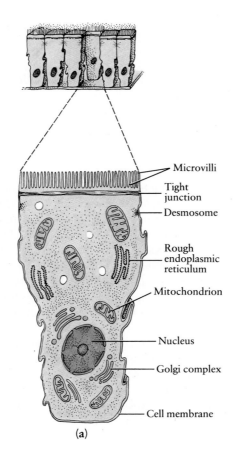

Microvilli

Tight junction

Desmosome

Rough endoplasmic reticulum

Mitochondrion

Nucleus

Golgi complex

Cell membrane

(a)

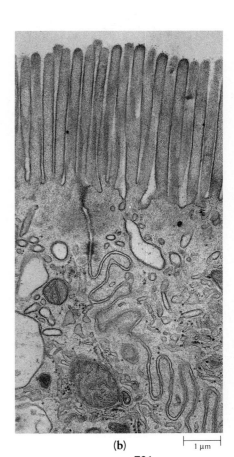

(b)

1 μm

721

TABLE 34-1   **Principal Digestive Enzymes**

ENZYME	SOURCE	SUBSTRATE	SITE OF ACTION
Salivary amylase	Salivary glands	Starches	Mouth
Pepsin	Stomach mucosa	Proteins, pepsinogen	Stomach
Pancreatic amylase	Pancreas	Starches	Small intestine
Lipase	Pancreas	Fats	Small intestine
Trypsin	Pancreas	Polypeptides, chymotrypsinogen	Small intestine
Chymotrypsin	Pancreas	Polypeptides	Small intestine
Carboxypeptidase	Pancreas	Polypeptides	Small intestine
Deoxyribonuclease	Pancreas	DNA	Small intestine
Enterokinase	Small intestine	Trypsinogen	Small intestine
Aminopeptidase	Small intestine	Polypeptides	Small intestine
Dipeptidase	Small intestine	Dipeptides	Small intestine
Maltase	Small intestine	Maltose	Small intestine
Lactase*	Small intestine	Lactose	Small intestine
Sucrase	Small intestine	Sucrose	Small intestine
Phosphatases	Small intestine	Nucleotides	Small intestine

* Often absent in adults, especially those of recent African origin.

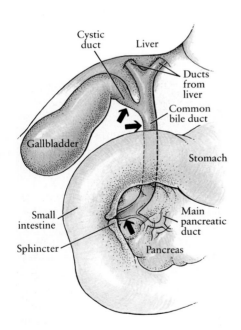

34-11 *Ducts from the liver, the gallbladder, and the pancreas merge just before emptying into the small intestine through a sphincter in its wall. The arrows indicate the locations in which gallstones typically lodge. Gallstones, which consist primarily of cholesterol, form when the delicate balance in the relative concentrations of bile ingredients is disturbed.*

A number of substances are involved in the digestive processes of the small intestine. Mucus is secreted by the goblet cells of the intestinal mucosa. Digestive enzymes are produced by epithelial cells of both the intestinal mucosa and the pancreas (Table 34–1). In addition to enzymes, the small intestine receives an alkaline fluid from the pancreas, which neutralizes the stomach acid, and bile, which is produced in the liver and stored in the gallbladder. Bile contains a mixture of salts that, like laundry detergents, emulsify fats, breaking them apart into droplets. In this form more surface area is available for attack by enzymes. Bile also contains some sodium bicarbonate, which, with the pancreatic fluid, neutralizes the stomach acid. Neutralizing the acidity is essential because the intestinal enzymes are optimally active at a pH of 7 to 8 and would be denatured by the acidic pH of the gastric juices that enter the intestine. Fluids from the pancreas, the liver, and the gallbladder travel toward the small intestine through a series of ducts that merge to form one large duct; this duct empties into the intestine about 10 centimeters below the pyloric sphincter (Figure 34–11).

In the small intestine, pancreatic amylases continue the breakdown of starch begun in the mouth, producing disaccharides. Lipases hydrolyze fats into glycerol and fatty acids. Three types of enzymes break down proteins. The members of one group break apart the long protein chains; each enzyme in this group acts only on the bonds linking particular amino acids, so that several enzymes are required to break a single large protein into shorter peptide fragments. A second type of enzyme acts only on the end of a chain, some on the amino end and some on the carboxyl end, splitting off dipeptides. A third group of enzymes then comes into action, breaking the remaining dipeptides into single amino acids.

The digestive activities of the small intestine are coordinated and regulated by hormones (Table 34–2). In the presence of acidic gastric juice, the duodenum releases secretin, a hormone that stimulates the pancreas and liver to secrete alkaline fluids. Fats and amino acids in the food stimulate the production of another hormone, cholecystokinin, which triggers the release of enzymes from the pancreas and the emptying of the gallbladder. At least eight other substances are suspected of being gastrointestinal hormones. In addition to hormonal influences, the intestinal tract is also regulated by the autonomic nervous system; stimulation by parasympathetic nerve fibers increases intestinal contractions, whereas inhibition by sympathetic fibers decreases contractions. Thus, a complex interplay of stimuli and checks and balances serves to activate digestive enzymes, to adjust the chemical environment, and to regulate movements of the intestines.

TABLE 34-2 **Major Gastrointestinal Hormones**

HORMONE	SOURCE	MAJOR STIMULUS FOR PRODUCTION	MAJOR ACTIONS
Gastrin	Stomach	Protein-containing food in stomach, also parasympathetic nerves to stomach	Stimulates secretion of gastric juices and muscular contractions of stomach and intestine
Secretin	Duodenum	HCl in duodenum	Stimulates secretion of alkaline pancreatic fluids and bile
Cholecystokinin	Duodenum	Fats and amino acids in duodenum	Stimulates release of enzymes from pancreas and of bile from gallbladder

Absorption of Nutrients

The food molecules released by the digestive processes are absorbed through the epithelial cells of the intestinal mucosa. Specialized enzymes—lactase, sucrase, and maltase—embedded in the epithelial cell membranes cleave disaccharides to monosaccharides, which are then rapidly absorbed by active transport and facilitated diffusion (page 136). Amino acids and dipeptides are absorbed by active transport. These molecules all enter the bloodstream by way of the capillaries of the villi.

Small fatty acids also enter the blood vessels of the intestine directly, but large fatty acids, glycerol, and cholesterol travel an indirect route. First, these molecules enter mucosal cells by passive diffusion. Within the cells, fatty acids and glycerol are resynthesized into fats (three molecules of fatty acid combined with one molecule of glycerol, as shown on page 68), which are then packaged into protein-coated droplets, known as chylomicrons. Similarly, cholesterol is packaged into low-density lipoprotein (LDL) complexes (see page 71). Both the chylomicrons and the LDLs are secreted into the lymph vessels and ultimately enter the bloodstream through ducts that empty into veins in the chest. In the bloodstream, the chylomicrons are gradually broken apart. Fats are delivered in the form of fatty acids to cells such as muscle cells, where they are oxidized for energy, or to fat cells, where they are stored. (One gram of fat, as we noted on page 67, contains more than twice as many calories as 1 gram of protein or glycogen; hence it is a very useful storage form for motile organisms.) The LDL particles are picked up by cells in the liver, where the cholesterol is stored, secreted in the bile, or repackaged for delivery to other body cells. It is used by all cells in the synthesis of cell membranes and, by specialized cells, for the production of steroid hormones.

Aids to Digestion

Compared to the microscopic photosynthetic cells of the open waters, plant material is tough and hard to digest. By logic, one would expect terrestrial herbivores to have sets of enzymes capable of breaking down cellulose and other structural polysaccharides found in plants. Not so. Terrestrial animals have evolved another solution: symbiotic relationships with bacteria and protists that perform these activities on their behalf, receiving, in return, a free food supply, protection (except from the host itself), and good housing.

In the warm, wet, airless interior of a mammalian digestive tract, symbionts break down plant products anaerobically. The glucose they use themselves; the host's share is mainly fatty acids, which are absorbed through the host's intestine. In addition, the microorganisms synthesize many vitamins, especially vitamin K and those of the B group, which are utilized by the host.

The fermentation process, the anaerobic oxidation of organic molecules, takes place in different areas in the intestinal tracts of different species. In the horse family, it occurs in the colon, which is enlarged, and in the cecum, a blind sac at the junction of the small and large intestines. (The cecum persists in humans in a vestigial form as the trouble-making appendix.) Rabbits and other lagomorphs have an enlarged cecum in which bacterial fermentation takes place and also a curious adaptation, peculiar to this group. At night, in their burrows, rabbits produce, from the cecum, a special type of feces, different from those produced in the daytime, that consist almost entirely of bacteria. They eat these feces, thus digesting and absorbing additional nutrients, obtaining vitamins produced by the bacteria, and recycling their beneficial symbionts.

The most elaborate specialization, anatomically speaking, is found among the ruminants, such as cattle, antelopes, sheep, and deer. It is probably not a coincidence that the ruminants are among the most successful terrestrial herbivores, with their rise in evolutionary prominence taking place at the time the grasslands were undergoing rapid expansion. Ruminants hastily shred grass blades and other plant parts and store them, regurgitating them as a cud and completing the chewing process at leisure. The actual fermentation takes place in the rumen, a large esophageal pouch. The contents of the rumen are then passed to the stomach where absorption begins. Bacteria and ciliates in the rumen not only break down the cellulose and other polysaccharides but also synthesize proteins, using ammonia and urea as nitrogen sources. Some of these microorganisms are also digested by the host, thus providing it with protein.

Ruminant digestion is a major activity. In cattle, the rumen-stomach complex makes up about 15 percent of the weight of the animal. To keep the mixture moist, a cow secretes 60 liters of saliva a day. She also produces about 2 liters of gas a minute, most of which escapes in sweet, chlorophyll-scented belches.

Omnivorous humans have no such symbionts and so are unable to utilize cellulose, which is eliminated as roughage. However, like other mammals, we are dependent on the bacteria of our digestive tracts for the synthesis of vitamins, and, as in the rabbit, a major component of our feces is symbiotic bacteria.

The Large Intestine: Further Absorption and Elimination

The absorption of water, sodium, and other minerals, a process that occurs primarily in the small intestine, continues in the large intestine. In the course of digestion, large amounts of water—approximately 7 liters per day—enter the stomach and small intestine as secretions of the glands emptying into and lining the digestive tract, by osmosis from the body fluids, and directly in the ingested food and drink. When the absorption of this water and the minerals it contains is disrupted, as in diarrhea, severe dehydration can result. Mortality from infant diarrhea, still the chief cause of infant death in many countries, is principally a consequence of water loss.

The large intestine harbors a considerable population of symbiotic bacteria (including the familiar *E. coli),* which break down food substances that escaped digestion and absorption in the small intestine. Living on these food substances, largely materials we lack the enzymes to digest, the bacteria synthesize amino acids and vitamins, some of which are absorbed into the bloodstream. These bacteria are our chief source of vitamin K.

A blind pouch off the large intestine, the appendix, is an evolutionary remnant from herbivorous ancestors (see essay). As many individuals know from personal experience, it may become irritated, inflamed, and then infected. If it ruptures, as a consequence of inflammation and swelling, it spills its bacterial contents into the abdominal cavity. Serious, and even fatal, infection can result. Although the appendix plays no known role in human digestion, it is one of the sites of interaction of cells involved in the immune response.

The bulk of the fecal matter consists of water, bacteria (mostly dead cells), and cellulose fibers, along with other indigestible substances. It is lubricated by mucus, which is secreted by some of the epithelial cells lining the large intestine, stored briefly in the rectum, and then eliminated through the anus as feces.

MAJOR ACCESSORY GLANDS

The Pancreas

As we have seen, the pancreas is the source of a number of substances essential to the digestive process. It is a specialized secretory organ, developing both in the embryo and in the course of evolutionary history from the upper portion of the small intestine. The bulk of the tissue of the pancreas (Figure 34–12) resembles salivary gland tissue and, like the salivary glands, secretes an amylase that plays a major role in the breakdown of starch. In addition, it secretes a number of other digestive enzymes.

The pancreas is also an endocrine (hormone-producing) gland. Clusters of pancreatic cells, known as the islets of Langerhans (named after their discoverer, not some romantic Pacific atoll), secrete insulin, glucagon, and somatostatin, which are released into the bloodstream and participate in the regulation of blood glucose.

The Liver

The liver also plays vital roles in digestion. The largest internal organ of the body, the liver is a three-pound chemical factory with an extraordinary variety of processes and products. It stores and releases carbohydrates, playing a central role in the regulation of blood glucose. It processes amino acids, converting them to carbohydrates, channeling them to other tissues of the body, and synthesizing essential proteins such as enzymes and clotting factors from them. It manufactures the plasma proteins that make the blood hypertonic in relation to the

34–12 *Portion of a pancreatic cell involved in the production of digestive enzymes. Notice the extensive rough endoplasmic reticulum characteristic of cells that synthesize proteins for export. The cell nucleus can be seen at the bottom of the micrograph, and, just outside the nuclear envelope, there are several mitochondria. The dark, spherical objects are secretory granules, moving toward the periphery of the cell.*

Other types of pancreatic cells synthesize the hormones insulin, glucagon, and somatostatin, which play major roles in the regulation of blood glucose.

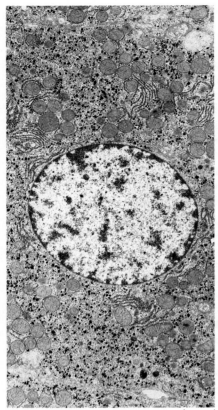

├─ 1 μm ─┤

34–13 *Electron micrograph of a liver cell. The dark granules are glycogen. Despite the variety of activities carried out by the liver, the cells of the liver all resemble one another in appearance and in function, with no division of labor among cell groups, as there is in the pancreas. Liver cells, as you would expect, have many mitochondria and abundant rough endoplasmic reticulum.*

interstitial ("between-the-tissues") fluids and so prevents the osmotic movement of water from bloodstream to tissues. It is the major source of the plasma lipoproteins, including LDLs and HDLs, that transport cholesterol, fats, and other water-insoluble substances in the bloodstream and is of central importance in the regulation of blood cholesterol (see page 71). It stores fat-soluble vitamins, such as A, D, and E. It produces bile (which is then stored in the gallbladder). It breaks down the hemoglobin from damaged and dead red blood cells to bilirubin, a yellow pigment; bilirubin is released into the bile and excreted through the intestinal tract. The liver inactivates a number of hormones, thus playing an important role in hormone regulation. It also breaks down a variety of foreign substances, some of which—alcohol, for instance—may form metabolic products that damage liver cells and interfere with their functions (see page 202).

REGULATION OF BLOOD GLUCOSE

The major function of digestion is, of course, to provide organic molecules that can serve as energy sources and raw materials for each cell of the body. Although vertebrates rarely eat 24 hours a day, their blood glucose—the major cellular energy supply and the fundamental building-block molecule—remains extraordinarily constant. As we have noted previously, the liver plays a central role in this critical process. Glucose and other monosaccharides are absorbed into the blood from the intestinal tract and are passed directly to the liver by way of the hepatic portal vein. The liver converts some of these monosaccharides to glycogen and fat, storing enough glycogen to satisfy the body's needs for about four hours. The fat is stored in fat cells, which can also form fat from glucose. Similarly, the liver breaks down excess amino acids (which are not stored) and converts them to glucose. The nitrogen from the amino acids is excreted in the form of urea, and the glucose is stored as glycogen.

Whether the liver takes up or releases glucose and the amount it takes up or releases are determined primarily by the concentration of glucose in the blood. As we shall see in Chapter 40, the concentration of glucose is, in turn, regulated by a number of hormones and is influenced by the autonomic nervous system. Among the hormones involved in this process are insulin, glucagon, and somatostatin, all produced by the pancreas. Insulin stimulates the uptake of glucose by cells, thus decreasing blood glucose. Glucagon promotes the breakdown of glycogen, thus increasing blood glucose. Somatostatin, which was the first vertebrate protein synthesized using recombinant DNA technology (page 351), has a variety of inhibitory effects that collectively help to regulate the rate at which glucose and other nutrients are absorbed from the digestive tract.

SOME NUTRITIONAL REQUIREMENTS

Because of the liver's activities in converting various types of food molecules into glucose and because most tissues can use fatty acids as an alternative fuel, the energy requirements of the body can be met by carbohydrates, proteins, or fats—the three principal types of food molecules. Energy requirements are ordinarily met by a combination of the three. Carbohydrates and proteins supply about the same number of calories per gram of dry weight, and fats supply about twice as many as either of them. However, because carbohydrates and proteins are always combined with water in the form in which they are eaten, the actual bulk of ingested material is usually nine or ten times greater.

Mother's Milk—It's the Real Thing

Throughout this chapter, we have been concerned with the digestive functions and nutritional requirements of mammalian adults. The digestive system of the mammalian infant, however, requires additional development and growth before it can perform its full range of functions. In the meantime, the infant is totally dependent on the complex, highly nutritive fluid known as milk, which is produced by the modified sweat glands from which the Mammalia derive their name.

In the course of evolution, the various species of mammals have developed milks especially suited to the nourishment of their own young. Human milk, for instance, contains about the same amount of fat (3.7 percent) as cow's milk, but it has less protein and more carbohydrate (in the form of lactose, milk sugar). The milk of a harp seal, on the other hand, is twice as concentrated as either cow's milk or human milk (45 percent water as compared to 87 percent water), has no carbohydrate, is 43 percent fat, and so fulfills the higher energy requirements of an animal that maintains a high body temperature in water. A kangaroo can provide an even more specialized diet; one mother carrying two joeys of different ages in her pouch supplies different milks (through different teats) to each.

In addition, each species produces milk with particular vitamins and minerals and, perhaps most important, a number of antibodies, other immunologically active proteins, and various white blood cells that protect the infant of that species against infection. For example, human milk contains antibodies that protect against *Staphylococcus, E. coli,* and polio, in particular. This immunological protection is especially important in the first few months of life, before the infant's own immune system is functioning fully. Another important feature of mother's milk is that it comes in a container that keeps it fresh, temperature-controlled, and free from bacterial contamination.

The length of time mammalian infants depend on milk for nourishment also varies with the species. Rabbits nurse for only a few days, cats for six to ten weeks, hyenas and humans for a year, elephants for four or five. However, with the exception of Western humans, once infancy is over, not only does milk consumption stop but there is also a decline in the level of lactase, the enzyme that breaks down lactose. This enzyme, present in most human infants, is absent in adults of most animal species and is much diminished in most human beings after the age of four. Lactose ingestion by individuals lacking the enzyme may cause intestinal cramps and diarrhea, as the lactose is broken down by bacterial fermentation. Natural selection for lactose digestion in adult humans is thought to have begun some 10,000 years ago when certain groups began to domesticate and milk mammals. Individuals of northern European descent are much more likely to have this enzyme as adults than are those of African or Mediterranean descent. This distribution of the allele (or of the gene that turns it on or off) is apparently correlated with the fact that northerners were more likely to drink milk (including reindeer milk) as adults, whereas people from hot climates, if they used milk at all, were likely to use fermented milk products, such as cheese or yogurt, in which lactose has already been broken down.

Various commercially produced formulas exist that, mixed with water, offer a substitute for human milk. Where families have access to clean water, sterilized bottles, good pediatric care, and money for an adequate amount of formula, milk substitutes can be used successfully by women who cannot or choose not to nurse their babies. However, at a World Health Organization assembly in 1981, 118 countries voted to regulate the promotion of infant formula (the United States cast the only vote against regulation). These regulations were aimed particularly at the promotion of infant formula in the Third World, where access to clean water or to facilities for boiling water is often simply not available. In fact, bottles are often not even washed between feedings. Another reason for the regulations is that the formula may be overdiluted to save money. Adding this to the absence in formulas of antibodies and other protective substances, it can be seen that formula-fed infants run a much higher risk of infectious disease than breast-fed babies. Many die of "bottle-baby disease," severe diarrhea, and some starve due to watered-down formula. Ironically, women in the Third World often use milk substitutes not as a matter of necessity but because bottle feeding is associated with progress, social status, and the adoption of Western conventions.

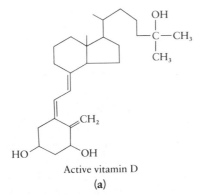

(b)

34–14 (a) *Vitamin D is a steroid-like substance (see page 70) that is produced in the skin by the action of ultraviolet rays from the sun on cholesterol (the absorbed energy opens the second ring of the molecule). Further chemical processing in the liver and kidneys produces the active form of the molecule shown here.*

According to one current hypothesis, the ancestors of modern Homo sapiens *originated in the tropics and were all dark-skinned. In those populations that moved northward, however, the screening effects of the darker pigment, which inhibits the production of vitamin D, caused selection for lighter skin color, whereas no such selection pressures occurred in the sun-drenched areas nearer the equator. In fact, in the tropics, selection would favor dark skin, which prevents carcinogenic ultraviolet rays from reaching the dermis. Nor did such selection occur among the Eskimos* (b), *who eat a diet rich in fish oils, a major source of vitamin D.*

In addition to calories, the cells of the body need the 20 different kinds of amino acids required for assembling proteins. When any one of the amino acids necessary for the synthesis of a particular protein is unavailable, the protein cannot be made, and the other amino acids are converted to carbohydrates and oxidized or stored. Vertebrates are not able to synthesize all 20 amino acids. Humans (and albino rats) can synthesize 12, either from a simple carbon skeleton or from another amino acid. The other eight, which must be obtained in the diet, are known as essential amino acids (see essay, page 76). Plants are the ultimate source of the essential amino acids, but it is difficult (although by no means impossible) to obtain sufficient quantities of them by eating a completely vegetable diet, largely because plant proteins are relatively deficient in lysine and tryptophan.

Mammals also require but cannot synthesize certain polyunsaturated fats that provide fatty acids needed for the synthesis of fats and a group of hormonelike compounds known as prostaglandins (see page 833). These essential fatty acids can be obtained by eating plants or insects (or eating other animals that have eaten plants or insects).

Vitamins are an additional group of molecules required by living cells that cannot be synthesized by animal cells. Many of them function as coenzymes, and they are characteristically required only in small amounts. Table 34–3 indicates some of the vitamins required in the human diet and their functions. Severe vitamin deficiencies, such as may occur in regions where malnutrition is chronic, can have appalling consequences. Studies triggered by an extraordinary incidence of blindness among children who were victims of the famines in Ethiopia in the mid-1980s have revealed that, in addition to its effects on vision, severe vitamin A deficiency dramatically increases deaths due to measles, diarrhea, and other childhood diseases. At the present time, the World Health Organization is mounting a massive campaign to supply high doses of supplementary vitamin A to children in poverty-stricken regions in Africa, Asia, and Latin America. Although such supplements are vital in cases of severe malnutrition, there is no clear evidence that the ingestion of amounts of any particular vitamin in excess of the amounts available in a well-balanced diet has any beneficial effect on a normally healthy individual. Some, including the fat-soluble vitamins A, D, and K, which can accumulate in body tissues, are toxic in large doses. One of the most concentrated sources of vitamin A known is polar bear liver, a half pound of which contains about 2,600 times the recommended daily allowance. For centuries, Eskimos and Arctic explorers have known that consumption of polar bear liver causes illness and can be fatal—a knowledge shared by sled dogs and Arctic birds, which also refuse to eat it.

The body also has a dietary requirement for a number of inorganic substances, or minerals. These include calcium and phosphorus for bone formation, iodine for thyroid hormone, iron for hemoglobin and cytochromes, and sodium, chloride, and other ions essential for ionic balance. Most of these are present in the ordinary diet or in drinking water. Like the vitamins, however, they must be given in supplementary form when the dietary intake is inadequate or when the individual is not able to assimilate them normally.

The Price of Affluence

The major nutritional problem among North Americans (and also Europeans) is obesity. In the United States, 30 percent of middle-aged women and 15 percent of middle-aged men are obese—that is, they weigh more than 120 percent of their appropriate weight. Obesity is correlated with a significant increase in coronary artery disease, diabetes, and other disorders.

TABLE 34-3 **Vitamins**

DESIGNATION: LETTER AND NAME	MAJOR SOURCES	FUNCTION	DEFICIENCY SYMPTOMS
A, carotene	Egg yolk, green or yellow vegetables, fruits, liver, butter	Formation of visual pigments, maintenance of normal epithelial structure	Night blindness; dry, flaky skin
B-complex vitamins:			
B_1, thiamine	Brain, liver, kidney, heart, pork, whole grains	Formation of coenzyme involved in Krebs cycle	Beri-beri, neuritis, heart failure
B_2, riboflavin	Milk, eggs, liver, whole grains	Part of electron carrier FAD	Photophobia, fissuring of skin
B_3, niacin (nicotinic acid)	Whole grains, liver and other meats, yeast	Part of electron carriers NAD, NADP, and of CoA	Pellagra, skin lesions, digestive disturbances
B_5, pantothenic acid	Present in most foods	Forms part of CoA	Neuromotor and cardiovascular disorders, gastrointestinal distress
B_6, pyridoxine	Whole grains, liver, kidney, fish, yeast	Coenzyme for amino acid metabolism and fatty acid metabolism	Dermatitis, nervous disorder
B_{12}, cyanocobalamin	Liver, kidney, brain, eggs, dairy products	Maturation of red blood cells, coenzyme in amino acid metabolism	Anemia, malformed red blood cells
Biotin	Egg yolk, synthesis by intestinal bacteria	Concerned with fatty acid synthesis, CO_2 fixation, and amino acid metabolism	Scaly dermatitis, muscle pains, weakness
Folic acid	Liver, leafy vegetables	Nucleic acid synthesis, formation of red blood cells	Failure of red blood cells to mature, anemia
C, ascorbic acid	Citrus fruits, tomatoes, green leafy vegetables, potatoes	Vital to collagen and ground (extracellular) substance	Scurvy, failure to form connective tissue fibers
D_3, calciferol	Fish oils, liver, fortified milk and other dairy products, action of sunlight on lipids in the skin	Increases Ca^{2+} absorption from gut, important in bone and tooth formation	Rickets (defective bone formation)
E, tocopherol	Green leafy vegetables, wheat germ, vegetable oils	Maintains resistance of red cells to hemolysis, cofactor in electron transport chain	Increased red blood cell fragility
K, naphthoquinone	Synthesis by intestinal bacteria, leafy vegetables	Enables synthesis of clotting factors by liver	Failure of blood coagulation

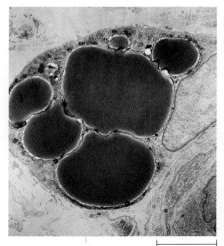

34-15 *A fat cell from the skin of a newborn rat. The nucleus is to the right. The large stored droplets of fat that nearly fill the cell have been fixed with osmium for electron microscopy. When excess calories are taken in in the diet, fat accumulates in these specialized cells, and when caloric intake is less than sufficient, fat is mobilized, broken down to glycerol and fatty acids, and released into the bloodstream.*

5 µm

In addition to excess calories, our diets appear to contain a number of health hazards as compared to those of other cultures. We consume, on the average, about twenty times as much salt as our bodies need. Excess salt has been correlated with hypertension (high blood pressure). Another hazard is animal fat, such as that present in beef and pork. It has been known for some time that diets high in animal fat interfere with the regulation of blood cholesterol (page 71), implicated in atherosclerosis and heart attacks. More recently, attention has been focused on the role of dietary fibers in decreasing colon cancer. As a result of a diet high in fat and protein and low in carbohydrate, the feces of North Americans have only about half the bulk of their Far Eastern or African counterparts, and are eliminated from the intestinal tract much more slowly. Also, the rate of colon cancer is far higher among North Americans, suggesting that the retention of a fat-laden, compact fecal mass in the lower bowel may contribute to cancer development.

A final nutritional hazard, also related to affluence, is our willingness to experiment with our own bodies and to adopt extreme diets and follow nutritional fads—fasting, liquid protein, "macrobiotic," high protein, low protein, high fiber, megavitamin—without consideration of our requirements as biological organisms and the dictates of our long, omnivorous evolutionary history.

SUMMARY

Digestion is the process by which food is broken down into molecules that can be taken up by the cells lining the intestine, transferred to the bloodstream, and so distributed to the individual cells of the body. It occurs in successive stages, regulated by an interplay of hormones and nervous stimuli. In mammals, food is processed initially in the mouth, where the breakdown of starch begins in humans. It moves through the esophagus to the stomach, where gastric juices destroy bacteria and begin to break down proteins.

Most of the digestion occurs in the upper portion of the small intestine, the duodenum; here, digestive activity, which is performed by enzymes, is almost completely under hormonal regulation. The breakdown of starch by amylases continues, fats are hydrolyzed by lipases, and proteins are reduced to dipeptides or single amino acids. Monosaccharides, amino acids, and dipeptides are absorbed into the blood vessels of the villi; fats are absorbed into the lymph vessels and ultimately enter the bloodstream. Hormones secreted by duodenal cells stimulate the functions of the pancreas and the liver. The pancreas releases an alkaline fluid containing digestive enzymes; the liver produces bile, which is also alkaline and emulsifies fats.

Much of the water that enters the stomach and small intestine in the course of digestion is reabsorbed in the small intestine itself. Most of the remaining water is reabsorbed from the residue of the food mass as it passes through the large intestine. The large intestine contains symbiotic bacteria, which are the source of certain vitamins. Undigested residues are eliminated from the large intestine.

The chief energy source for cells in the mammalian body is glucose circulating in the blood. The organ principally responsible for maintaining a steady supply of glucose is the liver, which stores glucose (in the form of glycogen) when glucose levels in the blood are high and breaks down glycogen, releasing glucose, when the levels drop. These activities of the liver are regulated by a number of different hormones.

Requirements for good nutrition include molecules for fuel (which can be obtained from carbohydrates, fats, or proteins), essential amino acids, essential fatty acids, vitamins, and certain minerals.

QUESTIONS

1. Distinguish, in terms of function, between: gastrointestinal hormones/gastrointestinal enzymes; gastric juice/bile; amylases/lipases; digestion/absorption; villi/microvilli; vitamins/essential amino acids.

2. Diagram the human digestive system, including all tubes, chambers, valves, and accessory organs. Describe the function of each structure.

3. Trace the chemical processing of a hamburger on a bun, with lettuce and tomato, as it passes through your digestive tract.

4. Most of the protein-digesting enzymes are secreted in an inactive form and are themselves activated by special enzymes secreted into the digestive tract. Explain the adaptive value of this two-step process.

5. If you oxidize a pound of fat, whether butter or the fat in your own body cells, about 4,200 kilocalories are released. Suppose that you go on a weight-reducing diet, limiting yourself to 1,000 kilocalories a day. You spend most of your time sitting in a well-heated library (studying, of course) and so you expend only about 3,000 kilocalories a day. How many pounds will you lose each week?

6. Four principal symptoms of cirrhosis of the liver and other liver diseases are jaundice (a yellowing of the eyes and skin), uncontrolled bleeding, increased sensitivity to drugs, and swelling of parts of the body, such as the legs and abdominal cavity. Explain each of these symptoms in terms of liver function.

7. Humans are omnivores rather than strict herbivores or carnivores or specialists that use a single organism as a food source. Humans have an unusually large number of substances specifically required in their diets. How might these two phenomena be related?

C H A P T E R **35**

Energy and Metabolism II: Respiration

Digestion, as we saw in the last chapter, is the process by which foodstuffs ingested by an animal are broken down into smaller organic molecules and absorbed into the bloodstream, through which they can be transported to the diverse cells of which the organism is composed. The release of energy from these molecules—the sole source of energy for heterotrophic cells—depends upon their oxidation. This process usually (although not always) requires oxygen, and when it does, it is called respiration. Actually, respiration has two meanings in biology. At the cellular level, it refers to the oxygen-requiring chemical reactions, discussed in Chapter 9, that take place in the mitochondria and are the chief source of energy for eukaryotic cells. At the level of a whole organism, it designates the process of taking in oxygen from the environment and returning carbon dioxide to it. The latter process—which is, of course, essential for the former—is the subject of this chapter.

Oxygen consumption is directly related to energy expenditure (Figure 35–1), and, in fact, energy requirements are usually calculated by measuring oxygen intake or the release of carbon dioxide. The energy expenditure at rest is known as basal metabolism. Metabolic rates increase sharply with exercise. A person exercising consumes 15 to 20 times the amount of oxygen that he or she consumes sitting still, with the oxygen consumption increasing in proportion to the energy expenditure.

35–1 *Oxygen is required for the energy-yielding reactions that take place in the mitochondria. The oxygen consumption of running animals—such as these female impalas—increases linearly with their speed.*

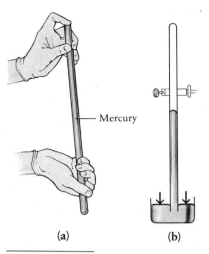

35–2 *Atmospheric pressure is usually measured by means of a mercury barometer. (a) To make a simple mercury barometer, put on a pair of protective gloves and then fill a long glass tube, open at one end, with mercury. Closing the tube with your finger, (b) invert it into a dish of mercury. Remove your finger and clamp the tube to a stand. The mercury level will drop until the pressure of its weight inside the tube is equal to the atmospheric pressure outside. At sea level, the height of the column will be about 760 millimeters (29.9 inches).*

DIFFUSION AND AIR PRESSURE

In every organism from an amoeba to an elephant, gas exchange—the exchange of oxygen and carbon dioxide between cells and the surrounding environment—takes place by diffusion. Diffusion, you will recall (page 129), is the net movement of particles from a region of higher concentration to a region of lower concentration as a result of their random motion. In describing gases, scientists speak of the pressure of a gas rather than its concentration. At sea level, the air around us exerts a pressure on our skin of 1 atmosphere (about 15 pounds per square inch). This pressure is enough to support a column of water about 10 meters high or a column of mercury 760 millimeters high (Figure 35–2). Atmospheric pressure is generally measured in terms of mercury simply because mercury is relatively heavy—so the column will not be inconveniently tall.

The total pressure of a mixture of gases, such as air, is the sum of the pressures of the separate gases in the mixture. The pressure of each gas is proportional to its concentration. Oxygen, for instance, makes up about 21 percent by volume of dry air (Table 35–1). Thus 21 percent of the total air pressure, or 160 millimeters of mercury (mm Hg), results from the pressure of oxygen in the air. This is known as the partial pressure of oxygen and is abbreviated P_{O_2}. In air that contains water vapor (also a gas), the volume of O_2 is proportionately less and so the P_{O_2} is less—around 155 mm Hg.

If a liquid containing no dissolved gases is exposed to air at atmospheric pressure, each of the gases in the air diffuses into the liquid until the partial pressure of each gas in the liquid is equal to the partial pressure of the gas in the air. Thus when one speaks of the P_{O_2} of a liquid, such as blood, one means the pressure of dry gas with which the dissolved O_2 in the liquid is in equilibrium. For example, blood with a P_{O_2} of 40 mm Hg would be in equilibrium with air in which the partial pressure of oxygen was 40 mm Hg. If, however, blood with a P_{O_2} of 40 mm Hg is exposed to the usual mixture of air, oxygen moves from the air into the blood until the blood P_{O_2} equals 155 mm Hg and equilibrium is reached. Conversely, if a liquid containing a dissolved gas is exposed to air in which the partial pressure of that gas is lower than in the liquid, the gas will leave the liquid until the partial pressures in the air and the liquid are equal. In short, gases move from a region of higher partial pressure to a region of lower partial pressure.

We are so accustomed to the pressure of the air around us that we are unaware of its presence or of its effects on us. However, if you visit a place, such as Mexico City, that is at a comparatively high altitude and therefore has a lower atmospheric pressure (and of course, a lower P_{O_2}), you will feel light-headed at first and will tire easily. As we shall see in an essay later in this chapter, a series of physiological adaptations are required for life at high altitude—and for successful mountain climbing.

The consequences of the opposite situation—higher gas pressures—can be seen in deep-sea divers. Early in the history of deep-sea diving, it was found that when divers come up from the bottom too quickly, they get the "bends," which are always painful and sometimes fatal. The bends develop as a result of breathing compressed air. Under high pressure, more nitrogen diffuses from the air in the lungs into solution in the blood and tissues. If the body is rapidly decompressed, the gas comes out of solution, and nitrogen bubbles form in the blood, like the carbon dioxide bubbles that appear in a bottle of soda when you decompress it by removing the top. The nitrogen bubbles lodge in the capillaries, stopping blood flow, or they invade nerves or other tissues. Diving mammals, such as whales and seals, have special adaptations that allow them to stay submerged at high pressures for relatively long periods of time. These adaptations—and their surprising human counterparts—will be the subject of another essay later in the chapter. First, however, we must consider the structure and function of respiratory systems.

TABLE 35–1 **Composition of Dry Air**	
GAS	% OF VOLUME
Oxygen	21
Nitrogen	77
Argon	1
Carbon dioxide	0.03
Other gases*	0.97

* Includes hydrogen, neon, krypton, helium, ozone, xenon, and now, unfortunately, in some environments, radon.

EVOLUTION OF RESPIRATORY SYSTEMS

Oxygen enters and moves within cells by diffusion. Within the cell, as we have noted, it takes part in the oxidation of organic compounds that serve as cellular energy sources. In this process, carbon dioxide is produced, which then diffuses out of the cell down the concentration (partial pressure) gradient. This is true of all cells, whether an amoeba, a *Paramecium*, a liver cell, or a brain cell. But, substances can move efficiently by diffusion only for very short distances (less than 1 millimeter). These limits pose no problem for very small animals, in which each cell is quite close to the surface, or for animals in which much of the body mass is not metabolically active—like the mesoglea ("jelly") of jellyfishes. Many eggs and embryos also obtain oxygen in this simple way, particularly in the early stages of development.

Diffusion, however, cannot possibly meet the needs of large organisms in which cells in the animal's interior may be many centimeters from the air or water serving as the oxygen source. As organisms increased in size in the course of evolution, there also evolved circulatory and respiratory systems that transport large numbers of gas molecules by bulk flow. (Remember that whereas diffusion is the result of the random movement of individual particles, bulk flow is the overall movement of a gas or liquid in response to pressure or gravity.)

An early stage in the evolution of gas-transport systems is exemplified by the earthworm (page 556). As is the case with most other kinds of worms, an earthworm has a network of capillaries just one cell layer beneath the surface of its body. Oxygen and carbon dioxide diffuse directly through the moist body surface into and out of the blood as it travels through these capillaries. The blood picks up oxygen by diffusion as it travels near the surface of the animal and releases oxygen by diffusion as it travels past the oxygen-poor cells in the interior of the earthworm's body. Conversely, blood picks up carbon dioxide from the cells and releases it by diffusion as it travels near the surface of the animal. Thus the gases move into and out of the earthworm by diffusion but are transported within the animal by bulk flow.

This system is particularly suitable for worms because their tube shape exposes a proportionately large surface area. Some worms can adjust their surface area in relation to oxygen supply. If you have an aquarium at home, you may be familiar with tubifex worms, which are frequently sold as fish food. When these worms are placed in water low in oxygen, such as a poorly aerated aquarium, they stretch out to as much as 10 times their normal length, increasing the surface area through which oxygen diffusion occurs and decreasing the diffusion distance.

Insects and some other arthropods, as we have seen (page 568), have evolved a different strategy. Air is piped directly into the tissues by a network of chitin-lined tubules (Figure 35–3). In large insects, diffusion is assisted by body movements, which propel the air into and out of the spiracles by bulk flow. This system is fine for small organisms but is a major limitation on the size that can be obtained by insects and other tracheal-system breathers. The planet may eventually be taken over by cockroaches or ants, but it is safe to bet they will not be the giant forms of science fiction.

Evolution of Gills

Gills and lungs are other ways of increasing the respiratory surface. Gills are usually outgrowths (Figure 35–4), whereas lungs are ingrowths, or cavities. The respiratory surface of the gill, like that of the earthworm, is a layer of cells, one cell thick, exposed to the environment on one side and to circulatory vessels on the other. The layers of gill tissue may be spread out flat, stacked, or convoluted in various ways. The gill of a clam, for instance, is shaped like a steam-heat radiator (which is also designed to provide a high surface-to-volume ratio).

2.5 μm

35–3 *Insects and some other terrestrial arthropods breathe by means of tracheae. These are inward tubular extensions of the exoskeleton, which are thickened in places, forming spiral supports that hold the tubes open. The tracheae are lined with chitin, which is molted when a new exoskeleton is formed. Gas exchange takes place at the thin, moist terminal ends of the tubules. This micrograph shows a portion of the tracheal system that services cells in the wall of the digestive tract of a cockroach.*

The insect respiratory tract, like that of humans, is a fertile breeding ground for infectious organisms. For example, the tracheae of honey bees are often parasitized by mites, leaving the bees somewhat lethargic and greatly reducing their production of honey. Scientists at the U.S. Department of Agriculture have found that treatment with menthol, the soothing ingredient in most cough drops and vapor rubs, kills the mites, enabling the bees to breathe easier and get back to work.

35–4 *Gills are essentially outpocketings of the epithelium that increase the surface area exposed to water. Often gills are covered by an exoskeleton, as in crustaceans, or a flaplike gill cover, as in fish. In the axolotl, the amphibian shown here, the external nature of the gills is clearly evident. The source of their bright red color is blood flowing through dense capillary networks just one cell layer beneath the gill surface.*

The vertebrate gill is believed to have originated primarily as a feeding device. Primitive vertebrates respired mostly through their skin. They filtered water into their mouths and out of what we now call their gill slits, extracting bits of organic matter from the water as it went through. *Branchiostoma* (page 591), which is believed to resemble closely the ancestral vertebrate, feeds in this way.

In the course of time, numerous selection pressures, chiefly involved with predation, came into operation. As one consequence, there was a trend toward an increasingly thick skin, even one armored or covered with scales. Such a skin is not, of course, useful for respiratory purposes. At the same time, related forces were operating to produce animals that were larger and swifter and so more efficient at capturing prey and escaping predators. Such animals had higher energy requirements—and consequently higher oxygen requirements. These problems were solved by the "capture" of the gill for a new purpose: respiratory exchange. The surface area and blood supply of the gill epithelium have slowly increased over the millennia. The modern gill is the result of this evolutionary process.

In most fishes, the water (in which oxygen is dissolved) is pumped in at the mouth by oscillations of the bony gill cover and flows out across the gills (Figure 35–5). In the gills of fish, the circulatory vessels are arranged so that the blood is pumped through them in a direction opposite to that of the oxygen-bearing water. This countercurrent arrangement (see page 130) results in a far more efficient transfer of oxygen to the blood than if the blood flowed in the same direction as the water. Also, the fish can regulate the rate of water flow, and sometimes assist it, by opening and closing its mouth. Fast swimmers, such as mackerel, obtain enough oxygen to meet their energy needs by keeping their mouths open as they swim. As a result, water moves rapidly over the gills. Such fish have become so dependent on this method of respiration that if they are kept in an aquarium or any other space where their motion is limited, they will suffocate.

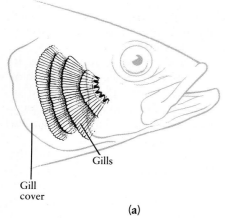

Gill cover

Gills

(a)

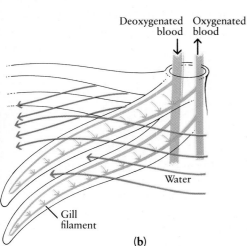

Deoxygenated blood   Oxygenated blood

Water

Gill filament

(b)

35–5 *In fish, oxygen enters the blood by diffusion from water flowing through the gills (**a**). The anatomical structure of the gills maximizes the rate of diffusion, which is proportional not only to the surface areas exposed but also to differences in concentration of the diffusing molecules. The greater the difference in concentration of a molecule, the more rapid its diffusion. (**b**) The circulatory vessels are arranged so that the blood is pumped through them in a direction opposite to that of the oxygen-bearing water. This countercurrent arrangement results in a far more complete transfer of oxygen to the blood than if the blood flowed in the same direction as the water.*

Evolution of Lungs

Lungs are internal cavities into which oxygen-containing air is taken. They have a disadvantage as compared with gills; it is more efficient from the point of view of diffusion to have a continuous flow across the respiratory surface. Air, however, is a far better source of oxygen than is water; 21 percent of the air of the modern atmosphere is oxygen, as compared to 0.5 percent by volume in water at 15°C. Not only must more water be processed to obtain a given amount of oxygen, but water also has a much higher viscosity than air. A fish spends up to 20 percent of its energy in the muscular work associated with respiration, whereas an air breather expends only 1 or 2 percent of its energy in respiration. Also, oxygen diffuses about 300,000 times more rapidly through air than through water, and so can be replenished much more quickly in air as it is used up by respiring organisms.

Lungs are not essential for air breathing. As we have seen, earthworms are air breathers. The overwhelming advantage of lungs, however, is that the respiratory surfaces can be kept moist without a large loss of water by evaporation. Although lungs are largely a vertebrate "invention," they are found in some invertebrates. Land-dwelling snails, for example, have independently evolved lungs that are remarkably similar to the lungs of some amphibians.

Some primitive fishes had lungs as well as gills, although, as we noted in Chapter 28, these lungs were not efficient enough to serve as more than accessory respiratory structures. They were probably a special adaptation to life in fresh water, which, unlike ocean water, may stagnate (become depleted of oxygen). A few species of lungfish still exist (Figure 28–17, page 595). By coming to the surface and gulping air into their lungs, they can live in water that does not have sufficient oxygen to support other fish life.

Amphibians and reptiles have relatively simple lungs, with small internal surface areas, although their lungs are far larger and more complex than those of the lungfish. The lungs of lungfish developed directly from the pharynx, the posterior portion of the mouth cavity, which leads to the digestive tract. In amphibians, reptiles, and other air-breathing vertebrates, we see the evolution of the windpipe, or **trachea,** guarded by a valve mechanism, the glottis, and of nostrils, which make it possible for the animal to breathe with its mouth closed. Amphibians still rely largely on their skin for gas exchange (Figure 35–6), but reptiles breathe almost entirely through their lungs.

An important feature of all vertebrate lungs is that the exchange of air with the atmosphere takes place by bulk flow as a result of changes in lung volume. Such lungs are known as ventilation lungs. Frogs gulp air and force it into their lungs in a swallowing motion; then they open the glottis and let the air out again. In reptiles, birds, and mammals, air is moved into and out of the lungs as a consequence of changes in the size of the thoracic cavity, brought about by muscular contractions and relaxations.

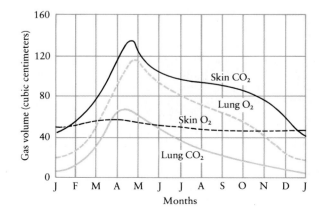

35–6 *Frogs have relatively small and simple lungs, and a major part of their respiration takes place through the skin. The chart shows the results of a study of a frog in which oxygen and carbon dioxide exchanges through the skin and lungs were measured simultaneously for one year. Can you explain why the respiratory activity peaks in April and May and is lowest in December and January?*

35–7 (a) *Although the lungs of birds are small, they are extraordinarily efficient. Each lung has several air sacs attached to it. With each cycle of expiration and inspiration, the air sacs empty and fill like balloons as they are compressed and expanded by movements of the body wall. No gas exchange takes place in the sacs. They appear rather to act as bellows, flushing fresh air through the lungs at every breath, always in the same direction. As a result, there is little residual "dead" air left in the lungs, as there is in mammals. (b) Scanning electron micrograph of lung tissue from a 14-day-old chicken. The tubes visible here are ventilated by air drawn in by the air sacs. Gas exchange takes place in the broad meshwork of air capillaries and blood capillaries, which make up the spongelike respiratory tissue seen here surrounding the tubes.*

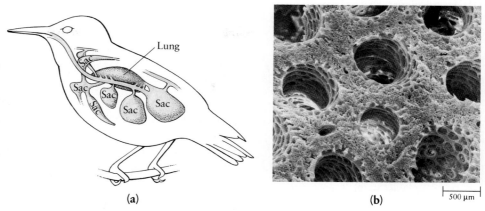

(a)

(b)

500 μm

Respiration in Large Animals: Some Principles

Figure 35–8 summarizes the types of respiratory systems found among the animals. In large animals, both diffusion and bulk flow move oxygen molecules between the external environment and actively metabolizing tissues. This movement occurs in four stages:

1. Movement by bulk flow of the oxygen-containing external medium (air or water) to a thin, moist epithelium close to small blood vessels in lungs or gills.

2. Diffusion of the oxygen across this epithelium into the blood.

3. Movement by bulk flow of the oxygen with the circulating blood to the tissues where it will be used.

4. Diffusion of the oxygen from the blood into the interstitial fluids, from whence it diffuses into the individual cells.

Carbon dioxide, which is produced in the tissue cells, follows the reverse path as it is eliminated from the body.

35–8 *Respiratory systems.* (a) *Gas exchange across the entire surface of the body is found in a wide range of small organisms from protists to earthworms.* (b) *Gas exchange across the surface of a flattened body is seen, for example, in flatworms. Flattening increases the surface-to-volume ratio and also decreases the distance over which diffusion has to occur within the body.*

(c) *External gills increase the surface area but are unprotected and therefore easily damaged. External gills are found in polychaete worms and some amphibians. Gas exchange usually takes place across the rest of the body surface as well.* (d) *With internal gills, a ventilation mechanism draws water over the highly vascularized gill surfaces, as in fish.*

(e) *Gas exchange at the terminal ends of fine tracheal tubes that branch through the body and penetrate all the tissues is characteristic of insects and some other arthropods.* (f) *Lungs are highly vascularized sacs into which air is drawn by a ventilation mechanism. Lungs are found in all air-breathing vertebrates and some invertebrates, such as terrestrial snails.*

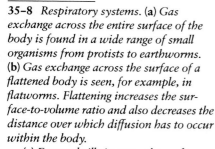

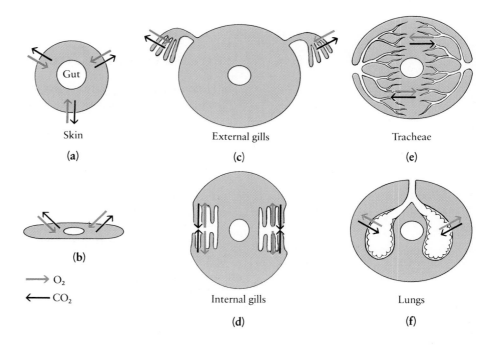

35–9 *The human respiratory system.* **(a)** *Air enters through the nose or mouth and passes into the pharynx, past the larynx, and down the trachea, bronchi, and bronchioles to the alveoli* **(b)** *in the lungs. The alveoli, of which there are some 300 million in a pair of lungs, are the sites of gas exchange. Oxygen and carbon dioxide diffuse into and out of the bloodstream through the capillaries in the walls of the alveoli.*

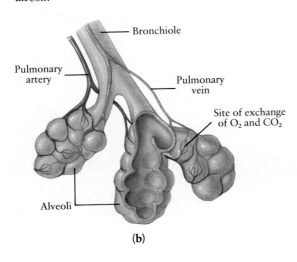

Bronchiole

Pulmonary artery

Pulmonary vein

Site of exchange of O_2 and CO_2

Alveoli

(b)

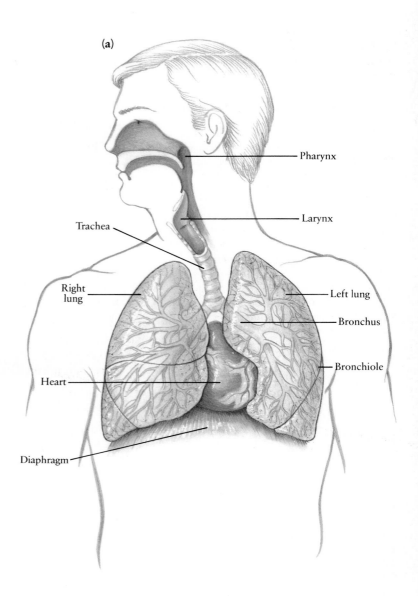

(a)

Pharynx

Larynx

Trachea

Right lung

Left lung

Bronchus

Bronchiole

Heart

Diaphragm

THE HUMAN RESPIRATORY SYSTEM

In *Homo sapiens,* inspiration (breathing in) and expiration (breathing out) usually take place through the nose. The nasal cavities are lined with hairs and cilia, both of which trap dust and other foreign particles. The epithelial cells that line the cavities secrete mucus, which humidifies the air and collects debris that can be removed by swallowing, sneezing, blowing the nose, or spitting. The cavities have a rich blood supply, which keeps their temperature high, warming the air before it reaches the lungs.

From the nasal passages, the air goes to the pharynx and from there to the larynx, located in the upper front part of the neck. An adult human larynx is shaped somewhat like a triangular box, with its point downward. Across it are stretched the vocal cords, which are two ligaments drawn taut across the lumen of the respiratory tract. Vibrations of these cords, by expired air, cause the sounds made in speech. Growth of the vocal cords is influenced by male sex hormones. At puberty, the cords in males become longer and thicker, sometimes so rapidly that the adolescent male temporarily loses control over them, occasionally emitting embarrassing squeaks. Laryngitis, which is simply an inflammation of the vocal cords, interferes with their vibration, and so you "lose your voice."

Cancer of the Lung

Lung cancer is the most rapidly increasing form of cancer in the United States. The death rate from lung cancer among U.S. men and women more than tripled during the period from 1950 to 1985. During that same period, the overall cancer death rate increased in the United States due entirely to an increase in the number of deaths from lung cancer. It is now the most common cause of death from cancer in both men and women, having recently surpassed breast cancer as the most common cause of death from cancer in women.

It is expected that well over 150,000 men and women will develop lung cancer in the United States this year, and more than 85 percent of them will die in less than five years. Most of these patients will be cigarette smokers.

The middle and lower lobes of a cancerous lung are shown in (a). The cancer is the solid grayish-white mass. Its rapid growth has replaced most of the normal lung tissue in the middle lobe. It has also probably begun to spread to other parts of the body. Notice the bronchial branch that leads into the cancer and is destroyed by it. The lung tissue remaining around the cancer is compressed, airless, and dark red. Away from the cancer, in the lower lobe, the lung is filled with air and is light pink.

The lung is normally protected by ciliated cells lining the trachea and bronchi. The cilia sweep out particles in the respired air that are caught in the mucus secreted by the goblet cells. These cilia are paralyzed by chemicals in cigarette smoke. Moreover, cancer cells arising in the bronchi generally do not have cilia. The scanning electron micrographs show (b) the normal ciliated surface of the bronchus and (c) the surface of a bronchus with cancer.

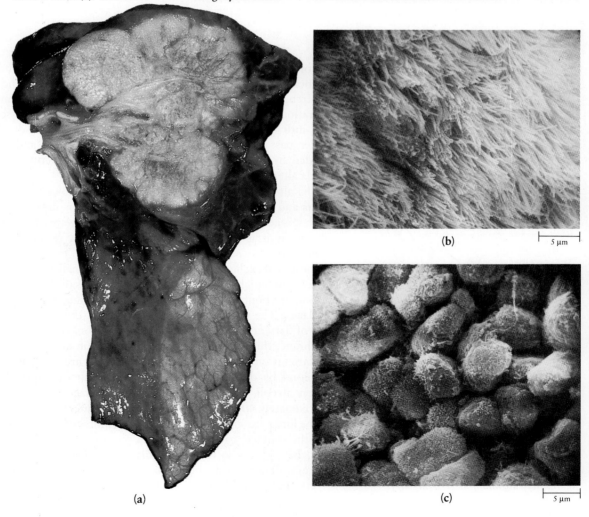

(a)

(b)   5 μm

(c)   5 μm

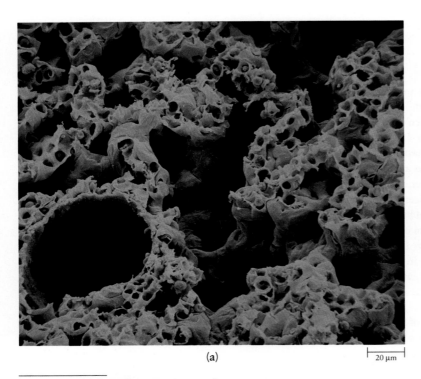

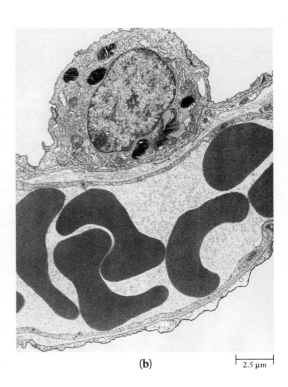

(a) |⊢⎯⎯⊣| 20 µm

(b) |⊢⊣| 2.5 µm

35–10 (a) *Scanning electron micrograph of lung tissue, showing numerous alveoli.* (b) *Longitudinal section of an alveolar capillary. The large, dark, irregularly shaped structures are red blood cells. An alveolar cell is visible at the top.*

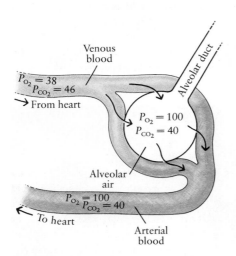

35–11 *Gases are exchanged by diffusion as a consequence of the different partial pressures of oxygen and carbon dioxide in the alveolus and the alveolar capillary. The figures indicate millimeters of mercury.*

From the larynx, inspired air travels through the trachea, which is a long membranous tube, also lined with ciliated epithelial cells. The walls of the trachea are strengthened by rings of cartilage that prevent it from collapsing during inspiration or when food from the adjacent esophagus presses on it. The trachea leads into the **bronchi** (singular, bronchus), which subdivide into smaller and smaller passageways, the **bronchioles.** The bronchi and bronchioles are surrounded by thin layers of smooth muscle. Contraction and relaxation of this muscle alters resistance to air flow.

Cilia along the trachea, bronchi, and bronchioles beat continuously, pushing mucus and foreign particles embedded in mucus up toward the pharynx, from which it is generally swallowed. We are usually aware of this production of mucus only when it is increased above normal as a result of an irritation of the membranes.

The actual exchange of gases takes place in small air sacs, the **alveoli,** which are clustered in bunches like grapes around the ends of the smallest bronchioles. Each alveolus is about 0.1 or 0.2 millimeter in diameter, and each is surrounded by capillaries. The walls of the capillaries and of the alveoli each consist of a single layer of flattened epithelial cells separated from one another by a thin basement membrane; thus the barrier between the air in an alveolus and the blood in its capillaries is only about 0.5 micrometer. Gases are exchanged between the air and the blood by diffusion (Figure 35–11). A pair of human lungs has about 300 million alveoli, providing a respiratory surface of some 70 square meters—approximately 40 times the external surface area of the entire human body.

The lungs are surrounded by a thin membrane known as the pleura, which also lines the thoracic cavity. The pleura secretes a small amount of fluid that lubricates the surfaces so that they slide past one another as the lungs expand and contract. Pleurisy is an inflammation of these membranes that causes them to secrete excess fluid that collects in the thoracic cavity.

35–12 *A model illustrating the way air is taken into and expelled from the lungs, showing the action of the diaphragm.*

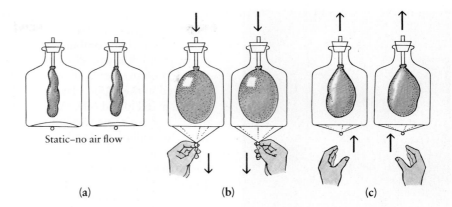

Static–no air flow

(a)   (b)   (c)

MECHANICS OF RESPIRATION

Air flows into or out of the lungs when the air pressure within the alveoli differs from the pressure of the external air (atmospheric pressure). When alveolar pressure is greater than atmospheric pressure, air flows out of the lungs, and expiration occurs. When alveolar pressure is less than atmospheric pressure, air flows into the lungs, and inspiration occurs.

The pressure in the lungs is varied by changes in the volume of the thoracic cavity, as illustrated by the model in Figure 35–12. These changes are brought about by the contraction and relaxation of the muscular diaphragm and of the intercostal ("between-the-ribs") muscles. We inhale by contracting the dome-shaped diaphragm, which flattens it and lengthens the thoracic cavity, and by contracting those intercostal muscles that pull the rib cage up and out. These movements enlarge the thoracic cavity; the pressure within it falls, and air moves into the lungs. Air is forced out of the lungs as the muscles relax, reducing the volume of the thoracic cavity. Usually, only about 10 percent of the air in the lung cavity is exchanged at every breath, but as much as 80 percent can be exchanged by deliberate deep breathing.

Whales and other large aquatic mammals suffocate on land because of the inability of their intercostal muscles to expand their massive chests to inhale when compressed under the weight of their bodies.

TRANSPORT AND EXCHANGE OF GASES

Hemoglobin and Its Function

Oxygen is relatively insoluble in blood plasma (the liquid part of blood)—only about 0.3 milliliter of oxygen will dissolve in 100 milliliters of plasma at normal atmospheric pressure. In insects, which do not depend on their blood to transport oxygen to the individual cells, this low solubility is of little consequence. In other animals, it would be a severe limitation were it not for the presence of special oxygen-carrying protein molecules, known as respiratory pigments, that raise the oxygen-transporting capacity of the blood as much as seventyfold. Such pigments are found in the blood of virtually all active animals except insects, including even the earthworm.

Hemoglobin is the respiratory pigment found among vertebrates and in a wide variety of invertebrate species representing many different phyla. A form of

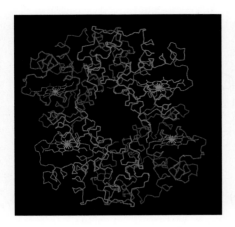

35-13 *A computer-generated model of the hemoglobin molecule. The turquoise lines represent the carbon backbones of the two alpha polypeptide chains, and the dark blue lines represent the carbon backbones of the two beta polypeptide chains. The porphyrin rings are shown in red, and the iron atoms as yellow starbursts.*

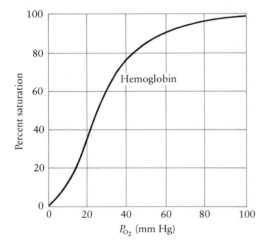

35-14 *Oxygen-hemoglobin association-dissociation curve. This curve represents figures for normal adult human hemoglobin at 38°C and at a normal pH. As the partial pressure of oxygen rises, hemoglobin picks up oxygen. When oxygen pressure reaches 100 mm Hg—the pressure usually present in the human lung—the hemoglobin becomes totally saturated with oxygen. As the P_{O_2} drops, the oxygen and the hemoglobin dissociate. Therefore, when oxygen-carrying blood reaches the capillaries, where pressure is only about 40 mm Hg or less, it gives up some of its oxygen to the tissues.*

oxygen carrier (hemocyanin), which contains copper rather than iron, is the most common respiratory pigment of mollusks and arthropods. (Unlike hemoglobin, which is red when combined with oxygen, hemocyanin is blue.) Other respiratory pigments are known, all of which are a combination of a metal-containing unit and a protein chain. In most invertebrates, respiratory pigments are simply dissolved in the blood plasma; in vertebrates and echinoderms, the pigments are carried in red blood cells. As we shall see in the next chapter, these cells are highly specialized for their transport function; a mature human red blood cell carries some 265 million molecules of hemoglobin.

Hemoglobin, as you will recall, is made up of four subunits, each of which comprises a heme unit and a polypeptide chain (Figure 35-13). The heme unit consists of a porphyrin ring with one atom of iron in the center. The iron in each heme unit can combine with one molecule of oxygen; thus each hemoglobin molecule can carry four molecules of oxygen. The oxygen molecules (O_2) are added one at a time:

$$Hb_4 + O_2 \rightleftharpoons Hb_4O_2$$

$$Hb_4O_2 + O_2 \rightleftharpoons Hb_4O_4$$

$$Hb_4O_4 + O_2 \rightleftharpoons Hb_4O_6$$

$$Hb_4O_6 + O_2 \rightleftharpoons Hb_4O_8$$

Combination of the first subunit (Hb) with O_2 increases the affinity of the second for O_2, and oxygenation of the second increases the affinity of the third, and so on. (As O_2 is taken up, the two beta polypeptide chains of hemoglobin move closer together, and this movement is apparently the reason the shift in affinity takes place.) As a consequence of the change in affinity of the hemoglobin molecule for oxygen, the curve relating the uptake of oxygen to P_{O_2} is not a straight line, but has a characteristic sigmoid shape (Figure 35-14). When hemoglobin is fully oxygenated, it enables our bloodstreams to carry about 65 times as much oxygen as could be transported by an equal volume of plasma alone.

Whether oxygen combines with hemoglobin or is released from it depends on the partial pressure of oxygen (P_{O_2}) in the surrounding blood plasma. Oxygen diffuses from the air into the alveolar capillaries. In these capillaries, where the P_{O_2} is high, most of the hemoglobin is combined with oxygen. In the tissues, however, where the P_{O_2} is lower, oxygen is released from the hemoglobin molecules into the plasma and diffuses into the tissues. This system compensates automatically for the oxygen requirements of the tissues. For example, in adult humans, the P_{O_2} as the blood leaves the lungs is about 100 mm Hg; at this pressure, the hemoglobin is saturated with oxygen. As the hemoglobin molecules travel through the tissue capillaries, the P_{O_2} drops, and as it drops, the oxygen bound to the hemoglobin molecules is given up. Little oxygen is yielded as the P_{O_2} drops from 100 mm Hg to 60 mm Hg. (This built-in safety factor ensures that oxygen is delivered to the tissues, even when the maximum blood P_{O_2} is lower than normal, as, for example, in individuals who live at high altitudes or who have heart and lung diseases.) However, as the P_{O_2} drops below 60 mm Hg, oxygen is given up much more readily (Figure 35-15).

The P_{O_2} of the blood in the tissue capillaries is normally about 40 mm Hg. As a consequence, even after the blood has passed through the tissue capillaries, its hemoglobin is still usually 70 percent saturated. The oxygen still carried by hemoglobin represents a reserve supply of this precious gas should the demand increase—as a result, for example, of exercise. In such a situation, the cells respire more rapidly, more oxygen is consumed, the P_{O_2} decreases, and more oxygen is released from the hemoglobin molecules.

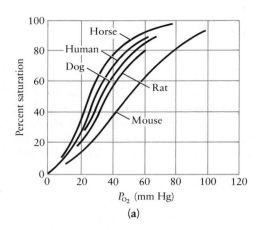

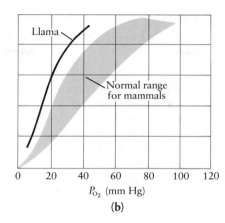

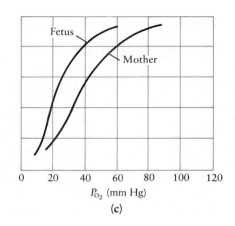

35–15 *These curves show how the amount of oxygen carried by the hemoglobin is related to oxygen pressure in a number of different mammals. A curve located to the right of another curve signifies that the oxygen is given up more readily at a given pressure. (a) Small animals have higher metabolic rates and so need more oxygen per gram of tissue than larger animals. Therefore, they have blood that gives up oxygen more readily. (b) The llama, which lives in the high Andes of South America, has a hemoglobin that enables its blood to take up oxygen more readily at the low atmospheric pressures. (c) The fetus must take up all its oxygen from the maternal blood. The hemoglobin of mammalian fetuses has a greater affinity for oxygen than does the hemoglobin of adult mammals, and so the oxygen tends to diffuse from the maternal blood to the fetal blood.*

Carbon dioxide is more soluble than oxygen in the blood, and a small amount of it is simply dissolved in the plasma. In addition, about 25 percent of it is bound to the amino groups of the hemoglobin molecules. However, most of the carbon dioxide (about 65 percent) is carried in the blood as bicarbonate ion (HCO_3^-). Bicarbonate ion is produced in a two-stage reaction. First, carbon dioxide combines with water to form carbonic acid. This reaction is catalyzed by the enzyme carbonic anhydrase found in red blood cells. Then carbonic acid, a weak acid, dissociates to yield bicarbonate and hydrogen ions:

$$CO_2 + H_2O \underset{\text{anhydrase}}{\overset{\text{Carbonic}}{\rightleftharpoons}} \underset{\substack{\text{Carbonic} \\ \text{acid}}}{H_2CO_3} \rightleftharpoons \underset{\substack{\text{Bicarbonate} \\ \text{ion}}}{HCO_3^-} + \underset{\substack{\text{Hydrogen} \\ \text{ion}}}{H^+}$$

As you can see by the arrows, this reaction can go in either direction. The direction it actually takes depends on the partial pressure of carbon dioxide in the blood. In the tissues, where the partial pressure of carbon dioxide is high, bicarbonate and hydrogen ions are formed. In the lungs, where the partial pressure of carbon dioxide is low, carbonic acid dissociates to form carbon dioxide and water. Once it is released, the carbon dioxide diffuses from the plasma into the alveoli and flows out of the lung with the expired air.

The partial pressure of carbon dioxide in the blood also affects hemoglobin's affinity for oxygen. In the tissues, where more carbon dioxide is taken up by the blood, the blood becomes increasingly acidic. As the acidity increases, hemoglobin's affinity for oxygen decreases and so it gives up its oxygen more readily.

Table 35–2 provides a quantitative illustration of the gas exchanges occurring in the lung during the respiratory process.

TABLE 35–2 **Composition of Respiratory Gas at Standard Atmospheric Pressure**

GAS	INSPIRED AIR		EXPIRED AIR		ALVEOLAR AIR	
	% OF VOLUME	PARTIAL PRESSURE (mm Hg)	% OF VOLUME	PARTIAL PRESSURE (mm Hg)	% OF VOLUME	PARTIAL PRESSURE (mm Hg)
O_2	20.71	157	14.6	111	13.2	100
CO_2	0.04	0.3	4.0	30	5.3	40
H_2O	1.25	9.5	5.9	45	5.9	45
N_2	78.00	593	75.5	574	75.6	574

Diving Mammals

Most mammals have about the same oxygen requirements as humans. Structurally, their respiratory organs are very similar. Yet marine mammals can dive to great depths and stay submerged for long periods of time. A sperm whale, for example, has been found at a depth of more than 1,000 meters and has been known to stay submerged for 75 minutes. A Weddell seal, a much smaller animal, can dive to 600 meters and stay submerged for 70 minutes, or can swim submerged for between 2,000 and 4,000 meters—1 to 2 miles. Are their lungs different from ours? Is their blood different? Do they have special oxygen reserves, like a submarine?

Studies in a variety of diving mammals have shown than none has lungs significantly larger, in proportion, than ours. In fact, seals (and perhaps other diving mammals as well) exhale before diving or early in the dive. Blood volume is greater, however. In humans, blood is about 7 percent of the body weight, whereas in diving marine mammals, it is 10 to 15 percent. The blood vessels are proportionately enlarged, and they appear to serve as a reservoir of oxygenated blood. Moreover, the proportion of red blood cells is higher and myoglobin is more concentrated, giving the muscles a very dark color. Still, physiologists calculate, these adaptations would not provide enough oxygen for long dives.

A major survival factor in diving mammals, studies have shown, is a group of automatic reactions known collectively as the diving reflex. During a dive, the heart rate slows and blood supply is reduced to one-tenth or one-twentieth of normal to tissues that are tolerant of oxygen deprivation, such as the digestive organs, skin, and muscles. Muscles obtain energy anaerobically, by glycolysis, producing large amounts of lactic acid and building up an oxygen debt (see page 190). Most of the oxygen is shunted to the heart and brain, whose cells would begin to die after about four minutes without oxygen. Blood supply is reduced only slightly to the adrenal glands, which produce hormones that are involved in regulation of the metabolic rate, the ionic balance of body fluids, and the rate of heartbeat. In a pregnant female, the fetus is also given high priority for the available oxygen.

Studies by Martin Nemiroff of the University of Michigan Medical School indicate that the diving reflex is also present in human beings. In humans, its primary survival value is during birth, when the infant may be cut off from an oxygen supply during the later stages of labor. Operation of the reflex can be demonstrated in a very simple way: immersing the face in cold water causes the heart rate to slow down.

The observations by Nemiroff came in the course of investigating 60 near-drownings. It has been generally assumed that a person who is submerged for four minutes or more will die, or, at best, suffer irreversible brain damage. In the group of 60, 15 had been rescued after four minutes or more in the chilly waters of Michigan lakes. Of these 15, 11 survived without brain damage. One of these, a college student who crashed through the ice in an automobile accident, was under water for 38 minutes; he finished the semester with a 3.2 average. Another survivor resumed his career as a practicing physician. All the survivors had been submerged in cold (below 21°C) water, which slowed the metabolic demands of their cells and therefore reduced their need for oxygen. Some of those rescued required assisted breathing for as long as 13 hours.

Moral: Even if a drowning victim is cold, blue, with no pulse, no heartbeat, and the eyes fixed in a glassy stare—as was the case with the student—if he or she has been in cold water, efforts at resuscitation should be started immediately and continued indefinitely in a hospital. We are not Weddell seals, but owing to an ancient survival strategy, we have unexpected powers of underwater survival.

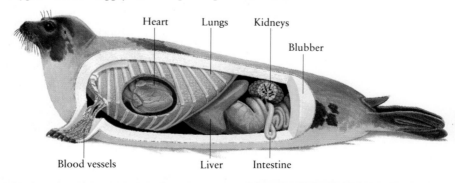

A seal exhales as it dives, reducing the likelihood of decompression sickness (the "bends"). It has 1.5 times as much blood as a land mammal of the same size and, when it dives, can reduce its heartbeat from 120 to 30 beats per minute.

Heart   Lungs   Kidneys

Blubber

Blood vessels   Liver   Intestine

35–16 *Tertiary structure of myoglobin, as deduced from x-ray diffraction analyses. Myoglobin closely resembles a single subunit of the four-subunit hemoglobin molecule. The heme group, to which the oxygen molecule binds, is shown in red.*

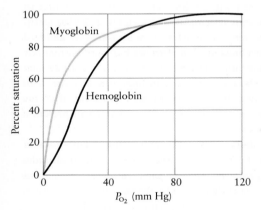

35–17 *Comparison of the oxygen associ-ation-dissociation curves of myoglobin and hemoglobin. Note that myoglobin remains almost 80 percent saturated with oxygen until the partial pressure of oxy-gen falls below 20 mm Hg. Therefore, myoglobin retains its oxygen in the resting cell and relinquishes it only when strenu-ous muscle activity uses up the available oxygen provided by hemoglobin.*

Myoglobin and Its Function

Myoglobin is a respiratory pigment found in skeletal muscle. Structurally, it resembles a single subunit of the hemoglobin molecule (Figure 35–16). Myoglo-bin's affinity for oxygen is greater than hemoglobin's and so it picks up oxygen from hemoglobin. It begins to release significant amounts of this oxygen only when the P_{O_2} in skeletal muscle falls below 20 mm Hg (Figure 35–17). Thus, when the muscle is at rest or engaged in only moderate activity, the myoglobin holds on to its oxygen. During strenuous exercise, however, when muscle cells are using oxygen rapidly and the partial pressure of oxygen in the muscle cells drops to zero, myoglobin gives up its oxygen. Thus myoglobin provides an additional reserve of oxygen for active muscles.

CONTROL OF RESPIRATION

The rate and depth of respiration are controlled by respiratory neurons in the brainstem. These neurons are responsible for normal breathing, which is rhythmic and automatic, like the beating of the heart. Unlike the beating of the heart, however, which few of us can control voluntarily, breathing may be brought under voluntary control within certain limits.

The respiratory neurons in the brain activate motor neurons in the spinal cord, causing the diaphragm and intercostal muscles to contract. This activity of the respiratory neurons is thought to occur spontaneously. Periodically, however, these neurons are inhibited, allowing expiration to occur. In addition to their own spontaneous activity, the respiratory neurons receive signals from receptors sensi-tive to carbon dioxide, oxygen, and hydrogen ions, as well as from receptors sensitive to the degree of stretch of the lungs and chest. Chemoreceptor cells located in the carotid arteries (Figure 35–18, page 747), which supply oxygen to the brain, signal the respiratory neurons when the concentration of oxygen in the blood decreases. The concentration of dissolved carbon dioxide and of hydrogen ion is simultaneously monitored by centers in the brain and also by chemorecep-tors in the carotid arteries. Thus, information is provided by a number of different, independent sensors.

High on Mt. Everest

High altitudes are harsh, dangerous environments, and their most dangerous feature is not the cold, the precipitous slopes, the threat of avalanche, or the blinding wind and snow; it is oxygen deprivation. For this reason, the ability of humans and other animals to adjust to high altitudes has long been of particular interest to physiologists.

Until very recently, about 6,000 meters was thought to be the limit for human survival. In 1978, however, two European climbers reached the summit of Mt. Everest (8,848 meters) without supplemental oxygen, raising new questions about physiological adaptability. Three years later, the American Medical Research Expedition to Everest set off with the goal of collecting the first data on human physiological function above 6,000 meters. The team included six highly experienced Himalayan climbers; a group of six "climbing scientists," all physicians with much climbing experience and an interest in high-altitude physiology; and eight physiologists who worked at the base camp at 5,400 meters and at a laboratory at 6,300 meters. Members of the expedition were able to make measurements at above 8,000 meters on human subjects (themselves), including continuous monitoring of the electrical activity of the heart and sampling of the gases in the alveoli—even at the summit itself, which two of the monitored "climbing scientists" reached.

Survival at this extreme altitude, the scientists found, depends largely on hyperventilation—extremely deep breathing. In fact, there seems to be a correlation between the capacity to hyperventilate and the capacity to be a mountain climber. This extremely deep breathing results in an astonishing decrease in the partial pressure of carbon dioxide in the lungs and bloodstream to less than one-fifth normal levels, with a corresponding increase in the pH of the blood. Even with hyperventilation, the partial pressure of oxygen in arterial blood is less than one-third the partial pressure at sea level, and work capacity is greatly diminished. The European climbers, for example, reported that they moved only 2 meters per minute as they approached the summit.

There were also changes in metabolism and brain function. At 6,300 meters, there was a striking loss of body weight, with two of the expedition members each losing 15 kilograms (33 pounds). The concentration of thyroid hormone, which increases the rate of cellular respiration, increased with increasing altitude. The levels of noradrenaline, a compound that functions both as a hormone and as a transmitter in the nervous system, were also elevated. Verbal learning and short-term memory, as measured by standard tests, declined at high altitudes but were normal one year later. In a simple brain function test, all 16 men tested showed a decrease in finger-tapping speed. In 15 of the 16, this abnormality persisted after the expedition, and in 13 of them, it was still present a year later.

In short, despite the remarkable adaptations of which the body is capable, extremely high altitudes have profound and, in some instances, permanent physiological effects. The summit of Mt. Everest is very close to the limits of human survival.

The main laboratory of the American Medical Research Expedition to Everest, at an altitude of 6,300 meters. An altitude of about 5,300 meters appears to be the limit for long-term human habitation.

This system is extremely sensitive to even the smallest change in the chemical composition of the blood, particularly to the concentration of hydrogen ion, which reflects the carbon dioxide concentration (P_{CO_2}), as shown in the equation on page 743. If P_{CO_2}—and therefore the concentration of H$^+$ ions—increases only slightly, breathing immediately becomes deeper and faster, permitting more carbon dioxide to leave the blood until the concentration of H$^+$ ions has returned to normal. If you deliberately hyperventilate (breathe deeply and rapidly) for a few moments, you will feel faint and dizzy because of the blood's (and, therefore, the brain's) increased alkalinity.

You can, as we have noted, deliberately increase your breathing rate by contracting and relaxing your diaphragm and chest muscles, but breathing is normally involuntary. It is impossible to commit suicide by deliberately holding your breath; as soon as you lose consciousness and P_{CO_2} rises, the involuntary controls take over once more and breathing resumes. The receptors sensitive to P_{O_2} provide a kind of backup system to the P_{CO_2} and H$^+$ sensors. In some cases of drug poisoning—for example, by morphine or barbiturates—the brainstem cells sensitive to H$^+$ become depressed. This causes a decrease in the breathing rate, leading ultimately to a reduction in the P_{O_2} of the blood. The oxygen sensors are then stimulated, and they maintain breathing. (Massive overdoses of these drugs, however, depress the activity of the respiratory neurons themselves.) This complex system of sensors, monitoring different factors in different locations, underlines the critical importance of an uninterrupted supply of oxygen to the cells of an animal's body—particularly those of the brain.

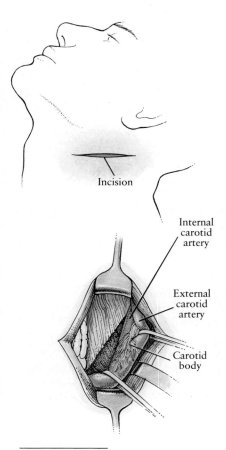

35–18 *Location of the carotid body, one of the receptors that monitor the concentration of dissolved oxygen* (P_{O_2}) *in the blood. Monitoring at this location is of critical importance, because the carotid arteries supply blood to the brain. To a lesser extent, the carotid body also monitors the concentrations of hydrogen ion* (H$^+$) *and carbon dioxide* (P_{CO_2}). *Hydrogen ion concentrations are also measured directly by neurons in the brain.*

Incision

Internal carotid artery

External carotid artery

Carotid body

SUMMARY

Heterotrophic cells obtain energy from the oxidation of carbon-containing compounds. This process releases carbon dioxide and, for maximum energy yields, requires oxygen. Respiration is the means by which an animal obtains oxygen for its cells and rids itself of carbon dioxide.

Oxygen is available in both water and air. It enters cells and body tissues by diffusion, moving from regions of higher partial pressure to regions of lower partial pressure. However, efficient movement of oxygen by diffusion requires a relatively large surface area exposed to the source of oxygen and a short distance over which the oxygen has to diffuse. Selection pressures for increasingly efficient means of gas exchange led to the evolution of gills and lungs. Both gills and lungs present enormously increased surface areas for the exchange of gases. They also have a rich blood supply for transporting these gases to and from other parts of the animal's body.

Respiration in large animals involves both diffusion and bulk flow. Bulk flow brings air or water to the lungs or gills and circulates oxygen and carbon dioxide in the bloodstream. Gases are exchanged by diffusion between the blood and the air in the lungs or the water around the gills and between the blood and the tissues.

In humans, air enters the lungs through the trachea, or windpipe, and goes from there into a network of increasingly smaller tubules, the bronchi and bronchioles, which terminate in small air sacs, the alveoli. Gas exchange actually takes place across the alveolar walls. Air moves into and out of the lungs as a result of changes in the pressure within the lungs, which, in turn, result from changes in the size of the thoracic cavity.

Respiratory pigments increase the oxygen-carrying capacity of the blood. In vertebrates, the respiratory pigment is hemoglobin, which is packed within red blood cells. Each hemoglobin molecule has four subunits, each of which can combine with one molecule of oxygen. The addition of each molecule of oxygen increases the affinity of the molecule for each subsequent molecule of oxygen.

Conversely, the loss of each molecule of oxygen facilitates the loss of the subsequent one. Carbon dioxide is transported in the blood plasma principally in the form of bicarbonate ion.

The rate and depth of respiration are controlled by respiratory neurons in the brainstem. These neurons, which activate motor neurons in the spinal cord that cause the diaphragm and intercostal muscles to contract, respond to signals caused by very slight changes in the hydrogen ion, carbon dioxide, and oxygen concentrations of the blood.

QUESTIONS

1. Distinguish among the following: gills/lungs; trachea/pharynx/larynx; bronchi/bronchioles/alveoli; hemoglobin/myoglobin/hemocyanin.

2. Why are calculations of partial pressure used in determining the movement of gases between liquids and the atmosphere rather than concentrations? When does a gas move from an area of lower concentration to one of higher concentration?

3. What are the advantages and disadvantages of obtaining oxygen from air rather than from water? You might be able to think of several of each besides those mentioned in the text.

4. Sketch and label a diagram of the human respiratory system. When you have finished, compare your drawing to Figure 35–9.

5. Explain the following statement: Frogs ventilate their lungs by positive pressure, whereas mammals, birds, and reptiles ventilate theirs by negative pressure.

6. One of the results of long-term smoking is the loss of bronchial cilia. What effects would you expect this to have on normal lung function?

7. Suppose a new cold remedy is guaranteed to suppress completely the secretion of mucus in the respiratory tract. Explain why you would or would not use this remedy during a cold.

8. Carbon monoxide (CO), which is extremely poisonous, has a greater affinity for hemoglobin than does oxygen. The resulting compound, which is a brighter red than normal hemoglobin, can no longer combine with oxygen. From these facts, suggest how you might recognize and give assistance to a victim of carbon monoxide poisoning.

Energy and Metabolism III: Circulation

In all animals except those of very small size or with an extremely simple body plan, blood is the chemical "highway" interconnecting the multitude of cells that form the organism's body. It is the medium in which the nutrient molecules processed by digestion and the oxygen molecules taken in by respiration are delivered to the individual cells. It also carries away the waste materials, including carbon dioxide and urea, produced by cells in the course of their metabolic activities. As we have seen, carbon dioxide picked up by the blood leaves the body by diffusion across the respiratory surfaces. In vertebrates, urea and other wastes are processed in the kidney and excreted from the body, as we shall see in the next chapter. In addition, the blood transports other important substances, such as hormones, enzymes, and antibodies, and has among its basic constituents the cells that defend the body against foreign invaders.

In this chapter, we shall first look at the composition of the blood and then consider the structure and dynamics of the systems by which blood and another fluid, lymph, are transported in the vertebrate body.

36–1 *The exchange of substances between the circulating blood and the body tissues takes place through the walls of the smallest blood vessels, the capillaries. The capillaries are so narrow that the red blood cells, with their oxygen-laden hemoglobin molecules, must move in single file.*

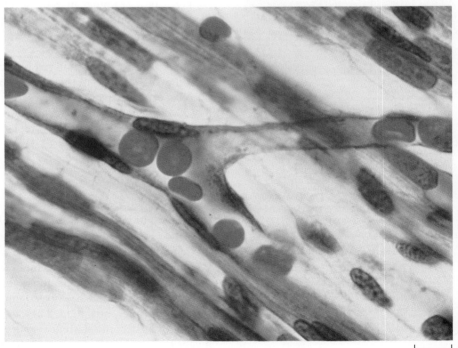

10 μm

THE BLOOD

An individual weighing 75 kilograms (165 pounds) has about 6 liters of blood, which is approximately 8 percent of body weight. About 60 percent of the blood is a straw-colored liquid called **plasma,** which is 90 percent water. The other 40 percent of the blood that is not plasma is made up of the red blood cells, white blood cells, and platelets. Blood, as we noted in Chapter 33, is classified as a type of connective tissue.

Plasma

With the exception of the oxygen and carbon dioxide carried by hemoglobin, most of the molecules needed by individual cells, as well as waste products from the cells, are carried along in the heavy traffic of the bloodstream dissolved in the plasma. In addition, the plasma contains protein molecules, known as plasma proteins, that are not nutrients or waste products but function instead in the bloodstream itself. Because of the dissolved plasma proteins, the osmotic potential (page 133) of the blood is greater than that of the surrounding interstitial fluid. Thus these proteins have the effect of preventing excessive loss of fluid from the bloodstream to the tissues. They also serve to bind certain ions and small molecules, preventing them from leaving the bloodstream, and to transport fats, cholesterol, and otherwise insoluble molecules in the bloodstream.

Plasma proteins are of three major types: albumin, fibrinogen, and globulins. The chief function of albumin is to maintain the high osmotic potential of the plasma in relation to the interstitial fluid, whereas fibrinogen is responsible for blood clotting. Among the most important of the globulins are the immunoglobulins, or antibodies (to be discussed in Chapter 39).

Red Blood Cells

Red blood cells, or **erythrocytes** (Figure 36–2), are specialized for the transport of oxygen. As a mammalian red blood cell matures, it extrudes its nucleus and mitochondria, and its other cellular structures dissolve. Almost the entire volume of a mature red blood cell is filled with hemoglobin, about 265 million molecules per cell in humans.

There are about 5 million red blood cells per cubic millimeter of blood—some 25 trillion (25×10^{12}) in the adult human body. Because red blood cells, lacking a nucleus, cannot repair themselves, their life span is comparatively short, some 120 to 130 days. At this moment, in your body, red blood cells are dying at a rate of about 2 million per second; to replace them, new ones are being formed in the bone marrow at the same rate.

White Blood Cells

For every 1,000 red blood cells in the human bloodstream, there are 1 or 2 white blood cells, or **leukocytes,** for a total of about 6,000 to 9,000 per cubic millimeter of blood. These cells are nearly colorless, are larger than red blood cells, contain no hemoglobin, and have a nucleus.

The chief function of the white blood cells is the defense of the body against invaders such as viruses, bacteria, and other foreign particles. Unlike red blood cells, white blood cells are not confined within the blood vessels but can migrate out into the interstitial fluid. They appear spherical in the bloodstream, but in the tissues they become flattened and amoeba-like. Like amoebas, they move by means of pseudopodia, and many are phagocytic (Figure 36–3). As we shall see in Chapter 39, certain types of white blood cells play key roles in the immune response.

10 µm

36–2 *In vertebrates, oxygen is transported in red blood cells, shown here in a scanning electron micrograph. Red blood cells are about 7 or 8 micrometers in diameter, significantly larger than some of the smallest capillaries, which are only 5 micrometers in diameter. The passage of the red blood cells through the capillaries is made possible by their "donut without a hole" shape, which enables them not only to twist and turn but also to fold. This triumph of biological form also provides a large surface area through which oxygen can diffuse to and from the hemoglobin molecules contained within the cell.*

36–3 *A scanning electron micrograph of a human white blood cell entrapping bacterial cells. This type of cell defends the body against pathogens and other harmful particles by extending pseudopodia to the foreign objects, which are then engulfed within phagocytic vacuoles and destroyed with the help of enzymes from the cell's lysosomes (page 118).*

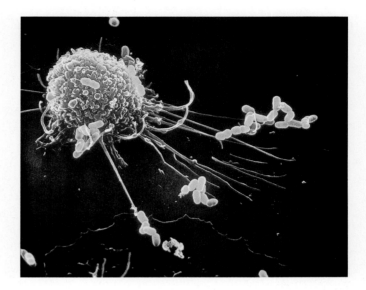

White blood cells are often destroyed in the course of fighting infection. Pus is composed largely of these dead cells. New white blood cells to take the place of those that are destroyed are formed constantly in the spleen, in bone marrow, and in certain other tissues.

Platelets

Platelets, so called because they look like little plates, are colorless, oval or irregularly shaped disks smaller than red blood cells (about 3 micrometers in diameter). Platelets are membrane-bound cytoplasmic fragments of unusually large cells, megakaryocytes, found in the bone marrow. They are, in effect, little bags of chemicals that play an essential role in initiating the clotting of blood and in plugging breaks in the blood vessels.

Blood Clotting

The clotting of blood is a complex phenomenon, requiring platelets and at least 15 factors normally present in the bloodstream or on cell membranes. The sequence of events begins when plasma encounters a rough surface or a protein molecule known as tissue factor. Tissue factor, which is thought to have many regulatory roles throughout the body, is found on the outer surface of many different cell types—but not on the cells of the inner lining of the blood vessels. When tissue factor reacts with a specific circulating plasma protein—as can occur when a blood vessel is broken—a cascade of chemical reactions is initiated. In this cascade, the product of each step of the reaction series acts as a catalyst for the next step, and the molecules involved are, like enzymes, used over and over. The result is that at each step in the series the number of molecules is amplified. Ultimately, a molecule called thromboplastin is activated. Thromboplastin acts to convert prothrombin, a plasma protein produced in the liver, to its active form, the enzyme thrombin:

$$\text{Prothrombin} \xrightarrow{\text{Thromboplastin}} \text{Thrombin}$$

Thrombin, in turn, converts fibrinogen, a soluble plasma protein, to fibrin:

$$\text{Fibrinogen} \xrightarrow{\text{Thrombin}} \text{Fibrin}$$

36-4 *An early stage in the formation of a blood clot. Note the fibers, composed of a protein known as fibrin, in which the red blood cells become enmeshed. Ultimately, the completed clot forms an impenetrable barrier, preventing both the loss of vital fluids and the entry of infectious organisms.*

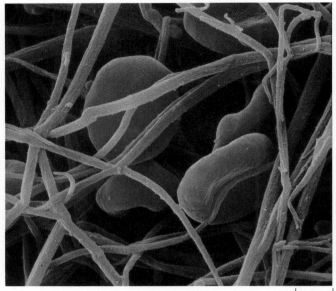

2.5 μm

The fibrin molecules clump together, forming an insoluble network that enmeshes red blood cells and platelets to form a clot (Figure 36–4). The clot contracts, pulling together the edges of the wound. In a typical clotting reaction, some 2 million thrombin molecules are produced from prothrombin, leading to the production of about 160 billion molecules of fibrin.

Clotting reactions also take place when blood is removed from the body and placed in a test tube, apparently triggered in some way by the contact with a foreign surface. The clear fluid that remains after the clot contracts is known as the blood serum. In a test tube, the clotting reactions always proceed inexorably to their conclusion unless anticoagulant chemicals are added. In the body, however, the reactions are precisely regulated by mechanisms that still remain mysterious. Were they not so regulated, we could, according to some calculations, clot from head to toe in 90 seconds after initiation of the reaction cascade.

Hemophilia is a group of genetically determined diseases that affect clotting (see page 392). In the most common type of hemophilia, one of the factors involved in the chain of reactions required to activate prothrombin, Factor VIII, occurs in a defective form. As a result, the blood does not clot easily. Hemophiliacs lacking Factor VIII are treated with Factor VIII extracted from normal blood. Such treatment, although lifesaving, also carries the risk of hepatitis, AIDS, and other blood-borne infectious diseases. Work is now under way, using recombinant DNA techniques, to develop a genetically engineered Factor VIII that will carry no risk of infection.

THE CARDIOVASCULAR SYSTEM

As we saw in Section 4, the systems through which blood is transported in animals vary in their structure and complexity. In the earthworm, for example, blood circulates in a closed system of continuous vessels; it is propelled by the contractions of "hearts" that are little more than expanded, muscular portions of particular blood vessels (see Figure 26–14, page 556). Mollusks, by contrast, have a much more complex heart, consisting of several chambers, but most of them have an open circulatory system: blood is pumped through vessels that open into spaces within the tissues, from which it returns to the gills and then to the heart.

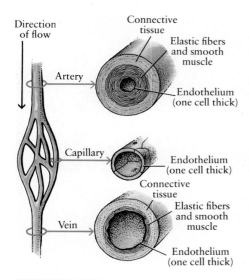

36-5 *Structure of blood vessels. Arteries have thick, tough, elastic walls that can withstand the high pressure of the blood as it leaves the heart. Capillaries have walls only one cell thick. Exchange of gases, nutrients, and wastes between the blood and the cells of the body takes place through these thin capillary walls. Veins usually have larger lumens (passageways) and always have thin, readily extensible walls that minimize resistance to the flow of blood on its return to the heart.*

The vertebrate heart, like that of the mollusks, is a muscular organ consisting of several chambers, but the blood vessels form a closed system, similar in principle to that of the earthworm, but considerably more elaborate. The heart and vessels together are known as the **cardiovascular system** (from *cardio,* meaning "heart," and *vascular,* meaning "vessel").

THE BLOOD VESSELS

In the cardiovascular system, the heart pumps blood into the large arteries, from which it travels to branching, smaller arteries, then to the smallest arteries (the **arterioles**), and then into networks of very small vessels, the **capillaries.** From the capillaries, the blood passes into small veins, the **venules,** then into larger veins, and through them, back to the heart.

In humans, the diameter of the opening of the largest artery, the **aorta,** is about 2.5 centimeters, that of the smallest capillary only 5 micrometers, and that of the largest vein, the **vena cava,** about 3 centimeters. Arteries, veins, and capillaries differ not only in their size but also in the structure of their walls (Figure 36–5). Of the three types of vessels, the arteries have the thickest, strongest walls, made up of three layers. The inner layer, or **endothelium** (a type of epithelial tissue), forms the lining of the vessels; the middle layer contains smooth muscle and elastic tissues; the outer layer, also elastic, is made of collagen and other supporting tissues. Because of their elasticity, the arteries stretch when blood is pumped into them and then recoil slowly. The pulsation felt when the fingertips are placed over an artery close to the body surface—as in the wrist—represents the alternating expansion and recoil of an elastic arterial wall. The walls of the veins, like those of the arteries, are three-layered, but they are thinner, less elastic, and more pliable (Figure 36–6). An empty vein collapses, whereas an empty artery remains open.

The Capillaries and Diffusion

The heart, arteries, and veins are, in essence, the means for getting the blood to and from the capillaries, where the actual function of the circulatory system is carried out. The walls of the capillaries consist of only one layer of cells (Figure 36–7), the endothelium, and the lumen (passageway) of the smallest capillaries is, as we have noted, 5 micrometers in diameter, just wide enough for twisted and folded red blood cells to move in single file (see Figure 36–1). The total length of the capillaries in a human adult is more than 80,000 kilometers (50,000 miles).

36-6 *In this scanning electron micrograph, a medium-sized vein is at the left and a smaller artery at the right. The artery can be identified by its thick, muscular wall. Connective tissue surrounds the two blood vessels and holds them in place.*

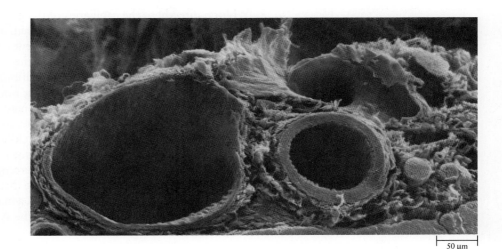

50 μm

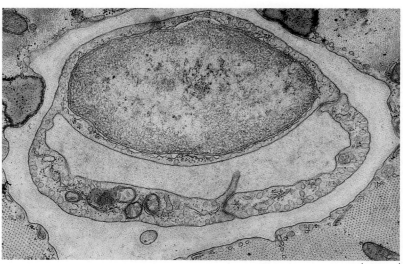

0.5 μm

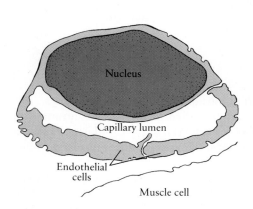

36–7 *Electron micrograph of a capillary from cardiac muscle. Portions of two endothelial cells can be seen, fitting together to form the capillary lumen. The size of the tiny gaps between the endothelial cells can be varied, and it is through these gaps that white blood cells migrate out into the tissues. The space between the capillary and the surrounding muscle contains interstitial fluid. If you look closely, you can see numerous pinocytic vesicles in the cytoplasm of the endothelial cells.*

Because of both of these facts, blood moves slowly through the capillary system. As it moves, gases (oxygen and carbon dioxide), hormones, and other materials are exchanged with the surrounding tissues by diffusion through junctions between the endothelial cells and through the cytoplasm of these cells. Organic molecules, such as glucose, are probably moved by transport systems of the endothelial cells. Pinocytic vesicles (page 138) have been observed in the endothelial cells; they presumably serve to ferry dissolved material into or out of the capillaries.

No cell in the human body is farther than 130 micrometers—a distance short enough for rapid diffusion—from a capillary. Even the cells in the walls of the large veins and arteries depend on capillaries for their blood supply, as does the heart itself.

THE HEART

Evolution of the Heart

In the course of vertebrate evolution, the heart has undergone some structural adaptations, as shown in Figure 36–8. Fish have a single heart divided into an **atrium,** the receiving area for the blood, and a **ventricle,** the pumping area from which the blood is expelled into the vessels. The ventricle of the fish heart pumps blood directly to the capillaries of the gills, where it picks up oxygen and releases carbon dioxide. From the gills, oxygenated blood is carried to the tissues. By this time, however, most of the propulsive force of the heartbeat has been dissipated by the resistance of the capillaries in the gills, so that the blood flow through the rest of the tissues (the systemic circulation) is relatively sluggish.

In amphibians, there are two atria; one receives oxygenated blood from the lungs, and the other receives deoxygenated blood from the systemic circulation. Both atria empty into a single ventricle. Although the ventricle is not divided, its internal structure ensures that the two kinds of blood remain relatively unmixed. The oxygenated blood is pumped into the systemic circulation under high pressure at the same time that the deoxygenated blood is pumped through the lungs. Branches of the vessels leading through the lungs also carry some of the deoxygenated blood to the moist skin, a major site of gas exchange in amphibians. Blood oxygenated in the skin enters the systemic circulation before returning to the heart.

36–8 *Vertebrate circulatory systems. Oxygen-rich blood is shown as red, and oxygen-poor blood as blue.* (**a**) *In the fish, the heart has only one atrium (A) and one ventricle (V). Blood oxygenated in the gill capillaries goes straight to the systemic capillaries without first returning to the heart.* (**b**) *In amphibians, the single primitive atrium has been divided into two separate chambers. Oxygen-rich blood from the lungs enters one atrium, and oxygen-poor blood from the tissues enters the other. Little mixing of the blood occurs in the ventricle, despite its lack of a structural division. From the ventricle, oxygen-rich blood is pumped to the body tissues at the same time that oxygen-poor blood is pumped to the lungs. Some of the oxygen-poor blood is diverted from the lungs to the skin, a major respiratory organ in amphibians.* (**c**) *In birds and mammals, both the atrium and the ventricle are divided into two separate chambers, so that there are, in effect, two hearts—one for pumping oxygen-poor blood through the lungs and one for pumping oxygen-rich blood through the body tissues.*

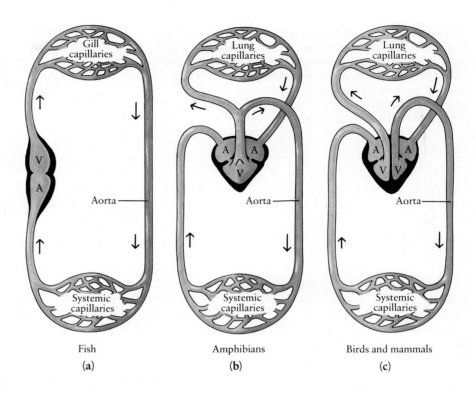

Fish
(**a**)

Amphibians
(**b**)

Birds and mammals
(**c**)

In birds and mammals, the heart is separated longitudinally into two functionally distinct organs, the right heart and the left heart, each with an atrium and a ventricle. The right heart receives blood from the tissues and pumps it into the lungs, where it becomes oxygenated. From the lungs, the oxygenated blood returns to the left heart, from which it is pumped at high pressure into the body tissues. This efficient, high-pressure circulatory system, with its full separation of oxygenated and deoxygenated blood, makes possible the high metabolic rate of both birds and mammals, with their constant body temperature and their generally high level of activity.

The Human Heart

Figure 36–9 shows a diagram of the human heart. Its walls are made up predominantly of a specialized type of muscle—cardiac muscle. Blood returning from the body tissues enters the right atrium through two large veins, the superior and inferior venae cavae. Blood returning from the lungs enters the left atrium through the pulmonary veins. The atria, which are thin-walled compared to the ventricles, expand as they receive the blood. Both atria then contract simultaneously, assisting the flow of blood through open valves into the ventricles. Then the ventricles contract simultaneously; the valves between the atria and ventricles are closed by the pressure of the blood in the ventricles. The right ventricle propels deoxygenated blood into the lungs through the pulmonary arteries; the left ventricle propels oxygenated blood into the aorta, from which it travels to the other body tissues. Valves between the ventricles and the pulmonary artery and the aorta close after the ventricles contract, thus preventing backflow of blood. In a healthy adult at rest, this rhythmic process takes place about 70 times a minute; under strenuous exercise, the rate more than doubles.

If you listen to a heartbeat, you hear "lubb-dup, lubb-dup." The deeper, first sound ("lubb") is the closing of the valves between the atria and the ventricles; the second sound ("dup") is the closing of the valves leading from the ventricles to the

36–9 *The human heart. Blood returning from the systemic circulation through the superior and inferior venae cavae enters the right atrium and passes to the right ventricle, which propels it through the pulmonary arteries to the lungs, where it is oxygenated. Blood from the lungs enters the left atrium through the pulmonary veins, passes to the left ventricle, and then is pumped through the aorta to the body tissues.*

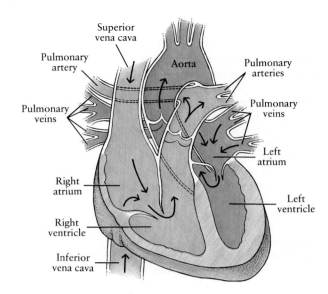

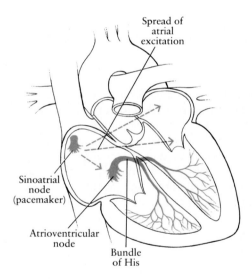

36–10 *The beat of the mammalian heart is controlled by a region of specialized muscle tissue in the right atrium, the sinoatrial node, which functions as the heart's pacemaker. Some of the nerves regulating the heart have their endings in this region. Excitation spreads from the pacemaker through the atrial muscle cells, causing both atria to contract almost simultaneously. When the wave of excitation reaches the atrioventricular node, its conducting fibers pass the stimulation to the bundle of His, which triggers almost simultaneous contraction of the ventricles. Because the fibers of the atrioventricular node conduct relatively slowly, the ventricles do not contract until after the atrial beat has been completed.*

arteries. If any one of the four valves is damaged, as from rheumatic fever, blood may leak back through the valve, producing the noise characterized as a "heart murmur" (a "ph-f-f-t" sound).

The total volume of blood pumped by the heart per minute is called the **cardiac output.** It is defined as:

$$\begin{matrix} \text{Cardiac output} \\ \text{(liters per minute)} \end{matrix} = \begin{matrix} \text{Heart rate} \\ \text{(beats per minute)} \end{matrix} \times \begin{matrix} \text{Stroke volume} \\ \text{(liters per beat)} \end{matrix}$$

Thus, if the heart beats 72 times per minute and ejects 0.07 liter of blood into the aorta with each beat, then

Cardiac output = 72 beats per minute × 0.07 liter of blood per beat
Cardiac output = 5 liters per minute

In addition to its function as a pump, the heart is also a hormone-secreting organ. In humans and other mammals, many of the cardiac muscle cells of the atria synthesize and release a peptide known as atrial natriuretic ("salt-excretion") factor or, more simply, as cardiac peptide. Receptors for cardiac peptide have been identified in the kidneys, blood vessels, adrenal glands, and brain. Although the full range of functions of this hormone—first sequenced in 1983—remain unknown, it appears to play a major role in the regulation of blood volume and blood pressure.

Regulation of the Heartbeat

Most muscle contracts only when stimulated by a motor nerve, but the stimulation of cardiac muscle cells originates in the muscle itself. A vertebrate heart will continue to beat even after it is removed from the body if it is kept in an oxygenated nutrient solution. In vertebrate embryos, the heart begins to beat very early in development, before the appearance of any nerve supply. In fact, isolated embryonic heart cells in a test tube will beat.

The contraction of cardiac muscle is initiated by a special area of the heart, the **sinoatrial node,** which is located in the right atrium (Figure 36–10). This region of tissue functions as the pacemaker. It is composed of specialized cardiac muscle cells that can spontaneously initiate their own impulse and contract. From the pacemaker the impulse spreads throughout the right and left atria. As it passes along the surface of the individual cardiac muscle cells, it activates their contractile machinery, and they contract. The impulse travels very quickly through gap junctions (page 142) that connect the cytoplasm of adjacent cardiac muscle cells.

36–11 *An electrocardiogram showing seven normal heartbeats. Each beat is denoted by a series of waves that record the electrical activity of the heart during contraction. Analysis of the rate and character of these waves can reveal abnormalities in heart function. The first small hump represents the current generated by the passage of the impulse through the atria, the spike by its passage through the ventricles, and the final hump by the return of the ventricles to their resting state. Atrial relaxation occurs during ventricular excitation; therefore the record of its electrical activity is obscured.*

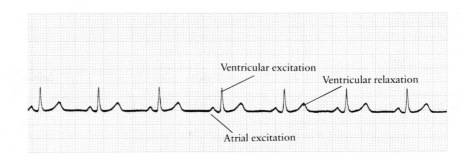

Ventricular excitation

Ventricular relaxation

Atrial excitation

As a consequence, many cells in both atria are activated almost simultaneously.

About 100 milliseconds after the pacemaker fires, impulses traveling through both special conducting fibers and the atrial muscle itself stimulate a second area of nodal tissue, the **atrioventricular node**. From the atrioventricular node, impulses are carried by special muscle fibers, the **bundle of His** (named after its discoverer), to the walls of the right and left ventricles, which then contract almost simultaneously. The bundle of His is the only electrical bridge between the atria and the ventricles. Although its fibers conduct impulses very rapidly, the atrioventricular node consists of slow-conducting fibers. As a consequence, a delay is imposed between the atrial and ventricular contractions, ensuring that the atrial beat is completed before the beat of the ventricles begins.

When the impulses from the conducting system travel across the heart, electric current generated on the heart's surface is transmitted to the body fluids, and from there some of it reaches the body surface. Appropriately placed electrodes on the surface connected to a recording instrument can measure this current. The output, an electrocardiogram (Figure 36–11), is important in assessing the heart's capability to initiate and transmit the impulses.

Although the autonomic nervous system does not initiate the vertebrate heartbeat, it does modify its rate (Figure 36–12). Fibers from the parasympathetic division travel in the vagus nerve (a large nerve that runs through the neck) to the pacemaker. Parasympathetic stimulation has a slowing effect on the pacemaker and thus decreases the rate of heartbeat. Sympathetic nerves stimulate the pacemaker, increasing the rate of heartbeat. Adrenaline from the adrenal medulla (the interior portion of the adrenal gland) affects the heart in the same way that the sympathetic nerves do.

36–12 *Autonomic regulation of the rate of heartbeat. Sympathetic fibers stimulate the sinoatrial node, whereas parasympathetic fibers, which are contained in the vagus nerve, inhibit it.*

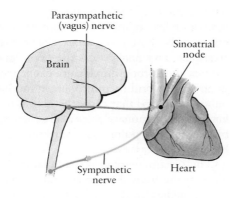

Parasympathetic (vagus) nerve

Brain

Sinoatrial node

Sympathetic nerve

Heart

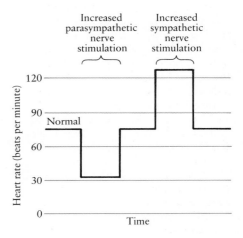

36–13 *Some of the principal circuits in the human cardiovascular system. Oxygenated blood is shown as red, and deoxygenated blood as blue. The portions of the lungs in which gas exchange occurs are served by the pulmonary circulation. Every other tissue of the body is served by the systemic circulation. Blood traveling through the capillaries supplies oxygen and nutrients to every cell of these tissues and carries off carbon dioxide and other wastes. At the venous ends of the capillary beds, blood passes into the venules (the smallest veins), then into larger veins, and finally back to the heart through either the superior (anterior) or the inferior (posterior) vena cava.*

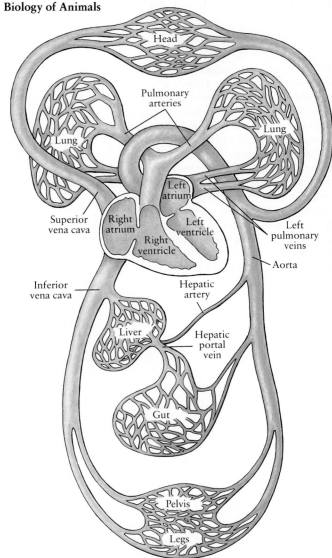

THE VASCULAR CIRCUITRY

As we have seen, there are two principal circuits in the cardiovascular system of an air-breathing vertebrate: the pulmonary circuit and the systemic circuit. Figure 36–13 shows, in more detail, these circuits as they occur in a representative mammal, *Homo sapiens*. In the pulmonary circuit, deoxygenated blood leaves the right ventricle of the heart through the pulmonary artery. This artery divides into right and left branches, which carry the blood to the right and left lungs, respectively. Within the lungs, the arteries divide into smaller arteries, then into arterioles, and finally into the even smaller alveolar capillaries through which oxygen and carbon dioxide are exchanged. Blood flows from the capillaries into small venules and then into larger and larger veins, finally draining into the four pulmonary veins that carry the blood, now oxygenated, to the left atrium of the heart. The pulmonary arteries are the only arteries that carry fully deoxygenated blood, and the pulmonary veins are the only veins that carry fully oxygenated blood.

The systemic circuit is much larger. Many major arteries, supplying different parts of the body, branch off the aorta after it leaves the left ventricle. The first two branches are the right and left coronary arteries, which bring oxygenated blood to the heart muscle itself. Another major subdivision of the systemic circulation supplies the brain. If the circulation of freshly oxygenated blood to the brain is cut off for even five seconds, unconsciousness results; after four to six minutes, brain cells are damaged irreversibly.

Among the features of the systemic circulation are several **portal systems,** in which blood flows through two distinct capillary beds, connected by either veins or arteries, before entering the veins that return it to the heart. For example, in the hepatic portal system, venous blood collected from the capillaries of the digestive tract is shunted via the hepatic portal vein through the liver. There it goes through a second capillary network before it is emptied into the inferior vena cava. In this way, the products of digestion can be directly processed by the liver. The liver also receives freshly oxygenated blood directly from a major artery, the hepatic artery. As we shall see in subsequent chapters, other portal systems play important roles in the chemical processing of blood in the kidneys and in the functions of the pituitary gland.

BLOOD PRESSURE

The contractions of the ventricles of the heart propel the blood into the arteries with considerable force. Blood pressure is a measure of the force per unit area with which blood pushes against the walls of the blood vessels. It is conventionally described in terms of how high it can push a column of mercury. For medical purposes, it is usually measured at the artery of the upper arm. Normal blood pressure in a young adult is generally about 120 millimeters of mercury (120 mm Hg) when the ventricles are contracting (the systolic blood pressure) and 80 mm Hg when the ventricles relax (diastolic pressure); this is stated as a blood pressure of 120/80. The pressure is generated by the pumping action of the heart and changes with the rate at which it contracts. The strength of these contractions, the elasticity of the arterial walls, and the rate at which blood flows from the arteries also play important roles in determining the blood pressure.

The rate of blood flow is directly proportional to blood pressure; the greater the pressure, the greater the rate of flow. The regulation of blood flow depends on a very simple physical principle: Fluid flow through a tube is proportional to the fourth power of the radius of the tube (r^4). As you may recall, the radius of a tube is one-half of its diameter. The diameter of the arterioles, which directly supply the capillaries, can be altered by rings of smooth muscle in the vessel walls. As the smooth muscle contracts, the opening of the arteriole gets smaller (vasoconstriction), and blood flow through the arteriole (and the capillary bed it feeds) decreases (Figure 36–14). Conversely, when the smooth muscle relaxes, the arteriole opens wider (vasodilation), and blood flow into the capillaries increases. These smooth muscles are influenced by autonomic nerves (chiefly sympathetic nerves), the hormones adrenaline and noradrenaline (norepinephrine), cardiac peptide (page 756), and the levels of other chemicals that are produced locally in the tissues themselves.

Constriction and dilation of the arterioles in different parts of the body regulate the blood flow—and thus the supply of oxygen and nutrients—according to the varying requirements of the animal. For example, blood flow through skeletal muscle increases during exercise, flow to the stomach and intestines increases during digestion, and flow through the skin increases at high temperatures and decreases at low temperatures. Of particular importance is a constant flow of blood to the brain. One of the mechanisms that prevents serious damage to human brain cells as a result of an inadequate blood supply is fainting, which causes the person to fall; as a consequence, the force of gravity does not have to be overcome for blood to flow to the brain. This response is often thwarted by well-meaning bystanders anxious to get the affected individual "back on his feet."

— Arteriole

— Venule

36–14 Diagram of a capillary bed. The muscular wall of the arteriole controls the blood flow through the capillaries. The arteriolar muscles, which are innervated by the sympathetic division of the autonomic nervous system, make it possible to regulate the blood supply to different regions of the body depending on the physiological state of the organism at different times.

Diseases of the Heart and Blood Vessels

Cardiovascular diseases cause almost as many deaths in the United States as accidents and all other diseases combined. According to recent estimates, about 65 million people in this country (more than 25 percent of the total population) have some form of cardiovascular disease.

About 55 percent of the cardiovascular deaths are caused by heart attacks. A heart attack is the result of an insufficient supply of blood (ischemia) to an area of heart muscle; with their oxygen supply cut off, the cardiac muscle cells may die. A heart attack can be caused by a blood clot—a thrombus—that forms in the blood vessels of the heart itself or by a clot that forms elsewhere in the body and travels to the heart and lodges in a vessel there. (A wandering clot such as this is known as an embolus.) A heart attack can also result from blockage of a blood vessel due to atherosclerosis. Recovery from a heart attack depends on how much of the heart tissue is damaged, where the damage occurs, and whether or not other blood vessels in the heart can enlarge their capacity and form new branches that supply these tissues, which then may recover to some extent. One-fourth of all deaths in the United States are caused by ischemic heart disease.

Angina pectoris is a related condition in which the heart muscle receives an insufficient blood supply (but not so little that muscle dies)—often a result of a narrowing of the vessels. Its symptoms, like those of a heart attack, are pain in the center of the chest and, often, in the left arm and shoulder.

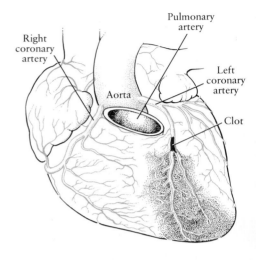

A heart attack. When a clot forms in a blood vessel, the cells in the area supplied by the vessel are deprived of oxygen and may die. The severity of the heart attack depends, in part, on the extent of damage to the heart muscle.

A stroke is caused by interference with the blood supply to the brain. This may be the result of a thrombus, an embolus, or the bursting of a blood vessel in the brain. Its effects depend on the extent of the damage and where it occurs in the brain.

(In fact, holding a fainting person upright can lead to severe shock and even death.) Psychological events also influence vasoconstriction and dilation and thus the distribution of blood flow. Familiar examples are blushing, turning pale with fear, angina pectoris (pain in the chest and left arm) precipitated by emotion, and erection of the penis or clitoris as a result of erotic stimulation.

As blood flows through the vascular circuitry, its pressure gradually drops as a consequence of both the damping caused by the recoil of the elastic arterial walls (see page 753) and the resistance of the arterioles and capillaries. Thus, blood pressure is not the same in the various parts of the cardiovascular system (Figure 36–15). As you would expect, it is highest in the aorta and other large systemic arteries, much lower in the veins, and lowest in the right atrium.

Atherosclerosis contributes to both heart attacks and strokes. In this disease, the linings of the arteries thicken due to the accumulation of abnormal smooth muscle cells, and their inner surfaces become roughened by deposits of cholesterol, fibrin, and cellular debris, as shown in the photos on page 71. Such arteries, becoming inelastic, no longer expand and contract, and blood moves with increasing difficulty through the narrowed vessels. Thrombi and emboli thus form more easily and are more likely to block the vessel.

The causes are not clear-cut. Atherosclerosis is associated with high blood pressure and with high levels of cholesterol in the blood. As we saw on page 71, a critical factor appears to be the relative proportions of the two principal forms in which cholesterol is carried in the bloodstream: high-density lipoproteins (HDLs), which appear to protect against atherosclerosis, and low-density lipoproteins (LDLs), which contribute to the disease process. Cigarette smoking, a lack of physical activity, a high dietary intake of cholesterol and saturated fats, and, in some cases, hereditary factors, are associated with increased levels of LDLs and decreased levels of HDLs.

Atherosclerosis is much less common in premenopausal women than it is among men of the same age, suggesting that female hormones may protect against the disease. Paradoxically, however, men with elevated levels of female hormones appear to be at an increased risk for atherosclerosis. Whatever the underlying cause may be, the difference in suscepti-bility to atherosclerosis is a major reason that, in the United States, women now live, on the average, almost 10 years longer than men.

In North America, the mortality rate from ischemic heart disease reached a peak in the mid-1960s and then, quite unexpectedly, began a decline that is still continuing; the total decrease in the mortality rate has, astonishingly, exceeded 30 percent. The increase, when it occurred, was attributed to increased cigarette smoking, increased stress, and an increasingly sedentary existence. The decreases, however, which are occurring at all age levels and in both sexes, do not correlate well with any known environmental factor. Perhaps if we could find out what it is that we are doing right, we could do it even better.

Hypertension—chronically increased arterial blood pressure—affects about 54 million people in the United States. It places additional strain on the arterial walls and increases the chances of emboli. Despite lack of knowledge about the causes, hypertension can be treated by drugs that act upon the autonomic nervous system to produce arteriolar dilation, and by measures that decrease blood volume, reducing the blood pressure to safer levels. In the United States, hypertension is about twice as common among blacks as among whites. Related to this disturbing statistic is the fact that the incidence of deaths among black men and women from cardiovascular disease is even higher than in the white population.

36–15 *In mammals, blood goes from the right heart to the lungs, from the lungs to the left heart, and, from the left heart, it enters the systemic circulation, moving from arteries to arterioles to capillaries to venules to veins. Blood pressure varies in the different areas of the cardiovascular system. The fluctuations in blood pressure produced by three heartbeats are shown in each section of the diagram. Note the fall in pressure as blood traverses the arterioles of the systemic circulation.*

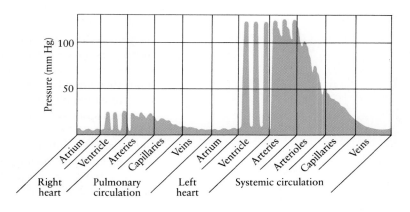

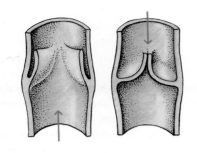

36–16 *Valves in the veins open to permit movement of blood toward the heart but close to prevent backflow.*

The veins, with their thin walls and relatively large diameters, offer little resistance to flow, making possible movement of the blood back to the heart despite its low pressure. Valves in the veins prevent backflow (Figure 36–16). The return of blood to the heart is enhanced by the contractions of skeletal muscles. For example, when you walk, your leg muscles bulge and squeeze the veins that lie between the contracting muscles, thus raising the pressure within the veins and increasing the flow. (If you have to stand still for long periods of time, try contracting your leg muscles periodically to move blood back toward the heart and prevent blood pooling.) Also, as the thoracic cavity expands on inspiration, the elastic walls of the veins in the chest dilate, venous pressure in the region of the heart decreases, and return of blood to the heart increases.

Cardiovascular Regulating Center

Activity of the nerves controlling the smooth muscle of the blood vessels is coordinated with the activity of nerves regulating heart rate and strength of heartbeat by the **cardiovascular regulating center.** This center is located in the medulla, a small part of the brain continuous with the spinal cord. It controls the sympathetic and parasympathetic nerves to the heart (see Figure 36–12) as well as the nerves to the smooth muscle in the arterioles. Thus, if there is a significant increase in blood flow due to dilation of blood vessels, the heart is simultaneously stimulated to beat faster, thus developing greater pressure to support the greater flow.

The cardiovascular regulating center integrates the reflexes that control blood pressure. It receives information about existing blood pressure from specialized stretch receptors in the carotid arteries (see Figure 35–18, page 747), the venae cavae, the aorta, and the heart. The effector organs of the reflex are, as we have indicated, the heart and blood vessels.

The blood-pressure reflex is an example of negative-feedback control. When pressure falls, the activity of the heart is increased and the blood vessels are constricted, raising the pressure again. Conversely, heart activity is decreased and the blood vessels dilated in response to high pressure.

THE LYMPHATIC SYSTEM

Earlier in this chapter, we discussed the role of plasma proteins in maintaining the high osmotic potential of the blood in relation to the interstitial fluid and thus in preventing excessive loss of fluid from the blood. However, as you may recall from Chapter 6, the osmotic potential of solutions separated by a selectively permeable membrane is only one of several factors influencing the movement of water across the membrane; another important factor is pressure. At the arteriolar end of the capillaries, the pressure of the circulating blood—the hydrostatic pressure—is greater than the osmotic potential, forcing some of the watery component of blood to leave the capillaries along with nutrient molecules and oxygen. By the time the blood reaches the venous end of the capillaries, the hydrostatic pressure has dropped, due to resistance in the capillary bed, to a point where it is less than the osmotic potential of the blood. Most—but not all—of the fluids then move back into the capillaries by osmosis. Fluids may also escape from the circulating blood into the tissues when the endothelium is damaged, as, for example, by a blow; damage to the endothelium allows the release not only of the fluids themselves but also of the plasma proteins that maintain the osmotic potential of the blood. When excessive fluid is lost, it tends to pool, producing the swelling known as edema.

In higher vertebrates, the fluids lost from the blood to the tissues are collected by the **lymphatic system** (Figure 36–17), which routes them back to the bloodstream. The lymphatic system is like the venous system in that it consists of an interconnecting network of progressively larger vessels. The larger vessels are, in fact, similar to veins in their structure, and the small vessels are much like the capillaries through which the blood is carried. An important difference, however, is that the lymph capillaries begin blindly in the tissues, rather than forming part of a continuous circuit. Interstitial fluid seeps into the lymph capillaries, from which it travels to large ducts that empty into the two subclavian veins; these veins, located below the two clavicles (collar bones), empty into the superior vena cava. The fluid carried in the lymphatic system is known as **lymph**. As you may recall, lymph is also the medium in which fats absorbed from the digestive tract are transported to the bloodstream (see Figure 34–9, page 721).

Some nonmammalian vertebrates have lymph "hearts," which help to move the fluid. In mammals, lymph is moved by contractions of the body muscles, with valves preventing backflow, as in the venous system. Also, recent studies have shown that lymph vessels contract rhythmically; these contractions may be the principal factor propelling the lymph.

Lymph nodes, which are masses of spongy tissue, are distributed throughout the lymphatic system. They have two functions: they are the sites of proliferation of lymphocytes, specialized white blood cells that are the effectors of the immune response (to be discussed in Chapter 39), and they remove cellular debris and foreign particles from the lymph before it enters the blood. The removal of chemical wastes, however, requires processing of the blood itself; this function is performed by the kidneys, as we shall see in the next chapter.

SUMMARY

Oxygen, nutrients, and other essential molecules, as well as waste products, are carried in the blood. The blood is composed of plasma, red blood cells (erythrocytes), white blood cells (leukocytes), and platelets. Plasma, the fluid part of the blood, is chiefly water in which nutrients, waste products, ions, antibodies, hormones, enzymes, plasma proteins, and other substances are dissolved or suspended. Red blood cells contain oxygen-bearing hemoglobin, and white blood cells defend the body against foreign invaders. Platelets participate in the clotting of blood, which occurs as the result of a reaction cascade involving at least 15 factors.

In vertebrates the blood is pumped by muscular contractions of the heart into a closed circuit of arteries, arterioles, capillaries, venules, and veins. This network, which includes both pulmonary and systemic circuits, ultimately services every cell in the body. The essential function of the circulatory system is performed by the capillaries, through which substances are exchanged with the interstitial fluid surrounding the individual cells of the body.

Evolutionary changes in the structure of the vertebrate heart can be correlated with changes in metabolic rates and the level of activity of the animals. Fish have a two-chambered heart, whereas amphibians have a three-chambered heart. Birds and mammals have a double circulatory system, made possible by a four-chambered heart that functions as two separate pumping organs. One side pumps deoxygenated blood to the lungs, and the other pumps oxygenated blood to the body tissues under high pressure.

Synchronization of the heartbeat is controlled by the sinoatrial node (the pacemaker), located in the right atrium, and by the atrioventricular node, which

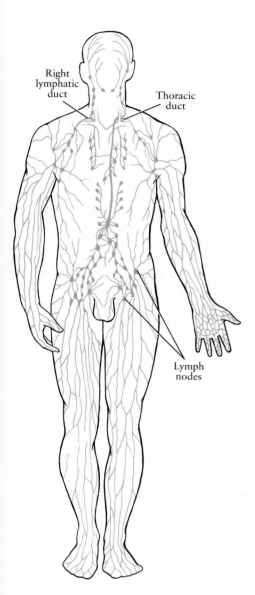

36–17 *The human lymphatic system consists of a network of lymph vessels and lymph nodes. Lymph reenters the bloodstream through the thoracic duct, which empties into the left subclavian vein, and through the right lymphatic duct, which empties into the right subclavian vein. These two veins empty into the superior vena cava.*

Right lymphatic duct

Thoracic duct

Lymph nodes

delays the stimulation of ventricular contraction until the atrial contraction is completed. The rate of heartbeat is secondarily under neural and hormonal regulation. Parasympathetic stimulation slows the heartbeat; sympathetic stimulation and adrenaline accelerate it.

Blood pressure is a measure of the force per unit area with which blood pushes against the walls of the blood vessels. It is generated by the pumping action of the heart and drops precipitously as the blood traverses the arterioles. The magnitude of the pressure within the arteries is influenced by the rate and strength of the heartbeat, the elasticity of the arterial walls, and the rate at which blood flows from the arteries into the arterioles. The rate of flow is, in turn, influenced by the degree of dilation or constriction of the arterioles. Activity of the nerves regulating the rate and strength of the heartbeat is coordinated with the activity of the nerves controlling the smooth muscle of the arterioles by the cardiovascular regulating center, located in the medulla of the brain.

Because the hydrostatic pressure in the arteries is greater than the osmotic potential of the blood, some fluids are forced out of the capillaries at the arteriolar end of capillary beds. These fluids either reenter the capillaries at the venous end of the beds (where the hydrostatic pressure is much lower) or are returned to the blood by the lymphatic system. The lymph also picks up cellular debris and foreign particles, which are filtered out by the lymph nodes.

QUESTIONS

1. Distinguish between the following: blood/plasma; aorta/vena cava; atrium/ventricle; right heart/left heart; sinoatrial node/atrioventricular node; systolic/diastolic.

2. Blood serum is the portion of the plasma remaining after a clot is formed. Name some of the components of the plasma that would not be present or would be present in a lesser amount in serum.

3. Give two reasons why atherosclerotic arteries are much more susceptible to clot formation and blockage than are normal, healthy arteries.

4. Label the diagram below.

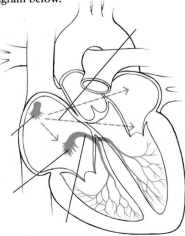

5. What is the advantage of the fibers of the atrioventricular node being slow-conducting?

6. The valves of the heart are not directly controlled by nerves. Yet, in most individuals, they open and shut at precisely the right points in the cardiac cycle for efficient heart operation. How is this precise timing possible? What does determine just when the valves will open and shut?

7. Trace the course of a single red blood cell from the right ventricle to the right atrium in a mammal. Trace the course of an oxygen molecule from the air to its arrival at a metabolizing cell.

8. How does the radius of a blood vessel affect the blood flow through it?

9. Explain the reasons for the changes in blood pressure shown in Figure 36–15.

10. When fair-skinned individuals are very frightened, they turn quite pale. What occurs to cause this change? Why is such an adaptation useful?

11. Individuals, particularly children, suffering from severe protein deprivation often have swollen, bloated bellies. What is the explanation for this phenomenon?

12. When an accident victim suffers blood loss, he or she is transfused with plasma rather than with whole blood. Why is plasma effective in meeting the immediate threat to life?

13. What is probably the immediate cause of death by crucifixion? (If you need a clue, see pages 759–760.)

Homeostasis I: Excretion and Water Balance

37-1 *The kangaroo rat, a common inhabitant of the American desert, may spend its entire life without drinking water, yet the composition of its body fluids remains essentially constant. As we shall see in this chapter, a number of physiological adaptations and mechanisms contribute to this remarkable feat.*

As we noted in Chapter 33, one of the advantages of multicellularity is the increased capacity for homeostasis—the maintenance of a controlled internal environment in which the cells can live and function. In animals, a great variety of activities contribute to homeostasis. Among the specific examples we have seen in previous chapters are the regulation of blood sugar, the uptake and distribution of oxygen to the cells, and the elimination of carbon dioxide from the body. In this chapter and the two following, we shall examine three particularly noteworthy homeostatic functions: regulation of the chemical composition of body fluids, regulation of temperature, and defense of the body against foreign invaders.

REGULATION OF THE CHEMICAL ENVIRONMENT

Animals are about 70 percent water. About two-thirds of this water is within the cells, and one-third is in the extracellular fluid that surrounds, bathes, and nourishes the cells. Thus the extracellular fluid serves the same purpose for the cells of an animal's body that the Precambrian seas served for the earliest unicellular organisms. As animals became multicellular in the course of evolution, they began to produce their own extracellular fluid, similar in composition to the salty fluid of the sea. As they did so, they also evolved mechanisms for regulating its composition.

Although the blood plasma constitutes only about 7 percent of total body fluids, the regulation of its composition is a key factor in regulation of the chemical environment throughout the vertebrate body. As we saw in the last chapter, blood is the supply line for chemicals taken up by the individual cells, and it carries away the wastes released by these cells. Blood can function as an efficient supply and sanitation medium only because cellular wastes are constantly removed from it. Such removal is quite different in principle from the elimination of feces from the intestinal tract. In the latter case, the bulk of what is eliminated is material, such as cellulose, that was never actually in the body, since it never passed through the epithelium of the digestive tract. Excretion of substances carried within the bloodstream, by contrast, is a very selective process of monitoring, analysis, selection, and rejection.

In many invertebrates and all vertebrates, the composition of the blood—and thus the internal chemical environment—is to a large extent regulated by special excretory organs. These organs include the nephridia of mollusks and annelids (page 547), the Malpighian tubules of insects (page 568), and the kidneys of vertebrates. Although other organs, especially the liver, play important roles in regulating the chemical environment, it is possible to relate major advances in vertebrate evolution—particularly the transition to land—to increasing efficiency of kidney function.

$$H_2N - \overset{\overset{\displaystyle CH_3}{|}}{\underset{\underset{\displaystyle COOH}{|}}{C}} - H + \tfrac{1}{2}O_2 \longrightarrow \overset{\overset{\displaystyle CH_3}{|}}{\underset{\underset{\displaystyle COOH}{|}}{C}} = O + NH_3$$

Alanine   Pyruvic
                                   acid

37–2 *The first step in the breakdown of amino acids is deamination—the removal of the amino group. The products of the reaction are ammonia and a carbon skeleton, which can be broken down to yield energy or converted to sugar or fat.*

Substances Regulated by the Kidneys

Regulation of the internal chemical environment of an animal involves solving three different—yet interwoven—problems: (1) excretion of metabolic wastes, (2) regulation of the concentrations of ions and other chemicals, and (3) maintenance of water balance.

The chief metabolic waste products that cells release into the bloodstream are carbon dioxide and nitrogenous compounds, mostly ammonia (NH_3), produced by the breakdown of amino acids (Figure 37–2). As we have seen, carbon dioxide diffuses out of the body across the respiratory surfaces, which may be skin, gills, or lungs. In simple aquatic animals, ammonia also leaves the body by diffusion into the surrounding water. Ammonia, however, is highly toxic, even in low concentrations. In more complex aquatic animals—and in all terrestrial animals—rapid diffusion of ammonia from the cells into an external water supply is not possible; thus it must be converted to nontoxic substances, which can be safely transported within the body to the excretory organs.

All birds, terrestrial reptiles, and insects convert their nitrogenous wastes into crystalline uric acid or uric acid salts, with the result that very little water is required for their excretion. In birds, the uric acid and uric acid salts are mixed with undigested wastes in the cloaca (the common exit chamber for the digestive, urinary, and reproductive tracts), and the combination is dropped as a semisolid paste, familiar to frequenters of public parks and admirers of outdoor statuary. This nitrogen-laden substance forms a rich natural fertilizer; guano, the excreta of seabirds, accumulates in such quantities on the small islands where great numbers of these birds gather that at one time it was harvested commercially.

In mammals, the ammonia resulting from the processing of nitrogenous wastes is quickly converted in the liver to urea (Figure 37–3), which diffuses into the bloodstream. This relatively nontoxic compound is then carried to the kidneys. Unlike uric acid, however, it must be dissolved in water for excretion.

Excretion is highly selective. For instance, although about half of the urea in the blood that enters a normal mammalian kidney is excreted, almost all the amino acids are retained. Glucose is not excreted unless it is present in high concentrations, as in diabetes mellitus. (The presence of glucose in the urine is, in fact, a basis for the diagnosis of this form of diabetes and the evaluation of its treatment.) Thus, although the kidneys have an excretory function, they are more accurately regarded as regulatory organs. Chemical regulation involves not only the retention of nutrient molecules such as glucose and amino acids but also the maintenance of closely controlled concentrations of ions. Such ions as Na^+, K^+, H^+, Mg^{2+}, Ca^{2+}, and HCO_3^- play vital roles in maintenance of protein structure, membrane permeability, and blood pH, in propagation of the nerve impulse, and in the contraction of muscles.

The concentration of a particular substance in the body depends not only on the absolute amount of the substance but also on the amount of water in which it is dissolved. Thus, regulation of the water content of body fluids is an important aspect of regulation of the chemical environment. The problem of water balance is such a universal one, biologically speaking, and is so important to the survival of

37–3 *Urea, the principal form in which nitrogen is excreted in most mammals. It is formed in the liver by the combination of two molecules of ammonia with one of carbon dioxide through a complex series of energy-requiring reactions. What would be the other product of this reaction?*

$$H_2N - \overset{\overset{\displaystyle }{}}{\underset{\underset{\displaystyle O}{\|}}{C}} - NH_2$$

Urea

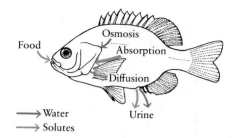

37-4 *Pathways by which water and solutes are gained and lost in freshwater fish. Because the body fluids are hypertonic to the surrounding environment, water enters the body of the fish by osmosis across the gill epithelium. Excess water is removed from the blood by the kidneys and excreted in the urine, which is much more dilute than the body fluids. Although the kidneys reabsorb the bulk of the essential solutes, some are nonetheless lost in the urine, and others leave the body by diffusion across the gills. These solutes are replaced principally by the action of specialized salt-absorbing cells in the gills, and, to a lesser extent, from the diet.*

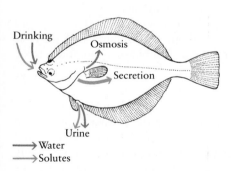

37-5 *Pathways by which water and solutes are gained and lost in bony saltwater fish. Because the body fluids are hypotonic to the surrounding environment, water leaves the body of the fish by osmosis across the gills. Water is also lost in the urine, where it is required to dissolve the urea removed from the blood by the kidneys. The fish maintains its internal fluid levels by drinking sea water, which entails the ingestion of solutes. Excess sodium and chloride ions are removed from the blood and excreted by specialized cells in the gills, while magnesium and sulfate ions are removed by the kidneys and excreted in the urine.*

the organism that we shall consider it in more detail before examining the structure and physiology of the kidney. Although the fundamental problem is always the same, the solution varies widely. A major factor is the availability of water to the organism.

WATER BALANCE

An Evolutionary Perspective

The earliest organisms probably had a salt and mineral composition much like that of the environment in which they lived. The early organisms and their surroundings were probably also isotonic; that is, each had the same total effective concentration of dissolved substances, so water did not tend to move either into or out of these organisms by osmosis. When organisms moved to fresh water (a hypotonic—less concentrated—environment), they had to develop systems for "bailing themselves out," since fresh water tended to move into their bodies; the contractile vacuole of *Paramecium* is an example of such a bailing device (see page 133).

If the earliest vertebrates—the fish—evolved in fresh water, as is generally believed, the first function of the kidneys, phylogenetically speaking, was probably to pump water out and to conserve salt and other desirable solutes, such as glucose. In freshwater fish today, the kidney works in just this way, primarily as a filter and reabsorber of solutes. The urine of these fish is hypotonic—that is, it has a concentration of solutes lower than that of body fluids. Some solutes are, however, inevitably lost both in the urine and by diffusion through the gills. This loss is counteracted by salt-absorbing cells in the gills, which actively transport salt back into the body (Figure 37-4).

When fish moved to the seas, they faced a different problem: the potential loss of water to their environment, principally by osmosis across the respiratory surfaces of the gills. The hagfish, a group of cartilaginous fish, have solved this problem by maintaining body fluids about as salty as the surrounding ocean waters. The body fluids of another group of cartilaginous fish, the sharks, are also isotonic with sea water, but this isotonicity is achieved in a different way. In the course of evolution, sharks developed an unusual tolerance for urea, so instead of constantly excreting it, as do most other fish, they retain a high concentration of it in their blood, which would otherwise be hypotonic to salt water. (This is a striking example of what has been called the "opportunism of evolution.")

The bony fish, which spread to the sea much later than the cartilaginous fish, have body fluids that are hypotonic to the marine environment, with a solute concentration only about one-third that of sea water. Thus they are constantly in danger of losing so much water to their environment that the solutes in their body fluids become so concentrated that the cells die. To compensate for their osmotic water loss, they drink sea water. This restores their water content but leads to a new problem—how to eliminate the excess salt ingested. This problem has been solved by the evolution of special gland cells in the gills that excrete excess salt. Magnesium and sulfate ions, also present in large quantities in sea water, are removed from the blood by the kidneys and excreted in the urine. Hence bony marine fish can drink freely from the water surrounding them and still remain hypotonic to it (Figure 37-5).

Sources of Water Gain and Loss in Terrestrial Animals

Since terrestrial animals do not always have ready access to either fresh or salt water, they must regulate water content in other ways, balancing gains and losses.

37–6 *Some marine animals, such as the turtle, have special glands in their heads that can excrete sodium chloride at a concentration about twice that of sea water. Since ancient times, turtle watchers have reported that these great armored reptiles come ashore, with tears in their eyes, to lay their eggs. It is only recently that biologists have learned that this is not caused by an excess of sentiment—as is the case with Lewis Carroll's mock turtle—but is, rather, a useful solution to the problem of excess salt from ingestion of sea water. Marine birds similarly excrete a salty fluid through their nostrils.*

They gain water by drinking fluids, by eating water-containing foods, and as an end product of certain metabolic reactions, such as the oxidative processes that take place in the mitochondria (see Figure 9–19, page 201). When 1 gram of glucose is oxidized, 0.6 gram of water is formed. When 1 gram of protein is oxidized, only about 0.3 gram of water is produced. Oxidation of 1 gram of fat, however, yields 1.1 grams of water because of the high hydrogen content of fat (the extra oxygen comes from the air).

Some animals can derive all their water from food and oxidation of nutrient molecules and therefore do not require fluids. The kangaroo rat (see Figure 37–1), for example, can live its entire existence without drinking water if it eats the right type of food. It is not surprising that it selects a diet of fatty seeds, which yield a large amount of water on oxidation. If it is fed high-protein seeds, such as soybeans—the oxidation of which produces a large amount of nitrogenous waste and a relatively small amount of water—the kangaroo rat will die of dehydration unless some other source of water is available. The kangaroo rat, like many other desert animals, is highly conservative in its water expenditures. It has no sweat glands, and, being nocturnal, it searches for food only when the surrounding air is relatively cool. Its feces have a very low water content, and its urine is highly concentrated (almost four times as concentrated as the most concentrated human urine). Its major water loss is through respiration, and even this loss is reduced by the animal's long nose in which some cooling of the expired air takes place, with condensation of water from it.

On the average, a human takes in about 2,300 milliliters of water a day in food and drink and gains an additional 200 milliliters a day by oxidation of nutrient molecules. Water is lost from the lungs in the form of moist exhaled air, is eliminated in the feces, is lost by evaporation from the skin, and is removed from the blood and excreted as urine. The latter is usually the major route of water loss. In a normal adult, the rate of water excretion in urine averages 1,500 milliliters a day. Although the actual amount of urine produced may vary from 500 to 2,300 milliliters per day, there will be a variation of less than 1 percent in the fluid content of the body. A minimum output of about 500 milliliters of water is necessary for health, since this much water is needed to remove potentially toxic waste products.

Water Compartments

The body has three principal water compartments: (1) the plasma (7 percent of body fluid), (2) the interstitial fluid and the lymph (28 percent of body fluid), and (3) the intracellular fluid, the fluid within the cells (65 percent of body fluid). Water is constantly moving from one compartment to another (Figure 37–8). The rates of exchange between compartments can be measured by administering a traceable substance, such as a harmless dye molecule, that passes readily through cell membranes, and then analyzing the concentration of the tracer material in each compartment.

37–7 *A terrestrial animal is in water balance when the total amount of water lost in expired air, in evaporation from the skin, and in the urine and feces equals the total amount of water gained by the intake of food and fluids and by the oxidation of food molecules.*

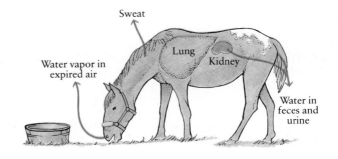

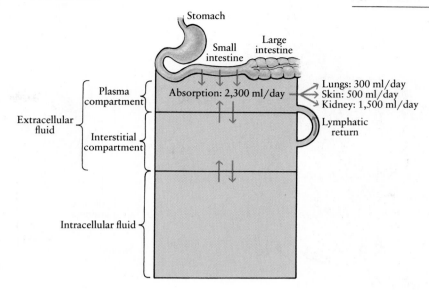

37–8 *The fluid compartments of the human body, showing the main routes of exchange among them. The interstitial fluid forms the environment in which the cells of the body live and multiply. The arrows indicate exchanges between various compartments. A relatively constant plasma volume is of extreme importance for maintenance of stable blood pressure and for normal cardiac function.*

Water absorbed from the digestive tract, the major source of water gain, passes largely into the intestinal capillaries and enters the plasma. This passage is mainly a result of osmosis. Because of the active transport of nutrient molecules and salts into the capillaries from the intestinal tract, the plasma becomes hypertonic in relation to the intestinal contents, and so water tends to follow the dissolved molecules into the plasma. Conversely, as blood moves through the systemic circulation, hydrostatic pressure forces watery fluid through the capillary walls into the interstitial fluid, as we saw on page 762. However, most of this fluid reenters the plasma osmotically or via the lymph vessels. Water remaining in the interstitial compartment comes in contact with the tissue cells. Since their membranes are permeable to water, water moves freely into the cells.

A number of factors affect the movement of water between compartments. Dehydration (water loss greater than water intake) increases the solute concentration of the extracellular fluid; water therefore moves out of the cells, including the cells of the mucous membranes of the oral cavity, producing the sensation of dryness that we associate with thirst. Physiological malfunctions, such as the retention of salts, which may occur as a consequence of kidney disease, or the loss of plasma proteins as a result of starvation, lead to the accumulation of interstitial fluids (edema).

Human sweat, unlike that of most other mammals, contains salt. The reason for salt excretion by the skin is not known, but it does tend to keep the solute concentration of the extracellular fluid relatively constant as water is lost through the skin. (It has also been suggested that the little crumbs of salt so produced clung to the fur of our primate ancestors and that these delicacies served as rewards for their companions who groomed them.) In cases of profuse sweating, if the water is replaced without the salt, water will move into body cells, diluting their contents. The effects of such dilution are particularly severe on the central nervous system, and water intoxication may produce disoriented behavior, convulsions, coma, and even death before the excess water can be excreted.

THE KIDNEY

In vertebrates, the complex functions involved in regulating the chemical composition of body fluids are performed chiefly by the kidney. The two human kidneys are dark red, bean-shaped organs about 10 centimeters long that lie at the back of the body, behind the stomach and liver.

The functional unit of the kidney is the **nephron** (Figure 37–9). It consists of a

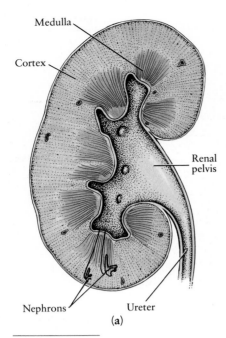

(a)

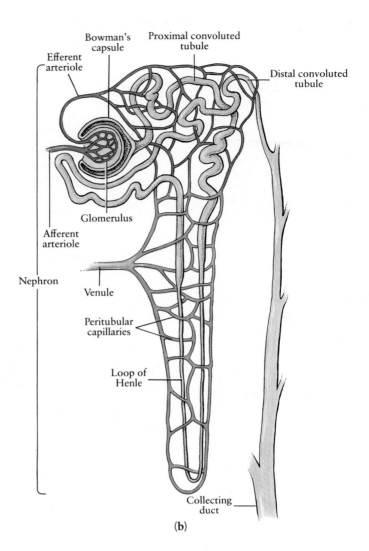

(b)

37–9 (a) *In longitudinal section, the human kidney is seen to be made up of two regions. The outer region, the cortex, contains the fluid-filtering mechanisms. The inner region, the medulla, is traversed by long loops of the renal tubules and by the collecting ducts carrying the urine. These ducts merge and empty into the funnel-shaped renal pelvis, which, in turn, empties into the ureter.*

(b) The nephron is the functional unit of the kidney. Blood enters the nephron through the afferent arteriole leading into the glomerulus. Fluid is forced out by the pressure of the blood through the thin capillary walls of the glomerulus into Bowman's capsule. The capsule connects with the long renal tubule, which has three regions: the proximal convoluted tubule; the loop of Henle, which extends into the medulla; and the distal convoluted tubule. As the fluid travels through the tubule, almost all the water, ions, and other useful substances are reabsorbed into the bloodstream through the peritubular capillaries. Other substances are secreted from the capillaries into the tubules. Waste materials and some water pass along the entire length of the tubule into the collecting duct and are excreted from the body as urine.

cluster of capillaries known as the **glomerulus** and a long narrow tube, the **renal tubule,** which originates as a bulb called **Bowman's capsule.** The renal tubule is made up of the **proximal** (near) and **distal** (far) **convoluted tubules,** which in humans and other mammals are connected by the **loop of Henle.** The nephron ends as the straight **collecting duct.** Each of the two human kidneys contains about a million nephrons with a total length of some 80 kilometers (50 miles) in an adult.

Urine is formed in the nephrons and passed from the collecting ducts into the renal pelvis, which is, in essence, a funnel. From this funnel, the urine trickles continuously through the ureter to the bladder, which stores the urine until it is passed out of the body through the urethra (Figure 37–10).

Function of the Kidney

Blood enters the kidney through the renal artery, which divides into progressively smaller arteries, leading ultimately to the arterioles, each of which feeds into a glomerulus. Unlike most other capillary beds, a glomerulus lies between two arterioles—the one leading in is the afferent arteriole, and the one leading out is the efferent arteriole. The efferent arteriole then divides again into capillaries, the peritubular ("around-the-tubes") capillaries, which surround the renal tubule and then merge to form a venule that empties into a small vein, leading ultimately to the renal vein. The efferent arteriole is thus a portal vessel (see page 759), providing a direct link between two distinct capillary beds.

Constriction of the afferent and efferent arterioles keeps the blood within the

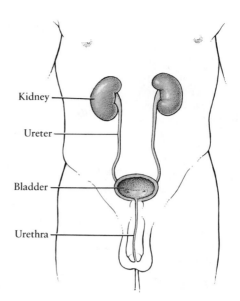

37–10 *Urinary system of the human male. Fluids are processed in the kidneys, which lie at the back of the body, behind the stomach and liver. Waste products and water—the urine—pass along a pair of tubes, the ureters, to the bladder, where they are stored. Urine leaves the body by way of the urethra. In the male mammal, a portion of the urethra also serves as the passageway for semen.*

Kidney

Ureter

Bladder

Urethra

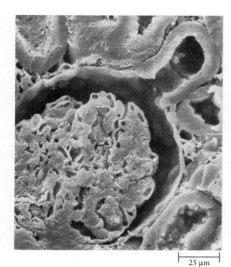

25 µm

37–11 *The section of kidney tissue shown in this scanning electron micrograph was prepared by the freeze-fracture technique. A glomerulus occupies the center of the micrograph, enclosed at the top and sides by Bowman's capsule. The space within the capsule is continuous with the lumen of the proximal tubule, the beginning of which you can see at the upper right.*

glomerulus at a pressure about twice that in other capillaries. As a consequence, about one-fifth of the blood plasma that enters the kidney is forced through the walls of the glomerular capillaries and the wall of Bowman's capsule into the lumen of the renal tubule (Figure 37–11). This crucial first process in the formation of urine is called **filtration**, and the fluid entering the capsule is the **filtrate**. Except for the absence of large molecules, such as proteins, which cannot cross the capillary wall, the filtrate has the same chemical composition as the plasma.

The filtrate then begins its long passage through the renal tubule, the wall of which is made up of a single layer of epithelial cells specialized for active transport. In a second process in the formation of urine, **secretion**, molecules remaining in the plasma after filtration are selectively removed from the peritubular capillaries and actively secreted into the filtrate. Penicillin, for example, is removed from the circulation in this way.

A third major process in urine formation, **reabsorption**, occurs simultaneously. Most of the water and solutes that initially entered the tubule during filtration are transported back into the peritubular capillaries. For example, glucose, most amino acids, and most vitamins are returned to the bloodstream. Secretion and reabsorption of many solutes take place by active transport; as a consequence, the kidney has a high energy requirement, higher on a per-gram basis than even the heart.

Finally, the remaining fluid—now the urine—leaves the nephron and passes into the renal pelvis. This process is **excretion**.

37–12 *The four basic components of renal function: (1) filtration through the glomerular capillaries into Bowman's capsule; (2) secretion from the peritubular capillary into the tubule; (3) reabsorption from the tubule to the peritubular capillary; (4) excretion. The pressure in the glomerular capillaries is regulated by constriction of the afferent and efferent arterioles.*

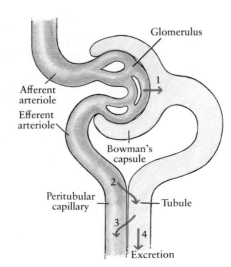

Glomerulus

Afferent arteriole

Efferent arteriole

Bowman's capsule

Peritubular capillary

Tubule

Excretion

37–13 *The formation of hypertonic urine in the human nephron. The filtrate entering the proximal convoluted tubule is isotonic with the blood plasma. Although sodium ions are pumped from the tubule here, with chloride ions following passively, the filtrate remains isotonic because water also moves out by osmosis. As the filtrate descends the loop of Henle, it becomes increasingly concentrated as water moves by osmosis into the surrounding zone of high solute concentration. This zone is created by the action of the wall of the ascending branch of the loop of Henle, which pumps out sodium and chloride ions, and by the diffusion of urea out of the lower portion of the collecting duct. Because the wall of the ascending branch of the loop is impermeable to water, the filtrate becomes less and less concentrated as sodium chloride is pumped out. By the time it reaches the distal convoluted tubule, it is hypotonic in relation to the blood plasma, and it remains hypotonic throughout the distal tubule. The filtrate then passes down the collecting duct, once more traversing the zone of high solute concentration.*

From this point onward, the urine concentration depends on antidiuretic hormone (ADH). If ADH is absent, the wall of the collecting duct is not permeable to water, no additional water is removed, and a less concentrated urine is excreted. If ADH is present, the cells of the collecting duct are permeable to water, which moves by osmosis into the surrounding fluid, as shown in the diagram. In this case a concentrated (hypertonic) urine is passed down the duct to the renal pelvis, the ureter, the bladder, and finally out the urethra.

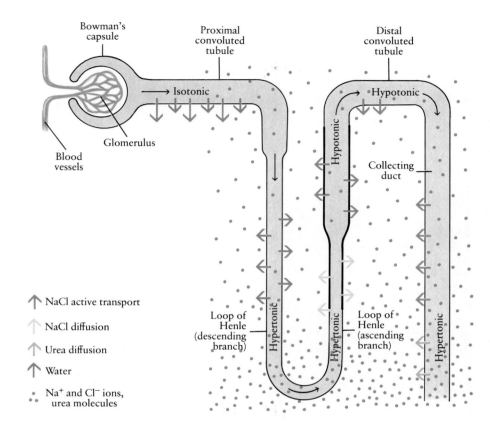

Water Conservation: The Loop of Henle

The control of urinary water loss is a major mechanism by which body water is regulated. Animals with free access to fresh water typically excrete a copious urine that is hypotonic in relation to their blood. The daily urinary output of a frog, for instance, totals 25 percent of its body weight. Most terrestrial animals cannot afford to be so profligate with water, however. In response to evolutionary pressures, birds and mammals have developed the ability to excrete hypertonic urines—that is, urines that are more concentrated than their body fluids. In mammals, this ability is associated with the hairpin-shaped section of the renal tubule known as the loop of Henle. By sampling fluids in and around different portions of the nephron, including different regions of the loop, and then analyzing the ions and molecules in each sample, physiologists have been able to determine how such a deceptively simple-looking structure makes this function possible.

Two structural features are the keys to the formation of a hypertonic urine. The first is the pathway formed by the basic structure of the nephron (Figure 37–13). From Bowman's capsule, the filtrate first enters the proximal convoluted tubule, descends the loop of Henle, ascends it, and then passes through the distal convoluted tubule into the collecting duct. The second feature is the differing permeability to water, salt, and urea of the wall in different parts of the nephron and the localized presence of membrane proteins for the active transport of salt. The wall of the proximal tubule is freely permeable to water and contains proteins that pump sodium ions (Na^+) out of the tubule by active transport, with chloride ions (Cl^-) following passively. The wall of the descending branch of the loop of Henle is also freely permeable to water but is relatively impermeable to salt. In the ascending branch of the loop the situation is quite different. The wall of the ascending branch is impermeable to water but permits the movement of salt: in the lower portion, diffusion of salt can occur, whereas in the upper portion, membrane proteins actively transport sodium and chloride ions out of the tubule. In the distal tubule, the wall is once more permeable to water and relatively permeable to salt. And, except under certain circumstances, so is the wall of the collecting duct. Moreover, the wall of the lower portion of the collecting duct is permeable to urea.

To understand the formation of a hypertonic urine, we must consider the consequences of the permeability of both the collecting duct and the ascending branch of the loop of Henle. First, because the lower portion of the collecting duct is permeable to urea, about half of the urea contained within the fluid passing through the duct diffuses into the surrounding interstitial fluid, from which it is ultimately reabsorbed into the peritubular capillaries. Second, in the upper portion of the ascending branch of the loop of Henle, sodium and chloride ions are actively pumped from the renal tubule into the interstitial fluid, through what is thought to be a cotransport process (see page 137). Water, however, cannot move through the wall of the ascending branch. As a result, sodium and chloride ions—as well as urea molecules—are present in high concentrations in the interstitial fluid bathing both branches of the loop of Henle and the collecting duct. Urea is most concentrated in the lower regions, and salt in the upper regions. With that in mind, let us now follow the filtrate as it moves through the nephron.

The fluid entering the proximal tubule from Bowman's capsule is isotonic with the blood plasma; that is, it has the same solute concentration as does the plasma. In the proximal tubule, sodium ions are pumped out, with chloride ions following passively. Water also moves out of the tubule by osmosis, following after these ions. Thus, as the fluid enters the descending branch of the loop of Henle, it is still isotonic with the blood plasma. Its volume is dramatically reduced, however; some 60 to 70 percent of the solutes and water contained within the original filtrate are removed from it during the passage through the proximal tubule. This water and the solutes it contains are taken up almost immediately by the peritubular capillaries and returned to the bloodstream.

The pathway of the loop then carries the remaining filtrate through a zone of high solute concentration—a result of the properties of both the ascending branch and the collecting duct. As we have noted, the walls of the descending branch are freely permeable to water, so large quantities of water move by osmosis from the tubule into the interstitial fluid (and from there into the peritubular capillaries), with only very small amounts of salt diffusing into the tubule. The fluid within the tubule is now hypertonic in relation to the plasma.

As the fluid travels through the ascending branch, some salt diffuses from the lower portion of the tubule and larger quantities are removed by the action of the transport proteins in the upper portion. As a consequence, the fluid within the tubule becomes hypotonic in relation to the plasma. In humans it remains hypotonic during its passage through the distal convoluted tubule. However, the fluid must again pass through the zone of high solute concentration as it flows down the collecting duct. Whether additional water is removed or not depends on the presence or absence of a hormone, antidiuretic hormone (ADH). If ADH is absent, the wall of the collecting duct is impermeable to water and a hypotonic urine is excreted. If, however, ADH is present, the wall is freely permeable to water. Water moves by osmosis through the wall, leaving within the duct a urine that is isotonic with the surrounding interstitial fluid but hypertonic in relation to the body fluids as a whole. In this way, mammals that need to conserve water are able to excrete a fluid, the urine, far more concentrated than the plasma from which it is derived.

37–14 *The cells of the kidney require large amounts of energy to power their active transport processes. This electron micrograph shows cells of the proximal tubule, where the active secretion and reabsorption of many solutes take place. Notice the numerous mitochondria and their position in relation to the brush border on the surface of the cells bordering the lumen of the tubule (the light area at the upper right of the micrograph). This border is made up of many very fine, hair-like extensions that, like microvilli, greatly increase the absorptive capacity of the cell. It is assumed that the mitochondria provide a sufficient nearby energy source to power active transport mechanisms residing in the convoluted cell surface. A red blood cell can be seen within the peri-tubular capillary, in the lower left of the micrograph.*

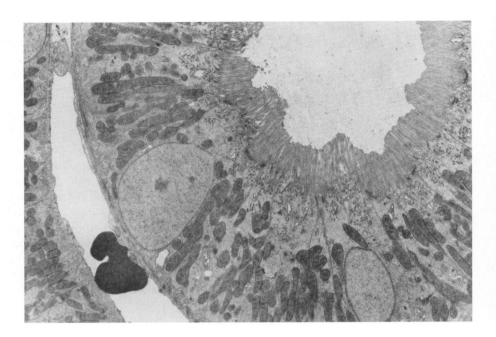

Although the ion pumps in the ascending branch of the loop of Henle and the permeability of the lower portion of the collecting duct to urea make possible the concentration differences on which this system depends, a key role is played by the hairpin structure of the loop as a whole. The flow of fluid through the loop provides another example of a countercurrent system. As we noted earlier (see page 130), a countercurrent flow can help to maintain differences in solute concentration from one end to the other of a system. When it is coupled to an active transport system, as in the loop of Henle, it can actually multiply the differences and is known as a countercurrent multiplier. The longer the loop, the greater the concentration differences that can be established. Since the primary factor limiting the concentration of the urine is the solute concentration surrounding the collecting duct, it should come as no surprise to learn that those mammals that excrete the most hypertonic urine also have the longest loops of Henle.

Control of Kidney Function: The Role of Hormones

In mammals, several different hormones act on the nephron to affect the composition of the urine. One of these is ADH, which is formed in the hypothalamus, a major regulatory center in the brain, and is stored in and released from the pituitary gland. As we have seen, ADH acts on the membranes of the collecting ducts of the nephrons and increases their permeability to water, so more water moves, by diffusion, back into the blood from the nephron. The amount of ADH released depends on the osmotic concentration of the blood and also on the blood pressure. Osmotic receptors that monitor the solute content of the blood are located in the hypothalamus. Pressure receptors that detect changes in blood volume are found in the walls of the heart, in the aorta, and in the carotid arteries. Stimuli received by these receptors are transmitted to the hypothalamus. Factors that increase the concentration of solutes in the blood or decrease blood pressure, or both, increase the production of ADH and conservation of water. Such factors include dehydration and hemorrhage. Factors that decrease blood concentration, such as the ingestion of large amounts of water, or that increase blood pressure—adrenaline, for instance—signal the hypothalamus to decrease ADH production, and so more water is excreted. Cold stress inhibits ADH secretion and thus increases urinary flow. Alcohol also suppresses ADH secretion and increases urinary flow, a phenomenon familiar to imbibers of beer and other alcoholic beverages. Pain and emotional stress trigger ADH secretion and thus decrease urinary flow.

A second hormone, aldosterone, which is produced by the adrenal cortex, stimulates reabsorption of sodium ions from the distal tubule and collecting duct and secretion of potassium ions into them. When the adrenal glands are removed, or when they function poorly (as in Addison's disease), excessive amounts of sodium chloride and water are lost in the urine, and the tissues of the body become depleted of them. Generalized weakness results, and if a patient with Addison's disease is not given hormone replacement therapy, the fluid loss can eventually be fatal. Aldosterone production is controlled by an extremely complex feedback circuit involving potassium ion levels in the bloodstream and processes initiated in the kidneys themselves.

Other hormones are also involved in the regulation of kidney function, particularly in response to increases in blood volume or blood pressure. The most intriguing of these substances is cardiac peptide (page 756), which inhibits the reabsorption of sodium from the distal tubule and thus increases the excretion of both sodium and water. This hormone, released from the atria of the heart, apparently exerts its effects both directly on the nephron itself and indirectly by inhibiting the release of aldosterone from the adrenal cortex. Receptors for cardiac peptide have been identified in both locations. There is also evidence that it may interact with other factors in the feedback circuit controlling aldosterone production. The complexity of the mechanisms regulating excretion provides additional evidence of the critical importance to the organism of maintaining a constant internal chemical environment.

SUMMARY

Homeostasis—the maintenance of a constant internal environment—is the result of a variety of processes within the animal body. One of the most critical homeostatic functions is regulation of the chemical composition of body fluids. This function, which in vertebrates is carried out primarily by the kidneys, involves (1) excretion of toxic waste products, especially the nitrogenous compounds produced by the breakdown of amino acids, (2) control of the levels of ions and other solutes in body fluids, and (3) maintenance of water balance.

Animals living in saltwater, freshwater, and terrestrial environments face different problems in maintaining the composition of body fluids. Terrestrial animals generally need to conserve water. Reptiles and birds excrete nitrogenous wastes in the form of crystalline uric acid, whereas mammals excrete them as urea, which must be dissolved in water.

Maintaining water balance involves equalizing gains and losses. The principal source of water gain in most mammals is in the diet; water is also formed as a result of oxidation of nutrient molecules. Water is lost in the feces and the urine, by respiration, and from the skin. Although the amount of water taken in and given off may vary widely from animal to animal and also from time to time in the same animal, depending largely on environmental circumstances, the volume of water in the body remains very nearly constant. The principal water compartments of the body are the plasma, the interstitial fluids, including the lymph, and the intracellular fluids. The major factor determining exchange of water among compartments of the body is osmotic potential.

The functional unit of the kidney is the nephron. Each nephron consists of a long tubule attached to a closed bulb (Bowman's capsule), which encloses a twisted cluster of capillaries, the glomerulus. Blood entering the glomerulus is under sufficient pressure to force plasma (minus the larger proteins) through the capillary walls into Bowman's capsule. As the filtrate makes its long passage through the nephron, cells of the renal tubule selectively reabsorb molecules from the filtrate and secrete other molecules into it. Glucose, amino acids, most ions, and a large amount of water are returned to the blood through the peritubular capillaries. Excess water and waste products, including about half of the urea present in the original filtrate, are excreted from the body as urine.

An important means of water conservation in mammals is the capacity for excreting a urine that is hypertonic in relation to the blood. The loop of Henle is the portion of the mammalian nephron that makes this possible.

The function of the nephron is influenced by hormones, chiefly antidiuretic hormone (ADH), which is produced by the hypothalamus and released from the pituitary gland; aldosterone, a hormone of the adrenal cortex; and cardiac peptide, released from the atria of the heart. ADH increases the return of water to the blood and so decreases water excretion. Aldosterone increases the reabsorption of sodium ions and water and the secretion of potassium ions. Cardiac peptide, by contrast, inhibits the reabsorption of sodium ions and water. All of these hormones play a role in regulation of both blood pressure and blood volume.

QUESTIONS

1. Distinguish among the following: afferent arteriole/efferent arteriole; Bowman's capsule/glomerulus; extracellular fluid/intracellular fluid; ADH/aldosterone/cardiac peptide.

2. Explain the following terms in relation to kidney function: filtration, secretion, reabsorption, and excretion.

3. Sketch a nephron, indicating the pathways of the blood and the glomerular filtrate.

4. Diagram the paths of glucose and urea through the human kidney.

5. Reexamine the micrographs in Figures 37–11 and 37–14. Describe the function of each structure visible in them in terms of the overall function of the nephron.

6. Why does a high-protein diet require an increased intake of water? (You should be able to think of two different reasons.) Why does a person lose some weight after shifting to a low-salt diet, even without reducing the caloric intake? Given the fact that amino acids in excess of the body's requirements are broken down by the liver, not stored, what is the advantage of a high-protein diet? What might be a disadvantage?

7. Sodium and chloride ions are actively pumped from the renal tubule in both the proximal convoluted tubule and the upper portion of the ascending branch of the loop of Henle. How does the mechanism of active transport in the proximal convoluted tubule differ from the mechanism in the ascending branch of the loop of Henle?

8. The text asserts that the primary factor limiting the concentration of the urine is the solute concentration surrounding the collecting duct. Why is this so?

9. In Figure 37–13, the labels within the tubule indicate the concentration of solutes in the filtrate in relation to their concentration in the blood plasma. From the information given about the movement of ions and of water, determine, for each labeled location, the concentration of the filtrate in relation to the concentration of the surrounding interstitial fluid.

10. Viewed as a whole, the kidney selectively removes substances from the bloodstream. A closer look, however, reveals that it does so by filtering out all small molecules through the glomerulus and then reabsorbing those that are needed as the filtrate passes along the renal tubule. Thus, the system need not identify wastes as such; rather, it must identify useful substances. Why would such an arrangement have been beneficial to early mammals? Why is it especially beneficial to modern mammals, including ourselves?

11. Could a human being on a life raft survive by drinking sea water? By catching and eating bony saltwater fish? Explain your answers.

Homeostasis II: Temperature Regulation

Some 200 years ago, Dr. Charles Blagden, then secretary of the Royal Society of London, went into a room that had been heated to a temperature of 126°C (260°F), taking with him a few friends, a small dog in a basket, and a steak. The entire group remained there for 45 minutes. Dr. Blagden and his friends emerged unaffected. So did the dog (the basket had kept its feet from being burned by the floor). But the steak was cooked. As this episode illustrates, the capacity of a living animal to regulate the internal temperature of its body is a critical factor in its well-being.

Physiological processes depend on both the maintenance of cellular structures and a multitude of biochemical reactions, virtually all controlled by enzymes. As we saw in Chapter 8, the rate at which an enzymatic reaction occurs is determined by a number of factors, one of the most important of which is temperature. In general, the reaction rate approximately doubles with every 10°C rise in temperature. However, for every protein, including every enzyme, there is a relatively narrow temperature range in which it maintains the three-dimensional conformation that is optimal for its particular functions. Above the upper limit of this range, the protein begins to lose its conformation, in the process known as denaturation. Once denaturation occurs, enzymes—and other proteins whose

38–1 Temperature regulation involves behavioral responses, as well as physiological and anatomical adaptations. In the cold, many small animals, such as these penguin chicks, huddle together for warmth. Huddling decreases the effective surface-to-volume ratio and thus the loss of heat from the body. A few of the chicks have lifted their heads from the warmth of the huddle to observe the photographer at work.

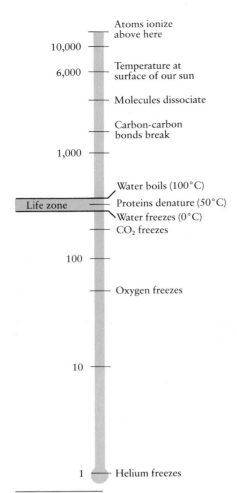

38–2 *Life processes can take place only within a very narrow range of temperature. The temperature scale shown here is the Kelvin, or absolute, scale. Absolute zero (0 K) is equivalent to −273.1°C, or −459°F, and is the temperature at which all molecular motion ceases.*

function depends on a specific shape—are inactivated. Low temperatures can also bring physiological processes to a halt. The slightly salty water that is the principal constituent of living tissue freezes at −1 or −2°C. Molecules that are not immobilized in the ice crystals are left in such a highly concentrated state that their normal interactions are completely disrupted. As a consequence of these biochemical facts, most animals must either occupy environments in which the temperature ranges from just below freezing to between 45 and 50°C, or through their own physiological processes create an internal environment within that range. The few exceptions, such as the fishes and arthropods of polar regions, have special adaptations, such as antifreeze molecules in their body fluids or the capacity to enter a state of near-total dormancy, that enable them to survive extreme conditions.

PRINCIPLES OF HEAT BALANCE

Water balance, as we saw in the preceding chapter, requires that the water gained equal the water lost through the body surface and in the urine and feces. Similarly, maintaining a constant temperature requires that heat gains equal heat losses. For animals, as for other living organisms, there are two primary sources of heat: the radiant energy of the sun and exothermic chemical reactions within the cells. However, heat can be lost from the body by any one of several different mechanisms (Figure 38–3).

In accordance with the second law of thermodynamics, heat moves from a warmer body to a colder body (see Figure 8–4, page 164). If the two bodies are in physical contact, so that kinetic energy can be transferred directly from molecule to molecule, the movement of heat is called **conduction.** Some materials are better conductors of heat than others. When you step out of bed barefoot on a cold morning, you probably prefer to step on a wool rug than on the bare floor. Although both are at the same temperature, the rug feels warmer. If you touch something metallic, such as a brass doorknob, it will feel even colder than the floor. These apparent differences in temperature are actually differences in the speed at which the different materials conduct heat away from your body.

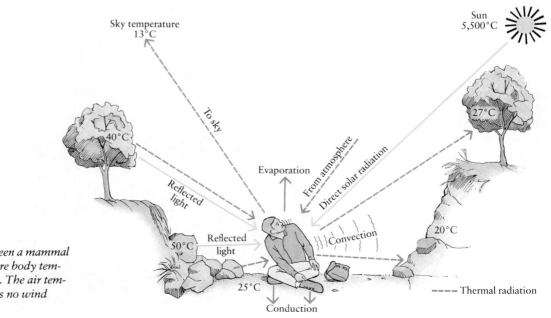

38–3 *Heat exchanges between a mammal and its environment. The core body temperature of the man is 37°C. The air temperature is 30°C and there is no wind movement.*

Water is an excellent conductor of heat, whereas air is a poor conductor. You are quite comfortable in air at 21°C (70°F) but may be uncomfortable in water at the same temperature. Conduction in fluids (air or water) is always influenced by **convection,** the movement of air or water in currents. Because both air and water become lighter as they get warmer, they move away from a heat source and are replaced by colder air or water, which again moves away as it warms. Fat, like air, is a poor conductor of heat, and both can serve as insulators. Animals that need to conserve heat are typically insulated with either fur or feathers, which trap air close to the body, or with fat or blubber.

Another route of heat loss is evaporation. As we saw in Chapter 2, every time a gram of water changes from a liquid to a gas, it takes more than 500 calories away with it. Many organisms, including ourselves, have exploited this property of water as a means for rapid adjustment of the heat balance.

Radiation, the transfer of energy by electromagnetic waves in the absence of direct contact, is a route not only of heat gain but also of heat loss. Depending on the wavelength of the radiation, energy may be transferred as light or heat (see Figure 10–3 on page 207). Light energy falling on an object is either absorbed as heat or reflected. As you undoubtedly know from your own experience, dark objects absorb heat more than do light-colored ones.

Body Size and the Transfer of Heat

Regardless of the mechanism, heat is transferred into or out of any object, animate or inanimate, across the body surface. Thus, the transfer of heat, like the diffusion of oxygen and carbon dioxide across the respiratory surface, is proportional to the surface area exposed. The smaller the object, as we saw in Chapter 5, the larger its surface-to-volume ratio. Ten pounds of ice, separated into individual ice cubes, will melt far faster than the equivalent volume in a solid block. For the same reason, it is much more difficult for a small animal to maintain a constant body temperature than it is for a large one. One of the ways in which many small animals deal with this problem in the face of extreme cold is togetherness, as illustrated in Figures 38–1 and 38–4.

38–4 *Within the wintering hive, bees maintain their temperature by clustering together in a dense ball; the lower the temperature, the denser the cluster. The clustered bees produce heat by constant muscular movements of their wings, legs, and abdomens. In very cold weather, the bees on the outside of the cluster keep moving toward the center, while those in the center move to the colder outside periphery. The entire cluster moves slowly about on the combs, eating the stored honey, which is their energy source.*

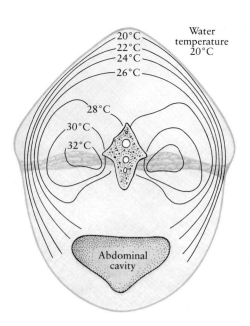

Water temperature 20°C

20°C
22°C
24°C
26°C
28°C
30°C
32°C
0°

Abdominal cavity

38–5 *The body of a 70-kilogram big-eye tuna, as shown in cross section. The internal temperature of the fish forms a gradient, ranging from a high of about 32°C deep within its body to 20°C (the temperature of the surrounding water) at its surface. Its body heat is produced by metabolic reactions and muscular activity, particularly in the hard-working, red, myoglobin-containing muscles that run the length of the body, parallel to the vertebral column.*

"COLD-BLOODED" VS. "WARM-BLOODED"

In common parlance, animals are often characterized as "cold-blooded" or "warm-blooded." This fits in with everyday experience: a snake is usually cold to the touch, and a bird feels warm. Actually, however, a "cold-blooded" animal may create an internal temperature for itself that is warmer than that of a "warm-blooded" animal. Another approach is to classify animals as ectotherms or endotherms. As we saw in Chapter 28, ectotherms are warmed from the outside in, and endotherms are warmed from the inside out. These categories correspond approximately but not completely with "cold-blooded" and "warm-blooded." Fish, for example, are considered "cold-blooded," yet some large fish, such as tuna, generate large amounts of metabolic heat and are, strictly speaking, endotherms (Figure 38–5).

When considering temperature regulation—rather than the actual temperature of an animal or the means by which its body is heated—the most useful classification of an animal is as a poikilotherm or a homeotherm. A **poikilotherm** (from the Greek word *poikilos*, meaning "changeable") has a variable temperature, and a **homeotherm,** a constant one. These terms, like ectotherm and endotherm, are generally used as synonymous with "cold-blooded" and "warm-blooded," though again the correspondence is not perfect. Fish that live deep in the sea, where the temperature of the water remains constant, have body temperatures far more constant than bats or hummingbirds, whose temperatures, as we shall see, fluctuate widely depending on their state of activity.

POIKILOTHERMS

Most aquatic animals are poikilotherms, maintaining a body temperature that is the same as the temperature of the surrounding water. Although the metabolic processes of such animals generate heat, it is usually quickly dissipated, even in large animals. In most fish, for example, heat is rapidly carried from the core of the body by the bloodstream and is lost by conduction into the water. A large proportion of this heat is lost from the gills. Exposure of a large, well-vascularized surface to the water is necessary in order to acquire oxygen, as you will recall from Chapter 35. This same process rapidly dissipates heat, so fish usually cannot maintain a body temperature significantly higher than that of the water. They also cannot maintain a temperature lower than that of the water, since they have no means of unloading heat.

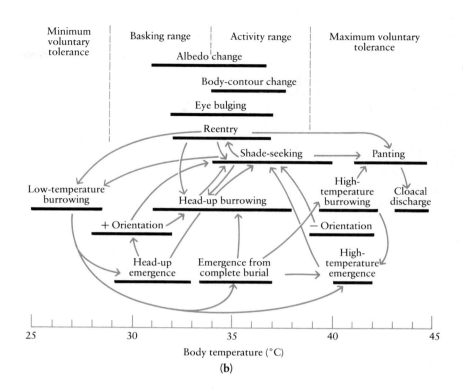

(a)

38–6 *In laboratory studies, the internal temperature of reptiles was shown to be almost the same as the temperature of the surrounding air. It was not until observations were made of these animals in their own environment that it was found that they have behavioral means of temperature regulation that give them a surprising degree of temperature independence. By absorbing solar energy, reptiles can raise their temperature well above that of the air around them.* (a) *Shown here is a horned lizard (often but not accurately known as a horned toad) that, having been overheated by the sun, has raised its body to allow cooling air currents to circulate across its underside.* (b) *This chart is based on field studies of the behavior of a horned lizard in response to temperature fluctuations. Changes in albedo (whiteness) are produced by pigment changes in the epithelial cells of the skin. Albedo changes affect the reflection versus absorption of light energy.*

In general, large bodies of water, for the reasons we discussed in Chapter 2, maintain a very stable temperature. At no place in the open ocean does the temperature vary more than 10°C in a year. In smaller bodies of water, where greater temperature changes occur, fish seek a level of optimal temperature, presumably the one to which their metabolic processes are adapted. However, because they can do almost nothing to make their own temperature different from that of the surrounding water, they can be quickly victimized by any rapid, drastic changes in water temperature.

Temperatures on land, in contrast to those in water, may vary annually in a given area by as much as 60 to 70°C. Thus temperature regulation, like water conservation, became a problem for animals when they moved from water to land. The first animals to become truly terrestrial, the reptiles (snakes, lizards, and tortoises), are, like fish, poikilotherms. Nonetheless, they are able to maintain remarkably stable body temperatures during their active hours by varying the amount of solar radiation they absorb (Figure 38–6). By careful selection of suitable sites, such as the slope of a hill facing the sun, and by orienting their bodies with a maximum surface exposed to sunlight, they can heat themselves rapidly. Such heating can occur as quickly as 1°C per minute, even on mornings when the air temperature, as on deserts or in the high mountains, is close to 0°C. By frequently changing position, these reptiles are able to keep their temperature within quite a narrow range as long as the sun is shining. When it is not, they seek the safety of their shelters. Although their body temperatures then drop and they become sluggish, they are not immobilized in an exposed position, where they are vulnerable to predators.

HOMEOTHERMS

Homeotherms are animals that maintain a constant body temperature despite fluctuations in their environment, and most maintain a body temperature well above that of their surroundings. Among modern animals, only birds and mam-

38–7 *Observed metabolic rates of mammals. Each division on the abscissa represents a tenfold increase in weight. The metabolic rate of very small mammals is much higher than that of larger mammals, owing principally to their greater surface-to-volume ratios.*

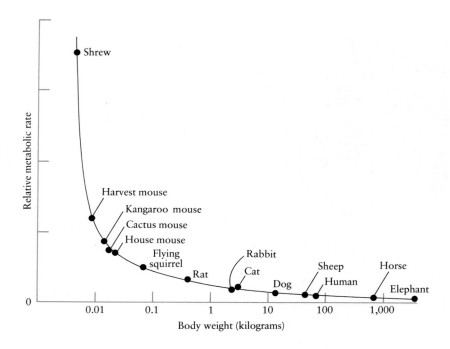

mals are true homeotherms. All homeotherms are endotherms, with the oxidation of glucose and other energy-yielding molecules within body cells as their primary source of heat.

Homeothermy brought with it a number of advantages, of which perhaps the most important was the capacity to function at peak efficiency even at low external temperatures. This capacity, for example, would have enabled the early mammals to be out and about at night, seeking food and mates, when the dominant animals of that era—the reptiles—were most likely inactive. It also made possible the invasion of less hospitable environments, particularly those with a relatively low temperature much of the year. In terms of energy requirements, however, the cost of homeothermy is high: the metabolic rate of a mammal is about 10 times that of a reptile of similar size at similar temperature. Moreover, the smaller the size, the higher the price (Figure 38–7). As a consequence of its high energy requirements, a mammal must absorb from its digestive tract about 10 times the quantity of nutrients as a reptile of comparable size. It does this by eating more and processing it faster, which is made possible by a more convoluted digestive tract with greater numbers of villi and microvilli.

Birds have even higher metabolic requirements. Although comparatively small, they maintain a higher body temperature—40 to 42°C—than most mammals. Also, unlike most small mammals, they spend much of their time exposed. Flight compounds their problems. To fly, birds must keep their weight down, and so they cannot store large amounts of fuel. Because of this and the high energy requirements of their way of life, birds need to eat constantly. A bird eating high-protein foods, such as seeds and insects, commonly consumes as much as 30 percent of its body weight per day. Bird migrations are dictated not so much by a need to seek warmer weather as by a need for longer days with more daylight hours for feeding themselves and their young.

Because it is heated from within, a homeotherm is warmer at the core of its body than at the periphery. Our temperature, for example, usually does not reach 37°C, or 98.6°F, until some distance below the skin surface. Heat is transported from the core to the periphery largely by the bloodstream. At the surface of the body, the heat is transferred to the air, as long as air temperature is less than body temperature. Temperature regulation involves increasing or decreasing heat production and increasing or decreasing heat loss at the body surface.

Avian Mechanical Engineers

Although we generally think of engineering as a profession invented and practiced only by representatives of *Homo sapiens,* birds have been practicing civil engineering in their design and construction of nests for far longer. Members of the family Megapodiidae, the mound-building birds of Australia, New Guinea, and other Pacific islands, are also accomplished mechanical engineers, harnessing the energy of external chemical reactions for their own purposes.

The developing young of homeotherms, both birds and mammals, must be maintained at a constant temperature, usually a temperature fairly close to that of the adult animal. In all mammals except the monotremes, this problem is solved automatically, as the young develop either within the body of the mother or, in the case of marsupials, in the mother's pouch. Birds generally keep their eggs warm by sitting on them, or in the case of penguins, tucking them on top of the feet, where they can be held close to the parent's body. In desert environments, however, there is a twofold problem: keeping the eggs cool during the day and yet warm at night, when the temperature drops dramatically. The mound-building mallee fowl of the Australian desert solve this problem by incubating their eggs in large compost heaps.

The parent birds, which are about the size of chickens, begin the process in the autumn by digging a pit, 3 meters wide by 1 meter deep, and raking plant litter into it. Next they build a huge mound of insulating sand around it, and then they wait for the spring rains. Following the rains, when the litter begins to decompose, the birds cover the fermenting heap with a layer of sand up to a meter in depth. Gradually, the fermentation reactions in the litter heat the mound from beneath. The male waits a few days and then begins checking its temperature with his beak, which contains a heat sensor. When the temperature is holding close to 33°C, he allows the female to begin laying her eggs. During the seven-week incubation period, he constantly checks and regulates the temperature in the mound by scraping away the sand around the eggs to expose them to cooling air currents or by piling up warm sand around them at night, maintaining a nearly constant 33°C, day and night.

The entire cycle from the beginning of mound building until the last egg hatches takes about a year. When the chicks hatch, completely feathered, their own internal temperature regulating mechanism is fully functional and they are able to feed and fend for themselves.

A pair of mallee fowl, surveying the result of their labors. All of the digging and moving of sand and litter is done with the large feet, from which this family of birds derive their name.

38–8 *Body temperature in mammals is regulated by a complex network of activities involving both the nervous and endocrine systems. The chief regulatory center is the hypothalamus, a brain center that, as we have seen, controls many physiological processes. In some animals (although not, apparently, in humans), the hormonal pathway—indicated by dashed lines—is of major importance. TRH is thyroid-releasing hormone, which is produced by the hypothalamus and which stimulates the production of TSH, thyroid-stimulating hormone, by the pituitary. TSH, in turn, stimulates the production of thyroxine, the thyroid hormone. Thyroxine increases cellular metabolism, apparently by acting directly on the mitochondria. Many behavioral responses, such as seeking sun or shelter, are also involved.*

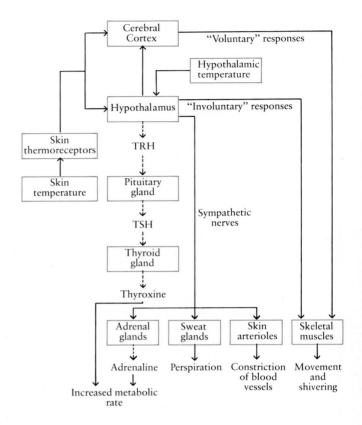

The Thermostat

The remarkable constancy of temperature characteristic of humans, and other homeotherms as well, is maintained by an automatic system—a thermostat—in the hypothalamus. This thermostat receives and integrates information from widely scattered temperature receptors, compares it to the set point of the thermostat, and, on the basis of this comparison, initiates appropriate responses. Unlike a furnace thermostat, which controls a simple on-off switch, the hypothalamic thermostat has a variety of complex responses at its command, as summarized in Figure 38–8.

Under ordinary conditions, the skin receptors for hot and cold are probably the most important sources of information about temperature change. However, the hypothalamus itself contains receptor cells that monitor the temperature of the blood flowing through it. As some interesting experiments have demonstrated, information received by these hypothalamic receptors can override that from other sources, ensuring that the core body temperature remains constant. For example, in a room in which the air is warmer than body temperature, if the blood circulating through a person's hypothalamus is cooled, he or she will stop perspiring, even though the skin temperature continues to rise.

The elevation of body temperature known as fever is due not to a malfunction of the hypothalamic thermostat but to a resetting. Thus, at the onset of fever, an individual typically feels cold and often has chills; although the body temperature is rising, it is still lower than the new thermostat setting. The substance primarily responsible for the resetting of the thermostat is a protein released by white blood cells in response to pathogens. The adaptive value of fever is not known with certainty, but current evidence indicates that moderate temperature increases stimulate the immune system, increasing not only the synthesis of antibodies but also the production of other substances and specialized cells that work together in the body's defenses against infection.

(a) (b) (c)

38–9 *The size of the extremities in a particular type of animal can often be correlated with the climate in which it lives.* (a) *The fennec fox of the North African desert has large ears, rich with blood vessels. As blood flows through the network of capillaries just below the skin surface, excess heat is dissipated from the body.* (b) *The red fox of the eastern United States has ears of intermediate size, and* (c) *the Arctic fox has relatively small ears. Similar correlations of characteristics such as size, weight, and color with environment can also sometimes be made among animals of a single species living over an extended geographic range.*

Regulating as Body Temperature Rises

As the body temperature of a mammal rises above its thermostat setting, the blood vessels near the skin surface are dilated, and the supply of blood to the skin increases. If the air is cooler than the body surface, heat can be transferred from the skin directly to the air. Animals that live in hot climates characteristically have larger exposed surface areas than animals that live in the cold (Figure 38–9).

Heat can also be lost from the surface by the evaporation of saliva or perspiration; the heat required for evaporation comes from blood traveling in vessels just below the skin surface. In humans and most other large mammals, when the external temperature rises above body temperature, perspiration begins. Horses and humans sweat from all over their body surfaces. Dogs and some other animals pant, so that air passes rapidly over their large, moist tongues, evaporating saliva. Cats lick themselves all over as the temperature rises, and evaporation of their saliva cools their body surface. For animals that dissipate heat through evaporation, temperature regulation at high temperatures necessarily involves water loss. This, in turn, stimulates thirst and water conservation by the kidneys. Dr. Blagden and his friends were probably very thirsty.

(a) (b) (c)

38–10 *Animals have various devices for utilizing the heat-absorbing properties of evaporating water.* (a) *Dogs unload heat by panting, which involves short, shallow breaths and the production of a copious saliva. When it is hot, a dog pants at a rate of about 300 to 400 times a minute,* compared with a respiration rate of from 10 to 40 times a minute in cool surroundings. Panting, in contrast to sweating, does not result in the loss of salt or other important ions. (b) Among the animals that can sweat over their entire body surface are hippopotamuses. Their sweat, unlike that of humans and horses, is pink. (c) Elephants, lacking sweat glands, wet down their thick, dry skins with mud or water. They also unload heat by flapping their ears, which are highly vascularized.

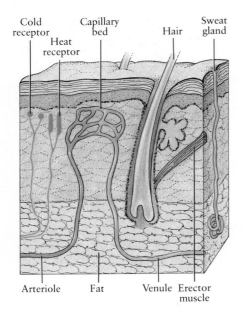

Cold receptor
Heat receptor
Capillary bed
Hair
Sweat gland

Arteriole   Fat   Venule   Erector muscle

38–11 *Cross section of human skin showing structures involved in temperature regulation. In cold, the arterioles constrict, reducing the flow of blood through the capillaries, and the hairs, each of which has a small erector muscle under nervous control, stand upright. Beneath the skin is a layer of subcutaneous fat that serves as insulation, retaining the heat in the underlying body tissues. With rising temperatures, the arterioles dilate and the sweat glands secrete a salty liquid. Evaporation of this liquid cools the skin surface, dissipating heat (approximately 540 calories for every gram of H_2O).*

Regulating as Body Temperature Falls

When the temperature of the circulating blood begins to fall below the thermostat setting, blood vessels near the skin surface are constricted, limiting heat loss from the skin. Metabolic processes increase. Part of this increase is due to increased muscular activity, either voluntary (shifting from foot to foot) or involuntary (shivering). Part is due to direct stimulation of metabolism by the endocrine and nervous systems. Adrenaline stimulates the release and oxidation of glucose. Autonomic nerves to fat increase its metabolic breakdown. In some mammals exposed to prolonged cold, the thyroid gland increases its release of thyroxine, the thyroid hormone. Thyroxine appears to exert its effects directly on the mitochondria.

Most mammals have a layer of subcutaneous fat that serves as insulation. Homeotherms also characteristically have hair or feathers that, as temperature falls, are pulled upright by erector muscles in the skin, trapping air, which insulates the surface. All we get, as our evolutionary legacy, are goose pimples.

Cutting Energy Costs

Less energy is consumed by a sleeping animal than by an active one. Bears, for example, nap for most of the winter, living on fat reserves, permitting their temperature to drop several degrees, and keeping their energy requirements down.

Turning down the thermostat saves more fuel. Some small animals have different day and night settings. A hummingbird, for instance, has an extremely high rate of fuel consumption, even for a bird. Its body temperature drops every evening when it is resting, thereby decreasing its metabolic requirements and its fuel consumption. Another effect of this resetting is a reduction in the loss of water across the respiratory surfaces.

Hibernation (from *hiber*, the Latin word for "winter") is another means of adjusting energy expenditures to food supplies. Hibernating animals do not stop regulating their temperatures altogether—like the hummingbird, they turn down their thermostats. Hibernators are mostly small animals, including a few insectivores, hamsters, ground squirrels, some South American opossums, some bats, and a few birds, such as the poor-will of the southwestern United States. In these animals, the thermostat is set very low, often close to the temperature of the surrounding air (if the air is above 0°C). The heartbeat slows; the heart of an active ground squirrel, for instance, beats 200 to 400 times a minute, whereas that of a hibernating one beats 7 to 10 times a minute. The metabolism, as measured by oxygen consumption, is reduced 20 to 100 times. Apparently, even aging stops; hibernators have much longer life spans than similar nonhibernators. However, despite these profound physiological changes, hibernators do not cease to monitor the external environment. If the animals are exposed to carbon dioxide, for example, they breathe more rapidly, and those of certain species wake up. Similarly, hibernating animals wake up if the temperature in their burrows drops to a life-threatening low of 0°C. Hibernators of some species can also be awakened by sound or by touch.

Arousal from hibernation can be rapid. In one experiment, bats kept in a refrigerator for 144 days without food were capable of sustained flight after 15 minutes at room temperature. As indicated, however, by the arousals at 0°C, arousal is a process of self-warming rather than of collecting environmental heat. Breathing becomes more regular and then more rapid, increasing the amount of oxygen available for consumption; subsequently, the animal "burns" its stored food supplies as it returns to its normal temperature and the fast pace of a homeothermic existence.

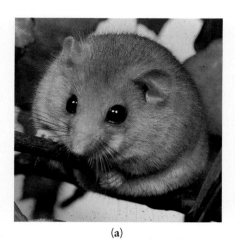

(a)

(b)

(c)

38–12 *Some energy conservation measures.* (a) *This dormouse has stored food reserves in body fat in preparation for hibernating. Hibernation saves energy by turning the thermostat down very low, slowing the heartbeat, and reducing the metabolic rate.* (b) *Note the heat-conserving reduction in surface-to-volume ratio in this hibernating golden-mantled ground squirrel.* (c) *Bats have a number of adaptations that enable them to conserve heat despite their small body size. Some hibernate. In others, the temperature drops when they are at rest. Members of the species shown here also roll up their ears when they get cold.*

ADAPTATIONS TO EXTREME TEMPERATURE

We rely mainly on our technology, rather than our physiology, to allow us to live in extreme climates, but many animals are comfortable in climates that we consider inhospitable or, indeed, uninhabitable.

Adaptations to Extreme Cold

Animals adapt to extreme cold largely by increased amounts of insulation. Fur and feathers, both of which trap air, provide insulation for Arctic land animals. Fur and feathers are usually shed to some extent in the spring and regrow in the fall.

Aquatic homeotherms, such as whales, walruses, seals, and penguins, are insulated with fat (neither fur nor feathers serve as effective insulators when wet). These marine animals, which survive in extremely cold water, can tolerate a very great drop in their skin surface temperature; measurements of skin temperature have shown that it is only a degree or so above that of the surrounding water. By permitting the skin temperature to drop, these animals, which maintain an internal temperature as high as that of a person, expend very little heat outside of their fat layer and so keep warm—like a diver in a wet suit.

Some large Arctic mammals, such as fur seals and polar bears, are so well insulated that, when they are on land, their principal physiological problem is unloading excess heat. One solution to this problem is to keep the fur moist; another is to sleep, which in some species of fur seals reduces heat production by almost 25 percent.

Countercurrent Exchange in Heat Conservation

Numerous animals conserve heat by permitting temperatures in the extremities to drop. The extremities of such animals are actually adapted to a different internal temperature than the rest of the animal. For example, the fat in the foot of an Arctic fox has a different thermal behavior than the fat in the rest of its body, so that its footpads are soft and resilient even at temperatures of −50°C. Also, for many of these animals, particularly those that stand on the ice, this capacity is essential for another reason. If, for example, the feet of an Arctic seabird were as warm as its body, they would melt the ice, which might then freeze over them, trapping the animal until the spring thaws set it free.

38–13 *The principle of countercurrent heat exchange as it occurs in the extremities of many animals is illustrated here by a hot-water pipe and a cold-water pipe placed side by side. In (a), the hot water and cold water flow in the same direction. Heat from the hot water warms the cold water until both temperatures equalize at 5 on this arbitrary scale. Thereafter, no further exchange takes place, and the outflowing water in both pipes is lukewarm. In (b), the flow is countercurrent, so heat transfer continues for the length of the pipes. The result is that the hot water transfers most of its heat as it travels through the pipe, and the cold water is warmed to almost the initial temperature of the hot water at its source.*

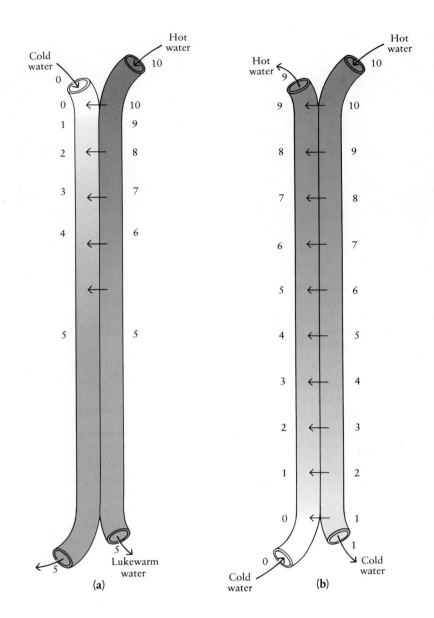

(a) (b)

In many Arctic animals, the lower internal temperature of the extremities is made possible by a countercurrent arrangement that keeps heat in the body and away from the extremities (Figure 38–13). The arteries and veins leading to and from the legs (or fins or tail) are juxtaposed in such a way that the chilled blood returning through the veins picks up heat from the blood entering through the arteries. The veins and arteries are closely apposed to give maximum surface for heat transfer. Thus the body heat carried by the blood is not wasted by being dissipated to the cold air at the extremities. Instead it serves the useful function of warming the chilled blood that would otherwise put a thermal burden on the body.

Adaptations to Extreme Heat

We are efficient homeotherms at high external temperatures, as Dr. Blagden's experiment showed. Our chief limitation in this regard is that we must evaporate a

38–14 *By facing the sun, a camel exposes as small an area of body surface as possible to the sun's radiation. Its body is insulated by fat on top, which minimizes heat gain by radiation. The underpart of its body, which has much less insulation, radiates heat out to the ground. Note the loose-fitting garments worn by the camel driver. When such garments are worn, sweat evaporates from the skin surface. With tight-fitting garments, the sweat evaporates from the surface of the garment instead, and much of the cooling effect is lost. The loose robes, which trap air, also serve to keep desert dwellers warm during the cold nights. Other adaptations of the camel to desert life include long eyelashes, which protect its eyes from the stinging sand, and flattened nostrils, which retard water loss.*

great deal of water in order to unload body heat, and so our water consumption is high. The camel, the philosophical-looking "ship of the desert," has several advantages over human desert dwellers. For one thing, a camel excretes a much more concentrated urine. Also, a camel can lose more water proportionally than a human and still continue to function. If a person loses 10 percent of the body weight in water, he or she becomes delirious, deaf, and insensitive to pain. If the loss is as much as 12 percent, the individual is unable to swallow and so cannot recover without assistance. Laboratory rats and many other common animals can tolerate dehydration of up to 12 to 14 percent of body weight. Camels can tolerate the loss of more than 25 percent of their body weight in water, going without drinking for as long as one week in the summer months, three weeks in the winter.

Probably most important, the camel can tolerate a fluctuation in internal temperature of as much as 6°C. This tolerance means that it can let its temperature rise during the daytime (which the human thermostat would never permit) and drop during the night. The camel begins the next day at below its normal temperature—storing up coolness, in effect. It is estimated that the camel saves as much as 5 liters of water a day as a result of these internal temperature fluctuations.

Camels' humps were once thought to be water-storage tanks, but actually they are localized fat deposits. Physiologists have suggested that the camel carries its fat in a dorsal hump, instead of distributed all over the body, because the hump, acting as an insulator, impedes heat flow into the body core. A uniform fat distribution, which is decidedly useful in Arctic animals, would not be in inhabitants of hot climates.

Most small desert animals are, like the earliest mammals, nocturnal. The avoidance of direct heat is their principal means of temperature regulation. Small desert animals usually do not unload heat by sweating or panting; because of their relatively large surface areas, such mechanisms would be extravagant in terms of water loss. As these animals remind us, physiological problems, even those that may seem quite distinct at first glance, are always interrelated. A successful solution to any specific physiological problem invariably involves the interactions of different systems, balancing the various demands on and needs of the animal as a whole.

SUMMARY

Life can exist only within a very narrow temperature range, from about 0°C to about 50°C, with few exceptions. Animals must either seek out environments with suitable temperatures or create suitable internal environments. Heat balance requires that the net heat loss from an animal equal the heat gain. The two primary sources of heat gain are the radiant energy of the sun and cellular metabolism. Heat is lost through conduction, which, in fluids, is aided by convection, through evaporation, and through radiation.

Water-dwelling animals usually maintain a body temperature that is the same as the relatively constant temperature of the water. On land, environmental temperatures are much more variable. Animals—whether aquatic or terrestrial—whose internal temperature fluctuates with that of the external environment are known as poikilotherms, whereas those that maintain a constant internal temperature despite external temperature changes are known as homeotherms. Terrestrial poikilotherms are generally ectotherms, taking in heat energy from the environment; through behavioral means, they are able to adjust the amount of heat taken in and thus regulate their internal temperature during daylight hours. All true homeotherms—birds and mammals—are endotherms, generating heat internally from the oxidation of glucose and other fuel molecules and by friction from muscular activity.

In mammals, temperature is regulated by a thermostat in the hypothalamus. This thermostat receives and integrates information from temperature-sensitive receptors in the skin and in the hypothalamus itself and triggers appropriate responses. As body temperature rises, blood vessels in the skin dilate, increasing the blood flow to the surface of the body. Evaporation of water from body surfaces increases heat loss. As body temperature falls, blood vessels in the skin constrict, reducing heat loss. Energy production is increased by muscular activity and by nervous and hormonal stimulation of metabolism.

The energy costs of homeothermy are high. Sleeping is one means of reducing energy consumption; in some small homeotherms, additional energy is conserved by a lowering of the thermostat setting when the animal is resting. For example, hummingbirds have different thermostat settings for day and night. Other homeotherms make seasonal adjustments, entering the state known as hibernation during the winter when temperatures are low and food is scarce.

Adaptations to extreme cold principally involve insulation by fur or feathers and heavy layers of subcutaneous fat, coupled with countercurrent mechanisms that prevent the loss of heat from the extremities. Adaptations to extreme heat in large animals such as the camel include a relatively wide tolerance for fluctuations in temperature and total water content. In smaller animals, water conservation measures and behavioral adaptations are of primary importance.

QUESTIONS

1. Distinguish among the following terms: endotherm, ectotherm, poikilotherm, and homeotherm.

2. Compare the surface-to-volume ratio of an Eskimo igloo with that of a California ranch house. In what way is the igloo well suited to the environment in which it is found?

3. Describe what happens in the human body as its temperature rises. As it falls.

4. In terms of temperature regulation, what is the advantage of having temperature monitors that are located externally, as on the outside of a building or in the skin? Why is it also important to monitor the internal temperature? Which should be the principal source of information for your hypothalamic thermostat?

5. Compare hibernation in animals with dormancy in plants. In what ways are they alike? In what ways are they different?

6. Plants, as well as animals, are warmed above air temperature by sunlight, and plants cannot actively seek shade. Why might the leaves in the sunlight at the top of an oak tree be smaller and more extensively lobed (more fingerlike) than the leaves in the shady areas lower on the tree?

7. Diagram a heat-conserving countercurrent mechanism in the leg of an Arctic animal.

8. In the first few patients to receive an artificial heart, a cardiac output (page 756) of 6 liters per minute was found to be optimal and was thus considered the standard to which the device should be adjusted. A later patient, weighing 220 pounds, in whom the artificial heart was used as a temporary measure while awaiting a heart transplant, seemed also to do well at this level. However, his core body temperature was above normal and the temperature of his skin and extremities was below normal. What adjustment did his doctors make to restore both core and surface temperatures to normal? Why was it needed?

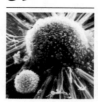

Homeostasis III: The Immune Response

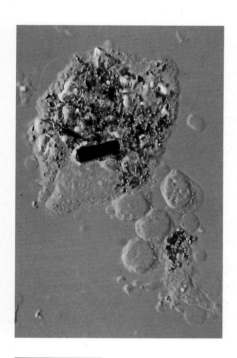

39–1 *The defense mechanisms of contemporary animals must deal with pollutants of human origin as well as with infectious microorganisms. This macrophage, a type of phagocytic white blood cell, was removed from a human lung. It had ingested a number of foreign particles, including a large rectangular fragment that is probably silicon, a principal component of computer chips. White blood cells are important in both nonspecific and specific responses to microorganisms and other possible pathogens.*

Every living thing, including yourself at this moment, is surrounded by potentially harmful microorganisms. Many of these microorganisms have the capacity not only to destroy individual cells but also to disrupt the numerous interrelated processes on which the continuing life of the organism depends; thus the defense against such microorganisms is an essential aspect of homeostasis.

Over the long course of evolution, organisms have developed a variety of defenses that function to exclude would-be invaders or to overcome them should they gain entry. In vertebrates, such defenses have evolved into an elaborate interacting network involving both nonspecific responses (in particular, inflammation) and the highly specific responses of the immune system, precisely tailored for each different invader. An understanding of these responses is of such practical importance for human health that most research has been specifically focused on mammals, with information derived from both laboratory animals, such as mice, and human patients. As a consequence, we now know a great deal about mammalian defensive mechanisms, but surprisingly little about those of other vertebrates.

NONSPECIFIC DEFENSES

Anatomic Barriers

The body's first line of defense against foreign invaders is its outer wrapping of skin and mucous membranes. The skin, with its tough layer of keratin, is an impregnable barrier as long as it is intact. When it is not, large numbers of microorganisms gain ready entry to the body. This is starkly illustrated in individuals who have suffered extensive burns; for such persons, the most immediate and greatest threats to life are the severe infections to which they are vulnerable.

The epithelium that forms mucous membranes is more fragile than the skin, but it is constantly flushed with fluids, such as mucus, saliva, and tears, that contain antimicrobial substances. The epithelium lining the respiratory tract is carpeted with cilia (page 122), which sweep away inhaled microorganisms, dirt, and debris trapped in the protective layer of mucus. The extremely acidic pH of the stomach contents creates an inhospitable environment for potential immigrants accompanying ingested food. In addition, the lower intestinal tract harbors resident populations of bacteria that defend their home territory against other microorganisms. Despite these defenses, the mucous membranes are the most common sites of entry of microorganisms or their toxins, typically through rips or tears in the epithelium.

The Inflammatory Response

If a microorganism penetrates the outer barrier, it encounters a second line of defense, consisting of a variety of agents carried by the circulating blood and

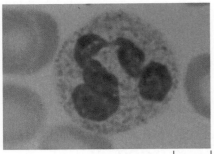

5 µm

39-2 *A neutrophil, the most common type of granulocyte; note the numerous small granules in the cytoplasm. The nucleus of a neutrophil is many-lobed; as a result, it appears to have many different shapes in different microscopic sections. Thus neutrophils are said to be polymorphonuclear, and they are often called "polymorphs" or "polys" for short.*

TABLE 39-1 **Types of White Blood Cells**

	PERCENT OF TOTAL
Granulocytes	
Neutrophils	50–70
Eosinophils	1–4
Basophils	0.1
Lymphocytes	20–40
Monocytes	2.8

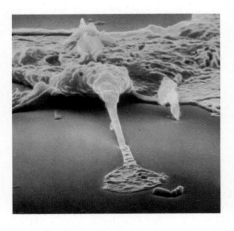

39-3 *Macrophages, as well as other types of white blood cells, move through the tissues by amoeboid motion, forming and retracting pseudopodia. They are also phagocytic, engulfing bacteria and other small particles. This patrolling macrophage has detected the chemical trail of an Escherichia coli cell, which it is about to capture with its precisely targeted pseudopod.*

lymph. Suppose, for example, you cut your skin. Cells in the area immediately release histamine and other chemicals that increase both blood flow into the area and the permeability of nearby capillaries. Circulating white blood cells, attracted by these chemicals, move through the capillary walls, crowding into the site of the injury. These cells engulf the foreign invaders, often literally eating themselves to death. Blood clots begin to form, walling off the injured area. The local temperature often rises, creating an environment unfavorable to the multiplication of microorganisms, while accelerating the motion of white blood cells. As a consequence of this series of events, known as the **inflammatory response,** the injured area becomes swollen, hot, red, and painful.

Both the inflammatory response and the more specific immune response depend on the interaction of a variety of types of white blood cells (Table 39–1). These cells, like the oxygen-carrying red blood cells, have a finite life span and must be continuously replenished. All of the different types of white blood cells, as well as the red blood cells, result from the differentiation and division of common, self-regenerating **stem cells** located in the marrow of long bones.

The principal cells involved in the inflammatory response are **granulocytes,** circulating white blood cells that are classified by their staining properties as neutrophils, eosinophils, or basophils. Neutrophils (Figure 39–2) are by far the most numerous, making up 50 to 70 percent of all white blood cells. When the first signs of inflammation appear, neutrophils, which are usually round and move freely through the bloodstream, begin to stick to the inner surface of the endothelium lining the blood vessels. They then develop amoeboid projections that enable them to push their way between the endothelial cells of the capillaries and move out into the infected tissues. Here they phagocytize microorganisms and other foreign particles, just as an amoeba engulfs its prey (page 474). The cytoplasmic granules from which granulocytes derive their name are actually lysosomes, and once the offending microparticle is within the neutrophil, lysosomes fuse with the phagocytic vacuole. Most phagocytized microbes are killed and digested within the vacuole by microbicidal proteins and lytic enzymes from the lysosomes. Some microbes—such as the encapsulated pneumococcus that played such a significant role in the earliest days of molecular genetics (page 282)—have evolved their own defenses against the defenses of their hosts and so remain virulent.

Basophils and eosinophils are also phagocytic. Basophils contain granules that rupture readily, releasing chemicals, such as histamine, that enhance the inflammatory response. Basophils are important components of allergic reactions; as we shall see later in this chapter, a key role in such reactions is played by mast cells, which are specialized, noncirculating basophils found in connective tissue. The role of eosinophils is not clear; they are found in increased numbers during infections involving internal parasites, such as worms.

Another type of circulating white blood cell that plays a major role in the inflammatory response is the **monocyte.** Monocytes, like neutrophils, are attracted to the site of an infection by chemicals released by both bacterial and host cells, generally arriving later than the neutrophils. Once on location, they are transformed into **macrophages,** becoming enlarged, amoeba-like, and phagocytic (Figure 39–3). Macrophages also lodge in the lymph nodes, spleen, liver, lungs, and connective tissues, where they entrap any microbes or foreign particles that may have penetrated the initial defenses (see Figure 39–1). They are also important, as we shall see, in activating the lymphocytes, other white blood cells that are the effectors of the immune response.

Inflammation can produce systemic as well as local effects. One common response to bacterial infection is as much as a fivefold increase in the proliferation and release of neutrophils. Another common response is fever (page 784), caused by a protein released by monocytes and macrophages in the course of the inflammatory response.

Interferons

Interferons differ from the other defense mechanisms of the body in two ways: (1) they are active only against viruses, and (2) they do not act directly on the invading viruses but rather stimulate the body's own cells to resist them.

The clue that led to the discovery of the interferons was the observation that an animal—human or mouse—infected with one virus is not usually susceptible to an infection with another virus. The source of this resistance was traced, with some difficulty, to a protein that was given the name of interferon. It is now known that at least three different classes of interferons exist, all small proteins that bind readily with other molecules and are active in very small amounts. As a result, the isolation of the first interferon took almost two decades.

When a cell is invaded by a virus, it releases interferon, which then interacts with receptor sites on the membranes of surrounding cells. Thus stimulated, these cells produce antiviral enzymes that block the translation of viral messenger RNA to protein. Only a very few molecules of interferon seem to be required to protect the surrounding cells from viral infection. Interferon molecules also interact with receptors on the surface of various types of white blood cells, stimulating both the inflammatory and immune responses.

Until recently, the only interferons available for research purposes were the minuscule amounts collected from mammalian cells exposed to virus in tissue culture. Now, as a result of recombinant DNA techniques, interferons are being produced in much larger quantities. These new supplies are being used to study clinical applications of interferons in the control of virus infections and to explore their effectiveness against certain forms of cancer. Among their many effects, interferons inhibit cell proliferation, which at one time led to hopes that they might be the long-sought "magic bullets" against this family of diseases. Thus far, however, only one extremely rare form of leukemia has been found to respond to an interferon.

THE IMMUNE SYSTEM

The immune response differs from the other defenses of the body in that it is highly specific, involving recognition of a particular invader and the tailoring of an attack against it. The response consists of two phases, a primary response to the initial attack of an invader and a rapid, secondary response to subsequent attacks by the same agent.

The specificity of the immune response derives from the actions and interactions of two remarkable groups of cells, known as **B lymphocytes** and **T lymphocytes** (or, more simply, as B cells and T cells). In mammals, the primary sites for the differentiation and proliferation of these cells are the bone marrow (B) and the thymus gland (T), a spongy, two-lobed organ that lies high in the chest.

The arena in which these cells operate is known as the immune system (Figure 39–4). As an organ system, it is more diffuse than the digestive system, for instance, or the excretory system, but it resembles these in being a functional, integrated unit. It includes not only the bone marrow and thymus gland, but also the lymph vessels, lymph nodes, spleen, and tonsils.

The lymph vessels, as you will recall, are the route for the return of interstitial fluid to the circulatory system. Strategically located within this system of vessels are the **lymph nodes**, which are masses of spongy tissue separated into compartments by connective tissue. Microorganisms, other foreign particles, and tissue

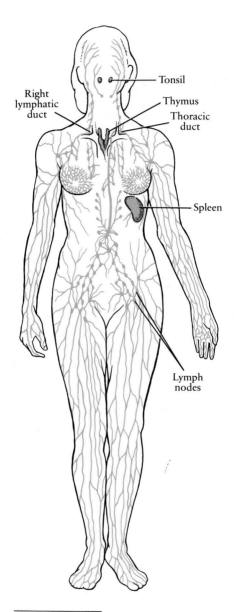

39–4 *The primary constituents of the human immune system are the bone marrow (not shown in this diagram) and the thymus, the sites of the initial proliferation of the B and T lymphocytes that are the effectors of the immune response. Other major components of the system are the lymph vessels, the numerous lymph nodes, the spleen, and the tonsils.*

Labels in figure: Right lymphatic duct, Tonsil, Thymus, Thoracic duct, Spleen, Lymph nodes

39–5 *Diagram of a lymph node. The cortical areas contain follicles, made up of reticulum cells, in which lymphocytes that have been exposed to foreign invaders are activated and undergo further proliferation. Thin, branched tubules, the medullary cords, traverse the node, adjacent to and within the central cistern. One cord is shown in longitudinal section, and the others are shown in cross section. Each cord consists of a small, central blood vessel and an outer network of cells, the mantle. The mantle, which is exposed to the lymphatic fluid on its outer surface, is typically distended with lymphocytes and other immunologically active cells that enter the bloodstream as the blood travels through the lymph nodes. Lymph nodes, which range in size from as small as a poppy seed to as large as a grape, typically increase in size during periods of exceptional immune system activity.*

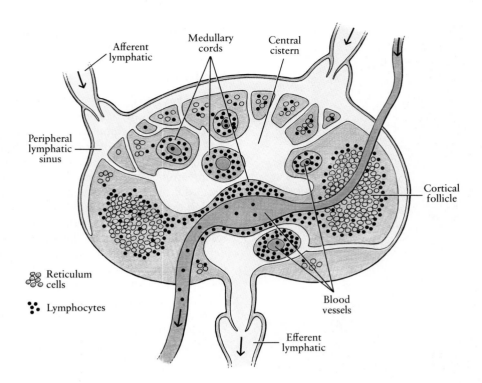

debris entering the extracellular spaces of any tissue are caught up in the interstitial fluid, swept into the channels of the lymphatic system, and trapped in the lymph nodes. Single lymph nodes are distributed throughout the body, but most are found clustered in particular areas, such as the neck, armpits, and groin. Lymph nodes serve as filters, removing microbes, foreign particles, tissue debris, and dead cells from the circulation. Those near the respiratory system, for example, are often filled with particles of soot or tobacco smoke. Lymph nodes near a cancer site may contain malignant cells that have broken loose from the primary growth; in cancer operations, such lymph nodes are often excised not only to remove any malignant cells they may contain but also to determine if the disease has spread from the primary site. Lymph nodes also trap bacterial cells and other microorganisms that have managed to make their way past the first line of defense, the skin and mucous membranes. Lymph nodes are densely populated by lymphocytes and macrophages, and it is within these structures that essential interactions among the cells involved in the immune response occur.

The structure of a lymph node is shown in Figure 39–5. Lymph enters the node by the afferent lymphatic vessels, filters through the peripheral lymphatic sinuses into the central cistern, and trickles out through the efferent lymphatic vessel. The valves within the vessels prevent backflow. The peripheral cortical areas of the node contain follicles where lymphocytes proliferate after exposure to a foreign substance. The incoming lymph carries bacteria, viruses, and other potential pathogens into the node, exposes them to the lymphocytes in the follicles, and collects lymphocytes and other immunologically active cells, which are then circulated throughout the body.

The spleen and the tonsils are also rich in lymphocytes and particle-trapping cells. In the spleen, foreign materials enter by way of the blood rather than the lymph; hence this organ is most important in the case of blood-borne infections. The tonsils trap airborne particles. Patches of lymphoid tissue with large lymphoid

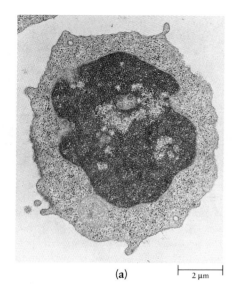

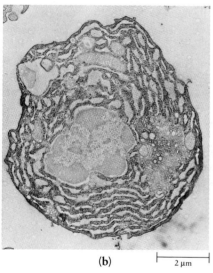

39–6 *As an activated B lymphocyte (a) differentiates into a plasma cell (b), it becomes highly specialized for the manufacture and secretion of antibodies. It grows larger, its nucleus becomes relatively smaller and less dense, and there is a large increase in endoplasmic reticulum and ribosomes.*

follicles—known as Peyer's patches—are also embedded in the wall of the intestine, lying between the inner lining of mucous membrane and the outer muscular coat and defending the body against the billions of microorganisms that inhabit the normal intestinal tract.

B LYMPHOCYTES AND THE FORMATION OF ANTIBODIES

B lymphocytes are the major protagonists in one type of immune response: the formation of **antibodies,** complex globular proteins that are also known as **immunoglobulins.** Antibodies make precise three-dimensional combinations with particular molecules or parts of molecules that the body recognizes as foreign, or "not-self." Any such molecular configuration that can trigger the synthesis of antibodies is known as an **antigen** (short for "antibody-generating substance"). Virtually all foreign proteins and most polysaccharides can act as antigens. The surface of a single cell, such as a bacterial cell, may have a number of different antigens, each of which can elicit production of a specific antibody.

The B Lymphocyte: A Life History

At any given time, about 2 trillion (2×10^{12}) B lymphocytes are on the alert in the human body. Many of these cells are on patrol, circulating with the bloodstream, squeezing out between the endothelial cells that form the walls of the capillaries, and migrating through the lymphatic system. Others are sessile, clustering in the lymph nodes, spleen, and other lymphoid tissues, where, as we have seen, they are exposed to circulating blood and lymph. Both the circulating and sessile B lymphocytes are small, round, nondividing, metabolically inactive cells. Embedded in the cell membrane of each B lymphocyte, and protruding from its surface, are antibodies with a specific three-dimensional structure. When a particular B lymphocyte meets its destiny in the form of an antigen with a three-dimensional structure complementary to the structure of the antibodies on its surface, the cell enlarges, its nucleolus swells, polysomes (page 314) form, and an increased synthesis of macromolecules begins. At the same time, microtubules form and the lymphocyte begins to divide. The proliferation of activated B lymphocytes often takes place in the follicles of the lymph nodes, which therefore enlarge during an infection.

The daughter cells resulting from B lymphocyte activation differentiate into two types, of which one is the **plasma cell** (Figure 39–6). Plasma cells, which rarely undergo further division, are, in essence, specialized antibody factories. A mature plasma cell can make 3,000 to 30,000 antibody molecules per second; these antibodies are released into the bloodstream and circulate throughout the body. However, it takes about five days to produce fully mature antibody-synthesizing cells working at this maximum capacity. Thus, if the microorganism is multiplying also, it may take about this long for the immune system to catch up. In the days before antibiotic therapy, about all the family doctor could do was to await the "break in the fever" that signaled this catching-up process and, often, forecast the patient's eventual recovery. Antibiotics, by suppressing the rate of multiplication of bacteria, enable antibody production to overtake the infection more rapidly.

The second type of cell produced by the antigen-stimulated B lymphocyte is the **memory cell.** Memory cells also produce antibodies, but they differ from plasma cells in their longevity: plasma cells live only a few days, whereas memory cells continue to circulate for long periods of time—up to a lifetime. Thus, the second time a particular pathogen gains entry to the body, large-scale production of antibodies to the invader begins immediately, often preventing any significant multiplication of the pathogen.

Death Certificate for Smallpox

In the early days of medicine, the human body was the laboratory and disease itself the instructor. Smallpox was one of the great teachers. Unlike other infections, which might go unreported or misdiagnosed, the pox left its unmistakable trace in history. It originated in the Far East and seems to have first been introduced into Europe by the returning Crusaders. Here it flourished and spread, until by the eighteenth century one in ten persons died of the pox, and 95 percent of those who survived their childhood had experienced it. About half had permanent scars and many were blinded. Young women studied their reflections in the mirror, waiting their almost inevitable turn. They learned to scratch their legs and feet at the first sign of the disease in the knowledge that the ugly lesions would then localize in these decorously concealed locations and so, perhaps, spare their faces and bosoms. No one could miss the fact that a person who had suffered an attack was thereby protected from a future one. Pox scars were required of domestic servants, particularly nursemaids, as a prime certificate of employability.

It came to be recognized that some outbreaks of pox were more severe than others (owing, we know now, to mutations of the virus). Since one had to have the disease some time, it was reasoned, it was advantageous to choose which pox and when. Intentional infection of children with material preserved from a mild attack was first practiced in the Far East. The Chinese did it in the form of powdered scabs, "heav-enly flowers," used as snuff. The Arabs carried matter from pox pustules around in nutshells and injected it under the skin on the point of a needle. Smallpox inoculation, known as variolation, was introduced into England in 1717 by Lady Mary Wortley Montagu, wife of the British Ambassador to Turkey. In 1746, a Hospital for the Inoculation against Smallpox had been established for the poor of London, where they could be confined during the course of the deliberate infection. Although the induced disease was usually mild, it produced serious illness in some persons. Moreover, since the pox caused by variolation was as contagious as normally contracted smallpox, it appears to have been responsible for some epidemics.

Edward Jenner was an English country doctor who, despite the derision of his colleagues, listened to the tales of country folk. They told him that smallpox never infected milkmaids or other persons who had previously had cowpox, a mild disease of farm animals that sometimes was transmitted to humans. Jenner tried variolation on several persons who had had cowpox and was unable to induce the customary infection. Then, in 1796, he performed his classic experiment on the farm boy Jamie Phipps. First he inoculated the eight-year-old with fluid taken from a pustule of a milkmaid with cowpox. Subsequently he inoculated the boy with material from a smallpox lesion. Fortunately—for himself, for Jamie, and for us all—the child did not contract smallpox. (The success of the cowpox inoculation depended, of

This rapid response by the memory cells is the source of immunity to many infectious diseases—such as smallpox, measles, mumps, and polio—following an infection. It is also the basis for vaccination against a number of diseases (see essay). Vaccines may be prepared using a closely related pathogen, as in the smallpox vaccine; a killed pathogen, as in the Salk vaccine against polio; or a strain of the pathogen that has been weakened by growing in another host organism or in tissue culture, as in the widely used oral (Sabin) vaccine against polio.

The use of recombinant DNA technology to synthesize purified antigenic proteins found on the surfaces of pathogens has now become a powerful new tool in the preparation of vaccines. This technology was first used to prepare a vaccine against the virus causing hoof-and-mouth disease, a serious problem in the livestock industry. The synthetic proteins that make up the vaccine are identical to components of the protein coat of the virus. More recently, recombinant DNA technology has yielded a safe and effective vaccine against the hepatitis B virus, a major cause of liver disease throughout the world. It is also the principal tool in

course, on the antigenic similarity between the two naturally occurring pathogens.) Jenner called the process vaccination, from *vacca,* the Latin word for cow. By 1800, at least 100,000 persons had been vaccinated, and smallpox began to lose its hold on the Western world.

It was not until almost 100 years later that Louis Pasteur discovered serendipitously that a virus or bacterium grown in tissues other than those of its normal host would often lose its virulence while retaining its immunogenicity. (He retained the term vaccination, in recognition of the earlier work of Jenner.) Pasteur's discovery provided the basis for most of our modern vaccines, such as those against poliomyelitis, diphtheria, and measles.

Despite the availability of an effective vaccine for smallpox, by the 1950s there were still an estimated 2 million new cases of smallpox every year. These were largely confined to the Far East, mostly among the urban poor of India, whose crowded living conditions provided as fertile a soil as the European cities had two centuries earlier. In 1973, the World Health Organization (WHO) declared war on smallpox, which involved the mass production of vaccine and, more important, search-and-destroy missions charged with finding each new case of the disease and vaccinating every susceptible person who might be exposed to it; 150,000 persons were in the task force. On April 23, 1977, an international commission declared India free of smallpox and so it has remained.

A child with smallpox, showing the pustules characteristic of the disease. This photograph was taken in a relief camp in Bangladesh in 1973.

In that same year, WHO recommended that smallpox research laboratories be closed and their supply of virus destroyed. One such laboratory was at the University of Birmingham in England. On July 25, 1978, during the last six months of the laboratory's existence, a woman medical photographer working on the floor below contracted smallpox; the virus, as it turned out, had made its way out of the laboratory through a duct system and the photographer was unvaccinated. The disease was diagnosed by the head of the laboratory, a prominent virologist, who was responsible for the woefully inadequate safety precautions. He committed suicide by cutting his throat. The photographer died five days later. Those appear to have been the last two deaths from smallpox on this planet.

current efforts to develop vaccines against malaria (page 473), schistosomiasis (page 537), and other parasitic diseases. The use of recombinant DNA technology to prepare vaccines has the advantage of eliminating any possibility of disease-causing infection or of introducing contaminants that might produce adverse side reactions.

The Action of Antibodies

Antibodies commonly act against invaders in one of three ways: (1) they may coat the foreign particles and cause them to clump together (Figure 39–7) in such a way that they can be taken up by phagocytic cells; (2) they may combine with them in such a way that they interfere with some vital activity—for example, by covering the protein coat of a virus at the site where the virus attaches to the host cell membrane, thereby preventing attachment; or (3) they may themselves, in combination with other blood components, known collectively as **complement,** actually lyse and destroy foreign cells.

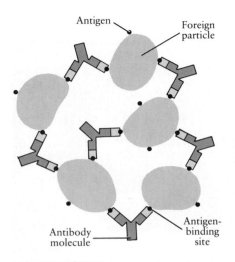

39-7 *As we saw in Figure 18-18 (page 371), each antibody molecule has two antigen-binding sites. Thus one antibody can bind to antigens on two different cells or particles, causing them to stick together (agglutinate). Phagocytic white blood cells then consume these larger masses of foreign particles and antibodies.*

Complement is a group of at least 11 different proteins found in blood. In particular combinations, these proteins function as lytic enzymes, acting at the point on the bacterial cell wall where antigen and antibody combine. By digesting holes in the foreign cells, complement causes them to burst. In addition to this lytic function, complement may itself coat the foreign cell and promote phagocytosis by other cells, and it mediates and enhances the inflammatory response.

The Structure of Antibodies

Determining the structure of antibodies was made possible by the peculiar properties of cancer cells. Virtually any type of cell in the body can turn cancerous. When it does, it produces a clone of rapidly multiplying malignant cells, all with similar properties. In the cancer known as multiple myeloma, the cancer cell is a plasma cell—in other words, a mature antibody-producing lymphocyte. As the cell multiplies, its descendants continue to turn out its one particular antibody, which, as the cells multiply, is produced in greater and greater quantities.

Using the antibodies produced by patients with multiple myeloma, Gerald Edelman and his colleagues at Rockefeller University were able to determine the amino acid sequence of a number of different antibody molecules. Each antibody, as we noted in Chapter 18 (page 371), is a complex protein consisting of four subunits, two identical light chains and two identical heavy chains (Figure 39–8). The light chains have about 214 amino acids each, and the heavy chains have about twice as many. Both the light chains and the heavy chains have constant (C) regions, in which the sequence of amino acids is identical from one molecule to the next. Each type of chain also has a variable (V) region, some 107 amino acids in

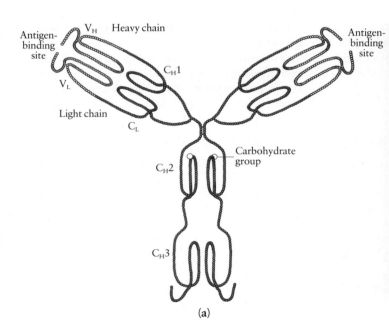

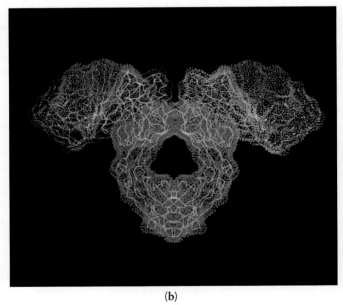

(a)   (b)

39-8 (a) *A drawing of an antibody molecule, based on a bead model made by Gerald Edelman and his colleagues, that suggests how the four chains may be folded. Each bead represents an amino acid, of which there are more than 1,200. The variable regions are shown in yellow, and the constant regions in brown. The*
labeled portions of the molecule are independently folded segments, known as domains. The C_H2 domain is important in interactions with complement, whereas the C_H3 domain enables phagocytic cells to bind to the agglutinated mass of antibodies and foreign particles.

(b) *A computer-generated model of an*
antibody molecule, showing the detailed structure of the surfaces of the heavy chains (blue dots) and light chains (green dots). The backbone of the constant regions is represented by the white and yellow lines, and the backbone of the variable regions by the red lines.

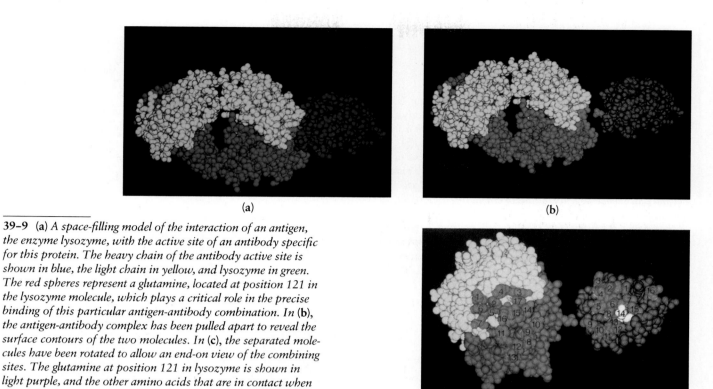

(a)

(b)

(c)

39–9 (a) *A space-filling model of the interaction of an antigen, the enzyme lysozyme, with the active site of an antibody specific for this protein. The heavy chain of the antibody active site is shown in blue, the light chain in yellow, and lysozyme in green. The red spheres represent a glutamine, located at position 121 in the lysozyme molecule, which plays a critical role in the precise binding of this particular antigen-antibody combination. In (b), the antigen-antibody complex has been pulled apart to reveal the surface contours of the two molecules. In (c), the separated molecules have been rotated to allow an end-on view of the combining sites. The glutamine at position 121 in lysozyme is shown in light purple, and the other amino acids that are in contact when the molecules are complexed are shown in red. The numbers on these amino acids are a computer code to their specific identity.*

length. Within this variable region, some of the sequences of amino acids are the same, but at approximately 40 positions the amino acids vary from one antibody to another. It is these variable amino acid sequences that, as the chains fold, come together to form the two active regions of the molecule, the sites that recognize and bind a specific antigen (Figure 39–9).

Biochemists have identified five distinct classes of immunoglobulins: IgG, IgA, IgD, IgM, and IgE. Structurally, these classes of antibodies are distinguished by the constant regions of their heavy chains; each class has a characteristic constant region, shared by all members of the class. There are functional differences as well. Antibodies of the IgG class, known as gamma globulin, are the principal type of circulating antibodies and are the type with which the great majority of the studies on antibody structure and function have been done. IgA antibodies are associated principally with mucosal immunity; they are found in such secretions as tears, saliva, milk, and the mucus on the interior lining of the digestive tract and the respiratory system. IgD molecules are found on the surface of B lymphocytes prior to activation and are the sites at which antigens initially bind to the cells. IgM molecules, which are also found on the surface of B lymphocytes, are the first to be secreted in the course of infection and last a relatively short time. IgE, according to present evidence, plays a major role in the expulsion of parasites, such as worms, from the intestinal tract. It also is involved in allergic reactions.

The Clonal Selection Theory of Antibody Formation

One of the most intriguing facts about the immune response is the great variety of antigens against which a single individual can produce antibodies. It is estimated that a mouse, for instance, can form antibodies against 10 million different antigens. Moreover, antibodies can be formed not only against the natural, common invaders that an individual organism might reasonably be expected to encounter in the course of its own life and those that its ancestors might have encountered, but also against synthetic antigens that are chemically unlike any substance found in nature.

39–10   *The clonal selection model of antibody formation.* (**a**) *An immature B lymphocyte, with one specific type of antibody displayed on its surface, encounters antigen molecules with a structure complementary to the binding site of its antibodies.* (**b**) *Antigens bind to the antibodies, setting in motion a series of changes within the cell.* (**c**) *The B lymphocyte begins to divide and differentiate* (**d**), *forming plasma cells and memory cells. Plasma cells secrete large quantities of circulating antibodies, all with a specificity identical to that of the antibodies on the surface of the original B lymphocyte. Memory cells bearing the same antibodies persist in the circulation indefinitely; they secrete antibodies only following a subsequent encounter with the same antigen.*

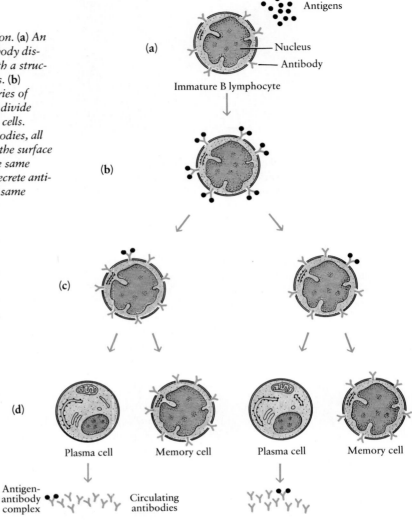

Our current understanding of antibody formation, first proposed in 1957 by the late Macfarlane Burnet of Australia, is known as the clonal selection theory. According to this model, each individual has a vast variety of different B lymphocytes, each genetically equipped with the capacity to synthesize only one type of antibody, which is displayed on its surface. Any given antigenic stimulus does not affect the great bulk of the B lymphocytes, but only those displaying an antibody that is able to bind that specific antigen. Thus, the antigen-antibody interaction "selects" particular lymphocytes. These cells then proliferate, producing clones of plasma cells and memory cells, all synthesizing the antibody displayed on the original B lymphocyte (Figure 39–10).

The clonal selection model predicted that (1) only a very small number of lymphocytes would respond to a given antigen, and (2) any one plasma cell would always form only one antibody. These predictions were initially confirmed by studies involving multiple myeloma cells. More dramatic confirmation has been provided by the development of monoclonal antibodies (see essay, page 802).

According to the clonal selection theory, antibodies are not "tailor-made" in response to an antigen. Rather, the antibodies displayed by the B lymphocytes are like the samples in the showroom of a huge "ready-to-wear" supplier; when a customer—an antigen—comes along that fits its antibodies, the manufacturer—the B lymphocyte—gears up its factory and begins mass production of that particular size and style.

The Genetics of Antibody Formation

In terms of genetics, the clonal selection theory would seem to predict that all the genes for making all the antibodies exist in each lymphocyte (indeed, if you follow the argument further, in every cell in the body), and that all but those coding for a single antibody are repressed in each lymphocyte's clone of plasma cells. Thus, for a number of years, the clonal selection model was in some difficulty because of simple mathematics. Each of us is apparently capable of making antibodies against 100 million different antigens. However, a human cell does not contain this many structural genes in its entire genome.

As we saw in Chapter 18, the seeming paradox was first resolved in work with mice, and further studies have indicated that the same principles apply in other mammals. In mice, the variable regions of both the light and heavy chains of antibodies are coded for by some 300 DNA sequences scattered through the genome, which are transposed and assembled into a variety of different arrangements as the B lymphocytes undergo their initial differentiation in the bone marrow. After transposition, the gene segments are ordered on the chromosome in the sequence of their appearance (Figure 18–19, page 372), separated by noncoding sequences (introns). The completed gene is then transcribed into RNA, and the introns are excised before the RNA is translated into the protein of the final antibody molecule. The number of possible combinations of these gene sequences has been calculated at 18 billion, enough to account for the enormous diversity of antibody molecules. Incredible as it may seem, this figure is actually low. Current evidence indicates that additional combinations are generated both by slight inaccuracies in the splicing and joining of the sequences that form the complete variable region genes and by somatic mutations that occur after the assembly of the genes.

T LYMPHOCYTES AND CELL-MEDIATED IMMUNITY

Circulating antibodies were long believed to be the sole effectors of immunity. It is now known, however, that there is another category of highly specific immune response that is effected by cell-to-cell interactions involving the other class of lymphocytes, the T lymphocytes. This is sometimes known as the **cell-mediated response.**

Unlike the circulating antibodies produced by B lymphocytes, which are primarily active against viruses and bacteria and the toxins they may produce, T lymphocytes interact with other eukaryotic cells—specifically, the body's own cells. Functionally, three classes of T lymphocytes are known. The cells of two of these classes, the helper T cells and the suppressor T cells, are the principal regulators of the immune response, including the activities of the B lymphocytes. Members of the third class, the cytotoxic T cells, act against foreign eukaryotic cells and against cells of the body that are infected by viruses or other microorganisms. When a virus, for example, is multiplying within a cell, it is protected from the action of antibodies. However, its presence is reflected by the appearance of new antigens on the surface of the infected cell, which make it possible for cytotoxic T cells to find the infected cell and lyse it, exposing the viruses to antibody action.

The T Lymphocyte: A Life History

T lymphocytes, like red blood cells, granulocytes, monocytes, and B lymphocytes, are the offspring of self-regenerating stem cells in the marrow of long bones. Because they are indistinguishable from B lymphocytes under the microscope, their identification as a distinct population of cells with unique functions was long delayed. It came about as a result of research directed at uncovering the function

├─ 2 μm ─┤

39–11 *Four T lymphocytes attacking a much larger cancer cell. Note the numerous folds and microvilli that greatly increase the surface area of the T lymphocytes. According to present evidence, cell-to-cell interactions involving lymphocytes, macrophages, and other immunologically active cells depend upon recognition of the pattern of antigenic glycoproteins on the surface of the cell membrane.*

Monoclonal Antibodies

When a microorganism, such as a bacterial cell, invades the body fluids (blood and lymph) of a mammal, a number of different B lymphocytes are activated, giving rise to plasma cell clones, each producing a different antibody. As we have noted previously, the surface of a bacterium bears a variety of proteins and polysaccharides, each of which is potentially antigenic. In addition, a particular antigen may bind to the antibodies displayed by different B lymphocytes. This occurs because the structural variations between the binding sites of different antibodies are sometimes quite subtle. The strength of the binding varies according to the precision of the antigen-antibody fit, and this, in turn, influences the strength and effectiveness of the response by the B lymphocyte.

Until recently, this responsiveness of numerous B lymphocytes to a given antigenic challenge created considerable difficulties for scientists seeking to unravel the details of the immune response. With a great deal of work, it was possible to separate different clones of plasma cells and to obtain partially purified antibodies. The yields, however, were low, and the plasma cells would rapidly die out in tissue culture. Although these problems did not occur with myeloma cells, which would survive indefinitely in tissue culture, producing vast amounts of their specific antibody, there was no way to manipulate the cells to produce other antibodies of potentially greater interest.

The solution to these difficulties was provided by Cesar Milstein and Georges Köhler, working in the Laboratory of Molecular Biology in Cambridge, England. In 1975, they developed a technique for fusing mouse myeloma cells with activated B lymphocytes from mice that had been immunized with antigenic substances of the scientists' choosing. The myeloma cells used in the fusion experiments were mutants that lacked the capacity to synthesize their own antibodies; all of the antibodies produced by the hybrid cells were thus those of the activated B lymphocytes. Separation of the hybrid cells from unfused myeloma cells and B lymphocytes depended on two factors: (1) the mutant myeloma cells were missing certain enzymes necessary for growth and replication in the special tissue culture medium in which the fusion was carried out, and (2) the B lymphocytes, which on their own would quickly die out, contained the missing enzymes. Thus, only the hybrid cells survived, with the necessary enzymes supplied by the B lymphocyte partner and the capacity for longevity supplied by the myeloma partner. Screening procedures, utilizing the differences in the antibodies produced by different hybrid cells, made possible the separation of the hybrids from one another. Maintained in tissue culture, each cell subsequently gave rise to a clone of identical cells, all producing the same pure antibody. Hence the term, monoclonal antibodies.

The fusion technique and the screening procedures required to separate the hybrid cells have now become so refined that monoclonal antibodies can be prepared against virtually any antigen of interest. Moreover, the technique has been extended to human B lymphocytes and myeloma cells, making possible the preparation of human monoclonal antibodies. An enormous variety of monoclonal antibodies are available for research, medical diagnosis, and therapeutic trials against various diseases. Monoclonal antibodies have made possible more detailed study of antibody structure and function, including the models shown in Figures 39–8b and 39–9; purification of naturally occurring substances present in very small quantities, such as the interferons; and elucidation of the detailed structure and function of the T lymphocytes, the other major actors in the immune response. They are used in pregnancy tests and to diagnose perplexing infectious diseases and certain types of cancer. In addition, monoclonal antibodies hold particular promise in the treatment of cancer. Malignant cells bear unusual antigens on their surface, unlike those found on normal cells; there is hope that by coupling chemotherapeutic agents with monoclonal antibodies to those antigens, it will be possible to target toxic chemicals directly to the cancer cells, while sparing the normal cells of the body.

All things considered, the technique for producing monoclonal antibodies has been one of the most significant developments in the history of immunology —an accomplishment for which Milstein and Köhler were awarded the Nobel Prize in 1984.

The preparation of monoclonal antibodies. When the immune system is challenged by foreign antigens, B lymphocytes bearing antibodies complementary to those antigens are activated and begin producing their specific antibodies. The result is (a) a mixture of circulating antibodies in the blood plasma. (b) Fusion of the antibody-producing lymphocytes (generally harvested from the spleen) with mutant myeloma cells produces hybrid cells, each of which synthesizes the antibody characteristic of its lymphocyte component. The hybrids can be separated from one another and maintained indefinitely in tissue culture. Each hybrid gives rise to a clone of cells, all producing the same specific antibody.

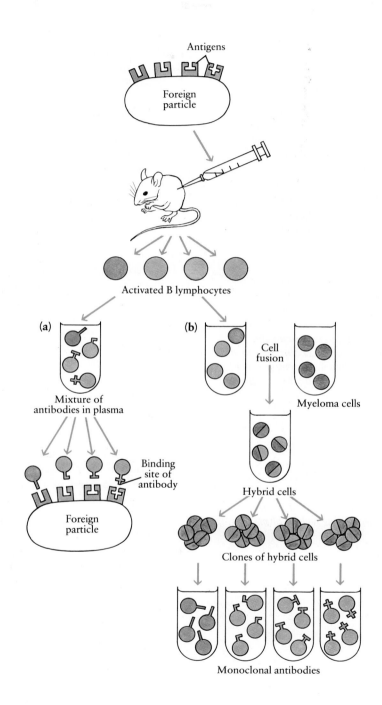

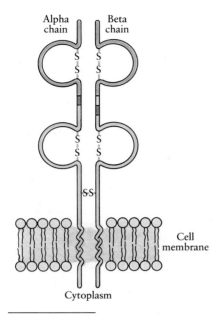

Alpha chain   Beta chain

Cell membrane

Cytoplasm

39–12 *A schematic representation of the structure of the T-cell receptor. Its polypeptide chains are, like those of an antibody molecule, coded by genes that are assembled from different components. Different portions of the chains are coded by variable genes (blue), diversity genes (yellow), joining genes (red), and constant genes (light green). The alpha chain of the resulting protein is characterized by acidic amino acids and the beta chain by basic amino acids (see Figure 3–18, page 73). The helical portions of the constant regions that anchor the receptor in the cell membrane are rich in hydrophobic amino acids. The binding site for antigens is a complex three-dimensional structure formed by the variable regions of the two chains.*

of the thymus gland. In humans, the thymus is located in the chest just behind the breastbone. The gland is large in infants and atrophies after puberty. In adults, surgical removal of the thymus has virtually no physiological effect. The first clue to its function came when a British investigator succeeded in removing the thymus gland in newborn mice—quite a feat when you consider that these infants are only 2 centimeters long and weigh about a gram. The animals survived, but they were sickly and runty due, further study showed, to their decreased resistance to infections. Following this lead, it was found that as early as the eleventh day of fetal life in the mouse (and the eighth week in the human) primitive blood cells creep into the embryonic thymus gland. These are the future T lymphocytes.

Within the thymus gland, the T-lymphocyte precursors go through a complex process of differentiation, selection, and maturation. Differentiation involves the synthesis of at least three different types of membrane glycoproteins, ultimately displayed on the surface of the mature T lymphocyte, that determine both its function and its antigenic specificity. The first type of membrane glycoprotein exists in one of two forms, known as T4 and T8,* and is correlated with function. Helper T cells bear the T4 molecule on their surface, whereas cytotoxic and suppressor T cells are characterized by the T8 molecule. The number of T4 cells formed is normally double or triple the number of T8 cells.

The second type of membrane glycoprotein is the receptor by which the T cell recognizes both the eukaryotic cells of the body itself and foreign antigens displayed on those cells. Although the T-cell receptor is not an antibody and is not secreted by the cell, there are important similarities. The receptor consists of two different protein chains, known as the alpha and beta chains (Figure 39–12). During the course of differentiation of each T lymphocyte, genes coding for these two chains are assembled by the selection and splicing together of variable, diversity, joining, and constant components—a process apparently identical to that occurring in differentiating B lymphocytes. The set of genes from which T-cell receptor genes are assembled is, however, distinct from the set from which antibody genes are assembled, and so are the resulting proteins.

Associated with the T-cell receptors on the surface of the differentiated cell are a set of five other proteins, known collectively as T3 (or as CD3). The structure and function of these molecules are still poorly understood, but they are thought to mediate the process of T-cell activation.

Differentiation of T lymphocytes is followed by a process of selection. As we have noted, the T-cell receptor recognizes foreign antigens displayed on the surface of the body's own cells. The process of gene rearrangement giving rise to the receptor is, however, random. As a consequence, some of the resulting T-cell receptors are incapable of recognizing "self"; others, by contrast, recognize it a little too well and thus have the potential to destroy healthy cells, wreaking havoc on the organism. Within the thymus, differentiated T lymphocytes with either of these characteristics are eliminated.

T lymphocytes that survive the selection process complete their maturation within the thymus and then are released to begin their duties throughout the body. Destiny for a T lymphocyte is like that for a B lymphocyte: the recognition and binding of antigens with a three-dimensional configuration that is complementary to the configuration of the T-cell receptors displayed on its surface. The result is also the same: cell division and differentiation, producing clones of active cells and memory cells. The functions of the active cells, however, are quite different from those of the plasma cells resulting from B lymphocyte activation. Before we consider these functions, we must pause to consider the mechanism by which T lymphocytes distinguish self from "not-self."

---

* In a more detailed system of nomenclature used to identify the membrane glycoproteins that are expressed during various stages of T-lymphocyte differentiation, these two types of molecules are known as CD4 and CD8; in this system, CD stands for "cluster of differentiation."

39-13 *Schematic representations of the structures of the antigens coded by genes of the major histocompatibility complex. (a) A Class I antigen consists of an alpha chain, coded by an MHC gene, combined with another protein, shown here in blue. This protein is coded by a gene that is not part of the major histocompatibility complex. (b) A Class II antigen consists of an alpha chain and a beta chain, both coded by MHC genes. Recent studies suggest that the three-dimensional structures formed by the outermost loops of these protein chains function as receptors to which foreign antigens bind. It is the combination of foreign antigen bound to MHC antigen that the T-cell receptor recognizes.*

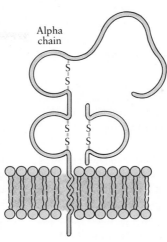

(a) Class I MHC antigen

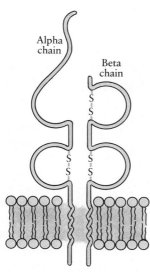

(b) Class II MHC antigen

The Major Histocompatibility Complex

Studies of T lymphocyte functions and of tissue transplant rejections have revealed that recognition of the body's own cells depends on a group of glycoprotein antigens found on the surface of nucleated cells. The protein components of these antigens are coded by a group of genes known as the **major histocompatibility complex,** or MHC. *(Histo is a Greek word signifying "tissue." The complex was first studied in the context of a search for compatible tissue grafts—that is, grafts that would not be rejected by the host.)*

The major histocompatibility complex consists of at least 20 different genes, and within the human population, each of these genes has as many as 8 to 10 alleles. The total number of different combinations is astronomical, and it is predicted that no two persons—except identical twins—will ever be found to have the same major histocompatibility complex. Each is as individual as a fingerprint. Thus, these antigens provide for extremely accurate recognition of self. They also provide for an extremely accurate determination of family relationships (see essay, page 806).

Among the cells displaying the antigens coded by the specific MHC genes that an individual has inherited are the cells of the thymus. During the selection process that occurs within the thymus, developing T lymphocytes are exposed to the MHC antigens displayed by the cells of the thymus. Those T lymphocytes whose receptors bind optimally to the displayed MHC antigens—tight enough, but not too tight—are the ones selected to complete their development and maturation.

Current evidence indicates that there are two classes of MHC antigens, known simply as Class I and Class II (Figure 39–13). These molecules differ in both structure and function. Class I molecules are found on cells throughout the body and are necessary for recognition by cytotoxic T cells. Class II molecules are present only on cells of the immune system and identify such cells to each other. Recently, the detailed three-dimensional structure of a Class I molecule has been revealed in an elegant series of x-ray crystallography studies carried out at Harvard University. The most interesting feature of the structure is a groove in the exterior surface of the molecule (Figure 39–14), just the right size to hold a peptide consisting of 12 to 20 amino acids. Although the detailed three-dimensional structure of a Class II molecule has not yet been determined, it is expected to be similar. With this information in mind, let us turn our attention to the activities of the T lymphocytes.

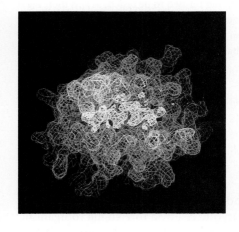

39-14 *A computer-generated representation of the surface of a Class I MHC antigen, based on x-ray crystallography studies. Note the deep groove that is thought to be the site at which antigenic peptides are bound.*

Children of the *Desaparecidos:*
An Application of MHC-Antigen Testing

Because we are diploid organisms, we each inherit two copies of the loci coding for our Class I MHC antigens and two copies of the loci for our Class II MHC antigens. Although it is theoretically possible that a given individual could be homozygous at all of the loci coding for either class of antigen, it is extremely improbable. As we have noted, at least 20 different genes are involved in the complete code for the two MHC antigens, and each of these genes has numerous alleles within the human population. Consequently, virtually all of us are heterozygous for our MHC antigens. Both of the completed genes for each class of MHC antigen are expressed, and thus each of us has two different Class I MHC antigens and two different Class II antigens, one of each class inherited from each parent. This fact, coupled with the enormous diversity of MHC antigens within the population as a whole, makes MHC antigens an extraordinarily valuable tool in determining biological relationships.

One of the most poignant applications of MHC testing has occurred in Argentina. From 1975 to 1983, Argentina was governed by a military junta that, in the name of national security, was responsible for the disappearance of more than 9,000 citizens, most of whom were ultimately murdered. Among the *desaparecidos* ("disappeared ones") were at least 195 children, some of whom were abducted with their parents or were abducted independently. Most of the children, however, were born to women who were pregnant at the time of their abduction and who were kept alive until after they gave birth. In the years since the return of democratic government to the country, an Argentinian human rights organization, the Grandmothers of the Plaza de Mayo, has located dozens of children thought to be *desaparecidos.* In most cases, these children have been living in the families of former members of the military and police units that carried out the abductions. In a few cases, a child has been in the custody of a passerby to whom the mother had frantically handed her baby as she was being abducted. The tool used to determine the biological identity of the children—and thus to facilitate their return to the care of their grandparents—is MHC-antigen testing.

When a child is located who is thought to be a *desaparecido,* a blood sample is obtained (under court order) and the particular MHC-antigen combinations he or she carries is determined. This is known as the haplotype, and it consists of two sets of numbers, representing the two inherited sets of antigens. Haplotypes are similarly obtained from the persons thought to be the child's biological grandparents. The chart below shows the result of the 1984 haplotype determinations of the surviving parents of Claudio Ernesto Logares and Monica Grinspon, *desaparecidos* who were abducted in May of 1978, and of the child thought to be their daughter, Paula Eva, who was two years old at the time of the abduction. Although Monica Grinspon's father was deceased, it was possible to reconstruct his haplotype from that of her siblings. The policeman and his wife with whom the child was living refused to be tested, as was their legal right. Even without this information, however, the haplotypes indicated a probability of more than 99.9 percent that the child was indeed Paula Eva Logares.

The Functions of the T Lymphocytes

Functionally, the simplest of the T lymphocytes are the cytotoxic T cells. As we have noted previously, infection of a eukaryotic cell by a foreign microorganism, such as a virus, results in the appearance of new antigens on the surface of that cell. These antigens, which are short peptides derived from the antigens originally on the surface of the infectious particle, are thought to be bound to and displayed in the surface groove of the individual's Class I MHC antigens. When a cytotoxic

The case was clinched when she was returned, by court order, to the home of her grandparents. The eight-year-old immediately went to her former room and asked for a doll with which she had last played at the age of two.

Some 39 children, orphaned during the years of the junta, have already been restored to their grandparents. Argentinian scientists and the Grandmothers of the Plaza de Mayo are now establishing a data bank containing both the haplotypes of persons whose grandchildren—or pregnant daughters—disappeared during the years of the junta and the haplotypes of children who are thought to be survivors. It is hoped that this data bank will make it possible to reunite many other families.

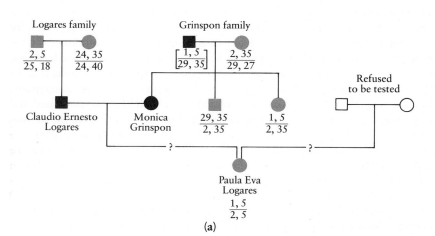

(a) *The haplotype determinations that indicated a 99.9 percent probability that the child being raised by an Argentinian policeman and his wife as their own daughter was actually Paula Eva Logares. The numbers above the lines represent one pair of Class I and Class II antigens, and the numbers below the line represent the other pair. Squares indicate males and circles females; individuals who were tested are shown in red, and those who were* deceased in black. (b) *A photograph of Liliana Pereyra, taken when she was 21. Abducted in 1977, when she was five months' pregnant, she was kept alive until February 1978, when she gave birth to a son. Forensic studies of her skeleton, found in 1985, indicate that she was killed by a shotgun blast to the head at close range. Her son has yet to be located.*

T cell encounters such a combination of Class I MHC antigen and foreign antigen to which its receptor can bind, it differentiates into active cells that attack and lyse the infected cells (Figure 39–15) and into memory cells that remain in the circulation indefinitely. In addition, the activated cytotoxic T cells release powerful chemicals, known as lymphokines, that attract macrophages and stimulate phagocytosis. Some of the cytotoxic T cells, known as "killer cells," secrete cytotoxins that destroy target cells directly; others secrete interferon. Thus, a whole battery of defenses is mobilized by activation of cytotoxic T cells.

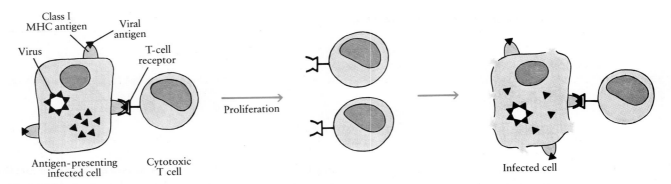

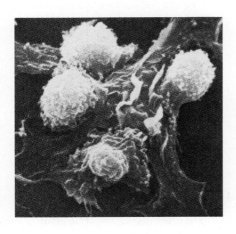

39–15 *When a virus invades a eukaryotic cell, its protein coat either remains on the surface of the cell or, as shown here, breaks apart within the cytoplasm. This is, of course, an essential step in the life of the virus, releasing its nucleic acid and* *enabling it to begin replication. As a consequence, however, telltale markers—viral antigens—appear on the surface of the infected cell and are displayed in conjunction with Class I MHC antigens. Cytotoxic T cells whose receptors are* *complementary to the specific antigenic combination that results bind to the cell and are thereby activated. This activation produces a clone of identical cytotoxic T cells, which then attack and destroy other infected cells.*

39–16 *This scanning electron micrograph shows four helper T cells attached to the surface of a macrophage (the large, flat cell). In the course of its duties in the inflammatory response, the macrophage has ingested cells of the bacterium* Listeria monocytogenes. *On its surface are antigens from the* Listeria *cells, displayed with Class II MHC antigens, a combination for which these particular helper T cells are specific. Thus macrophages constitute a key link between the inflammatory and immune responses. Macrophages also present foreign antigens to the membrane-bound antibodies of immature B lymphocytes, thereby participating in the first step in the production of circulating antibodies.*

The activities of the cytotoxic T cells, as well as those of the B lymphocytes, are regulated by the helper and suppressor T cells. Helper T cells are characterized by receptors that recognize and bind foreign antigens displayed in conjunction with the individual's Class II MHC antigens. Such combinations are found on the surface of both macrophages that have ingested foreign microorganisms (Figure 39–16) and activated B lymphocytes. A helper T cell encountering either type of cell bearing an antigenic combination to which it can bind becomes activated and begins producing proteins known as interleukins. These proteins act as hormones, stimulating the differentiation and proliferation of both B lymphocytes and cytotoxic T cells following activation. The actual binding of helper T cells to antigen-displaying B lymphocytes is also an essential step in the production of plasma cells and memory cells (Figure 39–17).

When an infection has been successfully eliminated, further activity of both B cells and T cells is suppressed. The mechanism of this suppression remains poorly understood. Moreover, although suppressor T cells are generally described as the "off" switch of the immune response, there are no reliable criteria by which to distinguish such cells from other T cells bearing the T8 glycoprotein. It is not yet known whether suppressor cells exist as a distinct class of T cells, or whether suppression is actually mediated by cytotoxic T cells and, perhaps, helper T cells.

Studies of T lymphocyte function have provided what is perhaps only the first glimpse of a vast cellular communication network, probably involving far more than the immune response. Various substances released by activated cells of the immune system appear to influence other cells of the body, particularly those of the endocrine and nervous systems. Conversely, the immune system seems to be influenced by the molecules responsible for cell-cell communication within the endocrine and nervous systems. The study of these complex interactions is one of the most intriguing areas of modern medical research, not only because of the basic biological knowledge that is being uncovered but also because of its potential bearing on many medical problems, a few of which we shall consider in the remainder of this chapter.

CANCER AND THE IMMUNE RESPONSE

Cancer cells resemble an individual's normal cells in many ways. Yet, within the body, they act like foreign organisms, invading and "choking off" or competing with normal tissues. Moreover, virtually all cancer cells have antigens on their cell surfaces that differ from the antigens of the normal cells of the individual and can be recognized as foreign. Does this mean that the body can mount an immune

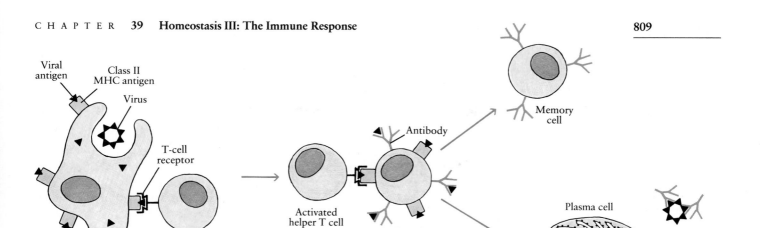

39–17 *A schematic representation of the interaction of macrophages, helper T cells, and B lymphocytes. Phagocytosis of a foreign microorganism or virus by a macrophage results in the display of foreign antigens on the surface of the macrophage. Helper T cells with a receptor that matches the combination of these specific antigens with Class II MHC antigens*

bind to the surface of the macrophage and are activated. Following activation, they bind to the same combination displayed on the surface of B lymphocytes that bear antibodies specific to the same foreign antigen and that have previously encountered the antigen. Binding of activated helper T cells to the activated B lymphocytes is an essential step in the differentia-

tion and proliferation of memory cells and plasma cells and the subsequent production of large quantities of circulating antibodies. The agglutinated masses of foreign particles and antibodies that result from a successful attack on the pathogen are cleared from the system by phagocytizing macrophages.

39–18 *The successful destruction of a cancer cell by cytotoxic T cells. (a) Recognition of the cancer cell as abnormal depends on the presence of unusual antigens displayed on its surface in conjunction with Class I MHC antigens. (b) As the cytotoxic T cells begin lysing the membrane of the cancer cell and consuming its contents, their shape changes. (c) When their work is completed, all that remains of the cancer cell is its cytoskeleton—a mass of lifeless protein fibers.*

response against its own cancers? A growing body of evidence suggests not only that cancer can induce an immune response (Figure 39–18) but also that it usually does so. In fact, according to this hypothesis, it usually does so successfully, overwhelming the cancer before it is ever detected; the cancers that are discovered represent occasional failures of the immune system. This conclusion suggests that bolstering the patient's immune response may provide a means for cancer prevention or control. It also suggests that the cell-mediated immune response may have had its evolutionary origins as a defense not only against parasitic invaders but also against the treacherous, malignant cells of the body itself.

(a)

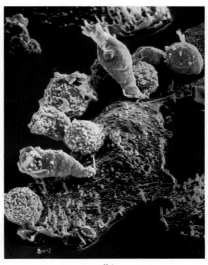

(b)

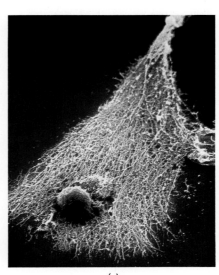

(c)

TISSUE TRANSPLANTS

Organ Transplants

People with extensive burns die from infection and because of the loss of body fluids from exposed areas. If skin is taken from one part of the patient's body and grafted to the burned area, the new tissue anneals to the exposed area, vascularization (the invasion of the transplanted tissue by blood vessels) takes place, and the tissue grows and spreads out. If a skin graft is taken from another individual (except the patient's identical twin), the initial stages of healing and vascularization take place, but then, on about the fifth to the seventh day, large numbers of white blood cells infiltrate the transplanted tissue, and by the tenth to twelfth day it has died. The infiltrating cells are mainly T lymphocytes and macrophages, and current evidence indicates that cytotoxic T cells and "killer cells" are responsible for transplant rejection.

The discovery and identification of the MHC antigens—a result of the search for compatible tissue grafts—is now making possible closer matches between donor and recipient in organ transplants. In the effort to further reduce rejection, transplant recipients are also generally given drugs to suppress the immune response. However, given that infection is a major complication among patients requiring skin grafts and is the leading cause of death among kidney transplant recipients, general suppression of the immune response is obviously not an ideal solution. Cyclosporin, a relatively new drug isolated from a soil fungus, acts selectively against the T cells involved in transplant rejection. Although the mechanism of its action is still poorly understood, cyclosporin has dramatically improved the success of organ transplants. Other cells of the immune system seem to be unaffected by this drug, with the result that the patient remains protected against infection.

Another approach to improving the success of organ transplants is to alter the transplanted tissue in some way to make it more acceptable to the recipient's natural defenses. The likelihood of rejection of a transplanted organ decreases greatly after several months have passed. Moreover, a second skin graft from the same donor is not more readily tolerated than the first, indicating that the ultimate acceptance of the transplant involves a change in the graft rather than in the immune response of the recipient. Identification of the exact nature of these changes may, in time, reduce the need for immunosuppressive drugs.

Blood Transfusions

The most frequently performed tissue transplants in modern medical practice are transfusions of blood. Blood transfusions are so routine that it is difficult to realize that in the past they often provoked severe—and sometimes fatal—immune responses.

At about the turn of the century, Karl Landsteiner set out to discover why blood transfusions between humans were sometimes safe and effective but, more often, led to serious complications. Mixing samples of blood taken from members of his laboratory staff, Landsteiner found that sometimes the red blood cells would agglutinate, and sometimes they would not. From these experiments, he came to realize that there were different categories of blood and that agglutination was caused by mixing blood of different categories. Soon after, the four major blood groups were determined: A, B, AB, and O.

Red blood cells, unlike nucleated cells, do not have MHC antigens on their surface. Instead, they display unique antigens, coded by an entirely different gene, which, in the human population, has three alleles (A, B, and O). The principal blood groups are defined by these antigens and by the presence of antibodies within the plasma (Table 39–2). The A and B alleles are codominant, whereas the

TABLE 39–2 **Blood Groups**

GROUP (ANTIGENS PRESENT)	GENOTYPE (ALLELES PRESENT)	REACTION WITH ANTIBODIES		ANTIBODIES IN BLOOD PLASMA
		ANTIBODY A	ANTIBODY B	
O	O/O	No	No	Antibody A, antibody B
A	A/A, O/A	Yes	No	Antibody B
B	B/B, O/B	No	Yes	Antibody A
AB	A/B	Yes	Yes	None

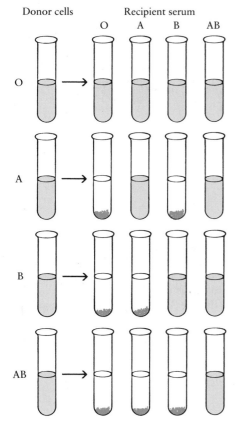

Donor cells   Recipient serum
                 O   A   B   AB

O

A

B

AB

39–19 *Severe and sometimes fatal reactions can occur following transfusions of blood of a different type from the recipient's. These reactions are the result of agglutination of the red blood cells caused by antibodies present in the recipient's blood. Antibodies in the donor's blood are generally of little consequence because they are so diluted in the recipient's blood. Blood-group reactions to transfusions can be demonstrated equally well in test tubes, as shown here. The blood serum (plasma from which the proteins involved in clotting have been removed) in which agglutination occurs has natural antibodies against the donor blood. Persons with type O blood, whose red blood cells have neither A nor B antigens, used to be called universal donors; similarly, those with AB blood (neither A nor B antibodies in the plasma) used to be called universal recipients. Now other factors are checked as well.*

O allele is recessive. Thus, individuals with type A blood have either two A alleles or one A and one O, and their red blood cells bear the A antigen. Their plasma does not contain antibodies against the A antigen (self) but does carry antibodies against the B antigen ("not-self"). Individuals with type B blood have the B antigen and antibodies against the A antigen. Individuals with type AB blood have both antigens but neither A nor B antibodies; conversely, type O individuals have neither antigen but both A and B antibodies. For reasons that are still not clear, circulating antibodies against the "not-self" antigens are always present—even though an individual has never been exposed to those antigens.

For blood transfusions to be performed safely, the blood types must be matched. If a person receives a transfusion containing red blood cells bearing a foreign antigen, the antibodies in his or her plasma will react with those cells, causing them to agglutinate (Figure 39–19). The resulting clumps of blood cells and antibodies can clog capillaries, blocking the vital flow of blood through the body.

Because blood types are inherited and are also easy to determine, they have been used to settle questions of paternity in legal actions. As you can see by studying Table 39–2, however, they can only demonstrate that someone is not the father of a particular child, not that he is. Evidence of a positive nature indicating the probability of a particular relationship can now be provided by MHC-antigen testing (see page 806).

The Rh Factor

Since Landsteiner's initial discoveries, additional antigens have been identified on the surface of red blood cells. Modern blood typing takes these into account also, thus minimizing the possibility of immune responses that could complicate recovery from the illness or injury that necessitates the blood transfusion. Among the most important of these antigens is the Rh factor, named after the rhesus monkeys in which the research leading to its discovery was carried out.

A once common medical problem caused by the Rh factor is hemolytic anemia of the newborn. During the last month before birth, the human fetus usually acquires antibodies from its mother. Most of these antibodies are beneficial. An important exception, however, is found in the antibodies formed against the Rh factor. Like the other surface antigens of red blood cells, the Rh factor is genetically determined. If a woman who lacks the Rh factor (that is, an Rh-negative woman) has children fathered by a man homozygous for the Rh factor, all of the children will be Rh positive; if he is a heterozygote, about half of the children will be Rh positive.

During the birth of an Rh-negative mother's first Rh-positive child, fetal red blood cells bearing the Rh antigen are likely to enter her bloodstream. The

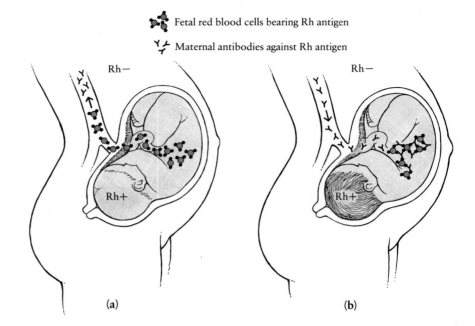

Fetal red blood cells bearing Rh antigen

Maternal antibodies against Rh antigen

Rh—

Rh+

(a)

Rh—

Rh+

(b)

39-20 *The events leading to Rh disease.
(a) Late in the first pregnancy—or during
the birth of the baby—fetal blood cells
spill across the barrier that separates the
maternal and fetal circulations. Antigens
on the red blood cells of the Rh-positive
fetus stimulate the production of antibodies by the immune system of the Rh-negative mother. These antibodies remain in
the mother's bloodstream indefinitely. (b)
Late in the second pregnancy, the antibodies pass through the barrier from the
maternal blood to the blood of the fetus. If
the fetus is Rh positive, the antibodies
react with the antigens on its red blood
cells, destroying them.*

consequences are the same as would occur with a transfusion of Rh-positive
blood: the mother's immune system produces antibodies against the foreign
antigens, and these antibodies persist in her blood. In subsequent pregnancies,
they may be transferred to the fetus. If that fetus is Rh positive, the antibodies will
react with its red blood cells, destroying them (Figure 39–20). This reaction can be
fatal either before or just after birth.

Now that its causes are recognized, Rh disease can be prevented by injecting
the Rh-negative mother, within 72 hours of her delivery, with antibodies against
the fetal Rh-positive red blood cells in her system. This destroys the cells and
prevents them from triggering antibody production.

DISORDERS OF THE IMMUNE SYSTEM

Autoimmune Diseases

The immune response is a powerful bulwark against disease, but it sometimes goes
awry. Usually the immune system can distinguish between self and "not-self."
Substances that are present during embryonic life, when the immune system is
developing, will not be antigenic in later life. This recognition occasionally breaks
down, however, and the immune system attacks cells of the body. Certain
disorders, among them myasthenia gravis, lupus erythematosus, and several types
of anemia, have been identified as autoimmune diseases—that is, diseases in
which an individual makes antibodies against his or her own cells. There is
growing evidence that other disorders, such as multiple sclerosis, juvenile-onset
diabetes, and some forms of rheumatoid arthritis, may have the same basis.

Allergies

Hay fever and other allergies are the result of immune responses to pollen, dust, or
other substances, such as some foods, that are weak antigens to which most
people do not react. When certain individuals are exposed to particular environmental antigens, the production of IgE antibodies by specific plasma cells is
stimulated, as is the formation of memory cells. Upon reexposure to the same
antigen, more IgE antibodies are formed. These antibodies circulate and attach
themselves to the mast cells and other noncirculating basophils found in connec-

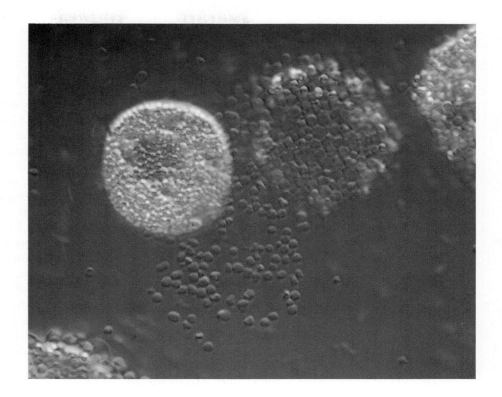

39–21 *Two mast cells, bearing receptors to which IgE antibodies to a specific antigen of birch pollen are bound. The cell on the left is "loaded" with granular inclusions, filled with histamine. The cell on the right, in response to the binding of the specific birch pollen antigen to the antibodies, is releasing its histamine-containing granules.*

tive tissue (page 792). Subsequent binding of the antigen to these attached antibodies triggers the release of histamine from the cells (Figure 39–21), which, in turn, induces an inflammatory response. This reaction typically occurs on an epithelial cell surface, producing increased mucus secretion, as in hay fever; hives or dermatitis; or cramps and diarrhea, as in the case of food allergies. Systemic reactions may result if the mast cells and other basophils release their chemicals into the circulation, causing dilation of the blood vessels, leading to a potentially dangerous fall in blood pressure, and constriction of the bronchioles (a syndrome known as anaphylactic shock). The evolutionary background of this decidedly maladaptive response—in which the response is far more dangerous than the antigen—remains a mystery.

Antihistamines, which counteract the histamine effect, suppress some of the symptoms of an allergic reaction. Most decongestants, on the other hand, promote the release of histamines, raising some questions about taking "allergy pills" that offer a combination of antihistamines and decongestants. Steroid hormones related to cortisone are used in more severe cases. These drugs act by suppressing the production of white blood cells and so of inflammation and the immune response in general.

Acquired Immune Deficiency Syndrome (AIDS)

Until fairly recently, epidemics of fatal infectious disease were a fact of life. For example, in 1918, at the close of World War I, a particularly deadly epidemic of influenza swept around the globe, leaving few families untouched. In the late 1940s and early 1950s, polio epidemics were a regular feature of summertime in the United States; most people who were alive at the time can remember more than one of their friends stricken by the disease. However, with the widespread introduction of antibiotics following World War II and, particularly, the development of polio vaccines in the mid-1950s, we entered an era in which it seemed that the conquest of infectious disease was virtually complete. That this is not so—and perhaps never can be so—has been powerfully demonstrated by the appearance of a virus that attacks the human immune system, leaving its victims susceptible to numerous types of opportunistic infections. The disease it causes is known as acquired immune deficiency syndrome (AIDS), and, thus far, it is invariably fatal.

AIDS was first identified in 1981, when epidemiologists noticed unusual clusters of a malignancy known as Kaposi's sarcoma, which affects the endothelial linings of the blood vessels. Previously, this rare cancer had been seen only in elderly men; its new victims, however, were young men. At about the same time—and also in young men—physicians began to observe an increasing incidence of fatal pneumonias and intestinal tract infections caused by ubiquitous but usually innocuous protists. In the past, such infections had been observed only in cancer patients and transplant recipients whose immune systems had been suppressed in the course of their treatment.

These facts suggested that the underlying cause of the new illnesses was a massive suppression of the immune system. Because most of the early victims were homosexual men, intravenous drug users, recipients of blood transfusions, or hemophiliacs who had received clotting factors prepared from donated blood, it appeared that the agent of this suppression was a microorganism transmitted sexually or through the exchange of blood. Within a remarkably short time—three years—the virus responsible was isolated and characterized. Since then, its principal effects on the immune system have been determined, its possible origins identified, and its modes of transmission traced.

The AIDS Virus and Its Effects

The AIDS virus is a retrovirus (page 342), known formally as human immunodeficiency virus (HIV). It is a particularly complex virus, as shown in Figure 39–22.

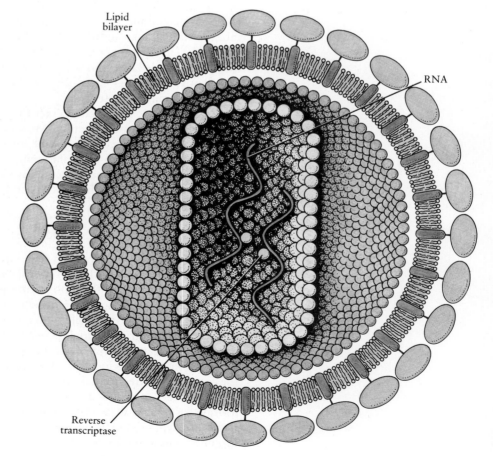

39-22 *The structure of the AIDS virus. The inner core consists of two molecules of RNA, accompanied by two or more molecules of the enzyme reverse transcriptase. Surrounding the core are envelopes formed of two distinct proteins. These are surrounded, in turn, by a lipid bilayer derived from the cell membrane of the host cell in which the virus previously replicated. Spanning this membrane are protein molecules to which are attached glycoprotein molecules that extend out from the surface.*

Lipid bilayer

RNA

Reverse transcriptase

39-23 *An electron micrograph of a degenerating helper T cell from which newly replicated AIDS virus particles are being released. The T cell is the irregular structure with the granular appearance, and the virus particles are the small disks containing a dark core. Numerous virus particles are visible in the three large vacuoles in the center of the cell. AIDS virus particles bud from the host cell in a process essentially identical to exocytosis, taking with them portions of the host cell membrane. Because large numbers of particles bud off at the same time, holes are torn in the membrane, allowing the contents of the cell to leak out. As you would expect, the result is the death of the cell.*

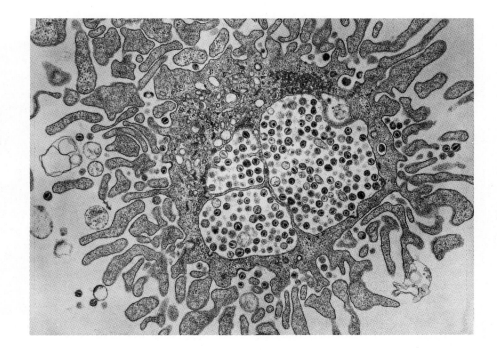

The inner core consists of two molecules of RNA and several copies of the enzyme reverse transcriptase. Surrounding the core are two distinct protein envelopes, which are, in turn, surrounded by a lipid bilayer studded with glycoproteins. The protein portions of the surface molecules contain both constant regions, identical from one strain of the virus to another, and variable regions, the genes for which have an unusually high rate of mutation. The unique—and deadly—feature of the outer glycoproteins is that they are a perfect three-dimensional match to the T4 molecules that characterize helper T cells. When the AIDS virus meets a helper T cell, the T4 molecule functions as its receptor, enabling the virus to enter the cell by receptor-mediated endocytosis (page 138). The virus is also able to infect macrophages, which are now thought to be an important reservoir for the virus within the body; the interactions between macrophages and helper T cells, described previously, may play a major role in the spread of the virus to helper T cells.

Once within a helper T cell, the RNA is released from its multilayered capsule, and reverse transcriptase catalyzes the transcription of complementary DNA and its incorporation into a chromosome of the host cell. In the host cell chromosome, this complementary DNA may lie latent as a provirus (page 373) for some time. Sooner or later, however, replication of new viral particles begins—at a rate much higher than that of other known viruses. In a relatively short time, enormous numbers of new viruses burst from the infected helper T cell, which is often destroyed in the process (Figure 39–23). These viruses invade other helper T cells, and the process is repeated. The victim is eventually left with few functional helper T cells, and, as we have seen, these cells are critical for the proliferation and activities of both B lymphocytes and cytotoxic T cells. The body cannot mount an effective immune response against cells harboring the AIDS virus, against the virus itself, against other invading microorganisms, or against any malignant cells that may be present or may develop. However, in the initial phases of the infection, before the population of helper T cells is severely depleted, B lymphocytes respond to the foreign antigens of the virus, producing circulating antibodies. Although these antibodies are apparently ineffectual in controlling the infection, they do remain in the bloodstream. Their detection is the basis of screening tests for the presence of the virus.

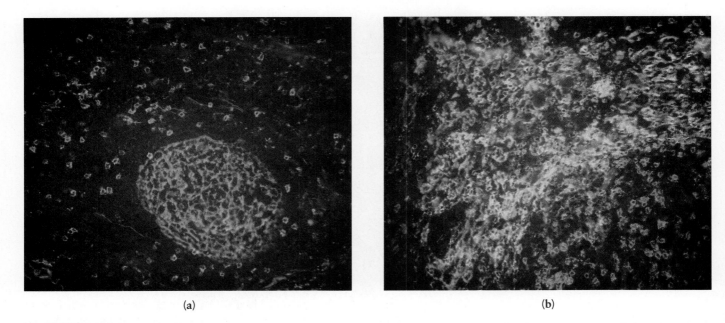

(a)   (b)

39–24 *The effect of the AIDS virus on the cortical follicles of the lymph nodes.* (a) *An immunofluorescence micrograph of the germinal center of a normal cortical follicle; this is the region in which B lymphocytes proliferate following activation. The reticulum cells of the follicle, shown in green, form a regular network in which antigen-presenting cells are trapped and can be exposed to B lymphocytes. Cytotoxic and suppressor T cells, shown in orange, populate the region around the germinal center but are not found within the center.* (b) *The same region of a lymph node from a person with AIDS-related complex (ARC), a condition that often precedes full development of the disease. As you can see, the regular structure of the germinal center has been destroyed, and cytotoxic and suppressor T cells have invaded the entire region.*

The consequences of infection with the AIDS virus are many. One of the earliest signs of this infection, as of so many others, is a swelling of the lymph nodes, the principal sites of the interactions among macrophages, helper T cells, B lymphocytes, and other T lymphocytes that are so essential in the immune response. In persons infected by the AIDS virus, the structure of the cortical follicles of the lymph nodes (see Figure 39–5, page 794) is completely disrupted, as shown in Figure 39–24. Although the reasons for this change remain unknown, it appears to constitute an additional assault on the capacity to mount an immune response. As the immune system becomes steadily more crippled, the victim becomes increasingly vulnerable to other diseases. Among the most common are *Pneumocystis carinii* pneumonia, parasitic gastrointestinal infections accompanied by severe diarrhea, and Kaposi's sarcoma and other cancers. Weight loss is extreme. Ultimately, the person dies, most often as a result of the opportunistic diseases or of heart failure.

In many AIDS victims, the central nervous system is affected, usually in the latter stages of the disease. The result is atrophy of the brain, dementia, and, in some cases, symptoms that resemble those of multiple sclerosis. Current evidence indicates that the nervous system is attacked not only by microorganisms thriving in the absence of an immune response but also by the AIDS virus itself, most likely carried into the nervous system by macrophages.

The time span from diagnosis of full-blown AIDS to death varies from a few months to as long as five years. The time span from initial infection to the appearance of the first symptoms varies even more and may be associated with the absence or presence of other infections that trigger an immune response—an event that appears to activate latent AIDS proviruses. About 30 percent of infected individuals develop active disease within five years, and it appears probable that everyone infected with the virus will eventually become ill; cases have now been documented in which more than 10 years have elapsed between infection and illness. (Large numbers of homosexual men participated in the clinical trials of the vaccine for hepatitis B in the 1970s, and throughout the years of the trials, blood samples were taken at intervals and frozen for future study; testing these samples for antibodies to the AIDS virus has made it possible to determine when many AIDS victims first became infected.)

39–25 *An African green monkey* (Cerco-pithecus aethiops). *These monkeys are host to a virus known as simian immuno-deficiency virus (SIV). Although SIV shares important similarities with human immunodeficiency virus (HIV), the cause of AIDS, it seldom produces illness in its monkey hosts.*

Transmission of the AIDS Virus

The AIDS virus is one of a family of primate retroviruses and appears to be most closely related to a relatively harmless virus that is endemic in the green monkeys of equatorial Africa. It has been hypothesized that this virus may have moved from the monkeys into the surrounding human population, and then undergone a series of mutations into its present deadly form. Tests of blood samples from individuals in equatorial Africa (saved for a variety of reasons) indicate that this may have occurred in the 1950s. According to this hypothesis, the virus gradually spread through the population, transmitted through sexual intercourse and, probably, through blood transfusions and other medical procedures in situations where sterilization was poor or nonexistent. Ultimately, it was acquired by visitors from Europe, North America, and the Caribbean and began its spread around the world.

Current evidence indicates that the virus can be transmitted through sexual intercourse, either vaginal or anal, through oral sex (perhaps including deep kissing), and through the exchange of blood. It is present at high levels in the semen and blood of infected individuals and can enter the body through any tear in the skin or mucous membranes—including those too small to be seen. Although the AIDS virus is one of the most virulent known, it is less readily transmitted than other viruses; without a surrounding environment of blood or semen or host cells, it quickly dies. There is no evidence that it can be transmitted through casual contact, hugs and light kisses, coughs or sneezes, on dishes used by an infected person, or on toilet seats. Thus far, no family member nursing an AIDS patient has become ill, and the few medical personnel who have tested positive for the virus were either accidentally pricked with a contaminated needle or exposed to large quantities of infected blood on portions of their body (for example, ungloved hands) where the skin was broken.

The initial spread of the virus in the West among homosexual men and intravenous drug users is thought to have occurred because these are relatively self-contained populations in which an individual might be repeatedly exposed. In equatorial Africa, AIDS is primarily a disease of heterosexuals, and its incidence among men and women is roughly equal. Continuing epidemiological studies suggest that members of the following groups in North America are presently at the greatest risk: (1) homosexual or bisexual men, (2) intravenous drug users, (3) sexual partners, whether male or female, of infected individuals, and (4) babies born to infected mothers. All available evidence indicates that once the virus has infected an individual, it remains for the rest of that person's life and can be transmitted in blood and semen—even if the person has no symptoms of illness.

The Prospects

As many as 2 million people in the United States alone are now thought to be infected by the AIDS virus. By November of 1988, 78,312 cases of AIDS had been documented in the United States, and 44,071 deaths had been attributed to the virus and its effects. It is estimated that by the end of 1991, an additional 270,000 cases will have been diagnosed, and another 178,760 deaths will have occurred. Although work is progressing on treatments and on the development of vaccines, any "quick fix" seems unlikely. Current treatments, which are directed at control of opportunistic infections and inhibition of reverse transcriptase (and thus of the replication of the virus), have serious side effects and are, at best, delaying actions.

The nucleotide sequences of the RNA of the AIDS virus have now been determined, and the codes for the various proteins of its coat have been identified. With this information, scientists are using recombinant DNA technology in the attempt to create synthetic vaccines against the virus. However, the fact that the natural antibodies synthesized by the B lymphocytes are ineffectual suggests that development of a truly protective vaccine may be extraordinarily difficult. The task is also compounded by the high mutation rate of the genes coding for key parts of the protein coat.

Tests for the AIDS antibody, which came into use in 1985, have provided a means of identifying donated blood carrying the virus and now provide a high level of protection for the blood supply. As a consequence, the number of new cases of AIDS in transfusion recipients and hemophiliacs has dropped dramatically. The spread of the virus through the homosexual population has also slowed significantly, in part because the most susceptible individuals have already contracted the disease and, perhaps even more important, because of changes in sexual practices. Presently, the greatest increase in new cases is occurring among drug users sharing needles, their sexual partners (many of whom are prostitutes), and their children. Whether a similar increase will subsequently occur among the population at large remains to be seen.

AIDS, like smallpox in another era (page 796), is one of the great teachers on the subject of immunology. The price of its lessons—in the suffering of its victims, their families and their friends, in lives cut short in their prime, in the strains on medical and social services, and in economic terms—is extraordinarily high. Our response will say much about us as a people.

SUMMARY

Animals have evolved a number of responses that exclude or destroy microorganisms, other foreign invaders, and cells not typically self. These responses depend on a variety of types of white blood cells, all the progeny of self-regenerating stem cells in the bone marrow.

The inflammatory response, which is nonspecific, involves the release of histamine and other chemicals, causing capillary distension, a local increase in temperature, and the mobilization of phagocytic granulocytes and macrophages at the site of infection. Interferons provide another and very different type of nonspecific defense, directed against viruses. Interferons are small proteins produced by virus-infected cells that stimulate nearby cells to defend themselves against viral infection and that also stimulate cells involved in the immune response.

The immune response is highly specific and involves two types of cells: B lymphocytes and T lymphocytes. B lymphocytes are the major protagonists in the formation of antibodies, large protein molecules whose binding sites are complementary to foreign molecules called antigens. The combination of antigen and antibody immobilizes the invader, destroying it or rendering it susceptible to phagocytosis.

Five classes of antibodies (immunoglobulins) are known, of which the circulating IgG antibodies are the most intensively studied. Antibodies consist of four subunits: two identical light chains and two identical heavy chains. Each of the four chains has a constant (C) region—a region common to all antibodies of its class—and a variable (V) region, which differs from one antibody to another. The antigen-binding sites—of which there are two on each antibody molecule—are formed by foldings of the variable regions of the light and heavy chains.

The accepted model of antibody formation is the clonal selection theory. According to this theory, differentiation of B lymphocyte precursors, which occurs in the bone marrow, produces a vast variety of different B lymphocytes, each capable of synthesizing antibodies with one particular three-dimensional structure for the binding site. Upon encountering an antigen that binds to the antibodies displayed on its surface, the B lymphocyte matures and divides, resulting in a clone of plasma cells all synthesizing circulating antibodies against that particular antigen. Memory cells are also produced, which persist in the bloodstream following infection and produce antibodies immediately upon subsequent exposure to the same antigen. This memory-cell response is the cause of the rapid and enhanced immunity following vaccination or many viral infections. The capacity to produce a tremendous variety of B lymphocytes, each able to synthesize one specific antibody, is accounted for by the large number of gene sequences (at least 300) coding for the variable regions of antibodies, by the transposition of these gene sequences in the course of lymphocyte differentiation, and by subsequent somatic mutations.

T lymphocytes, which undergo differentiation and maturation in the thymus, are responsible for cell-mediated immunity. Although these cells neither display nor produce antibodies, they do bear several types of glycoprotein molecules on their surface. One type of molecule, which exists in two forms, T4 and T8, identifies cells as helper T cells (T4) or as cytotoxic or suppressor T cells (T8). Cytotoxic T cells work principally against the body's own cells that are harboring intracellular viruses or other parasites. Helper T cells promote immune responses involving B lymphocytes and cytotoxic T cells, whereas suppressor T cells moderate the activities of B lymphocytes and other T cells.

The ability of the T lymphocytes to perform their functions depends upon another type of surface molecule, known as the T-cell receptor. The T-cell receptor consists of two polypeptide chains, each with constant and variable regions coded by genes that, like those for antibodies, are rearranged in the course of differentiation. The result is an enormous diversity of T lymphocytes, each bearing T-cell receptors with a single antigenic specificity. T-cell receptors recognize and bind to complementary foreign antigens presented in conjunction with genetically determined antigens found on the surface of the body's own nucleated cells. These "self" antigens are coded by a group of genes known as the major histocompatibility complex (MHC). Two classes of MHC antigens are known. Class I molecules, which are found on cells throughout the body, are essential in the identification of diseased cells by cytotoxic T cells. Class II molecules are found on the surface of macrophages and B lymphocytes. They are essential in the presentation of foreign antigens to the helper T cells, which are, in turn, essential for the activation and proliferation of both B lymphocytes and cytotoxic T cells.

Skin and other organs transplanted between individuals other than identical twins evoke an immune response by cytotoxic T cells that can lead to rejection of the transplanted organ. The success of such transplants has been dramatically improved by the development of MHC-antigen testing, ensuring a closer match between donor and recipient, and the use of the selectively immunosuppressive drug cyclosporin. Similarly, blood transfusions can evoke an immune response by circulating antibodies to the A and B antigens found on the surface of red blood cells. Blood typing involves not only the ABO blood groups but also other surface antigens of red blood cells, such as the Rh factor.

Disorders associated with the immune system include allergies, autoimmune diseases caused by an individual's immune responses to his or her own tissues, and AIDS, a fatal infectious disease. The retrovirus responsible for AIDS (human immunodeficiency virus, or HIV) invades and destroys helper T cells, leaving the victim's immune system incapable of responding to other infections or to malignancies. In the final stages of the disease, it invades other cells and tissues of the body, including those of the nervous system. The AIDS virus is present in high levels in the blood and semen of infected individuals and is transmitted by sexual contact (heterosexual or homosexual, oral, vaginal, or anal) and through the exchange of blood or blood products.

QUESTIONS

1. Distinguish among the following: granulocytes/monocytes/lymphocytes; B lymphocyte/T lymphocyte; antigen/antibody/T-cell receptor; antibody light chain/heavy chain; T-cell receptor alpha chain/beta chain; T4 molecule/T8 molecule; MHC antigens/ABO blood groups/Rh factor; Class I MHC antigens/Class II MHC antigens.

2. Describe the life history of a B lymphocyte, beginning with a precursor cell in the bone marrow and ending with a clone of plasma cells and a clone of memory cells. Include its interactions with other types of white blood cells.

3. Describe the life history of a cytotoxic T cell, beginning with a precursor cell in the bone marrow and ending with a clone of active cells and a clone of memory cells. Include its interactions with other types of white blood cells.

4. Explain the functions of the macrophages in the inflammatory and immune responses. How do these cells provide the essential link between the inflammatory response and the immune response?

5. What prominent feature in the structure of a virus makes it more susceptible than a bacterial cell to control by vaccination?

6. An early hypothesis of antibody formation, the instructive theory, proposed that the antibody was molded into its special configuration by its encounter with the antigen. Since then, data on the structure of proteins have invalidated this model. What are these data?

7. Predict the results of the following experiment: A mouse of strain A receives a skin graft from another strain A mouse and one from a strain B mouse. Two weeks later, the same mouse is grafted with skin from strain B and strain C. What are the fates of these grafts?

8. If you are of blood type O and neither of your parents is, what are their possible genotypes? What is the probability that one of your siblings is also of type O? (You may find it helpful to review the use of Punnett squares, page 240).

9. Although it can never be proved that someone is the father of a particular child, it is possible to prove that someone could *not* be the father. Fill in the table below to see how this is so. In the famous Charlie Chaplin paternity case in the 1940s, the baby's blood was B, the mother's A, and Chaplin's O. If you had been the judge, how would you have decided the case?*

PHENOTYPES OF PARENTS		CHILDREN POSSIBLE	CHILDREN NOT POSSIBLE
A	A		
A	B		
A	AB		
A	O		
B	B		
B	AB		
B	O		
AB	AB		
AB	O		
O	O		

* As a matter of fact, Chaplin was judged to be the father. Blood-group data are not admitted as evidence by some states in cases of disputed parentage.

Integration and Control I: The Endocrine System

As we have seen in the preceding chapters of this section, the maintenance of homeostasis in the body of an animal is a complex process, involving the regulation of diverse physiological activities. Internal and external conditions are constantly monitored, and the resulting information is transferred to centers in which it is integrated, leading to the initiation of appropriate responses. Throughout the living world, the integration and control necessary to maintain homeostasis are effected by chemical stimuli. Specific molecules interact with receptors that are either inside individual cells or embedded in their membranes, triggering changes in the cells that lead, directly or indirectly, to an appropriate response. In unicellular organisms, the response is direct and typically involves only the cell receiving the stimulus. In multicellular organisms, however, the response is more often indirect. Certain cells are specialized to receive and process information and to transmit it to other cells, thereby regulating the activities of the organism as a whole.

In small, simple animals and at many sites in larger, more complex animals, signalling molecules move from the cells in which they are produced to the cells on which they act by simple diffusion. Often, however, the target cells are a considerable distance away, and the signalling molecules are carried by bulk flow in the bloodstream, which can move them at a more rapid rate than would be possible by diffusion. This rate, although adequate for many of the processes involved in maintaining homeostasis, is nevertheless too slow for effective coordination of the numerous activities that characterize most animals. A much more rapid, direct channel of communication is provided by neurons, specialized cells that use an electrical signal—the nerve impulse—to conduct information over great distances.

For many years, the relatively slow, but long-acting regulation of the physiological activities of an animal by chemical means and the rapid conduction of information by electrical means were regarded as different phenomena, involving two distinct systems (Figure 40–1). Chemical regulation was considered the province of the endocrine system, the clusters of specialized secretory cells—glands—that synthesize and release chemical substances—hormones—that travel through the bloodstream. By contrast, the nervous system, the interconnected network of neurons and supporting tissues that attains its highest complexity in the vertebrates, was thought to be involved only in the conduction of nerve impulses. Recently, however, it has become apparent that there is enormous overlap between these two systems and they are more accurately viewed as different aspects of one overall system, the **neuroendocrine system.**

A variety of evidence supports this conception of one unified regulatory system. For example, although an individual neuron conducts information electrically, it transmits that information to other cells, including other neurons, through

40–1 *The classical distinction between endocrine control and nervous control is based on the means by which information is carried over long distances. (a) In endocrine control, specific molecules, known as hormones, diffuse into the bloodstream, which carries them through the body to target tissues. This transport process may take minutes to hours, and the effects are typically long-lasting. (b) In nervous control, electrical signals—nerve impulses—are conducted along a neuron to its terminus where specific molecules, known as neurotransmitters, are released and diffuse to the target tissues. The entire process takes only a fraction of a second, and the effect is similarly short-lived. Both hormones and neurotransmitters interact with specific receptors on or in the target cells, leading to a response.*

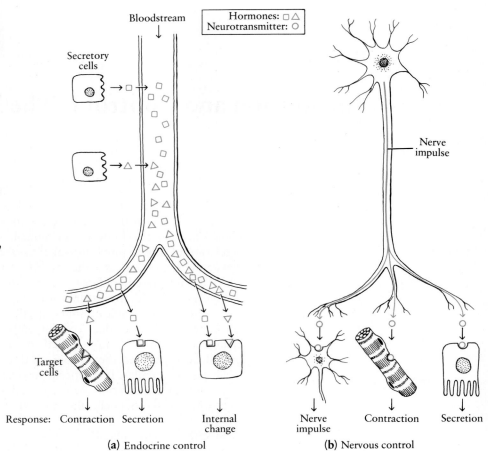

(a) Endocrine control   (b) Nervous control

the release of specific molecules known as **neurotransmitters.** A number of neurotransmitters are chemically identical to hormones. Moreover, some neurons, known as **neurosecretory cells,** release their signalling molecules into the bloodstream, which carries them—like hormones—to their target tissues. Although most endocrine glands are composed of columnar or cuboidal epithelial cells, some, such as the adrenal medulla, are clusters of neurosecretory cells. The molecules they release affect not only target tissues reached through the bloodstream but also other neurons that rapidly conduct the information to regulatory centers. In addition, certain substances that are, by all traditional definitions, hormones, such as insulin, have recently been found in the vertebrate brain.

Despite this evidence for a single, unified system, the older concepts of an endocrine system and a nervous system remain useful. By beginning our exploration of integration and control in the animal body with the endocrine system, as we shall do in this chapter, we can establish the basic principles of chemical communication among cells, which also apply to the nervous system. In Chapter 41, we shall focus our attention on the organization of the nervous system and the mechanisms by which information is transmitted through it. Chapter 42 will be devoted to the details of sensory perception, particularly of the external environment, and the subsequent motor response. In Chapter 43, all of these threads will come together again as we consider the culmination of vertebrate evolution, the brain.

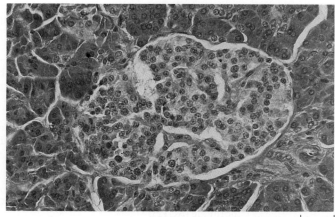

40–2 *A cross section of pancreatic tissue. The pancreas is both an endocrine and an exocrine gland. The group of small cells in the center of the micrograph are islet cells, which are the endocrine cells of the pancreas. Different types of islet cells secrete the hormones insulin, glucagon, and somatostatin. The surrounding exocrine cells produce digestive enzymes, which are carried through the pancreatic duct to the small intestine (see Figure 34–11, page 722).*

25 μm

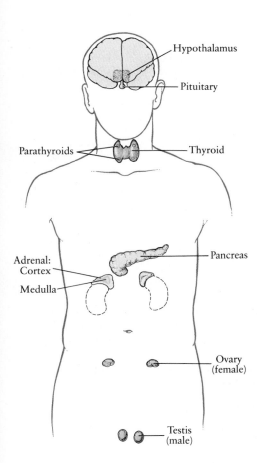

40–3 *Some of the hormone-producing (endocrine) organs. The pituitary releases hormones that, in turn, regulate the hormone secretions of the thyroid, the adrenal cortex (the outer layer of the adrenal gland), and the reproductive organs. The pituitary is itself under the regulatory control of a major brain center, the hypothalamus. The hypothalamus thus is the major link between the nervous system and the endocrine system.*

GLANDS AND THEIR PRODUCTS: AN OVERVIEW

Hormones are, by definition, organic molecules secreted in one part of an organism that diffuse or, in the case of vertebrates, are transported by the bloodstream to other parts of the organism, where they have specific effects on target organs or tissues. As we have seen in previous chapters, hormones are produced by a variety of different cell types: epithelial cells of the digestive tract (gastrin, secretin, cholecystokinin), cardiac muscle cells (cardiac peptide), white blood cells (histamine, lymphokines, interleukins), and even injured or infected cells (histamine, interferons). In these cell types and many others, the secretion of hormones is only one of several functions, although an essential one for the overall processes in which they—and the tissues of which they are a part—are involved. In glandular epithelial cells and neurosecretory cells, however, the secretion of hormones is the primary function, to which all other activities are subordinated.

Glandular epithelial cells and neurosecretory cells are often found in clusters, many of which are sufficiently large that they were readily identified by the early anatomists as distinct organs. With the discovery of their functions, which came much later, these organs were seen as constituting a distinct organ system. Still later, a distinction was drawn between exocrine glands and endocrine glands. **Exocrine glands** secrete their products into ducts; examples are digestive glands, the milk glands of the female mammal, and the sweat glands of the human skin. **Endocrine glands,** by contrast, secrete their products into the bloodstream (or, more precisely, into the extracellular fluids, from which they diffuse into the bloodstream); thus they are sometimes referred to as "ductless" glands.

The principal endocrine glands of the vertebrate body are shown in Figure 40–3. The hormones they secrete are of three general chemical types: steroids (see page 70), peptides or proteins, and amino acid derivatives. Hormones are characteristically active in very small amounts. It has been calculated, for example, that the concentration of adrenaline normally present in your bloodstream can be approximated by 8 milliliters (one teaspoonful) in a lake 2 meters deep and 100 meters in diameter. As befits such potent chemicals, playing key roles in the integration and control of the body's physiological functions, hormones are themselves under tight control. One aspect of this control is the regulation of their production. With very few exceptions, hormones are under negative feedback control, as described in Chapter 33. Another, equally important aspect is that they are rapidly degraded in the body; the steroids, peptides, and proteins are broken down by the liver, and the amines (amino acid derivatives) by enzymes in the blood.

Table 40–1 summarizes the principal endocrine glands of vertebrates. In this chapter, we shall discuss most of these glands, the hormones they secrete, and the regulation of their secretion. We shall, however, defer discussion of the glands and hormones involved in reproduction until Chapter 44.

823

TABLE 40-1 Some of the Principal Endocrine Glands of Vertebrates and the Hormones They Produce

GLAND	HORMONE	PRINCIPAL ACTION	MECHANISM CONTROLLING SECRETION	CHEMICAL COMPOSITION
Pituitary, anterior lobe	Growth hormone (somatotropin)	Stimulates growth of bone, inhibits oxidation of glucose, promotes breakdown of fatty acids	Hypothalamic hormone(s)	Protein
	Prolactin	Stimulates milk production and secretion in "prepared" gland	Hypothalamic hormone(s)	Protein
	Thyroid-stimulating hormone (TSH) (thyrotropin)	Stimulates thyroid	Thyroxine in blood; hypothalamic hormone(s)	Glycoprotein
	Adrenocorticotropic hormone (ACTH)	Stimulates adrenal cortex	Cortisol in blood; hypothalamic hormone(s)	Polypeptide (39 amino acids)
	Follicle-stimulating hormone (FSH)*	Stimulates ovarian follicle, spermatogenesis	Estrogen in blood; hypothalamic hormone(s)	Glycoprotein
	Luteinizing hormone (LH)*	Stimulates ovulation and formation of corpus luteum in female, interstitial cells in male	Progesterone or testosterone in blood; hypothalamic hormone(s)	Glycoprotein
Hypothalamus (via posterior pituitary)	Oxytocin	Stimulates uterine contractions, milk ejection	Nervous system	Peptide (9 amino acids)
	Antidiuretic hormone (ADH, vasopressin)	Controls water excretion	Osmotic concentration of blood; blood volume; nervous system	Peptide (9 amino acids)
Thyroid	Thyroxine, other thyroxinelike hormones	Stimulate and maintain metabolic activities	TSH	Iodinated amino acids
	Calcitonin	Inhibits release of calcium from bone	Concentration of Ca^{2+} ions in blood	Polypeptide (32 amino acids)
Parathyroid	Parathyroid hormone (parathormone)	Stimulates release of calcium from bone; stimulates conversion of vitamin D to active form, which promotes calcium uptake from gastrointestinal tract; inhibits calcium excretion	Concentration of Ca^{2+} ions in blood	Polypeptide (34 amino acids)
Adrenal cortex	Cortisol, other glucocorticoids	Affect carbohydrate, protein, and lipid metabolism	ACTH	Steroids
	Aldosterone	Affects salt and water balance	Processes initiated in the kidney; K^+ ions in blood	Steroid
Adrenal medulla	Adrenaline and noradrenaline	Increase blood sugar, dilate or constrict specific blood vessels, increase rate and strength of heartbeat	Nervous system	Catecholamines (amino acid derivatives)
Pancreas	Insulin	Lowers blood sugar, increases storage of glycogen	Concentration of glucose and amino acids in blood; somatostatin	Polypeptide (51 amino acids)
	Glucagon	Stimulates breakdown of glycogen to glucose in the liver	Concentration of glucose and amino acids in blood; somatostatin	Polypeptide (29 amino acids)
Pineal	Melatonin	Involved in regulation of circadian rhythms	Light-dark cycles	Catecholamine
Ovary, follicle	Estrogens*	Develop and maintain sex characteristics in females, initiate buildup of uterine lining	FSH	Steroids
Ovary, corpus luteum	Progesterone and estrogens*	Promote continued growth of uterine lining	LH	Steroids
Testis	Testosterone*	Supports spermatogenesis, develops and maintains sex characteristics of males	LH	Steroid

* These hormones will be discussed in Chapter 44.

THE PITUITARY GLAND

The pituitary gland was once considered the master gland of the body, as it is the source of hormones stimulating the reproductive organs, the adrenal cortex, and the thyroid. However, it is now known that this "master" gland is itself regulated by a key brain center, the hypothalamus. Hormones from the hypothalamus stimulate or, in some cases, inhibit the production of pituitary hormones. The pituitary, about the size of a kidney bean, is located at the base of the brain in the geometric center of the skull. It consists of three lobes: the anterior, the intermediate, and the posterior.

The Anterior Lobe

The anterior lobe of the pituitary is the source of at least six different hormones, each produced by recognizably different cells. One of these is growth hormone, sometimes called somatotropin, which stimulates protein synthesis and promotes the growth of bone. As is the case with most of the hormones, growth hormone is best known by the effects caused by too much or too little. If there is a deficit in growth hormone production in childhood, a midget results, the so-called "pituitary dwarf." An excess of growth hormone during childhood results in a giant; most circus giants are the result of such an excess. Excessive growth hormone in the adult does not lead to giantism, since growth of the long bones has ceased; it leads instead to acromegaly, an increase in the size of the jaw and the hands and feet, adult tissues that are still sensitive to the effects of growth hormone. Growth hormone also affects glucose metabolism, inhibiting the uptake and oxidation of glucose by some types of cells. It also stimulates the breakdown of fatty acids, thus conserving glucose. This hormone is now being produced by recombinant DNA techniques, which will extend its use in medical treatment and facilitate the study of its role in glucose metabolism.

A second hormone produced by the anterior pituitary is prolactin, which stimulates the secretion of milk in mammals. Its production is controlled by an inhibitory hormone produced by the hypothalamus. As long as the infant continues to nurse, the nerve impulses produced by the suckling of the breast are transmitted to the hypothalamus, which decreases production of prolactin-inhibiting hormone. The pituitary then releases prolactin, which, in turn, acts upon the breast to maintain the production of milk. Once suckling ceases, the synthesis and release of prolactin decrease and so milk production stops. Thus supply is regulated by demand.

Four of the hormones secreted by the anterior pituitary are **tropic hormones**—hormones that act upon other endocrine glands to regulate their secretions. One of these tropic hormones is TSH, the thyroid-stimulating hormone, also known as thyrotropin. TSH stimulates cells in the thyroid gland to increase their production and release of thyroxine, the thyroid hormone. In a negative feedback loop involving both the pituitary and the hypothalamus, the increased concentration of thyroxine inhibits the further secretion of TSH by the pituitary. Adrenocorticotropic hormone (ACTH) has a similar regulatory relationship with the production of cortisol, one of the hormones produced by the adrenal cortex (the outer layer of the adrenal gland).

The other two tropic hormones secreted by the anterior pituitary are gonadotropins—hormones that act upon the **gonads,** or gamete-producing organs (the testes and the ovaries). These hormones, follicle-stimulating hormone (FSH) and luteinizing hormone (LH), will be discussed in Chapter 44.

(a)

(b)

(c)

40–4 *When a green anole* (Anolis carolinensis) *moves from a light background to a dark background, or vice versa, its pigmentation changes. The color change is the result of the movement of pigment-containing granules within the pigmented cells of the skin. When the granules are concentrated in the center of the cells, the skin color is light; when they are dispersed into extensions of the cells, the color is dark. Melanocyte-stimulating hormone plays an important role in triggering the dispersal of these granules. The change from light to dark* (a–c) *takes 5 to 10 minutes, while the change from dark to light takes 20 to 30 minutes.*

The Intermediate and Posterior Lobes

In many vertebrates, the intermediate lobe of the pituitary gland is the source of melanocyte-stimulating hormone. In reptiles and amphibians, this hormone stimulates color changes associated with camouflage (Figure 40–4) or with behavior patterns such as aggression or courtship. In humans, in which the secretion of melanocyte-stimulating hormone is greatly reduced, its functions are unknown.

The posterior lobe of the pituitary gland stores the hormones produced by the hypothalamus.

THE HYPOTHALAMUS

The Pituitary-Hypothalamic Axis

The pituitary gland lies beneath the hypothalamus and is directly under its influence. The pituitary is also under the influence, by way of the hypothalamus, of other parts of the brain. The hypothalamus is the source of at least nine hormones that act either to stimulate or inhibit the secretion of hormones by the anterior pituitary. These hormones are small peptides, one only three amino acids in length. They are unusual not only for their small size but also for the way in which they reach their target gland. Produced by neurosecretory cells of the hypothalamus, they travel only a few millimeters to the pituitary, apparently never entering the general circulation. However, they make this brief passage by way of a portal system (Figure 40–5).

The first of these hypothalamic hormones to be discovered was TRH, thyrotropin-releasing hormone. As its name implies, it stimulates the release of thyrotropin (TSH) from the pituitary. The second was gonadotropin-releasing hormone (GnRH), which controls the release of the gonadotropic hormones LH and FSH.

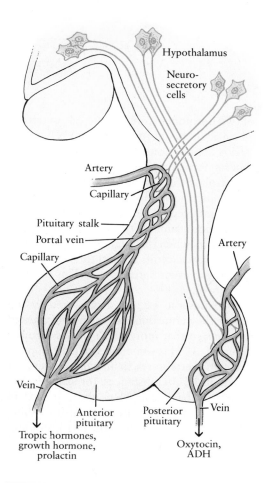

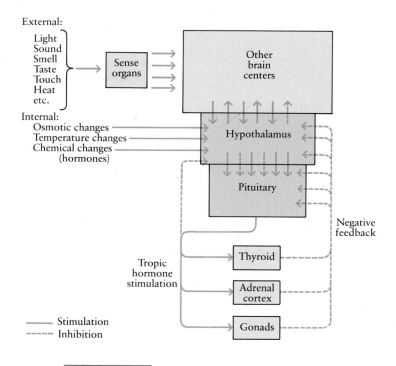

40–5 *Relationship between the hypothalamus and the pituitary. The hypothalamus communicates with the anterior lobe of the pituitary through a small portal system (see page 759). Neurosecretory cells of the hypothalamus secrete releasing or inhibiting hormones directly into capillaries that are linked by portal veins to a second capillary network in the anterior pituitary, where the hypothalamic hormones affect the production of pituitary hormones. Other hypothalamic neurosecretory cells produce oxytocin and ADH, which are transmitted to the posterior lobe of the pituitary through the nerve fibers. Following their release from the nerve endings in the posterior pituitary, these hormones diffuse into capillaries and thus enter the general circulation.*

40–6 *The production of many hormones is regulated by complex negative feedback systems involving the pituitary and the hypothalamus. The hypothalamus controls the pituitary's secretion of tropic hormones, and these, in turn, stimulate the secretion of hormones from the thyroid, adrenal cortex, and gonads (the testes or the ovaries). As the concentration of the hormones produced by these target glands rises in the blood, the hypothalamus decreases its production of releasing hormones, the pituitary decreases its hormone production, and production of hormones by the target glands also slows. By way of the hypothalamus, which receives information from many other parts of the brain, hormone production is also regulated in response to other changes in the external and internal environments.*

These discoveries were made by two groups of scientists, one headed by Roger Guillemin at the Salk Institute (requiring the brains of 5 million sheep) and the other by Andrew Schally in New Orleans (using a comparable supply of pigs). The third hormone found was not a releaser but, surprisingly, an inhibitor. Because it inhibited release of the growth hormone somatotropin, it was given the name of somatostatin. As you may recall, somatostatin was the first mammalian hormone to be synthesized using recombinant DNA technology (see page 351). Several more releasing hormones and one other inhibitory hormone have now been isolated, and the search is not over.

Figure 40–6 summarizes the negative feedback systems linking the hypothalamus and pituitary with the thyroid, adrenal cortex, and gonads. These feedback control systems, generally involving both the pituitary and the hypothalamus, provide for both homeostasis and response to changing conditions. The pituitary feedback system usually provides for constancy. However, it can be overridden by the hypothalamic system, which takes into account not only the balance between output and input but also the changes elsewhere in the body and in the external environment.

Other Hypothalamic Hormones

The hypothalamus is also the source of two hormones stored in and released from the posterior pituitary: oxytocin and antidiuretic hormone (ADH). Oxytocin accelerates childbirth by increasing uterine contractions during labor; these contractions also cause the uterus to regain its normal size and shape after delivery. The release of oxytocin is under the control of the nervous system and may be triggered by increasing pressure within the uterine wall or by movements of the fetus. In experimental animals, labor can be induced by mechanical stimulation of the uterus or electrical stimulation of the hypothalamus. Oxytocin is also responsible for the "letting down" of milk that occurs when the infant begins to suckle. The hormone promotes contraction of the muscle fibrils around the milk-secreting cells of the mammary glands. Oxytocin is present in males as well as females, but its function in the male, if any, is unknown.

ADH, as we saw in Chapter 37, decreases the excretion of water by the kidneys. It achieves this effect by increasing the permeability of the membranes of cells in the collecting ducts of the nephrons so that more water passes through them and back into the blood from the urine. ADH is sometimes called vasopressin because it increases blood pressure in many vertebrates; in humans, however, it has this effect only in response to certain unusual circumstances, for example, the blood loss of a severe hemorrhage. Oxytocin has some ADH effect and ADH some oxytocin effect. This cross action is not surprising because each of these hormones consists of only nine amino acids, and differences in the two hormones involve only two amino acids among the nine (Figure 40–7).

40-7 *Primary structures of antidiuretic hormone (ADH) and oxytocin, hormones produced in the hypothalamus and released from the posterior lobe of the pituitary. Note that each consists of only nine amino acids and that they differ by only two. This similarity of structure gives rise to some cross action by these hormones.*

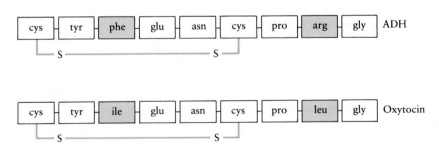

THE THYROID GLAND

The thyroid, under the influence of thyroid-stimulating hormone (TSH) from the pituitary, produces thyroxine, which is an amino acid combined with four atoms of iodine (Figure 40–8). Thyroxine (or, most likely, its metabolic product, triiodothyronine) accelerates the rate of cellular respiration. In some animals, it also plays a major role in temperature regulation (see Figure 38–8, page 784).

40-8 *Thyroxine, the principal hormone produced by the thyroid gland. Note the four iodine atoms in its structure. Triiodothyronine differs from thyroxine by having one less iodine atom on the OH-bearing ring. Because iodine is needed for thyroxine, it is an essential component of the human diet. Where iodine is present in the soil, it is available in minute quantities in drinking water and in plants. In the United States, table salt is ordinarily iodized or must be specifically labeled as being uniodized.*

Thyroxine

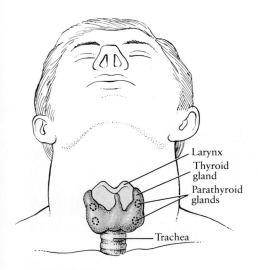

40–9 *The pea-sized parathyroid glands, the smallest of the known endocrine glands, are located behind or within the thyroid gland. They produce parathyroid hormone (parathormone), which increases concentrations of blood calcium. Calcitonin, a hormone produced by the thyroid gland, decreases blood calcium.*

Cortisol

Aldosterone

40–10 *The chemical structures of representatives of the two major groups of hormones secreted by the adrenal cortex. Cortisol is a glucocorticoid, and aldosterone is a mineralocorticoid. The hormones of both groups are steroids, identified by their characteristic four-ring structure. As you can see, the differences in the molecular structures of cortisol and aldosterone are minor, and yet their physiological roles are profoundly different.*

Hyperthyroidism, the overproduction of thyroxine, results in nervousness, insomnia, and excitability; increased heart rate and blood pressure; heat intolerance and excessive sweating; and weight loss. Hypothyroidism (too little thyroxine) in infancy affects development, particularly of the brain cells; if not treated in time, it can lead to permanent mental deficiency and dwarfism. In adults, hypothyroidism is associated with dry skin, intolerance to cold, and lack of energy. Hypothyroidism may be caused by insufficient iodine, which is needed to make thyroxine, and, in these cases, it is often associated with goiter, an enlargement of the thyroid gland.

The thyroid gland also secretes the hormone calcitonin in response to rising calcium levels in the fluid surrounding the thyroid cells, which reflects the levels in the blood. Calcitonin's major action is to inhibit the release of calcium ion from bone.

THE PARATHYROID GLANDS

The pea-sized parathyroid glands, the smallest of the known endocrine glands, are located behind or within the thyroid gland (Figure 40–9). They produce parathyroid hormone (parathormone), which plays an essential role in mineral metabolism, specifically in the regulation of calcium and phosphate ions, which exist in a reciprocal relationship in the blood. Calcium is normally present in mammalian blood in concentrations of about 6 milligrams per 100 milliliters of whole blood. A rise or fall of more than 2 or 3 milligrams per 100 milliliters can lead to such severe disturbances in blood coagulation, muscle contraction, and nerve function that death may follow within hours.

Parathyroid hormone increases the concentration of calcium ion in blood in several different ways. It stimulates the conversion of vitamin D into its active form (see Figure 34–14, page 728); active vitamin D, in turn, increases the absorption of calcium ion from the intestine. Parathyroid hormone also reduces excretion of calcium ion from the kidneys. In addition, it stimulates the release into the bloodstream of calcium from bone, which contains 99 percent of the body's total calcium. Thus, parathyroid hormone and calcitonin work as a fine-tuning mechanism, regulating blood calcium, with parathyroid hormone apparently playing the principal role. The production of both hormones is regulated directly by the concentration of calcium ions in the blood.

Hyperparathyroidism, caused by tumors of the parathyroids, occasionally occurs in humans. When there is too much parathyroid hormone, the bones lose large amounts of calcium, becoming soft and fragile, and the vertebrae may shrink, producing a loss of height. Removal of the parathyroid glands without hormone replacement therapy results in violent muscular contractions and spasms, leading to death.

ADRENAL CORTEX

The adrenal cortex—the outer layer of the adrenal gland—is the source of a number of steroid hormones. The adrenal (ad-renal) glands, as their name implies, are on top of the kidneys. About 50 different steroids, all very similar in structure, have been isolated from the adrenal cortex of various mammals. Some of these corticosteroids, as they are known, undoubtedly represent steps in the synthesis of various hormones, though most of them have some hormonal activity. In humans, there are two major groups of adrenocortical steroids, the glucocorticoids and the mineralocorticoids (Figure 40–10).

The Regulation of Bone Density

Calcitonin and parathyroid hormone play the principal roles in the regulation of calcium levels in the blood. The major reservoir of calcium on which their activities depend is, as we have seen, bone. The regulation of bone density, however, is a considerably more complex process than the regulation of blood calcium and involves a number of hormones in addition to calcitonin and parathyroid hormone.

Despite its apparent solidity, bone, like skin and the intestinal lining, is a turnover tissue. As we saw in Figure 33–6 (page 704), it is formed by cells known as osteoblasts, which secrete the collagen and other fibrils in which calcium compounds become crystallized. However, other cells, known as osteoclasts, break down bone, dissolving the crystalline compounds and releasing calcium to the blood. The maintenance of bone mass depends on a balance between the activities of these two types of cells. One of the contributing factors to osteoporosis, the severe loss of calcium from bone that afflicts many older persons, particularly women, is greater activity by osteoclasts than by osteoblasts.

The symptoms of osteoporosis are similar to those of hyperparathyroidism: the bones lose large amounts of calcium, becoming soft and fragile, and the vertebrae shrink, producing a loss of height. The consequences frequently include fractured vertebrae and hips, both of which can lead to serious complications. Current evidence indicates that hormones secreted by

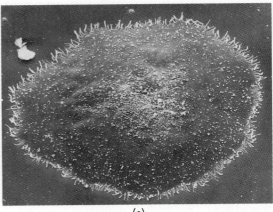

(a)

(a) An osteoclast, growing in tissue culture. These large cells, which are multinucleate, are produced by the fusion of many smaller cells. The precursors of osteoclasts are among the progeny of the bone marrow stem cells and may be related to monocytes and macrophages. The thin processes extending from the cell surface pump acid onto the bone, and the acid, in

stem cells in the bone marrow (page 792) play a major role in the activities of the osteoclasts. These substances, known as colony-stimulating factors, are the same molecules that stimulate the repeated divisions

Glucocorticoids

Cortisol is thought to be the most important glucocorticoid in humans. Cortisol and the other glucocorticoids promote the formation of glucose from protein and fat. At the same time, they decrease the utilization of glucose by most cells, with the notable exceptions of cells of the brain and the heart, thus favoring the activities of these vital organs at the expense of other body functions. Their release increases during periods of stress, such as facing new situations, engaging in athletic competition, and taking final exams. As we shall see in the next chapter, they work in concert with the sympathetic nervous system.

In addition to their effects on glucose metabolism, glucocorticoids suppress inflammatory and immune responses, an effect that may be a factor in the increased susceptibility to illness that often accompanies stress. Because of their immunosuppressive properties, cortisol and other glucocorticoids are sometimes used in the treatment of autoimmune diseases and severe allergic reactions.

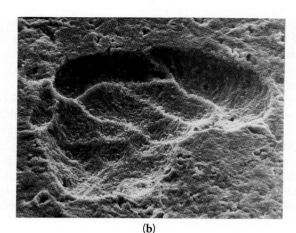

(b)

turn, leaches minerals from it. (b) *This crater, known as a resorption lacuna, was created on the surface of a bone by the activities of an osteoclast. Following the departure of the osteoclasts, osteoblasts migrate into resorption lacunae and begin their bone-creating activities to fill the lacunae.*

of the stem cells and regulate the differentiation of their progeny into red blood cells, granulocytes, monocytes, and lymphocytes. Stimulation of the osteoclasts also appears to involve molecules, particu-larly lymphokines and interleukins, produced in the course of the inflammatory and immune responses. There is evidence that prostaglandins (page 833) play a role as well. Damping the activity of the osteoclasts indirectly through the use of anti-inflammatory drugs is a component of medical therapy designed to halt the progress of osteoporosis.

Also contributing to the development of osteoporosis is a disruption of the pathways affecting absorption of calcium ion from the intestine, particularly those involved in the conversion of vitamin D into its active form. These pathways are influenced not only by parathyroid hormone but also by the female hormone estrogen, the production of which drops dramatically after menopause. Despite the claims of numerous television commercials, calcium supplements alone, in the absence of appropriate hormone replacement and anti-inflammatory agents, appear to be of little value in reversing or slowing the course of osteoporosis once it has begun. Accumulating evidence, however, indicates the importance of adequate calcium intake during childhood and adolescence, when bones are growing, and during the childbearing years. Drinking plenty of skim milk now, and even occasionally succumbing to that tempting chocolate milkshake, may help to provide the margin of safety —ample calcium reserves stored in the bones—that will enable you to be on the tennis court or golf course 50 years hence.

However, the serious side effects of the high doses often required limit their usefulness. Among these side effects are a reduced ability to combat infection; redistribution of body fat, resulting in "buffalo hump" and "moon face"; and mental disturbances, including both hyperactivity and depression.

The cortisol group of hormones are secreted in response to adrenocorticotropic hormone (ACTH). ACTH is secreted by the pituitary gland when it is stimulated by a releasing hormone from the hypothalamus. As with thyroxine, the secretion of the glucocorticoids is inhibited by negative feedback exerted on the pituitary and the hypothalamus. Recent studies indicate that positive feedback is also at work, involving an interleukin secreted by activated monocytes and macrophages. This molecule stimulates the secretion of ACTH, by acting either on the pituitary or the hypothalamus, or, perhaps, on both. Current evidence suggests that the immunosuppressive activities of corticosteroids may be part of the normal regulatory mechanism that turns off the inflammatory and immune responses when their work is done.

Mineralocorticoids

A second group of hormones secreted by the adrenal cortex comprises the mineralocorticoids, of which aldosterone is the primary example. These corticosteroids are involved in the regulation of ions, particularly sodium and potassium ions. The mineralocorticoids affect the transport of ions across the cell membranes of the nephrons and, as a consequence, have major effects both on ion concentrations in the blood and on water retention and loss. An increase in aldosterone secretion, as we noted in Chapter 37, results in greater reabsorption of sodium ions in the distal tubule and collecting duct of the nephron and increases the secretion of potassium ions into them. A deficiency in mineralocorticoids precipitates a critical loss of sodium ions from the body in the urine and, with it, a loss of water, leading, in turn, to a reduction in blood pressure.

In addition to glucocorticoids and mineralocorticoids, the adrenal cortex produces small amounts of male sex hormones in both males and females. An adrenal tumor may result in increased production of these hormones and, consequently, in women, the production of facial hair and other masculine characteristics. Bearded ladies in the circus were often the victims of such tumors. The adrenal cortex is also thought to secrete small amounts of female sex hormones.

ADRENAL MEDULLA

The adrenal medulla, the central portion of the adrenal gland, is a large cluster of neurosecretory cells whose nerve endings secrete adrenaline and noradrenaline (Figure 40–11) into the bloodstream. These hormones, known also as epinephrine and norepinephrine, increase the rate and strength of the heartbeat, raise blood pressure, stimulate respiration, and dilate the respiratory passages. They also increase the concentration of glucose in the bloodstream by promoting the activity of the enzyme that breaks down glycogen to glucose 6-phosphate. The adrenal medulla is stimulated by nerve fibers of the sympathetic division of the autonomic nervous system and so acts as an enforcer of sympathetic activity.

THE PANCREAS

The islet cells of the pancreas (Figure 40–2, page 823) are the source of insulin and glucagon, two hormones involved in the regulation of glucose metabolism. Insulin is secreted in response to a rise in blood sugar or amino acid concentration (as after a meal). It lowers the blood sugar by stimulating cellular uptake and utilization of glucose and by stimulating the conversion of glucose to glycogen.

When there is an insulin deficiency, as in persons with diabetes mellitus, the concentration of blood sugar rises so high that not all the glucose entering the kidney can be reabsorbed; the presence of glucose in the urine forms the basis of simple tests for diabetes. The loss of glucose is accompanied by loss of water. The resulting dehydration, which can lead to collapse of circulation, is one of the causes of death in an untreated diabetic.

Glucagon, produced by different islet cells of the pancreas, increases blood sugar. It stimulates the breakdown of glycogen to glucose in the liver and the breakdown of fats and proteins, which decreases glucose utilization.

Somatostatin, originally found in the hypothalamus, has now also been isolated from a third class of islet cells in the pancreas. It is released from the pancreas during digestion of a meal and exerts a variety of inhibitory effects on the digestive tract that, collectively, help to regulate the rate at which glucose and other nutrients are absorbed into the bloodstream. There is evidence that somatostatin also participates in controlling the synthesis of insulin and glucagon.

Adrenaline

Noradrenaline

40–11 *The chemical structures of adrenaline and noradrenaline, hormones secreted by the adrenal medulla. These compounds are modified amino acids known as catecholamines, characterized by a six-sided ring bearing two hydroxyl groups. They are synthesized from the amino acid tyrosine (see page 73). As we shall see in the next chapter, noradrenaline is also an important neurotransmitter.*

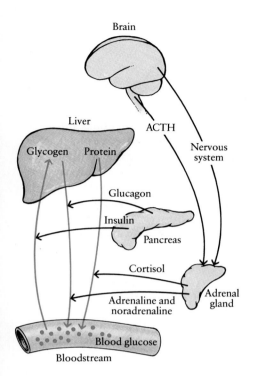

40–12 *Hormonal regulation of blood glucose. When blood sugar concentrations are low, the pancreas releases glucagon, which stimulates the breakdown of glycogen and the release of glucose from the liver. When blood sugar concentrations are high, the pancreas releases insulin, which removes glucose from the bloodstream by increasing its uptake by cells and promoting its conversion into glycogen, the storage form. Under conditions of stress, ACTH, produced by the pituitary, stimulates the adrenal cortex to produce cortisol and related hormones, which increase the breakdown of protein and its conversion to glucose in the liver. At the same time, the adrenal medulla releases adrenaline and noradrenaline, which also raise blood sugar.*

Growth hormone and somatostatin, not shown in this diagram, also affect blood glucose levels. Growth hormone inhibits the uptake and oxidation of glucose by many types of cells and stimulates the breakdown of fatty acids. Somatostatin influences the rate at which glucose is absorbed into the bloodstream from the digestive tract.

Thus, as we have seen, at least seven different hormones are involved in regulating blood sugar: growth hormone, cortisol, adrenaline, noradrenaline, insulin, glucagon, and somatostatin. This tight control over blood glucose, illustrated in Figure 40–12, ensures that glucose is always available for brain cells. Unlike other cells in the body, which can derive energy from the breakdown of amino acids and fats, brain cells can utilize only glucose under most circumstances. Thus they are immediately affected by low blood sugar. After several days of fasting, however, the brain begins to use fatty acids as an energy source. An insulin overdose, which causes a rapid drop in blood sugar, can result in coma and death.

THE PINEAL GLAND

The pineal gland is a small lobe in the forebrain, lying near the center of the brain in humans. In lower vertebrates, it contains light-sensitive cells and so is sometimes called the third eye. The pineal gland secretes the hormone melatonin. In many species, including chickens, rats, and humans, the production of melatonin rises sharply at night and falls rapidly in the daytime. Exposure to light during the dark cycle interrupts production of melatonin, recalling the studies on photoperiodism in plants (page 686).

In larval amphibians, melatonin causes pigment-containing granules in cells of the skin to aggregate near the center of the cells, causing a blanching of the skin associated with darkness. In sparrows, injections of melatonin induce roosting and the lowering of body temperature, both of which are characteristic nighttime events in these birds. Melatonin also inhibits the development of the gonads in species as disparate as chickens and hamsters; its decreased production during long-day periods is believed to be associated with the seasonal enlargement of the gonads in preparation for mating. In humans, the pineal gland may be involved in sexual maturation; tumors of the pineal have been associated with precocious puberty.

Thus there are a few tempting clues suggesting that the pineal gland may function as a biological timekeeper, but the way in which light affects the gland and the way in which melatonin alters physiological responses—if it does—are yet to be discovered.

PROSTAGLANDINS

Among the most potent of all substances produced by and released from cells are a group of related chemicals first detected in semen. These substances were thought at the time to be produced by the prostate gland, a structure of the male reproductive system, and so they were called prostaglandins. Later research revealed, however, that most of the prostaglandins in semen are synthesized in other structures, the seminal vesicles. Since their initial discovery, a large number of prostaglandins have been identified, all related structurally but with a variety of different, and sometimes directly opposite, effects.

Although prostaglandins have hormonelike properties, they differ from other hormones in several significant ways: (1) Unlike any other hormones, they are fatty acids. Most are formed by the oxygenation of a 20-carbon polyunsaturated fatty acid known as arachidonic acid. (2) They are produced by cell membranes in most—if not all—organs of the body, as opposed to other hormones, which are produced by glandular epithelium or neurosecretory cells. (3) Their target tissues are generally either the same tissues in which they are produced or the tissues of another individual. (4) They produce marked effects at extremely low concentra-

Circadian Rhythms

Virtually every living organism, from one-celled algae to bean plants (page 690) to *Homo sapiens,* exhibits circadian rhythms in many physiological functions. As we noted in Chapter 32, the chemical nature of the biological clock governing such rhythms remains unknown. Work with *Drosophila,* however, suggests that the products of one or more specific genes are involved. Mutations in a gene located on the *X* chromosome and known as *per* (for "period") have profound effects on the circadian rhythms of fruit flies. If the gene is deleted or rendered inoperative by a mutation, the flies become insomniacs, with no discernible sleep-wake cycle. Other mutations affect the length of the daily rhythms; for example, *per s* ("short periods") produces flies with daily cycles of 18 to 20 hours, and *per l* ("long periods") results in daily cycles of 28 to 30 hours. This same gene also affects the mating songs of male fruit flies, which beat their wings together in a rhythmic pattern that repeats at 60-second intervals. When the *per* gene is knocked out, the mating songs have no rhythms at all; *per s* results in songs at 40-second intervals, and *per l* in songs at 80-second intervals.

Investigators puzzled by how a single gene could affect both the daily 24-hour rhythm and the 60-second courtship song recently created mosaic flies in which cells in certain parts of the body contain the normal gene and cells in other parts of the body contain a mutant gene. They have found that the location in which *per* is expressed determines its effect; in cells of the head, it affects the daily rhythm, but in cells of the thorax, it affects the song rhythms. Moreover, the greater the quantities of the *per* gene product synthesized, the faster the clock runs. The gene has now been sequenced, and it turns out to contain repetitive sequences coding for alternating glycines and serines or threonines. Similar repetitive sequences have been identified in DNA from chickens, mice, and humans, but it is not yet known if these sequences are the portion of the gene controlling the rhythms.

Although the mechanism governing circadian rhythms remains mysterious, the effect of the rhythms in human physiology and behavior is not in doubt. For instance, we are more likely to be born between 3 and 4 A.M. and also to die in these same early morning hours. Body temperature fluctuates as much as 1°C (about 2°F) during the course of 24 hours, usually reaching a peak at about 4 P.M. and a low about 4 A.M. Alcohol tolerance is greatest, fortunately, at 5 P.M., while tolerance to pain is lowest at 6 P.M. in most persons. Respiration, heart rate, and urinary excretion of potassium, calcium, and sodium, all vary according to the time of day (heart rate by as much as 20 beats per minute). Secretion of various hormones follows circadian rhythms: serum levels of corticosteroids peak between 4 A.M. and 8 A.M. in people on a normal sleep-wake cycle; growth hormone levels rise about an hour after falling asleep; prolactin peaks around 3 A.M.; testosterone about 9 A.M.

Knowledge of circadian rhythms may have important practical consequences. A study by the Federal Aviation Agency showed that pilots flying from one time zone to another—from New York to Europe, for instance—exhibit "jet lag," a general decrease in mental alertness, an inability to concentrate, and an increase in decision time and physiological reaction time. Measurements of various functions show that the body may be "out of sync" for as much as a week after such a flight. This brings into question not only schedules for airline personnel but also present policies of speeding diplomats to foreign capitals at times of international crisis or of air transport of troops into combat. It may be significant that the accidents at the nuclear power plants at both Three Mile Island and Chernobyl occurred in the small hours of the morning local time; in the case of Three Mile Island, the workers on duty at the time of the accident had been alternating between the day shift and the night shift every week.

Medical researchers are finding that failure to take circadian fluctuations into account can lead to mistakes in diagnosis and treatment. For example, blood pressure may vary as much as 20 percent in the course of a day, so that a person might be found to be within the normal range at one time of day and diagnosed as hypertensive at another. Numbers of white blood cells may vary as much as 50 percent in a 24-hour period, rendering the same immunosuppressive therapy ineffectual at one time of day and life-threatening at another. Studies of cancer chemotherapy in mice indicate that proper timing of drug doses can mean a doubling of survival rate. Determination of a patient's chronobiology may someday become as routine in therapy as blood typing or taking a medical history.

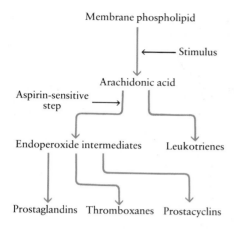

Membrane phospholipid

Stimulus

Arachidonic acid

Aspirin-sensitive step

Endoperoxide intermediates   Leukotrienes

Prostaglandins Thromboxanes Prostacyclins

40–13 *A number of different substances, known collectively as prostaglandins, are synthesized from the phospholipids of cell membranes. Among the stimuli known to trigger the production of arachidonic acid, the precursor of all of these substances, are neurotransmitters, other hormones, various drugs and toxic agents, and allergens. Depending on the cell type and the stimulus, arachidonic acid may be converted to leukotrienes, a category of prostaglandins involved in inflammatory and immune responses, or to intermediates known as endoperoxides. These molecules, in turn, may be converted to the prostaglandins that act on smooth muscle (the group first discovered, and for which the entire class of molecules is named), or to thromboxanes or prostacyclins, both of which affect platelet aggregation and the dilation or constriction of blood vessels. Aspirin inhibits the conversion of arachidonic acid to endoperoxides.*

tions, much lower than those of most hormones. A concomitant of their extraordinary potency is the fact that they are released in very small amounts and are rapidly broken down by enzyme systems in the body. If it were not for their presence in unusually large amounts in semen, they might never have been discovered.

Stimulation of Smooth Muscle

Among the many effects of prostaglandins, one of the most striking is their capacity to induce contractions in smooth muscle. This is believed to play several important roles in reproduction. The walls of the uterus, which are composed of smooth muscle, normally contract in continuous waves. After sexual intercourse, prostaglandins from semen are found in the female reproductive tract, where they increase the rhythmic contractions of the uterine wall and oviducts. It is believed that this action assists both the sperm on its journey to the oviduct and the oocyte as it travels from the oviduct to the uterus. The semen of some infertile males has been found to be poor in prostaglandins, and the uterus of infertile females is often unresponsive to prostaglandins.

The contractions of the uterus also increase during a menstrual period and reach their greatest strength when a woman is in labor. Prostaglandins produced in the uterine lining are believed to play a key role in triggering the onset of both menstruation and labor.

Increased understanding of prostaglandins and their capacity to stimulate smooth muscle has shed light on a long-perplexing medical problem. Between 30 and 50 percent of all women of childbearing age experience painful uterine cramps during the first day or two of each menstrual period, a condition known as dysmenorrhea. In most of these women, no abnormalities of the reproductive organs can be detected. Recent research has revealed, however, that the menstrual fluid of such women contains concentrations of prostaglandins two to three times higher than the levels found in the menstrual fluid of women without dysmenorrhea. Present evidence indicates that the increased prostaglandin levels not only cause stronger, more rapid contractions of the uterine walls but also reduce the blood supply to the tissue. As a result, less oxygen is available to the actively contracting muscles, creating an oxygen debt (page 190) and its accompanying pain. It is also hypothesized that prostaglandins act directly on pain nerve endings, causing them to fire more rapidly.

Clinical studies in this country and in Europe have shown that several compounds that inhibit prostaglandin synthesis are highly effective in reducing or, in some cases, completely eliminating the symptoms of dysmenorrhea. This alleviation is accompanied by a marked reduction in the levels of prostaglandins found in the menstrual fluid. Other studies are investigating the use of these compounds in preventing labor in women who are in danger of giving birth prematurely.

Other Prostaglandin Effects

Prostaglandins that are closely related chemically may differ widely in their physiological effects. Although many of them stimulate contractions of smooth muscle, as previously described, others inhibit smooth muscle contraction. Furthermore, one may affect the smooth muscle of bronchioles, another the smooth muscle of blood vessels. One, produced by platelets, is a potent stimulus for platelet aggregation (see page 751) and constriction of the blood vessels; another, produced by the endothelial cells that line the blood vessels, is a potent inhibitor of platelet aggregation and a dilator of the vessels. The way in which prostaglandins exert these multitudinous effects is not known.

Among the prostaglandins are a group of substances known as leukotrienes, which are produced principally by the various white blood cells involved in the inflammatory and immune responses. The leukotrienes include the interleukins released by activated helper T cells (page 808), as well as a variety of molecules released by stimulated macrophages and mast cells. As this information might lead you to expect, increased levels of prostaglandins have been implicated in disorders of the immune system such as rheumatoid arthritis, asthma, and severe allergies. Discovery of the prostaglandins and their involvement in inflammation has also provided a solution to another long-standing medical puzzle. Aspirin is one of the oldest and most effective medications known, but how it acts was, for many years, a mystery. Now it has been discovered that aspirin exerts its effects, at least in part, by inhibiting the synthesis of prostaglandins, thus producing its well-known soothing effects on inflammation and fever.

MECHANISMS OF ACTION OF HORMONES

Prostaglandins, like many of the regulatory chemicals in small organisms, travel only short distances to the cells they affect; the communication is rather like that of one person talking to another in the same room. Neurotransmitters released from the nerve endings of neurons similarly travel only a short distance; as shown in Figure 40–1 (page 822), neurons make close, anatomical connections with the cells or organs they influence, like a conversation on the telephone. Most hormones, by contrast, broadcast their messages. Whether or not these messages are received and acted upon depends upon the receptivity of the target tissue as much as on the chemical characteristics of the hormone. Moreover, target tissues may be receptive under some circumstances and not under others. For example, prolactin causes the production of milk—but only by the mammary glands and only when it is acting in concert with the female hormones estrogen and progesterone, thyroxine, adrenal steroids, and growth hormone.

The key to this specificity of hormone action lies in protein receptor molecules that, like enzymes, transport proteins, and the surface receptors of lymphocytes, have quite precise configurations that allow them to bind with one particular molecule but not with another that differs only slightly in structure. The study of hormones and their receptors has revealed two quite different mechanisms of action. In one, the receptor molecules are intracellular, in the nucleus or cytoplasm; in the other, the receptor molecules are embedded in the cell membrane. The steroid hormones and thyroid hormone utilize the first mechanism; the catecholamine, peptide, and protein hormones act through the second mechanism.

Intracellular Receptors

Steroid hormones are relatively small, lipid-soluble molecules. Thus they pass easily through cell membranes and freely enter all cells of the body. However, in the cytoplasm of their target cells—and only in their target cells—these hormones encounter a specific protein receptor molecule with which they combine (Figure 40–14). The hormone-receptor complex moves to the nucleus, where it binds to a specific chromosomal protein, initiating mRNA transcription from the particular DNA sequence regulated by that protein. After appropriate processing, the mRNA moves into the cytoplasm and protein synthesis occurs. The newly synthesized proteins may be structural proteins, enzymes, or other hormones; the result is a functional change in the cell, in the substances released from it, or in the receptors displayed on its surface.

Thyroid hormone, although not soluble in lipids, also readily passes through cell membranes, apparently by facilitated diffusion through a transport protein. Its receptor is not in the cytoplasm but rather in the nucleus. As with the steroid hormones, the hormone-receptor complex binds to the chromosome, initiating mRNA transcription leading to protein synthesis.

40-14 *The mechanism of action of a steroid hormone. The lipid-soluble hormone passes through the cell membrane into the cytoplasm. In its target cell, the hormone encounters a specific receptor to which it binds. The hormone-receptor complex then passes into the nucleus, where it combines with a chromosomal protein, triggering the transcription of a segment of DNA into mRNA. After processing, the mRNA is translated into protein. Depending on the hormone and the particular target cell, the newly synthesized protein may be an enzyme, another hormone or other product that will be secreted from the cell, or a structural protein, for example, a receptor for a different hormone. The altered functions of the cell as a consequence of the newly synthesized protein constitute the cell's response to the hormone.*

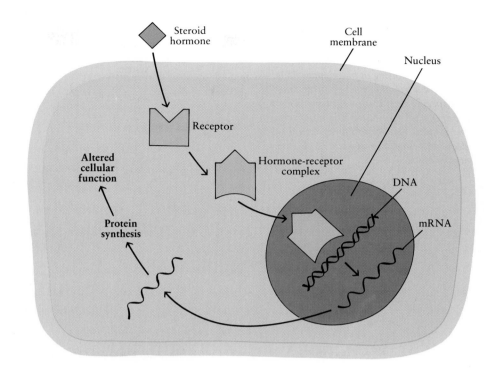

40-15 *Cyclic AMP (adenosine monophosphate) acts as a "second messenger" within a number of different types of vertebrate cells. Following stimulation by various hormones—the "first messengers"—cyclic AMP is formed from ATP. "Cyclic" refers to the fact that the atoms of the phosphate group form a ring. Cyclic AMP is also the chemical that attracts the amoebas of the cellular slime molds, causing them to aggregate into a slug-like body, which then behaves like a multicellular organism (see page 141).*

Membrane Receptors

The second group of hormones acts by combining with receptors in the membranes of target cells. The existence of these receptors has important medical implications. For example, it has long been known that juvenile diabetes—diabetes in young persons—is caused by a deficiency of the hormone insulin, which, as we have seen, promotes the uptake of glucose by cells. It was assumed by analogy that the diabetes found commonly in older persons and associated with obesity had the same cause. It has now been found, however, that adult diabetes often results from a decrease in the number of insulin receptor sites in target cell membranes rather than from a shortage of insulin. Such patients are treated most effectively by diet.

After a hormone combines with its membrane receptor, one of two events may follow, depending on the particular hormone. In some cases, the hormone-receptor complex is carried into the cytoplasm by receptor-mediated endocytosis (page 138). In other cases, the hormone never actually enters the cell; instead, its binding to the receptor sets in motion a "second messenger" that is responsible for the sequence of events inside the cell. The second messenger for many hormones is a chemical known as cyclic AMP (Figure 40–15).

An example of the second-messenger mechanism in action is the stimulation of the release of glucose from a liver cell by adrenaline (Figure 40–16). Adrenaline molecules bind to a receptor on the outer surface of the cell membrane. This event activates an enzyme, adenylate cyclase, that is bound on the inner surface of the membrane. Adenylate cyclase converts ATP to cyclic AMP (cAMP). The cyclic AMP binds to another enzyme, protein kinase, and activates it. This enzyme activates another enzyme, which, in turn, activates yet another. The final enzyme, phosphorylase *a*, then breaks down glycogen at a high rate, producing glucose 1-phosphate, which is further broken down to glucose and released from the cell. Since each enzyme greatly increases the rate of the particular reaction it catalyzes and can be used over and over again, the number of molecules involved in these

40–16 *Adrenaline triggers an amplification cascade in liver cells, similar in principle to the amplification cascade that occurs in the clotting of blood (page 751). Binding of a few molecules of adrenaline to its specific receptors on the outer surface of the cell membrane initiates a series of enzymatic reactions that result in the release of a very large amount of glucose into the blood.*

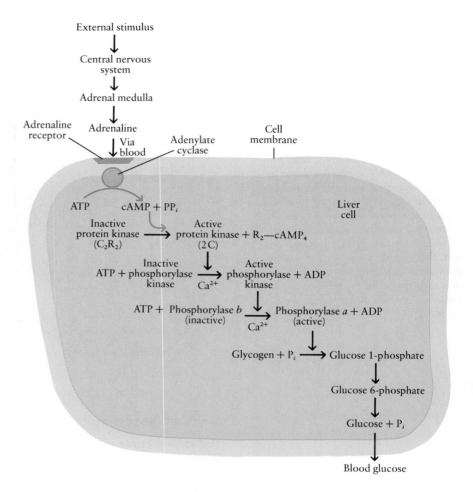

reactions is amplified at each step. Thus, the binding of a few molecules of adrenaline to cells in the liver leads to the activation of an estimated 25 million molecules of phosphorylase *a* and the consequent release of many grams of glucose into the blood.

Just about the same time that the role of cyclic AMP in mammalian cells was established—for which Earl W. Sutherland was awarded the Nobel Prize—biologists studying that peculiar group of organisms known as the cellular slime molds (see page 141) isolated a chemical of great importance in this biological system. As you will recall, the cells of the cellular slime mold begin as individual amoebas and then come together to form a single organism. The chemical that calls them together, which was named acrasin, was identified as cyclic AMP. You may also recall (page 325) that cyclic AMP plays a role in regulation of the *lac* operon in *Escherichia coli.*

More recently, it has been found that insulin is present in fruit flies, earthworms, protists, fungi, and even *E. coli.* Preliminary evidence indicates that other "mammalian" hormones, including ACTH, glucagon, and somatostatin, are present in unicellular organisms. Their function in these organisms is not known, but they may also, like acrasin, serve in cell-to-cell communication. These new discoveries of the universality of hormones provide other examples of the long thread of evolutionary history linking all organisms.

SUMMARY

Hormones are signalling molecules secreted in one part of an organism that diffuse or, in the case of vertebrates, are carried by the bloodstream to other tissues and organs, where they exert specific effects. Most hormones are secreted by glandular epithelial tissue or by neurosecretory cells. The principal endocrine glands of vertebrates include the pituitary, the hypothalamus, the thyroid, the parathyroids, the adrenal cortex and medulla, the pancreas (which is also an exocrine gland), the pineal, and the gonads (ovaries or testes).

The production of many hormones is regulated by negative feedback systems involving the anterior lobe of the pituitary gland and an area of the brain, the hypothalamus. Under the influence of hormones secreted by the hypothalamus, the pituitary produces tropic hormones that, in turn, stimulate the target glands to produce hormones. These hormones then act upon the pituitary or the hypothalamus (or both) to inhibit the production of the tropic hormones. Production of thyroid hormone and the steroid hormones of the adrenal cortex and gonads is regulated by the hypothalamus-pituitary system. The production of other hormones, such as calcitonin and parathyroid hormone, is regulated by the concentration in the bloodstream of other factors, such as ions.

Besides producing the tropic hormones, the anterior lobe of the pituitary also secretes somatotropin (growth hormone) and prolactin. In addition to producing at least nine peptide hormones (sometimes called releasing hormones) that act upon the anterior lobe of the pituitary, the hypothalamus produces the hormones ADH and oxytocin, which are stored in and released from the posterior lobe of the pituitary.

The islet cells of the pancreas are the source of three hormones involved in the regulation of blood glucose: insulin, which lowers blood sugar by stimulating cellular uptake of glucose; glucagon, which raises blood sugar by stimulating the breakdown of storage forms of glucose; and somatostatin. Blood sugar is also under the influence of adrenaline and noradrenaline, which are released from the adrenal medulla at times of stress; cortisol and other glucocorticoids, which are released from the adrenal cortex at times of stress; and somatotropin.

The pineal gland, located in the brain, is the source of melatonin and is believed to be involved with regulation of circadian and seasonal physiological changes. Its function in humans is not known.

Prostaglandins are a group of fatty acids that resemble other hormones in exerting effects on specific target tissues but that often act directly upon the tissues that produce them or, in some cases, on the tissues of another individual. Prostaglandins are formed in most, if not all, tissues of the body and affect such diverse functions as the contraction of smooth muscle, platelet aggregation, and the immune response.

Hormones act by at least two different mechanisms. Steroid hormones and thyroid hormone freely enter cells, where, after combining with an intracellular receptor, they exert a direct influence on the transcription of RNA. Catecholamine, peptide, and protein hormones, such as adrenaline, insulin, and glucagon, combine with receptor molecules on the surface of the target-cell membranes. The hormone-receptor combination may be carried into the cytoplasm by receptor-mediated endocytosis, or the combination may trigger the release of a "second messenger." The second messenger, in turn, sets off a series of events within the cell that is responsible for the end results of hormone activity. Cyclic AMP has been identified as the second messenger in many of these interactions.

QUESTIONS

1. Distinguish among the following: endocrine/exocrine; pituitary/hypothalamus; anterior pituitary/intermediate pituitary/posterior pituitary; thyroid gland/parathyroid glands; thyroxine/thyrotropin/triiodothyronine; thyroid-stimulating hormone/thyrotropin-releasing hormone; adrenal cortex/adrenal medulla; insulin/glucagon.

2. Diagram the feedback system regulating the production and release of thyroid hormone.

3. In terms of the feedback system diagrammed in Question 2, explain why a shortage of iodine produces goiter.

4. Describe how the concentration of calcium ion in the bloodstream is regulated by the thyroid and parathyroid glands.

5. Which hormones act to increase the level of blood glucose? To decrease it? How does each hormone exert its effects?

6. How does a steroid hormone exert its specific effects on a target cell? In what way is the action of thyroid hormone different?

7. How does a peptide or protein hormone exert its specific effects on a target cell?

8. What types of functions would you expect to be controlled by the endocrine system rather than by the nervous system? Does the reality, as described in this chapter, fulfill your expectations?

Integration and Control II: The Nervous System

As we saw in the last chapter, communication among cells—whether unicellular organisms or the component cells of a multicellular organism—depends upon chemical stimuli. Specific molecules are released from secretory cells and are transported to other cells, where they interact with protein receptors and trigger a response. The presence in prokaryotes, protists, and fungi of signalling molecules identical to those of vertebrates and the dual function of some vertebrate molecules as both hormones and neurotransmitters provide strong evidence that the endocrine and nervous systems had a common evolutionary origin in primitive cell-cell communication systems.

The specialization that distinguishes a nervous system from other communication systems is the neuron, a cell that converts appropriate stimuli into electrochemical signals that are rapidly conducted through the neuron itself, often over great distances. Neurons then transmit these signals to other neurons across junctions known as **synapses,** usually by the release of specific neurotransmitter molecules, and to effector cells, such as those of muscles and glands. The rapid communication provided by neurons and their arrangement in organized networks—nervous systems—are the keys to the integration and control underlying the active life style of animals.

EVOLUTION OF NERVOUS SYSTEMS

As we saw in Section 4, invertebrate nervous systems range from the very simple to the complex (Figure 41–1). In the cnidarian *Hydra,* the neurons form a diffuse network. They receive information from sensory receptor cells, which can be

41–1 *In* Hydra (a), *a cnidarian, the nerve impulse, which can travel in either direction, spreads out diffusely along the nerve net from the area of stimulation. In a planarian* (b), *there are two longitudinal nerve cords, with some aggregation of ganglia and sense organs at the anterior end. In annelids, such as the earthworm* (c), *the longitudinal nerve cords are fused together in a double ventral nerve cord. In the crayfish* (d), *an arthropod, the nerve cord is also double and ventral, with a series of ganglia, almost as large as the brain, that control particular segments of the body.*

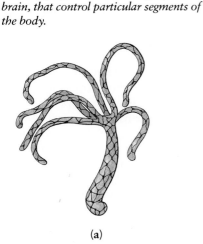

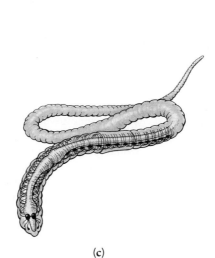

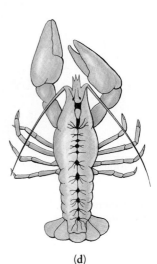

(a) (b) (c) (d)

found among the epithelial cells on the inner (feeding) and outer surfaces of the animal. The neurons stimulate epitheliomuscular cells that cause movements in the body wall.

The nervous system of the planarian, a flatworm, is more highly organized than that of *Hydra,* providing a more efficient coordination that makes possible increased motility. Some of the nerve net is condensed into two cords, and there are two clusters of nerve cell bodies at the anterior end of the body. As you will recall (page 535), such clusters of nerve cell bodies are known as ganglia (singular, ganglion).

In the earthworm, the two cords have come together in a fused, double nerve cord that runs along the ventral surface of the body. Along the nerve cord are ganglia, one for each segment of the body. The nerve cord forks just below the pharynx, and the two forks meet again in the head, terminating in two large dorsal ganglia.

Arthropods, such as the crayfish, also have a double ventral nerve cord and, in addition, may have sizable clusters of nerve cell bodies in the head region. Collectively, these ganglia are large enough to be called a brain. The nervous system also contains many other ganglia interconnected by nerve fibers that run along the ventral surface. In animals with this type of nervous system, many quite complicated activities—for example, the complex movements of the finely articulated appendages—are coordinated by the nearest ganglion.

In vertebrates, the nervous system, which is dorsal rather than ventral, has been greatly elaborated. Its central processing centers—the spinal cord and the brain—are enclosed and protected by the bones of the vertebral column and the skull. The trend in vertebrate evolution has been toward increased centralization of control in the brain—cephalization—a trend that may be still continuing. The precise integration that accompanies such centralization makes possible such complex behaviors as the dive of a kingfisher for a bass (see page 42) and the fashioning of a tiny lure by a fly fisherman.

ORGANIZATION OF THE VERTEBRATE NERVOUS SYSTEM

As we noted in Chapter 33, the vertebrate nervous system has a number of subdivisions (Figure 41–2) that can be distinguished by anatomical, physiological, and functional criteria. The primary and most obvious is the subdivision of the system into the central nervous system (the brain and spinal cord) and the peripheral nervous system (the sensory and motor pathways that carry information to and from the central nervous system). The motor pathways are further divided into the somatic system, which stimulates skeletal muscle, and the autonomic system, which relays signals to smooth muscle, cardiac muscle, and glands. The autonomic system is, in turn, subdivided into the sympathetic and parasympathetic divisions.

The functional unit of the vertebrate nervous system is, of course, the neuron, a nerve cell characterized by a cell body, an axon, and, often, many dendrites (see Figure 33–10, page 708). Neurons are surrounded and insulated by glial cells; in the central nervous system, these are neuroglia, and in the peripheral nervous system, they are Schwann cells. In vertebrates, as in invertebrates, nerve cell bodies are often found in clusters. Such clusters outside the central nervous system are called ganglia; inside the central nervous system, they are generally called **nuclei.** Axons (nerve fibers) are also grouped together, forming bundles; these bundles are known as **tracts** when they are in the central nervous system, and **nerves** when they are in the peripheral nervous system. The individual axons in tracts and nerves are often enveloped in insulating myelin sheaths formed by specialized glial

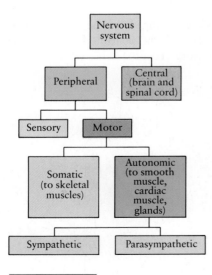

41–2 *A summary of the subdivisions of the vertebrate nervous system.*

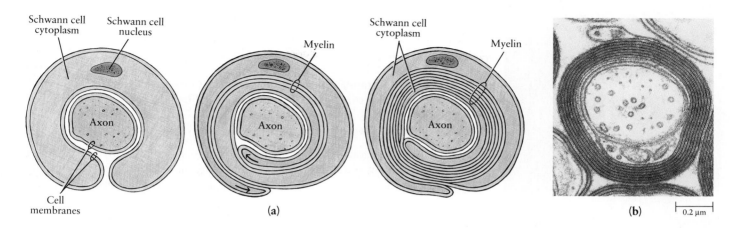

Schwann cell cytoplasm

Schwann cell nucleus

Axon

Cell membranes

(a)

Myelin

Axon

Schwann cell cytoplasm

Myelin

Axon

(b)

0.2 μm

41–3 (a) *Formation of a myelin sheath by a Schwann cell, a type of glial cell found in the peripheral nervous system. As the Schwann cell grows, it wraps itself around and around the axon and gradually extrudes its cytoplasm from between the layers. The myelin sheath, which consists of layers of lipid-containing cell membranes, insulates the nerve fiber. (b) Electron micrograph of a cross section of a mature myelin sheath. Its dark appearance is the result of chemicals used to "fix" the specimen for electron microscopy (see page 98); without such treatment, the myelin sheath appears white.*

cells (Figure 41–3). These sheaths are rich in lipids, giving nerves and tracts a glistening white appearance.

The Central Nervous System

The central nervous system is made up of the brain and the spinal cord, which provides the critical link between the brain and the rest of the body. Your spinal cord, which is a slim cylinder about as big around as your little finger, can be seen in cross section to be divided into a central area of gray matter and an outer area of white matter (Figure 41–4). The gray matter is mostly interneurons (which, as you may recall from page 709, conduct signals within localized regions of the central nervous system), the cell bodies of motor neurons, and neuroglia. The white matter consists of fiber tracts running longitudinally through the spinal cord; the fibers comprising these tracts are primarily the axons of relay neurons, which conduct signals between regions of the central nervous system.

41–4 *A portion of the human spinal cord and vertebral column. Each spinal nerve divides into two fiber bundles, the sensory root and the motor root, at the vertebral column. The sensory root connects with the cord dorsally; the cell bodies of the sensory neurons are in the dorsal root ganglia. The motor root connects ventrally with the spinal cord; the cell bodies of the motor neurons are in the spinal cord itself. The sympathetic ganglia, which form a chain, are part of the autonomic nervous system.*

The butterfly-shaped gray matter within the spinal cord is composed mostly of interneurons, cell bodies of motor neurons, and glial cells. The surrounding white matter consists of ascending and descending fiber tracts.

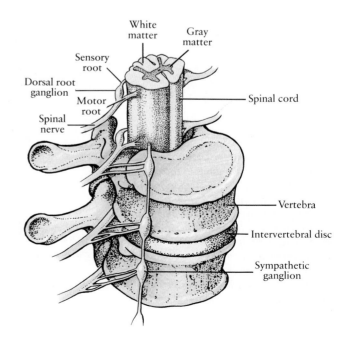

White matter

Gray matter

Sensory root

Dorsal root ganglion

Motor root

Spinal nerve

Spinal cord

Vertebra

Intervertebral disc

Sympathetic ganglion

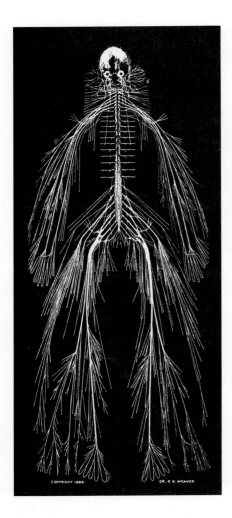

41-5 *The human nervous system, as dissected out in 1888 by Rufus B. Weaver, a Philadelphia physician. The subject, a maid employed by Dr. Weaver, had requested that her body be used to benefit science after her death. It took Dr. Weaver more than five months of daily work to remove the nerves.*

The spinal cord is continuous with the brainstem, which is the base of the brain. The brainstem contains fiber tracts conducting signals to and from the spinal cord and also the cell bodies of the neurons whose axons innervate the muscles and glands of the head. In addition, within the brainstem are centers for some of the important automatic regulatory functions, such as control of respiration and blood pressure, which are also influenced by other parts of the brain.

The Peripheral Nervous System

The peripheral nervous system is made up of neurons whose axons extend out of the central nervous system into the tissues and organs of the body. These include both motor (efferent) neurons, which carry signals out, and sensory (afferent) neurons, which carry signals in. The fibers of motor and sensory neurons are bundled together into nerves, which are classified as **cranial nerves,** those nerves that connect directly with the brain (such as the optic nerve), and **spinal nerves,** those that connect with the spinal cord. Pairs of spinal nerves enter and emerge from the cord through spaces between the vertebrae. The motor fibers of each pair innervate the muscles of a different area of the body, and the sensory fibers receive signals from sensory receptors in the same area. In humans there are 31 such pairs.

As you can see in Figure 41–4, the motor and sensory fibers of the spinal nerves separate from each other near the spinal cord. The cell bodies of the sensory neurons are in the dorsal root ganglia outside the spinal cord, and the sensory fibers feed into the dorsal side of the spinal cord. Here they may synapse with relay neurons, interneurons, or motor neurons, they may turn and ascend toward the brain, or they may do all of these. Fibers from the motor neurons emerge from the spinal cord on the ventral side. The cell bodies of motor neurons are located in the spinal cord, where they may receive signals from relay neurons, interneurons, and sensory neurons.

The four types of neurons are often interconnected in reflex arcs (Figure 41–6). A stimulus received by a sensory neuron is conducted to the central nervous system, where the neuron synapses with either a motor neuron (a monosynaptic reflex) or one or more interneurons (a polysynaptic reflex) that, in turn, synapse with a motor neuron that completes the arc, activating an effector that carries out the reflex action. Simultaneously, relay neurons conduct information concerning

41-6 *A polysynaptic reflex arc. In this example, free nerve endings in the skin, when appropriately stimulated, transmit signals along the sensory neuron to an interneuron in the spinal cord. The interneuron transmits the signal to a motor neuron. As a result of the stimulation of the motor neuron, muscle fibers contract. Relay neurons, not shown here, are also stimulated by the sensory neuron and carry the sensory information to the brain.*

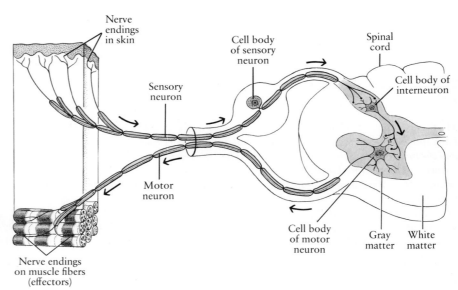

Nerve endings in skin

Cell body of sensory neuron

Spinal cord

Sensory neuron

Cell body of interneuron

Motor neuron

Cell body of motor neuron

Gray matter

White matter

Nerve endings on muscle fibers (effectors)

the event to other parts of the central nervous system. Thus, for instance, if your fingers touch a hot stove, your hand will automatically withdraw from the stove. Almost simultaneously, your brain will become aware of what has happened and you will take further action or make an appropriate comment.

Divisions of the Peripheral Nervous System: Somatic and Autonomic

As we noted earlier, there are two subdivisions of the motor pathways of the peripheral nervous system: the somatic and the autonomic. The autonomic ("involuntary") nervous system consists of the motor nerves that control cardiac muscle, glands, and smooth muscle (the type of muscle found in the walls of blood vessels and in the digestive, respiratory, excretory, and reproductive systems). The somatic ("voluntary") system controls the skeletal muscles—that is, the muscles that can be moved at will.

You will readily recognize that the distinction here between "voluntary" and "involuntary" is not clear-cut. Skeletal muscles—part of the somatic system—often move involuntarily, as in a reflex action. On the other hand, it has been reported that some individuals, such as practitioners of yoga or those who have had biofeedback training, can control their rate of heartbeat and the contractions of some smooth muscle, both of which are regulated by the autonomic system.

Anatomically, the motor neurons of the somatic system are distinct and separate from those of the autonomic nervous system, although axons of both types may be carried within the same nerve. The cell bodies of the motor neurons of the somatic system are located within the central nervous system, with long axons running without interruption all the way to the skeletal muscles. The pathways of the autonomic nervous system also include axons that originate in cell bodies inside the central nervous system, but these axons do not usually travel all the way to the target organs, or effectors. Instead, they synapse outside the central nervous system with motor neurons, which then innervate the effectors (see Figure 41–7 on the next page). These synapses occur within ganglia. Thus the neurons whose axons emerge from the central nervous system and terminate in the ganglia are known as **preganglionic,** while those whose axons emerge from the ganglia and terminate in the effectors are known as **postganglionic.** This two-neuron pathway constitutes a characteristic difference between the autonomic and somatic systems.

Another major difference between these systems is that, in vertebrates, the somatic system can only stimulate or not stimulate an effector; it cannot inhibit an effector. The autonomic system, by contrast, can stimulate or inhibit the activity of an effector. In the somatic system, sensory input often comes from neurons monitoring environmental change. The autonomic nervous system receives its sensory input from some of the same neurons as the somatic nervous system and also from sensory neurons monitoring changes in the interior of the body, such as the group signaling changes in blood pressure. These neurons are involved in reflexes similar to the one shown in Figure 41–6. One difference, however, is that, in reflex arcs involving the autonomic system, you are usually not conscious that the reflex action has taken place.

Divisions of the Autonomic Nervous System: Sympathetic and Parasympathetic

The autonomic nervous system itself has two divisions: the sympathetic division and the parasympathetic division. These two divisions are anatomically, physio-

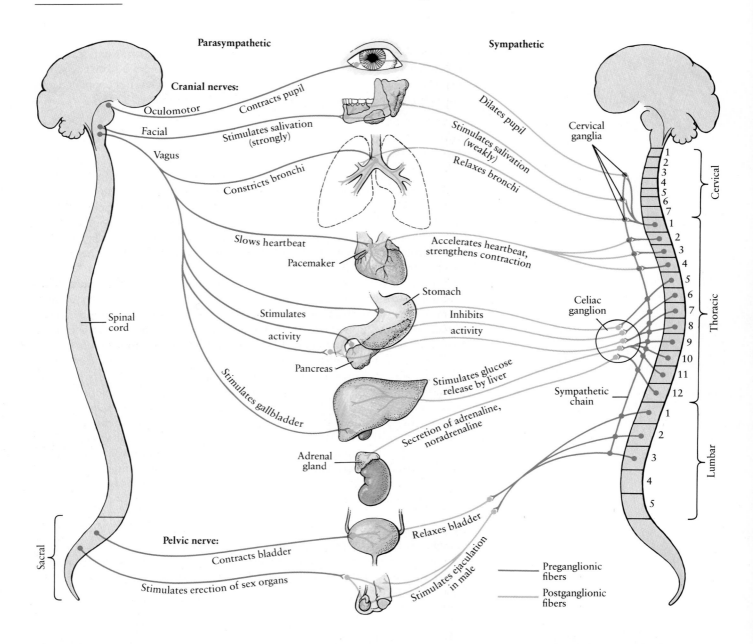

Parasympathetic

Cranial nerves:

Oculomotor — Contracts pupil

Facial — Stimulates salivation (strongly)

Vagus — Constricts bronchi

Slows heartbeat

Pacemaker

Stimulates activity

Pancreas

Stimulates gallbladder

Adrenal gland

Spinal cord

Sacral

Pelvic nerve:

Contracts bladder

Stimulates erection of sex organs

Sympathetic

Dilates pupil

Stimulates salivation (weakly)

Relaxes bronchi

Accelerates heartbeat, strengthens contraction

Stomach

Inhibits activity

Stimulates glucose release by liver

Secretion of adrenaline, noradrenaline

Cervical ganglia

Celiac ganglion

Sympathetic chain

Relaxes bladder

Stimulates ejaculation in male

Cervical 1 2 3 4 5 6 7

Thoracic 1 2 3 4 5 6 7 8 9 10 11 12

Lumbar 1 2 3 4 5

Preganglionic fibers

Postganglionic fibers

41–7 *The autonomic nervous system. It differs from the somatic system anatomically in that the axons emerging from the central nervous system do not travel without interruption to the effectors. Instead, they synapse outside the central nervous system with motor neurons, which then innervate the effectors. The fibers emerging from the central nervous system are known as preganglionic fibers, and those terminating in the effectors are known as postganglionic fibers.*

The autonomic nervous system con-sists of the sympathetic and the parasympathetic divisions. The preganglionic fibers of the parasympathetic division exit from the base of the brain and from the sacral region of the spinal cord and synapse with the postganglionic neurons at or near the target organs. The sympathetic division originates in the thoracic and lumbar regions. Preganglionic fibers of the sympathetic division synapse with postganglionic neurons in the chain of sympathetic ganglia or in other ganglia, such as the celiac ganglion, which is part of the

solar plexus. The neurotransmitter at the effector is usually noradrenaline in the sympathetic division and acetylcholine in the parasympathetic division.

Most, but not all, internal organs are innervated by both divisions, which usually function in opposition to each other. In general, the sympathetic division stimulates functions involved in "fight-or-flight" reactions, and the parasympathetic division stimulates more tranquil functions, such as digestion.

logically, and functionally distinct. The major anatomical and physiological differences between them are:

1. Axons of the sympathetic division originate in the thoracic (chest) and lumbar (lower back) regions of the spinal cord. Axons of the parasympathetic division emerge through the cranial (brain) region and the sacral ("tail") region of the spinal cord.

2. As previously stated, in the autonomic nervous system, there is always a relay system of two neurons connecting the central nervous system and the effector organ. These neurons synapse at a ganglion. In the sympathetic division, the ganglia are usually close to the central nervous system. Thus, characteristically, the preganglionic axon is short, and the postganglionic axon is long. In the parasympathetic division, the opposite is true: the ganglia are close to or embedded in the target organ. Thus the preganglionic axon is long, and the postganglionic axon is short.

3. Although the preganglionic nerve endings of both divisions release acetylcholine (Figure 41–8) as their neurotransmitter, the postganglionic nerve endings of the two divisions utilize different neurotransmitters. The neurotransmitter released by most postganglionic sympathetic nerve endings is noradrenaline. All postganglionic parasympathetic endings release acetylcholine.

Functionally, the two divisions are generally antagonistic. As you can see in Figure 41–7, most of the internal organs are innervated by axons from both divisions. They work in close cooperation with each other and with hormones secreted by the endocrine glands for the ultimate homeostatic regulation of the body. The parasympathetic division is involved primarily in the restorative activities of the body; it is particularly active, for example, after a heavy meal or following orgasm. Parasympathetic stimulation slows down the heartbeat, increases the movements of the smooth muscle of the intestinal wall, and stimulates secretions of the salivary glands and the digestive glands of the stomach.

The sympathetic division, by contrast, prepares the body for action. The physical characteristics of fear, for example, result from the increased discharge of neurons of the sympathetic division. Some of these cause the blood vessels in the skin and intestinal tract to contract; this contraction increases the return of the blood to the heart, raising the blood pressure and allowing more blood to be sent to the muscles. The heart beats both faster and stronger, and the respiratory rate increases. The pupils dilate. The muscles attached to the hair follicles in the skin contract; this is probably a legacy from our furry forebears, which looked larger and more ferocious with their hair standing on end (Figure 41–9). The rhythmic movement of the intestines stops, and the sphincters, muscles at the end of the intestines and the opening of the bladder, relax. These reactions inhibit digestive operations, but the relaxing of the sphincters may also, in extreme cases, have the disconcerting consequence of allowing involuntary defecation or urination. Sympathetic stimulation causes the adrenal medulla to pour out adrenaline, which, with other hormones, causes the release of large quantities of glucose from the liver into the bloodstream, as we saw in Figure 40–16. This glucose is the extra energy source for the muscles. As a consequence of this constellation of responses, the body as a whole is prepared for "fight or flight"—or, at least, for action that would have been appropriate at some earlier stage of our cultural evolution.

THE NERVE IMPULSE

Some 200 years ago, Luigi Galvani observed that the passage of an electric current along the nerve of a frog's leg made the muscle twitch. Since that time, it has been

H₃C—C—O—CH₂—CH₂—⁺N—CH₃

(chemical structure of Acetylcholine shown with O double-bonded to C, and three CH₃ groups on N)

Acetylcholine

41–8 *Acetylcholine, the principal neurotransmitter in motor fibers of the peripheral nervous system. It is released by both the preganglionic and postganglionic nerve endings of the parasympathetic division of the autonomic nervous system and by the preganglionic nerve endings of the sympathetic division. Noradrenaline (Figure 40–11, page 832) is usually the neurotransmitter at the postganglionic nerve endings of the sympathetic division. As we saw in the last chapter, noradrenaline and a related molecule, adrenaline, are also secreted by the adrenal medulla, the central portion of the adrenal gland.*

41–9 *A cat prepared for "fight or flight," showing the effects of stimulation by the sympathetic nervous system. This drawing is from Darwin's* The Expression of the Emotions in Man and Animals, *published in 1872.*

known that nerve conduction is associated with electrical phenomena. As you will recall, there are two types of electric charge, positive and negative; like charges repel one another, and unlike charges attract. Thus, negatively charged particles tend to move toward a region of positive charge, and vice versa. (In a metal, such as copper, only negative charges move, but the principle is the same.) Materials that permit the movement of charged particles, such as a copper wire or a solution containing ions, are known as conductors. Materials that do not permit the movement of charged particles, such as fat and rubber, are insulators.

The difference in the amount of electric charge between a region of positive charge and a region of negative charge is called the **electric potential.** As we saw in Chapter 9 (page 198), an electric potential is a form of potential energy, like a boulder at the top of a hill or water behind a dam. This potential energy is converted to electrical energy when charged particles are allowed to move through a solution or along a wire between the two regions of differing charge. The difference in potential energy between the two regions is measured in volts or, if it is very small, in millivolts.

For a time, it was assumed that a nerve impulse was an electric current—that is, a flow of ions—traveling along an axon in the same way that electrons flow along a wire. However, this model did not hold up under critical scrutiny. First, it was shown that the axon is a poor conductor of electricity—meaning that ions flow through it for only a short distance before the current dies out. By contrast, the nerve impulse is undiminished from one end of the axon to the other. Second, although the impulse is fast by biological criteria, it is much slower than an electric current. Finally, unlike an electric current, the strength of the impulse is always the same. There is either no nerve impulse in response to the stimulation of a nerve fiber, or there is a maximum response—all or nothing.

Major advances in understanding the nature of the nerve impulse occurred when it became possible to monitor changes in electric potential in an individual neuron. The organism that first made this possible was the squid, which has motor neurons with large, long axons (Figure 41–10). Measurements within an axon are made with microelectrodes tiny enough to penetrate a living cell without seriously injuring it. The microelectrodes are connected with a very sensitive voltmeter called an oscilloscope, which measures voltage (in millivolts) in relation to time (in milliseconds).

When both electrodes are outside the neuron, no voltage difference is recorded (Figure 41–11a). When one electrode penetrates the axon, a voltage difference of about 70 millivolts can be detected between the outside and the inside of the axon. The interior of the membrane is negatively charged with respect to the exterior (Figure 41–11b). This is the **resting potential** of the membrane. When the axon is stimulated, the oscilloscope records a very brief reversal of polarity—that is, the interior becomes positively charged in relation to the exterior (Figure 41–11c). This reversal in polarity is called the **action potential.** The action potential traveling along the membrane is the **nerve impulse.** It travels more slowly than an electric current, because it is actually a series of miniature electric circuits, but it is undiminished in intensity during its conduction.

The action potentials recorded from any one neuron are almost always the same. Figure 41–12 shows, for example, the action potentials produced by a single sensory neuron in the skin of a cat in response to pressure. All the impulses are the same size (all-or-nothing response). The only variation—and it is a critical variation—is the frequency with which the impulses are produced. The message carried by nerve impulses is, in effect, a Morse code with dashes but no dots.

The Ionic Basis of the Action Potential

The action potential depends on the electric potential of the axon, which is, in turn, made possible by differences in the concentration of ions on either side of

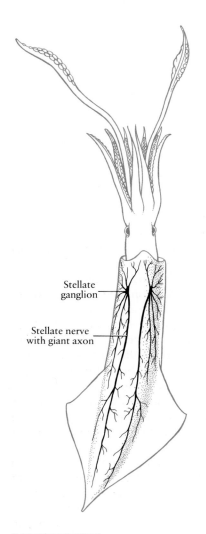

Stellate
ganglion

Stellate nerve
with giant axon

41–10 *The squid, one of the chief protagonists in research leading to an understanding of the nature of the nerve impulse. The stellate nerves contain the giant axons used in all the early studies of the nerve impulse. The giant axons innervate muscles in the wall of the mantle; powerful contractions of those muscles result in a rapid expulsion of water from the mantle cavity, producing the escape response described on page 553.*

41–11 *The electric potential across the membrane of the axon is measured by microelectrodes connected to an oscilloscope. (a) When both electrodes are outside the membrane, no potential is recorded. (b) When one electrode penetrates the membrane, the oscilloscope shows that the interior is negative with respect to the exterior and that the difference between the two is about 70 millivolts. This is the resting potential. (c) When the axon is stimulated and a nerve impulse passes along it, the oscilloscope shows a brief reversal of polarity—that is, the interior becomes positive in relation to the exterior. This brief reversal in polarity is the action potential.*

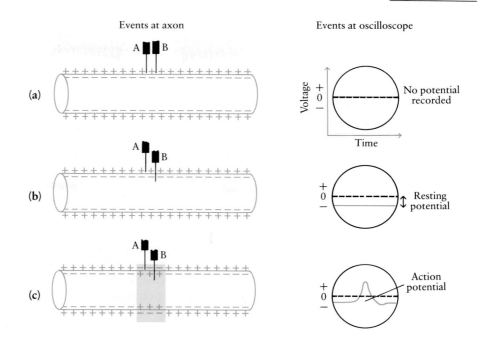

Events at axon   Events at oscilloscope

(a)

(b)

(c)

the membrane. Such differences in concentration—and thus in electric potential—are characteristic of all cells, and, as we have seen in earlier chapters, they can be used to power a variety of cellular processes. Their use to carry information through the body of an animal is perhaps the most sophisticated application to have evolved thus far.

In axons, the critical concentration differences involve potassium ions (K^+) and sodium ions (Na^+). In the resting state, the concentration of K^+ ions in the cytoplasm of an axon is about 30 times higher than in the fluid outside; conversely, the concentration of Na^+ ions is about 10 times higher in the extracellular fluid than in the cytoplasm. The distribution of ions on either side of the membrane is governed by three factors: (1) the diffusion of particles down a concentration gradient (page 129), (2) the attraction of particles with opposite charges and the repulsion of particles with like charges, and (3) the properties of the membrane itself.

The lipid bilayer of the axon membrane is, like the lipid bilayer of other cell membranes, impermeable to ions and most polar molecules. The movement of such particles through the membrane depends on the presence of integral membrane proteins that provide channels through which the particles can move, either by facilitated diffusion or active transport. The membrane of the axon is rich in proteins that provide channels for the movement of specific ions, particularly Na^+

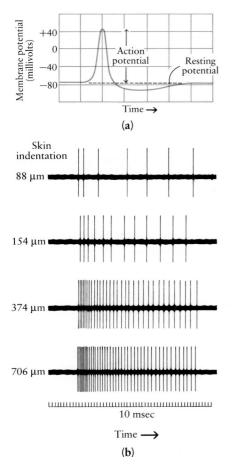

(a)

(b)

41–12 *(a) Nerve impulses can be monitored by electronic recording instruments. The impulses from any one neuron are all the same; that is, each impulse has the same duration and voltage change as any other.*

(b) Nerve impulses from a sensory neuron (touch receptor) in cat skin. The skin was touched and pressed in at varying depths, as indicated by the figures at the left. The more deeply the skin was pressed in, the more rapidly nerve impulses were produced. The vertical lines represent individual action potentials on a compressed time scale. As you can see, all the action potentials are the same size, but their frequency increases with the intensity of the stimulus.

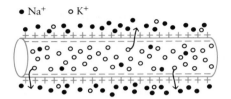

• Na⁺   ◦ K⁺

41-13 *Resting potential. The membrane of the axon is polarized, with a slight excess of negative charge inside the membrane caused by the outward diffusion of K⁺ ions. The movement of ions across the axon membrane depends on the presence of ion channels formed by integral membrane proteins. In the resting state, the K⁺ channels are open, but the Na⁺ channels are closed.*

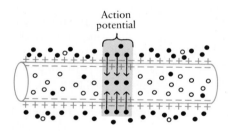

Action potential

41-14 *The action potential. A portion of the membrane becomes momentarily permeable to Na⁺ ions as Na⁺ channels open. The Na⁺ ions rush in, and the polarity of the membrane is reversed at that point.*

and K⁺. An important feature of these proteins is that changes in their conformation result in either the opening or the closing of the ion channels; thus, the channels are said to be **gated**. Depending on the specific ion channel, the changes in conformation that open or close its gate may be regulated by the voltage across the membrane or by the action of neurotransmitters (or drugs that mimic neurotransmitters). Another significant feature of the axon membrane is the presence of the integral membrane protein known as the sodium-potassium pump (page 137), which pumps Na⁺ ions out of the axon and K⁺ ions in.

When the axon membrane is in its resting state, the Na⁺ ion channels are mostly closed; as a consequence, the membrane is almost impermeable to Na⁺ ions. The few Na⁺ ions that diffuse in through open channels—moving down their concentration gradient—are promptly removed by the sodium-potassium pump. Many K⁺ ion channels, however, are open, and the membrane is thus relatively permeable to K⁺ ions. Because of the concentration gradient, K⁺ ions tend to move out of the cell; if no other forces were at work, they would, of course, move down the concentration gradient until their distribution were equal on either side of the membrane. However, because of the impermeability of the lipid bilayer, negatively charged ions cannot follow the K⁺ ions out of the cell. Thus, as K⁺ ions leave, an excess negative charge builds up inside the cell. This excess of negative charge attracts the positive K⁺ ions, impeding their further outward movement. As a result, an equilibrium is reached at which there is no net movement of K⁺ ions across the membrane. At the point of equilibrium, there is a slight excess of negative charge inside the cell, and the membrane is said to be polarized. This point of equilibrium is the resting potential (Figure 41-13).

When the membrane is stimulated, it suddenly becomes permeable to Na⁺ ions as Na⁺ channels open at the site of stimulation. The Na⁺ ions rush in, moving down their concentration gradient, attracted initially by the negative charge inside the axon. This influx of positively charged ions momentarily reverses the polarity of the membrane so that it becomes more positive on the inside than on the outside, producing the action potential (Figure 41-14). The change in Na⁺ permeability lasts for only about half a millisecond; then the Na⁺ channels close and the membrane regains its previous impermeability to Na⁺ ions. During this time, more K⁺ channels open, increasing the permeability to K⁺ ions. The result is an outward flow of K⁺ ions due to the concentration gradient and also to the positive charge inside the axon at the peak of the action potential. This outward flow of positive K⁺ ions counteracts the previous inward flow of positive Na⁺ ions, and the resting potential is quickly restored. The actual number of ions involved is very small. Only a very few Na⁺ ions need enter to reverse the polarity of the membrane, and only the same small number of K⁺ ions need move out of the cell to restore the resting potential. Subsequently, the sodium-potassium pump restores the Na⁺ and K⁺ concentrations to their original levels. As a consequence, action potentials can move along the axon in rapid fire without substantial changes occurring in the internal concentrations of Na⁺ and K⁺ ions.

Propagation of the Impulse

An important feature of the nerve impulse is that, once initiated, this transient reversal in polarity continues to move along the axon, renewing itself continuously, just as a flame traveling along a fuse ignites the fuse as it travels. The action potential is self-propagating because at its peak, when the inside of the membrane at the active region is comparatively positive, positively charged ions move from this region to the adjacent area inside the axon, which is still comparatively negative. As a result, the adjacent area becomes depolarized—that is, less negative (Figure 41-15). This depolarization opens Na⁺ channels, which are thus said to be

41–15 *Propagation of the nerve impulse. In advance of the action potential, a small segment of the membrane becomes slightly depolarized, owing to the flow of positively charged ions along the inside of the membrane. When the membrane becomes depolarized in this way, Na⁺ ion channels open and the permeability of the membrane to Na⁺ increases. Na⁺ ions move in through the open channels, creating an action potential in that region of the membrane and depolarizing the next adjacent segment of membrane. The nerve impulse is the action potential traveling along the membrane.*

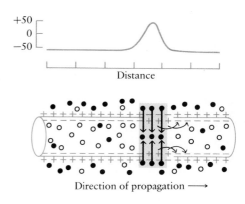

voltage-gated, and allows Na⁺ ions to rush in. The resulting increase in the internal concentration of Na⁺ ions depolarizes the next adjacent area of the membrane, causing its Na⁺ ion channels to open, and allowing the process to be repeated. As a consequence of this renewal process, repeating itself along the length of the membrane, the axon—a very poor conductor of an ordinary electric current—is capable of conducting a nerve impulse over a considerable distance with absolutely undiminished strength. The nerve impulse moves in one direction only because the segment of the axon behind the action potential has a brief refractory period during which its Na⁺ ion channels will not open; thus the action potential cannot go backwards.

The Role of the Myelin Sheath

As we saw in Figure 41–3 (page 843), long axons are generally enveloped in myelin sheaths formed by specialized glial cells. The myelin sheath, according to Nobel laureate John Eccles, is "a brilliant innovation." Because of it, the propagation of the nerve impulse is much more rapid in vertebrates than in invertebrates. The myelin sheath is not just an insulator. Its most important feature is that it is interrupted at regular intervals by openings, or nodes. Only at the nodes is it possible for Na⁺ and K⁺ ions to move into and out of the axon. Thus, in myelinated fibers—which include all the large nerve fibers of vertebrates—the impulse jumps from node to node (Figure 41–16), rather than moving continuously along the membrane like water along a wick. This saltatory (leaping) conduction greatly increases the velocity. Some large, myelinated nerve fibers, for example, conduct impulses as rapidly as 200 meters per second, compared with velocities of only a few millimeters per second in small, unmyelinated fibers. Also, because Na⁺ and K⁺ ions move across only a small portion of the axon's membrane, there is an enormous saving in energy expenditure by the sodium-potassium pump.

THE SYNAPSE

Signals travel from one neuron to another across the specialized junction known as the synapse, which may be electrical or chemical in nature (Figure 41–17). In the former, ions flow through gap junctions (page 142) connecting the cell membranes of closely juxtaposed neurons, and the nerve impulse moves directly from one neuron to the next. Such electrical synapses, which are common in lower invertebrates, have been identified at some sites in the mammalian brain. In chemical synapses, which make up the vast majority of connections between

41–16 *In an unmyelinated fiber* (a), *the action potential travels continuously along the axon, whereas in a myelinated fiber* (b), *the impulse jumps from node to node, greatly accelerating the conduction.*

Myelin

Node of Ranvier

(a)   (b)

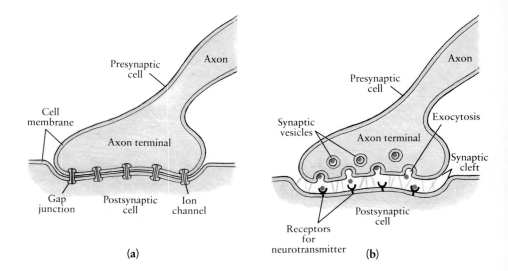

(a)

(b)

41–17 (a) *An electrical synapse. The arrival of an action potential at the axon terminal of the presynaptic cell is accompanied by changes in ion concentration. These changes are transmitted through gap junctions to the postsynaptic cell, where they depolarize the cell membrane and initiate a new action potential. (b) A chemical synapse. The arrival of an action potential at the axon terminal triggers the fusion of synaptic vesicles with the axon membrane, releasing neurotransmitter molecules into the synaptic cleft. These molecules diffuse to the postsynaptic cell, where they combine with specific receptors in the cell membrane. A network of protein fibers in the synaptic cleft anchors the presynaptic and postsynaptic membranes and, in some synapses, contains enzymes that rapidly degrade neurotransmitter molecules. Because of their shape, axon terminals are known as synaptic knobs or as "boutons" (French for "buttons").*

neurons in the mammalian nervous system, the two neurons never touch. As electron micrographs reveal (Figure 41–18), a space of about 20 nanometers, known as the **synaptic cleft,** separates the cell transmitting information (the presynaptic cell) from the cell receiving information (the postsynaptic cell). Information is transmitted across the synaptic cleft by means of the signalling molecules known as neurotransmitters. Unlike the nerve impulse along the axon —an all-or-nothing proposition—signals transmitted across chemical synapses are of varying strength and may have opposite effects. That is, some excite and some inhibit the postsynaptic cell.

Some neurotransmitters are synthesized in the cell body of the neuron and transported to the axon terminals, where they are packaged into synaptic vesicles and stored. Other neurotransmitters are both synthesized and packaged within the axon terminals. The release of neurotransmitter molecules is triggered by the arrival of an action potential at the axon terminal. The membrane in this region of the neuron is rich in membrane proteins that form channels for the transport of calcium ions (Ca^{2+}); these channels, like the Na^+ and K^+ channels, are regulated by the voltage across the axon membrane. Arrival of an action potential at the terminal alters the voltage, opening the channels and allowing Ca^{2+} ions to flow into the axon. This influx of Ca^{2+}, in turn, causes the synaptic vesicles to fuse with the cell membrane, emptying their contents into the synaptic cleft, in yet another example of exocytosis (page 138). The transmitter molecules diffuse from the presynaptic cell across the cleft and combine with receptor molecules on the membrane of the postsynaptic cell, setting in motion a series of events that, as we shall see shortly, may or may not trigger a nerve impulse in the postsynaptic cell.

41–18 (a) *Electron micrograph and* (b) *diagram of a cross section of a dendrite with which two axon terminals form synapses. Note the numerous synaptic vesicles, filled with neurotransmitter, in the axon terminals. Note also the fuzzy areas in the juxtaposed regions of the membranes of the presynaptic and postsynaptic cells. In presynaptic cells, these areas are specialized for exocytosis; in postsynaptic cells, they are rich in receptors for neurotransmitter molecules.*

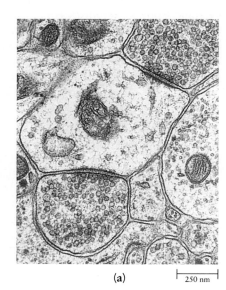

(a)

250 nm

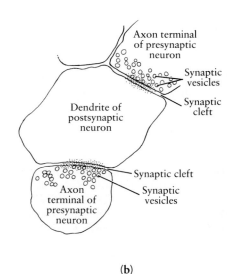

(b)

After their release, neurotransmitters are rapidly removed or destroyed, putting a halt to their effect; this is an essential feature in the control of the activities of the nervous system. The molecules may diffuse away or be broken down by specific enzymes, or they—or their breakdown products—may be taken up again by the axon terminal for recycling. At the same time, membrane from the synaptic vesicles, which fused with the cell membrane of the axon terminal, is apparently re-formed into vesicles by endocytosis, carried back into the cytoplasm, and recycled into new synaptic vesicles, filled with newly synthesized neurotransmitter. Membrane for the formation of new synaptic vesicles may also be supplied by smooth endoplasmic reticulum that extends from the cell body to the axon terminal.

The study of chemical synapses is one of the most active areas of contemporary neurobiological research, with a veritable tidal wave of new discoveries appearing in the scientific journals each week. Although the principal neurotransmitters have been known for many years, researchers are discovering an enormous number of chemicals that play a role in synaptic transmission, particularly in the central nervous system. Moreover, the specific receptors for different neurotransmitters are being identified, the details of their structure elucidated, and in some cases, their genes sequenced. Simultaneously, the biochemical events that occur when a neurotransmitter binds to its receptor are being revealed, as are the mechanisms by which a neuron integrates the information that it receives from the hundreds—or even thousands—of neurons that form synapses with it.

Neurotransmitters

A variety of chemical substances function as neurotransmitters. As we have seen, the principal neurotransmitters in the peripheral nervous system are acetylcholine (Figure 41–8) and noradrenaline. Acetylcholine is also found in the brain, although at relatively few synapses. Noradrenaline is a major neurotransmitter at some synapses in the hypothalamus and in other specific regions of the brain, where it is thought to play a role in arousal and attention. There is some evidence that severe depression may be related to an abnormally low level of noradrenaline at particular synapses; the two major types of antidepressants in clinical use apparently act by increasing the amount of noradrenaline at such synapses.

Many other neurotransmitters have been found in the central nervous system, including dopamine, serotonin (5-hydroxytryptamine), and gamma-aminobutyric acid (GABA), all of which, like noradrenaline, are amino acid derivatives (Figure 41–19). Dopamine is a transmitter for a relatively small group of neurons involved with muscular activity. Parkinson's disease, which is characterized by muscular tremors and weakness, is associated with a decrease in the number of dopamine-producing neurons and thus with the level of dopamine in certain areas of the brain. Serotonin is found in regions of the brain associated with arousal and attention; increasing levels of serotonin are associated with sleep. GABA is a major inhibitory transmitter in the central nervous system; the loss of GABA synapses is one of the features of Huntington's disease (page 395).

Almost all drugs that act in the brain to alter mood or behavior do so by enhancing or inhibiting the activity of neurotransmitter systems. Caffeine, nicotine, and the amphetamines, for example, stimulate brain activity by substituting for excitatory neurotransmitters at synapses. Chlorpromazine and related tranquilizers block dopamine receptors at many sites, whereas LSD inhibits brain serotonin.

Present evidence indicates two principal mechanisms by which neurotransmitters exert their effects on postsynaptic cells. In one mechanism, the binding of a neurotransmitter to its receptor triggers a change in the conformation of a membrane protein that functions as a channel for a specific ion. Depending on the

Noradrenaline

Dopamine

Serotonin (5-hydroxytryptamine)

gamma-Aminobutyric acid (GABA)

41–19 *The structures of some of the amino acid derivatives, known as biogenic amines, that function as neurotransmitters in the central nervous system. Biogenic amines are produced in neurons by slight chemical modifications of amino acids.*

Internal Opiates: The Endorphins

The word "opium" comes from the Greek *opion,* meaning "poppy juice." Since the time of the Greeks, poppy juice and its derivatives, such as morphine, have been used for the control of pain. They are the most potent painkillers known, and their physiological effects are greatly enhanced by the fact that they produce euphoria. They are also highly addictive, and although the pharmaceutical industry, urged on by the possibility of great profits, has made repeated attempts to develop an opium derivative that is non-addictive, their efforts have been uniformly unsuccessful.

All substances with opiate action are related chemically and have similarities in their three-dimensional structures. Thus, it was long suspected that opiates act upon the brain by binding to specific membrane receptors. Using opium derivatives labeled with radioactive isotopes, investigators were able to show that the central nervous system does, indeed, have receptors for opiates. These receptors are located primarily in the spinal cord, brainstem, and brain regions in which drives and emotions are thought to be translated into complex actions, such as seeking food or a mate. When opiates bind to neurons bearing their receptors, they act as inhibitory neuromodulators, causing decreased production of nerve impulses by the neurons. Such receptors have been found not just in humans but in all other vertebrates tested.

Why would vertebrate brains have opiate receptors? Only one answer seemed logical: Because vertebrate brains themselves must produce opiates. This rather startling conclusion triggered a search for naturally occurring substances with opiate activity. Many such internal opiates have now been isolated, all of which act as neuromodulators. They have been given the name of endorphins, for endogenous morphine-like substances.

Two types of endorphins are recognized. One group, known as the enkephalins, is widespread in the central nervous system and is also abundant in the adrenal medulla. The enkephalins that have been identified are two pentapeptides (peptides containing five amino acids) that differ by only one amino acid (see illustration). According to recent evidence, the two enkephalins are produced in multiple copies on a single polypeptide chain.

The other endorphins are produced primarily by the pituitary gland and perhaps by other tissues as well. Recently it has been found that the most common of these, beta-endorphin, is synthesized as part of a long peptide chain that also contains ACTH, the hormone that is released by the anterior pituitary and

receptor, binding of the neurotransmitter may open the channel, allowing ions to flow between the cytoplasm of the neuron and the surrounding fluid, or it may close the channel, shutting off a previously existing flow of ions. The consequence is a change in the degree of polarization across the membrane of the postsynaptic cell. In the second mechanism, you will not be surprised to learn, binding of the neurotransmitter to its receptor activates an enzyme in the cell membrane and sets in motion a second messenger, generally cyclic AMP (see page 837) or a related compound, cyclic GMP (guanosine monophosphate). The events that follow activation of the second messenger are complex, but the ultimate effect is a change in the degree of polarization of the postsynaptic cell. This change, however, takes place at a slower pace than the changes triggered by the opening or closing of ion channels. Although it appears that some neurotransmitters utilize only one of these mechanisms, others, including acetylcholine, have two different types of receptors, activating the two different mechanisms. Such neurotransmitters produce both a rapid, short-term effect and a slower, long-term effect.

In addition to the principal neurotransmitters—acetylcholine and the amino acid derivatives—other molecules, mostly small peptides, also play a role in synaptic transmission. These molecules, which may be released from the same axon terminals as the principal neurotransmitters or from other cells, are known

stimulates the adrenal cortex. Although there is a great deal of overlap in the primary structure of the various endorphins, the functional relationships among them are not yet known.

The endorphins are of great interest to medical researchers because of the insight they may afford into the relief of two extremely serious (and related) medical problems, opiate addiction and pain. The endorphins are believed to function as natural analgesics (pain relievers). Individuals in stressful situations —soldiers in battle, athletes at crucial moments in a contest—have often reported being unaware of what later proved to be an extremely painful injury and so being able to continue to function in a life-threatening (or victory-threatening) situation. The recent discovery that macrophages are among the cell types

bearing endorphin receptors suggests that these substances may also play a role in the stimulation of inflammatory and immune responses.

Morphine, heroin, and other exogenous opiates combine with the endorphin receptors, relieving stress, elevating mood, and soothing pain. However, it is hypothesized that these external opiates, acting by negative feedback, reduce the normal production of endorphins, resulting in ever-increasing dependence on the artificial source—or, in other words, in addiction.

*The amino acid sequences of three endorphins: (**a**) methionine-enkephalin, (**b**) leucine-enkephalin, and (**c**) beta-endorphin. The enkephalins are found in the brain tissue and adrenal medulla of vertebrates, and beta-endorphin is isolated from the pituitary gland. The first four amino acids of each sequence are identical.*

(a) tyr — gly — gly — phe — met
     Methionine-enkephalin

(b) tyr — gly — gly — phe — leu
     Leucine-enkephalin

(c) tyr — gly — gly — phe — met — thr — ser — glu — lys — ser — gln — thr — pro — leu — val — thr

     gln — gly — lys — lys — his — ala — asn — lys — val — ile — ala — asn — lys — phe — leu
                                        Beta-endorphin

as **neuromodulators.** Although neuromodulators may move directly across the synaptic cleft, they can also diffuse over a greater distance, affecting numerous cells within a local region of the central nervous system. Like neurotransmitters, they bind to specific membrane receptors and alter ion channels or set in motion second messengers; their effect is often to modulate the response of the cell to a principal neurotransmitter. Over 200 different substances that function as neuromodulators have been identified thus far. They include the endorphins (see essay), interferons and interleukins, hypothalamic releasing hormones, pituitary hormones, pancreatic hormones such as insulin, and even the digestive hormones gastrin and cholecystokinin (page 723).

The Integration of Information

As we noted previously, the dendrites and cell body of a single neuron may receive signals—in the form of neurotransmitter or neuromodulator molecules—from hundreds, or even thousands of synapses (see Figure 41–20 on the next page). The binding of each molecule to its receptor has some effect on the degree of polarization of the postsynaptic cell. If the effect is to make the interior of the cell less negative (depolarization), it is said to be excitatory. By contrast, if the effect is

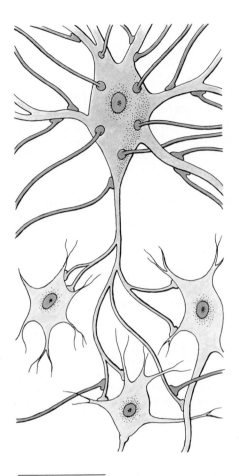

41–20 *A simplified representation of the many interconnections between neurons of the central nervous system. Note how numerous synapses, representing many different presynaptic neurons, converge on an individual neuron and how its axon, in turn, diverges to synapse on a number of other neurons. As you can see, most synapses are located on the dendrites and cell body of the postsynaptic neuron. Note, however, that there are also synapses on its axon, just before some of the axon terminals. These synapses are usually inhibitory or modulatory, influencing the response of the Ca²⁺ channels to the arrival of an action potential at the terminal and thus affecting the amount of neurotransmitter released.*

to maintain the membrane at or near the resting potential, or even to make the interior more negative (hyperpolarization), it is said to be inhibitory. The binding of acetylcholine, for example, is generally excitatory, causing a depolarization in the postsynaptic cell; GABA, on the other hand, is generally inhibitory.

The changes in polarity induced by neurotransmitters and neuromodulators spread from the synapses through the postsynaptic cell to a region known as the axon hillock. This is the region of the axon in which a nerve impulse can originate. If the collective effect is a sufficient depolarization to permit the influx of Na^+ ions that constitutes the beginning of an action potential, a nerve impulse is initiated in the axon of the postsynaptic cell and a new message is speeded on its way to the multitude of other neurons with which the axon synapses.

The processing of information that occurs within the cell body of each individual neuron plays a key role in the integration and control exercised jointly by the nervous and endocrine systems. It is affected not only by the specific neurotransmitters and neuromodulators received by the cell, but also by their quantity, the precise timing of their arrival, and the locations on the neuron of the various synapses and receptors. Each neuron is a tiny computer, summing an enormous quantity of information and issuing appropriate commands carried throughout the network with which we began this chapter.

SUMMARY

The nervous system, along with the endocrine system, integrates and controls the numerous functions that enable an animal to regulate its internal environment and to react to or deal with its external environment. The functional unit of the nervous system is the neuron, or nerve cell. A neuron consists of dendrites, which receive stimuli; a cell body, which contains the nucleus and metabolic machinery and also receives stimuli; and an axon, or nerve fiber, which relays stimuli to other cells.

The central nervous system consists of the brain and the spinal cord, which are encased, in vertebrates, in the skull and vertebral column. That part of the nervous system outside the central nervous system constitutes the peripheral nervous system. Cranial nerves enter and emerge from the brain in pairs, each pair consisting of motor fibers, sensory fibers, or both. Spinal nerves enter and emerge from the vertebral column, also in pairs. Each of these pairs innervates effectors and receives signals from sensory receptors of a different and distinct area of the body.

In vertebrates, the motor neurons of the peripheral nervous system are organized in two major divisions: (1) the somatic nervous system, which innervates the skeletal muscles, and (2) the autonomic nervous system, which controls cardiac muscle and the smooth muscles and glands involved in the digestive, circulatory, urinary, and reproductive functions. In the autonomic system, axons arising from neurons in the central nervous system synapse with motor neurons in ganglia outside the central nervous system. The postganglionic neurons stimulate or inhibit the effectors. The autonomic system has two divisions—sympathetic and parasympathetic—which are anatomically, physiologically, and functionally distinct (see Table 41–1).

Information received from the internal and external environments and instructions carried to effectors, such as muscles and glands, are transmitted in the nervous system as electrochemical signals. At rest, there is a difference in electric charge between the inside and outside of the axon membrane—the resting potential. With appropriate stimulation, an action potential, a transient reversal in membrane polarity, occurs. The action potential traveling along the axon membrane is the nerve impulse. Because all action potentials are the same size, the message carried by a particular axon can be varied only by a change in the

TABLE 41–1 **Comparison of Parasympathetic and Sympathetic Divisions**

	PARASYMPATHETIC DIVISION	SYMPATHETIC DIVISION
Preganglionic fibers	Long; emerge from cranial and sacral regions of spinal cord	Short; emerge from thoracic and lumbar regions of spinal cord
Location of synapse	In small local ganglia near or in innervated organs	In a series of ganglia close to the spinal cord or in ganglia halfway between spinal cord and innervated organs
Postganglionic fibers	Short	Long
Neurotransmitter	Preganglionic and postganglionic: acetylcholine	Preganglionic: acetylcholine; postganglionic usually noradrenaline
General effects	Promotes "vegetative" and restorative functions, such as digestion	Promotes "fight-or-flight" responses; inhibits "vegetative" functions

frequency or pattern of action potentials. In myelinated fibers, the nerve impulse leaps from node to node of the myelin sheath, thereby speeding conduction.

Neurons transmit signals to other neurons across a junction called a synapse. In most synapses, the signal crosses the synaptic cleft in the form of a chemical, a neurotransmitter, that binds to a specific receptor on the membrane of the postsynaptic cell. Other chemicals, known as neuromodulators, can diffuse through local areas of the central nervous system as well, binding to the membrane receptors of other neurons in the area. Binding of a neurotransmitter or neuromodulator to its receptor may open or close a membrane ion channel or set a second messenger in motion. The ultimate effect is a change in the membrane voltage of the postsynaptic cell. A single neuron may receive signals from many synapses, and, based on the summation of excitatory and inhibitory signals, an action potential will or will not be initiated in its axon. Thus individual neurons function as important relay and control centers in the integration of information by the nervous system.

QUESTIONS

1. Distinguish among the following: neuron/nerve/tract; ganglia/nuclei; central nervous system/peripheral nervous system; gray matter/white matter; afferent/efferent; preganglionic/postganglionic; resting potential/action potential/nerve impulse; presynaptic/postsynaptic; neurotransmitter/neuromodulator; adrenaline/noradrenaline/acetylcholine.

2. What are three significant differences between the somatic and autonomic nervous systems?

3. What are four significant differences between the sympathetic and parasympathetic divisions of the autonomic nervous system?

4. If a neuron is placed in a medium where ionic concentrations are the same as those of its own cytoplasm, what effect will this have on the resting potential?

5. Describe the way in which the nerve impulse propagates itself. Draw a diagram of this process.

6. Assume that three presynaptic neurons, A, B, and C, make adjacent synapses on the same postsynaptic neuron, D. No nerve impulse is initiated in the postsynaptic neuron as a result of single nerve impulses in A, B, or C, nor is one initiated if impulses arrive simultaneously in all three, in A and B, or in A and C. Only if impulses arrive together at the synapses of B and C will D fire an impulse. Explain these results in terms of excitatory and inhibitory neurotransmitters and their effects on the membrane voltage of the postsynaptic cell.

7. If you look closely at Figure 41–20, you will notice one neuron that has only very short dendrites and no axon at all. Axon terminals of other neurons form synapses on its cell body, but it forms no axonal synapses with other cells. Such neurons, which are located primarily in the brain, are incapable of initiating an action potential and are thus known as "nonspiking." Nevertheless, they affect the firing rate of other neurons. How do you think they might do this? Why might axons be superfluous for such neurons?

Integration and Control III: Sensory Perception and Motor Response

An animal's sensory equipment is the means by which it knows the world around it. Through sensory perception, animals, including ourselves, are informed of predator and prey, friend and foe, whether something is good to eat or not, and changes in the weather and the seasons. Sensory perception identifies infant to mother, mother to infant, and mate to mate. For us, as perhaps for many other animals as well, it is also a source of pleasure and esthetic enjoyment and provides essential tools for learning. The capacity of an animal to act on the information provided by its sensory equipment—to move toward desirable objects, to move away from harmful or potentially harmful objects, and to perform the multitudinous activities of its daily life—depends on its skeletal muscles. In this chapter, we shall first examine the structures and processes involved in sensory perception. Then we shall consider the marvelous molecular machinery that not only enables animals to jump, run, swim, or fly but also permits us to smile, to frown, and to perform precise manipulative activities.

SENSORY RECEPTORS AND THE INITIATION OF NERVE IMPULSES

The information received, processed, and transmitted by the neurons and synapses of the vertebrate brain and spinal cord is carried into the central nervous system by sensory neurons. The triggering of nerve impulses in a sensory neuron

42–1 *The acquisition of food by a predatory animal depends not only on the reception and processing of detailed sensory information regarding the potential food item and the difficulties in obtaining it but also on exquisite muscular coordination. For this young chacma baboon, one false step or one slip of the hand could mean the difference between feasting on a stolen ostrich egg and disaster—particularly if it attracted the attention of the adult ostriches responsible for guarding the nest.*

depends on transduction, the conversion of one form of energy—the energy of a stimulus—into another form of energy—the energy of an action potential. Stimuli come in a variety of forms—pressure, heat or cold, chemical substances, vibrations, and light. Different types of sensory receptors are specialized to respond to different types of stimuli. In all cases, however, when a sensory receptor has been sufficiently stimulated, its membrane permeability or that of a nearby sensory neuron is altered, initiating the action potentials that start information on its way through the nervous system. The more intense the stimulus, the greater the frequency of the action potentials (see Figure 41–12, page 849).

The differences among the senses lie not in the form in which the signals are coded and conducted—the action potential—but rather in their pattern and its reception and interpretation in the central nervous system. As we shall see in the next chapter, information from different sensory receptors is transmitted to different regions of the vertebrate brain; the particular sensation experienced—a sunset, the song of a whip-poor-will, or a cooling breeze across the face—depends on the region of the brain that is stimulated.

Types of Sensory Receptors

Sensory receptors are many and varied. Most animals, including ourselves, have mechanoreceptors (touch, position, and hearing), chemoreceptors (taste and smell), photoreceptors (vision), temperature receptors, and receptors for the sensation recognized as pain. Some animals, but apparently not *Homo sapiens,* also have electroreceptors and magnetoreceptors. Like the other structures of the body of an animal, sensory receptors are the products of adaptation and evolution, and by these long processes, they have been tailored to the animal's specific requirements. Although we think we see what is visible and hear what is audible, visible and audible are not actually properties of objects; they are instead properties of our particular sensory equipment. What we see is very different from what an insect sees (Figure 42–2), and a bat or a fish, acoustically speaking, might as well be living on another planet. Moreover, it is likely that an animal's insensitivity to a great number of the potential stimuli that surround it is also as useful as its sensitivity to others.

Functionally, sensory receptors can be categorized as interoceptors, proprioceptors, and exteroceptors. **Interoceptors** include the mechanoreceptors and chemoreceptors that are sensitive to blood pressure and O_2, CO_2, and H^+ concentrations in the carotid arteries. The temperature sensors of the hypothalamus are also interoceptors. We are usually not conscious of signals from these receptors, although sometimes the signals result in perceptions of, for example, pain, hunger, thirst, nausea, or the sensations, produced by stretch receptors, of having a full bladder or bowel.

Proprioceptors (from the Latin *proprius,* meaning "self"), which are sometimes considered a subset of interoceptors, provide information about the orientation of the body in space and the position of arms, legs, and other body parts. Because of proprioceptors, a praying mantis can strike unerringly at its victim (see Figure 27–22, page 583), and you can tie your shoelace in the dark, or, with your eyes shut, touch your nose with your fingers. The semicircular canals of the ear are major proprioceptive organs in many vertebrates, performing a function similar to that of the statocysts of jellyfish (see Figure 25–19a, page 529).

The most familiar sensory receptors are **exteroceptors**, which provide information about the external environment. Some exteroceptors are small and relatively simple in structure. Look again at the human skin (Figure 42–3) as an example. The simplest receptors are the free nerve endings, which are receptors for pain, temperature, and perhaps other sensations as well. Slightly more complex are the combinations of free nerve endings with a hair and its follicle. Each of these little

(a)

(b)

42–2 *An evening primrose, as perceived* (a) *by the human eye, which responds to light in the visible portion of the electromagnetic spectrum, and* (b) *by the honeybee eye, which responds to light in the ultraviolet portion of the spectrum.*

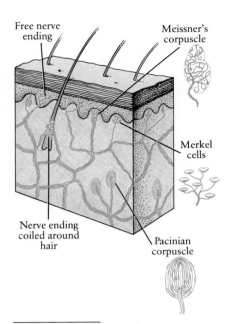

Free nerve
ending

Meissner's
corpuscle

Merkel
cells

Nerve ending
coiled around
hair

Pacinian
corpuscle

42–3 *Some sensory receptors present in human skin. The free nerve endings are largely pain and temperature receptors.*

The best understood of the skin receptors are the Pacinian corpuscles. The specialized nerve ending of a single myelinated fiber is encapsulated by the corpuscle, which is composed of many concentric layers of connective tissue. Pressure on these outer layers stimulates the firing of an action potential at this nerve ending.

Merkel cells and Meissner's corpuscles also respond to touch, as do the nerve endings surrounding the hair follicles. In regions where many hairs are present, the other touch receptors are sparse or absent.

organs is an exquisitely sensitive mechanoreceptor. When the hair is touched or bent, it causes changes in the nerve endings of a sensory neuron, setting off action potentials that are carried directly to the central nervous system. Three other types of mechanoreceptors are also shown, each a combination of one or more free nerve endings with an outer layer or layers of connective tissue. Meissner's corpuscles and Merkel cells are both involved with touch. They are found in particularly sensitive areas of the skin, such as the fingertips, palms, lips, and nipples, and are especially abundant where hairs are not present. They are responsible for the extraordinary cutaneous sensitivity of these parts of the human body and are associated with the ability, for example, to read Braille, do certain magic tricks, crack a safe, or enjoy a kiss. The Pacinian corpuscles, lying deeper within the tissues, respond to pressure and vibrations. The free nerve ending of the corpuscle is surrounded by layers of connective tissue and fluid. This layered structure is easily deformed, so it responds to even very slight pressure changes. However, a Pacinian corpuscle also adjusts quickly to pressure changes, in part because of its structure, and the nerve ending stops firing when the pressure is sustained. Of all the sensory receptors shown here, the simplest in structure—the free nerve endings associated with pain—are the least understood in terms of how they function. It seems likely that some of them are not mechanoreceptors, as are the others, but rather chemoreceptors, responding to very small amounts of some chemical released by cells when they are injured.

Chemoreception: Taste and Smell

If we were to lose our sense of taste or smell, our lives would lose many of their pleasures but we would not be greatly handicapped in our essential activities. Blindness or deafness is a far more serious threat. For many animals, however, the sense of smell is their window on the world. Many of the smaller mammals are nocturnal, and most of them live close to the ground in habitats in which their highly developed sense of smell provides the greatest amount of information about their environment. Presumably, it was the tree-dwelling habits of our immediate ancestors that made the sense of smell less useful for the primates.

Taste

Fish, particularly bottom feeders, have sensory cells for taste—the chemoreception of waterborne substances—scattered over the surface of their bodies. These cells play a major role in determining their behavior. In the carp, for example, the part of the brain that receives information from the taste cells is larger than all the other sensory centers combined. The catfish, also a bottom feeder, has taste cells on its body and, in addition, trails barbels, or "whiskers," along the bottom. These are richly supplied with gustatory nerve endings, and when they are stimulated, the fish turns and snaps at the source of the stimulus.

In terrestrial vertebrates, taste cells are located inside the mouth, where they act as sentinels, providing the information on which a final judgment can be made on what is and is not to be swallowed. The taste receptors and the supporting cells around them form the lemon-shaped clusters known as taste buds (see Figure 34–5, page 717). We are able to distinguish four primary tastes: sweet, sour, salty, and bitter. While each primary taste (or, more accurately, the molecules associated with the taste) appears to stimulate a different type of taste receptor, a single taste bud may respond to more than one category of substance.

Most animals seem to have the same general range of taste discrimination, although there is some difference in what animals "like" and "don't like." Birds, for example, will readily eat many seeds and insects with a bitter taste, whereas for most other vertebrates, bitterness serves as a warning signal. Cats are among the few animals that do not prefer substances with a sweet taste, and very few taste receptors that respond to sugar can be found in the cat's tongue.

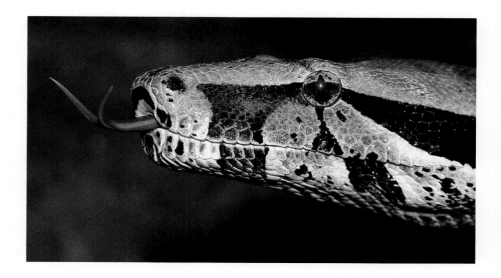

42–4 *Reptiles and amphibians have, in addition to nostrils, a special olfactory organ, the vomeronasal organ, in the roof of the mouth. Food substances are apparently tested in this organ. When a snake lashes its forked tongue in the air, it is collecting samples for testing in its vomeronasal cavities. The snake shown here is a boa constrictor.*

Smell

In fish, taste and smell are operationally very similar since both involve the detection of substances dissolved in the surrounding water. However, taste receptors and smell receptors differ anatomically in fish as in higher vertebrates, and the centers of taste and smell within the brain are entirely distinct.

In terrestrial animals, smell can be defined as the chemoreception of airborne substances. To be detected, however, these substances must first be dissolved in the watery layer of mucus overlying a specialized tissue, the **olfactory epithelium.** In humans, this tissue, which is located high within the nasal passages, is comparatively small. The part within each nasal passage is only about as large as a postage stamp. Each of these areas contains some 50 million receptor cells (see Figure 42–5 on the following page), characterized by numerous cilia whose membranes contain specific receptor molecules into which odor molecules of a complementary shape fit. Current evidence indicates that the binding of an odor molecule to its receptor triggers a series of events involving, once more, cyclic AMP. In the best-studied examples of olfactory transduction, binding of an odor molecule to its receptor activates an enzyme that phosphorylates GDP (guanosine diphosphate) to GTP. GTP, in turn, activates the enzyme adenylate cyclase, which, as we have seen, catalyzes the conversion of ATP to cyclic AMP. Cyclic AMP then binds to and opens a membrane channel through which Na^+ ions flow into the cilium, and from the cilium into the body of the olfactory cell. Other receptors on the cilia are thought to bind inhibitory and modulatory molecules, triggering other sequences of events that have their own effects on the membrane potential across the axon hillock. This membrane potential ultimately determines whether action potentials are or are not initiated by the olfactory cell and the frequency with which they occur.

Even with our relatively insensitive olfactory equipment, we are able to discriminate some 10,000 different odors. One of the current models of odor discrimination was developed in the 1960s by John Amoore while he was still an undergraduate at Oxford University in England. According to the present version of Amoore's model, all scents are made up of combinations of "primary" odors: the seven proposed by Amoore are camphoric, musky, floral, pepperminty, etherlike, pungent, and putrid. There are also, according to this hypothesis, different types of olfactory cells corresponding to the primary odors, with each type bearing membrane receptors for molecules of a particular shape. All substances that have a pungent odor, for example, have the same general molecular shape and fit into the same type of receptor. Some molecules, depending on which way they are oriented, can fit into more than one receptor and so can evoke in the brain two different types of signals. From the signals received from the various different types of cells, the brain constructs a "picture" of an odor.

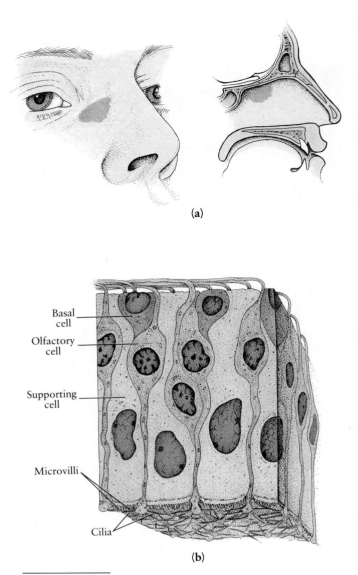

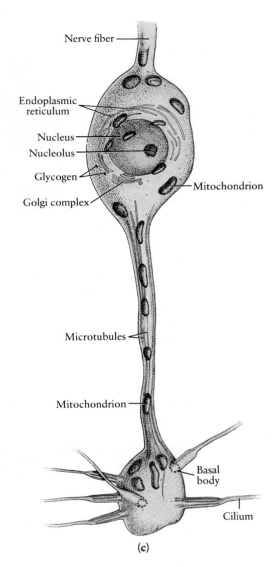

(a)

(b)

(c)

42–5 (a) *A patch of specialized tissue, the olfactory epithelium, arching over the roof of each nasal cavity, is responsible for our sense of smell. Also, much of what we call flavor in food is actually a result of volatile substances reaching this tissue.* (b) *The olfactory epithelium is composed of three types of cells: supporting cells, basal cells, and olfactory cells, which are the sensory receptors. Supporting cells are tall and columnar, wider near the surface than they are deep within the tissue, and their outer surfaces are covered with microvilli,* similar to the microvilli found on the surface of intestinal cells. Movements of these microvilli, in conjunction with proteins in the surrounding mucus, are thought to play a role in transporting odor molecules to the olfactory cells and, subsequently, in removing them. The basal cells, triangular in shape, are found along the innermost layer of the epithelium. Their function is unknown, but it is thought that they may give rise to new supporting cells when these are needed. (c) *The olfactory cell is long and narrow. The cilia protruding* from its exposed surface bear membrane receptors to which odor molecules, inhibitors, and modulators bind. This outermost part of the cell is connected with the cell body by a long stalk containing microtubules; changes in ion concentrations and second messengers set in motion by the binding of molecules to their receptors are transmitted to the cell body through this stalk. From the opposite end of the cell body, a nerve fiber extends through the surrounding tissue, conducting the computed sensory message to the brain.

Chemical Communication in Mammals

As we saw in Chapter 27 (page 585), the exchange of chemical messages—pheromones—among members of the same species is of great importance in the behavior of many insects. Chemoreception also plays a role in the behavior of most mammals. The males of many species—including domestic dogs and cats—scent-mark their territories with urine as a warning signal to other males. Males are attracted to females by special odors associated with estrus (being "in heat").

Among mice, males release substances in their urine that alter the reproductive cycles of the females. Juvenile females reach sexual maturity earlier when exposed to these substances, and the odor of urine from a strange male can cause a pregnant female to resorb her fetuses, leaving her free to mate with the newcomer.

In female rhesus monkeys, vaginal secretions of a volatile fatty acid have been shown to increase near the midpoint of the estrus cycle and to be attractive to male monkeys. Secretions of a similar compound also increase at the time of ovulation in the human female, but its attractiveness to human males has not been demonstrated. It has also been hypothesized that the chemicals (also fatty acids), which are the sources of "body odor" and are produced, beginning at puberty, by the apocrine sweat glands, originally played such a role. The pubic and axillary (underarm) hair that develops at about the same time would appear to have the function of retaining and amplifying these odors.

The most convincing evidence to date of pheromonal interactions in humans was assembled by Martha McClintock while still an undergraduate at Wellesley College. McClintock found a significant

A male dik-dik, a herbivore of the African plains, marking his territory by depositing scent on a grass stem. The glands from which the scent is secreted are located near the base of the eyes.

tendency toward synchronization of menstrual cycles over the school year among women who were roommates and close friends. The effect appeared to correlate directly with the time these women spent together and not with other external factors, such as food habits or photoperiodism. This finding was later extended by demonstrating that the odor from one woman's underarm secretions affected the timing of another woman's cycle, independent of any other contact.

Mechanoreception: Balance and Hearing

The mammalian ear is an organ containing two distinct mechanoreceptors, each of which transduces the energy of an enclosed, moving fluid into the energy of action potentials. One of these structures consists of three interlocking **semicircular canals** that provide information about the orientation of the head in space; this information is critical for the maintenance of physical balance. The other structure, the **cochlea** (snail), is a coiled chamber in which the sensory responses involved in hearing occur. Additional structures convert vibrations of the air (sound waves) into vibrations of the fluid contained within the cochlea. In both the semicircular canals and the cochlea, the primary sensory receptors are sensitive hair cells that vibrate in response to movements of the surrounding fluid.

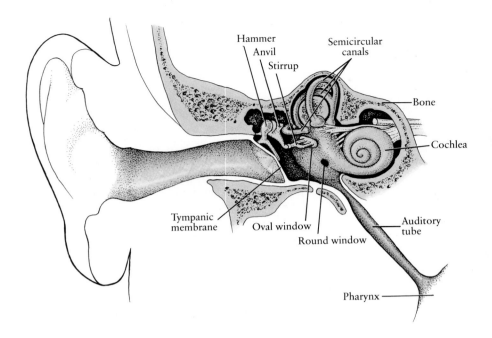

42–6 *The structure of the human ear. Sound waves entering the outer ear are funneled to the tympanic membrane, which they cause to vibrate. These vibrations are transmitted through three small bones of the middle ear—the hammer, anvil, and stirrup—to another membrane, the oval window membrane. Vibrations of this membrane, in turn, set off vibrations in the fluids in the cochlea, the structure of the inner ear concerned with hearing.*

The three semicircular canals are additional fluid-filled chambers within the bony labyrinth of the inner ear. Each one is in a plane perpendicular to the other two. Their function is to monitor the position of the head in space and to maintain equilibrium. Movements of the head set the fluid in these canals in motion, activating sensitive hair cells and triggering action potentials in sensory neurons with which they synapse.

The middle ear, as you can see, is connected with the upper pharynx by the auditory tube (also known as the Eustachian tube). This makes it possible to equalize the air pressure in the middle ear with atmospheric pressure; it also unfortunately makes the middle ear a fertile breeding ground for infectious microorganisms that enter the body through the nose or mouth.

Figure 42–6 diagrams the structures of the human ear. Sound travels through the ear canal to the tympanic membrane (the eardrum), in which it sets up a vibration. This vibration is transferred to a series of three very small and delicate bones in the middle ear, which are called, because of their shapes, hammer (malleus), anvil (incus), and stirrup (stapes). Vibrations in the tympanic membrane cause the stirrup to push gently and rapidly against the membrane covering the oval window, which leads to the fluid-filled cochlea, located in the inner ear.

The pressure must be amplified as it passes from the tympanic membrane to the cochlea because fluid is more difficult to move than air. This amplification is made possible by a difference in the sizes of the tympanic and oval window membranes. The oval window membrane is smaller than the tympanic membrane, resulting in more force per unit area—that is, more pressure—as the total force is transferred.

Figure 42–7 shows a diagram of the cochlea partially uncoiled. It consists essentially of three fluid-filled canals separated by membranes. The upper and lower canals are connected with one another at the far end of the spiral. At the near end of each of these two canals are movable membranes, the oval window membrane and the round window membrane. The stirrup vibrates against the oval window membrane at the near end of the upper canal, causing pressure waves in the fluid contained in these canals. The waves travel the length of the cochlea, around the far end, and back again to the round window membrane at the near end of the lower canal. As the oval window membrane moves in, the round window membrane moves out, keeping the pressure equalized.

The central canal contains the organ of Corti, which rests on the basilar membrane of the canal and contains the individual sensory cells, the hair cells. The movement of the fluid waves along the outer surface of the central canal causes vibrations in the basilar membrane, which, in turn, cause vibrations in the hair cells. Vibration of a hair cell opens ion channels in its membrane, causing a depolarization that, although it does not trigger an action potential, does alter the release of neurotransmitter molecules from the cell. Hair cells synapse with sensory neurons, which, when adequately stimulated at these synapses, initiate action potentials. The axons of these neurons form the auditory nerve, which links the ear to a region of the brain known as the cochlear nucleus. In the cochlear nucleus, neurons of the auditory nerve synapse with other neurons that synapse with still other neurons, ultimately carrying the information to the region of the brain associated with the conscious perception of sound.

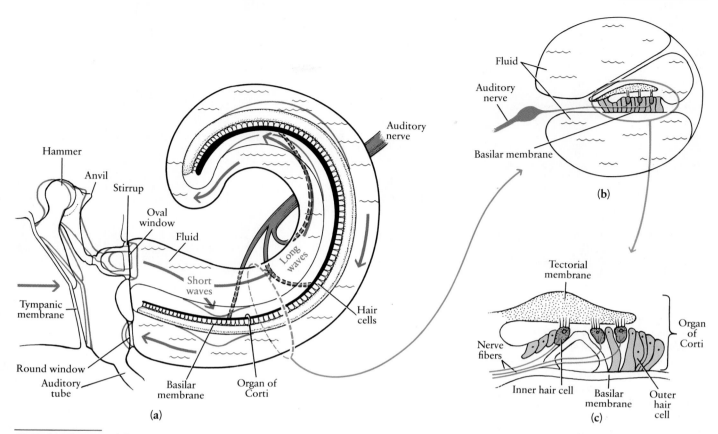

42–7 (a) *The part of the inner ear concerned with hearing is the cochlea, a coiled tube of 2.75 turns—shown here as if it were partially uncoiled. Vibrations transmitted from the tympanic membrane to the stirrup cause the stirrup to push against the oval window membrane, resulting in pressure waves in the fluid that fills the cochlear canals. Pressure waves in the fluid set up vibrations in the basilar membrane, stimulating the sensory cells in the organ of Corti, which rests on the basilar membrane. Sounds at different frequencies (or pitch) have their maximum effect on different areas of the membrane. The round window prevents the pressure from building up in the cochlea. A cross section of the cochlea is shown in* (b), *and a close-up of the organ of Corti in* (c).

The basilar membrane is narrower and less elastic at the end nearer the middle ear. Hence it does not vibrate uniformly along its length; instead different areas of the membrane respond to different frequencies of sound. Thus, different hair cells are stimulated by different frequencies. Human beings are capable of very fine discriminations of sound; such discriminations are made by the brain on the basis of the signals transmitted by the different sensory neurons with which the different hair cells synapse. In general, the human ear can detect sounds ranging from 16 to 20,000 cycles per second, although children can hear up to 25,000 cycles per second. (Middle C is 256 cycles per second.) From middle age onward, there is a progressive loss in ability to hear the higher frequencies.

Our inability to hear very low-pitched sounds (below 125 cycles per second at normal levels of intensity) is a useful adaptation. If we could hear at such low pitches, we would be barraged by sounds caused by internal movements within our own bodies, such as the flow of blood, as well as by sounds conducted through our bones. We do hear some sounds conducted directly through the skull—for example, the noise we make when we chew celery. We also hear our own voices mainly through the skull. This is why a recording of one's own voice sounds so startlingly unfamiliar.

Dogs, as all dog trainers know, can hear very high sounds (40,000 cycles per second) and mice squeak to one another at 80,000 cycles per second. An elephant named Lois, tested by investigators at the University of Kansas and the Ralph Mitchell Zoo in Independence, Kansas, could not hear sounds higher than 12,000 cycles per second but could hear sounds at 16 cycles per second at levels of intensity inaudible to human beings. (When Lois heard a sound, she pressed a lever with her trunk and was rewarded with half a cup of Kool-Aid.) Recent evidence indicates that elephants communicate with one another at pitches that would be the envy of any operatic basso—except that they are outside the range of human hearing.

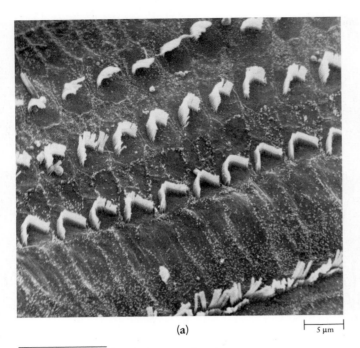

(a)

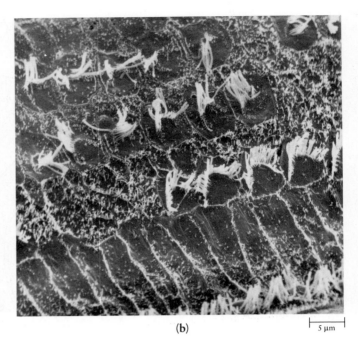

(b)

5 μm

5 μm

42–8 *Scanning electron micrographs of hair cells from the organ of Corti of a guinea pig before (a) and after (b) exposure to 24 hours of loud noise, comparable to that of a typical rock concert. On some of the hair cells, the orderly arrangement of cilia has been disrupted. Other hair cells have degenerated, losing their cilia.*

Photoreception: Vision

Eyes have evolved independently several times in the course of evolutionary history. Among the most highly developed of modern photoreceptor systems are the compound eye of arthropods (see page 579), the eye of the octopus (see page 552), and the vertebrate eye, of which the human eye, shown in Figure 42–9, is an example.

The vertebrate type of eye is often called a camera eye. In fact, it has a number of features in common with an ordinary camera equipped with several expensive accessories, such as a built-in cleaning and lubricating system, an exposure meter, and an automatic focus. Light from the object being viewed passes through the

42–9 *The eye is a complex organ composed of three layers of tissue, which form a fluid-filled sphere. The outer layer, the sclera, is white fibrous connective tissue that serves a protective function. The anterior portion of the sclera, the cornea, is transparent. The middle layer, the choroid, contains blood vessels. Its anterior portion is modified into the ciliary body, the suspensory ligament, and the iris. The ciliary body is a circle of smooth muscle from which extend the suspensory ligaments that hold the lens in place. The colored part of the eye, the iris, is a circular structure attached to the ciliary body. The pupil is a hole in the center of the iris, which regulates its size. The innermost layer of the eye, the retina, contains the photoreceptor cells, the rods and cones. The fovea, near the center of the retina, is the region of greatest visual acuity. Only the front of the eye is exposed; the rest of the eyeball is recessed in and protected by the bony socket of the skull.*

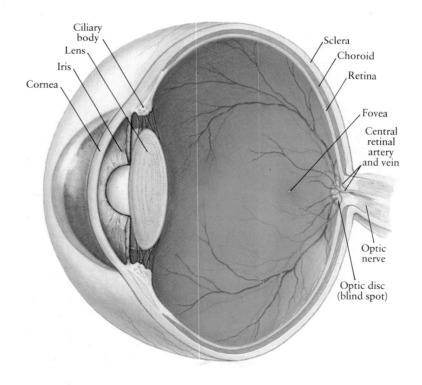

42–10 *Focusing the eye. Preliminary focusing is done by the cornea, and fine focusing by the lens, which is like a jelly-filled, transparent rubber balloon. The lens is held in place by suspensory ligaments attached to the ciliary body, which encircles the lens. The ciliary body contains the ciliary muscle, which is the muscle of accommodation. When it contracts, the lens is released, and, as it returns to its more nearly spherical shape, its curvature is increased.* (a) *The normal eye views distant objects with the lens stretched and flattened;* (b) *for viewing near objects, the lens is relaxed and more convex, bending the light rays more sharply.* (c) *Nearsightedness occurs when the eyeball is too long for the lens to focus a distant image on the retina;* (d) *it is corrected by a concave lens that bends the light rays out enough to focus the image.* (e) *Farsightedness is the result of an eyeball too short for the lens to focus a near object;* (f) *it is corrected by a convex lens that bends the light rays in before they reach the lens of the eye.*

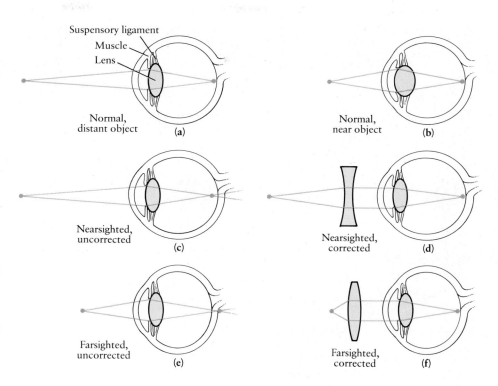

Suspensory ligament
Muscle
Lens

Normal, distant object **(a)**

Normal, near object **(b)**

Nearsighted, uncorrected **(c)**

Nearsighted, corrected **(d)**

Farsighted, uncorrected **(e)**

Farsighted, corrected **(f)**

transparent cornea and lens, which focus an inverted image of the object on the light-sensitive retina in the back of the eyeball. In mammals, fine focusing of the image on the retina is brought about by contracting or relaxing the ciliary muscles, thus changing the curvature, and hence the focal length, of the lens (Figure 42–10). Fish and amphibians, which lack ciliary muscles, focus their eyes the same way one focuses a camera with a lens of fixed focal length. Muscles within the eye change the position of the lens, drawing it back toward the retina in order to focus more distant objects.

Stereoscopic vision—that is, vision in three dimensions—depends on viewing the same visual field with both eyes simultaneously. When both eyes are trained on a distant object, both eyes see almost the same image. A nearby object, however, will present a slightly different image to each eye, a phenomenon known as parallax. The brain computes the disparity in the two images of a nearby object in such a way that we are able to judge its distance from us. When the eyes are trained on a very distant object, however, the disparity is too small; hence, we are not able to make such judgments for faraway objects. Instead, we estimate their distance on the basis of known sizes of the objects or such visual clues as the size and relative position of houses, people, or cars in the same field of vision. A young child viewing a distant object—an airplane, for example—may see it as little rather than as far away.

Tree-dwelling animals, such as the probable forerunners of *Homo sapiens,* usually have overlapping visual fields in the front that provide them with the stereoscopic vision essential for the distance judgments involved in moving from branch to branch. Predatory animals also tend to have stereoscopic vision, but animals that are more likely to be hunted than to hunt usually have an eye placed on each side of the head, giving a wide total visual field (Figure 42–11). Some birds with laterally placed eyes—the woodcock, the cuckoo, and certain species of crow—have binocular fields of vision both in front of them and behind them.

(a)

(b)

(c)

(d)

42–11 *Some vertebrate eyes. (a) The eyes of a mud skipper, a bottom-dwelling fish, are directed upward like periscopes. Mud skippers live in the shallow waters of mudflats and mangrove swamps in India and Asia. They can see objects in the air above the water, even when their bodies are totally submerged. (b) Eagles and other predatory birds often have eyes as large as ours (although the skull and brain are much smaller), two foveas in each retina, and strong powers of near and far accommodation. Shown here is a peregrine falcon. (c) Rabbits have eyes set on each side of their skulls, a placement that allows them to watch on both sides for predators while they are feeding. (d) The slow loris, a tree-dwelling primate of Southeast Asia, is nocturnal. Like many other nocturnal animals, it has large eyes. The photoreceptors of nocturnal animals are almost all rods, which are more light-sensitive than cones and are necessary for night vision.*

The Retina

The retina of the vertebrate eye contains the photoreceptor cells, which capture light energy and begin the process of transduction. These cells are of two types, named, because of their shapes, **rods** and **cones** (Figure 42–12). Rods are responsible for black-and-white vision; cones, for color vision.

Rods do not provide as great a degree of resolution as cones do, but they are more light-sensitive than cones. Dim light does not stimulate the cones, which is why the world becomes colorless to us at night. Nocturnal animals, which include most mammals, have retinas made up almost entirely of rods and thus have no color vision. Some diurnal animals, such as some reptiles and squirrels, have almost entirely cones. Higher primates, including humans, have both rods and cones.

As you can see in Figure 42–13, the retina of the vertebrate eye is anatomically inside out—that is, the photoreceptors of the eye are pointed toward the back of the eyeball. To reach them, light must pass through several layers of other neurons in the retina. Only about 10 percent of the light falling on the cornea reaches the retina. Of this light, that which is not captured by the photoreceptors is absorbed by pigmented epithelium that lines the back of the eyeball, just behind the photoreceptors. Some nocturnal vertebrates have a reflecting layer (tapetum) behind the photoreceptors that increases the likelihood of dim light stimulating the photoreceptors. This reflecting layer makes the animal's eyes seem to shine at night when light is directed into them.

The photoreceptor cells communicate, by way of intervening neurons, with the ganglion cells, whose axons form the optic nerve. When light is captured by the photoreceptor cells, a series of reactions occur that produce a change in their membrane polarity. This change influences their release of neurotransmitters at synapses they form with a group of cells known as bipolar cells. The release of neurotransmitters at these synapses causes, in turn, a change in the membrane polarity of the bipolar cells, and influences their release of neurotransmitters at synapses with the ganglion cells. The ultimate result of the stimulation by light is a change in the pattern in which action potentials fire in the axons of the ganglion cells.

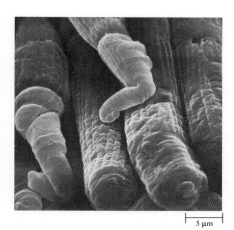

42–12 *Rods and cones as shown by the scanning electron microscope. A single photon of light (see page 208) is sufficient to cause a response in a photoreceptor; as few as five to eight photons in the blue-green frequency range can result in conscious perception.*

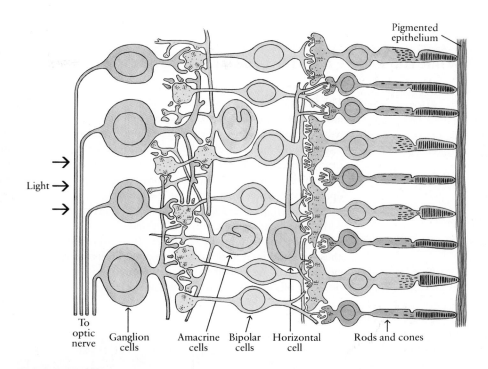

42–13 *The retina of the vertebrate eye. Light (shown here as entering from the left) must pass through several layers of cells to reach the photoreceptors (the rods and the cones) at the back of the eye. Signals from the photoreceptor cells are then transmitted through the bipolar cells to the* ganglion *cells, whose axons converge to become the optic nerve. Horizontal and amacrine cells, other neurons in the retina, also participate in the elaborate transmission paths. Some processing of information occurs in these pathways before nerve impulses leave the retina.*

Ganglion cell axons from all over the retina converge at the rear of the eyeball and bundle together like a cable to form the optic nerve, which connects the retina with the rest of the brain. The point at which the axons pass out of the retina is a blind spot, since photoreceptor cells are absent there. We generally are not aware of the existence of the blind spot since we usually see the same object with both eyes and the "missing piece" is always supplied by the other eye. You can demonstrate the existence of the blind spot with the help of Figure 42–14.

Information Processing in the Retina

There are about 125 million photoreceptors in the retina and about 1 million ganglion cell axons in the optic nerve, a reduction of 125 to 1. This reduction occurs principally with the rods. Many rods converge on each bipolar cell, and several bipolar cells, in turn, feed into each ganglion cell. Also, other neurons in the retina—horizontal and amacrine cells—participate in these connections and interactions. Therefore, a point-to-point representation is not transmitted from the rods—as with a television camera, for instance—but there is processing of the information before it even leaves the retina (see essay on the next page).

42–14 *With this diagram, you can prove to yourself that you have a blind spot in each eye. Hold the book about 30 centimeters (12 inches) from your face, cover your left eye, and gaze steadily at the X while slowly moving the book toward your face. Note that at a certain distance, the image of the dot becomes invisible. Then cover your right eye and gaze steadily at the dot while you move the book toward your face. What happens to the X?*

X   ●

What the Frog's Eye Tells the Frog's Brain

The frog's eye differs from the human eye in that it has no central fovea and the rods and cones are distributed uniformly over the surface of the retina. It resembles the human eye, however, in that these photoreceptor cells transmit their signals to a far fewer number of ganglion cells, whose axons make up the optic nerve leading from the eye to the brain.

By inserting microelectrodes into these axons while exposing the frog's eye to various kinds of stimuli, Jerome Lettvin and coworkers at the Massachusetts Institute of Technology found that different ganglion cells responded to different stimuli—light on, light off, or a big moving shadow, for instance.

Most interesting, one type of ganglion cell responded only to a small moving object; in other words, it was a bug detector. An object bigger than a bug would not stimulate these particular ganglion cells even if it was in motion, and a bug-sized object would not stimulate them if it was motionless. The existence of the bug detector corresponds nicely with certain well-known features of the animal's behavior. A frog will strike only—and virtually always—at a small moving object and will literally starve surrounded by dead insects. Thus, the information about an extremely important aspect of the frog's world is processed right in the retina itself.

*A European tree frog (**a**) observing and (**b**) capturing a blue-bottle fly.*

(a)

(b)

42–15 *Eye-hand coordination is a characteristic of the higher primates, such as the young olive baboon shown here grooming itself.*

The area of the retina in which the sharpest image is formed is known as the **fovea** (see Figure 42–9). In the fovea, the photoreceptor cells consist entirely of closely packed cones. These cones, instead of having the 125:1 relationship of the rods and ganglion cells, make one-to-one connections with the bipolar and ganglion cells. The one-to-one connections and the close packing of the cones provide greater resolution, giving a crisper picture.

Birds, which rely on vision above all other senses, may have two or three foveas. Also, the photoreceptors of birds tend to be more tightly packed. We have about 160,000 cones per square millimeter of retina, or some 5 to 10 million cones per eye. A hawk, with an eye of about the same size, has some 1 million cones per square millimeter and therefore a visual acuity about eight times that of a human being.

Among the mammals, only primates have a central fovea for sharp vision. In *Homo sapiens*, highly developed vision is undoubtedly closely correlated with the development of the use of the hands for fine manipulative movements. This reasonable hypothesis is supported by the fact that the human eye is so constructed that the sharpest images—images within the fovea—can be made of objects that are within the reach of the hand (Figure 42–15).

Visual Pigments and the Capture of Light

Both rods and cones contain light-sensitive compounds—visual pigments—embedded in a series of stacked membranes that are segregated in one portion of the cell (Figure 42–16). The pigments consist of a protein, called an opsin, and a carotenoid, called retinal, derived from vitamin A. (The relationships among retinal, vitamin A, and beta-carotene, the principal accessory photosynthetic pigment in plants, are summarized in Figure 10–7, page 212.) The opsin part of the molecule differs in different types of photoreceptors; the retinal is the same. As we noted previously, the eyes of arthropods, mollusks, and vertebrates all evolved separately, and yet each of the three types contains almost identical visual pigments. Since no animal can synthesize carotenoids, all vision depends on substances derived in the diet directly or indirectly from plants. Similar light-sensitive compounds occur in the cell membrane of the salt-loving halobacteria (see page 221), and they have also been identified in the eyespot of *Chlamydomonas*, suggesting a long evolutionary history in which the same, or closely related, light-sensitive molecules have been utilized for different functions.

42–16 (a) *Diagram of a rod from the human retina. The molecules of light-sensitive pigment, rhodopsin, are located on the lamellae. The foot makes synaptic contact with bipolar and horizontal cells.* (b) *An electron micrograph of a rod, showing the region in which the outer and inner segments connect. The outer segment of the rod (the portion of the cell pointing toward the back of the eyeball) consists of a stack of membranes, piled up one on top of the other like poker chips. The light-sensitive pigment is built into these membranes. In the inner segment of the cell, new pigment molecules and other substances required by the light-receptor area are synthesized. These molecules are transported through a narrow connecting stalk (indicated by the arrow) to the outer, membrane-packed segment, where the actual work of this highly specialized cell is done. A cross section of the stalk reveals, surprisingly, that it has the internal structure of a cilium, lacking only the two central microtubules. Cones exhibit the same general structure as rods.*

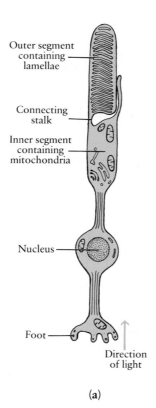

Outer segment containing lamellae

Connecting stalk

Inner segment containing mitochondria

Nucleus

Foot

Direction of light

(a)

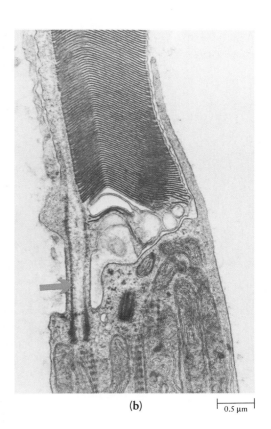

(b)

0.5 μm

In rods, the retinal-opsin combination is rhodopsin, sometimes called visual purple. When rhodopsin absorbs light, the retinal changes shape (Figure 42–17), triggering a change in shape in the opsin. This sets in motion a complex cascade of chemical reactions leading to a change in the membrane potential of the rod cell, and thus in its release of neurotransmitter to the bipolar cells. The reactions of the cascade involve at least two distinct enzymes and utilize cyclic GMP as a second messenger. Another participant is GTP, which, like ATP, is a ready source of

42-17 *When the retinal portion of the rhodopsin molecule is excited by light, rotation occurs at one bond (arrow), resulting in a significant change in its overall shape. The retinal is located in a cleft within the opsin portion of the molecule, and its change in shape causes a deformation in the protein. The resulting activated rhodopsin sets in motion the series of reactions that alter the membrane polarity of the rod cell and thus its release of neurotransmitter. One of the reactions in the series, catalyzed by an enzyme and powered by the hydrolysis of GTP to GDP, has the effect of restoring the opsin and retinal to their original conformations.*

Resting   Excited

cellular energy; energy supplied by GTP plays a key role in returning rhodopsin to its original shape, ready once more to respond to light and to initiate another series of reactions. The consequence of these reactions is not a depolarization of the rod cell membrane, but rather a hyperpolarization, resulting from the closure of membrane channels for Na^+. It is this hyperpolarization that alters the release of neurotransmitter from the cell.

Color vision in humans depends on the presence of three different types of cones, each containing one of three different visual pigments. Each of these pigments is made up of retinal and a slightly different opsin. As long ago as 1802, a three-receptor hypothesis for color vision was proposed; this hypothesis, supported by a number of psychological studies, was greatly bolstered by the identification of the three separate cone types and their pigments. Each pigment is most sensitive to the wavelength of one of the three fundamental colors—blue, green, or red. Different shades of color stimulate different combinations of these cones, and the signals, after being processed in the retina, are further processed by the brain and so become what we perceive as color. Most recently, the genes for each of the three opsins have been identified and sequenced, and the location of the genes for the pigments sensitive to red and green have been determined. As you would expect on the basis of your knowledge of the inheritance of color blindness, these two genes are located on the *X* chromosome, where multiple copies occur in tandem. Comparison of the sequences of all three genes strongly suggests that they, like the human globin genes, evolved from a single gene through processes of duplication and mutation (see page 367).

Visible light is only a small portion of the vast electromagnetic spectrum (see Figure 10–3, page 207). For the human eye, the visible spectrum ranges from violet light, which is made up of comparatively short light waves, to red light, the longest visible to us. Fortunately, we cannot see in the infrared (heat-wave) spectrum. Otherwise, we would see everything through an infrared glow emitted by our own bodies. Our photoreceptors are, however, sensitive to ultraviolet. Ordinarily, these waves are filtered out by yellow pigment in the lens, but persons who have had cataracts removed can read by ultraviolet light.

The perception of color often plays an important role in vertebrate behavior. Studies of the three-spined stickleback, a small freshwater fish, have shown that fighting behavior among the males is elicited by the red color that develops on their underbellies during the mating season. Niko Tinbergen, who kept a laboratory in England full of aquariums containing sticklebacks, found that the male sticklebacks would rush to the sides of their tanks and assume threatening postures every time a red mailtruck passed on the street outside. Similarly, it has been found that fighting behavior among English robins (see page 409) can be evoked by a small tuft of red feathers. Among birds, bright plumage, especially of the males, plays an important part in attracting the opposite sex and in courtship ceremonies.

THE RESPONSE TO SENSORY INFORMATION: MUSCLE CONTRACTION

The activities an animal undertakes as a result of the information received and processed by its sensory receptors and brain depend upon skeletal muscle, the effector of the somatic nervous system. Moreover, many of the adjustments in its internal environment depend upon cardiac muscle and smooth muscle, two of the effectors of the autonomic nervous system. Muscle is the principal tissue, by weight, in the vertebrate body. However, even if there were only a single muscle cell per organism, it would be worthy of our attention and admiration. Skeletal muscle is the one tissue of the biological world whose function on a macroscopic scale has been traced down to, and actually visualized in terms of, the very molecules that make it possible, a wonderful correspondence between structure and function.

The Structure of Skeletal Muscle

As we noted in Chapter 33, a skeletal muscle consists of bundles of muscle fibers—often hundreds of thousands of fibers—held together by connective tissue. Each fiber is a single, multinucleate cell 10 to 100 micrometers in diameter and, often, several centimeters long. Each muscle fiber is surrounded by an outer cell membrane called the **sarcolemma.** Like the membrane of the axon, the sarcolemma can propagate an action potential.

Embedded in the cytoplasm of each muscle fiber (cell) are some 1,000 to 2,000 smaller structural units, the **myofibrils** (from *myo,* the prefix for "muscle"). Tightly packed myofibrils run parallel for the length of the cell, crowding the nuclei to its periphery, where they are typically found just beneath the sarcolemma.

Each myofibril is encased in a sleevelike membrane structure, the **sarcoplasmic reticulum,** which is a specialized endoplasmic reticulum (Figure 42–18). The sacs of the sarcoplasmic reticulum contain calcium ions (Ca^{2+}), which, as we shall see, play an essential role in muscle contraction. Running perpendicular to the myofibrils is a system of transverse tubules, the T system. The membrane forming the T

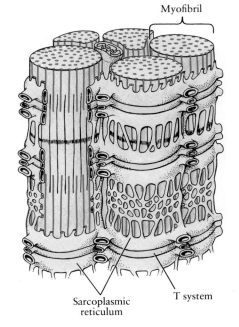

42–18 *A group of myofibrils, each of which is surrounded by the specialized endoplasmic reticulum of muscle cells, the sarcoplasmic reticulum. The sacs of the sarcoplasmic reticulum contain calcium ions, which, when released, trigger muscle contraction. Traversing the sarcoplasmic reticulum, perpendicular to the myofibrils, is a system of transverse tubules, the T system. The membrane forming the T system is continuous with the cell membrane; like the cell membrane, it can propagate an action potential.*

Myofibril

Sarcoplasmic reticulum

T system

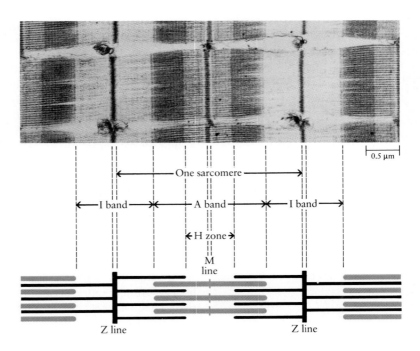

42-19 *Electron micrograph and diagram of a sarcomere, the contractile unit of muscle. Each sarcomere is composed of an array of thick and thin protein filaments arranged longitudinally. The Z line is where thin filaments from adjoining sarcomeres are anchored. The I band is a region that contains only thin filaments. The A band marks the extent of the thick filaments. The part of the A band where there are no thin filaments is called the H zone. The thick filaments are interconnected and held in place at the M line. Muscle contraction involves the sliding of the thin filaments between the thick ones.*

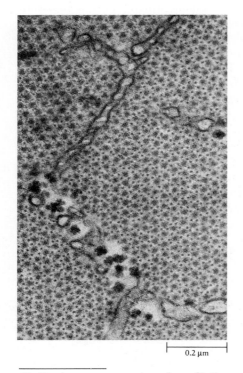

42-20 *A transverse section of myofibrils reveals a hexagonal array of filaments. As you can see, each thick filament is surrounded by six thin filaments.*

system is actually a complex invagination of the sarcolemma; it forms a series of channels (the interior of the transverse tubules) through which the solution of ions and molecules surrounding the muscle cell can flow. Despite the large diameter of the cell, no myofibril is separated from the extracellular fluid by more than the thickness of a lipid bilayer.

Myofibrils are composed of units called **sarcomeres,** each of which is about 2 or 3 micrometers in length. The repetition of these units gives the muscle its characteristic striated pattern. Figure 42-19 shows a sarcomere as seen in a longitudinal section of muscle. As the diagram shows, each sarcomere is composed of two types of filaments running parallel to one another. The thicker filaments in the central portion of the sarcomere are composed of the protein myosin; the thinner filaments are primarily actin, also a protein. The dense black line seen in the electron micrograph is the Z line, in which the thin filaments from adjacent sarcomeres interweave. The I band is the relatively clear, broad stripe that the Z line bisects. The large dense stripe in the center of the sarcomere, formed by thick filaments, is the A band; it is bisected by the central H zone, which contains thick filaments but no thin filaments. A transverse section of the region of the A band containing both types of filaments reveals that each thick filament is surrounded by six thin filaments (Figure 42-20).

The Contractile Machinery

The sarcomere is the functional unit of skeletal muscle, the mechanism by which contraction occurs. When the muscle is stimulated, the thin (actin) filaments of the sarcomere slide past the thick (myosin) filaments. Since the thin filaments are anchored into the Z line, this causes each sarcomere to shorten, and thus the myofibril as a whole to contract. As explained by the well documented sliding-filament model, cross bridges between the thick and thin filaments form, break, and re-form rapidly, as one filament "walks" along the other (Figure 42-21).

The actin strands of the thin filaments are composed of many globular actin molecules assembled in a long chain. As shown in Figure 42-21e, each thin filament consists primarily of two such actin chains wound around one another.

42–21 (a) *Skeletal muscle is composed of individual muscle cells, the muscle fibers. These are cylindrical cells, often many centimeters long, with numerous nuclei.* (b) *Each muscle fiber is made up of many cylindrical subunits, the myofibrils. These fibrils, which contain contractile proteins, run from one end of the cell to the other.* (c) *The myofibril is divided into segments, sarcomeres, by thin, dark partitions, the Z lines. The Z lines appear to run from myofibril to myofibril across the fiber. The sarcomeres of adjacent myofibrils are in line with each other, giving the muscle cell its striated appearance. Each sarcomere is made up of thick and thin filaments.* (d) *Chemical analysis shows that the thick filaments consist of bundles of a protein called myosin. Each individual myosin molecule is composed of two protein chains wound in a helix; the end of each chain is folded into a globular "head" structure.* (e) *Each thin filament consists primarily of two actin strands coiled about one another in a helical chain. Each strand is composed of globular actin molecules.* (f) *The globular myosin heads protruding from the thick filaments serve as hooks or levers, attaching to the actin molecules of the thin filaments and pulling them toward the center of the H zone, shortening the sarcomere and contracting the myofibril. When the myofibrils contract, the fiber shortens, and when enough fibers shorten, the entire muscle shortens, producing skeletal movements.*

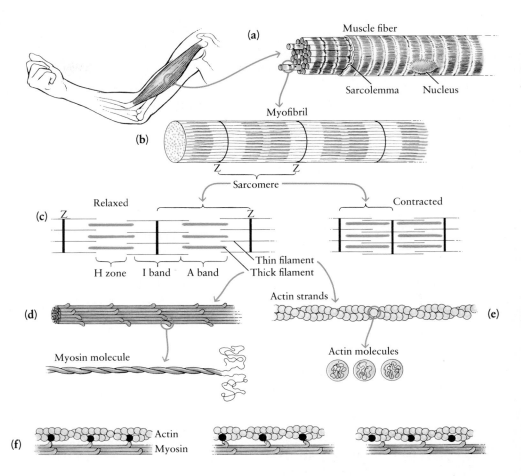

The thick filaments are composed of bundles of myosin molecules. A myosin molecule consists of two long protein chains, each containing some 1,800 amino acids and each with a globular "head" at one end. In the molecule these two chains are wound around each other, with the globular "heads" free. These heads have two crucial functions: they are the binding sites at which force is exerted on the thin filaments during contraction, and they also act as enzymes that split ATP to ADP, thus providing the energy for muscle contraction.

When a muscle fiber is stimulated, the heads of the myosin molecules move away from the thick (myosin) filament toward the thin (actin) filament, to which they attach themselves. The heads move with a swiveling, oarlike motion, pulling the thick filament, pushing the thin one. Thus, a repeated cycle of attachment, breaking away, and reattachment moves the two filaments, ratchetlike, past one another. The thin filaments on opposite sides of the H zone move toward one another, so the Z lines bordering the sarcomere are pulled together.

The contraction of the sarcomeres is dependent on ATP in two ways: hydrolysis of ATP by the myosin molecule provides the energy for the cycle, and combination of a new ATP molecule with the myosin molecule releases the myosin head from the binding site on the actin molecule. The stiffened muscles of a corpse—rigor mortis—are due to the absence of ATP and the subsequent locking of all the actin-myosin cross bridges.

The contractile machinery of cardiac muscle, like that of skeletal muscle, consists of sarcomeres and is thought to function in the same fashion. Contraction of smooth muscle also depends on assemblies of actin and myosin molecules, but the exact mechanism remains unknown. Recent studies of smooth muscle using tiny resin beads attached to the outside of the cells revealed a corkscrew pattern of contraction rather than the linear contraction observed in skeletal muscle. This suggests that the actin and myosin molecules, or perhaps elements of the cytoskeleton with which they interact, are arranged within the smooth muscle cell in the form of a helix.

The Regulation of Contraction

The regulation of contraction in skeletal muscle depends upon two other groups of organic molecules, troponin and tropomyosin, plus calcium (Ca^{2+}) ions. As shown in Figure 42–22, tropomyosin molecules are long, thin double cables that lie along the actin molecules of the thin filament, blocking the cross-bridge binding sites on those molecules. The troponin molecules are complexes of globular proteins that are located at regular intervals on the tropomyosin chains. When Ca^{2+} combines with the troponin molecules, they undergo conformational changes that result in shifting of the tropomyosin chains and exposure of the cross-bridge binding sites. The availability of Ca^{2+}, and thus the initiation of contraction, depend on stimulation of the muscle by a signal received from a motor neuron.

The Neuromuscular Junction

A motor neuron (Figure 42–23a) typically has a single long axon that branches as it reaches the muscle (Figure 42–23b). At the end of each branch, the axon emerges from the myelin sheath and becomes embedded in a groove on the surface of a muscle fiber, forming the **neuromuscular junction** (Figure 42–23c). As is the case with most synapses between neurons, the signal travels across the neuromuscular junction by means of a chemical transmitter—in this case, acetylcholine. However, unlike synaptic transmission between neurons, this is a direct, one-to-one relationship involving only excitation. The acetylcholine combines with receptors on the sarcolemma, depolarizing the muscle cell membrane and initiating an action potential that sweeps along the sarcolemma, including the invaginations that form the T system. As the electrochemical impulse moves through the T system, it alters the sarcoplasmic reticulum, which then releases Ca^{2+} ions. These ions continue to be released only as long as the fiber is stimulated; once stimulation stops, the ions are pumped back into the sacs of the sarcoplasmic reticulum by active transport. Thus it is the Ca^{2+} ions that turn the contractile machinery on and off.

42–22 *Tropomyosin and troponin molecules, both proteins, have a regulatory role in muscle contraction. When calcium ions are not present, tropomyosin molecules, long, thin double cables, block the cross-bridge binding sites on the thin (actin) filaments. Troponin molecules, which are globular proteins, are situated at regular intervals on the long tropomyosin chain. When calcium ion binds to troponin, the tropomyosin molecule shifts position, exposing the binding sites and permitting the cross bridges to form.*

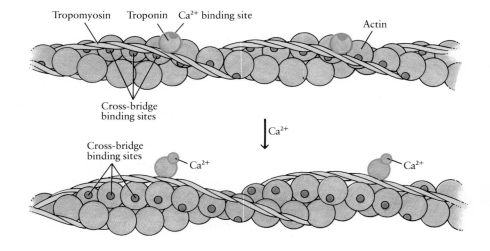

42–23 (a) *A motor neuron. The stimulus is received by the dendrites and cell body, which carry the signal to the axon hillock. The action potential initiated there travels along the axon, which is insulated by a myelin sheath composed of Schwann cell membranes.* (b) *The axon of each motor neuron divides into branches, each forming a neuromuscular junction with a different muscle fiber (cell). The motor neuron and the numerous muscle fibers that it innervates are known as a motor unit. Stimulation of a motor axon stimulates all of the fibers in that motor unit. Within a given muscle, fibers of different motor units are intermingled.* (c) *A neuromuscular junction. An action potential conducted along the axon of a motor neuron releases acetylcholine from synaptic vesicles into the synaptic cleft (see Figure 42–24). This neurotransmitter combines with receptor sites on the sarcolemma, altering the membrane permeability and initiating an action potential in the muscle fiber.*

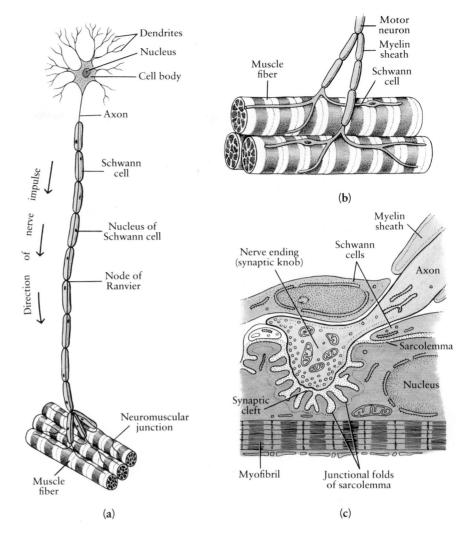

(a)

(b)

(c)

Stimulation of smooth and cardiac muscle also triggers the movement of Ca^{2+} ions into the cytoplasm, not only from the sarcoplasmic reticulum but also from the extracellular fluid. However, as you will recall (page 845), neurons of the autonomic system can either stimulate or inhibit their targets. A smooth muscle cell is typically postsynaptic to neurons of both the sympathetic and parasympathetic divisions; whether it contracts or not depends on the summed effect of the neurotransmitters released at excitatory, inhibitory, and modulatory synapses.

42–24 (a) *A motor neuron terminal. The larger, darker bodies are mitochondria. Below them are the spherical synaptic vesicles. At the bottom of the micrograph is a portion of a muscle cell. The light area immediately below the neuron terminal is the synaptic cleft, and below that is a junctional fold in the sarcolemma. The muscle fiber underlying the sarcolemma is shown in a transverse section, as in Figure 42–20.* (b) *Synaptic vesicles discharging into the synaptic cleft.*

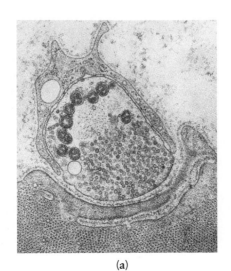

(a)

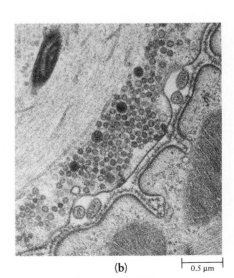

(b)

0.5 µm

Twitch Now, Pay Later

Vertebrate muscle fibers differ in their speed of contraction, which depends on the ATPase activity of their myosin heads, and in their energy metabolism. There are two general categories, red fibers and white fibers, most familiar to poultry fans as dark meat and light meat.

Red fibers are slow-twitch, pay-as-you-go fibers. They are relatively small, with large surface-to-volume ratios, a good blood supply, numerous mitochondria, and an abundance of myoglobin, which is mainly responsible for their color. As these characteristics would indicate, they have an abundant supply of oxygen, and oxidative phosphorylation provides most of their energy. In general, they use fatty acids as their fuel. (These fatty acids are stored in red fibers, which is why drumsticks tend to be greasy.) They are adapted for relatively continuous use that requires endurance. Thus in the domestic chicken, which walks some 14 hours a day, the leg muscles are mostly red fibers and the wing muscles—the "breast"—are white; whereas in a migratory bird—a wild duck, for instance—the "breast" is mostly red fibers.

White fibers, on the other hand, are fast-twitch, high-powered, twitch-now-pay-later muscles. They have fewer capillaries, fewer mitochondria, and less myoglobin. Their fuel supply is glucose or glycogen, which they break down by anaerobic glycolysis, resulting in lactic acid accumulation and oxygen debt. They are adapted for quick spurts of power. The strong, fast muscles of the legs of a rabbit or a frog are mostly white fibers.

The average human being has a fifty-fifty distribution of fast-twitch (white) and slow-twitch (red) fibers in the major skeletal muscles. However, recent studies have shown that the muscles of long-distance runners and swimmers average about 80 percent slow-twitch fibers, whereas sprinters tend to have about 75 percent fast-twitch fibers.

The proportion of slow-twitch fibers to fast-twitch fibers is genetically determined. (Identical twins have identical fiber patterns in their muscles, regardless of the type of muscular activity they engage in.) Training can cause fast-twitch fibers to behave more like slow-twitch fibers, but the reverse is not true. Some trainers recommend the use of fiber typing as part of a total body profile to determine what type of sport a would-be athlete should pursue. Russians and East Germans, who take their sports very seriously, are doing so already.

As we saw in Chapter 36, the contraction of cardiac muscle is initiated by the specialized cells of the pacemaker; signals transmitted to the pacemaker from the autonomic nervous system influence the rate at which the pacemaker fires but are not responsible for the actual initiation of contraction.

Many drugs act specifically on the neuromuscular junction. Curare, for instance, a plant extract that Indians of South America used to poison their arrow tips, blocks excitation of muscle by binding to acetylcholine receptors on the sarcolemma and produces paralysis. It is used medically as a muscle relaxant during surgery. The bacterial toxin of botulism—the most poisonous substance known—prevents nerve endings from liberating acetylcholine and so kills by paralysis of the muscles controlling breathing.

The Motor Unit

The axon of a single motor neuron and all the muscle fibers it innervates are known as a motor unit (Figure 42–25). The number of muscle fibers in a motor unit determines the fineness of control. In a muscle that moves the eyeball, for instance, a motor unit may contain as few as three muscle fibers, whereas in the biceps, each motor unit contains more than a thousand. Within a given muscle, fibers of different motor units are intermingled. A slight movement may involve the contraction of only a few motor units. The strength of contraction of a muscle

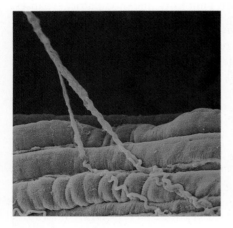

42–25 A scanning electron micrograph showing the axon of a motor neuron and three skeletal muscle fibers.

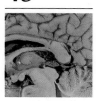

Integration and Control IV:
The Vertebrate Brain

43–1 *In this section of tissue from the gray matter of a human brain, you can see the cell bodies of a number of neurons, embedded in a dense meshwork of axons. All of the information processing that occurs in the brain depends on interconnected networks of neurons, of which there are estimated to be some 100 billion in the human brain.*

In the three preceding chapters, we have considered a number of the structures, molecules, and cellular processes by which relevant information is obtained and transmitted through the animal body, leading to adjustments in the internal environment and action in the external environment. As animals have become more complex in the course of their evolutionary history, the task of integration and control has similarly become increasingly complex. That it is accomplished so successfully in all vertebrates can be attributed to an accompanying evolutionary trend, the increased centralization of integration and control in one dominant processing center, the brain.

This trend has produced a structure that, in humans, weighs about 1,400 grams (3 pounds), has the consistency of semisoft cheese, and is, by far, the most complex and highly organized structure on this planet. Its functions are essential not only for the integration and control of the multitude of physiological activities that occur throughout the body but also for those processes that we identify as "mind"—consciousness, perception and understanding of information from the external environment, thought, memory, and the variety of emotions that characterize human experience. Although our knowledge of the structure and function of the vertebrate brain—and particularly of the human brain—is increasing rapidly, it is such a complex system that many think we shall never fully understand it.

The soft substance of the brain, upon which all of its functions depend, is well protected by the bones of the skull. Like the spinal cord, it consists of white matter—the fiber tracts, made white by their lipid-rich myelin sheaths—and gray matter. The gray matter contains not only the cell bodies of as many as 100 billion neurons but also numerous supporting cells of the neuroglia. In some areas of the human brain, neurons and neuroglial cells are so densely packed that a single cubic centimeter of gray matter contains some 6 million cell bodies, with each neuron making synaptic connections to as many as 80,000 others.

THE STRUCTURAL ORGANIZATION OF THE BRAIN: AN EVOLUTIONARY PERSPECTIVE

The vertebrate brain had its evolutionary beginning as a series of three bulges at the anterior end of the hollow, dorsal neural tube. In human embryonic development, as we shall see in Chapter 45, this history is repeated as a groove at the surface of the early embryo closes, giving rise to the tubular structure from which the central nervous system—the brain and spinal cord—will develop. The cavities, known as ventricles, within the three anterior bulges persist in the mature brain and are filled with the same cerebrospinal fluid that fills the interior of the spinal cord, which develops from the posterior portion of the neural tube. In lower vertebrates, the three anterior bulges retain their linear arrangement,

43-2 *A dorsal view showing the linear organization of the vertebrate brain. The brain has been cut horizontally to show the ventricles (cavities within the brain), which are continuous with the interior of the spinal cord. Like the central canal of the spinal cord, the ventricles are filled with fluid. The location of the cerebellum, which is a dorsal projection from the hindbrain, is indicated by the dashed line. The pons, another major structure of the hindbrain, is located ventrally and is not visible from this perspective.*

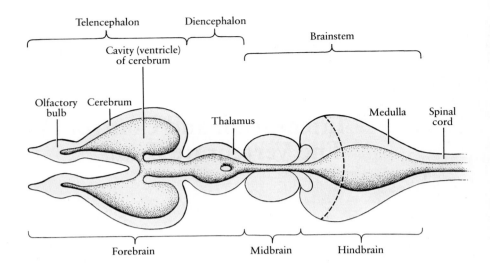

forming the hindbrain, the midbrain, and the forebrain (Figure 43–2). In birds and mammals, they become folded over one another in the course of development, but they can still be identified as distinct regions. Thus the terms hindbrain, midbrain, and forebrain are used to describe the principal regions of all vertebrate brains, including the human (Table 43–1).

TABLE 43-1 **Some Major Structures of the Human Brain**

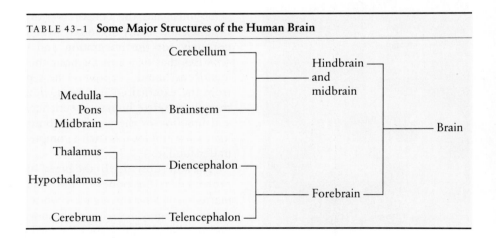

Hindbrain and Midbrain

The hindbrain and midbrain in birds and mammals can be seen as a knobby extension of the spinal cord, the **brainstem,** and a convoluted structure known as the **cerebellum.** The brainstem is the old brain, evolutionarily speaking, and is surprisingly similar from fish to *Homo sapiens* (Figure 43–3). Like the spinal cord, it contains nuclei (clusters of neuron cell bodies) involved with reflexes. Centers in the medulla, the posterior portion of the brainstem, control heartbeat and respiration, among other functions, which is why a blow to the base of the skull is so dangerous. The brainstem also contains sensory and motor neurons that serve the skin, muscles, and other structures of the head, as well as all the nerve fibers that pass between the spinal cord and the higher brain centers. Many of these fiber tracts cross over in the brainstem, so that the right side of the brain receives messages from and sends signals to the left side of the body, and vice versa. This crossing over is one of the major organizational features of all vertebrate nervous systems, but its evolutionary origin and significance remain unknown.

43–3 *The brains of five vertebrates. The brainstems (indicated in dark orange) include the medulla, pons, and midbrain. They are approximately the same in the different vertebrate groups. However, the cerebrum has become progressively larger in the course of evolution, and its two segments have folded upward, forming the two cerebral hemispheres. The cerebral cortex, the outer surface of the cerebral hemispheres, reaches its greatest development in the primates, particularly* Homo sapiens. *The olfactory bulb, readily visible in most vertebrate brains as a stalked knob at the anterior tip, is hidden in the human brain by the much more developed cerebrum.*

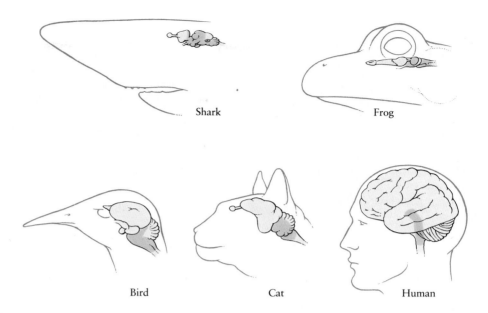

Shark

Frog

Bird

Cat

Human

The cerebellum, a dorsal outgrowth of the primitive hindbrain, is concerned with the execution and fine-tuning of complex patterns of muscular movement. It is much larger in homeotherms than in the more slow-moving fish and reptiles and reaches its greatest relative size in birds, in which it is associated with the exquisite coordination necessary for flight. A ventral enlargement of the hindbrain, the pons ("bridge") contains fibers that provide communication between the left and right portions of the cerebellum, as well as ascending and descending fiber tracts. Auditory information is also relayed through the pons.

In the lower vertebrates, a major part of the midbrain is made up of the optic lobes, which receive fibers of the optic nerves. In mammals, the analysis of visual information has become a function of the forebrain, and the midbrain serves primarily as a relay-and-reflex center.

Forebrain

As you can see in Figure 43–2, the primitive forebrain is divided into two major parts; these are called the **diencephalon** and the **telencephalon.** The diencephalon, which contains the thalamus and the hypothalamus, is a major coordinating center of the brain. The thalamus, two egg-shaped masses of gray matter tucked within the cerebrum, constitutes the main relay center between the brainstem and the higher brain centers. Its nuclei process and sort sensory information. The hypothalamus, lying just below the thalamus, contains nuclei responsible for coordinating the activities associated with sex, hunger, thirst, pleasure, pain, and anger. As we have seen in earlier chapters, it contains the mammalian thermostat (page 784) and is the source of the hormones ADH and oxytocin, which are stored in and released from the posterior lobe of the pituitary (page 828). Most important, it is the major center for integration of the nervous and endocrine systems, acting through its release of peptide hormones that regulate the secretion of tropic hormones from the anterior pituitary (page 826).

The telencephalon ("end brain") is the most anterior portion of the brain and the structure that has changed the most in the course of vertebrate evolution. In the most primitive vertebrates, the fishes, it is concerned almost entirely with olfactory information and is known as the rhinencephalon, or "smell brain." In

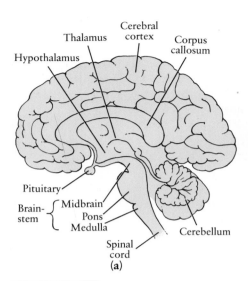

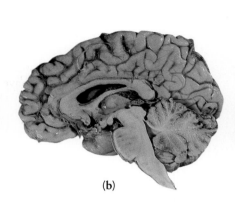

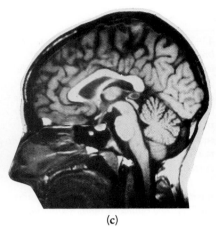

(b)   **(c)**

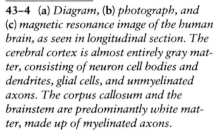

43–4 (a) *Diagram,* (b) *photograph, and* (c) *magnetic resonance image of the human brain, as seen in longitudinal section. The cerebral cortex is almost entirely gray matter, consisting of neuron cell bodies and dendrites, glial cells, and unmyelinated axons. The corpus callosum and the brainstem are predominantly white matter, made up of myelinated axons.*

The image in (c) *is that of an intact, living brain. Magnetic resonance imaging (MRI) is one of several new diagnostic tools available to neurologists. Like computerized axial tomography (CAT scans), MRI provides detailed images of soft tissues without requiring the injection of dyes or radioactive substances. The image results from a computerized summation of radio frequency signals emitted by the nuclei of hydrogen, sodium, and phosphorus atoms in the tissues when they are exposed to an intense magnetic field.*

reptiles, and especially birds, the most prominent structure of the telencephalon is the corpus striatum, which is involved in the control of complicated stereotyped behavior. In mammals, the **cerebrum,** the central portion of the telencephalon, is folded up into the two cerebral hemispheres and is greatly increased in size in relation to other parts of the brain. This increase reaches its greatest extent (so far) in the human brain, in which the many folds and convolutions of the surface of the cerebrum, the **cerebral cortex,** greatly increase its surface area. In humans, the cerebrum occupies 80 percent of the total brain volume. The wrinkling and folding of the cortex allows its enormous area of 2,500 square centimeters to fit within the confines of the skull. The cerebral hemispheres are connected to each other by a tightly packed, relatively large mass of fibers called the **corpus callosum.**

BRAIN CIRCUITS

In the brain, as elsewhere in the body, there is a division of labor, with different parts of the brain performing different, specific functions. However, integration and control of the multitude of processes occurring in an animal's body depend on coordination of all of the activities occurring in the different parts of the brain. Information is exchanged between different regions of the brain by way of diffuse tracts of bundled axons. Thus, a local network of neurons in one region of the brain can have its activity modified by—and can modify the activity of—networks of neurons located elsewhere in the brain. Two examples of such integration involve the reticular activating system and the limbic system.

43–5 *A brain, viewed from above. The many convolutions of the cerebral cortex are clearly visible. By these, you can immediately distinguish this brain as human. The brain and spinal cord are enclosed in three layers of membranous tissue, known as the meninges. The outermost layer, the dura mater, which has begun to peel away from the surface of the brain, is visible in this photograph.*

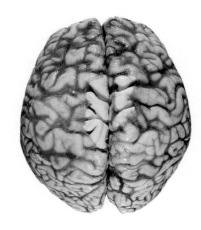

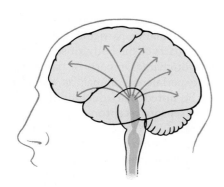

43-6 *The reticular formation is a diffuse network of neurons in the brainstem. Here, incoming stimuli are monitored and analyzed, leading to modulation of the activity of other areas of the brain. The reticular activating system includes the reticular formation and its thalamic extension. It is concerned with general alertness and the direction of attention.*

The Reticular Activating System

The reticular activating system is made up of the reticular formation, a core of tissue running through the brainstem (Figure 43-6), and neurons in the thalamus that function as an extension of this system. It is of particular interest because it is involved with arousal and with that hard-to-define state we know as consciousness. All the sensory systems have fibers that feed into this system, which apparently filters incoming stimuli and discriminates the important from the unimportant. Stimulation of the reticular activating system, either artificially or by incoming sensory impulses, results in increased electrical activity in other areas of the brain.

The existence of such a filtering system is well verified by ordinary experience. A person may sleep through the familiar blare of a subway train or a loud radio or TV program but wake instantly at the cry of the baby or the stealthy turn of a doorknob. Similarly, we may be unaware of the contents of a dimly overheard conversation until something important—our own name, for instance—is mentioned, and then our degree of attention increases.

The Limbic System

The limbic system is a network of mostly subcortical ("below the cortex") neurons that form a loop around the upper part of the diencephalon, linking the hypothalamus to the cerebral cortex and to other structures as well. It is thought to be the circuit by which drives and emotions, such as hunger, thirst, and desire for pleasure, are translated into complex actions, such as seeking food, drinking water, or courting a mate. As we shall see later in this chapter, it is also a principal circuit in the consolidation of memory. The structures of the limbic system are phylogenetically primitive, corresponding to the telencephalon (rhinencephalon) of reptiles.

THE CEREBRAL CORTEX

The cerebral cortex is a thin layer of gray matter about 1.5 to 4 millimeters thick, covering the surface of the cerebral hemispheres. It is the most recent development in the evolution of the vertebrate brain. Fish and amphibians have no cerebral cortex, and reptiles and birds have only a rudimentary indication of a cortex. More primitive mammals, such as rats, have a relatively smooth cortex. Among the primates, however, the cortex becomes increasingly complex. Of the approximately 100 billion nerve cells in the human brain, about 10 billion are estimated to be in the cerebral cortex.

In *Homo sapiens* and other primates, each of the cerebral hemispheres is divided into lobes by two deep fissures, or grooves, in the surface. The principal fissures are the central sulcus, which runs down the side of each hemisphere, and the lateral sulcus. There are four lobes—frontal, parietal, temporal, and occipital—on each hemisphere (Figure 43-7).

Although much of the functioning of the cerebral cortex remains poorly understood, it is the most thoroughly studied region of the human brain—and, inherently, the most fascinating. We shall therefore consider it in some detail, devoting particular attention to some of the more provocative findings about its divisions of labor.

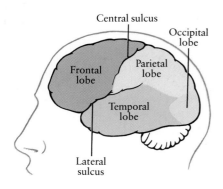

43-7 *The principal sulci (fissures) and lobes of the human cerebral cortex.*

Central sulcus

Occipital lobe

Parietal lobe

Frontal lobe

Temporal lobe

Lateral sulcus

Motor and Sensory Cortices

Certain areas of the cerebral cortex have been mapped in terms of the functions they perform (Figure 43-8). Some of the information is based upon observation of human patients in whom particular areas of the cortex have been destroyed by

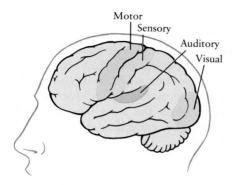

43–8 *The human cerebral cortex, showing the location of the motor and sensory areas, on either side of the central sulcus, and the auditory and visual zones. The motor and sensory cortices span the brain like earphones. Functionally, the left and right motor and sensory cortices are mirror images, with the left cortex receiving signals from and sending signals to the right side of the body, and vice versa.*

disease or accident, and some of it is derived from surgical procedures performed on experimental animals. Other studies involve stimulating particular areas of the cortex and observing what takes place in various parts of the body, or stimulating various sensory receptors and noting electrical discharges in parts of the cortex. Such investigations have been carried out in experimental animals as well as in humans undergoing brain surgery. It should be noted that touching the brain produces no sensation of pain because it has no sensory receptors.

These studies have shown that the area just anterior to the central sulcus, in the frontal lobe, contains neurons concerned with integration of activities performed by the skeletal muscles. Each point on the motor cortex, as it is called, is involved in the movement of a different part of the body. The relative amount of cortex allocated to a particular group of muscles varies from animal to animal. For instance, in humans, the areas controlling the hand and fingers are very large (Figure 43–9), whereas the area controlling a cat's paw is somewhat smaller and that for a horse's hoof is very small indeed.

Immediately posterior to the central sulcus, in the parietal lobe, is the sensory cortex. It is involved with the reception of tactile (touch) stimuli, as well as stimuli related to taste, temperature, and pain. Its greatest representation is for those parts of the body that are most richly endowed with sensory receptors: fingertips, tongue, lips, face, genitalia.

In the temporal lobe, partially buried within the lateral sulcus, is the auditory cortex (see Figure 43–8). This region of the cortex is the processing center for the

43–9 *A combined cross section of one hemisphere of the human cerebrum, indicating the functional areas of the motor and sensory cortices. The motor and sensory cortices are located on either side of the central sulcus in each hemisphere. The motor cortex is indicated in black; stimulation of these areas causes responses in corresponding parts of the opposite side of the body. The sensory cortex is in color; stimulation of various parts of the body produces electrical activity in corresponding parts of the opposite side of this cortex. Notice the relatively huge motor and sensory areas associated with the hand and the mouth. This map is based largely on studies done by neurosurgeon Wilder Penfield on patients undergoing surgical treatment for epilepsy.*

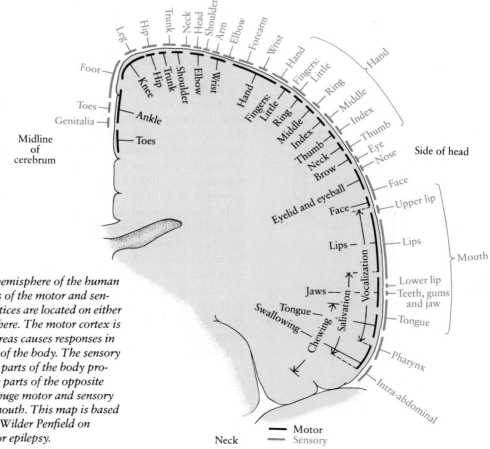

signals relayed from the sensory neurons of the ear. As you will recall from the last chapter, different hair cells of the ear, and thus different sensory neurons, are stimulated by different frequencies (pitches) of sound. The signals from the different sensory neurons are relayed to different regions of the auditory cortex, which can be mapped according to the frequencies eliciting the strongest response.

The visual cortex occupies the occipital lobe. By using a tiny point of light to stimulate very small regions of the retina, one after the other, investigators have been able to show that each region of the retina is represented by several corresponding but larger regions of the visual cortex. The fovea (page 866), which represents about 1 percent of the area of the human retina, projects to nearly 50 percent of the visual cortex. This enormous over-representation, combined with the very substantial portions of the motor and sensory cortices devoted to the hands, provide additional evidence of the importance of eye-hand coordination in primate evolution.

The Perception of Form

How do we perceive form? It was long thought that we perceive form simply because of the way the visual image is arranged on the visual cortex. However, numerous studies conducted over the past 30 years have revealed that a quite different mechanism is involved. Each of the regions of the visual cortex to which different regions of the retina project contains a variety of cells, different groups of which respond to different types of visual stimuli.

This feature of visual processing was first revealed in experiments performed by David H. Hubel and Torsten N. Wiesel at Harvard University. They used microelectrodes to record the responses of individual cells in the visual cortex of the cat to different visual stimuli and obtained some very surprising results (Figure 43–10). One class of cortical neurons, for example, responded only to a horizontal bar; as the bar was tipped away from the horizontal, discharges from the cells slowed and ceased. Another class responded best to a vertical bar, ceasing to fire

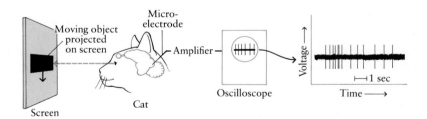

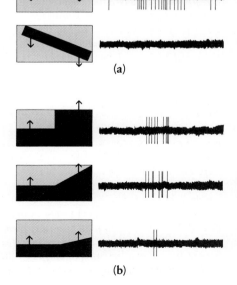

43–10 *The experimental method of Hubel and Wiesel. A cat with a small electrode recording from a single cell in its visual cortex observes a screen on which a moving shape is projected. (a) and (b) represent tracings from two different neurons; the arrows indicate the direction of movement of the shape on the screen. The first* *neuron (a) responds when the cat looks at a horizontal bar moving downward; if the moving bar is tilted rather than horizontal, the action potentials cease. The second neuron (b) responds to the upward movement of a right angle. Note the reduction in response as the angle increases.*

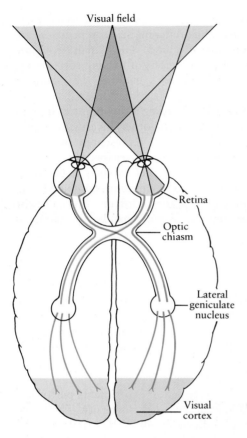

43–11 *The neural pathways leading from the retina to the visual cortex, as viewed from below. Axons of the ganglion cells travel from the retina to the lateral geniculate nuclei by way of the optic nerves. In the lateral geniculate nuclei, the ganglion cell axons synapse with other neurons that carry the signals to neurons of the visual cortex. The optic chiasm is a structure formed by the crossing over of axons conveying information from half of the visual field of each eye. Because of this crossing over, the right and left areas of the visual cortex each "see" only the opposite half of the visual field.*

as the bar was turned toward the horizontal. A third class fired only when the bar was moved from left to right; another only in response to a right-to-left movement. Another class of cells turned out to be a right-angle detector; such cells responded only to an angle moving across the visual field and were most excited when the angle was a right angle.

As we saw in the last chapter, considerable processing of visual information occurs in the retina before its transmission to the ganglion cells, which carry the information into the brain. In a region of the thalamus known as the lateral geniculate nucleus, the ganglion cells synapse with other neurons that relay the signals to the visual cortex (Figure 43–11). The lateral geniculate nucleus, however, is not merely a relay station; it is the site of further processing of the visual information. Its neurons apparently sort the signals representing different aspects of the visual image and then send them on to specific neurons of the visual cortex. What we see is not a direct image but rather a mental picture computed by the brain from a vast amount of coded information concerning the individual features that collectively form shapes, spatial relationships, and patterns of movement, light and dark, and color.

Left Brain/Right Brain

For more than 100 years it has been known that injury to the left side of the brain often results in impairment or loss of speech (aphasia), whereas a corresponding injury to the right side of the brain usually does not. Two areas in the left cerebral hemisphere concerned with speech have been mapped, primarily through work with patients who have had a stroke (an occlusion of the blood supply to a particular portion of the brain). These are known as Broca's area and Wernicke's area (Figure 43–12), each named for the nineteenth-century neurologist who first identified it. About 90 percent of all right-handed people and 65 percent of all left-handed people have these speech areas in the left cerebral cortex.

Broca's area is located just anterior to the region of the motor cortex that controls movements of the muscles of the lips, tongue, jaw, and vocal cords. Damage to this area results in slow and labored speech—if speech is possible at all—but does not affect comprehension. Wernicke's area is adjacent to and partially surrounds the auditory cortex. Localized damage in Wernicke's area results in speech that is fluent but often meaningless, and comprehension of both spoken and written words is impaired.

Because the left hemisphere in most people controls both language and the more capable hand, it has traditionally been regarded as dominant, with the right hemisphere described as mute, minor, or passive. However, severe perceptual and spatial disorders are now known to result from injury to the right hemisphere. For example, a person who has sustained such an injury may have difficulty orienting himself in space, may get lost easily and have difficulty finding his way in new or unfamiliar areas, and may have trouble recognizing familiar faces or voices.

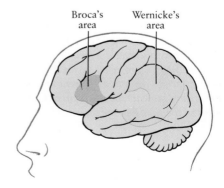

43–12 *The human cerebral cortex, showing the areas associated with language. Damage to Broca's area—the more anterior area—affects speech. Damage to Wernicke's area affects comprehension of language. These two areas of the left cerebral cortex are connected by a fiber tract.*

Musical talent also appears to reside in the right hemisphere: damage to the right hemisphere may result in loss of musical ability, leaving speech unimpaired. Alexander Luria, the eminent Russian neuropsychologist, described a composer whose best work was done after he became aphasic following a massive left hemisphere stroke.

The acquisition of different functions by the two cerebral hemispheres is seen as another way of increasing the functional capacity of the brain without increasing the size of the skull (which is calculated to be as large as possible in relation to the size of the birth canal). There is some evidence that this lateralization of function—that is, the differentiation of function between the two cerebral hemispheres—is part of the developmental process. It is well known, for instance, that in a young child areas of the right hemisphere can take over following damage to the left hemisphere, and completely normal speech may develop. The ability of the right hemisphere to assume this left-hemisphere function is closely correlated with the age at which the injury occurs.

It appears that songbirds also have some lateralization of function in their brains. Studies of songbirds indicate that their musical ability resides in the left hemisphere. Fernando Nottebohm at Rockefeller University has shown, for instance, that canaries with lesions in the left hemisphere sing a grossly distorted song. Canaries with right-hemisphere lesions show only minor changes in song production. Also, as in children with left-hemisphere damage, the right hemisphere of canaries can, with time, take over the functions of the damaged left hemisphere.

The song nuclei—the regions of the canary brain controlling song—can be readily identified anatomically. Nottebohm has found that the size of these regions varies from bird to bird and from season to season. Males with small song centers generally have small repertoires, and the song centers of females, which sing very little, are one-fourth the size of those of males. The size of the nuclei increases in the early spring and then decreases in the summer and fall.

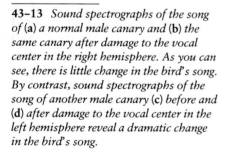

43–13 *Sound spectrographs of the song of* (a) *a normal male canary and* (b) *the same canary after damage to the vocal center in the right hemisphere. As you can see, there is little change in the bird's song. By contrast, sound spectrographs of the song of another male canary* (c) *before and* (d) *after damage to the vocal center in the left hemisphere reveal a dramatic change in the bird's song.*

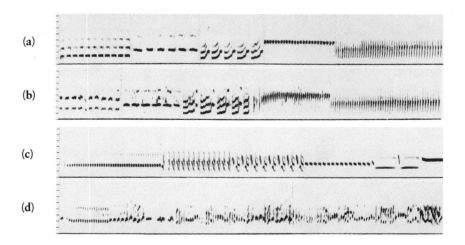

(a)

(b)

(c)

(d)

Split Brain

As we mentioned previously, the two cerebral hemispheres are connected by the corpus callosum. In some cases of epilepsy, severing the corpus callosum lessens the severity of the epileptic attacks. In the 1960s, Roger Sperry and his coworkers at the California Institute of Technology launched a series of studies on such patients, who came to be known as split-brain patients. The name arose because, as Sperry and his associates showed, once the corpus callosum was severed, the

Electrical Activity of the Brain

Many of the neurons of the brain fire action potentials spontaneously and continuously. This electrical activity can be monitored by measuring the difference in electric potential between an electrode placed on a specific area of the scalp and a "neutral" electrode placed elsewhere on the body or between pairs of electrodes on the head. The voltages that can be detected at the scalp are considerably less than the change in potential that occurs in a single action potential, which is about 110 millivolts. From the scalp of a normal adult, about 300 microvolts—only 0.3 millivolt—is the maximum that can be recorded, and thus extremely sensitive recording equipment is required.

An electroencephalogram (EEG) is a record of the electrical activity of the brain. EEG recordings represent the combined activity of large numbers of neurons. Electroencephalography is a useful tool for the diagnosis and monitoring of epilepsy (which is an electrical storm in the brain), cerebral tumors, and brain damage, because characteristic EEG patterns can be correlated with certain levels and types of brain activity. Electroencephalography is also used to detect brain death, which is indicated by the total cessation of electrical activity.

One of the characteristic EEG patterns is the alpha wave, a slow, fairly irregular wave with a frequency of about 8 to 12 cycles per second, recorded at the back of the head. Alpha waves are usually produced during periods of relaxation. In most persons they are conspicuous only when the eyes are shut. About two-thirds of the general population have alpha waves that are disrupted by attention. Of the remaining third, about one-half have almost no alpha waves at all and about one-half have persistent alpha waves that are not easily disrupted by attention. Another EEG pattern is the beta wave. Beta waves are lower amplitude (lower voltage) than alpha waves, but their frequency is greater, from 18 to 32 cycles per second. Beta waves occur in bursts and are associated with mental activity and excitement.

Sleep is characterized by repeated cycles in which the sleeping person passes through distinct stages during which the EEG waves get larger and slower. In the course of eight hours of sleep, a young adult typically passes through five such cycles. During each cycle there is usually a period of rapid, low-amplitude waves similar to those seen in alert persons. This stage, called paradoxical sleep, is associated with rapid eye movements (REMs), in which the eyes make rapid, coordinated movements.

REM sleep differs from the other sleep stages. For example, although there is electroencephalographic evidence of alertness, the muscles are more relaxed than in ordinary deep sleep. Further, although the sleeper is harder to awaken from REM than from deep sleep, once awakened, he or she is more alert. Fluctuations in heart rate, blood pressure, and respiration are also seen during REM sleep. Many male subjects experience erections of the penis during REM sleep, and women have similar erections of the clitoris with secretions of vaginal fluid. A person awakened during a period of REM sleep nearly always reports that he or she has been dreaming.

The significance of REM sleep is not known. At one time, REM sleep was thought to be the period when all dreaming occurred. On the basis of more recent work, it has been suggested that dreaming can

two hemispheres of the brain in these patients were functionally separated. In general, it was found that split-brain patients are able to carry out their normal activities, but under controlled experimental conditions, they behave as if they have two separate brains. If such patients are asked to identify by touch objects that they cannot see, for instance, they can name those they can feel with the right hand but not those they can feel only with the left hand. This is apparently because information from the right hand goes to the left cerebral hemisphere, where the speech centers of the brain are located, and information from the left hand goes to the right brain, which is generally mute. If, for instance, a patient's left hand is given a plastic object shaped like the number 2, which the patient cannot see, he or she is unable to identify the object verbally but can readily tell the experimenter what it is by extending two fingers.

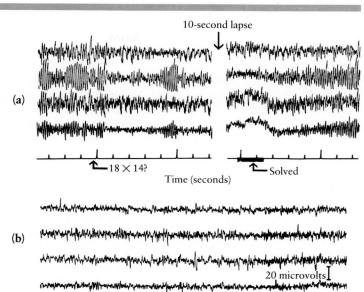

10-second lapse

(a)

18 × 14? Solved

Time (seconds)

(b)

20 microvolts

(a) *EEG of a student with conspicuous alpha rhythm. Electrodes were placed at four positions on the head. When the student was asked to multiply 18 × 14, the alpha rhythm was suppressed and then resumed after the problem was solved (10 seconds of the recording are omitted).* (b) *EEG of a student with a complete absence of alpha rhythms. According to a recent study, people who have almost no alpha rhythms under any conditions think almost exclusively by visual imagery, whereas people with persistent alpha rhythm tend to be abstract thinkers.*

occur at any time during sleep but that conditions for recalling dreams are most favorable when subjects are awakened during REM sleep. It has also been proposed that REM sleep serves as an information-processing period during which data from the previous waking cycle are sorted, processed, and stored. Supporting the hypothesis that this stage of sleep is essential to the memory process are studies showing that laboratory rats forget tasks they have learned if they are deprived of REM sleep. Similarly, evidence indicates that the student who stays up all night cramming for an exam will not have as good a recollection of the material studied as the student who studies and then sleeps.

The EEG measures the electrical activity of the cerebral cortex as recorded at the surface of the scalp. Within the last decade, technological advances have made possible the detection and localization of electrical activity deep within the human brain. The principle involved is well known to students of physics: any flow of electricity is accompanied by a magnetic field. The instrument used is called the SQUID (for Superconducting Quantum Interference Device). Brain tissue is essentially transparent to magnetic fields, so that by aiming two or more magnetic detectors at a source within the brain, it is possible to record and plot the source of activity from subcortical structures. This technique is presently used principally by clinical neurologists to locate the small areas within the brains of epileptics where their seizures originate. In addition, it appears to be an important new tool for correlating discrete areas of electrical activity in the brain with particular mental processes.

In split-brain patients, if a picture is briefly flashed before them while their eyes are held in a fixed position, the right hemisphere sees only the left side of the picture and the left hemisphere sees only the right side (see Figure 43–11). If a word is flashed to the left hemisphere, the patient is able to identify and write the word correctly. If the word is flashed to the right hemisphere, the patient is unable to speak or write the word. However, such a person is quite able to select an object corresponding to the word with his left hand from a group of objects hidden from the left hemisphere by a screen bisecting the visual field.

Sperry's findings created a flurry of interest in the psychological consequences of these "two realms of consciousness," but many authorities now believe that the popular speculations about them have far outstripped the hard evidence.

Intrinsic Processing Areas

The unmapped areas of the cerebral cortex were once known as the "association," or "silent" cortex. These terms were introduced in the course of the early mapping studies, when these areas did not respond in any specific way to stimulation and were thought to function as a sort of giant switchboard, interconnecting the motor and sensory cortices. More recent studies have shown, however, that the organization of the brain is more vertical than horizontal. For example, severing the motor from the sensory cortex by deep vertical cuts appears to have little, if any, effect on an animal's behavior. Anatomical studies support the interpretation that communication between the sensory and motor areas of the cortex takes place primarily via lower brain centers, particularly the thalamus.

The regions of the cortex traditionally considered to form the association cortex—all of the areas not mapped in Figures 43–8 and 43–12—are now more accurately described as intrinsic processing areas. This means that although they receive and process information from neurons in other areas of the brain, they do not receive directly relayed sensory information of the type, for example, that travels from the retina to the visual cortex by way of the lateral geniculate nuclei. Similarly, they transmit information to neurons in other areas of the brain but not directly to neurons leading out of the brain. Certain intrinsic processing areas do, however, receive information from the visual, auditory, or sensory cortices, and others transmit information to the motor cortex.

Mapping of the intrinsic processing areas is a considerably more complex undertaking than the mapping of the sensory and motor cortices, and this work is still in its earliest stages. However, using a variety of techniques, including the actual physical tracing of neurons and their connections, neurobiologists are beginning to gain some insight into the activities of these vast areas of the cortex. For example, it is now known that the posterior region of the parietal lobe and the lower portion of the temporal lobe receive signals transmitted by neurons of the adjacent visual cortex and are involved in the further processing of visual information (Figure 43–14). As you will recall (page 887), the visual cortex contains sets of very precisely tuned neurons, each receiving information about one specific feature of an object in the visual field. Current evidence indicates that the processing of all of these bits of information into a mental image of the whole object occurs in the lower portion of the temporal lobe. However, processing of the bits of information that locate the object in space—and identify its spatial relationships to other objects—occurs in the posterior region of the parietal lobe. A number of studies suggest that similar divisions of labor may also be involved in the processing of other types of sensory information.

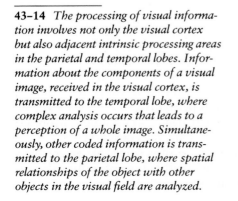

43–14 *The processing of visual information involves not only the visual cortex but also adjacent intrinsic processing areas in the parietal and temporal lobes. Information about the components of a visual image, received in the visual cortex, is transmitted to the temporal lobe, where complex analysis occurs that leads to a perception of a whole image. Simultaneously, other coded information is transmitted to the parietal lobe, where spatial relationships of the object with other objects in the visual field are analyzed.*

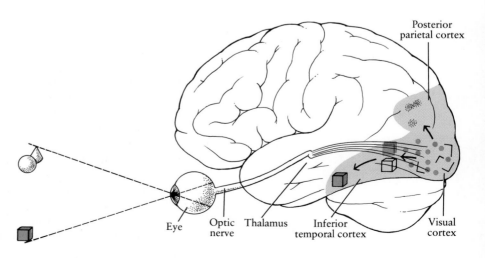

Posterior
parietal cortex

Eye   Optic
nerve   Thalamus   Inferior
temporal cortex   Visual
cortex

The proportion of the cerebral cortex devoted to intrinsic processing is much higher in primates than in other mammals and is very large in humans. This suggests that, in addition to their role in the processing of sensory information, these areas have something to do with what is special about the human mind. Also, about half of the total area involved is in the frontal lobes, the part of the brain that has developed most rapidly during the recent evolution of *Homo sapiens.* The frontal lobes are responsible for our high forehead, as compared with the beetle brow of our most immediate ancestors. Public appraisal of the function of the frontal lobes is reflected in the terms "high brow" and "low brow." The present scientific consensus is that the intrinsic processing areas are concerned with the integration of sensory information with emotion and its retention in memory, with the organization of ideas, which is, of course, a key component of learning, and with long-range planning and "intention." Although it will undoubtedly be a long time, if ever, before we understand the activities in these regions of the brain—which, of course, underlie our very capacity to understand—a variety of studies are beginning to shed light on the kinds of processes that may be occurring there.

LEARNING AND MEMORY

For neurobiologists, perhaps the greatest challenge is to understand the mechanisms of learning and memory. If we define learning as a change in behavior based on experience, and if the functions of the brain—the mind—are to be explained in terms of the structures of which it is composed, then learning—and the memory on which it depends—must involve changes in these structures. But what are these changes, and where and how do they take place?

Almost 60 years ago, the late Karl Lashley set out to locate the physical change, the trace of memory, which he called the engram. He taught rats and other animals to solve particular problems and then performed operations to see whether he could remove the portion of the cortex containing that particular engram. But he never found the engram. As long as he left enough brain tissue to enable the animal to respond to the test procedures at all, he left memory as well. And the amount of memory that remained was generally proportional to the amount of remaining brain tissue. Lashley concluded that memory is "nowhere and everywhere present." With the much more sophisticated techniques now available for brain research, contemporary neurobiologists are just beginning to unlock its hiding places.

There are two distinct types of memory, short-term and long-term. A simple example of short-term memory is looking up an unfamiliar number in a phone book; you usually remember it just long enough to dial it. (In fact, the capacity of short-term memory is about seven items, the number of digits necessary for a local call.) If you call the number enough times, it is transferred to long-term memory. This laying down of long-term memory is analogous to the establishment of a footpath. The more frequently the path is traveled, the better established it becomes. The analogy is strengthened by the familiar experience of consciously retrieving a name, for instance, by seeking out related information that puts one "on the right track." As we shall see, it is now thought that the establishment of such a pathway involves alterations in the synapses by which neurons communicate with one another.

The concept of two kinds of memory is supported by experience with patients with memory deficits; it is possible to lose one kind of memory and not the other. A blow to the head can result, for instance, in the soap-opera kind of amnesia for prior events while not interfering with short-term memory or the establishment of new long-term memories. (Typically, "lost" memories return, indicating that what has been lost is not the memory itself but rather the capacity to retrieve it.)

894

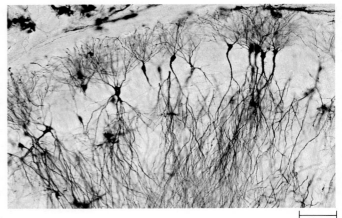

50 μm

43-15 *Neurons of the hippocampus, one of the principal structures involved in the consolidation of memory.*

Conversely, injury to the hippocampus, which is part of the limbic system, does not affect already established long-term memories but does interfere with the transfer of short-term memories to long-term memory. A patient with bilateral destruction of the hippocampus can remember where he lived as a child but not where he lives now, for example. He can carry on an apparently normal conversation, but if the person he was talking to leaves the room and returns a minute or two later, he will not recognize him.

Anatomical Pathways of Memory

Studies in experimental animals and with persons who have experienced memory loss as a result of disease, injury, or as a consequence of brain surgery required to treat disease or injury, have revealed that the hippocampus is only one of several structures involved in memory. Figure 43–16 shows the principal regions in the human brain presently thought to be involved in the consolidation and storage of memory. These regions include the hippocampus ("seahorse," given this name by early neuroanatomists because of its shape) and amygdala ("almond"), both located on the inner surface of the temporal lobe; the thalamus and another structure of the diencephalon, the mammillary body; the basal forebrain, one of the ancient parts of the telencephalon; and a portion of the frontal lobe known as the prefrontal cortex.

According to current hypotheses, information is transmitted along independent pathways from the various sensory cortical areas to the hippocampus and amygdala, from which independent pathways carry the information to the thalamus and the mammillary body. Neurons of the thalamus and mammillary body, in turn, conduct the information to the basal forebrain and to the prefrontal cortex. Parallel circuits transmit processed information in the opposite direction, in what is thought to be a positive feedback process. The basal forebrain, which degenerates in Alzheimer's disease, is a principal source of the neurotransmitter acetylcholine in the brain. The acetylcholine that its neurons release in the feedback circuit is apparently vital for the processes that occur in other parts of the circuit, particularly the amygdala and the hippocampus.

Experimental studies indicate that, in addition to its role as a relay station in these circuits, the amygdala is also the region in which information from different senses is linked—so that, for example, when you think of the seashore, you can remember not only the visual image of sand and waves but also the sound of the lapping waves, the smell and taste of salt spray, and the feel of sand between your toes. And, apparently because of connections between the amygdala and the hypothalamus, the memories have emotional content—the details of which depend on your particular experiences at the seashore. Other fibers of the amygdala appear to communicate back to the primary sensory cortical areas. It is hypothesized that the "storage" of sensory memories in the primary sensory areas is an important element in the subsequent recognition of similar sensory input.

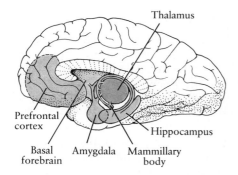

43-16 *Structures of the human brain involved in the consolidation and storage of memory, as shown in longitudinal section. Damage to any of these structures results in memory loss, the details of which vary according to the structure affected. For example, the memory loss associated with strokes typically involves damage to the prefrontal cortex, the thalamus, or the posterior portion of the hippocampus. Korsakoff's syndrome, an amnesia that develops in some chronic alcoholics, involves degeneration in the thalamus and mammillary body. The memory loss associated with Alzheimer's disease involves neurons in the basal forebrain. Inflammation or a temporary interruption of the oxygen supply to the brain can cause damage to the amygdala and the anterior portion of the hippocampus that also results in memory loss.*

Current evidence indicates that the traversing of all of these pathways, including the feedback pathways, is required for the consolidation of long-term memory. Although specific sensory memories appear to be stored in the sensory cortices, more complex memories may be stored elsewhere. The neurosurgeon Wilder Penfield, who was largely responsible for mapping the sensory and motor areas of the cortex, found that in a small percentage of his patients stimulation of certain areas of the cortex—especially in the temporal lobe—evoked particular experiences so vividly that patients reported that they felt they were actually reliving past events. Although there is currently some dispute as to whether Penfield actually evoked memories (as opposed to fantasies or dreams), it has been hypothesized that he may have made contact with memory pathways in the temporal lobe.

The pathways we have traced are those involved in "recognition" memory, which may be a distinct process from the memory that is involved in automatic motor responses to particular stimuli. This second type of memory, which has been termed "habit" or "procedural" memory, underlies such activities as the regurgitation of food by an adult bird in response to the open craw of its offspring or the shifting of your weight as you round a curve on a bicycle. It is thought to depend on a distinct set of pathways, involving the corpus striatum (page 884), an ancient part of the telencephalon.

Synaptic Modification

Although work with human patients and experimental animals is slowly elucidating the pathways through which information travels in the establishment of memory, it leaves unanswered the question of the changes at the cellular and molecular levels that form the "stuff" of memory. Possible answers to this question are beginning to emerge from studies performed by many groups of investigators in which invertebrate animals—or even isolated neurons or groups of neurons—are used as the model organisms. Invertebrates are useful because, first of all, their neurons are large and unmyelinated. Moreover, their nervous systems have far fewer neurons, numbering in the thousands rather than in the billions. Thus, invertebrate behavior circuits are relatively easy to trace—easy, that is, in contrast to the bewildering complexity of the circuits in the vertebrate nervous system and of the behaviors associated with their functioning.

A notable example of such work is that of Eric Kandel and his associates at Columbia University on the gill withdrawal reflex of the sea hare *(Aplysia)* (Figure 43–17). When this mollusk is touched gently on the underside, it quickly withdraws its siphon and delicate gills in a protective reaction. The stimulus, it has been found, causes 24 sensory neurons in the area to fire. These, in turn, activate interneurons and six motor neurons that control the movement of the gill and cause it to contract.

43–17 *The sea hare* Aplysia, *a shell-less mollusk that is shedding new light on the process of learning. The neurons of* Aplysia, *like those of the squid and other invertebrates, are quite large and their axons are unmyelinated. Moreover, its nervous system has far fewer neurons than a vertebrate nervous system. Individual neurons can be identified, their pattern of organization mapped, and microelectrodes inserted into them. Thus, investigators can trace the pathways followed by nerve impulses in response to particular stimuli and monitor the modifications in transmission associated with learning.*

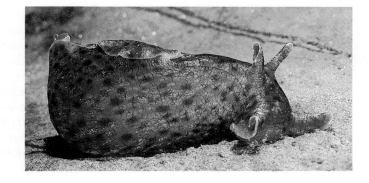

Alzheimer's Disease

Most of us take for granted the reliability of memory, day in and day out, as we engage in the many activities—both mental and physical—that make up our lives. Alzheimer's disease, which is estimated to afflict some 1.5 to 2 million older persons in this country alone, makes clear what a precious gift memory is and how devastating its loss can be.

Alzheimer's disease was first described in 1907 by the German neurologist Alois Alzheimer, who recognized it as a distinct pathology afflicting a very small number of people in their forties and fifties. Characterized by progressive memory loss, the disease ultimately leads to severe dementia, including the inability to think, speak, or perform even the most basic tasks of personal care. Death generally follows within 3 to 10 years. By the 1970s, it had become clear that many cases of "senility" in elderly persons, previously assumed to be an inevitable consequence of age, were actually the same disease first identified by Alzheimer.

The biological changes characteristic of Alzheimer's disease are revealed most strikingly in autopsy studies of brain tissue. Three abnormalities are found consistently: accumulations of tangled and twisted protein filaments within neuron cell bodies; structures known as neuritic plaques, which are clusters of degenerated axon terminals associated with a protein known as amyloid; and accumulations of this same protein adjacent to and within the walls of blood vessels. Although these abnormalities are found in various regions of the cerebral cortex, they are most apparent in the structures associated with memory—the hippocampus and the amygdala. In addition, there is a loss of neurons whose cell bodies are located in nuclei of the basal forebrain, which, as we have noted, are a major source of acetylcholine. The axons of these neurons extend not only to the hippocampus and amygdala but also into many areas of the cerebral cortex. The death of these neurons thus

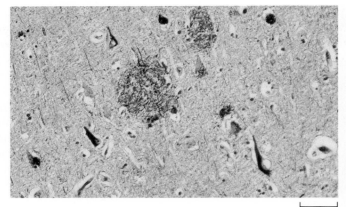

A section of tissue from the hippocampus of an Alzheimer's disease victim. The large round structure in the center (as well as the oval structure at the top) is a neuritic plaque, an aggregation of degenerating axons and amyloid protein. The dark, kite-shaped objects are the cell bodies of neurons containing abnormal tangles of protein filaments. These filaments are insoluble in water and impervious to chemical or enzymatic breakdown; they persist long after the cell containing them has died.

50 µm

If *Aplysia* is touched repeatedly, it becomes habituated to the stimulus and ceases to withdraw. Habituation, which is regarded as a very simple form of learning, is associated with a gradual decrease in the amount of neurotransmitter released by the repeatedly stimulated sensory neurons. This decrease is reflected, in turn, by a decline in the response of the motor neurons controlling the gill. However, if the stimulus is stronger—for example, a jab rather than a gentle stroking—*Aplysia* becomes sensitized to it; the motor response becomes more rapid and emphatic. At the synapse, the effect of sensitization is opposite to that of habituation; that is, there is a gradual increase in the amount of neurotransmitter released by the sensory neurons.

reduces the supply of acetylcholine in the regions of the brain where their axons terminate.

Whether the abnormalities observed with Alzheimer's are the cause of the disease or, instead, its consequence is not yet known. It is also unclear whether Alzheimer's is one disease, with one underlying cause, or whether it is a family of diseases, with several different causes that lead to the same set of pathological changes. There is strong evidence that genetic factors are involved in early-onset Alzheimer's disease and less conclusive evidence that they may also be involved in a significant proportion of the cases that occur much later in life. As a result, considerable research has been devoted to the identification of the gene coding for the amyloid protein associated with the disease. This protein, which consists of 42 amino acids, is synthesized as part of a much larger molecule that exhibits many of the characteristics of a membrane glycoprotein. Knowledge of the amino acid sequence has made possible the synthesis of gene probes, and these probes, along with the analysis of restriction enzyme fragments, have located this gene on chromosome 21. As you will recall (page 385), Down's syndrome is caused by an extra copy of chromosome 21, and perhaps significantly, many persons with Down's syndrome ultimately develop Alzheimer's disease.

Now that the gene coding for the amyloid protein has been identified, studies are underway to explore its function and regulation and to determine whether the protein is a normal or abnormal gene product. The role of external factors—toxic agents, infectious agents, and immunological responses to infection—in the disease process is also being explored. One of the most intriguing suggestions that other factors may be involved is provided by an outbreak of neurological disorders that occurred on the island of Guam in the early 1950s, causing the death of 20 percent of the native population over the age of 25. The disorders included amyotrophic lateral sclerosis (Lou Gehrig's disease), Parkinson's disease, and an Alzheimer-like dementia. Animal studies suggest that this outbreak was triggered by the ingestion of large amounts of an unusual amino acid present in high concentrations in a cycad (page 505) found on Guam. During the Japanese occupation of the island during World War II, food shortages were severe, and the native population subsisted principally on foods prepared from the cycad seeds. After the war ended, the cycad ceased to be a major food source, and since the 1950s, the incidence of neurological disorders has dropped dramatically.

The brain, as we have seen, is an enormously complex structure, in which numerous intricate, interdependent processes occur. To even begin to understand its normal functioning is extraordinarily difficult. To determine the cause or causes of a disease such as Alzheimer's is a staggering task. An article in the journal *Science* in April of 1986 estimated that it could take as long as 5 to 10 years to locate genetic markers for the hereditary form of Alzheimer's disease. And yet, a series of articles published in the same journal less than a year later reported not only the location of such genetic markers but also the cloning of the gene coding for the amyloid protein. It is therefore possible that a great deal more will be known by the time you read this.

Numerous studies with *Aplysia*, with another mollusk known as *Hermissenda*, and with preparations of tissue from the hippocampus of various mammals all support the hypothesis that alterations in the strength of synaptic transmission are critical in memory and learning. These alterations are thought to depend on changes in both the presynaptic and postsynaptic cells. One important element may be the opening or blocking of ion channels that influence the release of neurotransmitter by the presynaptic cell and the degree of depolarization or hyperpolarization of the postsynaptic cell in its resting state. Various models of the possible mechanisms by which these alterations occur and are made to endure are currently being intensively investigated. Among the factors that may be involved are second messengers, including both calcium ion and the ubiquitous cyclic AMP, changes in the pattern of protein synthesis, possibly affecting the numbers and kinds of membrane receptors, and shifts in the location of membrane receptors.

Although the problems of memory and learning are still as bewildering (and as fascinating) as those of human heredity were 40 years ago, neurobiologists appear to be on the threshold of new levels of understanding. Some scientists—Kandel is one of them—believe that the answers will come through a simple model, the equivalent of *Drosophila* and the bacteriophage T4. Others contend that the enormous complexities of the vertebrate brain will never be understood in terms of simple invertebrate models and single cells but rather that the secrets lie in the vast communication network itself. Stay tuned.

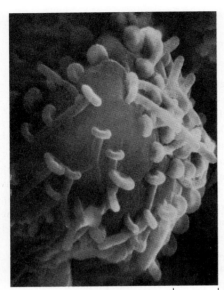

2.5 µm

43–18 *Synaptic knobs of the sea hare* Aplysia, *as shown by the scanning electron microscope. These knobs, which are axon terminals of a number of different presynaptic neurons, are all converging on the cell body of a single postsynaptic neuron. Research with* Aplysia *has demonstrated that changes in the strength of synaptic transmissions play a key role in learning.*

SUMMARY

One of the principal trends in vertebrate evolution has been the increasing centralization of control in one dominant processing center, the brain, which reaches its greatest complexity in *Homo sapiens*. The vertebrate brain develops from three bulges at the anterior end of the hollow, dorsal neural tube, giving rise to the hindbrain, midbrain, and forebrain. The hindbrain consists of the medulla, pons, and cerebellum, the brain structure associated with coordination of fine-tuned movements. The medulla, pons, and midbrain make up the brainstem, which controls vital functions such as heartbeat and respiration and serves as the relay between the spinal cord and the rest of the brain.

The posterior part of the forebrain, the diencephalon, comprises the thalamus, the main relay center between the brainstem and higher brain centers, and the hypothalamus, which contains nuclei associated with basic drives and emotions and is the center for the integration of the nervous and endocrine systems. The anterior part of the forebrain, the telencephalon, consists, in humans, of the two cerebral hemispheres connected by the corpus callosum and covered by the much convoluted cerebral cortex.

Coordination of the activities occurring in different parts of the brain depends on the exchange of information between local networks of neurons. Two major brain circuits involved in such exchanges are the reticular activating system, concerned with arousal and attention, and the limbic system, concerned with translating drives and emotions into actions.

In both experimental animals and humans, it has been possible to correlate particular areas of the cerebral cortex with particular functions. These areas include the motor cortex, sensory cortex, and parts of the cortex concerned with vision, hearing, and speech. In the motor and sensory cortices, the two cerebral hemispheres are mirror images of one another, with the right hemisphere controlling and receiving information from the left side of the body, and vice versa. However, the speech centers are found only in one hemisphere, nearly always the left one, and other faculties, such as spatial orientation and musical ability, appear to be associated with the right hemisphere. Usually the functions of the two hemispheres are integrated, but studies of patients whose corpus callosum has been severed indicate that the two hemispheres can function independently and confirm that they differ in their capacities.

Most of the human cortex has no direct sensory or motor function and consists of areas that receive signals from and transmit signals to neurons in other areas of the brain. Some of these areas participate in the further processing of information transmitted from the primary sensory, auditory, and visual cortices.

Others function in the integration of sensory information with memory and emotion, in the organization of ideas, and in long-term planning. About half of the total area devoted to such intrinsic processing is located in the frontal lobes, the part of the brain that developed most rapidly in human evolution.

Memory and learning are thought to involve both the processing of information through specific anatomical circuits and modifications of synaptic activity. Two types of memory have been identified: short-term memory and long-term memory. The consolidation of long-term "recognition" memory appears to depend on a circuit that includes the hippocampus, amygdala, diencephalon, basal forebrain, and prefrontal cortex. Feedback from this circuit to the primary sensory cortices is thought to play a role in subsequent recognition of similar sensory input. "Habit" or "procedural" memory appears to depend on a different circuit that includes the corpus striatum. Studies with invertebrate model systems indicate that, at the cellular level, memory and learning involve changes at synapses, including alterations in neurotransmitter release by presynaptic cells and the subsequent response by postsynaptic cells.

QUESTIONS

1. Distinguish among the following: brainstem/cerebellum/cerebrum; diencephalon/telencephalon; thalamus/hypothalamus; reticular activating system/limbic system; motor cortex/sensory cortex; Broca's area/Wernicke's area; left brain/right brain; hippocampus/amygdala.

2. Diagram the major structures of the vertebrate brain.

3. Relate the relative sizes of various portions of the brains of the animals in Figure 43–3 with the habitats in which they live and the ways in which they live, obtain food, and escape predators.

4. Diagram the major structures of the human brain.

5. Sketch the human cerebral cortex and indicate on it the areas that have been mapped.

6. What functions might be affected by damage (from stroke, accident, or disease) to the cerebellum? To the reticular formation? To the dorsal portion of the cerebral cortex anterior to the central sulcus?

7. Monkeys that have suffered damage to the amygdala are unable to remember whether an object, however familiar, is edible or inedible. Each time they encounter an object—for example, a banana—they not only look at it, but feel it, smell it, and taste it, before deciding whether to eat it or not. What is a probable explanation of this behavior?

8. Consider the axon of a presynaptic cell that, as a result of repeated stimulation, releases smaller quantities of excitatory neurotransmitter in response to arriving action potentials. What effect would you expect this to have on the initiation of action potentials in the postsynaptic cell?

9. Consider a postsynaptic cell in which membrane ion channels are blocked, with the result that the resting potential of the cell is less negative than it was when the ion channels were open. Will such a cell require a larger or smaller input of excitatory neurotransmitter in order to fire?

The Continuity of Life I: Reproduction

In the preceding chapters of this section, we have been primarily concerned with the ways in which the individual vertebrate body maintains itself. In this chapter and the next, we shall consider the process by which new individuals are produced, maintaining the continuity of the species, and how these individuals develop to become reasonable facsimiles of their parents.

As we saw in Chapters 25 through 28, there is enormous diversity in reproductive and life cycle patterns in the animal kingdom. Most vertebrates—and all mammals—reproduce sexually. As you will recall (page 249), sexual reproduction involves two events: meiosis and fertilization. In vertebrates, which are almost always diploid, meiosis produces gametes, the only haploid forms in the life cycle. The gametes are specialized for motility (sperm) or for production and storage of nutrients (eggs). They are produced in the gonads of individuals of the two separate sexes, male and female. In many invertebrates (insects, in particular), the generations are nonoverlapping, as is the case in annual plants. In vertebrates, however, parents not only survive after their young are produced but often are essential to the survival of the young. This evolutionary trend toward increasing parental care becomes pronounced among birds and reaches its fullest expression among certain of the mammals, ourselves included.

In most species of fish and in amphibians, as in many invertebrates, fertilization is external. Among organisms that lay amniote eggs (reptiles, birds, and monotreme mammals), fertilization is internal. The outer protective shell, produced as

44–1 Lioness and cub.

the egg moves through the female reproductive tract, is laid down after the egg cell is fertilized, enclosing the embryo and its membranes. Fertilization is also internal among marsupial and placental mammals, in which the embryo develops within the mother and is nourished by her.

In the following pages, we are going to describe sexual reproduction in mammals, using *Homo sapiens* as our representative organism. We shall first trace the development of the male gametes, then the development of the female gametes, and, finally, we shall describe the special structures and activities that provide for fertilization and the subsequent implantation of the developing embryo.

THE MALE REPRODUCTIVE SYSTEM

Sperm cells—the male gametes—are produced in the **testes** (singular, testis). The testes develop in the abdominal cavity of the male embryo and, in the human male, descend into an external sac, the scrotum. This descent usually occurs before birth. The function of the scrotum, apparently, is to keep the testes in an environment cooler than the abdominal cavity. A temperature 3°C lower than that of the body is necessary for the sperm to develop. Sperm are not produced in an undescended testis, and even temporary immersion of the testes in warm water—as in a hot bath—has been known to produce temporary sterility. (This effect is not reliable enough, however, to recommend its use as a birth control measure.) When the temperature outside the scrotal sac is warm, the sac is thin and hangs loose in multiple folds. When the outside temperature is cold, the muscles under the skin of the scrotum contract, drawing the testes close to the body. In this way, a fairly constant testicular temperature is maintained.

Spermatogenesis

Each testis (Figure 44–2) is subdivided into about 250 compartments (lobules), and each of these is packed with tightly coiled seminiferous ("seed-bearing") tubules. These are the sperm-producing regions of the testes. Between the tubules are the interstitial cells, the sources of the male sex hormone testosterone. Each seminiferous tubule is about 80 centimeters long, and the two testes together contain a total of about 500 meters of tubules. The sperm are produced continuously within the tubules.

44–2 *The testis* (a) *consists mostly of tightly packed coils of seminiferous tubules, containing* (b) *sperm cells in various stages of development. The entire developmental sequence takes eight to nine weeks. The interstitial cells, which are found in connective tissues between the tubules, are the source of the male hormone testosterone.* (c) *As shown in an idealized cross section of a portion of a seminiferous tubule, spermatogonia develop into cells known as primary spermatocytes. In the first meiotic division, these divide into two equal-sized cells, the secondary spermatocytes. In the second meiotic division, four equal-sized spermatids are formed. These differentiate into functional sperm cells. The Sertoli cells support and nourish the developing sperm. The sperm cells leave the testis through the epididymis and the vas deferens.*

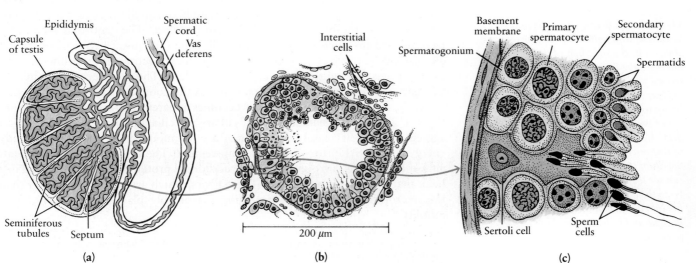

(a) (b) (c)

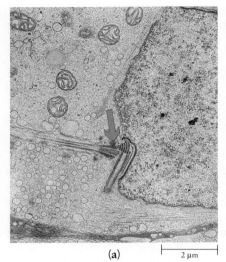

(a)   2 µm

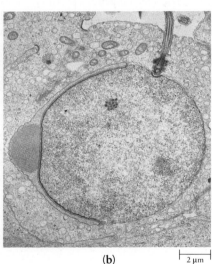

(b)   2 µm

44–3 *Mammalian spermatids in the process of differentiation.* (a) *The nucleus almost fills the right half of the electron micrograph. The two centrioles have moved to a position just behind the nucleus (see arrow), and the flagellum has begun to form from one of them. To the left of the nucleus are several mitochondria.* (b) *The nucleus occupies the center of this micrograph. At the upper right, the flagellum is forming and, at the left, you can see the beginning of the acrosome. Note the complex connecting piece that fits into a notch at the posterior end of the nucleus and links the flagellum to the nucleus.*

The tubules contain two types of cells: spermatogenic (sperm-producing) cells and Sertoli cells. The spermatogenic cells pass through several stages of differentiation. Once production of sperm begins at puberty in the human male, it goes on continuously; thus, in a single seminiferous tubule, it is possible to find cells in all the different stages of spermatogenesis. It is unusual, however, to find all of the stages in a single cross section, because spermatogenesis characteristically occurs in waves that travel down the tubules.

Cells in the first stage of spermatogenesis, the spermatogonia, line the basement membrane of each seminiferous tubule. Spermatogonia are diploid and have, in the human, 44 autosomes and 2 sex chromosomes, an *X* and a *Y*. Spermatogonia divide continuously. Some of the cells produced by these mitotic divisions remain undifferentiated, whereas others, in the course of their successive mitotic divisions, move away from the basement membrane and begin to differentiate, giving rise to primary spermatocytes. Primary spermatocytes undergo the first meiotic division* to produce two secondary spermatocytes, each of which contains 22 autosomes and either an *X* chromosome or a *Y* chromosome; each of the 23 chromosomes consists of two chromatids. The secondary spermatocytes undergo the second meiotic division to produce spermatids, each of which contains the haploid number of single chromosomes. Spermatids develop without further division into sperm cells, or spermatozoa. It takes eight to nine weeks for a spermatogonium to differentiate into four sperm cells. During this time the developing cells receive nutrients from adjacent Sertoli cells.

Differentiation of Spermatids

A spermatid is a small spherical or polygonal ("many-sided") cell that develops into a sperm cell. The sequence of changes by which the quite unremarkable spermatid becomes the highly specialized, very extraordinary sperm cell is an excellent example of cell differentiation. Differentiation is a process that, as we shall see in the next chapter, is an essential component of embryonic development.

The first visible sign of differentiation of a spermatid is the appearance, within the Golgi complex, of vesicles containing small, dark granules. These vesicles enlarge and coalesce into a single vesicle, the acrosome. The acrosome contains enzymes that will help the sperm to penetrate the protective layer surrounding the egg. The position of the acrosomal vesicle determines the polarity of the sperm; that is, it establishes where the anterior end, or "head," is going to be.

During the early stages of acrosome formation, the cell's pair of centrioles moves to the cell membrane at the end of the cell opposite the acrosome. One of these centrioles appears to initiate the assembly of tubulin dimers (page 78) into microtubules—the beginning of the sperm flagellum. The centrioles then move back toward the nucleus, carrying the cell membrane inward with them (Figure 44–3). One centriole eventually forms part of a connecting piece within the neck of the sperm, linking the flagellum to the nucleus. The other centriole, the one that gave rise to the flagellum, disintegrates as the connecting piece develops.

As the flagellum, or tail, grows, it becomes apparent that its axial filament has the 9 + 2 structure characteristic of eukaryotic cilia and flagella. Mitochondria aggregate about its basal end, forming a continuous spiral, providing a ready energy source (ATP) for the flagellar movement. The rest of the axial filament, almost to its tip, is surrounded by nine additional protein fibers tightly coiled in a helix that forms a fibrous sheath. These fibers, which are somewhat thicker than the microtubules within the flagellum, presumably play some role in sperm motility.

---

* For a review of meiosis, see pages 253 to 257. The stages of spermatogenesis are diagrammed in Figure 12–12 on page 258.

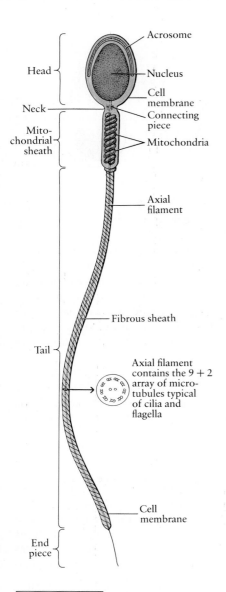

Head

Acrosome

Nucleus

Cell membrane

Neck

Connecting piece

Mito-chondrial sheath

Mitochondria

Axial filament

Fibrous sheath

Tail

Axial filament contains the 9 + 2 array of micro-tubules typical of cilia and flagella

Cell membrane

End piece

44–4 *Diagram of a human sperm cell. The mature cell consists primarily of the nucleus, carrying the "payload" of tightly condensed DNA and associated protein, the very powerful tail (flagellum), and mitochondria, which provide the power for sperm motility. The acrosome is a specialized lysosome (see page 119) containing enzymes that help the sperm penetrate the protective layers of an unfertilized secondary oocyte.*

During this period, the nucleus condenses, apparently by eliminating water. Once the tail has formed, the cell rapidly elongates. Longitudinal bundles of microtubules can be seen in the cell at this time, and they may play a role in changing the shape of the cell. As the cell lengthens, the bulk of the cytoplasm, together with the Golgi complex, is sloughed away.

In its final form, the sperm cell consists of the acrosome, the tightly condensed nucleus, the connecting piece in the neck, the mitochondrial sheath, and the long, powerful flagellum itself, all bounded by the cell membrane (Figure 44–4). In the fully differentiated sperm, all other functions have been subordinated to the task of providing motility for delivery of the "payload," the DNA and its associated protein, condensed and coiled in the sperm head. A young adult human male may produce several hundred million sperm per day; a ram may produce several billion.

Pathway of the Sperm

The pathway of the sperm can be traced in Figure 44–5. From the testis, the sperm are carried to the **epididymis,** which consists of a coiled tube 7 meters long, overlying the testis. It is surrounded by a thin, circular layer of smooth muscle fibers. The sperm are nonmotile when they enter the epididymis and gain motility only after some 18 hours there. (Maximum motility, however, occurs only after they have entered the female reproductive tract.)

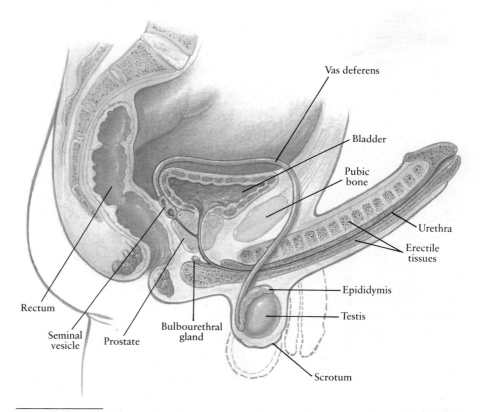

Vas deferens

Bladder

Pubic bone

Urethra

Erectile tissues

Epididymis

Testis

Scrotum

Rectum

Seminal vesicle

Prostate

Bulbourethral gland

44–5 *Diagram of the human male reproductive tract, showing the penis and scrotum before (dashed lines) and during erection. Sperm cells formed in the seminiferous tubules of the testis enter the epididymis. From there they move to the vas deferens, where most of them are stored. The vas deferens merges with a duct from the seminal vesicle and then, within the prostate gland, joins the urethra. The sperm cells are mixed with fluids, mostly from the seminal vesicles and prostate gland. The resulting mixture, the semen, is released through the urethra of the penis. The urethra is also the passageway for urine, which is stored in the bladder.*

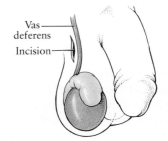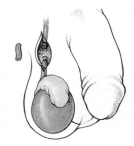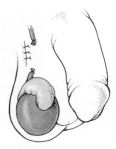

Vas deferens

Incision

44–6 *In a vasectomy, the vas deferens on each side is severed, and the cut ends are folded back and tied off, preventing the release of sperm from the testis. The sperm cells are reabsorbed by the body, and the ejaculate is normal except for the absence of sperm.*

This is a relatively safe and almost painless procedure that does not require a general anesthetic or hospitalization. Because the nerves and blood vessels between the testes and the rest of the body are left intact, the procedure does not affect hormone levels, sexual potency, or performance. Its chief drawback is that it is generally not reversible.

From the epididymis, the sperm pass to the **vas deferens** (plural, vasa deferentia), where most of them are stored. A vas deferens, an extension of the tightly coiled tubules of the epididymis, leads from each testis into the abdominal cavity. The vasa deferentia are covered with a heavy, three-layered coat of smooth muscle whose contractions propel the sperm along. Each vas deferens and its accompanying nerves, arteries, veins, and connective-tissue wrapping constitute a spermatic cord. The spermatic cords, one from each testis, are found along the same path the testes took during their descent in the embryo. Severing of the vasa deferentia (vasectomy) offers a relatively safe and almost painless means of birth control (Figure 44–6).

Within the posterior wall of the abdominal cavity, the vasa deferentia loop around the bladder, where they merge with the ducts of the **seminal vesicles.** The vas deferens from each testis then enters the **prostate gland** and merges with the urethra, which extends the length of the **penis.** The urethra serves both for the excretion of urine and the ejaculation of sperm.

Erection of the Penis and Orgasm in the Male

The function of the penis is to deposit sperm cells within the reproductive tract of the female. The penis, in various forms, has evolved independently in a number of species of insects and in other invertebrates. It is found among some reptiles and birds; all flightless birds have penes and so do all ducks, flightless or not. In most reptiles and birds, however, one opening, the cloaca, serves as the passage for eggs or sperm and also for the elimination of wastes. These animals mate by juxtaposition of their cloacae. Only among mammals is the penis found in all species.

The human penis is formed of three cylindrical masses of spongy erectile tissue, each of which contains a large number of small spaces, each about the size of a pinhead. Two of these masses are in the dorsal portion of the penis, and the third lies beneath them, surrounding the urethra (Figure 44–7). This third mass is enlarged at the distal (far) end to form the glans penis, which is a smooth protective cap over the spongy tissues. At the basal end, it is enlarged to form the bulb of the penis, which is embedded below the pelvic cavity and is surrounded by muscles that participate in orgasm. The exterior portion of the penis is covered by a loose, thin layer of skin, which at the end forms an encircling fold over the glans. This fold, the foreskin, is sometimes surgically removed (circumcision), usually shortly after birth. The urethra terminates in a slitlike opening in the glans.

Erection, which can be elicited by a variety of stimuli, occurs as a consequence of an increased flow of blood that fills the spongy, erectile tissues of the penis. The blood flow is controlled by parasympathetic nerve fibers to the blood vessels leading into the erectile tissues. As the tissues become distended, they compress the veins and so inhibit the flow of blood out of the tissues. With continued stimulation, the penis and the underlying bulb become hard and enlarged.

Erection is accompanied by the discharge into the urethra of a small amount of fluid from the bulbourethral glands, pea-shaped organs at the base of the penis. This fluid serves as a lubricant, facilitating the movement of spermatozoa along the male urethra and aiding penetration of the penis into the female. Continued stimulation of mechanoreceptors in the penis and scrotum (such as may be produced by repeated thrusting of the penis in the vagina) sends an escalating

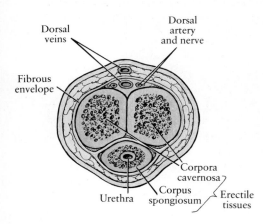

Dorsal veins

Dorsal artery and nerve

Fibrous envelope

Corpora cavernosa

Corpus spongiosum

Urethra

Erectile tissues

44–7 *A cross section of a human penis. The penis is formed of three cylindrical masses of spongy erectile tissue that contain a large number of small spaces, each about the size of a pinhead. Erection of the penis is caused by dilation of the blood vessels carrying blood to the spongy tissues.*

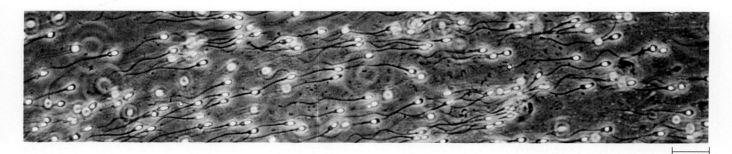

$\vdash$25 µm$\dashv$

44-8 *Human sperm. About 300 to 400 million sperm cells are present in the ejaculate of an average, healthy adult male.*

series of nerve impulses through reflex arcs in the lower spinal cord to motor neurons innervating different muscles of the reproductive system. Among the first muscles to contract are those in the scrotum, raising the testes close to the body, and those encircling each epididymis and vas deferens, moving the spermatozoa toward and into the urethra. As this occurs, the seminal vesicles secrete a fructose-rich fluid that nourishes the sperm cells. This fluid contains a high concentration of prostaglandins (page 833), which stimulate contractions in the musculature of the uterus and oviducts and so may assist the sperm in reaching the egg. The prostate gland adds a thin, milky, alkaline fluid that helps neutralize the normally acidic pH of the female reproductive tract. Finally, contractions are initiated in the muscles surrounding the bulb; these contractions propel the sperm and accompanying fluid out through the urethra (ejaculation) and produce some of the sensations associated with orgasm.

The sperm, along with the secretions from the seminal vesicles, the prostate gland, and the bulbourethral glands, constitute semen. The volume of semen measures from 3 to 4 milliliters per ejaculation. Even though sperm constitute less than 10 percent of the semen, about 300 to 400 million sperm cells are present in each ejaculate of a normal adult male. The mortality of sperm in the female reproductive tract is staggering—of the original 300 to 400 million cells, only a few hundred survive to reach the oviducts, where fertilization occurs. Although only one sperm cell needs to make contact with an oocyte for fertilization to occur, the odds are such that fertility requires the release of enormous numbers at one time. Males that produce fewer than 20 million sperm per milliliter of fluid are generally sterile.

The Role of Hormones

In addition to producing the sperm cells, the testes are also the major source of male hormones, known collectively as **androgens.** The principal androgen, testosterone, is necessary for the formation of sperm cells. It is a steroid, produced primarily by the interstitial cells of the testes (see Figure 44–2b). Other androgens are produced in the adrenal cortex (see page 832).

Androgens are first produced in early embryonic development, causing the male fetus to develop as a male rather than a female. After birth, androgen production continues at a very low level until the boy is about 10 years old. Then there is a surge in testosterone, resulting in the onset of sperm production (which marks the beginning of puberty) accompanied by enlargement of the penis and testes and also of the prostate and other accessory organs. In the healthy human male, a high level of testosterone production continues into the fourth decade of life, when it begins gradually to decline.

As can be readily observed, testosterone also has effects on other parts of the body not directly involved in the production and deposition of sperm. In the human male, these effects include growth of the larynx and an accompanying deepening of the voice, an increase in skeletal size, and a characteristic distribu-

(a)

(b)

(c)

(d)

44–9 *Secondary sex characteristics among vertebrates. The male and female animals in each of these photos are actively courting.*

tion of body hair. Androgens stimulate the biosynthesis of proteins and so of muscle tissue. They stimulate the apocrine sweat glands, whose secretions attract bacteria and so produce the body odors associated with sweat after puberty. And they may cause the sebaceous glands of the skin to become overactive, resulting in acne. Such characteristics, associated with sex hormones but not directly involved in reproduction, are known as **secondary sex characteristics.**

In other animals, testosterone is responsible for the lion's mane, the powerful musculature and fiery disposition of the stallion, the cock's comb and spurs, and the bright plumage of many adult male birds (Figure 44–9). It is also responsible for a variety of behavior patterns such as the scent marking of dogs, the courting behavior of sage grouse, and various forms of aggression toward other males found in a great many vertebrate species.

Since almost the beginnings of agriculture, domestic animals have been castrated in order to make them fatter, less tough, and more manageable. Eunuchs (human castrates) were traditionally, and for obvious reasons, used as harem guards, and as recently as the eighteenth century, selected boys were castrated before puberty in order to retain the purity of their soprano voices for church and opera choirs. As a consequence of these early practices, the effects of testosterone —or, more precisely, of its absence—were the first hormonal influences to be recognized and studied.

Regulation of Hormone Production

The production of testosterone is regulated by a negative feedback system involving, among other components, a gonadotropic hormone called luteinizing hormone (LH). LH is produced by the pituitary gland, under the influence of the hypothalamus. It is carried by the blood to the interstitial tissues of the testes, where it stimulates the output of testosterone. As the blood level of testosterone increases, the release of LH from the pituitary is slowed (Figure 44–10).

The testes are also under the influence of a second pituitary hormone, follicle-stimulating hormone (FSH). It acts on the Sertoli cells of the testes and, through them, on the developing sperm. Among the factors involved in the regulation of FSH production is a protein hormone, known as inhibin, that is secreted by the Sertoli cells. This hormone, as you might expect, inhibits FSH production. Its discovery is of more than academic interest; it would appear a likely candidate for the long-sought male contraceptive.

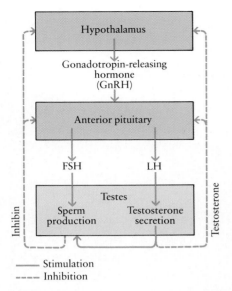

44–10 *Hormonal regulation of testicular function depends on negative feedback. The hypothalamus stimulates the anterior lobe of the pituitary to produce LH (luteinizing hormone), which, in turn, stimulates production and release of testosterone from the interstitial cells of the testes. The production of gonadotropin-releasing hormone (GnRH) by the hypothalamus is inhibited by the resulting increased concentration of testosterone and may also be inhibited by the increased concentration of LH. In addition, testosterone is thought to act directly on the pituitary to suppress the release of LH. As a consequence of these combined inhibitory effects, the secretion of LH by the pituitary is decreased.*

Similarly, in another negative feedback loop, FSH (follicle-stimulating hormone) acts on the Sertoli cells, which produce, in turn, the hormone inhibin. This hormone specifically inhibits FSH production. FSH itself may also act on the hypothalamus to reduce the production of GnRH.

The combined action of testosterone and FSH is required to initiate spermatogenesis.

44–11 *The female reproductive organs. Notice that the uterus lies at right angles to the vagina. This is one of the consequences of the bipedalism and upright posture of* Homo sapiens *and one of the reasons that childbirth is more difficult for the human female than for other mammals.*

In the human male, the rates of testosterone release are fairly constant. In many animals, however, male hormone production is triggered by changes in temperature, daylight, or other environmental cues and changes seasonally. Testosterone production may also be affected by social circumstances. Studies on bulls, for example, have shown that after a bull sees a cow, his blood level of LH rises as much as seventeenfold; within about half an hour, the blood level of testosterone reaches its peak. In many animal societies, including wolves and cape hunting dogs (the wild dogs of the African plains), socially inferior males may never become sexually mature, presumably because of depressed testosterone production. Human testosterone production also may vary according to the emotional climate. In a study made among Army GIs during the Vietnam War, testosterone levels in recruits in basic training and among combat troops were markedly lower than the normal levels of testosterone found in men in behind-the-lines assignments.

Testosterone production is also dramatically affected by the synthetic compounds known as anabolic steroids. These drugs, which are chemical variants of testosterone, were originally developed in Germany in the 1930s in an attempt to produce the muscle-building effects of the natural hormone without its masculinizing effects. Because of their chemical similarity to testosterone, anabolic steroids function as inhibitors in the negative feedback system regulating testosterone production. In adult males, their use can reduce testosterone levels by as much as 85 percent, causing shrinkage of the testes and growth of the breasts. Long-term use of these drugs also greatly increases the risk of kidney and liver damage, liver cancer, and heart disease. In adolescents, anabolic steroids can lead to premature baldness and failure to attain full height.

THE FEMALE REPRODUCTIVE SYSTEM

The female reproductive system is shown in Figure 44–11. The gamete-producing organs are the **ovaries,** each a solid mass of cells about 3 centimeters long. They are suspended in the abdominal cavity by ligaments (bands of connective tissue) and mesenteries. The oocytes, from which the eggs develop, are in the outer layer of the ovary.

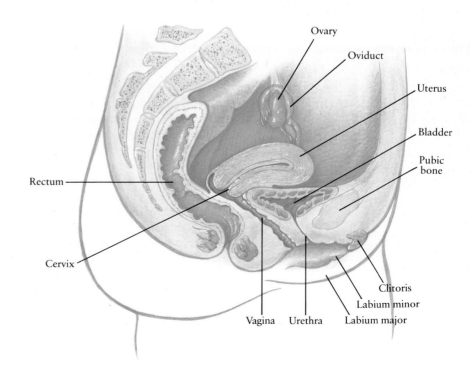

Sex and the Brain

As we saw in the last chapter, male songbirds have a song center that is located in the left cerebral hemisphere. Enlargement of the song center, which is associated with the establishment of territories and the initiation of courtship behavior, is produced by an increase in circulating testosterone. If a male songbird is castrated, the song center does not enlarge and the bird never sings. If the castrated bird is injected with testosterone, the song center enlarges and he begins to sing. Female birds normally do not sing, and both hemispheres are the same size. However, if adult female songbirds are given testosterone, their song centers enlarge and they begin to sing, although their musical repertoires never become as large or varied as those of the males. Thus, in songbirds, there are differences between the brains of male and female animals, and these differences can be traced, in part, to the influence of testosterone on the brain itself.

Rats don't sing, but they do have certain patterns of clearly defined behavior that differ between the sexes. The most obvious—and highly necessary—differences are in mating behavior. When a female rat in estrus (page 915) is exposed to a sexually competent male, she crouches and arches her back, exposing her genitals, in a posture called lordosis. The male mounts her, grabs her flanks, and makes pelvic thrusts that result in ejaculation. Lordosis is eliminated by removal of the ovaries and is restored by administration of estrogens. In fact, lordosis in a female rat depends on the presence of estrogens in cells of the hypothalamus.

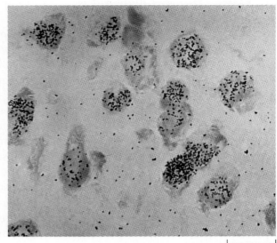

|—————| 50 μm

This autoradiograph of a section of brain tissue from an adult female rat shows the localization of tritium-labeled estradiol in the nuclei of cells of the hypothalamus. The black dots represent the presence of estradiol bound to an estrogen receptor. In rats and many other species, the hypothalamus has been shown to play a key role in the hormonal regulation of sexual behaviors.

Other important structures include the oviducts (sometimes called the Fallopian tubes or the uterine tubes), the uterus, the vagina, and the vulva. The **uterus** is a hollow, muscular, pear-shaped organ slightly smaller in size than a clenched fist (about 7.5 centimeters long and 5 centimeters wide) in the nonpregnant female. It lies almost horizontally in the abdominal cavity and is on top of the bladder. The uterus is lined by the endometrium, which has two principal layers, one of which is shed at menstruation and another from which the shed layer is regenerated. The smooth muscles in the walls of the uterus move in continuous waves. This motion possibly increases the motility of both the sperm on its journey to the oviduct and the oocyte, from which the egg develops, as it passes from the oviduct to the uterus. These contractions increase when the endometrium is shed during a menstrual period, and they are greatest when a woman is in labor. The muscular sphincter guarding the opening of the uterus is the **cervix**. The sperm pass through this opening on their way toward the oocyte. At the time of birth, the cervix dilates to allow the fetus to emerge.

In adult male rats, the normal repertoire of mating behavior ceases if the rat is castrated; the behavior is restored by testosterone injections. Studies using radioactive testosterone have shown that this hormone can also be localized in neurons in certain areas of the brain, particularly in a group of neurons in the hypothalamus. (This same area of the hypothalamus is the source of gonadotropin-releasing hormone.) Injection of testosterone to this area produces an increase in electrical activity of these neurons and an increase in sexual activity. Electrical stimulation of this area also produces an increase in sexual activity. However, administration of estrogens to these same castrated male rats does not produce lordosis or any other sex-specific female behavior. In short, there is a difference between the male and female brain in its response to hormones.

This differentiation between the male and female brain occurs during a critical period in early development. Rats have a 21-day gestation period. About eight days prior to birth, the embryonic male testes begin to secrete testosterone, and this secretion continues until the tenth day after birth. If male rats are castrated on the day of birth (thus depriving them of testosterone for about half the normal period), they never develop male sexual behavior. If they are given testosterone as adults, they do not respond to it. However, if these same adult rats are given estrogens, they show the lordosis behavior in the presence of normal males. By contrast, if male rats are castrated on the tenth day after birth or thereafter, they will not respond to the administration of estrogens. Thus, it is concluded that in rats sex hormones affect the brain in two stages: (1) an early period, during which differentiation of brain cells occurs, and (2) a sexually mature period, in which the differentiated brain cells respond to the hormones by evoking particular patterns of behavior.

Scientists are interested in what other types of behavior might be influenced by sex differences in the cellular organization of the brain. In rhesus monkeys, for instance, the critical burst of testosterone production occurs entirely in utero, around the middle of the 168-day gestation period. Female rhesus monkeys that are masculinized by high doses of testosterone given to their mothers during the critical period show an increase in rough-and-tumble play, an increase in aggressive behavior, and a decrease in maternal imitative behavior. The critical period in humans—the period of the surge in testosterone production—occurs about the second to third month in embryonic development. The question of what aspects of human behavior—if any—may be influenced by sex differences in the brain is a fascinating one to scientists and nonscientists alike. It is, however, generally agreed that even if such differences exist, they are largely overcome by the powerful forces of environment and culture.

The **vagina** is a muscular tube about 7.5 centimeters long that leads from the cervix of the uterus to the outside of the body. It is the receptive organ for the penis and is also the birth canal. Its exterior opening is between the urethra, the tube leading from the bladder, and the anus. The lining of the vagina is rich in glycogen, which bacteria normally present in the vagina convert to lactic acid. As a consequence, the vaginal tract is mildly acidic, with a pH between 4 and 5.

The external genital organs of the female are collectively known as the **vulva.** The **clitoris,** most of which is embedded in the surrounding tissue, is about 2 centimeters long and is homologous with the penis of the male (in the early embryo, the structures are identical). Like the penis, it is composed chiefly of erectile tissue. The clitoris has two bulbs (analogous to the penile bulb of the male) that lie on either side of the opening of the vagina. The **labia** (singular, labium) are folds of skin. The labia majora are fleshy and, in the adult, covered with pubic hair. They enclose and protect the underlying, more delicate structures. (Embryonically, they are homologous with the scrotum in the male.) The labia minora are thin and membranous.

44–12 *An unfertilized human secondary oocyte. The dark strands within the nucleus are chromosomes. A polar body is at the lower right.*

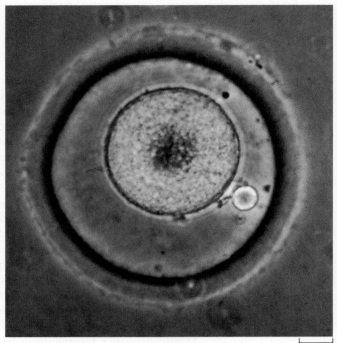

10 μm

Oogenesis

In human females, the primary oocytes begin to form about the third month of fetal development. By the time of birth, the two ovaries contain some 2 million primary oocytes, which have reached prophase of the first meiotic division. These primary oocytes remain in prophase until the female matures sexually. Then, under the influence of hormones, the first meiotic division of a primary oocyte resumes, resulting in a secondary oocyte and a polar body. The first meiotic division is completed at about the time of **ovulation** (the release of the oocyte from the ovary). Of the original 2 million primary oocytes, about 300 to 400 reach maturity, usually one at a time, about every 28 days from puberty to menopause, which typically occurs at about age 50. Given this timetable, a moment's reflection will reveal that more than 50 years may elapse between the beginning and the end of the first meiotic division in a particular oocyte.

Maturation of the oocyte involves both meiosis and a great increase in size. This size increase reflects the accumulation of stored food reserves and metabolic machinery, such as messenger RNA and enzymes, required for the early stages of development. As we saw in Figure 12–13 on page 258, oocytes do not divide into equal-sized cells, as spermatocytes do. Instead, one very large cell (100 micrometers in diameter in humans) is produced. The other nuclei are, in effect, discarded.

When a primary oocyte is ready to complete meiosis, the nuclear envelope fragments, and the chromosomes move to the surface of the cell. As the nucleus divides, the cytoplasm of the oocyte bulges out. One set of chromosomes moves into the bulge, which then pinches off into a small cell, the first polar body. The rest of the cellular material forms the large secondary oocyte. The first meiotic division is completed a few hours before ovulation. The second meiotic division does not take place until after fertilization. This division produces the ovum and another small polar body. As a consequence of these unequal cell divisions, most of the accumulated food reserves of the oocyte are passed on to a single ovum. The first polar body may also divide, although there is no functional reason for it to do so. All the polar bodies eventually die.

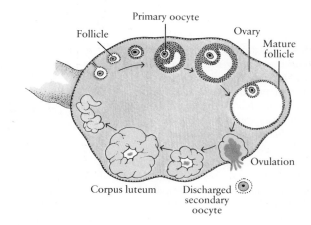

44–13 *Oocytes develop near the surface of the ovary within follicles. After a secondary oocyte is discharged from a follicle (ovulation), the remaining cells of the ruptured follicle give rise to the corpus luteum, which secretes estrogens and progesterone. If the ovum is not fertilized, the corpus luteum is reabsorbed within two weeks. If the ovum is fertilized, the corpus luteum persists for about three months, continuing its production of estrogens and progesterone. These hormones, which are later produced in large quantities by the placenta, maintain the uterus during pregnancy.*

Oocytes develop near the surface of the ovary. An oocyte and the specialized cells surrounding it are known as an **ovarian follicle** (Figure 44–13). The cells of the follicle supply nutrients to the growing oocyte and also secrete estrogens, the hormones that support the continued growth of the follicle and initiate the buildup of the endometrium. During the final stages of its growth, the follicle moves to the surface and produces a thin, blisterlike elevation that eventually bursts, releasing the oocyte (Figure 44–14). Usually a number of follicles begin to enlarge simultaneously, but only one becomes mature enough to release its oocyte, and the others regress.

44–14 *Ovulation. (a) A secondary oocyte, visible only as a tiny speck, begins to emerge from an ovarian follicle. (b, c) With explosive force, the oocyte bursts out of the follicle, surrounded by a halo of material known as the corona radiata. (d) Completely free of the follicle and much of the material it carried out with it, the oocyte begins its journey to an oviduct.*

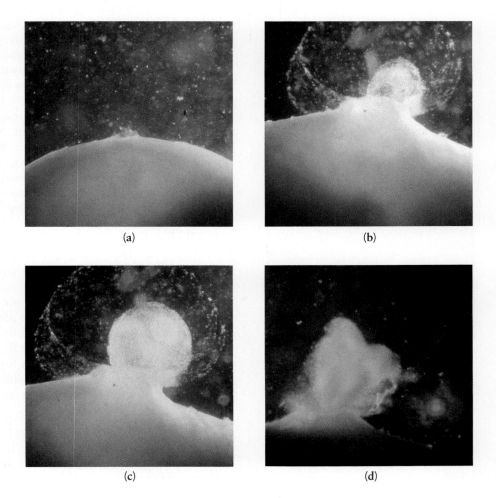

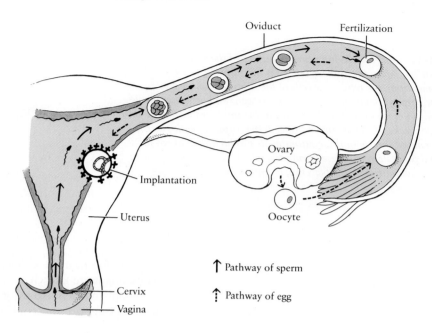

44–15 *Fertilization of the egg by sperm. About once a month in the nonpregnant female of reproductive age, an oocyte is ejected from an ovary and is swept into the adjacent oviduct. Fertilization, when it occurs, normally takes place within an oviduct, after which the young embryo passes down the oviduct and becomes implanted in the lining of the uterus. Muscular movements of the oviduct, plus the beating of the cilia that line it, propel the egg cell down the oviduct toward the uterus. If the oocyte is not fertilized within about 36 hours, it dies, usually by the time it reaches the uterus. A sperm cell has a life expectancy of about 48 hours within the female reproductive tract.*

Pathway of the Oocyte

When the oocyte is released from the follicle at ovulation, it is swept into the adjacent oviduct by the movement of the funnel-shaped opening of the oviduct over the surface of the ovary and by the beating of cilia that line the fingerlike projections surrounding this opening (Figure 44–15). This mechanism is so effective that women who have only one ovary and only one oviduct—and these on opposite sides of the body—have become pregnant.

The oocyte then moves slowly down the oviduct, propelled by peristaltic waves produced by the smooth muscles of the walls. The journey from the ovary to the uterus takes about three days. An unfertilized oocyte lives only about 72 hours after it is ejected from the follicle, but it is apparently capable of being fertilized for less than half of that time. So, fertilization, if it is to occur, must occur in an oviduct. If the egg cell is fertilized, the young embryo becomes implanted in the endometrium three to four days after it reaches the uterus, six or seven days after the egg cell was fertilized. If the oocyte is not fertilized, it dies, and the endometrial lining of the uterus is shed at menstruation. Fertilized eggs implanted in the endometrium are sometimes lost in abnormal menstrual flow, but it is difficult to estimate the number of such very short-lived pregnancies.

Sterilization in women is usually carried out by severing the oviducts, thus preventing sperm from meeting the oocyte. This procedure is known as tubal ligation. Passage of the oocyte and its fertilization are also prevented when the oviducts are blocked from natural causes, such as the formation of scar tissue following infection. Until recently, it was virtually impossible for women with such blockages to bear children. In the procedure known as in vitro fertilization, however, the oocyte is removed surgically from the ovary just prior to ovulation, fertilized with the husband's sperm in a laboratory dish, and inserted in the uterus at the time it would have arrived had it been fertilized naturally in the oviduct. In about 15 to 20 percent of the cases in which in vitro fertilization is attempted, the fertilized egg survives the procedure and subsequent development proceeds normally.

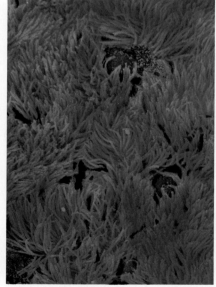

44–16 *Scanning electron micrograph of the inner lining of a mammalian oviduct. The cells of the lining of the oviduct have numerous microvilli, and the surface of the lining is carpeted with cilia. Interspersed among the ciliated cells are mucus-secreting goblet cells, a few of which are visible in this micrograph.*

Orgasm in the Female

Under the influence of a variety of stimuli, the clitoris and its bulbs become engorged and distended with blood, as does the penis of the male. This process is somewhat slower in women than in men, largely because the valves trapping the blood in the female sexual structures tend to leak. The distension of the tissues is accompanied by the secretion into the vagina of a fluid that both lubricates its walls and neutralizes its acidic, and therefore spermicidal, environment.

Orgasm in the female, as in the male, is marked by rhythmic muscular contractions, followed by expulsion into the veins of the blood trapped in the engorged tissues. Homologous muscles produce orgasm in the two sexes, but in the female there is no ejaculation of fluid through the urethra or the vagina. At orgasm, the cervix drops down into the upper portion of the vagina, where the semen tends to form a pool. The female orgasm also may produce contractions in the oviducts that propel the sperm upward. It has been calculated that it would take a sperm cell at least two hours to make its way up the oviducts under its own power, but sperm have been found in the oviducts as soon as five minutes after intercourse. Orgasm in the female, however, is not necessary for conception.

Hormonal Regulation in Females

The Menstrual Cycle

The production of oocytes in all vertebrate females is cyclic. It involves both the interplay of hormones and changes in the follicle cells and the lining of the uterus. This recurring pattern of varying hormone levels and tissue changes is known in humans as the **menstrual cycle.** It is timed and controlled by the hypothalamus. The hormones that participate in the extremely complex feedback system regulating the menstrual cycle include the estrogens and progesterone (the female sex hormones), the pituitary gonadotropins FSH and LH, and gonadotropin-releasing hormone (GnRH) from the hypothalamus. In low concentrations, estrogens act through negative feedback to inhibit the production of FSH and GnRH (and so of LH). In high concentrations, estrogens act through positive feedback to increase the sensitivity of the pituitary to GnRH and may also stimulate the secretion of GnRH; the result is an increase in the synthesis of LH and FSH by the pituitary. In high concentrations, progesterone, in the presence of estrogens, inhibits the secretion of GnRH and thus the production of LH and FSH.

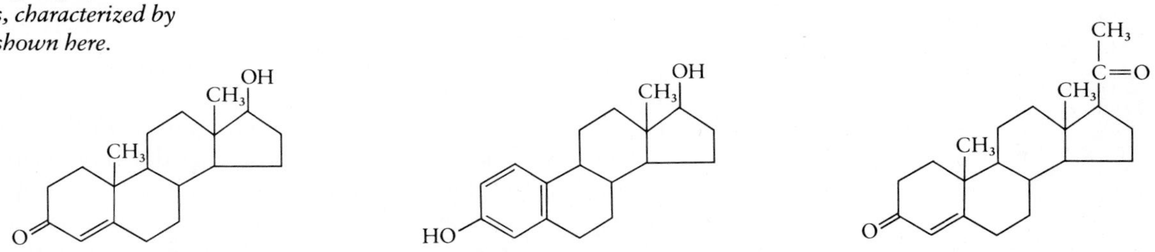

44–17 *The sex hormones testosterone, estradiol, and progesterone. Although there are a number of estrogens, estradiol is the most important. Note that the chemical structures of these three hormones differ only very slightly, in contrast to the great differences in their physiological effects—another example of the extreme specificity of biochemical actions. All of these belong to the group of molecules known as steroids, characterized by the four-ring structure shown here.*

Testosterone

Estradiol

Progesterone

At the beginning of the cycle, during menstruation, hormone levels are low (Figure 44–18). After a few days, an oocyte and its follicle begin to mature under the influence of the gonadotropins FSH and LH. As the follicle enlarges, it secretes increased amounts of estrogens, which stimulate the regrowth of the endometrium in preparation for implantation of a fertilized egg cell. The rapid rise in estrogen levels near the midpoint of the cycle triggers a sharp increase in the release of LH by the pituitary. The spurt of high LH stimulates the follicle to release the oocyte, which begins its passage to the uterus. Under the continued stimulus of LH, the cells of the emptied follicle grow larger and fill the cavity, producing the **corpus luteum** ("yellow body"). As the cells of the corpus luteum

44–18 *Diagram of the events that take place during the menstrual cycle, which involves changes in hormone levels, in structures at the surface of the ovary, and in the uterine lining. The cycle begins with the first day of menstrual flow, which is caused by the shedding of the endometrium, the lining of the uterine wall. The increase of FSH and LH at the beginning of the cycle promotes the growth of the ovarian follicle and its secretion of estrogens. Under the influence of estrogens, the endometrium regrows. The sudden rise in estrogens just before midcycle triggers a sharp increase in the release of LH from the pituitary, which stimulates the release of the oocyte (ovulation). (It is not known what role, if any, is played by the simultaneous increase in FSH.) Following ovulation, LH and FSH levels drop. The follicle is converted to the corpus luteum, which secretes estrogens and also progesterone. Progesterone further stimulates the endometrium, preparing it for implantation of the young embryo. If pregnancy does not occur, the corpus luteum degenerates, the production of progesterone and estrogens falls, the endometrium begins to slough off, FSH and LH concentrations increase once more, and the cycle begins anew.*

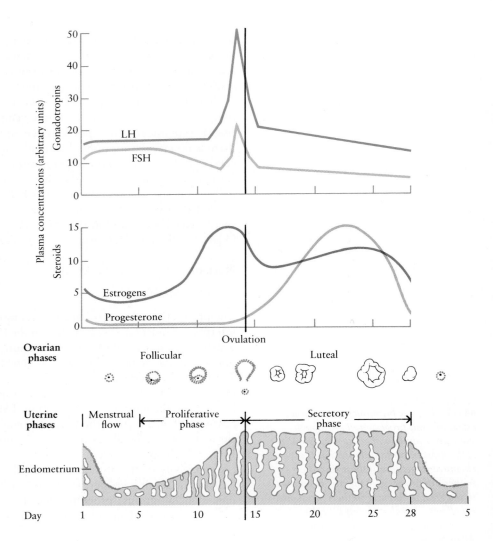

increase in size, they begin to synthesize significant amounts of progesterone as well as estrogens. As the progesterone levels increase, estrogens and progesterone together inhibit the production of GnRH and so of the gonadotropic hormones LH and FSH from the pituitary. If pregnancy does not occur, the corpus luteum is reabsorbed and the production of ovarian hormones drops. Without hormonal support, the endometrium can no longer sustain itself, and a portion of it is sloughed off in the menstrual fluid. Then, in response to the now low level of ovarian hormones, the level of pituitary gonadotropic hormones begins to rise again, followed by development of a new follicle and a rise in estrogens as the next monthly cycle begins.

The cycle usually lasts about 28 days, but individual variation is common. Even in women with cycles of average length, ovulation does not always occur at the same time in the cycle (which is the reason the "rhythm method" is an unreliable means of birth control). Although the menstrual cycle does not require an environmental cue, as do reproductive cycles in many other vertebrates, it is clearly under the influence of external factors to some extent. For example, some women find that emotional upset delays a menstrual period or eliminates it completely.

The onset of menstruation marks the beginning of puberty in the human female. The average age of puberty is 12.3 years, but the normal range is very wide. The increased production of female sex hormones preceding puberty induces the development of secondary sex characteristics, such as enlargement of the hips and breasts.

Estrus

Females of almost all mammalian species except *Homo sapiens* will mate only during **estrus,** the fertile period during which oocytes are released. Estrus may occur only once a year (as in wolves or deer), about once a month (as in cows and horses), or every few days (as in rats and mice).

The periods of estrus may last from only a few hours to three or four weeks. In animals, such as dogs, that produce eggs over a period of days, the eggs may be fertilized at different times and by different males, which explains in part why a mixed-breed litter can contain such an astonishing variety of siblings. In some mammals, such as cats, rabbits, and minks, although the egg is mature and the female receptive during estrus, ovulation occurs only under the stimulus of copulation. This is obviously a very efficient system, ensuring maximum economy in the utilization of gametes. There is suggestive evidence that in some women, also, ovulation may be triggered by sexual intercourse.

The human female appears to be one of the few female animals receptive to mating during infertile periods. Some anthropologists speculate that this receptivity coevolved with the establishment of strong pair-bond relationships between human or prehuman males and females. A consequence of this pair-bond relationship is a society based on a family unit, in contrast to most other primates, in which the social and breeding unit is a troop or band. The establishment of such family units is seen as the original basis for the traditional division of labor between the sexes, with the female concentrating on childbearing and the home and the male on hunting, protecting the family unit, and defending territory. If the anthropologists are right, this behavioral adaptation on the part of the human female has profoundly influenced the shape of human civilization.

CONTRACEPTIVE TECHNIQUES

Statistics show that 25 percent of women who conceive become pregnant within one month of trying to do so, 63 percent within six months, 75 percent within nine months, and 90 percent within 18 months. Using no contraceptive methods, 80 percent of women of childbearing age having regular sexual intercourse become pregnant within a year. However, a variety of contraceptive techniques is available for couples who wish to prevent or defer pregnancy. In Table 44–1, they are rated in order of effectiveness, in terms of the average number of pregnancies per year among women of childbearing age using them. In most cases, two figures are given for effectiveness. The first, lower, figure is an "ideal" figure, obtained when the method is used consistently and correctly. The second figure is an average figure, reflecting actual experience.

For many years the most widely used contraceptive techniques were barrier methods, such as the diaphragm and the condom. In the 1960s and 1970s, many couples abandoned barrier methods, and "the pill," a combination of synthetic estrogens and progesterone, came into wide use. When taken daily, it keeps the level of these hormones in the blood high enough to shut off production of the pituitary hormones FSH and LH. Without FSH the ovarian follicles do not ripen, and in the absence of LH no ovulation occurs, so pregnancy is not possible.

TABLE 44–1　Methods of Birth Control Currently Available

METHOD	MODE OF ACTION	EFFECTIVENESS (PREGNANCIES PER 100 WOMEN PER YEAR)	ACTION NEEDED AT TIME OF INTERCOURSE	REQUIRES INSTRUCTION IN USE	POSSIBLE UNDESIRABLE EFFECTS
Vasectomy	Prevents release of sperm	0	None	No	Usually produces irreversible sterility
Tubal ligation	Prevents passage of oocyte to uterus	0	None	No	Usually produces irreversible sterility
"The pill" (estrogens and progesterone)	Inhibits secretion of FSH and LH, thereby preventing follicle maturation and ovulation	0–10	None	Yes, timing	Early—some water retention, breast tenderness, nausea; late—increased risk of cardiovascular disease
"Minipill" (progesterone alone)	Causes changes in cervix and uterus that prevent conception	3–10	None	Yes, timing	Break-through bleeding, ectopic (tubal) pregnancy
"Morning-after pill" (50 × normal dose of estrogens)	Arrests pregnancy, probably by preventing implantation	?	None	Yes; prescribed only in emergencies, for example, for rape victims	Breast swelling, water retention, abdominal pain, nausea
Diaphragm with spermicidal jelly	Prevents sperm from entering uterus, jelly kills sperm	2–20	Yes, insertion before intercourse	Yes, must be inserted correctly each time; leave in at least 6 hours after	None usually, may cause irritation
Condom (worn by male)	Prevents sperm from entering vagina	3–36	Yes, male must put on after erection	Not usually	Some loss of sensation in male
Intrauterine device (wound with copper or containing progesterone that is slowly released)	Prevents fertilization or implantation	4–5	None; may be left in place for as long as 4 years	No, but must be inserted by a physician and periodically checked	Menstrual discomfort, displacement or loss of device, uterine infection, ectopic pregnancy
Cervical cap with spermicidal jelly	Prevents sperm from entering uterus, jelly kills sperm	4–?	Yes, insertion before intercourse	Yes, must be inserted correctly each time; more difficult to insert than diaphragm	None usually, may cause irritation
Contraceptive sponge (contains spermicide)	Prevents sperm from entering uterus; kills sperm	13–20	Yes, insertion before intercourse	Yes, must be inserted correctly each time; leave in at least 6 hours after	None usually, may cause irritation
Vaginal foam, jelly alone	Spermicidal, mechanical barrier to sperm	20–30	Yes, requires application before intercourse	Yes, must use within 30 minutes of intercourse; leave in at least 6 hours after	None usually, may cause irritation
Rhythm	Abstinence during probable time of ovulation	14–47	None	Yes, must know when to abstain, based on daily temperatures and/or cyclic changes in vaginal mucus	Requires abstinence during part of cycle
Douche	Washes out sperm that are still in the vagina	?–85	Yes, immediately after	No	May propel sperm through cervix into uterus
Withdrawal	Removes penis from vagina before ejaculation	?–93	Yes, withdrawal	No	Frustration in some

Recently, however, the diaphragm and the condom have once more become popular. In combination with spermicidal jellies, these two methods, and particularly the condom, provide a barrier not only against sperm but also against many infectious agents. As Table 44–1 reveals, however, their protection is far from absolute.

SUMMARY

In vertebrates, reproduction is characteristically sexual and involves two parents. The male and female gametes, sperm and eggs, are formed by meiosis in the gonads (the testes and the ovaries). Sperm cells are produced in the seminiferous tubules of the testes. The spermatogonia become primary spermatocytes; then, after the first meiotic division, secondary spermatocytes; and, following the second meiotic division, spermatids, which then differentiate into sperm cells. These sperm cells enter the epididymis, a tightly coiled tube overlying the testis, where they are partially mobilized. The epididymis is continuous with the vas deferens, which leads along the posterior wall of the abdominal cavity, around the bladder, and into the prostate gland. Just before entering the prostate, the two vasa deferentia merge with ducts of the seminal vesicles and then, within the prostate, with the urethra, which leads out through the penis.

The penis is composed largely of spongy erectile tissue that can become engorged with blood, enlarging and hardening. At the time of ejaculation, sperm are propelled along the vasa deferentia by contractions of a surrounding coat of smooth muscle. Secretions from the seminal vesicles, the prostate, and the bulbourethral glands are added to the sperm as they move toward the urethra. The resulting mixture, the semen, is expelled from the urethra by muscular contractions involving, among other structures, the base of the penis. These muscular contractions also contribute to the sensations of orgasm.

Production of sperm and the development of male secondary sex characteristics are under the control of hormones, including gonadotropin-releasing hormone (GnRH), the gonadotropins LH (luteinizing hormone) and FSH (follicle-stimulating hormone), and testosterone (the principal androgen). LH acts on the interstitial cells, located between the seminiferous tubules, to stimulate the production of testosterone. FSH and testosterone stimulate the production of sperm. Production of LH and FSH is regulated by a negative feedback system involving both the hypothalamus and the pituitary.

The female gamete-producing organs are the ovaries. The primary oocytes develop within nests of cells called follicles. The first meiotic division begins in the female fetus and is completed at ovulation. The second meiotic division is completed at fertilization. On the average, one secondary oocyte is released every 28 days and travels down the oviduct to the uterus. If it is fertilized (which usually takes place in an oviduct), the embryo becomes implanted in the lining of the uterus (the endometrium). If it is not fertilized, it degenerates and the endometrial lining is shed at menstruation.

The production of oocytes and the preparation of the endometrium for implantation of the embryo are cyclic. The reproductive cycle, which is known in humans as the menstrual cycle, is controlled by hormones, including gonadotropin-releasing hormone (GnRH), the gonadotropic hormones FSH and LH, and estrogens and progesterone (the female sex hormones). Before ovulation, FSH and LH from the pituitary stimulate the ripening of the follicle and the secretion of estrogens. After ovulation, the corpus luteum, which forms from the emptied follicle, produces both estrogens and progesterone. Progesterone and estrogens both stimulate the growth of the endometrium.

The part of the female reproductive cycle during which mating occurs in most mammals is known as estrus. Although ovulation is cyclic in human females, they, unlike most other animals, are continuously receptive to sexual intercourse, a behavior pattern that may have coevolved with strengthened pair bonding.

QUESTIONS

1. Distinguish among the following: spermatocytes/spermatids/sperm cells; oocyte/ovum; ovarian follicle/corpus luteum.

2. Consider the kinds of organisms that generally have external fertilization and those that generally have internal fertilization. What differences in their life styles require these differing mechanisms?

3. How is the shelled egg of a hen fertilized?

4. What would constitute the ejaculate of a man who has undergone a vasectomy? Would a vasectomy affect the structures associated with orgasm? Why or why not?

5. Describe the effects of luteinizing hormone (LH) and follicle-stimulating hormone (FSH) in the human male and female.

6. Explain the changes in the plasma concentrations of the gonadotropins and sex hormones during the menstrual cycle as shown on the graphs below.

7. During which days in the menstrual cycle is a woman most likely to become pregnant? (Include data on the longevity of eggs and sperm in making this calculation.) Why is the use of the calendar method of birth control much less effective than other methods?

8. What would be an appropriate method of contraception for a couple who have infrequent intercourse (once a month)? For a couple who have frequent intercourse (three times a week), with plans to have children eventually? For a couple who have frequent intercourse but do not wish to have any children? Under what circumstances, if any, would you elect to have a vasectomy or a tubal ligation?

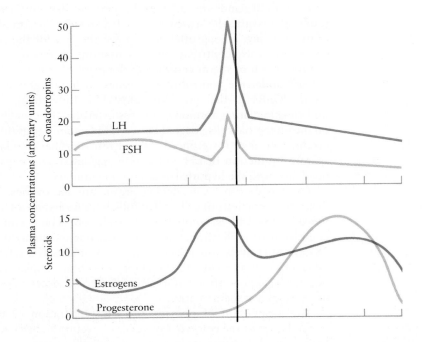

The Continuity of Life II: Development

Each of our lives began very modestly with the fusion of a tiny, motile sperm cell and a larger, but structurally simpler egg. If the newly fertilized egg of an animal—any animal—is examined under the electron microscope, there is no hint whatsoever of the awesome potential with which this single cell is endowed. How do the complex structures of the embryo and, subsequently, of the adult animal develop from this one, apparently simple cell? This question is one of the most fundamental in all of biology, and although it has commanded the attention of scientists for more than 100 years, a complete and comprehensive answer remains elusive.

During its development, the fertilized egg is transformed into a complete organism, closely resembling its parents and consisting, depending on the species, of hundreds to billions of cells. This process involves growth, differentiation, and morphogenesis—that is, increase in size, specialization of cells, tissues, and organs, and shaping of the adult body form. The course of development in many species has been closely observed and meticulously documented, illuminating the sequence of developmental events. Only in recent years, however, have the tools become available to begin deciphering the underlying cellular and molecular events that give rise to and control the observed processes.

Development, like other biological processes, is most clearly explained through concrete examples. Thus, in this chapter we shall first consider in some detail the course of embryonic development in three representative organisms—the sea urchin, the amphibian, and the chick—each of which has made major contributions to scientific understanding. With the general principles and overall pathways of development established, we shall then return to the fertilized human egg, which we left in the last chapter as it began its journey down the oviduct to the uterus, and consider the marvelous series of events that precede and lead to the birth of each new human infant.

45–1 *One of the distinguishing features of phylum Chordata, to which we and all other vertebrates belong, is a dorsal, hollow nerve cord. It develops from a groove, known as the neural groove, on the dorsal surface of the early embryo. As you can see in this scanning electron micrograph of the neural groove of a chick embryo, the tissue on either side of the groove curves upward, forming two arches that ultimately meet to form the hollow neural tube. As we shall see later in this chapter, the curvature is caused by changes in the shape of the individual cells.*

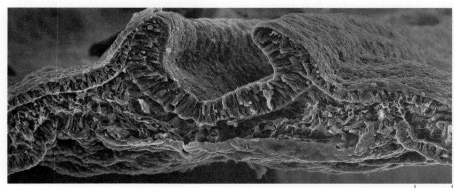

25 μm

DEVELOPMENT OF THE SEA URCHIN

Let us begin by following the course of development in the sea urchin, an echinoderm. As we saw in Chapter 26 (page 544), the earliest stages of embryonic development in echinoderms are, in important respects, similar to those observed in vertebrates and other chordates, indicating a closer evolutionary relationship than one would expect on the basis of the adult forms. Of all the invertebrates, echinoderms are the most useful as model organisms for the study of early vertebrate development. Sea urchins, in particular, have long been a favorite of embryologists because (1) their eggs, which are produced in great numbers, are fertilized and develop externally, making it possible to study their development under relatively simple laboratory conditions; (2) both egg and developing embryo are almost transparent, so that it is possible to observe many of the early events without disrupting them; and (3) the process is rapid. In about 48 hours, the zygote develops into a free-swimming larval form, known as the **pluteus.** Moreover, sea urchins are abundant in pleasant places such as Woods Hole, Massachusetts, where biologists like to spend the summer.

Fertilization and Activation of the Egg

Development begins with fertilization of the egg by the sperm (Figure 45–2). Like the mammalian sperm cell (page 903), the sperm cell of the sea urchin consists of an acrosome at its anterior tip, a highly condensed, tightly packed nucleus, a small amount of cytoplasm, and a long flagellum, all surrounded by a cell membrane. The egg cell, which is much larger than the sperm cell (see Figure 5–10b, page 109), is surrounded by an outer membrane, known as the **vitelline envelope,** attached to the cell membrane. Embedded in the vitelline envelope and displayed on its surface are species-specific protein receptors that participate in the binding of the sperm to the egg. Surrounding the vitelline envelope are additional layers of jelly.

When the sperm meets the egg, enzymes released from the acrosome dissolve the jelly layers at the point of contact, creating a path through which the sperm cell can move. Other molecules from the acrosome are the three-dimensional complement of the protein receptors displayed on the vitelline envelope. These molecules coat the outer surface of the sperm head; when they bind to their receptors, the sperm is able to penetrate the vitelline envelope. Ultimately, the cell membranes of the sperm and the egg make contact, they fuse, and the sperm nucleus enters the cytoplasm of the egg.

Fertilization—the fusion of sperm and egg—has at least four consequences. First, changes take place on the surface of the fertilized egg to prevent the entry of additional sperm. Within a second of sperm contact, the cell membrane of the

45–2 *Egg and sperm cells from a sea urchin:* (a) *sperm cells on the egg surface, and* (b) *a sperm penetrating the egg. The dark, torpedo-shaped body is the sperm nucleus.*

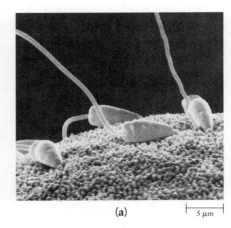

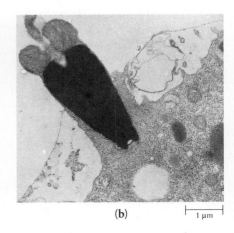

(a)   5 μm   (b)   1 μm

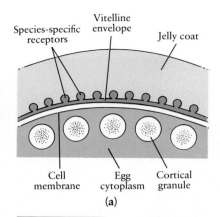

(a)

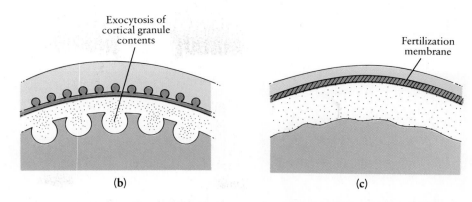

(b)   (c)

45–3 *Formation of the fertilization membrane.* **(a)** *The surface layers of an unfertilized egg consist of a jelly coat, the vitelline envelope, which bears species-specific protein receptors, and the cell membrane. Within the cytoplasm, near the cell membrane, are membrane-bound vesicles known as cortical granules. Contact of a sperm cell with the receptors on the vitelline envelope triggers an increase in the Ca^{2+} concentration in the cytoplasm. This causes the cortical granules to fuse with the cell membrane* **(b)**, *releasing their contents, including enzymes and polysaccharides, into the space between the cell membrane and the vitelline envelope. The enzymes break down the protein receptors on the vitelline envelope and the substances that hold it to the cell membrane. The polysaccharides raise the osmotic potential of the space beneath the vitelline envelope, causing water to flow inward, thus lifting the vitelline envelope off the surface of the egg* **(c)**. *The vitelline envelope "hardens" to form the fertilization membrane.*

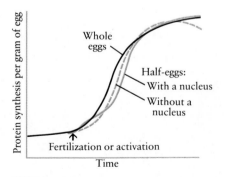

45–4 *Rates of protein synthesis during the first few hours after activation in whole eggs, half-eggs with a nucleus, and half-eggs without a nucleus. The initial pattern of protein synthesis is the same whether the activated egg contains a nucleus or not, indicating that the mRNA for early protein synthesis is present in the egg before fertilization.*

egg, which normally maintains an internal electric potential of about −60 millivolts, is depolarized. Sodium ions (Na^+) flow into the cytoplasm, temporarily reversing the polarity of the cell. This change in electric potential sweeps, wavelike, around the surface of the egg and makes the cell membrane suddenly unresponsive to the advances of other sperm. Simultaneously, calcium ions (Ca^{2+}) sequestered within the endoplasmic reticulum are released into the cytoplasm. The increase in the concentration of free Ca^{2+} ions triggers a series of reactions that cause the vitelline envelope to lift off the surface of the egg (Figure 45–3). The protective outer membrane that forms from the vitelline envelope is known as the **fertilization membrane.**

Second, the egg is activated metabolically, as evidenced by a dramatic increase in protein synthesis and a rise in oxygen consumption.

Third, the genetic material of the male is introduced into the female gamete—the haploid sperm nucleus fuses with the haploid egg nucleus to produce a diploid zygote nucleus. The genotype of the new individual is thus established.

Fourth, the egg begins to divide by mitosis, with the centrioles of the sperm participating in the organization of the mitotic spindle. The developmental chain of events is thus set in motion.

In many species, although activation of the egg and mitosis follow fertilization, they can also proceed without it. In the sea urchin egg, for instance, even exposure to a hypertonic solution can start the developmental chain of events. Frogs' eggs can be activated and will divide after being pricked with a glass needle or given a mild electric shock. In fact, in some species it has been possible to demonstrate that activation of the egg does not require any nucleus at all, male or female. Sea urchin eggs can be divided in two, half with a nucleus, half without. Each half can then be activated artificially. Protein production in the half without a nucleus is as great initially, on a gram-for-gram basis, as in the half with a nucleus or in the whole egg (Figure 45–4), although, of course, the enucleated half does not long survive. The egg, during the course of its maturation within the ovary, transcribes the genetic information needed for the early rounds of biosynthetic activity following fertilization, but the transcribed mRNA is inactive—stored in the cytoplasm of the egg—until fertilization initiates translation.

From Zygote to Pluteus

The initial stages of development in the sea urchin are shown in Figure 45–5, on the next page. The zygote divides about once an hour for 10 hours, in the process known as **cleavage.** The embryo at this stage is called a **morula.** As the cells continue to divide, Na^+ ions are pumped from them into the extracellular spaces, followed by an osmotic flow of water. This creates a fluid-filled cavity, known as the **blastocoel,** in the center of the embryo. When the blastocoel is fully formed, the embryo—now a fluid-filled sphere of a single layer of cells—is called the **blastula,** and its cells are known as blastomeres. The sea urchin blastula is about the same size as the egg cell from which it developed. Cleavage has not changed the total volume but has greatly altered the surface-to-volume ratio (page 103) and also the ratio of volume of the nucleus to that of the cytoplasm of the individual cells.

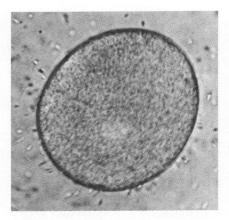

(a) *Numerous spermatozoa can be seen surrounding an unfertilized egg.*

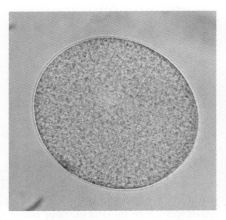

(b) *Fertilized egg; the fertilization membrane has just begun to form. The light area slightly above the center is the diploid nucleus.*

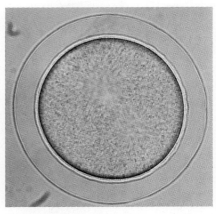

(c) *The fertilization membrane is fully formed. The egg has begun to divide; if you look closely, you can see that there are two nuclei.*

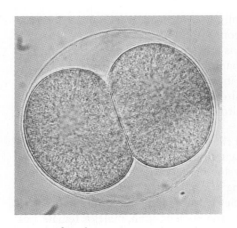

(d) *The first division.*

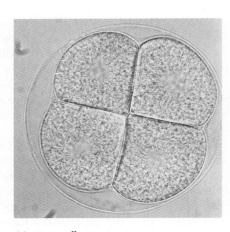

(e) *Four-cell stage.*

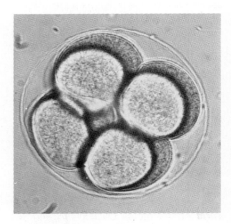

(f) *Eight-cell stage.*

45–5 *The early development of the sea urchin. Notice that as the egg divides, the cells become progressively smaller, so that by the blastula stage they are barely distinguishable, although the magnification has not changed.*

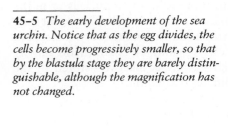

(g) *The blastocoel forms.*

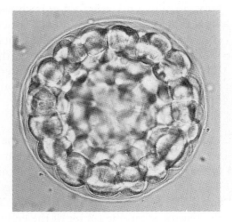

(h) *The mature blastula.*

0.1 mm

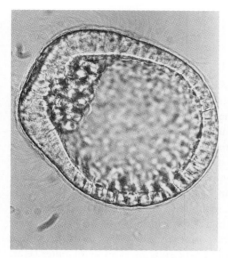

(a) *The beginning of gastrulation; the blastopore has begun to form at the upper left, and cells near the blastopore have begun to migrate across the blastocoel.*

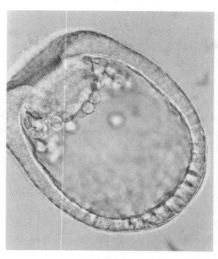

(b) *The outer cell layer begins to fold inward at the blastopore, forming the archenteron.*

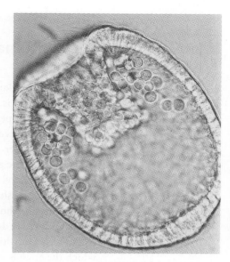

(c) *The outer layer of cells continues to move across the blastocoel.*

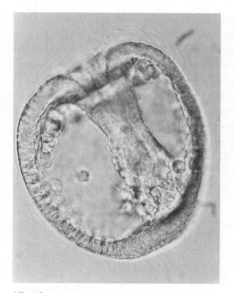

(d) *The mature gastrula.*

45–6 *Gastrulation in the sea urchin.*

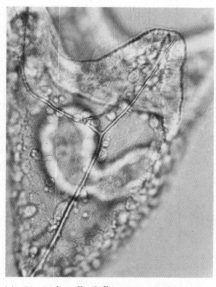

(e) *Gastrula cells differentiate and organize to form the pluteus larva.*

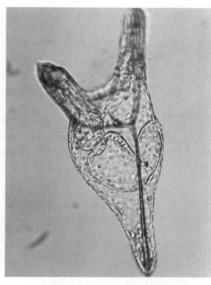

(f) *Within 48 hours after fertilization, the egg has developed into a free-swimming multicellular organism, the pluteus.*

The formation of the blastula is followed by a process known as **gastrulation** (from *gaster,* the Greek word for "stomach"), which gives rise to the primitive gut. Gastrulation in the sea urchin (Figure 45–6) begins with the formation of the **blastopore,** an opening into the blastula. Cells near the blastopore break loose and, with the aid of contractile pseudopodia with sticky tips, move over the interior surface of the blastula toward the opposite pole. These cells are known as the primary mesenchyme. Next, the entire cell layer closest to the blastopore turns inward, moving through the blastocoel to the opposite pole, forming a new cavity, the **archenteron.** The archenteron will ultimately develop into the digestive tract,

The Cytoplasmic Determination of Germ Cells

Early in the development of virtually all animals, the germ cells—that is, the cells that will ultimately differentiate into spermatogonia or oogonia—are set aside. Their fate is determined well before differentiation of any other cell types has begun. This early segregation of the germ cells is thought to reduce the number of cell divisions that they must undergo prior to spermatogenesis or oogenesis and, therefore, to reduce the likelihood of errors in DNA replication. At a later stage in development, the germ cells migrate from the site at which they were set aside to the site at which the gonads will form.

In insects that undergo a complete metamorphosis between a nonreproductive larval stage and a reproductive adult stage (see page 576), a total reorganization of the cells and tissues occurs. Consequently, early determination of the future germ cells and protection of their genetic integrity are of particular importance. The mechanism by which this is accomplished has been studied most extensively in *Drosophila*. As we saw in Section 3, more is known about the genetics of *Drosophila* than of any other animal. In laboratories throughout the world, populations of *Drosophila* are maintained for which the genetic makeup is known in extraordinary detail. Such populations make possible a variety of experiments producing subsequent generations in which the precise ancestry of the individual flies—or even portions of the flies—can be identified.

During cleavage in the fertilized egg of *Drosophila* and other insects, mitosis is not initially accompanied by cytokinesis. Numerous rounds of mitosis occur until the number of nuclei required for the cells of the blastula are produced. The nuclei then migrate to the periphery of the cell, followed by a wave of cytokinesis that produces the individual cells of the blastula. The blastula, like the egg from which it formed, is elongated, with a clear anterior-posterior axis. It consists of a single layer of cells—except at its posterior pole, where a group of cells visibly distinct from the other cells of the blastula form a second, outer layer. These cells, known as the pole cells, are the future germ cells. As the experiment diagrammed here shows, their fate is determined by a region of cytoplasm, known as the polar plasm, located at the posterior pole of the egg. In this experiment, polar plasm from an early *Drosophila* embryo of genotype A is injected into the anterior end of another embryo of a different genotype, B. When cytokinesis occurs in the genotype B embryo, pole cells form at both its anterior and posterior poles. Further development in this embryo is thwarted, but the pole cells at its anterior can be removed and injected into the posterior region of still another embryo, representing a third genotype, C. The fly that develops from this embryo is of genotype C. Breeding experiments, however, reveal that it produces gametes of both genotype C and genotype B. The gametes of genotype B are the pro-

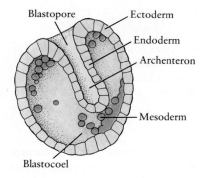

45–7 *The sea urchin gastrula. Gastrulation produces a three-layered embryo. The archenteron becomes the digestive tract, and the blastopore becomes the anus. Ultimately, the blastocoel is almost entirely obliterated. In this and subsequent illustrations, ectoderm is blue, endoderm is yellow, and mesoderm is red.*

and the blastopore will become the anus. As you will recall (page 545), formation of the anus at or near the blastopore is the defining characteristic of the deuterostomes, which include the echinoderms and the chordates.

As a result of the movements that take place at gastrulation, three embryonic tissue layers have been formed: an outside layer, the ectoderm; a middle layer, the mesoderm, which formed from the primary mesenchyme cells; and an inner layer, the endoderm (Figure 45–7). Also, the anterior-posterior axis of the embryo has become obvious.

Once gastrulation is completed, evidence of cell differentiation can be observed. Cells in the mesoderm begin to secrete calcium-containing granules that develop into tiny three-armed spicules. These become the supporting skeleton for the pluteus. At the point at which the endoderm touches the opposite surface of the blastocoel, ectodermal cells curve inward to form the mouth of the larva. The archenteron subdivides into stomach, intestine, and anus, and a ring of long cilia forms near the mouth region. The single fertilized egg cell has become a number of specialized, differentiated cells, performing specific functions, such as digestion; producing new organelles, such as cilia; and secreting new products, such as the calcium-containing skeleton.

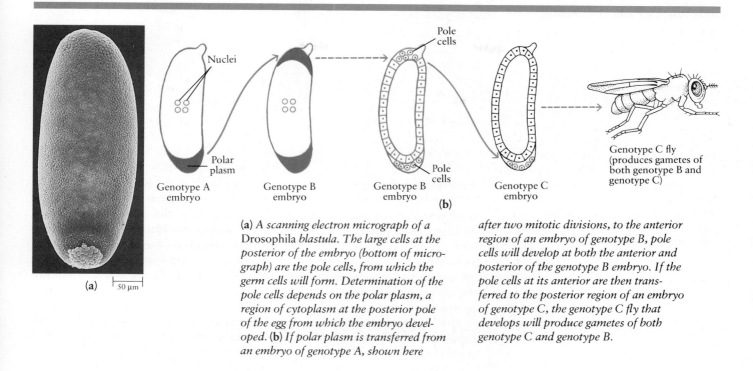

(a) *A scanning electron micrograph of a* Drosophila *blastula. The large cells at the posterior of the embryo (bottom of micrograph) are the pole cells, from which the germ cells will form. Determination of the pole cells depends on the polar plasm, a region of cytoplasm at the posterior pole of the egg from which the embryo developed.* (b) *If polar plasm is transferred from an embryo of genotype A, shown here*

after two mitotic divisions, to the anterior region of an embryo of genotype B, pole cells will develop at both the anterior and posterior of the genotype B embryo. If the pole cells at its anterior are then transferred to the posterior region of an embryo of genotype C, the genotype C fly that develops will produce gametes of both genotype C and genotype B.

geny of the pole cells that developed at the anterior end of the genotype B embryo in response to the injection of polar plasm from the genotype A embryo.

The substance or substances in the polar plasm of the *Drosophila* egg that determine the formation of pole cells and, subsequently, of the germ cells, are localized within small granules. These granules, which are present in the cytoplasm before fertiliza-

tion, are rich in RNA of maternal origin—coded by the genes of the female fly that laid the egg. Recent studies in amphibians suggest that a similar mechanism, involving RNA and other cytoplasmic determinants laid down in the unfertilized egg, may also be involved in the differentiation and setting apart of the germ cells of vertebrates.

The Influence of the Cytoplasm

The sea urchin egg contains a relatively small amount of stored food (yolk), which tends to be concentrated in the lower half of the egg. This lower half is called the vegetal half; the upper half is called the animal half. At cleavage, the first two cell divisions run longitudinally from the pole of the animal half to the pole of the vegetal half, perpendicular to one another—the way you would quarter an apple. If these four cells are separated, each can develop into a normal pluteus larva. During the third cleavage division, however, an important event takes place—the cytoplasm of the animal half and that of the vegetal half are separated as the four cells split across the equator, producing an eight-cell embryo (see Figure 45–5f). If the eight-cell embryo is separated into two halves longitudinally, normal development will occur, but if the embryo is divided along the third cleavage plane, the two halves develop abnormally. The top (animal) half has large tufts of cilia but no gut. The bottom (vegetal) half is almost all gut, with no mouth, shortened arms, no cilia, and only a few, if any, spicules (see Figure 45–8 on the next page).

The explanation for these observations lies in chemical factors, known as cytoplasmic determinants, that are distributed in two gradients, an animal gra-

45–8 *When an eight-cell sea urchin embryo is separated into two halves by moving a glass needle between the cells, the future development of the resulting halves depends on the plane of division. (a) If the division is longitudinal, producing embryos containing cells from both the animal and vegetal poles, subsequent development is normal. (b) If, however, the division is at the equator of the embryo, the four cells from the animal pole produce a ciliated embryo, permanently arrested at the blastula stage. The four cells from the vegetal pole give rise to an embryo that looks relatively normal but lacks critical structures, such as a mouth, cilia, and spicules.*

dient and a vegetal gradient, within the cytoplasm of the egg. Current evidence suggests that the cytoplasmic determinants are most likely messenger RNA or protein molecules, localized within the cytoplasm by the cytoskeleton. During cleavage, these molecules are "captured" within the cytoplasm of individual blastomeres. The particular cytoplasmic determinants contained within a cell then affect its subsequent pattern of development. As microsurgical experiments on sea urchin and other embryos make clear, it is not just the nucleus that controls the differentiation of cells. The cytoplasm surrounding the nucleus also has profound effects.

Other experiments have demonstrated that interactions among the individual blastomeres are also important in development. If you put the early blastula of a sea urchin in calcium-free sea water and agitate the water, the cells will separate. The cells still look the same, and neither their genetic material nor their cytoplasm has been altered in any crucial way. But they simply stop dividing and very shortly thereafter they deteriorate and die. Returning individual cells to normal sea water does not restore their developmental potential. Suppose, however, the cells are returned to normal sea water and, by gently stirring, are brought into contact with one another. Under these conditions the cells reaggregate, arrange themselves in their former pattern, and once more begin to divide.

DEVELOPMENT OF THE AMPHIBIAN

The eggs of frogs and many other amphibians are laid in shallow water and, like sea urchin eggs, are fertilized externally; hence, they too can be readily observed. The amphibian egg, however, contains a much larger amount of yolk. The animal half and the vegetal half of the egg differ markedly in appearance. For example, in frogs of the genus *Rana*, the yolk is massed in the lower hemisphere of the unfertilized egg, and the upper two-thirds of the egg is covered by a heavily pigmented layer of cytoplasm (Figure 45–9a). Immediately following fertilization, there is a massive reorganization of the cytoplasm. When the sperm penetrates the

45–9 *Formation of the gray crescent in a frog's egg. (a) Before fertilization, the upper two-thirds of the egg is "capped" by a heavily pigmented layer of cytoplasm. The bulk of the yolk is massed in the lower hemisphere. The nucleus of the egg is near the pole of the upper hemisphere. (b) The egg has been fertilized; the sperm nucleus has entered at the right and is moving toward the center of the egg. The trail that marks its path is caused by the disruption of pigment granules. The whole pigment cap has rotated toward the point of sperm penetration. The region on the other side of the egg, from which the pigmented layer has moved away, becomes the gray crescent. Accompanying these movements of the pigmented layer are rearrangements of the cytoplasm in more interior regions of the egg.*

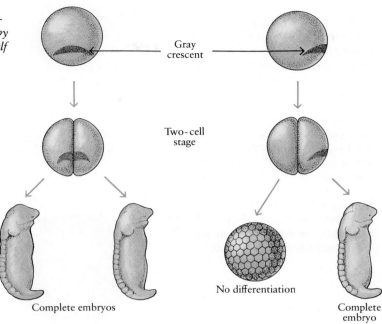

45–10 *The importance of the gray crescent cytoplasm in development was demonstrated by separating the two cells formed by the first cleavage of the egg. When the egg on the left divided, half of the gray crescent cytoplasm passed into each of the two new cells. When these cells were separated from each other, each formed a complete embryo. The first cleavage of the egg on the right resulted in all the gray crescent cytoplasm going to one cell and none to the other. When these cells were separated, the one without the crescent did not develop.*

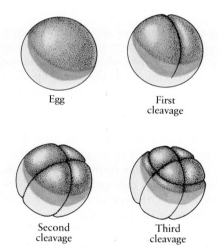

45–11 *Eggs, such as those of the frog, that contain a large amount of yolk concentrated in one hemisphere cleave unequally. The first longitudinal cleavage usually bisects the gray crescent. The second cleavage, also longitudinal, is at right angles to the first. Thus the egg is split through the poles to produce four cells shaped like the segments of an orange. The third cleavage separates the lower, yolkier (vegetal) part of the embryo from the upper, less yolky (animal) part. As you can see, the four cells in the animal hemisphere are much smaller than the four in the vegetal hemisphere.*

egg, the pigment cap rotates toward the point of sperm penetration, and a gray crescent appears on the side of the egg opposite the point of sperm entry (Figure 45–9b). Formation of the gray crescent is complete within 20 minutes.

In a series of brilliant experiments by Hans Spemann in the early 1900s, it was shown that the cytoplasm associated with the gray crescent is of critical importance in the later development of the embryo. For instance, if cells of an amphibian embryo are divided at the two-cell stage, each blastomere may develop into a normal embryo. Whether or not both cells develop normally depends on whether or not each has received some of the cytoplasm containing the gray crescent (Figure 45–10). These experiments provide further evidence that differences in the cytoplasm, and particularly in localized factors, play a major role in determining the course of early development.

Cleavage and Blastula Formation

Cleavage in the amphibian egg differs from that in the sea urchin egg chiefly because of differences in the yolk content. When yolk is absent or present only in small amounts, as in the sea urchin, cleavage is uniform, producing cells of similar size. When larger amounts of yolk are present, as in frogs, the egg divides unevenly (Figure 45–11), forming larger cells in the vegetal hemisphere. As with the sea urchin egg, cleavage results in the formation of a blastula, but the amphibian blastocoel is small and usually off-center. The blastopore appears as a crescent-shaped slit; it always forms at the boundary between the gray crescent and the vegetal hemisphere.

Gastrulation and Neural Tube Formation

Gastrulation in the amphibian differs from that in the sea urchin in detail but not in principle. In the sea urchin, one wall of the blastula moves inward, forming a tubular extension toward the other wall. In the yolky frog's egg, the process of inward movement is characterized by an extensive series of coordinated cellular movements—migration of single cells and sheets of cells—and the folding of cell layers.

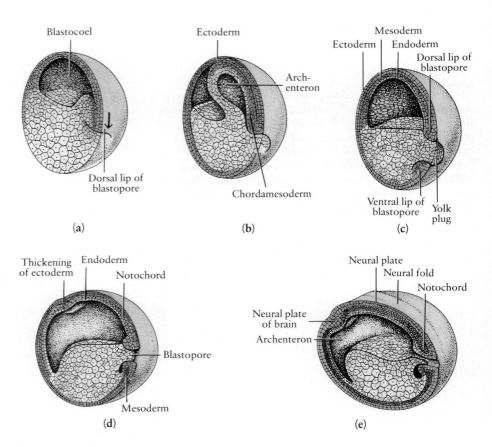

45–12 *Development of the frog gastrula. (a) The blastopore forms in the blastula, and cells from the outer surface begin moving across its dorsal lip to the interior. (b) As the cellular migrations progress, the blastocoel is obliterated and the archenteron created. (c) Three embryonic tissue layers are established: ectoderm, endoderm, and mesoderm. (d) The mesoderm directly over the roof of the archenteron, known as the chordamesoderm, differentiates to become the notochord (shown in green). The ectoderm on the dorsal surface overlying the notochord thickens and flattens, forming the neural ectoderm (shown in purple), from which the neural plate and neural folds begin to form (e).*

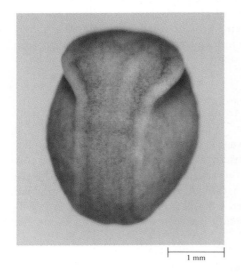

45–13 *The neural plate of a newt embryo. Newts are amphibians that, like other salamanders, have tails throughout their lives. The wide portion of the neural plate, at the anterior of the embryo, will give rise to the brain; the narrow portion will give rise to the spinal cord.*

The formation of the blastopore marks the initiation of gastrulation (Figure 45–12). Cells at the **dorsal lip** of the blastopore change shape and sink below the surface as they move to the interior. These cells are replaced by cells that move from their original positions on the surface of the embryo to the dorsal lip. As a result, the entire sheet of cells that forms the surface of the amphibian embryo appears to stream toward and over the lip of the blastopore as if it were being hauled over a pulley.

Once inside the embryo, the migrating cells move away from the blastopore and deeper into the interior. The direction in which they move is the future anterior-posterior axis of the animal. The advancing cells form the walls of an increasingly spacious archenteron, and the blastocoel disappears. The archenteron remains open to the outside at the blastopore, the site of the future anus. Yolk-laden cells are visible within the boundaries of the blastopore, forming the yolk plug.

In the course of gastrulation, the primary embryonic tissues—endoderm, mesoderm, and ectoderm—become arranged in a three-layered pattern. The floor of the archenteron is composed of yolk-laden endodermal cells. Its roof consists of endodermal cells that have been pushed and pulled into the interior by a sheet of mesodermal cells that lies above it along the axis of the embryo. This sheet of mesoderm includes cells destined to form the notochord (page 591) and is called the **chordamesoderm**. At the sides of the archenteron, other mesodermal cells have slipped between the ectoderm and endoderm, forming the **lateral plate mesoderm**. On the dorsal surface of the embryo, lying above the chordamesoderm, is a sheet of ectoderm that will ultimately give rise to the brain and spinal cord; this tissue is known as **neural ectoderm**. The ectoderm covering the rest of the gastrula, known as **epidermal ectoderm,** will give rise to the epidermis of the skin.

By the end of gastrulation, the first visible signs of differentiation have begun to appear. The chordamesoderm has formed the notochord, and the neural ectoderm has begun to thicken, forming the neural plate (Figure 45–13). The ridges of

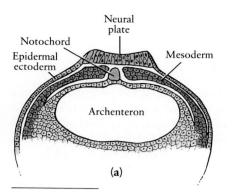

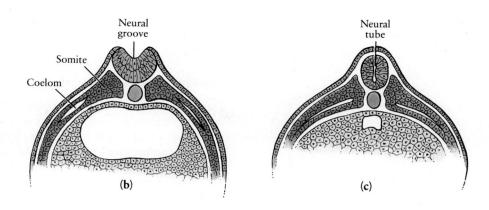

45–14 *The formation of the neural tube from the neural plate in the frog. (a, b) The thickened elevations of neural ectoderm on the right and left sides of the neural plate curve inward, forming the neural groove. The ridges bordering the neural groove then meet and fuse. (c) Finally, the resulting neural tube pinches off from the epidermal ectoderm.*

the neural plate curve upward and inward, forming the neural groove. Ultimately these ridges meet and fuse to form the **neural tube,** which pinches off from the rest of the ectoderm (Figure 45–14). At about the same time, the two longitudinal strips of chordamesoderm on either side of the differentiated notochord split into segments, forming the blocks of tissue called **somites.** In the lateral plate mesoderm, the coelom (page 545) forms between two layers of tissue. Thus the principal features of the vertebrate have been established. The endoderm, which contains the yolk, continues to take up much of the body mass until the stored nutrients are finally absorbed by the larva, the tadpole.

The movements of cells and tissues in gastrulation are completely regular and predictable, embryo after embryo. By applying harmless dyes to the surface of the late blastula, embryologists have developed fate maps (Figure 45–15a) that identify the groups of cells that give rise to the tissue layers of the mature gastrula. During subsequent development, these embryonic tissue layers differentiate to form the specialized tissues and organs of the adult animal (Figure 45–15b).

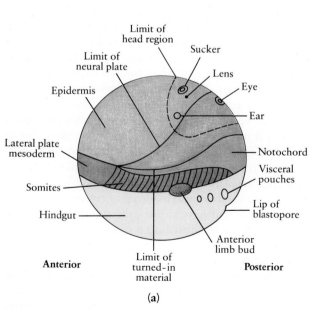

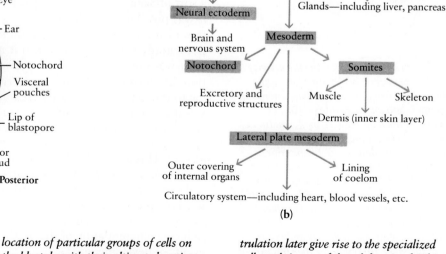

45–15 (a) *A fate map of the amphibian, showing the groups of cells on the surface of the blastula that, in the course of gastrulation and subsequent development, give rise to the tissues of the mature embryo (the tadpole). A comparison of the* location of particular groups of cells on the blastula with their ultimate location within the tadpole makes clear the extent of the cellular reorganization that occurs during amphibian development. (b) The tissue layers that form as a result of gastrulation later give rise to the specialized cells and tissues of the adult animal. This pattern of differentiation is characteristic of all vertebrates, ourselves included.

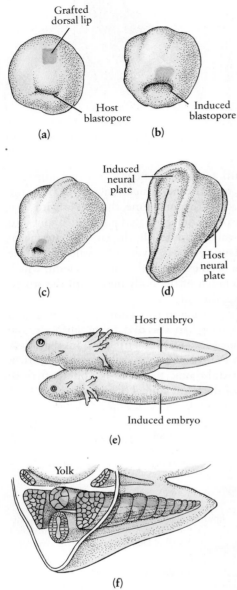

45–16 *Experiment showing the activity of the organizer.* **(a)** *The dorsal lip of the blastopore from one amphibian embryo is grafted onto another embryo.* **(b)** *The grafted dorsal lip induces the formation of a second blastopore, through which* **(c)** *the grafted tissue moves into the interior of the embryo at gastrulation. Note that in* **(b)** *and* **(c)** *the embryo is viewed from a different angle than in* **(a)**, *with the result that the host blastopore is not visible. As development proceeds, two neural plates are formed* **(d)**, *and a double embryo is produced* **(e)**. *In* **(f)** *you see the structure of the secondary embryo under the yolk of the host embryo. The dark brown cells are derived from the graft, and the light tan cells are host material that has been induced to undergo these differentiations.*

The Role of Tissue Interactions

Differentiation, as we saw in Chapter 18, is the result of the selective activation and inactivation of specific genes in the nucleus of a cell. Although molecular geneticists are gradually identifying the mechanisms that, at the level of the DNA molecule, control such activation and inactivation, the linkage of their discoveries to the observed events of development remains one of the major challenges of embryology. However, a variety of experiments, such as the transplantation of the nuclei of tadpole intestinal cells into enucleated eggs (see page 359), have demonstrated that, for some cell types, differentiation does not become irreversible until fairly late in the developmental process. Although the cells have become differentiated, they retain their full developmental potential. For many cell types, however, the developmental potential is gradually limited as gastrulation proceeds and sheets of cells become properly positioned in the embryo. This process, in which the fate of a cell becomes fixed, depends on a progressive series of interactions between different tissue types. In the amphibian, the first tissue to develop a unique character is the dorsal lip of the blastopore. It is the key to the events that follow.

The Organizer

The tissue that lies at the dorsal lip of the blastopore of the amphibian is one of the most thoroughly investigated of all embryonic tissues. Spemann, in a continuation of the studies described earlier (see Figure 45–10), discovered that he could divide the developing blastula in two and sometimes obtain two normal embryos. Once formation of the blastula was complete and gastrulation had begun, however, he found that he could obtain only one normal embryo from the operation. The embryo that was normal always developed from the half that contained the dorsal lip of the blastopore. (This was actually an extension of his previous experiments on the gray crescent; as you will recall, the dorsal lip forms in the blastula at the position once occupied in the egg by the gray crescent.)

Later, in 1924, Hilde Mangold, a student of Spemann, excised the dorsal lip from one salamander embryo and grafted it into the future belly region of a salamander embryo of a different species. When the embryo that had received the graft developed, it was found that a second embryo had formed within its tissues, a sort of Siamese twin to the first. Because the two salamanders differed in color, it was possible to distinguish the tissues of the host embryo from those of the transplant. The transplanted dorsal lip had given rise to chordamesoderm and a notochord, as it would have had it not been removed in the first place. This in itself was noteworthy because previous experiments had shown that other portions of an embryo of similar age, if transplanted, would develop strictly according to the site to which they were transplanted. Even more remarkable, however, was the fact that the rest of the second embryo was made up of tissues of the host. As countless repetitions of this experiment have shown, the transplanted dorsal lip has the effect of organizing tissues of the host into a second embryo (Figure 45–16). This is how the dorsal lip of the blastopore (the prospective chordamesoderm) came to be called the **organizer.**

Despite years of study, we still do not know the nature of the organizer or how it exerts its effect. However, it is known that in chordate embryos of every species examined, the chordamesoderm serves this same function of primary organizer, inducing differentiation of the overlying ectoderm and setting in motion the other events of development.

Induction and Inducers

The action of Spemann's organizer is just one of many tissue interactions that take place in development. This general process is known as **embryonic induction.**

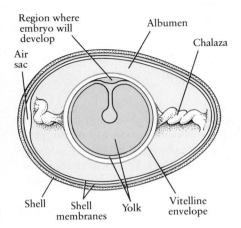

45–17 *Anatomy of a chicken egg. At the time of its release from the ovary in ovulation, the egg consists of the vitelline envelope and its contents. Fertilization, if it occurs, takes place in the upper end of the oviduct. As the egg moves through the oviduct, layers of albumen (egg white) synthesized by and released from cells of the oviduct are added to it. The first of these, the chalaza, is the ropelike structure often observed when an egg is opened. The chalaza suspends the egg cell and, much like a twisted rubber band, rotates it, ensuring that the embryo, developing on one side of the fertilized egg, is always upward, close to the warmth of the mother hen. Two shell membranes and the shell itself are added before the egg is laid.*

Embryonic induction occurs when two different types of tissue come in contact with one another in the course of development and one tissue induces the other to differentiate.

Numerous experiments have shown that induction usually takes place, even when direct contact is prevented between the two tissues, if chemical exchanges are allowed, as by placing a porous filter between them. If chemical exchanges are prevented, as by a piece of metal foil, induction does not occur. Although the chemical substance or substances exchanged have not yet been identified, they are apparently of a very general nature. Experiments have shown that the capacity of the organizer to induce neural plate formation is not species-specific. One can, for example, provoke the formation of a secondary neural tube in a chick embryo by means of a graft of tissue from a rabbit embryo. Moreover, a number of chemicals, some of which never occur in embryos, will induce neural plate formation in embryonic ectoderm. It is not, then, that the inducing substance endows the ectoderm with the capacity to form nerve tissue; rather, it merely evokes a potential that is already present in the responding tissue and initiates a new pattern of gene activity.

DEVELOPMENT OF THE CHICK

The hen's egg (Figure 45–17), familiar to us all, is different in many respects from the eggs previously examined. First, and most obvious, it is surrounded by a shell, which permits it to develop on land. Second, related to its terrestrial development, it is surrounded by a system of membranes. (Such eggs, as noted on page 596, are called amniote eggs.) Third, it contains a large amount of yolk. This large quantity of stored food makes possible a longer period of development before emergence of the immature animal. As a result, the chick, although still a juvenile at the time of hatching, is much farther along in development than the pluteus larva or the tadpole.

The yolk of the hen's egg, like that of the eggs of other birds and also of reptiles, is so large and dense that cleavage does not involve most of the egg mass. The only part of the fertilized egg that cleaves is a thin layer of cytoplasm that sits like a cap on top of the yolk and contains the nucleus. Cleavage of this thin layer produces a lozenge-shaped blastula known as a **blastodisc** (Figure 45–18). Birds, reptiles, and monotreme mammals develop from a blastodisc.

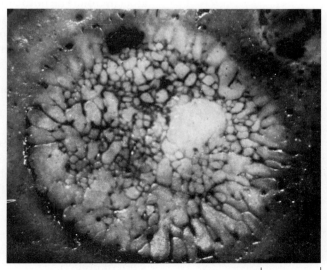

45–18 *The blastodisc of a chick embryo.*

0.5 mm

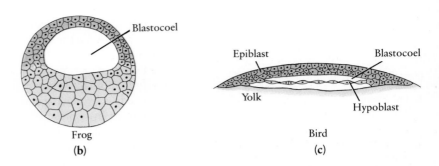

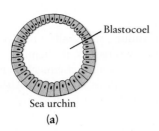

45–19 *Blastulas of* (a) *a sea urchin,* (b) *a frog, and* (c) *a bird. In the sea urchin and the frog, the blastula is a hollow sphere of cells. In the bird, it takes the form of a blastodisc—a flattened ball of cells—that sits on the surface of the yolk. The blastocoel is located between the epiblast and the hypoblast.*

If you break open a fertilized chick egg when it is first laid, you can see the blastodisc as a white mass about 2 millimeters in diameter on top of the yolk. Microscopic examination shows that this mass is made up of many cells (close to 100,000) and that the cells are in two layers: an upper layer, known as the **epiblast,** and a lower layer, the **hypoblast.** The space between them is the blastocoel (Figure 45–19).

If the egg is left intact and development continues, within a short period of time a visible line, known as the **primitive streak,** appears on the surface of the blastodisc. The primitive streak is, in effect, an elongated blastopore. At the beginning of gastrulation (Figure 45–20a), cells of the epiblast that will give rise to the mesoderm and endoderm of the embryo sink inward at the primitive streak and begin to spread out between the epiblast and the hypoblast. (Although some of the cells of the hypoblast contribute to the endoderm of the embryo proper, most of the hypoblast cells are subsequently involved in the formation of membranes surrounding the developing embryo.) As gastrulation continues (Figure 45–20b), the developing mesoderm spreads outward, forming a continuous middle layer. The notochord begins to form from a strip of chordamesoderm cells running from anterior to posterior in the middle of the mesodermal layer, and the coelom forms in the lateral plate mesoderm. Neural ectoderm begins to differentiate above the chordamesoderm. As the cells giving rise to the mesoderm move from the epiblast to the interior, they pull the future epidermal ectoderm and neural ectoderm from the sides of the blastodisc, forming a complete ectodermal covering over the surface of the embryo. As the process proceeds (Figure 45–20c), the embryo becomes tubular and lifts off the surface. At this stage, the similarity of the chick embryo to the frog embryo becomes apparent; both consist of concentric circles of ectoderm, mesoderm, and endoderm surrounding the archenteron. The notochord has differentiated, the somites have formed, and the sides of the neural plate are elevated. Gradually, the ridges of the neural plate come together, fusing to form the hollow, dorsal neural tube (Figure 45–20d). The basic body plan of the vertebrate-to-be is established.

Extraembryonic Membranes of the Chick

Eggs develop in water. In the amniote egg, this water is contained within a set of membranes formed from the tissues of the embryo, the **extraembryonic membranes.** These membranes begin as extensions of the blastodisc, and each is formed from a combination of two of the primary tissue types. The endoderm of the extraembryonic membranes is derived from cells of the hypoblast, and the mesoderm and ectoderm from cells originally located in the epiblast.

As Figure 45–21a (page 934) shows, the yolk gradually becomes surrounded by a membrane of mesoderm and endoderm, known as the **yolk sac.** The function of this membrane is nutritive; its endodermal cells digest the yolk, and blood vessels formed in its mesodermal component carry food molecules from the yolk into the embryo proper.

45-20 *Gastrulation and formation of the neural tube in the chick.* (**a**) *At the beginning of gastrulation, cells that will give rise to the mesoderm and endoderm of the embryo begin their migration through the primitive streak.* (**b**) *As gastrulation continues, the notochord begins to develop, and the lateral plate mesoderm splits into two layers to form the coelom. Neural ectoderm and epidermal ectoderm begin to differentiate.* (**c**) *The embryo becomes tubular and lifts off the surface of the yolk. The notochord has differentiated, the somites have formed, and the sides of the neural plate have begun to elevate.* (**d**) *The ridges of the neural plate come together, fusing to form the hollow, dorsal neural tube. As we shall see shortly, the portions of the three primary embryonic tissues located below the tubular embryo also undergo a series of foldings, producing a set of membranes that enclose and protect the embryo as it continues its development.*

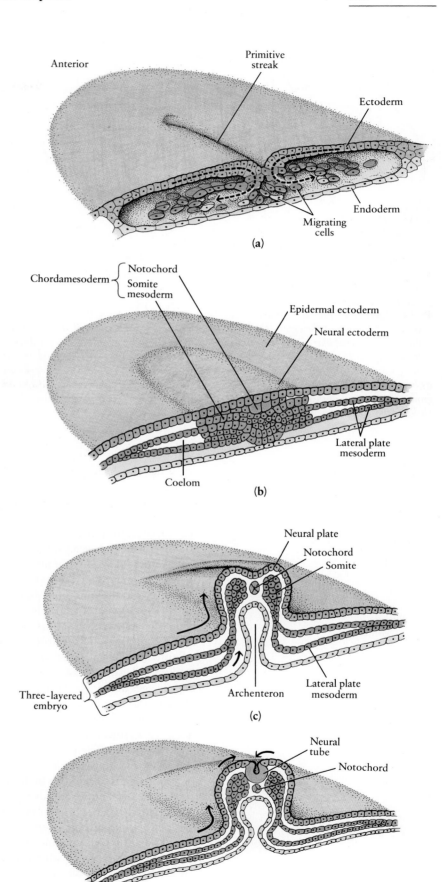

934

The four extraembryonic membranes:
Yolk sac (endoderm and mesoderm)
Allantois (endoderm and mesoderm)
Amnion (mesoderm and ectoderm)
Chorion (mesoderm and ectoderm)

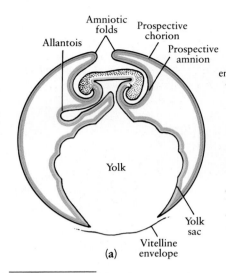

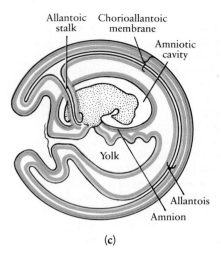

(a)   (b)   (c)

45-21 *Development of the extraembryonic membranes of the chick. As the embryo becomes tubular and lifts from the yolk (see Figure 45-20c), a series of membranes begin to form from the tissues at its base. (a) One membrane, the yolk sac, grows around and almost completely encloses the yolk. A second, the allantois,* *arises as an outgrowth of the rear of the gut. The third and fourth are elevated over the embryo by a folding process during which the membrane is doubled. When the folds fuse, two separate membranes are formed (b). The inner one is the amnion, and the outer one is the chorion. The* *chorion eventually fuses with the allantois to form the chorioallantoic membrane (c), which, in the later stages of development, encloses embryo, yolk, and all the other structures. Note that each membrane is composed of two primary embryonic tissues.*

When the developing embryo lifts from the surface of the blastodisc (see Figure 45-20c), the ectoderm and mesoderm located at its base begin to fold upward, around and over the tubular embryo. These folds of tissue, the amniotic folds, eventually fuse with each other over the embryo, enclosing it in a fluid-filled cavity, the **amniotic cavity** (Figure 45-21b). The saline fluid in the cavity serves to buffer the embryo against mechanical and thermal shocks. Fusion of the amniotic folds creates two membranes, each composed of a layer of ectoderm and mesoderm, separated by the extraembryonic coelom. The inner membrane is the **amnion,** and the outer membrane is the **chorion.**

As development proceeds and the embryo becomes increasingly demarcated from the yolk mass (although always attached to it by the yolk sac stalk), a ventral and posterior pouch is formed from the primitive hindgut. This pouch is the **allantois,** and, as you can see in Figure 45-21b, its wall is composed of endoderm and mesoderm. At first, the function of the allantois is excretion. Since bird and reptile eggs are closed off from their environment by the shell, there is no way to dispose of the products of nitrogen metabolism, most of which are toxic. Evolution has provided a dual answer to this problem. First, the nitrogenous wastes are converted to uric acid, which, as we saw in Chapter 37, is insoluble and of low toxicity. The uric acid is then voided from the embryo into the allantois, which essentially serves as an embryonic garbage bag. When a chick hatches, the accumulated uric acid can be found adhering to the abandoned shell.

As development proceeds further, the allantois expands in volume and pushes out into the extraembryonic coelom, gradually enveloping the embryo proper as it does so. Eventually, as Figure 45-21c shows, the allantois effectively obliterates the extraembryonic coelom, and its walls fuse with the chorion. Since the allantois is plentifully supplied with blood vessels, the **chorioallantoic membrane,** formed by the fusion of the allantoic wall with the chorion, acts as an efficient respiratory membrane for the embryo during its later development.

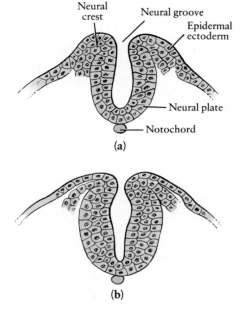

(a)

(b)

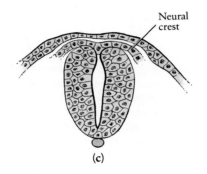

45–22 *During the formation of the neural tube, some of the neural ectoderm cells at the crests of the neural folds are pinched out as the folds come together. These cells form a variety of structures in the peripheral nervous system. Eventually, some of* them will migrate down toward the notochord and aggregate into ganglia. There they develop into sensory neurons, sending fibers up to connect with the dorsal part of the spinal cord and out into the surrounding tissues.

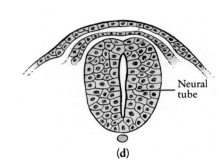

(c)

(d)

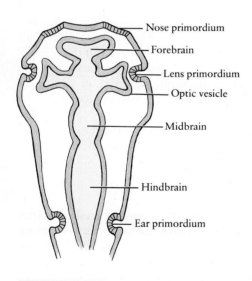

45–23 *The development of the brain and associated sense organs. At the anterior end of the neural tube, local swellings produce three distinct bulges—the forebrain, the midbrain, and the hindbrain. The forebrain then bulges laterally, and two saclike optic vesicles appear. At the same time, the epidermal ectoderm folds inward to meet the optic vesicles. The earliest stages—primordia—of ears and nostrils also appear first as infoldings of the epidermal ectoderm.*

Organogenesis: The Formation of Organ Systems

While the extraembryonic membranes are forming, the development of the embryo itself continues. These later stages of development, following cleavage and gastrulation, are generally termed **organogenesis,** the formation of the organ systems. Organogenesis begins with the inductive interaction between ectoderm and underlying chordamesoderm. Each of the three primary tissues formed during gastrulation then proceeds to undergo growth, differentiation, and morphogenesis. This process is essentially the same in all vertebrates.

Differentiation of the Ectoderm

The neural tube stretches out along the dorsal surface, growing and becoming longer and thinner as the embryo increases in size. The cells in the neural tube at first appear to be all alike, but some of them, the future motor neurons, begin to extend long processes that grow out beyond the neural tube and invade the peripheral organs and tissues.

When the neural tube is first forming, some of the ectodermal cells at the crests of the neural folds are left behind in the surrounding tissue, as shown in Figure 45–1 (page 919). Some of these cells from the neural crest, the future sensory neurons, migrate into positions around the neural tube (Figure 45–22) and form connections between the dorsal part of the tube and the surrounding tissues. Others become Schwann cells, which surround and insulate the nerve fibers of the peripheral nervous system, some become the pigment cells (melanocytes) found at the base of the epidermis, and some form the adrenal medulla. Meanwhile, mesodermal cells have begun to migrate toward the neural tube and underlying notochord and to aggregate around them. Here they differentiate into cartilage and give rise to the hollow vertebral column. If a portion of neural tube is transplanted to another region of the mesoderm, mesodermal cells will similarly aggregate around it as a result of tissue interaction.

As we noted in Chapter 43, the brain begins as three bulges in the foremost part of the neural tube (Figure 45–23). Almost immediately, two saclike protrusions, the optic vesicles, appear on the sides of the anterior region of the forming brain. These spherical vesicles enlarge until they come into contact with the epidermal ectoderm, while still remaining connected to the neural tube by the narrowing optic stalk. Within this stalk, the optic nerve will develop.

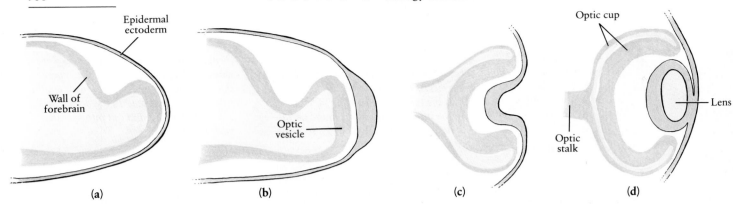

Epidermal
ectoderm

Wall of
forebrain

(a)

Optic
vesicle

(b)

(c)

Optic cup

Optic
stalk

Lens

(d)

45–24 *Eyes develop as a result of interactions between the epidermal ectoderm and the optic vesicles. When the optic vesicle reaches the epidermal ectoderm, the epidermal ectoderm begins to thicken and differentiate. The optic vesicle flattens out and invaginates, becoming the double-*

walled optic cup. The invaginated wall will become the retina; the outer, thinner wall develops into the pigment layer. The rim of the optic cup will later form the edge of the pupil. The thickened layer of epidermal ectoderm pinches off to become the transparent lens, and the overlying

epidermal ectoderm, which also becomes transparent, forms the cornea. The connection with the brain remains as the optic stalk, within which is the optic nerve. Note that the retina is thus a differentiated extension of the brain.

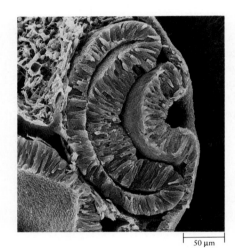

50 μm

45–25 *The developing eye of a chick embryo, as revealed by the scanning electron microscope. As the cells of the thick, invaginated layer of the optic cup differentiate into rods and cones and bipolar, horizontal, amacrine, and ganglion cells, the multiple connections that characterize the retina (see Figure 42–13, page 869) will become established.*

When the optic vesicle comes into contact with the inner surface of the epidermal ectoderm, the external surface of the vesicle flattens out and pushes inward. The vesicle thus becomes a double-walled cup. The rim of the cup becomes the edge of the pupil. The opening of the cup is large at first, but the rim bends inward and converges, so that the opening of the pupil becomes smaller. At the point where the optic vesicle touches the epidermal ectoderm, the epidermal ectoderm begins to differentiate into a lens, becoming transparent (Figure 45–24).

The differentiation of the lens of the eye is an example of a **secondary induction**. The inducing tissue (the optic vesicle) is itself the result of the primary induction of dorsal ectoderm by dorsal mesoderm. In fact, the formation of the complete eye involves at least six separate inductions, occurring in an orderly sequence, each one linked to the one before, with primary induction by the organizer as the starting point.

Other infoldings and differentiations of the epidermal ectoderm in the region of the head, associated with the presence of the embryonic brain, lead to the elaboration of the organs of olfaction and hearing.

Differentiation of the Mesoderm

The series of somites that came to lie on each side of the notochord shortly after gastrulation (Figure 45–27) now differentiate into three kinds of cells: (1) sclerotome cells, which later form skeletal elements; (2) dermatome cells, which become part of the developing skin; and (3) myotome cells, which form most of the musculature. In aquatic vertebrates, this segmental pattern, with its simple double series of back muscles, is retained in adulthood; it is well suited to the side-to-side motion of the body in swimming and crawling. In terrestrial vertebrates, however, this initial segmental arrangement is almost obliterated by modifications and additions of the muscles associated with locomotion on land.

Lateral plate mesoderm adjacent to the somites forms the kidneys, the gonads, and the ducts of the excretory and reproductive systems. Earlier in development, the lateral plate mesoderm in the ventral portion of the embryo split into two sheets, creating the coelom (see Figure 45–20b). One of these sheets forms the lining of the thoracic and abdominal cavities, and the other becomes the outer layer of the internal organs.

45–26 *Major stages of organogenesis and subsequent development in the chick embryo. (a) At 4 days, large blood vessels have grown out from the embryo into the yolk sac. Note the heart, above the junction of the blood vessels with the embryo. (b) At 5 days, the yolk sac has grown further around the yolk. Its endodermal cells secrete lipase (page 722), an enzyme that digests fat in the yolk. Nutrient molecules are carried to the embryo via the mesodermal blood vessels. (c) By 8 days, the large, pigmented eyes are clearly visible. (d) At 11 days, the limbs are developing. The brain is visible through the transparent skull. (e) By 14 days, feathers are beginning to appear. (f) By 21 days, just prior to hatching, a white knob is fully formed on the end of the beak; this is the egg tooth, which is used in pecking open the shell.*

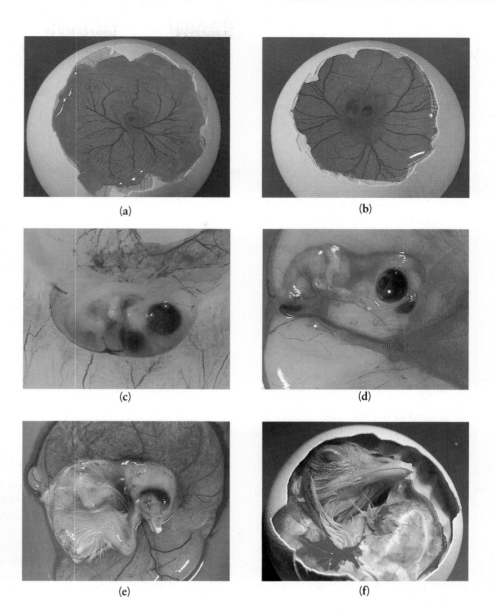

(a)

(b)

(c)

(d)

(e)

(f)

|——— 100 μm

45–27 *Somites of a chick embryo, flanking the neural tube. The formation of somites from the chordamesoderm on either side of the notochord proceeds from the anterior of the embryo (top of micrograph). Three pairs of somites are fully formed, and a fourth pair is beginning to separate from the posterior chordamesoderm. The wormlike extensions from the upper edge of the neural tube are groups of neural crest cells (Figure 45–22) beginning their downward migration.*

Differentiation of the Endoderm

The endoderm differentiates into tissues of the respiratory and digestive tracts and a number of related organs. Early in development, endodermal pouches develop at the anterior end of the archenteron. They push laterally until they meet the epidermal ectoderm, which folds inward to meet them. This produces a series of grooves on the surface of the embryo. In aquatic vertebrates, endoderm and ectoderm fuse and a perforation forms, around which gill filaments develop. In terrestrial vertebrates, including humans, portions of these endodermal pouches develop into the auditory tubes (see Figure 42–6, page 864), tonsils, parathyroid glands, and the thymus gland. Posterior to this region, the lungs develop as similar outpocketings, branching into two sacs. More posterior still, outpocketings from the primitive gut begin to differentiate into liver, gallbladder, and pancreas.

No organ system is derived from only one type of tissue. For example, the lining of the intestine is of endodermal origin; these lining cells secrete the digestive juices and absorb the digested materials. However, as we saw in Chapter 34, the functional structure of the intestine also includes muscles, connective tissue, blood vessels, nerves, and an outer wrapping. These are made up of tissues derived from mesoderm and ectoderm.

45–28 *Formation of the vertebrate neural tube depends on changes in the shape of individual cells brought about by movements of elements of the cytoskeleton.* **(a)** *Newly differentiated cells of neural ectoderm are cuboidal, and the microtubules within the cells appear to be randomly arranged.* **(b)** *As formation of the neural plate proceeds, the microtubules become arranged parallel to the dorsal-ventral axis of the embryo. As the microtubules lengthen through the addition of tubulin dimers (page 78), the cells become columnar. Just below the upper surface of the membrane of each cell is a band of actin filaments, running parallel to the left-right axis of the embryo. When this band contracts* **(c)***, the cells become wedge-shaped, forcing the entire sheet of cells to curve upward. A similar process is involved in the development of other curved or tubular structures that consist of sheets of cells.*

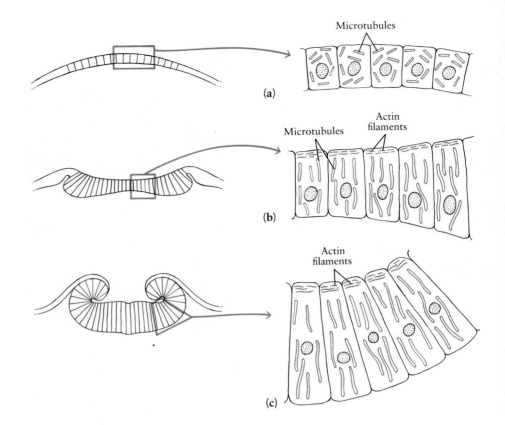

Morphogenesis: The Shaping of Body Form

Each part of the body has a characteristic shape and structure. For example, the spinal cord is basically a hollow tube, the liver is composed of distinctive lobules, the salivary glands consist of groups of secretory cells clustered around ducts, and the lungs consist of microscopic air spaces lined by extremely thin sheets of cells. Only a few cellular processes, repeated over and over in various permutations and combinations, appear to be responsible for shaping the various structures of the body. These processes include (1) increases or decreases in the rates of cell growth and division, (2) changes in the adhesion of cells to neighboring cells, (3) the deposition of extracellular materials, and (4) changes in cell shape brought about by extension or contraction. For example, formation of the neural tube from the flattened neural plate is accomplished by changes in the shape of the individual cells. As shown in Figure 45–28, cells in the neural plate first become elongated, a change that is associated with the alignment of microtubules parallel to the dorsal-ventral axis of the embryo. Following elongation, a layer of actin filaments, located just below the cell membrane in the dorsal portion of each cell, contracts. This causes the cells to become wedge-shaped, which, in turn, causes the edges of the neural plate to begin curving upward.

Such cellular processes give rise not only to the internal structures that are similar from one vertebrate to another but also to the distinctive structures that characterize different groups of vertebrates. Among the most striking features of terrestrial vertebrates are four limbs. As we saw in Figure 20–7 (page 414), the limbs of reptiles, birds, and mammals all have the same basic structure, and yet there are obvious—and significant—differences between, for example, the forelimbs of a bird (wings) and the forelimbs of a human (arms). Similar differences are apparent between the forelimbs of a bird and its hindlimbs (legs). The developmental process that gives rise to these differences is known as **pattern formation.** It has been studied principally through experimental manipulations of the developing chick wing.

The first visible sign of wing development is the appearance of a wing bud, which consists of a central mass of mesodermal cells covered by a thin layer of

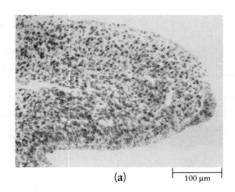

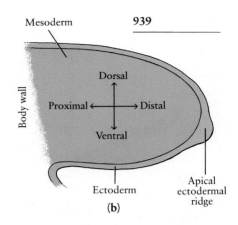

(a)   ⊢—— 100 μm ——⊣   (b)

45–29 (a) *The wing bud of a chick embryo, viewed from the posterior. This micrograph was taken three and a half days after the wing bud began its development. Note the thickened region, the apical ectodermal ridge, at its tip.* (b) *The tissues of the wing bud consist of an outer layer of ectoderm, from which the skin and feathers will form, and an inner layer of mesoderm, from which all of the bones, cartilage, tendons, and muscles of the wing will develop.*

ectoderm. As the wing bud begins to grow outward, a ridge of specialized tissue, the apical ectodermal ridge, forms at its tip (Figure 45–29). If this ridge is removed, elongation will not occur. Immediately beneath the apical ectodermal ridge is a region of actively dividing mesodermal cells, known as the progress zone, which moves outward with the apical ridge as the wing bud elongates. The progress zone produces the cells from which the structures of the completed wing are formed. As the wing bud elongates and the progress zone moves outward, newly divided cells are left behind and begin to differentiate into the characteristic structures of the wing from its base at the body wall (the proximal end) to its tip (the distal end).

Numerous experiments have been performed in which the apical ectodermal ridge, the mesodermal cells of the progress zone, or both have been grafted between wing buds at different stages of development. These experiments have shown that the length of time that cells spend near the apical ectodermal ridge before they are left behind is a critical factor in determining the structures into which they differentiate. The cells that are left behind first spend less time under the influence of substances secreted from the apical ectodermal ridge, and they become the proximal structures of the wing. Those that are left behind later spend more time in the region of the apical ectodermal ridge, and they become the distal structures of the wing.

A wing is characterized not only by a specific series of structures from its proximal end to its distal end but also by a specific series of structures along the axis from anterior to posterior (Figure 45–30). A variety of grafting experiments have shown that differentiation along this axis is controlled by chemical substances produced by cells at the posterior margin of the wing. For example, if a piece of tissue is transplanted from the posterior margin of one wing bud to the anterior margin of another, the wing that develops is a kind of Siamese twin in which the bones are duplicated in a mirror image of the normal wing structure (see Figure 45–31 on the next page).

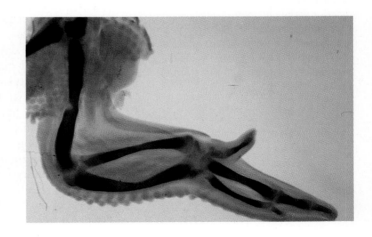

45–30 *A dorsal view of the wing of a chick embryo, 10 days after the wing bud first began its development. In the course of the evolution of the bird wing, two of the five digits of the basic vertebrate "hand" have been lost—the digits corresponding to the thumb and little finger. Thus, the digits of the wing are numbered, from anterior (top of micrograph) to posterior (bottom of micrograph), as 2, 3, and 4.*

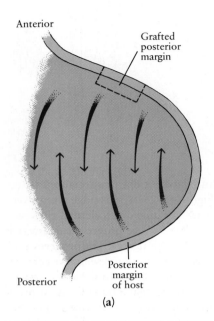

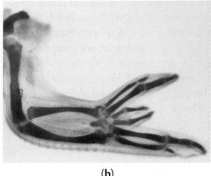

45–31 *Development of the correct structures on the anterior-to-posterior axis of the chick wing depends on a concentration gradient of a chemical substance, now known to be retinoic acid, released from cells at the posterior margin of the wing bud.* **(a)** *If cells from the posterior margin of one wing bud are grafted onto the anterior margin of another wing bud, and the posterior margin of the host wing bud is left intact, a dual gradient is established.* **(b)** *The result is a wing with six digits, arranged, from anterior to posterior, in the sequence 4, 3, 2, 2, 3, 4.*

The substance released from the posterior margin of the wing bud has recently been identified as retinoic acid, a member of the same chemical family that includes the carotenoids, vitamin A, and the retinal of the vertebrate eye. Retinoic acid diffuses from the posterior margin toward the anterior margin of the wing bud, creating a concentration gradient from high (posterior margin) to low (anterior margin). The receptors for retinoic acid have now been located on the surface of cells of the developing wing bud. Structurally the retinoic acid receptors are very similar to the receptors for steroid hormones and thyroid hormone, and it appears that their mechanism of action is also similar (see page 836). The particular sequence of cellular events that is initiated by the binding of retinoic acid to its receptors is thought to be influenced by the number of receptors activated, which is, of course, related to the location of the cell within the concentration gradient.

The identification of retinoic acid as a signalling molecule in wing development and the isolation and characterization of its receptor suggest that the integration and control of development are, in principle, similar to the integration and control of other physiological processes. A sense of high adventure now prevails as developmental biologists seek to identify other substances involved in the regulation of development, the mechanisms by which these substances turn genes on and off, the genes themselves, and the proteins for which they code.

DEVELOPMENT OF THE HUMAN EMBRYO

The basic patterns of development are remarkably similar throughout the animal kingdom, and particularly among vertebrates. Thus, experimental studies of the development of sea urchins, frogs, chicks, and mammals such as mice have provided a solid foundation for our understanding of human development—an area of biology in which experimental manipulation is, throughout the world, strictly prohibited.

Human oocytes, as we saw in the preceding chapter, mature in the ovary and are released at approximately 28-day intervals. Fertilization usually takes place in an oviduct. As in other species, fertilization results in (1) changes in the outer surface of the egg that prevent entry of other sperm cells, (2) metabolic activation of the egg, (3) introduction of the genetic material of the father, and (4) cleavage. After fertilization, the egg continues its passage down the oviduct, where the first cell divisions take place (Figure 45–32). At about 36 hours after fertilization, the fertilized egg divides to form two cells; at 60 hours, the two cells divide to form four cells. At three days, the four cells divide to form eight. In this early stage, all of the cells are of equal size, as they are in the sea urchin. During this period, the embryo is entirely on its own, following its own genetic program and supplying its own metabolic requirements. All it needs is a suitable fluid environment containing progesterone.

By about five days after fertilization, the blastula consists of some 120 cells, and a fluid-filled blastocoel has begun to form within it. The structure of the mammalian blastula, which is known as a **blastocyst,** is rather different from that of the blastulas we have examined previously. A cross section of the blastocyst looks somewhat like a ring with the stone on the inside. The inner cell mass (the stone) is a ball of cells at one pole of the blastula that will develop into the embryo proper. The ring is called the **trophoblast** (from the Greek word *trophe,* "to nourish"). It is composed of a double layer of cells and completely encloses the developing embryo. The trophoblast is the precursor of the chorion. (In the chick, as you will recall, the chorion develops from mesoderm and ectoderm formed at the base of the embryo during gastrulation. It is as if the mammalian embryo has skipped a step, arriving prepackaged.)

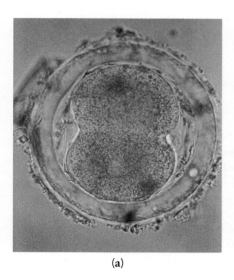

(a)

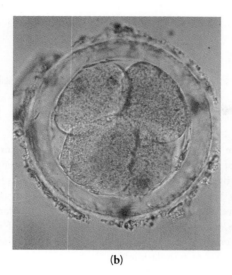

(b)

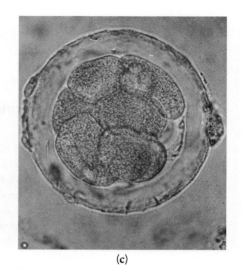

(c)

45–32 *A human embryo at (a) the two-cell stage, (b) the four-cell stage, and (c) the eight-cell stage. Although the number of cells has doubled with each cleavage, the volume of the embryo has not increased. Thus, the cells are still easily contained within the fertilization membrane. Sometimes the cells separate at the two-cell stage, resulting in identical twins.*

About three days after the embryo reaches the uterus (about six days after fertilization), the trophoblast makes contact with the tissues of the uterus and releases a protein hormone, chorionic gonadotropin. This hormone protects the pregnancy by stimulating the corpus luteum to continue its production of estrogens and progesterone, thus preventing menstruation (see page 913). Most pregnancy tests involve the detection of chorionic gonadotropin in blood or urine; with the development of monoclonal antibodies (page 802) against human chorionic gonadotropin, these tests have become much more sensitive and reliable.

The trophoblast cells, multiplying rapidly by now, chemically induce changes in the endometrium and invade it. As the embryo penetrates the endometrial tissues—the process known as **implantation** (Figure 45–33)—it becomes surrounded by ruptured blood vessels and the nutrient-filled blood escaping from them. The trophoblast thickens and develops fingerlike projections that invade the uterine lining to develop into the ectodermal component of the chorion.

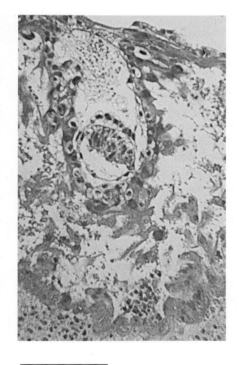

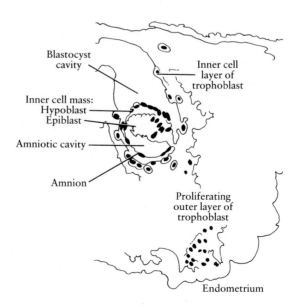

45–33 *Implantation of a human embryo. The tiny embryo invades the lining of the uterus within a week after fertilization. Subsequently, the placenta begins to form. This organ is the site at which oxygen,* nutrients, and other materials are exchanged between the mother and embryo; it is also the source of hormones that help to maintain pregnancy. Implantation usually occurs three to four days after the young embryo reaches the uterus. Once implantation has been achieved, the embryo is no longer independent but is parasitic on the mother.

941

Genetic Control of Development: The Homeobox

Underlying all of the developmental processes we have examined are, of course, genes—sequences of information encoded by DNA molecules. The cytoplasmic determinants that have such a profound influence on early development are the direct products of maternal genes, which are transcribed during the maturation of the oocyte and translated into protein after fertilization. These cytoplasmic factors, as well as all of the other substances involved in the induction of tissue differentiation and pattern formation, cause changes in the expression of specific genes in the nuclei of the embryonic cells they affect. The first steps in identifying these genes and in determining their functions in development are being made possible by the ubiquitous fruit fly, *Drosophila.*

Drosophila, like all insects, exhibits a striking segmentation, both in the larval stages and in the adult. Moreover, the adult is characterized by a series of complex appendages, including mouthparts, antennae, wings, and legs, precisely located on particular segments. Two distinct sets of genes controlling pattern formation in *Drosophila* have now been identified. One set, known as segmentation genes, controls the number and sequence of segments in the larva and, subsequently, in the three regions (head, thorax, and abdomen) of the adult fly. Mutations in these genes can result in larvae and flies in which certain segments are missing, or, alternatively, are duplicated. Another set, known as homeotic genes, controls the identity of segments; that is, they control the way in which a given segment develops and the appendages that it bears. Mutations of homeotic genes can result in bizarre disruptions of the normal developmental pattern. For example, mutations of the genes of the antennapedia complex, which controls the development of antennae on the head and of legs on the three segments of the thorax, can result in the development of legs on the head—where the antennae should have developed. Another group of homeotic genes, the bithorax complex, is involved in the control of the development of both legs and wings on the thoracic segments. In a normal fruit fly, the first thoracic segment bears a pair of legs, the second segment bears a pair of legs and a pair of wings, and the third segment bears a pair of legs and a pair of greatly reduced wings, known as halteres. Although the halteres play no role in the propulsion required for flight, they are critical in enabling the fly to maintain its balance while airborne. Mutations in two of the loci of the bithorax complex produce flies in which the third thoracic segment develops as a second thoracic segment; the result is a fly with four full-size wings and no halteres.

At present, we know very little about the functions of the proteins coded by the homeotic genes. Some of these proteins undoubtedly regulate the expression of other genes that have more specific roles in development. Others may be involved in the regulation of basic cellular activities, such as cell growth and division. For example, the gene known as *engrailed,* which is involved in the division of each segment into anterior and posterior compartments, appears to act by limiting the growth of populations of cells. In *engrailed* mutants, cells of the posterior part of each segment grow into the anterior part, which normally they would not penetrate.

Despite our lack of knowledge of the mechanisms by which homeotic genes exert their effects, many of the genes themselves have now been located on the *Drosophila* chromosomes and have been cloned and sequenced. In 1983, it was discovered that more than a dozen of the *Drosophila* homeotic genes contain a common DNA sequence of 180 nucleotides. This sequence is known as the homeobox. The nucleotide sequence of the homeobox has now been identified in the DNA of a variety of animals, ranging from worms to humans, all of which exhibit a segmental pattern at some stage of their development. The polypeptide strand dictated by the nucleotide sequence of the homeobox contains many basic amino acids, suggesting that it may function as a regulatory molecule that binds to DNA (which, as you will recall, is acidic), altering the course of gene expression. These discoveries suggest that even though the adult forms of different animals vary greatly, certain "master" genes play a critical role in the development of body organization and pattern throughout the animal kingdom.

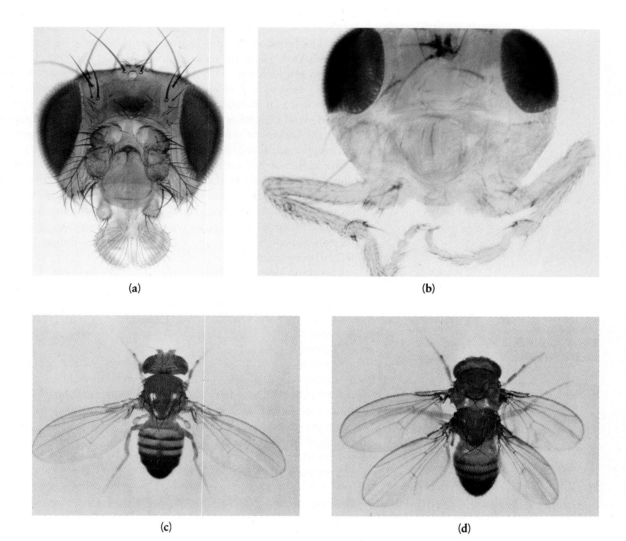

The effects of mutations in two groups of homeotic genes of Drosophila. (a) Head of a normal fly, and (b) head of a mutant fly in which the antennae are replaced by legs. (c) A normal fly with one pair of wings and one pair of halteres, and (d) a mutant fly with two pairs of wings and no halteres.

Extraembryonic Membranes

As the embryo becomes implanted, the extraembryonic membranes begin to develop. These have interesting similarities to and differences from the avian-reptilian membranes discussed previously. In the first place, the yolk sac, which develops first, has no yolk. This lack of yolk is a secondary evolutionary development. The monotremes produce eggs with yolks. The marsupials also produce eggs with yolks, but the yolks are discarded at the first cleavage. The placental mammals produce eggs with no yolks; instead there is a prominent cavity where the yolk "used to be." However, cleavage and the cellular migrations of gastrulation proceed as if the yolk were still there.

In humans and other mammals, the allantois buds off from the hindgut and eventually forms the primitive urinary bladder. Wastes are not segregated as uric acid and stored, as they are in the chick, but are transported as urea and ammonia to the maternal bloodstream. In the course of development, the endodermal component of the allantois becomes greatly reduced, but the allantoic mesoderm spreads out; it is the source of the major blood vessels on the embryonic side of the placenta. The allantoic stalk eventually becomes the umbilical cord, the embryo's principal connection with the uterine tissues. The blood vessels of the allantois transport oxygen and nutrients that have diffused in from the mother's circulatory system and carry off carbon dioxide and other wastes.

The third extraembryonic membrane is the amnion. The space between the amnion and the embryo is the amniotic cavity, and, as in the chick, it is filled with a saline fluid. The fourth membrane is the chorion, which is a combination of ectoderm from the trophoblast and mesoderm grown out from the embryo itself. By about the fourteenth day, chorionic villi begin to form: the formation of the villi represents the beginning of the mature placenta.

In amniocentesis (page 387), one of the two prenatal genetic testing procedures in current use, fluid samples are taken from the amniotic cavity. These samples contain cells that have been sloughed off by the embryo. The major drawback of amniocentesis is that it cannot be performed until the sixteenth week of pregnancy. A newer procedure, chorionic villus biopsy, can be performed as early as the eighth or ninth week of pregnancy. In this procedure, a small sample of tissue is removed from the chorion, which is, of course, genetically identical to the embryo itself.

The Placenta

By the end of the third week after conception, the placenta covers 20 percent of the uterus. It is a disc-shaped mass of spongy tissue through which all exchanges between mother and embryo take place. The placenta is formed as a result of the interactions of a maternal tissue, the endometrium, with the extraembryonic chorion, and has a rich blood supply from both. However, the embryonic and maternal circulatory systems are not directly connected (Figure 45–34), so maternal and embryonic blood cells do not mix. Molecules, including food and oxygen, diffuse from the maternal bloodstream through the placental tissue and into the blood vessels that carry them into the embryo. Similarly, carbon dioxide and other waste products from the embryo are picked up from the placenta by the maternal bloodstream and carried away for disposal through the mother's lungs and kidneys. The mother's blood volume begins to increase in response to these extra demands. Her appetite increases and the absorption of certain nutrients, such as calcium and iron, is increased.

From this point in development until birth, the embryo remains securely attached to the placenta by the umbilical cord. This permits the embryo to float freely in its sac of amniotic fluid.

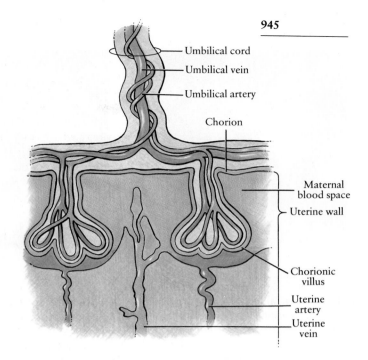

45-34 *From the placenta, numerous fingerlike chorionic villi project into the maternal blood space in the wall of the uterus. This space is kept charged with blood from branches of the uterine artery. Across the thin barrier separating fetal from maternal blood, exchange of materials takes place: soluble food substances, oxygen, water, and salts pass into the umbilical vein from the mother's blood; carbon dioxide and nitrogenous waste, brought to the placenta in the umbilical arteries, pass into the mother's blood. The placenta is thus the excretory organ of the embryo as well as its respiratory surface and its source of nourishment.*

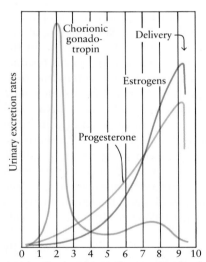

45-35 *Levels of estrogens, progesterone, and chorionic gonadotropin excreted in the urine during pregnancy. Urinary excretion rates reflect the concentrations of these hormones in the blood.*

The chorionic component of the placenta is, from the perspective of the mother's immune system, a graft of foreign tissue. As such, it has the potential to evoke a cell-mediated immune response (page 801), and this sometimes happens, leading to miscarriage. In most pregnancies, however, there is no immune response to the embryonic tissues. A selective immune suppression occurs that allows the embryo to develop unharmed while full protection of both mother and embryo from infectious agents is maintained. The mechanism by which this is accomplished is not yet understood.

As the placenta matures, it begins producing large amounts of estrogens and progesterone (Figure 45–35). By the end of the third month of pregnancy, the placenta has completely replaced the corpus luteum as a source of estrogens and progesterone. Chorionic gonadotropin is no longer produced by the chorion, and the corpus luteum degenerates. It is hypothesized that many of the miscarriages that occur in the third month result from the degeneration of the corpus luteum before the placenta is producing adequate levels of hormones to maintain the pregnancy.

The First Trimester

During the second week of its life, the embryo grows to 1.5 millimeters in length, and its major body axis begins to develop. (In this and subsequent measurements, the embryo is measured from crown to rump.) As it elongates, a primitive streak forms, very similar in appearance to the primitive streak of the chick. Cell migrations through the primitive streak establish the three-layered embryo.

During the third week, the embryo grows to 2.3 millimeters long, and most of its major organ systems begin to form. The neural groove is the beginning of the central nervous system, which is the first organ system to develop. By 22 days, the very rudimentary heart, still only a tube, begins to flutter and then to pulsate. From this time on, the heart will not stop its 100,000 beats per day until the death of the individual. Soon after, the eyes begin to form. Also, by this time, about 100 cells have been set aside in the yolk sac as germ cells, from which the egg or sperm cells of the individual will eventually develop. These cells begin to crawl, amoeba-like, toward the site at which the gonads will develop.

By the end of the first month, the embryo is 5 millimeters in length and has increased its mass 7,000 times. The neural groove has closed, and the embryo is now C-shaped. At this stage, it can be clearly seen that the tissues lateral to the

(a)

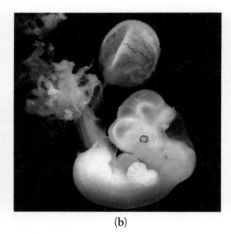

(b)

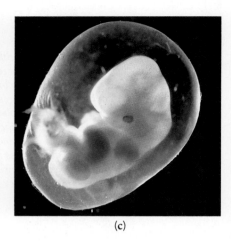

(c)

45–36 *Human embryos, early in their development. (a) In this four-week-old embryo, the neural groove is still open. The protuberances near the top are the limb buds of the future arms; the limb buds of the legs, which develop more slowly, are visible at the bottom. This embryo is 5 millimeters long. (b) By five weeks, the embryo is 10 millimeters in length. Its developing brain, an eye, hands, and a long tail can be seen. The yolk sac is at the top. (c) At six weeks, the embryo has grown to 15 millimeters in length. As it floats securely in the amniotic cavity, its heart beats rapidly. The brain continues to grow, and the eyes are more developed. The dark red object in the abdominal region is the liver. Skin folds have formed that will give rise to the outer ears.*

notochord are arranged in paired somites. The embryo typically has 40 pairs of somites, from which muscles, bones, and connective tissues will develop. The heart, even as it beats, develops from a simple set of paired, contracting tubes into a four-chambered vessel.

By 38 days, the germ cells reach their destination, the developing gonads. Although the rudimentary gonads have begun to form by this time, male and female embryos are still anatomically identical. Whether the infant develops as a male or as a female appears to be determined by a single gene situated on the Y chromosome. This gene, which was first identified late in 1987, codes for a DNA-binding regulatory protein. It is thought to be the master switch of sexual development, turning on other genes that lead to the differentiation of the primitive gonads into testes and of the germ cells into spermatogonia. In the absence of this gene, the gonads develop as ovaries and the germ cells become oogonia. More recently, a gene has been located on the X chromosome that, although not identical to the Y chromosome gene, is very similar in its nucleotide sequence. It is not yet known whether this is an inactive pseudogene (see page 374) or an active gene coding for a protein that acts either in conjunction with or in opposition to the product of the gene on the Y chromosome. In addition to the action of a master gene or genes, interactions between the germ cells and the primitive gonads are necessary for organogenesis to proceed. As the testes form, they begin their secretion of androgens, and under the influence of androgens, the external genitalia and other structures become masculinized. If a female embryo is exposed to androgens, it will become similarly masculinized, although, gonadally speaking, it will be female.

During the second month, the embryo increases in mass about 500 times. By the end of this period, it weighs about 1 gram, slightly less than the weight of an aspirin tablet, and is about 3 centimeters long. Despite its small size, it is almost human-looking, and from this time on it is generally referred to as a **fetus.** Its head is still relatively large, because of the early and rapid development of the brain, but the relative size of the head will continue to be reduced throughout gestation (and throughout childhood as well).

Arms, legs, elbows, knees, fingers, and toes are all forming during this time (Figure 45–37). As another reminder of our ancestry, there is a temporary tail. The tail reaches its greatest length in the second month and then, as its growth rate slows, becomes surrounded by other tissues and disappears. Gallbladder and pancreas are present at this stage, and there is clear differentiation of the divisions of the digestive tract. The liver now constitutes about 10 percent of the body of the fetus and is its main blood-forming organ. By the end of the second month, the major steps in organ development are more or less complete. In fact, the rest of development is mostly concerned with growth and the maturation of physiological processes.

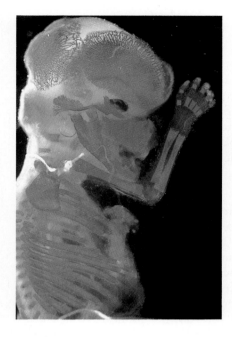

45–37 *By the seventh week of embryonic life, the skeleton begins to turn from cartilage to bone, as revealed by differential staining. A gradual layering of membranes, one on top of another, will give rise to the skull, which protects the developing brain.*

The first two months are the most sensitive period in human development as far as the possible influence of external factors is concerned. For example, when the arms and legs are mere rudiments (fourth and fifth weeks), a number of substances can upset the normal course of events and result in limb abnormalities. The experience with the presumably safe tranquilizer thalidomide in the early 1960s is a tragic and familiar example. Because of the widespread publicity about the "thalidomide babies," now adults, there was an increase in research into teratogens (substances that cause fetal deformities). A large number of drugs have been found to cause birth defects, and pregnant women and their doctors have become far more cautious about the use of any medication during these critical first months. Similarly, exposure to x-rays at doses that would not affect an adult or even an older fetus may produce permanent abnormalities. Heavy alcohol consumption and smoking also affect normal organogenesis; frequently, alcoholic mothers have miscarriages during the first trimester.

Infections may also affect the development of the embryo. Rubella (German measles) is a very mild disease in children and adults. Yet when contracted by the mother during the fourth through the twelfth weeks of pregnancy, it can have damaging effects on the formation of the heart, the lens of the eye, the inner ear, and the brain, depending on exactly when the infection occurs in relation to embryonic development.

During the third month, the fetus begins to move its arms and kick its legs, and the mother may become aware of its movements. Reflexes, such as the startle reflex and (by the end of the third month) sucking, first appear at this time. Its face becomes expressive; the fetus can squint, frown, or look surprised. Its respiratory organs are fairly well formed by this time but, of course, are not yet functional. The external sex organs begin to develop.

By the end of the third month, the fetus is about 9 centimeters long from the top of its head to its buttocks and weighs about 15 grams (0.5 ounce). It can suck

45–38 *Although the head of a 12-week-old fetus is disproportionately large, its features are clearly human. Its fingers and toes are fully developed, and outer ears and eyelids have formed. The lids are fused and will remain closed for the next three months. The umbilical cord, connecting the fetus to the placenta, contains a vein and two arteries.*

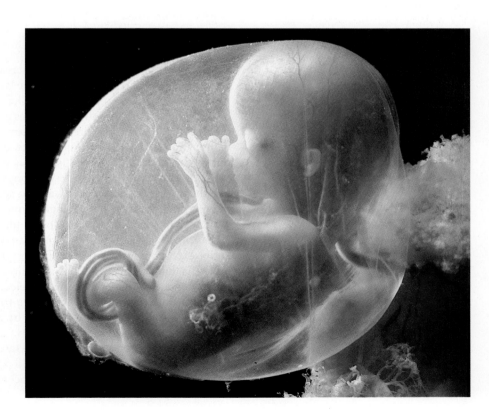

and swallow, and occasionally does swallow some of the fluid that surrounds it in the amniotic sac. The finger, palm, and toe prints are now so well developed that they can be clearly distinguished by ordinary fingerprinting methods. The kidneys and other structures of the urinary system develop rapidly, although waste products are still disposed of through the placenta. By the end of this period—the first trimester of development—all the major organ systems have been laid down.

The Second Trimester

During the fourth month, movements of the fetus become obvious to the mother. Its bony skeleton is forming and can be seen with x-rays. The body is becoming covered with a protective cheesy coating. The four-month-old fetus is about 14 centimeters long and weighs about 115 grams (4 ounces).

By the end of the fifth month, the placenta covers about 50 percent of the uterus. The fetus has grown to almost 20 centimeters and now weighs 250 grams. It has acquired hair on its head, and its body is covered with a fuzzy, soft hair called the lanugo, from the Latin word for "down." Its heart, which beats between 120 and 160 times per minute, can be heard with a stethoscope. The five-month-old fetus is already discarding some of its cells and replacing them with new ones, a process that will continue throughout its lifetime. Nevertheless, a five-month-old fetus cannot yet survive outside of the uterus. The youngest fetus on record to survive was about 23 weeks old and required continuous assistance in breathing, taking food, and maintaining its body temperature.

During the sixth month, the fetus has a sitting height of 30 to 36 centimeters and weighs about 680 grams. By the end of the sixth month, it could survive outside the mother's body, although probably only with respiratory assistance in an incubator. Its skin is red and wrinkled, and although teeth are only rarely visible at birth, they are already forming dentine. The cheesy body covering, which helps protect the fetus against abrasions, is now abundant. Reflexes are more vigorous. In the intestines is a pasty green mass of dead cells and bile, known as meconium, which will remain there until after birth.

The Final Trimester

During the final trimester, the fetus increases greatly in size and weight. In fact, it normally doubles in size just during the last two months. During this period, many nerve tracts are forming, and the number of brain cells is increasing at a very rapid rate. By the seventh month, brain waves can be recorded, through the abdomen of the mother, from the cerebral cortex of the fetus. Numerous studies have demonstrated that the protein intake of the mother is of critical importance during this period if the child is to have full development of its brain.

As the fetal period progresses, the physiology of the fetus becomes increasingly like the physiology of the adult, and so agents that affect the mother also threaten the maturing fetus. A depressingly familiar example is found in the infants born with cocaine, heroin, or methadone addiction.

During the last month of pregnancy, the baby usually begins to acquire antibodies from its mother, a process that continues after birth through the mother's milk (see page 727). Antibodies, which are far too large to diffuse across the placenta, are conveyed by highly selective active transport. The immunity they confer is only temporary. Within one to two months after birth, the maternal antibodies will be gradually replaced by antibodies manufactured by the baby's own immune system.

During this last month, the growth rate of the baby begins to slow down. (If it continued at the same rate, the child would weigh 90 kilograms—about 200 pounds—by its first birthday.) The placenta begins to regress and becomes tough and fibrous.

45–39 *A fetus at 18 weeks of age. When the thumb comes close to the mouth, the head turns, and the lips and tongue begin sucking motions—an important reflex for survival after birth.*

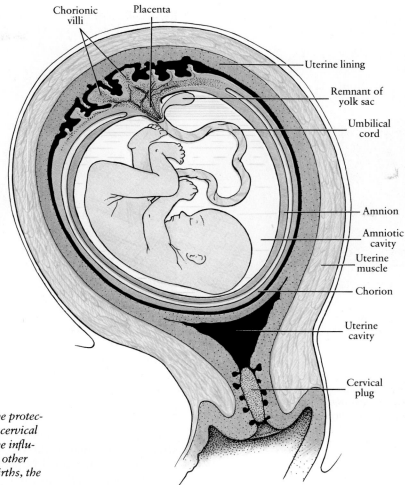

Chorionic villi

Placenta

Uterine lining

Remnant of yolk sac

Umbilical cord

Amnion

Amniotic cavity

Uterine muscle

Chorion

Uterine cavity

Cervical plug

45–40 *A human fetus, shortly before birth, showing the protective membranes and uterine tissues surrounding it. The cervical plug is composed largely of mucus. It develops under the influence of progesterone and serves to exclude bacteria and other infectious agents from the uterus. In 95 percent of all births, the fetus is in this head-down position.*

Weight at birth is the major factor in infant mortality. Infants weighing less than 2,000 grams (4 pounds, 6 ounces) at birth are at high risk of death or severe brain damage. Infants weighing less than 2,500 grams (5 pounds, 8 ounces) are considered to be low weight, and they are 40 times as likely to die within a month of birth as heavier infants. Two-thirds of all babies who die are low weight. From the 1950s until the early 1980s, infant mortality in the United States declined steadily, but it has now stabilized at 10.6 deaths for each 1,000 live births, placing the United States seventeenth in infant mortality worldwide. In other developed countries, however, infant mortality has continued to decline, reaching a level of 6.0 and 6.6 deaths per 1,000 births in Finland and Japan, respectively. The principal cause of low birth weight is inadequate maternal nutrition and prenatal care, which costs, on the average, $500 to $800 per pregnancy—less than the cost per day of hospitalization of a low-weight infant in an intensive care nursery. In the United States at the present time, a black infant is more than twice as likely to be born underweight as a white infant and twice as likely to die within four weeks of birth. This disturbing statistic reflects a lack of prenatal care and counseling that could, for the population group at highest risk—young, unmarried mothers —lead to heavier and healthier babies at birth.

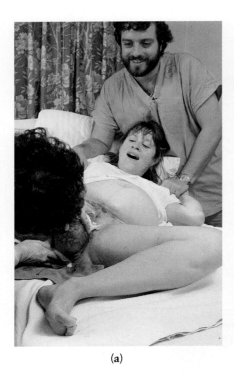

(a)

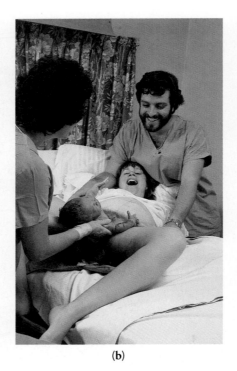

(b)

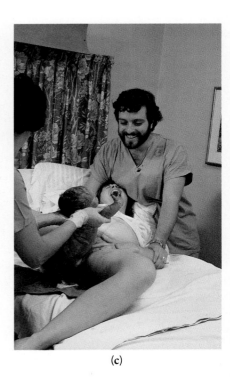

(c)

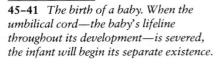

45–41 *The birth of a baby. When the umbilical cord—the baby's lifeline throughout its development—is severed, the infant will begin its separate existence.*

Birth

The date of birth is calculated as about 266 days after conception or 280 days after the beginning of the last menstrual period. Babies are rarely born on the scheduled day, but some 75 percent are born within two weeks of that day.

Labor is divided into three stages: dilation, expulsion, and placental stages. Dilation, which usually lasts from 2 to 16 hours (it is longer with the first baby than with subsequent births), begins with the onset of contractions of the uterus. It ends with the full dilation, or opening, of the cervix. At the beginning, uterine contractions occur at intervals of about 15 to 20 minutes and are relatively mild. By the end of the dilation stage, contractions are stronger and occur about every 1 to 2 minutes. At this point, the opening of the cervix is about 10 centimeters. Rupture of the amniotic sac, with the expulsion of fluids, usually occurs during this stage.

The second, or expulsion, stage lasts 2 to 60 minutes. It begins with the full dilation of the cervix and the appearance of the head in the cervix, called crowning. Contractions at this stage last from 50 to 90 seconds and are 1 or 2 minutes apart.

The third, or placental, stage begins immediately after the baby is born. It involves contractions of the uterus and the expelling of fluid, blood, and finally the placenta with the umbilical cord attached (also called the afterbirth). The placenta now weighs about 500 grams, about one-sixth of the weight of the infant. Minor uterine contractions continue; they help to stop the flow of blood and to return the uterus to its prepregnancy size and condition.

The baby emerges from the warm, protective enclosure in which it has been nourished and permitted to grow for nine months. The umbilical cord—until that moment, its lifeline—is severed. The baby cries as it takes its first breath, starts to breathe regularly, and so begins its independent existence.

45–42 *Within five to six minutes after vaginal delivery, the human infant is alert and its pupils are dilated, even in the presence of bright lights. This alertness, which may play a role in the bonding of infant and mother in the first hour of life, is the result of a surge of adrenaline and noradrenaline similar to that in "fight-or-flight" reactions (page 847). Birth puts great stress on an infant, as it is pushed and shoved by the uterine contractions, and temporary shortages of oxygen may occur as the blood vessels of the umbilical cord are pinched closed. Reflexes, similar to the diving reflex of marine mammals (page 744), alter the patterns of blood distribution, protecting vital organs, particularly the brain, heart, and lungs, which must begin functioning immediately after birth.*

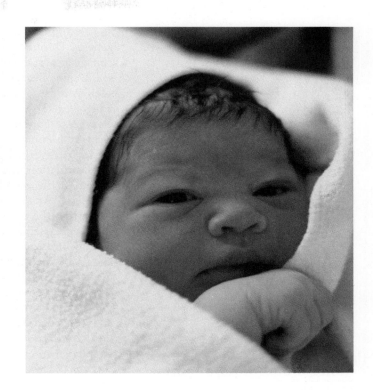

EPILOGUE

When did this particular human life begin? When the sperm encountered the egg? When the embryo became a fetus, visibly human? At the first heartbeat? The first brain wave? When the infant became viable as an independent entity? In the past, these matters were discussed by philosophers and theologians who were concerned with the question of when the soul enters the body. These issues have been revived in the ethical and legal controversies concerning abortion. In the evolutionary sense, however, none of these events marks the beginning of life. Life began more than 3 billion years ago and has been passed on since that time from organism to organism, generation after generation, to the present, and stretches on into the future, farther than the mind's eye can see. Each new organism is thus a temporary participant in the continuum of life. So is each sperm, each egg, indeed, in a sense, each living cell. Every individual, however, is a unique blend of heredity and experience, never to be duplicated and therefore irreplaceable; but from the perspective of the biological continuum, a human life lasts no longer than the blink of an eye.

SUMMARY

In the process of development, a fertilized egg becomes a complete organism, consisting of hundreds to billions of cells and closely resembling its parent organisms. This process involves growth, differentiation, and morphogenesis.

Development in most species of animals begins with fertilization—the fusion of egg and sperm. Fertilization results in (1) changes in the protective membranes of the egg that prevent fertilization by another sperm cell, (2) metabolic activation of the egg cell, (3) introduction of the genetic material of the male parent into the egg, and (4) cell division.

Development takes place in three stages: cleavage, gastrulation, and organogenesis. During cleavage, the original egg cell divides, with very slight, if any, change in overall volume. In eggs containing little or no yolk, such as those of the sea urchin, cleavage results in cells of approximately equal size. In yolkier eggs, such as those of the frog, cleavage is unequal, with fewer and larger cells in the yolky (vegetal) hemisphere. In eggs with a great deal of yolk, such as the hen's egg, cleavage is limited to a small, nonyolky disc at the top of the egg. In all cases, when cleavage is complete the embryo consists of a cluster of cells, the blastula, with a central cavity, the blastocoel.

Gastrulation involves the movement of cells into new relative positions and results in the establishment of three tissue layers: endoderm, mesoderm, and ectoderm. At the beginning of gastrulation, an opening forms, the blastopore. Cells from the outer surface of the embryo migrate through the blastopore into the blastocoel and form a new cavity, the archenteron, which will be the primitive gut. In birds and mammals, the homologue of the blastopore is the primitive streak. By the time the movements of the cells in gastrulation are complete, the mesoderm has differentiated into chordamesoderm and lateral plate mesoderm. The ectoderm on the dorsal surface overlying the chordamesoderm differentiates into neural ectoderm, flattening into the neural plate, which then curves upward and folds to form the neural tube. The chordamesoderm gives rise to the notochord and to the somites, which become embryonic muscle tissue. The lateral plate mesoderm splits into two layers, creating the fluid-filled coelom.

Each of the three primary tissue layers established during gastrulation gives rise to particular tissues and organs. The epidermis and the nervous system arise from ectoderm. The notochord, muscles, bones and cartilage, heart and blood vessels, excretory organs and gonads, the inner lining of the skin, and the outer linings of the digestive and respiratory tracts arise from mesoderm. The lungs, salivary glands, pancreas, liver, and inner lining of the digestive tract develop from endoderm.

The small area of tissue on the dorsal lip of the blastopore prior to gastrulation —the future chordamesoderm—is known as the organizer because, as shown by experiments on amphibians, it induces the cells overlying it to form the neural tube. If it is implanted in another embryo, it will induce the formation of a second neural tube in that embryo. Other organ systems also develop by this same process of embryonic induction, in which one tissue induces changes in the growth or migration of an adjacent tissue with which it comes in contact. A variety of factors are thought to be involved in the initial differentiation of the cells of the dorsal lip and in subsequent interactions. These include localized cytoplasmic determinants, particularly mRNA and proteins of maternal origin, substances that diffuse from specific cells and influence other cells, and cell adhesion molecules.

In amniote eggs, such as those of reptiles, birds, and mammals, the embryo, as it develops, forms four extraembryonic membranes: yolk sac, allantois, chorion, and amnion. In reptiles and birds, these membranes perform essential roles in supplying the developing embryo with food molecules and oxygen, in removing nitrogenous waste products, and in protecting the embryo from abrasion. In mammals, the yolk sac is the site in which germ cells are sequestered prior to their migration to the developing gonads, the allantois develops into the umbilical cord, the chorion forms the structures of the fetal side of the placenta, and the amnion, as in birds and reptiles, encloses the embryo in a fluid-filled cavity.

The mammalian embryo at the blastula stage is known as a blastocyst, and it consists of an inner cell mass and an outer layer of cells, the trophoblast. During the earliest stages of development, which take place in an oviduct, there is a large increase in the number of cells but little or no increase in the total size of the embryo. When the embryo descends into the uterus, at about the sixth day of development in humans, the trophoblast develops rapidly and invades the maternal tissues. The trophoblast becomes the chorion and interacts with tissues of the endometrium to form the placenta, through which the embryo receives its food and oxygen and excretes carbon dioxide and other waste products. The embryo is attached to the placenta by the umbilical cord.

When the human embryo is about 2 weeks old, a primitive streak forms, followed by the development of a neural plate and neural groove, which folds to form the neural tube. Although the embryo is still very small (about 2.5 millimeters long), most of the major organs have begun to form in these very early weeks, which is why damage caused to the embryo by viral infection, x-rays, or drugs during this period can be widespread. By the end of the second month, the embryo, now called a fetus, is almost human-looking, although it only weighs about 1 gram. By the end of the third month, all of the organ systems have been laid down. During the second trimester, development of the organ systems continues, and during the final trimester, there is a great increase in size and weight. Birth occurs, on the average, 266 days after fertilization.

QUESTIONS

1. Distinguish among the following: blastula/gastrula; blastopore/primitive streak; blastocoel/archenteron; ectoderm/mesoderm/endoderm; chordamesoderm/lateral plate mesoderm; fertilization/implantation; extraembryonic membranes/placenta; embryo/fetus; dilation/expulsion/placental stage.

2. Describe, in general terms, the end results of each of the following events: fertilization, cleavage, gastrulation, organogenesis.

3. Describe the similarities and differences in the blastulas of a sea urchin, an amphibian, a bird, and a mammal.

4. Cytoplasmic factors that influence the subsequent fate of cells may be localized or distributed in gradients. Give at least one example of each type, and describe its particular effects on the developing embryo.

5. The differentiation of functional sperm cells from spermatids, discussed in the preceding chapter, is under the control of genes on the Y chromosome. Sperm cells, however, are haploid and carry either an X chromosome or a Y chromosome. Explain, on the basis of principles discussed in this chapter, how it is possible for a spermatid containing an X chromosome (and thus no Y chromosome) to differentiate.

6. Follow the course of a single cell from its place of origin in the fertilized egg of a frog to its position in the eye cup of an early frog embryo. List, for each stage, all of the influences on this cell that might affect its history.

7. Follow the course of a single cell from its place of origin in the fertilized egg of a chick to its position in the third digit of the developing wing. List, for each stage, all of the influences on this cell that might affect its history.

8. In mammals, the fetus is basically feminine and fetal androgens are necessary to produce male characteristics. In birds, however, the fetus is basically masculine and fetal estrogens are necessary to produce female characteristics. Why would the arrangement in birds be unworkable in mammals?

9. If you look closely at Figure 45–34, you will notice that the umbilical arteries carry deoxygenated blood and the umbilical vein carries oxygenated blood. Why is this the case?

SUGGESTIONS FOR FURTHER READING

Books

BLOOM, FLOYD E., LAURA HOFSTADTER, and ARLYNE LAZERSON: *Brain, Mind, and Behavior,* W. H. Freeman and Company, New York, 1984.

   A readable companion to the television series, beautifully illustrated.

BRECHER, EDWARD M., et al.: *Licit and Illicit Drugs: The Consumers Union Report on Narcotics, Stimulants, Depressants, Inhalants, Hallucinogens, and Marijuana—Including Caffeine,* *Nicotine, and Alcohol,* Little, Brown and Company, Boston, 1973.*

   A well-balanced, straightforward, and nonsensational review.

BROWDER, L. W.: *Developmental Biology,* 2d ed., Saunders College/Holt, Rinehart and Winston, Philadelphia, 1984.

   A general, comprehensive coverage of basic embryology and modern developmental biology.

---

* Available in paperback.

ECKERT, ROGER, and DAVID RANDALL: *Animal Physiology: Mechanisms and Adaptations,* 3d ed., W. H. Freeman and Company, New York, 1988.

The emphasis is on basic principles. Well-written and handsomely illustrated.

GANONG, W. T.: *Review of Medical Physiology,* 12th ed., Lange Medical Publications, Los Angeles, 1987.*

A concise, accurate, up-to-date summary of human physiology. An excellent reference book.

GORDON, MALCOLM S., et al.: *Animal Physiology: Principles and Adaptations,* 4th ed., The Macmillan Company, New York, 1982.

A good animal physiology textbook, intended for an advanced course. The authors emphasize function as it relates to the survival of organisms in their natural environments.

KARP, GERALD, and N. J. BERRILL: *Development,* 2d ed., McGraw-Hill Book Company, New York, 1981.

A general textbook of developmental biology, with clear explanations and thorough coverage.

KESSEL, RICHARD G., and RANDY H. KARDON: *Tissues and Organs: A Text-Atlas of Scanning Electron Microscopy,* W. H. Freeman and Company, San Francisco, 1979.*

A collection of more than 700 outstanding scanning electron micrographs of vertebrate tissues and organs, beautifully reproduced. The accompanying text summarizes our knowledge of each organ system and its component parts.

LUCIANO, DOROTHY S., ARTHUR J. VANDER, and JAMES H. SHERMAN: *Human Anatomy and Physiology,* 2d ed., McGraw-Hill Book Company, New York, 1983.

An introductory anatomy and physiology textbook. The text is clearly written, and the detailed illustrations—particularly of the skeletal, muscular, nervous, and circulatory systems—are outstanding.

NEWSHOLME, ERIC, and TONY LEECH: *The Runner: Energy and Endurance,* Fitness Books, Roosevelt, N.J., 1984.*

A lively introduction to the energetics of the working human body. Equally useful as a primer for runners who want to know more about their own physiology and as an introduction to the biochemistry of carbohydrate and fat metabolism.

NILSSON, LENNART, AXEL INGLEMAN-SUNDBERG, and CLAES WIRSÉN: *A Child Is Born: The Drama of Life before Birth,* Dell Publishing, Inc., New York, 1986.*

This is an account of the history of life before birth. The book describes in detail the development of the unborn child from the moment of fertilization and also the changes in the mother during pregnancy. There are magnificent color photographs of the developing fetus.

NILSSON, LENNART: *The Body Victorious,* Delacorte Press, New York, 1987.

An account of the human immune system—its development, its normal functioning, and its disorders—filled with more of the stunning color photographs and micrographs for which Nilsson

is noted. Moreover, if you should be in any doubt as to the physiological consequences of cigarette smoking, you owe it to yourself to examine the relevant photographs in this book.

ROMER, ALFRED: *The Vertebrate Story,* 4th ed., The University of Chicago Press, Chicago, 1971.*

The history of vertebrate evolution, written by an expert but as readable as a novel.

ROMER, ALFRED, and THOMAS S. PARSONS: *The Vertebrate Body,* 6th ed., Saunders College/Holt, Rinehart and Winston, Philadelphia, 1985.

A thorough account of comparative vertebrate anatomy. The discussion of embryonic development is outstanding.

SCHMIDT-NIELSEN, KNUT: *Animal Physiology: Adaptation and Environment,* 3d ed., Cambridge University Press, New York, 1983.

Schmidt-Nielsen is concerned with underlying principles of animal physiology—the problems animals have to solve in order to survive. The emphasis is on comparative physiology, and the lucid exposition is illuminated by many interesting examples.

SCHMIDT-NIELSEN, KNUT: *Desert Animals,* Oxford University Press, New York, 1964.

Although considered the definitive work on the physiological problems relating to heat and water, this readable book also contains numerous anecdotes—such as that about Dr. Blagden—and many fascinating personal observations.

SCIENTIFIC AMERICAN: *The Brain,* W. H. Freeman and Company, New York, 1979.*

A reprint of the September 1979 issue of Scientific American. *Its 11 chapters by different authors are devoted to current knowledge of neurons and their organization and functions within the brain. The many drawings and micrographs are excellent.*

SCIENTIFIC AMERICAN: *Progress in Neuroscience,* W. H. Freeman and Company, New York, 1985.*

A collection of articles from Scientific American, *originally published between 1979 and 1984, covering such topics as the role of calcium ion at the synapse, the development of the nervous system, hearing, and the genetic control of behavior.*

SCIENTIFIC AMERICAN: *What Science Knows about AIDS,* W. H. Freeman and Company, New York, 1988.*

A reprint of the October 1988 issue of Scientific American. *Its 10 chapters by different authors provide a thorough review of our current knowledge of human immunodeficiency virus (HIV), the causative agent of AIDS; of the disease process and epidemiology of AIDS; and of the prospects and problems in the development of effective treatments and vaccines. Highly recommended.*

SHEPHERD, GORDON M.: *Neurobiology,* 2d ed., Oxford University Press, New York, 1988.

A general textbook that provides a broad coverage of contemporary neurobiology.

SMITH, HOMER W.: *From Fish to Philosopher,* Doubleday & Company, Inc., Garden City, N.Y., 1959.*

---

* Available in paperback.

Smith was an eminent specialist in the physiology of the kidney. Writing for the general public, he explains the role of this remarkable organ in the story of how, in the course of evolution, organisms have increasingly freed themselves from their environments.

THOMPSON, R. F.: *Introduction to Physiological Psychology,* 2d ed., Harper & Row, Publishers, Inc., New York, 1988.

Intended for the undergraduate student, this text presents an up-to-date survey of the biological foundations of psychology.

VANDER, ARTHUR J., JAMES H. SHERMAN, and DOROTHY S. LUCIANO: *Human Physiology: The Mechanisms of Body Function,* 4th ed., McGraw-Hill Book Company, New York, 1985.

Most highly recommended. The text is a model of clarity, and the diagrams, many of which we have borrowed, are splendid.

WALBOT, V., and N. HOLDER: *Developmental Biology,* Random House, Inc., New York, 1987.

A modern molecular and genetic approach to developmental biology. Highly recommended.

WEST, JOHN B.: *Everest—The Testing Place,* McGraw-Hill Book Company, New York, 1985.

An account, written for the general reader, of the adventures and findings of the American Medical Expedition to Everest (see page 746). Professor West, one of the leaders of the expedition, is a lucid and delightful writer on both human physiology and mountaineering.

Articles

ADA, GORDON L., and NOSSAL, GUSTAV: "The Clonal Selection Theory," *Scientific American,* August 1987, pages 62–69.

AOKI, CHIYE, and PHILIP SIEKEVITZ: "Plasticity in Brain Development," *Scientific American,* December 1988, pages 56–64.

AXELROD, JULIUS, and TERRY D. REISINE: "Stress Hormones: Their Interaction and Regulation," *Science,* vol. 224, pages 452–459, 1984.

BARNES, DEBORAH M.: "Steroids May Influence Changes in Mood," *Science,* vol. 232, pages 1344–1345, 1986.

BEACONSFIELD, PETER, GEORGE BIRDWOOD, and REBECCA BEACONSFIELD: "The Placenta," *Scientific American,* August 1980, pages 95–102.

BERRIDGE, MICHAEL J.: "The Molecular Basis of Communication within the Cell," *Scientific American,* October 1985, pages 142–152.

BRAMBLE, DENNIS M., and DAVID R. CARRIER: "Running and Breathing in Mammals," *Science,* vol. 219, pages 251–256, 1983.

BROWN, MICHAEL S., and JOSEPH L. GOLDSTEIN: "How LDL Receptors Influence Cholesterol and Atherosclerosis," *Scientific American,* November 1984, pages 58–66.

BUISSERET, PAUL D.: "Allergy," *Scientific American,* August 1982, pages 86–95.

CANTIN, MARC, and JACQUES GENEST: "The Heart as an Endocrine Gland," *Scientific American,* February 1986, pages 76–81.

CAPLAN, ARNOLD I.: "Cartilage," *Scientific American,* October 1984, pages 84–94.

CARMICHAEL, STEPHEN W., and HANS WINKLER: "The Adrenal Chromaffin Cell," *Scientific American,* August 1985, pages 40–49.

CERAMI, ANTHONY, HELEN VLASSARA, and MICHAEL BROWNLEE: "Glucose and Aging," *Scientific American,* May 1987, pages 90–96.

COHEN, IRUN R.: "The Self, the World and Autoimmunity," *Scientific American,* April 1988, pages 52–60.

COLLIER, R. JOHN, and DONALD A. KAPLAN: "Immunotoxins," *Scientific American,* July 1984, pages 56–64.

COTMAN, CARL W., and MANUEL NIETO-SAMPEDRO: "Cell Biology of Synaptic Plasticity," *Science,* vol. 225, pages 1287–1294, 1984.

CRAWSHAW, LARRY I., BRENDA P. MOFFITT, DANIEL E. LEMONS, and JOHN A. DOWNEY: "The Evolutionary Development of Vertebrate Thermoregulation," *American Scientist,* vol. 69, pages 543–550, 1981.

DOOLITTLE, RUSSELL F.: "Fibrinogen and Fibrin," *Scientific American,* December 1981, pages 126–135.

DUNANT, YVES, and MAURICE ISRAËL: "The Release of Acetylcholine," *Scientific American,* April 1985, pages 58–66.

EASTMAN, JOSEPH T., and ARTHUR L. DeVRIES: "Antarctic Fishes," *Scientific American,* November 1986, pages 106–114.

EDELMAN, GERALD M.: "Cell-Adhesion Molecules: A Molecular Basis for Animal Form," *Scientific American,* April 1984, pages 118–129.

EDELSON, RICHARD L., and JOSEPH M. FINK: "The Immunologic Function of Skin," *Scientific American,* June 1985, pages 46–53.

EISENBERG, EVAN, and TERRELL L. HILL: "Muscle Contraction and Free Energy Transduction in Biological Systems," *Science,* vol. 227, pages 999–1006, 1985.

FEDER, MARTIN E., and WARREN W. BURGGREN: "Skin Breathing in Vertebrates," *Scientific American,* November 1985, pages 126–142.

FINE, ALAN: "Transplantation in the Central Nervous System," *Scientific American,* August 1986, pages 52–58B.

FRENCH, ALAN R.: "The Patterns of Mammalian Hibernation," *American Scientist,* vol. 76, pages 568–575, 1988.

GALLO, ROBERT C.: "The AIDS Virus," *Scientific American*, January 1987, pages 46–56.

GARCIA-BELLIDO, ANTONIO, PETER A. LAWRENCE, and GINES MORATA: "Compartments in Animal Development," *Scientific American*, July 1979, pages 102–110.

GEHRING, WALTER J.: "The Molecular Basis of Development," *Scientific American*, October 1985, pages 152B-162.

GLICKSTEIN, MITCHELL: "The Discovery of the Visual Cortex," *Scientific American*, September 1988, pages 118–127.

GOLDE, DAVID W., and JUDITH C. GASSON: "Hormones that Stimulate the Growth of Blood Cells," *Scientific American*, July 1988, pages 62–70.

GOLDSTEIN, GARY W., and A. LORRIS BETZ: "The Blood-Brain Barrier," *Scientific American*, September 1986, pages 74–83.

GOTTLIEB, DAVID I.: "GABAergic Neurons," *Scientific American*, February 1988, pages 82–89.

HALL, BRIAN K.: "The Embryonic Development of Bone," *American Scientist*, vol. 76, pages 174–181, 1988.

HELLER, H. CRAIG, LARRY I. CRAWSHAW, and HAROLD T. HAMMEL: "The Thermostat of Vertebrate Animals," *Scientific American*, August 1978, pages 102–113.

HUDSPETH, A. J.: "The Hair Cells of the Inner Ear," *Scientific American*, January 1983, pages 54–64.

HYNES, RICHARD O.: "Fibronectins," *Scientific American*, June 1986, pages 42–51.

JACOBS, BARRY L.: "How Hallucinogenic Drugs Work," *American Scientist*, vol. 75, pages 386–392, 1987.

KOLATA, GINA: "Genes and Biological Clocks," *Science*, vol. 230, pages 1151–1152, 1985.

KORETZ, JANE F., and GEORGE H. HANDELMAN: "How the Human Eye Focuses," *Scientific American*, July 1988, pages 92–99.

LAGERCRANTZ, HUGO, and THEODORE A. SLOTKIN: "The 'Stress' of Being Born," *Scientific American*, April 1986, pages 100–107.

LAURENCE, JEFFREY: "The Immune System in AIDS," *Scientific American*, December 1985, pages 84–93.

MARRACK, PHILIPPA, and JOHN KAPPLER: "The T Cell and Its Receptor," *Scientific American*, February 1986, pages 36–45.

MARX, JEAN L.: "How the Brain Controls Birdsong," *Science*, vol. 217, pages 1125–1126, 1982.

MARX, JEAN L.: "The Immune System 'Belongs in the Body,' " *Science*, vol. 227, pages 1190–1192, 1985.

MASLAND, RICHARD H.: "The Functional Architecture of the Retina," *Scientific American*, December 1986, pages 102–111.

MILLER, C. ARDEN: "Infant Mortality in the U.S.," *Scientific American*, July 1985, pages 31–37.

MILSTEIN, CESAR: "Monoclonal Antibodies," *Scientific American*, October 1980, pages 66–74.

MISHKIN, MORTIMER, and TIM APPENZELLER: "The Anatomy of Memory," *Scientific American*, June 1987, pages 80–89.

MOOG, FLORENCE: "The Lining of the Small Intestine," *Scientific American*, November 1981, pages 154–176.

MORELL, PIERRE, and WILLIAM T. NORTON: "Myelin," *Scientific American*, May 1980, pages 89–116.

MORRISON, ADRIAN R.: "A Window on the Sleeping Brain," *Scientific American*, April 1983, pages 94–102.

NADEL, ETHAN R.: "Physiological Adaptations to Aerobic Training," *American Scientist*, vol. 73, pages 334–343, 1985.

ORCI, LELIO, JEAN-DOMINIQUE VASSALLI, and ALAIN PERRELET: "The Insulin Factory," *Scientific American*, September 1988, pages 85–94.

PERUTZ, M. F.: "Hemoglobin Structure and Respiratory Transport," *Scientific American*, December 1978, pages 92–125.

POGGIO, TOMASO, and CHRISTOF KOCH: "Synapses That Compute Motion," *Scientific American*, May 1987, pages 46–52.

REICHARDT, LOUIS F.: "Immunological Approaches to the Nervous System," *Science*, vol. 225, pages 1294–1299, 1984.

ROBERTS, LESLIE: "Zeroing in on the Sex Switch," *Science*, vol. 239, pages 21–23, 1988.

ROBINSON, THOMAS F., STEPHEN M. FACTOR, and EDMUND H. SONNENBLICK: "The Heart as a Suction Pump," *Scientific American*, June 1986, pages 84–91.

RODGER, JOHN C., and BELINDA L. DRAKE: "The Enigma of the Fetal Graft," *American Scientist*, vol. 75, pages 51–57, 1987.

ROSE, NOEL R.: "Autoimmune Diseases," *Scientific American*, February 1981, pages 80–103.

SCHMIDT-NIELSEN, KNUT: "Countercurrent Systems in Animals," *Scientific American*, May 1981, pages 118–128.

SCHNAPF, JULIE L., and DENIS A. BAYLOR: "How Photoreceptor Cells Respond to Light," *Scientific American*, April 1987, pages 40–47.

SNYDER, SOLOMON H.: "The Molecular Basis of Communication between Cells," *Scientific American*, October 1985, pages 132–141.

STALLONES, RUEL A.: "The Rise and Fall of Ischemic Heart Disease," *Scientific American*, November 1980, pages 53–59.

STRICKER, EDWARD M., and JOSEPH G. VERBALIS: "Hormones and Behavior: The Biology of Thirst and Sodium Appetite," *American Scientist*, vol. 76, pages 261–267, 1988.

STRYER, LUBERT: "The Molecules of Visual Excitation," *Scientific American*, July 1987, pages 42–50.

TAMARKIN, LAWRENCE, CURTIS J. BAIRD, and O. F. X. ALMEIDA: "Melatonin: A Coordinating Signal for Mammalian Reproduction?" *Science*, vol. 227, pages 714–720, 1985.

TONEGAWA, SUSUMU: "The Molecules of the Immune System," *Scientific American*, October 1985, pages 122–131.

UNANUE, EMIL R., and PAUL M. ALLEN: "The Basis for the Immunoregulatory Role of Macrophages and Other Accessory Cells," *Science,* vol. 236, pages 551–557, 1987.

WASSARMAN, PAUL M.: "The Biology and Chemistry of Fertilization," *Science,* vol. 235, pages 553–560, 1987.

WASSARMAN, PAUL M.: "Fertilization in Mammals," *Scientific American,* December 1988, pages 78–84.

WINGFIELD, JOHN C., et al.: "Testosterone and Aggression in Birds," *American Scientist,* vol. 75, pages 602–608, 1987.

WOLPERT, LEWIS: "Pattern Formation in Biological Development," *Scientific American,* October 1978, pages 154–164.

WURTMAN, RICHARD J.: "Alzheimer's Disease," *Scientific American,* January 1985, pages 62–74.

YOUNG, JOHN D., and ZANVIL A. COHN: "How Killer Cells Kill," *Scientific American,* January 1988, pages 38–44.

ZAPOL, WARREN M.: "Diving Adaptations of the Weddell Seal," *Scientific American,* June 1987, pages 100–105.

ZUCKER, MARJORIE B.: "The Functioning of Blood Platelets," *Scientific American,* June 1980, pages 86–103.

PART **3**

Biology of Populations

SECTION **7**

Evolution

Natural selection, as conceived by Darwin, is a two-stage process. First, variations, produced by chance, appear at random in a population. Second, organisms with favorable variations are more likely to survive, passing their characteristics, in some cases, to their offspring. Thus mimicry in insects, as in the case of this katydid, occurs by the slow accumulation, over generations, of slight, favorable variations, each adding to an individual's ability to escape predation and so to survive and leave offspring. Alfred Russel Wallace was among those calling Darwin's attention to such "wonderfully close" resemblances between insects and objects in their environment. These resemblances, as Darwin noted, are reinforced by the insects' patterns of behavior (also a result of natural selection).

Evolution: Theory and Evidence

As we have seen throughout the preceding chapters of this text, the thread that links together the enormous diversity of the living world is evolution, a process of change through time. A vast body of accumulated evidence indicates that the earth has had a long history and that all living organisms, including ourselves, arose in the course of that history from earlier, more primitive forms. This means that all species are descended from other species; in other words, all living things share common ancestors in the distant past. Organisms are what they are because of their history.

Evolution is inextricably intertwined with ecology in what G. E. Hutchinson of Yale University aptly called "the ecological theater and the evolutionary play." Separation of the two subjects is to a large extent arbitrary and artificial. For purposes of clarity, however, we have divided the study of populations of organisms, their interactions and their changes through time, into two sections. In this chapter, we shall consider the fundamental points of evolutionary theory and the major categories of evidence that pointed Darwin toward his theory and that provide the underpinnings of contemporary theory. In the subsequent chapters of this section, we shall examine in more detail the mechanisms by which the plot of the evolutionary play moves forward. In Section 8, we shall place ourselves in the ecological theater to observe the cast—all living things—in the variety of interactions with each other and the physical environment that constitute the ongoing play in action.

DARWIN'S THEORY

As we saw in the Introduction,* Charles Darwin was not the first to propose that the diversity of organisms is the result of historical processes, but credit for the theory of evolution is rightly his, for two reasons. First, his "long argument"—as *The Origin of Species* has been characterized—left little doubt that evolution had actually occurred and so marked a turning point in the science of biology. The second reason, which is closely related to the first, is that Darwin correctly perceived the general mechanism by which evolution occurs.

All of Darwin's contemporaries—whether amateurs, as was Darwin himself, strictly speaking, or academics—had access to much of the same information. The revolution in biological thought sparked by Darwin's theory would have eventually taken place even if Darwin had never existed, just as the structure of DNA would someday have been discovered in the absence of Watson and Crick. It is doubtful that Wallace alone (see page 8) could have persuaded the scientific world to accept his theory of evolution. He might well have been just another traveler with an odd tale that could not yet be made to fit. It was Darwin's dogged

46–1 *Polar bears, like many other Arctic birds and mammals, are notable for their cryptic coloration. Actually the hairs in a polar bear's coat are colorless; it appears white because of the refraction of light (the same phenomenon that makes a cloud look white). These hairs are hard, shiny, and hollow, and so they stay erect, do not mat when they are wet, and can be easily shaken free of water. Insulation is provided by a layer of blubber as much as 10 centimeters thick. Unlike their close relatives, the brown bears, which are omnivorous, polar bears have a digestive system specialized for an almost exclusively carnivorous diet—mostly seals. Polar bears diverged from brown bears only recently. The rapidness with which they became adapted to Arctic life is attributed to the intense selection pressure of this harsh environment.*

* If you have not read pages 1–9 of the Introduction, we suggest that you do so now.

46-2 *Within this long egg case, composed of a parchmentlike material, a whelk is depositing hundreds of eggs. When the baby mollusks hatch, they will break through thin areas in the individual capsules and emerge as miniature copies of their parents. In most species, once the egg case is produced and the eggs laid, the mother whelk does not participate further in raising the young. Despite the large number of young produced, there is, on the average, no increase in the number of whelks from generation to generation.*

intellectual pursuit of the subject and his vast body of personal knowledge that made it come about at the time and in the way that it did—as an intellectual avalanche.

Although it is now some 130 years since the first publication of *The Origin of Species,* Darwin's and Wallace's original concept of how evolution comes about still provides the basic framework for our understanding of the process. Their concept rested on five premises:

1. Organisms beget like organisms—in other words, there is stability in the process of reproduction.

2. In most species, the number of individuals in each generation that survive and reproduce is small compared with the number initially produced.

3. In any given population, there are chance variations among individual organisms—that is, variations that are not produced by the environment—and some of the variations are inheritable.

4. Which individuals will survive and reproduce and which will not are determined to a significant degree by the interaction between these chance variations and the environment. Some variations enable individuals to produce more offspring than other individuals. Darwin termed these "favorable" variations and argued that inherited favorable variations tend to become more and more common from one generation to the next. This is the process that Darwin called **natural selection.**

5. Given sufficient time, natural selection leads to the accumulation of changes that differentiate groups of organisms from one another.

EVIDENCE FOR MICROEVOLUTION

Darwin believed evolution to be such a slow process that it could never be observed directly. However, modern human civilization has produced such extremely strong selection pressures on some organisms that if one looks at small-scale evolutionary phenomena (known as microevolution), it is possible to observe not only the results but also the actual process of evolution by natural selection.

The Peppered Moth

One of the best-studied examples of natural selection in action is that of *Biston betularia,* the peppered moth. These moths were well known to British naturalists of the nineteenth century, who remarked that they were usually found on lichen-covered trees and rocks. Against this background, the light coloring of the moths made them practically invisible. Until 1845, all reported specimens of *Biston betularia* had been light-colored, but in that year one black moth of this species was captured in the growing industrial center of Manchester.

With the increasing industrialization of England, smoke particles began to pollute the foliage in the vicinity of industrial towns, killing the lichens and leaving the tree trunks bare. In heavily polluted districts, the trunks and even the rocks and ground became black. During this period, more and more black *Biston betularia* were found. Replacement of light-colored moths by dark ones proceeded briskly. By the 1950s, only a few of the light-colored population could be found, and these were far from industrial centers. Because of the prevailing westerly wind in England, pollutants were carried to the east of industrial towns, and the moths tended to be of the black variety right up to the east coast of England. The few light-colored populations were concentrated in the west, where lichens still grew. (This tendency for dark-colored forms to replace light-colored

46–3 *The two varieties of* Biston betularia, *the peppered moth, resting on* (**a**) *a lichen-covered tree trunk in an unpolluted English countryside and* (**b**) *a dark tree trunk, near Manchester. If you look carefully, you will notice that there are two moths in each photograph.*

Kettlewell demonstrated that on lichen-covered trees the black moths are more likely to be eaten by birds (which are visually oriented), whereas on polluted trees the light-colored moths are more likely to be eaten.

(a)

(b)

forms, known as industrial melanism, has been found among some 70 other moth species in England and some 100 species of moths in the area of Pittsburgh, Pennsylvania, a heavily industrialized city. It has also been observed in many species of butterflies.)

Where did the black *Biston betularia* come from? Eventually, it was demonstrated that the black color was the result of a rare, recurring mutation. The black moths had always been there, in very small numbers. But why had their numbers increased so dramatically? In the late 1950s, H. B. D. Kettlewell, a physician who was also an amateur collector of moths and butterflies, hypothesized that the color of the moths protected them from predators, notably birds (Figure 46–3). In the face of strong opposition from entomologists—all of whom claimed they had never seen a bird eat a *Biston betularia* of any color—he set out to test his hypothesis. He marked a sample of moths of each color by carefully putting a spot on the underside of the wings, where it could not be seen by a predator. Then he released known numbers of marked individuals into a bird reserve near Birmingham, an industrial area where 90 percent of the local *Biston betularia* population consisted of black moths. Another sample was released into an unpolluted Dorset countryside, where no black moths ordinarily occurred. He returned at night with light traps to recapture his marked moths. From the area around Birmingham, he recovered 40 percent of the black moths, but only 19 percent of the light ones. From the area in Dorset, 6 percent of the black moths and 12.5 percent of the light moths were retaken.

To clinch the argument, Kettlewell placed moths on tree trunks in both locations, focused hidden movie cameras on them, and was able to record birds actually selecting and eating the moths. As his films revealed, they do this so rapidly that it is not surprising it was not previously observed. When equal numbers of black and light moths were available near Birmingham, the birds seized 43 light-colored moths and only 15 black ones; in Dorset, they took 164 black moths but only 26 light-colored forms. Clearly, if you are a moth, it is advantageous to be black near Birmingham but light in Dorset.

More recently, strong controls have been instituted in Great Britain on the particulate content of smoke, and the heavy soot accumulation has begun to decrease. The light-colored moths are already increasing in proportion to the black forms, but it is not yet known whether a complete reversal either in pollution or in the selection process will come about. There is a moral to this story. Note that the black moth is not absolutely superior to the light one, or vice versa. It is, as Darwin realized, entirely a matter of time and place.

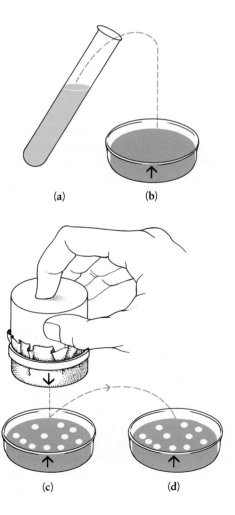

46–4 *A scale insect on a citrus tree. With their specialized mouthparts, these insects suck fluid from the phloem, debilitating the plant. If left unchecked, they eventually kill it.*

Insecticide Resistance

Another example of natural selection in action is the development of insecticide resistance. Chemicals poisonous to insects, such as DDT, were originally hailed as major saviors of human health and property. They have fallen into disfavor not only because of their tendency to accumulate in the environment (see page 1150), but also because of the extraordinary increase in resistant strains of insects. At least 225 species of insects are now resistant to one or more insecticides. One species is even able to remove a chlorine atom from a DDT molecule and use the remainder as food.

A particularly striking example of insecticide resistance has been found in the scale insects (Figure 46–4) that attack citrus trees in California. In the early 1900s, a concentration of hydrocyanic gas sufficient to kill nearly 100 percent of the insects was applied to orange groves at regular intervals with great success. By 1914, orange growers near Corona, California, began to notice that the standard dose of the fumigant was no longer sufficient to destroy one type of scale insect, the red scale. A concentration of the gas that had left fewer than 1 in 100 survivors in the nonresistant strain left 22 survivors out of 100 in the resistant strain. By crossing resistant and nonresistant strains, it was possible to show that the difference between the two involved a single gene locus. The mechanism for this resistance is not known, but one group of experiments has shown that the resistant individual can keep its spiracles (page 574) closed for 30 minutes under unfavorable conditions, whereas the nonresistant insect can do so for only 60 seconds.

Drug Resistance in Bacteria

A more recent example of natural selection concerns the development of drug resistance in bacteria. Following the period of rapid development of antibiotics after World War II, bacteria soon began to appear that were resistant to these agents, so widely hailed as "miracle drugs." What caused this resistance? Changes in the metabolism of individual bacteria or changes in the bacterial population—that is, evolution?

The question, important for medical as well as scientific reasons, was answered by a beautifully simple experiment performed by Joshua and Esther Lederberg. A set of petri dishes were prepared, each containing agar (a jellylike substance) and each marked on the side with an arrow. Bacterial cells grown in a nutrient broth were spread in a thin layer on one such dish. Within 24 hours, visible colonies began to appear, clones of the original scattered cells. The experimenters prepared a cylindrical block the same diameter as the inside of the petri dish, covered one end of it with a piece of velveteen and marked it on the side with an arrow. Then the arrow was lined up with the arrow on the petri dish, and the velveteen surface was pressed against the agar, picking up bacteria (Figure 46–5). The velveteen surface was then pressed against agar containing penicillin in another petri dish.

46–5 *The Lederbergs' replica-plating method for detecting and isolating drug-resistant bacteria. (a) The bacteria are cultured in a broth containing nutrient molecules. (b) A sample of the cell suspension is spread over the surface of a petri dish containing broth solidified with agar. (c) The plate is incubated until colonies appear. A piece of velveteen held by a ring that fits snugly around a cylindrical block is used to transfer a sample of each colony to a replicate plate, another petri dish containing a solidified medium (d).*

(a)   (b)

(c)   (d)

Again, the position was marked with an arrow. Each colony was thus placed in the new dish in exactly the same position it had occupied originally. Usually the bacterial colonies failed to survive in the penicillin medium, but eventually the Lederbergs found what they were looking for: bacteria growing on the penicillin plate.

In the next step of the experiment, they located the colony on the original, untreated petri dish that corresponded with the colony that had grown on the penicillin-laced agar. They transferred a sample of cells from this untreated colony to another petri dish containing no penicillin. The cells were allowed to grow, and when samples of these cells were tested in a medium containing penicillin, it was found that they too were penicillin-resistant, even though they had never been exposed to penicillin. Like the black moths and the cyanide-resistant scale insects, they were simply variants produced by chance in the original population and selected by their environment.

As we saw in Chapter 16, it is now known that the bacterial genes for drug resistance are carried on plasmids, small DNA molecules that can be transferred from one cell to another. Consequently, the spread within a bacterial population of mutations conferring resistance is much more rapid than would occur as a result of natural selection alone.

EVIDENCE FOR MACROEVOLUTION

The three examples we have just considered are among the many that support Darwin's proposal that natural selection is the mechanism of evolutionary change. In themselves, however, they do not provide evidence for macroevolution—that is, evolutionary change above the level of the species. The evidence for macroevolution, or the process that Darwin called "descent with modification," falls into five broad categories.

The Number of Species

One line of evidence, which had been building for more than a century before Darwin, was the existence of an enormous number of species. When observations had been confined to a limited area of the temperate zone, as they were before the great explorations of the eighteenth and nineteenth centuries, it was plausible that each kind of organism had been created separately. Even then, however, the existence of a score or more of very similar beetles (none of which seemed to serve a higher purpose) was something of an embarrassment.

It is no coincidence that Darwin and Wallace had come to doubt the doctrine of special creation while traveling in the tropics, where the widest variety of species is found and where, even today, many species new to science are encountered. Moreover, it became apparent that species are not as distinct as had previously been thought. For example, as Darwin traveled up the western coast of South America, he observed gradual changes in various characteristics of the plants and animals—evidence that organisms become modified, according to the different environments in which they live.

The creationists sought to accommodate doubters of fixed and unchanging species by tinkering with the definition of species. Some species had remained the same, they said; others represented altered forms of the originally created species and so were not really species at all in the special creation sense. Some creationists said it was the genera that were specially created and not the species at all, and so on. The species problem, made apparent by the nineteenth century explorations, was the first serious fissure in the monolith of special creation. (And, indeed, as we shall see in Section 8, the question of why there are so many species is still an unanswered one, although not for the same reasons.)

46–6 *In 1847, Henry Walter Bates, at the age of 22, formerly an apprentice to a hosiery manufacturer in Leicester, embarked on a trip up the River Amazon with his friend Alfred Russel Wallace. Their purpose: to gather facts "towards solving the problem of the origin of species." In his 11 years in Brazil, Bates collected 14,712 species of insects and other animals, of which some 8,000 were previously undescribed. Bates, like Wallace, was one of many scientists, naturalists, and explorers with whom Darwin corresponded. This drawing of Bates, reproduced from his book* The Naturalist on the River Amazon, *shows him collecting a curl-crested toucan.*

(a)

(b)

(c)

46–7 *Although these three mammals are similar in appearance and have similar life styles, they are not closely related.* (a) *The European rabbit, a placental mammal, is classified in the order Lagomorpha.* (b) *The Patagonian hare, or cavy, also a placental mammal, is classified in the order Rodentia.* (c) *The Australian "banded hare" is actually a wallaby, a member of the kangaroo family. Like the other native mammals of Australia, it is a marsupial.*

Biogeography

A second, related line of evidence in support of macroevolution comes from what is known as biogeography—the distribution of plants and animals in the various regions of the world. The naturalist explorers—Darwin and Wallace among them—were puzzled by the fact that places similar in climate and topography are often populated by very different organisms. According to creationist doctrine, each species was created specially for a particular way of life and placed in the locality for which it was suited—hence, no polar bears in the tropics, for instance.

Darwin began his voyage with this point of view, but many questions soon arose. Why, for example, did remote oceanic islands often have no terrestrial mammals at all but only peculiar species of bats? Why did England and Europe have rabbits galore, whereas similar areas in South America have only the Patagonian hare—taxonomically not a rabbit or hare at all but a rodent—and Australia a marsupial that resembled a hare?

Or, speaking of Australia, why did this island continent lack native placental mammals but contain instead a large array of marsupials, all clearly related to one another and found only rarely elsewhere on the planet? Was each of the 57 separate species of kangaroos created separately and dropped off in Australia? And why only in Australia? Or, more plausibly, was there perhaps an ancestral marsupial that gave rise to all these clearly related forms? The fossils of extinct marsupials had been described many years before Darwin visited Australia on the *Beagle;* in his later writings, he remarks on "this wonderful relationship on the same continent between the living and the dead."

With Darwin's observations on the Galapagos Islands, the question became still more finely focused. For example, the 13 species of finches on the Galapagos resembled South American species and also resembled one another. Had each been created separately and distributed among these volcanic islands (which the evidence indicated had been formed much more recently than the mainland)? Similar resemblances were also noted among the reptiles and plants of the different islands. In *The Voyage of the Beagle,* Darwin wrote:

> The distribution of the tenants of this archipelago would not be nearly so wonderful, if, for instance, one island had a mocking-thrush, and a second island some other quite distinct genus;—if one island had its genus of lizard, and a second island another distinct genus, or none whatever;—or if the different islands were inhabited, not by representative species of the same genera of plants, but by totally different genera . . . . But it is the circumstance, that several of the islands possess their own species of the tortoise, mocking-thrush, finches, and numerous plants, these species having the same general habits, occupying analogous situations, and obviously filling the same place in the natural economy of this archipelago, that strikes me with wonder.

These observations about geographic distribution did not, of course, disprove the possibility of special creation (which, because of its reliance on a supernatural agency, cannot be disproved). However, these examples and a multitude of other biogeographic examples provide strong evidence that living things are what and where they are because of events that occurred in the course of their previous history.

The Fossil Record

A third line of evidence in support of macroevolution is provided by the fossil record, which reveals a succession of living forms, with simpler forms generally preceding more complex forms. Geologic studies, as well as the collecting of specimens of plants and animals, were among Darwin's activities aboard the *Beagle.* The coasts of South America were of particular interest because they

(a)

(b)

(c)

46–8 *The rhea of South America (a), the emu of Australia (b), and the ostriches of Africa (c) are all very large, flightless birds found in similar habitats on the different continents. Was each a product of a separate act of special creation? For Darwin, biogeographic differences such as these cast doubts on the doctrine of special creation.*

showed evidence of widespread upheaval, with many geologic strata exposed. These strata, like those studied by William Smith in the British Isles, contained successive deposits of marine shells, some of them found high above sea level. Because of the gradually increasing percentage of modern species in the more recently deposited strata, Darwin was able to estimate their relative ages and to correlate strata from different localities, as Smith had before him (page 3).

In the course of his geologic studies Darwin came across many fossils of extinct mammals. Among the most interesting to him were those of giant armadillos. The fact that extinct armadillos were buried in the same South American plains where the only surviving species of these strange armored mammals lived provided tangible evidence of change and history.

Nowhere in the geologic record—either in his own observations or in the reports of others—did Darwin find exactly what he was seeking: evidence of a gradual transition between one species and another. Thus, as Darwin reiterates in *The Origin of Species,* the fossil record provided little evidence for him of how evolution came about. For Darwin's contemporaries, however, as for modern observers, the fossil record provided the overwhelming evidence that evolution had indeed occurred. During the decades immediately following the first publication of *The Origin of Species,* many new fossil finds were made. One of these was of *Archaeopteryx,* in 1862; Darwin describes this find in his later editions, noting its combination of avian and reptilian features (see Figure 28–22, page 598).

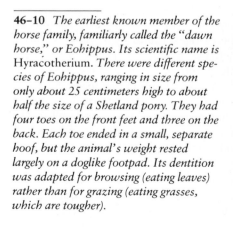

46-9 *Armadillos, mammals related to sloths and anteaters, were among the wholly new families of animals Darwin encountered in South America. He was particularly interested in the clear resemblance between the living and fossil forms he found there and later came to regard such resemblances as evidence of evolution. The nine-banded armadillo, shown here, is the one species that has spread northward into the United States, having crossed the Rio Grande into Texas in 1880.*

The most impressive early success in correlating Darwin's theory of evolution with fossil evidence stemmed from the discovery of a long series of fossil horses. In 1879, Othniel C. Marsh of Yale University published a description of horse genealogy, leading from the tiny dawn horse, Eohippus (Figure 46–10), through a number of stages of successively larger animals. These changes in size were accompanied by concomitant major evolutionary changes in teeth, legs, and feet.* Thomas H. Huxley, who characterized himself as "Darwin's bulldog," contending with Darwin's critics and arguing the case for evolution before the public, leaned heavily on horse genealogy in presenting the evidence for evolution. Describing Marsh's six stages of horses, spanning some 60 million years, before a meeting of The Zoological Society in 1880, Huxley stated:

> With respect to the interpretations of these facts, two, and only two, appear to be imaginable. The one assumes that these successive forms of equine animals have come into existence independently of one another. The other assumes that they are the result of the gradual modification undergone by the successive members of a continuous line of ancestry. As I am not aware that any zoologist maintains the first hypothesis, I do not feel called upon to discuss it. The adoption of the second, however, is equivalent to the acceptance of the doctrine of evolution so far as horses are concerned, and in the absence of evidence to the contrary, I shall suppose that it is accepted.

In the twentieth century, a steady stream of new discoveries has enormously increased our knowledge of the fossil record, which now extends back more than 3 billion years. In the case of many groups of organisms—vascular plants and vertebrates, for example, as we saw in Chapters 24 and 28—fossils have been found that exhibit a graded series of changes in anatomical characteristics, linking older forms with the modern forms and revealing pathways of divergence from common ancestors. The most dramatic of these fossil finds, which have shed light on the evolution of hominids, will be discussed in detail in Chapter 50.

---

* A more complete discussion of the evolutionary history of the horse will be given on pages 1025–1027.

46-10 *The earliest known member of the horse family, familiarly called the "dawn horse," or Eohippus. Its scientific name is* Hyracotherium. *There were different species of Eohippus, ranging in size from only about 25 centimeters high to about half the size of a Shetland pony. They had four toes on the front feet and three on the back. Each toe ended in a small, separate hoof, but the animal's weight rested largely on a doglike footpad. Its dentition was adapted for browsing (eating leaves) rather than for grazing (eating grasses, which are tougher).*

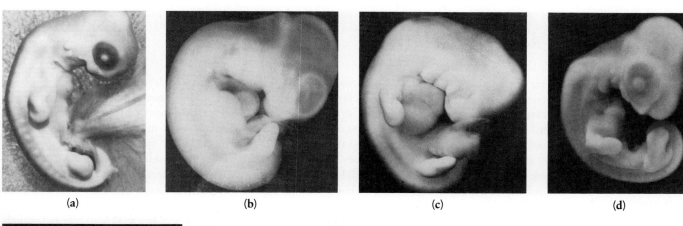

(a)   (b)   (c)   (d)

(e)

46-11 *One of these embryos is human; can you tell which? Homologies among vertebrates are clearly evident early in development, as the photos reveal. Embryo (a) is a turtle, (b) is a mouse, (c) is human, (d) is a chick, and (e) is a pig.*

Homology

A fourth major line of evidence comes from homologous structures and biochemical pathways. For example, as we saw in Figure 20–7 (page 414), the forelimbs of animals as diverse as crocodiles, birds, whales, horses, bats, and humans are all constructed of bones arranged in the same pattern. All vertebrates have four limbs—never six, or eight, or a hundred—and all have gill pouches, at least at some stage of their development (Figure 46–11). Whales and even some snakes retain the vestiges of pelvic and leg bones, for which they have no use (Figure 46–12). With a few unusual exceptions, virtually all mammals, ranging from mice to giraffes, have seven cervical vertebrae. If, however, one were starting from scratch, one might choose a somewhat different body plan for a giraffe, for instance, than for a meadow mouse. Yet the evidence confronting Darwin repeatedly suggested obsolete patterns built with hand-me-down materials.

The cellular basis of life was demonstrated within a decade of the publication of the first edition of *The Origin of Species.* Since that time, studies of cell ultrastructure and biochemistry have revealed innumerable new homologies at the submicroscopic and molecular levels of organization. The two-ply nature of cell membranes, the role of ribosomes in protein synthesis, the internal organization of eukaryotic cilia and flagella, glycolysis, the electron transport chain, the multitude of roles played by ATP, and, most important of all, the genetic code, make clear the historical unity of all living organisms.

46-12 *Many organisms retain traces of their evolutionary history. For example, the whale retains pelvic and leg bones as useless vestiges.*

Pelvis

Femur

The Record in the Rocks

The earth's long history is recorded in the rocks that lie at or near its surface, layer piled upon layer, like the chapters in a book. These layers, or strata, are formed as rocks in upland areas, are broken down to pebbles, sand, and clay, and are carried to the lowlands and the seas. Once deposited, they slowly become compacted and cemented into a solid form as new material is deposited above them. As continents and ocean basins change shape, some strata sink below the surface of an ocean or a lake, others are forced upward into mountain ranges, and some are worn away, in turn, by water, wind, or ice or are deformed by heat or pressure.

Individual strata may be paper-thin or many meters thick. They can be distinguished from one another by the types of parent material from which they were laid down, the way the material was transported, and the environmental conditions under which the strata were formed, all of which leave their traces in the rock. They can be distinguished, moreover, by the types of fossils they contain. Small marine fossils, in particular, can be associated with specific periods in the earth's history. The fossil record is seldom complete in any one place, but because of the identifying characteristics of the strata, it is possible to piece together the evidence from many different sources. It is somewhat like having many copies of the same book, all with chapters missing—but different chapters, so it is possible to reconstruct the whole.

The geologic eras—Precambrian, Paleozoic, Mesozoic, and Cenozoic—which are the major volumes of the geologic record, were identified and named in the early nineteenth century. These eras were subdivided into periods, many of which are named, quite simply, for the areas in which the particular strata were first studied or studied most completely: the Devonian for Devonshire in southern England, the Permian for the province of Perm in Russia, the Jurassic for the Jura Mountains between France and Switzerland, and so on.

Early attempts to date the various eras and periods were based simply on their relative ages compared to the age of the earth; obviously, a stratum occurring regularly above another was younger than the one below it. The first scientific estimate was made in the mid-1800s by the famous British physicist Lord Kelvin. On the basis of his calculations of the time necessary for the earth to have cooled from its original molten state, Kelvin maintained that the planet was about 100 million years old, a calculation that posed considerable difficulties for Darwin. (Kelvin was not aware of the existence under the earth's surface of radioactive materials that heat the planet from within.) In the last 40 years, however, new methods for determining the ages of strata have been developed involving measurements of the decay of radioactive isotopes. As a result, the estimated age of the earth has increased, in little more than a century, from 100 million years to about 4.6 billion years.

Geologic strata are now dated whenever possible by analysis of radioactive isotopes contained in crystals of igneous rock (rock formed from molten material) associated with particular strata. As we noted in

The Imperfection of Adaptation

Adaptation is a word with several meanings in biology. First, it can mean a state of being adjusted to the environment; every living organism is adapted in this sense, just as Abraham Lincoln's legs were, as he remarked, "just long enough to reach the ground." Second, adaptation can mean a process, which may take place either within the lifetime of an individual organism (physiological adaptation)—such as the production of more red blood cells in response to dwelling at high altitudes—or within a population as a whole over the course of many generations (evolutionary adaptation). Third, adaptation is commonly used to refer to a particular characteristic, that which is adapted, such as an eye or a hand. Adaptation in this sense—and its seeming perfection—has often appeared to provide strong support for the doctrine of special creation. It has been argued that it would be impossible for an organ of such perfection as the eye, for example, to come into existence from nothingness. Of what use would half an eye be?

In the badlands of South Dakota, the waters of the Missouri River and its tributaries have carved through the rock, revealing the geologic strata. These formations are of particular interest because they encompass the boundary between the Cretaceous and Tertiary periods—a time of mass extinction, including that of the dinosaurs (see page 1024).

Chapter 1, many naturally occurring isotopes are radioactive. All the heavier elements—atoms that have 84 or more protons in the nuclei—are unstable and, therefore, radioactive. All radioactive isotopes emit energy (as particles or rays) at a fixed rate; this process is known as radioactive "decay." The rate of decay is measured in terms of half-life: the half-life of a radioactive isotope is defined as the time in which half the atoms in a sample lose their radioactivity and become stable. Since the half-life of an isotope is constant, it is possible to calculate the fraction of decay that will take place for a given isotope in a given period of time. The radiometric clock starts to tick when the crystalline rock is formed.

Half-lives vary widely, depending on the isotope. The radioactive nitrogen isotope ^{13}N has a half-life of 10 minutes, and the most common isotope of uranium (^{238}U) has a half-life of 4.5 billion years. The uranium atom undergoes a series of decays, eventually being transformed to an isotope of lead (^{206}Pb). Thus the proportion of ^{238}U to ^{206}Pb in a given rock sample, for example, is a good indication of how long ago that rock was formed. Five different isotopes are now commonly employed as radiometric clocks, and in thousands of instances, rocks have been dated by three or more independent clocks.

Any theory of evolution requires, as Darwin knew well, that the earth have a long history. Thus, these radiometric clocks are doubly important to modern students of evolution. First, they demonstrate conclusively that the earth's age is close to 5 billion years; in other words, the earth is indeed old enough for evolution to have produced the observed diversity of organisms. Second, they provide the tools for estimating the relative ages of various rocks—and of the fossils within them—and so, for unraveling the details of the earth's biological past.

In the course of his career as a naturalist, Darwin amassed an enormous amount of information about living organisms; assembled in someone else's brain, much of it would have been charming, unrelated trivia. On the basis of this vast knowledge, Darwin knew that not all adaptations—"contrivances," he called them—are perfect. Adaptations are simply as good as they have to be, "long enough to reach the ground." The imperfection of many adaptations, as revealed on close examination, is a fifth strong line of evidence in support of macroevolution.

For instance, Darwin describes the highly ingenious workings of an orchid flower by which the flower "arranges" for a bee to fall into a pool of water, held in the highly modified petal that forms the lip, so it cannot fly out of the flower. Instead, the bee must crawl through narrow passageways and so deposit and pick up pollen in the course of its exit. However, Darwin also describes orchids and other types of plants with much less elaborate flowers, offering insects only bright

46–13 *A nesting woodcock, hidden among the brush on the woodland floor. As Darwin noted, in cases of mimicry or camouflage, each chance "improvement" in the disguise would be conserved by natural selection, with the end result representing the accumulation of such changes. Note that the behavior of the woodcock—sitting motionless—is as important as its protective coloration. Similarly, the orientation of the moths shown in Figure 46–3 plays an important role in concealing them.*

colors and a few drops of nectar, and finally plants with simple, inconspicuous flowers and windblown pollen. In short, there are gradations and varieties of adaptations, not merely one set of perfect solutions to a given problem.

From his correspondence with Wallace and Bates, Darwin also knew of remarkable examples of defensive mimicry among insects. He mentions a walking-stick insect that not only is shaped like a stick but also is overgrown with material that exactly resembles native mosses. This is an excellent example of how gradual "improvements," each carefully conserved, could be continuously advantageous to the individual organisms—as, of course, could a gradual improvement in the capacity to detect light and focus an image, leading ultimately to the eye.

THE THEORY TODAY

Since Darwin's time, massive additional evidence has accumulated supporting the fact of evolution—that all living organisms present on earth today have arisen from earlier forms in the course of the earth's long history. Indeed, all of modern biology is an affirmation of this relatedness of the many species of living things and of their gradual divergence from one another over the course of time. Since the publication of *The Origin of Species,* the important question, scientifically speaking, about evolution has not been whether it has taken place. That is no longer an issue among the vast majority of modern biologists. Today, the central and still fascinating questions for biologists concern the mechanisms by which evolution occurs.

A major weakness in the theory of evolution, as formulated by Darwin, was the absence of any valid mechanism to explain heredity. Although Mendel was at work on his experiments with pea plants at the time Darwin was writing *The Origin of Species,* his paper was not delivered until 1865 and did not enter the mainstream of biological thinking until after the turn of the twentieth century. The subsequent development of genetics made it possible to answer three questions that Darwin was never able to resolve: (1) how inherited characteristics are transmitted from one generation to the next; (2) why inherited characteristics are not "blended out" but can disappear and then reappear in later generations (like whiteness in pea flowers); and (3) how the variations arise on which natural selection acts.

The combination of Darwin's theory of evolution with the principles of Mendelian genetics is known as the neo-Darwinian synthesis, or the **synthetic theory** of evolution. (Here "synthetic" does not mean "artificial," which is the connotation it usually has for us, but has its original meaning of "the putting together of two or more different elements.") During the past 50 years, the synthetic theory has dominated scientific thinking about the process of evolution and has been enormously productive of new ideas and new experiments, as biologists have worked to unravel the details of how evolution takes place. As we shall see in subsequent chapters of this section, some aspects of the synthetic theory have recently come under challenge—partly as a result of new understandings of genetic mechanisms arising from rapid developments in molecular biology, and partly as a result of reevaluations of the fossil record. The current controversies, which primarily involve the rate of macroevolutionary change and the role played by chance in determining the direction of evolution, do not affect the basic principles of the synthetic theory. They do, however, promise to give us a more complete understanding of evolutionary mechanisms than we have at present.

SUMMARY

Darwin was not the first to propose a theory of evolution, but he was the first to describe a valid mechanism by which it might occur. His theory differed from others in that it envisioned evolution as a two-part process, depending upon (1) the existence in nature of inheritable variations among organisms and (2) the process of natural selection by which some organisms, by virtue of their inheritable variations, leave more offspring than others. Darwin's theory is rightfully regarded as the greatest unifying principle in biology.

Microevolution (small-scale change) caused by the pressures of human civilization on other organisms allows us to observe natural selection in action. Among the modern examples of natural selection acting on chance variations are the increase in number of black *Biston betularia* in industrial areas and the increase in insects resistant to insecticides and bacteria resistant to antibiotics.

There were five principal categories of evidence in support of macroevolution (evolution above the species level) available to Darwin at the time he formulated his theory of evolution: (1) the large number and diversity of species; (2) the fact that particular kinds of organisms were found in particular geographic areas but not in other areas similar in climate and topography; (3) the fossil record, which showed that organisms had a long history and had changed in the course of time; (4) homologies, or similarities, among structures of diverse organisms, implying a common ancestry; and (5) adaptations, which provided evidence not only that gradual changes occur in populations over time in response to selective forces in the environment but also that many adaptations are less than perfect. Since Darwin's time, large quantities of new evidence have accumulated in all of these categories.

A major weakness in Darwin's theory, unresolved for many years, was the absence of any valid mechanism to explain inheritance. In the 1930s, the work of a number of scientists led to the synthetic theory of evolution, which combines the principles of Mendelian genetics with Darwinian theory. The synthetic theory has provided—and continues to provide—the foundation from which biologists work as they attempt to unravel the details of evolutionary mechanisms.

46-14 *According to the geneticist Theodosius Dobzhansky (1900–1975), "Nothing in biology makes sense except in the light of evolution." The publication in 1937 of his* Genetics and the Origin of Species *is considered to mark the birth of the modern synthetic theory. Others who played major roles in the formulation of the synthetic theory included paleontologist George Gaylord Simpson (1902–1984), ornithologist Ernst Mayr, and botanist Ledyard Stebbins.*

QUESTIONS

1. What would have been the Lamarckian explanation (see page 4) for the development of drug resistance in bacteria? How did the Lederbergs' experiment disprove Lamarck's proposal as to how evolution takes place?

2. Darwin emphasized that when mammals (cats, goats, sheep, rodents, and so on) were introduced onto remote islands they often thrived, yet there were no indigenous mammals on the islands except bats. Can you explain why he considered this important?

3. Could the fossil record disprove that evolution has occurred? If so, how?

4. Could the doctrine of special creation be disproved? If so, how?

5. Since Darwin's time, biochemical analyses and electron microscopy have uncovered many new homologies. Name five.

6. Many of Darwin's arguments concerning evolution focused on imperfections rather than perfections. Can you explain why?

7. Modern medical research has given us not only vastly improved health but also numerous insights into human physiological processes that cannot, ethically or morally, be experimentally manipulated. The success of this research has depended heavily on the extrapolation from other organisms to *Homo sapiens*. On what foundation does this extrapolation rest?

The Genetic Basis of Evolution

47–1 *J. B. S. Haldane (1892–1964), whose comment on the plenitude of beetles has already been noted (page 574), was one of several mathematicians whose work provided the foundation on which the synthetic theory of evolution was built. Haldane once remarked that he would lay down his life for two brothers or four first cousins, thereby anticipating by some 40 years the concept of inclusive fitness, which we shall examine in Chapter 51.*

The concept of the gene, set forth by Mendel but unknown to Darwin, made it possible to understand, as Darwin could not, how variations could originate, be preserved, and be transmitted from one generation to another. It is instructive, however, for anyone interested in the development of scientific thought, to note that the early geneticists, including Bateson (page 263), de Vries (page 244), and even T. H. Morgan (page 263), did not accept Darwin's theory, with its concept of gradual change. Working with selected strains of organisms under laboratory conditions, they did not see the great range of variation in natural populations that Darwin had seen as a naturalist. The chief agent of evolution, in their eyes, had to be mutation. Most mutations they knew to be harmful; these were eliminated by natural selection. Occasionally, a useful mutation would come along and thus, suddenly and fortuitously, the species would be set on a new evolutionary course.

Darwinian evolution and Mendelian genetics became reconciled when biologists stopped thinking in terms of individual organisms and genotypes and began thinking in terms of populations, genes, and the frequencies of alleles. The branch of biology that emerged from the synthesis of Darwinian evolution and Mendelian principles is known as **population genetics.** A **population** can be defined as an interbreeding group of organisms of the same species, localized in space and time. For instance, all of the fish of one particular species in a pond constitute a population, as do all the fruit flies in one bottle. A population is unified and defined by its **gene pool,** which is simply the sum total of all the alleles of all the genes of all the individuals in the population. From the viewpoint of the population geneticist, the individual organism is only a temporary vessel, holding a small sampling of the gene pool for a moment in time. The subjects of interest to the population geneticist are gene pools, changes in their composition over time, and the forces causing these changes.

In natural populations, some alleles increase in frequency from generation to generation, and others decrease. If an individual has a favorable combination of alleles in its genotype, it is more likely to survive and reproduce; as a consequence, its alleles are more likely to be present in an increased proportion in the next generation. Conversely, if the combination of alleles is not favorable, the individual is less likely to survive and reproduce; thus representation of its alleles in the next generation will be reduced or perhaps eliminated. In fact, evolution can be precisely defined as any change in the frequency of alleles within a gene pool from one generation to the next.

Fitness, in the context of population genetics, does not mean physical well-being or optimal adaptation to the environment. The only criterion—the measurement—of an individual's fitness is the relative number of surviving offspring, that is, the extent to which the individual's genotype is present in succeeding generations.

Survival of the Fittest

The phrase "survival of the fittest" is often used in describing the Darwinian theory. In the early twentieth century, the doctrine of survival of the fittest in natural populations was used by some individuals to defend gross social inequalities and ruthless competitive tactics in industry on the grounds that they were merely in accord with the "laws of nature." This philosophy was referred to by some as social Darwinism. However, in fact, very little evolutionary change fits the concept of "nature red in tooth and claw." One fuchsia plant with flowers a little brighter than those of its neighbors and so better able to catch the attention of a passing hummingbird is a more pertinent model of the struggle for survival. Fitness, as measured by population geneticists, is determined solely by the relative number of descendants of an individual in a future population.

Natural selection in operation. A bee is making her choice.

THE EXTENT OF VARIATION

Like begets like, we know now, because of the remarkable precision with which the DNA is replicated and transmitted from each cell to its daughter cells at cell division. The DNA in the cells of any individual is, except for occasional mutations, a true replica of the DNA that individual received from its father and mother. Similarly, the DNA in the sperm or eggs of any individual is, except for occasional mutations, a true copy of one-half of the DNA that individual received from its father and mother. In fact, the mechanisms of DNA replication and transmission serve not only to link us to our immediate ancestors but also to all other living things.

This fidelity of duplication is, of course, essential to the survival of the individual organisms of which a population is composed. However, if evolution is to occur, there must be variations among individuals. Such variations make it possible for populations to change as conditions change; they are the raw material on which evolutionary forces act.

For Linnaeus and other creationists, each species was represented by a perfect type, and any variations from this type were considered to be imperfections. For Darwin, and for modern biologists as well, variation is a characteristic of the population; there is no ideal type, only a range of variants, changing in time and space. It was this concept, this shift in the view of nature from the ideal to the actual, that made possible the development of Darwin's theory.

One focus of research in modern population genetics has been to determine the extent of such variability (far greater than Darwin could have realized) and how variations are preserved and fostered in gene pools. The extent of variation has been revealed in a number of different ways.

47–2 As Darwin observed, variations exist among the members of any given species. Even in populations in which the individuals appear to us almost identical, such as the members of this group of king penguins, we know that variations exist because individuals have no difficulty in identifying their own mates or offspring. To penguins, all people probably look alike.

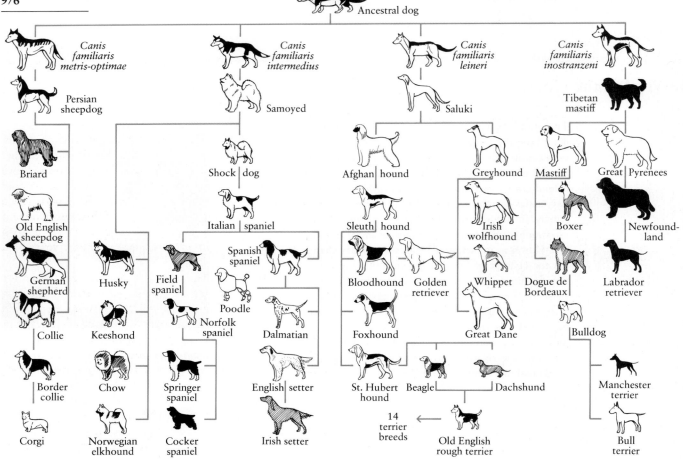

47–3 *All breeds of dogs have been pro-*
duced by artificial selection, a demonstra-
tion of the tremendous potential for
variability in a single species.

Breeding Experiments

Experiments in evolution were carried out by animal breeders and horticulturists for centuries before any concept of evolution was formulated. This process of choosing which individuals should be represented in the next generation—and which should not—was called **artificial selection** by Darwin, and he saw it as a direct analogy to the process of natural selection. First-time readers of *The Origin of Species* are sometimes surprised to find themselves, in the very first chapter, caught up in a treatise on pigeon breeding, one of the many subjects on which Darwin made himself an expert. By selecting birds with particular characteristics —such as a tail with more feathers or a larger beak—for breeding, pigeon fanciers had been able, over the years, to produce a number of exotic breeds: the short-faced tumbler, the pouter, the barb, the trumpeter, and the fan tail. These breeds of birds, all developed from the same wild species and still able to interbreed, differed widely in appearance, more widely in fact than many animals of different species. What the pigeon breeders showed, to put it in modern terms, is that there is a large amount of variability hidden in the gene pool and that this latent variability can be expressed under selection pressures.

The many breeds of dogs (Figure 47–3) provide another example of the variants present in a single species, an example more familiar to most modern readers and also well known to Darwin.

Bristle Number in Drosophila

Breeding experiments that demonstrate the extent of latent variability have also been carried out in the laboratory. In studies with *Drosophila melanogaster,* for example, an easily observable hereditary trait, the number of bristles on the ventral surface of the fourth and fifth abdominal segments (Figure 47–4), was

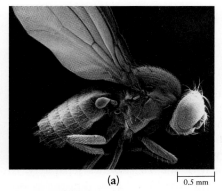

(a)

`0.5 mm`

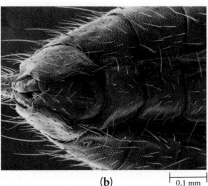

(b)

`0.1 mm`

47-4 *Scanning electron micrographs of* **(a)** *a fruit fly* (Drosophila melanogaster) *and* **(b)** *the ventral surface of its posterior abdominal segments. The results of selection experiments to increase and decrease the number of bristles on the ventral surface of the fourth and fifth abdominal segments are shown in Figure 47–5.*

47-5 *The results of experiments with* Drosophila melanogaster, *demonstrating the extent of latent variability in a population. From a single parental stock, one group was selected for an increase in the number of bristles on the ventral surface (high selection line) and one for a decrease in the bristle number (low selection line).*

As you can see, the high selection line rapidly reached a peak of 56, but then the stock began to become sterile. Selection was abandoned at generation 21 and begun again at generation 24. This time, the previous high bristle number was regained, and there was no apparent loss in reproductive capacity. Note that after generation 24 the stock interbreeding without selection was also continued, as indicated by the line of gray. After 60 generations, members of the freely breeding group from the high selection line had an average of 45 bristles. The low selection line died out owing to sterility.

chosen for selection. In the starting stock, the average number of bristles was 36. Two separate groups were interbred, with one group selected for increase of bristles and the other group for decrease of bristles. In every generation, those individuals with the fewest bristles were selected and crossbred with each other, and so were the individuals with the highest number of bristles.

Selection for low bristle number resulted in a drop over 30 generations from an average of 36 to an average of about 30 bristles. In the high-bristle-number line, progress was at first rapid and steady. In 21 generations, bristle number rose steadily from 36 to an average of about 56 (Figure 47–5). No new genetic material had been introduced; within the single population the potential for a wide range of bristle numbers already existed. Subsequent experiments with *Drosophila* and other organisms have shown that the choice of bristle number was not merely a fortunate accident. Many traits studied in breeding experiments have revealed a comparable range of natural variability.

There is a second part to the bristle-number story. The low-bristle-number line soon died out because it became sterile. Presumably, changes in factors affecting fertility had also taken place during selection. When sterility became severe in the high-bristle line, members of this line were permitted to interbreed without further selection. The average number of bristles fell sharply, and in five generations dropped from 56 to 40. Thereafter, as this line continued to breed without selection, the bristle number fluctuated up and down, usually between 40 and 45, which still was higher than the original 36. At generation 24, selection for high bristle number was begun again for a portion of this line. The previous high bristle number of 56 was regained, and this time there was no loss in reproductive capacity. Apparently, the genotype had become reorganized in such a way that alleles controlling bristle number were present in more favorable combinations with alleles affecting fertility.

Mapping studies have shown that bristle number is controlled by a large number of genes, at least one on every chromosome and sometimes several at different sites on the same chromosome. Although we do not know how important bristle number is to the survival of the animal, we do know that selection for this trait in some way disrupted the entire genotype. Livestock breeders are well aware of this consequence of artificial selection. Loss of fertility is a major problem in virtually all circumstances in which animals have been purposely inbred for particular characteristics. This result emphasizes the fact that in natural selection it is the entire phenotype that is selected, rather than certain isolated characteristics, as is usually the case in artificial selection.

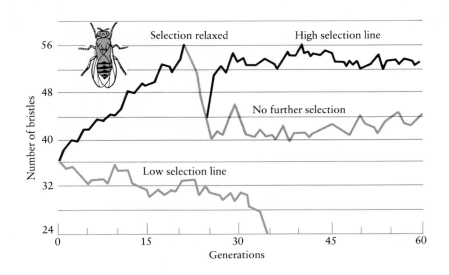

Quantifying Variability

Analysis at the molecular level provides a newer method for assessing variability. As we saw in Section 3, the amino acid sequences of proteins reflect the nucleotide sequences of the genes coding for them. J. L. Hubby and R. C. Lewontin ground up fruit flies from a natural population and extracted proteins from them. From these proteins, they were able to isolate 18 functionally different enzymes. Then they analyzed each of the 18 enzymes separately, to determine if its molecules existed in different structural forms (known as isozymes) or if the structure of the enzyme was uniform throughout the population. The method they used was electrophoresis—the same method used by Pauling to separate the variants of hemoglobin (see page 304).

Of the 18 enzymes studied in this way, nine were found to be composed of protein molecules that were indistinguishable by electrophoresis; in other words, the gene coding for each of these enzymes was the same throughout the entire population of fruit flies studied. However, each of the other nine enzymes was found to have two or more isozymes. Thus, without any direct analysis of the genes themselves, the investigators were able to conclude that among the fruit flies studied there were two or more alleles of the gene responsible for each of these nine enzymes. One enzyme had as many as six slightly different isozymes (Figure 47–6); that is, at least six alleles of the gene coding for that enzyme were present in the gene pool.

Each fruit-fly population examined was heterozygous for almost half of the genes tested. Each individual, it was estimated, was probably heterozygous for about 12 percent of its genes. Similar studies on humans, using accessible tissues such as blood or placenta, indicate that at least 25 percent of the genes in any given population are represented by two or more alleles, and individuals are heterozygous for at least 7 percent of their genes, on the average.

Now that methods for sequencing nucleotides in DNA have been developed, it is possible to make direct comparisons of the genetic material. Since all changes in nucleotides do not result in changes in amino acid sequences, and all changes in amino acid sequences are not detectable by electrophoresis, DNA sequencing is expected to reveal a greater extent of genetic variation.

Explaining the Extent of Variation

The findings of Hubby and Lewontin and others after them have, like most important scientific discoveries, raised major new questions. Many geneticists had previously thought that the individuals of a population should be close to genetic uniformity, as a result of a long history of selection for "optimal" genes. Yet, as these studies revealed, natural populations are far from uniform.

One school of geneticists, the selectionists, claim that even such small variations as those in enzyme structure are maintained by various forms of natural selection that favor some genotypes at some times and in some areas, and others at other times or in other localities. In the selectionist's view, all variation directly or indirectly affects fitness. An opposing school, the neutralists, claim that the observed variations in the protein molecules are so slight that they do not make any difference in the function of the organism and so are not affected by natural selection. (Such neutralism is, of course, supported by the evidence of molecular clocks, described on page 419.) Neutral alleles, according to the neutralist argument, accumulate as a result of random processes, including mutation. Although this difference of opinion cannot be resolved on the basis of present knowledge, we do have a growing understanding of the variety of processes that can act to maintain, increase, or decrease the variability in the gene pool of a population.

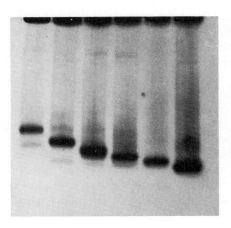

47–6 *The method used by Hubby and Lewontin to analyze* Drosophila *enzymes was electrophoresis. In this process, as you will recall, the proteins are dissolved, placed at the edge of a sheet of jellylike material, and exposed to a weak electric field. The rate at which the molecules move in the electric field is determined by their size and electric charge. As a result, proteins with even very slight structural differences can be separated.*

This photograph shows the separation of six different forms (isozymes) of one enzyme. The material in each column was obtained from flies homozygous for one of the six different alleles coding for the enzyme.

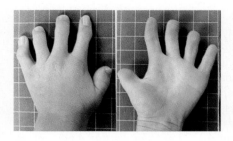

47-7 *A dominant allele is responsible for the characteristic known as brachydactylism (short fingers). In the brachydactylous hand shown here, the first bones of the fingers are of normal length, but the second and third bones are abnormally short. If brachydactylism is caused by a dominant allele, why is it such a rare characteristic?*

A STEADY STATE: THE HARDY-WEINBERG EQUILIBRIUM

In the early 1900s, biologists raised an important question about the maintenance of variability in populations. How, they asked, can both dominant and recessive alleles remain in populations? Why don't dominants simply drive out recessives? For example, given that brachydactylism (Figure 47-7) is caused by a dominant allele, why don't most or even all people have short, fat fingers? This question was answered in 1908 by G. H. Hardy, an English mathematician, and G. Weinberg, a German physician.

Working independently, Hardy and Weinberg showed that the genetic recombination that occurs at each generation in diploid organisms does not *by itself* change the overall composition of the gene pool. To demonstrate this, they examined the behavior of alleles in an idealized population in which five conditions hold:

1. No mutations occur.

2. There is no net movement of individuals—with their genes—into the population (immigration) or out of it (emigration).

3. The population is large enough that the laws of probability apply; that is, it is highly unlikely that chance alone can alter allele frequencies.

4. Mating is random.

5. All alleles are equally viable; in other words, there is no difference in reproductive success. The offspring of all possible matings are equally likely to survive to reproduce in the next generation.

Consider a single gene, which has only two alleles, *A* and *a*. Hardy and Weinberg demonstrated mathematically that if the five conditions listed above are met, the frequencies, or relative proportions, of alleles *A* and *a* in the population will not change from generation to generation. Moreover, the frequencies of the three possible combinations of these alleles—the genotypes *AA*, *Aa*, and *aa*—will not change from generation to generation. The gene pool will be in a steady state—an equilibrium—with respect to these alleles. This equilibrium is expressed by the following equation:

$$p^2 + 2pq + q^2 = 1$$

In this equation, the letter *p* designates the frequency of one allele, and the letter *q* designates the frequency of the other allele; *p* plus *q* must always equal 1 (that is, 100 percent of the alleles of that particular gene in the gene pool). The expression p^2 designates the frequency of individuals homozygous for one allele, q^2 the frequency of individuals homozygous for the other allele, and $2pq$ the frequency of heterozygotes.

Derivation of the Hardy-Weinberg Equation

To understand how Hardy and Weinberg arrived at their equation and demonstrated the equilibrium of allele and genotype frequencies in a population meeting the five stated conditions, let us look more closely at the single gene with only two alleles, *A* and *a*. We are interested in the relative proportions—that is, the frequencies—of *A* and *a* from one generation to the next. As we noted above, when there are only two alleles, *p* and *q* together must equal one: $p + q = 1$.

These proportions—*p* and *q*—could be expressed in terms of fractions, as done by Mendel, but since the proportions of the two alleles will probably not be equal, as they were in Mendel's carefully controlled experiments, it is more convenient to express the numbers as decimals. For example, suppose that in a

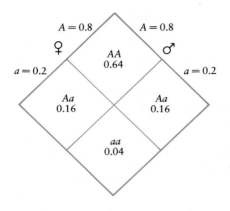

47–8 *Results of random mating in a population in which the frequency (p) of allele A equals 0.8 and the frequency (q) of allele a equals 0.2. In setting up this Punnett square, we are assuming that the gene is not sex-linked and that the alleles appear in the same frequencies in males as in females.*

particular population 80 percent of the alleles of the gene under study are allele *A*. The frequency of *A* is 0.8, or *p* = 0.8. Given that there are only two alleles, we then know that the frequency of allele *a* is 0.2 (*q* = 1 − *p*).

Let us assume that the relative frequencies of *A* and *a* are the same in both male and female (as they are for most alleles in natural populations). Now suppose that males and females mate at random with respect to alleles *A* and *a*. We can calculate the frequencies of the resulting genotypes by drawing a Punnett square. As you can see from Figure 47–8, the genotypes in the population produced by this random mating would consist of 64 percent *AA*, 32 percent *Aa*, and 4 percent *aa*.

Instead of drawing a Punnett square, we can do the same thing algebraically. Because *p* + *q* = 1, it follows that:

$$(p + q)(p + q) = 1 \times 1 = 1$$

Or, as you probably remember from algebra:

$$p^2 + 2pq + q^2 = 1$$

This algebraic expression of the genotype frequencies is the Hardy-Weinberg equation.

Let us apply this equation to the random mating that just occurred in our population. Taking the initial values for the frequencies of the two alleles, we obtain the following results:

$$p^2 = 0.8 \times 0.8 = 0.64 \text{ (the frequency of } AA \text{ genotypes)}$$
$$2pq = 2 \times 0.8 \times 0.2 = 0.32 \text{ (the frequency of } Aa \text{ genotypes)}$$
$$q^2 = 0.2 \times 0.2 = 0.04 \text{ (the frequency of } aa \text{ genotypes)}$$

What has happened to the frequencies of the two alleles in the gene pool as a result of this round of mating? We know from our calculations that the frequency of *AA* is 0.64. In addition, half of the alleles in the heterozygotes *(Aa)* are *A*, so the total frequency of allele *A* is 0.64 plus one-half of 0.32—that is, 0.64 plus 0.16, to give a total of 0.8. The frequency of allele *A (p)* has not changed. Similarly, the total frequency of allele *a* is 0.04 (in the homozygotes) plus 0.16 (half the alleles in the heterozygotes), or 0.2. The frequency of allele *a (q)* has also remained the same.

If another round of mating occurs, the proportion of *AA*, of *Aa*, and of *aa* genotypes in our population will again be 64 percent, 32 percent, and 4 percent respectively. Again the frequency of allele *A* will be 0.8 and of allele *a* 0.2. And so on, and so on, generation after generation. In an ideal population in which the five conditions are met, neither the allele frequencies nor the genotype frequencies change from generation to generation.

The Effect of Multiple Alleles

Although normally any one diploid individual has no more than two alleles of the same gene, there may, of course, be more than two alleles for a given gene in the gene pool. For instance, in the case of the gene for coat color in rabbits, there are at least four alleles. Different diploid combinations of these alleles give rise to coat colors ranging from dark gray through medium and light gray to white, as well as a variety in which most of the coat is white but the feet, ears, and muzzle are black (see Figure 13–9, page 269). The Hardy-Weinberg equilibrium applies equally well to situations such as this, in which there are multiple alleles of the same gene, although the equation representing the equilibrium is more complex. For example, the genotype frequencies of three alleles are expressed by the algebraic expansion of $(p + q + r)^2 = 1$, with *r* representing the frequency in the gene pool of the third allele.

The Significance of the Hardy-Weinberg Equilibrium

The Hardy-Weinberg equilibrium and its mathematical formulation have proved as valuable a foundation for population genetics as Mendel's principles have been for classical genetics. At first glance, this seems hard to understand, since the five conditions specified for the equilibrium are seldom likely to be met in a natural population. An analogy from physics may be useful. Newton's first law says that a body remains at rest or maintains a constant velocity when not acted upon by external force. In the real world, bodies are always acted upon by external forces, but this first law is an essential premise for examining the nature of such forces. Similarly, the frequencies of alleles in populations are always changing, but, in the absence of the Hardy-Weinberg equation, we would not know how to detect change, determine its magnitude and direction, or uncover the forces responsible for it. If, however, we can identify the individuals in a population that are homozygous for a particular allele of interest, we can then calculate the frequency of that allele—and all of the other terms of the Hardy-Weinberg equation. If we do this over a period of generations, we can chart precisely the changes that are taking place in the gene pool—and then look for the causes.

An Application of the Hardy-Weinberg Equation

As we have noted, the Hardy-Weinberg equilibrium holds only in populations in which the five conditions are met. What happens if one of these conditions is not met? Let us imagine another pair of alleles in which allele a has a damaging effect in the homozygous state, reducing the likelihood that a homozygous aa individual will survive to reproduce. We can estimate the number of aa genotypes in the population, perhaps by screening tests on newborn infants. Suppose, for instance, that the condition shows up in 1 in 10,000 infants. In other words, $q^2 = 1/10,000$, or 0.0001. Thus, $q = \sqrt{0.0001}$, or 0.01. If $q = 0.01$, then $p = 0.99$ and $2pq = 0.0198$, or almost 0.02. Thus about 2 percent of the population—one person in every 50—can be estimated to be a heterozygous carrier of this allele.

Now suppose that we do the same screening tests five years later and find that q is not 0.01 but 0.009, and then we repeat it a few years later and find that it has once again decreased very slightly, perhaps to 0.008. In other words, evolution is occurring. One allele is decreasing while the other is increasing (Figure 47–9). Using the Hardy-Weinberg equilibrium as our yardstick, we know not only that change has taken place and its direction but also that there must be some reason for it. We can then look for the factors causing the change.

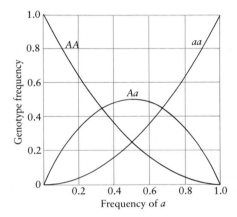

47–9 *The relationship between the frequency of allele* a *in the population and the frequency of the genotypes* AA, Aa, *and* aa. *Naturally, the more* AAs *there are, the lower the frequency of* a. *Because of the interrelationship of* AA, Aa, *and* aa, *a change in the frequency of either allele results in a corresponding and symmetrical change in the frequencies of the other allele and of the genotypes.*

THE AGENTS OF CHANGE

According to modern evolutionary theory, natural selection is the major force in changing allele frequencies. Because natural selection is of such great importance, we shall devote all of the next chapter to it. Here, let us look at some of the other agents that can change the frequencies of alleles in a population. There are four: mutations, gene flow, genetic drift, and nonrandom mating.

Mutations

Mutations, from the point of view of population genetics, are inheritable changes in the genotype. A mutation may involve deletion, transposition, or duplication of a portion of a DNA molecule, or the substitution of one or more nucleotides in the molecule. Mutations can occur not only in structural genes, such as the gene coding for the hemoglobin molecule, but also in regulatory genes, such as those responsible for turning on and off various processes in embryonic development.

47–10 *An example of the sometimes dramatic effects of mutation. The ewe in the middle is an Ancon, an unusually short-legged strain of sheep. The first Ancon on record was born in the late nineteenth century into the flock of a New England farmer. By inbreeding (the characteristic is transmitted as a recessive), it was possible to produce a strain of animals with legs too short to jump the low stone walls that traditionally enclosed New England sheep pastures. A similar strain was produced in northern Europe following an independent mutation there. At one time, it was proposed that evolution takes place in sudden, large jumps such as this—a concept sometimes referred to as the "hopeful monster" theory. One reason this concept has been abandoned is that nearly all mutations producing dramatic changes in the phenotype are harmful, as this one would be in a wild population.*

Among the agents known to cause mutations are x-rays, ultraviolet rays, radioactive compounds, and a variety of other chemical substances. Most mutations occur "spontaneously"—meaning simply that we do not know the chemical and physical forces that trigger them. Mutations are generally said to occur at random, or by chance. This does not mean that mutations occur without cause but rather that the events triggering them are independent of their subsequent effects. Although the rate of mutation can be influenced by environmental factors, the specific mutations produced are independent of the environment—and independent of their potential for subsequent benefit or harm to the organism and its offspring.

The rate of spontaneous mutation is generally low; for mutations detectable in the phenotype, it varies from 1 in 1,000 to 1 in 1,000,000 gametes per generation, depending on the allele involved. Different genes—and even different alleles of the same gene—have different rates of mutation. These differences probably have to do with the chemical composition of the gene (or allele) in question and its position in the chromosome. It is estimated that each new human individual, with approximately 100,000 genes (pairs of alleles), carries two new mutations. Thus, although the incidence of mutation in any given gene or any given individual is low, the number of new mutations per population generation is very high. Mutations are generally regarded as the raw material for evolutionary change. They provide variation for other evolutionary forces to act on, but they seldom, if ever, determine the direction of evolutionary change.

Gene Flow

Gene flow is the movement of alleles into or out of a population. It can occur as a result of the immigration or emigration of breeding individuals or, as in the case of plants and many aquatic invertebrates, the movement of gametes (for example, in the form of pollen) between populations. Gene flow can introduce new alleles into a population or can change allele frequencies. Its overall effect is to decrease the difference between populations, whereas natural selection is more likely to increase differences, producing populations more suited for different local conditions. Thus, gene flow often counteracts natural selection. As we shall see in Chapter 49, geographic barriers that prevent gene flow are very important in the formation of new species.

Genetic Drift

As we stated previously, the Hardy-Weinberg equilibrium holds true only if the population is large. This qualification is necessary because the equilibrium depends on the laws of probability. These laws—the laws of chance—apply equally well to flipping coins, rolling dice, or betting at roulette. In flipping coins,

47-11 *Among the Old Order Amish, a group founded by only a few couples some 200 years ago, there is an unusually high frequency of a rare allele. In its homozygous state, the allele results in extra fingers and dwarfism. This Amish child is a six-fingered dwarf.*

47-12 *Two northern elephant seal bulls fighting for supremacy of a harem, which may consist of as many as 50 females. Only a few males breed each year, and each breeding male fathers many offspring. This social system may be an additional factor contributing to the high degree of homozygosity found in the northern elephant seal population.*

it is possible for heads to show up six times in a row, but, on the average, heads will show up half the time and tails half the time. The more times the coin is flipped, the more closely the expected frequencies of half (0.50) and half (0.50) are approached.

Consider, for example, an allele, say *a*, that has a frequency of 1 percent. In a population of 1 million, 20,000 *a* alleles would be present in the gene pool. (Remember that each diploid individual carries two alleles for any given gene; in the gene pool of this population there are 2 million alleles for this particular gene, of which 1 percent, or 20,000, are allele *a*). But in a population of 50, only one copy of allele *a* would be present. If the individual carrying this allele failed to mate or were destroyed by chance before leaving offspring, allele *a* would be completely lost. Similarly, if 10 of the 49 individuals without allele *a* were lost, the frequency of *a* would jump from 1 in 100 to 1 in 80.

This phenomenon, a change in the gene pool that takes place as a result of chance, is **genetic drift.** Population geneticists and other evolutionary biologists generally agree that genetic drift plays a role in determining the evolutionary course of populations. Its relative importance, however, as compared to that of natural selection, is a matter of current debate. There are at least two situations in which it has been shown to be important.

The Founder Effect

A small population that branches off from a larger one may or may not be genetically representative of the larger population from which it was derived. Some rare alleles may be overrepresented or may be lost completely. As a consequence, even when and if the small population increases in size, it will have a different genetic composition—a different gene pool—from that of the parent group. This phenomenon, a type of genetic drift, is known as the **founder effect.**

An example of the founder effect is found in the Old Order Amish of Lancaster, Pennsylvania (Figure 47–11). Among these people, there is an unprecedented frequency of a recessive allele that, in the homozygous state, causes a combination of dwarfism and polydactylism (extra fingers). Since the group was founded in the early 1770s, some 61 cases of this rare congenital deformity have been reported, about as many as in all the rest of the world's population. Approximately 13 percent of the persons in the group, which numbers some 17,000, are estimated to carry this rare allele. The entire colony, which has kept itself virtually isolated from the rest of the world, is descended from only a few individuals. By chance, one of those must have been a carrier of the allele.

Population Bottleneck

Population bottleneck, another type of genetic drift, occurs when a population is drastically reduced in numbers by an event having little or nothing to do with the usual forces of natural selection. For example, from the 1820s to the 1880s, the northern elephant seal was hunted so heavily along the coast of California and Baja California that it was rendered almost extinct, with perhaps as few as 20 individuals remaining. Since 1884, when the seal was placed under the protection of the United States and Mexican governments, the population has increased to more than 30,000, all presumably descendants of this small group. Studies of blood samples taken from 124 seal pups showed them to be homozygous for some 21 gene loci, indicating (by comparison with other mammalian groups) a drastic loss of genetic variability. A population bottleneck is likely not only to eliminate some alleles entirely but also to cause others to become overrepresented in the gene pool. For example, the high rate of Tay-Sachs disease among Ashkenazic Jews (page 389) is attributed to a population bottleneck experienced by these people in the Middle Ages.

47-13 *At one time, the lesser white snow goose ("the goose from beyond the north wind") and the blue goose were believed to represent distinct species. More recently, it has been found that they actually represent one species, with individuals of two colors. The white is recessive; heterozygotes and homozygotes for blue are both the same dark blue color. However, birds mate preferentially with animals of their own color. As a result, there are more homozygotes than if mating were random.*

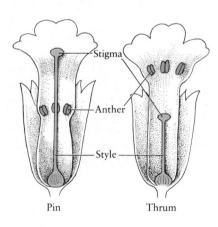

47-14 *Diagrams of two types of flowers ("pin" and "thrum") of the same species of primrose. Notice that the pollen-bearing anthers of the pin flower and the pollen-receiving stigma of the thrum flower are both situated about halfway up the length of the flower and that the pin stigma is level with the thrum anthers. An insect foraging for nectar in these plants would collect pollen on different areas of its body, so that thrum pollen would be deposited on pin stigmas, and vice versa.*

Nonrandom Mating

Disruption of the Hardy-Weinberg equilibrium can also be produced by nonrandom mating. A form of nonrandom mating particularly important in plants is self-pollination (as in the pea plants Mendel studied). In animals, nonrandom mating is often behavioral. For example, snow geese may be either white or blue (Figure 47–13). This is a type of variation known as **polymorphism,** in which two or more phenotypically distinct forms coexist in a population. White snow geese tend to mate preferentially with other white geese and blue with blue. Thus, assuming only two alleles are involved, there will be a decrease in the heterozygote (represented as $2pq$ in the Hardy-Weinberg equation), with concomitant increases in the two homozygotes (p^2 and q^2).

Note that nonrandom mating can cause changes in the genotype frequencies without necessarily producing any changes in the frequencies of the alleles in question. If, however, all female snow geese, white or blue, preferred to mate only with blue snow geese, and, as a consequence, some of the blue males mated with more than one female, and some of the white males with none, changes in the frequencies of the two alleles and also of the phenotypes would occur. In fact, as we shall see in the next chapter, such nonrandom mating produced by female choice is an important agent of natural selection in some species.

PRESERVATION AND PROMOTION OF VARIABILITY

Sexual Reproduction

By far the most important method by which eukaryotic organisms promote variation in their offspring is sexual reproduction. Sexual reproduction produces new genetic combinations in three ways: (1) by independent assortment at the time of meiosis—as diagrammed on page 259, (2) by crossing over with genetic recombination, and (3) by the combination of two different parental genomes at fertilization. At every generation, alleles are assorted into new combinations. In contrast, consider, for a moment, organisms that reproduce only asexually, by processes that involve mitosis and cytokinesis but not meiosis. Except when a mutation has occurred in the duplication process, the new organism will exactly resemble its only parent. In the course of time, various clones may form, each carrying one or more mutations, but unless the same mutations occur in the same clones, potentially favorable combinations are never going to accumulate in one genotype.

On the debit side, organisms that reproduce sexually can do so at only half the rate of asexually reproducing organisms. The only advantage to the organism of sexual reproduction, speaking strictly scientifically, is the promotion of variation, the production of new combinations of alleles among the offspring. Why such variation is advantageous to the individual organism is a matter of current controversy (see essay on page 986).

Mechanisms That Promote Outbreeding

Many ways have evolved by which new genetic combinations are promoted in sexually reproducing populations. Among plants, a variety of mechanisms ensure that the sperm-bearing pollen is from a different individual than the stigma it lights upon. Some plants, such as the holly and the date palm, have male flowers on one tree and female flowers on another. In others, such as the avocado, the pollen of a particular plant matures at a time when its own stigma is not receptive. In some species, anatomical arrangements inhibit self-pollination (Figure 47–14).

47–15 *Two banana slugs* (Ariolimax dolichophallus), *photographed on an oak tree near Santa Cruz, California, toward the end of a mating process that sometimes takes seven to eight hours. Courtship for these hermaphrodites begins when two neighboring slugs produce large amounts of slime, which is thought to contain chemical mating signals. Following several hours of ritualized circling, stroking, and biting, the animals become locked in a mutual copulatory embrace, in which each fertilizes the eggs of the other. Disengagement can sometimes be difficult, as in the case of these two slugs. Despite this problem, hermaphroditism is advantageous for banana slugs—as it is for other slow, solitary species—because it doubles the chances of finding a mate. Every mature member of the same species that a hermaphrodite meets will always be of the opposite sex—and of the same sex too. As a result of the mating, each individual can produce new offspring.*

Some plants have genes for self-sterility. Typically, such a gene has multiple alleles—s^1, s^2, s^3, and so on. A plant carrying the allele s^1 cannot pollinate a plant with an s^1 allele; one with an s^1/s^2 genotype cannot pollinate any plant with either of those alleles; and so forth. In one population of about 500 evening primrose plants, 37 different self-sterility alleles were found, and it has been estimated that there are more than 200 alleles for self-sterility in red clover. A plant with a rare self-sterility allele is more likely to be able to pollinate another plant than is a plant with a common self-sterility allele. As a consequence, the self-sterility system strongly encourages variability in a population; selection for the rare allele makes it more common, whereas more common alleles become rarer.

Among animals, even in those invertebrates that are hermaphrodites, such as earthworms, slugs, and many snails, an individual seldom fertilizes its own eggs. Among mammals, in particular, behavioral strategies promote outbreeding (mating with unrelated individuals). Often, for example, young males leave the family group as they reach reproductive age; this occurs among lions, gorillas, and baboons, to mention only a few. Among hunting dogs of the African plains, it is the young females who leave at reproductive age.

Most human cultures discourage inbreeding (breeding between closely related individuals). Many pretechnological societies, for example, demand that a young man choose a wife from another village rather than from his own community, and virtually all cultures have strong prohibitions against incest. Such prohibitions are particularly interesting when you consider that intermarriages of brother and sister or father and daughter would tend to keep property or power within the family and so could be socially and economically advantageous. Followers of Freud maintain that incest is forbidden by strong psychological taboos, as reflected in the Oedipus myth. Some biologists hold that the prohibition stems from the observed ill effects of incestuous matings (through greater chances of deleterious alleles becoming homozygous) and has been merely reinforced by cultural restraints. Others maintain that it is a genetically determined behavioral mechanism, analogous to those observed in other mammals.

47–16 *Male lions leave the pride in which they are born at the age of about 3 years. These young nomadic males are often found in small groups, such as the one shown here. They are characteristically of* *the same age and are probably brothers or, at least, from the same pride. They will probably eventually take over a pride of their own. This social system ensures outbreeding.*

Why Sex?

Sex is a very complicated and expensive way to reproduce. As we noted previously (page 259), a sexually reproducing population wastes half of its reproductive potential in producing males and can therefore increase in numbers only half as fast as an asexually reproducing population. The same genetic penalty is paid by individual females: a sexually reproducing female will propagate her genes only half as fast as an asexually reproducing female. Moreover, sexual reproduction often involves considerable expenditure of time and effort and may expose an organism to predators while it is seeking a mate, courting, and copulating.

If sexual reproduction is so inefficient and so risky, why has it not been eliminated by natural selection? The answer must lie in the genetic variation promoted by sexual reproduction, but it is not easy to see how producing genetically variable offspring could compensate for producing, effectively, only half as many of them. Several hypotheses have been proposed to attempt to account for the success and persistence of sexual reproduction.

According to the "best man" hypothesis, the environment is very changeable. As a result, the offspring are likely to grow up in conditions quite different from those experienced by their parents. Offspring that are genetically identical to their parents will be poorly adapted to the changed conditions. Sexual reproduction is a way of creating a great variety of new genotypes, a few of which will happen to be well adapted to the unpredictable conditions experienced by the next generation.

The "tangled bank" hypothesis, by contrast, emphasizes that the environment is very diverse. It offers a great variety of different opportunities, but each opportunity is limited—that is, it can be exploited successfully by only a limited number of individuals of a particular genotype. Producing many uniform offspring is futile because, being very similar, they will all attempt to exploit the same opportunity. If you have 10 children, all of whom would like to stay in the same small town in which they grew up, you wouldn't want them all to be doctors. Sexual reproduction is a way of avoiding competition between close relatives by suiting them to a variety of different ways of making a living.

Although often cited as correct, the "best man" hypothesis is, according to Graham Bell of McGill University, almost certainly wrong. If it were true, we would expect to find sexual reproduction in contin-

Diploidy

Another factor in the preservation of variability in eukaryotes is diploidy. In a haploid organism, genetic variations are immediately expressed in the phenotype and are therefore exposed to the selection process. In a diploid organism, however, such variations may be stored as recessives, as with the allele for white flowers in Mendel's pea plants. The extent to which a rare allele is protected is shown in Table 47–1.

TABLE 47–1 **Protection of Recessive Alleles by Diploidy**

FREQUENCY OF ALLELE a IN GENE POOL	GENOTYPE FREQUENCIES			PERCENTAGE OF ALLELE a IN HETEROZYGOTES
	AA	Aa	aa	
0.9	0.01	0.18	0.81	10
0.1	0.81	0.18	0.01	90
0.01	0.9801	0.0198	0.0001	99

As the table reveals, the lower the frequency of allele a, the smaller the proportion of it exposed in the aa homozygote becomes. The removal of the allele

ually changing, disturbed, or novel conditions, with asexual reproduction prevailing in stable, long-established environments. Exactly the reverse is now known to be true. For example, asexual reproduction is more common in freshwater environments and in Arctic regions, whereas sexual reproduction prevails in the more stable conditions provided by marine and tropical environments.

These ecological patterns are consistent with the "tangled bank" hypothesis, but they do not prove that it is correct. Other hypotheses may also be consistent with the observed pattern. For example, the "Red Queen" hypothesis—named after a character in *Through the Looking Glass*—points out that many species (such as predators and prey, or parasites and hosts) are continually running an evolutionary race, each trying to increase its efficiency at the expense of others. Success requires a means of rapid genetic change in order to keep pace with the adaptations and counter-adaptations of one's opponents. The continual reshuffling of alleles provided by sexual reproduction is such a means. If this hypothesis is correct, then sexual reproduction should be common wherever many species are found together, as in the sea and the tropics.

The Red Queen from Through the Looking Glass: *"You have to run faster than that to stay in the same place."*

Correlation of the occurrence of sexual reproduction with ecological information allows us to exclude the "best man" hypothesis, but different or more detailed comparisons are needed to decide between the "tangled bank" and "Red Queen" hypotheses. The evolution and function of sexual reproduction remain among the most challenging and perplexing problems in biology.

by natural selection slows down accordingly (see Figure 47–9, page 981). This result should be of special interest to proponents of eugenics, the science of improving the human gene pool through controlled breeding. For instance, consider a genetic disorder, such as PKU (page 388), that is expressed only in the homozygous recessive. If the frequency of allele a is about 0.01, individuals with the aa genotype make up 0.0001 of the population (1 child for every 10,000 born). It would take 100 generations, roughly 2,500 years, of a program of sterilization of homozygous individuals with this condition to halve the allele frequency (to 0.005) and reduce the number born with this genetic disorder to 1 in 40,000, which should be enough to discourage even the most zealous eugenicist.

Heterozygote Superiority

Recessive alleles, even ones that may be harmful in the homozygous state, may not only be sheltered in the heterozygous state, but sometimes may actually be selected for. This phenomenon, in which the heterozygotes have greater reproductive success than either type of homozygote, is known as **heterozygote superiority.** It is another way that genetic variability is preserved.

Sickle Cell Anemia

One of the best-studied examples of heterozygote superiority involving a single gene locus is found in association with sickle cell anemia. Until very recently, individuals homozygous for sickling almost never lived long enough to become parents. Therefore, almost every time two sickling alleles came together in a homozygous individual, two sickling alleles were removed from the gene pool. At one time, it was thought that the sickling allele was maintained in the population by a steady influx of new mutations. Yet in some African tribes, as much as 45 percent of the population is heterozygous for sickling, despite the loss of sickle alleles through homozygous individuals. To replace these alleles by mutations alone would require a rate of mutation about 1,000 times greater than any other known human mutation rate.

In the search for an alternative explanation, it was discovered that the sickling allele is maintained at high frequencies because the heterozygote has a selective advantage. In many regions of Africa, malaria is one of the leading causes of illness and death, especially among young children. Studies of the incidence of malaria among young children showed that susceptibility to malaria is significantly lower in individuals heterozygous for sickling than in normal homozygotes. Thus, although two alleles for sickle cell hemoglobin are eliminated virtually every time they appear in the homozygous state, positive selection for the heterozygote maintains the allele in the gene pool. Moreover, for reasons that are not known, women who carry the sickling allele are more fertile than normal homozygotes.

As this example makes clear, genetic variation not only is the raw material on which natural selection acts, but it can also be maintained by selection. In the next chapter, we shall consider other situations in which natural selection plays a key role in promoting and preserving genetic variation.

Heterosis, or Hybrid Vigor

When two different inbred strains of a crop plant are crossed with one another, the resultant strain is often found to be superior to either of the parent strains. Hybrid corn is a striking example. The development of hybrid corn caused a revolutionary improvement in the corn crop of the United States in the 1930s, because of the increased size and hardiness of the plants derived by crossing two different varieties to produce the seeds for each planting (Figure 47–17). This phenomenon is known as **heterosis**. Heterosis is believed to be one of the most widespread genetic mechanisms by which polymorphisms are maintained in populations.

Heterosis, also known as hybrid vigor, is apparently the result of the fact that hybrids are, by definition, heterozygous at far more loci than most natural varieties. The superiority of the hybrids may be because hybrids are less likely to be homozygous for deleterious alleles, or because of positive advantages conferred by greater heterozygosity.

THE ORIGIN OF GENETIC VARIATION

The newly won capacity to analyze the chromosomal DNA of eukaryotic organisms has, as we noted in Chapter 18, turned up a number of surprises. Large segments of DNA, it has been shown, have the capacity to produce duplicates of themselves and to disperse these duplicates to other locations in the chromosome and even to other chromosomes. These duplicate genes are then free to travel their own evolutionary course, leaving their functions to be carried out by the parent, original genes. The duplicate genes are therefore free of selective constraints, allowing mutations to accumulate.

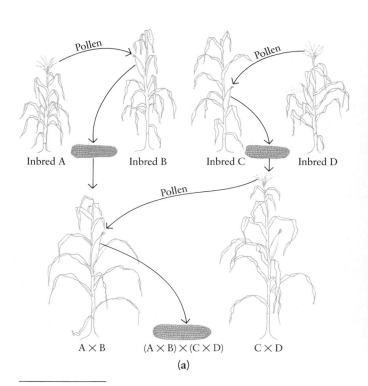

(a)

(b)

47-17 (a) *Hybrid corn is derived by first crossing strains A with B and C with D and then crossing the resulting single-cross plants to produce double-cross seed for planting. The increased size and hardiness of the hybrid are probably due to its increased heterozygosity.* (b) *Cross pollination of corn plants to produce hybrid seed.*

Evolutionary biologists speculate that all present-day structural genes had their beginnings in a very few protogenes that were then duplicated and modified, through the accumulation of mutations, over the last 4 billion years. Most important, there is clear evidence that this process of duplication and subsequent mutation is continuing at this very moment. Gene duplication and modification have undoubtedly played a major role in evolution; as our understanding of the details of these processes increases, revisions in some of the details of evolutionary theory may be required.

SUMMARY

Population genetics is a synthesis of the Darwinian theory of evolution with the principles of Mendelian genetics. A population, for the population geneticist, is an interbreeding group of organisms, defined and united by its gene pool (the sum of all the alleles of all the genes of all the individuals in the population). Evolution is the result of accumulated changes in the composition of the gene pool.

The extent of genetic variability in a population is a major determinant of its capacity for evolutionary change. Natural populations can be shown by breeding experiments—artificial selection—to harbor a wide spectrum of genetic variations. The extent of genetic variation can be quantified by comparison of protein structures and, more recently, the sequencing of DNA molecules. It has been estimated that at least 25 percent of the genes in any given human population are represented by two or more alleles.

The Hardy-Weinberg equilibrium describes the steady state in allele and genotype frequencies that would exist in an ideal population in which five conditions were met: (1) no mutations occurred, (2) there was no immigration or emigration, (3) the population was large, (4) mating was random, and (5) there was no difference in the reproductive success of the offspring. The Hardy-Weinberg equilibrium demonstrates that the genetic recombination that results from meiosis and fertilization cannot, in itself, change the frequencies of alleles in the gene pool. The mathematical expression of the Hardy-Weinberg equilibrium provides a quantitative method for determining the extent and direction of change in allele and genotype frequencies.

The principal agent of change in the composition of the gene pool is natural selection. Other agents of change include mutation, gene flow, genetic drift, and nonrandom mating. Mutations provide the raw material for change, but mutation rates are usually so low that mutations, in themselves, do not determine the direction of evolutionary change. Gene flow, the movement of alleles into or out of the gene pool, may introduce new alleles or alter the proportions of alleles already present; it often has the effect of counteracting natural selection. Genetic drift is the phenomenon in which certain alleles increase or decrease in frequency, and perhaps even disappear, as a result of chance events; examples of genetic drift, which is most likely to occur in small populations, are the founder effect and population bottleneck. Nonrandom mating causes changes in the proportions of genotypes but may or may not affect allele frequencies.

Sexual reproduction is the most important factor promoting genetic variability in populations. Mechanisms that promote outbreeding further promote variability. These mechanisms include self-sterility alleles in plants, anatomic adaptations that inhibit self-fertilization, and in animals, behavioral strategies. Variability is preserved by diploidy, which shelters rare, recessive alleles from selection. Natural selection can, however, act to promote and preserve variability. In cases of heterozygote superiority, for example, the heterozygote is selected over either homozygote, thus maintaining both alleles in the population. Heterosis, or hybrid vigor, is the result either of heterozygote superiority or of the masking of the effects of recessive alleles.

QUESTIONS

1. At the end of the experiment on bristle number in *Drosophila*, the flies in the high selection line had an average of 56 bristles. No fly at the beginning of the experiment had as many as 56, however. How do you explain this fact?

2. As noted in the text, the Hardy-Weinberg equation was first formulated in response to a question: Why, if some alleles are dominant and some are recessive, don't the dominants drive out the recessives? What is the fallacy in the reasoning underlying that question? How does the Hardy-Weinberg equation answer that question?

3. Suppose that in a breeding experiment, 7,000 *AA* individuals and 3,000 *aa* individuals mate at random. In the first generation of offspring, what would be the frequencies of the three genotypes (*AA*, *Aa*, and *aa*)? What would be the frequencies of the two alleles? What would be the values in the second generation, assuming that the Hardy-Weinberg assumptions hold?

4. Among black Americans, the frequency of sickle cell anemia is about 0.0025. What is the frequency of heterozygotes? When one black American marries another, what is the probability that both will be heterozygotes? If both are heterozygotes, what is the probability that one of their children will have sickle cell anemia?

5. How would the following affect the Hardy-Weinberg equilibrium and the phenotypes in a population: Increased mutation of *A* to *a* (assuming *A* is dominant)? Increased mutation of *a* to *A* (assuming *a* is recessive)?

6. What is the difference between gene flow and genetic drift? How does each affect the gene pool of a population?

7. What role might immigration play in promoting genetic variability? Compare the potential impact of immigration on the gene pool of a small island population versus that of a large mainland population.

8. How does self-fertilization, as in flowers, affect the Hardy-Weinberg equilibrium?

9. How do self-sterility alleles promote variability in a population?

10. In cases of heterozygote superiority involving a single gene with only two alleles, how does selection for the heterozygote affect the relative frequency of the two alleles? How would selection *against* the heterozygote affect this frequency?

11. Under what circumstances would two alleles be maintained in a population in exactly the same numbers?

12. In bacteria, the exchange of genetic information and reproduction take place as separate events. In eukaryotes, genetic recombination is generally linked with reproduction. Why do you think this might be so?

Natural Selection

According to Darwin's own account, the concept of natural selection came to him in 1838 upon reading Malthus's gloomy essay. Darwin realized that all populations—not just the human population—are potentially doomed to exceed their resources. Only a small fraction of the individuals that might exist are born and survive. According to Darwin, those that do survive are those that are "favoured," to use his term, by reason of slight, advantageous variations. This process of survival of the "favoured" he termed natural selection, by analogy with the artificial selection practiced by breeders of domestic animals and plants (Figure 48–1).

In terms of population genetics, natural selection is now defined more rigorously as the differential rate of reproduction of different genotypes in a population. This differential reproductive success, which is a result of interactions between the individual organisms and their environment (including other organisms), can result in changes in the gene pool of a population—that is, in evolution. According to the synthetic theory, natural selection is the major force in evolution.

NATURAL SELECTION AND THE MAINTENANCE OF VARIABILITY

In the course of the controversies that led to the formulation of the synthetic theory, some biologists argued that natural selection would serve only to eliminate the "less fit." As a consequence, it would tend to reduce the genetic variation in a

48–1 *Six vegetables produced from a single species of plant* (Brassica oleracea, *a member of the mustard family). They are the result of selection for leaves (kale), lateral buds (brussels sprouts), flowers and stem (broccoli), stem (kohlrabi), enlarged terminal buds (cabbage), and flower clusters (cauliflower). Kale most resembles the ancestral wild plant. Artificial selection, as practiced by animal breeders and horticulturists, gave Darwin the clue to the concept of natural selection.*

Kale   Brussels sprouts   Broccoli   Kohlrabi   Cabbage   Cauliflower

48–2 (a) *Four representative snails of the genus* Cepaea. *The shells of these snails exhibit a wide range of colors and banding patterns. Among the most common colors are yellow and pink. Shells of either color may be unbanded or may have as many as five bands of varying degrees of intensity. As shown here, banded snails living on dark, mottled backgrounds are less visible than unbanded snails and are therefore preyed upon less frequently. Conversely, in areas where the background is fairly uniform, unbanded snails have a survival advantage over the banded type.* **(b)** *A song thrush* (Turdus philomelos) *breaking open a yellow-shelled snail at an "anvil."*

population, acting in effect as an anti-evolutionary force. Modern population genetics has demonstrated this not to be true. As we saw in the example of the sickle cell allele (page 988), natural selection can be a critical factor in preserving and promoting variability in a population. There are many other examples of this effect; here we shall consider just one, which particularly illustrates the fact that a variety of selective forces can be at work simultaneously.

Balanced Polymorphism: Color and Banding in Snail Shells

As you will recall (page 984), polymorphism is the coexistence within a population of two or more phenotypically distinct forms. In some cases, one phenotype gradually replaces another, as in the case of the peppered moths (page 962), and the polymorphism is transient. In other cases, however, the phenotypes are maintained in fairly stable proportions by natural selection, and the polymorphism is said to be balanced.

One of the best-studied examples of a balanced polymorphism maintained by natural selection is found among land snails of the genus *Cepaea*. In one species, for instance, the shell of the snail may be yellow, brown, or any shade from pale fawn through pink and orange to red. The lip of the shell may be black or dark brown (normally) or pink or white (rarely). Up to five black or dark-brown longitudinal bands may decorate it (Figure 48–2a). Fossil evidence indicates that these different types of shells have coexisted for more than 10,000 years.

Studies among English colonies of *Cepaea* have revealed some of the selective forces at work on the snails, which occupy a variety of habitats, including rocks, bogs, and woodlands. These snails are preyed upon by birds, among which are song thrushes. Song thrushes select snails from the colonies and take them to nearby rocks, where they break them open and eat the soft parts, leaving the shells (Figure 48–2b). By comparing the proportions of types of shells around the thrush "anvils" with the proportions in the nearby colonies, investigators have been able to correlate the shell patterns of snails seized by the thrushes with the habitats of the snails. As Table 48–1 reveals, in habitats where the background is uniform (rocks, for instance) banded snails are more likely to be captured. Conversely, in habitats where the backgrounds are mottled (bogs and woodlands), unbanded snails are more likely to be the victims.

TABLE 48–1   A Comparison of Numbers of Snails with Banded Shells Collected Alive in a Uniform Habitat and Killed by Thrushes at Nearby Anvils

| | NUMBER OF SHELLS | | | PERCENT BANDED |
	BANDED	UNBANDED	TOTAL	
Living	264	296	560	47.1
Killed by thrushes	486	377	863	56.3

Studies of many different colonies have confirmed these correlations. In uniform environments, a higher proportion of snails is unbanded, whereas in rough, tangled habitats, such as woodland floors, far more tend to be banded. Similarly, the greenest habitats have the highest proportion of yellow shells, but among snails living on dark backgrounds, the yellow shells are much more visible and are clearly disadvantageous, judging from the evidence conveniently assembled by the thrushes.

Many of the snail colonies studied were at distances so great from one another that the possibility of movement between populations could be ruled out. Why, then, are both shell types still present in these colonies? One would expect populations living on uniform backgrounds to be composed almost entirely of unbanded snails, and colonies on dark, mottled backgrounds to lose most of their yellow-shelled individuals. The answer to this problem is not fully worked out, but it seems that there are physiological factors that are correlated with the particular shell patterns and that form a part of the same group of genes that control color and banding. Experiments have shown, for instance, that unbanded snails (especially yellow ones) are more heat-resistant and cold-resistant than banded snails. In other words, several different selection pressures are at work, and they appear to maintain the genetic variations of color and banding.

WHAT IS SELECTED?

As the studies with *Cepaea* illustrate, natural selection acts on the complete phenotype. In the days of fruit-fly genetics, when populations were being scanned for white eyes and short wings, phenotype became synonymous with physical appearance. In terms of evolutionary theory, however, it must be regarded as including such important characteristics as the optimum temperature at which a particular enzyme works, or the speed of response to a stimulus. In short, the phenotype includes all observable attributes of an organism.

A phenotype generally is the expression of many different genes. As a corollary, any particular phenotypic characteristic may be arrived at by a number of genotypic routes. For example, selection for high and low bristle number in *Drosophila* (page 976) led, in the first series of experiments, to sterility. The second attempt to select for high bristle number led to the same number of bristles (that is, the same phenotypic characteristic) but without a reduction in fertility. Obviously, different genotypes were responsible.

It is very rare that a single allele can determine a winning phenotype. In *Biston betularia*, black wing color is determined by a single allele. However, the peppered coloring may be the result of several different genes, and, along with the mottled appearance, there are other characteristics important for escaping hungry predators. These include selecting the right background, lying very still while on the tree trunk, and positioning oneself correctly so as to enhance the camouflage effect, all of which involve genetic factors. Such underlying genetic complexity is typical not only of the variety of mechanisms by which animals avoid predators (Figure 48–3) but also of most characteristics of the whole organism. Groups of genes that collectively produce coordinated phenotypic characteristics are known as **co-adaptive gene complexes,** or, when they are linked together on one chromosome, **supergenes.** Some of the chromosomal inversions seen in the giant chromosomes of *Drosophila* (page 278) may increase fitness by protecting supergenes from disassembly by recombination.

48–3 *Coordinated anatomical and behavioral characteristics protect the eyed hawkmoth from predators. (a) At rest, the moth positions itself on the bark of a tree in a way that enhances the camouflaging effect of its cryptically patterned wings. (b) Disturbed, the moth parts its wings, revealing a pair of "eyes," reported to be sufficiently menacing to scare away a bird.*

(a)   (b)

(a)

(b)

48-4 *Jeffrey pines usually grow tall and straight, as in (a). Environmental forces, however, can alter the normal growth patterns, as shown in (b). This tree is growing on a mountaintop in Yosemite National Park, California, where it is exposed to strong, constant winds.*

48-5 *The long and generally fruitless debate about the relative importance of heredity and environment in determining IQ has been enlivened in recent years by the discovery that in Japan the average IQ of schoolchildren has increased seven points in a single generation. Japanese schoolchildren now have an average IQ of 111 as compared to an average of 100 in the United States. This difference, seen among children as young as six, is most marked in tests of block designs, mazes, picture arrangement, and object assembly. The message on the headbands reads: "Strength in exams."*

As you will recall, the phenotype is not determined solely by the interactions of the multitude of alleles making up the genotype. It is also a product of the interaction of the genotype with the environment in the course of the individual's life (Figure 48-4). For example, among human identical twins (that is, twins produced from a single fertilized ovum and therefore having the same genotype), noticeable differences—such as body weight—between the two are often apparent even at birth, owing to differences in the intrauterine environment.

A particularly complex relationship between genotype, environment, and phenotype is illustrated by the IQ score. First, the IQ test itself is made up of a number of subtests of verbal comprehension, mathematical abilities, spatial concepts, and even value judgments (such as what you would do if you were the first to see a fire in a crowded theater). Scores achieved in the subtests by any individual may vary widely. Second, many variables are at work. Among those that can be measured are prenatal and perinatal nutrition, socioeconomic background, weight at birth, parental IQs, inborn errors of metabolism (PKU is one, but there are many others), chromosomal aberrations (as in Down's syndrome), oxygen deprivation at birth, and motivation and attitude. The IQ has a reality: the fact that an individual has an IQ of 95 or 125, or 145, or 165 predicts fairly accurately his or her performance in certain social situations. It also has genetic components, and some of these components, at least in past human history, probably had selective value. However, what was being selected, genotypically speaking, is virtually impossible to define.

Any particular phenotype lasts, in terms of evolutionary history, for as long as the blink of an eye. In the case of sexually reproducing organisms, the genotype is as unique and transient as the phenotype, shuffled and recombined at every generation. Only the individual genes survive. The great men and women of history have long since vanished, but each of us may carry one or two of their alleles as part of our human legacy.

TYPES OF SELECTION

As we have seen in the examples considered thus far, the results of natural selection in any given instance depend on the interaction of a variety of factors. Selection processes can be classified into five broad types on the basis of several different criteria. From the standpoint of its effect on the distribution of characteristics within a population, natural selection can be described as stabilizing, disruptive, or directional (Figure 48-6). When selection is influenced by the relative proportions of different phenotypes within a population, it is said to be

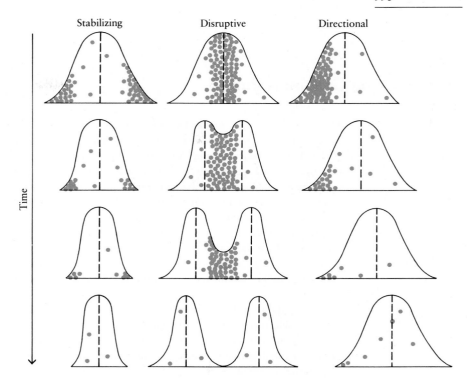

48-6 *A schematic representation of three types of natural selection acting on a trait, such as body size, that varies continuously throughout a population (see page 271 for a review of continuous variation). The horizontal axis of each small graph represents the varying dimensions of the trait being considered, with one extreme at the left, the other extreme at the right, and intermediate states in between. The black curve in each graph summarizes the proportion of individuals in the population exhibiting a particular variant of the trait. Initially, each of the populations displays a normal bell-shaped curve—most individuals exhibit intermediate variants of the trait, and only a few exhibit the extremes. The dots represent individuals in each generation that failed to reproduce or that left fewer than the average number of offspring.*

As you can see, the shapes of the curves change in succeeding generations. Stabilizing selection involves the elimination of extremes, producing a more uniform population. In disruptive selection, intermediate forms are eliminated, producing two divergent populations. Directional selection is the gradual elimination of one expression of the trait in favor of another; note how the bell-shaped curve moves to the right in each succeeding generation. Directional selection produces adaptive change, as may disruptive selection.

frequency-dependent. A fifth category, sexual selection, is defined by what is selected: characteristics of direct consequence in obtaining a mate and successfully reproducing.

Stabilizing Selection

Stabilizing selection, a process that is always in operation in all populations, is the elimination of extreme individuals. Many mutant forms are probably immediately weeded out in this way, often in the zygote or embryo. Clutch size in birds, for instance, is the result of stabilizing selection. Clutch size (the number of eggs a bird lays) is determined genetically, although it also appears to be influenced by ecological factors. In a study of Swiss starlings (Table 48–2), the percentage of birds surviving was found to increase for each clutch size up to five. With a clutch size larger than five, a smaller percentage of birds survived—apparently because of inadequate nutrition. Female Swiss starlings whose genotypes lead to a clutch size of four or five will have more surviving young, on the average, than members of the same species that lay more or fewer eggs. Although the number of eggs laid in a clutch may appear to be a simple trait, rather like pushing the right number for copies on the Xerox machine, it involves a number of coadapted physiological and behavioral characteristics. Among the factors involved are the synthesis of proteins for the yolk and albumen, the availability of calcium for the shells, and the length of time the female will mate.

TABLE 48-2 **Survival in Relation to Number of Young in Swiss Starlings**

Number of young in brood	1	2	3	4	5	6	7	8
Number of young marked	65	328	1,278	3,956	6,175	3,156	651	120
Number of marked birds recaptured after 3 months	0	6	26	82	128	53	10	1
Percentage of marked birds recaptured after 3 months	0	1.8	2.0	2.1	2.1	1.7	1.5	0.8

Human Blood Groups: A Puzzle

The blood types A, B, AB, and O represent the most thoroughly studied polymorphism found in human populations. Apparently, the three alleles associated with these blood types are part of our ancestral legacy, since the same blood types are also found in other primates. A great deal is known about the chemistry of the different blood groups and about the allele frequencies in different populations. Yet we know very little about how this polymorphism has been maintained.

Some population biologists regard the blood types as probably neutral in selective value. Others maintain that polymorphism in human blood groups is a result of selection. For example, among Caucasian males, life expectancy is greatest for those with group O and least with group B; exactly the opposite is true for Caucasian females with these blood types. People with type A blood run a relatively higher risk of cancer of the stomach and of pernicious anemia. People with type O blood have a higher risk of duodenal ulcers and are more likely to contract Asian flu. Most of these conditions, however, would not affect relative rates of reproduction—and thus act as selection forces—since they generally occur in individuals who are past reproductive age.

It has been suggested that there are correlations between the different blood types and susceptibility to such diseases as plague, leprosy, tuberculosis, syphilis, and smallpox—all diseases that, in the past, could have been powerful selection forces. However, such correlations have not been substantiated.

The geographic distributions of the A, B, AB, and O groups are irregular. For example, there is a large predominance of O in the Western Hemisphere and an increase in the B allele as one moves from Europe toward Central Asia. The B allele is totally absent in American Indians and Australian aborigines who have not mixed with Europeans. Although most American Indians are type O, the Blackfoot tribe has the highest frequency of type A blood found anywhere in the world (55 percent). Even within an area as small as the British Isles, there are significant variations in allele frequency, going both from north to south and from east to west. These differences may reflect some differences in the selective forces favoring particular blood groups under particular conditions, they may be the result of population migrations and genetic drift, or some combination of both. At the present time, we simply do not know.

Disruptive Selection

A second type of selection, **disruptive selection,** increases the two extreme types in a population at the expense of intermediate forms. Disruptive selection in the laboratory gave rise to the high- and low-bristle-number lines of *Drosophila* described earlier.

A particularly clear instance of disruptive selection in action has been demonstrated in studies of plants growing on soils that were previously contaminated by salts of heavy metals, such as lead and zinc, because of mining operations. The boundaries between contaminated and uncontaminated areas are often very sharp. Plants growing on uncontaminated soil are unable to survive on contaminated soil. Plants of the same species growing on contaminated soil are able to survive in the uncontaminated areas but cannot compete with those already growing there. Thus, the two extreme phenotypes have been "favoured" at the expense of intermediate forms, resulting in the development of very marked differences between the two groups in the 50 years or so since the mining operations were discontinued and the plants began to colonize the area. Disruptive selection of this sort may result in the formation of two new species, a subject we shall discuss in the next chapter.

(a)   (b)

48–7 (a) *In the coho salmon* (Oncorhynchus kisutch), *the hooknose male (left) and the female (right) both attain sexual maturity at age 3.* (b) *The jack male, however, attains sexual maturity at age 2 and is significantly smaller than either the female or the hooknose male.*

A more recently described example of disruptive selection is also an example of both frequency-dependent selection and sexual selection. As you may know, the coho salmon *(Oncorhynchus kisutch)* of the Pacific Northwest hatch in freshwater streams, in which they spend their first year before moving to the ocean, where they become sexually mature. Upon attaining maturity, the fish return to their home stream, breed, and die. Although females of this species attain sexual maturity and return to breed at age 3, the males may attain sexual maturity at either age 2 or age 3. The two-year-old males, known as "jacks," are about half the size of the three-year-old males, known as "hooknoses" (Figure 48–7). Whether a male will mature in two years and thus be a jack, or in three years and be a hooknose, is, in part, genetically determined. Studies by Mart R. Gross of Simon Fraser University in British Columbia have shown that disruptive selection not only maintains these two life-history characteristics in the population but favors the smallest jacks and the largest hooknose males. When a female spawns, the males closest to the nest in which her eggs are laid are the first to deposit sperm over them. The hooknose males jockey for proximity to the nests by fighting, and the largest males generally win. By contrast, the jacks sneak close to the nests by hiding among rocks or debris, or in shallow areas of the stream; the smaller the jack, the less likely it is to be discovered and chased away by a hooknose. Both the larger jacks and the smaller hooknoses seldom breed successfully, and the two extreme male types are thus maintained within the population.

Directional Selection

A third type of natural selection, **directional selection,** results in an increase in the proportion of individuals with an extreme phenotypic characteristic. It is therefore likely to result in the gradual replacement of one allele or group of alleles by another in the gene pool. Among the examples of directional selection we have already considered are industrial melanism in moths and butterflies, insecticide resistance, and drug resistance in bacteria. We shall see other examples in this chapter and the next.

Frequency-Dependent Selection

In the examples of natural selection we have considered thus far, we have assumed that the fitness of a phenotype is independent of its relative proportion within a population. In some situations, however, a type of natural selection known as

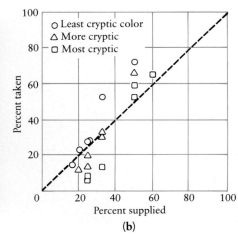

(a)   (b)

48–8 *Frequency-dependent selection has been demonstrated in studies utilizing corixid bugs, aquatic insects that are known more familiarly as water boatmen (a). These bugs, which are preyed upon by fish, often exist in different color forms, which provide different degrees of camouflage. (b) The results of experimental studies in which fish were provided with supplies of corixid bugs of the species Sigara distincta, which exists in three distinct color forms. When the three forms were present in equal proportions (33 percent of each type), the least cryptic form was the most likely to be eaten and the most cryptic form the least likely to be eaten. However, when the proportions of the three forms were unequal, the form present in the highest proportion was the most likely to be eaten, regardless of its degree of camouflage.*

frequency-dependent selection acts to decrease the frequency of more common phenotypes and to increase the frequency of less common ones. Such selection is, for example, a factor in maintaining hooknose and jack males in the coho salmon population. As the frequency of either type increases, the competition between males of that type becomes more intense, allowing greater opportunities for successful reproduction by males of the other type.

Predator-prey interactions can also lead to frequency-dependent selection. For instance, students of animal behavior report that many predators appear to form a "search image" that enables them to hunt a particular kind of prey most efficiently. If the prey individuals differ, for example, in color, the most common color will be preyed upon disproportionately (Figure 48–8). If, as a result, individuals with that color should become less common, selection pressure on them would be relaxed as predators switch their attention to individuals of alternative colors, which would now be more common. Thus frequency-dependent selection can be a factor in maintaining polymorphisms in prey populations.

A rather different type of frequency-dependent selection is the basis by which self-sterility alleles (page 985) maintain variability in plant populations.

Sexual Selection

As Darwin recognized, many of the conspicuous adaptations of animals have little to do with survival on a day-to-day basis but are instead the result of **sexual selection,** the "struggle between the members of one sex, generally the males, for the possession of the other sex." Sexual selection generally takes one of two forms: **intrasexual selection,** which is competition between members of one sex for matings with the opposite sex, and **intersexual selection,** in which members of either sex exert strong selective pressures on the characteristics of the opposite sex through their choice of mates. Examples of intrasexual selection are the competitions between elephant seal bulls (Figure 47–12, page 983) for control of a harem and between male salmon for proximity to the nests in which eggs will be laid. Intersexual selection comes into play in situations in which one sex, most often the female, makes an active choice of mate; as Darwin pointed out, it is the most likely explanation for the evolution of male ornamentation, such as the peacock's tail.

The relative intensities of intrasexual and intersexual selection on males versus females depend on the mating system of the particular species. Sexual reproduction requires a considerable investment of energy, which is allocated among at least three different components: competition for mates, the mating process itself

(a)

(b)

(c)

(d)

48–9 *Sexual selection is caused by competition for mates, usually among males, and often results in differences between the two sexes. (a) Even though male and female baboons eat the same foods, the males have large canines, as shown here, and are almost twice as large as the females, features of importance in male-male competition. (b) In impala, large antlers command respect from other males and are apparently attractive to females. (c) In many species of birds, such as the purple sugarbirds shown here, the females are colored in a way that blends with their surroundings, thus protecting them and their young, and the males have bright, conspicuous plumage. (d) In species in which mating couples form monogamous pairs, such as these albatrosses, males and females generally look alike.*

(including the production of gametes), and parental care. The way in which energy is allocated has a direct bearing on the nature of the sexual selection process at work. For example, in polygynous species, in which a few males sire most of the young, females invest more heavily in parental care and males in competition and mating. In such species, the males attempt to maximize the number of mates, whereas the females tend to be more discriminating in their mating habits. In monogamous species, in which both sexes generally contribute substantially to both mating and parental care, courtship rituals are often elaborate, with both males and females exhibiting "choosiness." In polyandrous species, in which females may have several mates but males usually have only one, males invest more heavily in parental care and females in mating. In such species, the females attempt to maximize the number of mates, while males tend to be more coy in their mating behavior.

These three mating systems—polygyny, monogamy, and polyandry—are idealized points along a continuum of varying energy investments by the two sexes. The observed mating systems of different species fall all along this continuum, and some may vary dramatically with ecological circumstances. For example, under conditions of abundant resources, Mormon crickets exhibit a polygynous mating system, but when resources are scarce, their mating system tends toward polyandry. It is interesting to note that human mating systems span almost the full range of this continuum and are broadly correlated with the economic conditions and social structure of local populations.

Male Ornamentation: The Role of Female Preference

Darwin's proposal that female preference plays a role in producing male ornamentation is only one of several hypotheses that could explain the sometimes extreme characteristics observed in many male animals. For example, another hypothesis is that male ornamentation evolved as a device for signaling between competing males, with female preference having little or no influence. However, a recent set of experiments by Malte Andersson, working with the long-tailed widowbirds of Kenya, has provided dramatic support for Darwin's hypothesis.

Long-tailed widowbirds exhibit a striking sexual dimorphism. The females, which are smaller than the males, are mottled brown and have short tails. The larger males, by contrast, are a brilliant black, with red epaulets, and have tails that average 50 centimeters (about 20 inches) in length. As you would expect on the basis of this degree of sexual dimorphism, long-tailed widowbirds have a polygynous mating system. Males compete for territories, in which the females subsequently nest. Andersson observed as many as six females nesting in the territories of some males, while the territories of other males contained no females at all. As the males fly over their territories, which are in the grasslands of the African savanna, their long tails and black plumage are read-ily visible, presumably to both predators and other widowbirds, as well as to human observers.

Darwin suggested that selection of males by females on the basis of an ornament, such as the widowbird's tail, would produce directional selection for further elaboration of the ornament until selection in the opposite direction—as, for example, by predators—balanced the effects of the sexual selection. On the basis of this hypothesis, one would predict that, given the opportunity, females would show preference for males with still greater elaboration of the ornament, even in species in which the males are already highly ornamented. Andersson tested this prediction by giving female widowbirds a choice between three groups of males: one group in which the tails were clipped to a length of 14 centimeters, one group in which the tails were the normal length (50 centimeters), and one group in which the tails were lengthened to 75 centimeters by gluing on feathers from the clipped tails of the first group. In the group in which the tails were the normal length, half of the birds were unaltered and half had their tails clipped and then reglued. Before the experimental alteration of the tails, similar numbers of females were nesting in the territories of all the males. Following the alterations of tail length, the males with

Sexual selection is thought to be the chief cause of sexual dimorphism, those differences between males and females that have to do not with the act of reproduction itself but rather with obtaining a mate. Examples of sexually selected characteristics are the extravagant plumage of many male birds, such as pheasants, the oversized antlers of deer and elk, and the larger body size, the mantle, and the huge canines of the male baboon. Sexual dimorphism is most marked among polygynous species. By contrast, in monogamous species, males and females tend to look very much alike; consider, for example, swans or geese. Among those few polyandrous bird species, in which the females do the courting and the males do the choosing, such as the phalarope, it is the female that is the gaudier, with the stay-at-home male the drab partner. For those of us weary of comparisons between the resplendent peacock and the dowdy peahen, it is comforting to realize that it is the female whose size, color, and general behavior are considered to be optimal for the environment, while the dimorphic characteristics of the male appear to be useful primarily for threat, display, and other bids for attention. In fact, these characteristics may well be maladaptive with respect to other traits, such as conspicuousness to predators.

lengthened tails attracted significantly more females than the males with either shortened tails or tails of normal length. Moreover, the ability of the males to compete for and successfully acquire territories appeared to be unaffected by the alterations in tail length. As this study demonstrates, at least in the case of the long-tailed widowbird, female mate choice is a dominant selective agent in the evolution of ornamentation.

Andersson's experimental results were published in 1982, more than 100 years after Darwin's hypothesis about the role of female choice. The length of time between that hypothesis and these elegant experiments is indicative of the difficulty often facing evolutionary biologists as they try to devise appropriate experimental or observational tests of key hypotheses. The challenge is not only in the asking of questions but also in the framing of questions in ways that can provide clear answers.

A male long-tailed widowbird, flying over his territory.

Darwin, recognizing that such males were not "fitter to survive in the struggle for existence," categorized sexual selection as a force separate from that of natural selection. However, with fitness stringently redefined in terms of relative numbers of surviving offspring, many investigators now believe that such a distinction is invalid and that sexual selection should be considered simply as one of the forms natural selection may take.

THE RESULT OF NATURAL SELECTION: ADAPTATION

Natural selection results in adaptation, with its several meanings and multiple manifestations. Consider, for example, a squirrel. Note how its tail serves as a counterbalance as the animal leaps and turns; in addition, the same marvelous structure serves as a parasol, a blanket, and an aerial rudder. Pluck a burr from your clothing and consider the artful contrivances by which it clings there. Consider the love and devotion characteristic of the domesticated dog; these are adaptations related to the procurement of food and shelter as stringently selected

(a)   (b)

48-10 (a) *The woodpecker has a number of adaptations that enable it to obtain food. These include two toes pointing backward with which the woodpecker clings to the tree bark, strong tail feathers that prop it up, a strong beak that can chisel holes in the bark, strong neck muscles that make the beak work as a hammer, air spaces in the skull that cushion the brain during hammering, and a very long tongue that can reach insects under the bark.*

(b) The woodpecker finch (Camarhynchus pallidus) is a rare phenomenon in the bird world because it is a tool user. Like the true woodpecker, it feeds on grubs, which it digs out of trees using its beak for a chisel. Lacking the woodpecker's long, barbed tongue, it resorts to an artificial probe, such as a twig or a cactus spine, to dislodge the grub. The woodpecker finch shown here selected a cactus spine, which it inserted into the grub hole. The bird has succeeded in prying out the grub, which it will eat. If the pick selected turns out to be an efficient tool, the bird will carry it from tree to tree in its search for grubs.

for as the beak of a woodpecker (Figure 48–10). Thread a needle; your capacity to do so represents the cumulative effect of millions of years of selection pressures for digital dexterity and eye-hand coordination. (The needle itself made its appearance a mere 10,000 years ago.)

As we noted earlier, natural selection involves interactions between individual organisms, their physical environment, and their biological environment—that is, other organisms. In many cases, the adaptations that result from natural selection can be clearly correlated with environmental factors or with the selective forces exerted by other organisms.

Adaptation to the Physical Environment: Clines and Ecotypes

Sometimes phenotypic variations within the same species follow a geographic distribution and can be correlated with gradual changes in temperature, humidity, or some other environmental condition. Such a graded variation in a trait or a complex of traits is known as a **cline**.

Many species exhibit north-south clines of various traits. House sparrows, for example, tend to have a smaller body size in the warmer parts of the range of the species, and a larger body size in the cooler parts (Figure 48–11). House sparrows were first introduced into North America between 1852 and 1860 in the form of small founder populations drawn principally from central England and Germany. From an evolutionary standpoint, the time that has elapsed since their introduction is very brief indeed, yet a dramatic degree of size differentiation has occurred. In cooler regions, a larger body size is advantageous for heat conservation, as we saw in Chapter 38; consequently, many homeothermic animals show geographic variations in size similar to those of the house sparrow. Conversely, among mammals, ears, tails, and other extremities are relatively longer in the warmer areas of a species' range; such adaptations allow for heat radiation from the animal. Plants growing in the south often have slightly different requirements for flowering or for ending dormancy than the same kind of plants growing in the north, although they may all belong to the same species.

A species that occupies many different habitats may appear to be slightly different in each one. Each group of distinct phenotypes is known as an **ecotype**. Are the differences among ecotypes determined entirely by the environment or do these differences represent adaptation resulting from the action of natural selection on genetic variation? Figure 48–12 illustrates a series of experiments carried out by Jens Clausen, David Keck, and William Hiesey, under the auspices of the Carnegie Institution of Washington. One of the plants studied was the perennial *Potentilla glandulosa*, a relative of the strawberry. Experimental gardens were established at various altitudes, and wild plants collected from near each of the

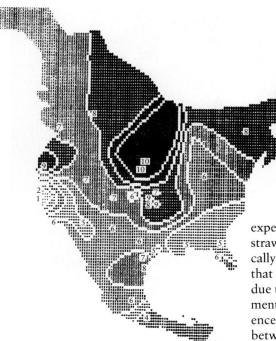

48–11 *The distribution by size of male house sparrows in North America. The higher numbers indicate larger sizes, based on a composite of 16 different measurements of the birds' skeletons. This map was generated and drawn by a computer.*

experimental sites were grown in all the gardens. *(Potentilla glandulosa,* like the strawberry, reproduces asexually by runners, making it possible to study genetically uniform replicates.) Under these conditions it was possible to demonstrate that many of the phenotypic differences among the ecotypes of *P. glandulosa* were due to genetic differences. It is not surprising that in these very different environments, different characteristics were selected for. Over time, the genetic differences between individual plants had been translated into genetic differences between subgroups of the *P. glandulosa* population. As we shall see in the next chapter, such a process is often the first step in the formation of new species.

Adaptation to the Biological Environment: Coevolution

When populations of two or more species interact so closely that each is a strong selective force on the other, simultaneous adjustments occur that result in **coevolution.** We have previously mentioned several examples of coevolution. One of the most important, in terms of sheer numbers of species and individuals involved, is the coevolution of flowers and their pollinators, described in Chapter 24. Here we shall consider another example, also involving plants and insects, those two ancient allies and enemies.

Milkweed, Monarchs, and Mimics

As we noted in Chapter 24, various families of plants have evolved chemical defenses, so-called "secondary substances," that, because they are toxic, bad-tasting, or both, deter predation by herbivores. The bitter white sap of plants of the milkweed family contains a cardiac glycoside, similar to digitalis, that acts as a deterrent to most herbivores. In the course of the evolutionary race to stay in the same place (see essay, page 986), some species of insects, including monarch butterflies, have evolved enzymes that enable the caterpillars to feed on the milkweeds without being poisoned.

48–12 *Ecotypes of* Potentilla glandulosa, *a relative of the strawberry. Notice the correlation between the height of the plant and the altitude at which it grows; there are other phenotypic differences among the plants as well. When plants from the four geographic areas are grown under identical conditions, many of these phenotypic differences persist and are passed on to the next generation, indicating that these plants are genotypically, as well as phenotypically, different.*

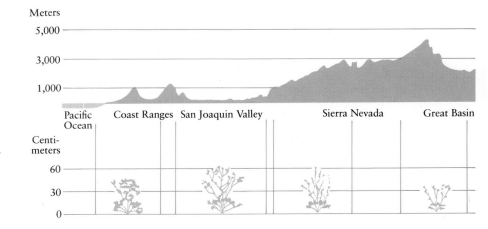

(a)

(c)

(d)

(b)

48–13 (a) *A monarch butterfly egg (center of photograph) clings to the bud of a milkweed.* (b) *When the egg hatches, two to four days after it is laid, a caterpillar emerges that feeds on the milkweed, ingesting and storing the toxic compounds produced by the plant. The monarch caterpillar and also the monarch butterfly* (c) *thus become unpalatable and poisonous. The conspicuous coloration of caterpillar and butterfly warns would-be predators. The viceroy butterfly* (d) *does not feed on milkweeds and is not poisonous, but it is protected by its similarity to the monarch.*

Monarch caterpillars not only utilize the plant tissues for food, but they also ingest and store up the toxic glycoside, which is then present in the adult forms, the butterflies. The butterflies, in turn, are distasteful and poisonous to their predators, which are insectivorous birds.

Like many small animals that taste bad, monarchs, as part of this same defensive strategy, came to have conspicuous warning colors that deter predators. Such warning colors are common in the animal world. They are found not only among insects but also among poisonous reptiles and amphibians. The skunk is a familiar mammalian example of an animal whose distinctive markings remind us to keep our distance. In many types of animals, especially insects, unrelated species often come to resemble one another in their warning characteristics—bees, wasps, and hornets are examples. This type of mimicry is known as Müllerian, after F. Müller, who first described the phenomenon. Müllerian mimicry is adaptive for all the individuals involved because each benefits from a predator's experience with another. As you might expect, other insects that feed on milkweeds have the same bright warning colors characteristic of the monarch.

Next to appear on the evolutionary scene, as events are reconstructed, were several species of butterflies that do not taste bad and that are not poisonous but that have coloration similar to that of the monarchs, and so escape predation. Such deceptive mimicry is known as Batesian because it was first described by Wallace's friend and traveling companion H. W. Bates (page 965) in 1862.

Laboratory experiments have confirmed the selective value of Batesian mimicry. Jane Brower, working at Oxford University, made artificial models by dipping mealworms in a solution of quinine, to give them a bitter taste, and then marking each one with a band of green cellulose paint. Other mealworms, which had first been dipped in distilled water, were painted green like the models, so as to produce mimics, and still others were painted orange to indicate another species. These colors were chosen deliberately: orange is a warning color, since it is clearly distinguishable, and green is usually found in species that are not repellent and that therefore benefit from being inconspicuous.

The painted mealworms were fed to caged starlings, which ordinarily eat mealworms voraciously. Each of the nine birds tested received models and mimics in varying proportions. After initial tasting and violent rejection, the models were generally recognized by their appearance and avoided. In consequence, their mimics were protected also.

Batesian mimicry obviously works to the advantage only of the mimic. The model, on the other hand, suffers from attacks not only by inexperienced predators but also by predators who have had their first experience with the mimic rather than with the model. The mimetic pattern will be at its greatest advantage if the mimic is rare—that is, less likely to be encountered than the model. However, even when as many as 60 percent of the green-banded worms were mimics, 80 percent of the mimics escaped predation. A mimic also has a

(a)

(b)

(c)

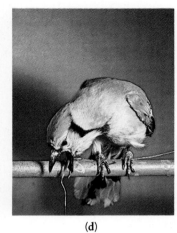
(d)

48-14 *A blue jay that has never before tasted a monarch butterfly is fed one in a controlled experiment. Shortly after ingesting the butterfly, the blue jay begins to show signs of being uncomfortable and then vomits. Upon subsequently being offered a monarch, the bird refused it.*

better chance if it times its emergence to appear after the model, thus reducing its chances of being encountered first. You will not be surprised to learn that in any given area, Batesian mimics do, indeed, generally emerge from the chrysalis after the model.

EVOLUTION AND THE IDEA OF PROGRESS

The concept of evolution has always carried with it some notion of progress. This vision is reinforced by the evidence that evolution has, over time, produced organisms increasingly larger, more complex in structure, and more sophisticated in design, culminating—as if it were preordained—in something as marvelous as ourselves. Natural selection does tend to push populations toward better solutions to the particular "problems in living" with which they are confronted.

(a)

(b)

(c)

48-15 *Model and mimics. (a) A yellow-jacket* (Vespula) *and some of its mimics: (b) a sand wasp, (c) a masarid wasp, (d) an anthidiine bee, and (e) a syrphid fly. All except the fly are Müllerian mimics; that is, they all sting. The fly is a Batesian mimic. Müllerian mimics have been compared to reputable tradesmen who share a common advertisement and divide its costs. Batesian mimics, by contrast, are like unscrupulous retailers who copy the advertisements of successful firms.*

(d)

(e)

1005

However, evolutionary change does not necessarily mean improvement by human criteria, nor does it necessarily even result in organisms that are better adapted to their immediate environment. Natural selection acts on the here and now; as a result, populations—particularly those engaged in intense evolutionary races—are often a generation behind. Their capacity to "keep up," and thus maintain adaptation, depends primarily on the existence of sufficient genetic variability in the population. Other factors are also involved.

Developmental and Structural Constraints

Natural selection has to work with what is available, not only in the chance variations in the genetic material but also in the three-dimensional phenotypic expression of the genotype. To take a simple example, all terrestrial vertebrates and their descendants (including ourselves) are limited structurally to a body with a long vertebral column and four limbs, one on each corner, with which our ancestors waddled up on shore. A structural engineer setting out to build a flying machine or a submarine would scrap these blueprints and start from scratch, but evolution can only build on past history. Many of our human ailments, such as unusual difficulty in childbirth and a propensity to lower back pain, can be traced directly to the conversion of this basically quadrupedal form to an upright stance. In *The Panda's Thumb,* Stephen Jay Gould points out that such absence of perfection and the presence of jerry-built contraptions (the panda's thumb, for example) are more convincing evidence that evolution has occurred than are the more usually cited examples of exquisite adaptation.

The conservatism of evolution is nowhere more evident than in embryonic development. Here the complexity of building an organism from a single cell and a meter or two of encoded instructions is apparently so great that, as we saw in Figure 46–11 (page 969), it is only possible to proceed along certain well-established, ancient pathways.

Eyeless Arthropods and Other Degenerates

Natural selection may lead to the loss of capacities rather than the increase of complexity. For example, cave animals that live in perpetual darkness often lack eyes (Figure 48–16); for an organism living in the dark, eyes are vulnerable and easily injured and there is no selection pressure to maintain them. Birds on islands where there are no predators may lose the ability to fly. In more extreme cases, as we saw in Chapter 25, parasitism can lead to degeneration of digestive organs and respiratory systems. Such organisms are highly evolved, in the sense of being extremely specialized—finely tuned to a particular way of life—although they may appear primitive and simple.

48–16 (a) *This cave-dwelling millipede* (Antriadesmus fragilis) *is, like other animals adapted for life in the lightless interiors of caves, sightless and unpigmented. Moving slowly and silently in limestone caves of the southern United States, millipedes of this species depend primarily on their long antennae for sensory information.* (b) *The Texas blind salamander* (Typhlomolge rathbuni) *is one of a number of cave-dwelling vertebrates.*

(a)

(b)

(a)
(b)

48-17 *An example of convergent evolution is provided by (a) the bull fur seal and (b) the king penguin shown here. Although one is a mammal and the other a bird, both have streamlined, fishlike bodies and a layer of insulating fat below the skin. The penguin swims by means of its flipperlike wings, using its webbed feet as rudders. The fur seal also swims primarily with its webbed forelimbs, using them like oars.*

PATTERNS OF EVOLUTION

Natural selection, as we have seen in numerous examples, is a complex process, operating continuously in all populations. Its outcome, in any particular instance, depends on a great many factors. Viewed from a broader perspective, however, it has the effect of producing different patterns of evolution. One of these patterns, which we have already considered, is coevolution, in which organisms of different species act as selective forces on each other. In other patterns of evolution, natural selection may produce remarkably similar phenotypes in distantly related organisms and, conversely, widely different phenotypes in closely related organisms.

Convergent Evolution

Organisms that occupy similar environments often come to resemble one another even though they may be only very distantly related phylogenetically. When they are subjected to similar selection pressures, they show similar adaptations. The whales, a mammalian group that includes the dolphins and porpoises, are similar to sharks and other large fish in their streamlined shape and other external features, but the fins of whales conceal the remnants of a tetrapod hand. Whales are homeotherms, like their land-dwelling ancestors, and they have lungs rather than gills. Similarly, two families of plants invaded deserts in different parts of the world, giving rise to the cacti and the euphorbs (Figure 48–18). Both evolved large fleshy stems with water-storage tissues and protective spines, and they appear superficially similar. However, their quite different flowers reveal their widely separate evolutionary origins.

48-18 *Members of (a) the euphorb family and (b) the cactus family have been separated for millennia of evolutionary history, with the cacti evolving in the deserts of the New World, and the euphorbs in the desert regions of Asia and Africa. Members of both families have fleshy stems adapted for water storage, protective spines, and greatly reduced leaves.*

(a)
(b)

48–19 *The polar bear is one example of divergent evolution. The white coloring of the polar bear is probably related to its need for camouflage while hunting rather than to a need to hide from predators, since it is one of the world's largest carnivores. (Its greatest enemies are commercial fishermen who do not appreciate the polar bear's fishing activities and hunters who, prizing the skins for trophies, have taken to tracking the bears by helicopter.) Unlike their southern cousins, the brown bears, polar bears (except for females with young) do not become lethargic in the winter but continue to roam the ice and icy waters in search of fish, seals, young walruses, and other prey.*

Divergent Evolution

Divergent evolution occurs when a population becomes isolated from the rest of the species and, owing to particular selection pressures, begins to follow a different evolutionary course. For example, *Ursus arctos,* the brown bear, is distributed throughout the Northern Hemisphere, ranging from the deciduous forests up through the coniferous forests and into the tundra, as it was some 1.5 million years ago. As is characteristic of such widespread species, there are many local ecotypes. During one of the massive glaciations of the Pleistocene, a population of *Ursus arctos* was split off from the main group, and, according to fossil evidence, this group, under selection pressure from the harsh environment, evolved into the polar bear, *Ursus maritimus* (Figure 48–19). Brown bears, although they are members of the order of carnivores and closely related to dogs, are mostly vegetarians, supplementing their diet only occasionally with fish and game. The polar bear, however, is almost entirely carnivorous, with seals being its staple diet. The polar bear differs physically from the other bears in a number of ways, including its white color, its carnivore-type teeth, its streamlined head and shoulders, and the stiff bristles that cover the soles of its feet, providing insulation and traction on the slippery ice.

As the example of the polar bear illustrates, divergent evolution can lead not only to the differentiation of locally adapted ecotypes but also to the formation of new species. In the next chapter, we shall examine this process in more detail, with a consideration of the other factors involved in speciation and macroevolutionary change.

SUMMARY

Natural selection is the differential reproduction of genotypes resulting from interactions between individual organisms and their environment. According to the synthetic theory, it is the major force in evolution. It can act both to produce change and to maintain variability within a population, as exemplified by the color and banding patterns in snails.

Natural selection can act only on characteristics expressed in the phenotype. The unit of selection is the entire phenotype—the whole organism. In extreme cases, a single allele may be decisive in selection. More typically, a successful phenotype is the result of the interaction of many genes. Groups of genes that together produce sets of coordinated characteristics in the phenotype are known as coadaptive gene complexes.

Three major categories of natural selection are stabilizing selection, in which extreme phenotypes are eliminated from the population; disruptive selection, in which the extreme phenotypes are selected at the expense of intermediate forms; and directional selection, in which one of the extremes is favored, pushing the population along a particular evolutionary pathway. Another type of selection is frequency-dependent selection, in which the fitness of a phenotype decreases as it becomes more common in the population and increases as it becomes less common. A fifth category, sexual selection, results from the competition for mates; it can greatly increase differential reproduction without improving adaptation to other environmental factors.

The result of natural selection is the adaptation—however imperfect—of populations to their environment. Evidence of adaptation to the physical environment can be seen in gradual variations that follow a geographic distribution (clines) and in distinct groups of phenotypes (ecotypes) of the same species occupying different habitats. Adaptation to the biological environment results from the selective forces exerted by interacting species of organisms on each other (coevolution). An example is the relationship between milkweeds, monarchs, and monarch mimics.

Evolution by natural selection does not necessarily produce a population with the best possible relationship to its environment. Because natural selection operates in the here and now, many members of a population may be at least one generation behind optimal adaptation. Moreover, the potential for evolutionary change is constrained by the extent of variability in the gene pool and the possibility of variation in structure and development of the individual organisms.

Observed patterns of evolution resulting from natural selection include coevolution, convergent evolution, and divergent evolution. In convergent evolution, dissimilar populations, only distantly related, come to resemble one another as a result of similar selection pressures. In divergent evolution, similar, related populations become more dissimilar, a process that can lead to the formation of new species.

QUESTIONS

1. Distinguish between genotypic variability and phenotypic variability. Which is acted upon by natural selection? Which is necessary for evolution?

2. What is the difference between artificial selection and natural selection? Under what circumstances might they become indistinguishable?

3. How does stabilizing selection affect the gene pool? Disruptive selection? Directional selection?

4. What characteristics in the contemporary human population are probably being maintained by stabilizing selection? Which might be subject to directional selection?

5. Imagine a rapidly expanding population in an environment that allows every organism that is born to survive and reproduce to its maximum extent. Which organisms will make the greatest contribution to the gene pool of future generations? Is natural selection acting on such a population?

6. From an evolutionary perspective, longevity of animals past reproductive age is generally useless. Why is this so? What situations might be exceptions to this statement? Can you think of any reason why longevity might be harmful from an evolutionary standpoint?

7. Distinguish between intrasexual selection and intersexual selection. Of which type is the disruptive selection occurring in male salmon? Of which type is the directional selection for increasing tail length in the long-tailed widowbird?

8. In his experiments with the long-tailed widowbird, Andersson clipped and reglued the tails of half of the males in the group with tails of normal length. Why did he perform this manipulation? Why was it important?

9. What tool-using animals other than the woodpecker finch can you think of?

10. If you were a poisonous butterfly, would you rather have a Batesian mimic or a Müllerian one? Why?

11. In her experiments with quinine-dipped mealworms and their Batesian mimics, Brower dipped the mimics in distilled water before painting on the distinctive green band. Similarly, the mealworms on which orange bands were painted were first dipped in distilled water. Why?

12. Distinguish between convergent evolution and divergent evolution. Which type of selection would you expect to be the major factor in convergent evolution? In divergent evolution?

On the Origin of Species

The central issue in the study of evolution concerns the source of the major changes that have taken place in the history of living organisms. Can microevolution, so well documented by the population geneticists, account for the great panorama of macroevolution revealed in the fossil record, including such events as the transition to land by both plants and vertebrates, the rise of the dinosaurs, and the appearance of the hominids? The bridge across this gulf is the process of speciation, a subject that Darwin, despite the title of his great work, did not really address. The origin of species is, however, of great current interest.

MODES OF SPECIATION

In Chapter 20, we defined a species as a group of natural populations whose members can interbreed with one another but cannot (or at least usually do not) interbreed with members of other such groups. The essential feature of this definition is reproductive isolation. We know that two separate species exist when the two can occupy the same space without interbreeding. In terms of population genetics, members of a species share a common gene pool effectively separated from the gene pools of other species. A central question, then, is how one pool of genes splits off from another to begin a separate evolutionary journey. A subsidiary question is how two species, often very similar to one another, inhabit the same place at the same time and yet remain reproductively isolated.

According to current perspectives, speciation is most commonly the result of the geographic separation of a population of organisms: this process is known as **allopatric** ("other country") **speciation.** Under certain circumstances, speciation may also occur without geographic isolation, in which case it is known as **sympatric speciation.**

Allopatric Speciation

Every widespread species that has been carefully studied has been found to contain geographically representative populations that differ from each other to a greater or lesser extent, as in the case of the house sparrow (page 1002) or the ecotypes of *Potentilla glandulosa* (page 1003). Such populations that differ somewhat genetically but still form part of the same species—that is, they can still interbreed—are known as races. Subspecies (Figure 49–1) are races that are sufficiently different phenotypically to have been formally recognized by a Latin name. A species composed of geographic races or subspecies is particularly susceptible to speciation if geographic barriers arise, preventing gene flow.

Geographic barriers are of many different types. Islands are frequent sites for the development of new species. The breakup of Pangaea (see essay on page 1012) profoundly altered the course of evolution, setting the island continent of Australia adrift as a veritable Noah's ark of marsupials. Populations of many

(a)

(b)

49–1 *Two subspecies that have formed on either side of a natural geographic barrier, the Tana River in Kenya, are* (a) *the common giraffe,* Giraffa camelopardalis *rothschildi, and* (b) *the reticulated giraffe,* Giraffa camelopardalis reticulata. *The two are considered subspecies because they are phenotypically distinct, but they* *interbreed readily in zoos, producing fertile hybrids, and hybrids are also found in nature.*

49–2 *For different species, islands, genetically speaking, may be* (a) *true islands,* (b) *mountaintops,* (c) *ponds, lakes, or even oceans, or* (d) *isolated clumps of vegetation.*

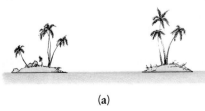

(a)

(b)

(c)

(d)

organisms can become cut off from one another by barriers less obvious than oceans (Figure 49–2). For a plant, an island may be a mist-veiled mountaintop, and for a fish, a freshwater lake. A forest grove may be an island for a small mammal. A few meters of dry ground can isolate two populations of snails. Islands may also form by the creation of barriers between formerly contiguous geographic zones. The Isthmus of Panama, for instance, has repeatedly submerged and reemerged in the course of geologic time. With each new emergence, the Atlantic and Pacific oceans became "islands," populations of marine organisms were isolated, and some new species formed. Then, when the oceans joined again (with the submergence of the Isthmus), the continents separated and became, in turn, the "islands."

If the isolated population is a small, peripheral one, it is more likely to differ from the parent population (because of the founder effect) and less likely to be subjected to gene flow. Once separated, the isolated population may begin to diverge genetically under the pressure of different selective forces. If enough time elapses and the differing selective forces are sufficiently great, the isolated population may diverge so much that, even if it were reunited with the parent population, interbreeding under natural conditions would no longer occur. At this point, speciation is said to have taken place. Every isolated population does not, of course, become a species. It may rejoin the parent group or it may perish, which is probably the fate of most small, isolated populations.

The Breakup of Pangaea

In the early 1900s, German geologist Alfred Wegener proposed that continents had migrated in the course of the earth's history and that this migration was the cause of mountain building and other geologic phenomena. Wegener's proposal was rejected because of the discovery of a thick crust below the ocean floor that seemed to render the movement of continents impossible. For decades the hypothesis of continental drift was viewed with as much suspicion as reports of flying saucers and extrasensory perception, despite the fact that numerous students of geography had noted how neatly the coastlines of South America and Africa complement one another.

In the past 25 years, continental drift has become respectable, having become firmly established as part of the new theory of plate tectonics. According to this theory, the outermost layer of the earth is divided into a number of segments, or plates. These plates, on which the continents rest, slide around the surface of the earth, moving in relation to one another. The geologic expression of this relative motion occurs mainly at the boundaries between the adjacent plates. Where plates collide, volcanic islands such as the Aleutians may be formed or mountain belts such as the Andes or Himalayas may be uplifted. At the boundaries where plates are separating, volcanic material wells up to fill the void. It is here that ocean basins are created. Plates may also move parallel to the boundary that joins them but in opposite directions, or in the same direction but at different speeds, as with the San Andreas fault.

About 200 million years ago, all the major continents were locked together in a supercontinent, Pangaea. Several reconstructions of Pangaea have been proposed, one of which is shown here. It is generally agreed that Pangaea began to break up about 190 million years ago, about the time the dinosaurs approached their zenith and the first mammals began to appear. First, the northern group of continents (Laurasia) split apart from the southern group (Gondwana). Subsequently, Gondwana broke into three parts: Africa–South America, Australia-Antarctica, and India. India drifted northward and collided with Asia about 50 million years ago. This collision initiated the uplift of the Himalayas, which continue to rise today as India still pushes northward into Asia.

By the end of the Cretaceous period, about 65 million years ago, South America and Africa had separated sufficiently to have formed half the South Atlantic, and Europe, North America, and Greenland had begun to drift apart. However, final separation between Europe and North America–Greenland did not occur until about 43 million years ago, during the Eocene epoch of the Tertiary period. Also during the Eocene, Australia finally split from Antarctica and moved northward to its present position. Later, dur-

Sympatric Speciation

Polyploidy

Polyploidy, an increase in the number of chromosomes beyond the typical diploid ($2n$) complement, is a well-documented mechanism by which new species are produced. Polyploidy may arise as a result of nondisjunction (page 385) during mitosis or meiosis, or may be generated when the chromosomes divide properly during mitosis or meiosis but cytokinesis does not subsequently occur. Polyploid individuals can be produced deliberately in the laboratory by the use of the drug colchicine, which prevents separation of chromosomes during mitosis. Polyploidy leading to the formation of new species sometimes occurs as a result of a doubling of the chromosome number within a species, a process known as **autopolyploidy** (Figure 49–3). More frequently, however, new species are generated by a doubling of the chromosome number in hybrid organisms, a process known as **allopolyploidy.**

A hybrid organism is the offspring of parents of different species. Hybrids can occur in animals (such as the mule), but they are far more common in plants. Kentucky bluegrass, for instance, is a promiscuous hybridizer and has crossbred with many related species, producing hundreds of hybrid races, each well adapted

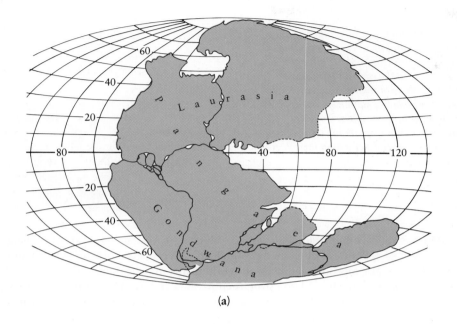

(a)

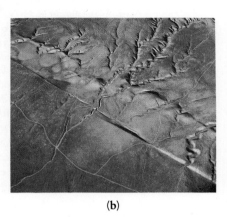

(b)

(a) *Pangaea.* (b) *The San Andreas fault is a boundary between two giant moving plates. The fault, running through San Francisco and continuing southeast of Los Angeles, is responsible for California's notorious earthquakes.*

ing the Pliocene, the two Americas were joined by the Isthmus of Panama, which was created by volcanic action.

From the outset, the theory of continental drift has been closely interwoven with that of evolution. One of the earliest and most impressive pieces of evidence in favor of continental drift was the discovery of fossil remains of a small, snaggle-toothed reptile, *Mesosaurus,* found in the coastal regions of Brazil and South Africa but nowhere else. Early in 1982, a team of American scientists returned from Antarctica with the first fossil of a land mammal ever found there—a marsupial. This find supports the theory that marsupials migrated by land from South America (where only some 40 marsupial species remain today) across Antarctica to Australia before the two separated, some 55 million years ago.

49–3 *The process of autopolyploidy. If nondisjunction occurs during meiosis, diploid (2n) gametes may result. Union of two such gametes, produced either by the same individual or by different individuals of the same species, will produce a tetraploid (4n) individual. Although this individual may be capable of sexual reproduction, it will be reproductively isolated from the parent species.*

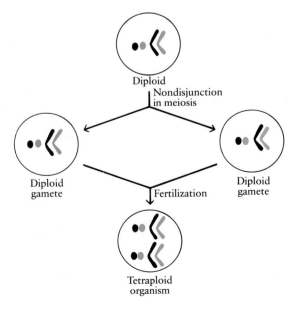

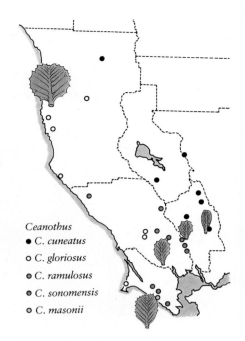

49–4 *The establishment of hybrid populations is an important evolutionary mechanism in many groups of plants, such as the mountain lilac. The map shows a portion of northern California (note San Francisco Bay at the bottom), a geologically complex region. Two relatively widespread and distinct species of mountain lilac, the coastal* Ceanothus gloriosus *and the interior* Ceanothus cuneatus *(which ranges far to the east out of the area depicted), have produced three series of hybrid populations (colored circles). Each hybrid is variable but stabilized and is able to grow better than either parent in the areas where it occurs. Leaves from representative populations are shown on the map. Clusters of species such as these tend to be common in geologically diverse areas.*

Ceanothus
- ● *C. cuneatus*
- ○ *C. gloriosus*
- ◐ *C. ramulosus*
- ◑ *C. sonomensis*
- ○ *C. masonii*

to the ecological conditions of the area in which it grows. Such hybrids, spreading by means of rhizomes, are sometimes able to outcompete both parents (Figure 49–4). Asexually reproducing races of hybrid plants may be considered species in that they are genetically isolated from their parents and from each other, but they do not conform to the working definition of a species because they do not interbreed.

Hybrids in both plants and animals are often sterile because the chromosomes cannot pair at meiosis (having no homologues), a necessary step for producing viable gametes (Figure 49–5a). If, however, polyploidy occurs in such a sterile hybrid and the resulting cells divide by further mitosis and cytokinesis so that they eventually produce a new individual asexually, that individual will have twice the number of chromosomes as its parent. As a consequence, it is reproductively isolated from the parental line. However, its chromosomes—now duplicated—can pair, meiosis can occur normally, and fertility is restored (Figure 49–5b). It is a new species, capable of sexual reproduction.

Sympatric speciation through polyploidy is an important, well-established phenomenon in plants. Approximately half of the 235,000 kinds of flowering plants have had a polyploid origin, and many important agricultural species, including wheat, are hybrid polyploids. This mechanism of speciation appears to have been less common among animals; however, several species of whiptail lizards (genus *Cnemidophorus*) of the American Southwest are allopolyploids.

Disruptive Selection

Sympatric speciation as a result of disruptive selection (page 996) is a more controversial issue. Cases of sympatric speciation are difficult to prove because of the time factor. Either (1) the two forms are in the process of diverging, as in the plants growing on the mine tailings (page 996), in which case it is not possible to prove that complete separation will ever occur, or (2) the two forms have separated completely, and it is not possible to prove that they were not formed allopatrically and have since been reunited.

One often-cited example of sympatric speciation produced by disruptive selection has been reported by Guy Bush of the University of Texas. The genus *Rhagoletis* comprises a group of species of small, brightly colored flies whose larvae feed on developing fruits. Each species of *Rhagoletis* feeds on fruits of only one plant family; the host fruits of the family serve not only as food but also as the rendezvous for courtship and mating, followed by the deposition of eggs. *Rhagoletis pomonella,* for example, is a species that feeds on hawthorns. In 1865, farmers in the Hudson River Valley reported that these flies had begun attacking their

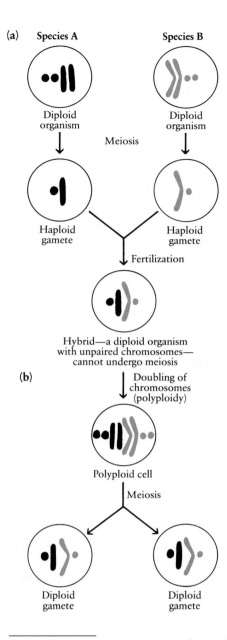

(a) Species A   Species B

Diploid organism   Diploid organism

Meiosis

Haploid gamete   Haploid gamete

Fertilization

Hybrid—a diploid organism with unpaired chromosomes— cannot undergo meiosis

(b)

Doubling of chromosomes (polyploidy)

Polyploid cell

Meiosis

Diploid gamete   Diploid gamete

49–5 (a) *An organism, such as a mule, that is a hybrid between two different species and is produced from two haploid (n) gametes, can grow normally because mitosis is normal. It cannot reproduce sexually, however, because the chromosomes cannot pair at meiosis.* (b) *If polyploidy occurs and the chromosome number doubles, the hybrid can produce viable gametes. Since each chromosome will have a partner, the chromosomes can pair at meiosis. The resultant gametes will be diploid (2n).*

apples. The infestation then spread rapidly to apple orchards in adjacent areas of Massachusetts and Connecticut. There are now two distinct sympatric races of *R. pomonella,* with a number of genetic differences; one feeds and reproduces on hawthorns, and the other on apples. They are isolated by their reproductive behavior and so may be regarded as on the path to speciation. Many biologists, however, question whether these races are truly sympatric. They point out that this may instead be an example of allopatric processes operating on a greatly reduced scale. For these flies—and other small organisms that are parasitic or highly specialized for feeding and reproducing in a particular microhabitat—environmental differences that appear small to us may represent significant barriers, creating "mini-islands."

MAINTAINING GENETIC ISOLATION

Once speciation has occurred, the now-separate species can live together without interbreeding, despite the fact that some are so similar phenotypically that only an expert with a microscope can tell them apart—*Drosophila* offers several examples. What factors operate to maintain genetic isolation of closely related species? Isolating mechanisms may be conveniently divided into two categories: premating mechanisms, which prevent mating between members of different species, and postmating mechanisms, which prevent the production of fertile offspring from such matings as are attempted or do occur. One of the most significant things about the postmating isolating mechanisms is that, in nature, they are rarely tested. The premating mechanisms alone usually prevent any interbreeding. This is an expected consequence of sexual selection; choosing the best mate or mates from among the members of one's own species requires a finer level of discrimination than is required to distinguish members of one's own species from those of a different species (Figure 49–6).

49–6 *A male frigate bird displays his crimson pouch. Throughout the courtship period, the pouch remains bright and inflated, even when the bird is flying or* sleeping. *The female will not mate with a male that does not have these species-specific characteristics.*

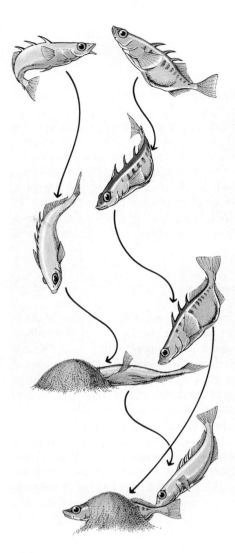

49-7 *Sticklebacks, small freshwater fish, have elaborate mating behavior. The male at breeding time, in response to increasing periods of sunlight, changes from dull brown to the radiant colors shown here. He builds a nest and begins to court females, zigging toward them and zagging away from them. A female ready to lay eggs responds by displaying her swollen belly. The male leads her down to the tunnel-like nest. He prods her tail and, in response, she lays her eggs and swims off. He follows her through, fertilizes the eggs, and stays to tend the brood. If either partner fails in any step of this quite elaborate ritual, no young are produced.*

Premating Isolating Mechanisms

Premating isolating mechanisms in vertebrates often involve elaborate behavioral rituals, visual signals, or frequently, a combination of the two (Figure 49–7). Visual behavioral signals may also be important in arthropods, as, for example, in the flashings of fireflies (see page 580).

Bird songs, frog calls, and the strident love notes of cicadas and crickets all serve to identify members of a species to one another. For example, studies of leopard frogs have revealed that some of the populations that can crossbreed under laboratory conditions do not do so in the wild. They are effectively isolated from one another by differences in their mating calls. (For this reason, artificial breeding experiments are not always a good indication of whether individuals belong to the same or different species.)

In many species of animals, pheromones (page 585) serve to bring the two sexes together. They may serve as signals to attract the male, as in the case of the cecropia moth, or to trigger the release of gametes by the female, as with oysters. Because such pheromones are species-specific, they also act, in effect, as isolating mechanisms.

Temporal differences also play an important role in reproductive isolation. Species differences in flowering times are important isolating mechanisms in plants. Most vertebrates—we are a notable exception—have seasons for mating, often controlled by temperature or by day length. Figure 49–8 shows the mating calendar of species of frogs near Ithaca, New York.

Postmating Isolating Mechanisms

On the rare occasions when members of two different species attempt to mate or succeed in doing so, anatomical or physiological incompatibilities often maintain genetic isolation. These postmating isolating mechanisms are of several different types. For instance:

1. Differences in the shape of the genitalia may prevent insemination, or differences in flower shape may prevent pollination.
2. The sperm may not be able to survive in the reproductive tract of the female, or the pollen tube may not be able to grow on the stigma.
3. The sperm cell may not be able to fuse with the ovum.
4. The ovum, once fertilized, may not develop.
5. The young may survive, but they may not become reproductively mature.
6. The offspring may be hardy but sterile—the mule, for example.

Postmating mechanisms reinforce premating ones. A female cricket that answers to the wrong song or a frog whose individual calendar is not synchronized with that of the rest of the species will contribute less (or nothing) to the gene pool. As a consequence, there is a steady selection for premating isolating mechanisms.

49-8 *Mating timetable for various frogs and toads that live near Ithaca, New York. In the two cases where two different species have mating seasons that coincide, the breeding sites differ. Peepers prefer woodland ponds and shallow water; leopard frogs breed in swamps; pickerel frogs mate in upland streams and ponds; and common toads use any ditch or puddle.*

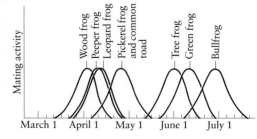

Creating Sexual Chaos

Among the most destructive of the agricultural pests plaguing U.S. farmers are cotton bollworms *(Heliothis zea)* and tobacco budworms *(Heliothis virescens)*. Their combined activities cause losses in cotton, corn, and soybeans totaling nearly $1 billion each year. They are currently controlled almost entirely by insecticides, all of which cause the death of other insect species, including those that prey on members of the genus *Heliothis.*

An alternative method for controlling harmful insects is the use of sex pheromones to lure males to their death. In 1982, scientists from the U.S. Department of Agriculture were experimenting with synthetic sex pheromones, looking for compounds that would either attract male budworms to traps or block their receptors so they could not find receptive females. While the scientists were testing one of these in the field, they discovered that it was particularly effective in an unexpected way: it lured male bollworms as well as male budworms, and even more surprising, it caused the male bollworms to cross the species barrier and mate with the female budworms.

The first isolating mechanism having failed owing to the duplicity and ingenuity of the scientists, a second was encountered. Members of the two species differ in the shapes of their genitalia. These minute differences do not prevent copulation, but once the mismatched pair copulate, they cannot separate again and so are locked in a deadly embrace.

The Department of Agriculture entomologists are cautious but hopeful that this "sexual chaos," as they

Tobacco budworms mating. One of the factors in maintaining genetic isolation among insect species is a lock-and-key fit of the genitalia. When bollworms attempt to mate with budworms, the key jams in the lock.

call it, could prove an effective means of insect control. One major problem is the existence of other isolating mechanisms—such as differences in times of mating—that may keep the two species apart despite the best scientific efforts.

AN EXAMPLE: DARWIN'S FINCHES

Admirers of Charles Darwin find it particularly appropriate that one of the best examples of speciation is provided by the finches observed by Darwin on his voyage to the Galapagos Islands. All the Galapagos finches are believed to have arisen from one common ancestral group—perhaps either a single pair or even a single female bearing a fertilized egg—transported from the South American mainland, some 950 kilometers away. How she or they got there is, of course, not known, but it may have been the result of some particularly severe storm. (Periodically, for example, some North American birds and insects appear on the coasts of Ireland and England after having been blown across the North Atlantic.) It is very likely that finches were the first land birds to colonize the islands.

We do have an idea of what greeted these unwilling adventurers. The Galapagos archipelago consists of 13 main volcanic islands, with many smaller islets and rocks. On some of the islands, craters rise to heights of more than a kilometer.

(a)

(b)

(c)

(d)

(e)

(f)

49-9 *Six of the 13 different species of Darwin's finches. Except for the warbler finch (a), which resembles a warbler more than a finch, the species look very much alike; the birds are all small and dusky-brown or blackish, with stubby tails. The most obvious differences among them are in their bills, which vary from small, thin beaks to large, thick ones.*

(b) The small ground finch (Geospiza fuliginosa) and (c) the large ground finch (Geospiza magnirostris) are both seed eaters. G. magnirostris, with a larger beak than G. fuliginosa, is able to crack larger seeds.

(d) The cactus ground finch (Geospiza conirostris) lives on cactus blooms and fruit. Notice that its beak is larger and more pointed than those of the other two ground finch species.

(e) The small tree finch (Camarhynchus parvulus) and (f) the large tree finch (Camarhynchus psittacula), both insectivorous, take prey of different sizes.

The islands were pushed up from the sea more than a million years ago, and most of them are still covered with black basaltic lava. The major vegetation is a dreary grayish-brown thornbush—making up vast areas of dense, leafless thicket—and a few tall tree cactuses. Inland and high up on the larger islands, the air is more humid, and there one can find rich, black soil and tall trees covered with ferns, orchids, lichens, and mosses, kept damp by a mist that forms around the volcanic peaks. During the rainy season, the area is dotted by sparkling shallow crater lakes. Thus the habitats available to the finches were highly diversified. Different groups of finches, occupying different habitats, would have been subjected to widely differing selective forces.

From the small ancestral group, 13 different species arose, plus one to the northeast on the Cocos Islands, 1,000 kilometers away. Apparently the various islands were near enough to one another that, over the years, small founding groups could emigrate. The islands were far enough apart, however, that once a group was established, there would be little or no gene flow between it and the parent group for a period of time long enough for significant differences to evolve as a result of natural selection and for genetic barriers to develop. The Cocos Islands are so far away that they were successfully invaded only once. (Finches are not very good long-distance fliers; if they were, this natural experiment in speciation would have been a failure.) Development of a new species in this way requires, it is estimated, at least 10,000 years of geographic isolation, but since there are many islands, several species could have been evolving at the same time.

The ancestral type was a finch, a smallish bird with a short, stout, conical bill especially adapted for seed-crushing. The ancestor is believed to have been a ground-feeding finch, and six of the Galapagos finches are ground finches (Figure 49–10). Four species of ground finches now live together on most of the islands. Three of them eat seeds and differ from one another mainly in the size of their beaks, which, in turn, of course, influences the size of seeds they eat. The fourth lives largely on the prickly pear and has a much longer and more pointed beak. The two other species of ground finch are usually found only on outlying islands, where some supplement their diet with cactus.

In addition to the ground finches, there are six species of tree finches, also differing from one another mainly in beak size and shape. One has a parrotlike beak, suited to its diet of buds and fruit. Four of these tree finches have insect-eating beaks, each adapted to a different size range of insects. The sixth, and most remarkable of the insect eaters, is the woodpecker finch (Figure 48–10b, page 1002)

49-10 *Evolutionary tree of the Galapagos finches. According to this hypothesis, all are derived from a single ancestral species. There are six ground finches (genus Geospiza), including the two cactus-feeding finches; six tree finches (genus Camarhynchus), including the woodpecker finch; and one warbler finch (genus Certhidea). Another warblerlike finch is found 1,000 kilometers away on the Cocos Islands.*

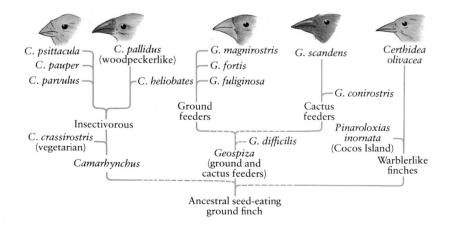

C. psittacula
C. pauper
C. parvulus
C. pallidus (woodpeckerlike)
C. heliobates
Insectivorous
C. crassirostris (vegetarian)
Camarhynchus

G. magnirostris
G. fortis
G. fuliginosa
Ground feeders
G. difficilis
Geospiza (ground and cactus feeders)

G. scandens
G. conirostris
Cactus feeders
Pinaroloxias inornata (Cocos Island)

Certhidea olivacea
Warblerlike finches

Ancestral seed-eating ground finch

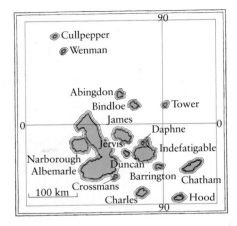

49–11 *The Galapagos Islands, some 950 kilometers west of the coast of Ecuador, have been called "a living laboratory of evolution." Species and subspecies of plants and animals that have been found nowhere else in the world inhabit these islands. "One is astonished," wrote Charles Darwin in 1837, "at the amount of creative force . . . displayed on these small, barren, and rocky islands. . . ."*

By all ordinary standards of external appearance and behavior, the thirteenth species of Galapagos finch would be classified as a warbler. Its beak is thin and pointed like a warbler's, it even has the warblerlike habit of flicking its wings partly open, and, warblerlike, it searches through leaves, twigs, and ground vegetation for small insects. However, its internal anatomy and other characteristics clearly place it among the finches. There is general agreement that it, too, is a descendant of the common ancestor or ancestors.

The differences among these three groups of finches—ground finches, tree finches, and the warbler finch—are sufficiently great that taxonomists classify them in three distinct genera. Recent work by Peter R. Grant and his associates has demonstrated that reproductive isolation of the individual species within the three genera depends on a combination of song characteristics and beak size and shape. It appears that young male and female finches learn to recognize their species-specific song from hearing the singing of their father and to recognize the beak of their species from the observation of the beaks of both parents during feeding. Subsequently, songs, which are sung by the males only, serve as long-range cues to species identity, and beak size and shape serve as short-range cues. Females enter the territory of a courting male in response to his song but appear, at least in part, to base their decisions on whether to mate or not on an inspection of his beak. The females usually enter the territories of several different males and inspect the territory owner's beak before mating. The males are also known to actively pursue females. Sometimes a male finch will mistakenly pursue a female of another species—the finches often look very much alike from the rear—only to lose interest as soon as he sees her beak. He will rarely court a female of another species following a close inspection of her beak.

The Effect of Natural Selection on Body and Beak Size

The Galapagos finches provide indirect evidence of past speciation resulting from geographic isolation coupled with differing selection forces in different environments. Until recently, however, there had been no definitive demonstration of natural selection at work on the finch populations. This gap has now been filled through detailed studies that Peter Grant and his colleagues have conducted on the Galapagos finches for more than 15 years. One series of studies of the ground finch *Geospiza fortis*, carried out on the small island of Daphne Major (0.4 square kilometers in area), has revealed the highest intensities of natural selection ever recorded for a vertebrate population. During these studies, which took place between 1975 and 1985, the investigators banded and monitored the population of *G. fortis*, kept continuous records of rainfall, and tracked both the abundance of food and its utilization by the finches.

During the early 1970s, rainfall on Daphne Major was normal and food was abundant. In 1976, the population of *G. fortis* on Daphne Major numbered between 1,200 and 1,400 birds. In 1977, however, a severe drought occurred. Production of the seeds upon which the finches feed dropped dramatically, and the food supply was rapidly depleted. Small seeds were the most quickly depleted, and, as a consequence, the average size and hardness of the seeds available to the finches increased. The effects on *G. fortis* were immediate. In 1977, no birds bred and the adult population declined by 85 percent, primarily as a result of mortality. The mortality, however, did not fall equally on all segments of the population. Smaller birds were lost from the population at a higher frequency, presumably because of their inability to crack large, hard seeds. As a result of this differential mortality, the average body and beak sizes of *G. fortis* on Daphne Major increased significantly—as Darwin's theory would predict. Higher rates of survival of larger birds were again observed during subsequent droughts in 1980 and 1982. Then, in

49–12 *These two finches illustrate the degree of variation in beak size in the* Geospiza fortis *population on Daphne Major. Individuals with large beaks, such as the bird on the left, had higher rates of survival during periods of drought and seed scarcity. Individuals with smaller beaks, such as the bird on the right, had higher rates of survival during periods of high rainfall and profuse seed production.*

1983, conditions were dramatically reversed as an El Niño (see page 1158) brought 1,359 millimeters of rain, some 10 to 20 times the normal annual range of rainfall. The result was a tremendous flush of seed production, particularly of small seeds. The *G. fortis* population responded by producing an average of eight broods that year, as compared to the normal one or two per year. Then, during 1984 and 1985, the smaller birds in the population survived at a higher rate than the larger birds, presumably because of their more efficient handling of the small, abundant seeds. This period of intense selection reversed the effects of the selection occurring during the drought years. If consistently sustained in one direction, such occasional periods of intense selection during the evolutionary history of the Galapagos finches would have been sufficient, in themselves, to have produced the impressive variation that we see within and between the species today.

THE EVIDENCE OF THE FOSSIL RECORD

Although we may make inferences about the course of macroevolution from the observation of living organisms, the proving ground of evolutionary theory is the fossil record itself. According to paleobiologists, the fossil record reveals that the diversity of organisms that have populated the earth is the result of three broad patterns of speciation, coupled with the effects of extinction—the complete demise not only of individual species but also of entire groups of species.

Phyletic Change

One pattern observed in the fossil record is change within a single lineage of organisms. This is known as **phyletic change.** Under the pressures of directional selection, a species gradually accumulates changes until eventually it is so distinct from its predecessors that it may properly be considered a new species—a new kind—of organism. Darwin's concept of evolution emphasized the slow, gradual accumulation of change, and phyletic change is, of course, a large-scale version of the microevolutionary events so ably documented by population geneticists.

Cladogenesis

The second pattern of evolutionary change is the splitting of lineages, or **cladogenesis** (the forming of branches). Species formed by cladogenesis are the contemporaneous descendants of a common ancestor—like Darwin's finches.

49-13 *The organisms of the Hawaiian Islands, like those of the Galapagos, provide examples of cladogenesis on volcanic islands. Consider, for instance, members of the genus* Cyanea, *found only in the Hawaiian Islands. These plants differ widely in leaf size and shape. All of the species are thought to have evolved from plants similar to* Cyanea lobata *(dark green).* Cyanea linearifolia *(upper left) is usually found in dry, sunny locations. Plants with fernlike leaves, like those on the right, are found in shady locations, where the divided, wide-bladed leaves are highly efficient in gathering light.*

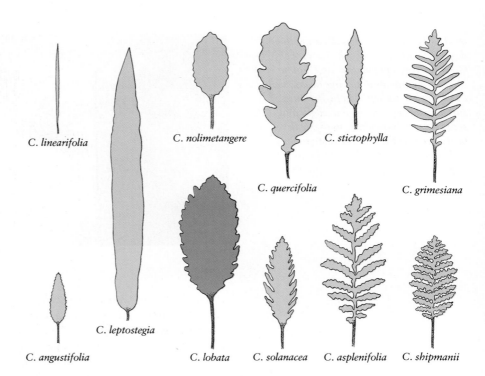

C. linearifolia

C. nolimetangere

C. stictophylla

C. quercifolia

C. grimesiana

C. leptostegia

C. angustifolia

C. lobata

C. solanacea

C. asplenifolia

C. shipmanii

Paleobiologists have tended to give more weight to the role of cladogenesis in evolution than to phyletic change. Ernst Mayr, an ornithologist and a leader in the formulation of modern evolutionary theory, agrees. He maintains that the formation of new species by the splitting off of small populations from the parent stock (the founder effect) is responsible for almost all major evolutionary change. In such small populations, favorable genetic combinations, if present, can increase rapidly in number and frequency without being diluted out by gene flow. Hence, evolution probably does not occur steadily and gradually but in spurts, accounting for the sudden increases of new species observed at many points in the fossil record. These two models—phyletic change and cladogenesis—have not been viewed as alternatives, however, but rather as complementary to one another. The question, until recently, was simply a matter of degree.

Adaptive Radiation

Another pattern of evolution seen in the fossil record is that of **adaptive radiation.** The late George Gaylord Simpson, a leading paleontologist, emphasized that adaptive radiation is the major pattern of macroevolution. In the fossil record, it is disclosed as the sudden (in geologic time) diversification of a group of organisms that share a common ancestor, often itself newly evolved. It is associated with the opening up of a new biological frontier that may be as vast as the land or the air, or, as in the case of the Galapagos finches, as small as an archipelago. Adaptive radiation is generally considered as combining both cladogenesis and phyletic change.

The fossil record contains many examples of adaptive radiation. For example, some 300 million years ago, the reptiles, liberated from an amphibian existence by the "invention" of the amniote egg (page 596), diversified rapidly into terrestrial environments. A similar, even more rapid burst of evolution later gave rise to the birds (page 598). As the dinosaurs became extinct, the mammals similarly burst forth upon the evolutionary scene, with many different kinds appearing simultaneously in the fossil record (Figure 49–14).

Placentals   **Marsupials**

Gray wolf
Canis lupus

Tasmanian wolf
Thylacinus cynocephalus

Ocelot
Felis pardalis

Tiger cat
Dasyurus maculatus

Honey glider
Petaurus breviceps

Flying squirrel
Glaucomys volans

Woodchuck
Marmota monax

Wombat
Vombatus ursinus platyrrhinus

Great anteater
Myrmecophaga jubata

Banded anteater
Myrmecobius fasciatus

Common mole
Scalopus aquaticus

Marsupial mole
Notoryctes tryphlops

House mouse
Mus musculus

Yellow-footed marsupial mouse
Antechinus flavipes

49–14 *Although the adaptive radiation of the marsupials in Australia and of the placental mammals on other continents were independent events, involving different immediate ancestors, striking similarities resulted in the descendant organisms. As shown in these examples, similar adaptations are exhibited by placentals and marsupials that occupy similar habitats and have similar life styles. All of the marsupials, however, are more closely related to each other than they are to any of the placentals, and vice versa.*

Extinction

A phenomenon particularly well documented by the fossil record is that of extinction. Only a small fraction of the species that have ever lived are presently in existence—certainly less than 1/10 of 1 percent, perhaps less than 1/1,000 of 1 percent. In a study that required six years, J. John Sepkoski, Jr., and David M. Raup of the University of Chicago gathered all available data on the extinction of marine organisms during the past 250 million years and analyzed the data statistically. Their analysis revealed a steady "background" extinction rate of about 180 to 300 species every one million years, interrupted every 26 million years by a period of mass extinctions. Some of these mass extinctions involved only a doubling or tripling of the "background" rate, while others resulted in the elimination of enormous numbers of species and, in some cases, of major taxonomic groups. The greatest extinction of all occurred at the end of the Permian period, some 248 million years ago (see Table 24–1, page 497). It is estimated that 80 to 85 percent of all species then living became extinct, with the rate rising as high as 96 percent for marine species living in shallow waters.

Of all the organisms that once lived but live no more, the dinosaurs are perhaps the most intriguing. For some 150 million years these giant reptiles and their relatives dominated the land, air, and waters—a success story that indicates exquisite adaptation to their environments. And then, about 65 million years ago, at the end of the Cretaceous period, they vanished. Many other species of terrestrial animals died out at about the same time, including virtually all of those weighing more than 25 kilograms (about 55 pounds), as did a large proportion of the plants in temperate regions. Marine life was particularly hard hit. Almost all of the planktonic forms in the oceans became extinct, along with several major groups of marine invertebrates. Paradoxically, tropical plants, small terrestrial animals, and freshwater organisms seem to have come through the general disaster relatively untouched.

In 1977, a group of scientists from the University of California at Berkeley, headed by geologist Walter Alvarez, made a most unexpected discovery. In the course of studies of sedimentary rocks in Gubbio, Italy, covering the transition from the Cretaceous period to the subsequent Tertiary period, they found that a layer of clay between the two sets of rocks contained unusually high levels of the metal iridium. This element, although relatively rare in the earth's crust, is abundant in meteorites. Similar iridium anomalies were found by the Berkeley group and by other scientists in clays at the Cretaceous-Tertiary boundary in other parts of Italy, in Denmark, Spain, and New Zealand, and in deep-sea cores from both the Atlantic and Pacific Oceans.

Subsequently, the Berkeley scientists proposed that the cause of the iridium anomaly—and the cause of the mass extinctions at the end of the Cretaceous period—was an asteroid, approximately 10 kilometers in diameter, that collided with the earth. They hypothesized that the impact and explosion of the asteroid kicked up a cloud of debris that circled the earth for a period of at least several months, producing continuous darkness, the cessation of photosynthesis, and a subsequent collapse of food supplies for heterotrophic organisms, ranging from planktonic forms to dinosaurs. Such darkness would be expected to cause significant climate changes, putting further stress on terrestrial organisms.

Many biologists, particularly many of those most familiar with the details of the fossil record on either side of the Cretaceous-Tertiary boundary, find serious flaws in this hypothesis. The available evidence indicates that different groups did not become extinct simultaneously, but that the extinctions occurred over a period of tens of thousands—perhaps hundreds of thousands—of years. This suggests more gradual processes at work. Nevertheless, continuing studies by a variety of workers indicate that the iridium anomaly, which does seem to be

49-15 *This photograph was taken at midday in Yakima, Washington, two days after the explosion of Mount St. Helens on May 18, 1980. As you can see, the volcanic ash hanging in the air, both at ground level and high above the city, effectively blocked most of the sunlight. According to the asteroid hypothesis, a similar, but more severe and long-lasting phenomenon triggered by the explosion of an asteroid colliding with the earth was responsible for months of darkness and the mass extinctions at the end of the Cretaceous period. A number of detailed models have indicated that even a limited nuclear exchange, which would throw vast quantities of debris and smoke high into the atmosphere, would create a comparable "nuclear winter," with devastating consequences for life on earth as we now know it.*

simultaneous with the planktonic extinctions, is real and worldwide. Of the various hypotheses advanced to explain it, only the asteroid hypothesis has held up well under extensive testing.

Enormous interest has been generated by the iridium anomaly, the asteroid hypothesis, and, more recently, the apparent periodicity of mass extinctions. As a consequence, scientists are now engaged in a thorough reexamination of the fossil record, geochemical studies of the boundary layers associated with the different mass extinctions, explorations of possible terrestrial sources of high levels of iridium (for example, certain types of volcanic eruptions), as well as new analyses of the geologic phenomena known to have occurred in the periods marked by high rates of extinction.

Whatever the cause or causes of mass extinctions, their effects on the subsequent course of evolutionary history are clear. When entire groups of organisms have died out, apparently without regard to the success of their particular solutions to the problems facing all living systems, new opportunities have opened up for the groups of organisms that survived. As the survivors have diversified in the course of exploiting newly available living space, new sets of solutions to the common problems have appeared, based on the survivors' genetic inheritance. At each mass extinction, the course of evolution has been dramatically altered, with some branches of the evolutionary tree permanently eliminated and others undergoing vast new diversifications.

EQUUS: A CASE STUDY

As a case study in macroevolution, let us consider horses, whose history is particularly well documented in the fossil record (Figure 49-16). At the beginning of the Tertiary period, shortly after the extinction of the dinosaurs, the horse lineage was represented by *Hyracotherium,* also known as Eohippus ("dawn horse"). It was a small herbivore (25 to 50 centimeters high at the shoulders) with three toes on its hind feet, four toes on its front feet, and doglike footpads on which its weight was carried. Its eyes were halfway between the top of its head and the tip of its nose. Its teeth had small grinding surfaces and low crowns, which probably could not have stood much wear. The teeth indicate that *Hyracotherium* did not eat grass; in fact, there was probably not much grass to eat then. It probably lived in forests and browsed on succulent leaves.

49–16 *The modern horse and some of its ancestors. Only one of the several branches represented in the fossil record is shown here. Over the past 60 million years, small several-toed browsers, such as* Hyracotherium, *were replaced in gradual stages by members of the genus* Equus, *characterized by, among other features, a larger size; broad molars adapted to grinding coarse grass blades; a single toe surrounded by a tough, protective keratin hoof; and a leg in which the bones of the lower leg had fused, with joints becoming more pulleylike and motion restricted to a single plane.*

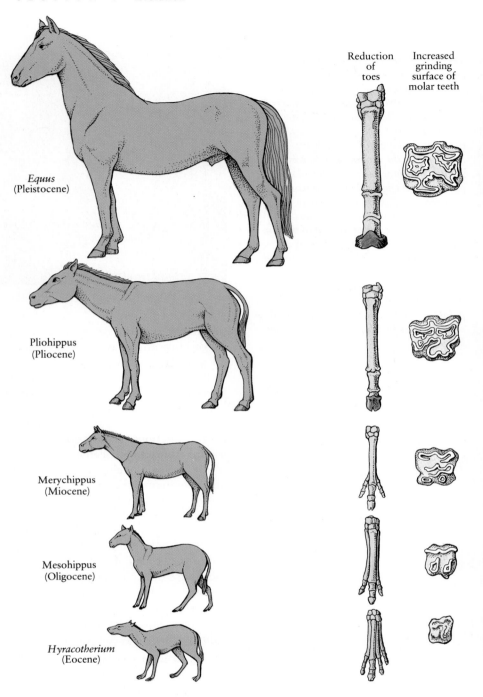

Reduction of toes

Increased grinding surface of molar teeth

Equus
(Pleistocene)

Pliohippus
(Pliocene)

Merychippus
(Miocene)

Mesohippus
(Oligocene)

Hyracotherium
(Eocene)

Slightly higher in the fossil-bearing strata, one finds larger horses, all three-toed, still browsing. Some of these clearly had molar teeth with large crowns that continued to grow as they were worn away (as do those of modern horses).

The strata of the Oligocene yield fossils of still larger horses. In these, the middle digit of each foot had expanded and was now the weight-bearing surface. In some, the two non-weight-bearing digits had become greatly reduced. These horses are thought to have been the direct ancestors of the genus *Equus,* which includes all modern horses.

These changes can be correlated with changes in the environment. In the time of *Hyracotherium,* much of the land was marshy and the chief vegetation was succulent leaves; the teeth of *Hyracotherium* were adapted for browsing. By the Miocene, the grasslands began to spread; horses whose teeth became adapted to

49-17 *The evolutionary history of the horse family, as summarized by George Gaylord Simpson in his classic book* Horses.

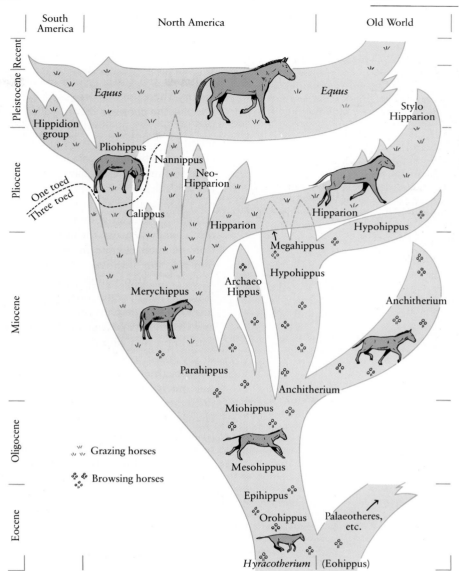

grinding grasses (which are very coarse and tough) survived, whereas those that remained browsers did not. The placement of the eye higher in the head may have facilitated watching for predators while grazing. The climate became drier, and the ground became harder; reduction of the number of toes, with the development of the spring-footed gait characteristic of modern horses, was an adaptation to harder ground and to larger size. An animal twice as high as another tends to weigh about eight times as much. Little *Hyracotherium* was probably as fast as the modern horse, but a larger, heavier horse with the foot and leg structure of *Hyracotherium* would have been too slow to escape from predators. (During this same period, predators were developing adaptations that rendered them better able to catch large herbivores, including horses.)

Thus the evolution of *Equus,* viewed from the long retrospective of the geologic record, represents a fairly straightforward accumulation of adaptive changes related to pressure for increased size and for grazing. It might be considered—as it was by Professors Marsh and Huxley (page 968)—as an example of phyletic change. As the number of fossil specimens has increased, however, it has become clear that at any given time many different species of horses coexisted, only some of which survived (Figure 49–17). Thus both phyletic change and cladogenesis are seen as contributing to the evolution of *Equus.*

PUNCTUATED EQUILIBRIA

Although the fossil record documents many important stages in evolutionary history, there are numerous gaps. As Darwin noted in Chapter X of *The Origin of Species,* the geologic record is ". . . a history of the world imperfectly kept, and written in a changing dialect; of this history, we possess the last volume alone, relating only to two or three countries. Of this volume, only here and there a short chapter has been preserved; and of each page, only here and there a few lines."

Many more fossils have, of course, been discovered in the 100 years since Darwin's death. We are no longer limited to the most recent volume, and some of the volumes are now far more complete. Nevertheless, fewer examples of gradual change have been found than might have been expected. Until recently, this discrepancy between the model of slow phyletic change and the poor documentation of such change in much of the fossil record has been ascribed to the imperfection of the record itself.

In 1972, two young scientists, Niles Eldredge of the American Museum of Natural History and Stephen Jay Gould of Harvard University, ventured the proposal that perhaps the fossil record is not so imperfect after all. Both Eldredge and Gould have backgrounds in geology and invertebrate paleontology, and both were impressed with the fact that there was very little evidence of phyletic change in the fossil species they studied. Typically, a species would appear abruptly in fossil-bearing strata, last 5 million to 10 million years, and disappear, apparently not much different than when it first appeared. Another species, related but distinctly different—"fully formed"—would take its place, persist with little change, and disappear equally abruptly. Suppose, Eldredge and Gould argued, that these long periods of no change ("stasis") punctuated by gaps are not flaws in the record but *are* the record, the evidence of what really happens.

How could it be that a new species would make such a sudden appearance? They found their answer in the model of allopatric speciation. If new species formed principally in small populations on the geographic periphery of the range of the species, if speciation occurred rapidly (by rapidly, paleobiologists mean in thousands rather than millions of years), and if the new species then outcompeted the old one, taking over its geographic range, the resulting fossil pattern would be the one observed. In the words of Gould: "Thus the fossil record is a faithful rendering of what evolutionary theory predicts, not a pitiful vestige of a once bountiful tale."

As first proposed by Eldredge and Gould, punctuated equilibria seemed to refer principally to the tempo of evolution. The population geneticists and Darwin before them had emphasized gradual change. There was clearly room, however, for the idea that populations would change more rapidly at some times than at others, particularly in periods of environmental stress, as in the case of the peppered moth (page 962).

As this model has become more fully developed, particularly by Steven M. Stanley of Johns Hopkins (also a paleontologist), it has become more radical and more controversial. Its proponents now argue that not only is cladogenesis the principal mode of evolutionary change (as Mayr stated some 40 years ago) but that selection occurs among species as well as among individuals. Thus, at any one time in evolutionary history it may be possible to find a number of related species coexisting, each departing in a different way from the ancestral type. For example, as the horses evolved, some species were larger, others remained small. Some became grazers, others remained browsers. The overall trends in horse evolution resulted from the differential survival of species, rather than from the phyletic change that occurred within species. Thus, in this new formulation, species take the place of individuals, and speciation and extinction substitute for birth and death. The role of speciation is analogous to the role of mutation in population

49–18 *A model of evolutionary change as a result of species selection. Speciation events moving to the left are equal in number to events moving to the right. However, the average rate of speciation increases to the right, while the average rate of extinction decreases in the same direction. Thus the direction of speciation itself plays no role in the direction of evolutionary change.*

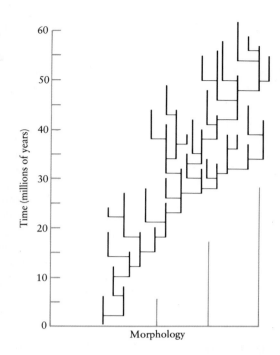

genetics; it provides the variation on which selection acts. According to this proposal, there are two levels of selection: in one, selection acts on the individual, and in the other, it acts on the species.

The punctuated equilibrium model, portions of which have now been assimilated into the synthetic theory, has stimulated vigorous debate among biologists, a reexamination of evolutionary mechanisms as currently understood, and a reappraisal of the evidence. These activities have at times been misinterpreted as a sign that Darwin's theory is "in trouble." In fact, they indicate that evolutionary biology is alive and well and that scientists are doing what they are supposed to be doing—asking questions. Darwin, we think, would have been delighted.

SUMMARY

A major question in evolutionary theory is whether microevolution (the gradual changes that take place within species) can account for macroevolution (the diversity among families, orders, classes, and phyla). The process of speciation—the formation of new species—is considered of central importance in answering this question.

A species is defined as a group of natural populations whose members can interbreed with one another but cannot (or at least usually do not) interbreed with members of other such groups. Two principal modes of speciation are recognized, allopatric and sympatric. Allopatric speciation occurs in geographically isolated populations. Sympatric speciation occurs principally in plants through polyploidy, often coupled with hybridization; it may also take place in some cases by disruptive selection.

The key event in speciation is genetic isolation. Once species have become genetically isolated, they can once again inhabit the same geographic area without interbreeding because of numerous elaborate behavioral, anatomical, and physiological mechanisms.

The fossil record discloses four components of evolutionary change: phyletic change, cladogenesis, adaptive radiation, and extinction. Phyletic change is gradual change within a single lineage over time. Cladogenesis, by contrast, is evolutionary change produced by the branching off of populations from one another to form new species. Adaptive radiation is the rapid formation of many new species from a single ancestral group, characteristically to fill a new ecological zone. Extinction is the disappearance of a species from the earth; the fossil record reveals a low, steady rate of extinction, interrupted periodically by mass extinctions involving enormous numbers of species. In the synthetic theory, macroevolution is regarded as the product of a combination of these patterns.

Paleontologists have presented evidence for an additional pattern of macroevolution known as punctuated equilibrium. They propose that new species are formed during bursts of rapid speciation among small peripheral populations, that the new species outcompete many of the then-existing species (which become extinct), persist for long periods with little change, and then, in turn, abruptly become extinct. The punctuationalists propose that major changes in evolution take place as a result of selection acting on species as well as on individuals.

QUESTIONS

1. Distinguish among the following: allopatric speciation/sympatric speciation; hybridization/polyploidy; premating isolating mechanisms/postmating isolating mechanisms; phyletic change/cladogenesis/adaptive radiation.

2. Define genetic isolation. Why is it such an important factor in speciation?

3. Give three possible reasons why sympatric speciation by hybridization and polyploidy is more common in plants than in animals.

4. Two species of plants, each capable of reproducing both sexually and asexually, became distinct from their parental lineages as a result of autopolyploidy. Individual plants of these two species then hybridized with each other. Would the resulting plants be capable of sexual reproduction? Would they constitute a new species? Explain your answers.

5. Describe the separate steps involved in the formation of the distinct species of Galapagos finches. How does this example of speciation illustrate both the founder principle and adaptive radiation?

6. What physical features of the Galapagos made possible the evolution of Darwin's finches?

7. The genetic isolation of long-established, closely related species is generally maintained almost entirely by premating isolating mechanisms. Explain, in terms of natural selection, why this is to be expected.

8. Recent evidence supporting the periodic occurrence of mass extinctions suggests that Cuvier and Agassiz (page 3) were correct in their assertion that a series of catastrophes led to the demise of many previously existing forms of life. Neither of these eminent scientists, however, could provide a testable explanation for the proliferation of new forms that followed each of the proposed catastrophes. How would modern evolutionary theory explain the proliferation of new forms after mass extinctions?

The Evolution of the Hominids

50–1 *A modern tree shrew, which the earliest primates probably resembled. If you look closely, you will see five-digited paws. Although clawed, they can be spread out and used for grasping. In some classification systems, tree shrews are grouped with the primates, and in others, with the insectivores, which indicates the closeness of the two evolutionary lines.*

Where does the story of human evolution begin? We might start with a chance combination of chemicals in some warm Precambrian sea. Or perhaps even with the formation of a small planet 150 million kilometers from a star. Or it might begin more than 4.5 billion years later, when some little tribe of hominids found they could sharpen a digging stick or create a sharp edge on a stone. In any case, it is a very long story, measured in human terms, and many of its details are lost to us, probably forever.

For present purposes, let us start the story about 200 million years ago, in the early Mesozoic era, at about the time of the first dinosaurs. In this same period of time—give or take a few million years—the first mammals appeared, arising from a primitive reptilian stock. Our information about these mammals is very slight. The entire length of the Jurassic and Cretaceous periods has left us with only a few fragments of skulls and some occasional teeth and jaws. From these scraps of evidence, we know that the first mammals were about the size of a mouse. They had sharp teeth, indicating that they were basically carnivorous. Since they were too small to attack most other vertebrates, however, they are assumed to have lived on insects and worms, supplementing their diet with tender buds, fruits, and perhaps eggs. These first mouse-sized mammals were probably nocturnal, and they were almost certainly warm-blooded. If such an animal were alive today, it would be classified as an insectivore, something like a ground shrew.

For about 130 million years, these small mammals led furtive existences in a land dominated by reptiles. In the course of this time, they diverged into three principal lineages: the monotremes, the marsupials, and the placentals (see page 600). Then suddenly, as geologic time is measured, the giant reptiles, the dinosaurs, disappeared, followed almost immediately by an explosive adaptive radiation of the mammals. This radiation gave rise to a variety of marsupials and about two dozen different lines of placentals. Among the placentals are carnivores, ranging in size from the saber-toothed cats down to small, weasel-like creatures; ungulates, which include not only the many wild grazing animals but also most of our domesticated farm animals; the omnipresent rodents; and such odd groups as the whales and dolphins, the bats, the modern insectivores, and the primates. We are placental mammals and members of the primate order, as are tarsiers, lemurs, monkeys, and apes, among others.

TRENDS IN PRIMATE EVOLUTION

Primate evolution began when a group of the small, shrewlike mammals took to the trees. Most trends in primate evolution seem to be related to various adaptations to arboreal life.

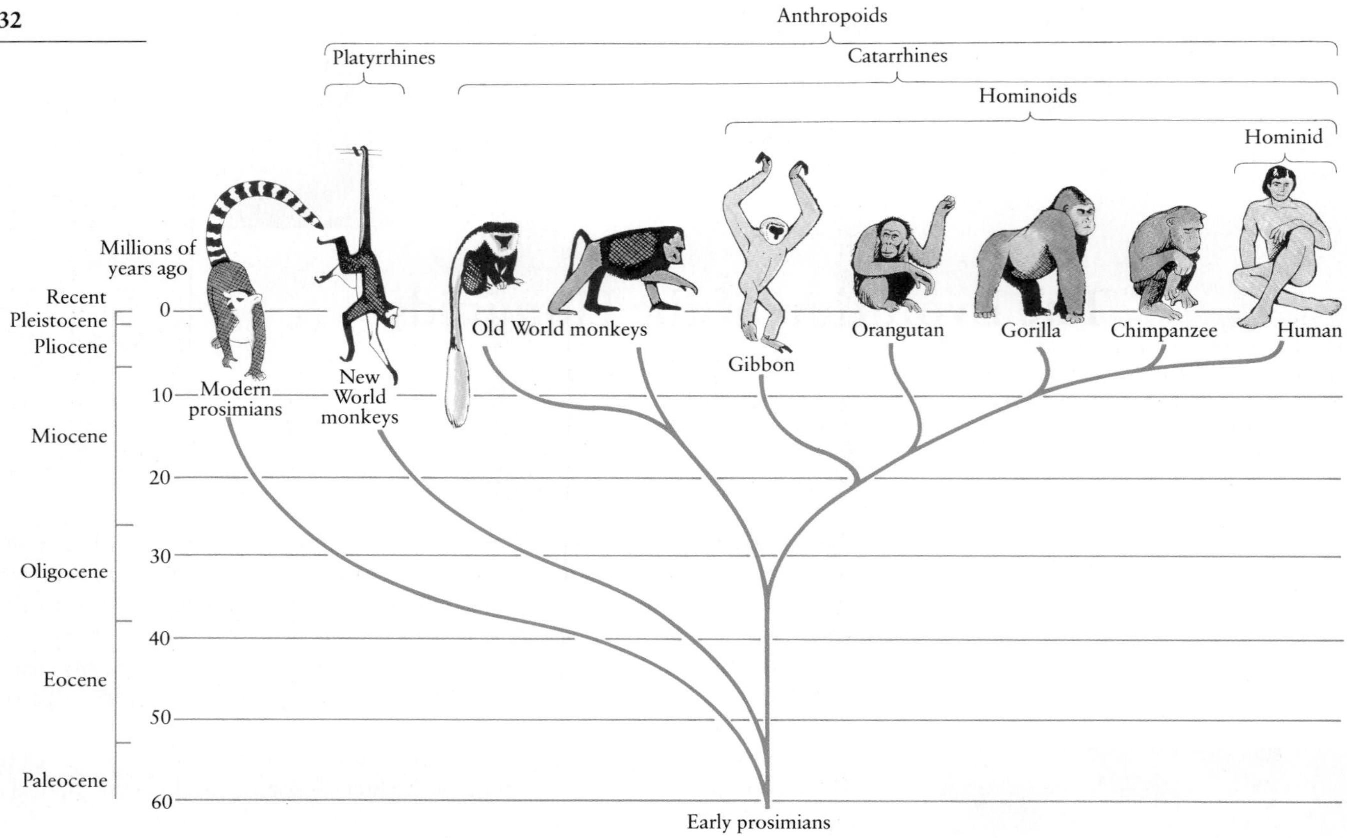

Anthropoids

Platyrrhines

Catarrhines

Hominoids

Hominid

Millions of years ago

Recent	0
Pleistocene	
Pliocene	
	10
Miocene	
	20
Oligocene	30
	40
Eocene	50
Paleocene	60

Old World monkeys

Gibbon

Orangutan

Gorilla

Chimpanzee

Human

Modern prosimians

New World monkeys

Early prosimians

50–2 *A tentative phylogenetic tree of the primates.*

The Primate Hand and Arm

The first four-legged mammals all had five separate digits on each hand and foot. Each digit, with the exception of the thumb and the first toe, had three separate segments, making it flexible and capable of independent movement. In the course of evolution, selection pressures for greater efficiency in running, digging, and seizing prey led to specialized hooves and paws in most mammals; in other mammals, selective forces led to the modification of limbs as flippers for swimming. The primates, however, retained and elaborated on the basic five-digit pattern. Notably, modern primates, with few exceptions, have a divergent thumb. The divergent thumb, which can be brought into opposition to the forefinger, greatly increases gripping power and dexterity. There is an evolutionary trend among the primates toward finer manipulative ability that reaches its culmination in humans (Figure 50–3).

In the basic quadrupedal structure of the early mammals and reptiles, the forelimb has two long bones (the radius and the ulna), a pattern that provides for flexibility. Among mammals, it is the primates, in particular, that can twist the radius, the bone on the thumb side, over the ulna so that the hand can be rotated through a full semicircle without moving the elbow or the upper arm. Similarly, only a few mammals have the ability to move the upper arm freely in the shoulder socket. A dog or horse, for instance, usually moves its legs in only one plane,

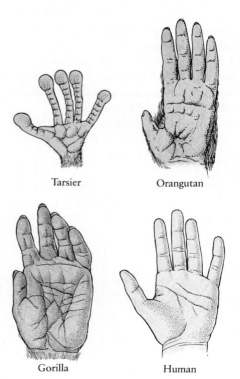

Tarsier

Orangutan

Gorilla

Human

50–3 *Some primate hands. The hand of the tarsier has enlarged adhesive skin pads for grasping branches. In the orangutan, the fingers are lengthened and the thumb reduced, which provide for efficient brachiating (swinging arm over arm through the trees). The gorilla's hand, which is used in walking as well as handling, has shortened fingers. The human thumb is larger proportionately than that of any other primate, and opposition of thumb and fingers, on which handling ability depends, is greatest in humans.*

(a)　　　　　　　　　　　　　　　　(b)

50–4 *Life in the treetops made maternal care a major factor in infant survival. Also the necessity for carrying the young for long periods resulted in strong selection pressures for reduced numbers of off-spring. (a) Anthropoids, such as this vervet monkey, usually have single births. (b) A mother chimpanzee with her infant. Field studies suggest that bonds between mother and offspring and perhaps also among siblings last well into adulthood, perhaps for a lifetime.*

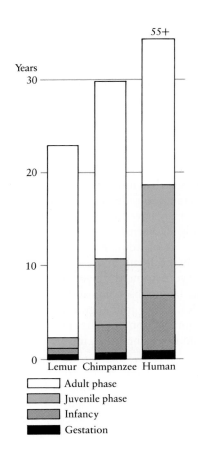

50–5 *In the course of primate evolution there has been a trend toward longer periods of juvenile dependency and thus longer periods of learning.*

forward and backward; some lemurs, South American monkeys, apes, and humans are among the few mammals that can rotate the arm widely in the socket, an advantageous characteristic for a tree-dwelling animal.

Most primates also have nails rather than claws. Nails leave the tactile surface of the digit free and so greatly increase the sensitivity of the digits for exploration and manipulation.

Visual Acuity

Another result of the move to the trees is the high premium placed on visual acuity, with a decreasing emphasis on the role of olfaction, the most important of the senses among many of the other mammalian orders. (Flying produced similar evolutionary pressures among the birds, which were also evolving rapidly during this same period.) This shift from dependence on smell to dependence on sight has anatomical consequences. Tree shrews, like many other animals, have eyes that are directed laterally, but among the primates there can be traced a steady evolutionary trend toward frontally directed eyes and stereoscopic vision.

Almost all primate retinas have cones as well as rods; cones, as we discussed on page 868, are concerned with color vision and with fine visual discrimination. Most primate retinas also have foveas, areas of closely packed cones that produce sharp visual images.

Care of the Young

Another principal trend in primate evolution is toward increased care of the young. Because mammals, by definition, nurse their young, they tend to have longer, stronger mother-child relationships than other vertebrates (with the exception, in some cases, of birds). In the larger primates, the young mature slowly and have long periods of dependency and learning (Figure 50–5).

Uprightness

Another adaptation to arboreal life is the ability to adopt an upright posture. Even quadrupedal primates, such as monkeys, can sit upright. One consequence of this posture is a change in the orientation of the head, allowing the animal to look straight ahead while in a vertical position; it is this characteristic, above all others, that makes our fellow primates look so "human" to us. Vertical posture was an important precondition for the eventual evolution of the upright stance characteristic of modern humans.

(a) (b)

50–6 *Two prosimians.* **(a)** *A ring-tailed lemur, combing its tail. The second digit of each foot has a special grooming claw.* **(b)** *A native of Indonesia, the little tarsier (about the size of a kitten) is specialized for leaping between vertical supports. It has stereoscopic vision and a relatively larger brain than most lemurs. Living entirely in trees, it has hands and feet with enlarged skin pads for grasping branches. As you may have guessed from the owl-like eyes, tarsiers are primarily nocturnal.*

MAJOR LINES OF PRIMATE EVOLUTION

Prosimians

Primates are generally divided into two major groups: the **prosimians** (lorises, bush babies, tarsiers, and lemurs) and the **anthropoids** (monkeys, apes, and humans). During the Paleocene and the Eocene (about 65 to 38 million years ago), a great abundance and variety of prosimians inhabited the tropical and subtropical forests that spread much farther north and south of the equator than they do today. Modern prosimians (Figure 50–6) are mostly small to medium-sized arboreal animals, and many are nocturnal. Insects typically form at least part of the diet of the smaller prosimians, while the larger ones eat varying combinations of leaves, fruits, and flowers.

Monkeys

Monkeys, along with the apes and humans, make up the higher primates, the anthropoids. Modern monkeys are generally larger than modern prosimians, their skulls are more rounded, and they are generally considered more intelligent, although this is an elusive quality to measure. They have full stereoscopic vision and also color discrimination. They are virtually all diurnal.

Monkeys typically move in bands composed of adult males and females, infants, and juveniles, but there is great variety in the details of social organization. The females protect and otherwise care for the young, and the males may perform protective functions for the group, ranging from observing and warning to attacking predators outright. In some species, males join together to mob a predator, harassing it by hooting and calling, jumping on branches until they fall on enemy heads, sometimes throwing sticks and branches, and, in some species, by group defecation and urination.

The monkeys probably arose from prosimian stock during the Eocene epoch, which ended 38 million years ago. Related fossil forms are found in the New World as early as the Oligocene. There are two principal groups (Figure 50–7): the New World monkeys, also known as platyrrhines (meaning "flat-nosed"), and the Old World monkeys, the catarrhines ("downward-nosed"). The separation of these groups took place with the breakup of Gondwana (see page 1012), with the platyrrhines evolving in South America and the catarrhines in Africa, quite possibly during the Oligocene, some 38 to 25 million years ago.

The New World monkeys, from South and Central America, are all strictly arboreal in their habits, and some of them use their tails as a fifth prehensile limb

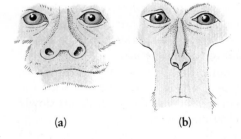

(a) (b)

50–7 *Early in their evolution, the anthropoids split into two main lines, the platyrrhine, or flat-nosed* **(a)**, *and the catarrhine, or downward-nosed* **(b)**. *New World monkeys are platyrrhines; Old World monkeys and the hominoids are catarrhines. There are many other characteristic anatomical differences between the two groups.*

50–8 *Only New World monkeys, such as the spider monkey, can hang by their tails. The platyrrhine monkeys are the only living primates native to the Americas.*

(Figure 50–8), which none of the Old World monkeys can do. Any monkey you see hanging by its tail is definitely a New World monkey. The New World monkeys include marmosets, howler monkeys, the spider monkey, and the capuchin—the familiar monkey of the organ-grinder.

The Old World monkeys include both arboreal and terrestrial species. Among the tree dwellers are the colobus monkeys, langurs, the mangabeys, and the guenons. Their tails are used for balance rather than as prehensile organs. The ground dwellers, which came from tree-dwelling ancestors and still make use of trees on occasion, include the macaques, or rhesus monkeys, and baboons. These primates walk on all fours, with their fingers flat against the ground.

Apes

The **hominoids,** a group now represented by the apes and ourselves, are cousins of the Old World monkeys and are grouped with them as catarrhines. Fossil apes are known in large numbers from deposits in Kenya and Uganda, ranging in age from 22 million to 14 million years ago. These Miocene apes have been given a variety of names, of which the best known is *Proconsul*. The landscape occupied by these early apes was largely forest and woodland. There were at least 10 species, with *Proconsul africanus* the best known. *P. africanus* was first discovered by Mary Leakey on an island in Lake Victoria; it was named after Consul, a favorite chimpanzee at the London Zoo. *P. africanus* and its relatives were tree-dwelling, probably fruit-eating, and about the size of baboons, with the males in some species much larger than the females. According to David Pilbeam of Harvard University, who has remained a voice of reason throughout the many controversies surrounding human evolution, "Its elbow and shoulder joints and feet are like a chimpanzee's, its wrist is like a monkey's, and its lumbar vertebrae are like a gibbon's. . . . *Proconsul* provides a salutary lesson for students of evolution: the relations inferred for the animal have depended on what part of the body was being studied."

The modern apes comprise four genera: *Hylobates* (gibbons), *Pongo* (orangutans), *Pan* (chimpanzees), and *Gorilla* (gorillas). Apes, with the exception of the gibbons, are larger than monkeys, and their brain is larger in proportion to their size. They are all capable of suspending their bodies from branches when in the trees, although among modern apes, only the gibbons move primarily by brachiation—swinging from one arm and then the other with their bodies upright (Figure 50–9). Upright suspension is thought to have played a role in the transition from the body structures associated with the horizontal position characteristic of the Old World monkeys and some lower primates to the body structure that led ultimately to our erect posture. Apes have relatively long arms and short legs, resting the weight of the front part of their bodies on their knuckles. As a result, even when they are on all fours, their bodies are partially erect.

50–9 *Brachiation, as exhibited by a gibbon, the smallest of the apes. Although gibbons can stand and walk upright, this is their usual and most efficient means of locomotion.*

50–10 *Orangutans, the least-studied of the primates, are found today only in the tropical rain forests of Borneo and Sumatra.*

Modern apes range widely in size. Gibbons, which are the smallest, weigh about 6 kilograms (about the size of a large house cat), with both sexes the same size. Male chimpanzees weigh about 40 kilograms (88 pounds) and females about 30 kilograms (66 pounds). Male orangutans can weigh up to 100 kilograms (220 pounds), with the females weighing about half that. A male gorilla is only about as tall as an average-sized man, but weighs two to three times as much, from 140 to 180 kilograms (300 to 400 pounds). Gorillas have strongly ridged brows, a strong heavy jaw and, on top of the skulls of adult males, a bony crest to which the strong jaw muscles are attached.

Social Behavior

Because of the evolutionary affinities between the other hominoids and ourselves, it is tempting to base interpretations of human behavior on that of other primates—and vice versa. However, social behavior, like size, varies greatly. Gibbons are the only monogamous hominoids, a status apparently reflected in their lack of sexual dimorphism (page 1000). They form permanent pairs, each couple living by itself with its young offspring. Male orangutans have ranges that generally overlap those of more than one female; adult females are often accompanied by one or two offspring.

50–11 *Researcher Dian Fossey with members of a mountain gorilla troop in Rwanda, in central Africa.*

Gorillas are highly social, living in troops consisting of 8 to 24 individuals, with more than twice as many females as males and a number of juveniles and infants. Each gorilla troop has a large, mature (silver-backed) male as a leader. Large males, in particular, have almost entirely abandoned the trees, but some of the smaller animals sleep in nests in the lower branches. Gorillas usually walk on all fours, but they can stand upright and walk erect for short distances. They also stand erect when challenging an enemy, threatening with broken branches, beating their chests, barking and roaring, even hitting themselves under the chin to make their teeth rattle.

A female gorilla of reproductive age mates only with the troop's highest ranking male that is not closely related to her. Once a female gives birth, she has a long period of nursing her infant, which lasts from two to four years, during which time she does not mate again.

Chimpanzees are the primates now thought to be the closest to humans. They usually move in groups, feeding on wild fruits, seeds, and pods of a large variety, and occasionally killing smaller animals for meat. Within a group there is a male dominance hierarchy; females are usually subordinate to adult males, but there is also a female dominance hierarchy. Temporary (two- or three-day) bonds may form between a male and a female in estrus, but otherwise there are no pair bonds, and receptive females typically mate with a series of males. However, there are often friendships among members of the group, and friends may spend as much as two hours a day in mutual grooming. Males sometimes cooperate in capturing prey, which consists most often of other primates, such as infant baboons and

50-12 *Chimpanzees are probably our closest cousins. They have a relatively fine precision grip and can use simple tools, such as a stick to dig out termites* (**a–c**), *a leaf as a blotter, or a branch as a weapon. They are* (**d**) *playful,* (**e**) *gregarious, and* (**f**) *noisy.* (**g**) *Because of their intelligence and their physiological resemblance to humans, chimpanzees are prize subjects for medical research, a practice that many find ethically troubling because of these very characteristics.*

colobus monkeys. Prey may be shared among the group. Offspring maintain strong bonds with their siblings and their mothers, apparently throughout their lifetimes. Chimps are gregarious, curious, boisterous, noisy, and extroverted. Although groups move through home ranges and some populations expand their range at the expense of others, there is generally not a strong sense of territoriality. Moreover, the bands are not well knit and exclusive, as they are in some primate groups. The composition of the group changes as newcomers are taken in and other members wander off.

Thus, as you can see, it is difficult to make correlations between human behavior and the diverse behavior of other, even closely related species.

THE EMERGENCE OF THE HOMINIDS

The First Hominids

In 1924, an explosion in a quarry that was being mined in Taung, South Africa, loosened a piece of rock containing a portion of the skull of a child. This specimen, along with many others, was sent to anatomist Raymond Dart in Johannesburg. After 73 days of chipping and picking at the crust of sand and limestone surrounding it, Dart was able to see that the little cranium had some humanlike features that distinguished it from both modern apes and their ancestors. These included the rounded appearance of the skull, the size and shape of the brain (Dart was a skilled neurologist), and the shape of the teeth. Also, the point of attachment of the vertebral column to the skull indicated that the young animal walked upright.

Dart reported his find in the British journal *Nature,* naming the new fossil *Australopithecus* ("southern ape"). The popular press proclaimed it momentarily as "the missing link," but he received little professional support for the Taung skull's having a place on the human family tree. Dart was not an anthropologist. The scientific community was not yet ready to recognize a human ancestor so small-brained and so apelike. Besides, it was generally agreed that Asia, with its cultural advantages, was the cradle of mankind, not darkest Africa. For more than 20 years, the Taung skull was virtually ignored.

Subsequent fossil discoveries, however, confirmed Dart's original findings and their interpretation. In 1985, Professor Dart celebrated both his ninety-second birthday and the diamond jubilee of his announcement to the world of the Taung child (Figure 50–13). The australopithecines are now generally accepted as **hominids**—members of the human family, Hominidae. A large number of australopithecine fossils have now been found; prominent among the fossil-finders have been Louis and Mary Leakey and their son Richard, of Kenya, and the American Donald Johanson and his colleagues, who have worked principally in Ethiopia.

How many species of *Australopithecus* there were and which of these, if any, are ancestors of the genus *Homo* are, however, matters on which hardly any two experts agree at the present moment. There are several reasons for these uncertainties. First, there is a paucity of hominid fossils, compared, for instance, to *Equus,* for which specimens abound. Second, they are often in fragments; one skull, for example, was in more than 300 pieces. In the words of one researcher: "It is like putting together a three-dimensional jigsaw puzzle, with some of the pieces missing and no picture on the box." Third, based on the evidence so far, the australopithecines seem to have been a heterogeneous group. Even those generally regarded as belonging to a single species vary in size and other characteristics. Adding to the complications, males appear to have been considerably larger than females, so it is not always clear if two sizes represent two species or two sexes. Finally, the search for human ancestors has, in recent years, become an exciting and intensely competitive undertaking, and discoverers of a new fossil are understandably predisposed to claim it as something distinctive and unique.

Current Status of the Australopithecines

The australopithecines, by present evidence, lived from more than 3.6 million years ago (a firm date for the earliest specimen) to at least 1.4 million years ago. The fossils found so far represent a number of different species. One is *Australopithecus africanus,* to which Dart's Taung child belongs. *A. africanus* was small

50–13 *Raymond Dart, with the skull of the Taung child, which was found in a limestone quarry in 1924 and to which he gave the name* Australopithecus africanus. *It was not until almost a quarter of a century later that the fossil was widely recognized by the paleoanthropological community as that of a hominid. In 1985, at a celebration of his ninety-second birthday and the sixtieth anniversary of his published report on his findings, Professor Dart said, "You know, I was never bitter about how I was treated. . . . I knew people wouldn't believe me. I wasn't in a hurry." Dart died in November of 1988, at the age of 95.*

The Footprints at Laetoli

The eruption of a volcano in the Rift Valley spewed a layer of ash over the plains of the southern Serengeti in what is now Tanzania. Soon after, there was a brief light shower and then, while the ash layer was still damp, some 20 different kinds of animals scurried, ran, and slithered over it, leaving their prints on the soft and slippery surface. These included hares, baboons, a rhinoceros, two types of giraffe, hyenas, many birds, a three-toed horse, a saber-toothed cat—and three hominids. At one point, one of the hominids stopped, paused, turned to the left and then, perhaps reassured, continued on. Under the heat of the equatorial sun, the ash dried, setting like concrete, and soon the footprints were covered by more ash and windblown silt.

In the same spot, by a happy coincidence, some 3.6 million years later, another group of young hominids were amusing themselves by hurling dried elephant dung at one another. One of these (a scientist from Harvard who was visiting Mary Leakey's excavations at Laetoli) dodged, slipped, fell, and found himself on eye level with some strange indentations. And so the Laetoli footprints, one of the most significant finds in all of paleoanthropology, were first discovered. Since that time, literally thousands of individual prints have been discovered in the same general area—probably more fossil animal tracks than have ever been found elsewhere in the world, an extraordinary panorama of the number and variety of ancient African animals. And, side by side with these are the arch, big toe, and heel marks that are clear proof that 3.6 million years ago there were hominids who walked fully upright with a bipedal human gait.

Hominid footprints in the volcanic ash of Laetoli. These fossil tracks extend in parallel for about 25 meters. The trail on the left was made by the smallest of the hominids, perhaps holding the hand of the one to the right. This one, the largest, was followed by another smaller hominid who walked in his or her footprints, partially obliterating them. The prints on the right are those of a three-toed horse.

(estimated at less than 30 kilograms, or 70 pounds), lightly built, with a brain case of about 440 cubic centimeters. (The average brain capacity of modern humans is 1,300 cubic centimeters.) The brain case is higher and more rounded than that of the apes, and the placement of the foramen magnum, the entry point of the spinal cord into the vertebral column, indicates an upright posture, as does the shape of the pelvis. The canines are small, not protruding like an ape's, and the incisors are as small as our own.

A second species is *Australopithecus robustus*, which, as its name implies, was larger (perhaps 40 kilograms, or 90 pounds) and much more massively built, with a much wider face. The front teeth are about the same size as ours, but the premolars and molars are huge, with some measuring 2.5 centimeters across. A gorilla-like bony crest, running fore and aft, supported powerful jaw muscles,

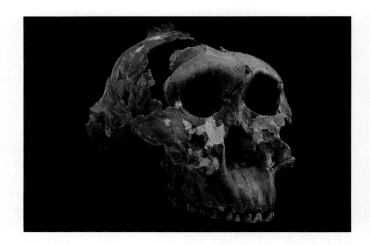

50-14 *The skull of Zinjanthropus, the "nutcracker man," now assigned to the species* Australopithecus boisei. *This discovery by the Leakeys focused attention on Africa as the cradle of mankind and marked the beginning of the present greatly expanded research interest in the early evolution of the hominids.*

which were accommodated in the large spaces inside the cheekbones. The size of the teeth and jaw muscles suggest that *A. robustus* ground large quantities of coarse plant material—nuts, hard-shelled fruits, roots, and tubers—between his (and her) massive molars. Although the facial skeleton is large, the brain case is small—a little larger than that of *A. africanus,* with an average capacity of about 500 cubic centimeters.

A third species, *Australopithecus boisei* (Figure 50–14), resembles *A. robustus* but is even more robust. *A. robustus* and *A. boisei* were geographically separated, with *A. robustus* being found only in southern Africa and *A. boisei* in East Africa. The first of these robust hominids to be discovered was originally named Zinjanthropus, the so-called "nutcracker man." It was found by Mary Leakey in the Olduvai Gorge in Tanzania in 1959 and dated by the then newly developed potassium-argon technique at 1.8 million years. The discovery of Zinjanthropus, the culmination of 20 years of exploration and preparation by Louis Leakey, marked the beginning of the great surge of interest in and support for research into human origins.

There have been various interpretations of the relationships among these three proposed species. For a number of years, one of the most widely accepted, because it fit the data then available, was that *A. africanus* evolved into *A. robustus,* which, in the course of phyletic evolution, became *A. boisei* and then became extinct. This scheme was rendered obsolete by a discovery in 1986, in northern Kenya, by Alan Walker of Johns Hopkins University, a member of Richard Leakey's team. The new fossil find, known as WT 17000 and dated at 2.5 million years ago, is the skull of a robust australopithecine more massive but older than any specimen found heretofore. WT 17000 shattered with a single blow the hypothesis of an orderly evolutionary progression toward greater robustness into extinction. *A. robustus* and *A. boisei* are now generally regarded as two separate branches on the family tree, with WT 17000 an early form of *A. boisei.* (To complicate things further, some have argued that WT 17000 should be regarded as a separate species.)

Yet to be resolved, and perhaps most interesting of all, is the place in the scheme of things of the oldest of the australopithecines, comprising fossils from Hadar in Ethiopia and from Laetoli, 1,600 kilometers to the south in Tanzania, dated from 3.6 to about 3 million years ago. Johanson, the discoverer of the Hadar fossils, known informally as the "First Family," has given the group the name of *Australopithecus afarensis.* The most famous *A. afarensis* is Lucy (Figure 50–15), named after the Beatles' song being played on the tape deck at the camp site. Based on Lucy, we can tell that adult females were only about 110 centimeters (3.5 feet) tall and weighed about 23 kilograms (50 pounds). Other individuals, interpreted by Johanson as males of the same species, were about twice as large, weighing upward of 45 kilograms (100 pounds). It is clear from various skeletal features that they walked upright (Figure 50–16). Some paleoanthropologists are

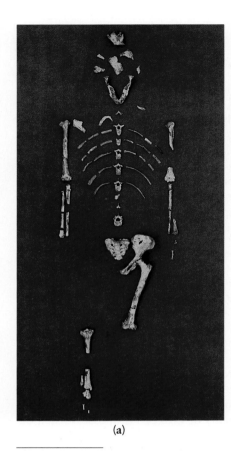

(a)

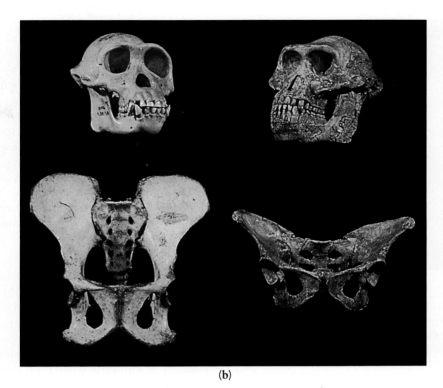

(b)

50–15 (a) *One of the most complete and oldest of the early hominids yet to be found, named Lucy by her discoverer, Donald Johanson. Lucy, the "First Family" (a remarkable fossil collection representing 13 individuals), and other* well-known hominid fossils were discovered in the Afar triangle in Ethiopia. Johanson claims that they represent a distinct species, Australopithecus afarensis. Others see them as members of A. africanus. Fossils ascribed to the same species were discovered by the Leakey group at Laetoli, 1,600 kilometers away, together with the haunting footprints shown on page 1039. (b) Comparison of the skull *and pelvis of a chimpanzee (left) and of a member of the "First Family" (right). Note that the skulls are similar but the pelvises are totally different, producing the differences in gait illustrated in Figure 50–16. The pelvis on the right is much closer to the modern human pelvis. The conclusion: hominids walked fully upright before there was any significant increase in brain size.*

50–16 *A comparison of the structures of the pelvis, legs, and feet of (a) a chimpanzee and (b) Lucy. Lucy's adaptation to upright walking was virtually complete; she was able to move forward smoothly, and the angle of her femur (red line) kept her legs under her body as she moved. As a chimpanzee walks, however, it must tilt from side to side (red arrow) in order to keep its center of gravity over the leading leg.*

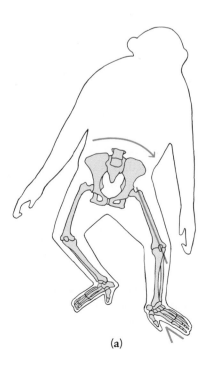

(a)

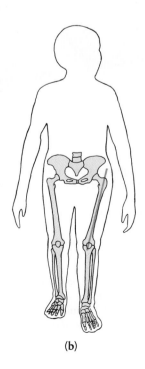

(b)

50–17 *This remarkably complete skull, known as 1470, was reconstructed from some 300 fossil fragments found by Richard Leakey's team in northern Kenya near the Ethiopian border. It is dated at 1.9 million years. Although skull 1470 was contemporaneous with the australopithecines, because of its much larger cranial capacity and the absence of a bony crest, it is assigned to the species* Homo habilis, *the most ancient human species.*

50–18 *Pebble tools such as these have been found in fossil strata at Olduvai and other sites in East Africa. From pebbles of lava and quartz, flakes were struck off in two directions at one end, making a somewhat pointed implement, or in a row on one side, making a chopper. These tools measure up to 10 centimeters in length and may have been used to prepare plant food and to butcher game. Recent evidence, however, suggests that the flakes struck from these pebbles were more commonly used as tools. The oldest have been dated at about 2.5 million years ago. They appear to be associated with* Homo habilis.

of the opinion that these fossils should be classified as early specimens of *A. africanus.* Others contend that the group is so heterogeneous—especially when it is extended to include the geographically distant Laetoli specimens—that it represents two or three different species, perhaps including one of the genus *Homo,* a subject we are about to explore. Here again, new data should resolve these differences and, very possibly, give rise to new ones. Unfortunately, the site in Ethiopia was closed to further excavations in 1982 for political reasons.

Thus, in short, four species are currently proposed: *A. afarensis,* the most ancient and the most recently discovered, plus *A. africanus, A. robustus,* and *A. boisei,* which may have been contemporaneous. And, WT 17000 may make five.

Homo habilis

Somewhere in the taxonomic tangle of *Australopithecus* is the first member of the genus *Homo,* but just exactly how the australopithecines and *Homo* are related is also at issue. The matter began in a storm of protest with the announcement by Louis Leakey in 1962 of the discovery, also at Olduvai, of a hominid that had lived about 1.75 million years ago, about the same time as *Zinjanthropus.* Because of its larger brain size, Leakey assigned it to the genus *Homo,* and because of its possible association with early stone tools dating from the same period, he called his find *H. habilis,* "handy man." The genus *Homo* suddenly tripled in age, and members of this genus were, according to Leakey, contemporaries of the australopithecines. This idea was not accepted easily. Opponents of Leakey's proposal insisted on classifying the new fossil as another example of *A. africanus.* Leakey, with his customary tact, accused his colleagues of "psychosclerosis." Finally, however, enough new specimens were found to grant this group of fossils a place on the family tree. The most convincing was skull 1470 (Figure 50–17), uncovered by Richard Leakey's group in northern Kenya. Skull 1470, dated at 1.9 million years ago, is large and lightly built, with a cranial capacity of about 775 cubic centimeters, about 50 percent larger than that of the robust australopithecines. The bony crest along the skull, so prominent in the robust australopithecines, is absent.

In 1986, at Olduvai, Tim White of Johanson's group found a specimen (OH 62) dated at about 1.8 million years ago. The fossils, which are those of an adult female, are unusual because limb fragments as well as skull fragments were found. The skull and dentition of OH 62 appear similar to other specimens of *H. habilis,* such as skull 1470. However, the limb fragments reveal that she was only 90 to 110 centimeters (3 to 3.5 feet) tall, about the same size as Lucy, and that she had long arms, like Lucy's, despite the fact that more than a million years separate the two.

As Louis Leakey first noted, stone tools occur in abundance in the same strata as fossils of *H. habilis.* These Oldowan tools (so called because they were first found in Olduvai) include small sharp slicers, flaked off from a larger stone, and simple pounders and choppers fashioned from the stone's core (Figure 50–18). These were used, it is speculated, for preparation of vegetable foods, hunting of small game, and perhaps the butchering of larger animals. Based on the analysis of animal remains and the lack of weapons for killing larger game, it is believed that such animals were scavenged rather than hunted.

Specimens attributed to *H. habilis* are found in strata dated at 2 million to 1.5 million years ago. The origin of this species is not yet clear. Proponents of *A. afarensis* believe that species to be directly ancestral to *H. habilis.* Others contend that *H. habilis* coexisted with other species at Hadar and Laetoli. A third possibility is that *A. africanus* was the progenitor. Some argue that *H. habilis* should be classified as an australopithecine, a point of view supported by the small, primitive skeleton of OH 62. New data and revised models may well have emerged by the time you finish this chapter.

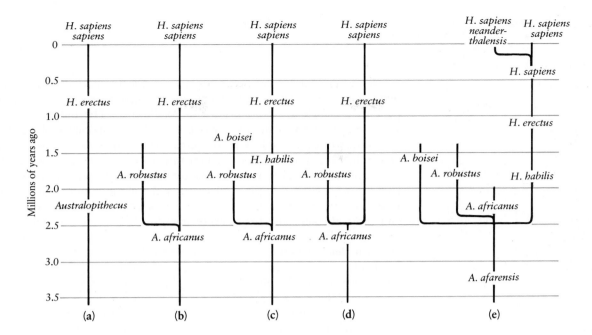

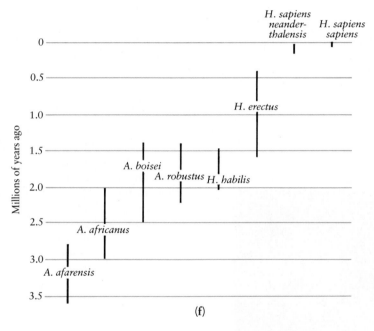

50–19 *Several models of human origins.* **(a)** *Until little more than 20 years ago, the hominid line was generally considered as a single lineage evolving gradually from* Australopithecus *through* Homo erectus *to* Homo sapiens. **(b)** *Next, based on the fossil evidence then available, it became generally accepted that there were two types of australopithecines, a robust and a gracile (slightly built) form, and that the robust form represented an evolutionary dead end.* **(c-e)** *The questions now remain of the status of* H. habilis *and of* A. afarensis. *Note that the models have become increasingly "bushy."*

(f) *The punctuationalists (page 1028) are pleased to point out that the more fossils that are discovered, the more different contemporaneous hominid species are found to have existed. Selection among these species fits the evidence better, they point out, than a gradual phyletic change from one to another.*

New Concepts in Hominid Evolution

Although the difficulties concerning the naming of the early species and their phyletic relationships have not improved in recent years and, indeed, may have grown more complex, three new and widely accepted ideas have emerged from the chaos, and three previous models of hominid evolution have been discarded.

One major concept that has been tested and found wanting is the single-species hypothesis, which held that only one species of hominid existed at any one time and that there was a straight phyletic progression from the first ape to walk upright directly to modern humans. This was maintained by some experts until as recently as 15 years ago. It is now clear that hominid evolution, like that of *Equus* and other lineages known from the fossil record, was not a ladder of progress but a bush with many branches, and most of these branches led to extinction (Figure 50–19). Our success was not preordained, and our survival was chancy. Our evolutionary history is much like that of other species.

The second major idea to emerge from the fossil evidence is that it was bipedalism—the capacity to walk on two feet rather than four—that set us on the path to humanness, not our superior intelligence. The old idea was hard to overcome. An example of its strength is found in one of science's biggest scandals, the Piltdown forgery. Piltdown is the name of a town in Sussex, England, near which, in 1912, fragments of a remarkably human cranium and a clearly apelike jaw were found in a gravel pit. It was subsequently proved, but not settled until half a century later, that the skull was that of a modern human only about 2,000 years old and the jaw was that of an orangutan, doctored to make it look older and more human. The fossils were uncritically accepted as authentic largely because the big-brained cranium conformed exactly to what was expected of our ancestor. And it was for that same reason that the Taung child was widely ignored.

Finally, and also contrary to long-held belief, the selective pressures for bipedalism did not involve the "freeing of the hands" for tool use. As we have seen, bipedalism and the splitting in the hominoid line occurred more than a million years before the appearance of simple stone tools in the fossil record.

A major question in paleoanthropology now—perhaps *the* major question—is what selective pressures gave rise to bipedalism. Clearly the transition would have involved a sacrifice in mobility. Even today, with our stride more fully evolved, we humans are hard put to keep up with baboons, chimpanzees, and other larger anthropoids. Dart and most others in the 1950s and 1960s (including Robert Ardrey, who popularized this point of view) attributed human evolution to carnivory, bloodthirstiness, and violence, with bipedalism freeing man's hands to hold weapons for killing prey and routing his enemies, including other humans. Although remnants of this point of view still linger, it is no longer compatible with the evidence. Owen Lovejoy of Kent State University has proposed a quite different scenario: the early appearance of bipedal locomotion in humans was due to the selective advantage it gave to males who could then procure food at a distance and carry it back to females and young, thus leading to a male-dominated, tightly knit nuclear family. Some say that bipedalism was triggered solely by the change in climate that set in at the end of the Miocene, shrinking the African forests and causing the spread of open savanna. Others suggest a female cooperative, with bipedalism favoring those members better suited to carry children, collect and gather food, and transport it for sharing. Females, according to the fossil record, were smaller and more numerous than males. "A social pattern like that of lions, with females working communally to support a few big, lazy territory-patrolling males would fit what we know about *Australopithecus*," according to Matt Cartmill of Yale University.

To quote David Pilbeam once more: "Our theories have often said far more about the theorists than they have about what actually happened."

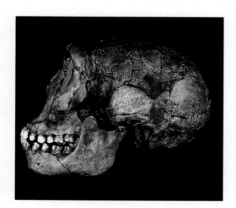

50–20 *The skull of the 12-year-old* Homo erectus *boy whose remarkably complete skeleton was discovered in 1984. Much larger than earlier hominids, H. erectus had a significantly larger brain. The skull walls are thick and heavy, and the brow ridges are prominent. The protruding occipital crest (clearly visible in this photograph) was the point of attachment of strong, heavy neck muscles. The jaw is relatively prognathous (protruding) and chinless.*

Homo erectus

The cause of his death is unknown, only that the body of the 12-year-old boy somehow found its way to the shores of a lake in northern Kenya. Here it was found 1.6 million years later by Kamoya Kimeu, long-time friend and colleague of Richard Leakey. It is both the oldest and the most complete specimen of *Homo erectus* yet discovered. The boy was surprisingly tall for his age, 165 centimeters (5 feet, 5 inches), and might well have reached 183 centimeters (6 feet) when fully grown. His skeleton was only subtly different from that of modern man. His skull, however, was much heavier, with the beetling brows and low forehead associated with other members of the species *H. erectus* (Figure 50–20).

Fossils of *H. erectus* were first found in Java in 1896 (Java man), and later, beginning in 1929, the fossils of more than 40 individuals were found in Peking

50–21 *The hand ax is a stone that has been worked on all its surfaces to provide what appears to be a gripping surface and various combinations of cutting edges, sometimes with a more or less sharp point. Hand axes came into use about 1.5 million years ago and are associated with* H. erectus.

(Peking man), followed by other discoveries in Africa, India, China, and Southeast Asia. (Fossils unequivocally assignable to *H. erectus* have not been found in Europe, although the distinctive tools associated with this species have.) Like the boy described above, these hominids had body skeletons much like our own and were about the same size that we are. The bones of their legs indicate that they had a stride similar to our own. The chief differences between *H. erectus* and *H. sapiens* are in the skull. *H. erectus* skull specimens are thick and massive with low foreheads. Their jaws and teeth are large (though smaller than those of *H. habilis*), and their chins are sloping. Brain capacity averages about 1,000 cubic centimeters, ranging from 700 to 1,200 cubic centimeters, overlapping that of modern humans.

Homo erectus had a new and highly distinctive tool, the hand ax (Figure 50–21). Tens of thousands of these have been found throughout Africa, Asia, and Europe. They all closely resemble one another, indicating the emergence of a cultural tradition in which skills and learning were passed from one generation to another. At some point, *H. erectus* also acquired the ability to control fire, which would have extended the range of their diet, not only making meat easier to chew but also, and perhaps more important, making it possible to eat plant parts that, uncooked, would have been too tough, too bitter, or too toxic. *H. erectus* was also the first of the hominids to inhabit the mouths of caves. Fire would have made such habitation safer, discouraging other cave dwellers, such as bears and saber-toothed cats, and also providing a social center.

Clearly, *H. erectus* was very different from the australopithecines and little *H. habilis*, which is what makes two of the latest fossil finds even more remarkable. Only 200,000 years separate the 12-year-old *H. erectus* from northern Kenya and the little adult female from Olduvai, yet they are astonishingly different. Was there an unprecedented spurt of evolution in those 200,000 years, as compared to the uneventfulness of the previous 1 million? Or did the little, long-armed *H. habilis* exist side by side with some other hominid ancestral to *H. erectus* and so to us? The answers to these questions may lie buried in the volcanic ash and dust of East Africa.

Homo sapiens

Homo sapiens is the name we give to the species of hominid that has not only a body skeleton much like ours but also a similar brain capacity (Figure 50–22). Three varieties, or subspecies, are commonly recognized: "archaic" *Homo sapiens*, which bears many resemblances to *H. erectus* and which some anthropologists believe should be a separate species; *Homo sapiens neanderthalensis*, another twig on the family bush; and *Homo sapiens sapiens*, the wisest of the wise, including, of course, ourselves.

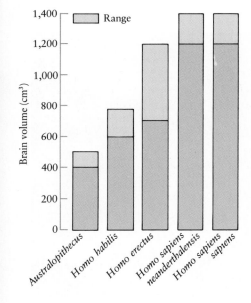

50–22 *Increase in brain volume in the course of hominid evolution. Some of the increase can be correlated with the increase in body size that was also taking place during this period, but most of it is believed to represent the results of strong selection pressures for intelligence.*

Archaic Homo sapiens

Fossils considered early, or archaic, *H. sapiens* are dated at 400,000 to 200,000 years ago. On the average, these individuals had more bulging foreheads, less prominent brow ridges, larger brains, and smaller teeth than the earlier *H. erectus*. Fossil materials of this group are scarce and fragmentary, and dates are uncertain. Most of the early *H. sapiens* fossils have been found in Europe (early Neanderthals), and some come from Asia and Africa. As we shall see, the place of these fossils in human evolutionary history is yet another matter under debate.

Homo sapiens neanderthalensis

The period from about 150,000 years ago to about 35,000 years ago abounds with specimens of what we have come to call the Neanderthals. They have been found largely in Europe but also in the Near East and central Asia, extending as far east as Afghanistan. The Neanderthals stood as erect as we do but were more heavily built and more muscular, with certain skeletal peculiarities. They had a brain capacity at least as large as ours, a long, low, massive skull, a prognathous (protruding) face, a low forehead, and heavy brow ridges. They are now usually classified as a variety of *H. sapiens*.

Neanderthals used hand-held stone tools that were much more sophisticated than those of *H. erectus*. Some of the stone tools appear to have been used for scraping hides, suggesting that Neanderthals wore clothing made of animal skins, which would certainly have been in keeping with the climate in which they lived.

Neanderthals buried their dead (Figure 50–23), often with food and weapons and, in at least one instance, with spring flowers. Formal burials such as these suggest a belief in life after death.

(a)

(b)

50–23 (a) *This large cave near the little village of Shanidar in northern Iraq has been continuously inhabited for more than 100,000 years. Nine Neanderthal skeletons have been found in the cave, including one who was, according to analysis of* fossil pollen, buried on a bed of woody branches and June flowers gathered from the hillside.

(b) *The grave of a young Neanderthal man, excavated in 1983 on Mount Carmel in Israel. After his death, estimated to* have occurred some 50,000 years ago, the flesh was allowed to decay and the cranium was removed before the remainder of the skeleton was covered with dirt.

50-24 *Cro-Magnon culture was characterized by a great increase in the types of specialized tools made from long and relatively thin flakes with parallel sides, called blades. This beautifully worked "laurel leaf" blade, fashioned from flint, was used as the point of a spear. Such blades are often more than 30 centimeters long and only 0.5 centimeter thick.*

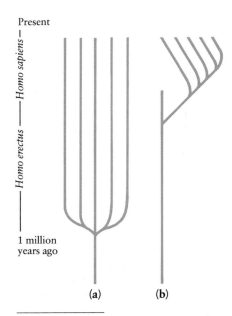

Present

Homo sapiens

Homo erectus

1 million
years ago

(a)   (b)

50-25 *Two models of the evolution of modern humans. (a) According to the candelabra hypothesis, populations of modern humans split off from earlier forms perhaps as much as 1 million years ago and evolved in parallel with one another. (b) The Noah's Ark model, by contrast, proposes a single, more recent source for modern humans with the extinction (for reasons unknown) of the earlier human forms.*

During the relatively short span of about 100,000 years, the Neanderthals spread all across Europe, the Middle East, and western and central Asia. They were contemporaries of forms of archaic *H. sapiens* found as far away as China and South Africa. Then they all disappeared abruptly, some 30,000 years ago.

Homo sapiens sapiens

All hominid fossils of the last 30,000 years are of anatomically modern humans, *H. sapiens sapiens*. Early Europeans of this type are commonly called Cro-Magnon, after the site in southwestern France where they were first discovered. However, much earlier modern populations are known from the Middle East; for example, representative fossils dated at 92,000 years ago have been found at Qafzeh, in Israel.

The Cro-Magnons, when they first appeared in Europe, came bearing a new, quite different, and far better tool kit (Figure 50–24). Their stone tools were essentially flakes—which, of course, had been in use for more than 2.5 million years—but they were struck from a carefully prepared core with the aid of a punch (a tool made to make another tool). These flakes, usually referred to as blades, were smaller, flatter, and narrower, and, most important, they could be and were shaped in a large variety of ways. They included, from the beginning, various scraping and piercing tools, flat-backed knives, awls, chisels, and a number of different engraving tools. Using these tools to work other materials, especially bone and ivory, Cro-Magnons made a variety of projectile points, barbed points for spears and harpoons, fishing hooks, and needles. They also produced some of the most exuberant and creative works of art in all of human history (see essay).

The Origin of Modern Humans

It is now generally accepted that the early hominids—all the australopithecines and *H. habilis*—evolved in Africa over a period of at least 3 million years. A major open question, however, concerns the events of the last few hundred thousand years, the period during which the evolution of modern humans took place. As we noted earlier, *H. erectus* fossils have been found not only in Africa but also abundantly, with a somewhat later appearance, in Asia and Europe. Until recently, on the basis of these fossil findings, many experts have subscribed to the so-called candelabra model of modern evolution (Figure 50–25a). According to this model, there were multiple early migrations from Africa, beginning perhaps as long as a million years ago, and these migrations established different populations of *H. erectus* that evolved separately into the different modern human races. Similarities among them were maintained, and speciation was prevented, by gene flow.

Opposed to the candelabra hypothesis is the Noah's Ark model (Figure 50–25b). According to this model, a small group of already modern humans, starting in one place, colonized the entire world, like the survivors on Noah's Ark. Some fossil findings support this point of view. For instance, the oldest unequivocally modern human skulls, dated at about 100,000 years ago, have been found in southern and eastern Africa. The next oldest are the group from Qafzeh in Israel, on the corridor to the Old World, which were found in the 1930s and have recently been dated at 92,000 years. The modern humans of Europe show up about 40,000 years ago. (Either we took a long time making our way northward, or earlier evidence has yet to be discovered.) Also, no fossils have been found that appear to be unequivocally intermediate between *H. erectus* or the Neanderthals and modern humans. Finally, the limb proportions of the first modern humans in Europe seem to be typical of equatorial people, not people adapted to cold climate, as the Neanderthals were.

The Art of the Caves

The cave paintings of western Spain and southern France, many surprisingly untouched by time, are part of a rich artistic tradition that endured for at least 20,000 years. Most depict animals, nearly all of them game animals, and the paintings are deep within the caves, so they must have been viewed (as they must have been painted) by the light of crude lamps or torches.

The meaning of these drawings and paintings has long been a matter of debate. Some of the animals are marked with darts or wounds (although very few appear to be seriously injured or dying). Such markings have led to the suggestion that the figures are examples of sympathetic magic, in which there is the notion that one can exert control over another creature by taking symbolic action against its image. The fact that some of the animals appear to be pregnant suggests that they may symbolize fertility. Many appear also to be in motion. Perhaps these animals, so vital to the hunters' welfare, were migratory in these areas, and they may have seemed to vanish at certain times in the year, mysteriously returning, heavy with young, in the springtime. This return of the animals might have been an event to be solicited or celebrated in much the same spirit as the rites of spring or Easter are celebrated by more recent peoples. Whatever their meaning, these images touch us, like the footprints at Laetoli and the burial at Shanidar, with a sense, across time, of sharing something of what it means to be

human. Paleolithic cave art in Europe came to an end perhaps 8,000 to 10,000 years ago, with the end of the last Ice Age. Not only were the tools and pigments laid aside, but the sacred places—for such they seem to have been—were no longer visited.

The Noah's Ark model is now receiving support from molecular biologists. The molecules in question are not the proteins and the chromosomal DNAs that are proving of such value in resolving other taxonomic puzzles. These accumulate change too slowly to provide any information about events so recent in evolutionary history. Rather, the molecules under scrutiny are the genes of the mitochondria. As we saw in Chapter 18, these organelles have their own DNA. Like chromosomal DNA, mitochondrial DNA slowly accumulates random changes over time, and these changes, like the changes in nuclear DNA, can serve as a measure of time and so of evolutionary distances. There are two important differences, however. First, according to studies of many diverse animal groups whose evolutionary backgrounds are well established, the mitochondrial clock ticks about 10 times faster than the nuclear clock. (The calibration of the clock is

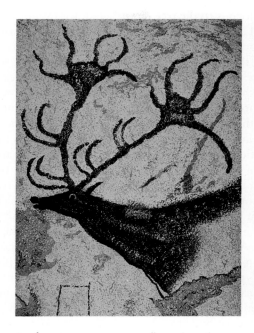

In almost every cave, a small number of the animals show wounds. Although the animals are shown in realistic detail, the weapons and the hunters, if shown at all, are abstractions, perhaps to keep their identity a secret. The illusion of movement is greatly enhanced by the patterns of light and shadow in the dark, narrow recesses of the cave.

still a matter of some dispute, however.) Because the clock ticks faster, it is more suitable for tracking more recent events, those measured in thousands of years rather than in millions of years. Second, in vertebrates, all mitochondria are of maternal origin, thus eliminating the possibility that changes in nucleotide sequence are the result of sexual recombination rather than the ticking of the clock. A pair of breeding individuals can transmit only one type of mitochondrial DNA, but they carry four haploid sets of transmissible nuclear genes. You have 16 great-great-grandparents, but all of your mitochondrial DNA comes from only one. As a consequence, mitochondrial lineages are easier to trace.

Rebecca Cann (Figure 50–26) and her colleagues from the University of California, Berkeley, collected mitochondrial DNA from 147 individuals in five different geographic populations—Africa, Asia, Europe, Australia, and New

50–26 *Analysis of the mitochondrial DNA of women from four continents has led to the conclusion that modern humans left Africa less than 200,000 years ago. This research was carried out by Rebecca Cann, shown here, and her colleagues at the University of California at Berkeley.*

Guinea. This DNA was cut with restriction enzymes, and fragment patterns were determined and compared. Based on these, the biologists constructed an evolutionary tree, analogous to the tree constructed, for example, from cytochrome *c* data (Figure 20–11, page 419). The important difference, however, was that it was based on variations within a species rather than among diverse taxa.

The results of the mitochondrial tree were unexpected on several counts. First, at the base of the tree was a single common ancestor, who came to be popularly known as "mitochondrial Eve." This does not mean that there was a single female from whom we are all descended, but rather that out of a population numbering perhaps several thousand, by chance, only one set of mitochondrial genes was passed on. (This finding, perhaps the most surprising to us, is the least disputed by population geneticists and others familiar with genetic drift and other manifestations of the laws of probability.) Second, this female lived only about 200,000 years ago (give or take 50,000 years); this finding is based on the known mutation rate for mitochondrial DNA. Third, she lived in Africa. The basis for this conclusion is that the African population is more diverse than the others, indicating simply that it has been around longer. Fourth, based on the degree of variation among the non-African groups, the original founder populations left Africa probably a little more than 100,000 years ago. Finally, there is no evidence of any introduction of new mitochondrial DNA into either the populations that remained in Africa or those that colonized the other continents.

If these new findings are supported by further data, we have a clear picture of a population of modern humans who were in existence in Africa about 200,000 years ago, who began migrating through Asia into Europe about 100,000 years ago, and who replaced the other resident human populations—*H. erectus* and *H. sapiens neanderthalensis*—upon their arrival.

The implications of these new findings are variously interpreted. According to Stephen Jay Gould, "It makes us realize that all human beings, despite differences in external appearance, are really members of a single entity that's had a very recent origin in one place. There is a kind of biological brotherhood that's much more profound than we ever realized." (Some of us might say sisterhood.) On the other hand, there is the unsettling question of the disappearance of *H. erectus* populations and of the Neanderthals. One possible explanation is that the migrating modern humans brought with them diseases to which the *H. erectus* and Neanderthal populations were not immune. However, Milford Wolpoff of the University of Michigan notes that if the new model is true—and he believes it is not—the appropriate biblical symbol of human evolution may not be Eve or Noah, but rather Cain.

In short, since its beginnings, the study of the history of human evolution has compelled us not only to look at the dry bones of our ancestors but also to examine what it means to be human. Depending upon our interpretations of the past and our views of the present, we can find in our biological history the justifications for our present shortcomings or the hopes for our salvation. Alternatively, we may conclude that, based on present knowledge, human behavior in the past appears to have been as complex and contradictory as we know it to be in the present and so has little bearing on our future. Under these circumstances, the best we can do is to rely on our recently increased cranial capacity and make our own choices among the bewildering alternatives.

SUMMARY

The first mammals arose from primitive reptilian stock about 200 million years ago and coexisted with the dinosaurs for 130 million years. The extinction of the dinosaurs was followed by a rapid adaptive radiation of the mammals. The primates are an order of mammals that became adapted to arboreal life. Primates are characterized by five-digited extremities adapted for grasping, nails rather than claws, and freely movable limbs. They are more dependent upon vision than upon smell, and the higher primates all have stereoscopic vision with foveas for fine focus and cones for color vision.

The two principal groups of living primates are the prosimians and the anthropoids. Prosimians were widespread and abundant during the Paleocene and Eocene epochs, some 65 to 38 million years ago. Modern prosimians include lorises, bush babies, lemurs, and tarsiers. The anthropoids include the New World monkeys, the Old World monkeys, and the hominoids (apes and humans). According to present evidence, the closest living relatives of the hominids—the members of the human family—are the chimpanzees.

From more than 3.6 million years ago to at least 1.4 million years ago, groups of hominids lived that, although they were small and their skulls were apelike, walked erect. Some used simple pebble tools. At least five species are now widely recognized: *Australopithecus afarensis, Australopithecus africanus, Australopithecus robustus, Australopithecus boisei,* and *Homo habilis.*

Homo erectus lived from at least 1.6 million to 300,000 years ago. Individuals were tall, with body skeletons closely resembling those of modern humans, but their skulls were much heavier, with flattened crania, heavy brows, large teeth and jaws, and a sloping chin. The hand ax is associated with *H. erectus.* Some groups at least occasionally occupied caves and, at later stages, certainly had fire, two developments that are probably related.

The species *Homo sapiens* comprises "archaic" *Homo sapiens, Homo sapiens neanderthalensis,* and *Homo sapiens sapiens.* Archaic *H. sapiens* fossils date from 400,000 to 200,000 years ago and indicate that these individuals had larger brains and smaller teeth than the earlier *H. erectus.* Neanderthal fossils date from about 150,000 to 35,000 years ago, the period of the last glacial advance. The majority of specimens have been found in Europe. Neanderthals had fire, inhabited caves, hunted large animals, and probably wore some sort of clothing. They used stone tools of a characteristic type. They buried their dead, sometimes with food and weapons. Neanderthals disappeared some 30,000 years ago.

The Cro-Magnons, anatomically modern humans, replaced the Neanderthals. The origin of these modern humans is under dispute. Modern biochemical evidence suggests that modern humans evolved in Africa and migrated from there only about 100,000 years ago, replacing previous populations of the genus *Homo* as they went.

QUESTIONS

1. Distinguish among the following: primate/prosimian; monkeys/apes; hominids/hominoids/anthropoids.

2. Name five evolutionary trends among primates, and discuss the probable selective value of each.

3. Suppose you were to meet a gathering of the following: *Homo erectus, Homo sapiens sapiens,* a Cro-Magnon, Peking Man, *Homo sapiens neanderthalensis,* Lucy, Zinjanthropus, *Australopithecus robustus,* and *Australopithecus africanus.* How would you distinguish one from the other?

Animal Behavior and Its Evolution

51–1 *A behavior pattern found in many young birds is a thrusting of the head upward toward the parent, with the mouth gaping open. A complementary behavior of the parent is the regurgitation of food into the gaping mouth of a nestling. When nestlings first hatch, their gaping response is stimulated by a light touch on the nest or a puff of air. As their vision improves, the response is stimulated by moving objects about the size and shape of the parent. In the bearded-tit, shown here, the parent's response also depends on conspicuous, species-specific markings in the mouth. These behavioral patterns are innate—that is, genetically determined—and require no learning process. As you might expect, they are subject to stringent natural selection: a baby bird that does not gape widely is not fed, and a parent that fails to respond to the gaping mouths of its young or that responds to the young of a different species will make no contribution to the gene pool of the next generation.*

As we have seen in numerous examples, organisms ranging from simple prokaryotes to complex animals *act:* they seek out suitable environments in which homeostasis can be maintained; they obtain nutrients and other essential molecules, such as water and oxygen; often they successfully avoid becoming nutrients for other organisms; and they produce offspring, a process that can involve complex patterns of courtship, mating, and parenting. These activities, along with many others, constitute behavior.

The behavioral characteristics of an organism—its sensitivity to particular stimuli and its patterns of response to those stimuli—are as much the products of natural selection as the shape of a tooth or a feedback loop that regulates blood pressure. An organism's behavior is vitally important for its survival and the successful production of offspring. The factors governing the evolution of behavioral characteristics are the same as those that apply to any other trait. First, there are variations among individual organisms in behavioral characteristics. Second, some of these variations are genetically determined; that is, they are influenced by the presence of particular alleles or combinations of alleles. Third, individuals with certain variations tend to have greater reproductive success than individuals with certain other variations. Fourth, as a result of this greater reproductive success, certain alleles tend to increase in the gene pool of a population, generation after generation.

The comparative study of patterns of behavior and the construction of hypotheses concerning their evolutionary origins is known as **ethology,** from the Greek word *ethos,* meaning "character" or "custom." The pioneers in this field have been the European zoologists Konrad Lorenz, Niko Tinbergen, and Karl von Frisch. At present, the study of behavior and its evolution is one of the most active and exciting areas of biological research. The scope of the field is enormous, ranging from laboratory studies of bacteria to long-term studies of animals in the wild. The students of behavior include not only ethologists but also biochemists, geneticists, developmental biologists, neurobiologists, psychologists, anthropologists, evolutionary biologists, and behavioral ecologists. An adequate introduction to the range of research under way and the knowledge gleaned thus far would require another book as long as this one. We shall, of necessity, limit ourselves to a few of the most fascinating areas of this vast subject.

THE GENETIC BASIS OF BEHAVIOR

Behavior—whether that of an *Escherichia coli* cell navigating toward a food source (see page 131) or yours in reading and reacting to this text—has its roots in the genetic program carried in the DNA molecules of the individual. The steps between a sequence of nucleotides in a DNA molecule and the behavior of an organism, even a very simple one, are many and complex. The process involves, at

51-2 *Four honey-bee pupae in different stages of their metamorphosis. Honey bees undergo their entire development from egg to adult in individual wax cells within the hive. During the early stages of development, the cells are open, enabling worker bees to feed the larvae continuously. When a larva has reached the size at which pupation begins, the workers add a wax cap to the cell, sealing in the pupa for its metamorphosis. A fully developed worker bee at the left has shed her pupal skin and is ready to emerge, which she will do by gnawing through the wax cap.*

51-3 *An "undertaker" honey bee, carrying the corpse of a dead bee away from the hive. Recent research suggests that in naturally occurring honey-bee colonies only about 1 percent of the workers exhibit this behavior.*

the least, the synthesis of specific molecules, their organization into particular structures capable of receiving and responding to stimuli, the development of pathways—either electrical or chemical—for the transmission of information within the organism, and the modification of those structures and pathways as a result of interactions of the organism with its environment.

For some behaviors, the role of genetic components can be clearly demonstrated. A classic example is the behavior of honey bees when a pupa (Figure 51-2), metamorphosing within its cell in the hive, becomes diseased and dies. In some honey-bee strains, known as hygienic, worker bees uncap the cell and remove the corpse. In other strains, known as unhygienic, this behavior does not occur. In the early 1960s, Walter Rothenbuhler, in a series of breeding experiments, first crossed unhygienic bees with hygienic bees. All of the offspring were unhygienic, indicating that this is a dominant characteristic. Rothenbuhler then made a testcross of these hybrid offspring with the original hygienic strain. Twenty-nine colonies of bees resulted. In eight colonies, the bees were unhygienic; they left the cells capped and did not remove dead pupae. Six colonies consisted of hygienic bees that uncapped the cells and removed dead pupae. In nine of the colonies, however, worker bees uncapped cells but left the dead pupae untouched, and in the remaining six colonies, they did not uncap cells but would remove the dead pupae if the cells were uncapped for them.

These results approximate the 1:1:1:1 ratio expected in a testcross involving two independently assorted genes (see page 243). They indicate that one gene, designated U, controls the uncapping behavior, whereas another gene, designated R, controls the removal behavior. Bees that are homozygous recessive for both genes (*uurr*) uncap cells and remove pupae. Bees that have at least one copy of the dominant allele for each gene (for example, *UuRr*) do neither. Bees that are *uuRR* or *uuRr* uncap the cells but do not remove the pupae, and bees that are *UUrr* or *Uurr* remove pupae after cells are uncapped for them.

As this example illustrates, a single allele may determine a particular behavioral characteristic of an organism. For most behaviors, however, the underlying genetics are considerably more complicated. For some behaviors, genes with multiple effects (that is, pleiotropic genes) play an important role. Fruit flies carrying the allele *Hk*, for instance, are more active than flies lacking this allele, jump violently when a shadow passes over them, exhibit abnormal mating behavior, and undergo rapid leg movements when anesthetized. As you might expect from their frenetic life style, they also have a shorter life span than flies without the allele. Most often, behavioral characteristics are the result of polygenic inheritance—that is, they depend on the integrated action of the alleles of a large number of genes.

PROXIMATE AND ULTIMATE CAUSATION

Imagine a pond on a spring day, with numerous tadpoles schooling near the water's edge. If you walk quietly along the bank of the pond and then make a sudden, loud noise, you will notice that the tadpoles immediately arch their bodies and shoot away into deeper water. Similarly, if you tap on the side of an aquarium in a pet shop, the fish also arch their backs, darting to the other side of the aquarium. Why do the tadpoles and the fish exhibit this behavior? This question, and similar questions about a vast diversity of behaviors exhibited by organisms, can be approached from two different points of view. One approach seeks an answer in terms of the **proximate cause**—that is, the immediate sequence of physiological events that lead to the observed behavior. Another approach seeks to explain the **ultimate cause** of the behavior—that is, its adaptive value and its evolutionary origins.

In the example of the startle behavior exhibited by many fish and larval amphibians, the proximate cause is stimulation of two giant neurons, known as Mauthner cells. The cell bodies of these neurons lie within the medulla, one on either side of the midline of the brain. Each neuron has two large dendrites that synapse with the acoustic nerve and a single giant axon that crosses to the opposite side of the brain and then descends the full length of the spinal cord. Each of these giant axons synapses with motor neurons in every segment on its side of the body. Stimulation of the Mauthner cells results in the firing first of one cell and then of the other, producing powerful contractions of the muscles of the body wall and tail that propel the animal forward.

The ultimate cause of the behavior, however, is to be found in its adaptive value in preserving the animal from predation. Individuals with a highly efficient response are more likely to avoid predation—and thus to survive to maturity and successfully reproduce—than those with a less efficient response.

FIXED ACTION PATTERNS

Some behavior patterns—for example, the gaping of a baby bird's mouth, the suckling of a human infant, or the tongue flick of a toad attempting to capture a fly—develop with a minimum of sensory experience. Although the behavior may be subsequently refined, the pattern appears, essentially complete, the first time the organism encounters the relevant stimulus. Such an innate behavioral pattern, which tends to be highly stereotyped, rigid, and predictable, is known as a **fixed action pattern.** The fixed action patterns of the members of a given species of the appropriate age, sex, and physiological condition are as specific and constant as their anatomical characteristics.

Fixed action patterns are initiated by external stimuli, known as **sign stimuli.** When these stimuli are communication signals exchanged between members of a species, they are known as **releasers.** Tinbergen hypothesized that certain specific areas within the brain, which he termed **innate releasing mechanisms,** respond to the sign stimuli. A fixed action pattern remains blocked until the organism encounters the appropriate sign stimulus, which then stimulates the innate releasing mechanism, setting in action the sequence of movements that constitute the behavior.

An example of a fixed action pattern studied by Lorenz and Tinbergen is the egg-rolling behavior of greylag geese. The nests of these geese are shallow depressions in the ground. If an egg rolls out of her nest, a goose will retrieve it using a stereotyped sequence of movements (Figure 51–4). First, she stretches her neck toward the egg and then begins rolling it back under her chin. Through a series of lateral movements of her head, she keeps the egg rolling back on a straight line toward the nest. Occasionally, however, the egg slips away to the side—but the goose continues the retrieval movements all the way back to the nest, even though the egg is no longer there to retrieve.

In most cases, once a fixed action pattern has been initiated, it cannot be altered but must be carried through to completion. With many fixed action patterns, the initial orientation of the organism is crucial. Consider a toad attempting to capture a fly; if it is to succeed, its body must be oriented correctly with respect to the position of the fly. If the toad is incorrectly oriented—or if the fly moves after the tongue flick has begun—the toad will miss.

Through the use of physical models, ethologists have been able to identify the releasers for a number of fixed action patterns. For example, the red bellies of male sticklebacks in breeding condition are releasers not only for the stereotyped mating behavior of stickleback females (see page 1016) but also for aggressive interactions between the males. Tinbergen constructed a series of models (Figure

51–4 *The fixed action pattern by which a greylag goose retrieves an egg that has rolled from her nest. Because an egg is not a perfect sphere, a series of lateral movements of the goose's head are required to keep the egg on a straight course back to the nest.*

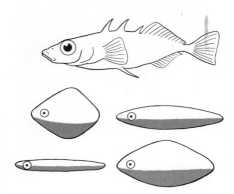

51–5 *Colors are the sign stimuli for many fixed action patterns in different animal species. In experiments by Niko Tinbergen with sticklebacks, the crude models, painted red on the ventral surface, elicited much stronger reactions in both male and female sticklebacks (aggressiveness in the males and attraction in the females) than did the exact replica of a male, which was not colored. Tinbergen, Lorenz, and von Frisch shared a Nobel Prize in 1973 for their work in animal behavior.*

51–6 *A female Belding's ground squirrel who has detected danger—most likely a potential predator—and is sounding the alarm. Ground squirrels who give alarm calls are more likely to be the victims of predators than those who hear the calls and retreat to the safety of their burrows. How such seemingly altruistic behavior could evolve is an intriguing question, to which we shall turn our attention later in this chapter.*

51–5), by which he was able to show that a male stickleback reacts much more aggressively to a very crude model of a male with a red belly than to an exact replica of a stickleback without the red coloration.

LEARNING

All patterns of behavior, even those that seem relatively complete on their first appearance, depend not only on environmental cues but also on the normal physiological development of the animal. An enormous amount of animal behavior also requires learning—a process in which the responses of the organism are modified as a result of experience. The capacity for learning appears to be loosely correlated with the length of the life span and with the size and complexity of the neural circuitry, particularly as concentrated in a brain. In small organisms, such as honey bees, with a short life span (and thus little available time for learning), most behavior appears to take the form of fixed action patterns. By contrast, in large organisms with a complex brain and a long life span, such as the primates, a large proportion of the behavioral repertoire is critically dependent on the prior experience of the individual.

One of the simplest forms of learning is habituation, in which an organism comes to ignore a persistent stimulus and go on about its other business, wasting neither time nor energy on a response. Examples of habituation that we have considered previously are the behavior of an amoeba when confronted by a light from which it cannot escape (page 474) and the response of *Aplysia* to repeated, gentle touching (page 895). Another example is provided by ground squirrels, which come to ignore the alarm calls of other ground squirrels if they are given frequently when there is, in fact, no danger. Other forms of learning are considerably more complex.

Associative Learning

A type of learning familiar to us all is association, in which one stimulus comes to be linked, through experience, with another one. If you keep pets in your home, you will be able to cite many examples of associative learning, such as goldfish coming to the corner of the aquarium to be fed as you walk toward the tank or your dog becoming excited at the sight of a leash. The first scientific studies of associative learning were performed in the 1920s by the Russian physiologist Ivan Pavlov. In his original **classical conditioning** experiments, Pavlov restrained a hungry dog in a harness and offered it small portions of food at regular intervals. When he signaled the delivery of food by preceding it with an external stimulus, such as the sound of a bell or a signal light, the dog began to respond to the external stimulus by salivating. After every few trials, the external stimulus was presented without the food, and the amount of saliva produced by the dog was measured. Pavlov found that the number of drops of saliva triggered by the external stimulus alone was directly proportional to the number of previous trials in which it had been followed by food. Pavlov referred to the food as an **unconditional stimulus,** since it triggered salivation—an **unconditional response** —in the absence of any other stimulus. The external signal, however, was a **conditional stimulus,** evoking a response—a **conditional response**—only after the dog had been conditioned to associate the signal with the reward of food.

In natural settings, associative learning often involves trial and error. A newly hatched chick, for example, will peck at any spots that contrast with the background; as it grows older, it learns which of these spots represent edible objects. Also, its pecking accuracy improves. When young chicks first peck at grain, only about 15 percent of the pecks lead to obtaining a morsel worth swallowing.

51–7 *Konrad Lorenz and followers. Many species of precocial birds (birds that are able to walk and feed as soon as they are born) follow the first moving object they see after hatching. The object is usually mother goose (for goslings, that is) but can be a matchbox on a string or a member of another species. In many species, this phenomenon, an example of the type of learning that occurs during a brief critical period and is known as imprinting, also influences mate selection in the adult bird.*

Accuracy improves with practice and also with maturity. Chicks that have been kept in the dark and fed by hand for a number of days peck more accurately than younger chicks, but they do not peck as accurately as chicks their own age that have had pecking experience. Trial-and-error learning is also known as **operant conditioning,** since the animal learns through operational experience to associate its behavior with the consequences of the behavior. For example, a laboratory rat that learns to run a maze to a goal box receives food as a reward, but if it makes any mistakes, it never obtains the food.

Imprinting

Closely related to associative learning is the development of discrimination. Of vital importance for the ultimate reproductive success of many animals is the discrimination of members of one's own species from members of all other species, a discrimination that may be based on a variety of cues. In many species, particularly birds, this learning occurs very rapidly during a specific **critical period** in the early life of the individual and depends on exposure to particular characteristics of the parent or parents. This type of learning is known as **imprinting.** The most familiar example of imprinting is the following response of many precocial birds, which keeps the young birds close behind and well within the protective range of the parent until the end of their juvenile period, when the response is lost. Imprinting is also involved in song learning in birds, a process of considerable complexity.

The Song of the White-Crowned Sparrow

About 150 days after hatching, a juvenile male white-crowned sparrow begins to utter a tentative, twittering call that has only a vague resemblance to the full song of the mature male. During the next 50 days or so, the juvenile's song gradually becomes more complex and sophisticated until, by the time the bird is 200 days old, he is singing a full song. His song not only identifies him as a white-crowned sparrow but also often provides information as to the specific locality from which he comes; in other words, the song patterns differ slightly from place to place, forming characteristic dialects. These dialects can be detected by human observers and reproduced by good whistlers such as ethologist Peter Marler of Rockefeller University.

Laboratory experiments undertaken by Marler and others have dissected the sequence of events in song learning in this species. Figure 51–8 shows sound spectrograms of (a) a wild bird, and (b) an adult, hand-reared bird that had never heard a song of its own species. Obviously some form of learning is required for production of the full song. Exposing birds to tape-recorded songs at various stages of their development revealed that exposure to the full song of the white-crowned male in a critical period of 10 to 50 days after hatching is required for development of the same song (in the same dialect) five to six months later.

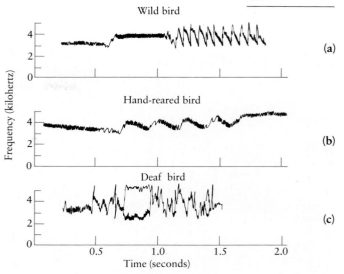

51–8 *Sound spectrograms for male white-crowned sparrows:* (a) *wild;* (b) *reared by hand but isolated from songs of other white-crowned sparrows;* (c) *deafened prior to five months of age (before he himself began to sing), after having heard normal white-crowned sparrow songs during the critical period. The unit in which the frequency (pitch) of the sound is expressed is the kilohertz; 1 kilohertz equals 1,000 cycles per second.*

This is true even if the bird is kept in complete sound isolation after two months of age. If, however, he does not hear the song until about the time he begins to sing himself, he can never produce a full song. Exposure to the song of other sparrow species during the critical period has no effect.

If a bird is deafened after he has heard songs of his own species during the critical period, the full song is not normal, though it is more complex than that of the sound-isolated bird (Figure 51–8c). Apparently, then, for normal song, three requirements must be met: (1) the bird must have the genetic capacity to recognize and reproduce the song, (2) he must hear the song during the critical period for imprinting, and (3) he must be able to hear himself sing the song. Apparently as he sings, he compares his own song to the song stored in his memory during the critical period, rehearsing until he gets it right. If a bird is deafened after he has mastered the full song, he continues to sing normally.

Imitative Learning

As the song of the white-crowned sparrow indicates, imitation is often a component of learning. Some of the most dramatic examples of learning through imitation have occurred in response to human activities. For instance, during the 1950s, titmice (small birds that are closely related to chickadees) began removing the lids (paper or foil) of milk bottles left on doorsteps in England and drinking off the cream. (At that time, milk was generally not homogenized, and the cream floated to the top.) The tearing of paper and paperlike bark is a common behavioral characteristic of this species, so one can guess that the first opened bottle may have been a happy accident. At least 11 species of birds subsequently took up the practice. The milk bottles were usually attacked within a few minutes after they were left at the door, and it was reported that troops of birds were following milk carts down the street.

A well-documented example of "monkey see, monkey do" is provided by the macaques (rhesus monkeys) of the Japanese island of Koshima. These primates lived and fed in the inland forest until about 30 years ago, when a group of Japanese researchers began throwing sweet potatoes on the beach for them. The group quickly got used to venturing onto the beach, brushing sand off the potatoes, and eating them. One year after the feeding started, a two-year-old female the scientists had named Imo was observed carrying a sweet potato to the water, dipping it in with one hand, and brushing off the sand with the other. Soon, other macaques began to wash their potatoes, too. Only macaques that had close associations with a potato washer took up the practice themselves. Thus it spread among close companions, siblings, and their mothers, but adult males, which were rarely part of these intimate groups, did not acquire the habit. However, when the

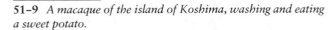

51–9 *A macaque of the island of Koshima, washing and eating a sweet potato.*

young females that learned potato washing matured and had offspring of their own, all of them learned potato washing from their mothers. Today all of the macaques of Koshima dip their potatoes in the salt water to rinse them off, and many of them, having acquired a taste for salt, dip them between bites.

This was only the beginning. Later, the scientists began scattering wheat kernels on the beach. Like the others, Imo, then four years old, had been picking the grains one by one out of the sand. One day she began carrying handfuls of sand and wheat to the shore and throwing them into the water. The sand sank, the wheat kernels floated to the top, and Imo collected the wheat and ate it. The researchers were particularly intrigued by this new behavior since it involved throwing away food once collected, much less a part of the macaques' normal behavioral repertoire than holding on to food and cleaning it off. Washing wheat spread through the group in much the same way that washing potatoes had. Now the macaques, which had never even been seen on the beaches before the feeding program began, have taken up swimming. The youngsters splash in the water on hot days. Some of them dive and bring up seaweed, and at least one has left Koshima and swum to a neighboring island, perhaps as a cultural missionary.

SOCIAL BEHAVIOR: AN INTRODUCTION

Of all behaviors, perhaps the most intriguing are the interactions that occur among animals, such as the macaques, living in structured societies. A **society** is a group of individuals of the same species, living together in an organized fashion, with divisions of resources, divisions of labor, and mutual dependence. Stimuli—that is, communications—exchanged among members of the group hold it together and maintain the social structure.

When biologists concerned with evolutionary theory began to analyze social behavior, some disturbing questions began to emerge. For instance, how can you explain, with a mechanism driven by differential reproductive success, the evolution of sterile castes—such as worker bees—in insects? How is it that in many vertebrate societies only a few of the males breed, with their right to breed seldom successfully challenged by other males? Why, among animals in groups, do certain ones utter warning cries or exhibit other forms of behavior that attract attention to the individual issuing the warning, thus threatening the warner's life? These behaviors are all examples of **altruism,** which is, by definition, behavior that benefits others and is performed at some risk or cost to the doer. How can acts of altruism be explained in terms of natural selection acting on the individual organism?

TABLE 51-1 **A Classification of Social Behaviors**

TYPE OF BEHAVIOR	EFFECT OF THE BEHAVIOR ON	
	THE DONOR	THE RECIPIENT
Selfish	Increases fitness	Decreases fitness
Cooperative	Increases fitness	Increases fitness
Altruistic	Decreases fitness	Increases fitness
Spiteful	Decreases fitness	Decreases fitness

Adapted from W. D. Hamilton, "The Evolution of Social Behavior," *Journal of Theoretical Biology,* vol. 7, pages 1–52, 1964.

Not all social behavior, of course, is altruistic. Table 51–1 shows a classification of social behaviors proposed by W. D. Hamilton, one of the leading theoreticians in this field. As you can see, each of these behaviors has a different potential effect on the direct fitness—that is, the reproductive success—of both the individual performing the behavior (the donor) and the recipient or target of the behavior. Selfish, cooperative, and altruistic behaviors have all been well documented in numerous species of animals living in a natural setting; spiteful behavior has thus far remained unobserved except in *Homo sapiens*. The evolution of selfish behavior through the action of natural selection poses no problems for evolutionary theory and, indeed, is what would be expected. Although the maintenance of cooperation by natural selection is relatively easy to understand, the mechanism by which it could have evolved initially is almost as perplexing as the mechanism underlying the evolution of altruism.

In our consideration of social behavior, we shall look first at several different types of animal societies, paying particular attention to their organization and the behavior of their individual members. Then we shall return to the question of how the observed behaviors could have evolved.

INSECT SOCIETIES

Insect societies are among the most ancient of all societies and, along with modern human societies, are among the most complex. Social insects include termites and hymenopterans (ants, wasps, and bees).

Stages of Socialization

As with other animals, the social insects evolved from forms that were originally solitary. Among bees, for example, true sociality appears to have evolved on at least eight separate occasions, and among wasps four times.

Most living species of bees and wasps are solitary and others show varying degrees of sociality. Thus it is possible to construct a scenario of the various stages of social evolution by the analysis of present-day species. Among the solitary species, the female builds a small nest, lays her eggs in it, stocks it with a food supply, and leaves it forever (Figure 51–10). She usually dies soon after, so there is no overlap between generations.

(a)

(b)

51–10 *Wasps, most of which are not truly social insects, do not tend their young, but often provide for them.* (a) *Wasps of the* Apanteles, *a solitary and parasitic genus, inject their eggs under the skins of caterpillars. The resultant larvae eat the internal tissues of the caterpillar, chew their way to the surface, and spin the cocoons shown here. One adult has just emerged from its cocoon.* (b) *Female wasps of the genus* Ammophila *hunt caterpillars, which they then deposit in previously prepared burrows in the soil. The egg is laid on the caterpillar, which, paralyzed but not dead, provides up to 40 days' food for the larva.*

Among subsocial or presocial species, the mother returns to feed the larvae for some period of time, and the emerging young may subsequently lay their eggs in the same nest or comb. However, the colony is not permanent (usually being destroyed over the winter), there is no division of labor, and all females are fertile.

Eusocial, or "truly social," insects are characterized by cooperation in caring for the young and a division of labor, with sterile individuals working on behalf of reproductive ones. All ants and termites and some species of wasps and bees—for example, honey bees—are eusocial.

Honey Bees

A honey-bee society usually has a population of 30,000 to 40,000 workers and one adult queen. Each worker, always a diploid female, begins life as a fertilized egg deposited by the queen in a separate wax cell. (Drones, or male bees, develop from unfertilized eggs and are therefore haploid.) The fertilized egg hatches to produce a white, grublike larva that is fed almost continuously by the nurse workers; each larval bee eats about 1,300 meals a day. After the larva has grown until it fills the cell, a matter of about six days, the nurses cover the cell with a wax lid, sealing it in. It pupates for about 12 days, after which the adult bee emerges (see Figure 51–2, page 1053).

The newly emerged adult worker rests for a day or two and then begins successive phases of employment. She is first a nurse, bringing honey and pollen from storage cells to the queen, drones, and larvae. This occupation usually lasts about a week, but it may be extended or shortened, depending on the conditions of the colony. Then she begins to produce wax, which is exuded from the abdomen, passed forward by the hind legs to the front legs, chewed thoroughly, and then used to enlarge the comb. During this stage of employment as a houseworking bee, she may also remove sick or dead comrades from the hive (see Figure 51–3, page 1053), clean emptied cells for reuse, or serve as a guard at the hive entrance. During this period, she begins to make brief trips outside, seemingly to become familiar with the immediate neighborhood. It is only in the third and final phase of her existence that the worker bee forages for nectar and pollen. At about six weeks of age, she dies.

The Queen

Queens are raised in special cells that are larger than the ordinary cells and shaped somewhat like a peanut shell. Although all fertilized eggs have the genetic potential to become queens, new queens develop only at certain times and under specific circumstances. According to the best recent evidence, queens become queens because of a generally more nutritious diet in the larval stage, especially rich in protein, as compared to the mostly carbohydrate (honey) diet fed the worker larvae.

51–11 *Honey-bee workers. (a) The first segment of each of the three pairs of legs has a patch of bristles on its inner surface. Those of the first and second pairs are pollen brushes, which gather the pollen that sticks to the bee's hairy body. On the third pair of legs, the bristles form a pollen comb that collects pollen from the brushes and the abdomen. From the comb, the pollen is forced up into the pollen basket, a concave surface fringed with hairs on the upper segment of the third pair of legs. Transfer of pollen to the pollen basket occurs in midflight. The sting is at the tip of the abdomen.*

(b) The mouthparts are fused into a sucking tube containing a tongue with which the bee obtains nectar. The antennae, attached to the head by a ball-and-socket joint, contain receptors for touch and for odors. The large compound eyes cannot see red (which is black, or colorless, to them) but can see ultraviolet, which is invisible to human eyes (see Figure 42–2, page 859).

(a)   (b)

51–12 *Honey bees.* (a) *Workers tending honey and pollen storage cells. The honey is made from nectar processed by special enzymes in the workers' bodies.* (b) *Workers feed the queen (upper right) and lick queen substance from her body. Queen substance, a pheromone, prevents sexual maturation in the workers.*

(a)   (b)

The queen exerts influences on her subjects by means of pheromones (page 585), of which there appear to be several. The influence of one of the pheromones, queen substance, inhibits ovarian development in the worker bees and prevents them from becoming queens or from producing rival queens.

If a hive loses its queen, workers will notice her absence very quickly and will become quite agitated. Very shortly, they begin enlarging worker cells to form emergency queen cells, and the larvae in the enlarged cells are then fed the special diet. Any diploid larva so treated will become a queen.

The Annual Cycle

One of the important differences between subsocial and eusocial bees is that colonies of eusocial bees survive the winter. Within the wintering hive, as we saw on page 779, bees maintain their temperature by clustering together in a dense ball; the lower the temperature, the denser the cluster.

In the spring, when the nectar supplies are at their peak, so many young may be raised that the group separates into two colonies. The new colony is always founded by the old queen, who leaves the hive, taking about half of the workers with her. The group stays together in a swarm for a few days, gathered around the queen, after which the swarm will settle in some suitable hollow tree or other shelter found by its scouts.

As the old queen is preparing to leave the hive, the new queens are getting ready to emerge. These two events are synchronized by sound signals transmitted through the comb. As these signals are exchanged, the workers remain motionless. During this period, ovarian development begins in some of the workers, a few of which lay eggs. The unfertilized eggs develop into males, or drones. After the old queen leaves the hive, a new young queen emerges, and any other developing queens are destroyed. The young queen then goes on her nuptial flight, exuding a pheromone (apparently also the queen substance) that entices the drones of neighboring colonies. She mates only on this one occasion (although she may mate with more than one male) and then returns to the hive to settle down to a life devoted to egg production.

During her nuptial flight, the queen receives enough sperm to last her entire life, which may be some five to seven years. These are stored in a special organ in her reproductive tract and are released, one at a time, to fertilize each egg as it is being laid. The queen usually lays unfertilized eggs only in the spring, at the time males are required to inseminate the new queens.

The drones' only contribution to the life of the hive is their participation in the nuptial flight. Since they are unable to feed themselves, they become an increasing liability to the social group. As nectar supplies decrease in the fall, they are stung to death by their sisters or are driven out.

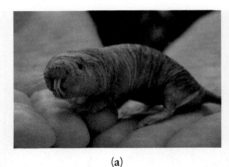

(a)

(b)

51–13 (a) *This tiny mammal, weighing less than 55 grams, is an adult naked mole rat. Naked mole rats, which live in underground tunnels in Kenya, Ethiopia, and Somalia, are the only vertebrates known to have a social system similar to that of the eusocial insects. In each colony, only a single dominant female (the queen rat) and one or a few males breed. The other members of the colony, both male and female, are workers who forage for food (mostly root vegetables) and dig and maintain the tunnels in which the colony lives.* (b) *When they are resting, naked mole rats huddle together to keep warm. As many workers as possible huddle under the queen rat, who, because of her large size, gives off large quantities of body heat.*

Naked mole rats can live more than 10 years; the queen rat usually breeds four times each year, producing an average of 10 pups per litter. It is thought that reproduction of the other females in the colony is suppressed by pheromones in the queen's urine.

VERTEBRATE SOCIETIES

With rare exceptions (Figure 51–13), vertebrate societies do not have the rigid caste systems characteristic of the truly social insects. Studies have revealed that many vertebrate societies are nonetheless highly structured, with social roles and access to resources determined by specific interactions that vary according to the species and the age and sex of the individuals.

Dominance Hierarchies

In many species of birds and mammals, dominance hierarchies, maintained by species-specific patterns of behavior, determine priority of access to resources and strongly influence relative reproductive success. One type of dominance hierarchy among vertebrates that has been studied in some detail is the pecking order in chickens.

A pecking order is established whenever a flock of hens is kept together over any period of time. In any one flock, one hen usually dominates all the others; she can peck any other hen without being pecked in return. A second hen can peck all hens but the first one; a third, all hens but the first two; and so on through the flock, down to the unfortunate pullet that is pecked by all and can peck none in return.

Hens that rank high in the pecking order have privileges such as first chance at the food trough, the roost, and the nest boxes. As a consequence, they can usually be recognized on sight by their sleek appearance and confident demeanor. Low-ranking hens tend to look dowdy and unpreened and to hover timidly on the fringes of the group.

During the period when a pecking order is being established, frequent and sometimes bloody battles may ensue, but once rank is fixed in the group, a mere raising or lowering of the head is sufficient to acknowledge the dominance or submission of one hen in relation to another. The overt fighting behavior has become **ritualized**, and life subsequently proceeds in harmony. If, however, new members are added to a flock, the entire pecking order must be reestablished. The resulting disorganization leads to more fighting, less eating, and less tending to the essential business of growth and egg laying.

Pecking orders have the effect of reducing the breeding population. Cocks and hens low in the pecking order copulate much less frequently than socially dominant chickens. Thus the final outcome is probably the same as it would be if

(a)

(b)

51–14 (a) *An adult wolf muzzling a young pup, who is begging for food. Similar muzzle-nuzzle gestures are used in maintaining dominance hierarchies of both males and females. Usually only the dominant male and dominant female breed, and the rest of the pack cooperate in caring for the young. Caring includes guarding the den and providing food, which is swallowed at the kill and then regurgitated for the pups. (Some domestic dogs regularly vomit at the sight of a puppy, which should be recognized not as a sign of disapproval but as a social reflex.)*

(b) *A subordinate baboon turns his buttocks toward a superior male. This gesture, known as presenting, is used between males and between females to signify submission or conciliation or to beg for special favors and by females to indicate their readiness to mate. The superior here appears to be reassuring the subordinate with a pat on the back.*

the social structure did not exist: the stronger and otherwise dominant animals eat better and leave the most offspring. However, because of the social hierarchy, this comes about with a minimum expenditure of lives and energy.

Territories and Territoriality

Many vertebrates stay close to their birthplaces, occupying a home range that is likely to be the same home range as that occupied by their parents. Even migratory birds that travel great distances are likely to return year after year to the same areas. Often these home ranges are defended, either by individuals or by groups, against other individuals or groups of the same species or closely related species that use the same resources. Areas so defended are known as **territories,** and the behavior of defending an area against rivals is known as **territoriality.**

Territoriality in Birds

Territoriality was first recognized by an English amateur naturalist and bird watcher, Eliot Howard. Howard observed that, in the spring, female birds were attracted to areas occupied by singing males of the same species, whereas other adult males of the species avoided those areas. In general, a breeding territory is established by a male. Courtship of the female, nest building, raising of the young, and often feeding are carried out within this territory. Frequently the female also participates in territory defense.

By virtue of territoriality, a mating pair has a better chance of obtaining food and nesting material in the area and a safe place to carry on all the activities associated with reproduction and care of the young. Some pairs carry out all their domestic activities within the territory. Others perform mating and nesting activities in the territories, which are defended vigorously, but gather food on a nearby communal feeding ground, where the birds congregate amicably together. A third type of territory functions only for courtship and mating, as in the bower of the bowerbird or the arena of the prairie chicken. In these territories, the males prance, strut, and posture—but very rarely fight—while the females look on and eventually indicate their choice of a mate by entering his territory. Males that have not been able to secure a territory for themselves are not able to reproduce; in fact, there is evidence from studies of some territorial species, such as the Australian magpie, that adults that do not secure territories do not mature sexually.

(a)

(b)

(c)

(d)

51–15 *Territories come in many shapes and sizes.* (a) *The male Uganda kob displays on his stamping ground, which is about 15 meters in diameter and is surrounded by similar stamping grounds on which other males display. A female signifies her choice by entering one of the stamping grounds and grazing there. Only a small proportion of males possess stamping grounds, and those that do are the only ones that breed.*

(b) *A fiddler crab's territory is a burrow, from which he signals with his large claw, beckoning females and warning off other males.*

(c) *Howler monkeys shift their territories as they move through the jungle canopy but maintain spacing between groups by chorusing.*

(d) *Territoriality is common among reef fish. For many species, a territory is a crevice in the coral, but for others, such as the clownfish shown here, the territory is a sea anemone. The fish is covered by thick slime that partially protects it from the poison of the tentacles, but its acceptance by the anemone is chiefly a consequence of behavioral adaptation of the fish, which even mates and raises its brood among the tentacles.*

Territorial Defense

Even though territorial boundaries may be invisible, they are clearly defined and recognized by the territory owner. With birds, for example, it is not the mere proximity of another bird of the same species that elicits aggression, but its presence within a particular area. The territory owner patrols his territory by flying from tree to tree. He will ignore a nearby rival outside his territory, but he will fly off to attack a more distant one that has crossed the border. Animals of other species are generally ignored unless they are prey or predators or are in competition for some limited resource.

Once an animal has taken possession of a territory, he or she is usually undefeatable on it. Among territory owners, prancing, posturing, scent marking, and singing and other types of calls usually suffice to repel intruders, which are at a great disadvantage. For example, a male cichlid, a tropical freshwater fish, will dart toward a rival male within his territory, but, as he chases the rival back into its

51–16 *Conflict between two male blue-striped grunts at their territorial borderline. Each grasps the other by the lips, and the loser is the one who lets go first. Although defeated, the loser swims back into his own territory unharmed.*

own territory, he begins to swim more slowly, the tail fin seemingly working harder and harder, just as if he were making his way against a current that increases in strength the farther he pushes into the other male's home ground. The fish know just where the boundaries are and, after chasing each other back and forth across them, will usually end up with each one trembling and victorious on his own side of the line.

Similarly, the expulsion from communal territories is typically accomplished by ritual rather than by force. For example, among the red grouse of Scotland, the males crow and threaten only very early in the morning, and then only when the weather is good. This ceremony may become so threatening that weaker members of the group leave the moor. Those that leave often starve or are killed by predators. Once the early-morning contest is over, the remaining birds flock together and feed side by side for the rest of the day.

KIN SELECTION

In 1962, V. C. Wynne-Edwards proposed that individuals that failed to reproduce were doing so for the benefit of the society to which they belonged. In this way, a society could maintain its population at a level always slightly below its resources, he argued, and so the entire group would benefit. Such behavior was perpetuated by the increased survival of groups whose members behaved with such altruistic self-restraint. Although Wynne-Edwards's hypothesis of group selection has now been rejected by almost all biologists, this proposal served to galvanize a whole series of extremely productive studies that have revolutionized the way modern biologists view social behavior.

Group selection was rejected on fairly simple grounds. If there were alleles for breeding and alleles for restraint-of-breeding, clearly the alleles for breeding would soon overwhelm the alleles for restraint-of-breeding in the gene pool. Rejection of group selection led to another proposal, put forward by W. D. Hamilton and based largely on studies of social insects. As Darwin himself realized, the evolution of sterile castes of insects posed a special problem for evolutionary theory. If evolution is based on the number of surviving offspring, how can natural selection acting on the individual organism result in the sterility of a large proportion of a population? Darwin concluded that, in some cases, natural selection might act not only on individuals but on families. Conceptually the idea is a simple one: members of families share inheritable characteristics.

A Tale of Two Territories

Although territorial defense is often ritualized, it can, on occasion, become bloody and violent, as illustrated by a series of incidents that occurred on Long Island, New York. Long Island is dotted with bays, inlets, lakes, and ponds, which are home to a great variety of birds, including a large population of mute swans. For some years, Lake Nowedonah, located on the South Fork of the island, has been the territory of a single mating pair of swans. Mating, nesting, and care of the newly hatched cygnets have occurred on the lake, with the entire family extending its range to neighboring ponds after the cygnets were a week or two old.

In the spring of 1986, a pair of young swans began nesting on one of these ponds but were quickly discovered by the older swans and driven away. The following year, the young swans tried again and were serenely incubating their eggs in an inconspicuous corner of the pond when they were spotted by the older swans, on an outing with their cygnets. The ensuing battle, which lasted more than three hours, led to the eviction of the younger swans and the death of one of the cygnets. While the older male engaged the younger male in combat, his mate attacked the younger female, who was on the nest incubating her eggs. The cygnets, of course, followed along and were soon within range of the younger female. She promptly began attacking those cygnets she could reach without abandoning the nest. This caused the older female to back off, with the cygnets huddled around her. One cygnet, however, was unable to escape the grasp of the younger female, who succeeded in drowning it. In the meantime, the males continued their battle—hissing, biting, and battering each other with their wings—until the younger male, a cringing mass of blood and feathers, abandoned the fight. The older male then turned his attention to the younger female, who was still dunking the now-dead cygnet. He went for her neck, driving her from the nest and the six eggs it contained. The exhausted, but triumphant, pair maintained patrol on the pond for some time before returning to the lake, where they reigned supreme for the remainder of the year. The younger male was later seen, wandering about on a road dazed and then huddled on a nearby lawn, before he and his mate disappeared.

In the spring of 1988, however, the younger pair returned again, laying their eggs earlier and successfully hatching seven cygnets some three weeks before the anniversary of the previous year's battle. The older swans on Lake Nowedonah, who had laid their eggs later, were fully occupied tending their own nest and did not discover the presence of the younger swans on the pond until a few weeks after their own seven cygnets had hatched. Once they had made the discovery, the conflict was resumed. In the course of the summer, three more bloody battles occurred. Two were on the pond, and one was on the lake, when the young male boldly ventured into the older swans' immediate territory. Although the young male lost each battle and was driven away by the older male, each time he returned, bruised and bloody, to the

---

Thus there are variations among families as well as among individuals, and families that have favorable variations are likely to leave more survivors than other families.

Hamilton elaborated Darwin's suggestion, modifying it to the concept of the gene pool. In population genetics, the measure of fitness is not the number of surviving offspring, but rather the increase or decrease in particular alleles in the gene pool. Viewed from this perspective, a mother takes risks and makes sacrifices for her offspring in order to increase the representation of her alleles in the gene pool. Or, to put it another way, she is programmed genetically to take certain risks or to make certain sacrifices for her offspring. To the extent that taking these risks or making these sacrifices increases her contribution to the gene pool, the alleles dictating this risk-taking program will increase in the next generation, and so on.

However, it is not just mothers and offspring that share alleles. For example,

The height of the battle in 1987. While the males were locked in combat, the younger female (left) was drowning a cygnet and the older female (right) was retreating to a safe distance with her remaining offspring.

pond and his mate and cygnets. During the final battle, in early August, all of his primary feathers were stripped off, rendering him, at least temporarily, unable to fly. Neither female was injured in any of the battles, nor were any cygnets lost as a result of the territorial dispute. Three cygnets were, however, lost from each family in the weeks immediately after hatching; it is thought that they fell victim to the snapping turtles that inhabit both the pond and the lake.

Toward the end of August of 1988, the young pair and their four surviving cygnets left the pond, head-ing for their wintering ground on a nearby bay. Although the male may have been able to fly by this time, he was walking, leading his family over the road and across the railroad tracks that lie between the pond and the bay. Later the family was spotted swimming near the edge of the bay. Given the scarcity of territories available in the area, and the young swans' success in raising their four cygnets, observers fully expect to see them back on the pond in the spring, prepared once more to defend their claim to the territory.

consider brothers—that is, sons of the same mother. Brothers with the same father share, on an average, half of their alleles with one another. Therefore, any allele that favorably influenced altruistic behavior among brothers could similarly increase its representation in the gene pool.

Hamilton's hypothesis, based on this principle, is called **kin selection**. Kin selection is the differential reproduction of lineages of related individuals—that is, different groups of related individuals of a species reproduce at different rates. The critical factor in kin selection is the effect of the individual on the reproductive success of its relatives. Tests of the hypothesis are based on degrees of relatedness of individuals. As a start, calculate your degree of relatedness to your parents and your siblings. Half your alleles came from your mother and half from your father; thus the probability that you received a particular allele from a particular parent is 50 percent, or 0.5. You and your sibling (ignoring the sex

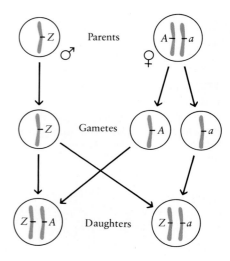

51–17 *In a haplodiploid species, daughters have a 50 percent chance of sharing an allele from the mother (A or a) and a 100 percent chance of sharing an allele from the father (Z).*

chromosome) each received half of your total alleles from your mother. Therefore, you each have a 50 percent chance of having received the same particular allele, so your total chance is 0.5×0.5, or 0.25. Your genetic relationship with regard to alleles from your father is the same, 0.25. Thus the total relatedness between two human siblings, taking into account the contribution from both parents, is, on the average, $0.25 + 0.25$, or 0.5. (This is often abbreviated as $r = 0.5$, with r standing for degree of relatedness.) Note that human siblings, like most other siblings with the same father, have the same average degree of relatedness to each other as to their parents and to their own offspring.

Now consider the honey bees. As we noted earlier, a fertilized (and therefore diploid) egg develops into a female, and an unfertilized, haploid egg develops into a male—a phenomenon known as **haplodiploidy.** Thus, for a female honey bee, her genetic relationship with her mother is the same as that of the usual vertebrate $(r = 0.5)$, but on the paternal side, since the father is haploid, each daughter gets an exact replica of the father's alleles (Figure 51–17). As a consequence, the average degree of relatedness between sister honey bees is not 0.5, but 0.75 $(0.25 + 0.5)$. If you are a female bee, more of your alleles will be represented in a reproductive sister (that is, a new queen) than in a daughter (Table 51–2). Hence, from this perspective, the hive's workers are not laboring for the queen. They are working for one another, and the queen is the machine by which they produce more of their own genotype, 75 percent at a time.

TABLE 51–2 Proportion of Alleles Shared by Relatives

NORMAL DIPLOID SEX DETERMINATION			HAPLOID SEX DETERMINATION		
ALLELES SHARED WITH HIS OR HER:	♂	♀	ALLELES SHARED WITH HIS OR HER:	♂	♀
Mother	0.5	0.5	Mother	1.0	0.5
Father	0.5	0.5	Father	0.0	0.5
Son	0.5	0.5	Son	0.0	0.5
Daughter	0.5	0.5	Daughter	1.0	0.5
Sister	0.5	0.5	Sister	0.5	0.75
Brother	0.5	0.5	Brother	0.5	0.25

51–18 (a) *This young silver-backed jackal, by helping raise his younger brother, probably increases his inclusive fitness more than if he raised his own family. Helpers are provided food by parents who, as the chart (b) clearly shows, have reasons to encourage their assistance. (c) Silver-backed jackal parents groom a helper. Being groomed is not only pleasurable but also removes insects and other parasites from the skin.*

(a)

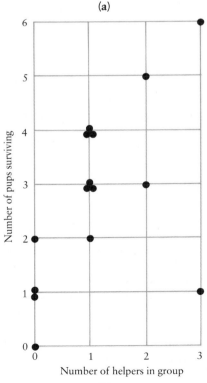

(b)

The workers' relationship with their brothers is quite different, however. The brother gets no alleles from his father, since he has no father. He gets half his mother's alleles, and that is all. His chance of getting the same half as any given sister is 0.5×0.5, or 0.25. However, if he mates, his daughters inherit all of his alleles. Moreover, such a mating would probably be with the new queen of a neighboring colony—not with one of his sisters. On this basis, one would predict that the male bee would be selfish, performing little in the way of service to the hive, and that the females would invest little energy in caring for their brothers. Both of these predictions turn out to be true.

Hamilton's example of kin selection among bees came under some criticism because all workers do not have the same father, since the queen may be inseminated by more than one male. (This would reduce the degree of relatedness between sisters and hence decrease the selective advantage of producing sisters rather than daughters.) Also, eusociality does not depend exclusively on haplodiploidy, since the highly social termites and naked mole rats, both of which are diploid rather than haplodiploid, also have sterile workers. However, Hamilton's hypothesis served to establish a new evolutionary perspective, that of **inclusive fitness**. The single criterion of Darwinian fitness is the relative number of an individual's offspring that survive to reproduce. The criterion of inclusive fitness is the relative number of an individual's alleles that are passed on from generation to generation, either as a result of his or her own reproductive success, or that of related individuals.

Tests of the Hypothesis

Hamilton's hypothesis is useful because it is testable. An example of such a test is the work of Patricia Moehlman, who studied 17 litters of silver-backed jackal pups near Ndutu Lodge in Tanzania. The family unit of the silver-backed jackal consists of a monogamous pair, young pups, and, often, one to three older siblings from a previous litter that serve as helpers (Figure 51–18). In litters with no helpers, Moehlman found that an average of only one pup was raised. With a single helper, three pups survived, and in one family with three helpers, six pups survived. Thus, by being helpers, young jackals (with an average relatedness of 0.5 to their siblings, just as they would have to their own offspring) produced more copies of their own alleles than if they had reproduced themselves with no helpers.

(c)

51-19 *A young scrub jay, surveying his surroundings.*

Among Florida scrub jays (Figure 51-19), the situation is somewhat different. Here again, breeding success is positively correlated with the number of helpers, which are usually close relatives of the breeding pair (Table 51-3). However, the success is somewhat less than would be achieved by the young jay's setting up its own nest and rearing its own young (Table 51-4). The reason that the young jays stay on as helpers is hypothesized, therefore, to be the shortage of available territories for young males. This hypothesis is supported by the fact that as soon as a territory becomes available, the male helper leaves. One of the most common ways to get a new territory is to inherit part of the parents' territory. Thus, one is led to predict that females will help less than males and that older, more dominant males will help more than the younger males, both predictions that have been shown to be true.

TABLE 51-3 Benefit of Helpers to Breeding Pairs of Florida Scrub Jays

	NUMBER OF OFFSPRING REARED	
	NO HELPERS	WITH HELPERS
Inexperienced pairs	1.03	2.06
Experienced pairs	1.62	2.20

From data of G. E. Woolfenden, analyzed by S. T. Emlen, in J. R. Krebs and N. B. Davies (eds.), *Behavioural Ecology: An Evolutionary Approach,* Blackwell Scientific Publications, Oxford, 1978.

TABLE 51-4 A Scrub Jay Helper Would Do Better By Setting up Its Own Breeding Territory, If It Could Find One, Than by Helping Its Parents at Home (results are calculated on the basis of the helper's being a full sibling to the young—that is, sharing one-half of its alleles)

OPTION		RESULT
1. Stay at home and help	Young produced by experienced parents with no help	1.62
	Young produced by pair with one helper	1.94
	Extra young due to presence of helper	0.32
	Genetic equivalents to helper ($r = 0.5$)	0.16
2. Go off and rear own young	Young reared by first-time breeders	1.36
	Genetic equivalents ($r = 0.5$)	0.68

From S. T. Emlen, in J. R. Krebs and N. B. Davies (eds.), *Behavioural Ecology: An Evolutionary Approach,* Blackwell Scientific Publications, Oxford, 1978.

THE SELFISH GENE

Many years ago the English satirist Samuel Butler remarked that a chicken was just an egg's way of making another egg. Updating this concept, some biologists see an organism as just a gene's way of making more genes. The argument is simple: The individual organism is transient. The genotype is fragmented at every generation. All that can survive from generation to generation is the gene. The way it survives is in the form of replicas. The more replicas, the better the chance of survival. The organism is the gene's survival machine, and it so programs the machine that it will turn out gene copies at a maximum rate, regardless of the personal cost to the organism. Thus, for example, there is a mite in which the female has a brood of one son and 20 daughters. The son mates with his sisters while still within the

51–20 *Two lionesses, probably sisters, with their cubs. In the plains of Africa, small groups of females (usually sisters or half-sisters) are the stable occupants of territories. The females of a pride hunt together cooperatively, come into estrus at about the same time, give birth at about the same time, and nurse each other's cubs. The males of a pride are transients, holding the territory and the females for only a few months, or, at most, a few years. New males taking over a pride often kill the young cubs. Following the death of the cubs, the females come into estrus again and so are available to perpetuate the genes of the newcomers, at least during the brief period of their reign.*

mother and dies before he is born—not much of a life for the male mite but a great low-risk, high-return strategy for his genes and a wise maternal investment.

The selfish-gene concept is admitted, even by its articulate sponsor, Richard Dawkins, to be an oversimplification. It has served, however, to bring into sharper focus questions on the evolution of behavior and of survival strategies.

Conflicts of Interest

In directing their attention to the survival of genes rather than of individuals, biologists realized that there were opportunities for serious conflicts of interest. These conflicts could occur not only between those individuals that were in obvious direct competition for some limited resource—such as territory, a food supply, or a mate—but also among individuals previously envisaged as working together for some common good—such as members of the nuclear family.

For example, a mammalian female invests in her offspring by providing nourishment in the form of milk. Nursing in mammals, however, delays the onset of estrus and so generally prevents new pregnancies. It is to the offspring's benefit to monopolize the milk supply for as long as possible; the mother, however, can maximize her fitness by weaning the offspring at an appropriate time and beginning a new investment.

51–21 *Mother langur holding an infant that has been fatally wounded by an infanticidal male at the time of takeover of her troop.*

Another example of intergenerational conflict occurs when a new male takes over a female with young. The new male's evolutionary goal, as programmed by his genes, is to produce his own young as rapidly as possible. The existing offspring (fathered by another male) may interfere with this objective, either by competing with their half-siblings, or by delaying estrus in the female. Sarah Blaffer Hrdy of Harvard University was the first to report the practice of infanticide by males at the time of takeover. Her observations, which were made on Hanuman langurs in India (Figure 51–21), were greeted with considerable skepticism when first reported in 1971, but similar acts have now been observed in other primates and also in birds, rodents, and lions. Infanticide can clearly represent a successful evolutionary strategy.

Male vs. Female

Among the most closely fought conflicts of interest are those that occur between the partners of a reproductive pair. A female, by definition, is the individual that produces the larger gametes. Female gametes are relatively expensive, metabolically speaking, and if the female must also carry the embryo in her body and care for the young after their birth, her investment in each reproductive effort may be, in terms of her total reproductive potential, very large indeed. Male gametes, by contrast, are usually cheap; a single insemination costs males of most species almost nothing. There are, however, exceptions. For example, the fertilization rate of the male lemon tetra, a common aquarium fish, is inversely proportional to the copulation rate, indicating that copulation can have a significant cost.

In most animal species, all that the male contributes to the next generation is his genes. In such situations, it is to the interest of the female's selfish genes to find themselves in the best possible company, thus promoting their survival in the next generation. Hence, intense competition may arise among the males to prove to the females that they are the best endowed. Such proof may take the form of elaborate courtship displays, in which strength, vigor, stamina, and other desirable traits are displayed, or it may take the form of dominance over other males. Since the same qualities are likely to be involved in both cases, the distinction is not always clear. Males also attract females by offers of food or other resources. This offer may be made in the form of a territory or a nesting site. In some species of birds and insects, males actually bring food offerings—a fly gift-wrapped in silk, for example. Because the male that commands the most resources, or the best resources, is likely also to be the male that is superior genetically, the offering or the territory may be less important as an actual resource than as a symbol of the male's desirability (see essay).

In some few species, males make an equal contribution to the care of the young. Selection for such a reproductive strategy would depend on at least two factors: first, that the additional parental investment would result in a significant increase in surviving young and, second, that the male has some way of being sure that the offspring he cares for are his own. Females in such situations may give up the opportunity of mating with the highest-ranking male in return for a male that will provide care for the young. Selection for such a reproductive strategy would include a means for the female to be sure that the male she has selected will remain with her after her eggs have been fertilized.

A long courtship, such as is seen among some species of birds, serves the purposes of both male and female. It gives each of them a chance to assess the qualities of the other; being choosey becomes equally important for a male under a monogamous system. For her, it symbolizes commitment and keeps him away from other females. It may also bring material benefits; for example, many female birds refuse to copulate until the nest is built. For him, the long courtship provides assurance that she has not already been inseminated. She loses nothing by cheating, whereas he loses his entire investment if she does.

51–22 *Displays, rituals, and ceremonies associated with mating are common among birds. These Japanese ground cranes, symbols of happiness, longevity, and marital fidelity, are performing their mating dance. The dance, which may be initiated by either the male or the female, is performed not only in the spring, as a prelude to mating, but also throughout the year. Pair bonding between the cranes characteristically endures for a lifetime, based, in part, on repeated performance of this ritual.*

Arts and Crafts of Bowerbirds

The bowerbirds are a family of fruit-eating songbirds found only in New Guinea and Australia. They are named for their elaborate nuptial structures. Their bowers include huts 1 to 2 meters in diameter, stick towers several meters high, walled avenues, decorated lawns, and moss platforms with parapets. The buildings are decorated with colored objects, including fruits, shells and rocks, flowers, and, if the birds live near human settlements, coins, marbles, false teeth, bottle tops, eyeglasses, and toothbrushes. At least six species use a tool, such as a wad of bark, to paint the bower with crushed plant matter, charcoal, or stolen blue laundry powder.

The male birds, as you can well imagine, spend much of their time building, adorning, and perfecting their bowers, and protecting them from raids by rival males. Young males watch adults. Adults steal ornaments from other adult males and damage their bowers. Some of the males are brightly colored with elaborate crests and plumes, and others are drab-colored and lack any special plumage; there is an inverse relationship between the splendor of the bower and the color and elaborateness of plumage of the birds themselves.

Male birds call from their nests and display to females if they approach. The male may carry one of the brightly colored objects during his display, and extend it toward the female, though she has never been seen to take it; observers interpret this part of the display as having originated in courtship feeding, just as the bower itself probably was once, in evolutionary history, a nest. Often several different adult males of the same species are competing within half a kilometer or less of one another. The female selects one, enters his bower, copulates briefly, and departs. Then she builds her own utilitarian nest some distance from the bower of her mate, lays her eggs, and cares for the young entirely on her own.

What are the rewards to the female of this activity? If she raises her brood, she will probably have sons who have inherited their father's artistic skills and so will be able to attract females whose activities will perpetuate her genes.

(a)

(a) *A pair of satin bowerbirds at a nest.* (b) *Great bowerbird, decorated with hibiscus.*

(b)

Studies at Duke University have shown that the togetherness of turtle doves—a traditional symbol of matrimonial bliss—may be more pragmatically interpreted as the male's ensuring that the female does not bear eggs fertilized by another male. The male keeps the female in sight all during nest-building activities. But once the eggs appear, he spends hours away from the nest. Such precaution on the part of the male may be warranted. Many birds—such as members of the heron family—share communal nesting areas, with nests within a meter or so of one another. Males and females take turns at the nest. Bird watchers have reported observing matings between the female on one nest and the male on another. To test whether or not such matings were actually successful, an ethologist captured some of the monogamously paired males, vasectomized them, and released them. The females paired with the sterile males were all able to lay fertile eggs and raise normal-sized clutches, with the help, of course, of their duped partners.

The Advantage of Waiting

Dominance hierarchies and territories can also be viewed from the perspective of the selfish gene. An animal that is not breeding because it is low in the dominance hierarchy or lacks a territory is not sacrificing itself for the good of the group or of the species. It is more likely waiting for its chance to move up the hierarchy, obtain a territory, or steal a copulation. If the individual is very low in a hierarchy, such an opportunity may never come, but still its chances of perpetuating its genotype are better than if it challenges a superior and is killed or badly injured.

Studies of bird territories confirm the hypothesis that individuals that lack territories are waiting for their turn. An empty territory is almost always immediately filled, as in the example of Florida scrub jays. Another example is provided by lions. As young male lions approach maturity, they are driven from the pride by the established adult males. The young males tend to stay together as nomads until they find a pride they can take over, often by defeating the resident males. Although their period of dominance may be limited (see Figure 51–20), their reproductive success is greater than if they had been killed or maimed resisting expulsion from the pride in which they were born. As we noted in Chapter 47, this social system also promotes outbreeding.

RECIPROCAL ALTRUISM

Seemingly altruistic acts may pay off, as we have seen, because they increase the probability of survival of genes shared by the performer of the act and the beneficiary. In 1971, Robert Trivers proposed another model for seemingly self-sacrificing behavior, which he termed **reciprocal altruism.** According to this model, an altruistic act is performed with the expectation that the favor will be returned. It might involve warning of danger, removing a tick from an inaccessible portion of the anatomy, or sharing food, to cite just a few possible examples. It also might involve acts of cooperation between organisms of different species, as well as between members of the same species. Reciprocal altruism may well have become established in early human societies, which is why this model is of special interest.

An example of reciprocal altruism is provided by vampire bats, which often have a difficult time locating suitable warm-blooded victims. Many individuals return each morning to the communal nesting sites with an empty stomach after an unsuccessful night of searching. A study by Gerald Wilkinson of wild vampire bats in Costa Rica showed that the bats regularly regurgitate blood to feed each other and that individuals in dire need are fed preferentially. A starved bat that receives blood reciprocates on a later day when its hunt has been successful. Not only related animals but also unrelated animals reciprocate with one another.

51–23 *Food sharing among nonrelated individuals is an example of reciprocal altruism. Here chimpanzees share a delicacy— the body of a red colobus monkey that one of the chimps has chanced upon. Meat is not actively sought as food by chimps but is highly prized when found.*

For reciprocal altruism to be a successful strategy that is resistant to exploitation by individuals who accept favors but never return them (that is, cheaters), two conditions are necessary. The first is that individuals meet more than once and are able to recognize each other. The second is that each individual cooperates on the first encounter and, on subsequent encounters, does whatever the other individual did on the preceding encounter. Under these particular conditions, cooperation pays off and cheating is punished.

One problem with reciprocal altruism is how to get it started; a sufficient number of cooperators must be present in the original population for the reciprocal behavior to become established. A possible, and reasonably probable, solution is that cooperation originates in a group of related individuals—that is, as kin selection—and spreads from the original cluster.

Another problem is that the individuals must have a way of recognizing one another. Thus, reciprocal altruism might not work in a large, impersonal society, such as that of honey bees. A variety of fascinating data, however, are accumulating that indicate that organisms of many different species, ranging from insects to amphibians to birds and mammals, are able to recognize and distinguish their own kin from other individuals of the same species (see essay). Recent psychological studies, some involving very young human infants, have demonstrated that the ability to recognize and remember the faces of others is highly developed among humans, which may be a related phenomenon.

THE BIOLOGY OF HUMAN BEHAVIOR

It is tempting—indeed almost irresistible—to draw parallels between human behavior and that observed in other species. The extent to which these concepts concerning the evolution of behavior can be extrapolated to the human species is a matter of current controversy. One group of biologists maintains that the human species is basically no different from any other species, that our genes are as selfish as any, and that if we seek to modify human behavior for the common good, we should understand its roots. An opposing group maintains that while early human ancestors may, in the past, have been governed by their genes, modern humans are so much a product of their culture and of their individual experience that such analyses are no longer valid. Moreover, they may be dangerous. The concept that biology determines human behavior lies at the roots of all notions of racial superiority, they point out. Thus it has provided the rationale for slavery, exploitation, and genocide. More commonly, the notion that our behavior is, to some extent, biologically determined allows us to forgive ourselves for, and even to justify, violence, aggressiveness, docility, and greed.

Kin Recognition in Tadpoles

For kin selection to be a factor in the evolution and maintenance of altruistic behavior, animals must be able to selectively direct their altruistic acts toward related individuals—thus increasing their own inclusive fitness. In many species, dispersal from the natal site is limited, and the probability is high that any individual of the same species in the vicinity—to whom assistance might be rendered—is a relative. In other species, however, dispersal is wider, and a group may consist of both related and unrelated individuals. How, then, can an animal distinguish relatives from nonrelatives, particularly if there has been no previous opportunity to get to know the relatives?

A series of illuminating studies conducted by Andrew Blaustein and Richard O'Hara of Oregon State University have revealed the mechanism involved in one organism, the Cascades frog *(Rana cascadae)*. Members of this species are highly gregarious, particularly in the tadpole stage, during which they maintain cohesive schools. The schooling behav-

ior is thought to aid in the defense against predators, principally the insect larvae that feed on these tadpoles. When a Cascades frog tadpole is captured and injured by an insect larva, it releases a chemical into the water that effectively warns the other tadpoles of the danger. Release of the chemical is followed by the flight of other members of the school.

In laboratory experiments, Blaustein and O'Hara allowed fertilized eggs to develop to the free-swimming tadpole stage under four different sets of conditions: (1) eggs from the same clutch—all laid by the same female and fertilized by the same male—developed together in the same tank; these individuals were full siblings; (2) eggs from one clutch developed at one end of a tank, with eggs from another, unrelated clutch developing at the other end of the tank; the two groups were separated by a mesh screen that allowed water to circulate freely throughout the tank; (3) a single egg developed within a mesh cylinder in a tank, while 12 eggs from an unrelated clutch devel-

Rana cascadae tadpoles schooling in a pond in the Cascade Range in Oregon. The preference of individuals of this species to associate with relatives rather than with unrelated individuals persists after their metamorphosis into adult frogs.

SUMMARY

The behavioral characteristics of organisms are the products of natural selection. Like anatomical and physiological adaptations, they affect the fitness of organisms. The study of behavior and its evolution is one of the most active areas of biological investigation, involving scientists from a diversity of disciplines. Ethology is the study of the behavior of animals in their natural environment, with particular emphasis on species-specific patterns of behavior and their evolutionary origins.

Although all behavior has its roots in the underlying genetic program of the individual, in only a few instances has it been possible to correlate a particular behavior with the presence or absence of specific alleles. Behaviors are typically

oped in the surrounding water; and (4) individual eggs developed in small tanks, in total isolation from any other individuals, related or unrelated. When the tadpoles hatched, they were given "preference tests," in which individual tadpoles were placed in tanks, sectioned by mesh screens, with related individuals at one end of the tank and unrelated individuals at the other. Regardless of the regimen under which the tadpole had developed, it spent most of its time at the end of the tank closest to its relatives. When tadpoles were mixed in tanks and allowed to school freely, time after time, they chose to school with relatives rather than with nonrelatives—even when they had been raised in a tank surrounded by nonrelatives. When, in field experiments, groups of related and unrelated individuals were released into ponds and lakes, they quickly sorted out into schools composed primarily of kin. Additional experiments, in which eggs were fertilized in the laboratory, with different males fertilizing the eggs of the same female and the same male fertilizing the eggs of different females, revealed that the tadpoles could distinguish between full siblings, half siblings, and nonsiblings, almost always preferring the company of the most closely related individuals available.

A second set of preference tests, in which the tested tadpoles were separated from the related and unrelated groups by glass barriers that allowed the tadpoles to see one another but prevented any water circulation between compartments, revealed that the cue that identifies its relatives to a Cascades frog tadpole is a chemical substance. In the absence of water bearing the identifying substance, determined by the investigators to be an odor, the tadpoles displayed no preference for either group.

The basis for the recognition of the odor of one's kin may be genetic or it may occur through a phenomenon known as phenotype matching. One hypothesis is that specific alleles code for the molecules responsible for the characteristic odor of an individual and that each individual bears receptors for the particular molecules it produces. Related individuals carry the same alleles, synthesize the same molecules, and bear the same receptors. According to this hypothesis, the recognition depends entirely on genetics, with no learning involved. With phenotype matching, however, learning is involved—and, the critical information to be learned, as indicated by the tadpoles that developed in isolation or in the presence of unrelated individuals, is one's own odor. Kin are recognized on the basis of the similarity of their odor to one's own odor. Thus far, no test has been devised that will enable investigators to determine which of these mechanisms is responsible for the kin recognition, and the possibility remains that both are involved.

Experiments by other investigators have demonstrated varying degrees of kin recognition in a vast diversity of animals, including social insects, ground squirrels, mice, birds, and monkeys, as well as tadpoles of other frog and toad species. In most species studied, the capacity to recognize kin appears early in development. It seems to function not only in determining the distribution of favors—such as warning signals—but also in enabling animals to avoid both fighting with kin and inbreeding.

the result of the interactions of a large number of genes, the influence of which may be further modified by interactions of the individual with its environment.

A full explanation as to why an animal performs a particular behavior involves both proximate and ultimate causation. The proximate cause of a behavior is typically a series of physiological responses to a particular stimulus; the ultimate cause, however, is its adaptive value for the survival and successful reproduction of the individual.

Patterns of behavior that appear essentially complete the first time the organism encounters the relevant stimulus are known as fixed action patterns. These patterns are highly stereotyped and rigid and, for members of a given species of a particular age, sex, and physiological condition, are as predictable and constant as the anatomical characteristics of the species. Communication signals among members of a species that act as stimuli for fixed action patterns are known as releasers, and the brain circuits that respond to releasers are known as innate releasing mechanisms.

Learning is the modification of behavior as a result of experience. Various categories of learning are recognized, of which habituation is one of the simplest. Associative learning, in which one stimulus comes to be associated with another, includes both classical conditioning, such as the experiments of Pavlov, and operant conditioning (trial-and-error learning). Imprinting is a form of learning of particular importance in the discrimination of members of one's own species from members of other species; it occurs quickly within a critical period in early life. Imitation of stimuli experienced during the critical period is a factor in song learning in birds, and other types of imitative learning have been observed in birds and mammals.

Of particular interest from an evolutionary standpoint is the behavior of animals in societies—groups of individuals of the same species, living together in an organized fashion, with divisions of resources and labor and with mutual dependence. Such behavior may be selfish, cooperative, altruistic, or spiteful—with differing consequences for the fitness of the donor and the recipient.

Insect societies are among the largest and most complex of animal societies. In the honey-bee colony (and among other eusocial insects as well), the queen is the only female reproductive form. She and her brood are tended by sterile workers. The behavior and physiology of the members of the hive are controlled by the exchange of chemical signals in the form of pheromones. This eusocial system is also exhibited by a mammal, the naked mole rat.

Vertebrate societies are often organized in terms of social dominance. Dominance hierarchies may take the form of pecking orders, in which higher-ranking animals have priority of access to food and other resources and reproduce more. Territoriality is a system of social dominance in which only animals with territories reproduce and animals with superior territories reproduce more. Territories may be "real"—that is, they may be actual areas containing food and nesting material to support a mating pair and young—or they may be symbolic, such as an arena or a bower.

A central question in the study of social behavior is the mechanism by which natural selection can result in behavior that limits the reproductive potential of individuals in the society. An early explanation was group selection, the hypothesis that animals were genetically programmed to refrain from breeding for the good of the society as a whole. This concept was rejected because there was no demonstrable way that group selection could be maintained against individual selection for increased reproduction.

This hypothesis has been largely supplanted by that of kin selection, which introduced the concept of inclusive fitness. Previous concepts of evolutionary fitness had focused on the relative number of surviving offspring in future generations. Inclusive fitness focuses on the relative number of an individual's alleles that are passed on from generation to generation, as a result of the reproductive success either of the individual or of related individuals.

From the viewpoint of inclusive fitness, social behavior is regulated by the "selfish gene," which programs the individual not necessarily for the individual's own well-being, or even survival, but only for the perpetuation of the allele by any means. Competition for representation in the future gene pool can occur not only between individuals obviously competing for the same resources—such as for food or mates—but also between parents and offspring and between members of a reproductive pair. The selfish-gene concept can also be used to explain the establishment of dominance hierarchies and territorial behavior. An individual that is low in a social hierarchy or that is without a territory is programmed by its genes to accept such an inferior status because the individual's chances of reproducing may be better if it waits than if it embarks on a probably futile challenge.

Altruism is behavior that carries a cost to the individual that performs it and benefits some other individual or individuals. Some acts of altruism are thought to be based on inclusive fitness, and others may be based on reciprocal altruism, the performance of an unselfish act with the expectation that it will be returned. For reciprocal altruism to be a successful strategy, resistant to cheating, it is necessary that individuals be able to recognize one another and that an individual cooperate with or cheat another individual on the basis of what that individual did the last time the two encountered each other. Reciprocal altruism is of particular interest in terms of the evolution of cooperative behavior in the ancestors of *Homo sapiens*.

QUESTIONS

1. Distinguish among the following: proximate cause/ultimate cause; fixed action pattern/innate releasing mechanism; sign stimulus/releaser; classical conditioning/operant conditioning; imprinting/imitation; subsocial/eusocial; queen/worker/drone; social dominance/territoriality; group selection/kin selection; Darwinian fitness/inclusive fitness; altruism/reciprocal altruism.

2. Define behavior. What distinguishes social behavior from other forms of behavior?

3. Territoriality and social dominance achieve the same results. What are they? How do the results differ in a population in which these forms of behavior are not present?

4. Among organisms in which both sexes are diploid, the degree of relatedness between parents and children is always 0.5, and the degree of relatedness between siblings has a theoretically possible range from 0 to 1 and averages 0.5. Explain.

5. As we noted on page 974, J. B. S. Haldane once remarked that he would lay down his life for two brothers or four cousins. On

what basis did he make such an offer? Actually, Haldane was being too generous. Explain why.

6. Animals that live within a dominance hierarchy may never breed, yet they may be as fit as those that do breed. How can this be explained?

7. How can one explain the evolution of eusociality in animals, such as termites and naked mole rats, that are not haplodiploid?

8. In order to protect their young against infanticidal males, females might refuse to mate with murderers. Would this be a successful strategy?

9. A hypothesis known as the "superterritory hypothesis" suggests that an animal should defend a territory larger than it really needs. Could this be spiteful behavior? Explain.

10. In what ways are human societies different from insect societies? How are they similar?

SUGGESTIONS FOR FURTHER READING

Books

ALCOCK, JOHN: *Animal Behavior: An Evolutionary Approach*, 4th ed., Sinauer Associates, Inc., Sunderland, Mass., 1989.

An outstanding text on animal behavior. Highly recommended.

ALTMANN, JEANNE: *Baboon Mother and Infants*, Harvard University Press, Cambridge, Mass., 1980.*

This book, the result of a nine-year study in Kenya, has been called the most significant contribution to the understanding of primate behavior to date. It is outstanding both for its quantitative data and for the author's ability to write clearly and interestingly.

BARASH, DAVID P.: *Sociobiology and Behavior*, 2d ed., American Elsevier Publishing Company, New York, 1982.

An excellent introduction to the evolution of social behavior, this text provides a clear outline of current sociobiological hypotheses and their tests. It is enriched by numerous examples of the behaviors discussed.

BATES, MARSTON, and PHILIP S. HUMPHREY, eds.: *The Darwin Reader*, Charles Scribner's Sons, New York, 1956.*

A collection of Darwin's writings, including The Autobiography *and excerpts from* The Voyage of the Beagle, The Origin of Species, The Descent of Man, *and* The Expression of the Emotions. *Darwin was a fine writer, and you can discover here the wide range of his interests and concerns at different periods of his life.*

CARLQUIST, SHERWIN: *Island Life: A Natural History of the Islands of the World*, Natural History Press, Garden City, N.Y., 1965.

An exploration of the nature of island life and the intricate and unexpected evolutionary patterns found in island plants and animals.

_____
* Available in paperback.

CLUTTON-BROCK, T. H., S. E. GUINNESS, and S. D. ALBON: *Red Deer: Behavior and Ecology of Two Sexes*, University of Chicago Press, Chicago, 1982.*

The aim of the authors, well realized, was "not merely to describe the ecology of the deer, but to interpret that ecology in terms of modern evolutionary theory and to use their study to illuminate evolution theory in general." This study, which follows more than a thousand deer over a 12-year time span, has already become a classic.

DARWIN, CHARLES: *The Origin of Species by Means of Natural Selection, or The Preservation of Favored Races in the Struggle for Life*, Doubleday & Company, Inc., Garden City, N.Y., 1960.*

Darwin's "long argument." Every student of biology should, at the very least, browse through this book to catch its special flavor and to begin to understand its extraordinary force.

DARWIN, CHARLES: *The Voyage of the Beagle*, Natural History Press, Garden City, N.Y., 1962.*

Darwin's own chronicle of the expedition on which he made the discoveries and observations that eventually led him to his theory of evolution. The sensitive, eager young Darwin that emerges from these pages is very unlike the image many of us have formed of him from his later portraits.

DAWKINS, RICHARD: *The Blind Watchmaker*, W. W. Norton & Company, New York, 1986.*

A thorough and delightfully written explication of modern evolutionary theory, the evidence in its support, and the controversies that have surrounded it.

DAWKINS, RICHARD: *The Selfish Gene*, Oxford University Press, New York, 1976.*

Dawkins argues that we are "survival machines" programmed to preserve the "selfish molecules" known as genes. A wit-sharpening account and also an entertaining one.

ELDREDGE, NILES (ed.): *The* Natural History *Reader in Evolution*, Columbia University Press, New York, 1987.*

An anthology of articles from Natural History, *with emphasis on adaptation through natural selection, speciation, and the rise and fall of major groups of organisms.*

FUTUYMA, DOUGLAS J.: *Evolutionary Biology*, 2d ed., Sinauer Associates, Inc., Sunderland, Mass., 1986.

An up-to-date textbook for an advanced course in evolution, focusing on controversial issues and problems.

FUTUYMA, DOUGLAS J.: *Science on Trial: The Case for Evolution*, Pantheon Books, New York, 1983.*

Futuyma presents a clear, straightforward analysis of the controversy between "scientific creationism" and modern biological science. Written for the general reader, the book also provides an excellent short course in the philosophy, history, and methodology of modern biology, as well as a thorough discussion of the major issues in evolutionary biology today.

GOODALL, JANE: *In the Shadow of Man*, Houghton Mifflin Company, Boston, 1983.*

A personal account, first published in 1971, of 11 years spent observing the complex social organization of a single chimpanzee community in Tanzania.

GOULD, STEPHEN JAY: *Ever Since Darwin: Reflections in Natural History*, W. W. Norton & Company, New York, 1977.*

GOULD, STEPHEN JAY: *The Flamingo's Smile: Reflections in Natural History*, W. W. Norton & Company, New York, 1985.*

GOULD, STEPHEN JAY: *Hen's Teeth and Horse's Toes: Further Reflections in Natural History*, W. W. Norton & Company, New York, 1983.*

GOULD, STEPHEN JAY: *The Panda's Thumb: More Reflections in Natural History*, W. W. Norton & Company, New York, 1980.*

Collections of thoughtful, witty, and well-written essays from Natural History. *Evolution is the central theme.*

GRANT, PETER R.: *Ecology and Evolution of Darwin's Finches*, Princeton University Press, Princeton, N.J., 1986.*

A new, major work on the Galapagos finches, based on field studies carried out by Grant and his students over a period of some 15 years. In the course of these studies, an enormous quantity of data was gathered, which is analyzed from the perspective of modern ecological and evolutionary theory. Clearly written and beautifully illustrated, this book is already considered a classic.

HRDY, SARAH BLAFFER: *The Langurs of Abu: Female and Male Strategies of Reproduction*, Harvard University Press, Cambridge, Mass., 1981.*

This book, which has been described as "spell-binding," is notable not only for its original observations but also for its contributions to the theoretical structure of sexual combat.

HRDY, SARAH BLAFFER: *The Woman that Never Evolved*, Harvard University Press, Cambridge, Mass., 1981.*

Both a sociobiologist and a feminist, Hrdy examines what it means to be female. Males are almost universally dominant over females in primate species, and Homo sapiens *is no exception. Yet in studying our own remote ancestors, the living primates, she discovers that the female is competitive, independent, sexually assertive, and has just as much at stake in the evolutionary game as her male counterpart.*

JOHANSON, DONALD, and MAITLAND EDEY: *Lucy: The Beginnings of Humankind*, Simon and Schuster, New York, 1981.*

A brash, exciting, firsthand account of the discovery of Lucy and of the conflicts and controversies among those seeking to uncover human origins.

KREBS, J. R., and N. B. DAVIES: *An Introduction to Behavioural Ecology*, 2d ed., Blackwell Scientific Publications, Inc., Boston, 1987.*

The best introduction to the general subject of social behavior. Well written with many good illustrations and an excellent balance between examples and theory.

LACK, DAVID: *Darwin's Finches*, Harper & Row, Publishers, Inc., New York, 1961.*

This short, readable book, first published in 1947, gives a marvelous account of the Galapagos, its finches and other inhabitants, and of the general process of evolution. Although in many respects superseded by Grant's book, it remains an excellent introduction to the Galapagos finches.

---

* Available in paperback.

LEAKEY, RICHARD E.: *The Making of Mankind*, E. P. Dutton & Co., Inc., New York, 1981.

LEAKEY, RICHARD E., and ROGER LEWIN: *Origins*, E. P. Dutton & Co., Inc., New York, 1977.*

Beautifully illustrated, thoughtful, and imaginative books about 3 million years of human history, and how we know what we do about it, with some speculations on human nature.

LEWIN, ROGER: *Bones of Contention*, Simon and Schuster, New York, 1988.*

A lively and highly recommended account of the personal and scientific disputes pervading the study of human evolution.

LEWIN, ROGER: *In the Age of Mankind: A Smithsonian Book of Human Evolution*, Smithsonian Books, Washington, D.C., 1988.

The best and most handsome introduction to this fascinating field of modern research.

LEWIN, ROGER: *Thread of Life: The Smithsonian Looks at Evolution*, Smithsonian Books, Washington, D.C., 1982.

Lewin is a clear, interesting, and well-informed writer on the subject of evolution. The book is beautifully illustrated.

LEWONTIN, RICHARD: *Human Diversity*, Scientific American Library, W. H. Freeman and Company, New York, 1984.

A leading population geneticist, Lewontin analyzes the biological bases of human variation and also explores their often controversial social ramifications—such as IQ, skin color, and sex-linked differences. This is a wise and beautiful book that, transcending politics and biology, becomes a celebration of human potential.

MAYR, ERNST: *The Growth of Biological Thought: Diversity, Evolution, and Inheritance*, Harvard University Press, Cambridge, Mass., 1982.*

This is the first of two projected volumes on the history of biology and its major ideas, written by one of the leading figures in the study of evolution. The introductory chapters provide an outstanding analysis of the philosophy and methodology of the biological sciences, followed by three sections on the history and current status of taxonomy, evolutionary biology, and genetics. This book, like Darwin's masterpiece, should at least be sampled by every serious student of biology.

MAYR, ERNST: *Populations, Species, and Evolution: An Abridgement of Animal Species and Evolution*, Harvard University Press, Cambridge, Mass., 1970.*

A masterly, authoritative, and illuminating statement of contemporary thinking about species—how they arise and their role as units of evolution.

MECH, L. DAVID: *The Wolf: The Ecology and Behavior of an Endangered Species*, University of Minnesota Press, Minneapolis, 1983.*

The definitive book on the social history of wolves, originally published in 1970. Mech, a wildlife biologist with the U.S. Fish and Wildlife Service, devoted more than 20 years to the study of wolves in their natural habitats.

MOOREHEAD, ALAN: *Darwin and the Beagle*, Harper & Row, Publishers, Inc., New York, 1972.*

A delightful narrative of Darwin's journey, beautifully illustrated with contemporary or near contemporary drawings, paintings, and lithographs.

MOSS, CYNTHIA: *Portraits in the Wild: Animal Behavior in East Africa*, 2d ed., University of Chicago Press, Chicago, 1982.*

A description of some of the studies carried out on various wild species. Full of fascinating bits of information and firsthand observations.

RAUP, DAVID M.: *The Nemesis Affair: A Story of the Death of Dinosaurs and the Ways of Science*, W. W. Norton & Company, New York, 1986.*

A popular account of mass extinctions, the various hypotheses that have been proposed to explain them, and the debates within the scientific community over those hypotheses.

SAGAN, CARL, and PAUL R. EHRLICH: *The Cold and the Dark: The World After Nuclear War*, W. W. Norton & Company, New York, 1985.*

A discussion, for the general reader, of the "nuclear winter" hypothesis and the biological consequences of a major nuclear exchange.

SANDERS, N. K.: *Prehistoric Art in Europe*, Penguin Books, Ltd., Harmondsworth, England, 1968.

An illustrated history of art's first 30,000 years.

SIMPSON, GEORGE GAYLORD: *Fossils and the History of Life*, Scientific American Library, W. H. Freeman and Company, New York, 1983.

A beautifully written and illustrated summary of the fossil evidence on which much of our knowledge of evolution rests, by one of the major figures in twentieth-century evolutionary biology.

SIMPSON, GEORGE GAYLORD: *Horses*, Oxford University Press, New York, 1951.

The story of the horse family in the modern world and through 60 million years of history. Simpson was not only an outstanding vertebrate paleontologist but also a graceful and delightful writer.

SIMPSON, GEORGE GAYLORD: *Penguins: Past and Present, Here and There*, Yale University Press, New Haven, Conn., 1983.*

As Simpson said, "Penguins are beautiful, interesting, inspiring, and funny." They are also excellent examples of evolution and adaptation, as this informal account demonstrates.

SIMPSON, GEORGE GAYLORD: *Splendid Isolation: The Curious History of South American Mammals*, Yale University Press, New Haven, Conn., 1983.*

Charles Darwin found in the South American fauna some of the major clues that led to the formulation of The Origin of Species. *Today, their history can be seen as an almost ideal natural experiment in evolution, told in Simpson's engaging style.*

* Available in paperback.

STANLEY, STEVEN M.: *Extinction, Scientific American* Library, W. H. Freeman and Company, New York, 1986.

A beautifully illustrated volume surveying the paleontological and geological evidence of the great extinctions of the past.

TINBERGEN, NIKO: *Curious Naturalists,* Natural History Library, Doubleday & Company, Inc., Garden City, N.Y., 1968.*

Some charming descriptions of the activities and discoveries of scientists studying the behavior of animals in their natural environment.

TOPOFF, HOWARD (ed.): *The* Natural History *Reader in Animal Behavior,* Columbia University Press, New York, 1987.*

A collection of Natural History *articles covering four broad topics: sensory processes and orientation, evolution of behavior, social organization, and behavioral development.*

TRIVERS, ROBERT: *Social Evolution,* The Benjamin/Cummings Publishing Company, Menlo Park, Calif., 1985.

An excellent introductory text on the evolution of social behavior, this book assumes a minimal background on the part of the reader. Well illustrated, with many examples, this is the most up-to-date introductory text currently available.

WALLACE, ALFRED RUSSEL: *The Malay Archipelago,* Oxford University Press, New York, 1987.

A reprint of Wallace's classic work, accompanied by an excellent introduction that provides much additional information about Wallace's life and scientific work.

WICKLER, WOLFGANG: *Mimicry in Plants and Animals,* World University Library, London, 1968.*

Many examples and illustrations of a delightful subject.

WILSON, E. O.: *The Insect Societies,* Harvard University Press, Cambridge, Mass., 1971.*

A comprehensive and fascinating account of the social insects. An unusual nominee for the National Book Award.

WILSON, E. O.: *Sociobiology: The New Synthesis,* Harvard University Press, Cambridge, Mass., 1975.

In this extremely interesting, beautifully written, and beautifully illustrated book, Wilson undertakes to set forth the biological principles that govern social behavior in all kinds of animals. The last chapter, which concerns human sociobiology, became a focus of controversy about the inheritance of behavioral traits in Homo sapiens. *Also available in an abridged, less technical version.*

WITTENBERGER, J. F.: *Animal Social Behavior,* Duxbury Press, Boston, 1981.

A text on the evolution of social behavior, written for advanced undergraduates. Strong on evolutionary logic, using many mathematical models, this is probably the most comprehensive book on the subject since Wilson's Sociobiology.

---

* Available in paperback.

Articles

ABELSON, PHILIP H.: "Creationism and the Age of the Earth," *Science,* vol. 215, page 119, 1982.

ALVAREZ, LUIS W., WALTER ALVAREZ, FRANK ASARO, and HELEN V. MICHEL: "Extraterrestrial Cause for the Cretaceous-Tertiary Extinction," *Science,* vol. 208, pages 1095–1108, 1980.

ARCHIBALD, J. DAVID, and WILLIAM A. CLEMENS: "Late Cretaceous Extinctions," *American Scientist,* vol. 70, pages 377–385, 1982.

BAMBACH, RICHARD K., CHRISTOPHER R. SCOTESE, and ALFRED M. ZIEGLER: "Before Pangea: The Geographies of the Paleozoic World," *American Scientist,* vol. 68, pages 26–38, 1980.

BEHRENSMEYER, ANNA K.: "Taphonomy and the Fossil Record," *American Scientist,* vol. 72, pages 558–566, 1984.

BLAUSTEIN, ANDREW R., and RICHARD K. O'HARA: "Kin Recognition in Tadpoles," *Scientific American,* January 1986, pages 108–116.

BORGIA, GERALD: "Sexual Selection in Bowerbirds," *Scientific American,* June 1986, pages 92–100.

BOUCOT, A. J., and JANE GRAY: "A Paleozoic Pangaea," *Science,* vol. 222, pages 571–581, 1983.

CARSON, HAMPTON L., et al.: "Hawaii: Showcase of Evolution," *Natural History,* December 1982, pages 16–72. A series of 10 articles on the evolution of the flora and fauna of the Hawaiian Islands.

CARTMILL, MATT, DAVID PILBEAM, and GLYNN ISAAC: "One Hundred Years of Paleoanthropology," *American Scientist,* vol. 74, pages 410–420, 1986.

COLE, CHARLES J.: "Unisexual Lizards," *Scientific American,* January 1984, pages 94–100.

COOK, L. M., G. S. MANI, and M. E. VARLEY: "Postindustrial Melanism in the Peppered Moth," *Science,* vol. 231, pages 611–613, 1986.

CREWS, DAVID, and MICHAEL C. MOORE: "Evolution of Mechanisms Controlling Mating Behavior," *Science,* vol. 231, pages 121–125, 1986.

DIAMOND, JARED M.: "Evolution of Bowerbirds' Bowers: Animal Origins of the Aesthetic Sense," *Nature,* vol. 297, pages 99–102, 1982.

DIAMOND, JARED M.: "Rapid Evolution of Urban Birds," *Nature,* vol. 324, pages 107–108, 1986.

EHRLICH, PAUL R., et al.: "Long-Term Biological Consequences of Nuclear War," *Science,* vol. 222, pages 1293–1300, 1983.

FAUL, HENRY: "A History of Geologic Time," *American Scientist,* vol. 66, pages 159–165, 1978.

GIBBS, H. L., and P. R. GRANT: "Oscillating Selection on Darwin's Finches," *Nature,* vol. 327, pages 511–513, 1987.

GOULD, STEPHEN JAY: "Bushes All the Way Down," *Natural History,* June 1987, pages 12–19.

GOULD, STEPHEN JAY: "Evolution and the Triumph of Homology, or Why History Matters," *American Scientist,* vol. 74, pages 60–69, 1986.

GOULD, STEPHEN JAY: "A Novel Notion of Neanderthal," *Natural History,* June 1988, pages 16–21.

GRANT, PETER R.: "Speciation and the Adaptive Radiation of Darwin's Finches," *American Scientist,* vol. 69, pages 653–663, 1981.

GROSS, MART R.: "Disruptive Selection for Alternative Life Histories in Salmon," *Nature,* vol. 313, pages 47–48, 1985.

HALLAM, ANTHONY: "End-Cretaceous Mass Extinction Event: Argument for Terrestrial Causation," *Science,* vol. 238, pages 1237–1242, 1987.

HAMILTON, W. D., and ROBERT AXELROD: "The Evolution of Cooperation," *Science,* vol. 211, pages 1390–1396, 1981.

HAY, RICHARD L., and MARY D. LEAKEY: "The Fossil Footprints of Laetoli," *Scientific American,* February 1982, pages 50–57.

HEYLER, DANIEL, and CECILE M. POPLIN: "The Fossils of Montceau-les-Mines," *Scientific American,* September 1988, pages 104–110.

HOLSINGER, JOHN R.: "Troglobites: The Evolution of Cave-Dwelling Organisms," *American Scientist,* vol. 76, pages 146–153, 1988.

HORN, HENRY S.: "Sociobiology," in Robert M. May (ed.): *Theoretical Ecology,* Sinauer Associates, Sunderland, Mass., 1981.

JARVIS, J. U. M.: "Eusociality in a Mammal: Cooperative Breeding in Naked Mole-Rat Colonies," *Science,* vol. 212, pages 571–573, 1981.

KOEHN, RICHARD K., and THOMAS J. HILBISH: "The Adaptive Importance of Genetic Variation," *American Scientist,* vol. 75, pages 134–141, 1987.

LEWIN, ROGER: "Africa: Cradle of Modern Humans," *Science,* vol. 237, pages 1292–1295, 1987.

LEWIN, ROGER: "The Surprising Genetics of Bottlenecked Flies," *Science,* vol. 235, pages 1325–1327, 1987.

LEWIN, ROGER: "The Unmasking of Mitochondrial Eve," *Science,* vol. 238, pages 24–26, 1987.

LOVEJOY, C. OWEN: "Evolution of Human Walking," *Scientific American,* November 1988, pages 118–125.

MARSHALL, LARRY G.: "Land Mammals and the Great American Interchange," *American Scientist,* vol. 76, pages 380–388, 1988.

MAY, ROBERT M.: "Evolution of Pesticide Resistance," *Nature,* vol. 315, pages 12–13, 1985.

MAYNARD SMITH, JOHN: "Evolution and the Theory of Games," *American Scientist,* vol. 64, pages 41–45, 1976.

MAYNARD SMITH, JOHN, and G. R. PRICE: "The Logic of Animal Conflict," *Nature,* vol. 246, pages 15–18, 1973.

MOEHLMAN, PATRICIA D.: "Social Organization in Jackals," *American Scientist,* vol. 75, pages 366–375, 1987.

NANCE, R. DAMIAN, THOMAS R. WORSLEY, and JUDITH B. MOODY: "The Supercontinent Cycle," *Scientific American,* July 1988, pages 72–79.

OFFICER, CHARLES B., and CHARLES L. DRAKE: "Terminal Cretaceous Environmental Events," *Science,* vol. 227, pages 1161–1167, 1985.

PILBEAM, DAVID: "The Descent of Hominoids and Hominids," *Scientific American,* March 1984, pages 84–96.

RAUP, DAVID M.: "Biological Extinction in Earth History," *Science,* vol. 231, pages 1528–1533, 1986.

RENSBERGER, BOYCE: "Bones of Our Ancestors," *Science 84,* April 1984, pages 28–39.

RUSSELL, DALE A.: "The Mass Extinctions of the Late Mesozoic," *Scientific American,* January 1982, pages 58–65.

SEELEY, THOMAS D.: "How Honeybees Find a Home," *Scientific American,* October 1982, pages 158–168.

SLATKIN, MONTGOMERY: "Gene Flow and the Geographic Structure of Natural Populations," *Science,* vol. 236, pages 787–792, 1987.

STANLEY, STEVEN M.: "Mass Extinctions in the Ocean," *Scientific American,* June 1984, pages 64–72.

STEBBINS, G. LEDYARD, and FRANCISCO J. AYALA: "The Evolution of Darwinism," *Scientific American,* July 1985, pages 72–82.

STEBBINS, G. LEDYARD, and FRANCISCO J. AYALA: "Is a New Evolutionary Synthesis Necessary?" *Science,* vol. 213, pages 967–971, 1981.

STRINGER, C. B., and P. ANDREWS: "Genetic and Fossil Evidence for the Origin of Modern Humans," *Science,* vol. 239, pages 1263–1268, 1988.

THORNHILL, RANDY, and DARRYL T. GWYNNE: "The Evolution of Sexual Differences in Insects," *American Scientist,* vol. 74, pages 382–389, 1986.

TURCO, RICHARD P., *et al.*: "The Climatic Effects of Nuclear War," *Scientific American,* August 1984, pages 33–43.

TURCO, RICHARD P., *et al.*: "Nuclear Winter: Global Consequences of Multiple Nuclear Explosions," *Science,* vol. 222, pages 1283–1292, 1983.

WALKER, ALAN, and MARK TEAFORD: "The Hunt for *Proconsul*," *Scientific American,* January 1989, pages 76–82.

WILKINSON, GERALD S.: "Reciprocal Food Sharing in the Vampire Bat," *Nature,* vol. 308, pages 181–184, 1984.

WILLIAMSON, PETER G.: "Morphological Stasis and Developmental Constraint: Real Problems for Neo-Darwinism," *Nature,* vol. 294, pages 214–215, 1981.

WILSON, E. O.: "The Sociogenesis of Insect Colonies," *Science,* vol. 228, pages 1489–1495, 1985.

WOOLFENDEN, G. E., and J. W. FITZPATRICK: "The Inheritance of Territory in Group Breeding Birds," *BioScience,* vol. 28, pages 104–108, 1978.

SECTION **8**

Ecology

Energy flows through an ecological system by way of a series of trophic (feeding) levels. At the first trophic level, in most ecosystems, are photosynthetic organisms, such as the grasses and other plants of this Serengeti plain. At the next trophic level are the primary consumers, the animals that derive their energy from plants, as, for example, zebras and wildebeests. The Serengeti is able to support many different kinds of herbivores because of slight differences in the type of vegetation each consumes.

Population Dynamics: The Numbers of Organisms

Ecology is the study of the interactions of organisms with their physical environment and with each other. As a science, it seeks to discover how organisms affect, and are affected by, their living and nonliving environment and to define how these interactions determine the kinds and numbers of organisms found in a particular place at a particular time.

Ecology is both the oldest and the youngest of the major subdivisions of biology. It is at least as old as the inquiring human mind—clearly not a subject any australopithecine could safely ignore. However, as a "hard" science it is young because it is only recently that biologists have been able to devise ways to analyze the multitude of variables that affect organisms in their natural environment, to study them quantitatively, and to construct models, pose hypotheses, and formulate and test predictions. The great ecologists of 40 years ago, who laid the foundations of modern ecology, were sensitive observers of nature; their successors, equally sensitive observers as they gather the data required to formulate and test hypotheses, are often also wizards at calculus, statistics, and computer modeling.

We are going to begin these chapters on ecology by looking at some of the factors affecting the size of populations. Subsequent chapters will be concerned with the interactions between and among the populations of different species in communities and with the interactions between communities and their physical environment in ecosystems. We shall end with a global overview of the biosphere.

A population, as you will recall, can be defined as a group of interbreeding organisms, members of the same species found in the same locality. An understanding of population dynamics is not only an essential foundation for studies of the diverse interactions among and between groups of organisms but also is of enormous practical importance. For example, fishery management aims to maximize yields of economically valuable fish. This requires, among other factors, a determination of the numbers of fish that can be caught each year while ensuring that the population will produce sufficient numbers for sizeable harvests in future years; this, in turn, depends on a knowledge of the population dynamics of the species in question. Identification of species in danger of extinction and of the types of intervention that may prevent their extinction also depends on a knowledge of population dynamics. Planning effective pest-control programs that will not have damaging consequences for other organisms and predicting the course of infectious diseases, including such human diseases as AIDS, are only a few of the other activities that depend on a similar knowledge.

52–1 At the next trophic level are the secondary consumers, the carnivores. Because only a fraction of the chemical energy is transferred at each trophic level, there are more plants than herbivores on the Serengeti and more herbivores than lions.

PROPERTIES OF POPULATIONS

As we noted in Chapter 1, different properties emerge at different levels of organization. A molecule has different properties from its constituent atoms, a

TABLE 52-1 Reproductive Capacity of the Housefly (Musca domestica)*

GENERATION	NUMBERS IF ALL SURVIVE
1	120
2	7,200
3	432,000
4	25,920,000
5	1,555,200,000
6	93,312,000,000
7	5,598,720,000,000

* In one year, about seven generations are produced. The numbers are based on the following: each female lays 120 eggs per generation, each fly survives just one generation, and half of the flies are females. Adapted from E. J. Kormondy, *Concepts of Ecology*, Prentice-Hall, Inc., Englewood Cliffs, N.J., 1969.

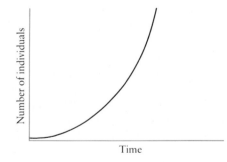

52-2 *An exponential growth curve. After an initial establishment phase, the population increases in the same fashion as a savings account with compound interest. Although the per capita rate of increase remains constant, the growth rate of the population increases rapidly as the number of reproducing individuals increases. Exponential growth is characteristic of small populations with access to abundant resources.*

cell different properties from its constituent molecules, and a multicellular organism different properties from its constituent cells. Similarly, a population has very different properties from the individuals it comprises. As a simple illustration, picture a patch of woods, or a field, or a lake shore to which you return from time to time, which perhaps your parents and grandparents visited and where, with any luck, your children may walk someday. If you go there in the spring, you will see, for the sake of the argument, squirrels, birds, wildflowers, and beetles—the same species you saw the preceding spring and the same species you expect to see in subsequent springs. Yet, year after year, it is unlikely that you see the same individual squirrels and birds, very unlikely that you encounter the same beetles, and seldom the same individual wildflowers. The individual is transitory, but the population endures, year after year.

Among the properties of populations that are not properties of individual organisms are growth patterns, mortality patterns, age structure, density, and dispersion.

Patterns of Population Growth

As Darwin noted some 150 years ago, the reproductive potential of all species is very high. It can be calculated, for instance, that a single female housefly producing 120 eggs (with half developing into females) can, within seven generations, give rise to almost 6×10^{12} houseflies (Table 52-1). With bacteria that have a generation time of 20 minutes, a single bacterium can give rise to eight bacteria within one hour, 512 in three hours, and 262,144 in six hours. Fortunately for us, few organisms ever attain their maximum rate of growth.

The rate of increase of a population is the increase in the number of individuals in a given unit of time per individual present. Thus, the annual rate of growth of the human population in the United States at present is about 9.4 per thousand, and in the world it is about 17.0 per thousand. In the absence of net immigration (the movement of other individuals of the species into the population from elsewhere) or emigration (the departure of individuals from the population), the increase is equal to the birth rate minus the death rate. Thus, the rate of increase can be equal to zero, or it can be a positive or a negative figure (as it now is for the human population of some countries). This property of a population is called its **per capita rate of increase,** and it is symbolized by the letter *r*.

The simplest model of population growth, in which the number of individuals increases at a constant rate, is known as **exponential growth.** It is described by the differential equation $dN/dt = rN$. The term dN/dt equals the growth rate of the population, that is, the change in the number of individuals over time. The equation states that the growth rate is equal to *r*, the per capita rate of increase, times *N*, the number of individuals *already present*. A graph of the equation is shown in Figure 52-2. As you can see, although the per capita rate of increase remains the same, the growth rate of the population—the rate of change in the number of individuals in the population—increases markedly over time. In other words, the slope of the growth curve is slight when the population is small, and then increases as the population gets larger. Exponential growth starts out slowly, but then it shoots up very rapidly as the number of reproducing individuals increases with each generation. The principle is the same as the compounding of interest on a savings account; the more you have, the more you get.

The exponential growth curve is most closely approximated by microorganisms cultivated in the laboratory, where resources are constantly renewed, by the initial stages of seasonal "blooms" of algae, and by the recent growth of the human population (see essay, page 1096). Under most circumstances, however, a population cannot long continue to increase exponentially without reaching some environmental limits imposed by shortages of food, space, oxygen, nesting or hiding places, accumulation of its own waste products, increased competition with other species, or increased predation on the population. In nature, short-term exponential growth is characteristic of so-called fugitive or opportunistic species that invade an area, rapidly use up local resources, and then either enter a phase of dormancy or move on. Weeds and some insects are examples of opportunists.

Sometimes, a population may hit an environmental limit prematurely and so "crash" to very low levels. An example can be found in the infestations of gypsy moths (Figure 52–3) that recurrently plague the northeastern United States. If a population grows rapidly enough, it can wipe out its food supply before the caterpillars complete metamorphosis and reach the reproductive stages. Or, alternatively, the population may reach such a high density that it provides a suitable environment for the exponential growth of an infectious microorganism (a virus, in the case of the gypsy moth) and is wiped out by disease.

The Effect of the Carrying Capacity

For many populations, the number of individuals is determined not by reproductive potential but by the environment. A given environment can support only a limited number of individuals of a particular population under any specific set of circumstances. Population size oscillates around this number, which is known as the **carrying capacity** of the environment. It is the average number of individuals of the population that the environment can support under a particular set of conditions. For animal species, the carrying capacity may be determined by food supply or access to sheltered sites. For plants, the determining factor may be access to sunlight or the availability of water.

Limiting factors may vary seasonally. When the wildebeests migrate through the Serengeti plain of East Africa, all the lions are well fed, but when the big herd animals are gone, the lion cubs often starve to death, and some of the less capable adults as well. The carrying capacity of a herd of deer is not the number counted in the springtime when the new fawns appear, but the number that can survive over several winters.

The patterns of population growth observed in nature are many and complex. One of the simplest patterns, which illustrates clearly the effect of the carrying capacity, is closely approximated by the following equation:

$$\frac{dN}{dt} = rN\left(\frac{K-N}{K}\right)$$

In this equation, r is the per capita rate of increase, as in the previous equation, and it is again multiplied by the number (N) of individuals present at any one time. K stands for the carrying capacity, the number of individuals the environment can support over a specified period of time. A graph of this equation is shown in Figure 52–4. Note that when N is very small, $(K-N)/K$ is close to 1, and the

52–3 A gypsy moth caterpillar. Infestations of these caterpillars can rapidly strip the leaves from trees. Spraying the caterpillars with insecticides often reduces the size of the population enough to prevent a population "crash" and thus maintains the population at a lower but constant level.

52–4 One of the simplest growth patterns observed in natural populations is known as logistic growth and is represented by a sigmoid, or S-shaped, curve. As with exponential growth, there is an initial establishment phase when population growth is relatively slow (1), followed by a phase of rapid acceleration (2). Then, as the population approaches the carrying capacity of the environment, the growth rate slows down (3 and 4), and finally stabilizes (5), though fluctuations around the carrying capacity may continue. Other growth patterns observed in natural populations are considerably more complex.

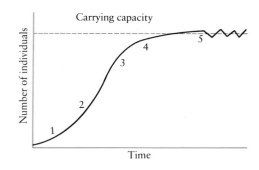

An Example of Ecological Modeling: Transmission of the AIDS Virus

Ecological systems are so complex that it is virtually impossible to examine all relevant factors simultaneously. Thus ecologists often focus their studies through the use of models, which are generally presented in mathematical or graphic form. Of necessity, all such models are simpler than the real world. However, by eliminating unnecessary details, they allow particular features of interest to be examined with more precision. In order to achieve the required simplification, many assumptions are made during the modeling process. These assumptions, based on the best information available at the time, may be changed later if they are shown to be incorrect. Models are powerful tools for gaining insight into both theoretical and practical ecological problems.

Recently, Robert May of Princeton University and Roy M. Anderson of the Imperial College, London, have developed models for the population dynamics of the AIDS virus (human immunodeficiency virus, or HIV). These models were formulated to predict the rates at which infection will spread through the population and the total number of cases that can be expected. This information is of vital importance to physicians, hospitals, public health officials, and various governmental agencies that must make predictions not only of the required financial outlays but also of the revenues that will be lost as previously employed individuals become ill and die. Among the factors considered in the models developed by May and Anderson are the length of time an HIV-positive individual is infectious (assumed to be throughout his or her life), the average number of other individuals infected by each previously infected individual during the early stages of the spread of the disease, and

the fraction of the total population that will eventually be infected.

Whether an infection can become established and spread within a population depends upon the average number of secondary infections—that is, new cases of the infection—produced by each infected individual. The infection will spread only if this quantity, designated as R_0, exceeds one. As May and Anderson point out, for a sexually transmitted disease, R_0 depends on three factors: β, the average probability that the infection is transmitted from infected to susceptible individuals during contact; c, the effective average rate at which new sexual partners are acquired; and D, the average length of time that an HIV-positive individual is infectious. The rate at which the virus would be expected to spread through a population during the early phases of its transmission is given by the following equation:

$$R_0 = \beta c D$$

This equation also provides a basis for designing programs to control the spread of the virus. Although D, the duration of infectiousness, is likely to be a constant, the magnitude of β and c may be lowered through public education programs that lead to a wider use of safer sexual practices and to a reduction in the rate at which new sexual partners are acquired.

According to May and Anderson, in a closed population, the fraction that is eventually infected depends upon both the magnitude of R_0 and the ratio of the variation in the rate of acquisition of new partners, designated as σ, to the mean rate of acquisition of new partners, m. As m is reduced, the ratio σ/m is

curve approximates the exponential growth curve of Figure 52–2. As N increases, $K - N$ decreases, and growth slows down, decreasing to zero when $N = K$. This slowing of population growth represents a decline in the population's rate of increase, perhaps due to competition among the individuals in the population for some limited resource. If the number of organisms exceeds the carrying capacity, the population's growth rate becomes negative and the population declines. Eventually, the population stabilizes and oscillates around the maximum size the environment can support. This model of population growth, represented by an S-shaped, or sigmoid (from the Greek letter sigma), curve, is called **logistic**.

increased. The effect of this ratio on the spread of the infection through the population is illustrated below. Note that the total fraction of the population eventually infected can be reduced by decreasing R_0, which depends on β and c, increasing the ratio σ/m, or both. The greatest decrease in m, and therefore the greatest increase in the ratio σ/m, is produced by a reduction in the activities of the most sexually active members of the population. Insofar as public education and awareness influence this factor, they can significantly alter the ultimate impact of the disease.

When AIDS first appeared among homosexual men in the United States, the spread of infection within that population followed an exponential growth curve. As we noted in Chapter 39, the rate of spread in that community has now slowed dramatically. Among intravenous drug users and their sexual partners, however, the growth curve of new infections remains exponential. Yet, as our examination of a portion of the May and Anderson model reveals, there is great potential for reducing both the rate of spread of the disease and the fraction of this second population that is eventually infected.

Seeking solutions to the AIDS epidemic—and even thinking about it—are difficult, not only because of its inherent biological complexity but also because of the tragic toll in human lives that it has already exacted and is likely to exact in the future. It is here that the modeling approach of May and Anderson may make its greatest contribution. For, while their analysis does not provide all—or even most—of the answers, it can, as May says, "help people think about the disease more clearly."

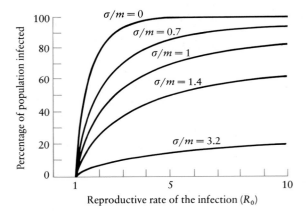

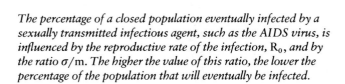

The percentage of a closed population eventually infected by a sexually transmitted infectious agent, such as the AIDS virus, is influenced by the reproductive rate of the infection, R_0, and by the ratio σ/m. The higher the value of this ratio, the lower the percentage of the population that will eventually be infected.

An understanding of the logistic model has practical applications. If, for instance, one wants to control a population of rats, killing half of them may merely reduce the population to the point at which it increases most rapidly. A more effective approach would be to reduce the carrying capacity, which, in the case of rats, usually means tighter control of garbage disposal. Similarly, if one wants to achieve maximum long-term productivity in the harvesting of a particular type of economically valuable fish, the population should not be harvested below the level of rapid growth unless one is willing to wait a long time for the population to recover.

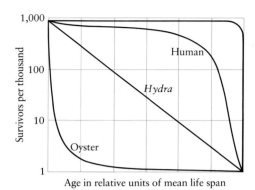

52–5 *Representative survivorship curves. In the oyster, mortality is extremely high during the free-swimming larval stage, but once the individual attaches itself to a favorable substrate, life expectancy levels off. Among* Hydra, *the mortality rate is the same at all ages.*

The curve at the top of the graph is for a hypothetical population in which all individuals live out the average life span of the species—a population, in other words, in which all individuals die at about the same age. The fact that the curve for humans approaches this hypothetical curve indicates that the human population as a whole is reaching some genetically programmed uniform age of mortality.

Mortality Patterns

Another important property, affecting both the size and the composition of a population, is its mortality pattern. Figure 52–5 shows three different patterns of mortality. Often natural populations show a combination of patterns. For instance, in a study made of the saltmarsh song sparrows of San Francisco Bay, it was estimated that of every 100 eggs laid, 26 are lost before hatching. Of the 74 live nestlings, only 52 leave the nest, and of these, 42 (80 percent) die the first year. The remaining 10 breed the following season, but during the next year, 43 percent of these die, leaving only 6 out of the original 100. Each subsequent year, mortality among the survivors amounts to 43 percent, apparently regardless of age. Once a bird survives its first, risk-laden year, the mortality rate for the subsequent years remains more or less constant. Thus, the early survivorship curve for the sparrows resembles that of the oyster, while the later curve is more *Hydra*-like. This situation, in which maximum longevity of the individual does not affect mortality patterns for the population as a whole, is characteristic of birds and of many other animals as well.

Age Structure

The mortality pattern of a population, in turn, affects another important property of the population, its age structure. The age structure of a population is the proportions of individuals of different ages in the population. In species in which the life span exceeds the reproductive span, knowledge of the age structure is useful in predicting future changes in population size. For example, if a large proportion of a population is of reproductive age or younger, as in the human population of India (Figure 52–6), population growth will continue at a high rate even though reproductive individuals only replace themselves. The age structure of the United States population (Figure 52–7) is one of the reasons why the population has continued to increase despite the fact that, on the average, young couples (between the ages of 20 and 30) are having slightly fewer than two children. (Another reason for the increase is continued immigration.) As population growth slows, however, the age structure gradually becomes more uniform. Ultimately, a population that is not increasing attains a stationary age structure.

52–6 *The age structure of the human population of India. This pyramidal shape is characteristic of developing nations, with half the population under 20 years of age. In the absence of emigration, the population can stay the same size only if death rates are as high as birth rates. Even if members of the present generation in India limit their family size to only two children per couple (which means cutting the current birth rate in half), population growth will not level off until about the year 2040—and at a level of well over a billion.*

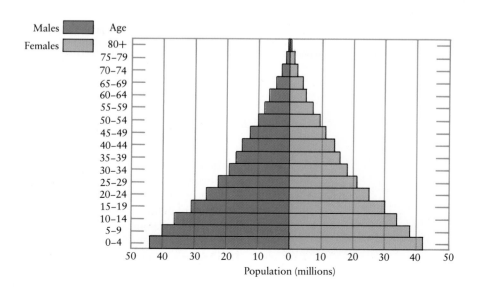
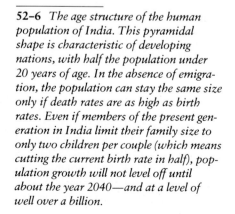

1910

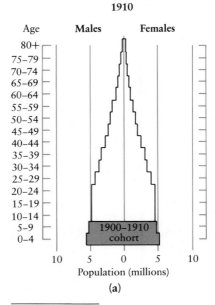

Age | Males | Females
80+
75–79
70–74
65–69
60–64
55–59
50–54
45–49
40–44
35–39
30–34
25–29
20–24
15–19
10–14
5–9
0–4

1900–1910
cohort

10   5   0   5   10
Population (millions)

(a)

1960

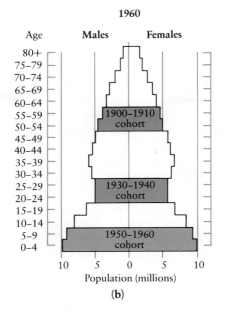

Age | Males | Females
80+
75–79
70–74
65–69
60–64
55–59
50–54
45–49
40–44
35–39
30–34
25–29
20–24
15–19
10–14
5–9
0–4

1900–1910
cohort

1930–1940
cohort

1950–1960
cohort

10   5   0   5   10
Population (millions)

(b)

2000
(projection)

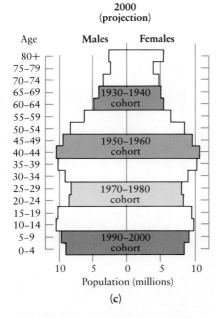

Age | Males | Females
80+
75–79
70–74
65–69
60–64
55–59
50–54
45–49
40–44
35–39
30–34
25–29
20–24
15–19
10–14
5–9
0–4

1930–1940
cohort

1950–1960
cohort

1970–1980
cohort

1990–2000
cohort

10   5   0   5   10
Population (millions)

(c)

52–7 *Age structure of the United States population.* **(a)** *In 1910, the graph of age structure was shaped like a pyramid, although its base—that is, the number of persons in the youngest age groups—was not as large as that of India's.* **(b, c)** *In subsequent years (actual and projected), the proportion of the population that is* more than 40 years of age has steadily increased. Note the decrease in the growth of the population during the depression years of 1930 to 1940 and the bulge produced by the "baby boom" of the 1950s. Although birth rates are now low in the United States, as they were through the 1970s, the increase in the number of individuals during the 1950s was reflected in an increase in population growth in the 1980s as the boom babies reached reproductive age. The term "cohort" refers to the group of individuals born during the decade indicated.

Density and Dispersion

Two additional, interrelated properties of populations are density and dispersion. Population density is the number of individual organisms per unit area or volume. For example, the number of *Paramecium aurelia* cells per cubic centimeter of pond water, of dandelions per square meter of lawn, and of field mice and oak trees per hectare are all measures of population density. A description of the dispersion provides additional information about the population: the pattern of distribution of the organisms within the two- or three-dimensional space.

The three basic dispersion patterns are shown in Figure 52–8. They are (1) random, in which the spacing between individuals is irregular, and the presence of one individual does not directly affect the location of another individual; (2) clumped, in which the individuals are aggregated in patches, and the presence of one individual increases the probability of finding another individual in close proximity; and (3) regular, in which individuals are evenly spaced within the area, and the presence of one individual decreases the probability of finding another individual in close proximity.

52–8 *The three basic dispersion patterns observed in natural populations are* **(a)** *random,* **(b)** *clumped, and* **(c)** *regular. The dots may represent individuals of the same species, populations of the same species, or populations of different species. Determination of dispersion patterns requires careful observation and precise mapping, repeated in a number of different areas and at different times.*

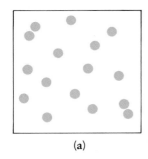

(a)

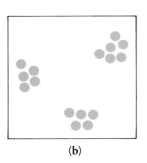

(b)

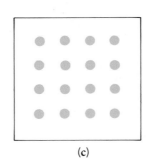

(c)

(a)

(b)

(c)

52–9 *Some examples of dispersion patterns.* (a) *Cliff swallows, like many other birds, exhibit a regular dispersion when perched on telephone wires or power lines.* (b) *Mussels and* (c) *quaking aspen typically exhibit clumped dispersion patterns.*

A number of factors, both nonliving and living, may affect the spatial distribution of a particular population. Dispersion patterns are often dependent upon the distribution of essential resources; if, for example, water is available only in patches, members of a population of a particular plant species may be found clumped in close proximity around these patches. In vertebrates, dispersion patterns often reflect social behavior. Individuals of a species that is highly territorial usually exhibit a regular spacing, whereas individuals of highly gregarious species tend to be clumped.

Dispersion patterns are not necessarily fixed and may vary either seasonally or at different stages in the life cycle. In many invertebrates, especially marine forms, the distribution of larvae is different from that of adults. Many bird species show a regular dispersion pattern during the breeding season, when territorial behavior is at its peak, but a clumped distribution during the rest of the year. Dispersion patterns may also vary according to fluctuations in important resources. For example, the density and distribution of creosote bushes, which grow in the deserts of the American southwest (see page 643), are correlated with the total annual rainfall. In areas of relatively high rainfall, the density of creosote plants is high, and they display a clumped dispersion pattern. In areas where rainfall is low, density of the plants is also low, and the dispersion patterns are regular. As you might expect, in areas with intermediate amounts of rainfall, the density of creosote bushes is intermediate between the two extremes. The dispersion pattern, however, is neither clumped nor regular but is instead random.

The dispersion pattern observed for a particular population or species depends upon the scale of the observation. On a local scale—a few square meters or kilometers, for example—a species may show a random or regular distribution. However, when viewed on a larger scale, such as the entire North American continent, the same species may show a clumped distribution.

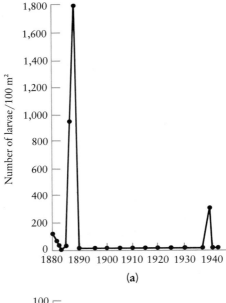

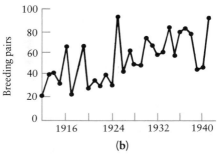

52–10 *Two examples of the sometimes extreme fluctuations that occur in population size and density.* (a) *The population density of dormant larvae of the moth* Dendrolimus pini, *recorded over a period of 60 years in a coniferous forest in Germany.* (b) *Variations over a span of 30 years in the size of the breeding population of the great tit* (Parus major), *a common European bird of the same genus as the titmice and chickadees of North America, recorded at a single location in the Netherlands.*

THE REGULATION OF POPULATION SIZE

Fifty years ago, the eminent ecologist Charles Elton noted that "no animal population remains the same for any great length of time, and . . . the numbers of most species are subject to violent fluctuations." The popular notion that "nature is in balance" and that populations generally reach an equilibrium state has come under severe criticism from contemporary ecologists. As the two examples illustrated in Figure 52–10 show, the size of a population may vary greatly over a period of years.

Although it is often difficult to understand why fluctuations in population size occur, such knowledge can be of great importance because population fluctuations in one species can have profound effects—for good or ill—on populations of other species, including ourselves. For example, earlier in this century an enormous increase occurred in the number of field mice in Kern County, California. The population density reached thousands of mice per hectare, leading not only to the devastation of crops but also to the infestation of playgrounds and living rooms. Abrupt declines in a population can also occur with staggering speed; a field plagued with mice in June may have practically none in July. A number of factors are thought to be involved in such fluctuations.

Limiting Factors

Among the influences affecting the size and density of a population are specific limiting factors that differ for different populations. Of critical importance is the organisms' range of tolerance for such factors as light, temperature, available water, salinity, nesting space, and shortages (or excesses) of required nutrients. If any essential requirement is in short supply or any environmental feature is too extreme, growth of the population is not possible, even though all other necessities are met (Figure 52–11).

The way limiting factors can affect the growth of a population is illustrated by a common pollution problem. When phosphorus is the limiting factor in the growth of freshwater algae in a lake or slow-moving stream, the release of phosphate-containing detergents into the water (as, for example, in the outflow from sewage systems) will produce a spectacular bloom of algae. This period of rapid growth continues until the available supply of some other essential element —perhaps calcium—is consumed. Then the algae begin to die, the decomposing bacteria take over, the respiration of the bacteria begins to use up the oxygen in the water, and oxygen becomes the limiting factor. Eventually, the concentration of oxygen drops below the tolerance levels of fish and other organisms, and these, too, begin to die. If the process is not interrupted, the lake becomes completely stagnant, and only anaerobic bacteria and other microorganisms can survive in it.

52–11 *The principle of limiting factors. Every species has a characteristic limiting factor curve for each factor in its environment. In the zones of intolerance, individual organisms cannot survive. In the zones of physiological stress, some individuals are able to survive, but the population cannot grow. In the optimum range, the population can flourish.*

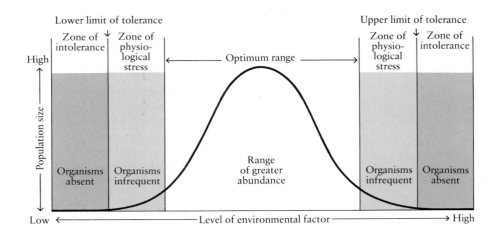

The Human Population Explosion

It is estimated that about 25,000 years ago there may have been as many as 3 million people. Some 15,000 years later, at the close of the Pleistocene epoch, the population probably numbered a little more than 5 million, spread over the entire world. At this point, some 10,000 years ago, the establishment of agricultural communities began. In the next 5,000 years, agriculture spread throughout the world and was accompanied by a twentyfold increase in the human population—to about 100 million in 3000 B.C., as shown in (a).

From 3000 B.C. until about 1650, human population growth slowed considerably. During this period —slightly less than 5,000 years—the population increased only fivefold, to about 500 million. At about this time, however, the rapid development of science, technology, and industrialization began, bringing profound changes in human life and its relationship to nature. In the 200 years between 1650 and 1850, the population doubled, to 1,000 million (1 billion), and then it doubled again by 1930, to 2 billion.

By 1989, there were 5.25 billion people on our planet, and the population is growing at a rate of 1.7 percent per year. This means a net increase in the world population of about 170 people every minute, more than 244,000 each day, and almost 90 million every year. If this rate of increase is sustained, there will be about 6.3 billion people on earth by the year 2000.

As we have seen, the rate at which a population increases is the result of both its birth rate and its death rate. The relative influence of these two factors has varied at different times in human history. In the period between 25,000 and 10,000 years ago, when humans lived as hunter-gatherers, the low rate of increase is believed to have resulted from a low birth rate, in turn the result of physiological factors. For example, in some primitive hunter-gatherer societies, a woman is unable to conceive a child until she is between 19 and 20; moreover, throughout the period when she is breast-feeding a child (which may be as long as three or four years), ovulation usually does not occur. Compared to other contemporary human societies, the age at first reproduction is later (and thus the total reproductive span of the woman is shorter) and the interval between children is longer; these factors combine to produce a much lower birth rate in the

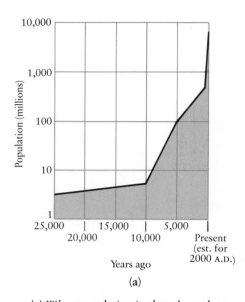

Population (millions) — Years ago — Present (est. for 2000 A.D.)

(a)

(a) *When population is plotted on a logarithmic scale (vertical axis) against time on a normal arithmetic scale (horizontal axis), the slope of the line between any two points represents the rate at which the population increased during that time interval. In the period from 25,000 to 10,000 years ago, when* Homo sapiens *lived as a hunter-gatherer, the rate of increase was low. During the next 5,000 years, when the agricultural revolution was spreading around the world, population growth increased dramatically. Population growth continued after an agricultural way of life had become established, but at a significantly lower rate. However, with the advent of the scientific and industrial revolution, about 1650, population growth again increased dramatically and is continuing at a very rapid rate.*

population. It has also been suggested that the difficulties of carrying small children in a nomadic society that lacked domesticated pack animals would have created social pressures for a long interval between children.

It is thought that the changes in culture and nutrition that accompanied the shift to an agricultural way of life may have led to a breakdown in the mechanisms that controlled the birth rate, resulting in the dramatic population increases that occurred between 10,000 and 5,000 years ago. The reasons for the subsequent slowing of population growth are not known with certainty. One contributing factor was probably an increased vulnerability to infectious diseases as people lived in closer proximity to each other within agricultural communities. Although the population continued to grow, increased mortality—particularly among the young—had a damping effect on the growth rate. Social customs, such as postpartum taboos, may have reduced the birth rate also.

The enormous recent increase in the world's population is primarily a result of the decline in death rates, especially among the young, as shown in (b). Despite the fact that this current phase of population increase is related to a reduction in death rates, some experts believe that further reductions in death rates and a general increase in the standard of living will reduce the rate of population growth. Other experts, however, disagree. They point out that for many countries a significant rise in the standard of living seems an almost unattainable goal and, in some cases, a rapidly receding one. Moreover, the age structure in many countries is such that even if birth rates were to drop dramatically, it would be many years before population size would begin to level off.

There is, however, a correlation between economic development and high birth rates. It is in the developing countries that the highest rates of population growth are found. In India, for example, which has struggled for years to feed its ever-surging population, the annual rate of population increase is 2.0 percent. Yet most Indian women do not seek help in birth control until they have three or four children.

The desire for large families is deeply rooted in the Indian culture. Moreover, in Indian tradition, children provide security for their parents in old age. Two studies of life expectancy in India have shown that, with the high death rate among children, it is necessary for a mother to bear five children if the couple is to be 95 percent certain that one son will survive the father's sixty-fifth birthday.

Within the last few years, the rate of population growth worldwide has begun to decline, but it is too early to know if this decline represents a long-term trend or is only temporary. What has become clear, however, is that the current growth of the human population poses urgent and complex problems, made more difficult by the unequal distribution of growth and available resources. Births are occurring at the greatest rates in precisely those areas where the new arrivals have the least chance of an adequate diet, good housing, schools, medical care, or future occupations. Moreover, because the affluent citizens of developed countries are not constantly reminded of the soaring population by the problems of hunger and crowding, so evident to the poor of other countries, they may feel it is not their problem. Yet a child born to a middle-class American will consume, in his or her lifetime, a far greater amount of the limited resources of the world—more than twice the amount of food, for instance—than a child born in a less developed country. Bringing human population growth under control is clearly an enormously complex undertaking, involving biological, economic, political, and social factors and requiring wisdom and commitment on the part of the diverse populations that constitute *Homo sapiens*.

(b) *In many tropical countries, the death rate has fallen rapidly since 1940, resulting in a rapid growth of the population. The drop in death rate is the result of increased medical services, control of malaria-bearing mosquitoes by DDT, and the availability of new antibacterial drugs, especially the antibiotics. Note that the birth rate has also begun to fall. The area marked in color indicates the population growth in the absence of net immigration or emigration. The data shown here are for Sri Lanka (formerly Ceylon).*

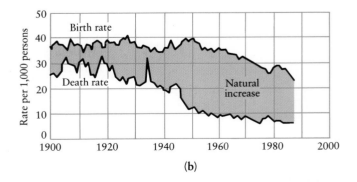

(b)

52–12 *A European meadow vole,* Microtus agrestis. *These rodents have a body length of about 10 centimeters, exclusive of the tail, which is typically about 4 centimeters long. They are found throughout Europe, including Great Britain, where they live in burrows. Within the order Rodentia, voles, lemmings, and muskrats are classified in family Muridae (rats and mice), subfamily Microtinae. Old World rats and mice, including dormice (page 787), are classified in subfamily Murinae; hamsters and New World rats and mice are members of subfamily Cricetinae.*

Density-Dependent and Density-Independent Factors

Ecologists often divide the factors that influence the growth of a population into **density-dependent** and **density-independent** factors. Factors that cause changes in either the birth rate or the mortality rate as the density of the population changes are said to be density-dependent. In our previous example, the limits on algal growth imposed by shortages of phosphorus and calcium were density-dependent. Birth rates and mortality rates, however, may also vary without regard to changes in population size or density. Factors that cause such changes are said to be density-independent, and they often involve weather-related events.

Numerous factors operate on populations in a density-dependent fashion. As a population increases in size, it may deplete its food supply, leading to increased competition among members of the population that ultimately leads to a higher mortality rate or a lower birth rate. Predators may be attracted to areas in which the density of prey organisms becomes high, capturing an increased proportion of the population. Similarly, diseases may spread more easily when population density is high.

Environmental disturbances often act as density-independent factors. For example, when Mount St. Helens erupted in 1980, populations of many species became locally extinct. Mortality was completely independent of the size and density of the populations. In many cases, however, the identification of density-independent factors is less clear-cut. The mortality resulting from a severe storm or cold spell may appear to be density-independent, yet animals that have secure places of shelter are more likely to survive than those that do not. And, the availability of good sheltering sites depends on the density of the population. As this example illustrates, there is often an interplay between density-dependent and density-independent factors, and populations are generally affected by both simultaneously.

Population Cycles

One of the most interesting and perplexing types of population regulation is the phenomenon of population cycles, exemplified most strikingly by voles *(Microtus).* Many species of voles, which include not only the meadow mice of the northeastern United States but also a variety of larger rodents in Europe, undergo population fluctuations that peak every three to four years. Although biologists have proposed numerous hypotheses to account for the cyclic periodicities, the phenomenon remains a mystery.

As you might expect, most of the proposed hypotheses involve density-dependent explanations. The most obvious hypothesis—that at peak density the voles exhaust their primary food supply—has been tested experimentally a number of times by providing additional food to high-density populations. In many of these experiments, the availability of adequate food has either had no effect at all or has only slightly slowed the subsequent decline. An alternative explanation suggests that when vole densities become very high, hormonal imbalances occur within individual voles, causing reproduction to cease. As a consequence, the population crashes to only a few individuals. With the density reduced, these individuals are once more able to reproduce, and the population gradually rebuilds. Another hypothesis proposes that as the number of voles increases, the number of predators also increases, leading to a reduction in the vole population as the predators feast. Other ecologists think that the explanation lies in emigration; according to this hypothesis, when population density is very high certain members of the population disperse to areas where the density is very low. Note that these hypotheses are not mutually exclusive; it is possible that all of the proposed factors are involved.

Density-independent hypotheses have also been proposed. Some investigators believe that vole population cycles are correlated with weather patterns, and a few have even suggested that there may be a relationship to sunspot cycles. Although this may seem farfetched, it is an indication of how baffling population cycles remain, despite years of study. This, in turn, is an indication of the complexity of population dynamics.

LIFE-HISTORY PATTERNS

Among the most interesting and variable properties of populations are **life-history patterns,** which are groups of coadapted traits affecting reproductive survival. Under what conditions, for example, will natural selection favor an organism that has two million almost microscopic young at one spawning, like an oyster, as opposed to an organism that has only one large infant, like an elephant? In view of the fact that physiological constraints prevent an oyster from producing a 10-kilogram offspring, this question may sound silly. However, as a result of studies such as those on clutch size in birds (page 995), biologists have become aware that life-history patterns vary from individual to individual within a population and also from one population to another population of related organisms. In other words, the patterns involve genetically determined variations subject to natural selection.

The Alternatives

Alternative life-history patterns have been given a variety of names. G. E. Hutchinson of Yale University called them prodigal and prudent, noting that despite the apparent value judgments of these terms, prodigal may be successful under some circumstances where prudent is not. Others have defined prodigal as opportunistic, and prudent as equilibrium. Generally speaking, prodigal or opportunistic patterns would appear more adaptive for "weedy" species, colonizers of open fields, for instance, whereas prudent or equilibrium patterns would appear to be more adaptive for a population at the carrying capacity. For this reason, Robert MacArthur and E. O. Wilson proposed that the patterns be classified as r-selected or K-selected. This proposal, however, has been criticized as an oversimplification. Many species clearly have both r-selected and K-selected characteristics. For example, some species of intertidal starfish have long lives (a K-selected characteristic) and produce numerous eggs (an r-selected characteristic). Other species exhibit predominantly r-selected patterns at some points in their life cycle and K-selected patterns at other points.

The characteristics associated with the alternative patterns are shown in Table 52–2. Each pair of characteristics represents the hypothetical extremes in a continuum of possibilities. The best choice in each continuum—the choice made by natural selection operating on a particular population—depends in large part

TABLE 52-2 Alternative Life-History Characteristics

PRODIGAL (r-SELECTED)	PRUDENT (K-SELECTED)
Many young	Few young
Small young	Large young
Rapid maturation	Slow maturation
Little or no parental care	Intensive parental care
Reproduction once ("Big Bang")	Reproduction many times

(a)

(b)

52–13 *Some examples of life-history patterns.* (a) *Annual plants, such as the zinnia, reproduce only once, typically setting large numbers of seeds. The parent plant dies soon after, and only the seeds survive.* (b) *In marsupial mice of the genus* Antechinus, *the males are "big bang" reproducers. They stake out territories, fight, cruise in search of females, and copulate for as long as 12 hours daily for up to 11 days, and then die, exhausted and sterile. The only male mice remaining are the embryos in the females' wombs.* (c) *A leatherback turtle returning to the sea after laying her eggs in a hole dug in the sand. The eggs, which she has covered over with sand, will be incubated by the warmth of the sun. Once the eggs are laid, the mother turtle does not participate further in raising the young.* (d) *A female water bug lays her eggs on the back of the male. The male then carries the eggs, aerating them, until they hatch. In the photograph, one egg can be seen hatching at the left.* (e) *Baby Canada geese remain with their parents through the summer after they are hatched and migrate with them to the winter feeding grounds. In proportion to life span, the time allotted to parental care by birds is longer than for any other animal group.* (f) *Elephant calves are suckled by their mothers for at least two years. The elephant social structure is a maternal hierarchy in which the young are zealously guarded by the mother and the sisters and the aunts.*

on other properties of the population (such as its mortality pattern and its age structure) and on properties of the environment it occupies.

Early or Late Reproduction

Suppose you are a plant with a very short life expectancy—a flowering annual, for example. The choice that has been made for you, as a result of the action of natural selection in the past, is to put as much energy as possible, as rapidly as possible, into one reproductive effort. The result is a large burst of seeds, increasing the probability that at least a few will survive your short life span and, in turn, reproduce. On the other hand, if you are an oak tree, once you have established a place for yourself, your life span is likely to be a very long one, more than 100 years. However, the likelihood of your progeny finding space on the forest floor is very small. You grow as tall as possible, for 20 years or more, since height aids in seed dispersal, and then put out moderate numbers of acorns per year, every year, over your entire life span. In this way, you are likely to produce more surviving offspring than if you had put the same amount of energy into one big burst of reproduction.

Bluefin tuna are like oak trees. Once a bluefin reaches a large size (about 4 meters in length, weighing about 750 kilograms), its chances of continued survival are extremely good. However, the mortality among eggs and larvae is always very high, and, in years in which the plankton growth is low, there are no survivors at all. So it is better for a bluefin to concentrate its efforts first on reaching a safe size and thereafter to expend a relatively small amount of reproductive effort each season over a long period of time.

Whether breeding is early or late can greatly influence the rate of population growth. Reconsider the housefly of Table 52–1. If the generation time (that is, the interval between the time an egg is laid and the time that individual, in turn, mates and lays eggs) were shortened by only two days (from a minimum of 13 to a minimum of 11), there would be time for one more generation each season. The total number of possible progeny per season would increase by 330×10^{12}.

Precocity does not always pay, however. Among many of the larger mammals, the rate of juvenile survival can be correlated with the experience of the mother, her size, or her social position, which are often determined by her age. In such situations, the action of natural selection favors delayed reproduction. A similar pattern is observed in the human population. Infant mortality is higher among teenage mothers in the United States than among women over 20, and higher still if the teenagers are from impoverished families, as they often are.

(c)

(d)

(e)

(f)

Small or Large Offspring

Among the reproductive characteristics particularly susceptible to natural selection is seed size in plants. Table 52–3 shows the average seed weights of populations of various species of goldenrod *(Solidago)* growing in a field (a new environment open to exploitation), and a prairie (an environment where populations are closer to the carrying capacity). Five of the six species examined were found in both habitats. As you can see from the table, there were remarkable differences both among species and among populations. When seed size was plotted against seed number, it was found, as expected, that the smaller the seeds on any given plant, the more numerous they were. Thus, in this situation and for these characteristics, the predictions of MacArthur and Wilson regarding life-history patterns (see Table 52–2) were fulfilled.

TABLE 52–3 **Mean Weight (in micrograms) per Seed of *Solidago* Species in Old Field and Prairie Communities**

	OLD FIELD	PRAIRIE
S. nemoralis	26.7 ± 2.1	104.0 ± 8.3
S. missouriensis	17.6 ± 0.6	39.3 ± 3.2
S. speciosa	19.5 ± 1.6	146.3 ± 11.7
S. canadensis	27.3 ± 2.3	58.3 ± 11.1
S. gigantea	—	50.8 ± 4.2
S. graminifolia	24.5 ± 2.7	10.6 ± 1.6

Adapted from P. A. Werner and W. J. Platt, "Ecological Relationships of Co-occurring Goldenrods *(Solidago:* Compositae)," *American Naturalist,* vol. 110, pages 959–971, 1976.

The Asexual Advantage

As we noted earlier (page 986), when we considered the evolution of sexual reproduction, an asexually reproducing population can increase in numbers much more rapidly than a sexually reproducing population. There are other advantages to asexual reproduction. Many plants, for example, reproduce by means of runners (see page 642), and by so doing, may be able to grow to cover a very large area. All of the plants produced represent a single genotype. A young plant that develops in this way has a continued supply of resources from the parent plant and thus a much higher probability of survival. Figure 52–14 shows the differences in survivorship in a species of buttercup among plants growing from seeds as compared to plants growing from runners. Note the very high mortality in the early stages of growth from seed—an oysterlike curve—compared to the uniform risk of death of the plantlets grown from runners.

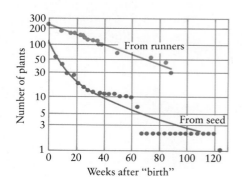

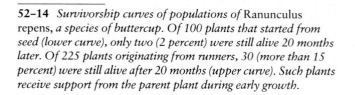

52–14 *Survivorship curves of populations of* Ranunculus repens, *a species of buttercup. Of 100 plants that started from seed (lower curve), only two (2 percent) were still alive 20 months later. Of 225 plants originating from runners, 30 (more than 15 percent) were still alive after 20 months (upper curve). Such plants receive support from the parent plant during early growth.*

52–15 *One of the most widely distributed of all the flowering plants is the dandelion. Its flowers produce enormous numbers of seeds parthenogenetically. The seeds are enclosed in fruits bearing a plumelike structure that is caught by the wind, thereby aiding dispersal. Dandelions also reproduce vegetatively; if the plant is lopped off down to the root by either hominid or herbivore, three more grow in its place.*

Parthenogenesis

Another form of asexual reproduction is parthenogenesis, the development of an organism from an unfertilized egg. In species in which the male gamete determines the sex of the offspring, parthenogenesis always results in all female offspring. Hence, it is far more efficient than sexual reproduction. If the houseflies of Table 52–1 had reproduced parthenogenetically, each female would have had twice as many female young in every generation, and the population would have reached 358×10^{12} at the end of seven generations.

Parthenogenesis in plants lacks the advantage of the parental support system supplied by vegetative growth, which is traded off for the possibilities of larger numbers and, usually, wider dispersal of the young. Dandelions reproduce parthenogenetically. They form conspicuous flowers and also some functionless pollen grains, which may be taken as evidence that the present asexual species of dandelions evolved from sexual ones. As a consequence of parthenogenetic reproduction, dandelions growing in a single locality often consist of several different populations, each composed of genetically identical individuals. Otto Solbrig compared two such populations of dandelions growing together in various localities near Ann Arbor, Michigan. One genotype, genotype D, outperformed the other under all environmental conditions, both in the number of plants that survived and in their total dry weight. On the other hand, genotype A always produced more seeds and produced them earlier, so it always got a head start whenever a newly disturbed area became available for occupation.

Completely asexual species are also found among small invertebrates—some rotifers, for example—as well as among plants. Recently, several species of fish, lizards, and frogs have been found that apparently reproduce only parthenogenetically. Many other organisms alternate sexual and asexual phases. Freshwater *Daphnia*, for instance, multiply by parthenogenesis when the plankton on which they feed is abundant. Then, in response to some environmental cue, they start producing both males and females. Typically, the asexual phase occurs when conditions are favorable for rapid local growth, and the sexual phase when the population is facing a less certain future and less homogeneous conditions.

Among the organisms that alternate sexual and asexual phases is the fungus-eating gall midge,* which should perhaps be awarded the prize for precocious development among multicellular animals. These small flies, which are found on mushroom beds, can reproduce sexually. However, when abundant food is available, a female can produce parthenogenetic eggs, which are retained inside her body. The larvae develop inside the mother, devour her tissues, and, completely skipping the usual metamorphosis, emerge with eggs inside their own tissues. Within two days, larvae emerge from these eggs, devour their own mother, and are soon devoured in turn. Eventually, in response to some environmental signal, the parthenogenetic cycle is broken, and the females produce normal males and females that fly off in search of new mushrooms.

Some Consequences of Life-History Patterns

Opportunistic organisms, which rapidly exploit an environment and then move on, would appear to lead risky lives, both as individuals and as species. There may not always be another mushroom. However, populations of such organisms typically possess remarkable recuperative powers, because a population can be built up quickly from only a few individuals. By contrast, populations composed of long-lived, slow-to-mature individuals, which would appear to have a higher probability of long-term survival, are very slow to recover when their size is

---

* This example of the gall midge was borrowed from Stephen Jay Gould, *Ever Since Darwin.*

52–16 *The natural habitat of the California condor, which soars over a huge territory in its search for food, has been destroyed by human invasion, and the species has been driven to the edge of extinction. As of 1988, only 28 individuals survived, all of which were in captivity, a setting in which the offspring have a chance at survival. Despite a high degree of parental care, the California condor cannot reproduce rapidly enough for a population this small to recover in the wild.*

reduced. Both the California condor and the whooping crane are in this category. Each does not begin to reproduce until it is about four years old, each raises only a single chick per season, and each is threatened with extinction.

SUMMARY

Ecology is the study of the interactions of organisms with their physical environment and with each other. Ecologists are seeking to quantify the variables that affect organisms in nature, to construct explanatory hypotheses for the observed distribution and abundance of organisms, and to make and test predictions based on their hypotheses.

A primary unit of ecological study is the population, a group of organisms of the same species living in the same place at the same time. Among the new properties that emerge at the level of the population are patterns of population growth, patterns of mortality, age structure, density, and dispersion.

The reproductive potential of most populations is high. When the full reproductive potential of a population is achieved (a relatively infrequent occurrence in nature), exponential growth can occur. It is represented by the equation $dN/dt = rN$. In this equation, r is the per capita rate of increase (which, in the absence of net immigration or emigration, equals the birth rate minus the death rate), N is the number of individuals in the population at any given time (t), and dN/dt is the growth rate of the population (the change in the number of individuals over time). A key aspect of exponential growth is that, although the per capita rate of increase remains the same, the growth rate increases as the size of the population increases. Exponential growth cannot long continue without a "crash" in population size.

The number of individuals in a particular population that the local environment can support over a specified period of time is known as the carrying capacity. The logistic model, which takes the carrying capacity into account, describes one of the simpler patterns of population growth observed in nature. Logistic growth is represented by the equation:

$$\frac{dN}{dt} = rN\left(\frac{K-N}{K}\right)$$

In this equation, K represents the carrying capacity. The graph of this equation at first resembles the exponential growth curve, rising slowly while N is still small and then shooting up rapidly as N increases. However, unlike exponential growth, logistic growth gradually slows as the population approaches the carrying capacity, and eventually the population stabilizes at or near the carrying capacity. The graphic result is an S-shaped curve.

Populations also have characteristic mortality patterns, with varying risk of death at varying ages. A related property is the age structure of the population, that is, the proportions of individuals of different ages. Age structure is an important factor in predicting the future growth of a population.

Two additional, interrelated properties of a population are its density and its dispersion pattern. Density is the number of individual organisms per unit area or volume, whereas the dispersion pattern describes the two- or three-dimensional arrangement of the organisms.

A complex variety of environmental factors, both living and nonliving, play a role in the regulation of population size. Factors that influence birth rates or death rates independently of the population density are said to be density-independent; they often involve severe environmental disturbances. Factors that cause changes in birth rates or death rates as the density of the population changes are said to be density-dependent; these factors include numerous resources that are available in limited supply. Some populations undergo regular cycles of growth and decline, phenomena that are still poorly understood.

Populations can be characterized by their life-history patterns, which are groups of coadapted traits affecting reproductive survival. These traits are generally genetically determined and thus subject to natural selection. The extremes of the alternative characteristics include many young versus few young, small young versus large young, rapid maturation versus slow maturation, little or no parental care versus intensive parental care, and reproducing once versus reproducing many times. In a wide variety of organisms, the alternatives also include asexual reproduction versus sexual reproduction. Some combinations of these characteristics are favorable for organisms of opportunistic species undergoing exponential growth, whereas others will be selected for in more stable populations living at or near the carrying capacity.

QUESTIONS

1. Distinguish between the following terms: exponential growth/logistic growth; limiting factors/carrying capacity; density-dependent factors/density-independent factors; opportunistic species/equilibrium species.

2. An old French riddle: "The pond lilies in a certain pond grow at a rate such that each day they cover twice as much of the pond as they did the day before. The pond is of a size that it will be completely covered at the end of 30 days. On what day is the pond half-covered? One-tenth covered? One-hundredth covered?"

What is the relevance of this riddle to human ecology?

3. Suppose that you have a "farm" on which you grow, harvest, and sell edible freshwater fish. The growth of the fish population is logistic. You, of course, wish to obtain maximum yields from your "farm" over a number of years. To ensure this, how large should you allow the population to become before you begin harvesting? Identify the point on the logistic growth curve (Figure 52–4) at which you should begin harvesting.

How large a population should you leave unharvested? Identify the point on the curve at which you should take no more fish from the population.

Factors in addition to the pattern of harvesting will affect the yields of fish obtained. What are some of these factors, and how might you adjust them to further increase the yields?

4. What would be the appearance of the survivorship curve of a

population of annual plants? Of automobiles? Of salmon? Of butterflies? Of dishes in a dishwasher?

5. Note that in the three graphs in Figure 52–7 there is a marked difference in the proportion of individuals more than 80 years old, but the vertical axis has not become appreciably longer. What do these data suggest? Does the age-structure graph of India (Figure 52–6) support your conclusion?

6. Distinguish among random, clumped, and regular dispersion patterns, and give an example of each. How does the scale from which organisms are viewed affect the dispersion pattern?

7. Describe the design of a laboratory experiment that would test the hypothesis that voles living in a high-density population reproduce at a lower rate than voles living in a low-density population.

8. Explain how each of the following factors would affect the growth rate of a population: age at first reproduction; time between generations; pre-reproductive mortality; post-reproductive mortality; length of period of parental care.

9. Imagine a hypothetical species in which a particular individual lives only 48 hours and produces only two offspring. How is it possible that this individual may achieve greater fitness than a longer-lived individual that produces 100 offspring? Explain in evolutionary terms.

CHAPTER **53**

Interactions in Communities

53-1 *Photographed in the Farallon Islands National Wildlife Refuge off the coast of California, this starfish of the genus* Pisaster *is opening a mussel, one of its favorite foods. Studies have shown that predation by* Pisaster, *particularly on mussels and barnacles, is an important factor in maintaining species diversity in the rocky coastal communities in which it lives.*

Populations are made up of individual organisms. Communities are made up of populations. Ecologically speaking, a **community** comprises all the populations of organisms inhabiting a common environment and interacting with one another. These interactions are, of course, major forces of natural selection. They also influence the number of individuals in each population and the number and kinds of species in the community. The interactions among different populations are enormously varied and complex, but they can generally be categorized as competitive, predatory, or symbiotic.

COMPETITION

Competition is the interaction between individual organisms of the same species (intraspecific competition) or of different species (interspecific competition) using the same resource, often present in limited supply. As a result of competition, the overall fitness—that is, the reproductive success—of the interacting individuals may be reduced. Among the many resources for which organisms may compete are food, water, light, or living space, including nesting sites or burrows. Competition that involves overt fighting or other face-to-face interaction is known as **interference competition.** Competition that, in the absence of direct interaction, involves the removal or preemption of a resource, leaving less for others, is said to be **exploitative.**

Competition is generally greatest among organisms that have similar requirements and life styles. Plants often compete with other plants for sunlight and water; herbivores, animals that eat plants and algae, may compete with other herbivores; carnivores, animals that eat animals, may compete with other carnivores. Moreover, within these categories, the more ecologically similar two species are, the more intense the competition between them is likely to be.

The Current Debate

For many years, competition has been invoked as a major force in determining the composition and structure of communities—that is, the number and kinds of species present and their spatial and temporal arrangement within the community. Recently, however, a number of ecologists have come to question the importance of competition as an influence on community composition and structure. The debate—at times acrimonious—that has ensued concerns not only the role of competition but also the methods to be used in testing ecological hypotheses.

As we noted in the last chapter, the investigators who laid the foundations of modern ecology were sensitive observers of nature. Many of these ecologists believed, as do a number of their successors, that careful observation of natural patterns of the distribution and abundance of organisms can—without experimental manipulation—yield the data required to test hypotheses about the role of

	New Guinea	Bougainville	Guadalcanal	Espirito Santo	Karkar	Kulambangra	Tolokiwa	Viti Levu	St. Matthias	Rennell
Island elevation (feet above sea level)	16,500	8,500	8,005	6,180	6,065	5,800	4,650	4,341	2,136	360
Lower limit of thrush range (feet above sea level)	9,000	4,000	4,000	0	3,850	3,400	2,460	0	0	0
Total number of bird species	513	98	102	50	53	80	44	48	43	42
Number of bird species in thrush habitat	31	28	36	33	23	34	23	36	35	31

53–2 *The altitudinal ranges of* Turdus poliocephalus, *a thrush of the southwestern Pacific that is related to the North American robin,* Turdus migratorius. *The range occupied by the thrush on each of 10 islands is shown in dark green; portions of each island not occupied by the thrush are in light green. The data below each island give its elevation, the lower altitudinal limit of the thrush, the total number of bird species on the island, and the number of species actually sharing the same habitat as the thrush.*

competition and other forces in shaping community structure. For these ecologists, the patterns themselves tell the story. An example of this approach is a study by Jared Diamond of the University of California at Los Angeles of the distribution patterns of the thrush *Turdus poliocephalus* on islands of the southwestern Pacific. Diamond determined the altitudes between which the thrush ranged on 10 different islands and then compared this information with data on the number of potential competitors (other species of birds). The data included not only the total number of species on each island but also the number of species within the habitat actually occupied by the thrushes. *T. poliocephalus* is omnivorous, consuming a diversity of food items, and thus is potentially in competition with species with widely varying food preferences. As you can see in Figure 53–2, on islands with a large number of potential competitors, the thrush occupies habitats at higher altitudes, whereas on islands with fewer potential competitors, it occupies habitats at all altitudes, from sea level to the top of the island. Diamond interprets these data as evidence that the thrushes, which can tolerate the physical conditions at higher altitudes, adjust their range in response to competition from other bird species. The number of potentially competing species within the altitudinal range occupied by the thrushes varies from a low of 23 on Karkar and Tolokiwa to a high of 36 on Guadalcanal and Viti Levu.

On the other side of the current debate are ecologists who believe that the testing of ecological hypotheses requires carefully structured and controlled experiments, conducted whenever possible under natural conditions. For example, although these ecologists allow that competition may play a role in the observed altitudinal range of *T. poliocephalus,* they believe that there are other possible explanations that should be considered before attributing a causal role to competition. They are uneasy about comparisons from one island to another and believe that such factors as differences in the terrain and other physical features of the islands, in the intensity of predation at different elevations, and in the kinds and amounts of food available should be analyzed. These ecologists maintain that experimental analyses, in which only one variable is manipulated at a time, are the only reliable means for identifying the processes that are actually occurring in nature.

We shall begin our consideration of competition—and of the controversies surrounding it—with the principle that underlies the study of the island thrush populations and that, until very recently, dominated the study of competition and the kinds of questions that ecologists asked about it.

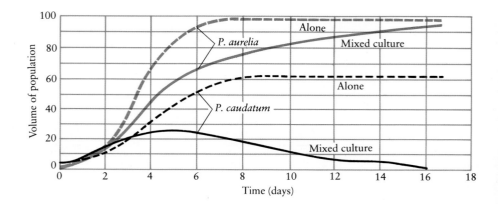

53–3 *The results of Gause's experiments with two species of* Paramecium *demonstrate the principle that if two species are in direct competition for the same limited resource—in this case, food—one eliminates the other.* Paramecium caudatum *and* Paramecium aurelia *were first grown separately under controlled conditions and with a constant food supply. As you can see,* P. aurelia *grew much more rapidly than* P. caudatum, *indicating that* P. aurelia *uses available food supplies more efficiently. When the two protists were grown together, the more rapidly growing species outmultiplied and eliminated the slower-growing species.*

The Principle of Competitive Exclusion

In 1934, the Russian biologist G. F. Gause formulated what became known as the principle of **competitive exclusion.** According to this principle, if two species are in competition for the same limited resource, one or the other will be more efficient at utilizing or controlling access to this resource and will eventually eliminate the other in situations in which they occur together. Gause demonstrated the validity of his principle in his own, now classic, experiments involving laboratory cultures of two species of *Paramecium: Paramecium aurelia* and *Paramecium caudatum.* When the two species were grown under identical conditions in separate containers, both grew well, but *P. aurelia* multiplied much more rapidly than *P. caudatum,* indicating that the former used the available food supply more efficiently than the latter. When the two were grown together, the former rapidly outmultiplied the latter, which soon died out (Figure 53–3).

In laboratory experiments, the fastest growing species is not always the most successful competitor, however, as observed with two species of duckweed, *Lemna gibba* and *Lemna polyrrhiza.* In pure culture, *L. gibba* grows more slowly than *L. polyrrhiza,* yet *L. gibba* always replaces *L. polyrrhiza* when they are grown together. The plant bodies of *L. gibba* have air-filled sacs that serve as little pontoons, so that these plants form a mass over the other species, cutting off the light. As a consequence, the shaded *L. polyrrhiza* dies out (Figure 53–4).

It is possible to devise different culture conditions so that the outcomes of both the *Paramecium* and *Lemna* experiments are reversed. However, as long as the conditions of the particular experiment are held constant, one species always wins and the other is always eventually eliminated.

53–4 *An experiment with two species of floating duckweed, tiny angiosperms found in ponds and lakes. One species,* Lemna polyrrhiza, *grows more rapidly in pure culture than the other species,* Lemna gibba. *But* L. gibba *has tiny air-filled sacs that float it on the surface, and so it shades the other species and is the victor in the competition for light.*

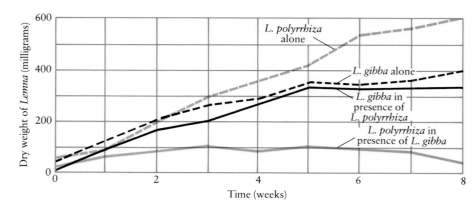

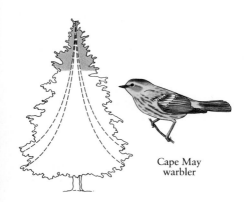

Cape May
warbler

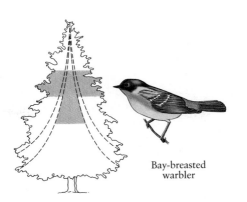

Bay-breasted
warbler

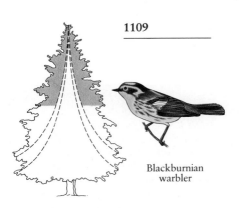

Blackburnian
warbler

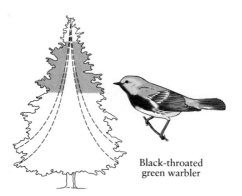

Black-throated
green warbler

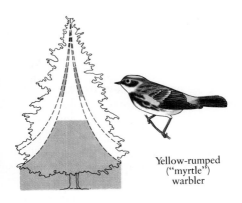

Yellow-rumped
("myrtle")
warbler

53–5 *The feeding zones in a spruce tree of five species of North American warblers. The colored areas in the tree indicate where each species spends at least half its feeding time. This partitioning of resources allows all five species to feed in the same trees.*

The Ecological Niche

Gause's competitive exclusion principle would lead one to expect that only dissimilar species would be found coexisting in natural communities. Yet, in fact, ecologically similar species are often found together in the same community. This observation raised the question of how similar two or more species can be and still continue to coexist in the same place at the same time, which led, in turn, to the concept of the **ecological niche**. This term is somewhat misleading because the word "niche" has a connotation of space, whereas an ecological niche is not the space occupied by an organism but rather the role that it plays. The simplest analogy is that the niche is an organism's profession, as distinct from its habitat, which is its address.

A working definition of the niche is more complex, however, and includes many more factors than the way in which an organism makes its living. A niche is, in fact, the total environment and way of life of all the members of a particular species of organism in the community. Its description includes physical factors, such as the temperature limits within which the organisms can survive and their requirements for moisture. It includes biological factors, such as the nature and amount of required food sources. It includes aspects of the behavior of the organism, such as patterns of movement and daily and seasonal activity cycles. Although only a few of these factors can be studied at any one time, all are likely to influence the interactions of the members of a species with the members of other species in the community.

Resource Partitioning

The concept of the ecological niche suggested that when similar species are found coexisting together, a close examination will reveal that their niches are, in fact, different. Although the organisms appear to the casual observer to be sharing and competing for the same resources, they are not. As numerous studies have revealed, resources are frequently partitioned among ecologically similar members of a community. The cause of the partitioning is currently a matter of considerable disagreement among ecologists, a topic to which we shall return shortly. First, however, let us consider some of the many examples of this phenomenon.

Woodland Warblers

Some New England forests are inhabited by five closely related species of warblers, all about the same size and all insect eaters. The late Robert MacArthur, an extraordinarily brilliant and innovative young ecologist, meticulously observed and timed where the warblers fed within the trees. His data showed that the five species—now sometimes referred to as MacArthur's warblers—each have different feeding zones in the trees (Figure 53–5). Because they exploit slightly different resources—that is, insects in different parts of the trees—the species can coexist. Observations of other groups of animals, including insects, fishes, desert rodents, and lizards, have revealed the same sort of partitioning of resources.

53–6 *A gerenuk, a member of the ante-lope family that is closely related to gazelles, feeding on the leaves of an acacia. Like many other plants of arid regions, this species of acacia has thorns that pro-tect it against vertebrate herbivores. The gerenuk, however, unlike other herbivores of the East African savanna, has special adaptations that enable it to circumvent these defenses. These graceful animals, also known as giraffe-necked gazelles, obtain water as well as food from the plants they eat and thus never need to drink.*

Bog Mosses

In bogs, mosses of the genus *Sphagnum* often appear to form a continuous cover, and several species are usually involved. How can these species, apparently very similar, continue to coexist? When the situation is examined in more detail, it is found that there are semiaquatic species growing along the bottoms of the wettest hollows; other species grow in drier places on the sides of the hummocks, which they help to form; and still other species grow only in the driest conditions on the tops of the hummocks, where they are eventually succeeded by one or more species of flowering plants. Therefore, although all the species of *Sphagnum* coexist, in the sense that they are all present in the same bog, they actually occupy different microhabitats and continually replace one another as the characteristics of each microhabitat change.

African Ungulates

East Africa is notable for the numbers and diversity of its large herbivores—both browsers (leaf eaters) and grazers (grass eaters)—which are often found living together in the same herds. The vegetation that supports these animals looks far less luxuriant and much more homogeneous than the animals that live upon it. Like the woodland warblers, however, the East African ungulates partition the resources.

The subdividing of resources by the browsers is the more obvious process. Only giraffes reach the taller branches, 3 or 4 meters off the ground. Gerenuks, long-legged, long-necked antelopes, stand on their hind legs to nibble delicately in the middle branches (Figure 53–6). The lower branches, a meter or so from ground level, are browsed upon by the rhinos; and the smaller antelopes, such as the dik-dik, hide in the thickets and browse on buds, new twigs, and seed pods.

The adaptations by which the grazers apportion the resources are less obvious, though equally effective. Zebras, which, unlike the browsers, have front teeth in both their upper and lower jaws, eat the older, tougher grass stems. Wildebeests specialize in younger stems and blades nearer the ground level. Gazelles, with their tiny muzzles, nip off the smallest green shoots of grass and also crop the little dicots that grow among the grasses and are almost entirely ignored by the larger herbivores. Thus, not only do the zebras inadvertently leave enough for the wildebeests and the wildebeests for the gazelles, but the tramping and cropping by the larger animals actually encourage the growth of the younger vegetation on which the smaller grazers depend (see page 642).

The Role of Past Competition in Resource Partitioning

The resource partitioning observed among woodland warblers, *Sphagnum* mosses, African ungulates, and many other organisms was long considered to be the result of competition. In some cases, such as the warblers, the competition was thought to be occurring in the present. In other cases, such as the mosses and the ungulates, it was thought to have occurred in the evolutionary past, leading to the differing adaptations that enable the organisms to coexist. This phenomenon, in which species that live together in the same environment tend to diverge in those characteristics that overlap, is known as **character displacement.**

One of the most frequently cited examples of character displacement is provided by the beaks of Darwin's finches. As we saw in Chapter 49, the large, medium, and small ground finch species are very similar except for differences in overall body size and in the sizes and shapes of their beaks. These differences in the beaks are correlated with the sizes of the seeds the birds eat. On islands such as Abingdon and Bindloe (see page 1020), where all three species of ground finch exist together, there are clear-cut differences in beak size. On Charles and

53–7 *Beak sizes in three species of ground finch found on the Galapagos Islands. Beak measurements are plotted horizontally, and the percentage of specimens of each species is shown vertically. Daphne and Crossmans, which are very small islands, each have only one species of ground finch. These species have beak sizes halfway between those of the medium-sized and small finches on the larger islands.*

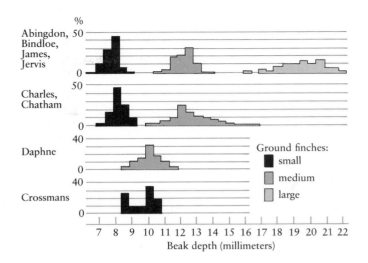

Chatham Islands, the large species is not found, and the beak sizes of the medium ground finches on these islands are larger and overlap the beak sizes of the large finches found on Abingdon and Bindloe. Daphne and Crossmans, which are very small islands, each have only one species; Daphne has the medium-sized finch and Crossmans, the small finch. These two populations have similar beak sizes, which are intermediate between those of the medium-sized and small finches on the larger islands (Figure 53–7).

These data have been interpreted in two different ways by ecologists. Some ecologists maintain that the observed differences in beak size are the result of the selection pressures exerted by interspecific competition. According to this interpretation, competition between organisms whose ecological niches overlap causes selection against individuals with overlapping characteristics, leading to the observed divergence between the species (Figure 53–8). Other ecologists, however, point out that it is impossible to determine if the differing beak sizes are the result of competitive interactions that occurred at times when the different species were coexisting on the same islands or if they are the result of adaptations to local conditions that occurred at times when the species were isolated from one another on different islands. These ecologists believe that conclusive evidence of character displacement requires a demonstration that the species in question are diverging—in the present—in areas where they are found together. According to Joseph H. Connell of the University of California at Santa Barbara, anything less is simply an invocation of "the ghost of competition past." Both groups of ecologists, however, agree that whatever their evolutionary cause, the differences in beak size and shape enable the different finch species to exploit different food sources and thus to coexist.

53–8 *One dimension of an ecological niche. The two bell-shaped curves represent resource utilization by two species in a community. The niche dimension might represent living space, as in the case of* Sphagnum *mosses; foraging space, as in the case of warblers; size of seeds eaten, as in the case of the Galapagos finches; and so on. Competition is potentially most intense in areas of niche overlap, leading to restriction of one or both species in living space, foraging space, or the size of seeds eaten, and so on. It is hypothesized that such competition results in selection against individuals with overlapping characteristics, leading to divergence in the niches of the two species.*

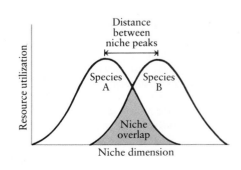

Experimental Approaches to the Study of Competition

Virtually all ecologists agree that competition does occur in nature, with the degree of intensity varying according to the particular species involved, the sizes of the interacting populations, and the abundance or scarcity of resources. Although analysis of distribution patterns and resource partitioning may provide clues as to the occurrence and importance of competition in a particular situation, experimental manipulations involving changes in community composition are required to demonstrate that competition is actually taking place. At their best, laboratory experiments, such as those performed by Gause, can only approximate natural conditions, which are invariably much more complex.

Field experiments of competition generally involve either the removal or addition of individuals of species hypothesized to be competitors. Any change in either the distribution or abundance of individuals of the species *not* manipulated experimentally is considered evidence of competition between the species. Although such experiments sound simple in principle, they are often difficult to perform. First, they require both experimental areas, in which the manipulations occur, and control areas, as similar as possible to the experimental areas, in which no manipulations occur. Such controls are essential to rule out the possibility that any changes observed are coincidental and would have occurred in the absence of the experimental manipulations. Second, replications, utilizing both experimental and control areas, must be performed at intervals of time that vary according to the system under study. Physical conditions (such as temperature), resource availability, and population densities all vary from season to season and from year to year, and as a consequence, the intensity of competition may vary. An experiment performed in June, for example, may yield different results than one performed in September.

Additional difficulties can arise because of the scale of the manipulations that would be required. How, for example, could you conduct removal or addition experiments to test the occurrence of competition among the woodland warblers without hampering the birds' freedom of movement to such an extent that their normal behavior patterns are disrupted? If you have observed warblers for any period of time, you know that they flit from tree to tree in their pursuit of insects. Do you tent an entire area of the forest and hope that you have not separated the birds from their nests? If you follow the example of contemporary ecologists, you will leave the question open, rather than create any additional stress for the warblers, which are already under pressure because of the destruction of the forests in Central America in which they winter.

Barnacles in Scotland

One of the clearest demonstrations of competition in a natural community was a study of barnacles performed by Joseph Connell, considered by many to be the founder of modern experimental ecology. Barnacles, as you will recall, are crustaceans. When they change from their free-swimming, larval form into their adult, sedentary form, they cement themselves to rocks and secrete a series of protective calcium-containing plates. Once attached, barnacles remain fixed, so that by making careful records, one can determine the history of a particular population. One can identify exactly which barnacles have died and which new ones have arrived between visits to the study site.

Connell studied two barnacle species, *Chthamalus stellatus* and *Semibalanus balanoides* (formerly known as *Balanus balanoides*), that live on the coasts of Scotland. *Chthamalus* is found on the high part of the intertidal seashore. As the tides go in and out, these barnacles are exposed to wide fluctuations of temperature and salinity and to the hazards of desiccation. The other species, *Semiba-*

53–9 *The zonation of barnacles at Millport, Isle of Cumbria, the site in Scotland where Connell performed his first experimental studies of competition. The* Chthamalus *barnacles, in the high part of the intertidal zone, are white; the* Semibalanus *barnacles, in the lower zone, are almost the same color as the rock to which they are attached. The grid, used for counting the barnacles, is 50 centimeters by 50 centimeters.*

lanus, occurs lower down, where the conditions are more benign. Although *Chthamalus* larvae, after their period of drifting in the plankton, often attach to rocks in the lower, *Semibalanus*-occupied zone, adults are rarely found there. Connell was able to show that in the lower zone, *Semibalanus,* which grows faster, ousts *Chthamalus* by crowding it off the rocks and growing over it or undercutting it. When Connell experimentally removed *Semibalanus* from the lower portion of the intertidal zone, *Chthamalus* invaded the area and thrived there. In the control areas, where removals were not conducted, each species remained in its own zone. As this experiment demonstrates, *Chthamalus* is not restricted to the upper intertidal zone by physiological tolerance limits. It is restricted by competition with *Semibalanus.* There is no evidence, however, that competition with *Chthamalus* keeps *Semibalanus* in the lower zone. Because *Semibalanus* lacks the physiological adaptations required for life in the upper zone, it cannot successfully invade it.

This study, and others like it, have generated the concepts of **fundamental niche** and **realized niche.** The fundamental niche describes the physiological limits of tolerance of the organism; it is the niche occupied by an organism in the absence of interactions with other organisms. The realized niche is that portion of the fundamental niche actually utilized; it is determined by physical factors and also by interactions with other organisms. In its fundamental niche, *Chthamalus,* for example, can occupy both the high and low intertidal zones, but because of niche overlap with *Semibalanus, Chthamalus* actually occupies a smaller area, its realized niche (Figure 53–10). Because *Semibalanus* is restricted by physiological limits, it can occupy only the lower intertidal zone. Its fundamental niche is narrower than that of *Chthamalus* and is totally included within it. In such a situation, the species with the narrow, included fundamental niche must be a superior competitor or it will be driven to extinction. *Semibalanus* can survive in the same intertidal community as *Chthamalus* because it is the superior competitor. *Chthamalus* can survive, despite its competitive inferiority, because its fundamental niche is broader, providing it with a refuge that cannot be invaded by *Semibalanus.*

The Prevalence of Competition

The results of Connell's studies of *Semibalanus* and *Chthamalus* provided strong support for the idea that competition can play a major role in determining the structure of communities. However, when Connell subsequently performed experimental studies of what appeared to be a similar situation involving barnacles on the coast of the state of Washington, he found that competition was not a factor. Recently, both Connell and Thomas Schoener of the University of California at Davis have conducted critical analyses of published experimental studies of competition. Connell's review, which covered 527 experiments involving 215 species, found that competition was indicated by about 40 percent of the experiments and involved more than half of the species studied. Schoener's review, which evaluated 150 experiments and used slightly different criteria, found that more than 75 percent of the species studied were involved in competitive interactions.

Winner Takes All

Most studies of competition have emphasized the adaptations and partitionings of resources that make possible the coexistence of similar species within a community. This is, however, a biased view, for it is difficult to study the interactions between species after one of the protagonists has left. Just as competition within species leads to the elimination of the great majority of individual

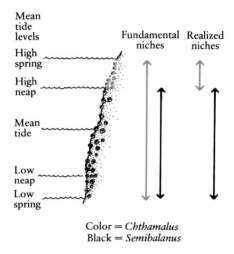

53–10 Interspecific competition between Semibalanus *and* Chthamalus *barnacles. The larvae of both species settle over a wide range, but the adults live in quite precisely restricted areas. The upward limits of the* Semibalanus *area are determined by physical factors such as desiccation. The* Chthamalus *barnacles, however, are prevented from living in the* Semibalanus *area not by physical factors (they would probably thrive there since the area is less physically limiting) but by the* Semibalanus *barnacles. The* Semibalanus *grows faster, and whenever it comes upon* Chthamalus *in the* Semibalanus *area, it either pries it off the rocks or grows right over it.*

53–11 *A male mountain bluebird at the entrance of a nesting box in Denver, Colorado. For more than 20 years, the U.S. Fish and Wildlife Service has been monitoring the breeding populations of the three bluebird species native to North America. A significant increase in the size of the populations began to occur in the five years ending in 1986, the most recent year for which data are available. This increase is attributed, in large part, to a concerted effort by the North American Bluebird Society and others to encourage people to provide suitable nesting boxes for the birds in potential bluebird habitats. An important factor in the design of the nesting boxes is that the entrance holes be large enough to admit bluebirds but small enough to exclude starlings.*

organisms, as Darwin observed, competition between species may lead to elimination of a species from the community. One example is the disappearance from many localities of bluebirds, which is thought to have been caused, in part, by the usurpation of their nesting sites by starlings. Starlings were introduced into Central Park, in New York City, in 1891, and are now found throughout the United States, whereas some of us have never seen a bluebird.

A similar example, also involving birds and human interference, is seen among the gull species of the northeastern United States. Herring gulls are such successful exploiters of garbage dumps that they have greatly increased in numbers at the expense of other species, notably the ring-billed gull and the laughing gull. These once plentiful birds have now also virtually disappeared, again perhaps as a result of competition for nesting sites.

A more recent example involves the California vole, which was accidentally introduced onto a small island in San Francisco Bay in the late 1950s. The vole rapidly colonized the island, and at the same time, the resident rodent on the island, the house mouse, began to decline in numbers. Eventually, it became extinct on the island. This would seem a fairly straightforward example of competition driving a species to local extinction. However, because the process was not studied in detail at the time it occurred, the possibility that predation or disease (perhaps introduced by the vole) caused extinction of the house mouse cannot be ruled out.

PREDATION

Predation is the eating of live organisms, including plants by animals, animals by animals, and even, as we have seen, animals by plants (page 666) or by fungi (page 489). Predators utilize a variety of techniques—known as foraging strategies—to obtain their food. Foraging strategies are under intense selection pressure; those individuals that forage most efficiently are likely to leave the most offspring. From the standpoint of potential prey, those individuals that are most successful at avoiding predation are likely to leave the most offspring. Thus, predation affects the evolution of both predator and prey. As we shall see, it also affects the number of organisms in a population and the diversity of species within a community.

The Arms Race

Predators and prey are engaged in an ever-escalating arms race, evidence of which is present in every community. For example, the beetle *Eleodes longicollis* has a gland located in the tip of its abdomen that contains a noxious secretion. When threatened, the beetle does a quick headstand and sprays the chemical. Grasshopper mice, which feed on insects, recognize *Eleodes longicollis*. They utilize them as prey by jamming them butt-end into the earth and eating them head first (Figure 53–12).

The seeds of legumes are the frequent prey of bruchid weevils, a kind of beetle. The adult lays its eggs on the developing pods, and the larvae, when they hatch, burrow into the seeds. Each larva occupies one seed, which it consumes as it grows. The legumes have countered with a variety of defenses. Some have evolved seeds so tiny that one will not sufficiently nourish a larva to pupation and so they are not preyed upon. Others produce a chemical that inhibits the protein-digesting enzymes. These defenses are still successful against most insects, but bruchids have evolved metabolic pathways that bypass the enzyme block. Soybeans appear to be a momentary victor: bruchid larvae laid in soybean seeds die. You will not be surprised to learn that chemicals isolated from the seeds have been found to inhibit bruchid development.

(a)

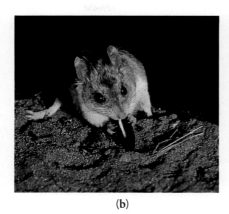

(b)

(c)

53–12 *Predator-prey interactions are frequently marked by an ever-increasing arms race.* (**a**) *The beetle* Eleodes longicollis *has glands in its abdomen that secrete a foul-smelling liquid. When disturbed,* Eleodes *stands on its head and sprays the liquid at the potential predator.* (**b**) *A grasshopper mouse that has solved*

the problem of how to eat Eleodes. *The mouse drives the posterior of the beetle into the ground and eats it head first.* (**c**) *In this photo,* Eleodes longicollis *is on the right. On the left is* Megasida obliterata, *a Batesian mimic (page 1004) of* Eleodes longicollis. *It resembles the latter physi-*

cally and, as you can see, emphasizes this resemblance by also standing on its head. However, Megasida *has no similar glands, no noxious secretion, and no spray. Its deceptive mimicry would probably do it little good with a grasshopper mouse.*

Bats, Moths, and Biosonar

In 1793, Italian scientist Lazzaro Spallanzani made a study of how animals find their way around in the darkness. He captured bats in the bell tower of the Cathedral of Pavia, blinded them, and turned them loose. Several days later, he captured the same bats again; not only had the blind animals found their way back to the bell tower but also their stomachs were full of the freshly caught flying insects that are their normal diet. Guessing that bats might "hear" their way through the darkness, Spallanzani plugged the ear canals of some captive bats and found that the deafened animals were wholly disoriented and bumped into obstacles at random.

In the 1930s, Donald Griffin, then an undergraduate at Harvard, repeated some of Spallanzani's experiments while monitoring bats with microphones sensitive to ultrasound. He found that bats emit cries above the frequency range of human hearing. When these sound waves hit a solid object, they echo back to the sensitive ears of the bat. The time required for the sound waves to return and the direction from which they return enable the bat to determine the distance and direction of the object. Moreover, the bat can tell if the object is moving toward it, moving away from it, or is stationary. If the frequency (that is, the pitch) of the returning sound is higher than the emitted sound, the distance between the bat and the object is decreasing; if the frequency is lower, the distance between the bat and the object is increasing; if the frequency is the same, the distance between the two is unchanging. (This is an example of the Doppler effect, which we experience whenever the whistle of a train or the wailing siren of an emergency vehicle approaches and then recedes from us.) Using the information provided by the echoes, bats can navigate skillfully through a dark room strung with wires little thicker than a human hair or can catch an insect as small as a mosquito.

Few other animals can hear the shrill cries of bats. Exceptions are the moths that are the normal prey of insectivorous bats. Recent studies by Canadian biologists M. Brock Fenton and James H. Fullard indicate that more than 95 percent of the moth species in Ontario have functioning ears, which may be located on the thorax, abdomen, or mouthparts. These ears are not sensitive to all

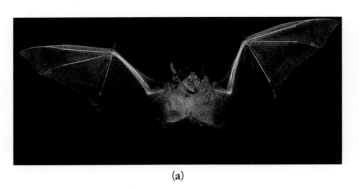

(a)

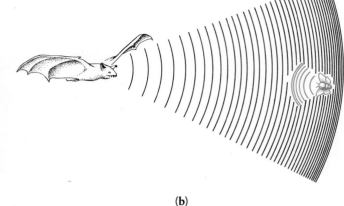

(b)

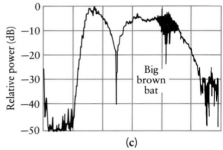

(c)

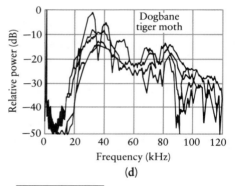

(d)

53–13 (a) *A big brown bat*, Eptesicus fuscus. (b) *As a bat flies, it emits shrill cries that are reflected back to it by any objects in the vicinity—for instance, as shown here, a moth.*

Graphs of the intensity (expressed in decibels) and frequency (expressed in kilohertz; 1 kHz = 1,000 cycles per second) of (c) *the echolocation calls of a big brown bat and* (d) *the clicks of its prey, a dogbane tiger moth. By mimicking the echolocation calls of the bat, the moth is able to jam the bat's information-processing system, allowing it to escape.*

frequencies of sound; instead, they are "tuned" to the frequencies most often emitted by the bats that are their predators. Moreover, the moths can detect the bats before the bats can detect them. Studies by the late Kenneth Roeder of Tufts University showed that some moths can detect loud bat cries at distances of up to 40 meters. By contrast, bats apparently cannot detect moths any farther than 20 meters away, and generally bats can echolocate only at distances ranging from 1 to 10 meters. On hearing the bat, the moth takes evasive action. If the bat is still at some distance, the moth flies in the opposite direction, often changing direction frequently. If the bat is close, the moth may suddenly collapse its wings and dive to the ground.

More recent studies have demonstrated that some moths have evolved an even more sophisticated defense. As the bat closes in on the moth, the moth emits a series of high-frequency clicks that mimic the echolocation cries of the bat, jamming the bat's information-processing system and so enabling the moth to make its escape (Figure 53–13).

Escape from Predation

Prey organisms often elude their predators without ever engaging them in combat. For example, members of some species, masters of camouflage, hide in plain sight (Figure 53–14). Physical hiding places, such as burrows, are important for many species, especially of small animals. Some animals use temporal strategies. Newborn wildebeest calves, for example, are extremely vulnerable to predation by hyenas, lions, and other large carnivores, and there are no hiding places for them on the African plains. Wildebeest herds often number in the thousands, and if wildebeest calves were produced at a steady rate throughout the year, as is the case with other species of antelopes that are able to conceal their young, most of the calves would die. However, the birth of calves are synchronized so almost all appear within the same short interval of time, with 80 percent being born within the same three-week period. Though many calves are lost to predation as the carnivores feast to satiety, the chance of survival of each individual is greatly increased, and the carrying capacity for the predators is minimized.

Predation and Population Dynamics

For many populations, predation is the major cause of death, yet, paradoxically, it is not at all clear that predation necessarily reduces the numbers of a prey population below the carrying capacity of its environment. However, when predation is heavier on certain age groups—juveniles versus adults, for instance—or certain life stages—such as caterpillars versus butterflies—it can alter the structure of a population and promote adjustments in life-history patterns.

Predation, especially on large herbivores, tends to cull animals in poor physical condition. Wolves, for instance, have great difficulty overtaking healthy adult

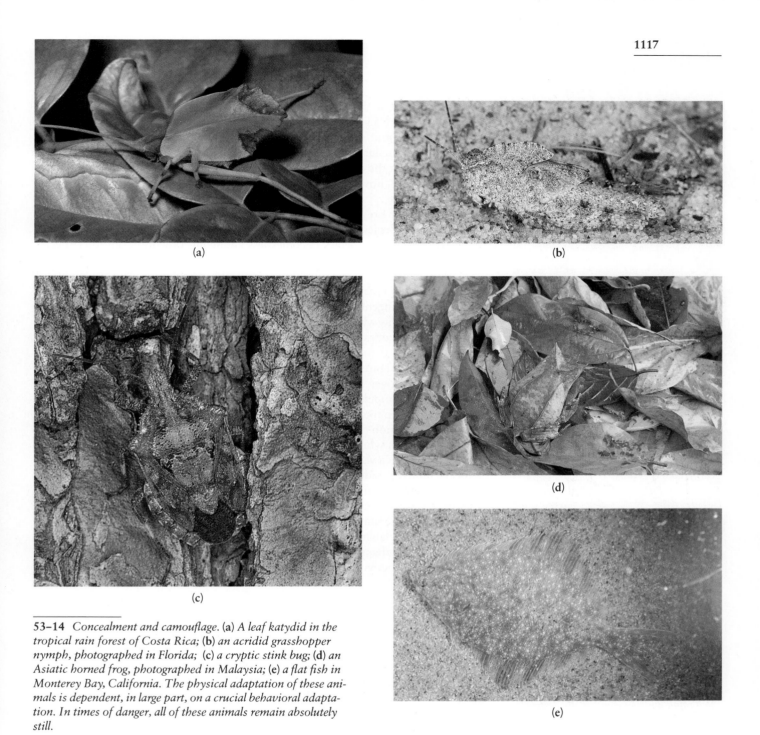

53–14 *Concealment and camouflage.* (a) *A leaf katydid in the tropical rain forest of Costa Rica;* (b) *an acridid grasshopper nymph, photographed in Florida;* (c) *a cryptic stink bug;* (d) *an Asiatic horned frog, photographed in Malaysia;* (e) *a flat fish in Monterey Bay, California. The physical adaptation of these animals is dependent, in large part, on a crucial behavioral adaptation. In times of danger, all of these animals remain absolutely still.*

(a) (b) (c) (d) (e)

moose or even healthy calves. A study of Isle Royale, an island in Lake Superior, showed that in some seasons more than 50 percent of the animals the wolves killed had lung disease, although the incidence of such individuals in the population was less than 2 percent. Thus, many of the animals killed by predators, according to this study, are animals that would have died anyway. (Modern human hunters, however, with their superior weapons and their desire for a "prize" specimen, are more likely to injure or destroy strong, healthy animals.)

In some situations, however, predators do limit their prey species. This has been most clearly demonstrated in cases involving the introduction of alien species. When prickly-pear cactus, for instance, was brought to Australia from South America, it escaped from the garden of the gentleman who imported it and spread into fields and pastureland until more than 12 million hectares were so densely covered with prickly pears that they could support almost no other vegetation (Figure 53–15). The cactus then began to take over the rest of Australia at the rate of about 400,000 hectares a year. It was not brought under control until a natural predator was imported—a South American moth, whose caterpillars feed only on the cactus. Now only an occasional cactus and a few moths can be found. (Note, however, that the introduction of the alien moth was, in itself, risky.)

Few predator-prey relationships are this simple. Most predators have more than one prey species, although one prey species may be preferred. Characteristically, when one prey species becomes less abundant, predation on other species increases so the proportions of each in the predator's diet fluctuate.

Although predators may not always limit prey populations, the availability of prey constitutes a major component of the carrying capacity of predator populations, often stringently affecting their size. This is evident in relatively simple situations, such as when a bloom of phytoplankton results from an upwelling of nutrients due to ocean currents and then is followed by a corresponding increase in zooplankton. A more complex example is that of the lynx and the snowshoe hare; the data (Figure 53–16) are based on pelts received yearly by the Hudson's Bay Company over a period of almost 100 years. As you can see, there are oscillations in population density that occur about every 10 years. Generally speaking, a rise in the hare population is followed by a rise in the lynx population; the hare population then plummets, and the lynx population follows.

This example, which has been studied by ecologists over the last 40 years, can be interpreted in a variety of ways. The traditional explanation is that overpredation by the lynx reduces the snowshoe hare population. The lynx population, heavily dependent on the snowshoe hare as prey, is reduced in turn. The reduction in predation then permits the snowshoe hares to increase in number, followed by an increase in the number of lynx, and so on.

A second explanation is that the hare population undergoes a regular 10-year cycle caused, perhaps, by diseases associated with crowding or the effects of its own predatory activities on the vegetation it consumes. This latter hypothesis is

(b)

53–15 (a) *Prickly-pear cactus* (Opuntia inermis) *growing on a pasture in Queensland, Australia, in November of 1926. Such rapid and environmentally destructive spread often occurs when alien organisms are introduced into a region where they have no competitors or predators.* (b) *The same pasture in October of 1929, slightly less than three years after the introduction of the cactus moth* (Cactoblastis).

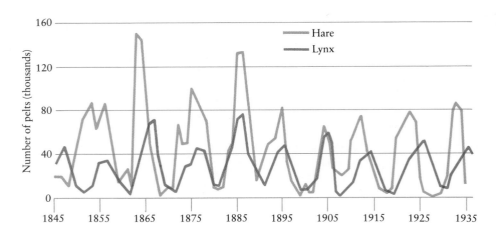

53–16 *The number of lynx and snowshoe hare pelts received yearly by the Hudson's Bay Company over a period of almost 100 years, indicating a pattern of 10-year oscillations in population density. The lynx reaches a population peak every* 9 or 10 years, and these peaks are followed in each case by several years of sharp decline. The snowshoe hare follows a similar cycle, with a peak abundance generally preceding that of the lynx by a year or more.

(a)

(b)

53-17 *An example of the effects of predation on species diversity.* (a) *In this high intertidal pool on the New England coast, the density of the herbivorous marine snail* Littorina littorea *is very low (between one and five individuals per square meter). The competitively superior green alga* Enteromorpha *dominates the pool, excluding other algae.* (b) *In a neighboring pool, less than a meter away, the density of* Littorina *is much higher—more than 250 individuals per square meter. The snails have grazed heavily on* Enteromorpha, *permitting the growth of other, inedible algal species. The aluminum grid at the top of the photograph was used to estimate the density of the snails.*

supported by two recent discoveries: (1) when overbrowsed, certain types of plants put out new shoots and leaves that contain chemicals toxic to hares, and (2) on an island where there are no lynx, the hare population undergoes similar cycles.

A third possibility is that the lynx undergo a regular cycle, independent of the hares, perhaps associated with some other factor, such as changes in the habits of their own predators, the human hunters. A decrease in the lynx population may permit growth of the hare population, or the two may oscillate independently. Thus, whereas the lynx and the hare used to serve as a simple model of predator-prey relationships, it is now perhaps even more instructive as an example of the difficulties in dealing with ecological variables.

Predation and Species Diversity

The number and kinds of species in a community can be greatly influenced by predation. Although predation may occasionally eliminate a prey species, many experimental studies have shown that it is often an important factor in maintaining species diversity in a community. R. T. Paine, for example, studied a community on the rocky coast of Washington. In this community, the principal predator was the starfish *Pisaster* (see Figure 53–1). At the beginning of the experiment, there were 15 prey species, including several species of barnacles and several kinds of mollusks, including mussels, limpets, and chitons. Paine systematically removed the starfish from one area of the community, 8 meters by 2 meters in size. By the end of the experiment, the number of prey species in the area from which the starfish were removed had declined to eight, and the community was dominated by one species of mussel. In the undisturbed community, starfish predation kept the densities of the prey populations low, reducing competition between the species and permitting all to survive. In the absence of the predator, the mussels were clear victors in the competition for living space.

Similarly, Jane Lubchenco, now of Oregon State University, showed in a series of experiments that the herbivorous marine snail *Littorina littorea* controls the abundance and type of algae in the higher intertidal pools on the New England coast. In such pools, the snails' preferred food (the green alga *Enteromorpha*) is competitively superior, and its removal by the snails permits the growth of other algal species (Figure 53–17). However, in areas exposed at high tide, where *Enteromorpha* is competitively inferior, its removal by the snails facilitates the growth of algal species that are competitively superior under exposed conditions. The result is domination by those species and a decrease in the total number of species in the exposed areas.

A terrestrial example of the role of predation in reducing competition and maintaining species diversity occurred a number of years ago when a virus epidemic struck the rabbit population in England. On the chalky soils characteristic of many parts of England, grasses grow more vigorously than the annual flowering plants, but the annuals are able to maintain their populations because rabbits crop the grasslands. When the epidemic severely reduced the rabbit population, the grasses took over, and the wildflowers disappeared.

SYMBIOSIS

As we saw in Chapter 21, symbiosis ("living together") is a close and long-term association between organisms of two different species. Long-continued symbiotic relationships can result in profound evolutionary changes in the organisms involved, as in the case of lichens (page 488), one of the oldest and most ecologically successful symbioses.

Although there is some disagreement among ecologists as to precisely what constitutes a symbiotic relationship, and the details of the relationship between

53–18 *Rabbits crowd a water hole in Australia. Imported from Europe as potentially valuable herbivores, they soon overran the countryside. The myxoma virus was introduced to control them, but now host and parasite are coexisting.*

two closely associated species are often difficult to determine, symbiotic relationships are generally considered to be of three kinds. If one species benefits and the other is harmed, the relationship is known as parasitism. If the relationship is beneficial to both species, it is called mutualism. Less common is commensalism, a relationship that is beneficial to one species and that neither benefits nor harms the other. An example of commensalism is the relationship between the marine annelid *Chaetopterus* and tiny crabs of the genus *Pinnixa*, which are residents of intertidal mud flats along the southeastern coast of the United States. Each *Chaetopterus* constructs a U-shaped tube in which it lives, and the tube usually also contains two crabs, one male and one female. Both the worm and the crabs feed on particles of food carried in water currents moved through the tube by fanlike appendages of the worm's body. The crabs obtain shelter and a steady food supply, and, as best anyone has been able to determine, their presence neither benefits nor harms the worms.

Parasitism

Parasitism may be considered as a special form of predation in which the predator is considerably smaller than the prey. The plants and animals in a natural community support hundreds of parasitic species—in fact, certainly thousands and perhaps millions if one were to count nematodes and viruses.

As with more obvious forms of predation, parasitic diseases are most likely to wipe out the very young, the very old, and the disabled—either directly or, more often, indirectly, by making them more susceptible to other predators or to the effects of climate or food shortages. It is predictable that a parasite-caused disease should not be too virulent or too efficient. If a parasite were to kill all the hosts for which it is adapted, it too would perish.

This principle is particularly well illustrated by a series of further misadventures on the continent of Australia. There were no rabbits in Australia until 1859, when an English gentleman imported a dozen from Europe to grace his estate. Six years later he had killed a total of 20,000 on his own property and estimated he had 10,000 remaining. In 1887, in New South Wales alone, Australians killed 20 million rabbits. By 1950, Australia was being stripped of its vegetation by the rabbit hordes. In that year, rabbits infected with myxoma virus were released on the continent. (Myxoma virus causes only a mild disease in South American rabbits, its normal hosts, but is usually fatal to the European rabbit.) At first, the effects were spectacular, and the rabbit population steadily declined, yielding a share of pastureland once more to the sheep herds on which much of the economy of the country depends. But then occasional rabbits began to survive the disease, and their litters also showed resistance to the myxoma virus.

A double process of selection had taken place. The original virus was so rapidly fatal that often a rabbit died before there was time for it to be bitten by a mosquito and thereby infect another rabbit; the virus strain then died with the rabbit. Strains less drastic in their effects, on the other hand, had a better chance of survival since they had a greater opportunity to spread to a new host. (A rabbit that survives an initial infection is immune to reinfection by that virus.) So selection began to work in favor of a less virulent strain of the myxoma virus. Simultaneously, rabbits that were resistant to the original virus began to proliferate (as in the Lederbergs' experiment; see page 964). Now, as a result of coevolution, the host-parasite relationship seems to be stabilizing.

Mutualism

If the current hypothesis as to the origin of eukaryotic cells (page 452) is correct, we owe our very origins to mutualism. Examples of present-day mutualistic associations are so abundant that we must forcibly limit ourselves to only a few.

(a)

(b)

(c)

53–19 *Mutualism.* (a) *Sea anemones on the back of a snail shell occupied by a hermit crab. The anemones protect and camouflage the crab and, in turn, gain mobility—and so a wider feeding range—from their association with the crab. Hermit crabs, which periodically move into new, larger shells, will coax their anemones to move with them.*

(b) *Cleaner fish are permitted to approach larger fish with impunity because they feed off algae, fungi, and other microorganisms on the fish's body. The fish recognize the cleaners by their distinctive markings. Other species of fish, by resembling the cleaners, are able to get close enough to the large fish to remove large bites of flesh. What would probably happen if the number of mimics began to approach the number of true cleaners?*

(c) *Aphids suck sap from the phloem, removing certain amino acids, sugars, and other nutrients from it and excreting most of it as "honeydew," or "sugar-lerp," as it is called in Australia, where it is harvested as food by the aborigines. Some species of aphids have been domesticated by some species of ants. These aphids do not excrete their honeydew at random, but only in response to caressing movements of the ants' antennae and forelimbs. The aphids involved in this symbiotic association have lost their own natural defenses, including their hard outer skeletons, relying upon their hosts for protection.*

(d) *Oxpeckers live on the ticks they remove from their hosts. An oxpecker forms an association with one particular animal, such as the young zebra shown here, conducting most of its activities, including courtship and mating, on the back of its host.*

(d)

Perhaps among the most significant are those that take place underground, between roots and nitrogen-fixing bacteria (page 665) and in mycorrhizae (page 490). Studies of underground root systems have revealed that intertwining roots often actually form grafts so that water, minerals, and organic materials are exchanged among plants of different species. As a result of such grafts, root stumps may be kept alive indefinitely. Some particularly colorful examples of mutualism are illustrated in Figure 53–19.

Ants and Acacias

Trees and shrubs of the genus *Acacia* grow throughout the tropical and subtropical regions of the world. In Africa and tropical America, where plants are preyed upon by large herbivores, acacia species are protected by thorns. (Acacias that evolved where there are no large browsers—Australia, for instance—lack thorns.)

1121

(a)

(b)

(c)   (d)

53–20 *Ants and acacias.* (a) *The beginning. A queen ant cuts an entrance into a thorn on a seedling bull's-horn acacia. She will hollow out the thorn and raise her first brood inside it.* (b) *The tip of an acacia leaf. The orange structures at the tips of the leaflets are Beltian bodies. They are a source of food for the ants.* (c) *A worker ant drinking from the nectary.* (d) *Warriors in a battle for possession of an acacia. Such battles occur when the branches of acacias grow to touch each other. The largest colony usually wins.*

On one of the African species of *Acacia,* ants of the genus *Crematogaster* gnaw entrance holes in the walls of the thorns and live permanently inside them. Each colony of ants inhabits the thorns on one or more trees. The ants obtain food from nectar-secreting glands on the leaves of the acacias and eat caterpillars and other herbivores that they find on the trees. Both the ants and the acacias appear to benefit from this association.

In the lowlands of Mexico and Central America, the ant-acacia relationship has been extended to even greater lengths. The bull's-horn acacia is found frequently in cutover or disturbed areas, where it grows extremely rapidly. This species of acacia has a pair of greatly swollen thorns, several centimeters in length, at the base of most leaves. The petioles bear nectaries, and at the very tip of each leaflet is a small structure, rich in oils and proteins, known as a Beltian body. Thomas Belt, the naturalist who first described these bodies, noted that their only apparent function was to nourish the ants. Ants live in the thorns, obtain sugars from the nectaries, eat the Beltian bodies, and feed them to their larvae.

Worker ants, which swarm over the surface of the plant, are very aggressive toward other insects and, indeed, toward animals of all sizes. They become alert at the mere rustle of a passing mammal, and when their tree is brushed by an animal, they swarm out and attack at once, inflicting painful, burning bites. The effect has been described as similar to walking into a large stinging nettle. Moreover, and even more surprising, other plants sprouting within as much as a meter of occupied acacias are chewed and mauled, and their bark is girdled; twigs and branches of other trees that touch an occupied acacia are similarly destroyed. Not surprisingly, acacias inhabited by these ants grow very rapidly, soon overtopping other vegetation.

Daniel Janzen, who first analyzed the ant-acacia relationship in detail, removed ants from acacias either by insecticides or by removing thorns or entire occupied branches. Acacias without their ants grew slowly and usually suffered severe damage from insect herbivores. Their stunted bodies were soon overshadowed by competing species of plants and vines. As for the ants, according to Janzen, these particular species live only on acacias.

There is an epilogue to the ant-acacia story. Three new species—a fly, a weevil, and a spider—have recently been discovered that mimic the ants that inhabit the acacias. So expert is their mimicry (probably involving chemical recognition signals as well as appearance) that the patrolling ants do not recognize them as interlopers, and so they enjoy the hospitality and protection of the ant-acacia complex.

COMMUNITY COMPOSITION AND THE QUESTION OF STABILITY

Viewed from a global perspective, ecological communities often seem to be at equilibrium, with many species persisting for many generations over large areas. However, when communities are examined on a local scale, it becomes apparent that they, like the individual populations of which they are composed, are often in a state of nonequilibrium. Two questions concerning community composition have long perplexed ecologists. First, what determines the number of species in a community? And, second, what factors underlie the changes in community composition that are revealed by close observation to occur with the passage of time?

The Island Biogeography Model

Because of their size and relative isolation, small islands are often excellent natural laboratories for the study of both evolutionary and ecological processes, as we have seen in numerous examples. In a classic study published in 1963, Robert MacArthur and E. O. Wilson used small islands as a model system to explore questions of community composition and stability. They hypothesized that the number of species on any given island remains relatively constant through time but that the identity of the species present is continually changing. According to their proposal, known as the **equilibrium hypothesis of island biogeography,** there is a balance between the rate at which new species immigrate to an island and the rate at which species already present become locally extinct. Although the number of species is in equilibrium, the species composition is in nonequilibrium, because when a species becomes extinct it is usually replaced by a different species.

The island biogeography model was tested in an ingenious way by Wilson and Daniel Simberloff. They selected a number of small mangrove islands off the Florida Keys and counted the number of species of arthropods on each. They then destroyed all the animal life by the simple expedient of covering the small islands with plastic tents and fumigating them. Their plant life intact, the islands were soon colonized again from the mainland, and the recolonization process was monitored. As predicted by the model, the total number of species present on an island after recolonization tended to be the same as the total number before the island was disturbed. However, and this is an important point, the species composition was often quite different from what it had been previously. Moreover, once the equilibrium number had been reached, the species composition continued to change, with extinction and immigration balancing one another out (Figure 53–21).

53–21 *Equilibrium model of diversity of species on an island. The immigration rate declines as more species reach the island because existing species will become better established and thus better able to compete against newly arriving species. Also, fewer immigrants will belong to new species. The extinction rate increases more rapidly at high species number because of increased competition between species. The equilibrium number of species (black line) is determined by the intersection of the immigration and extinction curves.*

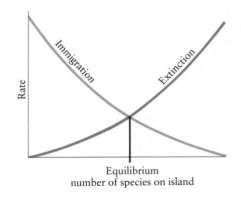

Equilibrium
number of species on island

According to the island biogeography model, the two most important variables influencing species diversity are the size of the island and its distance from a source, usually the mainland, that can provide colonists. Distant islands tend to have fewer species than islands closer to the mainland; this is thought to be a result of lower rates of immigration, which appear to be a function of the distance that potential colonists must travel. Smaller islands are thought to reach equilibrium with a smaller number of species than larger islands, primarily because extinction rates are higher on smaller islands. This may be because populations tend to be smaller on such islands and thus are more susceptible to the effects of both predation and environmental disturbances.

The Intermediate Disturbance Hypothesis

Different types of communities vary widely in the number and diversity of species present. Among the most diverse are tropical rain forests and coral reefs. Until recently, it was thought that the species composition of these communities was relatively constant, and they were frequently cited as prime examples of an equilibrium state. Their high species diversity was thought to be a function of their stability. It now appears, however, that their diversity may be a function not of their stability but rather of the frequency and magnitude of the disturbances to which they are subjected.

Disturbances can take many forms. In the tropical forest, for example, trees are killed or severely damaged by storms, landslides, lightning strikes, and insect outbreaks. The corals that form the basis of the coral reef community can be destroyed by predators, the severe waves that accompany tropical storms, and influxes of fresh water. Soon after a disturbance, open areas—of forest or reef—are invaded by immature forms—seeds, spores, larvae, or even gametes.

Initially, diversity in a newly colonized area is low. Only those species that are in close proximity to the disturbed area and that are reproducing at the time are able to exploit the newly available area. If disturbances are frequent, the community will consist only of those species that can invade, mature, and reproduce before the next disturbance occurs.

According to the **intermediate disturbance hypothesis,** as the interval between disturbances increases, so does species diversity. Species that are excluded by frequent disturbances (because they are slow to mature or have limited dispersal abilities) now have an opportunity to colonize. However, if the interval between disturbances increases still further, species diversity may begin to decline (Figure 53–22). The primary factor in this decline is thought to be interspecific competition, but even if all species were competitively equal, the species most resistant to ill effects from physical extremes, predation, or disease would eventually come to dominate the community.

Among the smallest self-contained communities are those found on boulders located in the intertidal zones of rocky coasts. These communities, which are dominated by the algae growing on the rock surface, are often subjected to massive disturbances as a result of severe waves, which may either strip the algae away or actually overturn the boulders. As a result, bare rock becomes available for colonization, either by vegetative regrowth of surviving individuals or by germination of spores. In a series of detailed observations in intertidal zones of southern California, Wayne Sousa of the University of California, Berkeley, found that large boulders, which are infrequently overturned, and small boulders, which are frequently overturned, are typically dominated by one or a few algal species. By contrast, medium-sized boulders, which are subjected to an intermediate number of disturbances, tend to have a greater diversity of species. Sousa extended his observations by experimentally stabilizing boulders, thus reducing the number of disturbances to which they were subjected. The algal communities

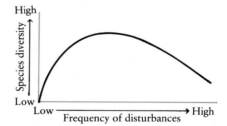

53–22 *According to the intermediate disturbance hypothesis, the diversity of species in a community is determined by the frequency of environmental disturbances. When disturbances are either very frequent or very infrequent, species diversity is low. By contrast, when the frequency of disturbances is intermediate, species diversity is high. Similar curves are obtained when species diversity is plotted against the time elapsed since the last disturbance and against the magnitude of the disturbance.*

on the stabilized boulders were less diverse than those on the unstabilized boulders, providing further support for the intermediate disturbance hypothesis.

Ecological Succession

Numerous observations have revealed that, if the interval between disturbances is relatively long, gradual changes occur in the composition of a community following the initial recolonization. The photosynthetic organisms that are usually the first colonists are generally replaced in time by other types, which gradually crowd out the earliest species and which may themselves be subsequently replaced. As the photosynthetic components of the system change, the accompanying animal life changes as well. This process is known as **ecological succession.**

An example of succession is provided by an abandoned field surrounded by other vegetation. Such an area is bombarded by the seeds of numerous plants and is rapidly covered by seedlings of those species whose seeds germinate the most rapidly. In an open field, the survivors among the seedlings will be those plants that can survive the sunlight and drying winds—weeds and grasses and, in many regions of North America, trees such as cedars, white pines, poplars, and birch. For a while, these plants dominate the community, but eventually they are replaced by other trees—oaks, red maples, white ash, and tulip trees. These trees are, in turn, replaced by hemlock, beech, and sugar maple, which, in the absence of major disturbances, dominate the forest indefinitely.

In many communities, the sequence of photosynthetic organisms is so regular and predictable that ecological succession was at one time viewed as analogous to the developmental processes of a single organism, with each stage "preparing the way" for the next by altering the local conditions of temperature, light, humidity, soil content, and so on. Ultimately, the community would, according to this view, reach a "mature," stable state, known as the climax community. This concept of succession, formulated in 1916 by the plant ecologist F. E. Clements, is known as **the facilitation hypothesis.** For many years, it dominated the study of ecological succession.

Recent studies have suggested that there are alternative mechanisms by which succession may occur. According to the **inhibition hypothesis,** the early species prevent—rather than assist—colonization by other species. Eventually, however, the earliest colonizers are replaced by later arrivals, and those species may, in turn, prevent colonization by others, until they too are replaced or a subsequent disturbance lowers their numbers. Another model, the **tolerance hypothesis,** suggests that the earliest species neither facilitate nor inhibit colonization by later species. The species dominant at any given time are those that can best tolerate the existing physical conditions and availability of resources.

53–23 (a) *When Mt. St. Helens exploded on May 18, 1980, shock waves leveled all of the fir trees in an area of about 18,000 hectares. Subsequent mudflows, moving at speeds of up to 80 kilometers an hour, buried the remaining vegetation. One of the few survivors was this* Anaphalis margaritacea *plant, photographed four months after the eruption, growing out of a crevice.* (b) *This small* Mullugo *plant is one of the few organisms eking out a living on a 92-year-old lava flow at Sullivan Bay, James Island, in the Galapagos. Differences in the material with which vegetation was covered (lava versus mud) may explain the much slower recovery time following the Galapagos eruption. What other factors might be involved?*

(a)   (b)

Conservation Biology and the Island Biogeography Model

As the demand for natural resources by expanding human populations increases, ecological communities are being fragmented at an ever accelerating pace. Formerly large and continuous natural communities are being reduced to isolated "islands," often surrounded by areas that are unsuitable for most of their constituent species. The most dramatic example of this destruction of natural communities is the unprecedented rate at which the tropical forests of the world are being cut down. However, the destruction of wildlife habitat is not unique to the tropical forests. Whenever a marsh is cut by a new roadway, a forest is cleared for agriculture or a housing development, or a river is interrupted by a dam, the result is a subdivision of ecological communities into smaller, increasingly isolated habitat islands.

In this situation, the equilibrium hypothesis of island biogeography has clear implications for conservation biology. As we have seen, this model predicts higher rates of extinction on and lower rates of immigration to small, isolated islands. According to this model, as natural communities are reduced to smaller and more isolated fragments, they can be expected to support fewer and fewer species.

This prediction of reduced species numbers in small, isolated habitat fragments has been supported by research in a number of different regions. One of the most thoroughly studied sites is Barro Colorado Island, an area of tropical rain forest located in Lake Gatun in Panama. The lake was created early in the twentieth century by the completion of the Panama Canal, isolating what is now Barro Colorado Island from the previously contiguous forest. Since that time, more than 50 species of birds have disappeared from the island, although they remain abundant on the mainland, only half a kilometer away. A similar pattern of extinction has been observed in the 86-hectare woodland of the Bogor Botanical Garden in Java, Indonesia. Fifty years ago, this woodland was isolated by the destruction of the surrounding woodlands. Since that time, it has lost 20 of the 62 species of birds originally breeding there—more than 30 percent of the original community—and another four species are close to extinction.

A similar process, involving natural events rather than human activities, is thought to have occurred on the mountains in the Great Basin region of the western United States. The communities on the upper regions of these mountains became isolated by the formation of deserts in the lowlands of the Great Basin following the last Ice Age. Today, the smallest and most isolated of these mountains support the fewest species of mammals, and it is estimated that most have lost at least 50 percent of their original mammalian species. On these same mountains, however, there has been less change in the number of bird species, presumably because of the birds' greater dispersal abilities.

It has been suggested that isolation of ecological communities may be reduced—and dispersal rates increased—by maintaining natural corridors between isolated communities. This possibility is being explored in the Netherlands, where studies are examining the role of hedgerows ("paths" of trees and shrubs, ranging from 1 to 10 meters in width) in increasing the rates of colonization of forest fragments. Some of the results of these studies have been surprising. For example, although hedgerows appear to increase the rates of dispersal of birds, they appear to reduce the rates of dispersal of some forest mammals. These studies indicate that all species cannot be managed in the same way and that the effects of landscape structure on animal movements are far from understood.

One of the most ambitious ecological projects carried out thus far, known as the Minimum Critical Size of Ecosystems Project, is now taking place in the Amazon rain forests of Brazil. This project, a cooperative investigation by Brazilian and American scientists, was inspired by the equilibrium hypothesis of island biogeography. It has already revealed that forest fragmentation can have substantial effects on the physical environment. Hot, dry winds blowing across surrounding cleared areas reduce the relative humidity along the edges of the forest fragments by as much as 20 percent. Moreover, increased light penetration at the edges of the forest fragments has elevated the temperatures there as much as 4.5°C (8.1°F) above the temperature in the forest interior. These physical changes may be responsible for some of the initial biological changes in the forest fragments: more trees die, more leaves drop from the remaining trees, and both the number of bird species and their population density decrease.

Although the equilibrium hypothesis of island biogeography began as a purely theoretical exercise to explain the maintenance of species diversity on islands, it has inspired research that is providing information vital for conservation biology, for the design and management of nature reserves, and for informed land-use planning. Some of the results of this research were not predicted by the original model. However, one of the most important predictions has held up: reductions in species diversity can be expected within nature reserves as they become increasingly isolated. For most groups of organisms, the numbers of species lost will depend upon the size of the reserves and their degree of isolation.

(a)

(b)

(c)

(a) *The tropical rain forest of Barro Colorado Island, Panama, as seen from an observation tower above the forest canopy (the island is in the foreground, and the Panamanian mainland is in the background). Biologists affiliated with the Smithsonian Tropical Research Institute, assisted by Earthwatch volunteers, are* conducting detailed studies of the forest and its occupants. (b) For North American visitors, one of the most familiar animals of the forest is the bay-breasted warbler, which summers in the forests of New England but winters on Barro Colorado Island—and nowhere else in Central or South America. (c) The forest interior.

(a)

(b)

53–24 (a) *Boulders in a low intertidal zone on the southern California coast. Those with bare upper surfaces have recently been overturned by the waves, while those covered with algae have remained undisturbed.* (b) *A collection of intertidal boulders covered with different successional stages of algae. The boulder in the center was overturned by wave action during the winter before this photograph was taken and has been colonized by the green alga* Ulva. *The surrounding boulders, which are larger, have not been disturbed as recently and are covered with later successional stages that are dominated by red algae.*

In a continuation of his earlier experiments, Wayne Sousa charted the patterns of algal succession on intertidal boulders by scraping boulders clean of algae and by adding bare rocks to the intertidal zone. The first colonists on the bare rock were the green algae *Ulva* (Figure 22–15c, page 463) and *Enteromorpha* (Figure 53–17a). These pioneer species grow rapidly and, within a short period of time, fully occupy the available space on the boulders (Figure 53–24). Later, large perennial brown and red algae become established, replacing the green algae; ultimately, the dominant organism is *Gigartina*, a red alga. To determine the mechanism by which this successional sequence occurred, Sousa removed algae at various stages of the process. For example, when he removed *Ulva,* he found that *Gigartina* was then able to colonize the boulders. Similarly, when he removed the middle successional species of red algae, *Gigartina* was also able to colonize. These results support the inhibition hypothesis, but they raise the question as to how the later species ever gain a foothold. The answer, it turns out, is that *Ulva* and the other early colonizers are subjected to heavy predation and have high mortality rates. As a consequence, small areas of space open up on the boulders, which are then seized by the later species.

Other experiments, however, have supported the facilitation hypothesis. In a study of successional patterns in intertidal areas in Oregon that are dominated by surfgrass, a perennial angiosperm with grasslike blades, Teresa Turner, now of the University of the Virgin Islands, found that colonization by surfgrass was dependent on earlier successional species of red algae. When Turner removed the red algae from experimental plots, the surfgrass was unable to become established (Figure 53–25). In plots from which the red algae were not removed, surfgrass colonized easily and soon became the dominant vegetation. As you may recall (see page 465), red algae are often filamentous. Surfgrass fruits, it turns out, bear small barbs with which they attach to these filaments. Without the red algae to anchor its fruits and the seedlings they produce, the surfgrass cannot gain a foothold.

These experiments suggest that, as with so many other ecological phenomena, there are no simple answers that apply in all situations. Each community is a

(a)

(b)

53-25 (a) *The two brown patches in the center of this photograph, each 1 meter square, are experimental plots in Turner's study of algal succession in low intertidal areas at Boiler Bay, Oregon. These plots were cleared of algae in September, before the winter dispersal of surfgrass fruits. They are in an early stage of succession, dominated by the brown alga* Petalonia. *Between the two experimental plots is a control plot in a later stage of succession, dominated by the red alga* Rhodomela larix. *The green patches in the photograph are surfgrass.* (b) *A surfgrass* (Phyllospadix scouleri) *seedling, attached to a red alga of the genus* Corallina.

unique assemblage of organisms, the product of a unique and ever-changing history involving both physical and biological factors. The world of living organisms is so vast and diverse—and ecologists are, relatively speaking, so few—that it may be many years before enough information is available to provide a thorough understanding of community ecology. Paradoxically, the processes that occur at the next level of ecological complexity—the ecosystem—are in many respects more clearly understood than those occurring within either populations or communities. These processes, which involve the flow of energy and the recycling of minerals, will be the subject of the next chapter.

SUMMARY

Populations live as part of a community—an assemblage of different organisms inhabiting a common environment and interacting with one another. Three major types of interspecific interaction in communities are competition, predation, and symbiosis.

The more similar organisms are in their requirements and life styles, the more intense competition is likely to be between them. As a result of competition, the overall fitness of the interacting individuals may be reduced. The importance of competition as an influence on community composition and structure is currently a matter of debate among ecologists, as are the methods to be used in studying competition and other aspects of community ecology.

The competitive exclusion principle, formulated by Gause, states that if two species are in competition for the same limited resource, one or the other will eventually be eliminated in situations in which they occur together. Similar species are able to coexist in the same community because they occupy different ecological niches. An ecological niche is defined by the resource requirements and total life style of members of a species, including their interactions with members of other species. Analyses of situations in which similar species coexist have demonstrated that resources are often subdivided, or partitioned, by the coexisting species.

Niche overlap describes the situation in which members of more than one species utilize the same limited resource. In communities in which niche overlap occurs, natural selection may result in an increase in the differences between the competing species, a phenomenon known as character displacement. Although the adaptations that enable various organisms to partition resources have often been attributed to character displacement, it is difficult, if not impossible, to distinguish between adaptations that occurred in response to competition and those that were the result of differing local conditions.

Competition has been demonstrated in numerous experimental studies. Experiments in which potential competitors were either removed from or added to a study site have given rise to the concepts of the fundamental niche and the realized niche. A fundamental niche represents the resources that would be utilized by a species in the absence of interactions with other organisms; a realized niche describes the resources actually utilized.

Predation is the eating of live organisms. It typically has a strong influence on the evolution of both predator and prey populations, as selection pressures favor ever-increasing efficiency in foraging by predators and ever-improving avoidance mechanisms by prey. Predator-prey interactions also influence population dynamics and may increase species diversity by reducing competition among prey species. The size of a predator population is often limited by the availability of prey. However, predation is not necessarily the major factor in regulating the population size of prey organisms, which may be influenced more by their own food supply.

Symbiosis is a close and long-term association between organisms of different species. It may be beneficial to one species and harmless to the other (commensalism), beneficial to one and harmful to the other (parasitism), or beneficial to both species (mutualism).

Two important questions regarding community composition and structure remain unanswered. First, what determines the number of species in a community? And, second, what factors underlie the changes in community composition with time? According to the island biogeography hypothesis, the number of species on islands reaches an equilibrium determined by the balance between immigration and extinction. Species composition may vary widely but the number remains approximately the same. For new species to gain a foothold, established species must become extinct, leading to a continual turnover in species composition. According to the intermediate disturbance hypothesis, the greatest species diversity is found in communities, such as tropical rain forests and coral reefs, that are subjected to environmental disturbances at an intermediate frequency. Communities in which disturbances are either infrequent or very frequent generally have a lower species diversity.

Following environmental disturbances, communities are recolonized by dispersal of immature forms from neighboring communities. If enough time elapses before the next major disturbance, a community typically goes through a process of ecological succession in which the earliest colonizers are replaced by other species that may, in turn, be replaced by still others. The mechanism of succession appears to vary from community to community. In some cases, the early-arriving species either facilitate or inhibit colonization by later-arriving species; in other cases, physical factors rather than biological interactions appear to be the major influence determining the successional stages of the community. Current evidence indicates that communities, like the populations of which they are composed, are dynamic, continually changing as conditions change.

QUESTIONS

1. Distinguish among the following: population/community; competitive exclusion/resource partitioning; habitat/niche; niche overlap/character displacement; fundamental niche/realized niche; symbiosis/commensalism/parasitism/mutualism.

2. Compare the effects of interspecific (between species) and intraspecific (within species) competition. What is the principal reason for the differences?

3. Compare the results of MacArthur's study of the warblers with Connell's experiment with *Semibalanus* and *Chthamalus* barnacles. What step did Connell perform that MacArthur did not? Why is the step important?

4. Consider the following chain of eater and eaten: plants-herbivores-carnivores. What are likely to be the critical resources for which organisms compete at each level? Are the levels different with respect to the mixture of resources, competition, and predation that might be involved in population regulation?

5. In the long and ruthless war between coyotes and sheep herders, studies have shown that (a) coyotes kill sheep, and (b) the percentage of sheep lost from herds in areas where coyotes have been exterminated is about the same as the percentage lost in areas where coyotes are still present. How could you explain this?

6. In the American southwest, grasses and mesquite compete with each other for dominance of the landscape. Mesquite, however, was rare before cattle were introduced into the western United States. How have cattle affected the competition between the two types of plant? Suppose all cattle were removed from a large area. What change would you predict in the competition between grasses and mesquite?

7. Introducing a new species into a community can have a number of possible effects. Name some of these possible consequences both to the community and to the species. What types of studies should be made before the importation of an "alien" organism? Some states and many countries have laws restricting such importations. Has your own state adopted any such laws? Are they, in your opinion, ecologically sound?

8. In the opinion of some ecologists, animals that eat seeds, such as the Galapagos ground finches, should be regarded as predators, while animals that eat leaves, such as deer, should be regarded as parasites. Justify this classification of herbivores as either predators or parasites.

9. Contrast the likely population dynamics of a predator-prey relationship versus a parasite-host relationship.

10. Compare the three hypotheses (facilitation, inhibition, and tolerance) that have been proposed to explain the process of ecological succession. In removal or addition experiments to identify the mechanism at work in a particular successional series, what type of results would constitute evidence in support of facilitation? Of inhibition? Of tolerance?

Ecosystems

As we saw in the last chapter, the populations within a community have numerous interactions with each other. Moreover, they interact with the abiotic (nonliving) environment. In all cases, these interactions have two consequences: (1) a one-way flow of energy through autotrophs (usually photosynthetic organisms) to heterotrophs, which eat either autotrophs or other heterotrophs; and (2) a cycling of materials, which move from the abiotic environment through the bodies of living organisms and back to the abiotic environment. This cycling of materials is dependent upon decomposers, organisms that break down organic materials into a form that can be used by autotrophs.

Such a combination of biotic and abiotic components through which energy flows and materials cycle is known as an ecological system, or **ecosystem.** Taking a global view, the entire surface of the earth can be seen as a single ecosystem. This view is useful when studying materials that are circulated on a worldwide basis, such as carbon dioxide, oxygen, and water. A suitably stocked aquarium or terrarium is also an ecosystem, and such models may be useful in studying certain ecological problems, such as the details of transfer of a particular mineral element. Most studies of ecosystems and their component communities have been made, however, on more or less self-contained natural units—for example, a pond, a swamp, or a meadow.

54–1 *Photograph of earth taken from the Apollo 17 spacecraft during the final lunar landing mission in 1972. Virtually the entire continent of Africa is visible, as is the Arabian Peninsula. The Mediterranean Sea is at the top left of the photograph. At the bottom, Antarctica is blanketed under a heavy cloud cover.*

Different parts of the earth's surface receive different amounts of solar energy. These differences determine worldwide patterns of climate, wind, and weather, which are, in turn, major factors affecting the structure of ecosystems.

We shall begin our consideration of ecosystems with the principal physical factor affecting them—the energy received from the caldron of hydrogen and helium atoms that we know as the sun.

SOLAR ENERGY

Life on earth is powered by the sun, which is also responsible for climate, wind, and weather. Every day, year in and year out, energy from the sun arrives at the upper surface of the earth's atmosphere at an average rate of 1.94 calories per square centimeter per minute, a total of about 1.3×10^{24} calories per year. This is known as the solar constant, and, although it is only a tiny fraction of the total energy radiated by the sun, it is a tremendous quantity of energy. However, because of the atmosphere, only a fraction of this energy reaches the surface of the earth and becomes available to living organisms.

The Influence of the Atmosphere

The atmosphere that lies between the surface of the earth and the incoming solar energy consists of four concentric shells, or layers, distinguished by temperature differences (Figure 54–2). The lowest layer, closest to the earth, is called the troposphere. It extends out about 10 kilometers and so covers all points on the earth's surface, although the tallest mountain peaks extend almost to its limits.

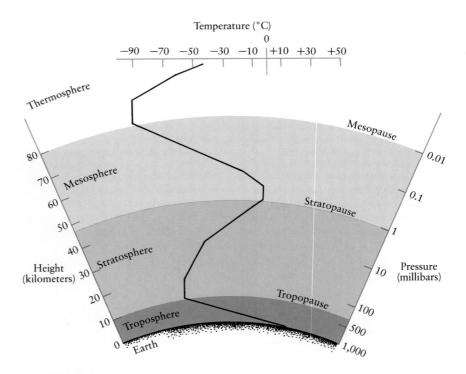

54-2 *Four main divisions of the atmosphere are the tropo-sphere, the stratosphere, the mesosphere, and the thermosphere. The boundaries between them are determined by abrupt changes in average temperature. The all-important ozone layer is located principally within the stratosphere.*

About 75 percent of all the molecules in the atmosphere are contained in this layer. At its outer boundary, the temperature is below −50°C. Almost all of the phenomena included under the broad general heading of weather take place within the troposphere.

The layer above the troposphere is the stratosphere, which extends to an altitude of about 50 kilometers. In the stratosphere, the temperature increases with altitude; the temperature at the outer boundary of the stratosphere is only slightly lower than the temperature at the earth's surface. The major reason for the warming of the stratosphere is its layer of ozone, which is proportionately densest near the outer boundary. Ozone (O_3) is formed when molecules of oxygen gas (O_2) are broken apart by radiant energy and recombine. Ozone molecules absorb most of the ultraviolet rays, converting their energy to heat; as a result, the temperature rises. About 99 percent of the atmosphere lies below the outer boundary of the stratosphere.

In the mesosphere, the third layer, there is a gradual decrease in temperature as the concentration of ozone decreases in the steadily thinning atmosphere. Then, above the mesosphere, we encounter another of those surprising temperature reversals: the temperature increases in the thermosphere. The individual molecules of the thermosphere, unshielded from the sun's rays, move at great speed, and, as you will recall (page 43), temperature is a measure of the average kinetic energy of molecules. Because there are few molecules, however, there is little heat. At the outer boundary of the thermosphere, the thin atmosphere blends with the hydrogen and helium atoms in the space between the stars.

Of the incoming solar energy, about 30 percent is reflected back into space from the clouds and dust, held closely to the earth's surface in the troposphere. Because of these reflections, earth, as seen from outer space, is a shining planet, brighter than Venus. Another 20 percent of the energy is absorbed by the atmosphere. Of this, 17 percent is taken up in the lower layers, mostly by water vapor, dust, and water droplets. This absorption of radiation warms the atmosphere slightly, though much of the energy is stored as latent heat in water vapor. An additional 1 to 3 percent is absorbed by oxygen and ozone in the stratosphere and mesosphere. This percentage, though small, is critically important because it represents most of the ultraviolet radiation. Ultraviolet and other high-energy radiations damage organic molecules, and should they reach the earth's surface in large quantities (as they were able to do before photosynthesis produced the protective oxygen-ozone layer), they would be lethal to most forms of terrestrial life.

The remaining 50 percent of the incoming radiation reaches the earth's surface. A small amount of this is reflected from bright areas, but most is absorbed. Energy absorbed by the oceans warms the surface of the water, evaporating water molecules and powering the water cycle (page 53). Solar energy absorbed by the ground is reradiated from the surface in longer (infrared) wavelengths—that is, as heat. The gases of the atmosphere are transparent to visible light, but carbon dioxide and water, in particular, are not transparent to infrared rays. As a result, the heat is held in the atmosphere, warming the surface of the earth.

Thus far, all of these factors have balanced out; heat loss and heat gain have remained in equilibrium. Yet the balance is a delicate one. Increase the earth's reflectivity, thicken its cloud cover, increase or decrease the CO_2 content of its atmosphere, or decrease its ozone layer, and the entire system would change in response. The nature and outcome of such changes—particularly the documented increases in the atmospheric CO_2 concentration (see page 226) and decreases in the ozone layer (see essay)—are currently matters of intense interest and concern.

The Threatened Ozone Layer

Approximately 99 percent of the ultraviolet radiation from the sun that reaches the stratosphere is converted to heat through a chemical reaction that continually recycles ozone (O_3) molecules. When ultraviolet radiation strikes an ozone molecule, the energy splits the molecule into highly reactive oxygen atoms; almost immediately, these atoms recombine, forming ozone once more and releasing energy in the form of heat. The chemical equilibrium is such that, for hundreds of millions of years, the quantity of ozone in the atmosphere has apparently remained constant, as has the very small percentage of the ultraviolet radiation reaching the earth's surface.

This is now changing, as a direct—although inadvertent—result of human activities. The most striking evidence of damage to the ozone layer has been the appearance of actual holes in the layer over both Antarctic and Arctic regions during their respective summers. Less dramatic, but of potentially far greater significance for living organisms, is growing evidence of a thinning of the ozone layer around the entire globe. The principal cause of the ozone loss is thought to be a group of synthetic chemicals known as chlorofluorocarbons. These chemicals are used throughout the world in refrigeration systems, including home refrigerators, freezers, and air-conditioners. They are also used in fire extinguishers, particularly small household extinguishers, and they were previously used as propellants in aerosol cans. For many years, chemists believed that chlorofluorocarbons released into the air diffused harmlessly into the upper atmosphere, where they were broken down by sunlight. In the 1970s, however, it was learned that, in the presence of ultraviolet radiation, chlorofluorocarbons react with ozone. Initially, the ultraviolet radiation causes a chlorine atom to break free of the chlorofluorocarbon molecule. The chlorine atom then reacts with an ozone molecule, breaking it apart in a reaction that is apparently not followed by a resynthesis of the ozone molecule. A single chlorine atom can react with—and destroy—as many as 100,000 ozone molecules. Analyses of this reaction strongly suggest that if the addition of chlorofluorocarbons to the atmosphere continues unchecked, the ozone layer will eventually be entirely destroyed.

Depletion of the ozone layer leads directly and inevitably to an increase in the amount of ultraviolet radiation reaching the earth's surface. For living systems, the consequences are many and serious. Ultraviolet radiation can trigger the dissociation of biologically important molecules, producing atoms and groups of atoms that contain unpaired electrons and are thus highly reactive. Especially sensitive to ultraviolet radiation are the phytoplankton that form the base of aquatic food chains; experimental studies have shown that even moderate levels of ultraviolet radiation are detrimental to these organisms and higher levels can cause mass mortality. Terrestrial photosynthetic organisms are also affected by increased levels of ultraviolet radiation; in a series of laboratory experiments in which some 100 crop plants were exposed to slightly increased levels of ultraviolet radiation, the yields were reduced in 20 percent. Other experiments have demonstrated that the nitrogen-fixing bacteria that form symbiotic associations with the roots of legumes are killed by high levels of ultraviolet radiation.

For humans, one of the most immediate consequences of increased levels of ultraviolet radiation is an increased incidence of skin cancer. Projections by the National Academy of Sciences indicate that a 1 percent reduction in ozone levels would result in 10,000 more cases of skin cancer each year in the United States alone. Of perhaps even greater concern is new evidence that ultraviolet radiation can severely damage the immune system, the body's bulwark against infectious disease.

On the basis of present data, it is estimated that a 5 percent reduction in the ozone layer will occur by the year 2000. This estimate is based on the amounts of chlorofluorocarbons that have already been released into the atmosphere. In the meantime, worldwide production of chlorofluorocarbons has continued to increase at a rate of 4.5 percent per year—leading to

A NASA satellite photograph of the South Pole taken on October 15, 1987, showing the most serious ozone depletion ever recorded. The pink area in the center is the ozone hole. The concentric areas of color surrounding the hole indicate successively more substantial concentrations of ozone, with red and orange representing the relatively small areas of high concentration.

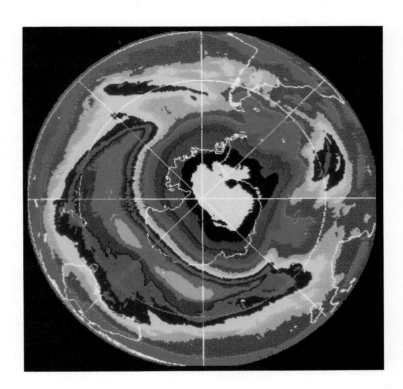

predictions of a 60 percent reduction in the ozone layer by the year 2050. Mathematical models predict that even a 16 percent reduction in the ozone layer would increase the amount of ultraviolet radiation reaching the earth's surface by about 40 percent, and a 30 percent reduction would double the amount.

Intense efforts to find replacements for chlorofluorocarbons, particularly for refrigeration systems, have only recently begun. A landmark treaty, drafted in Montreal in 1987 by 42 nations, binds its signatories to freeze chlorofluorocarbon production at 1986 levels by the end of 1989. Moreover, adherents

to the treaty are required to cut production and use of these chemicals by at least half by the end of the century. The countries that have now ratified the treaty, including not only the United States but also the 12 nations of the European Community, are responsible for more than two-thirds of world production of chlorofluorocarbons. Although this agreement will not halt the depletion of the ozone layer that is already underway, it is an important first step in slowing a process with potentially devastating effects for life on earth.

Climate, Wind, and Weather

The amount of energy received by various parts of the earth's surface is not uniform. This is the major factor determining the distribution of life on earth. In the vicinity of the equator, the sun's rays are almost perpendicular to the earth's surface, and this sector receives more energy per unit area than the regions to the north and south, with the polar regions receiving the least (Figure 54–3a). Moreover, because the earth, which is tilted on its axis, rotates once every 24 hours and completes an orbit around the sun about once every 365 days, the angle of the incident radiation, and so the amount of energy reaching different parts of the surface, changes hour by hour and season by season (Figure 54–3b).

54-3 (a) *A beam of solar energy striking the earth near one of the poles is spread over a wider area of the earth's surface than is a similar beam striking the earth near the equator.*

(b) *In the Northern and Southern Hemispheres, temperatures change in an annual cycle because the earth is slightly tilted on its axis in relation to its pathway around the sun. In winter in the Northern Hemisphere, the North Pole is tilted away from the sun, decreasing the angle at which the sun's rays strike the surface and also decreasing the duration of daylight, both of which result in lower temperatures. In summer in the Northern Hemisphere, the North Pole is tilted toward the sun. Note that the polar region of the Northern Hemisphere is continuously dark during the winter and continuously light during the summer.*

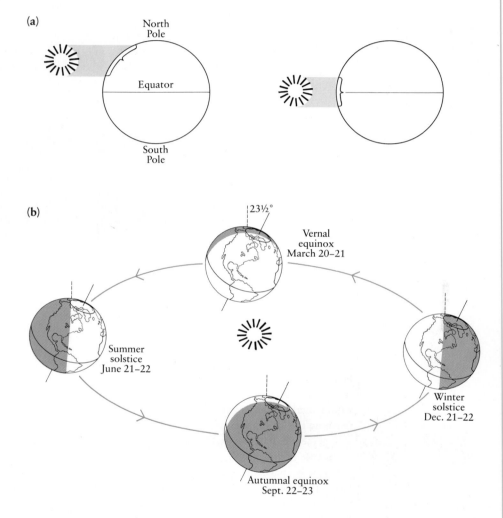

Temperature variations over the surface of the earth and the earth's rotation establish major patterns of air circulation and rainfall. These patterns depend, to a large extent, on the fact that cold air is denser than warm air. As a consequence, hot air rises and cold air falls. As air rises, it encounters lower pressure and consequently expands, and as a gas expands, it cools. Cooler air holds less moisture, so rising, cooling air tends to lose its moisture in the form of rain or snow.

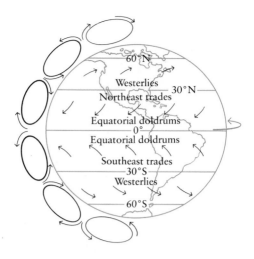

54-4 *The earth's surface is covered by belts of air currents, which determine the major patterns of wind and rainfall. Air rising at the equator loses moisture in the form of rain, and falling air at latitudes of 30° north and south is responsible for the great deserts found at these latitudes.*

The air is warmest along the equator, the region heated most intensely by the sun. This air rises, creating a low-pressure area (the doldrums) that draws air from north and south of the equator. As equatorial air rises, it cools, loses most of its moisture, and then falls again at latitudes of about 30° north and south, the regions where most of the great deserts of the world are found. This air warms, picks up moisture, and rises again at about 60° latitude (north and south); this is the polar front, another low-pressure area. A third, weaker belt rising at the polar front descends again at the poles, producing a region in which, as in other areas of descending air, there is virtually no rainfall. The earth's spinning movement twists the winds caused by this transfer of air from equator to poles, creating the major wind patterns (Figure 54-4).

The worldwide patterns are modified locally by a variety of factors. For example, along our own West Coast, where the winds are prevailing westerlies, the western slopes of the Sierra Nevada have abundant rainfall, while the eastern slopes are dry and desertlike (Figure 54-5). As the air from the ocean hits the western slope, it rises, is cooled, and releases its water. Then, after passing the crest of the mountain range, the air descends again, becomes warmer, and its water-holding capacity increases, resulting in a "rain shadow" on the eastern slope.

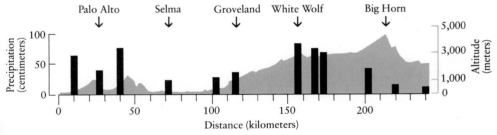

54-5 *The mean annual rainfall (vertical columns) in relation to altitude at a series of stations from Palo Alto on the Pacific Coast across the Coast Ranges and the* Sierra Nevada. *The prevailing winds are from the west, and there are rain shadows on the eastern slopes of the two mountain ranges.*

54-6 *The energy on which life depends enters the living world in the form of light. It is converted to chemical energy by photosynthetic organisms, such as the wild barley shown here. Such photosynthetic organisms are, in turn, the energy source for all heterotrophs, including ourselves.*

THE FLOW OF ENERGY

The flow of energy through ecosystems is the most important factor in their organization. Of the solar energy that reaches the earth's surface, only a very small fraction—an estimated 0.1 percent on a worldwide basis—is diverted into living systems. Even when light falls where vegetation is abundant, as in a forest, a cornfield, or a marsh, only 1 to 3 percent of that light (calculated on an annual basis) is used in photosynthesis. Yet this fraction, as small as it is, may result in the production—from carbon dioxide, water, and a few minerals—of several thousand grams (dry weight) of organic matter per year in a single square meter of field or forest, a total of about 120 billion metric tons of such organic matter per year on a worldwide basis.

Trophic Levels

The passage of energy from one organism to another takes place along a particular **food chain**—that is, a sequence of organisms related to one another as prey and

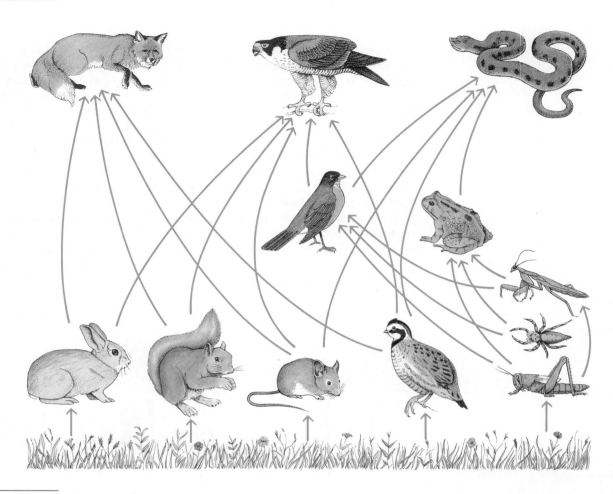

54–7 *Diagram of a food web. The arrows point in the direction of energy flow. This food web is much simplified; in reality, many more species of both plants and animals would be involved.*

predator. The first is eaten by the second, the second by the third, and so on, in a series of feeding levels, or **trophic levels.** In most ecosystems, food chains are linked together in complex **food webs,** with many branches and interconnections (Figure 54–7). Such a web may involve more than 100 different species, with predators characteristically taking more than one type of prey, and each type of prey being exploited by several different species of predators. The relation of any species to others in its food web is an important dimension of its ecological niche.

Producers

The first trophic level of a food web is always occupied by a primary producer. On land, the primary producer is usually a plant; in aquatic ecosystems, it is usually an alga. These photosynthetic organisms use light energy to make carbohydrates and other compounds, which then become sources of chemical energy. Producers far outweigh consumers; 99 percent of all the organic matter in the living world is made up of plants and algae. All heterotrophs combined account for only 1 percent of the organic matter.

Ecologists speak of the **productivity** of a trophic level, a community, or an ecosystem. **Gross productivity** is a measure of the rate at which energy is assimilated by the organisms in, for example, a particular trophic level. It might be considered analogous to the rate of gross income of a business. A more useful—and often more easily measured—quantity is the **net productivity** (Figure 54–8).

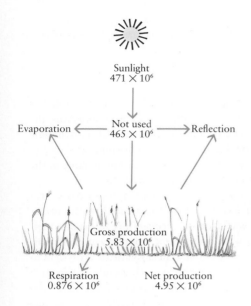

Sunlight
471×10^6

Evaporation ← Not used → Reflection
465×10^6

Gross production
5.83×10^6

Respiration
0.876×10^6

Net production
4.95×10^6

54–8 *Calculation of the productivity of a field in Michigan in which the vegetation was mostly perennial grasses and herbs. Measurements are in terms of calories per square meter per year. In this field, the net primary production—the amount of chemical energy stored in plant material—was 4,950,000 calories per square meter per year. Thus, slightly more than 1 percent of the 471,000,000 calories per square meter per year of sunlight reaching the field was converted to chemical energy and stored in the plant tissues.*

Net productivity is the gross productivity less the cost of all the metabolic activities of the organisms in question; this cost might be considered as equivalent to the cost of doing business. Net productivity is thus comparable to the rate of net profit. It is usually expressed as the amount of energy (measured in calories) stored in chemical compounds or as the increase in biomass (measured in grams or metric tons) in a particular period of time. (**Biomass** is a convenient shorthand term meaning the total dry weight of all the organisms being measured at any one time.) Net productivity is a measure of the rate at which organisms store energy, which then becomes available to the organisms of the next trophic level. In agricultural ecosystems, the standing crop at the end of the growing season represents the **net primary production** for that season.

As the data in Table 54–1 reveal, primary productivity varies enormously from one ecosystem type to another. In terrestrial ecosystems, the key factors influencing productivity are the duration and intensity of sunlight, temperature, and precipitation. In aquatic ecosystems, the availability of essential mineral elements is more often the principal factor affecting productivity.

TABLE 54–1 Average Primary Productivity of Selected Ecosystems

ECOSYSTEM	NET PRIMARY PRODUCTION (GRAMS PER METER² PER YEAR)	BIOMASS (KILOGRAMS PER METER²)
Open ocean	125	0.003
Algal beds and reefs	2,500	2
Estuaries	1,500	1
Lakes and streams	250	0.02
Swamps and marshes	2,000	15
Tropical forest	2,000	12
Temperate forest	1,250	32
Taiga	800	20
Woodland and shrubland	700	6
Tropical grassland	900	4
Temperate grassland	600	1.6
Tundra and alpine	140	0.6
Desert and semidesert scrub	90	0.7
Cultivated land	650	1

Adapted from R. H. Whittaker and G. E. Likens, Table 5.2, in H. Lieth and R. H. Whittaker (eds.), *Primary Productivity of the Biosphere*, Springer-Verlag, Inc., New York, 1975. Tropical forest biomass estimates based on Sandra Brown and Ariel E. Lugo, "Biomass of Tropical Forests: A New Estimate Based on Forest Volume," *Science*, vol. 223, pages 1290–1293, 1984.

Consumers

Energy enters the animal world through the activities of herbivores, animals that eat plants or algae. A herbivore may be a caterpillar, an elephant, a sea urchin, a snail, or a field mouse; each type of ecosystem has its characteristic complement of herbivores. Of the organic material consumed by herbivores, much is eliminated undigested. Most of the chemical energy from the digested food is used to maintain the metabolic processes of the animal and to power its daily activities: searching for food, eating and digesting it, mating and caring for the young, fleeing

Chemosynthetic Ecosystems

Throughout this text, we have emphasized that life on earth depends on radiant energy from the sun, which is converted by photosynthetic organisms—principally plants and algae—into chemical energy stored in the bonds of organic molecules. Although chemosynthetic bacteria (page 89) that are not dependent on the energy of sunlight to power their carbon-fixing reactions have long been known, until recently they were regarded as little more than oddities. As we noted previously (page 430), however, there is growing evidence that such bacteria may have been the earliest living systems on this planet. Moreover, a series of discoveries beginning in 1977 have revealed the existence of entire ecosystems in which the primary producers are chemosynthetic bacteria. Such ecosystems are dependent not on radiant energy from the sun but on geothermal energy—heat energy from within the earth itself.

The first known chemosynthetic ecosystem was discovered by the Alvin, the little research submarine from Woods Hole Oceanographic Institution, and its crew of scientific investigators, led by John B. Corliss of Oregon State University. The place was the Galapagos Rift, a boundary between adjacent tectonic plates (see page 1012). Sea water seeps into the fissures in the volcanic rocks erupting along this boundary, becomes heated, and rises again. As a consequence, oases of warmth are created in the near-freezing waters 2.5 kilometers below the surface. More important, the water reacts with the rocks deep within the earth's crust. Here, under extreme heat and pressure (300°C, 280 kilograms per square centimeter), chemical reactions take place. The crucial one, according to current evidence, is the reduction of sulfate in the sea water to hydrogen sulfide, with heat supplying the necessary energy for the reaction. Chemosynthetic bacteria oxidize the sulfide, obtaining the energy needed to extract carbon from carbon dioxide in the sea water and fix it in organic molecules.

More than a dozen of these deep-sea oases have now been located in the Pacific, each with chemosynthetic bacteria as the primary producers in a complex food web. The specific complement of animals present varies from one chemosynthetic ecosystem to another and is thought to be the result of whatever floating larvae happened to colonize the particular area during its early stages of development. Among the most spectacular inhabitants are the clams, measuring some 20 centimeters in diameter, which cover the lava floor. Other permanent inhabitants include mussels, crabs, and octopods. The bivalves feed on the bacteria, and the crustaceans and octopods on the bivalves. One oasis, known as the Garden of Eden, is dominated by huge tube worms. These tube worms, which lack mouths and digestive tracts, apparently obtain energy from organic molecules synthesized by chemosynthetic bacteria living within their tissues. Chemosynthetic bacteria have also been found living in the gills of some clams, suggesting that symbiosis plays an essential role in these ecosystems.

These discoveries have triggered not only a search for other deep-sea chemosynthetic ecosystems but also a closer look at the biological interactions of chemosynthetic bacteria found in more accessible surroundings. Such ecosystems have recently been found deep in the Gulf of Mexico in cold waters and off the Palos

predators, and so on. Although this energy is generally described as "lost" through respiration, it is important to realize that, for the individual organism, this is the essential energy on which its life depends. A fraction of the chemical energy consumed by the herbivore is converted to new animal biomass. The increase in animal biomass is the sum of the increase in weight of individual animals plus the weight of new offspring. It represents energy available to the next trophic level.

This next level, the secondary consumer level, is made up of carnivores, animals that eat other animals. The carnivore that devours the herbivore may be a lion, a minnow, a starfish, a robin, or a spider. In every case, only a small part of the

Giant tube worms, 1.5 meters tall, growing in the depths of the Pacific Ocean at the Garden of Eden oasis. These tube worms are dependent on chemosynthetic bacteria as their energy source. Also visible are yellow-shelled clams and an eyeless, colorless crab, other participants in the food web for which the bacteria are the primary producers. Unlike the near-freezing temperatures normally found at these depths, it was 17°C (63°F) in this area.

Verde peninsula of California in very shallow waters (1 to 10 meters deep). The latter discovery was made by a young, skin-diving scientist, Jeffrey Stein, who was puzzled by the black abalone he found grazing on mats of bacteria in sulfur-rich waters that would normally be toxic to animal life. Although the fissures through which the hot, sulfide-laden water escapes are only the size of teacups, they are surrounded by mats of chemosynthetic bacteria averaging 1 to 2 square meters in area. These mats are grazed not only by the black abalone but by other gastropods as well, including the California sea hare and the giant key-hole limpet.

Most recently, clams living in Los Angeles sewage outfalls have been found to contain symbiotic chemosynthetic bacteria, as have worms living in Bermuda. And, scientists have not yet examined the organisms of such sulfur-rich environments as pulpmill effluent zones, mangrove swamps, eel grass beds, and the Yellowstone geysers. It is already clear, however, that although most life on earth is based on radiant energy from the sun, there are highly viable alternatives—not only on this planet, but perhaps on others as well.

organic substance present in the body of the herbivore becomes incorporated into the body of the carnivore.

Some food chains have third and fourth consumer levels, but five links are usually the limit, regardless of the ecosystem. A study of 102 top predators (animals themselves free of predation) showed that there are usually only three links (four levels) in a food chain—plant to herbivore to carnivore to carnivore. For only one top predator were there more than five links (six species) involved. With each higher trophic level, there is a decrease in the total amount of energy stored in animal biomass and therefore available to other consumers.

(a)

(b)

54-9 *Representative consumers.* (**a**) *The hog deer* (Axis porcinus) *is a herbivore of the Indian subcontinent. Its common name derives from its relatively short stature; adult hog deer generally range from 60 to 72 centimeters in height.* (**b**) *An Indian python* (Python molurus), *a carnivore, begins the four-hour process of swallowing an adult hog deer that it has killed by constriction.*

Detritivores

Detritivores are organisms that live on the refuse, or detritus, of a community—dead leaves, branches, and tree trunks, the roots of annual plants, feces, carcasses, even the discarded exoskeletons of insects. They include scavengers, such as vultures, jackals, crabs, and earthworms, as well as decomposers, such as fungi and bacteria. Scavengers can be regarded as consumers that utilize dead prey rather than living prey. Decomposers are also consumers, but with a difference: they have evolved specializations that enable them to exploit sources of chemical energy, such as cellulose and nitrogenous waste products, that cannot be used by animals.

(a)

(b)

(c)

54-10 *Many birds, such as these vultures, are scavengers, living on dead animals. In East Africa, several species of vulture are often seen sharing the same carcass.* (**a**) *The lappet-faced vulture—the largest and most powerful, with a huge,* heavy beak—is the one that can first break into a carcass. Other vultures are often seen standing by a carcass awaiting its arrival. (**b**) Ruppell's griffon, also a vulture, specializes in reaching its long, snaky, featherless neck far inside the carcass to feed on the intestines and other internal organs. (**c**) The smaller Egyptian vulture, with its fine beak, can tear off scraps of flesh, such as from the skull, that the coarser beaks cannot grasp.

In a forest community, more than 90 percent of the net primary production is eventually consumed by detritivores rather than by herbivores. Some of this energy flows through the food web by way of consumers that feed on detritivores, while the rest of it is used in the metabolic processes of the detritivores themselves. As a result, essentially all of the energy stored by plants (and, in aquatic communities, by algae) is ultimately used to support life. Energy stored in organic matter goes unused only when it is trapped in an environment, such as a highly acidic peat bog, in which most detritivores cannot live (see Figure 8–18, page 176).

Efficiency of Energy Transfer

The shortness of food chains has long been attributed to the inefficiency involved in the transfer of energy from one trophic level to another, an explanation that, like so many others in ecology, is now undergoing renewed critical scrutiny. In general, however, only about 10 percent of the energy stored in a plant is converted to animal biomass in the herbivore that eats the plant. A similar relationship is found at each succeeding level. Thus, if an average of 1,500 kilocalories of light energy per square meter fall per day on a land surface covered with plants, about 15 kilocalories are converted to plant material (as noted earlier, only 1 to 3 percent of the incident radiant energy is used in photosynthesis). Of these 15 kilocalories, about 1.5 kilocalories are incorporated into the bodies of the herbivores that eat the plants, and about 0.15 kilocalorie is incorporated into the bodies of the carnivores that prey on the herbivores. Although meat is a more concentrated source of calories and nutrients than vegetation, carnivores must usually expend more energy in foraging than herbivores do, and so the net productivity of carnivores and herbivores may be roughly equivalent.

To give a concrete example, Lamont Cole of Cornell University, in his studies of Cayuga Lake, calculated that for every 1,000 calories of light energy utilized by algae in the lake, about 150 calories are reconstituted as small aquatic animals. Of these 150 calories, 30 calories are converted to smelt. If trout eat the smelt, about 6 calories are transferred, and if we eat the trout, we gain about 1.2 calories from the 1,000 calories of light energy originally stored in organic compounds. (Note that if we ate the smelt instead of the trout, we would derive about five times as much food energy from the original 1,000 calories. This, in short, is the argument for living lower on the food chain.)

As these figures suggest, the 10 percent "rule of thumb" is only a crude estimate. Actual measurements show wide variations in transfer efficiencies, from less than 1 percent to over 20 percent, depending on the species involved. The flow of energy with large losses at each successive level can be depicted as a pyramid such as that shown in Figure 54–11.

54–11 *Pyramid of energy flow for a river ecosystem in Florida. A relatively small proportion of the energy in the system is transferred at each trophic level. Much of the energy is used metabolically and is measured as calories lost in respiration.*

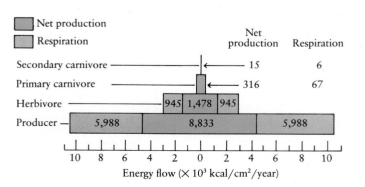

Energy Costs of Food Gathering

How much does a calorie cost, in terms of calories? For the expenditure of 1 calorie, an organism in most natural populations obtains from 2 to 20 calories in food energy. This is true for organisms whose expenditures are very high—such as the hummingbird, which spends up to 330 calories a minute—as well as for organisms whose expenditures are very low—such as the damselfly, which uses less than a calorie a day.

In simple human societies, in which individuals obtain their food without fossil-fuel energy subsidies, the ratio of food calories gained to calories invested is similar to that which prevails for the rest of the animal kingdom. Hunter-gatherers average 5 to 10 calories for each calorie spent; shifting agriculture (which requires no fertilizer) yields about 20 calories per calorie spent.

As is true in most societies (the social insects and civilized man are among the exceptions), almost the entire adult population has a share in the business of getting food. In the United States, about 20 percent of the population is involved in the food supply system. (Only about 2 percent are actually farmers; the rest are involved in food processing, transportation, and marketing.) Thus 80 percent are, for better or worse, free for other pursuits.

On the surface, it would seem that we expend less energy than most animals on the mundane work of food supply. Not so. At the turn of the century, for each calorie expended, including human labor, fuel for farm machinery and food transport, and the energy cost of fertilizer, we received about a calorie in return. Today, in the United States, as in other "advanced" technological societies, for every calorie invested we get a return of 0.1 calorie. This cost figure does not include the energy used for heating or lighting or running private automobiles (even those that bring the food home from the market) or electric can openers.

Obviously, other animals cannot live so profliga-

A Kalahari San (Bushman)—a hunter-gatherer—with a guinea hen he has snared. At his waist are a bow and quiver containing highly poisonous arrows.

tely; their energy income must exceed their expenditures. Like these other organisms, we are also dependent almost exclusively on solar energy. There is an important difference, however; because of our technology, we have been able to draw upon energy stored millions of years ago. It is only in the last decade that we have come to realize that not only are these resources finite but also they may soon be expended.

From Robert M. May, "Energy Costs of Food Gathering," *Nature*, vol. 225, page 669, 1975.

Energy Transfer and Ecosystem Structure

The energy relationships between the trophic levels determine the structure of an ecosystem in terms of both the numbers of organisms and the amount of biomass present. Figure 54–12a, for example, shows a pyramid of numbers for a grassland ecosystem. In this type of ecosystem, the primary producers (the individual grass plants) are small, and so a large number of them is required to support the primary consumers (the herbivores). In an ecosystem in which the primary producers are large (for instance, trees), one primary producer may support many herbivores, as indicated in Figure 54–12b.

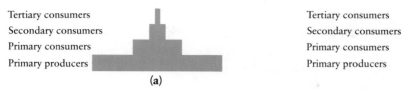

Tertiary consumers
Secondary consumers
Primary consumers
Primary producers

(a)

Tertiary consumers
Secondary consumers
Primary consumers
Primary producers

(b)

54–12 Pyramids of numbers for (a) a grassland ecosystem, in which the number of primary producers (grass plants) is large, and (b) a temperate forest, in which a single primary producer, a tree, can support a large number of herbivores.

Most pyramids of biomass take the form of an upright pyramid, as shown in Figure 54–13a, whether the producers are large or small. Pyramids of biomass are inverted only when the producers have very high reproduction rates. For example, in the ocean, the standing crop of phytoplankton may be smaller than the biomass of the zooplankton that feeds upon it (Figure 54–13b). Because the growth rate of the phytoplankton population is much more rapid than that of the zooplankton population, a small biomass of phytoplankton can supply food for a larger biomass of zooplankton. Like pyramids of numbers, pyramids of biomass indicate only the quantity of organic material present at one time; they do not give the total amount of material produced or, as do pyramids of energy, the rate at which it is produced.

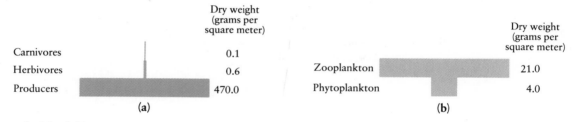

Dry weight
(grams per
square meter)

Carnivores — 0.1
Herbivores — 0.6
Producers — 470.0

(a)

Dry weight
(grams per
square meter)

Zooplankton — 21.0
Phytoplankton — 4.0

(b)

54–13 Pyramids of biomass for (a) a field in Georgia and (b) the English Channel. Such pyramids reflect the mass present at any one time; hence the seemingly paradoxical relationship between phytoplankton and zooplankton.

BIOGEOCHEMICAL CYCLES

Energy takes a one-way course through an ecosystem, but many substances cycle through the system. Such substances include water, nitrogen, carbon, phosphorus, potassium, sulfur, magnesium, calcium, sodium, chlorine, and also a number of other minerals, such as iron and cobalt, that are required by living systems in only very small amounts. The water cycle is shown on page 53, and the carbon cycle is on page 226. You may want to look at them again at this time.

Movements of inorganic substances are referred to as **biogeochemical cycles** because they involve geological as well as biological components of the ecosystem. The geological components are (1) the atmosphere, which is made up largely of gases, including water vapor; (2) the lithosphere, the solid crust of earth; and (3) the hydrosphere, comprising the oceans, lakes, and rivers, which cover three-fourths of the earth's surface.

The biological components of biogeochemical cycles include the producers, consumers, and detritivores (both scavengers and decomposers). The decomposers break down dead and discarded organic matter, completing the oxidation of the energy-rich compounds formed by photosynthesis. As we shall see in the case of the nitrogen cycle, the role of each decomposer may be very specialized—the performance of a single energy-yielding metabolic step. It has been calculated that for almost every reaction in the biosphere in which the conversion of one compound to another yields at least 15 kilocalories per mole, some organism has evolved that can exploit this energy to survive.

54–14 *The phosphorus cycle. Phosphorus is essential to all living systems as a component of energy-carrier molecules, such as ATP, and also of the nucleotides of DNA and RNA. Like other minerals, it is released from dead tissues by the activities of decomposers, taken up from soil and water by plants and algae, and cycled through the ecosystem.*

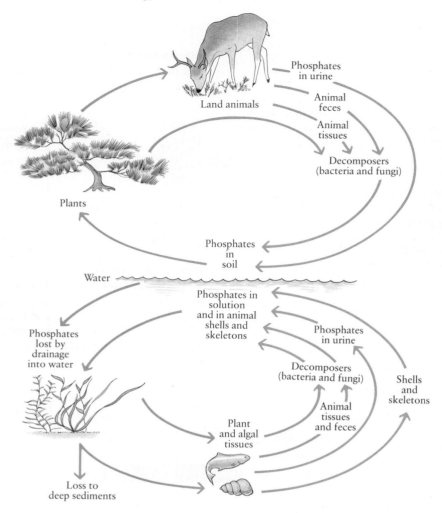

As a result of the metabolic work of the decomposers, inorganic substances are released from organic compounds and returned to the soil or water. From the soil or water, inorganic substances are taken back into the tissues of the primary producers, passed along to consumers, and then released to the detritivores, from which they enter the plants again, repeating the cycle.

The cycling through the ecosystem of one important mineral, phosphorus, is shown in Figure 54–14. Other minerals in the ecosystem undergo similar cycles, differing only in details.

The Nitrogen Cycle

The chief reservoir of nitrogen is the atmosphere; in fact, nitrogen makes up 78 percent of the gases in the atmosphere. Since most living things, however, cannot use elemental atmospheric nitrogen to make amino acids and other nitrogen-containing compounds, they are dependent on nitrogen present in soil minerals. So, despite the abundance of nitrogen in the atmosphere, a shortage of nitrogen in the soil is often the major limiting factor in plant growth. The process by which this limited amount of nitrogen is circulated and recirculated throughout the world of living organisms is known as the **nitrogen cycle.** The three principal stages of this cycle are (1) ammonification, (2) nitrification, and (3) assimilation.

Much of the nitrogen found in the soil is the result of the decomposition of organic materials and is in the form of complex organic compounds, such as proteins, amino acids, nucleic acids, and nucleotides. These nitrogenous com-

54–15 *The nitrogen cycle. Although the reservoir of nitrogen is in the atmosphere, where it makes up 78 percent of dry air, the movement of nitrogen in the ecosystem is more like that of a mineral than of a gas. Only a few microorganisms are capable of nitrogen fixation.*

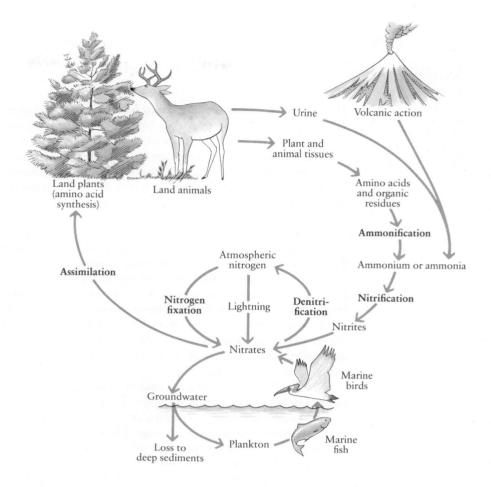

pounds are usually rapidly decomposed into simple compounds by soil-dwelling organisms, chiefly bacteria and fungi. These microorganisms use the proteins and amino acids to build their own needed proteins and release the excess nitrogen in the form of ammonia (NH_3) or ammonium ion (NH_4^+). This process is known as **ammonification.**

Several species of bacteria common in soils are able to oxidize ammonia or ammonium ion. This oxidation, known as **nitrification,** is an energy-yielding process, and the energy released is used by these bacteria as their primary energy source. One group of bacteria oxidizes ammonia (or ammonium) to nitrite (NO_2^-):

$$2NH_3 + 3O_2 \longrightarrow 2NO_2^- + 2H^+ + 2H_2O$$

Nitrite is toxic to many plants, but it rarely accumulates. Members of another genus of bacteria oxidize the nitrite to nitrate (NO_3^-), again with a release of energy:

$$2NO_2^- + O_2 \longrightarrow 2NO_3^-$$

Although plants can utilize ammonium directly, nitrate is the form in which most nitrogen moves from the soil into the roots. Hence, these bacteria play a crucial role in the recycling of nitrogen.

Once the nitrate is within the plant cell, it is reduced back to ammonium. In contrast to nitrification, this **assimilation** process requires energy. The ammonium ions thus formed are transferred to carbon-containing compounds to produce amino acids and other nitrogenous organic compounds needed by the plant.

Agricultural Ecosystems and a Hungry World

The earliest traces of agriculture, dating back about 11,000 years, are found in areas of the Near East that are now parts of Iran, Iraq, and Turkey. Here were the raw materials required by an agricultural economy: cereals, which are grasses with seeds capable of being stored for long periods of time without serious deterioration, and herbivorous herd animals, which can be readily domesticated. The grasses in these areas were wild wheats and barley, which still grow wild in the foothills. The animals were wild sheep and goats. By 8,100 years ago, agricultural communities were established in eastern Europe. By 7,000 years ago (about 5000 B.C.), agriculture had spread to the western Mediterranean and up the Danube River into central Europe and, by 6,000 years ago (about 4000 B.C.), to Britain. During this same period, agriculture originated independently in Central and South America, and perhaps slightly later in the Far East.

With the advent of widespread agriculture, *Homo sapiens* became a major actor in ecosystem dynamics. An area under cultivation is, in effect, an artificial ecosystem, maintained at a very early stage of ecological succession by human activities. Like other ecosystems subjected to frequent, large-scale disturbances, agricultural ecosystems are typically characterized by a small number of species, a relatively low total biomass, a high net productivity in relation to biomass, and a limited capacity to trap and retain nutrients.

In the earliest periods of agriculture, the most important of these features was undoubtedly the high net productivity, which provided the primitive farmers with a larger and more reliable food supply than they had previously known. As we saw in Chapter 52, an immediate consequence was a significant increase in the human population, a growth that has ultimately become exponential. The present-day consequences of this increase are many and serious—pollution, depletion of fuel supplies, destruction of natural resources, and extinction of other species as the human population expands into previously undisturbed ecosystems. Ironically, by far the most difficult problem and also the most urgent is hunger and starvation. Of the world's more than 5.25 billion people, at least 1 billion are inadequately nourished. It is estimated that some 20 million people die each year as the direct result of malnutrition, and many more deaths are thought to be caused by the indirect effects of malnutrition. Perhaps even more important, in terms of the world's future, are the effects, both physical and psychological, of the prolonged chronic hunger of so large a proportion of our population.

A major effort to increase world food supplies by increasing the productivity of agricultural ecosystems, known as the Green Revolution, has been under way for the past 40 years. The principal emphasis of this effort has been the development of new crop plants, especially grains. Enormous progress has been made. From 1950 to 1970, the production of wheat in Mexico increased from 270,000 metric tons per year to 2.35 million. In 1985, wheat production in Mexico reached an all-time high of 5.2 million metric tons, but by 1987, it had dropped back to 4.4 million. In the period from 1950 to 1970, the corn harvest in Mexico increased a more modest 250 percent, with yields per hectare almost doubling; in subsequent years, the corn harvest has continued to increase,

Although the nitrogen cycle appears complete and self-sustaining, nitrates are steadily lost from the soil and thus removed from the cycle. As we noted in Chapter 31, a variety of processes—the harvesting of plants, soil erosion, fire, and water percolating down through the soil—can reduce the amount of nitrate available to plants. Nitrates are also lost as a result of the activities of certain types of soil bacteria; in the absence of oxygen, these bacteria break down nitrates, releasing nitrogen back into the atmosphere and using the oxygen for their own respiration. This process, known as **denitrification,** takes place in poorly drained (hence, poorly aerated) soils. The cycle is maintained, despite these losses, primarily by the activities of the nitrogen-fixing bacteria (see pages 665–668), which incorporate gaseous nitrogen from the air into organic nitrogen-containing com-

reaching a record 14 million metric tons in 1985. Between 1950 and the present, India has increased its production of food grains about 2.8 percent a year. (Its population during this period has increased an average of 2.1 percent a year, which is significantly less.) Since 1971, China, the most populous nation in the world, has become agriculturally self-sustaining. Most of this success has come about as a result of improved varieties of crop plants, combined with better techniques of irrigation and fertilization.

Despite its acknowledged success, this massive effort has come under criticism in recent years. Modern agricultural ecosystems consist of huge stands of single species, an open invitation to insects, weeds, and disease organisms. Such opportunistic species can be kept at bay only by constant attention and the application of insecticides, herbicides, and fungicides. This susceptibility to invasion was tragically illustrated by the great potato famine of Ireland, which was caused by an oomycete (see page 467). The famine of 1845–1847 was responsible for more than a million deaths by starvation and initiated large-scale emigration from Ireland; within a decade, the population of Ireland dropped from 8 million to 4 million. Virtually the entire Irish potato crop was wiped out in a single week in the summer of 1846. A number of plant geneticists have warned that the new strains of wheat and rice, which have made major contributions toward feeding the growing human population, are particularly susceptible to such disasters because of their genetic uniformity and widespread distribution.

A related criticism concerns the enormous amounts of fertilizer required by the new plant varieties in order to achieve their high yields—fertilizer that must be repeatedly applied as nutrients are removed from the ecosystem by the harvesting of the crops. Fertilizer, pesticides, farm equipment, and fuel for the equipment are all increasingly expensive. Because large landowners are able to afford the necessary investment, whereas small-scale farmers are not, the Green Revolution is seen as accelerating the consolidation of farm lands into a few large holdings by the very wealthy.

Another problem is that although food production is still outstripping population growth, increased agricultural productivity will not be able to keep pace indefinitely with the rapid growth of the world's population. (If the population had remained at its 1950 levels, there would be more than enough food to feed everyone on the planet adequately today.)

Finally, there is a more fundamental though more elusive reason for the dissatisfaction with the Green Revolution. When it first began, it appeared to many to be an almost magical solution to problems so enormous and distressing that they had seemed insoluble. It is now clear, however, that poverty and famine and the unrest and violence they may bring will not be solved by a "technological fix." The Green Revolution must, of course, go forward. At the same time, we must recognize that the broader solutions are social, political, and ethical, involving not only the growth of crops but their distribution, not only the limiting of populations but the raising of living standards worldwide to tolerable levels.

pounds. Just as all organisms are ultimately dependent on photosynthesis for energy, they all depend on nitrogen fixation for their nitrogen.

Recycling in a Forest Ecosystem

Continuing studies of a deciduous forest ecosystem have shown that the plant life of a community plays a major role in its retention of nutrient elements. The studies have been carried out in the Hubbard Brook Experimental Forest in the White Mountain National Forest in New Hampshire. Investigators first established a procedure for determining the mineral budget—input and output, "profit" and "loss"—of areas in the forest. By analyzing the content of rain and snow, they were able to estimate input, and by constructing concrete weirs (Figure

54–16 *Weir at Hubbard Brook Experimental Forest in New Hampshire. Water from each of six experimental ecosystems was channeled through weirs, built where the water leaves the watershed, and was analyzed for chemical elements. The watershed behind the weir in this photograph has been stripped of vegetation. The experiments showed that such deforestation greatly increases the loss of nutrient elements from the system.*

54–16) that channeled the water flowing out of selected areas, they were able to calculate output. (A particular advantage of the site is that bedrock is present just below the soil surface, so that little material leaches downward.) They discovered, first, that the natural forest was extremely efficient in conserving its mineral elements. For example, the annual net loss of calcium from the ecosystem was 9.2 kilograms per hectare. This represents only about 0.3 percent of the calcium in the system. In the case of nitrogen, the ecosystem was actually accumulating this element at a rate of about 2 kilograms per hectare per year. There was a similar, though somewhat smaller, net gain of potassium.

In the winter of 1965–1966, all of the trees, saplings, and shrubs in one 15.6-hectare area of the forest were cut down completely. No organic materials were removed, however, and the soil was undisturbed. During the following summer, the area was sprayed with a herbicide to inhibit regrowth. During the four months from June to September, 1966, the runoff of water from the area was four times higher than in previous years. Net losses of calcium were 10 times higher than in the undisturbed forest and of potassium, 21 times higher. The most severe disturbance was seen in the nitrogen cycle. Plant and animal tissues continued to be decomposed to ammonia or ammonium, which then were acted upon by nitrifying bacteria to produce nitrates, the form in which nitrogen is usually taken up by plants. However, no plants were present, and the nitrate, a negatively charged ion, was not held in the soil. Net losses of nitrate nitrogen averaged 120 kilograms per hectare per year from 1966 to 1968. As a side effect, the stream draining the area became polluted with algal blooms, and its nitrate concentration exceeded the levels established by the U.S. Public Health Service for drinking water. Trees are now beginning to grow again on the devastated site, and the runoff of nutrients has dramatically decreased.

Studies in the tropical forests of Central and South America have further underlined the importance of plant life in mineral retention. In the tropical rain forest, in particular, virtually all mineral nutrients are held within the living components of the ecosystem—and none are retained by the soil for any length of time. When inorganic nutrients are released by the action of decomposers, they are immediately taken up again by the roots of plants. This has the effect of preventing leaching of nutrients from the soil by the frequent, and often torrential, rains. Another consequence is that when forest is cleared to make way for farming, it is often impossible to grow a second crop after the first has been harvested.

Concentration of Elements

The elements needed by living organisms often are present in their tissues in higher concentrations than in the surrounding air, soil, or water. This concentration of elements comes about as a result of the selective uptake of substances by living cells, amplified by the channeling effects of food chains. Under natural circumstances, this concentration effect is usually valuable; animals generally have a greater requirement for minerals than do plants because so much of the biomass of plants is cellulose. In some special cases, the effect of the accumulation may be quite dramatic. For example, the thousands of kilometers of coral reefs in the world's warmer waters are composed of calcium extracted from sea water over the millennia by one tiny polyp after another (see page 530).

Foreign substances can also get caught up in biogeochemical cycles and, as they are passed from one living organism to another, reach high concentrations as they approach the top of the food chain. DDT is probably the best known of the toxic substances whose effects were amplified in this way (Figure 54–17). Another was strontium-90, a radioactive element produced by nuclear testing in the 1950s. It is

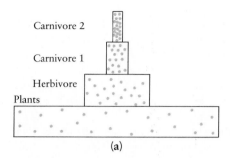

(a)

(b)

54–17 **(a)** *Concentration of DDT residues being passed along a simple food chain. As organic matter is transferred from one level to the next in the chain, about nine-tenths of it is usually respired or excreted; only the remaining 10 percent forms new biomass. The losses of DDT residue, however, are small in proportion to the loss of other organic matter through respiration and excretion. Consequently, the concentration of DDT increases as the material passes along the chain, and high concentrations occur in the carnivores.*

(b) *An eaglet and an egg that will never hatch, photographed in a nest near the Muskegon River in Michigan. DDT causes a bird's liver to break down the hormones that mobilize calcium at the time of egg production, resulting in thin-shelled, fragile eggs, like the one shown here. Birds at the top of food chains, such as the osprey, peregrine falcon, and bald eagle, were principal victims. With the banning of DDT in many countries, dramatic recoveries have been observed in populations of all three species. For example, more than 14,000 bald eagles were counted in 1982, as compared with about 3,000 in 1976.*

closely related to calcium and can take its place in many biochemical reactions. Strontium-90 from the fallout made its way through grasses into dairy cows and into milk; from there it became concentrated in the bones and teeth of children. By 1959 the bones of children in North America and Europe averaged an estimated 2.6 picocuries of strontium-90 per gram of bone calcium, compared with 0.4 picocurie per gram of calcium in the bones of adults. Although this level of radioactivity has not been proved dangerous, exposure to radioactive elements is known to cause leukemia, bone cancers, and genetic abnormalities and generally to shorten the life span. And, most important, the minimum exposure that can produce such effects is not established. Since the half-life of strontium-90 is 28 years (in other words, it takes 28 years for half of the atoms present to lose their radioactivity), this exposure still continues.

The channeling of strontium-90 along the food chain could have been anticipated. Other consequences of the atomic tests were less predictable. For example, in the Arctic only slight radiation exposure was expected because the amount of fallout that reaches the ground at the poles is much less than it is in the temperate zones. However, Eskimos in the Arctic were discovered to have concentrations of radioactivity in their bodies that were much higher than those found in the inhabitants of temperate regions. The key link in the chain was the lichens. Lichens, which obtain their minerals largely from the rain, had absorbed a large amount of fallout material, little of which had time to decay and none of which was dissipated by absorption into the soil. In the winter, caribou live almost exclusively on lichens, and, at the top of the food chain, Eskimos live largely on caribou.

The Lessons of Chernobyl

In the early hours of April 26, 1986, the operators of a nuclear power plant at Chernobyl in the Soviet Union began a series of tests of the operating capabilities of their newest reactor. According to official Soviet reports, without authorization the operators overrode the safety systems and deviated from standard procedures in order to conduct their tests. At some point, they lost control of the reactions occurring within the core of the reactor; its output suddenly rose to 120 times its rated capacity, rupturing portions of the core. The cooling water in the system was instantly turned to steam, producing a powerful explosion that blew the 1,000-ton reactor cover aside and spewed hot nuclear fuel into the sky. The reactor was immediately engulfed in fires that required more than 10 days to bring under control. During this period, an enormous quantity of radioactive material was released into the environment—more than 1,000 times the amount released in the accident at Three Mile Island in the United States in 1979.

Although the consequences of this accident were most severe in the areas closest to Chernobyl, they reached far beyond the Soviet Union, ultimately affecting some 100 million people in more than 20 European countries. The radioactive cloud from the accident was moved in a northwesterly direction by the wind, and, when the rains subsequently came, the radioactive material fell back to the earth. A substantial portion of the fallout was deposited on Norway, a country with no nuclear power plants. Had you been in the alpine tundra of Norway during the summer of 1986, you would have seen a landscape of low willows scattered on a thick mat of lichens, dotted with crystalline ponds and laced with rushing brooks. You would have seen ponies, descended from ancient Viking stocks, grazing on the upper slopes of the hills, and, with a little luck, herds of musk oxen and reindeer. The appearance was that of a wilderness little changed since the Pleistocene hunters had followed migrating herds northward. However, the Norwegians—well aware of the biological concentration of strontium-90 in Arctic regions during the 1950s—knew that the reality was otherwise.

54–18 *In the winter of 1986–1987, as in other winters, the Sami peoples of Norway rounded up their reindeer herds in preparation for the annual slaughter—a process that previously provided them not only with meat for their own use but also with income for the next year. In central Norway, the levels of radioactivity in the reindeer averaged more than 10 times the level allowed in meat that is to be marketed, and in some cases the levels were more than 20 times the allowed limit. The contaminated carcasses were buried in large pits—in effect, nuclear waste disposal sites—in uninhabited regions in the far north. Although the Norwegian government has provided compensation for the financial losses, preventing economic collapse, the Sami people face an interval of some 20 to 30 years before they can hope to use and sell the milk and meat of their reindeer once more.*

A principal component of the Chernobyl fallout was cesium-137, with a half-life of 30 years. As this element passed from the rainwater to the lichens and then to the reindeer, its concentration increased to levels far exceeding those considered safe for human consumption. The highest concentrations occurred in the milk, muscles, and bones of the reindeer—the traditional subsistence for the Sami, or Lapp, peoples of central and southern Norway. Given the half-life of cesium-137 and its concentration in the lichens, it appears that it will be a full generation before the Sami can once again rely upon meat and milk from their herds. By then, their unique culture may have disappeared forever—as a consequence of miscalculations and a lack of caution at a nuclear power plant more than 2,000 kilometers away.

The accident at Chernobyl and its aftermath remind us of several important lessons. The first, and most obvious, is that the biological concentration of elements is a very real phenomenon with potentially severe consequences, especially for organisms at the top of the food chain, including ourselves. The second is that we dare not become complacent about the safeguards surrounding the use of potentially hazardous materials or technologies; tragedies far worse than Chernobyl are possible. The third, and perhaps most important, lesson is that the consequences of our misadventures do not respect international borders or local environmental regulations, no matter how well conceived or faithfully followed. We—and all other living organisms—are interconnected in one global ecosystem.

SUMMARY

An ecosystem is a unit of biological organization made up of all the organisms in a given area and the environment in which they live. It is characterized by interactions between the living (biotic) and nonliving (abiotic) components that result in (1) a one-way flow of energy from the sun through autotrophs to heterotrophs, and (2) a cycling of mineral elements and other inorganic materials.

The ultimate source of energy for most ecosystems is the sun. Of the solar energy reaching the outer limits of the atmosphere, approximately half is reflected from or absorbed by the atmosphere. The distribution of the solar energy that reaches the earth's surface is affected by the position and movements of the earth in relation to the sun and by the movements of air and water over the surface. These factors can cause wide differences in temperature and rainfall from place to place and from season to season.

Within an ecosystem, there are trophic (feeding) levels. All ecosystems have at least three such levels: primary producers, which are usually plants or algae; primary consumers, which are usually animals; and detritivores, which live on animal wastes and dead plant and animal tissues. The primary producers (the autotrophs) convert a small proportion (about 1 to 3 percent) of the sun's energy into chemical energy. The primary consumers (herbivores) eat the primary producers. A carnivore that eats a herbivore is a secondary consumer, and so on. On the average, about 10 percent of the energy transferred at each trophic level is stored in body tissue; of the remaining 90 percent, part is used in the metabolism of the organism and part is unassimilated. This unassimilated energy is ultimately utilized by detritivores.

The movements of water, carbon, nitrogen, and mineral elements through ecosystems are known as biogeochemical cycles. In such cycles, inorganic materials from the air, water, or soil are taken up by primary producers, passed on to consumers, and eventually transferred to the detritivores known as decomposers —bacteria and fungi. In the course of their metabolism, decomposers release the inorganic materials to the soil or water in a form in which they can again be taken up by primary producers. Synthetic chemicals or radioactive elements released into the environment can be caught up in biogeochemical cycles, becoming concentrated in the tissues of organisms at higher trophic levels.

The nitrogen cycle is of critical importance to all organisms. It involves several stages: ammonification, the breakdown of nitrogenous organic compounds to ammonia or ammonium ion; nitrification, the oxidation of ammonia or ammonium to nitrates, which are taken up by plants; and assimilation, the conversion of nitrates to ammonia and its incorporation into organic compounds. Nitrogen-containing organic compounds are eventually returned to the soil or water, completing the cycle. Nitrogen lost from the ecosystem is replaced by nitrogen fixation, the incorporation of elemental nitrogen into organic compounds.

QUESTIONS

1. Distinguish among the following: community/ecosystem; biotic/abiotic; atmosphere/troposphere/stratosphere; solstice/equinox; food chain/food web; gross productivity/net productivity; producer/consumer/detritivore/decomposer; ammonification/nitrification/assimilation/denitrification.

2. What is the position of the earth in relation to the sun at the vernal equinox? At the winter solstice? What is the day-night cycle at the South Pole on these dates?

3. Describe what happens to the light energy striking a temperate forest ecosystem. What happens when it strikes a cornfield? A pond? A field on which cattle are grazing?

4. Describe what happens to a mineral nutrient in each of the environments in Question 3.

5. Consider each of the organisms below and list the effects of each on its ecosystem. Consider how the organism receives its inputs of energy and nutrients, where its outputs (metabolic wastes, offspring, dead carcasses) go, and its effects on other organisms.
   a. Earthworm
   b. Heterotrophic soil bacterium
   c. Oak tree or grass plant
   d. Deer or grasshopper
   e. Lion or wolf

6. Explain the different kinds of information provided by a pyramid of numbers, a pyramid of biomass, and a pyramid of energy flow. For what purposes might each type be more useful than the other two?

7. Among the highest energy-transfer efficiencies known occur when reptiles consume warm-blooded prey, such as birds or small mammals. Explain, in terms of characteristics of both prey and predator, why a high energy-transfer efficiency would be expected for this particular step in a food chain.

8. Although a carnivore at the top of a food chain is free of visible predation, it is—during its lifetime—a source of energy for many different species, representing as many as four kingdoms. Explain.

9. What are the implications for mineral cycling of human practices of fertilization of land and harvesting of crops? How are these implications different for nutrients whose major inorganic reservoir is the atmosphere rather than the soil?

CHAPTER 55

The Biosphere

The biosphere, that part of the earth in which life exists, is only a thin film on the surface of this little planet. It extends about 8 to 10 kilometers above sea level and a few meters down into the soil, as far as roots penetrate and microorganisms are found. It includes the surface waters and the ocean depths. It is patchy, differing in both depth and density.

In this, the final chapter of our text, we shall survey the characteristic patterns of life found in different regions of the biosphere.

LIFE IN THE WATERS

Life began in water, and although living organisms have long since conquered the land, by far the largest proportion of the biosphere consists of aquatic environments and their inhabitants. Freshwater environments can be conveniently classified as running water (rivers and streams) and standing water (lakes and ponds). Similarly, marine environments can be classified as oceans and seashores. Within these broad categories, there are, of course, a diversity of habitats, each with its own characteristic complement of organisms.

Rivers and Streams

Rivers and streams are characterized by continuously moving water. They may begin as outlets of ponds or lakes, as runoffs from melting ice or snow, or they may arise from springs (flows of groundwater emerging from bedrock).

55–1 *A fast-moving stream in Olympic National Park, Washington. The boulders within and beside the stream are constantly wetted by spray, making possible the luxuriant growth of mosses on their surfaces. What is a probable explanation for the total absence of mosses on some of the smaller boulders?*

(a)

(b)

(c)

55–2 *Some freshwater environments and their inhabitants.* **(a)** *A pond in the Pocono Mountains of Pennsylvania, covered with water lilies.* **(b)** *A great blue heron* (Ardea herodias), *with a bass that it has just caught.* **(c)** *A hungry muskrat.*

The character of life in a stream is determined to a large extent by the swiftness of the current, which generally changes as a stream moves downward from its source and, fed by tributaries, increases in volume and decreases in speed. In swift streams, most organisms live in the riffles, or shallows, where small photosynthetic organisms—algae and mosses—cling to the rock surfaces. Many insects, both adult and immature forms, live on the underside of rocks and gravel in the riffles. For those small organisms that can survive the swiftly moving current, there is an abundance of oxygen and nutrients swept along by the flowing waters.

As the stream travels along its course, the riffles are often interrupted by quieter pools, where organic materials may collect and be decomposed. Few plants can gain footholds on the shifting bottoms of stream pools, but some invertebrates, such as dragonflies and water striders, are typically found in or about the pools. Some organisms, notably trout, move back and forth between the riffles and the pools.

As streams broaden and become slower, they begin to take on the characteristics of lakes and ponds.

Lakes and Ponds

Bodies of standing water vary in size from small ponds to very large lakes, covering thousands of square kilometers. They contain three distinct zones: littoral, limnetic, and profundal.

The **littoral zone,** at the edge of the lake, is the most richly inhabited. Here the most conspicuous plants are angiosperms rooted to the bottom, such as cattails and rushes. Water lilies grow farther out from the shore. There is often a green blanket of duckweed, a small, free-floating angiosperm. Other pond weeds, which grow entirely beneath the water, lack a waxy outer cuticle and so can absorb minerals through their epidermis as well as through their roots. Their submerged plant surfaces harbor large numbers of small organisms. Snails, small arthropods, and mosquito larvae feed upon the plants and algae. Other insects that live among the submerged plants, such as the larvae of the dragonfly and damselfly, and the water scorpion, are carnivorous. Clams, worms, snails, and still other insect larvae burrow in the mud. Frogs, salamanders, water turtles, and water snakes are found almost exclusively in the littoral zone. The tadpoles of both frogs and salamanders feed primarily on algae, and the adults on insects; most water turtles feed on insects when young and on plants when fully grown. Water snakes are carnivores, feasting on amphibians and fish, which are also found in great numbers along the lake margins. Ducks, geese, and herons feed on the plants, insects, mollusks, fish, and amphibians abundant in this zone. The shallow margins of some lakes and ponds are marshy. Among the inhabitants of the marshes are such animals as snails, frogs, ducks, herons, bitterns, muskrats, otters, and beavers.

In the **limnetic zone,** the zone of open water, small, floating algae—phytoplankton—are usually the dominant photosynthetic organisms. This zone, which extends down to the limits of light penetration, is the habitat, for example, of smallmouth bass, bluegills, and, in colder waters, trout.

The deepwater **profundal zone,** extending down from the limnetic zone, has no plant life. Its principal occupants are detritivores—scavenging fish, aquatic worms, insect larvae, crustaceans, fungi, and bacteria—that consume the organic debris filtering down from the overlying water. The nutrients released by bottom-dwelling decomposers are recirculated to the upper levels of the lake by the twice-yearly overturn of the water (see page 46).

The Oceans

The oceans cover almost three-fourths of the surface of the earth. Life extends to their deepest portions, but photosynthesizing organisms are restricted to the

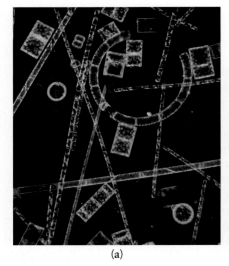

(a)

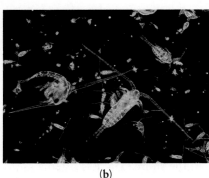

(b)

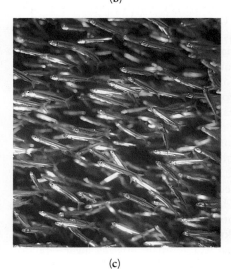

(c)

(d)

55–3 *Some ocean inhabitants.* (a) *A sample of living phytoplankton, taken from the English Channel in the spring. Most of these organisms are diatoms.* (b) *Zooplankton, the primary consumers in the sea, eat phytoplankton. Planktonic organisms form the base of the food webs in the pelagic division of the sea.* (c) *A school of dwarf herring, also known as "silversides."* (d) *A medusa of the hydrozoan* Gonionemus vertens, *photographed at Friday Harbor, Washington.*

upper, lighted zones. The sea has an average depth of more than 3 kilometers and, except for a relatively small surface fraction, is dark and cold. Most of its volume, therefore, is inhabited by bacteria, fungi, and animals, rather than by plants.

Sea water absorbs light readily. Even in clear water, less than 40 percent of the sunlight reaches a depth of 1 meter, and less than 1 percent of the sunlight that reaches the surface penetrates below 50 meters. Red, orange, and yellow wavelengths of light are absorbed first, so that only the shorter wavelengths, specifically blue and green, penetrate deeply. Thus, below depths of a few meters, only those photosynthetic organisms capable of utilizing the shorter wavelengths of light can grow.

There are two main divisions of life in the open ocean: **pelagic** (free-floating) and **benthic** (bottom-dwelling). A major component of the pelagic division is plankton. It is composed of rapidly dividing algae, intermingled with heterotrophic protists, small shrimp and other crustaceans, gelatinous invertebrates (such as jellyfish and ctenophores), and the eggs and larval forms of many fish and invertebrates. These planktonic forms provide food for fish and other relatively large pelagic animals. The benthic division contains the sessile animals, such as sponges, sea anemones, and clams, and many motile animals, such as worms, starfish, snails, crustaceans, and fish. A variety of fungi and bacteria also inhabit the benthic zone, subsisting on the accumulation of debris steadily drifting down from the more populated levels of the ocean.

Despite the fact that oceans cover three times more surface area of the planet than does the land, the total productivity of the open ocean—as measured by the amount of carbon converted to organic compounds by photosynthesis—is only about one-third as great. In fact, the open ocean is thought to be only slightly more productive per square meter than the desert (see Table 54–1, page 1139), presumably because of the low concentration of mineral nutrients in the areas of ocean where light penetrates and photosynthetic organisms can survive.

The major ocean currents, which are produced by a combination of winds and the earth's rotation, profoundly affect life in the oceans and alter the climate along the ocean coasts (see essay on page 1158). These patterns of water circulation—clockwise in the Northern Hemisphere and counterclockwise in the Southern Hemisphere—move currents of warm water north and south from the

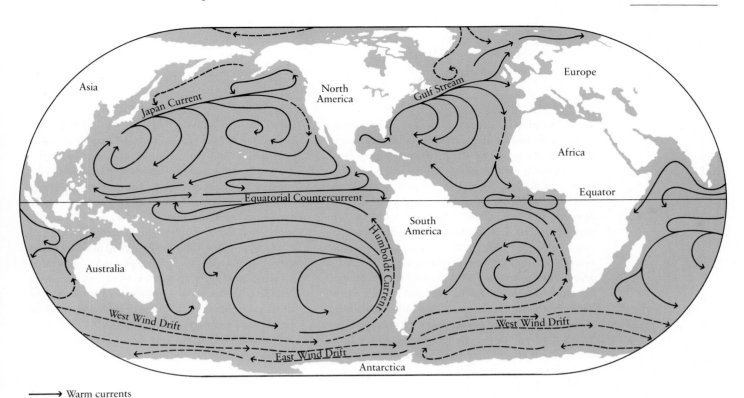

→ Warm currents

----→ Cold currents

55–4 *The major currents of the ocean have profound effects on climate. Because of the warming effects of the Gulf Stream, Europe is milder in temperature than is North America at similar latitudes. The eastern coast of South America is warmed by water from the equator, and the Humboldt Current brings cooler weather to the western coast of South America.*

equator (Figure 55–4). One such current, the Gulf Stream, warms a portion of the eastern coast of North America and the western shores of Europe, and another warms the eastern coast of South America. The same patterns of circulation bring cold waters to the western coasts of North and South America. Where the winds move the water continuously away from the shores, as off the coasts of Portugal and Peru, cold water rich in nutrients is brought to the surface from below, a process referred to as upwelling. Such areas contain high densities of pelagic life and traditionally have supported highly profitable fishing industries.

The Seashore

The edges of the continents extend 10 to 20 kilometers out into the sea. Along these edges, known as the continental shelves, nutrients are washed out from the land, and life is much denser than in the open seas. In temperate latitudes at the edge of the sea, where the large primary producers are brown algae (such as the rockweed *Fucus* and the kelp *Laminaria)*, net primary productivity is as high as anywhere else in the biosphere. Primary productivity is also extremely high in the coral reefs (page 530) of tropical regions.

Sessile animals, such as sponges and anemones, are found all over the ocean bottom, but they are most abundant in relatively shallow areas that are close to the shores. Predators, such as mollusks, echinoderms, crustaceans, and many kinds of fish, roam over the bottoms of the continental shelves. Eel grass, turtle grass, and seaweeds provide shelter for many animals and increase the supply of oxygen. Snails, sea slugs, and worms crawl over the surfaces of the plants and algae, feeding as they go.

Seashores—where the sea and the land join—are of three general types along most of the shores of the temperate zones: rocky, sandy, and muddy. Organisms that live on rocky coasts, like those that live in the riffles of fast-moving streams, often have special adaptations for clinging to rocks. The algae have strong holdfasts. The starfish of the rocky coast lies spread-eagled on the rocks, clinging with its tube feet. The abalone holds tight with its well-developed muscular foot. Mussels secrete coarse, ropelike strands that anchor them to rocky surfaces.

El Niño

The importance of ocean currents to weather patterns and to both marine and terrestrial organisms is dramatically illustrated by the disastrous consequences that can follow disturbance of the normal patterns of water circulation. The most frequent disturbance occurs in the Pacific Ocean and is known as El Niño (Spanish for "the Child") because its effects are usually first felt in South America around Christmastime. Major El Niño events occur approximately every 8 to 10 years, and smaller ones every 2 to 3 years. The most severe El Niño in the past century occurred between the fall of 1982 and the summer of 1983. Normally, the patterns of weather and ocean currents in the Pacific are determined by a large high-pressure system in the eastern Pacific and a large low-pressure system over Indonesia; air moves from the region of high pressure to the region of low pressure, creating winds that blow from east to west. These winds move warm surface water toward the western Pacific, and cold water moves in along the western coasts of North and South America as an undercurrent.

An El Niño is apparently triggered by a rise in the surface temperature of equatorial regions of the eastern Pacific, causing changes in the location and strength of the atmospheric pressure systems. These changes, in turn, cause a reversal in the wind direction and an accompanying reversal in the surface current. Over a period of two to three months, warm water surges back toward North and South America, trapping the cold undercurrent well below the surface. With higher water temperatures and a cutoff of nutrient upwelling, planktonic organisms die off—followed by the fish that feed on the plankton, fol-lowed by the seabirds that feed on the fish. The 1982–1983 El Niño not only caused a collapse of this food chain—and thus of the main fishing industries in Peru, Ecuador, and the Pacific Northwest—but also apparently disrupted weather patterns in many parts of the world. Massive rainstorms battered the usually dry coasts of California, Ecuador, and Peru that winter, and record snows fell in the Sierra Nevada and Rocky Mountains, followed by record flooding in the spring. Indonesia, the Philippines, India, Australia, Mexico, and southern Africa were stricken with droughts that destroyed vegetation and millions upon millions of animals, both domestic and wild, that had neither food nor water.

The complex, interacting events occurring in an El Niño are still poorly understood, but recent evidence suggests that they may be part of a natural, self-sustaining cycle of warming and cooling. In a 10-year international study begun in 1985, satellites, ships, and ground stations across the Pacific are continuously monitoring air pressure, water and air temperatures, and the direction and velocity of winds and water currents. This study made possible the accurate prediction, a year in advance, of a small El Niño that occurred in 1986–1987. If a major El Niño occurs during the period of the study, as is highly probable, our knowledge of the forces involved should increase greatly. The power of the oceans and the atmosphere is far beyond our meager capacity to control, but better understanding of El Niños may make possible further accurate predictions and thus action to minimize their impact on both human activities and natural ecosystems.

The organisms of the rocky coast face the additional problem of the rising and falling tides. The supratidal zone, which is wetted only by waves and spray, is a zone of dark algal and lichen growth. The intertidal zone, alternately submerged and exposed by the tides, is commonly characterized by *Fucus,* often intermixed with many other species of brown and red algae. Animal life includes barnacles, oysters, mussels, limpets, and periwinkles. The subtidal zone, always submerged, often contains forests of kelp *(Laminaria* and *Macrocystis),* sea squirts, starfish, and other invertebrates. The characteristic zonation is due, in part, to gradients of light, temperature, and wave action, partly to competitive interactions, and also to predation by both herbivores, such as sea urchins, and carnivores, such as starfish.

Sandy beaches have fewer inhabitants because of the constantly shifting sands. Clams, ghost crabs, sand fleas, lugworms, and other small invertebrates live below

(a)

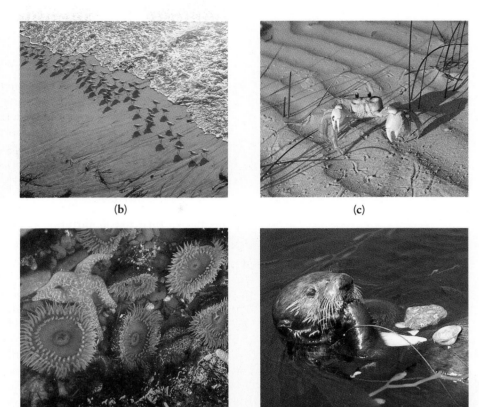

(b)

(c)

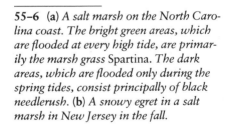

55–5 *Some examples of life at the seashore.* (a) *Sea oats, a common beach grass on the sandy dunes of the Atlantic coast.* (b) *Sanderlings, frequently seen on both the Atlantic and Pacific coasts of North America, race along the tide line, snatching up mollusks and crustaceans exposed as the waves retreat.* (c) *Common inhabitants of sandy tropical beaches are ghost crabs. These crabs, also known as racing crabs, run sideways at high speeds across the sand.* (d) *Among the inhabitants of this tidal pool on Rialto Beach, Washington, are kelp, sea anemones, and starfish.* (e) *A sea otter, eating clams while floating in a kelp bed in Monterey Bay, California. Note the large rock resting on the otter's abdomen; it is a tool, used as an anvil to break open the clam shells.*

(d)

(e)

the surface of the sand, feeding on the debris washed in and out by the tide. Above the sandy beaches, beach grasses, which spread by means of underground stems, are important for stabilizing the shifting dunes.

The mud flat, while not so rich or diverse in species as the rocky coast, supports a large number of organisms, with many animals living not only on but also beneath its surface. A mud flat can support tens of thousands of individuals per cubic meter.

Mud flats, salt marshes, and estuaries (areas where the fresh water of streams and rivers drains into the sea) are the receiving grounds for a constant flow of nutrients drained off from the land and so are extremely rich in animal life. They function as the spawning places and nurseries for many forms of marine life,

55–6 (a) *A salt marsh on the North Carolina coast. The bright green areas, which are flooded at every high tide, are primarily the marsh grass* Spartina. *The dark areas, which are flooded only during the spring tides, consist principally of black needlerush.* (b) *A snowy egret in a salt marsh in New Jersey in the fall.*

(a)

(b)

1159

including many commercially important species of fish and crustaceans. In the tropics and subtropics (including parts of Florida, Puerto Rico, and Hawaii), mangrove forests are important tideland communities, serving as spawning grounds for marine organisms and exporting minerals and nutrients. Because mud flats, salt marshes, and mangrove forests are often located in prime recreational and commercial areas and because they cannot be directly exploited for agriculture or lumbering, thousands of square kilometers of these wetlands are destroyed each year as they are filled and paved for human habitation—and thereby rendered sterile for many other species. Their protection is of special importance because of their role in nurturing life in the oceans.

LIFE ON THE LAND

The characteristic patterns of life on the land are determined principally by physical factors. Of the most immediate consequence to terrestrial organisms are temperature and precipitation (Figure 55–7). As we saw in the last chapter, these critical factors are influenced by the angle of the earth's axis in relation to its orbit around the sun and by the earth's rotation, which affect the duration and intensity of sunlight, the prevailing wind patterns, and the major ocean currents. Temperature and precipitation are also affected by the structure of the continents themselves.

The continents are composed largely of relatively light igneous rock (from the Latin word *ignis,* for "fire"), which is formed from molten material. The surfaces of the continents change constantly. They are crumpled by contractions and collisions as the continents rise, sink, and collide because of the motion of the plates on which they are carried (page 1012). As a consequence, the earth's surface is not at all uniform but varies widely from place to place in its composition and in its height above sea level. The mineral content of the earth's surface is a major factor affecting the growth of plants and other living organisms, and, as we saw in the example of the Sierra Nevada (page 1137), the mountain ranges of the continents do much to determine the patterns of rainfall.

55–7 *The principal determinants of the characteristic patterns of plant life in different regions of the biosphere are temperature, which decreases with increasing latitude, and precipitation.*

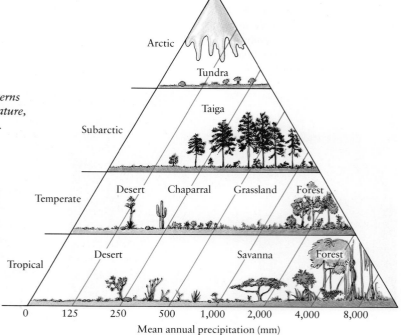

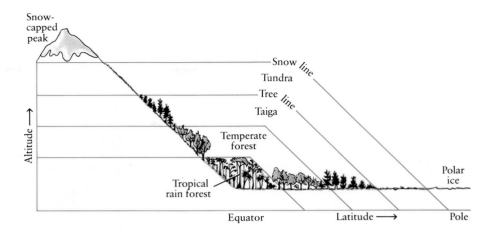

55–8 *In the Northern Hemisphere, we can experience a similar sequence of dominant plant life and its associated animal life by either traveling north for hundreds of kilometers or ascending a mountain. To experience a similar sequence in the Southern Hemisphere, we could ascend a mountain; however, by simply traveling south, we would never encounter vegetation corresponding to the taiga and the tundra of the Northern Hemisphere. Can you explain why?*

Mean atmospheric temperature decreases about 0.5°C for each degree of increase in latitude. Increases in elevation produce a similar effect, and in general, a change in mean atmospheric temperature corresponding to a 1° increase in latitude occurs with each rise of about 100 meters in elevation. This relationship has important consequences for the distribution of terrestrial organisms. For example, plants and animals characteristic of Arctic regions may be found near the equator in mountain ranges that run from north to south (Figure 55–8). There are, however, important differences between high-latitude habitats and high-altitude habitats. In mountains the air is clearer and the solar radiation is more intense. Most of the water vapor in the atmosphere—which plays a major role in preventing heat from radiating away from the earth at night—occurs below 2,000 meters. Consequently, nights in mountains are often much cooler than they are in regions with the same mean temperature but at higher latitudes and lower elevations. Moreover, there is pronounced seasonal variation both in day length and in temperature near the poles and relatively little variation near the equator. Those "Arctic" organisms that live in alpine environments near the equator must make physiological adjustments for such differences between high-latitude and high-altitude environments.

The Concept of the Biome

The land surface of the earth can be seen as being divided into a number of geographic areas distinguished by particular types of dominant plants. Thus each major continent has, for example, deserts, grasslands, and deciduous forests. These categories of characteristic plant life are called **biomes.**

As Figure 55–9 on pages 1162–1163 shows, the communities of plants and associated animal life that make up a biome are discontinuous, but a community may closely resemble another community on the other side of the planet. For instance, many of the deserts of the world look remarkably similar. However, when one looks more closely, one sees that although the physical features of the environment—temperature and rainfall—are the same, the particular organisms are not. But they look and act alike. The same is true of the Mediterranean scrub of California and Spain, the grasslands of North and South America, and so on.

A biome is a class, or category, not a place. When we speak of the tropical forest biome, we are not speaking of a particular geographic region, but rather of all the tropical forests on the planet. As with most abstractions, important details are omitted. For example, the boundaries are not as sharp as shown on maps nor are all areas of the world easy to categorize. However, the biome concept emphasizes one important truth: Where the climate is the same, the organisms are also very similar, even though these organisms are not related genetically and are far apart in their evolutionary history. Organisms of geographically separate patches of the same biome provide many examples of convergent evolution.

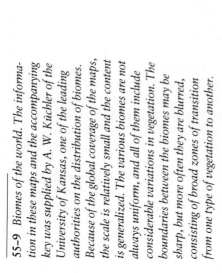

55-9 *Biomes of the world. The information in these maps and the accompanying key was supplied by A. W. Küchler, one of the leading authorities on the distribution of biomes. Because of the global coverage of the maps, the scale is relatively small and the content is generalized. The various biomes are not always uniform, and all of them include considerable variations in vegetation. The boundaries between the biomes may be sharp, but more often they are blurred, consisting of broad zones of transition from one type of vegetation to another.*

1 Temperate deciduous forests
2 Temperate mixed forests
3 Subtropical mixed forests
4 Taiga
5 Alpine forests and tundra
6 Mixed west-coast forests
7 Arctic tundra
8 Ice desert
9 Grasslands
10 Savannas
11 Mediterranean scrub
12 Deserts and semideserts
13 Juniper savanna
14 Southern woodland and scrub
15 Tropical mixed forests
16 Monsoon forests
17 Rain forests

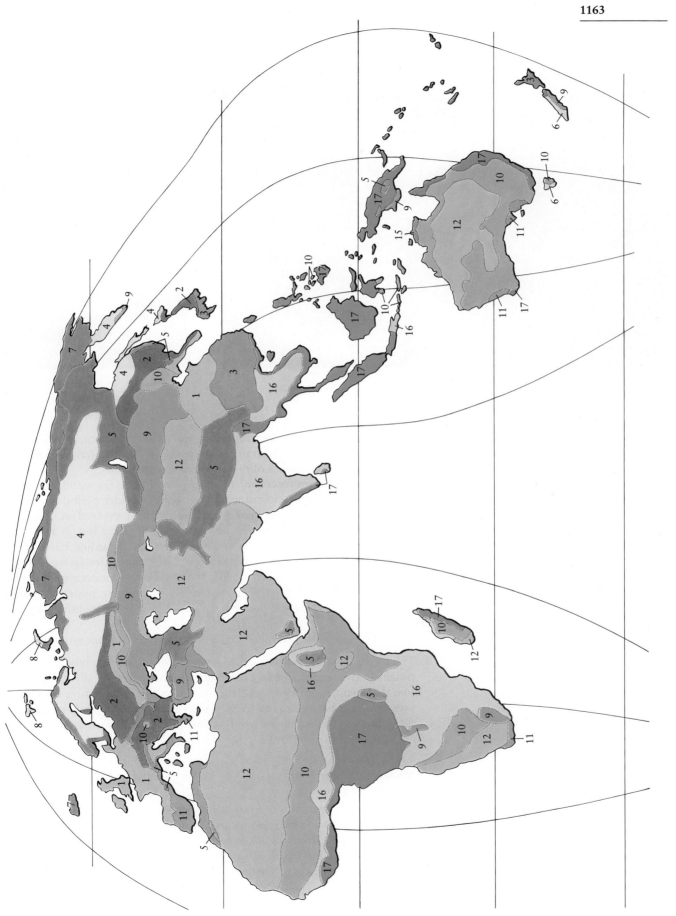

(a)

(b)

55–10 *The temperate deciduous forest of North America grades into temperate mixed forest in northern and upland regions and into subtropical mixed forest in the southeastern United States. (a) A beech and maple forest in Michigan, photographed in the spring. The forest floor is carpeted with large-flowered trillium (Trillium grandiflorum). (b) A beech and maple forest in the fall. (c) A rooting collybia mushroom in Hutcheson Forest, New Jersey. Primary consumers in the forest include (d) chipmunks, and (e) white-tailed deer. (f) This red fox, which has just killed a cottontail rabbit, is a secondary consumer. All of these organisms are common inhabitants of the North American deciduous forest.*

Temperate Forests

Temperate forests once covered most of eastern North America, as well as most of Europe, part of Japan and Australia, and the tip of South America. In the United States, only scattered patches of the original forest remain.

Temperate deciduous forests occupy areas where there is a warm, mild growing season with moderate precipitation, followed by a colder period less suited to plant growth. Leaf-shedding in the deciduous forest evolved as a protection against water loss. (As you will recall, most of the water lost in transpiration escapes through the stomata of the leaves.) However, discarding leaves is expensive, in terms of both energy and nutrients. For the water-saving benefits of defoliation to exceed the costs in energy and nutrients, deciduous trees must have a growing season of at least four months. Soils that are poor or leached of nutrients—such as those of pine barrens—are unable to support a deciduous forest.

In deciduous woodlands there are up to four layers of plant growth.

1. The tree layer, in which the crowns form a continuous canopy. The canopy is usually between 10 and 35 meters high. The tree layer usually has only one or two dominant species of trees, the victors in the competition for light. The presence of other species of trees in the canopy layer is likely to be due to a disturbance, such as fire, a wind storm, or cutting, that has permitted new colonists to gain a roothold.

2. The shrub layer, which grows to a height of about 5 meters. Shrubs and bushes resemble trees in that they are woody and deciduous, but they branch at or close to the ground.

3. The field layer, made up of grasses and other herbaceous (nonwoody) plants, including annual flowering plants, that typically bloom in the spring before the trees regain their leaves. Bracken and other ferns, whose large leaf areas make them efficient interceptors of light trickling through the canopy, are often conspicuous members of the field layer.

4. The ground layer, which consists of mosses and liverworts. The ground is also often covered with leaf litter.

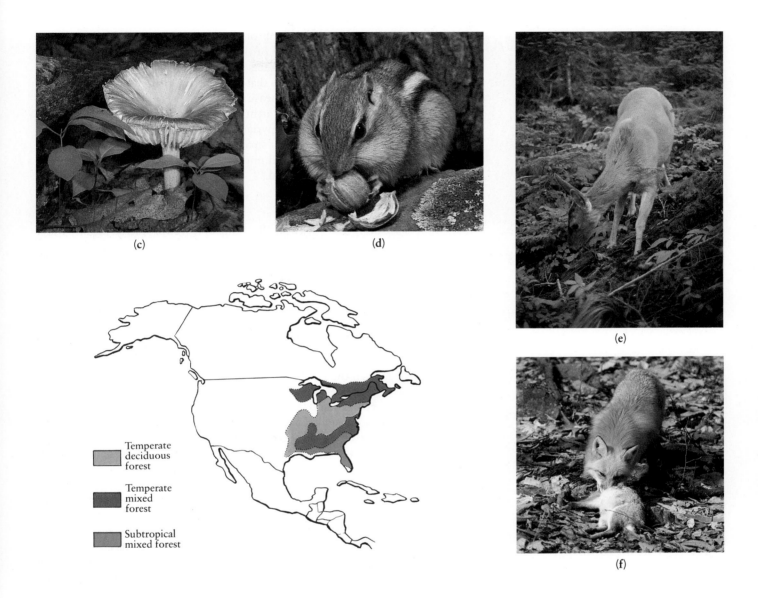

(c)

(d)

(e)

(f)

Temperate deciduous forest

Temperate mixed forest

Subtropical mixed forest

The dominant trees of temperate forests vary from region to region, depending largely on the local rainfall. In the northern and upland regions of the United States, oak, birch, beech, and maple are the most prominent trees. Before the chestnut blight struck North America, an oak-chestnut forest extended from Cape Ann, Massachusetts, and the Mohawk River valley of New York to the southern end of the Appalachian highland. Maple and basswood predominate in Wisconsin and Minnesota, and maple and beech in southern Michigan, becoming mixed with hemlock and white pine as the deciduous forest gives way to mixed forest. The southern and lowland regions have forests of oak and hickory. Along the southeastern coast of the United States, the warm, seasonally wet climate supports a mixed subtropical forest of pine, oak, and magnolia.

Temperate forests support an abundance of animal life. Smaller mammals, such as chipmunks, voles, squirrels, raccoons, opossums, and white-footed mice, live mainly on nuts and other fruits, mushrooms, and insects. The wolves, bobcats, gray foxes, and mountain lions, in the areas where they have not been driven out by the encroachments of civilization, feed on these smaller mammals. Deer live mainly on the forest borders, where they browse on shrubs and seedlings.

Beneath the ground layer is the soil of the forest, often a rich, gray-brown topsoil. Such soil is composed largely of organic material—decomposing leaves and other plant parts and decaying insects and other animals—and the bacteria, protists, fungi, worms, and arthropods that live on this organic matter. The roots of plants penetrate the soil to depths measurable in meters and add organic matter to the soil when they die. Carnivorous arthropods carry fragments of their prey to considerable depths in the soil. The myriad passageways left by dead roots and fungi and by the earthworms and other small animals that inhabit the forest make the soil a sponge that holds water and nutrients. Land where deciduous forests have stood often makes good farmland.

Coniferous Forests

Most conifers are evergreens, with small, compact leaves protected against water loss by a thick cuticle. The dropping of leaves is a more efficient adaptation to changing seasons, and conifers generally cannot compete with deciduous trees in temperate zones with adequate summer rainfall and rich soil. Coniferous forest biomes include the northern coniferous forest, also known as **taiga,** alpine forests, and the mixed west-coast forests found along portions of the coastlines of Chile and California.

The Taiga

The boundary between the temperate mixed forest and the northern coniferous forest occurs where summers become too short and winters too long for deciduous trees to grow well. The northern coniferous forest is characterized by long, severe winters and a constant cover of winter snow. It is composed chiefly of evergreen needle-leaved trees such as pine, fir, spruce, and hemlock. A thick layer of needles and dead twigs, matted together by fungal mycelia, covers the ground. Along the stream banks grow deciduous trees, such as tamarack, willow, birch, alder, and poplar. There are very few annual plants.

The principal large animals of the North American coniferous forest are elk, moose, mule deer, black bears, and grizzlies. Among the smaller animals are porcupines, red-backed voles, snowshoe hares, wolverines, lynxes, warblers, and grouse. The small animals use the dense growths of the evergreens for breeding and for shelter. Wolves feed upon the mammals, particularly the larger ones. The black bear and grizzly bear are omnivores—devouring leaves, buds, fruits, nuts, berries, fish, the supplies of campers, and occasionally the flesh of other mammals. Porcupines are bark eaters and may seriously damage trees by girdling them. Moose and mule deer are largely browsers. The ground layer of the coniferous forest is less richly populated by invertebrates than that of the deciduous forest, and the accumulated litter is slower to decompose.

The Pacific Northwest

The massive evergreen coniferous forests of the Pacific Northwest are adaptations to the winter-wet/summer-dry environment of that region. Because photosynthesis is limited by lack of moisture during the warm season, deciduous trees are at a disadvantage; they are usually found only along stream banks. The evergreen conifers, however, can synthesize carbohydrates all year round. Also, these trees, because of their massive size, can store water and nutrients for use during the dry season, and their thick barks and high crowns protect them from the fires characteristic of the region. These forests, which escaped the Pleistocene glaciation, are among the most ancient in North America, in terms of both individuals and populations.

(a)

(b)

(c)

(d)

(e)

(f)

55–11 *The coniferous forests of North America include the taiga, alpine forests, and mixed west-coast forest. (a) A coniferous forest of balsam spruce and white pine, photographed near the Canadian border. (b) Floor of a virgin northern coniferous forest, carpeted with red pine needles. Decay is slower than on the warmer, wetter floors of the deciduous forest, and, because of the shade cast by the mature trees, there is no undergrowth.*

Coniferous forests support a diversity of animals, including (c) grizzly bears, (d) great horned owls, (e) moose, and (f) mule deer.

Taiga

Mixed west-coast forest

Alpine forests and tundra

(a)

(b)

(c)

(d)

(e)

55–12 **(a)** *Tundra of North America on a long Arctic day. Kettle holes, formed by chunks of glacial ice, are a common sight.* **(b)** *The Arctic tern is one of a number of bird species that breed in the tundra, taking advantage of the long summer days to gather food for their nestlings. The terns winter in the Antarctic, following migra-* *tion routes of 13,000 to 18,000 kilometers. Within three months after hatching, the young must be ready to migrate.* **(c)** *A caribou, photographed in Alaska in the fall. Winter inhabitants of the tundra include* **(d)** *the Arctic fox and* **(e)** *the snowy owl.*

The Tundra

Where the climate is too cold and the winters too long even for conifers, coniferous forest grades into **tundra**. The Arctic tundra is a form of grassland that occupies one-tenth of the earth's land surface, forming a continuous belt across northern North America, Europe, and Asia. Similar vegetation is found above the tree line in mountainous regions.

The most characteristic feature of the Arctic tundra is permafrost, a layer of permanently frozen subsoil. During the summer the ground thaws to a depth of a few centimeters and becomes wet and soggy; in winter it freezes again. This freeze-thaw process, which tears and crushes the roots, keeps the plants small and stunted. Drying winter winds and abrasive, driven snow further reduce the growth of tundra plants. The virtually treeless vegetation of the tundra is dominated by herbaceous plants, such as grasses, sedges, and rushes, and woody shrubs like heather. Beneath these is a well-developed ground layer of mosses and lichens, particularly the lichen known as reindeer moss. All the flowering plants are perennials.

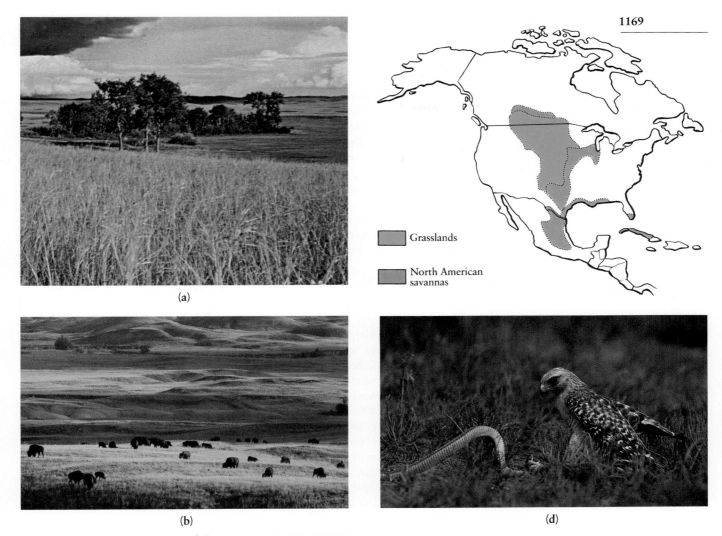

55–13 *The grasslands of North America include large regions of both temperate grassland and savanna.* (a) *A June day on a tall-grass prairie in North Dakota. The cottonwood grove by the prairie creek is characteristic of this biome. A thunderstorm is gathering on the horizon.*

(b) *Short-grass prairie with grazing bison, photographed in Sage Creek Wilderness, Badlands National Park, South Dakota. Among the other native animals of the grasslands are* (c) *the white-tailed prairie dog and* (d) *the red-shouldered hawk and its prey, the garter snake.*

The largest animals of the Arctic tundra are the musk oxen and caribou of North America and the reindeer of the Old World. Lemmings (small rodents with short tails) and ptarmigans (pigeon-sized grouse) are tundra herbivores. The white fox and the snowy owl of the Arctic are among the principal predators, feeding largely on lemmings.

During the brief Arctic summer, insects emerge in great numbers, and migratory birds visit, taking advantage of the insect hordes and the long periods of daylight to feed their young. The growing season in many areas of the tundra is less than two months.

Temperate Grasslands

Grasslands, which are transitional areas between temperate forests and deserts, are usually found in the interior areas of continents. They are characterized by rolling to flat terrain, hot-cold seasons (rather than the wet-dry seasons of the tropical grasslands), periodic droughts, and fires. The temperate grasslands of the world include the plains and prairies of North America, the steppes of the Soviet Union, the veld of South Africa, and the pampas of Argentina.

The vegetation is largely bunch or sod-forming grasses, often mixed with legumes (clovers and wild indigo) and a variety of annuals. In North America, there is a transition from the eastern temperate deciduous woodland, through the rich, moist, tall-grass prairie (the Corn Belt), to the more desertlike, western short-grass prairie (the Great Plains). Grasslands are drier to the west, where they are in the rain shadow of the Rockies. Periodic fires serve to maintain the nature of the grasslands, destroying tree seedlings and preventing their encroachment.

The temperate grasslands of the world support small, seed-eating rodents and also large herbivores, such as the bison of early America, the gazelles and zebras of the African veld, the wild horses, wild sheep, and ibex of the Asiatic steppes, and now the domestic herbivores. These large, grass-eating mammals, in turn, support carnivores, such as lions, tigers, and wolves, as well as omnivorous humans.

Tropical Grasslands: Savannas

Savannas are tropical grasslands with scattered clumps of trees (Figure 55–14). The transition from open forest with grassy undergrowth to savanna is gradual and is determined by the duration and severity of the dry season and, often, by fire and by grazing and browsing animals.

In the savanna, the critical competition is for water, and grasses are the victors. Grasses are well suited to a fine, sandy soil with seasonal rain because their roots form a dense network capable of extracting the maximum amount of water during the rainy period. During dry seasons, the aboveground portions of the plants die, but the deep roots are able to survive even many months of drought. The balance between woody plants and grasses is a delicate one. If rainfall decreases, the trees die. If rainfall increases, the trees increase in number until they shade the grasses, which, in turn, die. If the grasses are overgrazed (which often happens when people begin to use the savanna for agricultural purposes and introduce livestock), enough water is left in the soil so that the woody plants can increase in number, and the grassland is eventually destroyed.

The best-known savannas are those of Africa, which are inhabited by the most abundant and diverse group of large herbivores in the world, including the gazelles, the impala, the eland, the buffalo, the giraffe, the zebra, and the wildebeest.

55–14 *A savanna in East Africa. The giraffes are surrounded by a herd of impala. The trees in the background are acacias.*

(a)

(c)

(b)

55–15 *The Mediterranean scrub, or chaparral, of North America.* **(a)** *The bushy vegetation that characterizes the chaparral grows as dense as a mat on the foothills of southern California. It is the result of long, dry summers, during which much of the plant life is semidormant, followed by a cool, rainy season. The name comes from* chaparro, *the Indian word for the scrub oak that is one of the prominent components of the chaparral.* Chaps, *the leather leggings worn by cowboys making their way through this dense, dry growth, has the same derivation.* **(b)** *A cacomistle, or ring-tailed cat, a common inhabitant of the chaparral.*

(c) *Mediterranean scrub, or* maquis, *on Corfu, Greece. Note the limestone on which the plants are growing; it is characteristic of much of the Mediterranean coastline and of the islands within the Mediterranean.*

Mediterranean Scrub

Regions with mild winters and long, dry summers, such as the southern coast of California and southern Spain, are dominated by small trees or, often, by spiny shrubs with broad, thick evergreen leaves. This vegetation, known formally as Mediterranean scrub, has been given a variety of local names. In the United States, it is known as the **chaparral**, whereas in Mediterranean regions, it is the *maquis*, and in Chile it is the *matorral*. Mediterranean scrub is also found in southern Africa and along portions of the coast of Australia. Although the plants of these various areas are phylogenetically unrelated, they closely resemble one another in their growth patterns and characteristic appearance.

Mule deer live in the North American chaparral during the spring growing season, moving out to cooler regions during the summer. Many of the resident vertebrates—lizards, wren-tits, brown towhees, brush rabbits, and a diversity of rodents—are small and dull-colored, matching the dull-colored vegetation.

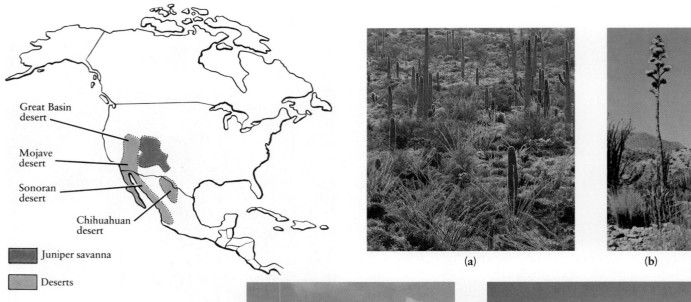

(a)

(b)

(c)

(d)

(e)

55–16 *The principal deserts of North America, adjacent to which are large areas of juniper savanna and semidesert.*

*The Sonoran stretches from southern California to western Arizona and down into Mexico. A dominant plant, the giant saguaro cactus (**a**) is often as much as 15 meters high, with a widespreading network of shallow roots. Water is stored in a thickened stem, which expands, accordionlike, after a rainfall.*

*To the east is the Chihuahuan desert, one of whose principal plants is the agave (**b**), or century plant, a monocot.*

*North of the Sonoran is the Mojave, whose characteristic plant is the Joshua tree (**c**). This plant was named by early Mormon colonists who thought that its strange, awkward form resembled a bearded patriarch gesticulating in prayer. The Mojave contains Death Valley, the lowest point on the continent (90 meters below sea level), only 130 kilometers from Mt. Whitney, whose elevation is more than 4,000 meters.*

*The Mojave blends into the Great Basin, a cold desert bounded by the Sierra Nevada to the west and the Rockies to the east. It is the largest and bleakest of the American deserts. The dominant plant form is sagebrush (**d**), shown here with the snow-covered Sierra Nevada mountains in the background.*

*Juniper savanna (**e**) is a high-altitude, cold biome named for its dominant plant species,* Juniperus osteosperma.

In the map legend:
Great Basin desert
Mojave desert
Sonoran desert
Chihuahuan desert
Juniper savanna
Deserts

The Desert

The great deserts of the world are located at latitudes of about 30°, both north and south, and extend poleward in the interiors of the continents. These are areas of falling, warming air and, consequently, little rainfall. Only about 5 percent of North America is desert. The Sahara, which stretches all the way from the Atlantic coast of Africa to the Arabian Peninsula, is the largest desert in the world (almost equal to the size of the United States) and is increasing in size, spreading along its southern boundaries. This spread is due in large part to the growth of the human population, resulting, in turn, in intensified grazing by domestic animals along the margins of the desert.

Desert regions are characterized by less than 25 centimeters of rain a year. Because there is little water vapor in the air to moderate the temperature, the nights are often extremely cold. The temperature may drop as much as 30°C at night, in contrast to the humid tropics, where day and night temperatures vary by only a few degrees.

Many desert plants are annuals that race from seed to flower to seed during periods when water is available; during the brief growing seasons, the desert may be carpeted with flowers. Many of the perennials are succulents (adapted for water storage). Some are drought-deciduous (dropping their leaves in dry seasons), some have small, leathery water-conserving leaves, and some, like cacti, have leaves that are modified as hard, dry spines. C_4 and CAM photosynthesis (pages 224 and 655), both of which conserve water, are common among desert plants, as are extensive root systems that are able to trap large amounts of water during the brief periods it is available.

The animals that live in the desert are also specially adapted to this extreme climate (see page 789). Reptiles and insects, for example, have waterproof outer coverings and dry, and therefore water-conserving, excretions. Many of the mammals of the desert are small and nocturnal and obtain what little water they require from the plants they eat.

55–17 *Inhabitants of the North American deserts:* **(a)** *a roadrunner, shown here resting during the hottest part of the day;* **(b)** *a collared lizard;* **(c)** *a jackrabbit, which is actually a hare;* **(d)** *a cougar, or mountain lion. Other familiar animals of the North American deserts include the kangaroo rat (Figure 37–1, page 765) and the horned lizard (Figure 38–6, page 781).*

(a)

(b)

(c)

(d)

(a)

(b)

(c)

55–18 *Tropical rain forest. (a) Lianas, such as this hummingbird-pollinated passion flower, are abundant in the tropical rain forest. (b) The diversity of the trees in the forest, which may reach several hundred species per hectare, is revealed when individual trees burst into bloom, like this guayacan tree in Panama. (c) A buttressed tree in the rain forest of Guatemala.*

Tropical Forests

In the equatorial zone, where most of the tropical forests of the world are found, the mean daily temperature is the same throughout the year, and the length of day varies by less than one hour. Rainfall, however, is often seasonal, and variations in total rainfall from one area to another are also caused by mountains and their rain shadows. Areas in which rainfall is limited all year are characterized by southern woodland and scrub, in which the trees have small, water-conserving leaves. In areas with distinct wet and dry seasons, tropical mixed forests and monsoon forests occur; these forests are dominated by trees that lose their leaves during the dry seasons. The tropical rain forest, the most complex of all ecosystems, is found in areas where rainfall is abundant all year round.

Tropical Rain Forest

In the tropical rain forest, total rainfall is between 200 and 400 centimeters per year, and a month with less than 10 centimeters of rain is considered relatively dry. There are more species of plants and animals in the tropical rain forest than in all the rest of the biomes of the world combined. As many as 100 species of trees can be counted on 1 hectare, but each species may be represented by only one tree (Figure 55–18). By contrast, a comparable area in a deciduous forest in the northeastern United States typically contains only a few tree species, but each species is represented by many individual trees.

The critical competition among plants of the tropical forest is for light. About 70 percent of all species of plants are trees. The upper tree story consists of solitary giants 50 to 60 meters tall. A lower story of trees characteristically forms a continuous canopy. The trees forming the canopy are remarkably similar in appearance. Their trunks are usually slender and branch only near the crown. The crowns are high up and relatively small as a result of crowding. Because the soil is perpetually wet, their roots do not reach deep into it, and trunks often end in thick buttresses that provide broad anchorage. Their leaves are large, leathery, and dark green; their bark is thin and smooth; and their flowers are often inconspicuous and greenish or whitish in color.

55–19 *An enormous variety of animals live in the tropical rain forest; for example, in Colombia alone, there are at least 1,400 species of birds.* (**a**) *An* Anartia amathea *butterfly, photographed in Ecuador.* (**b**) *A Colombian cone-headed grasshopper.* (**c**) *An orange-chinned parakeet in Panama.* (**d**) *A green tree viper, photographed in a forest in Malaysia.* (**e**) *Squirrel monkeys in the canopy of the Amazon forest.*

Woody vines, or lianas, are abundant, especially where an opening has appeared in the forest, as a result, for example, of a tree's falling; vines as long as 240 meters have been measured. There are also many epiphytes, which are plants that grow on other plants, often high above the forest floor. The epiphytes of the tropical rain forest germinate in the branches of trees and obtain water from the humid air of the canopy. Unlike the plants that have contact with the moist floor, epiphytes need to conserve water between rainfalls. Some epiphytes resemble desert succulents, having fleshy water-storing leaves and stems. Others have spongy roots or cup-shaped leaves that capture moisture and organic debris; many of these epiphytes can take up nutrients from decaying organisms in these storage tanks. A variety of plants, including ferns, orchids, mosses, and bromeliads, have exploited this life style.

An extraordinary variety of insects, birds, and other animals, including mammals, have moved into the treetops along with the vines and epiphytes to make it the most abundantly and diversely populated area of the tropical rain forest (Figure 55–19).

Little light reaches the forest floor (from 0.1 to 1 percent of the total), and the few plants that are found there are adapted to growing at low light intensities. Many of these, such as the African violet, are familiar to us as house plants.

Plants also compete for nutrients. The nutrient cycles are tight, and turnover is rapid. There is almost no accumulation of leaf litter on the forest floor, such as we find in our northern forests; decomposition is too rapid. Everything that touches the ground disappears almost immediately—carried off, consumed, or rapidly decomposed. As a consequence, the soils of tropical rain forests are relatively infertile. Many are chiefly composed of a red clay; these red soils are known as laterites, from the Latin word *later* or "brick." When laterite soils are cleared, in many cases they either erode rapidly or form thick, impenetrable crusts that cannot be cultivated after a season or two. Tropical soils are generally deficient in nutrients, and what nutrients are found there are likely to be leached out by heavy rainfall. Most of the nitrogen, phosphorus, calcium, and other nutrients are found in the plants rather than the soil, and trees that most effectively store these nutrients may be the ones that win the competition for light.

Tropical Forests, Mass Extinction, and Human Responsibility

Although tropical forests currently form about half of the forested area of the earth, they are being rapidly destroyed (Figure 55–20). If the current rate of destruction continues, almost all the tropical forests will have disappeared by the turn of the century, and with them the thousands of plants and animals found nowhere else in the biosphere. The destruction of the forests results directly from the rapid growth of the human population in the tropics and indirectly from the nature of tropical forest soils. Because such soils retain few nutrients, they are very poor for agricultural purposes and will support crops for only a few years after the trees have been cleared. Thus, the traditional agricultural practices in the tropics have been those of clearing and short-term cultivation. With a rapidly expanding, hungry human population, such practices have become immensely destructive as they are carried out on an ever-increasing scale. Commercial logging also contributes to the demise of the forests, as does the need of the local populations for firewood.

The loss of the organisms of the tropical forests would, in itself, represent a mass extinction on the scale of previous extinctions (see page 1024), but unlike the earlier extinctions, it would be the direct result of the activities of a single species—our own. Moreover, many biologists are concerned that the destruction

55–20 *Clearing tropical forests to provide land for agriculture or the grazing of cattle involves not only cutting the forest but also burning virtually all of the vegetation to release the stored nutrients into the soil. Within a few seasons, these precious nutrients are lost from the soil, rendering it sterile for most plant life.*

(a)

(b)

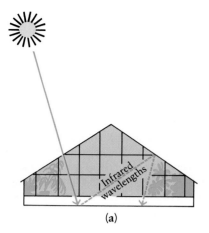

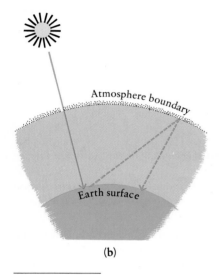

55–21 *The greenhouse effect.* (a) *Light rays penetrate the glass of the greenhouse, are absorbed by the plants and soil, and are then reradiated as longer-waved infrared radiation (heat). The glass does not permit these rays to escape, and so the heat remains within the greenhouse.*

(b) *Carbon dioxide, like glass, is transparent to light but absorbs the infrared rays, preventing their escape. An increase in atmospheric carbon dioxide, caused by the burning of fossil fuels and forest destruction, is expected to lead to an increase in the temperature of the biosphere as a whole. A study by the Environmental Protection Agency predicts that global temperatures could increase an average of 2°C by the year 2040 and 5°C by the year 2100.*

of the tropical forests will lead indirectly to extinctions in other biomes. As we noted previously (page 226), the concentration of carbon dioxide in the atmosphere has been rising steadily during the past 100 years, principally as a result of increased burning of fossil fuels. Recent analyses have shown that forest destruction is also a major contributor to increased atmospheric carbon dioxide levels. As forest is cleared, the oxidation of its biomass—either by burning or by the metabolic processes of decomposers—releases large amounts of carbon dioxide. Moreover, carbon dioxide that would previously have been used by the trees and other plants of the forest in photosynthesis remains in the atmosphere. As a result of the greenhouse effect (Figure 55–21), the increased carbon dioxide levels in the atmosphere are expected to result in a significant increase in global temperatures, beginning within the next 20 years. Among the expected consequences of this temperature increase are rising sea levels (as a result of the melting of polar ice), lengthened growing seasons and more precipitation in some areas, reduced precipitation and primary productivity in other areas, and the extinction of a variety of species as local environmental conditions change beyond the organisms' limits of adaptability.

Halting, or even slowing, the destruction of the tropical forests is an enormously complex undertaking that cannot be separated from the needs of the human population of the tropics. Setting aside areas of forest as undisturbed preserves and slowing the rate of human population growth are only part of the answer. There is a pressing need for developing more productive and appropriate agricultural practices for the tropics and devising techniques for harvesting the richness of the forest—providing a long-overdue improvement in the standard of living for local populations—without simultaneously destroying it.

In the age of the dinosaurs, the earliest primates survived, it would appear, largely by their wits. Now, if we and other species are to survive the consequences of our own evolution, we will have to use well the contents of our own skulls. For within the human brain—that complex collection of neurons and synapses—resides the uniquely human capacity to accumulate knowledge, to plan with foresight, and so to act with enlightened self-interest and even, on occasion, with compassion for others and reverence for the diversity of life on our planet.

SUMMARY

The biosphere is that part of the earth that contains living organisms. It is a thin film on the surface of the planet, irregular in its thickness and its density. The biosphere is affected by the position and movements of the earth in relation to the sun and by the movements of air and water over the earth's surface. These factors cause wide differences in temperature and rainfall from place to place and from season to season. There are also differences in the surfaces of the continents, both in composition and in altitude. These differences are reflected in differences in the kinds of plant and animal life found in different parts of the biosphere.

Distinctive communities of organisms are found in running fresh waters (rivers and streams), standing fresh waters (lakes and ponds), at different depths of the open oceans, and along rocky, sandy, and muddy seashores. On land, biomes are groupings of organisms with common patterns of climate and distinctive vegetation, distributed over a wide area. The major biomes include temperate forests (deciduous and mixed), coniferous forests (taiga, alpine, and west-coast mixed), tundra (Arctic and alpine), temperate grasslands, tropical grasslands (savannas), Mediterranean scrub, deserts, and tropical forests (southern woodland and scrub, mixed, monsoon, and rain forest). A major—and urgent—challenge for *Homo sapiens* is preservation of the tropical forests, the biomes with the greatest diversity of species found on this planet.

QUESTIONS

1. Mud flats are extremely rich in animal life, yet few plants are found there. What reasonable explanation can you give for the scarcity of plant life? How can the mud flats support such a profusion of animal life in the absence of plants?

2. What are the eight major biomes? Describe the principal abiotic features of each.

3. Name a plant and an animal associated with each of the eight major biomes, and describe their special adaptations.

4. Although each of the biomes we have considered is sufficiently different from all the others to warrant its identification as a distinct biome, there are important similarities among some of the biomes. Consider the following groups of biomes: tropical rain forest/monsoon forest; monsoon forest/temperate deciduous forest/taiga; savanna/temperate grasslands/tundra. Describe the essential similarities and the most significant differences in the

environmental factors affecting the members of each group. How do these factors affect the types of plants characterizing each biome?

5. The rate of decomposition of plant litter, animal wastes, and dead plants and animals varies from biome to biome. Describe the differences in decomposition rates in the following biomes: tropical rain forest, temperate deciduous forest, taiga. What factors in each biome are important in causing these differences? What are the consequences of these differing decomposition rates for nutrient recycling, soil quality, and the size and diversity of detritivore populations?

6. A leading ecologist has stated: "The plough is the most deadly weapon of extinction ever devised; not even thermonuclear weapons pose such a threat to the beauty and diversity of life on earth." Explain.

SUGGESTIONS FOR FURTHER READING

Books

BEGON, MICHAEL E., JOHN L. HARPER, and COLIN R. TOWNSEND: *Ecology: Individuals, Populations, and Communities*, Sinauer Associates, Sunderland, Mass., 1986.

An up-to-date, balanced, and thorough treatment of contemporary ecology.

CARSON, RACHEL: *Silent Spring*, Houghton Mifflin Company, Boston, 1962.*

This is the book that awakened the nation to the dangers of pesticides. Once seen as highly controversial, it is now regarded as a classic.

COLINVAUX, PAUL: *Why Big Fierce Animals Are Rare*, Princeton University Press, Princeton, N.J., 1978.*

A series of essays for the general reader.

EHRLICH, PAUL R., and J. ROUGHGARDEN: *The Science of Ecology*, The Macmillan Company, New York, 1987.

This recent textbook provides a solid introduction to the many facets of ecology. It gives a broad overview of the science and includes a great deal of information on behavioral ecology and sociobiology.

FORSYTH, ADRIAN, and KEN MIYATA: *Tropical Nature: Life and Death in the Rain Forests of Central and South America*, Charles Scribner's Sons, New York, 1984.*

This book, for the general reader, consists of 17 essays on different aspects of the complex—and still poorly understood—ecology of tropical rain forests. It is filled with a multitude of marvelous examples of the diverse organisms of the forests and their many interactions with each other. The authors, both specialists in tropical vertebrates, have spent seven years working in the forests, principally in Costa Rica and Peru, and bring vast personal experience to their essays.

HARPER, JOHN L.: *Population Biology of Plants*, Academic Press, Inc., New York, 1981.*

The world from a plant's perspective. A wonderful introduction to the ecology of populations and a must for any student of botany.

HEDRICK, PHILIP W.: *Population Biology: The Evolution and Ecology of Populations*, Jones and Bartlett, Boston, 1984.

A modern text on population biology that provides thorough coverage of the ecology, behavior, and evolution of populations. It is strong in genetics and contains clear explanations of the basic mathematical models used in population biology.

HUTCHINSON, G. EVELYN: *The Ecological Theater and the Evolutionary Play*, Yale University Press, New Haven, Conn., 1965.

By one of the great modern experts on freshwater ecology, this is a charming and sophisticated collection of essays on the influence of environment in evolution—and also on an astonishing variety of other subjects.

KREBS, CHARLES J.: *Ecology: The Experimental Analysis of Distribution and Abundance*, 3d ed., Harper & Row, Publishers, Inc., New York, 1985.

This excellent, modern text focuses on populations and on interactions among them. Requires some knowledge of mathematics.

LOPEZ, BARRY: *Arctic Dreams: Imagination and Desire in a Northern Landscape*, Bantam Books, New York, 1987.*

A brilliant evocation of the biogeography of the Arctic by a master nature writer. "This is a land," says Lopez, "where airplanes track icebergs the size of Cleveland and polar bears fly down out of the stars."

* Available in paperback.

MORAN, JOSEPH M., MICHAEL D. MORGAN, and JAMES H. WIERSMA: *An Introduction to Environmental Sciences,* 2d ed., W. H. Freeman and Company, San Francisco, 1985.

A clear, matter-of-fact presentation of the basic features of our planet and, in particular, of the biosphere. It provides a good background for the study of ecology in general and problems of environmental pollution and technological change in particular.

OWEN, DENIS: *Camouflage and Mimicry,* University of Chicago Press, Chicago, 1982.*

A short but thorough discussion of the diversity of ways that organisms deceive other organisms. A multitude of wonderful photographs accompany the text.

OWEN, JENNIFER: *Feeding Strategy,* University of Chicago Press, Chicago, 1982.*

This short, beautifully illustrated book explores the variety of ways in which animals obtain food. It considers not only herbivorous and carnivorous animals but also filter-feeders, parasites, and scavengers.

RICKLEFS, ROBERT E.: *Ecology,* 3d ed., Chiron Press, Inc., New York, 1987.

This outstanding textbook provides a comprehensive introduction to the basic concepts of modern ecology. Beautifully written and rich with examples, it is highly recommended.

SCHALLER, GEORGE: *The Serengeti Lion: A Study of Predator-Prey Relations,* University of Chicago Press, Chicago, 1976.*

One of the first, best, and most comprehensive of the modern field studies in animal ecology and behavior. Winner of the National Book Award.

SMITH, ROBERT L.: *Ecology and Field Biology,* 4th ed., Harper & Row Publishers, Inc., New York, 1989.

The outstanding sections of this text are those that deal with descriptions of biomes and communities and the plants and animals within them. Although meant to accompany a course in field biology, this book would be an asset and a pleasure to the amateur naturalist.

STORER, JOHN H.: *The Web of Life,* New American Library, Inc., New York, 1966.*

One of the first books ever written on ecology for the general public. In its simple presentation of the interdependence of living things, it remains a classic.

Articles

BERGERUD, ARTHUR T.: "Prey Switching in a Simple Ecosystem," *Scientific American,* December 1983, pages 130–141.

BURNHAM, CHARLES R.: "The Restoration of the American Chestnut," *American Scientist,* vol. 76, pages 478–487, 1988.

CASE, TED J., and MARTIN L. CODY: "Testing Theories of Island Biogeography," *American Scientist,* vol. 75, pages 402–411, 1987.

CHILDRESS, JAMES J., HORST FELBECK, and GEORGE N. SOMERO: "Symbiosis in the Deep Sea," *Scientific American,* May 1987, pages 114–120.

COCKBURN, ANDREW, and ANTHONY K. LEE: "Marsupial Femmes Fatales," *Natural History,* March 1988, pages 40–47.

CONNELL, JOSEPH H.: "Diversity and the Coevolution of Competitors, or the Ghost of Competition Past," *Oikos,* vol. 35, pages 131–138, 1980.

CONNELL, JOSEPH H.: "Diversity in Tropical Rain Forests and Coral Reefs," *Science,* vol. 199, pages 1302–1310, 1978.

CONNELL, JOSEPH H.: "On the Prevalence and Relative Importance of Interspecific Competition: Evidence from Field Experiments," *American Naturalist,* vol. 122, pages 661–696, 1983.

CONNELL, JOSEPH H., and RALPH O. SLATYER: "Mechanisms of Succession in Natural Communities and Their Role in Community Stability and Organization," *American Naturalist,* vol. 111, pages 1119–1144, 1977.

CONNOR, EDWARD F., and DANIEL SIMBERLOFF: "Competition, Scientific Method, and Null Models in Ecology," *American Scientist,* vol. 74, pages 155–162, 1986.

DETWILER, R. P., and C. A. S. HALL: "Tropical Forests and the Global Carbon Cycle," *Science,* vol. 239, pages 42–47, 1988.

DIAMOND, JARED M.: "Niche Shifts and the Rediscovery of Interspecific Competition," *American Scientist,* vol. 66, pages 322–331, 1978.

DIAMOND, JARED: "The Worst Mistake in the History of the Human Race," *Discover,* May 1987, pages 64–66.

DIAMOND, JARED M., K. DAVID BISHOP, and S. VAN BALEN: "Bird Survival in an Isolated Javan Woodland: Island or Mirror?" *Conservation Biology,* vol. 1, pages 132–142, 1987.

FENTON, M. BROCK, and JAMES H. FULLARD: "Moth Hearing and the Feeding Strategies of Bats," *American Scientist,* vol. 69, pages 266–275, 1981.

GOSZ, J. R., R. T. HOLMES, G. E. LIKENS, and B. H. BORMANN: "The Flow of Energy in a Forest Ecosystem," *Scientific American,* March 1978, pages 92–102.

HORN, HENRY S.: "Forest Succession," *Scientific American,* May 1975, pages 90–98.

HORN, MICHAEL S., and ROBIN H. GIBSON: "Intertidal Fishes," *Scientific American,* January 1988, pages 64–70.

JANZEN, DANIEL H.: "Tropical Ecological and Biocultural Restoration," *Science,* vol. 239, pages 243–244, 1988.

KARR, JAMES R.: "Avian Extinction on Barro Colorado Island, Panama: A Reassessment," *American Naturalist,* vol. 119, pages 220–237, 1982.

KERR, RICHARD A.: "Is the Greenhouse Here?" *Science,* vol. 239, pages 559–561, 1988.

LAWS, RICHARD M.: "The Ecology of the Southern Ocean," *American Scientist,* vol. 73, pages 26–40, 1985.

LEWIN, ROGER: "Ecological Invasions Offer Opportunities," *Science,* vol. 238, pages 752–753, 1987.

---

* Available in paperback.

LUBCHENCO, JANE: "Plant Species Diversity in a Marine Intertidal Community: Importance of Herbivore Food Preference and Algal Competitive Abilities," *American Naturalist,* vol. 112, pages 23–29, 1978.

MARES, MICHAEL A.: "Conservation in South America: Problems, Consequences, and Solutions," *Science,* vol. 233, pages 734–739, 1986.

MAY, ROBERT M.: "Parasitic Infections as Regulators of Animal Populations," *American Scientist,* vol. 71, pages 36–45, 1983.

MAY, ROBERT M., and ROY M. ANDERSON: "Transmission Dynamics of HIV Infection," *Nature,* vol. 326, pages 137–142, 1987.

MAY, ROBERT M., and JON SEGER: "Ideas in Ecology," *American Scientist,* vol. 74, pages 256–267, 1986.

MELLOR, JOHN W., AND SARAH GAVIAN: "Famine: Causes, Prevention, and Relief," *Science,* vol. 235, pages 539–545, 1987.

MOONEY, HAROLD A., PETER M. VITOUSEK, and PAMELA A. MATSON: "Exchange of Materials between Terrestrial Ecosystems and the Atmosphere," *Science,* vol. 238, pages 926–932, 1987.

PERRY, DONALD R.: "The Canopy of the Tropical Rain Forest," *Scientific American,* November 1984, pages 138–147.

POWER, J. F., and R. F. FOLLETT: "Monoculture," *Scientific American,* March 1987, pages 78–86.

RAMAGE, COLIN S.: "El Niño," *Scientific American,* June 1986, pages 76–83.

RICKLEFS, ROBERT E.: "Community Diversity: Relative Roles of Local and Regional Processes," *Science,* vol. 235, pages 167–171, 1987.

ROWLAND, F. SHERWOOD: "Chlorofluorocarbons and the Depletion of Stratospheric Ozone," *American Scientist,* vol. 77, pages 36–45, 1989.

SCHOENER, THOMAS W.: "The Controversy over Interspecific Competition," *American Scientist,* vol. 70, pages 586–595, 1982.

SCHOENER, THOMAS W.: "Field Experiments on Interspecific Competition," *American Naturalist,* vol. 122, pages 240–285, 1983.

SEBENS, KENNETH P.: "The Ecology of the Rocky Subtidal Zone," *American Scientist,* vol. 73, pages 548–557, 1985.

SIMONS, TED, STEVE K. SHERROD, MICHAEL W. COLLOPY, and M. ALAN JENKINS: "Restoring the Bald Eagle," *American Scientist,* vol. 76, pages 252–260, 1988.

SOUSA, WAYNE P.: "Disturbance in Marine Intertidal Boulder Fields: The Nonequilibrium Maintenance of Species Diversity," *Ecology,* vol. 60, pages 1225–1239, 1979.

SOUSA, WAYNE P.: "Experimental Investigations of Disturbance and Ecological Succession in a Rocky Intertidal Algal Community," *Ecological Monographs,* vol. 49, pages 227–254, 1979.

STEARNS, C.: "Life-History Tactics: A Review of the Ideas," *Quarterly Review of Biology,* vol. 51, pages 3–47, 1976.

STEPHENS, SHARON: "Lapp Life after Chernobyl," *Natural History,* December 1987, pages 32–41.

STOLARSKI, RICHARD S.: "The Antarctic Ozone Hole," *Scientific American,* January 1988, pages 30–36.

SUN, MARJORIE: "Costa Rica's Campaign for Conservation," *Science,* vol. 239, pages 1366–1369, 1988.

TURNER, TERESA: "Facilitation as a Successional Mechanism in a Rocky Intertidal Community," *American Naturalist,* vol. 121, pages 729–738, 1983.

WARING, RICHARD H.: "Land of the Giant Conifers," *Natural History,* October 1982, pages 54–63.

Appendixes

Metric Table

	QUANTITY	NUMERICAL VALUE	ENGLISH EQUIVALENT	CONVERTING ENGLISH TO METRIC
Length	kilometer (km)	1,000 (10^3) meters	1 km = 0.62 mile	1 mile = 1.609 km
	meter (m)	100 centimeters	1 m = 1.09 yards	1 yard = 0.914 m
			= 3.28 feet	1 foot = 0.305 m
	centimeter (cm)	0.01 (10^{-2}) meter	1 cm = 0.394 inch	1 foot = 30.5 cm
				1 inch = 2.54 cm
	millimeter (mm)	0.001 (10^{-3}) meter	1 mm = 0.039 inch	1 inch = 25.4 mm
	micrometer (μm)	0.000001 (10^{-6}) meter		
	nanometer (nm)	0.000000001 (10^{-9}) meter		
	angstrom (Å)	0.0000000001 (10^{-10}) meter		
Area	square kilometer (km²)	100 hectares	1 km² = 0.3861 square mile	1 square mile = 2.590 km²
	hectare (ha)	10,000 square meters	1 ha = 2.471 acres	1 acre = 0.4047 ha
	square meter (m²)	10,000 square centimeters	1 m² = 1.1960 square yards	1 square yard = 0.8361 m²
			= 10.764 square feet	1 square foot = 0.0929 m²
	square centimeter (cm²)	100 square millimeters	1 cm² = 0.155 square inch	1 square inch = 6.4516 cm²
Mass	metric ton (t)	1,000 kilograms = 1,000,000 grams	1 t = 1.103 tons	1 ton = 0.907 t
	kilogram (kg)	1,000 grams	1 kg = 2.205 pounds	1 pound = 0.4536 kg
	gram (g)	1,000 milligrams	1 g = 0.0353 ounce	1 ounce = 28.35 g
	milligram (mg)	0.001 gram		
	microgram (μg)	0.000001 gram		
Time	second (sec)	1,000 milliseconds		
	millisecond (msec)	0.001 second		
	microsecond (μsec)	0.000001 second		
Volume (solids)	1 cubic meter (m³)	1,000,000 cubic centimeters	1 m³ = 1.3080 cubic yards	1 cubic yard = 0.7646 m³
			= 35.315 cubic feet	1 cubic foot = 0.0283 m³
	1 cubic centimeter (cm³)	1,000 cubic millimeters	1 cm³ = 0.0610 cubic inch	1 cubic inch = 16.387 cm³
Volume (liquids)	kiloliter (kl)	1,000 liters	1 kl = 264.17 gallons	1 gal = 3.785 l
	liter (l)	1,000 milliliters	1 l = 1.06 quarts	1 qt = 0.94 l
				1 pt = 0.47 l
	milliliter (ml)	0.001 liter	1 ml = 0.034 fluid ounce	1 fluid ounce = 29.57 ml
	microliter (μl)	0.000001 liter		

Temperature Conversion Scale

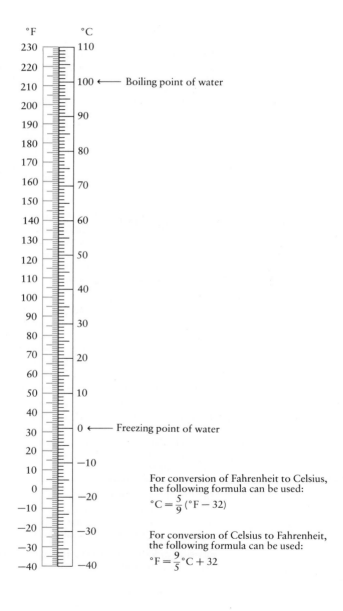

°F   °C

230 — 110

220

210 — 100 ← Boiling point of water

200

190 — 90

180 — 80

170

160 — 70

150

140 — 60

130

120 — 50

110

100 — 40

90 — 30

80

70 — 20

60

50 — 10

40

30 — 0 ← Freezing point of water

20

10 — −10

0

−10 — −20

−20 — −30

−30

−40 — −40

For conversion of Fahrenheit to Celsius, the following formula can be used:

$$°C = \frac{5}{9}(°F - 32)$$

For conversion of Celsius to Fahrenheit, the following formula can be used:

$$°F = \frac{9}{5}°C + 32$$

APPENDIX C

Classification of Organisms

There are several ways to classify organisms. The one presented here follows the overall scheme described at the end of Chapter 20. Organisms are divided into five major groups, or kingdoms: Monera, Protista, Fungi, Plantae, and Animalia.

The chief taxonomic categories are kingdom, division or phylum, class, order, family, genus, species. (The taxonomic categories of division and phylum are equivalent. The term "division" is generally used in the classification of prokaryotes, algae, fungi, and plants, whereas "phylum" is used in the classification of protozoa and animals.) The following classification includes all of the generally accepted divisions and phyla. Certain classes and orders, particularly those mentioned in this book, are also included, but the listing is far from complete. The number of species given for each group is an estimated number of living (that is, contemporary) species described and named to date.

KINGDOM MONERA

Monerans (prokaryotes) are cells that lack a nuclear envelope, chloroplasts and other plastids, mitochondria, and 9 + 2 flagella. Prokaryotes are unicellular but sometimes occur as filaments or other superficially multicellular bodies. Their predominant mode of nutrition is heterotrophic, by absorption, but some groups are autotrophic, either photosynthetic or chemosynthetic. Reproduction is primarily asexual, by binary fission or budding, but genetic exchanges occur in some as a result of conjugation, transformation, transduction, and exchanges of plasmids. Motile forms move by means of bacterial flagella or by gliding.

The classification of prokaryotes is not hierarchical, and the schemes that are most widely used do not reflect evolutionary relationships. Kingdom Monera contains representatives of two distinct lineages, the Archaebacteria and the Eubacteria. Archaebacteria include the methanogens, thermoacidophiles, and extreme halophiles. Among the principal lineages of Eubacteria are the green bacteria (sulfur and nonsulfur), purple bacteria (sulfur and nonsulfur) and related forms, spirochetes, cyanobacteria, and gram-positive bacteria. About 2,700 distinct species of prokaryotes are recognized.

KINGDOM PROTISTA

Eukaryotic organisms, including unicellular and multicellular photosynthetic autotrophs (algae), multinucleate or multicellular heterotrophs (slime molds and water molds), and unicellular or simple colonial heterotrophs (protozoa). Their modes of nutrition include photosynthesis, absorption, and ingestion. Reproduction is both sexual (in some forms) and asexual. They move by 9 + 2 flagella (or cilia) or pseudopodia or are nonmotile. About 60,000 living species and another 60,000 extinct species known only from their fossils.

DIVISION

DIVISION EUGLENOPHYTA: euglenoids. Unicellular photosynthetic (or sometimes secondarily heterotrophic) organisms with chlorophylls *a* and *b*. They store food as paramylon, an unusual carbohydrate. Euglenoids usually have a single apical flagellum and a contractile vacuole. Sexual reproduction is unknown. Euglenoids occur mostly in fresh water. There are some 1,000 species.

DIVISION CHRYSOPHYTA: diatoms, golden-brown algae, and yellow-green algae. Unicellular photosynthetic organisms with chlorophylls *a* and *c* and the accessory pigment fucoxanthin. Food stored as the carbohydrate laminarin or as large oil droplets. Cell walls consisting mainly of cellulose, sometimes heavily impregnated with siliceous materials. There may be as many as 13,000 living species.

CLASS

Class Bacillariophyceae: diatoms. Chrysophyta with double siliceous shells, the two halves of which fit together like a pillbox. They are sometimes motile by the secretion of mucilage fibrils along a specialized groove, the raphe. There are nearly 10,000 living species and at least 15,000 extinct species.

Class Chrysophyceae: golden-brown algae. A diverse group of organisms, including flagellated, amoeboid, and nonmotile forms, some naked and others with a cell wall that may be ornamented with siliceous scales. About 3,000 species.

DIVISION

DIVISION DINOFLAGELLATA: "spinning" flagellates. Unicellular photosynthetic organisms with chlorophylls *a* and *c*. Food is stored as starch. Cell walls contain cellulose. Most of the organisms in this division have two lateral flagella, one of which beats in a groove that encircles the organism. They probably have no form of sexual reproduction, and their mitosis is unique. Nearly 2,000 living species.

DIVISION CHLOROPHYTA: green algae. Unicellular, colonial, coenocytic, or multicellular, characterized by chlorophylls *a* and *b* and various carotenoids. The carbohydrate food reserve is starch. The cell walls consist of polysaccharides, including cellulose in some forms. Motile cells have two lateral or apical flagella. True multicellular genera do not exhibit complex patterns of differentiation. Multicellularity has arisen at least three times, and quite possibly more often. There are about 9,000 known species and possibly many more.

CLASS

Class Chlorophyceae: Unicellular, colonial, or multicellular green algae, found predominantly in fresh water. Cell division involves a system of microtubules parallel to the plane of cell division, and the nuclear envelope persists during mitosis. Sexual reproduction involves the formation of a dormant zygote that subsequently undergoes meiosis, producing the haploid cells in which the organisms spend most of the life cycle.

Class Charophyceae: Unicellular or multicellular green algae, found predominantly in fresh water. Cell division involves a system of microtubules perpendicular to the plane of cell division; the nuclear envelope breaks down during the course of mitosis. Sexual reproduction involves the formation of a dormant zygote that subsequently undergoes meiosis, producing the haploid cells in which the organisms spend most of the life cycle. Certain members of this class resemble plants more closely than do any other organisms.

Class Ulvophyceae: Coenocytic or multicellular green algae, found predominantly in salt water. Cell division involves a system of microtubules perpendicular to the plane of cell division, but the nuclear envelope persists during mitosis. Sexual reproduction often involves alternation of generations, with meiosis in the diploid sporophyte producing haploid spores that germinate to produce the haploid gametophyte.

DIVISION

DIVISION PHAEOPHYTA: brown algae. Multicellular marine organisms characterized by the presence of chlorophylls *a* and *c* and the pigment fucoxanthin. Their food reserve is the carbohydrate laminarin. Motile cells are biflagellate, with one forward flagellum and one trailing flagellum. A considerable amount of tissue differentiation is found in some of the kelps, with specialized conducting cells in some genera that transport photosynthate to dimly lighted regions of the alga. There is, however, no differentiation into leaves, roots, and stem, as in plants. About 1,500 species.

DIVISION RHODOPHYTA: red algae. Primarily marine organisms characterized by the presence of chlorophyll *a* and red pigments known as phycobilins. Their carbohydrate reserve is a special type of starch (floridean). No motile cells are present at any stage in the complex life cycle. The algal body is built up of closely packed filaments in a gelatinous matrix and is not differentiated into leaves, roots, and stem. It lacks specialized conducting cells. There are some 4,000 species.

DIVISION MYXOMYCOTA: plasmodial slime molds. Heterotrophic amoeboid organisms that form a coenocytic plasmodium that creeps along as a mass and eventually differentiates into sporangia, each of which is coenocytic and eventually gives rise to many spores. Predominant mode of nutrition is by ingestion. More than 550 species.

DIVISION ACRASIOMYCOTA: cellular slime molds. Heterotrophic organisms in which there are separate amoebas that eventually swarm together to form a mass but retain their identity within this mass, which eventually differentiates into a compound sporangium. Principal mode of nutrition is by ingestion. Seven genera and about 65 species.

DIVISION CHYTRIDIOMYCOTA: chytrids (water molds). Coenocytic aquatic heterotrophs with a vegetative body, or thallus, usually differentiated into rhizoids and a sporangium. Nutrition by absorption. Cell walls contain chitin. Asexual reproduction, with sexual reproduction in some forms. Both spores and gametes are flagellated. About 900 species.

DIVISION OOMYCOTA: oomycetes (water molds). Coenocytic filamentous heterotrophs, primarily aquatic. Nutrition by absorption. Their cell walls are composed of glucose polymers, including cellulose. Both asexual and sexual reproduction, with flagellated asexual spores and nonmotile gametes. The coenocytic filaments (hyphae) are diploid, and the haploid gametes are produced by meiosis. About 800 species.

PHYLUM

PHYLUM MASTIGOPHORA: mastigophorans (flagellates). Unicellular heterotrophs with flagella, most of which are symbiotic forms such as *Trichonympha* and *Trypanosoma,* the cause of sleeping sickness. Some free-living species. Reproduction usually asexual by binary fission. About 1,500 species.

PHYLUM SARCODINA: sarcodines. Unicellular heterotrophs with pseudopodia, such as amoebas. No stiffening pellicle; some produce shells. Reproduction may be asexual or sexual. About 11,500 living species and some 33,000 fossil species.

PHYLUM CILIOPHORA: ciliates. Unicellular heterotrophs with cilia, including *Paramecium* and *Stentor.* Reproduction is asexual, but genetic exchange through the phenomenon of conjugation is also common. About 8,000 species.

PHYLUM OPALINIDA: opalinids. Parasitic protists found mainly in the digestive tracts of frogs and toads. Covered uniformly with cilia or flagella. Reproduction asexual by fission or sexual, with flagellated gametes. About 400 species.

PHYLUM SPOROZOA: sporozoa. Parasitic protists; usually without locomotor organelles during a major part of their complex life cycle. Includes *Plasmodium,* several species of which cause malaria. About 5,000 species.

KINGDOM FUNGI

Eukaryotic filamentous or, rarely, unicellular organisms. The filamentous forms consist basically of a continuous mycelium; this mycelium becomes septate (partitioned off) in certain groups and at certain stages of the life cycle. Chitin is present in the cell walls of all fungi. Fungi are saprobic or parasitic heterotrophs, with nutrition by absorption. Reproductive cycles often include both sexual and asexual phases. Most fungi are haploid, with the zygote the only diploid stage in the life cycle. No flagella or cilia at any stage of the life cycle. Some 100,000 species of fungi have been named.

DIVISION

DIVISION ZYGOMYCOTA: terrestrial fungi, such as black bread mold, with the hyphae septate only during the formation of reproductive bodies. The division includes about 600 described species, of which about 30 occur as components of the endomycorrhizae that are found in about 80 percent of all vascular plants.

DIVISION ASCOMYCOTA: terrestrial and aquatic fungi, including *Neurospora,* powdery mildews, morels, and truffles. The hyphae are septate but the septa perforated; complete septa cut off the reproductive bodies, such as spores or gametangia. Sexual reproduction involves the formation of a characteristic cell, the ascus, in which meiosis takes place and within which spores are formed. The hyphae in many ascomycetes are packed together into complex "fruiting bodies." Yeasts are unicellular ascomycetes that reproduce asexually by budding. About 30,000 species, in addition to 25,000 species that occur in lichens.

DIVISION BASIDIOMYCOTA: terrestrial fungi, including the mushrooms and toadstools, with the hyphae septate but the septa perforated; complete septa cut off reproductive bodies. Sexual reproduction involves formation of basidia, in which meiosis takes place and on which the spores are borne. There are some 25,000 species.

DIVISION DEUTEROMYCOTA: Fungi Imperfecti. Mainly fungi in which the sexual cycle has not been observed. The deuteromycetes are classified by their asexual spore-bearing structures. There are some 25,000 species, including *Penicillium,* the original source of penicillin, fungi that cause athlete's foot and other skin diseases, and many of the molds that give cheeses, such as Roquefort and Camembert, their special flavor.

KINGDOM PLANTAE

Multicellular photosynthetic eukaryotes, primarily adapted to life on land. The photosynthetic pigment is chlorophyll *a,* with chlorophyll *b* and a number of carotenoids serving as accessory pigments. The cell walls contain cellulose. There is considerable differentiation of tissues and organs. Reproduction is primarily sexual with alternating gametophytic and sporophytic phases; the gametophytic phase has been progressively reduced in the course of evolution. The egg- and sperm-producing structures are multicellular and are surrounded by a sterile (nonreproductive) jacket layer; the zygote develops into an embryo, or young sporophyte, while encased in the archegonium (seedless plants) or embryo sac (seed plants). The living members of the plant kingdom include the bryophytes and nine divisions of vascular plants—plants with complex differentiation of the sporophyte into leaves, roots, and stems and with well-developed strands of conducting tissue for the transport of water and organic materials.

DIVISION

DIVISION BRYOPHYTA: liverworts, hornworts, and mosses. Multicellular plants with photosynthetic pigments and food reserves similar to those of the green algae. They have multicellular gametangia with a sterile jacket one-cell-layer thick. The sperm are biflagellate and motile. Gametophytes and sporophytes both exhibit complex multicellular patterns of development, but conducting tissues are usually completely absent and are not well differentiated when present; true roots, leaves, and stems are absent. Most of the photosynthesis is carried out by the gametophyte, upon which the sporophyte is nutritionally dependent, at least initially. There are some 16,000 species.

CLASS

Class Hepaticae: liverworts. The gametophytes are either thallose (not differentiated into roots, leaves, and stem) or "leafy," and the sporophytes are relatively simple in construction. About 6,000 species.

Class Anthocerotae: hornworts. The gametophytes are thallose. The sporophyte grows from a basal meristem for as long as conditions are favorable. Stomata are present on the sporophyte. About 100 species.

Class Musci: mosses. The gametophytes are "leafy." The sporophytes have complex patterns of spore discharge. Stomata are present on the sporophyte. About 9,500 species.

DIVISION

DIVISION PSILOPHYTA: whisk ferns. Homosporous vascular plants with or without microphylls. The sporophytes are extremely simple, and there is no differentiation between root and shoot. The sperm are motile. Two genera, with several species.

DIVISION LYCOPHYTA: club mosses. Homosporous and heterosporous vascular plants with microphylls; extremely diverse in appearance. All lycophytes have motile sperm. There are five genera, with about 1,000 species.

DIVISION SPHENOPHYTA: horsetails. Homosporous vascular plants with jointed stems marked by conspicuous nodes and elevated siliceous ribs. Sporangia are borne in a cone at the apex of the stem. Leaves are scalelike. Sperm are motile. Although now thought to have evolved from a megaphyll, the leaves of the horsetails are structurally indistinguishable from microphylls. There is one genus, *Equisetum,* with 15 living species.

DIVISION PTEROPHYTA: ferns. They are mostly homosporous, although some are heterosporous. All possess megaphylls. The gametophyte is more or less free-living and usually photosynthetic. Multicellular gametangia and free-swimming sperm are present. About 12,000 species.

DIVISION CONIFEROPHYTA: conifers. Seed plants with active cambial growth and simple, needle-like leaves; the ovules are not enclosed and the sperm are not flagellated. There are some 50 genera, with about 550 species; the most familiar group of gymnosperms.

DIVISION CYCADOPHYTA: cycads. Seed plants with sluggish cambial growth and pinnately compound, palmlike or fernlike leaves. The ovules are not enclosed. The sperm are flagellated and motile but are carried to the ovule in a pollen tube. Cycads are gymnosperms. There are 10 genera, with about 100 species.

DIVISION GINKGOPHYTA: ginkgo. Seed plants with active cambial growth and fan-shaped leaves with open dichotomous venation. The ovules are not enclosed and are fleshy at maturity. Sperm are carried to the ovule in a pollen tube but are flagellated and motile. They are gymnosperms. There is one species only.

DIVISION GNETOPHYTA: gnetophytes. Seed plants with many angiospermlike features. They are the only gymnosperms in which vessels are present in the xylem. Motile sperm are absent. There are three very distinctive genera, with about 70 species.

DIVISION ANTHOPHYTA: flowering plants (angiosperms). Seed plants in which the ovules are enclosed in a carpel (in all but a very few genera), and the seeds at maturity are borne within fruits. They are extremely diverse vegetatively but are characterized by the flower, which is basically insect-pollinated. Other modes of pollination, such as wind pollination, have been derived in a number of different lines. The gametophytes are much reduced, with the female gametophyte often consisting of only seven cells at maturity. Double fertilization involving the two nonmotile sperm nuclei of the mature male gametophyte gives rise to the zygote and to the primary endosperm nucleus; the former becomes the embryo and the latter a special nutritive tissue, the endosperm. About 235,000 species.

CLASS

Class Monocotyledones: monocots. Flower parts are usually in threes, leaf venation is usually parallel, vascular bundles in the young stem are scattered, true secondary growth is not present, and there is one cotyledon. About 65,000 species.

Class Dicotyledones: dicots. Flower parts are usually in fours or fives, leaf venation is usually netlike, the vascular bundles in the young stem are in a ring, there is true secondary growth with vascular cambium commonly present, and there are two cotyledons. About 170,000 species.

KINGDOM ANIMALIA

Eukaryotic multicellular organisms. Their principal mode of nutrition is by ingestion. Many animals are motile, and they generally lack the rigid cell walls characteristic of plants. Considerable cellular migration and reorganization of tissues often occur during the course of embryonic development. Their reproduction is primarily sexual, with male and female diploid organisms producing haploid gametes that fuse to form the zygote. More than 1.5 million living species have been described, and the actual number may be more than 50 million.

PHYLUM

PHYLUM PORIFERA: sponges. Simple multicellular animals, largely marine, with stiff skeletons, and bodies perforated by many pores that admit water containing food particles. All have choanocytes, "collar cells." About 5,000 species.

PHYLUM MESOZOA: extremely simple wormlike animals, all parasites of marine invertebrates. The body consists of 20 to 30 cells, organized in two layers. About 50 species.

PHYLUM CNIDARIA: polyps and jellyfishes. Radially symmetrical animals with a gastrovascular cavity and two-layered bodies of a jellylike consistency. Reproduction is asexual or sexual. They are the only organisms with cnidocytes, special stinging cells. All are aquatic and most are marine. About 9,000 species.

CLASS

Class Hydrozoa: Hydra, Obelia, and other *Hydra*-like animals. They are often colonial and frequently have a regular alternation of asexual and sexual forms. The polyp form is dominant.

Class Scyphozoa ("cup animals"): marine jellyfishes, including *Aurelia.* The medusa form is dominant. They have true muscle cells.

Class Anthozoa ("flower animals"): sea anemones, colonial corals, and related forms. They have no medusa stage.

PHYLUM

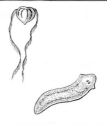

PHYLUM CTENOPHORA: comb jellies and sea walnuts. They are free-swimming, often almost spherical animals, with a gastrovascular cavity. They are translucent, gelatinous, delicately colored, and often bioluminescent. They possess eight bands of cilia, for locomotion. About 90 species.

PHYLUM PLATYHELMINTHES: flatworms. Bilaterally symmetrical with three embryonic tissue layers. The digestive cavity has only one opening. They have no coelom or pseudocoelom and no circulatory system. They have complex hermaphroditic reproductive systems and excrete by means of special flame cells. About 13,000 species.

CLASS

Class Turbellaria: planarians and other nonparasitic flatworms. They are ciliated, carnivorous, and have ocelli ("eyespots").

Class Trematoda: flukes. They are parasitic flatworms with digestive cavities.

Class Cestoda: tapeworms. They are parasitic flatworms with no digestive cavities; they absorb nourishment through body surfaces.

PHYLUM

PHYLUM GNATHOSTOMULIDA: tiny acoelomate marine worms, characterized by a unique pair of hard jaws. The digestive cavity has only one opening. About 80 species.

PHYLUM RHYNCHOCOELA: proboscis, nemertine, or ribbon worms. These acoelomate worms are nonparasitic, usually marine, and have a tubelike gut with mouth and anus, a retractile proboscis armed with a hook for capturing prey, and simple circulatory and reproductive systems. About 650 species.

PHYLUM NEMATODA: roundworms. The phylum includes minute free-living forms, such as vinegar eels, and plant and animal parasites, such as hookworms. They have a pseudocoelom and are characterized by elongated, cylindrical, bilaterally symmetrical bodies. About 12,000 species have been described and named, and there may be as many as 400,000 to 500,000 species.

PHYLUM NEMATOMORPHA: horsehair worms. They are extremely slender, brown or black pseudocoelomate worms up to 1 meter long. Adults are free-living, but the larvae are parasitic in arthropods. About 230 species.

PHYLUM ACANTHOCEPHALA: spiny-headed worms. They are parasitic pseudocoelomate worms with no digestive tract and a head armed with many recurved spines. About 500 species.

PHYLUM KINORHYNCHA: tiny pseudocoelomate worms that burrow in muddy ocean shores. They are short-bodied, covered with spines, and have a spiny retractile proboscis. About 100 species.

PHYLUM GASTROTRICHA: microscopic, wormlike pseudocoelomate animals that move by longitudinal bands of cilia. About 400 species.

PHYLUM LORICIFERA: microscopic pseudocoelomate animals of the ocean bottom, characterized by a plate-covered body, numerous spines projecting from the head, and a retractile tube for a mouth. The first of the three known species was discovered in 1982; there may be many more.

PHYLUM ROTIFERA: microscopic, wormlike or spherical pseudocoelomate animals, "wheel animalcules." They have a complete digestive tract, flame cells, and a circle of cilia on the head, the beating of which suggests a wheel; males are minute and either degenerate or unknown in many species. About 1,500 to 2,000 species.

PHYLUM ENTOPROCTA: microscopic, stalked, sessile animals that superficially resemble hydrozoans but are much more complex, having three embryonic tissue layers, a pseudocoelom, and a complete digestive tract. They were long misclassified with the coelomate bryozoans (ectoprocts), which they also resemble. About 75 species.

PHYLUM MOLLUSCA: unsegmented coelomate animals, with a head, a mantle, and a muscular foot, variously modified. They are mostly aquatic; soft-bodied, often with one or more hard shells, and a heart with three chambers. All mollusks, except bivalves, have a radula (rasplike organ used for scraping or marine drilling). At least 47,000 living species, and perhaps many more; an additional 35,000 species are known only from their fossils.

CLASS

Class Aplacophora: solenogasters. Wormlike marine animals, with no clearly defined shell, mantle, or foot. The presence of a radula identifies them as mollusks. About 250 species.

Class Polyplacophora: chitons. The mollusks with the closest resemblance to the hypothetical primitive form, they have an elongated body covered with a mantle in which are embedded eight dorsal shell plates. About 600 species.

Class Monoplacophora: Neopilina. Mostly deep-sea mollusks with a large, single dorsal shell and multiple pairs of gills, nephridia, and retractor muscles. Two genera with eight species.

Class Scaphopoda: tooth or tusk shells. They are marine mollusks with a conical tubular shell. About 350 species.

Class Bivalvia: two-shelled mollusks, including clams, oysters, mussels, scallops. They usually have a hatchet-shaped foot and no distinct head. Generally sessile. At least 7,500 species.

Class Gastropoda: asymmetrical mollusks, including snails, whelks, slugs. They usually have a spiral shell and a head with one or two pairs of tentacles. At least 37,500 living species and about 15,000 fossil species.

Class Cephalopoda: octopuses, squids, *Nautilus.* They are characterized by a "head-foot" with eight or ten arms or many tentacles, mouth with two horny jaws, and well-developed eyes and nervous system. The shell is external *(Nautilus),* internal (squid), or absent (octopus). All except *Nautilus* have ink glands. About 600 species.

PHYLUM

PHYLUM ANNELIDA: segmented worms. They usually have a well-developed coelom, a one-way digestive tract, head, circulatory system, nephridia, and well-defined nervous system. About 9,000 species.

CLASS

Class Oligochaeta: soil, freshwater, and marine annelids, including *Lumbricus* and other earthworms. They have scanty bristles and usually a poorly differentiated head. About 3,000 species.

Class Polychaeta: mainly marine worms, such as *Nereis.* They have a distinct head with tentacles, antennae, and specialized mouthparts. Parapodia are often brightly colored. About 6,000 species.

Class Hirudinea: leeches. They have a posterior sucker and usually an anterior sucker surrounding the mouth. They are freshwater, marine, and terrestrial; either free-living or parasitic. About 300 species.

PHYLUM

PHYLUM SIPUNCULA: peanut worms. Unsegmented marine worms with a stout body, a long, retractile proboscis, and a trochophore larva that resembles those of the polychaete annelids. About 300 species.

PHYLUM ECHIURA: spoon worms. Marine worms with a nonretractile proboscis that contracts to form a structure resembling a spoon. Their embryonic development and trochophore larvae resemble those of the polychaete annelids. About 100 species.

PHYLUM PRIAPULIDA: predatory burrowing marine worms, characterized by a retractile, spine-bearing proboscis. They resemble the pseudocoelomate kinorhynchs, but may have a true coelom. Only 15 species.

PHYLUM POGONOPHORA: beard worms. Unsegmented except at the posterior end, these slender marine worms live in long tubes in deep-sea sediments. Although they have no mouth or digestive tract, the anterior region of the body bears a crown of tentacles. About 100 species.

PHYLUM PENTASTOMIDA: wormlike parasites of vertebrate respiratory systems, sometimes with two pairs of short appendages at the anterior end of the body. They lack circulatory, respiratory, and excretory systems, but the nervous system resembles those of annelids and arthropods. About 70 species.

PHYLUM TARDIGRADA: water bears. Tiny segmented animals, with a thin cuticle and four pairs of stubby legs. They are common in fresh water and in the moisture film on mosses; when water is unavailable, they enter a state of suspended animation. About 350 species.

PHYLUM ONYCHOPHORA: *Peripatus.* Caterpillarlike animals with many short, unjointed pairs of legs. Their relatively soft bodies, segmentally arranged nephridia, muscular body walls, and ciliated reproductive tracts resemble those of the annelids; their antennae and eyes resemble those of both the polychaete annelids and the arthropods; their jaws, protective cuticle, brain, and circulatory and respiratory systems resemble those of arthropods. Most give birth to live young, and in some species the embryo is nourished through a placenta. About 70 species.

PHYLUM ARTHROPODA: The largest phylum in the animal kingdom, arthropods are segmented animals with paired, jointed appendages, a hard jointed exoskeleton, a complete digestive tract, reduced coelom, no nephridia, a dorsal brain, and a ventral nerve cord with paired ganglia in each segment. More than 1 million species have been classified to date.

CLASS

Class Merostomata: horseshoe crabs. Aquatic arthropods with chelicerae (pincers or fangs), pedipalps, compound eyes, four pairs of walking legs, and book gills. Only four species.

Class Pycnogonida: sea spiders. Aquatic chelicerates with slender bodies and four or, rarely, five pairs of legs, which are often very long. About 500 species.

Class Arachnida: spiders, mites, ticks, scorpions. Most members are terrestrial, air-breathing; four pairs of walking legs; chelicerae may be pincers or fangs; pedipalps are usually sensory. About 57,000 species.

Class Crustacea: lobsters, crayfish, crabs, shrimps. Crustaceans are mostly aquatic, with compound eyes, two pairs of antennae, one pair of mandibles, and typically two pairs of maxillae. The thoracic segments have appendages, and the abdominal segments are with or without appendages. About 25,000 species.

Class Chilopoda: centipedes. They have a head and 15 to 177 trunk segments, each with one pair of jointed appendages. About 3,000 species.

Class Diplopoda: millipedes. They have a head and a trunk with 20 to 200 body rings, each with two pairs of appendages. About 7,500 species.

Class Pauropoda: tiny, soft-bodied arthropods that resemble millipedes but have only 11 or 12 segments and 9 or 10 pairs of legs. They have branched antennae. About 300 species.

Class Symphyla: garden centipedes and their relatives. Soft-bodied arthropods with one pair of antennae, 12 pairs of jointed legs with claws, and a pair of unjointed posterior appendages. About 130 species.

Class Insecta: insects, including bees, ants, beetles, butterflies, fleas, lice, flies, etc. Most insects are terrestrial and breathe by means of tracheae. The body has three distinct parts (head, thorax, and abdomen); the head bears compound eyes and one pair of antennae; the thorax bears three pairs of legs and usually two pairs of wings. About 1 million species.

PHYLUM

PHYLUM BRACHIOPODA: lamp shells. Marine animals with two hard shells (one dorsal and one ventral), they superficially resemble clams. Fixed by a stalk or one shell in adult life, they feed by means of a lophophore. About 250 living species; 30,000 extinct species.

PHYLUM PHORONIDA: sedentary, elongated worms that secrete and live in a leathery tube. They have a U-shaped digestive tract and a ring of ciliated tentacles (the lophophore) with which they feed. Marine. Only 18 species.

PHYLUM BRYOZOA (ECTOPROCTA): "moss" animals. These microscopic aquatic organisms are characterized by the lophophore, a crown of hollow, ciliated tentacles, with which they feed and by a U-shaped digestive tract. They usually form fixed and branching colonies. Coelomates, they superficially resemble the pseudocoelomate entoprocts. Some species retain larvae in a special brood pouch. About 4,000 species.

PHYLUM ECHINODERMATA: starfish and sea urchins. Echinoderms are radially symmetrical in the adult stage, with well-developed coelomic cavities, an endoskeleton of calcareous ossicles and spines, and a unique water vascular system. They have tube feet. All are marine. About 6,000 living species and 20,000 species known only from their fossils.

CLASS

Class Crinoidea: sea lilies and feather stars. Sessile animals, they often have a jointed stalk for attachment, and they have 10 arms bearing many slender lateral branches. Most species are fossils; only 20 living species.

Class Stelleroidea: starfish and brittle stars. They have 5 to 50 arms, an oral surface directed downward, and rows of tube feet on each arm. Brittle stars have greatly elongated, highly flexible slender arms and rapid horizontal locomotion.

Class Echinoidea: sea urchins and sand dollars. Skeletal plates form a rigid external covering that bears many movable spines.

Class Concentricycloidea: sea daisies. The microscopic members of this newly established class have five plates on the dorsal surface, two rings of tube feet on the ventral surface, and no digestive tract.

Class Holothuroidea: sea cucumbers. They have a sausage-shaped or wormlike elongated body.

PHYLUM

PHYLUM CHAETOGNATHA: arrow worms. Free-swimming, planktonic marine worms, they have a coelom, a complete digestive tract, and a mouth with strong sickle-shaped hooks on each side. About 60 species.

PHYLUM HEMICHORDATA: acorn worms. The body is divided into three regions—proboscis, collar, and trunk. The coelomic cavities provide a hydrostatic skeleton similar to the water vascular system of echinoderms, and their larvae resemble starfish larvae. They have both ventral and dorsal nerve cords, and the anterior portion of the dorsal cord is hollow in some species. They also have a pharynx with gill slits. About 80 species.

PHYLUM CHORDATA: animals having at some stage a notochord, pharyngeal gill slits (or pouches), a hollow nerve cord on the dorsal side, and a tail. About 43,000 species.

SUBPHYLUM

SUBPHYLUM CEPHALOCHORDATA: lancelets. This small subphylum contains only *Branchiostoma* and related forms. They are somewhat fishlike marine animals with a permanent notochord the whole length of the body, a nerve cord, a pharynx with gill slits, and no cartilage or bone. About 28 species.

SUBPHYLUM UROCHORDATA: tunicates. Adults are saclike, usually sessile, often forming branching colonies. They feed by ciliary currents, have gill slits, a reduced nervous system, and no notochord. Larvae are active, with well-developed nervous system and notochord. They are marine. About 1,300 species.

SUBPHYLUM VERTEBRATA: vertebrates. The most important subphylum of Chordata. In the vertebrates the notochord is an embryonic structure; it is typically replaced in the course of development by cartilage or bone, forming the segmented vertebral column, or backbone. A cranium, or skull, surrounds a well-developed brain. Vertebrates usually have a tail. About 41,700 species.

CLASS

Class Agnatha: lampreys and hagfish. These are eel-like aquatic vertebrates without limbs, with a jawless sucking mouth, and no bones, scales, or fins. About 60 species.

Class Chondrichthyes: sharks, rays, skates, and other cartilaginous fishes. They have complicated copulatory organs, scales, and no air bladders. They are almost exclusively marine. About 625 species.

Class Osteichthyes: bony fishes, including nearly all modern freshwater fishes, such as sturgeon, trout, perch, anglerfish, lungfish, and some almost extinct groups. They usually have an air bladder or (rarely) a lung. More than 19,000 species.

Class Amphibia: salamanders, frogs, and toads. They usually respire by gills in the larval stage and by lungs in the adult stage. They have incomplete double circulation and a usually naked skin. The limbs are legs. They were the first vertebrates to inhabit the land and were ancestors of the reptiles. Their eggs are unprotected by a shell and embryonic membranes. About 2,500 species.

Class Reptilia: turtles, lizards, snakes, crocodiles; includes many extinct species such as the dinosaurs. Reptiles breathe by lungs and have incomplete double circulation. Their skin is usually covered with scales. The four limbs are legs (absent in snakes and some lizards). They are ectotherms. Most live and reproduce on land, although some are aquatic. The embryo is enclosed in an egg shell and has protective membranes. About 6,000 species.

Class Aves: birds. Birds are homeothermic animals with complete double circulation and a skin covered with feathers. The forelimbs are wings. The embryo is enclosed in an egg shell with protective membranes. About 9,000 living species; an estimated 14,000 species are extinct.

Class Mammalia: mammals. Mammals are homeothermic animals with complete double circulation. Their skin is usually covered with hair. The young are nourished with milk secreted by the mother. They have four limbs, usually legs (forelimbs, sometimes arms, wings, or fins), a diaphragm used in respiration, a lower jaw made up of a single pair of bones, three bones in each middle ear connecting eardrum and inner ear, and almost always seven vertebrae in the neck. About 4,500 species.

SUBCLASS

Subclass Prototheria: monotremes. These are the oviparous (egg-laying) mammals with imperfect temperature regulation. There are only three living species: the duckbilled platypus and spiny anteaters of Australia and New Guinea.

Subclass Metatheria: marsupials, including kangaroos, opossums, and others. Marsupials are viviparous mammals, usually with a yolk-sac placenta; the young are born in a very immature state and are carried in an external pouch of the mother for some time after birth. They are found chiefly in Australia and South America. About 260 species.

Subclass Eutheria: mammals with a well-developed chorioallantoic placenta. This subclass comprises the great majority of living mammals. There are 16 orders of Eutheria:

Order

Insectivora: shrews, moles, hedgehogs, etc.
Chiroptera: bats. Aerial mammals with the forelimbs wings.
Dermoptera: "flying" lemurs.
Edentata: toothless mammals—anteaters, sloths, armadillos, etc.
Lagomorpha: rabbits and hares.
Carnivora: carnivorous animals—cats, dogs, bears, weasels, seals, etc.
Tubulidentata: aardvarks.
Pholidota: pangolins.
Rodentia: rodents—rats, mice, squirrels, etc.
Artiodactyla: even-toed ungulates (hoofed mammals)—cattle, deer, camels, hippopotamuses, etc.
Perissodactyla: odd-toed ungulates—horses, zebras, rhinoceroses, etc.
Proboscidea: elephants.
Hyracoidea: hyraxes.
Sirenia: manatees, dugongs, and sea cows. Large aquatic mammals with the forelimbs finlike, the hind limbs absent.
Cetacea: whales, dolphins, and porpoises. Aquatic mammals with the forelimbs fins, the hind limbs absent.
Primates: prosimians, monkeys, apes, and humans.

Glossary

This list does not include units of measure or names of taxonomic groups, which can be found in Appendixes A and C, or terms that are used only once in the text and defined there.

abdomen: In vertebrates, the portion of the trunk containing visceral organs other than heart and lungs; in arthropods, the posterior portion of the body, made up of similar segments and containing the reproductive organs and part of the digestive tract.

abscisic acid (ABA) [L. *ab,* away, off + *scissio,* dividing]: A plant hormone with a variety of inhibitory effects; brings about dormancy in buds, maintains dormancy in many types of seeds, and effects stomatal closing; also known as the "stress hormone."

abscission [L. *ab,* away, off + *scissio,* dividing]: In plants, the dropping of leaves, flowers, fruits, or stems at the end of a growing season, as the result of formation of a two-layered zone of specialized cells (the abscission zone) and the action of a hormone (ethylene).

absorption [L. *absorbere,* to swallow down]: The movement of water and dissolved substances into a cell, tissue, or organism.

absorption spectrum: The characteristic pattern of the wavelengths (colors) of light that a particular pigment absorbs.

acetylcholine (asset-ill-**coal**-een): One of the principal chemicals (neurotransmitters) responsible for the transmission of nerve impulses across synapses.

acid [L. *acidus,* sour]: A substance that causes an increase in the number of hydrogen ions (H⁺) in a solution and a decrease in the number of hydroxide ions (OH⁻); having a pH of less than 7; the opposite of a base.

actin [Gk. *aktis,* a ray]: A protein, composed of globular subunits, that forms filaments that are among the principal components of the cytoskeleton. Also one of the two major proteins of muscle (the other is myosin); the principal constituent of the thin filaments.

action potential: A transient change in electric potential across a membrane; in nerve cells, results in conduction of a nerve impulse; in muscle cells, results in contraction.

action spectrum: The characteristic pattern of the wavelengths (colors) of light that elicit a particular reaction or response.

activation energy: The energy that must be possessed by atoms or molecules in order to react.

active site: The region of an enzyme surface that binds the substrate during the reaction catalyzed by the enzyme.

active transport: The energy-requiring transport of a solute across a cell membrane (or a membrane of an organelle) from a region of lower concentration to a region of higher concentration (that is, against a concentration gradient).

adaptation [L. *adaptare,* to fit]: (1) The evolution of features that make a group of organisms better suited to live and reproduce in their environment. (2) A peculiarity of structure, physiology, or behavior that aids the organism in its environment.

adaptive radiation: The evolution from a primitive and unspecialized ancestor of a number of divergent forms, each specialized to fill a distinct ecological niche; associated with the opening up of a new biological frontier.

adenosine diphosphate (ADP): A nucleotide consisting of adenine, ribose, and two phosphate groups; formed by the removal of one phosphate from an ATP molecule.

adenosine monophosphate (AMP): A nucleotide consisting of adenine, ribose, and one phosphate group; can be formed by the removal of two phosphates from an ATP molecule; in its cyclic form, functions as a "second messenger" for a number of vertebrate hormones and neurotransmitters.

adenosine triphosphate (ATP): The nucleotide that provides the energy currency for cell metabolism; composed of adenine, ribose, and three phosphate groups. On hydrolysis, ATP loses one phosphate group and one hydrogen ion to become adenosine diphosphate (ADP), releasing energy in the process. ATP is formed from ADP and inorganic phosphate in an enzymatic reaction that traps energy released by catabolism or energy captured in photosynthesis.

ADH: Abbreviation of antidiuretic hormone.

adhesion [L. *adhaerere,* to stick to]: The holding together of molecules of different substances.

ADP: Abbreviation of adenosine diphosphate.

adrenal gland [L. *ad,* near + *renes,* kidney]: A vertebrate endocrine gland. The cortex (outer surface) is the source of cortisol, aldosterone, and other steroid hormones; the medulla (inner core) secretes adrenaline and noradrenaline.

adrenaline: A hormone, produced by the medulla of the adrenal gland, that increases the concentration of sugar in the blood, raises blood pressure and heartbeat rate, and increases muscular power and resistance to fatigue; also a neurotransmitter across synaptic junctions. Also called epinephrine.

adventitious [L. *adventicius,* not properly belonging to]: Referring to a structure arising from an unusual place, such as roots growing from stems or leaves.

aerobic [Gk. *aēr,* air + *bios,* life]: Any biological process that can occur in the presence of molecular oxygen (O₂).

afferent [L. *ad,* near + *ferre,* to carry]: Bringing inward to a central part, applied to nerves and blood vessels.

agar: A gelatinous material prepared from certain red algae that is used to solidify nutrient media for growing microorganisms.

aldosterone [Gk. *aldainō,* to nourish + *stereō,* solid]: A hormone produced by the adrenal cortex that affects the concentration of ions in the blood; it stimulates the reabsorption of sodium and the excretion of potassium by the kidney.

alga, *pl.* **algae** (**al**-gah, **al**-jee): A unicellular or simple multicellular eukaryotic photosynthetic organism lacking multicellular sex organs.

alkaline: Pertaining to substances that increase the number of hydroxide ions (OH⁻) in a solution; having a pH greater than 7; basic; opposite of acidic.

allantois [Gk. *allant,* sausage]: One of the four extraembryonic membranes that form during the development of reptiles, birds, and mammals.

allele frequency: The proportion of a particular allele in a population.

alleles (al-eels) [Gk. *allelon,* of one another]: Two or more different forms of a gene. Alleles occupy the same position (locus) on homologous chromosomes and are separated from each other at meiosis.

allopatric speciation [Gk. *allos,* other + *patra,* fatherland, country]: Speciation that occurs as the result of the geographic separation of a population of organisms.

allosteric interaction [Gk. *allos,* other + *stereō,* solid, shape]: An interaction involving an enzyme that has two binding sites, the active site and a site into which another molecule, an allosteric effector, fits; the binding of the effector changes the shape of the enzyme and activates or inactivates it. Allosteric interactions also play a role in transport processes involving integral membrane proteins.

alternation of generations: A sexual life cycle in which a haploid (*n*) phase alternates with a diploid (2*n*) phase. The gametophyte (*n*) produces gametes (*n*) by mitosis. The fusion of gametes yields zygotes (2*n*). Each zygote develops into a sporophyte (2*n*) that forms haploid spores (*n*) by meiosis. Each haploid spore forms a new gametophyte, completing the cycle.

altruism: Self-sacrifice for the benefit of others; any form of behavior that increases the fitness of the recipient while reducing the fitness of the altruistic individual.

alveolus, *pl.* **alveoli** [L. dim. of *alveus,* cavity, hollow]: One of the many small air sacs within the lungs in which the bronchioles terminate. The thin walls of the alveoli contain numerous capillaries and are the site of gas exchange between the air in the alveoli and the blood in the capillaries.

amino acids (am-ee-no) [Gk. *Ammon,* referring to the Egyptian sun god, near whose temple ammonium salts were first prepared from camel dung]: Organic molecules containing nitrogen in the form of —NH_2 and a carboxyl group, —COOH, bonded to the same carbon atom; the "building blocks" of protein molecules.

ammonification: The process by which decomposers break down proteins and amino acids, releasing the excess nitrogen in the form of ammonia (NH_3) or ammonium ion (NH_4^+).

amnion (am-neon) [Gk. dim. of *amnos,* lamb]: One of the four extraembryonic membranes that form during the development of reptiles, birds, and mammals; it encloses a fluid-filled space, the amniotic cavity, that surrounds the developing embryo.

amniote egg: An egg that is isolated and protected from the environment during the period of its development by a series of extraembryonic membranes and, often, a more or less impervious shell; the amniote eggs of birds and many reptiles are completely self-sufficient, requiring only oxygen from the outside.

amoeboid [Gk. *amoibē,* change]: Moving or feeding by means of pseudopodia (temporary cytoplasmic protrusions from the cell body).

AMP: Abbreviation of adenosine monophosphate.

anabolism [Gk. *ana,* up + *-bolism* (as in metabolism)]: Within a cell or organism, the sum of all biosynthetic reactions (that is, chemical reactions in which larger molecules are formed from smaller ones).

anaerobe [Gk. *an,* without + *aēr,* air + *bios,* life]: Cell that can live without free oxygen; obligate anaerobes cannot live in the presence of oxygen; facultative anaerobes can live with or without oxygen.

anaerobic [Gk. *an,* without + *aēr,* air + *bios,* life]: Applied to a process that can occur without oxygen, such as fermentation; also applied to organisms that can live without free oxygen.

analogous [Gk. *analogos,* proportionate]: Applied to structures similar in function but different in evolutionary origin, such as the wing of a bird and the wing of an insect.

anaphase (anna-phase) [Gk. *ana,* up + *phasis,* form]: In mitosis and meiosis II, the stage in which the chromatids of each chromosome separate and move to opposite poles; in meiosis I, the stage in which homologous chromosomes separate and move to opposite poles.

androgens [Gk. *andros,* man + *genos,* origin, descent]: Male sex hormones; any chemical with actions similar to those of testosterone.

angiosperms (an-jee-o-sperms) [Gk. *angeion,* vessel + *sperma,* seed]: The flowering plants. Literally, a seed borne in a vessel; thus, any plant whose seeds are borne within a matured ovary (fruit).

anisogamy [Gk. *aniso,* unequal + *gamos,* marriage]: Sexual reproduction in which one gamete is larger than the other; both gametes are motile.

annual plant [L. *annus,* year]: A plant that completes its life cycle (from seed germination to seed production) and dies within a single growing season.

antennae: Long, paired sensory appendages on the head of many arthropods.

anterior [L. *ante,* before, toward, in front of]: The front end of an organism.

anther [Gk. *anthos,* flower]: In flowering plants, the pollen-bearing portion of a stamen.

antheridium, *pl.* **antheridia:** In bryophytes and some vascular plants, the multicellular sperm-producing organ.

anthropoid [Gk. *anthropos,* man, human]: A higher primate; includes monkeys, apes, and humans.

antibiotic [Gk. *anti,* against + *bios,* life]: An organic compound, inhibitory or toxic to other species, that is formed and secreted by an organism.

antibody [Gk. *anti,* against]: A globular protein, synthesized by a B lymphocyte, that is complementary to a foreign substance (antigen) with which it combines specifically.

anticodon: In a tRNA molecule, the three-nucleotide sequence that base pairs with the mRNA codon for the amino acid carried by that particular tRNA; the anticodon is complementary to the mRNA codon.

antidiuretic hormone (ADH) [Gk. *anti,* against + *diurgos,* thoroughly wet + *hormaein,* to excite]: A peptide hormone synthesized in the hypothalamus that inhibits urine excretion by inducing the reabsorption of water from the nephrons of the kidneys; also called vasopressin.

antigen [Gk. *anti,* against + *genos,* origin, descent]: A foreign substance, usually a protein or polysaccharide, that, when bound to a complementary antibody displayed on the surface of a B lymphocyte or to a complementary T-cell receptor, stimulates an immune response.

aorta (a-ore-ta) [Gk. *aeirein,* to lift, heave]: The major artery in blood-circulating systems; the aorta sends blood to the other body tissues.

apical dominance [L. *apex,* top]: In plants, the hormone-mediated influence of a terminal bud in suppressing the growth of axillary buds.

apical meristem [L. *apex,* top + Gk. *meristos,* divided]: In vascular plants, the growing point at the tip of the root or stem.

arboreal [L. *arbor,* tree]: Tree-dwelling.

archegonium, *pl.* **archegonia** [Gk. *archegonos,* first of a race]: In bryophytes and some vascular plants, the multicellular egg-producing organ.

archenteron [Gk. *arch,* first, or main + *enteron,* gut]: The main cavity within the early embryo (gastrula) of many animals; lined with endoderm, it opens to the outside by means of the blastopore and ultimately becomes the digestive tract.

artery: A vessel carrying blood from the heart to the tissues; arteries are usually thick-walled, elastic, and muscular. A small artery is known as an arteriole.

artificial selection: The breeding of selected organisms for the purpose of producing descendants with desired characteristics.

ascus, *pl.* **asci** (as-kus, as-i) [Gk. *askos,* wineskin, bladder]: In the fungi of division Ascomycota, a specialized cell within which two haploid nuclei fuse to produce a zygote that immediately divides by meiosis; at maturity, an ascus contains ascospores.

asexual reproduction: Any reproductive process, such as budding or the division of a cell or body into two or more approximately equal parts, that does not involve the union of gametes.

atmospheric pressure [Gk. *atmos,* vapor + *sphaira,* globe]: The weight of the earth's atmosphere over a unit area of the earth's surface.

atom [Gk. *atomos,* indivisible]: The smallest particle into which a chemical element can be divided and still retain the properties characteristic of the element; consists of a central core, the nucleus, containing protons and neutrons, and electrons that move around the nucleus.

atomic number: The number of protons in the nucleus of an atom; equal to the number of electrons in the neutral atom.

atomic weight: The average weight of all the isotopes of an element relative to the weight of an atom of the most common isotope of carbon (^{12}C), which is by convention assigned the integral value of 12; approximately equal to the number of protons plus neutrons in the nucleus of an atom.

ATP: Abbreviation of adenosine triphosphate, the principal energy-carrying compound of the cell.

ATP synthetase: The enzyme complex in the inner membrane of the mitochondrion and the thylakoid membrane of the chloroplast through which protons flow down the gradient established in the first stage of chemiosmotic coupling; the site of formation of ATP from ADP and inorganic phosphate during oxidative phosphorylation and photophosphorylation.

atrioventricular node [L. *atrium*, yard, court, hall + *ventriculus*, the stomach + *nodus*, knot]: A group of slow-conducting fibers in the atrium of the vertebrate heart that are stimulated by impulses originating in the sinoatrial node (the pacemaker) and that conduct impulses to the bundle of His, a group of fibers that stimulate contraction of the ventricles.

atrium, *pl.* **atria** (a-tree-um) [L., yard, court, hall]: A thin-walled chamber of the heart that receives blood and passes it on to a thick, muscular ventricle.

autonomic [Gk. *autos*, self + *nomos*, usage, law]: Self-controlling, independent of outside influences.

autonomic nervous system [Gk. *autos*, self + *nomos*, usage, law]: In the peripheral nervous system of vertebrates, the neurons and ganglia that are not ordinarily under voluntary control; innervates the heart, glands, visceral organs, and smooth muscle; subdivided into the sympathetic and parasympathetic divisions.

autosome [Gk. *autos*, self + *soma*, body]: Any chromosome other than the sex chromosomes. Humans have 22 pairs of autosomes and one pair of sex chromosomes.

autotroph [Gk. *autos*, self + *trophos*, feeder]: An organism that is able to synthesize all needed organic molecules from simple inorganic substances (e.g., H_2O, CO_2, NH_3) and some energy source (e.g., sunlight); in contrast to heterotroph. Plants, algae, and some groups of prokaryotes are autotrophs.

auxin [Gk. *auxein*, to increase + *in*, of, or belonging to]: One of a group of plant hormones with a variety of growth-regulating effects, including promotion of cell elongation.

auxotroph [L. *auxillium*, help + Gk. *trophos*, feeder]: A mutant with a defect in the enzymatic pathway for the synthesis of a particular molecule, which must therefore be supplied for normal growth.

axil [Gk. *axilla*, armpit]: The upper angle between a twig or leaf and the stem from which it grows.

axillary [Gk. *axilla*, armpit]: In botany, term applied to buds or branches occurring in the axil of a leaf.

axis: An imaginary line passing through a body or organ around which parts are symmetrically aligned.

axon [Gk. *axon*, axle]: A long process of a neuron, or nerve cell, that is capable of rapidly conducting nerve impulses over great distances.

B lymphocyte: A type of white blood cell capable of becoming an antibody-secreting plasma cell; a B cell.

bacteriophage [L. *bacterium* + Gk. *phagein*, to eat]: A virus that parasitizes a bacterial cell.

bark: In plants, all tissues outside the vascular cambium in a woody stem.

basal body [Gk. *basis*, foundation]: A cytoplasmic organelle of animals and some protists, from which cilia or flagella arise; identical in structure to the centriole, which is involved in mitosis and meiosis in animals and some protists.

base: A substance that causes an increase in the number of hydroxide ions (OH^-) in a solution and a decrease in the number of hydrogen ions (H^+); having a pH of more than 7; the opposite of an acid. *See* Alkaline.

base-pairing principle: In the formation of nucleic acids, the requirement that adenine must always pair with thymine (or uracil) and guanine with cytosine.

basidium, *pl.* **basidia** (ba-sid-ium) [L., a little pedestal]: A specialized reproductive cell of the fungi of division Basidiomycota, often club-shaped, in which nuclear fusion and meiosis occur; homologous with the ascus.

behavior: All of the acts an organism performs, as in, for example, seeking a suitable habitat, obtaining food, avoiding predators, and seeking a mate and reproducing.

biennial [L. *biennium*, a space of two years; *bi*, twice + *annus*, year]: Occurring once in two years; a plant that requires two years to complete its reproductive cycle; vegetative growth occurs in the first year, sexual reproduction and death in the second.

bilateral symmetry [L. *bi*, twice, two + *lateris*, side; Gk. *summetros*, symmetry]: A body form in which the right and left halves of an organism are approximate mirror images of each other.

bile: A yellow secretion of the vertebrate liver, temporarily stored in the gallbladder and composed of organic salts that emulsify fats in the small intestine.

binary fission [L. *binarius*, consisting of two things or parts + *fissus*, split]: Asexual reproduction by division of the cell or body into two equal, or nearly equal, parts.

binomial system [L. *bi*, twice, two + Gk. *nomos*, usage, law]: A system of naming organisms in which the name consists of two parts, with the first designating genus and the second, species; originated by Linnaeus.

biogeochemical cycle [Gk. *bios*, life + *geō*, earth + *chēmeia*, alchemy; *kyklos*, circle, wheel]: The cyclic path of an inorganic substance, such as carbon or nitrogen, through an ecosystem. Its geological components are the atmosphere, the crust of the earth, and the oceans, lakes, and rivers; its biological components are producers, consumers, and detritivores, including decomposers.

biological clock [Gk. *bios*, life + *logos*, discourse]: Proposed internal factor(s) in organisms that governs functions that occur rhythmically in the absence of external stimuli.

biomass [Gk. *bios*, life]: Total weight of all organisms (or some group of organisms) living in a particular habitat or place.

biome: One of the major types of distinctive plant formations; for example, the grassland biome, the tropical rain forest biome, etc.

biosphere [Gk. *bios*, life + *sphaira*, globe]: The zones of air, land, and water at the surface of the earth occupied by living things.

biosynthesis [Gk. *bios*, life + *synthesis*, a putting together]: Formation by living organisms of organic compounds from elements or simple compounds.

blade: (1) The broad, expanded part of a leaf. (2) The broad, expanded photosynthetic part of the thallus of a multicellular alga or a simple plant.

blastocoel [Gk. *blastos*, sprout + *koilos*, a hollow]: The fluid-filled cavity in the interior of a blastula.

blastocyst [Gk. *blastos*, sprout + *kystis*, sac]: The blastula stage of a developing mammal; consists of an inner cell mass that will give rise to the embryo proper and a double layer of cells, the trophoblast, that is the precursor of the chorion.

blastodisc [Gk. *blastos*, sprout + *discos*, a round plate]: Disklike area on the surface of a large, yolky egg that undergoes cleavage and gives rise to the embryo.

blastomere [Gk. *blastos*, sprout + *meris*, part of, portion]: One of many cells produced by cleavage of the fertilized egg.

blastopore [Gk. *blastos*, sprout + *poros*, a way, means, path]: In the gastrula stage of an embryo, the opening that connects the archenteron with the outside; represents the future mouth in some animals (protostomes), the future anus in others (deuterostomes).

blastula [Gk. *blastos*, sprout]: An animal embryo after cleavage and before gastrulation; usually consists of a fluid-filled sphere, the walls of which are composed of a single layer of cells.

bond strength: The strength with which a chemical bond holds two atoms together; conventionally measured in terms of the amount of energy, in kilocalories per mole, required to break the bond.

botany [Gk. *botanikos*, of herbs]: The study of plants.

Bowman's capsule: In the vertebrate kidney, the bulbous unit of the nephron, which surrounds the glomerulus. In filtration, the initial process in urine formation, blood plasma is forced from the glomerular capillaries into Bowman's capsule.

brainstem: The most posterior portion of the vertebrate brain; includes medulla, pons, and midbrain.

bronchus, *pl.* bronchi (bronk-us, bronk-eye) [Gk. *bronchos*, windpipe]: One of a pair of respiratory tubes branching into either lung at the lower end of the trachea; it subdivides into progressively finer passageways, the bronchioles, culminating in the alveoli.

bud: (1) In plants, an embryonic shoot, including rudimentary leaves, often protected by special bud scales. (2) In animals, an asexually produced outgrowth that develops into a new individual.

buffer: A combination of H^+-donor and H^+-acceptor forms of a weak acid or a weak base; a buffer prevents appreciable changes of pH in solutions to which small amounts of acids or bases are added.

bulb: A modified bud with thickened leaves adapted for underground food storage.

bulk flow: The overall movement of a fluid induced by gravity, pressure, or an interplay of both.

bundle of His: In the vertebrate heart, a group of muscle fibers that carry impulses from the atrioventricular node to the walls of the ventricles; the only electrical bridge between the atria and the ventricles.

C₃ pathway: *See* Calvin cycle.

C_3 pathway: *See* Calvin cycle.

C_4 pathway: The set of reactions by which some plants initially fix carbon in the four-carbon compound oxaloacetic acid; the carbon dioxide is later released in the interior of the leaf and enters the Calvin cycle. Also known as the Hatch-Slack pathway.

callus [L. *callos*, hard skin]: In plants, undifferentiated tissue; a term used in tissue culture, grafting, and wound healing.

calorie [L. *calor*, heat]: The amount of energy in the form of heat required to raise the temperature of 1 gram of water 1°C; in making metabolic measurements the kilocalorie (Calorie) is generally used. A Calorie is the amount of heat required to raise the temperature of 1 kilogram of water 1°C.

Calvin cycle: The set of reactions in which carbon dioxide is reduced to carbohydrate during the second stage of photosynthesis.

calyx [Gk. *kalyx*, a husk, cup]: Collectively, the sepals of a flower.

CAM photosynthesis: *See* Crassulacean acid metabolism.

capillaries [L. *capillaris*, relating to hair]: Smallest thin-walled blood vessels through which exchanges between blood and the tissues occur; connect arteries with veins.

capillary action: The movement of water or any liquid along a surface; results from the combined effect of cohesion and adhesion.

capsid: The protein coat surrounding the nucleic acid core of a virus.

capsule (kap-sul) [L. *capsula*, a little chest]: (1) A slimy layer around the cells of certain bacteria. (2) The sporangium of a bryophyte.

carbohydrate [L. *carbo*, charcoal + *hydro*, water]: An organic compound consisting of a chain or ring of carbon atoms to which hydrogen and oxygen are attached in a ratio of approximately 2:1; carbohydrates include sugars, starch, glycogen, cellulose, etc.

carbon cycle: Worldwide circulation and reutilization of carbon atoms, chiefly due to metabolic processes of living organisms. Inorganic carbon, in the form of carbon dioxide, is incorporated into organic compounds by photosynthetic organisms; when the organic compounds are broken down

in respiration, carbon dioxide is released. Large quantities of carbon are "stored" in the seas and the atmosphere, as well as in fossil fuel deposits.

carbon fixation: The second stage of photosynthesis; energy stored in ATP and NADPH by the energy-capturing reactions of the first stage is used to reduce carbon from carbon dioxide to simple sugars.

cardiovascular system [Gk. *kardio*, heart + L. *vasculum*, a small vessel]: In animals, the heart and blood vessels.

carnivore [L. *caro, carnis*, flesh + *voro*, to devour]: Predator that obtains its nutrients and energy by eating meat.

carotenoids [L. *carota*, carrot]: A class of pigments that includes the carotenes (yellows, oranges, and reds) and the xanthophylls (yellow); accessory pigments in photosynthesis.

carpel [Gk. *karpos*, fruit]: A leaflike floral structure enclosing the ovule or ovules of angiosperms, typically divided into ovary, style, and stigma; a flower may have one or more carpels, either single or fused. A single carpel or a group of fused carpels is also known as a pistil.

carrying capacity: In ecology, the average number of individuals of a particular population that the environment can support under a particular set of conditions.

cartilage [L. *cartilago*, gristle]: A connective tissue in skeletons of vertebrates; forms much of the skeleton of adult lower vertebrates and immature higher vertebrates.

Casparian strip (after Robert Caspary, German botanist): In the roots of plants, a thickened, waxy strip that extends around and seals the walls of endodermal cells, restricting the diffusion of solutes across the endodermis into the vascular tissues of the root.

catabolism [Gk. *katabole*, throwing down]: Within a cell or organism, the sum of all chemical reactions in which large molecules are broken down into smaller parts.

catalyst [Gk. *katalysis*, dissolution]: A substance that lowers the activation energy of a chemical reaction by forming a temporary association with the reacting molecules; as a result, the rate of the reaction is accelerated. Enzymes are catalysts.

category [Gk. *kategoria*, category]: In a hierarchical classification system, the level at which a particular group is ranked.

cell [L. *cella*, a chamber]: The structural unit of organisms, surrounded by a membrane and composed of cytoplasm and, in eukaryotes, one or more nuclei. In most plants, fungi, and bacteria, there is a cell wall outside the membrane.

cell cycle: A regular, timed sequence of the events of cell growth and division through which dividing cells pass.

cell membrane: The outer membrane of the cell; also called the plasma membrane.

cell plate: In the dividing cells of most plants (and in some algae), a flattened structure that forms at the equator of the mitotic spindle in early telophase; gives rise to the middle lamella.

cell theory: All living things are composed of cells; cells arise only from other cells. No exception has been found to these two principles since they were first proposed well over a century ago.

cellulose [L. *cellula*, a little cell]: The chief constituent of the cell wall in all plants and some protists; an insoluble complex carbohydrate formed of microfibrils of glucose molecules.

cell wall: A plastic or rigid structure, produced by the cell and located outside the cell membrane in most plants, algae, fungi, and prokaryotes; in plant cells, it consists mostly of cellulose.

central nervous system: In vertebrates, the brain and spinal cord; in invertebrates it usually consists of one or more cords of nervous tissue plus their associated ganglia.

centriole (sen-tree-ole) [Gk. *kentron*, center]: A cytoplasmic organelle identical in structure to a basal body; flagellated cells and all animal cells, including those without flagella, have centrioles at the spindle poles during division.

centromere (sen-tro-mere) [Gk. *kentron*, center + *meros*, a part]: Region of constriction of chromosome that holds sister chromatids together.

cerebellum [L. dim. of *cerebrum*, brain]: A subdivision of the vertebrate brain that lies above the brainstem and behind and below the cerebrum; functions in coordinating muscular activities and maintaining equilibrium.

cerebral cortex [L. *cerebrum*, brain]: A thin layer of neurons and glial cells forming the upper surface of the cerebrum, well developed only in mammals; the seat of conscious sensations and voluntary muscular activity.

cerebrum [L., brain]: The portion of the vertebrate brain occupying the upper part of the skull, consisting of two cerebral hemispheres united by the corpus callosum; coordinates most activities.

character displacement: A phenomenon in which species that live together in the same environment tend to diverge in those characteristics that overlap; exemplified by Darwin's finches.

chelicera, *pl.* **chelicerae** [Gk. *cheilos*, the edge, lips + *cheir*, arm]: First pair of appendages in horseshoe crabs, sea spiders, and arachnids; usually take the form of pincers or fangs.

chemical reaction: An interaction among atoms, ions, or molecules that results in the formation of new combinations of atoms, ions, or molecules; the making or breaking of chemical bonds.

chemiosmotic coupling: The mechanism by which ADP is phosphorylated to ATP in mitochondria and chloroplasts. The energy released as electrons pass down an electron transport chain is used to establish a proton gradient across an inner membrane of the organelle; when protons subsequently flow down this electrochemical gradient, the potential energy released is captured in the terminal phosphate bonds of ATP.

chemoreceptor: A sensory cell or organ that responds to the presence of a specific chemical stimulus; includes smell and taste receptors.

chemosynthetic: Applied to autotrophic bacteria that use the energy released by specific inorganic reactions to power their life processes, including the synthesis of organic molecules.

chemotactic [Gk. *cheimō*, storm + *taxis*, arrangement, order]: Of an organism, capable of responding to a chemical stimulus by moving toward or away from it.

chiasma, *pl.* **chiasmata (kye-az-ma)** [Gk., a cross]: Connection between paired homologous chromosomes at meiosis; the site at which crossing over and genetic recombination occur.

chitin (kye-tin) [Gk. *chitōn*, a tunic, undergarment]: A tough, resistant, nitrogen-containing polysaccharide present in the exoskeleton of arthropods, the epidermal cuticle or other surface structures of many other invertebrates, and in the cell walls of fungi.

chlorophyll [Gk. *chloros*, green + *phyllon*, leaf]: A class of green pigments that are the receptors of light energy in photosynthesis.

chloroplast [Gk. *chloros*, green + *plastos*, formed]: A membrane-bound, chlorophyll-containing organelle in eukaryotes (algae and plants) that is the site of photosynthesis.

chorion (core-ee-on) [Gk., skin, leather]: The outermost extraembryonic membrane of developing reptiles, birds, and mammals; in placental mammals it contributes to the structure of the placenta.

chromatid (crow-ma-tid) [Gk. *chrōma*, color]: Either of the two strands of a replicated chromosome, which are joined at the centromere.

chromatin [Gk. *chrōma*, color]: The deeply staining complex of DNA and histone proteins of which eukaryotic chromosomes are composed.

chromosome [Gk. *chrōma*, color + *soma*, body]: The structure that carries the genes. Eukaryotic chromosomes are visualized as threads or rods of chromatin, which appear in a contracted form during mitosis and meiosis, and are otherwise enclosed in a nucleus. Prokaryotic chromosomes consist of a closed circle of DNA with which a variety of proteins are associated. Viral chromosomes are linear or circular molecules of DNA or RNA.

chromosome map: A diagram of the linear order of the genes on a chromosome.

cilium, *pl.* **cilia (silly-um)** [L., eyelash]: A short, thin structure embedded in the surface of some eukaryotic cells, usually in large numbers and arranged in rows; has a highly characteristic internal structure of two inner microtu-

bules surrounded by nine pairs of outer microtubules; involved in locomotion and the movement of substances across the cell surface.

circadian rhythms [L. *circa*, about + *dies*, day]: Regular rhythms of growth or activity that occur on an approximately 24-hour cycle.

cladogenesis [Gk. *clados*, branch + *genesis*, origin]: The splitting of an evolutionary lineage into two or more separate lineages; one of the principal patterns of evolutionary change; also known as splitting evolution.

class: A taxonomic grouping of related, similar orders; category above order and below phylum.

cleavage: The successive cell divisions of a fertilized egg of an animal to form a multicellular blastula.

cline [Gk. *klinein*, to lean]: A graded series of changes in some characteristic within a species, correlated with some gradual change in temperature, humidity, or other environmental factor over the geographic range of the species.

cloaca [L., sewer]: The exit chamber from the digestive system in reptiles and birds; also may serve as the exit for the reproductive and urinary systems.

clone [Gk. *klon*, twig]: A line of cells, all of which have arisen from the same single cell by repeated cell divisions; a population of individuals derived by asexual reproduction from a single ancestor.

cnidocyte (ni-do-site) [Gk. *knide*, nettle + *kytos*, vessel]: A stinging cell containing a nematocyst; characteristic of cnidarians.

coadaptive gene complex: A group of genes that collectively produce coordinated phenotypic characteristics; when linked together on one chromosome, known as a supergene.

cochlea [Gk. *kochlias*, snail]: Part of the inner ear of mammals; concerned with hearing.

codominance: In genetics, the phenomenon in which the effects of both alleles at a particular locus are apparent in the phenotype of the heterozygote.

codon (code-on): Basic unit ("letter") of the genetic code; three adjacent nucleotides in a molecule of DNA or mRNA that form the code for a specific amino acid or for polypeptide chain termination.

coelom (see-loam) [Gk. *koilos*, a hollow]: A body cavity formed between layers of mesoderm and in which the digestive tract and other internal organs are suspended.

coenocytic (see-no-sit-ik) [Gk. *koinos*, shared in common + *kytos*, a vessel]: An organism or part of an organism consisting of many nuclei within a common cytoplasm.

coenzyme [L. *co*, together + Gk. *en*, in + *zyme*, leaven]: A nonprotein organic molecule that plays an accessory role in enzyme-catalyzed processes, often by acting as a donor or acceptor of a substance involved in the reaction. NAD$^+$, FAD, and coenzyme A are common coenzymes.

coevolution [L. *co*, together + *e-*, out + *volvere*, to roll]: The simultaneous evolution of adaptations in two or more populations that interact so closely that each is a strong selective force on the other.

cofactor: A nonprotein component that plays an accessory role in enzyme-catalyzed processes; some cofactors are ions, and others are coenzymes.

cohesion [L. *cohaerere*, to stick together]: The attraction or holding together of molecules of the same substance.

cohesion-tension theory: A theory accounting for the upward movement of water in plants. According to this theory, transpiration of a water molecule results in a negative (below 1 atmosphere) pressure in the leaf cells, inducing the entrance from the vascular tissue of another water molecule, which, because of the cohesive property of water, pulls with it a chain of water molecules extending up from the cells of the root tip.

coleoptile (coal-ee-op-tile) [Gk. *koleon*, sheath + *ptilon*, feather]: The sheath enclosing the apical meristem and leaf primordia of a germinating monocot.

collagen [Gk. *kolla*, glue]: A fibrous protein in bones, tendons, and other connective tissues.

collenchyma [Gk. *kolla*, glue]: In plants, a type of supporting cell with an irregularly thickened primary cell wall; alive at maturity.

colony: A group of organisms of the same species living together in close association.

commensalism [L. *com*, together + *mensa*, table]: *See* Symbiosis.

community: All of the populations of organisms inhabiting a common environment and interacting with one another.

companion cell: In angiosperms, a specialized parenchyma cell associated with a sieve-tube member and arising from the same mother cell as the sieve-tube member.

competition: Interaction between members of the same population or of two or more populations using the same resource, often present in limited supply; interference competition involves fighting or other direct interactions, whereas exploitative competition involves the removal or preemption of a resource.

competitive exclusion: The hypothesis that two species with identical ecological requirements cannot coexist stably in the same locality and the species that is more efficient in utilizing the available resources will exclude the other; also known as Gause's principle, after the Russian biologist G. F. Gause.

complementary DNA (cDNA): DNA molecules synthesized by reverse transcriptase from an RNA template.

compound [L. *componere*, to put together]: A chemical substance composed of two or more kinds of atoms in definite ratios.

compound eye: In arthropods, a complex eye composed of many separate elements, each with light-sensitive cells and a lens that can form an image.

condensation [L. *co*, together + *densare*, to make dense]: A type of chemical reaction in which two molecules join to form one larger molecule, simultaneously splitting out a molecule of water. The biosynthetic reactions in which monomers (e.g., monosaccharides, amino acids) are joined to form polymers (e.g., polysaccharides, polypeptides) are condensation reactions.

conditioning: A form of learning in which one stimulus (the conditional stimulus) comes to be associated with and elicit the same response as another stimulus (the unconditional stimulus).

cone: (1) In plants, the reproductive structure of a conifer. (2) In vertebrates, a type of photoreceptor cell in the retina, concerned with the perception of color and with the most acute discrimination of detail.

conjugation [L. *conjugatio*, a joining, connection]: The sexual process in some unicellular organisms by which genetic material is transferred from one cell to another by cell-to-cell contact.

connective tissues: Supporting or packing tissues that lie between groups of nerves, glands, and muscle cells, and beneath epithelial cells, in which the cells are irregularly distributed through a relatively large amount of extracellular material; include bone, cartilage, blood, and lymph.

consumer, in ecological systems: A heterotroph that derives its energy from living or freshly killed organisms or parts thereof. Primary consumers are herbivores; higher-level consumers are carnivores.

continental drift: The gradual movement of the earth's continents that has occurred over hundreds of millions of years.

continuous variation: A gradation of small differences in a particular trait, such as height, within a population; occurs in traits that are controlled by a number of genes.

convergent evolution [L. *convergere*, to turn together; *evolutio*, to unfold]: The independent development of similarities between unrelated groups, such as porpoises and sharks, resulting from adaptation to similar environments.

cork [L. *cortex*, bark]: A secondary tissue that is a major constituent of bark in woody and some herbaceous plants; made up of flattened cells, dead at maturity; restricts gas and water exchange and protects the vascular tissues from injury.

cork cambium [L. *cortex*, bark + *cambium*, exchange]: The lateral meristem that produces cork.

corolla (ko-role-a) [L. dim. of *corona*, wreath, crown]: Petals, collectively; usually the conspicuously colored flower parts.

corpus callosum [L., callous body]: In the vertebrate brain, a tightly packed mass of myelinated nerve fibers connecting the two cerebral hemispheres.

corpus luteum [L., yellowish body]: An ovarian structure that secretes estrogens and progesterone, which maintain the uterus during pregnancy. It develops from the remaining cells of the ruptured follicle following ovulation.

cortex [L., bark]: (1) The outer, as opposed to the inner, part of an organ, as in the adrenal gland. (2) In a stem or root, the primary tissue bounded externally by the epidermis and internally by the central cylinder of vascular tissue.

cosmid: In recombinant DNA technology, a vector constructed of a DNA segment flanked by cohesive regions (COS regions) of the bacteriophage lambda.

cotyledon (cottle-ee-don) [Gk. *kotyledon*, a cup-shaped hollow]: A leaflike structure of the embryo of a seed plant; contains stored food used during germination.

countercurrent exchange: An anatomical device for manipulating gradients so as to maximize uptake (or minimize loss) of O_2, heat, etc.

coupled reactions: In cells, the linking of endergonic (energy-requiring) reactions to exergonic (energy-releasing) reactions that provide enough energy to drive the endergonic reactions forward.

covalent bond [L. *con*, together + *valere*, to be strong]: A chemical bond formed as a result of the sharing of one or more pairs of electrons.

Crassulacean acid metabolism: A process by which some species of plants in hot, dry climates take in carbon dioxide during the night, fixing it in organic acids; the carbon dioxide is released during the day and used immediately in the Calvin cycle.

cristae: The "shelves" formed by the intricate folding of the inner membrane of the mitochondrion.

cross-fertilization: Fusion of gametes formed by different individuals; as opposed to self-fertilization.

crossing over: During meiosis, the exchange of genetic material between paired chromatids of homologous chromosomes.

cuticle (ku-tik-l) [L. *cuticula*, dim. of *cutis*, the skin]: (1) In plants, a layer of waxy substance (cutin) on the outer surface of epidermal cell walls. (2) In animals, the noncellular, outermost layer of many invertebrates.

cyclic AMP: A form of adenosine monophosphate (AMP) in which the atoms of the phosphate group form a ring; functions in chemical communication in slime molds, in positive regulation of operons, and as a "second messenger" for a number of vertebrate hormones and neurotransmitters.

cytochromes [Gk. *kytos*, vessel + *chrôma*, color]: Heme-containing proteins that participate in electron transport chains; involved in cellular respiration and photosynthesis.

cytokinesis [Gk. *kytos*, vessel + *kinesis*, motion]: Division of the cytoplasm of a cell following nuclear division.

cytokinin [Gk. *kytos*, vessel + *kinesis*, motion]: One of a group of chemically related plant hormones that promote cell division, among other effects.

cytoplasm (sight-o-plazm) [Gk. *kytos*, vessel + *plasma*, anything molded]: The living matter within a cell, excluding the genetic material.

cytoskeleton: A network of filamentous protein structures within the cytoplasm that maintains the shape of the cell, anchors its organelles, and is involved in cell motility; includes microtubules, actin filaments, and intermediate filaments.

deciduous [L. *decidere*, to fall off]: Refers to plants that shed their leaves at a certain season.

decomposers: Specialized detritivores, usually bacteria or fungi, that consume such substances as cellulose and nitrogenous waste products. Their metabolic processes release inorganic nutrients, which are then available for reuse by plants and other organisms.

denaturation: The loss of the native configuration of a macromolecule resulting, for instance, from heat treatment, extreme pH changes, chemical treatment, or other denaturing agents. It is usually accompanied by a loss of biological activity.

dendrite [Gk. *dendron*, tree]: A process of a neuron, typically branched, that receives stimuli from other cells.

denitrification: The process by which certain bacteria living in poorly aerated soils break down nitrates, using the oxygen for their own respiration and releasing nitrogen back into the atmosphere.

density-dependent factors: Factors affecting the birth rate or mortality rate of a population, the effects of which vary with the density (number of individuals per unit area or volume) of the population; include resources for which members of the same or different populations compete, predation, and disease.

density-independent factors: Factors affecting the birth rate or mortality rate of a population, the effects of which are independent of the density of the population; often involve weather-related events.

deoxyribonucleic acid (DNA) (dee-ox-y-rye-bo-new-**clay**-ick): The carrier of genetic information in cells, composed of two complementary chains of nucleotides wound in a double helix; capable of self-replication as well as coding for RNA synthesis.

dermis [Gk. *derma*, skin]: The inner layer of the skin, beneath the epidermis.

desmosome [Gk. *desmos*, bond + *soma*, body]: A type of cell-cell junction that provides mechanical strength in animal tissues; consists of a plaque of dense fibrous material between adjacent cells, with clusters of filaments looping in and out from the cytoplasm of the two cells.

detritivores [L. *detritus*, worn down, worn away + *voro*, to devour]: Organisms that live on dead and discarded organic matter; include large scavengers, smaller animals such as earthworms and some insects, as well as decomposers (fungi and bacteria).

deuterostome [Gk. *deuteros*, second + *stoma*, mouth]: An animal in which the anus forms at or near the blastopore in the developing embryo and the mouth forms secondarily elsewhere; echinoderms and chordates are deuterostomes. Deuterostomes are also characterized by radial cleavage during the earliest stages of development and by enterocoelous formation of the coelom.

development: The progressive production of the phenotypic characteristics of a multicellular organism, beginning with the fertilization of an egg.

diaphragm [Gk. *diaphrassein*, to barricade]: In mammals, a sheetlike tissue (tendon and muscle) forming the partition between the abdominal and thoracic cavities; functions in breathing.

dicotyledon (dye-cottle-ee-don) [Gk. *di*, double, two + *kotyledon*, a cup-shaped hollow]: A member of the class of flowering plants having two seed leaves, or cotyledons, among other distinguishing features; often abbreviated as dicot.

diencephalon [Gk. *di*, two + *enkephalos*, brain]: One of the two principal subdivisions of the vertebrate forebrain; the posterior portion of the forebrain, it contains the thalamus and the hypothalamus.

differentiation: The developmental process by which a relatively unspecialized cell or tissue undergoes a progressive (usually irreversible) change to a more specialized cell or tissue.

diffusion [L. *diffundere*, to pour out]: The net movement of suspended or dissolved particles down a concentration gradient as a result of the random spontaneous movements of individual particles; the process tends to distribute the particles uniformly throughout a medium.

digestion [L. *digestio*, separating out, dividing]: The breakdown of complex, usually insoluble foods into molecules that can be absorbed into the body and used by the cells.

dikaryon (dye-**care**-ee-on) [Gk. *di*, two + *karyon*, kernel]: A cell or organism with paired but not fused nuclei derived from different parents; found among the fungi of divisions Ascomycota and Basidiomycota.

dioecious (dye-ee-shus) [Gk. *di*, two + *oikos*, house]: In angiosperms, having the male (staminate) and female (carpellate) flowers on different individuals of the same species.

diploid [Gk. *di*, double, two + *ploion*, vessel]: The condition in which each autosome is represented twice (2*n*); in contrast to haploid (*n*).

disaccharide [Gk. *di*, two + *sakcharon*, sugar]: A carbohydrate molecule composed of two monosaccharide monomers; examples are sucrose, maltose, and lactose.

diurnal [L. *diurnus*, of the day]: Applied to organisms that are active during the daylight hours.

division: A taxonomic grouping of related, similar classes; a high-level category below kingdom and above class. Division is generally used in the classification of prokaryotes, algae, fungi, and plants, whereas an equivalent category, phylum, is used in the classification of protozoa and animals.

DNA: Abbreviation of deoxyribonucleic acid.

dominant allele: An allele whose phenotypic effect is the same in both the heterozygous and homozygous conditions.

dormancy [L. *dormire*, to sleep]: A period during which growth ceases and metabolic activity is greatly reduced; dormancy is broken when certain requirements, for example, of temperature, moisture, or day length, are met.

dorsal [L. *dorsum*, the back]: Pertaining to or situated near the back; opposite of ventral.

dorsal lip: The tissue on the dorsal edge of the blastopore of the vertebrate embryo; the prospective chordamesoderm, it functions as an organizer, inducing undifferentiated cells to follow a specific course of development.

double fertilization: A phenomenon unique to the angiosperms, in which the egg and one sperm nucleus fuse (resulting in a 2*n* fertilized egg, the zygote) and simultaneously the second sperm nucleus fuses with the two polar nuclei (resulting in a 3*n* endosperm nucleus).

duodenum (duo-dee-num) [L. *duodeni*, twelve each—from its length, about 12 fingers' breadth]: The upper portion of the small intestine in vertebrates, where food is digested into molecules that can be absorbed by intestinal cells.

ecological niche: A description of the roles and associations of a particular species in the community of which it is a part; the way in which an organism interacts with all of the biotic and abiotic factors in its environment.

ecological pyramid: A graphic representation of the quantitative relationships of numbers of organisms, biomass, or energy flow between the trophic levels of an ecosystem. Because large amounts of energy and biomass are dissipated at every trophic level, these diagrams nearly always take the form of pyramids.

ecological succession: The gradual process by which the species composition of a community changes.

ecology [Gk. *oikos*, home + *logos*, a discourse]: The study of the interactions of organisms with their physical environment and with each other and of the results of such interactions.

ecosystem [Gk. *oikos*, home + *systema*, that which is put together]: The organisms in a community plus the associated abiotic factors with which they interact.

ecotype [Gk. *oikos*, home + L. *typus*, image]: A locally adapted variant of a species, differing genetically from other ecotypes of the same species.

ectoderm [Gk. *ecto*, outside + *derma*, skin]: One of the three embryonic tissue layers of animals; it gives rise to the outer covering of the body, the sensory receptors, and the nervous system.

ectotherm [Gk. *ecto*, outside + *therme*, heat]: An organism, such as a reptile, that maintains its body temperature by taking in heat from the environment or giving it off to the environment. *See also* Poikilotherm.

effector [L. *ex*, out of + *facere*, to make]: Cell, tissue, or organ (such as muscle or gland) capable of producing a response to a stimulus.

efferent [L. *ex*, out of + *ferre*, to bear]: Carrying away from a center, applied to nerves and blood vessels.

egg: A female gamete, which usually contains abundant cytoplasm and yolk; nonmotile and often larger than a male gamete.

electric potential: The difference in the amount of electric charge between a region of positive charge and a region of negative charge. The establishment of electric potentials across cell and organelle membranes makes possible a number of phenomena, including the chemiosmotic synthesis of ATP, the conduction of nerve impulses, and muscle contraction.

electron: A subatomic particle with a negative electric charge equal in magnitude to the positive charge of the proton but with a much smaller mass; normally found within orbitals surrounding the atom's positively charged nucleus.

electron acceptor: Substance that accepts or receives electrons in an oxidation-reduction reaction, becoming reduced in the process.

electron carrier: A specialized molecule, such as a cytochrome, that can lose and gain electrons reversibly, alternately becoming oxidized and reduced.

electron donor: Substance that donates or gives up electrons in an oxidation-reduction reaction, becoming oxidized in the process.

electron transport: The movement of electrons down a series of electron-carrier molecules that hold electrons at slightly different energy levels; as electrons move down the chain, the energy released is used to form ATP from ADP and phosphate. Electron transport plays an essential role in the final stage of cellular respiration and in the energy-capturing reactions of photosynthesis.

element: A substance composed only of atoms of the same atomic number and that cannot be decomposed by ordinary chemical means.

embryo [Gk. *en*, in + *bryein*, to swell]: The early developmental stage of an organism produced from a fertilized egg; a young organism before it emerges from the seed, egg, or body of its mother. In humans, refers to the first two months of intrauterine life. *See* Fetus.

embryo sac: The female gametophyte of a flowering plant, contained within an ovule; typically consists of seven cells with a total of eight haploid nuclei.

endergonic [Gk. *endon*, within + *ergon*, work]: Energy-requiring, as in a chemical reaction; applied to an "uphill" process.

endocrine gland [Gk. *endon*, within + *krinein*, to separate]: Ductless gland whose secretions (hormones) are released into the extracellular spaces, from which they diffuse into the circulatory system; in vertebrates, includes pituitary, sex glands, adrenal, thyroid, and others.

endocytosis [Gk. *endon*, within + *kytos*, vessel]: A cellular process in which material to be taken into the cell induces the membrane to form a vacuole enclosing the material; the vacuole is released into the cytoplasm. Includes phagocytosis (endocytosis of solid particles), pinocytosis (endocytosis of liquids), and receptor-mediated endocytosis.

endoderm [Gk. *endon*, within + *derma*, skin]: One of the three embryonic tissue layers of animals; it gives rise to the epithelium that lines certain internal structures, such as most of the digestive tract and its outgrowths, most of the respiratory tract, and the urinary bladder, liver, pancreas, and some endocrine glands.

endodermis [Gk. *endon*, within + *derma*, skin]: In plants, a layer of specialized cells, one cell thick, that lies between the cortex and the vascular tissues in young roots. The Casparian strip of the endodermis prevents diffusion of solutes across the root.

endometrium [Gk. *endon*, within + *metrios*, of the womb]: The glandular lining of the uterus in mammals; thickens in response to secretion of estrogens and progesterone; one of its two principal layers is sloughed off in menstruation.

endoplasmic reticulum [Gk. *endon*, within + *plasma*, from cytoplasm; L. *reticulum*, network]: An extensive system of membranes present in most eukaryotic cells, dividing the cytoplasm into compartments and channels; often coated with ribosomes.

endorphin: One of a group of small peptides with morphine-like properties; produced by the vertebrate brain.

endosperm [Gk. *endon*, within + *sperma*, seed]: In plants, a *3n* tissue containing stored food that develops from the union of a sperm nucleus and the two nuclei of the central cell of the female gametophyte; found only in angiosperms.

endothelium [Gk. *endon*, + *thele*, nipple]: A type of epithelial tissue that forms the walls of the capillaries and the inner lining of arteries and veins.

endotherm [Gk. *endon*, within + *therme*, heat]: An organism, such as a bird or a mammal, that maintains its body temperature internally through metabolic processes. *See also* Homeotherm.

enterocoelous [Gk. *enteron*, gut + *koilos*, a hollow]: Formation of the coelom during embryonic development as cavities within mesoderm originating from outpocketings of the primitive gut; characteristic of deuterostomes.

entropy [Gk. *en*, in + *trope*, turning]: A measure of the randomness or disorder of a system.

enzyme [Gk. *en*, in + *zyme*, leaven]: A globular protein molecule that accelerates a specific chemical reaction.

epidermis [Gk. *epi*, on or over + *derma*, skin]: In plants and animals, the outermost layers of cells.

epinephrine: *See* Adrenaline.

episome: A plasmid that has become incorporated into a bacterial chromosome.

epistasis [Gk., a stopping]: Interaction between two nonallelic genes in which one of them interferes with or modifies the phenotypic expression of the other.

epithelial tissue [Gk. *epi*, on or over + *thele*, nipple]: In animals, a type of tissue that covers a body or structure or lines a cavity; epithelial cells form one or more regular layers with little intercellular material.

equilibrium [L. *aequus*, equal + *libra*, balance]: The state of a system in which no further net change is occurring; result of counterbalancing forward and backward processes.

erythrocyte (eh-rith-ro-site) [Gk. *erythros*, red + *kytos*, vessel]: Red blood cell, the carrier of hemoglobin.

estrogens [Gk. *oistros*, frenzy + *genos*, origin, descent]: Female sex hormones, which are the predominant secretions of the ovarian follicle during the preovulatory phase of the menstrual cycle; also produced by the corpus luteum and the placenta.

estrus [Gk. *oistros*, frenzy]: The mating period in female mammals, characterized by ovulation and intensified sexual activity.

ethology [Gk. *ethos*, habit, custom + *logos*, discourse]: The comparative study of patterns of animal behavior, with emphasis on their adaptive significance and evolutionary origin.

ethylene: A simple hydrocarbon ($H_2C = CH_2$) that functions as a plant hormone; plays a role in fruit ripening and leaf abscission.

eukaryote (you-car-ry-oat) [Gk. *eu*, good + *karyon*, nut, kernel]: A cell having a membrane-bound nucleus, membrane-bound organelles, and chromosomes in which DNA is combined with histone proteins; an organism composed of such cells.

eusocial [Gk. *eu*, good + L. *socius*, companion]: Applied to animal societies, such as those of certain insects, in which sterile individuals work on behalf of reproductive individuals.

evolution [L. *e-*, out + *volvere*, to roll]: Changes in the gene pool from one generation to the next as a consequence of processes such as mutation, natural selection, nonrandom mating, and genetic drift.

exergonic [Gk. *ex*, out of + *ergon*, work]: Energy-yielding, as in a chemical reaction; applied to a "downhill" process.

exocrine glands [Gk. *ex*, out of + *krinein*, to separate]: Glands, such as digestive glands and sweat glands, that secrete their products into ducts.

exocytosis [Gk. *ex*, out of + *kytos*, vessel]: A cellular process in which particulate matter or dissolved substances are enclosed in a vacuole and transported to the cell surface; there, the membrane of the vacuole fuses with the cell membrane, expelling the vacuole's contents to the outside.

exon: A segment of DNA that is transcribed into RNA and expressed; dictates the amino acid sequence of part of a polypeptide.

exoskeleton: The outer supporting covering of the body; common in arthropods.

exponential growth: In populations, the increasingly accelerated rate of growth due to the increasing number of individuals being added to the reproductive base. Exponential growth is very seldom approached or sustained in natural populations.

expressivity: In genetics, the degree to which a particular genotype is expressed in the phenotype of individuals with that genotype.

extinct [L. *exstinctus,* to be extinguished]: No longer existing.

extraembryonic membranes: In reptiles, birds, and mammals, membranes formed from embryonic tissues that lie outside the embryo proper, protecting it and aiding metabolism; include amnion, chorion, allantois, and yolk sac.

F₁ (first filial generation): The offspring resulting from the crossing of plants or animals of a parental generation.

F₂ (second filial generation): resulting from crossing members of the F₁ generation among themselves.

facilitated diffusion: The transport of substances across a cell or organelle membrane from a region of higher concentration to a region of lower concentration by protein molecules embedded in the membrane; driven by the concentration gradient.

Fallopian tube: *See* Oviduct.

family: A taxonomic grouping of related, similar genera; the category below order and above genus.

fatty acid: A molecule consisting of a —COOH group and a long hydrocarbon chain; fatty acids are components of fats, oils, phospholipids, glycolipids, and waxes.

feedback systems: Control mechanisms whereby an increase or decrease in the level of a particular factor inhibits or stimulates the production, utilization, or release of that factor; important in the regulation of enzyme and hormone levels, ion concentrations, temperature, and many other factors.

fermentation: The breakdown of organic compounds in the absence of oxygen; yields less energy than aerobic processes.

fertilization: The fusion of two haploid gamete nuclei to form a diploid zygote nucleus.

fetus [L., pregnant]: An unborn or unhatched vertebrate that has passed through the earliest developmental stages; a developing human from about the second month of gestation until birth.

fibril [L. *fibra,* fiber]: Any minute, threadlike structure within a cell.

fibrous protein: Insoluble structural protein in which the polypeptide chain is coiled along one dimension. Fibrous proteins constitute the main structural elements of many animal tissues.

filament [L. *filare,* to spin]: (1) A chain of cells. (2) In plants, the stalk of a stamen.

filtration: The first stage of kidney function; blood plasma is forced, under pressure, out of the glomerular capillaries into Bowman's capsule, through which it enters the renal tubule.

fission: *See* Binary fission.

fitness: The genetic contribution of an individual to succeeding generations relative to the contributions of other individuals in the population.

fixed action pattern: A behavior that appears substantially complete the first time the organism encounters the relevant stimulus; tends to be highly stereotyped, rigid, and predictable.

flagellum, *pl.* **flagella** (fla-jell-um) [L. *flagellum,* whip]: A long, threadlike organelle found in eukaryotes and used in locomotion and feeding; has an internal structure of nine pairs of microtubules encircling two central microtubules.

flower: The reproductive structure of angiosperms; a complete flower includes sepals, petals, stamens (male structures), and carpels (female structures).

food chain: A sequence of organisms related to one another as prey and predator.

food web: A set of interactions among organisms, including producers, consumers (herbivores and carnivores), and detritivores, through which energy and materials move within a community or ecosystem.

fossil [L. *fossilis,* dug up]: The remains of an organism, or direct evidence of its presence (such as tracks). May be an unaltered hard part (tooth or bone), a mold in a rock, petrification (wood or bone), unaltered or partially altered soft parts (a frozen mammoth).

founder effect: Type of genetic drift that occurs as the result of the founding of a population by a small number of individuals.

fovea [L., pit]: A small area in the center of the retina in which cones are concentrated; the area of sharpest vision.

free energy change: The total energy change that results from a chemical reaction or other process (such as evaporation); takes into account changes in both heat and entropy.

frequency-dependent selection: A type of natural selection that decreases the frequency of more common phenotypes in a population and increases the frequency of less common phenotypes.

fruit [L. *fructus,* fruit]: In angiosperms, a matured, ripened ovary or group of ovaries and associated structures; contains the seeds.

function [L. *fungor,* to busy oneself]: Characteristic role or action of a structure or process in the normal metabolism or behavior of an organism.

gametangium, *pl.* **gametangia** [Gk. *gamein,* to marry + L. *tangere,* to touch]: A unicellular or multicellular structure in which gametes are produced.

gamete (**gam**-meet) [Gk., wife]: A haploid reproductive cell whose nucleus fuses with that of another gamete of an opposite mating type or sex (fertilization); the resulting cell (zygote) may develop into a new diploid individual or, in some protists and fungi, may undergo meiosis to form haploid somatic cells.

gametophyte: In organisms that have alternation of haploid and diploid generations (all plants and some green algae), the haploid *(n)* gamete-producing generation.

ganglion, *pl.* **ganglia** (**gang**-lee-on) [Gk. *ganglion,* a swelling]: Aggregation of nerve cell bodies; in vertebrates, refers to an aggregation of nerve cell bodies located outside the central nervous system.

gap junction: A junction between adjacent animal cells that allows the passage of materials between the cells.

gastric [Gk. *gaster,* stomach]: Pertaining to the stomach.

gastrovascular cavity [Gk. *gaster,* stomach + L. *vasculum,* a small vessel]: A digestive cavity with only one opening, characteristic of the phyla Cnidaria (jellyfish, hydra, corals, etc.) and Ctenophora (comb jellies, sea walnuts); water circulating through the cavity supplies dissolved oxygen and carries away carbon dioxide and other waste products.

gastrula [Gk. *gaster,* stomach]: An animal embryo in the process of gastrulation; the stage of development during which the blastula, with its single layer of cells, turns into a three-layered embryo, made up of ectoderm, mesoderm, and endoderm, often enclosing an archenteron.

Gause's principle: *See* Competitive exclusion.

gene [Gk. *genos,* birth, race; L. *genus,* birth, race, origin]: A unit of heredity in the chromosome; a sequence of nucleotides in a DNA molecule that performs a specific function, such as coding for an RNA molecule or a polypeptide.

gene flow: The movement of alleles into or out of a population.

gene pool: All the alleles of all the genes of all the individuals in a population.

genetic code: The system of nucleotide triplets in DNA and RNA that carries genetic information; referred to as a code because it determines the amino acid sequence in the enzymes and other protein molecules synthesized by the organism.

genetic drift: Evolution (change in allele frequencies) owing to chance processes.

genetic isolation: The absence of genetic exchange between populations or species as a result of geographic separation or of premating or postmating mechanisms (behavioral, anatomical, or physiological) that prevent reproduction.

genome: The complete set of chromosomes, with their associated genes.

genomic DNA (gDNA): DNA fragments produced by the action of restriction enzymes on the DNA of a cell or organism.

genotype (jean-o-type): The genetic constitution of an individual cell or organism with reference to a single trait or a set of traits; the sum total of all the genes present in an individual.

genus, *pl.* **genera** (jean-us) [L. *genus*, race, origin]: A taxonomic grouping of closely related species.

geologic eras: See Table 24–1, pages 496–497.

germ cells [L. *germinare*, to bud]: Gametes or the cells that give rise directly to gametes

germination [L. *germinare*, to bud]: In plants, the resumption of growth or the development from seed or spore.

germ layer: A layer of distinctive cells in an embryo; an embryonic tissue layer. The majority of multicellular animals have three germ layers: ectoderm, mesoderm, and endoderm.

gibberellins (jibb-e-rell-ins) [Fr. *gibberella*, genus of fungi]: A group of chemically related plant growth hormones, whose most characteristic effect is stem elongation in dwarf plants and bolting.

gill: The respiratory organ of aquatic animals, usually a thin-walled projection from some part of the external body surface or, in vertebrates, from some part of the digestive tract.

gland [L. *glans, glandis*, acorn]: A structure composed of modified epithelial cells specialized to produce one or more secretions that are discharged to the outside of the gland.

globular protein [L. dim. of *globus*, a ball]: A polypeptide chain folded into a roughly spherical shape.

glomerulus (glom-mare-u-lus) [L. *glomus*, ball]: In the vertebrate kidney, a cluster of capillaries enclosed by Bowman's capsule; blood plasma minus large molecules filters through the walls of the glomerular capillaries into the renal tubule.

glucagon [Gk. *glukus*, sweet + *agō*, to lead toward]: Hormone produced in the pancreas that acts to raise the concentration of blood sugar.

glucose [Gk. *glukus*, sweet]: A six-carbon sugar ($C_6H_{12}O_6$); the most common monosaccharide in animals.

glycogen [Gk. *glukus*, sweet + *genos*, race or descent]: A complex carbohydrate (polysaccharide); one of the main stored food substances of most animals and fungi; it is converted into glucose by hydrolysis.

glycolipids [Gk. *glukus*, sweet + *lipos*, fat]: Organic molecules similar in structure to fats, but in which a short carbohydrate chain rather than a fatty acid is attached to the third carbon of the glycerol molecule; as a result, the molecule has a hydrophilic "head" and a hydrophobic "tail." Glycolipids are important constituents of cell and organelle membranes.

glycolysis (gly-coll-y-sis) [Gk. *glukus*, sweet + *lysis*, loosening]: The process by which a glucose molecule is changed anaerobically to two molecules of pyruvic acid with the liberation of a small amount of useful energy; catalyzed by cytoplasmic enzymes.

Golgi complex (goal-jee): An organelle present in many eukaryotic cells; consists of flat, membrane-bound sacs, tubules, and vesicles. It functions as a processing, packaging, and distribution center for substances that the cell manufactures.

gonad [Gk. *gone*, seed]: Gamete-producing organ of multicellular animals; ovary or testis.

granulocyte [L., *grānum*, grain or seed + Gk. *kytos*, vessel]: A type of phagocytic white blood cell involved in the inflammatory response; characterized by numerous lysosomes that give the cell a granular appearance under the light microscope. Granulocytes are classified on the basis of their staining properties as neutrophils, eosinophils, or basophils.

granum, *pl.* **grana** [L., grain or seed]: In chloroplasts, stacked membrane-bound disks (thylakoids) that contain chlorophylls and carotenoids and are the sites of the light-trapping reactions of photosynthesis.

gravitropism [L. *gravis*, heavy + Gk. *trope*, turning]: The direction of growth or movement in which the force of gravity is the determining factor; also called geotropism.

gross productivity: A measure of the rate at which energy is assimilated by the organisms in a trophic level, a community, or an ecosystem.

ground tissues: In leaves and young roots and stems, all tissues other than the epidermis and the vascular tissues.

guard cells: Specialized epidermal cells surrounding a pore, or stoma, in a leaf or green stem; changes in turgor of a pair of guard cells cause opening and closing of the pore.

gymnosperm [Gk. *gymnos*, naked + *sperma*, seed]: A seed plant in which the seeds are not enclosed in an ovary; the conifers are the most familiar group.

habitat [L. *habitare*, to live in]: The place in which individuals of a particular species can usually be found.

habituation [L. *habitus*, condition]: A response to a repeated stimulus in which the stimulus comes to be ignored and a previous behavior pattern is restored; one of the simplest forms of learning.

half-life: The average time required for the disappearance or decay of one-half of any amount of a given substance.

haploid [Gk. *haploos*, single + *ploion*, vessel]: Having only one set of chromosomes *(n)*, in contrast to diploid *(2n)*; characteristic of eukaryotic gametes, of gametophytes in plants, and of some protists and fungi.

Hardy-Weinberg equilibrium: The steady-state relationship between relative frequencies of two or more alleles in an idealized population; both the allele frequencies and the genotype frequencies will remain constant from generation to generation in a population breeding at random in the absence of evolutionary forces.

haustorium, *pl.* **haustoria** [L. *haustus*, from *haurire*, to drink, draw]: A projection from a parasitic oomycete, fungus, or plant that functions as a penetrating and absorbing organ.

heat of vaporization: The amount of heat required to change a given amount of a liquid into a gas; 540 calories are required to change 1 gram of liquid water into vapor.

heme [Gk. *haima*, blood]: The iron-containing group of heme proteins such as hemoglobin and the cytochromes.

hemocoel [Gk. *haima*, blood + *koilos*, a hollow]: A blood-filled space within the tissues; characteristic of animals with an incomplete circulatory system, such as mollusks and arthropods.

hemoglobin [Gk. *haima*, blood + L. *globus*, a ball]: The iron-containing protein in vertebrate blood that carries oxygen.

hemophilia [Gk. *haima*, blood + *philios*, friendly]: A group of hereditary diseases characterized by failure of the blood to clot and consequent excessive bleeding from even minor wounds.

hepatic [Gk. *hēpatikos*, liver]: Pertaining to the liver.

herbaceous (her-bay-shus) [L. *herba*, grass]: In plants, nonwoody.

herbivore [L. *herba*, grass + *vorare*, to devour]: A consumer that eats plants or other photosynthetic organisms to obtain its food and energy.

heredity [L. *herres, heredis*, heir]: The transmission of characteristics from parent to offspring.

hermaphrodite [Gk. *Hermes* and *Aphrodite*]: An organism possessing both male and female reproductive organs; hermaphrodites may or may not be self-fertilizing.

heterosis [Gk. *heteros*, other, different]: Hybrid vigor; the overall superiority of the hybrid over either parent.

heterotroph [Gk. *heteros*, other, different + *trophos*, feeder]: An organism that must feed on organic materials formed by other organisms in order to obtain energy and small building-block molecules; in contrast to autotroph. Animals, fungi, and many unicellular organisms are heterotrophs.

heterozygote [Gk. *heteros*, other + *zugōtos*, a pair]: A diploid organism that carries two different alleles at one or more genetic loci.

heterozygote superiority: The greater fitness of an organism heterozygous at a given genetic locus as compared with either homozygote.

hibernation [L. *hiberna*, winter]: A period of dormancy and inactivity, varying in length, depending on the species, and occurring in dry or cold seasons. During hibernation, metabolic processes are greatly slowed and, even in mammals, body temperature may drop to just above freezing.

histones: A group of five relatively small, basic polypeptide molecules found bound to the DNA of eukaryotic cells.

homeostasis (home-e-o-stay-sis) [Gk. *homos*, same or similar + *stasis*, standing]: Maintenance of a relatively stable internal physiological environment or internal equilibrium in an organism.

homeotherm [Gk. *homos*, same or similar + *therme*, heat]: An organism, such as a bird or mammal, capable of maintaining a stable body temperature independent of the environment.

hominid [L. *homo*, man]: Humans and closely related primates; includes modern and fossil forms, such as the australopithecines, but not the apes.

hominoid [L. *homo*, man]: Hominids and the apes.

homologues [Gk. *homologia*, agreement]: Chromosomes that carry corresponding genes and associate in pairs in the first stage of meiosis; each member of the pair is derived from a different parent.

homology [Gk. *homologia*, agreement]: Similarity in structure and/or position, assumed to result from a common ancestry, regardless of function, such as the wing of a bird and the foreleg of a mammal.

homozygote [Gk. *homos*, same or similar + *zugōtos*, a pair]: A diploid organism that carries identical alleles at one or more genetic loci.

hormone [Gk. *hormaein*, to excite]: An organic molecule secreted, usually in minute amounts, in one part of an organism that regulates the function of another tissue or organ.

host: (1) An organism on or in which a parasite lives. (2) A recipient of grafted tissue.

hybrid [L. *hybrida*, the offspring of a tame sow and a wild boar]: (1) Offspring of two parents that differ in one or more inheritable characteristics. (2) Offspring of two different varieties or of two different species.

hydrocarbon [L. *hydro*, water + *carbo*, charcoal]: An organic compound consisting of only carbon and hydrogen.

hydrogen bond: A weak molecular bond linking a hydrogen atom that is covalently bonded to another atom (usually oxygen, nitrogen, or fluorine) to another oxygen, nitrogen, or fluorine atom of the same or another molecule.

hydrolysis [L. *hydro*, water + Gk. *lysis*, loosening]: Splitting of one molecule into two by addition of H^+ and OH^- ions from water.

hydrophilic [L. *hydro*, water + Gk. *philios*, friendly]: Having an affinity for water; applied to polar molecules or polar regions of large molecules.

hydrophobic [L. *hydro*, water + Gk. *phobos*, fearing]: Having no affinity for water; applied to nonpolar molecules or nonpolar regions of molecules.

hypertonic [Gk. *hyper*, above + *tonos*, tension]: Of two solutions of different concentration, the solution that contains the higher concentration of solute particles; water moves across a selectively permeable membrane into a hypertonic solution.

hypha [Gk. *hyphe*, web]: A single tubular filament of a fungus or an oomycete; the hyphae together make up the mycelium, the matlike "body" of a fungus.

hypothalamus [Gk. *hypo*, under + *thalamos*, inner room]: The region of the vertebrate brain just below the thalamus; responsible for the integration of many basic behavioral patterns that involve correlation of neural and endocrine functions.

hypothesis [Gk. *hypo*, under + *tithenai*, to put]: A temporary working explanation or supposition based on accumulated facts and suggesting some general principle or relation of cause and effect; a postulated solution to a scientific problem that must be tested and if not validated, discarded.

hypotonic [Gk. *hypo*, under + *tonos*, tension]: Of two solutions of different concentration, the solution that contains the lower concentration of solute particles; water moves across a selectively permeable membrane from a hypotonic solution.

imbibition [L. *imbibere*, to drink in]: The capillary movement of water into germinating seeds and into substances such as wood and gelatin, which swell as a result.

immune response: A highly specific defensive reaction of the body to invasion by a foreign substance or organism; consists of a primary response in which the invader is recognized as foreign, or "not-self," and eliminated and a secondary response to subsequent attacks by the same invader. Mediated by two types of lymphocytes: B lymphocytes, which mature in the bone marrow and are responsible for antibody production, and T lymphocytes, which mature in the thymus and are responsible for cell-mediated immunity.

immunoglobulins: Complex, highly specific globular proteins synthesized by B lymphocytes; include both circulating antibodies and antibodies displayed on the surface of B lymphocytes prior to activation.

imprinting: A rapid and extremely narrow form of learning, common in birds and important in species recognition, that occurs during a very short critical period in the early life of an animal; depends on exposure to particular characteristics of the parent or parents.

inbreeding: The mating of individuals that are closely related genetically.

inclusive fitness: The relative number of an individual's alleles that are passed on from generation to generation, either as a result of his or her own reproductive success, or that of related individuals.

incomplete dominance: In genetics, the phenomenon in which the effects of both alleles at a particular locus are apparent in the phenotype of the heterozygote.

independent assortment: *See* Mendel's second law.

induction [L. *inducere*, to induce]: (1) In genetics, the phenomenon in which the presence of a substrate initiates transcription and translation of the genes coding for the enzymes required for its metabolism. (2) In embryonic development, the process in which one tissue or body part causes the differentiation of another tissue or body part.

inflammatory response: A nonspecific defensive reaction of the body to invasion by a foreign substance or organism; involves phagocytosis by white blood cells and is often accompanied by accumulation of pus and an increase in the local temperature.

innate releasing mechanism: In ethology, an area within an animal's brain that is hypothesized to respond to a specific stimulus, setting in motion, or "releasing," the sequence of movements that constitute a fixed action pattern.

insertion sequences: Relatively short sequences of DNA that can produce copies of themselves that become incorporated at other sites in the same chromosome or in other chromosomes; also known as simple transposons.

insulin: A peptide hormone, produced by the vertebrate pancreas, that acts to lower the concentration of glucose in the blood.

interferon: A protein made by virus-infected cells that inhibits viral multiplication.

intermediate filaments: Fibrous protein filaments that form part of the cytoskeleton; found in greatest density in cells subject to mechanical stress.

interneuron: Neuron that transmits signals from one neuron to another within a local region of the central nervous system; may receive signals from and transmit signals to many different neurons.

interphase: The portion of the cell cycle that occurs before mitosis or meiosis can take place; includes the G_1, S, and G_2 phases.

intron: A segment of DNA that is transcribed into RNA but is removed enzymatically from the RNA molecule before the mRNA enters the cytoplasm and is translated; also known as an intervening sequence.

invagination [L. *in*, in + *vagina*, sheath]: The local infolding of a layer of tissue, especially in animal embryos, so as to form a depression or pocket opening to the outside.

inversion: A chromosomal aberration in which a double break occurs and a segment is turned 180° before it is reincorporated into the chromosome.

ion (eye-on): Any atom or small molecule containing an unequal number of electrons and protons and therefore carrying a net positive or net negative charge.

ionic bond: A chemical bond formed as a result of the mutual attraction of ions of opposite charge.

isogamy [Gk. *isos,* equal + *gamos,* marriage]: Sexual reproduction in which both gametes are motile and are structurally alike.

isolating mechanisms: Mechanisms that prevent genetic exchange between individuals of different populations or species; they prevent mating or successful reproduction even when mating occurs; may be behavioral, anatomical, or physiological.

isotonic [Gk. *isos,* equal + *tonos,* tension]: Having the same concentration of solutes as another solution. If two isotonic solutions are separated by a selectively permeable membrane, there will be no net flow of water across the membrane.

isotope [Gk. *isos,* equal + *topos,* place]: Atom of an element that differs from other atoms of the same element in the number of neutrons in the atomic nucleus; isotopes thus differ in atomic weight. Some isotopes are unstable and emit radiation.

karyotype [Gk. *kara,* the head + *typos,* stamp or print]: The general appearance of the chromosomes of an organism with regard to number, size, and shape.

keratin [Gk. *karas,* horn]: One of a group of tough, fibrous proteins formed by certain epidermal tissues and especially abundant in skin, claws, hair, feathers, and hooves.

kidney: In vertebrates, the organ that regulates the balance of water and solutes in the blood and the excretion of nitrogenous wastes in the form of urine.

kinetic energy [Gk. *kinetikos,* putting in motion]: Energy of motion.

kinetochore [Gk. *kinetikos,* putting in motion + *choros,* chorus]: Disk-shaped protein structure within the centromere to which spindle fibers are attached during mitosis or meiosis.

kingdom: A taxonomic grouping of related, similar phyla or divisions; the highest-level category in biological classification.

kin selection: The differential reproduction of lineages of related individuals—that is, different groups of related individuals of a species reproduce at different rates; leads to an increase in the frequency of alleles shared by members of the groups with the greatest reproductive success.

Krebs cycle: Stage of cellular respiration in which acetyl groups are broken down into carbon dioxide; molecules reduced in the process can be used in ATP formation.

lagging strand: In DNA replication, the 3′ to 5′ strand of the DNA double helix, synthesized as a series of Okazaki fragments in the 5′ to 3′ direction; these segments are subsequently linked to one another in condensation reactions catalyzed by the enzyme DNA ligase.

lamella (lah-**mell**-ah) [L. dim. of *lamina,* plate or leaf]: Layer, thin sheet.

larva [L., ghost]: An immature animal that is anatomically very different from the adult; examples are caterpillars and tadpoles.

lateral meristem [L. *latus, lateris,* side + Gk. *meristos,* divided]: In vascular plants, one of the two rings of tissue (vascular cambium and cork cambium) that produce new cells for secondary growth.

leaching: The dissolving of minerals and other elements in soil or rocks by the downward movement of water.

leading strand: In DNA replication, the 5′ to 3′ strand of the DNA double helix, which is synthesized continuously.

learning: The process that leads to modification in individual behavior as the result of experience.

leucoplast [Gk. *leukos,* white + *plastes,* molder]: In plant cells, a colorless organelle that serves as a starch repository; usually found in cells not exposed to light, such as those in roots and the internal tissues of stems.

leukocyte [Gk. *leukos,* white + *kytos,* vessel]: White blood cell; principal types include granulocytes, monocytes and macrophages, and lymphocytes.

lichen: Organism composed of a fungus and a green alga or a cyanobacterium that are symbiotically associated.

life cycle: The entire span of existence of any organism from time of zygote formation (or asexual reproduction) until it itself reproduces.

limbic system [L. *limbus,* border]: Neuron network forming a loop around the inside of the brain and connecting the hypothalamus to the cerebral cortex; thought to be circuit by which drives and emotions are translated into complex actions and to play a role in the consolidation of memory.

linkage: The tendency for certain alleles to be inherited together because they are located on the same chromosome.

linkage group: A pair of homologous chromosomes.

lipid [Gk. *lipos,* fat]: One of a large variety of organic substances that are insoluble in polar solvents, such as water, but that dissolve readily in nonpolar organic solvents; includes fats, oils, waxes, steroids, phospholipids, glycolipids, and carotenes.

locus, *pl.* **loci** [L., place]: In genetics, the position of a gene in a chromosome. For any given locus, there may be a number of possible alleles.

logistic growth: A pattern of population growth in which growth is rapid when the population is small, gradually slows as the population approaches the carrying capacity of its environment, and then oscillates as the population stabilizes at or near its maximum size; it is one of the simplest growth patterns observed for populations in nature.

loop of Henle (after F. G. J. Henle, German pathologist): A hairpin-shaped portion of the renal tubule of mammals in which a hypertonic urine is formed by processes of diffusion and active transport.

lumen [L., light]: The cavity of a tubular structure, such as endoplasmic reticulum or a blood vessel.

lymph [L. *lympha,* water]: Colorless fluid derived from blood by filtration through capillary walls in the tissues; carried in special lymph ducts.

lymphatic system: The system through which lymph circulates; consists of lymph capillaries, which begin blindly in the tissues, and a network of progressively larger vessels that empty into the vena cava; also includes the lymph nodes, spleen, thymus, and tonsils.

lymph node [L. *lympha,* water + *nodus,* knot]: A mass of spongy tissues, separated into compartments; located throughout the lymphatic system, lymph nodes remove dead cells, debris, and foreign particles from the circulation; also are sites at which foreign antigens are displayed to immunologically active cells.

lymphocyte [L. *lympha,* water + Gk. *kytos,* vessel]: A type of white blood cell involved in the immune response; B lymphocytes differentiate into antibody-producing plasma cells, whereas cytotoxic T lymphocytes lyse diseased eukaryotic cells; other T lymphocytes interact with both cytotoxic T lymphocytes and with B lymphocytes.

lysis [Gk., a loosening]: Disintegration of a cell by rupture of its cell membrane.

lysogenic bacteria (lye-so-**jenn**-ick) [Gk. *lysis,* a loosening + *genos,* race or descent]: Bacteria carrying a bacteriophage integrated into the bacterial chromosome. The virus may subsequently set up an active cycle of infection, causing lysis of the bacterial cells.

lysosome [Gk. *lysis,* loosening + *soma,* body]: A membrane-bound organelle in which hydrolytic enzymes are segregated.

macromolecule [Gk. *makros,* large + L. dim. of *moles,* mass]: An extremely large molecule; refers specifically to proteins, nucleic acids, polysaccharides, and complexes of these.

macrophage [Gk. *makros,* large + *phagein,* to eat]: A type of phagocytic white blood cell important in both the inflammatory and immune responses.

major histocompatibility complex (MHC): In mammals, a group of at least 20 different genes, each with multiple alleles, coding for the protein components of the antigens that are displayed on nucleated cells and that serve to identify "self."

mandibles [L. *mandibula*, jaw]: In crustaceans, insects, and myriapods, the appendages immediately posterior to the antennae; used to seize, hold, bite, or chew food.

mantle: In mollusks, the outermost layer of the body wall or a soft extension of it; usually secretes a shell.

marine [L. *marini(us)*, from *mare*, the sea]: Living in salt water.

marsupial [Gk. *marsypos*, pouch, little bag]: A mammal in which the female has a ventral pouch or folds surrounding the nipples; the premature young leave the uterus and crawl into the pouch, where each one attaches itself by the mouth to a nipple until development is completed.

matrix: The dense solution in the interior of the mitochondrion, surrounding the cristae; contains enzymes, phosphates, coenzymes, and other molecules involved in cellular respiration.

mechanoreceptor: A sensory cell or organ that receives mechanical stimuli such as those involved in touch, pressure, hearing, and balance.

medulla (med-**dull**-a) [L., the innermost part]: (1) The inner, as opposed to the outer, part of an organ, as in the adrenal gland. (2) The most posterior region of the vertebrate brain; connects with the spinal cord.

medusa: The free-swimming, bell- or umbrella-shaped stage in the life cycle of many cnidarians; a jellyfish.

megaspore [Gk. *megas*, great, large + *spora*, a sowing]: In plants, a haploid (*n*) spore that develops into a female gametophyte.

meiosis (my-o-sis) [Gk. *meioun*, to make smaller]: The two successive nuclear divisions in which a single diploid (*2n*) cell forms four haploid (*n*) nuclei, and segregation, crossing over, and reassortment of the alleles occur; gametes or spores may be produced as a result of meiosis.

Mendel's first law: The factors for a pair of alternative characters are separate, and only one may be carried in a particular gamete (genetic segregation). In modern form: Alleles segregate in meiosis.

Mendel's second law: The inheritance of a pair of factors for one trait is independent of the simultaneous inheritance of factors for other traits, such factors "assorting independently" as though there were no other factors present (later modified by the discovery of linkage). Modern form: The alleles of unlinked genes assort independently.

menstrual cycle [L. *mensis*, month]: In humans and certain other primates, the cyclic, hormone-regulated changes in the condition of the uterine lining; marked by the periodic discharge of blood and disintegrated uterine lining through the vagina. Mammals with a menstrual cycle lack a well-defined period of estrus.

meristem [Gk. *merizein*, to divide]: The undifferentiated plant tissue, including a mass of rapidly dividing cells, from which new tissues arise.

mesenteries [Gk. *mesos*, middle + *enteron*, gut]: Double layers of mesoderm that suspend the digestive tract and other internal organs within the coelom.

mesoderm [Gk. *mesos*, middle + *derma*, skin]: In animals, the middle layer of the three embryonic tissue layers. In vertebrates, includes the chordamesoderm, which gives rise to the notochord and skeletal muscle, and the lateral plate mesoderm, which gives rise to the circulatory system, most of the excretory and reproductive systems, the lining of the coelom, and the outer covering of the internal organs.

mesophyll [Gk. *mesos*, middle + *phyllon*, leaf]: The internal tissue of a leaf, sandwiched between two layers of epidermal cells; consists of palisade parenchyma and spongy parenchyma cells.

messenger RNA (mRNA): A class of RNA molecules, each of which is complementary to one strand of DNA and which serves to carry the genetic information from the chromosome to the ribosomes, where it is translated into protein.

metabolism [Gk. *metabole*, change]: The sum of all chemical reactions occurring within a cell or organism.

metamere [Gk. *meta*, middle + *meros*, part]: One of a linear series of similar body segments.

metamorphosis [Gk. *metamorphoun*, to transform]: Abrupt transition from larval to adult form, such as the transition from tadpole to adult frog.

metaphase [Gk. *meta*, middle + *phasis*, form]: The stage of mitosis or meiosis during which the chromosomes lie in the equatorial plane of the spindle.

microbe [Gk. *mikros*, small + *bios*, life]: A microscopic organism.

micronutrient [Gk. *mikros*, small + L. *nutrire*, to nourish]: An inorganic nutrient required in only minute amounts for plant growth, such as iron, chlorine, copper, manganese, zinc, molybdenum, and boron.

microspore [Gk. *mikros*, small + *spora*, a sowing]: In plants, a haploid (*n*) spore that develops into a male gametophyte; in seed plants, it becomes a pollen grain.

microtubule [Gk. *mikros*, small + L. dim. of *tubus*, tube]: An extremely small hollow tube composed of two types of globular protein subunits. Among their many functions, microtubules make up the internal structure of cilia and flagella.

middle lamella: In plants, distinct layer between adjacent cell walls, rich in pectins and other polysaccharides; derived from the cell plate.

mimicry [Gk. *mimos*, mime]: The superficial resemblance in form, color, or behavior of certain organisms (mimics) to other more powerful or more protected ones (models), resulting in protection, concealment, or some other advantage for the mimic.

mineral: A naturally occurring element or inorganic compound.

mitochondrion, pl. **mitochondria** [Gk. *mitos*, thread + *chondros*, cartilage or grain]: An organelle, bound by a double membrane, in which the reactions of the Krebs cycle, terminal electron transport, and oxidative phosphorylation take place, resulting in the formation of CO_2, H_2O, and ATP from acetyl CoA and ADP. Mitochondria are the organelles in which most of the ATP of the eukaryotic cell is produced.

mitosis [Gk. *mitos*, thread]: Nuclear division characterized by chromosome replication and formation of two identical daughter nuclei.

mole [L. *moles*, mass]: The amount of an element equivalent to its atomic weight expressed in grams, or the amount of a substance equivalent to its molecular weight expressed in grams.

molecular weight: The sum of the atomic weights of the constituent atoms in a molecule.

molecule [L. dim. of *moles*, mass]: A particle consisting of two or more atoms held together by chemical bonds; the smallest unit of a compound that displays the properties of the compound.

molting: Shedding of all or part of an organism's outer covering; in arthropods, periodic shedding of the exoskeleton to permit an increase in size.

monocotyledon [Gk. *monos*, single + *kotyledon*, a cup-shaped hollow]: A member of the class of flowering plants having one seed leaf, or cotyledon, among other distinguishing features; often abbreviated as monocot.

monocyte [Gk. *monos*, single + *kytos*, vessel]: A type of circulating white blood cell that, in the presence of infectious organisms or other foreign invaders, becomes transformed into a macrophage.

monoecious (mo-**nee**-shus) [Gk. *monos*, single + *oikos*, house]: In angiosperms, having the male and female structures (the stamens and the carpels, respectively) on the same individual but on different flowers.

monomer [Gk. *monos*, single + *meros*, part]: A simple, relatively small molecule that can be linked to others to form a polymer.

monosaccharide [Gk. *monos*, single + *sakcharon*, sugar]: A simple sugar, such as glucose, fructose, ribose.

monotreme [Gk. *monos*, single + *trēma*, hole]: A nonplacental mammal, such as the duckbilled platypus, in which the female lays shelled eggs and nurses the young.

morphogenesis [Gk. *morphe*, form + *genesis*, origin]: The development of size, form, and other structural features of organisms.

morphological [Gk. *morphe,* form + *logos,* discourse]: Pertaining to form and structure, at any level of organization.

motor neuron: Neuron that conducts nerve impulses from the central nervous system to an effector, which is typically a muscle or a gland.

muscle fiber: Muscle cell; a long, cylindrical, multinucleated cell containing numerous myofibrils, which is capable of contraction when stimulated.

mutagen [L. *mutare,* to change + *genus,* source or origin]: A chemical or physical agent that increases the mutation rate.

mutant [L. *mutare,* to change]: An organism carrying a gene that has undergone a mutation.

mutation [L. *mutare,* to change]: The change of a gene from one allelic form to another; an inheritable change in the DNA sequence of a chromosome.

mutualism [L. *mutuus,* lent, borrowed]: *See* Symbiosis.

mycelium [Gk. *mykes,* fungus]: The mass of hyphae forming the body of a fungus.

mycorrhizae [Gk. *mykes,* fungus + *rhiza,* root]: Symbiotic associations between particular species of fungi and the roots of vascular plants.

myelin sheath [Gk. *myelinos,* full of marrow]: A lipid-rich layer surrounding the long axons of neurons in the vertebrate nervous system; in the peripheral nervous system, made up of the membranes of Schwann cells.

myofibril [Gk. *mys,* muscle + L. *fibra,* fiber]: Contractile element of a muscle fiber, made up of thick and thin filaments arranged in sarcomeres.

myoglobin [Gk. *mys,* muscle + L. *globus,* a ball]: An oxygen-binding, heme-containing globular protein found in muscles.

myosin [Gk. *mys,* muscle]: One of the principal proteins in muscle; makes up the thick filaments.

NAD: Abbreviation of nicotinamide adenine dinucleotide, a coenzyme that functions as an electron acceptor.

natural selection: A process of interaction between organisms and their environment that results in a differential rate of reproduction of different phenotypes in the population; can result in changes in the relative frequencies of alleles and genotypes in the population—that is, in evolution.

nectar [Gk. *nektar,* the drink of the gods]: A sugary fluid that attracts insects to plants.

negative feedback: A control mechanism whereby an increase in some substance inhibits the process leading to the increase; also known as feedback inhibition.

nematocyst [Gk. *nema, nematos,* thread + *kyst,* bladder]: A threadlike stinger, containing a poisonous or paralyzing substance, found in the cnidocyte of cnidarians.

nephridium, pl. **nephridia** [Gk. *nephros,* kidney]: A tubular excretory structure found in many invertebrates.

nephron [Gk. *nephros,* kidney]: The functional unit of the kidney in reptiles, birds, and mammals; a human kidney contains about 1 million nephrons.

nerve: A group or bundle of nerve fibers with accompanying connective tissue, located in the peripheral nervous system. A bundle of nerve fibers within the central nervous system is known as a tract.

nerve fiber: A filamentous process extending from the cell body of a neuron and conducting the nerve impulse; an axon.

nerve impulse: A rapid, transient, self-propagating change in electric potential across the membrane of an axon.

nervous system: All the nerve cells of an animal; the receptor-conductor-effector system; in humans, the nervous system consists of the central nervous system (brain and spinal cord) and the peripheral nervous system.

net productivity: In a trophic level, a community, or an ecosystem, the amount of energy (in calories) stored in chemical compounds or the increase in biomass (in grams or metric tons) in a particular period of time; it is the difference between gross productivity and the energy used by the organisms in respiration.

neural groove: Dorsal, longitudinal groove that forms in a vertebrate embryo; bordered by two neural folds; preceded by the neural-plate stage and followed by the neural-tube stage.

neural plate: Thickened strip of ectoderm in early vertebrate embryos that forms along the dorsal side of the body and gives rise to the central nervous system.

neural tube: Primitive, hollow, dorsal nervous system of the early vertebrate embryo; formed by fusion of neural folds around the neural groove.

neuromodulator: A chemical agent that is released by a neuron and diffuses through a local region of the central nervous system, acting on neurons within that region; generally has the effect of modulating the response to neurotransmitters.

neuron [Gk., nerve]: Nerve cell, including cell body, dendrites, and axon.

neurosecretory cell: A neuron that releases one or more hormones into the circulatory system.

neurotransmitter: A chemical agent that is released by a neuron at a synapse, diffuses across the synaptic cleft, and acts upon a postsynaptic neuron or muscle or gland cell and alters its electrical state or activity.

neutron (new-tron): An uncharged particle with a mass slightly greater than that of a proton. Found in the atomic nucleus of all elements except hydrogen, in which the nucleus consists of a single proton.

niche: *See* Ecological niche.

nitrification: The oxidation of ammonia or ammonium to nitrites and nitrates, as by nitrifying bacteria.

nitrogen cycle: Worldwide circulation and reutilization of nitrogen atoms, chiefly due to metabolic processes of living organisms; plants take up inorganic nitrogen and convert it into organic compounds (chiefly proteins), which are assimilated into the bodies of one or more animals; bacterial and fungal action on nitrogenous waste products and dead organisms return nitrogen atoms to the inorganic state.

nitrogen fixation: Incorporation of atmospheric nitrogen into inorganic nitrogen compounds available to plants, a process that can be carried out only by some soil bacteria, many free-living and symbiotic cyanobacteria, and certain symbiotic bacteria in association with legumes.

nitrogenous base: A nitrogen-containing molecule having basic properties (tendency to acquire an H^+ ion); a purine or pyrimidine.

nocturnal [L. *nocturnus,* of night]: Applied to organisms that are active during the hours of darkness.

node [L. *nodus,* knot]: In plants, a joint of a stem; the place where branches and leaves are joined to the stem.

nondisjunction [L. *non,* not + *disjungere,* to separate]: The failure of chromatids to separate during meiosis, resulting in one or more extra chromosomes in some gametes and correspondingly fewer in others.

noradrenaline: A hormone, produced by the medulla of the adrenal gland, that increases the concentration of sugar in the blood, raises blood pressure and heartbeat rate, and increases muscular power and resistance to fatigue; also one of the principal neurotransmitters; also called norepinephrine.

norepinephrine: *See* noradrenaline.

notochord [Gk. *noto,* back + L. *chorda,* cord]: A dorsal rodlike structure that runs the length of the body and serves as the internal skeleton in the embryos of all chordates; in most adult chordates the notochord is replaced by a vertebral column that forms around (but not from) the notochord.

nuclear envelope [L. *nucleus,* a kernel]: The double membrane surrounding the nucleus within a eukaryotic cell.

nucleic acid: A macromolecule consisting of nucleotides; the principal types are deoxyribonucleic acid (DNA) and ribonucleic acid (RNA).

nucleoid: In prokaryotic cells, the region of the cell in which the chromosome is localized.

nucleolus (new-klee-o-lus) [L., a small kernel]: A small, dense region visible in the nucleus of nondividing eukaryotic cells; consists of rRNA molecules, ribosomal proteins, and loops of chromatin from which the rRNA molecules are transcribed.

nucleosome [L. *nucleus,* a kernel + Gk. *soma,* body]: A complex of DNA and histone proteins that forms the fundamental packaging unit of eukaryotic DNA; its structure resembles a bead on a string.

nucleotide [L. *nucleus,* a kernel]: A molecule composed of a phosphate group, a five-carbon sugar (either ribose or deoxyribose), and a purine or pyrimidine base; nucleotides are the building blocks of nucleic acids.

nucleus [L., a kernel]: (1) The central core of an atom, containing protons and neutrons, around which electrons move. (2) The membrane-bound structure characteristic of eukaryotic cells that contains the genetic information in the form of DNA organized into chromosomes. (3) A group of nerve cell bodies within the central nervous system.

ocellus, *pl.* **ocelli** [L. dim. of *oculus,* eye]: A simple light receptor common among invertebrates.

Okazaki fragments (after R. Okazaki, Japanese geneticist): In DNA replication, the discontinuous segments in which the 3' to 5' strand (the lagging strand) of the DNA double helix is synthesized; typically 1,000 to 2,000 nucleotides long in prokaryotes, and 100 to 200 nucleotides long in eukaryotes.

olfactory [L. *olfacere,* to smell]: Pertaining to smell.

oligosaccharins [Gk. *oligo,* few + *sakcharon,* sugar]: Short carbohydrate chains released from plant cell walls in response to a variety of stimuli, including injury; hypothesized to play a role in regulation of plant growth and development.

ommatidium, *pl.* **ommatidia** [Gk. *ommos,* eye]: The single visual unit in the compound eye of arthropods; contains light-sensitive cells and a lens able to form an image.

omnivore [L. *omnis,* all + *vorare,* to devour]: An organism that "eats everything"; for example, an animal that eats both plants and meat.

oncogene [Gk. *onkos,* tumor + *genos,* birth, race]: One of a group of eukaryotic genes that closely resemble normal genes of the cells in which they are found and that are thought to play a role in the development of cancer; their gene products appear to be regulatory proteins, involved in the control of either cell growth or cell division.

oocyte (o-uh-sight) [Gk. *oion,* egg + *kytos,* vessel]: A cell that gives rise by meiosis to an ovum.

oogamy (oh-**og**-amy) [Gk. *oion,* egg + *gamos,* marriage]: Sexual reproduction in which one of the gametes, usually the larger, is not motile.

operator: A segment of DNA that interacts with a repressor protein to regulate the transcription of the structural genes of an operon.

operon [L. *opus, operis,* work]: In the bacterial chromosome, a segment of DNA consisting of a promoter, an operator, and a group of adjacent structural genes; the structural genes, which code for products related to a particular biochemical pathway, are transcribed onto a single mRNA molecule, and their transcription is regulated by a single repressor protein.

opportunistic species: Species characterized by high reproduction rates, rapid development, early reproduction, small body size, and uncertain adult survival.

orbital [L. *orbis,* circle, disk]: In the current model of atomic structure, the volume of space surrounding the atomic nucleus in which an electron will be found 90 percent of the time.

order: A taxonomic grouping of related, similar families; the category below class and above family.

organ [Gk. *organon,* tool]: A body part composed of several tissues grouped together in a structural and functional unit.

organelle [Gk. *organon,* instrument, tool]: A formed body in the cytoplasm of a cell.

organic [Gk. *organon,* instrument, tool]: Pertaining to (1) organisms or living things generally, or (2) compounds formed by living organisms, or (3) the chemistry of compounds containing carbon.

organism [Gk. *organon,* instrument, tool]: Any living creature, either unicellular or multicellular.

organizer [Gk. *organon,* instrument, tool]: In vertebrates, the part of an embryo capable of inducing undifferentiated cells to follow a specific course of development; in particular, the dorsal lip of the blastopore.

osmosis [Gk. *osmos,* impulse, thrust]: The diffusion of water across a selectively permeable membrane (a membrane that permits the free passage of water but prevents or retards the passage of a solute). In the absence of other factors that affect the water potential, the net movement of water is from the side containing a lower concentration of solute to the side containing a higher concentration.

osmotic potential [Gk. *osmos,* impulse, thrust]: The tendency of water to move across a selectively permeable membrane into a solution; it is determined by measuring the pressure required to stop the osmotic movement of water into the solution; the higher the solute concentration, the greater the osmotic potential of the solution.

ovary [L. *ovum,* egg]: (1) In animals, the egg-producing organ. (2) In flowering plants, the enlarged basal portion of a carpel or a fused carpel, containing the ovule or ovules; the ovary matures to become the fruit.

oviduct [L. *ovum,* egg + *ductus,* duct]: The tube serving to transport the eggs to the outside or to the uterus; also called uterine tube or Fallopian tube (in humans).

ovulation: In animals, release of an egg or eggs from the ovary.

ovule [L. dim. of *ovum,* egg]: In seed plants, a structure composed of a protective outer coat, a tissue specialized for food storage, and a female gametophyte with an egg cell; becomes a seed after fertilization.

ovum, *pl.* **ova** [L., egg]: The egg cell; female gamete.

oxidation: Gain of oxygen, loss of hydrogen, or loss of an electron by an atom, ion, or molecule. Oxidation and reduction take place simultaneously, with the electron lost by one reactant being transferred to another reactant.

oxidative phosphorylation: The process by which the energy released as electrons pass down the mitochondrial electron transport chain in the final stage of cellular respiration is used to phosphorylate (add a phosphate group to) ADP molecules, thereby yielding ATP molecules.

pacemaker: *See* Sinoatrial node.

paleontology [Gk. *palaios,* old + *onta,* things that exist + *logos,* discourse]: The study of the life of past geologic times, principally by means of fossils.

palisade cells [L. *palus,* stake + *cella,* a chamber]: In plant leaves, the columnar, chloroplast-containing parenchyma cells of the mesophyll.

pancreas (**pang**-kree-us) [Gk. *pan,* all + *kreas,* meat, flesh]: In vertebrates, a small, complex gland located between the stomach and the duodenum, which produces digestive enzymes and the hormones insulin and glucagon.

parasite [Gk. *para,* beside, akin to + *sitos,* food]: An organism that lives on or in an organism of a different species and derives nutrients from it.

parasitism: *See* Symbiosis.

parasympathetic division [Gk. *para,* beside, akin to]: A subdivision of the autonomic nervous system of vertebrates, with centers located in the brain and in the most anterior and most posterior parts of the spinal cord; stimulates digestion; generally inhibits other functions and restores the body to normal following emergencies.

parenchyma (pah-**renk**-ee-ma) [Gk. *para,* beside, akin to + *en,* in + *chein,* to pour]: A plant tissue composed of living, thin-walled, randomly arranged cells with large vacuoles; usually photosynthetic or storage tissue.

parthenogenesis [Gk. *parthenon,* virgin + *genesis,* birth]: The development of an organism from an unfertilized egg.

pellicle [L. dim. of *pellis,* skin]: A flexible series of protein strips inside the cell membrane of many protists.

penetrance: In genetics, the proportion of individuals with a particular genotype that show the phenotype ascribed to that genotype.

peptide bond [Gk. *pepto,* to soften, digest]: The type of bond formed when two amino acids are joined end to end; the acidic group (— COOH) of one amino acid is linked covalently to the basic group (—NH_2) of the next, and a molecule of water (H_2O) is removed.

perennial [L. *per*, through + *annus*, year]: A plant that persists in whole or in part from year to year and usually produces reproductive structures in more than one year.

pericycle [Gk. *peri*, around + *kyklos*, circle]: One or more layers of cells completely surrounding the vascular tissues of the root; branch roots arise from the pericycle.

peripheral nervous system [Gk. *peripherein*, to carry around]: All of the neurons and axons outside the central nervous system, including both motor neurons and sensory neurons; consists of the somatic nervous system and the autonomic nervous system.

peristalsis [Gk. *peristellein*, to wrap around]: Successive waves of muscular contraction in the walls of a tubular structure, such as the digestive tract or an oviduct; moves the contents, such as food or an egg cell, through the tube.

peritoneum [Gk. *peritonos*, stretched over]: A membrane that lines the body cavity and forms the external covering of the visceral organs.

permeable [L. *permeare*, to pass through]: Penetrable by molecules, ions, or atoms; usually applied to membranes that let given solutes pass through.

peroxisome: A membrane-bound organelle in which enzymes catalyzing peroxide-forming and peroxide-destroying reactions are segregated; in plant cells, the site of photorespiration.

petiole (pet-ee-ole) [Fr., from L. *petiolus*, dim. of *pes, pedis*, a foot]: The stalk of a leaf, connecting the blade of the leaf with the branch or stem.

pH: A symbol denoting the concentration of hydrogen ions in a solution; pH values range from 0 to 14; the lower the value, the more acidic a solution, that is, the more hydrogen ions it contains; pH 7 is neutral, less than 7 is acidic, more than 7 is alkaline.

phagocytosis [Gk. *phagein*, to eat + *kytos*, vessel]: Cell "eating." *See* Endocytosis.

phenotype [Gk. *phainein*, to show + *typos*, stamp, print]: Observable characteristics of an organism, resulting from interactions between the genotype and the environment.

pheromone (fair-o-moan) [Gk. *phero*, to bear, carry]: Substance secreted by an animal that influences the behavior or development of other animals of the same species, such as the sex attractants of moths, the queen substance of honey bees.

phloem (flow-em) [Gk. *phloos*, bark]: Vascular tissue of higher plants; conducts sugars and other organic molecules from the leaves to other parts of the plant; in angiosperms, composed of sieve-tube members, companion cells, other parenchyma cells, and fibers.

phospholipids: Organic molecules similar in structure to fats, but in which a phosphate group rather than a fatty acid is attached to the third carbon of the glycerol molecule; as a result, the molecule has a hydrophilic "head" and a hydrophobic "tail." Phospholipids form the basic structure of cell and organelle membranes.

phosphorylation: Addition of a phosphate group or groups to a molecule.

photon [Gk. *photos*, light]: The elementary particle of light and other electromagnetic radiations.

photoperiodism [Gk. *photos*, light]: The response to relative day and night length, a mechanism by which organisms measure seasonal change.

photophosphorylation [Gk. *photos*, light + *phosphoros*, bringing light]: The process by which the energy released as electrons pass down the electron transport chain between photosystems II and I during photosynthesis is used to phosphorylate ADP to ATP.

photoreceptor [Gk. *photos*, light]: A cell or organ capable of detecting light.

photorespiration [Gk. *photos*, light + L. *respirare*, to breathe]: The oxidation of carbohydrates in the presence of light and oxygen; occurs when the carbon dioxide concentration in the leaf is low in relation to the oxygen concentration.

photosynthesis [Gk. *photos*, light + *syn*, together + *tithenai*, to place]: The conversion of light energy to chemical energy; the synthesis of organic compounds from carbon dioxide and water in the presence of chlorophyll, using light energy.

phototropism [Gk. *photos*, light + *trope*, turning]: Movement in which the direction of the light is the determining factor, such as the growth of a plant toward a light source; a curving response to light.

phyletic change [Gk. *phylon*, race, tribe]: The changes taking place in a single lineage of organisms over a long period of time; one of the principal patterns of evolutionary change.

phylogeny [Gk. *phylon*, race, tribe]: Evolutionary history of a taxonomic group. Phylogenies are often depicted as "evolutionary trees."

phylum, pl. phyla [Gk. *phylon*, race, tribe]: A taxonomic grouping of related, similar classes; a high-level category below kingdom and above class. Phylum is generally used in the classification of protozoa and animals, whereas an equivalent category, division, is used in the classification of prokaryotes, algae, fungi, and plants.

physiology [Gk. *physis*, nature + *logos*, a discourse]: The study of function in cells, organs, or entire organisms; the processes of life.

phytochrome [Gk. *phyton*, plant + *chrōma*, color]: A plant pigment that is a photoreceptor for red or far-red light and is involved with a number of developmental processes, such as flowering, dormancy, leaf formation, and seed germination.

phytoplankton [Gk. *phyton*, plant + *planktos*, wandering]: Aquatic, free-floating, microscopic, photosynthetic organisms.

pigment [L. *pigmentum*, paint]: A colored substance that absorbs light over a narrow band of wavelengths.

pinocytosis [Gk. *pinein*, to drink + *kytos*, vessel]: Cell "drinking." *See* Endocytosis.

pituitary [L. *pituita*, phlegm]: Endocrine gland in vertebrates; the anterior lobe is the source of tropic hormones, growth hormone, and prolactin and is regulated by secretions of the hypothalamus; the posterior lobe stores and releases oxytocin and ADH produced by the hypothalamus.

placenta [Gk. *plax*, a flat object]: A tissue formed as the result of interactions between the inner lining of the mammalian uterus and the extraembryonic chorion; serves as the connection through which exchanges of nutrients and wastes occur between the blood of the mother and that of the embryo.

plankton [Gk. *planktos*, wandering]: Small (mostly microscopic) aquatic and marine organisms found in the upper levels of the water, where light is abundant; includes both photosynthetic (phytoplankton) and heterotrophic (zooplankton) forms.

planula [L. dim. of *planus*, a wanderer]: The ciliated, free-swimming type of larva formed by many cnidarians.

plasma [Gk., form or mold]: The clear, colorless fluid component of vertebrate blood, containing dissolved ions, molecules, and plasma proteins; blood minus the blood cells.

plasma cell: An antibody-producing cell resulting from the differentiation and proliferation of a B lymphocyte that has interacted with an antigen complementary to the antibodies displayed on its surface; a mature plasma cell can produce from 3,000 to 30,000 antibody molecules per second.

plasma membrane: The membrane surrounding the cytoplasm of a cell; the cell membrane.

plasmid: In prokaryotes, an extrachromosomal, independently replicating, small, circular DNA molecule.

plasmodesma, pl. plasmodesmata [Gk. *plassein*, to mold + *desmos*, band, bond]: In plants, a minute, cytoplasmic thread that extends through pores in cell walls and connects the cytoplasm of adjacent cells.

plastid [Gk. *plastos*, formed or molded]: A cytoplasmic, often pigmented, organelle in plant cells; includes leucoplasts, chromoplasts, and chloroplasts.

platelet (plate-let) [Gk. *platus*, flat]: In mammals, a round or biconcave disk suspended in the blood and involved in the formation of blood clots.

pleiotropy (plee-o-trope-ee) [Gk. *pleios*, more + *trope*, a turning]: The capacity of a gene to affect a number of different phenotypic characteristics.

poikilotherm [Gk. *poikilos*, changeable + *therme*, heat]: An organism with a body temperature that varies with that of the environment.

polar [L. *polus,* end of axis]: Having parts or areas with opposed or contrasting properties, such as positive and negative charges, head and tail.

polar body: Minute, nonfunctioning cell produced during those meiotic divisions that lead to egg cells; contains a nucleus but very little cytoplasm.

polar covalent bond: A covalent bond in which the electrons are shared unequally between the two atoms; the resulting polar molecule has regions of slightly negative and slightly positive charge.

pollen [L., fine dust]: In seed plants, spores consisting of an immature male gametophyte and a protective outer covering.

pollination [L. *pollen,* fine dust]: The transfer of pollen from the anther to a receptive surface of a flower.

polygenic inheritance [Gk. *polus,* many + *genos,* race, descent]: The determination of a given characteristic, such as weight or height, by the interaction of many genes.

polymer [Gk. *polus,* many + *meris,* part or portion]: A large molecule composed of many similar or identical molecular subunits.

polymorphism [Gk. *polus,* many + *morphe,* form]: The presence in a single population of two or more phenotypically distinct forms of a trait.

polyp [Gk. *polus,* many + *pous,* foot]: The sessile stage in the life cycle of cnidarians.

polypeptide [Gk. *polus,* many + *pepto,* to soften, digest]: A molecule consisting of a long chain of amino acids linked together by peptide bonds.

polyploid [Gk. *polus,* many + *ploion,* vessel]: Cell with more than two complete sets of chromosomes per nucleus.

polyribosome: Two or more ribosomes together with a molecule of mRNA that they are simultaneously translating; a polysome.

polysaccharide [Gk. *polus,* many + *sakcharon,* sugar]: A carbohydrate polymer composed of monosaccharide monomers in long chains; includes starch, cellulose.

polysome: *See* Polyribosome.

population: Any group of individuals of one species that occupy a given area at the same time; in genetic terms, an interbreeding group of organisms.

population bottleneck: Type of genetic drift that occurs as the result of a population being drastically reduced in numbers by an event having little to do with the usual forces of natural selection.

portal system [L. *porta,* gate]: In the circulatory system, a circuit in which blood flows through two distinct capillary beds, connected by either veins or arteries, before entering the veins that return it to the heart.

posterior: Of or pertaining to the rear, or tail, end.

potential energy: Energy in a potentially usable form that is not, for the moment, being used; often called "energy of position."

predator [L. *praedari,* to prey upon; from *prehendere,* to grasp, seize]: An organism that eats other living organisms.

pressure-flow hypothesis: A hypothesis accounting for sap flow through the phloem system. According to this hypothesis, the solution containing nutrient sugars moves through the sieve tubes by bulk flow, moving into and out of the sieve tubes by active transport and diffusion.

prey [L. *prehendere,* to grasp, seize]: An organism eaten by another organism.

primary growth: In plants, growth originating in the apical meristem of the shoots and roots, as contrasted with secondary growth; results in an increase in length.

primary structure of a protein: The amino acid sequence of a protein.

primate: A member of the order of mammals that includes anthropoids and prosimians.

primitive [L. *primus,* first]: Not specialized; at an early stage of evolution or development.

primitive streak [L. *primus,* first]: The thickened, dorsal, longitudinal strip of ectoderm and mesoderm in early avian, reptilian, and mammalian embryos; equivalent to the blastopore in other forms.

procambium [L. *pro,* before + *cambium,* exchange]: In plants, a primary meristematic tissue; gives rise to vascular tissues of the primary plant body and to the vascular cambium.

producer, in ecological systems: An autotrophic organism, usually a photosynthesizer, that contributes to the net primary productivity of a community.

progesterone [L. *progerere,* to carry forth or out + *steiras,* barren]: In mammals, a steroid hormone produced by the corpus luteum that prepares the uterus for implantation of the embryo; also produced by the placenta during pregnancy.

prokaryote [L. *pro,* before + Gk. *karyon,* nut, kernel]: A cell lacking a membrane-bound nucleus and membrane-bound organelles; a bacterium or a cyanobacterium.

promoter: Specific segment of DNA to which RNA polymerase attaches to initiate transcription of mRNA from an operon.

prophage: A bacterial virus (bacteriophage) integrated into a host chromosome.

prophase [Gk. *pro,* before + *phasis,* form]: An early stage in nuclear division, characterized by the condensing of the chromosomes and their movement toward the equator of the spindle. Homologous chromosomes pair up during meiotic prophase.

proprioceptor [L. *proprius,* one's own]: Receptor that senses movements, position of the body, or muscle strength.

prosimian [L. *pro,* before + *simia,* ape]: A lower primate; includes lemurs, lorises, tarsiers, and bush babies, as well as many fossil forms.

prostaglandins [Gk. *prostas,* a porch or vestibule + L. *glans,* acorn]: A group of fatty acids that function as chemical messengers; synthesized in most, possibly all, cells of the body; thought to play key roles in fertilization and in triggering the onset of both menstruation and labor.

prostate gland [Gk. *prostas,* a porch or vestibule + L. *glans,* acorn]: A mass of muscle and glandular tissue surrounding the base of the urethra in male mammals; the vasa deferentia merge with ducts from the seminal vesicles, enter the prostate gland, and there merge with the urethra. The prostate gland secretes an alkaline fluid that has a stimulating effect on the sperm as they are released.

protein [Gk. *proteios,* primary]: A complex organic compound composed of one or more polypeptide chains, each made up of many (about 100 or more) amino acids linked together by peptide bonds.

proton: A subatomic particle with a single positive charge equal in magnitude to the charge of an electron and with a mass slightly less than that of a neutron; a component of every atomic nucleus.

protoplasm [Gk. *protos,* first + *plasma,* anything molded]: Living matter.

protostome [Gk. *protos,* first + *stoma,* mouth]: An animal in which the mouth forms at or near the blastopore in the developing embryo; mollusks, annelids, and arthropods are protostomes. Protostomes are also characterized by spiral cleavage during the earliest stages of development and by schizocoelous formation of the coelom.

provirus: A virus of a eukaryote that has become integrated into a host chromosome.

pseudocoelom [Gk. *pseudes,* false + *koilos,* a hollow]: A body cavity consisting of a fluid-filled space between the endoderm and the mesoderm; characteristic of the nematodes.

pseudopodium [Gk. *pseudes,* false + *pous,* pod-, foot]: A temporary cytoplasmic protrusion from an amoeboid cell, which functions in locomotion or in feeding by phagocytosis.

pulmonary [L. *pulmonis,* lung]: Pertaining to the lungs.

pulmonary artery [L. *pulmonis,* lung]: In birds and mammals, an artery that carries deoxygenated blood from the right ventricle of the heart to the lungs, where it is oxygenated.

pulmonary vein [L. *pulmonis,* lung]: In birds and mammals, a vein that carries oxygenated blood from the lungs to the left atrium of the heart, from which blood is pumped into the left ventricle and from there to the body tissues.

punctuated equilibrium: A model of the mechanism of evolutionary change that proposes that long periods of no change ("stasis") are punctuated by periods of rapid speciation, with natural selection acting on species as well as on individuals.

Punnett square: The checkerboard diagram used for analysis of allele segregation.

pupa [L., girl, doll]: A developmental stage of some insects, in which the organism is nonfeeding, immotile, and sometimes encapsulated or in a cocoon; the pupal stage occurs between the larval and adult phases.

purine [Gk. *purinos,* fiery, sparkling]: A nitrogenous base, such as adenine or guanine, with a characteristic two-ring structure; one of the components of nucleic acids.

pyramid, ecological: *See* Ecological pyramid.

pyramid of energy: A diagram of the energy flow between the trophic levels of an ecosystem; plants or other autotrophs (at the base of the pyramid) represent the greatest amount of energy, herbivores next, then primary carnivores, secondary carnivores, etc.

pyrimidine: A nitrogenous base, such as cytosine, thymine, or uracil, with a characteristic single-ring structure; one of the components of nucleic acids.

quaternary structure of a protein: The overall structure of a globular protein molecule that consists of two or more polypeptide chains.

queen: In social insects (ants, termites, and some species of bees and wasps), the fertile, or fully developed, female whose function is to lay eggs.

radial symmetry [L. *radius,* a spoke of a wheel + Gk. *summetros,* symmetry]: The regular arrangement of parts around a central axis such that any plane passing through the central axis divides the organism into halves that are approximate mirror images; seen in cnidarians, ctenophorans, and adult echinoderms.

radiation [L. *radius,* a spoke of a wheel, hence, a ray]: Energy emitted in the form of waves or particles.

radioactive isotope: An isotope with an unstable nucleus that stabilizes itself by emitting radiation.

receptor: A protein or glycoprotein molecule with a specific three-dimensional structure, to which a substance (for example, a hormone, a neurotransmitter, or an antigen) with a complementary structure can bind; typically displayed on the surface of a membrane. Binding of a complementary molecule to a receptor may trigger a transport process or a change in processes occurring within the cell.

recessive allele [L. *recedere,* to recede]: An allele whose phenotypic effect is masked in the heterozygote by that of another, dominant allele.

reciprocal altruism: Performance of an altruistic act with the expectation that the favor will be returned.

recognition sequence: A specific sequence of nucleotides at which a restriction enzyme cleaves a DNA molecule.

recombinant DNA: DNA formed either naturally or in the laboratory by the joining of segments of DNA from different sources.

recombination: The formation of new gene combinations; in eukaryotes, may be accomplished by new associations of chromosomes produced during sexual reproduction or crossing over; in prokaryotes, may be accomplished through transformation, conjugation, or transduction.

reduction [L. *reducere,* to lead back]: Loss of oxygen, gain of hydrogen, or gain of an electron by an atom, ion, or molecule; oxidation and reduction take place simultaneously, with the electron lost by one reactant being transferred to another.

reflex [L. *reflectere,* to bend back]: Unit of action of the nervous system involving a sensory neuron, often one or more interneurons, and one or more motor neurons.

relay neuron: Neuron that transmits signals between different regions of the central nervous system.

releaser: In ethology, a stimulus that functions as a communication signal between members of the same species and that sets in motion, or

"releases," the sequence of movements that constitute a fixed action pattern.

renal [L. *renes,* kidneys]: Pertaining to the kidney.

replication fork: In DNA synthesis, the Y-shaped structure formed at the point where the two strands of the original molecule are being separated and the complementary strands are being synthesized.

repressor [L. *reprimere,* to press back, keep back]: In genetics, a protein that binds to the operator, preventing RNA polymerase from attaching to the promoter and transcribing the structural genes of the operon; coded by a segment of DNA known as the regulator.

resolving power [L. *resolvere,* to loosen, unbind]: The ability of a lens to distinguish two lines as separate.

respiration [L. *respirare,* to breathe]: (1) In aerobic organisms, the intake of oxygen and the liberation of carbon dioxide. (2) In cells, the oxygen-requiring stage in the breakdown and release of energy from fuel molecules.

resting potential: The difference in electric potential (about 70 millivolts) across the membrane of an axon at rest.

restriction enzymes: Enzymes that cleave the DNA double helix at specific nucleotide sequences.

reticular activating system [L. *reticulum,* a network]: A brain circuit involved with alertness and direction of attention to selected events; includes the reticular formation, a core of tissue that runs centrally through the brainstem, and neurons in the thalamus.

reticulum [L., network]: A fine network (e.g., endoplasmic reticulum).

retina [L. dim. of *rete,* net]: The light-sensitive layer of the vertebrate eye; contains several layers of neurons and photoreceptor cells (rods and cones); receives the image formed by the lens and transmits it to the brain via the optic nerve.

retrovirus [L., turning back]: An RNA virus that codes for an enzyme, reverse transcriptase, that transcribes the RNA into DNA.

reverse transcriptase: An enzyme that transcribes RNA into DNA; found only in association with retroviruses.

rhizoid [Gk. *rhiza,* root]: Rootlike anchoring structure in fungi and non-vascular plants.

rhizome [Gk. *rhizoma,* mass of roots]: In vascular plants, a horizontal stem growing along or below the surface of the soil; may be enlarged for storage or may function in vegetative reproduction.

ribonucleic acid (RNA) (rye-bo-new-**clay**-ick): A class of nucleic acids characterized by the presence of the sugar ribose and the pyrimidine uracil; includes mRNA, tRNA, and rRNA. RNA is the genetic material of many viruses.

ribosomal RNA (rRNA): A class of RNA molecules found, along with characteristic proteins, in ribosomes; transcribed from DNA of the chromatin loops that form the nucleolus.

ribosome: A small organelle composed of protein and ribonucleic acid; the site of translation in protein synthesis; in eukaryotic cells, often bound to the endoplasmic reticulum. Many ribosomes attached to a single strand of mRNA are called a polyribosome, or polysome.

RNA: Abbreviation of ribonucleic acid.

rod: Photoreceptor cell found in the vertebrate retina; sensitive to very dim light, responsible for "night vision."

root: The descending axis of a plant, normally below ground and serving both to anchor the plant and to take up and conduct water and minerals.

root hair: An extremely fine cytoplasmic extension of an epidermal cell of a young root; root hairs greatly increase the surface area for the uptake of water and minerals.

saprobe [Gk. *sapros,* rotten, putrid + *bios,* life]: An organism that feeds on nonliving organic matter.

sarcolemma [Gk. *sarx,* the flesh + *lemma,* husk]: The specialized cell membrane surrounding a muscle cell (muscle fiber); capable of propagating action potentials.

sarcomere [Gk. *sarx*, the flesh + *meris*, part of, portion]: Functional and structural unit of contraction in striated muscle.

sarcoplasmic reticulum [Gk. *sarx*, the flesh + *plasma*, from cytoplasm + L. *reticulum*, network]: The specialized endoplasmic reticulum that encases each myofibril of a muscle cell.

schizocoelous [Gk. *schizo*, to split + *koilos*, a hollow]: Formation of the coelom during embryonic development by a splitting of the mesoderm; characteristic of protostomes.

sclerenchyma [Gk. *skleros*, hard]: In plants, a type of supporting cell with thick, often lignified, secondary walls; may be alive or dead at maturity; includes fibers and sclereids.

secondary sex characteristics: Characteristics of animals that distinguish between the two sexes but that do not produce or convey gametes; includes facial hair of the human male and enlarged hips and breasts of the female.

secondary structure of a protein: The simple structure (often a helix, a sheet, or a cable) resulting from the spontaneous folding of a polypeptide chain as it is formed; maintained by hydrogen bonds and other weak forces.

secretion [L. *secernere*, to sever, separate]: (1) Product of any cell, gland, or tissue that is released through the cell membrane and that performs its function outside the cell that produced it. (2) The stage of kidney function in which, through active transport processes, molecules remaining in the blood plasma are selectively removed from the peritubular capillaries and pumped into the filtrate in the renal tubule.

seed: A complex structure formed by the maturation of the ovule of seed plants following fertilization; upon germination, a seed develops into a new sporophyte; generally consists of seed coat, embryo, and a food reserve.

segregation: *See* Mendel's first law.

selectively permeable [L. *seligere*, to gather apart + *permeare*, to go through]: Applied to membranes that permit passage of water and some solutes but block passage of most solutes; semipermeable.

self-fertilization: The union of egg and sperm produced by a single hermaphroditic organism.

self-pollination: The transfer of pollen from anther to stigma in the same flower or to another flower of the same plant, leading to self-fertilization.

semen [L., seed]: Product of the male reproductive system; includes sperm and the sperm-carrying fluids.

seminal vesicles [L. *semen*, seed + *vesicula*, a little bladder]: In male mammals, small vesicles, the ducts of which merge with the vasa deferentia as they enter the prostate gland; they produce an alkaline, fructose-containing fluid that suspends and nourishes the sperm cells.

sensory neuron: A neuron that conducts impulses from a sensory receptor to the central nervous system or central ganglion.

sensory receptor: A cell, tissue, or organ that detects internal or external stimuli.

septum [L., fence]: A partition, or cross wall, that divides a structure, such as a fungal hypha, into compartments.

sessile [L. *sedere*, to sit]: Attached; not free to move about.

sex chromosomes: Chromosomes that are different in the two sexes and that are involved in sex determination.

sex-linked trait: An inherited trait, such as color discrimination, determined by a gene located on a sex chromosome and that therefore shows a different pattern of inheritance in males and females.

sexual reproduction: Reproduction involving meiosis and fertilization.

sexual selection: A type of natural selection that acts on characteristics of direct consequence in obtaining a mate and successfully reproducing; thought to be the chief cause of sexual dimorphism, the striking phenotypic differences between the males and females of many species.

shoot: The aboveground portions, such as the stem and leaves, of a vascular plant.

sieve cell: A long, slender cell of the phloem of gymnosperms; involved in transport of sugars synthesized in the leaves to other parts of the plant.

sieve tube: A series of sugar-conducting cells (sieve-tube members) found in the phloem of angiosperms.

sinoatrial node [L. *sinus*, fold, hollow + *atrium*, yard, court, hall + *nodus*, knot]: Area of the vertebrate heart that initiates the heartbeat; located where the superior vena cava enters the right atrium; the pacemaker.

smooth muscle: Nonstriated muscle; lines the walls of internal organs and arteries and is under involuntary control.

social dominance: A hierarchical pattern of social organization involving domination of some members of a group by other members in a relatively orderly and long-lasting pattern.

society [L. *socius*, companion]: An organization of individuals of the same species in which there are divisions of resources, divisions of labor, and mutual dependence; a society is held together by stimuli exchanged among members of the group.

sociobiology: The study of the biological basis of social behavior.

solution: A homogeneous mixture of the molecules of two or more substances; the substance present in the greatest amount (usually a liquid) is called the solvent, and the substances present in lesser amounts are called solutes.

somatic cells [Gk. *soma*, body]: The differentiated cells composing body tissues of multicellular plants and animals; all body cells except those giving rise to gametes.

somatic nervous system [Gk. *soma*, body]: In vertebrates, the motor and sensory neurons of the peripheral nervous system that control skeletal muscle; the "voluntary" system, as contrasted with the "involuntary," or autonomic, nervous system.

somite: One of the blocks, or segments, of tissue into which the chorda-mesoderm is divided during differentiation of the vertebrate embryo.

specialized: (1) Of cells, having particular functions in a multicellular organism. (2) Of organisms, having special adaptations to a particular habitat or mode of life.

speciation: The process by which new species are formed.

species, *pl.* **species** [L., kind, sort]: A group of organisms that actually (or potentially) interbreed in nature and are reproductively isolated from all other such groups; a taxonomic grouping of anatomically similar individuals (the category below genus).

species-specific: Characteristic of (and limited to) a particular species.

specific: Unique; for example, the proteins in a given organism, the enzyme catalyzing a given reaction, or the antibody to a given antigen.

specific heat: The amount of heat (in calories) required to raise the temperature of 1 gram of a substance $1\,°C$. The specific heat of water is 1 calorie per gram.

sperm [Gk. *sperma*, seed]: A mature male sex cell, or gamete, usually motile and smaller than the female gamete.

spermatid [Gk. *sperma*, seed]: Each of four haploid *(n)* cells resulting from the meiotic divisions of a spermatocyte; each spermatid becomes differentiated into a sperm cell.

spermatocytes [Gk. *sperma*, seed + *kytos*, vessel]: The diploid *(2n)* cells formed by the enlargement of the spermatogonia; they give rise by meiotic division to the spermatids.

spermatogonia [Gk. *sperma*, seed + *gonos*, a child, the young]: The unspecialized diploid *(2n)* cells on the walls of the testes that, by meiotic division, become spermatocytes, then spermatids, then sperm cells.

sphincter [Gk. *sphinktēr*, a band]: A circular muscle surrounding the opening of a tubular structure or the juncture of different regions of a tubular structure (e.g., the pyloric sphincter, at the juncture of the stomach and the small intestine); contraction of the sphincter closes the passageway, and relaxation opens it.

spinal cord: Part of the vertebrate central nervous system; consists of a thick, dorsal, longitudinal bundle of nerve fibers extending posteriorly from the brain.

spindle: In dividing cells, the structure formed of microtubules that extends from pole to pole; the spindle fibers appear to maneuver the chromosomes into position during metaphase and to pull the newly separated chromosomes toward the poles during anaphase.

spiracle [L. *spirare,* to breathe]: One of the external openings of the respiratory system in terrestrial arthropods.

splitting evolution: *See* Cladogenesis.

sporangiophore (spo-ran-ji-o-for) [Gk. *spora,* seed + *phore,* from *phorein,* to bear]: A specialized hypha or a branch bearing one or more sporangia.

sporangium, *pl.* **sporangia** [Gk. *spora,* seed]: A unicellular or multicellular structure in which spores are produced.

spore [Gk. *spora,* seed]: An asexual reproductive or resting cell capable of developing into a new organism without fusion with another cell; in contrast to a gamete.

sporophyll [Gk. *spora,* seed + *phyllon,* the leaves]: Spore-bearing leaf. The carpels and stamens of flowers are modified sporophylls.

sporophyte [Gk. *spora,* seed + *phytos,* growing]: In organisms that have alternation of haploid and diploid generations (all plants and some green algae), the diploid (2*n*) spore-producing generation.

stamen [L., a thread]: The male structure of a flower, which produces microspores or pollen; usually consists of a stalk, the filament, bearing a pollen-producing anther at its tip.

starch [M.E. *sterchen,* to stiffen]: A class of complex, insoluble carbohydrates, the chief food-storage substances of plants; composed of 1,000 or more glucose units and readily broken down enzymatically into these units.

statocyst [Gk. *statos,* standing + *kystis,* sac]: An organ of balance, consisting of a vesicle containing granules of sand (statoliths) or some other material that stimulates sensory cells when the organism moves.

stem: The aboveground part of the axis of vascular plants, as well as anatomically similar portions below ground (such as rhizomes).

stem cells: The common, self-regenerating cells in the marrow of long bones that give rise, by differentiation and division, to red blood cells and all of the different types of white blood cells.

stereoscopic vision [Gk. *stereōs,* solid + *optikos,* pertaining to the eye]: Ability to perceive a single, three-dimensional image from the simultaneous but separate images delivered to the brain by each eye.

steroid: One of a group of lipids having four linked carbon rings and, often, a hydrocarbon tail; cholesterol, sex hormones, and the hormones of the adrenal cortex are steroids.

stigma [Gk. *stigme,* a prick mark, puncture]: In plants, the region of a carpel serving as a receptive surface for pollen grains, which germinate on it.

stimulus [L., goad, incentive]: Any internal or external change or signal that influences the activity of an organism or of part of an organism.

stoma, *pl.* **stomata** [Gk., mouth]: A minute opening in the epidermis of leaves and stems, bordered by guard cells, through which gases pass.

strategy [Gk. *strategein,* to maneuver]: A group of related traits, evolved under the influence of natural selection, that solve particular problems encountered by living organisms; often includes anatomical, physiological, and behavioral characteristics.

striated muscle [L., from *striare,* to groove]: Skeletal voluntary muscle and cardiac muscle. The name derives from the striped appearance, which reflects the arrangement of contractile elements.

stroma [Gk., a bed, from *stronnymi,* to spread out]: A dense solution that fills the interior of the chloroplast and surrounds the thylakoids.

structural gene: Any gene that codes for a protein; in distinction to regulatory genes.

style [L. *stilus,* stake, stalk]: In angiosperms, the stalk of a carpel, down which the pollen tube grows.

substrate [L. *substratus,* strewn under]: (1) The foundation to which an organism is attached. (2) A substance on which an enzyme acts.

succession: *See* Ecological succession.

sucrose: Cane sugar; a common disaccharide found in many plants; a molecule of glucose linked to a molecule of fructose.

sugar: Any monosaccharide or disaccharide.

supergene: *See* Coadaptive gene complex.

surface tension: A tautness of the surface of a liquid, caused by the cohesion of the molecules of liquid. Water has an extremely high surface tension.

symbiosis [Gk. *syn,* together with + *bioonai,* to live]: An intimate and protracted association between two or more organisms of different species. Includes mutualism, in which the association is beneficial to both; commensalism, in which one benefits and the other is neither harmed nor benefited; and parasitism, in which one benefits and the other is harmed.

sympathetic division: A subdivision of the autonomic nervous system, with centers in the midportion of the spinal cord; slows digestion; generally excites other functions.

sympatric speciation [Gk. *syn,* together with + *patra,* fatherland, country]: Speciation that occurs without geographic isolation of a population of organisms; usually occurs as the result of hybridization accompanied by polyploidy; may occur in some cases as a result of disruptive selection.

synapse [Gk. *synapsis,* a union]: A specialized junction between two neurons where the activity in one influences the activity in another; may be chemical or electrical, excitatory or inhibitory.

syngamy (sin-gamy) [Gk. *syn,* with + *gamos,* a marriage]: The union of gametes in sexual reproduction; fertilization.

synthesis [Gk. *syntheke,* a putting together]: The formation of a more complex substance from simpler ones.

synthetic theory: The currently prevailing theory of the mechanism of evolutionary change; combines the Darwinian two-step model of variation and selection with the principles of Mendelian genetics.

systematics [Gk. *systema,* that which is put together]: Scientific study of the kinds and diversity of organisms and of the relationships among them.

T lymphocyte: A type of white blood cell arising from precursors in the thymus gland and, upon maturation, involved in cell-mediated immunity and interactions with B lymphocytes; a T cell.

tagmosis [Gk. *tagma,* arrangement, order + *-osis,* process]: The formation of groups of segments (metameres) into body regions (tagmata) with functional differences.

taxon, *pl.* **taxa** [Gk. *taxis,* arrange, put in order]: A particular group, ranked at a particular categorical level, in a hierarchical classification scheme; for example, *Drosophila* is a taxon at the categorical level of genus.

taxonomy [Gk. *taxis,* arrange, put in order + *nomos,* law]: The study of the classification of organisms; the ordering of organisms into a hierarchy that reflects their essential similarities and differences.

telencephalon [Gk. *tēl,* far off + *enkephalos,* brain]: One of the two principal subdivisions of the vertebrate forebrain; the anterior portion of the forebrain, it contains the cerebrum and the olfactory bulbs.

telophase [Gk. *telos,* end + *phasis,* form]: The last stage in mitosis and meiosis, during which the chromosomes become reorganized into two new nuclei.

temperate bacteriophage: A bacterial virus that may become incorporated into the host-cell chromosome.

template: A pattern or mold guiding the formation of a negative or complement.

tentacles [L. *tentare,* to touch]: Long, flexible protrusions located about the mouth of many invertebrates; usually prehensile or tactile.

territory: An area or space occupied and defended by an individual or a group; trespassers are attacked (and usually defeated); may be the site of breeding, nesting, food gathering, or any combination thereof.

tertiary structure of a protein: A complex structure, usually globular, resulting from further folding of the secondary structure of a protein;

forms spontaneously due to attractions and repulsions among amino acids with different charges on their R groups.

testcross: A mating between a phenotypically dominant individual and a homozygous recessive "tester" to determine the genetic constitution of the dominant phenotype, that is, whether it is homozygous or heterozygous for the relevant gene.

testis, *pl.* **testes** [L., witness]: The sperm-producing organ; also the source of the male sex hormone testosterone.

testosterone [Gk. *testis,* testicle + *steiras,* barren]: A steroid hormone secreted by the testes in higher vertebrates and stimulating the development and maintenance of male sex characteristics and the production of sperm; the principal androgen.

tetrad [Gk. *tetras,* four]: In genetics, a pair of homologous chromosomes that have replicated and come together in prophase I of meiosis; consists of four chromatids.

thalamus [Gk. *thalamos,* chamber]: A part of the vertebrate forebrain just posterior to and tucked below the cerebrum; the main relay center between the brainstem and the higher brain centers.

thallus [Gk. *thallos,* a young twig]: A simple plant or algal body without true roots, leaves, or stems.

theory [Gk. *theorein,* to look at]: A generalization based on many observations and experiments; a verified hypothesis.

thermodynamics [Gk. *therme,* heat + *dynamis,* power]: The study of transformations of energy. The first law of thermodynamics states that, in all processes, the total energy of a system plus its surroundings remains constant. The second law states that all natural processes tend to proceed in such a direction that the disorder or randomness of the system increases.

thorax [Gk., breastplate]: (1) In vertebrates, that portion of the trunk containing the heart and lungs. (2) In crustaceans and insects, the fused, leg-bearing segments between head and abdomen.

thylakoid [Gk. *thylakos,* a small bag]: A flattened sac, or vesicle, that forms part of the internal membrane structure of the chloroplast; the site of the light-trapping reactions of photosynthesis and of photophosphorylation; stacks of thylakoids collectively form the grana.

thyroid [Gk. *thyra,* a door]: An endocrine gland of vertebrates, located in the neck; source of an iodine-containing hormone (thyroxine) that increases the metabolic rate and affects growth.

tight junction: A junction between adjacent animal cells that prevents materials from leaking through the tissue; for example, intestinal epithelial cells are surrounded by tight junctions.

tissue [L. *texere,* to weave]: A group of similar cells organized into a structural and functional unit.

tonoplast [Gk. *tonos,* stretching, tension + *plastos,* formed, molded]: In plant cells, the membrane surrounding the vacuole.

trachea, *pl.* **tracheae** (trake-ee-a) [Gk. *tracheia,* rough]: An air-conducting tube. (1) In insects and some other terrestrial arthropods, a system of chitin-lined air ducts. (2) In terrestrial vertebrates, the windpipe.

tracheid (tray-key-idd) [Gk. *tracheia,* rough]: In vascular plants, an elongated, thick-walled conducting and supporting cell of xylem, characterized by tapering ends and pitted walls without true perforations.

tract: A group or bundle of nerve fibers with accompanying connective tissue, located within the central nervous system.

transcription [L. *trans,* across + *scribere,* to write]: The enzymatic process by which the genetic information contained in one strand of DNA is used to specify a complementary sequence of bases in an RNA molecule.

transduction [L. *trans,* across + *ducere,* to lead]: (1) The transfer of genetic material (DNA) from one cell to another by a virus. (2) The conversion of one form of energy into another form of energy; for example, the conversion of the energy of a chemical stimulus into the energy of an action potential.

transfer RNA (tRNA) [L. *trans,* across + *ferre,* to bear or carry]: A class of small RNAs (about 80 nucleotides each) with two functional sites; one recognizes a specific activated amino acid; the other carries the nucleotide

triplet (anticodon) for that amino acid. Each type of tRNA accepts a specific activated amino acid and transfers it to a growing polypeptide chain as specified by the nucleotide sequence of the mRNA being translated.

transformation [L. *trans,* across + *formare,* to shape]: A genetic change produced by the incorporation into a cell of DNA from the external medium.

translation [L. *trans,* across + *latus,* that which is carried]: The process by which the genetic information present in a strand of mRNA directs the sequence of amino acids during protein synthesis.

translocation [L. *trans,* across + *locare,* to put or place]: (1) In plants, the transport of the products of photosynthesis from a leaf to another part of the plant. (2) In genetics, the breaking off of a piece of chromosome with its reattachment to a nonhomologous chromosome.

transpiration [L. *trans,* across + *spirare,* to breathe]: In plants, the loss of water vapor from the stomata.

transposon [L. *transponere,* to change the position of]: A DNA sequence carrying one or more genes that is capable of moving from one location in the chromosomes to another. Simple transposons, also known as insertion sequences, carry only the genes essential for transposition; complex transposons carry genes that code for additional proteins.

tritium: A radioactive isotope (^{3}H) of hydrogen with a half-life of 12.5 years.

trophic level [Gk. *trophos,* feeder]: The position of a species in the food web or chain, that is, its feeding level; a step in the movement of biomass or energy through an ecosystem.

trophoblast [Gk. *trophos,* feeder + *blastos,* sprout]: In the early mammalian embryo (the blastocyst), a double layer of cells that surrounds the inner cell mass and subsequently gives rise to the chorion.

tropic [Gk. *trope,* a turning]: Pertaining to behavior or action brought about by specific stimuli, for example, phototropic ("light-oriented") motion, gonadotropic ("stimulating the gonads") hormone.

tuber [L. *tuber,* bump, swelling]: A much-enlarged, short, fleshy underground stem, such as that of the potato.

turgor [L. *turgere,* to swell]: The pressure exerted on the inside of a plant cell wall by the fluid contents of the cell; the interior of the cell is hypertonic in relation to the fluids surrounding it and so gains water by osmosis.

urea [Gk. *ouron,* urine]: An organic compound formed in the vertebrate liver; principal form of disposal of nitrogenous wastes by mammals.

ureter [Gk. from *ourein,* to urinate]: The tube carrying urine from the kidney to the cloaca (in reptiles and birds) or to the bladder (in amphibians and mammals).

urethra [Gk. from *ourein,* to urinate]: The tube carrying urine from the bladder to the exterior of mammals.

uric acid [Gk. *ouron,* urine]: An insoluble nitrogenous waste product that is the principal excretory product in birds, reptiles, and insects.

urine [Gk. *ouron,* urine]: The liquid waste filtered from the blood by the kidney and stored in the bladder pending elimination through the urethra.

uterine tube: *See* Oviduct.

uterus [L., womb]: The muscular, expanded portion of the female reproductive tract modified for the storage of eggs or for housing and nourishing the developing embryo.

vacuole [L. *vacuus,* empty]: A membrane-bound, fluid-filled sac within the cytoplasm of a cell.

vagus nerve [L. *vagus,* wandering]: A nerve arising from the medulla of the vertebrate brain that innnervates the heart and visceral organs; carries parasympathetic fibers.

vaporization [L. *vapor,* steam]: The change from a liquid to a gas; evaporation.

vascular [L. *vasculum,* a small vessel]: Containing or concerning vessels that conduct fluid.

vascular bundle: In plants, a group of longitudinal supporting and conducting tissues (xylem and phloem).

vascular cambium [L. *vasculum,* a small vessel + *cambium,* exchange]: In plants, a cylindrical sheath of meristematic cells that divide mitotically, producing secondary phloem to one side and secondary xylem to the other, but always with a cambial cell remaining.

vas deferens, *pl.* **vasa deferentia** (vass **deff**-er-ens) [L. *vas,* a vessel + *deferre,* to carry down]: In mammals, the tube carrying sperm from a testis to the urethra.

vector [L., carrier]: In recombinant DNA, a small, self-replicating DNA molecule, or a portion thereof, into which a DNA segment can be spliced and introduced into a cell; generally a plasmid, a bacteriophage, or a cosmid.

vein [L. *vena,* a blood vessel]: (1) In plants, a vascular bundle forming part of the framework of the conducting and supporting tissue of a leaf. (2) In animals, a blood vessel carrying blood from the tissues to the heart. A small vein is known as a venule.

vena cava (vee-na **cah**-va) [L., blood vessel + hollow]: A large vein that brings blood from the tissues to the right atrium of the four-chambered mammalian heart. The superior vena cava collects blood from the forelimbs, head, and anterior or upper trunk; the inferior vena cava collects blood from the posterior body region.

ventral [L. *venter,* belly]: Pertaining to the undersurface of an animal that holds its body in a horizontal position; to the front surface of an animal that holds its body erect.

ventricle [L. *ventriculus,* the stomach]: A muscular chamber of the heart that receives blood from an atrium and pumps blood out of the heart, either to the lungs or to the body tissues.

vertebral column [L. *vertebra,* joint]: The backbone; in nearly all vertebrates, it forms the supporting axis of the body and protects the spinal cord.

vesicle [L. *vesicula,* a little bladder]: A small, intracellular membrane-bound sac.

vessel [L. *vas,* a vessel]: A tubelike element of the xylem of angiosperms; composed of dead cells (vessel members) arranged end to end. Its function is to conduct water and minerals from the soil.

viable [L. *vita,* life]: Able to live.

villus, *pl.* **villi** [L., a tuft of hair]: In vertebrates, one of the minute, fingerlike projections lining the small intestine that serve to increase the absorptive surface area of the intestine.

virus [L., slimy, liquid, poison]: A submicroscopic, noncellular particle composed of a nucleic acid core and a protein coat; parasitic; reproduces only within a host cell.

viscera [L., internal organs]: The collective term for the internal organs of an animal.

vitamin [L. *vita,* life]: Any of a number of unrelated organic substances that cannot be synthesized by a particular organism and are essential in minute quantities for normal growth and function.

water cycle: Worldwide circulation of water molecules, powered by the sun. Water evaporates from oceans, lakes, rivers, and, in smaller amounts, soil surfaces and bodies of organisms; water returns to the earth in the form of rain and snow. Of the water falling on land, some flows into rivers that pour water back into the oceans and some percolates down through the soil until it reaches a zone where all pores and cracks in the rock are filled with water (groundwater); the deep groundwater eventually reaches the oceans, completing the cycle.

water potential: The potential energy of water molecules; regardless of the reason (e.g., gravity, pressure, concentration of solute particles) for the water potential, water moves from a region where water potential is greater to a region where water potential is lower.

wild type: In genetics, the phenotype that is characteristic of the vast majority of individuals of a species in a natural environment.

worker: A member of the nonreproductive laboring caste of social insects.

xanthophyll [Gk. *xanthos,* yellow + *phyllon,* leaf]: In algae and plants, one of a group of yellow pigments; a member of the carotenoid group.

xylem [Gk. *xylon,* wood]: A complex vascular tissue through which most of the water and minerals are conducted from the roots to other parts of the plant; consists of tracheids or vessel members, parenchyma cells, and fibers; constitutes the wood of trees and shrubs.

yolk: The stored food in egg cells that nourishes the embryo.

yolk sac: In developing reptiles and birds, the extraembryonic membrane that surrounds and encloses the yolk; performs a nutritive function. In mammals, the extraembryonic membrane in which the germ cells are set aside very early in development.

zoology [Gk. *zoe,* life + *logos,* a discourse]: The study of animals.

zooplankton [Gk. *zoe,* life + *plankton,* wanderer]: A collective term for the nonphotosynthetic organisms present in plankton.

zygote (zi-goat) [Gk. *zygon,* yolk, pair]: The diploid (2*n*) cell resulting from the fusion of male and female gametes (fertilization); a zygote may either develop into a diploid individual by mitotic divisions or may undergo meiosis to form haploid *(n)* individuals that divide mitotically to form a population of cells.

Illustration Acknowledgments

Page ix © Rita Summers/Colorado Nature Photographic Studio; **Page xi** *(top)* © Stephen Dalton/NHPA; *(bottom, left to right)* © Jen & Des Bartlett/Bruce Coleman; © Michael Medford/Wheeler Pictures; Carolina Bilogical Supply Company; **Page xii** *(left to right)* © Larry West; © Grant Heilman Photography; © Paul W. Johnson/Biological Photo Service; **Page xiii** *(left to right)* Ripon Microslides; Nelson Max & Richard Dickerson; © Tony Mendoza/The Picture Cube; **Page xv** *(left to right)* © Brian Parker/Tom Stack & Associates; © Kim Taylor/Bruce Coleman Ltd.; © Larry West; **Page xvii** © E. R. Degginger/Earth Scenes; © James L. Castner; © Larry West; **Page xviii** (left to right) Lennart Nilsson, THE BODY VICTORIOUS. New York: Delacorte Press. Boehringer Ingelheim International GmbH; © E. R. Degginger/Animals Animals; Antone G. Jacobson; **Page xx** *(left to right)* © Frans Lanting; © Jen & Des Bartlett/ Bruce Coleman; © Jim Brandenburg; **Page xxii** *(left to right)* M. A. Chappell/Animals Animals; © James L. Castner; © Wolfgang Kaehler

I-1 © Francisco Erize/Bruce Coleman; **Page 1** © Raymond A. Mendez/ Animals Animals; **I-2** The Royal College of Surgeons of England; **I-3** © Chip & Rosa Maria Peterson; **I-4** (a), © J. Fennell/Bruce Coleman; (c) © W. H. Hodge/Peter Arnold; **I-5** American Museum of Natural History; **I-6** © Breck P. Kent/Animals Animals; **I-7** (a) Christopher Ralling; (b) Medical Illustration Unit, The Royal College of Surgeons of England; **I-9** © Frans Lanting/Bruce Coleman Ltd.; **I-10** (a) The Granger Collection; (b) Ann Ronan Picture Library; **Page 8** The Royal College of Surgeons of England; **I-11** (a) Field Museum, Photo Researchers; (b) © S. Robinson/NHPA; **I-12** (a) Rare Books Division, New York Public Library; (b) John Mais; **I-13** The Bettmann Archive; **I-14** © Laura Riley/Bruce Coleman; **I-15** (a) © Eric V. Gravé; (b) © J. Robert Waaland/Biological Photo Service; **I-16** © Larry West; **I-17** © Bruce Coleman; **I-18** © E. S. Ross; **I-19** Terry Erwin & Linda Sims, Smithsonian Institution; **I-20** (a) © Clem Haagner/Bruce Coleman; (b) © Raymond A. Mendez/Animals Animals

Page 20 National Optical Astronomy Observatories; **Page 21 & 1-1** © Sea Studio, Inc./Peter Arnold; **Page 23** © Jen & Des Bartlett/Bruce Coleman; **1-2** © John Cancalosi; **Page 26** (a) © John D. Cunningham/ Visuals Unlimited; (b) © M. Walker/NHPA; (c) © Mary M. Thacher/ Photo Researchers; **Page 27** (d) © Mitch Reardon/Photo Researchers; (e) © Jeff Foott; (f) © Dwayne M. Reed; (g) © Stephen Dalton/Photo Researchers; **1-5** © Bruce Coleman; **1-7** (b) © Charles M. Falco/Science Source, Photo Researchers; **1-8** © Jen & Des Bartlett/Bruce Coleman; **1-13** © Runk & Schoenberger/Grant Heilman Photography; **1-14** H. Berg & G. Forté; **1-15** (a) © Frieder Sauer/Bruce Coleman; (b) George I. Schwartz; (c) © Biology Media/Photo Researchers; (d) © Manfred Kage/ Peter Arnold

Page 40 © Michael Medford/Wheeler Pictures; **2-1** © E. R. Degginger/ Bruce Coleman; **2-4** Fritz Polking; **2-5** © Runk & Shoenberger/Grant Heilman Photography; **2-7** (b) © Brian Milne/Earth Scenes; **Page 51** (b) © John D. Cunningham/Visuals Unlimited; (c) © Michael Medford/ Wheeler Pictures; **2-11** Jeanne M. Riddle

Page 55 © Herbert B. Parsons/Photo NATS; **3-1** © E. R. Degginger/ Earth Scenes; **3-3** After DuPraw, E. J. (1968). *Cell and molecular biology.* New York: Academic Press, Inc.; **Page 58** John M. Sieburth; **3-7** © Charles & Elizabeth Schwartz/Animals Animals; **3-9** (a), (b), (c) After Lehninger, A. L. (1975). *Biochemistry,* 2d ed. New York: Worth Publishers, Inc.; (d)

L. M. Biedler; (e) J. C. Warren; **3-10** (c) R. D. Preston; **3-11** (b) © Herbert B. Parsons/Photo NATS; **3-13** © Caroline Kroeger/Animals Animals; **3-16** B. E. Juniper; **Page 71** (a), (b) © Sloop-Ober/Visuals Unlimited; **3-20** Sequence information from Lehninger, A. L. (1982). *Principles of biochemistry* (p. 135). New York: Worth Publishers, Inc.; **3-21** (b) After Wilson, E. O., *et al.* (1977). *Life, cells, organisms, populations.* Sunderland, MA: Sinauer Associates, Inc.; **3-22** (b) Computer graphics modeling and photography by Arthur J. Olson, Ph.D., Research Institute of Scripps Clinic, La Jolla, CA 92037, © 1988; **3-23** (a) After Alberts, B., Bray, D., Lewis, j., Raff, M., Roberts, K. & Watson, J. D. (1983). *Molecular biology of the cell.* New York: Garland Publishing Company; (b) Daniel Friend; **3-24** (a) © Anthony Bannister/NHPA; (b) © Robert L. Dunne/Bruce Coleman; **3-25** (a) After Karp, G. (1979). *Cell biology.* New York: McGraw-Hill Book Company; (b) © Manfred Kage/Peter Arnold; **3-27** Adapted from Dickerson, R. E. & Geis, I. (1969). *The structure and action of proteins.* Menlo Park, CA: W. A. Benjamin, Inc. Copyright 1969 by Dickerson & Geis; **3-28** (a), (b) Margaret Clark; **Page 83** After Lehninger, A. L. (1982). *op. cit.*

Page 84 © David M. Phillips; **4-1** Brent McCown; **4-2** Big Bear Solar Observatory; **Page 86** Pasteur Institute & The Rockefeller University Press; **4-4** © S. Johannson & Frank Lane/Bruce Coleman; **4-5** Sidney W. Fox; **4-6** © S. M. Awramik/Biological Photo Service; **4-8** (b) A. Ryter; **4-9** (b) Lang, N. J. (1965). *Journal of Phycology, 1,* 127–134; **4-10** (b) George Palade; **4-12** (b) Michael A. Walsh; **4-13** (b) Keith Porter; **4-14** After Alberts, *et al.* (1983). *op. cit.;* **4-15** (a), (b), (c) David M. Phillips; **4-17** After Alberts, *et al.* (1983). *op. cit.;* **4-18** *Ibid;* **4-19** (a)–(d) Keith Roberts & James Barnett

Page 102 Osborn, M. (October 1985). The molecules of life. *Scientific American;* **5-2** (a) © Eric V. Gravé/Photo Researchers; (b) © M. Schliwa/ Visuals Unlimited; **5-3** © J. Robert Waaland/Biological Photo Service; **5-4** J. D. Robertson; **5-6** Adapted in part from *Scientific American* (February 1984), p. 81, and in part from Darnell, J., Lodisch, H. & Baltimore, D. (1986). *Molecular cell biology.* New York: W. H. Freeman and Company; **5-7** (a, *photo*) Myron C. Ledbetter; (b) After Albershamm, P. (April 1975). *Scientific American;* **5-8** (a) Daniel Friend; (b) Nigel Unwin; (c) Barbara J. Stevens & Hewson Swift; **5-9** Ursula Goodenough; **5-10** (a) © Doug Wechsler; (b) Mia Tegner & David Epel; **5-13** Osborn, M. (October 1985). The molecules of life. *Scientific American;* **5-14** (a), (b), (c) *Ibid;* **Page 114** Adapted from Darnell, *et al.* (1986). *op. cit.;* **5-17** Peter Webster; **5-18** (a), (b) Don Fawcett; **5-19** Don Fawcett/Photo Researchers; **5-20** (b) Flickinger, C. J. (1975). *Journal of Cell Biology, 49,* 221.; **5-22** Birgit Satir; **5-23** (a) Don Fawcett; (b) G. Decker; **5-24** (b) Keith Porter; **5-25** (a) Roland R. Dute; (b) Myron C. Ledbetter; **5-26** David Stetler; **5-28** © Manfred Kage/Peter Arnold; **5-29** Don Fawcett; **5-30** Gregory Antipa; **5-31** (b) Peter Satir; **5-32** © David M. Phillips/Visuals Unlimited

Page 127 © M. I. Walker/NHPA; **6-2** Keith R. Porter; **6-3** © Frederick J. Dodd/Peter Arnold; **6-5** (a) G. M. Hughes; (b) After Schmidt-Nielson, K. (1979). *Animal physiology,* 2d ed. Cambridge: Cambridge University Press; **6-6** (a) © M. I. Walker/NHPA; (b), (c) © Thomas Eisner; **6-7** After Lehninger, A. L. (1975). *Biochemistry,* 2d ed. New York: Worth Publishers, Inc.; **6-9** (b) Daniel Branton; **6-10** Adapted from Raven, P. E., Eichhorn, S. E. & Evert, R. F. (1986). *Biology of plants,* 4th ed. (Fig. 4-9). New York: Worth Publishers, Inc.; **6-11** Adapted from Alberts, *et al.*

(1983). *Molecular biology of the cell* (p. 289). New York: Garland Publishing Company; **6-12** Keith R. Porter/Photo Researchers; **6-14** Adapted from Raven, *et. al.* (1986). *op. cit.;* **6-15** Adapted from *Scientific American* (May 1984), p. 54; **6-16** (a)-(d) Gregory Antipa; **6-17** (a)-(d) Perry, M. M. & Gilbert, A. B. (1979). *Journal of Cell Science, 39,* 257–272; **6-18** (a) Ray F. Evert; (b) Peter K. Hepler; **Page 141** (a), (b), (c) K. T. Raper; (d)-(g) © David Scharf/Peter Arnold; **6-19** (a) N. Bernard Gilula; (b) Adapted in part from Darnell *et al.* (1986). *Molecular cell biology* (Fig. 14-63). New York: W. H. Freeman and Company; in part from *Scientific American* (May 1978), p. 150; and in part from *Scientific American* (October 1985), p. 106.

Page 144 Carolina Biological Supply Company; **7-1** John Mais; **7-3** L. P. Wisniewski & K. Hirschhorn; **7-5** © David M. Phillips/Visuals Unlimited; **7-6** Carolina Biological Supply Company; **7-7** After Alberts, *et al.* (1983). *Molecular biology of the cell* (Fig. 11-11, p. 619). New York: Garland Publishing Company; **7-8** (a) Adapted from Alberts, *et al., op. cit.* (Fig. 11-47, p. 652); (b) M. J. Schibler; **7-9** (a) Andrew S. Bajer; (b), (c) Adapted from Alberts, *et al., op. cit.* (Fig. 11-48, p. 652); **7-10, 7-11,** and **7-12** Andrew S. Bajer; **7-13** (a)-(d) Carolina Biological Supply Company; **7-14** (a), (b) Beams, H. W. & Kessel, R. G. (1976). *American Scientist, 64,* 279; **7-15** James Cronshaw

Page 158 © Grant Heilman Photography; **Page 159** & **8-1** © Bruce Coleman; **Page 161** © Larry West; **8-2** © Zig Leszczynski/Animals Animals; **8-3** ©Alain Eurard/Photo Researchers; **Page 163** Lotte Jacobi; **8-5** (a)-(d) Miami Seaquarium; **8-7** Jeremy Pickett-Heaps; **8-8** After Lehninger, A. L. (1975). *Biochemistry,* 2d ed., New York: Worth Publishers, Inc.; **8-10** (a) *Ibid;* (b) William Goddard III; **8-16** Adapted from Lehninger, A. L. (1982). *Principles of biochemistry* (pp. 212 & 219). New York: Worth Publishers, Inc.; **8-18** V. Lennard; **8-19** After Watson, J. D., *et al.* (1970). *Molecular biology of the gene,* 2d ed., Menlo Park, CA: The Benjamin/Cummings Publishing Company; **8-22** After Lehninger (1975). *op. cit.;* **Page 179** (a) © Robert Pearcy/Animals Animals; (b) © Bill Curtsinger/Rapho, Photo Researchers; **8-24** (a) © Larry West; (b) © M. P. Price/Bruce Coleman; **Page 185** © R. D. Estes

Page 186 © Grant Heilman Photography; **9-1** (a) © Bob Evans/Peter Arnold; (b) Gary Robinson; (c) Bray, R. (1978). *Science, 200,* 333–334, © 1978 by AAAS; **9-5** (b) © Grant Heilman Photography; **9-7** Adapted from Darnell, *et al.* (1986). *Molecular cell biology.* New York: W. H. Freeman and Company; **Page 193** R. H. Kirschner; **9-8** (b) Lester J. Reed from Boyer, P. D., ed. (1970). *The enzymes, Vol. 1.* New York: Academic Press, Inc., **9-9** Adapted from Vander, A. J., Sherman, J. H. & Luciano, D. (1969). *Human physiology.* New York: McGraw-Hill Book Company; **9-10** After Lehninger, A. L. (1975). *Biochemistry,* 2d ed. New York: Worth Publishers, Inc.; **9-13** After Takano, T., Kallai, O. B., Swanson, R. and Dickerson, R. E. (1973). *Journal of biological chemistry, 248,* 5244; **9-14** Adapted from Lehninger, A. L. (1982). *Principles of biochemistry.* New York: Worth Publishers, Inc.; **9-15** After Alberts, B., *et al.* (1983). *Molecular biology of the cell.* New York: Garland Publishing Company; **9-16** (a) After Lehninger, A. L. (1982). *op. cit.;* (b) John N. Telford; **9-18** After Lehninger, A. L. (1975). *op. cit.;* **Page 202** (a), (b) Lieber, C. S. (March 1976). The metabolism of alcohol. *Scientific American;* **9-21** After Lehninger, A. L. (1975). *op. cit.*

Page 206 © Paul W. Johnson & J. McN. Sieburth/Biological Photo Service; **10-1** © J. Metzner/Peter Arnold; **10-4** (a), (b) Pearse, V. & Buchsbaum, R. (1987). *Living invertebrates.* Pacific Grove, CA: The Boxwood Press; photo by Karl J. Marschall; **10-8** Micrograph by Oxford Scientific Films/Bruce Coleman; **10-10** (a) A. D. Greenwood; (b) L. K. Shumway; **10-13** After Lehninger, A. L. (1975). *Biochemistry,* 2d ed. New York: Worth Publishers, Inc.; **Page 217** © Paul W. Johnson & J. McN. Sieburth/Biological Photo Service; **Page 221** Stoeckenius, W. (June 1976). The purple membrane of salt-loving bacteria. *Scientific American;* **10-17** Pallard, S. G. & Kozlowski, T. T. (1980). *New Phytologist, 85,* 363-368; **10-21** Ray F. Evert

Page 232 Edward Hicks, *Noah's Ark* (detail), The Philadelphia Museum of Art: Bequest of Lisa Norris Elkins; **Page 233** & **11-1** Computer Graphics Laboratory, University of California, San Francisco. © Regents of the University of California; **Page 235** Courtesy LKB Productions, Sweden; **11-2** (a), (b), (c) The Granger Collection; **11-3** The Bettmann

Archive; **11-4** The Granger Collection; **11-6** Adapted from von Frisch, K. (1964). *Biology.* translated by Jane Oppenheimer. New York: Harper & Row Publishers, Inc.; **11-12** Bill Ratcliffe; **11-13** (a) Dr. V. Orel, The Moravian Museum; (b) Courtesy LKB Productions, Sweden

Page 249 Ripon Microslides; **12-1** (a) © John Bova/Photo Researchers; (b) Arnold Sparrow, Brookhaven National Laboratory; **12-9** (a), (b) William Marks; **12-10** (a), (b) Mary E. Clutter; **12-12** After DuPraw, E. J. (1968). *Cell and molecular biology.* New York: Academic Press, Inc.; **Page 259** After Moore, J. L. (1963). *Heredity and development.* New York: Oxford University Press; **12-14** B. John; **Page 262** Ripon Microslides

Page 263 Ralph G. Somes, University of Connecticut; **13-1** Columbiana Collection, Rare Book and Manuscript Library, Columbia University; **13-4** © Jeremy Burgess/Science Photo Library, Photo Researchers; **Page 267** © George F. Godfrey/Animals Animals; **13-7** © John Chiasson/Gamma-Liaison; **13-9** (a) © Richard Kolar/Animals Animals; (b) © Grant Heilman Photography; (c), (d) © Jane Burton/Bruce Coleman; **13-10** (a)-(d) Ralph G. Somes, University of Connecticut **13-11** After Ayala, F. J. & Kiger, J. A., (1980). *Modern genetics.* Menlo Park, CA: The Benjamin/Cummings Publishing Company; **13-13** (a) After E. D. Merrell, 1964; (b) *Journal of Heredity, 15* (1914); **13-14** Photograph by F. B. Hutt (1930). *Journal of Genetics, 22,* 126; **13-15** B. John; **13-20** Beth Myers, courtesy of William Marks

Page 281 Nelson Max & Richard Dickerson; **14-1** Kleinschmidt, A. K., Land, D., Jacherts, D. & Zahn, R. K. (1962). *Biochemica Biophysica Acta, 61,* 857–864; **14-2** (a), (b) © Bruce Iverson; **14-3** After Koob, D. D. & Bogs, W. E. (1972). *The nature of life.* Reading, MA: Addison-Wesley Publishing Co.; **14-5** From Cairns, J., Stent, G. S., & Watson, J. D., eds. (1966). *Phage and the origins of molecular biology.* Cold Spring Harbor, NY: Cold Spring Harbor Laboratory of Quantitative Biology; **14-6** M. Wurtz, Biozentrum, University of Basel/Science Photo Library, Photo Researchers; **14-8** ©Lee D. Simon/Science Photo Library, Photo Researchers; **14-9** (a) From Watson, J. D. (1968). *The double helix.* New York: Atheneum Publishers; (b) Vittorio Luzatti; **14-10** (b) Nelson Max & Richard Dickerson; **Page 291** Watson (1968). *op. cit.;* **14-13** After Lehninger, A. L. (1975). *Biochemistry,* 2d ed. New York: Worth Publishers, Inc.; **14-14** *Ibid;* **14-16** (a-d) Bernhard Hirt; **14-17** Blumenthal, A. B., Kreigstein, H. J. & Hogness, D. S. (1973). *Cold Spring Harbor Symposium on Quantitative Biology, 38,* 205; **Page 300** C. M. Plork, Museum of Comparative Zoology, Harvard University

Page 301 © K. G. Murti/Visuals Unlimited; **15-1** © Gary R. Robinson/Visuals Unlimited; **15-6** (b) © K. G. Murti/Visuals Unlimited; **15-14** Hans Ris; **15-15** (b) Miller, O. L., Hamkalo, B. A. & Thomas, C. A. (1970). *Science, 69,* 392–395. © 1970 by AAAS

Page 319 Paris Match; **16-1** Jack Griffith; **16-5** Adapted from Darnell, J., Lodish, H. & Baltimore, G. (1986). *Molecular cell biology* (p. 278). New York: Scientific American Books; **16-6** Paris Match; **16-11** (a) Palchaudhuri, S., Bell, E. & Salton, M. R. J. (1975). *Infection and Immunity, 11,* 1141; (b), (c), (d) T. Kakefuda; **16-12** Judith Carnahan & Charles Brinton, Jr.; **16-13** After Ayala, F. J. & Kiger, J. A., (1980). *Modern genetics.* Menlo Park, CA: Benjamin/Cummings Publishing Co., Inc.; **16-14** *Ibid;* **16-15** Cohen, S. N. (December 27, 1969). *Nature;* **16-16** (a)-(d) B. Menge, J. V. D. Brock, H. Wunderli, K. Lickfield, M. Wurtz & E. Kellenberger; **16-20** Based on Alberts, *et al.* (1983). *Molecular biology of the cell.* New York: Garland Publishing Company; **16-22** Watson, *et al.,* (1987). *Molecular biology of the gene* (p. 335). Menlo Park, CA: Benjamin/Cummings Publishing Co., Inc.; **16-23** Stanley Falkow:

Page 340 Eli Lilly and Co.; **17-1** John C. Fiddes & Howard M. Goodman; **17-2** Adapted from Alberts, *et al.* (1983). *Molecular biology of the cell.* New York: Garland Publishing Company; **17-3** Adapted from Lehninger, A. L. (1982). *Principles of biochemistry* (p. 863, Fig. 28-4). New York: Worth Publishers, Inc.; **17-5** Adapted from Watson, J. D., Tooze, J. & Durtz, D. T. (1983). *Recombinant DNA* (p. 64, Fig. 5-5). New York: W. H. Freeman and Company; **17-6** (a) Stanley N. Cohen; **17-7** (a), (b), (c) Huntington Potter & David Dressler, *Life Magazine* © 1980 Time Inc.; **17-8** Adapted from Fristrom, J. W. & Spieth, P. T. (1980). *Principles of genetics* (p. 346, Fig. 12-20). New York & Concord: Chiron Press; **17-9** Adapted from Alberts, *et al.* (1983). *Molecular biology of the cell* (Fig.

4-56). New York: Garland Publishing Company; **17-10** (a) Jack Griffith; (b), (c) Adapted from Darnell, *et al.* (1986). *Molecular cell biology* (p. 247, Fig. 17-22). New York: W. H. Freeman and Company; (d) Daniel Nathans; **17-11** After Lehninger, A. L. (1982). *op. cit.;* **17-12** After *City of Hope Quarterly,* 7, 2. (Winter, 1978); **17-13** Eli Lilly and Co.; **17-14** Eugene W. Nester; **17-15** Keith Wood, University of California, San Diego

Page 355 Robert Noonan; **18-1** E. J. DuPraw; **18-2** Rich, A., *et al.* (1981). *Science,* 211, 171-176. © 1981 by AAAS; **18-3** (a) Victoria Foe; **18-4** (a) Barbara Hamkalo; (b), (c) Adapted from Alberts, *et al.* (1983). *Molecular biology of the cell* (Fig. 8-5). New York: Garland Publishing Company; **18-5** Victoria Foe; **18-6** After Alberts, *et al.* (1983). *op. cit.;* **18-8** James German; **18-9** George T. Rudkin; **18-10** (*photo*) James German; (*art*) After Chambon, P. (May 1981). Split genes. *Scientific American,* 60-71; **18-11** After Lewin, R. (1981). *Science,* 212, 28-32; **18-12** H. C. MacGregor; **18-13** (a), (b) Ullrich Scheer, W. W. Franke & M. F. Trendelenberg; **Page 366** Don Fawcett; **18-14** After Rahbar, S. (Winter 1982). Abnormalities of human hemoglobin. *City of Hope Quarterly,* 11, 2.; **18-15** After Goodenough, U. (1983). *Genetics,* 3rd ed. New York: Saunders College/Holt, Rinehart & Winston; **Page 370** Grabowski, P. J. & Cech, T. R. (1981). *Cell,* 23, 467-476; **18-18** After Lehninger, A. L. (1982). *Principles of biochemistry.* New York: Worth Publishers, Inc.; **18-19** *Ibid;* **18-20** UPI/Bettmann Newsphotos; **18-21** Adapted from Darnell, *et al.* (1986). *Molecular cell biology.* New York: W. H. Freeman and Company; **Page 375** Dr. Mary Clutter, Cold Spring Harbor Laboratory Research Archives; **18-22** (a), (b) R. D. Goldman; **18-23** (a) after Jon W. Gordon & Frank Ruddle; (b) Jon W. Gordon & Frank Ruddle; **18-24** R. L. Brinster; **18-25** Robert Noonan; **Page 381** R. Portman & M. L. Birnsteil

Page 382 © Tony Mendoza/The Picture Cube; **19-1** Gernsheim Collection, Humanities Research Center, University of Texas at Austin; **19-3** After Yunis, J. J. (1976). *Science,* 191, 1268-1270. © 1976 by AAAS; **19-4** (a) © Tony Mendoza/The Picture Cube; **19-7** (a), (b) Jorge J. Yunis; **19-8** (b) Courtesy Nell Ubbelohde & William Stryk; **19-9** After Stryer, L. (1988). *Biochemistry,* 3rd ed. (p. 512, Fig. 21-21). New York: W. H. Freeman and Company; **19-10** (a) Courtesy Jan Chalker; (b) John S. O'Brien; **19-11** Margaret Clark; **19-12** Scala/Art Resource; **19-13** Richmond Products, Boca Raton; **19-15** After Lerner, I. M. (1968). *Heredity, evolution, and society.* New York: W. H. Freeman and Company; **19-17** (a), (b) Steve Uzzell III; **Page 396** Lifecodes Corporation

Page 404 Donald R. Perry; **Page 405 & 20-1** © Michael Fogden/Animals Animals; **Page 407** © Brian Parker/Tom Stack & Associates; **20-2** (a) © Larry West; (b) © L. Campbell/NHPA; (c) John Shaw/NHPA; **20-3** (a) © John H. Gerard/DPI; (b) © Hans Reinhard/Bruce Coleman; **20-4** The Bettmann Archive; **20-6** (a) © Leonard Lee Rue III/Photo Researchers; (b) © Brian Parker/Tom Stack & Associates; (c) © Kevin Schafer/Tom Stack & Associates; (d) © Dale & Marian Zimmerman/Animals Animals; (e) © Tom McHugh/Photo Researchers; **20-8** R. M. Kristensen; **20-9 & 20-10** After Dobzhansky, T. (1977). *Evolution.* New York: W. H. Freeman and Company; **20-11** (a) After Lehninger, A. L. (1982). *Principles of biochemistry.* New York: Worth Publishers, Inc.; (b) Data provided by Lai-Su L. Yeh, Protein Identification Resource, National Biomedical Research Foundation, Georgetown University Medical Center; **20-12** Adapted from Sibley & Ahlquist (February 1986). *Scientific American,* 85; **Page 422** (a) © Ralph A. Reinhold/Animals Animals; (b) © Zig Leszczynski/Animals Animals; (c) Adapted from O'Brien, S. J., *et. al.* (September 12, 1985). *Nature,* 317, 141; **20-13** Adapted from Sibley & Ahlquist (February 1986). *Scientific American,* 92; **20-14** (a) © Manfred Kage/Peter Arnold; (b) © Eric V. Gravé; (c) © Robert P. Carr/Bruce Coleman; (d) © Marion Patterson/Black Star; (e) © Dwight R. Kuhn

Page 428 Hans Reichenbach; **21-1** © Paul Chesley/Photographers Aspen; **21-2** (a), (b), (c) David Greenwood; **21-3** Adapted from Fox, G. E. *et. al.* (1980). *Science,* 209, 459; **21-4** (*photo*) Lee D. Simon; **21-5** Victor Lorian & B. Atkinson; **21-6** (a) R. G. E. Murray; (b) I. D. J. Burdett & R. G. E. Murray; **21-7** John Swanson; **21-8** Hoeniger, J. F. M. (1965). *Journal of General Microbiology,* 40, 29; **21-9** (a) J. Adler; (b) After DePamphilis, M. L. & Adler, J. (1971). *Journal of Bacteriology,* 105, 395; **Page 435** D. L. Balkwill & D. Maratea; **21-10** (a) Eric V. Gravé; (b) R. S. Wolfe; **21-11** J. F. M. Hoeniger; **21-12** (a) Center for Disease Control; (b), (c) Listgarten, M. A. & Socransky, S. S. (1964). *Journal of Bacteriology,* 10, 127-138; **21-13** Virus Laboratory, Parke, Davis & Co.; **21-14** Hoeniger, J. F. M.

& Headly, C. L. (1968). *Journal of Bacteriology,* 96, 1835-1947; **21-15** Hans Reichenbach; **21-16** Stolp, H. & Starr, M. P. (1963). *Antoine von Leeuwenhoek,* 29, 217-248; **21-17** (a) Jerry D. Davis; (b) Germaine Cohen-Bazire; **21-18** (photo) M. Jost; **21-19** © Robert & Linda Mitchell; **Page 443** (a) Whately, J. & Lewin, R. A. (1977). *New Phytologist,* 79, 303-313; (b) F. R. Turner, Indiana University; **21-20** Jack Griffith; **21-21** (a), (b) Almeida, J. D. & Howatson, A. F. (1963). *Journal of Cell Biology,* 16, 616; (c) R. C. Williams & H. W. Fisher; (e) Lee D. Simon; **21-22** Jurgen Kartenbeck; **21-23** T. Koller & S. M. Sogo, Swiss Federal Institute of Technology, Zurich; **21-24** Stanley Prusiner; **21-25** Ny Carlsberg Glyptotek; **21-26** (a), (b) Computer graphics modeling and photography by Arthur J. Olson, Ph.D., Research Institute of Scripps Clinic, La Jolla, CA 92037 © 1988.

Page 452 © Kim Taylor/Bruce Coleman Ltd.; **22-1** © Eric V. Gravé; **22-2** (a) William J. Larsen; (b) D. A. Stetler & W. M. Laetsch; **22-3** (a), (b) R. W. Greene; **22-6** © D. P. Wilson/Eric & David Hosking; **22-9** (a) © Eric V. Gravé; (b) H. N. Guttmann; **22-10** (a) © Eric V. Gravé; (b) © D. P. Wilson/Eric & David Hosking; **22-11** (a) Kent Cambridge Scientific Instruments; (b) G. A. Fryxell; **22-12** (a), (b) © D. P. Wilson/Eric & David Hosking; **22-13** (a)-(d) Ronald Hoham, Colgate University; **22-14** (a) © M. I. Walker; (b) George I. Schwartz; (c) © Kim Taylor/Bruce Coleman Ltd.; **22-15** (a) © Grant Heilman Photography; (b) © Oxford Scientific Films, Animals Animals; (c) © D. P. Wilson/Eric & David Hosking; **22-17** (photo) Trond Braten, EM Laboratory for Biosciences, Oslo; **22-19** (a), (b) © D. P. Wilson/Eric & David Hosking; **22-20** (a) © Larry West; (b) © John Shaw/NHPA; **22-21** John W. Taylor; **22-22** Alma W. Barksdale; **22-23** After Niederhauser, J. S. & Cobb, W. (May 1959). *Scientific American;* **22-24** (a) © Eric V. Gravé; (b) Robert L. Owen; **22-25** (a), (b) © Eric V. Gravé/Photo Researchers; **22-26** © Eric V. Gravé; **Page 472** D. Kubai & H. Ris; **22-28** G. A. Horridge & S. L. Tamm; **22-30** (a)-(d) © Eric V. Gravé

Page 479 © John Shaw/NHPA; **23-1** © E. S. Ross; **23-3** © C. W. Perkins/Earth Scenes; **23-4** John H. Troughton; **23-5** H. C. Hoch & D. P. Maxwell; **23-6** (c) Carolina Biological Supply Company; **Page 484** John Hodgin from Buller, A. H. R. *Researches on Fungi, Vol. 6.* New York: Longman, Inc.; **23-7** (a) © Alvin E. Staffan; (b) © John Shaw; **23-9** (a) © C. Bracker; (b) Drawing by B. O. Dodge, from E. L. Tatum; **23-11** (a) © Agenzia Fotografica, Luisa Ricciarini, Milan; (b) © E. S. Ross; (c) Donald Simons; **23-12** G. L. Barron; **Page 489** (a) N. Allin & G. L. Barron, University of Guelph; (b) David Pramer; **23-13** (a) © E. S. Ross; (b) © J. N. A. Lott, McMaster Univ./Biological Photo Service; (c) © John Shaw/NHPA; **23-14** (a), (b) Vernon Ahmadjian; **23-15** (a) R. J. Molina, J. M. Trappe & G. S. Strickler; (b) B. Zak, U. S. Forest Service

Page 493 © Larry West; **24-1** © James L. Castner; **24-2** (a) L. E. Graham; (b) K. J. Nicklas; **24-3** L. E. Graham; **24-4** R. E. Magill, Botanical Research Institute, Pretoria; **24-5** (a), (b) © D. S. Neuberger; **24-6** (a) © E. S. Ross; (b), (c) © Robert A. Ross; **24-7** © D. S. Neuberger; **24-9** (a) © G. I. Bernard/Earth Scenes; (b), (c) © Ray F. Evert; **24-10 & 24-11** After Raven, P. H., Evert, R. F. & Eichhorn, S.E. (1987). *Biology of plants,* 4th ed. New York: Worth Publishers, Inc.; **24-12** (a., (b) © E. S. Ross; (c) © R. Carr; **24-13** (a) © Marilyn Wood/Photo NATS; (b) © B. Miller/Biological Photo Service; **24-14** © Liz & Tony Bomford/Survival Anglia Ltd., **Page 504** Kristine Rasmussen & Stuart Naquin; **24-15** (a) D. A. Steingraeber; (b) © G. R. Roberts; (c) © William H. Harlow/Photo Researchers; (d) © Michael Fogden/Earth Scenes; **24-17** (a) © Heather Angel; (b) © Photo Researchers; **24-19** © Grant Heilman Photography; **Page 510** © Stephen J. Krasemann/Photo Researchers; **24-24** © Larry West; **24-26** (a) © Robert & Linda Mitchell; (b), (c) E. S. Ross; (d) W. A. Calder; (e) Donna Howell; **24-27** (a) © Sandy Gregg/Imagery; (b) © G. I. Bernard/Oxford Scientific Films, Animals Animals; **24-28** © Patti Murray/Earth Scenes

Page 518 © H. Wes Pratt/Biological Photo Service; **25-1** © C. E. Mills/Biological Photo Service; **25-2 & 25-3** Nat Fain; **25-4** (a) Adolf Seilacher, University of Tübingen; (b) M. F. Glaessner; **25-5** (a), (b), (e) S. Conway Morris; (c), (d) H. B. Whittington; **25-7** © Marty Snyderman; **25-9** © Manfred Kage/Peter Arnold; **25-10** © Heather Angel; **25-11** Robert B. Short; **25-16** After Buchsbaum, R. & Milne, L. J. (1962). *The lower animals: Living invertebrates of the world.* Garden City, NY: Doubleday & Co., Inc.; **25-17** (a) © D. P. Wilson/Eric & David Hosking; (b) © Runk & Schoenberger/Grant Heilman Photography; **25-18** © Shaw Photos/

Bruce Coleman; **25-20** © H. Wes Pratt/Biological Photo Service; **Page 530** (a) © Allan Power/Bruce Coleman; (b) © Ian Took/Biofotos; **25-21** Runk & Schoenberger/Grant Heilman Photography; **25-24** (a) Maria Wimmer; **25-25** © Scott Johnson/Animals Animals; **25-27** After Roberts, M.B.V. (1971). *Biology: A functional approach.* New York: The Ronald Press Company; **25-29** (a) © J. H. Robinson/Photo Researchers; (b) © Eric V. Gravé/Photo Researchers; (c) © Walker England/Photo Researchers; **Page 537** (a) Ming Wong; (b) © Frank L. Lambrecht/Visuals Unlimited; **25-30** (a) W. Sterrer/Bermuda Biological Station; (b) © Kathie Atkinson/Oxford Scientific Films, Animals Animals; **25-31** © Kim Taylor/Bruce Coleman; **25-32** (c) © Larry Jensen/Visuals Unlimited; **25-33** (a) R. P. Higgins; (b) © M. I. Walker/Photo Researchers; (c) © Kim Taylor/Bruce Coleman; **25-34** © D. Wrobel/Biological Photo Service

Page 543 Doug Wechsler; **26-4** (a) After Roberts, M.B.V. (1971). *Biology: A functional approach.* New York: The Ronald Press Company; (b) A. Solem, Field Museum of Natural History; **26-5** (b), (d) Specimens courtesy of Edward I. Coher; (c) After a photograph supplied by Russel H. Jensen, Delaware Museum of Natural History; **26-6** © James Carmichael Jr./ NHPA; **26-7** (a), (b) © Doug Wechsler; **26-9** Carl Roessler/Animals Animals; **26-10** © G. I. Bernard/Oxford Scientific Films, Animals Animals; **26-11** (c) © D. P. Wilson/Eric & David Hosking; **26-15** © Oxford Scientific Films, Animals Animals; **26-16** © Sea Studios, Inc./ Peter Arnold; **26-17** © Russ Kinne/Photo Researchers; **26-18** (a) © Visuals Unlimited; (b) © Ralph Buchsbaum; (c) © Bradford Calloway; **26-19** © D. Foster, WHOI/Visuals Unlimited; **26-20** J. Teague Self; **26-21** Robert O. Schuster; **26-22** © Raymond A. Mendez/Animals Animals; **26-23** Russell Zimmer; **26-24** Specimen supplied by Edwad I. Coher; **26-25** © Dr. Steven K. Webster/Biological Photo Service; **26-26** Judith Winston, American Museum of Natural History

Page 565 © Robert & Linda Mitchell; **27-1** © Stephen Dalton/Oxford Scientific Films, Animals Animals; **27-2** © Ken Lucas/Biological Photo Service; **27-3** (a) © Doug Wechsler; (b) © Jack Dermid; **27-4** After Wilson, E. O. *et. al.* (1977). *Life: Cells, organisms, populations.* Sunderland, MA: Sinauer Associates, Inc.; **27-7** © Heather Angel; **27-9** (a) © Zig Leszczynski/Animals Animals; (b) © Karl Switak/NHPA; (c) © Michael Fogden/Animals Animals; (d) © Peter J. Bryant/Biological Photo Service; (e) © Stephen Dalton/NHPA; **27-11** (a) © A. Kerstitch/Taurus; (b) T. E. Adams; (c) © E. S. Ross; **27-12** © E. S. Ross; **27-13** (a) © Heather Angel; (b) © Robert & Linda Mitchell; (c) © Kjell B. Sandved/Bruce Coleman; **27-15** *(photo)* © Jane Burton/Bruce Coleman Ltd.; **27-16** (a) © E. S. Ross; (b) © John R. Macgregor/Peter Arnold; (c), (d) © E. S. Ross; **27-17** (a)–(h) © Peter J. Bryant/Biological Photo Service; **27-18** After Gordon, M. S. (1972). *Animal physiology: Principles and adaptations,* 2nd ed. New York: The Macmillan Company; **27-19** John Mais; **Page 581** James E. Lloyd; **27-20** After Roberts, M.B.V. (1971). *Biology: A functional approach.* New York: The Ronald Press Company; **27-21** © Don & Esther Phillips/Tom Stack & Associates; **27-23** Ross Hutchins; **27-25** © E. S. Ross; **27-26** © Alvin E. Staffan

Page 587 © Frans Lanting; **28-1** © Fred Bavendam/Peter Arnold; **28-2** © Jeff Rotman; **28-4** © Heather Angel; **28-5** (a) © Heather Angel; (b) © Fred Bavendam/Peter Arnold; **28-6** (a), (b) F. W. E. Rowe, Australian Museum; **28-7** © Steve Earley/Animals Animals; **28-8** (a), (b) © D. P. Wilson/Eric & David Hosking; **28-9** © Heather Angel; **28-10** (b) © Heather Angel; **28-11** (c) © R. L. Sefton/Bruce Coleman; **28-12** James Hanken; **28-13** © Breck P. Kent/Animals Animals; **28-14** © Zig Leszczynski/Animals Animals; **28-15** (a)–(d) Redrawn from Lewin, R. (1982). *The thread of life.* Washington, D.C.: Smithsonian Books; (e) © Oxford Scientific Films, Animals Animals; **28-16** American Museum of Natural History; **28-17** © Tom McHugh/Photo Researchers; **28-18** © Kim Taylor/Bruce Coleman; **28-20** (a) © Nielsen/Imagery; (b) © Gordon Langsbury/Bruce Coleman; **28-21** (a) © Zig Leszczynski/Animals Animals; (b) Lee R. Crist, Member of Golf Collectors' Society; **28-22** (a) American Museum of Natural History; (b) © Barbara Laing/Picture Group; **28-23** © Anthony Mercieca/Photo Researchers; **28-24** (a) The New York Zoological Society; (b) © Jack Dermid; **28-26** (a) © Michael Fogden/Bruce Coleman; (b) © Kenneth W. Fink/Bruce Coleman; (c) © Harry Engels/Photo Researchers; (d) © Jen & Des Bartlett/Bruce Coleman Ltd.; (e) © Frans Lanting; (f) © Jeff Rotman; (g) © C. Allan Morgan/Peter Arnold

Page 610 © Heather Angel; **Page 611 & 29-1** © Bruce D. Thomas; **Page 613** © E. R. Degginger/Earth Scenes; **29-3** (a)–(d) © Dennis Kunkel/ Phototake; **29-4** (a) © David L.Mulcahy; (b) Myron C. Ledbetter; **29-5** (a) © Louise K. Broman/Photo Researchers; (b) © James L. Castner; (c) © E. R. Degginger/Earth Scenes; (d) © John C. Coulter/Visuals Unlimited; (e) © Doug Wechsler; (f) © E. S. Ross; (g) © Virginia P. Weinland/Photo Researchers; **29-6** (c) © Ray F. Evert; **29-7** (a)–(d) © Ray F. Evert; **Page 622** © Ray F. Evert; **29-12** (a) © W. H. Hodge/Peter Arnold; (b) © Larry West; (c) © Renee Purse/Photo Researchers

Page 625 © James L. Castner; **30-4** © Manfred Kage/Peter Arnold; **30-5** © Ray F. Evert; **30-6** (a) © G. J. James/Biological Photo Service; (b) © James L. Castner; (c) © John D. Cunningham/Visuals Unlimited; **30-8** © Ray F. Evert; **30-9** © Robert Mitchell/Earth Scenes; **30-10** After Ray, P. (1971). *The living plant,* 2d ed. New York: Holt, Rinehart & Winston, Inc.; **30-11, 30-12 & 30-13** © Ray F. Evert; **30-14** After Ray (1971), *op. cit.*; **30-16** (a) © Jack Dermid; (b) © Robert Mitchell; **30-17** (a) © Biophoto Associates/Photo Researchers; (b), (c) © Ray F. Evert; (d) © Randy Moore/Visuals Unlimited; **30-18** (a), (b), (c) © Ray F. Evert; **30-19** (d) © J. N. A. Lott, McMaster University/Biological Photo Service; **30-20 & 30-21** © Ray F. Evert; **30-24** © G. I. Bernard/Oxford Scientific Films, Earth Scenes; **30-25** Frank C. Vasek; **30-27** From Ketchum, R. (1970). *The secret life of the forest,* American Heritage Press; **Page 646** (a) © Bruce Coleman; (b) C. W. Ferguson, Laboratory of Tree-Ring Research, University of Arizona

Page 650 © Breck P. Kent/Earth Scenes; **31-2** © James L. Castner; **31-3** John Shaw, National Audubon Society Collection/Photo Researchers; **31-4** After Richardson, M. (1968). *Translocation in plants.* New York: St. Martin's Press (After Stoat & Hoagland, 1939); **31-5** John L. Troughton; **31-6** Adapted from Northington, D. K. & Goodin, J. R. (1984). *The botanical world,* Mosby; **31-7** After Penny, M. G. & Bowling, D. J. F. (1974) *Planta,* 119, 17–25; **31-8** Jerry D. Davis; **31-9** © Roger Archibald/ Earth Scenes; **31-10** (a), © R. K. Burnard/BPS; (b) © C. W. May/BPS; **Page 658** (a) © Grant Heilman Photography; **Page 659** (b) Troughton, J. H. & Donaldson, L. (1972). *Probing plant structures.* New York: McGraw-Hill; **31-11** (a), (b) © Science Photo Library/Photo Researchers; (c) George A. Schaefers; **Page 661** (b), (c) Walter Eschrich & Eberhard Fritz; **31-15** © Robert P. Carr/Bruce Coleman; **31-16** S. A. Wilde; **31-17** © Runk & Schoenerger/Grant Heilman Photography; **Page 666** (a) © Jane Burton/Bruce Coleman; (b) © Breck P. Kent/Earth Scenes; **Page 667** (c), (d) © Runk & Schoenberger/Grant Heilman Photography; **31-18** (a) R. R. Herbert, R. I. Holsten & W. F. Hardy, E. I. duPont de Nemours & Co.; (b) R. R. Herbert, J. G. Griswell & R. W. F. Hardy, Central Research & Development Department, E. I. duPont de Nemours & Co.; (c) Mary Alice Webb; **31-19** © Karlene V. Schwartz

Page 671 © Larry West; **32-1** © Jack Dermid; **32-5** © Runk & Schoenberger/Grant Heilman Photography; **32-6** (a) © J. N. A. Lott, McMaster Univ./BPS; (b), (c) Nitsch, J. P. (1950). *American Journal of Botany,* 37(3); **32-9** H. R. Chen; **Page 678** Philip Harrington; **32-10** (a) © Ray F. Evert; (b) © Grant Heilman Photography; **32-12** © Larry West; **32-14** S. W. Wittwer & Michigan Agricultural Experiment Station; **32-15** After Wilson, E.O. *et. al.* (1977). *Life: Cells, organisms, populations.* Sunderland, MA: Sinauer Associates, Inc.; **32-16** D. E. Varner; **32-17** Kiem Trân Thanh Vân, CNRS, Laboratoire de Phytotron; **32-18 & 32-19** Randy Moore, **32-20**(a), (b) After Naylor, A. W. (May 1952). The control of flowering. *Scientific American;* **32-21** © John R. MacGregor/Peter Arnold; **32-24** © Breck P. Kent/EarthScenes; **32-25** (a), (b) © Jack Dermid; **32-27** © E. S. Ross; **32-28** (a) © Jeff Foott/Bruce Coleman; (b) © E. S. Ross; **32-29** (a), (b) © James L. Castner; **Page 696** After Wilson (1977), *op. cit.*

Page 698 Richard Clarkson/Time Magazine; **Page 699 & 33-1** © Biophoto Associates, Photo Researchers; **Page 701** © Lennart Nilsson, BEHOLD MAN. Boston: Little, Brown & Company; **33-4** © Veronika Burmeister/Visuals Unlimited; **33-5** (a), (b) L. A. Staehelin & B. E. Hull, University of Colorado, Boulder; **33-6** (b) Keith R. Porter; (c) © John D. Cunningham/Visuals Unlimited; **33-8** © Hugh E. Huxley; (b) Don Fawcett; **33-11** After Eckert, R. & Randall, D. (1978). *Animal physiology.* New York: W. H. Freeman & Company; **33-13** © Heather Angel; **33-16** © Lennart Nilsson, BEHOLD MAN. Boston: Little, Brown & Company

Page 919 Antone G. Jacobson; **45-1** Kathryn W. Tosney; **45-2** (a) William Byrd; (b) Everett Anderson; **45-3** Adapted from Alberts, *et al.* (1983). *Molecular biology of the cell.* New York: Garland Publishing Company; **45-4** After Baker, J. J. W. & Allen, G. E. (1977). *The study of biology,* 3rd ed. Reading, MA: Addison-Wesley Publishing Company, Inc.: **45-5 & 45-6** Tryggve Gustafson; **45-7** After Wilson, E. O. *et al.* (1978). *Life on earth,* 2d ed. Sunderland, MA: Sinauer Associates; **Page 925** (a) Turner, F. R. & Mahowald, A. P. (1976). Scanning electron microscopy of *Drosophila* embryogenesis: I. The structure of the egg envelopes and the formation of the cellular blastoderm. *Developmental Biology, 50,* 95-108; (b) Adapted from Alberts, *et al.* (1983), *op. cit.;* **45-11 & 45-12** After Huettner, 1949; **45-13** Antone G. Jacobson; **45-14** After Huettner, 1949; **45-16** After Waddington, 1966; **45-18** M. P. Olsen; **45-19 & 45-20** After Wilson, *et al.* (1978), *op. cit.;* **45-21** After Torrey, 1962; **45-23** After Sussman, M. (1964). *Animal growth and development,* 2d ed. Englewood Cliffs, NJ: Prentice-Hall, Inc.; **45-24** After Bodemer, C. W. (1968). *Modern embryology.* New York: Holt, Rinehart & Winston, Inc.; **45-25** S. R. Hilfer; **45-26** (a)-(f) © Runk & Schoenberger/Grant Heilman Photography; **45-27** Kathryn W. Tosney; **45-28** Adapted from Karp, G. & Berrill, N. J. (1981). *Development,* 2d ed. New York: McGraw-Hill; **45-29** (a) Martin Raff; (b) Adapted from Alberts, *et al.* (1983), *op. cit.;* **45-30 & 45-31** Lewis Wolpert; **45-32** (a), (b), (c) Petit Format, Nestle/Science Source, Photo Researchers; **45-33** Eugene M. Long; **Page 943** (a), (b) Walter J. Gehring, University of Basel; (c), (d) E. B. Lewis; **45-36** (a), (b), (c) © Lennart Nilsson, BEHOLD MAN. Boston: Little, Brown & Company; **45-37** © London Scientific Photos; **45-38 & 45-39** © Lennart Nilsson, BEHOLD MAN. Boston: Little, Brown & Company; **45-41** (a), (b), (c) © Jeffrey Reed/Medichrome; **45-42** © International Stock Photo

Page 958 © Bryn Campbell/Biofotos; **Page 959 & 46-1** © E. S. Ross; **Page 961** © Bruce D. Thomas; **46-2** © Zig Leszczynski/Animals Animals; **46-3** (a) © Michael W. F. Tweedie/Bruce Coleman; (b) © Michael W. F. Tweedie/Photo Researchers; **46-4** © R. E. Pelham/Bruce Coleman; **46-6** American Museum of Natural History; **46-7** (a) © Jane Burton/Bruce Coleman Ltd.; (b) © Leonard Lee Rue III/Bruce Coleman; (c) © G. R. Roberts; **46-8** (a) © Kenneth W. Fink/Photo Researchers; (b) © Steven C. Kaufman/Peter Arnold; (c) © A. B. Joyce/Photo Researchers; **46-9** © Jeff Foott; **46-10** Painting by Charles R. Knight, American Museum of Natural History; **46-11** (a)-(e) from Rugh, R. *Experimental embryology: Techniques and procedures,* 7th ed. Minneapolis: Burgess Publishing Co.; **46-12** Redrawn from Luria, Gould & Singer (1981). *A view of life.* Benjamin/Cummings; **Page 971** © Calvin Larsen/Photo Researchers; **46-13** © Bruce D. Thomas; **46-14** Office of Public Information, Columbia University

Page 974 © Jen & Des Bartlett/Bruce Coleman Ltd.; **47-1** Hulton Picture Company/The Bettmann Archive; **Page 975** © Stephen Dalton/Photo Researchers; **47-2** © George Holton/Photo Researchers; **47-4** (a), (b) John Mais; **47-5** After Mather & Harrison (1949). *Heredity, 3,* 977; **47-6** R. C. Lewontin; **47-7** Courtesy POSSUM; **47-10** Edmund B. Gerard; **47-11** Victor McKusick; **47-12** © Anthro-Photo; **47-13** © Jim Brandenburg; **47-15** © Daniel J. Harper; **47-16** © Jen & Des Bartlett/Bruce Coleman Ltd.; **Page 987** from Lewis Carroll. *Through the looking glass;* **47-17** (b) © John Colwell/Grant Heilman Photography

Page 991 © Frans Lanting; **48-1** After Harlow, J.R. (1976), *Scientific American;* **48-2** (a) © Heather Angel; (b) © Kim Taylor/Bruce Coleman Ltd.; **48-3** (a), (b) © Heather Angel; **48-4** (a) © Kenneth W. Fink/Photo Researchers; (b) © Kent & Donna Dannen/Photo Researchers; **48-5** Karu Kurita © 1982 DISCOVER Publications; **48-6** After Charles W. Brown, Santa Rosa Junior College; **48-7** (a), (b) © Mart R. Gross; **48-8** (a) © E. R. Degginger/Animals Animals; (b) Adapted from Futuyama (1979). *Evolutionary biology* (After Clarke 1962, based on data of Popham 1942); **48-9** (a) © A. J. Deane/Bruce Coleman Ltd.; (b) © Jonathan Scott/Seaphot Ltd.: Planet Earth Pictures; (c) © Avon & Tilford/Ardea Photographics; (d) © Frans Lanting; **Page 1001** © M. Andersson/VIREO; **48-10** (a) © Alan D. Cruikshank/Photo Researchers; (b) © Alan Root/Bruce Coleman; **48-11** From Gould, S. J. (1977). *Ever since Darwin.* New York: W. W. Norton & Company; **48-12** After Clausen, J. & Hiesey, W. M. (1950). *Publication 615,* Carnegie Institute of Washington; **48-13** (a) © Lincoln Brower; (b) © Frans Lanting; (c), (d) © Dwight R. Kuhn; **48-14** (a)-(d) © Lincoln Brower; **48-15** (a)-(d) © E. S. Ross; (e) © Hans Pfletschinger/Peter Arnold; **48-16** (a) © Chip Clark; (b) © Robert &

Linda Mitchell; **48-17** (a) © C. Haagner/Bruce Coleman; (b) M. P. Kahl/Bruce Coleman; **48-18** (a), (b) © E. S. Ross; **48-19** © E. R. Degginger/Bruce Coleman

Page 1010 © Jen & Des Bartlett/Bruce Coleman; **49-1** (a), (b) © Joan Baron; **49-2** After Wilson, E. O. & Bossert, W. H. (1971). *A primer of population biology.* Sunderland, MA: Sinauer Associates; **Page 1013** John B. Shelton; **49-4** After Raven, P. H. *et al.* (1986). *Biology of plants.* New York: Worth Publishers, Inc.; **49-6** © Jen & Des Bartlett/Bruce Coleman; **49-8** After Wallace, V. & Srb, A. (1964). *Adaptation.* Englewood Cliffs, NJ: Prentice-Hall, Inc.; **Page 1017** United States Department of Agriculture; **49-9** (a), (b), (e), (f) © Tui De Roy; (c) © Tui De Roy/Bruce Coleman; (d) © D. Cavagnaro/Peter Arnold; **49-10 & 49-11** After Lack, D. (1961). *Darwin's finches.* New York: Harper & Row, Inc.; **49-12** Peter R. Grant, *Ecology and evolution of Darwin's finches* © 1986 Princeton University Press. Plate 34 (Photo by P. T. Boag) reprinted with permission of Princeton University Press; **49-13** After Raven, P. H. *et al* (1986), *op. cit.;* **49-14** After Dobzhansky, T., Ayala, F. J., Stebbins, G. L. & Valentine, J.W. (1977). *Evolution.* New York: W. H. Freeman & Co. (after Simpson & Beck 1965); **49-15** © Bill Thompson/Woodfin Camp & Associates; **49-16 & 49-17** Redrawn from Simpson, G. G. (1951). *Horses.* New York: Oxford University Press; **49-18** After Stanley, S. M. (1979). *Macroevolution: Pattern and process.* New York: W. H. Freeman & Co.

Page 1031 Gallery of Prehistoric Art, Astoria, NY; **50-1** © A. Christiansen/Frank Lane Picture Agency; **50-4** (a) © E. S. Ross; (b) © H. Albrecht/Bruce Coleman; **50-5** *National Geographic* (November 1985); **50-6** (a), (b) © Tom McHugh/Photo Researchers; **50-7** After Howells, W. W. (1967). *Mankind in the making,* rev. ed. Garden City, NY: Doubleday & Co., Inc.; **50-8** © Townsend P. Dickinson/Comstock; **50-9** Ralph Morse, LIFE Magazine © 1965 Time Inc. **50-10** © Mickey Gibson/Tom Stack & Associates; **50-11** © Peter Veit/DRK Photo; **50-12** (a), (b), (c) © Peter Davey/Bruce Coleman; (d) Teleki/Baldwin; (e), (f) © Richard Wrangham/Anthro-Photo; (g) © Linda Koebner/Bruce Coleman; **50-13** © David L. Brill; **Page 1039** John Reader; **50-14** © David L. Brill, National Museum of Tanzania, Dar es Salaam; **50-15** (a), (b) The Cleveland Museum of Natural History; **50-16** *National Geographic* (November 1985); **50-17** © John Reader; **50-18** © Michael Holford; **50-20** © A. Walker; **50-21** © Michael Holford; **50-23** (a) Ralph S. Solecki; (b) © David L. Brill; **50-24** © Michael Holford; **Pages 1048-1049** Gallery of Prehistoric Art, Astoria, NY; **50-26** © James D. Wilson/Woodfin Camp & Associates

Page 1052 © Jim Brandenburg; **51-1** A. Christiansen/Frank Lane Picture Agency; **51-2** © Stephen Dalton/Photo Researchers; **51-3** P. Kirk Visscher; **51-4** Redrawn from Lorenz & Tinbergen, 1937; **51-6** © George D. Lepp/Comstock; **51-7** Nina Leen, LIFE Magazine © 1965 Time Inc.; **51-8** After Mazudazo Konish; **51-9** (a), (b) © C. R. Carpenter; **51-10** (a) © E. S. Ross; (b) © Alan Blank/Bruce Coleman; **51-11** (a) © E. S. Ross; (b) © Treat Davidson/NAS, Photo Researchers; **51-12** (a) © E. S. Ross; (b) © Colin G. Butler/Bruce Coleman; **51-13** (a) Christopher Springman © National Geographic Society; (b) © Raymond A. Mendez/Animals Animals; **51-14** (a) © Jim Brandenburg; (b) T. W. Ransom; **51-15** (a) © Leonard Lee Rue III/NAS, Photo Researchers; (b) © Robert Dunne/Photo Researchers; (c) © Sullivan & Rogers/Bruce Coleman; (d) © Zig Leszczynski/Animals Animals; **51-16** © Chesher/Photo Researchers; **Page 1067** Shirley Baty; **51-17** After Kreb, J. R. & Davies, N. B. (1981). *An introduction to behavioral ecology.* Sunderland, MA: Sinauer Associates; **51-18** (a), (c) © Patricia D. Moehlman; (b) After Krebs & Davies (1981), *op. cit.;* **51-19** © Charlie Heidecker/Visuals Unlimited; **51-20** © Joseph Popp/Anthro-Photo; **51-21** © Sarah Blaffer Hrdy/Anthro-Photo; **51-22** © Steven C. Kaufman/Peter Arnold; **Page 1073** (a), (b) © Hans & Judy Beste/Animals Animals; **51-23** © David Bygott/Anthro-Photo; **Page 1076** Andy Blaustein

Page 1084 © M. Philip Kahl, Jr./Photo Researchers; **Page 1085 & 52-1** © Günter Ziesler/Bruce Coleman Ltd.; **Page 1087** © M. A. Chappell/Animals Animals; **52-3** © John M. Burnley/Photo Researchers; **Page 1091** Adapted from May & Anderson (1987). *Nature, 326,* 140; **52-7** After *Nature, 298* (August 26, 1982); **52-8** Adapted from McNaughton & Wolfe (1979). *General ecology,* 2d ed. Holt, Rinehart, & Winston; **52-9** (a) © Joseph Van Wormer/Bruce Coleman Ltd.; (b) © Mark N. Boulton/Bruce Coleman Ltd.; (c) © Jeff Foott; **52-10** (a), (b) McNaughton & Wolfe

(1979), *op. cit.* (After Lack 1954); **52-11** After Kendeigh, S. C. (1961). *Animal ecology.* Englewood Cliffs, NJ: Prentice-Hall Inc. (After Shelford, 1911); **52-12** © Jane Burton/Bruce Coleman Ltd.; **52-13** (a) © Russ Kinne/Photo Researchers; (b) © C.A. Henley; (c) © Jane Burton/Bruce Coleman; (d) © Cosmos Blank/Photo Researchers; (e) © Eric Hosking/Bruce Coleman; (f) © E. S. Ross; **52-14** From Sarukhan & Harper, 1973; **52-15** © Jack Dermid; **52-16** © M. A. Chappell/Animals Animals

Page 1106 © James L. Castner; **53-1** © William E. Townsend, Jr./Photo Researchers; **53-2** Adapted from Jared Diamond (May-June 1978). *American Scientist,* 326; **53-5** After MacArthur, 1958; **53-6** © Joan Baron; **53-7** After Lack, D. (1961). *Darwin's finches.* New York: Harper & Row; **53-8** Adapted from Ricklefs, R. E. (1979). *Ecology,* 2d ed. New York: Chiron Press Inc.; **53-9** Joseph Connell; **53-11** © Maslowski/Photo Researchers; **53-12** (a), (b), (c) © Thomas Eisner; **53-13** (a) © Fred Whitehead/Animals Animals; (c), (d) After Fenton, M. B. & Fullard, J. H. (1981). Moth hearing and the feeding strategies of bats. *American Scientist,* 69, 266-275; **53-14** (a) © Michael Fogden/Bruce Coleman; (b), (c) © James L. Castner; (d) © Peter Ward/Bruce Coleman; (e) © E. S. Ross; **53-15** (a), (b) Australian Department of Lands; **53-17** (a), (b) Joe Lubchenco; **53-18** Australian News and Information Service; **53-19** (a) © R. Mariscal/Bruce Coleman; (b) © Douglas Faulkner; (c) © E. S. Ross; (d) © P. Ward/Bruce Coleman; **53-20** (a)-(d) © Daniel Janzen; **53-21** After Ricklefs (1979), *op. cit.;* **53-22** Adapted from Connell, J. H. (March 24, 1978). *Science,* 199, 1303; **53-23** (a) Roger Del Moral; (b) Shirley Baty; **Page 1127** (a), (b), (c) © Townsend Dickenson/Comstock; **53-24** (a), (b) Wayne Sousa; **53-25** (a), (b) Teresa Turner

Page 1131 © McDougal Tiger Tops/Ardea Photographics; **54-1** NASA; **Page 1135** NASA; **54-6** © Jack Dermid; **Page 1141** Robert R. Hessler; **54-9** (a), (b) © McDougal Tiger Tops/Ardea Photographics; **54-10** (a) © G. D. Plage/Bruce Coleman Ltd.; (b) © M. A. Chappell/Animals Animals; (c) © F. W. Lane/Bruce Coleman; **Page 1144** Irven de Vore/Anthro-Photo; **54-16** G. E. Likens; **54-17** (b) Michigan Department of Natural Resources; **54-18** © J. Donosa/Sygma

Page 1154 & 55-1 © Wolfgang Kaehler; **55-2** (a) © Michael P. Gadomski/Earth Scenes; (b) © Nielsen/Imagery; (c) © Alan G. Nelson/Animals Animals; **55-3** (a), (b) © D. P. Wilson/Eric & David Hosking; (c) © Tom Stack/Tom Stack & Associates; (d) © Claudia Mills/Biological Photo Service; **55-5** (a) © Jack Dermid; (b) © Frans Lanting; (c) © Stephen J. Krasemann/Photo Researchers; (d) © Breck P. Kent/Animals Animals; (e) © Jeff Foott; **55-6** (a) © Jack Dermid; (b) © Rip Griffith/Photo Researchers; **55-7 & 55-8** After Arms, K. & Camp, P.S. (1979). *Biology,* 2d ed. New York: Sanders College Publishing; **55-9** After A. W. Küchler; **55-10** (a) © Rod Planck/Tom Stack & Associates; (b) © Robert P. Carr/Bruce Coleman; (c) © Patti Murray/Earth Scenes; (d) © Breck P. Kent/Animals Animals; (e) © Wolfgang Kaehler; (f) © Leonard Lee Rue III/Animals Animals; **55-11** (a), (b) © Les Blacklock; (c) © Johnny Johnson/Animals Animals; (d) © Erwin & Peggy Bauer/Bruce Coleman; (e) © Leonard Lee Rue III/Animals Animals; (f) © Ralph A. Reinhold/Animals Animals; **55-12** (a) © Bill Ruth/Bruce Coleman; (b) © Bob & Clara Calhoun/Bruce Coleman; (c) © Kennan Ward; (d) © Bob & Clara Calhoun/Bruce Coleman; (e) © E. R. Degginger/Animals Animals; **55-13** (a) Pat Caulfield; (b) © Paul Rogers/Imagery; (c) © Franz J. Camenzind/Seaphot Ltd.: Planet Earth Pictures; (d) © Joe McDonald/Animals Animals; **55-14** © Peter Ward/Bruce Coleman; **55-15** (a) Jack Wilburn/Earth Scenes; (b) © R.J. Erwin/Photo Researchers; (c) © Ardea Photographics; **55-16** (a) © Bruce Coleman; (b) © Max Thompson/NAS, Photo Researchers; (c) © Ric Ergenbright; (d) © Greg Vaughn/Tom Stack & Associates; (e) J. Reveal; **55-17** (a) ©Jeff Foott/Bruce Coleman; (b), (d) © E. R. Degginger/Bruce Coleman; (c) © Stephen J. Krasemann/Peter Arnold; **55-18** (a) © E. S. Ross; (b) C. W. Rettenmeyer; (c) © Kevin Schafer/Tom Stack & Associates; **55-19** (a) © James H. Carmichael/Bruce Coleman; (b) © L. & D. Klein/Photo Researchers; (c) © Leonard Lee Rue III/Bruce Coleman; (d) © E. S. Ross; (e) © Wolfgang Kaehler; **55-20** (a), (b) © R. O. Bierregaard

Index

Facilitation hypothesis, **1125**, 1128, 1130
Factor VIII, in hemophilia, 392, 397, 752
FAD, 204, 729
  in Krebs cycle, 195, 200
FADH$_2$, 204
  and electron transport, 195–197
  in Krebs cycle, 195–196, 200–201
Fainting, 759–760
Fairy rings, 486
Fall overturn, 46
Family, 410, 412, 427, 915
Family tree, 423
Famine, 1149
Fangs, 566, 568, 585 (*see also* Chelicerae; Pincers)
Far-red light, and phytochrome, 686–687, 695
Farsightedness, 867
Fascicle, 505
Fast-twitch fibers, 878
Fat(s), **67–70, 228, 578** (*see also* Fatty acids, Lipids)
  absorption of, in lymphatic system, 763
  and calories, 68
  chemical bonding in, 34
  condensation reactions in, 67–68
  as connective tissue, 710
  in diet, 730
  digestion of, 720, 722–733, 730
  as energy storage, 67
  and fatty acids, 67
  and glucagon, 832
  and glucocorticoids, 830
  and heart disease, 761
  human requirements for, 726, 730
  insolubility of, 47
  as insulation, 68–69, 779, 786, 790
  as insulation, 68–69
  molecular structure of, 67
  saturated, 67–68
  and seasonal adaptation, **787–789**, 790
  subcutaneous, 710
  transport in bloodstream, 750
  unsaturated, 67–68
Fat cells, 723, 729
Fate maps, 929
Fatty acids, 55, 67–69, 202
  absorption of, 723
  breakdown of, 825
  in chemoreception, 863
  digestion of, 721
  essential, 730
  and fats, 67
  and growth hormone, 833
  in muscle fibers, 878
  properties, 82
  prostaglandins as, 833
  saturated, 67
  unsaturated, 67
Fear, 720, 760, 847
Feathers, 598, 599, 702, 779, 786–787, 790, 937, 938
  (*see also* Insulation)
Feather stars, 587, 589, 604
Feces, 547, 724–725, 768, 775
Feedback inhibition, **176**, 177, 184, 712–713
  and aldosterone, 775
  in blood pressure regulation, 762
  of hormone production, 823, 825, 827, 831, 839
  in liver regulation of glucose, 730
  and recognition memory, 899
  in vertebrates, 714
Feeding zones, 1109, 1110
Felidae, 412
Felis, 412
Female, 235–236, 243, 258
  biological role of, 235
  cell (*see* F⁻ cell, Conjugation)
  and color blindness, 391–392
  genetic disorders in, 400
Female choice, and sexual selection, 984, **998–1001**
Female cooperatives, in hominid evolution, 1044
Female hormones
  and atherosclerosis, 761
  regulation of, **913–915**
  reproductive system, **907–915**, 917
Fermentation, 11, 190–191, 204, 206
Femur, 705, 707, 1041
Fenton, M. Brock, 1115–1116
Fermentation, 11, 190–191, 204, 206
  in herbivore digestion, 724
  in urkaryotes, 453

and yeasts, 486
"Ferments," 190
Ferns, 256, 496–498, **502–503**, 504, 526, 1164
  fossils of, 502
  leaves of, 503
  life cycle of, **250–252**
  reproduction in, **502, 503**
  sporophyte of, 250
Fertility of soil, 663
Fertility factor (*see* F plasmid)
Fertilization, 236, 238, 426, 900, 901, **912**
  and allele frequency, 984, 989
  in angiosperms, 511, 615, **617**
  artificial, 236
  in chickens, 931
  and contraception, 915–917
  external, 519, 523, 538, 543, 550, 557, 558, 559, 588, 595, 900, 920, 926
  and gamete fusion, 252
  and genetic variability, 984
  in human beings, 940–941
  internal, 519, 523, 535, 542, 543, 550, 551, 553, 577, 595, 900
  and life cycle, 250
  and meiosis in oocyte, 917
  and nitrogen-fixing symbionts, 665
  in plants, 499, 502
  and sea urchins, 920–921, **922**, 951
    in sexual reproduction, 249–251, 261
  of self, 590
Fertilization membrane, **921–922**, 941
Fertilization tubes, in water molds, 467
Fertilizers (synthetic), 669, 1149
Fetus
  addictions in, 948
  amniocentesis of, 387–388, 400
  deformities of, 947
  heart of, 947–949
  and hemoglobin, 367, 743
  human, **946**
  male sex hormones in, 905
  prenatal detection of disease, 387
  spontaneous abortion of, 385
Feulgen, Robert, 282–283
Feulgen stain, 282–283
Fever, 784, 792–793, 795
F factor (*see* F plasmid)
F factor sequence, 327–328
Fiber (dietary), 730
Fiber cells (in plants), **637**, 647
Fibers
  in connective tissue, 704
  elastic, 704
  muscle, **706–708** (*see also* Muscle fibers)
  nerve (*see* Axons)
  reticular, 704
Fibrils, 77, 426
  contractile, 426
  of spirochetes, 437
Fibrin, **751–752**, 761
Fibrinogen, 750, **751**
Fibroblast, 100
  mouse, 146
Fibrosis, 394
Fibrous proteins, 77, 82, 111
Fibrous root systems, 635
Fibrous sheath, 902
Fiddle heads, 503
Fidelity
  and selfish gene, 1072
  as survival strategy, 1074
Field layer, 1164
"Fight or flight" reactions, 846–847, 857
  and infant alertness, 951
Filaments (*see* Actin filaments; Contractile filaments; Intermediate filaments; Myosin filaments)
  and bacterial shapes, 436
  in desmosomes, **703**
  in flowers, 511
  fungal, 480 (*see also* Hyphae)
  in prokaryote flagellum, 434
  of sarcomeres, 874
  of stamen, 614, 624
Filariasis, 540
Filter-feeding, 549
Filtration, in kidney, 771, 776
Fingers, 416
  development of, in human embryo, 946–948

evolution of, in primates, 1032
  and motor cortex, 886
Fire
  control of, and human evolution, 1045, 1051
  in grasslands, 1169–1170
Fire blight, 439
Fireflies, **580–581**
Firs, 505
"First family," 1040 (*see also* Hadar fossils)
First filial generation (*see* F$_1$ generation)
Fischer, Emil, 170
Fish(es), 587, **593–595**, 604 (*see also specific fishes*)
  bony, 593–595, 604
  brain of, 883
  cartilaginous, 593, 594, 595 (*see also* Dogfish; Sharks; Skates; Rays)
  in coral reefs, 530
  earliest, 497
  freshwater, 594
  heart, 763
  kidney in, 767
  lunged, 594–595
  respiration in, 734–737
  rhinencephalon, 883
  saltwater, 594
  startle behavior in, 1053–1054
  temperature regulation in, 780–781
  and transition to land, **593–595**
Fishing industries, and ocean currents, 1157
Fish oils, 728
Fission, 535
  binary, in protozoa, 568–569
"Fitness," 974–975, **1059**, 1066 (*see also* Inclusive fitness; Reproductive success)
  and behavioral evolution, 1076
  and competition, 1106, 1129
  of *Drosophila* for genetic research, 263
  of environment for photosynthesis, 211
  and frequency-dependent selection, 997, 1009
  of light, **209–211**
  of material, 238, 284
  and natural selection, 991–992
  and sexual selection, 1001
  and social behavior, 1059, 1078
  of water, 54
Fitz Roy, Captain, 1, 2, 5
Fixation
  of carbon, 216, 223–225, 228, 230
  of microscopic specimens, 96, 98, 101
  of nitrogen, 352
Fixatives, 96, 98
Fixed action patterns, **1054–1055**, 1077
Flagella, 92, 110, **122–123**, 125, 131, 149, 153, 251, 426
  algal, 456–457
  apical, 457
  ATPases in, 181
  vs. axial filaments, 437
  chemiosmotic power and, 200
  of dinoflagellates, 460
  of euglenoids, 458–459
  lateral, 457
  of opalinids, 473
  prokaryotic, **433–436**, 450
  protozoan, 468–470
  of slime molds, 469
  of sperm cell, 902–903
  of sponge, 522
  of water molds, 467, 469
Flagellated cells, 522
Flagellates, 455, 477
Flagellin, 433–434
Flakes, 1042, 1047
Flame cells, 537, 542
Flamingo, 423
Flatworms, 122, 520, 526, 532, **533–537**, 542 (*see also* Tapeworms; Trematodes; Turbellarians)
  free-living, 534
  life cycle of, 537
  marine, 534
  parasitic, 542
Flavin(s)
  and active transport in plants, 655
  in phototropism, 655, 673
Flavin adenine dinucleotide (*see* FAD)
Flavin mononucleotide (*see* FMN)
Flavoproteins, 195
Flemming, Walter, 109

gametes, 252–253, 258–259
genetics of, 266, **340–354**
genome, 380, 396–399
heredity and social organization, 235
inherited characteristics, 271
karyotype of, **383–388**
life cycle of, 252
number of chromosomes in, 264
origin of, **1047–1050**
and symbiosis, 537, 539
Human population
explosion of, **1096–1097**
and agricultural revolution, **1148–1149**
and destruction of rainforests, 1176–1177
and spread of Sahara, 1172
history of, 1096
theories of, 7 (*see also* Biosphere; Ecosystems;
Population dynamics; Populations)
Human society, and reciprocal altruism, 1074, 1079
Humboldt current, 1157
Humidity, and transpiration, 652
Hummingbird, 61, 67
Humulin, 354
Humus, 663–664, 670
Hunger
and limbic system, 885
and population explosion, 1097, **1148–1149**
sensory receptors for, 859
Hunter-gatherers, 1096, 1144
Hunting, 915
Huntington's disease, 391, 398, 853
as autosomal dominant, 391
diagnosis, 395, 397
Hutchinson, G. E., 961, 1099
Hutton, James, 2–3, 5
Huxley, Thomas H., 968, 1027
Hybridization, 308, 380
and cancer research, 378
denaturation, 363–364
DNA-DNA, **420–423**, 427
in genealogy studies, 420–421
and monoclonal antibodies, 802, 803
nucleic acid, 340, **345**, 346–347, 353–354
and repetitious DNA, 363–364, 365, 372
RNA and DNA, 308
Hybrid plant cultivation, 678
Hybrids
allopolyploidy in, 1012
RNA-DNA, 308
between subspecies, 1011, 1014
Hybrid vigor (*see* Heterosis)
Hydra, 259, 527
digestion in, 527
mortality rate of, 1092
nerve impulse in, 841
nervous system of, 527, 841–842
Hydrocarbons, **55**, 56, 61
Hydrochloric acid (HCl), 47, 720
Hydrogen
in anaerobic respiration, 439
in atmosphere, 733
atomic structure, 23–24
biological importance of, 35
bomb, 163
and cohesion-tension theory, 651
covalent bonds in, 32
ion, 48, 49 (*see* Hydrogen ions)
isotopes of, 35
and origin of life, 87
Hydrogen bonds, 210, 294, 296
and cohesion-tension theory, 651
and COS regions of lambda phage, 334
in DNA, 288–290, 299
and evaporation, 44
in insoluble compounds, 47
and protein structure, 74–75
in water, 33, **41–43**, 53
Hydrogen ions (H⁺), 48, 49, 204, 748
in blood, regulation of, 745–746
in electron transport, 197
in enzymatic reactions, 172
in glycolysis, 187–189
in Krebs cycle, 194
peroxide, 119
in photosynthesis, 218, 220, 229
in plant growth, 674–675
in plant movements, 693
in respiration, 743, 745

and stomatal movements, 653–654
Hydrogen sulfide, 217, 436, 439, 1140
Hydrolysis, 64–65, 722
of ATP, 181–182, 875
in enzyme regulation, 175
of sucrose, 170
Hydrolytic enzymes, in plants, **694–695**
and gibberellins, 681–682
Hydronium ion (H₃O⁺), 47, 54
Hydrophilic molecules, **46**, 54
and charge, 46
covalent bonds in, 46
and transport, 134, 136–137
Hydrophobic molecules, **47**, 54, 134, 136–137
Hydrostatic pressure, 128, 133–134, 604
in arteries, 764
of blood, vs. osmotic potential, 762
and lymphatic system, 762
and water gain, 769
Hydroxide ion (OH⁻), 47, 54, 56
Hydroxyl group, 56–57, 305
of nucleotides, 80
Hydrozoans, 526, **527–528**, 529, 541 (*see also* Hydra)
Hygienic honey bees, 1053
Hymenopterans, 573, 576, **1059–1061**, 1059 (*see also*
Ants; Bees; Wasps)
Hyperactivity, 831
Hyperparathyroidism, 829–830
Hyperpolarization
of postsynaptic cells, 856
of synaptic cells, and learning, 897
in vision, 872
Hypertension, **761**
and dietary salt, 730
Hypertonicity, 132–134, 142
of bacterial cells, 432
of blood, 725
of urine in mammals, 776
Hyperventilation, 746–747
Hyphae, 467–468, 481, 483, 485–487
cell walls of, 480
in ectomycorrhize, 491
in endomycorrhizae, 491
of fungi, 480–481, 489, 491
haustoria (*see* Haustoria)
of *Penicillium*, 488
rhizoids, 482
septate, 482
sporangiophores, 481–482
Hyphal sheath, 491
Hypoblast, 932
Hypocotyl, 631
Hypolimnion, 46
Hypothalamic releasing hormones, 855
Hypothalamus, **712**, 823–824, **826–828**, 831–832,
853, **883**, 898
and ADH, 774, 776, 827–828
and childbirth, 828
control of pituitary by, 825, **826–827**
and female sex hormones, 913
and limbic system, 885
and male sex hormones, **906**, 907–909, 917
negative feedback control in, 827
portal system of, **827**
and prolactin production, 825
and sexual behavior, regulation of, 914
and temperature regulation, 713, 784, 790
Hypotheses, in science, 14–15, 17, 85, 412 (*see also*
specific hypotheses)
of branching sequences, in cladistics, 417 (*see also*
Cladogram)
in ecology, 1087, 1104
and experimental design, 131, 1001
testing of, 417, 1106–1107
Hypothyroidism, 829
Hypotonicity, 132–134, 142
Hyracotherium, 409, **1025–1027** (*see also* Eohippus;
Horse)
Hyrax, 409

IAA (indoleacetic acid), 673–674, **676–677** (*see also*
Auxins)
I band, 874
Ibis, 423
Ice, 44, 1155
Ice Ages, and evolution of plants, **510**
Identical repeats, 335

Igneous rock, 1160
Ikatura, Keiichi, **350–351**
Imbibition, 42
Imitative learning, **1057–1058**, 1078
Immigration, 989, 1123–1124, 1130 (*see also* Island
biogeography model)
and community composition, 1123–1124, 1130
and destruction of habitat, 1126–1127
and extinction, 1123–1124
and gene flow, 982
and genetic variability, 989
and Hardy-Weinberg equation, 979
and population growth rate, 1088–1092
and speciation, in Darwin's finches, 1017
Immune disorders, 812–818, 819 (*see also* AIDS;
Allergies; Autoimmune diseases)
Immune response, **791–818**, 830–831, 835–836, 855
(*see also* Immune system)
and blood, 749, **750–751**
cancer and, **808–809**
and cellular communication, 808
evolution of, **819**
immune system, **793–819**
inflammatory response, **791–793**
and lymphatic system, 763
nonspecific defenses, **791–793**
primary, 793
and prostaglandins, 839
secondary, 793
Immune system, 711, **793–819**
AIDS, 813–819
allergies, 812–818
antibody coding in, 372
autoimmune diseases, 812–818
B lymphocytes and antibodies, **795–801**
and cancer, **809**
cell-mediated immunity, **801–808**, 819
and depletion of ozone layer, 1134
development of, in human infant, 948
disorders of, **812–818**
fever, 784
human, 793
immunoglobulins, 70, 72, 76
memory cell response, 819
mucosal immunity, 799
T lymphocytes, **801–808**
Immunity, **791–819**
cell-mediated, 801–808, 819
in fetus, 948
memory-cell response, 819
and mother's milk, 727
mucosal, 819
Immunization, 449, 797 (*see also* Inoculation;
Vaccination)
Immunofluorescence microscopy, 111
Immunoglobulins, 70, 72, 76, 750, **795–801**, 812,
818 (*see also* Antibodies)
Immunology, 797, 818
Immunosuppression, 831
Immunosuppressive drugs, 810, 813, 819, 834
Imperfect fungus, 488
Impetigo, 448
Implantation, of embryo, 901, 912, 916, 941
and progesterone, 914
Imprinting, **1056–1057**, 1078
in white-crowned sparrow, 1056–1057
Inbreeding, 977, 982
anatomical arrangements against, 984–985
and heterosis, 988
and kin selection, 1077
and mutation, 982
taboos against, 985
and sterility, 977, 985
Incest taboos, 985
Incisors, 601–602, **716–717**
Inclusive fitness, **974, 1069**, 1075, **1076–1077**, 1078
and altruism, 1079
and kin recognition, 1075, **1076–1077**
and selfish gene, **1070–1075**
Incomplete dominance, **268**, 279
Incomplete penetrance, 271, 279
Indehiscent fruits, 620–621 (*see also* Achenes; Nuts)
Independent assortment, **241–243**, 245, 247,
260–261, 270, 274, 276, 279
of behavioral genes, 1053
in meiosis, 260–261
principle of, **241–243**, 247
and sexual reproduction, 984

Independent effectors, 527
India, 1092, 1097
Indian pipe, 94
Indoleacetic acid (*see* IAA)
Induced-fit hypothesis, 170–171
Inducer protein, 323–325, 337
Inducers, embryonic, **930–931**, 952
Inducible enzymes, 321–322
Inducible operons, 323–325, 337
Induction, 337
Induction (embryonic), **930–931**, 952
   in chick embryo, 935–936
   primary, 930–931, 935–936
   secondary, 936
Industrialization, **962–965**, 973
Industrial melanism, **962–963**
   as directional selection, 997
Industrial revolution, and population growth, 1096–1097
Infant diarrhea, 449
Infanticide, 1071
Infant mortality rates, 949
Infection, 751
   by AIDS virus, **815–816**
   and antibodies, 819 (*see* Immune response; Immune system; Immunoglobulins)
   and birth defects, 947, 953
   and burns, 808
   and contraceptive techniques, 917
   and cytotoxic T cells, 808
   epidemics, 813
   and inflammatory response, 818 (*see also* Inflammatory response)
   and overcrowding, 437
   smallpox, 796–797
Infection cycle, 285, 286, 330–333, 338, 451
   in *E. coli*, 330
   and genetic recombination, 336
   lysogenic cells, **331–332**, 333
   lysogeny, 322, 330–331, **332**, 333
   lytic cycle, 330–335, 338
   microbial, 448
   phage, 344
   and prokaryotic nucleoid, 330
   retroviral, **342**
   susceptibility, 385
   and vacuoles, 330
   and viruses, 322, 330–335, 338, 432, 445–446, 451
Infection rate, of AIDS virus, **1090–1091**
Infection threads, 666–668
Infectious diseases, 1087, **1090–1091**
   epidemics, 813
   as limitation to population growth, 1084
   in plants, 678
Inflammatory response, **791–793**, 808, 818, 830–831, 835–836, 855
   and allergies, 813
   complement in, 798
   vs. immune response, 808
   and microbial infection, 448
Inflorescence, 616, 670
Influenza, 444–445, 813
Infrared light, 211, 872
Ingenhousz, Jan, 208, 217
Ingram, Vernon, 304
Ingrowths, 734
Inheritance, 238, 264, 413 (*see also* Genetics; Heredity)
   blending, **237–238**, 247
   early theories, 4–5
   and evolutionary theory, **962**, 973
   mechanism of, 247
   Mendelian, 361
   and natural selection, 7, 238
   non-Mendelian, 362
   polygenic, **271–272**, 279
   unit concept of, 247
   of variations, 247
Inhibin, 906–907
Inhibition
   of animal activity, 568
   of cell growth, 147–148
   in cerebral ganglia, 557
   competitive inhibition, **177**, 178, 184
   of enzyme activity, **177**
   irreversible inhibition, **177–178**, 184
   noncompetitive inhibition, **177–178**, 184
   reversible, 176–178

in seed coat, 623
of self-fertilization, 990
Inhibition hypothesis, **1125, 1128**, 1130
Inhibitor proteins, 177
Initiation, stage of mRNA translation, 312–313
Initiation codon, for DNA replication, 350–351
Initiation complex, 312, 314
Initiator proteins, for DNA replication, 294
Injury, response to (*see also* Immune response; Immune system; Inflammatory response)
   blood clotting, 753
   in cerebral hemispheres, 889
   edema in, 762
   in knee, 707
   and memory loss, 894
   in plants, 638–639, 647, 696
Innate releasing mechanisms, 1054, 1077
Inner cell membrane, of prokaryotes, 432
Inner ear, 865
Inoculation, 796
Inorganic molecules, 11 (*see also* Elements; Minerals)
   movement of in ecosystems, 1153 (*see* Biogeochemical cycles)
   oxidation of, in chemosynthetic autotrophs, 439
Input, of minerals, 1149
Insecticides, resistance to, **964**
   as directional selection, 997
   and malaria, 474
Insectivores, vs. early primates, 1031
Insects, 14, 554, 565, 566, 572, **573–578**, 586
   vs. animals, 14
   characteristics of, **574**
   chitin, 480
   determination of germ cells in, 924
   evolution of, 513
   and evolution of flowers, **513, 614**
   excretion in, 766
   exoskeleton of, 480
   life cycle of, **575–578**, 586
   microevolution in, **964**
   pollination by, 512–514
   reproduction in, 579
   resistance to insecticides in, **964**
   respiration in, 734
   social behavior in, **1059–1061**, 1078
Insertion sequences, 335, 374 (*see also* Transposons)
Insertion site, 329
   of prophage, 334
Insolubility, 47
Inspiration, 738, 741
Instars, 576, 577, 580
Insulation, 598, 599, **779, 786–787**, 790
   in camels, 789
   fats as, 68–69
Insulin, 72, 75–76, 138, 304, 349, 351, 822–824, 832–833, 837–839, 855 (*see also* Humulin)
   genetic code for, 359
   and regulation of blood glucose, 726
   secretion of, in pancreas, 725
   structure, 75
Integral membrane proteins, 104–106, 114, 124, 135, 178–179
   and cellular transport, 135, 142
   and electron transport, 200
   and membrane transport proteins, 178–179
   permeases, 135, 178–179
   in red blood cells, 114
Integration, 711–714
   of memory with sensory information, 898–899
   of muscular activities, 886
   of neuroendocrine system, at hypothalamus, 883
   in vertebrate brain, **881–899**
Integuments
   of angiosperm ovule, 619
   of gymnosperms, 506–507
Intelligence
   in chimpanzees, 1037
   evolution of, 1044
   in *Homo sapiens*, 702
Intelligence quotient (IQ), 994
Intentions, and intrinsic processing areas, 893
Interbreeding, 1014–1015, 1029
   and populations, 974, 989
   vs. reproductive isolation, 1010–1030
Intercellular spaces, 654
Intercostal muscles, in respiration, 741, 745, 748
Intercourse, 900–903, **904**, 905–918
   and contraception, **915–917**

and ovulation, 915
Interference competition, 1106
Interferons, **793**, 818, 823, 855
   of cytotoxic T cells, 807
   and monoclonal antibodies, 802
Intergenerational conflict, 1071
Interglacials, 510
Interior surface
   of cell membrane, 135, 138–139
Interleukins, 821, 823, 835–836, 855
Intermediate disturbance hypothesis, **1124–1125**, 1130
Intermediate filaments, 110, 111, **112**, 124
   and nucleolus, 111
Intermediate-repeat DNA, **365**, 380
Internal fertilization, 900
Internal membrane
   and cellular transport, 142, 197–198
Internal transport, 547
Interneurons, 708, **709**, 843–844
Internodes, 626, 641, 648
Interoceptos, 859
Interphase, **146**, 149, 155
   of meiosis, 253, 255, 261
   and DNA transcription, 359
Intersexual selection, 998–999
Interspecific competition, **1106–1114**
   in barnacles, 1113
   and ecological niches, 1111
   and intermediate disturbance hypothesis, 1124
Interspecific interactions, **1106–1122**
Interstitial cells, 824, **901**, 905, 906–907, 917
Interstitial fluid, 750, 754, 763, 775, 793–794
   as a body compartment, **768–769**
   and loop of Henle, 773–774
   and lymphatic system, 763
   and respiration, 737
   and substance exchange at capillaries, 763
Intertidal zones, 1113, 1119, 1158
   competition in, 1124–1125, 1127
Intervening sequence (*see* Intron)
Intestines
   absorption of calcium from, 831
   development
      in sea urchin embryo, 924
      in vertebrate embryo, 937
   in digestion, 730
   of earthworm, 555
   human, 544
   and immune system, 795
   motility, and nervous system, 723
Intoxication, 202
Intracellular fluids, 775
   as a body compartment, 768–769
Intracellular competition, 1106 (*see also* Dominance hierarchies; Inclusive fitness; Sexual selection)
Intraspecific discrimination, **1056**, 1078 (*see also* Kin selection)
Intrauterine devices (IUDs), **916**
Intravenous drug use, and AIDS, 817
Intrinsic processing centers, **892–893**, 898–899
Introns, **361–363**, 380, 420
   and alternative splicing, 369, 370
   and cellular nucleus, 380
   excision of, 369, 370
   and meiosis, 363
   in protein coding, 366
   and recombination, 363
   and viroids, 446–447
   and viruses, 362
Inversion, of chromosomal segments, 277, 279
   and protection of supergenes, 993
Invertebrates, 414, **519–697** (*see also specific invertebrates*)
   asexual reproduction in, 1103
   in coral reefs, 530
   embryonic development in, **920–926**
   evolution of, 538
   lower, **522–543**
   nervous systems of, 841–842
Inverted repeats, 335–336, 338, 374
In vitro fertilization, 912
Involuntary controls of respiration, 747
Involuntary muscles, 706 (*see also* Muscles)
Involuntary nervous system, 712 (*see* Autonomic nervous system)
Iodine, 728, 828, 829
Ions, **31**

inflammatory response, 791–793
interferons, 793
Nonspecific mineral requirements, 656
Nontemperate bacteriophages, 333–334
Noradrenaline, 746, 759, 824, 832–833, 839, 853, 857, 951
in infant alertness, 951
as neurotransmitter, 857
in sympathetic nervous system, 846–847
Northern coniferous forests (*see* Taiga)
Northern elephant seal, 983
Northern hemisphere, 1136, 1156–1157
North Pole, 1136
Nose, 738
Nostoc, 489
Nostrils, 736, 935
Notochord, 591–593, 604, 928–930, 932–936, 946, 952
Nottebohm, Fernando, 889
"Nuclear clock," 1048–1049
Nuclear envelope, 91, 108, 112, 116, 124, 426
in animal cell, 112
in cell division, 150–152
in cell variety, 125
in ciliates, 472
double membrane system, 127
in eukaryotes, 452–453
and exchange of substances, 127
and intermediate filaments, 111
and meiosis, 254–255, 258
in mitosis, 150–152
and nucleus, 366
in protozoa, 470
Nuclear family
conflicts of interest in, 1070
evolution of, 1044
Nuclear membrane, 108
Nuclear pores
and gene transfer, 380
and ribosome assembly, 366
Nuclear testing, **1150–1151**
Nuclear winter, 1025
Nuclei (in nervous system), 842 (*see also* Nervous system; Neurons)
Nucleic acid base-pairing
and recombinant DNA, 245
Nucleic acids, 55, 80, 420, 427
Nucleoid, 90, 102, 337
and infection cycle, 336
in prokaryotes, 431
and prokaryotic replication, 319–320
Nucleolus, 155, 472
in animal cell, 106, **112, 124**
in cell division, 150, 152
function, 366
during meiosis, 254, 368
during mitosis, 366
and nucleus, 366
structure, 366
Nucleosomes, **357, 379,** 444
Nucleotide(s), 55, 80, 82, 87, 305, 307, 309–312, 314, 316–317, 420
addition of, 315, 317
and behavior, 1052–1053
biosynthesis of, for DNA, 298
deletion of, 315, 317
disruption in sequence of, 315, 317
DNA and RNA, 308, 317
in DNA replication, 281, 283–284, 289–290, 292–299
and gene regulation, 380
sequence of, 307, 316
sequencing of, **340–341, 420,** 427, 429, 817, 942, 978
substitution of, 316
"synonym" sequences, 309
Nucleus (atomic), **23–24,** 38, 163
in atomic structure, 28
and oxidation-reduction reactions, 166
polar-covalent bonding and, 32
Nucleus (cellular), **90–92,** 93, 101, 102, **108–110,** 124, 127, 146, 155, 267, 383
in animal cell, **112**
in cell division, 155
chromosomes in, 356
in coenocytic organisms, 462
daughter, 145, 153–155
in meiosis, 252–253

diploid, 249, 253, 483
division of, 253, 462 (*see also* Meiosis; Mitosis)
envelope (*see* Nuclear envelope)
of eukaryotes, 452
and exchange of materials, 127
function, 109
of fungi, 485, 487
haploid, 253, 255–256, 258, 260
and heredity, 109–110
and introns, 380
membrane (*see* Nuclear membrane)
and messenger RNA tranbscription, 369
in mitosis, 150, 152–153
of muscle cells, 706
of neutrophils, 792
vs. nucleolus, 366
of oocyte, 910
of ovum, 922, 926
in plant cell, **113**
pores of, 366, 380
post-meiotic, 258
and rRNA genes, 365-**366**
of sperm cells, 902–903
Numerical phenetics, 415, **416,** 427
"Nurse" cells, 138
Nursing, 702
evolution of in primates, 1033
genetic conflicts of interest in, 1071
and monogamy, 1072
Nutrient canals, 705
Nutrient cycles, 1176
Nutrients, in forest ecosystems, 1147–1150
Nutrition
in arachnids, 570
in chemosynthetic autotrophs, **439** (*see also* Autotrophs)
in heterotrophs, 438–439 (*see also* Heterotrophs)
human, 70–71, 76, 730
of human embryos, 932, 937, 941, 944–945
in mammals, 601
modes of, 424, 426, 518, 555
of mothers, and birth weight, 949
in photosynthetic autotrophs, 440–443 (*see also* Autotrophs)
in plants, **650–670**
and population explosion, 1097
in prokaryotes, **438–442**
in reptiles, 597
Nutritional requirements, 726–729
Nutritive cell, 527–528
Nuts, 511, 621, 637
Nymphs, 576, 586

Oats, 226
Obelia, 527
Obesity, 728–729
Objectivity, in science, 17
Obligate anaerobes, 429, 439, 442
Obligate parasitism, 330
Observation
vs. experimentation in ecology, 1087, 1106–1107, 1112
in science, 14–17
Occipital lobe, 885
Ocean, 87
Ocean currents, 1156–1157, **1158,** 1160
Oceans, **1155–1157,** 1177 (*see also* Marine environments; Marine organisms)
currents of, 1156–1157, **1158,** 1160
and evolution, 1012
net primary productivity of, 1139
and sunlight, 1133
Ocelli, 528, 533, 535, 542, 582 (*see also* Eyes, simple)
evolution of, 529
Ochoa, Severo, 307
Octopuses, 545, 551, **552,** 553, 563–564
Odors, discrimination of, 861 (*see also* Smell)
Offspring
conflicts of interest with, 1071
and "fitness," 974–975
and Hardy-Weinberg equation, 989
live vs. eggs, 702
numbers of, in primates, 1033
size and number, and population growth, 1099, 1105
small vs. large, **1102**
variability in, 986
Offspring, care of, 561, 575, 589, 900, 917, 999, 1052

alternative patterns of, **1099–1104**
in birds, 598
in bivalves, 550
in California condor, 1104
among cephalopods, 553
evolution of in primates, 1033, 1034
in insects, 576
live vs. eggs, 702
in mammals, 600, 702
in sponges, 524
O'Hara, Richard, 1076
Oils, **67–70**
as algal food reserves, 457, 465
atomic bonding in, 34
Okazaki, Reiiji, **296–297**
Okazaki fragments, **296–297,** 299
and eukaryotic replication, 358
in prokaryotic replication, 320
Olduvai Gorge, 1040, 1042
Old World monkeys, 1034–1035, 1051
Oleic acid, 67–68
Olfaction (*see also* Odors; Smell)
development of, in embryo, 936
selection against, in primates, 1033, 1052
Olfactory bulb, 883
Olfactory cells, 862
Olfactory epithelium, 861, 862
Oligocene epoch, 496, 1026, 1034
Oligochaetes, 554, 555, 564 (*see also* Earthworms)
Oligonucleotides, synthetic, 343, 346, 353
Oligosaccharides, 673, **682,** 683, 688, 694–695
Ommatidia, 579, 580
response time of, 580
Omnivores, 550, 601, 1107, 1166
Onchocerca, 540
"Oncogene hypothesis" of cancer, 376
Oncogenes, 376–377, 381
One-celled organisms (*see also* Prokaryotes, Protists, Single-celled organisms)
algae, 460, 461
cell cycle in, 147
cell division in, 144
euglenoids, 458
in evolution, 461
green algae, 461
heterotrophs, **468–476**
protozoa, **468–476**
yeasts as, 486
"One-gene–one-enzyme" hypothesis, **301–304,** 316
One-way digestive tract, 538, 715
Onion root, 146, 150, 151
Onychophorans, 520, 560, 561, 564
Oocysts, 473
Oocytes, 294, **907, 908, 909,** 910–914, 940
and contraceptive techniques, 916
in estrus, 915
and hormones, 913–914
human, **907, 908, 909,** 910–914
pathway of, **912–913**
primary, 258, 917
production of, **910–911,** 924
and prostaglandins, 835
secondary, 258, 917
Oogamy, 463, 467, 469, 477, 494, 516
Oogenesis, **910–911,** 924
Oogonia, 258, 925, 946
Oomycetes, 455, 467–469, 477
vs. fungi, 482
and potato blight, 468, 678, 682, 1149
Opalina, 473
Opalinids, 455, 468, 469, 473, 477
Oparin, A. I., 87
Open circulatory system, vs. closed, 752–753
Operant conditioning, **1055–1056,** 1077
Operator gene, **323,** 324–325, 337
Operculum, 551, 569
Operons, **322–326,** 337, 368, 380
inducible, 323–325, 337
lac, 323, 325, 326, 337
operator gene, **323,** 324–325
prokaryotic, **322–326**
repressible, 323–325, 337
and RNA polymerase, 323–325
trp operon, 323, 325, 337
Opiates, **854–855**
addiction to, 854–855
Opisthonecta, 144
Opines, 352

Paleocene epoch, 496, 1034
Paleolithic art, 1042, **1048–1049**
Paleontology, 3–4, 15, 420, **1021–1025**, 1028, 1030
Paleozoic era, 497, 505, 510
Palisade cells, 214, 627, 648
Palmiter, Richard, 378
Palolo worms, 210
Palpi, 575, 585
Pampas, 1169
Pancreas, 116, 127, **715–716**, 730, 824, **832–833**
  and biosynthesis of chymotrypsin, 175
  in digestion, **725**
  as endocrine gland, 823
  as exocrine gland, 823
  and glucagon, 64
  hormones from, 855
  and regulation of blood glucose, 726
Pancreatic amylase, 722, **725**
Pandas, 422
Pandorina, 37, 461, 462
Pangaea, 1011, **1012–1013**
Panthera, 412
Panting, 785–786
Pantothenic acid, 194, 729
para-Aminobenzoic acid (*see* PABA)
Paradoxical sleep, 890
para-Hydroxyphenylpyruvate, 332
Parallax, 867
Paramecium, **13**, 102, 110, 122, 132–133, 139, 144, 471
  avoidance behavior in, **475–476**
  competition between, 1108
  locomotion in, 473
  macro- and micronuclei, 471
  temperature sensitivity in, 475, 476
Paramylon, 457–458, 476
Paraphyletic taxon, 418
Parapodia, 454, 558
Parasitism, 792
  and adaptation, 1006
  in dodder, 691–692
  in *E. coli*, 448
  in fungi, 480, 482, 486, 488, 492 (*see also* Mycorrhizae)
  in gastropods, 550
  in horsehair worms, 543
  in leeches, 558
  in mycoplasmas, 432
  in nematodes, 539–540, 542
  vs. predation, **1120**
  in prokaryotes, 432–439, 442, 451
  in protozoa, 468, 469, 470
  in tapeworms, 533, 536
  in viruses, 330
Parasympathetic nervous system, 764, 842, 845–847, 856, 857
  cardiovascular regulation in, 762
  and erection of penis, 904
  and regulation of digestion, 723
  regulation of heartbeat, **757**
  vs. sympathetic nervous system, 846, 847, 857
Parathyroid glands, 824, **829**, 937
Parathyroid hormone, 824, 829–831, 839
Parazoa, 520, 522
Parenchyma cells, 626, 627, 632, 636, 645, 648
  secondary growth and, 645
  and water movement, 652
Parent cell, in meiosis, 252
Parenting, 917
Parietal cells, 718
Parietal lobe, 885–886, 892
Parkinson's disease, 853, 897
Parotid glands, 718
Parthenogenesis, 541, 543, 557, 560, **1103**
Partial pressure, 733, 740, 742, 743, 745, 747
Partitioning, 529, 555, 562
  of resources (*see* Resource partitioning)
Partitions (*see also* Septae), 554
Passive diffusion
  and absorption in small intestine, 723
  in loop of Henle, 772–773
Pasteur, Louis, 11, **86–87**, 236, 797
Patagonian hare, 966
Patella, 707
Paternity, tests for, 811
Pathogenic microbes, 439, 448–451
  capsules in, 433
  toxins in, 448, 451

Pathways
  for oocyte, **903–905**
  for sperm, **903–905**
Pattern formation, in embryo, **938**
Patterns A–D, of Huntington's disease, 395, 397
Patterns of water circulation, 1156–1157
Pauling, Linus, 74, 288–289, 291, 304, 340
Pauropods, 573, 586
Pavlov, Ivan, 1055, 1078
P_{CO_2}, 740, 743, 746, 747
PCR, in diagnosis, 396
Peacock, 77
Peanut worms, 520, 559, 564
Peat, 504
Pecking
  and learning, 1055–1056
  order, 1062–1063, 1078
Pectin(s), 66, 154
  in cell walls, 457
  in collenchyma cells, 637
Pedicel, 562
Pedigree(s)
  for color blindness, 392
  for hemophilia, 393
  human, 382
  for Huntington's disease, 395
Pedipalps, 568–570, 585
Peking man, 1040–1041
Pelagic life, **1156–1157**
Pelecypoda (*see* Bivalves)
Pelican, 423
Pellagra, 729
Pellicle, 457
Pelvis, hominid vs. chimpanzee, 1041
Penetrance, **271**
  incomplete, 271, 279
Penfield, Wilder, 886, 895
Penicillin, 178, 449, 488
  resistance to, evolution of, 964–965
Penicillium, 481, 488
Penis, 535, 760, 903, **904**, 905, 909, 912
  vs. clitoris, 909, 912
  and contraception, 916
  erection and orgasm, **904–905**
Pentastomids, 520, 560, 564
Pentose(s), 60, 190
PEP, 223–224, 229
PEP carboxylase, 224–225, 230
Peppered moth, **962–965**, 973, 993, 1028
Pepsin, 176, 720, 722
Pepsinogen, 720, 722
Peptide bonds, **74**
  and carbonyl groups, 74
  and enzyme activity, 175
  and protein structure, 74
Peptide chain, 313
Peptide hormones, 883
Peptides, 826, 836, 839
  in digestion, 722
  hormones as, 823, 824
Peptidoglycans, 434, 450
  in prokaryotic cell walls, 107, 432
  in spore coat, 438
Per capita rate of increase, 1088, 1089, 1104
Perception, **887–889, 892**
Perch, 594
Perennials, 621, **622**, 624, 641, 1168, 1173
Perfect flower, 614
Per gene, 834
Pericycle, 647, 648
  and cell types in angiosperms, 647
  in roots, **633**, 634
  and vascular cambium, 644
Periods, geologic, 970
Periosteum, 705
Peripheral membrane proteins, 105–106, 124
Peripheral nervous system, 712, 842, **844–847**, 856, 935
  development of, in chick embryo, 935
  in digestive tract, 716, 718
  neurotransmitters in, 847
  in stomach, 720
  in yoga, 845
Peripheral proteins, 114
Perissodactyls, 602
Peristalsis, 912
Peritoneum, 716, **718**
Peritubular capillaries, 770–773, 776

Periwinkles, 550
Permafrost, 1168
Permeability
  of cell membrane, 127–128, 132–133, 142
  selective, 112, 124, 127–128, 142, 191, 197, 201
Permeases, 135, 178–179 (*see also* Carrier proteins, Transport proteins)
Permian period, 497, 503–504, 505, 509, 597, 1024
Peroxisomes, **118–119**, 125, 494
  in animal cell, 112
  of photosynthetic cells, 225
Perspiration, 703, 713, 784, 786
Pesticides, 1148
Petals, 238, 511, 513, 613–614, 619, 624
Petioles, 628, 679, 692
Petroleum, 504
Peyer's patches, 795
P_{fr}, **686–687** (*see* Phytochromes)
PGA, in three-carbon pathway, 222–224
pH, **48–49**, 52, 54, 184
  in biological systems, 52, 54
  of blood, and altitude, 746
  and cell growth in plants, 675
  and cell metabolism, 127
  and digestion, 715
  effect on cell division, 147
  and electron transport, 198
  and enzyme regulation, 175–176
  of female reproductive tract, 905, 909, 913
  of gastric juice, 720
  of intestinal enzymes, 722
  and oxygen debt, 190
  and rate of membrane transport, 178
  of soil, 664, 670
  of stomach, 791
Phacophyta, 455–457, 465, 477
Phage φX174, 349, **350**
Phagocytic cells, 797
Phagocytosis, 118, 121, **138–140**, 143, 281, 798, 818
  and cancer research, 378
  enzymes in, 138–139
  in heterotrophs, 138–139
  and inflammatory response, 792
  and lymphokines, 807
  and macrophages, 808
  and pathogenic bacteria, 433
  in protists, 138–139
  in white blood cells, 750–751
Pharynx, 534, 539, 541, 555, 591, 592, 604, 864
  in digestions, 715, 717, **718–720**
  in respiration, 736, 738
Phase-contrast microscopy, 99–101
Phases of meiosis, **253–257**
Phases of mitosis, 150–152
Phellem (*see* Cork cell)
Phenetics, 415, 418
Phenobarbitol, 117
Phenotype, **240–244**, 247, 266, 269, 271, 306, 315
  in balanced polymorphism, 992–993
  and behavior, 993
  and "central dogma" of molecular biology, 342
  and chromosome abnormalities, 386–387
  dominant, 241, 247
  effects of gene on, 268, 273
  and genetic interaction, 993–994
  vs. genotype, 993–994
  in learning, 1077
  and Mendelian genetics, **240–244,** 247
  and mutations, 982
  and natural selection, 977, **993–994, 995–1006**
  and nonrandom mating, 984
  novel, **269,** 279
  phenotype matching, 1077
  phenotypic ratio, 279
  recessive, 244, 247, 279
  and sexual reproduction, 994
  and speciation, 1010–1011
Phenotypic ratio, 239, 242–243, 247, 279
  independent assortment and, 243, 247
Phenylalanine, 73, 76, 309, 388–389 (*see also* Phenylketonuria)
Phenylketonuria, 388–389, 400
  and albinism, 388
  and alkaptonuria, 388
  incidence, 400
Pheromones, 585, 863
  in bees (*see* Queen substance)
  in naked mole rats, 1078

Tertiary structure, 184
  and enzyme activity, 170–171, 175
Testcross, 241, 243, 247, 265–266, 274, 276
Testes, 535, 824, 825, **901**, 903, 904, 905, 906, **907**, **917**
  development of in human embryo, 946
  hormone production, control of, 927
  and sex hormones, 70, 258
Testicles, 901
Testosterone, 70, 117, 824, **905**, 906–908, **913**, 917
  and interstitial cells, **901**
  in mammals, 906
  and sex differences in brain, 907–908
Tests (shells), 469, 470
Tetracycline, 344
  resistance, 329
Tetrad, 253–254
Tetrahydrocannabinol, 515
Tetrahymena furgasoni, 118, 370
Tetramers, 76
Tetranucleotide theory, 283–284
Tetraploidy, 1013
T-even bacteriophages, 285–286, 330, 341, 444–445
T4 bacteriophage, 330
T4 glycoprotein, 804, **815**, 819 (*see also* Helper T cells; Immune response; T lymphocytes)
Thalamus, **883**, **885**, 888, **892–893**, 898
Thalidomide, 947
Thallium, 25
Thallus, 463, 465, 467
Theca, 460
Thecodonts, 597
Theophrastus, 236
Theory, in science, 9, 14–17
Thermoacidophiles, 430
Thermocline, 46
Thermodynamics, **162–166**
  and cell transport, 180
  and diffusion, 180
  laws of, 12, **162–166**, 183, 710–711, 776
  and living systems, 166
Thermonuclear reaction, 84–85
  in sun, 161–163
Thermophilic bacteria, 428
Thermosphere, 1132, **1133**
Thermostat, 784–786, 790, 883
Theropods, 599
"Theta" chromosome, 295
Theta replication, 320
Thirst, 769, 785, 788–789, 859
30-nanometer fiber, of DNA, 357
32-cell stage, of embryo, 931
Thomas, Lewis, 16
Thoracic cavity, 701, 713, 718, 740–741, 747, 762, 936
Thoracic duct, 763
Thorax, 566, 571, 574, 584, 585
Thorns, 629, 1121
Three-carbon pathway, 222–223 (*see also* Calvin cycle)
Three-chambered heart, 754-**755**, 763
Threonine, 73, 76, 310
Thrombi, **760–761**
Thrombin, **751–752**
Thromboplastin, **835**
Thrush (bird), 423, 1107
Thrush (disease), 481, 488
Thumb, 601
  evolution of, 1032
Thylakoid membrane, **113**
Thylakoids, 120, 442
  inner compartments of, 213
  membrane, **113**
  in photosynthesis, **213–214**, 216, 218, 220–222, 224, 228–229
Thylakoid space, 213–214, 220, 229
Thymidine kinase, 378–379
Thymine, 80–81, 305–306, 310, 314, 351, 429
  in DNA, 283–284, 288–290, 292, 296, 299
Thymus, 819
  development in vertebrate embryo, 937
  and immune system, 793
  and MHC proteins, 804
  and T lymphocytes, 805, 819
Thyroid gland, 713, 784, 786, 828, 824, **828–829**
  hormone production, control of, 827
  and negative feedback control, 827
  and pituitary, 825

Thyroid hormone (*see* Thyroxine)
Thyroid-releasing hormone (*see* TSH)
Thyroid-stimulating hormone, 713, 784 (*see TSH*)
Thyrotropin, 825, 826
Thyrotropin-releasing hormone (*see* TRH)
Thyroxine (thyroid hormone), 713, 728, 784, 786, 824–825, 828, 836, 839, 940
Tibia, 707
Ticks, 570, 585
Tidal, 1158
Tideland communities, **1157–1160**
Tide line, 1159
Tiger, 412
Tight junctions, **703**, 721
Time, biological, 93
"Time's arrow," 164
Tinbergen, Niko, 872, 1052, 1054–1055
Ti plasmid, **352–353**, 354
Tips of chromosomes, 364
Tissue(s), 140
  animal, 142–143
  connective, 140
  embryonic, 525, 532, 542, **930–931** (*see also* Ectoderm; Endoderm; Mesoderm)
  epithelial, 140
  exchanges with blood, 749
  homogenization, 192–193
  human (*see* Bone; Connective Tissue; Epithelial Tissue; Muscle Tissue; Nerve Tissue)
  and human heart, 756
  interactions during development, **930–931**
  layers, 520, 532, 929
  muscle, 140
  nerve, 140
  organ, 140
  plant, 141–143 (*see also* Ground tissue; Phloem; Vascular tissue; Xylem)
  and respiration, 737
  systems, in angiosperms, **625–627**, 648
    transplants, 810–812
  turnover, 830
  in vertebrates, **702–710**, 713
Tissue factor, 751
Tissue systems, in angiosperms, **625–627**, 648
  dermal, 648
  differentiation of, 636
  ground, 648
  in roots, 634, 635
  vascular, 648
Titmice, 1057
T lymphocyte precursors, 804
T lymphocytes, **801**, **804–805**, 818, 836
  functions of, 806–808
  and immune system, **793**, **801–808**, 819
  and immunosuppressive drugs, 810
  life history of, 801, 804–805
  and monoclonal antibodies, 802
  selection of, 804–805
  and thymus gland, 804
Toads, 595, 1054
Toadstools, 481
Tobacco, 172, 307, 352–353, 678
Tobacco mosaic virus (TMV), 291, 305, 307, 444
Tocopherol, 729
Toes, 601, 946, 947, 948
Togetherness, 778–779
Tolerance hypothesis, **1125**
Tomatoes, 678
Tone deafness, 583
Tonegawa, Susumu, 372–373
Tongs, 565
Tongue, 717, 718
Tonicity
  of blood, 132
  hypertonicity, 132–134, 142
  hypotonicity, 132–134, 142
  isotonicity, 132–134, 142
Tonsils, 793–795, 937
Tonoplast, 115
Tool use, 1002, 1037, 1042, 1044, 1073, 1159
  in early *Homo sapiens*, 1046, 1047
  in *Homo erectus*, 1044–1045
Tooth shells, (*see* Tusk shells)
Top soil, 663, 1166
Topoisomerases, **294**, 297, 299
Torsion, **550**, 551
Touch, 588, 886
  response to, in plants, **691–693**

Touch receptors, 551, **582**, 586, 859, 860, 1060 (*see also* Mechanoreceptors; Sensilla)
Toxic waste, 488
Toxins, 72, 451
  in cdinarians, 526
  in dinoflagellates, 460
  in mushrooms, 487
  of octopus, 552, 553
"Tracers," 25
Trachea(e), 736
  of arthropods, 734, 737
  and evolution of lungs, 736
  human
    and cigarette smoke, 739
    in respiration, 738, 740, 747
  in mammals, 718–719
Tracheids, 107, 501, 509, **639**, 647–648
  in confier leaf, 508–509
  in secondary growth, 647
  water conduction in, 652, 660, 669
Tracheophytes, 497
Tracts, 842, 843
Trailer sequence, 320–321
Traits
  and clines, **1002**
  dominant, 260, 266, 268–271, 276
  hereditary, 249
  and life-history patterns, **1099**, 1105
  polygenic, 271–272, 279
  recessive, 260, 265–268, 274, 279
  sex-linked, 266–267, 278
  types of selection for, **995**
Transacetylase, 325
Transcription
  for amino acids, 320
  of DNA, 359
  efficiency of, 321–322
  in eukaryotes, 355, 358–360, 362, 368
  of Messenger RNA, **301–317**, **320–323**, 325, **368–370**, 372, 380
  and polypeptides, 369
  in prokaryotes, 320–322, 325, 337, 355, 358–360, 380
  and promoter sequences, 368
  and regulatory proteins, **360–361**
  reverse, 306, 340–342, 353, 373–376
  of RNA (*see* RNA)
  RNA polymerase, 368
  of structural genes, 320, 323, 325, 368
  units, 366
  of viral DNA, 330
Transcription units, 366
Transduction, 332–334, 336–338, 859
  vs. conjugation, 337
  general, **332–333**, 338
  and genetic recombination, 336–337
  olfactor, 861
  in prokaryotes, 437
  specialized, **333–334**, 338
Transfer, genetic, 331, 337–338
  and chromosomal insertions, 331–333, 338
  conjugation, **326–330**, 331, 337
  homologous recombination, 336–338
  and nuclear pores, 380
  in rolling-circle replication, **327–328**, 338
  "single-strand switch," 337
  transduction, **332–334**, 336–338
  transformation, 282, 329, 331, 336, 376
  and transposons, **337–338**, 379
Transfer RNA, 310, **311**, **312–317**, 420
  aminoacyl-tRNA synthetase, 311, 315, 317
  anticodons for, 312–313
  genetic code for, 362, 380
  initiator, 312
  linkage to amino acids, 312
  in prokaryote classification, 430
  shape of, 311, 317
  transcription for, 368
  varieties of, 311
Transformation, 282, 329, 331, 336, 376
  in bacteria, 282, 329, 331, 336
  and homologous recombination, 336
  in prokaryotes, 437
Transformations of energy (*see* Energy transformations)
Transforming factor, 281–284, 371
Transfusions, 810–812, 814, 817–818, 819
"Transgenic" mice, 378